Index of Applications

(continued)

Finite Mathematics
and Applied Calculus

Fourth Edition

STEFAN WANER Hofstra University

STEVEN R. COSTENOBLE Hofstra University

THOMSON
™
BROOKS/COLE

Australia • Brazil • Canada • Mexico • Singapore
Spain • United Kingdom • United States

Finite Mathematics and Applied Calculus, Fourth Edition
Stefan Waner and Steven R. Costenoble

Senior Acquisitions Editor: Carolyn Crockett

Executive Editor: Charlie Van Wagner

Development Editor: Danielle Derbenti

Assistant Editor: Ann Day

Editorial Assistant: Beth Gershman

Technology Project Manager: Donna Kelley

Marketing Manager: Joseph Rogove

Marketing Assistant: Jennifer Liang

Marketing Communications Manager: Darlene Amidon-Brent

Project Manager, Editorial Production: Cheryll Linthicum

Creative Director: Rob Hugel

Art Director: Lee Friedman

Print Buyer: Rebecca Cross

Permissions Editor: Roberta Broyer

Production Service: Interactive Composition Corporation

Text Designer: Tani Hasegawa

Art Editor: Tani Hasegawa

Photo Researcher: Kathleen Olson

Copy Editor: Randa Dubnick

Illustrator: Jade Myers

Cover Designer: MendeDesign

Cover Image: Brown Butterfly, overhead view. Giantstep Inc.

Cover Printer: Quebecor World/Taunton

Compositor: Interactive Composition Corporation

Printer: Quebecor World/Taunton

Thomson Higher Education
10 Davis Drive
Belmont, CA 94002-3098
USA

For more information about our products, contact us at:
Thomson Learning Academic Resource Center
1-800-423-0563
For permission to use material from this text or product, submit a request online at **http://www.thomsonrights.com**.
Any additional questions about permissions can be submitted by e-mail to **thomsonrights@thomson.com**.

Library of Congress Control Number: 2006925005

ISBN-13: 978-0-495-38427-4
ISBN-10: 0-495-38427-5

Finite Mathematics
and Applied Calculus

Fourth Edition

Brief Contents

Contents

Chapter 8   Random Variables and Statistics   543

Chapter 9   Nonlinear Models   615

Chapter 10   Introduction to the Derivative   685

Chapter 11   Techniques of Differentiation   797

Chapter 12   Applications of the Derivative   851

Chapter 13   The Integral   919

Preface

Finite Mathematics and Applied Calculus, Fourth Edition, is intended for a one- or two-term course for students majoring in business, the social sciences, or the liberal arts. Like the earlier editions, the Fourth Edition of *Finite Mathematics and Applied Calculus* is designed to address the challenge of generating enthusiasm and mathematical sophistication in an audience that is often under-prepared and lacks motivation for traditional mathematics courses. We meet this challenge by focusing on real-life applications that students can relate to, by presenting mathematical concepts intuitively and thoroughly, and by employing a writing style that is informal, engaging, and occasionally even humorous.

The Fourth edition goes further than earlier editions in implementing support for a wide range of instructional paradigms: from settings incorporating little or no technology to courses taught in computerized classrooms, and from classes in which a single form of technology is used exclusively to those incorporating several technologies. We fully support three forms of technology in this text: TI-83/84 graphing calculators, Excel spreadsheets, and the use of online utilities we have created for the book. In particular, our comprehensive support for Excel, both in the text and online, is highly relevant for students who are studying business and economics, where skill with spreadsheets may be vital to their future careers.

Our Approach to Pedagogy

Real World Orientation We are particularly proud of the diversity, breadth and abundance of examples and exercises included in this edition. A large number of these are based on real, referenced data from business, economics, the life sciences and the social sciences. Examples and exercises based on dated information have generally been replaced by more current versions; applications based on unique or historically interesting data have been kept.

Adapting real data for pedagogical use can be tricky; available data can be numerically complex, intimidating for students, or incomplete. We have modified and streamlined many of the real world applications, rendering them as tractable as any "made-up" application. At the same time, we have been careful to strike a pedagogically sound balance between applications based on real data and more traditional "generic" applications. Thus, the density and selection of real data-based applications has been tailored to the pedagogical goals and appropriate difficulty level for each section.

Readability We would like students to read this book. We would like students to *enjoy* reading this book. Thus, we have written the book in a conversational and student-oriented style, and have made frequent use of question-and-answer dialogues to encourage the development of the student's mathematical curiosity and intuition. We hope that this text will give the student insight into how a mathematician develops and thinks about mathematical ideas and their applications.

Five Elements of Mathematical Pedagogy to Address Different Learning Styles The "Rule of Four" is a common theme in many texts. Implementing this approach, we discuss many of the central concepts **numerically, graphically** and **algebraically,** and clearly delineate these distinctions. The fourth element, **verbal communication** of mathematical concepts, is emphasized through our discussions on translating English sentences into mathematical statements, and our Communication and Reasoning exercises at the end of each section. A fifth element, **interactivity,** is implemented through expanded use of question-and-answer dialogs, but is seen most dramatically within the student website. Using this resource, students can interact with the material in several ways: through interactive tutorials specific to concepts and examples covered in sections, and on-line utilities that automate a variety of tasks, from graphing to regression and matrix algebra.

Exercise Sets The substantial collection of exercises provides a wealth of material that can be used to challenge students at almost every level of preparation, and includes everything from straightforward drill exercises to interesting and rather challenging applications. The exercise sets have been carefully graded to move from straightforward "basic skills" exercises that mimic examples in the text to more interesting and challenging ones. With this edition, basic skills exercises and the most difficult exercises are marked for easy reference. We have also included, in virtually every section of every chapter, interesting applications based on real data, Communication and Reasoning exercises that help students articulate mathematical concepts, and exercises ideal for the use of technology.

Many of the scenarios used in application examples and exercises are revisited several times throughout the book. Thus, for instance, students will find themselves using a variety of techniques, from graphing through the use of derivatives and elasticity, to analyze the same application. Reusing scenarios and important functions provides unifying threads and shows students the complex texture of real-life problems.

New To This Edition

Content:

- Chapter 3 (page 173): We have included a new optional section on game theory as a modern and relevant illustration of the usefulness of matrix algebra. This gives instructors an alternative to input-output models as an applications topic.

- Chapter 4 (page 257): We have included an optional discussion of the use of the simplex method and duality in solving games, so students who have studied game theory in Chapter 3 can see yet another use of the simplex method.

- Chapter 7 (page 443): The sections on estimated probability (relative frequency) and theoretical probability are now combined for greater economy and conceptual clarity. Markov systems now appears as a single optional section at the end of the chapter illustrating an important application of both probability and matrix algebra.

- Appendix A (page A1): We have added an introductory section on logic up through truth tables, tautologies, contradictions, and an introduction to arguments. The material in this section is based on our highly successful and more extensive interactive logic supplement and extensive companion notes that continue to be available online for users of this edition.

- Chapter 10 (page 685): We now discuss limits and continuity in their traditional place before derivatives, but they may still be treated as optional if so desired.

- Chapter 13 (page 919): To help students better understand the definite integral and the Fundamental Theorem of Calculus, we have completely rewritten the chapter on the integral with greater emphasis on the computation of Riemann sums from graphs.

- Chapter 9 (page 615): We have placed more emphasis on half-life and doubling time in the chapter on exponential and logarithmic functions.

2.1 EXERCISES

● denotes basic skills exercises
◆ denotes challenging exercises
tech Ex indicates exercises that should be solved using technology

In Exercises 1–14, find all solutions of the given system of equations and check your answer graphically. **hint** [see Examples 1–4]

1. ● $x - y = 0$
$ x + y = 4$

2. ● $x - y = 0$
$ x + y = -6$

3. ● $x + y = 4$
$ x - y = 2$

4. ● $2x + y = 2$
$ -2x + y = 2$

5. ● $3x - 2y = 6$
$ 2x - 3y = -6$

6. ● $2x + 3y = 5$
$ 3x + 2y = 5$

7. ● $0.5x + 0.1y = 0.7$
$ 0.2x - 0.2y = 0.6$

8. ● $-0.3x + 0.5y = 0.1$
$ 0.1x - 0.1y = 0.4$

9. ● $\dfrac{x}{3} - \dfrac{y}{2} = 1$
$ \dfrac{x}{4} + y = -2$

10. ● $-\dfrac{2x}{3} + \dfrac{y}{2} = \dfrac{1}{6}$
$ \dfrac{x}{4} - y = -\dfrac{3}{4}$

Applications

25. ● **Resource Allocation** You manage an ice cream factory that makes two flavors: Creamy Vanilla and Continental Mocha. Into each quart of Creamy Vanilla go 2 eggs and 3 cups of cream. Into each quart of Continental Mocha go 1 egg and 3 cups of cream. You have in stock 500 eggs and 900 cups of cream. How many quarts of each flavor should you make in order to use up all the eggs and cream? **hint** [see Example 5]

26. ● **Class Scheduling** Enormous State University's Math Department offers two courses: Finite Math and Applied Calculus. Each section of Finite Math has 60 students, and each section of Applied Calculus has 50. The department will offer a total of 110 sections in a semester, and 6000 students would like to take a math course. How many sections of each course should the department offer in order to fill all sections and accommodate all of the students?

27. ● **Nutrition** Gerber Dutch Apple Dessert contains, in each serving, 60 calories and 11 grams of carbohydrates[1]. Gerber Mango Tropical Fruit Dessert contains, in each serving,

---

Chapter 12 Review

KEY CONCEPTS

12.1 Maxima and Minima
Relative maximum, relative minimum *p. 853*
Absolute maximum, absolute minimum *p. 853*
Stationary points, singular points, endpoints *p. 854*
Finding and classifying maxima and minima *p. 855*
Finding absolute extrema on a closed interval *p. 860*
Using technology to locate approximate extrema *p. 861*

12.2 Applications of Maxima and Minima
Minimizing average cost *p. 855*

12.3 The Second Derivative and Analyzing Graphs
The second derivative of a function f is the derivative of the derivative of f, written as f''. *p. 878*
The acceleration of a moving object is the second derivative of the position function. *p. 878*
Acceleration due to gravity *p. 879*
Acceleration of sales *p. 880*
Concave up, concave down, point of inflection *p. 881*
Locating points of inflection *p. 882*
The point of diminishing returns *p. 882*

12.4 Related Rates
If Q is a quantity changing over time t, then the derivative dQ/dt is the rate at which Q changes over time. *p. 893*
The expanding circle *p. 894*
Steps in solving related rates problems *p. 894*
The falling ladder *p. 895*
Average cost *p. 896*
Allocation of labor *p. 897*

12.5 Elasticity
Price elasticity of demand
$$E = -\frac{dq}{dp} \cdot \frac{p}{q}; \text{ demand is elastic}$$
if $E > 1$, inelastic if $E < 1$, has unit

---

mathematics At Work

Deb Farace

Working for the national accounts division for PepsiCo Beverages & Foods, I need to understand applied mathematics in order to control the variables associated with making profit, manufacturing, production, and most importantly selling our products to mass club channels. Examples of these large, "quality product at great value" outlets are Wal*Mart, Costco and Target. The types of products I handle include Gatorade, Tropicana, and Quaker foods.

Our studies show that the grocery store channels' sales are flattening or declining as a whole in lieu of large, national outlets like the above. So in order to maximize growth in this segment of our business, I meet with regional buying offices

TITLE Sr. National Accounts Manager
INSTITUTION PepsiCo Beverages & Foods

A number of factors must be taken into consideration in order to meet my company's financial forecasts. Precision using mathematical models is key here, since so many variables can impact last-minute decision-making. Extended variables of supply-and-demand include time of year, competitive landscape, special coupon distribution and other promotions, selling cycles and holidays, size of the outlets, and yes—even the weather.

For example, it's natural to assume that when it's hot outside people will buy more thirst-quenching products like Gatorade. But since our business is so precise, we need to understand mathematically how the weather affects sales. A mathematical model developed by Gatorade analyzes long-term data that impacts sales by geographic market due to the weather. Its findings include exponentially increased sales of Gatorade for each degree above the 90 degrees. I share our

---

TI-83/84 Technology Guide

Section 13.3

Example 5 Estimate the area under the graph of $f(x) = 1 - x^2$ over the interval $[0, 1]$ using $n = 100$, $n = 200$, and $n = 500$ partitions.

Solution with Technology There are several ways to compute Riemann sums with a graphing calculator. We illustrate one method. For $n = 100$, we need to compute the sum

$$\sum_{k=0}^{99} f(x_k)\Delta x = [f(0) + f(0.01) + \cdots + f(0.99)](0.01) \quad \text{See discussion in Example 5}$$

EXCEL Technology Guide

Section 13.3

Example 5 Estimate the area under the graph of $f(x) = 1 - x^2$ over the interval $[0, 1]$ using $n = 100$, $n = 200$, and $n = 500$ partitions.

Solution with Technology We need to compute various sums

FAQs Whether to Use Integration by Parts, and What Goes in the *D* and *I* Columns

Q: *Will integration by parts always work to integrate a product?*

A: No. While integration by parts often works for products in which one factor is a polynomial, it will almost *never* work in the examples of products we saw when discussing substitution in Section 13.2. For example, although integration by parts can be used to compute $\int (x^2 - x)e^{2x-1}\,dx$ (put $x^2 - x$ in the D column and e^{2x-1} in the I column), it *cannot* be used to compute $\int (2x - 1)e^{x^2-x}\,dx$ (put $u = x^2 - x$). Recognizing when to use integration by parts is best learned by experience. ∎

- Complete sections on Taylor Polynomials, the Chain Rule for Multivariate Calculus, and Calculus Applied to Probability are available as optional topics for custom published versions of this text.

- Throughout, we have improved the presentation of proofs and derivations of important results for clarity and accuracy.

← Exercises:

- Basic skills exercises at the beginning of each exercise set and challenging exercises near the end of some exercise sets are now highlighted to guide students and instructors.

- We have expanded the exercise sets themselves and carefully reorganized them to gradually increase in level and to include more basic skills exercises that carefully follow the examples.

← • We have annotated representative exercises with references to specific examples in the text where similar problems are solved.

- We have greatly expanded the chapter review exercise sections.

- A list of key concepts for review now appears at the end of each chapter.

Pedagogy:

← • **New Portfolios** are designed to convey to the student real-world experiences of professionals who have a background in mathematics and use it in their daily business interactions.

- **End-of-Chapter Technology Guides** We have placed detailed TI-83/84 and Microsoft® Excel Guides at the end of each chapter. This has allowed us to expand these instructions while not interrupting the flow of pedagogy in the text. These Guides are referenced liberally at appropriate points in the chapter, so instructors and students can easily use this material or not, as they prefer. Groups of exercises for which the use of technology is suggested or required appear throughout the exercise sets.

- **Question and Answer Dialogue** We frequently use informal question-and-answer dialogues that anticipate the kind of questions that may occur to the student and also guide the student through the development of new concepts. This feature has been streamlined, as has the "Frequently Asked Questions" feature at the end of each section.

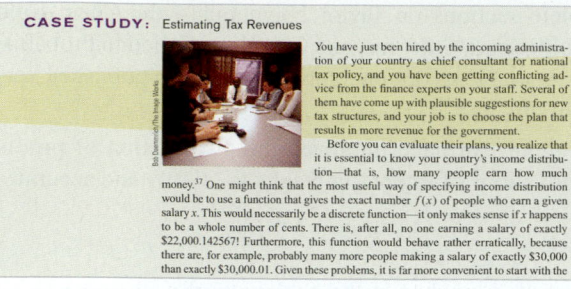

CASE STUDY: Estimating Tax Revenues

You have just been hired by the incoming administration of your country as chief consultant for national tax policy, and you have been getting conflicting advice from the finance experts on your staff. Several of them have come up with plausible suggestions for new tax structures, and your job is to choose the plan that results in more revenue for the government.

Before you can evaluate their plans, you realize that it is essential to know your country's income distribution—that is, how many people earn how much money.[37] One might think that the most useful way of specifying income distribution would be to use a function that gives the exact number $f(x)$ of people who earn a given salary x. This would necessarily be a discrete function—it only makes sense if x happens to be a whole number of cents. There is, after all, no one earning a salary of exactly $22,000.142567! Furthermore, this function would behave rather erratically, because there are, for example, probably many more people making a salary of exactly $30,000 than exactly $30,000.01. Given these problems, it is far more convenient to start with the

Continuing Features

- **Case Studies** Each chapter ends with a section titled "Case Study," an extended application that uses and illustrates the central ideas of the chapter, focusing on the development of mathematical models appropriate to the topics. These applications are ideal for assignment as projects, and to this end we have included groups of exercises at the end of each.

- **Before We Go On** Most examples are followed by supplementary discussions, which may include a check on the answer, a discussion of the feasibility and significance of a solution, or an in-depth look at what the solution means.

- **Quick Examples** Most definition boxes include quick, straightforward examples that a student can use to solidify each new concept.

- **Communication and Reasoning Exercises for Writing and Discussion** These are exercises designed to broaden the student's grasp of the mathematical concepts and develop modeling skills. They include exercises in which the student is asked to provide his or her own examples to illustrate a point or design an application with a given solution. They also include "fill in the blank" type exercises and exercises that invite discussion and debate. These exercises often have no single correct answer.

ThomsonNOW™ with Personalized Study

Help your students maximize study time, minimize stress, and get a better grade by finding out how they will do on a test before they take it! ThomsonNOW™ with Personalized Study is a customized learning tool for students that diagnoses their problem areas through a pre-test and then tailors individual solutions to fit their unique needs. Based on content from the author's widely recognized website, ThomsonNOW with Personalized Study is fully integrated with the text and features the following valuable assets:

- **Interactive Tutorials** Highly interactive tutorials are included on major topics, with guided exercises that parallel the text.

- **Detailed Chapter Summaries** Comprehensive summaries with interactive elements review all the basic definitions and problem solving techniques discussed in each chapter. These are a terrific pre-test study tool for students.

- **Downloadable Excel Tutorials** Detailed Excel tutorials are available for almost every section of the book. These interactive tutorials expand on the examples given in the text.

- **Online Utilities** Our collection of easy-to-use online utilities, written in Java™ and Javascript, allow students to solve many of the technology-based application exercises directly on the web page. The utilities available include function graphers and evaluators, regression tools, powerful matrix utilities and linear programming solvers. These utilities require nothing more than a standard, Java-capable web browser.

- **Downloadable Software** In addition to the web-based utilities the site offers a suite of free and intuitive stand-alone Macintosh® programs, including one for function graphing.

- **Supplemental Topics** We include complete interactive text and exercise sets for a selection of topics not ordinarily included in printed texts, but often requested by instructors.

Supplemental Material

For Students

Student Solutions Manual *by Waner and Costenoble*
ISBN: **0495016993**
The student solutions manual provides worked-out solutions to the odd-numbered exercises in the text (excluding Case Studies) as well as complete solutions to all the chapter review exercises.

Microsoft Excel Computer Laboratory Manual *by Anne D. Henriksen*
ISBN: 0495115010
This laboratory manual uses Microsoft Excel to solve real-world problems in a variety of scientific, technical, and business disciplines. It provides hands-on experience to demonstrate for students that calculus is a valuable tool for solving practical, real-world problems, while helping students increase their knowledge of Microsoft Excel. The manual is a set of self-contained computer exercises that are meant to be used over the course of a 15-week semester in a separate, 75-minute computer laboratory period. The weekly labs parallel the material in the text.

 **vMentor™** When students get stuck on a particular problem or concept, they need only log on to **vMentor**. Accessed through ThomsonNOW, vMentor allows students to talk (using their own computer microphones) to tutors who will skillfully guide them through the problem using an interactive whiteboard for illustration. Students who purchase access to vMentor have access to up to 40 hours of live tutoring a week!

For Instructors

Instructor's Solution Manual *by Waner and Costenoble*
ISBN: **0495016950**
The instructor's solutions manual provides worked-out solutions to all of the exercises in the text including Case Studies.

Test Bank *by James Ball*
ISBN: **0495016950**
This test bank contains numerous multiple choice and free response questions.

Instructor's Suite CD-ROM
ISBN: **0495016977**
The Instructor's Suite CD-ROM contains the instructor's solutions manual and test bank both in MS Word and as PDF files. There is also a multimedia library containing all of the art from the book in MS PowerPoint as well as individual jpeg files.

 **JoinIn on Turning Point** Thomson Brooks/Cole is pleased to offer you book-specific JoinIn content for electronic response systems tailored to *Finite Mathematics and Applied Calculus, 4th edition.* You can transform your classroom and assess your students' progress with instant in-class quizzes and polls. Turning Point software lets you pose book-specific questions and display students' answers seamlessly within Microsoft PowerPoint slides of your own lecture, in conjunction with the "clicker" hardware of your choice. Enhance how your students interact with you, your lecture, and each other.

Acknowledgments

This project would not have been possible without the contributions and suggestions of numerous colleagues, students and friends. We are particularly grateful to our colleagues at Hofstra and elsewhere who used and gave us useful feedback on previous editions. We are also grateful to everyone at Brooks/Cole for their encouragement and guidance throughout the project. Specifically, we would like to thank Curt Hinrichs for his unflagging enthusiasm, Danielle Derbenti for whipping the book into shape, Joe Rogove for telling the world about it, and Carolyn Crockett for stepping in at the last minute. In addition, we would like to thank Ann Day for coordinating our many ancillaries, Beth Gershman for her administrative support, and Cheryll Linthicum for shepherding the project through production.

We would also like to thank the numerous reviewers and accuracy checkers who provided many helpful suggestions that have shaped the development of this book.

Stefan Waner
Steven R. Costenoble

0

Algebra Review

Stewart Cohen/Getty Images

Get a better Grade!
www.thomsonedu.com/login

Logon to your Personalized Study plan to find:

• Section by section tutorials

• Graphers and other resources

Introduction

In this chapter we review some topics from algebra that you need to know to get the most out of this book. This chapter can be used either as a refresher course or as a reference.

There is one crucial fact you must always keep in mind: The letters used in algebraic expressions stand for numbers. All the rules of algebra are just facts about the arithmetic of numbers. If you are not sure whether some algebraic manipulation you are about to do is legitimate, try it first with numbers. If it doesn't work with numbers, it doesn't work.

0.1 Real Numbers

The **real numbers** are the numbers that can be written in decimal notation, including those that require an infinite decimal expansion. The set of real numbers includes all integers, positive and negative; all fractions; and the irrational numbers, those with decimal expansions that never repeat. Examples of irrational numbers are

$$\sqrt{2} = 1.414213562373\ldots$$

and

$$\pi = 3.141592653589\ldots$$

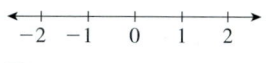

−2 −1 0 1 2

Figure **1**

It is very useful to picture the real numbers as points on a line. As shown in Figure 1, larger numbers appear to the right, in the sense that if $a < b$ then the point corresponding to b is to the right of the one corresponding to a.

Intervals

Some subsets of the set of real numbers, called **intervals,** show up quite often and so we have a compact notation for them.

Interval Notation

Here is a list of types of intervals along with examples.

	Interval	Description	Picture	Example
Closed	$[a, b]$	Set of numbers x with $a \le x \le b$	 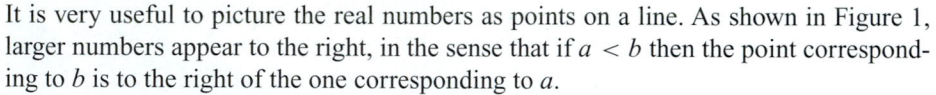 a b (includes end points)	$[0, 10]$
Open	(a, b)	Set of numbers x with $a < x < b$	a b (excludes end points)	$(-1, 5)$
Half-Open	$(a, b]$	Set of numbers x with $a < x \le b$	a b	$(-3, 1]$
	$[a, b)$	Set of numbers x with $a \le x < b$	 a b	$[0, 5)$

Infinite	$[a, +\infty)$	Set of numbers x with $a \leq x$		$[10, +\infty)$
	$(a, +\infty)$	Set of numbers x with $a < x$		$(-3, +\infty)$
	$(-\infty, b]$	Set of numbers x with $x \leq b$		$(-\infty, -3]$
	$(-\infty, b)$	Set of numbers x with $x < b$		$(-\infty, 10)$
	$(-\infty, +\infty)$	Set of all real numbers		$(-\infty, +\infty)$

Operations

There are five important operations on real numbers: addition, subtraction, multiplication, division, and exponentiation. "Exponentiation" means raising a real number to a power; for instance, $3^2 = 3 \cdot 3 = 9$; $2^3 = 2 \cdot 2 \cdot 2 = 8$.

A note on technology: Most graphing calculators and spreadsheets use an asterisk * for multiplication and a caret sign ˆ for exponentiation. Thus, for instance, 3×5 is entered as 3*5, $3x$ as 3*x, and 3^2 as 3ˆ2.

When we write an expression involving two or more operations, like

$$2 \cdot 3 + 4$$

or

$$\frac{2 \cdot 3^2 - 5}{4 - (-1)}$$

we need to agree on the order in which to do the operations. Does $2 \cdot 3 + 4$ mean $(2 \cdot 3) + 4 = 10$ or $2 \cdot (3 + 4) = 14$? We all agree to use the following rules for the order in which we do the operations.

Standard Order of Operations

Parentheses and Fraction Bars

First, calculate the values of all expressions inside parentheses or brackets, working from the innermost parentheses out, before using them in other operations. In a fraction, calculate the numerator and denominator separately before doing the division.

quick Examples

1. $6(2 + [3 - 5] - 4) = 6(2 + (-2) - 4) = 6(-4) = -24$.

2. $\dfrac{(4 - 2)}{3(-2 + 1)} = \dfrac{2}{3(-1)} = \dfrac{2}{-3} = -\dfrac{2}{3}$

3. $3/(2 + 4) = \dfrac{3}{2 + 4} = \dfrac{3}{6} = \dfrac{1}{2}$

4. $(x + 4x)/(y + 3y) = 5x/(4y)$

Exponents
Next, perform exponentiation.

quick Examples

1. $2 + 4^2 = 2 + 16 = 18$

2. $(2 + 4)^2 = 6^2 = 36$ Note the difference.

3. $2 \left(\dfrac{3}{4-5} \right)^2 = 2 \left(\dfrac{3}{-1} \right)^2 = 2(-3)^2 = 2 \times 9 = 18$

4. $2(1 + 1/10)^2 = 2(1.1)^2 = 2 \times 1.21 = 2.42$

Multiplication and Division
Next, do all multiplications and divisions, from left to right.

quick Examples

1. $2(3 - 5)/4 \cdot 2 = 2(-2)/4 \cdot 2$ Parentheses first

$= -4/4 \cdot 2$ Left-most product

$= -1 \cdot 2 = -2$ Multiplications and divisions, left to right

2. $2(1 + 1/10)^2 \times 2/10 = 2(1.1)^2 \times 2/10$ Parentheses first

$= 2 \times 1.21 \times 2/10$ Exponent

$= 4.84/10 = 0.484$ Multiplications and divisions, left to right

3. $4\dfrac{2(4-2)}{3(-2 \cdot 5)} = 4\dfrac{2(2)}{3(-10)} = 4\dfrac{4}{-30} = \dfrac{16}{-30} = -\dfrac{8}{15}$

Addition and Subtraction
Last, do all additions and subtractions, from left to right.

quick Examples

1. $2(3 - 5)^2 + 6 - 1 = 2(-2)^2 + 6 - 1 = 2(4) + 6 - 1 = 8 + 6 - 1 = 13$

2. $\left(\dfrac{1}{2} \right)^2 - (-1)^2 + 4 = \dfrac{1}{4} - 1 + 4 = -\dfrac{3}{4} + 4 = \dfrac{13}{4}$

3. $3/2 + 4 = 1.5 + 4 = 5.5$

4. $3/(2 + 4) = 3/6 = 1/2 = 0.5$ Note the difference.

5. $4/2^2 + (4/2)^2 = 4/2^2 + 2^2 = 4/4 + 4 = 1 + 4 = 5$

 tech Ex   ## Entering Formulas

Any good calculator or spreadsheet will respect the standard order of operations. However, we must be careful with division and exponentiation and use parentheses as necessary. The following table gives some examples of simple mathematical expressions and their equivalents in the functional format used in most graphing calculators, spreadsheets, and computer programs.

Mathematical Expression	Formula	Comments
$\dfrac{2}{3-x}$	`2/(3-x)`	Note the use of parentheses instead of the fraction bar. If we omit the parentheses, we get the expression shown next.
$\dfrac{2}{3}-x$	`2/3-x`	The calculator follows the usual order of operations.
$\dfrac{2}{3\times 5}$	`2/(3*5)`	Putting the denominator in parentheses ensures that the multiplication is carried out first. The asterisk is usually used for multiplication in graphing calculators and computers.
$\dfrac{2}{x}\times 5$	`(2/x)*5`	Putting the fraction in parentheses ensures that it is calculated first. Some calculators will interpret `2/3*5` as $\dfrac{2}{3\times 5}$, but `2/3(5)` as $\dfrac{2}{3}\times 5$.
$\dfrac{2-3}{4+5}$	`(2-3)/(4+5)`	Note once again the use of parentheses in place of the fraction bar.
2^3	`2^3`	The caret ^ is commonly used to denote exponentiation.
2^{3-x}	`2^(3-x)`	Be careful to use parentheses to tell the calculator where the exponent ends. Enclose the *entire exponent* in parentheses.
2^3-x	`2^3-x`	Without parentheses, the calculator will follow the usual order of operations: exponentiation and then subtraction.
3×2^{-4}	`3*2^(-4)`	On some calculators, the negation key is separate from the minus key.
$2^{-4\times 3}\times 5$	`2^(-4*3)*5`	Note once again how parentheses enclose the entire exponent.
$100\left(1+\dfrac{0.05}{12}\right)^{60}$	`100*(1+0.05/12)^60`	This is a typical calculation for compound interest.
$PV\left(1+\dfrac{r}{m}\right)^{mt}$	`PV*(1+r/m)^(m*t)`	This is the compound interest formula. *PV* is understood to be a single number (present value) and not the product of *P* and *V* (or else we would have used `P*V`).
$\dfrac{2^{3-2}\times 5}{y-x}$	`2^(3-2)*5/(y-x)` or `(2^(3-2)*5)/(y-x)`	Notice again the use of parentheses to hold the denominator together. We could also have enclosed the numerator in parentheses, although this is optional. (Why?)
$\dfrac{2^y+1}{2-4^{3x}}$	`(2^y+1)/(2-4^(3*x))`	Here, it is necessary to enclose both the numerator and the denominator in parentheses.
$2^y+\dfrac{1}{2}-4^{3x}$	`2^y+1/2-4^(3*x)`	This is the effect of leaving out the parentheses around the numerator and denominator in the previous expression.

Accuracy and Rounding

When we use a calculator or computer, the results of our calculations are often given to far more decimal places than are useful. For example, suppose we are told that a square has an area of 2.0 square feet and we are asked how long its sides are. Each side is the square root of the area, which the calculator tells us is

$$\sqrt{2} \approx 1.414213562$$

However, the measurement of 2.0 square feet is probably accurate to only two digits, so our estimate of the lengths of the sides can be no more accurate than that. Therefore, we round the answer to two digits:

$$\text{Length of one side} \approx 1.4 \text{ feet}$$

The digits that follow 1.4 are meaningless. The following guide makes these ideas more precise.

Significant Digits, Decimal Places, and Rounding

The number of **significant digits** in a decimal representation of a number is the number of digits that are not leading zeros after the decimal point (as in .0005) or trailing zeros before the decimal point (as in 5,400,000). We say that a value is **accurate to *n* significant digits** if only the first *n* significant digits are meaningful.

When to Round
After doing a computation in which all the quantities are accurate to no more than *n* significant digits, round the final result to *n* significant digits.

quick Examples

1. 0.00067 has two significant digits.   The 000 before 67 are leading zeros.
2. 0.000670 has three significant digits.   The 0 after 67 is significant.
3. 5,400,000 has two or more significant digits.   We can't say how many of the zeros are trailing.[*]
4. 5,400,001 has 7 significant digits.   The string of zeros is not trailing.
5. Rounding 63,918 to three significant digits gives 63,900.
6. Rounding 63,958 to three significant digits gives 64,000.
7. $\pi = 3.141592653...$ $\frac{22}{7} = 3.142857142...$ Therefore, $\frac{22}{7}$ is an approximation of π that is accurate to only three significant digits (3.14).
8. $4.02(1 + 0.02)^{1.4} \approx 4.13$ We rounded to three significant digits.

---

[*] If we obtained 5,400,000 by rounding 5,401,011, then it has three significant digits because the zero after the 4 is significant. On the other hand, if we obtained it by rounding 5,411,234, then it has only two significant digits. The use of scientific notation avoids this ambiguity: 5.40×10^6 (or 5.40 E6 on a calculator or computer) is accurate to three digits and 5.4×10^6 is accurate to two.

One more point, though: If, in a long calculation, you round the intermediate results, your final answer may be even less accurate than you think. As a general rule,

When calculating, don't round intermediate results. Rather, use the most accurate results obtainable or have your calculator or computer store them for you.

When you are done with the calculation, *then* round your answer to the appropriate number of digits of accuracy.

0.1 EXERCISES

Calculate each expression in Exercises 1–24, giving the answer as a whole number or a fraction in lowest terms.

1. $2(4+(-1))(2\cdot-4)$

2. $3+([4-2]\cdot 9)$

3. `20/(3*4)-1`

4. `2-(3*4)/10`

5. $\dfrac{3+([3+(-5)])}{3-2\times 2}$

6. $\dfrac{12-(1-4)}{2(5-1)\cdot 2-1}$

7. `(2-5*(-1))/1-2*(-1)`

8. `2-5*(-1)/(1-2*(-1))`

9. $2\cdot(-1)^2/2$

10. $2+4\cdot 3^2$

11. $2\cdot 4^2+1$

12. $1-3\cdot(-2)^2\times 2$

13. `3^2+2^2+1`

14. `2^(2^2-2)`

15. $\dfrac{3-2(-3)^2}{-6(4-1)^2}$

16. $\dfrac{1-2(1-4)^2}{2(5-1)^2\cdot 2}$

17. `10*(1+1/10)^3`

18. `121/(1+1/10)^2`

19. $3\left(\dfrac{-2\cdot 3^2}{-(4-1)^2}\right)$

20. $-\left(\dfrac{8(1-4)^2}{-9(5-1)^2}\right)$

21. $3\left(1-\left(-\dfrac{1}{2}\right)^2\right)^2+1$

22. $3\left(\dfrac{1}{9}-\left(\dfrac{2}{3}\right)^2\right)^2+1$

23. `(1/2)^2-1/2^2`

24. `2/(1^2)-(2/1)^2`

Convert each expression in Exercises 25–50 into its technology formula equivalent as in the table in the text.

25. $3\times(2-5)$

26. $4+\dfrac{5}{9}$

27. $\dfrac{3}{2-5}$

28. $\dfrac{4-1}{3}$

29. $\dfrac{3-1}{8+6}$

30. $3+\dfrac{3}{2-9}$

31. $3-\dfrac{4+7}{8}$

32. $\dfrac{4\times 2}{\left(\frac{2}{3}\right)}$

33. $\dfrac{2}{3+x}-xy^2$

34. $3+\dfrac{3+x}{xy}$

35. $3.1x^3-4x^{-2}-\dfrac{60}{x^2-1}$

36. $2.1x^{-3}-x^{-1}+\dfrac{x^2-3}{2}$

37. $\dfrac{\left(\frac{2}{3}\right)}{5}$

38. $\dfrac{2}{\left(\frac{3}{5}\right)}$

39. $3^{4-5}\times 6$

40. $\dfrac{2}{3+5^{7-9}}$

41. $3\left(1+\dfrac{4}{100}\right)^{-3}$

42. $3\left(\dfrac{1+4}{100}\right)^{-3}$

43. $3^{2x-1}+4^x-1$

44. $2^{x^2}-(2^{2x})^2$

45. 2^{2x^2-x+1}

46. $2^{2x^2-x}+1$

47. $\dfrac{4e^{-2x}}{2-3e^{-2x}}$

48. $\dfrac{e^{2x}+e^{-2x}}{e^{2x}-e^{-2x}}$

49. $3\left(1-\left(-\dfrac{1}{2}\right)^2\right)^2+1$

50. $3\left(\dfrac{1}{9}-\left(\dfrac{2}{3}\right)^2\right)^2+1$

0.2 | Exponents and Radicals

In Section 1 we discussed exponentiation, or "raising to a power"; for example, $2^3 = 2 \cdot 2 \cdot 2$. In this section we discuss the algebra of exponentials more fully. First, we look at *integer* exponents: cases in which the powers are positive or negative whole numbers.

Integer Exponents

Positive Integer Exponents

If a is any real number and n is any positive integer, then by a^n we mean the quantity $a \cdot a \cdot \cdots \cdot a$ (n times); thus, $a^1 = a$, $a^2 = a \cdot a$, $a^5 = a \cdot a \cdot a \cdot a \cdot a$. In the expression a^n the number n is called the **exponent,** and the number a is called the **base.**

quick Examples

$$3^2 = 9 \qquad\qquad 2^3 = 8$$
$$0^{34} = 0 \qquad\qquad (-1)^5 = -1$$
$$10^3 = 1000 \qquad\qquad 10^5 = 100,000$$

Negative Integer Exponents

If a is any real number *other than zero* and n is any positive integer, then we define

$$a^{-n} = \frac{1}{a^n} = \frac{1}{a \cdot a \cdot \cdots \cdot a} \quad (n \text{ } a\text{'s in the denominator})$$

quick Examples

$$2^{-3} = \frac{1}{2^3} = \frac{1}{8} \qquad\qquad 1^{-27} = \frac{1}{1^{27}} = 1$$

$$x^{-1} = \frac{1}{x^1} = \frac{1}{x} \qquad\qquad (-3)^{-2} = \frac{1}{(-3)^2} = \frac{1}{9}$$

$$y^7 y^{-2} = y^7 \frac{1}{y^2} = y^5 \qquad 0^{-2} \text{ is not defined}$$

Zero Exponent

If a is any real number other than zero, then we define

$$a^0 = 1$$

quick Examples

$$3^0 = 1 \qquad\qquad 1,000,000^0 = 1$$
$$0^0 \text{ is not defined}$$

When combining exponential expressions, we use the following identities.

Exponent Identity

1. $a^m a^n = a^{m+n}$

$2^3 2^2 = 2^{3+2} = 2^5 = 32$

$x^3 x^{-4} = x^{3-4} = x^{-1} = \dfrac{1}{x}$

$\dfrac{x^3}{x^{-2}} = x^3 \dfrac{1}{x^{-2}} = x^3 x^2 = x^5$

2. $\dfrac{a^m}{a^n} = a^{m-n}$ if $a \neq 0$

$\dfrac{4^3}{4^2} = 4^{3-2} = 4^1 = 4$

$\dfrac{x^3}{x^{-2}} = x^{3-(-2)} = x^5$

$\dfrac{3^2}{3^4} = 3^{2-4} = 3^{-2} = \dfrac{1}{9}$

3. $(a^n)^m = a^{nm}$

$(3^2)^2 = 3^4 = 81$

$(2^x)^2 = 2^{2x}$

4. $(ab)^n = a^n b^n$

$(4 \cdot 2)^2 = 4^2 2^2 = 64$

$(-2y)^4 = (-2)^4 y^4 = 16y^4$

5. $\left(\dfrac{a}{b}\right)^n = \dfrac{a^n}{b^n}$ if $b \neq 0$

$\left(\dfrac{4}{3}\right)^2 = \dfrac{4^2}{3^2} = \dfrac{16}{9}$

$\left(\dfrac{x}{-y}\right)^3 = \dfrac{x^3}{(-y)^3} = -\dfrac{x^3}{y^3}$

Caution

- In the first two identities, the bases of the expressions must be the same. For example, the first gives $3^2 3^4 = 3^6$, but does *not* apply to $3^2 4^2$.

- People sometimes invent their own identities, such as $a^m + a^n = a^{m+n}$, which is wrong! (Try it with $a = m = n = 1$.) If you wind up with something like $2^3 + 2^4$, you are stuck with it; there are no identities around to simplify it further. (You can factor out 2^3, but whether that is a simplification depends on what you are going to do with the expression next.)

Example 1 Combining the Identities

$\dfrac{(x^2)^3}{x^3} = \dfrac{x^6}{x^3}$ By (3)

$\quad = x^{6-3}$ By (2)

$\quad = x^3$

$$\frac{(x^4 y)^3}{y} = \frac{(x^4)^3 y^3}{y} \qquad \text{By (4)}$$

$$= \frac{x^{12} y^3}{y} \qquad \text{By (3)}$$

$$= x^{12} y^{3-1} \qquad \text{By (2)}$$

$$= x^{12} y^2$$

Example 2 Eliminating Negative Exponents

Simplify the following and express the answer using no negative exponents.

a. $\dfrac{x^4 y^{-3}}{x^5 y^2}$ **b.** $\left(\dfrac{x^{-1}}{x^2 y}\right)^5$

Solution

a. $\dfrac{x^4 y^{-3}}{x^5 y^2} = x^{4-5} y^{-3-2} = x^{-1} y^{-5} = \dfrac{1}{x y^5}$

b. $\left(\dfrac{x^{-1}}{x^2 y}\right)^5 = \dfrac{(x^{-1})^5}{(x^2 y)^5} = \dfrac{x^{-5}}{x^{10} y^5} = \dfrac{1}{x^{15} y^5}$

Radicals

If a is any nonnegative real number, then its **square root** is the nonnegative number whose square is a. For example, the square root of 16 is 4, because $4^2 = 16$. We write the square root of n as $\sqrt{n}$. (Roots are also referred to as **radicals.**) It is important to remember that $\sqrt{n}$ is never negative. Thus, for instance, $\sqrt{9}$ is 3, and not -3, even though $(-3)^2 = 9$. If we want to speak of the "negative square root" of 9, we write it as $-\sqrt{9} = -3$. If we want to write both square roots at once, we write $\pm\sqrt{9} = \pm 3$.

The **cube root** of a real number a is the number whose cube is a. The cube root of a is written as $\sqrt[3]{a}$ so that, for example, $\sqrt[3]{8} = 2$ (because $2^3 = 8$). Note that we can take the cube root of any number, positive, negative, or zero. For instance, the cube root of -8 is $\sqrt[3]{-8} = -2$ because $(-2)^3 = -8$. Unlike square roots, the cube root of a number may be negative. In fact, the cube root of a always has the same sign as a.

Higher roots are defined similarly. The **fourth root** of the *nonnegative* number a is defined as the nonnegative number whose fourth power is a, and written $\sqrt[4]{a}$. The **fifth root** of any number a is the number whose fifth power is a, and so on.

Note We cannot take an even-numbered root of a negative number, but we can take an odd-numbered root of any number. Even roots are always positive, whereas odd roots have the same sign as the number we start with. ■

Example 3 *n*th Roots

$\sqrt{4} = 2$	Because $2^2 = 4$
$\sqrt{16} = 4$	Because $4^2 = 16$
$\sqrt{1} = 1$	Because $1^2 = 1$
If $x \geq 0$, then $\sqrt{x^2} = x$	Because $x^2 = x^2$
$\sqrt{2} \approx 1.414213562$	$\sqrt{2}$ is not a whole number.
$\sqrt{1+1} = \sqrt{2} \approx 1.414213562$	First add, then take the square root.*
$\sqrt{9+16} = \sqrt{25} = 5$	Contrast with $\sqrt{9} + \sqrt{16} = 3 + 4 = 7$.
$\sqrt[3]{27} = 3$	Because $3^3 = 27$
$\sqrt[3]{-64} = -4$	Because $(-4)^3 = -64$
$\sqrt[4]{16} = 2$	Because $2^4 = 16$
$\sqrt[4]{-16}$ is not defined	Even-numbered root of a negative number
$\sqrt[5]{-1} = -1$, since $(-1)^5 = -1$	Odd-numbered root of a negative number
$\sqrt[n]{-1} = -1$ if n is any odd number	

---

*In general, $\sqrt{a+b}$ means the square root of the *quantity* $(a + b)$. The radical sign acts like a pair of parentheses or a fraction bar, telling us to evaluate what is inside before taking the root. (See the *Caution* on the next page.)

Q: *In the example we saw that $\sqrt{x^2} = x$ if x is nonnegative. What happens if x is negative* **?**

A: If x is negative, then x^2 is positive, and so $\sqrt{x^2}$ is still defined as the nonnegative number whose square is x^2. This number must be $|x|$, the **absolute value of x,** which is the nonnegative number with the same size as x. For instance, $|-3| = 3$, while $|3| = 3$, and $|0| = 0$. It follows that

$$\sqrt{x^2} = |x|$$

for every real number x, positive or negative. For instance,

$$\sqrt{(-3)^2} = \sqrt{9} = 3 = |-3|$$

and $\sqrt{3^2} = \sqrt{9} = 3 = |3|$

In general, we find that

$$\sqrt[n]{x^n} = x \text{ if } n \text{ is odd, and } \sqrt[n]{x^n} = |x| \text{ if } n \text{ is even} \qquad \blacksquare$$

We use the following identities to evaluate radicals of products and quotients.

Radicals of Products and Quotients

If a and b are any real numbers (nonnegative in the case of even-numbered roots), then

$$\sqrt[n]{ab} = \sqrt[n]{a}\,\sqrt[n]{b} \qquad \text{Radical of a product} = \text{Product of radicals}$$

$$\sqrt[n]{\frac{a}{b}} = \frac{\sqrt[n]{a}}{\sqrt[n]{b}} \quad \text{if } b \neq 0 \qquad \text{Radical of a quotient} = \text{Quotient of radicals}$$

Notes

- The first rule is similar to the rule $(a \cdot b)^2 = a^2 b^2$ for the square of a product, and the second rule is similar to the rule $\left(\dfrac{a}{b}\right)^2 = \dfrac{a^2}{b^2}$ for the square of a quotient.

- **Caution** There is no corresponding identity for addition:

$$\sqrt{a+b} \text{ is } not \text{ equal to } \sqrt{a} + \sqrt{b}$$

(consider $a = b = 1$, for example). Equating these expressions is a common error, so be careful!

quick Examples

1. $\sqrt{9 \cdot 4} = \sqrt{9}\sqrt{4} = 3 \times 2 = 6$ Alternatively, $\sqrt{9 \cdot 4} = \sqrt{36} = 6$

2. $\sqrt{\dfrac{9}{4}} = \dfrac{\sqrt{9}}{\sqrt{4}} = \dfrac{3}{2}$

3. $\sqrt{4(3+13)} = \sqrt{4(16)} = \sqrt{4}\sqrt{16} = 2 \times 4 = 8$

4. $\sqrt[3]{-216} = \sqrt[3]{(-27)8} = \sqrt[3]{-27}\sqrt[3]{8} = (-3)2 = -6$

5. $\sqrt{x^3} = \sqrt{x^2 \cdot x} = \sqrt{x^2}\sqrt{x} = x\sqrt{x}$ if $x \geq 0$

6. $\sqrt{\dfrac{x^2+y^2}{z^2}} = \dfrac{\sqrt{x^2+y^2}}{\sqrt{z^2}} = \dfrac{\sqrt{x^2+y^2}}{|z|}$ We can't simplify the numerator any further.

Rational Exponents

We already know what we mean by expressions such as x^4 and a^{-6}. The next step is to make sense of *rational* exponents: exponents of the form p/q with p and q integers as in $a^{1/2}$ and $3^{-2/3}$.

Q: *What should we mean by $a^{1/2}$* ?

A: The overriding concern here is that all the exponent identities should remain true. In this case the identity to look at is the one that says that $(a^m)^n = a^{mn}$. This identity tells us that

$$(a^{1/2})^2 = a^1 = a$$

That is, $a^{1/2}$, when squared, gives us a. But that must mean that $a^{1/2}$ is the *square root* of a, or

$$a^{1/2} = \sqrt{a}$$

A similar argument tells us that, if q is any positive whole number, then

$$a^{1/q} = \sqrt[q]{a}   \text{ the } q\text{th root of } a$$

Notice that if a is negative, this makes sense only for q odd. To avoid this problem, we usually stick to positive a. ∎

Q: *If p and q are integers (q positive), what should we mean by $a^{p/q}$* ?

A: By the exponent identities, $a^{p/q}$ should equal both $(a^p)^{1/q}$ and $(a^{1/q})^p$. The first is the qth root of a^p, and the second is the pth power of $a^{1/q}$, which gives us the following.

Conversion Between Rational Exponents and Radicals

If a is any nonnegative number, then

$$a^{p/q} = \sqrt[q]{a^p} = \left(\sqrt[q]{a}\right)^p$$

↑   ↑   ↑

Exponential form   Radical form

In particular,

$$a^{1/q} = \sqrt[q]{a} \quad \text{the } q\text{th root of } a$$

Notes

- If a is negative, all of this makes sense only if q is odd.
- All of the exponent identities continue to work when we allow rational exponents p/q. In other words, we are free to use all the exponent identities even though the exponents are not integers.

quick Examples

1. $4^{3/2} = (\sqrt{4})^3 = 2^3 = 8$ **2.** $8^{2/3} = (\sqrt[3]{8})^2 = 2^2 = 4$

3. $9^{-3/2} = \dfrac{1}{9^{3/2}} = \dfrac{1}{(\sqrt{9})^3} = \dfrac{1}{3^3} = \dfrac{1}{27}$ **4.** $\dfrac{\sqrt{3}}{\sqrt[3]{3}} = \dfrac{3^{1/2}}{3^{1/3}} = 3^{1/2-1/3} = 3^{1/6} = \sqrt[6]{3}$

5. $2^2 2^{7/2} = 2^2 2^{3+1/2} = 2^2 2^3 2^{1/2} = 2^5 2^{1/2} = 2^5 \sqrt{2}$

Example **4** **Simplifying Algebraic Expressions**

Simplify the following.

a. $\dfrac{(x^3)^{5/3}}{x^3}$ **b.** $\sqrt[4]{a^6}$ **c.** $\dfrac{(xy)^{-3}y^{-3/2}}{x^{-2}\sqrt{y}}$

Solution

a. $\dfrac{(x^3)^{5/3}}{x^3} = \dfrac{x^5}{x^3} = x^2$

b. $\sqrt[4]{a^6} = a^{6/4} = a^{3/2} = a \cdot a^{1/2} = a\sqrt{a}$

c. $\dfrac{(xy)^{-3}y^{-3/2}}{x^{-2}\sqrt{y}} = \dfrac{x^{-3}y^{-3}y^{-3/2}}{x^{-2}y^{1/2}} = \dfrac{1}{x^{-2+3}y^{1/2+3+3/2}} = \dfrac{1}{xy^5}$

Converting Between Rational, Radical, and Exponential Form

In calculus we must often convert algebraic expressions involving powers of x, such as $\dfrac{3}{2x^2}$, into expressions in which x does not appear in the denominator, such as $\dfrac{3}{2}x^{-2}$.

Also, we must often convert expressions with radicals, such as $\dfrac{1}{\sqrt{1+x^2}}$, into expressions with no radicals and all powers in the numerator, such as $(1+x^2)^{-1/2}$. In these cases, we are converting from **rational form** or **radical form** to **exponential form**.

Rational Form

An expression is in **rational form** if it is written with positive exponents only.

quick Examples

1. $\dfrac{2}{3x^2}$ is in rational form.

2. $\dfrac{2x^{-1}}{3}$ is not in rational form because the exponent of x is negative.

3. $\dfrac{x}{6} + \dfrac{6}{x}$ is in rational form.

Radical Form

An expression is in **radical form** if it is written with integer powers and roots only.

quick Examples

1. $\dfrac{2}{5\sqrt[3]{x}} + \dfrac{2}{x}$ is in radical form.

2. $\dfrac{2x^{-1/3}}{5} + 2x^{-1}$ is not in radical form because $x^{-1/3}$ appears.

3. $\dfrac{1}{\sqrt{1+x^2}}$ is in radical form, but $(1+x^2)^{-1/2}$ is not.

Exponential Form

An expression is in **exponential form** if there are no radicals and all powers of unknowns occur in the numerator. We usually write such expressions as sums or differences of terms of the form

$$\text{Constant} \times (\text{Expression with } x)^p \qquad \text{As in } \tfrac{1}{3}x^{-3/2}$$

quick Examples

1. $\dfrac{2}{3}x^4 - 3x^{-1/3}$ is in exponential form.

2. $\dfrac{x}{6} + \dfrac{6}{x}$ is not in exponential form because the second expression has x in the denominator.

3. $\sqrt[3]{x}$ is not in exponential form because it has a radical.

4. $(1+x^2)^{-1/2}$ is in exponential form, but $\dfrac{1}{\sqrt{1+x^2}}$ is not.

Example 5   Converting From One Form to Another

Convert the following to rational form:

a. $\dfrac{1}{2}x^{-2} + \dfrac{4}{3}x^{-5}$ 　　　　**b.** $\dfrac{2}{\sqrt{x}} - \dfrac{2}{x^{-4}}$

Convert the following to radical form:

c. $\dfrac{1}{2}x^{-1/2} + \dfrac{4}{3}x^{-5/4}$ 　　　　**d.** $\dfrac{(3+x)^{-1/3}}{5}$

Convert the following to exponential form:

e. $\dfrac{3}{4x^2} - \dfrac{x}{6} + \dfrac{6}{x}$ **f.** $\dfrac{2}{(x+1)^2} - \dfrac{3}{4\sqrt[5]{2x-1}}$

Solution

For (a) and (b), we eliminate negative exponents as we did in Example 2:

a. $\dfrac{1}{2}x^{-2} + \dfrac{4}{3}x^{-5} = \dfrac{1}{2} \cdot \dfrac{1}{x^2} + \dfrac{4}{3} \cdot \dfrac{1}{x^5} = \dfrac{1}{2x^2} + \dfrac{4}{3x^5}$

b. $\dfrac{2}{\sqrt{x}} - \dfrac{2}{x^{-4}} = \dfrac{2}{\sqrt{x}} - 2x^4$

For (c) and (d), we rewrite all terms with fractional exponents as radicals:

c. $\dfrac{1}{2}x^{-1/2} + \dfrac{4}{3}x^{-5/4} = \dfrac{1}{2} \cdot \dfrac{1}{x^{1/2}} + \dfrac{4}{3} \cdot \dfrac{1}{x^{5/4}}$

$$= \dfrac{1}{2} \cdot \dfrac{1}{\sqrt{x}} + \dfrac{4}{3} \cdot \dfrac{1}{\sqrt[4]{x^5}} = \dfrac{1}{2\sqrt{x}} + \dfrac{4}{3\sqrt[4]{x^5}}$$

d. $\dfrac{(3+x)^{-1/3}}{5} = \dfrac{1}{5(3+x)^{1/3}} = \dfrac{1}{5\sqrt[3]{3+x}}$

For (e) and (f), we eliminate any radicals and move all expressions involving x to the numerator:

e. $\dfrac{3}{4x^2} - \dfrac{x}{6} + \dfrac{6}{x} = \dfrac{3}{4}x^{-2} - \dfrac{1}{6}x + 6x^{-1}$

f. $\dfrac{2}{(x+1)^2} - \dfrac{3}{4\sqrt[5]{2x-1}} = 2(x+1)^{-2} - \dfrac{3}{4(2x-1)^{1/5}}$

$$= 2(x+1)^{-2} - \dfrac{3}{4}(2x-1)^{-1/5}$$

Solving Equations with Exponents

Example 6 Solving Equations

Solve the following equations:

a. $x^3 + 8 = 0$ **b.** $x^2 - \dfrac{1}{2} = 0$ **c.** $x^{3/2} - 64 = 0$

Solution

a. Subtracting 8 from both sides gives $x^3 = -8$. Taking the cube root of both sides gives $x = -2$.

b. Adding $\frac{1}{2}$ to both sides gives $x^2 = \frac{1}{2}$. Thus, $x = \pm\sqrt{\frac{1}{2}} = \pm\frac{1}{\sqrt{2}}$.

c. Adding 64 to both sides gives $x^{3/2} = 64$. Taking the reciprocal (2/3) power of both sides gives

$$(x^{3/2})^{2/3} = 64^{2/3}$$

$$x^1 = \left(\sqrt[3]{64}\right)^2 = 4^2 = 16$$

so $x = 16$

0.2 EXERCISES

Evaluate the expressions in Exercises 1–16.

1. 3^3 **2.** $(-2)^3$ **3.** $-(2\cdot 3)^2$ **4.** $(4\cdot 2)^2$

5. $\left(\dfrac{-2}{3}\right)^2$ **6.** $\left(\dfrac{3}{2}\right)^3$ **7.** $(-2)^{-3}$ **8.** -2^{-3}

9. $\left(\dfrac{1}{4}\right)^{-2}$ **10.** $\left(\dfrac{-2}{3}\right)^{-2}$ **11.** $2\cdot 3^0$ **12.** $3\cdot(-2)^0$

13. $2^3\,2^2$ **14.** $3^2 3$ **15.** $2^2 2^{-1} 2^4 2^{-4}$ **16.** $5^2 5^{-3} 5^2 5^{-2}$

Simplify each expression in Exercises 17–30, expressing your answer in rational form.

17. $x^3 x^2$ **18.** $x^4 x^{-1}$ **19.** $-x^2 x^{-3} y$

20. $-xy^{-1}x^{-1}$ **21.** $\dfrac{x^3}{x^4}$ **22.** $\dfrac{y^5}{y^3}$

23. $\dfrac{x^2 y^2}{x^{-1}y}$ **24.** $\dfrac{x^{-1}y}{x^2 y^2}$ **25.** $\dfrac{(xy^{-1}z^3)^2}{x^2 yz^2}$

26. $\dfrac{x^2 yz^2}{(xyz^{-1})^{-1}}$ **27.** $\left(\dfrac{xy^{-2}z}{x^{-1}z}\right)^3$ **28.** $\left(\dfrac{x^2 y^{-1}z^0}{xyz}\right)^2$

29. $\left(\dfrac{x^{-1}y^{-2}z^2}{xy}\right)^{-2}$ **30.** $\left(\dfrac{xy^{-2}}{x^2 y^{-1}z}\right)^{-3}$

Convert the expressions in Exercises 31–36 to rational form.

31. $3x^{-4}$ **32.** $\dfrac{1}{2}x^{-4}$ **33.** $\dfrac{3}{4}x^{-2/3}$

34. $\dfrac{4}{5}y^{-3/4}$ **35.** $1 - \dfrac{0.3}{x^{-2}} - \dfrac{6}{5}x^{-1}$

36. $\dfrac{1}{3x^{-4}} + \dfrac{0.1x^{-2}}{3}$

Evaluate the expressions in Exercises 37–56, rounding your answer to four significant digits where necessary.

37. $\sqrt{4}$ **38.** $\sqrt{5}$ **39.** $\sqrt{\dfrac{1}{4}}$

40. $\sqrt{\dfrac{1}{9}}$ **41.** $\sqrt{\dfrac{16}{9}}$ **42.** $\sqrt{\dfrac{9}{4}}$

43. $\dfrac{\sqrt{4}}{5}$ **44.** $\dfrac{6}{\sqrt{25}}$ **45.** $\sqrt{9}+\sqrt{16}$

46. $\sqrt{25}-\sqrt{16}$ **47.** $\sqrt{9+16}$ **48.** $\sqrt{25-16}$

49. $\sqrt[3]{8-27}$ **50.** $\sqrt[4]{81-16}$ **51.** $\sqrt[3]{27/8}$

52. $\sqrt[3]{8\times 64}$ **53.** $\sqrt{(-2)^2}$ **54.** $\sqrt{(-1)^2}$

55. $\sqrt{\dfrac{1}{4}(1+15)}$ **56.** $\sqrt{\dfrac{1}{9}(3+33)}$

Simplify the expressions in Exercises 57–64, given that x, y, a, b, and c are positive real numbers.

57. $\sqrt{a^2 b^2}$ **58.** $\sqrt{\dfrac{a^2}{b^2}}$ **59.** $\sqrt{(x+9)^2}$

60. $(\sqrt{x+9})^2$ **61.** $\sqrt[3]{x^3(a^3+b^3)}$ **62.** $\sqrt[4]{\dfrac{x^4}{a^4 b^4}}$

63. $\sqrt{\dfrac{4xy^3}{x^2 y}}$ **64.** $\sqrt{\dfrac{4(x^2+y^2)}{c^2}}$

Convert the expressions in Exercises 65–80 to exponential form.

65. $\sqrt{3}$ **66.** $\sqrt{8}$ **67.** $\sqrt{x^3}$

68. $\sqrt[3]{x^2}$ **69.** $\sqrt[3]{xy^2}$ **70.** $\sqrt{x^2 y}$

71. $\dfrac{x^2}{\sqrt{x}}$ **72.** $\dfrac{x}{\sqrt{x}}$ **73.** $\dfrac{3}{5x^2}$

74. $\dfrac{2}{5x^{-3}}$ **75.** $\dfrac{3x^{-1.2}}{2} - \dfrac{1}{3x^{2.1}}$

76. $\dfrac{2}{3x^{-1.2}} - \dfrac{x^{2.1}}{3}$ **77.** $\dfrac{2x}{3} - \dfrac{x^{0.1}}{2} + \dfrac{4}{3x^{1.1}}$

78. $\dfrac{4x^2}{3} + \dfrac{x^{3/2}}{6} - \dfrac{2}{3x^2}$ **79.** $\dfrac{1}{(x^2+1)^3} - \dfrac{3}{4\sqrt[3]{(x^2+1)}}$

80. $\dfrac{2}{3(x^2+1)^{-3}} - \dfrac{3\sqrt[3]{(x^2+1)^7}}{4}$

Convert the expressions in Exercises 81–92 to radical form.

81. $2^{2/3}$ **82.** $3^{4/5}$ **83.** $x^{4/3}$

84. $y^{7/4}$ **85.** $(x^{1/2}y^{1/3})^{1/5}$

86. $x^{-1/3}y^{3/2}$ **87.** $-\dfrac{3}{2}x^{-1/4}$

88. $\dfrac{4}{5}x^{3/2}$ **89.** $0.2x^{-2/3} + \dfrac{3}{7x^{-1/2}}$

90. $\dfrac{3.1}{x^{-4/3}} - \dfrac{11}{7}x^{-1/7}$ **91.** $\dfrac{3}{4(1-x)^{5/2}}$ **92.** $\dfrac{9}{4(1-x)^{-7/3}}$

Simplify the expressions in Exercises 93–102.

93. $4^{-1/2}4^{7/2}$ **94.** $2^{1/a}/2^{2/a}$ **95.** $3^{2/3}3^{-1/6}$

96. $2^{1/3}2^{-1}2^{2/3}2^{-1/3}$ **97.** $\dfrac{x^{3/2}}{x^{5/2}}$ **98.** $\dfrac{y^{5/4}}{y^{3/4}}$

99. $\dfrac{x^{1/2}y^2}{x^{-1/2}y}$ **100.** $\dfrac{x^{-1/2}y}{x^2 y^{3/2}}$

101. $\left(\dfrac{x}{y}\right)^{1/3}\left(\dfrac{y}{x}\right)^{2/3}$ **102.** $\left(\dfrac{x}{y}\right)^{-1/3}\left(\dfrac{y}{x}\right)^{1/3}$

Solve each equation in Exercises 103–116 for x, rounding your answer to four significant digits where necessary.

103. $x^2 - 16 = 0$ **104.** $x^2 - 1 = 0$ **105.** $x^2 - \dfrac{4}{9} = 0$

106. $x^2 - \dfrac{1}{10} = 0$ **107.** $x^2 - (1 + 2x)^2 = 0$

108. $x^2 - (2 - 3x)^2 = 0$ **109.** $x^5 + 32 = 0$

110. $x^4 - 81 = 0$ **111.** $x^{1/2} - 4 = 0$

112. $x^{1/3} - 2 = 0$ **113.** $1 - \dfrac{1}{x^2} = 0$ **114.** $\dfrac{2}{x^3} - \dfrac{6}{x^4} = 0$

115. $(x - 4)^{-1/3} = 2$ **116.** $(x - 4)^{2/3} + 1 = 5$

0.3 Multiplying and Factoring Algebraic Expressions

Multiplying Algebraic Expressions

Distributive Law

The **distributive law** for real numbers states that

$$a(b \pm c) = ab \pm ac$$
$$(a \pm b)c = ac \pm bc$$

for any real numbers a, b, and c.

quick Examples

1. $2(x - 3)$ is *not* equal to $2x - 3$ but is equal to $2x - 2(3) = 2x - 6$.
2. $x(x + 1) = x^2 + x$
3. $2x(3x - 4) = 6x^2 - 8x$
4. $(x - 4)x^2 = x^3 - 4x^2$
5. $(x + 2)(x + 3) = (x + 2)x + (x + 2)3 = (x^2 + 2x) + (3x + 6) = x^2 + 5x + 6$
6. $(x + 2)(x - 3) = (x + 2)x - (x + 2)3 = (x^2 + 2x) - (3x + 6) = x^2 - x - 6$

There is a quicker way of expanding expressions like the last two, called the "FOIL" method (First, Outer, Inner, Last). Consider, for instance, the expression $(x + 1)(x - 2)$. The FOIL method says: Take the product of the first terms: $x \cdot x = x^2$, the product of the outer terms: $x \cdot (-2) = -2x$, the product of the inner terms: $1 \cdot x = x$, and the product of the last terms: $1 \cdot (-2) = -2$, and then add them all up, getting $x^2 - 2x + x - 2 = x^2 - x - 2$.

Example 1   FOIL

a. $(x - 2)(2x + 5) = 2x^2 + 5x - 4x - 10 = 2x^2 + x - 10$

First   Outer   Inner   Last

b. $(x^2 + 1)(x - 4) = x^3 - 4x^2 + x - 4$

c. $(a - b)(a + b) = a^2 + ab - ab - b^2 = a^2 - b^2$

d $(a + b)^2 = (a + b)(a + b) = a^2 + ab + ab + b^2 = a^2 + 2ab + b^2$

e. $(a - b)^2 = (a - b)(a - b) = a^2 - ab - ab + b^2 = a^2 - 2ab + b^2$

The last three are particularly important and are worth memorizing.

Special Formulas

$$(a - b)(a + b) = a^2 - b^2 \qquad \text{Difference of two squares}$$
$$(a + b)^2 = a^2 + 2ab + b^2 \qquad \text{Square of a sum}$$
$$(a - b)^2 = a^2 - 2ab + b^2 \qquad \text{Square of a difference}$$

quick Examples

1. $(2 - x)(2 + x) = 4 - x^2$
2. $(1 + a)(1 - a) = 1 - a^2$
3. $(x + 3)^2 = x^2 + 6x + 9$
4. $(4 - x)^2 = 16 - 8x + x^2$

Here are some longer examples that require the distributive law.

Example 2 Multiplying Algebraic Expressions

a. $(x + 1)(x^2 + 3x - 4) = (x + 1)x^2 + (x + 1)3x - (x + 1)4$
$$= (x^3 + x^2) + (3x^2 + 3x) - (4x + 4)$$
$$= x^3 + 4x^2 - x - 4$$

b. $\left(x^2 - \dfrac{1}{x} + 1\right)(2x + 5) = \left(x^2 - \dfrac{1}{x} + 1\right)2x + \left(x^2 - \dfrac{1}{x} + 1\right)5$
$$= (2x^3 - 2 + 2x) + \left(5x^2 - \dfrac{5}{x} + 5\right)$$
$$= 2x^3 + 5x^2 + 2x + 3 - \dfrac{5}{x}$$

c. $(x - y)(x - y)(x - y) = (x^2 - 2xy + y^2)(x - y)$
$$= (x^2 - 2xy + y^2)x - (x^2 - 2xy + y^2)y$$
$$= (x^3 - 2x^2y + xy^2) - (x^2y - 2xy^2 + y^3)$$
$$= x^3 - 3x^2y + 3xy^2 - y^3$$

Factoring Algebraic Expressions

We can think of factoring as applying the distributive law in reverse—for example,

$$2x^2 + x = x(2x + 1)$$

which can be checked by using the distributive law. Factoring is an art that you will learn with experience and the help of a few useful techniques.

Factoring Using a Common Factor

To use this technique, locate a **common factor**—a term that occurs as a factor in each of the expressions being added or subtracted (for example, x is a common factor in $2x^2 + x$, because it is a factor of both $2x^2$ and x). Once you have located a common factor, "factor it out" by applying the distributive law.

quick Examples

1. $2x^3 - x^2 + x$ has x as a common factor, so
$$2x^3 - x^2 + x = x(2x^2 - x + 1)$$

2. $2x^2 + 4x$ has $2x$ as a common factor, so
$$2x^2 + 4x = 2x(x + 2)$$

3. $2x^2y + xy^2 - x^2y^2$ has xy as a common factor, so
$$2x^2y + xy^2 - x^2y^2 = xy(2x + y - xy)$$

4. $(x^2 + 1)(x + 2) - (x^2 + 1)(x + 3)$ has $x^2 + 1$ as a common factor, so
$$(x^2 + 1)(x + 2) - (x^2 + 1)(x + 3) = (x^2 + 1)[(x + 2) - (x + 3)]$$
$$= (x^2 + 1)(x + 2 - x - 3)$$
$$= (x^2 + 1)(-1) = -(x^2 + 1)$$

5. $12x(x^2 - 1)^5(x^3 + 1)^6 + 18x^2(x^2 - 1)^6(x^3 + 1)^5$ has $6x(x^2 - 1)^5(x^3 + 1)^5$ as a common factor, so
$$12x(x^2 - 1)^5(x^3 + 1)^6 + 18x^2(x^2 - 1)^6(x^3 + 1)^5$$
$$= 6x(x^2 - 1)^5(x^3 + 1)^5[2(x^3 + 1) + 3x(x^2 - 1)]$$
$$= 6x(x^2 - 1)^5(x^3 + 1)^5(2x^3 + 2 + 3x^3 - 3x)$$
$$= 6x(x^2 - 1)^5(x^3 + 1)^5(5x^3 - 3x + 2)$$

We would also like to be able to reverse calculations such as $(x + 2)(2x - 5) = 2x^2 - x - 10$. That is, starting with the expression $2x^2 - x - 10$, we would like to **factor** it to get the expression $(x + 2)(2x - 5)$. An expression of the form $ax^2 + bx + c$, where a, b, and c are real numbers, is called a **quadratic** expression in x. Thus, given a quadratic expression $ax^2 + bx + c$, we would like to write it in the form $(dx + e)(fx + g)$ for some real numbers d, e, f, and g. There are some quadratics, such as $x^2 + x + 1$, that cannot be factored in this form at all. Here, we consider only quadratics that do factor, and in such a way that the numbers d, e, f and g are integers (whole numbers; other cases are discussed in Section 5). The usual technique of factoring such quadratics is a "trial and error" approach.

Factoring Quadratics by Trial and Error

To factor the quadratic $ax^2 + bx + c$, factor ax^2 as $(a_1x)(a_2x)$ (with a_1 positive) and c as c_1c_2, and then check whether $ax^2 + bx + c = (a_1x \pm c_1)(a_2x \pm c_2)$. If not, try other factorizations of ax^2 and c.

quick Examples

1. To factor $x^2 - 6x + 5$, first factor x^2 as $(x)(x)$, and 5 as $(5)(1)$:

$(x + 5)(x + 1) = x^2 + 6x + 5$ No good

$(x - 5)(x - 1) = x^2 - 6x + 5$ Desired factorization

2. To factor $x^2 - 4x - 12$, first factor x^2 as $(x)(x)$, and -12 as $(1)(-12)$, $(2)(-6)$, or $(3)(-4)$. Trying them one-by-one gives

$(x + 1)(x - 12) = x^2 - 11x - 12$ No good

$(x - 1)(x + 12) = x^2 + 11x - 12$ No good

$(x + 2)(x - 6) = x^2 - 4x - 12$ Desired factorization

3. To factor $4x^2 - 25$, we can follow the above procedure, or recognize $4x^2 - 25$ as the difference of two squares:

$$4x^2 - 25 = (2x)^2 - 5^2 = (2x - 5)(2x + 5)$$

Note: Not all quadratic expressions factor. In Section 5 we look at a test that tells us whether or not a given quadratic factors.

Here are examples requiring either a little more work or a little more thought.

Example 3 Factoring Quadratics

Factor the following: **a.** $4x^2 - 5x - 6$ **b.** $x^4 - 5x^2 + 6$

Solution

a. Possible factorizations of $4x^2$ are $(2x)(2x)$ or $(x)(4x)$. Possible factorizations of -6 are $(1)(-6)$, $(2)(-3)$. We now systematically try out all the possibilities until we come up with the correct one.

$(2x)(2x)$ and $(1)(-6)$:	$(2x + 1)(2x - 6) = 4x^2 - 10x - 6$	No good
$(2x)(2x)$ and $(2)(-3)$:	$(2x + 2)(2x - 3) = 4x^2 - 2x - 6$	No good
$(x)(4x)$ and $(1)(-6)$:	$(x + 1)(4x - 6) = 4x^2 - 2x - 6$	No good
$(x)(4x)$ and $(2)(-3)$:	$(x + 2)(4x - 3) = 4x^2 + 5x - 6$	Almost!
Change signs:	$(x - 2)(4x + 3) = 4x^2 - 5x - 6$	Correct

b. The expression $x^4 - 5x^2 + 6$ is not a quadratic, you say? Correct, it's a quartic (a fourth degree expression). However, it looks rather like a quadratic. In fact, it is quadratic *in* x^2, meaning that it is

$$(x^2)^2 - 5(x^2) + 6 = y^2 - 5y + 6$$

where $y = x^2$. The quadratic $y^2 - 5y + 6$ factors as

$$y^2 - 5y + 6 = (y - 3)(y - 2)$$

so

$$x^4 - 5x^2 + 6 = (x^2 - 3)(x^2 - 2)$$

This is a sometimes useful technique.

Our last example is here to remind you why we should want to factor polynomials in the first place. We shall return to this in Section 5.

Example 4 Solving a Quadratic Equation by Factoring

Solve the equation $3x^2 + 4x - 4 = 0$.

Solution We first factor the left-hand side to get

$$(3x - 2)(x + 2) = 0$$

Thus, the product of the two quantities $(3x - 2)$ and $(x + 2)$ is zero. Now, if a product of two numbers is zero, one of the two must be zero. In other words, either $3x - 2 = 0$, giving $x = \frac{2}{3}$, or $x + 2 = 0$, giving $x = -2$. Thus, there are two solutions: $x = \frac{2}{3}$ and $x = -2$.

0.3 EXERCISES

Expand each expression in Exercises 1–22.

1. $x(4x + 6)$ **2.** $(4y - 2)y$

3. $(2x - y)y$ **4.** $x(3x + y)$

5. $(x + 1)(x - 3)$ **6.** $(y + 3)(y + 4)$

7. $(2y + 3)(y + 5)$ **8.** $(2x - 2)(3x - 4)$

9. $(2x - 3)^2$ **10.** $(3x + 1)^2$

11. $\left(x + \dfrac{1}{x}\right)^2$ **12.** $\left(y - \dfrac{1}{y}\right)^2$

13. $(2x - 3)(2x + 3)$

14. $(4 + 2x)(4 - 2x)$

15. $\left(y - \dfrac{1}{y}\right)\left(y + \dfrac{1}{y}\right)$

16. $(x - x^2)(x + x^2)$

17. $(x^2 + x - 1)(2x + 4)$

18. $(3x + 1)(2x^2 - x + 1)$

19. $(x^2 - 2x + 1)^2$

20. $(x + y - xy)^2$

21. $(y^3 + 2y^2 + y)(y^2 + 2y - 1)$

22. $(x^3 - 2x^2 + 4)(3x^2 - x + 2)$

In Exercises 23–30, factor each expression and simplify as much as possible.

23. $(x + 1)(x + 2) + (x + 1)(x + 3)$

24. $(x + 1)(x + 2)^2 + (x + 1)^2(x + 2)$

25. $(x^2 + 1)^5(x + 3)^4 + (x^2 + 1)^6(x + 3)^3$

26. $10x(x^2 + 1)^4(x^3 + 1)^5 + 15x^2(x^2 + 1)^5(x^3 + 1)^4$

27. $(x^3 + 1)\sqrt{x + 1} - (x^3 + 1)^2\sqrt{x + 1}$

28. $(x^2 + 1)\sqrt{x + 1} - \sqrt{(x + 1)^3}$

29. $\sqrt{(x + 1)^3} + \sqrt{(x + 1)^5}$

30. $(x^2 + 1)\sqrt[3]{(x + 1)^4} - \sqrt[3]{(x + 1)^7}$

*In Exercises 31–48, **(a)** factor the given expression; **(b)** set the expression equal to zero and solve for the unknown (x in the odd-numbered exercises and y in the even-numbered exercises).*

31. $2x + 3x^2$ **32.** $y^2 - 4y$

33. $6x^3 - 2x^2$ **34.** $3y^3 - 9y^2$

35. $x^2 - 8x + 7$ **36.** $y^2 + 6y + 8$

37. $x^2 + x - 12$ **38.** $y^2 + y - 6$

39. $2x^2 - 3x - 2$ **40.** $3y^2 - 8y - 3$

41. $6x^2 + 13x + 6$ **42.** $6y^2 + 17y + 12$

43. $12x^2 + x - 6$ **44.** $20y^2 + 7y - 3$

45. $x^2 + 4xy + 4y^2$ **46.** $4y^2 - 4xy + x^2$

47. $x^4 - 5x^2 + 4$ **48.** $y^4 + 2y^2 - 3$

0.4 Rational Expressions

Rational Expression

A **rational expression** is an algebraic expression of the form $\dfrac{P}{Q}$, where P and Q are simpler expressions (usually polynomials) and the denominator Q is not zero.

quick Examples

1. $\dfrac{x^2 - 3x}{x}$ $P = x^2 - 3x,\ Q = x$

2. $\dfrac{x + \frac{1}{x} + 1}{2x^2y + 1}$ $P = x + \dfrac{1}{x} + 1,\ Q = 2x^2y + 1$

3. $3xy - x^2$ $P = 3xy - x^2,\ Q = 1$

Algebra of Rational Expressions

We manipulate rational expressions in the same way that we manipulate fractions, using the following rules:

	Algebraic Rule	*quick* Example
Product:	$\dfrac{P}{Q} \cdot \dfrac{R}{S} = \dfrac{PR}{QS}$	$\dfrac{x+1}{x} \cdot \dfrac{x-1}{2x+1} = \dfrac{(x+1)(x-1)}{x(2x+1)} = \dfrac{x^2-1}{2x^2+x}$
Sum:	$\dfrac{P}{Q} + \dfrac{R}{S} = \dfrac{PS+RQ}{QS}$	$\dfrac{2x-1}{3x+2} + \dfrac{1}{x} = \dfrac{(2x-1)x + 1(3x+2)}{x(3x+2)}$ $= \dfrac{2x^2+2x+2}{3x^2+2x}$
Difference:	$\dfrac{P}{Q} - \dfrac{R}{S} = \dfrac{PS-RQ}{QS}$	$\dfrac{x}{3x+2} - \dfrac{x-4}{x} = \dfrac{x^2-(x-4)(3x+2)}{x(3x+2)}$ $= \dfrac{-2x^2+10x+8}{3x^2+2x}$
Reciprocal:	$\dfrac{1}{\left(\dfrac{P}{Q}\right)} = \dfrac{Q}{P}$	$\dfrac{1}{\left(\dfrac{2xy}{3x-1}\right)} = \dfrac{3x-1}{2xy}$
Quotient:	$\dfrac{\left(\dfrac{P}{Q}\right)}{\left(\dfrac{R}{S}\right)} = \dfrac{P}{Q} \cdot \dfrac{S}{R} = \dfrac{PS}{QR}$	$\dfrac{\left(\dfrac{x}{x-1}\right)}{\left(\dfrac{y-1}{y}\right)} = \dfrac{xy}{(x-1)(y-1)} = \dfrac{xy}{xy-x-y+1}$
Cancellation:	$\dfrac{P\cancel{R}}{Q\cancel{R}} = \dfrac{P}{Q}$	$\dfrac{(x-1)(xy+4)}{(x^2y-8)(x-1)} = \dfrac{xy+4}{x^2y-8}$

Caution Cancellation of summands is *invalid*. For instance,

$$\dfrac{\cancel{x} + (2xy^2 - y)}{\cancel{x} + 4y} = \dfrac{(2xy^2 - y)}{4y} \qquad \text{✗ } \textbf{\textit{WRONG!}} \qquad \text{Do } not \text{ cancel a summand.}$$

$$\dfrac{\cancel{x}(2xy^2 - y)}{4\cancel{x}y} = \dfrac{(2xy^2 - y)}{4y} \qquad \text{✔ } \textbf{\textit{CORRECT}} \qquad \text{Do cancel a factor.}$$

Here are some examples that require several algebraic operations.

Example 1 Simplifying Rational Expressions

a. $\dfrac{\left(\dfrac{1}{x+y} - \dfrac{1}{x}\right)}{y} = \dfrac{\left(\dfrac{x - (x+y)}{x(x+y)}\right)}{y} = \dfrac{\left(\dfrac{-y}{x(x+y)}\right)}{y} = \dfrac{-y}{xy(x+y)} = -\dfrac{1}{x(x+y)}$

b. $\dfrac{(x+1)(x+2)^2 - (x+1)^2(x+2)}{(x+2)^4} = \dfrac{(x+1)(x+2)[(x+2) - (x+1)]}{(x+2)^4}$

$= \dfrac{(x+1)(x+2)(x+2-x-1)}{(x+2)^4} = \dfrac{(x+1)(x+2)}{(x+2)^4} = \dfrac{x+1}{(x+2)^3}$

c. $\dfrac{2x\sqrt{x+1} - \dfrac{x^2}{\sqrt{x+1}}}{x+1} = \dfrac{\left(\dfrac{2x\left(\sqrt{x+1}\right)^2 - x^2}{\sqrt{x+1}}\right)}{x+1} = \dfrac{2x(x+1) - x^2}{(x+1)\sqrt{x+1}}$

$$= \dfrac{2x^2 + 2x - x^2}{(x+1)\sqrt{x+1}} = \dfrac{x^2 + 2x}{\sqrt{(x+1)^3}} = \dfrac{x(x+2)}{\sqrt{(x+1)^3}}$$

0.4 EXERCISES

Rewrite each expression in Exercises 1–16 as a single rational expression, simplified as much as possible.

1. $\dfrac{x-4}{x+1} \cdot \dfrac{2x+1}{x-1}$

2. $\dfrac{2x-3}{x-2} \cdot \dfrac{x+3}{x+1}$

3. $\dfrac{x-4}{x+1} + \dfrac{2x+1}{x-1}$

4. $\dfrac{2x-3}{x-2} + \dfrac{x+3}{x+1}$

5. $\dfrac{x^2}{x+1} - \dfrac{x-1}{x+1}$

6. $\dfrac{x^2-1}{x-2} - \dfrac{1}{x-1}$

7. $\dfrac{1}{\left(\frac{x}{x-1}\right)} + x - 1$

8. $\dfrac{2}{\left(\frac{x-2}{x^2}\right)} - \dfrac{1}{x-2}$

9. $\dfrac{1}{x}\left[\dfrac{x-3}{xy} + \dfrac{1}{y}\right]$

10. $\dfrac{y^2}{x}\left[\dfrac{2x-3}{y} + \dfrac{x}{y}\right]$

11. $\dfrac{(x+1)^2(x+2)^3 - (x+1)^3(x+2)^2}{(x+2)^6}$

12. $\dfrac{6x(x^2+1)^2(x^3+2)^3 - 9x^2(x^2+1)^3(x^3+2)^2}{(x^3+2)^6}$

13. $\dfrac{(x^2-1)\sqrt{x^2+1} - \dfrac{x^4}{\sqrt{x^2+1}}}{x^2+1}$

14. $\dfrac{x\sqrt{x^3-1} - \dfrac{3x^4}{\sqrt{x^3-1}}}{x^3-1}$

15. $\dfrac{\dfrac{1}{(x+y)^2} - \dfrac{1}{x^2}}{y}$

16. $\dfrac{\dfrac{1}{(x+y)^3} - \dfrac{1}{x^3}}{y}$

0.5 Solving Polynomial Equations

> **Polynomial Equation**
>
> A **polynomial equation** in one unknown is an equation that can be written in the form
>
> $$ax^n + bx^{n-1} + \cdots + rx + s = 0$$
>
> where $a, b, \ldots, r$ and s are constants.
>
> We call the largest exponent of x appearing in a nonzero term of a polynomial the **degree** of that polynomial.

quick Examples

1. $3x + 1 = 0$ has degree 1 because the largest power of x that occurs is $x = x^1$. Degree 1 equations are called **linear** equations.

2. $x^2 - x - 1 = 0$ has degree 2 because the largest power of x that occurs is x^2. Degree 2 equations are also called **quadratic equations,** or just **quadratics.**

3. $x^3 = 2x^2 + 1$ is a degree 3 polynomial (or **cubic**) in disguise. It can be rewritten as $x^3 - 2x^2 - 1 = 0$, which is in the standard form for a degree 3 equation.

4. $x^4 - x = 0$ has degree 4. It is called a **quartic.**

Now comes the question: How do we solve these equations for x? This question was asked by mathematicians as early as 1600 B.C. Let's look at these equations one degree at a time.

Solution of Linear Equations

By definition, a linear equation can be written in the form

$$ax + b = 0 \qquad \text{\textit{a} and \textit{b} are fixed numbers with } a \neq 0.$$

Solving this is a nice mental exercise: Subtract b from both sides and then divide by a, getting $x = -b/a$. Don't bother memorizing this formula; just go ahead and solve linear equations as they arise. If you feel you need practice, see the exercises at the end of the section.

Solution of Quadratic Equations

By definition, a quadratic equation has the form

$$ax^2 + bx + c = 0 \qquad \text{\textit{a}, \textit{b}, and \textit{c} are fixed numbers and } a \neq 0.^1$$

The solutions of this equation are also called the **roots** of $ax^2 + bx + c$. We're assuming that you saw quadratic equations somewhere in high school but may be a little hazy about the details of their solution. There are two ways of solving these equations—one works sometimes, and the other works every time.

Solving Quadratic Equations by Factoring (works sometimes)

If we can factor* a quadratic equation $ax^2 + bx + c = 0$, we can solve the equation by setting each factor equal to zero.

quick Examples

1. $x^2 + 7x + 10 = 0$

   $(x + 5)(x + 2) = 0$ Factor the left-hand side.

   $x + 5 = 0$ or $x + 2 = 0$ If a product is zero, one or both factors is zero.

   Solutions: $x = -5$ and $x = -2$

2. $2x^2 - 5x - 12 = 0$

   $(2x + 3)(x - 4) = 0$ Factor the left-hand side.

   $2x + 3 = 0$ or $x - 4 = 0$

   Solutions: $x = -3/2$ and $x = 4$

---
* See the section on factoring for a review of how to factor quadratics.

Test for Factoring

The quadratic $ax^2 + bx + c$, with a, b, and c being integers (whole numbers), factors into an expression of the form $(rx + s)(tx + u)$ with r, s, t and u integers precisely when the quantity $b^2 - 4ac$ is a perfect square (that is, it is the square of an integer). If this happens, we say that the quadratic **factors over the integers.**

quick Examples

1. $x^2 + x + 1$ has $a = 1$, $b = 1$, and $c = 1$, so $b^2 - 4ac = -3$, which is not a perfect square. Therefore, this quadratic does not factor over the integers.

2. $2x^2 - 5x - 12$ has $a = 2$, $b = -5$ and $c = -12$, so $b^2 - 4ac = 121$. Because $121 = 11^2$, this quadratic does factor over the integers (we factored it above).

---
[1] What happens if $a = 0$?

Solving Quadratic Equations with the Quadratic Formula (works every time)

The solutions of the general quadratic $ax^2 + bx + c = 0$ ($a \neq 0$) are given by

$$x = \frac{-b \pm \sqrt{b^2 - 4ac}}{2a}$$

We call the quantity $\Delta = b^2 - 4ac$ the **discriminant** of the quadratic (Δ is the Greek letter delta), and we have the following general rules:

- If Δ is positive, there are two distinct real solutions.
- If Δ is zero, there is only one real solution: $x = -\dfrac{b}{2a}$. (Why?)
- If Δ is negative, there are no real solutions.

quick Examples

1. $2x^2 - 5x - 12 = 0$ has $a = 2$, $b = -5$, and $c = -12$.

$$x = \frac{-b \pm \sqrt{b^2 - 4ac}}{2a} = \frac{5 \pm \sqrt{25 + 96}}{4} = \frac{5 \pm \sqrt{121}}{4} = \frac{5 \pm 11}{4}$$

$$= \frac{16}{4} \text{ or } \frac{6}{4}, = 4 \text{ or } -3/2 \qquad \color{red}{\Delta \text{ is positive in this example.}}$$

2. $4x^2 = 12x - 9$ can be rewritten as $4x^2 - 12x + 9 = 0$, which has $a = 4$, $b = -12$, and $c = 9$.

$$x = \frac{-b \pm \sqrt{b^2 - 4ac}}{2a} = \frac{12 \pm \sqrt{144 - 144}}{8} = \frac{12 \pm 0}{8} = \frac{12}{8} = \frac{3}{2}$$

$$\color{red}{\Delta \text{ is zero in this example.}}$$

3. $x^2 + 2x - 1 = 0$ has $a = 1$, $b = 2$, and $c = -1$.

$$x = \frac{-b \pm \sqrt{b^2 - 4ac}}{2a} = \frac{-2 \pm \sqrt{8}}{2} = \frac{-2 \pm 2\sqrt{2}}{2} = -1 \pm \sqrt{2}$$

The two solutions are $x = -1 + \sqrt{2} = 0.414\ldots$ and $x = -1 - \sqrt{2} = -2.414\ldots$.

$$\color{red}{\Delta \text{ is positive in this example.}}$$

4. $x^2 + x + 1 = 0$ has $a = 1$, $b = 1$, and $c = 1$. Since $\Delta = -3$ is negative, there are no real solutions. $\qquad \color{red}{\Delta \text{ is negative in this example.}}$

Q: *This is all very useful, but where does the quadratic formula come from?*

A: To see where it comes from, we will solve a general quadratic equation using "brute force." Start with the general quadratic equation.

$$ax^2 + bx + c = 0$$

First, divide out the nonzero number a to get

$$x^2 + \frac{bx}{a} + \frac{c}{a} = 0$$

Now we **complete the square:** Add and subtract the quantity $\dfrac{b^2}{4a^2}$ to get

$$x^2 + \frac{bx}{a} + \frac{b^2}{4a^2} - \frac{b^2}{4a^2} + \frac{c}{a} = 0$$

We do this to get the first three terms to factor as a perfect square:

$$\left(x + \frac{b}{2a}\right)^2 - \frac{b^2}{4a^2} + \frac{c}{a} = 0$$

(Check this by multiplying out.) Adding $\dfrac{b^2}{4a^2} - \dfrac{c}{a}$ to both sides gives:

$$\left(x + \frac{b}{2a}\right)^2 = \frac{b^2}{4a^2} - \frac{c}{a} = \frac{b^2 - 4ac}{4a^2}$$

Taking square roots gives

$$x + \frac{b}{2a} = \frac{\pm\sqrt{b^2 - 4ac}}{2a}$$

Finally, adding $-\dfrac{b}{2a}$ to both sides yields the result:

$$x = -\frac{b}{2a} + \frac{\pm\sqrt{b^2 - 4ac}}{2a}$$

or

$$x = \frac{-b \pm \sqrt{b^2 - 4ac}}{2a}$$ ■

Solution of Cubic Equations

By definition, a cubic equation can be written in the form

$$ax^3 + bx^2 + cx + d = 0 \qquad \text{\textcolor{red}{$a, b, c,$ and d are fixed numbers and $a \neq 0$.}}$$

Now we get into something of a bind. Although there is a perfectly respectable formula for the solutions, it is very complicated and involves the use of complex numbers rather heavily.[2] So we discuss instead a much simpler method that *sometimes* works nicely. Here is the method in a nutshell.

Solving Cubics by Finding One Factor

Start with a given cubic equation $ax^3 + bx^2 + cx + d = 0$.

Step 1 By trial and error, find one solution $x = s$. If $a, b, c,$ and d are integers, the only possible *rational* solutions[*] are those of the form $s = \pm(\text{factor of } d)/(\text{factor of } a)$.

Step 2 It will now be possible to factor the cubic as

$$ax^3 + bx^2 + cx + d = (x - s)(ax^2 + ex + f) = 0$$

To find $ax^2 + ex + f$, divide the cubic by $x - s$, using long division.[†]

Step 3 The factored equation says that either $x - s = 0$ or $ax^2 + ex + f = 0$. We already know that s is a solution, and now we see that the other solutions are the roots of the quadratic. Note that this quadratic may or may not have any real solutions, as usual.

quick Example

To solve the cubic $x^3 - x^2 + x - 1 = 0$, we first find a single solution. Here, $a = 1$ and $d = -1$. Because the only factors of ± 1 are ± 1, the only possible rational solutions are $x = \pm 1$. By substitution, we see that $x = 1$ is a solution. Thus $(x - 1)$ is a factor. Dividing by $(x - 1)$ yields the quotient $(x^2 + 1)$. Thus,

$$x^3 - x^2 + x - 1 = (x - 1)(x^2 + 1) = 0$$

so that either $x - 1 = 0$ or $x^2 + 1 = 0$.

Because the discriminant of the quadratic $x^2 + 1$ is negative, we don't get any real solutions from $x^2 + 1 = 0$, so the only real solution is $x = 1$.

---

[*] There may be *irrational* solutions, however; for example $x^3 - 2 = 0$ has the single solution $x = \sqrt[3]{2}$.

[†] Alternatively, use "synthetic division," a shortcut that would take us too far afield to describe.

---

[2] It was when this formula was discovered in the 16th century that complex numbers were first taken seriously. Although we would like to show you the formula, it is too large to fit in this footnote.

Possible Outcomes When Solving a Cubic Equation

If you consider all the cases, there are three possible outcomes when solving a cubic equation:

1. One real solution (as in the Quick Example on the previous page)

2. Two real solutions (try, for example, $x^3 + x^2 - x - 1 = 0$)

3. Three real solutions (see the next example)

Example **1** **Solving a Cubic**

Solve the cubic $2x^3 - 3x^2 - 17x + 30 = 0$.

Solution

First we look for a single solution. Here, $a = 2$ and $d = 30$. The factors of a are ± 1 and ± 2, and the factors of d are $\pm 1, \pm 2, \pm 3, \pm 5, \pm 6, \pm 10, \pm 15$ and ± 30. This gives us a large number of possible ratios: $\pm 1, \pm 2, \pm 3, \pm 5, \pm 6, \pm 10, \pm 15, \pm 30, \pm 1/2, \pm 3/2, \pm 5/2, \pm 15/2$. Undaunted, we first try $x = 1$ and $x = -1$, getting nowhere. So we move on to $x = 2$, and we hit the jackpot, because substituting $x = 2$ gives $16 - 12 - 34 + 30 = 0$. Thus, $(x - 2)$ is a factor. Dividing yields the quotient $2x^2 + x - 15$. Here is the calculation:

$$
\begin{array}{r}
2x^2 + x - 15 \\
x - 2 \enclose{longdiv}{2x^3 - 3x^2 - 17x + 30} \\
\underline{2x^3 - 4x^2} \\
x^2 - 17x \\
\underline{x^2 - 2x} \\
-15x + 30 \\
\underline{-15x + 30} \\
0
\end{array}
$$

Thus,

$$2x^3 - 3x^2 - 17x + 30 = (x - 2)(2x^2 + x - 15) = 0$$

Setting the factors equal to zero gives either $x - 2 = 0$ or $2x^2 + x - 15 = 0$. We could solve the quadratic using the quadratic formula, but luckily, we notice that it factors as

$$2x^2 + x - 15 = (x + 3)(2x - 5)$$

Thus, the solutions are $x = 2$, $x = -3$ and $x = 5/2$.

Solution of Higher-Order Polynomial Equations

Logically speaking, our next step should be a discussion of quartics, then quintics (fifth degree equations), and so on forever. Well, we've got to stop somewhere, and cubics may be as good a place as any. On the other hand, since we've gotten so far, we ought to at least tell you what is known about higher-order polynomials.

Quartics Just as in the case of cubics, there is a formula to find the solutions of quartics.[3]

---

[3] See, for example, *First Course in the Theory of Equations* by L. E. Dickson, (New York: Wiley, 1922), or *Modern Algebra* by B. L. van der Waerden, (New York: Frederick Ungar, 1953).

Quintics and Beyond All good things must come to an end, we're afraid. It turns out that there is no "quintic formula." In other words, there is no single algebraic formula or collection of algebraic formulas that gives the solutions to all quintics. This question was settled by the Norwegian mathematician Niels Henrik Abel in 1824 after almost 300 years of controversy about this question. (In fact, several notable mathematicians had previously claimed to have devised formulas for solving the quintic, but these were all shot down by other mathematicians—this being one of the favorite pastimes of practitioners of our art.) The same negative answer applies to polynomial equations of degree 6 and higher. It's not that these equations don't have solutions, it's just that they can't be found using algebraic formulas.[4] However, there are certain special classes of polynomial equations that can be solved with algebraic methods. The way of identifying such equations was discovered around 1829 by the French mathematician Évariste Galois.[5]

---

[4] What we mean by an "algebraic formula" is a formula in the coefficients using the operations of addition, subtraction, multiplication, division, and the taking of radicals. Mathematicians call the use of such formulas in solving polynomial equations "solution by radicals." If you were a math major, you would eventually go on to study this under the heading of Galois Theory.

[5] Both Abel (1802–1829) and Galois (1811–1832) died young. Abel died of tuberculosis at the age of 26, while Galois was killed in a duel at the age of 20.

0.5 EXERCISES

Solve the equations in Exercises 1–12 for x (mentally, if possible).

1. $x + 1 = 0$

2. $x - 3 = 1$

3. $-x + 5 = 0$

4. $2x + 4 = 1$

5. $4x - 5 = 8$

6. $\frac{3}{4}x + 1 = 0$

7. $7x + 55 = 98$

8. $3x + 1 = x$

9. $x + 1 = 2x + 2$

10. $x + 1 = 3x + 1$

11. $ax + b = c\,(a \neq 0)$

12. $x - 1 = cx + d\,(c \neq 1)$

By any method, determine all possible real solutions of each equation in Exercises 13–30. Check your answers by substitution.

13. $2x^2 + 7x - 4 = 0$

14. $x^2 + x + 1 = 0$

15. $x^2 - x + 1 = 0$

16. $2x^2 - 4x + 3 = 0$

17. $2x^2 - 5 = 0$

18. $3x^2 - 1 = 0$

19. $-x^2 - 2x - 1 = 0$

20. $2x^2 - x - 3 = 0$

21. $\frac{1}{2}x^2 - x - \frac{3}{2} = 0$

22. $-\frac{1}{2}x^2 - \frac{1}{2}x + 1 = 0$

23. $x^2 - x = 1$

24. $16x^2 = -24x - 9$

25. $x = 2 - \frac{1}{x}$

26. $x + 4 = \frac{1}{x - 2}$

27. $x^4 - 10x^2 + 9 = 0$

28. $x^4 - 2x^2 + 1 = 0$

29. $x^4 + x^2 - 1 = 0$

30. $x^3 + 2x^2 + x = 0$

Find all possible real solutions of each equation in Exercises 31–44.

31. $x^3 + 6x^2 + 11x + 6 = 0$

32. $x^3 - 6x^2 + 12x - 8 = 0$

33. $x^3 + 4x^2 + 4x + 3 = 0$

34. $y^3 + 64 = 0$

35. $x^3 - 1 = 0$

36. $x^3 - 27 = 0$

37. $y^3 + 3y^2 + 3y + 2 = 0$

38. $y^3 - 2y^2 - 2y - 3 = 0$

39. $x^3 - x^2 - 5x + 5 = 0$

40. $x^3 - x^2 - 3x + 3 = 0$

41. $2x^6 - x^4 - 2x^2 + 1 = 0$

42. $3x^6 - x^4 - 12x^2 + 4 = 0$

43. $(x^2 + 3x + 2)(x^2 - 5x + 6) = 0$

44. $(x^2 - 4x + 4)^2(x^2 + 6x + 5)^3 = 0$

0.6 Solving Miscellaneous Equations

Equations that are not polynomial equations of low degree often arise in calculus. Many of these complicated-looking equations can be solved easily if you remember the following, which we used in the previous section:

Solving an Equation of the Form $P \cdot Q = 0$

If a product is equal to 0, then at least one of the factors must be 0. That is, if $P \cdot Q = 0$, then either $P = 0$ or $Q = 0$.

quick Examples

1. $x^5 - 4x^3 = 0$

   $x^3(x^2 - 4) = 0$ Factor the left-hand side.

   Either $x^3 = 0$ or $x^2 - 4 = 0$ Either $P = 0$ or $Q = 0$.

   $x = 0, 2$ or -2 Solve the individual equations.

2. $(x^2 - 1)(x + 2) + (x^2 - 1)(x + 4) = 0$

   $(x^2 - 1)[(x + 2) + (x + 4)] = 0$ Factor the left-hand side.

   $(x^2 - 1)(2x + 6) = 0$

   Either $x^2 - 1 = 0$ or $2x + 6 = 0$ Either $P = 0$ or $Q = 0$.

   $x = -3, -1$, or 1 Solve the individual equations.

Example 1 Solving by Factoring

Solve $12x(x^2 - 4)^5(x^2 + 2)^6 + 12x(x^2 - 4)^6(x^2 + 2)^5 = 0$.

Solution

Again, we start by factoring the left-hand side:

$$12x(x^2 - 4)^5(x^2 + 2)^6 + 12x(x^2 - 4)^6(x^2 + 2)^5$$
$$= 12x(x^2 - 4)^5(x^2 + 2)^5[(x^2 + 2) + (x^2 - 4)]$$
$$= 12x(x^2 - 4)^5(x^2 + 2)^5(2x^2 - 2)$$
$$= 24x(x^2 - 4)^5(x^2 + 2)^5(x^2 - 1)$$

Setting this equal to 0, we get:

$$24x(x^2 - 4)^5(x^2 + 2)^5(x^2 - 1) = 0$$

which means that at least one of the factors of this product must be zero. Now it certainly cannot be the 24, but it could be the x: $x = 0$ is one solution. It could also be that

$$(x^2 - 4)^5 = 0$$

or

$$x^2 - 4 = 0$$

which has solutions $x = \pm 2$. Could it be that $(x^2 + 2)^5 = 0$? If so, then $x^2 + 2 = 0$, but this is impossible because $x^2 + 2 \geq 2$, no matter what x is. Finally, it could be that $x^2 - 1 = 0$, which has solutions $x = \pm 1$. This gives us five solutions to the original equation:

$$x = -2, -1, 0, 1, \text{ or } 2$$

Example 2 Solving by Factoring

Solve $(x^2 - 1)(x^2 - 4) = 10$.

Solution Watch out! You may be tempted to say that $x^2 - 1 = 10$ or $x^2 - 4 = 10$, but this does not follow. If two numbers multiply to give you 10, what must they be? There are lots of possibilities: 2 and 5, 1 and 10, $-500{,}000$ and -0.00002 are just a few. The fact that the left-hand side is factored is nearly useless to us if we want to solve this equation. What we will have to do is multiply out, bring the 10 over to the left, and hope that we can factor what we get. Here goes:

$$x^4 - 5x^2 + 4 = 10$$
$$x^4 - 5x^2 - 6 = 0$$
$$(x^2 - 6)(x^2 + 1) = 0$$

(Here we used a sometimes useful trick that we mentioned in Section 3: we treated x^2 like x and x^4 like x^2, so factoring $x^4 - 5x^2 - 6$ is essentially the same as factoring $x^2 - 5x - 6$.) *Now* we are allowed to say that one of the factors must be 0: $x^2 - 6 = 0$ has solutions $x = \pm\sqrt{6} = \pm 2.449\ldots$ and $x^2 + 1 = 0$ has no real solutions. Therefore, we get exactly two solutions, $x = \pm\sqrt{6} = \pm 2.449\ldots$.

To solve equations involving rational expressions, the following rule is very useful.

Solving an Equation of the Form *P/Q* = 0

$$\text{If } \frac{P}{Q} = 0, \text{ then } P = 0.$$

How else could a fraction equal 0? If that is not convincing, multiply both sides by Q (which cannot be 0 if the quotient is defined).

quick Example

$$\frac{(x + 1)(x + 2)^2 - (x + 1)^2(x + 2)}{(x + 2)^4} = 0$$

$(x + 1)(x + 2)^2 - (x + 1)^2(x + 2) = 0$ If $\frac{P}{Q} = 0$, then $P = 0$.

$(x + 1)(x + 2)[(x + 2) - (x + 1)] = 0$ Factor.

$(x + 1)(x + 2)(1) = 0$

Either $x + 1 = 0$ or $x + 2 = 0$

$x = -1$ or $x = -2$

$x = -1$ $x = -2$ does not make sense in the original equation: it makes the denominator 0. So it is not a solution and $x = -1$ is the only solution.

Example 3 Solving a Rational Equation

Solve $1 - \dfrac{1}{x^2} = 0$.

Solution Write 1 as $\frac{1}{1}$, so that we now have a difference of two rational expressions:

$$\frac{1}{1} - \frac{1}{x^2} = 0$$

To combine these we can put both over a common denominator of x^2, which gives

$$\frac{x^2 - 1}{x^2} = 0$$

Now we can set the numerator, $x^2 - 1$, equal to zero. Thus,

$$x^2 - 1 = 0$$

so

$$(x - 1)(x + 1) = 0$$

giving $x = \pm 1$.

+*Before we go on...* This equation could also have been solved by writing

$$1 = \frac{1}{x^2}$$

and then multiplying both sides by x^2. ∎

Example 4 **Another Rational Equation**

Solve $\dfrac{2x - 1}{x} + \dfrac{3}{x - 2} = 0$.

Solution

We *could* first perform the addition on the left and then set the top equal to 0, but here is another approach. Subtracting the second expression from both sides gives

$$\frac{2x - 1}{x} = \frac{-3}{x - 2}$$

Cross-multiplying [multiplying both sides by both denominators—that is, by $x(x - 2)$] now gives

$$(2x - 1)(x - 2) = -3x$$

so

$$2x^2 - 5x + 2 = -3x$$

Adding $3x$ to both sides gives the quadratic equation

$$2x^2 - 2x + 1 = 0$$

The discriminant is $(-2)^2 - 4 \cdot 2 \cdot 1 = -4 < 0$, so we conclude that there is no real solution.

+*Before we go on...* Notice that when we said that $(2x - 1)(x - 2) = -3x$, we were *not* allowed to conclude that $2x - 1 = -3x$ or $x - 2 = -3x$. ∎

Example 5 A Rational Equation with Radicals

Solve $\dfrac{\left(2x\sqrt{x+1} - \frac{x^2}{\sqrt{x+1}}\right)}{x+1} = 0.$

Solution

Setting the top equal to 0 gives

$$2x\sqrt{x+1} - \frac{x^2}{\sqrt{x+1}} = 0$$

This still involves fractions. To get rid of the fractions, we could put everything over a common denominator $(\sqrt{x+1})$ and then set the top equal to 0, or we could multiply the whole equation by that common denominator in the first place to clear fractions. If we do the second, we get

$$2x(x+1) - x^2 = 0$$
$$2x^2 + 2x - x^2 = 0$$
$$x^2 + 2x = 0$$

Factoring,

$$x(x+2) = 0$$

so either $x = 0$ or $x + 2 = 0$, giving us $x = 0$ or $x = -2$. Again, one of these is not really a solution. The problem is that $x = -2$ cannot be substituted into $\sqrt{x+1}$, because we would then have to take the square root of -1, and we are not allowing ourselves to do that. Therefore, $x = 0$ is the only solution.

0.6 EXERCISES

Solve the following equations:

1. $x^4 - 3x^3 = 0$

2. $x^6 - 9x^4 = 0$

3. $x^4 - 4x^2 = -4$

4. $x^4 - x^2 = 6$

5. $(x+1)(x+2) + (x+1)(x+3) = 0$

6. $(x+1)(x+2)^2 + (x+1)^2(x+2) = 0$

7. $(x^2+1)^5(x+3)^4 + (x^2+1)^6(x+3)^3 = 0$

8. $10x(x^2+1)^4(x^3+1)^5 - 10x^2(x^2+1)^5(x^3+1)^4 = 0$

9. $(x^3+1)\sqrt{x+1} - (x^3+1)^2\sqrt{x+1} = 0$

10. $(x^2+1)\sqrt{x+1} - \sqrt{(x+1)^3} = 0$

11. $\sqrt{(x+1)^3} + \sqrt{(x+1)^5} = 0$

12. $(x^2+1)\sqrt[3]{(x+1)^4} - \sqrt[3]{(x+1)^7} = 0$

13. $(x+1)^2(2x+3) - (x+1)(2x+3)^2 = 0$

14. $(x^2-1)^2(x+2)^3 - (x^2-1)^3(x+2)^2 = 0$

15. $\dfrac{(x+1)^2(x+2)^3 - (x+1)^3(x+2)^2}{(x+2)^6} = 0$

16. $\dfrac{6x(x^2+1)^2(x^2+2)^4 - 8x(x^2+1)^3(x^2+2)^3}{(x^2+2)^8} = 0$

17. $\dfrac{2(x^2-1)\sqrt{x^2+1} - \frac{x^4}{\sqrt{x^2+1}}}{x^2+1} = 0$

18. $\dfrac{4x\sqrt{x^3-1} - \frac{3x^4}{\sqrt{x^3-1}}}{x^3-1} = 0$

19. $x - \dfrac{1}{x} = 0$

20. $1 - \dfrac{4}{x^2} = 0$

21. $\dfrac{1}{x} - \dfrac{9}{x^3} = 0$

22. $\dfrac{1}{x^2} - \dfrac{1}{x+1} = 0$

23. $\dfrac{x-4}{x+1} - \dfrac{x}{x-1} = 0$

24. $\dfrac{2x-3}{x-1} - \dfrac{2x+3}{x+1} = 0$

25. $\dfrac{x+4}{x+1} + \dfrac{x+4}{3x} = 0$

26. $\dfrac{2x-3}{x} - \dfrac{2x-3}{x+1} = 0$

1

Functions and Linear Models

CASE STUDY Modeling Spending on Internet Advertising

You are the new director of *Impact Advertising Inc.'s* Internet division, which has enjoyed a steady 0.25% of the Internet advertising market. You have drawn up an ambitious proposal to expand your division in light of your anticipation that Internet advertising will continue to sky-rocket. The VP in charge of Financial Affairs feels that current projections (based on a linear model) do not warrant the level of expansion you propose. How can you persuade the VP that those projections do not fit the data convincingly?

Jeff Titcomb/Getty Images

Introduction

To analyze recent trends in spending on Internet advertising and to make reasonable projections, we need a mathematical model of this spending. Where do we start? To apply mathematics to real-world situations like this, we need a good understanding of basic mathematical concepts. Perhaps the most fundamental of these concepts is that of a function: a relationship that shows how one quantity depends on another. Functions may be described numerically and, often, algebraically. They can also be described graphically—a viewpoint that is extremely useful.

The simplest functions—the ones with the simplest formulas and the simplest graphs—are linear functions. Because of their simplicity, they are also among the most useful functions and can often be used to model real-world situations, at least over short periods of time. In discussing linear functions, we will meet the concepts of slope and rate of change, which are the starting point of the mathematics of change.

In the last section of this chapter, we discuss *simple linear regression*: construction of linear functions that best fit given collections of data. Regression is used extensively in applied mathematics, statistics, and quantitative methods in business. The inclusion of regression utilities in computer spreadsheets like Excel® makes this powerful mathematical tool readily available for anyone to use.

algebra Review

For this chapter you should be familiar with real numbers and intervals. To review this material, see Chapter 0.

1.1 Functions from the Numerical and Algebraic Viewpoints

The following table gives the weights of a particular child at various ages in her first year:

Age (months)	0	2	3	4	5	6	9	12
Weight (pounds)	8	9	13	14	16	17	18	19

Let's write $W(0)$ for the child's weight at birth (in pounds), $W(2)$ for her weight at 2 months, and so on (we read $W(0)$ as "W of 0"). Thus, $W(0) = 8$, $W(2) = 9$, $W(3) = 13$, ..., $W(12) = 19$. More generally, if we write t for the age of the child (in months) at any time during her first year, then we write $W(t)$ for the weight of the child at age t. We call W a **function** of the variable t, meaning that for each value of t between 0 and 12, W gives us a single corresponding number $W(t)$ (the weight of the child at that age).

In general, we think of a function as a way of producing new objects from old ones. The functions we deal with in this text produce new numbers from old numbers. The numbers we have in mind are the *real* numbers, including not only positive and negative integers and fractions but also numbers like $\sqrt{2}$ or π (see Chapter 0 for more on real numbers). For this reason, the functions we use are called **real-valued functions of a real variable.** For example, the function W takes the child's age in months and returns her weight in pounds at that age (Figure 1).

Age t → W → Weight $W(t)$

Figure **1**

The variable t is called the **independent** variable, while W is called the **dependent variable** as its value depends on t.

A function may be specified in several different ways. It may be specified **numerically,** by giving the values of the function for a number of values of the independent variable, as in the preceding table. It may be specified **verbally,** as in "Let $W(t)$ be the weight of the child at age t months in her first year."[1] In some cases we may be able to use an algebraic formula to calculate the function, and we say that the function is specified **algebraically.** In Section 1.2 we will see that a function may also be specified **graphically.**

Q: *For which values of t does it make sense to ask for W(t)? In other words, for which ages t is the function W defined?*

A: Since $W(t)$ refers to the weight of the child at age t months *in her first year*, $W(t)$ is defined when t is any number between 0 and 12, that is, when $0 \leq t \leq 12$. Using interval notation (see Appendix A), we can say that $W(t)$ is defined when t is in the interval $[0, 12]$. ∎

The set of values of the independent variable for which a function is defined is called its **domain** and is a necessary part of the definition of the function. Notice that the preceding table gives the values of $W(t)$ at only some of the infinitely many possible values in the domain $[0, 12]$.

The domain of a function is not always specified explicitly; if no domain is specified for a function f, we take the domain to be the largest set of numbers x for which $f(x)$ makes sense. This "largest possible domain" is sometimes called the **natural domain.**

Here is a summary of the terms we've just introduced.

Functions

A **real-valued function f of a real-valued variable x** assigns to each real number x in a specified set of numbers, called the **domain** of f, a unique real number $f(x)$, read "f of x." The variable x is called the **independent variable,** and f is called the **dependent variable.**

quick Examples

1. Let $W(t)$ be the weight (in pounds) at age t months of a particular child during her first year. The independent variable is t, and the dependent variable is W, the child's weight. The domain of W is $[0, 12]$ because it was specified that W gives the child's weight during her first year.

2. Let $f(x) = \dfrac{1}{x}$. The function f is specified algebraically. Some specific values of f are

$$f(2) = \frac{1}{2} \qquad f(3) = \frac{1}{3} \qquad f(-1) = \frac{1}{-1} = -1$$

Here, $f(0)$ is not defined because there is no such number as $1/0$. The natural domain of f consists of all real numbers except zero because $f(x)$ makes sense for all values of x other than $x = 0$.

---

[1] Specifying a function verbally in this way is useful for understanding what the function is doing, but it gives no numerical information.

Example 1   A Numerically Specified Function: Airline Profits

The following table[*] shows the cumulative net income of U.S. domestic airlines from January 2000 to the end of year x:

Year x (Since 2000)	0	1	2	3	4
Cumulative Net Income P ($ Billions)	12	2	−34	−51	−61

Viewing P as a function of x, give its domain and the values $P(0)$, $P(2)$, and $P(4)$. Compute $P(3) - P(2)$ and interpret the result. Also estimate and interpret the value $P(3.5)$.

Solution   The domain of P is the set of numbers x with $0 \leq x \leq 4$—that is, [0, 4].

From the table, we have:

$$P(0) = 12 \qquad \text{\$12 billion net income in 2000}$$

$$P(2) = -34 \qquad \text{\$34 billion cumulative loss from Jan. 2000 to Dec. 2002}$$

$$P(4) = -61 \qquad \text{\$61 billion cumulative loss from Jan. 2000 to Dec. 2004}$$

Also,

$$P(3) - P(2) = -51 - (-34) = -17$$

To interpret the result, notice that:

Cumulative net income through 2003 − Cumulative net income through 2002
= Net income in 2003

Thus, the net income in 2003 was −$17 billion. In other words, $17 billion was lost by the airline industry in 2003.

What about $P(3.5)$? Since $P(3) = -51$ and $P(4) = -61$, we estimate that

$$P(3.5) \approx -56 \qquad \text{−56 is midway between −51 and −61.}$$

The process of estimating values for a function between points where it is already known is called **interpolation.**

To interpret $P(3.5)$, note that $P(3)$ represents the accumulated net income through 2003, and $P(4)$ represents the accumulated net income through 2004. Thus, $P(3.5)$ represents the accumulated net income through June, 2003.

---

[*]2004 figure is an estimate based on first quarter results. SOURCE: Bureau of Transportation Statistics www.bts.gov/ Nov 15 2004.

+*Before we go on...*   In Example 1 we should not use the table to estimate $P(x)$ for values of x *outside* the domain—say, for $x = 10$. Estimating values for a function outside a range where it is already known is called **extrapolation.** As a general rule, extrapolation is far less reliable than interpolation: predicting the future from current data is difficult, especially given the vagaries of the marketplace. ∎

The two functions we have looked at so far were both specified numerically: we were given numerical values of the function evaluated at *certain* values of the independent variable. It would be more useful if we had a formula that would allow us to calculate the value of the function for *any* value of the independent variable we wished, that is, if the function were specified algebraically.

Example 2 An Algebraically Defined Function

Let f be the function specified by

$$f(x) = x^2 - 25x + 15$$

with domain $(-2, 10]$. When $0 \le x \le 4$, this formula gives an approximation of the airline cumulative net income function P in Example 1. Use the formula to calculate $f(0)$, $f(10)$, $f(-1)$, $f(a)$, and $f(x + h)$. Is $f(-2)$ defined?

Solution Let's check first that the values we are asked to calculate are all defined. Since the domain is stated to be $(-2, 10]$, the quantities $f(0)$, $f(10)$, and $f(-1)$ are all defined. The quantities $f(a)$ and $f(x + h)$ will also be defined if a and $x + h$ are understood to be in $(-2, 10]$. However, $f(-2)$ is not defined, since -2 is not in the domain $(-2, 10]$.

If we substitute 0 for x in the formula for $f(x)$, we get

$$f(0) = (0)^2 - 25(0) + 15 = 15$$

so $f(0) = 15$. Similarly,

$$f(10) = (10)^2 - 25(10) + 15 = 100 - 250 + 15 = -135$$
$$f(-1) = (-1)^2 - 25(-1) + 15 = 1 + 25 + 15 = 41$$
$$f(a) = a^2 - 25a + 15 \qquad \text{Substitute } a \text{ for } x.$$
$$f(x + h) = (x + h)^2 - 25(x + h) + 15 \qquad \text{Substitute } (x + h) \text{ for } x.$$
$$= x^2 + 2xh + h^2 - 25x - 25h + 15$$

Note how we placed parentheses around the number at which we are evaluating the function. If we omitted any of these parentheses, we would likely get errors:

$$f(-1) = (-1)^2 - 25(-1) + 15 \checkmark \qquad \text{NOT } -1^2 - 25(-1) + 15 \quad \times$$
$$f(x + h) = (x + h)^2 - 25(x + h) + 15 \checkmark \qquad \text{NOT } x + h^2 - 25(x + h) + 15 \quad \times$$

Note that there is nothing magical about the letter x. We might just as well say

$$f(t) = t^2 - 25t + 15$$

which defines *exactly the same function* as $f(x) = x^2 - 25x + 15$. For example, to calculate $f(10)$ from the formula for $f(t)$ we would substitute 10 for t, getting $f(10) = -135$, just as we did using the formula for $f(x)$.

+*Before we go on...* We said that the function f given in the Example 2 is an approximation of the cumulative net income function P of Example 1. The following table compares some of their values:[2]

x	0	2	3	4
$P(x)$	12	-34	-51	-61
$f(x)$	15	-31	-51	-69

---

[2] The function f is a "best-fit," or regression quadratic curve based on the data in Example 1 (coefficients are rounded). We will learn more about regression later in this chapter.

We call the algebraic function f an **algebraic model** of U.S. airlines' cumulative net income from Jan. 2000 because it uses an algebraic formula to model—or mathematically represent (approximately)—the cumulative net income. The particular kind of algebraic model we used is called a **quadratic model** (see the end of this section for the names of some commonly used models). ∎

Q: *The values of f(x) are close to but don't all equal those of P(x). Is this the best we can do with an algebraic model? Can't we get a formula that gives the cumulative net income data exactly?*

A: It is possible to find algebraic formulas that give the exact values of $P(x)$, but such formulas would be far more complicated than the one given, and quite possibly less useful.[3] ∎

Note Equation and Function Notation
Instead of using *function notation*

$$f(x) = x^2 - 25x + 15 \qquad \text{Function notation}$$

we could use equation notation

$$y = x^2 - 25x + 15 \qquad \text{Equation notation}$$

(the choice of the letter y is a convention) and we say that "y is a function of x." When we write a function in this form, the variable x is the independent variable and y is the dependent variable.

We could also write the above function as $f = x^2 - 25x + 15$, in which case the dependent variable would be f.

using *Technology*

Evaluating a function can be tedious to do by hand, but various technologies make this task easier. See the Technology Guides at the end of the chapter to find out how to create a table like the one in Example 3, using a TI-83/84 or Excel. Alternatively, go online and follow:

Chapter 1
→ Tools
   → Function Evaluator
     & Grapher

to find a utility you can use to evaluate functions like this.

Example 3 Evaluating a Function with Technology

Evaluate the function $f(x) = -0.4x^2 + 7x - 23$ for $x = 0, 1, 2, \ldots, 10$.

Solution

The first couple of evaluations go as follows:

$$f(0) = -0.4(0)^2 + 7(0) - 23 = -23$$
$$f(1) = -0.4(1)^2 + 7(1) - 23 = -0.4 + 7 - 23 = -16.4$$

Note that to evaluate $-0.4x^2$, we first compute x^2 and then multiply by -0.4. Continuing, we get the following table.

x	0	1	2	3	4	5	6	7	8	9	10
$f(x)$	−23	−16.4	−10.6	−5.6	−1.4	2	4.6	6.4	7.4	7.6	7

Sometimes, as in Example 4, we need to use several formulas to specify a single function.

---

[3] One reason that more complex formulas are often less realistic than simple ones is that it is often random phenomena in the real world, rather than algebraic relationships, that cause data to fluctuate. Attempting to model these random fluctuations using algebraic formulas amounts to imposing mathematical structure where structure does not exist.

using *Technology*

See the Technology Guides at
the end of the chapter to see
how to evaluate functions like
this using a TI-83/84 or Excel.
The techniques shown there
work for other technologies
as well, including the function
evaluator that you can find online.
Follow:

Chapter 1
→ Tools
   → Function Evaluator
      & Grapher

Example 4 A Piecewise-Defined Function: EBAY Stock

The price $V(t)$ in dollars of EBAY stock during the 10-week period starting July 1, 2004 can be approximated by the following function of time t in weeks ($t = 0$ represents July 1):[*]

$$V(t) = \begin{cases} 90 - 4t & \text{if } 0 \leq t \leq 5 \\ 60 + 2t & \text{if } 5 < t \leq 20 \end{cases}$$

What was the approximate price of EBAY stock after 1 week, after 5 weeks, and after 10 weeks?

Solution

We evaluate the given function at the corresponding value of t:

$t = 1$: $V(1) = 90 - 4(1) = 86$ We use the first formula since $0 \leq t \leq 5$.

$t = 5$: $V(5) = 90 - 4(5) = 70$ We use the first formula since $0 \leq t \leq 5$.

$t = 10$: $V(10) = 60 + 2(10) = 80$ We use the second formula since $5 < t \leq 20$.

Thus, the price of EBAY stock was $86 after 1 week, $70 after 5 weeks, and $80 after 10 weeks.

---

[*] Source for data: http://money.excite.com, November, 2004

The functions we used in Examples 1–4 above are **mathematical models** of real-life situations, because they model, or represent, situations in mathematical terms.

Mathematical Modeling

To mathematically model a situation means to represent it in mathematical terms. The particular representation used is called a **mathematical model** of the situation. Mathematical models do not always represent a situation perfectly or completely. Some (like Example 2) represent a situation only approximately, whereas others represent only some aspects of the situation.

quick Examples

Situation	Model
1. Albano's bank balance is twice Bravo's.	$a = 2b$ (a = Albano's balance, b = Bravo's)
2. The temperature is now 10°F and increasing by 20° per hour.	$T(t) = 10 + 20t$ (t = time in hours, T = temperature)
3. The volume of a rectangular solid with square base is obtained by multiplying the area of its base by its height.	$V = x^2 h$ (h = height, x = length of a side of the base)
4. U.S. airlines' cumulative net income	The table in Example 1 is a **numerical model** of U.S. airlines' income. The function in Example 2 is an **algebraic model** of U.S. airlines' income.
5. EBAY stock price	Example 4 gives a **piecewise algebraic model** of the EBAY stock price.

Table 1 lists some common types of functions that are often used to model real world situations.

Table **1** Common Types of Algebraic Functions

	Type of Function	*Example*
Linear	$f(x) = mx + b$ m, b constant	$f(x) = 3x - 2$ Technology format: 3*x – 2
Quadratic	$f(x) = ax^2 + bx + c$ a, b, c constant $(a \neq 0)$	$f(x) = -3x^2 + x - 1$ Technology format: –3*x^2 + x – 1
Cubic	$f(x) = ax^3 + bx^2 + cx + d$ a, b, c, d constant $(a \neq 0)$	$f(x) = 2x^3 - 3x^2 + x - 1$ Technology format: 2*x^3 – 3*x^2 + x – 1
Polynomial	$f(x) = ax^n + bx^{n-1} + \ldots + rx + s$ $a, b, \ldots, r, s$ constant (Includes all of the above functions)	All the above, and $f(x) = x^6 - x^4 + x - 3$ Technology format: x^6 – x^4 + x – 3
Exponential	$f(x) = Ab^x$ A, b constant (b positive)	$f(x) = 3(2^x)$ Technology format: 3*2^x
Rational	$f(x) = \dfrac{P(x)}{Q(x)}$ $P(x)$ and $Q(x)$ polynomials	$f(x) = \dfrac{x^2 - 1}{2x + 5}$ Technology format: (x^2 – 1)/(2*x + 5)

Functions and models other than linear ones are called **nonlinear.**

1.1 EXERCISES

● denotes basic skills exercises

tech Ex indicates exercises that should be solved using technology

In Exercises 1–4, evaluate or estimate each expression based on the following table. hint [see Example 1]

x	-3	-2	-1	0	1	2	3
$f(x)$	1	2	4	2	1	0.5	0.25

1. ● **a.** $f(0)$ **b.** $f(2)$

2. ● **a.** $f(-1)$ **b.** $f(1)$

3. ● **a.** $f(2) - f(-2)$ **b.** $f(-1)f(-2)$ **c.** $-2f(-1)$

4. ● **a.** $f(1) - f(-1)$ **b.** $f(1)f(-2)$ **c.** $3f(-2)$

5. ● Given $f(x) = 4x - 3$, find **a.** $f(-1)$ **b.** $f(0)$ **c.** $f(1)$ **d.** $f(y)$ **e.** $f(a + b)$ hint [see Example 2]

6. ● Given $f(x) = -3x + 4$, find **a.** $f(-1)$ **b.** $f(0)$ **c.** $f(1)$ **d.** $f(y)$ **e.** $f(a + b)$

7. ● Given $f(x) = x^2 + 2x + 3$, find **a.** $f(0)$ **b.** $f(1)$ **c.** $f(-1)$ **d.** $f(-3)$ **e.** $f(a)$ **f.** $f(x + h)$

8. ● Given $g(x) = 2x^2 - x + 1$, find **a.** $g(0)$ **b.** $g(-1)$ **c.** $g(r)$ **d.** $g(x + h)$

9. ● Given $g(s) = s^2 + \dfrac{1}{s}$, find **a.** $g(1)$ **b.** $g(-1)$ **c.** $g(4)$ **d.** $g(x)$ **e.** $g(s + h)$ **f.** $g(s + h) - g(s)$

10. ● Given $h(r) = \dfrac{1}{r + 4}$, find **a.** $h(0)$ **b.** $h(-3)$ **c.** $h(-5)$ **d.** $h(x^2)$ **e.** $h(x^2 + 1)$ **f.** $h(x^2) + 1$

11. ● Given $f(t) = \begin{cases} -t & \text{if } t < 0 \\ t^2 & \text{if } 0 \leq t < 4 \\ t & \text{if } t \geq 4 \end{cases}$

find **a.** $f(-1)$ **b.** $f(1)$ **c.** $f(4) - f(2)$ **d.** $f(3)f(-3)$ hint [see Example 4]

● basic skills   tech Ex technology exercise

12. ● Given $f(t) = \begin{cases} t-1 & \text{if } t \le 1 \\ 2t & \text{if } 1 < t < 5 \\ t^3 & \text{if } t \ge 5 \end{cases}$

find **a.** $f(0)$ **b.** $f(1)$ **c.** $f(4) - f(2)$
d. $f(5) + f(-5)$

In Exercises 13–16, say whether $f(x)$ is defined for the given values of x. If it is defined, give its value.

13. ● $f(x) = x - \dfrac{1}{x^2}$, with domain $(0, +\infty)$ **a.** $x = 4$

  b. $x = 0$ **c.** $x = -1$

14. ● $f(x) = \dfrac{2}{x} - x^2$, with domain $[2, +\infty)$

  a. $x = 4$ **b.** $x = 0$ **c.** $x = 1$

15. ● $f(x) = \sqrt{x + 10}$, with domain $[-10, 0)$
  a. $x = 0$ **b.** $x = 9$ **c.** $x = -10$

16. ● $f(x) = \sqrt{9 - x^2}$, with domain $(-3, 3)$
  a. $x = 0$ **b.** $x = 3$ **c.** $x = -3$

In Exercises 17–20, find and simplify (a) $f(x + h) - f(x)$
(b) $\dfrac{f(x + h) - f(x)}{h}$

17. $f(x) = x^2$ **18.** $f(x) = 3x - 1$
19. $f(x) = 2 - x^2$ **20.** $f(x) = x^2 + x$

In Exercises 21–24, first give the technology formula for the given function and then use technology to evaluate the function for the given values of x (when defined there).

21. tech Ex $f(x) = 0.1x^2 - 4x + 5; x = 0, 1, \ldots, 10$

22. tech Ex $g(x) = 0.4x^2 - 6x - 0.1; x = -5, -4, \ldots, 4, 5$

23. tech Ex $h(x) = \dfrac{x^2 - 1}{x^2 + 1}; x = 0.5, 1.5, 2.5, \ldots, 10.5$
  (Round all answers to four decimal places.)

24. tech Ex $r(x) = \dfrac{2x^2 + 1}{2x^2 - 1}; x = -1, 0, 1, \ldots, 9$ (Round all answers to four decimal places.)

Applications

25. ● *Employment* The following table lists the approximate number of people employed in the U.S. during the period 1995–2001, on July 1 of each year[4] ($t = 5$ represents 1995):

Year t	5	6	7	8	9	10	11
Employment $P(t)$ (Millions)	117	120	123	125	130	132	132

  a. Find or estimate $P(5)$, $P(10)$, and $P(9.5)$. Interpret your answers.
  b. What is the domain of P?

[4] The given values represent nonfarm employment, and are approximate. SOURCE: Bureau of Labor Statistics/*The New York Times*, December 17, 2001, p. C3.

26. ● *Cell Phone Sales* The following table lists the net sales (after-tax revenue) at the Finnish cell phone company Nokia for each year in the period 1995–2001[5] ($t = 5$ represents 1995):

Year t	5	6	7	8	9	10	11
Nokia Net Sales $P(t)$ (Billions of Dollars)	8	8	10	16	20	27	28

  a. Find or estimate $P(5)$, $P(10)$, and $P(7.5)$. Interpret your answers.
  b. What is the domain of P?

27. ● *Trade with China* The value of U.S. trade with China from 1994 through 2004 can be approximated by

$$C(t) = 3t^2 - 7t + 50 \text{ billion dollars}$$

(t is time in years since 1994).[6]

  a. Find an appropriate domain of C. Is $t \ge 0$ an appropriate domain? Why or why not?
  b. Compute $C(10)$. What does the answer say about trade with China?

28. ● *Scientific Research* The number of research articles in *Physics Review* that were written by researchers in the U.S. from 1983 through 2003 can be approximated by

$$A(t) = -0.01t^2 + 0.24t + 3.4 \text{ hundred articles}$$

(t is time in years since 1983).[7]

  a. Find an appropriate domain of A. Is $t \le 20$ an appropriate domain? Why or why not?
  b. Compute $A(10)$. What does the answer say about the number of research articles?

29. ● *Spending on Corrections in the 90s* The following table shows the annual spending by all states in the U.S. on corrections ($t = 0$ represents the year 1990):[8]

Year (t)	0	2	4	6	7
Spending ($ billion)	16	18	22	28	30

  a. Which of the following functions best fits the given data? (*Warning*: none of them fits exactly, but one fits more closely than the others.)

  (1) $S(t) = -0.2t^2 + t + 16$
  (2) $S(t) = 0.2t^2 + t + 16$
  (3) $S(t) = t + 16$

[5] SOURCE: Nokia/*New York Times*, February 6, 2002, p. A3.

[6] Based on a regression by the authors. Source for data: U.S. Census Bureau/*New York Times*, September 23, 2004, p. C1.

[7] Based on a regression by the authors. Source for data: The Americal Physical Society/*New York Times*, May 3, 2003, p. A1.

[8] Data are rounded. SOURCE: National Association of State Budget Officers/*The New York Times*, February 28, 1999, p. A1.

● basic skills tech Ex technology exercise

b. Use your answer to part (a) to "predict" spending on corrections in 1998, assuming that the trend continued.

30. ● ***Spending on Corrections in the 90s*** Repeat Exercise 29, this time choosing from the following functions:

(1) $S(t) = 16 + 2t$

(2) $S(t) = 16 + t + 0.5t^2$

(3) $S(t) = 16 + t - 0.5t^2$

31. ***Demand*** The demand for Sigma Mu Fraternity plastic brownie dishes is

$$q(p) = 361{,}201 - (p + 1)^2$$

where q represents the number of brownie dishes Sigma Mu can sell each month at a price of p. Use this function to determine

a. The number of brownie dishes Sigma Mu can sell each month if the price is set at 50¢,

b. The number of brownie dishes they can unload each month if they give them away,

c. The lowest price at which Sigma Mu will be unable to sell any dishes.

32. ***Revenue*** The total weekly revenue earned at Royal Ruby Retailers is given by

$$R(p) = -\frac{4}{3}p^2 + 80p$$

where p is the price (in dollars) RRR charges per ruby. Use this function to determine

a. The weekly revenue, to the nearest dollar, when the price is set at $20/ruby,

b. The weekly revenue, to the nearest dollar, when the price is set at $200/ruby (interpret your result).

c. The price RRR should charge in order to obtain a weekly revenue of $1200.

33. ● ***Processor Speeds*** The processor speed, in megahertz, of Intel processors could be approximated by the following function of time t in years since the start of 1995:[9]

$$P(t) = \begin{cases} 75t + 200 & \text{if } 0 \le t \le 4 \\ 600t - 1900 & \text{if } 4 < t \le 9 \end{cases}$$

a. Evaluate $P(0)$, $P(4)$, and $P(5)$ and interpret the results.

b. Use the model to estimate when processor speeds first hit 2.0 gigahertz (1 gigahertz = 1000 megahertz).

c. **tech** Ex Use technology to generate a table of values for $P(t)$ with $t = 0, 1, \ldots, 9$.

34. ● ***Leading Economic Indicators*** The value of the Conference Board Index of 10 economic indicators in the U.S. could be

approximated by the following function of time t in months since the end of December, 2002:[10]

$$E(t) = \begin{cases} 0.4t + 110 & \text{if } 6 \le t \le 15 \\ -0.2t + 119 & \text{if } 15 < t \le 20 \end{cases}$$

a. Estimate $E(10)$, $E(15)$, and $E(20)$ and interpret the results.

b. Use the model to estimate when—prior to March, 2004—the index was 115.

c. **tech** Ex Use technology to generate a table of values for $E(t)$ with $t = 6, 7, \ldots, 20$.

35. **tech** Ex ***Television Advertising*** The cost, in millions of dollars, of a 30-second television ad during the Super Bowl from 1990 to 2001 can be approximated by the following piecewise linear function ($t = 0$ represents 1990):[11]

$$C(t) = \begin{cases} 0.08t + 0.6 & \text{if } 0 \le t < 8 \\ 0.355t - 1.6 & \text{if } 8 \le t \le 11 \end{cases}$$

a. Give the technology formula for C and complete the following table of values of the function C.

t	0	1	2	3	4	5	6	7	8	9	10	11
$C(t)$												

b. Between 1998 and 2000, the cost of a Super Bowl ad was increasing at a rate of $_____ million per year

36. **tech** Ex ***Internet Purchases*** The percentage $p(t)$ of new car buyers who used the Internet for research or purchase since 1997 is given by the following function[12] ($t = 0$ represents 1997):

$$p(t) = \begin{cases} 10t + 15 & \text{if } 0 \le t < 1 \\ 15t + 10 & \text{if } 1 \le t \le 4 \end{cases}$$

a. Give the technology formula for p and complete the following table of values of the function p.

t	0	0.5	1	1.5	2	2.5	3	3.5	4
$p(t)$									

b. Between 1998 and 2000, the percentage of buyers of new cars who used the Internet for research or purchase was increasing at a rate of _____% per year

37. ***Income Taxes*** The U.S. Federal income tax is a function of taxable income. Write T for the tax owed on a taxable income

[10] SOURCE: The Conference Board/*New York Times*, November 19, 2004, p. C7.

[11] SOURCE: *New York Times*, January 26, 2001, p. C1.

[12] Model is based on data through 2000 (the 2000 value is estimated). SOURCE: J. D. Power Associates/*The New York Times*, January 25, 2000, p. C1.

[9] SOURCE: Sandpile.org/*New York Times*, May 17, 2004, p. C1.

● basic skills   **tech** Ex technology exercise

of I dollars. For tax year 2005, the function T for a single tax-payer was specified as follows:

If your taxable income was			of the
Over—	But not over—	Your tax is	amount over—
$0	7,300	.... 10%	$0
7,300	29,700	$730.00 + 15%	7,300
29,700	71,950	4,090.00 + 25%	29,700
71,950	150,150	14,652.50 + 28%	71,950
150,150	326,450	36,548,50 + 33%	150,150
326,450	....	94,727,50 + 35%	326,450

What was the tax owed by a single taxpayer on a taxable income of $26,000? On a taxable income of $65,000?

38. *Income Taxes* The income tax function T in Exercise 37 can also be written in the following form:

$$T(I) = \begin{cases} 0.10I & \text{if } 0 < I \le 7{,}300 \\ 730 + 0.15(I - 7{,}300) & \text{if } 7{,}300 < I \le 29{,}700 \\ 4{,}090.00 + 0.25(I - 29{,}700) & \text{if } 29{,}700 < I \le 71{,}950 \\ 14{,}652.50 + 0.28(I - 71{,}950) & \text{if } 71{,}950 < I \le 150{,}150 \\ 36{,}548.50 + 0.33(I - 150{,}150) & \text{if } 150{,}150 < I \le 326{,}450 \\ 94{,}727.50 + 0.35(I - 326{,}450) & \text{if } I > 326{,}450 \end{cases}$$

What was the tax owed by a single taxpayer on a taxable income of $25,000? On a taxable income of $125,000?

39. *Toxic Waste Treatment* The cost of treating waste by removing PCPs goes up rapidly as the quantity of PCPs removed goes up. Here is a possible model:

$$C(q) = 2000 + 100q^2$$

where q is the reduction in toxicity (in pounds of PCPs removed per day) and $C(q)$ is the daily cost (in dollars) of this reduction.

a. Find the cost of removing 10 pounds of PCPs per day.
b. Government subsidies for toxic waste cleanup amount to

$$S(q) = 500q$$

where q is as above and $S(q)$ is the daily dollar subsidy. Calculate the net cost function $N(q)$ (the cost of removing q pounds of PCPs per day after the subsidy is taken into account), given the cost function and subsidy above, and find the net cost of removing 20 pounds of PCPs per day.

40. *Dental Plans* A company pays for its employees' dental coverage at an annual cost C given by

$$C(q) = 1000 + 100\sqrt{q}$$

where q is the number of employees covered and $C(q)$ is the annual cost in dollars.

a. If the company has 100 employees, find its annual outlay for dental coverage.

b. Assuming that the government subsidizes coverage by an annual dollar amount of

$$S(q) = 200q$$

calculate the net cost function $N(q)$ to the company, and calculate the net cost of subsidizing its 100 employees. Comment on your answer.

41. `tech` Ex *Acquisition of Language* The percentage $p(t)$ of children who can speak at least single words by the age of t months can be approximated by the equation[13]

$$p(t) = 100 \left(1 - \frac{12{,}200}{t^{4.48}} \right) \quad (t \ge 8.5)$$

a. Give a technology formula for p.
b. Create a table of values of p for $t = 9, 10, \ldots, 20$ (rounding answers to one decimal place).
c. What percentage of children can speak at least single words by the age of 12 months?
d. By what age are 90% or more children speaking at least single words?

42. `tech` Ex *Acquisition of Language* The percentage $p(t)$ of children who can speak in sentences of five or more words by the age of t months can be approximated by the equation[14]

$$p(t) = 100 \left(1 - \frac{5.27 \times 10^{17}}{t^{12}} \right) \quad (t \ge 30)$$

a. Give a technology formula for p.
b. Create a table of values of p for $t = 30, 31, \ldots, 40$ (rounding answers to one decimal place).
c. What percentage of children can speak in sentences of five or more words by the age of 36 months?
d. By what age are 75% or more children speaking in sentences of five or more words?

Communication and Reasoning Exercises

43. ● If the market price m of gold varies with time t, then the independent variable is ___ and the dependent variable is ___.

44. ● Complete the following sentence: If weekly profit P is specified as a function of selling price s, then the independent variable is ___ and the dependent variable is ___.

45. ● Complete the following: The function notation for the equation $y = 4x^2 - 2$ is ___.

46. ● Complete the following: The equation notation for $C(t) = -0.34t^2 + 0.1t$ is ___.

---

[13] The model is the authors' and is based on data presented in the article *The Emergence of Intelligence* by William H. Calvin, *Scientific American*, October, 1994, pp. 101–107.
[14] Ibid.

● basic skills   `tech` Ex technology exercise

47. You now have 200 sound files on your hard drive, and this number is increasing by 10 sound files each day. Find a mathematical model for this situation.

48. The amount of free space left on your hard drive is now 50 gigabytes (GB) and is decreasing by 5 GB/month. Find a mathematical model for this situation.

49. Why is the following assertion false? "If $f(x) = x^2 - 1$, then $f(x + h) = x^2 + h - 1$."

50. Why is the following assertion false? "If $f(2) = 2$ and $f(4) = 4$, then $f(3) = 3$."

51. True or false: Every function can be specified numerically.

52. Which supplies more information about a situation: a numerical model or an algebraic model?

● basic skills   tech Ex technology exercise

1.2 Functions from the Graphical Viewpoint

Consider again the function W discussed in Section 1.1, giving a child's weight during her first year. If we represent the data given in Section 1.1 graphically by plotting the given pairs of numbers $(t, W(t))$, we get Figure 2. (We have connected successive points by line segments.)

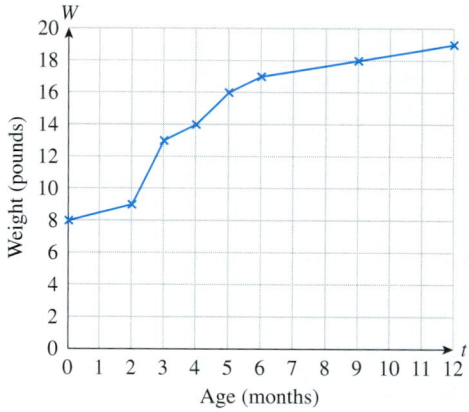

Figure **2**

Suppose now that we had only the graph without the table of data given in Section 1.1. We could use the graph to find values of W. For instance, to find $W(9)$ from the graph we do the following:

1. Find the desired value of t at the bottom of the graph ($t = 9$ in this case).

2. Estimate the height (W-coordinate) of the corresponding point on the graph (18 in this case).

Thus, $W(9) \approx 18$ pounds.[15]

We say that Figure 2 specifies the function W **graphically.** The graph is not a very accurate specification of W; the actual weight of the child would follow a smooth curve

---

[15] In a graphically defined function, we can never know the y-coordinates of points exactly; no matter how accurately a graph is drawn, we can only obtain *approximate* values of the coordinates of points. That is why we have been using the word *estimate* rather than *calculate* and why we say "$W(9) \approx 18$" rather than "$W(9) = 18$."

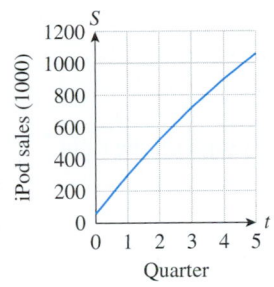

Figure **3**

Bartomeu Amengual/Index Stock Imagery

rather than a jagged line. However, the jagged line is useful in that it permits us to interpolate: for instance, we can estimate that $W(1) \approx 8.5$ pounds.

Example 1 A Function Specified Graphically: iPod Sales

Figure 3 shows the approximate quarterly sales of iPods for the second quarter in 2003 through the third quarter in 2004 ($t = 0$ represents the second quarter of 2003).[*]
Estimate and interpret $S(1)$, $S(4)$, and $S(5)$. What is the domain of S?

Solution We carefully estimate the S-coordinates of the points with t-coordinates 1, 4, and 5.

$$S(1) \approx 300$$

meaning that iPod sales in the third quarter of 2003 ($t = 1$) were approximately 300,000 units.

$$S(4) \approx 900$$

meaning that iPod sales in the second quarter of 2004 ($t = 4$) were approximately 900,000 units.

$$S(5) \approx 1050$$

meaning that iPod sales in the third quarter of 2004 ($t = 5$) were approximately 1,050,000 units.
The domain of S is the set of all values of t for which $S(t)$ is defined: $0 \leq t \leq 5$, or $[0, 5]$.

---

[*] Accurate sales figures are available from Apple financial statements, www.apple.com

Sometimes we are interested in drawing the graph of a function that has been specified in some other way—perhaps numerically or algebraically. We do this by plotting points with coordinates $(x, f(x))$.[16] Here is the formal definition of a graph.

Graph of a Function

The **graph of the function** f is the set of all points $(x, f(x))$ in the xy plane, where *we restrict the values of x to lie in the domain of f.*

quick Example

To sketch the graph of the function

$$f(x) = x^2 \qquad \text{Function notation}$$
$$\text{or} \qquad y = x^2 \qquad \text{Equation notation}$$

with domain the set of all real numbers, first choose some convenient values of x in the domain and compute the corresponding y-coordinates.

---

[16] Graphing utilities typically draw graphs by plotting and connecting a large number of points.

x	-3	-2	-1	0	1	2	3
$y = x^2$	9	4	1	0	1	4	9

Plotting these points gives the picture on the left, suggesting the graph on the right.[*]

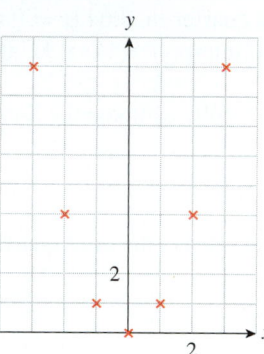 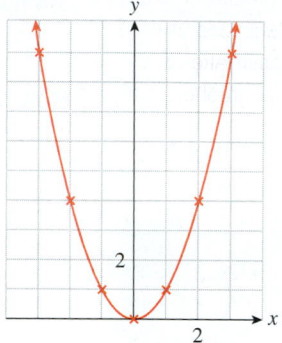

(This particular curve happens to be called a **parabola,** and its lowest point, at the origin, is called its **vertex.**)

---

[*] If you plot more points, you will find that they lie on a smooth curve as shown. That is why we did not use line segments to connect the points.

To draw the graph of a function, we often do as we did in the Quick Example above: We plot points of the form $(x, f(x))$ for several values of x in the domain of f, until we can get a good idea of the shape of the entire graph. (Calculus can give us information that allows us to draw a graph with relatively few points plotted.)

Example 2 Drawing the Graph of a Function: Web-Site Revenue

The monthly revenue[†] R from users logging on to your gaming site depends on the monthly access fee p you charge according to the formula

$$R(p) = -5600p^2 + 14{,}000p \qquad (0 \le p \le 2.5)$$

(R and p are in dollars.) Sketch the graph of R. Find the access fee that will result in the largest monthly revenue.

[†] The **revenue** resulting from one or more business transactions is the total payment received, sometimes called the gross proceeds.

Solution To sketch the graph of R by hand, we plot points of the form $(p, R(p))$ for several values of p in the domain $[0, 2.5]$ of R. First, we calculate several points:

p	0	0.5	1	1.5	2	2.5
$R(p) = -5600p^2 + 14{,}000p$	0	5600	8400	8400	5600	0

Graphing these points gives the graph shown in Figure 4(a), suggesting the parabola shown in Figure 4(b).

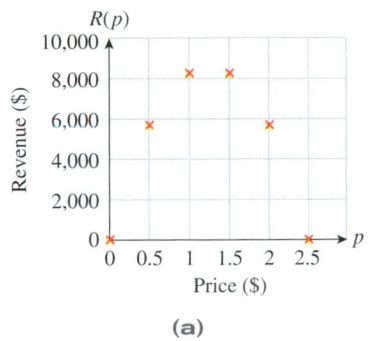
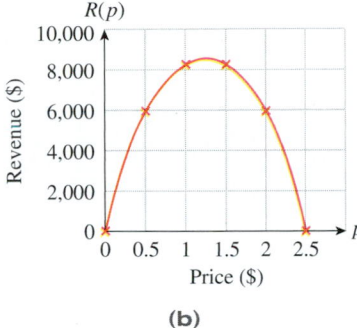

(a)   (b)

Figure **4**

The revenue graph appears to reach its highest point when $p = 1.25$, so setting the access fee at $1.25 appears to result in the largest monthly revenue.[*]

---

[*] We are hedging our language with words like *suggesting* and *appears* because the few points we have plotted don't, by themselves, allow us to draw these conclusions with certainty.

Note **Switching Between Equation and Function Notation**
As we discussed after Example 2 in Section 1.1, we can write the function R in Example 2 above in equation notation as

$$R = -5600p^2 + 14{,}000p \qquad \text{Equation notation}$$

The independent variable is p, and the dependent variable is R. Function notation and equation notation, using the same letter for the function name and the dependent variable, are often used interchangeably. It is important to be able to switch back and forth easily from function notation to equation notation.

Vertical Line Test

Every point in the graph of a function has the form $(x, f(x))$ for some x in the domain of f. Since f assigns a *single* value $f(x)$ to each value of x in the domain, it follows that, in the graph of f, there should be only one y corresponding to any such value of x—namely, $y = f(x)$. In other words, *the graph of a function cannot contain two or more points with the same x-coordinate—that is, two or more points on the same vertical line.*

On the other hand, a vertical line at a value of x not in the domain will not contain any points in the graph. This gives us the following rule:

Vertical-Line Test

For a graph to be the graph of a function, every vertical line must intersect the graph in *at most* one point.

quick Examples

As illustrated below, only graph B passes the vertical line test, so only graph B is the graph of a function.

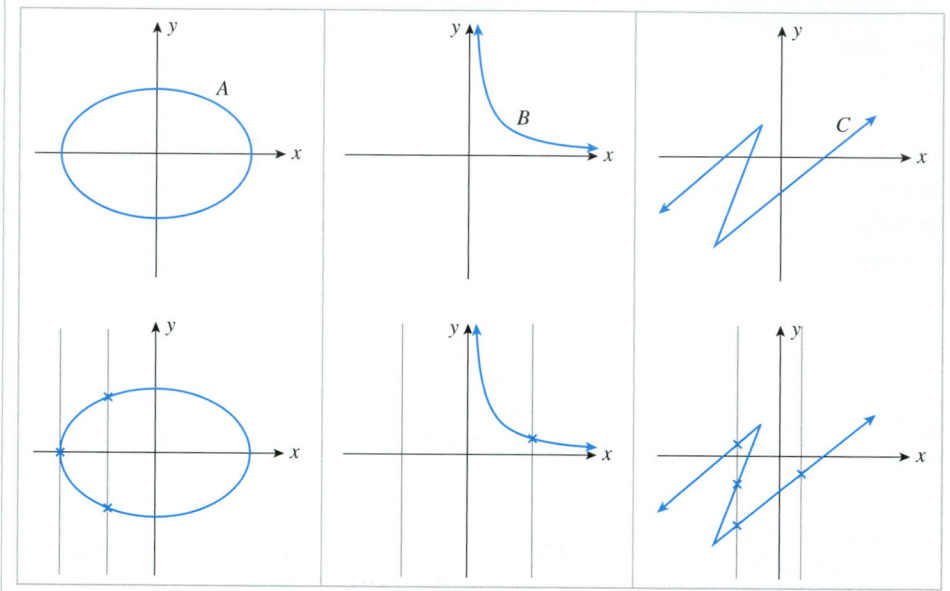

Graphing Piecewise-Defined Functions

Let us revisit the EBAY stock example from Section 1.1.

using *Technology*

To graph the function V using technology, consult the Technology Guides for Example 4 of Section 1.1 to see how to enter this piecewise-defined function. The Technology Guides for Example 2 of this section show how to then draw the graph.

Example 3   Graphing a Piecewise-Defined Function: EBAY Stock

The price $V(t)$ in dollars of EBAY stock during the 10-week period starting July 1, 2004 can be approximated by the following function of time t in weeks ($t = 0$ represents July 1):[*]

$$V(t) = \begin{cases} 90 - 4t & \text{if } 0 \le t \le 5 \\ 60 + 2t & \text{if } 5 < t \le 20 \end{cases}$$

Graph the function V.

Solution   As in Example 2, we can sketch the graph of V by hand by computing $V(t)$ for a number of values of t, plotting these points on the graph, and then connecting them.

_____

[*] Source for data: http://money.excite.com, November, 2004

t	0	5	10	15	20
$V(t)$	90	70	80	90	100

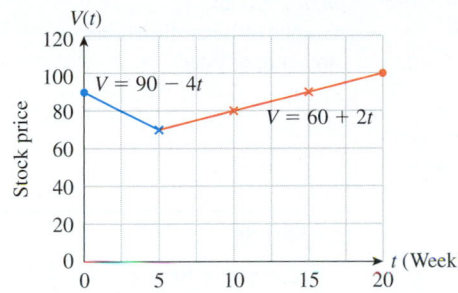

First formula   Second formula

Figure **5**

The graph (Figure 5) has the following features:

1. The first formula (the descending line) is used for $0 \leq t \leq 5$.

2. The second formula (the ascending line) is used for $5 < t \leq 20$.

3. The domain is [0, 20], so the graph is cut off at $t = 0$ and $t = 20$.

4. The heavy dots at the ends indicate the endpoints of the domain.

 using *Technology*

See the Technology Guide at the end of the chapter for comments on graphing this function using a TI-83/84 or Excel. The formula with inequalities used to graph this function in Excel also works on the various graphers online.

Example **4** Graphing More Complicated Piecewise-Defined Functions

Graph the function f specified by

$$f(x) = \begin{cases} -1 & \text{if } -4 \leq x < -1 \\ x & \text{if } -1 \leq x \leq 1 \\ x^2 - 1 & \text{if } 1 < x \leq 2 \end{cases}$$

Solution The domain of f is $[-4, 2]$, since $f(x)$ is only specified when $-4 \leq x \leq 2$. Further, the function changes formulas when $x = -1$ and $x = 1$.

To sketch the graph by hand, we first sketch the three graphs $y = -1$, $y = x$, and $y = x^2 - 1$, and then use the appropriate portion of each (Figure 6).

Note that solid dots indicate points on the graph, whereas the open dots indicate points *not* on the graph. For example, when $x = 1$, the inequalities in the formula tell us that we are to use the middle formula (x) rather than the bottom one ($x^2 - 1$). Thus, $f(1) = 1$, not 0, so we place a solid dot at $(1, 1)$ and an open dot at $(1, 0)$.

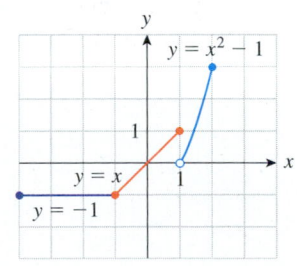

Figure **6**

We end this section with a list of some useful types of functions and their graphs (Table 2).

New Functions from Old: Scaled and Shifted Functions (Optional Section)

Online, follow:

Chapter 1
→ **Online Text**
  → **New Functions From Old: Scaled and Shifted Functions**

where you will find complete interactive text, examples, and exercises on scaling and translating the graph of a function by changing the formula.

Table 2 Functions and Their Graphs

Type of Function	Examples	
Linear $f(x) = mx + b$ m, b constant Graphs of linear functions are straight lines.	$y = x$	$y = -2x + 2$
Quadratic $f(x) = ax^2 + bx + c$ a, b, c constant ($a \neq 0$) Graphs of quadratic functions are called **parabolas.**	$y = x^2$	$y = -2x^2 + 2x + 4$
Technology formulas:	x^2	-2*x^2 + 2*x + 4
Cubic $f(x) = ax^3 + bx^2 + cx + d$ a, b, c, d constant ($a \neq 0$)	$y = x^3$	$y = -x^3 + 3x^2 + 1$
Technology formulas:	x^3	-x^3 + 3*x^2 + 1
Exponential $f(x) = Ab^x$ A, b constant ($b > 0$ and $b \neq 1$)	$y = 2^x$	$y = 4(0.5)^x$
Technology formulas:	2^x	4*0.5^x
Rational $f(x) = \dfrac{P(x)}{Q(x)};$ $P(x)$ and $Q(x)$ polynomials The graph of $y = 1/x$ is a **hyperbola.** The domain excludes zero since $1/0$ is not defined.	$y = \dfrac{1}{x}$	$y = \dfrac{x}{x - 1}$
Technology formulas:	1/x	x/(x - 1)

Table **2** (*Continued*)

Type of Function	Examples	
Absolute value For x positive or zero, the graph of $y = |x|$ is the same as that of $y = x$. For x negative or zero, it is the same as that of $y = -x$.	$y = |x|$	$y = |2x + 2|$
Technology formulas:	abs(x)	abs(2*x + 2)
Square Root The domain of $y = \sqrt{x}$ must be restricted to the nonnegative numbers, since the square root of a negative number is not real. Its graph is the top half of a horizontally oriented parabola.	$y = \sqrt{x}$	$y = \sqrt{4x - 2}$
Technology Formulas:	x^0.5 or √(x)	(4*x-2)^0.5 or √(4*x-2)

1.2 EXERCISES

● denotes basic skills exercises

tech Ex indicates exercises that should be solved using technology

In Exercises 1–4, use the graph of the function f to find approximations of the given values. hint [see Example 1]

1. ●

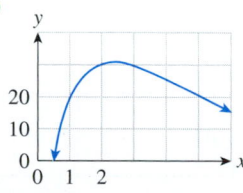

a. $f(1)$
b. $f(2)$
c. $f(3)$
d. $f(5)$
e. $f(3) - f(2)$

2. ●

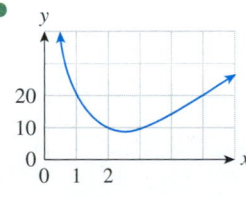

a. $f(1)$
b. $f(2)$
c. $f(3)$
d. $f(5)$
e. $f(3) - f(2)$

3. ●

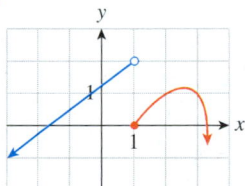

a. $f(-3)$ **b.** $f(0)$
c. $f(1)$ **d.** $f(2)$
e. $\dfrac{f(3) - f(2)}{3 - 2}$

4. ●

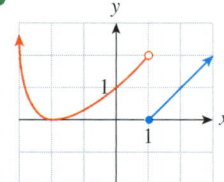

a. $f(-2)$ **b.** $f(0)$
c. $f(1)$ **d.** $f(3)$
e. $\dfrac{f(3) - f(1)}{3 - 1}$

In Exercises 5 and 6, match the functions to the graphs. Using technology to draw the graphs is suggested, but not required.

5. tech Ex
   a. $f(x) = x$ $(-1 \le x \le 1)$
   b. $f(x) = -x$ $(-1 \le x \le 1)$

c. $f(x) = \sqrt{x}$ $(0 < x < 4)$

d. $f(x) = x + \dfrac{1}{x} - 2$ $(0 < x < 4)$

e. $f(x) = |x|$ $(-1 \le x \le 1)$

f. $f(x) = x - 1$ $(-1 \le x \le 1)$

(I)

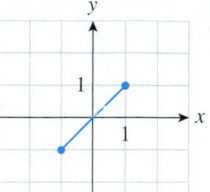

(II)

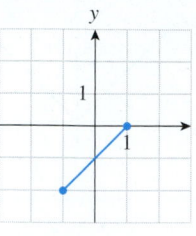

(III)

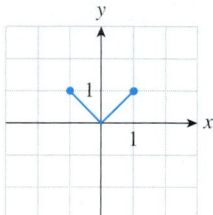

(IV)

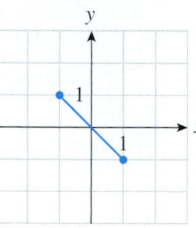

(V)

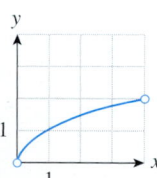

(VI)

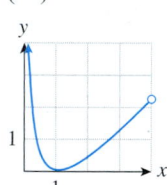

6. **tech** Ex

    a. $f(x) = -x + 4$ $(0 < x \le 4)$

    b. $f(x) = 2 - |x|$ $(-2 < x \le 2)$

    c. $f(x) = \sqrt{x + 2}$ $(-2 < x \le 2)$

    d. $f(x) = -x^2 + 2$ $(-2 < x \le 2)$

    e. $f(x) = \dfrac{1}{x} - 1$ $(0 < x \le 4)$

    f. $f(x) = x^2 - 1$ $(-2 < x \le 2)$

(I)

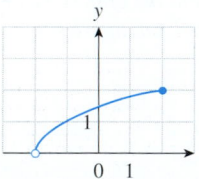

(II)

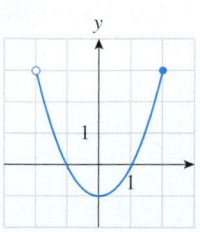

(III)

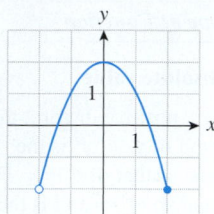

(IV)

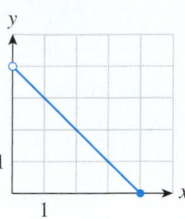

(V)

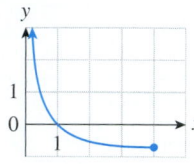

(VI)

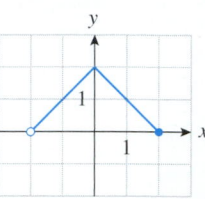

In Exercises 7–12, graph the given functions. Give the technology formula and use technology to check your graph. We suggest that you become familiar with these graphs, in addition to those in Table 2. *hint* [see Example 2]

7. ● $f(x) = -x^3$ (domain $(-\infty, +\infty)$)

8. ● $f(x) = x^3$ (domain $[0, +\infty)$)

9. ● $f(x) = x^4$ (domain $(-\infty, +\infty)$)

10. ● $f(x) = \sqrt[3]{x}$ (domain $(-\infty, +\infty)$)

11. ● $f(x) = \dfrac{1}{x^2}$ $(x \ne 0)$

12. ● $f(x) = x + \dfrac{1}{x}$ $(x \ne 0)$

In Exercises 13–18, sketch the graph of the given function, evaluate the given expressions, and then use technology to duplicate the graphs. Give the technology formula. *hint* [see Example 3]

13. ● $f(x) = \begin{cases} x & \text{if } -4 \le x < 0 \\ 2 & \text{if } 0 \le x \le 4 \end{cases}$

    a. $f(-1)$ **b.** $f(0)$ **c.** $f(1)$

14. ● $f(x) = \begin{cases} -1 & \text{if } -4 \le x \le 0 \\ x & \text{if } 0 < x \le 4 \end{cases}$

    a. $f(-1)$ **b.** $f(0)$ **c.** $f(1)$

15. ● $f(x) = \begin{cases} x^2 & \text{if } -2 < x \le 0 \\ 1/x & \text{if } 0 < x \le 4 \end{cases}$

    a. $f(-1)$ **b.** $f(0)$ **c.** $f(1)$

16. ● $f(x) = \begin{cases} -x^2 & \text{if } -2 < x \le 0 \\ \sqrt{x} & \text{if } 0 < x < 4 \end{cases}$

    a. $f(-1)$ **b.** $f(0)$ **c.** $f(1)$

17. ● $f(x) = \begin{cases} x & \text{if } -1 < x \le 0 \\ x + 1 & \text{if } 0 < x \le 2 \\ x & \text{if } 2 < x \le 4 \end{cases}$ *hint* [see Example 4]

    a. $f(0)$ **b.** $f(1)$ **c.** $f(2)$ **d.** $f(3)$

18. ● $f(x) = \begin{cases} -x & \text{if } -1 < x < 0 \\ x - 2 & \text{if } 0 \leq x \leq 2 \\ -x & \text{if } 2 < x \leq 4 \end{cases}$

   a. $f(0)$ **b.** $f(1)$ **c.** $f(2)$ **d.** $f(3)$

Applications

Sales of Sport Utility Vehicles Exercises 19–22 refer to the following graph, which shows the number $f(t)$ of sports utility vehicles (SUVs) sold in the U.S. each year from 1990 through 2003 ($t = 0$ represents 1990, and $f(t)$ represents sales in year t in thousands of vehicles).[17]

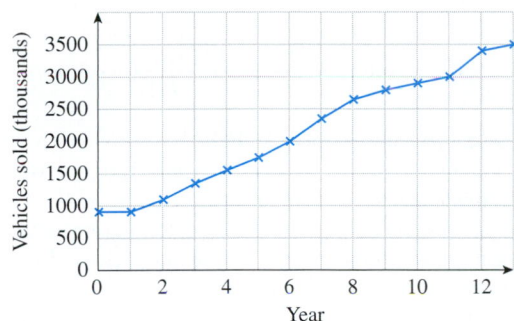

19. ● Estimate $f(6)$, $f(9)$, and $f(7.5)$. Interpret your answers.

20. ● Estimate $f(5)$, $f(11)$, and $f(1.5)$. Interpret your answers.

21. ● Which is larger: $f(6) - f(5)$ or $f(10) - f(9)$? Interpret the answer.

22. ● Which is larger: $f(10) - f(8)$ or $f(13) - f(11)$? Interpret the answer.

23. ● *Employment* The following graph shows the number $N(t)$ of people, in millions, employed in the U.S. (t is time in years, and $t = 0$ represents January 2000).[18]

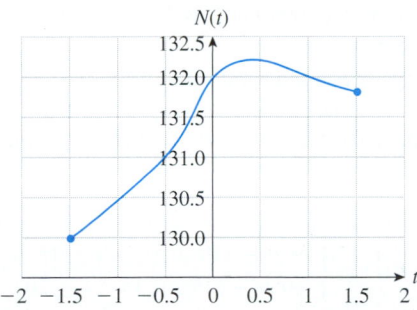

a. What is the domain of N?

b. Estimate $N(-0.5)$, $N(0)$, and $N(1)$. Interpret your answers.

c. On which interval is $N(t)$ falling? Interpret the result.

24. ● *Productivity* The following graph shows an index $P(t)$ of productivity in the U.S., where t is time in years, and $t = 0$ represents January 2000.[19]

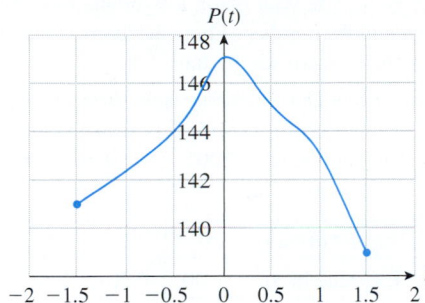

a. What is the domain of P?

b. Estimate $P(-0.5)$, $P(0)$, and $P(1.5)$. Interpret your answers.

c. On which interval is $P(t) \geq 144$? Interpret the result.

25. ● *Soccer Gear* The East Coast College soccer team is planning to buy new gear for its road trip to California. The cost per shirt depends on the number of shirts the team orders, as shown in the following table:

x (Shirts ordered)	5	25	40	100	125
$A(x)$ (Cost/shirt, \$)	22.91	21.81	21.25	21.25	22.31

a. Which of the following functions best models the data?

  (A) $A(x) = 0.005x + 20.75$

  (B) $A(x) = 0.01x + 20 + \dfrac{25}{x}$

  (C) $A(x) = 0.0005x^2 - 0.07x + 23.25$

  (D) $A(x) = 25.5(1.08)^{(x-5)}$

b. **tech** Ex Graph the model you chose in part (a) for $10 \leq x \leq 100$. Use your graph to estimate the lowest cost per shirt and the number of shirts the team should order to obtain the lowest price per shirt.

26. ● *Hockey Gear* The South Coast College hockey team wants to purchase wool hats for its road trip to Alaska. The cost per hat depends on the number of hats the team orders, as shown in the following table:

x (Hats ordered)	5	25	40	100	125
$A(x)$ (Cost/hat \$)	25.50	23.50	24.63	30.25	32.70

---

[17] 2000–2003 values were forecasts. Sources: Ford Motor Company/*The New York Times,* February 9, 1995, p. D17, Oak Ridge National Laboratory, Light Vehicle MPG and Market Shares System, AutoPacific, *The U.S. Car and Light Truck Market,* 1999, pp. 24, 120, 121.

[18] SOURCE: Haver Analytics: The Conference Board/*New York Times,* November 24, 2001.

[19] Ibid.

● basic skills   **tech** Ex technology exercise

a. Which of the following functions best models the data?

(A) $A(x) = 0.05x + 20.75$

(B) $A(x) = 0.1x + 20 + \dfrac{25}{x}$

(C) $A(x) = 0.0008x^2 - 0.07x + 23.25$

(D) $A(x) = 25.5(1.08)^{(x-5)}$

b. `tech` Ex Graph the model you chose in part (a) with $5 \le x \le 30$. Use your graph to estimate the lowest cost per hat and the number of hats the team should order to obtain the lowest price per hat.

27. ● *Value of Euro* The following table shows the approximate value V of one Euro in U.S. dollars from its introduction in January 2000 to January 2005. ($t = 0$ represents January, 2000.)[20]

t (Year)	0	2	5
V (Value in $)	1.00	0.90	1.30

Which of the following kinds of models would best fit the given data? Explain your choice of model. (A, a, b, c, and m are constants.)

(A) Linear: $V(t) = mt + b$

(B) Quadratic: $V(t) = at^2 + bt + c$

(C) Exponential: $V(t) = Ab^t$

28. ● *Household Income* The following table shows the approximate average household income in the U.S. in 1990, 1995, and 2003. ($t = 0$ represents 1990.)[21]

t (Year)	0	5	13
H (Household Income in $1000)	30	35	43

Which of the following kind of model would best fit the given data? Explain your choice of model. (A, a, b, c, and m are constants.)

(A) Linear: $H(t) = mt + b$

(B) Quadratic: $H(t) = at^2 + bt + c$

(C) Exponential: $H(t) = Ab^t$

29. `tech` Ex *Acquisition of Language* The percentage $p(t)$ of children who can speak at least single words by the age of t months can be approximated by the equation[22]

$$p(t) = 100 \left(1 - \frac{12{,}200}{t^{4.48}} \right) \quad (t \ge 8.5)$$

a. Give a technology formula for p.

b. Graph p for $8.5 \le t \le 20$ and $0 \le p \le 100$. Use your graph to answer parts (c) and (d).

c. What percentage of children can speak at least single words by the age of 12 months? (Round your answer to the nearest percentage point.)

d. By what age are 90% of children speaking at least single words? (Round your answer to the nearest month.)

30. `tech` Ex *Acquisition of Language* The percentage $p(t)$ of children who can speak in sentences of five or more words by the age of t months can be approximated by the equation[23]

$$p(t) = 100 \left(1 - \frac{5.27 \times 10^{17}}{t^{12}} \right) \quad (t \ge 30)$$

a. Give a technology formula for p.

b. Graph p for $30 \le t \le 45$ and $0 \le p \le 100$. Use your graph to answer parts (b) and (c).

c. What percentage of children can speak in sentences of five or more words by the age of 36 months? (Round your answer to the nearest percentage point.)

d. By what age are 75% of children speaking in sentences of five or more words? (Round your answer to the nearest month.)

31. ● *Processor Speeds* (Compare Exercise 33 in Section 1.1.) The processor speed, in megahertz, of Intel processors could be approximated by the following function of time t in years since the start of 1995:[24]

$$P(t) = \begin{cases} 75t + 200 & \text{if } 0 \le t \le 4 \\ 600t - 1900 & \text{if } 4 < t \le 9 \end{cases}$$

Sketch the graph of P and use your graph to estimate when processor speeds first reached 2.0 gigahertz (1 gigahertz = 1000 megahertz).

32. ● *Leading Economic Indicators* (Compare Exercise 34 in Section 1.1.) The value of the Conference Board Index of 10 economic indicators in the U.S. could be approximated by the following function of time t in months since the end of 2002:[25]

$$E(t) = \begin{cases} 0.4t + 110 & \text{if } 6 \le t \le 15 \\ -0.2t + 119 & \text{if } 15 < t \le 20 \end{cases}$$

Sketch the graph of E and use your graph to estimate when the index first reached 115.

---

[20] SOURCES: Bloomberg Financial Markets, International Monetary Fund/*New York Times*, May 18, 2003, p. I7

[21] In current dollars, unadjusted for inflation. SOURCE: U.S. Census Bureau; "Table H-5. Race and Hispanic Origin of Householder—Households by Median and Mean Income: 1967 to 2003;" published August 27, 2004; www.census.gov

[22] The model is the authors' and is based on data presented in the article *The Emergence of Intelligence* by William H. Calvin, *Scientific American*, October, 1994, pp. 101–107.

[23] Ibid.

[24] SOURCE: Sandpile.org/*New York Times*, May 17, 2004, p. C1.

[25] SOURCE: The Conference Board/*New York Times*, November 19, 2004, p. C7.

● basic skills   `tech` Ex technology exercise

33. `tech` Ex *Television Advertising* The cost, in millions of dollars, of a 30-second television ad during the Super Bowl in the years 1990–2001 can be approximated by the following piecewise linear function ($t = 0$ represents 1990):[26]

$$C(t) = \begin{cases} 0.08t + 0.6 & \text{if } 0 \le t < 8 \\ 0.355t - 1.6 & \text{if } 8 \le t \le 11 \end{cases}$$

a. Give a technology formula for C and use technology to graph the function C.

b. Based on the graph, a Superbowl ad first exceeded $2 million in what year?

34. `tech` Ex *Internet Purchases* The percentage $p(t)$ of buyers of new cars who used the Internet for research or purchase each year since 1997 is given by the following function[27] ($t = 0$ represents 1997):

$$p(t) = \begin{cases} 10t + 15 & \text{if } 0 \le t < 1 \\ 15t + 10 & \text{if } 1 \le t \le 4 \end{cases}$$

a. Give a technology formula for p and use technology to graph the function p.

b. Based on the graph, 50% or more of all new car buyers used the Internet for research or purchase in what years?

---

[26] SOURCE: *New York Times,* January 26, 2001, p. C1.

[27] Model is based on data through 2000 (the 2000 value is estimated). SOURCE: J.D. Power Associates/*The New York Times,* January 25, 2000, p. C1.

Communication and Reasoning Exercises

35. ● True or false: Every graphically specified function can also be specified numerically. Explain.

36. ● True or false: Every algebraically specified function can also be specified graphically. Explain.

37. ● True or false: Every numerically specified function with domain [0, 10] can also be specified graphically. Explain.

38. ● True or false: Every graphically specified function can also be specified algebraically. Explain.

39. ● How do the graphs of two functions differ if they are specified by the same formula but have different domains?

40. ● How do the graphs of two functions $f(x)$ and $g(x)$ differ if $g(x) = f(x) + 10$? (Try an example.)

41. How do the graphs of two functions $f(x)$ and $g(x)$ differ if $g(x) = f(x - 5)$? (Try an example.)

42. How do the graphs of two functions $f(x)$ and $g(x)$ differ if $g(x) = f(-x)$? (Try an example.)

● basic skills   `tech` Ex technology exercise

1.3 Linear Functions

Linear functions are among the simplest functions and are perhaps the most useful of all mathematical functions.

quick Example

Linear Function

A **linear function** is one that can be written in the form

$$f(x) = mx + b \qquad \text{Function form}$$

or

$$y = mx + b \qquad \text{Equation form}$$

where m and b are fixed numbers (the names m and b are traditional[*]).

$f(x) = 3x - 1$

$y = 3x - 1$

---

[*] Actually, c is sometimes used instead of b. As for m, there has even been some research lately into the question of its origin, but no one knows exactly why the letter m is used.

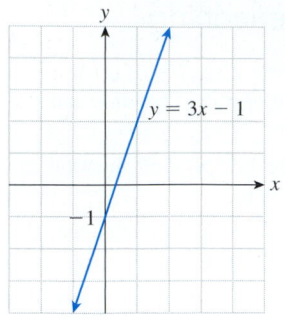

Figure **7**

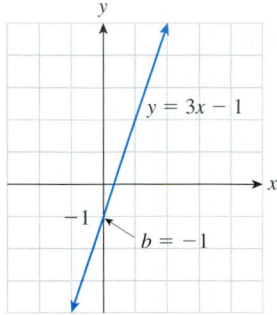

Figure **8**
y-intercept = b = −1
Graphically, b is the
y-intercept of the graph

Linear Functions from the Numerical and Graphical Point of View

The following table shows values of $y = 3x − 1$ ($m = 3$, $b = −1$) for some values of x:

x	−4	−3	−2	−1	0	1	2	3	4
y	−13	−10	−7	−4	−1	2	5	8	11

Its graph is shown in Figure 7.

Looking first at the table, notice that that setting $x = 0$ gives $y = −1$, the value of b.

Numerically, b is the value of y when x = 0

On the graph, the corresponding point $(0, −1)$ is the point where the graph crosses the y-axis, and we say that $b = −1$ is the **y-intercept** of the graph (Figure 8).

What about m? Looking once again at the table, notice that y increases by $m = 3$ units for every increase of 1 unit in x. This is caused by the term $3x$ in the formula: for every increase of 1 in x we get an increase of $3 \times 1 = 3$ in y.

Numerically, y increases by m units for every 1-unit increase of x

Likewise, for every increase of 2 in x we get an increase of $3 \times 2 = 6$ in y. In general, if x increases by some amount, y will increase by three times that amount. We write:

$$\text{Change in } y = 3 \times \text{Change in } x$$

The Change in a Quantity: Delta Notation

If a quantity q changes from q_1 to q_2, the **change in** q is just the difference:

$$\text{Change in } q = \text{Second value} − \text{First value}$$
$$= q_2 − q_1$$

Mathematicians traditionally use Δ (delta, the Greek equivalent of the Roman letter D) to stand for change, and write the change in q as Δq.

$$\Delta q = \text{Change in } q = q_2 − q_1$$

quick Examples

1. If x is changed from 1 to 3, we write

$$\Delta x = \text{Second value} − \text{First value} = 3 − 1 = 2$$

2. Looking at our linear function, we see that when x changes from 1 to 3, y changes from 2 to 8. So,

$$\Delta y = \text{Second value} − \text{First value} = 8 − 2 = 6$$

Using delta notation, we can now write, for our linear function,

$$\Delta y = 3\Delta x \qquad \text{Change in } y = 3 \times \text{Change in } x$$

or

$$\frac{\Delta y}{\Delta x} = 3$$

Because the value of y increases by exactly 3 units for every increase of 1 unit in x, the graph is a straight line rising by 3 units for every 1 unit we go to the right. We say that we have a **rise** of 3 units for each **run** of 1 unit. Because the value of y changes by $\Delta y = 3\Delta x$ units for every change of Δx units in x, in general we have a rise of $\Delta y = 3\Delta x$ units for each run of Δx units (Figure 9). Thus, we have a rise of 6 for a run of 2, a rise of 9 for a run of 3, and so on. So, $m = 3$ is a measure of the steepness of the line; we call m the **slope of the line:**

$$\text{Slope} = m = \frac{\Delta y}{\Delta x} = \frac{\text{Rise}}{\text{Run}}$$

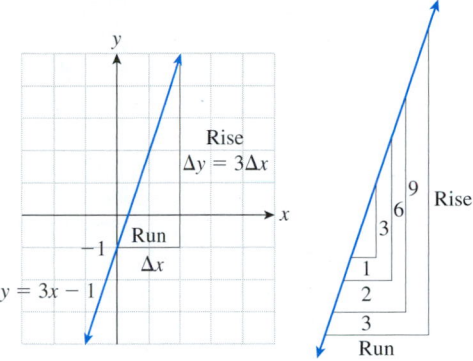

Figure **9**
Slope $= m = 3$
Graphically, m is the slope of the graph.

In general (replace the number 3 by a general number m), we can say the following.

The Roles of *m* and *b* in the Linear Function *f(x)* = *mx* + *b*

Role of *m*
Numerically If $y = mx + b$, then y changes by m units for every 1-unit change in x. A change of Δx units in x results in a change of $\Delta y = m\Delta x$ units in y. Thus,

$$m = \frac{\Delta y}{\Delta x} = \frac{\text{Change in } y}{\text{Change in } x}$$

Graphically m is the slope of the line $y = mx + b$:

$$m = \frac{\Delta y}{\Delta x} = \frac{\text{Rise}}{\text{Run}} = \text{Slope}$$

For positive m, the graph rises m units for every 1-unit move to the right, and rises $\Delta y = m\Delta x$ units for every Δx units moved to the right. For negative m, the graph drops $|m|$ units for every 1-unit move to the right, and drops $|m|\Delta x$ units for every Δx units moved to the right.

Graph of *y* = *mx* + *b*

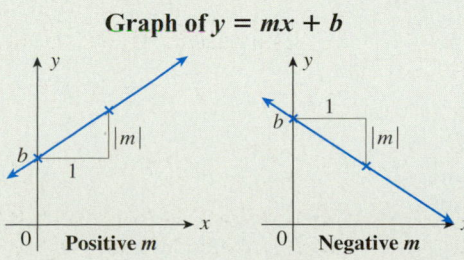

Role of b
Numerically When $x = 0$, $y = b$
Graphically b is the y-intercept of the line $y = mx + b$.

quick Examples

1. $f(x) = 2x + 1$ has slope $m = 2$ and y-intercept $b = 1$. To sketch the graph, we start at the y-intercept $b = 1$ on the y-axis, and then move 1 unit to the right and up $m = 2$ units to arrive at a second point on the graph. Now connect the two points to obtain the graph on the left.

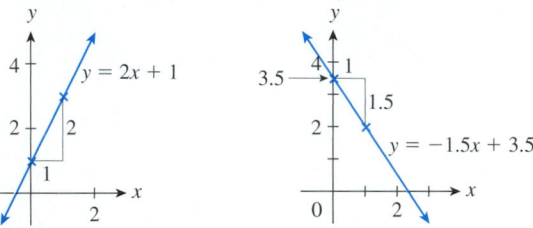

2. The line $y = -1.5x + 3.5$ has slope $m = -1.5$ and y-intercept $b = 3.5$. Since the slope is negative, the graph (above right) goes *down* 1.5 units for every 1 unit it moves to the right.

It helps to be able to picture what different slopes look like, as in Figure 10. Notice that the larger the absolute value of the slope, the steeper is the line.

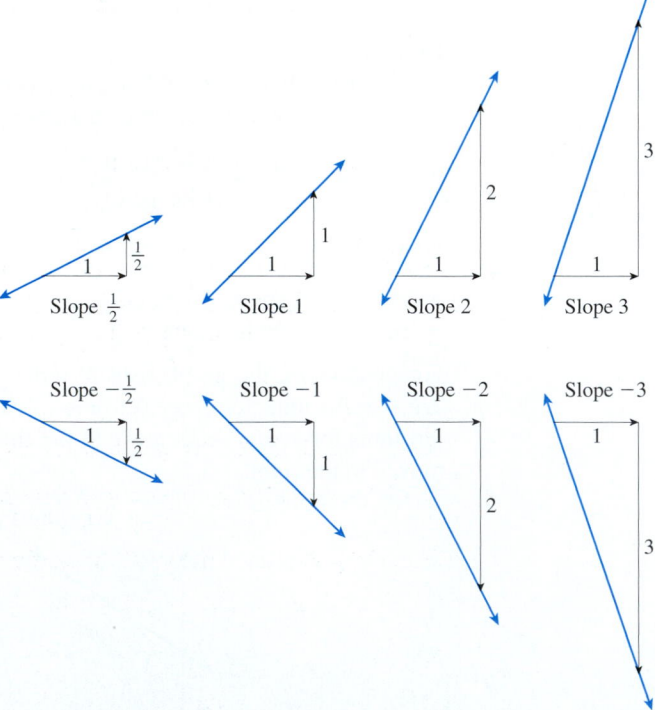

Figure **10**

Example 1 Recognizing Linear Data Numerically and Graphically

Which of the following two tables gives the values of a linear function? What is the formula for that function?

x	0	2	4	6	8	10	12
$f(x)$	3	-1	-3	-6	-8	-13	-15

x	0	2	4	6	8	10	12
$g(x)$	3	-1	-5	-9	-13	-17	-21

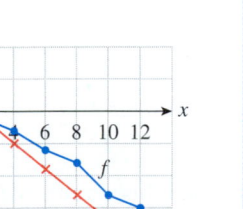
using *Technology*

Consult the Technology Guides at the end of the chapter to see how to generate tables showing the ratios $\Delta f / \Delta x$ and $\Delta g / \Delta x$. These tables show at a glance that f is not linear.

Solution The function f cannot be linear: If it were, we would have $\Delta f = m \Delta x$ for some fixed number m. However, although the change in x between successive entries in the table is $\Delta x = 2$ each time, the change in f is not the same each time. Thus, the ratio $\Delta f / \Delta x$ is not the same for every successive pair of points.

On the other hand, the ratio $\Delta g / \Delta x$ is the same each time, namely,

$$\frac{\Delta g}{\Delta x} = \frac{-4}{2} = -2$$

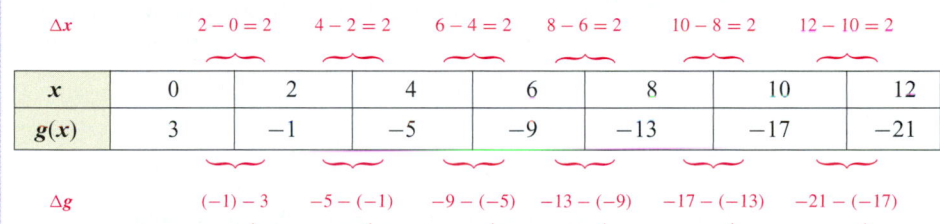

Δx		$2 - 0 = 2$	$4 - 2 = 2$	$6 - 4 = 2$	$8 - 6 = 2$	$10 - 8 = 2$	$12 - 10 = 2$
x	0	2	4	6	8	10	12
$g(x)$	3	-1	-5	-9	-13	-17	-21
Δg		$(-1) - 3$ $= -4$	$-5 - (-1)$ $= -4$	$-9 - (-5)$ $= -4$	$-13 - (-9)$ $= -4$	$-17 - (-13)$ $= -4$	$-21 - (-17)$ $= -4$

Thus, g is linear with slope $m = -2$. By the table, $g(0) = 3$, hence $b = 3$. Thus,

$$g(x) = -2x + 3 \qquad \text{Check that this formula gives the values in the table}$$

If you graph the points in the tables defining f and g above, it becomes easy to see that g is linear and f is not; the points of g lie on a straight line (with slope -2), whereas the points of f do not lie on a straight line (Figure 11).

Figure **11**

Example 2 Graphing a Linear Equation by Hand: Intercepts

Graph the equation $x + 2y = 4$. Where does the line cross the x- and y-axes?

Solution We first write y as a linear function of x by solving the equation for y.

$$2y = -x + 4$$

so

$$y = -\frac{1}{2}x + 2$$

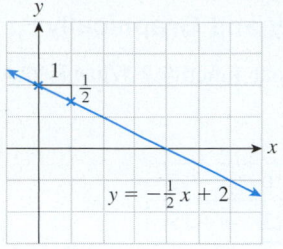

Figure 12

Now we can see that the graph is a straight line with a slope of $-1/2$ and a y-intercept of 2. We start at 2 on the y-axis and go down 1/2 unit for every 1 unit we go to the right. The graph is shown in Figure 12.

We already know that the line crosses the y-axis at 2. Where does it cross the x-axis? Wherever that is, we know that the y-coordinate will be 0 at that point. So, we set $y = 0$ and solve for x. It's most convenient to use the equation we were originally given:

$$x + 2(0) = 4 \qquad \text{\color{red}Original equation with } x = 0$$

so

$$x = 4$$

The line crosses the x-axis at 4.

➕ *Before we go on...* We could have graphed the equation in Example 2 another way, by first finding the intercepts. Once we know that the line crosses the y-axis at 2 and the x-axis at 4, we can draw those two points and then draw the line connecting them. ■

We now summarize the procedure for finding the intercepts of a line.

Finding the Intercepts

The x-**intercept** of a line is where it crosses the x-axis. To find it, set $y = 0$ and solve for x. The y-**intercept** is where it crosses the y-axis. If the equation of the line is written in as $y = mx + b$, then b is the y-intercept. Otherwise, set $x = 0$ and solve for y.

quick Example

Consider the equation $3x - 2y = 6$. To find its x-intercept, set $y = 0$ to find $x = 6/3 = 2$. To find its y-intercept, set $x = 0$ to find $y = 6/(-2) = -3$. The line crosses the x-axis at 2 and the y-axis at -3.

Computing the Slope of a Line

We know that the slope of a line is given by

$$\text{Slope} = m = \frac{\text{Rise}}{\text{Run}} = \frac{\Delta y}{\Delta x}$$

Recall that two points—say (x_1, y_1) and (x_2, y_2)—determine a line in the xy-plane. To find its slope, we need a run Δx and corresponding rise Δy. In Figure 13, we see that we can use $\Delta x = x_2 - x_1$, the change in the x-coordinate from the first point to the second, as our run, and $\Delta y = y_2 - y_1$, the change in the y-coordinate, as our rise. The resulting formula for computing the slope is given below.

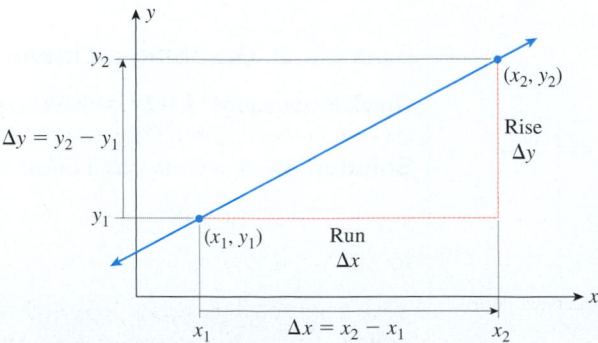

Figure 13

Computing the Slope of a Line

We can compute the slope m of the line through the points (x_1, y_1) and (x_2, y_2) using

$$m = \frac{\Delta y}{\Delta x} = \frac{y_2 - y_1}{x_2 - x_1}$$

quick Examples

1. The slope of the line through $(x_1, y_1) = (1, 3)$ and $(x_2, y_2) = (5, 11)$ is

$$m = \frac{\Delta y}{\Delta x} = \frac{y_2 - y_1}{x_2 - x_1} = \frac{11 - 3}{5 - 1} = \frac{8}{4} = 2$$

2. The slope of the line through $(x_1, y_1) = (1, 2)$ and $(x_2, y_2) = (2, 1)$ is

$$m = \frac{\Delta y}{\Delta x} = \frac{y_2 - y_1}{x_2 - x_1} = \frac{1 - 2}{2 - 1} = \frac{-1}{1} = -1$$

Q: *What if we had chosen to list the two points in Quick Example 1 in reverse order? That is, suppose we had taken $(x_1, y_1) = (5, 11)$ and $(x_2, y_2) = (1, 3)$. What would have been the effect on the computation of the slope?*

A: We would have found

$$m = \frac{\Delta y}{\Delta x} = \frac{y_2 - y_1}{x_2 - x_1} = \frac{3 - 11}{1 - 5} = \frac{-8}{-4} = 2$$

the same answer. The order in which we take the points is not important, *as long as we use the same order in the numerator and the denominator.* ∎

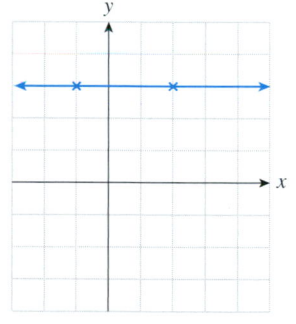

Figure **14**

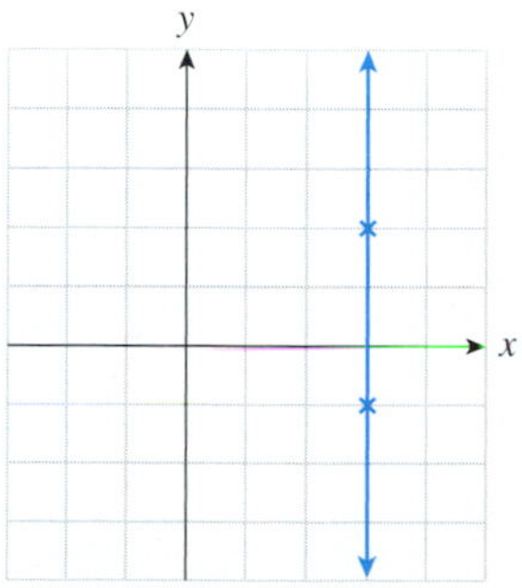

Example **3** Special Slopes

Find the slope of the line through $(2, 3)$ and $(-1, 3)$ and the slope of the line through $(3, 2)$ and $(3, -1)$.

Solution The line through $(2, 3)$ and $(-1, 3)$ has slope

$$m = \frac{\Delta y}{\Delta x} = \frac{3 - 3}{-1 - 2} = \frac{0}{-3} = 0$$

A line of slope 0 has 0 rise, so is a *horizontal* line, as shown in Figure 14. The line through $(3, 2)$ and $(3, -1)$ has slope

$$m = \frac{\Delta y}{\Delta x} = \frac{-1 - 2}{3 - 3} = \frac{-3}{0}$$

which is undefined. If we plot the two points, we see that the line passing through them is *vertical,* as shown in Figure 15.

Vertical lines have undefined slope.

Figure **15**

Finding a Linear Equation from Data: How to Make a Linear Model

If we happen to know the slope and y-intercept of a line, writing down its equation is straightforward. For example, if we know that the slope is 3 and the y-intercept is -1, then the equation is $y = 3x - 1$. Sadly, the information we are given is seldom so convenient. For instance, we may know the slope and a point other than the y-intercept, two points on the line, or other information.

We describe a straightforward method for finding the equation of a line: the **point-slope** method. As the name suggests, we need two pieces of information:

- The *slope m* (which specifies the direction of the line)
- A *point* (x_1, y_1) on the line (which pins down its location in the plane)

The equation of the line through the point (x_1, y_1) with slope m must have the form

$$y = mx + b$$

for some (unknown) number b. To determine b we use the fact that the line must pass through the point (x_1, y_1), so that (x_1, y_1) satisfies the equation $y = mx + b$. In other words,

$$y_1 = mx_1 + b$$

Solving for b gives

$$b = y_1 - mx_1$$

Summarizing:

The Point-Slope Formula

An equation of the line through the point (x_1, y_1) with slope m is given by

$$y = mx + b \qquad \text{Equation form}$$

where

$$b = y_1 - mx_1$$

quick Example   The line through $(2, 3)$ with slope 4 has equation

$$y = 4x + b, \quad \text{where } b = 3 - (4)(2) = -5, \text{ so } y = 4x - 5$$

Q: *When do we use the point-slope formula rather than the slope-intercept form* **?**

A: Use the point-slope formula to find the equation of a line when you are given information about a point and the slope of the line. The formula does not apply if the slope is undefined, as in a vertical line; see Example 4(d) below. The slope-intercept form is more useful for graphing a line whose equation you already have. ∎

Example 4 Using the Point-Slope Formula

Find equations for the following straight lines.

a. Through the points $(1, 2)$ and $(3, -1)$
b. Through $(2, -2)$ and parallel to the line $3x + 4y = 5$
c. Horizontal and through $(-9, 5)$
d. Vertical and through $(-9, 5)$

Solution In each case other than (d), we apply the point-slope formula.

a. To apply the point-slope formula, we need

- *Point* We have two to choose from, so we take the first, $(x_1, y_1) = (1, 2)$.
- *Slope* Not given directly, but we do have enough information to calculate it. Since we are given two points on the line, we can use the slope formula:

$$m = \frac{y_2 - y_1}{x_2 - x_1} = \frac{-1 - 2}{3 - 1} = -\frac{3}{2}$$

An equation of the line is therefore

$$y = -\frac{3}{2}x + b$$

where $b = y_1 - mx_1 = 2 - \left(-\frac{3}{2}\right)(1) = \frac{7}{2}$, so

$$y = -\frac{3}{2}x + \frac{7}{2}$$

b. Proceeding as before,

- *Point* Given here as $(2, -2)$.
- *Slope* We use the fact that *parallel lines have the same slope*. (Why?) We can find the slope of $3x + 4y = 5$ by solving for y and then looking at the coefficient of x:

$$y = -\frac{3}{4}x + \frac{5}{4} \qquad \textcolor{red}{\text{To find the slope, solve for } y.}$$

so the slope is $-3/4$.

An equation for the desired line is

$$y = -\frac{3}{4}x + b$$

where $b = y_1 - mx_1 = -2 - \left(-\frac{3}{4}\right)(2) = -\frac{1}{2}$

so $\qquad y = -\frac{3}{4}x - \frac{1}{2}$

c. We are given a point: $(-9, 5)$. Furthermore, we are told that the line is horizontal, which tells us that the slope is 0. Therefore, we get

$$y = 0x + b = b$$

where $b = y_1 - mx_1 = 5 - (0)(-9) = 5$

so $\qquad y = 5$

d. We are given a point: $(-9, 5)$. This time, we are told that the line is vertical, which means that the slope is undefined. Thus, we can't use the point-slope formula. (That formula makes sense only when the slope of the line is defined.) What can we do? Well, here are some points on the desired line:

$$(-9, 1), (-9, 2), (-9, 3), \ldots,$$

so $x = -9$ and $y = $ *anything*. If we simply say that $x = -9$, then these points are all solutions, so the equation is $x = -9$.

1.3 EXERCISES

● denotes basic skills exercises

tech Ex indicates exercises that should be solved using technology

In Exercises 1–6, a table of values for a linear function is given. Fill in the missing value and calculate m in each case.

1. ●

x	−1	0	1
y	5	8	

2. ●

x	−1	0	1
y	−1	−3	

3. ●

x	2	3	5
f(x)	−1	−2	

4. ●

x	2	4	5
f(x)	−1	−2	

5. ●

x	−2	0	2
f(x)	4		10

6. ●

x	0	3	6
f(x)	−1		−5

In Exercises 7–10, first find f(0), if not supplied, and then find the equation of the given linear function.

7. ●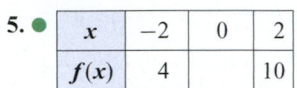

x	−2	0	2	4
f(x)	−1	−2	−3	−4

8. ●

x	−6	−3	0	3
f(x)	1	2	3	4

9. ●

x	−4	−3	−2	−1
f(x)	−1	−2	−3	−4

10. ●

x	1	2	3	4
f(x)	4	6	8	10

In each of Exercises 11–14, decide which of the two given functions is linear and find its equation. hint [see Example 1]

11. ●

x	0	1	2	3	4
f(x)	6	10	14	18	22
g(x)	8	10	12	16	22

12. ●

x	−10	0	10	20	30
f(x)	−1.5	0	1.5	2.5	3.5
g(x)	−9	−4	1	6	11

13. ●

x	0	3	6	10	15
f(x)	0	3	5	7	9
g(x)	−1	5	11	19	29

14. ●

x	0	3	5	6	9
f(x)	2	6	9	12	15
g(x)	−1	8	14	17	26

In Exercises 15–24, find the slope of the given line, if it is defined.

15. ● $y = -\dfrac{3}{2}x - 4$ **16.** ● $y = \dfrac{2x}{3} + 4$

17. ● $y = \dfrac{x+1}{6}$ **18.** ● $y = -\dfrac{2x-1}{3}$

19. ● $3x + 1 = 0$ **20.** ● $8x - 2y = 1$

21. ● $3y + 1 = 0$ **22.** ● $2x + 3 = 0$

23. ● $4x + 3y = 7$ **24.** ● $2y + 3 = 0$

In Exercises 25–38, graph the given equation.
hint [see Example 2]

25. ● $y = 2x - 1$ **26.** ● $y = x - 3$

27. ● $y = -\frac{2}{3}x + 2$ **28.** ● $y = -\frac{1}{2}x + 3$

29. ● $y + \frac{1}{4}x = -4$ **30.** ● $y - \frac{1}{4}x = -2$

31. ● $7x - 2y = 7$ **32.** ● $2x - 3y = 1$

33. ● $3x = 8$ **34.** ● $2x = -7$

35. ● $6y = 9$ **36.** ● $3y = 4$

37. ● $2x = 3y$ **38.** ● $3x = -2y$

In Exercises 39–54, calculate the slope, if defined, of the straight line through the given pair of points. Try to do as many as you can without writing anything down except the answer.
hint [see Example 3]

39. ● $(0, 0)$ and $(1, 2)$ **40.** ● $(0, 0)$ and $(−1, 2)$

41. ● $(−1, −2)$ and $(0, 0)$ **42.** ● $(2, 1)$ and $(0, 0)$

43. ● $(4, 3)$ and $(5, 1)$ **44.** ● $(4, 3)$ and $(4, 1)$

45. ● $(1, −1)$ and $(1, −2)$ **46.** ● $(−2, 2)$ and $(−1, −1)$

47. ● $(2, 3.5)$ and $(4, 6.5)$ **48.** ● $(10, −3.5)$ and $(0, −1.5)$

49. ● $(300, 20.2)$ and $(400, 11.2)$

50. ● $(1, −20.2)$ and $(2, 3.2)$

51. ● $(0, 1)$ and $\left(−\frac{1}{2}, \frac{3}{4}\right)$

52. ● $\left(\frac{1}{2}, 1\right)$ and $\left(−\frac{1}{2}, \frac{3}{4}\right)$

53. ● (a, b) and (c, d) $(a \neq c)$

54. ● (a, b) and (c, b) $(a \neq c)$

● basic skills   tech Ex technology exercise

55. ● In the following figure, estimate the slopes of all line segments.

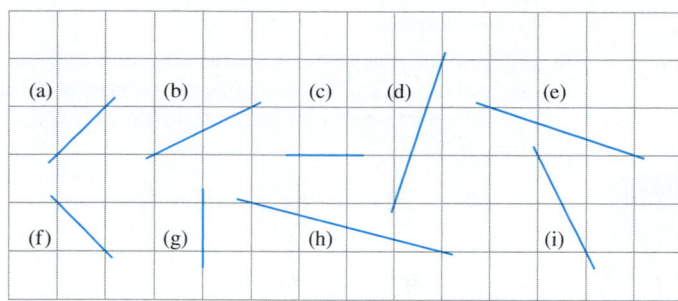

56. ● In the following figure, estimate the slopes of all line segments.

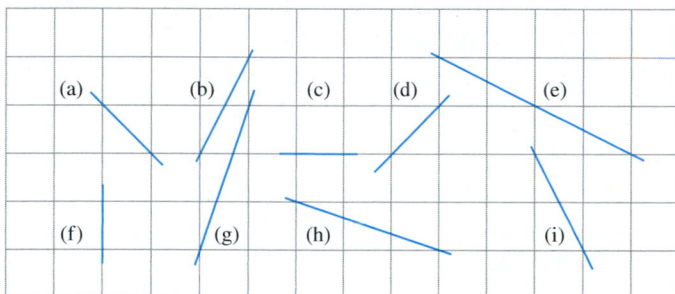

In Exercises 57–70, find a linear equation whose graph is the straight line with the given properties. hint [see Example 4]

57. ● Through $(1, 3)$ with slope 3

58. ● Through $(2, 1)$ with slope 2

59. ● Through $(1, -\frac{3}{4})$ with slope $\frac{1}{4}$

60. ● Through $(0, -\frac{1}{3})$ with slope $\frac{1}{3}$

61. ● Through $(20, -3.5)$ and increasing at a rate of 10 units of y per unit of x

62. ● Through $(3.5, -10)$ and increasing at a rate of 1 unit of y per 2 units of x.

63. ● Through $(2, -4)$ and $(1, 1)$

64. ● Through $(1, -4)$ and $(-1, -1)$

65. ● Through $(1, -0.75)$ and $(0.5, 0.75)$

66. ● Through $(0.5, -0.75)$ and $(1, -3.75)$

67. ● Through $(6, 6)$ and parallel to the line $x + y = 4$

68. ● Through $(1/3, -1)$ and parallel to the line $3x - 4y = 8$

69. ● Through $(0.5, 5)$ and parallel to the line $4x - 2y = 11$

70. ● Through $(1/3, 0)$ and parallel to the line $6x - 2y = 11$

Communication and Reasoning Exercises

71. ● How would you test a table of values of x and y to see if it comes from a linear function?

72. ● You have ascertained that a table of values of x and y corresponds to a linear function. How do you find an equation for that linear function?

73. ● To what linear function of x does the linear equation $ax + by = c$ $(b \neq 0)$ correspond? Why did we specify $b \neq 0$?

74. ● Complete the following. The slope of the line with equation $y = mx + b$ is the number of units that _____ increases per unit increase in _____.

75. ● Complete the following. If, in a straight line, y is increasing three times as fast as x, then its _____ is _____.

76. ● Suppose that y is decreasing at a rate of 4 units per 3-unit increase of x. What can we say about the slope of the linear relationship between x and y? What can we say about the intercept?

● basic skills   tech Ex technology exercise

77. ● If y and x are related by the linear expression $y = mx + b$, how will y change as x changes if m is positive? negative? zero?

78. ● Your friend April tells you that $y = f(x)$ has the property that, whenever x is changed by Δx, the corresponding change in y is $\Delta y = -\Delta x$. What can you tell her about f?

79. tech Ex Consider the following worksheet:

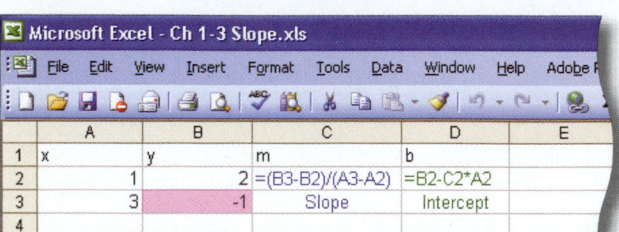

What is the effect on the slope of increasing the y-coordinate of the second point (the point whose coordinates are in Row 3)? Explain.

80. tech Ex Referring to the worksheet in Exercise 79, what is the effect on the slope of increasing the x-coordinate of the second point (the point whose coordinates are in row 3)? Explain.

● basic skills   tech Ex technology exercise

1.4 Linear Models

Using linear functions to describe or approximate relationships in the real world is called **linear modeling.** We start with some examples involving cost, revenue, and profit.

Cost, Revenue, and Profit Functions

Example 1 Linear Cost Function

As of January, 2005, Yellow Cab Chicago's rates were $1.90 on entering the cab plus $1.60 for each mile.[*]

a. Find the cost C of an x-mile trip.

b. Use your answer to calculate the cost of a 40-mile trip.

c. What is the cost of the second mile? What is the cost of the tenth mile?

d. Graph C as a function of x.

Solution

a. We are being asked to find how the cost C depends on the length x of the trip, or to find C as a function of x. Here is the cost in a few cases:

Cost of a 1-mile trip: $C = 1.60(1) + 1.90 = 3.50$ 1 mile @ $1.60 per mile plus $1.90

Cost of a 2-mile trip: $C = 1.60(2) + 1.90 = 5.10$ 2 miles @ $1.60 per mile plus $1.90

Cost of a 3-mile trip: $C = 1.60(3) + 1.90 = 6.70$ 3 miles @ $1.60 per mile plus $1.90

Do you see the pattern? The cost of an x-mile trip is given by the linear function:

$$C(x) = 1.60x + 1.90$$

Notice that the slope 1.60 is the incremental cost per mile. In this context we call 1.60 the **marginal cost;** the varying quantity $1.60x$ is called the **variable cost.** The

Photodisc/Getty Images

_____
[*]According to their website at www.yellowcabchicago.com/.

C-intercept 1.90 is the cost to enter the cab, which we call the **fixed cost.** In general, a linear cost function has the following form:

$$C(x) = \overbrace{mx}^{\text{Variable cost}} + b$$

Marginal cost   Fixed cost

b. We can use the formula for the cost function to calculate the cost of a 40-mile trip as:

$$C(40) = 1.60(40) + 1.90 = \$65.90$$

c. To calculate the cost of the second mile, we *could* proceed as follows:

Find the cost of a 1-mile trip: $C(1) = 1.60(1) + 1.90 = \$3.50$

Find the cost of a 2-mile trip: $C(2) = 1.60(2) + 1.90 = \$5.10$

Therefore, the cost of the second mile is $\$5.10 - \$3.50 = \$1.60$

But notice that this is just the marginal cost. In fact, the marginal cost is the cost of each additional mile, so we could have done this more simply:

Cost of second mile = Cost of tenth mile = Marginal cost = $1.60

d. Figure 16 shows the graph of the cost function, which we can interpret as a *cost vs. miles* graph. The fixed cost is the starting height on the left, while the marginal cost is the slope of the line.

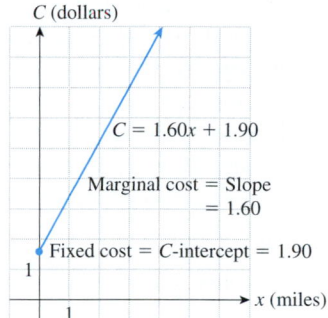

C (dollars)

$C = 1.60x + 1.90$

Marginal cost = Slope = 1.60

Fixed cost = *C*-intercept = 1.90

x (miles)

Figure **16**

✚ *Before we go on...* In general, the slope *m* measures the number of units of change in *y* per 1-unit change in *x*, so *we measure m in units of y per unit of x:*

Units of Slope = Units of *y* per unit of *x*

In Example 1, *y* is the cost *C*, measured in dollars, and *x* is the length of a trip, measured in miles. Hence,

Units of Slope = Units of *y* per Unit of *x* = Dollars per mile

The *y*-intercept *b*, being a value of *y*, is measured in the same units as *y*. In Example 1, *b* is measured in dollars. ∎

Here is a summary of the terms used in the preceding example, along with an introduction to some new terms.

Cost, Revenue, and Profit Functions

A **cost function** specifies the cost *C* as a function of the number of items *x*. Thus, $C(x)$ is the cost of *x* items. A cost function of the form

$$C(x) = mx + b$$

is called a **linear cost function.** The quantity *mx* is called the **variable cost** and the intercept *b* is called the **fixed cost.** The slope *m*, the **marginal cost,** measures the incremental cost per item.

Esteban Silva

TITLE Owner
INSTITUTION Regimen

Regimen is a retail shop and on-line merchant of high-end men's grooming products, a small business venture under my development. I came up with this concept in order fill the growing demand for men's grooming products from both graying baby boomers wanting to retain their competitive edge and young men who are increasingly accepting of the idea that is essential to be well styled and well groomed. The currently $3.5 billion a year men's grooming market is ever-expanding and there is tremendous opportunity for Regimen to take advantage of this untapped potential.

In the initial stages of this business venture I have relied on math to calculate the amount of capital needed to launch and sustain the business until it becomes profitable. Using spreadsheets I input projected sales figures and estimated monthly expenses to formulate if it is possible to realistically meet targets and achieve break-even in a timely matter. With assistance from a professional interior designer, I have drawn up plans which include space acquisition, contracting, and construction costs in order budget for build-out expenses.

I have teamed up with Yahoo! Small Business Solutions and devised an online advertising strategy which allows me to reach out to the niche customers my company's products are geared towards. Using a sponsored search method of advertising I pre-determine how much I am willing to spend for each combination of keywords which drive traffic onto my website via Yahoo! I can track on a daily basis the number of matches each combination of keywords are receiving and, therefore, determine if any of them need to be altered. It's very important that I analyze these figures frequently so I can redirect the limited marketing resources of this start-up company into the most effective channels available. Thankfully, the applied mathematics techniques I learned in college have helped me live the dream of owning my own business and being my own boss.

The **revenue** resulting from one or more business transactions is the total payment received, sometimes called the gross proceeds. If $R(x)$ is the revenue from selling x items at a price of m each, then R is the linear function $R(x) = mx$ and the selling price m can also be called the **marginal revenue.**

The **profit,** on the other hand, is the *net* proceeds, or what remains of the revenue when costs are subtracted. If the profit depends linearly on the number of items, the slope m is called the **marginal profit.** Profit, revenue, and cost are related by the following formula:

$$\text{Profit} = \text{Revenue} - \text{Cost}$$
$$P = R - C$$

If the profit is negative, say $-\$500$, we refer to a **loss** (of $500 in this case). To **break-even** means to make neither a profit nor a loss. Thus, break-even occurs when $P = 0$, or

$$R = C \qquad \text{Break-even}$$

The **break-even point** is the number of items x at which break-even occurs.

quick Example

If the daily cost (including operating costs) of manufacturing x T-shirts is $C(x) = 8x + 100$, and the revenue obtained by selling x T-shirts is $R(x) = 10x$, then the daily profit resulting from the manufacture and sale of x T-shirts is

$$P(x) = R(x) - C(x) = 10x - (8x + 100) = 2x - 100$$

Break-even occurs when $P(x) = 0$, or $x = 50$.

Example 2 Cost, Revenue, and Profit

The manager of the FrozenAir Refrigerator factory notices that on Monday it cost the company a total of $25,000 to build 30 refrigerators and on Tuesday it cost $30,000 to build 40 refrigerators.

a. Find a linear cost function based on this information. What is the daily fixed cost, and what is the marginal cost?

b. FrozenAir sells its refrigerators for $1500 each. What is the revenue function?

c. What is the profit function? How many refrigerators must FrozenAir sell in a day in order to break even for that day? What will happen if it sells fewer refrigerators? If it sells more?

Solution

a. We are seeking C as a linear function of x, the number of refrigerators sold:

$$C = mx + b$$

We are told that $C = 25,000$ when $x = 30$, and this amounts to being told that $(30, 25,000)$ is a point on the graph of the cost function. Similarly, $(40, 30,000)$ is another point on the line (Figure 17).

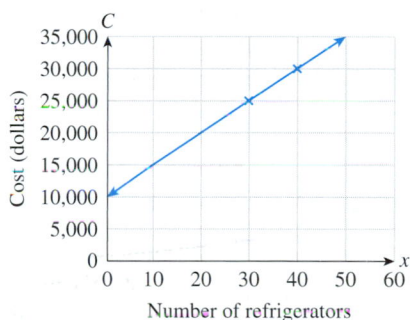

Figure **17**

We can now use the point-slope formula to construct a linear cost equation. Recall that we need two items of information: a point on the line and the slope:

- **Point** Let's use the first point: $(x_1, C_1) = (30, 25,000)$ C plays the role of y

- **Slope** $m = \dfrac{C_2 - C_1}{x_2 - x_1} = \dfrac{30,000 - 25,000}{40 - 30} = 500$ Marginal cost = $500

The cost function is therefore

$$C(x) = 500x + b$$

where $b = C_1 - mx_1 = 25,000 - (500)(30) = 10,000$ Fixed cost = $10,000

so $C(x) = 500x + 10,000$

Because $m = 500$ and $b = 10,000$ the factory's fixed cost is $10,000 each day, and its marginal cost is $500 per refrigerator.

b. The revenue FrozenAir obtains from the sale of a single refrigerator is $1500. So, if it sells x refrigerators, it earns a revenue of

$$R(x) = 1500x$$

c. For the profit, we use the formula

$$\text{Profit} = \text{Revenue} - \text{Cost}$$

For the cost and revenue, we can substitute the answers from parts (a) and (b) and obtain

$$P(x) = R(x) - C(x) \qquad \text{\textcolor{red}{Formula for profit}}$$

$$= 1500x - (500x + 10{,}000) \qquad \text{\textcolor{red}{Substitute } R(x) \text{ and } C(x)}$$

$$= 1000x - 10{,}000$$

Here, $P(x)$ is the daily profit FrozenAir makes by making and selling x refrigerators. Finally, to break even means to make zero profit. So, we need to find the x such that $P(x) = 0$. All we have to do is set $P(x) = 0$ and solve for x:

$$1000x - 10{,}000 = 0$$

giving

$$x = \frac{10{,}000}{1000} = 10$$

To break even, FrozenAir needs to manufacture and sell 10 refrigerators in a day.

For values of x less than the break-even point, 10, $P(x)$ is negative, so the company will have a loss. For values of x greater than the break-even point, $P(x)$ is positive, so the company will make a profit. This is the reason why we are interested in the point where $P(x) = 0$. Since $P(x) = R(x) - C(x)$, we can also look at the break-even point as the point where Revenue = Cost: $R(x) = C(x)$ (see Figure 18).

+ *Before we go on...* We can graph the cost and revenue functions from Example 2, and find the break-even point graphically (Figure 18):

$$\text{Cost: } C(x) = 500x + 10{,}000$$

$$\text{Revenue: } R(x) = 1500x$$

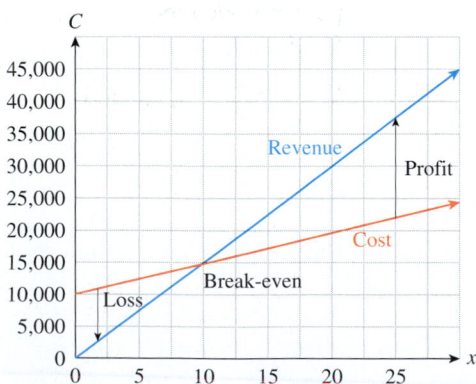

Break-even occurs at the point of intersection of the graphs of revenue and cost.

Figure **18**

The break-even point is the point where the revenue and cost are equal—that is, where the graphs of cost and revenue cross. Figure 18 confirms that break-even occurs when $x = 10$ refrigerators. Or, we can use the graph to find the break-even point in the first place. If we use technology, we can "zoom in" for an accurate estimate of the point of intersection.

   Excel has an interesting feature called "Goal Seek" that can be used to find the point of intersection of two lines numerically (rather than graphically). The downloadable Excel tutorial for this section contains detailed instructions on using Goal Seek to find break-even points. ■

Demand and Supply Functions

The demand for a commodity usually goes down as its price goes up. It is traditional to use the letter q for the (quantity of) demand, as measured, for example, in weekly sales. Consider the following example.

Example 3 Linear Demand Function

You run a small supermarket, and must determine how much to charge for Hot'n'Spicy brand baked beans. The following chart shows weekly sales figures for Hot'n'Spicy at two different prices.

Price/Can (*p*)	\$0.50	\$0.75
Demand (cans sold/week) (*q*)	400	350

a. Model the data by expressing the demand q as a linear function of the price p.

b. How do we interpret the slope? The q-intercept?

c. How much should you charge for a can of Hot'n'Spicy beans if you want the demand to increase to 410 cans per week?

Solution

a. A **demand equation** or **demand function** expresses demand q (in this case, the number of cans of beans sold per week) as a function of the unit price p (in this case, price per can). We model the demand using the two points we are given: (0.50, 400) and (0.75, 350).

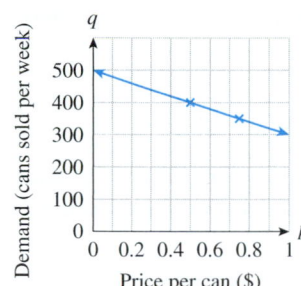

Figure **19**

$$\textit{Point: } (0.50, 400)$$

$$\textit{Slope: } m = \frac{q_2 - q_1}{p_2 - p_1} = \frac{350 - 400}{0.75 - 0.50} = \frac{-50}{0.25} = -200$$

Thus, the demand equation is

$$q = mp + b$$
$$= -200p + (400 - (-200)(0.50))$$

or $q = -200p + 500$

Figure 19 shows the data points and the linear model.

b. The key to interpreting the slope, $m = -200$, is to recall (see Example 1) that we measure the slope in units of y per unit of x. In this example, we mean units of q per unit of p, or the number of cans sold per dollar change in the price. Because m is negative, we see that the number of cans sold decreases as the price increases. We conclude that the weekly sales will drop by 200 cans per \$1-increase in the price.

To interpret the q-intercept, recall that it gives the q-coordinate when $p = 0$. Hence it is the number of cans the supermarket can "sell" every week if it were to give them away.[*]

c. If we want the demand to increase to 410 cans per week, we set $q = 410$ and solve for p:

$$410 = -200p + 500$$

$$200p = 90$$

$$p = \frac{90}{200} = \$0.45/\text{can}$$

using *Technology*

A graphing calculator or Excel can be used to find the answer to part (c) numerically; consult the Technology Guides at the end of the chapter to find out how.

+ *Before we go on...*

Q: *Just how reliable is the linear model used in Example 3* **?**

A: The *actual* demand graph could in principle be obtained by tabulating new sales figures for a large number of different prices. If the resulting points were plotted on the pq plane, they would probably suggest a curve and not a straight line. However, if you looked at a small enough portion of this curve, you could closely *approximate* it by a straight line. In other words, *over a small range of values of p, the linear model is accurate.* Linear models of real-world situations are generally reliable only for small ranges of the variables. (This point will come up again in some of the exercises.) ∎

Demand Function

A **demand equation** or **demand function** expresses demand q (the number of items demanded) as a function of the unit price p (the price per item). A **linear demand function** has the form

$$q(p) = mp + b$$

Interpretation of m
The (usually negative) slope m measures the change in demand per unit change in price. For instance, if p is measured in dollars and q in monthly sales and $m = -400$, then each \$1 increase in the price per item will result in a drop in sales of 400 items per month.

Interpretation of b
The y-intercept b gives the demand if the items were given away.

quick Example | If the demand for T-shirts, measured in daily sales, is given by $q(p) = -4p + 90$, where p is the sale price in dollars, then daily sales drop by four T-shirts for every \$1 increase in price. If the T-shirts were given away, the demand would be 90 T-shirts per day.

We have seen that a demand function gives the number of items consumers are willing to buy at a given price, and a higher price generally results in a lower demand. However, as the price rises, suppliers will be more inclined to produce these items (as opposed to spending their time and money on other products), so supply will generally rise. A **supply function** gives q, the number of items suppliers are willing to make available for sale[28], as a function of p, the price per item.

Example 4 Demand, Supply, and Equilibrium Price

Continuing with Example 3, consider the following chart, which shows weekly sales figures (the demand) for Hot'n'Spicy at two different prices, as well as the number of cans per week that you are prepared to place on sale (the supply) at these prices.

Price/Can	$0.50	$0.75
Demand (cans sold/week)	400	350
Supply (cans placed on sale/week)	300	500

a. Model these data with linear demand and supply functions.

b. How much should you charge per can of Hot'n'Spicy beans if you want the demand to equal the supply? How many cans will you sell at that price, known as the **equilibrium price?** What happens if you charge more than the equilibrium price? What happens if you charge less?

Solution

a. We have already modeled the demand function in Example 3:

$$q = -200p + 500$$

To model the supply, we use the first and third rows of the table. We are again given two points: $(0.50, 300)$ and $(0.75, 500)$:

Point: $(0.50, 300)$

Slope: $m = \dfrac{q_2 - q_1}{p_2 - p_1} = \dfrac{500 - 300}{0.75 - 0.50} = \dfrac{200}{0.25} = 800$

So, the supply equation is

$$q = mp + b$$
$$= 800p + [300 - (800)(0.50)]$$
$$= 800p - 100$$

b. To find where the demand equals the supply, we equate the two functions:

$$\text{Demand} = \text{Supply}$$
$$-200p + 500 = 800p - 100$$
$$-1000p = -600$$

so
$$p = \dfrac{-600}{-1000} = \$0.60$$

This is the equilibrium price. We can find the corresponding demand by substituting 0.60 for p in the demand (or supply) equation.

Equilibrium demand $= -200(0.60) + 500 = 380$ cans per week

---

[28] Although a bit confusing at first, it is traditional to use the same letter q for the quantity of supply and the quantity of demand, particularly when we want to compare them, as in the next example.

So, to balance supply and demand, you should charge $0.60 per can of Hot'n'Spicy beans and you should place 380 cans on sale each week.

If we graph supply and demand on the same set of axes, we obtain the graphs shown in Figure 20.

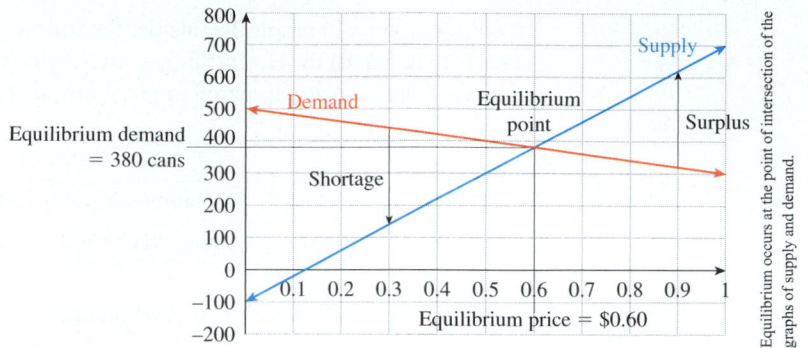

Figure 20

Figure 20 shows what happens if you charge prices other than the equilibrium price. If you charge, say, $0.90 per can ($p = 0.90$) then the supply will be larger than demand and there will be a weekly surplus. Similarly, if you charge less—say $0.30 per can— then the supply will be less than the demand, and there will be a shortage of Hot'n'Spicy beans.

+ *Before we go on...* We just saw in Example 4 that if you charge less than the equilibrium price, there will be a shortage. If you were to raise your price toward the equilibrium, you would sell more items and increase revenue, since it is the supply equation— and not the demand equation—that determines what you can sell below the equilibrium price. On the other hand, if you charge more than the equilibrium price, you will be left with a possibly costly surplus of unsold items (and will want to lower prices to reduce the surplus). Prices tend to move toward the equilibrium, so supply tends to equal demand. When supply equals demand, we say that the market **clears.** ■

Supply Function and Equilibrium Price

A **supply equation** or **supply function** expresses supply q (the number of items a supplier is willing to make available) as a function of the unit price p (the price per item). A **linear supply function** has the form

$$q(p) = mp + b$$

It is usually the case that supply increases as the unit price increases, so m is usually positive.

Demand and supply are said to be in **equilibrium** when demand equals supply. The corresponding values of p and q are called the **equilibrium price** and **equilibrium demand.** To find the equilibrium price, set demand equal to supply and solve for the unit price p. To find the equilibrium demand, evaluate the demand (or supply) function at the equilibrium price.

Change Over Time

Things around us change with time. Thus, there are many quantities, such as your income or the temperature in Honolulu, that it is natural to think of as functions of time.

Example 5 Growth of Sales

The U.S. Air Force's satellite-based Global Positioning System (GPS) allows people with radio receivers to determine their exact location anywhere on earth. The following table shows the estimated total sales of U.S.-made products that use the GPS.[*]

Year	1994	2000
Sales ($ Billions)	0.8	8.3

a. Use these data to model total sales of GPS-based products as a linear function of time t measured in years since 1994. What is the significance of the slope?

b. Use the model to predict when sales of GPS-based products will reach $13.3 billion, assuming they continue to grow at the same rate.

Solution

a. First, notice that 1994 corresponds to $t = 0$ and 2000 to $t = 6$. Thus, we are given the coordinates of two points on the graph of sales s as a function of time t: $(0, 0.8)$ and $(6, 8.3)$. The slope is

$$m = \frac{s_2 - s_1}{t_2 - t_1} = \frac{8.3 - 0.8}{6 - 0} = \frac{7.5}{6} = 1.25$$

Using the point $(0, 0.8)$, we get

$$s = mt + b$$
$$= 1.25t + 0.8 - (1.25)(0)$$
$$= 1.25t + 0.8$$

Notice that we calculated the slope as the ratio (change in sales)/(change in time). Thus, m is the *rate of change* of sales and is measured in units of sales per unit of time, or billions of dollars per year. In other words, to say that $m = 1.25$ is to say that sales are increasing by $1.25 billion per year.

b. Our model of sales as a function of time is

$$s = 1.25t + 0.8$$

Sales of GPS-based products will reach $13.3 billion when $s = 13.3$, or

$$13.3 = 1.25t + 0.8$$

Solving for t,

$$1.25t = 13.3 - 0.8 = 12.5$$
$$t = \frac{12.5}{1.25} = 10$$

In 2004 ($t = 10$), sales are predicted to reach $13.3 billion.

---

[*] Data estimated from published graph. SOURCE: U.S. Global Positioning System Industry Council/*New York Times,* March 5, 1996, p. D1.

Example 6 Velocity

You are driving down the Ohio Turnpike, watching the mileage markers to stay awake. Measuring time in hours after you see the 20-mile marker, you see the following markers each half hour:

Time (h)	0	0.5	1	1.5	2
Marker (mi)	20	47	74	101	128

Find your location s as a function of t, the number of hours you have been driving. (The number s is also called your **position** or **displacement.**)

Solution

If we plot the location s versus the time t, the five markers listed give us the graph in Figure 21.

These points appear to lie along a straight line. We can verify this by calculating how far you traveled in each half hour. In the first half hour, you traveled $47 - 20 = 27$ miles. In the second half hour you traveled $74 - 47 = 27$ miles also. In fact, you traveled exactly 27 miles each half hour. The points we plotted lie on a straight line that rises 27 units for every 0.5 unit we go to the right, for a slope of $27/0.5 = 54$.

To get the equation of that line, notice that we have the s-intercept, which is the starting marker of 20. From the slope intercept form (using s in place of y and t in place of x) we get:

$$s(t) = 54t + 20$$

Notice the significance of the slope: For every hour you travel, you drive a distance of 54 miles. In other words, you are traveling at a constant velocity of 54 mph. We have uncovered a very important principle:

In the graph of displacement versus time, velocity is given by the slope.

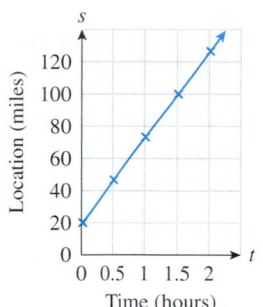

Figure 21

Linear Change Over Time
If a quantity q is a linear function of time t, so that

$$q(t) = mt + b$$

then the slope m measures the **rate of change** of q, and b is the quantity at time $t = 0$, the **initial quantity**. If q represents the position of a moving object, then the rate of change is also called the **velocity.**

Units of m and b
The units of measurement of m are units of q per unit of time; for instance, if q is income in dollars and t is time in years, then the rate of change m is measured in dollars per year.

The units of b are units of q; for instance, if q is income in dollars and t is time in years, then b is measured in dollars.

quick Example If the accumulated revenue from sales of your video game software is given by $R(t) = 2000t + 500$ dollars, where t is time in years from now, then you have earned $500 in revenue so far, and the accumulated revenue is increasing at a rate of $2000 per year.

Examples 1–6 share the following common theme.

General Linear Models

If $y = mx + b$ is a linear model of changing quantities x and y, then the slope m is the rate at which y is increasing per unit increase in x, and the y-intercept b is the value of y that corresponds to $x = 0$.

Units of m and b

The slope m is measured in units of y per unit of x, and the intercept b is measured in units of y.

quick Example

If the number n of spectators at a soccer game is related to the number g of goals your team has scored so far by the equation $n = 20g + 4$, then you can expect 4 spectators if no goals have been scored and 20 additional spectators per additional goal scored.

FAQs What to Use as x and y, and How to Interpret a Linear Model

Q: *In a problem where I must find a linear relationship between two quantities, which quantity do I use as x and which do I use as y?*

A: The key is to decide which of the two quantities is the independent variable, and which is the dependent variable. Then use the independent variable as x and the dependent variable as y. In other words, y depends on x.

Here are examples of phrases that convey this information, usually of the form *Find y [dependent variable] in terms of x [independent variable]*:

* Find the cost in terms of the number of items. $y = $ Cost, $x = $ # Items
* How does color depend on wavelength? $y = $ Color, $x = $ Wavelength

If no information is conveyed about which variable is intended to be independent, then you can use whichever is convenient. ∎

Q: *How do I interpret a general linear model $y = mx + b$?*

A: The key to interpreting a linear model is to remember the units we use to measure m and b:

The slope m is measured in units of y per unit of x; the intercept b is measured in units of y.

For instance, if $y = 4.3x + 8.1$ and you know that x is measured in feet and y in kilograms, then you can already say, "y is 8.1 kilograms when $x = 0$ feet, and increases at a rate of 4.3 kilograms per foot" without even knowing anything more about the situation! ∎

1.4 EXERCISES

● denotes basic skills exercises

◆ denotes challenging exercises

tech Ex indicates exercises that should be solved using technology

Applications

1. ● **Cost** A piano manufacturer has a daily fixed cost of $1200 and a marginal cost of $1500 per piano. Find the cost $C(x)$ of manufacturing x pianos in one day. Use your function to answer the following questions: *hint* [see Example 1]

a. On a given day, what is the cost of manufacturing 3 pianos?

b. What is the cost of manufacturing the 3rd piano that day?

c. What is the cost of manufacturing the 11th piano that day?

2. ● **Cost** The cost of renting tuxes for the Choral Society's formal is $20 down, plus $88 per tux. Express the cost C as a function of x, the number of tuxedos rented. Use your function to answer the following questions.

a. What is the cost of renting 2 tuxes?

b. What is the cost of the 2nd tux?

c. What is the cost of the 4098th tux?

d. What is the marginal cost per tux?

● basic skills   ◆ challenging   **tech** Ex technology exercise

3. ● *Cost* The RideEm Bicycles factory can produce 100 bicycles in a day at a total cost of $10,500 and it can produce 120 bicycles in a day at a total cost of $11,000. What are the company's daily fixed costs, and what is the marginal cost per bicycle? *hint* [see Example 2]

4. ● *Cost* A soft-drink manufacturer can produce 1000 cases of soda in a week at a total cost of $6000, and 1500 cases of soda at a total cost of $8500. Find the manufacturer's weekly fixed costs and marginal cost per case of soda.

5. ● *Break-even Analysis* Your college newspaper, *The Collegiate Investigator,* has fixed production costs of $70 per edition, and marginal printing and distribution costs of 40¢/copy. *The Collegiate Investigator* sells for 50¢/copy. *hint* [see Example 2]

  a. Write down the associated cost, revenue, and profit functions.
  b. What profit (or loss) results from the sale of 500 copies of *The Collegiate Investigator*?
  c. How many copies should be sold in order to break even?

6. ● *Break-even Analysis* The Audubon Society at Enormous State University (ESU) is planning its annual fund-raising "Eatathon." The society will charge students 50¢/serving of pasta. The only expenses the society will incur are the cost of the pasta, estimated at 15¢/serving, and the $350 cost of renting the facility for the evening.

  a. Write down the associated cost, revenue, and profit functions.
  b. How many servings of pasta must the Audubon Society sell in order to break even?
  c. What profit (or loss) results from the sale of 1500 servings of pasta?

7. ● *Demand* Sales figures show that your company sold 1960 pen sets each week when they were priced at $1/pen set, and 1800 pen sets each week when they were priced at $5/pen set. What is the linear demand function for your pen sets? *hint* [see Example 3]

8. ● *Demand* A large department store is prepared to buy 3950 per month of your neon-colored shower curtains for $5 each, but only 3700 shower curtains per month for $10 each. What is the linear demand function for your neon-colored shower curtains?

9. ● *Demand for Cell Phones* The following table shows worldwide sales of Nokia® cell phones and their average wholesale prices in 2004:[29]

Quarter	Second	Fourth
Wholesale Price ($)	111	105
Sales (millions)	45.4	51.4

  a. Use the data to obtain a linear demand function for (Nokia) cell phones, and use your demand equation to predict sales if Nokia lowered the price further to $103.
  b. Fill in the blanks: For every ____ increase in price, sales of cell phones decrease by ____ units.

10. ● *Demand for Cell Phones* The following table shows projected worldwide sales of (all) cell phones and wholesale prices:[30]

Year	2004	2008
Wholesale Price ($)	100	80
Sales (millions)	600	800

  a. Use the data to obtain a linear demand function for cell phones, and use your demand equation to predict sales if the price were set at $85.
  b. Fill in the blanks: For every ____ increase in price, sales of cell phones decrease by ____ units.

11. ● *Equilibrium Price* You can sell 90 pet chias per week if they are marked at $1 each, but only 30 each week if they are marked at $2/chia. Your chia supplier is prepared to sell you 20 chias each week if they are marked at $1/chia, and 100 each week if they are marked at $2 per chia. *hint* [see Example 4]

  a. Write down the associated linear demand and supply functions.
  b. At what price should the chias be marked so that there is neither a surplus nor a shortage of chias?

12. ● *Equilibrium Price* The demand for your college newspaper is 2000 copies each week if the paper is given away free of charge, and drops to 1000 each week if the charge is 10¢/copy. However, the university is prepared to supply only 600 copies per week free of charge, but will supply 1400 each week at 20¢ per copy.

  a. Write down the associated linear demand and supply functions.
  b. At what price should the college newspapers be sold so that there is neither a surplus nor a shortage of papers?

13. ● *Swimming Pool Sales* The following graph shows approximate annual sales of new in-ground swimming pools in the U.S.[31]

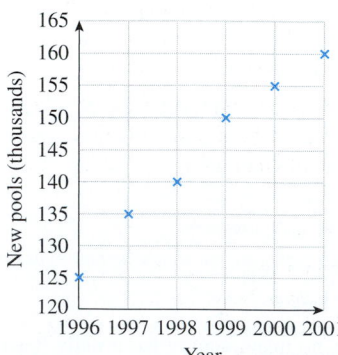

[30] Wholesale price projections are the authors'. Source for sales prediction: I-Stat/NDR December, 2004.

[31] 2001 figure is an estimate. SOURCE: PK Data/*New York Times*, July 5, 2001, p. C1.

[29] SOURCE: Embedded.com/Company reports December, 2004.

● basic skills   ◆ challenging   **tech** Ex technology exercise

a. Find two points on the graph such that the slope of the line segment joining them is the largest possible.

b. What does your answer tell you about swimming pool sales?

14. ● **Swimming Pool Sales** Repeat Exercise 13 using the following graph for sales of new above-ground swimming pools in the U.S.[32]

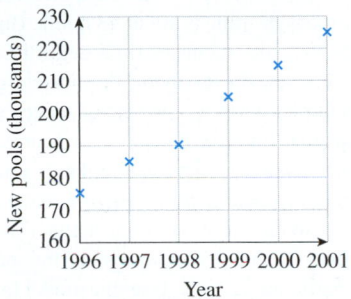

15. ● **Online Shopping** The number of online shopping transactions in the U.S. increased from 350 million transactions in 1999 to 450 million transactions in 2001.[33] Find a linear model for the number N of online shopping transactions in year t, with $t = 0$ corresponding to 2000. What is the significance of the slope? What are its units? *hint* [see Example 5]

16. ● **Online Shopping** The percentage of people in the U.S. who have ever purchased anything online increased from 20% in January, 2000 to 35% in January, 2002.[34] Find a linear model for the percentage P of people in the U.S. who have ever purchased anything online in year t, with $t = 0$ corresponding to January, 2000. What is the significance of the slope? What are its units?

17. ● **Medicare Spending** Annual federal spending on Medicare (in constant 2000 dollars) was projected to increase from $240 billion in 2000 to $600 billion in 2025.[35]

a. Use this information to express s, the annual spending on Medicare (in billions of dollars), as a linear function of t, the number of years since 2000. How fast is Medicare predicted to rise in the coming years?

b. Use your model to predict Medicare spending in 2040, assuming the spending trend continues.

18. ● **Pasta Imports** During the period 1990–2001, U.S. imports of pasta increased from 290 million pounds in 1990 ($t = 0$) by an average of 40 million pounds/year.[36]

---

[32] Ibid.

[33] Second half of 2001 data was an estimate. Source for data: Odyssey Research/*New York Times,* November 5, 2001, p. C1.

[34] January 2002 data was estimated. Source for data: Odyssey Research/*New York Times,* November 5, 2001, p. C1.

[35] Data are rounded. SOURCE: The Urban Institute's Analysis of the 1999 Trustee's Report www.urban.org

[36] Data are rounded. SOURCES: Department of Commerce/*New York Times,* September 5, 1995, p. D4, International Trade Administration (www.ita.doc.gov/) March 31, 2002.

a. Use these data to express q, the annual U.S. imports of pasta (in millions of pounds), as a linear function of t, the number of years since 1990.

b. Use your model to estimate U.S. pasta imports in 2005, assuming the import trend continued.

19. ● **Velocity** The position of a model train, in feet along a railroad track, is given by

$$s(t) = 2.5t + 10$$

after t seconds.

a. How fast is the train moving?

b. Where is the train after 4 seconds?

c. When will it be 25 feet along the track?

20. ● **Velocity** The height of a falling sheet of paper, in feet from the ground, is given by

$$s(t) = -1.8t + 9$$

after t seconds.

a. What is the velocity of the sheet of paper?

b. How high is it after 4 seconds?

c. When will it reach the ground?

21. **Fast Cars** A police car was traveling down Ocean Parkway in a high-speed chase from Jones Beach. It was at Jones Beach at exactly 10 PM ($t = 10$) and was at Oak Beach, 13 miles from Jones Beach, at exactly 10:06 PM.

a. How fast was the police car traveling?

b. How far was the police car from Jones Beach at time t? *hint* [see Example 6]

22. **Fast Cars** The car that was being pursued by the police in Exercise 21 was at Jones Beach at exactly 9:54 PM ($t = 9.9$) and passed Oak Beach (13 miles from Jones Beach) at exactly 10:06 PM, where it was overtaken by the police.

a. How fast was the car traveling?

b. How far was the car from Jones Beach at time t?

23. **Fahrenheit and Celsius** In the Fahrenheit temperature scale, water freezes at 32°F and boils at 212°F. In the Celsius scale, water freezes at 0°C and boils at 100°C. Assuming that the Fahrenheit temperature F and the Celsius temperature C are related by a linear equation, find F in terms of C. Use your equation to find the Fahrenheit temperatures corresponding to 30°C, 22°C, –10°C, and –14°C, to the nearest degree.

24. **Fahrenheit and Celsius** Use the information about Celsius and Fahrenheit given in Exercise 23 to obtain a linear equation for C in terms of F, and use your equation to find the Celsius temperatures corresponding to 104°F, 77°F, 14°F, and –40°F, to the nearest degree.

25. **Income** The well-known romance novelist Celestine A. Lafleur (a.k.a. Bertha Snodgrass) has decided to sell the screen rights to her latest book, *Henrietta's Heaving Heart,* to Boxoffice Success Productions for $50,000. In addition, the contract ensures Ms. Lafleur royalties of 5% of the net

profits.[37] Express her income I as a function of the net profit N, and determine the net profit necessary to bring her an income of $100,000. What is her marginal income (share of each dollar of net profit)?

26. Income Due to the enormous success of the movie *Henrietta's Heaving Heart*, based on a novel by Celestine A. Lafleur (see the Exercise 25), Boxoffice Success Productions decides to film the sequel, *Henrietta, Oh Henrietta*. At this point, Bertha Snodgrass (whose novels now top the best seller lists) feels she is in a position to demand $100,000 for the screen rights and royalties of 8% of the net profits. Express her income I as a function of the net profit N and determine the net profit necessary to bring her an income of $1,000,000. What is her marginal income (share of each dollar of net profit)?

Exercises 27–30 are based on the following data, showing the total amount of milk and cheese, in billions of pounds, produced in the 13 Western and 12 North-Central U.S. states in 1999 and 2000.[38]

Milk	1999	2000
Western States	56	60
North-Central States	57	59
Cheese	1999	2000
Western States	2.7	3.0
North-Central States	3.9	4.0

27. Use the data from both years to find a linear relationship giving the amount of milk w produced in Western states as a function of the amount of milk n produced in North-Central states. Use your model to predict how much milk Western states will produce if North-Central states produce 50 billion pounds.

28. Use the data from both years to find a linear relationship giving the amount of cheese w produced in Western states as a function of the amount of cheese n produced in North-Central states. Use your model to predict how much cheese Western states will produce if North-Central states produce 3.4 billion pounds.

29. Use the data to model cheese production c in Western states as a function of milk production m in Western states. According to the model, how many pounds of cheese are produced for every 10 pounds of milk?

30. Repeat Exercise 29 for North-Central states.

31. Biology The Snowtree cricket behaves in a rather interesting way: The rate at which it chirps depends linearly on the temperature. One summer evening you hear a cricket chirping at a rate of 140 chirps/minute, and you notice that the temperature is 80°F. Later in the evening the cricket has slowed down to 120 chirps/minute, and you notice that the temperature has dropped to 75°F. Express the temperature T as a function of the cricket's rate of chirping r. What is the temperature if the cricket is chirping at a rate of 100 chirps/ minute?

32. Muscle Recovery Time Most workout enthusiasts will tell you that muscle recovery time is about 48 hours. But it is not quite as simple as that; the recovery time ought to depend on the number of sets you do involving the muscle group in question. For example, if you do no sets of biceps exercises, then the recovery time for your biceps is (of course) zero. To take a compromise position, let's assume that if you do three sets of exercises on a muscle group, then its recovery time is 48 hours. Use these data to write a linear function that gives the recovery time (in hours) in terms of the number of sets affecting a particular muscle. Use this model to calculate how long it would take your biceps to recover if you did 15 sets of curls. Comment on your answer with reference to the usefulness of a linear model.

33. Profit Analysis—Aviation The operating cost of a Boeing 747-100, which seats up to 405 passengers, is estimated to be $5,132 per hour.[39] If an airline charges each passenger a fare of $100 per hour of flight, find the hourly profit P it earns operating a 747-100 as a function of the number of passengers x (be sure to specify the domain). What is the least number of passengers it must carry in order to make a profit?

34. Profit Analysis—Aviation The operating cost of a McDonnell Douglas DC 10-10, which seats up to 295 passengers, is estimated to be $3885 per hour.[40] If an airline charges each passenger a fare of $100 per hour of flight, find the hourly profit P it earns operating a DC 10-10 as a function of the number of passengers x (be sure to specify the domain). What is the least number of passengers it must carry in order to make a profit?

35. Break-even Analysis *(based on a question from a CPA exam)* The Oliver Company plans to market a new product. Based on its market studies, Oliver estimates that it can sell up to 5500 units in 2005. The selling price will be $2 per unit. Variable costs are estimated to be 40% of total revenue. Fixed costs are estimated to be $6000 for 2005. How many units should the company sell to break even?

36. Break-even Analysis *(based on a question from a CPA exam)* The Metropolitan Company sells its latest product at a unit price of $5. Variable costs are estimated to be 30% of the total revenue, while fixed costs amount to $7000 per month. How many units should the company sell per month in order to break even, assuming that it can sell up to 5000 units per month at the planned price?

37. ◆ **Break-even Analysis** *(from a CPA exam)* Given the following notations, write a formula for the break-even sales level.

---

[37] Percentages of net profit are commonly called "monkey points." Few movies ever make a net profit on paper, and anyone with any clout in the business gets a share of the *gross*, not the net.

[38] Figures are approximate. SOURCE: Department of Agriculture/*New York Times*, June 28, 2001, p. C1.

[39] In 1992. SOURCE: Air Transportation Association of America.
[40] Ibid.

● basic skills   ◆ challenging   **tech Ex** technology exercise

SP = Selling price per unit

FC = Total fixed cost

VC = Variable cost per unit

38. ◆ *Break-even Analysis (based on a question from a CPA exam)* Given the following notation, give a formula for the total fixed cost.

SP = Selling price per unit

VC = Variable cost per unit

BE = Break-even sales level in units

39. ◆ *Break-even Analysis—Organized Crime* The organized crime boss and perfume king Butch (Stinky) Rose has daily overheads (bribes to corrupt officials, motel photographers, wages for hit men, explosives, etc.) amounting to $20,000 per day. On the other hand, he has a substantial income from his counterfeit perfume racket: he buys imitation French perfume (Chanel Nº 22.5) at $20 per gram, pays an additional $30 per 100 grams for transportation, and sells it via his street thugs for $600 per gram. Specify Stinky's profit function, $P(x)$, where x is the quantity (in grams) of perfume he buys and sells, and use your answer to calculate how much perfume should pass through his hands per day in order that he break even.

40. ◆ *Break-even Analysis—Disorganized Crime* Butch (Stinky) Rose's counterfeit Chanel Nº 22.5 racket has run into difficulties; it seems that the *authentic* Chanel Nº 22.5 perfume is selling at only $500 per gram, whereas his street thugs have been selling the counterfeit perfume for $600 per gram, and his costs are $400 per gram plus $30 per gram transportation costs and commission. (The perfume's smell is easily detected by specially trained Chanel Hounds, and this necessitates elaborate packaging measures.) He therefore decides to price it at $420 per gram in order to undercut the competition. Specify Stinky's profit function, $P(x)$, where x is the quantity (in grams) of perfume he buys and sells, and use your answer to calculate how much perfume should pass through his hands per day in order that he break even. Interpret your answer.

41. ● *Television Advertising* The cost, in millions of dollars, of a 30-second television ad during the Super Bowl in the years 1990 to 2001 can be approximated by the following piecewise linear function ($t = 0$ represents 1990):[41]

$$C(t) = \begin{cases} 0.08t + 0.6 & \text{if } 0 \le t < 8 \\ 0.355t - 1.6 & \text{if } 8 \le t \le 11 \end{cases}$$

How fast and in what direction was the cost of an ad during the Super Bowl changing in 1999?

42. ● *Processor Speeds* The processor speed, in megahertz, of Intel processors could be approximated by the following function of time t in years since the start of 1995:[42]

$$P(t) = \begin{cases} 75t + 200 & \text{if } 0 \le t \le 4 \\ 600t - 1900 & \text{if } 4 < t \le 9 \end{cases}$$

How fast and in what direction was processor speed changing in 2002?

43. *Investment in Gold* Following are some approximate values of the Amex Gold BUGS Index[43]

Year	1995	2000	2004
Index	200	50	250

Take t to be the year since 1995 and y to be the BUGS index.

a. Model the 1995 and 2000 data with a linear equation.

b. Model the 2000 and 2004 data with a linear equation.

c. Use the results of parts (a) and (b) to obtain a piecewise linear model of the gold BUGS index for 1995–2004.

d. Use your model to estimate the index in 2002.

44. *Unemployment* The following table shows the number of unemployed persons in the U.S. from 1994, 2000, and 2004.[44]

Year	1994	2000	2004
Unemployment (Millions)	9	6	8

Take t to be the year since 1994 and y to be the number (in millions) of unemployed persons.

a. Model the 1994 and 2000 data with a linear equation.

b. Model the 2000 and 2004 data with a linear equation.

c. Use the results of parts (a) and (b) to obtain a piecewise linear model of the number (in millions) of unemployed persons for 1994–2004.

d. Use your model to estimate the number of unemployed persons in 2002.

45. *Career Choices* In 1989 approximately 30,000 college-bound high school seniors intended to major in computer and information sciences. This number decreased to approximately 23,000 in 1994, and rose to 60,000 in 1999.[45] Model this number C as a piecewise-linear function of the time t in years since 1989, and use your model to estimate the number of college-bound high school seniors who intended to major in computer and information sciences in 1992.

46. *Career Choices* In 1989 approximately 100,000 college-bound high school seniors intended to major in engineering. This number decreased to approximately 85,000 in 1994, and rose to 88,000 in 1999.[46] Model this number E as a piecewise-linear function of the time t in years since 1989, and use your model to estimate the number of college-bound high school seniors who intended to major in engineering in 1995.

---

[41] Source: *New York Times,* January 26, 2001, p. C1.

[42] Source: Sandpile.org/*New York Times,* May 17, 2004, p. C1.

[43] BUGS stands for "basket of unhedged gold stocks." Sources: www.321gold.com, Bloomberg Financial Markets/*New York Times,* Sept 7, 2003, p. BU8.

[44] Figures are seasonally adjusted and rounded. Source: U.S. Department of Labor, December, 2004. http://data.bls.gov

[45] Source: The College Board; National Science Foundation/*The New York Times,* September 2, 1999, p. C1.

[46] Ibid.

● basic skills   ◆ challenging   **tech Ex** technology exercise

47. **Divorce Rates** A study found that the divorce rate d appears to depend on the ratio r of available men to available women.[47] When the ratio was 1.3 (130 available men per 100 available women) the divorce rate was 22%. It rose to 35% if the ratio grew to 1.6, and rose to 30% if the ratio dropped to 1.1. Model these data by expressing d as a piecewise-linear function of r, and extrapolate your model to estimate the divorce rate if there are the same number of available men as women.

48. **Retirement** In 1950 the number N of retirees was approximately 150 per 1000 people aged 20–64. In 1990 this number rose to approximately 200 and is projected to rise to 275 in 2020.[48] Model N as a piecewise-linear function of the time t in years since 1950, and use your model to project the number of retirees per 1000 people aged 20–64 in 2010.

Communication and Reasoning Exercises

49. ● If y is measured in bootlags[49] and x is measured in $\overline{\overline{Z}}$ (zonars, the designated currency in Utarek, Mars)[50] and $y = mx + b$, then m is measured in _____ and b is measured in _____.

50. ● If the slope in a linear relationship is measured in miles per dollar, then the independent variable is measured in _____ and the dependent variable is measured in _____.

---

[47] The cited study, by Scott J. South and associates, appeared in the *American Sociological Review* (February, 1995). Figures are rounded. SOURCE: *The New York Times,* February 19, 1995, p. 40.

[48] Source: Social Security Administration/*The New York Times,* April 4, 1999, p. WK3.

[49] An ancient Martian unit of length; one bootlag is the mean distance from a Martian's foreleg to its rearleg.

[50] SOURCE: www.marsnext.com/comm/zonars.html

51. ● If a quantity is changing linearly with time, and it increases by 10 units in the first day, what can you say about its behavior in the third day?

52. ● The quantities Q and T are related by a linear equation of the form

$$Q = mT + b$$

When $T = 0$, Q is positive, but decreases to a negative quantity when T is 10. What are the signs of m and b? Explain your answers.

53. The velocity of an object is given by $v = 0.1t + 20$ m/sec, where t is time in seconds. The object is

(A) moving with fixed speed   (B) accelerating
(C) decelerating   (D) impossible to say from
                                    the given information

54. The position of an object is given by $x = 0.2t - 4$, where t is time in seconds. The object is

(A) moving with fixed speed   (B) accelerating
(C) decelerating   (D) impossible to say from
                                    the given information

55. Suppose the cost function is $C(x) = mx + b$ (with m and b positive), the revenue function is $R(x) = kx (k > m)$ and the number of items is increased from the break-even quantity. Does this result in a loss, a profit, or is it impossible to say? Explain your answer.

56. You have been constructing a demand equation, and you obtained a (correct) expression of the form $p = mq + b$, whereas you would have preferred one of the form $q = mp + b$. Should you simply switch p and q in the answer, should you start again from scratch, using p in the role of x and q in the role of y, or should you solve your demand equation for q? Give reasons for your decision.

● basic skills   ◆ challenging   **tech** Ex technology exercise

1.5 Linear Regression

We have seen how to find a linear model, given two data points: We find the equation of the line that passes through them. However, we often have more than two data points, and they will rarely all lie on a single straight line, but may often come close to doing so. The problem is to find the line coming *closest* to passing through all of the points.

Suppose, for example, that we are conducting research for a cable TV company interested in expanding into China and we come across the following figures showing the growth of the cable market there.[51]

---

[51] Data are approximate, and the 2001–2003 figures are estimates. SOURCES: HSBC Securities, Bear Sterns/*New York Times,* March 23, 2001, p. C1.

Year (t) ($t = 0$ represents 2000)	−4	−3	−2	−1	0	1	2	3
Households with Cable (y) (Millions)	50	55	57	60	68	72	80	83

A plot of these data suggests a roughly linear growth of the market. (Figure 22A).

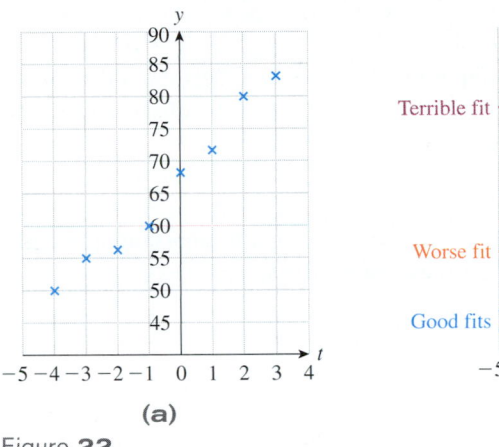

Figure **22**

These points suggest a line, although they clearly do not all lie on a single straight line. Figure 22B shows the points together with several lines, some fitting better than others. Can we precisely measure which lines fit better than others? For instance, which of the two lines labeled as "good" fits in Figure 22B models the data more accurately? We begin by considering, for each of the years 1996 through 2003, the difference between the actual number of households with cable (the **observed value**) and the number of households with cable predicted by a linear equation (the **predicted value**). The difference between the predicted value and the observed value is called the **residual.**

Residual = Observed Value − Predicted Value

On the graph, the residuals measure the vertical distances between the (observed) data points and the line (Figure 23) and they tell us how far the linear model is from predicting the number of households with cable.

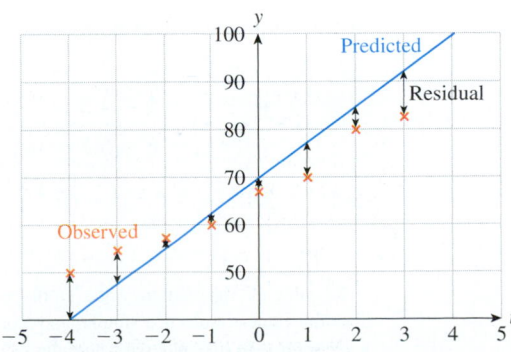

Figure **23**

The more accurate our model, the smaller the residuals should be. We can combine all the residuals into a single measure of accuracy by adding their *squares*. (We square the residuals in part to make them all positive.[52]) The sum of the squares of the residuals is called the **sum-of-squares error, SSE.** Smaller values of SSE indicate more accurate models.

Here are some definitions and formulas for what we have been discussing.

Observed and Predicted Values

Suppose we are given a collection of data points $(x_1, y_1), \ldots, (x_n, y_n)$. The n quantities $y_1, y_2, \ldots, y_n$ are called the **observed y-values.** If we model these data with a linear equation

$$\hat{y} = mx + b \qquad \text{\textcolor{red}{$\hat{y}$ stands for "estimated y" or "predicted y."}}$$

then the y-values we get by substituting the given x-values into the equation are called the **predicted y-values.**

$$\hat{y}_1 = mx_1 + b \qquad \text{Substitute x_1 for x.}$$

$$\hat{y}_2 = mx_2 + b \qquad \text{Substitute x_2 for x.}$$

$$\ldots$$

$$\hat{y}_n = mx_n + b \qquad \text{Substitute x_n for x.}$$

quick Example

Consider the three data points $(0, 2)$, $(2, 5)$, and $(3, 6)$. The observed y-values are $y_1 = 2$, $y_2 = 5$, and $y_3 = 6$. If we model these data with the equation $\hat{y} = x + 2.5$, then the predicted values are:

$$\hat{y}_1 = x_1 + 2.5 = 0 + 2.5 = 2.5$$

$$\hat{y}_2 = x_2 + 2.5 = 2 + 2.5 = 4.5$$

$$\hat{y}_3 = x_3 + 2.5 = 3 + 2.5 = 5.5$$

Residuals and Sum-of-Squares Error (SSE)

If we model a collection of data $(x_1, y_1), \ldots, (x_n, y_n)$ with a linear equation $\hat{y} = mx + b$, then the **residuals** are the n quantities (Observed Value − Predicted Value):

$$(y_1 - \hat{y}_1), (y_2 - \hat{y}_2), \ldots, (y_n - \hat{y}_n)$$

The **sum-of-squares error (SSE)** is the sum of the squares of the residuals:

$$\text{SSE} = (y_1 - \hat{y}_1)^2 + (y_2 - \hat{y}_2)^2 + (y_n - \hat{y}_n)^2$$

quick Example

For the data and linear approximation given above, the residuals are:

$$y_1 - \hat{y}_1 = 2 - 2.5 = -0.5$$

$$y_2 - \hat{y}_2 = 5 - 4.5 = 0.5$$

$$y_3 - \hat{y}_3 = 6 - 5.5 = 0.5$$

and $\quad \text{SSE} = (-0.5)^2 + (0.5)^2 + (0.5)^2 = 0.75$

---

[52] Why not add the absolute values of the residuals instead? Mathematically, using the squares rather than the absolute values results in a simpler and more elegant solution. Further, using the the squares always results in a *single* best-fit line in cases where the x-coordinates are all different, whereas this is not the case if we use absolute values.

Example 1 Computing the Sum-of-Squares Error

Using the data above on the cable television market in China, compute SSE, the sum-of-squares error, for the linear models $y = 8t + 72$ and $y = 5t + 68$. Which model is the better fit?

Solution We begin by creating a table showing the values of t, the observed (given) values of y, and the values predicted by the first model.

Year t	Observed y	Predicted $\hat{y} = 8t + 72$
-4	50	40
-3	55	48
-2	57	56
-1	60	64
0	68	72
1	72	80
2	80	88
3	83	96

We now add two new columns for the residuals and their squares.

 **using** *Technology*

Consult the Technology Guides at the end of the chapter to see how to generate these tables using a TI-83/84 or Excel. We can use either technology to graph the original data points and the lines.

Year t	Observed y	Predicted $\hat{y} = 8t + 72$	Residual $y - \hat{y}$	Residual^2 $(y - \hat{y})^2$
-4	50	40	$50 - 40 = 10$	$10^2 = 100$
-3	55	48	$55 - 48 = 7$	$7^2 = 49$
-2	57	56	$57 - 56 = 1$	$1^2 = 1$
-1	60	64	$60 - 64 = -4$	$(-4)^2 = 16$
0	68	72	$68 - 72 = -4$	$(-4)^2 = 16$
1	72	80	$72 - 80 = -8$	$(-8)^2 = 64$
2	80	88	$80 - 88 = -8$	$(-8)^2 = 64$
3	83	96	$83 - 96 = -13$	$(-13)^2 = 169$

SSE, the sum of the squares of the residuals, is then the sum of the entries in the last column,

$$\text{SSE} = 479$$

Repeating the process using the second model, $y = 5t + 68$, yields SSE $= 23$. Thus, the second model is a better fit.

+ Before we go on...

Q: *It seems clear from the figures that the second model in Example 1 gives a better fit. Why bother to compute SSE to tell me this?*

A: The difference between the two models we chose is so dramatic that it is clear from the graphs which is the better fit. However, if we used a third model with $m = 5$ and $b = 68.1$, then its graph would be almost indistinguishable from that of the second, but a better fit as measured by SSE $= 22.88$. ∎

Among all possible lines, there ought to be one with the least possible value of SSE—that is, the greatest possible accuracy as a model. The line (and there is only one such line) that minimizes the sum of the squares of the residuals is called the **regression line**, the **least-squares line**, or the **best-fit line**.

To find the regression line, we need a way to find values of m and b that give the smallest possible value of SSE. As an example, let us take the second linear model in the example above. We said in the "Before we go on . . ." discussion that increasing b from 68 to 68.1 had the desirable effect of decreasing SSE from 23 to 22.88. We could then decrease m to 4.9, further reducing SSE to 22.2. Imagine this as a kind of game: Alternately alter the values of m and b by small amounts until SSE is as small as you can make it. This works, but is extremely tedious and time-consuming.

Fortunately, there is an algebraic way to find the regression line. Here is the calculation. To justify it rigorously requires calculus of several variables or linear algebra.

Regression Line

The **regression line (least squares line, best-fit line)** associated with the points (x_1, y_1), $(x_2, y_2), \ldots, (x_n, y_n)$ is the line that gives the minimum sum-of-squares error (SSE). The regression line is

$$y = mx + b$$

where m and b are computed as follows:

$$m = \frac{n(\Sigma xy) - (\Sigma x)(\Sigma y)}{n(\Sigma x^2) - (\Sigma x)^2}$$

$$b = \frac{\Sigma y - m(\Sigma x)}{n}$$

$n =$ number of data points

The quantities m and b are called the **regression coefficients.**

Here, "Σ" means "the sum of." Thus, for example,

$$\Sigma x = \text{Sum of the } x\text{-values} = x_1 + x_2 + \cdots + x_n$$

$$\Sigma xy = \text{Sum of products} = x_1 y_1 + x_2 y_2 + \cdots + x_n y_n$$

$$\Sigma x^2 = \text{Sum of the squares of the } x\text{-values} = x_1^2 + x_2^2 + \cdots + x_n^2$$

On the other hand,

$$(\Sigma x)^2 = \text{Square of } \Sigma x = \text{Square of the sum of the } x\text{-values}$$

Example 2 The Cable Television Market in China

In Example 1 we considered the following data about the growth of the cable TV market in China.

Year (x) ($x = 0$ represents 2000)	−4	−3	−2	−1	0	1	2	3
Households with Cable (y) (Millions)	50	55	57	60	68	72	80	83

Find the best-fit linear model for these data and use the model to predict the number of Chinese households with cable in 2005.

Solution Let's organize our work in the form of a table, where the original data are entered in the first two columns and the bottom row contains the column sums.

using *Technology*

Consult the Technology Guides at the end of the chapter to see how to use built-in features of the TI-83/84 or Excel to find the regression line. Online, follow:

Chapter 1
→ Tools
    → Simple Regression

x	y	xy	x^2
−4	50	−200	16
−3	55	−165	9
−2	57	−114	4
−1	60	−60	1
0	68	0	0
1	72	72	1
2	80	160	4
3	83	249	9
$\sum$ (Sum) −4	525	−58	44

Since there are $n = 8$ data points, we get

$$m = \frac{n(\Sigma xy) - (\Sigma x)(\Sigma y)}{n(\Sigma x^2) - (\Sigma x)^2} = \frac{8(-58) - (-4)(525)}{8(44) - (-4)^2} \approx 4.87$$

and

$$b = \frac{\Sigma y - m(\Sigma x)}{n} \approx \frac{525 - (4.87)(-4)}{8} \approx 68.1$$

So, the regression line is

$$y = 4.87x + 68.1$$

To predict the number of Chinese households with cable in 2005 we substitute $x = 5$ and get $y \approx 92$ million households.

Coefficient of Correlation

If all the data points do not lie on one straight line, we would like to be able to measure how closely they can be approximated by a straight line. Recall that SSE measures the sum of the squares of the deviations from the regression line; therefore it constitutes a measurement of goodness of fit. (For instance, if SSE $= 0$, then all the points lie on a straight line.) However, SSE depends on the units we use to measure y, and also on the number of data points (the more data points we use, the larger SSE tends to be). Thus, while we can (and do) use SSE to compare the goodness of fit of two lines to the same data, we cannot use it to compare the goodness of fit of one line to one set of data with that of another to a different set of data.

To remove this dependency, statisticians have found a related quantity that can be used to compare the goodness of fit of lines to different sets of data. This quantity, called the **coefficient of correlation** or **correlation coefficient,** and usually denoted r, is between -1 and 1. The closer r is to -1 or 1, the better the fit. For an *exact* fit, we would have $r = -1$ (for a line with negative slope) or $r = 1$ (for a line with positive slope). For a bad fit, we would have r close to 0. Figure 24 shows several collections of data points with least squares lines and the corresponding values of r.

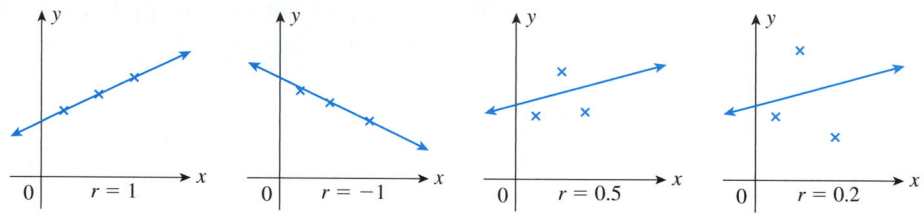

Figure **24**

Correlation Coefficient

The coefficient of correlation of the n data points $(x_1, y_1), (x_2, y_2), \dots, (x_n, y_n)$ is

$$r = \frac{n(\Sigma xy) - (\Sigma x)(\Sigma y)}{\sqrt{n(\Sigma x^2) - (\Sigma x)^2} \cdot \sqrt{n(\Sigma y^2) - (\Sigma y)^2}}$$

It measures how closely the data points $(x_1, y_1), (x_2, y_2), \dots, (x_n, y_n)$ fit the regression line.

Interpretation

If r is positive, the regression line has positive slope; if r is negative, the regression line has negative slope.

If $r = 1$ or -1, then all the data points lie exactly on the regression line; if it is close to ± 1, then all the data points are close to the regression line.

If r is close to 0, then y does not depend linearly on x.

Example **3** Computing the Coefficient of Correlation

Find the correlation coefficient for the data in Example 2. Is the regression line a good fit?

Solution

The formula for r requires Σx, Σx^2, Σxy, Σy, and Σy^2. We have all of these except for Σy^2, which we find in a new column, as shown.

x	y	xy	x^2	y^2
-4	50	-200	16	2500
-3	55	-165	9	3025
-2	57	-114	4	3249
-1	60	-60	1	3600
0	68	0	0	4624
1	72	72	1	5184
2	80	160	4	6400
3	83	249	9	6889
$\sum$ (Sum)	-4	525	-58	44	35,471

using *Technology*

Consult the Technology Guides at the end of the chapter to see how to use built-in features of the TI-83/84 or Excel to find the correlation coefficient. Alternatively, the online utility gives the correlation coefficient. Follow:

Chapter 1

→ Tools

→ Simple Regression

Substituting these values into the formula we get

$$r = \frac{n(\Sigma xy) - (\Sigma x)(\Sigma y)}{\sqrt{n(\Sigma x^2) - (\Sigma x)^2} \cdot \sqrt{n(\Sigma y^2) - (\Sigma y)^2}}$$

$$= \frac{8(-58) - (-4)(525)}{\sqrt{8(44) - (-4)^2} \cdot \sqrt{8(35,471) - 525^2}}$$

$$\approx 0.989$$

Thus, the fit is a fairly good one; that is, the original points lie nearly along a straight line, as can be confirmed by graphing the data in Example 2.

1.5 EXERCISES

● denotes basic skills exercises

techEx indicates exercises that should be solved using technology

In Exercises 1–4, compute the sum-of-squares error (SSE) by hand for the given set of data and linear model. hint [see Example 1]

1. ● $(1, 1), (2, 2), (3, 4)$; $y = x - 1$

2. ● $(0, 1), (1, 1), (2, 2)$; $y = x + 1$

3. ● $(0, -1), (1, 3), (4, 6), (5, 0)$; $y = -x + 2$

4. ● $(2, 4), (6, 8), (8, 12), (10, 0)$; $y = 2x - 8$

techEx *In Exercises 5–8, use technology to compute the sum-of-squares error (SSE) for the given set of data and linear models. Indicate which linear model gives the better fit.*

5. ● $(1, 1), (2, 2), (3, 4)$;
   a. $y = 1.5x - 1$ b. $y = 2x - 1.5$

6. ● $(0, 1), (1, 1), (2, 2)$;
   a. $y = 0.4x + 1.1$ b. $y = 0.5x + 0.9$

7. ● $(0, -1), (1, 3), (4, 6), (5, 0)$;
   a. $y = 0.3x + 1.1$ b. $y = 0.4x + 0.9$

8. ● $(2, 4), (6, 8), (8, 12), (10, 0)$;
   a. $y = -0.1x + 7$ b. $y = -0.2x + 6$

Find the regression line associated with each set of points in Exercises 9–12. Graph the data and the best-fit line. (Round all coefficients to 4 decimal places.) hint [see Example 2]

9. ● $(1, 1), (2, 2), (3, 4)$

10. ● $(0, 1), (1, 1), (2, 2)$

11. ● $(0, -1), (1, 3), (4, 6), (5, 0)$

12. ● $(2, 4), (6, 8), (8, 12), (10, 0)$

In the next two exercises, use correlation coefficients to determine which of the given sets of data is best fit by its associated regression line and which is fit worst. Is it a perfect fit for any of the data sets? hint [see Example 3]

13. ● a. $\{(1, 3), (2, 4), (5, 6)\}$
    b. $\{(0, -1), (2, 1), (3, 4)\}$
    c. $\{(4, -3), (5, 5), (0, 0)\}$

14. ● a. $\{(1, 3), (-2, 9), (2, 1)\}$
    b. $\{(0, 1), (1, 0), (2, 1)\}$
    c. $\{(0, 0), (5, -5), (2, -2.1)\}$

● basic skills   techEx technology exercise

Applications

15. ● *Worldwide Cell Phone Sales* Following are forecasts of worldwide annual cell phone handset sales:[53]

Year x	3	5	7
Sales y (Millions)	500	600	800

($x = 3$ represents 2003). Complete the following table and obtain the associated regression line. (Round coefficients to 2 decimal places.)

x	y	xy	x^2
3	500		
5	600		
7	800		
Totals			

Use your regression equation to project the 2008 sales.
hint [see Example 2]

16. ● *Investment in Gold* Following are approximate values of the Amex Gold BUGS Index:[54]

Year x	0	2	4
Index y	50	100	250

($x = 0$ represents 2000). Complete the following table and obtain the associated regression line. (Round coefficients to 2 decimal places.)

x	y	xy	x^2
0	50		
2	100		
4	250		
Totals			

Use your regression equation to estimate the 2003 index.

17. ● *E-Commerce* The following chart shows second quarter total retail e-commerce sales in the U.S. in 2000, 2002, and 2004 ($t = 0$ represents 2000):[55]

Year t	0	2	4
Sales ($ Billion)	6	10	16

Find the regression line (round coefficients to two decimal places) and use it to estimate second quarter retail e-commerce sales in 2003.

18. ● *Retail Inventories* The following chart shows total January retail inventories in U.S. department stores in 2000, 2002, and 2004 ($t = 0$ represents 2000):[56]

Year t	0	2	4
Inventory ($ Billion)	2.3	2.1	2.1

Find the regression line (round coefficients to two decimal places) and use it to estimate January retail inventories in 2001.

19. ● *Oil Recovery* In 2004 the Texas Bureau of Economic Geology published a study on the economic impact of using carbon dioxide enhanced oil recovery (EOR) technology to extract additional oil from fields that have reached the end of their conventional economic life. The following table gives the approximate number of jobs for the citizens of Texas that would be created at various levels of recovery.[57]

Percent Recovery (%)	20	40	80	100
Jobs Created (Millions)	3	6	9	15

Find the regression line and use it to estimate the number of jobs that would be created at a recovery level of 50%.

20. ● *Oil Recovery* (Refer to Exercise 19.) The following table gives the approximate economic value associated with various levels of oil recovery in Texas.[58]

Percent Recovery (%)	10	40	50	80
Economic Value ($ Billions)	200	900	1000	2000

---

[53] SOURCE: In-StatMDR, www.in-stat.com/ July, 2004.

[54] BUGS stands for "basket of unhedged gold stocks." SOURCES: www.321gold.com, Bloomberg Financial Markets/*New York Times,* Sept 7, 2003, p. BU8.

[55] SOURCE: U.S. Census Bureau www.census.gov December, 2004.

[56] SOURCE: U.S. Census Bureau www.census.gov December, 2004

[57] SOURCE: "CO_2 –Enhanced Oil Recovery Resource Potential in Texas: Potential Positive Economic Impacts" Texas Bureau of Economic Geology, April 2004 www.rrc.state.tx.us/tepc/CO2-EOR_white_paper.pdf

[58] SOURCE: "CO_2 –Enhanced Oil Recovery Resource Potential in Texas: Potential Positive Economic Impacts" Texas Bureau of Economic Geology, April 2004 www.rrc.state.tx.us/tepc/CO2-EOR_white_paper.pdf

● basic skills   **tech** Ex technology exercise

Find the regression line and use it to estimate the economic value associated with a recovery level of 70%.

21. tech Ex *Soybean Production* The following table shows soybean production, in millions of tons, in Brazil's *Cerrados* region, as a function of the cultivated area, in millions of acres.[59]

Area (Millions of Acres)	25	30	32	40	52
Production (Millions of Tons)	15	25	30	40	60

a. Use technology to obtain the regression line, and to show a plot of the points together with the regression line. (Round coefficients to two decimal places.)

b. Interpret the slope of the regression line.

22. tech Ex *Trade with Taiwan* The following table shows U.S. exports to Taiwan as a function of U.S. imports from Taiwan, based on trade figures in the period 1990–2003.[60]

Imports ($ Billions)	22	24	27	35	25
Exports ($ Billions)	12	15	20	25	17

a. Use technology to obtain the regression line, and to show a plot of the points together with the regression line. (Round coefficients to two decimal places.)

b. Interpret the slope of the regression line.

tech Ex *Exercises 23 and 24 are based on the following table comparing the median household income in the U.S. with the unemployment and poverty rates for various years from 1990 to 2003.*[61]

Year	Median Household Income ($1000)	Unemployment Rate (%)	Poverty Rate (%)
1990	39	5.6	13
1992	38	7.5	14
1994	39	6.1	14
1996	41	5.4	13
1998	44	4.5	12
2000	42	4	11
2002	41	5.8	12
2003	40	6	12

23. tech Ex *Poverty and Household Income*

a. Use x = median household income and y = poverty rate, and use technology to obtain the regression equation and graph the associated points and regression line. Round coefficients to two significant digits. Does the graph suggest a relationship between x and y?

b. What does the slope tell you about the relationship between the median household income and the poverty rate?

c. Use technology to obtain the coefficient of correlation r. Does the value of r suggest a strong correlation between x and y?

24. tech Ex *Poverty and Unemployment*

a. Use x = unemployment rate and y = poverty rate, and use technology to obtain the regression equation and graph the associated points and regression line. Round coefficients to two significant digits. Does the graph suggest a relationship between x and y?

b. What does the slope tell you about the relationship between the unemployment rate and the poverty rate?

c. Use technology to obtain the coefficient of correlation r. Does the value of r suggest a strong correlation between x and y?

25. tech Ex *New York City Housing Costs: Downtown* The following table shows the average price of a two-bedroom apartment in downtown New York City from 1994 to 2004. ($t = 0$ represents 1994)[62]

Year t	0	2	4	6	8	10
Price p ($ million)	0.38	0.40	0.60	0.95	1.20	1.60

a. Use technology to obtain the linear regression line, with regression coefficients rounded to two decimal places, and plot the regression line and the given points.

b. Does the graph suggest that a non-linear relationship between t and p would be more appropriate than a linear one? Why?

c. Use technology to obtain the residuals. What can you say about the residuals in support of the claim in part (b)?

26. tech Ex *Fiber-Optic Connections* The following table shows the number of fiber-optic cable connections to homes in the U.S. from 2000–2004 ($t = 0$ represents 2000):[63]

Year t	0	1	2	3	4
Connections c (Thousands)	0	10	25	65	150

---

[59] Source: Brazil Agriculture Ministry/*New York Times,* December 12, 2004, p. N32.

[60] Source: Taiwan Directorate General of Customs/*New York Times,* December 13, 2004, p. C7.

[61] Household incomes are in constant 2002 dollars. 2003 Figures for median income and poverty rate are estimates. The poverty threshold is approximately $18,000 for a family of four and $9200 for an individual. Sources: Census Bureau Current Population Survey/New York Times Sept 27, 2003, p. A10/ U.S. Department of Labor Bureau of Labor Statistics stats.bls.gov June 17, 2004.

[62] Data are rounded and 2004 figure is an estimte. Source: Miller Samuel/ *New York Times,* March 28, 2004, p. RE 11.

[63] Source: Render, Vanderslice & Associates/*New York Times,* October 11, 2004, p. C1.

● basic skills   tech Ex technology exercise

a. Use technology to obtain the linear regression line, with regression coefficients rounded to two decimal places, and plot the regression line and the given points.

b. Does the graph suggest that a non-linear relationship between t and p would be more appropriate than a linear one? Why?

c. Use technology to obtain the residuals. What can you say about the residuals in support of the claim in part (b)?

Communication and Reasoning Exercises

27. ● Why is the regression line associated with the two points (a, b) and (c, d) the same as the line that passes through both? (Assume that $a \neq c$.)

28. ● What is the smallest possible sum-of-squares error if the given points happen to lie on a straight line? Why?

29. ● If the points (x_1, y_1), (x_2, y_2), ..., (x_n, y_n) lie on a straight line, what can you say about the regression line associated with these points?

30. ● If all but one of the points (x_1, y_1), (x_2, y_2), ..., (x_n, y_n) lie on a straight line, must the regression line pass through all but one of these points?

31. Verify that the regression line for the points $(0, 0)$, $(-a, a)$, and (a, a) has slope 0. What is the value of r? (Assume that $a \neq 0$.)

32. Verify that the regression line for the points $(0, a)$, $(0, -a)$, and $(a, 0)$ has slope 0. What is the value of r? (Assume that $a \neq 0$.)

33. Must the regression line pass through at least one of the data points? Illustrate your answer with an example.

34. Why must care be taken when using mathematical models to extrapolate?

Chapter 1 Review

KEY CONCEPTS

1.1 Functions from the Numerical and Algebraic Viewpoints

Real-valued function f of a real-valued variable x, domain *p. 35*

Independent and dependent variables *p. 35*

Numerically specified function *p. 36*

Algebraically defined function *p. 37*

Piecewise-defined function *p. 39*

Mathematical model *p. 39*

Common types of algebraic functions *p. 40*

1.2 Functions from the Graphical Viewpoint

Graphically specified function *p. 45*

Graph of the function f *p. 45*

Drawing the graph of a function *p. 46*

Vertical line test *p. 48*

Graphing a piecewise-defined function *p. 48*

Graphs of common functions *p. 50*

1.3 Linear Functions

Linear function *p. 55*

Slope of a line: $m = \dfrac{\Delta y}{\Delta x} = \dfrac{\text{Change in } y}{\text{Change in } x}$

Interpretations of m *p. 57*

Interpretation of b: y-intercept *p. 58*

Recognizing linear data *p. 59*

Graphing a linear equation *p. 59*

x- and y-intercepts *p. 60*

Computing the slope: $m = \dfrac{y_2 - y_1}{x_2 - x_1}$ *p. 61*

Slopes of horizontal and vertical lines *p. 61*

Point-slope formula: $y = mx + b$ where $b = y_1 - mx_1$ *p. 62*

1.4 Linear Models

Linear modeling *p. 66*

Cost, revenue, and profit; marginal cost, revenue, and profit; break-even point *p. 67*

Demand *p. 72*

Supply; equilibrium *p. 74*

Linear change over time; rate of change; velocity: In the graph of displacement versus time, velocity is given by the slope. *p. 76*

General linear models *p. 77*

1.5 Linear Regression

Observed and predicted values *p. 84*

Residuals and sum-of-squares error (SSE) *p. 84*

Regression line (least squares line, best-fit line) *p. 86*

Correlation coefficient *p. 88*

REVIEW EXERCISES

In each of Exercises 1–4, use the graph of the function f to find approximations of the given values.

1.

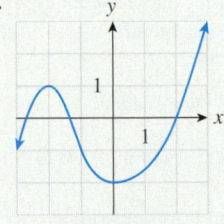

a. $f(-2)$ **b.** $f(0)$
c. $f(2)$ **d.** $f(2) - f(-2)$

2.

a. $f(-2)$ **b.** $f(0)$
c. $f(2)$ **d.** $f(2) - f(-2)$

3.

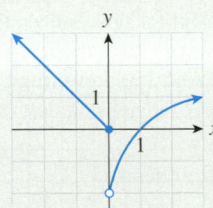

a. $f(-1)$ **b.** $f(0)$
c. $f(1)$ **d.** $f(1) - f(-1)$

4.

a. $f(-1)$ **b.** $f(0)$
c. $f(1)$ **d.** $f(1) - f(-1)$

In each of Exercises 5–8, graph the given function or equation.

5. $y = -2x + 5$

6. $2x - 3y = 12$

7. $y = \begin{cases} \frac{1}{2}x & \text{if } -1 \le x \le 1 \\ x - 1 & \text{if } 1 < x \le 3 \end{cases}$

8. $(x) = 4x - x^2$ with domain $[0, 4]$

In each of Exercises 9–13, decide whether the specified values come from a linear, quadratic, exponential, or absolute value function.

In each of Exercises 14–17, find the equation of the specified line.

	x	-2	0	1	2	4
9.	$f(x)$	4	2	1	0	2
10.	$g(x)$	-5	-3	-2	-1	1
11.	$h(x)$	1.5	1	0.75	0.5	0
12.	$k(x)$	0.25	1	2	4	16
13.	$u(x)$	0	4	3	0	-12

14. Through $(3, 2)$ with slope -3

15. Through $(-1, 2)$ and $(1, 0)$

16. Through $(1, 2)$ parallel to $x - 2y = 2$

17. With slope $1/2$ crossing $3x + y = 6$ at its x-intercept

In Exercises 18 and 19, determine which of the given lines better fits the given points.

18. $(-1, 1), (1, 2), (2, 0); y = -x/2 + 1$ or $y = -x/4 + 1$

19. $(-2, -1), (-1, 1), (0, 1), (1, 2), (2, 4), (3, 3); y = x + 1$ or $y = x/2 + 1$

In Exercises 20 and 21, find the line that best fits the given points and compute the correlation coefficient.

20. $(-1, 1), (1, 2), (2, 0)$

21. $(-2, -1), (-1, 1), (0, 1), (1, 2), (2, 4), (3, 3)$

Applications

22. As your online bookstore, OHaganBooks.com, has grown in popularity, you have been monitoring book sales as a function of the traffic at your site (measured in "hits" per day) and have obtained the following model:

$$n(x) = \begin{cases} 0.02x & \text{if } 0 \le x \le 1000 \\ 0.025x - 5 & \text{if } 1000 < x \le 2000 \end{cases}$$

where $n(x)$ is the average number of books sold in a day in which there are x hits at the site.

a. On average, how many books per day does your model predict you will sell when you have 500 hits in a day? 1000 hits in a day? 1500 hits in a day?

b. What does the coefficient 0.025 tell you about your book sales?

c. According to the model, how many hits per day will be needed in order to sell an average of 30 books per day?

23. Your monthly books sales have increased quite dramatically over the past few months, but now appear to be leveling off. Here are the sales figures for the past 6 months.

Month t	1	2	3	4	5	6
Daily Book Sales	12.5	37.5	62.5	72.0	74.5	75.0

a. Which of the following models best approximates the data?

(A) $S(t) = \dfrac{300}{4 + 100(5^{-t})}$ **(B)** $S(t) = 13.3t + 8.0$

(C) $S(t) = -2.3t^2 + 30.0t - 3.3$ **(D)** $S(t) = 7(3^{0.5t})$

b. What do each of the above models predict for the sales in the next few months: rising, falling, or leveling off?

24. To increase business at OHaganBooks.com, you plan to place more banner ads at well-known Internet portals. So far, you have the following data on the average number of hits per day at OHaganBooks.com versus your monthly advertising expenditures:

Advertising Expenditure ($/Month)	$2000	$5000
Website Traffic (Hits/Day)	1900	2050

You decide to construct a linear model giving the average number of hits h per day as a function of the advertising expenditure c.

a. What is the model you obtain?

b. Based on your model, how much traffic can you anticipate if you budget $6000 per month for banner ads?

c. Your goal is to eventually increase traffic at your site to an average of 2500 hits per day. Based on your model, how much do you anticipate you will need to spend on banner ads in order to accomplish this?

25. A month ago you increased expenditure on banner ads to $6000 per month, and you have noticed that the traffic at OHaganBooks.com has not increased to the level predicted by the linear model in Exercise 24. Fitting a quadratic function to the data you have gives the model

$$h = -0.000005c^2 + 0.085c + 1750$$

where h is the daily traffic (hits) at your website, and c is the monthly advertising expenditure.

a. According to this model, what is the current traffic at your site?

b. Does this model give a reasonable prediction of traffic at expenditures larger than $8500 per month? Why?

26. Besides selling books, you are generating additional revenue at OHaganBooks.com through your new online publishing service. Readers pay a fee to download the entire text of a novel. Author royalties and copyright fees cost you an average of $4 per novel, and the monthly cost of operating and maintaining the service amounts to $900 per month. You are currently charging readers $5.50 per novel.

a. What are the associated cost, revenue, and profit functions?

b. How many novels must you sell per month in order to break even?

c. If you lower the charge to $5.00 per novel, how many books will you need to sell in order to break even?

27. In order to generate a profit from your online publishing service, you need to know how the demand for novels depends on the price you charge. During the first month of the service, you were charging $10 per novel, and sold 350. Lowering the price to $5.50 per novel had the effect of increasing demand to 620 novels per month.

a. Use the given data to construct a linear demand equation.

b. Use the demand equation you constructed in part (a) to estimate the demand if you raised the price to $15 per novel.

c. Using the information on cost given in Exercise 26, determine which of the three prices ($5.50, $10 and $15) would result in the largest profit, and the size of that profit.

28. It is now several months later and you have tried selling your online novels at a variety of prices, with the following results:

Price	$5.50	$10	$12	$15
Demand (Monthly sales)	620	350	300	100

a. Use the given data to obtain a linear regression model of demand. (Round coefficients to four decimal places.)

b. Use the demand model you constructed in part (a) to estimate the demand if you charged $8 per novel. (Round the answer to the nearest novel.)

vMentor Do you need a live tutor for homework problems? Access vMentor on the ThomsonNOW! website at **www.thomsonedu.com** for one-on-one tutoring from a mathematics expert.

CASE STUDY: Modeling Spending on Internet Advertising

Jeff Titcomb/Getty Images

You are the new director of Impact Advertising Inc.'s Internet division, which has enjoyed a steady 0.25% of the Internet advertising market. You have drawn up an ambitious proposal to expand your division in light of your anticipation that Internet advertising will continue to skyrocket. However, upper management sees things differently and, based on the following e-mail, does not seem likely to approve the budget for your proposal.

TO: JCheddar@impact.com (J. R. Cheddar)

CC: CVODoylePres@impact.com. (C. V. O'Doyle, CEO)

FROM: SGLombardoVP@impact.com (S. G. Lombardo, VP Financial Affairs)

SUBJECT: Your Expansion Proposal

DATE: August 3, 2005

Hi John:

Your proposal reflects exactly the kind of ambitious planning and optimism we like to see in our new upper management personnel. Your presentation last week was most impressive, and obviously reflected a great deal of hard work and preparation.

I am in full agreement with you that Internet advertising is on the increase. Indeed, our Market Research department informs me that, based on a regression of the most recently available data, Internet advertising revenue in the U.S. will continue grow at a rate of approximately $1 billion per year. This translates into approximately $2.5 million in increased revenues per year for Impact, given our 0.25% market share. This rate of expansion is exactly what our planned 2006 budget anticipates. Your proposal, on the other hand, would require a budget of approximately *twice* the 2006 budget allocation, even though your proposal provides no hard evidence to justify this degree of financial backing.

At this stage, therefore, I am sorry to say that I am inclined not to approve the funding for your project, although I would be happy to discuss this further with you. I plan to present my final decision on the 2006 budget at next week's divisional meeting.

Regards, Sylvia

Refusing to admit defeat, you contact the Market Research department and request the details of their projections on Internet advertising. They fax you the following information:[64]

Year	1999	2000	2001	2002	2003	2004	2005
Spending on Advertising ($ Billion)	0	0.3	0.8	1.9	3	4.3	5.8

Regression Model: $y = 0.9857x - 0.6571$ (x = years since 1999)

Correlation Coefficient: $r = 0.9781$

Now you see where the VP got that $1 billion figure: The slope of the regression equation is close to 1, indicating a rate of increase of just under $1 billion per year. Also, the correlation coefficient is very high—an indication that the linear model fits the data well. In view of this strong evidence, it seems difficult to argue that revenues will increase by significantly more than the projected $1 billion per year.

To get a better picture of what's going on, you decide to graph the data together with the regression line in your spreadsheet program. What you get is shown in Figure 25.

You immediately notice that the data points seem to suggest a curve, and not a straight line. Then again, perhaps the suggestion of a curve is an illusion. Thus there are, you surmise, two possible interpretations of the data:

1. (Your first impression) As a function of time, spending on Internet advertising is non-linear, and is in fact accelerating (the rate of change is increasing), so a linear model is inappropriate.

2. (Devil's advocate) Spending on Internet advertising *is* a linear function of time; the fact that the points do not lie on the regression line is simply a consequence of random factors, such as the state of the economy, the stock market performance, etc.

You suspect that the VP will probably opt for the second interpretation and discount the graphical evidence of accelerating growth by claiming that it is an illusion: a "statistical fluctuation." That is, of course, a possibility, but you wonder how likely it really is.

For the sake of comparison, you decide to try a regression based on the simplest non-linear model you can think of—a quadratic function.

$$y = ax^2 + bx + c$$

Your spreadsheet allows you to fit such a function with a click of the mouse. The result is the following.

$$y = 0.1190x^2 + 0.2714x - 0.0619 \qquad (x = \text{number of years since 1999})$$

$$r = 0.9992 \qquad \qquad \text{See Note.}^*$$

Figure 26 shows the graph of the regression function together with the original data.

Aha! The fit is visually far better, and the correlation coefficient is even higher! Further, the quadratic model predicts 2006 spending as

$$y = 0.1190(7)^2 + 0.2714(7) - 0.0619 \approx \$7.67 \text{ billion}$$

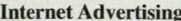

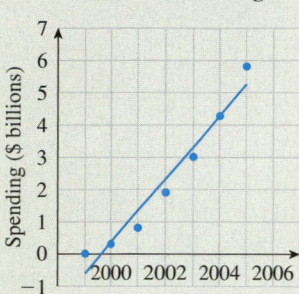

Internet Advertising

Figure **25**

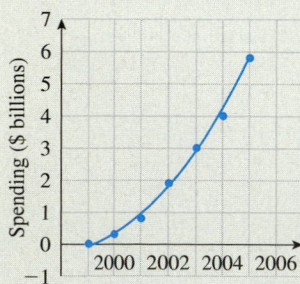

Internet Advertising

Figure **26**

---

[64] Figures are approximate, and the 1999–2001 figures are projected. SOURCE: Nielsen NetRatings/*The New York Times,* June 7, 1999, p. C17.

*Note that this r is *not* the linear correlation coefficient we defined on p. 92; what this r measures is how closely the *quadratic* regression model fits the data.

which is $1.87 billion above the 2005 spending figure in the table above. Given Impact Advertising's 0.25% market share, this translates into an increase in revenues of $4.7 million, which is almost double the estimate predicted by the linear model!

You quickly draft an e-mail to Lombardo, and are about to press "send" when you decide, as a precaution, to check with a statistician. He tells you to be cautious: the value of r will always tend to increase if you pass from a linear model to a quadratic one due to an increase in "degrees of freedom."[65] A good way to test whether a quadratic model is more appropriate than a linear one is to compute a statistic called the "p-value" associated with the coefficient of x^2. A low value of p indicates a high degree of confidence that the coefficient of x^2 cannot be zero (see below). Notice that if the coefficient of x^2 *is* zero, then you have a linear model.

You can, your friend explains, obtain the p-value using Excel as follows. First, set up the data in columns, with an extra column for the values of x^2.

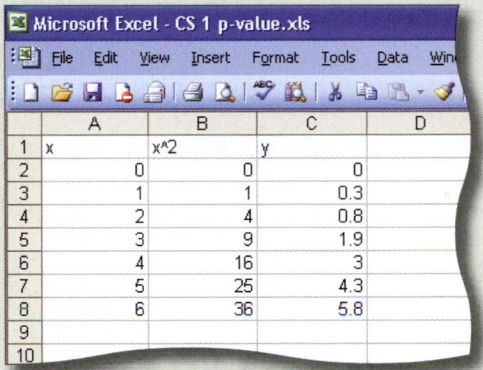

Next, choose "Data analysis" from the "Tools" menu, and choose "Regression." In the dialogue box, give the location of the data as shown in Figure 27.

Figure **27**

---

[65] The number of degrees of freedom in a regression model is 1 less than the number of coefficients. For a linear model, it is 1 (there are two coefficients: the slope m and the intercept b), and for a quadratic model it is 2. For a detailed discussion, consult a text on regression analysis.

Clicking "OK" then gives you a large chart of statistics. The p-value you want is in the very last row of the data: $p = 0.000478$.

Q: What does p actually measure?

A: Roughly speaking, $1 - p = .999522$ gives the degree of confidence you can have (99.9522%) in asserting that the coefficient of x^2 is not zero. (Technically, p is the probability—allowing for random fluctuation in the data—that, if the coefficient of x^2 were in fact zero, the "t-statistic" (10.4257, right next to the p-value on the spreadsheet) could be as large as it is. ∎

In short, you can go ahead and send your e-mail with 99% confidence!

Exercises

Suppose you are given the following data:

Year	1999	2000	2001	2002	2003	2004	2005
Spending on Advertising ($ Billion)	0	0.3	1.5	2.6	3.4	4.3	5.0

1. Obtain a linear regression model and the correlation coefficient r. According to the model, at what rate is spending on Internet Advertising increasing in the U.S.? How does this translate to annual revenues for Impact Advertising?

2. Use a spreadsheet or other technology to graph the data together with the best-fit line. Does the graph suggest a quadratic model (parabola)?

3. Test your impression in the preceding exercise by using technology to fit a quadratic function and graphing the resulting curve together with the data. Does the graph suggest that the quadratic model is appropriate?

4. Perform a regression analysis and find the associated p-value. What does it tell you about the appropriateness of a quadratic model?

Section **1.1**

Example 3 Evaluate the function $f(x) = -0.4x^2 + 7x - 23$ for $x = 0, 1, 2, \ldots, 10$.

Solution with Technology There are several ways to evaluate an algebraically defined function on a graphing calculator such as the TI-83/84.

1. Enter the function in the Y = screen, as

> Y₁ = -0.4*X^2+7*X-23
>
> Negative (-) and minus (−) are different keys on the TI-83/84.

or   Y₁ = -0.4X²+7X-23

(See Chapter 0 for a discussion of technology formulas.)

2. To evaluate $f(0)$, for example, enter the following in the home screen:

> Y₁(0)   This evaluates the function Y₁ at 0.

Alternatively, you can use the table feature:

1. After entering the function under Y₁, press 2ND TBLSET, and set Indpnt to Ask. (You do this once and for all; it will permit you to specify values for x in the table screen.)

2. Press 2ND TABLE, and you will be able to evaluate the function at several values of x. Here is a table showing some of the values requested:

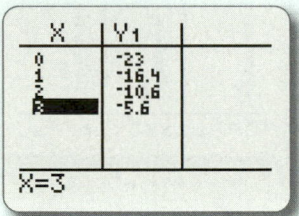

Example 4 The price $V(t)$ in dollars of EBAY stock during the 10-week period starting July 1, 2004 can be approximated by the following function of time t in weeks ($t = 0$ represents July 1):[66]

$$V(t) = \begin{cases} 90 - 4t & \text{if } 0 \le t \le 5 \\ 60 + 2t & \text{if } 5 < t \le 20 \end{cases}$$

What was the approximate price of EBAY stock after 1 week, after 5 weeks, and after 10 weeks?

[66] Source for data: http://money.excite.com, November, 2004

Solution with Technology The following formula defines the function V on the TI-83/84:

> (X≤5)*(90-4*X)+(X>5)*(60+2*X)

The logical operators (≤ and >, for example) can be found by pressing 2ND TEST.

When x is less than or equal to 5, the logical expression (x≤5) evaluates to 1 because it is true, and the expression (x>5) evaluates to 0 because it is false. The value of the function is therefore given by the expression (90-4*x). When x is greater than 5, the expression (x≤5) evaluates to 0 while the expression (x>5) evaluates to 1, so the value of the function is given by the expression (60+2*x).

As in Example 3, you can use the Table feature to compute several values of the function at once:

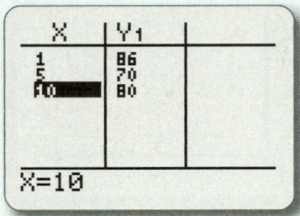

Section **1.2**

Example 2 The monthly revenue R from users logging on to your gaming site depends on the monthly access fee p you charge according to the formula

$$R(p) = -5600p^2 + 14{,}000p \qquad (0 \le p \le 2.5)$$

(R and p are in dollars.) Sketch the graph of R. Find the access fee that will result in the largest monthly revenue.

Solution with Technology You can reproduce the graph shown in Figure 4(b) in Section 1.2 as follows:

1. Enter

> Y₁ = -5600*X^2+14000*X

in the Y= screen.

2. Set the window coordinates: Xmin = 0, Xmax = 2.5, Ymin = 0, Ymax = 10000.

3. Press GRAPH.

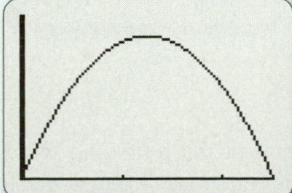

If you want to plot individual points (as in Figure 4(a) in Section 1.2) on the TI-83/84:

1. Enter the data in the stat list editor ($\boxed{\text{STAT}}$ EDIT) with the values of p in L_1, and the value of $R(p)$ in L_2.

2. Go to the $Y=$ window and turn $\texttt{Plot1}$ on by selecting it and pressing $\boxed{\text{ENTER}}$.

3. Now press ZoomStat ($\boxed{\text{GRAPH}}$ $\boxed{9}$) to obtain the plot.

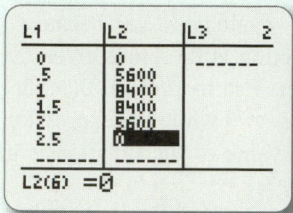

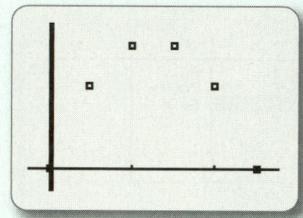

Example 4 Graph the function f specified by

$$f(x) = \begin{cases} -1 & \text{if } -4 \le x < -1 \\ x & \text{if } -1 \le x \le 1 \\ x^2 - 1 & \text{if } 1 < x \le 2 \end{cases}$$

Solution with Technology You can enter this function as

$$\underbrace{(X<-1)*(-1)}_{\text{First part}} + \underbrace{(-1\le X \text{ and } X\le1)*X}_{\text{Second part}} + \underbrace{(1<X)*(X^2-1)}_{\text{Third part}}$$

The logical operator $\texttt{and}$ is found in the TEST LOGIC menu. The following alternative formula will also work:

$$(X<-1)*(-1)+(-1\le X)*(X\le1)*X+(1<X)*(X^2-1)$$

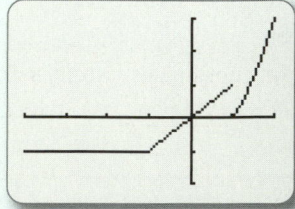

Section 1.3

Example 1 Which of the following two tables gives the values of a linear function? What is the formula for that function?

x	0	2	4	6	8	10	12
$f(x)$	3	−1	−3	−6	−8	−13	−15

x	0	2	4	6	8	10	12
$f(x)$	3	−1	−5	−9	−13	−17	−21

Solution with Technology You can compute the successive quotients $m = \Delta y / \Delta x$ as follows, using the TI-83/84.

1. Enter the values of x and $f(x)$ in the lists L_1 and L_2, which is most easily done using the stat list editor ($\boxed{\text{STAT}}$ EDIT).

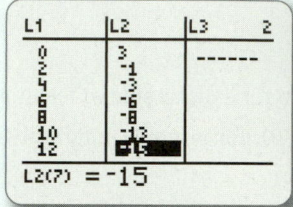

2. Highlight the heading L_3 and enter the following formula (with the quotes, as explained below):

   $$\texttt{"}\Delta\texttt{List(L}_2\texttt{)/}\Delta\texttt{List(L}_1\texttt{)"}$$

The $\texttt{"}\Delta\texttt{List"}$ function is found under $\boxed{\text{LIST}}$ OPS and computes the differences between successive elements of a list, returning a list with one less element. The formula above then computes the quotients $\Delta y / \Delta x$ in the list L_3.

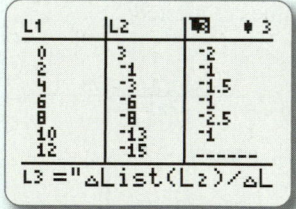

As you can see in the third column, $f(x)$ is not linear. To redo the computation for $g(x)$, all you need to do is edit the values of L_2 in the stat list editor. By putting quotes around the formula we used for L_3, we "attached" the formula to L_3 so that it would update automatically.

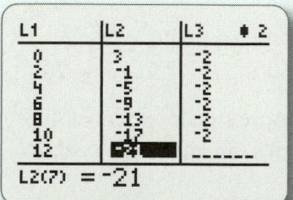

Section 1.4

Example 3(c) The following chart shows weekly sales figures for Hot'n'Spicy brand baked beans at two different prices.

Price/Can (p)	$0.50	$0.75
Demand (cans sold/week) (q)	400	350

How much should you charge for a can of Hot'n'Spicy beans if you want the demand to increase to 410 cans per week?

Solution with Technology In part (c) we can find p numerically using the Table feature in the TI-83/84.

1. Make sure that the table settings permit you to enter values of x: Press $\boxed{\text{2ND}}$ $\boxed{\text{TBLSET}}$ and set `Indpnt` to `Ask`.

2. Enter

   $$\text{Y}_1 = -200*\text{X} + 500 \qquad \text{\color{red}Demand equation}$$

   and press $\boxed{\text{2ND}}$ $\boxed{\text{TABLE}}$.

3. You will now be able to adjust the price (x) until you find the value at which the demand equals 410.

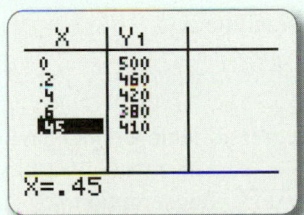

Section 1.5

Example 1 Using the data at the beginning of the section on the cable television market in China, compute SSE, the sum-of-squares error, for the linear models $y = 8t + 72$ and $y = 5t + 68$. Which model is the better fit?

Solution with Technology We can use the "List" feature in the TI-83/84 to automate the computation of SSE.

1. Use the stat list editor ($\boxed{\text{STAT}}$ EDIT) to enter the given data in the first two columns, called L_1 and L_2. (If there is already data in a column you want to use, you can clear it by highlighting the column heading (e.g., L_1), using the arrow key, and pressing $\boxed{\text{CLEAR}}$ $\boxed{\text{ENTER}}$.)

2. To compute the predicted values, highlight the heading L_3 using the arrow keys, and enter the formula for the predicted values:

   $$8*\text{L}_1+72 \qquad \text{\color{red}L_1 is $\boxed{\text{2ND}}$ [1]}$$

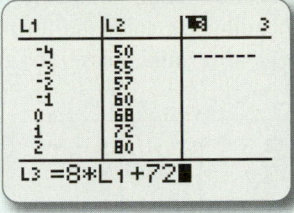

Pressing $\boxed{\text{ENTER}}$ again will fill column 3 with the predicted values. (Note that only seven of the eight data points can be seen on the screen at one time.)

3. Highlight the heading L_4 and enter the following formula (with the quotes):

   $$\text{"}(\text{L}_2-\text{L}_3)\text{^}2\text{"} \qquad \text{\color{red}Squaring the residuals}$$

Pressing $\boxed{\text{ENTER}}$ will fill column 4 with the squares of the residuals. (Putting quotes around the formula will allow us to easily check the second model, as we shall see.)

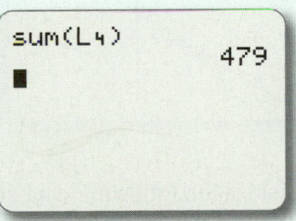

4. To compute SSE, the sum of the entries in L_4, go to the home screen and enter `sum(L_4)`. The sum function is found by pressing $\boxed{\text{2ND}}$ $\boxed{\text{LIST}}$ and selecting MATH.

5. To check the second model, go back to the List screen, highlight the heading L_3, enter the formula for the second model, `5*L_1+68`, and press ENTER.

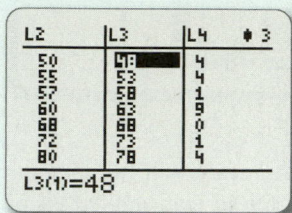

Because we put quotes around the formula for the residuals in L_4, the TI-83/84 remembered the formula and automatically recalculated the values. On the home screen we can again calculate `sum(L_4)` to get SSE for the second model.

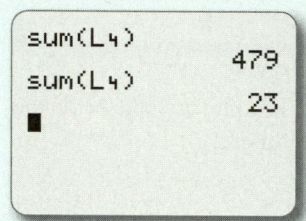

The second model gives a much smaller SSE, so is the better fit.

  You can also use the TI-83/84 to plot both the original data points and the two lines.

1. Turn PLOT1 on in the STAT PLOTS window, obtained by pressing 2ND STAT PLOT.

2. To show the lines, enter them in the "Y=" screen as usual.

3. To obtain a convenient window showing all the points and the lines, press ZOOM and choose option #9: `ZoomStat`.

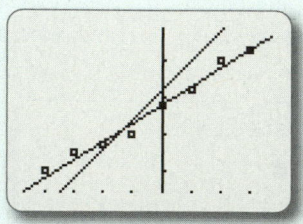

Example 2 Consider the following data about the growth of the cable TV market in China.

Year (t) ($t = 0$ represents 2000)	−4	−3	−2	−1	0	1	2	3
Households with Cable (y) (Millions)	50	55	57	60	68	72	80	83

Find the best-fit linear model for these data.

Solution with Technology

1. Enter the data in the TI-83/84 using the List feature, putting the x-coordinates in L_1 and the y-coordinates in L_2, just as in Example 1.

2. Press STAT, select CALC, and choose option #4: `LinReg(ax+b)`. Pressing ENTER will cause the equation of the regression line to be displayed in the home screen:

So, the regression line is

$$y \approx 4.87x + 68.1$$ Coefficients rounded to 3 significant digits[67]

3. To graph the regression line without having to enter it by hand in the "Y=" screen, press Y =, clear the contents of Y_1, press VARS, choose option #5: `Statistics`, select `EQ, and then choose #1: RegEQ`. The regression equation will then be entered under Y_1.

4. To simultaneously show the data points, press 2ND STAT PLOT and turn PLOT1 on as in Example 1.

5. To obtain a convenient window showing all the points and the line, press ZOOM and choose option #9: `ZoomStat`.

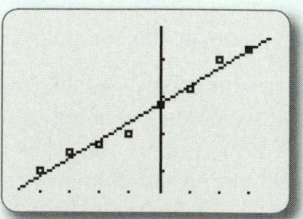

---

[67] The original data only gives 2 significant digits, so we rounded the coefficients to one more digit than that to give us a "safety margin" of one digit. Further digits are probably meaningless.

Example 3 Find the correlation coefficient for the data in Example 2.

Solution with Technology A utility that will calculate linear regression lines will also calculate the correlation coefficient. To find the correlation coefficient using a TI-83/84, you need to tell the calculator to show you the coefficient at the same time that it shows you the regression line. To do this, press $\boxed{\text{CATALOG}}$ and select "Diagnosticon" from the list. The command will be pasted to the home screen, and you should then press $\boxed{\text{ENTER}}$ to execute the command. Once you have done this, the "LinReg

(ax+b)" command will show you not only a and b, but r and r^2 as well.

```
LinReg
 y=ax+b
 a=4.869047619
 b=68.05952381
 r²=.9782343
 r=.9890572784
```

EXCEL Technology Guide

Section **1.1**

Example 3 Evaluate the function $f(x) = -0.4x^2 + 7x - 23$ for $x = 0, 1, 2, \ldots, 10$.

Solution with Technology To create a table of values of f using Excel:

1. Set up two columns—one for the values of x and one for the values of $f(x)$.

2. To enter the sequence of values $0, 1, \ldots, 10$ in the x column, start by entering the first two values, 0 and 1, highlight both of them, and drag the **fill handle** (the little dot at the lower right-hand corner of the selection) down until you reach Row 12. (Why 12?)

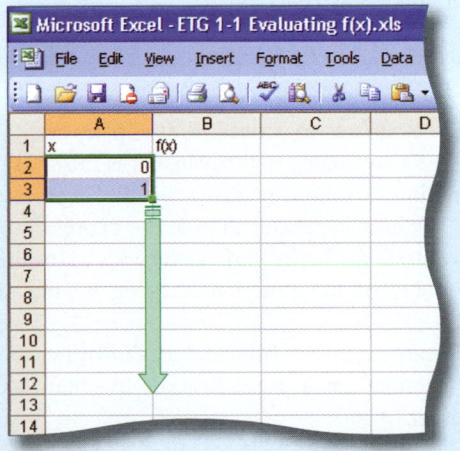

3. Enter the formula for f in cell B2. The technology formula (see Chapter 0) for f is

   $\quad -0.4*x\text{^}2+7*x-23$ Technology formula

   To get the formula to use for Excel, replace each occurrence of x by the name of the cell holding the value of x (cell A2 in this case) and obtain

   $\quad =-0.4*A2\text{^}2+7*A2-23$ A2 refers to the cell containing the value of x

Note: Instead of typing in the name of the cell "A2" each time, you can simply click on the cell A2, and "A2" will be automatically inserted. The formula

   $\quad =-0.4*x\text{^}2+7*x-23$

will also work in many versions of Excel, provided you have entered the heading x in cell A1 as shown. Try it!

    Enter this formula in cell B2, press Enter, and then drag the resulting value down (using the fill handle) to cell B12, as shown below (center), to obtain the result shown at the bottom.

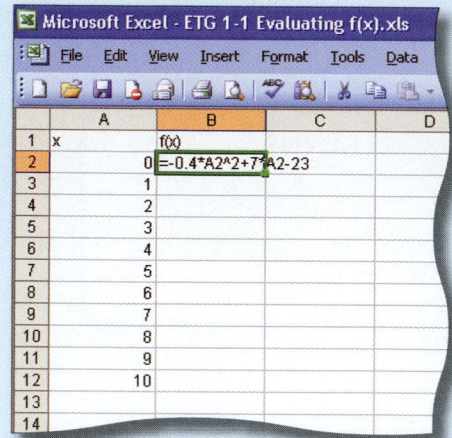

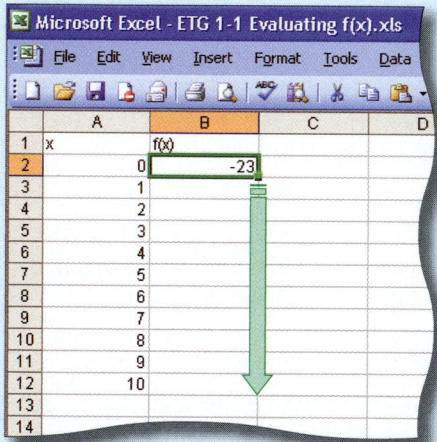

Warning: In interpreting a negative sign at the start of an expression, Excel uses a different convention from the usual mathematical one and the one used by almost all other technology and programming languages:

Excel Formula	Usual Interpretation	Excel Interpretation	
-x^2	$-x^2$	$(-x)^2$	Same as x^2 Different
2-x^2	$2 - x^2$	$2 - x^2$	The same
-1*x^2	$-x^2$	$-x^2$	The same
-(x^2)	$-x^2$	$-x^2$	The same

Thus, if a formula begins with $-x^2$, you should enter it in Excel as

=-1*x^2   or   -(x^2)   With x replaced by the cell holding x

For example:

1. To enter $-x^2 + 4x - 3$ in Excel, type = -1*x^2+4*x-3 .

2. To enter $-3x^2 + 4x - 3$ in Excel, type =-3*x^2+4*x-3 .

3. To enter $4x - x^2$ in Excel, type =4*x-x^2 .

In short, you need to be careful only when the expression you want to use begins with a negative sign in front of an x.

Example 4 The price $V(t)$ in dollars of EBAY stock during the 10-week period starting July 1, 2004 can be approximated by the following function of time t in weeks ($t = 0$ represents July 1):[68]

$$V(t) = \begin{cases} 90 - 4t & \text{if } 0 \le t \le 5 \\ 60 + 2t & \text{if } 5 < t \le 20 \end{cases}$$

What was the approximate price of EBAY stock after 1 week, after 5 weeks, and after 10 weeks?

Solution with Technology The following formula defines the function V in Excel:

(x<=5)*(90-4*x)+(x>5)*(60+2*x)

When x is less than or equal to 5, the logical expression (x<=5) evaluates to 1 because it is true, and the expression (x>5) evaluates to 0 because it is false. The value of the function is therefore given by the expression (90-4*x). When x is greater than 5, the expression (x<=5) evaluates to 0 while the expression (x>5) evaluates to 1, so the value of the function is given by the expression (60+2*x).

We can set up a worksheet as shown (see Example 3 for instructions on using a formula to obtain values of a function in Excel).

---

[68] Source for data: http://money.excite.com, November, 2004.

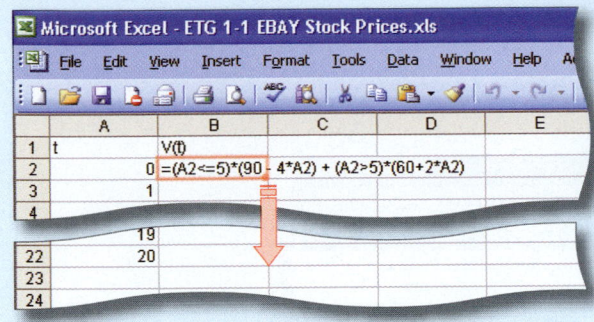

Using the IF Function in Excel

The following worksheet shows how we can get the same result using the IF function in Excel:

The IF function evaluates its first argument, which tests to see if the value of *t* is in the range $t \le 5$. If the first argument is true, IF returns the result of evaluating its second argument: 90-4*A2; if not, it returns the result of evaluating its third argument: 60+2*A2.

In either case, the final result will look something like this:

Section 1.2

Example 2 The monthly revenue *R* from users logging on to your gaming site depends on the monthly access fee *p*

you charge according to the formula

$$R(p) = -5600p^2 + 14,000p \qquad (0 \le p \le 2.5)$$

(*R* and *p* are in dollars.) Sketch the graph of *R*. Find the access fee that will result in the largest monthly revenue.

Solution with Technology

1. Begin by creating a table of values for the function, as in Example 3 of Section 1.1, by entering the values of the independent variable *p* in column A and the formula for the function in cell B2, then copying the formula into the cells beneath it as shown.

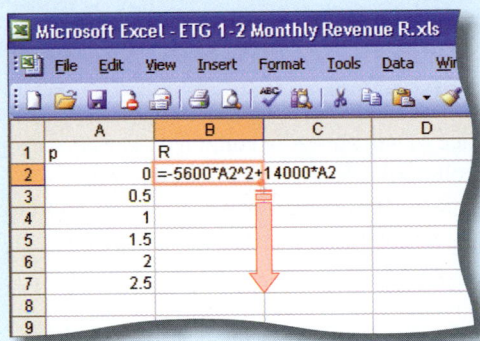

2. To draw the graph shown in Figure 4(a) of Section 1.2, select (highlight) both columns of data and then ask Excel to insert a chart.

3. When it asks you to specify what type of chart, select the XY (Scatter) option. This tells the program that your data specify the *x*- and *y*-coordinates of a sequence of points. In the same dialogue box, select the option that shows points connected by lines.

4. Press Next to bring up a new dialogue called Data Type, where you should make sure that the Series in Columns option is selected, telling the program that the *x*- and *y*-coordinates are arranged vertically, down columns.

5. You can then set various other options, add labels, and otherwise fiddle with it until it looks nice.

To get a smoother curve, you need to plot many more points. Here is a method of plotting 100 points (in addition to the starting point), similar to the Excel worksheet posted online that you can find by following:

Chapter 1 → Excel Tutorials → Section 1.2: Functions from the Graphical Viewpoint

Moreover, if you decide to follow this method, you can save the resulting spreadsheet and use it as an "Excel graphing calculator" to graph any other function (see below).

1. Set up your worksheet as follows:

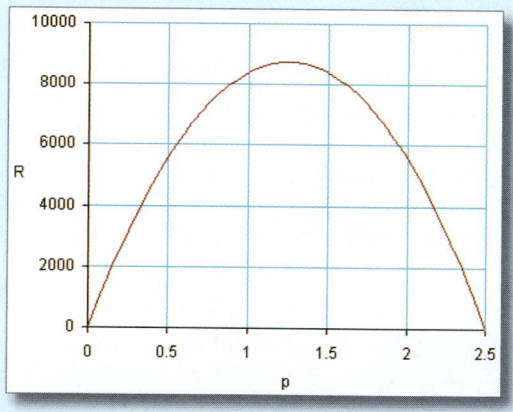

(Columns C and D are empty in case you want to add additional functions to graph.) The 101 values of the *x*-coordinate (the price *p*) will appear in column A. The corresponding values of the *y*-coordinate (the revenue *R*) will appear in column B. In column F you see some settings: Xmin = 0 and Xmax = 2.5 (see Figure 4). Delta X (in cell F5) is the amount by which the *x*-coordinate is increased as you go from one *x*-value in column A to the next, starting with Xmin in A2. Enter the formula for $R(p)$ in Cell B2 and copy it into the other cells in column B as shown.

2. When done, graph the data in columns A and B, choosing the scatter plot option with subtype `Points connected by lines with no markers`.

Save this worksheet with its graph and you can use it to graph new functions as follows:

1. Enter the new values for Xmin and Xmax in column F.

2. Enter the new function in cell B2 (using A2 in place of *x*).

3. Copy the contents of cell B2 to cells B3–B102.

The graph will be updated automatically.

Example 4 Graph the function *f* specified by

$$f(x) = \begin{cases} -1 & \text{if } -4 \leq x < -1 \\ x & \text{if } -1 \leq x \leq 1 \\ x^2 - 1 & \text{if } 1 < x \leq 2 \end{cases}$$

Solution with Technology You can use either of the following formulas:

$$= (x<-1)*(-1) + (-1<=x)*(x<=1)*x + (1<x)*(x^2-1)$$

First part · Second part · Third part

or

$$=IF(x<-1,-1,IF(x<=1,x,x^2-1))$$

Since the third part of the formula specifying *f* is not linear, we need to plot many points in Excel to get a smooth graph. Here is one possible setup (for a smoother curve, plot more points):

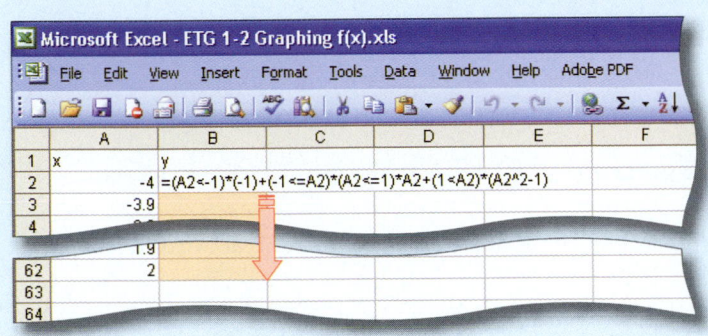
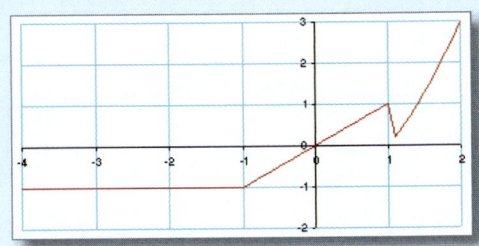

Notice that Excel does not handle the transition at $x = 1$ correctly and connects the two parts of the graph with a spurious line segment.

Section 1.3

Example 1 Which of the following two tables gives the values of a linear function? What is the formula for that function?

x	0	2	4	6	8	10	12
f(x)	3	−1	−3	−6	−8	−13	−15

x	0	2	4	6	8	10	12
f(x)	3	−1	−5	−9	−13	−17	−21

Solution with Technology The following worksheet shows how you can compute the successive quotients $m = \Delta y/\Delta x$, and hence check whether a given set of data shows a linear relationship, in which case all the quotients will be the same. (The shading indicates that the formula is to be copied down only as far as cell C7. Why not cell C8?)

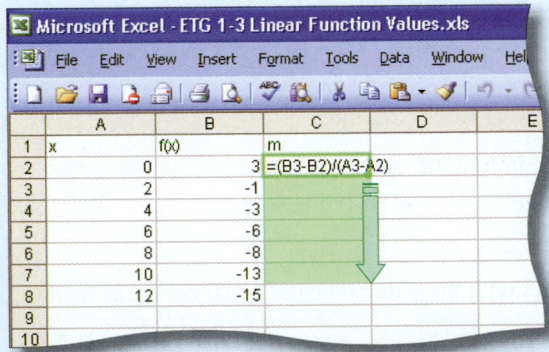

Here are the results for both $f(x)$ and $g(x)$:

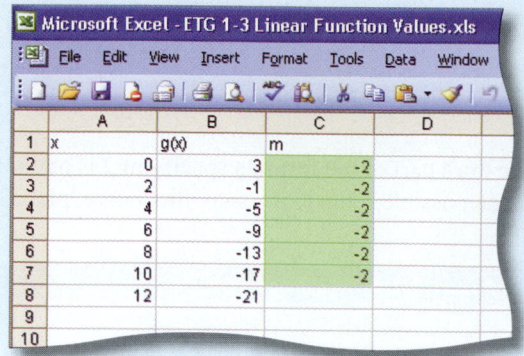

Section 1.4

Example 3(c) The following chart shows weekly sales figures for Hot'n'Spicy brand baked beans at two different prices.

Price/Can (p)	$0.50	$0.75
Demand (cans sold/week) (q)	400	350

How much should you charge for a can of Hot'n'Spicy beans if you want the demand to increase to 410 cans per week?

Solution with Technology In part (c), we can find p numerically by using a worksheet like the following to compute the demand for many prices starting at $p = \$0.00$, until we find the price at which the demand equals 410.

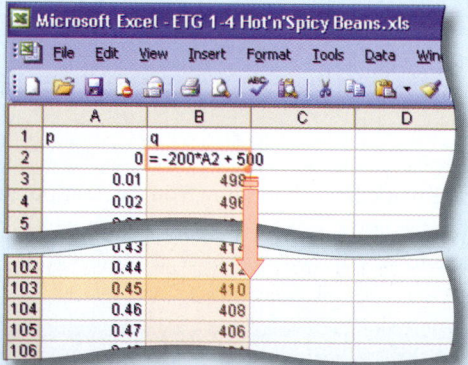

Section 1.5

Example 1 Using the data at the beginning of Section 1.5 on the cable television market in China, compute SSE, the sum-of-squares error, for the linear models $y = 8t + 72$ and $y = 5t + 68$. Which model is the better fit?

Solution with Technology

1. Begin by setting up your worksheet with the observed data in two columns, t and y, and the predicted data for the first model in the third.

Notice that instead of using the numerical equation for the first model in column C, we used absolute references to the cells containing the slope m and the intercept b. This way, we can switch from one linear model to the next by changing only m and b in cells E2 and F2. (We have deliberately left column D empty in anticipation of the next step.)

2. In column D we compute the squares of the residuals using the Excel formula

$$=(B2-C2)^2$$

Here is the completed worksheet, with SSE in cell F4.

Changing *m* to 5 and *b* to 68 gives the sum of squares error for the second model, SSE = 23.

	A	B	C	D	E	F	G
1	t	y (Observed)	y (Predicted)	Residual^2	m	b	
2	-4	50	48	4	5	68	
3	-3	55	53	4			
4	-2	57	58	1	SSE:	23	
5	-1	60	63	9			
6	0	68	68	0			
7	1	72	73	1			
8	2	80	78	4			
9	3	83	83	0			
10							
11							

Microsoft Excel - ETG 1-5 Cable Television Market in China.xls
File   Edit   View   Insert   Format   Tools   Data   Window   Help   Adobe PDF

Thus, the second model is a better fit.

You can also use Excel to plot both the original data points and each of the two lines. Use a scatter plot to graph the data in columns A through C in each of the last two worksheets above.

$$y = 8t + 72$$

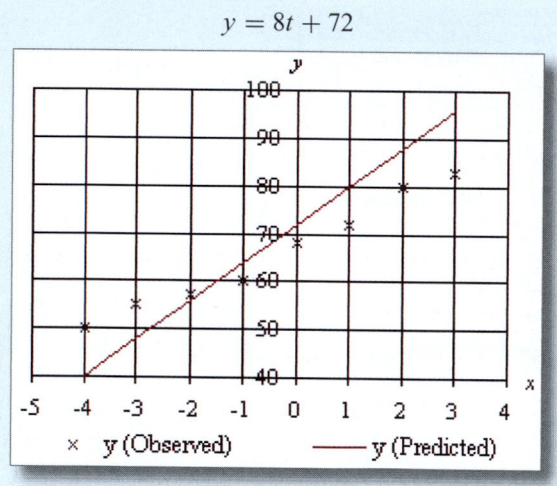

$$y = 5t + 68$$

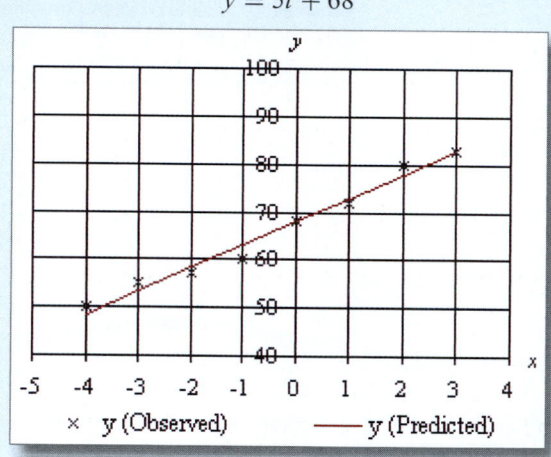

Example 2 Consider the following data about the growth of the cable TV market in China.

Year (*t*) (*t* = 0 represents 2000)	−4	−3	−2	−1	0	1	2	3
Households with Cable (*y*) (Millions)	50	55	57	60	68	72	80	83

Find the best-fit linear model for these data.

Solution with Technology Here are two Excel shortcuts for linear regression; one graphical and one based on an Excel formula.

Using the Trendline

1. Start with the original data and a "scatter plot."

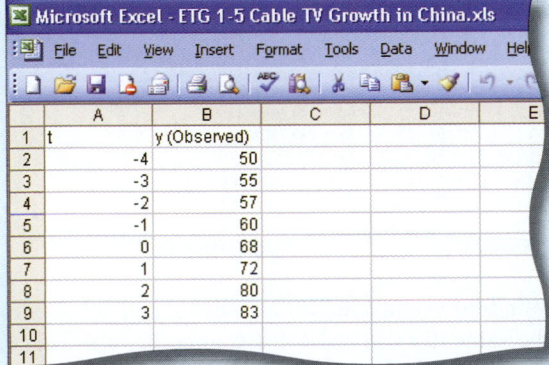

2. Click on the chart, and select "Add Trendline . . ." from the Chart menu. Then select a "Linear" type (the default) and, under "Options", check the option "Display equation on chart".

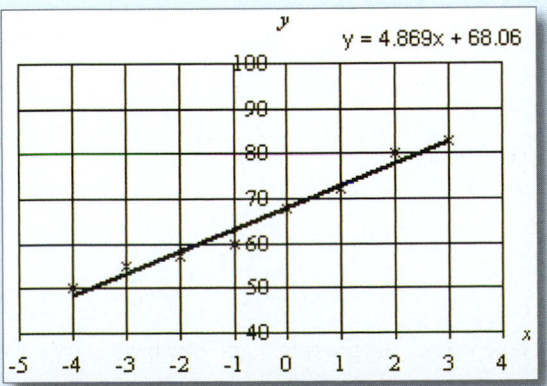

Using a Formula

Alternatively, you can use the "LINEST" function (for "linear estimate"):

1. Enter your data as above, and select a block of unused cells two wide and one tall; for example C2:D2.

2. Enter the formula

    =LINEST(B2:B9,A2:A9)

as shown, and press Control-Shift-Enter.

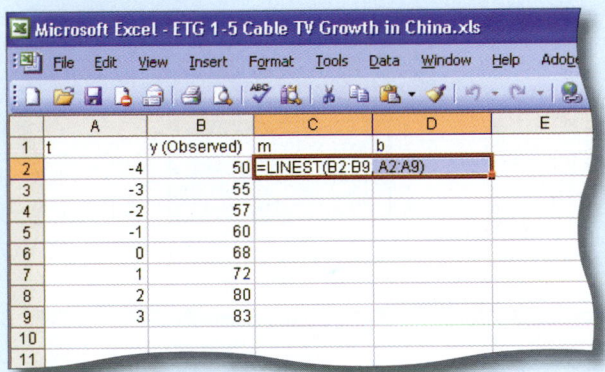

The result should look like this.

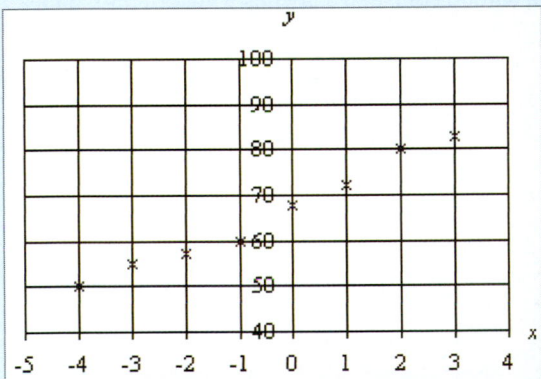

The values of *m* and *b* appear in cells C2 and D2 as shown,

Example 3 Find the correlation coefficient for the data in Example 2.

Solution with Technology In Excel, when you add a trend line to a chart you can select the option "Display r-squared value on chart" to show the value of r^2 on the chart (it is common to examine r^2, which takes on values between 0

and 1, instead of r). Alternatively, the LINEST function we used in Example 2 can be used to display quite a few statistics about a best fit line, including r^2:

1. Instead of selecting a block of cells two wide and one tall as we did in Example 2, we select one two wide and *five* tall.

2. We now enter the requisite LINEST formula with two additional arguments set to "TRUE" as shown, and press Control-Shift-Enter.

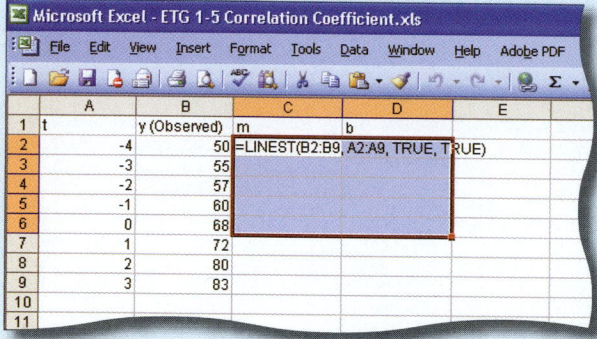

The result should look something like this:

	A	B	C	D	E
1	t	y (Observed)	m	b	
2	-4	50	4.869047619	68.05952381	
3	-3	55	0.296505855	0.695367867	
4	-2	57	0.9782343	1.92157756	
5	-1	60	269.6630844	6	
6	0	68	995.7202381	22.1547619	
7	1	72			
8	2	80			
9	3	83			
10					
11					

The values of m and b appear in cells C2 and D2 as before, and the value of r^2 in cell C4. (Among the other numbers shown is SSE in cell D6. For the meanings of the remaining numbers shown, see the on-line help for LINEST in Excel; a good course in statistics wouldn't hurt, either.)

2 Systems of Linear Equations and Matrices

CASE STUDY The Impact of Regulating Sulfur Emissions

You have been hired as a consultant to the Environmental Protection Agency (EPA). In an effort to curb the effects of acid rain on the ecosystem, the agency is considering regulations that will require a 15-million-ton reduction in sulfur emissions. You have been asked to estimate the cost of the proposed regulations to the major utility companies and also their effect on jobs in the coal mining industry. The data you have available show the annual cost to utilities and the cost in jobs for emission reductions of up to 12 million tons. Your assignment is to use these figures to compute projections for a 15-million-ton reduction.

David Woodfall/Getty Images

Introduction

In Chapter 1 we studied single functions and equations. In this chapter we seek solutions to **systems** of two or more equations. For example, suppose we need to *find two numbers whose sum is* 3 *and whose difference is* 1. In other words, we need to find two numbers x and y such that $x + y = 3$ and $x - y = 1$. The only solution turns out to be $x = 2$ and $y = 1$, a solution you might easily guess. But, how do we know that this is the only solution, and how do we find solutions systematically? When we restrict ourselves to systems of *linear* equations, there is a very elegant method for determining the number of solutions and finding them all. Moreover, as we will see, many real-world applications give rise to just such systems of linear equations.

We begin in Section 2.1 with systems of two linear equations in two unknowns and some of their applications. In Section 2.2 we study a powerful matrix method, called *row reduction,* for solving systems of linear equations in any number of unknowns. In Section 2.3 we look at more applications.

Computers have been used for many years to solve the large systems of equations that arise in the real world. You probably already have access to devices that will do the row operations used in row reduction. Many graphing calculators can do them, as can spreadsheets and various special-purpose applications, including utilities available at the website. Using such a device or program makes the calculations quicker and helps avoid arithmetic mistakes. Then there are programs (and calculators) into which you simply feed the system of equations and out pop the solutions. We can think of what we do in this chapter as looking inside the "black box" of such a program. More important, we talk about how, starting from a real-world problem, to get the system of equations to solve in the first place. This conversion no computer will yet do for us.

2.1 Systems of Two Equations in Two Unknowns

Suppose you have $3 in your pocket to spend on snacks and a drink. If x represents the amount you'll spend on snacks and y represents the amount you'll spend on a drink, you can say that $x + y = 3$. On the other hand, if for some reason you want to spend $1 more on snacks than on your drink, you can also say that $x - y = 1$. These are simple examples of **linear equations in two unknowns.**

Linear Equations in Two Unknowns

A **linear equation in two unknowns** is an equation that can be written in the form

$$ax + by = c$$

with a, b, and c being real numbers. The number a is called the **coefficient of x** and b is called the **coefficient of y.** A **solution** of an equation consists of a pair of numbers: a value for x and a value for y that satisfy the equation.

quick Example

In the linear equation $3x - y = 15$, the coefficients are $a = 3$ and $b = -1$. The point $(x, y) = (5, 0)$ is a solution, since $3(5) - (0) = 15$.

In fact, a single linear equation like $3x - y = 15$ has infinitely many solutions: We could solve for $y = 3x - 15$ and then, for every value of x we choose, we can get the corresponding value of y, giving a solution (x, y). As we saw in Chapter 1, these solutions are the points on a straight line, the *graph* of the equation.

In this section we are concerned with pairs (x, y) that are solutions of *two* linear equations at the same time. For example, $(2, 1)$ is a solution of both of the equations $x + y = 3$ and $x - y = 1$, since substituting $x = 2$ and $y = 1$ into these equations gives $2 + 1 = 3$ (true) and $2 - 1 = 1$ (also true), respectively. So, in the simple example we began with, you could spend \$2 on snacks and \$1 on a drink.

In the following set of examples, you will see how to graphically and algebraically solve a system of two linear equations in two unknowns. Then we'll return to some more interesting applications.

Example 1 Two Ways of Solving a System: Graphically and Algebraically

Find all solutions (x, y) of the following system of two equations:

$$x + y = 3$$
$$x - y = 1$$

Solution We will see how to find the solution(s) in two ways: graphically and algebraically. Remember that a solution is a pair (x, y) that simultaneously satisfies *both* equations.

Method 1: Graphical We already know that the solutions of a single linear equation are the points on its graph, which is a straight line. For a point to represent a solution of two linear equations, it must lie simultaneously on both of the corresponding lines. In other words, it must be a point where the two lines cross, or intersect. A look at Figure 1 should convince us that the lines cross only at the point $(2, 1)$, so this is the only possible solution.

Method 2: Algebraic In the algebraic approach, we try to combine the equations in such a way as to eliminate one variable. In this case, notice that if we add the left-hand sides of the equations, the terms with y are eliminated. So, we add the first equation to the second (that is, add the left-hand sides and add the right-hand sides[*]):

$$x + y = 3$$
$$\underline{x - y = 1}$$
$$2x + 0 = 4$$
$$2x \quad\;\; = 4$$
$$x \quad\;\;\; = 2$$

Now that we know that x has to be 2, we can substitute back into either equation to find y. Choosing the first equation (it doesn't matter which we choose), we have

$$2 + y = 3$$
$$y = 3 - 2$$
$$= 1$$

We have found that the only possible solution is $x = 2$ and $y = 1$, or

$$(x, y) = (2, 1)$$

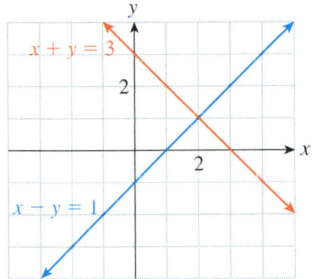

Figure **1**

using *Technology*

See the Technology Guides at the end of the chapter to find out how to use a TI-83/84 or Excel to help solve a system graphically.

---

[*] We can add these equations because when we add equal amounts to both sides of an equation, the results are equal. That is, if $A = B$ and $C = D$, then $A + C = B + D$.

+ *Before we go on...* In our discussion of the break-even point in Section 1.4, we found the intersection of two straight lines by setting equal two linear functions of x (cost and revenue). We could find the solution in Example 1 here in a similar way: First we solve our two equations for y to obtain

$$y = -x + 3$$
$$y = x - 1$$

and then we equate the right-hand sides and solve to find $x = 2$. The value for y is then found by substituting $x = 2$ in either equation. ∎

Q: *So why don't we solve all systems of equations this way* **?**

A: The elimination method extends more easily to systems with more equations and unknowns. It is the basis for the matrix method of solving systems—a method we discuss in Section 2.2. So, we shall use it exclusively for the rest of this section. ∎

Example 2 illustrates the drawbacks of the graphical method.

Example **2** Solving a System: Algebraically vs. Graphically

Solve the system

$$3x + 5y = 0$$
$$2x + 7y = 1$$

Solution

Method 1: Graphical First, solve for y, obtaining $y = -\dfrac{3}{5}x$ and $y = -\dfrac{2}{7}x + \dfrac{1}{7}$. Graphing these equations, we get Figure 2.

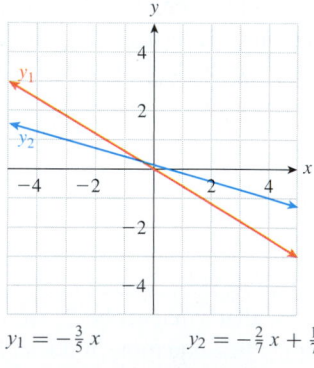

$$y_1 = -\tfrac{3}{5}x \qquad y_2 = -\tfrac{2}{7}x + \tfrac{1}{7}$$

Figure **2**

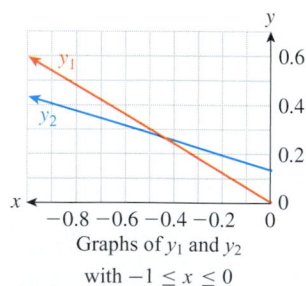

Graphs of y_1 and y_2
with $-1 \le x \le 0$

Figure **3**

The lines appear to intersect slightly above and to the left of the origin. Redrawing with a finer scale (or zooming in using graphing technology), we can get the graph in Figure 3.

If we look carefully at Figure 3, we see that the graphs intersect near $(-0.45, 0.27)$. Is the point of intersection *exactly* $(-0.45, 0.27)$? (Substitute these values into the equations to find out.) In fact, it is impossible to find the exact solution of this system graphically, but we now have a ballpark answer that we can use to help check the following algebraic solution.

Method 2: Algebraic We first see that adding the equations is not going to eliminate either x or y. Notice, however, that if we multiply (both sides of) the first equation by 2 and the second by -3, the coefficients of x will become 6 and -6. *Then* if we add them, x will be eliminated. So we proceed as follows:

$$2(3x + 5y) = 2(0)$$
$$-3(2x + 7y) = -3(1)$$

gives

$$6x + 10y = 0$$
$$-6x - 21y = -3$$

Adding these equations we get

$$-11y = -3$$

so that

$$y = \frac{3}{11} = 0.\overline{27}$$

Substituting $y = \dfrac{3}{11}$ in the first equation gives

$$3x + 5\left(\frac{3}{11}\right) = 0$$
$$3x = -\frac{15}{11}$$
$$x = -\frac{5}{11} = -0.\overline{45}$$

The solution is $(x, y) = \left(-\frac{5}{11}, \frac{3}{11}\right) = (-0.\overline{45}, 0.\overline{27})$.

Notice that the algebraic method gives us the exact solution that we could not find with the graphical method. Still, we can check that the graph and our algebraic solution agree to the accuracy with which we can read the graph. To be absolutely sure that our answer is correct we should check it:

$$3\left(-\frac{5}{11}\right) + 5\left(\frac{3}{11}\right) = -\frac{15}{11} + \frac{15}{11} = 0 \qquad ✔$$

$$2\left(-\frac{5}{11}\right) + 7\left(\frac{3}{11}\right) = -\frac{10}{11} + \frac{21}{11} = 1 \qquad ✔$$

Get in the habit of checking your answers.

+*Before we go on...*

Q: *In solving the system in Example 2 algebraically, we multiplied (both sides of) the equations by numbers. How does that affect their graphs?*

A: Since multiplying both sides of an equation by a nonzero number has no effect on its solutions, the graph (which represents the set of all solutions) is unchanged. ∎

Before doing some more examples, let's summarize what we've said about solving systems of equations.

> ## Graphical Method for Solving a System of Two Equations in Two Unknowns
>
> Graph both equations on the same graph. (For example, solve each for y to find the slope and y-intercept.) A point of intersection gives the solution to the system. To find the point, you may need to adjust the range of x-values you use. To find the point accurately you may need to use a smaller range (or zoom in if using technology).

> ## Algebraic Method for Solving a System of Two Equations in Two Unknowns
>
> Multiply each equation by a nonzero number so that the coefficients of x are the same in absolute value but opposite in sign. Add the two equations to eliminate x; this gives an equation in y that we can solve to find its value. Substitute this value of y into one of the original equations to find the value of x. (Note that we could eliminate y first instead of x if it's more convenient.)

Sometimes, something appears to go wrong with these methods. The following examples show what can happen.

Example 3 Inconsistent System

Solve the system

$$x - 3y = 5$$
$$-2x + 6y = 8$$

Solution   To eliminate x, we multiply the first equation by 2 and then add:

$$2x - 6y = 10$$
$$-2x + 6y = 8$$

Adding gives

$$0 = 18$$

But this is absurd! This calculation shows that if we had two numbers x and y that satisfied both equations, it would be true that $0 = 18$. Since 0 is *not* equal to 18, there can be no such numbers x and y. In other words, *the system has no solutions,* and is called an **inconsistent system.**

In slope-intercept form these lines are $y = \frac{1}{3}x - \frac{5}{3}$ and $y = \frac{1}{3}x + \frac{4}{3}$. Notice that they have the same slope but different y-intercepts. This means that they are parallel, but different lines. Plotting them confirms this fact (Figure 4).

Since they are parallel, they do not intersect. Since a solution must be a point of intersection, we again conclude that there is no solution.

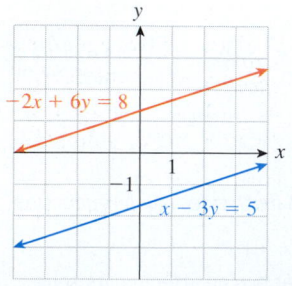

Figure **4**

Example 4 Redundant System

Solve the system

$$x + y = 2$$
$$2x + 2y = 4$$

Solution Multiplying the first equation by -2 gives

$$-2x - 2y = -4$$
$$2x + 2y = 4$$

Adding gives the not-very-enlightening result

$$0 = 0$$

Now what has happened? Looking back at the original system, we note that the second equation is really the first equation in disguise (it is the first equation multiplied by 2). Put another way, if we solve both equations for y, we find that, in slope-intercept form, both equations become the same:

$$y = -x + 2$$

Since the second equation gives us the same information as the first, we say that this is a **redundant,** or **dependent system.** In other words, we really have only one equation in two unknowns. From Chapter 1, we know that a single linear equation in two unknowns has infinitely many solutions, one for each value of x. (Recall that to get the corresponding solution for y, we solve the equation for y and substitute the x-value.) The entire set of solutions can be summarized as follows:

x is arbitrary

$y = 2 - x$ Solve the first equation for y.

This set of solutions is called the **general solution** because it includes all possible solutions. When we write the general solution this way, we say that we have a **parameterized solution** and that x is the **parameter.**

We can also write the general solution as

$$(x, 2 - x) \quad x \text{ arbitrary}$$

Different choices of the parameter x lead to different **particular solutions.** For instance, choosing $x = 3$ gives the particular solution

$$(x, y) = (3, -1)$$

Because there are infinitely many values of x from which to choose, there are infinitely many solutions.

What does this system of equations look like graphically? Since the two equations are really the same, their graphs are identical, each being the line with x-intercept 2 and y-intercept 2. The "two" lines intersect at every point, so there is a solution for each point on the common line. In other words, we have a "whole line of solutions" (Figure 5).

We could also have solved the first equation for x instead and used y as a parameter, obtaining another form of the general solution:

$$(2 - y, y) \quad y \text{ arbitrary} \qquad \text{Alternate form of the general solution}$$

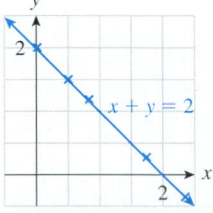

The line of solutions $(y = 2 - x)$ and some particular solutions

Figure 5

We summarize the three possible outcomes we have encountered.

> **Possible Outcomes for a System of Two Linear Equations in Two Unknowns**
>
> 1. **A single (or *unique*) solution** This happens when the lines corresponding to the two equations are distinct and not parallel so that they intersect at a single point. (See Example 1.)
>
> 2. **No solution** This happens when the two lines are parallel. We say that the system is **inconsistent.** (See Example 3.)
>
> 3. **An infinite number of solutions** This occurs when the two equations represent the same straight line, and we say that such a system is **redundant,** or **dependent.** In this case, we can represent the solutions by choosing one variable arbitrarily and solving for the other. (See Example 4.)
>
> In cases 1 and 3, we say that the system of equations is **consistent** because it has at least one solution.

You should think about straight lines and convince yourself that these are the only three possibilities.

Applications

Example 5 Blending

Acme Baby Foods mixes two strengths of apple juice. One quart of Beginner's juice is made from 30 fluid ounces of water and 2 fluid ounces of apple juice concentrate. One quart of Advanced juice is made from 20 fluid ounces of water and 12 fluid ounces of concentrate. Every day Acme has available 30,000 fluid ounces of water and 3600 fluid ounces of concentrate. If the company wants to use all the water and concentrate, how many quarts of each type of juice should it mix?

Solution In all applications we follow the same general strategy.

1. ***Identify and label the unknowns.*** What are we asked to find? To answer this question, it is common to respond by saying, "The unknowns are Beginner's juice and Advanced juice." Quite frankly, this is a baffling statement. Just what is unknown about juice? We need to be more precise:

   *The unknowns are (1) the **number of quarts** of Beginner's juice and (2) the **number of quarts** of Advanced juice made each day.*

   So, we label the unknowns as follows: Let

   x = number of quarts of Beginner's juice made each day
   y = number of quarts of Advanced juice made each day

2. ***Use the information given to set up equations in the unknowns.*** This step is trickier, and the strategy varies from problem to problem. Here, the amount of juice the company can make is constrained by the fact that they have limited amounts of water and concentrate. This example shows a kind of application we will often see, and it is helpful in these problems to use a table to record the amounts of the resources used.

	Beginner's (x)	Advanced (y)	Available
Water (fl oz)	30	20	30,000
Concentrate (fl oz)	2	12	3600

We can now set up an equation for each of the items listed on the left.

Water: We read across the first row. If Acme mixes x quarts of Beginner's juice, each quart using 30 fluid ounces of water, and y quarts of Advanced juice, each using 20 fluid ounces of water, it will use a total of $30x + 20y$ fluid ounces of water. But we are told that the total has to be 30,000 fluid ounces. Thus, $30x + 20y = 30,000$. This is our first equation.

Concentrate: We read across the second row. If Acme mixes x quarts of Beginner's juice, each using 2 fluid ounces of concentrate, and y quarts of Advanced juice, each using 12 fluid ounces of concentrate, it will use a total of $2x + 12y$ fluid ounces of concentrate. But we are told that the total has to be 3600 fluid ounces. Thus, $2x + 12y = 3600$.

Now we have two equations:

$$30x + 20y = 30,000$$
$$2x + 12y = 3600$$

To make the numbers easier to work with, let's divide (both sides of) the first equation by 10 and the second by 2:

$$3x + 2y = 3000$$
$$x + 6y = 1800$$

We can now eliminate x by multiplying the second equation by -3 and adding:

$$
\begin{array}{r}
3x + 2y = 3000 \\
-3x - 18y = -5400 \\
\hline
-16y = -2400
\end{array}
$$

So, $y = 2400/16 = 150$. Substituting this into the equation $x + 6y = 1800$ gives $x + 900 = 1800$, and so $x = 900$. The solution is $(x, y) = (900, 150)$. In other words, the company should mix 900 quarts of Beginner's juice and 150 quarts of Advanced juice.

Example 6 Blending

A medieval alchemist's love potion calls for a number of eyes of newt and toes of frog, the total being 20, but with twice as many newt eyes as frog toes. How many of each is required?

Solution As in the preceding example, the first step is to identify and label the unknowns. Let

$$x = \text{number of newt eyes}$$
$$y = \text{number of frog toes}$$

As for the second step—setting up the equations—a table is less appropriate here than in the preceding example. Instead, we translate each phrase of the problem into an equation. The first sentence tells us that the total number of eyes and toes is 20. Thus,

$$x + y = 20$$

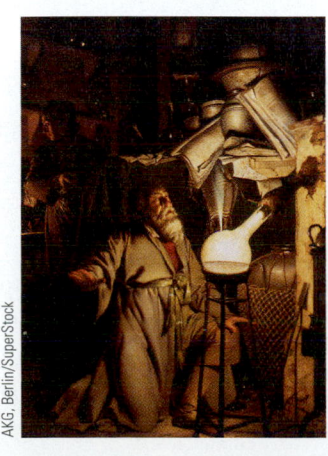

The end of the first sentence gives us more information, but the phrase "twice as many newt eyes as frog toes" is a little tricky: does it mean that $2x = y$ or that $x = 2y$? We can decide which by rewording the statement using the phrases "the *number of* newt eyes," which is x, and "the *number of* frog toes," which is y. Rephrased, the statement reads:

*The **number of** newt eyes is twice the **number of** frog toes.*

(Notice how the word "twice" is forced into a different place.) With this rephrasing, we can translate directly into algebra:

$$x = 2y$$

In standard form $(ax + by = c)$, this equation reads

$$x - 2y = 0$$

Thus, we have the two equations:

$$x +\ y = 20$$
$$x - 2y = 0$$

To eliminate x, we multiply the second equation by -1 and then add:

$$x +\ y = 20$$
$$-x + 2y = 0$$

We'll leave it to you to finish solving the system and find that $x = 13\frac{1}{3}$ and $y = 6\frac{2}{3}$.

So, the recipe calls for exactly $13\frac{1}{3}$ eyes of newt and $6\frac{2}{3}$ toes of frog. The alchemist needs a very sharp scalpel and a very accurate balance (not to mention a very strong stomach).

We saw in Chapter 1 that the *equilibrium price* of an item (the price at which supply equals demand) and the *break-even point* (the number of items that must be sold to break even) can both be described as the intersection points of two lines. In terms of the language of this chapter, what we were really doing was solving a system of two linear equations in two unknowns, as illustrated in the following problem.

Example 7 Equilibrium Price

The demand for refrigerators in West Podunk is given by

$$q = -\frac{p}{10} + 100$$

where q is the number of refrigerators that the citizens will buy each year if the refrigerators are priced at p dollars each. The supply is

$$q = \frac{p}{20} + 25$$

where now q is the number of refrigerators the manufacturers will be willing to ship into town each year if they are priced at p dollars each. Find the equilibrium price and the number of refrigerators that will be sold at that price.

Solution Figure 6 shows the demand and supply curves. The equilibrium price occurs at the point where these two lines cross, which is where demand equals supply.

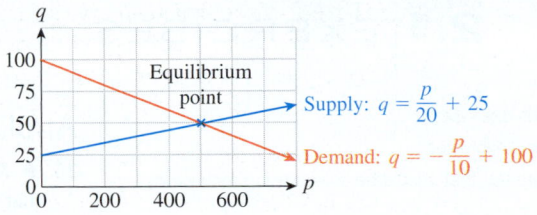

Figure 6

The graph suggests that the equilibrium price is $500, and zooming in confirms this. To solve this system algebraically, first write both equations in standard form:

$$\frac{p}{10} + q = 100$$

$$-\frac{p}{20} + q = 25$$

We can clear fractions and also prepare to eliminate p if we multiply the first equation by 10 and the second by 20:

$$p + 10q = 1000$$

$$-p + 20q = 500$$

and so:

$$30q = 1500$$

$$q = 50$$

Substituting this value of q into either equation gives us $p = 500$. Thus, the equilibrium price is $500, and 50 refrigerators will be sold at this price.

We could also have solved this system of equations by setting the two expressions for q (the supply and the demand) equal to each other:

$$-p/10 + 100 = p/20 + 25$$

and then solving for p. (See the *Before we go on* discussion at the end of Example 1.)

FAQs Setting Up the Equations

Q: *Looking through these examples, I notice that in some, we can tabulate the information given and read off the equations (as in Example 5), whereas in others (like Example 6), we have to reword each sentence to turn it into an equation. How do I know what approach to use?*

A: There is no hard-and-fast rule, and indeed some applications might call for a bit of each approach. However, it is generally not hard to see when it would be useful to tabulate values: Lists of the numbers of ingredients or components generally lend themselves to tabulation, whereas phrases like "twice as many of these as those" generally require direct translation into equations (after rewording if necessary). ∎

2.1 | EXERCISES

● denotes basic skills exercises

◆ denotes challenging exercises

tech Ex indicates exercises that should be solved using technology

In Exercises 1–14, find all solutions of the given system of equations and check your answer graphically. hint [*see Examples 1–4*]

1. ● $x - y = 0$
  $x + y = 4$

2. ● $x - y = 0$
  $x + y = -6$

3. ● $x + y = 4$
  $x - y = 2$

4. ● $2x + y = 2$
  $-2x + y = 2$

5. ● $3x - 2y = 6$
  $2x - 3y = -6$

6. ● $2x + 3y = 5$
  $3x + 2y = 5$

7. ● $0.5x + 0.1y = 0.7$
  $0.2x - 0.2y = 0.6$

8. ● $-0.3x + 0.5y = 0.1$
  $0.1x - 0.1y = 0.4$

9. ● $\dfrac{x}{3} - \dfrac{y}{2} = 1$
  $\dfrac{x}{4} + y = -2$

10. ● $-\dfrac{2x}{3} + \dfrac{y}{2} = -\dfrac{1}{6}$
  $\dfrac{x}{4} - y = -\dfrac{3}{4}$

11. ● $2x + 3y = 1$
  $-x - \dfrac{3y}{2} = -\dfrac{1}{2}$

12. ● $2x - 3y = 1$
  $6x - 9y = 3$

13. ● $2x + 3y = 2$
  $-x - \dfrac{3y}{2} = -\dfrac{1}{2}$

14. ● $2x - 3y = 2$
  $6x - 9y = 3$

tech Ex *In Exercises 15–24, use technology to obtain approximate solutions graphically. All solutions should be accurate to one decimal place. (Zoom in for improved accuracy.)*

15. tech Ex $2x + 8y = 10$
  $x + y = 5$

16. tech Ex $2x - y = 3$
  $x + 3y = 5$

17. tech Ex $3.1x - 4.5y = 6$
  $4.5x + 1.1y = 0$

18. tech Ex $0.2x + 4.5y = 1$
  $1.5x + 1.1y = 2$

19. tech Ex $10.2x + 14y = 213$
  $4.5x + 1.1y = 448$

20. tech Ex $100x + 4.5y = 540$
  $1.05x + 1.1y = 0$

21. tech Ex Find the intersection of the line through (0, 1) and (4.2, 2) and the line through (2.1, 3) and (5.2, 0).

22. tech Ex Find the intersection of the line through (2.1, 3) and (4, 2) and the line through (3.2, 2) and (5.1, 3).

23. tech Ex Find the intersection of the line through (0, 0) and (5.5, 3) and the line through (5, 0) and (0, 6).

24. tech Ex Find the intersection of the line through (4.3, 0) and (0, 5) and the line through (2.1, 2.2) and (5.2, 1).

Applications

25. ● *Resource Allocation* You manage an ice cream factory that makes two flavors: Creamy Vanilla and Continental Mocha. Into each quart of Creamy Vanilla go 2 eggs and 3 cups of cream. Into each quart of Continental Mocha go 1 egg and 3 cups of cream. You have in stock 500 eggs and 900 cups of cream. How many quarts of each flavor should you make in order to use up all the eggs and cream? hint [*see Example 5*]

26. ● *Class Scheduling* Enormous State University's Math Department offers two courses: Finite Math and Applied Calculus. Each section of Finite Math has 60 students, and each section of Applied Calculus has 50. The department will offer a total of 110 sections in a semester, and 6000 students would like to take a math course. How many sections of each course should the department offer in order to fill all sections and accommodate all of the students?

27. ● *Nutrition* Gerber Mixed Cereal for Baby contains, in each serving, 60 calories and 11 grams of carbohydrates[1]. Gerber Mango Tropical Fruit Dessert contains, in each serving, 80 calories and 21 grams of carbohydrates[2]. If you want to provide your child with 200 calories and 43 grams of carbohydrates, how many servings of each should you use?

28. ● *Nutrition* Anthony Altino is mixing food for his young daughter and would like the meal to supply 1 gram of protein and 5 milligrams of iron. He is mixing together cereal, with 0.5 grams of protein and 1 milligram of iron per ounce, and fruit, with 0.2 grams of protein and 2 milligrams of iron per ounce. What mixture will provide the desired nutrition?

29. ● *Nutrition* One serving of Campbell's® Pork & Beans contains 5 grams of protein and 21 grams of carbohydrates[3]. A typical slice of white bread provides 2 grams of protein and 11 grams of carbohydrates per slice. The U.S. RDA (Recommended Daily Allowance) is 60 grams of protein each day.

  a. I am planning a meal of "beans on toast" and wish to have it supply one-half of the RDA for protein and 139 grams of carbohydrates. How should I prepare my meal?

  b. Is it possible to have my meal supply the same amount of protein as in part (a) but only 100 g of carbohydrates?

30. ● *Nutrition* One serving of Campbell's® Pork & Beans contains 5 grams of protein and 21 grams of carbohydrates[4]. A typical slice of "lite" rye bread contains 4 grams of protein and 12 grams of carbohydrates.

---

[1] SOURCE: nutrition information printed on the box.

[2] SOURCE: nutrition information printed on the jar.

[3] According to the label information on a 16 oz. can.

[4] Ibid.

● basic skills   ◆ challenging   tech Ex technology exercise

a. I am planning a meal of "beans on toast" and wish to have it supply one third of the U.S. RDA for protein (see the preceding exercise), and 80 grams of carbohydrates. How should I prepare my meal?

b. Is it possible to have my meal supply the same amount of protein as in part (a) but only 60 grams of carbohydrates?

Creatine Supplements Exercises 31 and 32 are based the following data on three popular bodybuilding supplements. (Figures shown correspond to a single serving.)[5]

	Creatine (g)	Carbohydrates (g)	Alpha Lipoic Acid (mg)	Cost ($)
Cell-Tech® (MuscleTech)	10	75	200	2.20
RiboForce HP® (EAS)	5	15	0	1.60
GlutaCell Xtreme® (Global/Image)	5	3	100	1.00

31. ● You are thinking of combining Cell-Tech and RiboForce HP to obtain a 10-day supply that provides exactly 80 grams of creatine and 1000 milligrams of alpha lipoic acid. How many servings of each supplement should you combine in order to meet your requirements? What will it cost?

32. ● You are thinking of combining Cell-Tech and GlutaCell Xtreme to obtain a 10-day supply that provides exactly 125 grams of creatine and 420 grams of carbohydrates. How many servings of each supplement should you combine in order to meet your requirements? What will it cost?

33. ● *Investments: Tech Stocks* In November 2003, Cisco (CSCO) stock rose from $20 to $22.50 per share, and Nokia (NOK) rose from $17 to $18 per share.[6] If you invested a total of $4550 in these stocks at the beginning of November and sold them for $4950 at the end of November, how many shares of each stock did you buy?

34. ● *Investments: Tech Stocks* In November 2003, Nortel (NT) began and ended the month at $4.50 per share, and Altria Group (MO) increased from $46 to $52 per share.[7] If you invested a total of $5500 in these stocks at the beginning of November and sold them for $6100 at the end of November, how many shares of each stock did you buy?

35. *Investments: Utility Stocks* In July 2004, Consolidated Edison (ED) cost $40 per share and yielded 5.5% per year in dividends, while Keyspan Corp (KSE) cost $36 per share and yielded 5% per year in dividends.[8] If you invested a total of $11,200 in these stocks and earned $580 in dividends in a year, how many shares of each stock did you purchase? (Assume the dividend rate was unchanged for the year.)

36. *Investments: Bank Stocks* In December, 2003, Bank of America (BAC) cost $80 per share and yielded 4% per year in dividends, while Royal Bank of Scotland (RBS-K) cost $28 per share and yielded 7.5% per year in dividends.[9] If you invested a total of $6800 in these stocks and earned $370 in dividends in a year, how many shares of each stock did you purchase? (Assume the dividend rate was unchanged for the year.)

37. ● *Voting* The U.S. House of Representatives has 435 members. If an appropriations bill passes the House with 49 more members voting in favor than against, how many voted in favor and how many voted against?

38. ● *Voting* The U.S. Senate has 100 members. For a bill to pass with a supermajority, at least twice as many senators must vote for the bill as vote against it. If all 100 senators vote, how many must vote for a bill for it to pass with a supermajority?

39. ● *Intramural Sports* The best sports dorm on campus, Lombardi House, has won a total of 12 games this semester. Some of these games were soccer games, and the others were football games. According to the rules of the university, each win in a soccer game earns the winning house 2 points, whereas each win in a football game earns them 4 points. If the total number of points Lombardi House earned was 38, how many of each type of game did they win?

40. ● *Law* Enormous State University's campus publication, *The Campus Inquirer*, ran a total of 10 exposés five years ago, dealing with alleged recruiting violations by the football team and with theft by the student treasurer of the film society. Each exposé dealing with recruiting violations resulted in a $4 million libel suit, and the treasurer of the film society sued the paper for $3 million as a result of each exposé concerning his alleged theft. Unfortunately for *The Campus Inquirer*, all the lawsuits were successful, and the paper wound up being ordered to pay $37 million in damages. (It closed down shortly thereafter.) How many of each type of exposé did the paper run?

41. ● *Purchasing* (from the GMAT) Elena purchased brand *X* pens for $4.00 apiece and brand *Y* pens for $2.80 apiece. If Elena purchased a total of 12 of these pens for $42.00, how many brand *X* pens did she purchase?

---

[5] SOURCE: Nutritional information supplied by the manufacturers. www.netrition.com. Cost per serving is approximate.

[6] Share prices are approximate at or close to the dates cited. Quotes provided by Comstock, Dec 2, 2003. http://money.excite.com.

[7] Share prices are approximate. Quotes provided by Comstock, Dec 2, 2003. http://money.excite.com.

[8] Share prices and dividend rates are approximate. Quotes provided by Comstock, August 1, 2004. http://money.excite.com.

[9] Share prices and dividend rates are approximate. Quotes provided by Comstock, Dec 19, 2003. http://money.excite.com.

● basic skills   ◆ challenging   **tech** Ex technology exercise

42. ● Purchasing (based on a question from the GMAT) Earl is ordering supplies. Yellow paper costs $5.00 per ream while white paper costs $6.50 per ream. He would like to order 100 reams total, and has a budget of $560. How many reams of each color should he order?

43. ● Equilibrium Price The demand and supply functions for pet chias are, respectively, $q = -60p + 150$ and $q = 80p - 60$, where p is the price in dollars. At what price should the chias be marked so that there is neither a surplus nor a shortage of chias? *hint* [see Example 7]

44. ● Equilibrium Price The demand and supply functions for your college newspaper are, respectively, $q = -10,000p + 2000$ and $q = 4000p + 600$, where p is the price in dollars. At what price should the newspapers be sold so that there is neither a surplus nor a shortage of papers?

45. ● Supply and Demand (from the GRE Economics Test) The demand curve for widgets is given by $D = 85 - 5P$, and the supply curve is given by $S = 25 + 5P$, where P is the price of widgets. When the widget market is in equilibrium, what is the quantity of widgets bought and sold?

46. ● Supply and Demand (from the GRE Economics Test) In the market for soybeans, the demand and supply functions are $Q_D = 100 - 10P$ and $Q_S = 20 + 5P$, where Q_D is quantity demanded, Q_S is quantity supplied, and P is price in dollars. If the government sets a price floor of $7, what will be the resulting surplus or shortage?

47. ● Equilibrium Price In June 2001, the retail price of a 25-kg bag of cornmeal was $8 in Zambia; by December, the price had risen to $11. The result was that one retailer reported a drop in sales from 15 bags/day to 3 bags/day.[10] Assume that the retailer is prepared to sell 3 bags/day at $8 and 15 bags/day at $11. Find linear demand and supply equations and then compute the retailer's equilibrium price.

48. ● Equilibrium Price At the start of December 2001, the retail price of a 25 kg bag of cornmeal was $10 in Zambia, while by the end of the month, the price had fallen to $6.[11] The result was that one retailer reported an increase in sales from 3 bags/day to 5 bags/day. Assume that the retailer is prepared to sell 18 bags/day at $8 and 12 bags/day at $6. Obtain linear demand and supply equations, and hence compute the retailer's equilibrium price.

49. ● Pollution Joe Slo, a college sophomore, neglected to wash his dirty laundry for 6 weeks. By the end of that time, his roommate had had enough and tossed Joe's dirty socks and T-shirts into the trash, counting a total of 44 items. (A pair of dirty socks counts as one item.) The roommate noticed that there were three times as many pairs of dirty socks as T-shirts. How many of each item did he throw out?

50. ● Diet The local sushi bar serves 1-ounce pieces of raw salmon (consisting of 50% protein) and $1\frac{1}{4}$-ounce pieces of raw tuna (40% protein). A customer's total intake of protein amounts to $1\frac{1}{2}$ ounces after consuming a total of three pieces. How many of each did the customer consume? (Fractions of pieces are permitted.)

51. Management (from the GMAT) A manager has $6000 budgeted for raises for 4 full-time and 2 part-time employees. Each of the full-time employees receives the same raise, which is twice the raise that each of the part-time employees receives. What is the amount of the raise that each full-time employee receives?

52. Publishing (from the GMAT) There were 36,000 hardback copies of a certain novel sold before the paperback version was issued. From the time the first paperback copy was sold until the last copy of the novel was sold, 9 times as many paperback copies as hardback copies were sold. If a total of 441,000 copies of the novel were sold in all, how many paperback copies were sold?

Communication and Reasoning Exercises

53. ● A system of three equations in two unknowns corresponds to three lines in the plane. Describe how these lines might be positioned if the system has a unique solution.

54. ● A system of three equations in two unknowns corresponds to three lines in the plane. Describe several ways that these lines might be positioned if the system has no solutions.

55. ● Both the supply and demand equations for a certain product have negative slope. Can there be an equilibrium price? Explain.

56. ● You are solving a system of equations with x representing the number of rocks and y representing the number of pebbles. The solution is $(200, -10)$. What do you conclude?

57. Referring to Exercise 25, suppose that the solution of the corresponding system of equations was 198.7 gallons of vanilla and 100.89 gallons of mocha. If your factory can produce only whole numbers of gallons, would you recommend rounding the answers to the nearest whole number? Explain.

58. Referring to Exercise 25 but using different data, suppose that the general solution of the corresponding system of equations was $(200 - y, y)$, where x = number of gallons vanilla, and y = number of gallons of mocha. Your factory can produce only whole numbers of gallons. There are infinitely many ways of making the ice cream, mixes, right? Explain.

59. Select one: Multiplying both sides of a linear equation by a nonzero constant results in a linear equation whose graph is
(A) parallel to   **(B)** the same as
(C) not always parallel to   **(D)** not the same as
the graph of the original equation.

---

[10] The prices quoted are approximate. (Actual prices varied from retailer to retailer.) SOURCE: *New York Times,* December 24, 2001, p. A4.

[11] Ibid.

● basic skills   ◆ challenging   **tech Ex** technology exercise

60. Select one: If the addition or subtraction of two linear equations results in the equation $3 = 3$, then the graphs of those equations are

(A) equal   **(B)** parallel

(C) perpendicular   **(D)** none of the above

61. Select one: If the addition or subtraction of two linear equations results in the equation $0 = 3$, then the graphs of those equations are

(A) equal   **(B)** parallel

(C) perpendicular   **(D)** not parallel

62. Select one: If adding two linear equations gives $x = 3$, and subtracting them gives $y = 3$, then the graphs of those equations are

(A) equal   **(B)** parallel

(C) perpendicular   **(D)** not parallel

63. Invent an interesting application that leads to a system of two equations in two unknowns with a unique solution.

64. Invent an interesting application that leads to a system of two equations in two unknowns with no solution.

65. ◆ How likely do you think it is that a "random" system of two equations in two unknowns has a unique solution? Give some justification for your answer.

66. ◆ How likely do you think it is that a "random" system of three equations in two unknowns has a unique solution? Give some justification for your answer.

● basic skills   ◆ challenging   _tech_ Ex technology exercise

2.2 Using Matrices to Solve Systems of Equations

In this section we describe a systematic method for solving systems of equations that makes solving large systems of equations straightforward. Although this method may seem a little cumbersome at first, it will prove _immensely_ useful in this and the next several chapters.

First of all, notice that a linear equation (for example, $2x - y = 3$) is entirely determined by its _coefficients_ (here, the numbers 2 and -1) and its _constant term_ or _right-hand side_ (here, 3). In other words, if we were simply given the row of numbers

$$[2 \quad -1 \quad 3]$$

we could easily reconstruct the original linear equation by multiplying the first number by x, the second by y, and inserting a plus sign and an equals sign, as follows:

$$2 \cdot x + (-1) \cdot y = 3$$

or $2x - y = 3$

Similarly, the equation

$$-4x + 2y = 0$$

is represented by the row

$$[-4 \quad 2 \quad 0]$$

and the equation

$$-3y = \frac{1}{4}$$

is represented by

$$\begin{bmatrix} 0 & -3 & \dfrac{1}{4} \end{bmatrix}$$

As the last example shows, the first number is always the coefficient of x and the second is the coefficient of y. If an x or a y is missing, we write a zero for its coefficient. We shall call such a row the **coefficient row** of an equation.

If we have a system of equations, for example the system

$$\begin{aligned} 2x - \;\; y &= 3 \\ -x + 2y &= -4 \end{aligned}$$

we can put the coefficient rows together like this:

$$\begin{bmatrix} 2 & -1 & 3 \\ -1 & 2 & -4 \end{bmatrix}$$

We call this the **augmented matrix** of the system of equations. The term *augmented* means that we have included the right-hand sides 3 and -4. We will often drop the word *augmented* and simply refer to the matrix of the system. A **matrix** (plural: **matrices**) is nothing more than a rectangular array of numbers as above.

Matrix

A **matrix** is a rectangular array of numbers.

Augmented Matrix

The **augmented matrix** of a system of linear equations is the matrix whose rows are the coefficient rows of the equations.

quick Example   The augmented matrix of the system

$$\begin{aligned} x + y &= 3 \\ x - y &= 1 \end{aligned}$$

is

$$\begin{bmatrix} 1 & 1 & 3 \\ 1 & -1 & 1 \end{bmatrix}$$

We'll be studying matrices in their own right more carefully in Chapter 3.

Q: *What good are coefficient rows and matrices?*

A: Think about what we do when we multiply both sides of an equation by a number. For example, consider multiplying both sides of the equation $2x - y = 3$ by -2 to get $-4x + 2y = -6$. All we are really doing is multiplying the coefficients and the right-hand side by -2. This corresponds to *multiplying the row* $[2 \;\; -1 \;\; 3]$ *by* -2, that is, multiplying every number in the row by -2. We shall see that any manipulation we want to do with equations can be done instead with rows, and this fact leads to a method of solving equations that is systematic and generalizes easily to larger systems. ■

Here is the same operation both in the language of equations and the language of rows. (We refer to the equation here as *Equation* 1, or simply E_1 for short, and to the row as *Row* 1, or R_1.)

Equation		Row	
E_1: $2x - y = 3$		$[\;2 \quad -1 \quad 3]$	R_1
Multiply by -2: $(-2)E_1$: $-4x + 2y = -6$		$[-4 \quad 2 \quad -6]$	$(-2)R_1$

Multiplying both sides of an equation by the number a corresponds to multiplying the coefficient row by a.

Now look at what we do when we add two equations:

Equation		Row	
E_1: $2x - y = 3$		$[\;2 \quad -1 \quad 3]$	R_1
E_2: $-x + 2y = -4$		$[-1 \quad 2 \quad -4]$	R_2
Add: $E_1 + E_2$: $x + y = -1$		$[\;1 \quad 1 \quad -1]$	$R_1 + R_2$

All we are really doing is *adding the corresponding entries in the rows*, or *adding the rows*. In other words,

Adding two equations corresponds to adding their coefficient rows.

In short, the manipulations of equations that we saw in the preceding section can be done more easily with rows in a matrix, since we don't have to carry x, y, and other unnecessary notation along with us; x and y can always be inserted at the end if desired.

The manipulations we are talking about are known as **row operations.** In particular, we use three **elementary row operations.**

Elementary Row Operations[*]

Type 1: Replacing R_i by aR_i (where $a \neq 0$)[†]
In words: multiplying or dividing a row by a nonzero number.

Type 2: Replacing R_i by $aR_i \pm bR_j$ (where $a \neq 0$)
Multiplying a row by a nonzero number and adding or subtracting a multiple of another row.

Type 3: Switching the order of the rows
This corresponds to switching the order in which we write the equations; occasionally this will be convenient.

[*] We are using the term *elementary row operations* a little more freely than most books. Some mathematicians insist that $a = 1$ in an operation of Type 2, but the less restrictive version is very useful.

[†] Multiplying an equation or row by zero gives us the not very surprising result $0 = 0$. In fact, we lose any information that that equation provided, which usually means that the resulting system has more solutions than the original system.

For Types 1 and 2, we write the instruction for the row operation *next to the row we wish to replace* (see the quick examples below).

quick Examples

Type 1: $\begin{bmatrix} 1 & 3 & -4 \\ 0 & 4 & 2 \end{bmatrix} 3R_2 \rightarrow \begin{bmatrix} 1 & 3 & -4 \\ 0 & 12 & 6 \end{bmatrix}$ Replace R_2 by $3R_2$

Type 2: $\begin{bmatrix} 1 & 3 & -4 \\ 0 & 4 & 2 \end{bmatrix} 4R_1 - 3R_2 \rightarrow \begin{bmatrix} 4 & 0 & -22 \\ 0 & 4 & 2 \end{bmatrix}$ Replace R_1 by $4R_1 - 3R_2$

Type 3: $\begin{bmatrix} 1 & 3 & -4 \\ 0 & 4 & 2 \\ 1 & 2 & 3 \end{bmatrix} R_1 \leftrightarrow R_2 \rightarrow \begin{bmatrix} 0 & 4 & 2 \\ 1 & 3 & -4 \\ 1 & 2 & 3 \end{bmatrix}$ Switch R_1 and R_2

using *Technology*

See the Technology Guides at the end of the chapter to see how to do row operations using a TI-83/84 or Excel.

One very important fact about the elementary row operations is that they do not change the solutions of the corresponding system of equations. In other words, the new system of equations that we get by applying any one of these operations will have exactly the same solutions as the original system: It is easy to see that numbers that make the original equations true will also make the new equations true, because each of the elementary row operations corresponds to a valid operation on the original equations. That any solution of the new system is a solution of the old system follows from the fact that these row operations are *reversible*: The effects of a row operation can be reversed by applying another row operation, called its **inverse.** Here are some examples of this reversibility. (Try them out in the above Quick Examples.)

Operation	Inverse Operation
Replace R_2 by $3R_2$.	Replace R_2 by $\frac{1}{3}R_2$.
Replace R_1 by $4R_1 - 3R_2$.	Replace R_1 by $\frac{1}{4}R_1 + \frac{3}{4}R_2$.
Switch R_1 and R_2.	Switch R_1 and R_2.

Our objective, then, is to use row operations to change the system we are given into one with exactly the same set of solutions, in which it is easy to see what the solutions are.

Solving Systems of Equations Using Row Operations

Now we put rows to work for us in solving systems of equations. Let's start with a complicated looking system of equations:

$$-\frac{2x}{3} + \frac{y}{2} = -3$$
$$\frac{x}{4} - y = \frac{11}{4}$$

We begin by writing the matrix of the system:

$$\begin{bmatrix} -\frac{2}{3} & \frac{1}{2} & -3 \\ \frac{1}{4} & -1 & \frac{11}{4} \end{bmatrix}$$

Now what do we do with this matrix?

Step 1 *Clear the fractions and/or decimals (if any) using operations of Type 1.* To clear the fractions, we multiply the first row by 6 and the second row by 4. We record the operations by writing the symbolic form of an operation next to the row it will change, as follows.

$$\begin{bmatrix} -\frac{2}{3} & \frac{1}{2} & -3 \\ \frac{1}{4} & -1 & \frac{11}{4} \end{bmatrix} \begin{matrix} 6R_1 \\ 4R_2 \end{matrix}$$

By this we mean that we will replace the first row by $6R_1$ and the second by $4R_2$. Doing these operations gives

$$\begin{bmatrix} -4 & 3 & -18 \\ 1 & -4 & 11 \end{bmatrix}$$

Step 2 *Designate the first nonzero entry in the first row as the* **pivot.** In this case we designate the entry -4 in the first row as the "pivot" by putting a box around it:

$$\begin{bmatrix} \boxed{-4} & 3 & -18 \\ 1 & -4 & 11 \end{bmatrix} \leftarrow \text{Pivot row}$$

$\uparrow$
Pivot column

Q: What is a "pivot"?

A: A **pivot** is an entry in a matrix that is used to "clear a column" (see Step 3). In this procedure, we will always select the first nonzero entry of a row as our pivot. In Chapter 4, when we study the simplex method, we will select our pivots differently. ■

Step 3 *Use the pivot to clear its column using operations of Type 2.* By **clearing a column**, we mean changing the matrix so that the pivot is the only nonzero number in its column. The procedure of clearing a column using a designated pivot is also called **pivoting.**

$$\begin{bmatrix} \boxed{-4} & 3 & -18 \\ 0 & \# & \# \end{bmatrix} \leftarrow \text{Desired row 2 (the "\#"s stand for as yet unknown numbers)}$$

$\uparrow$
Cleared pivot column

We want to replace R_2 by a row of the form $aR_2 \pm bR_1$ to get a zero in column 1. Moreover—and this will be important when we discuss the simplex method in Chapter 4—*we are going to choose positive values for both a and b.*[12] We need to choose a and b so that we get the desired cancellation. We can do this quite mechanically as follows:

a. Write the name of the row you need to change on the left and that of the pivot row on the right.

$$R_2 \qquad R_1$$

$\uparrow$ $\qquad$ $\uparrow$
Row to change   Pivot row

[12] Thus, the only place a negative sign may appear is between aR_2 and bR_1 as indicated in the formula $aR_2 \pm bR_1$.

b. Focus on the pivot column, $\begin{bmatrix} -4 \\ 1 \end{bmatrix}$. Multiply each row by the *absolute value* of the entry currently in the other (we are not permitting a or b to be negative).

$$\mathbf{4}R_2 \qquad\qquad \mathbf{1}R_1$$

↑   ↑
From Row 1   From Row 2

The effect is to make the two entries in the pivot column numerically the same. Sometimes, you can accomplish this by using smaller values of a and b.

c. If the entries in the pivot column have opposite signs, insert a plus ($+$). If they have the same sign, insert a minus ($-$). Here, we get the instruction

$$4R_2 + 1R_1$$

d. Write the operation next to the row you want to change, and then replace that row using the operation:

$$\begin{bmatrix} \boxed{-4} & 3 & -18 \\ 1 & -4 & 11 \end{bmatrix} 4R_2 + 1R_1 \rightarrow \begin{bmatrix} -4 & 3 & -18 \\ 0 & -13 & 26 \end{bmatrix}$$

We have cleared the pivot column and completed Step 3.

Note In general, the row operation you use should always have the following form:[13]

$$a R_c \qquad \pm \qquad b R_p$$

↑   ↑
Row to change   Pivot row

with a and b both positive.

The next step is one that can be performed at any time. ∎

Simplification Step (Optional) *If, at any stage of the process, all the numbers in a row are multiples of an integer, divide by that integer—a Type 1 operation.*

This is an optional but extremely helpful step: It makes the numbers smaller and easier to work with. In our case, the entries in R_2 are divisible by 13, so we divide that row by 13. (Alternatively, we could divide by -13. Try it.)

$$\begin{bmatrix} -4 & 3 & -18 \\ 0 & -13 & 26 \end{bmatrix} \tfrac{1}{13}R_2 \rightarrow \begin{bmatrix} -4 & 3 & -18 \\ 0 & -1 & 2 \end{bmatrix}$$

Step 4 *Select the first nonzero number in the second row as the pivot, and clear its column.* Here we have combined two steps in one: selecting the new pivot and clearing the column (pivoting). The pivot is shown below, as well as the desired result when the column has been cleared:

$$\begin{bmatrix} -4 & 3 & -18 \\ 0 & \boxed{-1} & 2 \end{bmatrix} \rightarrow \begin{bmatrix} \# & 0 & \# \\ 0 & -1 & 2 \end{bmatrix} \leftarrow \text{desired row}$$

↑   ↑
pivot column   cleared pivot column

---

[13] We are deviating somewhat from the traditional procedure here. It is traditionally recommended first to divide the pivot row by the pivot, turning the pivot into a 1. This allows us to always use $a = 1$. The procedure we use here is easier for hand calculations since it avoids the use of fractions. See the end of this section for an example using the traditional procedure.

We now wish to get a 0 in place of the 3 in the pivot column. Let's run once again through the mechanical steps to get the row operation that accomplishes this.

a. Write the name of the row you need to change on the left and that of the pivot row on the right:

$$R_1 \qquad\qquad R_2$$
$$\uparrow \qquad\qquad \uparrow$$

Row to change Pivot Row

b. Focus on the pivot column, $\begin{bmatrix} 3 \\ -1 \end{bmatrix}$. Multiply each row by the absolute value of the entry currently in the other:

$$\mathbf{1}R_1 \qquad\qquad \mathbf{3}R_2$$
$$\uparrow \qquad\qquad \uparrow$$

From Row 2 From Row 1

c. If the entries in the pivot column have opposite signs, insert a plus ($+$). If they have the same sign, insert a minus ($-$). Here, we get the instruction

$$1R_1 + 3R_2$$

d. Write the operation next to the row you want to change and then replace that row using the operation.

$$\begin{bmatrix} -4 & 3 & -18 \\ 0 & \boxed{-1} & 2 \end{bmatrix} \begin{array}{c} R_1 + 3R_2 \\ \\ \end{array} \rightarrow \begin{bmatrix} -4 & 0 & -12 \\ 0 & -1 & 2 \end{bmatrix}$$

Now we are essentially done, except for one last step.

Final Step *Using operations of Type* **1,** *turn each pivot (the first nonzero entry in each row) into a* **1.** We can accomplish this by dividing the first row by -4 and multiplying the second row by -1:

$$\begin{bmatrix} -4 & 0 & -12 \\ 0 & -1 & 2 \end{bmatrix} \begin{array}{c} -\frac{1}{4}R_1 \\ -R_2 \end{array} \rightarrow \begin{bmatrix} 1 & 0 & 3 \\ 0 & 1 & -2 \end{bmatrix}$$

The matrix now has the following nice form:

$$\begin{bmatrix} \boxed{1} & 0 & \# \\ 0 & \boxed{1} & \# \end{bmatrix}$$

(This is the form we will always obtain with two equations in two unknowns when there is a unique solution.) This form is nice because, when we translate back into equations, we get

$$1x + 0y = 3$$
$$0x + 1y = -2$$

In other words,

$$x = 3 \quad \text{and} \quad y = -2$$

and so we have found the solution, which we can also write as $(x, y) = (3, -2)$.

The procedure we've just demonstrated is called **Gauss-Jordan**[14] **reduction** or **row reduction.** It may seem too complicated a way to solve a system of two equations in two unknowns, and it is. However, for systems with more equations and more unknowns, it is very efficient.

In Example 1 below we use row reduction to solve a system of linear equations in *three* unknowns: x, y, and z. Just as for a system in two unknowns, a **solution** of a system in any number of unknowns consists of values for each of the variables that, when substituted, satisfy all of the equations in the system. Again just as for a system in two unknowns, any system of linear equations in any number of unknowns has either no solution, exactly one solution, or infinitely many solutions. There are no other possibilities.

Solving a system in three unknowns graphically would require the graphing of planes (flat surfaces) in three dimensions. (The graph of a linear equation in three unknowns is a flat surface.) The use of row reduction makes three-dimensional graphing unnecessary.

Example 1 Solving a System by Gauss-Jordan Reduction

Solve the system

$$x - y + 5z = -6$$
$$3x + 3y - z = 10$$
$$x + 3y + 2z = 5$$

Solution The augmented matrix for this system is

$$\begin{bmatrix} 1 & -1 & 5 & -6 \\ 3 & 3 & -1 & 10 \\ 1 & 3 & 2 & 5 \end{bmatrix}$$

Note that the columns correspond to x, y, z, and the right-hand side, respectively. We begin by selecting the pivot in the first row and clearing its column. Remember that clearing the column means that we turn *all* other numbers in the column into zeros. Thus, to clear the column of the first pivot, we need to change two rows, setting up the row operations in exactly the same way as above.

$$\begin{bmatrix} \boxed{1} & -1 & 5 & -6 \\ 3 & 3 & -1 & 10 \\ 1 & 3 & 2 & 5 \end{bmatrix} \begin{matrix} \\ R_2 - 3R_1 \to \\ R_3 - R_1 \end{matrix} \begin{bmatrix} 1 & -1 & 5 & -6 \\ 0 & 6 & -16 & 28 \\ 0 & 4 & -3 & 11 \end{bmatrix}$$

Notice that both row operations have the required form

$$a R_c \pm b R_1$$

↑   ↑

Row to change   Pivot row

with a and b both positive.

---

[14] Carl Friedrich Gauss (1777–1855) was one of the great mathematicians, making fundamental contributions to number theory, analysis, probability and statistics, as well as many fields of science. He developed a method of solving systems of linear equations by row reduction, which was then refined by Wilhelm Jordan (1842–1899) into the form we are showing you here.

Now we use the optional simplification step to simplify R_2:

$$\begin{bmatrix} 1 & -1 & 5 & -6 \\ 0 & 6 & -16 & 28 \\ 0 & 4 & -3 & 11 \end{bmatrix} \frac{1}{2}R_2 \rightarrow \begin{bmatrix} 1 & -1 & 5 & -6 \\ 0 & 3 & -8 & 14 \\ 0 & 4 & -3 & 11 \end{bmatrix}$$

Next, we select the pivot in the second row and clear its column:

$$\begin{bmatrix} 1 & -1 & 5 & -6 \\ 0 & \boxed{3} & -8 & 14 \\ 0 & 4 & -3 & 11 \end{bmatrix} \begin{array}{l} 3R_1 + R_2 \\ \\ 3R_3 - 4R_2 \end{array} \rightarrow \begin{bmatrix} 3 & 0 & 7 & -4 \\ 0 & 3 & -8 & 14 \\ 0 & 0 & 23 & -23 \end{bmatrix}$$

R_1 and R_3 are to be changed.

R_2 is the pivot row.

We simplify R_3.

$$\begin{bmatrix} 3 & 0 & 7 & -4 \\ 0 & 3 & -8 & 14 \\ 0 & 0 & 23 & -23 \end{bmatrix} \frac{1}{23}R_3 \rightarrow \begin{bmatrix} 3 & 0 & 7 & -4 \\ 0 & 3 & -8 & 14 \\ 0 & 0 & 1 & -1 \end{bmatrix}$$

Now we select the pivot in the third row and clear its column:

$$\begin{bmatrix} 3 & 0 & 7 & -4 \\ 0 & 3 & -8 & 14 \\ 0 & 0 & \boxed{1} & -1 \end{bmatrix} \begin{array}{l} R_1 - 7R_3 \\ R_2 + 8R_3 \end{array} \rightarrow \begin{bmatrix} 3 & 0 & 0 & 3 \\ 0 & 3 & 0 & 6 \\ 0 & 0 & 1 & -1 \end{bmatrix}$$

R_1 and R_2 are to be changed.

R_3 is the pivot row.

Finally, we turn all the pivots into 1s:

using *Technology*

See the Technology Guides at the end of the chapter to see how to use a TI-83/84 or Excel to solve this system of equations.

$$\begin{bmatrix} 3 & 0 & 0 & 3 \\ 0 & 3 & 0 & 6 \\ 0 & 0 & 1 & -1 \end{bmatrix} \begin{array}{l} \frac{1}{3}R_1 \\ \frac{1}{3}R_2 \end{array} \rightarrow \begin{bmatrix} 1 & 0 & 0 & 1 \\ 0 & 1 & 0 & 2 \\ 0 & 0 & 1 & -1 \end{bmatrix}$$

The matrix is now reduced to a simple form, so we translate back into equations to obtain the solution:

$$x = 1, \ y = 2, \ z = -1, \ \text{or} \ (x, y, z) = (1, 2, -1)$$

Notice the form of the very last matrix in the example:

$$\begin{bmatrix} 1 & 0 & 0 & \# \\ 0 & 1 & 0 & \# \\ 0 & 0 & 1 & \# \end{bmatrix}$$

The 1s are on the **(main) diagonal** of the matrix; the goal in Gauss-Jordan reduction is to reduce our matrix to this form. If we can do so, then we can easily read off the solution, as we saw in Example 1. However, as we will see in several examples in this section, it is not always possible to achieve this ideal state. After Example 6, we will give a form that is always possible to achieve.

Example 2 Solving a System by Gauss-Jordan Reduction

Solve the system:

$$2x + y + 3z = 1$$
$$4x + 2y + 4z = 4$$
$$x + 2y + z = 4$$

Solution

$$\begin{bmatrix} \boxed{2} & 1 & 3 & 1 \\ 4 & 2 & 4 & 4 \\ 1 & 2 & 1 & 4 \end{bmatrix} \begin{matrix} \\ R_2 - 2R_1 \\ 2R_3 - R_1 \end{matrix} \rightarrow \begin{bmatrix} 2 & 1 & 3 & 1 \\ 0 & 0 & -2 & 2 \\ 0 & 3 & -1 & 7 \end{bmatrix}$$

Now we have a slight problem: The number in the position where we would like to have a pivot—the second column of the second row—is a zero and thus cannot be a pivot. There are two ways out of this problem. One is to move on to the third column and pivot on the −2. Another is to switch the order of the second and third rows so that we can use the 3 as a pivot. We will do the latter.

$$\begin{bmatrix} 2 & 1 & 3 & 1 \\ 0 & 0 & -2 & 2 \\ 0 & 3 & -1 & 7 \end{bmatrix} \begin{matrix} \\ \\ R_2 \leftrightarrow R_3 \end{matrix} \rightarrow \begin{bmatrix} 2 & 1 & 3 & 1 \\ 0 & \boxed{3} & -1 & 7 \\ 0 & 0 & -2 & 2 \end{bmatrix} \begin{matrix} 3R_1 - R_2 \\ \\ \end{matrix}$$

$$\rightarrow \begin{bmatrix} 6 & 0 & 10 & -4 \\ 0 & 3 & -1 & 7 \\ 0 & 0 & -2 & 2 \end{bmatrix} \begin{matrix} \\ \\ -\frac{1}{2}R_3 \end{matrix} \rightarrow \begin{bmatrix} 6 & 0 & 10 & -4 \\ 0 & 3 & -1 & 7 \\ 0 & 0 & \boxed{1} & -1 \end{bmatrix} \begin{matrix} R_1 - 10R_3 \\ R_2 + R_3 \\ \end{matrix}$$

$$\rightarrow \begin{bmatrix} 6 & 0 & 0 & 6 \\ 0 & 3 & 0 & 6 \\ 0 & 0 & 1 & -1 \end{bmatrix} \begin{matrix} \frac{1}{6}R_1 \\ \frac{1}{3}R_2 \\ \end{matrix} \rightarrow \begin{bmatrix} 1 & 0 & 0 & 1 \\ 0 & 1 & 0 & 2 \\ 0 & 0 & 1 & -1 \end{bmatrix}$$

Thus, the solution is $(x, y, z) = (1, 2, -1)$, as you can check in the original system.

Example 3 Inconsistent System

Solve the system:

$$x + y + z = 1$$
$$2x - y + z = 0$$
$$4x + y + 3z = 3$$

Solution

$$\begin{bmatrix} \boxed{1} & 1 & 1 & 1 \\ 2 & -1 & 1 & 0 \\ 4 & 1 & 3 & 3 \end{bmatrix} \begin{matrix} \\ R_2 - 2R_1 \\ R_3 - 4R_1 \end{matrix} \rightarrow \begin{bmatrix} 1 & 1 & 1 & 1 \\ 0 & \boxed{-3} & -1 & -2 \\ 0 & -3 & -1 & -1 \end{bmatrix} \begin{matrix} 3R_1 + R_2 \\ \\ R_3 - R_2 \end{matrix}$$

$$\rightarrow \begin{bmatrix} 3 & 0 & 2 & 1 \\ 0 & -3 & -1 & -2 \\ 0 & 0 & 0 & 1 \end{bmatrix}$$

Stop. That last row translates into $0 = 1$, which is nonsense, and so, as in Example 3 in Section 2.1, we can say that this system has no solution. We also say, as we did for systems with only two unknowns, that a system with no solution is **inconsistent.** A system with at least one solution is **consistent.**

+ *Before we go on...*

Q: How, exactly, does the nonsensical equation $0 = 1$ tell us that there is no solution of the system in Example 3?

A: Here is an argument similar to that in Example 3 in Section 2.1: If there *were* three numbers x, y, and z satisfying the original system of equations, then manipulating the equations according to the instructions in the row operations above leads us to conclude that $0 = 1$. Since 0 is *not* equal to 1, there can be no such numbers x, y, and z. ■

Example 4 Infinitely Many Solutions

Solve the system:

$$
\begin{aligned}
x + y + z &= 1 \\
\frac{1}{4}x - \frac{1}{2}y + \frac{3}{4}z &= 0 \\
x + 7y - 3z &= 3
\end{aligned}
$$

Solution

$$
\begin{bmatrix}
1 & 1 & 1 & 1 \\
\frac{1}{4} & -\frac{1}{2} & \frac{3}{4} & 0 \\
1 & 7 & -3 & 3
\end{bmatrix}
\begin{matrix} \\ 4R_2 \end{matrix}
\rightarrow
\begin{bmatrix}
1 & 1 & 1 & 1 \\
1 & -2 & 3 & 0 \\
1 & 7 & -3 & 3
\end{bmatrix}
\begin{matrix} \\ R_2 - R_1 \\ R_3 - R_1 \end{matrix}
$$

$$
\rightarrow
\begin{bmatrix}
1 & 1 & 1 & 1 \\
0 & -3 & 2 & -1 \\
0 & 6 & -4 & 2
\end{bmatrix}
\begin{matrix} \\ \\ \frac{1}{2}R_3 \end{matrix}
\rightarrow
\begin{bmatrix}
1 & 1 & 1 & 1 \\
0 & -3 & 2 & -1 \\
0 & 3 & -2 & 1
\end{bmatrix}
\begin{matrix} 3R_1 + R_2 \\ \\ R_3 + R_2 \end{matrix}
$$

$$
\rightarrow
\begin{bmatrix}
3 & 0 & 5 & 2 \\
0 & -3 & 2 & -1 \\
0 & 0 & 0 & 0
\end{bmatrix}
$$

There are no nonzero entries in the third row, so there can be no pivot in the third row. We skip to the final step and turn the pivots we did find into 1s.

$$
\begin{bmatrix}
3 & 0 & 5 & 2 \\
0 & -3 & 2 & -1 \\
0 & 0 & 0 & 0
\end{bmatrix}
\begin{matrix} \frac{1}{3}R_1 \\ -\frac{1}{3}R_2 \\ \end{matrix}
\rightarrow
\begin{bmatrix}
1 & 0 & \frac{5}{3} & \frac{2}{3} \\
0 & 1 & -\frac{2}{3} & \frac{1}{3} \\
0 & 0 & 0 & 0
\end{bmatrix}
$$

Now we translate back into equations and obtain:

$$
\begin{aligned}
x + \tfrac{5}{3}z &= \tfrac{2}{3} \\
y - \tfrac{2}{3}z &= \tfrac{1}{3} \\
0 &= 0
\end{aligned}
$$

But how does this help us find a solution? The last equation doesn't tell us anything useful, so we ignore it. The thing to notice about the other equations is that we can easily solve the first equation for x and the second for y, obtaining

$$x = \tfrac{2}{3} - \tfrac{5}{3}z$$
$$y = \tfrac{1}{3} + \tfrac{2}{3}z$$

This is the solution! We can choose z to be any number and get corresponding values for x and y from the formulas above. This gives us infinitely many different solutions. Thus, the general solution (see Example 4 in Section 2.1) is

$$x = \tfrac{2}{3} - \tfrac{5}{3}z$$
$$y = \tfrac{1}{3} + \tfrac{2}{3}z \qquad\qquad \textcolor{magenta}{\text{General solution}}$$
$$z \text{ is arbitrary}$$

We can also write the general solution as

$$\left(\tfrac{2}{3} - \tfrac{5}{3}z, \tfrac{1}{3} + \tfrac{2}{3}z, z\right) \ z \text{ arbitrary} \qquad\qquad \textcolor{magenta}{\text{General solution}}$$

This general solution has z as the parameter. Specific choices of values for the parameter z give particular solutions. For example, the choice $z = 6$ gives the particular solution

$$x = \tfrac{2}{3} - \tfrac{5}{3}(6) = -\tfrac{28}{3}$$
$$y = \tfrac{1}{3} + \tfrac{2}{3}(6) = \tfrac{13}{3} \qquad\qquad \textcolor{magenta}{\text{Particular solution}}$$
$$z = 6$$

while the choice $z = 0$ gives the particular solution $(x, \ y, \ z) = \left(\tfrac{2}{3}, \tfrac{1}{3}, 0\right)$.

Note that, unlike the system given in preceding example, the system given in this example does have solutions, and is thus *consistent*.

+*Before we go on...* Why were there infinitely many solutions to Example 4? The reason is that the third equation was really a combination of the first and second equations to begin with, so we effectively had only two equations in three unknowns.[15] Choosing a specific value for z (say, $z = 6$) has the effect of supplying the "missing" equation.

Q: How do we know when there are have infinitely many solutions?

A: When there are solutions (we have a consistent system, unlike the one in Example 3), and when the matrix we arrive at by row reduction has fewer pivots than there are unknowns. In Example 4 we had three unknowns but only two pivots.

Q: How do we know which variables to use as parameters in a parameterized solution?

A: The variables to use as parameters are those in the columns without pivots. In Example 4 there were pivots in the x and y columns, but no pivot in the z column, and it was z that we used as a parameter. ■

---

[15] In fact, you can check that the third equation, E_3, is equal to $3E_1 - 8E_2$. Thus, the third equation could have been left out, since it conveys no more information than the first two. The process of row reduction always eliminates such a redundancy by creating a row of zeros.

Example 5 Four Unknowns

Solve the system:

$$\begin{aligned} x + 3y + 2z - \quad w &= 6 \\ 2x + 6y + 6z + \quad 3w &= 16 \\ x + 3y - 2z - 11w &= -2 \\ 2x + 6y + 8z + \quad 8w &= 20 \end{aligned}$$

Solution

$$\begin{bmatrix} 1 & 3 & 2 & -1 & 6 \\ 2 & 6 & 6 & 3 & 16 \\ 1 & 3 & -2 & -11 & -2 \\ 2 & 6 & 8 & 8 & 20 \end{bmatrix} \begin{matrix} \\ R_2 - 2R_1 \\ R_3 - R_1 \\ R_4 - 2R_1 \end{matrix} \rightarrow \begin{bmatrix} 1 & 3 & 2 & -1 & 6 \\ 0 & 0 & 2 & 5 & 4 \\ 0 & 0 & -4 & -10 & -8 \\ 0 & 0 & 4 & 10 & 8 \end{bmatrix}$$

There is no pivot available in the second column, so we move on to the third column.

$$\begin{bmatrix} 1 & 3 & 2 & -1 & 6 \\ 0 & 0 & 2 & 5 & 4 \\ 0 & 0 & -4 & -10 & -8 \\ 0 & 0 & 4 & 10 & 8 \end{bmatrix} \begin{matrix} R_1 - R_2 \\ \\ R_3 + 2R_2 \\ R_4 - 2R_2 \end{matrix} \rightarrow \begin{bmatrix} 1 & 3 & 0 & -6 & 2 \\ 0 & 0 & 2 & 5 & 4 \\ 0 & 0 & 0 & 0 & 0 \\ 0 & 0 & 0 & 0 & 0 \end{bmatrix} \begin{matrix} \\ \frac{1}{2}R_2 \\ \\ \end{matrix}$$

$$\rightarrow \begin{bmatrix} 1 & 3 & 0 & -6 & 2 \\ 0 & 0 & 1 & \frac{5}{2} & 2 \\ 0 & 0 & 0 & 0 & 0 \\ 0 & 0 & 0 & 0 & 0 \end{bmatrix}$$

Translating back into equations, we get:

$$\begin{aligned} x + 3y - 6w &= 2 \\ z + \tfrac{5}{2}w &= 2 \end{aligned}$$

(We have not written down the equations corresponding to the last two rows, each of which is $0 = 0$.) Since there are no pivots in the y or w columns, we use these two variables as parameters. We bring them over to the right-hand sides of the equations above and write the general solution as

$$\begin{aligned} x &= 2 - 3y + 6w \\ y &\text{ is arbitrary} \\ z &= 2 - \tfrac{5}{2}w \\ w &\text{ is arbitrary} \end{aligned}$$

or

$$(x, y, z, w) = (2 - 3y + 6w, y, 2 - 5w/2, w) \ y, w \text{ arbitrary}$$

As in Example 4, you can check that the general solution is correct by substituting back into the original equations. (Since y and w are arbitrary, substitute nothing for them, but check that, as in Example 4, they cancel out when we substitute the formulas for x and z.)

+_Before we go on..._ In Examples 4 and 5, you might have noticed an interesting phenomenon: If at any time in the process, two rows are equal or one is a multiple of the other, then one of those rows (eventually) becomes all zero. ∎

Up to this point, we have always been given as many equations as there are unknowns. However, we shall see in the next section that some applications lead to systems where the number of equations is not the same as the number of unknowns. As the following example illustrates, such systems can be handled the same way as any other.

Example 6 Number of Equations ≠ Number of Unknowns

Solve the system:

$$x + y = 1$$
$$13x - 26y = -11$$
$$26x - 13y = 2$$

Solution We proceed exactly as before and ignore the fact that there is one more equation than unknown.

$$
\begin{bmatrix}
\boxed{1} & 1 & 1 \\
13 & -26 & -11 \\
26 & -13 & 2
\end{bmatrix}
\begin{matrix} \\ R_2 - 13R_1 \\ R_3 - 26R_1 \end{matrix}
\rightarrow
\begin{bmatrix}
1 & 1 & 1 \\
0 & -39 & -24 \\
0 & -39 & -24
\end{bmatrix}
\begin{matrix} \\ \frac{1}{3}R_2 \\ \frac{1}{3}R_3 \end{matrix}
$$

$$
\rightarrow
\begin{bmatrix}
1 & 1 & 1 \\
0 & \boxed{-13} & -8 \\
0 & -13 & -8
\end{bmatrix}
\begin{matrix} 13R_1 + R_2 \\ \\ R_3 - R_2 \end{matrix}
\rightarrow
\begin{bmatrix}
13 & 0 & 5 \\
0 & -13 & -8 \\
0 & 0 & 0
\end{bmatrix}
\begin{matrix} \frac{1}{13}R_1 \\ -\frac{1}{13}R_2 \\ \\ \end{matrix}
$$

$$
\rightarrow
\begin{bmatrix}
1 & 0 & \frac{5}{13} \\
0 & 1 & \frac{8}{13} \\
0 & 0 & 0
\end{bmatrix}
$$

Thus, the solution is $(x, y) = \left(\frac{5}{13}, \frac{8}{13}\right)$.

If, instead of a row of zeros, we had obtained, say, [0 0 6] in the last row, we would immediately have concluded that the system was inconsistent.

The fact that we wound up with a row of zeros indicates that one of the equations was actually a combination of the other two; you can check that the third equation can be obtained by multiplying the first equation by 13 and adding the result to the second. Since the third equation therefore tells us nothing that we don't already know from the first two, we call the system of equations **redundant,** or **dependent** (compare Example 4 in Section 2.1).

+*Before we go on...* If you now take another look at Example 5, you will find that we could have started with the following smaller system of two equations in four unknowns

$$x + 3y + 2z - w = 6$$
$$2x + 6y + 6z + 3w = 16$$

and obtained the same general solution as we did with the larger system. Verify this by solving the smaller system. ■

The preceding examples illustrated that we cannot always reduce a matrix to the form shown before Example 2, with pivots going all the way down the diagonal. What we *can* always do is reduce a matrix to the following form:

Reduced Row Echelon Form

A matrix is said to be in **reduced row echelon form** or to be **row-reduced** if it satisfies the following properties.

P1. The first nonzero entry in each row (called the **leading entry** of that row) is a 1.

P2. The columns of the leading entries are **clear** (i.e., they contain zeros in all positions other than that of the leading entry).

P3. The leading entry in each row is to the right of the leading entry in the row above, and any rows of zeros are at the bottom.

quick Examples

$$\begin{bmatrix} 1 & 0 & 0 & 2 \\ 0 & 1 & 0 & 4 \\ 0 & 0 & 1 & -3 \end{bmatrix}, \begin{bmatrix} 0 & 1 & -3 \\ 0 & 0 & 0 \end{bmatrix}, \text{ and } \begin{bmatrix} 1 & 0 & 0 & -2 \\ 0 & 0 & 1 & 4 \\ 0 & 0 & 0 & 0 \end{bmatrix} \text{ are row-reduced.}$$

$$\begin{bmatrix} 1 & 1 & 0 & 2 \\ 0 & 1 & 0 & 4 \\ 0 & 0 & 1 & -3 \end{bmatrix}, \begin{bmatrix} 0 & 1 & -3 \\ 0 & 0 & 1 \end{bmatrix}, \text{ and } \begin{bmatrix} 0 & 0 & 1 & 4 \\ 1 & 0 & 0 & -2 \\ 0 & 0 & 0 & 0 \end{bmatrix} \text{ are not row-reduced.}$$

You should check in the examples we did that the final matrices were all in reduced row echelon form.

It is an interesting and useful fact, though not easy to prove, that any two people who start with the same matrix and row-reduce it will reach exactly the same row-reduced matrix, even if they use different row operations.

The Traditional Gauss-Jordan Method (Optional)

In the version of the Gauss-Jordan method we have presented, we eliminated fractions and decimals in the first step and then worked with integer matrices, partly to make hand computation easier and partly for mathematical elegance. However, complicated fractions and decimals present no difficulty when we use technology. The following example illustrates the more traditional approach to Gauss-Jordan reduction used in many of the computer programs that solve the huge systems of equations that arise in practice.[16]

Example **7** **Solving a System with the Traditional Gauss-Jordan Method**

Solve the following system using the traditional Gauss-Jordan method:

$$\begin{aligned} 2x + y + 3z &= 5 \\ 3x + 2y + 4z &= 7 \\ 2x + y + 5z &= 10 \end{aligned}$$

---

[16] Actually, for reasons of efficiency and accuracy, the methods used in commercial programs are close to, but not exactly, the method presented here. To learn more, consult a text on numerical methods.

Solution We make two changes in our method. First, there is no need to get rid of decimals (since computers and calculators can handle decimals as easily as they can integers). Second, after selecting a pivot, *divide the pivot row by the pivot value, turning the pivot into a* 1. It is easier to determine the row operations that will clear the pivot column if the pivot is a 1.

If we use technology to solve this system of equations, the sequence of matrices might look like this:

$$\begin{bmatrix} \boxed{2} & 1 & 3 & 5 \\ 3 & 2 & 4 & 7 \\ 2 & 1 & 5 & 10 \end{bmatrix} \begin{matrix} \frac{1}{2}R_1 \\ \\ \end{matrix} \rightarrow \begin{bmatrix} \boxed{1} & 0.5 & 1.5 & 2.5 \\ 3 & 2 & 4 & 7 \\ 2 & 1 & 5 & 10 \end{bmatrix} \begin{matrix} \\ R_2 - 3R_1 \\ R_3 - 2R_1 \end{matrix}$$

$$\rightarrow \begin{bmatrix} 1 & 0.5 & 1.5 & 2.5 \\ 0 & \boxed{0.5} & -0.5 & -0.5 \\ 0 & 0 & 2 & 5 \end{bmatrix} \begin{matrix} \\ 2R_2 \\ \end{matrix} \rightarrow \begin{bmatrix} 1 & 0.5 & 1.5 & 2.5 \\ 0 & \boxed{1} & -1 & -1 \\ 0 & 0 & 2 & 5 \end{bmatrix} \begin{matrix} R_1 - 0.5R_2 \\ \\ \end{matrix}$$

$$\rightarrow \begin{bmatrix} 1 & 0 & 2 & 3 \\ 0 & 1 & -1 & -1 \\ 0 & 0 & \boxed{2} & 5 \end{bmatrix} \begin{matrix} \\ \\ \frac{1}{2}R_3 \end{matrix} \rightarrow \begin{bmatrix} 1 & 0 & 2 & 3 \\ 0 & 1 & -1 & -1 \\ 0 & 0 & \boxed{1} & 2.5 \end{bmatrix} \begin{matrix} R_1 - 2R_3 \\ R_2 + R_3 \\ \end{matrix}$$

$$\rightarrow \begin{bmatrix} 1 & 0 & 0 & -2 \\ 0 & 1 & 0 & 1.5 \\ 0 & 0 & 1 & 2.5 \end{bmatrix}$$

The solution is $(x, y, z) = (-2, 1.5, 2.5)$.

Q: The solution to Example 7 looked quite easy. Why didn't we use the traditional method from the start*?*

A: It looked easy because we deliberately chose an example with simple numbers. In all but the most contrived examples, the decimals or fractions involved get very complicated very quickly.

FAQs   **Getting Unstuck, Going Round in Circles, and Knowing When to Stop**

Q: Help! I have been doing row operations on this matrix for half an hour. I have filled two pages, and I am getting nowhere. What do I do*?*

A: Here is a way of keeping track of where you are *at any stage of the process* and also deciding what to do next.

Starting at the top row of your current matrix:

1. Scan along the row until you get to the leading entry: the first nonzero entry. If there is none—that is, the row is all zero—go to the next row.
2. Having located the leading entry, scan up and down its *column*. If its column is not clear (that is, it contains other nonzero entries), use your leading entry as a pivot to clear its column as in the examples in this section.
3. Now go to the next row and start again at Step 1.

When you have scanned all the rows and find that all the columns of the leading entries are clear, it means you are done (except possibly for reordering the rows so that the leading entries go from left to right as you read down the matrix, and zero rows are at the bottom). ∎

Q: No good. I have been following these instructions, but every time I try to clear a column, I unclear a column I had already cleared. What is going on?

A: Are you using *leading entries* as pivots? Also, are you *using the pivot* to clear its column? That is, are your row operations all of the following form?

$$aR_c \pm bR_p$$

↑ Row to change   ↑ Pivot row

The instruction next to the row you are changing should involve only that row and the pivot row, even though you might be tempted to use some other row instead.

Q: Must I continue until I get a matrix that has 1s down the leading diagonal and 0s above and below?

A: Not necessarily. You are completely done when your matrix is row-reduced: each leading entry is a 1, the column of each leading entry is clear, and the leading entries go from left to right. You are done *pivoting* when the column of each leading entry is clear. After that, all that remains is to turn each pivot into a 1 (the "Final Step") and, if necessary, rearrange the rows. ∎

Online, follow:

Chapter 2

→ **Tools**

to find the following resources:
- A page that pivots and does row operations automatically
- A TI-83/84 program that pivots and does other row operations
- An Excel worksheet that pivots and does row operations automatically
- Macintosh pivoting software available for download

2.2 EXERCISES

● denotes basic skills exercises

tech Ex indicates exercises that should be solved using technology

In Exercises 1–42, use Gauss-Jordan row reduction to solve the given systems of equation. We suggest doing some by hand, and others using technology. hint [see Examples 1–6]

1. ● $x + y = 4$
  $x - y = 2$

2. ● $2x + y = 2$
  $-2x + y = 2$

3. ● $3x - 2y = 6$
  $2x - 3y = -6$

4. ● $2x + 3y = 5$
  $3x + 2y = 5$

5. ● $2x + 3y = 1$
  $-x - \dfrac{3y}{2} = -\dfrac{1}{2}$

6. ● $2x - 3y = 1$
  $6x - 9y = 3$

7. ● $2x + 3y = 2$
  $-x - \dfrac{3y}{2} = -\dfrac{1}{2}$

8. ● $2x - 3y = 2$
  $6x - 9y = 3$

9. ● $x + y = 1$
  $3x - y = 0$
  $x - 3y = -2$

10. ● $x + y = 1$
  $3x - 2y = -1$
  $5x - y = \dfrac{1}{5}$

11. ● $x + y = 0$
  $3x - y = 1$
  $x - y = -1$

12. ● $x + 2y = 1$
  $3x - 2y = -2$
  $5x - y = \dfrac{1}{5}$

13. ● $0.5x + 0.1y = 1.7$
  $0.1x - 0.1y = 0.3$
  $x + y = \dfrac{11}{3}$

14. ● $-0.3x + 0.5y = 0.1$
  $x - y = 4$
  $\dfrac{x}{17} + \dfrac{y}{17} = 1$

15. ● $-x + 2y - z = 0$
  $-x - y + 2z = 0$
  $2x - z = 4$

16. ● $x + 2y = 4$
  $y - z = 0$
  $x + 3y - 2z = 5$

● basic skills   tech Ex technology exercise

17. ● $x + y + 6z = -1$
$\frac{1}{3}x - \frac{1}{3}y + \frac{2}{3}z = 1$
$\frac{1}{2}x \quad\quad + z = 0$

18. ● $x - \frac{1}{2}y \quad\quad = 0$
$\frac{1}{3}x + \frac{1}{3}y + \frac{1}{3}z = 2$
$\frac{1}{2}x \quad\quad - \frac{1}{2}z = -1$

19. ● $-\frac{1}{2}x + y - \frac{1}{2}z = 0$
$-\frac{1}{2}x - \frac{1}{2}y + z = 0$
$x - \frac{1}{2}y - \frac{1}{2}z = 0$

20. ● $x - \frac{1}{2}y \quad\quad = 0$
$\frac{1}{2}x - \quad \frac{1}{2}z = -1$
$3x - y - z = -2$

21. ● $x + y + 2z = -1$
$2x + 2y + 2z = 2$
$\frac{3}{5}x + \frac{3}{5}y + \frac{3}{5}z = \frac{2}{5}$

22. ● $x + y - z = -2$
$x - y - 7z = 0$
$\frac{2}{7}x \quad\quad - \frac{8}{7}z = 14$

23. ● $-0.5x + 0.5y + 0.5z = 1.5$
$4.2x + 2.1y + 2.1z = 0$
$0.2x \quad\quad + 0.2z = 0$

24. ● $0.25x - 0.5y \quad\quad = 0$
$0.2x + 0.2y - 0.2z = -0.6$
$0.5x - 1.5y + z = 0.5$

25. ● $2x - y + z = 4$
$3x - y + z = 5$

26. ● $3x - y - z = 0$
$x + y + z = 4$

27. ● $0.75x - 0.75y - z = 4$
$x - y + 4z = 0$

28. ● $2x - y + z = 4$
$-x + 0.5y - 0.5z = 1.5$

29. $3x + y - z = 12$

30. $x + y - 3z = -21$
(Yes: One equation in three unknowns!)

31. $x + y + 2z = -1$
$2x + 2y + 2z = 2$
$0.75x + 0.75y + z = 0.25$
$-x \quad\quad - 2z = 21$

32. $x + y - z = -2$
$x - y - 7z = 0$
$0.75x - 0.5y + 0.25z = 14$
$x + y + z = 4$

33. $x + y + 5z \quad\quad = 1$
$y + 2z + w = 1$
$x + 3y + 7z + 2w = 2$
$x + y + 5z + w = 1$

34. $x + y \quad\quad + 4w = 1$
$2x - 2y - 3z + 2w = -1$
$4y + 6z + w = 4$
$2x + 4y + 9z \quad\quad = 6$

35. $x + y + 5z \quad\quad = 1$
$y + 2z + w = 1$
$x + y + 5z + w = 1$
$x + 2y + 7z + 2w = 2$

36. $x + y + \quad 4w = 1$
$2x - 2y - 3z + 2w = -1$
$4y + 6z + w = 4$
$3x + 3y + 3z + 7w = 4$

37. $x - 2y + z - 4w = 1$
$x + 3y + 7z + 2w = 2$
$2x + y + 8z - 2w = 3$

38. $x - 3y - 2z - w = 1$
$x + 3y + z + 2w = 2$
$2x \quad\quad - z + w = 3$

39. $x + y + z + u + v = 15$
$y - z + u - v = -2$
$z + u + v = 12$
$u - v = -1$
$v = 5$

40. $x - y + z - u + v = 1$
$y + z + u + v = 2$
$z - u + v = 1$
$u + v = 1$
$v = 1$

41. $x - y + z - u + v = 0$
$y - z + u - v = -2$
$x \quad\quad - 2v = -2$
$2x - y + z - u - 3v = -2$
$4x - y + z - u - 7v = -6$

42. $x + y + z + u + v = 15$
$y + z + u + v = 3$
$x + 2y + 2z + 2u + 2v = 18$
$x - y - z - u - v = 9$
$x - 2y - 2z - 2u - 2v = 6$

In Exercises 43–46, use technology to solve the systems of equations. Express all solutions as fractions.

43. `tech`Ex $x + 2y - z + w = 30$
$2x \quad\quad - z + 2w = 30$
$x + 3y + 3z - 4w = 2$
$2x - 9y \quad\quad + w = 4$

44. `tech`Ex $4x - 2y + z + w = 20$
$3y + 3z - 4w = 2$
$2x + 4y \quad\quad - w = 4$
$x + 3y + 3z \quad\quad = 2$

45. `tech`Ex $x + 2y + 3z + 4w + 5t = 6$
$2x + 3y + 4z + 5w + t = 5$
$3x + 4y + 5z + w + 2t = 4$
$4x + 5y + z + 2w + 3t = 3$
$5x + y + 2z + 3w + 4t = 2$

● basic skills   `tech`Ex technology exercise

46. tech Ex

$$x - 2y + 3z - 4w = 0$$
$$-2x + 3y - 4z + t = 0$$
$$3x - 4y + w - 2t = 0$$
$$-4x + z - 2w + 3t = 0$$
$$y - 2z + 3w - 4t = 1$$

tech Ex *In Exercises 47–50, use technology to solve the systems of equations. Express all solutions as decimals, rounded to one decimal place.*

47. tech Ex
$$1.6x + 2.4y - 3.2z = 4.4$$
$$5.1x - 6.3y + 0.6z = -3.2$$
$$4.2x + 3.5y + 4.9z = 10.1$$

48. tech Ex
$$2.1x + 0.7y - 1.4z = -2.3$$
$$3.5x - 4.2y - 4.9z = 3.3$$
$$1.1x + 2.2y - 3.3z = -10.2$$

49. tech Ex
$$-0.2x + 0.3y + 0.4z - t = 4.5$$
$$2.2x + 1.1y - 4.7z + 2t = 8.3$$
$$9.2y - 1.3t = 0$$
$$3.4x + 0.5z - 3.4t = 0.1$$

50. tech Ex
$$1.2x - 0.3y + 0.4z - 2t = 4.5$$
$$1.9x - 0.5z - 3.4t = 0.2$$
$$12.1y - 1.3t = 0$$
$$3x + 2y - 1.1z = 9$$

Communication and Reasoning Exercises

51. ● What is meant by a pivot? What does pivoting do?

52. ● Give instructions to check whether or not a matrix is row-reduced.

53. ● You are row-reducing a matrix and have chosen a −6 as a pivot in Row 4. Directly above the pivot, in Row 1, is a 15. What row operation can you use to clear the 15?

54. ● You are row-reducing a matrix and have chosen a −4 as a pivot in Row 2. Directly below the pivot, in Row 4, is a −6. What row operation can you use to clear the −6?

55. ● In the matrix of a system of linear equations, suppose that two of the rows are equal. What can you say about the row-reduced form of the matrix?

56. ● In the matrix of a system of linear equations, suppose that one of the rows is a multiple of another. What can you say about the row-reduced form of the matrix?

57. Your friend Frans tells you that the system of linear equations you are solving cannot have a unique solution because the reduced matrix has a row of zeros. Comment on his claim.

58. Your other friend Hans tells you that, since he is solving a consistent system of 5 linear equations in 6 unknowns, he will get infinitely many solutions. Comment on his claim.

59. If the reduced matrix of a consistent system of linear equations has 5 rows, 3 of which are zero, and 5 columns, how many parameters does the general solution contain?

60. If the reduced matrix of a consistent system of linear equations has 5 rows, 2 of which are zero, and 7 columns, how many parameters does the general solution contain?

61. Suppose a system of equations has a unique solution. What must be true of the number of pivots in the reduced matrix of the system? Why?

62. Suppose a system has infinitely many solutions. What must be true of the number of pivots in the reduced matrix of the system? Why?

63. Give an example of a system of three linear equations with the general solution $x = 1$, $y = 1 + z$, z arbitrary. (Check your system by solving it.)

64. Give an example of a system of three linear equations with the general solution $x = y - 1$, y arbitrary, $z = y$. (Check your system by solving it.)

● basic skills   tech Ex technology exercise

2.3 Applications of Systems of Linear Equations

In the examples and the exercises of this section, we consider scenarios that lead to systems of linear equations in three or more unknowns. Some of these applications will strike you as a little idealized or even contrived compared with the kinds of problems

Kate Laepple

TITLE Director of Programs and Member Services
INSTITUTION Delaware Valley Grantmakers

Patrick Farace

Delaware Valley Grantmakers (DVG) is the regional associate of grantmakers in the Philadelphia area. As a professional association for people, organizations and businesses that make charitable grants in Greater Philadelphia and neighboring regions, our mission is to promote and enhance regional philanthropy by providing our members with opportunities for networking, knowledge sharing, and professional development that add critical value to their work.

As the Director of Programs and Member Services I focus on planning and implementing approximately 40 educational programs annually including a full-day conference for both grantmakers and nonprofit organizations, as well as overseeing all member recruitment and retention activities. DVG has a membership of over 150 organizations and 600 individuals within those organizations. In my capacity I need to provide those individuals and organizations with information they can use to be more effective in their grantmaking dollars which fund the arts, health programs, the environment, community development, etc.

From the outside, it wouldn't necessarily seem as if math would be a necessity for someone in my position at an organization like DVG. However, I quickly learned that there is virtually no job that doesn't include using math. In fact I have always been someone who was intimidated by math and pursued a career which wouldn't require applying it everyday! But as a nonprofit organization, we need to understand, manage, and most importantly *stretch* a small budget. Whether planning a daylong conference, publishing a quarterly newsletter, or implementing a grant-funded project—an understanding of multiple budgets as part of a bigger whole is crucial.

For example, throughout the course of a fiscal year, I need to assess my budget and make decisions about managing money for various programs. During our annual conference (which is the most costly event that we run throughout the year) we set a budget ahead of time but unforeseen costs always arise. Consequently, there needs to be consistent review of the expenses we're incurring while balancing it against the sponsorship income that we collect. Simple arithmetic isn't enough which is why the applied math courses I took in college come in handy here.

My job is one of relationships. I need to provide members with consistent programs and services that they value and want to fund. Without an understanding of the math needed to manage a complex budget, I would not be able to offer the members of DVG the educational opportunities and information they need to be informed in their funding decisions which positively impact communities around the world.

you might encounter in the real world.[17] One reason is that we will not have tools to handle more realistic versions of these applications until we have studied linear programming in Chapter 4.

In each example that follows, we set up the problem as a linear system and then give the solution. For practice, you should do the row reduction necessary to get the solution.

Example 1 Blending

The Arctic Juice Company makes three juice blends: PineOrange, using 2 quarts of pineapple juice and 2 quarts of orange juice per gallon; PineKiwi, using 3 quarts of pineapple juice and 1 quart of kiwi juice per gallon; and OrangeKiwi, using 3 quarts of orange juice and 1 quart of kiwi juice per gallon. Each day the company has 800 quarts of pineapple juice, 650 quarts of orange juice, and 350 quarts of kiwi juice available. How many gallons of each blend should it make each day if it wants to use up all of the supplies?

---

[17] See the discussion at the end of the first example below.

Solution We take the same steps to understand the problem that we took in Section 2.1. The first step is to identify and label the unknowns. Looking at the question asked in the last sentence, we see that we should label the unknowns like this:

$$x = \text{number of gallons of PineOrange made each day}$$
$$y = \text{number of gallons of PineKiwi made each day}$$
$$z = \text{number of gallons of OrangeKiwi made each day}$$

Next, we can organize the information we are given in a table:

	PineOrange (x)	PineKiwi (y)	OrangeKiwi (z)	Total Available
Pineapple Juice (qt)	2	3	0	800
Orange Juice (qt)	2	0	3	650
Kiwi Juice (qt)	0	1	1	350

Notice how we have arranged the table; we have placed headings corresponding to the unknowns along the top, rather than down the side, and we have added a heading for the available totals. This gives us a table that is essentially the matrix of the system of linear equations we are looking for. (However, read the caution in the Before we go on . . . section.)

Now we read across each row of the table. The fact that we want to use exactly the amount of each juice that is available leads to the following three equations:

$$\begin{aligned} 2x + 3y \quad\quad &= 800 \\ 2x \quad\quad + 3z &= 650 \\ y + z &= 350 \end{aligned}$$

The solution of this system is $(x, y, z) = (100, 200, 150)$, so Arctic Juice should make 100 gallons of PineOrange, 200 gallons of PineKiwi, and 150 gallons of OrangeKiwi each day.

+ Before we go on...

Caution
We do not recommend relying on the coincidence that the table we created to organize the information in Example 1 happened to be the matrix of the system; it is too easy to set up the table "sideways" and get the wrong matrix. You should always write down the system of equations *and be sure you understand each equation.* For example, the equation $2x + 3y = 800$ in Example 1 indicates that the number of quarts of pineapple juice that will be used ($2x + 3y$) is equal to the amount available (800 quarts). By thinking of the reason for each equation, you can check that you have the correct system. If you have the wrong system of equations to begin with, solving it won't help you. ∎

Q: *Just how realistic is the scenario in Example 1?*

A: This is a very unrealistic scenario, for several reasons:

1. Isn't it odd that we happened to end up with exactly the same number of equations as unknowns? Real scenarios are rarely so considerate. If there had been four equations, there would in all likelihood have been no solution at all. However, we need to understand these idealized problems before we can tackle the real world.

2. Even if a real-world scenario does give the same number of equations as unknowns, there is still no guarantee that there will be a unique solution consisting of positive values. What, for instance, would we have done in this example if x had turned out to be negative?

3. The requirement that we use exactly all the ingredients would be an unreasonable constraint in real life. When we discuss linear programming, we will be able to substitute the more reasonable constraint that you use no more than is available, and we will add the more reasonable objective that you maximize profit. ∎

Example 2 Aircraft Purchasing

An airline is considering the purchase of aircraft to meet an estimated demand for 3200 seats. The airline has decided to buy Boeing 747s, which seat 400 passengers and are priced at \$200 million each; Boeing 777s, which seat 300 passengers and are priced at \$160 million; and Airbus A330s, which seat 300 passengers and are priced at \$120 million.[*] Assuming that the airline wishes to buy three times as many 777s as 747s, how many of each should it order to meet the demand for seats, given a \$1600 million spending target?

Solution We label the unknowns as follows:

$$x = \text{number of 747s ordered from Boeing}$$
$$y = \text{number of 777s ordered from Boeing}$$
$$z = \text{number of A330s ordered from Airbus}$$

We must now set up the equations. We can organize some (but not all) of the given information in a table:

	Boeing 747s (x)	Boeing 777s (y)	Airbus A330s (z)	Total
Passengers	400	300	300	3200
Cost (\$ millions)	200	160	120	1600

Reading across, we get the equations expressing the facts that the airline needs to seat 3000 passengers and that it will spend \$1600 million:

$$400x + 300y + 300z = 3200$$
$$200x + 160y + 120z = 1600$$

There is an additional piece of information we have not yet used: the airline wishes to purchase three times as many 777s as 747s. As we said in Section 2.1, it is easiest to translate a statement like this into an equation if we first reword it using the phrase "the number of." Thus, we say: "The number of 777s ordered is three times the number of 747s ordered," or

$$y = 3x$$
$$3x - y = 0$$

---

[*] The prices are approximate 2003 prices; seating capacities have been rounded. Prices and seating capacities found at the companies' websites and in an August 2002 press release announcing a Deutsche Lufthansa order for ten A330-300s to be delivered in 2004–2005.

We now have a system of three equations in three unknowns:

$$400x + 300y + 300z = 3200$$
$$200x + 160y + 120z = 1600$$
$$3x - \quad y \qquad = 0$$

Solving the system, we get the solution $(x, y, z) = (2, 6, 2)$. Thus, the airline should order two 747s, six 777s, and two A330s.

Example 3 Traffic Flow

Traffic through downtown Urbanville flows through the one-way system shown in Figure 7.

Traffic counting devices installed in the road (shown as boxes) count 200 cars entering town from the west each hour, 150 leaving town on the north each hour, and 50 leaving town on the south each hour.

a. From this information, is it possible to determine how many cars drive along Allen, Baker, and Coal streets every hour?

b. What is the maximum possible traffic flow along Baker Street?

c. What is the minimum possible traffic along Allen Street?

d. What is the maximum possible traffic flow along Coal Street?

Solution

a. Our unknowns are:

x = number of cars per hour on Allen Street
y = number of cars per hour on Baker Street
z = number of cars per hour on Coal Street

Assuming that, at each intersection, cars do not fall into a pit or materialize out of thin air, the number of cars entering each intersection has to equal the number exiting. For example, at the intersection of Allen and Baker Streets there are 200 cars entering and $x + y$ cars exiting:

Traffic In = Traffic Out
$$200 = x + y$$

At the intersection of Allen and Coal Streets, we get

Traffic In = Traffic Out
$$x = z + 150$$

and at the intersection of Baker and Coal Streets, we get

Traffic In = Traffic Out
$$y + z = 50$$

We now have the following system of equations:

$$x + y \qquad = 200$$
$$x \qquad - z = 150$$
$$y + z = 50$$

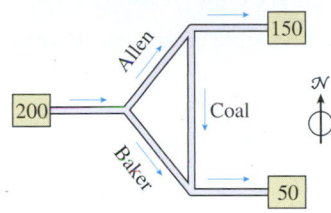

Mediolmages/Getty Images

Figure **7**

If we solve this system using the methods of the preceding section, we find that it has infinitely many solutions. The general solution is

$$x = z + 150$$
$$y = -z + 50$$
$$z \text{ is arbitrary}$$

Since we do not have a unique solution, it is *not* possible to determine how many cars drive along Allen, Baker and Coal Streets every hour.

b. The traffic flow along Baker Street is measured by y. From the general solution,

$$y = -z + 50$$

where z is arbitrary. How arbitrary is z? It makes no sense for any of the variables x, y, or z to be negative in this scenario, so $z \geq 0$. Therefore, the largest possible value y can have is

$$y = -0 + 50 = 50 \text{ cars per hour}$$

c. The traffic flow along Allen Street is measured by x. From the general solution,

$$x = z + 150$$

where $z \geq 0$, as we saw in part (b). Therefore, the smallest possible value x can have is

$$x = 0 + 150 = 150 \text{ cars per hour}$$

d. The traffic flow along Coal Street is measured by z. Referring to the general solution, we see that z shows up in the expressions for both x and y:

$$x = z + 150$$
$$y = -z + 50$$

In the first of these equations, there is nothing preventing z from being as big as we like; the larger we make z, the larger x becomes. However, the second equation places a limit on how large z can be: if $z > 50$, then y is negative, which is impossible. Therefore, the largest value z can take is 50 cars per hour.

From the discussion above, we see that z is not completely arbitrary: we must have $z \geq 0$ and $z \leq 50$. Thus, z has to satisfy $0 \leq z \leq 50$ for us to get a realistic answer.

+ *Before we go on...* Here are some questions to think about in Example 3: If you wanted to nail down x, y, and z to see where the cars are really going, how would you do it with only one more traffic counter? Would it make sense for z to be fractional? What if you interpreted x, y, and z as *average* numbers of cars per hour over a long period of time?

Traffic flow is only one kind of flow in which we might be interested. Water and electricity flows are others. In each case, to analyze the flow, we use the fact that the amount entering an intersection must equal the amount leaving it. ∎

Example 4 Transportation

A car rental company has four locations in the city: Northside, Eastside, Southside, and Westside. The Westside location has 20 more cars than it needs, and the Eastside location has 15 more cars than it needs. The Northside location needs 10 more cars than it has, and the Southside location needs 25 more cars than it has. It costs $10 (in salary and gas) to have an employee drive a car from Westside to Northside. It costs $5 to drive a car from Eastside to Northside. It costs $20 to drive a car from Westside to Southside, and it costs $10 to drive a car from Eastside to Southside. If the company will spend a total of $475 rearranging its cars, how many cars will it drive from each of Westside and Eastside to each of Northside and Southside?

Solution

Figure 8 shows a diagram of this situation. Each arrow represents a route along which the rental company can drive cars. Next to each location is written the number of extra cars the location has or the number it needs. Next to each route is written the cost of driving a car along that route.

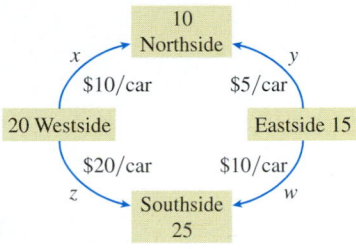

     The unknowns are the number of cars the company will drive along each route, so we have the following four unknowns, as indicated in the figure:

x = number of cars driven from Westside to Northside
y = number of cars driven from Eastside to Northside
z = number of cars driven from Westside to Southside
w = number of cars driven from Eastside to Southside

Figure 8

Consider the Northside location. It needs 10 more cars, so the total number of cars being driven to Northside should be 10. This gives us the equation

$$x + y = 10$$

Similarly, the total number of cars being driven to Southside should be 25, so

$$z + w = 25$$

Considering the number of cars to be driven out of the Westside and Eastside locations, we get the following two equations as well:

$$x + z = 20$$
$$y + w = 15$$

There is one more equation that we should write down, the equation that says that the company will spend $475:

$$10x + 5y + 20z + 10w = 475$$

Thus, we have the following system of five equations in four unknowns:

$$
\begin{aligned}
x + y & & & = 10 \\
& z + & w & = 25 \\
x & + z & & = 20 \\
y & + & w & = 15 \\
10x + 5y + 20z + 10w & & & = 475
\end{aligned}
$$

Solving this system, we find that $(x, y, z, w) = (5, 5, 15, 10)$. In words, the company will drive 5 cars from Westside to Northside, 5 from Eastside to Northside, 15 from Westside to Southside, and 10 from Eastside to Southside.

+*Before we go on...* A very reasonable question to ask in Example 4 is, Can the company rearrange its cars for less than $475? Even better, what is the least possible cost? In general, a question asking for the optimal cost may require the techniques of linear programming that we will discuss in Chapter 4. However, in this case we can approach the problem directly. If we remove the equation that says that the total cost is $475 and solve the system consisting of the other four equations, we find that there are infinitely many solutions and that the general solution may be written as

$$x = w - 5$$
$$y = 15 - w$$
$$z = 25 - w$$
$$w \text{ is arbitrary}$$

This allows us to write the total cost as a function of w:

$$10x + 5y + 20z + 10w = 10(w - 5) + 5(15 - w) + 20(25 - w) + 10w$$
$$= 525 - 5w$$

So, the larger we make w, the smaller the total cost will be. The largest we can make w is 15 (why?), and if we do so we get $(x, y, z, w) = (10, 0, 10, 15)$ and a total cost of $450. ∎

2.3 EXERCISES

● denotes basic skills exercises

◆ denotes challenging exercises

tech Ex indicates exercises that should be solved using technology

Applications

1. ● *Resource Allocation* You manage an ice cream factory that makes three flavors: Creamy Vanilla, Continental Mocha, and Succulent Strawberry. Into each batch of Creamy Vanilla go 2 eggs, 1 cup of milk, and 2 cups of cream. Into each batch of Continental Mocha go 1 egg, 1 cup of milk, and 2 cups of cream, while into each batch of Succulent Strawberry go 1 egg, 2 cups of milk, and 1 cup of cream. You have in stock 350 eggs, 350 cups of milk, and 400 cups of cream. How many batches of each flavor should you make in order to use up all of your ingredients? *hint* [see Example 1]

2. ● *Resource Allocation* You own a hamburger franchise and are planning to shut down operations for the day, but you are left with 13 bread rolls, 19 defrosted beef patties, and 15 opened cheese slices. Rather than throw them out, you decide to use them to make burgers that you will sell at a discount. Plain burgers each require 1 beef patty and 1 bread roll, double cheeseburgers each require 2 beef patties, 1 bread roll, and 2 slices of cheese, while regular cheeseburgers each require 1 beef patty, 1 bread roll, and 1 slice of cheese. How many of each should you make?

3. ● *Resource Allocation* Urban Community College is planning to offer courses in Finite Math, Applied Calculus, and Computer Methods. Each section of Finite Math has 40 students and earns the college $40,000 in revenue. Each section of Applied Calculus has 40 students and earns the college $60,000, while each section of Computer Methods has 10 students and earns the college $20,000. Assuming the college wishes to offer a total of six sections, to accommodate 210 students, and to bring in $260,000 in revenues, how many sections of each course should it offer? *hint* [see Example 2]

4. ● *Resource Allocation* The Enormous State University History Department offers three courses, Ancient, Medieval, and Modern History, and the chairperson is trying to decide how many sections of each to offer this semester. The department is allowed to offer 45 sections total, there are 5000 students who would like to take a course, and there are 60 professors to teach them. Sections of Ancient History have 100 students each, sections of Medieval History hold 50 students each, and sections of Modern History have 200 students each. Modern History sections are taught by a team of 2 professors, while Ancient and Medieval History need only 1 professor per section. How many sections of each course should the chair schedule in order to offer all the sections that they are allowed to, accommodate all of the students, and give one teaching assignment to each professor?

5. ● **Purchasing; Aircraft** In Example 2 we saw that Boeing 747s seat 400 passengers and are priced at $200 million each, Boeing 777s seat 300 passengers and are priced at $160 million, and the European Airbus A330s seat 300 passengers and are priced at $120 million. You are the purchasing manager of an airline company and have a spending goal of $2400 million for the purchase of new aircraft to seat a total of 5000 passengers. Your company has a policy of supporting U.S. industries, and you have been instructed to buy twice as many U.S. manufactured aircraft (Boeing) as foreign aircraft (Airbus). Given the selection of three aircraft, how many of each should you order?

6. ● **Purchasing; Aircraft** Repeat the preceding exercise, but with the following data: Your spending goal is $6400 million, you must seat a total of 13,000 passengers, and you are required to purchase twice as many Boeing 777s as Boeing 747s.

7. ● **Supply** A bagel store orders cream cheese from three suppliers, Cheesy Cream Corp. (CCC), Super Smooth & Sons (SSS), and Bagel's Best Friend Co. (BBF). One month, the total order of cheese came to 100 tons (they do a booming trade). The costs were $80, $50 and $65 per ton from the three suppliers respectively, with total cost amounting to $5990. Given that the store ordered the same amount from CCC and BBF, how many tons of cream cheese were ordered from each supplier?

8. ● **Supply** Refer back to Exercise 7. The bagel store's outlay for cream cheese the following month was $2310, when it purchased a total of 36 tons. Two more tons of cream cheese came from Bagel's Best Friend Co. than from Super Smooth & Sons. How many tons of cream cheese came from each supplier?

9. ● **Pest Control** Conan the Great has boasted to his hordes of followers that many a notorious villain has fallen to his awesome sword: his total of 560 victims consists of evil sorcerers, trolls and orcs. These he has slain with a total of 620 mighty thrusts of his sword; evil sorcerers and trolls each requiring two thrusts (to the chest) and orcs each requiring one thrust (to the neck). When asked about the number of trolls he has slain, he replies, "I, the mighty Conan, despise trolls five times as much as I despise evil sorcerers. Accordingly, five times as many trolls as evil sorcerers have fallen to my sword!" How many of each type of villain has he slain?

10. ● **Manufacturing: Perfume** The Fancy French Perfume Company recently had its secret formula divulged. It turned out that it was using, as the three ingredients, rose oil, oil of fermented prunes, and alcohol. Moreover, each 22-ounce econo-size bottle contained 4 more ounces of alcohol than oil of fermented prunes, while the amount of alcohol was equal to the combined volume of the other two ingredients. How much of each ingredient did it use in an econo-size bottle?[18]

11. ● **Music CD Sales** In 2000, total revenues from sales of recorded rock, rap, and classical music amounted to $5.8 billion. Rock music brought in twice as much revenue as rap music and

900% the revenue of classical music.[19] How much revenue was earned in each of the three categories of recorded music?

12. **Music CD Sales** In 2000, combined revenues from sales of "oldies" and soundtracks amounted to $0.23 billion. Country music brought in 15 times as much revenue as soundtracks, and revenue from the sale of oldies was 30% above that of soundtracks.[20] How much revenue was earned in each of the three categories of recorded music?

13. **Donations** The Enormous State University Good Works Society recently raised funds for three worthwhile causes; the Math Professors' Benevolent Fund (MPBF), the Society of Computer Nerds (SCN), and the NY Jets. Because the society's members are closet jocks, the society donated twice as much to the NY Jets as to the MPBF, and equal amounts to the first two funds (it is unable to distinguish between mathematicians and nerds). Further, for every $1 it gave to the MPBF, it decided to keep $1 for itself; for every $1 it gave to the SCN, it kept $2, and for every $1 to the Jets, it also kept $2. The treasurer of the Society, Johnny Treasure, was required to itemize all donations for the Dean of Students, but discovered to his consternation that he had lost the receipts! The only information available to him was that the society's bank account had swelled by $4200. How much did the society donate to each cause?

14. **Tenure** Professor Walt is up for tenure, and wishes to submit a portfolio of written student evaluations as evidence of his good teaching. He begins by grouping all the evaluations into four categories: good reviews; bad reviews—a typical one being "GET RID OF WALT! THE MAN CAN'T TEACH!"—mediocre reviews—such as "I suppose he's OK, given the general quality of teaching here at _____ University," and reviews left blank. When he tallies up the piles, Walt gets a little worried: There are 280 more bad reviews than good ones and only half as many blank reviews as bad ones. The good reviews and blank reviews together total 170. On an impulse, he decides to even up the piles a little by removing 280 of the bad reviews (including the memorable "GET RID OF WALT" review), and this leaves him with a total of 400 reviews of all types. How many of each category of reviews were there originally?

Airline Costs (Use of technology recommended) Exercises 15 and 16 are based on the following chart, which shows the amount spent by five U.S. airlines to fly one available seat one mile in the fourth quarter of 2000.[21]

Airline	United	American	Continental	SkyWest	SouthWest
Cost	11.3¢	11.0¢	9.8¢	20.3¢	7.8¢

[19] "Rap" includes "Hip Hop." Revenues are based on total manufacturers' shipments at suggested retail prices, and are rounded to the nearest $0.1 billion. SOURCE: Recording Industry Association of America, www.riaa.com, March, 2002.

[20] SOURCE: Ibid.

[21] SOURCE: Company financial statements, obtained from their respective websites. The cost per available seat-mile is a commonly used operating statistic in the airline industry.

[18] Most perfumes consist of 10 to 20% perfume oils dissolved in alcohol. This may or may not be reflected in this company's formula!

● basic skills   ◆ challenging   **tech**Ex technology exercise

15. `tech` Ex Suppose that, on a 3000-mile New York-Los Angeles flight, United, American and SouthWest flew a total of 210 empty seats, costing them a total of $65,580. If United had three times as many empty seats as American, how many empty seats did each of these three airlines carry on its flight?

16. `tech` Ex Suppose that, on a 2000-mile Miami-Memphis flight, Continental, SkyWest, and SouthWest flew a total of 200 empty seats, costing them a total of $58,200. If SkyWest had twice as many empty seats as SouthWest, how many empty seats did each of these three airlines carry on its flight?

Investing: Municipal Bond Funds Exercises 17 and 18 are based on the following data on three tax-exempt municipal bond funds.[22]

	2003 Yield
PNF (Pimco NY)	6%
FDMMX (Fidelity Spartan Mass)	5%
FFLIX (Fidelity Spartan Florida)	7%

17. ● You invested a total of $9000 in the three funds at the beginning of 2003, including an equal amount in FDMMX and FFLIX. Your 2003 yield for the year from the first two funds amounted to $400. How much did you invest in each of the three funds?

18. ● You invested a total of $6000 in the three funds at the beginning of 2003, including an equal amount in PNF and FDMMX. Your total yields for 2003 amounted to $360. How much did you invest in each of the three funds?

Investing: Stocks Exercises 19 and 20 are based on the following data on three computer stocks.[23]

	Price	*Dividend Yield*
APPL (Apple Computer)	$25	0.6%
HPQ (Hewlett Packard)	25	1.2
DELL	35	0

19. You invested a total of $5800 on Apple, Hewlett Packard, and Dell shares at the above prices, and expected to earn $21 in annual dividends. If you purchased a total of 200 shares, how many shares of each stock did you purchase?

20. You invested a total of $6750 on Apple, Hewlett Packard, and Dell shares at the above prices, and expected to earn $45 in annual dividends. If you purchased a total of 250 shares, how many shares of each stock did you purchase?

21. `tech` Ex *Internet Audience* In November 2003, the four companies with the largest number of home Internet users in the US were Microsoft, Time Warner, Yahoo, and Google, with a combined audience of 284 million users.[24] Taking x to be the Microsoft audience in millions, y the Time Warner audience in millions, z the Yahoo audience in millions, and u the Google audience in millions, it was observed that

$$z - u = 3(x - y) + 6$$
$$x + y = 50 + z + u$$

and $x - y + z - u = 42$

How large was the audience of each of the four companies in November, 2003?

22. `tech` Ex *Internet Audience* In November, 2003, the four organizations ranking 5 through 8 in home Internet users in the US were eBay, the U.S. Government, Amazon, and Lycos, with a combined audience of 112 million users.[25] Taking x to be the eBay audience in millions, y the U.S. Government audience in millions, z the Amazon audience in millions, and u the Lycos audience in millions, it was observed that

$$y - z = z - u$$
$$x - y = 3(y - u) + 5$$

and $x - y + z - u = 12$

How large was the audience of each of the four companies in November, 2003?

23. `tech` Ex *Market Share: Homeowners Insurance* The three market leaders in homeowners insurance in Missouri are State Farm, American Family Insurance Group, and Zurich Insurance. Based on data from 1999 through 2002, two relationships between the Missouri homeowners insurance market shares are found to be

$$x = 0.48 - 0.85y - 0.71z$$
$$z = 0.95 - 1.9w$$

where x, y, z, and w are, respectively, the fractions of the market held by State Farm, American Family, Zurich Insurance, and Other.[26] Given that the four groups account for the entire market, obtain a third equation relating x, y, z, and w, and solve the associated system of three linear equations to show how the market shares of State Farm, American Family and Zurich Insurance depend on the share held by Other. (Round all coefficients to two decimal places.) Which of the three companies' market share is most impacted by the share held by Other?

[22] Yields are rounded. SOURCES: www.pimcofunds.com, www.fidelity.com. January 18, 2004.

[23] Stocks were trading at or near the given prices in January, 2004. Dividends are rounded, based on values in January 18, 2004. SOURCE: http://money.excite.com, January 18, 2004.

[24] SOURCE: Nielsen/NetRatings www.nielsen-netratings.com/. January 1, 2004.

[25] SOURCE: Nielsen/NetRatings www.nielsen-netratings.com/. January 1, 2004.

[26] SOURCE: Missouri Dept. of Insurance, http://insurance.mo.gov. January 1, 2004.

● basic skills   ◆ challenging   `tech` Ex technology exercise

24. tech Ex *Market Share: Auto Insurance* The three market leaders in auto insurance in Missouri are State Farm, American Family Insurance Group, and Zurich Insurance, Based on data from 1999 through 2002, two relationships between the Missouri auto insurance market shares are found to be

$$x = 0.30 - 0.12y - 0.70z$$
$$z = 0.35 - 0.53w$$

where x, y, z, and w are, respectively, the fractions of the market held by State Farm, American Family, Zurich Insurance, and Other.[27] Given that the four groups account for the entire market, obtain a third equation relating x, y, z, and w, and solve the associated system of three linear equations to show how the market shares of State Farm, American Family and Zurich Insurance depend on the share held by Other. (Round all coefficients to two decimal places.) Which of the three companies' market share is most impacted by the share held by Other?

25. ● *Inventory Control* Big Red Bookstore wants to ship books from its warehouses in Brooklyn and Queens to its stores, one on Long Island and one in Manhattan. Its warehouse in Brooklyn has 1000 books and its warehouse in Queens has 2000. Each store orders 1500 books. It costs $1 to ship each book from Brooklyn to Manhattan and $2 to ship each book from Queens to Manhattan. It costs $5 to ship each book from Brooklyn to Long Island and $4 to ship each book from Queens to Long Island.

a. If Big Red has a transportation budget of $9000 and is willing to spend all of it, how many books should Big Red ship from each warehouse to each store in order to fill all the orders?

b. Is there a way of doing this for less money? *hint* [see Example 4]

26. ● *Inventory Control* The Tubular Ride Boogie Board Company has manufacturing plants in Tucson, AZ and Toronto, Ontario. You have been given the job of coordinating distribution of the latest model, the Gladiator, to outlets in Honolulu and Venice Beach. The Tucson plant, when operating at full capacity, can manufacture 620 Gladiator boards per week, while the Toronto plant, beset by labor disputes, can produce only 410 boards per week. The outlet in Honolulu orders 500 Gladiator boards per week, while Venice Beach orders 530 boards per week. Transportation costs are as follows:

Tucson to Honolulu: $10 per board; Tucson to Venice Beach: $5 per board;

Toronto to Honolulu: $20 per board; Toronto to Venice Beach: $10 per board.

a. Assuming that you wish to fill all orders and assure full capacity production at both plants, is it possible to meet a total transportation budget of $10,200? If so, how many Gladiator boards are shipped from each manufacturing plant to each distribution outlet?

b. Is there a way of doing this for less money?

27. *Tourism* In the 1990s, significant numbers of tourists traveled from North America and Europe to Australia and South Africa. In 1998, a total of 1,390,000 of these tourists visited Australia, while 1,140,000 of them visited South Africa. Further, 630,000 of them came from North America and 1,900,000 of them came from Europe.[28] (Assume no single tourist visited both destinations or traveled from both North America and Europe.)

a. The given information is not sufficient to determine the number of tourists from each region to each destination. Why?

b. If you were given the additional information that a total of 2,530,000 tourists traveled from these two regions to these two destinations, would you now be able to determine the number of tourists from each region to each destination? If so, what are these numbers?

c. If you were given the additional information that the same number of people from Europe visited South Africa as visited Australia, would you now be able to determine the number of tourists from each region to each destination? If so, what are these numbers?

28. *Tourism* In the 1990s, significant numbers of tourists traveled from North America and Asia to Australia and South Africa. In 1998, a total of 2,230,000 of these tourists visited Australia, while 390,000 of them visited South Africa. Also, 630,000 of these tourists came from North America, and a total of 2,620,000 tourists traveled from these two regions to these two destinations.[29] (Assume no single tourist visited both destinations or traveled from both North America and Asia.)

a. The given information is not sufficient to determine the number of tourists from each region to each destination. Why?

b. If you were given the additional information that a total of 1,990,000 tourists came from Asia, would you now be able to determine the number of tourists from each region to each destination? If so, what are these numbers?

c. If you were given the additional information that 200,000 tourists visited South Africa from Asia, would you now be able to determine the number of tourists from each region to each destination? If so, what are these numbers?

29. *Alcohol* The following table shows some data from a 2000 study on substance use among 10th graders in the U.S. and Europe.[30]

	Used Alcohol	*Alcohol-Free*	*Totals*
U.S.	x	y	14,000
Europe	z	w	95,000
Totals	63,550	45,450	

---

[28] Figures are rounded to the nearest 10,000. SOURCES: South African Dept. of Environmental Affairs and Tourism; Australia Tourist Commission/ *The New York Times,* January 15, 2000, p. C1.

[29] Ibid.

[30] "Used Alcohol" indicates consumption of alcohol at least once in the past 30 days. SOURCE: Council of Europe/University of Michigan "Monitoring the Future"/*New York Times* February 21, 2001, p. A10.

[27] SOURCE: Missouri Dept. of Insurance, http://insurance.mo.gov. January 1, 2004.

● basic skills   ◆ challenging   tech Ex technology exercise

a. The table leads to a linear system of four equations in four unknowns. What is the system? Does it have a unique solution? What does this indicate about the given and the missing data?

b. *tech* Ex Given that the number of U.S. 10th graders who were alcohol-free was 50% more than the number who had used alcohol, find the missing data.

30. *Tobacco* The following table shows some data from the same study cited in Exercise 29.[31]

	Smoked Cigarettes	**Cigarette-Free**	**Totals**
U.S.	x	y	14,000
Europe	z	w	95,000
Totals		70,210	109,000

a. The table leads to a linear system of four equations in four unknowns. What is the system? Does it have a unique solution? What does this indicate about the missing data?

b. *tech* Ex Given that 31,510 more European 10th graders smoked cigarettes than U.S. 10th graders, find the missing data.

31. ● *Traffic Flow* One-way traffic through Enormous State University is shown in the figure, where the numbers indicate daily counts of vehicles.

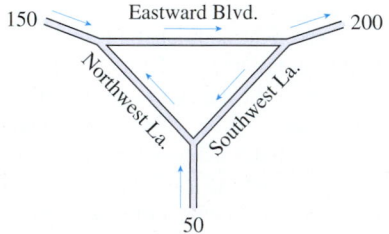

a. Is it possible to determine the daily flow of traffic along each of the three streets from the information given? If your answer is *yes*, what is the traffic flow along each street? If your answer is *no*, what additional information would suffice?

b. Is a flow of 60 vehicles per day along Southwest Lane consistent with the information given?

c. What is the minimum traffic flow possible along Northwest Lane consistent with the information given? *hint* [see Example 3]

32. ● *Traffic Flow* The traffic through downtown East Podunk flows through the one-way system shown below.

---

[31] "Smoked cigarettes" indicates that at least one cigarette was smoked in the past 30 days. SOURCE: Ibid.

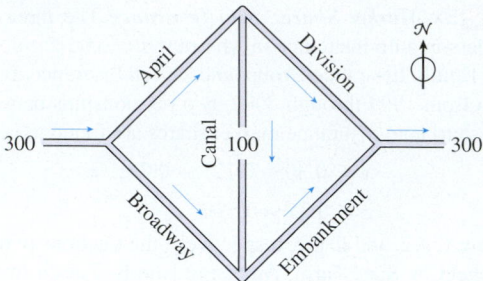

Traffic counters find that 300 vehicles enter town from the west each hour, and 300 leave town toward the east each hour. Also, 100 cars drive down Canal Street each hour.

a. Write down the general solution of the associated system of linear equations. Is it possible to determine the number of vehicles on each street per hour?

b. On which street could you put another traffic counter in order to determine the flow completely?

c. What is the minimum traffic flow along April Street consistent with the information given?

33. *Traffic Management* The Outer Village Town Council has decided to convert its (rather quiet) main street, Broadway, to a one-way street, but is not sure of the direction of most of the traffic. The accompanying diagram illustrates the downtown area of Outer Village, as well as the *net* traffic flow along the intersecting streets (in vehicles per day). (There are no one-way streets; a net traffic flow in a certain direction is defined as the traffic flow in that direction minus the flow in the opposite direction.)

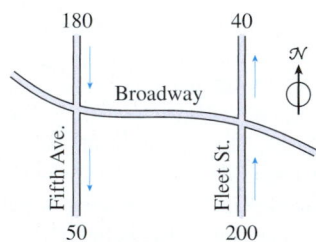

a. Is the given information sufficient to determine the net traffic flow along the three portions of Broadway shown? If your answer is *yes*, give the traffic flow along each stretch. If your answer is *no*, what additional information would suffice? [*Hint:* For the direction of net traffic flow, choose either east or west. If a corresponding value is negative, it indicates net flow in the opposite direction.]

b. Assuming that there is little traffic (less than 160 vehicles per day) east of Fleet Street, in what direction is the net flow of traffic along the remaining stretches of Broadway?

34. *Electric Current* Electric current measures (in **amperes** or **amps**) the flow of electrons through wires. Like traffic flow, the current entering an intersection of wires must equal the

● basic skills   ◆ challenging   *tech* Ex technology exercise

current leaving it.[32] Here is an electrical circuit known as a **Wheatstone bridge.**

a. If the currents in two of the wires are 10 amps and 5 amps as shown, determine the currents in the unlabeled wires in terms of suitable parameters.

b. In which wire should you measure the current in order to know all of the currents exactly?

35. *Econometrics* **(From the GRE Economics Test)** This and the next exercise are based on the following simplified model of the determination of the money stock.

$$M = C + D$$

$$C = 0.2D$$

$$R = 0.1D$$

$$H = R + C$$

where

$M =$ money stock
$C =$ currency in circulation
$R =$ bank reserves
$D =$ deposits of the public
$H =$ high-powered money

If the money stock were \$120 billion, what would bank reserves have to be?

36. *Econometrics* **(From the GRE Economics Test)** With the model in the previous exercise, if H were equal to \$42 billion, what would M equal?

CAT Scans CAT (Computerized Axial Tomographic) scans are used to map the exact location of interior features of the human body. CAT scan technology is based on the following principles: (1) different components of the human body (water, gray matter, bone, etc.) absorb X-rays to different extents; and (2) to measure the X-ray absorption by a specific region of, say, the brain, it suffices to pass a number of line-shaped pencil beams of X-rays through the brain at different angles and measure the total absorption for each beam, which is the sum of the absorptions of the regions through which it passes. The accompanying diagram illustrates a simple example. (The number in each region shows its absorption, and the number on each X-ray beam shows the total absorption for that beam.)[33]

---

[32] This is known as **Kirchhoff's Current Law,** named after Gustav Robert Kirchhoff (1824–1887). Kirchhoff made important contributions to the fields of geometric optics, electromagnetic radiation and electrical network theory.

[33] Based on a COMAP article *Geometry, New Tools for New Technologies,* by J. Malkevitch, Video Applications Library, COMAP, 1992. The absorptions are actually calibrated on a logarithmic scale. In real applications, the size of the regions is very small, and very large numbers of beams must be used.

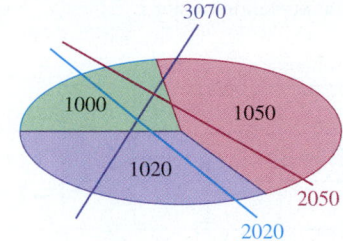

In Exercises 37–42, use the table and the given X-ray absorption diagrams to identify the composition of each of the regions marked by a letter.

Type	Air	Water	Gray Matter	Tumor	Blood	Bone
Absorption	0	1000	1020	1030	1050	2000

37.

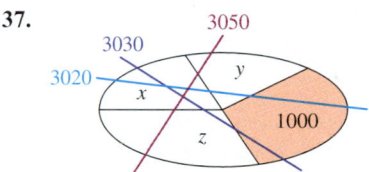

38.

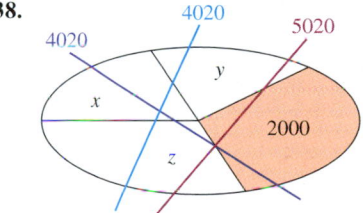

39.

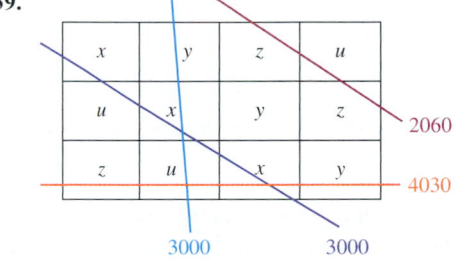

40.

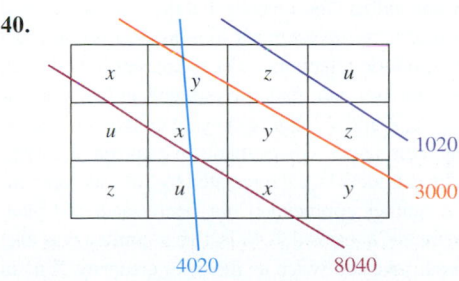

● basic skills   ◆ challenging   **tech Ex** technology exercise

41. Identify the composition of site x.

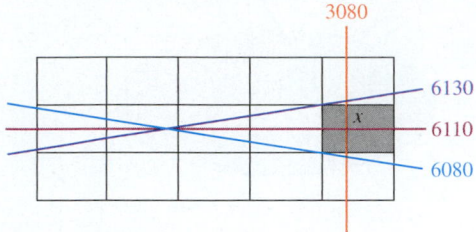

(The horizontal and slanted beams each pass through five regions.)

42. Identify the composition of site x.

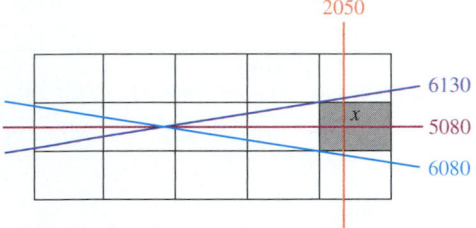

(The horizontal and slanted beams each pass through five regions.)

43. ◆ *Voting* In the 75th Congress (1937–39) the U.S. House of Representatives had 333 Democrats, 89 Republicans, and 13 members of other parties. Suppose that a bill passed the House with 31 more votes in favor than against, with 10 times as many Democrats voting for the bill as Republicans, and with 36 more non-Democrats voting against the bill than for it. How many Democrats, how many Republicans, and how many members of other parties voted in favor of the bill?

44. ◆ *Voting* In the 75th Congress (1937–39) there were in the Senate 75 Democrats, 17 Republicans, and 4 members of other parties. Suppose that a bill passed the Senate with 16 more votes in favor than against, with three times as many Democrats voting in favor as non-Democrats voting in favor, and 32 more Democrats voting in favor than Republicans voting in favor. How many Democrats, how many Republicans, and how many members of other parties voted in favor of the bill?

45. ◆ *Investments* Things have not been going too well here at Accurate Accounting, Inc. since we hired Todd Smiley. He has a tendency to lose important documents, especially around April, when tax returns of our business clients are due. Today Smiley accidentally shredded Colossal Conglomerate Corp's investment records. We must therefore reconstruct them based on the information he can gather. Todd recalls that the company earned an $8 million return on investments totaling $65 million last year. After a few frantic telephone calls to sources in Colossal, he learned that Colossal had made investments in four companies last year: X, Y, Z and W. (For reasons of confidentiality, we are withholding their names.) Investments in company X earned 15% last year, investments in Y depreciated by 20% last year, investments in Z neither appreciated nor depreciated last year, while investments in W earned 20% last year. Smiley was also told that Colossal invested twice as much in company X as in

company Z, and three times as much in company W as in company Z. Does Smiley have sufficient information to piece together Colossal's investment portfolio before its tax return is due next week? If so, what does the investment portfolio look like?

46. ◆ *Investments* Things are going from bad to worse here at Accurate Accounting, Inc.! Colossal Conglomerate Corp's tax return is due tomorrow and the accountant Todd Smiley seems to have no idea how Colossal earned a return of $8 million on a $65 million investment last year. It appears that, although the returns from companies X, Y, Z and W were as listed in Exercise 45, the rest of the information there was wrong. What Smiley is now being told is that Colossal only invested in companies X, Y and Z and that the investment in X amounted to $30 million. His sources in Colossal still maintain that twice as much was invested in company X as in company Z. What should Smiley do?

Communication and Reasoning Exercises

47. ● Are Exercises 1 and 2 realistic in their expectation of using up all the ingredients? What does your answer have to do with the solution(s) of the associated system of equations?

48. ● Suppose that you obtained a solution for Exercise 3 or 4 consisting of positive values that were not all whole numbers. What would such a solution signify about the situation in the exercise? Should you round these values to the nearest whole numbers?

In Exercises 49–54, x, y and z represent the weights of the three ingredients X, Y and Z in a gasoline blend. Say which of the following is represented by a linear equation in x, y and z, and give a form of the equation when it is:

49. ● The blend consists of 100 pounds of ingredient X.

50. ● The blend is free of ingredient X.

51. ● The blend contains 30% ingredient Y by weight.

52. ● The weight of ingredient X is the product of the weights of ingredients Y and Z.

53. There is at least 30% ingredient Y by weight.

54. There is twice as much ingredient X by weight as Y and Z combined.

55. Make up an entertaining word problem leading to the following system of equations.

$$10x + 20y + 10z = 100$$
$$5x + 15y = 50$$
$$x + y + z = 10$$

56. Make up an entertaining word problem leading to the following system of equations.

$$10x + 20y = 300$$
$$10z + 20w = 400$$
$$20x + 10z = 400$$
$$10y + 20w = 300$$

● basic skills   ◆ challenging   **tech** Ex technology exercise

Chapter 2 Review

KEY CONCEPTS

2.1 Systems of Two Equations in Two Unknowns
Linear equation in two unknowns *p. 114*
Coefficient *p. 114*
Solution of an equation in two unknowns *p. 116*
Graphical method for solving a system of two linear equations *p. 116*
Algebraic method for solving a system of two linear equations *p. 117*
Possible outcomes for a system of two linear equations *p. 117*

Redundant or dependent system *p. 119*
Consistent system *p. 120*

2.2 Using Matrices to Solve Systems of Equations
Matrix *p. 128*
Augmented matrix of a system of linear equations *p. 128*
Elementary row operations *p. 129*
Pivot *p. 131*
Clearing a column; pivoting *p. 131*

Gauss-Jordan or row reduction *p. 134*
Reduced row echelon form *p. 141*

2.3 Applications of Systems of Linear Equations
Resource allocation *p. 146*
(Traffic) flow *p. 149*
Transportation *p. 150*

REVIEW EXERCISES

In each of Exercises 1–6, graph the equations and determine how many solutions the system has, if any.

1. $x + 2y = 4$
$2x - y = 1$

2. $0.2x - 0.1y = 0.3$
$0.2x + 0.2y = 0.4$

3. $\frac{1}{2}x - \frac{3}{4}y = 0$
$6x - 9y = 0$

4. $2x + 3y = 2$
$-x - 3y/2 = 1/2$

5. $x + y = 1$
$2x + y = 0.3$
$3x + 2y = \frac{13}{10}$

6. $3x + 0.5y = 0.1$
$6x + y = 0.2$
$\frac{3x}{10} - 0.05y = 0.01$

Solve each of the systems of linear equations in Exercises 7–18.

7. $x + 2y = 4$
$2x - y = 1$

8. $0.2x - 0.1y = 0.3$
$0.2x + 0.2y = 0.4$

9. $\frac{1}{2}x - \frac{3}{4}y = 0$
$6x - 9y = 0$

10. $2x + 3y = 2$
$-x - 3y/2 = 1/2$

11. $x + y = 1$
$2x + y = 0.3$
$3x + 2y = \frac{13}{10}$

12. $3x + 0.5y = 0.1$
$6x + y = 0.2$
$\frac{3x}{10} - 0.05y = 0.01$

13. $x + 2y = -3$
$x - z = 0$
$x + 3y - 2z = -2$

14. $x - y + z = 2$
$7x + y - z = 6$
$x - \frac{1}{2}y + \frac{1}{3}z = 1$
$x + y + z = 6$

15. $x - \frac{1}{2}y + z = 0$
$\frac{1}{2}x - \frac{1}{2}z = -1$
$\frac{3}{2}x - \frac{1}{2}y + \frac{1}{2}z = -1$

16. $x + y - 2z = -1$
$-2x - 2y + 4z = 2$
$0.75x + 0.75y - 1.5z = -0.75$

17. $x = \frac{1}{2}y$
$\frac{1}{2}x = -\frac{1}{2}z + 2$
$z = -3x + y$

18. $x - y + z = 1$
$y - z + w = 1$
$x + z - w = 1$
$2x + z = 3$

19. The Fahrenheit and Celsius (or centigrade) temperature scales are related by the equation

$$5F - 9C = 160$$

where F is the Fahrenheit temperature of an object and C is its Celsius temperature.

  a. What temperature should an object be if its Fahrenheit and Celsius temperatures are the same?

  b. What temperature should an object be if its Celsius temperature is half its Fahrenheit temperature?

  c. Is it possible for the Fahrenheit temperature of an object to be 1.8 times its Celsius temperature? Explain.

20. Let x, y, z, and w represent the population in millions of four cities A, B, C, and D. Express each of the following as an equation in x, y, z, and w. If the equation is linear, say so and express it in the standard form $ax + by + cz + dw = k$.

  a. The total population of the four cities is 10 million people

b. City A has three times as many people as cities B and C combined.

c. City D is actually a ghost town; there are no people living in it.

d. The population of city A is the sum of the squares of the populations of the other three cities

e. City C has 30% more people than City B.

f. City C has 30% fewer people than City B.

Applications

Purchasing In Exercises 21–24, you are the buyer for OHagan-Books.com, and are considering increasing stocks of romance and horror novels at the new OHaganBooks.com warehouse in Texas. You have offers from two publishers: Duffin House and Higgins Press. Duffin offers a package of 5 horror novels and 5 romance novels for $50, and Higgins offers a package of 5 horror and 11 romance novels for $150.

21. How many packages should you purchase from each publisher to get exactly 4500 horror novels and 6600 romance novels?

22. You want to spend a total of $50,000 on books and have promised to buy twice as many packages from Duffin as from Higgins. How many packages should you purchase from each publisher?

23. Your accountant tells you that you can now afford to spend a total of $60,000 on romance and horror books. She also reminds you that you had signed an agreement to spend the same amount of money at both publishers. How many packages should you purchase from each publisher?

24. Your accountant tells you that you can now afford to spend a total of $90,000 on romance and horror books. She also reminds you that you had signed an agreement to spend twice as much money for books from Duffin as from Higgins. How many packages should you purchase from each publisher?

25. *Equilibrium* The demand per year for "Finite Math the OHagan Way" is given by $q = -1000p + 140,000$, where p is the price per book in dollars. The supply is given by $q = 2000p + 20,000$. Find the price at which supply and demand balance.

26. *Equilibrium* OHaganbooks.com CEO John O'Hagan announces to a stunned audience at the annual board meeting that he is considering expanding into the jumbo jet airline manufacturing business. The demand per year for jumbo jets is given by $q = -2p + 18$ where p is the price per jet in millions of dollars. The supply is given by $q = 3p + 3$. Find the price the envisioned O'Hagan jumbo jet division should charge to balance supply and demand.

27. *Feeding Schedules* Billy-Sean O'Hagan's 36-gallon tropical fish tank contains three types of carnivorous creatures: baby sharks, piranhas and squids, and he feeds them three types of delicacies: goldfish, angelfish and butterfly fish. Each baby shark can consume 1 goldfish, 2 angelfish and 2 butterfly fish per day; each piranha can consume 1 goldfish and 3 butterfly fish per day (the piranhas are rather large as a result of their diet) while each squid can consume 1 goldfish and 1 angelfish per day. After a trip to the local pet store, he was able to feed his creatures to capacity, and noticed that 21 goldfish, 21 angelfish and 35 butterfly fish were eaten. How many of each type of creature does he have?

28. *Resource Allocation* Duffin House is planning its annual Song Festival, when it will serve three kinds of delicacies: granola treats, nutty granola treats, and nuttiest granola treats. The following table shows the ingredients required (in ounces) for a single serving of each delicacy, as well as the total amount of each ingredient available.

	Granola	Nutty granola	Nuttiest granola	Total available
Toasted oats	1	1	5	1500
Almonds	4	8	8	10,000
Raisins	2	4	8	4000

The Song Festival planners at Duffin House would like to use up all the ingredients. Is this possible? If so, how many servings of each kind of delicacy can they make?

29. *Website Traffic* OHaganBooks.com has two principal competitors: JungleBooks.com and FarmerBooks.com. Combined website traffic at the three sites is estimated at 10,000 hits per day. Only 10% of the hits at OHaganBooks.com result in orders, whereas JungleBooks.com and FarmerBooks.com report that 20% of the hits at their sites result in book orders. Together, the three sites process 1500 book orders per day. FarmerBooks.com appears to be the most successful of the three, and gets as many book orders as the other two combined. What is the traffic (in hits per day) at each of the sites?

30. *Investing in Stocks* Billy-Sean O'Hagan is the treasurer at his college fraternity, which recently earned $12,400 in its annual carwash fundraiser. Billy-Sean decided to invest all the proceeds in the purchase of three computer stocks: HAL, POM, and WELL.

	Price per Share	Dividend Yield
HAL	$100	0.5 %
POM	20	1.50
WELL	25	0

If the investment was expected to earn $56 in annual dividends, and he purchased a total of 200 shares, how many shares of each stock did he purchase?

31. *IPOs* Duffin House, Higgins Press, and Sickle Publications all went public on the same day recently. John O'Hagan had the opportunity to participate in all three initial public offerings (partly because he and Marjory Duffin are good friends). He made a considerable profit when he sold all of the stock two days later on the open market. The following table shows the purchase price and percentage yield on the investment in each company.

	Purchase Price per Share	Yield
Duffin House (DHS)	$8	20%
Higgins Press (HPR)	$10	15%
Sickle Publications (SPUB)	$15	15%

He invested $20,000 in a total of 2000 shares, and made a $3400 profit from the transactions. How many shares in each company did he purchase?

32. Degree Requirements During his lunch break, John O'Hagan decides to devote some time to assisting his son Billy-Sean, who is having a terrible time coming up with a college course schedule. One reason for this is the very complicated Bulletin of Suburban State University. It reads as follows:

All candidates for the degree of Bachelor of Arts at SSU must take a total of 124 credits from the Sciences, Fine Arts, Liberal Arts, and Mathematics,[34] including an equal number of Science and Fine Arts credits, and twice as many Mathematics credits as Science credits and Fine Arts credits combined, but with Liberal Arts credits exceeding Mathematics credits by exactly one third of the number of Fine Arts credits.

What are all the possible degree programs for Billy-Sean?

33. Network Traffic All book orders received at the Order Department at OHaganBooks.com are transmitted through a small computer network to the Shipping Department. The following diagram shows the network (which uses two intermediate computers as routers), together with some of the average daily traffic measured in book orders.

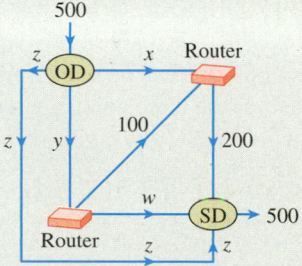

OD = Order department
SD = Shipping department

a. Set up a system of linear equations in which the unknowns give the average traffic along the paths labeled x, y, z, w, and find the general solution.
b. What is the minimum volume of traffic along y?
c. What is the maximum volume of traffic along w?
d. If there is no traffic along z, find the volume of traffic along all the paths.
e. If there is the same volume of traffic along y and z, what is the volume of traffic along w?

34. Business Retreats Marjory Duffin is planning a joint business retreat for Duffin House and OHaganBooks.com at Laguna Surf City, but is concerned about traffic conditions (she feels that too many cars tend to spoil the ambiance of a seaside retreat). She managed to obtain the following map from the Laguna Surf City Engineering Department (all the streets are one-way as indicated). The counters show traffic every five minutes.

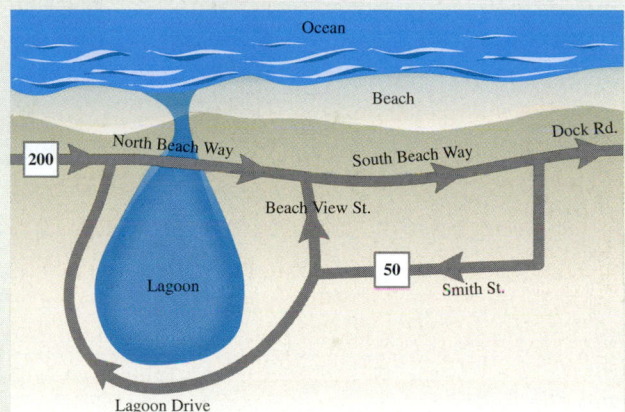

a. Set up and solve the associated system of linear equations. *Be sure to give the general solution.* (Take x = traffic along North Beach Way, y = traffic along South Beach Way, z = traffic along Beach View St., u = traffic along Lagoon Drive and v = traffic along Dock Road.)
b. Assuming that all roads are one-way in the directions shown, what, if any, is the maximum possible traffic along Lagoon Drive?
c. The Laguna Surf City Traffic Department is considering opening up Beach View Street to two-way traffic, but an environmentalist group is concerned that this will result in increased traffic on Lagoon Drive. What, if any, is the maximum possible traffic along Lagoon Drive assuming that Beach View Street is two-way and the traffic counter readings are as shown?

35. Shipping On the same day that the sales department at Duffin House received an order for 600 packages from the OHaganBooks.com Texas headquarters, it received an additional order for 200 packages from FantasyBooks.com, based in California. Duffin House has warehouses in New York and Illinois. The Illinois warehouse is closing down, and must

---

[34] Strictly speaking, mathematics is not a science; it is the Queen of the Sciences, although we like to think of it as the Mother of all Sciences.

clear all 300 packages it has in stock. Shipping costs per package of books are as follows:

New York to Texas: $20   New York to California: $50
Illinois to Texas: $30   Illinois to California: $40

Is it possible to fill both orders and clear the Illinois warehouse at a cost of $22,000? If so, how many packages should be sent from each warehouse to each online bookstore?

vMentor Do you need a live tutor for homework problems? Access vMentor on the ThomsonNOW! website at **www.thomsonedu.com** for one-on-one tutoring from a mathematics expert.

CASE STUDY: The Impact of Regulating Sulfur Emissions

David Woodfall/Getty Images

Your consulting company has been hired by the EPA. The EPA is considering regulations requiring a 15 million-ton rollback in sulfur emissions in an effort to curb the effects of acid rain on the ecosystem. The EPA would like to have an estimate of the cost to the major utility companies and the effect on jobs in the coal mining industry. The following data are available:[35]

Strategy	Annual Cost to Utilities ($ billions)	Cost in Jobs (number of jobs lost)
8 million-ton rollback	20.4	14,100
10 million-ton rollback	34.5	21,900
12 million-ton rollback	93.6	13,400[*]

Your assignment is to use these data to give projections of the annual cost to utilities and the cost in jobs if the regulations were to be enacted.

You decide to consider the annual cost C to utilities and the job loss J separately. After giving the situation some thought, you decide to have two equations, one giving C in terms of the rollback tonnage t and the other giving J in terms of t. Your first inclination is to try linear equations—that is, an equation of the form

$$C = at + b \quad (a \text{ and } b \text{ constants})$$

and a similar one for J, but you quickly discover that the data simply won't fit, no matter what the choice of the constants. The reason for this can be seen graphically by plotting C and J versus t (Figure 9). In neither case do the three points lie on a straight line. In fact, the job data are not even *close* to being linear. Thus, you will need curves to model these data.

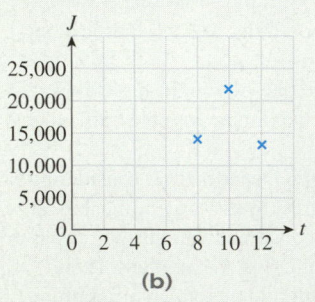

(a)

(b)

Figure **9**

---

[35] SOURCE: Congress of the United States, Congressional Budget Office, *Curbing Acid Rain: Cost, Budget, and Coal Market Effects* (Washington, DC: U.S. Government Printing Office, 1986): xx, xxii, 23, 80.

[*] The reason that job losses drop when the rollback is increased to 12 million tons is that a rollback of this magnitude requires that expensive scrubbers be installed to filter emissions, even if a utility company has switched from coal to other energy sources. Once the scrubbers are installed, it pays the utility companies to switch back to (cheaper) coal as a primary source of energy. A 10-million-ton reduction, on the other hand, results in a massive move away from coal—this being cheaper than installing scrubbers—and hence a dramatic job loss in the coal mining industry.

After giving the matter further thought, you remember something your mathematics instructor once said: the simplest curve passing through any three points not all on the same line is a parabola. Since you are looking for a simple model of the data, you decide to try a parabola. A general parabola has the equation

$$C = at^2 + bt + c$$

where a, b and c are constants. The problem now is: what are a, b, and c? You decide to try substituting the values of C and t into the general equation, and you get the following.

$$t = 8, C = 20.4 \quad \text{gives} \quad 20.4 = 64a + 8b + c$$
$$t = 10, C = 34.5 \quad \text{gives} \quad 34.5 = 100a + 10b + c$$
$$t = 12, C = 93.6 \quad \text{gives} \quad 93.6 = 144a + 12b + c$$

Now you notice that you have three linear equations in three unknowns! You solve the system:

$$a = 5.625, b = -94.2, c = 414$$

Thus your cost equation becomes

$$C = 5.625t^2 - 94.2t + 414$$

You now substitute $t = 15$ to get $C = 266.625$. In other words, you are able to predict an annual cost to the utility industry of $266.625 billion.

Forging ahead, you turn to the jobs equation, and write

$$J = at^2 + bt + c$$

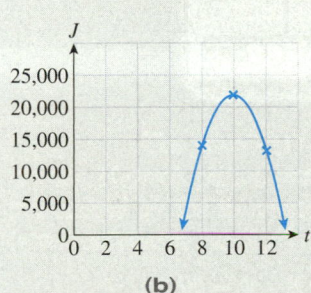

(a)

Substituting the data from the second column of the chart, you obtain

$$14,100 = 64a + 8b + c$$
$$21,900 = 100a + 10b + c$$
$$13,400 = 144a + 12b + c$$

You solve this system and find that

$$a = -2,037.5, b = 40,575, c = -180,100$$

Thus, your jobs equation is

$$J = -2,037.5t^2 + 40,575t - 180,100$$

Substituting $t = 15$, you get $J = -29,912.5$. Uh-oh! A negative number! Just what does this mean? Well, if there were to be $-29,913$ jobs lost, then there would be 29,913 new jobs *created*![36] Figure 10 shows the parabolas superimposed on the data points.

You thus submit the following projection: a 15-million-ton rollback will result in an annual cost of approximately $266.6 billion to the utility industry, but will also result in the creation of approximately 29,900 new jobs in the coal mining industry.

(b)

Figure 10

---

[36] The logic is this: Since the stringent new emission standards would require that even more scrubbers be installed, the utility industry might as well increase its use of coal (which tends to be far cheaper than other energy sources), thus creating a boom for the coal industry.

Exercises

1. Repeat the computations above using the following rounded figures.

Strategy	Annual Cost to Utilities ($ billions)	Cost in Jobs (number of jobs lost)
8-million-ton rollback	20	14,000
10-million-ton rollback	35	22,000
12-million-ton rollback	94	13,000

2. If the 8 million ton rollback data had not been available, what projections would you have given? (Use the original data.)

3. If the 10 million ton rollback data had not been available, what projections would you have given? (Use the original data.)

4. Find the equation of the parabola that passes through the points $(1, 2)$, $(2, 9)$ and $(3, 19)$.

5. Is there a parabola that passes through the points $(1, 2)$, $(2, 9)$ and $(3, 16)$?

6. Is there a parabola that passes though the points $(1, 2)$, $(2, 9)$, $(3, 19)$ and $(-1, 2)$?

7. `tech` Ex Use a spreadsheet to estimate the size of the rollback that results in the lowest annual cost to utilities.

8. `tech` Ex You submit your projections on the cost of regulating sulfur emissions, and the EPA tells you, "Thank you very much, but we made a small mistake: a 15-million-ton rollback has in fact already been in effect for the past year, and has resulted in a cost of $250 billion to the utilities and the creation of 8,000 jobs in the coal mining industry." (Apparently, utilities have been switching back to coal at a greater rate than anticipated.) In view of this, the EPA is now considering a 20-million-ton rollback. You are to come up with projections by tomorrow. [*Hint:* Since you now have four data points on each graph, try a general cubic instead: $C = at^3 + bt^2 + ct + d$.]

Section 2.1

Example 1 Find all solutions (x, y) of the following system of two equations:

$$x + y = 3$$
$$x - y = 1$$

Solution with Technology You can use a graphing calculator to draw the graphs of the two equations on the same set of axes and to check the solution. First, solve the equations for y, obtaining $y = -x + 3$ and $y = x - 1$. On the TI-83/84:

1. Set

    Y₁=-X+3
    Y₂=X-1

2. Decide on the range of x-values you want to use. As in Figure 1, let us choose the range $[-4, 4]$.[38]

3. In the WINDOW menu set Xmin $= -4$ and Xmax $= 4$.

4. Press ZOOM and select Zoomfit to set the y range.

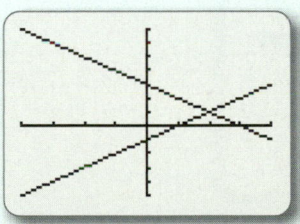

You can now zoom in for a more accurate view by choosing a smaller x-range that includes the point of intersection, like $[1.5, 2.5]$, and using Zoomfit again. You can also use the trace feature to see the coordinates of points near the point of intersection.

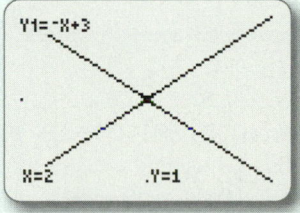

---

[38] How did we come up with this interval? Trial and error. You might need to try several intervals before finding one that gives a graph showing the point of intersection clearly.

To check that $(2, 1)$ is the correct solution, use the table feature to compare the two values of y corresponding to $x = 2$:

1. Press 2ND TABLE.

2. Set X $= 2$, and compare the corresponding values of Y₁ and Y₂; they should each be 1.

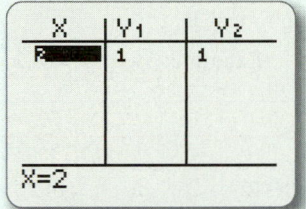

Q: Is it always possible to find the exact answer using a TI-83/84 this way?

A: No. Consider the system:

$$2x + 3y = 1$$
$$-3x + 2y = 1$$

Its solution is $(x, y) = \left(-\frac{1}{13}, \frac{5}{13}\right)$. (Check it!) If we write this solution in decimal form, we get repeating decimals: $(-0.0769\ 230769230\ldots, 0.384615384615\ldots)$. Since the graphing calculator gives the coordinates of points only to several decimal places, it can show only an approximate answer. ∎

Q: How accurate is the answer shown using the trace feature?

A: That depends. We can increase the accuracy up to a point by zooming in on the point of intersection of the two graphs. But there is a limit to this: Most graphing calculators are capable of giving an answer correct to about 13 decimal places. This means, for instance, that, in the eyes of the TI-83/84, 2.000 000 000 000 1 is exactly the same as 2 (subtracting them yields 0). It follows that if you attempt to use a window so narrow that you need approximately 13 significant digits to distinguish the left and right edges, you will run into accuracy problems. ∎

Section 2.2

Row Operations with a TI-83/84 Start by entering the matrix into [A] using MATRX EDIT. You can then do row operations on [A] using the following instructions (*row, *row+, and rowSwap are found in the MATRX MATH menu).

Row Operation	TI-83/84 Instruction (Matrix name is [A])
$R_i \rightarrow kR_i$	`*row(k,[A],i)→[A]`
Example: $R_2 \rightarrow 3R_2$	`*row(3,[A],2)→[A]`
$R_i \rightarrow R_i + kR_j$	`*row+(k,[A],j,i)→[A]`
Examples: $R_1 \rightarrow R_1 - 3R_2$	`*row+(-3,[A],2,1)→[A]`
$R_1 \rightarrow 4R_1 - 3R_2$	`*row(4,[A],1)→[A]` `*row+(-3,[A],2,1)→[A]`
Swap R_i and R_j	`rowSwap([A],i,j)→[A]`
Example: Swap R_1 and R_2	`rowSwap([A],1,2)→[A]`

Example 1 Solve the system

$$x - y + 5z = -6$$
$$3x + 3y - z = 10$$
$$x + 3y + 2z = 5$$

Solution with Technology

1. Begin by entering the matrix into `[A]` using $\boxed{\text{MATRX}}$ EDIT. Only three columns can be seen at a time; you can see the rest of the matrix by scrolling left or right.

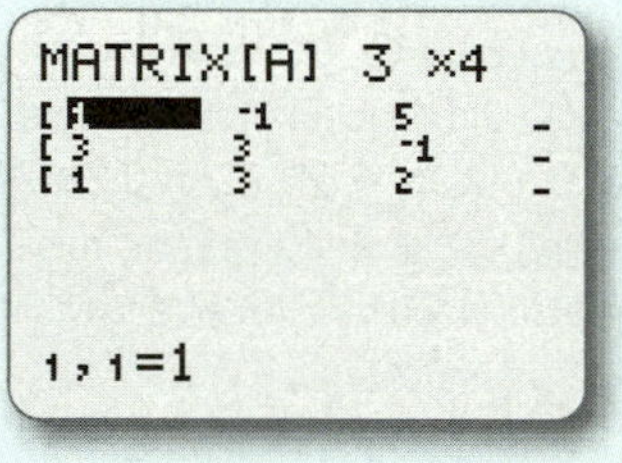

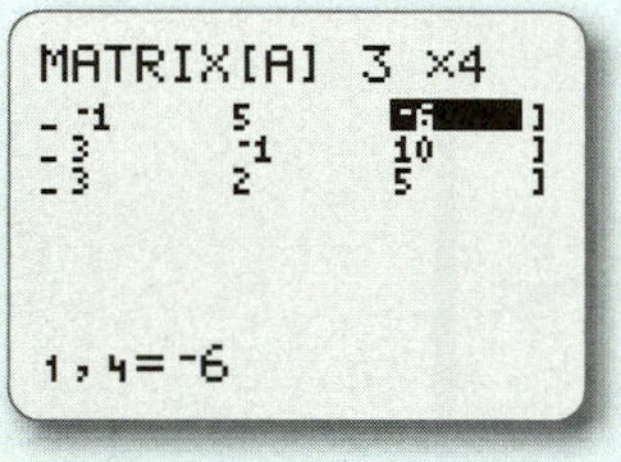

```
MATRIX[A] 3 ×4
[1     -1    5    -
[3     3     -1   -
[1     3     2    -


1,1=1
```

```
MATRIX[A] 3 ×4
-  -1    5    -6  ]
-  3     -1   10  ]
-  3     2    5   ]


1,4=-6
```

2. Now perform the operations given in Example 1:

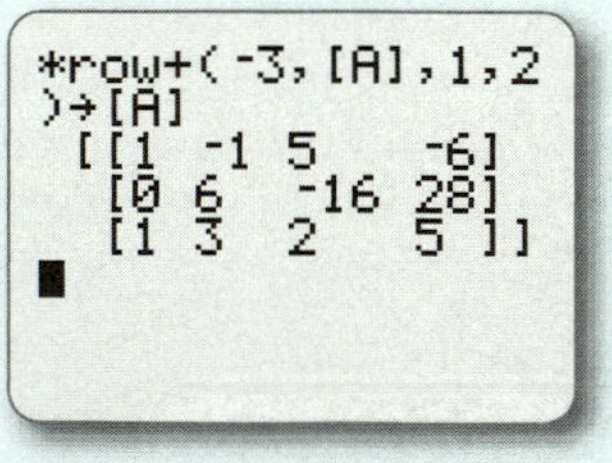

```
*row+(-3,[A],1,2
)→[A]
  [[1  -1  5   -6]
   [0  6   -16 28]
   [1  3   2   5 ]]
■
```

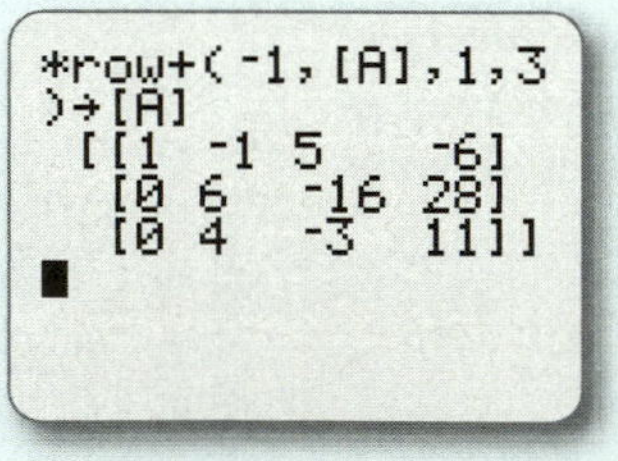

```
*row+(-1,[A],1,3
)→[A]
  [[1  -1  5   -6]
   [0  6   -16 28]
   [0  4   -3  11]]
■
```

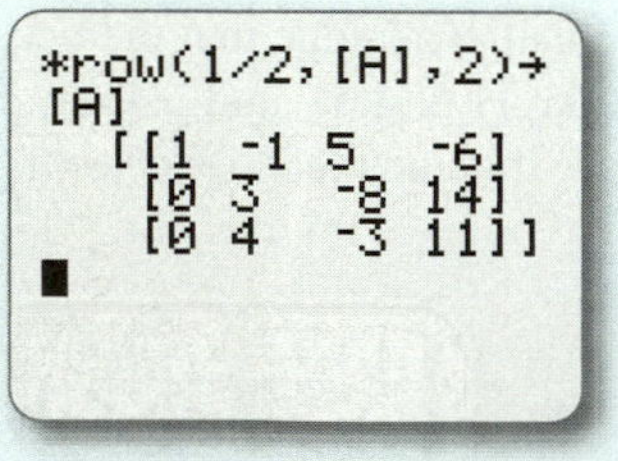

```
*row(1/2,[A],2)→
[A]
  [[1  -1  5   -6]
   [0  3   -8  14]
   [0  4   -3  11]]
■
```

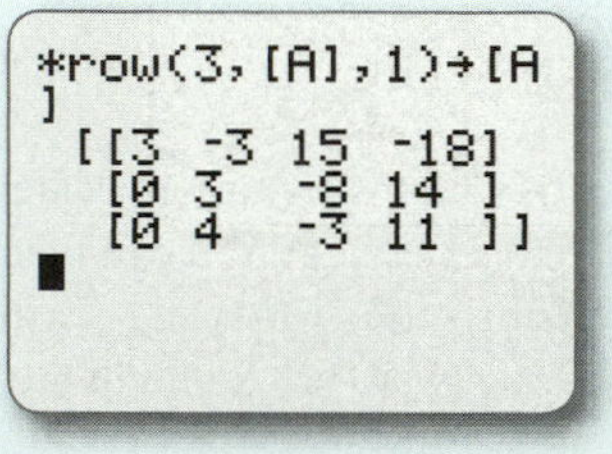

```
*row(3,[A],1)→[A
]
  [[3  -3  15  -18]
   [0  3   -8  14 ]
   [0  4   -3  11 ]]
■
```

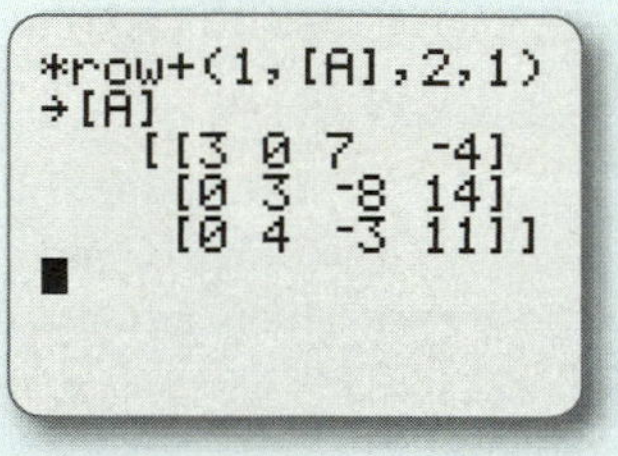

```
*row+(1,[A],2,1)
→[A]
  [[3  0   7   -4]
   [0  3   -8  14]
   [0  4   -3  11]]
■
```

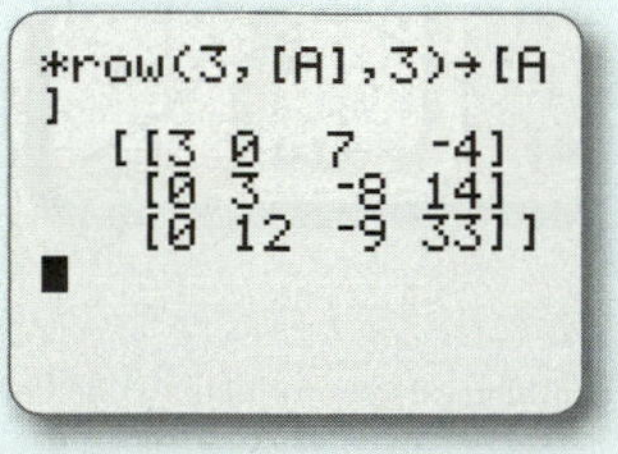

```
*row(3,[A],3)→[A
]
  [[3  0   7   -4]
   [0  3   -8  14]
   [0  12  -9  33]]
■
```

```
*row+(-4,[A],2,3
)→[A]
  [[3 0 7  -4 ]
   [0 3 -8 14 ]
   [0 0 23 -23]]
■
```

```
*row(1/23,[A],3)
→[A]
  [[3 0 7  -4]
   [0 3 -8 14]
   [0 0 1  -1]]
■
```

```
*row+(-7,[A],3,1
)→[A]
  [[3 0 0  3 ]
   [0 3 -8 14]
   [0 0 1  -1]]
■
```

```
*row+(8,[A],3,2)
→[A]
  [[3 0 0 3 ]
   [0 3 0 6 ]
   [0 0 1 -1]]
■
```

```
*row(1/3,[A],1)→
[A]
  [[1 0 0 1 ]
   [0 3 0 6 ]
   [0 0 1 -1]]
■
```

```
*row(1/3,[A],2)→
[A]
  [[1 0 0 1 ]
   [0 1 0 2 ]
   [0 0 1 -1]]
■
```

As in Example 1, we can now read the solution from the right hand column: $x = 1$, $y = 2$, $z = -1$.

Note: The TI-83/84 has a function, `rref`, that gives the reduced row echelon form of a matrix in one step. (See the text for the definition of reduced row echelon form.) Internally it uses a variation of Gauss-Jordan reduction to do this. ■

EXCEL Technology Guide

Section **2.1**

Example **1** Find all solutions (x, y) of the following system of two equations:

$$x + y = 3$$
$$x - y = 1$$

Solution with Technology You can use Excel to draw the graphs of the two equations on the same set of axes, and to check the solution.

1. Solve the equations for y, obtaining $y = -x + 3$ and $y = x - 1$.

2. To graph these lines, you *could* use the Excel graphing template for smooth curves developed in Chapter 1. However, there is no need to plot a large number of points when graphing a straight line; two points suffice. The following worksheet shows all you really need for a plot of the two lines in question:

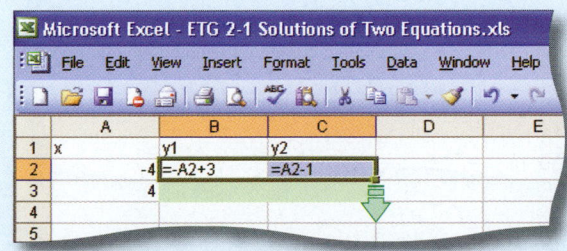

The two values of x give the x-coordinates of the two points we will use as endpoints of the lines. (We have—somewhat arbitrarily—chosen the range $[-4, 4]$ for x.) The formula for the first line, $y = -x + 3$, is in cell B2, and the formula for the second line, $y = x - 1$, is in C2.

3. Copy these two cells as shown to yield the following result:

	A	B	C	D	E
1	x	y1	y2		
2	-4	7	-5		
3	4	-1	3		
4					
5					

Microsoft Excel - ETG 2-1 Solutions of Two Equations.xls
File   Edit   View   Insert   Format   Tools   Data   Window   Help

4. For the graph, select all 9 cells and create a scatter graph with line segments joining the data points. Instruct Excel to insert a chart and select the "XY (Scatter)" option. In the same dialogue box, select the option that shows points connected by lines. Press "Next" to bring up a new dialogue called "Data Type," where you should make sure that the "Series in Columns" option is selected, telling the program that the x- and y-coordinates are arranged vertically, down columns. Your graph should appear as shown in Figure 11.

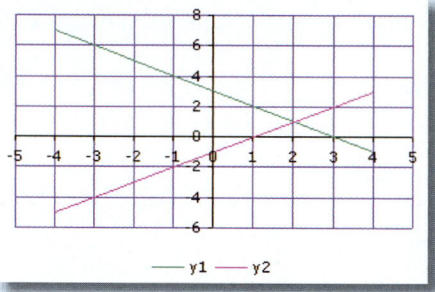

Figure **11**

To zoom in:

1. First decide on a new x range, say, $[1, 3]$.

2. Change the value in cell A2 to 1 and the value in cell A3 to 3, and Excel will automatically update the y-values and the graph.[39]

To check that $(2, 1)$ is the correct solution:

1. Enter the value 2 in cell A4 in your spreadsheet.

2. Copy the formulas in B3 and C3 down to row 4 to obtain the corresponding values of y.

Microsoft Excel - ETG 2-1 Solutions of Two Equations.xls
File   Edit   View   Insert   Format   Tools   Data   Window   Help

	A	B	C	D	E
1	x	y1	y2		
2	-4	7	-5		
3	4	-1	3		
4	2	1	1		
5					
6					

---

[39] Excel sets the y range of the graph automatically, and the range it chooses may not always be satisfactory. It may sometimes be necessary to "narrow down" the x or y range by double-clicking on the axis in question and specifying a new range.

Since the values of *y* agree, we have verified that (2, 1) is a solution of both equations.

Q: *Is it always possible to find the* exact *answer using Excel this way*?

A: No. Consider the system:

$$2x + 3y = 1$$
$$-3x + 2y = 1$$

Its solution is $(x, y) = \left(-\frac{1}{13}, \frac{5}{13}\right)$. (Check it!) If we write this solution in decimal form, we get repeating decimals: $(-0.076923\ 0769230\ldots, 0.384615384615\ldots)$. Since Excel rounds all numbers to 15 significant digits (so that, in the eyes of Excel, 2.000 000 000 000 001 is exactly the same as 2), it can show only an approximate answer. ■

Section **2.2**

Row Operations with Excel To Excel, a row is a block of data, or an **array,** and Excel has built-in the capability to handle arrays in much the same way as it handles single cells. Consider the following example:

$$\begin{bmatrix} 1 & 3 & -4 \\ 0 & 4 & 2 \end{bmatrix} 3R_2 \rightarrow \begin{bmatrix} 1 & 3 & -4 \\ 0 & 12 & 6 \end{bmatrix} \qquad \text{Replace } R_2 \text{ by } 3R_2$$

1. Enter the original matrix in a convenient location, say the cells A1 through C2.

2. To get the first row of the new matrix, which is simply a copy of the first row of the old matrix, decide where you want to place the new matrix, and highlight *the whole row of the new matrix,* say A4:C4.

3. Enter the formula =A1:C1 (the easiest way to do this is to type "=" and then use the mouse to select cells A1 through C1, i.e., the first row of the old matrix).

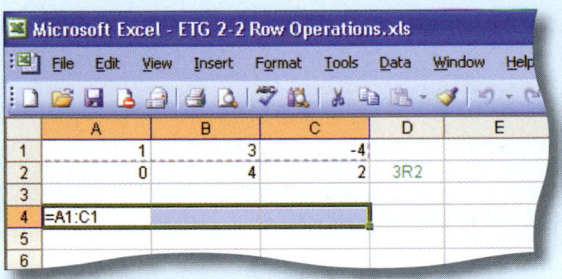

Enter formula for new first row   Press Control-Shift-Enter

4. Press **Control-Shift-Enter** (instead of "Enter" alone) and the whole row will be copied.[40] Pressing Control-Shift-Enter tells Excel that your formula is an *array formula,* one that returns an array rather than a single number. Once entered, Excel will show an array formula enclosed in "curly braces." (Note that you must also use Control-Shift-Enter to delete any array you create: select the block you wish to delete and press Delete followed by Control-Shift-Enter.)

5. Similarly, to get the second row, select cells A5 through C5, where the new second row will go, enter the formula =3*A2:C2, and press Control-Shift-Enter. (Again, the easiest way to enter the formula is to type "=3*" and then select the cells A2 through C2 using the mouse.)

---

[40] Note that on a Macintosh, Command-Enter and Command-Shift-Enter have the same effect as Control-Shift-Enter.

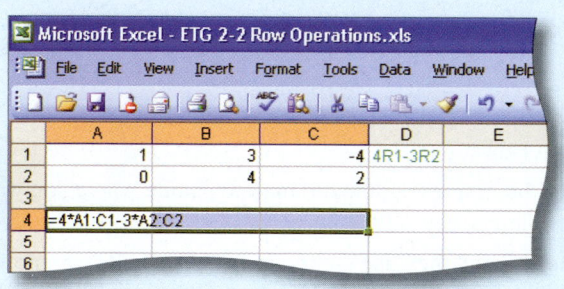

Enter formula for new second row

$\rightarrow$

Press Control-Shift-Enter

We can perform the following operation in a similar way:

$$\begin{bmatrix} 1 & 3 & -4 \\ 0 & 4 & 2 \end{bmatrix} 4R_1 - 3R_2 \rightarrow \begin{bmatrix} 4 & 0 & -22 \\ 0 & 4 & 2 \end{bmatrix}$$

Replace R_1 by $4R_1 - 3R_2$

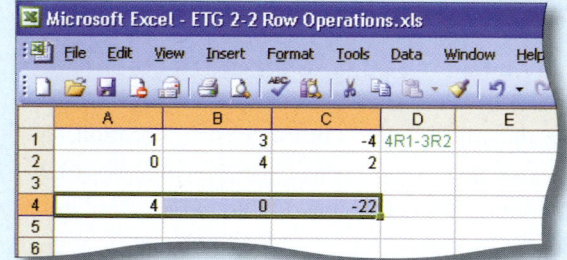

$\rightarrow$

(To easily enter the formula for $4R_1 - 3R_2$, type "=4*," select the first row A1:C1 using the mouse, type "-3*," and then select the second row A2:C2 using the mouse.)

Example 1 Solve the system:

$$\begin{aligned} x - \quad y + 5z &= -6 \\ 3x + 3y - \quad z &= 10 \\ x + 3y + 2z &= 5 \end{aligned}$$

Solution with Technology Here is the complete row reduction as it would appear in Excel.

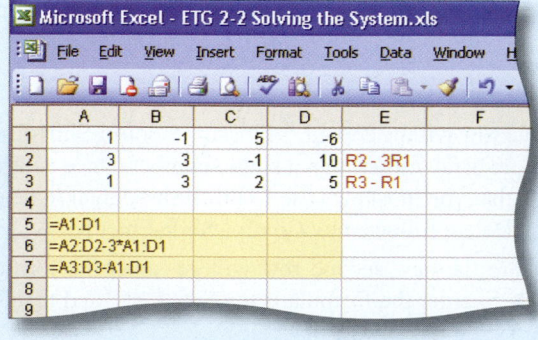

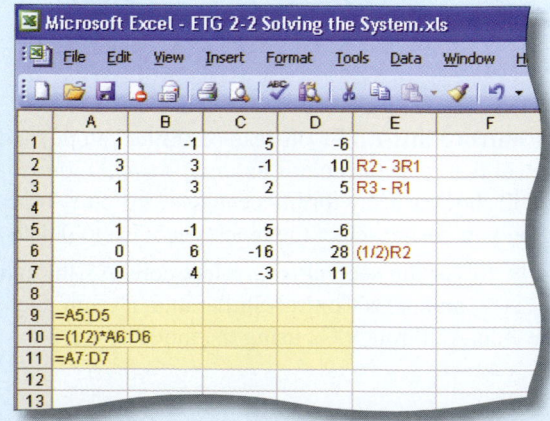

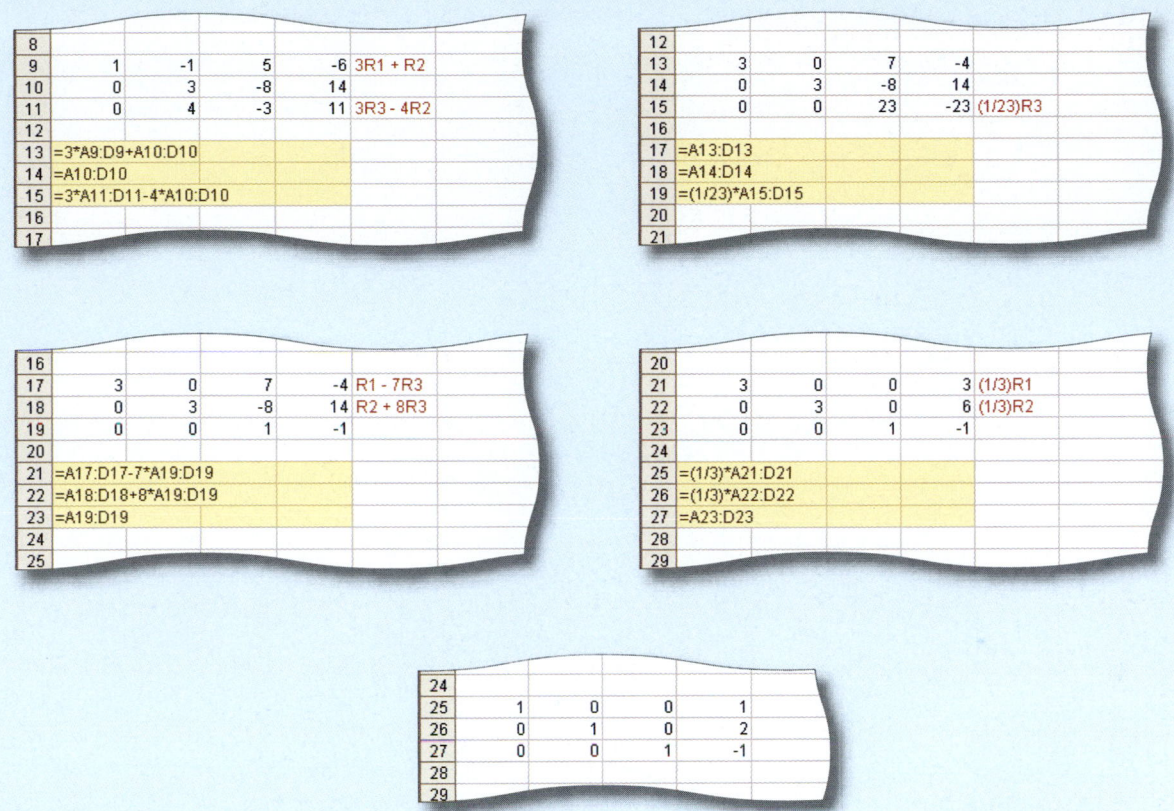

	A	B	C	D	E
8					
9	1	-1	5	-6	3R1 + R2
10	0	3	-8	14	
11	0	4	-3	11	3R3 - 4R2
12					
13	=3*A9:D9+A10:D10				
14	=A10:D10				
15	=3*A11:D11-4*A10:D10				
16					
17					

	A	B	C	D	E
12					
13	3	0	7	-4	
14	0	3	-8	14	
15	0	0	23	-23	(1/23)R3
16					
17	=A13:D13				
18	=A14:D14				
19	=(1/23)*A15:D15				
20					
21					

	A	B	C	D	E
16					
17	3	0	7	-4	R1 - 7R3
18	0	3	-8	14	R2 + 8R3
19	0	0	1	-1	
20					
21	=A17:D17-7*A19:D19				
22	=A18:D18+8*A19:D19				
23	=A19:D19				
24					
25					

	A	B	C	D	E
20					
21	3	0	0	3	(1/3)R1
22	0	3	0	6	(1/3)R2
23	0	0	1	-1	
24					
25	=(1/3)*A21:D21				
26	=(1/3)*A22:D22				
27	=A23:D23				
28					
29					

	A	B	C	D
24				
25	1	0	0	1
26	0	1	0	2
27	0	0	1	-1
28				
29				

As in Example 1, we can now read the solution from the right hand column: $x = 1$, $y = 2$, $z = -1$. What do you notice if you change the entries in cells D1, D2, and D3?

3

Matrix Algebra and Applications

CASE STUDY The Japanese Economy

A senator walks into your cubicle in the Congressional Budget Office. "Look here," she says, "I don't see why the Japanese trade representative is getting so upset with my proposal to cut down on our use of Japanese finance and insurance. He claims that it'll hurt Japan's mining operations. But just look at Japan's input-output table. The finance sector doesn't use any input from the mining sector. How can our cutting down demand for finance and insurance hurt mining?" How should you respond?

Jose Fuste Raga/Zefa/Corbis

Introduction

We used matrices in Chapter 2 simply to organize our work. It is time we examined them as interesting objects in their own right. There is much that we can do with matrices besides row operations: We can add, subtract, multiply, and even, in a sense, "divide" matrices. We use these operations to study game theory and input-output models in this chapter, and Markov chains in a later chapter.

Many calculators, electronic spreadsheets, and other computer programs can do these matrix operations, which is a big help in doing calculations. However, we need to know how these operations are defined to see why they are useful and to understand which to use in any particular application.

3.1 Matrix Addition and Scalar Multiplication

Let's start by formally defining what a matrix is and introducing some basic terms.

Matrix, Dimension, and Entries

An $m \times n$ **matrix** A is a rectangular array of real numbers with m rows and n columns. We refer to m and n as the **dimensions** of the matrix. The numbers that appear in the matrix are called its **entries.** We customarily use capital letters $A, B, C, \ldots$ for the names of matrices.

quick Examples

1. $A = \begin{bmatrix} 2 & 0 & 1 \\ 33 & -22 & 0 \end{bmatrix}$ is a 2 × 3 matrix because it has 2 rows and 3 columns.

2. $B = \begin{bmatrix} 2 & 3 \\ 10 & 44 \\ -1 & 3 \\ 8 & 3 \end{bmatrix}$ is a 4 × 2 matrix because it has 4 rows and 2 columns.

The entries of A are 2, 0, 1, 33, -22, and 0. The entries of B are the numbers 2, 3, 10, 44, -1, 3, 8, and 3.

Hint: Remember that the number of rows is given first and the number of columns second. An easy way to remember this is to think of the acronym "RC" for "Row then Column."

Referring to the Entries of a Matrix

There is a systematic way of referring to particular entries in a matrix. If i and j are numbers, then the entry in the ith row and jth column of the matrix A is called the ij**th entry** of A. We usually write this entry as a_{ij} or A_{ij}. (If the matrix was called B, we would write its ijth entry as b_{ij} or B_{ij}.) Notice that this follows the "RC" convention: The row number is specified first and the column number second.

quick Example

With $A = \begin{bmatrix} 2 & 0 & 1 \\ 33 & -22 & 0 \end{bmatrix}$,

$a_{13} = 1$ First row, third column

$a_{21} = 33$ Second row, first column

using *Technology*

See the Technology Guides at the end of the chapter to see how matrices are entered and used in a TI-83/84 or Excel. For the authors' web-based utility, follow:

Chapter 3
→ Tools
→ Matrix Algebra Tool

There you will find a computational tool that allows you to do matrix algebra. Use the following format to enter the matrix A on the previous page (spaces are optional):

```
A = [2, 0, 1
33, −22, 0]
```

To display the matrix A, type A in the formula box and press "Compute."

According to the labeling convention, the entries of the matrix A above are

$$A = \begin{bmatrix} a_{11} & a_{12} & a_{13} \\ a_{21} & a_{22} & a_{23} \end{bmatrix}$$

In general, the $m \times n$ matrix A has its entries labeled as follows:

$$A = \begin{bmatrix} a_{11} & a_{12} & a_{13} & \cdots & a_{1n} \\ a_{21} & a_{22} & a_{23} & \cdots & a_{2n} \\ \vdots & \vdots & \vdots & \ddots & \vdots \\ a_{m1} & a_{m2} & a_{m3} & \cdots & a_{mn} \end{bmatrix}$$

We say that two matrices A and B are **equal** if they have the same dimensions and the corresponding entries are equal. Note that a 3×4 matrix can never equal a 3×5 matrix because they do not have the same dimensions.

Example 1 Matrix Equality

Let $A = \begin{bmatrix} 7 & 9 & x \\ 0 & -1 & y+1 \end{bmatrix}$ and $B = \begin{bmatrix} 7 & 9 & 0 \\ 0 & -1 & 11 \end{bmatrix}$. Find the values of x and y such that $A = B$.

Solution For the two matrices to be equal, we must have corresponding entries equal, so

$$\begin{aligned} x &= 0 & a_{13} = b_{13} \\ y+1 &= 11 \quad \text{or} \quad y = 10 & a_{23} = b_{23} \end{aligned}$$

+*Before we go on...* Note in Example 1 that the matrix equation

$$\begin{bmatrix} 7 & 9 & x \\ 0 & -1 & y+1 \end{bmatrix} = \begin{bmatrix} 7 & 9 & 0 \\ 0 & -1 & 11 \end{bmatrix}$$

is really six equations in one: $7 = 7, 9 = 9, x = 0, 0 = 0, -1 = -1$. and $y + 1 = 11$. We used only the two that were interesting. ∎

Row Matrix, Column Matrix, and Square Matrix

A matrix with a single row is called a **row matrix,** or **row vector.** A matrix with a single column is called a **column matrix** or **column vector.** A matrix with the same number of rows as columns is called a **square matrix.**

quick Examples

The 1×5 matrix $C = \begin{bmatrix} 3 & -4 & 0 & 1 & -11 \end{bmatrix}$ is a row matrix.

The 4×1 matrix $D = \begin{bmatrix} 2 \\ 10 \\ -1 \\ 8 \end{bmatrix}$ is a column matrix.

The 3×3 matrix $E = \begin{bmatrix} 1 & -2 & 0 \\ 0 & 1 & 4 \\ -4 & 32 & 1 \end{bmatrix}$ is a square matrix.

Matrix Addition and Subtraction

The first matrix operations we discuss are matrix addition and subtraction. The rules for these operations are simple.

Matrix Addition and Subtraction

Two matrices can be added (or subtracted) if and only if they have the same dimensions. To add (or subtract) two matrices of the same dimensions, we add (or subtract) the corresponding entries. More formally, if A and B are $m \times n$ matrices, then $A + B$ and $A - B$ are the $m \times n$ matrices whose entries are given by:

$$(A + B)_{ij} = A_{ij} + B_{ij} \qquad ij\text{th entry of the sum} = \text{sum of the } ij\text{th entries}$$
$$(A - B)_{ij} = A_{ij} - B_{ij} \qquad ij\text{th entry of the difference} = \text{difference of the } ij\text{th entries}$$

Visualizing Matrix Addition

$$\begin{bmatrix} 2 & -3 \\ 1 & 0 \end{bmatrix} + \begin{bmatrix} 1 & 1 \\ -2 & 1 \end{bmatrix} = \begin{bmatrix} 3 & -2 \\ -1 & 1 \end{bmatrix}$$

quick Examples

1. $\begin{bmatrix} 2 & -3 \\ 1 & 0 \\ -1 & 3 \end{bmatrix} + \begin{bmatrix} 9 & -5 \\ 0 & 13 \\ -1 & 3 \end{bmatrix} = \begin{bmatrix} 11 & -8 \\ 1 & 13 \\ -2 & 6 \end{bmatrix}$ Corresponding entries added

2. $\begin{bmatrix} 2 & -3 \\ 1 & 0 \\ -1 & 3 \end{bmatrix} - \begin{bmatrix} 9 & -5 \\ 0 & 13 \\ -1 & 3 \end{bmatrix} = \begin{bmatrix} -7 & 2 \\ 1 & -13 \\ 0 & 0 \end{bmatrix}$ Corresponding entries subtracted

Example 2 Sales

The A-Plus auto parts store chain has two outlets, one in Vancouver and one in Quebec. Among other things, it sells wiper blades, windshield cleaning fluid, and floor mats. The monthly sales of these items at the two stores for two months are given in the following tables:

January Sales

	Vancouver	Quebec
Wiper Blades	20	15
Cleaning Fluid (bottles)	10	12
Floor Mats	8	4

February Sales

	Vancouver	Quebec
Wiper Blades	23	12
Cleaning Fluid (bottles)	8	12
Floor Mats	4	5

using *Technology*

See the Technology Guides at the end of the chapter to see how to add and subtract matrices using a TI-83/84 or Excel. Alternatively, use the Matrix Algebra Tool at

Chapter 3
→ Tools
→ Matrix Algebra Tool

There, first enter the two matrices you wish to add or subtract (subtract, in this case) as shown:

    J = [20, 15, 10
    12, 8, 4]

    F = [23, 12, 8
    12, 4, 5]

To compute their difference, type F-J in the formula box and press "Compute." (You can enter multiple formulas separated by commas in the formula box. For instance, F+J, F-J will compute both the sum and difference.)

Use matrix arithmetic to calculate the change in sales of each product in each store from January to February.

Solution The tables suggest two matrices:

$$J = \begin{bmatrix} 20 & 15 \\ 10 & 12 \\ 8 & 4 \end{bmatrix} \quad \text{and} \quad F = \begin{bmatrix} 23 & 12 \\ 8 & 12 \\ 4 & 5 \end{bmatrix}$$

To compute the change in sales of each product for both stores, we want to subtract corresponding entries in these two matrices. In other words, we want to compute the difference of the two matrices:

$$F - J = \begin{bmatrix} 23 & 12 \\ 8 & 12 \\ 4 & 5 \end{bmatrix} - \begin{bmatrix} 20 & 15 \\ 10 & 12 \\ 8 & 4 \end{bmatrix} = \begin{bmatrix} 3 & -3 \\ -2 & 0 \\ -4 & 1 \end{bmatrix}$$

Thus, the change in sales of each product is the following:

	Vancouver	Quebec
Wiper Blades	3	−3
Cleaning Fluid (bottles)	−2	0
Floor Mats	−4	1

Scalar Multiplication

A matrix A can be added to itself because the expression $A + A$ is the sum of two matrices that have the same dimensions. When we compute $A + A$, we end up doubling every entry in A. So we can think of the expression $2A$ as telling us to *multiply every element in A by* 2.

In general, to multiply a matrix by a number, multiply every entry in the matrix by that number. For example,

$$6 \begin{bmatrix} \frac{5}{2} & -3 \\ 1 & 0 \\ -1 & \frac{5}{6} \end{bmatrix} = \begin{bmatrix} 15 & -18 \\ 6 & 0 \\ -6 & 5 \end{bmatrix}$$

It is traditional when talking about matrices to call individual numbers **scalars.** For this reason, we call the operation of multiplying a matrix by a number **scalar multiplication.**

Example 3 Sales

The revenue generated by sales in the Vancouver and Quebec branches of the A-Plus auto parts store (see Example 2) was as follows:

January Sales in Canadian Dollars

	Vancouver	Quebec
Wiper Blades	140.00	105.00
Cleaning Fluid	30.00	36.00
Floor Mats	96.00	48.00

using *Technology*

See the Technology Guides at the end of the chapter to see how to compute scalar multiples using a TI-83/84 or Excel. Alternatively, go to the online Matrix Algebra Tool at

  Chapter 3

  → Tools

    → Matrix Algebra Tool

There, enter the January sales in U.S. Dollars:

    A = [140, 105
    30, 36
    96, 48]

Then type 0.65*A in the formula box and press "Compute."

If the Canadian dollar was worth \$0.65 U.S. at the time, compute the revenue in U.S. dollars.

Solution We need to multiply each revenue figure by 0.65. Let A be the matrix of revenue figures in Canadian dollars:

$$A = \begin{bmatrix} 140.00 & 105.00 \\ 30.00 & 36.00 \\ 96.00 & 48.00 \end{bmatrix}$$

The revenue figures in U.S. dollars are then given by the scalar multiple

$$0.65A = 0.65 \begin{bmatrix} 140.00 & 105.00 \\ 30.00 & 36.00 \\ 96.00 & 48.00 \end{bmatrix} = \begin{bmatrix} 91.00 & 68.25 \\ 19.50 & 23.40 \\ 62.40 & 31.20 \end{bmatrix}$$

In other words, in U.S. dollars, \$91 worth of wiper blades was sold in Vancouver, \$68.25 worth of wiper blades was sold in Quebec, and so on.

Formally, scalar multiplication is defined as follows:

Scalar Multiplication

If A is an $m \times n$ matrix and c is a real number, then cA is the $m \times n$ matrix obtained by multiplying all the entries of A by c. (We usually use lowercase letters $c, d, e, \ldots$ to denote scalars.) Thus, the ijth entry of cA is given by

$$(cA)_{ij} = c(A_{ij})$$

In words, this rule is: To get the ijth entry of cA, multiply the ijth entry of A by c.

Example 4 Combining Operations

Let $A = \begin{bmatrix} 2 & -1 & 0 \\ 3 & 5 & -3 \end{bmatrix}$, $B = \begin{bmatrix} 1 & 3 & -1 \\ 5 & -6 & 0 \end{bmatrix}$, and $C = \begin{bmatrix} x & y & w \\ z & t+1 & 3 \end{bmatrix}$

Evaluate the following: $4A$, xB, and $A + 3C$.

Solution First, we find $4A$ by multiplying each entry of A by 4:

$$4A = 4 \begin{bmatrix} 2 & -1 & 0 \\ 3 & 5 & -3 \end{bmatrix} = \begin{bmatrix} 8 & -4 & 0 \\ 12 & 20 & -12 \end{bmatrix}$$

Similarly, we find xB by multiplying each entry of B by x:

$$xB = x \begin{bmatrix} 1 & 3 & -1 \\ 5 & -6 & 0 \end{bmatrix} = \begin{bmatrix} x & 3x & -x \\ 5x & -6x & 0 \end{bmatrix}$$

We get $A + 3C$ in two steps as follows:

$$A + 3C = \begin{bmatrix} 2 & -1 & 0 \\ 3 & 5 & -3 \end{bmatrix} + 3 \begin{bmatrix} x & y & w \\ z & t+1 & 3 \end{bmatrix}$$

$$= \begin{bmatrix} 2 & -1 & 0 \\ 3 & 5 & -3 \end{bmatrix} + \begin{bmatrix} 3x & 3y & 3w \\ 3z & 3t+3 & 9 \end{bmatrix}$$

$$= \begin{bmatrix} 2+3x & -1+3y & 3w \\ 3+3z & 3t+8 & 6 \end{bmatrix}$$

Addition and scalar multiplication of matrices have nice properties, reminiscent of the properties of addition and multiplication of real numbers. Before we state them, we need to introduce some more notation.

If A is any matrix, then $-A$ is the matrix $(-1)A$. In other words, $-A$ is A multiplied by the scalar -1. This amounts to changing the signs of all the entries in A. For example,

$$-\begin{bmatrix} 4 & -2 & 0 \\ 6 & 10 & -6 \end{bmatrix} = \begin{bmatrix} -4 & 2 & 0 \\ -6 & -10 & 6 \end{bmatrix}$$

For any two matrices A and B, $A - B$ is the same as $A + (-B)$. (Why?)

Also, a **zero matrix** is a matrix all of whose entries are zero. Thus, for example, the 2×3 zero matrix is

$$O = \begin{bmatrix} 0 & 0 & 0 \\ 0 & 0 & 0 \end{bmatrix}$$

Now we state the most important properties of the operations that we have been talking about:

Properties of Matrix Addition and Scalar Multiplication

If A, B, and C are any $m \times n$ matrices and if O is the zero $m \times n$ matrix, then the following hold:

$A + (B + C) = (A + B) + C$	*Associative law*
$A + B = B + A$	*Commutative law*
$A + O = O + A = A$	*Additive identity law*
$A + (-A) = O = (-A) + A$	*Additive inverse law*
$c(A + B) = cA + cB$	*Distributive law*
$(c + d)A = cA + dA$	*Distributive law*
$1A = A$	*Scalar unit*
$0A = O$	*Scalar zero*

These properties would be obvious if we were talking about addition and multiplication of *numbers,* but here we are talking about addition and multiplication of *matrices.* We are using "+" to mean something new: matrix addition. There is no reason why matrix addition has to obey *all* the properties of addition of numbers. It happens that it does obey many of them, which is why it is convenient to call it *addition* in the first place. This means that we can manipulate equations involving matrices in much the same way that we manipulate equations involving numbers. One word of caution: We haven't yet discussed how to multiply matrices, and it probably isn't what you think. It will turn out that multiplication of matrices does *not* obey all the same properties as multiplication of numbers.

Transposition

We mention one more operation on matrices:

Transposition

If A is an $m \times n$ matrix, then its **transpose** is the $n \times m$ matrix obtained by writing its rows as columns, so that the ith row of the original matrix becomes the ith column of the transpose. We denote the transpose of the matrix A by A^T.

Visualizing Transposition

$$\begin{bmatrix} 2 & -3 \\ 1 & 0 \\ 5 & 1 \end{bmatrix} \qquad \begin{bmatrix} 2 & 1 & 5 \\ -3 & 0 & 1 \end{bmatrix}$$

quick Examples

1. Let $B = \begin{bmatrix} 2 & 3 \\ 10 & 44 \\ -1 & 3 \\ 8 & 3 \end{bmatrix}$. Then $B^T = \begin{bmatrix} 2 & 10 & -1 & 8 \\ 3 & 44 & 3 & 3 \end{bmatrix}$.

$\qquad$ 4 × 2 matrix $\qquad\qquad\qquad$ 2 × 4 matrix

2. $\begin{bmatrix} -1 & 1 & 2 \end{bmatrix}^T = \begin{bmatrix} -1 \\ 1 \\ 2 \end{bmatrix}$

$\quad$ 1 × 3 matrix $\qquad$ 3 × 1 matrix

 using *Technology*

See the Technology Guides at the end of the chapter to see how to transpose a matrix using a TI-83/84 or Excel. Alternatively, go to the online Matrix Algebra Tool at

Chapter 3
→ Tools
$\quad$ → Matrix Algebra Tool

There, first enter the matrix you wish to transpose:

```
A = [2, 0, 1
33, -22, 0]
```

Then type A^T in the formula box and press "Compute."

Properties of Transposition

If A and B are $m \times n$ matrices, then the following hold:

$$(A + B)^T = A^T + B^T$$
$$(cA)^T = c(A^T)$$
$$(A^T)^T = A$$

To see why the laws of transposition are true, let us consider the first one: $(A + B)^T = A^T + B^T$. The left-hand side is the transpose of $A + B$, and so is obtained by first adding A and B, and then writing the rows as columns. This is the same as first writing the rows of A and B individually as columns before adding, which gives the right-hand side. Similar arguments can be used to establish the other laws of transposition.

3.1 EXERCISES

● denotes basic skills exercises

tech Ex indicates exercises that should be solved using technology

In each of Exercises 1–10, find the dimensions of the given matrix and identify the given entry.

1. ● $A = [1 \quad 5 \quad 0 \quad \frac{1}{4}]; a_{13}$

2. ● $B = [44 \quad 55]; b_{12}$

3. ● $C = \begin{bmatrix} \frac{5}{2} \\ 1 \\ -2 \\ 8 \end{bmatrix}; C_{11}$

4. ● $D = \begin{bmatrix} 15 & -18 \\ 6 & 0 \\ -6 & 5 \\ 48 & 18 \end{bmatrix}; d_{31}$

5. ● $E = \begin{bmatrix} e_{11} & e_{12} & e_{13} & \cdots & e_{1q} \\ e_{21} & e_{22} & e_{23} & \cdots & e_{2q} \\ \vdots & \vdots & \vdots & \ddots & \vdots \\ e_{p1} & e_{p2} & e_{p3} & \cdots & e_{pq} \end{bmatrix}; E_{22}$

6. ● $A = \begin{bmatrix} 2 & -1 & 0 \\ 3 & 5 & -3 \end{bmatrix}; A_{21}$

7. ● $B = \begin{bmatrix} 1 & 3 \\ 5 & -6 \end{bmatrix}; b_{12}$

8. ● $C = \begin{bmatrix} x & y & w & e \\ z & t+1 & 3 & 0 \end{bmatrix}; C_{23}$

9. ● $D = [d_1 \quad d_2 \quad \cdots \quad d_n]; D_{1r}$ (any r)

10. ● $E = [d \quad d \quad d \quad d]; E_{1r}$ (any r)

11. ● Solve for $x, y, z,$ and w. *hint* [see Example 1]

$$\begin{bmatrix} x+y & x+z \\ y+z & w \end{bmatrix} = \begin{bmatrix} 3 & 4 \\ 5 & 4 \end{bmatrix}$$

12. ● Solve for $x, y, z,$ and w.

$$\begin{bmatrix} x-y & x-z \\ y-w & w \end{bmatrix} = \begin{bmatrix} 0 & 0 \\ 0 & 6 \end{bmatrix}$$

In Exercises 13–20, let $A = \begin{bmatrix} 0 & -1 \\ 1 & 0 \\ -1 & 2 \end{bmatrix}, B = \begin{bmatrix} 0.25 & -1 \\ 0 & 0.5 \\ -1 & 3 \end{bmatrix},$

and $C = \begin{bmatrix} 1 & -1 \\ 1 & 1 \\ -1 & -1 \end{bmatrix}.$ *Evaluate:*

hint [see Quick Examples on pp. 176 and 180, and Example 4]

13. ● $A + B$ 14. ● $A - C$ 15. ● $A + B - C$

16. ● $12B$ 17. ● $2A - C$ 18. ● $2A + 0.5C$

19. ● $2A^T$ 20. ● $A^T + 3C^T$

In Exercises 21–28, let $A = \begin{bmatrix} 1 & -1 & 0 \\ 0 & 2 & -1 \end{bmatrix},$

$B = \begin{bmatrix} 3 & 0 & -1 \\ 5 & -1 & 1 \end{bmatrix},$ *and* $C = \begin{bmatrix} x & 1 & w \\ z & r & 4 \end{bmatrix}.$ *Evaluate:*

21. ● $A + B$ 22. ● $B - C$ 23. ● $A - B + C$

24. ● $\frac{1}{2}B$ 25. ● $2A - B$ 26. ● $2A - 4C$

27. ● $3B^T$ 28. ● $2A^T - C^T$

tech Ex *Use technology in Exercises 29–36.*

Let $A = \begin{bmatrix} 1.5 & -2.35 & 5.6 \\ 44.2 & 0 & 12.2 \end{bmatrix}, B = \begin{bmatrix} 1.4 & 7.8 \\ 5.4 & 0 \\ 5.6 & 6.6 \end{bmatrix},$

$C = \begin{bmatrix} 10 & 20 & 30 \\ -10 & -20 & -30 \end{bmatrix}$ *Evaluate:*

29. ● $A - C$ 30. ● $C - A$ 31. ● $1.1B$

32. ● $-0.2B$ 33. ● $A^T + 4.2B$ 34. ● $(A + 2.3C)^T$

35. ● $(2.1A - 2.3C)^T$ 36. ● $(A - C)^T - B$

Applications

37. ● *Cost of Real Estate* The following table shows the cost of one square foot of residential real estate, in dollars per square foot, at the start of 2001, and the changes in 2002 and 2003.[1]

	New York	London	Hong Kong
2001	600	620	400
Change in 2002	120	60	−50
Change in 2003	40	120	−50

Use matrix algebra to find the cost of residential real estate in each city in **a.** 2002 and **b.** 2003. *hint* [see Example 2]

38. ● *Army Personnel* The following table shows the number of personnel, in thousands, in three branches of the U.S. Army in 2001, and the changes in 2002 and 2003.[2]

	Active Duty	Reserve	National Guard
2001	75	35	60
Change in 2002	5	−15	2
Change in 2003	−12	5	−17

Use matrix algebra to find to find the number of personnel in each branch in **a.** 2002 and **b.** 2003.

39. ● *Inventory* The Left Coast Bookstore chain has two stores, one in San Francisco and one in Los Angeles. It stocks three kinds of book: hardcover, softcover, and plastic (for infants).

---

[1] Figures are rounded. SOURCE: CLSA, Honk Kong Monetary Authority/ *New York Times,* August 15, 2003, p. C1.

[2] Figures are rounded. SOURCE: Army Recruiting Command, Army National Guard Bureau/*New York Times,* September 22, 2003, p. A14.

● basic skills   tech Ex technology exercise

At the beginning of January, the central computer showed the following books in stock:

	Hard	**Soft**	**Plastic**
San Francisco	1000	2000	5000
Los Angeles	1000	5000	2000

Suppose its sales in January were as follows: 700 hardcover books, 1300 softcover books, and 2000 plastic books sold in San Francisco, and 400 hardcover, 300 softcover, and 500 plastic books sold in Los Angeles. Write these sales figures in the form of a matrix, and then show how matrix algebra can be used to compute the inventory remaining in each store at the end of January.

40. ● *Inventory* The Left Coast Bookstore chain discussed in Exercise 39 actually maintained the same sales figures for the first 6 months of the year. Each month, the chain restocked the stores from its warehouse by shipping 600 hardcover, 1500 softcover, and 1500 plastic books to San Francisco and 500 hardcover, 500 softcover, and 500 plastic books to Los Angeles.

a. Use matrix operations to determine the total sales over the 6 months, broken down by store and type of book.

b. Use matrix operations to determine the inventory in each store at the end of June.

41. ● *Profit* Annual revenues and production costs at Luddington's Wellington Boots & Co. are shown in the following spreadsheet.

Use matrix algebra to compute the profits from each sector each year.

42. ● *Revenue* The following spreadsheet gives annual production costs and profits at Gauss Jordan Sneakers, Inc.

Use matrix algebra to compute the revenues from each sector each year.

43. ● *Population Distribution* In 1980 the U.S. population, broken down by regions, was 49.1 million in the Northeast, 58.9 million in the Midwest, 75.4 million in the South, and 43.2 million in the West.[3] In 1990 the population was 50.8 million in the Northeast, 59.7 million in the Midwest, 85.4 million in the South, and 52.8 million in the West. Set up the population figures for each year as a row vector, and then show how to use matrix operations to find the net increase or decrease of population in each region from 1980 to 1990.

44. ● *Population Distribution* In 1990 the U.S. population, broken down by regions, was 50.8 million in the Northeast, 59.7 million in the Midwest, 85.4 million in the South, and 52.8 million in the West.[4] Between 1990 and 2000, the population in the Northeast grew by 2.8 million, the population in the Midwest grew by 4.7 million, the population in the South grew by 14.8 million, and the population in the West grew by 10.4 million. Set up the population figures for 1990 and the growth figures for the decade as row vectors. Assuming that the population will grow by the same numbers from 2000 to 2010 as they did from 1990 to 2000, show how to use matrix operations to find the population in each region in 2010.

Bankruptcy Filings Exercises 45–48 are based on the following chart, which shows the numbers of personal bankruptcy filings in three New York/New Jersey regions during various months of 2001–2002.[5]

	Jan 01	**Apr**	**Jul**	**Oct**	**Jan 02**
Manhattan	150	250	150	100	150
Brooklyn	300	400	300	200	250
Newark	250	400	250	200	200

45. ● Use matrix algebra to determine the total number of bankruptcy filings in each of the given months.

46. ● Use matrix algebra to determine the total number of bankruptcy filings in each of the given regions.

47. Use matrix algebra to determine in which month the difference between the number of bankruptcy filings in Brooklyn and in Newark was greatest.

48. Use matrix algebra to determine in which region the difference between the bankruptcy filings in April 2001 and January 2001 was greatest.

49. *Inventory* Microbucks Computer Company makes two computers, the Pomegranate II and the Pomegranate Classic, at

---

[3] Source: U.S. Census Bureau. Statistical Abstract of the United States: 2001. http://www.census.gov.

[4] Ibid.

[5] Data are approximate four-week moving averages. Source: Lundquist Consulting/*New York Times,* February 10, 2002, p. L1.

● basic skills　tech Ex technology exercise

two different factories. The Pom II requires 2 processor chips, 16 memory chips, and 20 vacuum tubes, while the Pom Classic requires 1 processor chip, 4 memory chips, and 40 vacuum tubes. Microbucks has in stock at the beginning of the year 500 processor chips, 5000 memory chips, and 10,000 vacuum tubes at the Pom II factory, and 200 processor chips, 2000 memory chips, and 20,000 vacuum tubes at the Pom Classic factory. It manufactures 50 Pom II's and 50 Pom Classics each month.

a. Find the company's inventory of parts after two months, using matrix operations.

b. When (if ever) will the company run out of one of the parts?

50. *Inventory* Microbucks Computer Company, besides having the stock mentioned in Exercise 49, gets shipments of parts every month in the amounts of 100 processor chips, 1000 memory chips, and 3000 vacuum tubes at the Pom II factory, and 50 processor chips, 1000 memory chips, and 2000 vacuum tubes at the Pom Classic factory.

a. What will the company's inventory of parts be after six months?

b. When (if ever) will the company run out of one of the parts?

51. *Tourism* The following table gives the number of people (in thousands) who visited Australia and South Africa in 1998:[6]

	To	Australia	South Africa
From	**North America**	440	190
	Europe	950	950
	Asia	1790	200

You predict that in 2008, 20,000 fewer people from North America will visit Australia and 40,000 more will visit South Africa, 50,000 more people from Europe will visit each of Australia and South Africa, and 100,000 more people from Asia will visit South Africa, but there will be no change in the number visiting Australia.

a. Use matrix algebra to predict the number of visitors from the three regions to Australia and South Africa in 2008.

b. Take A to be the 3×2 matrix whose entries are the 1998 tourism figures and take B to be the 3×2 matrix whose entries are the 2008 tourism figures. Give a formula (in terms of A and B) that predicts the average of the numbers of visitors from the three regions to Australia and South Africa in 1998 and 2008. Compute its value.

---

[6] Figures are rounded to the nearest 10,000. Sources: South African Dept. of Environmental Affairs and Tourism; Australia Tourist Commission/ *The New York Times*, January 15, 2000, p. C1.

52. *Tourism* Referring to the 1998 tourism figures given in the preceding exercise, assume that the following (fictitious) figures represent the corresponding numbers from 1988.

	To	Australia	South Africa
From	**North America**	500	100
	Europe	900	800
	Asia	1400	50

Take A to be the 3×2 matrix whose entries are the 1998 tourism figures and take B to be the 3×2 matrix whose entries are the 1988 tourism figures.

a. Compute the matrix $A - B$. What does this matrix represent?

b. Assuming that the changes in tourism over 1988–1998 are repeated in 1998–2008, give a formula (in terms of A and B) that predicts the number of visitors from the three regions to Australia and South Africa in 2008.

Communication and Reasoning Exercises

53. ● What does it mean when we say that $(A + B)_{ij} = A_{ij} + B_{ij}$?

54. ● What does it mean when we say that $(cA)_{ij} = c(A_{ij})$?

55. ● What would a 5×5 matrix A look like if $A_{ii} = 0$ for every i?

56. ● What would a matrix A look like if $A_{ij} = 0$ whenever $i \neq j$?

57. Give a formula for the ijth entry of the transpose of a matrix A.

58. A matrix is **symmetric** if it is equal to its transpose. Give an example of **a.** a nonzero 2×2 symmetric matrix and **b.** a nonzero 3×3 symmetric matrix.

59. A matrix is **skew-symmetric** or **antisymmetric** if it is equal to the negative of its transpose. Give an example of **a.** a nonzero 2×2 skew-symmetric matrix and **b.** a nonzero 3×3 skew-symmetric matrix.

60. Referring to Exercises 58 and 59, what can be said about a matrix that is both symmetric and skew-symmetric?

61. Why is matrix addition associative?

62. Describe a scenario (possibly based on one of the preceding examples or exercises) in which you might wish to compute $A - 2B$ for certain matrices A and B.

63. Describe a scenario (possibly based on one of the preceding examples or exercises) in which you might wish to compute $A + B - C$ for certain matrices A, B and C.

3.2 | Matrix Multiplication

Suppose we buy 2 CDs at $3 each and 4 Zip disks at $5 each. We calculate our total cost by computing the products' price × quantity and adding:

$$\text{Cost} = 3 \times 2 + 5 \times 4 = \$26$$

Let us instead put the prices in a row vector

$$P = [3 \quad 5] \qquad \text{The price matrix}$$

and the quantities purchased in a column vector,

$$Q = \begin{bmatrix} 2 \\ 4 \end{bmatrix} \qquad \text{The quantity matrix}$$

Q: *Why a row and a column?*

A: It's rather a long story, but mathematicians found that it works best this way . . .

Because P represents the prices of the items we are purchasing and Q represents the quantities, it would be useful if the product PQ represented the total cost, a *single number* (which we can think of as a 1×1 matrix). For this to work, PQ should be calculated the same way we calculated the total cost:

$$PQ = [3 \quad 5]\begin{bmatrix} 2 \\ 4 \end{bmatrix} = [3 \times 2 + 5 \times 4] = [26]$$

Notice that we obtain the answer by multiplying each entry in P (going from left to right) by the corresponding entry in Q (going from top to bottom) and then adding the results. ∎

The Product *Row* × *Column*

The **product** AB of a row matrix A and a column matrix B is a 1×1 matrix. The length of the row in A must match the length of the column in B for the product to be defined. To find the product, multiply each entry in A (going from left to right) by the corresponding entry in B (going from top to bottom) and then add the results.

Visualizing

$$\begin{bmatrix} 2 \\ 10 \\ -1 \end{bmatrix} \qquad \begin{matrix} 2 \times 2 & = & 4 \\ 4 \times 10 & = & 40 \\ 1 \times (-1) = & & \underline{-1} \\ & & 43 \end{matrix}$$

Product of first entries = 4
Product of second entries = 40
Product of third entries = −1
Sum of products = 43

$[2 \quad 4 \quad 1]$

quick Examples

1. $[2 \quad 1]\begin{bmatrix} -3 \\ 1 \end{bmatrix} = [2 \times (-3) + 1 \times 1] = [-6 + 1] = [-5]$

2. $[2 \quad 4 \quad 1]\begin{bmatrix} 2 \\ 10 \\ -1 \end{bmatrix} = [2 \times 2 + 4 \times 10 + 1 \times (-1)] = [4 + 40 + (-1)] = [43]$

Notes

1. In the discussion so far, *the row is on the left and the column is on the right* (RC again). (Later we will consider products where the column matrix is on the left and the row matrix is on the right.)

2. The row size has to match the column size. This means that, if we have a 1×3 row on the left, then the column on the right must be 3×1 in order for the product to make sense. For example, the product

$$[a \quad b \quad c]\begin{bmatrix} x \\ y \end{bmatrix}$$

is not defined. ■

Example 1 Revenue

The A-Plus auto parts store mentioned in examples in the previous section had the following sales in its Vancouver store:

	Vancouver
Wiper Blades	20
Cleaning Fluid (bottles)	10
Floor Mats	8

The store sells wiper blades for $7.00 each, cleaning fluid for $3.00 per bottle, and floor mats for $12.00 each. Use matrix multiplication to find the total revenue generated by sales of these items.

Solution We need to multiply each sales figure by the corresponding price and then add the resulting revenue figures. We represent the sales by a column vector, as suggested by the table.

$$Q = \begin{bmatrix} 20 \\ 10 \\ 8 \end{bmatrix}$$

We put the selling prices in a row vector.

$$P = [7.00 \quad 3.00 \quad 12.00]$$

We can now compute the total revenue as the product

$$R = PQ = [7.00 \quad 3.00 \quad 12.00]\begin{bmatrix} 20 \\ 10 \\ 8 \end{bmatrix}$$

$$= [140.00 + 30.00 + 96.00] = [266.00]$$

So, the sale of these items generated a total revenue of $266.00.

Note We could also have written the quantity sold as a row vector (which would be Q^T) and the prices as a column vector (which would be P^T) and then multiplied them in the opposite order ($Q^T P^T$). Try this. ■

using *Technology*

See the Technology Guides at the end of the chapter to see how to multiply matrices using a TI-83/84 or Excel. To use the online Matrix Algebra Tool at

Chapter 3
→ Tools
→ Matrix Algebra Tool

enter the matrices *P* and *Q* as follows:

```
P = [7, 3, 12]
Q = [20
10
8]
```

Then type P*Q in the formula box and press "Compute."

Example 2 Relationship with Linear Equations

a. Represent the matrix equation

$$[2 \quad -4 \quad 1] \begin{bmatrix} x \\ y \\ z \end{bmatrix} = [5]$$

as an ordinary equation.

b. Represent the linear equation $3x + y - z + 2w = 8$ as a matrix equation.

Solution

a. If we perform the multiplication on the left, we get the 1×1 matrix $[2x - 4y + z]$. Thus the equation may be rewritten as

$$[2x - 4y + z] = [5]$$

Saying that these two 1×1 matrices are equal means that their entries are equal, so we get the equation

$$2x - 4y + z = 5$$

b. This is the reverse of part (a):

$$[3 \quad 1 \quad -1 \quad 2] \begin{bmatrix} x \\ y \\ z \\ w \end{bmatrix} = [8]$$

+*Before we go on...* The row matrix $[3 \quad 1 \quad -1 \quad 2]$ in Example 2 is the row of **coefficients** of the original equation (see Section 2.1). ∎

Now to the general case of matrix multiplication:

The Product of Two Matrices: General Case

In general for matrices A and B, we can take the product AB only if the number of columns of A equals the number of rows of B (so that we can multiply the rows of A by the columns of B as above). The product AB is then obtained by taking its ijth entry to be:

ijth entry of $AB = $ Row i of $A \times$ Column j of B As defined above

quick Examples (R stands for row; C stands for column)

$$
\begin{array}{c}
\qquad \quad C_1 \quad C_2 \quad C_3 \\
\qquad \quad \downarrow \quad\;\; \downarrow \quad\;\; \downarrow \\
1.\ R_1 \to [2 \quad 0 \quad -1 \quad 3] \begin{bmatrix} 1 & 1 & -8 \\ 1 & -6 & 0 \\ 0 & 5 & 2 \\ -3 & 8 & 1 \end{bmatrix} = [R_1 \times C_1 \quad R_1 \times C_2 \quad R_1 \times C_3] \\
= [-7 \quad 21 \quad -15]
\end{array}
$$

$$
\begin{array}{c}
\qquad\qquad\quad C_1 \quad C_2 \\
\qquad\qquad\quad \downarrow \quad\;\; \downarrow \\
2.\ \begin{matrix} R_1 \to \\ R_2 \to \end{matrix} \begin{bmatrix} 1 & -1 \\ 0 & 2 \end{bmatrix} \begin{bmatrix} 3 & 0 \\ 5 & -1 \end{bmatrix} = \begin{bmatrix} R_1 \times C_1 & R_1 \times C_2 \\ R_2 \times C_1 & R_2 \times C_2 \end{bmatrix} = \begin{bmatrix} -2 & 1 \\ 10 & -2 \end{bmatrix}
\end{array}
$$

In matrix multiplication we always take

Rows on the left $\times$ Columns on the right

Look at the dimensions in the two Quick Examples above.

Match

$(1 \times 4)(4 \times 3) \to 1 \times 3$ $(2 \times 2)(2 \times 2) \to 2 \times 2$

The fact that the number of columns in the left-hand matrix equals the number of rows in the right-hand matrix amounts to saying that the middle two numbers must match as above. If we "cancel" the middle matching numbers, we are left with the dimensions of the product.

Before continuing with examples, we state the rule for matrix multiplication formally.

Multiplication of Matrices: Formal Definition

If A is an $m \times n$ matrix and B is an $n \times k$ matrix, then the product AB is the $m \times k$ matrix whose ijth entry is the product

Row i of A $\times$ Column j of B

$$(AB)_{ij} = [a_{i1} \quad a_{i2} \quad a_{i3} \dots a_{in}] \begin{bmatrix} b_{1j} \\ b_{2j} \\ b_{3j} \\ \vdots \\ b_{nj} \end{bmatrix} = a_{i1}b_{1j} + a_{i2}b_{2j} + a_{i3}b_{3j} + \cdots + a_{in}b_{nj}$$

Example 3 Matrix Product

Calculate:

a. $\begin{bmatrix} 2 & 0 & -1 & 3 \\ 1 & -1 & 2 & -2 \end{bmatrix} \begin{bmatrix} 1 & 1 & -8 \\ 1 & 0 & 0 \\ 0 & 5 & 2 \\ -2 & 8 & -1 \end{bmatrix}$ **b.** $\begin{bmatrix} -3 \\ 1 \end{bmatrix} [2 \quad 1]$

Solution

a. Before we start the calculation, we check that the dimensions of the matrices match up.

Match

2×4 4×3

$$\begin{bmatrix} 2 & 0 & -1 & 3 \\ 1 & -1 & 2 & -2 \end{bmatrix} \begin{bmatrix} 1 & 1 & -8 \\ 1 & 0 & 0 \\ 0 & 5 & 2 \\ -2 & 8 & -1 \end{bmatrix}$$

The product of the two matrices is defined, and the product will be a 2×3 matrix (we remove the matching 4s: $(2 \times 4)(4 \times 3) \to 2 \times 3$). To calculate the product, we

using *Technology*

See the Technology Guides at the end of the chapter for more on multiplying matrices using a TI-83/84 or Excel. To use the online Matrix Algebra Tool at

Chapter 3
→ Tools
→ Matrix Algebra Tool

enter the matrices A and B as follows:

```
A = [2, 0, -1, 3
1, -1, 2, -2]

B = [1, 1, -8
1, 0, 0
0, 5, 2
-2, 8, -1]
```

Then type A*B in the formula box and press "Compute." If you try to multiply two matrices whose product is not defined, you will get an error alert box telling you that.

follow the prescription above:

$$
\begin{array}{c} \begin{array}{ccc} C_1 & C_2 & C_3 \\ \downarrow & \downarrow & \downarrow \end{array} \end{array}
$$

$$
\begin{array}{c} R_1 \to \\ R_2 \to \end{array}
\begin{bmatrix} 2 & 0 & -1 & 3 \\ 1 & -1 & 2 & -2 \end{bmatrix}
\begin{bmatrix} 1 & 1 & -8 \\ 1 & 0 & 0 \\ 0 & 5 & 2 \\ -2 & 8 & -1 \end{bmatrix}
=
\begin{bmatrix} R_1 \times C_1 & R_1 \times C_2 & R_1 \times C_3 \\ R_2 \times C_1 & R_2 \times C_2 & R_2 \times C_3 \end{bmatrix}
$$

$$
=
\begin{bmatrix} -4 & 21 & -21 \\ 4 & -5 & -2 \end{bmatrix}
$$

b. The dimensions of the two matrices given are 2×1 and 1×2. Because the 1s match, the product is defined, and the result will be a 2×2 matrix.

$$
\begin{array}{c} R_1 \to \\ R_2 \to \end{array}
\begin{bmatrix} -3 \\ 1 \end{bmatrix}
\begin{array}{c} \begin{array}{cc} C_1 & C_2 \\ \downarrow & \downarrow \end{array} \\ [2 \quad 1] \end{array}
=
\begin{bmatrix} R_1 \times C_1 & R_1 \times C_2 \\ R_2 \times C_1 & R_2 \times C_2 \end{bmatrix}
=
\begin{bmatrix} -6 & -3 \\ 2 & 1 \end{bmatrix}
$$

Note In part (a) we *cannot* multiply the matrices in the opposite order—the dimensions do not match. We say simply that the product in the opposite order is **not defined.** In part (b) we *can* multiply the matrices in the opposite order, but we would get a 1×1 matrix if we did so. Thus, order is important when multiplying matrices. In general, if AB is defined, then BA need not even be defined. If BA is also defined, it may not have the same dimensions as AB. And even if AB and BA have the same dimensions, they may have different entries (see the next example). ∎

Example 4 *AB* versus *BA*

Let $A = \begin{bmatrix} 1 & -1 \\ 0 & 2 \end{bmatrix}$ and $B = \begin{bmatrix} 3 & 0 \\ 5 & -1 \end{bmatrix}$. Find AB and BA.

Solution Note first that A and B are both 2×2 matrices, so the products AB and BA are both defined and are both 2×2 matrices—unlike the case in Example 3(b). We first calculate AB:

$$
AB = \begin{bmatrix} 1 & -1 \\ 0 & 2 \end{bmatrix}\begin{bmatrix} 3 & 0 \\ 5 & -1 \end{bmatrix} = \begin{bmatrix} -2 & 1 \\ 10 & -2 \end{bmatrix}
$$

Now let's calculate BA.

$$
BA = \begin{bmatrix} 3 & 0 \\ 5 & -1 \end{bmatrix}\begin{bmatrix} 1 & -1 \\ 0 & 2 \end{bmatrix} = \begin{bmatrix} 3 & -3 \\ 5 & -7 \end{bmatrix}
$$

Notice that BA has no resemblance to AB! Thus, we have discovered that, even for square matrices:

Matrix multiplication is not commutative.

In other words, $AB \neq BA$ in general, even when AB and BA both exist and have the same dimensions. (There are instances when $AB = BA$ for particular matrices A and B, but this is an exception, not the rule.)

Example 5 Revenue

January sales at the A-Plus auto parts stores in Vancouver and Quebec are given in the following table.

	Vancouver	Quebec
Wiper Blades	20	15
Cleaning Fluid (bottles)	10	12
Floor Mats	8	4

The usual selling prices for these items are $7.00 each for wiper blades, $3.00 per bottle for cleaning fluid, and $12.00 each for floor mats. The discount prices for A-Plus Club members are $6.00 each for wiper blades, $2.00 per bottle for cleaning fluid, and $10.00 each for floor mats. Use matrix multiplication to compute the total revenue at each store, assuming first that all items were sold at the usual prices, and then that they were all sold at the discount prices.

Solution We can do all of the requested calculations at once with a single matrix multiplication. Consider the following two labeled matrices.

$$Q = \begin{matrix} & \mathbf{V} & \mathbf{Q} \\ \mathbf{Wb} \\ \mathbf{Cf} \\ \mathbf{Fm} \end{matrix} \begin{bmatrix} 20 & 15 \\ 10 & 12 \\ 8 & 4 \end{bmatrix}$$

$$P = \begin{matrix} & \mathbf{Wb} & \mathbf{Cf} & \mathbf{Fm} \\ \mathbf{Usual} \\ \mathbf{Discount} \end{matrix} \begin{bmatrix} 7.00 & 3.00 & 12.00 \\ 6.00 & 2.00 & 10.00 \end{bmatrix}$$

The first matrix records the quantities sold, while the second records the sales prices under the two assumptions. To compute the revenue at both stores under the two different assumptions, we calculate $R = PQ$.

$$R = PQ = \begin{bmatrix} 7.00 & 3.00 & 12.00 \\ 6.00 & 2.00 & 10.00 \end{bmatrix} \begin{bmatrix} 20 & 15 \\ 10 & 12 \\ 8 & 4 \end{bmatrix}$$

$$= \begin{bmatrix} 266.00 & 189.00 \\ 220.00 & 154.00 \end{bmatrix}$$

We can label this matrix as follows.

$$R = \begin{matrix} & \mathbf{V} & \mathbf{Q} \\ \mathbf{Usual} \\ \mathbf{Discount} \end{matrix} \begin{bmatrix} 266.00 & 189.00 \\ 220.00 & 154.00 \end{bmatrix}$$

In other words, if the items were sold at the usual price, then Vancouver had a revenue of $266 while Quebec had a revenue of $189, and so on.

+ Before we go on... In Example 5 we were able to calculate PQ because the dimensions matched correctly: $(2 \times 3)(3 \times 2) \rightarrow 2 \times 2$. We could also have multiplied them in the opposite order and gotten a 3×3 matrix. Would the product QP be meaningful? In an application like this, not only do the dimensions have to match, but also the *labels* have to match for the result to be meaningful. The labels on the three columns of P are the parts that were sold, and these are also the labels on the three rows of Q. Therefore, we can "cancel labels" at the same time that we cancel the dimensions in the product. However, the labels on the two columns of Q do not match the labels on the two rows of P, and there is no useful interpretation of the product QP in this situation. ∎

There are very special square matrices of every size: $1 \times 1, 2 \times 2, 3 \times 3$, and so on, called the **identity** matrices.

Identity Matrix

The $n \times n$ identity matrix I is the matrix with 1s down the **main diagonal** (the diagonal starting at the top left) and 0s everywhere else. In symbols,

$$I_{ii} = 1, \quad \text{and}$$
$$I_{ij} = 0 \quad \text{if } i \neq j$$

quick Examples

1. 1×1 identity matrix $I = [1]$

2. 2×2 identity matrix $I = \begin{bmatrix} 1 & 0 \\ 0 & 1 \end{bmatrix}$

3. 3×3 identity matrix $I = \begin{bmatrix} 1 & 0 & 0 \\ 0 & 1 & 0 \\ 0 & 0 & 1 \end{bmatrix}$

4. 4×4 identity matrix $I = \begin{bmatrix} 1 & 0 & 0 & 0 \\ 0 & 1 & 0 & 0 \\ 0 & 0 & 1 & 0 \\ 0 & 0 & 0 & 1 \end{bmatrix}$

Note Identity matrices are always square matrices, meaning that they have the same number of rows as columns. There is no such thing, for example, as the "2×4 identity matrix."

The next example shows why I is interesting. ∎

Example 6 Identity Matrix

Evaluate the products AI and IA, where $A = \begin{bmatrix} a & b & c \\ d & e & f \\ g & h & i \end{bmatrix}$ and I is the 3×3 identity matrix.

Solution

First notice that A is arbitrary; it could be any 3×3 matrix.

$$AI = \begin{bmatrix} a & b & c \\ d & e & f \\ g & h & i \end{bmatrix} \begin{bmatrix} 1 & 0 & 0 \\ 0 & 1 & 0 \\ 0 & 0 & 1 \end{bmatrix} = \begin{bmatrix} a & b & c \\ d & e & f \\ g & h & i \end{bmatrix}$$

using *Technology*

See the Technology Guides at the end of the chapter to see how to get an identity matrix using a TI-83/84 or Excel. In the online Matrix Algebra Tool at

   Chapter 3
   → Tools
      → Matrix Algebra Tool

use I in a formula to refer to the identity matrix of any dimension. The program will choose the correct dimension in the context of the formula. For example, if *A* is a 3 x 3 matrix, then the expression I-A uses the 3 x 3 identity matrix for I.

and

$$IA = \begin{bmatrix} 1 & 0 & 0 \\ 0 & 1 & 0 \\ 0 & 0 & 1 \end{bmatrix} \begin{bmatrix} a & b & c \\ d & e & f \\ g & h & i \end{bmatrix} = \begin{bmatrix} a & b & c \\ d & e & f \\ g & h & i \end{bmatrix}$$

In both cases, the answer is the matrix *A* we started with. In symbols,

$$AI = A$$

and

$$IA = A$$

no matter which 3×3 matrix *A* you start with. Now this should remind you of a familiar fact from arithmetic:

$$a \cdot 1 = a$$

and

$$1 \cdot a = a$$

That is why we call the matrix *I* the 3×3 *identity* matrix, because it appears to play the same role for 3×3 matrices that the identity 1 does for numbers.

+ *Before we go on...* Try a similar calculation using 2×2 matrices: Let $A = \begin{bmatrix} a & b \\ c & d \end{bmatrix}$, let *I* be the 2×2 identity matrix, and check that $AI = IA = A$. In fact, the equation

$$AI = IA = A$$

works for square matrices of every dimension. It is also interesting to notice that $AI = A$ if *I* is the 2×2 identity matrix and *A* is any 3×2 matrix (try one). In fact, if *I* is any identity matrix, then $AI = A$ whenever the product is defined, and $IA = A$ whenever this product is defined. ∎

We can now add to the list of properties we gave for matrix arithmetic at the end of Section 3.1 by writing down properties of matrix multiplication. In stating these properties, we shall assume that all matrix products we write are defined—that is, that the matrices have correctly matching dimensions. The first eight properties are the ones we've already seen; the rest are new.

Properties of Matrix Addition and Multiplication

If *A, B* and *C* are matrices, if *O* is a zero matrix, and if *I* is an identity matrix, then the following hold:

$A + (B + C) = (A + B) + C$	*Additive associative law*
$A + B = B + A$	*Additive commutative law*
$A + O = O + A = A$	*Additive identity law*
$A + (-A) = O = (-A) + A$	*Additive inverse law*
$c(A + B) = cA + cB$	*Distributive law*
$(c + d)A = cA + dA$	*Distributive law*

$1A = A$	*Scalar unit*
$0A = O$	*Scalar zero*
$A(BC) = (AB)C$	*Multiplicative associative law*
$c(AB) = (cA)B$	*Multiplicative associative law*
$c(dA) = (cd)A$	*Multiplicative associative law*
$AI = IA = A$	*Multiplicative identity law*
$A(B + C) = AB + AC$	*Distributive law*
$(A + B)C = AC + BC$	*Distributive law*
$OA = AO = O$	*Multiplication by zero matrix*

Note that we have not included a multiplicative commutative law for matrices, because the equation $AB = BA$ does not hold in general. In other words, matrix multiplication is *not* exactly like multiplication of numbers. (You have to be a little careful because it is easy to apply the commutative law without realizing it.)

We should also say a bit more about transposition. Transposition and multiplication have an interesting relationship. We write down the properties of transposition again, adding one new one.

Properties of Transposition

$$(A + B)^T = A^T + B^T$$
$$(cA)^T = c(A^T)$$
$$(AB)^T = B^T A^T$$

Notice the change in order in the last one. The order is crucial.

quick Examples

1. $\left(\begin{bmatrix} 1 & -1 \\ 0 & 2 \end{bmatrix}\begin{bmatrix} 3 & 0 \\ 5 & -1 \end{bmatrix}\right)^T = \begin{bmatrix} -2 & 1 \\ 10 & -2 \end{bmatrix}^T = \begin{bmatrix} -2 & 10 \\ 1 & -2 \end{bmatrix}$ $(AB)^T$

2. $\begin{bmatrix} 3 & 0 \\ 5 & -1 \end{bmatrix}^T\begin{bmatrix} 1 & -1 \\ 0 & 2 \end{bmatrix}^T = \begin{bmatrix} 3 & 5 \\ 0 & -1 \end{bmatrix}\begin{bmatrix} 1 & 0 \\ -1 & 2 \end{bmatrix} = \begin{bmatrix} -2 & 10 \\ 1 & -2 \end{bmatrix}$ $B^T A^T$

3. $\begin{bmatrix} 1 & -1 \\ 0 & 2 \end{bmatrix}^T\begin{bmatrix} 3 & 0 \\ 5 & -1 \end{bmatrix}^T = \begin{bmatrix} 1 & 0 \\ -1 & 2 \end{bmatrix}\begin{bmatrix} 3 & 5 \\ 0 & -1 \end{bmatrix} = \begin{bmatrix} 3 & 5 \\ -3 & -7 \end{bmatrix}$ $A^T B^T$

These properties give you a glimpse of the field of mathematics known as **abstract algebra.** Algebraists study operations like these that resemble the operations on numbers but differ in some way, such as the lack of commutativity for multiplication seen here.

We end this section with more on the relationship between linear equations and matrix equations, which is one of the important applications of matrix multiplication.

Example 7 Matrix Form of a System of Linear Equations

a. If

$$A = \begin{bmatrix} 1 & -2 & 3 \\ 2 & 0 & -1 \\ -3 & 1 & 1 \end{bmatrix}, X = \begin{bmatrix} x \\ y \\ z \end{bmatrix}, \text{ and } B = \begin{bmatrix} 3 \\ -1 \\ 0 \end{bmatrix}$$

rewrite the matrix equation $AX = B$ as a system of linear equations.

b. Express the following system of equations as a matrix equation of the form $AX = B$:

$$2x + y = 3$$
$$4x - y = -1$$

Solution

a. The matrix equation $AX = B$ is

$$\begin{bmatrix} 1 & -2 & 3 \\ 2 & 0 & -1 \\ -3 & 1 & 1 \end{bmatrix} \begin{bmatrix} x \\ y \\ z \end{bmatrix} = \begin{bmatrix} 3 \\ -1 \\ 0 \end{bmatrix}$$

As in Example 2(a), we first evaluate the left-hand side and then set it equal to the right-hand side.

$$\begin{bmatrix} 1 & -2 & 3 \\ 2 & 0 & -1 \\ -3 & 1 & 1 \end{bmatrix} \begin{bmatrix} x \\ y \\ z \end{bmatrix} = \begin{bmatrix} x - 2y + 3z \\ 2x - z \\ -3x + y + z \end{bmatrix}$$

$$\begin{bmatrix} x - 2y + 3z \\ 2x - z \\ -3x + y + z \end{bmatrix} = \begin{bmatrix} 3 \\ -1 \\ 0 \end{bmatrix}$$

Because these two matrices are equal, their corresponding entries must be equal.

$$x - 2y + 3z = 3$$
$$2x \quad\quad - z = -1$$
$$-3x + y + z = 0$$

In other words, the matrix equation $AX = B$ is equivalent to this system of linear equations. Notice that the coefficients of the left-hand sides of these equations are the entries of the matrix A. We call A the **coefficient matrix** of the system of equations. The entries of X are the unknowns and the entries of B are the right-hand sides $3, -1$, and 0.

b. As we saw in part (a), the coefficient matrix A has entries equal to the coefficients of the left-hand sides of the equations. Thus,

$$A = \begin{bmatrix} 2 & 1 \\ 4 & -1 \end{bmatrix}$$

X is the column matrix consisting of the unknowns, while B is the column matrix consisting of the right-hand sides of the equations, so

$$X = \begin{bmatrix} x \\ y \end{bmatrix} \quad \text{and} \quad B = \begin{bmatrix} 3 \\ -1 \end{bmatrix}$$

The system of equations can be rewritten as the matrix equation $AX = B$ with this A, X, and B.

This translation of systems of linear equations into matrix equations is really the first step in the method of solving linear equations discussed in Chapter 2. There we worked with the **augmented matrix** of the system, which is simply A with B adjoined as an extra column.

$Q\!:$ *When we write a system of equations as $AX = B$, couldn't we solve for the unknown X by dividing both sides by A ?*

$A\!:$ If we interpret division as multiplication by the inverse (for example, $2 \div 3 = 2 \times 3^{-1}$), we shall see in the next section that *certain* systems of the form $AX = B$ can be solved in this way, by multiplying both sides by A^{-1}. We first need to discuss what we mean by A^{-1} and how to calculate it. ∎

3.2 EXERCISES

● denotes basic skills exercises
◆ denotes challenging exercises
tech Ex indicates exercises that should be solved using technology

In Exercises 1–28, compute the products. Some of these may be undefined. Exercises marked tech *Ex should be done using technology. The others should be done two ways: by hand and by using technology where possible.* hint *[see Example 3]*

1. ● $\begin{bmatrix} 1 & 3 & -1 \end{bmatrix} \begin{bmatrix} 9 \\ 1 \\ -1 \end{bmatrix}$
2. ● $\begin{bmatrix} 4 & 0 & -1 \end{bmatrix} \begin{bmatrix} -4 \\ 1 \\ 8 \end{bmatrix}$

3. ● $\begin{bmatrix} -1 & \frac{1}{2} \end{bmatrix} \begin{bmatrix} -\frac{1}{3} \\ 1 \end{bmatrix}$
4. ● $\begin{bmatrix} -1 & 1 \end{bmatrix} \begin{bmatrix} \frac{3}{4} \\ \frac{1}{4} \end{bmatrix}$

5. ● $\begin{bmatrix} 0 & -2 & 1 \end{bmatrix} \begin{bmatrix} x \\ y \\ z \end{bmatrix}$
6. ● $\begin{bmatrix} 4 & -1 & 1 \end{bmatrix} \begin{bmatrix} -x \\ x \\ y \end{bmatrix}$

7. ● $\begin{bmatrix} 1 & 3 & 2 \end{bmatrix} \begin{bmatrix} 1 \\ -1 \end{bmatrix}$
8. ● $\begin{bmatrix} 3 & 2 \end{bmatrix} \begin{bmatrix} 1 & -2 \end{bmatrix}$

9. ● $\begin{bmatrix} -1 & 1 \end{bmatrix} \begin{bmatrix} -3 & 1 & 4 & 3 \\ 0 & 1 & -2 & 1 \end{bmatrix}$

10. ● $\begin{bmatrix} 2, & -1 \end{bmatrix} \begin{bmatrix} -3 & 1 & 4 & 3 \\ 4 & 0 & 1 & 3 \end{bmatrix}$

11. ● $\begin{bmatrix} 1 & -1 & 2 & 3 \end{bmatrix} \begin{bmatrix} -1 & 2 & 0 \\ 2 & -1 & 0 \\ 0 & 5 & 2 \\ -1 & 8 & 1 \end{bmatrix}$

12. ● $\begin{bmatrix} 0 & 1 & -1 & 2 \end{bmatrix} \begin{bmatrix} 1 & -2 & 1 \\ 0 & 1 & 3 \\ 6 & 0 & 0 \\ -1 & -2 & 11 \end{bmatrix}$

13. ● $\begin{bmatrix} 1 & 0 & -1 \\ 1 & 1 & 2 \end{bmatrix} \begin{bmatrix} 0 & 1 & -1 \\ 1 & 0 & 1 \\ 4 & 8 & 0 \end{bmatrix}$

14. ● $\begin{bmatrix} 0 & 1 & -1 \\ 3 & 1 & -1 \end{bmatrix} \begin{bmatrix} 1 & 1 \\ 4 & 2 \\ 0 & 1 \end{bmatrix}$

15. ● $\begin{bmatrix} 1 & 0 \\ 1 & -1 \end{bmatrix} \begin{bmatrix} 0 & 1 \\ 0 & 1 \end{bmatrix}$
16. ● $\begin{bmatrix} 1 & -1 \\ 1 & -1 \end{bmatrix} \begin{bmatrix} 3 & -3 \\ 5 & -7 \end{bmatrix}$

17. ● $\begin{bmatrix} 0 & 1 \\ 0 & 1 \end{bmatrix} \begin{bmatrix} 1 & 0 \\ 1 & -1 \end{bmatrix}$
18. ● $\begin{bmatrix} 3 & -3 \\ 5 & -7 \end{bmatrix} \begin{bmatrix} 1 & -1 \\ 1 & -1 \end{bmatrix}$

19. ● $\begin{bmatrix} 1 & -1 \\ 1 & -1 \end{bmatrix} \begin{bmatrix} 2 & 3 \\ 2 & 3 \end{bmatrix}$
20. ● $\begin{bmatrix} 0 & 1 \\ 1 & 0 \end{bmatrix} \begin{bmatrix} 3 & -3 \\ 2 & -1 \end{bmatrix}$

21. ● $\begin{bmatrix} 1 & -1 \\ -1 & 1 \end{bmatrix} \begin{bmatrix} 2 & 3 \\ 2 & 3 \\ 1 & 1 \end{bmatrix}$

22. ● $\begin{bmatrix} 0 & 1 & -1 \\ 0 & -1 & 1 \end{bmatrix} \begin{bmatrix} 3 & -3 \\ 2 & -1 \end{bmatrix}$

23. ● $\begin{bmatrix} 1 & 0 & -1 \\ 2 & -2 & 1 \\ 0 & 0 & 1 \end{bmatrix} \begin{bmatrix} 1 & -1 & 4 \\ 1 & 1 & 0 \\ 0 & 4 & 1 \end{bmatrix}$

24. ● $\begin{bmatrix} 1 & 2 & 0 \\ 4 & -1 & 1 \\ 1 & 0 & 1 \end{bmatrix} \begin{bmatrix} 1 & 2 & -4 \\ 4 & 1 & 0 \\ 0 & -2 & 1 \end{bmatrix}$

25. ● $\begin{bmatrix} 1 & 0 & 1 & 0 \\ -1 & 1 & 0 & 1 \\ -2 & 0 & 1 & 4 \\ 0 & -1 & 0 & 1 \end{bmatrix} \begin{bmatrix} 1 \\ -3 \\ 2 \\ 0 \end{bmatrix}$

26. ● $\begin{bmatrix} 1 & 1 & -7 & 0 \\ -1 & 0 & 2 & 4 \\ -1 & 0 & -2 & 1 \\ 1 & -1 & 1 & 1 \end{bmatrix} \begin{bmatrix} 1 \\ -3 \\ 2 \\ 1 \end{bmatrix}$

27. tech Ex $\begin{bmatrix} 1.1 & 2.3 & 3.4 & -1.2 \\ 3.4 & 4.4 & 2.3 & 1.1 \\ 2.3 & 0 & -2.2 & 1.1 \\ 1.2 & 1.3 & 1.1 & 1.1 \end{bmatrix} \begin{bmatrix} -2.1 & 0 & -3.3 \\ -3.4 & -4.8 & -4.2 \\ 3.4 & 5.6 & 1 \\ 1 & 2.2 & 9.8 \end{bmatrix}$

28. `tech` Ex $\begin{bmatrix} 1.2 & 2.3 & 3.4 & 4.5 \\ 3.3 & 4.4 & 5.5 & 6.6 \\ 2.3 & -4.3 & -2.2 & 1.1 \\ 2.2 & -1.2 & -1 & 1.1 \end{bmatrix} \begin{bmatrix} 9.8 & 1 & -1.1 \\ 8.8 & 2 & -2.2 \\ 7.7 & 3 & -3.3 \\ 6.6 & 4 & -4.4 \end{bmatrix}$

29. ● Find[*] $A^2 = A \cdot A$, $A^3 = A \cdot A \cdot A$, A^4, and A^{100}, given that

$$A = \begin{bmatrix} 0 & 1 & 1 & 1 \\ 0 & 0 & 1 & 1 \\ 0 & 0 & 0 & 1 \\ 0 & 0 & 0 & 0 \end{bmatrix}.$$

30. ● Repeat the preceding exercise with $A = \begin{bmatrix} 0 & 2 & 0 & -1 \\ 0 & 0 & 2 & 0 \\ 0 & 0 & 0 & 2 \\ 0 & 0 & 0 & 0 \end{bmatrix}$.

Exercises 31–38 should be done two ways: (a) by hand; (b) using technology where possible.

Let

$$A = \begin{bmatrix} 0 & -1 & 0 & 1 \\ 10 & 0 & 1 & 0 \end{bmatrix}, B = \begin{bmatrix} 0 & -1 \\ 1 & 1 \\ -1 & 3 \\ 5 & 0 \end{bmatrix}, C = \begin{bmatrix} 1 & -1 \\ 1 & 1 \\ 1 & 1 \\ 1 & 1 \end{bmatrix}.$$

Evaluate:

31. ● AB **32.** ● AC **33.** ● $A(B - C)$ **34.** ● $(B - C)A$

Let

$$A = \begin{bmatrix} 1 & -1 \\ 0 & 2 \\ 0 & -2 \end{bmatrix}, B = \begin{bmatrix} 3 & 0 & -1 \\ 5 & -1 & 1 \end{bmatrix}, C = \begin{bmatrix} x & 1 & w \\ z & r & 4 \end{bmatrix}.$$

Evaluate:

35. ● AB **36.** ● AC **37.** ● $A(B + C)$ **38.** ● $(B + C)A$

In Exercises 39–44, calculate (a) $P^2 = P \cdot P$ (b) $P^4 = P^2 \cdot P^2$ and (c) P^8. (Round all entries to four decimal places.) (d) Without computing it explicitly, find P^{1000}.

39. $P = \begin{bmatrix} 0.2 & 0.8 \\ 0.2 & 0.8 \end{bmatrix}$ **40.** $P = \begin{bmatrix} 0.1 & 0.9 \\ 0.1 & 0.9 \end{bmatrix}$

41. $P = \begin{bmatrix} 0.1 & 0.9 \\ 0 & 1 \end{bmatrix}$ **42.** $P = \begin{bmatrix} 1 & 0 \\ 0.8 & 0.2 \end{bmatrix}$

43. $P = \begin{bmatrix} 0.25 & 0.25 & 0.50 \\ 0.25 & 0.25 & 0.50 \\ 0.25 & 0.25 & 0.50 \end{bmatrix}$

44. $P = \begin{bmatrix} 0.3 & 0.3 & 0.4 \\ 0.3 & 0.3 & 0.4 \\ 0.3 & 0.3 & 0.4 \end{bmatrix}$

---

[*] $A \cdot A \cdot A$ is $A(A \cdot A)$, or the equivalent $(A \cdot A)A$ by the associative law. Similarly, $A \cdot A \cdot A \cdot A = A(A \cdot A \cdot A) = (A \cdot A \cdot A)A = (A \cdot A)(A \cdot A)$; it doesn't matter where we place parentheses.

In Exercises 45–48, translate the given matrix equations into systems of linear equations. *hint* [see Example 7]

45. ● $\begin{bmatrix} 2 & -1 & 4 \\ -4 & \frac{3}{4} & \frac{1}{3} \\ -3 & 0 & 0 \end{bmatrix} \begin{bmatrix} x \\ y \\ z \end{bmatrix} = \begin{bmatrix} 3 \\ -1 \\ 0 \end{bmatrix}$

46. ● $\begin{bmatrix} 1 & -1 & 4 \\ -\frac{1}{3} & -3 & \frac{1}{3} \\ 3 & 0 & 1 \end{bmatrix} \begin{bmatrix} x \\ y \\ z \end{bmatrix} = \begin{bmatrix} -3 \\ -1 \\ 2 \end{bmatrix}$

47. ● $\begin{bmatrix} 1 & -1 & 0 & 1 \\ 1 & 1 & 2 & 4 \end{bmatrix} \begin{bmatrix} x \\ y \\ z \\ w \end{bmatrix} = \begin{bmatrix} -1 \\ 2 \end{bmatrix}$

48. ● $\begin{bmatrix} 0 & 1 & 6 & 1 \\ 1 & -5 & 0 & 0 \end{bmatrix} \begin{bmatrix} x \\ y \\ z \\ w \end{bmatrix} = \begin{bmatrix} -2 \\ 9 \end{bmatrix}$

In Exercises 49–52, translate the given systems of equations into matrix form. *hint* [see Example 7]

49. ● $x - y = 4$
    $2x - y = 0$

50. ● $2x + y = 7$
    $-x = 9$

51. ● $x + y - z = 8$
    $2x + y + z = 4$
    $\dfrac{3x}{4} + \dfrac{z}{2} = 1$

52. ● $x + y + 2z = -2$
    $4x + 2y - z = -8$
    $\dfrac{x}{2} + \dfrac{y}{3} = 1$

Applications

53. ● **Revenue** Your T-shirt operation is doing a booming trade. Last week you sold 50 tie-dye shirts for $15 each, 40 Suburban State University Crew shirts for $10 each, and 30 Lacrosse T-shirts for $12 each. Use matrix operations to calculate your total revenue for the week. *hint* [see Example 1]

54. ● **Revenue** Karen Sandberg, your competitor in Suburban State U's T-shirt market, has apparently been undercutting your prices and outperforming you in sales. Last week she sold 100 tie dye shirts for $10 each, 50 (low quality) Crew shirts at $5 apiece, and 70 Lacrosse T-shirts for $8 each. Use matrix operations to calculate her total revenue for the week.

55. ● **Revenue** Recall the Left Coast Bookstore chain from the preceding section. In January, it sold 700 hardcover books, 1300 softcover books, and 2000 plastic books in San Francisco; it sold 400 hardcover, 300 softcover, and 500 plastic books in Los Angeles. Now, hardcover books sell for $30 each, softcover books sell for $10 each, and plastic books sell for $15 each. Write a column matrix with the price data and show how matrix multiplication (using the sales and price data matrices) may be used to compute the total revenue at the two stores. *hint* [see Example 5]

56. ● **Profit** Refer back to Exercise 55, and now suppose that each hardcover book costs the stores $10, each softcover book

● basic skills   ◆ challenging   `tech` Ex technology exercise

costs \$5, and each plastic book costs \$10. Use matrix operations to compute the total *profit* at each store in January.

57. ● *Publishing* Editors' workloads were increasing during the 1990's, as the following table shows.[7]

	1993	1994	1995	1996
Books per Editor	3	3.5	5	5.2
Number of Editors	16,000	15,000	12,500	13,000

Use matrix multiplication to estimate the total number of books published during the years 1993–1996.

58. ● *Real Estate* The following table shows the cost of one square foot of residential real estate, in dollars per square foot, at the start of 2003[8] together with the number of square feet your development company intends to purchase in each city.

	New York	London	Hong Kong
Cost per sq. foot	760	800	300
Number of sq. ft	500	800	1000

Use matrix multiplication to estimate the total cost of the real estate.

tech Ex *Income Exercises 59–62 are based on the following Excel sheet, which shows the 2004 and 2005 male and female population in various age groups, as well as per capita incomes.[9]*

59. ● Use matrix algebra to estimate the total income for females in 2004. (Round the answer to two significant digits.)

60. ● Use matrix algebra to estimate the total income for males in 2005.

61. ● Give a single matrix formula that expresses the difference in total income between males and females in 2005, and compute its value, rounded to two significant digits.

62. ● Give a single matrix formula that expresses the total income in 2005, and compute its value, rounded to two significant digits.

63. ● *Cheese Production* The total amount of cheese, in billions of pounds, produced in the 13 western and 12 north central U.S. states in 1999 and 2000 was as follows.[10]

	1999	2000
Western States	2.7	3.0
North Central States	3.9	4.0

Thinking of this table as a (labeled) 2×2 matrix P, compute the matrix product $[-1 \quad 1]P$. What does this product represent?

64. ● *Milk Production* The total amount of milk, in billions of pounds, produced in the 13 western and 12 north central U.S. states in 1999 and 2000 was as follows.[11]

	1999	2000
Western States	56	60
North Central States	57	59

Thinking of this table as a (labeled) 2×2 matrix P, compute the matrix product $P \begin{bmatrix} -1 \\ 1 \end{bmatrix}$. What does this product represent?

Bankruptcy Filings Exercises 65–70 are based on the following chart, which shows the number of personal bankruptcy filings in three New York/New Jersey regions during various months of 2001–2002.[12]

	Jan 01	Jul 01	Jan 02
Manhattan	150	150	150
Brooklyn	300	300	250
Newark	250	250	200

65. Each month, your law firm handles 10% of all bankruptcy filings in Manhattan, 5% of all filings in Brooklyn, and 20% of all filings in Newark. Use matrix multiplication to compute the total number of bankruptcy filings handled by your firm in each of the months shown.

---

[7] Books per Editor refers to new titles. Data are rounded. SOURCES: R.R. Bowker, EEOC/*The New York Times,* June 29, 1998, p. E1.

[8] Figures are rounded. SOURCE: CLSA, Honk Kong Monetary Authority/*New York Times,* August 15, 2003, p. C1.

[9] The population figures are Census Bureau estimates, and the income figures are 2001 mean per capita incomes. All figures are rounded. SOURCE: U.S. Census Bureau www.census.gov.

[10] Ibid.

[11] Figures are approximate. SOURCE: Department of Agriculture/*New York Times,* June 28, 2001, p. C1.

[12] Data are approximate four-week moving averages. SOURCE: Lundquist Consulting/*New York Times,* February 10, 2002, p. L1.

● basic skills   ◆ challenging   tech Ex technology exercise

66. Your law firm handled 10% of all bankruptcy filings in each region in January 2001, 30% of all filings in July 2001, and 20% of all filings in January 2002. Use matrix multiplication to compute the total number of bankruptcy filings handled by your firm in each of the regions shown.

67. Let A be the 3×3 matrix whose entries are the figures in the table, and let $B = [\,1 \quad 1 \quad 0\,]$. What does the matrix BA represent?

68. Let A be the 3×3 matrix whose entries are the figures in the table, and let $B = [\,1 \quad 1 \quad 0\,]^T$. What does the matrix AB represent?

69. Write a matrix product whose computation gives the total number by which the combined filings in Manhattan and Newark exceeded the filings in Brooklyn.

70. Write a matrix product whose computation gives the total number by which bankruptcy filings in January, 2001, exceeded filings in January, 2002.

71. *Costs* Microbucks Computer Co. makes two computers, the Pomegranate II and the Pomegranate Classic. The Pom II requires 2 processor chips, 16 memory chips, and 20 vacuum tubes, while the Pom Classic requires 1 processor chip, 4 memory chips, and 40 vacuum tubes. There are two companies that can supply these parts: Motorel can supply them at $100 per processor chip, $50 per memory chip, and $10 per vacuum tube, while Intola can supply them at $150 per processor chip, $40 per memory chip, and $15 per vacuum tube. Write down all of this data in two matrices, one showing the parts required for each model computer, and the other showing the prices for each part from each supplier. Then show how matrix multiplication allows you to compute the total cost for parts for each model when parts are bought from either supplier.

72. *Profits* Refer back to Exercise 71. It actually costs Motorel only $25 to make each processor chip, $10 for each memory chip, and $5 for each vacuum tube. It costs Intola $50 per processor chip, $10 per memory chip, and $7 per vacuum tube. Use matrix operations to find the total profit Motorel and Intola would make on each model.

73. *Tourism* The following table gives the number of people (in thousands) who visited Australia and South Africa in 1998.[13]

	To	**Australia**	**South Africa**
From	**North America**	440	190
	Europe	950	950
	Asia	1790	200

You estimate that 5% of all visitors to Australia and 4% of all visitors to South Africa decide to settle there permanently. Take A to be the 3×2 matrix whose entries are the 1998 tourism figures in the above table and take

$$B = \begin{bmatrix} 0.05 \\ 0.04 \end{bmatrix} \quad \text{and} \quad C = \begin{bmatrix} 0.05 & 0 \\ 0 & 0.04 \end{bmatrix}$$

Compute the products AB and AC. What do the entries in these matrices represent?

74. ● *Tourism* Referring to the tourism figures in the preceding exercise, you estimate that from 1998 to 2008, tourism from North America to each of Australia and South Africa will have increased by 20%, tourism from Europe by 30%, and tourism from Asia by 10%. Take A to be the 3×2 matrix whose entries are the 1998 tourism figures and take

$$B = [\,1.2 \quad 1.3 \quad 1.1\,] \text{ and } C = \begin{bmatrix} 1.2 & 0 & 0 \\ 0 & 1.3 & 0 \\ 0 & 0 & 1.1 \end{bmatrix}$$

Compute the products AB and AC. What do the entries in these matrices represent?

75. ◆ *tech Ex Population Movement* In 2003, the population of the U.S., broken down by regions, was 53.3 million in the Northeast, 64.0 million in the Midwest, 101.6 million in the South, and 65.4 million in the West.[14] The matrix P below shows the population movement during the period 2003–2004. (Thus, 98.79% of the population in the Northeast stayed there, while 0.20% of the population in the Northeast moved to the Midwest, and so on.)

	To NE	To MW	To S	To W
From NE	0.9879	0.0020	0.0081	0.0019
From MW	0.0014	0.9895	0.0063	0.0028
From S	0.0027	0.0025	0.9927	0.0022
From W	0.0010	0.0030	0.0050	0.9909

$P = $ (bracket enclosing the matrix above)

Set up the 2003 population figures as a row vector. Then use matrix multiplication to compute the population in each region in 2004. (Round all answers to the nearest 0.1 million.)

76. ◆ *tech Ex Population Movement* Assume that the percentages given in the preceding exercise also describe the population movements from 2004–2005. Use two matrix multiplications to predict from the data in the preceding exercise the population in each region in 2005. (Round the *final* answers to the nearest 0.1 million, but do not round the intermediate answer.)

---

[13] Figures are rounded to the nearest 10,000. SOURCES: South African Dept. of Environmental Affairs and Tourism; Australia Tourist Commission/ *The New York Times,* January 15, 2000, p. C1.

[14] Note that this exercise ignores migration into or out of the country. The internal migration figures and 2004 population figures are accurate. SOURCE: U.S. Census Bureau, Current Population Survey, 2004 Annual Social and Economic Supplement, Internet release date June 23, 2005. www.census.gov.

● basic skills   ◆ challenging   *tech* Ex technology exercise

Communication and Reasoning Exercises

77. ● Give an example of two matrices A and B such that AB is defined but BA is not defined.

78. ● Give an example of two matrices A and B of different dimensions such that both AB and BA are defined.

79. ● Compare addition and multiplication of 1×1 matrices to the arithmetic of numbers.

80. ● In comparing the algebra of 1×1 matrices, as discussed so far, to the algebra of real numbers (see Exercise 79), what important difference do you find?

81. ● Comment on the following claim: Every matrix equation represents a system of equations.

82. ● When is it true that both AB and BA are defined, even though neither A nor B is a square matrix?

83. Find a scenario in which it would be useful to "multiply" two row vectors according to the rule

$$[\,a \quad b \quad c\,][\,d \quad e \quad f\,] = [\,ad \quad be \quad cf\,]$$

84. Make up an application whose solution reads as follows.

$$\text{"Total revenue} = [\,10 \quad 100 \quad 30\,] \begin{bmatrix} 10 & 0 & 3 \\ 1 & 2 & 0 \\ 0 & 1 & 40 \end{bmatrix}\text{"}$$

85. What happens in Excel if, instead of using the function MMULT, you use "ordinary multiplication" as shown here?

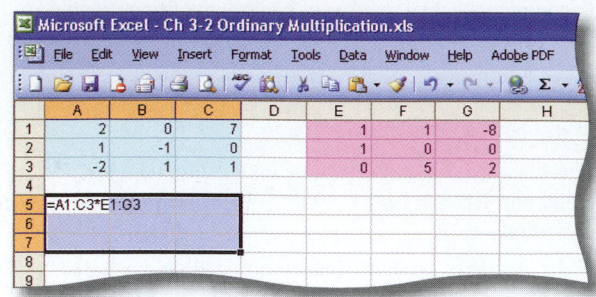

86. Define the *naïve product* $A \,\square\, B$ of two $m \times n$ matrices A and B by

$$(A \,\square\, B)_{ij} = A_{ij} B_{ij}$$

(This is how someone who has never seen matrix multiplication before might think to multiply matrices.) Referring to Example 1 in this section, compute and comment on the meaning of $P \,\square\, (Q^T)$.

● basic skills   ◆ challenging   **tech** Ex technology exercise

3.3 Matrix Inversion

Now that we've discussed matrix addition, subtraction, and multiplication, you may well be wondering about matrix *division*. In the realm of real numbers, division can be thought of as a form of multiplication: Dividing 3 by 7 is the same as multiplying 3 by $1/7$, the inverse of 7. In symbols, $3 \div 7 = 3 \times (1/7)$, or 3×7^{-1}. In order to imitate division of real numbers in the realm of matrices, we need to discuss the multiplicative **inverse**, A^{-1}, of a matrix A.

Note Because multiplication of real numbers is commutative, we can write, for example, $\frac{3}{7}$ as either 3×7^{-1} or $7^{-1} \times 3$. In the realm of matrices, multiplication is not commutative, so from now on we shall *never* talk about "division" of matrices (by "$\frac{B}{A}$" should we mean $A^{-1}B$ or BA^{-1}?). ■

Before we try to find the inverse of a matrix, we must first know exactly what we *mean* by the inverse. Recall that the inverse of a number a is the number, often written a^{-1}, with the property that $a^{-1} \cdot a = a \cdot a^{-1} = 1$. For example, the inverse of 76 is the number $76^{-1} = 1/76$, because $(1/76) \cdot 76 = 76 \cdot (1/76) = 1$. This is the number calculated by the x^{-1} button found on most calculators. Not all numbers have an inverse.

For example—and this is the only example—the number 0 has no inverse, because you cannot get 1 by multiplying 0 by anything.

The inverse of a matrix is defined similarly. To make life easier, we shall restrict attention to **square** matrices, matrices that have the same number of rows as columns.[15]

Inverse of a Matrix

The **inverse** of an $n \times n$ matrix A is that $n \times n$ matrix A^{-1} which, when multiplied by A on either side, yields the $n \times n$ identity matrix I. Thus,

$$AA^{-1} = A^{-1}A = I$$

If A has an inverse, it is said to be **invertible.** Otherwise, it is said to be **singular.**

quick Examples

1. The inverse of the 1×1 matrix $[3]$ is $[1/3]$, because $[3][1/3] = [1] = [1/3][3]$.

2. The inverse of the $n \times n$ identity matrix I is I itself, because $I \times I = I$. Thus, $I^{-1} = I$.

3. The inverse of the 2×2 matrix $A = \begin{bmatrix} 1 & -1 \\ -1 & -1 \end{bmatrix}$ is $A^{-1} = \begin{bmatrix} \frac{1}{2} & -\frac{1}{2} \\ -\frac{1}{2} & -\frac{1}{2} \end{bmatrix}$, because

$$\begin{bmatrix} 1 & -1 \\ -1 & -1 \end{bmatrix}\begin{bmatrix} \frac{1}{2} & -\frac{1}{2} \\ -\frac{1}{2} & -\frac{1}{2} \end{bmatrix} = \begin{bmatrix} 1 & 0 \\ 0 & 1 \end{bmatrix} \qquad AA^{-1} = I$$

and $\qquad \begin{bmatrix} \frac{1}{2} & -\frac{1}{2} \\ -\frac{1}{2} & -\frac{1}{2} \end{bmatrix}\begin{bmatrix} 1 & -1 \\ -1 & -1 \end{bmatrix} = \begin{bmatrix} 1 & 0 \\ 0 & 1 \end{bmatrix} \qquad A^{-1}A = I$

Notes

1. It is possible to show that if A and B are square matrices with $AB = I$, then it must also be true that $BA = I$. In other words, once we have checked that $AB = I$, we know that B is the inverse of A. The second check, that $BA = I$, is unnecessary.

2. If B is the inverse of A, then we can also say that A is the inverse of B (why?). Thus, we sometimes refer to such a pair of matrices as an **inverse pair** of matrices. ∎

Example **1 Singular Matrix**

Can $A = \begin{bmatrix} 1 & 1 \\ 0 & 0 \end{bmatrix}$ have an inverse?

Solution No. To see why not, notice that both entries in the second row of AB will be 0, no matter what B is. So AB cannot equal I, no matter what B is. Hence, A is singular.

+ *Before we go on...* If you think about it, you can write down many similar examples of singular matrices. There is only one number with no multiplicative inverse (0), but there are many matrices having no inverses. ∎

---

[15] Nonsquare matrices *cannot* have inverses in the sense that we are talking about here. This is not a trivial fact to prove.

Finding the Inverse of a Square Matrix

Q: *In the box, it was stated that the inverse of $\begin{bmatrix} 1 & -1 \\ -1 & -1 \end{bmatrix}$ is $\begin{bmatrix} \frac{1}{2} & -\frac{1}{2} \\ -\frac{1}{2} & -\frac{1}{2} \end{bmatrix}$. How was that obtained?*

A: We can think of the problem of finding A^{-1} as a problem of finding four unknowns, the four unknown entries of A^{-1}:

$$A^{-1} = \begin{bmatrix} x & y \\ z & w \end{bmatrix}$$

These unknowns must satisfy the equation $AA^{-1} = I$, or

$$\begin{bmatrix} 1 & -1 \\ -1 & -1 \end{bmatrix}\begin{bmatrix} x & y \\ z & w \end{bmatrix} = \begin{bmatrix} 1 & 0 \\ 0 & 1 \end{bmatrix}$$

If we were to try to find the first column of A^{-1}, consisting of x and z, we would have to solve

$$\begin{bmatrix} 1 & -1 \\ -1 & -1 \end{bmatrix}\begin{bmatrix} x \\ z \end{bmatrix} = \begin{bmatrix} 1 \\ 0 \end{bmatrix}$$

or

$$x - z = 1$$
$$-x - z = 0$$

To solve this system by Gauss-Jordan reduction, we would row-reduce the augmented matrix, which is A with the column $\begin{bmatrix} 1 \\ 0 \end{bmatrix}$ adjoined.

$$\begin{bmatrix} 1 & -1 & | & 1 \\ -1 & -1 & | & 0 \end{bmatrix} \rightarrow \begin{bmatrix} 1 & 0 & | & x \\ 0 & 1 & | & z \end{bmatrix}$$

To find the second column of A^{-1} we would similarly row-reduce the augmented matrix obtained by tacking on to A the second column of the identity matrix.

$$\begin{bmatrix} 1 & -1 & | & 0 \\ -1 & -1 & | & 1 \end{bmatrix} \rightarrow \begin{bmatrix} 1 & 0 & | & y \\ 0 & 1 & | & w \end{bmatrix}$$

The row operations used in doing these two reductions would be exactly the same. We could do both reductions simultaneously by "doubly augmenting" A, putting both columns of the identity matrix to the right of A.

$$\begin{bmatrix} 1 & -1 & | & 1 & 0 \\ -1 & -1 & | & 0 & 1 \end{bmatrix} \rightarrow \begin{bmatrix} 1 & 0 & | & x & y \\ 0 & 1 & | & z & w \end{bmatrix}$$

We carry out this reduction in the following example. ∎

Example 2 Computing Matrix Inverse

Find the inverse of each matrix.

a. $P = \begin{bmatrix} 1 & -1 \\ -1 & -1 \end{bmatrix}$ **b.** $Q = \begin{bmatrix} 1 & 0 & 1 \\ 2 & -2 & -1 \\ 3 & 0 & 0 \end{bmatrix}$

Solution

a. As described above, we put the matrix P on the left and the identity matrix I on the right to get a 2×4 matrix.

$$\begin{bmatrix} 1 & -1 & | & 1 & 0 \\ -1 & -1 & | & 0 & 1 \end{bmatrix}$$
$$PI$$

using *Technology*

See the Technology Guides at the end of the chapter to see how to invert a matrix using a TI-83/84 or Excel. You can also invert matrices with the Matrix Algebra Tool. Follow:

Chapter 3
→ Tools
  → Matrix Algebra Tool

To invert a matrix A you have entered, type A^-1 or A^(-1) in the formula box and press "Compute." To invert a matrix using the row-reduction method in Example 2, you can use either the browser-based Pivot and Gauss-Jordan Tool or the Excel-based Excel Pivot and Gauss-Jordan Tool, also accessible at

Chapter 3 → Tools

We now row reduce the whole matrix:

$$\begin{bmatrix} 1 & -1 & 1 & 0 \\ -1 & -1 & 0 & 1 \end{bmatrix} \begin{matrix} \\ R_2 + R_1 \end{matrix} \rightarrow \begin{bmatrix} 1 & -1 & 1 & 0 \\ 0 & -2 & 1 & 1 \end{bmatrix} \begin{matrix} 2R_1 - R_2 \\ \\ \end{matrix}$$

$$\rightarrow \begin{bmatrix} 2 & 0 & 1 & -1 \\ 0 & -2 & 1 & 1 \end{bmatrix} \begin{matrix} \frac{1}{2}R_1 \\ -\frac{1}{2}R_2 \end{matrix} \rightarrow \begin{bmatrix} 1 & 0 & \frac{1}{2} & -\frac{1}{2} \\ 0 & 1 & -\frac{1}{2} & -\frac{1}{2} \end{bmatrix}$$
$$\quad\quad\quad\quad\quad\quad\quad\quad\quad\quad\quad\quad I \quad\quad\quad P^{-1}$$

We have now solved the systems of linear equations that define the entries of P^{-1}. Thus,

$$P^{-1} = \begin{bmatrix} \frac{1}{2} & -\frac{1}{2} \\ -\frac{1}{2} & -\frac{1}{2} \end{bmatrix}$$

b. The procedure to find the inverse of a 3×3 matrix (or larger) is just the same as for a 2×2 matrix. We place Q on the left and the identity matrix (now 3×3) on the right, and reduce.

$$\begin{matrix} Q & & & I & & \end{matrix}$$
$$\begin{bmatrix} 1 & 0 & 1 & 1 & 0 & 0 \\ 2 & -2 & -1 & 0 & 1 & 0 \\ 3 & 0 & 0 & 0 & 0 & 1 \end{bmatrix} \begin{matrix} \\ R_2 - 2R_1 \\ R_3 - 3R_1 \end{matrix} \rightarrow \begin{bmatrix} 1 & 0 & 1 & 1 & 0 & 0 \\ 0 & -2 & -3 & -2 & 1 & 0 \\ 0 & 0 & -3 & -3 & 0 & 1 \end{bmatrix} \begin{matrix} 3R_1 + R_3 \\ R_2 - R_3 \rightarrow \\ \end{matrix}$$

$$\begin{bmatrix} 3 & 0 & 0 & 0 & 0 & 1 \\ 0 & -2 & 0 & 1 & 1 & -1 \\ 0 & 0 & -3 & -3 & 0 & 1 \end{bmatrix} \begin{matrix} \frac{1}{3}R_1 \\ -\frac{1}{2}R_2 \rightarrow \\ -\frac{1}{3}R_3 \end{matrix} \begin{bmatrix} 1 & 0 & 0 & 0 & 0 & \frac{1}{3} \\ 0 & 1 & 0 & -\frac{1}{2} & -\frac{1}{2} & \frac{1}{2} \\ 0 & 0 & 1 & 1 & 0 & -\frac{1}{3} \end{bmatrix}$$
$$\quad\quad\quad\quad\quad\quad\quad\quad\quad\quad\quad\quad\quad I \quad\quad\quad\quad Q^{-1}$$

Thus,

$$Q^{-1} = \begin{bmatrix} 0 & 0 & \frac{1}{3} \\ -\frac{1}{2} & -\frac{1}{2} & \frac{1}{2} \\ 1 & 0 & -\frac{1}{3} \end{bmatrix}$$

We have already checked that P^{-1} is the inverse of P. You should also check that Q^{-1} is the inverse of Q.

The method we used in Example 2 can be summarized as follows:

Inverting an *n* × *n* Matrix

In order to determine whether an $n \times n$ matrix A is invertible or not, and to find A^{-1} if it does exist, follow this procedure:

1. Write down the $n \times 2n$ matrix $[A\,|\,I]$ (this is A with the $n \times n$ identity matrix set next to it).

2. Row-reduce $[A\,|\,I]$.

3. If the reduced form is $[I\,|\,B]$ (i.e., has the identity matrix in the left part), then A is invertible and $B = A^{-1}$. If you cannot obtain I in the left part, then A is singular. (See Example 3 below.)

Although there is a general formula for the inverse of a matrix, it is not a simple one. In fact, using the formula for anything larger than a 3×3 matrix is so inefficient that the row-reduction procedure is the method of choice even for computers. However, the general formula is very simple for the special case of 2×2 matrices:

Formula for the Inverse of a 2 × 2 Matrix

The inverse of a 2×2 matrix is

$$\begin{bmatrix} a & b \\ c & d \end{bmatrix}^{-1} = \frac{1}{ad - bc} \begin{bmatrix} d & -b \\ -c & a \end{bmatrix}, \quad \text{provided } ad - bc \neq 0$$

If the quantity $ad - bc$ is zero, then the matrix is singular (noninvertible). The quantity $ad - bc$ is called the **determinant** of the matrix $\begin{bmatrix} a & b \\ c & d \end{bmatrix}$.

---

quick Examples

1. $\begin{bmatrix} 1 & 2 \\ 3 & 4 \end{bmatrix}^{-1} = \frac{1}{(1)(4) - (2)(3)} \begin{bmatrix} 4 & -2 \\ -3 & 1 \end{bmatrix} = -\frac{1}{2} \begin{bmatrix} 4 & -2 \\ -3 & 1 \end{bmatrix} = \begin{bmatrix} -2 & 1 \\ \frac{3}{2} & -\frac{1}{2} \end{bmatrix}.$

2. $\begin{bmatrix} 1 & -1 \\ 2 & -2 \end{bmatrix}$ has determinant $ad - bc = (1)(-2) - (-1)(2) = 0$ and so is singular.

The formula for the inverse of a 2×2 matrix can be obtained using the technique of row reduction. (See the Communication and Reasoning Exercises at the end of the section.)

As we have mentioned above, not every square matrix has an inverse, as we see in the next example.

Example 3 Singular 3 × 3 Matrix

Find the inverse of the matrix $S = \begin{bmatrix} 1 & 1 & 2 \\ -2 & 0 & 4 \\ 3 & 1 & -2 \end{bmatrix}$, if it exists.

Solution We proceed as before.

$$\begin{array}{cc} S & I \end{array}$$

$$\begin{bmatrix} 1 & 1 & 2 & | & 1 & 0 & 0 \\ -2 & 0 & 4 & | & 0 & 1 & 0 \\ 3 & 1 & -2 & | & 0 & 0 & 1 \end{bmatrix} \begin{array}{c} \\ R_2 + 2R_1 \\ R_3 - 3R_1 \end{array} \rightarrow \begin{bmatrix} 1 & 1 & 2 & | & 1 & 0 & 0 \\ 0 & 2 & 8 & | & 2 & 1 & 0 \\ 0 & -2 & -8 & | & -3 & 0 & 1 \end{bmatrix} \begin{array}{c} 2R_1 - R_2 \\ \\ R_3 + R_2 \end{array}$$

$$\rightarrow \begin{bmatrix} 2 & 0 & -4 & | & 0 & -1 & 0 \\ 0 & 2 & 8 & | & 2 & 1 & 0 \\ 0 & 0 & 0 & | & -1 & 1 & 1 \end{bmatrix}$$

We stopped here, even though the reduction is incomplete, because there is *no hope* of getting the identity on the left-hand side. Completing the row reduction will not change the three zeros in the bottom row. So what is wrong? Nothing. As in Example 1, we have here a singular matrix. Any square matrix that, after row reduction, winds up with a row of zeros is singular (see Exercise 77).

+*Before we go on...* In practice, deciding whether a given matrix is invertible or singular is easy: simply try to find its inverse. If the process works, then the matrix is invertible, and we get its inverse. If the process fails, then the matrix is singular. If you try to invert a singular matrix using a spreadsheet, calculator, or computer program, you should get an error. Sometimes, instead of an error, you will get a spurious answer due to round-off errors in the device. ∎

Using the Inverse to Solve a System of n Linear Equations in n Unknowns

Having used systems of equations and row reduction to find matrix inverses, we will now use matrix inverses to solve systems of equations. Recall that, at the end of the previous section, we saw that a system of linear equations could be written in the form

$$AX = B$$

where A is the coefficient matrix, X is the column matrix of unknowns, and B is the column matrix of right-hand sides. Now suppose that there are as many unknowns as equations, so that A is a square matrix, and suppose that A is invertible. The object is to solve for the matrix X of unknowns, so we multiply both sides of the equation by the inverse A^{-1} of A, getting

$$A^{-1}AX = A^{-1}B$$

Notice that we put A^{-1} on the left on both sides of the equation. Order matters when multiplying matrices, so we have to be careful to do the same thing to both sides of the equation. But now $A^{-1}A = I$, so we can rewrite the last equation as

$$IX = A^{-1}B$$

Also, $IX = X$ (I being the identity matrix), so we really have

$$X = A^{-1}B$$

and we have solved for X!

Moreover, we have shown that, if A is invertible and $AX = B$, then the only *possible* solution is $X = A^{-1}B$. We should check that $A^{-1}B$ is actually a solution by substituting back into the original equation.

$$AX = A(A^{-1}B) = (AA^{-1})B = IB = B$$

Thus, $X = A^{-1}B$ is a solution and is the only solution. Therefore, if A is invertible, $AX = B$ has exactly one solution.

On the other hand, if $AX = B$ has no solutions or has infinitely many solutions, we can conclude that A is not invertible (why?). To summarize:

Solving the Matrix Equation $AX = B$

If A is an invertible matrix, then the matrix equation $AX = B$ has the unique solution

$$X = A^{-1}B$$

quick Example

The system of linear equations

$$2x \qquad + z = 9$$
$$2x + y - z = 6$$
$$3x + y - z = 9$$

can be written as $AX = B$, where

$$A = \begin{bmatrix} 2 & 0 & 1 \\ 2 & 1 & -1 \\ 3 & 1 & -1 \end{bmatrix}, X = \begin{bmatrix} x \\ y \\ z \end{bmatrix} \quad \text{and} \quad B = \begin{bmatrix} 9 \\ 6 \\ 9 \end{bmatrix}$$

The matrix A is invertible with inverse

$$A^{-1} = \begin{bmatrix} 0 & -1 & 1 \\ 1 & 5 & -4 \\ 1 & 2 & -2 \end{bmatrix} \qquad \text{You should check this}$$

Thus,

$$X = A^{-1}B = \begin{bmatrix} 0 & -1 & 1 \\ 1 & 5 & -4 \\ 1 & 2 & -2 \end{bmatrix} \begin{bmatrix} 9 \\ 6 \\ 9 \end{bmatrix} = \begin{bmatrix} 3 \\ 3 \\ 3 \end{bmatrix}$$

so that $(x, y, z) = (3, 3, 3)$ is the (unique) solution to the system.

Example 4 Solving Systems of Equations Using an Inverse

Solve the following three systems of equations.

a. $\begin{aligned} 2x \quad\quad + z &= 1 \\ 2x + y - z &= 1 \\ 3x + y - z &= 1 \end{aligned}$
b. $\begin{aligned} 2x \quad\quad + z &= 0 \\ 2x + y - z &= 1 \\ 3x + y - z &= 2 \end{aligned}$
c. $\begin{aligned} 2x \quad\quad + z &= 0 \\ 2x + y - z &= 0 \\ 3x + y - z &= 0 \end{aligned}$

using *Technology*

See the Technology Guides at the end of the chapter to see how to do these calculations using a TI-83/84 or Excel. If, instead, you use the online Matrix Algebra Tool, you can use the formulas `A^-1*B`, `A^-1*C`, and `A^-1*D` to compute these products after entering the matrices *A*, *B* and *C* as usual (See p. 175.)

Solution We *could* go ahead and row-reduce all three augmented matrices as we did in Chapter 2, but this would require a lot of work. Notice that the coefficients are the same in all three systems. In other words, we can write the three systems in matrix form as

a. $AX = B$ **b.** $AX = C$ **c.** $AX = D$

where the matrix A is the same in all three cases:

$$A = \begin{bmatrix} 2 & 0 & 1 \\ 2 & 1 & -1 \\ 3 & 1 & -1 \end{bmatrix}$$

Now the solutions to these systems are

a. $X = A^{-1}B$ **b.** $X = A^{-1}C$ **c.** $X = A^{-1}D$

so the main work is the calculation of the single matrix A^{-1}, which we have already noted (Quick Example above) is

$$A^{-1} = \begin{bmatrix} 0 & -1 & 1 \\ 1 & 5 & -4 \\ 1 & 2 & -2 \end{bmatrix}$$

Thus, the three solutions are

a. $X = A^{-1}B = \begin{bmatrix} 0 & -1 & 1 \\ 1 & 5 & -4 \\ 1 & 2 & -2 \end{bmatrix} \begin{bmatrix} 1 \\ 1 \\ 1 \end{bmatrix} = \begin{bmatrix} 0 \\ 2 \\ 1 \end{bmatrix}$

b. $X = A^{-1}C = \begin{bmatrix} 0 & -1 & 1 \\ 1 & 5 & -4 \\ 1 & 2 & -2 \end{bmatrix} \begin{bmatrix} 0 \\ 1 \\ 2 \end{bmatrix} = \begin{bmatrix} 1 \\ -3 \\ -2 \end{bmatrix}$

c. $X = A^{-1}D = \begin{bmatrix} 0 & -1 & 1 \\ 1 & 5 & -4 \\ 1 & 2 & -2 \end{bmatrix} \begin{bmatrix} 0 \\ 0 \\ 0 \end{bmatrix} = \begin{bmatrix} 0 \\ 0 \\ 0 \end{bmatrix}$

+*Before we go on...* We have been speaking of *the* inverse of a matrix A. Is there only one? It is not hard to prove that a matrix A cannot have more than one inverse: If B and C were both inverses of A, then

$$
\begin{aligned}
B &= BI && \text{Property of the identity} \\
&= B(AC) && \text{Since } C \text{ is an inverse of } A \\
&= (BA)C && \text{Associative law} \\
&= IC && \text{Since } B \text{ is an inverse of } A \\
&= C && \text{Property of the identity}
\end{aligned}
$$

In other words, if B and C were both inverses of A, then B and C would have to be equal. ∎

FAQs   Which Method to Use in Solving a System

Q: *Now we have two methods to solve a system of linear equations $AX = B$:   (1) Compute $X = A^{-1}B$, or (2) row-reduce the augmented matrix. Which is the better method?*

A: Each method has its advantages and disadvantages. Method (1), as we have seen, is very efficient when you must solve several systems of equations with the same coefficients, but it works only when the coefficient matrix is *square* (meaning that you have the same number of equations as unknowns) *and invertible* (meaning that there is a unique solution). The row-reduction method will work for all systems. Moreover, for all but the smallest systems, the most efficient way to find A^{-1} is to use row reduction. Thus, in practice, the two methods are essentially the same when both apply. ∎

3.3 EXERCISES

● denotes basic skills exercises
◆ denotes challenging exercises
tech Ex indicates exercises that should be solved using technology

In Exercises 1–6, determine whether or not the given pairs of matrices are inverse pairs. hint [see Quick Examples p. 199]

1. ● $A = \begin{bmatrix} 0 & 1 \\ 1 & 0 \end{bmatrix}$, $B = \begin{bmatrix} 0 & 1 \\ 1 & 0 \end{bmatrix}$

2. ● $A = \begin{bmatrix} 2 & 0 \\ 0 & 3 \end{bmatrix}$, $B = \begin{bmatrix} \frac{1}{2} & 0 \\ 0 & \frac{1}{2} \end{bmatrix}$

3. ● $A = \begin{bmatrix} 2 & 1 & 1 \\ 0 & 1 & 1 \\ 0 & 0 & 1 \end{bmatrix}$, $B = \begin{bmatrix} \frac{1}{2} & -\frac{1}{2} & 0 \\ 0 & 1 & -1 \\ 0 & 0 & 1 \end{bmatrix}$

4. ● $A = \begin{bmatrix} 1 & 1 & 1 \\ 0 & 1 & 1 \\ 0 & 0 & 1 \end{bmatrix}$, $B = \begin{bmatrix} 1 & -1 & 0 \\ 0 & 1 & -1 \\ 0 & 0 & 1 \end{bmatrix}$

5. ● $A = \begin{bmatrix} a & 0 & 0 \\ 0 & b & 0 \\ 0 & 0 & 0 \end{bmatrix}$, $B = \begin{bmatrix} a^{-1} & 0 & 0 \\ 0 & b^{-1} & 0 \\ 0 & 0 & 0 \end{bmatrix}$ $(a, b \neq 0)$

6. ● $A = \begin{bmatrix} a & 0 & 0 \\ 0 & b & 0 \\ 0 & 0 & c \end{bmatrix}$, $B = \begin{bmatrix} a^{-1} & 0 & 0 \\ 0 & b^{-1} & 0 \\ 0 & 0 & c^{-1} \end{bmatrix}$ $(a, b, c \neq 0)$

In Exercises 7–26, use row reduction to find the inverses of the given matrices if they exist, and check your answers by multiplication. hint [see Example 2]

7. ● $\begin{bmatrix} 1 & 1 \\ 2 & 1 \end{bmatrix}$ **8.** ● $\begin{bmatrix} 0 & 1 \\ 1 & 1 \end{bmatrix}$ **9.** ● $\begin{bmatrix} 0 & 1 \\ 1 & 0 \end{bmatrix}$ **10.** ● $\begin{bmatrix} 4 & 0 \\ 0 & 2 \end{bmatrix}$

11. ● $\begin{bmatrix} 2 & 1 \\ 1 & 1 \end{bmatrix}$ **12.** ● $\begin{bmatrix} 3 & 0 \\ 0 & \frac{1}{2} \end{bmatrix}$ **13.** ● $\begin{bmatrix} 2 & 1 \\ 4 & 2 \end{bmatrix}$ **14.** ● $\begin{bmatrix} 1 & 1 \\ 6 & 6 \end{bmatrix}$

15. ● $\begin{bmatrix} 1 & 1 & 1 \\ 0 & 1 & 1 \\ 0 & 0 & 1 \end{bmatrix}$ **16.** ● $\begin{bmatrix} 1 & 2 & 3 \\ 0 & 1 & 2 \\ 0 & 0 & 1 \end{bmatrix}$ **17.** ● $\begin{bmatrix} 1 & 1 & 1 \\ 1 & 0 & 2 \\ 1 & -1 & 1 \end{bmatrix}$

18. ● $\begin{bmatrix} 1 & 2 & 3 \\ 0 & 2 & 3 \\ 1 & 0 & 1 \end{bmatrix}$ **19.** ● $\begin{bmatrix} 1 & 1 & 1 \\ 1 & -1 & 0 \\ 1 & 2 & 3 \end{bmatrix}$ **20.** ● $\begin{bmatrix} 1 & -1 & 3 \\ 0 & 1 & 3 \\ 1 & 1 & 1 \end{bmatrix}$

● basic skills   ◆ challenging   **tech** Ex technology exercise

21. $\begin{bmatrix} 1 & 1 & 1 \\ 1 & 0 & 1 \\ 1 & -1 & 1 \end{bmatrix}$ **22.** $\begin{bmatrix} 1 & 1 & 1 \\ 0 & 1 & 1 \\ 1 & 0 & 0 \end{bmatrix}$

23. $\begin{bmatrix} 1 & 0 & 1 & 0 \\ -1 & 1 & 0 & 1 \\ -1 & 0 & 0 & 1 \\ 0 & -1 & 0 & 1 \end{bmatrix}$ **24.** $\begin{bmatrix} 0 & 1 & 1 & 0 \\ -1 & 1 & 1 & 1 \\ -1 & 1 & 0 & 1 \\ 0 & -1 & 0 & 1 \end{bmatrix}$

25. $\begin{bmatrix} 1 & 2 & 3 & 4 \\ 0 & 1 & 2 & 3 \\ 0 & 0 & 1 & 2 \\ 0 & 0 & 0 & 1 \end{bmatrix}$ **26.** $\begin{bmatrix} 0 & 0 & 0 & 1 \\ 0 & 0 & 1 & 0 \\ 0 & 1 & 0 & 0 \\ 1 & 0 & 0 & 0 \end{bmatrix}$

In Exercises 27–34, compute the determinant of the given matrix. If the determinant is nonzero, use the formula for inverting a 2 × 2 matrix to calculate the inverse of the given matrix. hint [see Quick Examples p. 202]

27. $\begin{bmatrix} 1 & 1 \\ 1 & -1 \end{bmatrix}$ **28.** $\begin{bmatrix} 4 & 1 \\ 0 & 2 \end{bmatrix}$ **29.** $\begin{bmatrix} 1 & 2 \\ 3 & 4 \end{bmatrix}$

30. $\begin{bmatrix} 1 & 0 \\ 0 & 1 \end{bmatrix}$ **31.** $\begin{bmatrix} \frac{1}{6} & -\frac{1}{6} \\ 0 & \frac{1}{6} \end{bmatrix}$

32. $\begin{bmatrix} 2 & 1 \\ 4 & 2 \end{bmatrix}$ **33.** $\begin{bmatrix} 1 & 0 \\ \frac{3}{4} & 0 \end{bmatrix}$ **34.** $\begin{bmatrix} 1 & 1 \\ 1 & 1 \end{bmatrix}$

tech Ex *In Exercises 35–42, use technology to find the inverse of the given matrix (when it exists). Round all entries in your answer to two decimal places. [Caution: Because of rounding errors, technology sometimes produces an "inverse" of a singular matrix. These can be often recognized by their huge entries.]*

35. $\begin{bmatrix} 1.1 & 1.2 \\ 1.3 & -1 \end{bmatrix}$ **36.** $\begin{bmatrix} 0.1 & -3.2 \\ 0.1 & -1.5 \end{bmatrix}$

37. $\begin{bmatrix} 3.56 & 1.23 \\ -1.01 & 0 \end{bmatrix}$ **38.** $\begin{bmatrix} 9.09 & -5.01 \\ 1.01 & 2.20 \end{bmatrix}$

39. $\begin{bmatrix} 1.1 & 3.1 & 2.4 \\ 1.7 & 2.4 & 2.3 \\ 0.6 & -0.7 & -0.1 \end{bmatrix}$ **40.** $\begin{bmatrix} 2.1 & 2.4 & 3.5 \\ 6.1 & -0.1 & 2.3 \\ -0.3 & -1.2 & 0.1 \end{bmatrix}$

41. $\begin{bmatrix} 0.01 & 0.32 & 0 & 0.04 \\ -0.01 & 0 & 0 & 0.34 \\ 0 & 0.32 & -0.23 & 0.23 \\ 0 & 0.41 & 0 & 0.01 \end{bmatrix}$

42. $\begin{bmatrix} 0.01 & 0.32 & 0 & 0.04 \\ -0.01 & 0 & 0 & 0.34 \\ 0 & 0.32 & -0.23 & 0.23 \\ 0.01 & 0.96 & -0.23 & 0.65 \end{bmatrix}$

In Exercises 43–48, use matrix inversion to solve the given systems of linear equations. (You previously solved all of these systems using row reduction in Chapter 2.) hint [see Quick Example p. 203]

43. $x + y = 4$
$x - y = 1$

44. $2x + y = 2$
$2x - 3y = 2$

45. $\dfrac{x}{3} + \dfrac{y}{2} = 0$
$\dfrac{x}{2} + y = -1$

46. $\dfrac{2x}{3} - \dfrac{y}{2} = \dfrac{1}{6}$
$\dfrac{x}{2} - \dfrac{y}{2} = -1$

47. $-x + 2y - z = 0$
$-x - y + 2z = 0$
$2x \quad - z = 6$

48. $x + 2y \quad = 4$
$y - z = 0$
$x + 3y - 2z = 5$

In Exercises 49 and 50, use matrix inversion to solve each collection of systems of linear equations. hint [see Example 4]

49. a. $-x - 4y + 2z = 4$
$x + 2y - z = 3$
$x + y - z = 8$

b. $-x - 4y + 2z = 0$
$x + 2y - z = 3$
$x + y - z = 2$

c. $-x - 4y + 2z = 0$
$x + 2y - z = 0$
$x + y - z = 0$

50. a. $-x - 4y + 2z = 8$
$x \quad - z = 3$
$x + y - z = 8$

b. $-x - 4y + 2z = 8$
$x \quad - z = 3$
$x + y - z = 2$

c. $-x - 4y + 2z = 0$
$x \quad - z = 0$
$x + y - z = 0$

Applications

Some of the following exercises are similar or identical to exercises and examples in Chapter 2. Use matrix inverses to find the solutions. We suggest you invert some of the matrices by hand, and others using technology.

51. ● *Nutrition* A four-ounce serving of Campbell's® Pork & Beans contains 5 grams of protein and 21 grams of carbohydrates.[16] A typical slice of "lite" rye bread contains 4 grams of protein and 12 grams of carbohydrates.

a. I am planning a meal of "beans-on-toast" and I want it to supply 20 grams of protein and 80 grams of carbohydrates. How should I prepare my meal?

b. If I require *A* grams of protein and *B* grams of carbohydrates, give a formula that tells me how many slices of bread and how many servings of Pork & Beans to use.

52. ● *Nutrition* According to the nutritional information on a package of Honey Nut Cheerios® brand cereal, each 1-ounce serving of Cheerios contains 3 grams protein and 24 grams carbohydrates.[17] Each half-cup serving of enriched skim milk contains 4 grams protein and 6 grams carbohydrates.

a. I am planning a meal of cereal and milk and I want it to supply 26 grams of protein, and 78 grams of carbohydrates. How should I prepare my meal?

---

[16] According to the label information on a 16-oz can.

[17] Actually, it is 23 grams carbohydrates. We made it 24 grams to simplify the calculation.

● basic skills   ◆ challenging   tech Ex technology exercise

b. If I require A grams of protein and B grams of carbohydrates, give a formula that tells me how many servings of milk and Cheerios to use.

53. ● *Resource Allocation* You manage an ice cream factory that makes three flavors: Creamy Vanilla, Continental Mocha, and Succulent Strawberry. Into each batch of Creamy Vanilla go two eggs, one cup of milk, and two cups of cream. Into each batch of Continental Mocha go one egg, one cup of milk, and two cups of cream. Into each batch of Succulent Strawberry go one egg, two cups of milk, and one cup of cream. Your stocks of eggs, milk, and cream vary from day to day. How many batches of each flavor should you make in order to use up all of your ingredients if you have the following amounts in stock?

a. 350 eggs, 350 cups of milk, and 400 cups of cream
b. 400 eggs, 500 cups of milk, and 400 cups of cream
c. A eggs, B cups of milk, and C cups of cream

54. ● *Resource Allocation* The Arctic Juice Company makes three juice blends: PineOrange, using 2 quarts of pineapple juice and 2 quarts of orange juice per gallon; PineKiwi, using 3 quarts of pineapple juice and 1 quart of kiwi juice per gallon; and OrangeKiwi, using 3 quarts of orange juice and 1 quart of kiwi juice per gallon. The amount of each kind of juice the company has on hand varies from day to day. How many gallons of each blend can it make on a day with the following stocks?

a. 800 quarts of pineapple juice, 650 quarts of orange juice, 350 quarts of kiwi juice.
b. 650 quarts of pineapple juice, 800 quarts of orange juice, 350 quarts of kiwi juice.
c. A quarts of pineapple juice, B quarts of orange juice, C quarts of kiwi juice.

Investing in Municipal Bond Funds *Exercises 55 and 56 are based on the following data on three tax-exempt municipal bond funds.*[18]

	2003 Yield
PNF (Pimco NY)	6%
FDMMX (Fidelity Spartan Mass)	5%
FFLIX (Fidelity Spartan Florida)	7%

55. ● You invested a total of $9000 in the three funds at the beginning of 2003, including an equal amount in FDMMX and FFLIX. Your 2003 yield for the year from the first two funds amounted to $400. How much did you invest in each of the three funds?

56. ● You invested a total of $6000 in the three funds at the beginning of 2003, including an equal amount in PNF and FDMMX. Your total yields for 2003 amounted to $360. How much did you invest in each of the three funds?

[18] Yields are rounded. Sources: www.pimcofunds.com, www.fidelity.com. January 18, 2004.

Investing in Stocks *Exercises 57 and 58 are based on the following data on three computer stocks.*[19]

	Price	Dividend Yield
APPL (Apple Computer)	$25	0.6%
HPQ (Hewlett Packard)	25	1.2
DELL	35	0

57. You invested a total of $5800 in Apple, Hewlett Packard, and Dell shares at the above prices, and expected to earn $21 in annual dividends. If you purchased a total of 200 shares, how many shares of each stock did you purchase?

58. You invested a total of $6750 in Apple, Hewlett Packard, and Dell shares at the above prices, and expected to earn $45 in annual dividends. If you purchased a total of 250 shares, how many shares of each stock did you purchase?

59. ◆ tech Ex *Population Movement* In 2003, the population of the U.S., broken down by regions, was 53.3 million in the Northeast, 64.0 million in the Midwest, 101.6 million in the South, and 65.4 million in the West.[20] The matrix P below shows the population movement during the period 2003–2004. (Thus, 98.79% of the population in the Northeast stayed there, while 0.20% of the population in the Northeast moved to the Midwest, and so on.)

$$P = \begin{array}{c} \\ \text{From NE} \\ \text{From MW} \\ \text{From S} \\ \text{From W} \end{array} \begin{array}{cccc} \overset{\text{To}}{\underset{\text{NE}}{}} & \overset{\text{To}}{\underset{\text{MW}}{}} & \overset{\text{To}}{\underset{\text{S}}{}} & \overset{\text{To}}{\underset{\text{W}}{}} \\ \begin{bmatrix} 0.9879 & 0.0020 & 0.0081 & 0.0019 \\ 0.0014 & 0.9895 & 0.0063 & 0.0028 \\ 0.0027 & 0.0025 & 0.9927 & 0.0022 \\ 0.0010 & 0.0030 & 0.0050 & 0.9909 \end{bmatrix} \end{array}$$

Set up the 2003 population figures as a row vector. Assuming that these percentages also describe the population movements from 2002 to 2003, show how matrix inversion and multiplication allow you to compute the population in each region in 2002. (Round all answers to the nearest 0.1 million.)

60. ◆ tech Ex *Population Movement* Assume that the percentages given in the preceding exercise also describe the population movements from 2001 to 2002. Show how matrix inversion and multiplication allow you to compute the population in each region in 2001. (Round all answers to the nearest 0.1 million.)

61. ◆ tech Ex *Rotations* If a point (x, y) in the plane is rotated counterclockwise through an angle of $45°$, its new coordinates

[19] Stocks were trading at or near the given prices in January, 2004. Dividends are rounded, based on values in January 18, 2004. Source: http://money.excite.com, January 18, 2004.

[20] Note that this exercise ignores migration into or out of the country. The internal migration figures and 2004 population figures are accurate. Source: U.S. Census Bureau, Current Population Survey, 2004 Annual Social and Economic Supplement, Internet release date June 23, 2005. www.census.gov.

● basic skills   ◆ challenging   tech Ex technology exercise

(x', y') are given by

$$\begin{bmatrix} x' \\ y' \end{bmatrix} = R \begin{bmatrix} x \\ y \end{bmatrix}$$

where R is the 2×2 matrix $\begin{bmatrix} a & -a \\ a & a \end{bmatrix}$ and $a = \sqrt{1/2} \approx 0.7071$.

a. If the point $(2, 3)$ is rotated counterclockwise through an angle of $45°$, what are its (approximate) new coordinates?

b. Multiplication by what matrix would result in a counter-clockwise rotation of $90°$? $135°$? (Express the matrices in terms of R. Hint: Think of a rotation through $90°$ as two successive rotations through $45°$.)

c. Multiplication by what matrix would result in a *clockwise* rotation of $45°$?

62. ◆ `tech` Ex **Rotations** If a point (x, y) in the plane is rotated counterclockwise through an angle of $60°$, its new coordinates are given by

$$\begin{bmatrix} x' \\ y' \end{bmatrix} = S \begin{bmatrix} x \\ y \end{bmatrix}$$

where S is the 2×2 matrix $\begin{bmatrix} a & -b \\ b & a \end{bmatrix}$ and $a = 1/2$ and $b = \sqrt{3/4} \approx 0.8660$.

a. If the point $(2, 3)$ is rotated counterclockwise through an angle of $60°$, what are its (approximate) new coordinates?

b. Referring to Exercise 61, multiplication by what matrix would result in a counterclockwise rotation of $105°$? (Express the matrices in terms of S and the matrix R from Exercise 61. Hint: Think of a rotation through $105°$ as a rotation through $60°$ followed by one through $45°$.)

c. Multiplication by what matrix would result in a *clockwise* rotation of $60°$?

`tech` Ex **Encryption** Matrices are commonly used to encrypt data. Here is a simple form such an encryption can take. First, we represent each letter in the alphabet by a number. For example, if we take <space> = 0, A = 1, B = 2, and so on,

"ABORT MISSION" becomes

[1　2　15　18　20　0　13　9　19　19　9　15　14]

To encrypt this coded phrase, we use an invertible matrix of any size with integer entries. For instance, let us take A to be the 2×2 matrix $\begin{bmatrix} 1 & 2 \\ 3 & 4 \end{bmatrix}$. We can first arrange the coded sequence of numbers in the form of a matrix with two rows (using zero in the last place if we have an odd number of characters) and then multiply on the left by A.

$$\text{Encrypted Matrix} = \begin{bmatrix} 1 & 2 \\ 3 & 4 \end{bmatrix} \begin{bmatrix} 1 & 15 & 20 & 13 & 19 & 9 & 14 \\ 2 & 18 & 0 & 9 & 19 & 15 & 0 \end{bmatrix}$$

$$= \begin{bmatrix} 5 & 51 & 20 & 31 & 57 & 39 & 14 \\ 11 & 117 & 60 & 75 & 133 & 87 & 42 \end{bmatrix}$$

.ich we can also write as

[5　11　51　117　20　60　31　75　57　133　39　87　14　42].

To decipher the encoded message, multiply the encrypted matrix by A^{-1}.

63. Use the matrix A to encode the phrase "GO TO PLAN B."

64. Use the matrix A to encode the phrase "ABANDON SHIP."

65. Decode the following message, which was encrypted using the matrix A.

[33　69　54　126　11　27　20　60　29　59　65　149　41　87]

66. Decode the following message, which was encrypted using the matrix A.

[59　141　43　101　7　21　29　59　65　149　41　87]

Communication and Reasoning Exercises

67. ● Multiple Choice: If A and B are square matrices with $AB = I$ and $BA = I$, then
(A) B is the inverse of A.　　**(B)** A and B must be equal.
(C) A and B must both be singular.
(D) At least one of A and B is singular.

68. ● Multiple Choice: If A is a square matrix with $A^3 = I$, then
(A) A must be the identity matrix.　**(B)** A is invertible.
(C) A is singular.　**(D)** A is both invertible and singular.

69. ● What can you say about the inverse of a 2×2 matrix of the form $\begin{bmatrix} a & b \\ a & b \end{bmatrix}$?

70. ● If you think of numbers as 1×1 matrices, which numbers are invertible 1×1 matrices?

71. Use matrix multiplication to check that the inverse of a general 2×2 matrix is given by

$$\begin{bmatrix} a & b \\ c & d \end{bmatrix}^{-1} = \frac{1}{ad - bc} \begin{bmatrix} d & -b \\ -c & a \end{bmatrix} \quad (\text{provided } ad - bc \neq 0)$$

72. ◆ Derive the formula in Exercise 71 using row reduction. (Assume that $ad - bc \neq 0$.)

73. A **diagonal** matrix D has the following form.

$$D = \begin{bmatrix} d_1 & 0 & 0 & \cdots & 0 \\ 0 & d_2 & 0 & \cdots & 0 \\ 0 & 0 & d_3 & \cdots & 0 \\ \vdots & \vdots & \vdots & \ddots & \vdots \\ 0 & 0 & 0 & \cdots & d_n \end{bmatrix}$$

When is D singular? Why?

74. If a square matrix A row-reduces to the identity matrix, must it be invertible? If so, say why, and if not, give an example of such a (singular) matrix.

75. If A and B are invertible, check that $B^{-1}A^{-1}$ is the inverse of AB.

76. Solve the matrix equation $A(B + CX) = D$ for X. (You may assume that all the matrices are square and invertible.)

77. ◆ In Example 3 we said that, if a square matrix A row-reduces to a matrix with a row of zeros is singular. Why?

78. ◆ Your friend has two square matrices A and B, neither of them the zero matrix, with the property that AB is the zero matrix. You immediately tell him that neither A nor B can possibly be invertible. How can you be so sure?

● basic skills　　◆ challenging　　`tech` Ex technology exercise

3.4 | Game Theory

It frequently happens that you are faced with making a decision or choosing a best strategy from several possible choices. For instance, you might need to decide whether to invest in stocks or bonds, whether to cut prices of the product you sell, or what offensive play to use in a football game. In these examples, the result depends on something you cannot control. In the first case, your success depends on the future behavior of the economy. In the second case, it depends in part on whether your competitors also cut prices, and in the third case, it depends on the defensive strategy chosen by the opposing team.

We can model situations like these using **game theory.** We represent the various options and payoffs in a matrix and can then calculate the best single strategy or combination of strategies using matrix algebra and other techniques.

Game theory is very new compared with most of the mathematics you learn. It was invented in the 1920s by the noted mathematicians Emile Borel (1871–1956) and John von Neumann (1903–1957). Game theory's connection with linear programming was discovered even more recently, in 1947, by von Neumann, and further advances were made by the mathematician John Nash[21] (1928–), for which he received the 1994 Nobel Prize for Economics.

The Payoff Matrix and Expected Value of a Game

We have probably all played the simple game "Paper, Scissors, Rock" at some time in our lives. It goes as follows: There are two players—let us call them A and B—and at each turn, both players produce, by a gesture of the hand, either paper, a pair of scissors, or a rock. Rock beats scissors (since a rock can crush scissors), but is beaten by paper (since a rock can be covered by paper), while scissors beat paper (since scissors can cut paper). The round is a draw if both A and B show the same item. We could turn this into a betting game if, at each turn, we require the loser to pay the winner 1¢. For instance, if A shows a rock and B shows paper, then A pays B 1¢.

Paper, Scissors, Rock is an example of a **two-person zero-sum game.** It is called a zero-sum game because each player's loss is equal to the other player's gain.[22] We can represent this game by a matrix, called the **payoff matrix.**

$$
\mathbf{A} \begin{array}{c} \\ p \\ s \\ r \end{array}
\begin{array}{c} \mathbf{B} \\ \begin{array}{ccc} p & s & r \end{array} \\ \left[\begin{array}{ccc} 0 & -1 & 1 \\ 1 & 0 & -1 \\ -1 & 1 & 0 \end{array} \right] \end{array}
$$

or just

$$
P = \left[\begin{array}{ccc} 0 & -1 & 1 \\ 1 & 0 & -1 \\ -1 & 1 & 0 \end{array} \right]
$$

if we choose to omit the labels. In the payoff matrix, Player A's options, or **moves,** are listed on the left, while Player B's options are listed on top. We think of A as playing the rows and B as playing the columns. Positive numbers indicate a win for the row player,

---

[21] Nash's turbulent life is the subject of the biography "A Beautiful Mind" by Sylvia Nasar (Simon & Schuster, 1998). The 2001 Academy Award-winning movie of the same title is a somewhat fictionalized account.

[22] An example of a *nonzero sum game* would be one in which the government taxed the earnings of the winner. In that case the winner's gain would be less than the loser's loss.

while negative numbers indicate a loss for the row player. Thus, for example, the p, s entry represents the outcome if A plays p (paper) and B plays s (scissors). In this event, B wins, and the -1 entry there indicates that A loses 1¢. (If that entry were -2 instead, it would have meant that A loses 2¢.)

Two-Person Zero-Sum Game

A **two-person zero-sum game** is one in which one player's loss equals the other's gain. We assume that the outcome is determined by each player's choice from among a fixed, finite set of moves. If Player A has m moves to choose from and Player B has n, we can represent the game using the **payoff matrix,** the $m \times n$ matrix showing the result of each possible pair of choices of moves.

In each round of the game, the way a player chooses a move is called a **strategy.** A player using a **pure strategy** makes the same move each round of the game. For example, if a player in the above game chooses to play scissors at each turn, then that player is using the pure strategy s. A player using a **mixed strategy** chooses each move a certain percentage of the time in a random fashion; for instance, Player A might choose to play p 50% of the time, and each of s and r 25% of the time.

Our ultimate goal is to be able to determine which strategy is best for each player to use. To do that, we need to know how to evaluate strategies. The fundamental calculation we need is that of the **expected value** of a pair of strategies. Let's look at a simple example.

Example 1 Expected Value

Consider the following game:

$$\begin{array}{cc} & \mathbf{B} \\ & \begin{array}{cc} a & b \end{array} \\ \mathbf{A} \begin{array}{c} p \\ q \end{array} & \begin{bmatrix} 3 & -1 \\ -2 & 3 \end{bmatrix} \end{array}$$

Player A decides to pick moves at random, choosing to play p 75% of the time and q 25% of the time. Player B also picks moves at random, choosing a 20% of the time and b 80% of the time. On average, how much does A expect to win or lose?

Solution Suppose they play the game 100 times. Each time they play there are four possible outcomes:

Case 1: A plays p, B plays a.
Because A plays p only 75% of the time and B plays a only 20% of the time, we expect this case to occur $0.75 \times 0.20 = 0.15$, or 15% of the time, or 15 times out of 100. Each time this happens, A gains 3 points, so we get a contribution of $15 \times 3 = 45$ points to A's total winnings.

Case 2: A plays p, B plays b.
Because A plays p only 75% of the time and B plays b only 80% of the time, we expect this case to occur $0.75 \times 0.80 = 0.60$, or 60 times out of 100. Each time this happens, A loses 1 point, so we get a contribution of: $60 \times -1 = -60$ to A's total winnings.

Case 3: A plays q, B plays a.
This case occurs $0.25 \times 0.20 = 0.05$, or 5 out of 100 times, with a loss of 2 points to A each time, giving a contribution of $5 \times -2 = -10$ to A's total winnings.

Case 4: *A plays q, B plays b.*

This case occurs $0.25 \times 0.80 = 0.20$, or 20 out of 100 times, with a gain of 3 points to A each time, giving a contribution of $20 \times 3 = 60$ to A's total winnings.

Summing to get A's total winnings and then dividing by the number of times the game is played gives the average value of

$$(45 - 60 - 10 + 60)/100 = 0.35$$

so that A can expect to win an average of 0.35 points per play of the game. We call 0.35 the **expected value** of the game when A and B use these particular strategies.

This calculation was somewhat tedious and it would only get worse if A and B had many moves to choose from. There is a far more convenient way of doing exactly the same calculation, using matrix multiplication: We start by representing the player's strategies as matrices. For reasons to become clear in a moment, we record A's strategy as a row matrix:

$$R = [0.75 \quad 0.25]$$

We record B's strategy as a column matrix:

$$C = \begin{bmatrix} 0.20 \\ 0.80 \end{bmatrix}$$

(We will sometimes write column vectors using transpose notation, writing, for example, $[0.20 \quad 0.80]^T$ for the column above, to save space.) Now: *the expected value is the matrix product RPC where P is the payoff matrix!*

$$\text{Expected value} = RPC = [0.75 \quad 0.25] \begin{bmatrix} 3 & -1 \\ -2 & 3 \end{bmatrix} \begin{bmatrix} 0.20 \\ 0.80 \end{bmatrix}$$

$$= [1.75 \quad 0] \begin{bmatrix} 0.20 \\ 0.80 \end{bmatrix} = [0.35]$$

Why does this work? Write out the arithmetic involved in the matrix product RPC to see what we calculated:

$$[0.75 \times 3 + 0.25 \times (-2)] \times 0.20 + [0.75 \times (-1) + 0.25 \times 3] \times 0.80$$

$$= 0.75 \times 3 \times 0.20 + 0.25 \times (-2) \times 0.20 + 0.75 \times (-1) \times 0.80 + 0.25 \times 3 \times 0.80$$
$$= \quad \text{Case 1} \quad + \quad \text{Case 3} \quad + \quad \text{Case 2} \quad + \quad \text{Case 4}$$

So, the matrix product does all at once the various cases we considered above.

To summarize what we just saw:

▦ using *Technology*

The use of technology becomes indispensable when we need to do several calculations or when the matrices involved are big. See the end-of-chapter Technology Guide notes on Example 3 in Section 3.2 for instructions on multiplying matrices using a TI-83/84 or Excel. Online, use the Matrix Algebra Tool.

The Expected Value of a Game for Mixed Strategies *R* and *C*

The **expected value of a game for given mixed strategies** is the average payoff that occurs if the game is played a large number of times with the row and column players using the given strategies.

To compute the expected value of a game:

1. Write the row player's mixed strategy as a row matrix R.

2. Write the column player's mixed strategy as a column matrix C.

3. Calculate the product RPC, where P is the payoff matrix. This product is a 1×1 matrix whose entry is the expected value e.

quick Example

Consider "Paper, Scissors, Rock,"

$$\mathbf{A} \begin{array}{c} \\ p \\ s \\ r \end{array} \begin{array}{ccc} & \mathbf{B} & \\ p & s & r \\ \left[\begin{array}{ccc} 0 & -1 & 1 \\ 1 & 0 & -1 \\ -1 & 1 & 0 \end{array}\right] \end{array}$$

Suppose that the row player plays *paper* half the time and each of the other two strategies a quarter of the time, and the column player always plays *scissors*. We write

$$R = \begin{bmatrix} \dfrac{1}{2} & \dfrac{1}{4} & \dfrac{1}{4} \end{bmatrix}, \quad C = \begin{bmatrix} 0 \\ 1 \\ 0 \end{bmatrix}$$

so,

$$e = RPC = \begin{bmatrix} \dfrac{1}{2} & \dfrac{1}{4} & \dfrac{1}{4} \end{bmatrix} \begin{bmatrix} 0 & -1 & 1 \\ 1 & 0 & -1 \\ -1 & 1 & 0 \end{bmatrix} \begin{bmatrix} 0 \\ 1 \\ 0 \end{bmatrix} = \begin{bmatrix} \dfrac{1}{2} & \dfrac{1}{4} & \dfrac{1}{4} \end{bmatrix} \begin{bmatrix} -1 \\ 0 \\ 1 \end{bmatrix} = -\dfrac{1}{4}$$

Thus, player A loses an average of once every four plays.

Solving a Game

Now that we know how to evaluate particular strategies, we want to find the *best* strategy. The next example takes us another step toward that goal.

Example 2 Television Ratings Wars

Commercial TV station RTV and cultural station CTV are competing for viewers in the Tuesday prime-time 9–10 PM time slot. RTV is trying to decide whether to show a sitcom, a docudrama, a reality show, or a movie, while CTV is thinking about either a nature documentary, a symphony concert, a ballet, or an opera. A television rating company estimates the payoffs for the various alternatives as follows. (Each point indicates a shift of 1000 viewers from one channel to the other; thus, for instance, –2 indicates a shift of 2000 viewers from RTV to CTV.)

		CTV			
		Nature Doc.	Symphony	Ballet	Opera
RTV	Sitcom	2	1	−2	2
	Docudrama	−1	1	−1	2
	Reality Show	−2	0	0	1
	Movie	3	1	−1	1

a. If RTV notices that CTV is showing nature documentaries half the time and symphonies the other half, what would RTV's best strategy be, and how many viewers would it gain if it followed this strategy?

b. If, on the other hand, CTV notices that RTV is showing docudramas half the time and reality shows the other half, what would CTV's best strategy be, and how many viewers would it gain or lose if it followed this strategy?

Solution

a. We are given the matrix of the game, P, in the table above, and we are given CTV's strategy $C = [0.50 \quad 0.50 \quad 0 \quad 0]^T$. We are not given RTV's strategy R. To say that RTV is looking for its best strategy is to say that it wants the expected value $e = RPC$ to be as large as possible. So, we take $R = [x \quad y \quad z \quad t]$ and look for values for x, y, z, and t that make RPC as large as possible. First, we calculate e in terms of these unknowns:

$$e = RPC = [x \quad y \quad z \quad t] \begin{bmatrix} 2 & 1 & -2 & 2 \\ -1 & 1 & -1 & 2 \\ -2 & 0 & 0 & 1 \\ 3 & 1 & -1 & 1 \end{bmatrix} \begin{bmatrix} 0.50 \\ 0.50 \\ 0 \\ 0 \end{bmatrix}$$

$$= [x \quad y \quad z \quad t] \begin{bmatrix} 1.5 \\ 0 \\ -1 \\ 2 \end{bmatrix} = 1.5x - z + 2t$$

Now, the unknowns x, y, z, and t must be nonnegative and add up to 1 (why?). Because t has the largest coefficient, 2, we'll get the best result by making it as large as possible, namely, $t = 1$, leaving $x = y = z = 0$. Thus, RTV's best strategy is $R = [0 \quad 0 \quad 0 \quad 1]$. In other words, RTV should use the pure strategy of showing a movie every Tuesday evening. If it does so, the expected value will be

$$e = 1.5(0) - 0 + 2(1) = 2$$

so RTV can expect to gain 2000 viewers.

b. Here, we are given $R = [0 \quad 0.50 \quad 0.50 \quad 0]$ and are not given CTV's strategy C, so this time we take $C = [x \quad y \quad z \quad t]^T$ and calculate e:

$$e = RPC = [0 \quad 0.50 \quad 0.50 \quad 0] \begin{bmatrix} 2 & 1 & -2 & 2 \\ -1 & 1 & -1 & 2 \\ -2 & 0 & 0 & 1 \\ 3 & 1 & -1 & 1 \end{bmatrix} \begin{bmatrix} x \\ y \\ z \\ t \end{bmatrix}$$

$$= [-1.5 \quad 0.5 \quad -0.5 \quad 1.5] \begin{bmatrix} x \\ y \\ z \\ t \end{bmatrix} = -1.5x + 0.5y - 0.5z + 1.5t$$

Now, CTV wants e to be as *small* as possible (why?). Because x has the largest negative coefficient, CTV would like it to be as large as possible: $x = 1$, so the rest of the unknowns must be zero Thus, CTV's best strategy is $C = [1 \quad 0 \quad 0 \quad 0]^T$, that is, show a nature documentary every night. If it does so, the expected value will be

$$e = -1.5(1) + 0.5(0) - 0.5(0) + 1.5(0) = -1.5$$

so CTV can expect to gain 1500 viewers.

This example illustrates the fact that, no matter what mixed strategy one player selects, the other player can choose an appropriate *pure* counterstrategy in order to maximize its gain. How does this affect what decisions you should make as one of the players? If you were on the board of directors of RTV, you might reason as follows: Since for every mixed strategy you try, CTV can find a best counterstrategy (as in part (b)) it is in your company's best interest to select a mixed strategy that *minimizes* the effect of CTV's best counterstrategy. This is called the **minimax criterion.**

Minimax Criterion

A player using the **minimax criterion** chooses a strategy that, among all possible strategies, minimizes the effect of the other player's best counterstrategy. That is, an optimal (best) strategy according to the minimax criterion is one that minimizes the maximum damage the opponent can cause.

This criterion assumes that your opponent is determined to win. More precisely, it assumes the following.

Fundamental Principle of Game Theory

Each player tries to use its best possible strategy, and assumes that the other player is doing the same.

This principle is not always followed by every player. For example, one of the players may be nature and may choose its move at random, with no particular purpose in mind. In such a case, criteria other than the minimax criterion may be more appropriate. For example, there is the "maximax" criterion, which maximizes the maximum possible payoff (also known as the "reckless" strategy), or the criterion that seeks to minimize "regret" (the difference between the payoff you get and the payoff you *would have gotten* if you had known beforehand what was going to happen).[23] But, we shall assume here the fundamental principle and try to find optimal strategies under the minimax criterion.

Finding the optimal strategy is called **solving the game.** In general, solving a game can be done using linear programming, as we shall see in the next chapter. However, we can solve 2×2 games "by hand" as we shall see in the next example. First, we notice that some large games can be reduced to smaller games.

Consider the game in the preceding example, which had the following matrix:

$$P = \begin{bmatrix} 2 & 1 & -2 & 2 \\ -1 & 1 & -1 & 2 \\ -2 & 0 & 0 & 1 \\ 3 & 1 & -1 & 1 \end{bmatrix}$$

Compare the second and third columns through the eyes of the column player, CTV. Every payoff in the third column is as good as or better, from CTV's point of view, than the corresponding entry in the second column. Thus, no matter what RTV does, CTV will do better showing a ballet (third column) than a symphony (second column). We say that the third column **dominates** the second column. As far as CTV is concerned, we might as well forget about symphonies entirely, so we remove the second column. Similarly, the third column dominates the fourth, so we can remove the fourth column, too. This gives us a smaller game to work with:

$$P = \begin{bmatrix} 2 & -2 \\ -1 & -1 \\ -2 & 0 \\ 3 & -1 \end{bmatrix}$$

---

[23] See *Location in Space: Theoretical Perspectives in Economic Geography,* 3rd Edition, by Peter Dicken and Peter E. Lloyd, HarperCollins Publishers, 1990, pp. 276 ff.

Now compare the first and last rows. Every payoff in the last row is larger than the corresponding payoff in the first row, so the last row is always better to RTV. Again, we say that the last row dominates the first row, and we can discard the first row. Similarly, the last row dominates the second row, so we discard the second row as well. This reduces us to the following game:

$$P = \begin{bmatrix} -2 & 0 \\ 3 & -1 \end{bmatrix}$$

In this matrix, neither row dominates the other and neither column dominates the other. So, this is as far as we can go with this line of argument. We call this **reduction by dominance.**

Reduction by Dominance

One *row* **dominates** another if every entry in the former is greater than or equal to the corresponding entry in the latter. Put another way, one row dominates another if it is always at least as good for the row player.

One *column* dominates another if every entry in the former is less than or equal to the corresponding entry in the latter. Put another way, one column dominates another if it is always at least as good for the column player.

Procedure for Reducing by Dominance:
1. Check whether there is any row in the (remaining) matrix that is dominated by another row. Remove all dominated rows.
2. Check whether there is any column in the (remaining) matrix that is dominated by another column. Remove all dominated columns.
3. Repeat steps 1 and 2 until there are no dominated rows or columns.

Let us now go back to the "television ratings wars" example and see how we can solve a game using the minimax criterion once we are down to a 2×2 payoff matrix.

Example **3 Solving a 2 × 2 Game**

Continuing the preceding example:

a. Find the optimal strategy for RTV.

b. Find the optimal strategy for CTV.

c. Find the expected value of the game if RTV and CTV use their optimal strategies.

Solution As in the text, we begin by reducing the game by dominance, which brings us down to the following 2×2 game:

		CTV	
		Nature Doc.	**Ballet**
RTV	**Reality Show**	−2	0
	Movie	3	−1

a. Now let's find RTV's optimal strategy. Because we don't yet know what it is, we write down a general strategy:

$$R = [x \quad y]$$

Because $x + y = 1$, we can replace y by $1 - x$:

$$R = [x \quad 1 - x]$$

We know that CTV's best counterstrategy to R will be a pure strategy (see the discussion after Example 2), so let's compute the expected value of the game for each of CTV's possible pure strategies:

$$e = [x \quad 1 - x] \begin{bmatrix} -2 & 0 \\ 3 & -1 \end{bmatrix} \begin{bmatrix} 1 \\ 0 \end{bmatrix}$$

$$= (-2)x + 3(1 - x) = -5x + 3$$

$$f = [x \quad 1 - x] \begin{bmatrix} -2 & 0 \\ 3 & -1 \end{bmatrix} \begin{bmatrix} 0 \\ 1 \end{bmatrix}$$

$$= 0x - (1 - x) = x - 1$$

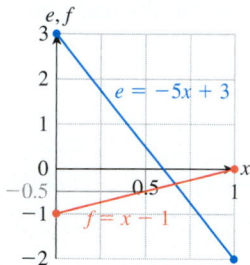

Figure **1**

Because both e and f depend on x, we can graph them as in Figure 1.

If, for instance, RTV happened to choose $x = 0.5$, then the expected values for CTV's two pure strategies are $e = -5(1/2) + 3 = 1/2$ and $f = 1/2 - 1 = -1/2$. The worst outcome for RTV is the lower of the two, f, and this will be true wherever the graph of f is below the graph of e. On the other hand, if RTV chose $x = 1$, the graph of e would be lower and the worst possible expected value would be $e = -5(1) + 3 = -2$. Since RTV can choose x to be any value between 0 and 1, the worst possible outcomes are those shown by the orange portion of the graph in Figure 2.

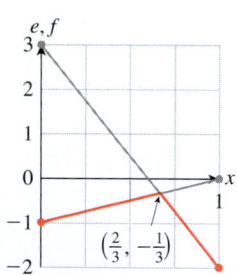

Figure **2**

Because RTV is trying to make the worst possible outcome as large as possible (that is, to minimize damages), it is seeking the point on the orange portion of the graph that is highest. This is the intersection point of the two lines. To calculate its coordinates, it's easiest to equate the two functions of x:

$$-5x + 3 = x - 1,$$

$$-6x = -4,$$

or $\qquad x = \dfrac{2}{3}$

The e (or f) coordinate is then obtained by substituting $x = 2/3$ into the expression for e (or f) giving:

$$e = -5\left(\frac{2}{3}\right) + 3$$

$$= -\frac{1}{3}$$

We conclude that RTV's best strategy is to take $x = 2/3$, giving an expected value of $-1/3$. In other words, RTV's optimal mixed strategy is:

$$R = \begin{bmatrix} \dfrac{2}{3} & \dfrac{1}{3} \end{bmatrix}$$

Going back to the original game, RTV should show reality shows 2/3 of the time and movies 1/3 of the time. It should not bother showing any sitcoms or docudramas. It expects to lose, on average, 333 viewers to CTV, but all of its other options are worse.

b. To find CTV's optimal strategy, we must reverse roles and start by writing its unknown strategy as follows.

$$C = \begin{bmatrix} x \\ 1-x \end{bmatrix}$$

We calculate the expected values for the two pure row strategies:

$$e = \begin{bmatrix} 1 & 0 \end{bmatrix} \begin{bmatrix} -2 & 0 \\ 3 & -1 \end{bmatrix} \begin{bmatrix} x \\ 1-x \end{bmatrix}$$

$$= -2x$$

and

$$f = \begin{bmatrix} 0 & 1 \end{bmatrix} \begin{bmatrix} -2 & 0 \\ 3 & -1 \end{bmatrix} \begin{bmatrix} x \\ 1-x \end{bmatrix}$$

$$= 3x - (1-x)$$

$$= 4x - 1$$

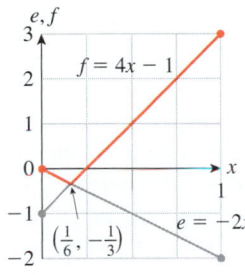

Figure **3**

As with the row player, we know that the column player's best strategy will correspond to the intersection of the graphs of e and f (Figure 3).

(Why is the upper edge orange, rather than the lower edge?) The graphs intersect when

$$-2x = 4x - 1$$

or

$$x = \frac{1}{6}$$

The corresponding value of e (or f) is

$$e = -2\left(\frac{1}{6}\right) = -\frac{1}{3}$$

Thus CTV's optimal mixed strategy is $\begin{bmatrix} \frac{1}{6} & \frac{5}{6} \end{bmatrix}^T$ and the expected value is $-1/3$. So, CTV should show nature documentaries $1/6$ of the time and ballets $5/6$ of the time. It should not bother to show symphonies or operas. It expects to gain, on average, 333 viewers from RTV.

c. We can now calculate the expected value of the game as usual, using the optimal strategies we found in parts (a) and (b).

$$e = RPC$$

$$= \begin{bmatrix} \frac{2}{3} & \frac{1}{3} \end{bmatrix} \begin{bmatrix} -2 & 0 \\ 3 & -1 \end{bmatrix} \begin{bmatrix} \frac{1}{6} \\ \frac{5}{6} \end{bmatrix}$$

$$= -\frac{1}{3}$$

+*Before we go on...* In Example 3, it is no accident that the expected value with the optimal strategies equals the expected values we found in (a) and (b). If we call the expected value with the optimal strategies the **expected value of the game,** the row player's optimal strategy guarantees an average payoff no smaller than the expected value, while the column player's optimal strategy guarantees an average payoff no larger. Together they force the average value to be the expected value of the game. ∎

Expected Value of a Game

The **expected value of a game** is its expected value when the row and column players use their optimal (minimax) strategies. By using its optimal strategy, the row player guarantees an expected value no lower than the expected value of the game, no matter what the column player does. Similarly, by using its optimal strategy, the column player guarantees an expected value no higher than the expected value of the game, no matter what the row player does.

Q: *What about games that don't reduce to 2 × 2 matrices? Are these solved in a similar way?*

A: The method illustrated in the above example cannot easily be generalized to solve bigger games (i.e., games that cannot be reduced to 2 × 2 matrices); solving even a 2 × 3 game using this approach would require us to consider graphs in three dimensions. To be able to solve games of arbitrary size, we need to wait until the next chapter (Section 4.5) where we describe a method for solving a game, using the simplex method, that works for all payoff matrices. ∎

Strictly Determined Games

The following example illustrates an interesting special case.

Example 4 Strictly Determined Game

Solve the following game:

$$
\mathbf{A} \begin{array}{c} \\ s \\ t \\ u \end{array} \overset{\overset{\textstyle \mathbf{B}}{\begin{array}{ccc} p & q & r \end{array}}}{\left[\begin{array}{ccc} -4 & -3 & 3 \\ 2 & -1 & -2 \\ 1 & 0 & 2 \end{array} \right]}
$$

Solution If we look carefully at this matrix, we see that no row dominates another and no column dominates another, so we can't reduce it. Nor do we know how to solve a 3×3 game, so it looks as if we're stuck. However, there is a way to understand this particular game. With the minimax criterion in mind, let's begin by considering the worst possible outcomes for the row player for each possible move. We do this by circling the smallest payoff in each row, the **row minima:**

$$
\mathbf{A} \begin{array}{c} \\ s \\ t \\ u \end{array} \overset{\overset{\textstyle \mathbf{B}}{\begin{array}{ccc} p & q & r \end{array}}}{\left[\begin{array}{ccc} \boxed{-4} & -3 & 3 \\ 2 & -1 & \boxed{-2} \\ 1 & \boxed{0} & 2 \end{array} \right]} \begin{array}{c} \text{Row minima} \\ -4 \\ -2 \\ 0 \leftarrow \text{largest} \end{array}
$$

So, for example, if A plays move s, the worst possible outcome is to lose 4. Player A takes the least risk by using move u, which has the largest row minimum.

We do the same thing for the column player, remembering that smaller payoffs are better for B and larger payoffs worse. We draw a box around the largest payoff in each column, the **column maxima:**

$$\mathbf{A} \begin{array}{c} \\ s \\ t \\ u \end{array} \overset{\begin{array}{ccc} \quad p & \quad q & \quad r \end{array}}{\overset{\mathbf{B}}{\left[\begin{array}{ccc} -4 & -3 & \boxed{3} \\ \boxed{2} & -1 & -2 \\ 1 & \boxed{0} & 2 \end{array}\right]}}$$

Column maxima   2   0   3
                                  ↑
                             smallest

Player B takes the least risk by using move q, which has the smallest column maximum.

Now put the circles and boxes together:

$$\mathbf{A} \begin{array}{c} \\ s \\ t \\ u \end{array} \overset{\begin{array}{ccc} \quad p & \quad q & \quad r \end{array}}{\overset{\mathbf{B}}{\left[\begin{array}{ccc} \boxed{-4} & -3 & \boxed{3} \\ \boxed{2} & -1 & \boxed{-2} \\ 1 & \boxed{0} & 2 \end{array}\right]}}$$

Notice that the uq entry is both circled and boxed: it is both a row minimum and a column maximum. We call such an entry a **saddle point.**

Now we claim that the optimal strategy for A is to always play u while the optimal strategy for B is to always play q. By playing u, A guarantees that the payoff will be 0 or higher, no matter what B does, so the expected value of the game has to be at least 0. On the other hand, by playing q, B guarantees that the payoff will be 0 or less, so the expected value of the game has to be no more than 0. That means that the expected value of the game must be 0, and A and B have no strategies that could do any better for them than the pure strategies u and q.

+ *Before we go on...* You should consider what happens in an example like the television rating wars game of Example 3. In that game, the largest row minimum is –1 while the smallest column maximum is 0; there is no saddle point. The row player can force a payoff of at least –1 by playing a pure strategy (always showing movies, for example), but can do better, forcing an average payoff of $-1/3$, by playing a mixed strategy. Similarly, the column player can force the payoff to be 0 or less with a pure strategy, but can do better, forcing an average payoff of $-1/3$, with a mixed strategy. Only when there is a saddle point will pure strategies be optimal. ∎

Strictly Determined Game

A **saddle point** is a payoff that is simultaneously a row minimum and a column maximum (both boxed and circled in our approach). If a game has a saddle point, the corresponding row and column strategies are the optimal ones, and we say that the game is **strictly determined.**

FAQs   Solving a Game

using Technology

To play a strictly determined game against your computer, and also to automatically reduce by dominance and solve arbitrary games up to 5×5, go online and follow the path

Chapter 3
→ Tools
   → Game Theory Tool

Q: We've seen several ways of trying to solve a game. What should I do and in what order?

A: Here are the steps you should take when trying to solve a game:

1. Reduce by dominance. This should always be your first step.
2. If you were able to reduce to a 1×1 game, you're done. The optimal strategies are the corresponding pure strategies, as they dominate all the others.
3. Look for a saddle point in the reduced game. If it has one, the game is strictly determined, and the corresponding pure strategies are optimal.
4. If your reduced game is 2×2 and has no saddle point, use the method of Example 3 to find the optimal mixed strategies.
5. If your reduced game is larger than 2×2 and has no saddle point, you have to use linear programming to solve it, but that will have to wait until the following chapter. ∎

3.4 EXERCISES

● denotes basic skills exercises
◆ denotes challenging exercises
tech Ex indicates exercises that should be solved using technology

Reduce the payoff matrices in Exercises 1–6 by dominance.

1. ●

$$\mathbf{A} \begin{array}{c} a \\ b \end{array} \begin{array}{ccc} p & q & r \\ \left[\begin{array}{ccc} 1 & 1 & 10 \\ 2 & 3 & -4 \end{array} \right] \end{array}$$

2. ●

$$\mathbf{A} \begin{array}{c} a \\ b \end{array} \begin{array}{ccc} p & q & r \\ \left[\begin{array}{ccc} 2 & 0 & 10 \\ 15 & -4 & -5 \end{array} \right] \end{array}$$

3. ●

$$\mathbf{A} \begin{array}{c} 1 \\ 2 \\ 3 \end{array} \begin{array}{ccc} a & b & c \\ \left[\begin{array}{ccc} 2 & -4 & -9 \\ -1 & -2 & -3 \\ 5 & 0 & -1 \end{array} \right] \end{array}$$

4. ●

$$\mathbf{A} \begin{array}{c} 1 \\ 2 \\ 3 \end{array} \begin{array}{ccc} a & b & c \\ \left[\begin{array}{ccc} 0 & -1 & -5 \\ -3 & -10 & 10 \\ 2 & 3 & -4 \end{array} \right] \end{array}$$

5. ●

$$\mathbf{A} \begin{array}{c} p \\ q \\ r \\ s \end{array} \begin{array}{ccc} a & b & c \\ \left[\begin{array}{ccc} 1 & -1 & -5 \\ 4 & 0 & 2 \\ 3 & -3 & 10 \\ 3 & -5 & -4 \end{array} \right] \end{array}$$

6. ●

$$\mathbf{A} \begin{array}{c} p \\ q \\ r \\ s \end{array} \begin{array}{ccc} a & b & c \\ \left[\begin{array}{ccc} 2 & -4 & 9 \\ 1 & 1 & 0 \\ -1 & -2 & -3 \\ 1 & 1 & -1 \end{array} \right] \end{array}$$

In Exercises 7–12, decide whether the game is strictly determined. If it is, give the players' optimal pure strategies and the value of the game. hint [see Example 4]

7. ●

$$\mathbf{A} \begin{array}{c} a \\ b \end{array} \begin{array}{cc} p & q \\ \left[\begin{array}{cc} 1 & 1 \\ 2 & -4 \end{array} \right] \end{array}$$

8. ●

$$\mathbf{A} \begin{array}{c} a \\ b \end{array} \begin{array}{cc} p & q \\ \left[\begin{array}{cc} -1 & 2 \\ 10 & -1 \end{array} \right] \end{array}$$

9. ●

$$\mathbf{A} \begin{array}{c} a \\ b \end{array} \begin{array}{ccc} p & q & r \\ \left[\begin{array}{ccc} 2 & 0 & -2 \\ -1 & 3 & 0 \end{array} \right] \end{array}$$

10. ●

$$\mathbf{A} \begin{array}{c} a \\ b \end{array} \begin{array}{ccc} p & q & r \\ \left[\begin{array}{ccc} -2 & 1 & -3 \\ -2 & 3 & -2 \end{array} \right] \end{array}$$

11. ●

$$\mathbf{A} \begin{array}{c} P \\ Q \\ R \\ S \end{array} \begin{array}{ccc} a & b & c \\ \left[\begin{array}{ccc} 1 & -1 & -5 \\ 4 & -4 & 2 \\ 3 & -3 & -10 \\ 5 & -5 & -4 \end{array} \right] \end{array}$$

12. ●

$$\mathbf{A} \begin{array}{c} P \\ Q \\ R \\ S \end{array} \begin{array}{ccc} a & b & c \\ \left[\begin{array}{ccc} -2 & -4 & 9 \\ 1 & 1 & 0 \\ -1 & -2 & -3 \\ 1 & 1 & -1 \end{array} \right] \end{array}$$

In Exercises 13–16, calculate the expected value of the game with payoff matrix

$$P = \begin{bmatrix} 2 & 0 & -1 & 2 \\ -1 & 0 & 0 & -2 \\ -2 & 0 & 0 & 1 \\ 3 & 1 & -1 & 1 \end{bmatrix}$$

using the mixed strategies supplied. hint [see Example 1]

● basic skills   ◆ challenging   **tech Ex** technology exercise

13. ● $R = [0 \quad 1 \quad 0 \quad 0], C = [1 \quad 0 \quad 0 \quad 0]^T$

14. ● $R = [0 \quad 0 \quad 0 \quad 1], C = [0 \quad 1 \quad 0 \quad 0]^T$

15. ● $R = [0.5 \quad 0.5 \quad 0 \quad 0], C = [0 \quad 0 \quad 0.5 \quad 0.5]^T$

16. ● $R = [0 \quad 0.5 \quad 0 \quad 0.5], C = [0.5 \quad 0.5 \quad 0 \quad 0]^T$

In Exercises 17–20, either a mixed column or row strategy is given. In each case, use

$$P = \begin{bmatrix} 0 & -1 & 5 \\ 2 & -2 & 4 \\ 0 & 3 & 0 \\ 1 & 0 & -5 \end{bmatrix}$$

and find the optimal pure strategy (or strategies) the other player should use. Express the answer as a row or column matrix. Also determine the resulting expected value of the game.
hint [see Example 2]

17. ● $C = [0.25 \quad 0.75 \quad 0]^T$ 18. ● $C = \left[\dfrac{1}{3} \quad \dfrac{1}{3} \quad \dfrac{1}{3}\right]^T$

19. ● $R = \left[\dfrac{1}{2} \quad 0 \quad \dfrac{1}{4} \quad \dfrac{1}{4}\right]$ 20. ● $R = [0.8 \quad 0.2 \quad 0 \quad 0]$

*In Exercises 21–24, find: **a.** the optimal mixed row strategy;*
***b.** the optimal mixed column strategy, and **c.** the expected value of the game in the event that each player uses his or her optimal mixed strategy.* *hint* [see Example 3]

21. ● $P = \begin{bmatrix} -1 & 2 \\ 0 & -1 \end{bmatrix}$ 22. ● $P = \begin{bmatrix} -1 & 0 \\ 1 & -1 \end{bmatrix}$

23. ● $P = \begin{bmatrix} -1 & -2 \\ -2 & 1 \end{bmatrix}$ 24. ● $P = \begin{bmatrix} -2 & -1 \\ -1 & -3 \end{bmatrix}$

Applications

Set up the payoff matrix in each of Exercises 25–32.

25. ● **Games to Pass the Time** You and your friend have come up with the following simple game to pass the time: at each round, you simultaneously call "heads" or "tails." If you have both called the same thing, your friend wins one point; if your calls differ, you win one point.

26. ● **Games to Pass the Time** Bored with the game in Exercise 25, you decide to use the following variation instead: If you both call "heads" your friend wins two points; if you both call "tails" your friend wins 1 point; if your calls differ, then you win two points if you called "heads," and one point if you called "tails."

27. ● **War Games** You are deciding whether to invade France, Sweden or Norway, and your opponent is simultaneously deciding which of these three countries to defend. If you invade a country that your opponent is defending, you will be defeated (payoff: -1), but if you invade a country your opponent is not defending, you will be successful (payoff: $+1$).

28. ● **War Games** You must decide whether to attack your opponent by sea or air, and your opponent must simultaneously decide whether to mount an all-out air defense, an all-out coastal defense (against an attack from the sea) or a combined air and coastal defense. If there is no defense for your mode of attack, you win 100 points. If your attack is met by a shared air and coastal defense, you win 50 points. If your attack is met by an all-out defense, you lose 200 points.

29. **Marketing** Your fast-food outlet, Burger Queen, has obtained a license to open branches in three closely situated South African cities: Brakpan, Nigel, and Springs. Your market surveys show that Brakpan and Nigel each provide a potential market of 2000 burgers a day, while Springs provides a potential market of 1000 burgers per day. Your company can only finance an outlet in one of those cities at the present time. Your main competitor, Burger Princess, has also obtained licenses for these cities, and is similarly planning to open only one outlet. If you both happen to locate at the same city, you will share the total business from all three cities equally, but if you locate in different cities, you will each get all the business in the cities in which you have located, plus half the business in the third city. The payoff is the number of burgers you will sell per day minus the number of burgers your competitor will sell per day.

30. **Marketing** Repeat Exercise 29, given that the potential sales markets in the three cities are: Brakpan, 2500 per day; Nigel, 1500 per day; Springs, 1200 per day.

31. **Betting** When you bet on a racehorse with odds of m–n, you stand to win m dollars for every bet of n dollars if your horse wins; for instance, if the horse you bet is running at 5–2 and wins, you will win \$5 for every \$2 you bet. (Thus a \$2 bet will return \$7.). Here are some actual odds from a 1992 race at Belmont Park, NY.[24] The favorite at 5–2 was Pleasant Tap. The second choice was Thunder Rumble at 7–2, while the third choice was Strike the Gold at 4–1. Assume you are making a \$10 bet on one of these horses. The payoffs are your winnings. (If your horse does not win, you lose your entire bet. Of course, it is possible for none of your horses to win.)

32. **Betting** Referring to Exercise 31, suppose that just before the race, there has been frantic betting on Thunder Rumble, with the result that the odds have dropped to 2–5. The odds on the other two horses remain unchanged.

33. ● **Price Wars** Computer Electronics, Inc. (CE) and the Gigantic Computer Store (GCS) are planning to discount the price they charge for the HAL Laptop Computer, of which they are the only distributors. Because Computer Electronics provides a free warranty service, they can generally afford to charge more. A market survey provides the following data on the

---

[24] SOURCE: *The New York Times,* September 18, 1992, p.B14.

● basic skills   ◆ challenging   **tech Ex** technology exercise

gains to CE's market share that will result from different pricing decisions:

GCS

		$900	$1000	$1200
	$1000	15%	60%	80%
CE	$1200	15%	60%	60%
	$1300	10%	20%	40%

a. Use reduction by dominance to determine how much each company should charge. What is the effect on CE's market share?

b. CE is aware that GCS is planning to use reduction of dominance to determine its pricing policy, and wants its market share to be as large as possible. What effect, if any, would the information about GCS have on CE's best strategy?

34. ● *More Price Wars* (Refer to Exercise 33) A new market survey results in the following revised data:

GCS

		$900	$1000	$1200
	$1000	20%	60%	60%
CE	$1200	15%	60%	60%
	$1300	10%	20%	40%

a. Use reduction by dominance to determine how much each company should charge. What is the effect on CE's market share?

b. In general, why do price wars tend to force prices down?

35. ● *Wrestling Tournaments* City Community College (CCC) plans to host Midtown Military Academy (MMA) for a wrestling tournament. Each school has three wrestlers in the 190 lb. weight class: CCC has Pablo, Sal, and Edison, while MMA has Carlos, Marcus and Noto. Pablo can beat Carlos and Marcus, Marcus can beat Edison and Sal, Noto can beat Edison, while the other combinations will result in an even match. Set up a payoff matrix, and use reduction by dominance to decide which wrestler each team should choose as their champion. Does one school have an advantage over the other?

36. ● *Wrestling Tournaments* (Refer to Exercise 35) One day before the wrestling tournament discussed in Exercise 35, Pablo sustains a hamstring injury, and is replaced by Hans, who (unfortunately for CCC) can be beaten by both Carlos and Marcus. Set up the payoff matrix, and use reduction by dominance to decide which wrestler each team should choose as their champion. Does one school have an advantage over the other?

37. ● *The Battle of Rabaul-Lae*[25] In the Second World War, during the struggle for New Guinea, intelligence reports revealed that the Japanese were planning to move a troop and supply convoy from the port of Rabaul at the Eastern tip of New Britain to Lae, which lies just west of New Britain on New Guinea. It could either travel via a northern route which was plagued by poor visibility, or by a southern route, where the visibility was clear. General Kenney, who was the commander of the Allied Air Forces in the area, had the choice of concentrating reconnaissance aircraft on one route or the other, and bombing the Japanese convoy once it was sighted. Kenney's

Japanese Commander's Strategies

		Northern Route	Southern Route
Kenney's Strategies	Northern Route	2	2
	Southern Route	1	3

staff drafted the following outcomes for his choices, where the payoffs are estimated days of bombing time:

What would you have recommended to General Kenney? What would you have recommended to the Japanese Commander?[26] How much bombing time results if these recommendations are followed?

38. ● *The Battle of Rabaul-Lae* Referring to Exercise 37, suppose that General Kenney had a third alternative: Splitting his reconnaissance aircraft between the two routes, and resulting in the following estimates:

Japanese Commander's Strategies

		Northern Route	Southern Route
	Northern Route	2	2
Kenney's Strategies	Split Reconnaissance	1.5	2.5
	Southern Route	1	3

What would you have recommended to General Kenney? What would you have recommended to the Japanese Commander? How much bombing time results if these recommendations are followed?

39. *The Prisoner's Dilemma* Slim Lefty and Joe Rap have been arrested for grand theft auto, having been caught red-handed driving away in a stolen 2005 Porsche. Although the police have more than enough evidence to convict them both, they feel that a confession would simplify the work of the prosecution. They decide to interrogate the prisoners separately. Slim and Joe are both told of the following plea-bargaining arrangement: if both confess, they will each receive a two-year

---

[25] As discussed in *Games and Decisions* by R.D. Luce and H. Raiffa Section 11.3 (New York; Wiley, 1957). This is based on an article in the *Journal of the Operations Research Society of America* **2** (1954) 365–385.

[26] The correct answers to parts (a) and (b) correspond to the actual decisions both commanders made.

● basic skills ◆ challenging *tech*Ex technology exercise

sentence; if neither confesses, they will both receive five-year sentences, and if only one confesses (and thus squeals on the other), he will receive a suspended sentence, while the other will receive a 10-year sentence. What should Slim do?

40. *More Prisoners' Dilemmas* Jane Good and Prudence Brown have been arrested for robbery, but the police lack sufficient evidence for a conviction, and so decide to interrogate them separately in the hope of extracting a confession. Both Jane and Prudence are told the following: if they both confess, they will each receive a five-year sentence; if neither confesses, they will be released; if one confesses, she will receive a suspended sentence, while the other will receive a ten-year sentence. What should Jane do?

41. *Campaign Strategies*[27] Florida and Ohio are "swing states" that have a large bounty of electoral votes and are therefore highly valued by presidential campaign strategists. Suppose it is now the weekend before Election Day 2004, and each candidate (Bush and Kerry) can visit only one more state. Further, to win the election, Bush needs to win both of these states. Currently Bush has a 40% chance of winning Ohio and a 60% chance of winning Florida. Therefore, he has a $0.40 \times 0.60 = 0.24$, or 24% chance of winning the election. Assume that each candidate can increase his probability of winning a state by 10% if he, and not his opponent, visits that state. If both candidates visit the same state, there is no effect.

a. Set up a payoff matrix with Bush as the row player and Kerry as the column player, where the payoff for a specific set of circumstances is the probability (expressed as a percentage) that Bush will win both states.

b. Where should each candidate visit under the circumstances?

42. *Campaign Strategies* Repeat Exercise 41, this time assuming that Bush has an 80% chance of winning Ohio and a 90% chance of winning Florida.

43. ● *Retail Discount Wars* Just one week after your Abercrom B men's fashion outlet has opened at a new location opposite Burger King® in the Mall, your rival, Abercrom A, opens up directly across from you, You have been informed that Abercrom A is about to launch either a 30% off everything sale or a 50% off everything sale. You, on the other hand, have decided to either *increase* prices (to make your store seem more exclusive) or do absolutely nothing. You construct the following payoff matrix, where the payoffs represent the number of customers your outlet can expect to gain from Abercrom A:

Abercrom A

	30% Off	50% Off
Abercrom B Do Nothing	−60	−40
Increase prices	30	−50

There is a 20% chance that Abercrom A will opt for the "30% off" sale and an 80% chance that they will opt for the "50% Off" sale. Your sense from upper management at Abercrom B is that there is a 50% chance you will be given the go-ahead to raise prices. What is the expected resulting effect on your customer base?

44. ● *More Retail Discount Wars* Your Abercrom B mens fashion outlet has a 30% chance of launching an expensive new line of used auto-mechanic dungarees (complete with grease stains) and a 70% chance of staying instead with its traditional torn military-style dungarees. Your rival across from you in the mall, Abercrom A, appears to be deciding between a line of torn gym shirts and a more daring line of "empty shirts" (that is, empty shirt boxes). Your corporate spies reveal that there is a 20% chance that Abercrom A will opt for the empty shirt option. The following payoff matrix gives the number of customers your outlet can expect to gain from Abercrom A in each situation:

Abercrom A

	Torn Shirts	Empty Shirts
Abercrom B Mechanics	10	−40
Military	−30	50

What is the expected resulting effect on your customer base?

45. ● *Factory Location*[28] A manufacturer of electrical machinery is located in a cramped, though low-rent, factory close to the center of a large city. The firm needs to expand, and it could do so in one of three ways: (1) remain where it is and install new equipment, (2) move to a suburban site in the same city, or (3) relocate in a different part of the country where labor is cheaper. Its decision will be influenced by the fact that one of the following will happen: (I) the government may introduce a program of equipment grants, (II) a new suburban highway may be built, or (III) the government may institute a policy of financial help to companies who move into regions of high unemployment. The value to the company of each combination is given in the following payoff matrix.

Government's Options

Manufacturer's Options		I	II	III
	1	200	150	140
	2	130	220	130
	3	110	110	220

[27] Based on *Game Theory for Swingers: What states should the candidates visit before Election Day?* by Jordan Ellenberg. Source: www.slate.com.

[28] Adapted from an example in *Location in Space: Theoretical Perspectives in Economic Geography* by P. Dicken and P.E. Lloyd, Harper & Row, 1990.

● basic skills ◆ challenging **techEx** technology exercise

If the manufacturer judges that there is a 20% probability that the government will go with option I, a 50% probability that they will go with option II, and a 30% probability that they will go with option III, what is the manufacturer's best option?

46. ● *Crop Choice*[29] A farmer has a choice of growing wheat, barley, or rice. Her success will depend on the weather, which could be dry, average, or wet. Her payoff matrix is as follows.

Weather

		Dry	Average	Wet
	Wheat	20	20	10
Crop Choices	Barley	10	15	20
	Rice	10	20	20

If the probability that the weather will be dry is 10%, the probability that it will be average is 60%, and the probability that it will be wet is 30%, what is the farmer's best choice of crop?

47. ● *Study Techniques* Your mathematics test is tomorrow, and will cover the following topics: game theory, linear programming, and matrix algebra. You have decided to do an "all-nighter" and must determine how to allocate your eight hours of study time among the three topics. If you were to spend the entire eight hours on any one of these topics (thus using a pure strategy) you feel confident that you would earn a 90% score on that portion of the test, but would not do so well on the other topics. You have come up with the following table, where the entries are your expected scores. (The fact that linear programming and matrix algebra are used in game theory is reflected in these numbers).

Test

Your Strategies	Game Theory	Linear Programming	Matrix Algebra
Game Theory	90	70	70
Linear Programming	40	90	40
Matrix Algebra	60	40	90

You have been told that the test will be weighted as follows: game theory: 25%; linear programming: 50%; matrix algebra: 25%.

a. If you spend 25% of the night on game theory, 50% on linear programming, and 25% on matrix algebra, what score do you expect to get on the test?

b. Is it possible to improve on this by altering your study schedule? If so, what is the highest score you can expect on the test?

c. If your study schedule is according to part (a) and your teacher decides to forget her promises about how the test will be weighted and instead base it all on a single topic, which topic would be worst for you, and what score could you expect on the test?

48. ● *Study Techniques* Your friend Joe has been spending all of his time in fraternity activities, and thus knows absolutely nothing about any of the three topics on tomorrow's math test. (See Exercise 47.) Because you are recognized as an expert on the use of game theory to solve study problems, he has turned to you for advice as to how to spend his "all-nighter." As the following table shows, his situation is not so rosy. (Since he knows no linear programming or matrix algebra, the table shows, for instance, that studying game theory all night will not be much use in preparing him for this topic.)

Test

Joe's Strategies	Game Theory	Linear Programming	Matrix Algebra
Game Theory	30	0	20
Linear Programming	0	70	0
Matrix Algebra	0	0	70

Assuming that the test will be weighted as described in Exercise 47, what are the answers to parts (a), (b), and (c) as they apply to Joe?

49. *Staff Cutbacks* Frank Tempest manages a large snowplow service in Manhattan, Kansas, and is alarmed by the recent weather trends; there have been no significant snowfalls since 1993. He is therefore contemplating laying off some of his workers, but is unsure about whether to lay off 5, 10 or 15 of his 50 workers. Being very methodical, he estimates his annual net profits based on four possible annual snowfall figures: 0 inches, 20 inches, 40 inches and 60 inches. (He takes into account the fact that, if he is running a small operation in the face of a large annual snowfall, he will lose business to his competitors because he will be unable to discount on volume.)

	0 inches	20 inches	40 inches	60 inches
5 laid off	−$500,000	−$200,000	$10,000	$200,000
10 laid off	−$200,000	$0	$0	$0
15 laid off	−$100,000	$10,000	−$200,000	−$300,000

a. During the past 10 years, the region has had 0 inches twice, 20 inches twice, 40 inches 3 times, and 60 inches 3 times. Based on this information, how many workers should Tempest lay off, and how much would it cost him?

b. There is a 50% chance that Tempest will lay off 5 workers and a 50% chance that he will lay off 15 workers. What is the worst thing Nature can do to him in terms of snowfall? How much would it cost him?

[29] Ibid.

● basic skills   ◆ challenging   **tech Ex** technology exercise

c. The Gods of Chaos (who control the weather) know that Tempest is planning to use the strategy in part (a), and are determined to hurt Tempest as much as possible. Tempest, being somewhat paranoid, suspects it too. What should he do?

50. **Textbook Writing** You are writing a college-level textbook on finite mathematics, and are trying to come up with the best combination of word problems. Over the years, you have accumulated a collection of amusing problems, serious applications, long complicated problems, and "generic" problems.[30] Before your book is published, it must be scrutinized by several reviewers who, it seems, are never satisfied with the mix you use. You estimate that there are three kinds of reviewers: the "no-nonsense" types who prefer applications and generic problems, the "dead serious" types, who feel that a college-level text should be contain little or no humor and lots of long complicated problems, and the "laid-back" types, who believe that learning best takes place in a light-hearted atmosphere bordering on anarchy. You have drawn up the following chart, where the payoffs represent the reactions of reviewers on a scale of -10 (ballistic) to $+10$ (ecstatic):

Reviewers

	No-Nonsense	Dead Serious	Laid-Back
Amusing	-5	-10	10
Serious	5	3	0
Long	-5	5	3
Generic	5	3	-10

You (label to the left of table, rows)

a. Your first draft of the book contained no generic problems, and equal numbers of the other categories. If half the reviewers of your book were "dead serious" and the rest were equally divided between the "no-nonsense" and "laid-back" types, what score would you expect?

b. In your second draft of the book, you tried to balance the content by including some generic problems and eliminating several amusing ones, and wound up with a mix of which one eighth were amusing, one quarter were serious, three eighths were long, and a quarter were generic. What kind of reviewer would be *least* impressed by this mix?

c. What kind of reviewer would be *most* impressed by the mix in your second draft?

51. ● *Advertising* The Softex Shampoo Company is considering how to split its advertising budget between ads on two radio stations: WISH and WASH. Its main competitor, Splish Shampoo, Inc. has found out about this, and is considering countering Softex's ads with its own, on the same radio stations. (Proposed jingle: *Softex, Shmoftex; Splash with Splish*) Softex has calculated that, were it to devote its entire adver-

tising budget to ads on WISH, it would increase revenues in the coming month by $100,000 in the event that Splish was running all its ads on the less popular WASH, but would lose $20,000 in revenues if Splish ran its ads on WISH. If, on the other hand, it devoted its entire budget to WASH ads, it would neither increase nor decrease revenues in the event that Splish was running all its ads on the more popular WISH, and would in fact lose $20,000 in revenues if Splish ran its ads on WASH. What should Softex do, and what effect will this have on revenues?

52. ● *Labor Negotiations* The management team of the Abstract Concrete Company is negotiating a three-year contract with the labor unions at one of its plants, and is trying to decide on its offer for a salary increase. If it offers a 5% increase and the unions accept the offer, Abstract Concrete will gain $20 million in projected profits in the coming year, but if labor rejects the offer, the management team predicts that it will be forced to increase the offer to the union demand of 15%, thus halving the projected profits. If Abstract Concrete offers a 15% increase, the company will earn $10 million in profits over the coming year if the unions accept. If the unions reject, they will probably go out on strike (because management has set 15% as its upper limit) and management has decided that it can then in fact gain $12 million in profits by selling out the defunct plant in retaliation. What intermediate percentage should the company offer, and what profit should it project?

Communication and Reasoning Exercises

53. ● Why is a saddle point called a "saddle point"?

54. ● Can the payoff in a saddle point ever be larger than all other payoffs in a game? Explain.

55. ◆ One day, while browsing through an old *Statistical Abstract of the United States,* you came across the following data which show the number of females employed (in thousands) in various categories according to their educational attainment.[31]

	Managerial/ Professional	Technical/Sales/ Administrative	Service	Precision Production	Operators/ Fabricators
Less than 4 years high school	260	1080	2020	260	1400
4 years of high school only	2430	9510	3600	570	2130
1 to 3 years of college	2690	5080	1080	160	350
At least 4 years of college	7210	2760	380	70	110

---

[30] of the following type: "An oil company has three refineries: A, B and C, each of which uses three processes: P_1, P_2, and P_3. Process P_1 uses 100 units of chemical C_1 and costs $100 per day . . ."

[31] SOURCE: *Statistical Abstract of the United States 1991* (111th Ed.) U.S. Department of Commerce, Economics and Statistics Administration, and Bureau of the Census.

● basic skills   ◆ challenging   **tech**Ex technology exercise

Because you had been studying game theory that day, the first thing you did was to search for a saddle point. Having found one, you conclude that, as a female, your best strategy in the job market is to forget about a college career. Find the flaw in this reasoning.

56. ◆ Exercises 47 and 48 seem to suggest that studying a single topic prior to an exam is better than studying all the topics in that exam. Comment on this discrepancy between the game theory result and common sense.

57. ◆ Explain what is wrong with a decision to play the mixed strategy [0.5 0.5] by alternating the two strategies: play the first strategy on the odd-numbered moves and the second strategy on the even-numbered moves. Illustrate your argument by devising a game in which your best strategy is [0.5 0.5].

58. ◆ Describe a situation in which a both a mixed strategy and a pure strategy are equally effective.

● basic skills   ◆ challenging   *tech* Ex technology exercise

3.5 Input-Output Models

In this section we look at an application of matrix algebra developed by Wassily Leontief (1906–1999) in the middle of the twentieth century. In 1973, he won the Nobel Prize in Economics for this work. The application involves analyzing national and regional economies by looking at how various parts of the economy interrelate. We'll work out some of the details by looking at a simple scenario.

First, we can think of the economy of a country or a region as being composed of various **sectors,** or groups of one or more industries. Typical sectors are the manufacturing sector, the utilities sector, and the agricultural sector. To introduce the basic concepts, we shall consider two specific sectors: the coal-mining sector (Sector 1) and the electric utilities sector (Sector 2). Both produce a commodity: the coal-mining sector produces coal, and the electric utilities sector produces electricity. We measure these products by their dollar value. By **one unit** of a product, we mean $1 worth of that product.

Here is the scenario.

1. To produce one unit ($1 worth) of coal, assume that the coal-mining sector uses 50¢ worth of coal (to power mining machinery, say) and 10¢ worth of electricity.

2. To produce one unit ($1 worth) of electricity, assume that the electric utilities sector uses 25¢ worth of coal and 25¢ worth of electricity.

These are *internal* usage figures. In addition to this, assume that there is an *external* demand (from the rest of the economy) of 7000 units ($7,000 worth) of coal and 14,000 units ($14,000 worth) of electricity over a specific time period (one year, say). Our basic question is: How much should each of the two sectors supply in order to meet both internal and external demand?

The key to answering this question is to set up equations of the form:

Total supply = Total demand

The unknowns, the values we are seeking, are

x_1 = the total supply (in units) from Sector 1 (coal) and
x_2 = the total supply (in units) from Sector 2 (electricity)

Our equations then take the following form:

Total supply from Sector 1 = Total demand for Sector 1 products

$$x_1 = 0.50x_1 \qquad + \qquad 0.25x_2 \qquad + \qquad 7000$$

Coal required by Sector 1   Coal required by Sector 2   External demand for coal

Total supply from Sector 2 = Total demand for Sector 2 products

$$x_2 = 0.10x_1 \qquad + \qquad 0.25x_2 \qquad + \qquad 14{,}000$$

Electricity required by Sector 1   Electricity required by Sector 2   External demand for electricity

This is a system of two linear equations in two unknowns:

$$x_1 = 0.50x_1 + 0.25x_2 + 7000$$
$$x_2 = 0.10x_1 + 0.25x_2 + 14{,}000$$

We can rewrite this system of equations in matrix form as follows:

$$\underbrace{\begin{bmatrix} x_1 \\ x_2 \end{bmatrix}}_{\text{Production}} = \underbrace{\begin{bmatrix} 0.50 & 0.25 \\ 0.10 & 0.25 \end{bmatrix} \begin{bmatrix} x_1 \\ x_2 \end{bmatrix}}_{\text{Internal demand}} + \underbrace{\begin{bmatrix} 7000 \\ 14{,}000 \end{bmatrix}}_{\text{External demand}}$$

In symbols,

$$X = AX + D$$

Here,

$$X = \begin{bmatrix} x_1 \\ x_2 \end{bmatrix}$$

is called the **production vector.** Its entries are the amounts produced by the two sectors. The matrix

$$D = \begin{bmatrix} 7000 \\ 14{,}000 \end{bmatrix}$$

is called the **external demand** vector, and

$$A = \begin{bmatrix} 0.50 & 0.25 \\ 0.10 & 0.25 \end{bmatrix}$$

is called the **technology matrix.** The entries of the technology matrix have the following meanings:

$$a_{11} = \text{units of Sector 1 needed to produce one unit of Sector 1}$$
$$a_{12} = \text{units of Sector 1 needed to produce one unit of Sector 2}$$
$$a_{21} = \text{units of Sector 2 needed to produce one unit of Sector 1}$$
$$a_{22} = \text{units of Sector 2 needed to produce one unit of Sector 2}$$

You can remember this order by the slogan "In the side, out the top."

Now that we have the matrix equation

$$X = AX + D$$

we can solve it as follows. First, subtract AX from both sides:

$$X - AX = D$$

Because $X = IX$, where I is the 2×2 identity matrix, we can rewrite this as

$$IX - AX = D$$

Now factor out X:

$$(I - A)X = D$$

If we multiply both sides by the inverse of $(I - A)$, we get the solution

$$X = (I - A)^{-1}D$$

Input-Output Model

In an input-output model, an economy (or part of one) is divided into n **sectors.** We then record the $n \times n$ **technology matrix** A, whose ijth entry is the number of units from Sector i used in producing one unit from Sector j ("in the side, out the top"). To meet an **external demand** of D, the economy must produce X, where X is the **production vector.** These are related by the equations

$$X = AX + D$$

or

$$X = (I - A)^{-1}D \qquad \text{Provided } (I - A) \text{ is invertible}$$

quick Example

In the scenario above, $A = \begin{bmatrix} 0.50 & 0.25 \\ 0.10 & 0.25 \end{bmatrix}, X = \begin{bmatrix} x_1 \\ x_2 \end{bmatrix}$, and $D = \begin{bmatrix} 7{,}000 \\ 14{,}000 \end{bmatrix}$.

The solution is

$$X = (I - A)^{-1}D$$

$$\begin{bmatrix} x_1 \\ x_2 \end{bmatrix} = \left(\begin{bmatrix} 1 & 0 \\ 0 & 1 \end{bmatrix} - \begin{bmatrix} 0.50 & 0.25 \\ 0.10 & 0.25 \end{bmatrix} \right)^{-1} \begin{bmatrix} 7{,}000 \\ 14{,}000 \end{bmatrix} \qquad \text{Calculate } I - A$$

$$= \begin{bmatrix} 0.50 & -0.25 \\ -0.10 & 0.75 \end{bmatrix}^{-1} \begin{bmatrix} 7{,}000 \\ 14{,}000 \end{bmatrix} \qquad \text{Calculate } (I - A)^{-1}$$

$$= \begin{bmatrix} \frac{15}{7} & \frac{5}{7} \\ \frac{2}{7} & \frac{10}{7} \end{bmatrix} \begin{bmatrix} 7{,}000 \\ 14{,}000 \end{bmatrix}$$

$$= \begin{bmatrix} 25{,}000 \\ 22{,}000 \end{bmatrix}$$

In other words, to meet the demand, the economy must produce \$25,000 worth of coal and \$22,000 worth of electricity.

The next example uses actual data from the U.S. economy (we have rounded the figures to make the computations less complicated). It is rare to find input-output data already packaged for you as a technology matrix. Instead, the data commonly found in statistical sources come in the form of "input-output tables," from which we will have to construct the technology matrix.

Dynamic Graphics, Inc./Jupiterimages

Example 1 Petroleum and Natural Gas

Consider two sectors of the U.S. economy: crude petroleum and natural gas (*crude*) and petroleum refining and related industries (*refining*). According to government figures,* in 1998 the crude sector used $27,000 million worth of its own products and $750 million worth of the products of the refining sector to produce $87,000 million worth of goods (crude oil and natural gas). The refining sector in the same year used $59,000 million worth of the products of the crude sector and $15,000 million worth of its own products to produce $140,000 million worth of goods (refined oil and the like). What was the technology matrix for these two sectors? What was left over from each of these sectors for use by other parts of the economy or for export?

Solution First, for convenience, we record the given data in the form of a table, called the **input-output table.** (All figures are in millions of dollars.)

	To	**Crude**	**Refining**
From	**Crude**	27,000	59,000
	Refining	750	15,000
	Total Output	87,000	140,000

The entries in the top portion are arranged in the same way as those of the technology matrix: The ijth entry represents the number of units of Sector i that went to Sector j. Thus, for instance, the 59,000 million entry in the 1, 2 position represents the number of units of Sector 1, crude, that were used by Sector 2, refining. ("In the side, out the top".)

We now construct the technology matrix. The technology matrix has entries $a_{ij} =$ units of Sector i used to produce *one* unit of Sector j. Thus,

$a_{11} =$ units of crude to produce one unit of crude. We are told that 27,000 million units of crude were used to produce 87,000 million units of crude. Thus, to produce *one* unit of crude, $27,000/87,000 \approx 0.31$ units of crude were used, and so $a_{11} \approx 0.31$. (We have rounded this value to two significant digits; further digits are not reliable due to rounding of the original data.)

$a_{12} =$ units of crude to produce one unit of refined:
$$a_{12} = 59,000/140,000 \approx 0.42$$

$a_{21} =$ units of refined to produce one unit of crude:
$$a_{21} = 750/87,000 \approx 0.0086$$

$a_{22} =$ units of refined to produce one unit of refined:
$$a_{22} = 15,000/140,000 \approx 0.11$$

This gives the technology matrix

$$A = \begin{bmatrix} 0.31 & 0.42 \\ 0.0086 & 0.11 \end{bmatrix}$$
    Technology Matrix

In short *we obtained the technology matrix from the input-output table by dividing the Sector 1 column by the Sector 1 total, and the Sector 2 column by the Sector 2 total.*

---

* The data have been rounded to two significant digits. SOURCE: *Survey of Current Business*, December, 2001, U.S. Department of Commerce. The *Survey of Current Business* and the input-output tables themselves are available at the website of the Department of Commerce's Bureau of Economic Analysis, www.bea.gov.

Now we also know the total output from each sector, so *we have already been given the production vector:*

$$X = \begin{bmatrix} 87,000 \\ 140,000 \end{bmatrix}$$ Production Vector

What we are asked for is the external demand vector D, the amount available for the outside economy. To find D, we use the equation

$$X = AX + D$$ Relationship of X, A, and D

where, this time, we are given A and X, and must solve for D. Solving for D gives

$$D = X - AX$$

$$= \begin{bmatrix} 87,000 \\ 140,000 \end{bmatrix} - \begin{bmatrix} 0.31 & 0.42 \\ 0.0086 & 0.11 \end{bmatrix} \begin{bmatrix} 87,000 \\ 140,000 \end{bmatrix}$$

$$\approx \begin{bmatrix} 87,000 \\ 140,000 \end{bmatrix} - \begin{bmatrix} 86,000 \\ 16,000 \end{bmatrix} = \begin{bmatrix} 1000 \\ 124,000 \end{bmatrix}$$ We rounded to 2 digits[†]

The first number, $1000 million, is the amount produced by the crude sector that is available to be used by other parts of the economy or to be exported. (In fact, because something has to happen to all that crude petroleum and natural gas, this is the amount actually used or exported, where use can include stockpiling.) The second number, $124,000 million, represents the amount produced by the refining sector that is available to be used by other parts of the economy or to be exported. Complete the example using a TI-83/84 or Excel.

Note that we could have calculated D more simply from the input-output table. The internal use of units from the crude sector was the sum of the outputs from that sector:

$$27,000 + 59,000 = 86,000$$

Because 87,000 units were actually produced by the sector, that left a surplus of $87,000 - 86,000 = 1000$ units for export. We could compute the surplus from the refining sector similarly. (The two calculations actually come out slightly different, because we rounded the intermediate results.) The calculation in Example 2 below cannot be done as trivially, however.

[†]Why?

using *Technology*

See the Technology Guides at the end of the chapter to see how to compute the technology matrix and then calculate the external demand vector using a TI 83/84 or Excel.

Input-Output Table

National economic data are often given in the form of an **input-output table.** The ijth entry in the top portion of the table is the number of units that go from Sector i to Sector j. The "Total outputs" are the total numbers of units produced by each sector. We obtain the technology matrix from the input-output table by dividing the Sector 1 column by the Sector 1 total, the Sector 2 column by the Sector 2 total, and so on.

quick Example

Input-Output Table:

	To	Skateboards	Wood
From	**Skateboards**	20,000*	0
	Wood	100,000	500,000
	Total Output	200,000	5,000,000

*The production of skateboards required skateboards due to the fact that skateboard workers tend to commute to work on (what else?) skateboards!

Technology Matrix:

$$A = \begin{bmatrix} \dfrac{20,000}{200,000} & \dfrac{0}{5,000,000} \\ \dfrac{100,000}{200,000} & \dfrac{500,000}{5,000,000} \end{bmatrix} = \begin{bmatrix} 0.1 & 0 \\ 0.5 & 0.1 \end{bmatrix}$$

Example 2 Rising Demand

Suppose that external demand for refined petroleum rises to $200,000 million, but the demand for crude remains $1000 million (as in Example 1). How do the production levels of the two sectors considered in Example 1 have to change?

Solution We are being told that now

$$D = \begin{bmatrix} 1000 \\ 200,000 \end{bmatrix}$$

and we are asked to find X. Remember that we can calculate X from the formula

$$X = (I - A)^{-1}D$$

Now

$$I - A = \begin{bmatrix} 1 & 0 \\ 0 & 1 \end{bmatrix} - \begin{bmatrix} 0.31 & 0.42 \\ 0.0086 & 0.11 \end{bmatrix} = \begin{bmatrix} 0.69 & -0.42 \\ -0.0086 & 0.89 \end{bmatrix}$$

We take the inverse using our favorite technique and find that, to four significant digits,[*]

$$(I - A)^{-1} \approx \begin{bmatrix} 1.458 & 0.6880 \\ 0.01409 & 1.130 \end{bmatrix}$$

Now we can compute X:

$$X = (I - A)^{-1}D = \begin{bmatrix} 1.458 & 0.6880 \\ 0.01409 & 1.130 \end{bmatrix} \begin{bmatrix} 1000 \\ 200,000 \end{bmatrix} \approx \begin{bmatrix} 140,000 \\ 230,000 \end{bmatrix}$$

(As in Example 1, we have rounded all the entries in the answer to two significant digits.) Comparing this vector to the production vector used in Example 1, we see that production in the crude sector has to increase from $87,000 million to $140,000 million, while production in the refining sector has to increase from $140,000 million to $230,000 million.

---

[*] Since A is accurate to two digits, we should use more than two significant digits in intermediate calculations so as not to lose additional accuracy. We must, of course, round the final answer to two digits.

Note Using the matrix $(I - A)^{-1}$, we have a slightly different way of solving Example 2. We are asking for the effect on production of a *change* in the final demand of 0 for crude and $200,000 - 124,000 = \$76,000$ million for refined products. If we multiply $(I - A)^{-1}$ by the matrix representing this *change,* we obtain

$$\begin{bmatrix} 1.458 & 0.6880 \\ 0.01409 & 1.130 \end{bmatrix} \begin{bmatrix} 0 \\ 76,000 \end{bmatrix} \approx \begin{bmatrix} 53,000 \\ 90,000 \end{bmatrix}$$

$(I - A)^{-1} \times$ Change in Demand = Change in Production

We see the changes required in production: an increase of $53,000 million in the crude sector and an increase of $90,000 million in the refining sector.

Notice that the increase in external demand for the products of the refining sector requires the crude sector to also increase production, even though there is no increase in the *external* demand for its products. The reason is that, in order to increase production, the refining sector needs to use more crude oil, so that the *internal* demand for crude oil goes up. The inverse matrix $(I - A)^{-1}$ takes these **indirect effects** into account in a nice way.

By replacing the $76,000 by $1 in the computation we just did, we see that a $1 increase in external demand for refined products will require an increase in production of $0.6880 in the crude sector, as well as an increase in production of $1.130 in the refining sector. This is how we interpret the entries in $(I - A)^{-1}$, and this is why it is useful to look at this matrix inverse rather than just solve $(I - A)X = D$ for X using, say, Gauss-Jordan reduction. Looking at $(I - A)^{-1}$, we can also find the effects of an increase of $1 in external demand for crude: an increase in production of $1.458 in the crude sector and an increase in production of $0.01409 in the refining sector.

Here are some questions to think about: Why are the diagonal entries of $(I - A)^{-1}$ (slightly) larger than 1? Why is the entry in the lower left so small compared to the others? ∎

Interpreting $(I - A)^{-1}$: Indirect Effects

If A is the technology matrix, then the *ij*th entry of $(I - A)^{-1}$ is the change in the number of units Sector i must produce in order to meet a one-unit increase in external demand for Sector j products. To meet a rising external demand, the necessary change in production for each sector is given by

$$\text{Change in production} = (I - A)^{-1}D^+$$

where D^+ is the change in external demand.

quick Example

Take Sector 1 to be skateboards, and Sector 2 to be wood, and assume that

$$(I - A)^{-1} = \begin{bmatrix} 1.1 & 0 \\ 0.6 & 1.1 \end{bmatrix}$$

Then

$a_{11} = 1.1 =$ number of additional units of skateboards that must be produced to meet a one-unit increase in the demand for skateboards (Why is this number larger than 1?)

$a_{12} = 0 =$ number of additional units of skateboards that must be produced to meet a one-unit increase in the demand for wood (why is this number 0?)

$a_{21} = 0.6 =$ number of additional units of wood that must be produced to meet a one-unit increase in the demand for skateboards

$a_{22} = 1.1 =$ number of additional units of wood that must be produced to meet a one-unit increase in the demand for wood

To meet an increase in external demand of 100 skateboards and 400 units of wood, the necessary change in production is

$$(I - A)^{-1}D^+ = \begin{bmatrix} 1.1 & 0 \\ 0.6 & 1.1 \end{bmatrix} \begin{bmatrix} 100 \\ 400 \end{bmatrix} = \begin{bmatrix} 110 \\ 500 \end{bmatrix}$$

so 110 additional skateboards and 500 additional units of wood will need to be produced.

In the preceding examples, we used only two sectors of the economy. The data used in Examples 1 and 2 were taken from an input-output table published by the U.S. Department of Commerce, in which the whole U.S. economy was broken down into 85 sectors. This in turn was a simplified version of a model in which the economy was broken into about 500 sectors. Obviously, computers are required to make a realistic input-output analysis possible. Many governments collect and publish input-output data as part of their national planning. The United Nations collects these data and publishes collections of national statistics. The U.N. also has a useful set of links to government statistics at the following URL:

www.un.org/Depts/unsd/sd_natstat.htm

Example 3 Kenya Economy

Consider four sectors of the economy of Kenya:* (1) the traditional economy, (2) agriculture, (3) manufacture of metal products and machinery, and (4) wholesale and retail trade. The input-output table for these four sectors for 1976 looks like this (all numbers are 1000s of K£):

	To	1	2	3	4
From	1	8600	0	0	0
	2	0	20,000	24	0
	3	1500	530	15,000	660
	4	810	8500	5800	2900
Total Output		87,000	530,000	110,000	180,000

using *Technology*

See the Technology Guides at the end of the chapter for more on how to do the computations in this example using a TI-83/84 or Excel. Alternatively, go to the Matrix Algebra Tool at

Chapter 3
→ Tools
→ Matrix Algebra Tool

There, type the entries of A as the quotients (column entry/ column total—you need not calculate them first), enter D, and then use the format
(I-A)^(-1)*D.

Suppose that external demand for agriculture increased by K£50,000,000 and that external demand for metal products and machinery increased by K£10,000,000. How would production in these four sectors have to change to meet this rising demand?

Solution To find the change in production necessary to meet the rising demand, we need to use the formula

$$\text{Change in production} = (I - A)^{-1}D^+$$

where A is the technology matrix and D^+ is the change in demand:

$$D^+ = \begin{bmatrix} 0 \\ 50,000 \\ 10,000 \\ 0 \end{bmatrix}$$

With entries shown rounded to two significant digits, the matrix A is

$$A = \begin{bmatrix} 0.099 & 0 & 0 & 0 \\ 0 & 0.038 & 0.00022 & 0 \\ 0.017 & 0.001 & 0.14 & 0.0037 \\ 0.0093 & 0.016 & 0.053 & 0.016 \end{bmatrix}$$

Entries shown rounded to 2 significant digits

---

* Figures are rounded. SOURCE: *Input-Output Tables for Kenya 1976,* Central Bureau of Statistics of the Ministry of Economic Planning and Community Affairs, Kenya.

The next calculation is best done using technology:

$$\text{Change in production} = (I - A)^{-1}D^+ = \begin{bmatrix} 0 \\ 52{,}000 \\ 12{,}000 \\ 1500 \end{bmatrix}$$ Entries shown rounded to 2 significant digits

Looking at this result, we see that the changes in external demand will leave the traditional economy unaffected, production in agriculture will rise by K£52 million, production in the manufacture of metal products and machinery will rise by K£12 million, and activity in wholesale and retail trade will rise by K£1.5 million.

+*Before we go on...* Can you see why the traditional economy was unaffected in Example 3? Although it takes inputs from other parts of the economy, it is not itself an input to any other part. In other words, there is no intermediate demand for the products of the traditional economy coming from any other part of the economy, and so an increase in production in any other sector of the economy will require no increase from the traditional economy. On the other hand, the wholesale and retail trade sector does provide input to the agriculture and manufacturing sectors, so increases in those sectors do require an increase in the trade sector.

One more point: Notice how small the off-diagonal entries in $(I - A)^{-1}$ are. This says that increases in each sector have relatively small effects on the other sectors. We say that these sectors are **loosely coupled.** Regional economies, where many products are destined to be shipped out to the rest of the country, tend to show this phenomenon even more strongly. Notice in Example 2 that those two sectors are **strongly coupled**, because a rise in demand for refined products requires a comparable rise in the production of crude.∎

3.5 EXERCISES

● denotes basic skills exercises

tech Ex indicates exercises that should be solved using technology

1. ● Let A be the technology matrix $A = \begin{bmatrix} 0.2 & 0.05 \\ 0.8 & 0.01 \end{bmatrix}$, where Sector 1 is paper and Sector 2 is wood. Fill in the missing quantities.

    a. ___ units of wood are needed to produce one unit of paper.

    b. ___ units of paper are used in the production of one unit of paper.

    c. The production of each unit of wood requires the use of ___ units of paper.

2. ● Let A be the technology matrix $A = \begin{bmatrix} 0.01 & 0.001 \\ 0.2 & 0.004 \end{bmatrix}$, where Sector 1 is computer chips and Sector 2 is silicon. Fill in the missing quantities.

    a. ___ units of silicon are required in the production of one unit of silicon.

    b. ___ units of computer chips are used in the production of one unit of silicon.

    c. The production of each unit of computer chips requires the use of ___ units of silicon.

3. ● Each unit of television news requires 0.2 units of television news and 0.5 units of radio news. Each unit of radio news requires 0.1 units of television news and no radio news. With Sector 1 as television news and Sector 2 as radio news, set up the technology matrix A.

4. ● Production of one unit of cologne requires no cologne and 0.5 units of perfume. Into one unit of perfume go 0.1 units of cologne and 0.3 units of perfume. With Sector 1 as cologne and Sector 2 as perfume, set up the technology matrix A.

In each of Exercises 5–12, you are given a technology matrix A and an external demand vector D. Find the corresponding production vector X. hint [see Quick Example on p. 228]

5. ● $A = \begin{bmatrix} 0.5 & 0.4 \\ 0 & 0.5 \end{bmatrix}$, $D = \begin{bmatrix} 10,000 \\ 20,000 \end{bmatrix}$

6. ● $A = \begin{bmatrix} 0.5 & 0.4 \\ 0 & 0.5 \end{bmatrix}$, $D = \begin{bmatrix} 20,000 \\ 10,000 \end{bmatrix}$

7. ● $A = \begin{bmatrix} 0.1 & 0.4 \\ 0.2 & 0.5 \end{bmatrix}$, $D = \begin{bmatrix} 25,000 \\ 15,000 \end{bmatrix}$

8. ● $A = \begin{bmatrix} 0.1 & 0.2 \\ 0.4 & 0.5 \end{bmatrix}$, $D = \begin{bmatrix} 24,000 \\ 14,000 \end{bmatrix}$

9. ● $A = \begin{bmatrix} 0.5 & 0.1 & 0 \\ 0 & 0.5 & 0.1 \\ 0 & 0 & 0.5 \end{bmatrix}$, $D = \begin{bmatrix} 1000 \\ 1000 \\ 2000 \end{bmatrix}$

10. ● $A = \begin{bmatrix} 0.5 & 0.1 & 0 \\ 0.1 & 0.5 & 0.1 \\ 0 & 0 & 0.5 \end{bmatrix}$, $D = \begin{bmatrix} 3000 \\ 3800 \\ 2000 \end{bmatrix}$,

11. ● $A = \begin{bmatrix} 0.2 & 0.2 & 0 \\ 0.2 & 0.4 & 0.2 \\ 0 & 0.2 & 0.2 \end{bmatrix}$, $D = \begin{bmatrix} 16,000 \\ 8000 \\ 8000 \end{bmatrix}$

12. ● $A = \begin{bmatrix} 0.2 & 0.2 & 0.2 \\ 0.2 & 0.4 & 0.2 \\ 0.2 & 0.2 & 0.2 \end{bmatrix}$, $D = \begin{bmatrix} 7000 \\ 14,000 \\ 7000 \end{bmatrix}$

13. ● Given $A = \begin{bmatrix} 0.1 & 0.4 \\ 0.2 & 0.5 \end{bmatrix}$, find the changes in production required to meet an increase in demand of 50 units of Sector 1 products and 30 units of Sector 2 products.

14. ● Given $A = \begin{bmatrix} 0.5 & 0.4 \\ 0 & 0.5 \end{bmatrix}$, find the changes in production required to meet an increase in demand of 20 units of Sector 1 products and 10 units of Sector 2 products.

15. ● Let $(I - A)^{-1} = \begin{bmatrix} 1.5 & 0.1 & 0 \\ 0.2 & 1.2 & 0.1 \\ 0.1 & 0.7 & 1.6 \end{bmatrix}$ and assume that the external demand for the products in Sector 1 increases by 1 unit. By how many units should each sector increase production? What do the columns of the matrix $(I - A)^{-1}$ tell you? hint [see Quick Example on p. 232]

16. ● Let $(I - A)^{-1} = \begin{bmatrix} 1.5 & 0.1 & 0 \\ 0.1 & 1.1 & 0.1 \\ 0 & 0 & 1.3 \end{bmatrix}$, and assume that the external demand for the products in each of the sectors increases by 1 unit. By how many units should each sector increase production?

In Exercises 17 and 18, obtain the technology matrix from the given input-output table. hint [see Example 1]

17. ●

	To	A	B	C
From	A	1000	2000	3000
	B	0	4000	0
	C	0	1000	3000
Total Output		5000	5000	6000

18. ●

	To	A	B	C
From	A	0	100	300
	B	500	400	300
	C	0	0	600
Total Output		1000	2000	3000

Applications

19. ● *Campus Food* The two campus cafeterias, the Main Dining Room and Bits & Bytes, typically use each other's food in doing business on campus. One weekend, the input-output table was as follows.[32]

	To	Main DR	Bits & Bytes
From	**Main DR**	$10,000	$20,000
	Bits & Bytes	5000	0
	Total Output	50,000	40,000

Given that the demand for food on campus last weekend was $45,000 from the Main Dining Room and $30,000 from Bits & Bytes, how much did the two cafeterias have to produce to meet the demand last weekend? hint [see Example 1]

20. ● *Plagiarism* Two student groups at Enormous State University, the Choral Society and the Football Club, maintain files of term papers that they write and offer to students for research purposes. Some of these papers they use themselves in generating more papers. In order to avoid suspicion of plagiarism by faculty members (who seem to have astute memories), each paper is given to students or used by the clubs only once (no copies are kept). The number of papers that were used in the production of new papers last year is shown in the following input-output table:

	To	Choral Soc.	Football Club
From	**Choral Soc.**	20	10
	Football Club	10	30
	Total Output	100	200

---

[32] For some reason, the Main Dining Room consumes a lot of its own food!

● basic skills   *tech* Ex technology exercise

Given that 270 Choral Society papers and 810 Football Club papers will be used by students outside of these two clubs next year, how many new papers do the two clubs need to write?

21. tech Ex *Communication Equipment*[33] Two sectors of the U.S. economy are (1) audio, video, and communication equipment and (2) electronic components and accessories. In 1998, the input-output table involving these two sectors was as follows (all figures are in millions of dollars):

	To	Equipment	Components
From	Equipment	6000	500
	Components	24,000	30,000
	Total Output	90,000	140,000

Determine the production levels necessary in these two sectors to meet an external demand for $80,000 million of communication equipment and $90,000 million of electronic components. Round answers to two significant digits.

22. tech Ex *Wood and Paper*[34] Two sectors of the U.S. economy are (1) lumber and wood products and (2) paper and allied products. In 1998 the input-output table involving these two sectors was as follows (all figures are in millions of dollars).

	To	Wood	Paper
From	Wood	36,000	7000
	Paper	100	17,000
	Total Output	120,000	120,000

If external demand for lumber and wood products rises by $10,000 million and external demand for paper and allied products rises by $20,000 million, what increase in output of these two sectors is necessary? Round answers to two significant digits.

23. ● *Australia Economy*[35] Two sectors of the Australian economy are (1) textiles and (2) clothing and footwear. The 1977 input-output table involving these two sectors results in the following value for $(I - A)^{-1}$:

$$(I - A)^{-1} = \begin{bmatrix} 1.228 & 0.182 \\ 0.006 & 1.1676 \end{bmatrix}$$

Complete the following sentences.

a. ____ additional dollars worth of clothing and footwear must be produced to meet a $1 increase in the demand for textiles.

b. 0.182 additional dollars worth of ____ must be produced to meet a one-dollar increase in the demand for ____.

24. *Australia Economy*[36] Two sectors of the Australian economy are (1) community services and (2) recreation services. The 1978–79 input-output table involving these two sectors results in the following value for $(I - A)^{-1}$:

$$(I - A)^{-1} = \begin{bmatrix} 1.0066 & 0.00576 \\ 0.00496 & 1.04206 \end{bmatrix}$$

Complete the following sentences.

a. 0.00496 additional dollars worth of ____ must be produced to meet a $1 increase in the demand for ____.

b. ____ additional dollars worth of community services must be produced to meet a one-dollar increase in the demand for community services.

tech Ex *Exercises 25–28 require the use of technology.*

25. tech Ex *United States Input-Output Table*[37] Four sectors of the U.S. economy are (1) livestock and livestock products, (2) other agricultural products, (3) forestry and fishery products, and (4) agricultural, forestry, and fishery services. In 1977 the input-output table involving these four sectors was as follows (all figures are in millions of dollars):

	To	1	2	3	4
From	1	11,937	9	109	855
	2	26,649	4285	0	4744
	3	0	0	439	61
	4	5423	10,952	3002	216
Total Output		97,795	120,594	14,642	47,473

Determine how these four sectors would react to an increase in demand for livestock (Sector 1) of $1000 million, how they would react to an increase in demand for other agricultural products (Sector 2) of $1000 million, and so on.

26. tech Ex *United States Input-Output Table*[38] Four sectors of the U.S. economy are (1) motor vehicles, (2) truck and bus bodies, trailers, and motor vehicle parts, (3) aircraft and parts, and (4) other transportation equipment. In 1998 the input-output table involving these four sectors was (all figures in millions of dollars):

	To	1	2	3	4
From	1	75	1092	0	1207
	2	64,858	13,081	7	1070
	3	0	0	21,782	0
	4	0	0	0	1375
Total Output		230,676	135,108	129,376	44,133

---

[33] The data have been rounded. SOURCE: *Survey of Current Business*, December, 2001, U.S. Department of Commerce.

[34] Ibid.

[35] SOURCE: *Australian National Accounts and Input-Output Tables 1978–1979*, Australian Bureau of Statistics.

[36] Ibid.

[37] SOURCE: *Survey of Current Business* December, 2001, U.S. Department of Commerce.

[38] Ibid.

● basic skills   tech Ex technology exercise

Determine how these four sectors would react to an increase in demand for motor vehicles (Sector 1) of $1000 million, how they would react to an increase in demand for truck and bus bodies (Sector 2) of $1000 million, and so on.

27. *Australia Input-Output Table* Four sectors of the Australian economy are (1) agriculture, (2) forestry, fishing, and hunting, (3) meat and milk products, and (4) other food products. In 1978–79 the input-output table involving these four sectors was as follows (all figures are in millions of Australian dollars).[39]

	To	1	2	3	4
From	1	678.4	3.7	3341.5	1023.5
	2	15.5	6.9	17.1	124.5
	3	47.3	4.3	893.1	145.8
	4	312.5	22.1	83.2	693.5
Total Output		9401.3	685.8	6997.3	4818.3

a. How much additional production by the meat and milk sector is necessary to accommodate a $100 increase in the demand for agriculture?

b. Which sector requires the most of its own product in order to meet a $1 increase in external demand for that product?

28. *Australia Input-Output Table* Four sectors of the Australian economy are (1) petroleum and coal products, (2) nonmetallic mineral products, (3) basic metals and products, and (4) fabricated metal products. In 1978–79 the input-output table involving these four sectors was as follows (all figures are in millions of Australian dollars).[40]

	To	*1*	*2*	*3*	*4*
From	1	174.1	30.5	120.3	14.2
	2	0	190.1	55.8	12.6
	3	2.1	40.2	1418.7	1242.0
	4	0.1	7.3	40.4	326.0
Total Output		3278.0	2188.8	6541.7	4065.8

a. How much additional production by the petroleum and coal products sector is necessary to accommodate a $1000 increase in the demand for fabricated metal products?

b. Which sector requires the most of the product of some other sector in order to meet a $1 increase in external demand for that product?

Communication and Reasoning Exercises

29. ● What would it mean if the technology matrix A were the zero matrix?

30. ● Can an external demand be met by an economy whose technology matrix A is the identity matrix? Explain.

31. What would it mean if the total output figure for a particular sector of an input-output table were equal to the sum of the figures in the row for that sector?

32. What would it mean if the total output figure for a particular sector of an input-output table were less than the sum of the figures in the row for that sector?

33. What does it mean if an entry in the matrix $(I - A)^{-1}$ is zero?

34. Why do we expect the diagonal entries in the matrix $(I - A)^{-1}$ to be slightly larger than 1?

35. Why do we expect the off-diagonal entries of $(I - A)^{-1}$ to be less than 1?

36. Why do we expect all the entries of $(I - A)^{-1}$ to be non-negative?

[39]SOURCE: *Australian National Accounts and Input-Output Tables 1978–1979,* Australian Bureau of Statistics.
[40] Ibid.

Chapter **3** Review

KEY CONCEPTS

3.1 Matrix Addition and Scalar Multiplication

$m \times n$ matrix, dimensions, entries. *p. 174*
Referring to the entries of a matrix *p. 174*
Matrix equality *p. 175*
Row, column, and square matrices *p. 175*
Addition and subtraction of matrices *p. 176*
Scalar multiplication *p. 177*
Properties of matrix addition and scalar multiplication *p. 179*
The transpose of a matrix *p. 180*
Properties of transposition *p. 180*

3.2 Matrix Multiplication

Multiplying a row by a column *p. 184*
Linear equation as a matrix equation *p. 186*
The product of two matrices: general case *p. 186*
Identity matrix *p. 190*
Properties of matrix addition and multiplication *p. 191*

Properties of transposition and multiplication *p. 192*
A system of linear equations can be written as a single matrix equation *p. 192*

3.3 Matrix Inversion

The inverse of a matrix, singular matrix *p. 199*
Procedure for finding the inverse of a matrix *p. 200*
Formula for the inverse of a 2×2 matrix; determinant of a 2×2 matrix *p. 201*
Using an inverse matrix to solve a system of equations. *p. 203*

3.4 Game Theory

Two-person zero sum game, payoff matrix *p. 210*
A strategy specifies how a player chooses a move *p. 210*
The expected value of a game for given mixed strategies R and C *p. 210*
An optimal strategy, according to the minimax criterion, is one that mini-

mizes the maximum damage your opponent can cause you. *p. 214*
The Fundamental Principle of Game Theory *p. 214*
Procedure for reducing by dominance *p. 215*
Procedure for solving a 2×2 game *p. 215*
The expected value of a game is its expected value when the players use their optimal strategies. *p. 218*
A strictly determined game is one with a saddle point. *p. 219*
Steps to follow in solving a game *p. 220*

3.5 Input-Output Models

An input-output model divides an economy into sectors. The technology matrix records the interactions of these sectors and allows us to relate external demand to the production vector. *p. 228*
Procedure for finding a technology matrix from an input-output table *p. 230*
The entries of $(I - A)^{-1}$ *p. 232*

REVIEW EXERCISES

For Exercises 1–10, let

$$A = \begin{bmatrix} 1 & 2 & 3 \\ 4 & 5 & 6 \end{bmatrix}, B = \begin{bmatrix} 1 & -1 \\ 0 & 1 \end{bmatrix},$$

$$C = \begin{bmatrix} -1 & 0 \\ 1 & 1 \\ 0 & 1 \end{bmatrix} \text{ and } D = \begin{bmatrix} -3 & -2 & -1 \\ 1 & 2 & 3 \end{bmatrix}$$

For each of the following, determine whether the expression is defined, and if it is, evaluate it.

1. $A + B$ **2.** $A - D$

3. $2A^T + C$ **4.** AB

5. $A^T B$ **6.** A^2

7. B^2 **8.** B^3

9. $AC + B$ **10.** $CD + B$

For each matrix in Exercises 11–16, find the inverse or determine that the matrix is singular.

11. $\begin{bmatrix} 1 & -1 \\ 0 & 1 \end{bmatrix}$ **12.** $\begin{bmatrix} 1 & 2 \\ 0 & 0 \end{bmatrix}$

13. $\begin{bmatrix} 1 & 2 & 3 \\ 0 & 4 & 1 \\ 0 & 0 & 1 \end{bmatrix}$ **14.** $\begin{bmatrix} 1 & 2 & 3 & 4 \\ 1 & 3 & 4 & 2 \\ 0 & 1 & 2 & 3 \\ 0 & 0 & 1 & 2 \end{bmatrix}$

15. $\begin{bmatrix} 1 & 2 & 3 & 4 \\ 2 & 3 & 3 & 3 \\ 0 & 1 & 2 & 3 \\ 0 & 0 & 1 & 2 \end{bmatrix}$ **16.** $\begin{bmatrix} 0 & 1 & 0 & 0 \\ 1 & 0 & 0 & 0 \\ 0 & 0 & 0 & 1 \\ 0 & 0 & 1 & 0 \end{bmatrix}$

Write each system of linear equations in Exercises 17–20 as a matrix equation, and solve by inverting the coefficient matrix.

17. $\begin{aligned} x + 2y &= 0 \\ 3x + 4y &= 2 \end{aligned}$ **18.** $\begin{aligned} x + y + z &= 3 \\ y + 2z &= 4 \\ y - z &= 1 \end{aligned}$

19. $\begin{aligned} x + y + z &= 2 \\ x + 2y + z &= 3 \\ x + y + 2z &= 1 \end{aligned}$ **20.** $\begin{aligned} x + y &= 0 \\ y + z &= 1 \\ z + w &= 0 \\ x - w &= 3 \end{aligned}$

In each of Exercises 21–24, solve the game with the given matrix and give the expected value of the game.

21. $P = \begin{bmatrix} 2 & 1 & 3 & 2 \\ -1 & 0 & -2 & 1 \\ 2 & 0 & 1 & 3 \end{bmatrix}$ **22.** $P = \begin{bmatrix} 3 & -3 & -2 \\ -1 & 3 & 0 \\ 2 & 2 & 1 \end{bmatrix}$

23. $P = \begin{bmatrix} -1 & -3 & -2 \\ -1 & 3 & 0 \\ 3 & 3 & -1 \end{bmatrix}$ **24.** $P = \begin{bmatrix} 1 & 4 & 3 & 3 \\ 0 & -1 & 2 & 3 \\ 2 & 0 & -1 & 2 \end{bmatrix}$

In each of Exercises 25–28, find the production vector X corresponding to the given technology matrix A and external demand vector D.

25. $A = \begin{bmatrix} 0.3 & 0.1 \\ 0 & 0.3 \end{bmatrix}$, $D = \begin{bmatrix} 700 \\ 490 \end{bmatrix}$

26. $A = \begin{bmatrix} 0.7 & 0.1 \\ 0.1 & 0.7 \end{bmatrix}$, $D = \begin{bmatrix} 1000 \\ 2000 \end{bmatrix}$

27. $A = \begin{bmatrix} 0.2 & 0.2 & 0.2 \\ 0 & 0.2 & 0.2 \\ 0 & 0 & 0.2 \end{bmatrix}$, $D = \begin{bmatrix} 32,000 \\ 16,000 \\ 8,000 \end{bmatrix}$

28. $A = \begin{bmatrix} 0.5 & 0.1 & 0 \\ 0.1 & 0.5 & 0.1 \\ 0 & 0.1 & 0.5 \end{bmatrix}$, $D = \begin{bmatrix} 23,000 \\ 46.000 \\ 23,000 \end{bmatrix}$

Applications

It is now July 1 and online sales of romance, science fiction, and horror novels at OHaganBooks.com were disappointingly slow over the past month. Exercises 29 and 30 are based on the following tables:

Inventory of books in stock on June 1 at the OHaganBooks.com warehouses in Texas and Nevada:

Books in Stock (June 1)

	Romance	Sci Fi	Horror
Texas	2500	4000	3000
Nevada	1500	3000	1000

Online sales during June:

Online Sales (June)

	Romance	Sci Fi	Horror
Texas	300	500	100
Nevada	100	450	200

29. *Inventory* Use matrix algebra to compute the inventory at each warehouse at the end of June.

30. *Inventory* It is now July 15. Based on July sales to date, the e-commerce manager projects July sales as follows:

July Sales (Projected)

	Romance	Sci Fi	Horror
Texas	280	550	100
Nevada	50	500	120

Assuming that sales continue at this level for the next few months, write down a matrix equation showing the inventory N at each warehouse x months after July 1 (assuming no new books are ordered during that period).

31. *Revenue* It is now the end of July and OHaganBooks.com's e-commerce manager bursts into the CEO's office. "I thought you might want to know, John, that our sales figures are exactly what I projected two weeks ago (see Exercise 30). Is that good market analysis or what?" OHaganBooks.com has charged an average of $5 for romance novels, $6 for science fiction novels, and $5.50 for horror novels. Use the projected sales figures from Exercise 30 and matrix arithmetic to compute the total revenue OHaganBooks.com earned at each warehouse in July.

32. *Revenue* OHaganBooks.com pays an average of $2 for romance novels, $3.50 for science fiction novels, and $1.50 for horror novels. Use this information together with the information in Exercise 31 to compute the profit OHaganBooks.com earned from sales of these books at each warehouse in July.

OHaganBooks.com has two main competitors: JungleBooks.com and FarmerBooks.com, and no other competitors of any significance on the horizon. Exercises 33–35 are based on the following table, which shows the movement of customers during July.[41] (Thus, for instance, the first row tells us that 80 percent of OHaganBooks.com's customers remained loyal, 10 percent of them went to JungleBooks.com, and the remaining 10 percent to FarmerBooks.com.)

	To OHagan	To Jungle	To Farmer
From OHagan	0.8	0.1	0.1
From Jungle	0.4	0.6	0
From Farmer	0.2	0	0.8

At the beginning of July, OHaganBooks.com had an estimated 2000 customers, while its two competitors had 4000 each.

33. *Competition* Set up the July 1 customer numbers in a row matrix, and use matrix arithmetic to estimate the number of customers each company has at the end of July.

---

[41] By a "customer" of one of the three e-commerce sites, we mean someone who purchases more at that site than at any of the two competitors.

34. *Competition* Assuming the July trends continue in August, predict the number of customers each company will have at the end of August.

35. *Competition* Name one or more important factors that the model we have used in this question does not take into account.

Acting on a "tip" from Marjory Duffin, John O'Hagan decided that his company should invest a significant sum in the stock of publisher Duffin House. Exercises 36 and 37 are based on the following table, which shows what information he pieced together later, after some of the records had been lost.

Date	Number of Shares	Price per Share
July 1	?	$20
August 1	?	$10
September 1	?	$5
Total	5000	

Over the three months shown, the company invested a total of $50,000 in Duffin stock, and, on August 15, was paid dividends of 10¢ per share, for a total of $300.

36. *Investments* Use matrix inversion to determine how many shares OHaganBooks.com purchased on each of the three dates shown.

37. *Investments* (Refer to Exercise 36.) On October 1, the shares purchased on July 1 were sold at $3 per share. The remaining shares were sold one month later at $1 per share. Use matrix algebra to determine the total loss (taking into account the dividends paid on August 15) incurred as a result of the Duffin stock fiasco.

Publisher Marjory Duffin reveals that JungleBooks may be launching a promotional scheme in which it will offer either two books for the price of one, or three books for the price of two (Marjory can't quite seem to remember which, and is not certain whether they will go with the scheme at all). Your marketing advisers Floody and O'Lara seem to have different ideas as to how to respond. Floody suggests you counter by offering *three* books for the price of one, while O'Lara suggests that you instead offer a free copy of the "Finite Mathematics Student Solutions Manual" with every purchase. After a careful analysis, you come up with the following payoff matrix, where the payoffs represent the number of customers, in thousands, you expect to gain from JungleBooks.

	JungleBooks		
	No Promo	2 for Price of 1	3 for Price of 2
O'Hagan No Promo	0	−60	−40
3 for Price of 1	30	20	10
Finite Math	20	0	15

Use the above information in Exercises 38–40.

38. *Competition* Determine whether the game is strictly determined. (If so, what should you do?)

39. *Competition* After a very expensive dinner at an exclusive restaurant, Marjory suddenly "remembers" that the Jungle-Books CEO mentioned to her (at a less expensive restaurant) that there is only a 20 percent chance JungleBooks will launch a "2 for the price of 1" promotion, and a 40 percent chance that it will launch a "3 for the price of 2" promotion. What should you do in view of this new information, and what will the expected effect be on your customer base?

40. *Competition* You are about to go with the option chosen in the preceding exercise when one of your corporate spies reveals that Marjory Duffin has just been seen at the lavish Donald Club 8000 in earnest conversation with JungleBooks CEO François Dubois. All bets are off; you can assume now that JungleBooks knows what options you're considering and knows the payoff matrix above. *Now* what should you do, and how many customers should you expect to gain or lose?

Some of the books sold by OHaganBooks.com are printed at Bruno Mills, Inc., a combined paper mill and printing company. Exercises 41–44 are based on the following typical monthly input-output table for Bruno Mills' paper and book printing sectors.

To	Paper	Books
From **Paper**	$20,000	$50,000
Books	2,000	5,000
Total Output	200,000	100,000

41. *Production* Find the technology matrix for Bruno Mills' paper and book printing sectors.

42. *Production* Compute $(I - A)^{-1}$. What is the significance of the $(1, 2)$ entry?

43. *Production* Approximately $1700 worth of the books sold each month by OHaganBooks.com are printed at Bruno Mills, Inc., and OHaganBooks.com uses approximately $170 worth of Bruno Mills' paper products each month. What is the total value of paper and books that must be produced by Bruno Mills, Inc. in order to meet demand from OHaganBooks.com?

44. *Production* Currently, Bruno Mills, Inc. has a monthly capacity of $500,000 of paper products and $200,000 of books. What level of external demand would cause Bruno to meet the capacity for both products?

Mentor Do you need a live tutor for homework problems? Access vMentor on the ThomsonNOW! website at **www.thomsonedu.com** for one-on-one tutoring from a mathematics expert.

CASE STUDY: The Japanese Economy

Jose Fuste Raga/zefa/Corbis

A senator walks into your cubicle in the Congressional Budget Office. "Look here," she says, "I don't see why the Japanese trade representative is getting so upset with my proposal to cut down on our use of Japanese finance and insurance. He claims that it'll hurt Japan's mining operations. But just look at Japan's input-output table. The finance sector doesn't use any input from the mining sector. How can our cutting back on finance and insurance hurt mining?" Indeed, the senator is right about the input-output table, which you have hanging on your wall. Here is what it looks like (all figures are in 100 million yen):[42]

	1	2	3	4	5	6	7	8	9	10	11	12	13
1	19,221	8	99,417	1,610	0	100	0	1	23	0	20	12,491	0
2	0	41	53,006	8,187	13,189	0	0	0	0	0	7	45	7
3	25,376	959	1,247,342	259,049	14,488	38,247	13,316	1,635	54,426	3,827	26,614	267,704	4,963
4	503	105	13,909	2,242	11,664	5,924	1,340	22,788	4,712	1,591	4,627	11,793	0
5	716	471	59,111	6,203	25,035	11,654	1,940	2,265	8,764	1,811	8,528	46,068	915
6	6,559	291	171,655	61,848	3,146	11,242	2,225	1,066	18,053	762	4,683	78,454	1,123
7	5,303	733	43,394	9,533	7,238	58,662	35,348	32,706	30,879	2,252	824	53,827	9,003
8	42	157	11,353	2,731	2,533	38,416	6,772	4,790	8,308	2,445	498	27,639	735
9	7,273	4,021	93,244	46,994	6,781	53,416	7,046	1,623	52,905	4,151	8,369	38,765	1,416
10	135	76	8,682	4,913	1,118	19,014	6,740	435	3,465	9,167	3,826	37,155	98
11	0	0	0	0	0	0	0	0	0	0	0	0	4,614
12	1,780	688	207,289	69,945	24,464	53,225	37,826	10,267	65,577	19,902	18,595	142,920	3,327
13	1,507	228	23,154	1,788	1,702	5,874	1,456	5,112	2,286	1,196	4,263	11,511	0

The sectors are

1. Agriculture, forestry, and fishery
2. Mining
3. Manufacturing
4. Construction
5. Electric power, gas & water supply
6. Commerce
7. Finance and insurance
8. Real estate
9. Transport
10. Communication and broadcasting

---

[42] SOURCE: *1995 Input-Output Tables for Japan,* Management and Coordination Agency, Government of Japan, March 2000. Obtained from www.stat.go.jp.

11. Public administration

12. Services

13. Activities not elsewhere classified

The total output from each sector is given in the following table:

1	2	3	4	5	6	7	8	9	10	11	12	13
158,178	16,595	3,145,585	881,493	264,635	1,023,216	363,346	641,852	501,138	147,628	262,170	1,909,996	55,176

"If I look at just the mining and finance sectors," says the senator, "I'm looking at this input-output table."

	To	Mining (2)	Finance (7)
From	**Mining (2)**	41	0
	Finance (7)	733	35,348
	Total Output	16,595	363,346

"That gives me

$$A = \begin{bmatrix} 0.0025 & 0 \\ 0.0442 & 0.0973 \end{bmatrix}$$

and so

$$(I - A)^{-1} = \begin{bmatrix} 1.0025 & 0 \\ 0.0491 & 1.1078 \end{bmatrix}$$

That last column tells me that any change in demand for finance will have no effect on demand for mining."

Now you have to explain to the senator a point that we fudged a bit in Section 3.5. What she said assumes that changing the external demand for finance (that is, the demand from outside of these two sectors) will not change the external demand for mining. But in fact, that is unlikely to be true. Changing the demand for finance will change the demand for other sectors in the economy directly (the manufacturing sector, for example), which in turn may change the demand for mining. To see these indirect effects properly, you tell the senator that she must look at the whole Japanese economy. The technology matrix A is then

$$\begin{bmatrix}
0.121515 & 0.000482 & 0.031605 & 0.001826 & 0 & 0.000098 & 0 & 0.000002 & 0.000046 & 0 & 0.000076 & 0.006540 & 0 \\
0 & 0.002471 & 0.016851 & 0.009288 & 0.049838 & 0 & 0 & 0 & 0 & 0 & 0.000027 & 0.000024 & 0.000127 \\
0.160427 & 0.057788 & 0.396537 & 0.293875 & 0.054747 & 0.037379 & 0.036648 & 0.002547 & 0.108605 & 0.025923 & 0.101514 & 0.140159 & 0.089949 \\
0.003180 & 0.006327 & 0.004422 & 0.002543 & 0.044076 & 0.005790 & 0.003688 & 0.035504 & 0.009403 & 0.010777 & 0.017649 & 0.006174 & 0 \\
0.004527 & 0.028382 & 0.018792 & 0.007037 & 0.094602 & 0.011390 & 0.005339 & 0.003529 & 0.017488 & 0.012267 & 0.032529 & 0.024119 & 0.016583 \\
0.041466 & 0.017535 & 0.054570 & 0.070163 & 0.011888 & 0.010987 & 0.006124 & 0.001661 & 0.036024 & 0.005162 & 0.017862 & 0.041075 & 0.020353 \\
0.033526 & 0.044170 & 0.013795 & 0.010815 & 0.027351 & 0.057331 & 0.097285 & 0.050956 & 0.061618 & 0.015255 & 0.003143 & 0.028182 & 0.163169 \\
0.000266 & 0.009461 & 0.003609 & 0.003098 & 0.009572 & 0.037544 & 0.018638 & 0.007463 & 0.016578 & 0.016562 & 0.001900 & 0.014471 & 0.013321 \\
0.045980 & 0.242302 & 0.029643 & 0.053312 & 0.025624 & 0.052204 & 0.019392 & 0.002529 & 0.105570 & 0.028118 & 0.031922 & 0.020296 & 0.025663 \\
0.000853 & 0.004580 & 0.002760 & 0.005573 & 0.004225 & 0.018583 & 0.018550 & 0.000678 & 0.006914 & 0.062095 & 0.014594 & 0.019453 & 0.001776 \\
0 & 0 & 0 & 0 & 0 & 0 & 0 & 0 & 0 & 0 & 0 & 0 & 0.083623 \\
0.011253 & 0.041458 & 0.065898 & 0.079348 & 0.092444 & 0.052017 & 0.104105 & 0.015996 & 0.130856 & 0.134812 & 0.070927 & 0.074827 & 0.060298 \\
0.009527 & 0.013739 & 0.007361 & 0.002028 & 0.006431 & 0.005741 & 0.004007 & 0.007964 & 0.004562 & 0.008101 & 0.016260 & 0.006027 & 0
\end{bmatrix}$$

and $(I - A)^{-1}$ is

$$
\begin{bmatrix}
1.151615 & 0.008820 & 0.064373 & 0.023826 & 0.008161 & 0.004945 & 0.005439 & 0.001751 & 0.011777 & 0.005486 & 0.009378 & 0.019111 & 0.009187 \\
0.006998 & 1.008375 & 0.032070 & 0.020476 & 0.059556 & 0.003058 & 0.002844 & 0.001357 & 0.006714 & 0.003262 & 0.006507 & 0.007120 & 0.005696 \\
0.354517 & 0.202960 & 1.754282 & 0.569941 & 0.190926 & 0.114240 & 0.119857 & 0.039317 & 0.281756 & 0.113835 & 0.232270 & 0.296456 & 0.228125 \\
0.008384 & 0.014417 & 0.013625 & 1.010041 & 0.053425 & 0.010797 & 0.007977 & 0.037093 & 0.016950 & 0.015866 & 0.023000 & 0.012313 & 0.007258 \\
0.018575 & 0.046643 & 0.046090 & 0.028805 & 1.117416 & 0.020504 & 0.014828 & 0.006923 & 0.035840 & 0.023719 & 0.046783 & 0.039408 & 0.032861 \\
0.075374 & 0.047784 & 0.112144 & 0.115655 & 0.038606 & 1.026034 & 0.022400 & 0.009075 & 0.068981 & 0.023330 & 0.041195 & 0.067975 & 0.044748 \\
0.065086 & 0.085917 & 0.054440 & 0.045494 & 0.055681 & 0.080239 & 1.122213 & 0.062448 & 0.098977 & 0.034915 & 0.023922 & 0.053391 & 0.199228 \\
0.008625 & 0.021395 & 0.016655 & 0.015301 & 0.018399 & 0.044074 & 0.025858 & 1.010222 & 0.028489 & 0.023738 & 0.008612 & 0.022987 & 0.023259 \\
0.081959 & 0.291054 & 0.084629 & 0.098878 & 0.065802 & 0.071023 & 0.036062 & 0.010081 & 1.143621 & 0.046935 & 0.055252 & 0.046613 & 0.053068 \\
0.007101 & 0.013239 & 0.012988 & 0.015212 & 0.011790 & 0.024942 & 0.026509 & 0.003336 & 0.017217 & 1.072079 & 0.020839 & 0.027392 & 0.011983 \\
0.001291 & 0.001568 & 0.001402 & 0.000791 & 0.000964 & 0.000731 & 0.000602 & 0.000760 & 0.000851 & 0.000969 & 1.001677 & 0.000902 & 0.084107 \\
0.067590 & 0.122810 & 0.160730 & 0.160074 & 0.153207 & 0.093094 & 0.148358 & 0.034321 & 0.206047 & 0.181389 & 0.117705 & 1.129124 & 0.127110 \\
0.015436 & 0.018751 & 0.016768 & 0.009458 & 0.011522 & 0.008746 & 0.007199 & 0.009085 & 0.010175 & 0.011591 & 0.020055 & 0.010782 & 1.005784 \\
\end{bmatrix}
$$

You tell the senator to look at the 7th column to see the effects of a change in demand for finance. There is indeed an effect on mining: Every 1 yen increase in demand for finance produces a 0.002844 yen increase in demand for mining. This also means that every 1 yen *decrease* in demand for finance produces a 0.002844 yen *decrease* in demand for mining. So the Japanese trade representative is right to complain that the senator's plan will hurt their mining companies.

Exercises

1. What does the $(2, 7)$ entry in the matrix A tell you?

2. What does the $(2, 7)$ entry in the matrix $(I - A)^{-1}$ tell you?

3. Why are none of the entries of $(I - A)^{-1}$ negative?

4. Why are the diagonal entries of $(I - A)^{-1}$ close to, and larger than, 1?

5. An increase in demand for the products of which sector of the Japanese economy would have the least impact on the mining sector?

6. An increase in demand for the products of which sector of the Japanese economy would have the most impact on the mining sector?

7. Which sector of the Japanese economy has the greatest percentage of its total output available for external consumption?

8. Referring to the conclusion of this section, try to account for most of the 0.002844 yen by looking at the technology matrix A. For example, a 1 yen increase in demand for finance produces directly a 0.036648 yen increase in demand for manufacturing, which in turn produces a $(0.036648)(0.016851) = 0.000618$ yen increase in demand for mining. What other two-step effects are there? Do they account for all of the 0.002844 yen?

9. Explain why A^2 gives you the total two-step effects of increases in each sector on the others. What about three-step effects? How do you see *all* of the direct and indirect effects?

On the TI-83/84, matrices are referred to as [A], [B], and so on through [J]. To enter a matrix, press MATRX to bring up the matrix menu, select EDIT, select a matrix, and press ENTER. Then enter the dimensions of the matrix followed by its entries. When you want to use a matrix, press MATRX, select the matrix and press ENTER.

Section **3.1**

Example 2 The A-Plus auto parts store chain has two outlets, one in Vancouver and one in Quebec. Among other things, it sells wiper blades, windshield cleaning fluid, and floor mats. The monthly sales of these items at the two stores for two months are given in the following tables:

January Sales

	Vancouver	Quebec
Wiper Blades	20	15
Cleaning Fluid (bottles)	10	12
Floor Mats	8	4

February Sales

	Vancouver	Quebec
Wiper Blades	23	12
Cleaning Fluid (bottles)	8	12
Floor Mats	4	5

Use matrix arithmetic to calculate the change in sales of each product in each store from January to February.

Solution with Technology On the TI-83/84, adding matrices is similar to adding numbers. The sum of the matrices [A] and [B] is [A]+[B]; their difference, of course, is [A]-[B]. As in the text, for this example,

1. Create two matrices, [J] and [F].
2. Compute their difference, [F]-[J] using [F]-[J]→[D].

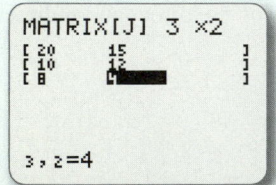

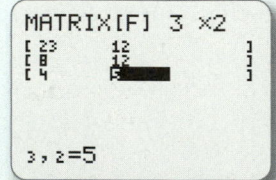

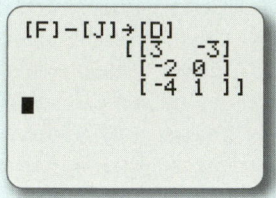

Note that we have stored the difference in the matrix [D] in case we need it for later use.

Example 3 The revenue generated by sales in the Vancouver and Quebec branches of the A-Plus auto parts store was as follows:

January Sales in Canadian Dollars

	Vancouver	Quebec
Wiper Blades	140.00	105.00
Cleaning Fluid (bottles)	30.00	36.00
Floor Mats	96.00	48.00

If the Canadian dollar was worth $0.65US at the time, compute the revenue in U.S. dollars.

Solution with Technology Scalar multiplication on the TI-83/84 is similar to multiplication of numbers.

1. Enter the matrix A using the matrix editor.

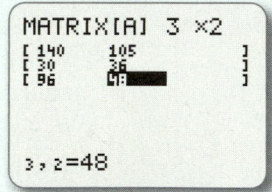

2. Calculate $0.65A$ using the formula 0.65*[A] or just 0.65[A].

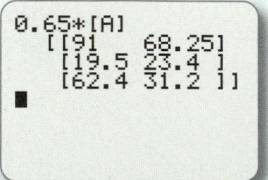

Transposition On the TI-83/84 you can access the symbol "T" by pressing [MATRX], selecting MATH, selecting T, and then pressing [ENTER]. Here is an example:

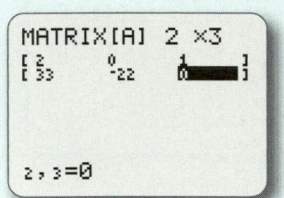

The matrix A

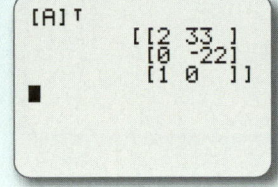

Computation of [A]^T

Section 3.2

Example 1 The A-Plus auto parts store had the following sales in its Vancouver store.

	Vancouver
Wiper Blades	20
Cleaning Fluid (bottles)	10
Floor Mats	8

The store sells wiper blades for $7.00 each, cleaning fluid for $3.00 per bottle, and floor mats for $12.00 each. Use matrix multiplication to find the total revenue generated by sales of these items.

Solution with Technology On the TI-83/84, the format for multiplying matrices is the same as for multiplying numbers: [A][B] or [A] ∗ [B] will give the product. Because we can't use P and Q as names of matrices on the TI-83/84, we'll use [A] for P and [B] for Q. We enter the matrices and then multiply them:

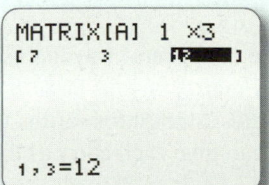

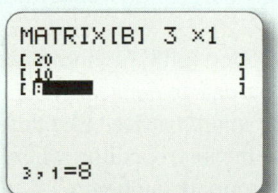

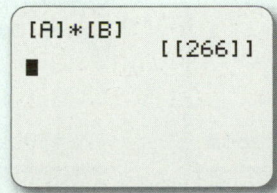

Example 3 Calculate

$$\begin{bmatrix} 2 & 0 & -1 & 3 \\ 1 & -1 & 2 & -2 \end{bmatrix} \begin{bmatrix} 1 & 1 & -8 \\ 1 & 0 & 0 \\ 0 & 5 & 2 \\ -2 & 8 & -1 \end{bmatrix}$$

Solution with Technology We can do this computation just as we did the one in Example 1. Note that if you try to multiply two matrices whose product is not defined, you will get the error "DIM MISMATCH" (dimension mismatch).

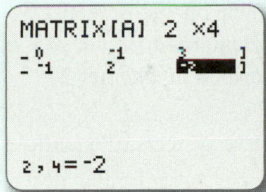

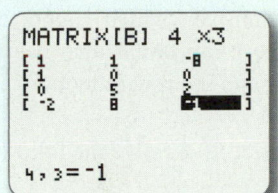

Note that, while editing, you can see only three columns of [A] at a time.

Example 6 —Identity Matrix On the TI-83/84, the function identity(n) (in the ([MATRX] MATH menu) returns the $n \times n$ identity matrix.

Section 3.3

Example 2 Find the inverse of

$$Q = \begin{bmatrix} 1 & 0 & 1 \\ 2 & -2 & -1 \\ 3 & 0 & 0 \end{bmatrix}$$

Solution with Technology On a TI-83/84, you can invert the square matrix [A] by entering [A] $\boxed{x^{-1}}$ $\boxed{\text{ENTER}}$.

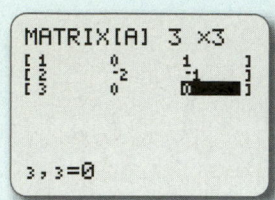
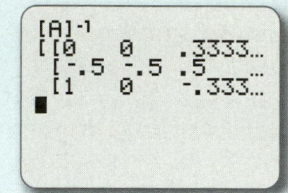

(Note that you can use the right and left arrow keys to scroll the answer, which cannot be shown on the screen all at once.) You could also use the calculator to help you go through the row reduction, as described in Chapter 2.

Example 4 Solve the following three systems of equations.

a.
$$2x \quad + z = 1$$
$$2x + y - z = 1$$
$$3x + y - z = 1$$

b.
$$2x \quad + z = 0$$
$$2x + y - z = 1$$
$$3x + y - z = 2$$

c.
$$2x \quad + z = 0$$
$$2x + y - z = 0$$
$$3x + y - z = 0$$

Solution with Technology

1. Enter the four matrices A, B, C, and D:

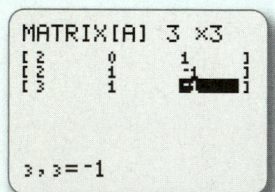
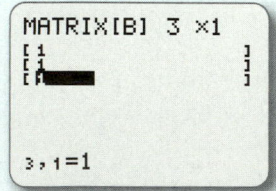

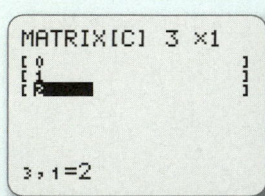
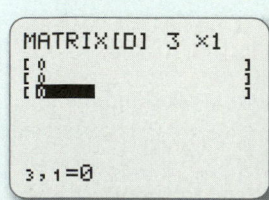

2. Compute the solutions $A^{-1}B$, $A^{-1}C$. and $A^{-1}D$:

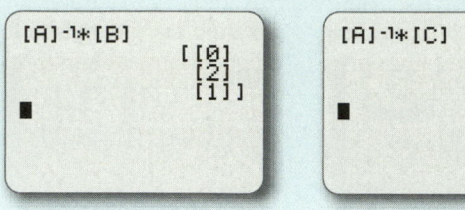

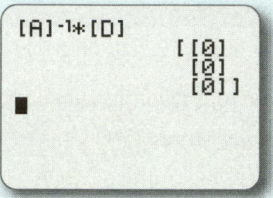

Section **3.5**

Example 1   Recall that the input-output table in Example 1 looks like this:

	To	Crude	Refining
From	**Crude**	27,000	59,000
	Refining	750	15,000
	Total Output	87,000	140,000

What was the technology matrix for these two sectors? What was left over from each of these sectors for use by other parts of the economy or for export?

Solution with Technology There are several ways to use these data to create the technology matrix in your TI-83/84. For small matrices like this, the most straightforward is to use the matrix editor, where you can give each entry as the appropriate quotient:

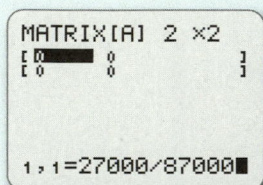
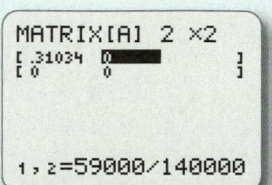

and so on. Once we have the technology matrix [A] and the production vector [B] (remember that we can't use [X] as a matrix name), we can calculate the external demand vector:

```
[B]-[A]*[B]
      [[1000 ]
       [124250]]
■
```

Of course, when interpreting these numbers, we must remember to round to two significant digits, because our original data were accurate to only that many digits.

Here is an alternative way to calculate the technology matrix that may be better for examples with more sectors.

1. Begin by entering the columns of the input-output table as lists in the list editor (STAT EDIT).

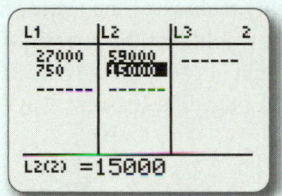

2. We now want to divide each column by the total output of its sector and assemble the results into a matrix. We can do this using the List ▶ matr function (under the MATRX MATH menu) as follows:

```
List▶matr(L1/870
00,L2/140000,[A]
)
                Done
[A]
[[.3103448276 .…
 [.0086206897 .…
■
```

Example 3 Consider four sectors of the economy of Kenya:[43] (1) the traditional economy, (2) agriculture, (3) manufacture of metal products and machinery, and (4) wholesale and retail trade. The input-output table for these

four sectors for 1976 looks like this (all numbers are 1000s of K£):

	To	**1**	**2**	**3**	**4**
From	**1**	8600	0	0	0
	2	0	20,000	24	0
	3	1500	530	15,000	660
	4	810	8500	5800	2900
Total Output		87,000	530,000	110,000	180,000

Suppose that external demand for agriculture increased by K£50,000,000 and that external demand for metal products and machinery increased by K£10,000,000. How would production in these four sectors have to change to meet this rising demand?

Solution with Technology

1. Enter the technology matrix [A] using one of the techniques above.

```
MATRIX[A]  4 ×4
[ 0        0       0      ]
[.03774  2.2E-4   0      ]
[.001    .13636  .00367 ]
[.01604  .05273  ■■■■■  ]

4,4=.0161111111…
```

2. Enter the matrix D^+ as [D].

```
MATRIX[D]  4 ×1
[ 0     ]
[ 50000 ]
[ 10000 ]
[ ■■■■  ]

4,1=0
```

3. You can then compute the change in production with the following formula:

```
(identity(4)-[A])⁻¹ [D]
```

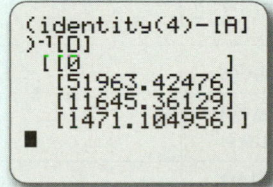

---

[43] Figures are rounded. SOURCE: *Input-Output Tables for Kenya 1976*, Central Bureau of Statistics of the Ministry of Economic Planning and Community Affairs, Kenya.

EXCEL Technology Guide

Section 3.1

To enter a matrix in a spreadsheet, we put its entries in any convenient block of cells. For example, the matrix A in the first Quick Example of this section might look like this:

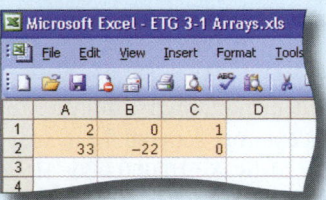

Excel refers to such blocks of data as **arrays**, which it can handle in much the same way as it handles single cells of data. For instance, when typing a formula, just as clicking on a cell creates a reference to that cell, selecting a whole array of cells will create a reference to that array. An array is referred to using an **array range** consisting of the top-left and bottom-right cell coordinates, separated by a colon. For example, the array range A1:C2 refers to the 2×3 matrix above, with top-left corner A1 and bottom-right corner C2.

Example 2 The A-Plus auto parts store chain has two outlets, one in Vancouver and one in Quebec. Among other things, it sells wiper blades, windshield cleaning fluid, and floor mats. The monthly sales of these items at the two stores for two months are given in the following tables:

January Sales

	Vancouver	Quebec
Wiper Blades	20	15
Cleaning Fluid (bottles)	10	12
Floor Mats	8	4

February Sales

	Vancouver	Quebec
Wiper Blades	23	12
Cleaning Fluid (bottles)	8	12
Floor Mats	4	5

Use matrix arithmetic to calculate the change in sales of each product in each store from January to February.

Solution with Technology

1. To add or subtract two matrices in Excel, first input their entries in two separate arrays in the spreadsheet (we've also added labels as in the tables above, which you might do if you wanted to save the spreadsheet for later use):

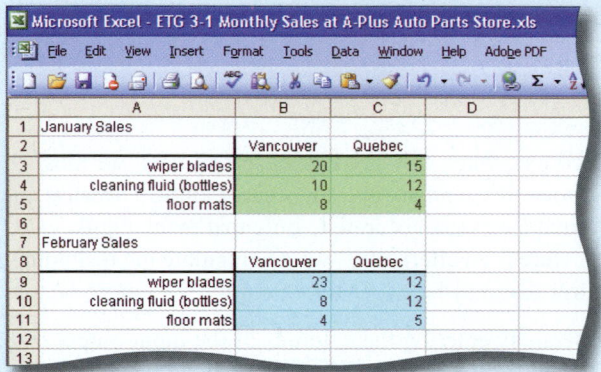

2. Select (highlight) a block of the same size (3×2 in this case) where you would like the answer, $F - J$, to appear, enter the formula =B9:C11-B3:C5, and then type Control+Shift+Enter. The easiest way to do this is as follows:

- Highlight cells B15:C17.   Where you want the answer to appear

- Type "=".

- Highlight the matrix F.   Cells B9 through C11

- Type "-".

- Highlight the matrix J.   Cells B3 through C5

- Type Control+Shift+Enter.   Not just Enter

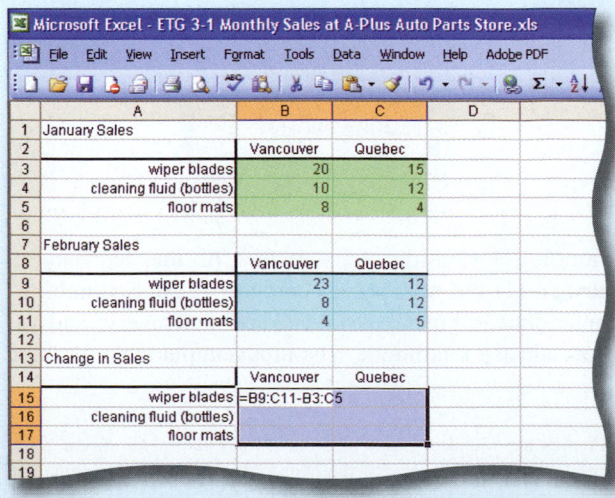

Typing Control+Shift+Enter (instead of Enter) tells Excel that your formula is an *array formula,* one that returns a matrix rather than a single number.[44] Once entered, the formula bar will show the formula you entered enclosed in "curly braces," indicating that it is an array formula. Note that you must use Control+Shift+Enter to delete any array you create: select the block you wish to delete and press Delete followed by Control+Shift+Enter.

Example 3 The revenue generated by sales in the Vancouver and Quebec branches of the A-Plus auto parts store was as follows:

January Sales in Canadian Dollars

	Vancouver	Quebec
Wiper Blades	140.00	105.00
Cleaning Fluid (bottles)	30.00	36.00
Floor Mats	96.00	48.00

If the Canadian dollar was worth $0.65U.S. at the time, compute the revenue in U.S. dollars.

Solution with Technology Scalar multiplication is done in Excel in much the same way we did matrix addition above:

As with matrix addition, we use Control+Shift+Enter to evaluate scalar multiplication (or to evaluate any formula that uses an array).

Transposition To transpose a matrix in Excel, first highlight the block where you would like the transpose to appear (shaded in the spreadsheet below). (Note that it should have the correct dimensions for the transpose: 3×2 in the case

shown below.) Then type the formula =TRANSPOSE (A1:C2) (use the array range appropriate for the matrix you want to transpose) in the formula bar and press Control+Shift+Enter. The easiest way to do this is as follows:

1. Highlight Cells A4:B6. The 3×2 block for the answer
2. Type "=TRANSPOSE(".
3. Highlight the matrix you want to transpose. Cells A1:C2
4. Type ")".
5. Type Control+Shift+Enter.   Not just Enter

The transpose will appear in the region you highlighted.

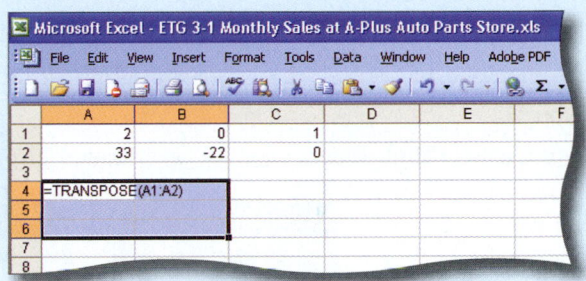

Section **3.2**

Example 1 The A-Plus auto parts store had the following sales in its Vancouver store.

	Vancouver
Wiper Blades	20
Cleaning Fluid (bottles)	10
Floor Mats	8

The store sells wiper blades for $7.00 each, cleaning fluid for $3.00 per bottle, and floor mats for $12.00 each. Use matrix multiplication to find the total revenue generated by sales of these items.

Solution with Technology In Excel, the formula for matrix multiplication is MMULT. (Ordinary multiplication, *, will *not* work.)

1. To find the product PQ of a row and a column in Excel, first enter the matrices P and Q anywhere in the spreadsheet.
2. Select the cell where you would like the result to appear (G1 in the example shown below), and use MMULT as shown. (Because the product occupies only a single cell, it is not necessary to press Control+Shift+Enter, but it won't hurt.)

---

[44] Note that on a Macintosh, Command-Enter has the same effect as Control+Shift+Enter.

TECHNOLOGY GUIDE

As usual, you can use the mouse to avoid typing the array ranges:

1. Click on cell G1.

2. Type "=MMULT(".

3. Highlight the matrix P.   Cells A1 through C1

4. Type ",".

5. Highlight the matrix Q.   Cells E1 through E3

6. Type ")".

7. Type Control+Shift+Enter.   "Enter" also works in this case

Online, follow

Chapter 3 → Excel Tutorials → 3.3 Matrix Inversion

for a downloadable Excel tutorial that covers this and other examples in this section.

Example 3 Calculate the product $\begin{bmatrix} 2 & 0 & -1 & 3 \\ 1 & -1 & 2 & -2 \end{bmatrix} \begin{bmatrix} 1 & 1 & -8 \\ 1 & 0 & 0 \\ 0 & 5 & 2 \\ -2 & 8 & -1 \end{bmatrix}$.

Solution with Technology

1. Enter the two matrices as shown in the spreadsheet and highlight a block where you want the answer to appear. (Note that it should have the correct dimensions for the product: 2×3.)

2. Enter the formula =MMULT(A1:D2,F1:H4) (using the mouse to avoid typing the array ranges if you like) and press Control+Shift+Enter. The product will appear in the region you highlighted. If you try to multiply two matrices whose product is not defined, you will get the error "!VALUE#".

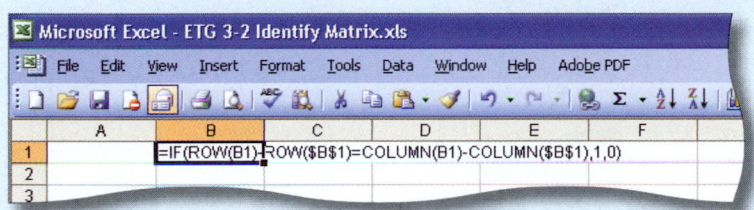

Example 6—Identity Matrix There is no function in Excel that returns an identity matrix. If you need a small identity matrix, it's simplest to just enter the 1s and 0s by hand. If you need a large identity matrix, here is one way to get it quickly.

1. Say we want a 4×4 identity matrix in the cells B1:E4. Enter the following formula in cell B1:

=IF(ROW(B1)-ROW($B$1)=COLUMN(B1)-COLUMN($B$1),1,0)

2. Press Enter, then copy cell B1 to cells B1:E4. The formula will return 1 along the diagonal of the matrix and 0s elsewhere, giving you the identity matrix. Why does this formula work?

Section **3.3**

Example 2 Find the inverse of

$$Q = \begin{bmatrix} 1 & 0 & 1 \\ 2 & -2 & -1 \\ 3 & 0 & 0 \end{bmatrix}$$

Solution with Technology In Excel, the function MINVERSE computes the inverse of a matrix.

1. Enter Q somewhere convenient, for example, in cells A1:C3.

TECHNOLOGY GUIDE

2. Choose the block where you would like the inverse to appear, highlight the whole block.

3. Enter the formula =MINVERSE(A1:C3) and press Control+Shift+Enter.

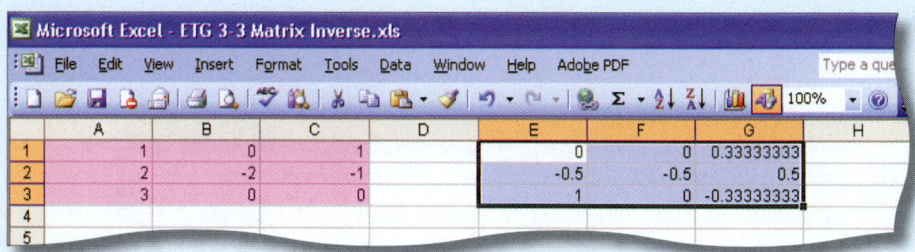

The inverse will appear in the region you highlighted. (To convert the answer to fractions, format the cells as fractions.)

If a matrix is singular, Excel will register an error by showing #NUM! in each cell.

Although Excel appears to invert the matrix in one step, it is going through the procedure in the text or some variation of it to find the inverse. Of course, you could also use Excel to help you go through the row reduction, just as in Chapter 2.

Online, follow

Chapter 3 → Excel Tutorials → 3.3 Matrix Inversion

for a downloadable Excel tutorial that covers this and other examples in this section.

Example 4 Solve the following three systems of equations.

a. $2x \quad + z = 1$	**b.** $2x \quad + z = 0$	**c.** $2x \quad + z = 0$
$2x + y - z = 1$	$2x + y - z = 1$	$2x + y - z = 0$
$3x + y - z = 1$	$3x + y - z = 2$	$3x + y - z = 0$

Solution with Technology Spreadsheets like Excel instantly update calculated results every time the contents of a cell are changed. We can take advantage of this to solve the three systems of equations given above using the same worksheet as follows.

1. Enter the matrices A and B from the matrix equation $AX = B$.

2. Select a 3×1 block of cells for the matrix X.

3. The Excel formula we can use to calculate X is

=MMULT(MINVERSE(A1:C3),E1:E3) $A^{-1}B$

(As usual, use the mouse to select the ranges for *A* and *B* while typing the formula, and don't forget to press Control+Shift+Enter.) Having obtained the solution to part (a), you can now simply modify the entries in Column E to see the solutions for parts (b) and (c).

Note Your spreadsheet for part (a) may look like this:

What is that strange number doing in cell A6? "E-16" represents "$\times 10^{-16}$", so the entry is really

$$1.11022 \times 10^{-16} = 0.000\,000\,000\,000\,000\,111022 \approx 0$$

Mathematically, it is supposed to be *exactly* zero (see the solution to part (a) in the text) but Excel made a small error in computing the inverse of *A*, resulting in this spurious value. Note, however, that it is accurate (agrees with zero) to 15 decimal places! In practice, when we see numbers arise in matrix calculations that are far smaller than all the other entries, we can usually assume they are supposed to be zero. ■

Section **3.5**

Example 1 Recall that the input-output table in Example 1 looks like this:

	To	**Crude**	**Refining**
From	**Crude**	27,000	59,000
	Refining	750	15,000
	Total Output	87,000	140,000

What was the technology matrix for these two sectors? What was left over from each of these sectors for use by other parts of the economy or for export?

Solution with Technology

1. Enter the input-output table in a spreadsheet:

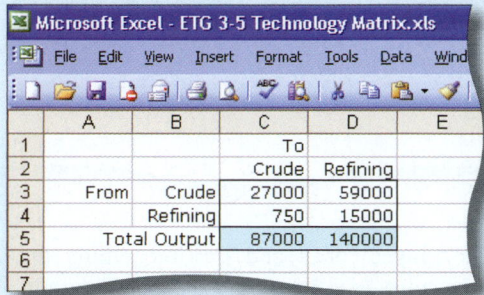

2. To obtain the technology matrix, we divide each column by the total output of its sector:

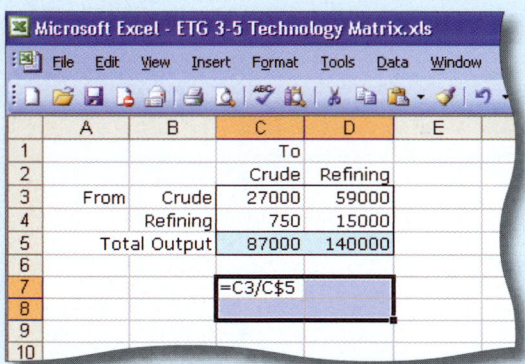

The formula `=C3/C$5` is copied into the shaded 2 × 2 block shown \ on the left. (The $ sign in front of the 5 forces the program to always divide by the total in Row 5 even when the formula is copied from Row 7 to Row 8. The result is the technology matrix shown on the right.

3. Using the techniques discussed in the second section, we can now compute $D = X - AX$ to find the demand vector.

Example 3 Consider four sectors of the economy of Kenya:[45] (1) the traditional economy, (2) agriculture, (3) manufacture of metal products and machinery, and (4) wholesale and retail trade. The input-output table for these four sectors for 1976 looks like this (all numbers are 1000s of K£):

	To	1	2	3	4
From	**1**	8600	0	0	0
	2	0	20,000	24	0
	3	1500	530	15,000	660
	4	810	8500	5800	2900
Total Output		87,000	530,000	110,000	180,000

---

[45] Figures are rounded. SOURCE: *Input-Output Tables for Kenya 1976*, Central Bureau of Statistics of the Ministry of Economic Planning and Community Affairs, Kenya.

Suppose that external demand for agriculture increased by K£50,000,000 and that external demand for metal products and machinery increased by K£10,000,000. How would production in these four sectors have to change to meet this rising demand?

Solution with Technology

1. Enter the input-output table in the spreadsheet.

2. Compute the technology matrix by dividing each column by the column total.

3. Insert the identity matrix I in preparation for the next step.

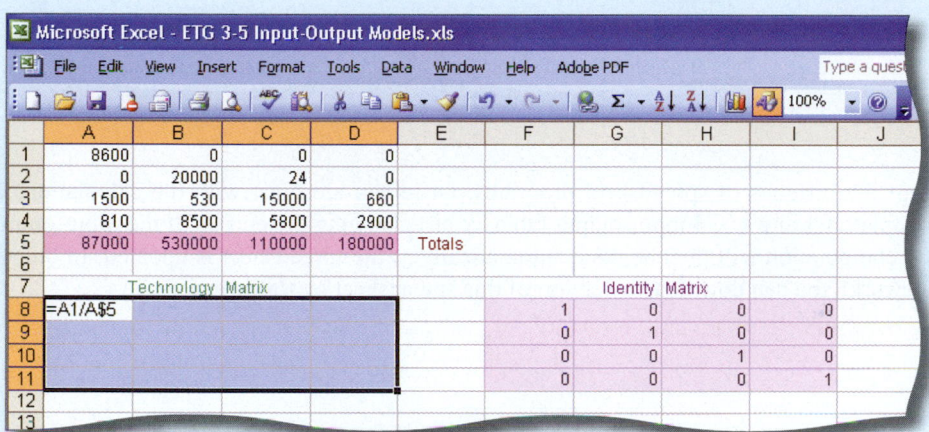

4. To see how each sector reacts to rising external demand, you must calculate the inverse matrix $(I - A)^{-1}$, as shown below (remember to use Control+Shift+Enter each time):

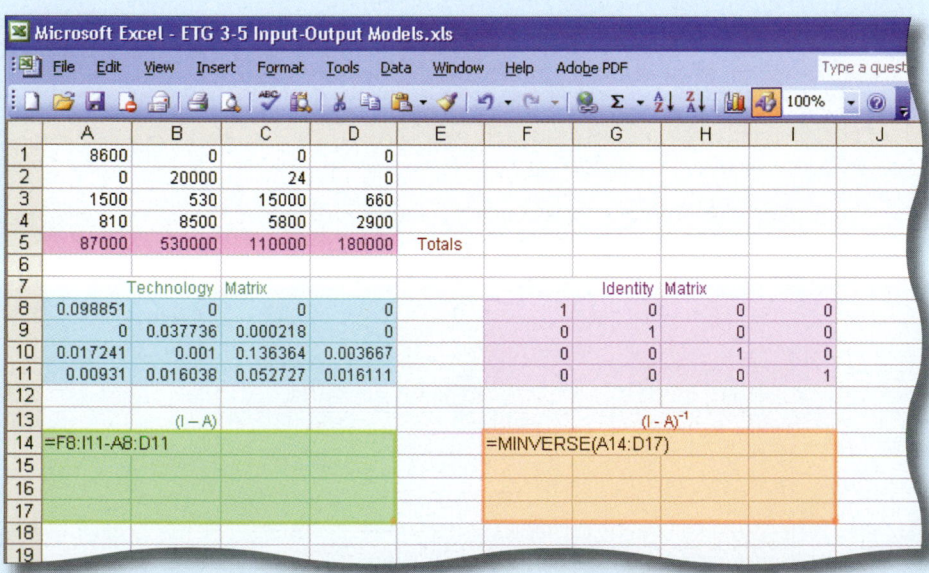

5. To compute $(I - A)^{-1}D^+$, enter D^+ as a column and use the MMULT operation (see Example 3 in Section 3.2).

		$(I - A)$				$(I - A)^{-1}$			
12									
13		$(I - A)$				$(I - A)^{-1}$			
14	0.901149	0	0	0	1.109694	0	0	0	
15	0	0.962264	-0.000218	0	5.03E-06	1.039216	0.000263	9.79E-07	
16	-0.017241	-0.001	0.863636	-0.003667	0.022203	0.001276	1.158159	0.004316	
17	-0.00931	-0.016038	-0.052727	0.983889	0.011691	0.017008	0.062071	1.016606	
18									
19	D		$(I - A)^{-1}D$						
20	0		0						
21	50000		51963.42						
22	10000		11645.36						
23	0		1471.105						
24									
25									

Note Here is one of the beauties of spreadsheet programs: Once you are done with the calculation, you can use the spreadsheet as a template for any 4×4 input-output table by just changing the entries of the input-output matrix and/or D^+. The rest of the computation will then be done automatically as the spreadsheet is updated. In other words, you can use it do your homework! You can download a version of this spreadsheet by following

Chapter 3 → Excel Tutorials → 3.5 Input-Output Models ∎

4

Linear Programming

CASE STUDY Airline Scheduling

Fly-by-Night Airlines flies airplanes from five cities: Los Angeles, Chicago, Atlanta, New York, and Boston. Due to a strike, the airline has not had sufficient crews to fly its planes, so some planes have been stranded in Chicago and Atlanta. On the other hand, Los Angeles, New York, and Boston need more planes than they have. The airline knows how much it costs to move a plane from one city to another. To get back on schedule at the lowest cost, how should Fly-by-Night rearrange its planes?

Austin Brown/Getty Images

Introduction

In this chapter we begin to look at one of the most important types of problems for business and the sciences: finding the largest or smallest possible value of some quantity (such as profit or cost) under certain constraints (such as limited resources). We call such problems **optimization** problems because we are trying to find the best, or optimum, value. The optimization problems we look at in this chapter involve linear functions only and are known as **linear programming** (LP) problems. One of the main purposes of calculus, which you may study later, is to solve nonlinear optimization problems.

Linear programming problems involving only two unknowns can usually be solved by a graphical method which we discuss in Sections 4.1 and 4.2. When there are three or more unknowns, we must use an algebraic method, as we had to do for systems of linear equations. The method we use is called the **simplex method.** Invented in 1947 by George B. Dantzig[1] (1914–2005), the simplex method is still the most commonly used technique to solve LP problems in real applications, from finance to the computation of trajectories for guided missiles.

The simplex method can be used for hand calculations when the numbers are fairly small and the unknowns are few. Practical problems often involve large numbers and many unknowns, however. Problems like routing telephone calls or airplane flights, or allocating resources in a manufacturing process can involve tens of thousands of unknowns. Solving such problems by hand is obviously impractical, and so computers are regularly used. Although computer programs most often use the simplex method, mathematicians are always seeking faster methods. The first radically different method of solving LP problems was the **ellipsoid algorithm** published in 1979 by the Soviet mathematician Leonid G. Khachiyan[2] (1952–2005). In 1984, Narendra Karmarkar (1957–), a researcher at Bell Labs, created a more efficient method now known as **Karmarkar's algorithm.** Although these methods (and others since developed) can be shown to be faster than the simplex method in the worst cases, it seems to be true that the simplex method is still the fastest in the applications that arise in practice.

Calculators and spreadsheets are very useful aids in the simplex method. In practice, software packages do most of the work, so you can think of what we teach you here as a peek inside a "black box." What the software cannot do for you is convert a real situation into a mathematical problem, so the most important lessons to get out of this chapter are (1) how to recognize and set up a linear programming problem, and (2) how to interpret the results.

4.1 Graphing Linear Inequalities

By the end of the next section, we will be solving linear programming (LP) problems with two unknowns. We use inequalities to describe the constraints in a problem, such as limitations on resources. Recall the basic notation for inequalities.

---

[1] Dantzig is the real-life source of the story of the student who, walking in late to a math class, copies down two problems on the board, thinking they're homework. After much hard work he hands in the solutions, only to discover that he's just solved two famous unsolved problems. This actually happened to Dantzig in graduate school in 1939. SOURCES: D. J. Albers and C. Reid, "An Interview of George B. Dantzig: The Father of Linear Programming," *College Math. Journal,* v. 17 (1986), pp. 293–314. Quoted and discussed in the context of the urban legends it inspired at http://www.snopes.com/college/homework/unsolvable.asp.

[2] Dantzig and Khachiyan died approximately two weeks apart in 2005. The *New York Times* ran their obituaries together on May 23, 2005.

quick Examples

Nonstrict Inequalities

$a \le b$ means that a **is less than or equal to** b.

$3 \le 99, -2 \le -2, 0 \le 3$

$a \ge b$ means that a **is greater than or equal to** b.

$3 \ge 3, 1.78 \ge 1.76, \dfrac{1}{3} \ge \dfrac{1}{4}$

There are also the inequalities $<$ and $>$, called **strict** inequalities because they do not permit equality. We do not use them in this chapter.

Following are some of the basic rules for manipulating inequalities. Although we illustrate all of them with the inequality $\le$, they apply equally well to inequalities with $\ge$ and to the strict inequalities $<$ and $>$.

quick Examples

Rules for Manipulating Inequalities

1. The same quantity can be added to or subtracted from both sides of an inequality:

   If $x \le y$, then $x + a \le y + a$ for any real number a.

   $x \le y$ implies $x - 4 \le y - 4$

2. Both sides of an inequality can be multiplied or divided by a positive constant:

   If $x \le y$ and a is positive, then $ax \le ay$.

   $x \le y$ implies $3x \le 3y$

3. Both sides of an inequality can be multiplied or divided by a negative constant if the inequality is *reversed*:

   If $x \le y$ and a is negative, then $ax \ge ay$.

   $x \le y$ implies $-3x \ge -3y$

4. The left and right sides of an inequality can be switched if the inequality is *reversed*:

   If $x \le y$, then $y \ge x$; if $y \ge x$, then $x \le y$.

   $3x \ge 5y$ implies $5y \le 3x$

Here are the particular kinds of inequalities in which we're interested:

Linear Inequalities and Solving Inequalities

An **inequality in the unknown x** is the statement that one expression involving x is less than or equal to (or greater than or equal to) another. Similarly, we can have an **inequality in x and y**, which involves expressions that contain x and y; an **inequality in x, y, and z**; and so on. A **linear inequality** in one or more unknowns is an inequality of the form

$$ax \le b \text{ (or } ax \ge b\text{),} \quad a \text{ and } b \text{ real constants}$$
$$ax + by \le c \text{ (or } ax + by \ge c\text{),} \quad a, b, \text{ and } c \text{ real constants}$$
$$ax + by + cz \le d, \quad a, b, c, \text{ and } d \text{ real constants}$$
$$ax + by + cz + dw \le e, \quad a, b, c, d, \text{ and } e \text{ real constants}$$

and so on.

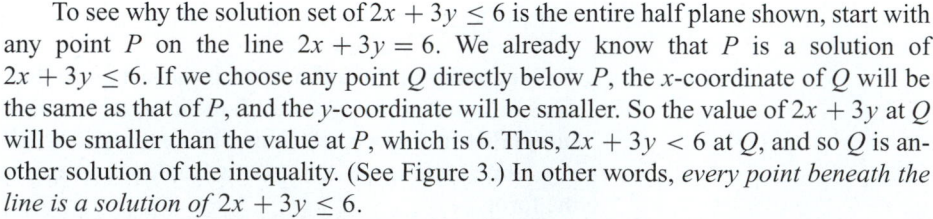

quick Examples

$2x + 8 \geq 89$	Linear inequality in x
$2x^3 \leq x^3 + y$	Nonlinear inequality in x and y
$3x - 2y \geq 8$	Linear inequality in x and y
$x^2 + y^2 \leq 19z$	Nonlinear inequality in x, y, and z
$3x - 2y + 4z \leq 0$	Linear inequality in x, y, and z

A **solution** of an inequality in the unknown x is a value for x that makes the inequality true. For example, $2x + 8 \geq 89$ has a solution $x = 50$ because $2(50) + 8 \geq 89$. Of course, it has many other solutions as well. Similarly, a solution of an inequality in x and y is a pair of values (x, y) making the inequality true. For example, $(5, 1)$ is a solution of $3x - 2y \geq 8$ because $3(5) - 2(1) \geq 8$. To **solve** an inequality is to find the set of *all* solutions.

Solving Linear Inequalities in Two Variables

Our first goal is to solve linear inequalities in two variables—that is, inequalities of the form $ax + by \leq c$. As an example, let's solve

$$2x + 3y \leq 6$$

We already know how to solve the *equation* $2x + 3y = 6$. As we saw in Chapter 1, the solution of this equation may be pictured as the set of all points (x, y) on the straight-line graph of the equation. This straight line has x-intercept 3 (obtained by putting $y = 0$ in the equation) and y-intercept 2 (obtained by putting $x = 0$ in the equation) and is shown in Figure 1.

Notice that, if (x, y) is any point on the line, then x and y not only satisfy the *equation* $2x + 3y = 6$, but they also satisfy the *inequality* $2x + 3y \leq 6$, because being equal to 6 qualifies as being less than or equal to 6.

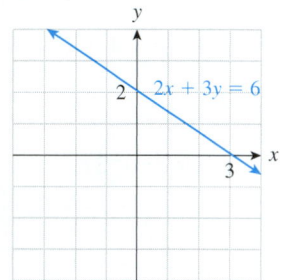

Figure **1**

Q: *Do the points on the line give all possible solutions to the inequality?*

A: No. For example, try the origin, $(0, 0)$. Since $2(0) + 3(0) = 0 \leq 6$, the point $(0, 0)$ is a solution that does not lie on the line. In fact, here is a possibly surprising fact: The solution to any linear inequality in two unknowns is represented by an entire **half plane:** the set of all points on one side of the line (including the line itself). Thus, because $(0, 0)$ is a solution of $2x + 3y \leq 6$ and is not on the line, every point on the same side of the line as $(0, 0)$ is a solution as well (the blue region below the line in Figure 2 shows which half plane constitutes the solution set). ■

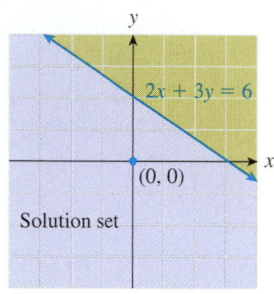

Figure **2**

To see why the solution set of $2x + 3y \leq 6$ is the entire half plane shown, start with any point P on the line $2x + 3y = 6$. We already know that P is a solution of $2x + 3y \leq 6$. If we choose any point Q directly below P, the x-coordinate of Q will be the same as that of P, and the y-coordinate will be smaller. So the value of $2x + 3y$ at Q will be smaller than the value at P, which is 6. Thus, $2x + 3y < 6$ at Q, and so Q is another solution of the inequality. (See Figure 3.) In other words, *every point beneath the line is a solution of* $2x + 3y \leq 6$.

On the other hand, any point above the line is directly above a point on the line, and so $2x + 3y > 6$ for such a point. Thus, *no point above the line is a solution of* $2x + 3y \leq 6$.

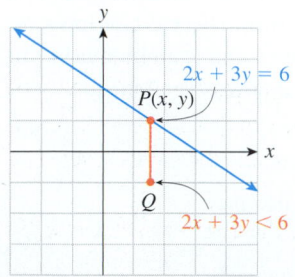

Figure **3**

The same kind of argument can be used to show that the solution set of every inequality of the form $ax + by \leq c$ or $ax + by \geq c$ consists of the half plane above or below the line $ax + by = c$. The "test-point" procedure we describe below gives us an

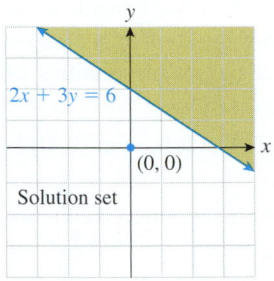

Figure 4

easy method for deciding whether the solution set includes the region above or below the corresponding line.

Now we are going to do something that will appear backward at first (but makes it simpler to sketch sets of solutions of *systems* of linear inequalities). For our standard drawing of the region of solutions of $2x + 3y \leq 6$, we are going to *shade only the part that we do not want and leave the solution region blank*. Think of covering over or "blocking out" the unwanted points, leaving those that we do want in full view (but remember that the points on the boundary line are also points that we want). The result is Figure 4. The reason we do this should become clear in Example 2.

> ### Sketching the Region Represented by a Linear Inequality in Two Variables
>
> **1.** Sketch the straight line obtained by replacing the given inequality with an equality.
>
> **2.** Choose a test point not on the line; $(0, 0)$ is a good choice if the line does not pass through the origin.
>
> **3.** If the test point satisfies the inequality, then the set of solutions is the entire region on the same side of the line as the test point. Otherwise, it is the region on the other side of the line. In either case, shade (block out) the side that does *not* contain the solutions, leaving the solution set unshaded.

quick Example

Here are the three steps used to graph the inequality $x + 2y \geq 5$.

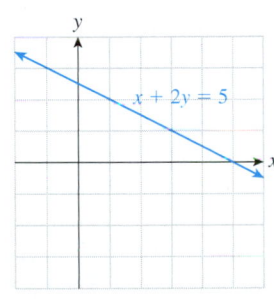
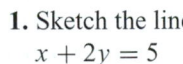

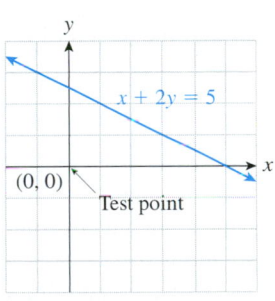

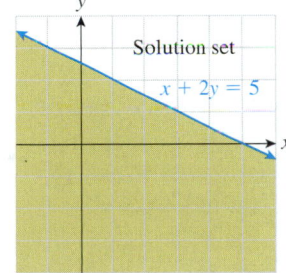

using *Technology*

See the Technology Guides at the end of the chapter to see how to graph inequalities using a TI-83/84 or Excel.

1. Sketch the line
$x + 2y = 5$

2. Test the point $(0, 0)$
$0 + 2(0) \not\geq 5$
Inequality is not satisfied.

3. Since the inequality is not satisfied, shade the region containing the test point.

Example 1 Graphing Single Inequalities

Sketch the regions determined by each of the following inequalities:

a. $3x - 2y \leq 6$ **b.** $6x \leq 12 + 4y$ **c.** $x \leq -1$ **d.** $y \geq 0$ **e.** $x \geq 3y$

Solution

a. The boundary line $3x - 2y = 6$ has x-intercept 2 and y-intercept -3 (Figure 5). We use $(0, 0)$ as a test point (because it is not on the line). Because $3(0) - 2(0) \leq 6$, the inequality is satisfied by the test point $(0, 0)$, and so it lies inside the solution set. The solution set is shown in Figure 5.

b. The given inequality, $6x \leq 12 + 4y$, can be rewritten in the form $ax + by \leq c$ by subtracting $4y$ from both sides:

$$6x - 4y \leq 12$$

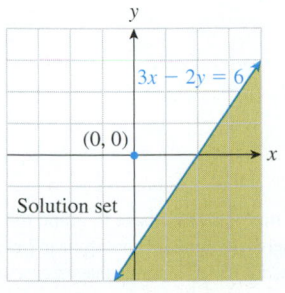

Figure 5

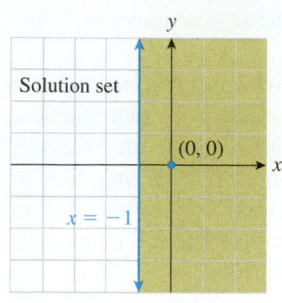

Figure **6**

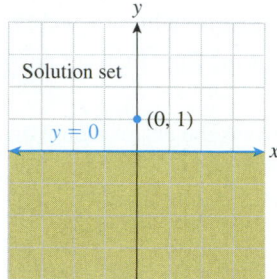

Figure **7**

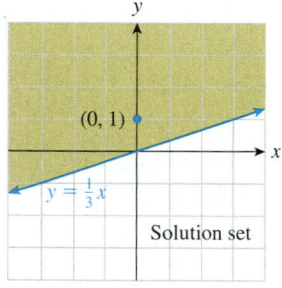

Figure **8**

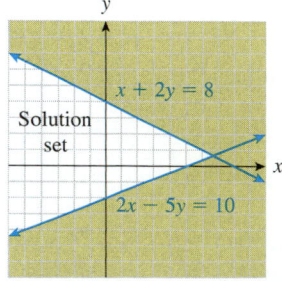

Figure **9**

Dividing both sides by 2 gives the inequality $3x - 2y \leq 6$, which we considered in part (a). Now, *applying the rules for manipulating inequalities does not affect the set of solutions.* Thus, the inequality $6x \leq 12 + 4y$ has the same set of solutions as $3x - 2y \leq 6$ (see Figure 5).

c. The region $x \leq -1$ has as boundary the vertical line $x = -1$. The test point $(0, 0)$ is not in the solution set, as shown in Figure 6.

d. The region $y \geq 0$ has as boundary the horizontal line $y = 0$ (that is, the x-axis). We cannot use $(0, 0)$ for the test point because it lies on the boundary line. Instead, we choose a convenient point not on the line $y = 0$—say, $(0, 1)$. Because $1 \geq 0$, this point is in the solution set, giving us the region shown in Figure 7.

e. The line $x \geq 3y$ has as boundary the line $x = 3y$ or, solving for y,

$$y = \frac{1}{3}x$$

This line passes through the origin with slope $1/3$, so again we cannot choose the origin as a test point. Instead, we choose $(0, 1)$. Substituting these coordinates in $x \geq 3y$ gives $0 \geq 3(1)$, which is false, so $(0, 1)$ is not in the solution set, as shown in Figure 8.

Example **2** Graphing Simultaneous Inequalities

Sketch the region of points that satisfy both inequalities:

$$2x - 5y \leq 10$$
$$x + 2y \leq 8$$

Solution Each inequality has a solution set that is a half plane. If a point is to satisfy *both* inequalities, it must lie in both sets of solutions. Put another way, if we cover the points that are not solutions to $2x - 5y \leq 10$ and then also cover the points that are not solutions to $x + 2y \leq 8$, the points that remain uncovered must be the points we want, those that are solutions to both inequalities. The result is shown in Figure 9, where the unshaded region is the set of solutions.[*]

As a check, we can look at points in various regions in Figure 9. For example, our graph shows that $(0, 0)$ should satisfy both inequalities, and it does;

$$2(0) - 5(0) = 0 \leq 10 \qquad ✔$$
$$0 + 2(0) = 0 \leq 8 \qquad ✔$$

On the other hand, $(0, 5)$ should fail to satisfy one of the inequalities.

$$2(0) - 5(5) = -25 \leq 10 \qquad ✔$$
$$0 + 2(5) = 10 > 8 \qquad ✗$$

One more: $(5, -1)$ should fail one of the inequalities:

$$2(5) - 5(-1) = 15 > 10 \qquad ✗$$
$$5 + 2(-1) = 3 \leq 8 \qquad ✔$$

---

[*] ***Technology Note*** Although these graphs are quite easy to do by hand, the more lines we have to graph the more difficult it becomes to get everything in the right place, and this is where graphing technology can become important. This is especially true when, for instance, three or more lines intersect in points that are very close together and hard to distinguish in hand-drawn graphs.

Example 3 Corner Points

Sketch the region of solutions of the following system of inequalities and list the coordinates of all the corner points.

$$3x - 2y \le 6$$
$$x + y \ge -5$$
$$y \le 4$$

Solution Shading the regions that we do not want leaves us with the triangle shown in Figure 10. We label the corner points A, B, and C as shown.

Each of these corner points lies at the intersection of two of the bounding lines. So, to find the coordinates of each corner point, we need to solve the system of equations given by the two lines. To do this systematically, we make the following table:

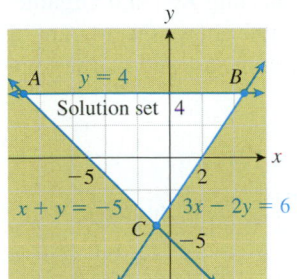

Figure **10**

Point	Lines through Point	Coordinates
A	$y = 4$ $x + y = -5$	$(-9, 4)$
B	$y = 4$ $3x - 2y = 6$	$\left(\dfrac{14}{3}, 4\right)$
C	$x + y = -5$ $3x - 2y = 6$	$\left(-\dfrac{4}{5}, -\dfrac{21}{5}\right)$

Here, we have solved each system of equations in the middle column to get the point on the right, using the techniques of Chapter 2. You should do this for practice.[*]

As a partial check that we have drawn the correct region, let us choose any point in its interior—say, $(0, 0)$. We can easily check that $(0, 0)$ satisfies all three given inequalities. It follows that all of the points in the triangular region containing $(0, 0)$ are also solutions.

---

[*] ***Technology Note*** Using the trace feature makes it easy to locate corner points graphically. Remember to zoom in for additional accuracy when appropriate. Of course, you can also use technology to help solve the systems of equations, as we discussed in Chapter 2.

Take another look at the regions of solutions in Examples 2 and 3 (Figure 11).

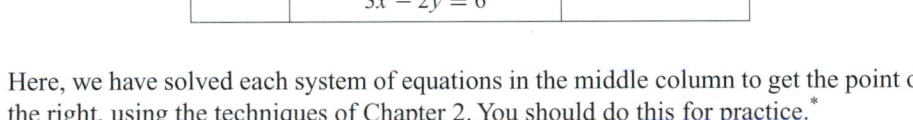

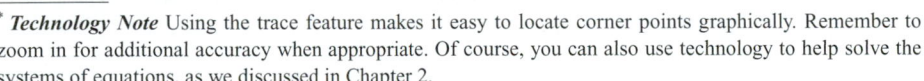

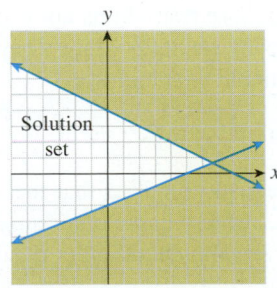

(a) Example 2 Solution Set   (b) Example 3 Solution Set

Figure **11**

Notice that the solution set in Figure 11(a) extends infinitely far to the left, whereas the one in Figure 11(b) is completely enclosed by a boundary. Sets that are completely enclosed are called **bounded,** and sets that extend infinitely in one or more directions are **unbounded.** For example, all the solution sets in Example 1 are unbounded.

Example 4 Resource Allocation

Socaccio Pistachio Inc. makes two types of pistachio nuts: Dazzling Red and Organic. Pistachio nuts require food color and salt, and the following table shows the amount of food color and salt required for a 1-kilogram batch of pistachios, as well as the total amount of these ingredients available each day.

	Dazzling Red	*Organic*	*Total Available*
Food Color (g)	2	1	20
Salt (g)	10	20	220

Use a graph to show the possible numbers of batches of each type of pistachio Socaccio can produce each day. This region (the solution set of a system of inequalities) is called the **feasible region.**

Solution As we did in Chapter 2, we start by identifying the unknowns: Let x be the number of batches of Dazzling Red manufactured per day and let y be the number of batches of Organic manufactured each day.

Now, because of our experience with systems of linear equations, we are tempted to say: for food color $2x + y = 20$ and for salt, $10x + 20y = 220$. However, no one is saying that Socaccio has to use all available ingredients; the company might choose to use fewer than the total available amounts if this proves more profitable. Thus, $2x + y$ can be anything *up to a total of* 20. In other words,

$$2x + y \le 20$$

Similarly,

$$10x + 20y \le 220$$

There are two more restrictions not explicitly mentioned: Neither x nor y can be negative (the company cannot produce a negative number of batches of nuts). Therefore, we have the additional restrictions

$$x \ge 0 \quad y \ge 0$$

These two inequalities tell us that the feasible region (solution set) is restricted to the first quadrant, because in the other quadrants, either x or y or both x and y are negative. So instead of shading out all other quadrants, we can simply restrict our drawing to the first quadrant.

The (bounded) feasible region shown in Figure 12 is a graphical representation of the limitations the company faces.

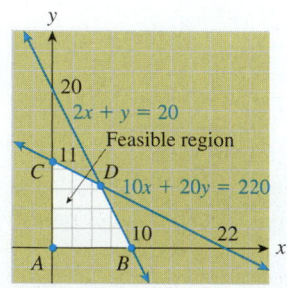

Figure **12**

mathematics At Work

Amy Bredbenner

TITLE **Toxicologist**

INSTITUTION **International Flavors & Fragrances, Inc.**

Haejo Hwang

International Flavors and Fragrances is a leading creator and manufacturer of fragrances and flavors. Our mission is to develop new fragrance and flavor molecules for a variety of everyday products such as cosmetics, detergents, food, and beverages. Several different departments at my company work together to create these flavors and fragrances, including expert perfumers and flavorists blending a variety of unique molecules together to create different types of tastes and scents. But as a toxicologist my goal is to perform a safety evaluation of each new molecule before it can be sold to the public. This safety evaluation includes testing the product to determine a safe use level and using scientific models to predict how the material will act on the human body and the environment.

How does mathematics play a part in my daily tasks? As an IFF toxicologist (a.k.a. Regulatory Compliance Specialist), I incorporate mathematics into my job by using a chemical program called "Persistent, Bioaccumulative and Toxic Profiler." PBT uses the structure of the chemical and mathematical equations to determine if that chemical structure may pose a hazard to the environment.

For example, when a chemist in another department develops a new fragrance chemical she sends it to me. I will then input the structure of that chemical in to the PBT Profile and it will calculate various physical and chemical properties of the new chemical. PBT and I calculate how long a material will persist in the environment, whether or not it is likely to bioaccumulate in the body, and how toxic the material will be to aquatic species. Based on these predictions, I determine what type of further information is necessary in order to evaluate the safety of the new chemical and whether or not it may be hazardous to people and the environment.

So, as you can see, mathematics helps toxicologists determine the safe use level of everyday products in order to protect the consumer and the environment from harm. You may consider this when you read this chapter and take your first bite of your dinner tonight!

+ *Before we go on...* Every point in the feasible region in Example 4 represents a value for x and a value for y that do not violate any of the company's restrictions. For example, the point $(5, 6)$ lies well inside the region, so the company can produce 5 batches of Dazzling Red nuts and 6 batches of Organic without exceeding the limitations on ingredients [that is, $2(5) + 6 = 16 \leq 20$ and $10(5) + 20(6) = 170 \leq 220$]. The corner points A, B, C, and D are significant if the company wishes to realize the greatest profit, as we will see in Section 4.2. We can find the corners as in the following table:

Point	Lines through Point	Coordinates
A		$(0, 0)$
B		$(10, 0)$
C		$(0, 11)$
D	$2x + y = 20$ $10x + 20y = 220$	$(6, 8)$

(We have not listed the lines through the first three corners because their coordinates can be read easily from the graph.) Points on the line segment DB represent use of all the food color (because the segment lies on the line $2x + y = 20$), and points on the line segment CD represent use of all the salt (because the segment lies on the line $10x + 20y = 220$). Note that the point D is the only solution that uses all of both ingredients. ∎

> ## FAQs   Recognizing Whether to Use a Linear Inequality or a Linear Equation
>
> **Q:** *How do I know whether to model a situation by a linear inequality like y: $3x + 2y \leq 10$ or by a linear equation like $3x + 2y = 10$?*
>
> **A:** Here are some key phrases to look for: *at most, up to, no more than, at least, or more, exactly.* Suppose, for instance, that nuts cost 3¢, bolts cost 2¢, *x* is the number of nuts you can buy, and *y* is the number of bolts you can buy.
>
> - If you have *up to* 10¢ to spend, then $3x + 2y \leq 10$.
> - If you must spend *exactly* 10¢, then $3x + 2y = 10$.
> - If you plan to spend *at least* 10¢, then $3x + 2y \geq 10$.
>
> The use of inequalities to model a situation is often more realistic than the use of equations; for instance, one cannot always expect to exactly fill all orders, spend the exact amount of one's budget, or keep a plant operating at exactly 100% capacity. ∎

4.1 EXERCISES

● denotes basic skills exercises

tech Ex indicates exercises that should be solved using technology

In Exercise 1–26, sketch the region that corresponds to the given inequalities, say whether the region is bounded or unbounded, and find the coordinates of all corner points (if any). hint *[see Example 1]*

1. ● $2x + y \leq 10$

2. ● $4x - y \leq 12$

3. ● $-x - 2y \leq 8$

4. ● $-x + 2y \geq 4$

5. ● $3x + 2y \geq 5$

6. ● $2x - 3y \leq 7$

7. ● $x \leq 3y$

8. ● $y \geq 3x$

9. ● $\dfrac{3x}{4} - \dfrac{y}{4} \leq 1$

10. ● $\dfrac{x}{3} + \dfrac{2y}{3} \geq 2$

11. ● $x \geq -5$

12. ● $y \leq -4$

13. ● $4x - \ y \leq 8$
$x + 2y \leq 2$ hint
[see Examples 2 and 3]

14. ● $2x + \ y \leq 4$
$x - 2y \geq 2$

15. ● $3x + 2y \geq 6$
$3x - 2y \leq 6$
$x \qquad \geq 0$

16. ● $3x + 2y \leq 6$
$3x - 2y \geq 6$
$-y \geq 2$

17. ● $x + y \geq 5$
$x \qquad \leq 10$
$y \leq 8$
$x \geq 0, y \geq 0$

18. ● $2x + 4y \geq 12$
$x \qquad \leq 5$
$y \leq 3$
$x \geq 0, y \geq 0$

19. ● $20x + 10y \leq 100$
$10x + 20y \leq 100$
$10x + 10y \leq 60$
$x \geq 0, y \geq 0$

20. ● $30x + 20y \leq 600$
$10x + 40y \leq 400$
$20x + 30y \leq 450$
$x \geq 0, y \geq 0$

21. ● $20x + 10y \geq 100$
$10x + 20y \geq 100$
$10x + 10y \geq 80$
$x \geq 0, y \geq 0$

22. ● $30x + 20y \geq 600$
$10x + 40y \geq 400$
$20x + 30y \geq 600$
$x \geq 0, y \geq 0$

23. ● $-3x + 2y \leq 5$
$3x - 2y \leq 6$
$x \qquad \leq 2y$
$x \geq 0, y \geq 0$

24. ● $-3x + 2y \leq 5$
$3x - 2y \geq 6$
$y \leq x/2$
$x \geq 0, y \geq 0$

25. ● $2x - \ y \geq 0$
$x - \ 3y \leq 0$
$x \geq 0, y \geq 0$

26. ● $-x + \ y \geq 0$
$4x - 3y \geq 0$
$x \geq 0, y \geq 0$

tech Ex *In Exercises 27–32, we suggest you use technology. Graph the regions corresponding to the inequalities, and find the coordinates of all corner points (if any) to two decimal places:*

27. tech Ex $2.1x - 4.3y \geq 9.7$

28. tech Ex $-4.3x + 4.6y \geq 7.1$

● basic skills   tech Ex technology exercise

29. tech Ex $-0.2x + 0.7y \geq 3.3$
    $1.1x + 3.4y \geq 0$

30. tech Ex $0.2x + 0.3y \geq 7.2$
    $2.5x - 6.7y \leq 0$

31. tech Ex $4.1x - 4.3y \leq 4.4$
    $7.5x - 4.4y \leq 5.7$
    $4.3x + 8.5y \leq 10$

32. tech Ex $2.3x - 2.4y \leq 2.5$
    $4.0x - 5.1y \leq 4.4$
    $6.1x + 6.7y \leq 9.6$

Applications

33. ● **Resource Allocation** You manage an ice cream factory that makes two flavors: Creamy Vanilla and Continental Mocha. Into each quart of Creamy Vanilla go 2 eggs and 3 cups of cream. Into each quart of Continental Mocha go 1 egg and 3 cups of cream. You have in stock 500 eggs and 900 cups of cream. Draw the feasible region showing the number of quarts of vanilla and number of quarts of mocha that can be produced. Find the corner points of the region. *hint* [see Example 4]

34. ● **Resource Allocation** Podunk Institute of Technology's Math Department offers two courses: Finite Math and Applied Calculus. Each section of Finite Math has 60 students, and each section of Applied Calculus has 50. The department is allowed to offer a total of up to 110 sections. Furthermore, no more than 6000 students want to take a math course (no student will take more than one math course). Draw the feasible region that shows the number of sections of each class that can be offered. Find the corner points of the region.

35. ● **Nutrition** Ruff Inc. makes dog food out of chicken and grain. Chicken has 10 grams of protein and 5 grams of fat per ounce, and grain has 2 grams of protein and 2 grams of fat per ounce. A bag of dog food must contain at least 200 grams of protein and at least 150 grams of fat. Draw the feasible region that shows the number of ounces of chicken and number of ounces of grain Ruff can mix into each bag of dog food. Find the corner points of the region.

36. ● **Purchasing** Enormous State University's Business School is buying computers. The school has two models to choose from, the Pomegranate and the iZac. Each Pomegranate comes with 400 MB of memory and 80 GB of disk space, and each iZac has 300 MB of memory and 100 GB of disk space. For reasons related to its accreditation, the school would like to be able to say that it has a total of at least 48,000 MB of memory and at least 12,800 GB of disk space. Draw the feasible region that shows the number of each kind of computer it can buy. Find the corner points of the region.

37. ● **Nutrition** Each serving of Gerber Mixed Cereal for Baby contains 60 calories and 11 grams of carbohydrates, and each

serving of Gerber Mango Tropical Fruit Dessert contains 80 calories and 21 grams of carbohydrates.[3] You want to provide your child with at least 140 calories and at least 32 grams of carbohydrates. Draw the feasible region that shows the number of servings of cereal and number of servings of dessert that you can give your child. Find the corner points of the region.

38. ● **Nutrition** Each serving of Gerber Mixed Cereal for Baby contains 60 calories, 11 grams of carbohydrates, and no Vitamin C. Each serving of Gerber Apple Banana Juice contains 60 calories, 15 grams of carbohydrates, and 120 percent of the U.S. Recommended Daily Allowance (RDA) of Vitamin C for infants.[4] You want to provide your child with at least 120 calories, at least 26 grams of carbohydrates, and at least 50 percent of the U.S. RDA of Vitamin C for infants. Draw the feasible region that shows the number of servings of cereal and number of servings of juice that you can give your child. Find the corner points of the region.

39. ● **Municipal Bond Funds** The Pimco New York Municipal Bond Fund (PNF) and the Fidelity Spartan Mass Fund (FDMMX) are tax-exempt municipal bond funds. In 2003, the Pimco fund yielded 6% while the Fidelity fund yielded 5%.[5] You would like to invest a total of up to $80,000 and earn at least $4200 in interest in the coming year (based on the given yields). Draw the feasible region that shows how much money you can invest in each fund. Find the corner points of the region.

40. ● **Municipal Bond Funds** In 2003, the Fidelity Spartan Mass Fund (FDMMX) yielded 5%, and the Fidelity Spartan Florida Fund (FFLIX) yielded 7%.[6] You would like to invest a total of up to $60,000 and earn at least $3500 in interest. Draw the feasible region that shows how much money you can invest in each fund (based on the given yields). Find the corner points of the region.

41. **Investments** Your portfolio manager has suggested two high-yielding stocks: The Altria Group (MO) and Reynolds American (RAI). MO shares cost $50 and yield 4.5% in dividends. RAI shares cost $55 and yield 5% in dividends.[7] You have up to $12,100 to invest, and would like to earn at least $550 in dividends. Draw the feasible region that shows how many shares in each company you can buy. Find the corner points of the region. (Round each coordinate to the nearest whole number.)

42. **Investments** Your friend's portfolio manager has suggested two high-yielding stocks: Consolidated Edison (ED) and

---

[3] SOURCE: Nutrition information printed on the jar.

[4] Ibid.

[5] Yields are rounded. SOURCES: www.pimcofunds.com, www.fidelity.com. January 18, 2004.

[6] Yields are rounded. SOURCE: www.fidelity.com. January 18, 2004.

[7] Share prices and yields are approximate (2004). SOURCE: http://money.excite.com.

● basic skills   tech Ex technology exercise

Royal Bank of Scotland (RBS-K). ED shares cost $40 and yield 5.5% in dividends. RBS-K shares cost $25 and yield 7.5% in dividends.[8] You have up to $30,000 to invest, and would like to earn at least $1650 in dividends. Draw the feasible region that shows how many shares in each company you can buy. Find the corner points of the region. (Round each coordinate to the nearest whole number.)

43. *Advertising* You are the marketing director for a company that manufactures bodybuilding supplements and you are planning to run ads in *Sports Illustrated* and *GQ Magazine*. Based on readership data, you estimate that each one-page ad in *Sports Illustrated* will be read by 650,000 people in your target group, while each one-page ad in *GQ* will be read by 150,000.[9] You would like your ads to be read by at least three million people in the target group and, to ensure the broadest possible audience, you would like to place at least three full-page ads in each magazine Draw the feasible region that shows how many pages you can purchase in each magazine. Find the corner points of the region. (Round each coordinate to the nearest whole number.)

44. *Advertising* You are the marketing director for a company that manufactures bodybuilding supplements and you are planning to run ads in *Sports Illustrated* and *Muscle and Fitness*. Based on readership data, you estimate that each one-page ad in *Sports Illustrated* will be read by 650,000 people in your target group, while each one-page ad in *Muscle and Fitness* will be read by 250,000 people in your target group.[10] You would like your ads to be read by at least four million people in the target group and, to ensure the broadest possible audience, you would like to place at least three full-page ads in each magazine during the year. Draw the feasible region showing how many pages you can purchase in each magazine. Find the corner points of the region. (Round each coordinate to the nearest whole number.)

Communication and Reasoning Exercises

45. ● Find a system of inequalities whose solution set is unbounded.

46. ● Find a system of inequalities whose solution set is empty.

47. ● How would you use linear inequalities to describe the triangle with corner points $(0, 0)$, $(2, 0)$, and $(0, 1)$?

[8] Share prices and yields are approximate (2004). SOURCE: http://money.excite.com.

[9] The readership data for *Sports Illustrated* is based, in part, on the results of a readership survey taken in March 2000. The readership data for *GQ* is fictitious. SOURCE. Mediamark Research Inc./*New York Times,* May 29, 2000, p. C1.

[10] The readership data for both magazines are based on the results of a readership survey taken in March 2000. SOURCE: Mediamark Research Inc./*New York Times,* May 29, 2000, p. C1.

48. ● Explain the advantage of shading the region of points that do not satisfy the given inequalities. Illustrate with an example.

49. ● Describe at least one drawback to the method of finding the corner points of a feasible region by drawing its graph, when the feasible region arises from real-life constraints.

50. ● Draw several bounded regions described by linear inequalities. For each region you draw, find the point that gives the greatest possible value of $x + y$. What do you notice?

In Exercises 51–54, you are mixing x grams of ingredient A and y grams of ingredient B. Choose the equation or inequality that models the given requirement.

51. ● There should be at least 3 more grams of ingredient A than ingredient B.
   (A) $3x - y \le 0$ **(B)** $x - 3y \ge 0$
   (C) $x - y \ge 3$ **(D)** $3x - y \ge 0$

52. ● The mixture should contain at least 25% of ingredient A by weight.
   (A) $4x - y \le 0$ **(B)** $x - 4y \ge 0$ **(C)** $x - y \ge 4$
   (D) $3x - y \ge 0$

53. There should be at least 3 parts (by weight) of ingredient A to 2 parts of ingredient B.
   (A) $3x - 2y \ge 0$ **(B)** $2x - 3y \ge 0$
   (C) $3x + 2y \ge 0$ **(D)** $2x + 3y \ge 0$

54. There should be no more of ingredient A (by weight) than ingredient B.
   (A) $x - y = 0$ **(B)** $x - y \le 0$
   (C) $x - y \ge 0$ **(D)** $x + y \ge y$

55. You are setting up a system of inequalities in the unknowns x and y. The inequalities represent constraints faced by Fly-by-Night Airlines, where x represents the number of first-class tickets it should issue for a specific flight and y represents the number of business-class tickets it should issue for that flight. You find that the feasible region is empty. How do you interpret this?

56. In the situation described in the preceding exercise, is it possible instead for the feasible region to be unbounded? Explain your answer.

57. Create an interesting scenario that leads to the following system of inequalities:

$$20x + 40y \le 1000$$
$$30x + 20y \le 1200$$
$$x \ge 0, y \ge 0$$

58. Create an interesting scenario that leads to the following system of inequalities:

$$20x + 40y \ge 1000$$
$$30x + 20y \ge 1200$$
$$x \ge 0, y \ge 0$$

● basic skills   **tech** Ex technology exercise

4.2 Solving Linear Programming Problems Graphically

As we saw in Example 4 in Section 4.1, in some scenarios the possibilities are restricted by a system of linear inequalities. In that example, it would be natural to ask which of the various possibilities gives the company the largest profit. This is a kind of problem known as a *linear programming problem* (commonly referred to as an LP problem).

Linear Programming (LP) Problems

A **linear programming problem** in two unknowns x and y is one in which we are to find the maximum or minimum value of a linear expression

$$ax + by$$

called the **objective function,** subject to a number of linear **constraints** of the form

$$cx + dy \leq e \quad \text{or} \quad cx + dy \geq e$$

The largest or smallest value of the objective function is called the **optimal value,** and a pair of values of x and y that gives the optimal value constitutes an **optimal solution.**

quick Example

$$
\begin{aligned}
\text{Maximize} \quad & p = x + y && \text{Objective function} \\
\text{subject to} \quad & \left. \begin{array}{l} x + 2y \leq 12 \\ 2x + y \leq 12 \\ x \geq 0, y \geq 0 \end{array} \right\} && \text{Constraints}
\end{aligned}
$$

See Example 1 for a method of solving this LP problem (that is, finding an optimal solution and value).

The set of points (x, y) satisfying all the constraints is the **feasible region** for the problem. Our methods of solving LP problems rely on the following facts:

Fundamental Theorem of Linear Programming

- If an LP problem has optimal solutions, then at least one of these solutions occurs at a corner point of the feasible region.
- Linear programming problems with bounded, nonempty feasible regions always have optimal solutions.

Let's see how we can use this to solve an LP problem, and then we'll discuss why it's true.

Example 1 Solving an LP Problem

$$
\begin{aligned}
\text{Maximize} \quad & p = x + y \\
\text{subject to} \quad & x + 2y \leq 12 \\
& 2x + y \leq 12 \\
& x \geq 0, y \geq 0
\end{aligned}
$$

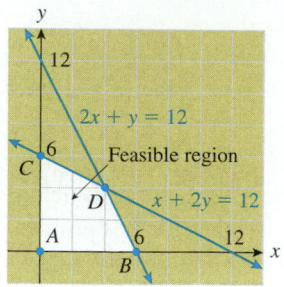

Figure **13**

Solution We begin by drawing the feasible region for the problem. We do this using the techniques of Section 4.1, and we get Figure 13.

Each **feasible point** (point in the feasible region) gives an x and a y satisfying the constraints. The question now is, which of these points gives the largest value of the objective function $p = x + y$? The Fundamental Theorem of Linear Programming tells us that the largest value must occur at one (or more) of the corners of the feasible region. In the following table, we list the coordinates of each corner point and we compute the value of the objective function at each corner.

Corner Point	Lines through Point	Coordinates	$p = x + y$
A		$(0, 0)$	0
B		$(6, 0)$	6
C		$(0, 6)$	6
D	$x + 2y = 12$ $2x + y = 12$	$(4, 4)$	8

Now we simply pick the one that gives the largest value for p, which is D. Therefore, the optimal value of p is 8, and an optimal solution is $(4, 4)$.

Now we owe you an explanation of why one of the corner points should be an optimal solution. The question is, which point in the feasible region gives the largest possible value of $p = x + y$?

Consider first an easier question: Which points result in a *particular value* of p? For example, which points result in $p = 2$? These would be the points on the line $x + y = 2$, which is the line labeled $p = 2$ in Figure 14.

Now suppose we want to know which points make $p = 4$: These would be the points on the line $x + y = 4$, which is the line labeled $p = 4$ in Figure 14. Notice that this line is parallel to but higher than the line $p = 2$. (If p represented profit in an application, we would call these **isoprofit lines**, or **constant-profit lines**.) Imagine moving this line up or down in the picture. As we move the line down, we see smaller values of p, and as we move it up, we see larger values. Several more of these lines are drawn in Figure 14. Look, in particular, at the line labeled $p = 10$. This line does not meet the feasible region, meaning that no feasible point makes p as large as 10. Starting with the line $p = 2$, as we move the line up, increasing p, there will be a last line that meets the feasible region. In the figure it is clear that this is the line $p = 8$, and this meets the feasible region in only one point, which is the corner point D. Therefore D gives the greatest value of p of all feasible points.

If we had been asked to maximize some other objective function, such as $p = x + 3y$, then the optimal solution might be different. Figure 15 shows some of the isoprofit lines for this objective function. This time, the last point that is hit as p increases is C, not D. This tells us that the optimal solution is $(0, 6)$, giving the optimal value $p = 18$.

This discussion should convince you that the optimal value in an LP problem will always occur at one of the corner points. By the way, it is possible for the optimal value to occur at *two* corner points and at all points along an edge connecting them. (Do you see why?) We will see this in Example 3(b).

Here is a summary of the method we have just been using.

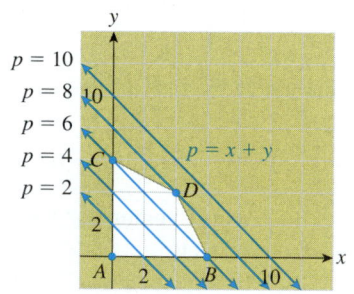

Figure **14**

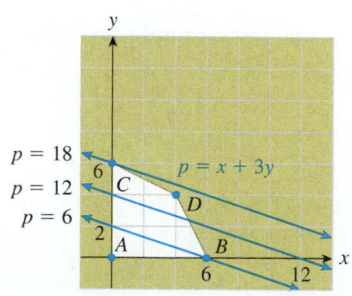

Figure **15**

Graphical Method for Solving Linear Programming Problems in Two Unknowns (Bounded Feasible Regions)

1. Graph the feasible region and check that it is bounded.

2. Compute the coordinates of the corner points.

3. Substitute the coordinates of the corner points into the objective function to see which gives the maximum (or minimum) value of the objective function.

4. Any such corner point is an optimal solution.

Note If the feasible region is unbounded, this method will work only if there are optimal solutions; otherwise, it will not work. We will show you a method for deciding this on p. 275. ■

Applications

Example 2 Resource Allocation

Acme Babyfoods mixes two strengths of apple juice. One quart of Beginner's juice is made from 30 fluid ounces of water and 2 fluid ounces of apple juice concentrate. One quart of Advanced juice is made from 20 fluid ounces of water and 12 fluid ounces of concentrate. Every day Acme has available 30,000 fluid ounces of water and 3600 fluid ounces of concentrate. Acme makes a profit of 20¢ on each quart of Beginner's juice and 30¢ on each quart of Advanced juice. How many quarts of each should Acme make each day to get the largest profit? How would this change if Acme made a profit of 40¢ on Beginner's juice and 20¢ on Advanced juice?

Solution Looking at the question that we are asked, we see that our unknown quantities are

$$x = \text{number of quarts of Beginner's juice made each day}$$
$$y = \text{number of quarts of Advanced juice made each day}$$

(In this context, x and y are often called the **decision variables,** because we must decide what their values should be in order to get the largest profit.) We can write down the data given in the form of a table (the numbers in the first two columns are amounts per quart of juice):

	Beginner's, x	*Advanced, y*	*Available*
Water (ounces)	30	20	30,000
Concentrate (ounces)	2	12	3600
Profit (¢)	20	30	

Because nothing in the problem says that Acme must use up all the water or concentrate, just that it can use no more than what is available, the first two rows of the table give us two inequalities:

$$30x + 20y \leq 30,000$$
$$2x + 12y \leq 3600$$

Dividing the first inequality by 10 and the second by 2 gives

$$3x + 2y \leq 3000$$
$$x + 6y \leq 1800$$

We also have that $x \geq 0$ and $y \geq 0$ because Acme can't make a negative amount of juice. To finish setting up the problem, we are asked to maximize the profit, which is

$$p = 20x + 30y \qquad \textcolor{red}{\text{Expressed in ¢}}$$

This gives us our LP problem:

$$\begin{aligned}
\text{Maximize} \quad & p = 20x + 30y \\
\text{subject to} \quad & 3x + 2y \leq 3000 \\
& x + 6y \leq 1800 \\
& x \geq 0, \, y \geq 0
\end{aligned}$$

The (bounded) feasible region is shown in Figure 16.

The corners and the values of the objective function are listed in the following table:

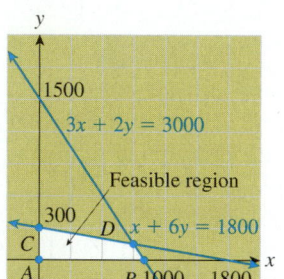

Figure **16**

Point	Lines through point	Coordinates	$p = 20x + 30y$
A		(0, 0)	0
B		(1000, 0)	20,000
C		(0, 300)	9000
D	$3x + 2y = 3000$ $x + 6y = 1800$	(900, 150)	22,500

We are seeking to maximize the objective function p, so we look for corner points that give the maximum value for p. Because the maximum occurs at the point D, we conclude that the (only) optimal solution occurs at D. Thus, the company should make 900 quarts of Beginner's juice and 150 quarts of Advanced juice, for a largest possible profit of 22,500¢, or $225.

If, instead, the company made a profit of 40¢ on each quart of Beginner's juice and 20¢ on each quart of Advanced juice, then we would have $p = 40x + 20y$. This gives the following table:

Point	Lines through Point	Coordinates	$p = 40x + 20y$
A		(0, 0)	0
B		(1000, 0)	40,000
C		(0, 300)	6000
D	$3x + 2y = 3000$ $x + 6y = 1800$	(900, 150)	39,000

We can see that, in this case, Acme should make 1000 quarts of Beginner's juice and no Advanced juice, for a largest possible profit of 40,000¢, or $400.

✛*Before we go on...* Notice that, in the first version of the problem in Example 2, the company used all the water and juice concentrate:

Water: $30(900) + 20(150) = 30{,}000$

Concentrate: $2(900) + 12(150) = 3600$

In the second version, it used all the water but not all the concentrate:

Water: $30(100) + 20(0) = 30{,}000$

Concentrate: $2(100) + 12(0) = 200 < 3600$ ■

Example 3 Investments

The Solid Trust Savings & Loan Company has set aside $25 million for loans to home buyers. Its policy is to allocate at least $10 million annually for luxury condominiums. A government housing development grant it receives requires, however, that at least one-third of its total loans be allocated to low-income housing.

a. Solid Trust's return on condominiums is 12% and its return on low-income housing is 10%. How much should the company allocate for each type of housing to maximize its total return?

b. Redo part (a), assuming that the return is 12% on both condominiums and low-income housing.

Solution

a. We first identify the unknowns: Let x be the annual amount (in millions of dollars) allocated to luxury condominiums and let y be the annual amount allocated to low-income housing.

We now look at the constraints. The first constraint is mentioned in the first sentence: The total the company can invest is $25 million. Thus,

$$x + y \leq 25$$

(The company is not required to invest all of the $25 million; rather, it can invest *up to* $25 million.) Next, the company has allocated at least $10 million to condos. Rephrasing this in terms of the unknowns, we get

The amount allocated to condos is at least $10 million.

The phrase "is at least" means $\geq$. Thus, we obtain a second constraint:

$$x \geq 10$$

The third constraint is that at least one-third of the total financing must be for low-income housing. Rephrasing this, we say:

The amount allocated to low-income housing is at least one-third of the total.

Because the total investment will be $x + y$, we get

$$y \geq \frac{1}{3}(x + y)$$

We put this in the standard form of a linear inequality as follows:

$$3y \geq x + y \qquad \textcolor{red}{\text{Multiply both sides by 3.}}$$

$$-x + 2y \geq 0 \qquad \textcolor{red}{\text{Subtract } x + y \text{ from both sides.}}$$

There are no further constraints.

Now, what about the return on these investments? According to the data, the annual return is given by

$$p = 0.12x + 0.10y$$

We want to make this quantity p as large as possible. In other words, we want to

$$\begin{aligned}
\text{Maximize} \quad & p = 0.12x + 0.10y \\
\text{subject to} \quad & x + y \le 25 \\
& x \ge 10 \\
& -x + 2y \ge 0 \\
& x \ge 0, y \ge 0
\end{aligned}$$

(Do you see why the inequalities $x \ge 0$ and $y \ge 0$ are slipped in here?) The feasible region is shown in Figure 17.

We now make a table that gives the return on investment at each corner point:

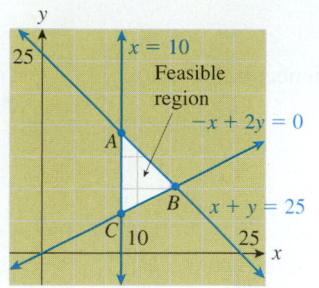

Figure **17**

Point	Lines through Point	Coordinates	$p = 0.12x + 0.10y$
A	$x = 10$ $x + y = 25$	$(10, 15)$	2.7
B	$x + y = 25$ $-x + 2y = 0$	$(50/3, 25/3)$	2.833
C	$x = 10$ $-x + 2y = 0$	$(10, 5)$	1.7

From the table, we see that the values of x and y that maximize the return are $x = 50/3$ and $y = 25/3$, which give a total return of \$2.833 million. In other words, the most profitable course of action is to invest \$16.667 million in loans for condominiums and \$8.333 million in loans for low-income housing, giving a maximum annual return of \$2.833 million.

b. The LP problem is the same as for part (a) except for the objective function:

$$\begin{aligned}
\text{Maximize} \quad & p = 0.12x + 0.12y \\
\text{subject to} \quad & x + y \le 25 \\
& x \ge 10 \\
& -x + 2y \ge 0 \\
& x \ge 0, y \ge 0
\end{aligned}$$

Here are the values of p at the three corners:

Point	Coordinates	$p = 0.12x + 0.12y$
A	$(10, 15)$	3
B	$(50/3, 25/3)$	3
C	$(10, 5)$	1.8

Looking at the table, we see that a curious thing has happened: We get the same maximum annual return at both A and B. Thus, we could choose either option to maximize the annual return. In fact, any point along the line segment AB will yield an annual return of $3 million. For example, the point $(12, 13)$ lies on the line segment AB and also yields an annual revenue of $3 million. This happens because the "isoreturn" lines are parallel to that edge.

+ *Before we go on...* What breakdowns of investments would lead to the *lowest* return for parts (a) and (b)? ∎

The preceding examples all had bounded feasible regions. If the feasible region is unbounded, then, *provided there are optimal solutions,* the fundamental theorem of linear programming guarantees that the above method will work. The following procedure determines whether or not optimal solutions exist and finds them when they do.

Solving Linear Programming Problems in Two Unknowns (Unbounded Feasible Regions)

If the feasible region of an LP problem is unbounded, proceed as follows:

1. Draw a rectangle large enough so that all the corner points are inside the rectangle (and not on its boundary):

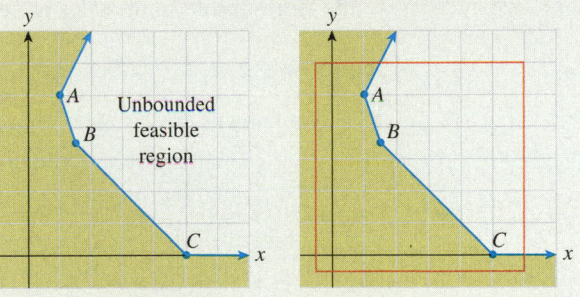

Corner points: A, B, C Corner points inside the rectangle

2. Shade the outside of the rectangle so as to define a new bounded feasible region, and locate the new corner points:

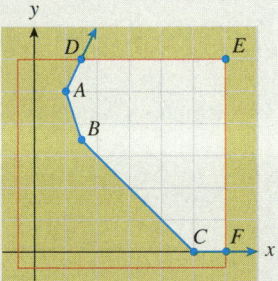

New corner points: D, E, and F

3. Obtain the optimal solutions using this bounded feasible region.

4. If any optimal solutions occur at one of the original corner points (A, B, and C in the figure), then the LP problem has that corner point as an optimal solution. Otherwise, the LP problem has no optimal solutions. When the latter occurs, we say that the **objective function is unbounded,** because it can assume arbitrarily large (positive or negative) values.

In the next two examples, we work with unbounded feasible regions.

Example 4 Cost

You are the manager of a small store that specializes in hats, sunglasses, and other accessories. You are considering a sales promotion of a new line of hats and sunglasses. You will offer the sunglasses only to those who purchase two or more hats, so you will sell at least twice as many hats as pairs of sunglasses. Moreover, your supplier tells you that, due to seasonal demand, your order of sunglasses cannot exceed 100 pairs. To ensure that the sale items fill out the large display you have set aside, you estimate that you should order at least 210 items in all.

a. Assume that you will lose $3 on every hat and $2 on every pair of sunglasses sold. Given the constraints above, how many hats and pairs of sunglasses should you order to lose the least amount of money in the sales promotion?

b. Suppose instead that you lose $1 on every hat sold but make a profit of $5 on every pair of sunglasses sold. How many hats and pairs of sunglasses should you order to make the largest profit in the sales promotion?

c. Now suppose that you make a profit of $1 on every hat sold but lose $5 on every pair of sunglasses sold. How many hats and pairs of sunglasses should you order to make the largest profit in the sales promotion?

Solution

a. The unknowns are:

$$x = \text{number of hats you order}$$
$$y = \text{number of pairs of sunglasses you order}$$

The objective is to minimize the total loss:

$$c = 3x + 2y$$

Now for the constraints. The requirement that you will sell at least twice as many hats as sunglasses can be rephrased as:

The number of hats is at least twice the number of pairs of sunglasses

or

$$x \geq 2y$$

which, in standard form, is

$$x - 2y \geq 0$$

Next, your order of sunglasses cannot exceed 100 pairs, so

$$y \leq 100$$

Finally, you would like to sell at least 210 items in all, giving

$$x + y \geq 210$$

Thus, the LP problem is the following:

$$\begin{aligned}
\text{Minimize} \quad & c = 3x + 2y \\
\text{subject to} \quad & x - 2y \geq 0 \\
& y \leq 100 \\
& x + y \geq 210 \\
& x \geq 0, y \geq 0
\end{aligned}$$

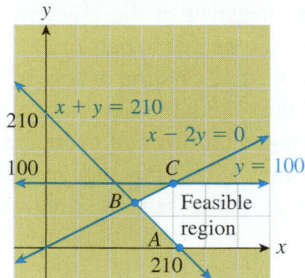

Figure **18**

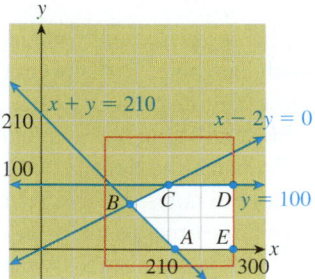

Figure **19**

The feasible region is shown in Figure 18.

This region is unbounded, so there is no guarantee that there are any optimal solutions. Following the procedure described above, we enclose the corner points in a rectangle as shown in Figure 19. (There are many infinitely many possible rectangles we could have used. We chose one that gives convenient coordinates for the new corners.)

We now list all the corners of this bounded region along with the corresponding values of the objective function c:

Point	Lines through Point	Coordinates	$c = 3x + 2y$ ($)
A		$(210, 0)$	630
B	$x + y = 210$ $x - 2y = 0$	$(140, 70)$	560
C	$x - 2y = 0$ $y = 100$	$(200, 100)$	800
D		$(300, 100)$	1100
E		$(300, 0)$	900

The corner point that gives the minimum value of the objective function c is B. Because B is one of the corner points of the original feasible region, we conclude that our linear programming problem has an optimal solution at B. Thus, the combination that gives the smallest loss is 140 hats and 70 pairs of sunglasses.

b. The LP problem is the following:

$$\text{Maximize} \quad p = -x + 5y$$
$$\text{subject to} \quad x - 2y \geq 0$$
$$y \leq 100$$
$$x + y \geq 210$$
$$x \geq 0, y \geq 0$$

Because most of the work is already done for us in part (a), all we need to do is change the objective function in the table that lists the corner points:

Point	Lines through Point	Coordinates	$p = -x + 5y$ ($)
A		$(210, 0)$	-210
B	$x + y = 210$ $x - 2y = 0$	$(140, 70)$	210
C	$x - 2y = 0$ $y = 100$	$(200, 100)$	300
D		$(300, 100)$	200
E		$(300, 0)$	-300

The corner point that gives the maximum value of the objective function p is C. Because C is one of the corner points of the original feasible region, we conclude that our LP problem has an optimal solution at C. Thus, the combination that gives the largest profit ($300) is 200 hats and 100 pairs of sunglasses.

c. The objective function is now $p = x - 5y$, which is the negative of the objective function used in part (b). Thus, the table of values of p is the same as in part (b), except that it has opposite signs in the p column. This time we find that the maximum value of p occurs at E. However, E is not a corner point of the original feasible region, so the LP problem has no optimal solution. Referring to Figure 18, we can make the objective p as large as we like by choosing a point far to the right in the unbounded feasible region. Thus, the objective function is unbounded; that is, it is possible to make an arbitrarily large profit.

Example 5 Resource Allocation

You are composing a very avant-garde ballade for violins and bassoons. In your ballade, each violinist plays a total of two notes and each bassoonist only one note. To make your ballade long enough, you decide that it should contain at least 200 instrumental notes. Furthermore, after playing the requisite two notes, each violinist will sing one soprano note, while each bassoonist will sing three soprano notes.[*] To make the ballade sufficiently interesting, you have decided on a minimum of 300 soprano notes. To give your composition a sense of balance, you wish to have no more than three times as many bassoonists as violinists. Violinists charge $200 per performance and bassoonists $400 per performance. How many of each should your ballade call for in order to minimize personnel costs?

Solution First, the unknowns are x = number of violinists and y = number of bassoonists. The constraint on the number of instrumental notes implies that

$$2x + y \geq 200$$

because the total number is to be *at least* 200. Similarly, the constraint on the number of soprano notes is

$$x + 3y \geq 300$$

The next one is a little tricky. As usual, we reword it in terms of the quantities x and y.

The number of bassoonists should be no more than three times the number of violinists.

Thus, $y \leq 3x$

or $3x - y \geq 0$

Finally, the total cost per performance will be

$$c = 200x + 400y$$

We wish to minimize total cost. So, our linear programming problem is as follows:

$$
\begin{aligned}
\text{Minimize} \quad & c = 200x + 400y \\
\text{subject to} \quad & 2x + y \geq 200 \\
& x + 3y \geq 300 \\
& 3x - y \geq 0 \\
& x \geq 0, y \geq 0
\end{aligned}
$$

---

[*] Whether or not these musicians are capable of singing decent soprano notes will be left to chance. You reason that a few bad notes will add character to the ballade.

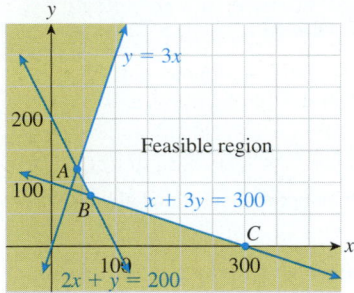

Figure **20**

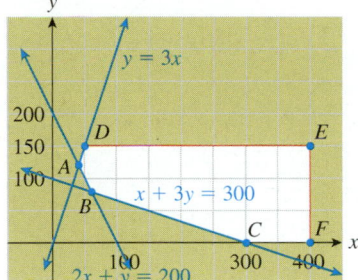

Figure **21**

We get the feasible region shown in Figure 20.[*] The feasible region is unbounded, and so we add a convenient rectangle as before (Figure 21).

Point	Lines through Point	Coordinates	$c = 200x + 400y$
A	$2x + y = 200$ $3x - y = 0$	(40, 120)	56,000
B	$2x + y = 200$ $x + 3y = 300$	(60, 80)	44,000
C		(300, 0)	60,000
D	$3x - y = 0$ $y = 150$	(50, 150)	70,000
E		(400, 150)	140,000
F		(400, 0)	80,000

From the table we see that the minimum cost occurs at B, a corner point of the original feasible region. The linear programming problem thus has an optimal solution, and the minimum cost is \$44,000 per performance, employing 60 violinists and 80 bassoonists. (Quite a wasteful ballade, one might say.)

---

[*] Here is an example where graphing technology would help in determining the corner points. Unless you are very confident in the accuracy of your sketch, how do you know that the line $y = 3x$ falls to the left of the point B, for example? If it were to fall to the right, then B would not be a corner point and the solution would be different. You could (and should) check that B satisfies the inequality $3x - y \geq 0$, so that the line falls to the left of B as shown. However, if you use a graphing calculator or computer, you can be fairly confident of the picture produced without doing further calculations.

FAQs Recognizing a Linear Programming Problem, Setting Up Inequalities, and Dealing with Unbounded Regions

Q: *How do I recognize when an application leads to an LP problem as opposed to a system of linear equations?*

A: Here are some cues that suggest an LP problem:

- Key phrases suggesting inequalities rather than equalities, like *at most, up to, no more than, at least*, and *or more*.
- A quantity that is being maximized or minimized (this will be the objective). Key phrases are *maximum, minimum, most, least, largest, greatest, smallest, as large as possible*, and *as small as possible*. ∎

Q: *How do I deal with tricky phrases like "there should be no more than twice as many nuts as bolts" or "at least 50% of the total should be bolts"?*

A: The easiest way to deal with phrases like this is to use the technique we discussed in Chapter 2: reword the phrases using "the number of . . .", as in

The number of nuts (x) is no more than twice the number of bolts (y) $x \leq 2y$
The number of bolts is at least 50% of the total $y \geq 0.50(x + y)$ ∎

Online, follow:

Chapter 4

→ Tools

→ Linear Programming
  Grapher

for a utility that does everything
(solves linear programming
problems with two unknowns and
even draws the feasible region).

Q: *Do I always have to add a rectangle to deal with unbounded regions*?

A: Under some circumstances, you can tell right away whether optimal solutions exist, even when the feasible region is unbounded.

Note that the following apply only when we have the constraints $x \geq 0$ and $y \geq 0$.

1. If you are minimizing $c = ax + by$ with a and b nonnegative, then optimal solutions always exist. (Examples 4(a) and 5 are of this type.)

2. If you are maximizing $p = ax + by$ with a and b nonnegative (and not both zero), then there is no optimal solution unless the feasible region is bounded.

Do you see why statements (1) and (2) are true? ■

4.2 EXERCISES

● denotes basic skills exercises

◆ denotes challenging exercises

In Exercises 1–20, solve the LP problems. If no optimal solution exists, indicate whether the feasible region is empty or the objective function is unbounded. hint [see Example 1]

1. ● Maximize $p = x + y$
subject to $x + 2y \leq 9$
$2x + y \leq 9$
$x \geq 0, y \geq 0$

2. ● Maximize $p = x + 2y$
subject to $x + 3y \leq 24$
$2x + y \leq 18$
$x \geq 0, y \geq 0$

3. ● Minimize $c = x + y$
subject to $x + 2y \geq 6$
$2x + y \geq 6$
$x \geq 0, y \geq 0$

4. ● Minimize $c = x + 2y$
subject to $x + 3y \geq 30$
$2x + y \geq 30$
$x \geq 0, y \geq 0$

5. ● Maximize $p = 3x + y$
subject to $3x - 7y \leq 0$
$7x - 3y \geq 0$
$x + y \leq 10$
$x \geq 0, y \geq 0$

6. ● Maximize $p = x - 2y$
subject to $x + 2y \leq 8$
$x - 6y \leq 0$
$3x - 2y \geq 0$
$x \geq 0, y \geq 0$

7. ● Maximize $p = 3x + 2y$
subject to $0.2x + 0.1y \leq 1$
$0.15x + 0.3y \leq 1.5$
$10x + 10y \leq 60$
$x \geq 0, y \geq 0$

8. ● Maximize $p = x + 2y$
subject to $30x + 20y \leq 600$
$0.1x + 0.4y \leq 4$
$0.2x + 0.3y \leq 4.5$
$x \geq 0, y \geq 0$

9. ● Minimize $c = 0.2x + 0.3y$
subject to $0.2x + 0.1y \geq 1$
$0.15x + 0.3y \geq 1.5$
$10x + 10y \geq 80$
$x \geq 0, y \geq 0$

10. ● Minimize $c = 0.4x + 0.1y$
subject to $30x + 20y \geq 600$
$0.1x + 0.4y \geq 4$
$0.2x + 0.3y \geq 4.5$
$x \geq 0, y \geq 0$

11. ● Maximize and minimize $p = x + 2y$
subject to $x + y \geq 2$
$x + y \leq 10$
$x - y \leq 2$
$x - y \geq -2$

12. ● Maximize and minimize $p = 2x - y$
subject to $x + y \geq 2$
$x - y \leq 2$
$x - y \geq -2$
$x \leq 10, y \leq 10$

13. ● Maximize $p = 2x + 3y$
subject to $0.1x + 0.2y \geq 1$
$2x + y \geq 10$
$x \geq 0, y \geq 0$

14. ● Maximize $p = 3x + 2y$
subject to $0.1x + 0.1y \geq 0.2$
$y \leq 10$
$x \geq 0, y \geq 0$

● basic skills   ◆ challenging

15. ● Minimize $c = 2x + 4y$

subject to $0.1x + 0.1y \geq 1$

$x + 2y \geq 14$

$x \geq 0, y \geq 0$

16. ● Maximize $p = 2x + 3y$

subject to $-x + y \geq 10$

$x + 2y \leq 12$

$x \geq 0, y \geq 0$

17. Minimize $c = 3x - 3y$

subject to $\dfrac{x}{4} \leq y$

$y \leq \dfrac{2x}{3}$

$x + y \geq 5$

$x + 2y \leq 10$

$x \geq 0, y \geq 0$

18. Minimize $c = -x + 2y$

subject to $y \leq \dfrac{2x}{3}$

$x \leq 3y$

$y \geq 4$

$x \geq 6$

$x + y \leq 16$

19. Maximize $p = x + y$

subject to $x + 2y \geq 10$

$2x + 2y \leq 10$

$2x + y \geq 10$

$x \geq 0, y \geq 0$

20. Minimize $c = 3x + y$

subject to $10x + 20y \geq 100$

$0.3x + 0.1y \geq 1$

$x \geq 0, y \geq 0$

Applications

21. ● *Resource Allocation* You manage an ice cream factory that makes two flavors: Creamy Vanilla and Continental Mocha. Into each quart of Creamy Vanilla go 2 eggs and 3 cups of cream. Into each quart of Continental Mocha go 1 egg and 3 cups of cream. You have in stock 500 eggs and 900 cups of cream. You make a profit of $3 on each quart of Creamy Vanilla and $2 on each quart of Continental Mocha. How many quarts of each flavor should you make in order to earn the largest profit? *hint* [see Example 2]

22. ● *Resource Allocation* Podunk Institute of Technology's Math Department offers two courses: Finite Math and Applied Calculus. Each section of Finite Math has 60 students,

and each section of Applied Calculus has 50. The department is allowed to offer a total of up to 110 sections. Furthermore, no more than 6000 students want to take a math course (no student will take more than one math course). Suppose the university makes a profit of $100,000 on each section of Finite Math and $50,000 on each section of Applied Calculus (the profit is the difference between what the students are charged and what the professors are paid). How many sections of each course should the department offer to make the largest profit?

23. ● *Nutrition* Ruff, Inc. makes dog food out of chicken and grain. Chicken has 10 grams of protein and 5 grams of fat per ounce, and grain has 2 grams of protein and 2 grams of fat per ounce. A bag of dog food must contain at least 200 grams of protein and at least 150 grams of fat. If chicken costs 10¢ per ounce and grain costs 1¢ per ounce, how many ounces of each should Ruff use in each bag of dog food in order to minimize cost? *hint* [see Example 4]

24. ● *Purchasing* Enormous State University's Business School is buying computers. The school has two models from which to choose, the Pomegranate and the iZac. Each Pomegranate comes with 400 MB of memory and 80 GB of disk space; each iZac has 300 MB of memory and 100 GB of disk space. For reasons related to its accreditation, the school would like to be able to say that it has a total of at least 48,000 MB of memory and at least 12,800 GB of disk space. If the Pomegranate and the iZac cost $2000 each, how many of each should the school buy to keep the cost as low as possible?

25. ● *Nutrition* Each serving of Gerber Mixed Cereal for Baby contains 60 calories and 11 grams of carbohydrates. Each serving of Gerber Mango Tropical Fruit Dessert contains 80 calories and 21 grams of carbohydrates.[11] If the cereal costs 30¢ per serving and the dessert costs 50¢ per serving, and you want to provide your child with at least 140 calories and at least 32 grams of carbohydrates, how can you do so at the least cost? (Fractions of servings are permitted.)

26. ● *Nutrition* Each serving of Gerber Mixed Cereal for Baby contains 60 calories, 10 grams of carbohydrates, and no Vitamin C. Each serving of Gerber Apple Banana Juice contains 60 calories, 15 grams of carbohydrates, and 120 percent of the U.S. Recommended Daily Allowance (RDA) of Vitamin C for infants. The cereal costs 10¢ per serving and the juice costs 30¢ per serving. If you want to provide your child with at least 120 calories, at least 25 grams of carbohydrates, and at least 60 percent of the U.S. RDA of Vitamin C for infants, how can you do so at the least cost? (Fractions of servings are permitted.)

27. ● *Energy Efficiency* You are thinking of making your home more energy efficient by replacing some of the light bulbs with compact fluorescent bulbs, and insulating part or all of your exterior walls. Each compact fluorescent light bulb costs $4 and saves you an average of $2 per year in energy costs, and each

---

[11] SOURCE: Nutrition information printed on the containers.

● basic skills   ◆ challenging

square foot of wall insulation costs $1 and saves you an average of $0.20 per year in energy costs.[12] Your home has 60 light fittings and 1100 sq. ft. of uninsulated exterior wall. You can spend no more than $1200 and would like to save as much per year in energy costs as possible. How many compact fluorescent light bulbs and how many square feet of insulation should you purchase? How much will you save in energy costs per year?

28.● *Energy Efficiency* (Compare with the preceding exercise.) You are thinking of making your mansion more energy efficient by replacing some of the light bulbs with compact fluorescent bulbs, and insulating part or all of your exterior walls. Each compact fluorescent light bulb costs $4 and saves you an average of $2 per year in energy costs, and each square foot of wall insulation costs $1 and saves you an average of $0.20 per year in energy costs.[13] Your mansion has 200 light fittings and 3000 sq. ft. of uninsulated exterior wall. To impress your friends, you would like to spend as much as possible, but save no more than $800 per year in energy costs (you are proud of your large utility bills). How many compact fluorescent light bulbs and how many square feet of insulation should you purchase? How much will you save in energy costs per year?

Creatine Supplements Exercises 29 and 30 are based on the following data on three bodybuilding supplements. (Figures shown correspond to a single serving.)[14]

	Creatine (g)	Carbohydrates (g)	Taurine (g)	Alpha Lipoic Acid (mg)	Cost ($)
Cell-Tech® (MuscleTech)	10	75	2	200	2.20
RiboForce HP® (EAS)	5	15	1	0	1.60
Creatine Transport® (Kaizen)	5	35	1	100	0.60

29.● You are thinking of combining Cell-Tech and Riboforce HP to obtain a 10-day supply that provides at least 80 grams of creatine and at least 10 grams of taurine, but no more than 750 grams of carbohydrates and no more than 1000 milligrams of alpha lipoic acid. How many servings of each supplement should you combine to meet your specifications at the least cost?

30.● You are thinking of combining Cell-Tech and Creatine Transport to obtain a 10-day supply that provides at least 80 grams of creatine and at least 10 grams of taurine, but no more than 600 grams of carbohydrates and no more than 2000 milligrams of alpha lipoic acid. How many servings of each supplement should you combine to meet your specifications at the least cost?

plement should you combine to meet your specifications at the least cost?

31.● *Resource Allocation.* Your salami manufacturing plant can order up to 1000 pounds of pork and 2400 pounds of beef per day for use in manufacturing its two specialties: "Count Dracula Salami" and "Frankenstein Sausage." Production of the Count Dracula variety requires 1 pound of pork and 3 pounds of beef for each salami, while the Frankenstein variety requires 2 pounds of pork and 2 pounds of beef for every sausage. In view of your heavy investment in advertising Count Dracula Salami, you have decided that at least one-third of the total production should be Count Dracula. On the other hand, due to the health-conscious consumer climate, your Frankenstein Sausage (sold as having less beef) is earning your company a profit of $3 per sausage, while sales of the Count Dracula variety are down and it is earning your company only $1 per salami. Given these restrictions, how many of each kind of sausage should you produce to maximize profits, and what is the maximum possible profit? *hint* [see Example 3]

32.● *Project Design* The Megabuck Hospital Corp. is to build a state-subsidized nursing home catering to homeless patients as well as high-income patients. State regulations require that every subsidized nursing home must house a minimum of 1000 homeless patients and no more than 750 high-income patients in order to qualify for state subsidies. The overall capacity of the hospital is to be 2100 patients. The board of directors, under pressure from a neighborhood group, insists that the number of homeless patients should not exceed twice the number of high-income patients. Due to the state subsidy, the hospital will make an average profit of $10,000 per month for every homeless patient it houses, whereas the profit per high-income patient is estimated at $8000 per month. How many of each type of patient should it house in order to maximize profit?

Investing Exercises 33 and 34 are based on the following data on four U.S. computer-related stocks.[15]

	Price	Dividend Yield	Earnings Per Share
IBM	$90	1%	$5.00
HPQ (Hewlett Packard)	20	2	1.20
AAPL (Apple)	70	0	1.20
MSFT (Microsoft)	25	1	0.70

33. You are planning to invest up to $10,000 in IBM and HPQ shares. You want your investment to yield at least $120 in dividends and you want to maximize the total earnings. How many shares of each company should you purchase?

---

[12] SOURCE: American Council for an Energy-Efficient Economy/*New York Times,* December 1, 2003, p. C6.

[13] Ibid.

[14] SOURCE: Nutritional information supplied by the manufacturers at www.netrition.com. Cost per serving is approximate.

---

[15] Stocks were trading at or near the given prices in January, 2005. Earnings per share are rounded. SOURCE: http://money.excite.com.

● basic skills   ◆ challenging

34. You are planning to invest up to $43,000 in AAPL and MSFT shares. For tax reasons, you want your investment to yield no more than $10 in dividends. You want to maximize the total earnings. How many shares of each company should you purchase?

35. *Investments* Your portfolio manager has suggested two high-yielding stocks: The Altria Group (MO) and Reynolds American (RAI). MO shares cost $50, yield 4.5% in dividends, and have a risk index of 2.0 per share. RAI shares cost $55, yield 5% in dividends, and have a risk index of 3.0 per share.[16] You have up to $12,100 to invest and would like to earn at least $550 in dividends. How many shares (to the nearest tenth of a unit) of each stock should you purchase to meet your requirements and minimize the total risk index for your portfolio? What is the minimum total risk index?

36. *Investments* Your friend's portfolio manager has suggested two high-yielding stocks: Consolidated Edison (ED) and Royal Bank of Scotland (RBS-K). ED shares cost $40, yield 5.5% in dividends, and have a risk index of 1.0 per share. RBS-K shares cost $25, yield 7.5% in dividends, and have a risk index of 1.5 per share.[17] You have up to $30,000 to invest and would like to earn at least $1650 in dividends. How many shares (to the nearest tenth of a unit) of each stock should you purchase in order to meet your requirements and minimize the total risk index for your portfolio?

37. *Television Advertising* In February, 2002, each episode of "Becker" was typically seen in 8.3 million homes, while each episode of "The Simpsons" was seen in 7.5 million homes.[18] Your marketing services firm has been hired to promote Bald No More, Inc.'s, hair replacement process by buying at least 30 commercial spots during episodes of "Becker" and "The Simpsons." The cable company running "Becker" has quoted a price of $2000 per spot, while the cable company showing "The Simpsons" has quoted a price of $1500 per spot. Bald No More's advertising budget for TV commercials is $70,000, and it would like no more than 50% of the total number of spots to appear on "The Simpsons." How many spots should you purchase on each show to reach the most homes?

38. *Television Advertising* In February, 2002, each episode of "Boston Public" was typically seen in 7.0 million homes, while each episode of "NYPD Blue" was seen in 7.8 million homes.[19] Your marketing services firm has been hired to promote Gauss Jordan Sneakers by buying at least 30 commercial spots during episodes of "Boston Public" and "NYPD Blue." The cable company running "Boston Public" has quoted a price of $2000 per spot, while the cable company showing "NYPD Blue" has quoted a price of $3000 per spot. Gauss Jordan Sneakers' advertising budget for TV commercials is $70,000, and it would like at least 75% of the total number of spots to appear on "Boston Public." How many spots should you purchase on each show to reach the most homes?

39. *Planning* My friends: I, the mighty Brutus, have decided to prepare for retirement by instructing young warriors in the arts of battle and diplomacy. For each hour spent in battle instruction, I have decided to charge 50 ducats. For each hour in diplomacy instruction I shall charge 40 ducats. Due to my advancing years, I can spend no more than 50 hours per week instructing the youths, although the great Jove knows that they are sorely in need of instruction! Due to my fondness for physical pursuits, I have decided to spend no more than one-third of the total time in diplomatic instruction. However, the present border crisis with the Gauls is a sore indication of our poor abilities as diplomats. As a result, I have decided to spend at least 10 hours per week instructing in diplomacy. Finally, to complicate things further, there is the matter of Scarlet Brew: I have estimated that each hour of battle instruction will require 10 gallons of Scarlet Brew to quench my students' thirst, and that each hour of diplomacy instruction, being less physically demanding, requires half that amount. Because my harvest of red berries has far exceeded my expectations, I estimate that I'll have to use at least 400 gallons per week in order to avoid storing the fine brew at great expense. Given all these restrictions, how many hours per week should I spend in each type of instruction to maximize my income?

40. *Planning* Repeat the preceding exercise with the following changes: I would like to spend no more than half the total time in diplomatic instruction, and I must use at least 600 gallons of Scarlet Brew.

41. *Resource Allocation* One day, Gillian the magician summoned the wisest of her women. "Devoted followers," she began, "I have a quandary: As you well know, I possess great expertise in sleep spells and shock spells, but unfortunately, these are proving to be a drain on my aural energy resources; each sleep spell costs me 500 pico-shirleys of aural energy, while each shock spell requires 750 pico-shirleys. Clearly, I would like to hold my overall expenditure of aural energy to a minimum, and still meet my commitments in protecting the Sisterhood from the ever-present threat of trolls. Specifically, I have estimated that each sleep spell keeps us safe for an average of two minutes, while every shock spell protects us for about three minutes. We certainly require enough protection to last 24 hours of each day, and possibly more, just to be safe. At the same time, I have noticed that each of my sleep spells can immobilize three trolls at once, while one of my typical shock spells (having a narrower range) can immobilize only two trolls at once. We are faced, my sisters, with an onslaught of 1200 trolls per day! Finally, as you are no doubt aware, the

[16] Share prices and yields are approximate (2004). SOURCE: http://money.excite.com. Risk indices are fictitious.

[17] Ibid.

[18] Ratings are for February 18–24, 2002. SOURCE: Nielsen Media Research/www.ptd.net. February 28, 2002.

[19] Ibid.

● basic skills ◆ challenging

bylaws dictate that for a Magician of the Order to remain in good standing, the number of shock spells must be between one-quarter and one-third the number of shock and sleep spells combined. What do I do, oh Wise Ones?"

42. *Risk Management* The Grand Vizier of the Kingdom of Um is being blackmailed by numerous individuals and is having a very difficult time keeping his blackmailers from going public. He has been keeping them at bay with two kinds of payoff: gold bars from the Royal Treasury and political favors. Through bitter experience, he has learned that each payoff in gold gives him peace for an average of about 1 month, while each political favor seems to earn him about a month and a half of reprieve. To maintain his flawless reputation in the Court, he feels he cannot afford any revelations about his tainted past to come to light within the next year. Thus it is imperative that his blackmailers be kept at bay for 12 months. Furthermore, he would like to keep the number of gold payoffs at no more than one-quarter of the combined number of payoffs because the outward flow of gold bars might arouse suspicion on the part of the Royal Treasurer. The Grand Vizier feels that he can do no more than seven political favors per year without arousing undue suspicion in the Court. The gold payoffs tend to deplete his travel budget. (The treasury has been subsidizing his numerous trips to the Himalayas.) He estimates that each gold bar removed from the treasury will cost him four trips. On the other hand because the administering of political favors tends to cost him valuable travel time, he suspects that each political favor will cost him about two trips. Now, he would obviously like to keep his blackmailers silenced and lose as few trips as possible. What is he to do? How many trips will he lose in the next year?

43. ◆ ***Management***[20] You are the service manager for a supplier of closed-circuit television systems. Your company can provide up to 160 hours per week of technical service for your customers, although the demand for technical service far exceeds this amount. As a result, you have been asked to develop a model to allocate service technicians' time between new customers (those still covered by service contracts) and old customers (whose service contracts have expired). To ensure that new customers are satisfied with your company's service, the sales department has instituted a policy that at least 100 hours per week be allocated to servicing new customers. At the same time, your superiors have informed you that the company expects your department to generate at least $1200 per week in revenues. Technical service time for new customers generates an average of $10 per hour (because much of the service is still under warranty) and for old customers generates $30 per hour. How many hours per week

should you allocate to each type of customer to generate the most revenue?

44. ◆ ***Scheduling***[21] The Scottsville Textile Mill produces several different fabrics on eight dobby looms which operate 24 hours per day and are scheduled for 30 days in the coming month. The Scottsville Textile Mill will produce only Fabric 1 and Fabric 2 during the coming month. Each dobby loom can turn out 4.63 yards of either fabric per hour. Assume that there is a monthly demand of 16,000 yards of Fabric 1 and 12,000 yards of Fabric 2. Profits are calculated as 33¢ per yard for each fabric produced on the dobby looms.

a. Will it be possible to satisfy total demand?
b. In the event that total demand is not satisfied, the Scottsville Textile Mill will need to purchase the fabrics from another mill to make up the shortfall. Its profits on resold fabrics ordered from another mill amount to 20¢ per yard for Fabric 1 and 16¢ per yard for Fabric 2. How many yards of each fabric should it produce to maximize profits?

Communication and Reasoning Exercises

45. ● Multiple Choice: If a linear programming problem has a bounded, nonempty feasible region, then optimal solutions **(A)** must exist   **(B)** may or may not exist   **(C)** cannot exist

46. ● Multiple Choice: If a linear programming problem has an unbounded, nonempty feasible region, then optimal solutions **(A)** must exist   **(B)** may or may not exist   **(C)** cannot exist

47. ● What can you say if the optimal value occurs at two adjacent corner points?

48. ● Describe at least one drawback to using the graphical method to solve a linear programming problem arising from a real-life situation.

49. ● Create a linear programming problem in two variables that has no optimal solution.

50. ● Create a linear programming problem in two variables that has more than one optimal solution.

51. ● Create an interesting scenario leading to the following linear programming problem:

$$\text{Maximize} \quad p = 10x + 10y$$
$$\text{subject to} \quad 20x + 40y \le 1000$$
$$30x + 20y \le 1200$$
$$x \ge 0, y \ge 0$$

---

[20] Loosely based on a similiar problem in *An Introduction to Management Science* (6th Ed.) by D. R. Anderson, D. J. Sweeny and T. A. Williams (West, 1991).

[21] Adapted from *The Calhoun Textile Mill Case* by J. D. Camm, P. M. Dearing and S. K. Tadisina as presented for case study in *An Introduction to Management Science* (6th Ed.) by D. R. Anderson, D. J. Sweeney and T. A. Williams (West, 1991). Our exercise uses a subset of the data given in the cited study.

● basic skills   ◆ challenging

52. ● Create an interesting scenario leading to the following linear programming problem:

Minimize $c = 10x + 10y$

subject to $20x + 40y \geq 1000$
$30x + 20y \geq 1200$
$x \geq 0, y \geq 0$

53. Use an example to show why there may be no optimal solution to a linear programming problem if the feasible region is unbounded.

54. Use an example to illustrate why, in the event that an optimal solution does occur despite an unbounded feasible region, that solution corresponds to a corner point of the feasible region.

55. You are setting up an LP problem for Fly-by-Night Airlines with the unknowns x and y, where x represents the number of first-class tickets it should issue for a specific flight and y represents the number of business-class tickets it should issue for that flight, and the problem is to maximize profit. You find that there are two different corner points that maximize the profit. How do you interpret this?

56. In the situation described in the preceding exercise, you find that there are no optimal solutions. How do you interpret this?

57. ◆ Consider the following example of a *nonlinear* programming problem: Maximize $p = xy$ subject to $x \geq 0$, $y \geq 0$, $x + y \leq 2$. Show that p is zero on every corner point, but is greater than zero at many noncorner points.

58. ◆ Solve the nonlinear programming problem in Exercise 57.

● basic skills   ◆ challenging

4.3 The Simplex Method: Solving Standard Maximization Problems

The method discussed in Section 4.2 works quite well for LP problems in two unknowns, but what about three or more unknowns? Because we need an axis for each unknown, we would need to draw graphs in three dimensions (where we have x-, y- and z-coordinates) to deal with problems in three unknowns, and we would have to draw in hyperspace to answer questions involving four or more unknowns. Given the state of technology as this book is being written, we can't easily do this. So we need another method for solving LP problems that will work for any number of unknowns. One such method, called the **simplex method,** has been the method of choice since it was invented by George Dantzig in 1947. To illustrate it best, we first use it to solve only so-called standard maximization problems.

Standard Maximization Problem

A **standard maximization problem in *n* unknowns** is an LP problem in which we are required to *maximize* (not minimize) an objective function of the form[*]

$$p = ax + by + cz + \cdots \ (n \text{ terms})$$

where a, b, c, ... are numbers, subject to the constraints

$$x \geq 0, \ y \geq 0, \ z \geq 0, \ \ldots$$

---

[*] As in the chapter on linear equations we will seldom use the traditional subscripted variables $x_1, x_2, \ldots$. These are very useful names when you start running out of letters of the alphabet, but we should not find ourselves in that predicament.

and further constraints of the form

$$Ax + By + Cz + \cdots \leq N$$

where A, B, C, $\ldots$, N are numbers with N *nonnegative*. Note that the inequality here must be $\leq$, not $=$ or $\geq$.

quick Examples

1. Maximize $p = 2x - 3y + 3z$

   subject to $2x \qquad + z \leq 7$

                $-x + 3y - 6z \leq 6$ This is a standard maximization problem.

                $x \geq 0, y \geq 0, z \geq 0$

2. Maximize $p = 2x - 3y + 3z$

   subject to $2x \qquad + z \geq 7$

                $-x + 3y - 6z \leq 6$ This is *not* a standard maximization problem.

                $x \geq 0, y \geq 0, z \geq 0$

The inequality $2x + z \geq 7$ cannot be written in the required form. If we reverse the inequality by multiplying both sides by -1, we get $-2x - z \leq -7$, but a negative value on the right side is not allowed.

The idea behind the simplex method is this: In any linear programming problem, there is a feasible region. If there are only two unknowns, we can draw the region; if there are three unknowns, it is a solid region in space; and if there are four or more unknowns, it is an abstract higher-dimensional region. But it is a faceted region with corners (think of a diamond), and it is at one of these corners that we will find the optimal solution. Geometrically, what the simplex method does is to start at the corner where all the unknowns are 0 (possible because we are talking of standard maximization problems) and then walk around the region, from corner to adjacent corner, always increasing the value of the objective function, until the best corner is found. In practice, we will visit only a small number of the corners before finding the right one. Algebraically, as we are about to see, this walking around is accomplished by matrix manipulations of the same sort as those used in the chapter on systems of linear equations.

We describe the method while working through an example.

Example 1   Introducing the Simplex Method

Maximize $p = 3x + 2y + z$

subject to $2x + 2y + \ z \leq 10$

             $x + 2y + 3z \leq 15$

       $x \geq 0, \ y \geq 0, \ z \geq 0$

Solution

Step 1 ***Convert to a system of linear equations.*** The inequalities $2x + 2y + z \leq 10$ and $x + 2y + 3z \leq 15$ are less convenient than equations. Look at the first inequality. It says that the left-hand side, $2x + 2y + z$, must have some positive number (or zero) *added to it* if it is to equal 10. Because we don't yet know what x, y, and z are, we are not yet sure what number to add to the left-hand side. So we invent a new unknown, $s \geq 0$, called a **slack variable,** to "take up the slack," so that

$$2x + 2y + z + s = 10$$

Turning to the next inequality, $x + 2y + 3z \leq 15$, we now add a slack variable to its left-hand side, to get it up to the value of the right-hand side. We might have to add a different number than we did the last time, so we use a new slack variable, $t \geq 0$, and obtain

$$x + 2y + 3z + t = 15 \qquad \text{Use a different slack variable for each constraint.}$$

Now we write the system of equations we have (including the one that defines the objective function) in standard form.

$$
\begin{aligned}
2x + 2y + z + s \phantom{{}+t} \phantom{{}+p} &= 10 \\
x + 2y + 3z \phantom{{}+s} + t \phantom{{}+p} &= 15 \\
-3x - 2y - z \phantom{{}+s} \phantom{{}+t} + p &= 0
\end{aligned}
$$

Note three things: First, all the variables are neatly aligned in columns, as they were in Chapter 2. Second, in rewriting the objective function $p = 3x + 2y + z$, we have left the coefficient of p as $+1$ and brought the other variables over to the same side of the equation as p. This will be our standard procedure from now on. *Don't* write $3x + 2y + z - p = 0$ (even though it means the same thing) because the negative coefficients will be important in later steps. Third, the above system of equations has fewer equations than unknowns, and hence cannot have a unique solution.

Step 2 *Set up the initial tableau.* We represent our system of equations by the following table (which is simply the augmented matrix in disguise), called **the initial tableau:**

	x	y	z	s	t	p	
	2	2	1	1	0	0	10
	1	2	3	0	1	0	15
	-3	-2	-1	0	0	1	0

The labels along the top keep track of which columns belong to which variables.

Now notice a peculiar thing. If we rewrite the matrix using the variables s, t, and p first, we get the following matrix

$$
\begin{array}{ccccccc}
s & t & p & x & y & z & \\
\left[\begin{array}{cccccc}
1 & 0 & 0 & 2 & 2 & 1 \\
0 & 1 & 0 & 1 & 2 & 3 \\
0 & 0 & 1 & -3 & -2 & -1
\end{array}\right. & & & & & & \left.\begin{array}{c}
10 \\
15 \\
0
\end{array}\right]
\end{array}
\qquad \text{Matrix with } s, t, \text{ and } p \text{ columns first}
$$

which is already in reduced form. We can therefore read off the general solution (see Section 2.2) to our system of equations as

$$
\begin{aligned}
s &= 10 - 2x - 2y - z \\
t &= 15 - x - 2y - 3z \\
p &= 0 + 3x + 2y + z
\end{aligned}
$$

x, y, z arbitrary

Thus, we get a whole family of solutions, one for each choice of x, y, and z. One possible choice is to set x, y, and z all equal to 0. This gives the particular solution

$$s = 10, t = 15, p = 0, x = 0, y = 0, z = 0 \qquad \text{Set } x = y = z = 0 \text{ above.}$$

This solution is called the **basic solution** associated with the tableau. The variables s and t are called the **active** variables, and x, y, and z are the **inactive** variables. (Other terms used are **basic** and **nonbasic** variables.)

We can obtain the basic solution directly from the tableau as follows.

- The active variables correspond to the cleared columns (columns with only one nonzero entry).
- The values of the active variables are calculated as shown below.
- All other variables are inactive, and set equal to zero.

Inactive $x=0$	Inactive $y=0$	Inactive $z=0$	Active $s=\frac{10}{1}$	Active $t=\frac{15}{1}$	Active $p=\frac{0}{1}$	
x	y	z	s	t	p	
2	2	1	1	0	0	10
1	2	3	0	1	0	15
−3	−2	−1	0	0	1	0

As an additional aid to recognizing which variables are active and which are inactive, we label each row with the name of the corresponding active variable. Thus, the complete initial tableau looks like this.

	x	y	z	s	t	p	
s	2	2	1	1	0	0	10
t	1	2	3	0	1	0	15
p	−3	−2	−1	0	0	1	0

This basic solution represents our starting position $x = y = z = 0$ in the feasible region in xyz space.

We now need to move to another corner point. To do so, we choose a pivot[*] in one of the first three columns of the tableau and clear its column. Then we will get a different basic solution, which corresponds to another corner point. Thus, in order to move from corner point to corner point, all we have to do is choose suitable pivots and clear columns in the usual manner.

The next two steps give the procedure for choosing the pivot.

Step 3 *Select the pivot column* (the column that contains the pivot we are seeking).

Selecting the Pivot Column

Choose the negative number with the largest magnitude on the left-hand side of the bottom row (that is, don't consider the last number in the bottom row). Its column is the pivot column. (If there are two candidates, choose either one.) If all the numbers on the left-hand side of the bottom row are zero or positive, then we are done, and the basic solution is the optimal solution.

---

[*] Also see Section 2.2 for a discussion of pivots and pivoting.

Simple enough. The most negative number in the bottom row is -3, so we choose the x column as the pivot column:

	x	y	z	s	t	p	
s	**2**	2	1	1	0	0	10
t	**1**	2	3	0	1	0	15
p	**−3**	−2	−1	0	0	1	0

↑
pivot column

Q: *Why choose the pivot column this way?*

A: The variable labeling the pivot column is going to be increased from 0 to something positive. In the equation $p = 3x + 2y + z$, the fastest way to increase p is to increase x because p would increase by 3 units for every 1-unit increase in x. (If we chose to increase y, then p would increase by only 2 units for every 1-unit increase in y, and if we increased z instead, p would grow even more slowly.) In short, choosing the pivot column this way makes it likely that we'll increase p as much as possible. ■

Step 4 *Select the pivot in the pivot column.*

Selecting the Pivot

1. The pivot must always be a positive number. (This rules out zeros and negative numbers, such as the −3 in the bottom row.)

2. For each positive entry b in the pivot column, compute the ratio a/b, where a is the number in the rightmost column in that row. We call this a **test ratio.**

3. Of these ratios, choose the smallest one. The corresponding number b is the pivot.

In our example, the test ratio in the first row is $10/2 = 5$, and the test ratio in the second row is $15/1 = 15$. Here, 5 is the smallest, so the 2 in the upper left is our pivot.

	x	y	z	s	t	p		Test Ratios
s	**2**	2	1	1	0	0	10	$\dfrac{10}{2} = 5$
t	1	2	3	0	1	0	15	$\dfrac{15}{1} = 15$
p	−3	−2	−1	0	0	1	0	

Q: *Why select the pivot this way?*

A: The rule given above guarantees that, after pivoting, all variables will be nonnegative in the basic solution. In other words, it guarantees that we will remain in the feasible region. We will explain further after finishing this example. ■

Step 5 *Use the pivot to clear the column in the normal manner and then relabel the pivot row with the label from the pivot column.* It is important to follow the exact prescription described in Section 2.2 for formulating the row operations:

$$a R_c \pm b R_p \qquad \qquad a \text{ and } b \text{ both positive}$$

↑   ↑

Row to change   Pivot row

All entries in the last column should remain nonnegative after pivoting. Furthermore, because the x column (and no longer the s column) will be cleared, x will become an active variable. In other words, the s on the left of the pivot will be replaced by x. We call s the **departing,** or **exiting variable** and x the **entering variable** for this step.

Entering variable
↓

Departing
variable →

		x	y	z	s	t	p		
	s	2	2	1	1	0	0	10	
	t	1	2	3	0	1	0	15	$2R_2 - R_1$
	p	-3	-2	-1	0	0	1	0	$2R_3 + 3R_1$

This gives

	x	y	z	s	t	p	
x	2	2	1	1	0	0	10
t	0	2	5	-1	2	0	20
p	0	2	1	3	0	2	30

This is the second tableau.

Step 6 ***Go to Step 3*** But wait! According to Step 3, we are finished because there are no negative numbers in the bottom row. Thus, we can read off the answer. Remember, though, that the solution for x, the first active variable, is not just $x = 10$, but is $x = 10/2 = 5$ because the pivot has not been reduced to a 1. Similarly, $t = 20/2 = 10$ and $p = 30/2 = 15$. All the other variables are zero because they are inactive. Thus, the solution is as follows: p has a maximum value of 15, and this occurs when $x = 5$, $y = 0$ and $z = 0$. (The slack variables then have the values $s = 0$ and $t = 10$.)

Q: *Why can we stop when there are no negative numbers in the bottom row? Why does this tableau give an optimal solution?*

A: The bottom row corresponds to the equation $2y + z + 3s + 2p = 30$, or

$$p = 15 - y - \frac{1}{2}z - \frac{3}{2}s$$

Think of this as part of the general solution to our original system of equations, with y, z, and s as the parameters. Because these variables must be nonnegative, *the largest possible value of p in any feasible solution of the system comes when all three of the parameters are 0.* Thus, the current basic solution must be an optimal solution.[*] ■

_____

[*] Calculators or spreadsheets could obviously be a big help in the calculations here, just as in Chapter 2. We'll say more about that after the next couple of examples.

We owe some further explanation for Step 4 of the simplex method. After Step 3, we knew that x would be the entering variable, and we needed to choose the departing variable. In the next basic solution, x was to have some positive value and we wanted this value to be as large as possible (to make p as large as possible) without making any other variables negative. Look again at the equations written in Step 2:

$$s = 10 - 2x - 2y - z$$
$$t = 15 - x - 2y - 3z$$

We needed to make either s or t into an inactive variable and hence zero. Also, y and z were to remain inactive. If we had made s inactive, then we would have had $0 = 10 - 2x$, so $x = 10/2 = 5$. This would have made $t = 15 - 5 = 10$, which would be fine. On the other hand, if we had made t inactive, then we would have had $0 = 15 - x$, so $x = 15$, and this would have made $s = 10 - 2 \cdot 15 = -20$, which would *not* be fine, because slack variables must be nonnegative. In other words, we had a choice of making $x = 10/2 = 5$ or $x = 15/1 = 15$, but making x larger than 5 would have made another variable negative. We were thus compelled to choose the smaller ratio, 5, and make s the departing variable. Of course, we do not have to think it through this way every time. We just use the rule stated in Step 4. (For a graphical explanation, see Example 3.)

Example 2 Simplex Method

Find the maximum value of $p = 12x + 15y + 5z$, subject to the constraints:

$$2x + 2y + z \le 8$$
$$x + 4y - 3z \le 12$$
$$x \ge 0, \ y \ge 0, \ z \ge 0$$

Solution Following Step 1, we introduce slack variables and rewrite the constraints and objective function in standard form:

$$
\begin{aligned}
2x + 2y + z + s &= 8 \\
x + 4y - 3z + t &= 12 \\
-12x - 15y - 5z + p &= 0
\end{aligned}
$$

We now follow with Step 2, setting up the initial tableau:

	x	y	z	s	t	p	
s	2	2	1	1	0	0	8
t	1	4	−3	0	1	0	12
p	−12	−15	−5	0	0	1	0

For Step 3, we select the column over the negative number with the largest magnitude in the bottom row, which is the y column. For Step 4, finding the pivot, we see that the test ratios are 8/2 and 12/4, the smallest being $12/4 = 3$. So we select the pivot in the t row and clear its column:

	x	y	z	s	t	p		
s	2	2	1	1	0	0	8	$2R_1 - R_2$
t	1	[4]	−3	0	1	0	12	
p	−12	−15	−5	0	0	1	0	$4R_3 + 15R_2$

The departing variable is t and the entering variable is y. This gives the second tableau.

	x	y	z	s	t	p	
s	3	0	5	2	−1	0	4
y	1	4	−3	0	1	0	12
p	−33	0	−65	0	15	4	180

We now go back to Step 3. Because we still have negative numbers in the bottom row, we choose the one with the largest magnitude (which is −65), and thus our pivot column is the z column. Because negative numbers can't be pivots, the only possible choice for the pivot is the 5. (We need not compute the test ratios because there would only be one from which to choose.) We now clear this column, remembering to take care of the departing and entering variables.

	x	y	z	s	t	p		
s	3	0	$\boxed{5}$	2	−1	0	4	
y	1	4	−3	0	1	0	12	$5R_2 + 3R_1$
p	−33	0	−65	0	15	4	180	$R_3 + 13R_1$

This gives

	x	y	z	s	t	p	
z	3	0	5	2	−1	0	4
y	14	20	0	6	2	0	72
p	6	0	0	26	2	4	232

Notice how the value of p keeps climbing: It started at 0 in the first tableau, went up to $180/4 = 45$ in the second, and is currently at $232/4 = 58$. Because there are no more negative numbers in the bottom row, we are done and can write down the solution: p has a maximum value of $\frac{232}{4} = 58$, and this occurs when

$$x = 0$$

$$y = \frac{72}{20} = \frac{18}{5} \quad \text{and}$$

$$z = \frac{4}{5}$$

The slack variables are both zero.

As a partial check on our answer, we can substitute these values into the objective function and the constraints:

$$58 = 12(0) + 15(18/5) + 5(4/5) \qquad ✔$$

$$2(0) + 2(18/5) + (4/5) = 8 \le 8 \qquad ✔$$

$$0 + 4(18/5) − 3(4/5) = 12 \le 12 \qquad ✔$$

We say that this is only a partial check, because it shows only that our solution is feasible, and that we have correctly calculated p. It does not show that we have the optimal solution. This check will *usually* catch any arithmetic mistakes we make, but it is not foolproof.

Applications

In the next example (further exploits of Acme Baby Foods—compare Example 2 in Section 2) we show how the simplex method relates to the graphical method.

Example 3 Resource Allocation

Acme Baby Foods makes two puddings, vanilla and chocolate. Each serving of vanilla pudding requires 2 teaspoons of sugar and 25 fluid ounces of water, and each serving of chocolate pudding requires 3 teaspoons of sugar and 15 fluid ounces of water. Acme has available each day 3600 teaspoons of sugar and 22,500 fluid ounces of water. Acme makes no more than 600 servings of vanilla pudding because that is all that it can sell each day. If Acme makes a profit of 10¢ on each serving of vanilla pudding and 7¢ on each serving of chocolate, how many servings of each should it make to maximize its profit?

Solution We first identify the unknowns. Let

$x =$ the number of servings of vanilla pudding

$y =$ the number of servings of chocolate pudding

The objective function is the profit $p = 10x + 7y$, which we need to maximize. For the constraints, we start with the fact that Acme will make no more than 600 servings of vanilla: $x \le 600$. We can put the remaining data in a table as follows:

	Vanilla	*Chocolate*	*Total Available*
Sugar (teaspoons)	2	3	3600
Water (ounces)	25	15	22,500

Because Acme can use no more sugar and water than is available, we get the two constraints:

$$2x + 3y \le 3600$$
$$25x + 15y \le 22{,}500 \qquad \text{Note that all the terms are divisible by 5.}$$

Thus our linear programming problem is this:

Maximize $\quad p = 10x + 7y$

subject to $\quad x \le 600$
$\qquad\qquad 2x + 3y \le 3600$
$\qquad\qquad 5x + 3y \le 4500 \qquad$ We divided $25x + 15y \le 22{,}500$ by 5.
$\qquad\qquad x \ge 0, \ y \ge 0$

Next, we introduce the slack variables and set up the initial tableau.

$$\begin{aligned}
x + s   &= 600 \\
2x + 3y + t &= 3600 \\
5x + 3y   + u &= 4500 \\
-10x - 7y   + p &= 0
\end{aligned}$$

Note that we have had to introduce a third slack variable, u. There need to be as many slack variables as there are constraints (other than those of the $x \geq 0$ variety).

Q: What do the slack variables say about Acme puddings ?

A: The first slack variable, s, represents the number you must add to the number of servings of vanilla pudding actually made to obtain the maximum of 600 servings. The second slack variable, t, represents the amount of sugar that is left over once the puddings are made, and u represents the amount of water left over. ■

We now use the simplex method to solve the problem:

	x	y	s	t	u	p		
s	[1]	0	1	0	0	0	600	
t	2	3	0	1	0	0	3600	$R_2 - 2R_1$
u	5	3	0	0	1	0	4500	$R_3 - 5R_1$
p	-10	-7	0	0	0	1	0	$R_4 + 10R_1$

	x	y	s	t	u	p		
x	1	0	1	0	0	0	600	
t	0	3	-2	1	0	0	2400	$R_2 - R_3$
u	0	[3]	-5	0	1	0	1500	
p	0	-7	10	0	0	1	6000	$3R_4 + 7R_3$

	x	y	s	t	u	p		
x	1	0	1	0	0	0	600	$3R_1 - R_2$
t	0	0	[3]	1	-1	0	900	
y	0	3	-5	0	1	0	1500	$3R_3 + 5R_2$
p	0	0	-5	0	7	3	28,500	$3R_4 + 5R_2$

	x	y	s	t	u	p	
x	3	0	0	-1	1	0	900
s	0	0	3	1	-1	0	900
y	0	9	0	5	-2	0	9000
p	0	0	0	5	16	9	90,000

Thus, the solution is as follows: The maximum value of p is $90{,}000/9 = 10{,}000¢ = \$100$, which occurs when $x = 900/3 = 300$, and $y = 9000/9 = 1000$. (The slack variables are $s = 900/3 = 300$ and $t = u = 0$.)

using *Technology*

See the Technology Guides at the end of the chapter for a discussion of using a TI-83/84 or Excel to help with the simplex method. Or, go online and follow

Chapter 4

→ Tools

  → Pivot and Gauss-Jordan Tool

to find a utility that allows you to avoid doing the calculations in each pivot step: Just highlight the entry you wish to use as a pivot, and press "Pivot on Selection."

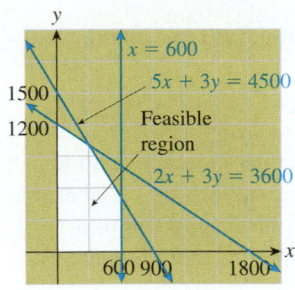

Figure **22**

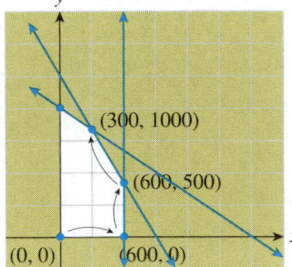

Figure **23**

+ *Before we go on...* Because the problem in Example 3 had only two variables, we could have solved it graphically. It is interesting to think about the relationship between the two methods. Figure 22 shows the feasible region. Each tableau in the simplex method corresponds to a corner of the feasible region, given by the corresponding basic solution. In this example, the sequence of basic solutions is

$$(x, y) = (0, 0), (600, 0), (600, 500), (300, 1000)$$

This is the sequence of corners shown in Figure 23. In general, we can think of the simplex method as walking from corner to corner of the feasible region, until it locates the optimal solution. In problems with many variables and many constraints, the simplex method usually visits only a small fraction of the total number of corners.

We can also explain again, in a different way, the reason we use the test ratios when choosing the pivot. For example, when choosing the first pivot we had to choose among the test ratios 600, 1800, and 900 (look at the first tableau). In Figure 22, you can see that those are the three x-intercepts of the lines that bound the feasible region. If we had chosen 1800 or 900, we would have jumped along the x-axis to a point outside of the feasible region, which we do not want to do. In general, the test ratios measure the distance from the current corner to the constraint lines, and we must choose the smallest such distance to avoid crossing any of them into the unfeasible region.

It is also interesting in an application like this to think about the values of the slack variables. We said above that s is the difference between the maximum 600 servings of vanilla that might be made and the number that is actually made. In the optimal solution, $s = 300$, which says that 300 fewer servings of vanilla were made than the maximum possible. Similarly, t was the amount of sugar left over. In the optimal solution, $t = 0$, which tells us that all of the available sugar is used. Finally, $u = 0$, so all of the available water is used as well. ∎

Online, follow:

Chapter 4

→ Tools

to find utilities that automate the simplex method to varying extents:

- Pivot and Gauss-Jordan Tool (Pivots and does row operations)
- Simplex Method Tool (Solves entire LP problems; shows all tableaus)

Summary: The Simplex Method for Standard Maximization Problems

To solve a standard maximization problem using the simplex method, we take the following steps:

1. Convert to a system of equations by introducing **slack variables** to turn the constraints into equations and by rewriting the objective function in standard form.

2. Write down the initial **tableau.**

3. Select the pivot column: Choose the negative number with the largest magnitude in the left-hand side of the bottom row. Its column is the pivot column. (If there are two or more candidates, choose any of them.) If all the numbers in the left-hand side of the bottom row are zero or positive, then we are finished, and the basic solution maximizes the objective function. (See below for the basic solution.)

4. Select the pivot in the pivot column: The pivot must always be a positive number. For each positive entry b in the pivot column, compute the ratio a/b, where a is the number in the last column in that row. Of these **test ratios,** choose the smallest one. The corresponding number b is the pivot.

5. Use the pivot to clear the column in the normal manner (taking care to follow the exact prescription for formulating the row operations described in Chapter 2) and then relabel the pivot row with the label from the pivot column. The variable

originally labeling the pivot row is the **departing,** or **exiting, variable,** and the variable labeling the column is the **entering variable.**

6. Go to step 3.

To get the **basic solution** corresponding to any tableau in the simplex method, set to zero all variables that do not appear as row labels. The value of a variable that does appear as a row label (an **active variable**) is the number in the rightmost column in that row divided by the number in that row in the column labeled by the same variable.

FAQs Troubleshooting the Simplex Method

Q: *What if there is no candidate for the pivot in the pivot column? For example, what do we do with a tableau like the following?*

	x	y	z	s	t	p	
z	0	0	5	2	0	0	4
y	-8	20	0	6	5	0	72
p	-20	0	0	26	15	4	232

A: Here, the pivot column is the x column, but there is no suitable entry for a pivot (because zeros and negative numbers can't be pivots). This happens when the feasible region is unbounded and there is also no optimal solution. In other words, p can be made as large as we like without violating the constraints. ∎

Q: *What should we do if there is a negative number in the rightmost column?*

A: A negative number will not appear above the bottom row in the rightmost column if we follow the procedure correctly. (The bottom right entry is allowed to be negative if the objective takes on negative values as in a negative profit, or loss.) Following are the most likely errors leading to this situation:

• The pivot was chosen incorrectly (don't forget to choose the *smallest* test ratio). When this mistake is made, one or more of the variables will be negative in the corresponding basic solution.
• The row operation instruction was written backwards or performed backwards (for example, instead of $R_2 - R_1$, it was $R_1 - R_2$). This mistake can be corrected by multiplying the row by -1.
• An arithmetic error occurred. (We all make those annoying errors from time to time.) ∎

Q: *What about zeros in the rightmost column?*

A: Zeros are permissible in the rightmost column. For example, the constraint $x - y \leq 0$ will lead to a zero in the rightmost column. ∎

Q: *What happens if we choose a pivot column other than the one with the most negative number in the bottom row?*

A: There is no harm in doing this as long as we choose the pivot in that column using the smallest test ratio. All it might do is slow the whole calculation by adding extra steps. ∎

One last suggestion: If it is possible to do a simplification step (dividing a row by a positive number) *at any stage,* we should do so. As we saw in Chapter 2, this can help prevent the numbers from getting out of hand.

4.3 EXERCISES

● denotes basic skills exercises

◆ denotes challenging exercises

tech Ex indicates exercises that should be solved using technology

1. ● Maximize $p = 2x + y$

subject to $x + 2y \le 6$

$-x + y \le 4$

$x + y \le 4$

$x \ge 0, y \ge 0$ *hint* [see Examples 1 and 2]

2. ● Maximize $p = x$

subject to $x - y \le 4$

$-x + 3y \le 4$

$x \ge 0, y \ge 0$

3. ● Maximize $p = x - y$

subject to $5x - 5y \le 20$

$2x - 10y \le 40$

$x \ge 0, y \ge 0$

4. ● Maximize $p = 2x + 3y$

subject to $3x + 8y \le 24$

$6x + 4y \le 30$

$x \ge 0, y \ge 0$

5. ● Maximize $p = 5x - 4y + 3z$

subject to $5x + 5z \le 100$

$5y - 5z \le 50$

$5x - 5y \le 50$

$x \ge 0, y \ge 0, z \ge 0$

6. ● Maximize $p = 6x + y + 3z$

subject to $3x + y \le 15$

$2x + 2y + 2z \le 20$

$x \ge 0, y \ge 0, z \ge 0$

7. ● Maximize $p = 7x + 5y + 6z$

subject to $x + y - z \le 3$

$x + 2y + z \le 8$

$x + y \le 5$

$x \ge 0, y \ge 0, z \ge 0$

8. ● Maximize $p = 3x + 4y + 2z$

subject to $3x + y + z \le 5$

$x + 2y + z \le 5$

$x + y + z \le 4$

$x \ge 0, y \ge 0, z \ge 0$

9. ● Maximize $z = 3x_1 + 7x_2 + 8x_3$

subject to $5x_1 - x_2 + x_3 \le 1500$

$2x_1 + 2x_2 + x_3 \le 2500$

$4x_1 + 2x_2 + x_3 \le 2000$

$x_1 \ge 0, x_2 \ge 0, x_3 \ge 0$

10. ● Maximize $z = 3x_1 + 4x_2 + 6x_3$

subject to $5x_1 - x_2 + x_3 \le 1500$

$2x_1 + 2x_2 + x_3 \le 2500$

$4x_1 + 2x_2 + x_3 \le 2000$

$x_1 \ge 0, x_2 \ge 0, x_3 \ge 0$

11. ● Maximize $p = x + y + z + w$

subject to $x + y + z \le 3$

$y + z + w \le 4$

$x + z + w \le 5$

$x + y + w \le 6$

$x \ge 0, y \ge 0, z \ge 0, w \ge 0$

12. ● Maximize $p = x - y + z + w$

subject to $x + y + z \le 3$

$y + z + w \le 3$

$x + z + w \le 4$

$x + y + w \le 4$

$x \ge 0, y \ge 0, z \ge 0, w \ge 0$

13. Maximize $p = x + y + z + w + v$

subject to $x + y \le 1$

$y + z \le 2$

$z + w \le 3$

$w + v \le 4$

$x \ge 0, y \ge 0, z \ge 0, w \ge 0, v \ge 0$

14. Maximize $p = x + 2y + z + 2w + v$

subject to $x + y \le 1$

$y + z \le 2$

$z + w \le 3$

$w + v \le 4$

$x \ge 0, y \ge 0, z \ge 0, w \ge 0, v \ge 0$

tech Ex *In Exercises 15–20 we suggest the use of technology. Round all answers to two decimal places.*

15. tech Ex Maximize $p = 2.5x + 4.2y + 2z$
subject to $0.1x + y - 2.2z \leq 4.5$
$2.1x + y + z \leq 8$
$x + 2.2y \leq 5$
$x \geq 0, y \geq 0, z \geq 0$

16. tech Ex Maximize $p = 2.1x + 4.1y + 2z$
subject to $3.1x + 1.2y + z \leq 5.5$
$x + 2.3y + z \leq 5.5$
$2.1x + y + 2.3z \leq 5.2$
$x \geq 0, y \geq 0, z \geq 0$

17. tech Ex Maximize $p = x + 2y + 3z + w$
subject to $x + 2y + 3z \leq 3$
$y + z + 2.2w \leq 4$
$x + z + 2.2w \leq 5$
$x + y + 2.2w \leq 6$
$x \geq 0, y \geq 0, z \geq 0, w \geq 0$

18. tech Ex Maximize $p = 1.1x - 2.1y + z + w$
subject to $x + 1.3y + z \leq 3$
$1.3y + z + w \leq 3$
$x + z + w \leq 4.1$
$x + 1.3y + w \leq 4.1$
$x \geq 0, y \geq 0, z \geq 0, w \geq 0$

19. tech Ex Maximize $p = x - y + z - w + v$
subject to $x + y \leq 1.1$
$y + z \leq 2.2$
$z + w \leq 3.3$
$w + v \leq 4.4$
$x \geq 0, y \geq 0, z \geq 0, w \geq 0, v \geq 0$

20. tech Ex Maximize $p = x - 2y + z - 2w + v$
subject to $x + y \leq 1.1$
$y + z \leq 2.2$
$z + w \leq 3.3$
$w + v \leq 4.4$
$x \geq 0, y \geq 0, z \geq 0, w \geq 0, v \geq 0$

Applications

21. ● *Purchasing* You are in charge of purchases at the student-run used-book supply program at your college, and you must decide how many introductory calculus, history, and marketing texts should be purchased from students for resale. Due to budget limitations, you cannot purchase more than 650 of these textbooks each semester. There are also shelf-space limitations: Calculus texts occupy 2 units of shelf space each, history books 1 unit each, and marketing texts 3 units each, and you can spare at most 1000 units of shelf space for the texts. If the used book program makes a profit of $10 on each calculus text, $4 on each history text, and $8 on each marketing text, how many of each type of text should you purchase

to maximize profit? What is the maximum profit the program can make in a semester? *hint* [see Example 3]

22. ● *Sales* The Marketing Club at your college has decided to raise funds by selling three types of T-shirt: one with a single-color "ordinary" design, one with a two-color "fancy" design, and one with a three-color "very fancy" design. The club feels that it can sell up to 300 T-shirts. "Ordinary" T-shirts will cost the club $6 each, "fancy" T-shirts $8 each, and "very fancy" T-shirts $10 each, and the club has a total purchasing budget of $3000. It will sell "ordinary" T-shirts at a profit of $4 each, "fancy" T-shirts at a profit of $5 each, and "very fancy" T-shirts at a profit of $4 each. How many of each kind of T-shirt should the club order to maximize profit? What is the maximum profit the club can make?

23. ● *Resource Allocation* Arctic Juice Company makes three juice blends: PineOrange, using 2 portions of pineapple juice and 2 portions of orange juice per gallon; PineKiwi, using 3 portions of pineapple juice and 1 portions of kiwi juice per gallon; and OrangeKiwi, using 3 portions of orange juice and 1 portions of kiwi juice per gallon. Each day the company has 800 portions of pineapple juice, 650 portions of orange juice, and 350 portions of kiwi juice available. Its profit on PineOrange is $1 per gallon, its profit on PineKiwi is $2 per gallon, and its profit on OrangeKiwi is $1 per gallon. How many gallons of each blend should it make each day to maximize profit? What is the largest possible profit the company can make?

24. ● *Purchasing* Trans Global Tractor Trailers has decided to spend up to $1,500,000 on a fleet of new trucks, and it is considering three models: the Gigahaul, which has a capacity of 6000 cubic feet and is priced at $60,000; the Megahaul, with a capacity of 5000 cubic feet and priced at $50,000; and the Picohaul, with a capacity of 2000 cubic feet, priced at $40,000. The anticipated annual revenues are $500,000 for each new truck purchased (regardless of size). Trans Global would like a total capacity of up to 130,000 cubic feet, and feels that it cannot provide drivers and maintenance for more than 30 trucks. How many of each should it purchase to maximize annual revenue? What is the largest possible revenue it can make?

25. ● *Resource Allocation* The Enormous State University History Department offers three courses, Ancient, Medieval, and Modern History, and the department chairperson is trying to decide how many sections of each to offer this semester. They may offer up to 45 sections total, up to 5000 students would like to take a course, and there are 60 professors to teach them (no student will take more than one history course, and no professor will teach more than one section). Sections of Ancient History have 100 students each, sections of Medieval History have 50 students each, and sections of Modern History have 200 students each. Modern History sections are taught by a team of two professors, while Ancient and Medieval History need only one professor per section. Ancient History nets the university $10,000 per section, Medieval nets $20,000, and Modern History nets $30,000 per section. How many sections of each course should the department offer in order to generate

● basic skills   ◆ challenging   tech Ex technology exercise

the largest profit? What is the largest profit possible? Will there be any unused time slots, any students who did not get into classes, or any professors without anything to teach?

26. ● *Resource Allocation* You manage an ice cream factory that makes three flavors: Creamy Vanilla, Continental Mocha, and Succulent Strawberry. Into each batch of Creamy Vanilla go 2 eggs, 1 cup of milk, and 2 cups of cream. Into each batch of Continental Mocha go 1 egg, 1 cup of milk, and 2 cups of cream. Into each batch of Succulent Strawberry go 1 egg, 2 cups of milk, and 2 cups of cream. You have in stock 200 eggs, 120 cups of milk, and 200 cups of cream. You make a profit of $3 on each batch of Creamy Vanilla, $2 on each batch of Continental Mocha and $4 on each batch of Succulent Strawberry.

   a. How many batches of each flavor should you make to maximize your profit?

   b. In your answer to part (a), have you used all the ingredients?

   c. Due to the poor strawberry harvest this year, you cannot make more than 10 batches of Succulent Strawberry. Does this affect your maximum profit?

27. ● *Agriculture* Your small farm encompasses 100 acres, and you are planning to grow tomatoes, lettuce, and carrots in the coming planting season. Fertilizer costs per acre are: $5 for tomatoes, $4 for lettuce, and $2 for carrots. Based on past experience, you estimate that each acre of tomatoes will require an average of 4 hours of labor per week, while tending to lettuce and carrots will each require an average of 2 hours per week. You estimate a profit of $2000 for each acre of tomatoes, $1500 for each acre of lettuce, and $500 for each acre of carrots. You can afford to spend no more than $400 on fertilizer, and your farm laborers can supply up to 500 hours per week. How many acres of each crop should you plant to maximize total profits? In this event, will you be using all 100 acres of your farm?

28. ● *Agriculture* Your farm encompasses 500 acres, and you are planning to grow soy beans, corn, and wheat in the coming planting season. Fertilizer costs per acre are: $5 for soy beans, $2 for corn, and $1 for wheat. You estimate that each acre of soy beans will require an average of 5 hours of labor per week, while tending to corn and wheat will each require an average of 2 hours per week. Based on past yields and current market prices, you estimate a profit of $3000 for each acre of soy beans, $2000 for each acre of corn, and $1000 for each acre of wheat. You can afford to spend no more than $3000 on fertilizer, and your farm laborers can supply 3000 hours per week. How many acres of each crop should you plant to maximize total profits? In this event, will you be using all the available labor?

29. ● *Resource Allocation* (Note that the following exercise is almost identical to Exercise 8 in Section 2.3, except for one important detail. Refer back to your solution of that problem—if you did it—and then attempt this one.) The Enormous State University Choral Society is planning its annual Song Festival, when it will serve three kinds of delicacies: granola treats, nutty granola treats, and nuttiest granola treats. The following table shows some of the ingredients required

(in ounces) for a single serving of each delicacy, as well as the total amount of each ingredient available.

	Granola	Nutty Granola	Nuttiest Granola	Total Available
Toasted Oats	1	1	5	1500
Almonds	4	8	8	10,000
Raisins	2	4	8	4000

The society makes a profit of $6 on each serving of granola, $8 on each serving of nutty granola, and $3 on each serving of nuttiest granola. Assuming that the Choral Society can sell all that they make, how many servings of each will maximize profits? How much of each ingredient will be left over?

30. ● *Resource Allocation* Repeat the preceding exercise, but this time assume that the Choral Society makes a $3 profit on each of its delicacies.

31. ● *Recycling* Safety-Kleen operates the world's largest oil refinery at Elgin, Illinois. You have been hired by the company to determine how to allocate its intake of up to 50 million gallons of used oil to its three refinery processes: A, B, and C. You are told that electricity costs for process A amount to $150,000 per million gallons treated, while for processes B and C, the costs are respectively $100,000 and $50,000 per million gallons treated. Process A can recover 60 percent of the used oil, process B can recover 55 percent, and process C can recover only 50 percent. Assuming a revenue of $4 million per million gallons of recovered oil and an annual electrical budget of $3 million, how much used oil would you allocate to each process in order to maximize total revenues?[22]

32. ● *Recycling* Repeat the preceding exercise, but this time assume that process C can handle only up to 20 million gallons per year.

Creatine Supplements Exercises 33 and 34 are based on the following data on four popular bodybuilding supplements. (Figures shown correspond to a single serving.)[23]

	Creatine (g)	Carbohydrates (g)	Taurine (g)	Alpha Lipoic Acid (mg)
Cell-Tech® (MuscleTech)	10	75	2	200
RiboForce HP® (EAS)	5	15	1	0
Creatine Transport® (Kaizen)	5	35	1	100
Pre-Load Creatine (Optimum)	6	35	1	25

[22] These figures are realistic: Safety-Kleen's actual 1993 capacity was 50 million gallons, its recycled oil sold for approximately $4 per gallon, its recycling process could recover approximately 55 percent of the used oil, and its electrical bill was $3 million. SOURCE: *Oil Recycler Greases Rusty City's Economy, Chicago Tribune,* May 30 1993, Section 7, p.1.

[23] SOURCE: Nutritional information supplied by the manufacturers at www. netrition.com. Cost per serving is approximate.

● basic skills   ◆ challenging   **tech** Ex technology exercise

33. ● You are thinking of combining the first three supplements in the table above to obtain a 10-day supply that gives you the maximum possible amount of creatine, but no more than 1000 milligrams of alpha lipoic acid and 225 grams of carbohydrates. How many servings of each supplement should you combine to meet your specifications, and how much creatine will you get?

34. ● Repeat Exercise 33, but use the last three supplements in the table instead.

Investing Exercises 35 and 36 are based the following data on four U.S. computer-related stocks.[24]

	Price	*Dividend Yield*	*Earnings per share*
IBM	$80	1%	$5
HPQ (Hewlett Packard)	20	2	1
AAPL (Apple)	70	0	1

35. You are planning to invest up to $10,000 in IBM, HPQ, and AAPL shares. You desire to maximize your share of the companies' earnings but, for tax reasons, want to earn no more than $200 in dividends. Your broker suggests that because AAPL stocks pay no dividends, you should invest only in AAPL. Is she right?

36. Repeat Exercise 35 under the assumption that IBM stocks have climbed to $120 on speculation, but its dividend yield and EPS (earnings per share) are unchanged.

37. ❚tech❙Ex *Loan Planning*[25] Enormous State University's employee credit union has $5 million available for loans in the coming year. As VP in charge of finances, you must decide how much capital to allocate to each of four different kinds of loans, as shown in the following table.

Type of Loan	*Annual Rate of Return*
Automobile	8%
Furniture	10
Signature	12
Other secured	10

State laws and credit union policies impose the following restrictions:

- Signature loans may not exceed 10 percent of the total investment of funds.
- Furniture loans plus other secured loans may not exceed automobile loans.
- Other secured loans may not exceed 200 percent of automobile loans.

How much should you allocate to each type of loan to maximize the annual return?

38. ❚tech❙Ex *Investments* You have $100,000 and are considering investing in three municipal bond funds, Pimco, Fidelity Spartan, and Columbia Oregon. You have the following data:[26]

Company	*Yield*
PNF (Pimco NY)	7%
FDMMX (Fidelity Spartan Mass)	5
CMBFX (Columbia Oregon)	4

Your broker has made the following suggestions:

- At least 50 percent of your total investment should be in CMBFX.
- No more than 10 percent of your total investment should be in PNF.

How much should you invest in each fund to maximize your anticipated returns while following your broker's advice?

39. *Portfolio Management* If x dollars is invested in a company that controls, say, 30 percent of the market with 5 brand-names, then $0.30x$ is a measure of market exposure and $5x$ is a measure of brand-name exposure. Now suppose you are a broker at a large securities firm, and one of your clients would like to invest up to $100,000 in recording industry stocks. You decide to recommend a combination of stocks in four of the world's largest companies: Warner Music, Universal Music, Sony, and EMI. (See the table.[27])

	Warner Music	*Universal Music*	*Sony*	*EMI*
Market Share	12%	20%	20%	15%
Number of Labels (Brands)	8	20	10	15

You would like your client's brand-name exposure to be as large as possible but his total market exposure to be $15,000 or less. (This would reflect an average of 15 percent.) Furthermore,

[24] NOK, HWP and DELL were trading at or near the given prices in January, 2005. IBM was trading near the given price in August 2004. Earnings per share are rounded. SOURCE: http://money.excite.com.

[25] Adapted from an exercise in *An Introduction to Management Science* (6th. ed.) by D. R. Anderson, D. J. Sweeny and T. A. Williams (West).

[26] Yields are as of January 2004. SOURCE: http://finance.yahoo.com.

[27] The number of labels includes only major labels. Market shares are approximate, and represent the period 2000–2002. SOURCES: Various, including www.emigroup.com, http://finance.vivendi.com/discover/financial, http://business2.com, March, 2002.

● basic skills   ◆ challenging   ❚tech❙Ex technology exercise

you would like at least 20 percent of the investment to be in Universal because you feel that its control of the DGG and Phillips labels is advantageous for its classical music operations. How much should you advise your client to invest in each company?

40. *Portfolio Management* Referring to Exercise 39, suppose instead that you wanted your client to maximize his total market exposure but limit his brand-name exposure to 1.5 million or less (representing an average of 15 labels or fewer per company), and still invest at least 20 percent of the total in Universal. How much should you advise your client to invest in each company?

41. [tech] Ex *Transportation Scheduling* (This exercise is almost identical to Exercise 24 in Section 2.3 but is more realistic; one cannot always expect to fill all orders exactly, and keep all plants operating at 100 percent capacity.) The Tubular Ride Boogie Board Company has manufacturing plants in Tucson, Arizona, and Toronto, Ontario. You have been given the job of coordinating distribution of the latest model, the Gladiator, to their outlets in Honolulu and Venice Beach. The Tucson plant, when operating at full capacity, can manufacture 620 Gladiator boards per week, while the Toronto plant, beset by labor disputes, can produce only 410 boards per week. The outlet in Honolulu orders 500 Gladiator boards per week, while Venice Beach orders 530 boards per week. Transportation costs are as follows: Tucson to Honolulu: $10 per board; Tucson to Venice Beach: $5 per board; Toronto to Honolulu: $20 per board; Toronto to Venice Beach: $10 per board. Your manager has informed you that the company's total transportation budget is $6550. You realize that it may not be possible to fill all the orders, but you would like the total number of boogie boards shipped to be as large as possible. Given this, how many Gladiator boards should you order shipped from each manufacturing plant to each distribution outlet?

42. [tech] Ex *Transportation Scheduling* Repeat the preceding exercise, but use a transportation budget of $5050.

43. [tech] Ex *Transportation Scheduling* Your publishing company is about to start a promotional blitz for its new book, *Physics for the Liberal Arts*. You have 20 salespeople stationed in Chicago and 10 in Denver. You would like to fly at most 10 into Los Angeles and at most 15 into New York. A round-trip plane flight from Chicago to LA costs $195;[28] from Chicago to NY costs $182; from Denver to LA costs $395; and from Denver to NY costs $166. You want to spend at most $4520 on plane flights. How many salespeople should you fly

from each of Chicago and Denver to each of LA and NY to have the most salespeople on the road?

44. [tech] Ex *Transportation Scheduling* Repeat the preceding exercise, but this time, spend at most $5320.

Communication and Reasoning Exercises

45. ● Can the following linear programming problem be stated as a standard LP problem? If so, do it; if not, explain why.

Maximize $p = 3x - 2y$

subject to $x - y + z \geq 0$

$x - y - z \leq 6$

46. ● Can the following linear programming problem be stated as a standard LP problem? If so, do it; if not, explain why.

Maximize $p = -3x - 2y$

subject to $x - y + z \geq 0$

$x - y - z \geq -6$

47. ● Why is the simplex method useful? (After all, we do have the graphical method for solving LP problems.)

48. ● Are there any types of linear programming problems that cannot be solved with the methods of this section but that can be solved using the methods of the preceding section? Explain.

49. Your friend Janet is telling everyone that if there are only two constraints in a linear programming problem, then, in any optimal basic solution, at most two unknowns (other than the objective) will be nonzero. Is she correct? Explain.

50. Your other friend Jason is telling everyone that if there is only one constraint in a standard linear programming problem, then you will have to pivot at most once to obtain an optimal solution. Is he correct? Explain.

51. What is a "basic solution"? How might one find a basic solution of a given system of linear equations?

52. In a typical simplex method tableau, there are more unknowns than equations, and we know from the chapter on systems of linear equations that this typically implies the existence of infinitely many solutions. How are the following types of solutions interpreted in the simplex method?

a. Solutions in which all the variables are positive.
b. Solutions in which some variables are negative.
c. Solutions in which the inactive variables are zero.

53. ◆ Can the value of the objective function decrease in passing from one tableau to the next? Explain.

54. ◆ Can the value of the objective function remain unchanged in passing from one tableau to the next? Explain.

---

[28] Prices from Travelocity, at www.travelocity.com, for the week of June 3, 2002, as of May 5, 2002.

4.4 The Simplex Method: Solving General Linear Programming Problems

As we saw in Section 4.2, not all LP problems are standard maximization problems. We might have constraints like $2x + 3y \geq 4$ or perhaps $2x + 3y = 4$. Or, you might have to minimize, rather than maximize, the objective function. General problems like this are almost as easy to deal with as the standard kind: There is a modification of the simplex method that works very nicely. The best way to illustrate it is by means of examples. First, we discuss nonstandard maximization problems.

Nonstandard Maximization Problems

Example 1 Maximizing with Mixed Constraints

$$\text{Maximize} \quad p = 4x + 12y + 6z$$
$$\text{subject to} \quad x + y + z \leq 100$$
$$4x + 10y + 7z \leq 480$$
$$x + y + z \geq 60$$
$$x \geq 0, y \geq 0, z \geq 0$$

Solution We begin by turning the first two inequalities into equations as usual because they have the standard form. We get

$$x + y + z + s \quad = 100$$
$$4x + 10y + 7z \quad + t = 480$$

We are tempted to use a slack variable for the third inequality, $x + y + z \geq 60$, but *adding* something positive to the left-hand side will not make it equal to the right: It will get even bigger. To make it equal to 60, we must *subtract* some nonnegative number. We will call this number u (because we have already used s and t) and refer to u as a **surplus variable** rather than a slack variable. Thus, we write

$$x + y + z - u = 60$$

Continuing with the setup, we have

$$x + y + z + s \qquad\qquad = 100$$
$$4x + 10y + 7z \quad + t \qquad\qquad = 480$$
$$x + y + z \qquad\quad - u \qquad = 60$$
$$-4x - 12y - 6z \qquad\qquad + p = 0$$

This leads to the initial tableau:

	x	y	z	s	t	u	p	
s	1	1	1	1	0	0	0	100
t	4	10	7	0	1	0	0	480
*u	1	1	1	0	0	−1	0	60
p	−4	−12	−6	0	0	0	1	0

We put a star next to the third row because the basic solution corresponding to this tableau is

$$x = y = z = 0, s = 100, t = 480, u = 60/(-1) = -60$$

Several things are wrong here. First, the values $x = y = z = 0$ do not satisfy the third inequality $x + y + z \geq 60$. Thus, this basic solution is *not feasible*. Second—and this is really the same problem—the surplus variable u is negative, whereas we said that it should be nonnegative. The star next to the row labeled u alerts us to the fact that the present basic solution is not feasible and that the problem is located in the starred row, where the active variable u is negative.

Whenever an active variable is negative, we star the corresponding row.

In setting up the initial tableau, we star those rows coming from $\geq$ inequalities.

The simplex method as described in the preceding section assumed that we began in the feasible region, but now we do not. Our first task is to get ourselves into the feasible region. In practice, we can think of this as getting rid of the stars on the rows. Once we get into the feasible region, we go back to the method of the preceding section.

There are several ways to get into the feasible region. The method we have chosen is one of the simplest to state and carry out. (We will see why this method works at the end of the example.)

The Simplex Method for General Linear Programming Problems

Star all rows that give a negative value for the associated active variable (except for the objective variable, which is allowed to be negative). If there are starred rows, you will need to begin with Phase I.

Phase I: Getting into the Feasible Region (Getting Rid of the Stars)
In the first starred row, find the largest positive number. Use test ratios as in Section 4.3 to find the pivot in that column (exclude the bottom row), and then pivot on that entry. (If the lowest ratio occurs both in a starred row and an un-starred row, pivot in a starred row rather than the un-starred one.) Check to see which rows should now be starred. Repeat until no starred rows remain, and then go on to Phase II.

Phase II: Use the Simplex Method for Standard Maximization Problems
If there are any negative entries on the left side of the bottom row after Phase I, use the method described in the preceding section.

Because there is a starred row, we need to use Phase I. The largest positive number in the starred row is 1, which occurs three times. Arbitrarily select the first, which is in the first column. In that column, the smallest test ratio happens to be given by the 1 in the u row, so this is our first pivot.

pivot column
↓

	x	y	z	s	t	u	p		
s	1	1	1	1	0	0	0	100	$R_1 - R_3$
t	4	10	7	0	1	0	0	480	$R_2 - 4R_3$
*u	1	1	1	0	0	−1	0	60	
p	−4	−12	−6	0	0	0	1	0	$R_4 + 4R_3$

This gives

	x	y	z	s	t	u	p	
s	0	0	0	1	0	1	0	40
t	0	6	3	0	1	4	0	240
x	1	1	1	0	0	−1	0	60
p	0	−8	−2	0	0	−4	1	240

Notice that we removed the star from row 3. To see why, look at the basic solution given by this tableau:

$$x = 60, y = 0, z = 0, s = 40, t = 240, u = 0$$

None of the variables is negative anymore, so there are no rows to star. The basic solution is therefore feasible—it satisfies all the constraints.

Now that there are no more stars, we have completed Phase I, so we proceed to Phase II, which is just the method of the preceding section.

	x	y	z	s	t	u	p		
s	0	0	0	1	0	1	0	40	
t	0	6	3	0	1	4	0	240	
x	1	1	1	0	0	−1	0	60	$6R_3 - R_2$
p	0	−8	−2	0	0	−4	1	240	$3R_4 + 4R_2$

	x	y	z	s	t	u	p	
s	0	0	0	1	0	1	0	40
y	0	6	3	0	1	4	0	240
x	6	0	3	0	−1	−10	0	120
p	0	0	6	0	4	4	3	1680

And we are finished. Thus the solution is

$$p = 1680/3 = 560, x = 120/6 = 20, y = 240/6 = 40, z = 0$$

The slack and surplus variables are

$$s = 40, t = 0, u = 0$$

+*Before we go on...* We owe you an explanation of why this method works. When we perform a pivot in Phase I, one of two things will happen. As in Example 1, we may pivot in a starred row. In that case, the negative active variable in that row will become inactive (hence zero) and some other variable will be made active with a positive value because we are pivoting on a positive entry. Thus, at least one star will be eliminated. (We will not introduce any new stars because pivoting on the entry with the smallest test ratio will keep all nonnegative variables nonnegative.)

The second possibility is that we may pivot on some row other than a starred row. Choosing the pivot via test ratios again guarantees that no new starred rows are created. A little bit of algebra shows that the value of the negative variable in the first starred row must increase toward zero. (Choosing the *largest* positive entry in the starred row will make it a little more likely that we will increase the value of that variable as much as possible; the rationale for choosing the largest entry is the same as that for choosing the most negative entry in the bottom row during Phase II.) Repeating this procedure as necessary, the value of the variable must eventually become zero or positive, assuming that there are feasible solutions to begin with.

So, one way or the other, we can eventually get rid of all the stars. ∎

Here is an example that begins with two starred rows.

Example 2 More Mixed Constraints

Maximize $p = 2x + y$
subject to $\quad x + y \geq 35$
$\qquad\qquad x + 2y \leq 60$
$\qquad\qquad 2x + y \geq 60$
$\qquad\qquad x \qquad\quad \leq 25$
$\qquad\qquad x \geq 0, y \geq 0$

Solution We introduce slack and surplus variables, and write down the initial tableau:

$$x + y - s \qquad\qquad\qquad = 35$$
$$x + 2y \qquad + t \qquad\qquad = 60$$
$$2x + y \qquad\qquad - u \qquad = 60$$
$$x \qquad\qquad\qquad + v \quad = 25$$
$$-2x - y \qquad\qquad\qquad + p = 0$$

	x	y	s	t	u	v	p	
*s	1	1	−1	0	0	0	0	35
t	1	2	0	1	0	0	0	60
*u	2	1	0	0	−1	0	0	60
v	1	0	0	0	0	1	0	25
p	−2	−1	0	0	0	0	1	0

We locate the largest positive entry in the first starred row (row 1). There are two to choose from (both 1s); let's choose the one in the x column. The entry with the smallest test ratio in that column is the 1 in the v row, so that is the entry we use as the pivot:

Pivot column
↓

	x	y	s	t	u	v	p		
*s	1	1	−1	0	0	0	0	35	$R_1 - R_4$
t	1	2	0	1	0	0	0	60	$R_2 - R_4$
*u	2	1	0	0	−1	0	0	60	$R_3 - 2R_4$
v	[1]	0	0	0	0	1	0	25	
p	−2	−1	0	0	0	0	1	0	$R_5 + 2R_4$

	x	y	s	t	u	v	p	
*s	0	1	−1	0	0	−1	0	10
t	0	2	0	1	0	−1	0	35
*u	0	1	0	0	−1	−2	0	10
x	1	0	0	0	0	1	0	25
p	0	−1	0	0	0	2	1	50

Notice that both stars are still there because the basic solutions for s and u remain negative (but less so). The only positive entry in the first starred row is the 1 in the y column, and that entry also has the smallest test ratio in its column (actually, it is tied with the 1 in the u column, so we could choose either one).

	x	y	s	t	u	v	p		
*s	0	[1]	−1	0	0	−1	0	10	
t	0	2	0	1	0	−1	0	35	$R_2 - 2R_1$
*u	0	1	0	0	−1	−2	0	10	$R_3 - R_1$
x	1	0	0	0	0	1	0	25	
p	0	−1	0	0	0	2	1	50	$R_5 + R_1$

	x	y	s	t	u	v	p	
y	0	1	−1	0	0	−1	0	10
t	0	0	2	1	0	1	0	15
u	0	0	1	0	−1	−1	0	0
x	1	0	0	0	0	1	0	25
p	0	0	−1	0	0	1	1	60

The basic solution is $x = 25$, $y = 10$, $s = 0$, $t = 15$, $u = 0/(-1) = 0$, and $v = 0$. Because there are no negative variables left (even u has become 0), we are in the feasible region, so we can go on to Phase II, shown next. (Filling in the instructions for the row operations is an exercise.)

	x	y	s	t	u	v	p	
y	0	1	−1	0	0	−1	0	10
t	0	0	2	1	0	1	0	15
u	0	0	[1]	0	−1	−1	0	0
x	1	0	0	0	0	1	0	25
p	0	0	−1	0	0	1	1	60

	x	y	s	t	u	v	p	
y	0	1	0	0	−1	−2	0	10
t	0	0	0	1	[2]	3	0	15
s	0	0	1	0	−1	−1	0	0
x	1	0	0	0	0	1	0	25
p	0	0	0	0	−1	0	1	60

	x	y	s	t	u	v	p	
y	0	2	0	1	0	−1	0	35
u	0	0	0	1	2	3	0	15
s	0	0	2	1	0	1	0	15
x	1	0	0	0	0	1	0	25
p	0	0	0	1	0	3	2	135

The optimal solution is

$$x = 25, \, y = 35/2 = 17.5, \, p = 135/2 = 67.5 \quad (s = 7.5, t = 0, u = 7.5)$$

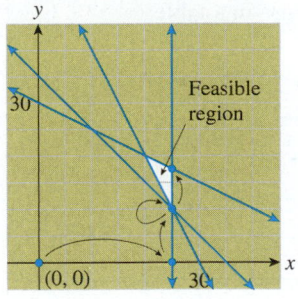

Figure **24**

+ *Before we go on...* Because Example 2 had only two unknowns, we can picture the sequence of basic solutions on the graph of the feasible region. This is shown in Figure 24.

You can see that there was no way to jump from (0, 0) in the initial tableau directly into the feasible region because the first jump must be along an axis. (Why?) Also notice that the third jump did not move at all. To which step of the simplex method does this correspond? ■

Minimization Problems

Now that we know how to deal with nonstandard constraints, we consider **minimization problems**, problems in which we have to minimize, rather than maximize, the objective function. The idea is to *convert a minimization problem into a maximization problem*, which we can then solve as usual.

Suppose, for instance, we want to minimize $c = 10x - 30y$ subject to some constraints. The technique is as follows: define a new variable p by taking p to be the negative of c, so that $p = -c$. Then, the larger we make p, the smaller c becomes. For example, if we can make p increase from -10 to -5, then c will decrease from 10 to 5. So, if we are looking for the smallest value of c, we might as well look for the largest value of p instead. More concisely,

Minimizing c is the same as maximizing $p = -c$.

Now because $c = 10x - 30y$, we have $p = -10x + 30y$, and the requirement that we "minimize $c = 10x - 30y$" is now replaced by "maximize $p = -10x + 30y$."

Minimization Problems

We convert a minimization problem into a maximization problem by taking the negative of the objective function. All the constraints remain unchanged.

quick Example

Minimization Problem	$\rightarrow$	**Maximization Problem**
Minimize $\quad c = 10x - 30y$		Maximize $\quad p = -10x + 30y$
subject to $\quad 2x + y \leq 160$		subject to $\quad 2x + y \leq 160$
$x + 3y \geq 120$		$x + 3y \geq 120$
$x \geq 0, y \geq 0$		$x \geq 0, y \geq 0$

Example 3 Purchasing

You are in charge of ordering furniture for your company's new headquarters. You need to buy at least 200 tables, 500 chairs, and 300 computer desks. Wall-to-Wall Furniture (WWF) is offering a package of 20 tables, 25 chairs, and 18 computer desks for $2000, whereas rival Acme Furniture (AF) is offering a package of 10 tables, 50 chairs, and 24 computer desks for $3000. How many packages should you order from each company to minimize your total cost?

Solution The unknowns here are

$x =$ number of packages ordered from WWF
$y =$ number of packages ordered from AF

We can put the information about the various kinds of furniture in a table:

	WWF	AF	Needed
Tables	20	10	200
Chairs	25	50	500
Computer Desks	18	24	300
Cost ($)	2000	3000	

From this table we get the following LP problem:

$$\text{Minimize} \quad c = 2000x + 3000y$$
$$\text{subject to} \quad 20x + 10y \geq 200$$
$$25x + 50y \geq 500$$
$$18x + 24y \geq 300$$
$$x \geq 0, y \geq 0$$

Before we start solving this problem, notice that all inequalities may be simplified. The first is divisible by 10, the second by 25, and the third by 6. (However, this affects the meaning of the surplus variables; see *Before we go on* below.) Dividing gives the following simpler problem:

$$\text{Minimize} \quad c = 2000x + 3000y$$
$$\text{subject to} \quad 2x + \ y \geq 20$$
$$x + 2y \geq 20$$
$$3x + 4y \geq 50$$
$$x \geq 0, y \geq 0$$

Following the discussion that preceded this example, we convert to a maximization problem:

$$\text{Maximize} \quad p = -2000x - 3000y$$
$$\text{subject to} \quad 2x + \ y \geq 20$$
$$x + 2y \geq 20$$
$$3x + 4y \geq 50$$
$$x \geq 0, y \geq 0$$

We introduce surplus variables.

$$2x + \quad y - s \qquad\qquad = 20$$
$$x + \quad 2y \quad - t \qquad = 20$$
$$3x + \quad 4y \qquad - u \quad = 50$$
$$2000x + 3000y \qquad\qquad + p = 0$$

The initial tableau is then

	x	y	s	t	u	p	
*s	2	1	−1	0	0	0	20
*t	1	2	0	−1	0	0	20
*u	3	4	0	0	−1	0	50
p	2000	3000	0	0	0	1	0

The largest entry in the first starred row is the 2 in the upper left, which happens to give the smallest test ratio in its column.

	x	y	s	t	u	p		
*s	2	1	-1	0	0	0	20	
*t	1	2	0	-1	0	0	20	$2R_2 - R_1$
*u	3	4	0	0	-1	0	50	$2R_3 - 3R_1$
p	2000	3000	0	0	0	1	0	$R_4 - 1000R_1$

	x	y	s	t	u	p		
x	2	1	-1	0	0	0	20	$3R_1 - R_2$
*t	0	3	1	-2	0	0	20	
*u	0	5	3	0	-2	0	40	$3R_3 - 5R_2$
p	0	2000	1000	0	0	1	$-20{,}000$	$3R_4 - 2000R_2$

	x	y	s	t	u	p		
x	6	0	-4	2	0	0	40	$5R_1 - R_3$
y	0	3	1	-2	0	0	20	$5R_2 + R_3$
*u	0	0	4	10	-6	0	20	
p	0	0	1000	4000	0	3	$-100{,}000$	$R_4 - 400R_3$

	x	y	s	t	u	p		
x	30	0	-24	0	6	0	180	$R_1/6$
y	0	15	9	0	-6	0	120	$R_2/3$
t	0	0	4	10	-6	0	20	$R_3/2$
p	0	0	-600	0	2400	3	$-108{,}000$	$R_4/3$

This completes Phase I. We are not yet at the optimal solution, so after performing the simplifications indicated we proceed with Phase II.

	x	y	s	t	u	p		
x	5	0	-4	0	1	0	30	$R_1 + 2R_3$
y	0	5	3	0	-2	0	40	$2R_2 - 3R_3$
t	0	0	2	5	-3	0	10	
p	0	0	-200	0	800	1	$-36{,}000$	$R_4 + 100R_3$

	x	y	s	t	u	p	
x	5	0	0	10	-5	0	50
y	0	10	0	-15	5	0	50
s	0	0	2	5	-3	0	10
p	0	0	0	500	500	1	$-35{,}000$

The optimal solution is

$x = 50/5 = 10$, $y = 50/10 = 5$, $p = -35{,}000$, so $c = 35{,}000$ ($s = 5, t = 0, u = 0$)

You should buy 10 packages from Wall-to-Wall Furniture and 5 from Acme Furniture, for a minimum cost of \$35,000.

+*Before we go on...* The surplus variables in the preceding example represent pieces of furniture over and above the minimum requirements. The order you place will give you 50 extra tables ($s = 5$, but s was introduced after we divided the first inequality by 10, so the actual surplus is $10 \times 5 = 50$), the correct number of chairs ($t = 0$), and the correct number of computer desks ($u = 0$). ∎

The preceding LP problem is an example of a **standard minimization problem**—in a sense the opposite of a standard maximization problem: We are *minimizing* an objective function, where all the constraints have the form $Ax + By + Cz + \cdots \geq N$. We will discuss standard minimization problems more fully in Section 4.5, as well as another method of solving them.

FAQs **When to Switch to Phase II, Equality Constraints, and Troubleshooting**

Q: How do I know when to switch to Phase II?

A: After each step, check the basic solution for starred rows. You are not ready to proceed with Phase II until all the stars are gone. ∎

Q: How do I deal with an equality constraint, such as $2x + 7y - z = 90$?

A: Although we haven't given examples of equality constraints, they can be treated by the following trick: *Replace an equality by two inequalities.* For example, replace the equality $2x + 7y - z = 90$ by the two inequalities $2x + 7y - z \leq 90$ and $2x + 7y - z \geq 90$. A little thought will convince you that these two inequalities amount to the same thing as the original equality! ∎

Q: What happens if it is impossible to choose a pivot using the instructions in Phase I?

A: In that case, the LP problem has no solution. In fact, the feasible region is empty. If it is impossible to choose a pivot in phase II, then the feasible region is unbounded and there is no optimal solution. ∎

4.4 EXERCISES

● denotes basic skills exercises

tech Ex indicates exercises that should be solved using technology

1. ● Maximize $p = x + y$
subject to $x + 2y \geq 6$
$-x + y \leq 4$
$2x + y \leq 8$
$x \geq 0, y \geq 0$ *hint* [see Examples 1 and 2]

2. ● Maximize $p = 3x + 2y$
subject to $x + 3y \geq 6$
$-x + y \leq 4$
$2x + y \leq 8$
$x \geq 0, y \geq 0$

3. ● Maximize $p = 12x + 10y$
subject to $x + y \leq 25$
$x \geq 10$
$-x + 2y \geq 0$
$x \geq 0, y \geq 0$

4. ● Maximize $p = x + 2y$
subject to $x + y \leq 25$
$y \geq 10$
$2x - y \geq 0$
$x \geq 0, y \geq 0$

● basic skills   tech Ex technology exercise

5. ● Maximize $p = 2x + 5y + 3z$
subject to
$$x + y + z \le 150$$
$$x + y + z \ge 100$$
$$x \ge 0, y \ge 0, z \ge 0$$

6. ● Maximize $p = 3x + 2y + 2z$
subject to
$$x + y + 2z \le 38$$
$$2x + y + z \ge 24$$
$$x \ge 0, y \ge 0, z \ge 0$$

7. ● Maximize $p = 2x + 3y + z + 4w$
subject to
$$x + y + z + w \le 40$$
$$2x + y - z - w \ge 10$$
$$x + y + z + w \ge 10$$
$$x \ge 0, y \ge 0, z \ge 0, w \ge 0$$

8. ● Maximize $p = 2x + 2y + z + 2w$
subject to
$$x + y + z + w \le 50$$
$$2x + y - z - w \ge 10$$
$$x + y + z + w \ge 20$$
$$x \ge 0, y \ge 0, z \ge 0, w \ge 0$$

9. ● Minimize $c = 6x + 6y$
subject to
$$x + 2y \ge 20$$
$$2x + y \ge 20$$
$$x \ge 0, y \ge 0$$ *hint* [see Example 3]

10. ● Minimize $c = 3x + 2y$
subject to
$$x + 2y \ge 20$$
$$2x + y \ge 10$$
$$x \ge 0, y \ge 0$$

11. ● Minimize $c = 2x + y + 3z$
subject to
$$x + y + z \ge 100$$
$$2x + y \ge 50$$
$$y + z \ge 50$$
$$x \ge 0, y \ge 0, z \ge 0$$

12. ● Minimize $c = 2x + 2y + 3z$
subject to
$$x + z \ge 100$$
$$2x + y \ge 50$$
$$y + z \ge 50$$
$$x \ge 0, y \ge 0, z \ge 0$$

13. ● Minimize $c = 50x + 50y + 11z$
subject to
$$2x + z \ge 3$$
$$2x + y - z \ge 2$$
$$3x + y - z \le 3$$
$$x \ge 0, y \ge 0, z \ge 0$$

14. ● Minimize $c = 50x + 11y + 50z$
subject to
$$3x + z \ge 8$$
$$3x + y - z \ge 6$$
$$4x + y - z \le 8$$
$$x \ge 0, y \ge 0, z \ge 0$$

15. ● Minimize $c = x + y + z + w$
subject to
$$5x - y + w \ge 1000$$
$$z + w \le 2000$$
$$x + y \le 500$$
$$x \ge 0, y \ge 0, z \ge 0, w \ge 0$$

16. ● Minimize $c = 5x + y + z + w$
subject to
$$5x - y + w \ge 1000$$
$$z + w \le 2000$$
$$x + y \le 500$$
$$x \ge 0, y \ge 0, z \ge 0, w \ge 0$$

tech Ex *In Exercises 17–22, we suggest the use of technology. Round all answers to two decimal places.*

17. **tech** Ex Maximize $p = 2x + 3y + 1.1z + 4w$
subject to
$$1.2x + y + z + w \le 40.5$$
$$2.2x + y - z - w \ge 10$$
$$1.2x + y + z + 1.2w \ge 10.5$$
$$x \ge 0, y \ge 0, z \ge 0, w \ge 0$$

18. **tech** Ex Maximize $p = 2.2x + 2y + 1.1z + 2w$
subject to
$$x + 1.5y + 1.5z + w \le 50.5$$
$$2x + 1.5y - z - w \ge 10$$
$$x + 1.5y + z + 1.5w \ge 21$$
$$x \ge 0, y \ge 0, z \ge 0, w \ge 0$$

19. **tech** Ex Minimize $c = 2.2x + y + 3.3z$
subject to
$$x + 1.5y + 1.2z \ge 100$$
$$2x + 1.5y \ge 50$$
$$1.5y + 1.1z \ge 50$$
$$x \ge 0, y \ge 0, z \ge 0$$

20. **tech** Ex Minimize $c = 50.3x + 10.5y + 50.3z$
subject to
$$3.1x + 1.1z \ge 28$$
$$3.1x + y - 1.1z \ge 23$$
$$4.2x + y - 1.1z \le 28$$
$$x \ge 0, y \ge 0, z \ge 0$$

21. **tech** Ex Minimize $c = 1.1x + y + 1.5z - w$
subject to
$$5.12x - y + w \le 1000$$
$$z + w \ge 2000$$
$$1.22x + y \le 500$$
$$x \ge 0, y \ge 0, z \ge 0, w \ge 0$$

22. **tech** Ex Minimize $c = 5.45x + y + 1.5z + w$
subject to
$$5.12x - y + w \ge 1000$$
$$z + w \ge 2000$$
$$1.12x + y \le 500$$
$$x \ge 0, y \ge 0, z \ge 0, w \ge 0$$

Applications

23. ● *Agriculture*[29] Your small farm encompasses 100 acres, and you are planning to grow tomatoes, lettuce, and carrots in the coming planting season. Fertilizer costs per acre are: $5 for tomatoes, $4 for lettuce, and $2 for carrots. Based on past experience, you estimate that each acre of tomatoes will require an average of 4 hours of labor per week, while tending to lettuce and carrots will each require an average of 2 hours per week. You estimate a profit of $2000 for each acre of tomatoes, $1500 for each acre of lettuce and $500 for each acre of carrots. You would like to spend at least $400 on fertilizer (your niece owns the

_____
[29] Compare Exercise 27 in Section 4.3.

● basic skills   **tech** Ex technology exercise

company that manufactures it) and your farm laborers can supply up to 500 hours per week. How many acres of each crop should you plant to maximize total profits? In this event, will you be using all 100 acres of your farm? *hint* [see Example 3]

24. ● *Agriculture*[30] Your farm encompasses 900 acres, and you are planning to grow soy beans, corn, and wheat in the coming planting season. Fertilizer costs per acre are: $5 for soy beans, $2 for corn, and $1 for wheat. You estimate that each acre of soy beans will require an average of 5 hours of labor per week, while tending to corn and wheat will each require an average of 2 hours per week. Based on past yields and current market prices, you estimate a profit of $3000 for each acre of soy beans, $2000 for each acre of corn, and $1000 for each acre of wheat. You can afford to spend no more than $3000 on fertilizer, but your labor union contract stipulates at least 2000 hours per week of labor. How many acres of each crop should you plant to maximize total profits? In this event, will you be using more than 2000 hours of labor?

25. ● *Politics* The political pollster Canter is preparing for a national election. It would like to poll at least 1500 Democrats and 1500 Republicans. Each mailing to the East Coast gets responses from 100 Democrats and 50 Republicans. Each mailing to the Midwest gets responses from 100 Democrats and 100 Republicans. And each mailing to the West Coast gets responses from 50 Democrats and 100 Republicans. Mailings to the East Coast cost $40 each to produce and mail, mailings to the Midwest cost $60 each, and mailings to the West Coast cost $50 each. How many mailings should Canter send to each area of the country to get the responses it needs at the least possible cost? What will it cost?

26. ● *Purchasing* Bingo's Copy Center needs to buy white paper and yellow paper. Bingos can buy from three suppliers. Harvard Paper sells a package of 20 reams of white and 10 reams of yellow for $60, Yale Paper sells a package of 10 reams of white and 10 reams of yellow for $40, and Dartmouth Paper sells a package of 10 reams of white and 20 reams of yellow for $50. If Bingo's needs 350 reams of white and 400 reams of yellow, how many packages should it buy from each supplier to minimize the cost? What is the least possible cost?

27. ● *Resource Allocation* Succulent Citrus produces orange juice and orange concentrate. This year the company anticipates a demand of at least 10,000 quarts of orange juice and 1000 quarts of orange concentrate. Each quart of orange juice requires 10 oranges, and each quart of concentrate requires 50 oranges. The company also anticipates using at least 200,000 oranges for these products. Each quart of orange juice costs the company 50¢ to produce, and each quart of concentrate costs $2.00 to produce. How many quarts of each product should Succulent Citrus produce to meet the demand and minimize total costs?

28. ● *Resource Allocation* Fancy Pineapple produces pineapple juice and canned pineapple rings. This year the company anti-

cipates a demand of at least 10,000 pints of pineapple juice and 1000 cans of pineapple rings. Each pint of pineapple juice requires 2 pineapples, and each can of pineapple rings requires 1 pineapple. The company anticipates using at least 20,000 pineapples for these products. Each pint of pineapple juice costs the company 20¢ to produce, and each can of pineapple rings costs 50¢ to produce. How many pints of pineapple juice and cans of pineapple rings should Fancy Pineapple produce to meet the demand and minimize total costs?

29. ● *Music CD Sales* In 2000 industry revenues from sales of recorded music amounted to $3.6 billion for rock music, $1.8 billion for rap music, and $0.4 billion for classical music.[31] You would like the selection of music in your music store to reflect, in part, this national trend, so you have decided to stock at least twice as many rock music CDs as rap CDs. Your store has an estimated capacity of 20,000 CDs, and, as a classical music devotee, you would like to stock at least 5000 classical CDs. Rock music CDs sell for $12 on average, rap CDs for $15, and classical CDs for $12. How many of each type of CD should you stock to get the maximum retail value?

30. ● *Music CD Sales* Your music store's main competitor, Nuttal Hip Hop Classic Store, also wishes to stock at most 20,000 CDs, with at least half as many rap CDs as rock CDs and at least 2000 classical CDs. It anticipates an average sale price of $15/rock CD, $10/rap CD and $10/classical CD. How many of each type of CD should it stock to get the maximum retail value, and what is the maximum retail value?

31. **tech** Ex *Nutrition* Each serving of Gerber Mixed Cereal for Baby contains 60 calories and no Vitamin C. Each serving of Gerber Mango Tropical Fruit Dessert contains 80 calories and 45 percent of the U.S. Recommended Daily Allowance (RDA) of Vitamin C for infants. Each serving of Gerber Apple Banana Juice contains 60 calories and 120 percent of the RDA of Vitamin C for infants.[32] The cereal costs 10¢/serving, the dessert costs 53¢/serving, and the juice costs 27¢/serving. If you want to provide your child with at least 120 calories and at least 120 percent of the RDA of Vitamin C, how can you do so at the least cost?

32. **tech** Ex *Nutrition* Each serving of Gerber Mixed Cereal for Baby contains 60 calories, no Vitamin C, and 11 grams of carbohydrates. Each serving of Gerber Mango Tropical Fruit Dessert contains 80 calories, 45 percent of the RDA of Vitamin C for infants, and 21 grams of carbohydrates. Each serving of Gerber Apple Banana Juice contains 60 calories, 120 percent of the RDA of Vitamin C for infants, and 15 grams of carbohydrates.[33] Assume that the cereal costs 11¢/serving, the dessert

---

[30] Compare Exercise 28 in Section 4.3.

[31] "Rap" includes "Hip Hop." Revenues are based on total manufacturers' shipments at suggested retail prices and are rounded to the nearest $0.1 billion. SOURCE: Recording Industry Association of America, www.riaa.com, March, 2002.

[32] SOURCE: Nutrition information supplied by Gerber.

[33] Ibid.

● basic skills   **tech** Ex technology exercise

costs 50¢/serving, and the juice costs 30¢/serving. If you want to provide your child with at least 180 calories, at least 120 percent of the RDA of Vitamin C, and at least 37 grams of carbohydrates, how can you do so at the least cost?

33. tech Ex *Purchasing* Cheapskate Electronics Store needs to update its inventory of stereos, TVs, and DVD players. There are three suppliers it can buy from: Nadir offers a bundle consisting of 5 stereos, 10 TVs, and 15 DVD players for $3000. Blunt offers a bundle consisting of 10 stereos, 10 TVs, and 10 DVD players for $4000. Sonny offers a bundle consisting of 15 stereos, 10 TVs, and 10 DVD players for $5000. Cheapskate Electronics needs at least 150 stereos, 200 TVs, and 150 DVD players. How can it update its inventory at the least possible cost? What is the least possible cost?

34. tech Ex *Purchasing* Federal Rent-a-Car is putting together a new fleet. It is considering package offers from three car manufacturers. Fred Motors is offering 5 small cars, 5 medium cars, and 10 large cars for $500,000. Admiral Motors is offering 5 small, 10 medium, and 5 large cars for $400,000. Chrysalis is offering 10 small, 5 medium, and 5 large cars for $300,000. Federal would like to buy at least 550 small cars, at least 500 medium cars, and at least 550 large cars. How many packages should it buy from each car maker to keep the total cost as small as possible? What will be the total cost?

tech Ex *Creatine Supplements Exercises 35 and 36 are based on the following data on four bodybuilding supplements. (Figures shown correspond to a single serving.)[34]*

	Creatine (g)	Carbohydrates (g)	Taurine (g)	Alpha Lipoic Acid (mg)	Cost ($)
Cell-Tech® (MuscleTech)	10	75	2	200	2.20
RiboForce HP® (EAS)	5	15	1	0	1.60
Creatine Transport® (Kaizen)	5	35	1	100	0.60
Pre-Load Creatine (Optimum)	6	35	1	25	0.50

35. tech Ex (Compare Exercise 29 in Section 4.2.) You are thinking of combining Cell-Tech, RiboForce HP, and Creatine Transport to obtain a 10-day supply that provides at least 80 grams of creatine and at least 10 grams of taurine, but no more than 750 grams of carbohydrates and 1000 milligrams of alpha lipoic acid. How many servings of each supplement should you combine to meet your specifications for the least cost?

36. tech Ex (Compare Exercise 30 in Section 4.2.) You are thinking of combining RiboForce HP, Creatine Transport,

and Pre-Load Creatine to obtain a 10-day supply that provides at least 80 grams of creatine and at least 10 grams of taurine, but no more than 600 grams of carbohydrates and 2000 milligrams of alpha lipoic acid. How many servings of each supplement should you combine to meet your specifications for the least cost?

37. *Subsidies* The Miami Beach City Council has offered to subsidize hotel development in Miami Beach, and it is hoping for at least two hotels with a total capacity of at least 1400. Suppose that you are a developer interested in taking advantage of this offer by building a small group of hotels in Miami Beach. You are thinking of three prototypes: a convention-style hotel with 500 rooms costing $100 million, a vacation-style hotel with 200 rooms costing $20 million, and a small motel with 50 rooms costing $4 million. The City Council will approve your plans, provided you build at least one convention-style hotel and no more than two small motels.

a. How many of each type of hotel should you build to satisfy the city council's wishes and stipulations while minimizing your total cost?

b. Now assume that the city council will give developers 20 percent of the cost of building new hotels in Miami Beach, up to $50 million.[35] Will the city's $50 million subsidy be sufficient to cover 20 percent of your total costs?

38. *Subsidies* Refer back to the preceding exercise. You are about to begin the financial arrangements for your new hotels when the city council informs you that it has changed its mind and now requires at least two vacation-style hotels and no more than four small motels.

a. How many of each type of hotel should you build to satisfy the city council's wishes and stipulations while minimizing your total costs?

b. Will the city's $50 million subsidy limit still be sufficient to cover 20 percent of your total costs?

39. *Transportation Scheduling* We return to your exploits coordinating distribution for the Tubular Ride Boogie Board Company.[36] You will recall that the company has manufacturing plants in Tucson, Arizona and Toronto, Ontario, and you have been given the job of coordinating distribution of their latest model, the Gladiator, to their outlets in Honolulu and Venice Beach. The Tucson plant can manufacture up to 620 boards per week, while the Toronto plant, beset by labor disputes, can produce no more than 410 Gladiator boards per week. The outlet in Honolulu orders 500 Gladiator boards per week, while Venice Beach orders 530 boards per week. Transportation costs are as follows: Tucson to Honolulu: $10/board; Tucson to Venice Beach: $5/board; Toronto to Honolulu: $20/board; Toronto to Venice Beach: $10/board.

---

[34] Source: Nutritional information supplied by the manufacturers at www.netrition.com. Cost per serving is approximate.

[35] The Miami Beach City Council made such an offer in 1993. (*Chicago Tribune*, June 20, 1993, Section 7, p.8).

[36] See Exercise 24 in Section 2.3 and also Exercise 41 in Section 4.3. This time, we will use the simplex method to solve the version of this problem we first considered in the chapter on systems of equations.

● basic skills   tech Ex technology exercise

Your manager has said that you are to be sure to fill all orders and ship the boogie boards at a minimum total transportation cost. How will you do it?

40. *Transportation Scheduling* In the situation described in the preceding exercise, you have just been notified that workers at the Toronto boogie board plant have gone on strike, resulting in a total work stoppage. You are to come up with a revised delivery schedule by tomorrow with the understanding that the Tucson plant can push production to a maximum of 1000 boards per week. What should you do?

41. *Finance* Senator Porkbarrel habitually overdraws his three bank accounts, at the Congressional Integrity Bank, Citizens' Trust, and Checks R Us. There are no penalties because the overdrafts are subsidized by the taxpayer. The Senate Ethics Committee tends to let slide irregular banking activities as long as they are not flagrant. At the moment (due to Congress' preoccupation with a Supreme Court nominee), a total overdraft of up to $10,000 will be overlooked. Porkbarrel's conscience makes him hesitate to overdraw accounts at banks whose names include expressions like "integrity" and "citizens' trust." The effect is that his overdrafts at the first two banks combined amount to no more than one-quarter of the total. On the other hand, the financial officers at Integrity Bank, aware that Senator Porkbarrel is a member of the Senate Banking Committee, "suggest" that he overdraw at least $2500 from their bank. Find the amount he should overdraw from each bank in order to avoid investigation by the Ethics Committee and overdraw his account at Integrity by as much as his sense of guilt will allow.

42. *Scheduling* Because Joe Slim's brother was recently elected to the State Senate, Joe's financial advisement concern, Inside Information Inc., has been doing a booming trade, even though the financial counseling he offers is quite worthless. (None of his seasoned clients pays the slightest attention to his advice.) Slim charges different hourly rates to different categories of individuals: $5000/hour for private citizens, $50,000/hour for corporate executives, and $10,000/hour for presidents of universities. Due to his taste for leisure, he feels that he can spend no more than 40 hours/week in consultation. On the other hand, Slim feels that it would be best for his intellect were he to devote at least 10 hours of consultation each week to university presidents. However, Slim always feels somewhat uncomfortable dealing with academics, so he would prefer to spend no more than half his consultation time with university presidents. Furthermore, he likes to think of himself as representing the interests of the common citizen, so he wishes to offer at least 2 more hours of his time each week to private citizens than to corporate executives and university presidents combined. Given all these restrictions, how many hours each week should he spend with each type of client in order to maximize his income?

43. tech Ex ***Transportation Scheduling*** Your publishing company is about to start a promotional blitz for its new book, *Physics for the Liberal Arts.* You have 20 salespeople stationed

in Chicago and 10 in Denver. You would like to fly at least 10 into Los Angeles and at least 15 into New York. A round-trip plane flight from Chicago to LA costs $195;[37] from Chicago to NY costs $182; from Denver to LA costs $395; and from Denver to NY costs $166. How many salespeople should you fly from each of Chicago and Denver to each of LA and NY to spend the least amount on plane flights?

44. tech Ex ***Transportation Scheduling*** Repeat Exercise 43, but now suppose that you would like at least 15 salespeople in Los Angeles.

45. tech Ex ***Hospital Staffing*** As the staff director of a new hospital, you are planning to hire cardiologists, rehabilitation specialists, and infectious disease specialists. According to recent data, each cardiology case averages $12,000 in revenue, each physical rehabilitation case $19,000, and each infectious disease case, $14,000.[38] You judge that each specialist you employ will expand the hospital caseload by about 10 patients per week. You already have 3 cardiologists on staff, and the hospital is equipped to admit up to 200 patients per week. Based on past experience, each cardiologist and rehabilitation specialist brings in one government research grant per year, while each infectious disease specialist brings in three. Your board of directors would like to see a total of at least 30 grants per year and would like your weekly revenue to be as large as possible. How many of each kind of specialist should you hire?

46. tech Ex ***Hospital Staffing*** Referring to Exercise 45, you completely misjudged the number of patients each type of specialist would bring to the hospital per week. It turned out that each cardiologist brought in 120 new patients per *year,* each rehabilitation specialist brought in 90 per year, and each infectious disease specialist brought in 70 per year.[39] It also turned out that your hospital could deal with no more than 1960 new patients per year. Repeat the preceding exercise in the light of this corrected data.

Communication and Reasoning Exercises

47. ● Explain the need for Phase I in a nonstandard LP problem.

48. ● Explain the need for Phase II in a nonstandard LP problem.

49. ● Explain briefly why we would need to use Phase I in solving a linear programming problem with the constraint $x + 2y - z \geq 3$.

---

[37] Prices from Travelocity, at www.travelocity.com, for the week of June 3, 2002, as of May 5, 2002.

[38] These (rounded) figures are based on an Illinois survey of 1.3 million hospital admissions (*Chicago Tribune,* March 29, 1993, Section 4, p.1). SOURCE: Lutheran General Health System, Argus Associates, Inc.

[39] These (rounded) figures were obtained from the survey referenced in the preceding exercise by dividing the average hospital revenue per physician by the revenue per case.

● basic skills   tech Ex technology exercise

50. ● Which rows do we star, and why?

51. ● Multiple choice: Consider the following linear programming problem:

$$\text{Maximize} \quad p = x + y$$
$$\text{subject to} \quad x - 2y \geq 0$$
$$2x + y \leq 10$$
$$x \geq 0, y \geq 0$$

This problem

(A) Must be solved using the techniques of Section 4.4.

(B) Must be solved using the techniques of Section 4.3.

(C) Can be solved using the techniques of either section.

52. ● Multiple choice: Consider the following linear programming problem:

$$\text{Maximize} \quad p = x + y$$
$$\text{subject to} \quad x - 2y \geq 1$$
$$2x + y \leq 10$$
$$x \geq 0, y \geq 0$$

This problem

(A) Must be solved using the techniques of Section 4.4.

(B) Must be solved using the techniques of Section 4.3.

(C) Can be solved using the techniques of either section.

53. Find a linear programming problem in three variables that requires one pivot in Phase I.

54. Find a linear programming problem in three variables that requires two pivots in Phase I.

55. Find a linear programming problem in two or three variables with no optimal solution, and show what happens when you try to solve it using the simplex method.

56. Find a linear programming problem in two or three variables with more than one optimal solution, and investigate which solution is found by the simplex method.

● basic skills   **tech Ex** technology exercise

4.5 The Simplex Method and Duality (OPTIONAL)

We mentioned **standard minimization problems** in the last section. These problems have the following form.

Standard Minimization Problem

A **standard minimization problem** is an LP problem in which we are required to *minimize* (not maximize) a linear objective function

$$c = as + bt + cu + \cdots$$

of the variables $s, t, u, \ldots$ (in this section, we will always use the letters $s, t, u, \ldots$ for the unknowns in a standard minimization problem) subject to the constraints

$$s \geq 0, t \geq 0, u \geq 0, \ldots$$

and further constraints of the form

$$As + Bt + Cu + \cdots \geq N$$

where $A, B, C, \ldots$ and N are numbers with N nonnegative.

A **standard linear programming problem** is an LP problem that is either a standard maximization problem or a standard minimization problem. An LP problem satisfies the **nonnegative objective condition** if all the coefficients in the objective function are nonnegative.

---

quick Examples   **Standard Minimization and Maximization Problems**

1. Minimize $c = 2s + 3t + 3u$
   subject to $2s \qquad + \; u \geq 10$ This is a standard minimization problem
   $\qquad\qquad s + 3t - 6u \geq 5$ satisfying the nonnegative objective
   $\qquad\qquad s \geq 0, t \geq 0, u \geq 0$ condition.

2. Maximize $p = 2x + 3y + 3z$
   subject to $2x \qquad + \; z \leq 7$ This is a standard maximization problem
   $\qquad\qquad x + 3y - 6z \leq 6$ satisfying the nonnegative objective
   $\qquad\qquad x \geq 0, y \geq 0, z \geq 0$ condition.

3. Minimize $c = 2s - 3t + 3u$
   subject to $2s \qquad + \; u \geq 10$ This is a standard minimization problem
   $\qquad\qquad s + 3t - 6u \geq 5$ that does *not* satisfy the nonnegative
   $\qquad\qquad s \geq 0, t \geq 0, u \geq 0$ objective condition.

We saw a way of solving minimization problems in Section 4.4, but a mathematically elegant relationship between maximization and minimization problems gives us another way of solving minimization problems that satisfy the nonnegative objective condition. This relationship is called **duality.**

To describe duality, we must first represent an LP problem by a matrix. This matrix is *not* the first tableau but something simpler: Pretend you forgot all about slack variables and also forgot to change the signs of the objective function.[40] As an example, consider the following two standard[41] problems.

Problem 1

Maximize $p = 20x + 20y + 50z$
subject to $2x + \; y + 3z \leq 2000$
   $\qquad x + 2y + 4z \leq 3000$
   $\qquad x \geq 0, y \geq 0, z \geq 0$

We represent this problem by the matrix

$$\begin{bmatrix} 2 & 1 & 3 & 2000 \\ 1 & 2 & 4 & 3000 \\ 20 & 20 & 50 & 0 \end{bmatrix} \quad \begin{matrix} \text{Constraint 1} \\ \text{Constraint 2} \\ \text{Objective} \end{matrix}$$

Notice that the coefficients of the objective function go in the bottom row and we place a zero in the bottom right corner.

Problem 2 (from Example 3 in Section 4.4)

Minimize $c = 2000s + 3000t$
subject to $2s + \; t \geq 20$
   $\qquad s + 2t \geq 20$
   $\qquad 3s + 4t \geq 50$
   $\qquad s \geq 0, t \geq 0$

---

[40] Forgetting these things is exactly what happens to many students under test conditions!

[41] Although duality does not require the problems to be standard, it does require them to be written in so-called *standard form*: in the case of a maximization problem all constraints need to be (re)written using $\leq$, while for a minimization problem all constraints need to be (re)written using $\geq$. It is least confusing to stick with standard problems, which is what we will do in this section.

Problem 2 is represented by

$$
\begin{bmatrix}
2 & 1 & 20 \\
1 & 2 & 20 \\
3 & 4 & 50 \\
\hline
2000 & 3000 & 0
\end{bmatrix}
\begin{array}{l}
\text{Constraint 1} \\
\text{Constraint 2} \\
\text{Constraint 3} \\
\text{Objective}
\end{array}
$$

These two problems are related: the matrix for Problem 1 is the transpose of the matrix for Problem 2. (Recall that the transpose of a matrix is obtained by writing its rows as columns; see Section 3.1.) When we have a pair of LP problems related in this way, we say that the two are *dual* LP problems.

Dual Linear Programming Problems

Two LP problems, one a maximization and one a minimization problem, are **dual** if the matrix that represents one is the transpose of the matrix that represents the other.

Finding the Dual of a Given Problem
Given an LP problem, we find its dual as follows:

1. Represent the problem as a matrix (see above).

2. Take the transpose of the matrix.

3. Write down the dual, which is the LP problem corresponding to the new matrix. If the original problem was a maximization problem, its dual will be a minimization problem, and vice versa.

The original problem is called the **primal problem,** and its dual is referred to as the **dual problem.**

quick Examples

Primal problem
Minimize $c = s + 2t$
subject to $5s + 2t \geq 60$
$\qquad\quad 3s + 4t \geq 80$
$\qquad\quad\; s + \; t \geq 20$
$\qquad\quad\; s \geq 0, t \geq 0$

$\xrightarrow{\;\mathbf{1}\;}$

$$
\begin{bmatrix}
5 & 2 & 60 \\
3 & 4 & 80 \\
1 & 1 & 20 \\
1 & 2 & 0
\end{bmatrix}
$$

Dual problem
Maximize $p = 60x + 80y + 20z$
subject to $5x + 3y + z \leq 1$
$\qquad\quad 2x + 4y + z \leq 2$
$\qquad\quad\; x \geq 0, y \geq 0, z \geq 0$

$\xrightarrow{\;\mathbf{2}\;}$

$$
\begin{bmatrix}
5 & 3 & 1 & 1 \\
2 & 4 & 1 & 2 \\
60 & 80 & 20 & 0
\end{bmatrix}
$$

$\xrightarrow{\;\mathbf{3}\;}$

The following theorem justifies what we have been doing, and says that solving the dual problem of an LP problem is equivalent to solving the original problem.

Fundamental Theorem of Duality

a. If an LP problem has an optimal solution, then so does its dual. Moreover, the primal problem and the dual problem have the same optimal value for their objective functions.

b. Contained in the final tableau of the simplex method applied to an LP problem is the solution to its dual problem: It is given by the bottom entries in the columns associated with the slack variables, divided by the entry under the objective variable.

The theorem[42] gives us an alternative way of solving minimization problems that satisfy the nonnegative objective condition. Let's illustrate by solving Problem 2 above.

Example 1 Solving by Duality

Minimize $c = 2000s + 3000t$
subject to $2s + \ t \geq 20$
$\qquad\qquad s + 2t \geq 20$
$\qquad\quad 3s + 4t \geq 50$
$\qquad\quad s \geq 0, t \geq 0$

Solution

Step 1 *Find the dual problem.* Write the primal problem in matrix form and take the transpose:

$$
\begin{bmatrix} 2 & 1 & 20 \\ 1 & 2 & 20 \\ 3 & 4 & 50 \\ 2000 & 3000 & 0 \end{bmatrix} \rightarrow \begin{bmatrix} 2 & 1 & 3 & 2000 \\ 1 & 2 & 4 & 3000 \\ 20 & 20 & 50 & 0 \end{bmatrix}
$$

The dual problem is:

Maximize $p = 20x + 20y + 50z$
subject to $2x + \ y + 3z \leq 2000$
$\qquad\qquad x + 2y + 4z \leq 3000$
$\qquad\quad x \geq 0, y \geq 0, z \geq 0$

Step 2 *Use the simplex method to solve the dual problem.* Because we have a standard maximization problem, we do not have to worry about Phase I but go straight to Phase II.

	x	y	z	s	t	p	
s	2	1	[3]	1	0	0	2000
t	1	2	4	0	1	0	3000
p	−20	−20	−50	0	0	1	0

	x	y	z	s	t	p	
z	2	1	3	1	0	0	2000
t	−5	[2]	0	−4	3	0	1000
p	40	−10	0	50	0	3	100,000

	x	y	z	s	t	p	
z	9	0	6	6	−3	0	3000
y	−5	2	0	−4	3	0	1000
p	15	0	0	30	15	3	105,000

Note that the maximum value of the objective function is $p = 105{,}000/3 = 35{,}000$. By the theorem, this is also the optimal value of c in the primal problem!

---

[42] The proof of the theorem is beyond the scope of this book but can be found in a textbook devoted to linear programming, like *Linear Programming* by Vašek Chvátal (San Francisco: W. H. Freeman and Co., 1983,) which has a particularly well-motivated discussion.

Step 3 ***Read off the solution to the primal problem by dividing the bottom entries in the columns associated with the slack variables by the entry in the p column.*** Here is the final tableau again with the entries in question highlighted.

	x	y	z	s	t	p	
z	9	0	6	6	−3	0	3000
y	−5	2	0	−4	3	0	1000
p	15	0	0	**30**	**15**	**3**	**105,000**

The solution to the primal problem is

$$s = 30/3 = 10, t = 15/3 = 5, c = 105{,}000/3 = 35{,}000$$

(Compare this with the method we used to solve Example 3 in the preceding section. Which method seems more efficient?)

+ *Before we go on...* Can you now see the reason for using the variable names $s, t, u, \ldots$ in standard minimization problems? ■

Q: *Is the theorem also useful for solving problems that do not satisfy the nonnegative objective condition* ***?***

A: Consider a standard minimization problem that does not satisfy the nonnegative objective condition, such as

$$\text{Minimize} \quad c = 2s - t$$
$$\text{subject to} \quad 2s + 3t \geq 2$$
$$s + 2t \geq 2$$
$$s \geq 0, t \geq 0$$

Its dual would be

$$\text{Maximize} \quad p = 2x + 2y$$
$$\text{subject to} \quad 2x + \ y \leq 2$$
$$3x + 2y \leq -1$$
$$x \geq 0, y \geq 0$$

This is not a standard maximization problem because the right-hand side of the second constraint is negative. In general, if a problem does not satisfy the nonnegative condition, its dual is not standard. Therefore, to solve the dual by the simplex method will require using Phase I as well as Phase II, and we may as well just solve the primal problem that way to begin with. Thus, duality helps us solve problems only when the primal problem satisfies the nonnegative objective condition. ■

In many economic applications, the solution to the dual problem also gives us useful information about the primal problem, as we will see in the following example.

Example 2 Shadow Costs

You are trying to decide how many vitamin pills to take. SuperV brand vitamin pills each contain 2 milligrams of vitamin X, 1 milligram of vitamin Y, and 1 milligram of vitamin Z. Topper brand vitamin pills each contain 1 milligram of vitamin X, 1 milligram of

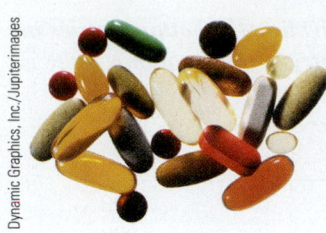

vitamin Y, and 2 milligrams of vitamin Z. You want to take enough pills daily to get at least 12 milligrams of vitamin X, 10 milligrams of vitamin Y, and 12 milligrams of vitamin Z. However, SuperV pills cost 4¢ each and Toppers cost 3¢ each, and you would like to minimize the total cost of your daily dosage. How many of each brand of pill should you take? How would changing your daily vitamin requirements affect your minimum cost?

Solution   This is a straightforward minimization problem. The unknowns are

$$s = \text{number of SuperV brand pills}$$
$$t = \text{number of Topper brand pills}$$

The linear programming problem is

$$
\begin{aligned}
\text{Minimize} \quad & c = 4s + 3t \\
\text{subject to} \quad & 2s + t \geq 12 \\
& s + t \geq 10 \\
& s + 2t \geq 12 \\
& s \geq 0, t \geq 0
\end{aligned}
$$

We solve this problem by using the simplex method on its dual, which is

$$
\begin{aligned}
\text{Maximize} \quad & p = 12x + 10y + 12z \\
\text{subject to} \quad & 2x + y + z \leq 4 \\
& x + y + 2z \leq 3 \\
& x \geq 0, y \geq 0, z \geq 0
\end{aligned}
$$

After pivoting three times, we arrive at the final tableau:

	x	y	z	s	t	p	
x	6	0	−6	6	−6	0	6
y	0	1	3	−1	2	0	2
p	0	0	6	2	8	1	32

Therefore, the answer to the original problem is that you should take 2 SuperV vitamin pills and 8 Toppers at a cost of 32¢ per day.

Now, the key to answering the last question, which asks you to determine how changing your daily vitamin requirements would affect your minimum cost, is to look at the solution to the dual problem. From the tableau we see that $x = 1$, $y = 2$, and $z = 0$. To see what x, y, and z might tell us about the original problem, let's look at their units. In the inequality $2x + y + z \leq 4$, the coefficient 2 of x has units "mg of vitamin X/SuperV pill," and the 4 on the right-hand side has units "¢/SuperV pill." For $2x$ to have the same units as the 4 on the right-hand side, x must have units "¢/mg of vitamin X." Similarly, y must have units "¢/mg of vitamin Y" and z must have units "¢/mg of vitamin Z." One can show (although we will not do it here) that x gives the amount that would be added to the minimum cost for each increase[*] of 1 milligram of vitamin X in our daily requirement. For example, if we were to increase our requirement from 12 milligrams to 14 milligrams, an increase of 2 milligrams, the minimum cost would change by $2x = 2$¢, from 32¢ to 34¢. (Try it; you'll end up taking 4 SuperV pills and 6 Toppers.) Similarly, each increase of 1 milligram of vitamin Y in the requirements would increase the cost by $y = 2$¢. These costs are called the **marginal costs** or the **shadow costs** of the vitamins.

---

[*] To be scrupulously correct, this works only for relatively small changes in the requirements, not necessarily for very large ones.

What about $z = 0$? The shadow cost of vitamin Z is 0¢/mg, meaning that you can increase your requirement of vitamin Z without changing your cost. In fact, the solution $s = 2$ and $t = 8$ provides you with 18 milligrams of vitamin Z, so you can increase the required amount of vitamin Z up to 18 milligrams without changing the solution at all.

We can also interpret the shadow costs as the effective cost to you of each milligram of each vitamin in the optimal solution. You are paying 1¢/milligram of vitamin X, 2¢/milligram of vitamin Y, and getting the vitamin Z for free. This gives a total cost of $1 \times 12 + 2 \times 10 + 0 \times 12 = 32$¢, as we know. Again, if you change your requirements slightly, these are the amounts you will pay per milligram of each vitamin.

Game Theory

We return to a topic we discussed in Section 3.4: solving two-person zero-sum games. In that section, we described how to solve games that could be reduced to 2×2 games or smaller. It turns out that we can solve larger games using linear programming and duality. We summarize the procedure, work through an example, and then discuss why it works.

Solving a Matrix Game

Step 1 Reduce the payoff matrix by dominance.

Step 2 Add a fixed number k to each of the entries so that they all become nonnegative and no column is all zero.

Step 3 Write 1s to the right of and below the matrix, and then write down the associated standard maximization problem. Solve this primal problem using the simplex method.

Step 4 Find the optimal strategies and the expected value as follows:

Column Strategy

1. Express the solution to the primal problem as a column vector.

2. Normalize by dividing each entry of the solution vector by p (which is also the sum of all the entries).

3. Insert zeros in positions corresponding to the columns deleted during reduction.

Row Strategy

1. Express the solution to the dual problem as a row vector.

2. Normalize by dividing each entry by p, which will once again be the sum of all the entries.

3. Insert zeros in positions corresponding to the rows deleted during reduction.

Value of the Game: $e = \dfrac{1}{p} - k$.

Example 3 Restaurant Inspector

You manage two restaurants, Tender Steaks Inn (TSI) and Break for a Steak (BFS). Even though you run the establishments impeccably, the Department of Health has been sending inspectors to your restaurants on a daily basis and fining you for minor infractions.

You've found that you can head off a fine if you're present, but you can cover only one restaurant at a time. The Department of Health, on the other hand, has two inspectors, who sometimes visit the same restaurant and sometimes split up, one to each restaurant. The average fines you have been getting are shown in the following matrix.

Health Inspectors

You go to	Both at BFS	Both at TSI	One at Each
TSI	$8000	0	$2000
BFS	0	$10,000	$4000

How should you choose which restaurant to visit to minimize your expected fine?

Solution This matrix is not quite the payoff matrix because fines, being penalties, should be negative payoffs. Thus, the payoff matrix is the following:

$$P = \begin{bmatrix} -8000 & 0 & -2000 \\ 0 & -10,000 & -4000 \end{bmatrix}$$

We follow the steps above to solve the game using the simplex method.

Step 1 There are no dominated rows or columns, so this game does not reduce.

Step 2 We add $k = 10,000$ to each entry so that none are negative, getting the following new matrix (with no zero column):

$$\begin{bmatrix} 2000 & 10,000 & 8000 \\ 10,000 & 0 & 6000 \end{bmatrix}$$

Step 3 We write 1s to the right and below this matrix:

$$\begin{bmatrix} 2000 & 10,000 & 8000 & 1 \\ 10,000 & 0 & 6000 & 1 \\ 1 & 1 & 1 & 0 \end{bmatrix}$$

The corresponding standard maximization problem is the following.

$$\begin{aligned} \text{Maximize} \quad & p = x + y + z \\ \text{subject to} \quad & 2000x + 10,000y + 8000z \le 1 \\ & 10,000x \qquad\qquad + 6000z \le 1 \\ & x \ge 0, y \ge 0, z \ge 0 \end{aligned}$$

Step 4 We use the simplex method to solve this problem. After pivoting twice, we arrive at the final tableau:

	x	y	z	s	t	p	
y	0	50,000	34,000	5	−1	0	4
x	10,000	0	6000	0	1	0	1
p	0	0	14,000	5	4	50,000	9

Column Strategy The solution to the primal problem is

$$\begin{bmatrix} x \\ y \\ z \end{bmatrix} = \begin{bmatrix} \frac{1}{10,000} \\ \frac{4}{50,000} \\ 0 \end{bmatrix}$$

We divide each entry by $p = 9/50,000$, which is also the sum of the entries. This gives the optimal column strategy:

$$C = \begin{bmatrix} \frac{5}{9} \\ \frac{4}{9} \\ 0 \end{bmatrix}$$

Thus, the inspectors' optimal strategy is to stick together, visiting BFS with probability 5/9 and TSI with probability 4/9.

Row Strategy   The solution to the dual problem is

$$[s \quad t] = \begin{bmatrix} \dfrac{1}{10,000} & \dfrac{4}{50,000} \end{bmatrix}$$

Once again, we divide by $p = 9/50,000$ to find the optimal row strategy:

$$R = \begin{bmatrix} \dfrac{5}{9} & \dfrac{4}{9} \end{bmatrix}$$

Thus, you should visit TSI with probability 5/9 and BFS with probability 4/9.

Value of the Game   Your expected average fine is

$$e = \frac{1}{p} - k = \frac{50,000}{9} - 10,000 = -\frac{40,000}{9} \approx -\$4,444$$

✚*Before we go on...* We owe you an explanation of why the procedure we used in Example 3 works. The main point is to understand how we turn a game into a linear programming problem. It's not hard to see that adding a fixed number k to all the payoffs will change only the payoff, increasing it by k, and not change the optimal strategies. So let's pick up Example 3 from the point where we were considering the following game:

$$P = \begin{bmatrix} 2000 & 10,000 & 8000 \\ 10,000 & 0 & 6000 \end{bmatrix}$$

We are looking for the optimal strategies R and C for the row and column players, respectively; if e is the value of the game, we will have $e = RPC$. Let's concentrate first on the column player's strategy $C = [u \quad v \quad w]^T$, where u, v, and w are the unknowns we want to find. Because e is the value of the game, if the column player uses the optimal strategy C and the row player uses any old strategy S, the expected value with these strategies has to be e or better for the column player, so $SPC \le e$. Let's write that out for two particular choices of S. First, consider $S = [1 \quad 0]$:

$$[1 \quad 0] \begin{bmatrix} 2000 & 10,000 & 8000 \\ 10,000 & 0 & 6000 \end{bmatrix} \begin{bmatrix} u \\ v \\ w \end{bmatrix} \le e$$

Multiplied out, this gives

$$2000u + 10,000v + 8000w \le e$$

Next, do the same thing for $S = [0 \quad 1]$:

$$[0 \quad 1] \begin{bmatrix} 2000 & 10,000 & 8000 \\ 10,000 & 0 & 6000 \end{bmatrix} \begin{bmatrix} u \\ v \\ w \end{bmatrix} \le e$$

$$10,000u + 6000w \le e$$

It turns out that if these two inequalities are true, then $SPC \leq e$ for any S at all, which is what the column player wants. These are starting to look like constraints in a linear programming problem, but the e appearing on the right is in the way. We get around this by dividing by e, which we know to be positive because all of the payoffs are nonnegative and no column is all zero (so the column player can't force the value of the game to be 0; here is where we need these assumptions). We get the following inequalities:

$$2000\left(\frac{u}{e}\right) + 10{,}000\left(\frac{v}{e}\right) + 8000\left(\frac{w}{e}\right) \leq 1$$

$$10{,}000\left(\frac{u}{e}\right) \qquad\qquad + 6000\left(\frac{w}{e}\right) \leq 1$$

Now we're getting somewhere. To make these look even more like linear constraints, we replace our unknowns u, v, and w with new unknowns, $x = u/e$, $y = v/e$, and $z = w/e$. Our inequalities then become:

$$2000x + 10{,}000y + 8000z \leq 1$$

$$10{,}000x \qquad\qquad + 6000z \leq 1$$

What about an objective function? From the point of view of the column player, the objective is to find a strategy that will minimize the expected value e. In order to write e in terms of our new variables x, y, and z, we use the fact that our original variables, being the entries in the column strategy, have to add up to 1: $u + v + w = 1$. Dividing by e gives

$$\frac{u}{e} + \frac{v}{e} + \frac{w}{e} = \frac{1}{e}$$

or

$$x + y + z = \frac{1}{e}$$

Now we notice that, if we *maximize* $p = x + y + z = 1/e$, it will have the effect of minimizing e, which is what we want. So, we get the following linear programming problem:

$$\begin{aligned} \text{Maximize} \quad & p = x + y + z \\ \text{subject to} \quad & 2000x + 10{,}000y + 8000z \leq 1 \\ & 10{,}000x \qquad\qquad + 6000z \leq 1 \\ & x \geq 0,\, y \geq 0,\, z \geq 0 \end{aligned}$$

Why can we say that x, y, and z should all be nonnegative? Because the unknowns u, v, w, and e must all be nonnegative.

So now, if we solve this linear programming problem to find x, y, z, and p, we can find the column player's optimal strategy by computing $u = xe = x/p$, $v = y/p$, and $w = z/p$. Moreover, the value of the game is $e = 1/p$. (If we added k to all the payoffs, we should now adjust by subtracting k again to find the correct value of the game.)

Turning now to the row player's strategy, if we repeat the above type of argument from the row player's viewpoint, we'll end up with the following linear programming problem to solve:

$$\begin{aligned} \text{Minimize} \quad & c = s + t \\ \text{subject to} \quad & 2000s + 10{,}000t \geq 1 \\ & 10{,}000s \qquad\qquad \geq 1 \\ & 8000s + \quad 6000t \geq 1 \\ & s \geq 0,\, t \geq 0 \end{aligned}$$

This is, of course, the dual to the problem we solved to find the column player's strategy, so we know that we can read its solution off of the same final tableau. The optimal value of c will be the same as the value of p, so $c = 1/e$ also. The entries in the optimal row strategy will be s/c and t/c. ∎

FAQs   When to Use Duality

Q: *Given a minimization problem, when should I use duality, and when should I use the two-phase method in Section 4.4?*

A: If the original problem satisfies the nonnegative objective condition (none of the coefficients in the objective function are negative), then you can use duality to convert the problem to a standard maximization one, which can be solved with the one-phase method. If the original problem does not satisfy the nonnegative objective condition, then dualizing results in a non-standard LP problem, so dualizing may not be worthwhile. ∎

Q: *When is it absolutely necessary to use duality?*

A: Never. Duality gives us an efficient but not necessary alternative for solving standard minimization problems. ∎

4.5 EXERCISES

● denotes basic skills exercises

◆ denotes challenging exercises

tech Ex indicates exercises that should be solved using technology

In Exercises 1–8, write down (without solving) the dual LP problem. *hint* [see Quick Example on p. 317]

1. ● Maximize $p = 2x + y$
subject to $x + 2y \leq 6$
$-x + y \leq 2$
$x \geq 0, y \geq 0$

2. ● Maximize $p = x + 5y$
subject to $x + y \leq 6$
$-x + 3y \leq 4$
$x \geq 0, y \geq 0$

3. ● Minimize $c = 2s + t + 3u$
subject to $s + t + u \geq 100$
$2s + t \geq 50$
$s \geq 0, t \geq 0, u \geq 0$

4. ● Minimize $c = 2s + 2t + 3u$
subject to $s + u \geq 100$
$2s + t \geq 50$
$s \geq 0, t \geq 0, u \geq 0$

5. ● Maximize $p = x + y + z + w$
subject to $x + y + z \leq 3$
$y + z + w \leq 4$
$x + z + w \leq 5$
$x + y + w \leq 6$
$x \geq 0, y \geq 0, z \geq 0, w \geq 0$

6. ● Maximize $p = x + y + z + w$
subject to $x + y + z \leq 3$
$y + z + w \leq 3$
$x + z + w \leq 4$
$x + y + w \leq 4$
$x \geq 0, y \geq 0, z \geq 0, w \geq 0$

7. ● Minimize $c = s + 3t + u$
subject to $5s - t + v \geq 1000$
$u - v \geq 2000$
$s + t \geq 500$
$s \geq 0, t \geq 0, u \geq 0, v \geq 0$

8. ● Minimize $c = 5s + 2u + v$
subject to $s - t + 2v \geq 2000$
$u + v \geq 3000$
$s + t \geq 500$
$s \geq 0, t \geq 0, u \geq 0, v \geq 0$

In Exercises 9–22, solve the standard minimization problems using duality. (You may already have seen some of them in earlier sections, but now you will be solving them using a different method.) *hint* [see Example 1]

9. ● Minimize $c = s + t$
subject to $s + 2t \geq 6$
$2s + t \geq 6$
$s \geq 0, t \geq 0$

10. ● Minimize $c = s + 2t$
subject to $s + 3t \geq 30$
$2s + t \geq 30$
$s \geq 0, t \geq 0$

11. ● Minimize $c = 6s + 6t$
    subject to $s + 2t \geq 20$
                  $2s + t \geq 20$
                  $s \geq 0, t \geq 0$

12. ● Minimize $c = 3s + 2t$
    subject to $s + 2t \geq 20$
                  $2s + t \geq 10$
                  $s \geq 0, t \geq 0$

13. ● Minimize $c = 0.2s + 0.3t$
    subject to $0.2s + 0.1t \geq 1$
                  $0.15s + 0.3t \geq 1.5$
                  $10s + 10t \geq 80$
                  $s \geq 0, t \geq 0$

14. ● Minimize $c = 0.4s + 0.1t$
    subject to $30s + 20t \geq 600$
                  $0.1s + 0.4t \geq 4$
                  $0.2s + 0.3t \geq 4.5$
                  $s \geq 0, t \geq 0$

15. ● Minimize $c = 2s + t$
    subject to $3s + t \geq 30$
                  $s + t \geq 20$
                  $s + 3t \geq 30$
                  $s \geq 0, t \geq 0$

16. ● Minimize $c = s + 2t$
    subject to $4s + t \geq 100$
                  $2s + t \geq 80$
                  $s + 3t \geq 150$
                  $s \geq 0, t \geq 0$

17. ● Minimize $c = s + 2t + 3u$
    subject to $3s + 2t + u \geq 60$
                  $2s + t + 3u \geq 60$
                  $s \geq 0, t \geq 0, u \geq 0$

18. ● Minimize $c = s + t + 2u$
    subject to $s + 2t + 2u \geq 60$
                  $2s + t + 3u \geq 60$
                  $s \geq 0, t \geq 0, u \geq 0$

19. ● Minimize $c = 2s + t + 3u$
    subject to $s + t + u \geq 100$
                  $2s + t \geq 50$
                  $t + u \geq 50$
                  $s \geq 0, t \geq 0, u \geq 0$

20. ● Minimize $c = 2s + 2t + 3u$
    subject to $s + u \geq 100$
                  $2s + t \geq 50$
                  $t + u \geq 50$
                  $s \geq 0, t \geq 0, u \geq 0$

21. ● Minimize $c = s + t + u$
    subject to $3s + 2t + u \geq 60$
                  $2s + t + 3u \geq 60$
                  $s + 3t + 2u \geq 60$
                  $s \geq 0, t \geq 0, u \geq 0$

22. ● Minimize $c = s + t + 2u$
    subject to $s + 2t + 2u \geq 60$
                  $2s + t + 3u \geq 60$
                  $s + 3t + 6u \geq 60$
                  $s \geq 0, t \geq 0, u \geq 0$

In Exercises 23–28, solve the games with the given payoff matrices. *hint* [see Example 3]

23. ● $P = \begin{bmatrix} -1 & 1 & 2 \\ 2 & -1 & -2 \end{bmatrix}$

24. ● $P = \begin{bmatrix} 1 & -1 & 2 \\ 1 & 2 & 0 \end{bmatrix}$

25. ● $P = \begin{bmatrix} -1 & 1 & 2 \\ 2 & -1 & -2 \\ 1 & 2 & 0 \end{bmatrix}$

26. ● $P = \begin{bmatrix} 1 & -1 & 2 \\ 1 & 2 & 0 \\ 0 & 1 & 1 \end{bmatrix}$

27. ● $P = \begin{bmatrix} -1 & 1 & 2 & -1 \\ 2 & -1 & -2 & -3 \\ 1 & 2 & 0 & 1 \\ 0 & 2 & 3 & 3 \end{bmatrix}$

28. ● $P = \begin{bmatrix} 1 & -1 & 2 & 0 \\ 1 & 2 & 0 & 1 \\ 0 & 1 & 1 & 0 \\ 2 & 0 & -2 & 2 \end{bmatrix}$

Applications

The following applications are similar to ones in preceding exercise sets. Use duality to answer them. *hint* [see Example 2]

29. ● **Resource Allocation** Meow makes cat food out of fish and cornmeal. Fish has 8 grams of protein and 4 grams of fat per ounce, and cornmeal has 4 grams of protein and 8 grams of fat. A jumbo can of cat food must contain at least 48 grams of protein and 48 grams of fat. If fish and cornmeal both cost 5¢/ounce, how many ounces of each should Meow use in each can of cat food to minimize costs? What are the shadow costs of protein and of fat?

30. ● **Resource Allocation** Oz makes lion food out of giraffe and gazelle meat. Giraffe meat has 18 grams of protein and 36 grams of fat per pound, while gazelle meat has 36 grams of protein and 18 grams of fat per pound. A batch of lion food must contain at least 36,000 grams of protein and 54,000 grams of fat. Giraffe meat costs $2/pound and gazelle meat costs $4/pound. How many pounds of each should go into each batch of lion food in order to minimize costs? What are the shadow costs of protein and fat?

31. ● **Nutrition** Ruff makes dog food out of chicken and grain. Chicken has 10 grams of protein and 5 grams of fat/ounce, and grain has 2 grams of protein and 2 grams of fat/ounce. A bag of dog food must contain at least 200 grams of protein and at least 150 grams of fat. If chicken costs 10¢/ounce and grain costs 1¢/ounce, how many ounces of each should Ruff use in each bag of dog food in order to minimize cost? What are the shadow costs of protein and fat?

32. ● **Purchasing** The Enormous State University's Business School is buying computers. The school has two models to choose from, the Pomegranate and the iZac. Each Pomegranate

● basic skills   ◆ challenging   **tech**Ex technology exercise

comes with 400 MB of memory and 80 GB of disk space, while each iZac has 300 MB of memory and 100 GB of disk space. For reasons related to its accreditation, the school would like to be able to say that it has a total of at least 48,000 MB of memory and at least 12,800 GB of disk space. If both the Pomegranate and the iZac cost $2000 each, how many of each should the school buy to keep the cost as low as possible? What are the shadow costs of memory and disk space?

33. ● *Nutrition* Each serving of Gerber Mixed Cereal for Baby contains 60 calories and no Vitamin C. Each serving of Gerber Mango Tropical Fruit Dessert contains 80 calories and 45 percent of the U.S. Recommended Daily Allowance (RDA) of Vitamin C for infants. Each serving of Gerber Apple Banana Juice contains 60 calories and 120 percent of the U.S. RDA of Vitamin C for infants.[43] The cereal costs 10¢/serving, the dessert costs 53¢/serving, and the juice costs 27¢/serving. If you want to provide your child with at least 120 calories and at least 120 percent of the U.S. RDA of Vitamin C, how can you do so at the least cost? What are your shadow costs for calories and Vitamin C?

34. ● *Nutrition* Each serving of Gerber Mixed Cereal for Baby contains 60 calories, no Vitamin C, and 11 grams of carbohydrates. Each serving of Gerber Mango Tropical Fruit Dessert contains 80 calories, 45 percent of the U.S. Recommended Daily Allowance (RDA) of Vitamin C for infants, and 21 grams of carbohydrates. Each serving of Gerber Apple Banana Juice contains 60 calories, 120 percent of the U.S. RDA of Vitamin C for infants, and 15 grams of carbohydrates.[44] Assume that the cereal costs 11¢/serving, the dessert costs 50¢/serving, and the juice costs 30¢/serving. If you want to provide your child with at least 180 calories, at least 120 percent of the U.S. RDA of Vitamin C, and at least 37 grams of carbohydrates, how can you do so at the least cost? What are your shadow costs for calories, Vitamin C, and carbohydrates?

35. ● *Politics* The political pollster Canter is preparing for a national election. It would like to poll at least 1500 Democrats and 1500 Republicans. Each mailing to the East Coast gets responses from 100 Democrats and 50 Republicans. Each mailing to the Midwest gets responses from 100 Democrats and 100 Republicans. And each mailing to the West Coast gets responses from 50 Democrats and 100 Republicans. Mailings to the East Coast cost $40 each to produce and mail, mailings to the Midwest cost $60 each, and mailings to the West Coast cost $50 each. How many mailings should Canter send to each area of the country to get the responses it needs at the least possible cost? What will it cost? What are the shadow costs of a Democratic response and a Republican response?

36. ● *Purchasing* Bingo's Copy Center needs to buy white paper and yellow paper. Bingo's can buy from three suppliers. Harvard Paper sells a package of 20 reams of white and 10 reams of yellow for $60; Yale Paper sells a package of 10 reams of white and 10 reams of yellow for $40, and Dartmouth Paper sells a package of 10 reams of white and 20 reams of yellow

for $50. If Bingo's needs 350 reams of white and 400 reams of yellow, how many packages should it buy from each supplier so as to minimize the cost? What is the lowest possible cost? What are the shadow costs of white paper and yellow paper?

37. *Resource Allocation* One day Gillian the Magician summoned the wisest of her women. "Devoted followers," she began, "I have a quandary: As you well know, I possess great expertise in sleep spells and shock spells, but unfortunately, these are proving to be a drain on my aural energy resources; each sleep spell costs me 500 pico-shirleys of aural energy, while each shock spell requires 750 pico-shirleys. Clearly, I would like to hold my overall expenditure of aural energy to a minimum, and still meet my commitments in protecting the Sisterhood from the ever-present threat of trolls. Specifically, I have estimated that each sleep spell keeps us safe for an average of 2 minutes, while every shock spell protects us for about 3 minutes. We certainly require enough protection to last 24 hours of each day, and possibly more, just to be safe. At the same time, I have noticed that each of my sleep spells can immobilize 3 trolls at once, while one of my typical shock spells (having a narrower range) can immobilize only 2 trolls at once. We are faced, my sisters, with an onslaught of 1200 trolls per day! Finally, as you are no doubt aware, the bylaws dictate that for a Magician of the Order to remain in good standing, the number of shock spells must be between one-quarter and one-third the number of shock and sleep spells combined. What do I do, oh Wise Ones?"

38. *Risk Management* The Grand Vizier of the Kingdom of Um is being blackmailed by numerous individuals and is having a very difficult time keeping his blackmailers from going public. He has been keeping them at bay with two kinds of payoff: gold bars from the Royal Treasury and political favors. Through bitter experience, he has learned that each payoff in gold gives him peace for an average of about one month, and each political favor seems to earn him about a month and a half of reprieve. To maintain his flawless reputation in the court, he feels he cannot afford any revelations about his tainted past to come to light within the next year. Thus, it is imperative that his blackmailers be kept at bay for 12 months. Furthermore, he would like to keep the number of gold payoffs at no more than one-quarter of the combined number of payoffs because the outward flow of gold bars might arouse suspicion on the part of the Royal Treasurer. The gold payoffs tend to deplete the Grand Vizier's travel budget. (The treasury has been subsidizing his numerous trips to the Himalayas.) He estimates that each gold bar removed from the treasury will cost him four trips. On the other hand, because the administering of political favors tends to cost him valuable travel time, he suspects that each political favor will cost him about two trips. Now, he would obviously like to keep his blackmailers silenced and lose as few trips as possible. What is he to do? How many trips will he lose in the next year?

39. *Game Theory—Politics* Incumbent Tax N. Spend and challenger Trick L. Down are running for county executive, and polls show them to be in a dead heat. The election hinges on three cities: Littleville, Metropolis, and Urbantown. The candidates have decided to spend the last weeks before the election campaigning in

---

[43] SOURCE: Nutrition information provided by Gerber.

[44] Ibid.

● basic skills   ◆ challenging   **tech**Ex technology exercise

those three cities; each day each candidate will decide in which city to spend the day. Pollsters have determined the following payoff matrix, where the payoff represents the number of votes gained or lost for each one-day campaign trip.

T. N. Spend

		Littleville	Metropolis	Urbantown
	Littleville	−200	−300	300
T. L. Down	Metropolis	−500	500	−100
	Urbantown	−500	0	0

What percentage of time should each candidate spend in each city in order to maximize votes gained? If both candidates use their optimal strategies, what is the expected vote?

40. Game Theory—Marketing Your company's new portable phone/music player/PDA/bottle washer, the RunMan, will compete against the established market leader, the iNod, in a saturated market. (Thus, for each device you sell, one fewer iNod is sold.) You are planning to launch the RunMan with a traveling road show, concentrating on two cities, New York and Boston. The makers of the iNod will do the same to try to maintain their sales. If, on a given day, you both go to New York, you will lose 1000 units in sales to the iNod. If you both go to Boston, you will lose 750 units in sales. On the other hand, if you go to New York and your competitor to Boston, you will gain 1500 units in sales from them. If you go to Boston and they to New York, you will gain 500 units in sales. What percentage of time should you spend in New York and what percentage in Boston, and how do you expect your sales to be affected?

41. Game Theory—Morra Games A three-finger Morra game is a game in which two players simultaneously show one, two, or three fingers at each round. The outcome depends on a predetermined set of rules. Here is an interesting example: If the number of fingers shown by A and B differ by 1, then A loses one point. If they differ by more than 1, the round is a draw. If they show the same number of fingers, A wins an amount equal to the sum of the fingers shown. Determine the optimal strategy for each player and the expected value of the game.

42. Game Theory—Morra Games Referring to the preceding exercise, consider the following rules for a three-finger Morra game: If the sum of the fingers shown is odd, then A wins an amount equal to that sum. If the sum is even, B wins the sum. Determine the optimal strategy for each player and the expected value of the game.

43. ♦ tech Ex Game Theory—Military Strategy Colonel Blotto is a well-known game in military strategy.[45] Here is a version of this game: Colonel Blotto has four regiments under his command, while his opponent, Captain Kije, has three. The armies are to try to occupy two locations, and each commander must decide how many regiments to send to each location.

The army that sends more regiments to a location captures that location as well as the other army's regiments. If both armies send the same number of regiments to a location, then there is a draw. The payoffs are one point for each location captured and one point for each regiment captured. Find the optimum strategy for each commander and also the value of the game.

44. ♦ tech Ex Game Theory—Military Strategy Referring to the preceding exercise, consider the version of Colonel Blotto with the same payoffs given there except that Captain Kije earns two points for each location captured, while Colonel Blotto continues to earn only one point. Find the optimum strategy for each commander and also the value of the game.

Communication and Reasoning Exercises

45. ● Give one possible advantage of using duality to solve a standard minimization problem.

46. ● Multiple choice: To ensure that the dual of a minimization problem will result in a standard maximization problem,

(A) The primal problem should satisfy the nonnegative objective condition.

(B) The primal problem should be a standard minimization problem.

(C) The primal problem should not satisfy the nonnegative objective condition.

47. ● Give an example of a standard minimization problem whose dual is *not* a standard maximization problem. How would you go about solving your problem?

48. ● Give an example of a nonstandard minimization problem whose dual is a standard maximization problem.

49. Solve the following nonstandard minimization problem using duality. Recall from a footnote in the text that to find the dual you must first rewrite all of the constraints using "≥." The Miami Beach City Council has offered to subsidize hotel development in Miami Beach, and is hoping for at least two hotels with a total capacity of at least 1400. Suppose that you are a developer interested in taking advantage of this offer by building a small group of hotels in Miami Beach. You are thinking of three prototypes: a convention-style hotel with 500 rooms costing $100 million, a vacation-style hotel with 200 rooms costing $20 million, and a small motel with 50 rooms costing $4 million. The city council will approve your plans provided you build at least one convention-style hotel and no more than two small motels. How many of each type of hotel should you build to satisfy the city council's wishes and stipulations while minimizing your total cost?

50. Given a minimization problem, when would you solve it by applying the simplex method to its dual, and when would you apply the simplex method to the minimization problem itself?

51. Create an interesting application that leads to a standard maximization problem. Solve it using the simplex method and note the solution to its dual problem. What does the solution to the dual tell you about your application?

---

[45] See Samuel Karlin, Mathematical Methods and Theory in Games, Programming and Economics (Addison-Wesley, 1959).

● basic skills   ♦ challenging   tech Ex technology exercise

Chapter 4 Review

KEY CONCEPTS

4.1 Graphing Linear Inequalities

Inequalities, strict and nonstrict *p. 259*
Linear inequalities *p. 259*
Solution of an inequality *p. 259*
Sketching the region represented by a linear inequality in two variables *p. 261*
Bounded and unbounded regions *p. 264*
Feasible region *p. 264*

4.2 Solving Linear Programming Problems Graphically

Linear programming (LP) problem; objective function; constraints; optimal value; optimal solution *p. 269*
Feasible region *p. 269*
Fundamental Theorem of Linear Programming *p. 269*
Graphical method for solving an LP problem *p. 271*

Decision variables *p. 271*
Procedure for solving an LP problem with an unbounded feasible region *p. 275*

4.3 The Simplex Method: Solving Standard Maximization Problems

Standard maximization problem *p. 285*
Slack variable *p. 287*
Tableau *p. 287*
Active (or basic) variables; inactive (or nonbasic) variables; basic solution *p. 288*
Rules for selecting the pivot column *p. 288*
Rules for selecting the pivot; test ratios *p. 289*
Departing or exiting variable, entering variable *p. 290*

4.4 The Simplex Method: Solving General Linear Programming Problems

Surplus variable *p. 302*
Phase I and Phase II for solving general LP problems *p. 303*
Using the simplex method to solve a minimization problem *p. 307*

4.5 The Simplex Method and Duality

Standard minimization problem *p. 315*
Standard LP problem *p. 315*
Nonnegative objective condition *p. 315*
Dual LP problems; primal problem; dual problem *p. 317*
Fundamental Theorem of Duality *p. 317*
Shadow costs *p. 320*
LP problem associated to a two-person zero-sum game *p. 321*

REVIEW EXERCISES

In each of Exercises 1–4, sketch the region corresponding to the given inequalities, say whether it is bounded, and give the coordinates of all corner points.

1. $2x - 3y \leq 12$

2. $x \leq 2y$

3. $x + 2y \leq 20$
$3x + 2y \leq 30$
$x \geq 0, y \geq 0$

4. $3x + 2y \geq 6$
$2x - 3y \leq 6$
$3x - 2y \geq 0$
$x \geq 0, y \geq 0$

In each of Exercises 5–8, solve the given linear programming problem graphically.

5. Maximize $p = 2x + y$
subject to $3x + y \leq 30$
$x + y \leq 12$
$x + 3y \leq 30$
$x \geq 0, y \geq 0$

6. Maximize $p = 2x + 3y$
subject to $x + y \geq 10$
$2x + y \geq 12$
$x + y \leq 20$
$x \geq 0, y \geq 0$

7. Minimize $c = 2x + y$
subject to $3x + y \geq 30$
$x + 2y \geq 20$
$2x - y \geq 0$
$x \geq 0, y \geq 0$

8. Minimize $c = 3x + y$
subject to $3x + 2y \geq 6$
$2x - 3y \leq 0$
$3x - 2y \geq 0$
$x \geq 0, y \geq 0$

In each of Exercises 9–16, solve the given linear programming problem using the simplex method. If no optimal solution exists, indicate whether the feasible region is empty or the objective function is unbounded.

9. Maximize $p = x + y + 2z$
subject to $x + 2y + 2z \leq 60$
$2x + y + 3z \leq 60$
$x \geq 0, y \geq 0, z \geq 0$

10. Maximize $p = x + y + 2z$
subject to $x + 2y + 2z \leq 60$
$2x + y + 3z \leq 60$
$x + 3y + 6z \leq 60$
$x \geq 0, y \geq 0, z \geq 0$

11. Maximize $p = x + y + 3z$
subject to $x + y + z \geq 100$
$y + z \leq 80$
$x \quad + z \leq 80$
$x \geq 0, y \geq 0, z \geq 0$

12. Maximize $p = 2x + y$
subject to $x + 2y \geq 12$
$2x + y \leq 12$
$x + y \leq 5$
$x \geq 0, y \geq 0$

13. Minimize $c = x + 2y + 3z$
subject to $3x + 2y + z \geq 60$
$2x + y + 3z \geq 60$
$x \geq 0, y \geq 0, z \geq 0$

14. `tech` Ex Minimize $c = x + y - z$
subject to $3x + 2y + z \geq 60$
$2x + y + 3z \geq 60$
$x + 3y + 2z \geq 60$
$x \geq 0, y \geq 0, z \geq 0$

15. Minimize $c = x + y + z + w$
subject to $x + y \qquad\quad \geq 30$
$x \quad + z \qquad \geq 20$
$x + y \quad - w \leq 10$
$y + z - w \leq 10$
$x \geq 0, y \geq 0, z \geq 0, w \geq 0$

16. Minimize $c = 4x + y + z + w$
subject to $x + y \qquad\quad \geq 30$
$y - z \qquad \leq 20$
$z - w \leq 10$
$x \geq 0, y \geq 0, z \geq 0, w \geq 0$

In each of Exercises 17–20, solve the given linear programming problem using duality.

17. Minimize $c = 2x + y$
subject to $3x + 2y \geq 60$
$2x + y \geq 60$
$x + 3y \geq 60$
$x \geq 0, y \geq 0$

18. Minimize $c = 2x + y + 2z$
subject to $3x + 2y + z \geq 100$
$2x + y + 3z \geq 200$
$x \geq 0, y \geq 0, z \geq 0$

19. Minimize $c = 2x + y$
subject to $3x + 2y \geq 10$
$2x - y \leq 30$
$x + 3y \geq 60$
$x \geq 0, y \geq 0$

20. Minimize $c = 2x + y + 2z$
subject to $3x - 2y + z \geq 100$
$2x + y - 3z \leq 200$
$x \geq 0, y \geq 0, z \geq 0$

In each of Exercises 21–24, solve the game with the given payoff matrix.

21. $P = \begin{bmatrix} -1 & 2 & -1 \\ 1 & -2 & 1 \\ 3 & -1 & 0 \end{bmatrix}$ **22.** $P = \begin{bmatrix} -3 & 0 & 1 \\ -4 & 0 & 0 \\ 0 & -1 & -2 \end{bmatrix}$

23. $P = \begin{bmatrix} -3 & -2 & 3 \\ 1 & 0 & 0 \\ -2 & 2 & 1 \end{bmatrix}$ **24.** $P = \begin{bmatrix} -4 & -2 & -3 \\ 1 & -3 & -2 \\ -3 & 1 & -4 \end{bmatrix}$

Exercises 25–27 are adapted from the Actuarial Exam on Operations Research.

25. You are given the following linear programming problem:

Minimize $c = x + 2y$
subject to $-2x + y \geq 1$
$x - 2y \geq 1$
$x \geq 0, y \geq 0$

Which of the following is true?
(A) The problem has no feasible solutions.
(B) The objective function is unbounded.
(C) The problem has optimal solutions.

26. Repeat the preceding exercise with the following linear programming problem:

Maximize $p = x + y$
subject to $-2x + y \leq 1$
$x - 2y \leq 2$
$x \geq 0, y \geq 0$

27. Determine the optimal value of the objective function.

You are given the following linear programming problem.

Maximize $Z = x_1 + 4x_2 + 2x_3 - 10$
subject to $4x_1 + x_2 + x_3 \leq 45$
$-x_1 + x_2 + 2x_3 \leq 0$
$x_1, x_2, x_3 \geq 0$

Applications

Purchases *In Exercises 28–31, you are the buyer for OHagan-Books.com and are considering increasing stocks of romance and horror novels at the new OHaganBooks.com warehouse in Texas. You have offers from two publishers: Duffin House and Higgins Press. Duffin offers a package of 5 horror novels and 5 romance novels for $50, and Higgins offers a package of 5 horror and 10 romance novels for $80.*

28. How many packages should you purchase from each publisher to obtain at least 4000 horror novels and 6000 romance novels at minimum cost? What is the minimum cost?

29. As it turns out, John O'Hagan promised Marjory Duffin that OHaganBooks.com would buy at least 20 percent more packages from Duffin as from Higgins, but you still want to obtain at least 4000 horror novels and 6000 romance novels at minimum cost. *Without solving the problem,* say which of the following statements are possible:

(A) The cost will stay the same.
(B) The cost will increase.
(C) The cost will decrease.
(D) It will be impossible to meet all the conditions.
(E) The cost will become unbounded.

30. If you wish to meet all the requirements in the preceding exercise at minimum cost, how many packages should you purchase from each publisher? What is the minimum cost?

31. You are just about to place your orders when your sales manager reminds you that, in addition to all your commitments above, you had also assured Sean Higgins that you would spend at least as much on purchases from Higgins as from Duffin. *Now* what should you do?

32. *Profit* Duffin Press, which has become the largest publisher of books sold at the OHaganBooks.com site, prints three kinds of books: paperback, quality paperback, and hardcover. The amounts of paper, ink, and time on the presses required for each kind of book are given in this table:

	Paperback	Quality Paperback	Hardcover	Total Available
Paper (pounds)	3	2	1	6000
Ink (gallons)	2	1	3	6000
Time (minutes)	10	10	10	22,000
Profit	$1	$2	$3	

The table also lists the total amounts of paper, ink, and time available in a given day and the profits made on each kind of book. How many of each kind of book should Duffin print to maximize profit?

33. *Purchases* You are just about to place book orders from Duffin and Higgins when Ewing Books enters the fray and offers its own package of horror and romance novels. To complicate things further, Duffin and Higgins both change their offers, you decide to renege on the commitments you had made to them, and your sales manager has changed her mind about the number of books OHaganBooks.com will require. Taking all of this (and other factors too complicated to explain) into account, you arrive at the following LP problem:

Minimize $c = 50x + 150y + 100z$
subject to $5x + 10y + 5z \geq 4000$
$2x + 10y + 5z \geq 6000$
$x - y + z \leq 0$
$x \geq 0, y \geq 0, z \geq 0$

What is the solution to this linear programming problem?

Degree Requirements During his lunch break, John OHagan decides to devote some time to assisting his son Billy Sean, who continues to have a terrible time planning his college course

schedule. *The latest Bulletin of Suburban State University claims to have added new flexibility to its course requirements, but it remains as complicated as ever. It reads as follows:*

> All candidates for the degree of Bachelor of Arts at SSU must take at least 120 credits from the Sciences, Fine Arts, Liberal Arts and Mathematics combined, including at least as many Science credits as Fine Arts credits, and at most twice as many Mathematics credits as Science credits, but with Liberal Arts credits exceeding Mathematics credits by no more than one-third of the number of Fine Arts credits.

Science and fine arts credits cost $300 each, and liberal arts and mathematics credits cost $200 each. John would like to have Billy Sean meet all the requirements at a minimum total cost.

34. Set up (without solving) the associated linear programming problem.

35. tech Ex How many of each type of credit should Billy Sean take? What will the total cost be?

36. *Shipping* On the same day that the sales department at Duffin House received an order for 600 packages from the OHaganBooks.com Texas headquarters, it received an additional order for 200 packages from FantasyBooks.com, based in California. Duffin House has warehouses in New York and Illinois. The New York warehouse has 600 packages in stock, but the Illinois warehouse is closing down and has only 300 packages in stock. Shipping costs per package of books are as follows: New York to Texas: $20; New York to California: $50; Illinois to Texas: $30; Illinois to California: $40. What is the lowest total shipping cost for which Duffin House can fill the orders? How many packages should be sent from each warehouse to each online bookstore at a minimum shipping cost?

37. *Marketing* Marjory Duffin, head of Duffin House, reveals to John O'Hagan that FantasyBooks.com is considering several promotional schemes: It may offer two books for the price of one, three books for the price of two, or possibly a free copy of *Brain Surgery for Klutzes* with each order. O'Hagan's marketing advisers Floody and O'Lara seem to have different ideas as to how to respond. Floody suggests offering *three* books for the price of one, while O'Lara suggests instead offering a free copy of the *Finite Mathematics Student Solutions Manual* with every purchase. After a careful analysis, O'Hagan comes up with the following payoff matrix, where the payoffs represent the number of customers, in thousands, O'Hagan expects to gain from FantasyBooks.

		FantasyBooks			
		No Promo	2 for Price of 1	3 for Price of 2	Brain Surgery
	No Promo	0	−60	−40	10
O'Hagan	3 for Price of 1	30	20	10	15
	Finite Math	20	0	15	10

Find the optimal strategies for both companies and the expected shift in customers.

CASE STUDY   Airline Scheduling

Austin Brown/Getty Images

You are the traffic manager for Fly-by-Night Airlines, which flies airplanes from five cities: Los Angeles, Chicago, Atlanta, New York, and Boston. Due to a strike, you have not had sufficient crews to fly your planes, so some of your planes have been stranded in Chicago and in Atlanta. In fact, 15 extra planes are in Chicago and 30 extra planes are in Atlanta. Los Angeles needs at least 10 planes to get back on schedule, New York needs at least 20, and Boston needs at least 10. It costs $50,000 to fly a plane from Chicago to Los Angeles, $10,000 to fly from Chicago to New York, and $20,000 to fly from Chicago to Boston. It costs $70,000 to fly a plane from Atlanta to Los Angeles, $20,000 to fly from Atlanta to New York, and $50,000 to fly from Atlanta to Boston. How should you rearrange your planes to get back on schedule at the lowest cost?

As always, you remember to start by identifying the unknowns. You need to decide how many planes you will fly from each of Chicago and Atlanta to each of Los Angeles, New York, and Boston. This gives you six unknowns:

$$x = \text{number of planes to fly from Chicago to LA}$$
$$y = \text{number of planes to fly from Chicago to NY}$$
$$z = \text{number of planes to fly from Chicago to Boston}$$
$$u = \text{number of planes to fly from Atlanta to LA}$$
$$v = \text{number of planes to fly from Atlanta to NY}$$
$$w = \text{number of planes to fly from Atlanta to Boston}$$

Los Angeles needs at least 10 planes, so you know that

$$x + u \geq 10$$

Similarly, New York and Boston give you the inequalities

$$y + v \geq 20$$
$$z + w \geq 10$$

Because Chicago has only 15 extra planes, you have

$$x + y + z \leq 15$$

Because Atlanta has 30 extra planes,

$$u + v + w \leq 30$$

Finally, you want to minimize the cost, which is given by

$$C = 50{,}000x + 10{,}000y + 20{,}000z + 70{,}000u + 20{,}000v + 50{,}000w$$

This gives you your LP problem:

Minimize $C = 50{,}000x + 10{,}000y + 20{,}000z + 70{,}000u + 20{,}000v + 50{,}000w$

subject to

$$
\begin{aligned}
x \qquad\quad + u \qquad\qquad\quad &\geq 10\\
y \qquad\qquad + v \qquad\quad &\geq 20\\
z \qquad\qquad\quad + w &\geq 10\\
x + y + z \qquad\qquad\qquad\quad &\leq 15\\
u + v + w &\leq 30\\
\end{aligned}
$$

$x \geq 0,\, y \geq 0,\, z \geq 0,\, u \geq 0,\, v \geq 0,\, w \geq 0$

Because you have six unknowns, you know that you will have to use the simplex method, so you start converting into the proper form. You change the minimization problem into the problem of maximizing $P = -C$. At the same time, you notice that it would actually be enough to maximize

$$P = -5x - y - 2z - 7u - 2v - 5w$$

(this will keep the numbers small). You subtract the surplus variables p, q, and r from the left-hand sides of the first three inequalities, and you add the slack variables s and t to the last two. Then it's time to go to work. Remembering all the rules for pivoting, and using a computer to help with the calculations, you get the following sequence of tableaux:

	x	y	z	u	v	w	p	q	r	s	t	P	
p	1	0	0	1	0	0	−1	0	0	0	0	0	10
q	0	1	0	0	1	0	0	−1	0	0	0	0	20
r	0	0	1	0	0	1	0	0	−1	0	0	0	10
s	1	1	1	0	0	0	0	0	0	1	0	0	15
t	0	0	0	1	1	1	0	0	0	0	1	0	30
P	5	1	2	7	2	5	0	0	0	0	0	1	0

	x	y	z	u	v	w	p	q	r	s	t	P	
x	1	0	0	1	0	0	−1	0	0	0	0	0	10
q	0	1	0	0	1	0	0	−1	0	0	0	0	20
r	0	0	1	0	0	1	0	0	−1	0	0	0	10
s	0	1	1	−1	0	0	1	0	0	1	0	0	5
t	0	0	0	1	1	1	0	0	0	0	1	0	30
P	0	1	2	2	2	5	5	0	0	0	0	1	−50

	x	y	z	u	v	w	p	q	r	s	t	P	
x	1	0	0	1	0	0	−1	0	0	0	0	0	10
v	0	1	0	0	1	0	0	−1	0	0	0	0	20
r	0	0	1	0	0	1	0	0	−1	0	0	0	10
s	0	1	1	−1	0	0	1	0	0	1	0	0	5
t	0	−1	0	1	0	1	0	1	0	0	1	0	10
P	0	−1	2	2	0	5	5	2	0	0	0	1	−90

	x	y	z	u	v	w	p	q	r	s	t	P	
x	1	0	0	1	0	0	−1	0	0	0	0	0	10
v	0	1	0	0	1	0	0	−1	0	0	0	0	20
w	0	0	1	0	0	1	0	0	−1	0	0	0	10
s	0	1	$\boxed{1}$	−1	0	0	1	0	0	1	0	0	5
t	0	−1	−1	1	0	0	0	1	1	0	1	0	0
P	0	−1	−3	2	0	0	5	2	5	0	0	1	−140

	x	y	z	u	v	w	p	q	r	s	t	P	
x	1	0	0	1	0	0	−1	0	0	0	0	0	10
v	0	1	0	0	1	0	0	−1	0	0	0	0	20
w	0	−1	0	$\boxed{1}$	0	1	−1	0	−1	−1	0	0	5
z	0	1	1	−1	0	0	1	0	0	1	0	0	5
t	0	0	0	0	0	0	1	1	1	1	1	0	5
P	0	2	0	−1	0	0	8	2	5	3	0	1	−125

	x	y	z	u	v	w	p	q	r	s	t	P	
x	1	1	0	0	0	−1	0	0	1	1	0	0	5
v	0	1	0	0	1	0	0	−1	0	0	0	0	20
u	0	−1	0	1	0	1	−1	0	−1	−1	0	0	5
z	0	0	1	0	0	1	0	0	−1	0	0	0	10
t	0	0	0	0	0	0	1	1	1	1	1	0	5
P	0	1	0	0	0	1	7	2	4	2	0	1	−120

Now you see what to do: You should fly 5 planes from Chicago to LA, none from Chicago to NY, and 10 from Chicago to Boston. You should fly 5 planes from Atlanta to LA, 20 from Atlanta to NY, and none from Atlanta to Boston. The total cost will be $1,200,000 to get the airline back on schedule.

Exercises

1. Your boss calls you on the carpet for not flying any planes from Chicago to New York, which is the cheapest run. Come up with a convincing reason for having made that choice. Your job may be on the line.

2. Your boss *insists* that you fly at least 5 planes from Chicago to New York. What is your best option now?

3. If LA needed 15 planes rather than 10, what would be your best option?

4. If NY needed 25 planes rather than 20, what would be your best option?

5. If Boston needed 15 planes rather than 10, what would be your best option?

6. If there were only 10 planes in Chicago, what would be your best option?

7. If there were 35 planes in Atlanta, what would be your best option?

8. Suppose that the Boston airport was socked in by snow, so you only needed to send planes to LA and NY. What would be your best option?

9. Suppose that the NY airport was socked in by snow, so you only needed to send planes to LA and Boston. What would be your best option?

10. A higher cost per flight from Atlanta to NY would change the answer reached in the text. By how much would the cost have to rise to change the answer?

11. The linear programming problem here satisfies the nonnegative objective condition. Solve it using duality. What do the values of the dual variables tell you about the situation?

Section 4.1

Some calculators, including the TI-83/84, will shade one side of a graph, but you need to tell the calculator which side to shade. For instance, to obtain the solution set of $2x + 3y \leq 6$ shown in Figure 4:

1. Solve the corresponding equation $2x + 3y = 6$ for y and use the following input:

2. The icon to the left of "Y_1" tells the calculator to shade above the line. You can cycle through the various shading options by positioning the cursor to the left of Y_1 and pressing ENTER until you see the one you want. Here's what the graph will look like:

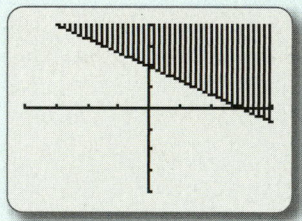

Section 4.3

Example 3 The Acme Baby Foods example in the text leads to the following linear programming problem:

$$\text{Maximize} \quad p = 10x + 7y$$
$$\text{subject to} \quad x \qquad\qquad \leq 600$$
$$2x + 3y \leq 3600$$
$$5x + 3y \leq 4500$$
$$x \geq 0, \; y \geq 0$$

Solve it using technology.

Solution with Technology When we introduce slack variables, we get the following system of equations:

$$
\begin{aligned}
x \quad\quad\; + s \qquad\qquad\quad\; &= 600 \\
2x + 3y \quad\;\; + t \qquad\quad\; &= 3600 \\
5x + 3y \qquad\quad + u \quad\; &= 4500 \\
-10x - 7y \qquad\qquad + p &= 0
\end{aligned}
$$

We use the PIVOT program for the TI-83/84 to help with the simplex method. This program is available at

Chapter 4 → Tools

Because the calculator handles decimals as easily as integers, there is no need to avoid them, except perhaps to save limited screen space. If we don't need to avoid decimals, we can use the traditional Gauss-Jordan method (see the discussion at the end of Section 2.2): After selecting your pivot, and prior to clearing the pivot column, *divide the pivot row by the value of the pivot, thereby turning the pivot into a 1*.

The main drawback to using the TI-83/84 is that we can't label the rows and columns. We can mentally label them as we go, but we can do without them entirely if we wish. We begin by entering the initial tableau as the matrix [A]. (Another drawback to using the TI-83/84 is that it can't show the whole tableau at once. Here and below we show tableaux across several screens. Use the TI-83/84's arrow keys to scroll a matrix left and right so you can see all of it.)

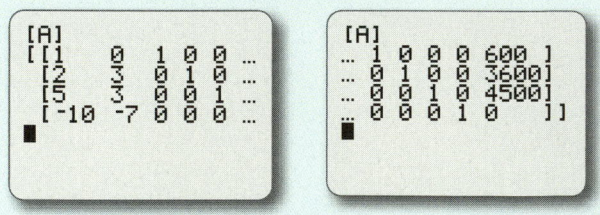

The following is the sequence of tableaux we get while using the simplex method with the help of the PIVOT program.

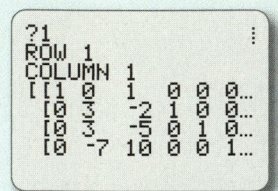
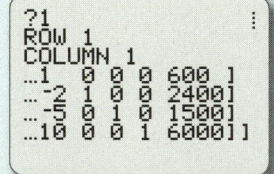

After determining that the next pivot is in the third row and second column, we divide the third row by the pivot, 3, and then pivot:

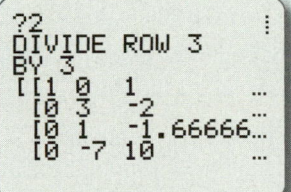

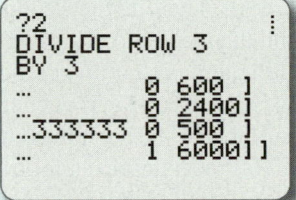

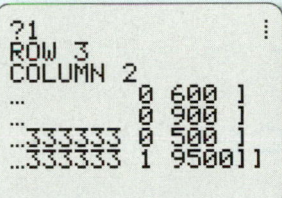

The next pivot is the 3 in the second row, third column. We divide its row by 3 and pivot:

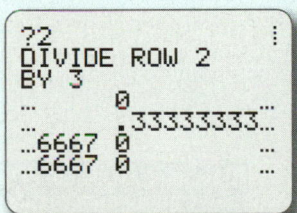

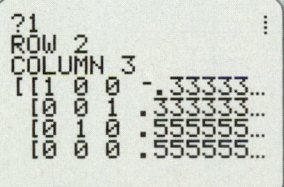

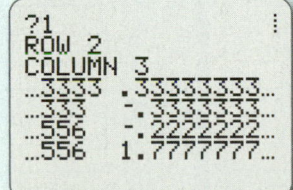

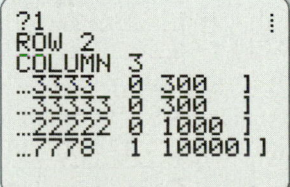

There are no negative numbers in the bottom row, so we're finished. How do we read off the optimal solutions if we don't have labels, though? Look at the columns containing one 1 and three 0s. They are the x column, the y column, the s column, and the p column. Think of the 1 that appears in each of these columns as a pivot whose column has been cleared. If we had labels, the row containing a pivot would have the same label as the column containing that pivot. We can now read off the solution as follows:

x column: The pivot is in the first row, so row 1 would have been labeled with x. We look at the rightmost column to read off the value $x = 300$.

y column: The pivot is in row 3, so we look at the rightmost column to read off the value $y = 1000$.

s column: The pivot is in row 2, so we look at the rightmost column to read off the value $s = 300$.

p column: The pivot is in row 3, so we look at the rightmost column to read off the value $p = 10{,}000$.

Thus, the maximum value of p is $10{,}000¢ = \$100$, which occurs when $x = 300$ and $y = 1000$. The values of the slack variables are $s = 300$ and $t = u = 0$. (Look at the t and u columns to see that they must be inactive.)

EXCEL Technology Guide

Section 4.1

Excel is not a particularly good tool for graphing linear inequalities because it cannot easily shade one side of a line. One solution is to use the "error bar" feature to indicate which side of the line *should* be shaded. For example, here is how we might graph the inequality $2x + 3y \leq 6$.

1. Create a scatter graph using two points to construct a line segment (as in Chapter 1). (Notice that we had to solve the equation $2x + 3y = 6$ for y.)

Microsoft Excel - ETG 4-1 Cunstruct a Line Segment.xls

File Edit View Insert Format Tools Data Window Help

	A	B	C	D
1	x	y		
2		−10	=-(2/3)*A2 + 2		
3		10			
4				
5				

2. Double-click on the line-segment, and use the "X-Error Bars" feature to obtain a diagram similar to the one on the left below, where the error bars indicate the direction of shading.

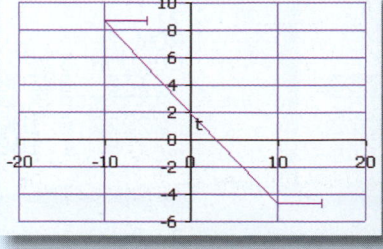
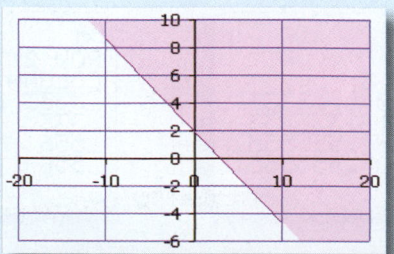

Alternatively, you can use the Drawing Palette to create a polygon with a semi-transparent fill, as shown on the right.

Section **4.3**

Example 3 The Acme Baby Foods example in the text leads to the following linear programming problem:

$$\text{Maximize} \quad p = 10x + 7y$$
$$\text{subject to} \quad x \quad\quad\quad \le 600$$
$$2x + 3y \le 3600$$
$$5x + 3y \le 4500$$
$$x \ge 0, \ y \ge 0$$

Solve it using technology.

Solution with Technology When we introduce slack variables, we get the following system of equations:

$$x \quad\quad + s \quad\quad\quad\quad = 600$$
$$2x + 3y \quad + t \quad\quad\quad = 3600$$
$$5x + 3y \quad\quad + u \quad = 4500$$
$$-10x - 7y \quad\quad\quad + p = 0$$

We use the Excel pivot tool to help with the simplex method. This spreadsheet is available at

Chapter 4 → Tools → Excel Pivot and Gauss-Jordan Tool

Because Excel handles decimals as easily as integers, there is no need to avoid them. If we don't need to avoid decimals, we can use the traditional Gauss-Jordan method (see the discussion at the end of Section 2.2): After selecting your pivot, and prior to clearing the pivot column, *divide the pivot row by the value of the pivot, thereby turning the pivot into a 1.*

The main drawback to using a tool like Excel is that it will not automatically keep track of row and column labels. We could enter and modify labels by hand as we go, or we can do without them entirely if we wish. The following is the sequence of tableaux we get while using the simplex method with the help of the pivoting tool.

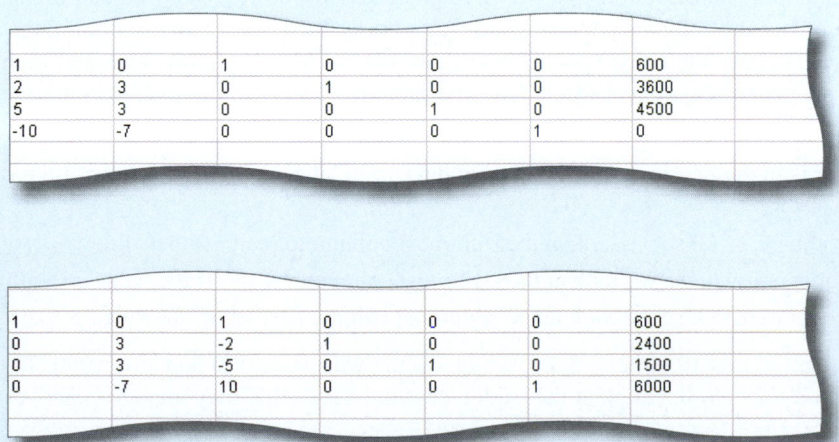

After determining that the next pivot is in the third row and second column, we divide the third row by the pivot, 3, before actually pivoting:

1	0	1	0	0	0	600
0	3	-2	1	0	0	2400
0	1	-1.6666667	0	0.33333333	0	500
0	-7	10	0	0	1	6000

1	0	1	0	0	0	600
0	0	3	1	1	0	900
0	1	-1.6666667	0	0.33333333	0	500
0	0	-1.6666667	0	2.33333333	1	9500

Again, we divide by the pivot, this time in the second row, third column.

1	0	1	0	0	0	600
0	0	1	0.33333333	-0.3333333	0	300
0	1	-1.6666667	0	0.33333333	0	500
0	0	-1.6666667	0	2.33333333	1	9500

1	0	0	-0.3333333	0.33333333	0	300
0	0	1	0.33333333	-0.3333333	0	300
0	1	0	0.55555556	-0.2222222	0	1000
0	0	0	0.55555556	1.77777778	1	10000

There are no negative numbers in the bottom row, so we're finished. How do we read off the optimal solutions if we don't have labels, though? Look at the columns containing one 1 and three 0s. They are the x column, the y column, the s column, and the p column. Think of the 1 that appears in each of these columns as a pivot whose column has been cleared. If we had labels, the row containing a pivot would have the same label as the column containing that pivot. We can now read off the solution as follows:

x **column:** The pivot is in the first row, so row 1 would have been labeled with x. We look at the rightmost column to read off the value $x = 300$.

y **column:** The pivot is in row 3, so we look at the rightmost column to read off the value $y = 1000$.

s **column:** The pivot is in row 2, so we look at the rightmost column to read off the value $s = 300$.

p **column:** The pivot is in row 3, so we look at the rightmost column to read off the value $p = 10,000$.

Thus, the maximum value of p is $10,000¢ = \$100$, which occurs when $x = 300$ and $y = 1000$. The values of the slack variables are $s = 300$ and $t = u = 0$. (Look at the t and u columns to see that they must be inactive.)

5

The Mathematics of Finance

CASE STUDY Saving for College

Tuition costs at colleges and universities increased at an average rate of 5.8% per year from 1991 through 2001. The 2001–2002 average cost for a private college was $23,578. How much would the parents of a newborn have to invest each month in a mutual fund expected to yield 6% per year to pay for their child's college education?

Cindy Charles/PhotoEdit

Introduction

A knowledge of the mathematics of investments and loans is important not only for business majors but also for everyone who deals with money, which is all of us. This chapter is largely about *interest*: interest paid by an investment, interest paid on a loan, and variations on these.

We focus on three forms of investment: investments that pay simple interest, investments in which interest is compounded, and annuities. An investment that pays *simple interest* periodically gives interest directly to the investor, perhaps in the form of a monthly check. If instead, the interest is reinvested, the interest is *compounded,* and the value of the account grows as the interest is added. An *annuity* is an investment earning compound interest into which periodic payments are made or from which periodic withdrawals are made; in the case of periodic payments, such an investment is more commonly called a *sinking fund*. From the point of view of the lender, a loan is a kind of annuity.

We also look at bonds, the primary financial instrument used by companies and governments to raise money. Although bonds nominally pay simple interest, determining their worth, particularly in the secondary market, requires an annuity calculation.

5.1   Simple Interest

You deposit $1000, called the **principal** or **present value,** into a savings account. The bank pays you 5% interest, in the form of a check, each year. How much interest will you earn each year? Because the bank pays you 5% interest each year, your annual (or yearly) interest will be 5% of $1000, or $0.05 \times 1000 = \$50$.

Generalizing this calculation, call the present value PV and the interest rate (expressed as a decimal) r. Then INT, the annual interest paid to you, is given by

$$INT = PVr$$

If the investment is made for a period of t years, then the total interest accumulated is t times this amount, which gives us the following:

Simple Interest

The **simple interest** on an investment (or loan) of PV dollars at an annual interest rate of r for a period of t years is

$$INT = PVrt$$

quick Example

The simple interest over a period of 4 years on a $5000 investment earning 8% per year is

$$INT = PVrt$$
$$= (5000)(0.08)(4) = \$1600$$

Note on Multiletter Variables

Multiletter variables like PV and INT are traditional in finance textbooks, calculators (like the TI-83/84), and spreadsheets (like Excel). Just watch out for expressions like PVr, which is the product of two things, PV and r, not three.

Given your $1000 investment at 5% simple interest, how much money will you have after 2 years? To find the answer, we need to add the accumulated interest to the principal to get the **future value** (FV) of your deposit.

$$FV = PV + INT = \$1000 + (1000)(0.05)(2) = \$1100$$

In general, we can compute the future value as follows:

$$FV = PV + INT = PV + PVrt = PV(1 + rt)$$

Future Value for Simple Interest

The **future value** of an investment of PV dollars at an annual simple interest rate of r for a period of t years is

$$FV = PV(1 + rt)$$

quick Examples

1. The value, at the end of 4 years, of a $5000 investment earning 8% simple interest per year is

$$FV = PV(1 + rt)$$
$$= 5000[1 + (0.08)(4)] = \$6600$$

2. Writing the future value in Quick Example 1 as a function of time, we get

$$FV = 5000(1 + 0.08t)$$
$$= 5000 + 400t$$

which is a linear function of time t. The intercept is $PV = \$5000$, and the slope is the annual interest, $400 per year.

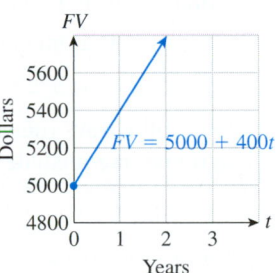

In general, *Simple interest growth is a linear function of time, with intercept given by the present value and slope given by annual interest.*

Example 1 Savings Accounts

In August 2005, the Amex Bank of Canada was paying 2.40% interest on savings accounts.[*] If the interest is paid as simple interest, find the future value of a $2000 deposit in 6 years. What is the total interest paid over the period?

Solution We use the future value formula:

$$FV = PV(1 + rt)$$
$$= 2000[1 + (0.024)(6)] = 2000[1.144] = \$2288$$

---

[*] SOURCE: Canoe Money, August 16, 2005. http://money.canoe.ca/rates/savings.html.

The total interest paid is given by the simple interest formula.

$$INT = PVrt$$
$$= (2000)(0.024)(6) = \$288$$

Note To find the interest paid, we could also have computed

$$INT = FV - PV = 2288 - 2000 = \$288 \ \blacksquare$$

+*Before we go on...* In the preceding example, we could look at the future value as a function of time:

$$FV = 2000(1 + 0.024t) = 2000 + 48t$$

Thus, the future value is growing linearly at a rate of \$48 per year. $\blacksquare$

Example 2 Bridge Loans

Fisher/Thatcher/Getty Images

When "trading up," homeowners sometimes have to buy a new house before they sell their old house. One way to cover the costs of the new house until they get the proceeds from selling the old house is to take out a short-term *bridge loan*. Suppose a bank charges 12% simple annual interest on such a loan. How much will be owed at the maturation (the end) of a 90-day bridge loan of \$90,000?

Solution We use the future value formula

$$FV = PV(1 + rt)$$

with $t = 90/365$, the fraction of a year represented by 90 days:

$$FV = 90{,}000[1 + (0.12)(90/365)]$$
$$= \$92{,}663.01$$

(We will always round our answers to the nearest cent after calculation. Be careful not to round intermediate results.)

+*Before we go on...* Many banks use 360 days for this calculation rather than 365, which makes a "year" for the purposes of the loan slightly shorter than a calendar year. The effect is to increase the amount owed:

$$FV = 90{,}000[1 + (0.12)(90/360)] = \$92{,}700 \ \blacksquare$$

One of the primary ways companies and governments raise money is by selling **bonds.** At its most straightforward, a corporate bond promises to pay simple interest, usually twice a year, for a length of time until it **matures,** at which point it returns the original investment to the investor (U.S. Treasury Notes and Bonds are similar). Things get more complicated when the selling price is negotiable, as we will see in the following sections.

Example 3 Corporate Bonds

The Megabucks Corporation is issuing 10-year bonds paying an annual rate of 6.5%. If you buy \$10,000 worth of bonds, how much interest will you earn every 6 months, and how much interest will you earn over the life of the bonds?

Solution Using the simple interest formula, every 6 months you will receive

$$INT = PVrt$$

$$= (10{,}000)(0.065)\left(\frac{1}{2}\right) = \$325$$

Over the 10-year life of the bonds, you will earn

$$INT = PVrt$$

$$= (10{,}000)(0.065)(10) = \$6500$$

in interest. So, at the end of 10 years, when your original investment is returned to you, your \$10,000 will have turned into \$16,500.

We often want to turn an interest calculation around: Rather than starting with the present value and finding the future value, there are times when we know the future value and need to determine the present value. Solving the future value formula for PV gives us the following.

Present Value for Simple Interest

The present value of an investment at an annual simple interest rate of r for a period of t years, with future value FV, is

$$PV = \frac{FV}{1 + rt}$$

quick Example

If an investment earns 5% simple interest and will be worth \$1000 in 4 years, then its present value (its initial value) is

$$PV = \frac{FV}{1 + rt}$$

$$= \frac{1000}{1 + (0.05)(4)} = \$833.33$$

Here is a typical example. U.S. Treasury bills (T-bills) are short-term investments (up to 1 year) that pay you a set amount after a period of time.

Example 4 Treasury Bills

A U.S. Treasury bill paying \$10,000 after 6 months earns 3.67% simple annual interest.[*] How much would it cost to buy?

Solution The future value of the T-bill is \$10,000; the price we pay is its present value. We know that

$$FV = \$10{,}000$$

$$r = 0.0367$$

---

[*] Rate on August 12, 2005. SOURCE: *Federal Reserve Statistical Release H.15: Selected Interest Rates,* dated August 15, 2005, obtained from the White House's Economics Statistics Briefing Room, www.whitehouse.gov/ fsbr/esbr.html. By the way, because interest rates are often reported to 0.01%, you will sometimes see the term **basis point,** which means one one-hundredth of a percent. Thus, for example, 25 basis points is the same as 0.25%, or 0.0025.

and

$$t = 0.5$$

Substituting into the present value formula, we have

$$PV = \frac{10,000}{1 + (0.0367)(0.5)} = 9819.81$$

so we will pay $9819.81 for the T-bill.

+ *Before we go on...* The simplest way to find the interest earned on the T-bill is by subtraction:

$$INT = FV - PV = 10,000 - 9819.81 = \$180.19 \ \blacksquare$$

Example **5 Treasury Bills**

A T-bill paying $10,000 after 6 months sells at a discount rate of 3.6%.[*] What does it cost? What is the annual **yield;** that is, what simple annual interest rate does it pay?

Solution To say that a *1-year* T-bill sells at a **discount rate** of 3.6% is to say that its selling price is 3.6% lower than its maturity value of $10,000. The discount rates for T-bills of other lengths are adjusted to give annual figures. Thus, the selling price for this 6-month T-bill will be $3.6/2 = 1.8\%$ below its maturity value. This makes the selling price, or present value,

$$PV = 10,000 - (10,000)(0.036/2) \qquad \text{Maturity value – Discount}$$
$$= 10,000(1 - 0.018) = \$9820$$

To find the annual interest rate, notice that

$$FV = \$10,000$$
$$PV = \$9820$$

and

$$t = 0.5$$

and we wish to find r. Substituting in the future value formula, we get

$$FV = PV(1 + rt)$$
$$10,000 = 9820(1 + 0.5r)$$

so

$$1 + 0.5r = 10,000/9820$$

and

$$r = (10,000/9820 - 1)/0.5 \approx 0.0367$$

Thus, the bond is paying 3.67% simple annual interest.

---

[*] Ibid.

$+$ *Before we go on...* The T-bill in Example 5 is the same one as in Example 4 (with a bit of rounding). The interest rate and the discount rate are two different ways of telling what the investment pays. One of the Communication and Reasoning Exercises asks you to find a formula for the interest rate in terms of the discount rate. ∎

Fees on loans can also be thought of as a form of interest.

Example 6 Tax Refunds

You are expecting a tax refund of $800. Because it may take up to 6 weeks to get the refund, your tax preparation firm offers, for a fee of $40, to give you an "interest-free" loan of $800 to be paid back with the refund check. If we think of the fee as interest, what simple annual interest rate is the firm charging?

Solution If we view the $40 as interest, then the future value of the loan (the value of the loan to the firm, or the total you will pay the firm) is $840. Thus, we have

$$FV = 840$$
$$PV = 800$$
$$t = 6/52 \qquad \text{Using 52 weeks in a year}$$

and we wish to find r. Substituting, we get

$$FV = PV(1 + rt)$$
$$840 = 800(1 + 6r/52) = 800 + \frac{4800r}{52}$$

so

$$\frac{4800r}{52} = 840 - 800 = 40$$
$$r = \frac{40 \times 52}{4800} \approx 0.43$$

In other words, the firm is charging you 43% annual interest! Save your money and wait 6 weeks for your refund.

5.1 EXERCISES

● denotes basic skills exercises

In Exercises 1–6, compute the simple interest for the specified period and the future value at the end of the period. Round all answers to the nearest cent. hint [see Example 1]

1. ● $2000 is invested for 1 year at 6% per year.

2. ● $1000 is invested for 10 years at 4% per year.

3. ● $20,200 is invested for 6 months at 5% per year.

4. ● $10,100 is invested for 3 months at 11% per year.

5. ● You borrow $10,000 for 10 months at 3% per year.

6. ● You borrow $6000 for 5 months at 9% per year.

In Exercises 7–12, find the present value of the given investment. hint [see Quick Example on p. 345]

7. ● An investment earns 2% per year and is worth $10,000 after 5 years.

8. ● An investment earns 5% per year and is worth $20,000 after 2 years.

● basic skills

9. ● An investment earns 7% per year and is worth $1,000 after 6 months.

10. ● An investment earns 10% per year and is worth $5,000 after 3 months.

11. ● An investment earns 3% per year and is worth $15,000 after 15 months.

12. ● An investment earns 6% per year and is worth $30,000 after 20 months.

Applications

In Exercises 13–26, compute the specified quantity. Round all answers to the nearest month, the nearest cent, or the nearest 0.001%, as appropriate.

13. ● **Simple Loans** You take out a 6-month, $5,000 loan at 8% simple interest. How much would you owe at the end of the 6 months? *hint* [see Example 2]

14. ● **Simple Loans** You take out a 15-month, $10,000 loan at 11% simple interest. How much would you owe at the end of the 15 months?

15. ● **Savings** How much would you have to deposit in an account earning 4.5% simple interest if you wanted to have $1000 after 6 years? *hint* [see Example 4]

16. ● **Simple Loans** Your total payment on a 4-year loan, which charged 9.5% simple interest, amounted to $30,360. How much did you originally borrow?

17. ● **Bonds** A 5-year bond costs $1000 and will pay a total of $250 interest over its lifetime. What is its annual interest rate?

18. ● **Bonds** A 4-year bond costs $10,000 and will pay a total of $2800 in interest over its lifetime. What is its annual interest rate?

19. **Simple Loans** A $4000 loan, taken now, with a simple interest rate of 8% per year, will require a total repayment of $4640. When will the loan mature?

20. **Simple Loans** The simple interest on a $1000 loan at 8% per year amounted to $640. When did the loan mature?

21. **Treasury Bills** At auction on August 18, 2005, 6-month T-bills were sold at a discount of 3.705%.[1] What was the simple annual yield? *hint* [see Example 5]

22. **Treasury Bills** At auction on August 18, 2005, 3-month T-bills were sold at a discount of 3.470%.[2] What was the simple annual yield?

23. **Fees** You are expecting a tax refund of $1000 in 4 weeks. A tax preparer offers you an "interest-free" loan of $1000 for a fee of $50 to be repaid by your refund check when it arrives in 4 weeks. Thinking of the fee as interest, what simple interest rate would you be paying on this loan? *hint* [see Example 6]

24. **Fees** You are expecting a tax refund of $1500 in 3 weeks. A tax preparer offers you an "interest-free" loan of $1500 for a fee of $60 to be repaid by your refund check when it arrives in 3 weeks. Thinking of the fee as interest, what simple interest rate would you be paying on this loan?

25. **Fees** You take out a 2-year, $5000 loan at 9% simple annual interest. The lender charges you a $100 fee. Thinking of the fee as additional interest, what is the actual annual interest rate you will pay?

26. **Fees** You take out a 3-year, $7000 loan at 8% simple annual interest. The lender charges you a $100 fee. Thinking of the fee as additional interest, what is the actual annual interest rate you will pay?

Stock Investments *Exercises 27–32 are based on the following chart, which shows monthly figures for Apple Computer, Inc., stock[3] (adjusted for two stock splits that took place in this time period).*

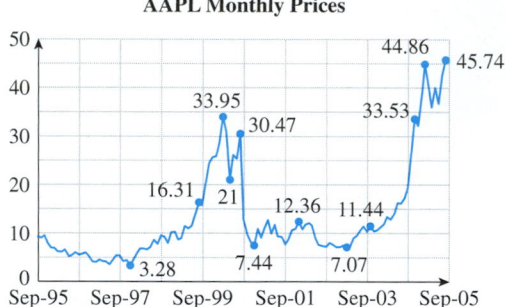

AAPL Monthly Prices

Marked are the following points on the chart:

Dec. 1997	Aug. 1999	Mar. 2000	May 2000	Aug. 2000	Dec. 2000
3.28	16.31	33.95	21	30.47	7.44
Jan. 2002	Mar. 2003	Oct. 2003	Nov. 2004	Feb. 2005	Aug. 2005
12.36	7.07	11.44	33.53	44.86	45.74

27. Calculate to the nearest 0.01% your annual return (on a simple interest basis) if you had bought Apple stock in December, 1997, and sold in August, 2005.

28. Calculate to the nearest 0.01% your annual loss (on a simple interest basis) if you had bought Apple stock in March, 2000, and sold in January, 2002.

29. Suppose you bought Apple stock in January, 2002. If you later sold at one of the marked dates on the chart, which of those dates would have given you the largest annual return (on a simple interest basis), and what was that return?

---

[1] SOURCE: The Bureau of the Public Debt's website: www.publicdebt.treas.gov.

[2] Ibid.

[3] SOURCE: Yahoo! Finance, http://finance.yahoo.com.

30. Suppose you bought Apple stock in August, 1999. If you later sold at one of the marked dates on the chart, which of those dates would have given you the largest annual loss (on a simple interest basis), and what was that loss?

31. Did Apple's stock undergo simple interest increase in the period December, 1997, through March, 2000? (Give a reason for your answer.)

32. If Apple's stock underwent simple interest increase from February, 2005, through August, 2005 and into 2006, what would the price be in December, 2006?

Population Exercises 33–38 are based on the following graph, which shows the population of San Diego County from 1950 to 2000.[4]

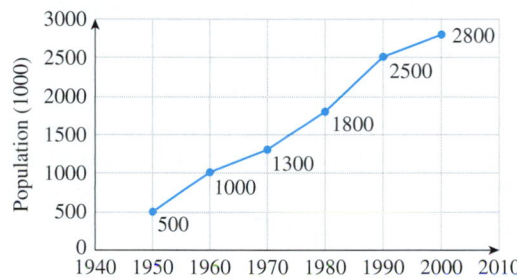

33. ● At what annual (simple interest) rate did the population of San Diego County increase from 1950 to 2000?

34. ● At what annual (simple interest) rate did the population of San Diego County increase from 1950 to 1990?

35. If you used your answer to Exercise 33 as the annual (simple interest) rate at which the population was growing since 1950, what would you predict the San Diego County population to be in 2010?

36. If you used your answer to Exercise 34 as the annual (simple interest) rate at which the population was growing since 1950, what would you predict the San Diego County population to be in 2010?

37. Use your answer to Exercise 33 to give a linear model for the population of San Diego County from 1950 to 2000. Draw the graph of your model over that period of time.

38. Use your answer to Exercise 34 to give a linear model for the population of San Diego County from 1950 to 2000. Draw the graph of your model over that period of time.

Communication and Reasoning Exercises

39. ● One or more of the following graphs represent the future value of an investment earning simple interest. Which one(s)? Give the reason for your choice(s).

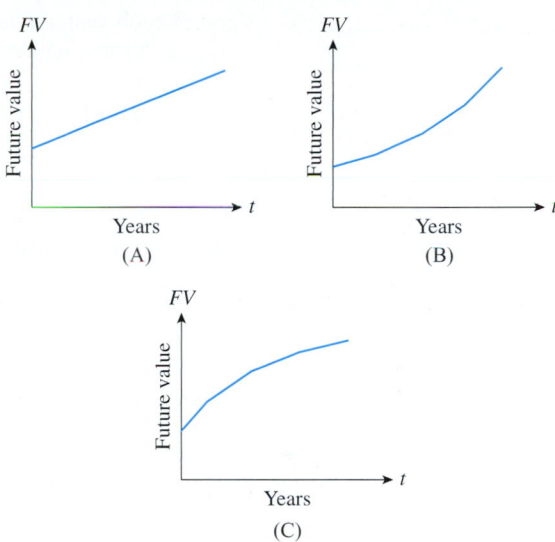

40. ● Given that $FV = 5t + 400$, for what interest rate is this the equation of future value (in dollars) as a function of time t (in years)?

41. *Interpreting the News* You hear the following on your local radio station's business news: "The economy last year grew by 1%. This was the second year in a row in which the economy showed a 1% growth." Because the rate of growth was the same two years in a row, this represents simple interest growth, right? Explain your answer.

42. *Interpreting the News* You hear the following on your local radio station's business news: "The economy last year grew by 1%. This was the second year in a row in which the economy showed a 1% growth." This means that, in dollar terms, the economy grew more last year than the year before. Why?

43. Explain why simple interest is not the appropriate way to measure interest on a savings account that pays interest directly into your account.

44. Suppose that a one-year T-bill sells at a discount rate d. Find a formula, in terms of d, for the simple annual interest rate r the bill will pay.

---

[4] SOURCE: Census Bureau/*New York Times,* April 2, 2002, p. F3.

5.2 Compound Interest

You deposit $1000 into a savings account. The bank pays you 5% interest, which it deposits into your account, or **reinvests,** at the end of each year. At the end of 5 years, how much money will you have accumulated? Let us compute the amount you have at the end of each year. At the end of the first year, the bank will pay you simple interest of 5% on your $1000, which gives you

$$PV(1 + rt) = 1000(1 + 0.05)$$
$$= \$1050$$

At the end of the second year, the bank will pay you another 5% interest, but this time computed on the total in your account, which is $1050. Thus, you will have a total of

$$1050(1 + 0.05) = \$1102.50$$

If you were being paid simple interest on your original $1000, you would have only $1100 at the end of the second year. The extra $2.50 is the interest earned on the $50 interest added to your account at the end of the first year. Having interest earn interest is called **compounding** the interest. We could continue like this until the end of the fifth year, but notice what we are doing: Each year we are multiplying by $1 + 0.05$. So, at the end of 5 years, you will have

$$1000(1 + 0.05)^5 \approx \$1276.28$$

It is interesting to compare this to the amount you would have if the bank paid you simple interest:

$$1000(1 + 0.05 \times 5) = \$1250.00$$

The extra $26.28 is again the effect of compounding the interest.

Banks often pay interest more often than once a year. Paying interest quarterly (four times per year) or monthly is common. If your bank pays interest monthly, how much will your $1000 deposit be worth after 5 years? The bank will not pay you 5% interest every month, but will give you 1/12 of that,[5] or 5/12% interest each month. Thus, instead of multiplying by $1 + 0.05$ every year, we should multiply by $1 + 0.05/12$ each month. Because there are $5 \times 12 = 60$ months in 5 years, the total amount you will have at the end of 5 years is

$$1000\left(1 + \frac{0.05}{12}\right)^{60} \approx \$1283.36$$

Compare this to the $1276.28 you would get if the bank paid the interest every year. You earn an extra $7.08 if the interest is paid monthly because interest gets into your account and starts earning interest earlier. The amount of time between interest payments is called the **compounding period.**

---

[5] This is approximate. They will actually give you 31/365 of the 5% at the end of January and so on, but it's simpler and reasonably accurate to call it 1/12.

The following table summarizes the results above.

Time in Years	Amount with Simple Interest	Amount with Annual Compounding	Amount with Monthly Compounding
0	$1000	$1000	$1000
1	$1000(1 + 0.05)$ $= \$1050$	$1000(1 + 0.05)$ $= \$1050$	$1000(1 + 0.05/12)^{12}$ $= \$1051.16$
2	$1000(1 + 0.05 \times 2)$ $= \$1100$	$1000(1 + 0.05)^2$ $= \$1102.50$	$1000(1 + 0.05/12)^{24}$ $= \$1104.94$
5	$1000(1 + 0.05 \times 5)$ $= \$1250$	$1000(1 + 0.05)^5$ $= \$1276.28$	$1000(1 + 0.05/12)^{60}$ $= \$1283.36$

The preceding calculations generalize easily to give the general formula for future value when interest is compounded.

Future Value for Compound Interest

The future value of an investment of PV dollars earning interest at an annual rate of r compounded (reinvested) m times per year for a period of t years is

$$FV = PV\left(1 + \frac{r}{m}\right)^{mt}$$

or

$$FV = PV(1 + i)^n$$

where $i = r/m$ is the interest paid each compounding period and $n = mt$ is the total number of compounding periods.

quick Examples

1. To find the future value after 5 years of a $10,000 investment earning 6% interest, with interest reinvested every month, we set $PV = 10,000, r = 0.06, m = 12,$ and $t = 5$. Thus,

$$FV = PV\left(1 + \frac{r}{m}\right)^{mt} = 10,000\left(1 + \frac{0.06}{12}\right)^{60} \approx \$13,488.50$$

2. Writing the future value in Quick Example 1 as a function of time, we get

$$FV = 10,000\left(1 + \frac{0.06}{12}\right)^{12t}$$

$$= 10,000(1.005)^{12t}$$

which is an **exponential** function of time t.

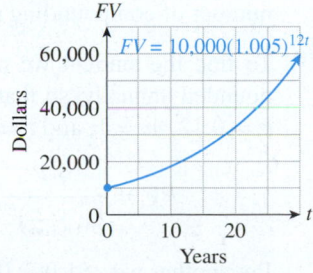

In general, *Compound interest growth is an exponential function of time.*

using *Technology*

See the Technology Guides at the end of the chapter for a discussion of using a TI-83/84 or Excel for compound interest calculations. Online, follow:

Chapter 5
→ Online Utilities
   → Time Value of Money
      Utility

to find a utility very similar to the TI-83/84's TVM Solver, and instructions for its use.

Example 1   Savings Accounts

In August 2005, the Amex Bank of Canada was paying 2.40% interest on savings accounts.[*] If the interest is compounded quarterly, find the future value of a $2000 deposit in 6 years. What is the total interest paid over the period?

Solution   We use the future value formula with $m = 4$:

$$FV = PV\left(1 + \frac{r}{m}\right)^{mt}$$

$$= 2000\left(1 + \frac{0.0240}{4}\right)^{(4)(6)} \approx \$2308.77$$

The total interest paid is

$$INT = FV - PV = 2308.77 - 2000 = \$308.77$$

---

[*] Source: Canoe Money, August 16, 2005. http://money.canoe.ca/rates/savings.html.

Example 1 illustrates the concept of the **time value of money:** A given amount of money received now will usually be worth a different amount to us than the same amount received some time in the future. In the example above, we can say that $2000 received now is worth the same as $2308.77 received six years from now, because if we receive $2000 now, we can turn it into $2308.77 by the end of six years.

We often want to know, for some amount of money in the future, what is the equivalent value at present. As we did for simple interest, we can solve the future value formula for the present value and obtain the following formula.

Present Value for Compound Interest

The present value of an investment earning interest at an annual rate of r compounded m times per year for a period of t years, with future value FV, is

$$PV = \frac{FV}{\left(1 + \frac{r}{m}\right)^{mt}}$$

or

$$PV = \frac{FV}{(1 + i)^n} = FV(1 + i)^{-n}$$

where $i = r/m$ is the interest paid each compounding period and $n = mt$ is the total number of compounding periods.

quick Example

To find the amount we need to invest in an investment earning 12% per year, compounded annually, so that we will have $1 million in 20 years, use $FV = \$1,000,000$, $r = 0.12$, $m = 1$, and $t = 20$:

$$PV = \frac{FV}{\left(1 + \frac{r}{m}\right)^{mt}} = \frac{1,000,000}{(1 + 0.12)^{20}} \approx \$103,666.77$$

Put another way, $1,000,000 20 years from now is worth only $103,666.77 to us now, if we have a 12% investment available.

In the preceding section, we mentioned that a bond pays interest until it reaches maturity, at which point it pays you back an amount called its **maturity value** or **par value.** The two parts, the interest and the maturity value, can be separated and sold and traded by themselves. A **zero coupon bond** is a form of corporate bond that pays no interest during its life but, like U.S. Treasury bills, promises to pay you the maturity value when it reaches maturity. Zero coupon bonds are often created by removing or *stripping* the interest coupons from an ordinary bond, so are also known as **strips.** Zero coupon bonds sell for less than their maturity value, and the return on your investment is the difference between what the investor pays and the maturity value. Although no interest is actually paid, we measure the return on investment by thinking of the interest rate that would make the selling price (the present value) grow to become the maturity value (the future value).[6]

 using *Technology*

See the Technology Guides at the end of the chapter for a discussion of using a TI-83/84 or Excel for this example.

Example 2 Zero Coupon Bonds

Megabucks Corporation is issuing 10-year zero coupon bonds. How much would you pay for bonds with a maturity value of $10,000 if you wish to get a return of 6.5% compounded annually?[*]

Solution As we said earlier, we think of a zero coupon bond as if it were an account earning (compound) interest. We are asked to calculate the amount you will pay for the bond—the present value PV. We have

$$FV = \$10,000$$
$$r = 0.065$$
$$t = 10$$
$$m = 1$$

We can now use the present value formula:

$$PV = \frac{FV}{\left(1 + \frac{r}{m}\right)^{mt}}$$

$$PV = \frac{10,000}{\left(1 + \frac{0.065}{1}\right)^{10 \times 1}} \approx \$5327.26$$

Thus, you should pay $5327.26 to get a return of 6.5% annually.

---

[*] The return investors look for depends on a number of factors, including risk (the chance that the company will go bankrupt and you will lose your investment); the higher the risk, the higher the return. U.S. Treasuries are considered risk free because the federal government has never defaulted on its debts. On the other hand, so-called junk bonds are high-risk investments (below investment grade) and have correspondingly high yields.

+*Before we go on...* Particularly in financial applications, you will hear the word **"discounted"** in place of "compounded" when discussing present value. Thus, the result of Example 2 might be phrased, "The present value of $10,000 to be received 10 years from now, with an interest rate of 6.5% discounted annually, is $5,327.26." ■

---

[6] The IRS refers to this kind of interest as **original issue discount (OID)** and taxes it as if it were interest actually paid to you each year.

Time value of money calculations are often done to take into account inflation, which behaves like compound interest. Suppose, for example, that inflation is running at 5% per year. Then prices will increase by 5% each year, so if PV represents the price now, the price one year from now will be 5% higher, or $PV(1 + 0.05)$. The price a year from then will be 5% higher still, or $PV(1 + 0.05)^2$. Thus, the effects of inflation are compounded just as reinvested interest is.

Example 3 Inflation

Inflation in East Avalon is 5% per year. TruVision television sets cost $200 today. How much will a comparable set cost 2 years from now?

Solution To find the price of a television set 2 years from now, we compute the future value of $200 at an inflation rate of 5% compounded yearly:

$$FV = 200(1 + 0.05)^2 = \$220.50$$

Example 4 Constant Dollars

Inflation in North Avalon is 6% per year. Which is really more expensive, a car costing $20,000 today or one costing $22,000 in 3 years?

Solution We cannot compare the two costs directly because inflation makes $1 today worth more (it buys more) than a dollar three years from now. We need the two prices expressed in comparable terms, so we convert to **constant dollars.** We take the car costing $22,000 three years from now and ask what it would cost in today's dollars. In other words, we convert the future value of $22,000 to its present value:

$$
\begin{aligned}
PV &= FV(1 + i)^{-n} \\
&= 22{,}000(1 + 0.06)^{-3} \\
&\approx \$18{,}471.62
\end{aligned}
$$

Thus, the car costing $22,000 in 3 years actually costs less than the one costing $20,000 now.

+*Before we go on...* In the presence of inflation, the only way to compare prices at different times is to convert all prices to constant dollars. We pick some fixed time and compute future or present values as appropriate to determine what things would have cost at that time. ■

There are some other interesting calculations related to compound interest, besides present and future values.

Example 5 Effective Interest Rate

You have just won $1 million in the lottery and are deciding what to do with it during the next year before you move to the South Pacific. Bank Ten offers 10% interest, compounded annually, while Bank Nine offers 9.8% compounded monthly. In which should you deposit your money?

Solution Let's calculate the future value of your $1 million after one year in each of the banks:

Bank Ten: $FV = 1(1 + 0.10)^1 = \$1.1$ million

Bank Nine: $FV = 1\left(1 + \dfrac{0.098}{12}\right)^{12} = \$1.1025$ million

Bank Nine turns out to be better: It will pay you a total of $102,500 in interest over the year, whereas Bank Ten will pay only $100,000 in interest.

Another way of looking at the calculation in Example 5 is that Bank Nine gave you a total of 10.25% interest on your investment over the year. We call 10.25% the **effective interest rate** of the investment (also referred to as the **annual percentage yield,** or **APY** in the banking industry); the stated 9.8% is called the **nominal** interest rate. In general, to best compare two different investments, it is wisest to compare their *effective*—rather than nominal—interest rates.

Notice that we got 10.25% by computing

$$\left(1 + \frac{0.098}{12}\right)^{12} = 1.1025$$

and then subtracting 1 to get 0.1025, or 10.25%. Generalizing, we get the following formula.

Effective Interest Rate

The effective interest rate r_{eff} of an investment paying a nominal interest rate of r_{nom} compounded m times per year is

$$r_{\text{eff}} = \left(1 + \frac{r_{\text{nom}}}{m}\right)^m - 1$$

To compare rates of investments with different compounding periods, always compare the effective interest rates rather than the nominal rates.

quick Example

To calculate the effective interest rate of an investment that pays 8% per year, with interest reinvested monthly, set $r_{\text{nom}} = 0.08$ and $m = 12$, to obtain

$$r_{\text{eff}} = \left(1 + \frac{0.08}{12}\right)^{12} - 1 \approx 0.0830, \quad \text{or} \quad 8.30\%$$

Example 6 How Long to Invest

You have $5000 to invest at 6% interest compounded monthly. How long will it take for your investment to grow to $6000?

Solution If we use the future value formula, we already have the values

$$FV = 6000$$
$$PV = 5000$$
$$r = 0.06$$
$$m = 12$$

using *Technology*

See the Technology Guides at the end of the chapter for a discussion of using a TI-83/84 or Excel for this example.

Substituting, we get

$$6000 = 5000 \left(1 + \frac{0.06}{12}\right)^{12t}$$

If you are familiar with logarithms, you can solve explicitly for t as follows:

$$\left(1 + \frac{0.06}{12}\right)^{12t} = \frac{6000}{5000} = 1.2$$

$$\log\left(1 + \frac{0.06}{12}\right)^{12t} = \log 1.2$$

$$12t \log\left(1 + \frac{0.06}{12}\right) = \log 1.2$$

$$t = \frac{\log 1.2}{12 \log\left(1 + \frac{0.06}{12}\right)} \approx 3.046 \approx 3 \text{ years}$$

Another approach is to use a bit of trial and error to find the answer. Let's see what the future value is after 2, 3, and 4 years:

$$5000 \left(1 + \frac{0.06}{12}\right)^{12 \times 2} = 5635.80 \qquad \text{Future value after 2 years}$$

$$5000 \left(1 + \frac{0.06}{12}\right)^{12 \times 3} = 5983.40 \qquad \text{Future value after 3 years}$$

$$5000 \left(1 + \frac{0.06}{12}\right)^{12 \times 4} = 6352.45 \qquad \text{Future value after 4 years}$$

From these calculations, it looks as if the answer should be a bit more than 3 years, but is certainly between 3 and 4 years. We could try 3.5 years next and then narrow it down systematically until we have a pretty good approximation of the correct answer.

Graphing calculators and spreadsheets give us alternative methods of solution.

FAQs   Recognizing When to Use Compound Interest and the Meaning of Present Value

Q: *How do I distinguish a problem that calls for compound interest from one that calls for simple interest?*

A: Study the scenario to ascertain whether the interest is being withdrawn as it is earned or reinvested (deposited back into the account). If the interest is being withdrawn, the problem is calling for simple interest because the interest is not itself earning interest. If it is being reinvested, the problem is calling for compound interest. ∎

Q: *How do I distinguish present value from future value in a problem?*

A: The present value always refers to the value of an investment before any interest is included (or, in the case of a depreciating investment, before any depreciation takes place). As an example, the future value of a bond is its maturity value. The value of $1 today in constant 1990 dollars is its present value (even though 1990 is in the past). ∎

5.2 EXERCISES

- ● denotes basic skills exercises
- ◆ denotes challenging exercises

tech Ex indicates exercises that should be solved using technology

In Exercises 1–8, calculate, to the nearest cent, the future value of an investment of $10,000 at the stated interest rate after the stated amount of time. hint [see Example 1]

1. ● 3%/year, compounded annually, after 10 years

2. ● 4%/year, compounded annually, after 8 years

3. ● 2.5%/year, compounded quarterly (4 times/year), after 5 years

4. ● 1.5%/year, compounded weekly (52 times/year), after 5 years

5. ● 6.5%/year, compounded daily (assume 365 days/year), after 10 years

6. ● 11.2%/year, compounded monthly, after 12 years

7. ● 0.2% per month, compounded monthly, after 10 years

8. ● 0.45% per month, compounded monthly, after 20 years

In Exercises 9–14, calculate the present value of an investment that will be worth $1000 at the stated interest rate after the stated amount of time. hint [see Example 2]

9. ● 10 years, at 5%/year, compounded annually

10. ● 5 years, at 6%/year, compounded annually

11. ● 5 years, at 4.2%/year, compounded weekly (assume 52 weeks per year)

12. ● 10 years, at 5.3%/year, compounded quarterly

13. ● 4 years, depreciating 5% each year

14. ● 5 years, depreciating 4% each year

In Exercises 15–20, find the effective annual interest rates of the given annual interest rates. Round your answers to the nearest 0.01%. hint [see Quick Example on p. 355]

15. ● 5% compounded quarterly

16. ● 5% compounded monthly

17. ● 10% compounded monthly

18. ● 10% compounded daily (assume 365 days per year)

19. ● 10% compounded hourly (assume 365 days per year)

20. ● 10% compounded every minute (assume 365 days per year)

Applications

21. ● **Savings** You deposit $1000 in an account at the Lifelong Trust Savings and Loan that pays 6% interest compounded quarterly. By how much will your deposit have grown after 4 years?

22. ● **Investments** You invest $10,000 in Rapid Growth Funds, which appreciate by 2% per year, with yields reinvested quarterly. By how much will your investment have grown after 5 years?

23. ● **Depreciation** During the year ending April, 2002, the S&P 500 index depreciated by approximately 6%.[7] Assuming that this trend were to continue, how much would a $3000 investment in an S&P index fund be worth in 3 years?

24. ● **Depreciation** During the year ending April, 2002, the MSCI World Index depreciated by approximately 3.4%.[8] Assuming that this trend were to continue, how much would a $3000 investment in an MSCI index fund be worth in 3 years?

25. ● **Bonds** You want to buy a 10-year zero-coupon bond with a maturity value of $5000 and a yield of 5.5% annually. How much will you pay?

26. ● **Bonds** You want to buy a 15-year zero-coupon bond with a maturity value of $10,000 and a yield of 6.25% annually. How much will you pay?

27. **Investments** When I was considering what to do with my $10,000 Lottery winnings, my broker suggested I invest half of it in gold, the value of which was growing by 10% per year, and the other half in CDs, which were yielding 5% per year, compounded every 6 months. Assuming that these rates are sustained, how much will my investment be worth in 10 years?

28. **Investments** When I was considering what to do with the $10,000 proceeds from my sale of technology stock, my broker suggested I invest half of it in municipal bonds, whose value was growing by 6% per year, and the other half in CDs, which were yielding 3% per year, compounded every 2 months. Assuming that these interest rates are sustained, how much will my investment be worth in 10 years?

29. **Depreciation** During a prolonged recession, property values on Long Island depreciated by 2% every 6 months. If my house cost $200,000 originally, how much was it worth 5 years later?

30. **Depreciation** Stocks in the health industry depreciated by 5.1% in the first nine months of 1993.[9] Assuming that this trend were to continue, how much would a $40,000 investment be worth in 9 years? [Hint: 9 years corresponds to 12 nine-month periods.]

31. **Retirement Planning** I want to be earning an annual salary of $100,000 when I retire in 15 years. I have been offered a job that guarantees an annual salary increase of 4% per year, and the starting salary is negotiable. What salary should I request in order to meet my goal?

---

[7] Values were compared in mid-April. SOURCE: *New York Times,* April 17, 2002.

[8] Ibid.

[9] SOURCE: *The New York Times,* October 9, 1993, p.37.

32. *Retirement Planning* I want to be earning an annual salary of $80,000 when I retire in 10 years. I have been offered a job that guarantees an annual salary increase of 5% per year, and the starting salary is negotiable. What salary should I request in order to meet my goal?

33. *Present Value* Determine the amount of money, to the nearest dollar, you must invest at 6% per year, compounded annually, so that you will be a millionaire in 30 years.

34. *Present Value* Determine the amount of money, to the nearest dollar, you must invest now at 7% per year, compounded annually, so that you will be a millionaire in 40 years.

35. *Stocks* Six years ago, I invested some money in Dracubunny Toy Co. stock, acting on the advice of a "friend." As things turned out, the value of the stock decreased by 5% every 4 months, and I discovered yesterday (to my horror) that my investment was worth only $297.91. How much did I originally invest?

36. *Sales* My recent marketing idea, the *Miracle Algae Growing Kit,* has been remarkably successful, with monthly sales growing by 6% every 6 months over the past 5 years. Assuming that I sold 100 kits the first month, what is the present rate of sales?

37. *Inflation* Inflation has been running 2% per year. A car now costs $30,000. How much would it have cost five years ago? *hint* [see Example 3]

38. *Inflation* (Compare Exercise 37) Inflation has been running 1% every six months. A car now costs $30,000. How much would it have cost five years ago?

39. *Inflation* Housing prices have been rising 6% per year. A house now costs $200,000. What would it have cost 10 years ago?

40. *Inflation* (Compare Exercise 39) Housing prices have been rising 0.5% each month. A house now costs $200,000. What would it have cost 10 years ago?

41. *Constant Dollars* Inflation is running 3% per year when you deposit $1000 in an account earning interest of 5% per year compounded annually. In *constant dollars*, how much money will you have two years from now? [Hint: First calculate the value of your account in two years' time, and then find its present value based on the inflation rate.] *hint* [see Example 4]

42. *Constant Dollars* Inflation is running 1% per month when you deposit $10,000 in an account earning 8% compounded monthly. In *constant dollars*, how much money will you have two years from now? [See the hint for Exercise 41.]

43. *Investments* You are offered two investments. One promises to earn 12% compounded annually. The other will earn 11.9% compounded monthly. Which is the better investment? *hint* [see Example 5]

44. *Investments* You are offered three investments. The first promises to earn 15% compounded annually, the second will earn 14.5% compounded quarterly, and the third will earn 14% compounded monthly. Which is the best investment?

45. *History* Legend has it that the Manhattan Indians sold Manhattan Island to the Dutch in 1626 for $24. In 2001, the total value of Manhattan real estate was estimated to be $136,106 million.[10] Suppose that the Manhattan Indians had taken that $24 and invested it at 6.2% compounded annually (a relatively conservative investment goal). Could the Manhattan Indians have bought back the island in 2001?

46. *History* Repeat Exercise 45, assuming that the Manhattan Indians had invested the $24 at 6.1% compounded annually.

Inflation Exercises 47–54 are based on the following table, which shows the 2002 annual inflation rates in several Latin American countries.[11] Assume that the rates shown continue indefinitely.

Country	Argentina	Brazil	Colombia	Chile	Ecuador	Mexico	Uruguay
Currency	Peso	Real	Peso	Peso	Sucre	Peso	Peso
Inflation Rate (%)	8	8	6	3	15	5	4

47. ● If an item in Brazil now costs 100 reals, what do you expect it to cost 5 years from now? (Answer to the nearest real.)

48. ● If an item in Argentina now costs 1000 pesos, what do you expect it to cost 5 years from now? (Answer to the nearest peso.)

49. ● If an item in Chile will cost 1000 pesos in 10 years, what does it cost now? (Answer to the nearest peso.)

50. ● If an item in Mexico will cost 20,000 pesos in 10 years, what does it cost now? (Answer to the nearest peso.)

51. You wish to invest 1000 pesos in Colombia at 8% annually, compounded twice a year. Find the value of your investment in 10 years, expressing the answer in constant pesos. (Answer to the nearest peso.)

52. You wish to invest 1000 pesos in Uruguay at a 8% annually, compounded twice a year. Find the value of your investment in 10 years, expressing the answer in constant pesos. (Answer to the nearest peso.)

53. Which is the better investment: an investment in Chile yielding 4.3% per year, compounded annually, or an investment in Ecuador yielding 16.2% per year, compounded every six months? Support your answer with figures that show the future value of an investment of one unit of currency in constant units.

54. Which is the better investment: an investment in Argentina yielding 10% per year, compounded annually, or an investment in Uruguay, yielding 5% per year, compounded every six months? Support your answer with figures that show the

[10] SOURCE: Queens County Overall Economic Development Corporation, 2001. www.queensny.org/outlook/Outlook_Sp_7.pdf.

[11] Consumer Price Indices as of March 2002. SOURCE: Latin Focus, March 2002. www.latin-focus.com.

● basic skills   ◆ challenging   **tech Ex** technology exercise

future value of an investment of one unit of currency in constant units.

Stock Investments *Exercises 55–60 are based on the following chart, which shows monthly figures for Apple Computer, Inc., stock[12] (adjusted for two stock splits that took place in this time period).*

AAPL Monthly Prices

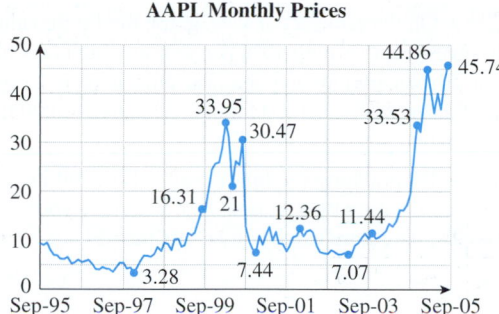

Marked are the following points on the chart:

Dec. 1997	Aug. 1999	Mar. 2000	May 2000	Aug. 2000	Dec. 2000
3.28	16.31	33.95	21	30.47	7.44
Jan. 2002	Mar. 2003	Oct. 2003	Nov. 2004	Feb. 2005	Aug. 2005
12.36	7.07	11.44	33.53	44.86	45.74

55. Calculate to the nearest 0.01% your annual return (assuming annual compounding) if you had bought Apple stock in December, 1997, and sold in August, 2005.

56. Calculate to the nearest 0.01% your annual loss (assuming annual compounding) if you had bought Apple stock in March, 2000, and sold in January, 2002.

57. Suppose you bought Apple stock in January, 2002. If you later sold at one of the marked dates on the chart, which of those dates would have given you the largest annual return (assuming annual compounding), and what was that return?

58. Suppose you bought Apple stock in August, 1999. If you later sold at one of the marked dates on the chart, which of those dates would have given you the largest annual loss (assuming annual compounding), and what was that loss?

59. Did Apple's stock undergo a compound interest increase in the period December, 1997, through March, 2000? (Give a reason for your answer.)

60. If Apple's stock underwent compound interest increase from February, 2005, through August, 2005, and into 2006, what would the price be in December, 2006?

61. tech Ex ***Competing Investments*** I just purchased $5000 worth of municipal funds which are expected to yield 5.4% per year, compounded every six months. My friend has just

purchased $6000 worth of CDs which will earn 4.8% per year, compounded every six months. Determine when, to the nearest year, the value of my investment will be the same as hers, and what this value will be. [Hint: You can either graph the values of both investments or make tables of the values of both investments.]

62. tech Ex ***Investments*** Determine when, to the nearest year, $3000 invested at 5% per year, compounded daily, will be worth $10,000.

63. tech Ex ***Epidemics*** At the start of 1985, the incidence of AIDS was doubling every six months and 40,000 cases had been reported in the U.S. Assuming this trend would have continued, determine when, to the nearest tenth of a year, the number of cases would have reached 1 million.

64. tech Ex ***Depreciation*** My investment in Genetic Splicing, Inc., is now worth $4354 and is depreciating by 5% every six months. For some reason, I am reluctant to sell the stock and swallow my losses. Determine when, to the nearest year, my investment will drop below $50.

65. ◆ ***Bonds*** Once purchased, bonds can be sold in the secondary market. The value of a bond depends on the prevailing interest rates, which vary over time. Suppose that, in January, 1982, you bought a 30-year zero coupon U.S. Treasury bond with a maturity value of $100,000 and a yield of 15% annually.
   a. How much did you pay for the bond?
   b. In January, 1999, your bond had 13 years remaining until maturity. Rates on U.S. Treasury bonds of comparable length were about 4.75%. If you sold your bond to an investor looking for a return of 4.75% annually, how much money would you have received?
   c. Using your answers to parts (a) and (b), what was the annual yield on your 17-year investment?

66. ◆ ***Bonds*** Suppose that, in January, 1999, you bought a 30-year zero coupon U.S. Treasury bond with a maturity value of $100,000 and a yield of 5% annually.
   a. How much did you pay for the bond?
   b. Suppose that, 15 years later, interest rates have risen again, to 12%. If you sell your bond to an investor looking for a return of 12%, how much money will you receive?
   c. Using your answers to parts (a) and (b), what will be the annual yield on your 15 year investment?

Communication and Reasoning Exercises

67. ● Why is the graph of the future value of a compound interest investment as a function of time not a straight line (assuming a nonzero rate of interest)?

68. ● An investment that earns 10% (compound interest) every year is the same as an investment that earns 5% (compound interest) every six months, right?

69. If a bacteria culture is currently 0.01g and increases in size by 10% each day, then its growth is linear, right?

---

[12] SOURCE: Yahoo! Finance, http://finance.yahoo.com.

● basic skills   ◆ challenging   tech Ex technology exercise

70. At what point is the future value of a compound interest investment the same as the future value of a simple interest investment at the same annual rate of interest?

71. If two equal investments have the same effective interest rate and you graph the future value as a function of time for each of them, are the graphs necessarily the same? Explain your answer.

72. For what kind of compound interest investments is the effective rate the same as the nominal rate? Explain your answer.

73. For what kind of compound interest investments is the effective rate greater than the nominal rate? When is it smaller? Explain your answer.

74. If an investment appreciates by 10% per year for five years (compounded annually) and then depreciates by 10% per year (compounded annually) for five more years, will it have the same value as it had originally? Explain your answer.

75. You can choose between two investments in zero coupon bonds: one maturing in 10 years, and the other maturing in 15 years. If you knew the rate of inflation, how would you decide which is the better investment?

76. If you knew the various inflation rates for the years 1995 through 2001, how would you convert $100 in 2002 dollars to 1995 dollars?

77. `tech` Ex On the same set of axes, graph the future value of a $100 investment earning 10% per year as a function of time over a 20-year period, compounded once a year, 10 times a year, 100 times a year, 1000 times a year, and 10,000 times a year. What do you notice?

78. `tech` Ex By graphing the future value of a $100 investment that is depreciating by 1% each year, convince yourself that, eventually, the future value will be less than $1.

● basic skills   ◆ challenging   `tech` Ex technology exercise

5.3 Annuities, Loans, and Bonds

A typical defined-contribution pension fund works as follows:[13] Every month while you work, you and your employer deposit a certain amount of money in an account. This money earns (compound) interest from the time it is deposited. When you retire, the account continues to earn interest, but you may then start withdrawing money at a rate calculated to reduce the account to zero after some number of years. This account is an example of an **annuity,** an account earning interest into which you make periodic deposits or from which you make periodic withdrawals. In common usage, the term "annuity" is used for an account from which you make withdrawals. There are various terms used for accounts into which you make payments, based on their purpose. Examples include **savings account, pension fund,** or **sinking fund.** A sinking fund is generally used by businesses or governments to accumulate money to pay off an anticipated debt, but we'll use the term to refer to any account into which you make periodic payments.

Sinking Funds

Suppose you make a payment of $100 at the end of every month into an account earning 3.6% interest per year, compounded monthly. This means that your investment is earning $3.6\%/12 = 0.3\%$ per month. We write $i = 0.036/12 = 0.003$. What will be the value of the investment at the end of two years (24 months)?

Think of the deposits separately. Each earns interest from the time it is deposited, and the total accumulated after two years is the sum of these deposits and the interest they earn. In other words, the accumulated value is the sum of the future values of the deposits, taking into account how long each deposit sits in the account. Figure 1 shows a timeline with the deposits and the contribution of each to the final value.

---

[13] Defined-contribution pensions are increasingly common, replacing the defined-benefit pensions that were once the norm. In a defined-benefit pension, the size of your pension is guaranteed; it is typically a percentage of your final working salary. In a defined-contribution plan the size of your pension depends on how well your investments do.

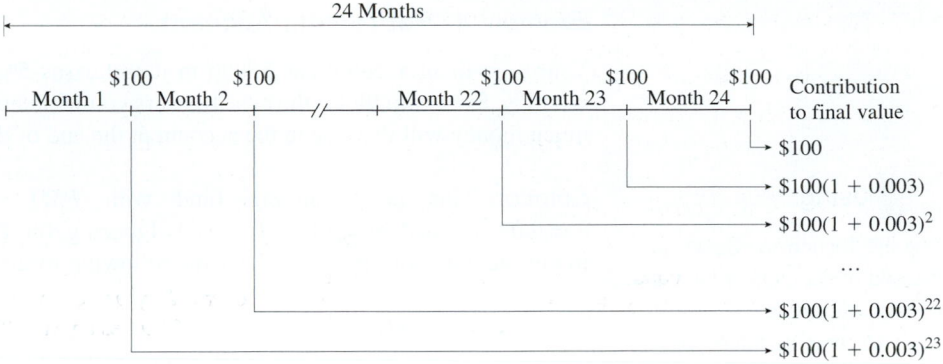

Figure **1**

For example, the very last deposit (at the end of month 24) has no time to earn interest, so it contributes only $100. The very first deposit, which earns interest for 23 months, by the future value formula for compound interest contributes $100(1 + 0.003)^{23}$ to the total. Adding together all of the future values gives us the total future value:

$$FV = 100 + 100(1 + 0.003) + 100(1 + 0.003)^2 + \cdots + 100(1 + 0.003)^{23}$$
$$= 100[1 + (1 + 0.003) + (1 + 0.003)^2 + \cdots + (1 + 0.003)^{23}]$$

Fortunately, this sort of sum is well-known (to mathematicians, anyway[14]) and there is a convenient formula for its value:[15]

$$1 + x + x^2 + \cdots + x^{n-1} = \frac{x^{n} - 1}{x - 1}$$

In our case, with $x = 1 + 0.003$, this formula allows us to calculate the future value:

$$FV = 100\frac{(1 + 0.003)^{24} - 1}{(1 + 0.003) - 1} = 100\frac{(1.003)^{24} - 1}{0.003} \approx \$2484.65$$

It is now easy to generalize this calculation.

Future Value of a Sinking Fund

A **sinking fund** is an account earning compound interest into which you make periodic deposits. Suppose that the account has an annual rate of r compounded m times per year, so that $i = r/m$ is the interest rate per compounding period. If you make a payment of PMT at the end of each period, then the future value after t years, or $n = mt$ periods, will be

$$FV = PMT\frac{(1 + i)^n - 1}{i}$$

quick Example

At the end of each month you deposit $50 into an account earning 2% annual interest compounded monthly. To find the future value after five years, we use $i = 0.02/12$ and $n = 12 \times 5 = 60$ compounding periods, so

$$FV = 50\frac{(1 + 0.02/12)^{60} - 1}{0.02/12} = \$3152.37$$

---

[14] It is called a **geometric series.**

[15] The quickest way to convince yourself that this formula is correct is to multiply out $(x - 1)(1 + x + x^2 + \cdots + x^{n-1})$ and see that you get $x^n - 1$. You should also try substituting some numbers. For example, $1 + 3 + 3^2 = 13 = (3^3 - 1)/(3 - 1)$.

using *Technology*

See the Technology Guides at the end of the chapter for a discussion of using a TI-83/84 or Excel to compute future values of sinking funds.

Example 1 Retirement Account

Your retirement account has $5000 in it and earns 5% interest per year compounded monthly. Every month for the next 10 years you will deposit $100 into the account. How much money will there be in the account at the end of those 10 years?

Solution This is a sinking fund with $PMT = \$100$, $r = 0.05$, $m = 12$, so $i = 0.05/12$, and $n = 12 \times 10 = 120$. Ignoring for the moment the $5000 already in the account, your payments have the following future value:

$$FV = PMT\frac{(1+i)^n - 1}{i}$$

$$= 100\frac{\left(1 + \frac{0.05}{12}\right)^{120} - 1}{\frac{0.05}{12}}$$

$$\approx \$15,528.23$$

What about the $5000 that was already in the account? That sits there and earns interest, so we need to find its future value as well, using the compound interest formula:

$$FV = PV(1+i)^n$$
$$= 5000(1 + 0.05/12)^{120}$$
$$= \$8235.05$$

Hence, the total amount in the account at the end of 10 years will be

$$\$15,528.23 + 8235.05 = \$23,763.28$$

Sometimes we know what we want the future value to be and need to determine the payments necessary to achieve that goal. We can simply solve the future value formula for the payment.

Payment Formula for a Sinking Fund

Suppose that an account has an annual rate of r compounded m times per year, so that $i = r/m$ is the interest rate per compounding period. If you want to accumulate a total of FV in the account after t years, or $n = mt$ periods, by making payments of PMT at the end of each period, then each payment must be

$$PMT = FV\frac{i}{(1+i)^n - 1}$$

Example 2 Education Fund

Tony and Maria have just had a son, José Phillipe. They establish an account to accumulate money for his college education, in which they would like to have $100,000 after 17 years. If the account pays 4% interest per year compounded quarterly, and they make deposits at the end of every quarter, how large must each deposit be for them to reach their goal?

using *Technology*

See the Technology Guides at the end of the chapter for a discussion of using a TI-83/84 or Excel to compute payments for sinking funds.

Solution This is a sinking fund with $FV = \$100{,}000$, $m = 4$, $n = 4 \times 17 = 68$, and $r = 0.04$, so $i = 0.04/4 = 0.01$. From the payment formula, we get

$$PMT = 100{,}000 \frac{0.01}{(1 + 0.01)^{68} - 1} \approx \$1033.89$$

So, Tony and Maria must deposit \$1033.89 every quarter in order to meet their goal.

Annuities

Suppose we deposit an amount PV now in an account earning 3.6% interest per year, compounded monthly. Starting one month from now, the bank will send us monthly payments of \$100. What must PV be so that the account will be drawn down to \$0 in exactly 2 years?

As before, we write $i = r/m = 0.036/12 = 0.003$, and we have $PMT = 100$. The first payment of \$100 will be made one month from now, so its present value is

$$\frac{PMT}{(1 + i)^n} = \frac{100}{1 + 0.003} = 100(1 + 0.003)^{-1} \approx \$99.70$$

In other words, that much of the original PV goes towards funding the first payment. The second payment, two months from now, has a present value of

$$\frac{PMT}{(1 + i)^n} = \frac{100}{(1 + 0.003)^2} = 100(1 + 0.003)^{-2} \approx \$99.40$$

That much of the original PV funds the second payment. This continues for two years, at which point we receive the last payment, which has a present value of

$$\frac{PMT}{(1 + i)^n} = \frac{100}{(1 + 0.003)^{24}} = 100(1 + 0.003)^{-24} \approx \$93.06$$

and that exhausts the account. Figure 2 shows a timeline with the payments and the present value of each.

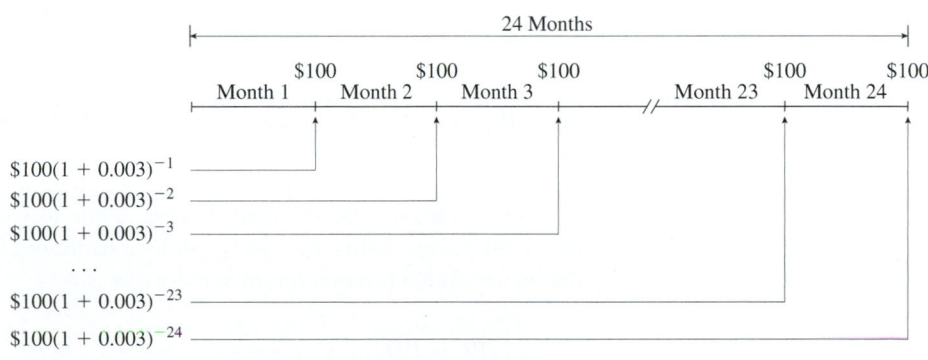

Figure **2**

Because PV must be the sum of these present values, we get

$$PV = 100(1 + 0.003)^{-1} + 100(1 + 0.003)^{-2} + \cdots + 100(1 + 0.003)^{-24}$$
$$= 100[(1 + 0.003)^{-1} + (1 + 0.003)^{-2} + \cdots + (1 + 0.003)^{-24}]$$

We can again find a simpler formula for this sum:

$$x^{-1} + x^{-2} + \cdots + x^{-n} = \frac{1}{x^n}(x^{n-1} + x^{n-2} + \cdots + 1)$$

$$= \frac{1}{x^n} \cdot \frac{x^n - 1}{x - 1} = \frac{1 - x^{-n}}{x - 1}$$

So, in our case,

$$PV = 100\frac{1 - (1 + 0.003)^{-24}}{(1 + 0.003) - 1}$$

or

$$PV = 100\frac{1 - (1.003)^{-24}}{0.003} \approx \$2312.29$$

If we deposit $2312.29 initially and the bank sends us $100 per month for two years, our account will be exhausted at the end of that time.

Generalizing, we get the following formula:

Present Value of an Annuity

An **annuity** is an account earning compound interest from which periodic withdrawals are made. Suppose that the account has an annual rate of r compounded m times per year, so that $i = r/m$ is the interest rate per compounding period. Suppose also that the account starts with a balance of PV. If you receive a payment of PMT at the end of each compounding period, and the account is down to $0 after t years, or $n = mt$ periods, then

$$PV = PMT\frac{1 - (1 + i)^{-n}}{i}$$

quick Example   At the end of each month you want to withdraw $50 from an account earning 2% annual interest compounded monthly. If you want the account to last for five years (60 compounding periods), it must have the following amount to begin with:

$$PV = 50\frac{1 - (1 + 0.02/12)^{-60}}{0.02/12} = \$2852.62$$

Note If you make your withdrawals at the end of each compounding period, as we've discussed so far, you have an **ordinary annuity.** If, instead, you make withdrawals at the beginning of each compounding period, you have an **annuity due.** Because each payment occurs one period earlier, there is one less period in which to earn interest, hence the present value must be larger by a factor of $(1 + i)$ to fund each payment. So, the present value formula for an annuity due is

$$PV = PMT(1 + i)\frac{1 - (1 + i)^{-n}}{i}$$

In this book, we will concentrate on ordinary annuities. ∎

Example 3 Trust Fund

You wish to establish a trust fund from which your niece can withdraw $2000 every six months for 15 years, at the end of which time she will receive the remaining money in the trust, which you would like to be $10,000. The trust will be invested at 7% per year compounded every six months. How large should the trust be?

 using *Technology*

See the Technology Guides at the end of the chapter for a discussion of using a TI-83/84 or Excel to compute the present value of an annuity.

Solution We view this account as having two parts, one funding the semiannual payments and the other funding the $10,000 lump sum at the end. The amount of money necessary to fund the semiannual payments is the present value of an annuity, with $PMT = 2000$, $r = 0.07$ and $m = 2$, so $i = 0.07/2 = 0.035$, and $n = 2 \times 15 = 30$. Substituting gives

$$PV = 2000\frac{1 - (1 + 0.035)^{-30}}{0.035}$$

$$= \$36{,}784.09$$

To fund the lump sum of $10,000 after 15 years, we need the present value of $10,000 under compound interest:

$$PV = 10{,}000(1 + 0.035)^{-30}$$

$$= \$3562.78$$

Thus the trust should start with $36,784.09 + 3562.78 = $40,346.87.

Sometimes we know how much money we begin with and for how long we want to make withdrawals. We then want to determine the amount of money we can withdraw each period. For this, we simply solve the present value formula for the payment.

Payment Formula for an Ordinary Annuity

Suppose that an account has an annual rate of r compounded m times per year, so that $i = r/m$ is the interest rate per compounding period. Suppose also that the account starts with a balance of PV. If you want to receive a payment of PMT at the end of each compounding period, and the account is down to $0 after t years, or $n = mt$ periods, then

$$PMT = PV\frac{i}{1 - (1 + i)^{-n}}$$

Example 4 Education Fund

Tony and Maria (see Example 2), having accumulated $100,000 for José Phillipe's college education, would now like to make quarterly withdrawals over the next four years. How much money can they withdraw each quarter in order to draw down the account to zero at the end of the four years? (Recall that the account pays 4% interest compounded quarterly.)

 using *Technology*

See the Technology Guides at the end of the chapter for a discussion of using a TI-83/84 or Excel to compute the payment for an annuity.

Solution Now Tony and Maria's account is acting as an annuity with a present value of $100,000. So, $PV = \$100{,}000$, $r = 0.04$ and $m = 4$, giving $i = 0.04/4 = 0.01$, and $n = 4 \times 4 = 16$. We use the payment formula to get

$$PMT = 100{,}000\frac{0.01}{1 - (1 + 0.01)^{-16}} \approx \$6794.46$$

So, they can withdraw $6794.46 each quarter for four years, at the end of which time their account balance will be 0.

Paul Seheult/Eye Ubiquitous/Corbis

Example 5 Saving for Retirement

Jane Q. Employee has just started her new job with Big Conglomerate, Inc., and is already looking forward to retirement. BCI offers her as a pension plan an annuity that is guaranteed to earn 6% annual interest compounded monthly. She plans to work for 40 years before retiring and would then like to be able to draw an income of $7000 per month for 20 years. How much do she and BCI together have to deposit per month into the fund to accomplish this?

Solution Here we have the situation we described at the beginning of the section: a sinking fund accumulating money to be used later as an annuity. We know the desired payment out of the annuity, so we work backward. The first thing we need to do is calculate the present value of the annuity required to make the pension payments. We use the annuity present value formula with $PMT = 7000$, $i = r/m = 0.06/12 = 0.005$ and $n = 12 \times 20 = 240$.

$$PV = PMT\frac{1 - (1 + i)^{-n}}{i}$$

$$= 7000\frac{1 - (1 + 0.005)^{-240}}{0.005} \approx \$977{,}065.40$$

This is the total that must be accumulated in the sinking fund during the 40 years she plans to work. In other words, this is the *future* value, *FV*, of the sinking fund. (Thus, the present value in the first step of our calculation is the future value in the second step.) To determine the payments necessary to accumulate this amount, we use the sinking fund payment formula with $FV = 977{,}065.40$, $i = 0.005$, and $n = 12 \times 40 = 480$.

$$PMT = FV\frac{i}{(1 + i)^n - 1}$$

$$= 977{,}065.40\frac{0.005}{1.005^{480} - 1}$$

$$\approx \$490.62$$

So, if she and BCI collectively deposit $490.62 per month into her retirement fund, she can retire with the income she desires.

Installment Loans

In a typical installment loan, such as a car loan or a home mortgage, we borrow an amount of money and then pay it back with interest by making fixed payments (usually every month) over some number of years. From the point of view of the lender, this is an annuity. Thus, loan calculations are identical to annuity calculations.

Example 6 Home Mortgages

Marc and Mira are buying a house, and have taken out a 30-year, $90,000 mortgage at 8% interest per year. What will their monthly payments be?

Solution From the bank's point of view, a mortgage is an annuity. In this case, the present value is $PV = \$90{,}000$, $r = 0.08$, $m = 12$, and $n = 12 \times 30 = 360$. To find the payments, we use the payment formula:

$$PMT = 90{,}000\frac{0.08/12}{1 - (1 + 0.08/12)^{-360}} \approx \$660.39$$

The word "mortgage" comes from the French for "dead pledge." The process of paying off a loan is called **amortizing** the loan, meaning to kill the debt owed.

Example 7 Amortization Schedule

Continuing Example 6: Mortgage interest is tax-deductible, so it is important to know how much of a year's mortgage payments represents interest. How much interest will Marc and Mira pay in the first year of their mortgage?

Solution Let us calculate how much of each month's payment is interest and how much goes to reducing the outstanding principal. At the end of the first month Marc and Mira must pay one month's interest on $90,000, which is

$$\$90,000 \times \frac{0.08}{12} = \$600$$

The remainder of their first monthly payment, $660.39 - 600 = \$60.39$ goes to reducing the principal. Thus, in the second month the outstanding principal is $90,000 - 60.39 = \$89,939.61$, and part of their second monthly payment will be for the interest on this amount, which is

$$\$89,939.61 \times \frac{0.08}{12} = \$599.60$$

The remaining $\$660.39 - \$599.60 = \$60.79$ goes to further reduce the principal. If we continue this calculation for the 12 months of the first year, we get the beginning of the mortgage's **amortization schedule.**

 using *Technology*

See the Technology Guides at the end of the chapter for a discussion of using a TI-83/84 or Excel to construct an amortization schedule.

Month	Interest Payment	Payment on Principal	Outstanding Principal
0			$90,000.00
1	$600.00	$60.39	89,939.61
2	599.60	60.79	89,878.82
3	599.19	61.20	89,817.62
4	598.78	61.61	89,756.01
5	598.37	62.02	89,693.99
6	597.96	62.43	89,631.56
7	597.54	62.85	89,568.71
8	597.12	63.27	89,505.44
9	596.70	63.69	89,441.75
10	596.28	64.11	89,377.64
11	595.85	64.54	89,313.10
12	595.42	64.97	89,248.13
Total	$7172.81	$751.87	

As we can see from the totals at the bottom of the columns, Marc and Mira will pay a total of $7172.81 in interest in the first year.

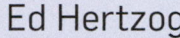

mathematics At Work

Ed Hertzog

TITLE Web Developer

INSTITUTION Renga Five, LLC

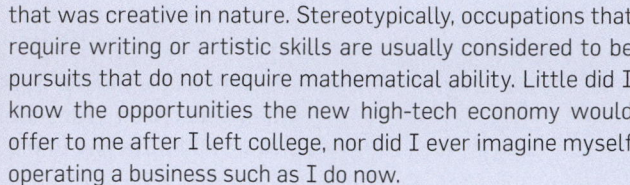

Patrick Farace

Renga Five, LLC is a web design firm based in Philadelphia, serving small- and medium-sized business' graphic design, programming, Internet marketing, and e-commerce needs.

As the President of a small consulting company, I am involved in all facets of the business, from a financial, creative and technical standpoint. Whether I am calculating revenue projections for the next three months or writing a piece of software to automate business processes for an accountant, math is an integral part of my day-to-day activities as a small business owner.

While I was obtaining my college degree in Political Science I assumed I would most likely find myself in a career

that was creative in nature. Stereotypically, occupations that require writing or artistic skills are usually considered to be pursuits that do not require mathematical ability. Little did I know the opportunities the new high-tech economy would offer to me after I left college, nor did I ever imagine myself operating a business such as I do now.

In my day-to-day work life I may have to perform any number of tasks that involve diverse mathematical concepts. One week I may be writing an application to graph financial statistics. The next I may be building an e-commerce shopping cart. In any of these circumstances, my math skills have limitless value to me. My career path will only involve more of this skill in the future.

Bonds

Suppose that a corporation offers a 10-year bond paying 6.5% with payments every six months. As we saw in Example 3 of Section 5.1, this means that if we pay $10,000 for bonds with a maturity value of $10,000, we will receive $6.5/2 = 3.25\%$ of $10,000, or $325, every six months for 10 years, at the end of which time the corporation will give us the original $10,000 back. But bonds are rarely sold at their maturity value. Rather, they are auctioned off and sold at a price the bond market determines they are worth.

For example, suppose that bond traders are looking for an investment that has a **rate of return** or **yield** of 7% rather than the stated 6.5% (sometimes called the **coupon interest rate** to distinguish it from the rate of return). How much would they be willing to pay for the bonds above with a maturity value of $10,000? Think of the bonds as an investment that will pay the owner $325 every six months for 10 years, at the end of which time it will pay $10,000. We can treat the $325 payments as if they come from an annuity and determine how much an investor would pay for such an annuity if it earned 7% compounded semiannually. Separately, we determine the present value of an investment worth $10,000 ten years from now, if it earned 7% compounded semiannually. For the first calculation, we use the annuity present value formula, with $i = 0.07/2$ and $n = 2 \times 10 = 20$.

$$PV = PMT\frac{1 - (1 + i)^{-n}}{i}$$
$$= 325\frac{1 - (1 + 0.07/2)^{-20}}{0.07/2}$$
$$= \$4619.03$$

For the second calculation, we use the present value formula for compound interest:

$$PV = 10,000(1 + 0.07/2)^{-20}$$
$$= \$5025.66$$

Thus, an investor looking for a 7% return will be willing to pay $4619.03 for the semi-annual payments of $325 and $5025.66 for the $10,000 payment at the end of 10 years, for a total of $4619.03 + 5025.66 = $9644.69 for the $10,000 bond.

Example 8 Bonds

Suppose that bond traders are looking for only a 6% yield on their investment. How much would they pay per $10,000 for the 10-year bonds above, which have a coupon interest rate of 6.5% and pay interest every six months?

Solution We redo the calculation with $r = 0.06$. For the annuity calculation we now get

$$PV = 325 \frac{1 - (1 + 0.06/2)^{-20}}{0.06/2} = \$4835.18$$

For the compound interest calculation we get

$$PV = 10,000(1 + 0.06/2)^{-20} = \$5536.76$$

Thus, traders would be willing to pay a total of $4835.18 + $5536.76 = $10,371.94 for bonds with a maturity value of $10,000.

+ *Before we go on...* Notice how the selling price of the bonds behaves as the desired yield changes. As desired yield goes up, the price of the bonds goes down, and as desired yield goes down, the price of the bonds goes up. When the desired yield equals the coupon interest rate, the selling price will equal the maturity value. Therefore, when the yield is higher than the coupon interest rate, the price of the bond will be below its maturity value, and when the yield is lower than the coupon interest rate, the price will be above the maturity value.

As we've mentioned before, the desired yield depends on many factors, but it generally moves up and down with prevailing interest rates. And interest rates have historically gone up and down cyclically. The effect on the value of bonds can be quite dramatic (see Exercises 65 and 66 in the preceding section). Because bonds can be sold again once bought, someone who buys bonds while interest rates are high and then resells them when interest rates decline can make a healthy profit. ∎

Example 9 Rate of Return on a Bond

Suppose that a 5%, 20-year bond sells for $9800 per $10,000 maturity value. What rate of return will investors get?

using *Technology*

See the Technology Guides at the end of the chapter for a discussion of using a TI-83/84 or Excel to find the yield using built-in functions.

Solution Assuming the usual semiannual payments, we know the following about the annuity calculation:

$$PMT = 0.05 \times 10,000/2 = 250$$
$$n = 20 \times 2 = 40$$

What we do not know is r or i, the annual or semiannual interest rate. So, we write

$$PV = 250 \frac{1 - (1 + i)^{-40}}{i}$$

For the compound interest calculation, we know $FV = 10,000$ and $n = 40$ again, so we write

$$PV = 10,000(1 + i)^{-40}$$

Adding these together should give the selling price of $9800:

$$250\frac{1 - (1 + i)^{-40}}{i} + 10,000(1 + i)^{-40} = 9800$$

This equation cannot be solved for i directly. The best we can do by hand is a sort of trial and error approach, substituting a few values for i in the left-hand side of the above equation to get an estimate:

i	0.01	0.02	0.03
$250\dfrac{1 - (1 + i)^{-40}}{i} + 10,000(1 + i)^{-40}$	14,925	11,368	8844

Because we want the value to be 9800, we see that the correct answer is somewhere between $i = 0.02$ and $i = 0.03$. Let us try the value midway between 0.02 and 0.03; namely $i = 0.025$.

i	0.02	0.025	0.03
$250\dfrac{1 - (1 + i)^{-40}}{i} + 10,000(1 + i)^{-40}$	11,368	10,000	8,844

Now we know that the correct value of i is somewhere between 0.025 and 0.03, so we choose for our next estimate of i the number midway between them: 0.0275. We could continue in this fashion to obtain i as accurately as we like. In fact, $i \approx 0.02581$, corresponding to an annual rate of return of approximately 5.162%.

FAQs   Which Formula to Use

Q: *We have retirement accounts, trust funds, loans, bonds, and so on. Some are sinking funds, others are annuities. How do we distinguish among them, so we can tell which formula to use*?

A: In general, remember that a sinking fund is an interest-bearing fund into which payments are made, while an annuity is an interest-bearing fund from which money is withdrawn. Here is a list of some of the accounts we have discussed in this section:

- *Retirement Accounts* A retirement account is a sinking fund while payments are being made into the account (prior to retirement) and an annuity while a pension is being withdrawn (after retirement).
- *Education Funds* These are similar to retirement accounts.
- *Trust Funds* A trust fund is an annuity if periodic withdrawals are made.
- *Installment Loans* We think of an installment loan as an investment a bank makes in the lender. In this way, the lender's payments can be viewed as the bank's withdrawals, and so a loan is an annuity.
- *Bonds* A bond pays regular fixed amounts until it matures, at which time it pays its maturity value. We think of the bond as an annuity coupled with a compound interest investment funding the payment of the maturity value. We can then determine its present value based on the current market interest rate.

From a mathematical point of view, sinking funds and annuities are really the same thing. See the Communication and Reasoning Exercises for more about this. ∎

5.3 EXERCISES

● denotes basic skills exercises
◆ denotes challenging exercises
tech Ex indicates exercises that should be solved using technology

Find the amount accumulated in the sinking funds in Exercises 1–6. (Assume end-of-period deposits and compounding at the same intervals as deposits.) hint [see Example 1]

1. ● $100 deposited monthly for 10 years at 5% per year

2. ● $150 deposited monthly for 20 years at 3% per year

3. ● $1000 deposited quarterly for 20 years at 7% per year

4. ● $2000 deposited quarterly for 10 years at 7% per year

5. $100 deposited monthly for 10 years at 5% per year in an account containing $5000 at the start

6. $150 deposited monthly for 20 years at 3% per year in an account containing $10,000 at the start

Find the periodic payments necessary to accumulate the amounts given in Exercises 7–12 in a sinking fund. (Assume end-of-period deposits and compounding at the same intervals as deposits.) hint [see Example 2]

7. ● $10,000 in a fund paying 5% per year, with monthly payments for 5 years

8. ● $20,000 in a fund paying 3% per year, with monthly payments for 10 years

9. ● 75,000 in a fund paying 6% per year, with quarterly payments for 20 years

10. ● $100,000 in a fund paying 7% per year, with quarterly payments for 20 years

11. $20,000 in a fund paying 5% per year, with monthly payments for 5 years, if the fund contains $10,000 at the start

12. $30,000 in a fund paying 3% per year, with monthly payments for 10 years, if the fund contains $10,000 at the start

Find the present value of the annuity necessary to fund the withdrawals given in Exercises 13–18. (Assume end-of-period withdrawals and compounding at the same intervals as withdrawals.) hint [see Example 3]

13. ● $500 per month for 20 years, if the annuity earns 3% per year

14. ● $1000 per month for 15 years, if the annuity earns 5% per year

15. ● $1500 per quarter for 20 years, if the annuity earns 6% per year

16. ● $2000 per quarter for 20 years, if the annuity earns 4% per year

17. $500 per month for 20 years, if the annuity earns 3% per year and if there is to be $10,000 left in the annuity at the end of the 20 years

18. $1000 per month for 15 years, if the annuity earns 5% per year and if there is to be $20,000 left in the annuity at the end of the 15 years

Find the periodic withdrawals for the annuities given in Exercises 19–24. (Assume end-of-period withdrawals and compounding at the same intervals as withdrawals.) hint [see Example 4]

19. ● $100,000 at 3%, paid out monthly for 20 years

20. ● $150,000 at 5%, paid out monthly for 15 years

21. ● $75,000 at 4%, paid out quarterly for 20 years

22. ● $200,000 at 6%, paid out quarterly for 15 years

23. $100,000 at 3%, paid out monthly for 20 years, leaving $10,000 in the account at the end of the 20 years

24. $150,000 at 5%, paid out monthly for 15 years, leaving $20,000 in the account at the end of the 15 years

Determine the periodic payments on the loans given in Exercises 25–28. hint [see Example 6]

25. ● $10,000 borrowed at 9% for 4 years, with monthly payments

26. ● $20,000 borrowed at 8% for 5 years, with monthly payments

27. ● $100,000 borrowed at 5% for 20 years, with quarterly payments

28. ● $1,000,000 borrowed at 4% for 10 years, with quarterly payments

Determine the selling price, per $1000 maturity value, of the bonds[16] in Exercises 29–32. (Assume twice-yearly interest payments.) hint [see Example 8]

29. 10 year, 4.875% bond, with a yield of 4.880%

30. 30 year, 5.375% bond, with a yield of 5.460%

31. 2 year, 3.625% bond, with a yield of 3.705%

32. 5 year, 4.375% bond, with a yield of 4.475%

Determine the yield on the bonds[17] in Exercises 33–36. (Assume twice-yearly interest payments.) hint [see Example 9]

33. tech Ex 5 year, 3.5% bond, selling for $994.69 per $1000 maturity value

34. tech Ex 10 year, 3.375% bond, selling for $991.20 per $1000 maturity value

35. tech Ex 2 year, 3% bond, selling for $998.86 per $1000 maturity value

---

[16] These are actual U.S. Treasury notes and bonds auctioned in 2001 and 2002. SOURCE: The Bureau of the Public Debt's website, at www.publicdebt.treas.gov.

[17] Ibid.

● basic skills   ◆ challenging   tech Ex technology exercise

36. `tech` Ex 5 year, 4.625% bond, selling for $998.45 per $1000 maturity value

Applications

37. *Pensions* Your pension plan is an annuity with a guaranteed return of 3% per year (compounded monthly). You would like to retire with a pension of $5000 per month for 20 years. If you work 40 years before retiring, how much must you and your employer deposit each month into the fund? *hint* [see Example 5]

38. *Pensions* Meg's pension plan is an annuity with a guaranteed return of 5% per year (compounded quarterly). She would like to retire with a pension of $12,000 per quarter for 25 years. If she works 45 years before retiring, how much money must she and her employer deposit each quarter?

39. *Pensions* Your pension plan is an annuity with a guaranteed return of 4% per year (compounded quarterly). You can afford to put $1200 per quarter into the fund, and you will work for 40 years before retiring. After you retire, you will be paid a quarterly pension based on a 25-year payout. How much will you receive each quarter?

40. *Pensions* Jennifer's pension plan is an annuity with a guaranteed return of 5% per year (compounded monthly). She can afford to put $300 per month into the fund, and she will work for 45 years before retiring. If her pension is then paid out monthly based on a 20-year payout, how much will she receive per month?

41. ● *Car Loans* While shopping for a car loan, you get the following offers: Solid Savings & Loan is willing to loan you $10,000 at 9% interest for four years. Fifth Federal Bank & Trust will loan you the $10,000 at 7% interest for three years. Both require monthly payments. You can afford to pay $250 per month. Which loan, if either, can you take?

42. ● *Business Loans* You need to take out a loan of $20,000 to start up your T-shirt business. You have two possibilities: One bank is offering a 10% loan for five years, and another is offering a 9% loan for four years. Which will have the lower monthly payments? On which will you end up paying more interest total?

Note: In Exercises 43–48 we suggest the use of the finance functions in the TI-83/84 or amortization tables in Excel. *hint* [see Example 7]

43. `tech` Ex *Mortgages* You take out a 15-year mortgage for $50,000, at 8%, to be paid off monthly. Construct an amortization table showing how much you will pay in interest each year, and how much goes towards paying off the principal.

44. `tech` Ex *Mortgages* You take out a 30-year mortgage for $95,000 at 9.75%, to be paid off monthly. If you sell your house after 15 years, how much will you still owe on the mortgage?

45. `tech` Ex *Mortgages* This exercise describes a popular kind of mortgage. You take out a $75,000, 30-year mortgage. For the first five years the interest rate is held at 5%, but for the remaining 25 years it rises to 9.5%. The payments for the first five years are calculated as if the 5% rate were going to remain in effect for all 30 years, and then the payments for the last 25 years are calculated to amortize the debt remaining at the end of the fifth year. What are your monthly payments for the first five years, and what are they for the last 25 years?

46. `tech` Ex *Adjustable Rate Mortgages* You take out an adjustable rate mortgage for $100,000 for 20 years. For the first 5 years, the rate is 4%. It then rises to 7% for the next 10 years, and then 9% for the last 5 years. What are your monthly payments in the first 5 years, the next 10 years, and the last 5 years? (Assume that each time the rate changes, the payments are recalculated to amortize the remaining debt if the interest rate were to remain constant for the remaining life of the mortgage.)

47. `tech` Ex *Refinancing* Your original mortgage was a $96,000, 30-year 9.75% mortgage. After 4 years you refinance the remaining principal for 30 years at 6.875%. What was your original monthly payment? What is your new monthly payment? How much will you save in interest over the course of the loan by refinancing?

48. `tech` Ex *Refinancing* Kara and Michael take out a $120,000, 30-year, 10% mortgage. After 3 years they refinance the remaining principal with a 15 year, 6.5% loan. What were their original monthly payments? What is their new monthly payment? How much did they save in interest over the course of the loan by refinancing?

49. `tech` Ex *Fees* You take out a 2-year, $5000 loan at 9% interest with monthly payments. The lender charges you a $100 fee that can be paid off, interest free, in equal monthly installments over the life of the loan. Thinking of the fee as additional interest, what is the actual annual interest rate you will pay?

50. `tech` Ex *Fees* You take out a 3-year, $7000 loan at 8% interest with monthly payments. The lender charges you a $100 fee that can be paid off, interest free, in equal monthly installments over the life of the loan. Thinking of the fee as additional interest, what is the actual annual interest rate you will pay?

51. `tech` Ex *Savings* You wish to accumulate $100,000 through monthly payments of $500. If you can earn interest at an annual rate of 4% compounded monthly, how long (to the nearest year) will it take to accomplish your goal?

52. `tech` Ex *Retirement* Alonzo plans to retire as soon as he has accumulated $250,000 through quarterly payments of $2500. If Alonzo invests this money at 5.4% interest, compounded quarterly, when (to the nearest year) can he retire?

53. `tech` Ex *Loans* You have a $2000 credit card debt, and you plan to pay it off through monthly payments of $50. If you are being charged 15% interest per year, how long (to the nearest 0.5 years) will it take you to repay your debt?

54. `tech` Ex *Loans* You owe $2000 on your credit card, which charges you 15% interest. Determine, to the nearest 1¢, the minimum monthly payment that will allow you to eventually repay your debt.

● basic skills   ◆ challenging   `tech` Ex technology exercise

55. tech Ex *Savings* You are depositing $100 per month in an account that pays 4.5% interest per year (compounded monthly), while your friend Lucinda is depositing $75 per month in an account that earns 6.5% interest per year (compounded monthly). When, to the nearest year, will her balance exceed yours?

56. tech Ex *Car Leasing* You can lease a $15,000 car for $300 per month. For how long (to the nearest year) should you lease the car so that your monthly payments are lower than if you were purchasing it with an 8%-per-year loan?

Communication and Reasoning Exercises

57. ● Your cousin Simon claims that you have wasted your time studying annuities: If you wish to retire on an income of $1000 per month for 20 years, you need to save $1000 per month for 20 years. Explain why he is wrong.

58. Your other cousin Trevor claims that you will earn more interest by accumulating $10,000 through smaller payments than through larger payments made over a shorter period. Is he correct? Give a reason for your answer.

59. A real estate broker tells you that doubling the period of a mortgage halves the monthly payments. Is he correct? Support your answer by means of an example.

60. Another real estate broker tells you that doubling the size of a mortgage doubles the monthly payments. Is he correct? Support your answer by means of an example.

61. ◆ Consider the formula for the future value of a sinking fund with given payments. Show algebraically that the present value of that future value is the same as the present value of the annuity required to fund the same payments.

62. ◆ Give a nonalgebraic justification for the result from the preceding exercise.

● basic skills   ◆ challenging   tech Ex technology exercise

Chapter 5 Review

KEY CONCEPTS

5.1 Simple Interest

Simple interest *p. 342*
Future value *p. 343*
Bond, maturity value *p. 344*
Present value *p. 345*

5.2 Compound Interest

Compound interest *p. 350*
Future value for compound interest
p. 351

Present value for compound interest
p. 352
Zero coupon bond or strip *p. 353*
Inflation, constant dollars *p. 354*
Effective interest rate, annual percentage
yield (APY) *p. 355*

5.3 Annuities, Loans, and Bonds

Annuity, sinking fund *p. 360*
Future value of a sinking fund *p. 361*

Payment formula for a sinking fund
p. 362
Present value of an annuity *p. 364*
Ordinary annuity, annuity due *p. 364*
Withdrawal formula for an annuity *p. 365*
Installment loan *p. 366*
Amortization schedule *p. 367*
Bond *p. 368*

REVIEW EXERCISES

In each of Exercises 1–6, find the future value of the investment.

1. $6000 for 5 years at 4.75% simple annual interest

2. $10,000 for 2.5 years at 5.25% simple annual interest

3. $6000 for 5 years at 4.75% compounded monthly

4. $10,000 for 2.5 years at 5.25% compounded semiannually

5. $100 deposited at the end of each month for 5 years, at 4.75% interest compounded monthly

6. $2000 deposited at the end of each half-year for 2.5 years, at 5.25% interest compounded semiannually

In each of Exercises 7–12, find the present value of the investment.

7. Worth $6000 after 5 years at 4.75% simple annual interest

8. Worth $10,000 after 2.5 years at 5.25% simple annual interest

9. Worth $6000 after 5 years at 4.75% compounded monthly

10. Worth $10,000 after 2.5 years at 5.25% compounded semiannually

11. Funding $100 withdrawals at the end of each month for 5 years, at 4.75% interest compounded monthly

12. Funding $2000 withdrawals at the end of each half-year for 2.5 years, at 5.25% interest compounded semiannually

In each of Exercises 13–18, find the amounts indicated.

13. The monthly deposits necessary to accumulate $12,000 after 5 years in an account earning 4.75% compounded monthly

14. The semiannual deposits necessary to accumulate $20,000 after 2.5 years in an account

15. The monthly withdrawals possible over 5 years from an account earning 4.75% compounded monthly and starting with $6000

16. The semiannual withdrawals possible over 2.5 years from an account earning 5.25% compounded semiannually and starting with $10,000

17. The monthly payments necessary on a 5-year loan of $10,000 at 4.75%

18. The semiannual payments necessary on a 2.5-year loan of $15,000 at 5.25%

19. How much would you pay for a $10,000, 5-year, 6% bond if you want a return of 7%? (Assume that the bond pays interest every 6 months.)

20. How much would you pay for a $10,000, 5-year, 6% bond if you want a return of 5%? (Assume that the bond pays interest every 6 months.)

21. *tech* Ex A $10,000, 7-year, 5% bond sells for $9800. What return does it give you? (Assume that the bond pays interest every 6 months.)

22. *tech* Ex A $10,000, 7-year, 5% bond sells for $10,200. What return does it give you? (Assume that the bond pays interest every 6 months.)

In each of Exercises 23–28, find the time requested, to the nearest 0.1 year.

23. The time it would take $6000 to grow to $10,000 at 4.75% simple annual interest

24. The time it would take $10,000 to grow to $15,000 at 5.25% simple annual interest

25. *tech* Ex The time it would take $6000 to grow to $10,000 at 4.75% interest compounded monthly

26. *tech* Ex The time it would take $10,000 to grow to $15,000 at 5.25% interest compounded semiannually

27. *tech* Ex The time it would take to accumulate $10,000 by depositing $100 at the end of each month in an account earning 4.75% interest compounded monthly

28. **tech** Ex The time it would take to accumulate $15,000 by depositing $2000 at the end of each half-year in an account earning 5.25% compounded semiannually

Applications

Total online revenues at OHaganBooks.com during 1999, its first year of operation, amounted to $150,000.

29. After December, 1999, revenues increased by a steady 20% each year. Track OHaganBooks.com's revenues for the subsequent 5 years, assuming that this rate of growth continued. During which year did the revenue surpass $300,000?

30. Unfortunately, the picture for net income was not so bright: The company lost $20,000 in the fourth quarter of 1999. However, the quarterly loss decreased at an annual rate of 15%. How much did the company lose during the third quarter of 2001?

31. In order to finance anticipated expansion, you are considering making a public offering of OHaganBooks.com shares at $3.00 a share. You are not sure how many shares to offer, but you would like the total value of the shares to reach a market value of at least $500,000 six months after the initial offering. Your financial adviser, Sally McCormack, tells you that you can expect the value of the stock to double in the first day of trading, and then you should expect an appreciation rate of around 8% per month for the first six months. How many shares should you sell?

Unfortunately, the stock market takes a dive just as you are concluding plans for your stock offering in Question 31, so you decide to postpone the offering until the market shows renewed vigor. In the meantime, there is an urgent need to finance the continuing losses at OHaganBooks.com, so you decide to seek a $250,000 loan.

32. The two best deals you are able to find are Industrial Bank, which offers a 10-year 9.5% loan, and Expansion Loans, offering an 8-year 6.5% loan. What are the monthly payments for each loan?

33. You have estimated that OHaganBooks.com can afford to pay only $3000 per month to service the loan. What, to the nearest dollar, is the largest amount the company can borrow from Expansion Loans?

34. **tech** Ex What interest rate would Expansion Loans have to offer in order to meet the company's loan requirements at a price it can afford?

OHaganBooks.com has just introduced a retirement package for the employees. Under the annuity plan operated by Sleepy Hollow Retirement, Inc., the monthly contribution by the company on behalf of each employee is $800. Each employee can then supplement that amount through payroll deductions.

35. The current rate of return of Sleepy Hollow Inc.'s retirement fund is 7.3%, and James Callahan, your site developer, plans to retire in 10 years. He contributes $1000 per month to the plan (in addition to the company contribution of $800). Currently, there is $50,000 in his retirement annuity. How much (to the nearest dollar) will it be worth when he retires?

36. How much of that retirement fund was from the company contribution? (The company did not contribute toward the $50,000 Callahan now has.)

37. Callahan actually wants to retire with $500,000. How much should he contribute each month to the annuity?

38. On second thoughts, Callahan wants to be in a position to draw at least $5000 per month for 30 years after his retirement. He feels he can invest the proceeds of his retirement annuity at 8.7% per year in perpetuity. Given the information in part (a), how much will he need to contribute to the plan starting now?

Actually, James Callahan is quite pleased with himself; one year ago he purchased a $50,000 government bond paying 7.2% per year (with interest paid every six months) and maturing in 10 years, and interest rates have come down since then.

39. The current interest rate on 10-year government bonds is 6.3%. If he were to auction the bond at the current interest rate, how much would he get?

40. If he holds on to the bond for six more months and the interest rate drops to 6%, how much will the bond be worth then?

41. **tech** Ex James suspects that interest rates will come down further during the next six months. If he hopes to auction the bond for $54,000 in six months' time, what will the interest rate need to be at that time?

vMentor Do you need a live tutor for homework problems? Access vMentor on the ThomsonNOW! website at **www.thomsonedu.com** for one-on-one tutoring from a mathematics expert.

CASE STUDY: Saving for College

Cindy Charles/PhotoEdit

Tuition costs at public and private four-year colleges and universities increased at an average rate of about 5.8% per year from 1991 through 2001. The 2001–2002 costs for tuition, fees, room, and board were $9,008 for a public in-state college and $23,578 for a private college.[18]

Mr. and Mrs. Wong have an appointment tomorrow with you, their investment counselor, to discuss a plan to save for their newborn child's college education. You have already recommended that they invest a fixed amount each month in mutual funds expected to yield 6% per year (with earnings reinvested each month) until their child begins college at age 18. They have indicated that they are not sure when they will start making the monthly payments, however, and of course they don't know which type of college or university their child will attend.

You decide that you should create an Excel worksheet that will compute the monthly payment *PMT* for various possibilities for the age of the child at the time the Wongs begin making payments and the current cost of tuition at the college. The figures you have are for 2001 and it is now 2006, so you update your figures using the future value formula for compound interest, assuming that the 5.8% rate of increase continued.

$$9008(1 + 0.058)^5 = \$11,941$$

and

$$23,578(1 + 0.058)^5 = \$31,256$$

You begin to create your worksheet by entering various possible ages and the two tuition figures you just computed.

Microsoft Excel - CS 5 Compute the Monthly Payment PMT.xls

File   Edit   View   Insert   Format   Tools   Data   Window   Help   Adobe PDF   Type a question for

	A	B	C	D	E	F	G
1		**Tuition**					
2	**Age**	$11,941.00	$31,256.00				
3	0						
4	1						
5	2						
6	3						
7	4						
8	5						
9	6						
10	7						
11	8						
12	9						
13	10						
14							
15							

Ultimately, you would like to compute in cells B3 through C13 the monthly payments corresponding to the various possible combinations of age and tuition. You decide that it

[18] Figures are enrollment-weighted averages. Source: *Trends in College Pricing 2001,* The College Board, 2001, available Online at www.collegeboard.com.

would also be good to enter the rate of inflation of tuition costs and the assumed return on the mutual fund in their own cells.

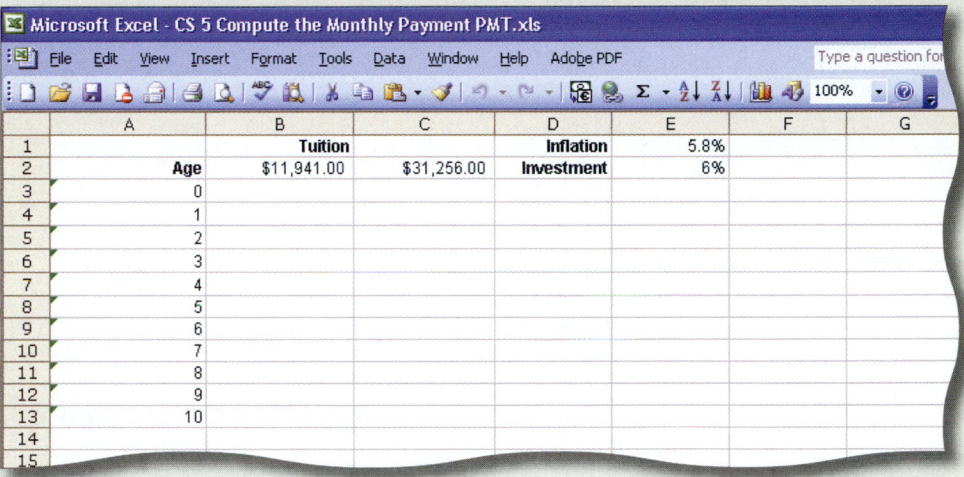

You realize that you need to do a two-stage calculation: First, you need to find the amount of money required to fund the tuition payments. Second, you will need to find the monthly payments that will lead to that amount of money.

The first calculation is not a simple annuity because the amount the Wongs need to withdraw for tuition will continue to rise over the course of the four years their child is in college. You begin by computing the 8 semiannual tuition payments that will be required. The first occurs 18 years hence and pays for half a year, so will be half of the future value of the current tuition, with the assumed annual inflation rate. The second will pay for the second half of that year, so it will be identical. You can compute the future value using the FV worksheet function or directly. The next 6 payments also occur in pairs. You leave room above the tuition payments to compute the required accumulation.

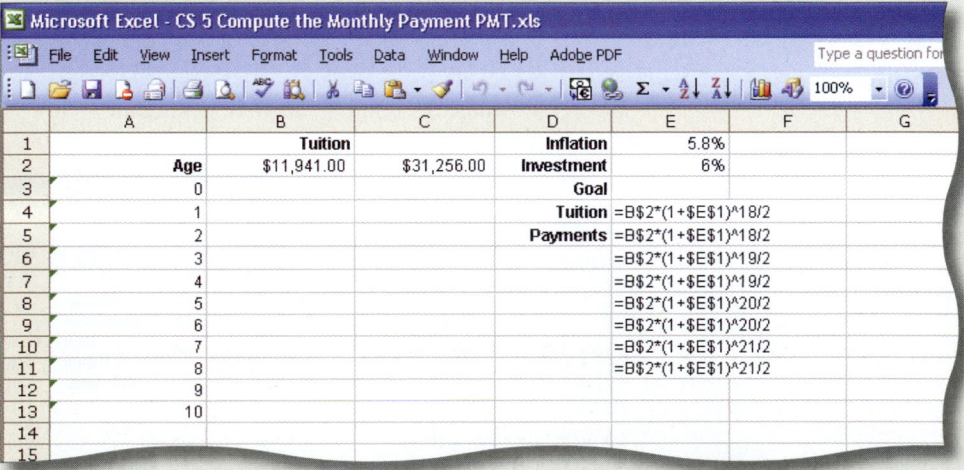

Copying these formulas into column F gives you two columns of figures—the one in column E being the tuition payments that will be required for a public college and the one in column F being the payments that will be required for a private college.

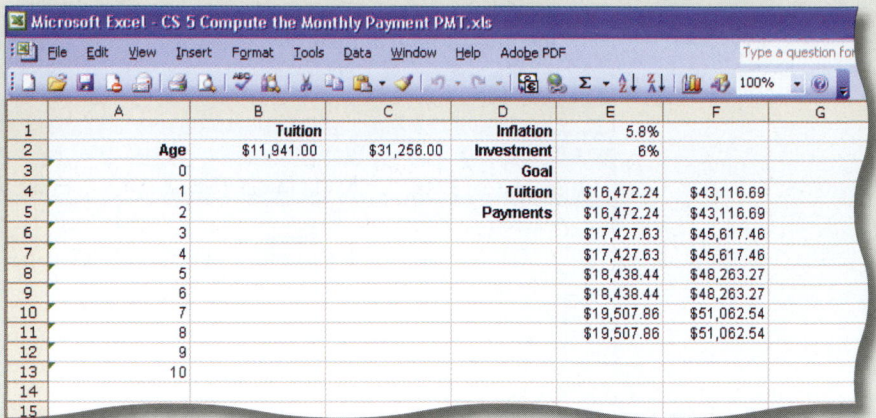

Now, enough money needs to be accumulated in the account to fund each of these payments. Because the account continues to earn interest, the amount required to fund each payment should be discounted back to the beginning of the child's college career. Thus, for column E (public college), you want a formula to compute

$$E4 + E5(1 + i)^{-1} + E6(1 + i)^{-2} + \cdots + E11(1 + i)^{-7}$$

where i is the semiannual interest rate for the mutual fund. Fortunately, Excel has a worksheet function just for this purpose, called NPV (Net Present Value). It takes an interest rate and a series of payments and computes the sum of the present values of those payments, assuming that the payments are made regularly at the end of each period. Because the first payment (in E4) will be made right away, you need to apply NPV to only E5 through E11 and add in E4. One more thing you need to do is determine the semiannual interest rate for the mutual fund. This is similar to the effective rate calculation: Because the fund earns (6%)/12 each month, over 6 months it will earn interest at a rate of

$$\left(1 + \frac{0.06}{12}\right)^{6} - 1$$

Thus, you enter the formula

```
=E4+NPV((1+$E$2/12)^6-1,E5:E11)
```

into cell E3 and copy it into F3 as well, getting the following result:

![Excel spreadsheet screenshot: Microsoft Excel - CS 5 Compute the Monthly Payment PMT.xls with columns A through G. Row 1: B=Tuition, D=Inflation, E=5.8%. Row 2: A=Age, B=$11,941.00, C=$31,256.00, D=Investment, E=6%. Row 3: A=0, D=Goal, E=$129,162.31, F=$338,087.02. Row 4: A=1, D=Tuition, E=$16,472.24, F=$43,116.69. Row 5: A=2, D=Payments, E=$16,472.24, F=$43,116.69. Row 6: A=3, E=$17,427.63, F=$45,617.46. Row 7: A=4, E=$17,427.63, F=$45,617.46. Row 8: A=5, E=$18,438.44, F=$48,263.27. Row 9: A=6, E=$18,438.44, F=$48,263.27. Row 10: A=7, E=$19,507.86, F=$51,062.54. Row 11: A=8, E=$19,507.86, F=$51,062.54. Row 12: A=9. Row 13: A=10.]

You have now completed the calculation of the amount that needs to be accumulated in the account to pay for college. What remains is to compute the monthly payment needed to reach this goal, but this is a straightforward sinking fund calculation. You enter the following formula in cell B3 and copy it to all the cells in the rectangle B3:C13:

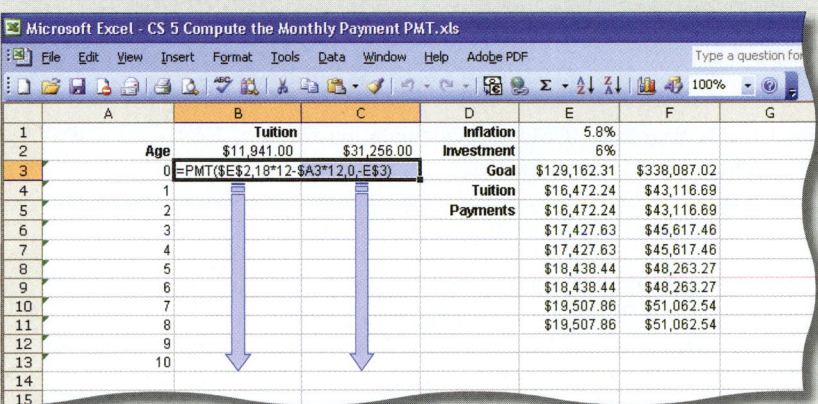

Finally, this gives you the computations you were looking for.

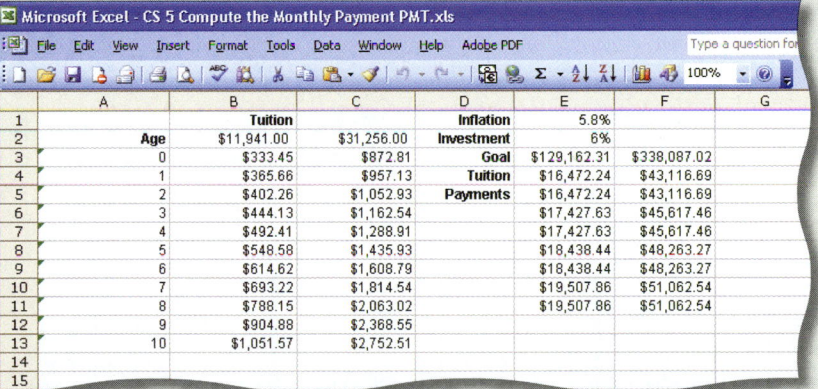

Thus, for example, if Mr. and Mrs. Wong start saving for a private college next year, when their child is one year old, they will need to invest $957.13 per month to save enough money to pay fully for private college.

Exercises

1. tech Ex Graph the monthly payments needed to save for a four-year public college as a function of the age of the child when payments start, from birth through age 17.

2. tech Ex Graph the monthly payments needed to save for a four-year private college as a function of the age of the child when payments start, from birth through age 17.

3. tech Ex Redo the calculations in the discussion above for the case where college costs are increasing by only 4% per year and the Wongs can invest their money at 10% per year.

4. Which strategy results in a lower monthly payment: waiting a year longer to begin college or starting to save for college a year earlier?

Section 5.2

Example 1 In August, 2005, the Amex Bank of Canada was paying 2.40% interest on savings accounts. If the interest is compounded quarterly, find the future value of a $2000 deposit in 6 years. What is the total interest paid over the period?

Solution with Technology We could calculate the future value using the TI-83/84 by entering

```
2000(1+0.0240/4)^(4*6)
```

However, the TI-83/84 has this and other useful calculations built into its TVM (Time Value of Money) Solver.

1. On the TI-83/84, press 2nd FINANCE and then choose item 1:TVM Solver, from the menu. (On a TI-84 Plus, press APPS then choose item 1:Finance and then choose item 1:TVM Solver.) This brings up the TVM Solver window.

 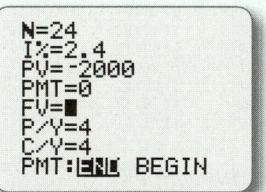

The second screen shows the values you should enter for this example. The various variables are:

N   Number of compounding periods
I%   Annual interest rate, as percent, not decimal
PV   Negative of present value
PMT   Payment per period (0 in this section)
FV   Future value
P/Y   Payments per year
C/Y   Compounding periods per year
PMT: Not used in this section

Several things to notice:

- *I%* is the *annual* interest rate, corresponding to *r*, not *i*, in the compound interest formula.

380

- The present value, *PV*, is entered as a negative number. In general, when using the TVM Solver, any amount of money you give to someone else (such as the $2000 you deposit in the bank) will be a negative number, whereas any amount of money someone gives to you (such as the future value of your deposit, which the bank will give back to you) will be a positive number.

- *PMT* is not used in this example (it will be used in the next section) and should be 0.

- *FV* is the future value, which we shall compute in a moment; it doesn't matter what you enter now.

- *P/Y* and *C/Y* stand for payments per year and compounding periods per year, respectively: They should both be set to the number of compounding periods per year for compound interest problems (setting *P/Y* automatically sets *C/Y* to the same value).

- *PMT*: *END* or *BEGIN* is not used in this example and it doesn't matter which you select.

2. To compute the future value, use the up or down arrow to put the cursor on the *FV* line, then press ALPHA SOLVE.

Example 2 Megabucks Corporation is issuing 10-year zero coupon bonds. How much would you pay for bonds with a maturity value of $10,000 if you wish to get a return of 6.5% compounded annually?

Solution with Technology To compute the present value using a TI-83/84:

1. Enter the numbers shown on the left in the TVM Solver window.

2. Put the cursor on the PV line, and press ALPHA SOLVE.

Why is the present value given as negative?

Example 6 You have $5000 to invest at 6% interest compounded monthly. How long will it take for your investment to grow to $6000?

Solution with Technology

1. Enter the numbers shown on the left in the TVM Solver window.

2. Put the cursor on the N line, and press ALPHA SOLVE.

Recall that I% is the annual interest rate, corresponding to r in the formula, but N is the number of compounding periods, so number of months in this example. Thus, you will need to invest your money for about 36.5 months, or just over three years, before it grows to $6000.

Section 5.3

Example 1 Your retirement account has $5000 in it and earns 5% interest per year compounded monthly. Every month for the next 10 years, you will deposit $100 into the account. How much money will there be in the account at the end of those 10 years?

Solution with Technology
We can use the TVM Solver in the TI-83/84 to calculate future values like these:

1. Enter the values shown on the left.

2. With the cursor on the FV line, press ALPHA SOLVE to find the future value.

Note that the TVM Solver allows you to put the $5000 already in the account as the present value of the account.

3. Following the TI-83/84's usual convention, set *PV* to the *negative* of the present value because this is money you paid into the account.

4. Likewise, set *PMT* to −100 because you are paying $100 each month.

5. Set the number of payment and compounding periods to 12 per year.

6. Finally, set the payments to be made at the end of each period.

Example 2 Tony and Maria have just had a son, José Phillipe. They establish an account to accumulate money for his college education. They would like to have $100,000 in this account after 17 years. If the account pays 4% interest per year compounded quarterly, and they make deposits at the end of every quarter, how large must each deposit be for them to reach their goal?

Solution with Technology

1. In the TVM Solver in the TI-83/84, enter the values shown below.

2. Solve for *PMT*.

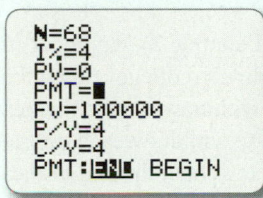 

Why is *PMT* negative?

Example 3 You wish to establish a trust fund from which your niece can withdraw $2000 every six months for 15 years, at which time she will receive the remaining money in the trust, which you would like to be $10,000. The trust will be invested at 7% per year compounded every six months. How large should the trust be?

Solution with Technology

1. In the TVM Solver in the TI-83/84, enter the values shown below.

2. Solve for *PV*.

The payment and future value are positive because you (or your niece) will be receiving these amounts from the investment.

TECHNOLOGY GUIDE

Note We have assumed that your niece receives the withdrawals at the end of each compounding period, so that the trust fund is an ordinary annuity. If, instead, she receives the payments at the beginning of each compounding period, it is an annuity due. You switch between the two types of annuity by changing PMT: END at the bottom to PMT: BEGIN. ■

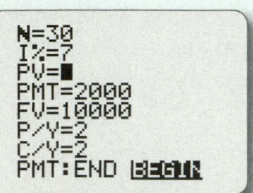 

As mentioned in the text, the present value must be higher to fund payments at the beginning of each period, because the money in the account has less time to earn interest.

Example 4 Tony and Maria (see Example 2), having accumulated $100,000 for José Phillipe's college education, would now like to make quarterly withdrawals over the next 4 years. How much money can they withdraw each quarter in order to draw down the account to zero at the end of the 4 years? (Recall that the account pays 4% interest compounded quarterly.)

Solution with Technology

1. Enter the values shown below in the TI-83/84 TVM Solver.

2. Solve for *PMT*.

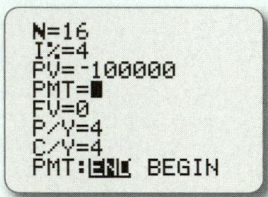 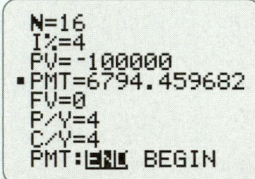

The present value is negative because Tony and Maria do not possess it; the bank does.

Example 7 Marc and Mira are buying a house, and have taken out a 30-year, $90,000 mortgage at 8% interest per year. Mortgage interest is tax-deductible, so it is important to know how much of a year's mortgage payments represents interest. How much interest will Marc and Mira pay in the first year of their mortgage?

Solution with Technology The TI-83/84 has built-in functions to compute the values in an amortization schedule.

1. First, use the TVM Solver to find the monthly payment.

Three functions correspond to the last three columns of the amortization schedule given in the text: ΣInt, ΣPrn, and bal (found in the menu accessed through [2ND] [FINANCE]). They all require that the values of *I%*, *PV*, and *PMT* be entered or calculated ahead of time; calculating the payment in the TVM Solver in Step 1 accomplishes this.

2. Use ΣInt($m, n, 2$) to compute the sum of the interest payments from payment *m* through payment *n*. For example,

$$\Sigma\text{Int}(1, 12, 2)$$

will return −7172.81, the total paid in interest in the first year, which answers the question asked in this example. (The last argument, 2, tells the calculator to round all intermediate calculations to two decimal places—that is, the nearest cent—as would the mortgage lender.)

3. Use ΣPrn($m, n, 2$) to compute the sum of the payments on principal from payment *m* through payment *n*. For example,

$$\Sigma\text{Prn}(1, 12, 2)$$

will return −751.87, the total paid on the principal in the first year.

4. Finally, bal($n, 2$) finds the balance of the principal outstanding after *n* payments. For example,

$$\text{bal}(12, 2)$$

will return the value 89248.13, the balance remaining at the end of one year.

5. To construct an amortization schedule as in the text, make sure that FUNC is selected in the [MODE] window; then enter the functions in the [Y=] window as shown below.

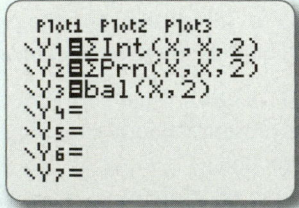

6. Press ⌐2ND⌐ ⌐TBLSET⌐ and enter the values shown here.

7. Press ⌐2ND⌐ ⌐TABLE⌐, to get the table shown here.

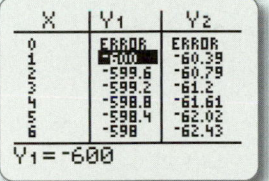 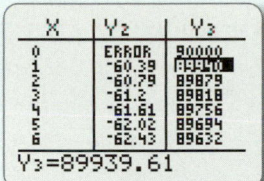

The column labeled X gives the month, the column labeled Y_1 gives the interest payment for each month, the column labeled Y_2 gives the payment on principal for each month, and the column labeled Y_3 (use the right arrow button to make it visible) gives the outstanding principal.

8. To see later months, use the down arrow. As you can see, some of the values will be rounded in the table, but by selecting a value (as the outstanding principal at the end of the first month is selected in the second screen) you can see its exact value at the bottom of the screen.

Example 9 Suppose that a 5%, 20-year bond sells for $9800 per $10,000 maturity value. What rate of return will investors get?

Solution with Technology We can use the TVM Solver in the TI-83/84 to find the interest rate just as we use it to find any other one of the variables.

1. Enter the values shown in the TVM Solver window.

2. Solve for *I%*. (Recall that *I%* is the annual interest rate, corresponding to *r* in the formula.)

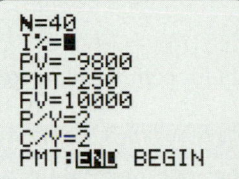 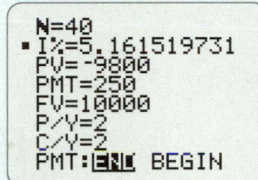

Thus, at $9800 per $10,000 maturity value, these bonds yield 5.162% compounded semiannually.

EXCEL Technology Guide

Section **5.2**

Example 1 In August 2005, the Amex Bank of Canada was paying 2.40% interest on savings accounts. If the interest is compounded quarterly, find the future value of a $2000 deposit in 6 years. What is the total interest paid over the period?

Solution with Technology You can either compute compound interest directly or use financial functions built into Excel. The following worksheet has more than we need for this example, but will be useful for other examples in this and the next section.

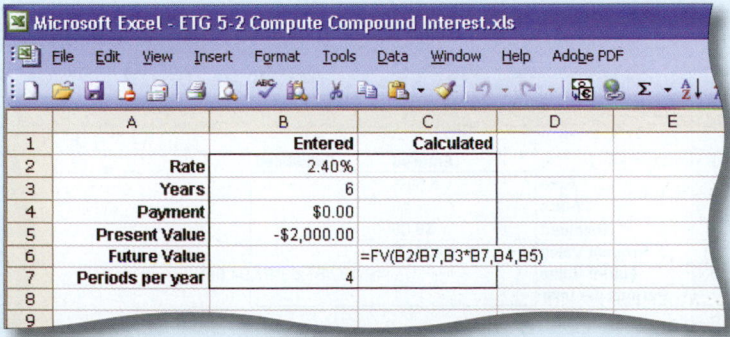

TECHNOLOGY GUIDE

For this example, the payment amount in B4 should be 0 (we shall use it in the next section).

1. Enter the other numbers as shown. As with other technologies, like the TVM Solver in the TI-83/84 calculator, money that you pay to others (such as the $2000 you deposit in the bank) should be entered as negative, whereas money that is paid to you is positive.

2. The formula entered in C6 uses the built-in FV function to calculate the future value based on the entries in column B. This formula has the following format:

$$FV(i, n, PMT, PV)$$

i = interest per period	We use B2/B7 for the interest
n = number of periods	We use B3*B7 for the number of periods
PMT = payment per period	The payment is 0 (cell B4)
PV = present value	The present value is in cell B5

Instead of using the built-in FV function, we could use

```
=-B5*(1+B2/B7)^(B3*B7)
```

based on the future value formula for compound interest. After calculation the result will appear in cell C6.

Microsoft Excel - ETG 5-2 Compute Compound Interest.xls

File   Edit   View   Insert   Format   Tools   Data   Window   Help   Adobe PDF

	A	B	C	D	E
1		Entered	Calculated		
2	Rate	2.40%			
3	Years	6			
4	Payment	$0.00			
5	Present Value	-$2,000.00			
6	Future Value		$2,308.77		
7	Periods per year	4			
8					
9					

Note that we have formatted the cells B4:C6 as currency with two decimal places. If you change the values in column B, the future value in column C will be automatically recalculated.

Example 2 Megabucks Corporation is issuing 10-year zero coupon bonds. How much would you pay for bonds with a maturity value of $10,000 if you wish to get a return of 6.5% compounded annually?

Solution with Technology   You can compute present value in Excel using the PV worksheet function. The following worksheet is similar to the one in the preceding example, except that we have entered a formula for computing the present value from the entered values.

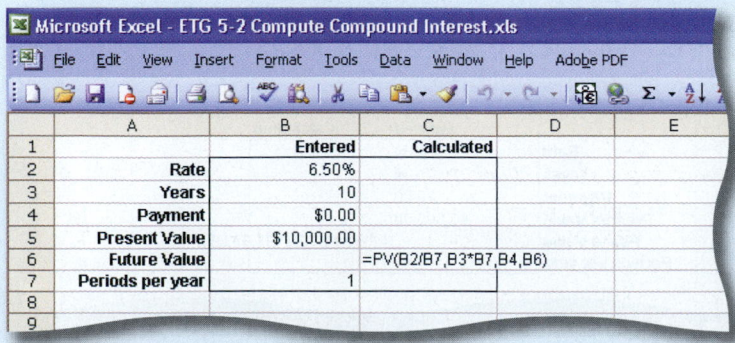

Microsoft Excel - ETG 5-2 Compute Compound Interest.xls

File   Edit   View   Insert   Format   Tools   Data   Window   Help   Adobe PDF

	A	B	C	D	E
1		Entered	Calculated		
2	Rate	6.50%			
3	Years	10			
4	Payment	$0.00			
5	Present Value	$10,000.00			
6	Future Value		=PV(B2/B7,B3*B7,B4,B6)		
7	Periods per year	1			
8					
9					

The next worksheet shows the calculated value.

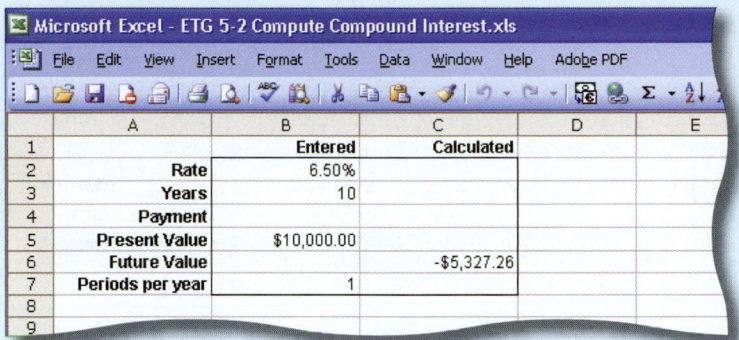

Why is the present value negative?

Example 6 You have $5000 to invest at 6% interest compounded monthly. How long will it take for your investment to grow to $6000?

Solution with Technology You can compute the requisite length of an investment in Excel using the NPER worksheet function. The following worksheets show the calculation.

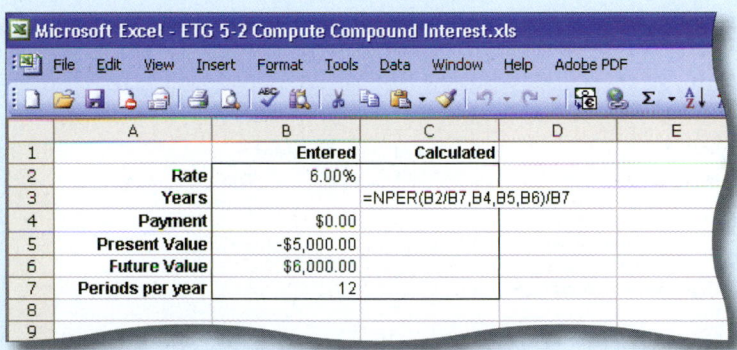

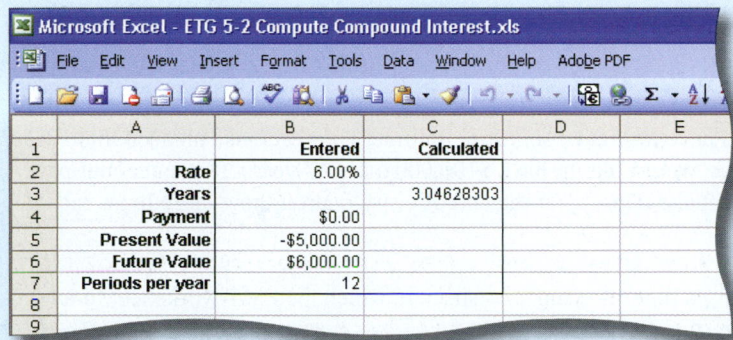

The NPER function computes the number of compounding periods, months in this case, so we divide by B7, the number of periods per year, to calculate the number of years, which appears as 3.046. So, you need to invest your money for just over 3 years for it to grow to $6000.

Section **5.3**

Example 1 Your retirement account has $5000 in it and earns 5% interest per year compounded monthly. Every month for the next 10 years you will deposit $100 into the account. How much money will there be in the account at the end of those 10 years?

Solution with Technology We can use exactly the same worksheet that we used in Example 1 in the preceding section. In fact, we included the "Payment" row in that worksheet just for this purpose.

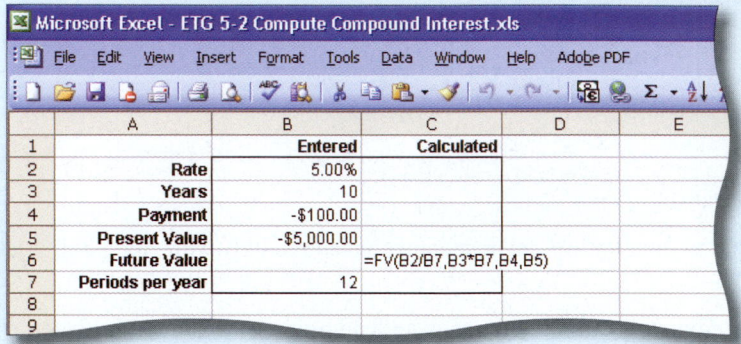

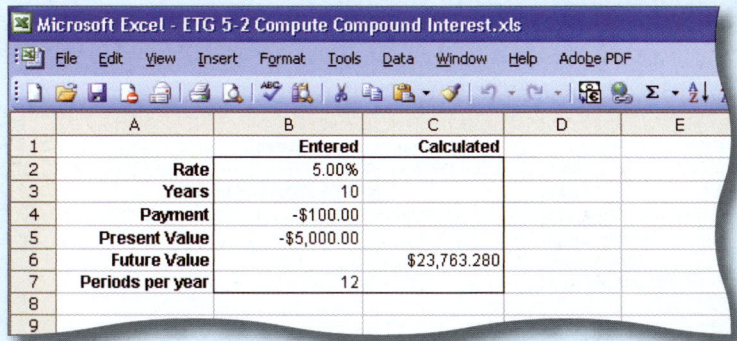

Note that the FV function allows us to enter, as the last argument, the amount of money already in the account. Following the usual convention, we enter the present value and the payment as *negative,* because these are amounts you pay into the account.

Example 2 Tony and Maria have just had a son, José Phillipe. They establish an account to accumulate money for his college education, in which they would like to have $100,000 after 17 years. If the account pays 4% interest per year compounded quarterly, and they make deposits at the end of every quarter, how large must each deposit be for them to reach their goal?

Solution with Technology Use the following worksheet, in which the PMT worksheet function is used to calculate the required payments.

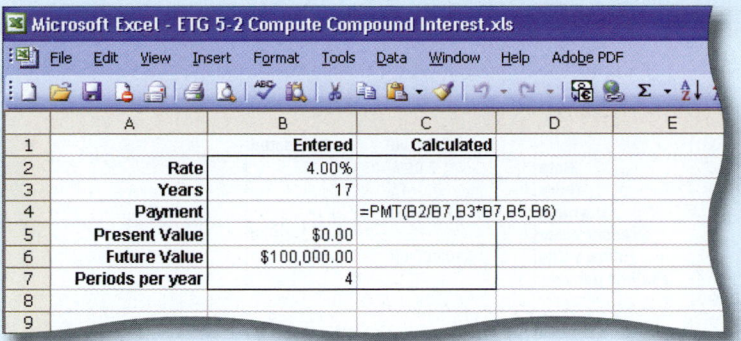

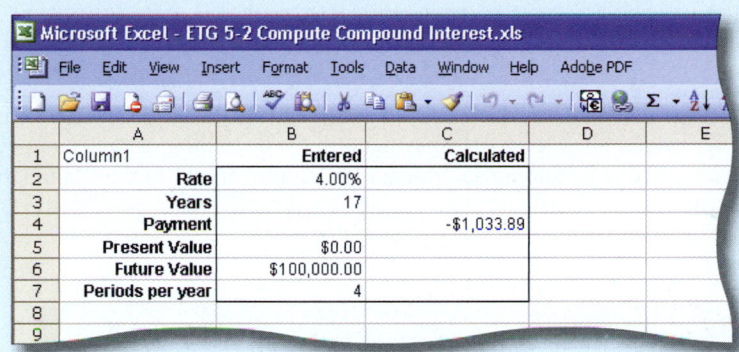

Why is the payment negative?

Example **3** You wish to establish a trust fund from which your niece can withdraw $2000 every six months for 15 years, at which time she will receive the remaining money in the trust, which you would like to be $10,000. The trust will be invested at 7% per year compounded every six months. How large should the trust be?

Solution with Technology You can use the same worksheet as in Example 2 in Section 5.2.

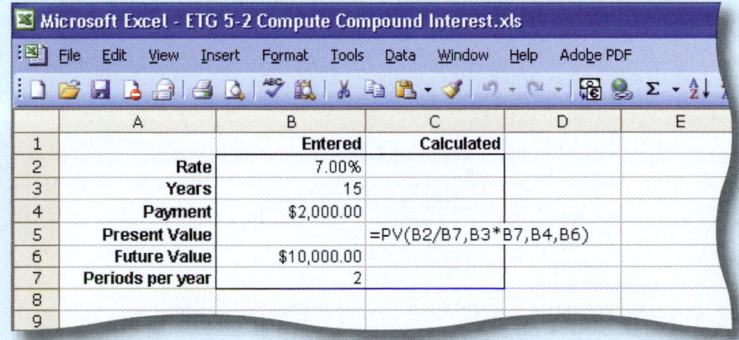

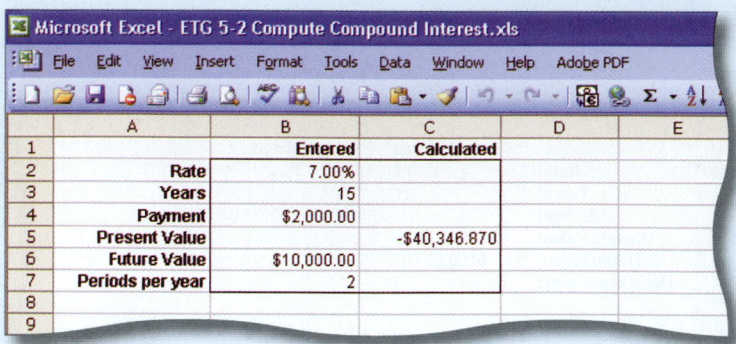

The payment and future value are positive because you (or your niece) will be receiving these amounts from the investment.

Note We have assumed that your niece receives the withdrawals at the end of each compounding period, so that the trust fund is an ordinary annuity. If, instead, she receives the payments at the beginning of each compounding period, it is an annuity due. You switch to an annuity due by adding an optional last argument of 1 to the PV function (and similarly for the other finance functions in Excel).

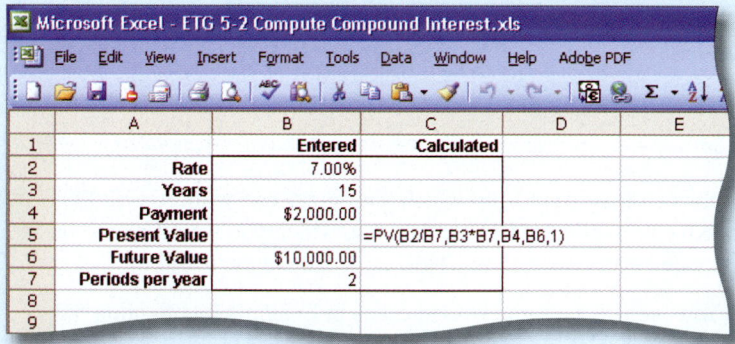

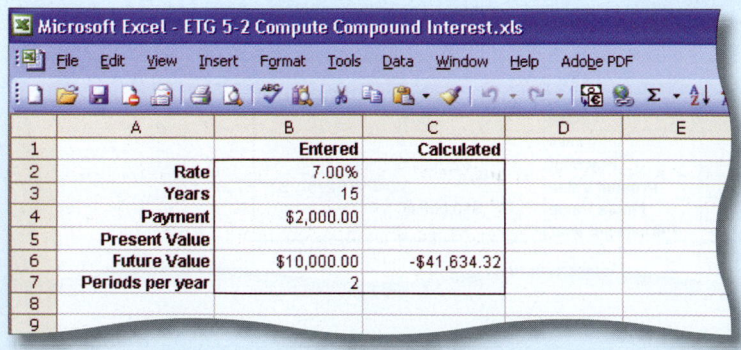

As mentioned in the text, the present value must be higher to fund payments at the beginning of each period, because the money in the account has less time to earn interest. ▪

Example 4 Tony and Maria (see Example 2), having accumulated $100,000 for José Phillipe's college education, would now like to make quarterly withdrawals over the next four years. How much money can they withdraw each quarter in order to draw down the account to zero at the end of the four years? (Recall that the account pays 4% interest compounded quarterly.)

Solution with Technology You can use the same worksheet as in Example 2.

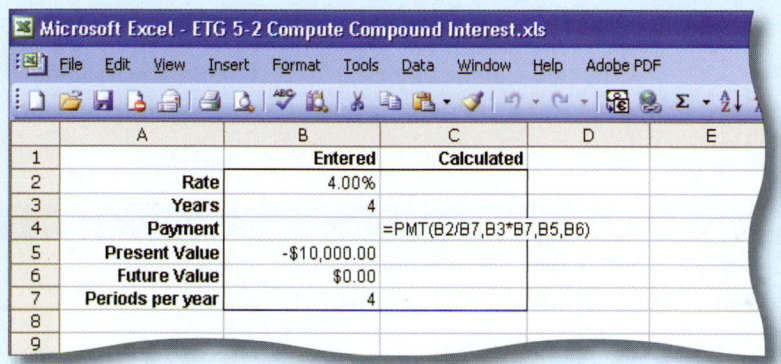

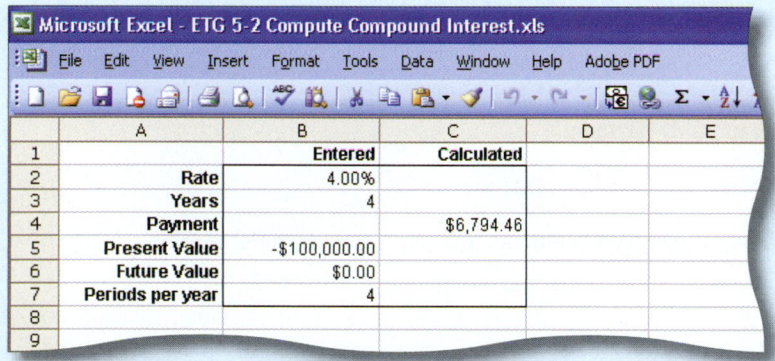

The present value is negative since Tony and Maria do not possess it; the bank does.

Example 7 Marc and Mira are buying a house, and have taken out a 30-year, $90,000 mortgage at 8% interest per year. Mortgage interest is tax-deductible, so it is important to know how much of a year's mortgage payments represents interest. How much interest will Marc and Mira pay in the first year of their mortgage?

TECHNOLOGY GUIDE

Solution with Technology We construct an amortization schedule with which we can answer the question.

1. Begin with the worksheet below.

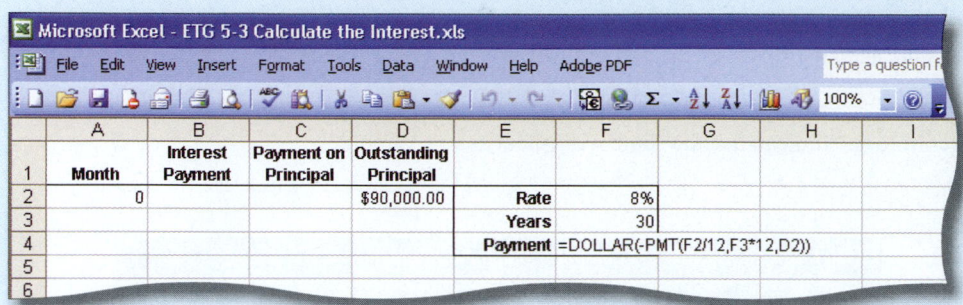

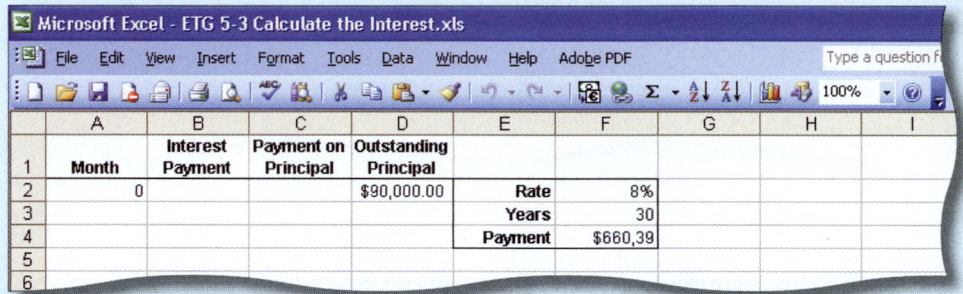

Note the formula for the monthly payment:

    =DOLLAR(-PMT(F2/12,F3*12,D2))

The function `DOLLAR` rounds the payment to the nearest cent, as the bank would.

2. Calculate the interest owed at the end of the first month using the formula

    =DOLLAR(D2*F$2/12)

in cell B3.

3. The payment on the principal is the remaining part of the payment, so enter

    =F$4-B3

in cell C3.

4. Calculate the outstanding principal by subtracting the payment on the principal from the previous outstanding principal, by entering

    =D2-C3

in cell D3.

5. Copy the formulas in cells B3, C3, and D3 into the cells below them to continue the table.

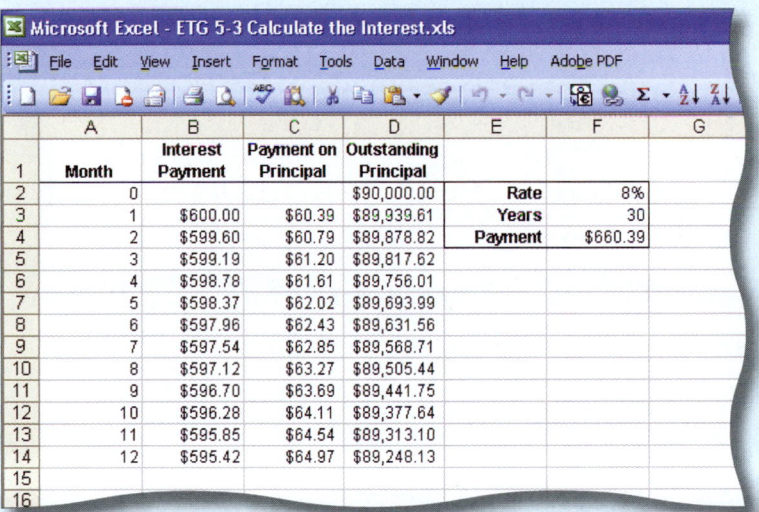

The result should be something like the following:

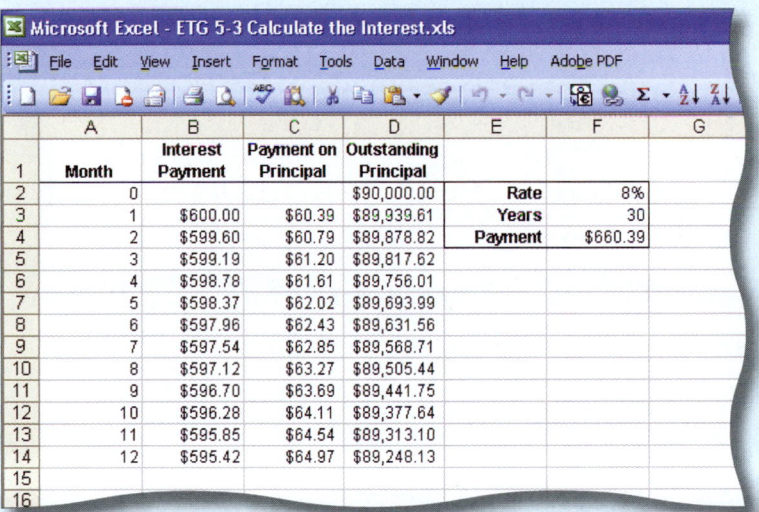

6. Adding the calculated interest payments gives us the total interest paid in the first year: $7172.81.

Note Excel has built-in functions that compute the interest payment (IPMT) or the payment on the principle (PPMT) in a given period. We could also have used the built-in future value function (FV) to calculate the outstanding principal each month. The main problem with using these functions is that, in a sense, they are too accurate. They do not take into account the fact that payments and interest are rounded to the nearest cent. Over time, this rounding causes the actual value of the outstanding principal to differ from what the FV function would tell us. In fact, because the actual payment is rounded slightly upwards (to $660.39 from 660.38811...), the principal is reduced slightly faster than necessary and a last payment of $660.39 would be $2.95 larger than needed to clear out the debt. The lender would reduce the last payment by $2.95 for this reason; Marc and Mira will pay only $657.44 for their final payment. This is common: The last payment on an install-ment loan is usually slightly larger or smaller than the others, to compensate for the rounding of the monthly payment amount. ∎

Example 9 Suppose that a 5%, 20-year bond sells for $9800 per $10,000 maturity value. What rate of return will investors get?

Solution with Technology Use the following worksheet, in which the RATE worksheet function is used to calculate the interest rate.

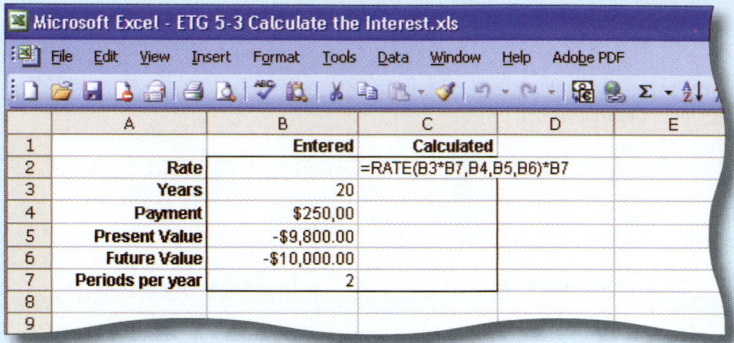

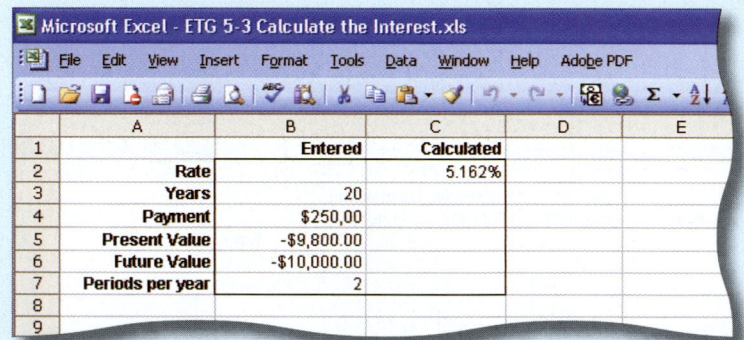

6

Sets and Counting

CASE STUDY You're the Expert—Designing a Puzzle

As Product Design Manager for Cerebral Toys, Inc., you are constantly on the lookout for ideas for intellectually stimulating yet inexpensive toys. Your design team recently came up with an idea for a puzzle consisting of a number of plastic cubes. Each cube will have two faces colored red, two white, and two blue, and there will be exactly two cubes with each possible configuration of colors. The goal of the puzzle is to seek out the matching pairs, thereby enhancing a child's geometric intuition and three-dimensional manipulation skills. If the kit is to include every possible configuration of colors, how many cubes will the kit contain?

David Young-Wolff/PhotoEdit

Introduction

The theory of sets is the foundation for most of mathematics. It also has direct applications—for example in searching computer databases. We will use set theory extensively in the chapter on probability, and thus much of this chapter revolves around the idea of a **set of outcomes** of a procedure such as rolling a pair of dice or choosing names from a database. Also important in probability is the theory of **counting** the number of elements in a set, which is called **combinatorics.**

Counting elements is not a trivial proposition; for example, the betting game Lotto (used in many state lotteries) has you pick six numbers from some range—say, 1–55. If your six numbers match the six numbers chosen in the "official drawing," you win the top prize. How many Lotto tickets would you need to buy to guarantee that you will win? That is, how many Lotto tickets are possible? By the end of this chapter, we will be able to answer these questions.

6.1 Sets and Set Operations

In this section we review the basic ideas of set theory with examples in applications that recur throughout the rest of the chapter.

Sets

Sets and Elements

A **set** is a collection of items, referred to as the **elements** of the set.

Visualizing a Set

Set

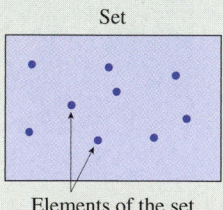

Elements of the set

We usually use a capital letter to name a set and braces to enclose the elements of a set.

quick Examples

$W = \{$Amazon, eBay, Apple$\}$
$N = \{1, 2, 3, \ldots\}$

$x \in A$ means that x **is an element of** the set A. If x is not an element of A, we write $x \notin A$.

Amazon $\in W$ (W as above)
Microsoft $\notin W$ $2 \in N$

$B = A$ means that A and B have the same elements. The order in which the elements are listed does not matter.

$\{5, -9, 1, 3\} = \{-9, 1, 3, 5\}$
$\{1, 2, 3, 4\} \neq \{1, 2, 3, 6\}$

$B \subseteq A$ means that B is a **subset** of A; every element of B is also an element of A.

$\{$eBay, Apple$\} \subseteq W$
$\{1, 2, 3, 4\} \subseteq \{1, 2, 3, 4\}$

$B \subset A$ means that B is a **proper subset** of A: $B \subseteq A$, but $B \neq A$.	$\{\text{eBay, Apple}\} \subset W$ $\{1, 2, 3\} \subset \{1, 2, 3, 4\}$ $\{1, 2, 3\} \subset N$
$\emptyset$ is the **empty set,** the set containing no elements. It is a subset of every set.	$\emptyset \subseteq W$ $\emptyset \subset W$
A **finite** set has finitely many elements. An **infinite** set does not have finitely many elements.	$W = \{\text{Amazon, eBay, Apple}\}$ is a finite set. $N = \{1, 2, 3, \ldots\}$ is an infinite set.

One type of set we'll use often is the **set of outcomes** of some activity or experiment. For example, if we toss a coin and observe which side faces up, there are two possible outcomes, heads (H) and tails (T). The set of outcomes of tossing a coin once can be written

$$S = \{\text{H, T}\}$$

As another example, suppose we roll a die that has faces numbered 1 through 6, as usual, and observe which number faces up. The set of outcomes *could* be represented as

$$S = \left\{ \boxed{\cdot}, \boxed{\cdot\,\cdot}, \boxed{\cdot\cdot\cdot}, \boxed{::}, \boxed{\cdot::}, \boxed{:::} \right\}$$

However, we can much more easily write

$$S = \{1, 2, 3, 4, 5, 6\}$$

Example 1 Two Dice: Distinguishable vs. Indistinguishable

a. Suppose we have two dice that we can distinguish in some way—say, one is blue and one is green. If we roll both dice, what is the set of outcomes?

b. Describe the set of outcomes if the dice are indistinguishable.

Solution

a. A systematic way of laying out the set of outcomes for a distinguishable pair of dice is shown in Figure 1.

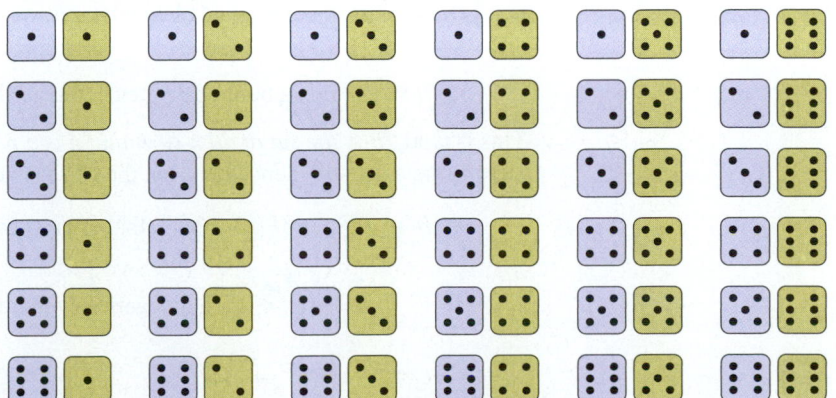

Figure **1**

In the first row all the blue dice show a 1, in the second row a 2, in the third row a 3, and so on. Similarly, in the first column all the green dice show a 1, in the second column a 2, and so on. The diagonal pairs (top left to bottom right) show all the "doubles." Using the picture as a guide, we can write the set of 36 outcomes as follows.

$$S = \begin{cases} (1,1),\ (1,2),\ (1,3),\ (1,4),\ (1,5),\ (1,6), \\ (2,1),\ (2,2),\ (2,3),\ (2,4),\ (2,5),\ (2,6), \\ (3,1),\ (3,2),\ (3,3),\ (3,4),\ (3,5),\ (3,6), \\ (4,1),\ (4,2),\ (4,3),\ (4,4),\ (4,5),\ (4,6), \\ (5,1),\ (5,2),\ (5,3),\ (5,4),\ (5,5),\ (5,6), \\ (6,1),\ (6,2),\ (6,3),\ (6,4),\ (6,5),\ (6,6) \end{cases} \qquad \text{Distinguishable dice}$$

Notice that S is also the set of outcomes if we roll a single die twice, if we take the first number in each pair to be the outcome of the first roll and the second number the outcome of the second roll.

b. If the dice are truly indistinguishable, we will have no way of knowing which die is which once they are rolled. Think of placing two identical dice in a closed box and then shaking the box. When we look inside afterward, there is no way to tell which die is which. (If we make a small marking on one of the dice or somehow keep track of it as it bounces around, we are *distinguishing* the dice.) We regard two dice as **indistinguishable** if we make no attempt to distinguish them. Thus, for example, the two different outcomes $(1,3)$ and $(3,1)$ from part (a) would represent the same outcome in part (b) (one die shows a 3 and the other a 1). Because the set of outcomes should contain each outcome only once, we can remove $(3,1)$. Following this approach gives the following smaller set of outcomes:

$$S = \begin{cases} (1,1),\ (1,2),\ (1,3),\ (1,4),\ (1,5),\ (1,6), \\ (2,2),\ (2,3),\ (2,4),\ (2,5),\ (2,6), \\ (3,3),\ (3,4),\ (3,5),\ (3,6), \\ (4,4),\ (4,5),\ (4,6), \\ (5,5),\ (5,6), \\ (6,6) \end{cases} \qquad \text{Indistinguishable dice}$$

Example 2 Set-Builder Notation

Let $B = \{0, 2, 4, 6, 8\}$. B is the set of all nonnegative even integers less than 10. If we don't want to list the individual elements of B, we can instead use "set-builder notation," and write

$$B = \{n \mid n \text{ is a nonnegative even integer less than } 10\}$$

This is read "*B is the set of all n such that n is a nonnegative even integer less than 10.*" Here is the correspondence between the words and the symbols:

B is the set of all n such that n is a nonnegative even integer less than 10

$$B = \{n \mid n \text{ is a nonnegative even integer less than } 10\}$$

+ *Before we go on...* Note that the **nonnegative** integers *include* 0, whereas the **positive** integers *exclude* 0. ∎

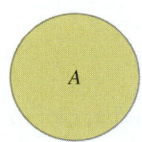

Figure **2**

Venn Diagrams

We can visualize sets and relations between sets using **Venn diagrams.** In a Venn diagram, we represent a set as a region, often a disk (Figure 2).

The elements of A are the points inside the region. The following Venn diagrams illustrate the relations we've discussed so far.

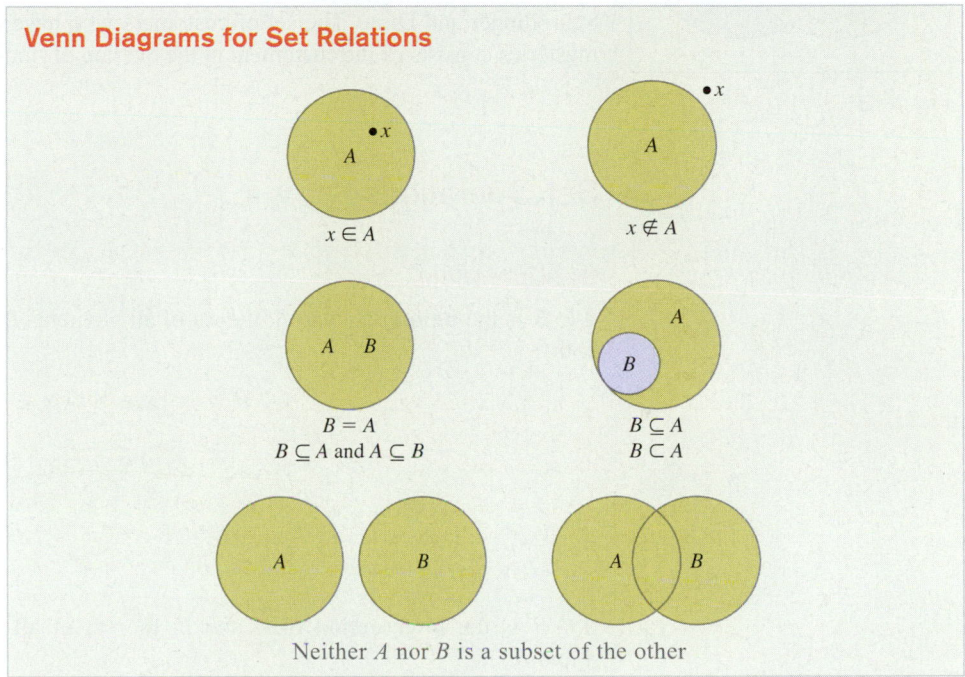

Venn Diagrams for Set Relations

$x \in A$

$x \notin A$

$B = A$
$B \subseteq A$ and $A \subseteq B$

$B \subseteq A$
$B \subset A$

Neither A nor B is a subset of the other

Note Although the diagram for $B \subseteq A$ suggests a proper subset, it is customary to use the same diagram for both subsets and proper subsets. ∎

Example **3 Customer Interests**

NobelBooks.com (a fierce competitor of OHaganBooks.com) maintains a database of customers and the types of books they have purchased. In the company's database is the set of customers

$$S = \{\text{Einstein, Bohr, Millikan, Heisenberg, Schrödinger, Dirac}\}$$

A search of the database for customers who have purchased cookbooks yields the subset

$$A = \{\text{Einstein, Bohr, Heisenberg, Dirac}\}$$

Another search, this time for customers who have purchased mysteries, yields the subset

$$B = \{\text{Bohr, Heisenberg, Schrödinger}\}$$

NobelBooks.com wants to promote a new combination mystery/cookbook, and wants to target two subsets of customers: those who have purchased either cookbooks or

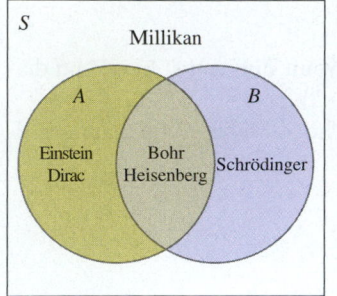

Figure **3**

mysteries (or both) and, for additional promotions, those who have purchased both cookbooks and mysteries. Name the customers in each of these subsets.

Solution We can picture the database and the two subsets using the Venn diagram in Figure 3.

The set of customers who have purchased either cookbooks or mysteries (or both) consists of the customers who are in A or B or both: Einstein, Bohr, Heisenberg, Schrödinger and Dirac. The set of customers who have purchased both cookbooks and mysteries consists of the customers in the overlap of A and B, Bohr and Heisenberg.

Set Operations

Set Operations

$A \cup B$ is the **union** of A and B, the set of all elements that are either in A or in B (or in both).

$$A \cup B = \{x \mid x \in A \text{ or } x \in B\}$$

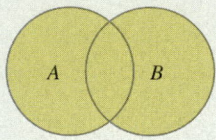

$A \cap B$ is the **intersection** of A and B, the set of all elements that are common to A and B.

$$A \cap B = \{x \mid x \in A \text{ and } x \in B\}$$

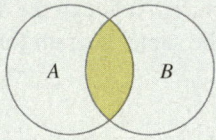

Logical Equivalents
Union: For an element to be in $A \cup B$, it must be in A **or** in B.
Intersection: For an element to be in $A \cap B$, it must be in A **and** in B.

quick Examples

If $A = \{a, b, c, d\}$, and $B = \{c, d, e, f\}$ then

$$A \cup B = \{a, b, c, d, e, f\}$$
$$A \cap B = \{c, d\}$$

Note Mathematicians always use "or" in its *inclusive* sense: one thing or another *or both*. ∎

There is one other operation we use, called the **complement** of a set A, which, roughly speaking, is the set of things *not* in A.

Q: Why only "roughly"? Why not just form the set of things not in A*?*

A: This would amount to assuming that there is a set of *all things*. (It would be the complement of the empty set.) Although tempting, talking about entities such as the "set of all things" leads to paradoxes.[1] Instead, we first need to fix a set *S* of *all objects under consideration*, or the *universe of discourse*, which we generally call the **universal set** for the discussion. For example, when we search the web, we take *S* to be the set of all web pages. When talking about integers, we take *S* to be the set of all integers. In other words, our choice of universal set depends on the context. The complement of a set $A \subseteq S$ is then the set of *things in S* that are not in *A*. ∎

Complement

If *S* is the universal set and $A \subseteq S$, then A' is the **complement** of *A* (in *S*), the set of all elements of *S* not in *A*.

$$A' = \{x \in S \mid x \notin A\} = \text{Gold Region Below}$$

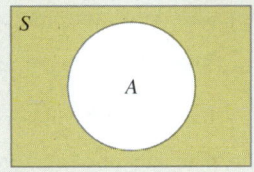

Logical Equivalent

For an element to be in A', it must be in *S* but **not** in *A*.

quick Example

If $S = \{a, b, c, d, e, f, g\}$ and $A = \{a, b, c, d\}$ then

$$A' = \{e, f, g\}$$

In the following example we use set operations to describe the sets we found in Example 3, as well as some others.

Example 4 Customer Interests

NobelBooks.com maintains a database of customers and the types of books they have purchased. In the company's database is the set of customers

$$S = \{\text{Einstein, Bohr, Millikan, Heisenberg, Schrödinger, Dirac}\}$$

A search of the database for customers who have purchased cookbooks yields the subset

$$A = \{\text{Einstein, Bohr, Heisenberg, Dirac}\}$$

---

[1] The most famous such paradox is called "Russell's Paradox," after the mathematical logician (and philosopher and pacifist) Bertrand Russell. It goes like this: If there were a set of all things, then there would also be a (smaller) set of all sets. Call it *S*. Now, because *S* is the *set* of *all* sets, it must contain itself as a member. In other words, $S \in S$. Let *P* be the subset of *S* consisting of all sets that are *not* members of themselves. Now we pose the following question: Is *P* a member of itself? If it is, then, because it is the set of all sets that are *not* members of themselves, it is not a member of itself. On the other hand, if it is *not* a member of itself, then it qualifies as an element of *P*. In other words, it *is* a member of itself! Because neither can be true, something is wrong. What is wrong is the assumption that there is such a thing as the set of all sets or the set of all things.

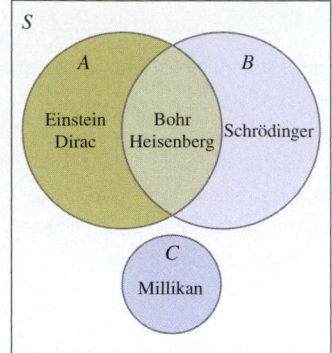

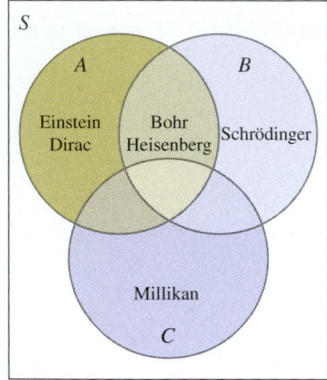

Figure **4**

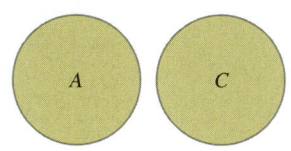

Figure **5**

Another search, this time for customers who have purchased mysteries, yields the subset

$$B = \{\text{Bohr, Heisenberg, Schrödinger}\}$$

A third search, for customers who had registered with the site but not used their first-time customer discount yields the subset

$$C = \{\text{Millikan}\}$$

Use set operations to describe the following subsets:

a. The subset of customers who have purchased either cookbooks or mysteries

b. The subset of customers who have purchased both cookbooks and mysteries

c. The subset of customers who have not purchased cookbooks

d. The subset of customers who have purchased cookbooks but have not used their first-time customer discount

Solution Figure 4 shows two alternative Venn diagram representations of the database. Although the second version shows C overlapping A and B, the placement of the names inside shows that there are no customers in those overlaps.

a. The subset of customers who have bought either cookbooks *or* mysteries is

$$A \cup B = \{\text{Einstein, Bohr, Heisenberg, Schrödinger, Dirac}\}$$

b. The subset of customers who have bought both cookbooks *and* mysteries is

$$A \cap B = \{\text{Bohr, Heisenberg}\}$$

c. The subset of customers who have *not* bought cookbooks is

$$A' = \{\text{Millikan, Schrödinger}\}$$

Note that, for the universal set, we are using the set S of all customers in the database.

d. The subset of customers who have bought cookbooks but have not used their first-time purchase discount is the empty set

$$A \cap C = \emptyset$$

When the intersection of two sets is empty, we say that the two sets are **disjoint.** In a Venn diagram, disjoint sets are drawn as regions that don't overlap, as in Figure 5.[*]

---

[*] People new to set theory sometimes find it strange to consider the empty set a valid set. Here is one of the times where it is very useful to do so. If we did not, we would have to say that $A \cap C$ was defined only when A and C had something in common. Having to deal with the fact that this set operation was not always defined would quickly get tiresome.

+*Before we go on...* Computer databases and the web can be searched using so-called "Boolean searches." These are search requests using "and," "or," and "not." Using "and" gives the intersection of separate searches, using "or" gives the union, and using "not" gives the complement. In the next section we'll see how web search engines allow such searches. ∎

Cartesian Product

There is one more set operation we need to discuss.

Cartesian Product

The **Cartesian product** of two sets, A and B, is the set of all ordered pairs (a, b) with $a \in A$ and $b \in B$.

$$A \times B = \{(a, b) \mid a \in A \text{ and } b \in B\}$$

In words, $A \times B$ is the set of all ordered pairs whose first component is in A and whose second component is in B.

quick Examples

1. If $A = \{a, b\}$ and $B = \{1, 2, 3\}$, then

$$A \times B = \{(a, 1), (a, 2), (a, 3), (b, 1), (b, 2), (b, 3)\}$$

Visualizing $A \times B$

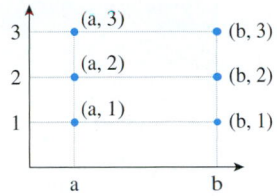

2. If $S = \{H, T\}$, then

$$S \times S = \{(H, H), (H, T), (T, H), (T, T)\}$$

In other words, if S is the set of outcomes of tossing a coin once, then $S \times S$ is the set of outcomes of tossing a coin twice.

3. If $S = \{1, 2, 3, 4, 5, 6\}$, then

$$S \times S = \left\{ \begin{array}{llllll} (1, 1), & (1, 2), & (1, 3), & (1, 4), & (1, 5), & (1, 6), \\ (2, 1), & (2, 2), & (2, 3), & (2, 4), & (2, 5), & (2, 6), \\ (3, 1), & (3, 2), & (3, 3), & (3, 4), & (3, 5), & (3, 6), \\ (4, 1), & (4, 2), & (4, 3), & (4, 4), & (4, 5), & (4, 6), \\ (5, 1), & (5, 2), & (5, 3), & (5, 4), & (5, 5), & (5, 6), \\ (6, 1), & (6, 2), & (6, 3), & (6, 4), & (6, 5), & (6, 6) \end{array} \right\}$$

In other words, if S is the set of outcomes of rolling a die once, then $S \times S$ is the set of outcomes of rolling a die twice (or rolling two distinguishable dice).

4. If $A = \{\text{red, yellow}\}$ and $B = \{\text{Mustang, Firebird}\}$ then

$A \times B = \{(\text{red, Mustang}), (\text{red, Firebird}), (\text{yellow, Mustang}), (\text{yellow, Firebird})\}$
which we might also write as

$A \times B = \{\text{red Mustang, red Firebird, yellow Mustang, yellow Firebird}\}$

Example 5 Representing Cartesian Products

The manager of an automobile dealership has collected data on the number of pre-owned Acura, Infiniti, Lexus, and Mercedes cars the dealership has from the 2004, 2005, and 2006 model years. In entering this information on a spreadsheet, the manager would like to have each spreadsheet cell represent a particular year and make. Describe this set of cells.

Solution Because each cell represents a year and a make, we can think of the cell as a pair (year, make), as in (2004, Acura). Thus, the set of cells can be thought of as a Cartesian product:

$$Y = \{2004, 2005, 2006\} \qquad \text{Year of car}$$

$$M = \{\text{Acura, Infiniti, Lexus, Mercedes}\} \qquad \text{Make of car}$$

$$Y \times M = \left\{ \begin{array}{llll} (2004, \text{Acura}) & (2004, \text{Infiniti}) & (2004, \text{Lexus}) & (2004, \text{Mercedes}) \\ (2005, \text{Acura}) & (2005, \text{Infiniti}) & (2005, \text{Lexus}) & (2005, \text{Mercedes}) \\ (2006, \text{Acura}) & (2006, \text{Infiniti}) & (2006, \text{Lexus}) & (2006, \text{Mercedes}) \end{array} \right\} \text{Cells}$$

Thus, the manager might arrange the spreadsheet as follows:

	A	B	C	D	E	F
		Acura	**Infiniti**	**Lexus**	**Mercedes**	
1		**Acura**	**Infiniti**	**Lexus**	**Mercedes**	
2	**2004**	(2004 Acura)	(2004 Infiniti)	(2004 Lexus)	(2004 Mercedes)	
3	**2005**	(2005 Acura)	(2005 Infiniti)	(2005 Lexus)	(2005 Mercedes)	
4	**2006**	(2006 Acura)	(2006 Infiniti)	(2006 Lexus)	(2006 Mercedes)	
5						
6						

Microsoft Excel - Ch 6-1 Pre-Owned Cars.xls — File Edit View Insert Format Tools Data Window Help Adobe PDF

The highlighting shows the 12 cells to be filled in, representing the numbers of cars of each year and make. For example, in cell B2 should go the number of 2004 Acuras the dealership has.

✛*Before we go on...* The arrangement in the spreadsheet in Example 5 is consistent with the matrix notation in Chapter 2. We could also have used the elements of Y as column labels along the top and the elements of M as row labels down the side. Along those lines, we can also visualize the Cartesian product $Y \times M$ as a set of points in the *xy*-plane ("Cartesian plane") as shown in Figure 6.

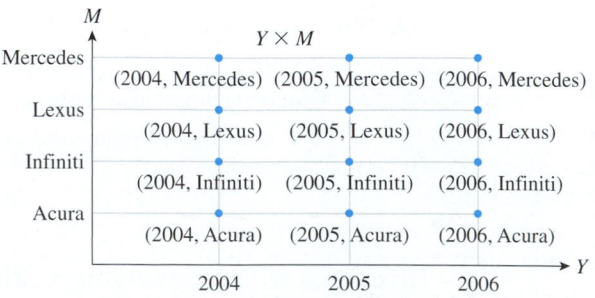

Figure **6**

FAQs   The Many Meanings of "And"

Q: *Suppose A is the set of actors and B is the set of all baseball players. Then the set of all actors* and *baseball players is $A \cap B$, right?*

A: Wrong. The fact that the word "and" appears in the description of a set does not always mean that the set is an intersection; the word "and" can mean different things in different contexts. $A \cap B$ refers to the set of elements that are in both A and B, hence to actors who are also baseball players. On the other hand, the set of all actors and baseball players is the set of people who are either actors or baseball players (or both), which is $A \cup B$. We can use the word "and" to describe both sets:

$A \cap B$ = {people who are both actors *and* baseball players}
$A \cup B$ = {people who are actors *or* baseball players} = {all actors *and* baseball players}

6.1 EXERCISES

● denotes basic skills exercises

tech Ex indicates exercises that should be solved using technology

List the elements in each of the sets in Exercises 1–16.

1. ● The set F consisting of the four seasons. *hint* [see Example 1]

2. ● The set A consisting of the authors of this book.

3. ● The set I of all positive integers no greater than 6.

4. ● The set N of all negative integers greater than -3.

5. ● $A = \{n \mid n$ is a positive integer and $0 \leq n \leq 3\}$. *hint* [see Example 2]

6. ● $A = \{n \mid n$ is a positive integer and $0 < n < 8\}$.

7. ● $B = \{n \mid n$ is an even positive integer and $0 \leq n \leq 8\}$.

8. ● $B = \{n \mid n$ is an odd positive integer and $0 \leq n \leq 8\}$.

9. ● The set of all outcomes of tossing a pair of **(a)** distinguishable coins **(b)** indistinguishable coins.

10. ● The set of outcomes of tossing three **(a)** distinguishable coins **(b)** indistinguishable coins.

11. ● The set of all outcomes of rolling two distinguishable dice such that the numbers add to 6.

12. ● The set of all outcomes of rolling two distinguishable dice such that the numbers add to 8.

13. ● The set of all outcomes of rolling two indistinguishable dice such that the numbers add to 6.

14. ● The set of all outcomes of rolling two indistinguishable dice such that the numbers add to 8.

15. ● The set of all outcomes of rolling two distinguishable dice such that the numbers add to 13.

16. ● The set of all outcomes of rolling two distinguishable dice such that the numbers add to 1.

In each of Exercises 17–20, draw a Venn diagram that illustrates the relationships among the given sets. *hint* [see Example 3]

17. ● $S = $ {eBay, Google, Amazon, OHaganBooks, Hotmail}, $A = $ {Amazon, OHaganBooks}, $B = $ {eBay, Amazon}, $C = $ {Amazon, Hotmail}.

18. ● $S = $ {Apple, Dell, Gateway, Pomegranate, Compaq}, $A = $ {Gateway, Pomegranate, Compaq}, $B = $ {Dell, Gateway, Pomegranate, Compaq}, $C = $ {Apple, Dell, Compaq}.

19. ● $S = $ {eBay, Google, Amazon, OHaganBooks, Hotmail}, $A = $ {Amazon, Hotmail}, $B = $ {eBay, Google, Amazon, Hotmail}, $C = $ {Amazon, Hotmail}.

20. ● $S = $ {Apple, Dell, Gateway, Pomegranate, Compaq}, $A = $ {Apple, Dell, Pomegranate, Compaq}, $B = $ {Pomegranate}, $C = $ {Pomegranate}.

Let $A = $ {June, Janet, Jill, Justin, Jeffrey, Jello}, $B = $ {Janet, Jello, Justin}, and $C = $ {Sally, Solly, Molly, Jolly, Jello}. Find each set in Exercises 21–34. *hint* [see Quick Examples on p. 398]

21. ● $A \cup B$
22. ● $A \cup C$
23. ● $A \cup \emptyset$
24. ● $B \cup \emptyset$
25. ● $A \cup (B \cup C)$
26. ● $(A \cup B) \cup C$
27. ● $C \cap B$
28. ● $C \cap A$
29. ● $A \cap \emptyset$
30. ● $\emptyset \cap B$
31. ● $(A \cap B) \cap C$
32. ● $A \cap (B \cap C)$
33. ● $(A \cap B) \cup C$
34. ● $A \cup (B \cap C)$

In Exercises 35–42, $A = $ {small, medium, large}, $B = $ {blue, green}, and $C = $ {triangle, square}. *hint* [see Quick Examples on p. 401]

35. ● List the elements of $A \times C$.

● basic skills   tech Ex technology exercise

36. ● List the elements of $B \times C$.

37. ● List the elements of $A \times B$.

38. ● The elements of $A \times B \times C$ are the ordered triples (a, b, c) with $a \in A, b \in B$, and $c \in C$. List all the elements of $A \times B \times C$.

39. ● Represent $B \times C$ as cells in a spreadsheet. *hint* [see Example 5]

40. ● Represent $A \times C$ as cells in a spreadsheet.

41. ● Represent $A \times B$ as cells in a spreadsheet.

42. ● Represent $A \times A$ as cells in a spreadsheet.

Let $A = \{H, T\}$ be the set of outcomes when a coin is tossed, and let $B = \{1, 2, 3, 4, 5, 6\}$ be the set of outcomes when a die is rolled. Write each set in Exercises 43–46 in terms of A and/or B and list its elements.

43. ● The set of outcomes when a die is rolled and then a coin tossed.

44. ● The set of outcomes when a coin is tossed twice.

45. ● The set of outcomes when a coin is tossed three times.

46. ● The set of outcomes when a coin is tossed twice and then a die is rolled.

Let S be the set of outcomes when two distinguishable dice are rolled, let E be the subset of outcomes in which at least one die shows an even number, and let F be the subset of outcomes in which at least one die shows an odd number. List the elements in each subset given in Exercises 47–52.

47. ● E' **48.** ● F'

49. ● $(E \cup F)$ **50.** ● $(E \cap F)'$

51. ● $E' \cup F'$ **52.** ● $E' \cap F'$

Use Venn diagrams to illustrate the following identities for subsets A, B, and C of S.

53. $(A \cup B)' = A' \cap B'$ De Morgan's Law

54. $(A \cap B)' = A' \cup B'$ De Morgan's Law

55. $(A \cap B) \cap C = A \cap (B \cap C)$ Associative Law

56. $(A \cup B) \cup C = A \cup (B \cup C)$ Associative Law

57. $A \cup (B \cap C) = (A \cup B) \cap (A \cup C)$ Distributive Law

58. $A \cap (B \cup C) = (A \cap B) \cup (A \cap C)$ Distributive Law

59. $S' = \emptyset$ **60.** $\emptyset' = S$

Applications

Databases A free-lance computer consultant keeps a database of her clients, which contains the names

$S = \{$Acme, Brothers, Crafts, Dion, Effigy, Floyd, Global, Hilbert$\}$

The following clients owe her money:

$A = \{$Acme, Crafts, Effigy, Global$\}$

The following clients have done at least \$10,000 worth of business with her:

$B = \{$Acme, Brothers, Crafts, Dion$\}$

The following clients have employed her in the last year:

$C = \{$Acme, Crafts, Dion, Effigy, Global, Hilbert$\}$

In Exercises 61–68, a subset of clients is described that the consultant could find using her database. Write each subset in terms of A, B, and C and list the clients in that subset. hint [see Example 4]

61. ● The clients who owe her money and have done at least \$10,000 worth of business with her.

62. ● The clients who owe her money or have done at least \$10,000 worth of business with her.

63. ● The clients who have done at least \$10,000 worth of business with her or have employed her in the last year.

64. ● The clients who have done at least \$10,000 worth of business with her and have employed her in the last year.

65. ● The clients who do not owe her money and have employed her in the last year.

66. ● The clients who do not owe her money or have employed her in the last year.

67. The clients who owe her money, have not done at least \$10,000 worth of business with her, and have not employed her in the last year.

68. The clients who either do not owe her money, have done at least \$10,000 worth of business with her, or have employed her in the last year.

69. ● ***Boat Sales*** You are given data on revenues from sales of sail boats, motor boats, and yachts for each of the years 2003 through 2006. How would you represent these data in a spreadsheet? The cells in your spreadsheet represent elements of which set?

70. ● ***Health-Care Spending*** Spending in most categories of health care in the U.S. increased dramatically in the last 30 years of the 1900s.[2] You are given data showing total spending on prescription drugs, nursing homes, hospital care, and professional services for each of the last three decades of the 1900s. How would you represent these data in a spreadsheet? The cells in your spreadsheet represent elements of which set?

Communication and Reasoning Exercises

71. ● You sell iPods® and jPods. Let I be the set of all iPods you sold last year, and let J be the set of all jPods you sold last year. What set represents the collection of all iPods and jPods you sold combined?

---

[2] Source: Department of Health and Human Services/*New York Times*, January 8, 2002, p. A14.

● basic skills   **tech**Ex technology exercise

72. ● You sell two models of music players: the *yoVaina Grandote* and the *yoVaina Minúsculito,* and each comes in three colors: Infraroja, Ultravioleta, and Radiografía. Let *M* be the set of models and let *C* be the set of colors. What set represents the different choices a customer can make?

73. ● You are searching online for techno music that is neither European nor Dutch. In set notation, which set of music files are you searching for?
(A) Techno ∩ (European ∩ Dutch)′
(B) Techno ∩ (European ∪ Dutch)′
(C) Techno ∪ (European ∩ Dutch)′
(D) Techno ∪ (European ∪ Dutch)′

74. ● You would like to see either a World War II movie, or one that that is based on a comic book character but does not feature aliens. Which set of movies are you interested in seeing?
(A) WWII ∩ (Comix ∩ Aliens′)
(B) WWII ∩ (Comix ∪ Aliens′)
(C) WWII ∪ (Comix ∩ Aliens′)
(D) WWII ∪ (Comix ∪ Aliens′)

75. Explain, illustrating by means of an example, why $(A \cap B) \cup C \neq A \cap (B \cup C)$.

76. Explain, making reference to operations on sets, why the statement "He plays soccer or rugby and cricket" is ambiguous.

77. Explain the meaning of a universal set, and give two different universal sets that could be used in a discussion about sets of positive integers.

78. Is the set of outcomes when two indistinguishable dice are rolled (Example 1) a Cartesian product of two sets? If so, which two sets; if not, why not?

79. Design a database scenario that leads to the following statement: To keep the factory operating at maximum capacity, the plant manager should select the suppliers in $A \cap (B \cup C')$.

80. Design a database scenario that leads to the following statement: To keep her customers happy, the bookstore owner should stock periodicals in $A \cup (B \cap C')$.

81. Rewrite in set notation: She prefers movies that are not violent, are shorter than two hours, and have neither a tragic ending nor an unexpected ending.

82. Rewrite in set notation: He will cater for any event as long as there are no more than 1000 people, it lasts for at least three hours, and it is within a 50 mile radius of Toronto.

● basic skills   **tech** Ex technology exercise

6.2 Cardinality

In this section, we begin to look at a deceptively simple idea: the size of a set, which we call its **cardinality.**

Cardinality

If *A* is a finite set, then its **cardinality** is

$$n(A) = \text{number of elements in } A$$

Visualizing Cardinality

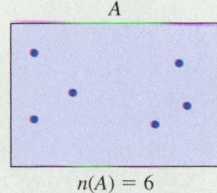

$n(A) = 6$

quick Examples   **1.** Let $S = \{a, b, c\}$. Then $n(S) = 3$.

2. Let *S* be the set of outcomes when two distinguishable dice are rolled. Then $n(S) = 36$ (see Example 1 in Section 6.1).

3. $n(\emptyset) = 0$ because the empty set has no elements.

Counting the elements in a small, simple set is straightforward. To count the elements in a large, complicated set, we try to describe the set as built of simpler sets using the set operations. We then need to know how to calculate the number of elements in, for example, a union, based on the number of elements in the simpler sets whose union we are taking.

The Cardinality of a Union

How can we calculate $n(A \cup B)$ if we know $n(A)$ and $n(B)$? Our first guess might be that $n(A \cup B)$ is $n(A) + n(B)$. But consider a simple example. Let

$$A = \{a, b, c\}$$

and

$$B = \{b, c, d\}$$

Then $A \cup B = \{a, b, c, d\}$ so $n(A \cup B) = 4$, but $n(A) + n(B) = 3 + 3 = 6$. The calculation $n(A) + n(B)$ gives the wrong answer because the elements b and c are counted twice, once for being in A and again for being in B. To correct for this overcounting, we need to subtract the number of elements that get counted twice, which is the number of elements that A and B have in common, or $n(A \cap B) = 2$ in this case. So, we get the right number for $n(A \cup B)$ from the following calculation:

$$n(A) + n(B) - n(A \cap B) = 3 + 3 - 2 = 4$$

This argument leads to the following general formula.

Cardinality of a Union

If A and B are finite sets, then

$$n(A \cup B) = n(A) + n(B) - n(A \cap B)$$

In particular, if A and B are disjoint (meaning that $A \cap B = \emptyset$), then

$$n(A \cup B) = n(A) + n(B)$$

(When A and B are disjoint we say that $A \cup B$ is a **disjoint union**.)

Visualizing Cardinality of a Union

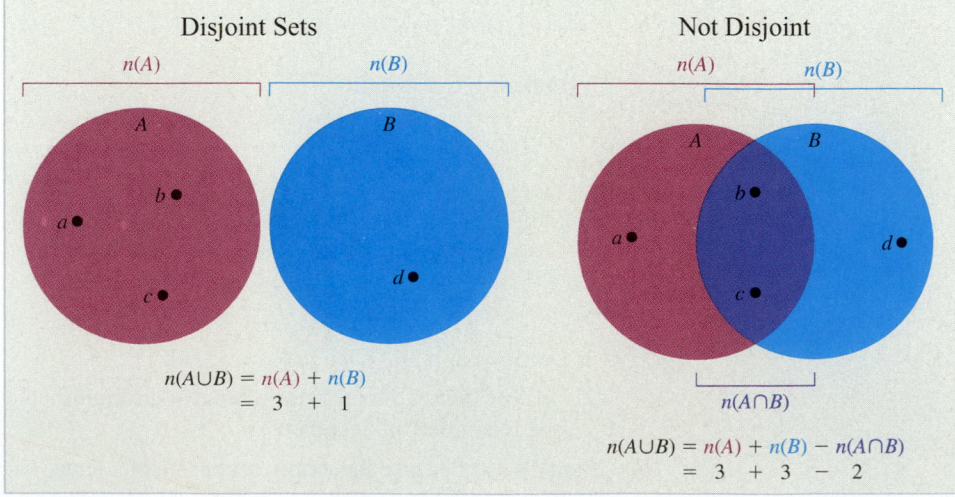

quick Examples

1. If $A = \{a, b, c, d\}$ and $B = \{b, c, d, e, f\}$, then

$$n(A \cup B) = n(A) + n(B) - n(A \cap B) = 4 + 5 - 3 = 6$$

In fact, $A \cup B = \{a, b, c, d, e, f\}$.

2. If $A = \{a, b, c\}$ and $B = \{d, e, f\}$, then $A \cap B = \emptyset$, so

$$n(A \cup B) = n(A) + n(B) = 3 + 3 = 6$$

Example 1 Web Searches

In February, 2005, a search using Google® for "manned Jupiter mission" yielded 32 websites containing that phrase, and a search for "Venus terraforming project" yielded 36 sites. A search for sites containing both phrases yielded only 4 websites. How many websites contained either "manned Jupiter mission," "Venus terraforming project," or both?

Solution Let A be the set of sites containing "manned Jupiter mission" and let B be the set of sites containing "Venus terraforming project." We are told that

$$n(A) = 32$$
$$n(B) = 36$$
$$n(A \cap B) = 4 \qquad \text{"manned Jupiter mission" AND "Venus terraforming project"}$$

The formula for the cardinality of the union tells us that

$$n(A \cup B) = n(A) + n(B) - n(A \cap B) = 32 + 36 - 4 = 64$$

So, 64 sites in the Google database contained one or both of the phrases "manned Jupiter mission" or "Venus terraforming project." (This was subsequently confirmed by doing a search for "manned Jupiter mission" OR "Venus terraforming project.")

+*Before we go on...* Each search engine has a different way of specifying a search for a union or an intersection. At Google or AltaVista®, you can use "OR" for the union and "AND" for the intersection.

Although the formula $n(A \cup B) = n(A) + n(B) - n(A \cap B)$ always holds mathematically, you may sometimes find that, in an actual search, the numbers don't add up. The reason? Search engines are not perfect. AltaVista, for example, explains that the number reported may be only an estimate of the actual number of sites in their database containing the search words.[3] ∎

Q: Is there a similar formula for $n(A \cap B)$?

A: The formula for the cardinality of a union can also be thought of as a formula for the cardinality of an intersection. We can solve for $n(A \cap B)$ to get

$$n(A \cap B) = n(A) + n(B) - n(A \cup B)$$

In fact, we can think of this formula as an equation relating four quantities. If we know any three of them, we can use the equation to find the fourth (see Example 2 below). ∎

---

[3] E-mail correspondence with the authors, January, 2002.

Q: Is there a similar formula for $n(A')$?

A: We can get a formula for the cardinality of a complement as follows: If S is our universal set and $A \subseteq S$, then S is the disjoint union of A and its complement. That is,

$$S = A \cup A' \quad \text{and} \quad A \cap A' = \varnothing$$

Applying the cardinality formula for a disjoint union, we get

$$n(S) = n(A) + n(A')$$

We can solve for $n(A')$ or for $n(A)$ to get the following formulas. ∎

Cardinality of a Complement

If S is a finite universal set and A is a subset of S, then

$$n(A') = n(S) - n(A)$$

and

$$n(A) = n(S) - n(A')$$

Visualizing Cardinality of a Complement

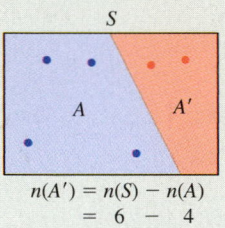

$$n(A') = n(S) - n(A)$$
$$= 6 - 4$$

quick Example   If $S = \{a, b, c, d, e, f\}$ and $A = \{a, b, c, d\}$, then

$$n(A') = n(S) - n(A) = 6 - 4 = 2$$

In fact, $A' = \{e, f\}$.

Example 2 Cookbooks

In February, 2005, a search at Amazon.com found 53,000 books on cooking.* Of these, 6000 were on American cooking, 2000 were on vegetarian cooking, and 7000 were on either American or vegetarian cooking (or both). How many of these books were not on both American and vegetarian cooking?

Solution Let S be the set of all 53,000 books on cooking, let A be the set of books on American cooking, and let B be the set of books on vegetarian cooking. We wish to find the size of the complement of the set of books on both American and vegetarian cooking— that is, $n((A \cap B)')$. Using the formula for the cardinality of a complement, we have

$$n((A \cap B)') = n(S) - n(A \cap B) = 53,000 - n(A \cap B)$$

---

* Precisely, it found that many books under the subject "cooking." Figures are rounded to the nearest 1000.

To find $n(A \cap B)$, we use the formula for the cardinality of a union:

$$n(A \cup B) = n(A) + n(B) - n(A \cap B)$$

Substituting the values we were given, we find

$$7000 = 6000 + 2000 - n(A \cap B)$$

which we can solve to get

$$n(A \cap B) = 1000$$

Therefore,

$$n((A \cap B)') = 53,000 - n(A \cap B) = 53,000 - 1000 = 52,000$$

So, 52,000 of the books on cooking were not on both American and vegetarian cooking.

Example 3 iPods, iMacs, and Powerbooks

The following table shows sales, in thousands of units, of iPods®, Powerbooks®, and iMacs® in the fourth quarter of 2003 and in the third and fourth quarters of 2004.[*]

	iPods	Powerbooks	iMacs	Total
2003 Q4	340	180	250	770
2004 Q3	860	220	240	1320
2004 Q4	1030	210	230	1470
Total	2230	610	720	3560

Let S be the set of all these iPods, Powerbooks, and iMacs. Let D be the set of all iPods sold during the stated quarters, let M be the set of iMacs sold during the stated quarters, and let A be the set of all items sold in the fourth quarter of 2003. Describe the following sets and compute their cardinality:

a. A' **b.** $D \cap A'$ **c.** $(D \cap A)'$ **d.** $M \cup A$

Solution Because all the figures are stated in thousands of units, we'll give our calculations and results in thousands of units as well.

a. A' is the set of all items not sold in the fourth quarter of 2003. To compute its cardinality, we could add the totals for all the other quarters listed in the rightmost column:

$$n(A') = 1320 + 1470 = 2790 \text{ thousand items}$$

Alternatively, we can use the formula for the cardinality of a complement (referring again to the totals in the table):

$$n(A') = n(S) - n(A) = 3560 - 770 = 2790 \text{ thousand items}$$

b. $D \cap A'$ is the intersection of the set of all iPods and the set of all items not sold in the fourth quarter of 2003. In other words, it is the set of all iPods not sold in the fourth

---

[*] Figures are rounded. SOURCE: Company report www.apple.com January, 2005.

quarter of 2003. Here is the table with the corresponding sets D and A' shaded ($D \cap A'$ is the overlap):

	iPods	Powerbooks	iMacs	Total
2003 Q4	340	180	250	770
2004 Q3	860	220	240	1320
2004 Q4	1030	210	230	1470
Total	2230	610	720	3560

From the figure:

$$n(D \cap A') = 860 + 1030 = 1890 \text{ thousand iPods}$$

c. $D \cap A$ is the set of all iPods sold in the fourth quarter of 2003, and so $(D \cap A)'$ is the set of all items *other than* iPods sold in the fourth quarter of 2003:

	iPods	Powerbooks	iMacs	Total
2003 Q4	340	180	250	770
2004 Q3	860	220	240	1320
2004 Q4	1030	210	230	1470
Total	2230	610	720	3560

From the formula for the cardinality of a complement:

$$n(D \cap A)' = n(S) - n(D \cap A)$$
$$= 3560 - 340 = 3220 \text{ thousand items}$$

d. $M \cup A$ is the set of items that were either iMacs or sold in the fourth quarter of 2003:

	iPods	Powerbooks	iMacs	Total
2003 Q4	340	180	250	770
2004 Q3	860	220	240	1320
2004 Q4	1030	210	230	1470
Total	2230	610	720	3560

To compute it, we can use the formula for the cardinality of a union:

$$n(M \cup A) = n(M) + n(A) - n(M \cap A)$$
$$= 720 + 770 - 250 = 1240 \text{ thousand items}$$

To determine the cardinality of a union of three or more sets, like $n(A \cup B \cup C)$, we can think of $A \cup B \cup C$ as a union of two sets, $(A \cup B)$ and C, and then analyze each piece using the techniques we already have. Alternatively, there are formulas for the cardinalities of unions of any number of sets, but these formulas get more and more complicated as the number of sets grows. In many applications, like the following example, we can use Venn diagrams instead.

Example 4 Reading Lists

A survey of 300 college students found that 100 had read *War and Peace,* 120 had read *Crime and Punishment,* and 100 had read *The Brothers Karamazov*. It also found that 40 had read only *War and Peace,* 70 had read *War and Peace* but not *The Brothers Karamazov,* and 80 had read *The Brothers Karamazov* but not *Crime and Punishment*. Only 10 had read all three novels. How many had read none of these three novels?

Solution There are four sets mentioned in the problem: the universe S consisting of the 300 students surveyed, the set W of students who had read *War and Peace,* the set C of students who had read *Crime and Punishment,* and the set K of students who had read *The Brothers Karamazov*. Figure 7 shows a Venn diagram representing these sets.

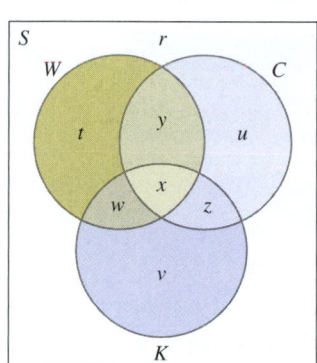

Figure **7**

We have put labels in the various regions of the diagram to represent the number of students in each region. For example, x represents the number of students in $W \cap C \cap K$, which is the number of students who have read all three novels. We are told that this number is 10, so

$$x = 10$$

(You should draw the diagram for yourself and fill in the numbers as we go along.) We are also told that 40 students had read only *War and Peace,* so

$$t = 40$$

We are given none of the remaining regions directly. However, because 70 had read *War and Peace* but not *The Brothers Karamazov,* we see that t and y must add up to 70. Because we already know that $t = 40$, it follows that $y = 30$. Further, because a total of 100 students had read *War and Peace,* we have

$$x + y + t + w = 100$$

Substituting the known values of x, y, and t gives

$$10 + 30 + 40 + w = 100$$

so $w = 20$. Because 80 students had read *The Brothers Karamazov* but not *Crime and Punishment,* we see that $v + w = 80$, so $v = 60$ (because we know $w = 20$). We can now calculate z using the fact that a total of 100 students had read *The Brothers Karamazov:*

$$60 + 20 + 10 + z = 100$$

giving $z = 10$. Similarly, we can now get u using the fact that 120 students had read *Crime and Punishment:*

$$10 + 30 + 10 + u = 120$$

giving $u = 70$. Of the 300 students surveyed, we've now found $x + y + z + w + t + u + v = 240$. This leaves

$$r = 60$$

who had read none of the three novels.

The Cardinality of a Cartesian Product

We've covered all the operations except Cartesian product. To find a formula for $n(A \times B)$, consider the following simple example:

$$A = \{H, T\}$$
$$B = \{1, 2, 3, 4, 5, 6\}$$

so that

$$A \times B = \{H1, H2, H3, H4, H5, H6, T1, T2, T3, T4, T5, T6\}$$

As we saw in Example 5 in Section 6.1, the elements of $A \times B$ can be arranged in a table or spreadsheet with $n(A) = 2$ rows and $n(B) = 6$ elements in each row.

In a region with 2 rows and 6 columns, there are $2 \times 6 = 12$ cells. So,

$$n(A \times B) = n(A)n(B)$$

in this case. There is nothing particularly special about this example, however, and that formula holds true in general.

Cardinality of a Cartesian Product

If A and B are finite sets, then

$$n(A \times B) = n(A)n(B)$$

quick Example

If $A = \{a, b, c\}$ and $B = \{x, y, z, w\}$, then

$$n(A \times B) = n(A)n(B) = 3 \times 4 = 12$$

Example 5 Coin Tosses

a. If we toss a coin twice and observe the sequence of heads and tails, how many possible outcomes are there?

b. If we toss a coin three times, how many possible outcomes are there?

c. If we toss a coin ten times, how many possible outcomes are there?

Solution

a. Let $A = \{H, T\}$ be the set of possible outcomes when a coin is tossed once. The set of outcomes when a coin is tossed twice is $A \times A$, which has

$$n(A \times A) = n(A)n(A) = 2 \times 2 = 4$$

possible outcomes.

b. When a coin is tossed three times, we can think of the set of outcomes as the product of the set of outcomes for the first two tosses, which is $A \times A$, and the set of outcomes for the third toss, which is just A. The set of outcomes for the three tosses is then $(A \times A) \times A$, which we usually write as $A \times A \times A$ or A^3. The number of outcomes is

$$n((A \times A) \times A) = n(A \times A)n(A) = (2 \times 2) \times 2 = 8$$

c. Considering the result of part (b), we can easily see that the set of outcomes here is $A^{10} = A \times A \times \cdots \times A$ (10 copies of A), or the set of ordered sequences of ten Hs and Ts. It's also easy to see that

$$n(A^{10}) = [n(A)]^{10} = 2^{10} = 1024$$

+ Before we go on... We can start to see the power of these formulas for cardinality. In Example 5, we were able to calculate that there are 1024 possible outcomes when we toss a coin 10 times without writing out all 1024 possibilities and counting them. ∎

6.2 EXERCISES

● denotes basic skills exercises

◆ denotes challenging exercises

Let A = {Dirk, Johan, Frans, Sarie}, B = {Frans, Sarie, Tina, Klaas, Henrika}, C = {Hans, Frans}. Find the numbers indicated in Exercises 1–6. hint [see Quick Examples on p. 407]

1. ● $n(A) + n(B)$

2. ● $n(A) + n(C)$

3. ● $n(A \cup B)$

4. ● $n(A \cup C)$

5. ● $n(A \cup (B \cap C))$

6. ● $n(A \cap (B \cup C))$

7. ● Verify that $n(A \cup B) = n(A) + n(B) - n(A \cap B)$ with A and B as above.

8. ● Verify that $n(A \cup C) = n(A) + n(C) - n(A \cap C)$ with A and C as above.

Let A = {H, T}, B = {1, 2, 3, 4, 5, 6}, and C = {red, green, blue}. Find the numbers indicated in Exercise 9–14. hint [see Example 5]

9. ● $n(A \times A)$

10. ● $n(B \times B)$

11. ● $n(B \times C)$

12. ● $n(A \times C)$

13. ● $n(A \times B \times B)$

14. ● $n(A \times B \times C)$

15. ● If $n(A) = 43$, $n(B) = 20$, and $n(A \cap B) = 3$, find $n(A \cup B)$.

16. ● If $n(A) = 60$, $n(B) = 20$, and $n(A \cap B) = 1$, find $n(A \cup B)$.

17. ● If $n(A \cup B) = 100$ and $n(A) = n(B) = 60$, find $n(A \cap B)$.

18. ● If $n(A) = 100$, $n(A \cup B) = 150$, and $n(A \cap B) = 40$, find $n(B)$.

Let S = {Barnsley, Manchester United, Southend, Sheffield United, Liverpool, Maroka Swallows, Witbank Aces, Royal Tigers, Dundee United, Lyon} be a universal set, A = {Southend, Liverpool, Maroka Swallows, Royal Tigers}, and B = {Barnsley, Manchester United, Southend}. Find the numbers indicated in Exercises 19–24. hint [see Quick Example on p. 408]

19. ● $n(A')$

20. ● $n(B')$

21. ● $n((A \cap B)')$

22. ● $n((A \cup B)')$

23. ● $n(A' \cap B')$

24. ● $n(A' \cup B')$

25. ● With S, A, and B as above, verify that $n((A \cap B)') = n(A') + n(B') - n((A \cup B)')$.

26. ● With S, A, and B as above, verify that $n(A' \cap B') + n(A \cup B) = n(S)$.

In Exercises 27–30, use the given information to complete the solution of each partially solved Venn diagram. hint [see Example 4]

27. ●

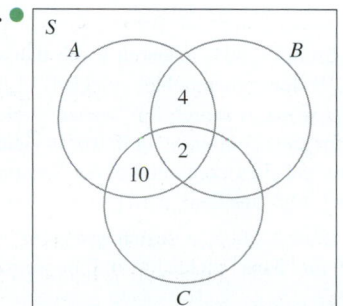

$n(A) = 20$, $n(B) = 20$, $n(C) = 28$, $n(B \cap C) = 8$, $n(S) = 50$

● basic skills   ◆ challenging

28.

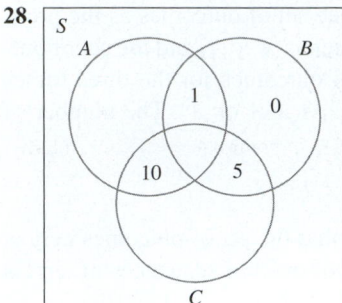

$n(A) = 16, n(B) = 11, n(C) = 30, n(S) = 40$

29.

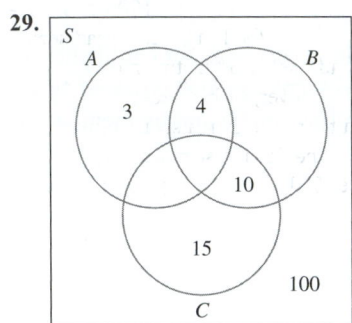

$n(A) = 10, n(B) = 19, n(S) = 140$

30.

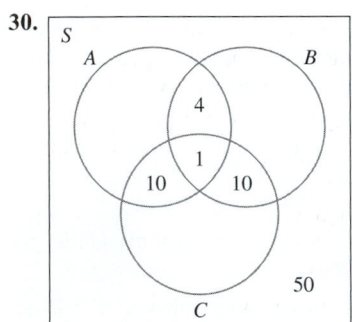

$n(A \cup B) = 30, n(B \cup C) = 30, n(A \cup C) = 35$

Applications

31. ● **Web Searches** In February, 2005, a search using the web search engine Google™ for "costenoble" yielded 11,000 websites containing that word. A search for "waner" yielded 69,000 sites. A search for sites containing both words yielded 4000 sites.[4] How many websites contained either "costenoble" or "waner" or both? *hint* [see Example 1]

32. ● **Web Searches** In February, 2005, a search using the web search engine Google™ for "hans" yielded 41 million websites containing that word. A search for "franz" yielded 18 million sites. A search for sites containing both words yielded 3 million sites.[5] How many websites contained either "hans" or "franz" or both?

33. ● **Amusement** On a particularly boring transatlantic flight, one of the authors amused himself by counting the heads of the people in the seats in front of him. He noticed that all 37 of them either had black hair or had a whole row to themselves (or both). Of this total, 33 had black hair and 6 were fortunate enough to have a whole row of seats to themselves. How many black-haired people had whole rows to themselves?

34. ● **Restaurant Menus** Your favorite restaurant offers a total of 14 desserts, of which 8 have ice cream as a main ingredient and 9 have fruit as a main ingredient. Assuming that all of them have either ice cream or fruit or both as a main ingredient, how many have both?

Publishing *Exercises 35–40 are based on the following table, which shows the results of a survey of authors by a fictitious publishing company.*

	New Authors	Established Authors	Total
Successful	5	25	30
Unsuccessful	15	55	70
Total	20	80	100

Consider the following subsets of the set S of all authors represented in the table: C, the set of successful authors; U, the set of unsuccessful authors; N, the set of new authors; and E, the set of established authors. *hint* [see Example 3]

35. ● Describe the sets $C \cap N$ and $C \cup N$ in words. Use the table to compute $n(C), n(N), n(C \cap N)$ and $n(C \cup N)$. Verify that $n(C \cup N) = n(C) + n(N) - n(C \cap N)$.

36. ● Describe the sets $N \cap U$ and $N \cup U$ in words. Use the table to compute $n(N), n(U), n(N \cap U)$ and $n(N \cup U)$. Verify that $n(N \cup U) = n(N) + n(U) - n(N \cap U)$.

37. ● Describe the set $C \cap N'$ in words, and find the number of elements it contains.

38. ● Describe the set $U \cup E'$ in words, and find the number of elements it contains.

39. What percentage of established authors are successful? What percentage of successful authors are established?

40. What percentage of new authors are unsuccessful? What percentage of unsuccessful authors are new?

---

[4] Figures are rounded to the nearest 1000.

[5] Figures are rounded to the nearest million.

● basic skills   ◆ challenging

Recreational Boat Sales Exercises 41–46 are based on the following table, which shows the amount spent, in billions of dollars, on recreational boats and accessories in the U.S. during the period 1999–2001[6] (Take S to be the set of all dollars represented in the table.)

	Used Boats U	New Boats N	Accessories R	Total
1999 A	7	7	4	18
2000 B	7	7.5	5	19.5
2001 C	8	8	5	21
Total	22	22.5	14	58.5

In Exercises 41–46, use symbols to describe each set, and compute its cardinality.

41. ● The set of dollars spent on new boats in 2001.

42. ● The set of dollars spent on accessories or spent in 2001.

43. ● The set of dollars spent in 2001 excluding new boats.

44. ● The set of dollars spent on new boats excluding 2001.

45. The set of dollars spent in 1999 on new and used boats.

46. The set of dollars spent in years other than 1999 on new boats and accessories.

Exercises 47–52 are based on the following table, which shows the performance of a selection of 100 stocks after one year. (Take S to be the set of all stocks represented in the table.)

	Companies			
	Pharmaceutical P	Electronic E	Internet I	Total
Increased V	10	5	15	30
Unchanged* N	30	0	10	40
Decreased D	10	5	15	30
Total	50	10	40	100

*If a stock stayed within 20% of its original value, it is classified as "unchanged."

47. ● Use symbols to describe the set of non-Internet stocks that increased. How many elements are in this set?

48. ● Use symbols to describe the set of Internet stocks that did not increase. How many elements are in this set?

49. ● Compute $n(P' \cup N)$. What does this number represent?

50. ● Compute $n(P \cup N')$. What does this number represent?

51. ● Calculate $\dfrac{n(V \cap I)}{n(I)}$. What does the answer represent?

52. ● Calculate $\dfrac{n(D \cap I)}{n(D)}$. What does the answer represent?

53. *Medicine* In a study of Tibetan children,[7] a total of 1556 children were examined. Of these, 1024 had rickets. Of the 243 urban children in the study, 93 had rickets.

a. How many children living in nonurban areas had rickets?

b. How many children living in nonurban areas did not have rickets?

54. *Medicine* In a study of Tibetan children,[8] a total of 1556 children were examined. Of these, 615 had caries (cavities). Of the 1313 children living in nonurban areas, 504 had caries.

a. How many children living in urban areas had caries?

b. How many children living in urban areas did not have caries?

55. ● *Entertainment* According to a survey of 100 people regarding their movie attendance in the last year, 40 had seen a science fiction movie, 55 had seen an adventure movie, and 35 had seen a horror movie. Moreover, 25 had seen a science fiction movie and an adventure movie, 5 had seen an adventure movie and a horror movie, and 15 had seen a science fiction movie and a horror movie. Only 5 people had seen a movie from all three categories.
hint [see Example 4]

a. Use the given information to set up a Venn diagram and solve it.

b. Complete the following sentence: The survey suggests that __ % of science fiction movie fans are also horror movie fans.

56. ● *Athletics* Of the 4700 students at Medium Suburban College (MSC), 50 play collegiate soccer, 60 play collegiate lacrosse, and 96 play collegiate football. Only 4 students play both collegiate soccer and lacrosse, 6 play collegiate soccer and football, and 16 play collegiate lacrosse and football. No students play all three sports.

a. Use the given information to set up a Venn diagram and solve it.

b. Complete the following sentence: __ % of the college soccer players also play one of the other two sports at the collegiate level.

57. ● *Entertainment* In a survey of 100 Enormous State University students, 21 enjoyed classical music, 22 enjoyed rock music, and 27 enjoyed house music. Five of the students enjoyed both classical and rock. How many of those that enjoyed rock did not enjoy classical music?

58. ● *Entertainment* Refer back to the preceding exercise. You are also told that 5 students enjoyed all three kinds of music while 53 enjoyed music in none of these categories. How many students enjoyed both classical and rock but disliked house music?

[6] Figures are approximate. SOURCE: National Marine Manufacturers Association/*New York Times*, January 10. 2002, p. C1.

[7] SOURCE: N. S. Harris et al., Nutritional and Health Status of Tibetan Children Living at High Altitudes, *New England Journal of Medicine*, 344(5), February 1, 2001, pp. 341–347.

[8] Ibid.

● basic skills   ◆ challenging

Communication and Reasoning Exercises

59. ● Why is the Cartesian product referred to as a "product"? [*Hint:* Think about cardinality.]

60. ● Refer back to your answer to Exercise 59. What set operation could one use to represent the *sum* of two disjoint sets A and B? Why?

61. ● Formulate an interesting application whose answer is $n(A \cap B) = 20$.

62. ● Formulate an interesting application whose answer is $n(A \times B) = 120$.

63. When is $n(A \cup B) \neq n(A) + n(B)$?

64. When is $n(A \times B) = n(A)$?

65. When is $n(A \cup B) = n(A)$?

66. When is $n(A \cap B) = n(A)$?

67. ◆ Use a Venn diagram or some other method to obtain a formula for $n(A \cup B \cup C)$ in terms of $n(A), n(B), n(C)$, $n(A \cap B), n(A \cap C), n(B \cap C)$ and $n(A \cap B \cap C)$.

68. ◆ Suppose that A and B are nonempty sets with $A \subset B$ and $n(A)$ at least two. Arrange the following numbers from smallest to largest (if two numbers are equal, say so): $n(A)$, $n(A \times B), n(A \cap B), n(A \cup B), n(B \times A), n(B), n(B \times B)$.

● basic skills   ◆ challenging

6.3 The Addition and Multiplication Principles

Let's start with a really simple example. You walk into an ice cream parlor and find that you can choose between ice cream, of which there are 15 flavors, and frozen yogurt, of which there are 5 flavors. How many different selections can you make? Clearly, you have $15 + 5 = 20$ different desserts from which to choose. Mathematically, this is an example of the formula for the cardinality of a disjoint union: If we let A be the set of ice creams you can choose from, and B the set of frozen yogurts, then $A \cap B = \emptyset$ and we want $n(A \cup B)$. But the formula for the cardinality of a disjoint union is $n(A \cup B) = n(A) + n(B)$, which gives $15 + 5 = 20$ in this case.

This example illustrates a very useful general principle.

Addition Principle

When choosing among r disjoint alternatives, if

    alternative 1 has n_1 possible outcomes,

    alternative 2 has n_2 possible outcomes,

    . . .

    alternative r has n_r possible outcomes,

then you have a total of $n_1 + n_2 + \cdots + n_r$ possible outcomes.

quick Example

At a restaurant you can choose among 8 chicken dishes, 10 beef dishes, 4 seafood dishes, and 12 vegetarian dishes. This gives a total of $8 + 10 + 4 + 12 = 34$ different dishes to choose from.

Here is another simple example. In that ice cream parlor, not only can you choose from 15 flavors of ice cream, but you can also choose from 3 different sizes of cone. How many different ice cream cones can you select from? If we let A again be the set of ice cream flavors and now let C be the set of cone sizes, we want to pick a flavor *and* a size. That is, we want to pick an element of $A \times C$, the Cartesian product. To find the number of choices we have, we use the formula for the cardinality of a Cartesian product: $n(A \times C) = n(A)n(C)$. In this case, we get $15 \times 3 = 45$ different ice cream cones we can select.

This example illustrates another general principle.

> **Multiplication Principle**
>
> When making a sequence of choices with r steps, if
>
> > step 1 has n_1 possible outcomes
> >
> > step 2 has n_2 possible outcomes
> >
> > . . .
> >
> > step r has n_r possible outcomes
>
> then you have a total of $n_1 \times n_2 \times \cdots \times n_r$ possible outcomes.

quick Example   At a restaurant you can choose among 5 appetizers, 34 main dishes, and 10 desserts. This gives a total of $5 \times 34 \times 10 = 1700$ different meals (each including one appetizer, one main dish, and one dessert) from which you can choose.

Things get more interesting when we have to use the addition and multiplication principles in tandem.

Example 1 Desserts

You walk into an ice cream parlor and find that you can choose between ice cream, of which there are 15 flavors, and frozen yogurt, of which there are 5 flavors. In addition, you can choose among 3 different sizes of cones for your ice cream or 2 different sizes of cups for your yogurt. How many different desserts can you choose from?

Solution It helps to think about a definite procedure for deciding which dessert you will choose. Here is one we can use:

Alternative 1: An ice cream cone
   Step 1 Choose a flavor.
   Step 2 Choose a size.

Alternative 2: A cup of frozen yogurt
   Step 1 Choose a flavor.
   Step 2 Choose a size.

That is, we can choose between alternative 1 and alternative 2. If we choose alternative 1, we have a sequence of two choices to make: flavor and size. The same is true of alternative 2. We shall call a procedure in which we make a sequence of decisions a **decision algorithm.**[*] Once we have a decision algorithm, we can use the addition and multiplication principles to count the number of possible outcomes.

Alternative 1: An ice cream cone
   Step 1 Choose a flavor; 15 choices
   Step 2 Choose a size; 3 choices
   There are $15 \times 3 = 45$ possible choices in alternative 1.   Multiplication Principle

Alternative 2: A cup of frozen yogurt
   Step 1 Choose a flavor; 5 choices
   Step 2 Choose a size; 2 choices
   There are $5 \times 2 = 10$ possible choices in alternative 2.   Multiplication Principle

   So, there are $45 + 10 = 55$ possible choices of desserts.   Addition Principle

_____
[*] An algorithm is a procedure with definite rules for what to do at every step.

✛*Before we go on...* Decision algorithms can be illustrated by **decision trees.** To simplify the picture, suppose we had fewer choices in Example 1—say, only 2 choices of ice cream flavor: vanilla and chocolate, and 2 choices of yogurt flavor: banana and raspberry. This gives us a total of $2 \times 3 + 2 \times 2 = 10$ possible desserts. We can illustrate the decisions we need to make when choosing what to buy in the diagram in Figure 8, called a decision tree.

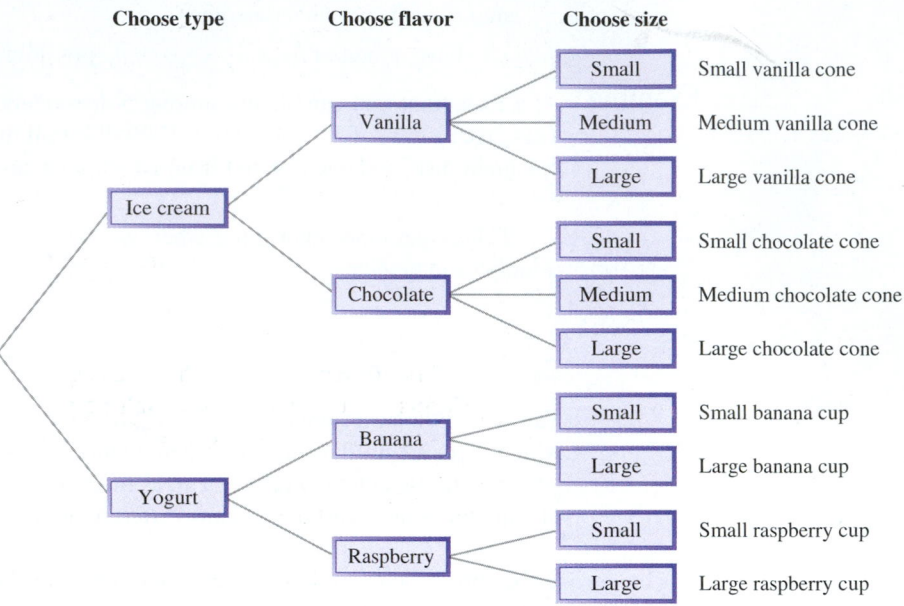

Figure **8**

To count the number of possible gadgets, *pretend you are designing a gadget,* and list the decisions to be made at each stage.

We do not use decision trees much in this chapter because, while they provide a good way of thinking about decision algorithms, they're not really practical for counting large sets. Similar diagrams will be very useful, however, in the chapter on probability. ■

Decision Algorithm

A **decision algorithm** is a procedure in which we make a sequence of decisions. We can use decision algorithms to determine the number of possible items by pretending we are *designing* such an item (for example, an ice-cream cone) and listing the decisions or choices we should make at each stage of the process.

quick Example

An iPod is available in two sizes. The larger size comes in 2 colors and the smaller size (the Mini) comes in 4 colors. A decision algorithm for "designing" an iPod is:

Alternative 1: Select Large:
    Step 1 Choose a color: 2 choices
    (so, there are 2 choices for Alternative 1)

Alternative 2: Select a Mini:
    Step 1 Choose a color: 4 choices
    (so, there are 4 choices for Alternative 2)

Thus, there are $2 + 4 = 6$ possible choices of iPods.

Caution

For a decision algorithm to give the correct number of possible items, it must be the case that *changing one or more choices results in a different item* (see Example 3).

Example 2 Exams

An exam is broken into two parts, Part A and Part B, both of which you are required to do. In Part A you can choose between answering 10 true-false questions or answering 4 multiple-choice questions, each of which has 5 answers to choose from. In Part B you can choose between answering 8 true-false questions and answering 5 multiple-choice questions, each of which has 4 answers to choose from. How many different collections of answers are possible?

Solution While deciding what answers to write down, we use the following decision algorithm:

Step 1 Do Part A.
   Alternative 1: Answer the 10 true-false questions.
       Steps 1–10 Choose true or false for each question; 2 choices each.
       There are $2 \times 2 \times \cdots \times 2 = 2^{10} = 1024$ choices in alternative 1.

   Alternative 2: Answer the 4 multiple-choice questions.
       Steps 1–4 Choose one answer for each question; 5 choices each.
       $5 \times 5 \times 5 \times 5 = 5^4 = 625$ choices in alternative 2.
   $1024 + 625 = 1649$ choices in step 1

Step 2 Do Part B.
   Alternative 1: Answer the 8 true-false questions; 2 choices each.
       $2^8 = 256$ choices in alternative 1

   Alternative 2: Answer the 5 multiple-choice questions, 4 choices each.
       $4^5 = 1024$ choices in alternative 2
   $256 + 1024 = 1280$ choices in step 2

There are $1649 \times 1280 = 2{,}110{,}720$ different collections of answers possible.

The next example illustrates the need to select your decision algorithm with care.

Example 3 Scrabble®

You are playing Scrabble® and have the following letters to work with: k, e, r, e. Because you are losing the game, you would like to use all your letters to make a single word, but you can't think of any four-letter words using all these letters. In desperation, you decide to list *all* the four-letter sequences possible to see if there are any valid words among them. How large is your list?

Solution It may first occur to you to try the following decision algorithm.

Step 1 Select the first letter; 4 choices.
Step 2 Select the second letter; 3 choices.

Step 3 Select the third letter; 2 choices.
Step 4 Select the last letter; 1 choice.

This gives $4 \times 3 \times 2 \times 1 = 24$ choices. However, something is wrong with the algorithm.

Q: What is wrong with this decision algorithm?

A: We didn't take into account the fact that there are two "e"s;[*] different decisions in Steps 1–4 can produce the same sequence. Suppose, for example, that we selected the first "e" in Step 1, the second "e" in Step 2, and then the "k" and the "r." This would produce the sequence "eekr." If we selected the *second* "e" in Step 1, the *first* "e" in Step 2, and then the "k" and "r," we would obtain the *same* sequence: "eekr." In other words, the decision algorithm produces two copies of the sequence "eekr" in the associated decision tree. (In fact, it produces two copies of each possible sequence of the letters.)

For a decision algorithm to be valid, each sequence of choices must produce a different result. ∎

Because our original algorithm is not valid, we need a new one. Here is a strategy that works nicely for this example. Imagine, as before, that we are going to construct a sequence of four letters. This time we are going to imagine that we have a sequence of four empty slots: ☐☐☐☐, and instead of selecting letters to fill the slots from left to right, we are going to select *slots* in which to place each of the letters. Remember that we have to use these letters: k, e, r, e. We proceed as follows, leaving the "e"s until last.

Step 1 Select an empty slot for the k; 4 choices (e.g. ☐☐k☐)
Step 2 Select an empty slot for the r; 3 choices (e.g. r☐k☐).
Step 3 Place the "e"s in the remaining two slots; 1 choice!

Thus the multiplication principle yields $4 \times 3 \times 1 = 12$ choices.

---

[*] Consider the following extreme case: If all four letters were "e," then there would be only a single sequence: "eeee" and not the 24 predicted by the decision algorithm.

+*Before we go on...* You should try constructing a decision tree for Example 3, and you will see that each sequence of four letters is produced exactly once when we use the correct (second) decision algorithm. ∎

FAQs Creating and Testing a Decision Algorithm

Q: How do I set up a decision algorithm to count how many items there are in a given scenario?

A: Pretend that you are *designing* such an item (for example, pretend that you are designing an ice-cream cone) and come up with a step-by-step procedure for doing so, listing the decisions you should make at each stage. ∎

Q: Once I have my decision algorithm, how do I check if it is valid?

A: Ask yourself the following question: "Is it possible to get the same item (the exact same ice-cream cone, say) by making different decisions when applying the algorithm?" If the answer is "yes," then your decision algorithm is invalid. Otherwise, it is valid. ∎

6.3 EXERCISES

● denotes basic skills exercises

◆ denotes challenging exercises

1. ● An experiment requires a choice among 3 initial setups. The first setup can result in 2 possible outcomes, the second in 3 possible outcomes, and the third in 5 possible outcomes. What is the total number of outcomes possible? *hint* [see Quick Example on p. 416]

2. ● A surgical procedure requires choosing among 4 alternative methodologies. The first can result in 4 possible outcomes, the second in 3 possible outcomes, and the remaining methodologies can each result in 2 possible outcomes. What is the total number of outcomes possible?

3. ● An experiment requires a sequence of 3 steps. The first step can result in 2 possible outcomes, the second in 3 possible outcomes, and the third in 5 possible outcomes. What is the total number of outcomes possible? *hint* [see Quick Example on p. 417]

4. ● A surgical procedure requires 4 steps. The first can result in 4 possible outcomes, the second in 3 possible outcomes, and the remaining two can each result in 2 possible outcomes. What is the total number of outcomes possible?

For the decision algorithms in Exercises 5–12, find how many outcomes are possible. *hint* [see Example 1]

5. ● Alternative 1: Alternative 2:
  Step 1: 1 outcome   Step 1: 2 outcomes
  Step 2: 2 outcomes   Step 2: 2 outcomes
          Step 3: 1 outcome

6. ● Alternative 1: Alternative 2:
  Step 1: 1 outcome   Step 1: 2 outcomes
  Step 2: 2 outcomes   Step 2: 2 outcomes
  Step 3: 2 outcomes

7. ● Step 1: Step 2:
  Alternative 1: 1 outcome   Alternative 1: 2 outcomes
  Alternative 2: 2 outcomes   Alternative 2: 2 outcomes
          Alternative 3: 1 outcome

8. ● Step 1: Step 2:
  Alternative 1: 1 outcome   Alternative 1: 2 outcomes
  Alternative 2: 2 outcomes   Alternative 2: 2 outcomes
  Alternative 3: 2 outcomes

9. ● Alternative 1: Alternative 2: 5 outcomes
  Step 1:
    Alternative 1: 3 outcomes
    Alternative 2: 1 outcome
  Step 2: 2 outcomes

10. ● Alternative 1: 2 outcomes Alternative 2:
          Step 1:
            Alternative 1: 4 outcomes
            Alternative 2: 1 outcome
          Step 2: 2 outcomes

11. ● Step 1: Step 2: 5 outcomes
  Alternative 1:
    Step 1: 3 outcomes
    Step 2: 1 outcome
  Alternative 2: 2 outcomes

12. ● Step 1: 2 outcomes Step 2:
          Alternative 1:
            Step 1: 4 outcomes
            Step 2: 1 outcome
          Alternative 2: 2 outcomes

13. ● How many different four-letter sequences can be formed from the letters a, a, a, b? *hint* [see Example 3]

14. ● How many different five-letter sequences can be formed from the letters a, a, a, b, c?

Applications

15. ● *Ice Cream* When Baskin-Robbins® was founded in 1945, it made 31 different flavors of ice cream.[9] If you had a choice of having your ice cream in a cone, a cup, or a sundae, how many different desserts could you have?

16. ● *Ice Cream* At the beginning of 2002, Baskin-Robbins® claimed to have "nearly 1000 different ice cream flavors.[10] Assuming that you could choose from 1000 different flavors, that you could have your ice cream in a cone, a cup, or a sundae, and that you could choose from a dozen different toppings, how many different desserts could you have?

17. ● *Binary Codes* A binary digit, or "bit," is either 0 or 1. A nybble is a four-bit sequence. How many different nybbles are possible?

18. ● *Ternary Codes* A ternary digit is either 0, 1, or 2. How many sequences of six ternary digits are possible?

19. ● *Ternary Codes* A ternary digit is either 0, 1, or 2. How many sequences of six ternary digits containing a single 1 and a single 2 are possible?

20. ● *Binary Codes* A binary digit, or "bit," is either 0 or 1. A nybble is a four-bit sequence. How many different nybbles containing a single 1 are possible?

21. ● *Reward* While selecting candy for students in his class, Professor Murphy must choose between gummy candy and

---

[9] SOURCE: Company website www.baskinrobbins.com.

[10] Ibid.

licorice nibs. Gummy candy packets come in three sizes, while packets of licorice nibs come in two. If he chooses gummy candy, he must select gummy bears, gummy worms, or gummy dinos. If he chooses licorice nibs, he must choose between red and black. How many choices does he have? *hint* [see Example 2]

22. ● *Productivity* Professor Oger must choose between an extra writing assignment and an extra reading assignment for the upcoming spring break. For the writing assignment, there are two essay topics to choose from and three different mandatory lengths (30 pages, 35 pages, or 40 pages). The reading topic would consist of one scholarly biography combined with one volume of essays. There are five biographies and two volumes of essays to choose from. How many choices does she have?

23. ● *Zip™ Disks* Zip™ disks come in two sizes (100 MB and 250 MB), packaged singly, in boxes of five, or in boxes of ten. When purchasing singly, you can choose from five colors; when purchasing in boxes of five or ten you have two choices, black or an assortment of colors. If you are purchasing Zip disks, how many possibilities do you have to choose from?

24. ● *Radar Detectors* Radar detectors are either powered by their own battery or plug into the cigarette lighter socket. All radar detectors come in two models: no-frills and fancy. In addition, detectors powered by their own batteries detect either radar or laser, or both, whereas the plug-in types come in models that detect either radar or laser, but not both. How many different radar detectors can you buy?

25. ● *Multiple-Choice Tests* Professor Easy's final examination has 10 true-false questions followed by 2 multiple-choice questions. In each of the multiple-choice questions, you must select the correct answer from a list of 5. How many answer sheets are possible?

26. ● *Multiple-choice Tests* Professor Tough's final examination has 20 true-false questions followed by 3 multiple-choice questions. In each of the multiple-choice questions, you must select the correct answer from a list of 6. How many answer sheets are possible?

27. ● *Tests* A test requires that you answer either Part A or Part B. Part A consists of 8 true-false questions, and Part B consists of 5 multiple-choice questions with 1 correct answer out of 5. How many different completed answer sheets are possible?

28. ● *Tests* A test requires that you answer first Part A and then either Part B or Part C. Part A consists of 4 true-false questions, Part B consists of 4 multiple-choice questions with 1 correct answer out of 5, and Part C consists of 3 questions with one correct answer out of 6. How many different completed answer sheets are possible?

29. ● *Stock Portfolios* Your broker has suggested that you diversify your investments by splitting your portfolio among mutual funds, municipal bond funds, stocks, and precious metals. She suggests four good mutual funds, three municipal bond funds, eight stocks, and three precious metals (gold, silver, and platinum).

a. Assuming your portfolio is to contain one of each type of investment, how many different portfolios are possible?

b. Assuming your portfolio is to contain three mutual funds, two municipal bond funds, one stock, and two precious metals, how many different portfolios are possible?

30. ● *Menus* The local diner offers a meal combination consisting of an appetizer, a soup, a main course, and a dessert. There are five appetizers, two soups, four main courses, and five desserts. Your diet restricts you to choosing between a dessert and an appetizer. (You cannot have both.) Given this restriction, how many three-course meals are possible?

31. ● *Computer Codes* A computer "byte" consists of eight "bits," each bit being either a 0 or a 1. If characters are represented using a code that uses a byte for each character, how many different characters can be represented?

32. ● *Computer Codes* Some written languages, like Chinese and Japanese, use tens of thousands of different characters. If a language uses roughly 50,000 characters, a computer code for this language would have to use how many bytes per character? (See Exercise 31.)

33. ● *Symmetries of a Five-Pointed Star* A five-pointed star will appear unchanged if it is rotated through any one of the angles 0°, 72°, 144°, 216°, or 288°. It will also appear unchanged if it is flipped about the axis shown in the figure. A *symmetry* of the five-pointed star consists of either a rotation, or a rotation followed by a flip. How many different symmetries are there altogether?

34. ● *Symmetries of a Six-Pointed Star* A six-pointed star will appear unchanged if it is rotated through any one of the angles 0°, 60°, 120°, 180°, 240°, or 300°. It will also appear unchanged if it is flipped about the axis shown in the figure. A *symmetry* of the six-pointed star consists of either a rotation, or a rotation followed by a flip. How many different symmetries are there altogether?

35. ● *Variables in BASIC*[11] A variable name in the programming language BASIC can be either a letter or a letter followed by a decimal digit—that is, one of the numbers 0, 1, . . . , 9. How many different variables are possible?

36. ● *Employee IDs* A company assigns to each of its employees an ID code that consists of one, two, or three letters followed by a digit from 0 through 9. How many employee codes does the company have available?

37. ● *Tournaments* How many ways are there of filling in the blanks for the following (fictitious) soccer tournament?

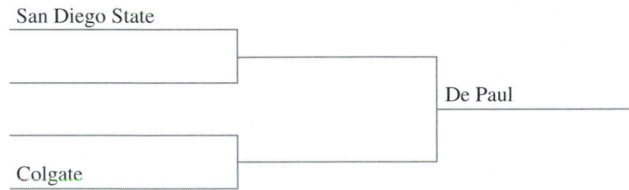

North Carolina
Central Connecticut
Virginia
Syracuse

38. ● *Tournaments* How many ways are there of filling in the blanks for a (fictitious) soccer tournament involving the four teams San Diego State, De Paul, Colgate, and Hofstra?

San Diego State
De Paul
Colgate

39. *Telephone Numbers* A telephone number consists of a sequence of 7 digits not starting with 0 or 1.

a. How many telephone numbers are possible?
b. How many of them begin with either 463, 460, or 400?
c. How many telephone numbers are possible if no two adjacent digits are the same? (For example, 235-9350 is permitted, but not 223-6789.)

40. *Social Security Numbers* A Social Security Number is a sequence of 9 digits.

a. How many Social Security Numbers are possible?
b. How many of them begin with either 023 or 003?
c. How many Social Security Numbers are possible if no two adjacent digits are the same? (For example, 235-93-2345 is permitted, but not 126-67-8189.)

41. *DNA Chains* DNA (deoxyribonucleic acid) is the basic building block of reproduction in living things. A DNA chain is a sequence of chemicals called *bases*. There are four possible bases: thymine (T), cytosine (C), adenine (A), and guanine (G).

a. How many three-element DNA chains are possible?
b. How many *n*-element DNA chains are possible?

c. A human DNA chain has 2.1×10^{10} elements. How many human DNA chains are possible?

42. *Credit Card Numbers*[12] Each customer of Mobil Credit Corporation is given a 9-digit number for computer identification purposes.

a. If each digit can be any number from 0 to 9, are there enough different account numbers for 10 million credit-card holders?
b. Would there be if the digits were only 0 or 1?

43. *HTML* Colors in HTML (the language in which many web pages are written) can be represented by 6-digit hexadecimal codes: sequences of six integers ranging from 0 to 15 (represented as 0, . . . , 9, A, B, . . . , F).

a. How many different colors can be represented?
b. Some monitors can only display colors encoded with pairs of repeating digits (such as 44DD88). How many colors can these monitors display?
c. Grayscale shades are represented by sequences $xyxyxy$ consisting of a repeated pair of digits. How many grayscale shades are possible?
d. The pure colors are pure red: $xy0000$; pure green: $00xy00$; and pure blue: $0000xy$. ($xy = FF$ gives the brightest pure color, while $xy = 00$ gives the darkest: black). How many pure colors are possible?

44. *Telephone Numbers* In the past, a local telephone number in the U.S. consisted of a sequence of two letters followed by five digits. Three letters were associated with each number from 2 to 9 (just as in the standard telephone layout shown in the figure) so that each telephone number corresponded to a sequence of 7 digits. How many different sequences of 7 digits were possible?

45. *Romeo and Juliet* Here is a list of the main characters in Shakespeare's *Romeo and Juliet.* The first 7 characters are male and the last 4 are female.

Escalus, *prince of Verona*
Paris, *kinsman to the prince*
Romeo, *of Montague Household*
Mercutio, *friend of Romeo*

---

[11] From *Applied Combinatorics* by F.S. Roberts (Prentice Hall, New Jersey, 1984)

[12] Ibid.

● basic skills   ◆ challenging

Benvolio, *friend of Romeo*
Tybalt, *nephew to Lady Capulet*
Friar Lawrence, *a Franciscan*
Lady Montague, *of Montague Household*
Lady Capulet, *of Capulet Household*
Juliet, *of Capulet Household*
Juliet's Nurse

A total of 10 male and 8 female actors are available to play these roles. How many possible casts are there? (All roles are to be played by actors of the same gender as the character.)

46. *Swan Lake* The Enormous State University's Accounting Society has decided to produce a version of the ballet *Swan Lake,* in which all the female roles (including all of the swans) will be danced by men, and vice versa. Here are the main characters:

Prince Siegfried
Prince Siegfried's Mother
Princess Odette, *the White Swan*
The Evil Duke Rotbart
Odile, *the Black Swan*
Cygnet #1, *young swan*
Cygnet #2, *young swan*
Cygnet #3, *young swan*

The ESU Accounting Society has on hand a total of 4 female dancers and 12 male dancers who are to be considered for the main roles. How many possible casts are there?

47. *License Plates* Many U.S. license plates display a sequence of 3 letters followed by 3 digits.

a. How many such license plates are possible?

b. In order to avoid confusion of letters with digits, some states do not issue standard plates with the last letter an I, O, or Q. How many license plates are still possible?

c. Assuming that the letter combinations VET, MDZ and DPZ are reserved for disabled veterans, medical practitioners and disabled persons, respectively, how many license plates are possible, also taking the restriction in part (b) into account?

48. *License Plates*[13] License plates in Montana have a sequence consisting of: (1) a digit from 1 to 9, (2) a letter, (3) a dot, (4) a letter, and (5) a four-digit number.

a. How many different license plates are possible?

b. How many different license plates are available for citizens if numbers that end with 0 are reserved for official state vehicles?

49. *Mazes*

a. How many four-letter sequences are possible that contain only the letters R and D, with D occurring only once?

b. Use part (a) to calculate the number of possible routes from Start to Finish in the maze shown in the figure, where each move is either to the right or down.

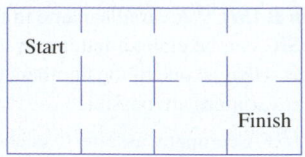

c. Comment on what would happen if we also allowed left and up moves.

50. *Mazes*

a. How many six-letter sequences are possible that contain only the letters R and D, with D occurring only once?

b. Use part (a) to calculate the number of possible routes from Start to Finish in the maze shown in the figure, where each move is either to the right or down.

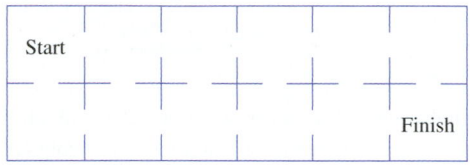

c. Comment on what would happen if we also allowed left and up moves.

51. *Car Engines*[14] In a six-cylinder V6 engine, the even-numbered cylinders are on the left and the odd-numbered cylinders are on the right. A good firing order is a sequence of the numbers 1 through 6 in which right and left sides alternate.

a. How many possible good firing sequences are there?

b. How many good firing sequences are there that start with a cylinder on the left?

52. *Car Engines* Repeat the preceding exercise for an eight-cylinder V8 engine.

53. *Minimalist Art* You are exhibiting your collection of minimalist paintings. Art critics have raved about your paintings, each of which consists of 10 vertical colored lines set against a white background. You have used the following rule to produce your paintings: Every second line, starting with the first, is to be either blue or grey, while the remaining five lines are to be either all orange, all pink, or all purple. Your collection is complete: Every possible combination that satisfies the rules occurs. How many paintings are you exhibiting?

54. *Combination Locks* Dripping wet after your shower, you have clean forgotten the combination of your lock. It is one of

---

[13] SOURCE: The License Plates of the World website: www.worldlicenseplates.com.

[14] Adapted from an exercise in *Basic Techniques of Combinatorial Theory* by D.I.A. Cohen (New York: John Wiley, 1978).

● basic skills ◆ challenging

those "standard" combination locks, which uses a three-number combination with each number in the range 0 through 39. All you remember is that the second number is either 27 or 37, while the third number ends in a 5. In desperation, you decide to go through all possible combinations using the information you remember. Assuming that it takes about 10 seconds to try each combination, what is the longest possible time you may have to stand dripping in front of your locker?

55. *Product Design* Your company has patented an electronic digital padlock that a user can program with his or her own four-digit code. (Each digit can be 0 through 9.) The padlock is designed to open if either the correct code is keyed in or—and this is helpful for forgetful people—if exactly one of the digits is incorrect.

   a. How many incorrect codes will open a programmed padlock?

   b. How many codes will open a programmed padlock?

56. *Product Design* Your company has patented an electronic digital padlock which has a telephone-style keypad. Each digit from 2 through 9 corresponds to three letters of the alphabet (see the figure for Exercise 44). How many different four-letter sequences correspond to a single four-digit sequence using digits in the range 2 through 9?

57. *Calendars* The World Almanac[15] features a "perpetual calendar," a collection of 14 possible calendars. Why does this suffice to be sure there is a calendar for every conceivable year?

58. *Calendars* How many possible calendars are there that have February 12 falling on a Sunday, Monday, or Tuesday?

59. *Programming in BASIC* (Some programming knowledge assumed for this exercise.) How many iterations will be carried out in the following routine?

```
For i = 1 to 10
    For j = 2 to 20
        For k = 1 to 10
            Print i, j, k
        Next k
    Next j
Next i
```

60. *Programming in JAVASCRIPT* (Some programming knowledge assumed for this exercise.) How many iterations will be carried out in the following routine?

```
for (i = 1; i <= 2; i++) {
    for (j = 1; j <= 2; j++) {
        for (k = 1; k <= 2; k++)
            sum += i+j+k;
    }
}
```

[15] SOURCE: *The World Almanac and Book of Facts* 1992 (New York: Pharos Books, 1992).

61. ◆ ***Building Blocks*** Use a decision algorithm to show that a rectangular solid with dimensions $m \times n \times r$ can be constructed with $m \cdot n \cdot r$ cubical $1 \times 1 \times 1$ blocks (see the figure).

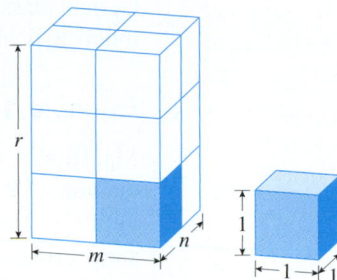

Rectangular solid made up of $1 \times 1 \times 1$ cubes

62. ◆ ***Matrices*** (Some knowledge of matrices is assumed for this exercise.) Use a decision algorithm to show that an $m \times n$ matrix must have $m \cdot n$ entries.

63. ◆ ***Morse Code*** In Morse Code, each letter of the alphabet is encoded by a different sequence of dots and dashes. Different letters may have sequences of different lengths. How long should the longest sequence be in order to allow for every possible letter of the alphabet?

64. ◆ ***Numbers*** How many odd numbers between 10 and 99 have distinct digits?

Communication and Reasoning Exercises

65. ● Complete the following sentences: The multiplication principle is based on the cardinality of the _____ of two sets.

66. ● Complete the following sentences: The addition principle is based on the cardinality of the _____ of two disjoint sets.

67. ● You are packing for a short trip and want to take 2 of the 10 shirts you have hanging in your closet. Critique the following decision algorithm and calculation of how many different ways you can choose two shirts to pack: Step 1, choose one shirt, 10 choices. Step 2, choose another shirt, 9 choices. Hence there are 90 possible choices of two shirts.

68. ● You are designing an advertising logo that consists of a tower of five squares. Three are yellow, one is blue, and one is green. Critique the following decision algorithm and calculation of the number of different five-square sequences: Step 1: choose the first square, 5 choices. Step 2: choose the second square, 4 choices. Step 3: choose the third square: 3 choices. Step 4: choose the fourth square, 2 choices. Step 5: choose the last square: 1 choice. Hence there are 120 possible five-square sequences.

69. Construct a decision algorithm that gives the correct number of five-square sequences in Exercise 68.

70. Find an interesting application that requires a decision algorithm with two steps in which each step has two alternatives.

● basic skills   ◆ challenging

6.4 Permutations and Combinations

Certain classes of counting problems come up frequently, and it is useful to develop formulas to deal with them.

Example 1 Casting

Ms. Birkitt, the English teacher at Brakpan Girls High School, wanted to stage a production of R. B. Sheridan's play, *The School for Scandal*. The casting was going well until she was left with five unfilled characters and five seniors who were yet to be assigned roles. The characters were Lady Sneerwell, Lady Teazle, Mrs. Candour, Maria, and Snake; while the unassigned seniors were April, May, June, Julia and Augusta. How many possible assignments are there?

Solution To decide on a specific assignment, we use the following algorithm:

Step 1 Choose a senior to play Lady Sneerwell; 5 choices.

Step 2 Choose one of the remaining seniors to play Lady Teazle; 4 choices.

Step 3 Choose one of the now remaining seniors to play Mrs. Candour; 3 choices.

Step 4 Choose one of the now remaining seniors to play Maria; 2 choices.

Step 5 Choose the remaining senior to play Snake; 1 choice.

Thus, there are $5 \times 4 \times 3 \times 2 \times 1 = 120$ possible assignments of seniors to roles.

What the situation in Example 1 has in common with many others is that we start with a set—here the set of seniors—and we want to know how many ways we can put the elements of that set in order in a list. In this example, an ordered list of the five seniors—say,

1. May

2. Augusta

3. June

4. Julia

5. April

corresponds to a particular casting:

Cast

Lady Sneerwell	May
Lady Teazle	Augusta
Mrs. Candour	June
Maria	Julia
Snake	April

We call an ordered list of items a **permutation** of those items.

If we have n items, how many permutations of those items are possible? We can use a decision algorithm similar to the one we used in the preceding example to select a permutation.

Step 1 Select the first item; n choices.

Step 2 Select the second item; $n - 1$ choices.

Step 3 Select the third item; $n - 2$ choices.

· · ·

Step $n - 1$ Select the next-to-last item; 2 choices.

Step n Select the last item; 1 choice.

Thus, there are $n \times (n - 1) \times (n - 2) \times \cdots \times 2 \times 1$ possible permutations. We call this number **n factorial,** which we write as $n!$.

Permutations

A **permutation of n items** is an ordered list of those items. The number of possible permutations of n items is given by **n factorial,** which is

$$n! = n \times (n - 1) \times (n - 2) \times \cdots \times 2 \times 1$$

for n a positive integer, and

$$0! = 1$$

Visualizing Permutations
Permutations of 3 colors in a flag:

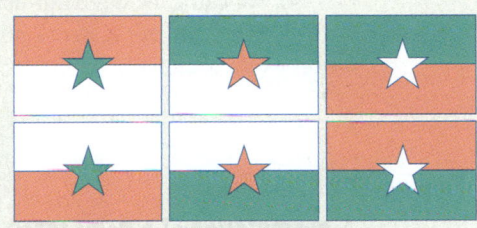

$3! = 3 \times 2 \times 1 = 6$ possible flags

quick Examples

1. The number of permutations of 5 items is $5! = 5 \times 4 \times 3 \times 2 \times 1 = 120$.

2. The number of ways 4 CDs can be played in sequence is $4! = 4 \times 3 \times 2 \times 1 = 24$.

3. The number of ways 3 cars can be matched with 3 drivers is $3! = 6$.

Sometimes, instead of constructing an ordered list of *all* the items of a set, we might want to construct a list of only *some* of the items, as in the next example.

Scott Barrow/SuperStock

Example **2** Corporations*

At the end of 2001, the 10 largest companies (by market capitalization) listed on the New York Stock Exchange were, in alphabetical order, American International Group, AOL Time Warner, Citigroup, Exxon Mobil Corp., General Electric Co., IBM, Johnson & Johnson, Merck & Co., Pfizer, and Wal-Mart Stores. You would like to apply to six of these companies for a job and you would like to list them in order of job preference. How many such ordered lists are possible?

Solution We want to count ordered lists, but we can't use the permutation formula because we don't want all 10 companies in the list, just 6 of them. So we fall back to a decision algorithm.

Step 1 Choose the first company; 10 choices.

Step 2 Choose the second company; 9 choices.

---

* SOURCE: New York Stock Exchange website: www.nyse.com.

Step 3 Choose the third one; 8 choices;

Step 4 Choose the fourth one; 7 choices.

Step 5 Choose the fifth one; 6 choices.

Step 6 Choose the sixth one; 5 choices.

Thus, there are $10 \times 9 \times 8 \times 7 \times 6 \times 5 = 151{,}200$ possible lists of six. We call this number the **number of permutations of 6 items chosen from 10,** or the **number of permutations of 10 items taken 6 at a time.**

✛ *Before we go on...* We wrote the answer as the product $10 \times 9 \times 8 \times 7 \times 6 \times 5$. But it is useful to notice that we can write this number in a more compact way:

$$10 \times 9 \times 8 \times 7 \times 6 \times 5 = \frac{10 \times 9 \times 8 \times 7 \times 6 \times 5 \times 4 \times 3 \times 2 \times 1}{4 \times 3 \times 2 \times 1}$$

$$= \frac{10!}{4!} = \frac{10!}{(10-6)!}$$

So, we can generalize our definition of permutation to allow for the case in which we use only some of the items, not all. Check that, if $r = n$ below, this is the same definition we gave above.

Permutations of *n* items taken *r* at a time

A **permutation of *n* items taken *r* at a time** is an ordered list of r items chosen from a set of n items. The number of permutations of n items taken r at a time is given by

$$P(n, r) = n \times (n - 1) \times (n - 2) \times \cdots \times (n - r + 1)$$

(there are r terms multiplied together). We can also write

$$P(n, r) = \frac{n!}{(n - r)!}$$

quick Example

The number of permutations of 6 items taken 2 at a time is

$$P(6, 2) = 6 \times 5 = 30$$

which we could also calculate as

$$P(6, 2) = \frac{6!}{(6 - 2)!} = \frac{6!}{4!} = \frac{720}{24} = 30$$

What if we don't care about the order of the items we're choosing? Consider the following example:

Example 3 Corporations

Suppose we simply wanted to pick two of the 10 companies listed in Example 2 to apply to, without regard to order. How many possible choices do we have? What if we wanted to choose six to apply to, without regard to order?

Solution Our first guess might be $P(10, 2) = 10 \times 9 = 90$. However, that is the number of *ordered lists* of two companies. We said that we don't care which is first and which second. For example, we consider the list

1. Merck   **2.** Pfizer

to be the same as

1. Pfizer   **2.** Merck.

Because every set of two companies occurs twice in the 90 lists, once in one order and again in the reverse order, we would count every set of two twice. Thus, there are $90/2 = 45$ possible choices of two companies.

Now, if we wish to pick six companies, again we might start with $P(10, 6) = 151,200$. But now, every set of six companies appears as many times as there are different orders in which they could be listed. Six things can be listed in $6! = 720$ different orders, so the number of ways of choosing six companies is $151,200/720 = 210$.

In Example 3 we were concerned with counting not the number of ordered lists, but the number of *unordered sets* of companies. For ordered lists we used the word *permutation*; for unordered sets we use the word **combination.**

Permutations and Combinations

A **permutation** of n items taken r at a time is an *ordered list* of r items chosen from n. A **combination** of n items taken r at a time is an *unordered set* of r items chosen from n.

Visualizing

1. ●
2. ●
3. ●

Permutation   Combination

Note Because lists are usually understood to be ordered, when we refer to a list of items, we will always mean an *ordered* list. Similarly, because sets are understood to be unordered, when we refer to a set of items we will always mean an *unordered* set. In short:

Lists are ordered. Sets are unordered. ∎

---

quick Example

There are six permutations of the three letters a, b, c taken two at a time:

1. a, b;   **2.** b, a;   **3.** a, c;   **4.** c, a;   **5.** b, c;   **6.** c, b.

There are six lists containing two of the letters a, b, c

There are three combinations of the three letters a, b, c taken two at a time:

1. {a, b};   **2.** {a, c};   **3.** {b, c}.

There are three sets containing two of the letters a, b, c

How do we count the number of possible combinations of n items taken r at a time? We generalize the calculation done in Example 3. The number of permutations is $P(n, r)$, but each set of r items occurs $r!$ times because this is the number of ways in which those r items can be ordered. So, the number of combinations is $P(n, r)/r!$.

Combinations of *n* items taken *r* at a time

The number of **combinations of *n* items taken *r* at a time** is given by

$$C(n, r) = \frac{P(n, r)}{r!} = \frac{n \times (n-1) \times (n-2) \times \cdots \times (n-r+1)}{r!}$$

We can also write

$$C(n, r) = \frac{n!}{r!(n-r)!}$$

quick Examples

1. The number of combinations of 6 items taken 2 at a time is

$$C(6, 2) = \frac{6 \times 5}{2 \times 1} = 15$$

which we can also calculate as

$$C(6, 2) = \frac{6!}{2!(6-2)!} = \frac{6!}{2!4!} = \frac{720}{2 \times 24} = 15$$

2. The number of sets of 4 marbles chosen from 6 is

$$C(6, 4) = \frac{6 \times 5 \times 4 \times 3}{4 \times 3 \times 2 \times 1} = 15$$

Note There are other common notations for $C(n, r)$. Calculators often have $_nC_r$. In mathematics we often write $\binom{n}{r}$ which is also known as a **binomial coefficient.** Because $C(n, r)$ is the number of ways of choosing a set of r items from n, it is often read "*n* choose *r*." ∎

Example 4 Calculating Combinations

Calculate   **a.** $C(11, 3)$ **b.** $C(11, 8)$.

Solution The easiest way to calculate $C(n, r)$ by hand is to use the formula

$$C(n, r) = \frac{n \times (n-1) \times (n-2) \times \cdots \times (n-r+1)}{r!}$$

$$= \frac{n \times (n-1) \times (n-2) \times \cdots \times (n-r+1)}{r \times (r-1) \times (r-2) \times \cdots \times 1}$$

Both the numerator and the denominator have r factors, so we can begin with n/r and then continue multiplying by decreasing numbers on the top and the bottom until we hit 1 in the denominator. When calculating, it helps to cancel common factors from the numerator and denominator before doing the multiplication in either one.

a. $C(11, 3) = \dfrac{11 \times 10 \times 9}{3 \times 2 \times 1} = \dfrac{11 \times \overset{5}{\cancel{10}} \times \overset{3}{\cancel{9}}}{\cancel{3} \times \cancel{2} \times 1} = 165$

b. $C(11, 8) = \dfrac{11 \times 10 \times 9 \times 8 \times 7 \times 6 \times 5 \times 4}{8 \times 7 \times 6 \times 5 \times 4 \times 3 \times 2 \times 1}$

$$= \dfrac{11 \times \overset{5}{\cancel{10}} \times \overset{3}{\cancel{9}}}{\cancel{3} \times \cancel{2} \times 1} = 165$$

+*Before we go on...* It is no coincidence that the answers for parts (a) and (b) of Example 4 are the same. Consider what each represents. $C(11, 3)$ is the number of ways of choosing 3 items from 11—for example, electing 3 trustees from a slate of 11. Electing those 3 is the same as choosing the 8 who do not get elected. Thus, there are exactly as many ways to choose 3 items from 11 as there are ways to choose 8 items from 11. So, $C(11, 3) = C(11, 8)$. In general,

$$C(n, r) = C(n, n - r)$$

We can also see this equality by using the formula

$$C(n, r) = \frac{n!}{r!(n - r)!}$$

If we substitute $n - r$ for r, we get exactly the same formula.

Use the equality $C(n, r) = C(n, n - r)$ to make your calculations easier. Choose the one with the smaller denominator to begin with. ■

Example 5 Calculating Combinations

Calculate **a.** $C(11, 11)$ **b.** $C(11, 0)$

Solution

a. $C(11, 11) = \dfrac{11 \times 10 \times 9 \times 8 \times 7 \times 6 \times 5 \times 4 \times 3 \times 2 \times 1}{11 \times 10 \times 9 \times 8 \times 7 \times 6 \times 5 \times 4 \times 3 \times 2 \times 1} = 1$

b. What do we do with that 0? What does it mean to multiply 0 numbers together? We know from above that $C(11, 0) = C(11, 11)$, so we must have $C(11, 0) = 1$. How does this fit with the formulas? Go back to the calculation of $C(11, 11)$:

$$1 = C(11, 11) = \frac{11!}{11!(11 + 11)!} = \frac{11!}{11!0!}$$

This equality is true only if we agree that $0! = 1$, which we do. Then

$$C(11, 0) = \frac{11!}{0!11!} = 1$$

+*Before we go on...* There is nothing special about 11 in the calculation in Example 5. In general,

$$C(n, n) = C(n, 0) = 1$$

After all, there is only one way to choose n items out of n: choose them all. Similarly, there is only one way to choose 0 items out of n: choose none of them. ■

Now for a few more complicated examples that illustrate the applications of the counting techniques we've discussed.

Example 6 Lotto

In the betting game Lotto, used in many state lotteries, you choose six different numbers in the range 1–55 (the upper number varies). The order in which you choose them is irrelevant. If your six numbers match the six numbers chosen in the "official drawing," you win the top prize. If Lotto tickets cost $1 for two sets of numbers and you decide to buy tickets that cover every possible combination, thereby guaranteeing that you will win the top prize, how much money will you have to spend?

Solution We first need to know how many sets of numbers are possible. Because order does not matter, we are asking for the number of combinations of 55 numbers taken 6 at a time. This is

$$C(55, 6) = \frac{55 \times 54 \times 53 \times 52 \times 51 \times 50}{6 \times 5 \times 4 \times 3 \times 2 \times 1} = 28,989,675$$

Because $1 buys you two of these, you need to spend $28,989,675/2 = \$14,494,838$ (rounding up to the nearest dollar) to be assured of a win!

+*Before we go on...* The calculation in Example 6 shows that you should not bother buying all these tickets if the winning prize is less than about $14.5 million. Even if the prize is higher, you need to account for the fact that many people will play and the prize may end up split among several winners, not to mention the impracticality of filling out millions of betting slips. ∎

Example **7** Marbles

A bag contains 3 red marbles, 3 blue ones, 3 green ones, and 2 yellow ones (all distinguishable from one another).

a. How many sets of 4 marbles are possible?

b. How many sets of 4 are there such that each one is a different color?

c. How many sets of 4 are there in which at least 2 are red?

d. How many sets of 4 are there in which none are red, but at least one is green?

Solution

a. We simply need to find the number of ways of choosing 4 marbles out of 11, which is

$$C(11, 4) = 330 \text{ possible sets of 4 marbles}$$

b. We use a decision algorithm for choosing such a set of marbles.

Step 1 Choose one red one from the 3 red ones; $C(3, 1) = 3$ choices.
Step 2 Choose one blue one from the 3 blue ones; $C(3, 1) = 3$ choices.
Step 3 Choose one green one from the 3 green ones; $C(3, 1) = 3$ choices.
Step 4 Choose one yellow one from the 2 yellow ones; $C(2, 1) = 2$ choices.

This gives a total of $3 \times 3 \times 3 \times 2 = 54$ possible sets.

c. We need another decision algorithm. To say that at least 2 must be red means that either 2 are red or 3 are red (with a total of 3 red ones). In other words, we have two *alternatives*.

Alternative 1: Exactly 2 red marbles
Step 1 Choose 2 red ones; $C(3, 2) = 3$ choices.
Step 2 Choose 2 nonred ones. There are 8 of these, so we get $C(8, 2) = 28$ possible choices.

Thus, the total number of choices for this alternative is $3 \times 28 = 84$.

Alternative 2: Exactly 3 red marbles.
Step 1 Choose the 3 red ones; $C(3, 3) = 1$ choice.
Step 2 Choose 1 nonred one; $C(8, 1) = 8$ choices.
Thus, the total number of choices for this alternative is $1 \times 8 = 8$.

By the addition principle, we get a total of $84 + 8 = 92$ sets.

d. The phrase "at least 1 green" tells us that we again have some alternatives.

Alternative 1: 1 green marble
Step 1 Choose 1 green marble from the 3; $C(3, 1) = 3$ choices.
Step 2 Choose 3 nongreen, nonred marbles; $C(5, 3) = 10$ choices.
Thus, the total number of choices for alternative 1 is $3 \times 10 = 30$.

Alternative 2: 2 green marbles
Step 1 Choose 2 green marbles from the 3; $C(3, 2) = 3$ choices.
Step 2 Choose 2 nongreen, nonred marbles; $C(5, 2) = 10$ choices.
Thus, the total number of choices for alternative 2 is $3 \times 10 = 30$.

Alternative 3: 3 green marbles
Step 1 Choose 3 green marbles from the 3; $C(3, 3) = 1$ choice.
Step 2 Choose 1 nongreen, nonred marble; $C(5, 1) = 5$ choices.
Thus, the total number of choices for alternative 3 is $1 \times 5 = 5$.

The addition principle now tells us that the number of sets of four marbles with none red, but at least one green, is $30 + 30 + 5 = 65$.

+*Before we go on...* Here is an easier way to answer part (d) of Example 7. First, the total number of sets having *no* red marbles is $C(8, 4) = 70$. Next, of those, the number containing no green marbles is $C(5, 4) = 5$. This leaves $70 - 5 = 65$ sets that contain no red marbles but having at least one green marble. (We have really used here the formula for the cardinality of the complement of a set.) ∎

The last example concerns poker hands. For those unfamiliar with playing cards, here is a short description. A standard deck consists of 52 playing cards. Each card is in one of 13 denominations: Ace, 2, 3, 4, 5, 6, 7, 8, 9, 10, Jack (J), Queen (Q), and King (K), and in one of four suits: hearts (♥), diamonds (♦), clubs (♣), and spades (♠). Thus, for instance, the Jack of spades, J♠, refers to the denomination of Jack in the suit of spades. The entire deck of cards is thus

A♥	2♥	3♥	4♥	5♥	6♥	7♥	8♥	9♥	10♥	J♥	Q♥	K♥
A♦	2♦	3♦	4♦	5♦	6♦	7♦	8♦	9♦	10♦	J♦	Q♦	K♦
A♣	2♣	3♣	4♣	5♣	6♣	7♣	8♣	9♣	10♣	J♣	Q♣	K♣
A♠	2♠	3♠	4♠	5♠	6♠	7♠	8♠	9♠	10♠	J♠	Q♠	K♠

Example 8 Poker Hands

In the card game poker, a hand consists of a set of five cards from a standard deck of 52. A **full house** is a hand consisting of three cards of one denomination ("three of a kind"—e.g., three 10s) and two of another ("two of a kind"—e.g., two Queens). Here is an example of a full house: 10♣, 10♦, 10♠, Q♥, Q♣.

a. How many different poker hands are there?

b. How many different full houses are there that contain three 10s and two Queens?

c. How many different full houses are there altogether?

Solution

a. Because the order of the cards doesn't matter, we simply need to know the number of ways of choosing a set of 5 cards out of 52, which is

$$C(52, 5) = 2{,}598{,}960 \text{ hands}$$

b. Here is a decision algorithm for choosing a full house with three 10s and two Queens.

Step 1 Choose three 10s. Because there are four 10s to choose from, we have $C(4, 3) = 4$ choices.

Step 2 Choose 2 Queens; $C(4, 2) = 6$ choices.

Thus, there are $4 \times 6 = 24$ possible full houses with three 10s and two Queens.

c. Here is a decision algorithm for choosing a full house.

Step 1 Choose a denomination for the three of a kind; 13 choices.

Step 2 Choose 3 cards of that denomination. Because there are 4 cards of each denomination (one for each suit), we get $C(4, 3) = 4$ choices.

Step 3 Choose a different denomination for the two of a kind. There are only 12 denominations left, so we have 12 choices.

Step 4 Choose 2 of that denomination; $C(4, 2) = 6$ choices.

Thus, by the multiplication principle, there are a total of $13 \times 4 \times 12 \times 6 = 3744$ possible full houses.

FAQs   **Recognizing When to Use Permutations or Combinations**

Q: *How can I tell whether a given application calls for permutations or combinations?*

A: Decide whether the application calls for ordered lists (as in situations where order is implied) or for unordered sets (as in situations where order is not relevant). Ordered lists are permutations, whereas unordered sets are combinations. ∎

6.4 EXERCISES

● denotes basic skills exercises

◆ denotes challenging exercises

Evaluate each number in Exercises 1–16. hint [see Quick Examples on pp. 427, 428, & 430]

1. ● 6!

2. ● 7!

3. ● 8!/6!

4. ● 10!/8!

5. ● $P(6, 4)$

6. ● $P(8, 3)$

7. ● $P(6, 4)/4!$

8. ● $P(8, 3)/3!$

9. ● $C(3, 2)$

10. ● $C(4, 3)$

11. ● $C(10, 8)$

12. ● $C(11, 9)$

13. ● $C(20, 1)$

14. ● $C(30, 1)$

15. ● $C(100, 98)$

16. ● $C(100, 97)$

17. ● How many ordered lists are there of 4 items chosen from 6?

18. ● How many ordered sequences are possible that contain 3 objects chosen from 7?

19. ● How many unordered sets are possible that contain 3 objects chosen from 7?

20. ● How many unordered sets are there of 4 items chosen from 6?

21. ● How many five-letter sequences are possible that use the letters b, o, g, e, y once each?

22. ● How many six-letter sequences are possible that use the letters q, u, a, k, e, s once each?

23. ● How many three-letter sequences are possible that use the letters q, u, a, k, e, s at most once each?

● basic skills   ◆ challenging

24. ● How many three-letter sequences are possible that use the letters b, o, g, e, y at most once each?

25. ● How many three-letter (unordered) sets are possible that use the letters q, u, a, k, e, s at most once each?

26. ● How many three-letter (unordered) sets are possible that use the letters b, o, g, e, y at most once each?

27. How many six-letter sequences are possible that use the letters a, u, a, a, u, k? [Hint: Use the decision algorithm discussed in Example of Section 6.3.]

28. How many six-letter sequences are possible that use the letters f, f, a, a, f, f? [See the hint for the preceding exercise.]

Marbles For Exercises 29–42, a bag contains 3 red marbles, 2 green ones, 1 lavender one, 2 yellows, and 2 orange marbles. *hint* [see Example 7]

29. ● How many possible sets of four marbles are there?

30. ● How many possible sets of three marbles are there?

31. ● How many sets of four marbles include all the red ones?

32. ● How many sets of three marbles include all the yellow ones?

33. ● How many sets of four marbles include none of the red ones?

34. ● How many sets of three marbles include none of the yellow ones?

35. ● How many sets of four marbles include one of each color other than lavender?

36. ● How many sets of five marbles include one of each color?

37. ● How many sets of five marbles include at least two red ones?

38. ● How many sets of five marbles include at least one yellow one?

39. ● How many sets of five marbles include at most one of the yellow ones?

40. ● How many sets of five marbles include at most one of the red ones?

41. How many sets of five marbles include either the lavender one or exactly one yellow one but not both colors?

42. How many sets of five marbles include at least one yellow one but no green ones?

Dice If a die is rolled 30 times, there are 6^{30} different sequences possible. Exercises 43–46 ask how many of these sequences satisfy certain conditions. [Hint: Use the decision algorithm discussed in Example 3 of Section 6.3.]

43. What fraction of these sequences have exactly 5 ones?

44. What fraction of these sequences have exactly 5 ones and 5 twos?

45. What fraction of these sequences have exactly 15 even numbers?

46. What fraction of these sequences have exactly 10 numbers less than or equal to 2?

Applications

47. ● *Itineraries* Your international diplomacy trip requires stops in Thailand, Singapore, Hong Kong, and Bali. How many possible itineraries are there? *hint* [see Examples 1 & 2]

48. ● *Itineraries* Refer back to the preceding exercise. How many possible itineraries are there in which the last stop is Thailand?

Poker Hands A poker hand consists of five cards from a standard deck of 52. (See the chart preceding Example 8.) In Exercises 49–54, find the number of different poker hands of the specified type. *hint* [see Example 8]

49. ● Two pairs (two of one denomination, two of another denomination, and one of a third).

50. ● Three of a kind (three of one denomination, one of another denomination, and one of a third).

51. ● Two of a kind (two of one denomination and three of different denominations).

52. ● Four of a kind (all four of one denomination and one of another).

53. Straight (five cards of consecutive denominations: A, 2, 3, 4, 5 up through 10, J, Q, K, A, not all of the same suit.) (Note that the Ace counts either as a 1 or as the denomination above King.)

54. Flush (five cards all of the same suit, but not consecutive denominations)

Dogs of the Dow The "Dogs of the Dow" refer to the stocks listed on the Dow with the highest dividend yield. Exercises 55 and 56 are based on the following table, which shows the top ten stocks of the "Dogs of the Dow" list in February, 2005:[16]

Symbol	Company	Price	Yield
SBC	SBC Communications	23.62	5.46%
GM	General Motors	36.64	5.46%
MRK	Merck	28.02	5.42%
MO	Altria	63.20	4.62%
VZ	Verizon	35.68	4.32%
JPM	JP Morgan Chase	37.00	3.68%
C	Citigroup	48.38	3.64%
PFE	Pfizer	24.35	3.12%
DD	DuPont	47.23	2.96%
GE	General Electric	35.75	2.46%

55. ● You decide to make a small portfolio consisting of a collection of five of the top ten Dogs of the Dow.

   a. How many portfolios are possible?

---

[16] SOURCE: www.dogsofthedow.com.

● basic skills   ◆ challenging

b. How many of these portfolios contain MO and GM but neither GE nor DD?

c. How many of these portfolios contain at least four stocks with yields above 4%?

56. ● You decide to make a small portfolio consisting of a collection of six of the top ten Dogs of the Dow.

a. How many portfolios are possible?

b. How many of these portfolios contain SBC but not MO?

c. How many of these portfolios contain at most one stock priced above $40?

Day Trading Day traders typically buy and sell stocks (or other investment instruments) during the trading day and sell all investments by the end of the day. Exercises 57 and 58 are based on the following table, which shows the closing prices on February 10, 2005, of 12 stocks selected by your broker, Prudence Swift, as well as the change that day.

Tech Stocks	Close	Change
NT (Nortel)	$3.05	−0.09
MSFT (Microsoft)	$26.06	−0.01
CSCO (Cisco)	$17.58	−0.05
NOK (Nokia)	$15.81	0.14
INTC (Intel)	$23.50	0.20
EBAY (eBay)	$81.22	2.23
Non-Tech Stocks		
ATVI (Activision)	$20.41	−3.49
DUK (Duke)	$26.84	0.09
KSE (Keyspan)	$40.06	−0.01
ED (Con Ed)	$44.60	0.39
MO (Altria Group)	$66.40	1.12
AACE (Ace Cash Express)	$26.45	−0.25

57. ● On the morning of February 10, 2005, Swift advised you to purchase a collection of three tech stocks and two nontech stocks, all chosen at random from those listed in the table. You were to sell all the stocks at the end of the trading day.

a. How many possible collections are possible?

b. You tend to have bad luck with stocks—they usually start going down the moment you buy them. How many of the collections in part (a) consist entirely of stocks that declined in value by the end of the day?

c. Using the answers to parts (a) and (b), what would you say your chances were of choosing a collection consisting entirely of stocks that declined in value by the end of the day?

58. ● On the morning of February 10, 2005, Swift advised your friend to purchase a collection of three stocks chosen at random from those listed in the table. Your friend was to sell all the stocks at the end of the trading day.

a. How many possible collections are possible?

b. How many of the collections in part (a) included exactly two tech stocks that increased in value by the end of the day?

c. Using the answers to parts (a) and (b), what would you say the chances were that your friend chose a collection that included exactly two tech stocks that increased in value by the end of the day?

Movies Exercises 59 and 60 are based on the following list of top movie rentals (based on revenue) for the weekend ending February 6, 2005:[17]

Name	Revenue ($ million)
The Grudge	$9.24
Ray	$6.54
The Forgotten	$6.01
Friday Night Lights	$5.99
Cellular	$5.85
Shall We Dance	$5.82
The Village	$5.58
AVP: Alien Vs. Predator	$5.46
Catwoman	$5.10
Sky Captain and the World of Tomorrow	$5.09

59. Rather than study for math, you and your buddies decide to get together for a marathon movie-watching popcorn-guzzling event on Saturday night. You decide to watch four movies selected at random from the above list.

a. How many sets of four movies are possible?

b. Your best friends, the Lara twins, refuse to see "Shall We Dance" on the grounds that it is "for girliemen" and also insist that at least one of "Alien Vs. Predator" or "Sky Captain" be among the movies selected. How many of the possible groups of four will satisfy the twins?

c. Comparing the answers in parts (a) and (b), would you say the Lara twins are more likely than not to be satisfied with your random selection?

60. Rather than study for astrophysics, you and your friends decide to get together for a marathon movie-watching gummy-bear munching event on Saturday night. You decide to watch three movies selected at random from the above list.

a. How many sets of three movies are possible?

b. Your best friends, the Pelogrande twins, refuse to see either "Alien Vs. Predator" or "Sky Captain" on the grounds that they are "for idiots" and also insist that no more than one of "The Grudge" and "Ray" should be among the movies selected. How many of the possible groups of four will satisfy the twins?

---

[17] Source: www.imdb.com/Charts/videolast

● basic skills ◆ challenging

c. Comparing the answers in parts (a) and (b), would you say the Pelogrande twins are more likely than not to be satisfied with your random selection?

61. *Traveling Salesperson* Suppose you are a salesperson who must visit the following 23 cities: Dallas, Tampa, Orlando, Fairbanks, Seattle, Detroit, Chicago, Houston, Arlington, Grand Rapids, Urbana, San Diego, Aspen, Little Rock, Tuscaloosa, Honolulu, New York, Ithaca, Charlottesville, Lynchville, Raleigh, Anchorage, and Los Angeles. Leave all your answers in factorial form.

a. How many possible itineraries are there that visit each city exactly once?

b. Repeat part (a) in the event that the first five stops have already been determined.

c. Repeat part (a) in the event that your itinerary must include the sequence Anchorage, Fairbanks, Seattle, Chicago, and Detroit, in that order.

62. *Traveling Salesperson* Refer back to the preceding exercise (and leave all your answers in factorial form).

a. How many possible itineraries are there that start and end at Detroit, and visit every other city exactly once?

b. How many possible itineraries are there that start and end at Detroit, visit Chicago twice and every other city once?

c. Repeat part (a) in the event that your itinerary must include the sequence Anchorage, Fairbanks, Seattle, Chicago, and New York, in that order.

In Exercises 63–68, calculate how many different sequences can be formed that use the letters of each given word. [Decide where, for example, all the s's will go, rather than what will go in each position. Leave your answer as a product of terms of the form $C(n, r)$.]

63. Mississippi

64. Mesopotamia

65. Megalomania

66. Schizophrenia

67. Casablanca

68. Desmorelda

69. (*From the GMAT*) Ben and Ann are among 7 contestants from which 4 semifinalists are to be selected. Of the different possible selections, how many contain neither Ben nor Ann?

(A) 5   **(B)** 6   **(C)** 7   **(D)** 14   **(E)** 21

70. (*Based on a Question from the GMAT*) Ben and Ann are among 7 contestants from which 4 semifinalists are to be selected. Of the different possible selections, how many contain Ben but not Ann?

(A) 5   **(B)** 8   **(C)** 9   **(D)** 10   **(E)** 20

71. ◆ (*From the GMAT Exam*) If 10 persons meet at a reunion and each person shakes hands exactly once with each of the others, what is the total number of handshakes?

(A) $10 \cdot 9 \cdot 8 \cdot 7 \cdot 6 \cdot 5 \cdot 4 \cdot 3 \cdot 2 \cdot 1$ **(B)** $10 \cdot 10$
(C) $10 \cdot 9$ **(D)** 45   **(E)** 36

72. ◆ (*Based on a question from the GMAT Exam*) If 12 businesspeople have a meeting and each pair exchanges business cards, how many business cards, total, get exchanged?

(A) $12 \cdot 11 \cdot 10 \cdot 9 \cdot 8 \cdot 7 \cdot 6 \cdot 5 \cdot 4 \cdot 3 \cdot 2 \cdot 1$ **(B)** $12 \cdot 12$
(C) $12 \cdot 11$ **(D)** 66   **(E)** 72

73. ◆ *Product Design* The Honest Lock Company plans to introduce what it refers to as the "true combination lock." The lock will open if the correct set of three numbers from 0 through 39 is entered in any order.

a. How many different combinations of three different numbers are possible?

b. If it is allowed that a number appear twice (but not three times), how many more possibilities are created?

c. If it is allowed that any or all of the numbers may be the same, what is the total number of combinations that will open the lock?

74. ◆ *Product Design* Repeat Exercise 73 for a lock based on selecting from the numbers 0 through 19.

75. ◆ *Theory of Linear Programming* (Some familiarity with linear programming is assumed for this exercise.) Suppose you have a linear programming problem with two unknowns and 20 constraints. You decide that graphing the feasible region would take a lot of work, but then you recall that corner points are obtained by solving a system of two equations in two unknowns obtained from two of the constraints. Thus, you decide that it might pay instead to locate all the possible corner points by solving all possible combinations of two equations and then checking whether each solution is a feasible point.

a. How many systems of two equations in two unknowns will you be required to solve?

b. Generalize this to n constraints.

76. ◆ *More Theory of Linear Programming* (Some familiarity with linear programming is assumed for this exercise.) Before the advent of the simplex method for solving linear programming problems, the following method was used: Suppose you have a linear programming problem with three unknowns and 20 constraints. You locate corner points as follows: Selecting three of the constraints, you turn them into equations (by replacing the inequalities with equalities), solve the resulting system of three equations in three unknowns, and then check to see whether the solution is feasible.

a. How many systems of three equations in three unknowns will you be required to solve?

b. Generalize this to n constraints.

Communication and Reasoning Exercises

77. ● If you were hard pressed to study for an exam on counting and had only enough time to study one topic, would you

● basic skills   ◆ challenging

choose the formula for the number of permutations or the multiplication principle? Give reasons for your choice.

78. ● Which of the following represent permutations?

   (A) An arrangement of books on a shelf.
   (B) A group of 10 people in a bus.
   (C) A committee of 5 senators chosen from 100.
   (D) A presidential cabinet of 5 portfolios chosen from 20.

79. ● You are tutoring your friend for a test on sets and counting and she asks the question: "How do I know what formula to use for a given problem?" What is a good way to respond?

80. ● Complete the following. If a counting procedure has 5 alternatives, each of which has 4 steps of 2 choices each, then there are ___ outcomes. On the other hand, if there are 5 steps, each of which has 4 alternatives of 2 choices each, then there are ___ outcomes.

81. A textbook has the following exercise. "Three students from a class of 50 are selected to take part in a play. How many casts are possible?" Comment on this exercise.

82. Explain why the coefficient of a^2b^4 in $(a + b)^6$ is $C(6, 2)$ (this is a consequence of the **binomial theorem**). [Hint: In the product $(a + b)(a + b) \cdots (a + b)$ (6 times), in how many different ways can you pick 2 a's and 4 b's to multiply together?]

Chapter 6 Review

KEY CONCEPTS

6.1 Sets and Set Operations

Sets, elements, subsets, proper
  subsets, empty set, finite and infinite
  sets *p. 394*

Visualizing sets and relations between
  sets using Venn diagrams *p. 397*

Union:
  $A \cup B = \{x \mid x \in A \text{ or } x \in B\}$
  p. 398

Intersection: $A \cap B =$
  $\{x \mid x \in A \text{ and } x \in B\}$ *p. 398*

Universal sets, complements *p. 399*

Disjoint sets: $A \cap B = \emptyset$ *p. 400*

Cartesian product: $A \times B =$
  $\{(a, b) \mid a \in A \text{ and } b \in B\}$ *p. 400*

6.2 Cardinality

Cardinality: $n(A) =$
  number of elements in A. *p. 405*

If A and B are finite sets, then $n(A \cup B) =$
  $n(A) + n(B) - n(A \cap B)$. *p. 406*

If A and B are disjoint finite sets, then
  $n(A \cup B) = n(A) + n(B)$. In this
  case, we say that $A \cup B$ is a **disjoint
  union.** *p. 406*

If S is a finite universal set and A is a
  subset of S, then $n(A') = n(S) - n(A)$
  and $n(A) = n(S) - n(A')$. *p. 408*

If A and B are finite sets, then
  $n(A \times B) = n(A)n(B)$. *p. 412*

6.3 The Addition and Multiplication Principles

Addition Principle *p. 416*

Multiplication Principle *p. 417*

Decision Algorithm: a procedure for
  making a sequence of decisions
  to choose an element of a set *p. 418*

6.4 Permutations and Combinations

n factorial:
  $n! = n \times (n-1) \times (n-2) \times \cdots \times 2 \times 1$
  p. 427

Permutation of n items taken r at a time:
  $$P(n,r) = \frac{n!}{(n-r)!} \quad \text{p. 428}$$

Combination of n items taken r at a time:
  $$C(n,r) = P(n,r)/r! = \frac{n!}{r!(n-r)!}$$
  p. 430

Using the equality
  $C(n,r) = C(n, n-r)$ to simplify
  calculations of combinations.
  p. 431

REVIEW EXERCISES

In each of Exercises 1–5, list the elements of the given set.

1. The set N of all negative integers greater than or equal to -3.

2. The set of all outcomes of tossing a coin 5 times.

3. The set of all outcomes of tossing 2 distinguishable dice such that the numbers are different.

4. The sets $(A \cap B) \cup C$ and $A \cap (B \cup C)$ where $A = \{1, 2, 3, 4, 5\}$, $B = \{3, 4, 5\}$ and $C = \{1, 2, 5, 6, 7\}$

5. The sets $A \cup B'$ and $A \times B'$ where $A = \{a, b\}$, $B = \{b, c\}$ and $S = \{a, b, c, d\}$

In each of Exercises 6–9, write the indicated set in terms of the given sets A and B.

6. S: the set of all customers; A: the set of all customers who owe money; B: the set of all customers who owe at least \$1000. The set of all customers who owe money but owe less than \$1000

7. A: the set of outcomes when a day in August is selected; B: the set of outcomes when a time of day is selected. The set of outcomes when a day in August and a time of that day are selected

8. S: the set of outcomes when two dice are rolled; E: those outcomes in which at most one die shows an even number, F:

those outcomes in which the sum of the numbers is 7. The set of outcomes in which both dice show an even number or sum to 7.

9. S: the set of all integers; P: the set of all positive integers; E: the set of all even integers; Q: the set of all integers that are perfect squares ($Q = \{0, 1, 4, 9, 16, 25, \dots\}$). The set of all integers that are not positive odd perfect squares

In each of Exercises 10–13, give a formula for the cardinality rule or rules needed to answer the question, and then give the solution.

10. There are 32 students in categories A and B combined; 24 are in A and 24 are in B. How many are in both A and B?

11. You have read 150 of the 400 novels in your home, but your sister Roslyn has read 200, of which only 50 are novels you have read as well. How many have neither of you read?

12. The *Apple iMac®* comes in 3 models, each with 5 colors to choose from. How many combinations are possible?

13. You roll 2 dice, 1 red and 1 green. Losing combinations are doubles (both dice show the same number) and outcomes in which the green die shows an odd number and the red die shows an even number. The other combinations are winning ones. How many winning combinations are there?

Recall that a poker hand consists of 5 cards from a standard deck of 52. In each of Exercises 14–17, find the number of different poker hands of the specified type. Leave your answer in terms of combinations.

14. A full house with either 2 Kings and 3 Queens or 2 Queens and 3 Kings

15. Two of a kind with no Aces

16. Three of a kind with no Aces

17. Straight Flush (5 cards of the same suit with consecutive denominations: A, 2, 3, 4, 5 up through 10, J, Q, K, A)

In each of Exercises 18–22, consider a bag containing 4 red marbles, 2 green ones, 1 transparent one, 3 yellow ones, and 2 orange ones.

18. How many possible sets of 5 marbles are there?

19. How many sets of 5 marbles include all the red ones?

20. How many sets of 5 marbles do not include all the red ones?

21. How many sets of 5 marbles include at least 2 yellow ones?

22. How many sets of 5 marbles include at most one of the red ones but no yellow ones?

Applications

OHaganBooks.com currently operates three warehouses: one in Washington, one in California, and the new one in Texas. Book inventories are shown in the following table:

	Sci Fi	Horror	Romance	Other	Total
Washington	10,000	12,000	12,000	30,000	64,000
California	8000	12,000	6000	16,000	42,000
Texas	15,000	15,000	20,000	44,000	94,000
Total	33,000	39,000	38,000	90,000	200,000

Take the first letter of each category to represent the corresponding set of books; for instance, S is the set of sci fi books in stock, W is the set of books in the Washington warehouse, and so on. In each of Exercises 23–28, describe the given set in words and compute its cardinality.

23. $S \cup T$

24. $H \cap C$

25. $C \cup S'$

26. $(R \cap T) \cup H$

27. $R \cap (T \cup H)$

28. $(S \cap W) \cup (H \cap C')$

OHaganBooks.com has two main competitors: JungleBooks.com, and FarmerBooks.com. At the beginning of August, OHaganBooks.com had 3500 customers. Of these, a total of 2000 customers were shared with JungleBooks.com and 1500 with FarmerBooks.com. Furthermore, 1000 customers were shared with both. Use these data for Exercises 29–33,

29. How many of all these customers are exclusive OHaganBooks.com customers?

30. JungleBooks.com has a total of 3600 customers, FarmerBooks.com has 3400, and they share 1100 customers between them.

How many of their customers are not customers of OHaganBooks.com?

31. Which of the three companies has the largest number of exclusive customers?

32. OHaganBooks.com is interested in merging with one of its two competitors. Which merger would give it the largest combined customer base, and how large would that be?

33. Referring to the preceding exercise, which merger would give OHaganBooks.com the largest *exclusive* customer base, and how large would that be?

As the customer base at OHaganBooks.com grows, software manager Ruth Nabarro is thinking of introducing identity codes for all the online customers. Help her with the questions in Exercises 34–37.

34. If she uses 3-letter codes, how many different customers can be identified?

35. If she uses codes with 3 different letters, how many different customers can be identified?

36. It appears that Nabarro has finally settled on codes consisting of two letters followed by two digits. For technical reasons, the letters must be different and the first digit cannot be a zero. How many different customers can be identified?

37. The CEO sends Nabarro the following memo:

```
To: Ruth Nabarro, Software Manager
From: John O'Hagan, CEO
Subject: Customer Identity Codes

I have read your proposal for the cus-
tomer ID codes. However, due to our am-
bitious expansion plans, I would like
our system software to allow for at
least 500,000 customers. Please adjust
your proposal accordingly.
```

Nabarro is determined to have a sequence of letters followed by some digits, and, for reasons too complicated to explain, there cannot be more than two letters, the letters must be different, the digits must all be different, and the first digit cannot be a zero. What is the form of the shortest code she can use to satisfy the CEO, and how many different customers can be identified?

After an exhausting day at the office (you are CEO of OHagan Books.com), you return home and find yourself having to assist your son Billy Sean, who continues to have a terrible time planning his first-year college course schedule. The latest Bulletin of Suburban State University reads as follows:

All candidates for the degree of Bachelor of Arts at SSU must take, in their first year, at least 10 courses in the Sciences, Fine Arts, Liberal Arts and Mathematics combined, of which at least 2 must be in each of the Sciences and Fine Arts, and exactly 3 must be in each of the Liberal Arts and Mathematics.

Help him with the answers to Exercises 38–40.

38. If the Bulletin lists exactly 5 first-year-level science courses and 6 first-year-level courses in each of the other categories, how many course combinations are possible that meet the minimum requirements?

39. Reading through the course descriptions in the bulletin a second time, you notice that Calculus I (listed as one of the mathematics courses) is a required course for many of the other courses, and you decide that it would be best if Billy Sean included Calculus I. Further, two of the Fine Arts courses cannot both be taken in the first year. How many course combinations are possible that meet the minimum requirements and include Calculus I?

40. To complicate things further, in addition to the requirement in part (b), Physics II has Physics I as a prerequisite (both are listed as first-year science courses, but it is not necessary to take both). How many course combinations are possible that include Calculus I and meet the minimum requirements?

Mentor Do you need a live tutor for homework problems? Access vMentor on the ThomsonNOW! website at **www.thomsonedu.com** for one-on-one tutoring from a mathematics expert.

CASE STUDY: Designing a Puzzle

David Young-Wolff/PhotoEdit

As Product Design Manager for Cerebral Toys Inc., you are constantly on the lookout for ideas for intellectually stimulating yet inexpensive toys. You recently received the following memo from Felix Frost, the developmental psychologist on your design team.

To: Felicia

From: Felix

Subject: Crazy Cubes

We've hit on an excellent idea for a new educational puzzle (which we are calling "Crazy Cubes" until Marketing comes up with a better name). Basically, Crazy Cubes will consist of a set of plastic cubes. Two faces of each cube will be colored red, two will be colored blue, and two white, and there will be exactly two cubes with each possible configuration of colors. The goal of the puzzle is to seek out the matching pairs, thereby enhancing a child's geometric intuition and three-dimensional manipulation skills. The kit will include every possible configuration of colors. We are, however, a little stumped on the following question: How many cubes will the kit contain? In other words, how many possible ways can one color the faces of a cube so that two faces are red, two are blue, and two are white?

Looking at the problem, you reason that the following three-step decision algorithm ought to suffice:

> **Step 1** Choose a pair of faces to color red; $C(6, 2) = 15$ choices.
>
> **Step 2** Choose a pair of faces to color blue; $C(4, 2) = 6$ choices.
>
> **Step 3** Choose a pair of faces to color white; $C(2, 2) = 1$ choice.

This algorithm appears to give a total of $15 \times 6 \times 1 = 90$ possible cubes. However, before sending your reply to Felix, you realize that something is wrong, because there are different choices that result in the same cube. To describe some of these choices,

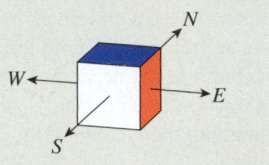

Figure 9

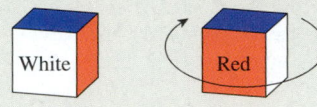

Figure 10

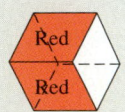

Figure 11

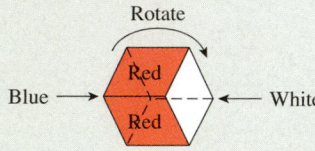

Figure 12

imagine a cube oriented so that four of its faces are facing the four compass directions (Figure 9). Consider choice 1 with the top and bottom faces blue, north and south faces white, and east and west faces red and choice 2 with the top and bottom faces blue, north and south faces red, and east and west faces white. These cubes are actually the same, as you see by rotating the second cube 90 degrees (Figure 10).

You therefore decide that you need a more sophisticated decision algorithm. Here is one that works.

Alternative 1: Faces with the same color opposite each other. Place one of the blue faces down. Then the top face is also blue. The cube must look like the one drawn in Figure 10. Thus there is only one choice here.

Alternative 2: Red faces opposite each other and the other colors on adjacent pairs of faces. Again there is only one choice, as you can see by putting the red faces on the top and bottom and then rotating.

Alternative 3: White faces opposite each other and the other colors on adjacent pairs of faces; one possibility.

Alternative 4: Blue faces opposite each other and the other colors on adjacent pairs of faces; one possibility.

Alternative 5: Faces with the same color adjacent to each other. Look at the cube so that the edge common to the two red faces is facing you and horizontal (Figure 11). Then the faces on the left and right must be of different colors because they are opposite each other. Assume that the face on the right is white. (If it's blue, then rotate the die with the red edge still facing you to move it there, as in Figure 12.) This leaves two choices for the other white face, on the upper or the lower of the two back faces. This alternative gives 2 choices.

It follows that there are $1 + 1 + 1 + 1 + 2 = 6$ choices. Because the Crazy Cubes kit will feature two of each cube, the kit will require 12 different cubes.[18]

Exercises

In all of the following exercises, there are three colors to choose from: red, white, and blue.

1. In order to enlarge the kit, Felix suggests including two each of two-colored cubes (using two of the colors red, white, and blue) with three faces one color and three another. How many additional cubes will be required?

2. If Felix now suggests adding two each of cubes with two faces one color, one face another color, and three faces the third color, how many additional cubes will be required?

3. Felix changes his mind and suggests the kit use tetrahedral blocks with two colors instead (see the figure). How many of these would be required?

4. Once Felix finds the answer to the preceding exercise, he decides to go back to the cube idea, but this time insists that all possible combinations of up to three colors should be included. (For instance, some cubes will be all one color, others will be two colors.) How many cubes should the kit contain?

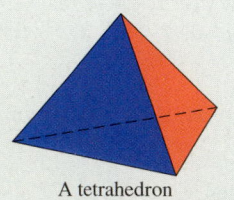

A tetrahedron

[18] There is a beautiful way of calculating this and similar numbers, called Polya enumeration, but it requires a discussion of topics well outside the scope of this book. Take this as a hint that counting techniques can use some of the most sophisticated mathematics.

7 Probability

CASE STUDY The Monty Hall Problem

On the game show *Let's Make a Deal,* you are shown three doors, A, B, and C, and behind one of them is the Big Prize. After you select one of them—say, door A—to make things more interesting the host (Monty Hall) opens one of the other doors—say, door B—revealing that the Big Prize is not there. He then offers you the opportunity to change your selection to the remaining door, door C. Should you switch or stick with your original guess? Does it make any difference?

The Everett Collection

Introduction

What is the probability of winning the lottery twice? What are the chances that a college athlete whose drug test is positive for steroid use is actually using steroids? You are playing poker and have been dealt two jacks. What is the likelihood that one of the next three cards you are dealt will also be a jack? These are all questions about probability.

Understanding probability is important in many fields, ranging from risk management in business through hypothesis testing in psychology to quantum mechanics in physics. The goal of this chapter is to familiarize you with the basic concepts of probability theory and to give you a working knowledge that you can apply in a variety of situations.

In the first two sections, the emphasis is on translating real-life situations into the language of sample spaces, events, and probability. Once we have mastered the language of probability, we spend the rest of the chapter studying some of its theory and applications. The last section gives an interesting application of both probability and matrix arithmetic.

7.1 Sample Spaces and Events

Sample Spaces

At the beginning of a football game, to ensure fairness, the referee tosses a coin to decide who will get the ball first. When the ref tosses the coin and observes which side faces up, there are two possible results: heads (H) and tails (T). These are the *only* possible results, ignoring the (remote) possibility that the coin lands on its edge. The act of tossing the coin is an example of an **experiment.** The two possible results, H and T, are possible **outcomes** of the experiment, and the set $S = \{H, T\}$ of all possible outcomes is the **sample space** for the experiment.

Experiments, Outcomes, and Sample Spaces

An **experiment** is an occurrence with a result, or **outcome,** that is uncertain before the experiment takes place. The set of all possible outcomes is called the **sample space** for the experiment.

quick Examples

1. *Experiment:* Flip a coin and observe the side facing up.
   Outcomes: H, T
   Sample Space: $S = \{H, \ T\}$

2. *Experiment:* Select a student in your class.
   Outcomes: The students in your class
   Sample Space: The set of students in your class

3. *Experiment:* Select a student in your class and observe the color of his or her hair.
   Outcomes: red, black, brown, blond, green, . . .
   Sample Space: {red, black, brown, blond, green, . . .}

4. *Experiment:* Cast a die and observe the number facing up.
   Outcomes: 1, 2, 3, 4, 5, 6
   Sample Space: $S = \{1, 2, 3, 4, 5, 6\}$

5. *Experiment:* Cast two distinguishable dice and observe the numbers facing up.
Outcomes: $(1, 1), (1, 2), \ldots, (6, 6)$ (36 outcomes)

$$
\textit{Sample Space: } S = \left\{ \begin{array}{llllll}
(1, 1), & (1, 2), & (1, 3), & (1, 4), & (1, 5), & (1, 6), \\
(2, 1), & (2, 2), & (2, 3), & (2, 4), & (2, 5), & (2, 6), \\
(3, 1), & (3, 2), & (3, 3), & (3, 4), & (3, 5), & (3, 6), \\
(4, 1), & (4, 2), & (4, 3), & (4, 4), & (4, 5), & (4, 6), \\
(5, 1), & (5, 2), & (5, 3), & (5, 4), & (5, 5), & (5, 6), \\
(6, 1), & (6, 2), & (6, 3), & (6, 4), & (6, 5), & (6, 6)
\end{array} \right\}
$$

$n(S) = $ the number of outcomes in $S = 36$

6. *Experiment:* Cast two indistinguishable dice and observe the numbers facing up.
Outcomes: $(1, 1), (1, 2), \ldots, (6, 6)$ (21 outcomes)

$$
\textit{Sample Space: } S = \left\{ \begin{array}{llllll}
(1, 1), & (1, 2), & (1, 3), & (1, 4), & (1, 5), & (1, 6), \\
 & (2, 2), & (2, 3), & (2, 4), & (2, 5), & (2, 6), \\
 & & (3, 3), & (3, 4), & (3, 5), & (3, 6), \\
 & & & (4, 4), & (4, 5), & (4, 6), \\
 & & & & (5, 5), & (5, 6), \\
 & & & & & (6, 6)
\end{array} \right\}
$$

$n(S) = 21$

7. *Experiment:* Cast two dice and observe the *sum* of the numbers facing up.
Outcomes: 2, 3, 4, 5, 6, 7, 8, 9, 10, 11, 12
Sample Space: $S = \{2, 3, 4, 5, 6, 7, 8, 9, 10, 11, 12\}$

8. *Experiment:* Choose 2 cars (without regard to order) at random from a fleet of 10.
Outcomes: Collections of 2 cars chosen from 10
Sample Space: The set of all collections of 2 cars chosen from 10

$n(S) = C(10, 2) = 45$

The following example introduces a sample space that we'll use in several other examples.

Example 1 School and Work

In a survey[*] conducted by the Bureau of Labor Statistics, the high school graduating class of 2003 was divided into those who went on to college and those who did not. Those who went on to college were further divided into those who went to two-year colleges and those who went to four-year colleges. All graduates were also asked whether they were working or not. Find the sample space for the experiment "Select a member of the high school graduating class of 2003 and classify his or her subsequent school and work activity."

---

[*] "College Enrollment and Work Activity of High School Graduates," U.S. Bureau of Labor Statistics, April 2004, available at www.bls.gov/schedule/archives/all_nr.htm.

Solution The tree in Figure 1 shows the various possibilities.

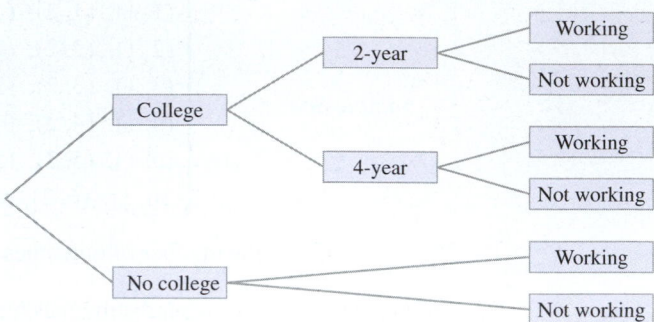

Figure **1**

The sample space is

$S = \{$2-year college & working, 2-year college & not working,
4-year college & working, 4-year college & not working,
no college & working, no college & not working$\}$.

Events

In Example 1, suppose we are interested in the event that a 2003 high school graduate was working. In mathematical language, we are interested in the *subset* of the sample space consisting of all outcomes in which the graduate was working.

Events

Given a sample space S, an **event** E is a subset of S. The outcomes in E are called the **favorable** outcomes. We say that E **occurs** in a particular experiment if the outcome of that experiment is one of the elements of E—that is, if the outcome of the experiment is favorable.

Visualizing an Event
In the following figure, the favorable outcomes (events in E) are shown in blue.

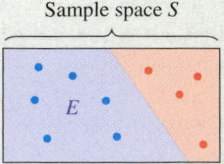

Sample space S

quick Examples

1. *Experiment:* Roll a die and observe the number facing up.

$S = \{1, 2, 3, 4, 5, 6\}$

Event: E: The number observed is odd.

$E = \{1, 3, 5\}$

2. *Experiment:* Roll two distinguishable dice and observe the numbers facing up.

$$S = \{(1, 1), (1, 2), \ldots, (6, 6)\}$$

Event: F: The dice show the same number.

$$F = \{(1, 1), (2, 2), (3, 3), (4, 4), (5, 5), (6, 6)\}$$

3. *Experiment:* Roll two distinguishable dice and observe the numbers facing up.

$$S = \{(1, 1), (1, 2), \ldots, (6, 6)\}$$

Event: G: The sum of the numbers is 1.

$$G = \emptyset \qquad \text{There are no favorable outcomes.}$$

4. *Experiment:* Select a city beginning with "J."
Event: E: The city is Johannesburg.

$$E = \{\text{Johannesburg}\} \qquad \text{An event can consist of a single outcome.}$$

5. *Experiment:* Roll a die and observe the number facing up.
Event: E: The number observed is either even or odd.

$$E = S = \{1, 2, 3, 4, 5, 6\} \qquad \text{An event can consist of all possible outcomes.}$$

6. *Experiment:* Select a student in your class.
Event: E: The student has red hair.

$$E = \{\text{red-haired students in your class}\}$$

7. *Experiment:* Draw a hand of two cards from a deck of 52.
Event: H: Both cards are diamonds.

H is the set of all hands of 2 cards chosen from 52 such that both cards are diamonds.

Here are some more examples of events.

Example 2 Dice

We roll a red die and a green die and observe the numbers facing up. Describe the following events as subsets of the sample space.

a. *E:* The sum of the numbers showing is 6.

b. *F:* The sum of the numbers showing is 2.

Solution Here (again) is the sample space for the experiment of throwing two dice.

$$S = \begin{Bmatrix} (1, 1), & (1, 2), & (1, 3), & (1, 4), & (1, 5), & (1, 6), \\ (2, 1), & (2, 2), & (2, 3), & (2, 4), & (2, 5), & (2, 6), \\ (3, 1), & (3, 2), & (3, 3), & (3, 4), & (3, 5), & (3, 6), \\ (4, 1), & (4, 2), & (4, 3), & (4, 4), & (4, 5), & (4, 6), \\ (5, 1), & (5, 2), & (5, 3), & (5, 4), & (5, 5), & (5, 6), \\ (6, 1), & (6, 2), & (6, 3), & (6, 4), & (6, 5), & (6, 6) \end{Bmatrix}$$

a. In mathematical language, E is the subset of S that consists of all those outcomes in which the sum of the numbers showing is 6. Here is the sample space once again, with the outcomes in question shown in color:

$$S = \begin{Bmatrix} (1,1), & (1,2), & (1,3), & (1,4), & (1,5), & (1,6), \\ (2,1), & (2,2), & (2,3), & (2,4), & (2,5), & (2,6), \\ (3,1), & (3,2), & (3,3), & (3,4), & (3,5), & (3,6), \\ (4,1), & (4,2), & (4,3), & (4,4), & (4,5), & (4,6), \\ (5,1), & (5,2), & (5,3), & (5,4), & (5,5), & (5,6), \\ (6,1), & (6,2), & (6,3), & (6,4), & (6,5), & (6,6) \end{Bmatrix}$$

Thus, $E = \{(1,5), (2,4), (3,3), (4,2), (5,1)\}$.

b. The only outcome in which the numbers showing add to 2 is (1,1). Thus,

$$F = \{(1,1)\}$$

Example 3 School and Work

Let S be the sample space of Example 1. List the elements in the following events:

a. The event E that a 2003 high school graduate was working.

b. The event F that a 2003 high school graduate was not going to a two-year college.

Solution

a. We had this sample space:

$S = \{$2-year college & working, two-year college & not working,
4-year college & working, four-year college & not working,
no college & working, no college & not working$\}$

We are asked for the event that a graduate was working. Whenever we encounter a phrase involving "the event that . . . ," we mentally translate this into mathematical language by changing the wording.

Replace the phrase "the event that . . . " by the phrase "the subset of the sample space consisting of all outcomes in which . . . "

Thus we are interested in the subset of the sample space consisting of all outcomes in which the graduate was working. This gives

$E = \{$two-year college & working, four-year college & working, no college & working$\}$

The outcomes in E are illustrated by the shaded cells in Figure 2.

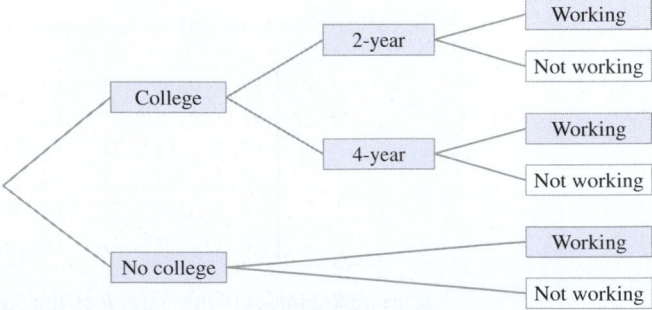

Figure **2**

b. We are looking for the event that a graduate was not going to a two-year college, that is, the subset of the sample space consisting of all outcomes in which the graduate was not going to a two-year college. Thus,

$$F = \{\text{four-year college \& working, four-year college \& not working,}$$
$$\text{no college \& working, no college \& not working}\}$$

The outcomes in F are illustrated by the shaded cells in Figure 3.

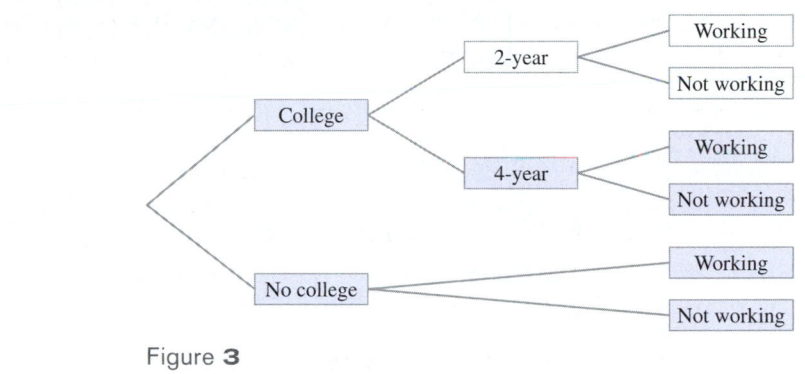

Figure **3**

Complement, Union, and Intersection of Events

Events may often be described in terms of other events, using set operations such as complement, union, and intersection.

Complement of an Event

The **complement** of an event E is the set of outcomes not in E. Thus, the complement of E represents the event that E *does not occur*.

Visualizing the Complement

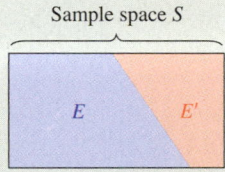

Sample space S

quick Examples

1. You take four shots at the goal during a soccer game and record the number of times you score. Describe the event that you score at least twice, and also its complement.

$S = \{0, 1, 2, 3, 4\}$ Set of outcomes

$E = \{2, 3, 4\}$ Event that you score at least twice

$E' = \{0, 1\}$ Event that you do not score at least twice

2. You roll a red die and a green die and observe the two numbers facing up. Describe the event that the sum of the numbers is not 6.

$$S = \{(1, 1), (1, 2), \ldots, (6, 6)\}$$

$$F = \{(1, 5), (2, 4), (3, 3), (4, 2), (5, 1)\} \qquad \text{Sum of numbers is 6.}$$

$$F' = \begin{cases} (1, 1), & (1, 2), & (1, 3), & (1, 4), & & (1, 6), \\ (2, 1), & (2, 2), & (2, 3), & & (2, 5), & (2, 6), \\ (3, 1), & (3, 2), & & (3, 4), & (3, 5), & (3, 6), \\ (4, 1), & & (4, 3), & (4, 4), & (4, 5), & (4, 6), \\ & (5, 2), & (5, 3), & (5, 4), & (5, 5), & (5, 6), \\ (6, 1), & (6, 2), & (6, 3), & (6, 4), & (6, 5), & (6, 6) \end{cases} \qquad \text{Sum of numbers is not 6.}$$

Union of Events

The **union** of the events E and F is the set of all outcomes in E or F (or both). Thus, $E \cup F$ represents the event that E occurs *or* F occurs (or both).[*]

quick Example

Roll a die.

E: The outcome is a 5; $E = \{5\}$

F: The outcome is an even number; $F = \{2, 4, 6\}$

$E \cup F$: The outcome is either a 5 *or* an even number; $E \cup F = \{2, 4, 5, 6\}$

_____

[*] As in the preceding chapter, when we use the word *or*, we agree to mean one or the other *or both*. This is called the **inclusive or** and mathematicians have agreed to take this as the meaning of *or* to avoid confusion.

Intersection of Events

The **intersection** of the events E and F is the set of all outcomes common to E and F. Thus, $E \cap F$ represents the event that both E *and* F occur.

quick Example

Roll two dice; one red and one green.

E: The red die is 2

F: The green die is odd

$E \cap F$: The red die is 2 and the green die is odd; $E \cap F = \{(2, 1), (2, 3), (2, 5)\}$

Example 4 Weather

Let R be the event that it will rain tomorrow, let P be the event that it will be pleasant, let C be the event that it will be cold, and let H be the event that it will be hot.

a. Express in words: $R \cap P'$, $R \cup (P \cap C)$.

b. Express in symbols: Tomorrow will be either a pleasant day or a cold and rainy day; it will not, however, be hot.

Solution The key here is to remember that intersection corresponds to *and* and union to *or*.

a. $R \cap P'$ is the event that it will rain *and* it will not be pleasant.

$R \cup (P \cap C)$ is the event that either it will rain, or it will be pleasant and cold.

b. If we rephrase the given statement using *and* and *or* we get "Tomorrow will be either a pleasant day or a cold and rainy day, and it will not be hot."

$$[P \cup (C \cap R)] \cap H' \qquad \text{Pleasant, or cold and rainy, and not hot.}$$

The nuances of the English language play an important role in this formulation. For instance, the effect of the pause (comma) after "rainy day" suggests placing the preceding clause $P \cup (C \cap R)$ in parentheses. In addition, the phrase "cold and rainy" suggests that C and R should be grouped together in their own parentheses.

Example 5 iPods, iMacs and Powerbooks

(Compare Example 3 in Section 6.2.) The following table shows sales, in thousands of units, of iPods®, Powerbooks®, and iMacs® in the fourth quarter of 2003 and in the third and fourth quarters of 2004.[*]

	iPods	Powerbooks	iMacs	Total
2003 Q4	340	180	250	770
2004 Q3	860	220	240	1320
2004 Q4	1030	210	230	1470
Total	2230	610	720	3560

Consider the experiment in which an item is selected at random from those in the table. Let D be the event that it was an iPod, let M be the event that it was an iMac, and let A be the event that it was sold in the fourth quarter of 2003. Describe the following events and compute their cardinality:

a. A' **b.** $D \cap A$ **c.** $M \cup A'$

Solution Before we answer the questions, note that S is the set of all items represented in the table, so S has a total of 3,560,000 outcomes.

a. A' is the event that the item was not sold in the fourth quarter of 2003. Its cardinality is

$$n(A') = n(S) - n(A) = 3560 - 770 = 2790 \text{ thousand items}$$

b. $D \cap A$ is the event that it was an iPod and was sold in the fourth quarter of 2003. Referring to the table, we find

$$n(D \cap A) = 340 \text{ thousand items}$$

---

[*] Figures are rounded. SOURCE: Company report www.apple.com January 2005.

c. $M \cup A'$ is the event that either it was an iMac or it was not sold in the fourth quarter of 2003:

	iPods	Powerbooks	iMacs	Total
2003 Q4	340	180	250	770
2004 Q3	860	220	240	1320
2004 Q4	1030	210	230	1470
Total	2230	610	720	3560

To compute its cardinality, we use the formula for the cardinality of a union:

$$n(M \cup A') = n(M) + n(A') - n(M \cap A')$$
$$= 720 + 2790 - (240 + 230) = 3040 \text{ thousand items}$$

+ *Before we go on...* We could shorten the calculation in part (c) of Example 5 even further using De Morgan's law to write $n(M \cup A') = n((M' \cap A)') = 3560 - (340 + 180) = 3040$ thousand items, a calculation suggested by looking at the table. ∎

The case where $E \cap F$ is empty is interesting, and we give it a name.

Mutually Exclusive Events

If E and F are events, then E and F are said to be **disjoint** or **mutually exclusive** if $E \cap F$ is empty. (Hence, they have no outcomes in common.)

Visualizing Mutually Exclusive Events

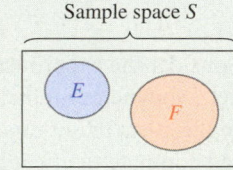

Sample space S

Interpretation

It is impossible for mutually exclusive events to occur simultaneously.

quick Examples In each of the following examples, E and F are mutually exclusive events.

1. Roll a die and observe the number facing up. E: The outcome is even; F: The outcome is odd.

$$E = \{2, 4, 6\}, \ F = \{1, 3, 5\}$$

2. Toss a coin three times and record the sequence of heads and tails. E: All three tosses land the same way up, F: One toss shows heads and the other two show tails.

$$E = \{HHH, TTT\}, \ F = \{HTT, THT, TTH\}$$

3. Observe tomorrow's weather. E: It is raining; F: There is not a cloud in the sky.

FAQs   Specifying the Sample Space

Q: *How do I determine the sample space in a given application?*

A: Strictly speaking, an experiment should include a description of what kinds of objects are in the sample space, as in:

Cast a die and observe the number facing up.
Sample space: the possible numbers facing up, [1, 2, 3, 4, 5, 6].

Choose a person at random and record her social security number and whether she is blonde.
Sample space: pairs (9-digit number, Y/N).

However, in many of the scenarios discussed in this chapter and the next, an experiment is specified more vaguely, as in "Select a student in your class." In cases like this, the nature of the sample space should be determined from the context. For example, if the discussion is about grade-point averages and gender, the sample space can be taken to consist of pairs (grade-point average, M/F). ∎

7.1 | EXERCISES

● denotes basic skills exercises

In Exercises 1–18, describe the sample space S of the experiment and list the elements of the given event. (Assume that the coins are distinguishable and that what is observed are the faces or numbers that face up.) hint [see Examples 1–3]

1. ● Two coins are tossed; the result is at most one tail.

2. ● Two coins are tossed; the result is one or more heads.

3. ● Three coins are tossed; the result is at most one head.

4. ● Three coins are tossed; the result is more tails than heads.

5. ● Two distinguishable dice are rolled; the numbers add to 5.

6. ● Two distinguishable dice are rolled; the numbers add to 9.

7. ● Two indistinguishable dice are rolled; the numbers add to 4.

8. ● Two indistinguishable dice are rolled; one of the numbers is even and the other is odd.

9. ● Two indistinguishable dice are rolled; both numbers are prime.[1]

10. ● Two indistinguishable dice are rolled; neither number is prime.

11. ● A letter is chosen at random from those in the word *Mozart*; the letter is a vowel.

12. ● A letter is chosen at random from those in the word *Mozart*; the letter is neither *a* nor *m*.

13. ● A sequence of two different letters is randomly chosen from those of the word *sore*; the first letter is a vowel.

14. ● A sequence of two different letters is randomly chosen from those of the word *hear*; the second letter is not a vowel.

15. ● A sequence of two different digits is randomly chosen from the digits 0–4; the first digit is larger than the second.

16. ● A sequence of two different digits is randomly chosen from the digits 0–4; the first digit is twice the second.

17. ● You are considering purchasing either a domestic car, an imported car, a van, an antique car, or an antique truck; you do not buy a car.

18. ● You are deciding whether to enroll for Psychology 1, Psychology 2, Economics 1, General Economics, or Math for Poets; you decide to avoid economics.

19. ● A packet of gummy candy contains 4 strawberry gums, 4 lime gums, 2 black currant gums, and 2 orange gums. April May sticks her hand in and selects 4 at random. Complete the following sentences:
    a. The sample space is the set of . . .
    b. April is particularly fond of combinations of 2 strawberry and 2 black currant gums. The event that April will get the combination she desires is the set of . . .

20. ● A bag contains 3 red marbles, 2 blue ones, and 4 yellow ones, and Alexandra pulls out 3 of them at random. Complete the following sentences:
    a. The sample space is the set of . . .
    b. The event that Alexandra gets one of each color is the set of . . .

21. President George W. Bush's cabinet consisted of the Secretaries of State, Treasury, Defense, Interior, Agriculture, Commerce,

---

[1] A positive integer is **prime** if it is neither 1 nor a product of smaller integers.

Labor, Health and Human Services, Housing and Urban Development, Transportation, Energy, Education, Veterans Affairs, and the Attorney General.[2] Assuming that President Bush had 20 candidates, including Colin Powell, to fill these posts (and wished to assign no one to more than one post), complete the following sentences:

a. The sample space is the set of . . .
b. The event that Colin Powell is the Secretary of State is the set of . . .

22. A poker hand consists of a set of 5 cards chosen from a standard deck of 52 playing cards. You are dealt a poker hand. Complete the following sentences:

a. The sample space is the set of . . .
b. The event "a full house" is the set of . . . (Recall that a full house is three cards of one denomination and two of another.)

Suppose two dice (one red, one green) are rolled. Consider the following events. A: the red die shows 1; B: the numbers add to 4; C: at least one of the numbers is 1; and D: the numbers do not add to 11. In Exercises 23–30, express the given event in symbols and say how many elements it contains. hint [see Example 4]

23. ● The red die shows 1 and the numbers add to 4.

24. ● The red die shows 1 but the numbers do not add to 11.

25. ● The numbers do not add to 4.

26. ● The numbers add to 11.

27. ● The numbers do not add to 4, but they do add to 11.

28. ● Either the numbers add to 11 or the red die shows a 1.

29. ● At least one of the numbers is 1 or the numbers add to 4.

30. ● Either the numbers add to 4, or they add to 11, or at least one of them is 1.

_____
[2] Source: The Whitehouse website, www.whitehouse.gov.

Let W be the event that you will use the website tonight, let I be the event that your math grade will improve, and let E be the event that you will use the website every night. In Exercises 31–38, express the given event in symbols.

31. ● You will use the website tonight and your math grade will improve.

32. ● You will use the website tonight or your math grade will not improve.

33. ● Either you will use the website every night, or your math grade will not improve.

34. ● Your math grade will not improve even though you use the website every night.

35. Either your math grade will improve, or you will use the website tonight but not every night.

36. You will use the website either tonight or every night, and your grade will improve.

37. (Compare Exercise 35.) Either your math grade will improve or you will use the website tonight, but you will not use it every night.

38. (Compare Exercise 36.) You will either use the website tonight, or you will use it every night and your grade will improve.

Applications

Housing Prices *Exercises 39–44 are based on the following map, which shows the percentage increase in housing prices from September 30, 2003 to September 30, 2004 in each of nine regions (U.S. Census divisions):[3]*

_____
[3] Source: Third Quarter 2004 House Price Index, released December 1, 2004 by the Office of Federal Housing Enterprise Oversight; available online at www.ofheo.gov/media/pdf/3q04hpi.pdf.

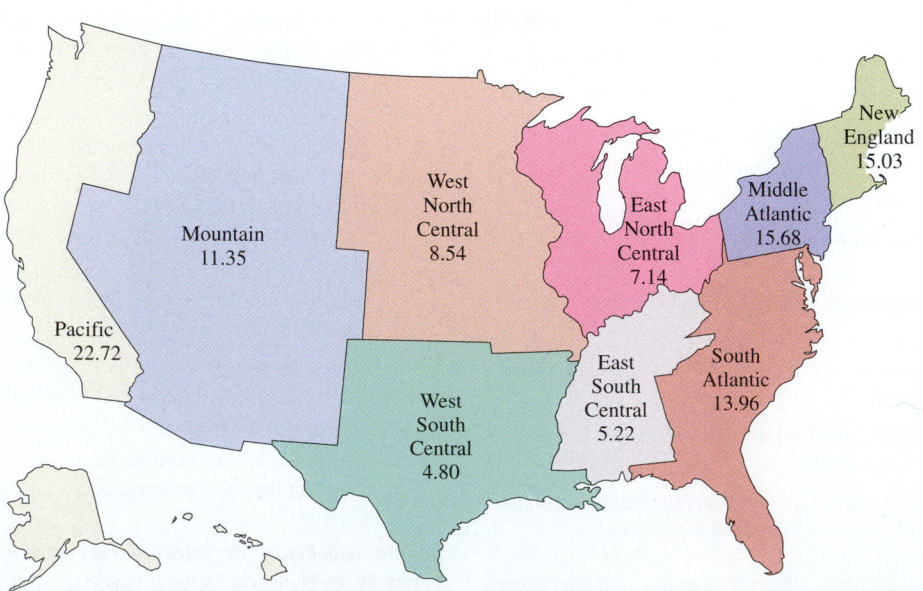

39. ● You are choosing a region of the country to move to. Describe the event E that the region you choose saw an increase in housing prices of 15% or more.

40. ● You are choosing a region of the country to move to. Describe the event F that the region you choose saw an increase in housing prices of less than 10%.

41. ● You are choosing a region of the country to move to. Let E be the event that the region you choose saw an increase in housing prices of 15% or more, and let F be the event that the region you choose is on the east coast. Describe the events $E \cup F$ and $E \cap F$ both in words and by listing the outcomes of each.

42. ● You are choosing a region of the country to move to. Let E be the event that the region you choose saw an increase in housing prices of less than 10%, and let F be the event that the region you choose is not on the east coast. Describe the events $E \cup F$ and $E \cap F$ both in words and by listing the outcomes of each.

43. You are choosing a region of the country to move to. Which of the following pairs of events are mutually exclusive?

  a. E: You choose a region from among the three with the highest percentage increase in housing prices;
  F: You choose a region that is not on the east or west coast.

  b. E: You choose a region from among the three with the highest percentage increase in housing prices;
  F: You choose a region that is not on the west coast.

44. You are choosing a region of the country to move to. Which of the following pairs of events are mutually exclusive?

  a. E: You choose a region from among the three with the lowest percentage increase in housing prices;
  F: You choose a region from among the central divisions.

  b. E: You choose a region from among the three with the lowest percentage increase in housing prices;
  F: You choose a region from among the noncentral divisions.

Publishing *Exercises 45–52 are based on the following table, which shows the results of a survey of authors by a (fictitious) publishing company.*

	New Authors	Established Authors	Total
Successful	5	25	30
Unsuccessful	15	55	70
Total	20	80	100

Consider the following events: S: an author is successful; U: an author is unsuccessful; N: an author is new; and E: an author is established. hint *[see Example 5]*

45. ● Describe the events $S \cap N$ and $S \cup N$ in words. Use the table to compute $n(S \cap N)$ and $n(S \cup N)$.

46. ● Describe the events $N \cap U$ and $N \cup U$ in words. Use the table to compute $n(N \cap U)$ and $n(N \cup U)$.

47. ● Which of the following pairs of events are mutually exclusive: N and E; N and S; S and E?

48. ● Which of the following pairs of events are mutually exclusive: U and E; U and S; S and N?

49. ● Describe the event $S \cap N'$ in words and find the number of elements it contains.

50. ● Describe the event $U \cup E'$ in words and find the number of elements it contains.

51. What percentage of established authors are successful? What percentage of successful authors are established?

52. What percentage of new authors are unsuccessful? What percentage of unsuccessful authors are new?

Exercises 53–60 are based on the following table, which shows the performance of a selection of 100 stocks after one year. (Take S to be the set of all stocks represented in the table.)

	Companies			
	Pharmaceutical P	Electronic E	Internet I	Total
Increased V	10	5	15	30
Unchanged* N	30	0	10	40
Decreased D	10	5	15	30
Total	50	10	40	100

* If a stock stayed within 20% of its original value, it is classified as "unchanged."

53. ● Use symbols to describe the event that a stock's value increased but it was not an Internet stock. How many elements are in this event?

54. ● Use symbols to describe the event that an Internet stock did not increase. How many elements are in this event?

55. ● Compute $n(P' \cup N)$. What does this number represent?

56. ● Compute $n(P \cup N')$. What does this number represent?

57. ● Find all pairs of mutually exclusive events among the events P, E, I, V, N, and D.

58. ● Find all pairs of events that are not mutually exclusive among the events P, E, I, V, N, and D.

59. Calculate $\dfrac{n(V \cap I)}{n(I)}$. What does the answer represent?

60. Calculate $\dfrac{n(D \cap I)}{n(D)}$. What does the answer represent?

Animal Psychology *Exercises 61–66 concern the following chart, which shows the way in which a dog moves its facial*

● basic skills

muscles when torn between the drives of fight and flight.[4] The "fight" drive increases from left to right; the "flight" drive increases from top to bottom. (Notice that an increase in the "fight" drive causes its upper lip to lift, while an increase in the "flight" drive draws its ears downwards.)

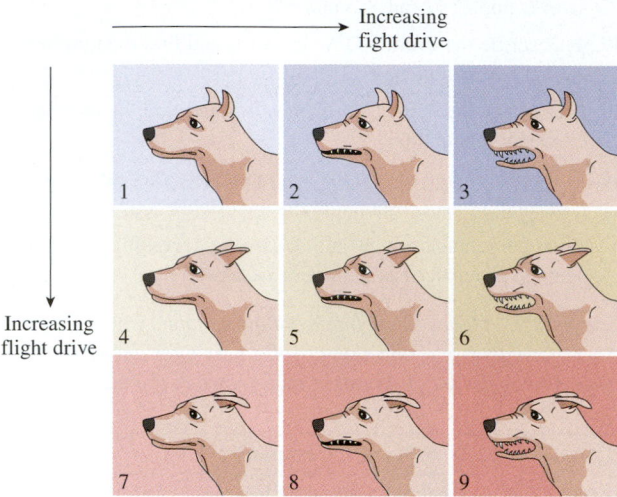

Increasing fight drive →

↓ Increasing flight drive

61. Let E be the event that the dog's flight drive is the strongest, let F be the event that the dog's flight drive is weakest, let G be the event that the dog's fight drive is the strongest, and let H be the event that the dog's fight drive is weakest. Describe the following events in terms of E, F, G and H using the symbols $\cap$, $\cup$ and $'$.

a. The dog's flight drive is not strongest and its fight drive is weakest.

b. The dog's flight drive is strongest or its fight drive is weakest.

c. Neither the dog's flight drive nor its fight drive is strongest.

62. Let E be the event that the dog's flight drive is the strongest, let F be the event that the dog's flight drive is weakest, let G be the event that the dog's fight drive is the strongest, and let H be the event that the dog's fight drive is weakest. Describe the following events in terms of E, F, G and H using the symbols $\cap$, $\cup$ and $'$.

a. The dog's flight drive is weakest and its fight drive is not weakest.

b. The dog's flight drive is not strongest or its fight drive is weakest.

c. Either the dog's flight drive or its fight drive fails to be strongest.

63. Describe the following events explicitly (as subsets of the sample space):

a. The dog's fight and flight drives are both strongest.

b. The dog's fight drive is strongest, but its flight drive is neither weakest nor strongest.

64. Describe the following events explicitly (as subsets of the sample space):

a. Neither the dog's fight drive nor its flight drive is strongest.

b. The dog's fight drive is weakest, but its flight drive is neither weakest nor strongest.

65. Describe the following events in words:

a. $\{1, 4, 7\}$

b. $\{1, 9\}$

c. $\{3, 6, 7, 8, 9\}$

66. Describe the following events in words:

a. $\{7, 8, 9\}$

b. $\{3, 7\}$

c. $\{1, 2, 3, 4, 7\}$

Exercises 67–74 use counting arguments from the preceding chapter.

67. *Gummy Bears* A bag contains 6 gummy bears. Noel picks 4 at random. How many possible outcomes are there? If one of the gummy bears is raspberry, how many of these outcomes include the raspberry gummy bear?

68. *Chocolates* My couch potato friend enjoys sitting in front of the TV and grabbing handfuls of 5 chocolates at random from his snack jar. Unbeknownst to him, I have replaced one of the 20 chocolates in his jar with a cashew. (He hates cashews with a passion.) How many possible outcomes are there the first time he grabs 5 chocolates? How many of these include the cashew?

69. *Horse Races* The 7 contenders in the fifth horse race at Aqueduct on February 18, 2002, were: Pipe Bomb, Expect a Ship, All That Magic, Electoral College, Celera, Cliff Glider, and Inca Halo.[5] You are interested in the first 3 places (winner, second place, and third place) for the race.

a. Find the cardinality $n(S)$ of the sample space S of all possible finishes of the race. (A finish for the race consists of a first, second, and third place winner.)

b. Let E be the event that Electoral College is in second or third place, and let F be the event that Celera is the winner. Express the event $E \cap F$ in words, and find its cardinality.

70. *Intramurals* The following five teams will be participating in Urban University's hockey intramural tournament: the Independent Wildcats, the Phi Chi Bulldogs, the Gate Crashers, the Slide Rule Nerds, and the City Slickers. Prizes will be awarded for the winner and runner-up.

a. Find the cardinality $n(S)$ of the sample space S of all possible outcomes of the tournament. (An outcome of the tournament consists of a winner and a runner-up.)

b. Let E be the event that the City Slickers are runners-up, and let F be the event that the Independent Wildcats are neither the winners nor runners-up. Express the event $E \cup F$ in words, and find its cardinality.

---

[4] SOURCE: *On Aggression* by Konrad Lorenz (Fakenham, Norfolk: University Paperback Edition, Cox & Wyman Limited , 1967).

[5] SOURCE: *Newsday*, Feb. 18, 2002, p. A36.

● basic skills

In Exercises 71–74, Pablo randomly picks 3 marbles from a bag of 8 marbles (4 red ones, 2 green ones, and 2 yellow ones).

71. How many outcomes are there in the sample space?

72. How many outcomes are there in the event that Pablo grabs 3 red marbles?

73. How many outcomes are there in the event that Pablo grabs one marble of each color?

74. How many outcomes are there in the event that Pablo's marbles are not all the same color?

Communication and Reasoning Exercises

75. ● Complete the following sentence. An event is a ____.

76. ● Complete the following sentence. Two events E and F are mutually exclusive if their intersection is _____.

77. ● If E and F are events, then $(E \cap F)'$ is the event that ____.

78. ● If E and F are events, then $(E' \cap F')$ is the event that ____.

79. True or false: Every set S is the sample space for some experiment. Explain.

80. True or false: Every sample space S is a finite set. Explain.

81. Describe an experiment in which a die is cast and the set of outcomes is $\{0, 1\}$.

82. Describe an experiment in which two coins are flipped and the set of outcomes is $\{0, 1, 2\}$.

83. Two distinguishable dice are rolled. Could there be two mutually exclusive events that both contain outcomes in which the numbers facing up add to 7?

84. Describe an experiment in which two dice are rolled and describe two mutually exclusive events that both contain outcomes in which both dice show a 1.

● basic skills

7.2 Estimated and Theoretical Probability

Estimated Probability

Suppose you have a coin that you think is not fair and you would like to determine the likelihood that heads will come up when it is tossed. You could estimate this likelihood by tossing the coin a large number of times and counting the number of times heads comes up. Suppose, for instance, that in 100 tosses of the coin, heads comes up 58 times. The fraction of times that heads comes up, $58/100 = .58$, is the **estimated probability** or **relative frequency** of heads coming up when the coin is tossed. In other words, saying that the estimated probability of heads coming up is .58 is the same as saying that heads came up 58% of the time in your series of experiments.

Now let's think about this example in terms of sample spaces and events. First of all, there is an experiment that has been repeated $N = 100$ times: Toss the coin and observe the side facing up. The sample space for this experiment is $S = \{H, T\}$. Also, there is an event E in which we are interested: the event that heads comes up, which is $E = \{H\}$. The number of times E has occurred, or the **frequency** of E, is $fr(E) = 58$. The estimated probability of the event E is then

$$P(E) = \frac{fr(E)}{N} \qquad \text{Frequency of event } E \over \text{Number of repetitions } N$$
$$= \frac{58}{100} = .58$$

Notes

1. The estimated probability gives us an *estimate* of the likelihood that heads will come up when that particular coin is tossed.

2. The larger the number of times the experiment is performed, the more accurate an estimate we expect the estimated probability to be. ■

Estimated Probability

When an experiment is performed a number of times, the **estimated probability** or **relative frequency** of an event E is the fraction of times that the event E occurs. If the experiment is performed N times and the event E occurs $fr(E)$ times, then the estimated probability is given by

$$P(E) = \frac{fr(E)}{N} \qquad \text{Fraction of times } E \text{ occurs.}$$

The number $fr(E)$ is called the **frequency** of E. N, the number of times that the experiment is performed, is called the number of **trials** or the **sample size**. If E consists of a single outcome s, then we refer to $P(E)$ as the estimated probability or relative frequency of the outcome s, and we write $P(s)$.

Visualizing Estimated Probability

$$P(E) = \frac{fr(E)}{N} = \frac{4}{10} = .4$$

The collection of the estimated probabilities of *all* the outcomes is the **estimated probability distribution** or **relative frequency distribution.**

Note In this text, we almost always use the term *estimated probability* rather than *relative frequency.* ∎

quick Examples

1. **Experiment:** Roll a pair of dice and add the numbers that face up.

   Event: E: The sum is 5.

   If the experiment is repeated 100 times and E occurs on 10 of the rolls, then the estimated probability of E is

   $$P(E) = \frac{fr(E)}{N} = \frac{10}{100} = .10$$

2. If 10 rolls of a single die resulted in the outcomes 2, 1, 4, 4, 5, 6, 1, 2, 2, 1, then the associated estimated probability distribution is shown in the following table:

Outcome	1	2	3	4	5	6
Probability	.3	.3	0	.2	.1	.1

Example 1   Sales of Light Trucks

In a survey of 75 light truck sales in 2004, 20 were found to be small SUVs, 25 were large SUVs, 5 were small pickups, and 25 were large pickups.[*] What is the estimated probability that a light truck sold in the U.S. is not a large pickup?

Solution The experiment consists of choosing a light truck sold in the U.S. and determining its type. The sample space suggested by the information given is

$$S = \{\text{Small SUV, Large SUV, Small Pickup, Large Pickup}\}$$

---

[*] The proportions are based on 2004 sales data. SOURCE: Environmental Protection Agency, Ward's AutoInfoBank/ *New York Times,* February 8, 2005, p. C1.

and we are interested in the event

$$E = \{\text{Small SUV, Large SUV, Small Pickup}\}$$

The sample size is $N = 75$ and the frequency of E is $fr(E) = 20 + 25 + 5 = 50$. Thus, the estimated probability of E is

$$P(E) = \frac{fr(E)}{N} = \frac{50}{75} \approx .67$$

+ *Before we go on...* In Example 1, you might ask how accurate the estimate of .67 is or how well it reflects *all* of the light trucks sold in the U.S. absent any information about national sales figures. The field of statistics provides the tools needed to say to what extent the estimated probability can be trusted. ■

Example 2 Auctions on eBay

Chris Ware/The Image Works

The following chart shows the results of a survey of 50 paintings on eBay whose listings were close to expiration, examining the number of bids each had received.[*]

Bids	0	1–5	6–10	>10
Frequency	10	17	13	10

Consider the experiment in which a painting whose listing is close to expiration is chosen and the number of bids is observed.

a. Find the estimated probability distribution.

b. Find the estimated probability that a painting whose listing was close to expiration had received no more than 10 bids.

Solution

a. The following table shows the estimated probability of each outcome, which we find by dividing each frequency by the sum $N = 50$:

Probability Distribution

Bids	0	1–5	6–10	>10
Probability	$\frac{10}{50} = .20$	$\frac{17}{50} = .34$	$\frac{13}{50} = .26$	$\frac{10}{50} = .20$

b. $E = \{0 \text{ bids}, 1\text{–}5 \text{ bids}, 6\text{–}10 \text{ bids}\}$; thus,

$$P(E) = \frac{fr(E)}{N} = \frac{10 + 17 + 13}{50} = \frac{40}{50} = .80$$

Alternatively, notice that we can obtain the same answer from the probability distribution in part (a) by simply adding the probabilities of the outcomes in E:

$$P(E) = .20 + .34 + .26 = .80$$

This property of probability distributions is discussed below.

_____

[*] The 50 paintings whose listings were closest to expiration in the category "Original-Listed by Artist" on February 19, 2005. www.eBay.com.

Following are some important properties of estimated probability that we can observe in Example 2.

Some Properties of Estimated Probability Distributions

Let $S = \{s_1, s_2, \ldots, s_n\}$ be a sample space and let $P(s_i)$ be the estimated probability of the event $\{s_i\}$. Then

1. $0 \le P(s_i) \le 1$

2. $P(s_1) + P(s_2) + \cdots + P(s_n) = 1$

3. If $E = \{e_1, e_2, \ldots, e_r\}$, then $P(E) = P(e_1) + P(e_2) + \cdots + P(e_r)$.

In words:

1. The estimated probability of each outcome is a number between 0 and 1 (inclusive).

2. The estimated probabilities of all the outcomes add up to 1.

3. The estimated probability of an event E is the sum of the estimated probabilities of the individual outcomes in E.

 **using** *Technology*

See the Technology Guides at the end of the chapter to see how to use a TI-83/84 or Excel to simulate multiple coin tosses.

Estimated Probability and Increasing Sample Size

A "fair" coin is one that is as likely to come up heads as it is to come up tails. In other words, we expect heads to come up 50% of the time if we toss such a coin many times. Put more precisely, we expect the estimated probability to approach .5 as the number of trials gets larger. Figure 4 shows how the estimated probability behaved for one sequence of coin tosses. For each N we have plotted what fraction of times the coin came up heads in the first N tosses.

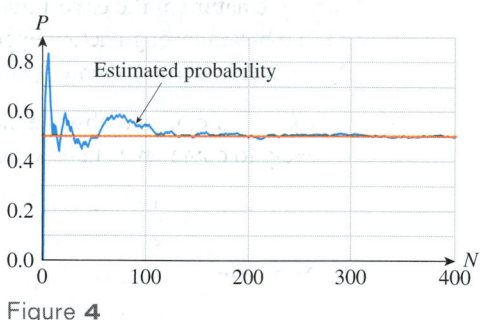

Figure **4**

Notice that the estimated probability graph meanders as N increases, sometimes getting closer to .5, and sometimes drifting away again. However, the graph tends to meander within smaller and smaller distances of .5 as N increases.[6]

In general, this is how estimated probability behaves; as N gets large, the estimated probability approaches some fixed value: the *theoretical probability*.

Theoretical Probability

It is understandable if you are a little uncomfortable with estimated probability because it does not always agree with what you intuitively feel to be true. For instance, if you toss

---

[6] This can be made precise by the concept of "limit" used in calculus.

a fair coin 100 times and heads happen to come up 62 times, the experiment seems to suggest that the probability of heads is .62, even though you *know* that the "actual" probability is .50 (because the coin is fair).

Q: *So what do we mean by the "actual" probability*?

A: First, some terminology: We shall refer to the "actual" probability as the **theoretical probability.** We can define it in two ways: one more intuitive (but not very precise) and one more precise (but not very intuitive). ∎

Theoretical Probability

Intuitive Definition of Theoretical Probability

The **theoretical probability,** or **probability,** $P(E)$, of an event E is the fraction of times we *expect* E to occur if we repeat the same experiment over and over.

More Precise Definition of Theoretical Probability

The **theoretical probability,** or **probability,** $P(E)$, of an event E is the *limiting value* of the estimated probability as the number of trials gets larger and larger. That is, the estimated probability approaches the theoretical probability as the number of trials gets larger and larger.[*] Thus,

Estimated probability is an approximation, or estimate, of theoretical probability. The larger the number of trials, the more accurate we expect this approximation to be.

If E consists of a single outcome s, we refer to $P(E)$ as the probability of the outcome s, and write $P(s)$ for $P(E)$. The collection of the probabilities of all the outcomes is the **probability distribution.**

Determining Theoretical Probability

Theoretical probability is determined *analytically*—that is, by using our knowledge about the nature of the experiment rather than through actual experimentation. The best we can obtain through actual experimentation is an *estimate* of the theoretical probability (hence the term *estimated probability*).

quick Examples

1. Toss a fair coin and observe the side that faces up. Because we expect that heads is as likely to come up as tails, we conclude that the theoretical probability distribution is

$$P(\text{H}) = \frac{1}{2}, \quad P(\text{T}) = \frac{1}{2}$$

2. Roll a fair die. Because we expect to roll a 1 one sixth of the time,

$$P(1) = \frac{1}{6}$$

Similarly, $P(2) = \frac{1}{6}$, $P(3) = \frac{1}{6}$, ..., $P(6) = \frac{1}{6}$.

3. Roll a pair of fair dice (recall that there are a total of 36 outcomes if the dice are distinguishable). If E is the event that the sum of the numbers that face up is 5, then $E = \{(1, 4), (2, 3), (3, 2), (4, 1)\}$. Because all 36 outcomes are equally likely,

$$P(E) = \frac{4}{36} = \frac{1}{9}$$

---

[*] See Figure 4 in the preceding section. In calculus, the idea of a "limiting value" can be made very precise, and we say that $P(E)$ is the limit of the estimated probability of E as the number of trials $N \to +\infty$.

Notes

1. Another distinction between estimated and theoretical probability is that the estimated probability of an event E is the fraction of times E *actually occurs* in a repeated experiment, whereas the theoretical probability is the fraction of times we *expect E to occur* in the long term.

2. We use the same letter P for theoretical probability as we used for estimated probability. Whether we are talking about estimated or theoretical probability will either be stated or be clear from the context. In the same vein, we shall often simply speak of "probability" in reference to either estimated or theoretical probability.

Now, how do we calculate theoretical probability "using our knowledge about the nature of the experiment"? There is no easy way to calculate theoretical probability in general, and sometimes there is no way at all. For instance, the probability that it will snow in Toronto on November 1 is (we suspect) impossible to determine theoretically, so we must rely on estimated probabilities for weather prediction. However, notice that, in the Quick Examples above, all the outcomes were equally likely and, for an event E all we did was compute the ratio

$$\frac{\text{Number of favorable outcomes}}{\text{Total number of outcomes}} = \frac{n(E)}{n(S)}$$

Computing Theoretical Probability: Equally Likely Outcomes

In an experiment in which all outcomes are equally likely, the theoretical probability of an event E is given by

$$P(E) = \frac{\text{Number of favorable outcomes}}{\text{Total number of outcomes}} = \frac{n(E)}{n(S)}$$

Visualizing Theoretical Probability with Equally Likely Outcomes

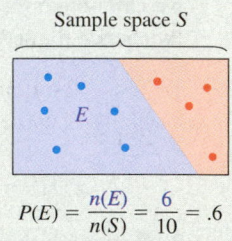

Sample space S

$$P(E) = \frac{n(E)}{n(S)} = \frac{6}{10} = .6$$

Note Remember that this formula will work *only* when the outcomes are equally likely. If, for example, a die is *weighted,* then the outcomes may not be equally likely, and the formula above will not apply.

---

quick Examples

1. Toss a fair coin three times. The probability that we throw exactly two heads is

$$P(E) = \frac{n(E)}{n(S)} = \frac{3}{8} \qquad \text{There are 8 equally likely outcomes and } E = \{\text{HHT, HTH, THH}\}.$$

2. Roll a pair of fair dice. The probability that we roll a double (both dice show the same number) is

$$P(E) = \frac{n(E)}{n(S)} = \frac{6}{36} = \frac{1}{6} \qquad E = \{(1, 1), (2, 2), (3, 3), (4, 4), (5, 5), (6, 6)\}$$

3. Randomly choose a person from a class of 40, in which 6 have red hair. If E is the event that a randomly selected person in the class has red hair, then

$$P(E) = \frac{n(E)}{n(S)} = \frac{6}{40} = .15$$

Example 3 Sales of Light Trucks

(Compare Example 1) Approximately 750,000 light trucks were sold in the U.S. in 2004. Of these, 200,000 were small SUVs, 250,000 were large SUVs, 50,000 were small pickups, and 250,000 were large pickups.[*]

a. What is the probability that a light truck sold in the U.S. in 2004 was an SUV?

b. What is the probability that a light truck sold in the U.S. in 2004 was not a large pickup?

Solution

a. The experiment suggested by the question consists of randomly choosing a light truck sold in the U.S. and determining its type. We are interested in the event E that the light truck was an SUV. So,

$S =$ the set of light trucks sold in the U.S. in 2004; $n(S) = 750,000$.

$E =$ the set of SUVs sold in the U.S. in 2004; $n(E) = 200,000 + 250,000 = 450,000$.

Are the outcomes equally likely in this experiment? Yes, because we are as likely to choose one light truck as another. Thus,

$$P(E) = \frac{n(E)}{n(S)} = \frac{450,000}{750,000} = .6$$

b. Let the event F consist of those light trucks sold in 2004 that were not large pickups.

$$n(F) = 750,000 - 250,000 = 500,000$$

Hence,

$$P(F) = \frac{n(F)}{n(S)} = \frac{500,000}{750,000} \approx .67$$

---

[*] The proportions are based on 2004 sales data. SOURCE: Environmental Protection Agency, Ward's AutoInfoBank/*The New York Times*, February 8, 2005, p. C1.

Q: In Example 1 we had a similar example about trucks, but we called the probabilities calculated there estimated. Here they are theoretical. What is the difference?

A: In Example 1, the data were based on the results of a survey, or sample, of only 75 light trucks (out of a total of about 750,000 sold in the U.S.), and were therefore incomplete. (A statistician would say that we were given *sample statistics*.) It follows that any inference we draw from the 75 surveyed, such as the probability that a light truck sold in the U.S. is not a large pickup, is uncertain, and this is the cue that tells us that we are working with estimated probability. Think of the survey as an experiment (for instance, select a graduating senior) repeated 75 times—exactly the setting for estimated probability.

In Example 3 above, on the other hand, the data do not describe how *some* light truck sales are broken down into the categories described, but they describe how *all 750,000* light truck sales in the U.S. are broken down. (The statistician would say that we were given *population statistics* in this case, because the data describe the entire "population" of light trucks sold in the U.S. in 2004.) ∎

Example 4 iPods, iMacs and Powerbooks

(Compare Example 5 in Section 7.1.) The following table shows sales, in thousands of units, of iPods®, Powerbooks®, and iMacs® in the fourth quarter of 2003 and in the third and fourth quarters of 2004.[*]

	iPods	Powerbooks	iMacs	Total
2003 Q4	340	180	250	770
2004 Q3	860	220	240	1320
2004 Q4	1030	210	230	1470
Total	2230	610	720	3560

If one of the items is selected at random, find the (theoretical) probabilities of the following events:

a. It is an iPod.
b. It was sold in the third quarter of 2004.
c. It is an iPod sold in the third quarter of 2004.
d. It is either an iPod or was sold in the third quarter of 2004.
e. It is not an iPod.

Solution

a. The sample space S is the set of all these machines, and the outcomes consist of all the items sold in the given quarters. Because an item is being selected at random, all the outcomes are equally likely. If D is the event that the selected item is an iPod, then

$$P(D) = \frac{n(D)}{n(S)} = \frac{2230}{3560} \approx .626$$

The event D is represented by the blue shaded region in the table:

	iPods	Powerbooks	iMacs	Total
2003 Q4	340	180	250	770
2004 Q3	860	220	240	1320
2004 Q4	1030	210	230	1470
Total	2230	610	720	3560

b. If A is the event that the selected item was sold in the third quarter of 2004, then

$$P(A) = \frac{n(A)}{n(S)} = \frac{1320}{3560} \approx .371$$

---

[*] Figures are rounded. SOURCE: Company report www.apple.com January 2005.

In the table, A is represented as shown:

	iPods	Powerbooks	iMacs	Total
2003 Q4	340	180	250	770
2004 Q3	860	220	240	1320
2004 Q4	1030	210	230	1470
Total	2230	610	720	3560

c. The event that the selected item is an iPod sold in the third quarter of 2004 is the event $D \cap A$, and

$$P(D \cap A) = \frac{n(D \cap A)}{n(S)} = \frac{860}{3560} \approx .242$$

In the table, $D \cap A$ is represented by the overlap of the regions representing D and A:

	iPods	Powerbooks	iMacs	Total
2003 Q4	340	180	250	770
2004 Q3	860	220	240	1320
2004 Q4	1030	210	230	1470
Total	2230	610	720	3560

d. The event that the selected item is either an iPod or was sold in the third quarter of 2004 is the event $D \cup A$, and is represented by the blue shaded area in the table:

	iPods	Powerbooks	iMacs	Total
2003 Q4	340	180	250	770
2004 Q3	860	220	240	1320
2004 Q4	1030	210	230	1470
Total	2230	610	720	3560

$$P(D \cup A) = \frac{n(D \cup A)}{n(S)} = \frac{1320 + 2230 - 860}{3560} \approx .756$$

(Notice that we used the formula
$$n(D \cup A) = n(D) + n(A) - n(D \cap A)$$
$$= 1320 + 2230 - 860 = 2690$$
to compute the cardinality of $D \cup A$.)

e. The event that the selected item is not an iPod is the event D'.

	iPods	Powerbooks	iMacs	Total
2003 Q4	340	180	250	770
2004 Q3	860	220	240	1320
2004 Q4	1030	210	230	1470
Total	2230	610	720	3560

$$P(D') = \frac{n(D')}{n(S)} = \frac{610 + 720}{3560} = \frac{1330}{3560} \approx .374$$

+*Before we go on...* How are the probabilities $P(D)$ and $P(D')$ in Example 4 related? Notice that they add to 1. We shall explore this, and general facts about probability, in Section 7.3. ■

We saw that estimated probability has the following properties for a sample space $S = \{s_1, s_2, \ldots, s_n\}$:

a. $0 \le P(s_i) \le 1$

b. $P(s_1) + P(s_2) + \cdots + P(s_n) = 1$

c. We can obtain the probability of an event E by adding up the probabilities of the outcomes in E.

These properties hold for theoretical probability as well. Because the probability of an outcome is the fraction of times we expect it to occur, probabilities are always numbers between 0 and 1, inclusive. To get the fraction of times an event will occur, we clearly need to add the fractions of times each outcome in the event will occur. And, because the event $E = S$ *must* occur every time, $P(S) = P(s_1) + \cdots + P(s_n) = 1$.

Example 5 Indistinguishable Dice

We recall from Section 7.1 that the sample space when we roll a pair of indistinguishable dice is

$$S = \begin{Bmatrix} (1, 1), & (1, 2), & (1, 3), & (1, 4), & (1, 5), & (1, 6), \\ & (2, 2), & (2, 3), & (2, 4), & (2, 5), & (2, 6), \\ & & (3, 3), & (3, 4), & (3, 5), & (3, 6), \\ & & & (4, 4), & (4, 5), & (4, 6), \\ & & & & (5, 5), & (5, 6), \\ & & & & & (6, 6) \end{Bmatrix}$$

Compute the probabilities of all the outcomes.

Solution Because there are 21 outcomes, it is tempting to say that the probability of each outcome is 1/21. However, the outcomes are not all equally likely. For instance, the outcome (2, 3) is twice as likely as (2, 2), because (2, 3) can occur in two ways (it corresponds to the event $\{(2, 3), (3, 2)\}$ for distinguishable dice). For purposes of calculating probability, it is easiest to use calculations for distinguishable dice.[*] Here are some examples.

Outcome (indistinguishable dice)	(1, 1)	(1, 2)	(2, 2)	(1, 3)	(2, 3)	(3, 3)
Corresponding event (distinguishable dice)	$\{(1, 1)\}$	$\{(1, 2), (2, 1)\}$	$\{(2, 2)\}$	$\{(1, 3), (3, 1)\}$	$\{(2, 3), (3, 2)\}$	$\{(3, 3)\}$
Probability	$\dfrac{1}{36}$	$\dfrac{2}{36} = \dfrac{1}{18}$	$\dfrac{1}{36}$	$\dfrac{2}{36} = \dfrac{1}{18}$	$\dfrac{2}{36} = \dfrac{1}{18}$	$\dfrac{1}{36}$

If we continue this process for all 21 outcomes, we will find that they add to 1.

---

[*] Note that any pair of real dice can be distinguished in principle because they possess slight differences, although we may regard them as indistinguishable by not attempting to distinguish them. Thus, the probabilities of events must be the same as for the corresponding events for distinguishable dice.

Example 6 Weighted Dice

In order to impress your friends with your die-rolling skills, you have surreptitiously weighted your die in such a way that 6 is three times as likely to come up as any one of the other numbers. (All the other outcomes are equally likely.) Find the probability distribution for a roll of the die and use it to calculate the probability of an even number coming up.

Solution Let us label our unknowns (there appear to be two of them):

x = probability of rolling a 6.
y = probability of rolling any one of the other numbers.

We are first told that "6 is three times as likely to come up as any other number." If we rephrase this in terms of our unknown probabilities we get, "the probability of rolling a 6 is three times the probability of rolling any other number." In symbols,

$$x = 3y$$

We must also use a piece of information not given to us, but one we know must be true: The sum of the probabilities of all the outcomes is 1:

$$x + y + y + y + y + y = 1$$

or

$$x + 5y = 1$$

We now have two linear equations in two unknowns, and we solve for x and y. Substituting the first equation ($x = 3y$) in the second ($x + 5y = 1$) gives

$$8y = 1$$

or

$$y = \frac{1}{8}$$

To get x, we substitute the value of y back into either equation and find

$$x = \frac{3}{8}$$

Thus, the probability distribution is the one shown in the following table.

Outcome	1	2	3	4	5	6
Probability	$\frac{1}{8}$	$\frac{1}{8}$	$\frac{1}{8}$	$\frac{1}{8}$	$\frac{1}{8}$	$\frac{3}{8}$

We can use the distribution to calculate the probability of an even number coming up by adding the probabilities of the favorable outcomes.

$$P(\{2,\ 4,\ 6\}) = \frac{1}{8} + \frac{1}{8} + \frac{3}{8} = \frac{5}{8}$$

Thus there is a $5/8 = .625$ chance that an even number will come up.

$+$*Before we go on...* We should check that the probability distribution in Example 6 satisfies the requirements: 6 is indeed three times as likely to come up as any other number. Also, the probabilities we calculated do add up to 1:

$$\frac{1}{8} + \frac{1}{8} + \frac{1}{8} + \frac{1}{8} + \frac{1}{8} + \frac{3}{8} = 1 \quad\blacksquare$$

FAQs   Distinguishing Theoretical Probability from Estimated Probability

Q: *How do I know whether a given probability is estimated or theoretical*?

A: Ask yourself this: Has the probability been arrived at experimentally, by performing a number of trials and counting the number of times the event occurred? If so, the probability is estimated. If, on the other hand, the probability was computed by analyzing the experiment under consideration rather than by performing actual trials of the experiment, it is theoretical. ∎

Q: *Out of every 100 homes, 22 have broadband Internet service. Thus, the probability that a house has broadband service is .22. Is this probability estimated or theoretical*?

A: That depends on how the ratio 22 out of 100 was arrived at. If it is based on a poll of all homes, then the probability is theoretical. If it is based on a survey of only a sample of homes, it is estimated (see the Q/A following Example 3). ∎

7.2 | EXERCISES

● denotes basic skills exercises

◆ denotes challenging exercises

tech Ex indicates exercises that should be solved using technology

In Exercises 1–4, calculate the estimated probability $P(E)$ using the given information.

1. ● $N = 100$, $fr(E) = 40$ **2.** ● $N = 500$, $fr(E) = 300$

3. ● 800 adults are polled and 640 of them support universal health-care coverage. E is the event that an adult supports universal health coverage. *hint* [see Example 1]

4. ● 800 adults are polled and 640 of them support universal health-care coverage. E is the event that an adult does not support universal health coverage.

Exercises 5–10 are based on the following table, which shows the frequency of outcomes when two distinguishable coins were tossed 4000 times and the uppermost faces were observed.

Outcome	HH	HT	TH	TT
Frequency	1100	950	1200	750

5. ● Determine the estimated probability distribution.
hint [see Example 2]

6. ● What is the estimated probability that heads comes up at least once?

7. ● What is the estimated probability that the second coin lands with heads up?

8. ● What is the estimated probability that the first coin lands with heads up?

9. ● Would you judge the second coin to be fair? Give a reason for your answer.

10. ● Would you judge the first coin to be fair? Give a reason for your answer.

Exercises 11–14 require the use of a calculator or computer with a random number generator.

11. ● Simulate 100 tosses of a fair coin, and compute the estimated probability that heads comes up.

12. ● Simulate 100 throws of a fair die, and calculate the estimated probability that the result is a 6.

13. ● Simulate 50 tosses of two coins, and compute the estimated probability that the outcome is one head and one tail (in any order).

14. ● Simulate 100 throws of two fair dice, and calculate the estimated probability that the result is a double 6.

● basic skills   ◆ challenging   **tech Ex** technology exercise

In Exercises 15–20, calculate the (theoretical) probability $P(E)$ using the given information, assuming that all outcomes are equally likely. hint [see Quick Examples on p. 462]

15. ● $n(S) = 20$, $n(E) = 5$ **16.** ● $n(S) = 8$, $n(E) = 4$

17. ● $n(S) = 10$, $n(E) = 10$ **18.** ● $n(S) = 10$, $n(E) = 0$

19. ● $S = \{a, b, c, d\}$, $E = \{a, b, d\}$

20. ● $S = \{1, 3, 5, 7, 9\}$, $E = \{3, 7\}$

In Exercises 21–30, an experiment is given together with an event. Find the (theoretical) probability of each event, assuming that the coins and dice are distinguishable and fair, and that what is observed are the faces or numbers uppermost. (Compare with Exercises 1–10 in Section 7.1.)

21. ● Two coins are tossed; the result is at most one tail.

22. ● Two coins are tossed; the result is one or more heads.

23. ● Three coins are tossed; the result is at most one head.

24. ● Three coins are tossed; the result is more tails than heads.

25. ● Two dice are rolled; the numbers add to 5.

26. ● Two dice are rolled; the numbers add to 9.

27. ● Two dice are rolled; the numbers add to 1.

28. ● Two dice are rolled; one of the numbers is even, the other is odd.

29. ● Two dice are rolled; both numbers are prime.[7]

30. ● Two dice are rolled; neither number is prime.

31. ● If two indistinguishable dice are rolled, what is the probability of the event {(4, 4), (2, 3)}? What is the corresponding event for a pair of distinguishable dice? hint [see Example 5]

32. ● If two indistinguishable dice are rolled, what is the probability of the event {(5, 5), (2, 5), (3, 5)}? What is the corresponding event for a pair of distinguishable dice?

33. A die is weighted in such a way that each of 2, 4, and 6 is twice as likely to come up as each of 1, 3, and 5. Find the probability distribution. What is the probability of rolling less than 4? hint [see Example 6]

34. Another die is weighted in such a way that each of 1 and 2 is three times as likely to come up as each of the other numbers. Find the probability distribution. What is the probability of rolling an even number?

35. A tetrahedral die has 4 faces, numbered 1–4. If the die is weighted in such a way that each number is twice as likely to land facing down as the next number (1 twice as likely as 2, 2 twice as likely as 3, and so on) what is the probability distribution for the face landing down?

36. A dodecahedral die has 12 faces, numbered 1–12. If the die is weighted in such a way that 2 is twice as likely to land facing up as 1, 3 is three times as likely to land facing up as 1, and so on, what is the probability distribution for the face landing up?

Applications

37. ● **Fast Food** In 2004, the U.S. produced approximately 25 billion pounds of beef and 46 billion pounds of potatoes.[8] McDonald's purchased approximately 1 billion pounds of each.

  a. Find the estimated probability that a pound of beef produced in the U.S. is purchased by McDonald's.

  b. Find the estimated probability that a pound of potatoes produced in the U.S. is not purchased by McDonald's.

38. ● **Fast Food** In 2005, McDonald's purchased approximately 225 million pounds of fruits and vegetables, which included 110 million pounds of lettuce and 50 million of tomatoes.[9]

  a. Find the estimated probability that a randomly selected pound of fruits and vegetables purchased by McDonald's is tomatoes.

  b. Find the estimated probability that a randomly selected pound of fruits and vegetables purchased by McDonald's is not lettuce.

39. ● **Motor Vehicle Safety** The following table shows crashworthiness ratings for 10 small SUVs.[10] (3=Good, 2=Acceptable, 1=Marginal, 0=Poor)

Frontal Crash Test Rating	3	2	1	0
Frequency	1	4	4	1

  a. Find the estimated probability distribution for the experiment of choosing a small SUV at random and determining its frontal crash rating.

  b. What is the estimated probability that a randomly selected small SUV will have a crash test rating of "Acceptable" or better?

40. ● **Motor Vehicle Safety** The following table shows crashworthiness ratings for 16 small cars.[11] (3=Good, 2=Acceptable, 1=Marginal, 0=Poor)

Frontal Crash Test Rating	3	2	1	0
Frequency	1	11	2	2

---

[7] A positive integer is prime if it is neither 1 nor a product of smaller integers.

[8] Source: Department of Agriculture/*New York Times,* February 20, 2005, p. BU8.

[9] Figures are in part estimates for 2005 and in part 2004 figures. Source: McDonald's/*New York Times,* February 20, 2005, p. BU8.

[10] Ratings by the Insurance Institute for Highway Safety. Sources: Oak Ridge National Laboratory: "An Analysis of the Impact of Sport Utility Vehicles in the United States" Stacy C. Davis, Lorena F. Truett, (August 2000)/Insurance Institute for Highway Safety http://www-cta.ornl.gov/Publications/Final SUV report.pdf.

[11] Ratings by the Insurance Institute for Highway Safety. Sources: Oak Ridge National Laboratory: "An Analysis of the Impact of Sport Utility Vehicles in the United States" Stacy C. Davis, Lorena F. Truett, (August 2000)/Insurance Institute for Highway Safety http://www-cta.ornl.gov/Publications/Final SUV report.pdf http://www.highwaysafety.org/vehicle_ratings.

● basic skills   ◆ challenging   **tech Ex** technology exercise

a. Find the estimated probability distribution for the experiment of choosing a small car at random and determining its frontal crash rating.

b. What is the estimated probability that a randomly selected small car will have a crash test rating of "Marginal" or worse?

41. ● *Internet Connections* The following pie chart shows results of a survey of 1000 U.S. households with Internet connections.[12]

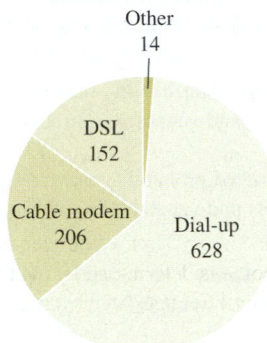

Other
14

DSL
152

Cable modem
206

Dial-up
628

a. Determine the estimated probability of each type of Internet connection for the experiment of choosing a household with an Internet connection and determining the type of connection. (Round probabilities to two decimal places.)

b. What is the estimated probability that a household with an Internet connection has either a cable modem or DSL connection?

42. ● *Internet Connections* The following pie chart shows results of a survey of 1000 rural U.S. households with Internet connections.[13]

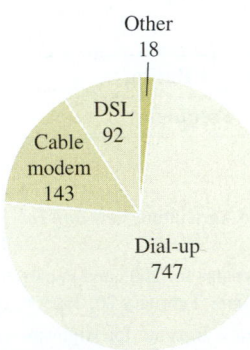

Other
18

DSL
92

Cable modem
143

Dial-up
747

a. Determine the estimated probability of each type of Internet connection for the experiment of choosing a rural

household with an Internet connection and determining the type of connection. (Round probabilities to two decimal places.)

b. What is the estimated probability that a rural household with an Internet connection has neither a cable modem nor DSL connection?

43. ● *Stock Index* The following table shows the closing values of the Dow Jones Industrial Average at the end of each month over the 10-month period beginning May 1, 2004.[14]

Month	1	2	3	4	5	6	7	8	9	10
Dow	10,300	10,200	10,300	10,200	10,200	10,200	10,100	10,500	10,800	10,600

Use these data to construct the estimated probability distribution using the following three outcomes. Low: the Dow is at or below 10,200; Middle: the Dow is above 10,200 but not above 10,500; High: the Dow is above 10,500.

44. ● *Stock Index* The following table shows the closing values of the NASDAQ Composite Index at the end of each month over the 10-month period beginning May 1, 2004.[15]

Month	1	2	3	4	5	6	7	8	9	10
Index	1940	1980	2020	1860	1860	1900	1980	2120	2160	2080

Use these data to construct the estimated probability distribution using the following three outcomes. Low: the NASDAQ is at or below 1900; Middle: the NASDAQ is above 1900 but not above 2000; High: the NASDAQ is above 2000.

Publishing Exercises 45–54 are based on the following table, which shows the results of a survey of 100 authors by a publishing company.

	New Authors	Established Authors	Total
Successful	5	25	30
Unsuccessful	15	55	70
Total	20	80	100

Compute the estimated probabilities of the given events if an author as specified is chosen at random.

45. An author is established and successful.

46. An author is unsuccessful and new.

47. An author is a new author.

48. An author is successful.

49. An author is unsuccessful.

50. An author is established.

51. A successful author is established.

52. An unsuccessful author is established.

[12] Based on a 2003 survey. SOURCE: A Nation Online: Entering the Broadband Age, US Department of Commerce, September 2004 www.ntia.doc.gov/reports/anol/index.html.

[13] Ibid.

[14] Values are rounded. SOURCE: money.excite.com.

[15] Ibid.

● basic skills   ◆ challenging   **tech** Ex technology exercise

53. An established author is successful.

54. A new author is unsuccessful.

55. *Public Health* A random sampling of chicken in supermarkets revealed that approximately 80% was contaminated with the organism *Campylobacter*.[16] Of the contaminated chicken, 20% had the strain resistant to antibiotics. Construct an estimated probability distribution showing the following outcomes when chicken is purchased at a supermarket: *U:* the chicken is not infected with *Campylobacter*; *C:* the chicken is infected with nonresistant *Campylobacter*; *R:* the chicken is infected with resistant *Campylobacter*.

56. *Public Health* A random sampling of turkey in supermarkets found 58% to be contaminated with *Campylobacter*, and 84% of those to be resistant to antibiotics.[17] Construct an estimated probability distribution showing the following outcomes when turkey is purchased at a supermarket: *U:* the turkey is not infected with *Campylobacter*; *C:* the turkey is infected with nonresistant *Campylobacter*; and *R:* the turkey is infected with resistant *Campylobacter*.

57. *Organic Produce* A 2001 Agriculture Department study of more than 94,000 samples from more than 20 crops showed that 73% of conventionally grown foods had residues from at least one pesticide. Moreover, conventionally grown foods were six times as likely to contain multiple pesticides as organic foods. Of the organic foods tested, 23% had pesticide residues, which includes 10% with multiple pesticide residues.[18] Compute two probability distributions: one for conventional produce and one for organic produce, showing the estimated probabilities that a randomly selected product has no pesticide residues, has residues from a single pesticide, and has residues from multiple pesticides.

58. *Organic Produce* Repeat Exercise 57 using the following information for produce from California: 31% of conventional food and 6.5% of organic food had residues from at least one pesticide. Assume that, as in Exercise 57, conventionally grown foods were six times as likely to contain multiple pesticides as organic foods. Also assume that 3% of the organic food has residues from multiple pesticides.

59. *Steroids Testing* A pharmaceutical company is running trials on a new test for anabolic steroids. The company uses the test on 400 athletes known to be using steroids and 200 athletes known not to be using steroids. Of those using steroids, the new test is positive for 390 and negative for 10. Of those not using steroids, the test is positive for 10 and negative for 190. What is the estimated probability of a **false negative** result

(the probability that an athlete using steroids will test negative)? What is the estimated probability of a **false positive** result (the probability that an athlete not using steroids will test positive)?

60. *Lie Detectors* A manufacturer of lie detectors is testing its newest design. It asks 300 subjects to lie deliberately and another 500 to tell the truth. Of those who lied, the lie detector caught 200. Of those who told the truth, the lie detector accused 200 of lying. What is the estimated probability of the machine wrongly letting a liar go, and what is the probability that it will falsely accuse someone who is telling the truth?

61. ◆ techEx *Public Health* Refer back to Exercise 55. Simulate the experiment of selecting chicken at a supermarket and determining the following outcomes: *U:* The chicken is not infected with *campylobacter; C:* The chicken is infected with nonresistant *campylobacter; R:* The chicken is infected with resistant *campylobacter.* [Hint: Generate integers in the range 1–100. The outcome is determined by the range. For instance if the number is in the range 1–20, regard the outcome as *U,* etc.]

62. ◆ techEx *Public Health* Repeat the preceding exercise, but use turkeys and the data given in Exercise 56.

Sales The following table shows sales, in thousands of units, of iPods®, Powerbooks®, and iMacs® in the fourth quarter of 2004.[19]

	iPods	Powerbooks	iMacs
Units (1000)	1030	210	230

In Exercises 63–68, use this information to compute the probabilities of the given events, rounded to the nearest .01. hint [see Example 3]

63. ● An item sold in 2004 Q4 was an iPod.

64. ● An item sold in 2004 Q4 was an iMac.

65. ● An item sold in 2004 Q4 was either an iPod or an iMac.

66. ● An item sold in 2004 Q4 was either Powerbook or an iMac.

67. ● An item sold in 2004 Q4 was not a Powerbook.

68. ● An item sold in 2004 Q4 was not an iPod.

69. ● *Ethnic Diversity* The following pie-chart shows the ethnic makeup of California schools in the 2000–2001 academic year.[20]

---

[16] *Campylobacter* is one of the leading causes of food poisoning in humans. Thoroughly cooking the meat kills the bacteria. SOURCE: *The New York Times,* October 20, 1997, p. A1. Publication of this article first brought *Campylobacter* to the attention of a wide audience.

[17] Ibid.

[18] The 10% figure is an estimate SOURCE: *New York Times,* May 8, 2002, p. A29.

---

[19] Figures are rounded. SOURCE: Company report www.apple.com January 2005.

[20] SOURCE: CBEDS data collection, Educational Demographics, October 2000. www.cde.ca.gov/resrc/factbook/ethnicpies.htm.

● basic skills   ◆ challenging   techEx technology exercise

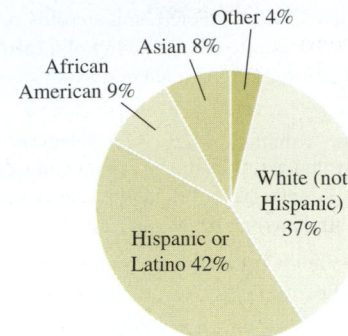

Write down the probability distribution showing the probability that a randomly selected California student in 2000–2001 belonged to one of the ethnic groups named. What is the probability that a student is neither white nor Asian?

70. ● ***Ethnic Diversity*** (Compare Exercise 69.) The following pie-chart shows the ethnic makeup of California schools in the 1981–1982 academic year.[21]

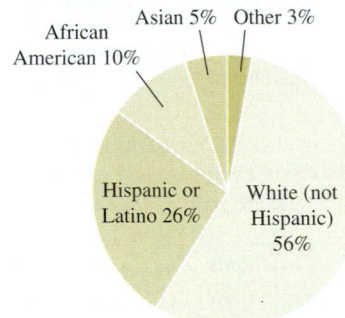

Write down the probability distribution showing the probability that a randomly selected California student in 1981–1982 belonged to one of the ethnic groups named. What is the probability that a student is neither Hispanic, Latino, nor African American?

71. ***Internet Investments in the 90s*** The following excerpt is from an article in *The New York Times* in July, 1999.[22]

> While statistics are not available for Web entrepreneurs who fail, the venture capitalists that finance such Internet start-up companies have a rule of thumb. For every 10 ventures that receive financing—and there are plenty who do not—2 will be stock market successes, which means spectacular profits for early investors; 3 will be sold to other concerns, which translates into more modest profits; and the rest will fail.

a. What is a sample space for the scenario?
b. Write down the associated probability distribution.

c. What is the probability that a start-up venture that receives financing will realize profits for early investors?

72. ◆ ***Internet Investments in the 90s*** The following excerpt is from an article in *The New York Times* in July, 1999.[23]

> Right now, the market for Web stocks is sizzling. Of the 126 initial public offerings of Internet stocks priced this year, 73 are trading above the price they closed on their first day of trading ... Still, 53 of the offerings have failed to live up to their fabulous first-day billings, and 17 [of these] are below the initial offering price.

Assume that, on the first day of trading, all stocks closed higher than their initial offering price.

a. What is a sample space for the scenario?
b. Write down the associated probability distribution. (Round your answers to two decimal places.)
c. What is the probability that an Internet stock purchased during the period reported ended either below its initial offering price or above the price it closed on its first day of trading? *hint* [see Example 3]

73. ● ***Market Share: Light Vehicles*** In 2003, 25% of all light vehicles sold (SUVs, pickups, passenger cars, and minivans) in the U.S. were SUVs and 15% were pickups. Moreover, a randomly chosen vehicle sold that year was five times as likely to be a passenger car as a minivan.[24] Find the associated probability distribution.

74. ● ***Market Share: Light Vehicles*** In 2000, 15% of all light vehicles (SUVs, pickups, passenger cars, and minivans) sold in the U.S. were pickups and 55% were passenger cars. Moreover, a randomly chosen vehicle sold that year was twice as likely to be an SUV as a minivan.[25] Find the associated probability distribution.

Gambling In Exercises 75–82 are detailed some of the nefarious dicing practices of the Win Some/Lose Some Casino. In each case, find the probabilities of all the possible outcomes and also the probability that an odd number or an odd sum faces up.

75. ● Some of the dice are specially designed so that 1 and 6 never come up, and all the other outcomes are equally likely.

---

[21] Ibid.
[22] Article: "Not All Hit It Rich in the Internet Gold Rush," *The New York Times*, July 20, 1999, p. A1.

[23] Article: Ibid. Source for data: Comm-Scan/*The New York Times*, July 20, 1999, p. A1.
[24] Source: Environmental Protection Agency/*The New York Times*, Jun 28, 2003.
[25] Ibid.

● basic skills   ◆ challenging   **tech** Ex technology exercise

76. ● Other dice are specially designed so that 1 comes up half the time, 6 never comes up, and all the other outcomes are equally likely.

77. ● Some of the dice are cleverly weighted so that each of 2, 3, 4, and 5 is twice as likely to come up as 1 is, and 1 and 6 are equally likely.

78. ● Other dice are weighted so that each of 2, 3, 4, and 5 is half as likely to come up as 1 is, and 1 and 6 are equally likely.

79. Some pairs of dice are magnetized so that each pair of mismatching numbers is twice as likely to come up as each pair of matching numbers.

80. Other pairs of dice are so strongly magnetized that mismatching numbers never come up.

81. Some dice are constructed in such a way that deuce (2) is five times as likely to come up as 4 and three times as likely to come up as each of 1, 3, 5, and 6.

82. Other dice are constructed in such a way that deuce (2) is six times as likely to come up as 4 and four times as likely to come up as each of 1, 3, 5, and 6.

Communication and Reasoning Exercises

83. ● Complete the following. The estimated probability of an event E is defined to be _____.

84. ● Interpret the popularity rating of the student council president as an estimated probability by specifying an appropriate experiment and also what is observed.

85. ● Ruth tells you that when you roll a pair of fair dice, the theoretical probability of obtaining a pair of matching numbers is 1/6. To test this claim, you roll a pair of fair dice 20 times, and never once get a pair of matching numbers. This proves that either Ruth is wrong or the dice are not fair, right?

86. ● Design an experiment based on rolling a fair die for which there are at least three outcomes with different probabilities.

87. How would you measure the estimated probability that the weather service accurately predicts the next day's temperature?

88. Suppose that you toss a coin 100 times and get 70 heads. If you continue tossing the coin, the estimated probability of heads overall should approach 50% if the coin is fair. Will you have to get more tails than heads in subsequent tosses to "correct" for the 70 heads you got in the first 100 tosses?

89. Tony has had a "losing streak" at the casino—the chances of winning the game he is playing are 40%, but he has lost 5 times in a row. Tony argues that, because he should have won 2 times, the game must obviously be "rigged." Comment on his reasoning.

90. Maria is on a winning streak at the casino. She has already won four times in a row and concludes that her chances of winning a fifth time are good. Comment on her reasoning.

● basic skills   ◆ challenging   *tech* Ex technology exercise

7.3 | Properties of Probability Distributions

Estimated and theoretical probability can be thought of as two different ways, appropriate in different situations, to find a **probability distribution,** an assignment of probabilities to all the outcomes in a sample space. We record the common properties of probability distributions.

Probability Distribution

A (finite) **probability distribution** is an assignment of a number $P(s_i)$, the **probability of s_i,** to each outcome of a finite sample space $S = \{s_1, s_2, \ldots, s_n\}$. The probabilities must satisfy

1. $0 \le P(s_i) \le 1$

   and

2. $P(s_1) + P(s_2) + \cdots + P(s_n) = 1$.

We find the probability of an event E, written $P(E)$, by adding up the probabilities of the outcomes in E.

If $P(E) = 0$, we call E an **impossible event.** The event $\emptyset$ is always impossible, since *something* must happen.

quick Examples

1. All the examples of estimated and theoretical probability distributions we have considered are examples of probability distributions.

2. Let us take $S = \{H, T\}$ and make the assignments $P(H) = .2$, $P(T) = .8$. Because these numbers are between 0 and 1 and add to 1, they specify a probability distribution.

3. With $S = \{H, T\}$ again, we could also take $P(H) = 1$, $P(T) = 0$, so that T is an impossible event.

4. The following table gives a probability distribution for the sample space

$$S = \{1, 2, 3, 4, 5, 6\}$$

Outcome	1	2	3	4	5	6
Probability	.3	.3	0	.1	.2	.1

It follows that

$P(\{1, 6\}) = .3 + .1 = .4$

$P(\{2, 3\}) = .3 + 0 = .3$

$P(3) = 0$ An impossible event

The distribution in Quick Example 4 could be the probability distribution of a weighted die. Whether or not a die with these exact (theoretical) probabilities for outcomes can actually be manufactured is a question we leave for die engineers to ponder. As mathematicians, we can certainly *conceive* of such a die, and that is all the justification we need to consider its probability distribution. By studying probability distributions in the abstract, without regard for how they originate, we can find properties that all distributions must satisfy, whether estimated or theoretical.

The Probability of the Union of Two Events

For example, so far, all we know about computing the probability of an event E is that $P(E)$ is the sum of the probabilities of the individual outcomes in E. Suppose, though, that we do not know the probabilities of the individual outcomes in E but we do know that $E = A \cup B$, where we happen to know $P(A)$ and $P(B)$. How do we compute the probability of $A \cup B$? We might be tempted to say that $P(A \cup B)$ is $P(A) + P(B)$, but let us look at an example using the probability distribution in Quick Example 4. For A let us take the event $\{2, 4, 5\}$, and for B let us take $\{2, 4, 6\}$. $A \cup B$ is then the event $\{2, 4, 5, 6\}$. We know that we can find the probabilities $P(A)$, $P(B)$, and $P(A \cup B)$ by adding the probabilities of all the outcomes in these events, so

$P(A) = P(\{2, 4, 5\}) = .3 + .1 + .2 = .6$

$P(B) = P(\{2, 4, 6\}) = .3 + .1 + .1 = .5$, and

$P(A \cup B) = P(\{2, 4, 5, 6\}) = .3 + .1 + .2 + .1 = .7$

Our first guess was wrong: $P(A \cup B) \neq P(A) + P(B)$. Notice, however, that the outcomes in $A \cap B$ are counted twice in computing $P(A) + P(B)$, but only once in computing $P(A \cup B)$:

$$P(A) + P(B) = P(\{2, 4, 5\}) + P(\{2, 4, 6\}) \qquad A \cap B = \{2, 4\}$$
$$= (.3 + .1 + .2) + (.3 + .1 + .1) \qquad P(A \cap B) \text{ counted twice}$$
$$= 1.1$$

whereas

$$P(A \cup B) = P(\{2, 4, 5, 6\}) = .3 + .1 + .2 + .1 \qquad P(A \cap B) \text{ counted once}$$
$$= .7$$

Thus, if we take $P(A) + P(B)$ and then subtract the surplus $P(A \cap B)$, we get $P(A \cup B)$. In symbols,

$$P(A \cup B) = P(A) + P(B) - P(A \cap B)$$
$$.7 = .6 + .5 - .4$$

(See Figure 5.) We call this formula the **addition principle.** One more thing: Notice that our original guess $P(A \cup B) = P(A) + P(B)$ would have worked if we had chosen A and B with no outcomes in common; that is, if $A \cap B = \emptyset$. When $A \cap B = \emptyset$, recall that we say that A and B are mutually exclusive.

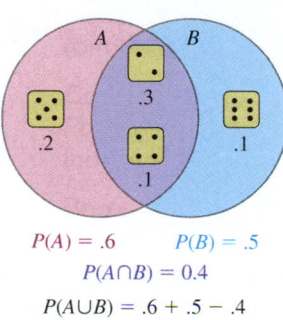

$P(A) = .6 \qquad P(B) = .5$
$P(A \cap B) = 0.4$
$P(A \cup B) = .6 + .5 - .4$

Figure **5**

Addition Principle

If A and B are any two events, then

$$P(A \cup B) = P(A) + P(B) - P(A \cap B)$$

Visualizing the Addition Principle
In the figure, the area of the union is obtained by adding the areas of A and B and then subtracting the overlap (because it is counted twice when we add the areas).

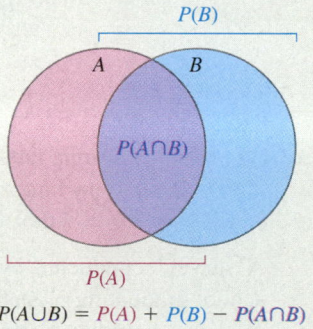

$P(A \cup B) = P(A) + P(B) - P(A \cap B)$

Addition Principle for Mutually Exclusive Events
If $A \cap B = \emptyset$, we say that A and B are **mutually exclusive,** and we have

$$P(A \cup B) = P(A) + P(B) \qquad \text{Because } P(A \cap B) = 0$$

Visualizing the Addition Principle for Mutually Exclusive Events
If A and B do not overlap, then the area of the union is obtained by adding the areas of A and B.

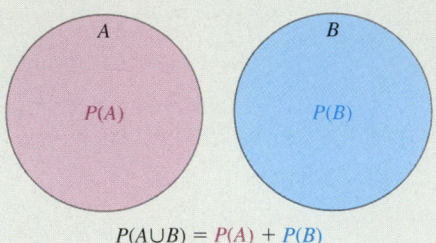

$$P(A \cup B) = P(A) + P(B)$$

This holds true also for more than two events: If $A_1, A_2, \ldots, A_n$ are mutually exclusive events (that is, the intersection of every pair of them is empty), then

$$P(A_1 \cup A_2 \cup \ldots \cup A_n) = P(A_1) + P(A_2) + \cdots + P(A_n)$$

Addition principle for many mutually exclusive events.

quick Examples

1. There is a 10% chance of rain (R) tomorrow, a 20% chance of high winds (W), and a 5% chance of both. The probability of either rain or high winds (or both) is

$$P(R \cup W) = P(R) + P(W) - P(R \cap W)$$
$$= .10 + .20 - .05 = .25$$

2. The probability that you will be in Cairo at 6 A.M. tomorrow (C) is .3, while the probability that you will be in Alexandria at 6 A.M tomorrow (A) is .2. Thus, the probability that you will be either in Cairo or Alexandria at 6 A.M. tomorrow is

$$P(C \cup A) = P(C) + P(A) \qquad \text{\textit{A} and \textit{C} are mutually exclusive.}$$
$$= .3 + .2 = .5$$

3. When a pair of fair dice is rolled, the probability of the numbers that face up adding to 7 is 6/36, the probability of their adding to 8 is 5/36, and the probability of their adding to 9 is 4/36. Thus, the probability of the numbers adding to 7, 8, or 9 is

$$P(7 \cup 8 \cup 9) = P(7) + P(8) + P(9) \qquad \text{The events are mutually exclusive.*}$$
$$= \frac{6}{36} + \frac{5}{36} + \frac{4}{36} = \frac{15}{36} = \frac{5}{12}$$

---

*The sum of the numbers that face up cannot equal two different numbers at the same time.

Example 1 School and Work

A survey[†] conducted by the Bureau of Labor Statistics found that 64% of the high school graduating class of 2003 went on to college the following year, while 55% of the class was working. Furthermore, 92% were either in college or working, or both.

a. What percentage went on to college and work at the same time?

b. What percentage went on to college but not work?

[†] "College Enrollment and Work Activity of High School Graduates," U.S. Bureau of Labor Statistics, April 2004, available at www.bls.gov/schedule/archives/all_nr.htm.

Solution We can think of the experiment of choosing a member of the high school graduating class of 2003 at random. The sample space is the set of all these graduates.

a. We are given information about two events:

A: A graduate went on to college; $P(A) = .64$

B: A graduate went on to work; $P(B) = .55$

We are also told that $P(A \cup B) = .92$. We are asked for the probability that a graduate went on to both college and work, $P(A \cap B)$. To find $P(A \cap B)$, we take advantage of the fact that the formula

$$P(A \cup B) = P(A) + P(B) - P(A \cap B)$$

can be used to calculate any one of the four quantities that appear in it as long as we know the other three. Substituting the quantities we know, we get

$$.92 = .64 + .55 - P(A \cap B)$$

so

$$P(A \cap B) = .64 + .55 - .92 = .27$$

Thus, 27% of the graduates went on to college and work at the same time.

b. We are asked for the probability of a new event:

C: A graduate went on to college but not work.

C is the part of A outside of $A \cap B$, so $C \cup (A \cap B) = A$, and C and $A \cap B$ are mutually exclusive. (See Figure 6.)

Thus, applying the addition principle, we have

$$P(C) + P(A \cap B) = P(A)$$

From part (a), we know that $P(A \cap B) = .27$, so

$$P(C) + .27 = .64$$

giving

$$P(C) = .37$$

In other words, 37% of the graduates went on to college but not work.

You can use the formula $P(A \cup B) = P(A) + P(B) - P(A \cap B)$ to calculate any of the 4 quantities in the formula if you know the other 3.

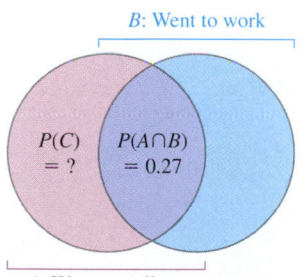

B: Went to work

$P(C)$ = ? $P(A \cap B)$ = 0.27

A: Went to college
$P(A) = 0.64$

Figure **6**

We can use the addition principle to deduce other useful properties of a probability distribution.

More Principles of Probability Distributions

The following rules hold for any sample space S and any event A:

$$P(S) = 1 \qquad \text{The probability of } \textit{something} \text{ happening is 1.}$$

$$P(\emptyset) = 0 \qquad \text{The probability of } \textit{nothing} \text{ happening is 0.}$$

$$P(A') = 1 - P(A) \qquad \text{The probability of } A \textit{ not} \text{ happening is 1 minus the probability of } A.$$

Note

We can also write the third equation as

$$P(A) = 1 - P(A')$$

or

$$P(A) + P(A') = 1$$

Visualizing the Rule for Complements

Think of A' as the portion of S outside of A. Adding the two areas gives the area of all of S, equal to 1.

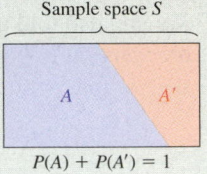

Sample space S

$P(A) + P(A') = 1$

quick Examples

1. There is a 10% chance of rain (R) tomorrow. Therefore, the probability that it will *not* rain is

$$P(R') = 1 - P(R) = 1 - .10 = .90$$

2. The probability that Eric Ewing will score at least two goals is .6. Therefore, the probability that he will score at most one goal is $1 - .6 = .4$.

Q: Can you persuade me that all of these principles are true?

A: Let us take them one at a time. ∎

We know that $S = \{s_1, s_2, \ldots, s_n\}$ is the set of all outcomes, and so

$$P(S) = P(\{s_1, s_2, \ldots, s_n\})$$
$$= P(s_1) + P(s_2) + \cdots + P(s_n)$$
$$= 1$$

We add the probabilities of the outcomes to obtain the probability of an event.
By the definition of a probability distribution

Now, note that $S \cap \emptyset = \emptyset$, so that S and $\emptyset$ are mutually exclusive. Applying the addition principle gives

$$P(S) = P(S \cup \emptyset) = P(S) + P(\emptyset)$$

Subtracting $P(S)$ from both sides gives $0 = P(\emptyset)$.

If A is any event in S, then we can write

$$S = A \cup A'$$

where A and A' are mutually exclusive. (Why?) Thus, by the addition principle,

$$P(S) = P(A) + P(A')$$

Because $P(S) = 1$, we get

$$1 = P(A) + P(A')$$

or $P(A') = 1 - P(A)$

Example 2 Employment

In February, 2005, the probability that a randomly selected U.S. resident of working age (16 and over) was employed was approximately .62.[*] The probability that the resident of

* SOURCE: Bureau of Labor Statistics press release, January 4, 2002, obtained from www.bls.gov.

working age was unemployed but actively searching for a job was approximately .05. Calculate the probabilities of the following events.

a. A resident of working age was unemployed.

b. A resident was unemployed and not actively searching for a job.

Solution

a. Let us write E for the event that a resident (of working age) was employed. We are given that $P(E) = .62$. The event that the resident was *un*employed is the complement of E, and its probability is given by

$$P(E') = 1 - P(E) = 1 - .62 = .38$$

b. We are given the probability that a resident was unemployed and searching for a job, and are asked to find the probability that a resident was unemployed and *not* searching for a job. The two corresponding events

J: A resident was unemployed and searching for a job; $P(J) = .05$

N: A resident was unemployed and not searching for a job

are mutually exclusive events whose union is E', the set of all unemployed residents. Hence,

$$P(E') = P(N) + P(J)$$
$$.38 = P(N) + .05$$

giving

$$P(N) = .38 - .05 = .33$$

Thus, there was a 33% chance that a U.S. resident of working age was unemployed and not searching for a job.

The next example is identical to Example 4 in Section 7.2. What is different is the way in which we calculate some of the answers.

Example 3 iPods, iMacs and Powerbooks

The following table shows sales, in thousands of units, of iPods®, Powerbooks®, and iMacs® in the fourth quarter of 2003 and in the third and fourth quarters of 2004.[*]

	iPods	Powerbooks	iMacs	Total
2003 Q4	340	180	250	770
2004 Q3	860	220	240	1320
2004 Q4	1030	210	230	1470
Total	2230	610	720	3560

[*] Figures are rounded. SOURCE: Company report www.apple.com January 2005.

If one of the items is selected at random, find the probabilities of the following events:

a. It is an iPod.

b. It was sold in the third quarter of 2004.

c. It is an iPod sold in the third quarter of 2004.

d. It is either an iPod or was sold in the third quarter of 2004.

e. It is not an iPod.

Solution The answers to parts (a), (b), and (c) are calculated exactly as in Example 4 in Section 7.2.

a. If D is the event that the selected item is an iPod, then

$$P(D) = \frac{n(D)}{n(S)} = \frac{2230}{3560} \approx .626$$

b. If A is the event that the selected item was sold in the third quarter of 2004, then

$$P(A) = \frac{n(A)}{n(S)} = \frac{1320}{3560} \approx .371$$

c. The event that the selected item is an iPod sold in the third quarter of 2004 is the event $D \cap A$, and

$$P(D \cap A) = \frac{n(D \cap A)}{n(S)} = \frac{860}{3560} \approx .242$$

d. The event that the selected item is either an iPod or was sold in the third quarter of 2004 is the event $D \cup A$, and may be computed using the formula for the probability of the union:

$$P(D \cup A) = P(D) + P(A) - P(D \cap A)$$
$$\approx .626 + .371 - .242 = .755$$

(A more precise answer is .756. The last decimal place in the above computation is off due to the fact that we rounded $P(D)$, $P(A)$, and $P(D \cap A)$.)

e. The event that the selected item is not an iPod is the event D'; its probability may be computed using the formula for the probability of the complement:

$$P(D') = 1 - P(D) \approx 1 - .626 = .374$$

7.3 EXERCISES

● denotes basic skills exercises

◆ denotes challenging exercises

In Exercises 1–16, use the given information to find the indicated probability. hint [see Quick Examples on p. 476]

1. ● $P(A) = .1$, $P(B) = .6$, $P(A \cap B) = .05$. Find $P(A \cup B)$.

2. ● $P(A) = .3$, $P(B) = .4$, $P(A \cap B) = .02$. Find $P(A \cup B)$.

3. ● $A \cap B = \emptyset$, $P(A) = .3$, $P(A \cup B) = .4$. Find $P(B)$.

4. ● $A \cap B = \emptyset$, $P(B) = .8$, $P(A \cup B) = .8$. Find $P(A)$.

5. ● $A \cap B = \emptyset$, $P(A) = .3$, $P(B) = .4$. Find $P(A \cup B)$.

6. ● $A \cap B = \emptyset$, $P(A) = .2$, $P(B) = .3$. Find $P(A \cup B)$.

7. ● $P(A \cup B) = .9$, $P(B) = .6$, $P(A \cap B) = .1$. Find $P(A)$.

8. ● $P(A \cup B) = 1.0$, $P(A) = .6$, $P(A \cap B) = .1$. Find $P(B)$.

● basic skills   ◆ challenging

9. ● $P(A) = .75$. Find $P(A')$.

10. ● $P(A) = .22$. Find $P(A')$.

11. ● A, B and C are mutually exclusive. $P(A) = .3$, $P(B) = .4$, $P(C) = .3$. Find $P(A \cup B \cup C)$.

12. ● A, B and C are mutually exclusive. $P(A) = .2$, $P(B) = .6$, $P(C) = .1$. Find $P(A \cup B \cup C)$.

13. ● A and B are mutually exclusive. $P(A) = .3$, $P(B) = .4$. Find $P((A \cup B)')$.

14. ● A and B are mutually exclusive. $P(A) = .4$, $P(B) = .4$. Find $P((A \cup B)')$.

15. ● $A \cup B = S$ and $A \cap B = \emptyset$. Find $P(A) + P(B)$.

16. ● $P(A \cup B) = .3$ and $P(A \cap B) = .1$. Find $P(A) + P(B)$.

In Exercises 17–22, determine whether the information shown is consistent with a probability distribution. If not, say why.

17. ● $P(A) = .2$; $P(B) = .1$; $P(A \cup B) = .4$

18. ● $P(A) = .2$; $P(B) = .4$; $P(A \cup B) = .2$

19. ● $P(A) = .2$; $P(B) = .4$; $P(A \cap B) = .2$

20. ● $P(A) = .2$; $P(B) = .4$; $P(A \cap B) = .3$

21. ● $P(A) = .1$; $P(B) = 0$; $P(A \cup B) = 0$

22. ● $P(A) = .1$; $P(B) = 0$; $P(A \cap B) = 0$

23. Complete the following probability distribution table and then calculate the stated probabilities.

Outcome	a	b	c	d	e
Probability	.1	.05	.6	.05	

   a. $P(\{a, c, e\})$.
   b. $P(E \cup F)$, where $E = \{a, c, e\}$ and $F = \{b, c, e\}$,
   c. $P(E')$, where E is as in part (b),
   d. $P(E \cap F)$, where E and F are as in part (b).

24. Repeat the preceding exercise using the following table.

Outcome	a	b	c	d	e
Probability	.1		.65	.1	.05

Applications

25. ● **Astrology** The astrology software package, Turbo Kismet,[26] works by first generating random number sequences and then interpreting them numerologically. When I ran it yesterday, it informed me that there was a 1/3 probability that I would meet a tall dark stranger this month, a 2/3 probability that I would travel this month, and a 1/6 probability that I would meet a tall dark stranger and also travel this month. What is

the probability that I will either meet a tall dark stranger or that I will travel this month? *hint* [see Quick Example 1 on p. 476 and Example 1]

26. ● **Astrology** Another astrology software package, Java Kismet, is designed to help day traders choose stocks based on the position of the planets and constellations. When I ran it yesterday, it informed me that there was a .5 probability that Amazon.com will go up this afternoon, a .2 probability that Yahoo.com will go up this afternoon, and a .2 chance that both will go up this afternoon. What is the probability that either Amazon.com or Yahoo.com will go up this afternoon?

27. ● **Polls** According to a *New York Times*/CBS poll released in March 2005, 61% of those polled ranked jobs or health care as the top domestic priority.[27] What is the probability that a randomly selected person polled did not rank either as the top domestic priority? *hint* [see Example 2]

28. ● **Polls** According to the *New York Times*/CBS poll of March, 2005, referred to in Exercise 27, 72% of those polled ranked neither Iraq nor North Korea as the top foreign policy issue.[28] What is the probability that a randomly selected person polled ranked either Iraq or North Korea as the top foreign policy issue?

29. ● **Resources** In 2003, the probability that a randomly chosen bag of cement from the world supply was consumed in the U.S. was .06, while the probability that it was consumed in China was .40.[29] What is the probability that a randomly chosen bag of cement was consumed in neither country?

30. ● **Resources** In 2003, the probability that a randomly chosen barrel of oil from the world supply was consumed in the U.S. was .25, while the probability that it was consumed in China was .12. What is the probability that a randomly chosen barrel of oil was consumed in neither country?

Student Admissions *Exercises 31–46 are based on the following table, which shows the profile, by Math SAT I scores, of admitted students at UCLA for the Fall 2004 semester.[30]*

Math SAT I

	200–399	400–499	500–599	600–699	700–800	Total
Admitted	7	212	1124	2882	5309	9534
Not Admitted	687	3512	8689	12,230	5150	30,268
Total Applicants	694	3724	9813	15,112	10,459	39,802

Determine the theoretical probabilities of the following events. (Round your answers to the nearest .01). *hint* [see Example 3]

31. ● An applicant was admitted.

32. ● An applicant had a Math SAT below 400.

33. ● An applicant had a Math SAT below 400 and was admitted.

---

[26] The name and concept were borrowed from a hilarious (as yet unpublished) novel by the science-fiction writer William Orr, who also happens to be a faculty member at Hofstra University.

[27] SOURCE: *New York Times*, March 3, 2005, p. A20.

[28] Ibid.

[29] SOURCE: Goldman Sachs/*New York Times*, December 6, 2004, p. C4.

[30] SOURCE: University of California website, February, 2004. www.admissions.ucla.edu/Prospect/Adm_fr/Frosh_Prof04.htm.

● basic skills   ◆ challenging

34. ● An applicant had a Math SAT of 700 or above and was admitted.

35. ● An applicant was not admitted.

36. ● An applicant did not have a Math SAT below 400.

37. ● An applicant had a Math SAT in the range 500–599 or was admitted.

38. ● An applicant had a Math SAT of 700 or above or was admitted.

39. ● An applicant was neither admitted nor had a Math SAT in the range 500–599.

40. ● An applicant neither had a Math SAT of 700 or above nor was admitted.

41. An applicant who had a Math SAT below 400 was admitted.

42. An applicant who had a Math SAT of 700 or above was admitted.

43. An admitted student had a Math SAT of 700 or above.

44. An admitted student had a Math SAT below 400.

45. A rejected applicant had a Math SAT below 600.

46. A rejected applicant had a Math SAT of at least 600.

47. *Social Security* According to the *New York Times*/CBS poll of March, 2005, referred to in Exercise 27, 79% agreed that it should be the government's responsibility to provide a decent standard of living for the elderly, and 43% agreed that it would be a good idea to invest part of their Social Security taxes on their own.[31] What is the smallest percentage of people who could have agreed with both statements? What is the largest percentage of people who could have agreed with both statements?

48. *Social Security* According to the *New York Times*/CBS poll of March, 2005, referred to in Exercise 27, 49% agreed that Social Security taxes should be raised if necessary to keep the system afloat, and 43% agreed that it would be a good idea to invest part of their Social Security taxes on their own.[32] What is the largest percentage of people who could have agreed with at least one of these statements? What is the smallest percentage of people who could have agreed with at least one of these statements?

49. *Greek Life* The ΤΦΦ Sorority has a tough pledging program—it requires its pledges to master the Greek alphabet forward, backward, and "sideways." During the last pledge period, two-thirds of the pledges failed to learn it backward and three quarters of them failed to learn it sideways; 5 of the 12 pledges failed to master it either backward or sideways. Because admission into the sisterhood requires both backward and sideways mastery, what fraction of the pledges were disqualified on this basis?

50. *Swords and Sorcery* Lance the Wizard has been informed that tomorrow there will be a 50% chance of encountering the evil

---

[31] Source: *New York Times,* March 3, 2005, p. A20.
[32] Ibid.

Myrmidons and a 20% chance of meeting up with the dreadful Balrog. Moreover, Hugo the Elf has predicted that there is a 10% chance of encountering both tomorrow. What is the probability that Lance will be lucky tomorrow and encounter neither the Myrmidons nor the Balrog?

51. *Public Health* A study shows that 80% of the population has been vaccinated against the Venusian flu, but 2% of the vaccinated population gets the flu anyway. If 10% of the total population gets this flu, what percent of the population either gets the vaccine or gets the disease?

52. *Public Health* A study shows that 75% of the population has been vaccinated against the Martian ague, but 4% of this group gets this disease anyway. If 10% of the total population gets this disease, what is the probability that a randomly selected person has been neither vaccinated nor has contracted Martian ague?

Communication and Reasoning Exercises

53. ● Complete the following sentence. The probability of the union of two events is the sum of the probabilities of the two events if _____.

54. ● A friend of yours asserted at lunch today that, according to the weather forecast for tomorrow, there is a 52% chance of rain and a 60% chance of snow. "But that's impossible!" you blurted out, "the percentages add up to more than 100%." Explain why you were wrong.

55. A certain experiment is performed a large number of times, and the event E has estimated probability equal to zero. This means that it has theoretical probability zero, right?

56. ◆ Your friend tells you that once lightning has struck a particular spot, it is bound to do so again within the next five thunderstorms. Last night, lightning struck the first hole green at the golf course. What can you say about the theoretical probability that it will strike there again? Explain.

57. Explain how the addition principle for mutually exclusive events follows from the general addition principle.

58. Explain how the property $P(A') = 1 - P(A)$ follows directly from the properties of a probability distribution.

59. ◆ It is said that lightning never strikes twice in the same spot. Assuming this to be the case, what is the theoretical probability that lightning will strike your favorite dining spot during a thunderstorm? Explain.

60. ◆ A certain event has theoretical probability equal to zero. This means it will never occur, right?

61. ◆ Find a formula for the probability of the union of three (not necessarily mutually exclusive) events A, B, and C.

62. ◆ Four events A, B, C, and D have the following property: If any two events have an outcome in common, that outcome is common to all four events. Find a formula for the probability of their union.

● basic skills   ◆ challenging

7.4 Probability and Counting Techniques (OPTIONAL)

We saw in the preceding section that, when all outcomes in a sample space are equally likely, we can use the following formula.

Computing Theoretical Probability: Equally Likely Outcomes

In an experiment in which all outcomes are equally likely, the theoretical probability of an event E is given by

$$P(E) = \frac{\text{Number of favorable outcomes}}{\text{Total number of outcomes}} = \frac{n(E)}{n(S)}$$

This formula is simple, but calculating $n(E)$ and $n(S)$ may not be. In this section, we look at some examples in which we need to use the counting techniques discussed in the preceding chapter.

Example 1 Marbles

A bag contains 4 red marbles and 2 green ones. Upon seeing the bag, Suzy (who has compulsive marble-grabbing tendencies) sticks her hand in and grabs 3 at random. Find the probability that she will get both green marbles.

Solution

According to the formula, we need to know these numbers:

- The number of elements in the sample space S.
- The number of elements in the event E.

S is the set of all outcomes that can occur, and has nothing to do with having green marbles.

First of all, what is the sample space? The sample space is the set of all possible outcomes, and each outcome consists of a set of 3 marbles (in Suzy's hand). So, the set of outcomes is the set of all sets of 3 marbles chosen from a total of 6 marbles (4 red and 2 green). Thus,

$$n(S) = C(6, 3) = 20$$

Now what about E? This is the event that Suzy gets both green marbles. We must *rephrase this as a subset of S* in order to deal with it: "E is the collection of sets of three marbles such that one is red and two are green." Thus, $n(E)$ is the *number* of such sets, which we determine using a decision algorithm.

Step 1 Choose a red marble; $C(4, 1) = 4$ possible outcomes.

Step 2 Choose the two green marbles; $C(2, 2) = 1$ possible outcome.

We get $n(E) = 4 \times 1 = 4$. Now,

$$P(E) = \frac{n(E)}{n(S)} = \frac{4}{20} = \frac{1}{5}$$

Thus, there is a one in five chance of Suzy's getting both the green marbles.

Example 2 Investment Lottery

In order to "spice up" your investment portfolio, you decided to ignore your broker's cautious advice and select three stocks at random from the six most active stocks listed on the New York Stock Exchange on March 1, 2005:[*]

Company	Symbol	Close	Change
Elan	ELN	$7.99	+$0.02
Titan	TTN	$18.15	+$0.80
Nokia	NOK	$10.02	−$0.22
Lear	LEA	$44.80	−$7.96
British Petroleum	BP	$63.59	−$0.52
Pfizer	PFE	$26.77	+$0.16

Find the probabilities of the following events:

a. Your portfolio included ELN and NOK.

b. At most two of the stocks in your portfolio declined in value.

Solution First, the sample space is the set of all collections of 3 stocks chosen from the 6. Thus,

$$n(S) = C(6, 3) = 20$$

a. The event E of interest is the event that your portfolio includes ELN and TTN. Thus, E is the set of all groups of 3 stocks that include ELN and TTN. Because there is only one more stock left to choose,

$$n(E) = C(4, 1) = 4$$

We now have

$$P(E) = \frac{n(E)}{n(S)} = \frac{4}{20} = \frac{1}{5} = .2$$

b. Let F be the event that at most two of the stocks in your portfolio declined in value. Thus, F is the set of all portfolios of three stocks of which at most two declined in value. To calculate $n(F)$, we use the following decision algorithm, noting that three of the stocks declined in value and three did not.

Alternative 1: None of the stocks declined in value.
  Step 1 Choose three stocks that did not decline in value; $C(3, 3) = 1$ possibility.

Alternative 2: One of the stocks declined in value.
  Step 1 Choose one stock that declined in value; $C(3, 1) = 3$ possibilities.
  Step 2 Choose two stocks that did not decline in value; $C(3, 2) = 3$ possibilities.
  This gives $3 \times 3 = 9$ possibilities for this alternative.

Alternative 3: Two of the stocks declined in value.
  Step 1 Choose two stocks that declined in value; $C(3, 2) = 3$ possibilities.
  Step 2 Choose one stock that did not decline in value; $C(3, 1) = 3$ possibilities.
  This gives $3 \times 3 = 9$ possibilities for this alternative.

---

[*] Changes based on pre-market close. Source: Yahoo! Finance (http://finance.yahoo.com), March 2, 2005.

So, we have a total of $1 + 9 + 9 = 19$ possible portfolios. Thus,

$$n(F) = 19$$

and

$$P(F) = \frac{n(F)}{n(S)} = \frac{19}{20} = .95$$

At most 2 of the stocks declined in value.

Complementary events

At least 3 of the stocks declined in value.

+ *Before we go on...* When counting the number of outcomes in an event, the calculation is sometimes easier if we look at the *complement* of that event. In the case of part (b) of the Example 2, the complement of the event F is

F': At least three of the stocks in your portfolio declined in value.

Because there are only three stocks in your portfolio, this is the same as the event that all three stocks in your portfolio declined in value. The decision algorithm for $n(F')$ is far simpler:

Step 1 Choose 3 stocks that declined in value: $C(3, 3) = 1$ possibility.

So, $n(F') = 1$, giving

$$n(F) = n(S) - n(F') = 20 - 1 = 19$$

as we calculated above. ∎

Example 3 Poker Hands

You are dealt 5 cards from a well-shuffled standard deck of 52. Find the probability that you have a full house. (Recall that a full house consists of 3 cards of one denomination and 2 of another.)

Solution The sample space S is the set of all possible 5-card hands dealt from a deck of 52. Thus,

$$n(S) = C(52, 5) = 2,598,960$$

If the deck is thoroughly shuffled, then each of these 5-card hands is equally likely. Now consider the event E, the set of all possible 5-card hands that constitute a full house. To calculate $n(E)$, we use a decision algorithm, which we show in the following compact form.

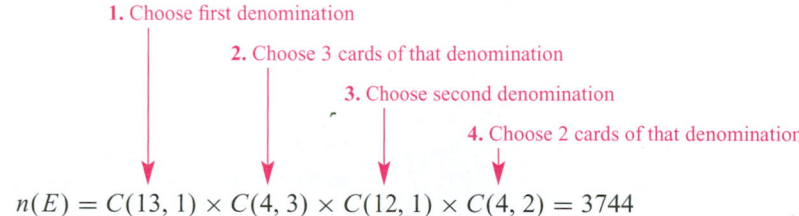

1. Choose first denomination

2. Choose 3 cards of that denomination

3. Choose second denomination

4. Choose 2 cards of that denomination

$$n(E) = C(13, 1) \times C(4, 3) \times C(12, 1) \times C(4, 2) = 3744$$

Thus,

$$P(E) = \frac{n(E)}{n(S)} = \frac{3744}{2,598,960} \approx .00144$$

In other words, there is an approximately 0.144% chance that you will be dealt a full house.

Example 4 More Poker Hands

You are playing poker, and you have been dealt the following hand:

$$J\spadesuit, J\diamondsuit, J\heartsuit, 2\clubsuit, 10\spadesuit$$

You decide to exchange the last two cards. The exchange works as follows: The two cards are discarded (not replaced in the deck) and you are dealt two new cards.

a. Find the probability that you end up with a full house.

b. Find the probability that you end up with four jacks.

c. What is the probability that you end up with either a full house or four jacks?

Solution

a. In order to get a full house, you must be dealt two of a kind. The sample space S is the set of all pairs of cards selected from what remains of the original deck of 52. You were dealt 5 cards originally, so there are $52 - 5 = 47$ cards left in the deck. Thus, $n(S) = C(47, 2) = 1081$. The event E is the set of all pairs of cards that constitute two of a kind. Note that you cannot get two jacks because only one is left in the deck. Also, only three 2s and three 10s are left in the deck. We have

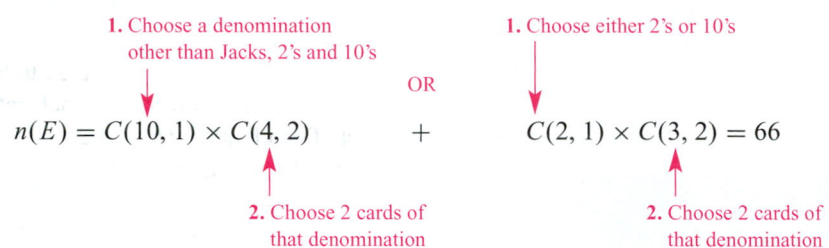

Thus,

$$P(E) = \frac{n(E)}{n(S)} = \frac{66}{1081} \approx .0611$$

b. We have the same sample space as in part (a). Let F be the set of all pairs of cards that include the missing jack of clubs. So,

1. Choose the jack of Clubs

2. Choose 1 card from the remaining 46

$$n(F) = C(1, 1) \times C(46, 1) = 46$$

Thus,

$$P(F) = \frac{n(F)}{n(S)} = \frac{46}{1081} \approx .0426$$

c. We are asked to calculate the probability of the event $E \cup F$. From the addition principle, we have

$$P(E \cup F) = P(E) + P(F) - P(E \cap F)$$

Because $E \cap F$ means "E and F," $E \cap F$ is the event that the pair of cards you are dealt are two of a kind and include the Jack of clubs. But this is impossible because only one jack is left. Thus $E \cap F = \emptyset$, and so $P(E \cap F) = 0$. This gives us

$$P(E \cup F) = P(E) + P(F) \approx .0611 + .0426 = .1037$$

In other words, there is slightly better than a one in ten chance that you will wind up with either a full house or four of a kind, given the original hand.

+ *Before we go on...* A more accurate answer to part (c) of Example 4 is $(66 + 46)/1081 \approx$ .1036; we lost some accuracy in rounding the answers to parts (a) and (b). ∎

Example **5 Committees**

The University Senate bylaws at Hofstra University[*] state the following:

> The Student Affairs Committee shall consist of one elected faculty senator, one faculty senator-at-large, one elected student senator, five student senators-at-large (including one from the graduate school), two delegates from the Student Government Association, and the president of the Student Government Association or his/her designate. It shall be chaired by the elected student senator on the Committee and it shall be advised by the Dean of Students or his/her designate.

You are an undergraduate student and, even though you are not an elected student senator, you would very much like to serve on the Student Affairs Committee. The senators-at-large as well as the Student Government delegates are chosen by means of a random drawing from a list of candidates. There are already 13 undergraduate candidates for the position of senator-at-large, and 6 candidates for Student Government delegates, and you have been offered a position on the Student Government Association by the chairperson (who happens to be a good friend of yours), should you wish to join it. (This would make you ineligible for a senator-at-large position.) What should you do?

Solution You have two options. Option 1 is to include your name on the list of candidates for the senator-at-large position. Option 2 is to join the Student Government Association (SGA) and add your name to its list of candidates. Let us look at the two options separately.

Option 1: Add your name to the senator-at-large list.
This will result in a list of 14 undergraduates for 4 undergraduate positions. The sample space is the set of all possible outcomes of the random drawing. Each outcome consists of a set of 4 lucky students chosen from 14. Thus,

$$n(S) = C(14, 4) = 1001$$

We are interested in the probability that you are among the chosen four. Thus, E is the set of sets of 4 that include you.

1. Choose yourself

2. Choose 3 from the remaining 13

$$n(E) = C(1, 1) \times C(13, 3) = 286$$

---

[*] As of 1999. SOURCE: Hofstra University Faculty Policy Series.

So,

$$P(E) = \frac{n(E)}{n(S)} = \frac{286}{1001} = \frac{2}{7} \approx .2857$$

Option 2: Join the SGA and add your name to its list.
This results in a list of 7 candidates from which 2 are selected. For this case, the sample space consists of all sets of 2 chosen from 7, so

$$n(S) = C(7, 2) = 21$$

and

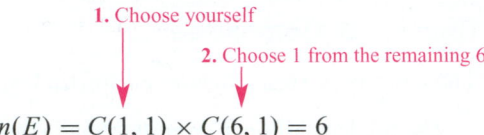

1. Choose yourself

2. Choose 1 from the remaining 6

$$n(E) = C(1, 1) \times C(6, 1) = 6$$

Thus,

$$P(E) = \frac{n(E)}{n(S)} = \frac{6}{21} = \frac{2}{7} \approx .2857$$

In other words, the probability of being selected is exactly the same for Option 1 as it is for Option 2! Thus, you can choose either option, and you will have slightly less than a 29% chance of being selected.

7.4 EXERCISES

● denotes basic skills exercises

◆ denotes challenging exercises

Recall from Example 1 that whenever Suzy sees a bag of marbles, she grabs a handful at random. In Exercises 1–10, she has seen a bag containing 4 red marbles, 3 green ones, 2 white ones, and 1 purple one. She grabs 5 of them. Find the probabilities of the following events, expressing each as a fraction in lowest terms. hint [see Example 1]

1. ● She has all the red ones.

2. ● She has none of the red ones.

3. ● She has at least 1 white one.

4. ● She has at least 1 green one.

5. ● She has 2 red ones and 1 of each of the other colors.

6. ● She has 2 green ones and 1 of each of the other colors.

7. ● She has at most 1 green one.

8. ● She has no more than 1 white one.

9. ● She does not have all the red ones.

10. ● She does not have all the green ones.

Dogs of the Dow The phrase "Dogs of the Dow" refers to the stocks listed on the Dow with the highest dividend yield. Exercises 11–16 are based on the following table, which shows the top ten stocks of the "Dogs of the Dow" list in February, 2005:[33]

Symbol	Company	Yield
SBC	SBC Communications	5.46%
GM	General Motors	5.46%
MRK	Merck	5.42%
MO	Altria	4.62%
VZ	Verizon	4.32%
JPM	JP Morgan Chase	3.68%
C	Citigroup	3.64%
PFE	Pfizer	3.12%
DD	DuPont	2.96%
GE	General Electric	2.46%

hint [see Example 2]

[33] SOURCE: http://www.dogsofthedow.com.

11. ● If you selected two of these stocks at random, what is the probability that both the stocks in your selection had yields of 5% or more?

12. ● If you selected three of these stocks at random, what is the probability that all three of the stocks in your selection had yields of 5% or more?

13. ● If you selected four of these stocks at random, what is the probability that your selection included the company with the highest yield and excluded the company with the lowest yield?

14. ● If you selected four of these stocks at random, what is the probability that your selection included SBC and GM but excluded DD and GE?

15. If your portfolio included 100 shares of PFE and you then purchased 100 shares each of any two companies on the list at random, find the probability that you ended up with a total of 200 shares of PFE.

16. If your portfolio included 100 shares of PFE and you then purchased 100 shares each of any three companies on the list at random, find the probability that you ended up with a total of 200 shares of PFE.

17. ● *Tests* A test has 3 parts. Part A consists of 8 true-false questions, Part B consists of 5 multiple choice questions with 5 choices each, and Part C requires you to match 5 questions with 5 different answers one-to-one. Assuming that you make random guesses in filling out your answer sheet, what is the probability that you will earn 100% on the test? (Leave your answer as a formula.)

18. ● *Tests* A test has 3 parts. Part A consists of 4 true-false questions, Part B consists of 4 multiple choice questions with 5 choices each, and Part C requires you to match 6 questions with 6 different answers one-to-one. Assuming that you make random choices in filling out your answer sheet, what is the probability that you will earn 100% on the test? (Leave your answer as a formula.)

Poker In Exercises 19–24, you are asked to calculate the probability of being dealt various poker hands. (Recall that a poker player is dealt 5 cards at random from a standard deck of 52.) Express each of your answers as a decimal rounded to four decimal places, unless otherwise stated. hint [see Example 3]

19. ● **Two of a kind:** 2 cards with the same denomination and 3 cards with other denominations (different from each other and that of the pair) Example: K♣, K♥, 2♠, 4♦, J♠

20. ● **Three of a kind:** 3 cards with the same denomination and 2 cards with other denominations (different from each other and that of the 3) Example: Q♣, Q♥, Q♠, 4♦, J♠

21. ● **Two pair:** 2 cards with one denomination, 2 with another, and 1 with a third. Example: 3♣, 3♥, Q♠, Q♥, 10♠

22. ● **Straight Flush:** 5 cards of the same suit with consecutive denominations but not a royal flush (a royal flush consists of the 10, J, Q, K, and A of one suit). Round the answer to one significant digit. Examples: A♣, 2♣, 3♣, 4♣, 5♣, or

9♦, 10♦, J♦, Q♦, K♦, or A♥, 2♥, 3♥, 4♥, 5♥, but *not* 10♦, J♦, Q♦, K♦, A♦

23. ● **Flush:** 5 cards of the same suit, but not a straight flush or royal flush. Example: A♣, 5♣, 7♣, 8♣, K♣

24. ● **Straight:** five cards with consecutive denominations, but not all of the same suit. Examples: 9♦, 10♦, J♣, Q♥, K♦, and 10♥, J♦, Q♦, K♦, A♦.

25. ● *The Monkey at the Typewriter* Suppose that a monkey is seated at a computer keyboard and randomly strikes the 26 letter keys and the space bar. Find the probability that its first 39 characters (including spaces) will be "to be or not to be that is the question." (Leave your answer as a formula.)

26. ● *The Cat on the Piano* A standard piano keyboard has 88 different keys. Find the probability that a cat, jumping on 4 keys in sequence and at random (possibly with repetition), will strike the first 4 notes of Beethoven's Fifth Symphony. (Leave your answer as a formula.)

27. ● *(Based on a question from the GMAT)* Tyler and Gebriella are among 7 contestants from which 4 semifinalists are to be selected at random. Find the probability that neither Tyler nor Gebriella is selected.

28. ● *(Based on a question from the GMAT)* Tyler and Gebriella are among 7 contestants from which 4 semifinalists are to be selected at random. Find the probability that Tyler but not Gebriella is selected.

29. *Lotteries* The Sorry State Lottery requires you to select 5 different numbers from 0 through 49. (Order is not important.) You are a Big Winner if the 5 numbers you select agree with those in the drawing, and you are a Small-Fry Winner if 4 of your 5 numbers agree with those in the drawing. What is the probability of being a Big Winner? What is the probability of being a Small-Fry Winner? What is the probability that you are either a Big Winner or a Small-Fry winner?

30. *Lotto* The Sad State Lottery requires you to select a sequence of three different numbers from zero through 49. (Order is important.) You are a winner if your sequence agrees with that in the drawing, and you are a booby prize winner if your selection of numbers is correct, but in the wrong order. What is the probability of being a winner? What is the probability of being a booby prize winner? What is the probability that you are either a winner or a booby prize winner?

31. *Transfers* Your company is considering offering 400 employees the opportunity to transfer to its new headquarters in Ottawa and, as personnel manager, you decide that it would be fairest if the transfer offers are decided by means of a lottery. Assuming that your company currently employs 100 managers, 100 factory workers, and 500 miscellaneous staff, find the following probabilities, leaving the answers as formulas:

a. All the managers will be offered the opportunity.
b. You will be offered the opportunity.

32. *Transfers* (Refer back to the preceding exercise.) After thinking about your proposed method of selecting employees for the opportunity to move to Ottawa, you decide it might be a better idea to select 50 managers, 50 factory workers, and 300 miscellaneous staff, all chosen at random. Find the probability that you will be offered the opportunity. (Leave your answer as a formula.)

33. *Lotteries* In a New York State daily lottery game, a sequence of 3 digits (not necessarily different) in the range 0–9 are selected at random. Find the probability that all 3 are different.

34. *Lotteries* Refer back to the preceding exercise. Find the probability that 2 of the 3 digits are the same.

35. *Sports* The following table shows the results of the Big Eight Conference for the 1988 college football season.[34]

Team	Won	Lost
Nebraska (NU)	7	0
Oklahoma (OU)	6	1
Oklahoma State (OSU)	5	2
Colorado (CU)	4	3
Iowa State (ISU)	3	4
Missouri (MU)	2	5
Kansas (KU)	1	6
Kansas State (KSU)	0	7

This is referred to as a "perfect progression." Assuming that the "Won" score for each team is chosen at random in the range 0–7, find the probability that the results form a perfect progression.[35] (Leave your answer as a formula.)

36. *Sports* Refer back to Exercise 35. Find the probability of a perfect progression with Nebraska scoring 7 wins and 0 losses. (Leave your answer as a formula.)

37. *Graph Searching* A graph consists of a collection of **nodes** (the dots in the figure) connected by **edges** (line segments from one node to another). A **move on a graph** is a move from one node to another along a single edge. Find the probability of going from Start to Finish in a sequence of two

random moves in the graph shown. (All directions are equally likely.)

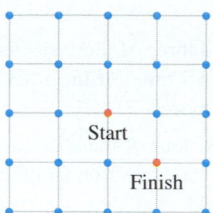

38. *Graph Searching* Refer back to Exercise 37. Find the probability of going from Start to one of the Finish nodes in a sequence of two random moves in the following figure. (All directions are equally likely.)

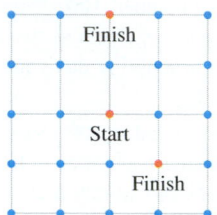

39. *Tournaments* What is the probability that North Carolina will beat Central Connecticut but lose to Virginia in the following (fictitious) soccer tournament? (Assume that all outcomes are equally likely.)

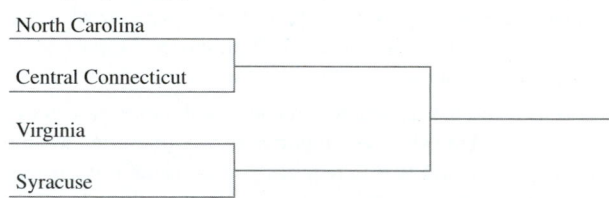

40. *Tournaments* In a (fictitious) soccer tournament involving the four teams San Diego State, De Paul, Colgate, and Hofstra, find the probability that Hofstra will play Colgate in the finals and win. (Assume that all outcomes are equally likely and that the teams not listed in the first round slots are placed at random.)

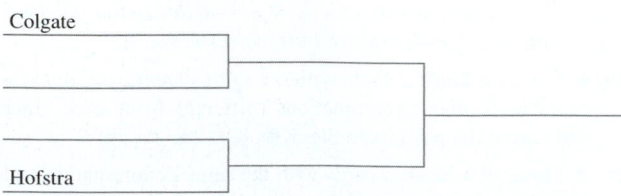

41. ◆ *Product Design* Your company has patented an electronic digital padlock which a user can program with his or her own 4-digit code. (Each digit can be 0 through 9, and repetitions are allowed.) The padlock is designed to open either if the correct code is keyed in or—and this is helpful for forgetful people—if exactly one of the digits is incorrect. What is the probability that a randomly chosen sequence of 4 digits will open a programmed padlock?

---

[34] Source: On the probability of a perfect progression, *The American Statistician,* August 1991, vol. 45, no. 3, p. 214.

[35] Even if all the teams are equally likely to win each game, the chances of a perfect progression actually coming up are a little more difficult to estimate, because the number of wins by one team impacts directly on the number of wins by the others. For instance, it is impossible for all eight teams to show a score of 7 wins and 0 losses at the end of the season—someone must lose! It is, however, not too hard to come up with a counting argument to estimate the total number of win-lose scores actually possible.

42. ◆ *Product Design* Assume that you already know the first digit of the combination for the lock described in Exercise 41. Find the probability that a random guess of the remaining three digits will open the lock. *hint* [see Example 5]

43. ◆ *Committees* An investigatory Committee in the Kingdom of Utopia consists of a chief investigator (a Royal Party member), an assistant investigator (a Birthday Party member), 2 at-large investigators (either party), and 5 ordinary members (either party). Royal Party member Larry Sifford is hoping to avoid serving on the committee, unless he is the Chief Investigator and Otis Taylor, a Birthday Party member, is the Assistant Investigator. The committee is to be selected at random from a pool of 12 candidates (including Larry Sifford and Otis Taylor), half of whom are Royal Party and half of whom are Birthday Party.

a. How many different committees are possible?

b. How many committees are possible in which Larry's hopes are fulfilled? (This includes the possibility that he's not on the committee at all.)

c. What is the probability that he'll be happy with a randomly selected committee?

44. ◆ *Committees* A committee is to consist of a chair, 3 hagglers, and 4 do-nothings. The committee is formed by choosing randomly from a pool of 10 people and assigning them to the various "jobs."

a. How many different committees are possible?

b. Norman is eager to be the chair of the committee. What is the probability that he will get his wish?

c. Norman's girl friend Norma is less ambitious and would be happy to hold any position on the committee provided Norman is also selected as a committee member. What is the probability that she will get her wish and serve on the committee?

d. Norma does not get along with Oona (who is also in the pool of prospective members) and would be most unhappy if Oona were to chair the committee. Find the probability that all her wishes will be fulfilled: she and Norman are on the committee and it is not chaired by Oona.

Communication and Reasoning Exercises

45. ● What is wrong with the following argument? A bag contains 2 blue marbles and 2 red ones; 2 are drawn at random. Because there are 4 possibilities—(red, red), (blue, blue), (red, blue) and (blue, red)—the probability that both are red is 1/4.

46. ● What is wrong with the following argument? When we roll two indistinguishable dice, the number of possible outcomes (unordered groups of two not necessarily distinct numbers) is 21 and the number of outcomes in which both numbers are the same is 6. Hence, the probability of throwing a double is $6/21 = 2/7$.

47. Suzy grabs 2 marbles out of a bag of 5 red marbles and 4 green ones. She could do so in two ways: She could take them out one at a time, so that there is a first and a second marble, or she could grab 2 at once so that there is no order. Does the method she uses to grab the marbles affect the probability that she gets 2 red marbles?

48. If Suzy grabs 2 marbles, one at a time, out of a bag of 5 red marbles and 4 green ones, find an event with a probability that depends on the order in which the two marbles are drawn.

49. ● Create an interesting application whose solution requires finding a probability using combinations.

50. ● Create an interesting application whose solution requires finding a probability using permutations.

● basic skills   ◆ challenging

7.5 Conditional Probability and Independence

Cyber Video Games, Inc., ran a television ad in advance of the release of its latest game, "Ultimate Hockey." As Cyber Video's director of marketing, you would like to assess the ad's effectiveness, so you ask your market research team to survey video game players. The results of its survey of 2000 video game players are summarized in the following table:

	Saw Ad	Did Not See Ad	Total
Purchased Game	100	200	300
Did Not Purchase Game	200	1500	1700
Total	300	1700	2000

The market research team concludes in its report that the ad is highly persuasive, and recommends using the company that produced the ad for future projects.

But wait, how could the ad possibly have been persuasive? Only 100 people who saw the ad purchased the game, while 200 people purchased the game without seeing the ad at all! At first glance, it looks as though potential customers are being *put off* by the ad. However, let us analyze the figures a little more carefully.

First, let us restrict attention to those players who saw the ad (first column of data: "Saw Ad") and compute the estimated probability that a player *who saw the ad* purchased Ultimate Hockey.

	Saw Ad
Purchased Game	100
Did Not Purchase Game	200
Total	300

To compute this probability, we calculate

Probability that someone who saw the ad purchased the game

$$= \frac{\text{Number of people who saw the ad and bought the game}}{\text{Total number of people who saw the ad}} = \frac{100}{300} \approx .33$$

In other words, 33% of game players who saw the ad went ahead and purchased the game. Let us compare this with the corresponding probability for those players who did *not* see the ad (second column of data "Did Not See Ad"):

	Did Not See Ad
Purchased Game	200
Did Not Purchase Game	1500
Total	1700

Probability that someone who did not see the ad purchased the game

$$= \frac{\text{Number of people who did not see the ad and bought the game}}{\text{Total number of people who did not see the ad}} = \frac{200}{1700} \approx .12$$

In other words, only 12% of game players who did not see the ad purchased the game, whereas 33% of those who *did* see the ad purchased the game. Thus, it appears that the ad *was* highly persuasive.

Here's some terminology. In this example there were two related events of importance:

A: A video game player purchased Ultimate Hockey

B: A video game player saw the ad

The first probability we computed was the estimated probability that a video game player purchased Ultimate Hockey *given that* he or she saw the ad. We call the latter probability the (estimated) **probability of *A*, given *B***, and we write it as $P(A \mid B)$. We call $P(A \mid B)$ a **conditional probability**—it is the probability of *A* under the condition that B occurred. Put another way, it is the probability of A occurring if the sample space is reduced to just those outcomes in B.

$$P(\text{Purchased Game } \textit{Given That } \text{Saw the Ad}) = P(A \mid B) \approx .33$$

The second probability we computed was the estimated probability that a video game player purchased Ultimate Hockey *given that* he or she did not see the ad, or the **probability of A, given B'.**

$$P(\text{Purchased Game } \textit{Given That } \text{Did Not See the Ad}) = P(A \mid B') \approx .12$$

Calculating Conditional Probabilities

How do we calculate conditional probabilities? In the example above we used the ratio

$$P(A \mid B) = \frac{\text{Number of people who saw the ad and bought the game}}{\text{Total number of people who saw the ad}}$$

The numerator is the frequency of $A \cap B$, and the denominator is the frequency of B:

$$P(A \mid B) = \frac{fr(A \cap B)}{fr(B)}$$

Now, we can write this formula in another way:

$$P(A \mid B) = \frac{fr(A \cap B)}{fr(B)} = \frac{fr(A \cap B)/N}{fr(B)/N} = \frac{P(A \cap B)}{P(B)}$$

We therefore have the following definition, which applies to general probability distributions.

Conditional Probability

If A and B are events with $P(B) \neq 0$, then the probability of A given B is

$$P(A \mid B) = \frac{P(A \cap B)}{P(B)}$$

Visualizing Conditional Probability

In the figure, $P(A \mid B)$ is represented by the fraction of B that is covered by A.

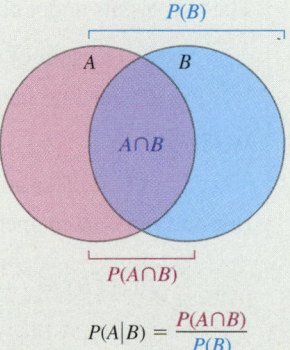

quick Examples

1. If there is a 50% chance of rain (R) and a 10% chance of both rain and lightning (L), then the probability of lightning, given that it rains, is

$$P(L \mid R) = \frac{P(L \cap R)}{P(R)} = \frac{.10}{.50} = .20$$

Here are two more ways to express the result:

• If it rains, the probability of lightning is .20.

• Assuming that it rains, there is a 20% chance of lightning.

2. Referring to the Cyber Video data above, the probability that a video game player did not purchase the game (A'), given that she did not see the ad (B'), is

$$P(A' \mid B') = \frac{P(A' \cap B')}{P(B')} = \frac{1500/2000}{1700/2000} = \frac{15}{17} \approx .88$$

Q: *Returning to the video game sales survey, how do we compute the* ordinary *probability of A, not "given" anything?*

A: We look at the event A that a randomly chosen game player purchased Ultimate Hockey *regardless of whether or not he or she saw the ad.* In the "Purchased Game" row we see that a total of 300 people purchased the game out of a total of 2000 surveyed. Thus, the (estimated) probability of A is

$$P(A) = \frac{fr(A)}{N} = \frac{300}{2000} = .15$$

We sometimes refer to $P(A)$ as the **unconditional** probability of A to distinguish it from conditional probabilities like $P(A \mid B)$ and $P(A \mid B')$. ∎

Now, let's see some more examples involving conditional probabilities.

Example 1 Dice

If you roll a fair die twice and observe the numbers that face up, find the probability that the sum of the numbers is 8, given that the first number is 3.

Solution We begin by recalling that the sample space when we roll a fair die twice is the set $S = \{(1, 1), (1, 2), \ldots, (6, 6)\}$ containing the 36 different equally likely outcomes.

The two events under consideration are

A: The sum of the numbers is 8.

B: The first number is 3.

We also need

$A \cap B$: The sum of the numbers is 8 and the first number is 3.

But this can only happen in one way: $A \cap B = \{(3, 5)\}$. From the formula, then,

$$P(A \mid B) = \frac{P(A \cap B)}{P(B)} = \frac{1/36}{6/36} = \frac{1}{6}$$

+*Before we go on...* There is another way to think about Example 1. When we say that the first number is 3, we are restricting the sample space to the six outcomes (3, 1), (3, 2), . . . , (3, 6), all still equally likely. Of these six, only one has a sum of 8, so the probability of the sum being 8, given that the first number is 3, is 1/6. ∎

Notes

1. Remember that, in the expression $P(A \mid B)$, A is the event whose probability you want, given that you know the event B has occurred.
2. From the formula, notice that $P(A \mid B)$ is not defined if $P(B) = 0$. Could $P(A \mid B)$ make any sense if the event B were impossible? ∎

Example 2 School and Work

A survey[*] of the high school graduating class of 2003, conducted by the Bureau of Labor Statistics, found that, if a graduate went on to college, there was a 42% chance that he or she would work at the same time. On the other hand, there was a 64% chance that a randomly selected graduate would go on to college. What is the probability that a graduate went to college and work at the same time?

Solution To understand what the question asks and what information is given, it is helpful to rephrase everything using the standard wording "*the probability that ___* " and "*the probability that ___ given that ___.*" Now we have, "The probability that a graduate worked, given that the graduate went on to college, equals .42. (See Figure 7.) The probability that a graduate went on to college is .64." The events in question are as follows:

W: A high school graduate went on to work

C: A high school graduate went on to college

From our rephrasing of the question we can write:

$$P(W \mid C) = .42. \qquad P(C) = .64. \qquad \text{Find } P(W \cap C)$$

The definition

$$P(W \mid C) = \frac{P(W \cap C)}{P(C)}$$

can be used to find $P(W \cap C)$:

$$P(W \cap C) = P(W \mid C)P(C)$$
$$= (.42)(.64) \approx .27$$

Thus there is a 27% chance that a member of the high school graduating class of 2003 went on to college and work at the same time.

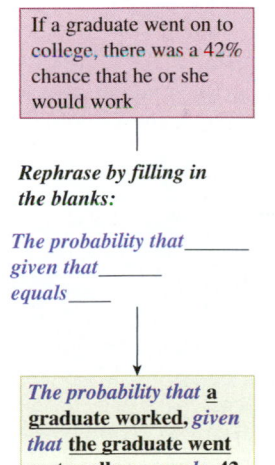

If a graduate went on to college, there was a 42% chance that he or she would work

Rephrase by filling in the blanks:

The probability that_____ given that_____ equals____

The probability that a graduate worked, *given that* **the graduate went on to college,** *equals* **.42**

$P(\text{Worked} \mid \text{Went to college}) = .42$

Figure **7**

---

[*] "College Enrollment and Work Activity of High School Graduates," U.S. Bureau of Labor Statistics, April, 2004, available at www.bls.gov/schedule/archives/all_nr.htm.

The Multiplication Principle and Trees

In Example 2, we saw that the formula

$$P(A \mid B) = \frac{P(A \cap B)}{P(B)}$$

can be used to calculate $P(A \cap B)$ if we rewrite the formula in the following form, known as the **multiplication principle for conditional probability:**

Multiplication Principle for Conditional Probability

If A and B are events, then

$$P(A \cap B) = P(A \mid B)P(B)$$

quick Example

If there is a 50% chance of rain (R) and a 20% chance of a lightning (L) if it rains, then the probability of both rain and lightning is

$$P(R \cap L) = P(L \mid R)P(R) = (.20)(.50) = .10$$

The multiplication principle is often used in conjunction with **tree diagrams.** Let's return to Cyber Video Games, Inc., and its television ad campaign. Its marketing survey was concerned with the following events:

A: A video game player purchased Ultimate Hockey

B: A video game player saw the ad

We can illustrate the various possibilities by means of the two-stage "tree" shown in Figure 8.

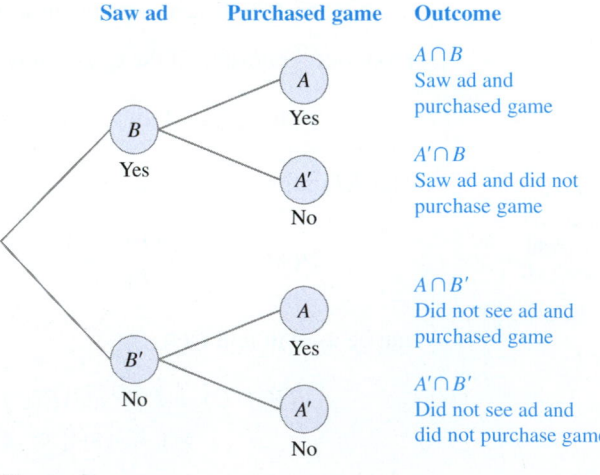

Figure **8**

Consider the outcome $A \cap B$. To get there from the starting position on the left, we must first travel up to the B node. (In other words, B must occur.) Then we must travel up the branch from the B node to the A node. We are now going to associate a probability with each branch of the tree: the probability of traveling along that branch *given that we have gotten to its beginning node.* For instance, the probability of traveling up the branch from the starting position to the *B* node is $P(B) = 300/2000 = .15$ (see the data in the survey). The probability of going up the branch from the *B* node to the *A* node is the probability that *A* occurs, given that *B* has occurred. In other words, it is the *conditional* probability $P(A \mid B) \approx .33$. (We calculated this probability at the beginning of the section.) The probability of the outcome $A \cap B$ can then be computed using the multiplication principle:

$$P(A \cap B) = P(B)P(A \mid B) \approx (.15)(.33) \approx .05$$

In other words, *to obtain the probability of the outcome $A \cap B$, we multiply the probabilities on the branches leading to that outcome* (Figure 9).

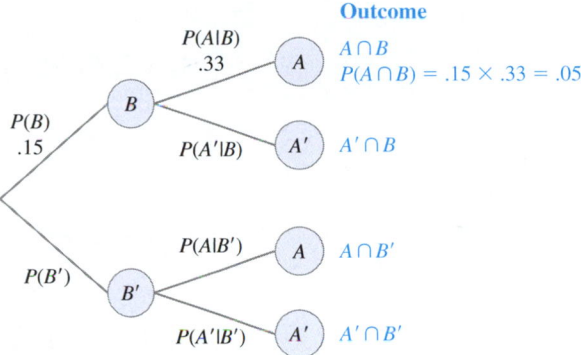

Figure **9**

The same argument holds for the remaining three outcomes, and we can use the table given at the beginning of this section to calculate all the conditional probabilities shown in Figure 10.

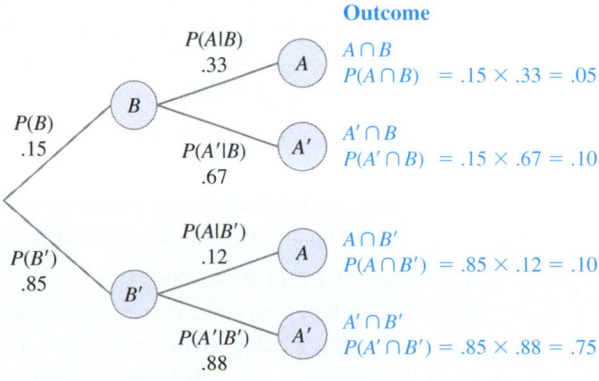

Figure **10**

Note The sum of the probabilities on the branches leaving any node is always 1 (why?). This observation often speeds things up because after we have labeled one branch (or, all but one, if a node has more than two branches leaving it), we can easily label the remaining one. ∎

Example 3 Unfair Coins

An experiment consists of tossing two coins. The first coin is fair, while the second coin is twice as likely to land with heads facing up as it is with tails facing up. Draw a tree diagram to illustrate all the possible outcomes, and use the multiplication principle to compute the probabilities of all the outcomes.

Solution A quick calculation shows that the probability distribution for the second coin is $P(H) = 2/3$ and $P(T) = 1/3$. (How did we get that?) Figure 11 shows the tree diagram and the calculations of the probabilities of the outcomes.

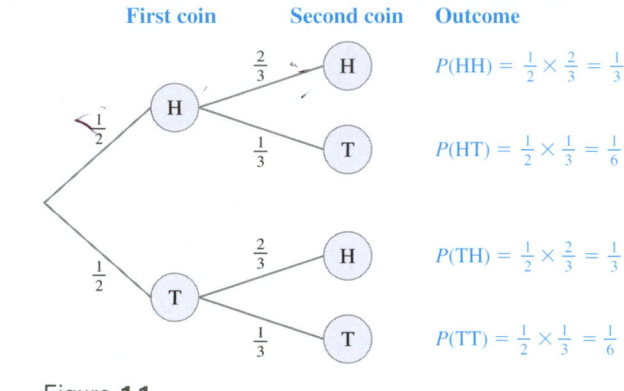

Figure **11**

Independence

Let us go back once again to Cyber Video Games, Inc., and its ad campaign. How did we assess the ad's effectiveness? We considered the following events.

A: A video game player purchased Ultimate Hockey

B: A video game player saw the ad

We used the survey data to calculate $P(A)$, the probability that a video game player purchased Ultimate Hockey, and $P(A \mid B)$, the probability that a video game player *who saw the ad* purchased Ultimate Hockey. When these probabilities are compared, one of three things can happen.

Case 1 $P(A \mid B) > P(A)$
This is what the survey data actually showed: A video game player was more likely to purchase Ultimate Hockey if he or she saw the ad. This indicates that the ad is effective; seeing the ad had a positive effect on a player's decision to purchase the game.

Case 2 $P(A \mid B) < P(A)$
If this had happened, then a video game player would have been *less* likely to purchase Ultimate Hockey if he or she saw the ad. This would have indicated that the ad had "backfired"; it had, for some reason, put potential customers off. In this case, just as in the first case, the event B would have had an effect—a negative one—on the event A.

Case 3 $P(A \mid B) = P(A)$
In this case seeing the ad would have had absolutely no effect on a potential customer's buying Ultimate Hockey. Put another way, the probability of A occurring *does not depend* on whether B occurred or not. We say in a case like this that the events A and B are **independent**.

In general, we say that two events A and B are independent if $P(A \mid B) = P(A)$. When this happens, we have

$$P(A) = P(A \mid B) = \frac{P(A \cap B)}{P(B)}$$

so

$$P(A \cap B) = P(A)P(B)$$

Conversely, if $P(A \cap B) = P(A)P(B)$, then, assuming $P(B) \neq 0,$[†] $P(A) = P(A \cap B)/P(B) = P(A \mid B)$. Thus, saying that $P(A) = P(A \mid B)$ is the same as saying that $P(A \cap B) = P(A)P(B)$. Also, we can switch A and B in this last formula and conclude that saying that $P(A \cap B) = P(A)P(B)$ is the same as saying that $P(B \mid A) = P(B)$.

[†] We shall only discuss the independence of two events in cases where their probabilities are both nonzero.

Independent Events

The events A and B are **independent** if

$$P(A \cap B) = P(A)P(B)$$

Equivalent formulas (assuming neither A nor B is impossible) are

$$P(A \mid B) = P(A)$$

and $P(B \mid A) = P(B)$

If two events A and B are not independent, then they are **dependent.**

The property $P(A \cap B) = P(A)P(B)$ can be extended to three or more independent events. If, for example, A, B and C are three mutually independent events (that is, each one of them is independent of each of the other two and of their intersection), then, among other things,

$$P(A \cap B \cap C) = P(A)P(B)P(C)$$

Testing for Independence

To check whether two events A and B are independent, we compute $P(A)$, $P(B)$, and $P(A \cap B)$. If $P(A \cap B) = P(A)P(B)$, the events are independent; otherwise, they are dependent. Sometimes it is obvious that two events, by their nature, are independent, so a test is not necessary. For example, the event that a die you roll comes up 1 is clearly independent of whether or not a coin you toss comes up heads.

To test for independence, calculate the three quantities $P(A)$, $P(B)$ and $P(A \cap B)$ separately, and then see if $P(A \cap B) = P(A) \cdot P(B)$.

quick Examples

1. Roll two distinguishable dice (one red, one green) and observe the numbers that face up.

A: The red die is even; $P(A) = \dfrac{18}{36} = \dfrac{1}{2}$

B: The dice have the same parity[*]; $P(B) = \dfrac{18}{36} = \dfrac{1}{2}$

---

[*] Two numbers have the **same parity** if both are even or both are odd. Otherwise, they have **opposite parity**.

$A \cap B$: Both dice are even; $P(A \cap B) = \dfrac{9}{36} = \dfrac{1}{4}$

$P(A \cap B) = P(A)P(B)$, and so A and B are independent.

2. Roll two distinguishable dice and observe the numbers that face up.

A: The sum of the numbers is 6; $P(A) = \dfrac{5}{36}$

B: Both numbers are odd; $P(B) = \dfrac{9}{36} = \dfrac{1}{4}$

$A \cap B$: The sum is 6, and both are odd; $P(A \cap B) = \dfrac{3}{36} = \dfrac{1}{12}$

$P(A \cap B) \neq P(A)P(B)$, and so A and B are dependent.

Example 4 Weather Prediction

According to the weather service, there is a 50% chance of rain in New York and a 30% chance of rain in Honolulu. Assuming that New York's weather is independent of Honolulu's, find the probability that it will rain in at least one of these cities.

Solution We take A to be the event that it will rain in New York and B to be the event that it will rain in Honolulu. We are asked to find the probability of $A \cup B$, the event that it will rain in at least one of the two cities. We use the addition principle:

$$P(A \cup B) = P(A) + P(B) - P(A \cap B)$$

We know that $P(A) = .5$ and $P(B) = .3$. But what about $P(A \cap B)$? Because the events A and B are independent, we can compute

$$
\begin{aligned}
P(A \cap B) &= P(A)P(B) \\
&= (.5)(.3) = .15
\end{aligned}
$$

Thus,

$$
\begin{aligned}
P(A \cup B) &= P(A) + P(B) - P(A \cap B) \\
&= .5 + .3 - .15 \\
&= .65
\end{aligned}
$$

So, there is a 65% chance that it will rain either in New York or in Honolulu (or in both).

Example 5 Roulette

You are playing roulette, and have decided to leave all 10 of your $1 chips on black for five consecutive rounds, hoping for a sequence of five blacks which, according to the rules, will leave you with $320. There is a 50% chance of black coming up on each spin, ignoring the complicating factor of zero or double zero. What is the probability that you will be successful?

Solution Because the roulette wheel has no memory, each spin is independent of the others. Thus, if A_1 is the event that black comes up the first time, A_2 the event that it comes up the second time, and so on, then

$$P(A_1 \cap A_2 \cap A_3 \cap A_4 \cap A_5) = P(A_1)P(A_2)P(A_3)P(A_4)P(A_5) = \left(\frac{1}{2}\right)^5 = \frac{1}{32}$$

The next example is a version of a well known "brain teaser" that forces one to think carefully about conditional probability.

Example 6 Legal Argument

A man was arrested for attempting to smuggle a bomb on board an airplane. During the subsequent trial, his lawyer claimed that, by means of a simple argument, she would prove beyond a shadow of a doubt that her client was not only innocent of any crime, but was in fact contributing to the safety of the other passengers on the flight. This was her eloquent argument: "Your Honor, first of all, my client had absolutely no intention of setting off the bomb. As the record clearly shows, the detonator was unarmed when he was apprehended. In addition—and your Honor is certainly aware of this—there is a small but definite possibility that there will be a bomb on any given flight. On the other hand, the chances of there being *two* bombs on a flight are so remote as to be negligible. There is in fact no record of this having *ever* occurred. Thus, because my client had already brought one bomb on board (with no intention of setting it off) and because we have seen that the chances of there being a second bomb on board were vanishingly remote, it follows that the flight was far safer as a result of his action! I rest my case." This argument was so elegant in its simplicity that the judge acquitted the defendant. Where is the flaw in the argument? (Think about this for a while before reading the solution.)

Solution   The lawyer has cleverly confused the phrases "two bombs on board" and "a second bomb on board." To pinpoint the flaw, let us take B to be the event that there is one bomb on board a given flight, and let A be the event that there are two independent bombs on board. Let us assume for argument's sake that $P(B) = 1/1,000,000 = .000\,001$. Then the probability of the event A is

$$(.000\,001)(.000\,001) = .000\,000\,000\,001$$

This *is* vanishingly small, as the lawyer contended. It was at this point that the lawyer used a clever maneuver: She assumed in concluding her argument that the probability of having two bombs on board was the same as the probability of having a *second* bomb on board. But to say that there is a *second* bomb on board is to imply that there already is one bomb on board. This is therefore a *conditional* event: the event that there are two bombs on board, *given that there is already one bomb on board.* Thus, the probability that there is a second bomb on board is the probability that there are two bombs on board, given that there is already one bomb on board, which is

$$P(A \mid B) = \frac{P(A \cap B)}{P(B)} = \frac{.000\,000\,000\,001}{.000\,001} = .000\,001$$

In other words, it is the same as the probability of there being a single bomb on board to begin with! Thus the man's carrying the bomb onto the plane did not improve the flight's safety at all.[*]

---

[*] If we want to be picky, there was a *slight* decrease in the probability of a second bomb because there was one less seat for a potential second bomb bearer to occupy. In terms of our analysis, this is saying that the event of one passenger with a bomb and the event of a second passenger with a bomb are not completely independent.

7.5 | EXERCISES

● denotes basic skills exercises

◆ denotes challenging exercises

In Exercises 1–10, compute the indicated quantity.

1. ● $P(B) = .5$, $P(A \cap B) = .2$. Find $P(A \mid B)$.

2. ● $P(B) = .6$, $P(A \cap B) = .3$. Find $P(A \mid B)$.

3. ● $P(A \mid B) = .2$, $P(B) = .4$. Find $P(A \cap B)$.

4. ● $P(A \mid B) = .1$, $P(B) = .5$. Find $P(A \cap B)$.

5. ● $P(A \mid B) = .4$, $P(A \cap B) = .3$. Find $P(B)$.

6. ● $P(A \mid B) = .4$, $P(A \cap B) = .1$. Find $P(B)$.

7. ● $P(A) = .5$, $P(B) = .4$. A and B are independent. Find $P(A \cap B)$.

8. ● $P(A) = .2$, $P(B) = .2$. A and B are independent. Find $P(A \cap B)$.

9. ● $P(A) = .5$, $P(B) = .4$. A and B are independent. Find $P(A \mid B)$.

10. ● $P(A) = .3$, $P(B) = .6$. A and B are independent. Find $P(B \mid A)$.

In Exercises 11–14, supply the missing quantities.

11. ●

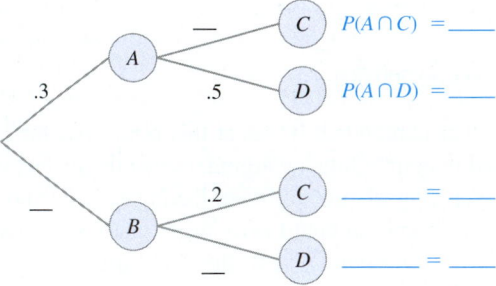

12. ●

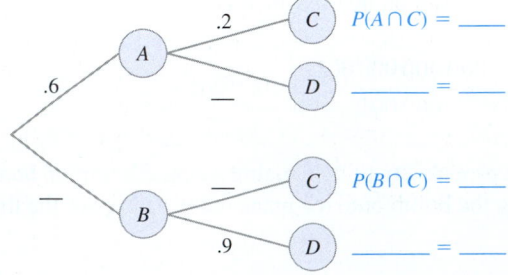

13. ●

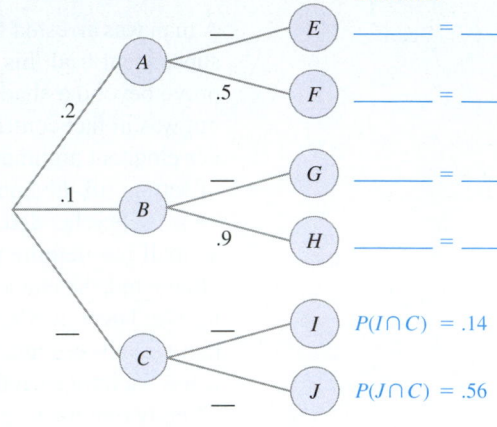

14. ●

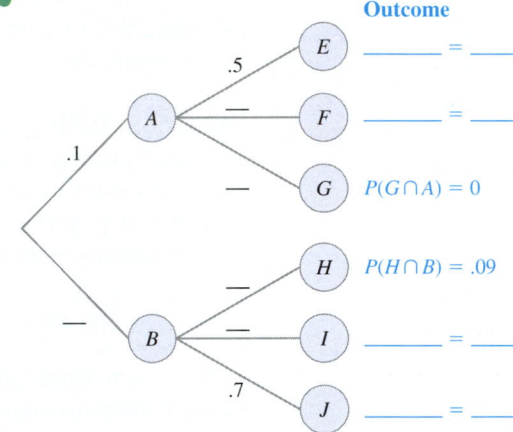

In Exercises 15–20, find the conditional probabilities of the indicated events when two fair dice (one red and one green) are rolled. hint [see Example 1]

15. ● The sum is 5, given that the green one is not a 1.

16. ● The sum is 6, given that the green one is either 4 or 3.

17. ● The red one is 5, given that the sum is 6.

18. ● The red one is 4, given that the green one is 4.

19. ● The sum is 5, given that the dice have opposite parity.

20. ● The sum is 6, given that the dice have opposite parity.

Exercises 21–26 require the use of counting techniques from the last chapter. A bag contains 3 red marbles, 2 green ones, 1 fluorescent pink one, 2 yellow ones, and 2 orange ones. Suzy grabs 4 at random. Find the probabilities of the indicated events.

21. ● She gets all the red ones, given that she gets the fluorescent pink one.

22. ● She gets all the red ones, given that she does not get the fluorescent pink one.

23. ● She gets none of the red ones, given that she gets the fluorescent pink one.

24. ● She gets one of each color other than fluorescent pink, given that she gets the fluorescent pink one.

25. ● She gets one of each color other than fluorescent pink, given that she gets at least one red one.

26. ● She gets at least two red ones, given that she gets at least one green one.

In Exercises 27–30, say whether the given pairs of events are (A) independent, (B) mutually exclusive, or (C) neither.

27. ● A: Your new skateboard design is a success.

B: Your new skateboard design is a failure.

28. ● A: Your new skateboard design is a success.

B: There is life in the Andromeda galaxy.

29. ● A: Your new skateboard design is a success.

B: Your competitor's new skateboard design is a failure.

30. ● A: Your first coin flip results in heads.

B: Your second coin flip results in heads.

In Exercises 31–36, two dice (one red and one green) are rolled, and the numbers that face up are observed. Test the given pairs of events for independence. hint [see Quick Examples on pp. 495–500]

31. ● A: The red die is 1, 2, or 3; B: The green die is even.

32. ● A: The red die is 1; B: The sum is even.

33. ● A: Exactly one die is 1; B: The sum is even.

34. ● A: Neither die is 1 or 6; B: The sum is even.

35. ● A: Neither die is 1; B: Exactly one die is 2.

36. ● A: Both dice are 1; B: Neither die is 2.

37. ● If a coin is tossed 11 times, find the probability of the sequence H, T, T, H, H, H, T, H, H, T, T. hint [see Example 5]

38. ● If a die is rolled 4 times, find the probability of the sequence 4, 3, 2, 1.

Applications

39. ● In 2004, the probability that a person in the U.S. would declare personal bankruptcy was .006. The probability that a person in the U.S. would declare personal bankruptcy and had recently experienced a "big three" event (loss of job, medical problem, or divorce or separation) was .005.[36] What was the probability that a person had recently experienced one of the "big three" events, given that she had declared personal bankruptcy? (Round your answer to one decimal place.)

40. ● In 2004, the probability that a person in the U.S. would declare personal bankruptcy was .006. The probability that a person in the U.S. would declare personal bankruptcy and had recently overspent credit cards was .0004.[37] What was the probability that a person had recently overspent credit cards given that he had declared personal bankruptcy?

41. ● By the end of 2004, approximately 136 million people were registered at eBay. Of these, 56 million were active users. 95 million, including 41 million of the active users, had registered by the end of 2003.[38] Find the probability that a registered eBay user was an active user, given that the user had registered by the end of 2003. (Round your answer to two decimal places.)

42. ● Refer to the data given in Exercise 41. Find the probability that a registered eBay user had registered by 2003, given that the user was an active user. (Round your answer to two decimal places.)

43. ● *Social Security* According to a *New York Times*/CBS poll released in March, 2005, 79% agreed that it should be the government's responsibility to provide a decent standard of living for the elderly, and 43% agreed that it would be a good idea to invest part of their Social Security taxes on their own.[39] If agreement with one of these propositions is independent of agreement with the other, what is the probability that a person agreed with both propositions? (Round your answer to two decimal places.) hint [see Example 2]

44. ● *Social Security* According to the *New York Times*/CBS poll of March, 2005, referred to in Exercise 43, 49% agreed that Social Security taxes should be raised if necessary to keep the system afloat, and 43% agreed that it would be a good idea to invest part of their Social Security taxes on their own.[40] If agreement with one of these propositions is independent of agreement with the other, what is the probability that a person agreed with both propositions? (Round your answer to two decimal places.)

45. ● *Marketing* A market survey shows that 40% of the population used Brand X laundry detergent last year, 5% of the population gave up doing its laundry last year, and 4% of the population used Brand X and then gave up doing laundry last year. Are the events of using Brand X and giving up doing laundry independent? Is a user of Brand X detergent more or less likely to give up doing laundry than a randomly chosen person?

46. ● *Marketing* A market survey shows that 60% of the population used Brand Z computers last year, 5% of the population quit their jobs last year, and 3% of the population used Brand Z

[36] Probabilities are approximate. SOURCE: *New York Times*, March 13, 2005, p, WK3.

[37] The .0004 figure is an estimate by the authors. SOURCE: *New York Times*, March 13, 2005, p, WK3.

[38] Users who bid, bought, or listed an item within the previous 12 month period.

[39] SOURCE: *New York Times*, March 3, 2005, p. A20.

[40] Ibid.

computers and then quit their jobs. Are the events of using Brand Z computers and quitting your job independent? Is a user of Brand Z computers more or less likely to quit a job than a randomly chosen person?

47. ● *Road Safety* In 1999, the probability that a randomly selected vehicle would be involved in a deadly tire-related accident was approximately 3×10^{-6}, whereas the probability that a tire-related accident would prove deadly was .02.[41] What was the probability that a vehicle would be involved in a tire-related accident?

48. ● *Road Safety* In 1998, the probability that a randomly selected vehicle would be involved in a deadly tire-related accident was approximately 2.8×10^{-6}, while the probability that a tire-related accident would prove deadly was .016.[42] What was the probability that a vehicle would be involved in a tire-related accident?

Publishing Exercises 49–56 are based on the following table, which shows the results of a survey of 100 authors by a publishing company.

	New Authors	Established Authors	Total
Successful	5	25	30
Unsuccessful	15	55	70
Total	20	80	100

Compute the following conditional probabilities:

49. ● An author is established, given that she is successful.

50. ● An author is successful, given that he is established.

51. ● An author is unsuccessful, given that he is a new author.

52. ● An author is a new author, given that she is unsuccessful.

53. ● An author is unsuccessful, given that she is established.

54. ● An author is established, given that he is unsuccessful.

55. ● An unsuccessful author is established.

56. ● An established author is successful.

In Exercises 57–60, draw an appropriate tree diagram and use the multiplication principle to calculate the probabilities of all the outcomes. hint [see Example 3]

57. ● *Sales* Each day, there is a 40% chance that you will sell an automobile. You know that 30% of all the automobiles you sell are two-door models, and the rest are four-door models.

58. ● *Product Reliability* You purchase Brand X floppy discs one quarter of the time and Brand Y floppies the rest of the time. Brand X floppy discs have a 1% failure rate, while Brand Y floppy discs have a 3% failure rate.

59. ● *Car Rentals* Your auto rental company rents out 30 small cars, 24 luxury sedans, and 46 slightly damaged "budget" vehicles. The small cars break down 14% of the time, the luxury sedans break down 8% of the time, and the "budget" cars break down 40% of the time.

60. ● *Travel* It appears that there is only a 1 in 5 chance that you will be able to take your spring vacation to the Greek Islands. If you are lucky enough to go, you will visit either Corfu (20% chance) or Rhodes. On Rhodes, there is a 20% chance of meeting a tall dark stranger, while on Corfu, there is no such chance.

61. ● *Weather Prediction* There is a 50% chance of rain today and a 50% chance of rain tomorrow. Assuming that the event that it rains today is independent of the event that it rains tomorrow, draw a tree diagram showing the probabilities of all outcomes. What is the probability that there will be no rain today or tomorrow?

62. ● *Weather Prediction* There is a 20% chance of snow today and a 20% chance of snow tomorrow. Assuming that the event that it snows today is independent of the event that it snows tomorrow, draw a tree diagram showing the probabilities of all outcomes. What is the probability that it will snow by the end of tomorrow?

Education and Employment Exercises 63–72 are based on the following table, which shows U.S. employment figures in February, 2005, broken down by educational attainment.[43] All numbers are in millions, and represent civilians aged 16 years and over. Those classed as "not in labor force" were not employed nor actively seeking employment. Round all answers to 2 decimal places.

	Employed	Unemployed	Not in Labor Force	Total
Less Than High School Diploma	11.6	1.0	15.5	28.1
High School Diploma Only	36.2	1.9	21.9	60
Some College or Associate's Degree	33.4	1.5	12.9	47.8
Bachelor's Degree or Higher	39.6	1.0	11.2	51.8
Total	120.8	5.4	61.5	187.7

63. ● Find the probability that a person was employed, given that the person had a bachelor's degree or higher.

---

[41] The original data reported 3 tire-related deaths per million vehicles. SOURCE: *New York Times* analysis of National Traffic Safety Administration crash data/Polk Company vehicle registration data/*New York Times*, Nov. 22, 2000, p. C5.

[42] The original data reported 2.8 tire-related deaths per million vehicles. SOURCE: Ibid.

[43] Figures are seasonally adjusted and rounded. SOURCE: Bureau of Labor Statistics press release, March 4, 2005, obtained from www.bls.gov.

● basic skills   ◆ challenging

64. ● Find the probability that a person was employed, given that the person had attained less than a high school diploma.

65. ● Find the probability that a person had a bachelor's degree or higher, given that the person was employed.

66. ● Find the probability that a person had attained less than a high school diploma, given that the person was employed.

67. Find the probability that a person who had not completed a bachelor's degree or higher was not in the labor force.

68. Find the probability that a person who had completed at least a high school diploma was not in the labor force.

69. Find the probability that a person who had completed a bachelor's degree or higher and was in the labor force was employed.

70. Find the probability that a person who had completed less than a high school diploma and was in the labor force was employed.

71. Your friend claims that an unemployed person is more likely to have a high school diploma only than an employed person. Respond to this claim by citing actual probabilities.

72. Your friend claims that a person not in the labor force is more likely to have less than a high school diploma than an employed person. Respond to this claim by citing actual probabilities.

73. *Airbag Safety* According to a study conducted by the Harvard School of Public Health, a child seated in the front seat who was wearing a seatbelt was 31% more likely to be killed in an accident if the car had an air bag that deployed than if it did not.[44] Let the sample space S be the set of all accidents involving a child seated in the front seat wearing a seatbelt. Let K be the event that the child was killed and let D be the event that the airbag deployed. Fill in the missing terms and quantities: $P(\_\_\_ \mid \_\_\_) = \_\_\_ \times P(\_\_\_ \mid \_\_\_)$.

74. *Airbag Safety* According to the study cited in Exercise 73, a child seated in the front seat not wearing a seatbelt was 84% more likely to be killed in an accident if the car had an air bag that deployed than if it did not.[45] Let the sample space S be the set of all accidents involving a child seated in the front seat not wearing a seatbelt. Fill in the missing terms and quantities: $P(\_\_\_ \mid \_\_\_) = \_\_\_ \times P(\_\_\_ \mid \_\_\_)$.

75. *Productivity* A company wishes to enhance productivity by running a one-week training course for its employees. Let T be the event that an employee participated in the course, and let I be the event that an employee's productivity improved the week after the course was run.

   a. Assuming that the course has a positive effect on productivity, how are $P(I|T)$ and $P(I)$ related?

   b. If T and I are independent, what can one conclude about the training course?

76. *Productivity* Consider the events T and I in the preceding exercise.

   a. Assuming that everyone who improved took the course but that not everyone took the course, how are $P(T \mid I)$ and $P(T)$ related?

   b. If half the employees who improved took the course and half the employees took the course, are T and I independent?

77. *Internet Use in 2000* The following pie chart shows the percentage of the population that used the Internet in 2000, broken down further by family income, and based on a survey taken in August, 2000.[46]

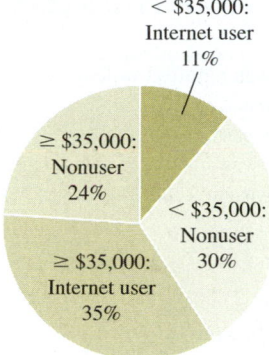

   a. Determine the probability that a randomly chosen person was an Internet user, given that his or her family income was at least $35,000.

   b. Based on the data, was a person more likely to be an Internet user if his or her family income was less than $35,000 or $35,000 or more? (Support your answer by citing the relevant conditional probabilities.)

78. *Internet Use in 2001* Repeat Exercise 77 using the following pie chart, which shows the results of a similar survey taken in September, 2001.[47]

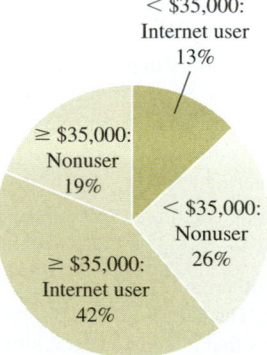

---

[44] The study was conducted by Dr. Segul-Gomez at the Harvard School of Public Health. SOURCE: *New York Times*, December 1, 2000, p. F1.

[45] Ibid.

[46] SOURCE: *Falling Through the Net: Toward Digital Inclusion, A Report on Americans' Access to Technology Tools*, U.S. Department of Commerce, October, 2000. Available at www.ntia.doc.gov/ntiahome/fttn00/contents00.html.

[47] SOURCE: *A Nation Online: How Americans Are Expanding Their Use of the Internet*, U.S. Department of Commerce, February, 2002. Available at www.ntia.doc.gov/ntiahome/dn/index.html.

● basic skills   ◆ challenging

Auto Theft Exercises 79–84 are based on the following table, which shows the probability that an owner of the given model would report his or her vehicle stolen a one-year period.[48]

Brand	Jeep Wrangler	Suzuki Sidekick (2 door)	Toyota Land Cruiser	Geo Tracker (2 door)	Acura Integra (2 door)
Probability	.0170	.0154	.0143	.0142	.0123
Brand	Mitsubishi Montero	Acura Integra (4 door)	BMW 3-series (2 door)	Lexus GS300	Honda Accord (2 door)
Probability	.0108	.0103	.0077	.0074	.0070

In an experiment in which a vehicle is selected, consider the following events:

R: The vehicle was reported stolen.
J: The vehicle was a Jeep Wrangler.
$A2$: The vehicle was an Acura Integra (two-door).
$A4$: The vehicle was an Acura Integra (four-door).
A: The vehicle was an Acura Integra (either two-door or four-door).

79. Fill in the blanks: $P(\underline{\quad} \mid \underline{\quad}) = .0170$.

80. Fill in the blanks: $P(\underline{\quad} \mid A4) = \underline{\quad}$.

81. Which of the following is true?
   (A) There is a 1.43% chance that a vehicle reported stolen was a Toyota Land Cruiser.
   (B) Of all the vehicles reported stolen, 1.43% of them were Toyota Land Cruisers.
   (C) Given that a vehicle was reported stolen, there is a .0143 probability that it was a Toyota Land Cruiser.
   (D) Given that a vehicle was a Toyota Land Cruiser, there was a 1.43% chance that it was reported stolen.

82. Which of the following is true?
   (A) $P(R \mid A) = .0123 + .0103 = .0226$.
   (B) $P(R' \mid A2) = 1 - .0123 = .9877$.
   (C) $P(A2 \mid A) = .0123/(.0123 + .0103) \approx .544$.
   (D) $P(R \mid A2') = 1 - .0123 = .9877$.

83. It is now January, and I own a BMW 3-series and a Lexus GS300. Because I house my vehicles in different places, the event that one of my vehicles gets stolen does not depend on the event that the other gets stolen. Compute each probability to 6 decimal places.

   a. Both my vehicles will get stolen this year.
   b. At least one of my vehicles will get stolen this year.

84. It is now December, and I own a Mitsubishi Montero and a Jeep Wrangler. Because I house my vehicles in different places, the event that one of my vehicles gets stolen does not depend on the event that the other gets stolen. I have just returned from a one-year trip to the Swiss Alps.

   a. What is the probability that my Montero, but not my Wrangler, has been stolen?
   b. Which is more likely: the event that my Montero was stolen or the event that *only* my Montero was stolen?

85. *Drug Tests* If 90% of the athletes who test positive for steroids in fact use them, and 10% of all athletes use steroids and test positive, what percentage of athletes test positive?

86. *Fitness Tests* If 80% of candidates for the soccer team pass the fitness test, and only 20% of all athletes are soccer team candidates who pass the test, what percentage of the athletes are candidates for the soccer team?

87. *Food Safety* According to a University of Maryland study of 200 samples of ground meats,[49] the probability that a sample was contaminated by salmonella was .20. The probability that a salmonella-contaminated sample was contaminated by a strain resistant to at least three antibiotics was .53. What was the probability that a ground meat sample was contaminated by a strain of salmonella resistant to at least three antibiotics?

88. *Food Safety* According to the study mentioned in Exercise 87,[50] the probability that a ground meat sample was contaminated by salmonella was .20. The probability that a salmonella-contaminated sample was contaminated by a strain resistant to at least one antibiotic was .84. What was the probability that a ground meat sample was contaminated by a strain of salmonella resistant to at least one antibiotic?

89. ◆ *Food Safety* According to a University of Maryland study of 200 samples of ground meats,[51] the probability that one of the samples was contaminated by salmonella was .20. The probability that a salmonella-contaminated sample was contaminated by a strain resistant to at least one antibiotic was .84, and the probability that a salmonella-contaminated sample was contaminated by a strain resistant to at least three antibiotics was .53. Find the probability that a ground meat sample that was contaminated by an antibiotic-resistant strain was contaminated by a strain resistant to at least three antibiotics.

90. ◆ *Food Safety* According to a University of Maryland study of 200 samples of ground meats,[52] the probability that a

---

[48] Data are for insured vehicles, for 1995 to 1997 models except Wrangler, which is for 1997 models only. SOURCE: Highway Loss Data Institute/*The New York Times*, March 28, 1999, p. WK3.

[49] As cited in the *New York Times*, October 16, 2001, p. A12.
[50] Ibid.
[51] Ibid.
[52] Ibid.

● basic skills   ◆ challenging

ground meat sample was contaminated by a strain of salmonella resistant to at least three antibiotics was .11. The probability that someone infected with any strain of salmonella will become seriously ill is .10. What is the probability that someone eating a randomly-chosen ground meat sample will not become seriously ill with a strain of salmonella resistant to at least three antibiotics?

Communication and Reasoning Exercises

91. ● Name three events, each independent of the others, when a fair coin is tossed four times.

92. ● Name three pairs of independent events when a pair of distinguishable and fair dice is rolled and the numbers that face up are observed.

93. ● You wish to ascertain the probability of an event E, but you happen to know that the event F has occurred. Is the probability you are seeking $P(E)$ or $P(E|F)$? Give the reason for your answer.

94. ● Your television advertising campaign seems to have been very successful: 10,000 people who saw the ad purchased your product, while only 2000 people purchased the product without seeing the ad. Explain how additional data could show that your ad campaign was, in fact, a failure.

95. If $A \subseteq B$ and $P(B) \neq 0$, why is $P(A | B) = \dfrac{P(A)}{P(B)}$?

96. If $B \subseteq A$ and $P(B) \neq 0$, why is $P(A | B) = 1$?

97. Your best friend thinks that it is impossible for two mutually exclusive events with nonzero probabilities to be independent. Establish whether or not he is correct.

98. Another of your friends thinks that two mutually exclusive events with nonzero probabilities can never be dependent. Establish whether or not she is correct.

99. ◆ Show that if A and B are independent, then so are A' and B' (assuming none of these events has zero probability). [Hint: $A' \cap B'$ is the complement of $A \cup B$.]

100. ◆ Show that if A and B are independent, then so are A and B' (assuming none of these events has zero probability). [Hint: $P(B' | A) + P(B | A) = 1$.]

● basic skills   ◆ challenging

7.6 Bayes' Theorem and Applications

Should schools test their athletes for drug use? A problem with drug testing is that there are always false positive results, so one can never be certain that an athlete who tests positive is in fact using drugs. Here is a typical scenario.

Example 1 Steroids Testing

Gamma Chemicals advertises its anabolic steroid detection test as being 95% effective at detecting steroid use, meaning that it will show a positive result on 95% of all anabolic steroid users. It also states that its test has a false positive rate of 6%. This means that the probability of a nonuser testing positive is .06. Estimating that about 10% of its athletes are using anabolic steroids, Enormous State University (ESU) begins testing its football players. The quarterback, Hugo V. Huge, tests positive and is promptly dropped from the team. Hugo claims that he is not using anabolic steroids. How confident can we be that he is not telling the truth?

Solution There are two events of interest here: the event T that a person tests positive, and the event A that the person tested uses anabolic steroids. Here are the probabilities we are given:

$$P(T | A) = .95$$
$$P(T | A') = .06$$
$$P(A) = .10$$

We are asked to find $P(A \mid T)$, the probability that someone who tests positive is using anabolic steroids. We can use a tree diagram to calculate $P(A \mid T)$. The trick to setting up the tree diagram is to use as the first branching the events with *unconditional* probabilities we know. Because the only unconditional probability we are given is $P(A)$, we use A and A' as our first branching (Figure 12).

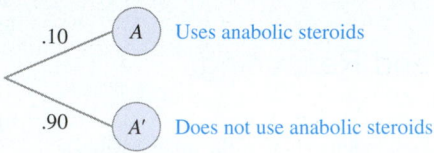

Figure **12**

For the second branching, we use the outcomes of the drug test: positive (T) or negative (T'). The probabilities on these branches are conditional probabilities because they depend on whether or not an athlete uses steroids (see Figure 13). (We fill in the probabilities that are not supplied by remembering that the sum of the probabilities on the branches leaving any node must be 1.)

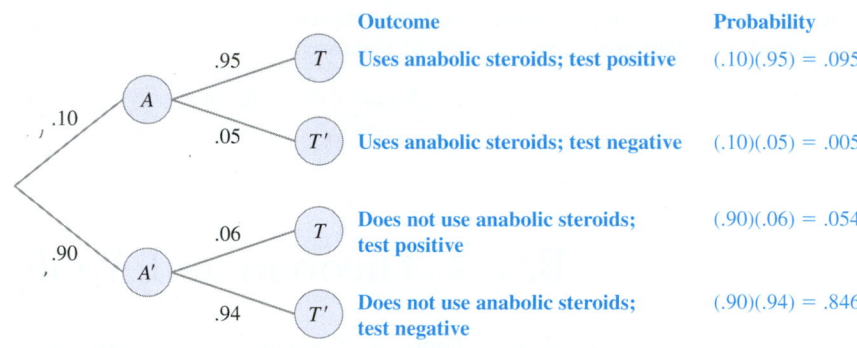

Figure **13**

We can now calculate the probability we are asked to find:

$$P(A \mid T) = \frac{P(A \cap T)}{P(T)} = \frac{P(\text{Uses anabolic steroids and tests positive})}{P(\text{Tests positive})}$$

$$\frac{P(\text{Using } A \text{ and } T \text{ branches})}{\text{Sum of } P(\text{Using branches ending in } T)}$$

From the tree diagram, we see that $P(A \cap T) = .095$. To calculate $P(T)$, the probability of testing positive, notice that there are two outcomes on the tree diagram that reflect a positive test result. The probabilities of these events are .095 and .054. Because these two events are mutually exclusive (an athlete either uses steroids or does not, but not both), the probability of a test being positive (ignoring whether or not steroids are used) is the sum of these probabilities, .149. Thus,

$$P(A \mid T) = \frac{.095}{.095 + .054} = \frac{.095}{.149} \approx .64$$

Thus there is a 64% chance that a randomly selected athlete who tests positive, like Hugo, is using steroids. In other words, we can be 64% confident that Hugo is lying.

✛*Before we go on...* Note that the correct answer in Example 1 is 64%, *not* the 94% we might suspect from the test's false positive rating. In fact, we can't answer the question asked without knowing the percentage of athletes who actually use steroids. For instance, if *no* athletes at all use steroids, then Hugo must be telling the truth, and so the test result has no significance whatsoever. On the other hand, if *all* athletes use steroids, then Hugo is definitely lying, regardless of the outcome of the test.

False positive rates are determined by testing a large number of samples known not to contain drugs and computing estimated probabilities. False negative rates are computed similarly by testing samples known to contain drugs. However, the accuracy of the tests depends also on the skill of those administering them. False positives were a significant problem when drug testing started to become common, with estimates of false positive rates for common immunoassay tests ranging from 10–30% on the high end,[53] but the accuracy has improved since then. Because of the possibility of false positive results, positive immunoassay tests need to be confirmed by the more expensive and much more reliable gas chromatograph/mass spectrometry (GC/MS) test. See also the NCAA's drug testing policy, available at http://www1.ncaa.org/membership/ed_outreach/health-safety/drug_testing/index.html (the section on Institutional Drug Testing addresses the problem of false positives). ■

Bayes' Theorem

The calculation we used to answer the question in Example 1 can be recast as a formula known as **Bayes' theorem.** Figure 14 shows a general form of the tree we used in Example 1.

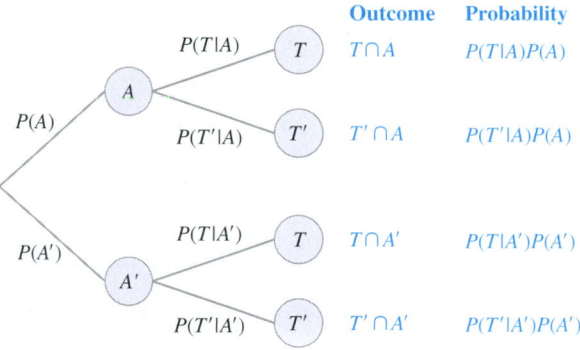

Figure **14**

We first calculated

$$P(A \mid T) = \frac{P(A \cap T)}{P(T)}$$

as follows. We first calculated the numerator $P(A \cap T)$ using the multiplication principle:

$$P(A \cap T) = P(T \mid A)P(A)$$

---

[53] *Drug Testing in the Workplace,* ACLU Briefing Paper, 1996.

We then calculated the denominator $P(T)$ by using the addition principle for mutually exclusive events together with the multiplication principle:

$$P(T) = P(A \cap T) + P(A' \cap T)$$
$$= P(T \mid A)P(A) + P(T \mid A')P(A')$$

Substituting gives

$$P(A \mid T) = \frac{P(T \mid A)P(A)}{P(T \mid A)P(A) + P(T \mid A')P(A')}$$

This is the short form of Bayes' theorem.

Bayes' Theorem (Short Form)

If A and T are events, then

Bayes' Formula

$$P(A \mid T) = \frac{P(T \mid A)P(A)}{P(T \mid A)P(A) + P(T \mid A')P(A')}$$

Using a Tree

$$P(A \mid T) = \frac{P(\text{Using } A \text{ and } T \text{ branches})}{\text{Sum of } P(\text{Using branches ending in } T)}$$

---

quick Example

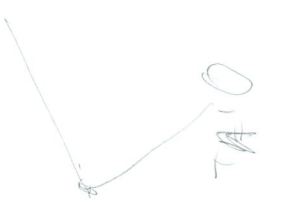

Let us calculate the probability that an athlete from Example 1 who tests positive is actually using steroids if only 5% of ESU athletes are using steroids. Thus,

$$P(T \mid A) = .95$$
$$P(T \mid A') = .06$$
$$P(A) = .05$$
$$P(A') = .95$$

and so

$$P(A \mid T) = \frac{P(T \mid A)P(A)}{P(T \mid A)P(A) + P(T \mid A')P(A')} = \frac{(.95)(.05)}{(.95)(.05) + (.06)(.95)} \approx .45$$

In other words, it is actually more likely that such an athlete does *not* use steroids than he does.[*]

---

[*] Without knowing the results of the test we would have said that there was a probability of $P(A) = 0.05$ that the athlete is using steroids. The positive test result raises the probability to $P(A \mid T) = 0.45$, but the test gives too many false positives for us to be any more certain than that that athlete is actually using steroids.

Remembering the Formula Although the formula looks complicated at first sight, it is not hard to remember if you notice the pattern. Or, you could re-derive it yourself by thinking of the tree diagram.

The next example illustrates that we can use either a tree diagram or the Bayes' theorem formula.

Example 2 Lie Detectors

The Sherlock Lie Detector Company manufactures the latest in lie detectors, and the Count-Your-Pennies (CYP) store chain is eager to use them to screen their employees for theft. Sherlock's advertising claims that the test misses a lie only once in every 100 instances. On the other hand, an analysis by a consumer group reveals 20% of people who are telling the truth fail the test anyway.[*] The local police department estimates that 1 out of every 200 employees has engaged in theft. When the CYP store first screened their employees, the test indicated Mrs. Prudence V. Good was lying when she claimed that she had never stolen from CYP. What is the probability that she was lying and had in fact stolen from the store?

Solution We are asked for the probability that Mrs. Good was lying, and in the preceding sentence we are told that the lie detector test showed her to be lying. So, we are looking for a conditional probability: the probability that she is lying, given that the lie detector test is positive. Now we can start to give names to the events:

L: A subject is lying

T: The test is positive (indicated that the subject was lying)

We are looking for $P(L \mid T)$. We know that 1 out of every 200 employees engages in theft; let us assume that no employee admits to theft while taking a lie detector test, so the probability $P(L)$ that a test subject is lying is $1/200$. We also know the false negative and false positive rates $P(T' \mid L)$ and $P(T \mid L')$.

Using a tree diagram
Figure 15 shows the tree diagram.

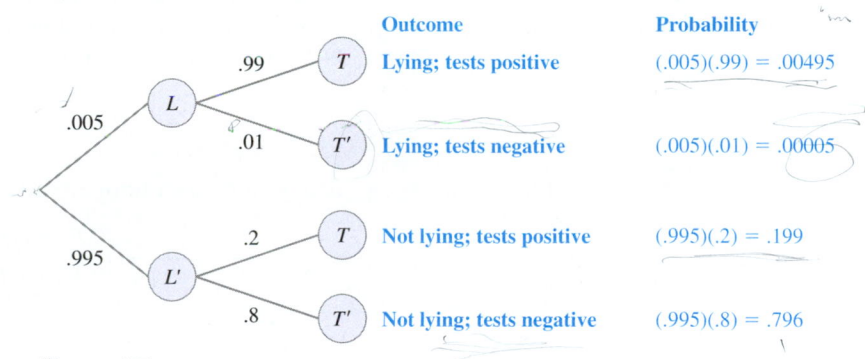

Figure **15**

We see that

$$P(L \mid T) = \frac{P(\text{Using } L \text{ and } T \text{ branches})}{\text{Sum of } P(\text{Using branches ending in } T)}$$

$$= \frac{.00495}{.00495 + .199} \approx .024$$

This means that there was only a 2.4% chance that poor Mrs. Good was lying and had stolen from the store!

---

[*] The reason for this is that many people show physical signs of distress when asked accusatory questions. Many people are nervous around police officers even if they have done nothing wrong.

Using Bayes' Theorem

We have

$$P(L) = .005$$
$$P(T' \mid L) = .01, \text{ from which we obtain}$$
$$P(T \mid L) = .99$$
$$P(T \mid L') = .2$$

and so

$$P(L \mid T) = \frac{P(T \mid L)P(L)}{P(T \mid L)P(L) + P(T \mid L')P(L')} = \frac{(.99)(.005)}{(.99)(.005) + (.2)(.995)} \approx .024$$

Expanded Form of Bayes' Theorem

We have seen the "short form" of Bayes' theorem. What is the "long form?" To motivate an expanded form of Bayes' theorem, look again at the formula we've been using:

$$P(A \mid T) = \frac{P(T \mid A)P(A)}{P(T \mid A)P(A) + P(T \mid A')P(A')}$$

The events A and A' form a **partition** of the sample space S; that is, their union is the whole of S and their intersection is empty (Figure 16).

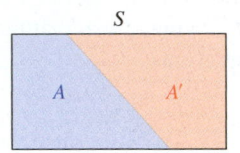

S

A A'

A and A' form a partition of S.

Figure **16**

The expanded form of Bayes' theorem applies to a partition of S into three or more events, as shown in Figure 17.

By saying that the events A_1, A_2 and A_3 form a partition of S, we mean that their union is the whole of S and the intersection of any two of them is empty, as in the figure. When we have a partition into three events as shown, the formula gives us $P(A_1 \mid T)$ in terms of $P(T \mid A_1)$, $P(T \mid A_2)$, $P(T \mid A_3)$, $P(A_1)$, $P(A_2)$, and $P(A_3)$.

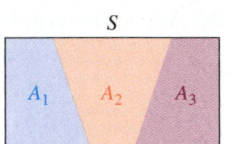

S

A_1 A_2 A_3

A_1, A_2, and A_3 form a partition of S.

Figure **17**

Bayes' Theorem (Expanded Form)

If the events A_1, A_2, and A_3 form a partition of the sample space S, then

$$P(A_1 \mid T) = \frac{P(T \mid A_1)P(A_1)}{P(T \mid A_1)P(A_1) + P(T \mid A_2)P(A_2) + P(T \mid A_3)P(A_3)}$$

As for why this is true, and what happens when we have a partition into *four or more* events, we will wait for the exercises. In practice, as was the case with a partition into two events, we can often compute $P(A_1 \mid T)$ by constructing a tree diagram.

Example **3** School and Work

A survey[*] conducted by the Bureau of Labor Statistics found that approximately 22% of the high school graduating class of 2004 went on to a two-year college, 44% went on to a four-year college, and the remaining 34% did not go on to college. Of those who went on to a two-year college, 52% worked at the same time, 32% of those going on to a

---

[*] "College Enrollment and Work Activity of High School Graduates," U.S. Bureau of Labor Statistics, October 2004, available at www.bls.gov/schedule/archives/all_nr.htm.

four-year college worked, and 62% of those who did not go on to college worked. What percentage of those working had not gone on to college?

Solution We can interpret these percentages as probabilities if we consider the experiment of choosing a member of the high school graduating class of 2004 at random. The events we are interested in are these:

R_1: A graduate went on to a two-year college

R_2: A graduate went on to a four-year college

R_3: A graduate did not go to college

A: A graduate went on to work

The three events R_1, R_2, and R_3 partition the sample space of all graduates into three events. We are given the following probabilities:

$$P(R_1) = .22 \qquad P(R_2) = .44 \qquad P(R_3) = .34$$
$$P(A \mid R_1) = .52 \quad P(A \mid R_2) = .32 \quad P(A \mid R_3) = .62$$

We are asked to find the probability that a graduate who went on to work did not go to college, so we are looking for $P(R_3 \mid A)$. Bayes' formula for these events is

$$P(R_3 \mid A) = \frac{P(A \mid R_3)P(R_3)}{P(A \mid R_1)P(R_1) + P(A \mid R_2)P(R_2) + P(A \mid R_3)P(R_3)}$$

$$= \frac{(.62)(.34)}{(.52)(.22) + (.32)(.44) + (.62)(.34)} \approx .45$$

Thus we conclude that 45% of all those working had not gone on to college.

✚ *Before we go on...* We could also solve Example 3 using a tree diagram. As before, the first branching corresponds to the events with unconditional probabilities that we know: R_1, R_2, and R_3. You should complete the tree and check that you obtain the same result as above. ■

7.6 EXERCISES

● denotes basic skills exercises

◆ denotes challenging exercises

In Exercises 1–8, use Bayes' theorem or a tree diagram to calculate the indicated probability. Round all answers to four decimal places. *hint* [see Quick Examples on p. 510]

1. ● $P(A \mid B) = .8$, $P(B) = .2$, $P(A \mid B') = .3$. Find $P(B \mid A)$.

2. ● $P(A \mid B) = .6$, $P(B) = .3$, $P(A \mid B') = .5$. Find $P(B \mid A)$.

3. ● $P(X \mid Y) = .8$, $P(Y') = .3$, $P(X \mid Y') = .5$. Find $P(Y \mid X)$.

4. ● $P(X \mid Y) = .6$, $P(Y') = .4$, $P(X \mid Y') = .3$. Find $P(Y \mid X)$.

5. ● Y_1, Y_2, Y_3 form a partition of S. $P(X \mid Y_1) = .4$, $P(X \mid Y_2) = .5$, $P(X \mid Y_3) = .6$, $P(Y_1) = .8$, $P(Y_2) = .1$. Find $P(Y_1 \mid X)$.

6. ● Y_1, Y_2, Y_3 form a partition of S. $P(X \mid Y_1) = .2$, $P(X \mid Y_2) = .3$, $P(X \mid Y_3) = .6$, $P(Y_1) = .3$, $P(Y_2) = .4$. Find $P(Y_1 \mid X)$.

7. ● Y_1, Y_2, Y_3 form a partition of S. $P(X \mid Y_1) = .4$, $P(X \mid Y_2) = .5$, $P(X \mid Y_3) = .6$, $P(Y_1) = .8$, $P(Y_2) = .1$. Find $P(Y_2 \mid X)$.

8. ● Y_1, Y_2, Y_3 form a partition of S. $P(X \mid Y_1) = .2$, $P(X \mid Y_2) = .3$, $P(X \mid Y_3) = .6$, $P(Y_1) = .3$, $P(Y_2) = .4$. Find $P(Y_2 \mid X)$.

Applications

9. ● *Music Downloading* According to a study on the effect of music downloading on spending on music, 11% of all

● basic skills   ◆ challenging

Internet users had decreased their spending on music.[54] We estimate that 40% of all music fans used the Internet at the time of the study.[55] If 20% of non-Internet users had decreased their spending on music, what percentage of those who had decreased their spending on music were Internet users? *hint* [see Examples 1 and 2]

10. ● *Music Downloading* According to the study cited in the preceding exercise, 36% of experienced file-sharers with broadband access had decreased their spending on music. Let us estimate that 3% of all music fans were experienced file-sharers with broadband access at the time of the study.[56] If 20% of the other music fans had decreased their spending on music, what percentage of those who had decreased their spending on music were experienced file-sharers with broadband access?

11. ● *Weather* It snows in Greenland an average of once every 25 days, and when it does, glaciers have a 20% chance of growing. When it does not snow in Greenland, glaciers have only a 4% chance of growing. What is the probability that it is snowing in Greenland when glaciers are growing?

12. ● *Weather* It rains in Spain an average of once every 10 days, and when it does, hurricanes have a 2% chance of happening in Hartford. When it does not rain in Spain, hurricanes have a 1% chance of happening in Hartford. What is the probability that it rains in Spain when hurricanes happen in Hartford?

13. ● *Side Impact Hazard* In 2004, 45.4% of all light vehicles were cars, and the rest were pickups or SUVs. The probability that a severe side-impact crash would prove deadly to a driver depended on the type of vehicle he or she was driving at the time, as shown in the table:[57]

Car	1.0
Light Truck or SUV	.3

What is the probability that the victim of a deadly side-impact accident was driving a car?

14. ● *Side Impact Hazard* In 2004, 27.3% of all light vehicles were light trucks, and the rest were cars or SUVs. The probability that a severe side-impact crash would prove deadly to a driver depended on the type of vehicle he or she was driving

at the time, as shown in the table:[58]

Light Truck	.2
Car or SUV	.7

What is the probability that the victim of a deadly side-impact accident was driving a car or SUV?

15. ● *Athletic Fitness Tests* Any athlete who fails the Enormous State University's women's soccer fitness test is automatically dropped from the team. Last year, Mona Header failed the test, but claimed that this was due to the early hour. (The fitness test is traditionally given at 5 AM on a Sunday morning.) In fact, a study by the ESU Physical Education Department suggested that 50% of athletes fit enough to play on the team would fail the soccer test, although no unfit athlete could possibly pass the test. It also estimated that 45% of the athletes who take the test are fit enough to play soccer. Assuming these estimates are correct, what is the probability that Mona was justifiably dropped?

16. ● *Academic Testing* Professor Frank Nabarro insists that all senior physics majors take his notorious physics aptitude test. The test is so tough that anyone *not* going on to a career in physics has no hope of passing, whereas 60% of the seniors who do go on to a career in physics still fail the test. Further, 75% of all senior physics majors in fact go on to a career in physics. Assuming that you fail the test, what is the probability that you will not go on to a career in physics?

17. ● *Side Impact Hazard* (Compare Exercise 13.) In 2004, 27.3% of all light vehicles were light trucks, 27.3% were SUVs, and 45.4% were cars. The probability that a severe side-impact crash would prove deadly to a driver depended on the type of vehicle he or she was driving at the time, as shown in the table:[59]

Light Truck	.210
SUV	.371
Car	1.000

What is the probability that the victim of a deadly side-impact accident was driving an SUV? *hint* [see Example 3]

18. ● *Side Impact Hazard* In 1986, 23.9% of all light vehicles were pickups, 5.0% were SUVs, and 71.1% were cars. Refer to Exercise 17 for the probabilities that a severe side-impact crash would prove deadly. What is the probability that the victim of a deadly side-impact accident was driving a car?

19. ● *University Admissions* In fall 2004, UCLA admitted 24% of its California resident applicants, 22% of its applicants from

[54] Regardless of whether they used the Internet to download music. SOURCE: *New York Times,* May 6, 2002, p. C6.

[55] According to the U.S. Department of Commerce, 51% of all U.S. households had computers in 2001.

[56] Around 15% of all online households had broadband access in 2001 according to a *New York Times* article (Dec 24, 2001, p. C1).

[57] A "serious" side-impact accident is defined as one in which the driver of a car would be killed. SOURCE: National Highway Traffic Safety Administration/*New York Times,* May 30, 2004, p. BU 9.

[58] Ibid.

[59] Ibid.

● basic skills   ◆ challenging

other U.S. states, and 20% of its international student applicants. Of all its applicants, 87% were California residents, 9% were from other U.S. states, and 4% were international students.[60] What percentage of all admitted students were California residents? (Round your answer to the nearest 1%.)

20. ● *University Admissions* In fall 2002, UCLA admitted 26% of its California resident applicants, 18% of its applicants from other U.S. states, and 13% of its international student applicants. Of all its applicants, 86% were California residents, 11% were from other U.S. states, and 4% were international students.[61] What percentage of all admitted students were California residents? (Round your answer to the nearest 1%.)

21. ● *Internet Use* In 2000, 86% of all Caucasians in the U.S., 77% of all African-Americans, 77% of all Hispanics, and 85% of residents not classified into one of these groups used the Internet for e-mail.[62] At that time, the U.S. population was 69% Caucasian, 12% African-American, and 13% Hispanic. What percentage of U.S. residents who used the Internet for e-mail were Hispanic?

22. ● *Internet Use* In 2000, 59% of all Caucasians in the U.S., 57% of all African-Americans, 58% of all Hispanics, and 54% of residents not classified into one of these groups used the Internet to search for information.[63] At that time, the U.S. population was 69% Caucasian, 12% African-American, and 13% Hispanic. What percentage of U.S. residents who used the Internet for information search were African-American?

23. *Market Surveys* A *New York Times* survey[64] of homeowners showed that 86% of those with swimming pools were married couples, and the other 14% were single. It also showed that 15% of all homeowners had pools.

   a. Assuming that 90% of all homeowners without pools are married couples, what percentage of homes owned by married couples have pools?

   b. Would it better pay pool manufacturers to go after single homeowners or married homeowners? Explain.

24. *Crime and Preschool*. Another *New York Times* survey[65] of needy and disabled youths showed that 51% of those who had no preschool education were arrested or charged with a crime by the time they were 19, whereas only 31% who had preschool education wound up in this category. The survey

did not specify what percentage of the youths in the survey had preschool education, so let us take a guess at that and estimate that 20% of them had attended preschool.

   a. What percentage of the youths arrested or charged with a crime had no preschool education?

   b. What would this figure be if 80% of the youths had attended preschool? Would youths who had preschool education be more likely to be arrested or charged with a crime than those who did not? Support your answer by quoting probabilities.

25. ◆ *Grade Complaints* Two of the mathematics professors at Enormous State are Professor A (known for easy grading) and Professor F (known for tough grading). Last semester, roughly three quarters of Professor F's class consisted of former students of Professor A; these students apparently felt encouraged by their (utterly undeserved) high grades. (Professor F's own former students had fled in droves to Professor A's class to try to shore up their grade point averages.) At the end of the semester, as might have been predicted, all of Professor A's former students wound up with a C– or lower. The rest of the students in the class—former students of Professor F who had decided to "stick it out"—fared better, and two-thirds of them earned higher than a C–. After discovering what had befallen them, all the students who earned C– or lower got together and decided to send a delegation to the Department Chair to complain that their grade point averages had been ruined by this callous and heartless beast! The contingent was to consist of 10 representatives selected at random from among them. How many of the 10 would you estimate to have been former students of Professor A?

26. *Weather Prediction* A local TV station employs Desmorelda, "Mistress of the Zodiac," as its weather forecaster. Now, when it rains, Sagittarius is in the shadow of Jupiter one-third of the time, and it rains on 4 out of every 50 days. Sagittarius falls in Jupiter's shadow on only one in every five rainless days. The powers that be at the station notice a disturbing pattern to Desmorelda's weather predictions. It seems that she always predicts that it will rain when Sagittarius is in the shadow of Jupiter. What percentage of the time is she correct? Should they replace her?

27. *Employment in the 1980s* In a 1987 survey of married couples with earnings, 95% of all husbands were employed. Of all employed husbands, 71% of their wives were also employed.[66] Noting that either the husband or wife in a couple with earnings had to be employed, find the probability that the husband of an employed woman was also employed.

28. *Employment in the 1980s* Repeat the preceding exercise in the event that 50% of all husbands were employed.

[60] SOURCE: UCLA Web site, May, 2005. www.admissions.ucla.edu/Prospect/Adm_fr/Frosh_Prof04.htm.

[61] SOURCE: UCLA Web site, May, 2002. www.admissions.ucla.edu/Prospect/Adm_fr/Frosh_Prof.htm.

[62] SOURCE: NTIA and ESA, U.S. Department of Commerce, using August 2000 U.S. Bureau of The Census Current Population Survey Supplement.

[63] Ibid.

[64] SOURCE: *All about Swimming Pools, The New York Times,* September 13, 1992.

[65] SOURCE: *Governors Develop Plan to Help Preschool Children, The New York Times,* August 2, 1992.

[66] SOURCE: *Statistical Abstract of the United States*, 111th Ed., 1991, U.S. Dept. of Commerce/U.S. Bureau of Labor Statistics. Figures rounded to the nearest 1%.

● basic skills   ◆ challenging

29. *Juvenile Delinquency* According to a study at the Oregon Social Learning Center, boys who had been arrested by age 14 were 17.9 times more likely to become chronic offenders than those who had not.[67] Use these data to estimate the percentage of chronic offenders who had been arrested by age 14 in a city where 0.1% of all boys have been arrested by age 14. (Hint: Use Bayes' formula rather than a tree.)

30. *Crime* According to the same study at the Oregon Social Learning Center, chronic offenders were 14.3 times more likely to commit violent offenses than people who were not chronic offenders.[68] In a neighborhood where 2 in every 1000 residents is a chronic offender, estimate the probability that a violent offender is also a chronic offender. (Hint: Use Bayes' formula rather than a tree.)

31. *Benefits of Exercise* According to a study in *The New England Journal of Medicine*,[69] 202 of a sample of 5990 middle-aged men had developed diabetes. It also found that men who were very active (burning about 3500 calories daily) were half as likely to develop diabetes compared with men who were sedentary. Assume that one-third of all middle-aged men are very active, and the rest are classified as sedentary. What is the probability that a middle-aged man with diabetes is very active?

32. *Benefits of Exercise* Repeat Exercise 31, assuming that only one in ten middle-aged men is very active.

33. ◆ *Airbag Safety* According to a study conducted by the Harvard School of Public Health, a child seated in the front seat who was wearing a seatbelt was 31% more likely to be killed in an accident if the car had an air bag that deployed than if it did not.[70] Airbags deployed in 25% of all accidents. For a child seated in the front seat wearing a seatbelt, what is the probability that the airbag deployed in an accident in which the child was killed? (Round your answer to two decimal places.)

34. ◆ *Airbag Safety* According to the study cited in Exercise 33, a child seated in the front seat who was not wearing a seatbelt was 84% more likely to be killed in an accident if the car had an air bag that deployed than if it did not.[71] Airbags deployed in 25% of all accidents. For a child seated in the front seat not wearing a seatbelt, what is the probability that the airbag deployed in an accident in which the child was killed? (Round your answer to two decimal places.)

Communication and Reasoning Exercises

35. ● Your friend claims that the probability of *A* given *B* is the same as the probability of *B* given A. How would you convince him that he is wrong?

36. ● Complete the following sentence. To use Bayes' formula to compute $P(E \mid F)$, you need to be given _____.

37. Give an example in which a steroids test gives a false positive only 1 percent of the time, and yet if an athlete tests positive, the chance that he or she has used steroids is under 10%.

38. Give an example in which a steroids test gives a false positive 30% of the time, and yet if an athlete tests positive, the chance that he or she has used steroids is over 90%.

39. Use a tree to derive the expanded form of Bayes' Theorem for a partition of the sample space *S* into three events R_1, R_2, and R_3.

40. Write down an expanded form of Bayes' Theorem that applies to a partition of the sample space *S* into four events R_1, R_2, R_3, and R_4.

41. ◆ *Politics* The following letter appeared in the *New York Times*.[72]

> To the Editor:
>
>   It stretches credulity when William Safire contends (column, Jan. 11) that 90 percent of those who agreed with his Jan. 8 column, in which he called the First Lady, Hillary Rodham Clinton, "a congenital liar," were men and 90 percent of those who disagreed were women.
>
>   Assuming these percentages hold for Democrats as well as Republicans, only 10 percent of Democratic men disagreed with him. Is Mr. Safire suggesting that 90 percent of Democratic men supported him? How naive does he take his readers to be?
>
> A. D.
> New York, Jan. 12, 1996

Comment on the letter writer's reasoning.

42. ◆ *Politics* Refer back to the preceding exercise. If the letter writer's conclusion was correct, what percentage of all Democrats would have agreed with Safire's column?

---

[67] Based on a study, by Marion S. Forgatch, of 319 boys from high-crime neighborhoods in Eugene, Oregon. SOURCE: W, Wayt Gibbs, Seeking the Criminal Element, *Scientific American,* March 1995, pp. 101–107.

[68] Ibid.

[69] As cited in an article in *The New York Times* on July 18, 1991.

[70] The study was conducted by Dr. Segul-Gomez at the Harvard School of Public Health. SOURCE: *The New York Times,* December 1, 2000, p. F1.

[71] Ibid.

---

[72] The original letter appeared in *The New York Times*, January 16, 1996, p. A16. The authors have edited the first phrase of the second paragraph slightly for clarity; the original sentence read: "Assuming the response was equally divided between Democrats and Republicans, . . . "

7.7 Markov Systems

Many real-life situations can be modeled by processes that pass from state to state with given probabilities. A simple example of such a **Markov system** is the fluctuation of a gambler's fortune as he or she continues to bet. Other examples come from the study of trends in the commercial world and the study of neural networks and artificial intelligence. The mathematics of Markov systems is an interesting combination of probability and matrix arithmetic.

Here is a basic example we shall use many times: A market analyst for Gamble Detergents is interested in whether consumers prefer powdered laundry detergents or liquid detergents. Two market surveys taken one year apart revealed that 20% of powdered detergent users had switched to liquid one year later, while the rest were still using powder. Only 10% of liquid detergent users had switched to powder one year later, with the rest still using liquid.

We analyze this example as follows: Every year a consumer may be in one of two possible **states:** He may be a powdered detergent user or a liquid detergent user. Let us number these states: A consumer is in state 1 if he uses powdered detergent and in state 2 if he uses liquid. There is a basic **time step** of one year. If a consumer happens to be in state 1 during a given year, then there is a probability of 20% = .2 (the chance that a randomly chosen powder user will switch to liquid) that he will be in state 2 the next year. We write

$$p_{12} = .2$$

to indicate that the probability of going *from* state 1 *to* state 2 in one time step is .2. The other 80% of the powder users are using powder the next year. We write

$$p_{11} = .8$$

to indicate that the probability of *staying* in state 1 from one year to the next is .8.[†] What if a consumer is in state 2? Then the probability of going to state 1 is given as 10% = .1, so the probability of remaining in state 2 is .9. Thus,

$$p_{21} = .1$$

and

$$p_{22} = .9$$

We can picture this system as in Figure 18, which shows the **state transition diagram** for this example. The numbers p_{ij}, which appear as labels on the arrows, are the **transition probabilities.**

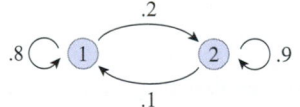

Figure **18**

[†] Notice that these are actually *conditional* probabilities. For instance, p_{12} is the probability that the system (the consumer in this case) will go into state 2, *given that the system* (the consumer) *is in state 1.*

Markov System, States, and Transition Probabilities

A **Markov system**[*] (or **Markov process** or **Markov chain**) is a system that can be in one of several specified **states.** There is specified a certain **time step,** and at each step the system will randomly change states or remain where it is. The probability of going from state i to state j is a fixed number p_{ij}, called the **transition probability.**

---

[*] Named after the Russian mathematician A.A. Markov (1856–1922) who first studied these "non-deterministic" processes.

quick Example

The Markov system depicted in Figure 18 has two states: state 1 and state 2. The transition probabilities are as follows:

$$p_{11} = \text{Probability of going from state 1 to state 1} = .8$$
$$p_{12} = \text{Probability of going from state 1 to state 2} = .2$$
$$p_{21} = \text{Probability of going from state 2 to state 1} = .1$$
$$p_{22} = \text{Probability of going from state 2 to state 2} = .9$$

Notice that, because the system must go somewhere at each time step, the transition probabilities originating at a particular state always add up to 1. For example, in the transition diagram above, when we add the probabilities originating at state 1, we get $.8 + .2 = 1$.

The transition probabilities may be conveniently arranged in a matrix.

Transition Matrix

The **transition matrix** associated with a given Markov system is the matrix P whose ijth entry is the transition probability p_{ij}, the transition probability of going *from* state *i to* state *j*. In other words, the entry in position *ij* is the *label on the arrow going from state i to state j* in a state transition diagram.

Thus, the transition matrix for a system with two states would be set up as follows:

To:

$$\text{From: } \begin{matrix} & 1 & 2 \\ 1 & \begin{bmatrix} p_{11} & p_{12} \\ p_{21} & p_{22} \end{bmatrix} \end{matrix}$$

Arrows Originating in State 1
Arrows Originating in State 2

quick Example

In the system pictured in Figure 18, the transition matrix is

$$P = \begin{bmatrix} .8 & .2 \\ .1 & .9 \end{bmatrix}$$

Note Notice that because the sum of the transition probabilities that originate at any state is 1, *the sum of the entries in any row of a transition matrix is 1.* ∎

Now, let's start doing some calculations.

Example 1 Laundry Detergent Switching

Consider the Markov system found by Gamble Detergents at the beginning of this section. Suppose that 70% of consumers are now using powdered detergent, while the other 30% are using liquid.

a. What will be the distribution one year from now? (That is, what percentage will be using powdered and what percentage liquid detergent?)

b. Assuming that the probabilities remain the same, what will be the distribution two years from now? Three years from now?

Solution

a. First, let us think of the statement that 70% of consumers are using powdered detergent as telling us a probability: The probability that a randomly chosen consumer uses powdered detergent is .7. Similarly, the probability that a randomly chosen consumer uses liquid detergent is .3. We want to find the corresponding probabilities one year from now. To do this, consider the tree diagram in Figure 19.

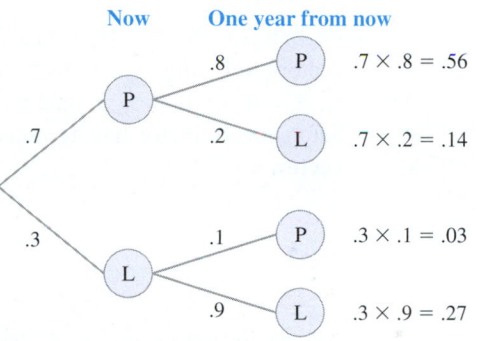

Figure **19**

The first branching shows the probabilities now, while the second branching shows the (conditional) transition probabilities. So, if we want to know the probability that a consumer is using powdered detergent one year from now, it will be

Probability of using powder after one year $= .7 \times .8 + .3 \times .1 = .59$.

On the other hand, we have:

Probability of using liquid after one year $= .7 \times .2 + .3 \times .9 = .41$.

Now, here's the crucial point: *These are exactly the same calculations as in the matrix product*

$$[.7 \quad .3]\begin{bmatrix} .8 & .2 \\ .1 & .9 \end{bmatrix} = [.59 \quad .41]$$

↑   ↑   ↑
Initial distribution   Transition matrix   Distribution after 1 step

Thus, to get the distribution of detergent users after one year, all we have to do is multiply the **initial distribution vector** [.7 .3] by the transition matrix P. The result is [.59 .41], the **distribution vector after one step.**

b. Now what about the distribution after *two* years? If we assume that the same fraction of consumers switch or stay put in the second year as in the first, we can simply repeat the calculation we did above, using the new distribution vector:

$$[.59 \quad .41]\begin{bmatrix} .8 & .2 \\ .1 & .9 \end{bmatrix} = [.513 \quad .487]$$

↑   ↑   ↑
Distribution after 1 step   Transition matrix   Distribution after 2 steps

using *Technology*

See the Technology Guides at the end of the chapter for convenient ways to use a graphing calculator or Excel to find distribution vectors after several steps. Online, follow:

Chapter 7

→ Online Utilities

  → Matrix Algebra Tool

You can then enter the transition matrix P and the initial distribution vector v in the input area, and compute the various distribution vectors using the following formulas:

v*P   1 step
v*P^2   2 steps
v*P^3   3 steps

Thus, after two years we can expect 51.3% of consumers to be using powdered detergent and 48.7% to be using liquid detergent. Similarly, after three years we have

$$[.513 \quad .487] \begin{bmatrix} .8 & .2 \\ .1 & .9 \end{bmatrix} = [.4591 \quad .5409]$$

So, after three years 45.91% of consumers will be using powdered detergent and 54.09% will be using liquid. Slowly but surely, liquid detergent seems to be winning.

+*Before we go on...* Note that the sum of the entries is 1 in each of the distribution vectors in Example 1. In fact, these vectors are giving the probability distributions for each year of finding a randomly chosen consumer using either powdered or liquid detergent. A vector having nonnegative entries adding up to 1 is called a **probability vector.** ∎

Distribution Vector after *m* Steps

A **distribution vector** is a probability vector giving the probability distribution for finding a Markov system in its various possible states. If v is a distribution vector, then the distribution vector one step later will be vP. The distribution m steps later will be

$$\text{Distribution after } m \text{ steps} = v \cdot P \cdot P \cdot \ldots \cdot P (m \text{ times}) = vP^m$$

quick Example

If $P = \begin{bmatrix} 0 & 1 \\ .5 & .5 \end{bmatrix}$ and $v = [.2 \quad .8]$, then we can calculate the following distribution vectors:

$$vP = [.2 \quad .8] \begin{bmatrix} 0 & 1 \\ .5 & .5 \end{bmatrix} = [.4 \quad .6] \qquad \text{Distribution after one step.}$$

$$vP^2 = (vP)P = [.4 \quad .6] \begin{bmatrix} 0 & 1 \\ .5 & .5 \end{bmatrix} = [.3 \quad .7] \qquad \text{Distribution after two steps.}$$

$$vP^3 = (vP^2)P = [.3 \quad .7] \begin{bmatrix} 0 & 1 \\ .5 & .5 \end{bmatrix} = [.35 \quad .65] \qquad \text{Distribution after three steps.}$$

What about the matrix P^m that appears above? Multiplying a distribution vector v times P^m gives us the distribution m steps later, so we can think of P^m and the m-step transition matrix. More explicitly, consider the following example.

Example 2 Powers of the Transition Matrix

Continuing the example of detergent switching, suppose that a consumer is now using powdered detergent. What are the probabilities that the consumer will be using powdered or liquid detergent two years from now? What if the consumer is now using liquid detergent?

Solution To record the fact that we know that the consumer is using powdered detergent, we can take as our initial distribution vector $v = [1 \ 0]$. To find the distribution two years from now, we compute vP^2. To make a point, we do the calculation slightly differently:

$$vP^2 = [1 \quad 0] \begin{bmatrix} .8 & .2 \\ .1 & .9 \end{bmatrix} \begin{bmatrix} .8 & .2 \\ .1 & .9 \end{bmatrix}$$

$$= [1 \quad 0] \begin{bmatrix} .66 & .34 \\ .17 & .83 \end{bmatrix}$$

$$= [.66 \quad .34]$$

So, the probability that our consumer is using powdered detergent two years from now is .66, while the probability of using liquid detergent is .34. The point to notice is that these are the entries in the first row of P^2. Similarly, if we consider a consumer now using liquid detergent, we should take the initial distribution vector to be $v = [0 \quad 1]$ and compute

$$vP^2 = [0 \quad 1] \begin{bmatrix} .66 & .34 \\ .17 & .83 \end{bmatrix} = [.17 \quad .83]$$

Thus, the bottom row gives the probabilities that a consumer, now using liquid detergent, will be using either powdered or liquid detergent two years from now.

In other words, the ijth entry of P^2 gives the probability that a consumer, starting in state i, will be in state j after two time steps.

What is true in Example 2 for two time steps is true for any number of time steps:

Powers of the Transition Matrix

$P^m (m = 1, 2, 3, \ldots)$ is the **m-step transition matrix.** The ijth entry in P^m is the probability of a transition from state i to state j in m steps.

quick Example

If $P = \begin{bmatrix} 0 & 1 \\ .5 & .5 \end{bmatrix}$ then

$$P^2 = P \cdot P = \begin{bmatrix} .5 & .5 \\ .25 & .75 \end{bmatrix} \qquad \text{2-Step transition matrix}$$

$$P^3 = P \cdot P^2 = \begin{bmatrix} .25 & .75 \\ .375 & .625 \end{bmatrix} \qquad \text{3-Step transition matrix}$$

The probability of going from state 1 to state 2 in 2 steps = (1, 2)-entry of P^2 = .5.
The probability of going from state 1 to state 2 in 3 steps = (1, 2)-entry of P^3 = .75.

What happens if we follow our laundry detergent-using consumers for many years?

Example 3 Long-Term Behavior

Suppose that 70% of consumers are now using powdered detergent while the other 30% are using liquid. Assuming that the transition matrix remains valid the whole time, what will be the distribution 1, 2, 3, . . . , and 50 years later?

Solution Of course, to do this many matrix multiplications, we're best off using technology. We already did the first three calculations in an earlier example.

Distribution after 1 year:
$$[.7 \quad .3]\begin{bmatrix} .8 & .2 \\ .1 & .9 \end{bmatrix} = [.59 \quad .41]$$

Distribution after 2 years:
$$[.59 \quad .41]\begin{bmatrix} .8 & .2 \\ .1 & .9 \end{bmatrix} = [.513 \quad .487]$$

Distribution after 3 years:
$$[.513 \quad .487]\begin{bmatrix} .8 & .2 \\ .1 & .9 \end{bmatrix} = [.4591 \quad .5409]$$

. . .

Distribution after 48 years: $\quad [.33333335 \quad .66666665]$

Distribution after 49 years: $\quad [.33333334 \quad .66666666]$

Distribution after 50 years: $\quad [.33333334 \quad .66666666]$

Thus, the distribution after 50 years is approximately $[.33333334 \quad .66666666]$.

Something interesting seems to be happening in Example 3. The distribution seems to be getting closer and closer to

$$[.333333\ldots \quad .666666\ldots] = \begin{bmatrix} \dfrac{1}{3} & \dfrac{2}{3} \end{bmatrix}$$

Let's call this distribution vector v_∞. Notice two things about v_∞:

• v_∞ is a probability vector.
• If we calculate $v_\infty P$, we find

$$v_\infty P = \begin{bmatrix} \dfrac{1}{3} & \dfrac{2}{3} \end{bmatrix}\begin{bmatrix} .8 & .2 \\ .1 & .9 \end{bmatrix} = \begin{bmatrix} \dfrac{1}{3} & \dfrac{2}{3} \end{bmatrix} = v_\infty$$

In other words,

$$v_\infty P = v_\infty$$

We call a probability vector v with the property that $vP = v$ a **steady-state (probability) vector.**

Q: Where does the name steady-state *vector come from?*

A: If $vP = v$, then v is a distribution that will not change from time step to time step. In the example above, because [1/3 2/3] is a steady-state vector, if 1/3 of consumers use powdered detergent and 2/3 use liquid detergent one year, then the proportions will be the same the next year. Individual consumers may still switch from year to year, but as many will switch from powder to liquid as switch from liquid to powder, so the number using each will remain constant. ∎

But how do we find a steady-state vector?

Example 4 Calculating the Steady-State Vector

Calculate the steady-state probability vector for the transition matrix in the preceding examples:

$$P = \begin{bmatrix} .8 & .2 \\ .1 & .9 \end{bmatrix}$$

Solution We are asked to find

$$v_\infty = [x \quad y]$$

This vector must satisfy the equation

$$v_\infty P = v_\infty$$

or

$$[x \quad y]\begin{bmatrix} .8 & .2 \\ .1 & .9 \end{bmatrix} = [x \quad y]$$

Doing the matrix multiplication gives

$$[.8x + .1y \quad .2x + .9y] = [x \quad y]$$

Equating corresponding entries gives

$$.8x + .1y = x$$
$$.2x + .9y = y$$

or

$$-.2x + .1y = 0$$
$$.2x - .1y = 0$$

Now these equations are really the same equation. (Do you see that?) There is one more thing we know, though: Because $[x \quad y]$ is a probability vector, its entries must add up to 1. This gives one more equation:

$$x + y = 1$$

Taking this equation together with one of the two equations above gives us the following system:

$$x + \quad y = 1$$
$$-.2x + .1y = 0$$

We now solve this system using any of the techniques we learned for solving systems of linear equations. We find that the solution is $x = 1/3$, and $y = 2/3$, so the steady-state vector is

$$v_\infty = [x \quad y] = \begin{bmatrix} \dfrac{1}{3} & \dfrac{2}{3} \end{bmatrix}$$

as suggested in Example 3.

The method we just used works for any size transition matrix and can be summarized as follows.

Calculating the Steady-State Distribution Vector

To calculate the steady-state probability vector for a Markov System with transition matrix P, we solve the system of equations given by

$$x + y + z + \cdots = 1$$
$$[x \quad y \quad z \ldots]P = [x \quad y \quad z \ldots]$$

where we use as many unknowns as there are states in the Markov system. The steady-state probability vector is then

$$v_\infty = [x \quad y \quad z \ldots]$$

using *Technology*

See the Technology Guides at the end of section to see how to use a graphing calculator or Excel to find the steady-state probability vector. Online, you can use the Pivot and Gauss-Jordan Tool to solve the system of equations that gives you v_∞ by following the path

Chapter 7
→ Online Utilities
  → Pivot and Gauss-Jordan Tool

Q: *Is there always a steady-state distribution vector* **?**

A: Yes, although the explanation why is more involved than we can give here. ■

Q: *In Example 3, we started with a distribution vector v and found that vP^m got closer and closer to v_∞ as m got larger. Does that always happen* **?**

A: It does if the Markov system is **regular,** as we define below, but may not for other kinds of systems. Again, we shall not prove this fact here. ■

Regular Markov Systems

A **regular** Markov system is one for which some power of its transition matrix P has no zero entries. If a Markov system is regular, then

1. It has a unique steady-state probability vector v_∞, and

2. If v is any probability vector whatsoever, then vP^m approaches v_∞ as m gets large. We say that the **long-term behavior** of the system is to have distribution (close to) v_∞.

Interpreting the Steady-State Vector
In a regular Markov system, the entries in the steady-state probability vector give the long-term probabilities that the system will be in the corresponding states, or the fractions of time one can expect to find the Markov system in the corresponding states.

quick Examples

1. The system with transition matrix $P = \begin{bmatrix} .8 & .2 \\ .1 & .9 \end{bmatrix}$ is regular because $P(= P^1)$ has no zero entries.

2. The system with transition matrix $P = \begin{bmatrix} 0 & 1 \\ .5 & .5 \end{bmatrix}$ is regular because $P^2 = \begin{bmatrix} .5 & .5 \\ .25 & .75 \end{bmatrix}$ has no zero entries.

3. The system with transition matrix $P = \begin{bmatrix} 0 & 1 \\ 1 & 0 \end{bmatrix}$ is *not* regular: $P^2 = \begin{bmatrix} 1 & 0 \\ 0 & 1 \end{bmatrix}$ and $P^3 = P$ again, so the powers of P alternate between these two matrices. Thus, every power of P has zero entries. Although this system has a steady-state vector, namely $[.5 \quad .5]$, if we take $v = [1 \quad 0]$, then $vP = [0 \quad 1]$ and $vP^2 = v$, so the distribution vectors vP^m just alternate between these two vectors, not approaching v_∞.

We finish with one more example.

Example 5 Gambler's Ruin

A timid gambler, armed with her annual bonus of $20, decides to play roulette using the following scheme. At each spin of the wheel, she places $10 on red. If red comes up, she wins an additional $10; if black comes up, she loses her $10. For the sake of simplicity, assume that she has a probability of 1/2 of winning. (In the real game, the probability is slightly lower—a fact that many gamblers forget.) She keeps playing until she has either gotten up to $30 or lost it all. In either case, she then packs up and leaves. Model this situation as a Markov system and find the associated transition matrix. What can we say about the long-term behavior of this system?

Solution We must first decide on the states of the system. A good choice is the gambler's financial state, the amount of money she has at any stage of the game. According to her rules, she can be broke, have $10, $20, or $30. Thus, there are four states: $1 = \$0$, $2 = \$10$, $3 = \$20$, and $4 = \$30$. Because she bets $10 each time, she moves down $10 if she loses (with probability $1/2$) and up $10 if she wins (with probability also $1/2$), until she reaches one of the extremes. The transition diagram is shown in Figure 20.

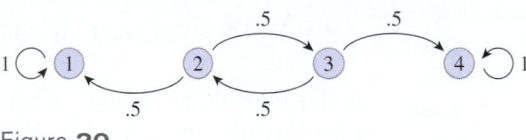

Figure **20**

Note that once the system enters state 1 or state 4, it does not leave; with probability 1, it stays in the same state.[*] We call such states **absorbing states.** We can now write down the transition matrix:

$$P = \begin{bmatrix} 1 & 0 & 0 & 0 \\ .5 & 0 & .5 & 0 \\ 0 & .5 & 0 & .5 \\ 0 & 0 & 0 & 1 \end{bmatrix}$$

(Notice all the 0 entries, corresponding to possible transitions that we did not draw in the transition diagram because they have 0 probability of occurring. We usually leave out such arrows.) Is this system regular? Take a look at P^2:

$$P^2 = \begin{bmatrix} 1 & 0 & 0 & 0 \\ .5 & .25 & 0 & .25 \\ .25 & 0 & .25 & .5 \\ 0 & 0 & 0 & 1 \end{bmatrix}$$

Notice that the first and last rows haven't changed. After two steps, there is still no chance of leaving states 1 or 4. In fact, no matter how many powers we take, no matter how many steps we look at, there will still be no way to leave either of those states, and the first and last rows will still have plenty of zeros. This system is not regular.

Nonetheless, we can try to find a steady-state probability vector. If we do this (and you should set up the system of linear equations and solve it), we find that there are infinitely many steady-state probability vectors, namely all vectors of the form $[x\ \ 0\ \ 0\ \ 1-x]$ for $0 \le x \le 1$. (You can check directly that these are all steady-state vectors.) As with a regular system, if we start with any distribution, the system will tend toward one of these steady-state vectors. In other words, eventually the gambler will either lose all her money or leave the table with $30.

But which outcome is more likely, and with what probability? One way to approach this question is to try computing the distribution after many steps. The distribution that represents the gambler starting with $20 is $v = [0\ \ 0\ \ 1\ \ 0]$. Using technology, it's easy to compute vP^n for some large values of n:

$$vP^{10} \approx [.333008\ \ 0\ \ .000976\ \ .666016]$$

$$vP^{50} \approx [.333333\ \ 0\ \ 0\ \ .666667]$$

---

[*] These states are like "Roach Motels®" ("Roaches check in, but they don't check out")!

Online, follow:

Chapter 7

→ Online Utilities

→ Markov Simulation

where you can enter the transition matrix and watch a visual simulation showing the system hopping from state to state. Run the simulation a number of times to get an experimental estimate of the "time to absorption." (Take the average time to absorption over many runs.) Also, try varying the transition probabilities to study the effect on the time to absorption.

So, it looks like the probability that she will leave the table with $30 is approximately 2/3, while the probability that she loses it all is 1/3.

What if she started with only $10? Then our initial distribution would be $v = [0 \ 1 \ 0 \ 0]$ and

$$vP^{10} \approx [.666016 \quad .000976 \quad 0 \quad .333008]$$

$$vP^{50} \approx [.666667 \quad 0 \quad 0 \quad .333333]$$

So, this time, the probability of her losing everything is about 2/3 while the probability of her leaving with $30 is 1/3. There is a way of calculating these probabilities exactly using matrix arithmetic; however, it would take us too far afield to describe it here.

✛*Before we go on...* Another interesting question is, how long will it take the gambler in Example 5 to get to $30 or lose it all? Again, this can be calculated using matrix arithmetic. ∎

7.7 EXERCISES

● denotes basic skills exercises

◆ denotes challenging exercises

tech Ex indicates exercises that should be solved using technology

In Exercises 1–10, write down the transition matrix associated with each state transition diagram.

1. ●

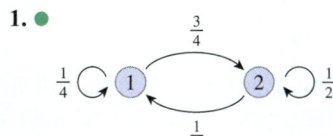

2. ●

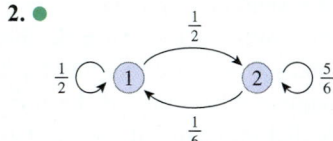

3. ● **4.** ●

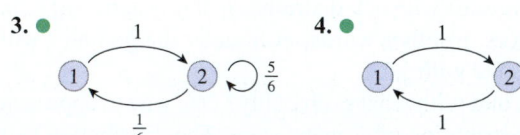

5. ●
6. ●

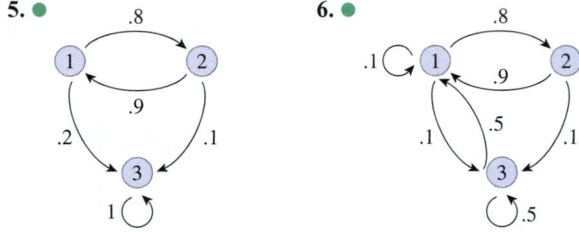

7. ●
8. ●

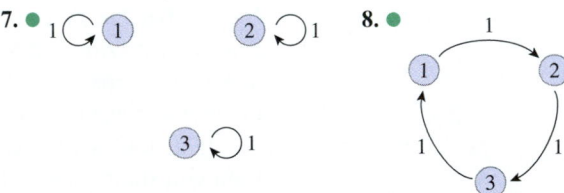

9. ●

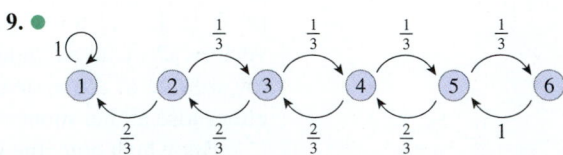

10. ●

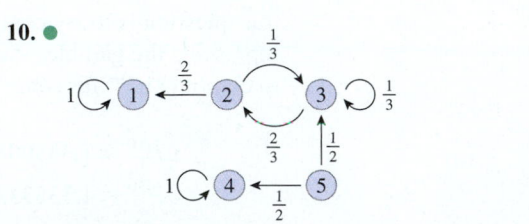

● basic skills ◆ challenging tech Ex technology exercise

In each of Exercises 11–24, you are given a transition matrix P and initial distribution vector v. Find

    a. the two-step transition matrix and

    b. the distribution vectors after one, two, and three steps.

    hint [see Quick Examples on pp. 520 and 521]

11. ● $P = \begin{bmatrix} .5 & .5 \\ 0 & 1 \end{bmatrix}$, $v = [1 \quad 0]$

12. ● $P = \begin{bmatrix} 1 & 0 \\ .5 & .5 \end{bmatrix}$, $v = [0 \quad 1]$

13. ● $P = \begin{bmatrix} .2 & .8 \\ .4 & .6 \end{bmatrix}$, $v = [.5 \quad .5]$

14. ● $P = \begin{bmatrix} 1/3 & 2/3 \\ 1/2 & 1/2 \end{bmatrix}$, $v = [1/4 \quad 3/4]$

15. ● $P = \begin{bmatrix} 1/2 & 1/2 \\ 1 & 0 \end{bmatrix}$, $v = [2/3 \quad 1/3]$

16. ● $P = \begin{bmatrix} 0 & 1 \\ 1/4 & 3/4 \end{bmatrix}$, $v = [1/5 \quad 4/5]$

17. ● $P = \begin{bmatrix} 3/4 & 1/4 \\ 3/4 & 1/4 \end{bmatrix}$, $v = [1/2 \quad 1/2]$

18. ● $P = \begin{bmatrix} 2/3 & 1/3 \\ 2/3 & 1/3 \end{bmatrix}$, $v = [1/7 \quad 6/7]$

19. ● $P = \begin{bmatrix} .5 & .5 & 0 \\ 0 & 1 & 0 \\ 0 & .5 & .5 \end{bmatrix}$, $v = [1 \quad 0 \quad 0]$

20. ● $P = \begin{bmatrix} .5 & 0 & .5 \\ 1 & 0 & 0 \\ 0 & .5 & .5 \end{bmatrix}$, $v = [0 \quad 1 \quad 0]$

21. ● $P = \begin{bmatrix} 0 & 1 & 0 \\ 1/3 & 1/3 & 1/3 \\ 1 & 0 & 0 \end{bmatrix}$, $v = [1/2 \quad 0 \quad 1/2]$

22. ● $P = \begin{bmatrix} 1/2 & 1/2 & 0 \\ 1/2 & 1/2 & 0 \\ 1/2 & 0 & 1/2 \end{bmatrix}$, $v = [0 \quad 0 \quad 1]$

23. ● $P = \begin{bmatrix} .1 & .9 & 0 \\ 0 & 1 & 0 \\ 0 & .2 & .8 \end{bmatrix}$, $v = [.5 \quad 0 \quad .5]$

24. ● $P = \begin{bmatrix} .1 & .1 & .8 \\ .5 & 0 & .5 \\ .5 & 0 & .5 \end{bmatrix}$, $v = [0 \quad 1 \quad 0]$

In each of Exercises 25–36, you are given a transition matrix P. Find the steady-state distribution vector. hint [see Example 4]

25. $P = \begin{bmatrix} 1/2 & 1/2 \\ 1 & 0 \end{bmatrix}$

26. $P = \begin{bmatrix} 0 & 1 \\ 1/4 & 3/4 \end{bmatrix}$

27. $P = \begin{bmatrix} 1/3 & 2/3 \\ 1/2 & 1/2 \end{bmatrix}$

28. $P = \begin{bmatrix} .2 & .8 \\ .4 & .6 \end{bmatrix}$

29. $P = \begin{bmatrix} .1 & .9 \\ .6 & .4 \end{bmatrix}$

30. $P = \begin{bmatrix} .2 & .8 \\ .7 & .3 \end{bmatrix}$

31. $P = \begin{bmatrix} .5 & 0 & .5 \\ 1 & 0 & 0 \\ 0 & .5 & .5 \end{bmatrix}$

32. $P = \begin{bmatrix} 0 & .5 & .5 \\ .5 & .5 & 0 \\ 1 & 0 & 0 \end{bmatrix}$

33. $P = \begin{bmatrix} 0 & 1 & 0 \\ 1/3 & 1/3 & 1/3 \\ 1 & 0 & 0 \end{bmatrix}$

34. $P = \begin{bmatrix} 1/2 & 1/2 & 0 \\ 1/2 & 1/2 & 0 \\ 1/2 & 0 & 1/2 \end{bmatrix}$

35. $P = \begin{bmatrix} .1 & .9 & 0 \\ 0 & 1 & 0 \\ 0 & .2 & .8 \end{bmatrix}$

36. $P = \begin{bmatrix} .1 & .1 & .8 \\ .5 & 0 & .5 \\ .5 & 0 & .5 \end{bmatrix}$

Applications

37. ● *Marketing* A market survey shows that half the owners of Sorey State Boogie Boards became disenchanted with the product and switched to C&T Super Professional Boards the next surf season, while the other half remained loyal to Sorey State. On the other hand, three quarters of the C&T Boogie Board users remained loyal to C&T, while the rest switched to Sorey State. Set these data up as a Markov transition matrix, and calculate the probability that a Sorey State Board user will be using the same brand two seasons later.
hint [see Example 1]

38. ● *Major Switching* At Suburban Community College, 10% of all business majors switched to another major the next semester, while the remaining 90% continued as business majors. Of all nonbusiness majors, 20% switched to a business major the following semester, while the rest did not. Set up these data as a Markov transition matrix, and calculate the probability that a business major user will no longer be a business major in two semesters' time.

39. ● *Pest Control* In an experiment to test the effectiveness of the latest roach trap, the "Roach Resort," 50 roaches were placed in the vicinity of the trap and left there for an hour. At the end of the hour, it was observed that 30 of them had "checked in," while the rest were still scurrying around. (Remember that "once a roach checks in, it never checks out.")

    a. Set up the transition matrix P for the system with decimal entries, and calculate P^2 and P^3.

    b. If a roach begins outside the "Resort," what is the probability of it "checking in" by the end of 1 hour? 2 hour? 3 hours?

    c. What do you expect to be the long-term impact on the number of roaches? *hint* [see Example 5]

40. ●*Employment* You have worked for the Department of Administrative Affairs (DAA) for 27 years, and you still have little or no idea exactly what your job entails. To make your life a little more interesting, you have decided on the following course of action. Every Friday afternoon, you will use your desktop computer to generate a random digit from 0 to 9 (inclusive). If the digit is a zero, you will immediately quit your job, never to return. Otherwise, you will return to work the following Monday.

a. Use the states (1) employed by the DAA and (2) not employed by the DAA to set up a transition probability matrix P with decimal entries, and calculate P^2 and P^3.

b. What is the probability that you will still be employed by the DAA after each of the next three weeks?

c. What are your long-term prospects for employment at the DAA?

41. *Risk Analysis* An auto insurance company classifies each motorist as "high-risk" if the motorist has had at least one moving violation during the past calendar year and "low risk" if the motorist has had no violations during the past calendar year. According to the company's data, a high-risk motorist has a 50 percent chance of remaining in the high-risk category the next year and a 50 percent chance of moving to the low-risk category. A low-risk motorist has a 10 percent chance of moving to the high-risk category the next year and a 90 percent chance of remaining in the low-risk category. In the long term, what percentage of motorists fall in each category?

42. *Debt Analysis* A credit card company classifies its cardholders as falling into one of two credit ratings: "good" and "poor." Based on its rating criteria, the company finds that a cardholder with a good credit rating has an 80 percent chance of remaining in that category the following year and a 20 percent chance of dropping into the poor category. A cardholder with a poor credit rating has a 40 percent chance of moving into the good rating the following year and a 60 percent chance of remaining in the poor category. In the long term, what percentage of cardholders fall in each category?

43. *Textbook Adoptions* College instructors who adopt this book are (we hope!) twice as likely to continue to use the book the following semester as they are to drop it, whereas nonusers are nine times as likely to remain nonusers the following year as they are to adopt this book.

a. Determine the probability that a nonuser will be a user in two years.

b. In the long term, what proportion of college instructors will be users of this book?

44. *Confidence Level* Tommy the Dunker's performance on the basketball court is influenced by his state of mind: If he scores, he is twice as likely to score on the next shot as he is to miss, whereas if he misses a shot, he is three times as likely to miss the next shot as he is to score.

a. If Tommy has missed a shot, what is the probability that he will score two shots later?

b. In the long term, what percentage of shots are successful?

45. *Debt Analysis* As the manager of a large retailing outlet, you have classified all credit customers as falling into one of the following categories: Paid Up, Outstanding 0-90 Days, Bad

Debts. Based on an audit of your company's records, you have come up with the following table, which gives the probabilities that a single credit customer will move from one category to the next in the period of one month.

	Paid up	To 0–90 days	Bad debts
From Paid up	.5	.5	0
0-90 days	.5	.3	.2
Bad debts	0	.5	.5

How do you expect the company's credit customers to be distributed in the long term?

46. *Debt Analysis* Repeat the preceding exercise using the following table:

	Paid up	To 0–90 days	Bad debts
From Paid up	.8	.2	0
0-90 days	.5	.3	.2
Bad debts	0	.5	.5

47. *Income Brackets* The following diagram shows the movement of U.S. households among three income groups—affluent, middle class, and poor—over the 11-year period 1980–1991.[73]

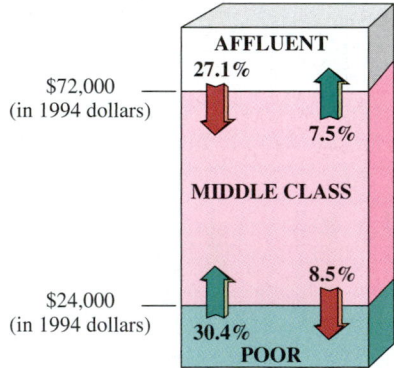

a. Use the transitions shown in the diagram to construct a transition matrix (assuming zero probabilities for the transitions between affluent and poor).

b. Assuming that the trend shown was to continue, what percent of households classified as affluent in 1980 were predicted to become poor in 2002? (Give your answer to the nearest 0.1%.)

c. tech Ex According to the model, what percentage of all U.S. households will be in each income bracket in the long term? (Give your answer to the nearest 0.1 percent.)

---

[73] The figures were based on household after-tax income. The study was conducted by G. J. Duncan of Northwestern University and T. Smeeding of Syracuse University and based on annual surveys of the personal finances of 5,000 households since the late 1960s. (The surveys were conducted by the University of Michigan.) SOURCE: *The New York Times,* June 4, 1995, p. E4.

● basic skills   ◆ challenging   tech Ex technology exercise

48. *Income Brackets* The following diagram shows the movement of U.S. households among three income groups—affluent, middle class, and poor—over the 12-year period 1967–1979.[74]

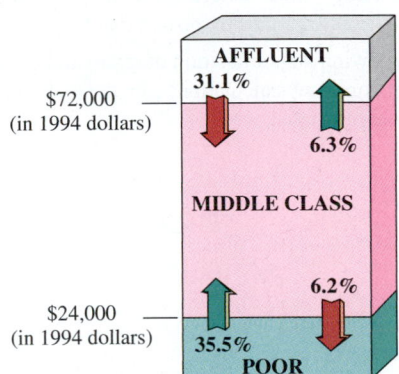

$72,000
(in 1994 dollars)

$24,000
(in 1994 dollars)

a. Use the transitions shown in the diagram to construct a transition matrix (assuming zero probabilities for the transitions between affluent and poor).

b. Assuming that the trend shown had continued, what percent of households classified as affluent in 1967 would have been poor in 1991? (Give your answer to the nearest 0.1 percent.)

c. **tech**Ex According to the model, what percentage of all U.S. households will be in each income bracket in the long term? (Give your answer to the nearest 0.1 percent.)

49. tech Ex ***Income Distribution*** A University of Michigan study shows the following one-generation transition probabilities among four major income groups.[75]

Father's Income	Eldest son's income			
	Bottom 10%	10–50%	50–90%	Top 10%
Bottom 10%	.30	.52	.17	.01
10–50%	.10	.48	.38	.04
50–90%	.04	.38	.48	.10
Top 10%	.01	.17	.52	.30

In the long term, what percentage of male earners would you expect to find in each category? Why are the long-range figures not necessarily 10% in the lowest 10% income bracket, 40% in the 10-50% range, 40% in the 50-90% range, and 10% in the top 10% range?

50. tech Ex ***Income Distribution*** Repeat Exercise 49, using the following data:

Father's Income	Son's income			
	Bottom 10%	10–50%	50–90%	Top 10%
Bottom 10%	.50	.32	.17	.01
10–50%	.10	.48	.38	.04
50–90%	.04	.38	.48	.10
Top 10%	.01	.17	.32	.50

Market Share: Cell phones *Three of the largest cellular phone companies in 2004 were Verizon, Cingular, and AT&T Wireless. Exercises 51 and 52 are based on the following figure, which shows percentages of subscribers who switched from one company to another during the third quarter of 2003:[76]*

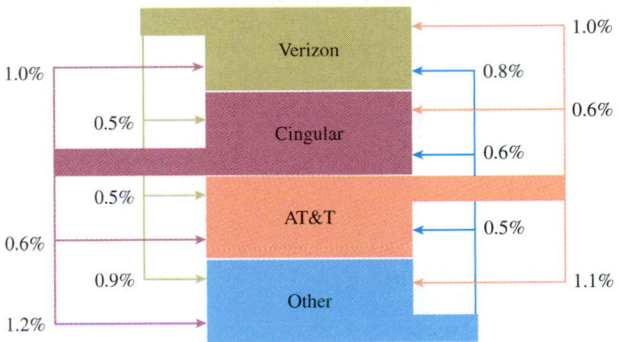

51. a. tech Ex Use the diagram to set up an associated transition matrix.

b. At the end of the third quarter of 2003, the market shares were: Verizon: 29.7 percent, Cingular: 19.3 percent, AT&T: 18.1 percent, and Other: 32.9 percent. Use your Markov system to estimate the percentage shares at the *beginning* of the third quarter of 2003.

c. Using the information from part (b), estimate the market shares at the end of 2005. Which company is predicted to gain the most in market share?

52. a. tech Ex Use the diagram to set up an associated transition matrix.

b. At the end of the third quarter of 2003, the market shares were: Verizon: 29.7 percent, Cingular: 19.3 percent, AT&T: 18.1 percent, Other: 32.9 percent. Use your Markov system to estimate the percentage shares one year earlier.

[74] Ibid.

[75] SOURCE: Gary Solon, University of Michigan/*New York Times,* May 18, 1992, p. D5. We have adjusted some of the figures so that the probabilities add to 1; they did not do so in the original table due to rounding.

[76] Published market shares and "churn rate" (percentage drops for each company) were used to estimate the individual transition percentages. "Other" consists of Sprint, Nextel, and T Mobile. No other cellular companies are included in this analysis. Source: The Yankee Group/*New York Times,* January 21, 2004, p. C1.

● basic skills   ◆ challenging   **tech**Ex technology exercise

c. Using the information from part (b), estimate the market shares at the end of 2010. Which company is predicted to lose the most in market share?

53. ● **Dissipation** Consider the following five-state model of one-dimensional dissipation without drift:

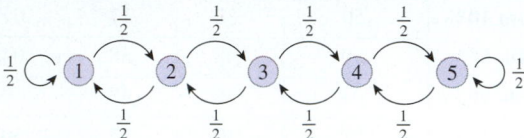

Find the steady-state distribution vector.

54. ● **Dissipation with Drift** Consider the following five-state model of one-dimensional dissipation with drift:

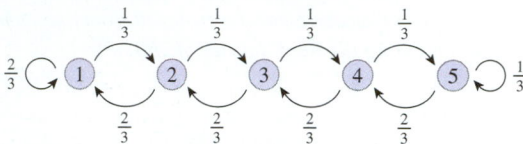

Find the steady-state distribution vector.

Communication and Reasoning Exercises

55. Describe an interesting situation that can be modeled by the transition matrix

$$P = \begin{bmatrix} .2 & .8 & 0 \\ 0 & 1 & 0 \\ .4 & .6 & 0 \end{bmatrix}$$

56. Describe an interesting situation that can be modeled by the transition matrix

$$P = \begin{bmatrix} .8 & .1 & .1 \\ 1 & 0 & 0 \\ .3 & .3 & .4 \end{bmatrix}$$

57. Describe some drawbacks to using Markov processes to model the behavior of the stock market with states (1) bull market (2) bear market.

58. Can the repeated toss of a fair coin be modeled by a Markov process? If so, describe a model; if not, explain the reason.

59. Explain: If Q is a matrix whose rows are steady-state distribution vectors, then $QP = Q$.

60. Construct a four-state Markov system so that both [.5 .5 0 0] and [0 0 .5 .5] are steady-state vectors. [*hint:* Try one in which no arrows link the first two states to the last two.]

61. ◆ Refer to the following state transition diagram and explain in words (without doing any calculation) why the steady-state vector has a zero in position 1.

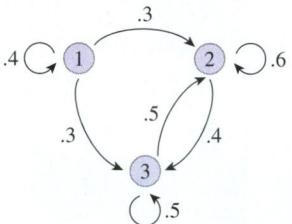

62. ◆ Without doing any calculation, find the steady-state distribution of the following system and explain the reasoning behind your claim.

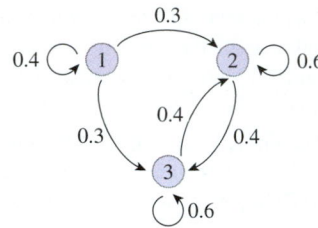

63. ◆ Construct a regular state transition diagram that possesses the steady-state vector [.3, .3 .4].

64. ◆ Construct a regular state transition diagram possessing a steady-state vector [.6 .3 0 .1].

65. ◆ Show that if a Markov system has two distinct steady-state distributions v and w, then $\dfrac{v+w}{2}$ is another steady-state distribution.

66. ◆ If higher and higher powers of P approach a fixed matrix Q, explain why the rows of Q must be steady-state distributions vectors.

Chapter 7 Review

KEY CONCEPTS

7.1 Sample Spaces and Events

Experiment, outcome, sample space *p. 444*

Event *p. 446*

The complement of an event *p. 449*

Unions of events *p. 450*

Intersections of events *p. 450*

Mutually exclusive events *p. 452*

7.2 Estimated and Theoretical Probability

Estimated probability or relative
frequency *p. 458*

Estimated probability distribution or
relative frequency distribution *p. 458*

Properties of estimated probability distribution *p. 460*

Theoretical probability *p. 461*

Equally likely outcomes: $P(E) = n(E)/n(S)$ *p. 462*

7.3 Properties of Probability Distributions

Finite probability distribution: $0 \le P(s_i) \le 1$ and
$P(s_1) + \cdots + P(s_n) = 1$ *p. 473*

If $P(E) = 0$, we call E an impossible event *p. 474*

Addition principle: $P(A \cup B) = P(A) + P(B) - P(A \cap B)$
p. 475

If A and B are mutually exclusive, then
$P(A \cup B) = P(A) + P(B)$ *p. 475*

If S is the sample space, then $P(S) = 1$, $P(\emptyset) = 0$, and
$P(A') = 1 - P(A)$ *p. 477*

7.4 Probability and Counting Techniques

Use counting techniques from Chapter 6 to calculate
probability *p. 483*

7.5 Conditional Probability and Independence

Conditional probability: $P(A \mid B) = P(A \cap B)/P(B)$ *p. 493*

Multiplication principle for conditional probability:
$P(A \cap B) = P(A \mid B)P(B)$ *p. 496*

Independent events: $P(A \cap B) = P(A)P(B)$ *p. 499*

7.6 Bayes' Theorem and Applications

Bayes' Theorem (short form):

$$P(A \mid T) = \frac{P(T \mid A)P(A)}{P(T \mid A)P(A) + P(T \mid A')P(A')} \quad p.\ 510$$

Bayes' Theorem (partition of sample space into 3 events):

$$P(A_1 \mid T) = \frac{P(T \mid A_1)P(A_1)}{P(T \mid A_1)P(A_1) + P(T \mid A_2)P(A_2) + P(T \mid A_3)P(A_3)}$$

p. 512

7.7 Markov Systems

Markov system, Markov process, states, transition
probabilities *p. 517*

Transition matrix associated with a given Markov system *p. 518*

Distribution vector *p. 520*

If v is a distribution vector, then the distribution vector one
step later will be vP. The distribution after m steps will
be vP^m *p. 520*

P^m is the m-step transition matrix. The ijth entry in P^m is
the probability of a transition from state i to state j in m
steps *p. 521*

A steady-state (probability) vector is a probability vector v such
that $vP = v$ *p. 522*

Calculation of the steady-state probability vector *p. 523*

A regular Markov system, long-term behavior of a regular
Markov system *p. 524*

An absorbing state is one for which the probability of staying in
the state is 1 (and the probability of leaving it for any other
state is 0) *p. 525*

REVIEW EXERCISES

In each of Exercises 1–6, say how many elements are in the sample space S, list the elements of the given event E, and compute the probability of E.

1. Three coins are tossed; the result is one or more tails.

2. Four coins are tossed; the result is fewer heads then tails.

3. Two distinguishable dice are rolled; the numbers facing up add to 7.

4. Three distinguishable dice are rolled; the number facing up add to 5.

5. A die is weighted so that each of 2, 3, 4, and 5 is half as likely to come up as either 1 or 6; however, 2 comes up.

6. Two indistinguishable dice are rolled; the numbers facing up add to 7.

In each of Exercises 7–10, calculate the estimated probability P(E).

7. Two coins are tossed 50 times, and two heads come up 12 times. E is the event that at least one tail comes up.

8. Ten stocks are selected at random from a portfolio. Seven of them have increased in value since their purchase, and the rest have decreased. Eight of them are Internet stocks and two of those have decreased in value. E is the event that a stock has either increased in value or is an Internet stock.

9. You have read 150 of the 400 novels in your home, but your sister Roslyn has read 200, of which only 50 are novels you have read as well. E is the event that a novel has been read by neither you nor your sister.

10. You roll two dice 10 times. Both dice show the same number three times, and on two rolls, exactly one number is odd. E is the event that the sum of the numbers is even.

In each of Exercises 11–14, calculate the theoretical probability $P(E)$.

11. There are 32 students in categories A and B combined. Some are in both, 24 are in A, and 24 are in B. E is the event that a randomly selected student (among the 32) is in both categories.

12. You roll two dice, one red and one green. Losing combinations are doubles (both dice showing the same number) and outcomes in which the green die shows an odd number and the red die shows an even number. The other combinations are winning ones. E is the event that you roll a winning combination.

13. The *jPlay* portable music/photo/video player and bottle opener comes in three models: A, B, and C, each with five colors to choose from, and there are equal numbers of each combination. E is the event that a randomly selected *jPlay* is either orange (one of the available colors), a Model A, or both.

14. The Heavy Weather Service predicts that for tomorrow there is a 50 percent chance of tornadoes, a 20 percent chance of a monsoon, and a 10 percent chance of both. What is the probability that we will be lucky tomorrow and encounter neither tornadoes nor a monsoon?

A bag contains 4 red marbles, 2 green ones, 1 transparent one, 3 yellow ones, and 2 orange ones. You select 5 at random. In each of Exercises 15–20, compute the probability of the given event.

15. You have selected all the red ones.

16. You have selected all the green ones.

17. All are different colors.

18. At least 1 is not red.

19. At least 2 are yellow.

20. None are yellow and at most 1 is red.

In each of Exercises 21–26, find the probability of being dealt the given type of 5-card hand from a standard deck of 52 cards. (None of these is a recognized poker hand.) Express your answer in terms of combinations.

21. **Kings and Queens:** Each of the five cards is either a king or a queen

22. **Five Pictures:** All picture cards (J, Q, K)

23. **Fives and Queens:** Three fives, the queen of spades and one other queen

24. **Prime Full House:** A full house (three cards of one denomination, two of another) with the face value of each card a prime number (Ace = 1, J = 11, Q = 12, K = 13)

25. **Full House of Commons:** A full house (three cards of one denomination, two of another) with no royal cards (that is, no J, Q, K, or Ace).

26. **Black Two Pair:** Five black cards (spades or clubs), two with one denomination, two with another and one with a third

Two dice, one green and one yellow, are rolled. In each of Exercises 27–32, find the conditional probability, and also say whether the indicated pair of events is independent.

27. The sum is 5, given that the green one is not 1 and the yellow one is 1.

28. That the sum is 6, given that the green one is either 1 or 3 and the yellow one is 1.

29. The yellow one is 4, given that the green one is 4.

30. The yellow one is 5, given that the sum is 6.

31. The dice have the same parity, given that both of them are odd.

32. The sum is 7, given that the dice do not have the same parity.

A poll shows that half the consumers who use Brand A switched to Brand B the following year, while the other half stayed with Brand A. Three quarters of the Brand B users stayed with Brand B the following year, while the rest switched to Brand A. Use this information to answer Exercises 33–36.

33. Give the associated Markov state distribution matrix with state 1 representing using Brand A, and state 2 represented by using Brand B.

34. Compute the associated two- and three-step transition matrices. What is the probability that a Brand A user will be using Brand B three years later?

35. If two-thirds of consumers are presently using Brand A and one-third are using Brand B, how are these consumers distributed in three years' time?

36. In the long term, what fraction of the time will a user spend using each of the two brands?

Applications

OHaganBooks.com currently operates three warehouses: one in Washington, one in California, and the new one in Texas. Book inventories are shown in the following table.

	Sci-Fi	Horror	Romance	Other	Total
Washington	10,000	12,000	12,000	30,000	64,000
California	8000	12,000	6000	16,000	42,000
Texas	15,000	15,000	20,000	44,000	94,000
Total	33,000	39,000	38,000	90,000	200,000

A book is selected at random. In each of Exercises 37–42, compute the probability of the given event.

37. That it is either a sci-fi book or stored in Texas (or both).

38. That it is a sci-fi book stored in Texas.

39. That it is a sci-fi book, given that it is stored in Texas.

40. That it was stored in Texas, given that it was a sci-fi book.

41. That it was stored in Texas, given that it was not a sci-fi book.

42. That it was not stored in Texas, given that it was a sci-fi book.

In order to gauge the effectiveness of the OHaganBooks.com site, you recently commissioned a survey of online shoppers. According to the results, 2 percent of online shoppers visited OHaganBooks.com during a one-week period, while 5 percent of them visited at least one of OHaganBooks.com's two main competitors: JungleBooks.com and FarmerBooks.com. Use this information to answer Exercises 43–47.

43. What percentage of online shoppers never visited OHaganBooks.com?

44. Assuming that visiting OHaganBooks.com was independent of visiting a competitor, what percentage of online shoppers visited either OHaganBooks.com or a competitor?

45. Under the assumption of Exercise 44, what is the estimated probability that an online shopper will visit none of the three sites during a week?

46. Actually, the assumption in Exercise 44 is not what was found by the survey, because an online shopper visiting a competitor was in fact more likely to visit OHaganBooks.com than a randomly selected online shopper. Let H be the event that an online shopper visits OHaganBooks.com, and let C be the event that he visits a competitor. Which is greater: $P(H \cap C)$ or $P(H)P(C)$? Why?

47. What the survey found is that 25 percent of online shoppers who visited a competitor also visited OHaganBooks.com. Given this information, what percentage of online shoppers visited OHaganBooks.com and neither of its competitors?

Not all visitors to the OHaganBooks.com site actually purchase books, and not all OHaganBooks.com customers buy through the website (some call them in and others use a mail-order catalog). According to statistics gathered at the website, 8 percent of online shoppers who visit the site during the course of a single week will purchase books. However, the survey mentioned prior to Exercise 43 revealed that 2 percent of online shoppers visited the site during the course of a week. Another survey estimated that 0.5 percent of online shoppers who did not visit the site during the course of a week nonetheless purchased books at OHaganBooks.com. Use this information to answer Exercises 48–50.

48. Complete the following sentence: Online shoppers who visit the OHaganBooks.com website are ___ times as likely to purchase books than shoppers who do not.

49. What is the probability that an online shopper will not visit the site during the course of a week but still purchase books?

50. What is the probability that an online shopper who purchased books during a given week also visited the site?

As mentioned earlier, OHaganBooks.com has two main competitors, JungleBooks.com and FarmerBooks.com, and no other competitors of any significance. The following table shows the movement of customers during July.[77] (Thus, for instance, the first row tells us that 80 percent of OHaganBooks.com's customers remained loyal, 10 percent of them went to JungleBooks.com and the remaining 10 percent to FarmerBooks.com.)

		To		
		OHaganBooks	**JungleBooks**	**FarmerBooks**
From	**OHaganBooks**	80%	10%	10%
	JungleBooks	40%	60%	0%
	FarmerBooks	20%	0%	80%

At the beginning of July, OHaganBooks.com had an estimated market share of one-fifth of all customers, while its two competitors had two-fifths each. Use this information to answer Exercises 51–54.

51. Estimate the market shares each company had at the end of July.

52. Assuming the July trends continue in August, predict the market shares of each company at the end of August.

53. Name one or more important factors that the Markov model does not take into account.

54. Assuming the July trend were to continue indefinitely, predict the market share enjoyed by each of the three e-commerce sites.

vMentor Do you need a live tutor for homework problems? Access vMentor on the ThomsonNOW! website at **www.thomsonedu.com** for one-on-one tutoring from a mathematics expert.

---

[77] By a "customer" of one of the three e-commerce sites, we mean someone who purchases more at that site than at any of the two competitors' sites.

CASE STUDY: The Monty Hall Problem

The Everett Collection

Here is a famous "paradox" that even mathematicians find counterintuitive. On the game show *Let's Make a Deal*, you are shown three doors, A, B, and C, and behind one of them is the Big Prize. After you select one of them—say, door A—to make things more interesting the host (Monty Hall), who knows what is behind each door, opens one of the other doors—say, door B—to reveal that the Big Prize is not there. He then offers you the opportunity to change your selection to the remaining door, door C. Should you switch or stick with your original guess? Does it make any difference?

Most people would say that the Big Prize is equally likely to be behind door A or door C, so there is no reason to switch.[78] In fact, this is wrong: The prize is more likely to be behind door C! There are several ways of seeing why this is so. Here is how you might work it out using Bayes' theorem.

Let A be the event that the Big Prize is behind door A, B the event that it is behind door B, and C the event that it is behind door C. Let F be the event that Monty has opened door B and revealed that the Prize is not there. You wish to find $P(C \mid F)$ using Bayes' theorem. To use that formula you need to find $P(F \mid A)$ and $P(A)$ and similarly for B and C. Now, $P(A) = P(B) = P(C) = 1/3$ because at the outset, the prize is equally likely to be behind any of the doors. $P(F \mid A)$ is the probability that Monty will open door B if the Prize is actually behind door A, and this is $1/2$ because we assume that he will choose either B or C randomly in this case. On the other hand, $P(F \mid B) = 0$, because he will never open the door that hides the prize. Also, $P(F \mid C) = 1$ because if the prize is behind door C, he must open door B to keep from revealing that the prize is behind door C. Therefore,

$$P(C \mid F) = \frac{P(F \mid C)P(C)}{P(F \mid A)P(A) + P(F \mid B)P(B) + P(F \mid C)P(C)}$$

$$= \frac{1 \cdot \frac{1}{3}}{\frac{1}{2} \cdot \frac{1}{3} + 0 \cdot \frac{1}{3} + 1 \cdot \frac{1}{3}} = \frac{2}{3}$$

You conclude from this that you *should* switch to door C because it is more likely than door A to be hiding the Prize.

Here is a more elementary way you might work it out. Consider the tree diagram of possibilities shown in Figure 21.

The top two branches of the tree give the cases in which the Prize is behind door A, and there is a total probability of $1/3$ for that case. The remaining two branches with nonzero probabilities give the cases in which the prize is behind the door that you did not choose, and there is a total probability of $2/3$ for that case. Again, you conclude that you should switch your choice of doors because the one you did not choose is twice as likely as door A to be hiding the Big Prize.

---

[78] This problem caused quite a stir in late 1991 when this problem was discussed in Marilyn vos Savant's column in *Parade* magazine. Vos Savant gave the answer that you should switch. She received about 10,000 letters in response, most of them disagreeing with her, including several from mathematicians.

Prize behind   **Monty opens**

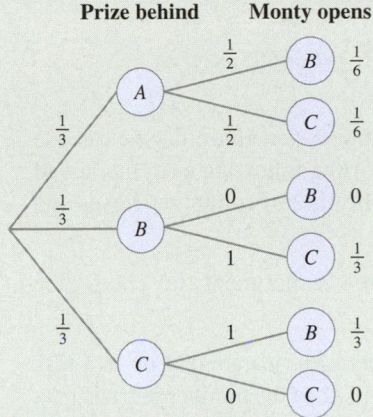

Figure **21**

Exercises

1. The answer you came up with, to switch to the other door, depends on the strategy Monty Hall uses in picking the door to open. Suppose that he actually picks one of doors B and C at random, so that there is a chance that he will reveal the Big Prize. If he opens door B and it happens that the Prize is not there, should you switch or not?

2. What if you know that Monty's strategy is always to open door B if possible (i.e., it does not hide the Big prize) after you choose A?

   a. If he opens door B, should you switch?

   b. If he opens door C, should you switch?

3. Repeat the analysis of the original game, but suppose that the game uses four doors instead of three (and still only one prize).

4. Repeat the analysis of the original game, but suppose that the game uses 1000 doors instead of three (and still only one prize).

TECHNOLOGY GUIDE

Section 7.2

The TI-83 has a random number generator that we can use to simulate experiments. For the following example, recall that a fair coin has probability 1/2 of coming up heads and 1/2 of coming up tails.

Example Use a simulated experiment to check the following.

a. The estimated probability of heads coming up in a toss of a fair coin approaches 1/2 as the number of trials gets large.

b. The estimated probability of heads coming up in two consecutive tosses of a fair coin approaches 1/4 as the number of trials gets large.[79]

Solution with Technology

a. Let us use 1 to represent heads and 0 to represent tails. We need to generate a list of **random binary digits** (0 or 1). One way to do this—and a method that works for most forms of technology—is to generate a random number between 0 and 1 and then round it to the nearest whole number, which will be either 0 or 1.

We generate random numbers on the TI-83 using the "rand" function. To round the number X to the nearest whole number on the TI-83, follow [MATH] →NUM, select "round," and enter round(X,0). This instruction rounds X to zero decimal places—that is, to the nearest whole number. Since we wish to round a random number we need to enter

> round(rand,0)   **To obtain** rand,
> **follow** [MATH] →PRB

The result will be either 0 or 1. Each time you press [ENTER] you will now get another 0 or 1. The TI-83 can also generate a random integer directly (without the need for rounding) through the instruction

> randInt(0,1)   **To obtain** randInt,
> **follow** [MATH] →PRB

In general, the command randInt(*m*, *n*) generates a random integer in the range [*m*, *n*]. The following sequence of 100 random binary digits was produced using technology.[80]

0	1	0	0	1	1	0	1	0	0
0	1	0	0	0	0	0	0	1	0
1	1	0	0	0	1	0	0	1	1
1	1	1	0	1	0	0	0	1	0
1	1	1	1	1	1	1	0	0	1
1	0	1	1	1	0	0	1	1	0
0	1	0	1	1	1	0	1	1	1
1	0	0	0	0	0	0	1	1	1
1	1	1	1	0	0	1	1	1	0
1	1	1	0	1	1	0	1	0	0

If we use only the first row of data (corresponding to the first ten tosses), we find

$$P(\text{H}) = \frac{fr(1)}{N} = \frac{4}{10} = .4$$

Using the first two rows ($N = 20$) gives

$$P(\text{H}) = \frac{fr(1)}{N} = \frac{6}{20} = .3$$

Using all ten rows ($N = 100$) gives

$$P(\text{H}) = \frac{fr(1)}{N} = \frac{54}{100} = .54$$

This is somewhat closer to the theoretical probability of 1/2 and supports our intuitive notion that the larger the number of trials, the more closely the estimated probability should approximate the theoretical value.[81]

b. We need to generate pairs of random binary digits and then check whether they are both 1s. Although the TI-83 will generate a pair of random digits if you enter round(rand(2),0), it would be a lot more convenient if the calculator could tell you right away whether both digits are 1s (corresponding to two consecutive heads in a coin toss). Here is a simple way of accomplishing this. Notice that if we *add* the two random binary digits, we obtain either 0, 1, or 2, telling us the number of heads that result from the two consecutive throws. Therefore, all we need to do is add the pairs of random digits and then count the number of times 2 comes up. A formula we can use is

> randInt(0,1)+randInt(0,1)

What would be even *more* convenient is if the result of the calculation would be either 0 or 1, with 1 signifying

---

[79] Since the set of outcomes of a pair of coin tosses is {HH, HT, TH, TT}, we expect HH to come up once in every four trials, on average.

[80] The instruction randInt(0,1,100)→L₁ will generate a list of 100 random 0s and 1s and store it in L₁, where it can be summed with Sum(L₁) (under [2ND] [LIST] →MATH).

[81] Do not expect this to happen every time. Compare, for example, *P*(H) for the first five rows and for all ten rows.

success (two consecutive heads) and 0 signifying failure. Then, we could simply add up all the results to obtain the number of times two heads occurred. To do this, we first divide the result of the calculation above by 2 (obtaining 0, .5, or 1, where now 1 signifies success) and then round *down* to an integer using a function called "int":

```
int(0.5*(randInt(0,1)+randInt(0,1)))
```

Following is the result of 100 such pairs of coin tosses, with 1 signifying success (two heads) and 0 signifying failure (all other outcomes). The last column records the number of successes in each row and the total number at the end.

1	1	0	0	0	0	0	0	0	0	2
0	1	0	0	0	0	0	1	0	1	3
0	1	0	0	1	1	0	0	0	1	4
0	0	0	0	0	0	0	0	1	0	1
0	1	0	0	1	0	0	1	0	0	3
1	0	1	0	0	0	0	0	0	0	2
0	0	0	0	0	0	0	0	0	1	1
0	1	1	1	1	0	0	0	0	1	5
1	1	0	1	0	0	1	1	0	0	5
0	0	0	0	0	0	0	0	1	0	1
										27

Now, as in part (a), we can compute estimated probabilities, with D standing for the outcome "two heads":

First 10 trials: $P(D) = \dfrac{fr(1)}{N} = \dfrac{2}{10} = .2$

First 20 trials: $P(D) = \dfrac{fr(1)}{N} = \dfrac{5}{20} = .25$

First 50 trials: $P(D) = \dfrac{fr(1)}{N} = \dfrac{13}{50} = .26$

100 trials: $P(D) = \dfrac{fr(1)}{N} = \dfrac{27}{100} = .27$

Q: *What is happening with the data? The probabilities seem to be getting less accurate as N increases!*

A: Quite by chance, exactly 5 of the first 20 trials resulted in success, which matches the theoretical probability. Figure 22 shows an Excel plot of estimated probability versus N (for N a multiple of 10). Notice that, as N increases, the graph seems to meander within smaller distances of .25. ∎

Q: *The above techniques work fine for simulating coin tosses. What about rolls of a fair die, where we want outcomes between 1 and 6?*

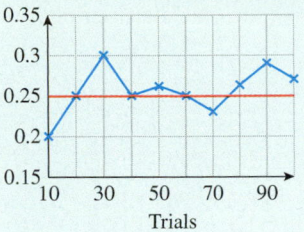

Figure 22

A: We can simulate a roll of a die by generating a random integer in the range 1 through 6. The following formula accomplishes this:

$$1 + \text{int}(5.99999*\text{rand})$$

(We used 5.99999 instead of 6 to avoid the outcome 7.) ∎

Section **7.7**

Example 1 Consider the Markov system found by Gamble Detergents at the beginning of this section. Suppose that 70 percent of consumers are now using powdered detergent while the other 30 percent are using liquid. What will be the distribution one year from now? Two years from now? Three years from now?

Solution with Technology In Chapter 3 we saw how to set up and multiply matrices. For this example, we can use the matrix editor to define [A] as the initial distribution and [B] as the transition matrix (remember that the only names we can use are [A] through [J]).

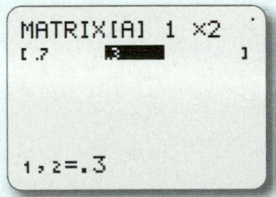   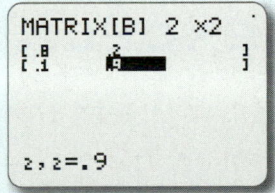

Entering [A] (obtained by pressing [MATRX] [1] [ENTER]) will show you the initial distribution. To obtain the distribution after 1 step, press [X] [MATRX] [2] [ENTER], which has the effect of multiplying the previous answer by the transition matrix [B]. Now, just press [ENTER] repeatedly to continue multiplying by the transition matrix and obtain the distribution after any number of steps. The screenshot shows the initial distribution `[A]` and the distributions after 1, 2, and 3 steps.

```
[A]
          [[.7 .3]]
Ans*[B]
       [[.59 .41]]
     [[.513 .487]]
   [[.4591 .5409]]
```

Example 4 Calculate the steady-state probability vector for the transition matrix in the preceding examples.

Solution with Technology Finding the steady-state probability vector comes down to solving a system of equations. As discussed in Chapters 2 and 3, there are several ways to use a calculator to help. The most straightforward is to use matrix inversion to solve the matrix form of the system. In this case, as in the text, the system of equations we need to solve is

$$x + y = 1$$
$$-.2x + .1y = 0$$

We write this as the matrix equation $AX = B$ with

$$A = \begin{bmatrix} 1 & 1 \\ -.2 & .1 \end{bmatrix} \qquad B = \begin{bmatrix} 1 \\ 0 \end{bmatrix}$$

To find $X = A^{-1}B$ using the TI-83, we first use the matrix editor to enter these matrices as `[A]` and `[B]`, then compute `[A]`$^{-1}$`[B]` on the home screen.

```
[A]⁻¹[B]
   [[.3333333333]
    [.6666666667]]
```

To convert the entries to fractions, we can follow this by the command

➤ Frac $\boxed{\text{MATH}}$ $\boxed{\text{ENTER}}$ $\boxed{\text{ENTER}}$

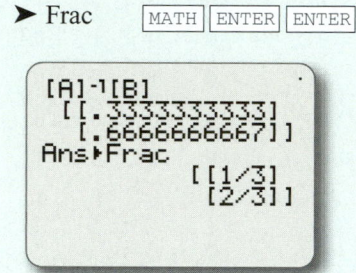

```
[A]⁻¹[B]
   [[.3333333333]
    [.6666666667]]
Ans▶Frac
          [[1/3]
           [2/3]]
```

EXCEL Technology Guide

Section 7.2

Excel has a random number generator that we can use to simulate experiments. For the following example, recall that a fair coin has probability $1/2$ of coming up heads and $1/2$ of coming up tails.

Example Use a simulated experiment to check the following.

a. The estimated probability of heads coming up in a toss of a fair coin approaches $1/2$ as the number of trials gets large.

b. The estimated probability of heads coming up in two consecutive tosses of a fair coin approaches $1/4$ as the number of trials gets large.[82]

Solution with Technology

a. Let us use 1 to represent heads and 0 to represent tails. We need to generate a list of **random binary digits** (0 or 1). One way to do this—and a method that works for most forms of technology—is to generate a random

---

[82] Since the set of outcomes of a pair of coin tosses is {HH, HT, TH, TT}, we expect HH to come up once in every four trials, on average.

number between 0 and 1 and then round it to the nearest whole number, which will be either 0 or 1.

In Excel, the formula `RAND()` gives a random number between 0 and 1.[83] The function `ROUND(X,0)` rounds X to zero decimal places—that is, to the nearest integer. Therefore, to obtain a random binary digit in any cell just enter the following formula.

```
=ROUND(RAND(),0)
```

Excel can also generate a random integer directly (without the need for rounding) through the formula

```
=RANDBETWEEN(0,1)
```

To obtain a whole array of random numbers, just drag this formula into the cells you wish to use.

b. We need to generate pairs of random binary digits and then check whether they are both 1s. It would be convenient if the spreadsheet could tell you right away whether both digits are 1s (corresponding to two consecutive heads in a coin toss). Here is a simple way of accomplishing this. Notice that if we *add* two random binary digits, we obtain either 0, 1, or 2, telling us the number of heads that result from the two consecutive throws. Therefore, all we need to do is add pairs of random digits and then count the number of times 2 comes up. Formulas we can use are

```
=RANDBETWEEN(0,1)+RANDBETWEEN(0,1)
```

What would be even *more* convenient is if the result of the calculation would be either 0 or 1, with 1 signifying success (two consecutive heads) and 0 signifying failure. Then, we could simply add up all the results to obtain the number of times two heads occurred. To do this, we first divide the result of the calculation above by 2 (obtaining 0, .5, or 1, where now 1 signifies success) and then round *down* to an integer using a function called "int":

```
=INT(0.5*(RANDBETWEEN(0,1)+
   RANDBETWEEN(0,1)))
```

Following is the result of 100 such pairs of coin tosses, with 1 signifying success (two heads) and 0 signifying failure (all other outcomes). The last column records the number of successes in each row and the total number at the end.

1	1	0	0	0	0	0	0	0	0	2
0	1	0	0	0	0	0	1	0	1	3
0	1	0	0	1	1	0	0	0	1	4
0	0	0	0	0	0	0	0	1	0	1
0	1	0	0	1	0	0	1	0	0	3
1	0	1	0	0	0	0	0	0	0	2
0	0	0	0	0	0	0	0	0	1	1
0	1	1	1	1	0	0	0	0	1	5
1	1	0	1	0	0	1	1	0	0	5
0	0	0	0	0	0	0	0	1	0	1
										27

Now, as in part (a), we can compute estimated probabilities, with D standing for the outcome "two heads":

First 10 trials: $\quad P(D) = \dfrac{fr(1)}{N} = \dfrac{2}{10} = .2$

First 20 trials: $\quad P(D) = \dfrac{fr(1)}{N} = \dfrac{5}{20} = .25$

First 50 trials: $\quad P(D) = \dfrac{fr(1)}{N} = \dfrac{13}{50} = .26$

100 trials: $\quad P(D) = \dfrac{fr(1)}{N} = \dfrac{27}{100} = .27$

Q: What is happening with the data? The probabilities seem to be getting less accurate as N increases!

A: Quite by chance, exactly 5 of the first 20 trials resulted in success, which matches the theoretical probability. Figure 23 shows an Excel plot of estimated probability versus N (for N a multiple of 10). Notice that, as N increases, the graph seems to meander within smaller distances of .25.

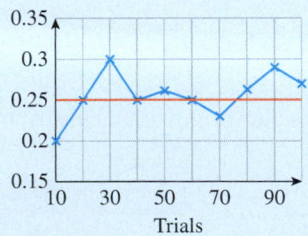

Figure 23

Q: The above techniques work fine for simulating coin tosses. What about rolls of a fair die, where we want outcomes between 1 and 6?

A: We can simulate a roll of a die by generating a random integer in the range 1 through 6. The following formula

---

[83] The parentheses after RAND are necessary even though the function takes no arguments.

accomplishes this:

```
=1 + INT(5.99999*RAND())
```

(We used 5.99999 instead of 6 to avoid the outcome 7.) ∎

Section 7.7

Example 1 Consider the Markov system found by Gamble Detergents at the beginning of this section. Suppose that 70 percent of consumers are now using powdered detergent while the other 30 percent are using liquid. What will be the distribution one year from now? Two years from now? Three years from now?

Solution with Technology In Excel, enter the initial distribution vector in cells A1 and B1 and the transition matrix to the right of that, as shown.

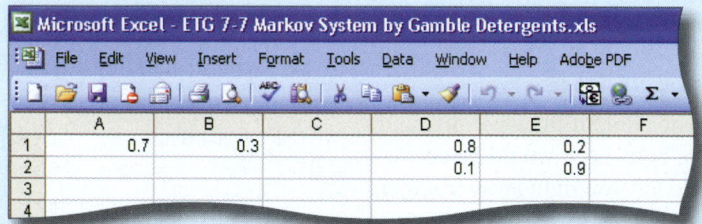

To calculate the distribution after one step, use the array formula

```
=MMULT(A1:B1,$D$1:$E$2)
```

The absolute cell references (dollar signs) ensure that the formula always refers to the same transition matrix, even if we copy it into other cells. To use the array formula, select cells A2 and B2, where the distribution vector will go, enter this formula, then press Control+Shift+Enter.[84]

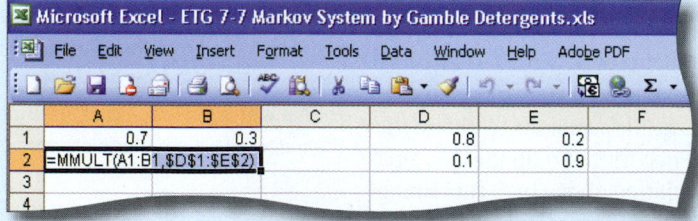

The result is the following, with the distribution after one step highlighted.

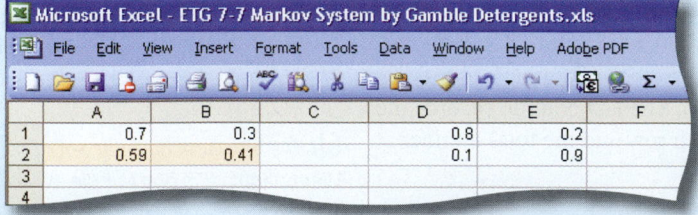

---

[84] On a Macintosh, you can also use Command+Enter.

To calculate the distribution after two steps, select cells A2 and B2 and drag the fill handle down to copy the formula to cells A3 and B3. Note that the formula now takes the vector in A2:B2 and multiplies it by the transition matrix to get the vector in A3:B3. To calculate several more steps, drag down as far as desired.

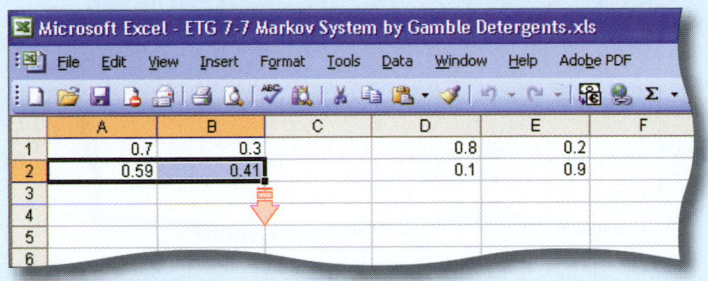

Example 4 Calculate the steady-state probability vector for the transition matrix in the preceding examples.

Solution with Technology Finding the steady-state probability vector comes down to solving a system of equations. As discussed in Chapters 2 and 3, there are several ways to use Excel to help. The most straightforward is to use matrix inversion to solve the matrix form of the system. In this case, as in the text, the system of equations we need to solve is

$$x + \ \ y = 1$$
$$-.2x + .1y = 0$$

We write this as the matrix equation $AX = B$ with

$$A = \begin{bmatrix} 1 & 1 \\ -.2 & .1 \end{bmatrix} \qquad B = \begin{bmatrix} 1 \\ 0 \end{bmatrix}$$

We enter A in cells A1:B2, B in cells D1:D2, and the formula for $X = A^{-1}B$ in a convenient location, say B4:B5.

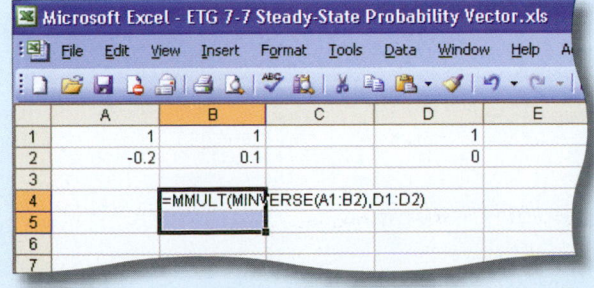

When we press Control+Shift+Enter we see the result:

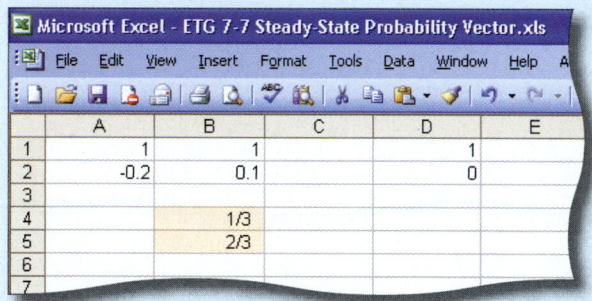

If we want to see the answer in fraction rather than decimal form, we format the cells as fractions.

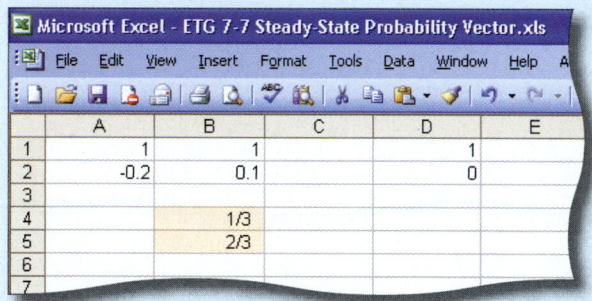

8

Random Variables and Statistics

CASE STUDY Spotting Tax Fraud with Benford's Law

You are a tax fraud specialist working for the Internal Revenue Service (IRS), and you have just been handed a portion of the tax return from Colossal Conglomerate. The IRS suspects that the portion you were handed may be fraudulent and would like your opinion. Is there any mathematical test, you wonder, that can point to a suspicious tax return based on nothing more than the numbers entered?

age fotostock/SuperStock

Introduction

Statistics is the branch of mathematics concerned with organizing, analyzing, and interpreting numerical data. For example, given the current annual incomes of 1000 lawyers selected at random, you might wish to answer some questions: If I become a lawyer, what income am I likely to earn? Do lawyers' salaries vary widely? How widely?

To answer questions like these, it helps to begin by organizing the data in the form of tables or graphs. This is the topic of the first section of the chapter. The second section describes an important class of examples that are applicable to a wide range of situations, from tossing a coin to product testing.

Once the data are organized, the next step is to apply mathematical tools for analyzing the data and answering questions like those posed above. Numbers such as the **mean** and the **standard deviation** can be computed to reveal interesting facts about the data. These numbers can then be used to make predictions about future events.

The chapter ends with a section on one of the most important distributions in statistics, the **normal distribution.** This distribution describes many sets of data and also plays an important role in the underlying mathematical theory.

8.1 Random Variables and Distributions

Random Variables

In many experiments we can assign numerical values to the outcomes. For instance, if we roll a die, each outcome has a value from 1 through 6. If you select a lawyer and ascertain his or her annual income, the outcome is again a number. We call a rule that assigns a number to each outcome of an experiment a **random variable.**

Random Variable

A **random variable** X is a rule that assigns a number, or **value,** to each outcome in the sample space of an experiment.[*]

Visualizing a Random Variable

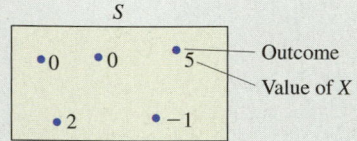

---

quick Examples

1. Roll a die; X = the number facing up.

2. Select a mutual fund; X = the number of companies in the fund portfolio.

3. Select a computer; X = the number of megabytes of memory it has.

4. Survey a group of 20 college students; X = the mean SAT.

---

[*] In the language of functions (Chapter 1), a random variable is a *real-valued function* whose domain is the sample space.

Discrete and Continuous Random Variables
A **discrete** random variable can take on only specific, isolated numerical values, like the outcome of a roll of a die, or the number of dollars in a randomly chosen bank account. A **continuous** random variable, on the other hand, can take on any values within a continuum or an interval, like the temperature in Central Park, or the height of an athlete in centimeters. Discrete random variables that can take on only finitely many values (like the outcome of a roll of a die) are called **finite** random variables.

quick Examples

Random Variable	Values	Type
1. Select a mutual fund; $X =$ the number of companies in the fund portfolio.	$\{1, 2, 3, \ldots\}$	Discrete Infinite
2. Take 5 shots at the goal during a soccer match; $X =$ the number of times you score.	$\{0, 1, 2, 3, 4, 5\}$	Finite
3. Measure the length of an object; $X =$ its length in centimeters.	Any positive real number	Continuous
4. Roll a die until you get a 6; $X =$ the number of times you roll the die.	$\{1, 2, \ldots\}$	Discrete Infinite
5. Bet a whole number of dollars in a race where the betting limit is $100; $X =$ the amount you bet.	$\{0, 1, \ldots, 100\}$	Finite
6. Bet a whole number of dollars in a race where there is no betting limit; $X =$ the amount you bet.	$\{0, 1, \ldots, 100, 101, \ldots\}$	Discrete Infinite

Notes
1. In Chapter 7, the only sample spaces we considered in detail were finite sample spaces. However, in general, sample spaces can be infinite, as in many of the experiments mentioned above.

2. There are some "borderline" situations. For instance, if X is the salary of a factory worker, then X is, strictly speaking, discrete. However, the values of X are so numerous and close together that in some applications it makes sense to model X as a continuous random variable. ■

For the moment, we shall consider only finite random variables.

Example 1 **Finite Random Variable**

Let X be the number of heads that come up when a coin is tossed three times. List the value of X for each possible outcome. What are the possible values of X?

Solution First, we describe X as a random variable.

X is the rule that assigns to each outcome the number of heads that come up.

We take as the outcomes of this experiment all possible sequences of three heads and tails. Then, for instance, if the outcome is HTH, the value of X is 2. An easy way to list the values of X for all the outcomes is by means of a table.

2 Heads ($X = 2$)

Online, follow:

Chapter 8

to find an interactive simulation based on Example 1.

Outcome	HHH	HHT	HTH	HTT	THH	THT	TTH	TTT
Value of X	3	2	2	1	2	1	1	0

From the table, we also see that the possible values of X are 0, 1, 2, and 3.

✚ *Before we go on...* Remember that X is just a rule we decide on. In Example 1, we could have taken X to be a different rule, such as the number of tails or perhaps the number of heads minus the number of tails. These different rules are examples of different random variables associated with the same experiment. ■

Example 2 Stock Prices

You have purchased $10,000 worth of stock in a biotech company whose newest arthritis drug is awaiting approval by the F.D.A. If the drug is approved this month, the value of the stock will double by the end of the month. If the drug is rejected this month, the stock's value will decline by 80%. If no decision is reached this month, its value will decline by 10%. Let X be the value of your investment at the end of this month. List the value of X for each possible outcome.

Solution   There are three possible outcomes: the drug is approved this month, it is rejected this month, and no decision is reached. Once again, we express the random variable as a rule.

The random variable X is the rule that assigns to each outcome the value of your investment at the end of this month.

We can now tabulate the values of X as follows:

Outcome	Approved this month	Rejected this month	No decision
Value of X	$20,000	$2000	$9000

Probability Distribution of a Finite Random Variable

Given a random variable X, it is natural to look at certain *events*—for instance, the event that $X = 2$. By this, we mean the event consisting of all outcomes that have an assigned X-value of 2. Looking once again at the chart in Example 1, with X being the number of heads that face up when a coin is tossed three times, we find the following events:

The event that $X = 0$ is {TTT}.
The event that $X = 1$ is {HTT, THT, TTH}.
The event that $X = 2$ is {HHT, HTH, THH}.
The event that $X = 3$ is {HHH}.
The event that $X = 4$ is Ø.   There are no outcomes with four heads.

Each of these events has a certain probability. For instance, the probability of the event that $X = 2$ is 3/8 because the event in question consists of three of the eight possible (equally likely) outcomes. We shall abbreviate this by writing

$$P(X = 2) = \frac{3}{8}$$ The probability that $X = 2$ is 3/8.

Similarly,

$$P(X = 4) = 0$$ The probability that $X = 4$ is 0.

When X is a finite random variable, the collection of the probabilities of X equaling each of its possible values is called the **probability distribution** of X. Because the probabilities in a probability distribution can be estimated or theoretical, we shall discuss both *estimated probability distributions* (or *relative frequency distributions*) and *theoretical probability distributions* of random variables. (See the next two examples.)

Probability Distribution of a Finite Random Variable

If X is a finite random variable, with values $n_1, n_2, \ldots$ then its **probability distribution** lists the probabilities that $X = n_1$, $X = n_2$, $\ldots$ The sum of these probabilities is always 1.

Visualizing the Probability Distribution of a Random Variable

If each outcome in S is equally likely, we get the probability distribution shown for the random variable X.

S

$\bullet 0 \quad \bullet 0 \quad \bullet 5$

$\bullet 2 \quad \bullet -1$

Probability Distribution of X

x	-1	0	2	5
$P(X = x)$	$\frac{1}{5} = .2$	$\frac{2}{5} = .4$	$\frac{1}{5} = .2$	$\frac{1}{5} = .2$

Here $P(X = x)$ means "the probability that the random variable X has the specific value x."

quick Example

Roll a fair die; $X =$ the number facing up. Then, the (theoretical) probability that any specific value of X occurs is $\frac{1}{6}$. So, the probability distribution of X is the following (notice that the probabilities add up to 1):

x	1	2	3	4	5	6
$P(X = x)$	$\frac{1}{6}$	$\frac{1}{6}$	$\frac{1}{6}$	$\frac{1}{6}$	$\frac{1}{6}$	$\frac{1}{6}$

Note The distinction between X (upper case) and x (lower case) in the tables above is important; X stands for the random variable in question, whereas x stands for a specific *value* of X (so that x is always a number). Thus, if, say $x = 2$, then $P(X = x)$ means $P(X = 2)$, the probability that X is 2. Similarly, if Y is a random variable, then $P(Y = y)$ is the probability that Y has the specific value y. ∎

using *Technology*

See the Technology Guides at the end of the chapter to find out how to draw histograms using a TI-83/84 or Excel. Alternatively, go online and follow:

Chapter 8

→ Online Utilities

→ Histogram Maker

to find a utility for drawing histograms.

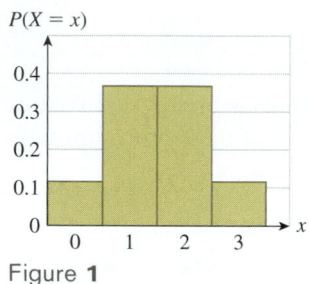

$P(X = x)$

Figure 1

The online simulation for Example 1 gives estimated probabilities for the random variable of Example 3.

Example 3 Theoretical Probability Distribution

Let X be the number of heads that face up in three tosses of a coin. Give the probability distribution of X.

Solution X is the random variable of Example 1, so its values are 0, 1, 2, and 3. The probability distribution of X is given in the following table:

x	0	1	2	3
$P(X = x)$	$\frac{1}{8}$	$\frac{3}{8}$	$\frac{3}{8}$	$\frac{1}{8}$

Notice that the probabilities add to 1, as we might expect.

We can use a bar graph to visualize a probability distribution. Figure 1 shows the bar graph for the probability distribution we obtained. Such a graph is sometimes called a **histogram.**

➕ *Before we go on...* The probabilities in the table in Example 3 are *theoretical* probabilities. To obtain a similar table of *estimated* probabilities, we would have to repeatedly toss a coin three times and calculate the fraction of times we got 0, 1, 2, and 3 heads. ▪

Note The table of probabilities in Example 3 looks like the probability distribution associated with an experiment, as we studied in Section 7.2. In fact, the probability distribution of a random variable is not really new. Consider the following experiment: toss three coins and count the number of heads. The associated probability distribution (as per Section 7.2) would be this:

Outcome	0	1	2	3
Probability	$\frac{1}{8}$	$\frac{3}{8}$	$\frac{3}{8}$	$\frac{1}{8}$

The difference is that in this chapter we are thinking of 0, 1, 2, and 3 not as the outcomes of the experiment, but as values of the random variable X. ▪

Example 4 Estimated Probability Distribution

The following table shows the (fictitious) income brackets of a sample of 1000 lawyers in their first year out of law school.

Income Bracket	$20,000–$29,999	$30,000–$39,999	$40,000–$49,999	$50,000–$59,999	$60,000–$69,999	$70,000–$79,999	$80,000–$89,999
Number	20	80	230	400	170	70	30

Think of the experiment of choosing a first-year lawyer at random (all being equally likely) and assign to each lawyer the number X that is the midpoint of his or her income bracket. Find the probability distribution of X.

using *Technology*

See the Technology Guides at
the end of the chapter to see
how to use a TI-83/84 or Excel
to convert frequencies into
probabilities.

Solution Statisticians refer to the income brackets as **measurement classes.** Because the first measurement class contains incomes that are at least $20,000, but less than $30,000, its midpoint is $25,000.[*] Similarly the second measurement class has midpoint $35,000, and so on. We can rewrite the table with the midpoints, as follows:

x	25,000	35,000	45,000	55,000	65,000	75,000	85,000
Frequency	20	80	230	400	170	70	30

We have used the term *frequency* rather than *number,* although it means the same thing. This table is called a **frequency table.** It is *almost* the probability distribution for X, except that we must replace frequencies by probabilities. (We did this in calculating estimated probabilities in the preceding chapter.) We start with the lowest measurement class. Because 20 of the 1000 lawyers fall in this group, we have

$$P(X = 25,000) = \frac{20}{1000} = .02$$

We can calculate the remaining probabilities similarly to obtain the following distribution:

x	25,000	35,000	45,000	55,000	65,000	75,000	85,000
$P(X = x)$	.02	.08	.23	.40	.17	.07	.03

Note again the distinction between X and x: X stands for the random variable in question, whereas x stands for a specific value (25,000, 35,000, . . . , or 85,000) of X.

---

[*] One might argue that the midpoint should be $(20,000 + 29,999)/2 = 24,999.50$, but we round this to 25,000. So, technically we are using "rounded" midpoints of the measurement classes.

Example 5 Probability Distribution: Greenhouse Gases

The following table shows per capita emissions of greenhouse gases for various countries, rounded to the nearest 5 metric tons.[†]

Country	Per Capita Emissions (metric tons)	Country	Per Capita Emissions (metric tons)
Australia	30	Czech Rep.	15
Austria	10	Denmark	15
Belgium	15	Estonia	10
Britain	10	Finland	15
Bulgaria	10	France	10
Canada	20	Germany	10

(continued)

[†] Figures are measured in carbon dioxide equivalent and are based on 1998 data. SOURCE: "Comprehensive Emissions Per Capita for Industrialized Countries," Hal Turton and Clive Hamilton, The Australia Institute, 2001. www.tai.org.au/Publications_Files/Papers&Sub_Files/Per%20Capita.pdf.

Country	Per Capita Emissions (metric tons)	Country	Per Capita Emissions (metric tons)
Greece	10	Poland	10
Hungary	10	Portugal	5
Iceland	10	Romania	5
Ireland	15	Russian Fed.	5
Italy	10	Spain	10
Japan	10	Sweden	5
Netherlands	15	Switzerland	5
New Zealand	15	Ukraine	10
Norway	10	U.S.A.	20

Consider the experiment in which a country is selected at random from this list, and let X be the per capita greenhouse gas emissions for that country. Find the probability distribution of X and graph it with a histogram. Use the probability distribution to compute $P(X \geq 20)$ (the probability that X is 20 or more) and interpret the result.

Solution The values of X are the possible emissions figures, which we take to be 0, 5, 10, 15, 20, 25, and 30. In the table below, we first compute the frequency of each value of X by counting the number of countries that produce that per capita level of greenhouse gases. For instance, there are 7 countries that have $X = 15$. Then, we divide each frequency by the sample size $N = 30$ to obtain the probabilities.

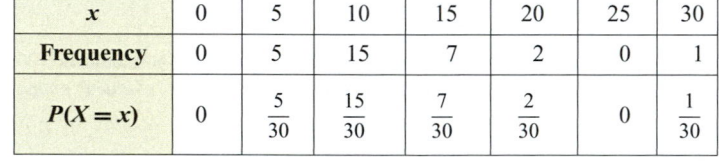

x	0	5	10	15	20	25	30
Frequency	0	5	15	7	2	0	1
$P(X = x)$	0	$\dfrac{5}{30}$	$\dfrac{15}{30}$	$\dfrac{7}{30}$	$\dfrac{2}{30}$	0	$\dfrac{1}{30}$

Figure 2 shows the resulting histogram.

Finally, we compute $P(X \geq 20)$, the probability of all events with an X-value of 20 or more. From the table, we obtain

$$P(X \geq 20) = \frac{2}{30} + 0 + \frac{1}{30} = \frac{3}{30} = .1$$

Thus, there is a 10% chance that a country randomly selected from the given list produces 20 or more metric tons per capita of greenhouse gases.

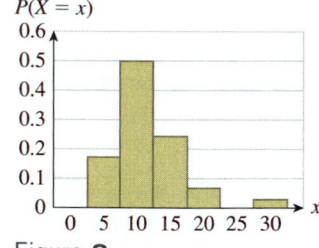

Figure **2**

FAQs   Recognizing What to Use as a Random Variable and Deciding on Its Values

Q: *In an application, how, exactly, do I decide what to use as a random variable X?*

A: Be as systematic as possible: First, decide what the experiment is and what its sample space is. Then, based on what is asked for in the application, complete the following sentence: "X assigns ___ to each outcome." For instance, "X assigns <u>the number of flavors</u> to each packet of gummy bears selected," or "X assigns <u>the average faculty salary</u> to each college selected." ∎

> Q: *Once I have decided what X should be, how do I decide what values to assign it?*
>
> A: Ask yourself: What are the conceivable values I could get for X? Then choose a collection of values that includes all of these. For instance, if X is the number of heads obtained when a coin is tossed 5 times, then the possible values of X are 0, 1, 2, 3, 4, and 5. If X is the average faculty salary in dollars, rounded to the nearest $5000, then possible values of X could be 20,000, 25,000, 30,000, and so on, up to the highest salary in your data. ∎

8.1 EXERCISES

● denotes basic skills exercises
◆ denotes challenging exercises

In Exercises 1–10, classify each random variable X as finite, discrete infinite, or continuous, and indicate the values that X can take. hint [see Quick Examples on p. 545]

1. ● Roll two dice; $X =$ the sum of the numbers facing up.

2. ● Open a 500-page book on a random page; $X =$ the page number.

3. ● Select a stock at random; $X =$ your profit, to the nearest dollar, if you purchase one share and sell it one year later.

4. ● Select an electric utility company at random; $X =$ the exact amount of electricity, in gigawatt hours, it supplies in a year.

5. ● Look at the second hand of your watch; X is the time it reads in seconds.

6. ● Watch a soccer game; $X =$ the total number of goals scored.

7. ● Watch a soccer game; $X =$ the total number of goals scored, up to a maximum of 10.

8. ● Your class is given a mathematics exam worth 100 points; X is the average score, rounded to the nearest whole number

9. ● According to quantum mechanics, the energy of an electron in a hydrogen atom can assume only the values $k/1, k/4, k/9, k/16, \ldots$ for a certain constant value k. $X =$ the energy of an electron in a hydrogen atom

10. ● According to classical mechanics, the energy of an electron in a hydrogen atom can assume any positive value. $X =$ the energy of an electron in a hydrogen atom

In Exercises 11–18, (a) say what an appropriate sample space is; (b) complete the following sentence: "X is the rule that assigns to each . . . "; (c) list the values of X for all the outcomes. hint [see Example 1]

11. ● X is the number of tails that come up when a coin is tossed twice.

12. ● X is the largest number of consecutive times heads comes up in a row when a coin is tossed three times.

13. ● X is the sum of the numbers that face up when two dice are rolled.

14. ● X is the value of the larger number when two dice are rolled.

15. ● X is the number of red marbles that Tonya has in her hand after she selects 4 marbles from a bag containing 4 red marbles and 2 green ones and then notes how many there are of each color.

16. ● X is the number of green marbles that Stej has in his hand after he selects 4 marbles from a bag containing 3 red marbles and 2 green ones and then notes how many there are of each color.

17. ● The mathematics final exam scores for the students in your study group are 89%, 85%, 95%, 63%, 92%, and 80%.

18. ● The capacities of the hard drives of your dormitory suite mates' computers are 10GB, 15GB, 20GB, 25GB, 30GB, and 35GB.

19. ● The random variable X has this probability distribution table:

x	2	4	6	8	10
$P(X = x)$	.1	.2	—	—	.1

a. Assuming $P(X = 8) = P(X = 6)$, find each of the missing values.
b. Calculate $P(X \geq 6)$.
hint [see Quick Example on p. 547]

20. ● The random variable X has the probability distribution table shown below:

x	−2	−1	0	1	2
$P(X = x)$	—	—	.4	.1	.1

a. Calculate $P(X \geq 0)$ and $P(X < 0)$.
b. Assuming $P(X = -2) = P(X = -1)$, find each of the missing values.

In Exercises 21–28, give the probability distribution for the indicated random variable and draw the corresponding histogram. hint [see Example 3]

21. ● A fair die is rolled, and X is the number facing up.

22. ● A fair die is rolled, and X is the square of the number facing up.

● basic skills   ◆ challenging

23. ● Three fair coins are tossed, and X is the square of the number of heads showing.

24. ● Three fair coins are tossed, and X is the number of heads minus the number of tails.

25. ● A red and a green die are rolled, and X is the sum of the numbers facing up.

26. ● A red and a green die are rolled, and
$$X = \begin{cases} 0 & \text{If the numbers are the same} \\ 1 & \text{If the numbers are different.} \end{cases}$$

27. A red and a green die are rolled, and X is the larger of the two numbers facing up.

28. A red and a green die are rolled, and X is the smaller of the two numbers facing up.

Applications

29. ● *Sport Utility Vehicles—Tow Ratings* The following table shows tow ratings (in pounds) for some popular sports utility vehicles:[1]

Mercedes Grand Marquis V8	2000
Jeep Wrangler I6	2000
Ford Explorer V6	3000
Dodge Dakota V6	4000
Mitsubishi Montero V6	5000
Ford Explorer V8	6000
Dodge Durango V8	6000
Dodge Ram 1500 V8	8000
Ford Expedition V8	8000
Hummer 2-door Hardtop	8000

Let X be the tow rating of a randomly chosen popular SUV from the list above.

a. What are the values of X?

b. Compute the frequency and probability distributions of X.

c. What is the probability that an SUV (from the list above) is rated to tow no more than 5000 pounds?
hint [see Example 5]

30. ● *Housing Prices* The following table shows the average percentage increase in the price of a house from 1980 to 2001 in 9 regions of the U.S.[2]

New England	300
Pacific	225
Middle Atlantic	225
South Atlantic	150
Mountain	150
West North Central	125
West South Central	75
East North Central	150
East South Central	125

Let X be the percentage increase in the price of a house in a randomly selected region.

a. What are the values of X?

b. Compute the frequency and probability distribution of X.

c. What is the probability that, in a randomly selected region, the percentage increase in the cost of a house exceeded 200%?

31. ● *Mold Counts* The following histogram shows the mold counts in Atlanta, GA, on 16 different days in March and April of 2005.[3]

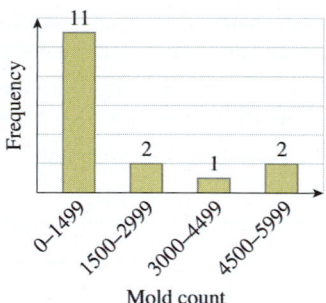

What is the associated random variable? Represent the data as an estimated probability distribution using the (rounded) midpoints of the given measurement classes. *hint* [see Example 4]

32. ● *Pollen Counts* Repeat the preceding exercise, using the following histogram showing the pollen counts in Miami, FL, on 20 different days in March and April of 2005.[4]

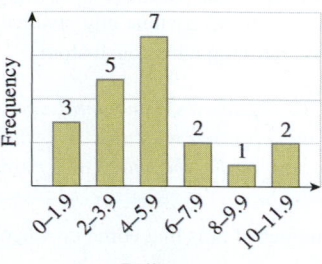

---

[1] Tow rates are for 2000 models and vary considerably within each model. Figures cited are rounded. For more detailed information, consult www.rvsafety.com/towrate2k.htm.

[2] Percentages are rounded to the nearest 25%. SOURCE: Third Quarter 2001 House Price Index, released November 30, 2001 by the Office of Federal Housing Enterprise Oversight; available online at www.ofheo.gov/house/3q01hpi.pdf.

[3] SOURCE: American Academy of Allergy Asthma & Immunology, April, 2005 www.aaaai.org.

[4] Ibid.

● basic skills   ◆ challenging

33. ● *Income Distribution up to $100,000* The following table shows the distribution of household incomes for a sample of 1000 households in the U.S. with incomes up to $100,000.[5]

Income Bracket	0–19,999	20,000–39,999	40,000–59,999	60,000–79,999	80,000–99,999
Households	270	280	200	150	100

a. Let X be the midpoint of a bracket in which a household falls. Find the probability distribution of X and graph its histogram.

b. Shade the area of your histogram corresponding to the probability that a U.S. household has a value of X of more than 50,000. What is this probability?

34. ● *Income Distribution up to $100,000* Repeat Exercise 33, using the following data for a sample of 1000 Hispanic households in the U.S.[6]

Income Bracket	0–19,999	20,000–39,999	40,000–59,999	60,000–79,999	80,000–99,999
Households	300	340	190	110	60

35. ● *Grade Point Averages* The grade point averages of the students in your mathematics class are

3.2, 3.5, 4.0, 2.9, 2.0, 3.3, 3.5, 2.9, 2.5, 2.0,
2.1, 3.2, 3.6, 2.8, 2.5, 1.9, 2.0, 2.2, 3.9, 4.0

Use these raw data to construct a frequency table with the following measurement classes: 1.1–2.0, 2.1–3.0, 3.1–4.0, and find the probability distribution using the (rounded) midpoint values as the values of X. *hint* [see Example 5]

36. ● *Test Scores* Your scores for the 20 surprise math quizzes last semester were (out of 10)

4.5, 9.5, 10.0, 3.5, 8.0, 9.5, 7.5, 6.5, 7.0, 8.0,
8.0, 8.5, 7.5, 7.0, 8.0, 9.0, 10.0, 8.5, 7.5, 8.0.

Use these raw data to construct a frequency table with the following brackets: 2.1–4.0, 4.1–6.0, 6.1–8.0, 8.1–10.0, and find the probability distribution using the (rounded) midpoint values as the values of X.

37. *Car Purchases* To persuade his parents to contribute to his new car fund, Carmine has spent the last week surveying the ages of 2000 cars on campus. His findings are reflected in the following frequency table:

Age of Car (years)	0	1	2	3	4	5	6	7	8	9	10
Number of Cars	140	350	450	650	200	120	50	10	5	15	10

Carmine's jalopy is 6 years old. He would like to make the following claim to his parents: "x percent of students have cars newer than mine." Use a probability distribution to find x.

38. *Car Purchases* Carmine's parents, not convinced of his need for a new car, produced the following statistics showing the ages of cars owned by students on the Dean's List:

Age of Car (years)	0	1	2	3	4	5	6	7	8	9	10
Number of Cars	0	2	5	5	10	10	15	20	20	20	40

They then claimed that if he kept his 6-year old car for another year, his chances of getting on the Dean's List would be increased by x percent. Use a probability distribution to find x.

Highway Safety Exercises 39–48 are based on the following table, which shows crashworthiness ratings for several categories of motor vehicles.[7] In all of these exercises, take X as the crash-test rating of a small car, Y as the crash-test rating for a small SUV, and so on, as shown in the table.

		Overall Frontal Crash Test Rating			
	Number Tested	3 (Good)	2 (Acceptable)	1 (Marginal)	0 (Poor)
Small Cars X	16	1	11	2	2
Small SUVs Y	10	1	4	4	1
Medium SUVs Z	15	3	5	3	4
Passenger Vans U	13	3	0	3	7
Midsize Cars V	15	3	5	0	7
Large Cars W	19	9	5	3	2

39. ● Compute the probability distribution for X.

40. ● Compute the probability distribution for Y.

41. ● Compute $P(X \geq 2)$ and interpret the result.

42. ● Compute $P(Y \leq 1)$ and interpret the result.

43. ● Compare $P(Y \geq 2)$ and $P(Z \geq 2)$. What does the result suggest about SUVs?

44. ● Compare $P(V \geq 2)$ and $P(Z \geq 2)$. What does the result suggest?

45. Which of the six categories shown has the *lowest* probability of a Good rating?

46. Which of the six categories shown has the *highest* probability of a Poor rating?

47. You choose, at random, a small car and a small SUV. What is the probability that both will be rated at least 2?

---

[5] Based on the actual household income distribution in 2003. SOURCE: U.S. Census Bureau, Current Population Survey, 2004 Annual Social and Economic Supplement. http://www.bls.census.gov/cps/asec/2004/sdata.htm.
[6] Ibid.

[7] Ratings are by the Insurance Institute for Highway Safety. SOURCES: Oak Ridge National Laboratory: "An Analysis of the Impact of Sport Utility Vehicles in the United States" Stacy C. Davis, Lorena F. Truett, (August 2000)/Insurance Institute for Highway Safety www-cta.ornl. gov/Publications/Final SUV report.pdf www.highwaysafety.org/ vehicle_ratings.

● basic skills   ◆ challenging

48. You choose, at random, a small car and a midsize car. What is the probability that both will be rated at most 1?

Exercises 49 and 50 assume familiarity with counting arguments and probability (Section 7.4).

49. Camping Kent's Tents has 4 red tents and 3 green tents in stock. Karin selects 4 of them at random. Let X be the number of red tents she selects. Give the probability distribution and find $P(X \geq 2)$.

50. Camping Kent's Tents has 5 green knapsacks and 4 yellow ones in stock. Curt selects 4 of them at random. Let X be the number of green knapsacks he selects. Give the probability distribution and find $P(X \leq 2)$.

51. ◆ **Testing Your Calculator** Use your calculator or computer to generate a sequence of 100 random digits in the range 0–9, and test the random number generator for uniformness by drawing the distribution histogram.

52. ◆ **Testing Your Dice** Repeat Exercise 51, but this time, use a die to generate a sequence of 50 random numbers in the range 1–6.

Communication and Reasoning Exercises

53. ● Are all infinite random variables necessarily continuous? Explain.

54. ● Are all continuous random variables necessarily infinite? Explain.

55. ● If you are unable to compute the (theoretical) probability distribution for a random variable X, how can you estimate the distribution?

56. ● What do you expect to happen to the probabilities in a probability distribution as you make the measurement classes smaller?

57. Give an example of a real-life situation that can be modeled by a random variable with a probability distribution whose histogram is highest on the left.

58. How wide should the bars in a histogram be so that the probability $P(a \leq X \leq b)$ equals the area of the corresponding portion of the histogram?

59. ◆ Give at least one scenario in which you might prefer to model the number of pages in a randomly selected book using a continuous random variable rather than a discrete random variable.

60. ◆ Give at least one reason why you might prefer to model a temperature using a discrete random variable rather than a continuous random variable.

● basic skills   ◆ challenging

8.2 Bernoulli Trials and Binomial Random Variables

Your electronic production plant produces video game joysticks. Unfortunately, quality control at the plant leaves much to be desired, and 10% of the joysticks the plant produces are defective. A large corporation has expressed interest in adopting your product for its new game console, and today an inspection team will be visiting to test video game joysticks as they come off the assembly line. If the team tests five joysticks, what is the probability that none will be defective? What is the probability that more than one will be defective?

In this scenario we are interested in the following, which is an example of a particular type of finite random variable called a **binomial random variable:** Think of the experiment as a sequence of five "trials" (in each trial the inspection team chooses one joystick at random and tests it) each with two possible outcomes: "success" (a defective joystick) and "failure" (a non-defective one).[8] If we now take X to be the number of

_____
[8] These are customary names for the two possible outcomes, and often do not indicate actual success or failure at anything. "Success" is the label we give the outcome of interest—in this case, finding a defective joystick.

successes (defective joysticks) the inspection team finds, we can recast the questions above as follows: Find $P(X = 0)$ and $P(X > 1)$.

Bernoulli Trial; Binomial Random Variable

A **Bernoulli**[*] **trial** is an experiment that has two possible outcomes, called **success** and **failure.** If the probability of success is p then the probability of failure is $q = 1 - p$.

Visualizing a Bernoulli Trial:

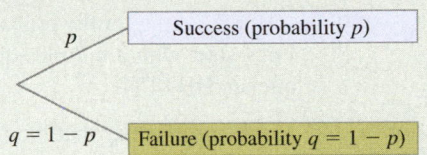

Tossing a coin three times is an example of a **sequence of independent Bernoulli trials:** a sequence of Bernoulli trials in which the outcomes in any one trial are independent (in the sense of the preceding chapter) of those in any other trial, and in which the probability of success is the same for all the trials.

A **binomial random variable** is one that counts the number of successes in a sequence of independent Bernoulli trials, where the number of trials is fixed.

Visualizing a Binomial Random Variable:

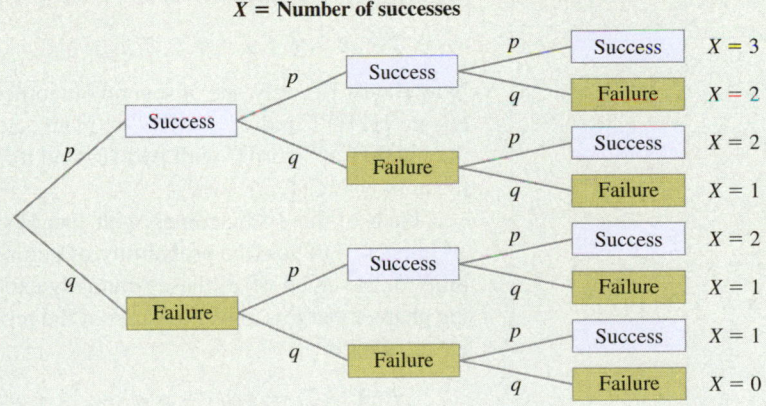

quick Examples **Binomial Random Variables**

1. Roll a die ten times; X is the number of times you roll a 6.

2. Provide a property with flood insurance for 20 years; X is the number of years, during the 20-year period, during which the property is flooded.[†]

3. You know that 60% of all bond funds will depreciate in value next year, and you randomly select four from a very large number of possible choices; X is the number of bond funds you hold that will depreciate next year. (X is approximately binomial.[‡])

---

[*] Jakob Bernoulli (1654–1705) was one of the pioneers of probability theory.

[†] Assuming that the occurrence of flooding one year is independent of whether there was flooding in earlier years.

[‡] Since the number of bond funds is extremely large, choosing a "loser" (a fund that will depreciate next year) does not significantly deplete the pool of "losers," and so the probability that the next fund you choose will be a "loser," is hardly affected. Hence, p is very nearly constant and we can think of X as being a binomial variable.

Example 1 Probability Distribution of a Binomial Random Variable

Suppose that we have a possibly unfair coin with the probability of heads p and the probability of tails $q = 1 - p$.

a. Let X be the number of heads you get in a sequence of 5 tosses. Find $P(X = 2)$.

b. Let X be the number of heads you get in a sequence of n tosses. Find $P(X = x)$.

Solution

a. We are looking for the probability of getting exactly 2 heads in a sequence of 5 tosses. Let's start with a simpler question: What is the probability that we will get the sequence HHTTT?

The probability that the first toss will come up heads is p.

The probability that the second toss will come up heads is also p.

The probability that the third toss will come up tails is q.

The probability that the fourth toss will come up tails is q.

The probability that the fifth toss will come up tails is q.

The probability that the first toss will be heads *and* the second will be heads *and* the third will be tails *and* the fourth will be tails *and* the fifth will be tails equals the probability of the *intersection* of these five events. Because these are independent events, the probability of the intersection is the product of the probabilities, which is

$$p \times p \times q \times q \times q = p^2 q^3$$

Now HHTTT is only one of several outcomes with two heads and three tails. Two others are HTHTT and TTTHH. How many such outcomes are there altogether? This is the number of "words" with two H's and three T's, and we know from Chapter 6 that the answer is $C(5, 2) = 10$.

Each of the 10 outcomes with two H's and three T's has the same probability: $p^2 q^3$. (Why?) Thus, the probability of getting one of these 10 outcomes is the probability of the union of all these (mutually exclusive) events, and we saw in the preceding chapter that this is just the sum of the probabilities. In other words, the probability we are after is

$$P(X = 2) = p^2 q^3 + p^2 q^3 + \cdots + p^2 q^3 \qquad \textcolor{magenta}{C(5, 2) \text{ times}}$$
$$= C(5, 2) p^2 q^3$$

The structure of this formula is as follows:

$$\underset{\textcolor{magenta}{\underset{\text{Number of heads}}{\uparrow\ \uparrow\ \uparrow\ \uparrow}}}{P(X = 2) = C(5, 2)\, p^{\overset{\textcolor{magenta}{\text{Number of heads}}}{2}} q^{\overset{\textcolor{magenta}{\text{Number of tails}}}{3}}}$$

Number of heads   Number of tails

Number of tosses | | Probability of tails
Number of heads   Probability of heads

b. What we did using the numbers 5 and 2 in part (a) works as well in general. For the general case, with n tosses and x heads, replace 5 with n and replace 2 with x to get:

$$P(X = x) = C(n, x)\, p^x q^{n-x}$$

(Note that the coefficient of q is the number of tails, which is $n - x$.)

The calculation in Example 1 applies to any binomial random variable, so we can say the following:

Probability Distribution of Binomial Random Variable

If X is the number of successes in a sequence of n independent Bernoulli trials, then

$$P(X = x) = C(n, x)p^x q^{n-x}$$

where

n = number of trials
p = probability of success
q = probability of failure = $1 - p$

quick Example

If you roll a fair die 5 times, the probability of throwing exactly 2 sixes is

$$P(X = 2) = C(5, 2)\left(\frac{1}{6}\right)^2 \left(\frac{5}{6}\right)^3 = 10 \times \frac{1}{36} \times \frac{125}{216} \approx .1608$$

Here, we used $n = 5$ and $p = 1/6$, the probability of rolling a six on one roll of the die.

Example 2 Aging

The probability that a randomly chosen person in Cape Coral, FL, is 65 years old or older[*] is approximately .2.

a. What is the probability that, in a randomly selected sample of 6 Cape Coral Floridians, exactly 4 of them are 65 or older?

b. If X is the number of people aged 65 or older in a sample of 6, construct the probability distribution of X and plot its histogram.

c. Compute $P(X \leq 2)$.

d. Compute $P(X \geq 2)$.

Solution

a. The experiment is a sequence of Bernoulli trials; in each trial we select a person and ascertain his or her age. If we take "success" to mean selection of a person aged 65 or older, then the probability distribution is

$$P(X = x) = C(n, x)p^x q^{n-x}$$

where n = number of trials = 6
p = probability of success = .2
q = probability of failure = .8

So,

$$P(X = 4) = C(6, 4)(.2)^4(.8)^2$$
$$= 15 \times .0016 \times .64 = .01536$$

---

[*] The actual figure in 2000 was .196 SOURCE: U.S. Census Bureau, Census 2000 Summary File 1. www.census.gov/prod/2001pubs/c2kbr01-10.pdf.

b. We have already computed $P(X = 4)$. Here are all the calculations:

$$P(X = 0) = C(6, 0)(.2)^0(.8)^6$$
$$= 1 \times 1 \times .262144 = .262144$$

$$P(X = 1) = C(6, 1)(.2)^1(.8)^5$$
$$= 6 \times .2 \times .32768 = .393216$$

$$P(X = 2) = C(6, 2)(.2)^2(.8)^4$$
$$= 15 \times .04 \times .4096 = .24576$$

$$P(X = 3) = C(6, 3)(.2)^3(.8)^3$$
$$= 20 \times .008 \times .512 = .08192$$

$$P(X = 4) = C(6, 4)(.2)^4(.8)^2$$
$$= 15 \times .0016 \times .64 = .01536$$

$$P(X = 5) = C(6, 5)(.2)^5(.8)^1$$
$$= 6 \times .00032 \times .8 = .001536$$

$$P(X = 6) = C(6, 6)(.2)^6(.8)^0$$
$$= 1 \times .000064 \times 1 = .000064$$

The probability distribution is therefore as follows:

x	0	1	2	3	4	5	6
$P(X = x)$	.262144	.393216	.24576	.08192	.01536	.001536	.000064

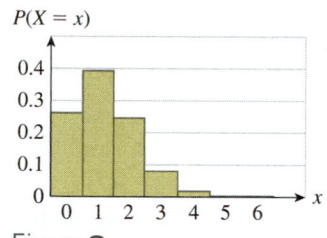

$P(X = x)$

Figure **3**

Figure 3 shows its histogram.

c. $P(X \leq 2)$—the probability that the number of people selected who are at least 65 years old is either 0, 1, or 2—is the probability of the union of these events and is thus the sum of the three probabilities:

$$P(X \leq 2) = P(X = 0) + P(X = 1) + P(X = 2)$$
$$= .262144 + .393216 + .24576$$
$$= .90112$$

d. To compute $P(X \geq 2)$, we *could* compute the sum

$$P(X \geq 2) = P(X = 2) + P(X = 3) + P(X = 4) + P(X = 5) + P(X = 6)$$

but it is far easier to compute the probability of the complement of the event:

$$P(X < 2) = P(X = 0) + P(X = 1)$$
$$= .262144 + .393216 = .65536$$

and then subtract the answer from 1:

$$P(X \geq 2) = 1 - P(X < 2)$$
$$= 1 - .65536 = .34464$$

using *Technology*

See the Technology Guides at the end of the chapter to find out how to compute the probability distribution of a binomial random variable using a TI-83/84 or Excel. Alternatively, go online and follow:

Chapter 8

→ Online Utilities

→ Binomial Distribution Utility

to find a utility for computing this distribution and drawing its histogram.

FAQs   Terminology and Recognizing When to Use the Binomial Distribution

Q: What is the difference between Bernoulli trials and a binomial random variable?

A: A Bernoulli trial is a type of experiment, whereas a binomial random variable is the resulting kind of random variable. More precisely, if your experiment consists of performing a sequence of *n* Bernoulli trials (think of throwing a dart *n* times at random points on a dartboard hoping to hit the bull's eye), then the random variable *X* that counts the number of successes (the number of times you actually hit the bull's eye) is a binomial random variable. ∎

Q: How do I recognize when a situation gives a binomial random variable?

A: Make sure that the experiment consists of a sequence of independent Bernoulli trials; that is, a sequence of a fixed number of trials of an experiment that has two outcomes, where the outcome of each trial does not depend on the outcomes in previous trials, and where the probability of success is the same for all the trials. For instance, repeatedly throwing a dart at a dartboard hoping to hit the bull's eye does not constitute a sequence of Bernoulli trials if you adjust your aim each time depending on the outcome of your previous attempt. This dart-throwing experiment can be modeled by a sequence of Bernoulli trials if you make no adjustments after each attempt and your aim does not improve (or deteriorate) with time. ∎

8.2 EXERCISES

● denotes basic skills exercises

◆ denotes challenging exercises

tech Ex indicates exercises that should be solved using technology

In Exercises 1–10, you are performing 5 independent Bernoulli trials with $p = .1$ and $q = .9$. Calculate the probability of each of the stated outcomes. Check your answer using technology.
hint [see Quick Examples on p. 557]

1. ● 2 successes

2. ● 3 successes

3. ● No successes

4. ● No failures

5. ● All successes

6. ● All failures

7. ● At most 2 successes

8. ● At least 4 successes

9. ● At least 3 successes

10. ● At most 3 successes

In Exercises 11–18, X is a binomial variable with $n = 6$ and $p = .4$. Compute the given probabilities. Check your answer using technology. *hint* [see Example 2]

11. ● $P(X = 3)$

12. ● $P(X = 4)$

13. ● $P(X \leq 2)$

14. ● $P(X \leq 1)$

15. ● $P(X \geq 5)$

16. ● $P(X \geq 4)$

17. ● $P(1 \leq X \leq 3)$

18. ● $P(3 \leq X \leq 5)$

In Exercises 19 and 20, graph the histogram of the given binomial distribution. Check your answer using technology.

19. ● $n = 5, p = \frac{1}{4}, q = \frac{3}{4}$.

20. ● $n = 5, p = \frac{1}{3}, q = \frac{2}{3}$.

In Exercises 21 and 22, graph the histogram of the given binomial distribution and compute the given quantity, indicating the corresponding region on the graph.

21. ● $n = 4, p = \frac{1}{3}, q = \frac{2}{3}; P(X \leq 2)$

22. ● $n = 4, p = \frac{1}{4}, q = \frac{3}{4}; P(X \leq 1)$

Applications

23. ● *Retirement* The probability that a randomly chosen person in Britain is of pension age[9] is approximately .25. What is the probability that, in a randomly selected sample of 5 people, 2 are of pension age? *hint* [see Example 2]

24. ● *Alien Retirement* The probability that a randomly chosen citizen-entity of Cygnus is of pension age[10] is approximately .8. What is the probability that, in a randomly selected sample of 4 citizen-entities, all of them are of pension age?

25. ● *90s Internet Stock Boom* According to a July, 1999, article[11] in the *New York Times,* venture capitalists had this "rule of thumb": The probability that an Internet start-up company will

---

[9] SOURCE: Carnegie Center, Moscow/*The New York Times,* March 15, 1998, p. 10.

[10] 12,000 bootlags, which translates to approximately 20 minutes Earth time.

[11] Not All Hit It Rich in the Internet Gold Rush, *The New York Times,* July 20, 1999, p. A1.

● basic skills   ◆ challenging   **tech** Ex technology exercise

be a "stock market success" resulting in "spectacular profits for early investors" is .2. If you were a venture capitalist who invested in 10 Internet start-up companies, what was the probability that at least 1 of them would be a stock market success? (Round your answer to four decimal places.)

26. ● *90s Internet Stock Boom* According to the article cited in Exercise 25, 13.5% of Internet stocks that entered the market in 1999 ended up trading below their initial offering prices. If you were an investor who purchased 5 Internet stocks at their initial offering prices, what was the probability that at least 4 of them would end up trading at or above their initial offering price? (Round your answer to four decimal places.)

27. ● *Job Training* *(from the GRE Exam in Economics)* In a large on-the-job training program, half of the participants are female and half are male. In a random sample of 3 participants, what is the probability that an investigator will draw at least 1 male?

28. ● *Job Training* *(based on a question from the GRE Exam in Economics)* In a large on-the-job training program, half of the participants are female and half are male. In a random sample of 5 participants, what is the probability that an investigator will draw at least 2 males?

29. ● *Manufacturing* Your manufacturing plant produces air bags, and it is known that 10% of them are defective. Five air bags are tested.

a. Find the probability that 3 of them are defective.
b. Find the probability that at least 2 of them are defective.

30. ● *Manufacturing* Compute the probability distribution of the binomial variable described in the preceding exercise, and use it to compute the probability that if 5 air bags are tested, at least 1 will be defective and at least 1 will not.

31. ● *Teenage Pastimes* According to a study,[12] the probability that a randomly selected teenager watched a rented video at least once during a week was .71. What is the probability that at least 8 teenagers in a group of 10 watched a rented movie at least once last week?

32. ● *Other Teenage Pastimes* According to the study cited in the preceding exercise, the probability that a randomly selected teenager studied at least once during a week was only .52. What is the probability that less than half of the students in your study group of 10 have studied in the last week?

33. ● *Fast Food* The probability that a randomly selected pound of beef produced in the U.S. is purchased by McDonald's is .04.[13] Ten pounds of beef are chosen at random.

a. What is the probability that exactly two of them are purchased by McDonald's?
b. *tech* Ex Use technology to generate the probability distribution for the associated binomial random variable.

c. Fill in the blank: If 10 pounds of beef produced in the U.S. are selected at random, the number of pounds most likely to have been purchased by McDonald's is _____.

34. ● *Fast Food* There is a .4 probability that a randomly selected pound of vegetables purchased by McDonald's is tomatoes.[14] Eight pounds of vegetables purchased by McDonald's are chosen at random.

a. What is the probability that exactly 4 of them are tomatoes?
b. *tech* Ex Use technology to generate the probability distribution for the associated binomial random variable.
c. Fill in the blank: If 8 pounds of vegetables purchased by McDonald's are selected at random, it is most likely that the number of pounds of tomatoes is _____.

35. *Triple Redundancy* In order to ensure reliable performance of vital computer systems, aerospace engineers sometimes employ the technique of "triple redundancy," in which 3 identical computers are installed in a space vehicle. If 1 of the 3 computers gives results different from the other 2, it is assumed to be malfunctioning and it is ignored. This technique will work as long as no more than one computer malfunctions. Assuming that an onboard computer is 99% reliable (that is, the probability of its failing is 0.01), what is the probability that at least 2 of the 3 computers will malfunction?

36. *IQ Scores* Mensa is a club for people who have high IQ scores. To qualify, your IQ must be at least 132, putting you in the top 2% of the general population. If a group of 10 people are chosen at random, what is the probability that at least 2 of them qualify for Mensa?

37. *tech* Ex *Standardized Tests* Assume that on a standardized test of 100 questions, a person has a probability of 80% of answering any particular question correctly. Find the probability of answering between 75 and 85 questions, inclusive, correctly. (Assume independence, and round your answer to four decimal places.)

38. *tech* Ex *Standardized Tests* Assume that on a standardized test of 100 questions, a person has a probability of 80% of answering any particular question correctly. Find the probability of answering at least 90 questions correctly. (Assume independence, and round your answer to four decimal places.)

39. *tech* Ex *Product Testing* It is known that 43% of all the ZeroFat hamburger patties produced by your factory actually contain more than 10 grams of fat. Compute the probability distribution for $n = 50$ Bernoulli trials.

a. What is the most likely value for the number of burgers in a sample of 50 that contain more than 10 grams of fat?
b. Complete the following sentence: There is an approximately 71% chance that a batch of 50 ZeroFat patties contains ____ or more patties with more than 10 grams of fat.
c. Compare the graphs of the distributions for $n = 50$ trials and $n = 20$ trials. What do you notice?

---

[12] SOURCES: Rand Youth Poll/Teen-age Research Unlimited/*The New York Times,* March 14, 1998, p. D1.

[13] SOURCE: Department of Agriculture/*New York Times,* February 20, 2005, p. BU8.

---

[14] Ibid.

● basic skills   ◆ challenging   *tech* Ex technology exercise

40. `tech` Ex *Product Testing* It is known that 65% of all the ZeroCal hamburger patties produced by your factory actually contain more than 1000 calories. Compute the probability distribution for $n = 50$ Bernoulli trials.

   a. What is the most likely value for the number of burgers in a sample of 50 that contain more than 1000 calories?

   b. Complete the following sentence: There is an approximately 73% chance that a batch of 50 ZeroCal patties contains _____ or more patties with more than 1000 calories.

   c. Compare the graphs of the distributions for $n = 50$ trials and $n = 20$ trials. What do you notice?

41. `tech` Ex *Quality Control* A manufacturer of light bulbs chooses bulbs at random from its assembly line for testing. If the probability of a bulb's being bad is .01, how many bulbs do they need to test before the probability of finding at least one bad one rises to more than .5? (You may have to use trial and error to solve this.)

42. `tech` Ex *Quality Control* A manufacturer of light bulbs chooses bulbs at random from its assembly line for testing. If the probability of a bulb's being bad is .01, how many bulbs do they need to test before the probability of finding at least two bad ones rises to more than .5? (You may have to use trial and error to solve this.)

43. *Highway Safety* According to a study,[15] a male driver in the U.S. will average 562 accidents per 100 million miles. Regard an n-mile trip as a sequence of n Bernoulli trials with "success" corresponding to having an accident during a particular mile. What is the probability that a male driver will have an accident in a one-mile trip?

44. *Highway Safety*: According to the study cited in the preceding exercise, a female driver in the U.S. will average 611 accidents per 100 million miles. Regard an n-mile trip as a sequence of n Bernoulli trials with "success" corresponding to having an accident during a particular mile. What is the probability that a female driver will have an accident in a one-mile trip?

45. ◆ *Mad Cow Disease* In March, 2004, the U.S. Agriculture Department announced plans to test approximately 243,000 slaughtered cows per year for mad cow disease (bovine spongiform encephalopathy).[16] When announcing the plan, the Agriculture Department stated that "by the laws of probability, that

many tests should detect mad cow disease even if it is present in only 5 cows out of the 45 million in the nation."[17] Test the Department's claim by computing the probability that, if only 5 out of 45 million cows had mad cow disease, at least 1 cow would test positive in a year (assuming the testing was done randomly).

46. ◆ *Mad Cow Disease* According to the article cited in Exercise 45, only 223,000 of the cows being tested for bovine spongiform encephalopathy were to be "downer cows;" cows unable to walk to their slaughter. Assuming that just one downer cow in 500,000 is infected on average, use a binomial distribution to find the probability that 2 or more cows would test positive in a year. Your associate claims that "by the laws of probability, that many tests should detect at least two cases of mad cow disease even if it is present in only 2 cows out of a million downers." Comment on that claim.

Communication and Reasoning Exercises

47. ● A soccer player is more likely to score on his second shot if he was successful on his first. Can we model a succession of shots a player takes as a sequence of Bernoulli trials? Explain.

48. ● A soccer player takes repeated shots on goal. What assumption must we make if we want to model a succession of shots by a player as a sequence of Bernoulli trials?

49. ● Your friend just told you that "misfortunes always occur in threes." If life is just a sequence of Bernoulli trials, is this possible? Explain.

50. ● Suppose an experiment consists of repeatedly (every week) checking whether your graphing calculator battery has died. Is this a sequence of Bernoulli trials? Explain.

51. Why is the following not a binomial random variable? Select, without replacement, 5 marbles from a bag containing 6 red marbles and 2 blue ones, and let X be the number of red marbles you have selected.

52. By contrast with Exercise 51, why can the following be modeled by a binomial random variable? Select, without replacement, 5 electronic components from a batch of 10,000 in which 1000 are defective, and let X be the number of defective components you select.

---

[15] Data are based on a report by National Highway Traffic Safety Administration released in January, 1996. SOURCE for data: U.S. Department of Transportation/*The New York Times,* April 9, 1999, p. F1.

[16] SOURCE: *The New York Times,* March 17, 2004, p. A19.

[17] As stated in the *New York Times* article.

● basic skills   ◆ challenging   `tech` Ex technology exercise

8.3 | Measures of Central Tendency

Mean, Median, and Mode of a Set of Data

One day you decide to measure the popularity rating of your statistics instructor, Mr. Pelogrande. Ideally, you should poll all of Mr. Pelogrande's students, which is what statisticians would refer to as the **population.** However, it would be difficult to poll all the members of the population in question (Mr. Pelogrande teaches more than 400 students). Instead, you decide to survey 10 of his students, chosen at random, and ask them to rate Mr. Pelogrande on a scale of 0–100. The survey results in the following set of data:

$$60, 50, 55, 0, 100, 90, 40, 20, 40, 70$$

Such a collection of data is called a **sample,** because the 10 people polled represent only a (small) sample of Mr. Pelogrande's students. We should think of the individual scores 60, 50, 55, . . . as values of a random variable: Choose one of Mr. Pelogrande's students at random and let X be the rating the student gives to Mr. Pelogrande.

How do we distill a single measurement, or **statistic,** from this sample that would describe Mr. Pelogrande's popularity? Perhaps the most commonly used statistic is the **average,** or **mean,** which is computed by adding the scores and dividing the sum by the number of scores in the sample:

$$\text{Sample Mean} = \frac{60 + 50 + 55 + 0 + 100 + 90 + 40 + 20 + 40 + 70}{10} = \frac{525}{10} = 52.5$$

We might then conclude, based on the sample, that Mr. Pelogrande's average popularity rating is about 52.5. The usual notation for the sample mean is $\bar{x}$, and the formula we use to compute it is

$$\bar{x} = \frac{x_1 + x_2 + \cdots + x_n}{n}$$

where $x_1, x_2, \ldots, x_n$ are the values of X in the sample.

A convenient way of writing the sum that appears in the numerator is to use **summation** or **sigma notation.** We write the sum $x_1 + x_2 + \cdots + x_n$ as

$$\sum_{i=1}^{n} x_i$$

$\displaystyle\sum_{i=1}^{n}$ by itself stands for "the sum, from $i = 1$ to n."

$\displaystyle\sum_{i=1}^{n} x_i$ stands for "the sum of the x_i, from $i = 1$ to n."

We think of i as taking on the values $1, 2, \ldots, n$ in turn, making x_i equal $x_1, x_2, \ldots, x_n$ in turn, and we then add up these values.

Sample and Mean

A **sample** is a sequence of values (or scores) of a random variable X. (The process of collecting such a sequence is sometimes called **sampling** X.) The **sample mean** is the average of the values, or **scores,** in the sample. To compute the sample mean, we use the following formula:

$$\bar{x} = \frac{x_1 + x_2 + \cdots + x_n}{n} = \frac{\sum_{i=1}^{n} x_i}{n}$$

or simply

$$\bar{x} = \frac{\sum_i x_i}{n}$$ $\sum_i$ stands for "sum over all i"*

Here, n is the **sample size** (number of scores), and $x_1, x_2, \ldots, x_n$ are the individual values.

If the sample $x_1, x_2, \ldots, x_n$ consists of all the values of X from the entire population[†] (for instance, the ratings given Mr. Pelogrande by *all* of his students), we refer to the mean as the **population mean,** and write it as μ (Greek "mu") instead of $\bar{x}$.

Visualizing the Mean

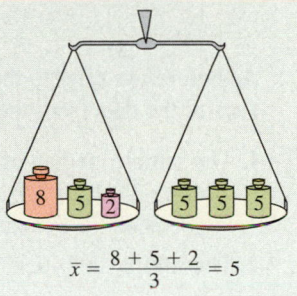

$$\bar{x} = \frac{8 + 5 + 2}{3} = 5$$

quick Examples

1. The mean of the sample 1, 2, 3, 4, 5 is $\bar{x} = 3$.

2. The mean of the sample −1, 0, 2 is $\bar{x} = \dfrac{-1 + 0 + 2}{3} = \dfrac{1}{3}$.

3. The mean of the population −3, −3, 0, 0, 1 is $\mu = \dfrac{-3 - 3 + 0 + 0 + 1}{5} = -1$.

---

* In Section 1.5 we simply wrote $\sum x$ for the sum of all the x_i, but here we will use the subscripts to make it easier to interpret formulas in this and the next section.

† When we talk about *populations,* the understanding is that the underlying experiment consists of selecting a member of a given population and ascertaining the value of X.

Note: Sample Mean versus Population Mean

Determining a population mean can be difficult or even impossible. For instance, computing the mean household income for the U.S. would entail recording the income of every single household in the U.S. Instead of attempting to do this, we usually use sample means instead. The larger the sample used, the more accurately we expect the sample mean to approximate the population mean. Estimating how accurately a sample mean based on a given sample size approximates the population mean is possible, but we will not go into that in this book. ∎

The mean $\bar{x}$ is an attempt to describe where the "center" of the sample is. It is therefore called a **measure of central tendency.** There are two other common measures of central tendency: the "middle score," or **median,** and the "most frequent score," or **mode.** These are defined as follows.

Median and Mode

The **sample median** m is the middle score (in the case of an odd-size sample), or average of the two middle scores (in the case of an even-size sample) when the scores in a sample are arranged in ascending order.

A **sample mode** is a score that appears most often in the collection. (There may be more than one mode in a sample.)

Visualizing the Median and Mode

Median = Middle score = 4

2 2 2 4 4 8 8

Mode = Most frequent score = 2

As before, we refer to the **population median** and **population mode** if the sample consists of the data from the entire population.

---

quick Examples

1. The sample median of 2, –3, –1, 4, 2 is found by first arranging the scores in ascending order: –3, –1, 2, 2, 4 and then selecting the middle (third) score: $m = 2$. The sample mode is also 2 because this is the score that appears most often.

2. The sample 2, 5, 6, –1, 0, 6 has median $m = (2 + 5)/2 = 3.5$ and mode 6.

The mean tends to give more weight to scores that are further away from the center than does the median. For example, if you take the largest score in a collection of more than two numbers and make it larger, the mean will increase but the median will remain the same. For this reason the median is often preferred for collections that contain a wide range of scores. The mode can sometimes lie far from the center and is thus used less often as an indication of where the "center" of a sample lies.

Example 1   Teenage Spending

A 10-year survey of spending patterns of U.S. teenagers yielded the following figures[*] (in billions of dollars spent in a year): 90, 90, 85, 80, 80, 80, 80, 85, 90, 100. Compute and interpret the mean, median, and mode, and illustrate the data on a graph.

Solution The *mean* is given by

$$\bar{x} = \frac{\sum_i x_i}{n}$$

$$= \frac{90 + 90 + 85 + 80 + 80 + 80 + 80 + 85 + 90 + 100}{10} = \frac{860}{10} = 86$$

Thus, spending by teenagers averaged $86 billion per year.

Grace/zefa/Corbis

---

[*] Spending figures are rounded, and cover the years 1988 through 1997. SOURCE: Rand Youth Poll/Teenage Research Unlimited/*The New York Times,* March 14, 1998, p. D1.

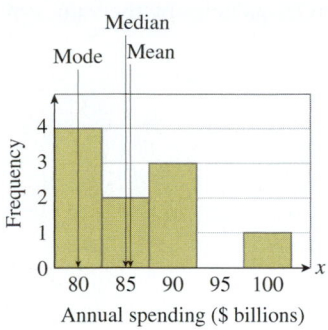

Figure **4**

For the *median,* we arrange the sample data in ascending order:

$$80, 80, 80, 80, 85, 85, 90, 90, 90, 100$$

We then take the average of the two middle scores:

$$m = \frac{85 + 85}{2} = 85$$

This means that in half the years in question, teenagers spent $85 billion or less, and in half they spent $85 billion or more.

For the *mode* we choose the score (or scores) that occurs most frequently: $80 billion. Thus, teenagers spent $80 billion per year more often than any other amount.

The frequency histogram in Figure 4 illustrates these three measures.

+ *Before we go on...* There is a nice geometric interpretation of the difference between the median and mode: The median line shown in Figure 4 divides the total area of the histogram into two equal pieces, whereas the mean line passes through its "center of gravity"; if you placed the histogram on a knife-edge along the mean line, it would balance. ∎

Expected Value of a Finite Random Variable

Now, instead of looking at a sample of values of a given random variable, let us look at the probability distribution of the random variable itself and see if we can predict the sample mean without actually taking a sample. This prediction is what we call the *expected value* of the random variable.

Example **2 Expected Value of a Random Variable**

Suppose you roll a fair die a large number of times. What do you expect to be the average of the numbers that face up?

Solution Suppose we take a sample of n rolls of the die (where n is large). Because the probability of rolling a 1 is 1/6, we would expect that we would roll a 1 one sixth of the time, or $n/6$ times. Similarly, each other number should also appear $n/6$ times. The frequency table should then look like this:

x	1	2	3	4	5	6
Number of Times x is Rolled (frequency)	$\dfrac{n}{6}$	$\dfrac{n}{6}$	$\dfrac{n}{6}$	$\dfrac{n}{6}$	$\dfrac{n}{6}$	$\dfrac{n}{6}$

Note that we would not really expect the scores to be evenly distributed in practice, although for very large values of n we would expect the frequencies to vary only by a small percentage. To calculate the sample mean, we would add up all the scores and divide by the sample size. Now, the table tells us that there are $n/6$ ones, $n/6$ twos, $n/6$ threes, and so on, up to $n/6$ sixes. Adding these all up gives

$$\sum_i x_i = \frac{n}{6} \cdot 1 + \frac{n}{6} \cdot 2 + \frac{n}{6} \cdot 3 + \frac{n}{6} \cdot 4 + \frac{n}{6} \cdot 5 + \frac{n}{6} \cdot 6$$

(Notice that we can obtain this number by multiplying the frequencies by the values of X and then adding.) Thus, the mean is

$$\bar{x} = \frac{\sum_i x_i}{n}$$

$$= \frac{\frac{n}{6} \cdot 1 + \frac{n}{6} \cdot 2 + \frac{n}{6} \cdot 3 + \frac{n}{6} \cdot 4 + \frac{n}{6} \cdot 5 + \frac{n}{6} \cdot 6}{n}$$

$$= \frac{1}{6} \cdot 1 + \frac{1}{6} \cdot 2 + \frac{1}{6} \cdot 3 + \frac{1}{6} \cdot 4 + \frac{1}{6} \cdot 5 + \frac{1}{6} \cdot 6 \qquad \text{\textcolor{red}{Divide top and bottom by } n.}$$

$$= 3.5$$

This is the average value we expect to get after a large number of rolls or, in short, the **expected value** of a roll of the die. More precisely, we say that this is the expected value of the random variable X whose value is the number we get by rolling a die. Notice that n, the number of rolls, does not appear in the expected value. In fact, we could redo the calculation more simply by dividing the frequencies in the table by n *before* adding. Doing this replaces the frequencies with the *probabilities,* $1/6$. That is, it *replaces the frequency distribution with the probability distribution.*

x	1	2	3	4	5	6
$P(X = x)$	$\frac{1}{6}$	$\frac{1}{6}$	$\frac{1}{6}$	$\frac{1}{6}$	$\frac{1}{6}$	$\frac{1}{6}$

The expected value of X is then the sum of the products $x \cdot P(X = x)$. This is how we shall compute it from now on.

To obtain the expected value, multiply the values of X by their probabilities, and then add the results.

Expected Value of a Finite Random Variable

If X is a finite random variable that takes on the values $x_1, x_2, \ldots, x_n$, then the **expected value** of X, written $E(X)$ or μ, is

$$\mu = E(X) = x_1 \cdot P(X = x_1) + x_2 \cdot P(X = x_2) + \cdots + x_n \cdot P(X = x_n)$$
$$= \sum_i x_i \cdot P(X = x_i)$$

In Words

To compute the expected value from the probability distribution of X, we multiply the values of X by their probabilities and add up the results.

Interpretation

We interpret the expected value of X as a *prediction* of the mean of a large random sample of measurements of X; in other words, it is what we "expect" the mean of a large number of scores to be. (The larger the sample, the more accurate this prediction will tend to be.)

quick Example   If X has the distribution shown,

x	-1	0	4	5
$P(X = x)$	.3	.5	.1	.1

then $\mu = E(X) = -1(.3) + 0(.5) + 4(.1) + 5(.1) = -.3 + 0 + .4 + .5 = .6$.

Example 3 Sports Injuries

According to historical data, the number of injuries that a member of the Enormous State University women's soccer team will sustain during a typical season is given by the following probability distribution table:

Injuries	0	1	2	3	4	5	6
Probability	.2	.2	.22	.2	.15	.01	.02

If X denotes the number of injuries sustained by a player during one season, compute $E(X)$ and interpret the result.

 using *Technology*

See the Technology Guides at the end of the chapter to see how to use a TI-83/84 or Excel to automate the computation of expected values.

Solution We can compute the expected value using the following tabular approach: take the probability distribution table, add another row in which we compute the product $xP(X = x)$, and then add these products together.

x	0	1	2	3	4	5	6	
$P(X = x)$	.2	.2	.22	.2	.15	.01	.02	**Total:**
$xP(X = x)$	0	.2	.44	.6	.6	.05	.12	2.01

The total of the entries in the bottom row is the expected value. Thus,

$$E(X) = 2.01$$

We interpret the result as follows: If many soccer players are observed for a season, we predict that the average number of injuries each will sustain is about 2.

Example 4 Roulette

A roulette wheel (of the kind used in the U.S.) has the numbers 1 through 36, 0 and 00. A bet on a single number pays 35 to 1. This means that if you place a $1 bet on a single number and win (your number comes up), you get your $1 back plus $35 (that is, you gain $35). If your number does not come up, you lose the $1 you bet. What is the expected gain from a $1 bet on a single number?

Solution The probability of winning is 1/38, so the probability of losing is 37/38. Let X be the gain from a $1 bet. X has two possible values: $X = -1$ if you lose and $X = 35$ if you win. $P(X = -1) = 37/38$ and $P(X = 35) = 1/38$. This probability distribution and the calculation of the expected value are given in the following table:

x	-1	35	
$P(X = x)$	$\frac{37}{38}$	$\frac{1}{38}$	**Total:**
$xP(X = 9\ x)$	$-\frac{37}{38}$	$\frac{35}{38}$	$-\frac{2}{38}$

So, we expect to average a small loss of $2/38 \approx \$0.0526$ on each spin of the wheel.

+*Before we go on...* Of course, you cannot actually lose the expected $0.0526 on one spin of the roulette wheel in Example 4. However, if you play many times, this is what you expect your *average* loss per bet to be. For example, if you played 100 times, you could expect to lose about $100 \times 0.0526 = \$5.26$. ∎

A betting game in which the expected value is zero is called a **fair game.** For example, if you and I flip a coin, and I give you $1 each time it comes up heads but you give me $1 each time it comes up tails, then the game is fair. Over the long run, we expect to come out even. On the other hand, a game like roulette in which the expected value is not zero is **biased.** Most casino games are slightly biased in favor of the house.[18] Thus, most gamblers will lose only a small amount and many gamblers will actually win something (and return to play some more). However, when the earnings are averaged over the huge numbers of people playing, the house is guaranteed to come out ahead. This is how casinos make (lots of) money.

Expected Value of a Binomial Random Variable

Suppose you guess all the answers to the questions on a multiple-choice test. What score can you expect to get? This scenario is an example of a sequence of Bernoulli trials (see the preceding section), and the number of correct guesses is therefore a binomial random variable whose expected value we wish to know. There is a simple formula for the expected value of a binomial random variable.

Expected Value of Binomial Random Variable

If X is the binomial random variable associated with n independent Bernoulli trials, each with probability p of success, then the expected value of X is

$$\mu = E(X) = np$$

quick Example

If X is the number of successes in 20 Bernoulli trials with $p = .7$, then the expected number of successes is $\mu = E(X) = (20)(.7) = 14$.

Where does this formula come from? We *could* use the formula for expected value and compute the sum

$$E(X) = 0C(n, 0)p^0 q^n + 1C(n, 1)p^1 q^{n-1} + 2C(n, 2)p^2 q^{n-2} + \cdots + nC(n, n)p^n q^0$$

directly (using the binomial theorem), but this is one of the many places in mathematics where a less direct approach is much easier. X is the number of successes in a sequence of n Bernoulli trials, each with probability p of success. Thus, p is the fraction of time we expect a success, so out of n trials we expect np successes. Because X counts successes, we expect the value of X to be np. (With a little more effort, this can be made into a formal proof that the sum above equals np.)

---

[18] Only rarely are games not biased in favor of the house. However, blackjack played without continuous shuffle machines can be beaten by card counting.

Example 5 Guessing on an Exam

An exam has 50 multiple-choice questions, each having 4 choices. If a student randomly guesses on each question, how many correct answers can he or she expect to get?

Solution Each guess is a Bernoulli trial with probability of success 1 in 4, so $p = .25$. Thus, for a sequence of $n = 50$ trials,

$$\mu = E(X) = np = (50)(.25) = 12.5$$

Thus, the student can expect to get about 12.5 correct answers.

Q: *Wait a minute. How can a student get a fraction of a correct answer?*

A: Remember that the expected value is the average number of correct answers a student will get if he or she guesses on a large number of such tests. Or, we can say that if many students use this strategy of guessing, they will average about 12.5 correct answers each. ■

Estimating the Expected Value from a Sample

It is not always possible to know the probability distribution of a random variable. For instance, if we take X to be the income of a randomly selected lawyer, one could not be expected to know the probability distribution of X. However, we can still obtain a good *estimate* of the expected value of X (the average income of all lawyers) by using the estimated probability distribution based on a large random sample.

Example 6 Estimating an Expected Value

The following table shows the (fictitious) incomes of a random sample of 1000 lawyers in their first year out of law school.

Income Bracket	\$20,000–\$29,999	\$30,000–\$39,999	\$40,000–\$49,999	\$50,000–\$59,999	\$60,000–\$69,999	\$70,000–\$79,999	\$80,000–\$89,999
Number	20	80	230	400	170	70	30

Estimate the average of the incomes of all lawyers in their first year out of law school.

Solution We first interpret the question in terms of a random variable. Let X be the income of a lawyer selected at random from among all currently practicing first-year lawyers in the U.S. We are given a sample of 1000 values of X, and we are asked to find the expected value of X. First, we use the midpoints of the income brackets to set up an (estimated) probability distribution for X:

x	25,000	35,000	45,000	55,000	65,000	75,000	85,000
$P(X = x)$	.02	.08	.23	.40	.17	.07	.03

Our estimate for $E(X)$ is then

$$E(X) = \sum_i x_i \cdot P(X = x_i)$$

$$= (25,000)(.02) + (35,000)(.08) + (45,000)(.23) + (55,000)(.40)$$
$$+ (65,000)(.17) + (75,000)(.07) + (85,000)(.03) = \$54,500$$

Thus, $E(X)$ is approximately $54,500. That is, the average income of all currently practicing first-year lawyers in the U.S. is approximately $54,500.

FAQs   Recognizing When to Compute the Mean and When to Compute the Expected Value

Q: *When am I supposed to compute the mean (add the values of X and divide by n) and when am I supposed to use the expected value formula?*

A: The formula for the mean (adding and dividing by the number of observations) is used to compute the mean of a sequence of random scores, or sampled values of X. If, on the other hand, you are given the probability distribution for X (even if it is only an estimated probability distribution) then you need to use the expected value formula. ■

8.3 EXERCISES

● denotes basic skills exercises

◆ denotes challenging exercises

tech Ex indicates exercises that should be solved using technology

Compute the mean, median, and mode of the data samples in Exercises 1–8. hint [see Quick Examples on pp. 563, 564]

1. ● −1, 5, 5, 7, 14

2. ● 2, 6, 6, 7, −1

3. ● 2, 5, 6, 7, −1, −1

4. ● 3, 1, 6, −3, 0, 5

5. ● $\frac{1}{2}, \frac{3}{2}, -4, \frac{5}{4}$

6. ● $-\frac{3}{2}, \frac{3}{8}, -1, \frac{5}{2}$

7. ● 2.5, −5.4, 4.1, −0.1, −0.1

8. ● 4.2, −3.2, 0, 1.7, 0

9. Give a sample of 6 scores with mean 1 and with median $\neq$ mean. (Arrange the scores in ascending order.)

10. Give a sample of 5 scores with mean 100 and median 1. (Arrange the scores in ascending order.)

In Exercises 11–16, calculate the expected value of X for the given probability distribution. hint [see Quick Examples on p. 566]

11. ●

x	0	1	2	3
P(X = x)	.5	.2	.2	.1

12. ●

x	1	2	3	4
P(X = x)	.1	.2	.5	.2

13. ●

x	10	20	30	40
P(X = x)	$\frac{15}{50}$	$\frac{20}{50}$	$\frac{10}{50}$	$\frac{5}{50}$

14. ●

x	2	4	6	8
P(X = x)	$\frac{1}{20}$	$\frac{15}{20}$	$\frac{2}{20}$	$\frac{2}{20}$

15. ●

x	−5	−1	0	2	5	10
P(X = x)	.2	.3	.2	.1	.2	0

16. ●

x	−20	−10	0	10	20	30
P(X = x)	.2	.4	.2	.1	0	.1

In Exercises 17–28, calculate the expected value of the given random variable X. [Exercises 23, 24, 27, and 28 assume familiarity with counting arguments and probability (Section 7.4).] hint [see Example 2]

17. ● X is the number that faces up when a fair die is rolled.

18. ● X is the number selected at random from the set $\{1, 2, 3, 4\}$.

19. ● X is the number of tails that come up when a coin is tossed twice.

20. ● X is the number of tails that come up when a coin is tossed three times.

● basic skills   ◆ challenging   **tech Ex** technology exercise

21. X is the higher number when two dice are rolled.

22. X is the lower number when two dice are rolled.

23. X is the number of red marbles that Suzy has in her hand after she selects 4 marbles from a bag containing 4 red marbles and 2 green ones.

24. X is the number of green marbles that Suzy has in her hand after she selects 4 marbles from a bag containing 3 red marbles and 2 green ones.

25. 20 darts are thrown at a dartboard. The probability of hitting a bull's-eye is .1. Let X be the number of bull's-eyes hit.

26. 30 darts are thrown at a dartboard. The probability of hitting a bull's-eye is $\frac{1}{5}$. Let X be the number of bull's-eyes hit.

27. tech Ex Select five cards without replacement from a standard deck of 52, and let X be the number of queens you draw.

28. tech Ex Select five cards without replacement from a standard deck of 52, and let X be the number of red cards you draw.

Applications

29. ● **Sports Utility Vehicles—Tow Ratings** Following is a sample of tow ratings (in pounds) for some popular 2000 model light trucks:[19] *hint* [see Quick Examples on pp. 563, 564]

    2000, 2000, 3000, 4000, 5000, 6000, 6000, 6000, 8000, 8000

    Compute the mean and median of the given sample. Fill in the blank: There were as many light trucks with a tow rating of more than _____ pounds as there were with tow ratings below that.

30. ● **Housing Prices** Following is a sample of the percentage increases in the price of a house from 1980 to 2001 in 8 regions of the U.S.:[20]

    75, 130, 145, 150, 150, 225, 225, 300

    Compute the mean and median of the given sample. Fill in the blank: As many regions in the U.S. had housing increases averaging more than ____ % as below that.

31. ● **Gold** The following figures show the price of gold in dollars per ounce for the 10-business day period April 4–April 15, 2004:[21]

    424, 425, 426, 428, 425, 429, 427, 428, 423, 425

    Find the sample mean, median, and mode(s). What do your answers tell you about the price of gold?

32. ● **Silver** The following figures show the price of silver in dollars per ounce for the 10-business day period April 4–April 15, 2004:[22]

    7.0, 6.9, 7.0, 7.1, 7.0, 7.2, 7.2, 7.2, 7.1, 7.1

    Find the sample mean, median, and mode(s). What do your answers tell you about the price of silver?

33. ● **Supermarkets** A survey of 52 U.S. supermarkets yielded the following relative frequency table, where X is the number of checkout lanes at a randomly chosen supermarket.[23] *hint* [see Example 3]

x	1	2	3	4	5	6	7	8	9	10
$P(X=x)$	.01	.04	.04	.08	.10	.15	.25	.20	.08	.05

   a. Compute $\mu = E(X)$ and interpret the result.
   b. Which is larger, $P(X < \mu)$ or $P(X > \mu)$? Interpret the result.

34. ● **Video Arcades** Your company, Sonic Video, Inc., has conducted research that shows the following probability distribution, where X is the number of video arcades in a randomly chosen city with more than 500,000 inhabitants.

x	0	1	2	3	4	5	6	7	8	9
$P(X=x)$	.07	.09	.35	.25	.15	.03	.02	.02	.01	.01

   a. Compute $\mu = E(X)$ and interpret the result.
   b. Which is larger, $P(X < \mu)$ or $P(X > \mu)$? Interpret the result.

35. ● **School Enrollment** The following table shows the approximate numbers of school goers in the U.S. (residents who attended some educational institution) in 1998, broken down by age group.[24]

Age	3–6.9	7–12.9	13–16.9	17–22.9	23–26.9	27–42.9
Population (millions)	12	24	15	14	2	5

Use the rounded midpoints of the given measurement classes to compute the probability distribution of the age X of a school goer. (Round probabilities to two decimal places.) Hence compute the expected value of X. What information does the expected value give about residents enrolled in schools?

---

[19] Tow rates vary considerably within each model. Figures cited are rounded. For more detailed information, consult www.rvsafety.com/towrate2k.htm.

[20] Percentages are rounded to the nearest 25%. SOURCE: Third Quarter 2001 House Price Index, released November 30, 2001 by the Office of Federal Housing Enterprise Oversight; available online at www.ofheo.gov/house/3q01hpi.pdf.

[21] Prices rounded to the nearest $1. SOURCE: www.kitco.com/gold.londonfix.html.

[22] Prices rounded to the nearest 10¢. SOURCE: Ibid.

[23] SOURCES: J.T. McClave, P.G. Benson, T. Sincich, Statistics for Business and Economics, 7th Ed. (Prentice Hall, 1998) p. 177, W. Chow et al. "A model for predicting a supermarket's annual sales per square foot," Graduate School of Management, Rutgers University.

[24] Data are approximate, SOURCE: Statistical Abstract of the United States: 2000.

● basic skills   ◆ challenging   tech Ex technology exercise

36. ● *School Enrollment* Repeat Exercise 35, using the following data from 1980.[25]

Age	3–6.9	7–12.9	13–16.9	17–22.9	23–26.9	27–42.9
Population (millions)	8	20	11	13	1	3

37. `tech` Ex *Mold Counts* The following histogram shows the mold counts in Atlanta, GA, on 16 different days in March and April of 2005.[26] *hint* [see Example 6]

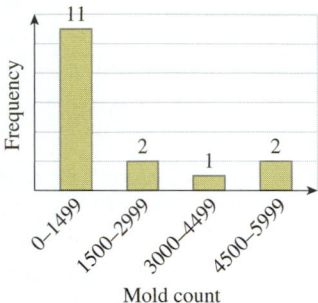

Mold count

Use the estimated probability distribution based on the (rounded) midpoints of the given measurement classes to obtain an estimate of the expected mold count (rounded to the nearest whole number) in Atlanta.

38. `tech` Ex *Pollen Counts* Repeat the preceding exercise using the following histogram showing the pollen counts in Miami, FL, on 20 different days in March and April of 2005.[27]

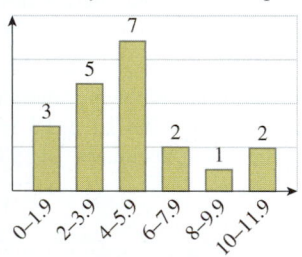

Pollen count

39. `tech` Ex *Income Distribution up to $100,000* The following table shows the distribution of household incomes for a sample of 1000 households in the U.S. with incomes up to $100,000.[28]

Income Bracket	0–19,999	20,000–39,999	40,000–59,999	60,000–79,999	80,000–99,999
Households	270	280	200	150	100

Use this information to estimate, to the nearest $1000, the average household income for such households.

40. `tech` Ex *Income Distribution up to $100,000* Repeat Exercise 39, using the following data for a sample of 1000 Hispanic households in the U.S.[29]

Income Bracket	0–19,999	20,000–39,999	40,000–59,999	60,000–79,999	80,000–99,999
Households	300	340	190	110	60

Highway Safety Exercises 41–44 are based on the following table, which shows crashworthiness ratings for several categories of motor vehicles.[30] In all of these exercises, take X as the crash-test rating of a small car, Y as the crash-test rating for a small SUV, and so on as shown in the table.

	Number Tested	Overall Frontal Crash-Test Rating			
		3 (Good)	2 (Acceptable)	1 (Marginal)	0 (Poor)
Small cars X	16	1	11	2	2
Small SUVs Y	10	1	4	4	1
Medium SUVs Z	15	3	5	3	4
Passenger vans U	13	3	0	3	7
Midsize cars V	15	3	5	0	7
Large cars W	19	9	5	3	2

41. Compute the probability distributions and expected values of X and Y. Based on the results, which of the two types of vehicle performed better in frontal crashes?

42. Compute the probability distributions and also expected values of Z and V. Based on the results, which of the two types of vehicle performed better in frontal crashes?

43. `tech` Ex Based on expected values, which of the following categories performed best in crash tests: small cars, midsize cars, or large cars?

44. `tech` Ex Based on expected values, which of the following categories performed best in crash tests: small SUVs, medium SUVs, or passenger vans?

45. *Roulette* A roulette wheel has the numbers 1 through 36, 0, and 00. Half of the numbers from 1 through 36 are red, and a bet on red pays even money (that is, if you win, you will get back your $1 plus another $1). How much do you expect to win with a $1 bet on red? *hint* [see Example 4]

---

[25] Ibid.

[26] SOURCE: American Academy of Allergy Asthma & Immunology, April, 2005 www.aaaai.org.

[27] Ibid.

[28] Based on the actual household income distribution in 2003. SOURCE: U.S. Census Bureau, Current Population Survey, 2004 Annual Social and Economic Supplement. http://www.bls.census.gov/cps/asec/2004/sdata.htm.

[29] Ibid.

[30] Ratings are by the Insurance Institute for Highway Safety. SOURCES: Oak Ridge National Laboratory: "An Analysis of the Impact of Sport Utility Vehicles in the United States" Stacy C. Davis, Lorena F. Truett, (August 2000)/ Insurance Institute for Highway Safety www-cta.ornl.gov/Publications/Final SUV report.pdf www.highwaysafety.org/vehicle_ratings.

● basic skills   ◆ challenging   `tech` Ex technology exercise

46. *Roulette* A roulette wheel has the numbers 1 through 36, 0, and 00. A bet on two numbers pays 17 to 1 (that is, if one of the two numbers you bet comes up, you get back your $1 plus another $17). How much do you expect to win with a $1 bet on two numbers?

47. ● *Teenage Pastimes* According to a study,[31] the probability that a randomly selected teenager shopped at a mall at least once during a week was .63. How many teenagers in a randomly selected group of 40 would you expect to shop at a mall during the next week? *hint* [see Example 5]

48. ● *Other Teen-Age Pastimes* According to the study referred to in the preceding exercise, the probability that a randomly selected teenager played a computer game at least once during a week was .48. How many teenagers in a randomly selected group of 30 would you expect to play a computer game during the next seven days?

49. *Manufacturing* Your manufacturing plant produces air bags, and it is known that 10% of them are defective. A random collection of 20 air bags is tested.

   a. How many of them would you expect to be defective?
   b. In how large a sample would you expect to find 12 defective airbags?

50. *Spiders* Your pet tarantula, Spider, has a .12 probability of biting an acquaintance who comes into contact with him. Next week, you will be entertaining 20 friends (all of whom will come into contact with Spider).

   a. How many guests should you expect Spider to bite?
   b. At your last party, Spider bit 6 of your guests. Assuming that Spider bit the expected number of guests, how many guests did you have?

Exercises 51 and 52 assume familiarity with counting arguments and probability (Section 7.4).

51. *Camping* Kent's Tents has 4 red tents and 3 green tents in stock. Karin selects 4 of them at random. Let X be the number of red tents she selects. Give the probability distribution of X and find the expected number of red tents selected.

52. *Camping* Kent's Tents has 5 green knapsacks and 4 yellow ones in stock. Curt selects 4 of them at random. Let X be the number of green knapsacks he selects. Give the probability distribution of X, and find the expected number of green knapsacks selected.

53. tech Ex *Stock Portfolios* You are required to choose between two stock portfolios, FastForward Funds and SolidState Securities. Stock analysts have constructed the following probability distributions for next year's rate of return for the two funds.

FastForward Funds

Rate of return	−.4	−.3	−.2	−.1	0	.1	.2	.3	.4
Probability	.015	.025	.043	.132	.289	.323	.111	.043	.019

SolidState Securities

Rate of return	−.4	−.3	−.2	−.1	0	.1	.2	.3	.4
Probability	.012	.023	.050	.131	.207	.330	.188	.043	.016

Which of the two funds gives the higher expected rate of return?

54. tech Ex *Risk Management* Before making your final decision whether to invest in FastForward Funds or SolidState Securities (see Exercise 53), you consult your colleague in the risk management department of your company. She informs you that, in the event of a stock market crash, the following probability distributions for next year's rate of return would apply:

FastForward Funds

Rate of return	−.8	−.7	−.6	−.5	−.4	−.2	−.1	0	.1
Probability	.028	.033	.043	.233	.176	.230	.111	.044	.102

SolidState Securities

Rate of return	−.8	−.7	−.6	−.5	−.4	−.2	−.1	0	.1
Probability	.033	.036	.038	.167	.176	.230	.211	.074	.035

Which of the two funds offers the lowest risk in case of a market crash?

55. ◆ *Insurance Schemes* The Acme Insurance Company is launching a drive to generate greater profits, and it decides to insure racetrack drivers against wrecking their cars. The company's research shows that, on average, a racetrack driver races four times a year and has a 1 in 10 chance of wrecking a vehicle, worth an average of $100,000, in every race. The annual premium is $5000, and Acme automatically drops any driver who is involved in an accident (after paying for a new car), but does not refund the premium. How much profit (or loss) can the company expect to earn from a typical driver in a year? [Hint: Use a tree diagram to compute the probabilities of the various outcomes.]

56. ◆ *Insurance* The Blue Sky Flight Insurance Company insures passengers against air disasters, charging a prospective passenger $20 for coverage on a single plane ride. In the event of a fatal air disaster, it pays out $100,000 to the named beneficiary. In the event of a nonfatal disaster, it pays out an average of $25,000 for hospital expenses. Given that the probability of a plane's crashing on a single trip[32] is 0.00000087, and that a passenger involved in a plane crash has a 0.9 chance of being killed, determine the profit (or loss) per passenger that the insurance company expects to make on each trip. [Hint: Use a tree to compute the probabilities of the various outcomes.]

---

[31] SOURCE: Rand Youth Poll/Teen-age Research Unlimited/*The New York Times,* March 14, 1998, p. D1.

[32] This was the probability of a passenger plane crashing per departure in 1990. (SOURCE: National Transportation Safety Board.)

● basic skills   ◆ challenging   tech Ex technology exercise

Communication and Reasoning Exercises

57. ● In a certain set of scores, there are as many values above the mean as below it. It follows that
(A) The median and mean are equal.
(B) The mean and mode are equal.
(C) The mode and median are equal.
(D) The mean, mode, and median are all equal.

58. ● In a certain set of scores, the median occurs more often than any other score. It follows that
(A) The median and mean are equal.
(B) The mean and mode are equal.
(C) The mode and median are equal.
(D) The mean, mode, and median are all equal.

59. ● Your friend Charlesworth claims that the median of a collection of data is always close to the mean. Is he correct? If so, say why; if not, give an example to prove him wrong.

60. ● Your other friend Imogen asserts that Charlesworth is wrong and that it is the mode and the median that are always close to each other. Is she correct? If so, say why; if not, give an example to prove her wrong.

61. ● Must the expected number of times you hit a bull's-eye after 50 attempts always be a whole number? Explain.

62. ● Your statistics instructor tells you that the expected score of the upcoming midterm test is 75%. That means that 75% is the most likely score to occur, right?

63. Your grade in a recent midterm was 80%, but the class average was 83%. Most people in the class scored better than you, right?

64. Your grade in a recent midterm was 80%, but the class median was 100%. Your score was lower than the average score, right?

65. Slim tells you that the population mean is just the mean of a suitably large sample. Is he correct? Explain.

66. Explain how you can use a sample to estimate an expected value.

67. Following is an excerpt from a full-page ad by MoveOn.org in the *New York Times* criticizing President G.W. Bush:[33]

On Tax Cuts:

George Bush: "... Americans will keep, this year, an average of almost $1000 more of their own money."

The Truth: Nearly half of all taxpayers get less than $100. And 31% of all taxpayers get nothing at all.

The statements referred to as "The Truth" contradict the statement attributed to President Bush, right? Explain.

68. Following is an excerpt from a five-page ad by WeissneggerForGov.org in *The Martian Enquirer* criticizing Supreme Martian Administrator, Gov. Red Davis

On Worker Accommodation:

Gov. Red Davis: "The median size of Government worker habitats in Valles Marineris is at least 400 square feet."

Weissnegger: "The average size of a Government worker habitat in Valles Marineris is a mere 150 square feet."

The statements attributed to Weissnegger do not contradict the statement attributed to Gov. Davis, right? Explain.

69. Sonia has just told you that the expected household income in the U.S. is the same as the population mean of all U.S. household incomes. Clarify her statement by describing an experiment and an associated random variable X so that the expected household income is the expected value of X.

70. If X is a random variable, what is the difference between a sample mean of measurements of X and the expected value of X? Illustrate by means of an example.

---

[33] SOURCE: Full-page ad in the *New York Times,* September 17, 2003, p. A25.

● basic skills   ◆ challenging   **tech Ex** technology exercise

---

8.4 Measures of Dispersion

Variance and Standard Deviation of a Set of Scores

Your grade on a recent midterm was 68%; the class average was 72%. How do you stand in comparison with the rest of the class? If the grades were widely scattered, then your grade may be close to the mean and a fair number of people may have done a lot worse than you (Figure 5a). If, on the other hand, almost all the grades were within a few points of the average, then your grade may not be much higher than the lowest grade in the class (Figure 5b).

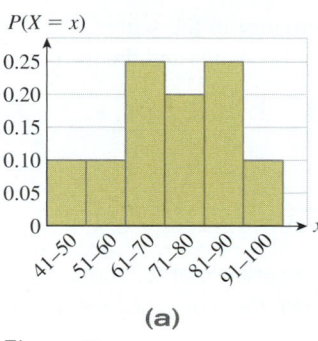
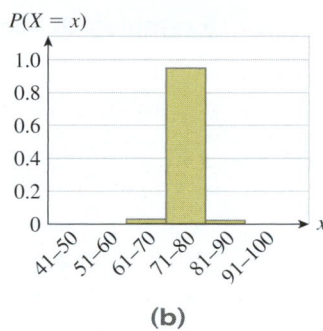

(a)   **(b)**

Figure **5**

This scenario suggests that it would be useful to have a way of measuring not only the central tendency of a set of scores (mean, median or mode) but also the amount of "scatter" or "dispersion" of the data.

If the scores in our set are $x_1, x_2, \ldots, x_n$ and their mean is $\bar{x}$ (or μ in the case of a population mean), we are really interested in the distribution of the differences $x_i - \bar{x}$. We could compute the *average* of these differences, but this average will always be 0. (Why?) It is really the *sizes* of these differences that interest us, so we might try computing the average of the absolute values of the differences. This idea is reasonable, but it leads to technical difficulties that are avoided by a slightly different approach: The statistic we use is based on the average of the *squares* of the differences, as explained in the following definitions.

Sample Variance and Standard Deviation

The **sample variance** of the sample $x_1, x_2, \ldots, x_n$ of n values of X is given by

$$s^2 = \frac{(x_1 - \bar{x})^2 + (x_2 - \bar{x})^2 + \cdots + (x_n - \bar{x})^2}{n - 1} = \frac{\sum_{i=1}^{n} (x_i - \bar{x})^2}{n - 1}$$

The **sample standard deviation** is the square-root of the sample variance:

$$s = \sqrt{s^2}$$

Visualizing Small and Large Variance

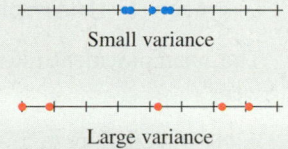

Small variance

Large variance

Population Variance and Standard Deviation

If the values $x_1, x_2, \ldots, x_n$ are all the measurements of X in the entire population, then the **population variance** is given by

$$\sigma^2 = \frac{(x_1 - \mu)^2 + (x_2 - \mu)^2 + \cdots + (x_n - \mu)^2}{n} = \frac{\sum_{i=1}^{n} (x_i - \mu)^2}{n}$$

(Remember that μ is the symbol we use for the *population* mean.) The **population standard deviation** is the square root of the population variance:

$$\sigma = \sqrt{\sigma^2}$$

quick Examples

1. The sample variance of the scores 1, 2, 3, 4, 5 is the sum of the squares of the differences between the scores and the mean $\bar{x} = 3$, divided by $n - 1 = 4$:

$$s^2 = \frac{(1-3)^2 + (2-3)^2 + (3-3)^2 + (4-3)^2 + (5-3)^2}{4} = \frac{10}{4} = 2.5$$

so

$$s = \sqrt{2.5} \approx 1.58$$

2. The population variance of the scores 1, 2, 3, 4, 5 is the sum of the squares of the differences between the scores and the mean $\mu = 3$, divided by $n = 5$:

$$\sigma^2 = \frac{(1-3)^2 + (2-3)^2 + (3-3)^2 + (4-3)^2 + (5-3)^2}{5} = \frac{10}{5} = 2$$

so

$$\sigma = \sqrt{2} \approx 1.41$$

Q: *The population variance is the average of the squares of the differences between the values and the mean. But why do we divide by $n - 1$ instead of n when calculating the sample variance?*

A: In real-life applications, we would like the sample variance to approximate the population variance. In statistical terms, we would like the expected value of the sample variance s^2 to be the same as the population variance σ^2. The sample variance s^2 as we have defined it is the "unbiased estimator" of the population variance σ^2 that accomplishes this task; if, instead, we divided by n in the formula for s^2, we would, on average, tend to underestimate the population variance. (See the online text on Sampling Distributions for further discussion of unbiased estimators.) Note that as the sample size gets larger and larger, the discrepancy between the formulas for s^2 and σ^2 becomes negligible; dividing by n gives almost the same answer as dividing by $n - 1$. ∎

Here's a simple example of calculating standard deviation.

Example 1 Unemployment

The unemployment rates (in percentage points) in the U.S. for the years 1999–2004 were[*]

4, 4, 5, 6, 6, 5

Compute the sample mean and standard deviation, rounded to 1 decimal place. What percentage of the scores fall within one standard deviation of the mean? What percentage fall within two standard deviations of the mean?

Solution The sample mean is

$$\bar{x} = \frac{\sum_i x_i}{n} = \frac{4+4+5+6+6+5}{6} = \frac{30}{6} = 5$$

---

[*] Figures are rounded. SOURCES: SWA, 222; Bureau of Labor Statistics (BLS), http://www.bls.gov.

using *Technology*

See the Technology Guides at the end of the chapter to see how to compute means and standard deviations using a TI-83/84 or Excel.

The sample variance is

$$s^2 = \frac{\sum\limits_{i=1}^{n}(x_i - \bar{x})^2}{n-1}$$

$$= \frac{(4-5)^2 + (4-5)^2 + (5-5)^2 + (6-5)^2 + (6-5)^2 + (5-5)^2}{5}$$

$$= \frac{1+1+0+1+1+0}{5} = \frac{4}{5} = 0.8$$

Thus, the sample standard deviation is

$$s = \sqrt{0.8} \approx 0.9 \qquad \text{Rounded to 1 decimal place}$$

To ask which scores fall "within one standard deviation of the mean" is to ask which scores fall in the interval $[\bar{x} - s, \bar{x} + s]$, or $[5 - 0.9, 5 + 0.9] = [4.1, 5.9]$. Only 2 of the 6 scores (the two 5s) fall in this interval, so the percentage of scores that fall within one standard deviation of the mean is $2/6 \approx 33\%$.

For two standard deviations, the interval in question is $[\bar{x} - 2s, \bar{x} + 2s] = [5 - 1.8, 5 + 1.8] = [3.2, 6.8]$, which includes all of the scores. In other words, 100% of the scores fall within two standard deviations of the mean.

Q: In Example 1, 33 percent of the scores fell within one standard deviation of the mean and all of them fell within two standard deviations of the mean. Is this typical?

A: Actually, this is atypical due to the small number of scores involved. There are two useful methods for *estimating* the percentage of scores that fall within any number of standard deviations of the mean. The first method applies to any set of data, and is due to P.L. Chebyshev (1821–1894), while the second applies to "nice" sets of data, and is based on the "normal distribution," which we shall discuss in Section 5. ∎

Chebyshev's Rule

For any set of data, the following statements are true:

At least 3/4 of the scores fall within 2 standard deviations of the mean (within the interval $[\bar{x} - 2s, \bar{x} + 2s]$ for samples or $[\mu - 2\sigma, \mu + 2\sigma]$ for populations).

At least 8/9 of the scores fall within 3 standard deviations of the mean (within the interval $[\bar{x} - 3s, \bar{x} + 3s]$ for samples or $[\mu - 3\sigma, \mu + 3\sigma]$ for populations).

At least 15/16 of the scores fall within 4 standard deviations of the mean (within the interval $[\bar{x} - 4s, \bar{x} + 4s]$ for samples or $[\mu - 4\sigma, \mu + 4\sigma]$ for populations).

. . .

At least $1 - 1/k^2$ of the scores fall within k standard deviations of the mean (within the interval $[\bar{x} - ks, \bar{x} + ks]$ for samples or $[\mu - k\sigma, \mu + k\sigma]$ for populations).

Visualizing Chebyshev's Rule

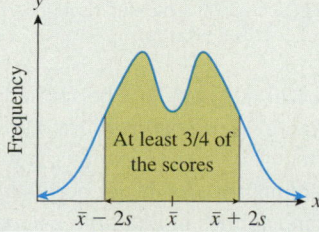

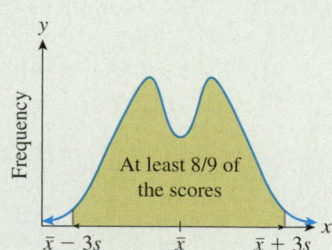

Empirical Rule*

For a set of data whose frequency distribution is "bell-shaped" and symmetric (see Figure 6), the following is true:

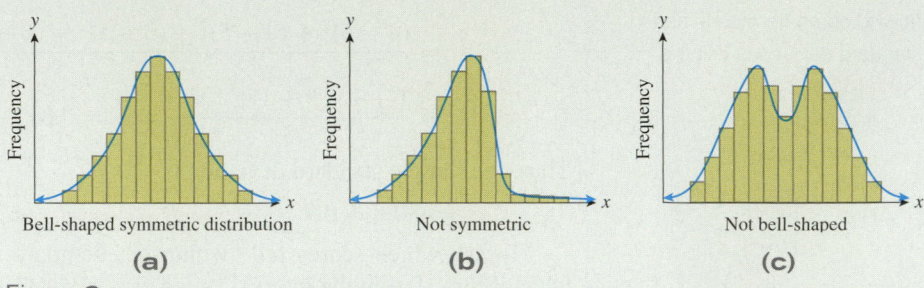

Bell-shaped symmetric distribution

(a)

Not symmetric

(b)

Not bell-shaped

(c)

Figure **6**

Approximately 68 percent of the scores fall within 1 standard deviation of the mean (within the interval $[\bar{x} - s, \bar{x} + s]$ for samples or $[\mu - \sigma, \mu + \sigma]$ for populations).

Approximately 95 percent of the scores fall within 2 standard deviations of the mean (within the interval $[\bar{x} - 2s, \bar{x} + 2s]$ for samples or $[\mu - 2\sigma, \mu + 2\sigma]$ for populations).

Approximately 99.7 percent of the scores fall within 3 standard deviations of the mean (within the interval $[\bar{x} - 3s, \bar{x} + 3s]$ for samples or $[\mu - 3\sigma, \mu + 3\sigma]$ for populations).

Visualizing the Empirical Rule

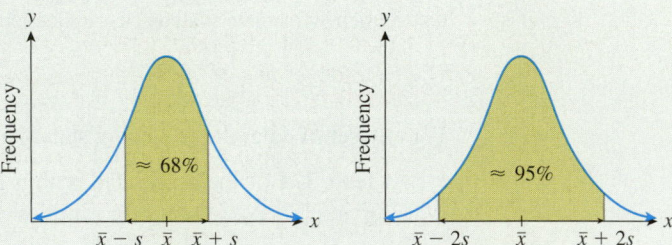

quick Examples

1. If the mean of a sample is 20 with standard deviation $s = 2$, then at least $15/16 = 93.75\%$ of the scores lie within 4 standard deviations of the mean—that is, in the interval [12, 28].

2. If the mean of a sample with a bell-shaped symmetric distribution is 20 with standard deviation $s = 2$, then approximately 95 percent of the scores lie in the interval [16, 24].

_____

* Unlike Chebyshev's rule, which is a precise theorem, the empirical rule is a "rule of thumb" that is intentionally vague about what exactly is meant by a "bell-shaped distribution" and "approximately such-and-such percent." (As a result, the rule is often stated differently in different textbooks.) We will see in Section 8.5 that if the distribution is a *normal* one, the empirical rule translates to a precise statement.

The empirical rule could not be applied in Example 1. The distribution was not bell-shaped (sketch it to see for yourself) and the fact that there were only 6 scores limits the accuracy further. The empirical rule is, however, accurate in distributions that are bell-shaped and symmetric, even if not perfectly so. Chebyshev's rule, on the other hand, is always valid (and applies in Example 1 in particular) but tends to be "overcautious" and in practice underestimates how much of a distribution lies in a given interval.

Steve Glanz

TITLE Senior Project Engineer, IMS Development

INSTITUTION Amtrak

In my position as Senior Project Engineer on Amtrak's nascent Infrastructure Management System (IMS), the team I work with is using mathematics to ensure you have a safe, comfortable ride every time you step on the train. The IMS is essentially an enterprise asset management system that will allow Amtrak to more effectively maintain its infrastructure, manage its workforce, and control costs associated with the maintenance of the railroad.

Amtrak's Northeast Corridor is unique in that it is the most heavily traversed stretch of railroad in the United States. In addition, it is the only railroad that supports heavy freight rail, commuter operations, and high-speed rail. Because of these unique traffic patterns it is imperative to ensure that the rail geometry is maintained to very high standards. One of the ways in which we use mathematics to improve ride quality and safety while reducing maintenance costs is through the development of a "roughness profile" for the Amtrak owned and operated rail along the Northeast Corridor. The roughness profile is an excellent indicator of where track work must be done in order to maintain a safe and comfortable railroad. Using the latest in testing equipment, we are able to record many aspects of the rail geometry including the measurement of deformations in the rail caused by the varied weight and speed of the trains running on the Corridor. From these measurements, we then calculate the standard deviation from construction and safety standards to create the aforementioned "roughness profile"; we can then generate a work plan to improve safety and comfort as well as optimize our resources.

While this is one explicit example of a real-world application of mathematics that I've experienced in my career, it is certainly not the only one. From computer programming to project management to financial analysis, I have used virtually every mathematical discipline at one time or another. In my experience, mathematics is everywhere you look.

Figure **7**

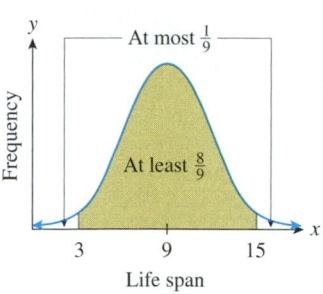

Figure **8**

Example **2 Automobile Life**

The average life span of a Batmobile® is 9 years, with a standard deviation of 2 years. My own Batmobile lasted less than 3 years before being condemned to the bat-junkyard.

a. Without any further knowledge about the distribution of Batmobile life spans, what can one say about the percentage of Batmobiles that last less than 3 years?

b. Refine the answer in part (a), assuming that the distribution of Batmobile life spans is bell-shaped and symmetric.

Solution

a. If we are given no further information about the distribution of Batmobile life spans, we need to use Chebyshev's rule. Because the life span of my Batmobile was more than 6 years (or 3 standard deviations) shorter than the mean, it lies outside the range $[\mu - 3\sigma, \mu + 3\sigma] = [3, 15]$. Because *at least* 8/9 of the life spans of all Batmobiles lie in this range, *at most* 1/9, or 11 percent, of the life spans lie outside this range (see Figure 7). Some of these, like the life span of my own Batmobile, are less than 3 years, while the rest are more than $\mu + 3\sigma = 15$ years.

b. Because we know more about the distribution now than we did in part (a), we can use the empirical rule and obtain sharper results. The empirical rule predicts that approximately 99.7 percent of the life spans of Batmobiles lie in the range $[\mu - 3\sigma, \mu + 3\sigma] = [3, 15]$. Thus, approximately $1 - 99.7\% = 0.3\%$ of them lie outside that range. Because the distribution is symmetric, however, more can be said: half of that 0.3 percent, or 0.15 percent of Batmobiles will last longer than 15 years, while the other 0.15 percent are, like my own ill-fated Batmobile, doomed to a life span of less than 3 years (see Figure 8).

Variance and Standard Deviation of a Finite Random Variable

We calculated the population variance by taking the mean of the quantities $(x_i - \mu)^2$. The x_i are the values of a finite random variable X. Thus, the population variance of a set of data can be written as the mean of all the values of $(X - \mu)^2$. But for a population, the mean of all the values of $(X - \mu)^2$ is the same as the expected value of $(X - \mu)^2$, which we can write as $E([X - \mu]^2)$. In general, we make the following definition.

Variance and Standard Deviation of a Finite Random Variable

If X is a finite random variable taking on values $x_1, x_2, \ldots, x_n$, then the **variance** of X is

$$\sigma^2 = E([X - \mu]^2)$$
$$= (x_1 - \mu)^2 P(X = x_1) + (x_2 - \mu)^2 P(X = x_2) + \cdots + (x_n - \mu)^2 P(X = x_n)$$
$$= \sum_i (x_i - \mu)^2 P(X = x_i)$$

The **standard deviation** of X is then the square root of the variance:

$$\sigma = \sqrt{\sigma^2}$$

To compute the variance from the probability distribution of X, first compute the expected value μ and then compute the expected value of $(X - \mu)^2$.

quick Example

The following distribution has expected value $\mu = E(X) = 2$:

x	-1	2	3	10
$P(X = x)$	.3	.5	.1	.1

The variance of X is

$$\sigma^2 = (x_1 - \mu)^2 P(X = x_1) + (x_2 - \mu)^2 P(X = x_2) + \cdots + (x_n - \mu)^2 P(X = x_n)$$
$$= (-1 - 2)^2(.3) + (2 - 2)^2(.5) + (3 - 2)^2(.1) + (10 - 2)^2(.1) = 9.2$$

The standard deviation of X is

$$\sigma = \sqrt{9.2} \approx 3.03$$

We can calculate the variance and standard deviation of a random variable using a tabular approach just as when we calculated the expected value in Example 3 in the preceding section.

Example 3 Variance of a Random Variable

Compute the variance and standard deviation for the following probability distribution.

x	10	20	30	40	50	60
$P(X = x)$	.2	.2	.3	.1	.1	.1

Solution

We first compute the expected value, μ, in the usual way:

x	10	20	30	40	50	60	
$P(X = x)$	.2	.2	.3	.1	.1	.1	
$xP(X = x)$	2	4	9	4	5	6	$\mu = 30$

Next, we add an extra three rows:

- a row for the differences $(x - \mu)$, which we get by subtracting μ from the values of X
- a row for the squares $(x - \mu)^2$, which we obtain by squaring the values immediately above
- a row for the products $(x - \mu)^2 P(X = x)$, which we obtain by multiplying the values in the second and the fifth rows.

x	10	20	30	40	50	60	
$P(X = x)$	.2	.2	.3	.1	.1	.1	
$xP(X = x)$	2	4	9	4	5	6	$\mu = 30$
$x - \mu$	-20	-10	0	10	20	30	
$(x - \mu)^2$	400	100	0	100	400	900	
$(x - \mu)^2 P(X = x)$	80	20	0	10	40	90	$\sigma^2 = 240$

using *Technology*

See the Technology Guides at the end of the chapter to find out how to use a TI-83/84 or Excel to automate the calculations of Exercise 3.

The sum of the values in the last row is the variance. The standard deviation is then the square root of the variance:

$$\sigma = \sqrt{240} \approx 15.49$$

Note Chebyshev's rule and the empirical rule apply to random variables just as they apply to samples and populations, as we illustrate in the following example. ∎

Example 4 Internet Commerce

Your newly launched company, CyberPromo, Inc., sells computer games on the Internet.

a. Statistical research indicates that the lifespan of an Internet marketing company such as yours is symmetrically distributed with an expected value of 30 months and standard deviation of 4 months. Complete the following sentence:

There is (at least/at most/approximately)_____ a _____ percent chance that CyberPromo will still be around for more than three years.

b. How would the answer to part (a) be affected if the distribution of lifespans was not known to be symmetric?

Solution

a. Do we use Chebyshev's rule or the empirical rule? Because the empirical rule requires that the distribution be both symmetric and bell-shaped—not just symmetric—we cannot conclude that it applies here, so we are forced to use Chebyshev's rule instead.

Let X be the lifespan of an Internet commerce site. The expected value of X is 30 months, and the hoped-for lifespan of CyberPromo, Inc., is 36 months, which is 6 months, or $6/4 = 1.5$ standard deviations, above the mean. Chebyshev's rule tells us that X is within $k = 1.5$ standard deviations of the mean at least $1 - 1/k^2$ of the time; that is,

$$P(24 \leq X \leq 36) \geq 1 - \frac{1}{k^2} = 1 - \frac{1}{1.5^2} \approx .56$$

In other words, at least 56 percent of all Internet marketing companies have life spans in the range of 24 to 36 months. Thus, *at most* 44 percent have life spans outside this range. Because the distribution is symmetric, at most 22 percent have life spans longer than 36 months. Thus we can complete the sentence as follows:

There is at most a 22 percent chance that CyberPromo will still be around for more than three years.

b. If the given distribution was not known to be symmetric, how would this affect the answer? We saw above that regardless of whether the distribution is symmetric or not, at most 44 percent have lifespans outside the range 24 to 36 months. Because the distribution is not symmetric, we cannot conclude that at most half of the 44 percent have lifespans longer than 36 months, and all we can say is that *no more than 44 percent can possibly have life spans longer than 36 years.* In other words:

There is at most a 44 percent chance that CyberPromo will still be around for more than three years.

Variance and Standard Deviation of a Binomial Random Variable

We saw that there is an easy formula for the expected value of a binomial random variable: $\mu = np$, where n is the number of trials and p is the probability of success. Similarly, there is a simple formula for the variance and standard deviation.

Variance and Standard Deviation of a Binomial Random Variable

If X is a binomial random variable associated with n independent Bernoulli trials, each with probability p of success, then the variance and standard deviation of X are given by

$$\sigma^2 = npq \quad \text{and} \quad \sigma = \sqrt{npq}$$

where $q = 1 - p$ is the probability of failure.

quick Example

If X is the number of successes in 20 Bernoulli trials with $p = .7$, then the standard deviation is $\sigma = \sqrt{npq} = \sqrt{(20)(.7)(.3)} \approx 2.05$.

Note For values of p near $1/2$ and large values of n, a binomial distribution is bell-shaped and (nearly) symmetric, hence the empirical rule applies. One rule of thumb is that we can use the empirical rule when both $np \geq 10$ and $nq \geq 10$.[34] ∎

---

[34] Remember that the empirical rule only gives an *estimate* of probabilities. In Section 8.5 we give a more accurate approximation that takes into account the fact that the binomial distribution is not continuous.

Example 5 Internet Commerce

You have calculated that there is a 40 percent chance that a hit on your Web page results in a fee paid to your company CyberPromo, Inc. Your Web page receives 25 hits per day. Let X be the number of hits that result in payment of the fee ("successful hits").

a. What are the expected value and standard deviation of X?

b. Complete the following: On approximately 95 out of 100 days, I will get between ___ and ___ successful hits.

Solution

a. The random variable X is binomial with $n = 25$ and $p = .4$. To compute μ and σ, we use the formulas

$$\mu = np = (25)(.4) = 10 \text{ successful hits}$$

$$\sigma = \sqrt{npq} = \sqrt{(25)(.4)(.6)} \approx 2.45 \text{ hits}$$

b. Because $np = 10 \geq 10$ and $nq = (25)(.6) = 15 \geq 10$, we can use the empirical rule, which tells us that there is an approximately 95 percent probability that the number of successful hits is within two standard deviations of the mean—that is, in the interval

$$[\mu - 2\sigma, \mu + 2\sigma] = [10 - 2(2.45), 10 + 2(2.45)] = [5.1, 14.9]$$

Thus, on approximately 95 out of 100 days, I will get between 5.1 and 14.9 successful hits.

FAQs Recognizing When to Use the Empirical Rule or Chebyshev's Rule

Q: *How do I decide whether to use Chebyshev's rule or the empirical rule?*

A: Check to see whether the probability distribution you are considering is both symmetric and bell-shaped. If so, you can use the empirical rule. If not, then you must use Chebyshev's rule. Thus, for instance, if the distribution is symmetric but not known to be bell-shaped, you must use Chebyshev's rule. ∎

8.4 EXERCISES

● denotes basic skills exercises

◆ denotes challenging exercises

tech Ex indicates exercises that should be solved using technology

Compute the (sample) variance and standard deviation of the data samples given in Exercises 1–8. (You calculated the means in the last exercise set. Round all answers to two decimal places.) hint [see Quick Examples on p. 576]

1. ● $-1, 5, 5, 7, 14$

2. ● $2, 6, 6, 7, -1$

3. ● $2, 5, 6, 7, -1, -1$

4. ● $3, 1, 6, -3, 0, 5$

5. ● $\dfrac{1}{2}, \dfrac{3}{2}, -4, \dfrac{5}{4}$

6. ● $-\dfrac{3}{2}, \dfrac{3}{8}, -1, \dfrac{5}{2}$

7. ● $2.5, -5.4, 4.1, -0.1, -0.1$

8. ● $4.2, -3.2, 0, 1.7, 0$

In Exercises 9–14, calculate the standard deviation of X for each probability distribution. (You calculated the expected values in the last exercise set. Round all answers to two decimal places.) hint [see Quick Example on p. 580]

9. ●

x	0	1	2	3
$P(X = x)$	.5	.2	.2	.1

● basic skills   ◆ challenging   tech Ex technology exercise

10. ●

x	1	2	3	4
$P(X = x)$	.1	.2	.5	.2

11. ●

x	10	20	30	40
$P(X = x)$	$\frac{3}{10}$	$\frac{2}{5}$	$\frac{1}{5}$	$\frac{1}{10}$

12. ●

x	2	4	6	8
$P(X = x)$	$\frac{1}{20}$	$\frac{15}{20}$	$\frac{2}{20}$	$\frac{2}{20}$

13. ●

x	−5	−1	0	2	5	10
$P(X = x)$	.2	.3	.2	.1	.2	0

14. ●

x	−20	−10	0	10	20	30
$P(X = x)$	.2	.4	.2	.1	0	.1

In Exercises 15–24, calculate the expected value, the variance, and the standard deviation of the given random variable X. (You calculated the expected values in the last exercise set. Round all answers to two decimal places.)

15. ● X is the number that faces up when a fair die is rolled.

16. ● X is the number selected at random from the set $\{1, 2, 3, 4\}$.

17. ● X is the number of tails that come up when a coin is tossed twice.

18. ● X is the number of tails that come up when a coin is tossed three times.

19. X is the higher number when two dice are rolled.

20. X is the lower number when two dice are rolled.

21. X is the number of red marbles that Suzy has in her hand after she selects 4 marbles from a bag containing 4 red marbles and 2 green ones.

22. X is the number of green marbles that Suzy has in her hand after she selects 4 marbles from a bag containing 3 red marbles and 2 green ones.

23. Twenty darts are thrown at a dartboard. The probability of hitting a bull's-eye is .1. Let X be the number of bull's-eyes hit.

24. Thirty darts are thrown at a dartboard. The probability of hitting a bull's-eye is $\frac{1}{5}$. Let X be the number of bull's-eyes hit.

Applications

25. ● *Popularity Ratings* In your bid to be elected class representative, you have your election committee survey 5 randomly chosen students in your class and ask them to rank you on a scale of 0–10. Your rankings are 3, 2, 0, 9, 1. *hint* [see Example 1 and Quick Examples on p. 578]

    a. Find the sample mean and standard deviation. (Round your answers to two decimal places.)

    b. Assuming the sample mean and standard deviation are indicative of the class as a whole, in what range does the empirical rule predict that approximately 68 percent of the class will rank you? What other assumptions must we make to use the rule?

26. ● *Popularity Ratings* Your candidacy for elected class representative is being opposed by Slick Sally. Your election committee has surveyed six of the students in your class and had them rank Sally on a scale of 0–10. The rankings were 2, 8, 7, 10, 5, 8.

    a. Find the sample mean and standard deviation. (Round your answers to two decimal places.)

    b. Assuming the sample mean and standard deviation are indicative of the class as a whole, in what range does the empirical rule predict that approximately 95 percent of the class will rank Sally? What other assumptions must we make to use the rule?

27. ● *Unemployment* Following is a sample of unemployment rates (in percentage points) in the U.S. sampled from the period 1990–2004:[35]

    4.2, 4.7, 5.4, 5.8, 4.9

    a. Compute the mean and standard deviation of the given sample. (Round your answers to one decimal place.)

    b. Assuming the distribution of unemployment rates in the population is symmetric and bell-shaped, 95 percent of the time, the unemployment rate is between _____ and _____ percent.

28. ● *Unemployment* Following is a sample of unemployment rates among Hispanics (in percentage points) in the US sampled from the period 1990–2004:[36]

    7.7, 7.5, 9.3, 6.9, 8.6

    a. Compute the mean and standard deviation of the given sample. (Round your answers to one decimal place.)

    b. Assuming the distribution of unemployment rates in the population of interest is symmetric and bell-shaped, 68 percent of the time, the unemployment rate is between _____ and _____ percent.

29. ● *Sport Utility Vehicles—Tow Ratings* Following is a sample of tow ratings (in pounds) for some popular 2000 model light trucks:[37]

    2000, 2000, 3000, 4000, 5000, 6000, 6000, 6000, 8000, 8000

    a. Compute the mean and standard deviation of the given sample. (Round your answers to the nearest whole number.)

    b. Assuming the distribution of tow ratings for all popular light trucks is symmetric and bell-shaped, 68 percent of

---

[35] SOURCE for data: Bureau of Labor Statistics (BLS), www.bls.gov.

[36] Ibid.

[37] Tow rates vary considerably within each model. Figures cited are rounded. For more detailed information, consult www.rvsafety.com/towrate2k.htm.

● basic skills   ◆ challenging   *tech* Ex technology exercise

all light trucks have tow ratings between ____ and ____. What is the percentage of scores in the sample that fall in this range?

30. ● *Housing Prices* Following is a sample of the percentage increases in the price of a house from 1980 to 2001 in eight regions of the U.S.[38]

75, 125, 150, 150, 150, 225, 225, 300

   a. Compute the mean and standard deviation of the given sample. (Round answers to the nearest whole number.)
   b. Assuming the distribution of percentage housing price increases for all regions is symmetric and bell-shaped, 68 percent of all regions in the U.S. reported housing increases between ____ and ____. What is the percentage of scores in the sample that fall in this range?

31. tech Ex *Sport Utility Vehicles* Following are highway driving gas mileages of a selection of medium-sized sport utility vehicles (SUVs):[39]

17, 18, 17, 18, 21, 16, 21, 18, 16, 14, 15, 22, 17, 19, 17, 18

   a. Find the sample standard deviation (rounded to 2 decimal places).
   b. In what gas mileage range does Chebyshev's inequality predict that at least $8/9$ (approximately 89 percent) of the selection will fall?
   c. What is the actual percentage of SUV models of the sample that fall in the range predicted in part (b)? Which gives the more accurate prediction of this percentage: Chebyshev's rule or the empirical rule?

32. tech Ex *Sport Utility Vehicles* Following are the city driving gas mileages of a selection of sport utility vehicles (SUVs):[40]

14, 15, 14, 15, 13, 16, 12, 14, 19, 18, 16, 16, 12, 15, 15, 13

   a. Find the sample standard deviation (rounded to 2 decimal places).
   b. In what gas mileage range does Chebyshev's inequality predict that at least 75 percent of the selection will fall?
   c. What is the actual percentage of SUV models of the sample that fall in the range predicted in part (b)? Which gives the more accurate prediction of this percentage: Chebyshev's rule or the empirical rule?

33. ● *Shopping Malls* A survey of all the shopping malls in your region yields the following probability distribution, where X is the number of movie theater screens in a selected mall:

Number of Movie Screens	0	1	2	3	4
Probability	.4	.1	.2	.2	.1

Compute the expected value μ and the standard deviation σ of X. (Round answers to two decimal places.) What percentage of malls have a number of movie theater screens within two standard deviations of μ?

34. ● *Pastimes* A survey of all the students in your school yields the following probability distribution, where X is the number of movies that a selected student has seen in the past week:

Number of Movies	0	1	2	3	4
Probability	.5	.1	.2	.1	.1

Compute the expected value μ and the standard deviation σ of X. (Round answers to two decimal places.) For what percentage of students is X within two standard deviations of μ?

35. tech Ex *Income Distribution up to $100,000* The following table shows the distribution of household incomes for a sample of 1000 households in the U.S. with incomes up to $100,000.[41]

2000 Income (thousands)	$10	$30	$50	$70	$90
Households	270	280	200	150	100

Compute the expected value μ and the standard deviation σ of the associated random variable X. If we define a "lower income" family as one whose income is more than one standard deviation below the mean, and a "higher income" family as one whose income is at least one standard deviation above the mean, what is the income gap between higher- and lower-income families in the U.S.? (Round your answers to the nearest $1000.)

36. tech Ex *Income Distribution up to $100,000* Repeat Exercise 35, using the following data for a sample of 1000 Hispanic households in the U.S.[42]

2000 Income (thousands)	$10	$30	$50	$70	$90
Households	300	340	190	110	60

---

[38] Percentages are rounded to the nearest 25 percent. SOURCE: Third Quarter 2001 House Price Index, released November 30, 2001 by the Office of Federal Housing Enterprise Oversight; available online at www.ofheo.gov/house/3q01hpi.pdf.

[39] Figures are the low-end of ranges for 1999 models tested. SOURCE: Oak Ridge National Laboratory: "An Analysis of the Impact of Sport Utility Vehicles in the United States" Stacy C. Davis, Lorena F. Truett, (August 2000)/Insurance Institute for Highway Safety http://cta.ornl.gov/cta/Publications/pdf/ORNL_TM_2000_147.pdf.

[40] Ibid.

[41] Based on the actual household income distribution in 2003. SOURCE: U.S. Census Bureau, Current Population Survey, 2004 Annual Social and Economic Supplement. http://pubdb3.census.gov/macro/032004/hhinc/new06_000.htm.

[42] Ibid.

● basic skills   ◆ challenging   tech Ex technology exercise

37. ● *Hispanic Employment: Male* The following table shows the approximate number of males of Hispanic origin employed in the U.S. in 2005, broken down by age group.[43]

Age	15–24.9	25–54.9	55–64.9
Employment (thousands)	16,000	13,000	1600

a. Use the rounded midpoints of the given measurement classes to compute the expected value and the standard deviation of the age X of a male Hispanic worker in the U.S. (Round all probabilities and intermediate calculations to two decimal places.)

b. In what age interval does the empirical rule predict that 68 percent of all male Hispanic workers will fall? (Round answers to the nearest year.)

38. ● *Hispanic Employment: Female* Repeat Exercise 37, using the corresponding data for females of Hispanic origin.[44]

Age	15–24.9	25–54.9	55–64.9
Employment (thousands)	1200	5000	600

a. Use the rounded midpoints of the given measurement classes to compute the expected value and the standard deviation of the age X of a female Hispanic worker in the U.S. (Round all probabilities and intermediate calculations to two decimal places.)

b. In what age interval does the empirical rule predict that 68 percent of all female Hispanic workers will fall? (Round answers to the nearest year.)

39. ● *Commerce* You have been told that the average life span of an Internet-based company is 2 years, with a standard deviation of 0.15 years. Further, the associated distribution is highly skewed (not symmetric). Your Internet company is now 2.6 years old. What percentage of all Internet-based companies have enjoyed a life span at least as long as yours? Your answer should contain one of the following phrases: *At least; At most; Approximately.* *hint* [see Example 2]

40. ● *Commerce* You have been told that the average life span of a car-compounding service is 3 years, with a standard deviation of 0.2 years. Further, the associated distribution is symmetric but not bell-shaped. Your car-compounding service is exactly 2.6 years old. What fraction of car-compounding services last at most as long as yours? Your answer should contain one of the following phrases: *At least; At most; Approximately.*

41. ● *Batmobiles* The average life span of a Batmobile® is 9 years, with a standard deviation of 2 years.[45] Further, the probability distribution of the life spans of Batmobiles is symmetric, but not known to be bell-shaped.

Because my old Batmobile has been sold as bat-scrap, I have decided to purchase a new one. According to the above information, there is

(A) At least   **(B)** At most   **(C)** Approximately

a ____ percent chance that my new Batmobile will last 13 years or more.

42. ● *Spiderman Coupés* The average life span of a Spiderman Coupé is 8 years, with a standard deviation of 2 years. Further, the probability distribution of the life spans of Spiderman Coupés is not known to be bell-shaped or symmetric. I have just purchased a brand-new Spiderman Coupé. According to the above information, there is

(A) At least   **(B)** At most   **(C)** Approximately

a ____ percent chance that my new Spiderman Coupé will last for less than 4 years.

43. ● *Teenage Pastimes* According to a study,[46] the probability that a randomly selected teenager shopped at a mall at least once during a week was .63. Let X be the number of students in a randomly selected group of 40 that will shop at a mall during the next week.

a. Compute the expected value and standard deviation of X. (Round answers to two decimal places.)

b. Fill in the missing quantity: There is an approximately 2.5 percent chance that ____ or more teenagers in the group will shop at a mall during the next week. *hint* [see Example 5]

44. ● *Other Teenage Pastimes* According to the study referred to in the preceding exercise, the probability that a randomly selected teenager played a computer game at least once during a week was .48. Let X be the number of teenagers in a randomly selected group of 30 who will play a computer game during the next seven days.

a. Compute the expected value and standard deviation of X. (Round answers to 2 decimal places.)

b. Fill in the missing quantity: There is an approximately 16 percent chance that ____ or more teenagers in the group will play a computer game during the next seven days.

45. *Teenage Marketing* In 2000, 22 percent of all teenagers in the U.S. had checking accounts.[47] Your bank, TeenChex Inc., is

---

[43] Figures are rounded. Bounds for the age groups for the first and third categories were adjusted for computational convenience. SOURCE: Bureau of Labor Statistics ftp://ftp.bls.gov/pub/suppl/empsit.cpseed15.txt.

[44] Ibid.

[45] See Example 2.

[46] SOURCE: Rand Youth Poll/Teenage Research Unlimited/*The New York Times,* March 14, 1998, p. D1.

[47] SOURCE: Teenage Research Unlimited, January 25, 2001 www.teenresearch.com.

● basic skills   ◆ challenging   **tech** Ex technology exercise

interested in targeting teenagers who do not already have a checking account.

a. If TeenChex selects a random sample of 1000 teenagers, what number of teenagers *without* checking accounts can it expect to find? What is the standard deviation of this number? (Round the standard deviation to one decimal place.)

b. Fill in the missing quantities: There is an approximately 95 percent chance that between ___ and ___ teenagers in the sample will not have checking accounts. (Round answers to the nearest whole number.)

46. *Teenage Marketing* In 2000, 18 percent of all teenagers in the U.S. owned stocks or bonds.[48] Your brokerage company, TeenStox Inc., is interested in targeting teenagers who do not already own stocks or bonds.

a. If TeenStox selects a random sample of 2000 teenagers, what number of teenagers who *do not* own stocks or bonds can it expect to find? What is the standard deviation of this number? (Round the standard deviation to one decimal place.)

b. Fill in the missing quantities: There is an approximately 99.7 percent chance that between ___ and ___ teenagers in the sample will not own stocks or bonds. (Round answers to the nearest whole number.)

47. tech Ex *Supermarkets* A survey of supermarkets in the U.S. yielded the following relative frequency table, where X is the number of checkout lanes at a randomly chosen supermarket:[49]

x	1	2	3	4	5	6	7	8	9	10
$P(X = x)$	.01	.04	.04	.08	.10	.15	.25	.20	.08	.05

a. Compute the mean, variance and standard deviation (accurate to one decimal place).

b. As financial planning manager at Express Lane Mart, you wish to install a number of checkout lanes that is in the range of at least 75 percent of all supermarkets. What is this range according to Chebyshev's inequality? What is the *least* number of checkout lanes you should install so as to fall within this range?

48. tech Ex *Video Arcades* Your company, Sonic Video, Inc., has conducted research that shows the following probability distribution, where X is the number of video arcades in a randomly chosen city with more than 500,000 inhabitants:

x	0	1	2	3	4	5	6	7	8	9
$P(X = x)$	.07	.09	.35	.25	.15	.03	.02	.02	.01	.01

a. Compute the mean, variance and standard deviation (accurate to one decimal place).

b. As CEO of Startrooper Video Unlimited, you wish to install a chain of video arcades in Sleepy City, U.S.A. The city council regulations require that the number of arcades be within the range shared by at least 75 percent of all cities. What is this range? What is the *largest* number of video arcades you should install so as to comply with this regulation?

Distribution of Wealth If we model after-tax household income by a normal distribution, then the figures of a 1995 study imply the information in the following table, which should be used for Exercises 49–60.[50] Assume that the distribution of incomes in each country is bell-shaped and symmetric.

Country	U.S.	Canada	Switzerland	Germany	Sweden
Mean Household Income	$38,000	$35,000	$39,000	$34,000	$32,000
Standard Deviation	$21,000	$17,000	$16,000	$14,000	$11,000

49. ● If we define a "poor" household as one whose after-tax income is at least 1.3 standard deviations below the mean, what is the household income of a poor family in the U.S.?

50. ● If we define a "poor" household as one whose after-tax income is at least 1.3 standard deviations below the mean, what is the household income of a poor family in Switzerland?

51. ● If we define a "rich" household as one whose after-tax income is at least 1.3 standard deviations above the mean, what is the household income of a rich family in the U.S.?

52. ● If we define a "rich" household as one whose after-tax income is at least 1.3 standard deviations above the mean, what is the household income of a rich family in Sweden?

53. Refer to Exercise 49. Which of the five countries listed has the poorest households (i.e., the lowest cutoff for considering a household poor)?

54. Refer to Exercise 52. Which of the five countries listed has the wealthiest households (i.e., the highest cutoff for considering a household rich)?

55. Which of the five countries listed has the largest gap between rich and poor?

56. Which of the five countries listed has the smallest gap between rich and poor?

57. ● What percentage of U.S. families earned an after-tax income of $17,000 or less?

---

[48] Ibid.

[49] Source: J.T. McClave, P.G. Benson, T. Sincich, *Statistics for Business and Economics,* 7th Ed. (Prentice Hall, 1998) p. 177, W. Chow et al. "A model for predicting a supermarket's annual sales per square foot," Graduate School of Management, Rutgers University.

[50] The data are rounded to the nearest $1000 and based on a report published by the Luxembourg Income Study. The report shows after-tax income, including government benefits (such as food stamps) of households with children. Our figures were obtained from the published data by assuming a normal distribution of incomes. All data were based on constant 1991 U.S. dollars and converted foreign currencies (adjusted for differences in buying power). Source: Luxembourg Income Study/*The New York Times,* August 14, 1995, p. A9.

● basic skills   ◆ challenging   tech Ex technology exercise

58. ● What percentage of U.S. families earned an after-tax income of $80,000 or more?

59. ● What is the after-tax income range of approximately 99.7 percent of all Germans?

60. ● What is the after-tax income range of approximately 99.7 percent of all Swedes?

tech Ex *Aging Exercises 61–66 are based on the following list, which shows the percentage of aging population (residents of age 65 and older) in each of the 50 states in 1990 and 2000:*[51]

2000

> 6, 9, 10, 10, 10, 11, 11, 11, 11, 11,
> 11, 11, 12, 12, 12, 12, 12, 12, 12, 12,
> 12, 12, 12, 13, 13, 13, 13, 13, 13, 13,
> 13, 13, 13, 13, 13, 13, 13, 14, 14, 14,
> 14, 14, 14, 14, 15, 15, 15, 15, 16, 18

1990

> 4, 9, 10, 10, 10, 10, 10, 11, 11, 11,
> 11, 11, 11, 11, 11, 12, 12, 12, 12, 12,
> 12, 13, 13, 13, 13, 13, 13, 13, 13, 13,
> 13, 13, 13, 13, 14, 14, 14, 14, 14,
> 14, 14, 14, 15, 15, 15, 15, 15, 15, 18

61. **tech Ex** Compute the population mean and standard deviation for the 2000 data.

62. **tech Ex** Compute the population mean and standard deviation for the 1990 data.

63. **tech Ex** Compare the actual percentage of states whose aging population in 2000 was within one standard deviation of the mean to the percentage predicted by the empirical rule. Comment on your answer.

64. **tech Ex** Compare the actual percentage of states whose aging population in 1990 was within one standard deviation of the mean to the percentage predicted by the empirical rule. Comment on your answer.

65. **tech Ex** What was the actual percentage of states whose aging population in 1990 was within two standard deviations of the mean? Is Chebyshev's rule valid? Explain.

66. **tech Ex** What was the actual percentage of states whose aging population in 2000 was within two standard deviations of the mean? Is Chebyshev's rule valid? Explain.

67. *Electric Grid Stress* The following chart shows the approximate standard deviation of the power grid frequency, in

1/1000 cycles per second, taken over six-month periods. (0.9 is the average standard deviation.)[52]

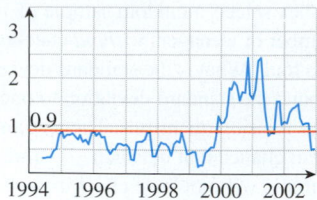

Which of the following statements are true? (More than one may be true.)

(A) The power grid frequency was at or below the mean until late 1999.

(B) The power grid frequency was more stable in mid-1999 than in 1995.

(C) The power grid frequency was more stable in mid-2002 than in mid-1999.

(D) The greatest fluctuations in the power grid frequency occurred in 2000–2001.

(E) The power grid frequency was more stable around January 1995 than around January 1999.

68. *Electric Grid Stress* The following chart shows the approximate monthly means of the power grid frequency, in 1/1000 cycles per second. 0.0 represents the desired frequency of exactly 60 cycles per second.[53]

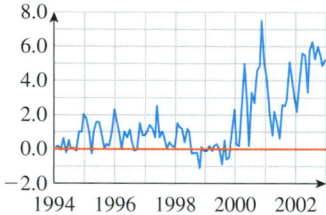

Which of the following statements are true? (More than one may be true.)

(A) Both the mean and the standard deviation show an upward trend from 2000 on.

(B) The mean, but not the standard deviation, shows an upward trend from 2000 on.

(C) The demand for electric power peaked in the second half of 2001.

(D) The standard deviation was larger in the second half of 2002 than in the second half of 1999.

(E) The mean of the monthly means in 2000 was lower than that for 2002, but the standard deviation of the monthly means was higher.

---

[51] Percentages are rounded and listed in ascending order. SOURCE: U.S. Census Bureau, Census 2000 Summary File 1. www.census.gov/prod/2001pubs/c2kbr01-10.pdf.

[52] SOURCE: Robert Blohm, energy consultant and adviser to the North American Electric Reliability Council/*The New York Times,* August 20, 2003, p. A16.

[53] Ibid.

● basic skills   ◆ challenging   **tech Ex** technology exercise

Communication and Reasoning Exercises

69. ● Which is greater: the sample standard deviation or the population standard deviation? Explain.

70. ● Suppose you take larger and larger samples of a given population. Would you expect the sample and population standard deviations to get closer or further apart? Explain.

71. ● In one Finite Math class, the average grade was 75 and the standard deviation of the grades was 5. In another Finite Math class, the average grade was 65 and the standard deviation of the grades was 20. What conclusions can you draw about the distributions of the grades in each class?

72. ● You are a manager in a precision manufacturing firm and you must evaluate the performance of two employees. You do so by examining the quality of the parts they produce. One particular item should be 50.0 ± 0.3 mm long to be usable. The first employee produces parts that are an average of 50.1 mm long with a standard deviation of 0.15 mm. The second employee produces parts that are an average of 50.0 mm long with a standard deviation of 0.4 mm. Which employee do you rate higher? Why? (Assume that the empirical rule applies.)

73. If a finite random variable has an expected value of 10 and a standard deviation of 0, what must its probability distribution be?

74. If the values of X in a population consist of an equal number of 1s and -1s, what is its standard deviation?

75. ◆ Find an algebraic formula for the population standard deviation of a sample $\{x, y\}$ of two scores ($x \le y$).

76. ◆ Find an algebraic formula for the sample standard deviation of a sample $\{x, y\}$ of two scores ($x \le y$).

● basic skills   ◆ challenging   *tech* Ex technology exercise

8.5 Normal Distributions

Continuous Random Variables

Figure 9 shows the probability distributions for the number of successes in sequences of 10 and 15 independent Bernoulli trials, each with probability of success $p = .5$.

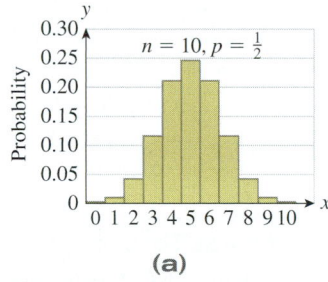
(a)

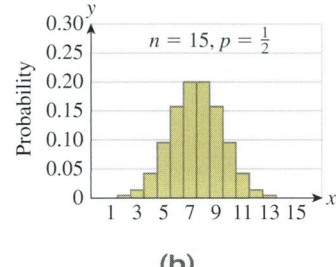
(b)

Figure **9**

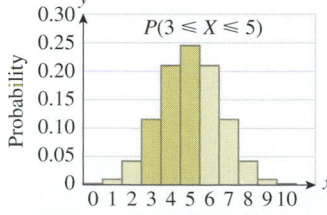

Figure **10**

Because each column is 1 unit wide, its area is numerically equal to its height. Thus, the area of each rectangle can be interpreted as a probability. For example, in Figure 9(a) the area of the rectangle over $X = 3$ represents $P(X = 3)$. If we want to find $P(3 \le X \le 5)$, we can add up the areas of the three rectangles over 3, 4 and 5, shown shaded in Figure 10. Notice that if we add up the areas of *all* the rectangles in Figure 9(a), the total is 1 because $P(0 \le X \le 10) = 1$. We can summarize these observations.

Properties of the Probability Distribution Histogram

In a probability distribution histogram where each column is one unit wide:

- The total area enclosed by the histogram is 1 square unit.
- $P(a \leq X \leq b)$ is the area enclosed by the rectangles lying between and including $X = a$ and $X = b$.

This discussion is motivation for considering another kind of random variable, one whose probability distribution is specified not by a bar graph, as above, but by the graph of a function.

Continuous Random Variable; Probability Density Function

A **continuous random variable** X may take on any real value whatsoever. The probabilities $P(a \leq X \leq b)$ are defined by means of a **probability density function,** a function whose graph lies above the x-axis with the total area between the graph and the x-axis being 1. The probability $P(a \leq X \leq b)$ is defined to be the area enclosed by the curve, the x-axis, and the lines $x = a$ and $x = b$ (see Figure 11).

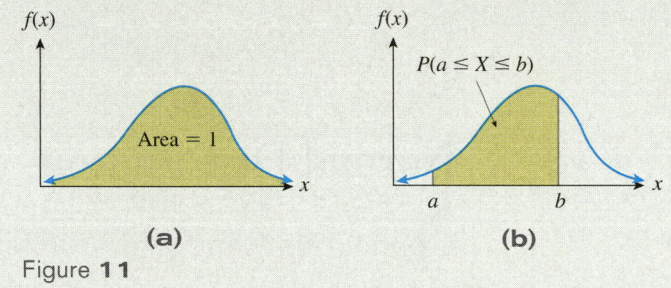

(a)   **(b)**

Figure **11**

Notes

1. In Chapter 7, we defined probability distributions only for *finite* sample spaces. Because continuous random variables have infinite sample spaces, we need the definition above to give meaning to $P(a \leq X \leq b)$ if X is a continuous random variable.

2. If $a = b$, then $P(X = a) = P(a \leq X \leq a)$ is the area under the curve between the lines $x = a$ and $x = a$—no area at all! Thus, when X is a continuous random variable, $P(X = a) = 0$ for every value of a.

3. Whether we take the region on the right in Figure 11 to include the boundary or not does not affect the area. The probability $P(a < X < b)$ is defined as the area strictly between the vertical lines $x = a$ and $x = b$, but is, of course, the same as $P(a \leq X \leq b)$, because the boundary contributes nothing to the area. When we are calculating probabilities associated with a continuous random variable,

$$P(a \leq X \leq b) = P(a < X \leq b) = P(a \leq X < b) = P(a < X < b) \blacksquare$$

Normal Density Functions

Among all the possible probability density functions, there is an important class of functions called **normal density functions**, or **normal distributions**. The graph of a normal density function is bell-shaped and symmetric, as the figure below shows. The formula for a normal density function is rather complicated looking:

$$f(x) = \frac{1}{\sigma\sqrt{2\pi}}e^{-\frac{(x-\mu)^2}{2\sigma^2}}$$

The quantity μ is called the **mean** and can be any real number. The quantity σ is called the **standard deviation** and can be any positive real number. The number $e = 2.7182\ldots$ is a useful constant that shows up many places in mathematics, much as the constant π does. Finally, the constant $1/(\sigma\sqrt{2\pi})$ that appears in front is there to make the total area come out to be 1. We rarely use the actual formula in computations; instead, we use tables or technology.

Normal Density Function; Normal Distribution

A **normal density function,** or **normal distribution,** is a function of the form

$$f(x) = \frac{1}{\sigma\sqrt{2\pi}}e^{-\frac{(x-\mu)^2}{2\sigma^2}}$$

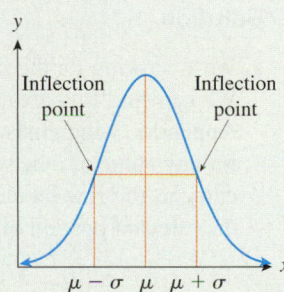

where μ is the mean and σ is the standard deviation. The "inflection points" are the points where the curve changes from bending in one direction to bending in another.[*]

---

[*] Pretend you were driving along the curve in a car. Then the points of inflection are the points where you would change the direction in which you are steering (from left to right or right to left).

Figure 12 shows the graph of several normal density functions. The third of these has mean 0 and standard deviation 1, and is called the **standard normal distribution.** We use Z rather than X to refer to the standard normal variable.

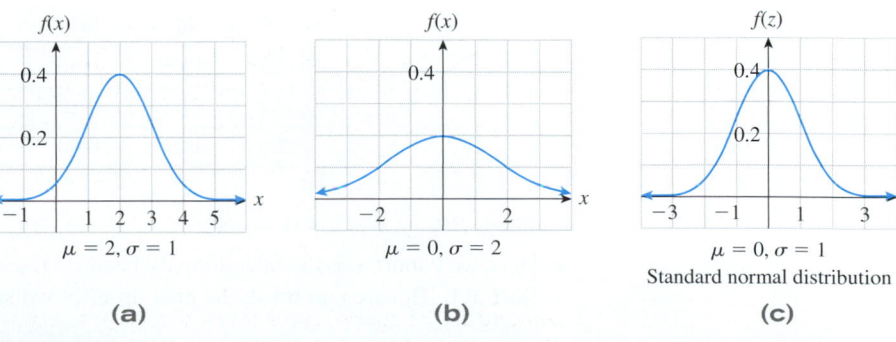

Figure **12**

Calculating Probabilities Using the Standard Normal Distribution

The standard normal distribution has $\mu = 0$ and $\sigma = 1$. The corresponding variable is called the **standard normal variable,** which we always denote by Z. Recall that to calculate the probability $P(a \leq Z \leq b)$, we need to find the area under the distribution curve between the vertical lines $z = a$ and $z = b$. We can use the table in the Appendix to look up these areas, or we can use technology. Here is an example.

Example 1 Standard Normal Distribution

Let Z be the standard normal variable. Calculate the following probabilities:

a. $P(0 \leq Z \leq 2.4)$

b. $P(0 \leq Z \leq 2.43)$

c. $P(-1.37 \leq Z \leq 2.43)$

d. $P(1.37 \leq Z \leq 2.43)$

 using *Technology*

See the Technology Guides at the end of the chapter for details on using a TI-83/84 or Excel to calculate the area under the standard normal curve. Alternatively, go online and follow:

Chapter 8

→ Online Utilities

   → Normal Distribution Utility

for a utility for this purpose.

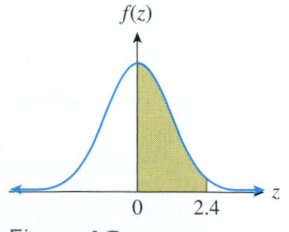

Figure **13**

Solution

a. We are asking for the shaded area under the standard normal curve shown in Figure 13. We can find this area, correct to four decimal places, by looking at the table in the Appendix, which lists the area under the standard normal curve from $Z = 0$ to $Z = b$ for any value of b between 0 and 3.09. To use the table, write 2.4 as 2.40, and read the entry in the row labeled 2.4 and the column labeled .00. $(2.4 + .00 = 2.40)$ Here is the relevant portion of the table:

Z	.00	.01	.02	.03
2.3	.4893	.4896	.4898	.4901
→ **2.4**	.4918	.4920	.4922	.4925
2.5	.4938	.4940	.4941	.4943

Thus, $P(0 \leq Z \leq 2.40) = .4918$.

b. The area we require can be read from the same portion of the table shown above. Write 2.43 as $2.4 + .03$, and read the entry in the row labeled 2.4 and the column labeled .03:

Z	.00	.01	.02	.03
2.3	.4893	.4896	.4898	.4901
→ **2.4**	.4918	.4920	.4922	.4925
2.5	.4938	.4940	.4941	.4943

Thus, $P(0 \leq Z \leq 2.43) = .4925$.

c. Here we cannot use the table directly because the range $-1.37 \leq Z \leq 2.43$ does not start at 0. But we can break the area up into two smaller areas that start or end at 0:

$$P(-1.37 \leq Z \leq 2.43) = P(-1.37 \leq Z \leq 0) + P(0 \leq Z \leq 2.43)$$

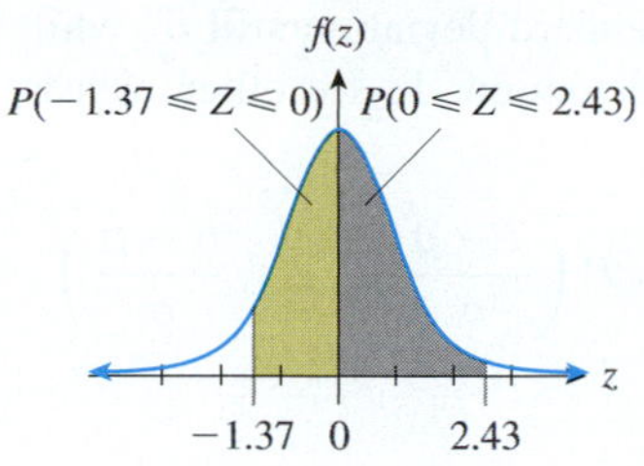

$P(-1.37 \leqslant Z \leqslant 0)$ $P(0 \leqslant Z \leqslant 2.43)$

Figure **14**

In terms of the graph, we are splitting the desired area into two smaller areas (Figure 14).

We already calculated the area of the right-hand piece in part (b):

$$P(0 \leq Z \leq 2.43) = .4925$$

For the left-hand piece, the symmetry of the normal curve tells us that

$$P(-1.37 \leq Z \leq 0) = P(0 \leq Z \leq 1.37)$$

This we can find on the table. Look at the row labeled 1.3 and the column labeled .07, and read

$$P(-1.37 \leq Z \leq 0) = P(0 \leq Z \leq 1.37) = .4147$$

Thus,

$$P(-1.37 \leq Z \leq 2.43) = P(-1.37 \leq Z \leq 0) + P(0 \leq Z \leq 2.43)$$
$$= .4147 + .4925$$
$$= .9072$$

d. The range $1.37 \leq Z \leq 2.43$ does not contain 0, so we cannot use the technique of part (c). Instead, the corresponding area can be computed as the *difference* of two areas:

$$P(1.37 \leq Z \leq 2.43) = P(0 \leq Z \leq 2.43) - P(0 \leq Z \leq 1.37)$$
$$= .4925 - .4147$$
$$= .0778$$

Calculating Probabilities for Any Normal Distribution

Although we have tables to compute the area under the *standard* normal curve, there are no readily available tables for nonstandard distributions. For example, If $\mu = 2$ and $\sigma = 3$, then how would we calculate $P(0.5 \leq X \leq 3.2)$? The following conversion formula provides a method for doing so:

Standardizing a Normal Distribution

If X has a normal distribution with mean μ and standard deviation σ, and if Z is the standard normal variable, then

$$P(a \leq X \leq b) = P\left(\frac{a-\mu}{\sigma} \leq Z \leq \frac{b-\mu}{\sigma}\right)$$

quick Example   If $\mu = 2$ and $\sigma = 3$, then

$$P(0.5 \leq X \leq 3.2) = P\left(\frac{0.5-2}{3} \leq Z \leq \frac{3.2-2}{3}\right)$$
$$= P(-0.5 \leq Z \leq 0.4) = .1915 + .1554 = .3469$$

To completely justify the above formula requires more mathematics than we shall discuss here. However, here is the main idea: if X is normal with mean μ and standard deviation σ, then $X - \mu$ is normal with mean 0 and standard deviation still σ, while $(X - \mu)/\sigma$ is normal with mean 0 and standard deviation 1. In other words, $(X - \mu)/\sigma = Z$. Therefore,

$$P(a \leq X \leq b) = P\left(\frac{a - \mu}{\sigma} \leq \frac{X - \mu}{\sigma} \leq \frac{b - \mu}{\sigma}\right) = P\left(\frac{a - \mu}{\sigma} \leq Z \leq \frac{b - \mu}{\sigma}\right)$$

Example 2 Quality Control

Pressure gauges manufactured by Precision Corp. must be checked for accuracy before being placed on the market. To test a pressure gauge, a worker uses it to measure the pressure of a sample of compressed air known to be at a pressure of exactly 50 pounds per square inch. If the gauge reading is off by more than 1 percent (0.5 pounds), it is rejected. Assuming that the reading of a pressure gauge under these circumstances is a normal random variable with mean 50 and standard deviation 0.4, find the percentage of gauges rejected.

Solution If X is the reading of the gauge, then X has a normal distribution with $\mu = 50$ and $\sigma = 0.4$. We are asking for $P(X < 49.5 \text{ or } X > 50.5) = 1 - P(49.5 \leq X \leq 50.5)$. We calculate

$$P(49.5 \leq X \leq 50.5) = P\left(\frac{49.5 - 50}{0.4} \leq Z \leq \frac{50.5 - 50}{0.4}\right) \qquad \text{Standardize}$$

$$= P(-1.25 \leq Z \leq 1.25)$$
$$= 2 \cdot P(0 \leq Z \leq 1.25)$$
$$= 2(.3944) = .7888$$

So, $P(X < 49.5 \text{ or } X > 50.5) = 1 - P(49.5 \leq X \leq 50.5)$
$$= 1 - .7888 = .2112$$

In other words, about 21 percent of the gauges will be rejected.

In many applications, we need to know the probability that a value of a normal random variable will lie within one standard deviation of the mean, or within two standard deviations, or within some number of standard deviations. To compute these probabilities, we first notice that, if X has a normal distribution with mean μ and standard deviation σ, then

$$P(\mu - k\sigma \leq X \leq \mu + k\sigma) = P(-k \leq Z \leq k)$$

by the standardizing formula. We can compute these probabilities for various values of k using the table in the Appendix, and we obtain the following results.

Probability of a normal distribution being within k standard deviations of its mean

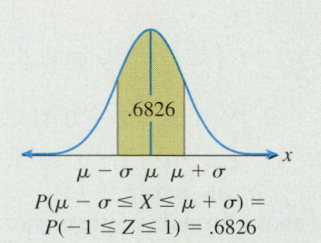

$P(\mu - \sigma \leq X \leq \mu + \sigma) =$
$P(-1 \leq Z \leq 1) = .6826$

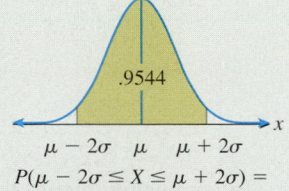

$P(\mu - 2\sigma \leq X \leq \mu + 2\sigma) =$
$P(-2 \leq Z \leq 2) = .9544$

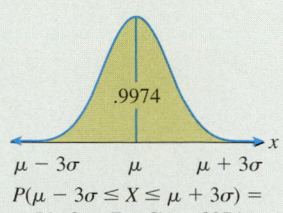

$P(\mu - 3\sigma \leq X \leq \mu + 3\sigma) =$
$P(-3 \leq Z \leq 3) = .9974$

---

using _Technology_

See the Technology Guides at the end of the chapter for details on using a TI-83/84 or Excel to calculate the area under a general normal curve. Alternatively, go online and follow:

Chapter 8

→ Online Utilities

→ Normal Distribution Utility

for a utility for this purpose.

Now you can see where the empirical rule comes from! Notice also that the probabilities above are a good deal larger than the lower bounds given by Chebyshev's rule. Chebyshev's rule must work for distributions that are skew or any shape whatsoever.

Example 3 Loans

The values of mortgage loans made by a certain bank one year were normally distributed with a mean of $120,000 and a standard deviation of $40,000.

a. What is the probability that a randomly selected mortgage loan was in the range of $40,000–$200,000?

b. You would like to state in your annual report that 50 percent of all mortgage loans were in a certain range with the mean in the center. What is that range?

Solution

a. We are asking for the probability that a loan was within 2 standard deviations ($80,000) of the mean. By the calculation done above, this probability is .9544.

b. We look for the k such that

$$P(120,000 - k \cdot 40,000 \leq X \leq 120,000 + k \cdot 40,000) = .5$$

Because

$$P(120,000 - k \cdot 40,000 \leq X \leq 120,000 + k \cdot 40,000) = P(-k \leq Z \leq k)$$

we look in the Appendix to see for which k we have

$$P(0 \leq Z \leq k) = .25$$

so that $P(-k \leq Z \leq k) = .5$. That is, we look *inside* the table to see where 0.25 is, and find the corresponding k. We find

$$P(0 \leq Z \leq 0.67) = .2486$$
and $\quad P(0 \leq Z \leq 0.68) = .2517$

Therefore, the k we want is about half-way between 0.67 and 0.68, call it 0.675. This tells us that 50 percent of all mortgage loans were in the range

$$120,000 - 0.675 \cdot 40,000 = \$93,000$$
to $\quad 120,000 + 0.675 \cdot 40,000 = \$147,000$

Normal Approximation to a Binomial Distribution

You might have noticed that the histograms of some of the binomial distributions we have drawn (for example, those in Figure 1) have a very rough bell shape. In fact, in many cases it is possible to draw a normal curve that closely approximates a given binomial distribution.

Normal Approximation to a Binomial Distribution

If X is the number of successes in a sequence of n independent Bernoulli trials, with probability p of success in each trial, and if the range of values of X within 3 standard deviations of the mean lies entirely within the range 0 to n (the possible values of X), then

$$P(a \leq X \leq b) \approx P(a - 0.5 \leq Y \leq b + 0.5)$$

where Y has a normal distribution with the same mean and standard deviation as X; that is, $\mu = np$ and $\sigma = \sqrt{npq}$, where $q = 1 - p$.

Notes

1. The condition that $0 \leq \mu - 3\sigma < \mu + 3\sigma \leq n$ is satisfied if n is sufficiently large and p is not too close to 0 or 1; it ensures that most of the normal curve lies in the range 0 to n.

2. In the formula $P(a \leq X \leq b) \approx P(a - 0.5 \leq Y \leq b + 0.5)$, we assume that a and b are integers. The use of $a - 0.5$ and $b + 0.5$ is called the **continuity correction.** To see that it is necessary, think about what would happen if you wanted to approximate, say, $P(X = 2) = P(2 \leq X \leq 2)$. Should the answer be 0?

Figures 15 and 16 show two binomial distributions with their normal approximations superimposed, and illustrate how closely the normal approximation fits the binomial distribution.

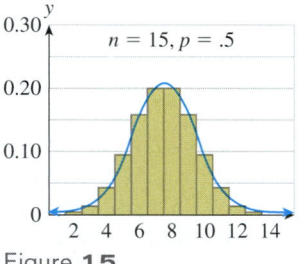

Figure **15**

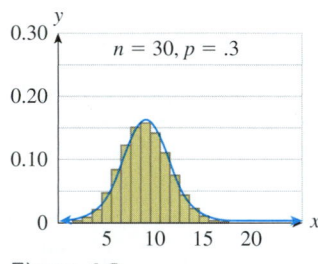

Figure **16**   ■

Example **4** Coin Tosses

a. If you flip a fair coin 100 times, what is the probability of getting more than 55 heads or fewer than 45 heads?

b. What number of heads (out of 100) would make you suspect that the coin is not fair?

Solution

a. We are asking for

$$P(X < 45 \text{ or } X > 55) = 1 - P(45 \leq X \leq 55)$$

We *could* compute this by calculating

$$1 - [C(100, 45)(.5)^{45}(.5)^{55} + C(100, 46)(.5)^{46}(.5)^{54} + \cdots + C(100, 55)(.5)^{55}(.5)^{45}]$$

but we can much more easily *approximate* it by looking at a normal distribution with mean $\mu = 50$ and standard deviation $\sigma = \sqrt{(100)(.5)(.5)} = 5$. (Notice that three standard deviations above and below the mean is the range 35 to 65, which is well within the range of possible values for X, which is 0 to 100, so the approximation should be a good one.) Let Y have this normal distribution. Then

$$P(45 \le X \le 55) \approx P(44.5 \le Y \le 55.5)$$
$$= P(-1.1 \le Z \le 1.1)$$
$$= .7286$$

Therefore,

$$P(X < 45 \text{ or } X > 55) \approx 1 - .7286 = .2714$$

b. This is a deep question that touches on the concept of **statistical significance:** What evidence is strong enough to overturn a reasonable assumption (the assumption that the coin is fair)? Statisticians have developed sophisticated ways of answering this question, but we can look at one simple test now. Suppose we tossed a coin 100 times and got 66 heads. If the coin were fair, then $P(X > 65) \approx P(Y > 65.5) = P(Z > 3.1) \approx .001$. This is small enough to raise a reasonable doubt that the coin is fair. However, we should not be too surprised if we threw 56 heads because we can calculate $P(X > 55) \approx .1357$, which is not such a small probability. As we said, the actual tests of statistical significance are more sophisticated than this, but we shall not go into them.

FAQs   When to Subtract From .5 and When Not To

Q: *When computing probabilities like, say $P(Z \le 1.2)$, $P(Z \ge 1.2)$, or $P(1.2 \le Z \le 2.1)$ using a table, just looking up the given values (1.2, 2.1 or whatever) is not enough. Sometimes you have to subtract from .5, sometimes not. Is there a simple rule telling me what to do when?*

A: The simplest—and also most instructive—way of knowing what to do is to draw a picture of the standard normal curve, and shade in the area you are looking for. Drawing pictures also helps one come up with the following mechanical rules:

1. To compute $P(a \le Z \le b)$, look up the areas corresponding to $|a|$ and $|b|$ in the table. If a and b have opposite sign, add these areas. Otherwise, subtract the smaller area from the larger.
2. To compute $P(Z \le a)$, look up the area corresponding to $|a|$. If a is positive, add .5. Otherwise, subtract from .5.
3. To compute $P(Z \ge a)$, look up the area corresponding to $|a|$. If a is positive, subtract from .5. Otherwise, add .5. ∎

8.5 EXERCISES

● denotes basic skills exercises
◆ denotes challenging exercises

Note: Answers for Section 8.5 were computed using the 4-digit table in the appendix, and may differ slightly from the more accurate answers generated using technology.

In Exercises 1–8, Z is the standard normal distribution. Find the indicated probabilities. *hint* [see Example 1]

1. ● $P(0 \le Z \le 0.5)$
2. ● $P(0 \le Z \le 1.5)$
3. ● $P(-0.71 \le Z \le 0.71)$
4. ● $P(-1.71 \le Z \le 1.71)$

● basic skills   ◆ challenging

5. ● $P(-0.71 \leq Z \leq 1.34)$ **6.** ● $P(-1.71 \leq Z \leq 0.23)$

7. ● $P(0.5 \leq Z \leq 1.5)$ **8.** ● $P(0.71 \leq Z \leq 1.82)$

In Exercises 9–14, X has a normal distribution with the given mean and standard deviation. Find the indicated probabilities.
hint [see Quick Example on p. 593]

9. ● $\mu = 50$, $\sigma = 10$, find $P(35 \leq X \leq 65)$

10. ● $\mu = 40$, $\sigma = 20$, find $P(35 \leq X \leq 45)$

11. ● $\mu = 50$, $\sigma = 10$, find $P(30 \leq X \leq 62)$

12. ● $\mu = 40$, $\sigma = 20$, find $P(30 \leq X \leq 53)$

13. ● $\mu = 100$, $\sigma = 15$, find $P(110 \leq X \leq 130)$

14. ● $\mu = 100$, $\sigma = 15$, find $P(70 \leq X \leq 80)$

15. Find the probability that a normal variable takes on values within 0.5 standard deviations of its mean.

16. Find the probability that a normal variable takes on values within 1.5 standard deviations of its mean.

17. Find the probability that a normal variable takes on values more than $\frac{2}{3}$ standard deviations away from its mean.

18. Find the probability that a normal variable takes on values more than $\frac{5}{3}$ standard deviations away from its mean.

19. ● If you roll a die 100 times, what is the approximate probability that you will roll between 10 and 15 ones, inclusive? (Round your answer to two decimal places.) *hint* [see Example 4]

20. ● If you roll a die 100 times, what is the approximate probability that you will roll between 15 and 20 ones, inclusive? (Round your answer to two decimal places.)

21. ● If you roll a die 200 times, what is the approximate probability that you will roll fewer than 25 ones, inclusive? (Round your answer to two decimal places.)

22. ● If you roll a die 200 times, what is the approximate probability that you will roll more than 40 ones? (Round your answer to two decimal places.)

Applications

23. ● *SAT Scores* SAT test scores are normally distributed with a mean of 500 and a standard deviation of 100. Find the probability that a randomly chosen test-taker will score between 450 and 550. *hint* [see Example 3]

24. ● *SAT Scores* SAT test scores are normally distributed with a mean of 500 and a standard deviation of 100. Find the probability that a randomly chosen test-taker will score 650 or higher.

25. ● *LSAT Scores* LSAT test scores are normally distributed with a mean of 500 and a standard deviation of 100. Find the probability that a randomly chosen test-taker will score between 300 and 550.

26. ● *LSAT Scores* LSAT test scores are normally distributed with a mean of 500 and a standard deviation of 100. Find the

probability that a randomly chosen test-taker will score 250 or lower.

27. ● *IQ Scores* IQ scores (as measured by the Stanford-Binet intelligence test) are normally distributed with a mean of 100 and a standard deviation of 16. What percentage of the population has an IQ score between 110 and 140? (Round your answer to the nearest percentage point.)

28. ● *IQ Scores* Refer to Exercise 27. What percentage of the population has an IQ score between 80 and 90? (Round your answer to the nearest percentage point.)

29. ● *IQ Scores* Refer to Exercise 27. Find the approximate number of people in the U.S. (assuming a total population of 280,000,000) with an IQ higher than 120.

30. ● *IQ Scores* Refer to Exercise 27. Find the approximate number of people in the U.S. (assuming a total population of 280,000,000) with an IQ higher than 140.

31. ● *Baseball* The mean batting average in major league baseball is about 0.250. Supposing that batting averages are normally distributed, that the standard deviation in the averages is 0.03, and that there are 250 batters, what is the expected number of batters with an average of at least 0.400?

32. ● *Baseball* The mean batting average in major league baseball is about 0.250. Supposing that batting averages are normally distributed, that the standard deviation in the averages is 0.05, and that there are 250 batters, what is the expected number of batters with an average of at least 0.400?[54]

33. ● *Marketing* Your pickle company rates its pickles on a scale of spiciness from 1 to 10. Market research shows that customer preferences for spiciness are normally distributed, with a mean of 7.5 and a standard deviation of 1. Assuming that you sell 100,000 jars of pickles, how many jars with a spiciness of 9 or above do you expect to sell?

34. ● *Marketing* Your hot sauce company rates its sauce on a scale of spiciness of 1 to 20. Market research shows that customer preferences for spiciness are normally distributed, with a mean of 12 and a standard deviation of 2.5. Assuming that you sell 300,000 bottles of sauce, how many bottles with a spiciness below 9 do you expect to sell?

Distribution of Wealth If we model after-tax household income with a normal distribution, then the figures of a 1995 study imply the information in the following table, which should be

---

[54] The last time that a batter ended the year with an average above 0.400 was in 1941. The batter was Ted Williams of the Boston Red Sox, and his average was 0.406. Over the years, as pitching and batting have improved, the standard deviation in batting averages has declined from around 0.05 when professional baseball began to around 0.03 by the end of the twentieth century. For a very interesting discussion of statistics in baseball and in evolution, see Stephen Jay Gould, *Full House: The Spread of Excellence from Plato to Darwin,* Random House, 1997.

● basic skills   ◆ challenging

used for Exercises 35–40.[55] Assume that the distribution of incomes in each country is normal, and round all percentages to the nearest whole number.

Country	U.S.	Canada	Switzerland	Germany	Sweden
Mean household income	$38,000	$35,000	$39,000	$34,000	$32,000
Standard deviation	$21,000	$17,000	$16,000	$14,000	$11,000

35. ● What percentage of U.S. households had an income of $50,000 or more?

36. ● What percentage of German households had an income of $50,000 or more?

37. ● What percentage of Swiss households are either very wealthy (income at least $100,000) or very poor (income at most $12,000)?

38. ● What percentage of Swedish households are either very wealthy (income at least $100,000) or very poor (income at most $12,000)?

39. ● Which country has a higher proportion of very poor families (income $12,000 or less): the U.S. or Canada?

40. ● Which country has a higher proportion of very poor families (income $12,000 or less): Canada or Switzerland?

41. *Comparing IQ Tests* IQ scores as measured by both the Stanford-Binet intelligence test and the Wechsler intelligence test have a mean of 100. The standard deviation for the Stanford-Binet test is 16, while that for the Wechsler test is 15. For which test do a smaller percentage of test-takers score less than 80? Why?

42. *Comparing IQ Tests* Referring to Exercise 41, for which test do a larger percentage of test-takers score more than 120?

43. *Product Repairs* The new copier your business bought lists a mean time between failures of 6 months, with a standard deviation of 1 month. One month after a repair, it breaks down again. Is this surprising? (Assume that the times between failures are normally distributed.)

44. *Product Repairs* The new computer your business bought lists a mean time between failures of 1 year, with a standard deviation of 2 months. Ten months after a repair, it breaks down again. Is this surprising? (Assume that the times between failures are normally distributed.)

Software Testing Exercises 45–50 are based on the following information, gathered from student testing of a statistical software package called MODSTAT.[56] Students were asked to complete certain tasks using the software, without any instructions. The results were as follows. (Assume that the time for each task is normally distributed.)

Task	Mean Time (minutes)	Standard Deviation
Task 1: Descriptive Analysis of Data	11.4	5.0
Task 2: Standardizing Scores	11.9	9.0
Task 3: Poisson Probability Table	7.3	3.9
Task 4: Areas Under Normal Curve	9.1	5.5

45. ● Find the probability that a student will take at least 10 minutes to complete Task 1.

46. ● Find the probability that a student will take at least 10 minutes to complete Task 3.

47. Assuming that the time it takes a student to complete each task is independent of the others, find the probability that a student will take at least 10 minutes to complete each of Tasks 1 and 2.

48. Assuming that the time it takes a student to complete each task is independent of the others, find the probability that a student will take at least 10 minutes to complete each of Tasks 3 and 4.

49. ◆ It can be shown that if X and Y are independent normal random variables with means μ_X and μ_Y and standard deviations σ_X and σ_Y respectively, then their sum $X + Y$ is also normally distributed and has mean $\mu = \mu_X + \mu_Y$ and standard deviation $\sigma = \sqrt{\sigma_X^2 + \sigma_Y^2}$. Assuming that the time it takes a student to complete each task is independent of the others, find the probability that a student will take at least 20 minutes to complete both Tasks 1 and 2.

50. ◆ Referring to Exercise 49, compute the probability that a student will take at least 20 minutes to complete both Tasks 3 and 4.

51. ● *Computers* In 2001, 51 percent of all households in the U.S. had a computer.[57] Find the probability that in a small town with 800 households, at least 400 had a computer in 2001. *hint* [see Example 4]

52. ● *Television Ratings* Based on data from Nielsen Research, there is a 15 percent chance that any television that is turned on during the time of the evening newscasts will be tuned to

---

[55] The data is rounded to the nearest $1000 and is based on a report published by the Luxembourg Income Study. The report shows after-tax income, including government benefits (such as food stamps) of households with children. Our figures were obtained from the published data by assuming a normal distribution of incomes. All data was based on constant 1991 U.S. dollars and converted foreign currencies (adjusted for differences in buying power). SOURCE: Luxembourg Income Study/*The New York Times,* August 14, 1995, p. A9.

[56] Data are rounded to one decimal place. SOURCE: *Student Evaluations of MODSTAT,* by Joseph M. Nowakowski, Muskingum College, New Concord, OH, 1997. http://members.aol.com/rcknodt/pubpage.htm.

[57] SOURCE: NTIA and ESA, U.S. Department of Commerce, using U.S. Bureau of the Census Current Population Survey supplements.

ABC's evening news show.[58] Your company wishes to advertise on a small local station carrying ABC that serves a community with 2500 households that regularly tune in during this time slot. Find the approximate probability that at least 400 households will be tuned in to the show.

53. ● *Aviation* The probability of a plane crashing on a single trip in 1989 was .00000165.[59] Find the approximate probability that in 100,000,000 flights, there will be fewer than 180 crashes.

54. ● *Aviation* The probability of a plane crashing on a single trip in 1990 was .00000087. Find the approximate probability that in 100,000,000 flights, there will be more than 110 crashes.

55. *Insurance* Your company issues flight insurance. You charge $2 and in the event of a plane crash, you will pay out $1 million to the victim or his or her family. In 1989, the probability of a plane crashing on a single trip was .00000165. If ten people per flight buy insurance from you, what was your approximate probability of losing money over the course of 100 million flights in 1989? [Hint: First determine how many crashes there must be for you to lose money.]

56. *Insurance* Refer back to the preceding exercise. What is your approximate probability of losing money over the course of 10 million flights?

57. ◆ *Polls* In a certain political poll, each person polled has a 90 percent probability of telling his or her real preference. Suppose that 55 percent of the population really prefer candidate Goode, and 45 percent prefer candidate Slick. First find the probability that a person polled will say that he or she prefers Goode. Then find the approximate probability that, if 1000 people are polled, more than 52 percent will say they prefer Goode.

58. ◆ *Polls* In a certain political poll, each person polled has a 90 percent probability of telling his or her real preference. Suppose that 1000 people are polled and 51 percent say that they prefer candidate Goode, while 49 percent say that they prefer candidate Slick. Find the approximate probability that Goode could do at least this well if, in fact, only 49 percent prefer Goode.

59. ◆ *IQ Scores* Mensa is a club for people with high IQs. To qualify, you must be in the top 2 percent of the population. One way of qualifying is by having an IQ of at least 148, as measured by the Cattell intelligence test. Assuming that scores on this test are normally distributed with a mean of

100, what is the standard deviation? [Hint: Use the table in the Appendix "backwards."]

60. ◆ *SAT Scores* Another way to qualify for Mensa (see the previous exercise) is to score at least 1250 on the SAT [combined Critical Reading (Verbal, before March 2005) and Math scores], which puts you in the top 2 percent. Assuming that SAT scores are normally distributed with a mean of 1000, what is the standard deviation? [See the hint for the previous exercise.]

Communication and Reasoning Exercises

61. ● Under what assumptions are the estimates in the empirical rule exact?

62. ● If X is a continuous random variable, what values can the quantity $P(X = a)$ have?

63. ● Which is larger for a continuous random variable, $P(X \le a)$ or $P(X < a)$?

64. ● Which of the following is greater: $P(X \le b)$ or $P(a \le X \le b)$?

65. A uniform continuous distribution is one with a probability density curve that is a horizontal line. If X takes on values between the numbers a and b with a uniform distribution, find the height of its probability density curve.

66. Which would you expect to have the greater variance: the standard normal distribution or the uniform distribution taking values between -1 and 1? Explain.

67. ◆ Which would you expect to have a density curve that is higher at the mean: the standard normal distribution, or a normal distribution with standard deviation 0.5? Explain.

68. ◆ Suppose students must perform two tasks: Task 1 and Task 2. Which of the following would you expect to have a smaller standard deviation?
   (A) The time it takes a student to perform both tasks if the time it takes to complete Task 2 is independent of the time it takes to complete Task 1.
   (B) The time it takes a student to perform both tasks if students will perform similarly in both tasks.

   Explain.

---

[58] SOURCE: Nielsen Media Research/ABC Network/*New York Times*, March 18, 2002, p. C1.

[59] Source for this exercise and the following three: National Transportation Safety Board.

● basic skills   ◆ challenging

Chapter 8 Review

KEY CONCEPTS

8.1 Random Variables and Distributions

Random variable; discrete vs continuous random variable *p. 545*

Probability distribution of a random variable *p. 546*

Using measurement classes *p. 549*

8.2 Bernoulli Trials and Binomial Random Variables

Bernoulli trial; binomial random variable *p. 555*

Probability distribution of binomial random variable:

$P(X = x) = C(n, x)p^x q^{n-x}$ *p. 557*

8.3 Measures of Central Tendency

Sample, sample mean; population, population mean *p. 563*

Sample median, sample mode *p. 564*

Expected value of a random variable:

$$\mu = E(X) = \sum_i x_i \cdot P(X = x_i)$$

p. 565

Mean of a binomial random variable:

$\mu = E(X) = np$ *p. 565*

8.4 Measures of Dispersion

Sample variance:

$$s^2 = \frac{\sum_{i=1}^{n}(x_i - \bar{x})^2}{n - 1}$$

Sample standard deviation: $s = \sqrt{s^2}$ *p. 575*

Population variance:

$$\sigma^2 = \frac{\sum_{i=1}^{n}(x_i - \mu)^2}{n}$$

Population standard deviation:

$\sigma = \sqrt{\sigma^2}$ *p. 575*

Chebyshev's Rule *p. 577*

Empirical Rule *p. 578*

Variance of a random variable:

$$\sigma^2 = \sum_i (x_i - \mu)^2 P(X = x_i)$$

Standard deviation of X: $\sigma = \sqrt{\sigma^2}$ *p. 580*

Variance and standard deviation of a binomial random variable:

$\sigma^2 = npq, \sigma = \sqrt{npq}$ *p. 582*

8.5 Normal Distributions

Probability density function *p. 590*

Normal density function; normal distribution; standard normal distribution *p. 590*

Computing probabilities based on the standard normal distribution *p. 593*

Standardizing a normal distribution *p. 593*

Computing probabilities based on non-standard normal distributions *p. 593*

Normal approximation to a binomial distribution *p. 596*

REVIEW EXERCISES

In Exercises 1–5, find the probability distribution for the given random variable and draw a histogram.

1. A couple has two children; $X =$ the number of boys. (Assume an equal likelihood of a child being a boy or a girl.)

2. A four-sided die (with sides numbered 1 through 4) is rolled twice in succession; $X =$ the sum of the two numbers.

3. 48.2 percent of *Xbox*® players are in their teens, 38.6 percent are in their twenties, 11.6 percent are in their thirties, and the rest are in their forties; $X =$ age of an *Xbox* player. (Use the midpoints of the measurement classes.)

4. From a bin that contains 20 defective joysticks and 30 good ones, 3 are chosen at random; $X =$ the number of defective joysticks chosen. (Round all probabilities to four decimal places.)

5. Two dice are weighted so that each of 2, 3, 4 and 5 is half as likely to face up as each of 1 and 6; $X =$ the number of ones that face up when both are thrown.

6. Use any method to calculate the sample mean, median, and standard deviation of the following sample of scores: -1, 2, 0, 3, 6

7. Give an example of a sample of four scores with mean 1, and median 0. (Arrange them in ascending order.)

8. Give an example of a sample of six scores with sample standard deviation 0 and mean 2.

9. Give an example of a population of six scores with mean 0 and population standard deviation 1.

10. Give an example of a sample of five scores with mean 0 and sample standard deviation 1.

A die is constructed in such a way that rolling a 6 is twice as likely as rolling each other number. That die is rolled four times. Let X be the number of times a 6 is rolled. Evaluate the probabilities in Exercises 11–15.

11. $P(X = 1)$

12. The probability that 6 comes up at most twice

13. The probability that X is at least 2

14. $P(X > 3)$

15. $P(1 \leq X \leq 3)$

16. A couple has three children; $X =$ the number of girls. (Assume an equal likelihood of a child being a boy or a girl.) Find the probability distribution, expected value and standard deviation of X, and complete the following sentence with the smallest possible whole number: All values of X lie within ___ standard deviations of the expected value.

17. A random variable X has the following frequency distribution.

x	-3	-2	-1	0	1	2	3
$fr(X = x)$	1	2	3	4	3	2	1

Find the probability distribution, expected value and standard deviation of X, and complete the following sentence: 87.5 percent (or 14/16) of the time, X is within ___ (round to one decimal place) standard deviations of the expected value.

18. A random variable X has expected value $\mu = 100$ and standard deviation $\sigma = 16$. Use Chebyshev's rule to find an interval in which X is guaranteed to lie with a probability of at least 90 percent.

19. A random variable X has a bell-shaped, symmetric distribution, with expected value $\mu = 100$ and standard deviation $\sigma = 30$. Use the empirical rule to give an interval in which X lies approximately 95 percent of the time

In Exercises 20–25 the mean and standard deviation of a normal variable X are given. Find the indicated probability.

20. X is the standard normal variable Z; $P(0 \leq Z \leq 1.5)$

21. X is the standard normal variable Z; $P(Z \leq -1.5)$

22. X is the standard normal variable Z; $P(|Z| \geq 2.1)$

23. $\mu = 100$, $\sigma = 16$; $P(80 \leq X \leq 120)$

24. $\mu = 0$, $\sigma = 2$; $P(X \leq -1)$

25. $\mu = -1$, $\sigma = 0.5$; $P(X \geq 1)$

Applications

26. Marketing As a promotional gimmick, OHaganBooks.com has been selling copies of the Encyclopædia Galactica at an extremely low price that is changed each week at random in a nationally televised drawing. The following table summarizes the anticipated sales.

Price	$5.50	$10	$12	$15
Frequency (weeks)	1	2	3	4
Weekly sales	6200	3500	3000	1000

a. What is the expected value of the price of Encyclopædia Galactica?

b. What are the expected weekly sales of Encyclopædia Galactica?

27. Marketing Refer to Exercise 26.

a. What is the expected weekly revenue? (Revenue = Price per copy sold × Number of copies sold).

b. True or false? If X and Y are two random variables, then $E(XY) = E(X)E(Y)$ (the expected value of the product of two random variables is the product of the expected values). Support your claim by referring to the answers of part (a), and Exercise 26.

Online Sales *Exercises 28–30 are based on the following table, which shows the number of online orders at OHaganBooks.com per million residents in 100 U.S. cities during one month:*

Orders (per million residents)	1–2.9	3–4.9	5–6.9	7–8.9	9–10.9
Number of Cities	25	35	15	15	10

28. Let X be the number of orders per million residents in a randomly chosen U.S. city (use rounded midpoints of the given measurement classes). Construct the probability distribution for X and hence compute the expected value μ of X and standard deviation σ. (Round answers to four decimal places.)

29. What range of orders per million residents does the empirical rule predict from approximately 68 percent of all cities? Would you judge that the empirical rule applies? Why?

30. The actual percentage of cities from which you obtain between 3 and 8 orders per million residents is (choose the correct answer that gives the most specific information)

(A) Between 50 percent and 65 percent

(B) At least 65 percent

(C) At least 50 percent

(D) 57.5 percent

Mac vs. Windows *On average, 5 percent of all hits by Mac OS® users and 10 percent of all hits by Windows® users result in orders for books at OHaganBooks.com. Due to online promotional efforts, the site traffic is approximately 10 hits per hour by Mac OS users, and 20 hits per hour by Windows users. Compute the probabilities in Exercises 31–35. (Round all answers to three decimal places.)*

31. What is the probability that exactly 3 Windows users will order books in the next hour?

32. What is the probability that at most 3 Windows users will order books in the next hour?

33. What is the probability that exactly 1 Mac OS user and 3 Windows users will order books in the next hour?

34. What assumption must you make to justify your calculation in Exercise 33?

35. How many orders for books can OHaganBooks.com expect in the next hour?

Online Cosmetics *OHaganBooks.com has launched a subsidiary, GnuYou.com, which sells beauty products online. Most products sold by GnuYou.com are skin creams and hair products. Exercises 36–38 are based on the following table, which shows monthly revenues earned through sales of these products. (Assume a normal distribution. Round all answers to three decimal places.)*

Product	Skin Creams	Hair Products
Mean Monthly Revenue	$38,000	$34,000
Standard Deviation	$21,000	$14,000

36. What is the probability that GnuYou.com will sell *at least* $50,000 worth of skin cream next month?

37. What is the probability that GnuYou.com will sell *at most* $50,000 worth of hair products next month?

38. Which type of product is most likely to yield sales of less than $12,000 next month? (Support your conclusion numerically.)

39. *Intelligence* Billy-Sean O'Hagan, now a senior at Suburban State University, has done exceptionally well and has just joined Mensa, a club for people with high IQs. Within Mensa is a group called the Three Sigma Club because their IQ scores are at least 3 standard deviations higher than the U.S. mean. Assuming a U.S. population of 280,000,000, how many people in the U.S. are qualified for the Three Sigma Club? Round your answer to the nearest 1000 people.)

40. *Intelligence* To join Mensa (not necessarily the Three Sigma Club), one needs an IQ of at least 132, corresponding to the top 2 percent of the population. Assuming that scores on this test are normally distributed with a mean of 100, what is the standard deviation? (Round your answer to the nearest whole number.)

41. *Intelligence* Based on the information given in Exercises 39 and 40, what score must Billy Sean have to get into the Three Sigma Club? (Assume that IQ scores are normally distributed with a mean of 100, and use the rounded standard deviation.)

vMentor Do you need a live tutor for homework problems? Access vMentor on the ThomsonNOW! website at **www.thomsonedu.com** for one-on-one tutoring from a mathematics expert.

CASE STUDY: Spotting Tax Fraud with Benford's Law[60]

age fotostock/SuperStock

You are a tax fraud specialist working for the Internal Revenue Service (IRS), and you have just been handed a portion of the tax return from Colossal Conglomerate. The IRS suspects that the portion you were handed may be fraudulent, and would like your opinion. Is there any mathematical test, you wonder, that can point to a suspicious tax return based on nothing more than the numbers entered?

You decide, on an impulse, to make a list of the first digits of all the numbers entered in the portion of the Colossal Conglomerate tax return (there are 625 of them). You reason that, if the tax return is an honest one, the first digits of the numbers should be uniformly distributed. More precisely, if the experiment consists of selecting a number at random from the tax return, and the random variable X is defined to be the first digit of the selected number, then X should have the following probability distribution:

x	1	2	3	4	5	6	7	8	9
$P(X = x)$	$\frac{1}{9}$	$\frac{1}{9}$	$\frac{1}{9}$	$\frac{1}{9}$	$\frac{1}{9}$	$\frac{1}{9}$	$\frac{1}{9}$	$\frac{1}{9}$	$\frac{1}{9}$

You then do a quick calculation based on this probability distribution, and find an expected value of $E(X) = 5$. Next, you turn to the Colossal Conglomerate tax return data and calculate the relative frequency (estimated probability) of the actual numbers in the tax return. You find the following results.

Colossal Conglomerate Return

y	1	2	3	4	5	6	7	8	9
$P(Y = y)$	.29	.1	.04	.15	.31	.08	.01	.01	.01

---

[60] The discussion is based on the article "Following Benford's Law, or Looking Out for No. 1" by Malcolm W. Browne, *The New York Times,* August 4, 1998, p. F4. The use of Benford's Law in detecting tax evasion is discussed in a Ph.D. dissertation by Dr. Mark J. Nigrini (Southern Methodist University, Dallas).

It certainly does look suspicious! For one thing, the digits 1 and 5 seem to occur a lot more often than any of the other digits, and roughly 3 times what you predicted. Moreover, when you compute the expected value, you obtain $E(Y) = 3.48$, considerably lower than the value of 5 you predicted. Gotcha! you exclaim.

You are about to file a report recommending a detailed audit of Colossal Conglomerate when you recall an article you once read about first digits in lists of numbers. The article dealt with a remarkable discovery in 1938 by Dr. Frank Benford, a physicist at General Electric. What Dr. Benford noticed was that the pages of logarithm tables that listed numbers starting with the digits 1 and 2 tended to be more soiled and dog-eared than the pages that listed numbers starting with higher digits—say, 8. For some reason, numbers that start with low digits seemed more prevalent than numbers that start with high digits. He subsequently analyzed more than 20,000 sets of numbers, such as tables of baseball statistics, listings of widths of rivers, half-lives of radioactive elements, street addresses, and numbers in magazine articles. The result was always the same: Inexplicably, numbers that start with low digits tended to appear more frequently than those that start with high ones, with numbers beginning with the digit 1 most prevalent of all.[61] Moreover, the expected value of the first digit was not the expected 5, but 3.44.

Because the first digits in Colossal Conglomerate's return have an expected value of 3.48, very close to Benford's value, it might appear that your suspicion was groundless after all. (Back to the drawing board . . .)

Out of curiosity, you decide to investigate Benford's discovery more carefully. What you find is that Benford did more than simply observe a strange phenomenon in lists of numbers. He went further and derived the following formula for the probability distribution of first digits in lists of numbers:

$$P(X = x) = \log(1 + 1/x) \qquad (x = 1, 2, \ldots, 9)$$

You compute these probabilities, and find the following distribution (the probabilities are all rounded, and thus do not add to exactly 1).

x	1	2	3	4	5	6	7	8	9
$P(X = x)$	.30	.18	.12	.10	.08	.07	.06	.05	.05

You then enter these data along with the Colossal Conglomerate tax return data in your spreadsheet program and obtain the graph shown in Figure 17.

The graph shows something awfully suspicious happening with the digit 5. The percentage of numbers in the Colossal Conglomerate return that begin with 5 far exceeds Benford's prediction that approximately 8 percent of all numbers should begin with 5.

Now it seems fairly clear that you are justified in recommending Colossal Conglomerate for an audit, after all.

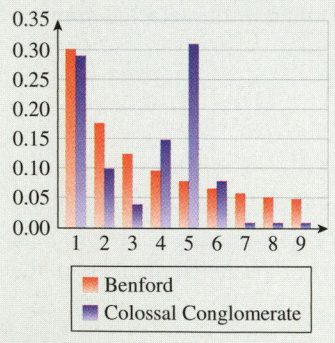

Benford

Colossal Conglomerate

Figure **17**

Q: Because no given set of data can reasonably be expected to satisfy Benford's Law exactly, how can I be certain that the Colossal Conglomerate data is not simply due to chance?

A: You can never be 100 percent certain. It is certainly conceivable that the tax figures just happen to result in the "abnormal" distribution in the Colossal Conglomerate tax return. However—and this is the subject of "inferential statistics"—there is a method for deciding whether you can be, say "95 percent certain" that the anomaly reflected in the data is not due to chance. To check, you must first compute a statistic that determines how far a given set of data deviates from

---

[61] This does not apply to all lists of numbers. For instance, a list of randomly chosen numbers between 100 and 999 will have first digits uniformly distributed between 1 and 9.

satisfying a theoretical prediction (Benford's Law, in this case). This statistic is called a **sum-of-squares error,** and given by the following formula (reminiscent of the variance):

$$\text{SSE} = n \left[\frac{[P(y_1) - P(x_1)]^2}{P(x_i)} + \frac{[P(y_2) - P(x_2)]^2}{P(x_2)} + \cdots + \frac{[P(y_9) - P(x_9)]^2}{P(x_9)} \right]$$

Here, n is the sample size: 625 in the case of Colossal Conglomerate. The quantities $P(x_i)$ are the theoretically predicted probabilities according to Benford's Law, and the $P(y_i)$ are the probabilities in the Colossal Conglomerate return. Notice that if the Colossal Conglomerate return probabilities had exactly matched the theoretically predicted probabilities, then SSE would have been zero. Notice also the effect of multiplying by the sample size n: The larger the sample, the more likely that the discrepancy between the $P(x_i)$ and the $P(y_i)$ is not due to chance. Substituting the numbers gives

$$\text{SSE} \approx 625 \left[\frac{[.29 - .30]^2}{.30} + \frac{[.1 - .18]^2}{.18} + \cdots + \frac{[.01 - .05]^2}{.05} \right]$$

$$\approx 552^{62}$$ ∎

Q: *The value of SSE does seem quite large. But how can I use this figure in my report? I would like to say something impressive, such as "Based on the portion of the Colossal Conglomerate tax return analyzed, one can be 95 percent certain that the figures are anomalous."?*

A: The error SSE is used by statisticians to answer exactly such a question. What they would do is compare this figure to the largest SSE we would have expected to get by chance in 95 out of 100 selections of data that *do* satisfy Benford's law. This "biggest error" is computed using a "Chi-Squared" distribution and can be found in Excel by entering

    =CHIINV(0.05, 8)

Here, the 0.05 is $1 - 0.95$, encoding the "95 percent certainty," and the 8 is called the "number of degrees of freedom" = number of outcomes (9) minus 1.

   You now find, using Excel, that the chi-squared figure is 15.5, meaning that the largest SSE that you could have expected purely by chance is 15.5. Because Colossal Conglomerate's error is much larger at 552, you can now justifiably say in your report that there is a 95% certainty that the figures are anomalous.[63] ∎

Exercises

Which of the following lists of data would you expect to follow Benford's law? If the answer is "no," give a reason.

1. Distances between cities in France, measured in kilometers.

2. Distances between cities in France, measured in miles.

3. The grades (0–100) in your math instructor's grade book.

4. The Dow Jones averages for the past 100 years.

5. Verbal SAT scores of college-bound high school seniors.

6. Life spans of companies.

---

[62] If you use more accurate values for the probabilities in Benford's distribution, the value is approximately 560.

[63] What this actually means is that, if you were to do a similar analysis on a large number of tax returns, and you designated as "not conforming to Benford's Law" all of those whose value of SSE was larger than 15.5, you would be justified in 95 percent of the cases.

tech Ex Use technology to determine whether the given distribution of first digits fails, with 95 percent certainty, to follow Benford's law.

7. Good Neighbor Inc.'s tax return ($n = 1000$)

y	1	2	3	4	5	6	7	8	9
$P(Y = y)$	.31	.16	.13	.11	.07	.07	.05	.06	.04

8. Honest Growth Funds Stockholder Report ($n = 400$)

y	1	2	3	4	5	6	7	8	9
$P(Y = y)$	.28	.16	.1	.11	.07	.09	.05	.07	.07

Section 8.1

Example 3 Let X be the number of heads that face up in three tosses of a coin. We obtained the following probability distribution of X in the text:

x	0	1	2	3
$P(X = x)$	$\frac{1}{8}$	$\frac{3}{8}$	$\frac{3}{8}$	$\frac{1}{8}$

Use technology to obtain the corresponding histogram.

Solution with Technology

1. In the TI-83/84, you can enter a list of probabilities as follows: press $\boxed{\text{STAT}}$, choose EDIT, and then press $\boxed{\text{ENTER}}$. Clear columns L_1 and L_2 if they are not already cleared. (Select the heading of a column and press $\boxed{\text{CLEAR}}$ $\boxed{\text{ENTER}}$ to clear it.) Enter the values of X in the column under L_1 (pressing $\boxed{\text{ENTER}}$ after each entry) and enter the frequencies in the column under L_2.

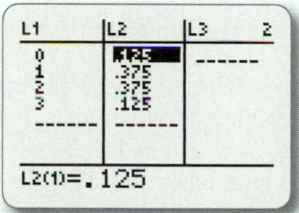

2. To graph the data as in Figure 1, first set the $\boxed{\text{WINDOW}}$ to $0 \leq X \leq 4$, $0 \leq Y \leq 0.5$, and Xscl = 1 (the width of the bars). Then turn STAT PLOT on ([2nd] $\boxed{\text{Y=}}$), and configure it by selecting the histogram icon, setting Xlist = L_1 and Freq = L_2. Then hit $\boxed{\text{GRAPH}}$.

 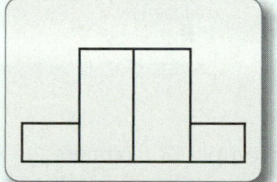

Example 4 We obtained the following frequency table in the text:

x	25,000	35,000	45,000	55,000	65,000	75,000	85,000
Frequency	20	80	230	400	170	70	30

Find the probability distribution of X.

Solution with Technology We need to divide each frequency by the sum. Although the computations in this example (dividing the 7 frequencies by 1000) are simple to do by hand, they could become tedious in general, so technology is helpful.

1. On the TI-83/84, press $\boxed{\text{STAT}}$, select EDIT, enter the values of X in the L_1 list, and enter the frequencies in the L_2 list as in Example 3.

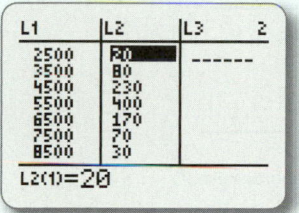

2. Then, on the home screen, enter

$$L_2/1000 \rightarrow L_3 \qquad \text{L_2 is } \boxed{\text{2nd}}\ \boxed{2}\text{, L_3 is } \boxed{\text{2nd}}\ \boxed{3}$$

or, better yet,

$$L_2/\text{sum}(L_2) \rightarrow L_3 \qquad \text{Sum is found in } \boxed{\text{2nd}}\ \boxed{\text{STAT}},$$
$$\text{under MATH}$$

3. After pressing $\boxed{\text{ENTER}}$ you can now go back to the $\boxed{\text{STAT}}$ EDIT screen, and you will find the probabilities displayed in L_3 as shown.

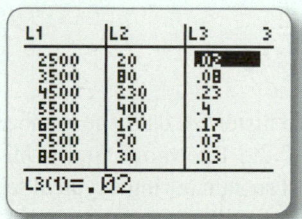

Section 8.2

Example 2(b) The probability that a randomly chosen person in Cape Coral, FL, is 65 years old or older[64] is approximately .2. If X is the number of people aged 65 or older in a sample of 6, construct the probability distribution of X.

---

[64] The actual figure in 2000 was .196 SOURCE: U.S. Census Bureau, Census 2000 Summary File 1. www.census.gov/prod/2001pubs/c2kbr01-10.pdf.

Solution with Technology In the "Y=" screen, you can enter the binomial distribution formula

$$Y_1 = 6 \ nCr \ X*0.2^X*0.8^(6-X)$$

directly (to get nCr, press MATH and select PRB), and hit TABLE. You can then replicate the table in the text by choosing $X = 0, 1, \ldots, 6$ (use the TBLSET screen to set "Indpnt" to "Ask" if you have not already done so).

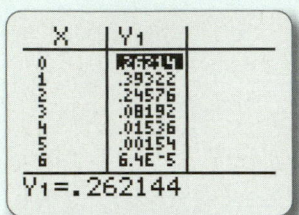

The TI-83/84 also has a built-in binomial distribution function that you can use in place of the explicit formula:

$Y_1 = binompdf(6, 0.2, X)$ Press 2nd VARS 0

The TI-83/84 function binompcf (directly following binompdf) gives the value of the *cumulative* distribution function, $P(0 \leq X \leq x)$.

To graph the resulting probability distribution on your calculator, follow the instructions for graphing a histogram in Section 8.1.

Section **8.3**

Example 3 According to historical data, the number of injuries that a member of the Enormous State University women's soccer team will sustain during a typical season is given by the following probability distribution table:

Injuries	0	1	2	3	4	5	6
Probability	.2	.2	.22	.2	.15	.01	.02

If X denotes the number of injuries sustained by a player during one season, compute $E(X)$.

Solution with Technology To obtain the expected value of a probability distribution on the TI-83/84, press STAT, select EDIT, and then press ENTER, and enter the values of X in the L_1 list and the probabilities in the column in the L_2

list. Then, on the home screen, you can obtain the expected value as

$$\text{sum } (L_1 * L_2)$$

L_1 is 2nd 1 L_2 is 2nd 2

Sum is found in 2nd STAT , under MATH

Section **8.4**

Example 1 The unemployment rates (in percentage points) in the U.S. for the years 1999–2004 were*

$$4, \ 4, \ 5, \ 6, \ 6, \ 5$$

Compute the sample mean and standard deviation.

Solution with Technology On the TI-83/84, enter the sample scores in list L_1 on the STAT/EDIT screen, then go to STAT/CALC, select 1-Var Stats, and hit ENTER. The resulting display shows, among other statistics, the sample standard deviation s as "Sx" as well as the population standard deviation σ as "σx."

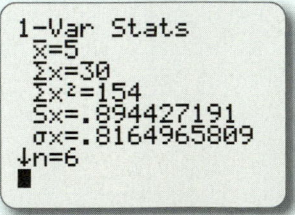

Example 3 Compute the variance and standard deviation for the following probability distribution.

x	10	20	30	40	50	60
$P(X=x)$	.2	.2	.3	.1	.1	.1

---

* Figures are rounded. SOURCES: SWA, 222; Bureau of Labor Statistics (BLS), www.bls.gov.

Solution with Technology

1. As in Example 3 in the preceding section, begin by entering the probability distribution of X into columns L_1 and L_2 in the LIST screen (press STAT and select EDIT).

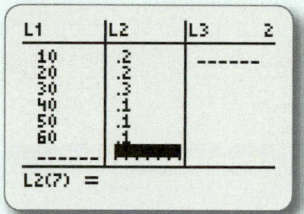

2. Then, on the home screen, enter

   sum(L_1*L_2) → M   Stores the value of μ as M
   Sum is found in 2nd STAT, under MATH

3. To obtain the variance, enter the following.

   sum((L_1-M)^2*L_2)   Computation of $\sum(x-\mu)^2 P(X=x)$

```
sum(L1*L2)+M
                 30
sum((L1-M)^2*L2)
                240
■
```

Section 8.5

Example 1 (b), (c) Let Z be the standard normal variable. Calculate the following probabilities.
b. $P(0 \le Z \le 2.43)$

c. $P(-1.37 \le Z \le 2.43)$

Solution with Technology On the TI-83/84, press 2nd VARS to obtain the selection of distribution functions. The first function, normalpdf, gives the values of the normal density function (whose graph is the normal curve). The second, normalcdf, gives $P(a \le Z \le b)$. For example, to compute $P(0 \le Z \le 2.43)$, enter

   normalcdf(0, 2.43)

To compute $P(-1.37 \le Z \le 2.43)$, enter

   normalcdf(-1.37, 2.43)

```
normalcdf(0,2.43
)
        .4924505885
normalcdf(-1.37,
2.43)
        .9071070809
■
```

Example 2 Pressure gauges manufactured by Precision Corp. must be checked for accuracy before being placed on the market. To test a pressure gauge, a worker uses it to measure the pressure of a sample of compressed air known to be at a pressure of exactly 50 pounds per square inch. If the gauge reading is off by more than 1 percent (0.5 pounds), it is rejected. Assuming that the reading of a pressure gauge under these circumstances is a normal random variable with mean 50 and standard deviation 0.4, find the percentage of gauges rejected.

Solution with Technology As seen in the text, we need to compute $1 - P(49.5 \le X \le 50.5)$ with $\mu = 50$ and $\sigma = 0.4$. On the TI-83/84, the built-in normalcdf function permits us to compute $P(a \le X \le b)$ for nonstandard normal distributions as well. The format is

   normalcdf(a, b, μ, σ) $P(a \le X \le b)$

For example, we can compute $P(49.5 \le X \le 50.5)$ by entering

   normalcdf(49.5, 50.5, 50, 0.4)

Then subtract it from 1 to obtain the answer:

```
normalcdf(49.5,5
0.5,50,.4)
        .7887003221
1-Ans
        .2112996779
■
```

EXCEL Technology Guide

Section 8.1

Example 3 Let X be the number of heads that face up in three tosses of a coin. We obtained the following probability distribution of X in the text:

x	0	1	2	3
$P(X = x)$	$\frac{1}{8}$	$\frac{3}{8}$	$\frac{3}{8}$	$\frac{1}{8}$

Use technology to obtain the corresponding histogram.

Solution with Technology

1. In Excel, enter the values of X in one column and the probabilities in another.

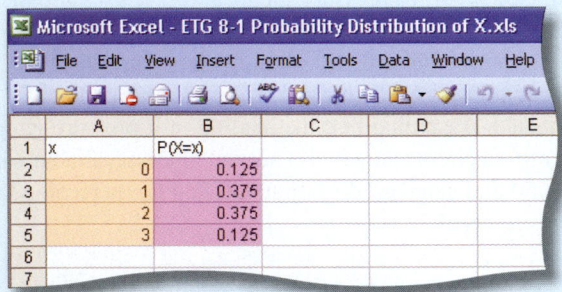

2. Next, select *only* the column of probabilities (B2–B5) and then choose Insert→Chart. In the Chart dialog, select "Column" under "Standard Types" and then press "Next" and select "Series." At the bottom of the resulting dialogue box, click in the blank area to the right of "Category (X) Axis Labels" and use your mouse to select the column of X-values (A2–A5). You can then press "Finish." To adjust the width of the bars, double-click on a bar, select "Options" in the menu that appears, and adjust the gap width to your liking. Here is a possible result (gap width zero):

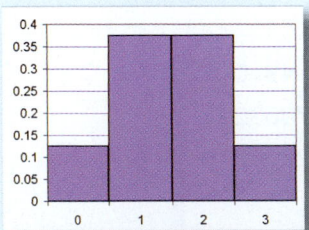

Example 4 We obtained the following frequency table in the text:

x	25,000	35,000	45,000	55,000	65,000	75,000	85,000
Frequency	20	80	230	400	170	70	30

Find the probability distribution of X.

Solution with Technology We need to divide each frequency by the sum. Although the computations in this example (dividing the 7 frequencies by 1000) are simple to do by hand, they could become tedious in general, so technology is helpful. Excel manipulates lists with ease. Set up your spreadsheet as shown.

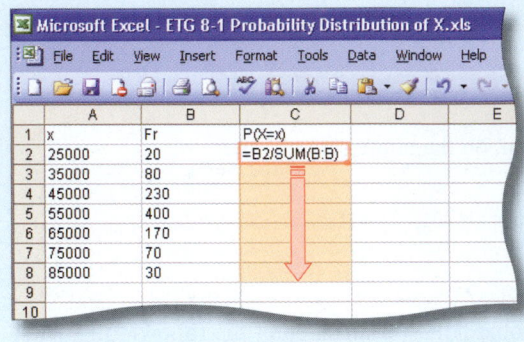

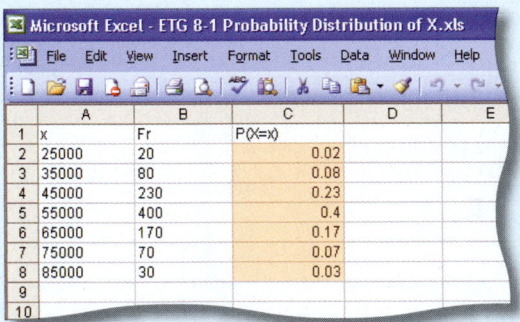

The formula `SUM(B:B)` gives the sum of all the numerical entries in Column B. You can now change the frequencies to see the effect on the probabilities. You can also add new values and frequencies to the list if you copy the formula in column C further down the column.

Section 8.2

Example 2(b) The probability that a randomly chosen person in Cape Coral, FL, is 65 years old or older[65] is approximately .2. If X is the number of people aged 65 or

---

[65] The actual figure in 2000 was .196 SOURCE: U.S. Census Bureau, Census 2000 Summary File 1. www.census.gov/prod/2001pubs/c2kbr01-10 .pdf.

older in a sample of 6, construct the probability distribution of X.

Solution with Technology You can generate the binomial distribution as follows in Excel:

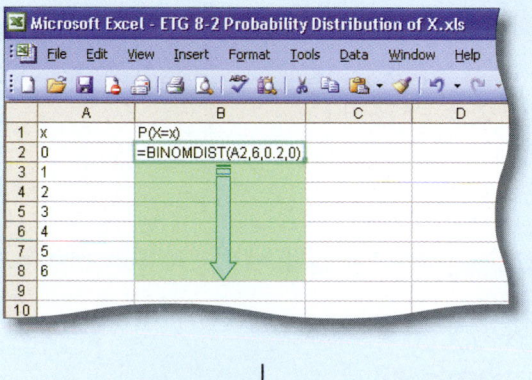

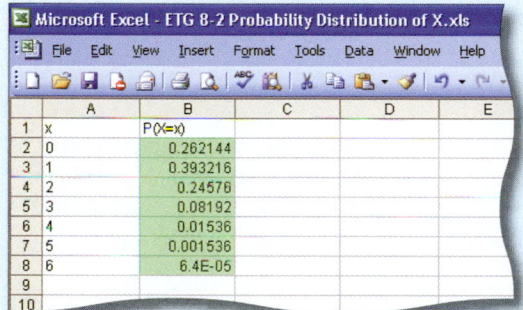

The values of X are shown in column A, and the probabilities are computed in column B. The arguments of the BINOMDIST function are as follows:

$$\text{BINOMDIST}(x, n, p, \text{Cumulative } (0 = \text{no}, 1 = \text{yes}).)$$

Setting the last argument to 0 (as shown) gives $P(X = x)$. Setting it to 1 gives $P(X \le x)$.

To graph the resulting probability distribution using Excel, follow the instructions for graphing a histogram in Section 8.1.

Section **8.3**

Example 3 According to historical data, the number of injuries that a member of the Enormous State University women's soccer team will sustain during a typical season is given by the following probability distribution table:

Injuries	0	1	2	3	4	5	6
Probability	.2	.2	.22	.2	.15	.01	.02

If X denotes the number of injuries sustained by a player during one season, compute $E(X)$.

Solution with Technology As the method we used suggests, the calculation of the expected value from the probability distribution is particularly easy to do using a spreadsheet program such as Excel. The following worksheet shows one way to do it. (The first two columns contain the probability distribution of X; the quantities $xP(X = x)$ are summed in cell C9.)

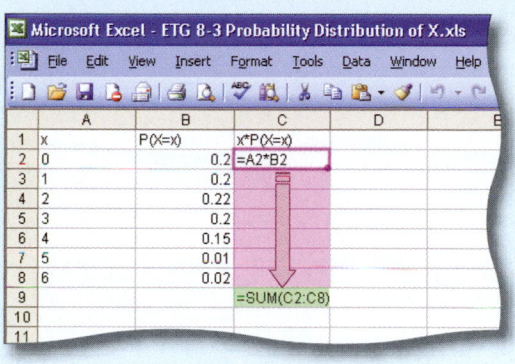

An alternative is to use the SUMPRODUCT function in Excel: Once we enter the first two columns above, the formula

$$\texttt{=SUMPRODUCT(A2:A8,B2:B8)}$$

computes the sum of the products of corresponding entries in the columns, giving us the expected value.

Section **8.4**

Example 1 The unemployment rates (in percentage points) in the U.S. for the years 1999–2004 were[66]

$$4, \ 4, \ 5, \ 6, \ 6, \ 5$$

Compute the sample mean and standard deviation.

[66] Figures are rounded. SOURCES: SWA, 222; Bureau of Labor Statistics (BLS), www.bls.gov.

Solution with Technology To compute the sample mean and standard deviation of a collection of scores in Excel, set up your spreadsheet as follows:

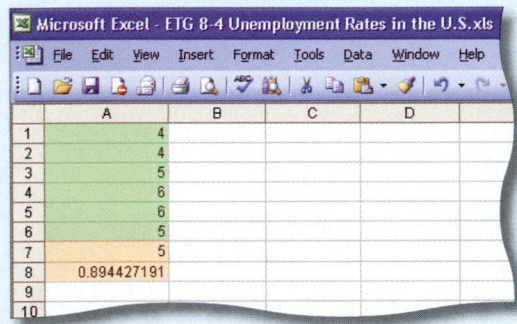

For the population standard deviation, use

    =STDEVP(A1:A6)

Example 3 Compute the variance and standard deviation for the following probability distribution.

x	10	20	30	40	50	60
$P(X = x)$	.2	.2	.3	.1	.1	.1

Solution with Technology As in Example 3 in the preceding section, begin by entering the probability distribution into columns A and B, and then proceed as shown:

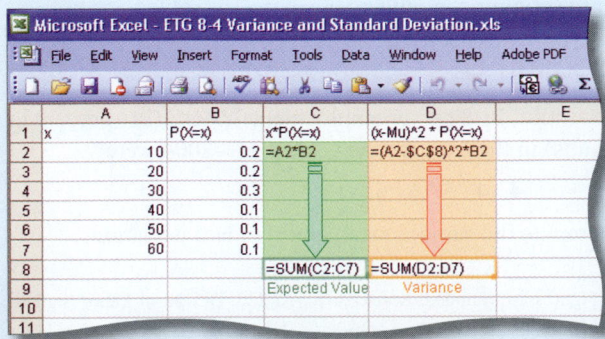

The variance then appears in cell D8:

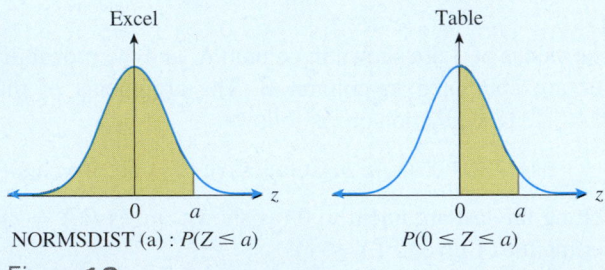

Section **8.5**

Example 1(b), (c) Let Z be the standard normal variable. Calculate the following probabilities.
b. $P(0 \leq Z \leq 2.43)$
c. $P(-1.37 \leq Z \leq 2.43)$

Solution with Technology In Excel, the function NORMS-DIST (**N**ormal **S**tandard **Dist**ribution) gives the area shown on the left in Figure 18. (Tables such as the one in the Appendix give the area shown on the right.)

Excel

NORMSDIST (a) : $P(Z \leq a)$

Table

$P(0 \leq Z \leq a)$

Figure 18

To compute a general area, $P(a \leq Z \leq b)$ in Excel, subtract the cumulative area to a from that to b:

    =NORMSDIST(b)-NORMSDIST(a) $P(a \leq Z \leq b)$

In particular, to compute $P(0 \leq Z \leq 2.43)$, use

    =NORMSDIST(2.43)-NORMSDIST(0)

and to compute $P(-1.37 \leq Z \leq 2.43)$, use

    =NORMSDIST(2.43)-NORMSDIST(-1.37)

TECHNOLOGY GUIDE

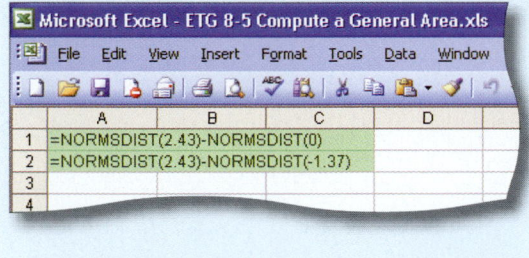

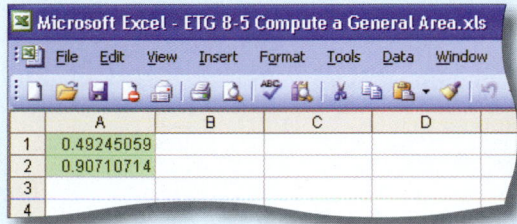

Example 2 Pressure gauges manufactured by Precision Corp. must be checked for accuracy before being placed on the market. To test a pressure gauge, a worker uses it to measure the pressure of a sample of compressed air known to be at a pressure of exactly 50 pounds per square inch. If the gauge reading is off by more than 1 percent (0.5 pounds), it is rejected. Assuming that the reading of a pressure gauge under these circumstances is a normal random variable with mean 50 and standard deviation 0.4, find the percentage of gauges rejected.

Solution with Technology In Excel, we use the function NORMDIST instead of NORMSDIST. Its format is similar to NORMSDIST, but includes extra arguments as shown.

$$\texttt{=NORMDIST(a, } \mu \texttt{, } \sigma \texttt{, 1)} \quad P(X \leq a)$$

(The last argument, set to 1, tells Excel that you want the cumulative distribution.) To compute $P(a \leq X \leq b)$ we enter the following in any vacant cell:

$$\texttt{=NORMDIST(b, } \mu \texttt{, } \sigma \texttt{, 1)}$$
$$\texttt{- NORMDIST(a, } \mu \texttt{, } \sigma \texttt{, 1)} \quad P(a \leq X \leq b)$$

For example, we can compute $P(49.5 \leq X \leq 50.5)$ by entering

$$\texttt{=NORMDIST(50.5,50,0.4,1)}$$
$$\texttt{-NORMDIST(49.5,50,0.4,1)}$$

We then subtract it from 1 to obtain the answer:

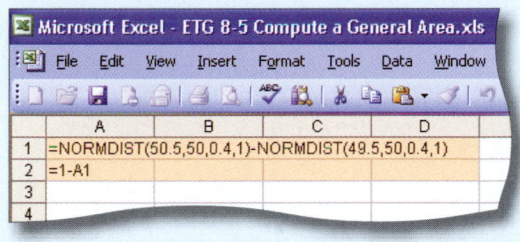

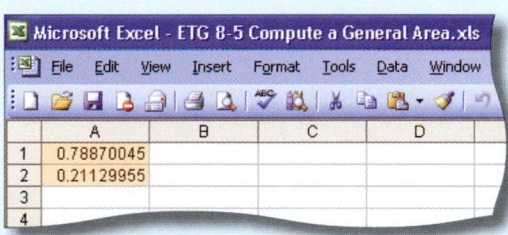

9

Nonlinear Models

CASE STUDY Checking up on Malthus

In 1798 Thomas R. Malthus (1766–1834) published an influential pamphlet, later expanded into a book, titled *An Essay on the Principle of Population as It Affects the Future Improvement of Society*. One of his main contentions was that population grows geometrically (exponentially), while the supply of resources such as food grows only arithmetically (linearly). Some 200 years later, you have been asked to check the validity of Malthus's contention. How do you go about doing so?

Park Street/PhotoEdit

Introduction

To see if Malthus was right, we need to see if the data fit the models (linear and exponential) that he suggested or if other models would be better. We saw in Chapter 1 how to fit a linear model. In this chapter we discuss how to construct models using various *nonlinear* functions.

The nonlinear functions we consider in this chapter are the *quadratic* functions, the simplest nonlinear functions; the *exponential* functions, essential for discussing many kinds of growth and decay, including the growth (and decay) of money in finance and the initial growth of an epidemic; the *logarithmic* functions, needed to fully understand the exponential functions; and the *logistic* functions, used to model growth with an upper limit, such as the spread of an epidemic.

algebra Review

For this chapter, you should be familiar with the algebra reviewed in Chapter 0, Section 2.

9.1 Quadratic Functions and Models

In Chapter 1 we studied linear functions. Linear functions are useful, but in real-life applications, they are often accurate for only a limited range of values of the variables. The relationship between two quantities is often best modeled by a curved line rather than a straight line. The simplest function with a graph that is not a straight line is a *quadratic* function.

Quadratic Function

A **quadratic function** of the variable x is a function that can be written in the form

$$f(x) = ax^2 + bx + c \qquad \text{Function form}$$

or

$$y = ax^2 + bx + c \qquad \text{Equation form}$$

where a, b, and c are fixed numbers (with $a \neq 0$).

quick Examples

1. $f(x) = 3x^2 - 2x + 1$ $\qquad\qquad$ $a = 3, b = -2, c = 1$

2. $g(x) = -x^2$ $\qquad\qquad$ $a = -1, b = 0, c = 0$

3. $R(p) = -5600p^2 + 14{,}000p$ $\qquad\qquad$ $a = -5600, b = 14{,}000, c = 0$

Every quadratic function $f(x) = ax^2 + bx + c$ $(a \neq 0)$ has a **parabola** as its graph. Following is a summary of some features of parabolas that we can use to sketch the graph of any quadratic function.[1]

---

[1] We shall not fully justify the formula for the vertex and the axis of symmetry until we have studied some calculus, although it is possible to do so with just algebra.

Features of a Parabola

The graph of $f(x) = ax^2 + bx + c \, (a \neq 0)$ is a **parabola.** If $a > 0$ the parabola opens upward (concave up) and if $a < 0$ it opens downward (concave down):

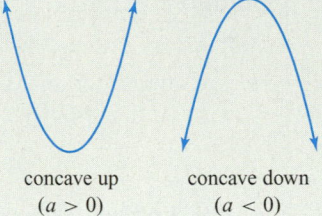

concave up   concave down
$(a > 0)$ $(a < 0)$

Vertex, Intercepts, and Symmetry

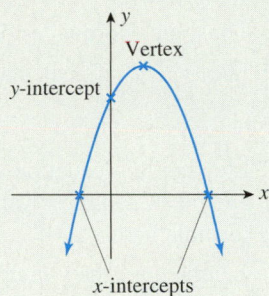

Vertex   The vertex is the turning point of the parabola (see the above figure). Its x-coordinate is $-\dfrac{b}{2a}$. Its y-coordinate is $f\left(-\dfrac{b}{2a}\right)$.

x-Intercepts   (if any) These occur when $f(x) = 0$; that is, when

$$ax^2 + bx + c = 0$$

Solve this equation for x by either factoring or using the quadratic formula. The x-intercepts are

$$x = \frac{-b \pm \sqrt{b^2 - 4ac}}{2a}$$

If the **discriminant** $b^2 - 4ac$ is positive, there are two x-intercepts. If it is zero, there is a single x-intercept (at the vertex). If it is negative, there are no x-intercepts (so the parabola doesn't touch the x-axis at all).

y-Intercept   This occurs when $x = 0$, so

$$y = a(0)^2 + b(0) + c = c$$

Symmetry   The parabola is symmetric with respect to the vertical line through the vertex, which is the line $x = -\dfrac{b}{2a}$.

Note that the x-intercepts can also be written as

$$x = -\frac{b}{2a} \pm \frac{\sqrt{b^2 - 4ac}}{2a}$$

making it clear that they are located symmetrically on either side of the line $x = -b/(2a)$. This partially justifies the claim that the whole parabola is symmetric with respect to this line.

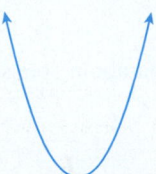

Figure **1**

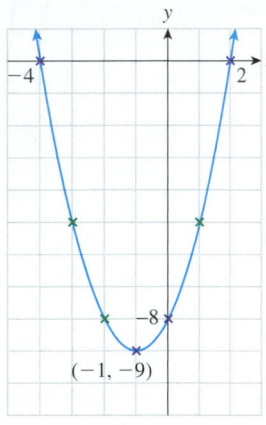

$$y = x^2 + 2x - 8$$

Figure **2**

Example **1** Sketching the Graph of a Quadratic Function

Sketch the graph of $f(x) = x^2 + 2x - 8$ by hand.

Solution Here, $a = 1$, $b = 2$, and $c = -8$. Because $a > 0$, the parabola is concave up (Figure 1).

Vertex: The x-coordinate of the vertex is

$$x = -\frac{b}{2a} = -\frac{2}{2} = -1$$

To get its y-coordinate, we substitute the value of x back into $f(x)$ to get

$$y = f(-1) = (-1)^2 + 2(-1) - 8 = 1 - 2 - 8 = -9$$

Thus, the coordinates of the vertex are $(-1, -9)$.

x-Intercepts: To calculate the x-intercepts (if any), we solve the equation

$$x^2 + 2x - 8 = 0$$

Luckily, this equation factors as $(x + 4)(x - 2) = 0$. Thus, the solutions are $x = -4$ and $x = 2$, so these values are the x-intercepts. (We could also have used the quadratic formula here.)

y-Intercept: The y-intercept is given by $c = -8$.

Symmetry: The graph is symmetric around the vertical line $x = -1$.

Now we can sketch the curve as in Figure 2. (As we see in the figure, it is helpful to plot additional points using the equation $y = x^2 + 2x - 8$, and to use symmetry to obtain others.)

Example **2** One *x*-Intercept and No *x*-Intercepts

Sketch the graph of each quadratic function, showing the location of the vertex and intercepts.

a. $f(x) = 4x^2 - 12x + 9$

b. $g(x) = -\dfrac{1}{2}x^2 + 4x - 12$

using *Technology*

See the Technology Guides at the end of the chapter to find out how to do the calculations and the graph in part (a) using a TI-83/84 or Excel.

Solution

a. We have $a = 4$, $b = -12$, and $c = 9$. Because $a > 0$, this parabola is concave up.

Vertex: $x = -\dfrac{b}{2a} = \dfrac{12}{8} = \dfrac{3}{2}$ *x*-coordinate of vertex

$$y = f\left(\frac{3}{2}\right) = 4\left(\frac{3}{2}\right)^2 - 12\left(\frac{3}{2}\right) + 9 = 0$$ *y*-coordinate of vertex

Thus, the vertex is at the point $(3/2, 0)$.

x-Intercepts: $4x^2 - 12x + 9 = 0$

$$(2x - 3)^2 = 0$$

The only solution is $2x - 3 = 0$, or $x = 3/2$. Note that this coincides with the vertex, which lies on the x-axis.

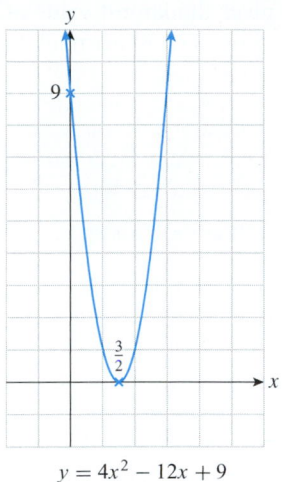

$$y = 4x^2 - 12x + 9$$

Figure **3**

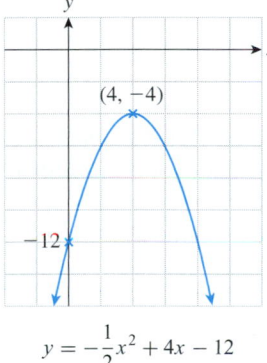

$$y = -\frac{1}{2}x^2 + 4x - 12$$

Figure **4**

y-Intercept: $c = 9$

Symmetry: The graph is symmetric around the vertical line $x = 3/2$.

The graph is the narrow parabola shown in Figure 3. (As we remarked in Example 1, plotting additional points and using symmetry helps us obtain an accurate sketch.)

b. Here, $a = -1/2$, $b = 4$, and $c = -12$. Because $a < 0$, the parabola is concave down. The vertex has x-coordinate $-b/(2a) = 4$, with corresponding y-coordinate $f(4) = -\frac{1}{2}(4)^2 + 4(4) - 12 = -4$. Thus, the vertex is at $(4, -4)$.

For the x-intercepts, we must solve $-\frac{1}{2}x^2 + 4x - 12 = 0$. If we try to use the quadratic formula, we discover that the discriminant is $b^2 - 4ac = 16 - 24 = -8$. Because the discriminant is negative, there are no solutions of the equation, so there are no x-intercepts.

The y-intercept is given by $c = -12$, and the graph is symmetric around the vertical line $x = 4$.

Because there are no x-intercepts, the graph lies entirely below the x-axis, as shown in Figure 4. (Again, you should plot additional points and use symmetry to ensure that your sketch is accurate.)

Applications

Recall that the **revenue** resulting from one or more business transactions is the total payment received. Thus, if q units of some item are sold at p dollars per unit, the revenue resulting from the sale is

$$\text{revenue} = \text{price} \times \text{quantity}$$
$$R = pq$$

Example **3** Demand and Revenue

Alien Publications, Inc., predicts that the demand equation for the sale of its latest illustrated sci-fi novel *Episode 93: Yoda vs. Alien* is

$$q = -2000p + 150{,}000$$

where q is the number of books it can sell each year at a price of \$$p$ per book. What price should Alien Publications, Inc., charge to obtain the maximum annual revenue?

Solution The total revenue depends on the price, as follows:

$$R = pq \qquad \text{\textcolor{red}{Formula for revenue}}$$
$$= p(-2000p + 150{,}000) \qquad \text{\textcolor{red}{Substitute for q from demand equation.}}$$
$$= -2000p^2 + 150{,}000p \qquad \text{\textcolor{red}{Simplify.}}$$

We are after the price p that gives the maximum possible revenue. Notice that what we have is a quadratic function of the form $R(p) = ap^2 + bp + c$, where $a = -2000$, $b = 150{,}000$, and $c = 0$. Because a is negative, the graph of the function is a parabola, concave down, so its vertex is its highest point (Figure 5). The p-coordinate of the vertex is

$$p = -\frac{b}{2a} = -\frac{150{,}000}{-4000} = 37.5$$

This value of p gives the highest point on the graph and thus gives the largest value of $R(p)$. We may conclude that Alien Publications, Inc., should charge \$37.50 per book to maximize its annual revenue.

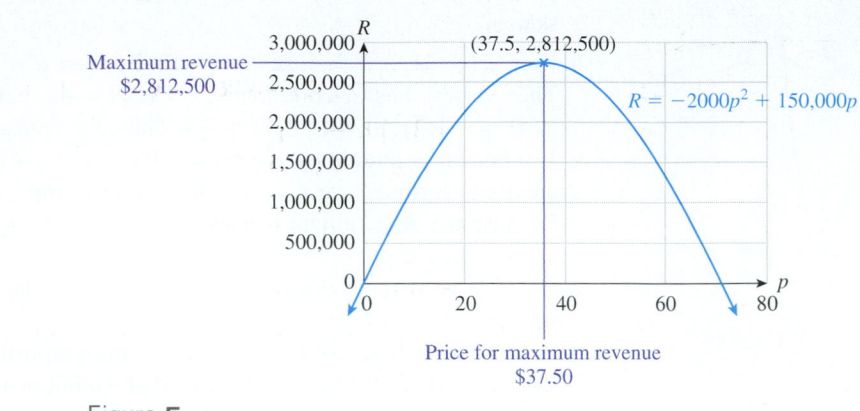

Figure **5**

+ *Before we go on...* You might ask what the maximum annual revenue is for the publisher in Example 3. Because $R(p)$ gives us the revenue at a price of \$$p$, the answer is $R(37.5) = -2000(37.5)^2 + 150,000(37.5) = 2,812,500$. In other words, the company will earn total annual revenues from this book amounting to \$2,812,500. ▪

Example **4** Demand, Revenue, and Profit

As the operator of Workout Fever Health Club, you calculate your demand equation to be

$$q = -0.06p + 84$$

where q is the number of members in the club and p is the annual membership fee you charge.

a. Your annual operating costs are a fixed cost of \$20,000 per year plus a variable cost of \$20 per member. Find the annual revenue and profit as functions of the membership price p.

b. At what price should you set the annual membership fee to obtain the maximum revenue? What is the maximum possible revenue?

c. At what price should you set the annual membership fee to obtain the maximum profit? What is the maximum possible profit? What is the corresponding revenue?

Solution

a. The annual revenue is given by

$$R = pq \qquad \text{Formula for revenue}$$
$$= p(-0.06p + 84) \qquad \text{Substitute for } q \text{ from demand equation.}$$
$$= -0.06p^2 + 84p \qquad \text{Simplify.}$$

The annual cost C is given by

$$C = 20{,}000 + 20q \qquad \text{\color{magenta}\$20,000 plus \$20 per member}$$

However, this is a function of q, and not p. To express C as a function of p we substitute for q using the demand equation $q = -0.06p + 84$:

$$\begin{aligned} C &= 20{,}000 + 20(-0.06p + 84) \\ &= 20{,}000 - 1.2p + 1680 \\ &= -1.2p + 21{,}680 \end{aligned}$$

Thus, the profit function is

$$\begin{aligned} P &= R - C & &\text{\color{magenta}Formula for profit} \\ &= -0.06p^2 + 84p - (-1.2p + 21{,}680) & &\text{\color{magenta}Substitute for revenue and cost.} \\ &= -0.06p^2 + 85.2p - 21{,}680 \end{aligned}$$

b. From part (a) the revenue function is given by

$$R = -0.06p^2 + 84p$$

This is a quadratic function ($a = -0.06$, $b = 84$, $c = 0$) whose graph is a concave-down parabola (Figure 6).

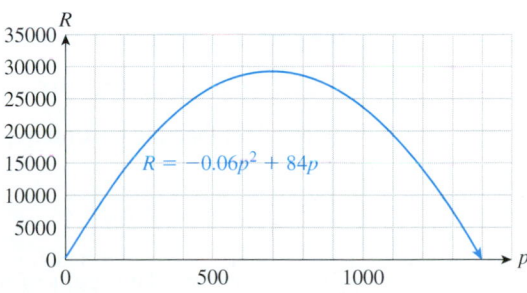

Figure **6**

The maximum revenue corresponds to the highest point of the graph: the vertex, of which the p-coordinate is

$$p = -\frac{b}{2a} = -\frac{84}{2(-0.06)} \approx \$700$$

This is the membership fee you should charge for the maximum revenue. The corresponding maximum revenue is given by the y-coordinate of the vertex in Figure 6:

$$R(700) = -0.06(700)^2 + 84(700) = \$29{,}400$$

c. From part (a), the profit function is given by

$$P = -0.06p^2 + 85.2p - 21{,}680$$

Like the revenue function, the profit function is quadratic ($a = -0.06$, $b = 85.2$, $c = -21,680$). Figure 7 shows both the revenue and profit functions.

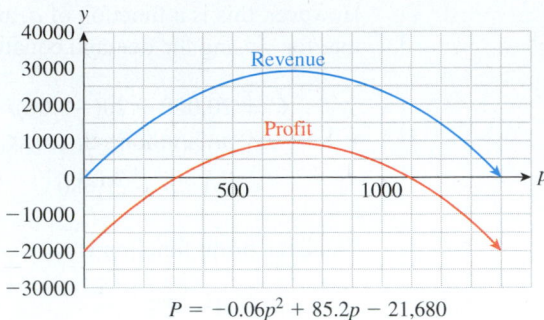

$$P = -0.06p^2 + 85.2p - 21,680$$

Figure **7**

The maximum profit corresponds to the vertex, whose p-coordinate is

$$p = -\frac{b}{2a} = -\frac{85.2}{2(-0.06)} \approx \$710$$

This is the membership fee you should charge for the maximum profit. The corresponding maximum profit is given by the y-coordinate of the vertex of the profit curve in Figure 7:

$$P(710) = -0.06(710)^2 + 85.2(710) - 21,680 = \$8,566$$

The corresponding revenue is

$$R(710) = -0.06(710)^2 + 84(710) = \$29,394$$

slightly less than the maximum possible revenue of $29,400.

+ *Before we go on...* The result of part (c) of Example 4 tells us that the vertex of the profit curve in Figure 7 is slightly to the right of the vertex in the revenue curve. However, the difference is tiny compared to the scale of the graphs, so the graphs appear to be parallel. ∎

Q: *Charging $710 membership brings in less revenue than charging $700. So why charge $710*?

A: A membership fee of $700 does bring in slightly larger revenue than a fee of $710, but it also brings in a slightly larger membership which in turn raises the operating expense and has the effect of *lowering* the profit slightly (to $8560). In other words, the slightly higher fee, while bringing in less revenue, also lowers the cost, and the net result is a larger profit. ∎

Fitting a Quadratic Function to Data: Quadratic Regression

In Section 1.5 we saw how to fit a regression line to a collection of data points. Here, we see how to use technology to obtain the **quadratic regression curve** associated with a set of points. The quadratic regression curve is the quadratic curve $y = ax^2 + bx + c$ that best fits the data points in the sense that the associated sum-of-squares error (SSE—see Section 1.5) is a minimum. Although there are algebraic methods for obtaining the quadratic regression curve, it is normal to use technology to do this.

Example 5 Currency

The following table shows the value of the euro (€) in U.S. dollars since it began trading in January, 1999 ($t = 0$ represents January 2000).[*]

Year t	-1	0	1	2	3	4	4.5
Value ($)	1.2	1	0.9	0.9	1.1	1.3	1.2

a. Is a linear model appropriate for these data?

b. Find the quadratic model

$$V(t) = at^2 + bt + c$$

that best fits the data.

Solution

a. To see whether a linear model is appropriate, we plot the data points and the regression line using one of the methods of Example 3 in Section 1.5 (Figure 8).

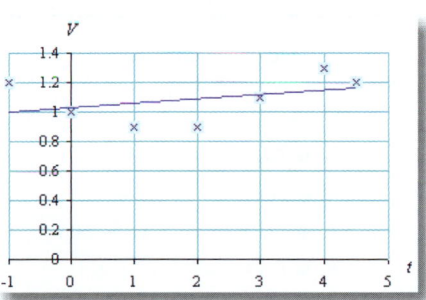

Figure **8**

From the graph, we can see that the given data suggest a curve and not a straight line: The observed points are above the regression line at the ends but below in the middle. (We would expect the data points from a linear relation to fall randomly above and below the regression line.)

b. The quadratic model that best fits the data is the quadratic regression model. As with linear regression, there are algebraic formulas to compute a, b, and c, but they are rather involved. However, we exploit the fact that these formulas are built into graphing calculators, spreadsheets, and other technology, and obtain the regression curve using technology (see Figure 9):

$$V(t) = 0.0399t^2 - 0.1150t + 1.0154 \qquad \text{Coefficients rounded to four decimal places}$$

_____
[*]Prices are rounded 20-day averages. Source for data: http://finance.yahoo.com/.

---

using *Technology*

See the Technology Guides at the end of the chapter for detailed instructions on using a TI-83/84 or Excel to find quadratic regression curves. Alternatively, go online and follow:

  Chapter 9

  → Tools

    → Simple Regression

to find a utility for finding regression curves of various sorts.

Figure **9**

Notice from the graphs above that the quadratic regression model appears to give a far better fit than the linear regression model. This impression is supported by the values of SSE. For the linear regression model, SSE ≈ 0.13. For the quadratic regression model, SSE is much smaller, approximately 0.03, indicating a much better fit.

9.1 EXERCISES

● denotes basic skills exercises

tech Ex indicates exercises that should be solved using technology

In Exercises 1–10, sketch the graphs of the quadratic functions, indicating the coordinates of the vertex, the y-intercept, and the x-intercepts (if any). hint [see Example 1]

1. ● $f(x) = x^2 + 3x + 2$ **2.** ● $f(x) = -x^2 - x$

3. ● $f(x) = -x^2 + 4x - 4$ **4.** ● $f(x) = x^2 + 2x + 1$

5. ● $f(x) = -x^2 - 40x + 500$ **6.** ● $f(x) = x^2 - 10x - 600$

7. ● $f(x) = x^2 + x - 1$ **8.** ● $f(x) = x^2 + \sqrt{2}x + 1$

9. ● $f(x) = x^2 + 1$ **10.** ● $f(x) = -x^2 + 5$

In Exercises 11–14, for each demand equation, express the total revenue R as a function of the price p per item, sketch the graph of the resulting function, and determine the price p that maximizes total revenue in each case. hint [see Example 3]

11. ● $q = -4p + 100$ **12.** ● $q = -3p + 300$

13. ● $q = -2p + 400$ **14.** ● $q = -5p + 1200$

tech Ex *In Exercises 15–18, use technology to find the quadratic regression curve through the given points. (Round all coefficients to four decimal places.) hint [see Example 5]*

15. *tech* Ex $\{(1, 2), (3, 5), (4, 3), (5, 1)\}$

16. *tech* Ex $\{(-1, 2), (-3, 5), (-4, 3), (-5, 1)\}$

17. *tech* Ex $\{(-1, 2), (-3, 5), (-4, 3)\}$

18. *tech* Ex $\{(2, 5), (3, 5), (5, 3)\}$

Applications

19. ● ***Trade with China*** The following chart shows the value of U.S. trade with China for the period 1994–2004 ($t = 0$ represents 1994).[2] *hint* [see "Features of a Parabola" p. 617]

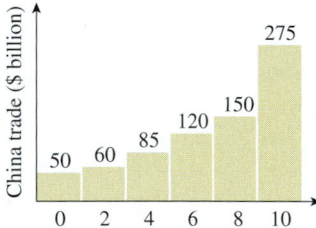

a. If you want to model the trade figures with a function of the form,

$$f(t) = at^2 + bt + c$$

would you expect the coefficient a to be positive or negative? Why?

b. Which of the following models best approximates the data given? (Try to answer this without actually computing values.)

(A) $f(t) = 3t^2 - 7t - 50$
(B) $f(t) = -3t^2 - 7t + 50$
(C) $f(t) = 3t^2 - 7t + 50$
(D) $f(t) = -3t^2 - 7t - 50$

---

[2] 2004 figure is an estimate. SOURCE: U.S. Census Bureau/*New York Times,* September 23, 2004, p. C1.

● basic skills   *tech* Ex technology exercise

c. What is the nearest year that would correspond to the vertex of the graph of the correct model from part (b)? What is the danger of extrapolating the data backwards?

20. ● **Scientific Research** The following chart shows the number of research articles in the prominent journal *Physics Review* that were written by researchers in the U.S. for the period 1983–2003 ($t = 0$ represents 1983).[3]

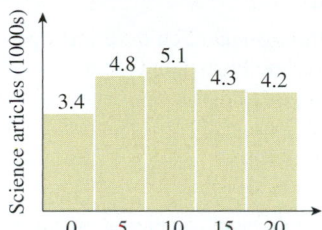

a. If you want to model the publication figures with a function of the form,

$$f(t) = at^2 + bt + c$$

would you expect the coefficient a to be positive or negative? Why?

b. Which of the following models best approximates the data given? (Try to answer this without actually computing values.)
(A) $f(t) = -0.01t^2 + 0.24t + 3.4$
(B) $f(t) = 0.01t^2 + 0.24t + 3.4$
(C) $f(t) = -0.01t^2 + 0.24t - 3.4$
(D) $f(t) = 0.01t^2 + 0.24t - 3.4$

c. In which year or years does the model in part (b) predict the number of articles published was greatest? What is the danger of extrapolating the data forwards?

21. ● **Sport Utility Vehicles** The average weight of an SUV could be approximated by

$$W = 3t^2 - 90t + 4200 \quad (5 \le t \le 27)$$

where t is its year of manufacture ($t = 0$ represents 1970) and W is the average weight of an SUV in pounds.[4] Sketch the graph of W as a function of t. According to the model, in what year were SUVs the lightest? What was their average weight in that year? *hint* [see Example 1]

22. ● **Sedans** The average weight of a sedan could be approximated by

$$W = 6t^2 - 240t + 4800 \quad (5 \le t \le 27)$$

where t is its year of manufacture ($t = 0$ represents 1970) and W is the average weight of a sedan in pounds.[5] Sketch the

graph of W as a function of t. According to the model, in what year were sedans the lightest? What was their average weight in that year?

23. ● **Fuel Efficiency** The fuel efficiency (in miles per gallon) of an SUV depends on its weight according to the formula[6]

$$E = 0.000\,001\,6x^2 - 0.016x + 54 \quad (1800 \le x \le 5400)$$

where x is the weight of an SUV in pounds. According to the model, what is the weight of the least fuel-efficient SUV? Would you trust the model for weights greater than the answer you obtained? Explain.

24. ● **Global Warming** The amount of carbon dioxide (in pounds per 15,000 miles) released by a typical SUV depends on its fuel efficiency according to the formula[7]

$$W = 32x^2 - 2080x + 44,000 \quad (12 \le x \le 33)$$

where x is the fuel efficiency of an SUV in miles per gallon. According to the model, what is the fuel efficiency of the SUV with the least carbon dioxide pollution? Comment on the reliability of the model for fuel efficiencies that exceed your answer.

25. ● **Revenue** The market research department of the Better Baby Buggy Co. predicts that the demand equation for its buggies is given by $q = -0.5p + 140$, where q is the number of buggies it can sell in a month if the price is p per buggy. At what price should it sell the buggies to get the largest revenue? What is the largest monthly revenue?

26. ● **Revenue** The Better Baby Buggy Co. has just come out with a new model, the Turbo. The market research department predicts that the demand equation for Turbos is given by $q = -2p + 320$, where q is the number of buggies it can sell in a month if the price is p per buggy. At what price should it sell the buggies to get the largest revenue? What is the largest monthly revenue? *hint* [see Example 3]

27. ● **Revenue** Pack-Em-In Real Estate is building a new housing development. The more houses it builds, the less people will be willing to pay, due to the crowding and smaller lot sizes. In fact, if it builds 40 houses in this particular development, it can sell them for $200,000 each, but if it builds 60 houses, it will only be able to get $160,000 each. Obtain a linear demand equation and hence determine how many houses Pack-Em-In should build to get the largest revenue. What is the largest possible revenue?

28. ● **Revenue** Pack-Em-In has another development in the works. If it builds 50 houses in this development, it will be

---

[3] Source: The American Physical Society/*New York Times* May 3, 2003, p. A1.

[4] The quadratic model is based on data published in *The New York Times,* November 30, 1997, p. 43.

[5] Ibid.

[6] Fuel efficiency assumes 50% city driving and 50% highway driving. The model is based on a quadratic regression using data from 18 models of SUV. Source for data: Environmental Protection Agency, National Highway Traffic Safety Administration, American Automobile Manufacturers' Association, Ford Motor Company/*The New York Times,* November 30, 1997, p. 43.

[7] Ibid.

● basic skills   **tech** Ex technology exercise

able to sell them at $190,000 each, but if it builds 70 houses, it will get only $170,000 each. Obtain a linear demand equation and hence determine how many houses it should build to get the largest revenue. What is the largest possible revenue?

29. ● *Website Profit* You operate a gaming website, www.mudbeast.net, where users must pay a small fee to log on. When you charged $2 the demand was 280 log-ons per month. When you lowered the price to $1.50, the demand increased to 560 log-ons per month. *hint* [see Example 4]

   a. Construct a linear demand function for your website and hence obtain the monthly revenue R as a function of the log-on fee x.

   b. Your Internet provider charges you a monthly fee of $30 to maintain your site. Express your monthly profit P as a function of the log-on fee x, and hence determine the log-on fee you should charge to obtain the largest possible monthly profit. What is the largest possible monthly profit?

30. ● *T-Shirt Profit* Two fraternities, Sig Ep and Ep Sig, plan to raise money jointly to benefit homeless people on Long Island. They will sell Yoda vs. Alien T-shirts in the student center, but are not sure how much to charge. Sig Ep treasurer Augustus recalls that they once sold 400 shirts in a week at $8 per shirt, but Ep Sig treasurer Julius has solid research indicating that it is possible to sell 600 per week at $4 per shirt.

   a. Based on this information, construct a linear demand equation for Yoda vs. Alien T-shirts, and hence obtain the weekly revenue R as a function of the unit price x.

   b. The university administration charges the fraternities a weekly fee of $500 for use of the Student Center. Write down the monthly profit P as a function of the unit price x, and hence determine how much the fraternities should charge to obtain the largest possible weekly profit. What is the largest possible weekly profit?

31. ● *Website Profit* The latest demand equation for your gaming website, www.mudbeast.net, is given by

$$q = -400x + 1200$$

where q is the number of users who log on per month and x is the log-on fee you charge. Your Internet provider bills you as follows:

   Site maintenance fee: $20 per month
   High-volume access fee: 50¢ per log-on

Find the monthly cost as a function of the log-on fee x. Hence, find the monthly profit as a function of x and determine the log-on fee you should charge to obtain the largest possible monthly profit. What is the largest possible monthly profit?

32. ● *T-Shirt Profit* The latest demand equation for your Yoda vs. Alien T-shirts is given by

$$q = -40x + 600$$

where q is the number of shirts you can sell in one week if you charge $x per shirt. The Student Council charges you $400 per week for use of their facilities, and the T-shirts cost you $5 each. Find the weekly cost as a function of the unit price x. Hence, find the weekly profit as a function of x and determine the unit price you should charge to obtain the largest possible weekly profit. What is the largest possible weekly profit?

33. *Nightclub Management* You have just opened a new nightclub, Russ' Techno Pitstop, but are unsure of how high to set the cover charge (entrance fee). One week you charged $10 per guest and averaged 300 guests per night. The next week you charged $15 per guest and averaged 250 guests per night.

   a. Find a linear demand equation showing the number of guests q per night as a function of the cover charge p.

   b. Find the nightly revenue R as a function of the cover charge p.

   c. The club will provide two free nonalcoholic drinks for each guest, costing the club $3 per head. In addition, the nightly overheads (rent, salaries, dancers, DJ, etc.) amount to $3000. Find the cost C as a function of the cover charge p.

   d. Now find the profit in terms of the cover charge p, and hence determine the cover charge you should charge for a maximum profit.

34. *Television Advertising* As Sales Manager for Montevideo Productions, Inc., you are planning to review the prices you charge clients for television advertisement development. You currently charge each client an hourly development fee of $2500. With this pricing structure, the demand, measured by the number of contracts Montevideo signs per month, is 15 contracts. This is down 5 contracts from the figure last year, when your company charged only $2000.

   a. Construct a linear demand equation giving the number of contracts q as a function of the hourly fee p Montevideo charges for development.

   b. On average, Montevideo bills for 50 hours of production time on each contract. Give a formula for the total revenue obtained by charging $p per hour

   c. The costs to Montevideo Productions are estimated as follows:

   Fixed costs: $120,000 per month
   Variable costs: $80,000 per contract

   Express Montevideo Productions' monthly cost (**i**) as a function of the number q of contracts and (**ii**) as a function of the hourly production charge p.

   d. Express Montevideo Productions' monthly profit as a function of the hourly development fee p and find the price it should charge to maximize the profit.

● basic skills   **tech** Ex technology exercise

35. `tech` Ex **Trade with China** The following table shows the value of U.S. trade with China in 1994, 1999, and 2004 (see Exercise 19; $t = 0$ represents 1994).[8]

Year t	0	5	10
China Trade ($ Billion)	50	95	275

Find a quadratic model for these data, and use your model to estimate the value of U.S. trade with China in 2000. Compare your answer with the actual figure shown in Exercise 19. *hint* [see Example 5]

36. `tech` Ex **Scientific Research** The following table shows the number of research articles in *Physics Review* that were written by researchers in the U.S. in 1983, 1993, and 2003 (See Exercise 20; $t = 0$ represents 1983).[9]

Year t	0	10	20
Science Articles (Hundreds)	34	51	42

Find a quadratic model for these data, and use your model to estimate the number of articles published in 1998. Compare your answer with the actual figure shown in Exercise 20.

37. `tech` Ex **iPod Sales** The following table shows Apple iPod sales from the second quarter in 2003 through the third quarter in 2004 ($t = 0$ represents the second quarter of 2003):[10]

Quarter t	0	1	2	3	4	5
iPod Sales (Thousands)	80	304	336	733	807	860

a. Find a quadratic regression model for these data. (Round coefficients to two decimal places.) Graph the model together with the data.
b. Assuming the trend had continued, estimate iPod sales in the fourth quarter of 2004 ($t = 6$) to the nearest 1000 units.
c. Actual sales of iPods in the fourth quarter of 2004 exceeded 2 million units. What does this fact suggest about using regression curves to predict sales?

38. `tech` Ex **Fiber-Optic Connections** Phone companies have been scrambling to install fiber-optic cable in order to compete with television cable companies. The following table shows the number of fiber-optic cable connections to homes in the U.S. from 2000 to 2004 ($t = 0$ represents 2000):[11]

Year t	0	1	2	3	4
Connections (Thousands)	0	10	25	65	150

a. Find a quadratic regression model for these data. Graph the model, together with the data.
b. Assuming the trend had continued, estimate the number of connections in 2005 to the nearest 1000 homes.
c. Is the quadratic model appropriate for long-term prediction of the number of network connections? Why?

Communication and Reasoning Exercises

39. ● Suppose the graph of revenue as a function of unit price is a parabola that is concave down. What is the significance of the coordinates of the vertex, the x-intercepts, and the y-intercept?

40. ● Suppose the height of a stone thrown vertically upward is given by a quadratic function of time. What is the significance of the coordinates of the vertex, the (possible) x-intercepts, and the y-intercept?

41. ● How might you tell, roughly, whether a set of data should be modeled by a quadratic rather than by a linear equation?

42. ● A member of your study group tells you that, because the following set of data does not suggest a straight line, the data are best modeled by a quadratic.

x	0	2	4	6	8
y	1	2	1	0	1

Comment on her suggestion.

43. Explain why, if demand is a linear function of unit price p (with negative slope) then there must be a *single value of p* that results in the maximum revenue.

44. Explain why, if the average cost of a commodity is given by $y = 0.1x^2 - 4x - 2$, where x is the number of units sold, there is a single choice of x that results in the lowest possible average cost.

45. If the revenue function for a particular commodity is $R(p) = -50p^2 + 60p$, what is the (linear) demand function? Give a reason for your answer.

46. If the revenue function for a particular commodity is $R(p) = -50p^2 + 60p + 50$, can the demand function be linear? What is the associated demand function?

---

[8] 2004 figure is an estimate. SOURCE: U.S. Census Bureau/*New York Times*, September 23, 2004, p. C1.

[9] SOURCE: The American Physical Society/*New York Times* May 3, 2003, p. A1.

[10] SOURCE: Apple financial statements, www.apple.com

[11] SOURCE: Render, Vanderslice & Associates/*New York Times*, October 11, 2004, p. C1.

● basic skills   `tech` Ex technology exercise

9.2 Exponential Functions and Models

The quadratic functions we discussed in Section 9.1 can be used to model many nonlinear situations. However, exponential functions give better models in some applications, including population growth, radioactive decay, the growth or depreciation of financial investments, and many other phenomena.

To work effectively with exponential functions, we need to know the laws of exponents. The following list, similar to the one in the algebra review in Chapter 0, gives the laws of exponents we will be using.

quick Examples

The Laws of Exponents

If b and c are positive and x and y are any real numbers, then the following laws hold:

Law

1. $b^x b^y = b^{x+y}$ $\qquad$ $2^3 2^2 = 2^5 = 32$ $\qquad$ $2^{3-x} = 2^3 2^{-x}$

2. $\dfrac{b^x}{b^y} = b^{x-y}$ $\qquad$ $\dfrac{4^3}{4^2} = 4^{3-2} = 4^1 = 4$ $\qquad$ $3^{x-2} = \dfrac{3^x}{3^2} = \dfrac{3^x}{9}$

3. $\dfrac{1}{b^x} = b^{-x}$ $\qquad$ $9^{-0.5} = \dfrac{1}{9^{0.5}} = \dfrac{1}{3}$ $\qquad$ $2^{-x} = \dfrac{1}{2^x}$

4. $b^0 = 1$ $\qquad$ $(3.3)^0 = 1$ $\qquad$ $x^0 = 1$ if $x \neq 0$

5. $(b^x)^y = b^{xy}$ $\qquad$ $(2^3)^2 = 2^6 = 64$ $\qquad$ $\left(\dfrac{1}{2}\right)^x = (2^{-1})^x = 2^{-x}$

6. $(bc)^x = b^x c^x$ $\qquad$ $(4 \cdot 2)^2 = 4^2 2^2 = 64$ $\qquad$ $10^x = 5^x 2^x$

7. $\left(\dfrac{b}{c}\right)^x = \dfrac{b^x}{c^x}$ $\qquad$ $\left(\dfrac{4}{3}\right)^2 = \dfrac{4^2}{3^2} = \dfrac{16}{9}$ $\qquad$ $\left(\dfrac{1}{2}\right)^x = \dfrac{1^x}{2^x} = \dfrac{1}{2^x}$

Here are the functions we will study in this section.

Exponential Function

An **exponential function** has the form

$$f(x) = Ab^x$$

Technology: A*b^x

where A and b are constants with $A \neq 0$ and b positive. We call b the **base** of the exponential function.

quick Examples

1. $f(x) = 2^x$ $\qquad$ $A = 1, b = 2$; Technology: 2^x

$\quad f(1) = 2^1 = 2$ $\qquad$ 2^1

$\quad f(-3) = 2^{-3} = \dfrac{1}{8}$ $\qquad$ 2^(-3)

$\quad f(0) = 2^0 = 1$ $\qquad$ 2^0

2. $g(x) = 20(3^x)$ $A = 20, b = 3$; Technology: `20*3^x`

$g(2) = 20(3^2) = 20(9) = 180$ `20*3^2`

$g(-1) = 20(3^{-1}) = 20\left(\dfrac{1}{3}\right) = 6\dfrac{2}{3}$ `20*3^(-1)`

3. $h(x) = 2^{-x} = \left(\dfrac{1}{2}\right)^x$ $A = 1, b = \frac{1}{2}$; Technology: `2^(-x)` or `(1/2)^x`

$h(1) = 2^{-1} = \dfrac{1}{2}$ `2^(-1)` or `(1/2)^1`

$h(2) = 2^{-2} = \dfrac{1}{4}$ `2^(-2)` or `(1/2)^2`

4. $k(x) = 3 \cdot 2^{-4x} = 3(2^{-4})^x$ $A = 3, b = 2^{-4}$; Technology: `3*2^(-4*x)`

$k(-2) = 3 \cdot 2^{-4(-2)}$ `3*2^(-4*(-2))`

$\qquad = 3 \cdot 2^8 = 3 \cdot 256 = 768$

Exponential Functions from the Numerical and Graphical Points of View

The following table shows values of $f(x) = 3(2^x)$ for some values of x ($A = 3, b = 2$):

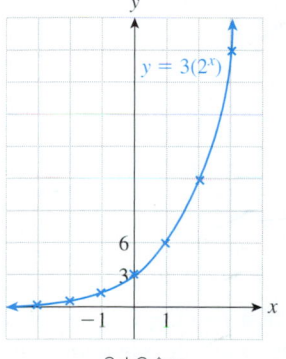

$y = 3(2^x)$

$3*2^x$

Figure 10

x	-3	-2	-1	0	1	2	3
$f(x)$	$\frac{3}{8}$	$\frac{3}{4}$	$\frac{3}{2}$	3	6	12	24

Its graph is shown in Figure 10.

Notice that the y-intercept is $A = 3$ (obtained by setting $x = 0$). In general:

In the graph of $f(x) = Ab^x$, A is the y-intercept, or the value of y when $x = 0$.

What about b? Notice from the table that the value of y is multiplied by $b = 2$ for every increase of 1 in x. If we decrease x by 1, the y-coordinate gets *divided* by $b = 2$.

The value of y is multiplied by b for every one-unit increase of x.

x	-3	-2	-1	0	1	2	3
$f(x)$	$\frac{3}{8}$	$\frac{3}{4}$	$\frac{3}{2}$	3	6	12	24

Multiply by 2

On the graph, if we move one unit to the right from any point on the curve, the y-coordinate doubles. Thus, the curve becomes dramatically steeper as the value of x increases. This phenomenon is called **exponential growth.**

Exponential Function Numerically and Graphically

For the exponential function $f(x) = Ab^x$:

Role of A

$f(0) = A$, so A is the y-intercept of the graph of f.

Role of b

If x increases by 1, $f(x)$ is multiplied by b.
If x increases by 2, $f(x)$ is multiplied by b^2.
$\vdots$

If x increases by Δx, $f(x)$ is multiplied by $b^{\Delta x}$.

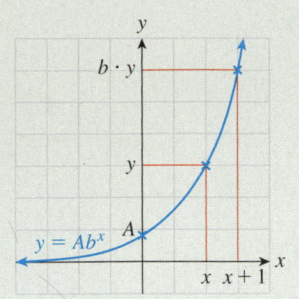

If x increases by 1, y is multiplied by b.

quick Examples

1. $f_1(x) = 2^x$, $f_2(x) = \left(\dfrac{1}{2}\right)^x = 2^{-x}$

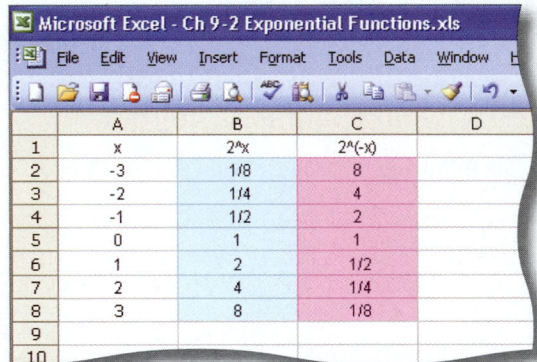

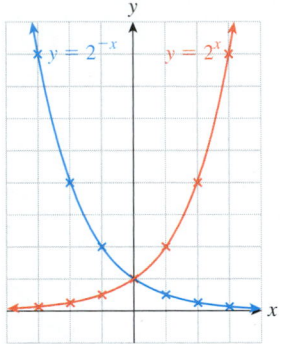

When x increases by 1, $f_2(x)$ is multiplied by $\frac{1}{2}$. The function $f_1(x) = 2^x$ illustrates exponential growth, while $f_2(x) = \left(\frac{1}{2}\right)^x$ illustrates the opposite phenomenon: **exponential decay.**

2. $f_1(x) = 2^x$, $f_2(x) = 3^x$, $f_3(x) = 1^x$

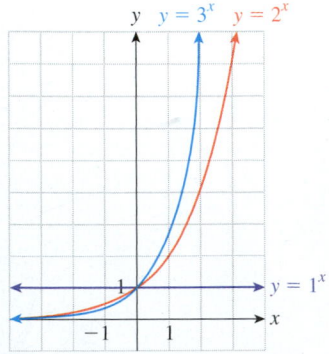

If x increases by 1, 3^x is multiplied by 3. Note also that all the graphs pass through $(0, 1)$. (Why?)

Example 1 Recognizing Exponential Data Numerically and Graphically

Some of the values of two functions, f and g, are given in the following table:

x	-2	-1	0	1	2
$f(x)$	-7	-3	1	5	9
$g(x)$	$\frac{2}{9}$	$\frac{2}{3}$	2	6	18

One of these functions is linear, and the other is exponential. Which is which?

Solution

Remember that a linear function increases (or decreases) by the same amount every time x increases by 1. The values of f behave this way: Every time x increases by 1, the value of $f(x)$ increases by 4. Therefore, f is a linear function with a *slope* of 4. Since $f(0) = 1$, we see that

$$f(x) = 4x + 1$$

is a linear formula that fits the data.

On the other hand, every time x increases by 1, the value of $g(x)$ is *multiplied* by 3. Since $g(0) = 2$, we find that

$$g(x) = 2(3^x)$$

is an exponential function fitting the data.

We can visualize the two functions f and g by plotting the data points (Figure 11).

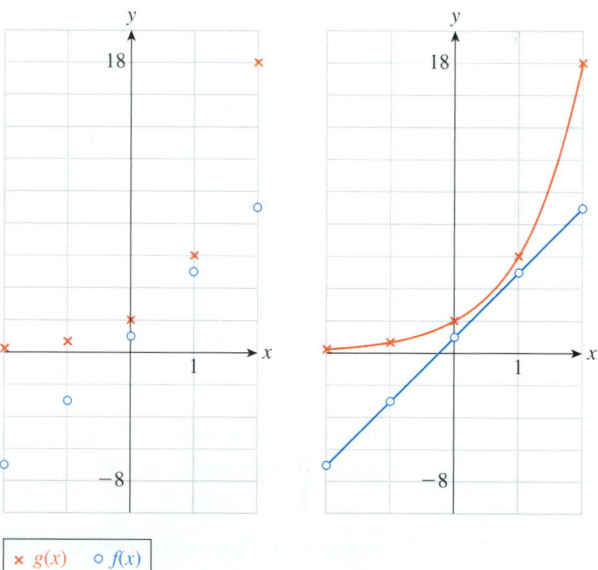

| ✕ $g(x)$ ○ $f(x)$ |

Figure **11**

The data points for $f(x)$ clearly lie along a straight line, whereas the points for $g(x)$ lie along a curve, as shown in the graph on the right. The y-coordinate of each point for $g(x)$ is three times the y-coordinate of the preceding point, demonstrating that the curve is an exponential one.

In Section 1.3, we discussed a method for calculating the equation of the line that passes through two given points. In the following example, we show a method for calculating the equation of the exponential curve through two given points.

Example 2 Finding the Exponential Curve through Two Points

Find an equation of the exponential curve through $(1, 6)$ and $(3, 24)$.

Solution

We want an equation of the form

$$y = Ab^x \quad (b > 0)$$

Substituting the coordinates of the given points, we get

$$6 = Ab^1 \qquad \text{Substitute } (1, 6).$$
$$24 = Ab^3 \qquad \text{Substitute } (3, 24).$$

If we now divide the second equation by the first, we get

$$\frac{24}{6} = \frac{Ab^3}{Ab} = b^2$$
$$b^2 = 4$$
$$b = 2 \qquad b \text{ is positive in an exponential function}$$

Now that we have b, we can substitute its value into the first equation to obtain

$$6 = 2A \qquad \text{Substitute } b = 2 \text{ into the equation } 6 = Ab.$$
$$A = 3$$

We have both constants, $A = 3$ and $b = 2$, so the model is

$$y = 3(2^x)$$

Example 7 will show how to use technology to fit an exponential function to two or more data points.

Applications

Recall some terminology we mentioned earlier: A quantity y experiences **exponential growth** if $y = Ab^t$ with $b > 1$. (Here we use t for the independent variable, thinking of time.) It experiences **exponential decay** if $y = Ab^t$ with $0 < b < 1$. Here we discuss examples of the numerous applications of exponential growth and decay. Our first application is to public health.

Example 3 Exponential Growth: Epidemics

In the early stages of the AIDS epidemic during the 1980s, the number of cases in the U.S. was increasing by about 50% every six months. By the start of 1983, there were approximately 1600 AIDS cases in the U.S.[*]

---

[*]Data based on regression of the 1982–1986 figures. Source for data: Centers for Disease Control and Prevention. HIV/AIDS Surveillance Report, 2000;12 (No. 2).

a. Assuming an exponential growth model, find a function that predicts the number of people infected t years after the start of 1983.

b. Use the model to estimate the number of people infected by October 1, 1986, and also by the end of that year.

Solution

a. One way of finding the desired exponential function is to reason as follows: At time $t = 0$ (January 1, 1983), the number of people infected was 1600, so $A = 1600$. Every six months, the number of cases increased to 150% of the number six months earlier—that is, to 1.50 times that number. Each year, it therefore increased to $(1.50)^2 = 2.25$ times the number one year earlier. Hence, after t years, we need to multiply the original 1600 by 2.25^t, so the model is

$$y = 1600(2.25^t) \text{ cases}$$

Alternatively, if we wish to use the method of Example 2, we need two data points. We are given one point: $(0, 1600)$. Since y increased by 50% every six months, six months later it reached $1600 + 800 = 2400$ ($t = 0.5$). This information gives a second point: $(0.5, 2400)$. We can now apply the method in Example 2 to find the model above.

b. October 1, 1986, corresponds to $t = 3.75$ (because October 1 is 9 months, or $9/12 = 0.75$ of a year after January 1). Substituting this value of t in the model gives

$$y = 1600(2.25^{3.75}) \approx 33{,}481 \text{ cases} \qquad \texttt{1600*2.25\^3.75}$$

By the end of 1986, the model predicts that

$$y = 1600(2.25^4) = 41{,}006 \text{ cases}$$

(The actual number of cases was around 41,700.)

+*Before we go on...* Increasing the number of cases by 50% every six months couldn't continue for very long and this is borne out by observations. If increasing by 50% every six months did continue, then by January 2003 ($t = 20$), the number of infected people would have been

$$1600(2.25^{20}) \approx 17{,}700{,}000{,}000$$

a number that is more than 50 times the size of the U.S. population! Thus, although the exponential model is fairly reliable in the early stages of the epidemic, it is unreliable for predicting long-term trends. ■

Epidemiologists use more sophisticated models to measure the spread of epidemics, and these models predict a leveling-off phenomenon as the number of cases becomes a significant part of the total population. We discuss such a model, the **logistic function,** in Section 9.4.

Exponential functions arise in finance and economics mainly through the idea of **compound interest.** Suppose you invest $500 (the **present value**) in an investment account with an annual yield of 15%, and the interest is reinvested at the end of every year (we say that the interest is **compounded** once a year). Let t represent the number of years since you made the initial $500 investment. Each year, the investment is worth 115% (or 1.15 times) its value the previous year. The **future value** A of your investment changes over time t, so we think of A as a function of t. The following table illustrates how we can calculate the future value for several values of t:

t	0	1	2	3
Future Value $A(t)$	500	575	661.25	760.44

A $500(1.15)$ $500(1.15)^2$ $500(1.15)^3$

$\times 1.15$ $\times 1.15$ $\times 1.15$

Thus, $A(t) = 500(1.15)^t$. A traditional way to write this formula is

$$A(t) = P(1+r)^t$$

where P is the present value ($P = 500$) and r is the annual interest rate ($r = 0.15$).

If, instead of compounding the interest once a year, we compound it every three months (four times a year), we would earn one quarter of the interest ($r/4$ of the current investment) every three months. Because this would happen $4t$ times in t years, the formula for the future value becomes

$$A(t) = P\left(1+\frac{r}{4}\right)^{4t}$$

Compound Interest

If an amount (**present value**) P is invested for t years at an annual rate of r, and if the interest is compounded (reinvested) m times per year, then the **future value** A is

$$A(t) = P\left(1+\frac{r}{m}\right)^{mt}$$

A special case is **interest compounded once a year:**

$$A(t) = P(1+r)^t$$

quick Example

If $2000 is invested for two and a half years in a mutual fund with an annual yield of 12.6% and the earnings are reinvested each month, then $P = 2000, r = 0.126, m = 12$, and $t = 2.5$, which gives

$$A(2.5) = 2000\left(1+\frac{0.126}{12}\right)^{12\times2.5}$$

    `2000*(1+0.126/12)^(12*2.5)`

$$= 2000(1.0105)^{30} = \$2736.02$$

Example 4 Compound Interest: Investments

Consider the scenario in the preceding Quick Example: you invest $2000 in a mutual fund with an annual yield of 12.6% and the interest is reinvested each month.

a. Find the associated exponential model.

b. Use the model to estimate the year when the value of your investment will reach $5000.

Solution

a. Apply the formula

$$A(t) = P\left(1+\frac{r}{m}\right)^{mt}$$

using *Technology*

See the Technology Guides at the end of the chapter for detailed instructions on using a TI-83/84 or Excel to create the tables shown here. Alternatively, go online and follow:

Chapter 9
→ Tools
→ Function Evaluator
& Grapher

to use the function evaluator to create a table.

with $P = 2000$, $r = 0.126$, and $m = 12$. We get

$$A(t) = 2000\left(1 + \frac{0.126}{12}\right)^{12t} \quad \text{\textcolor{red}{2000*(1+0.126/12)\^(12*t)}}$$

$$A(t) = 2000(1.0105)^{12t}$$

This is the exponential model. (What would happen if we left out the last set of parentheses in the technology formula?)

b. We need to find the value of t for which $A(t) = \$5000$, so we need to solve the equation

$$5000 = 2000(1.0105)^{12t}$$

In Section 9.3 we will learn how to use logarithms to do this algebraically, but we can answer the question now using a graphing calculator, a spreadsheet, or the Function Evaluator and Grapher tool at the website. Just enter the model and compute the balance at the end of several years. Here are examples of tables obtained using various forms of technology:

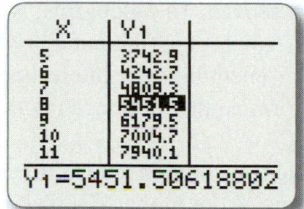

TI-83/84

Microsoft Excel - Ch 9-2 Compound Interest of Investments.xls

	A	B	C	D	E
1	t	A			
2	0	$2,000.00			
3	1	$2,267.07			
4	2	$2,569.81			
5	3	$2,912.98			
6	4	$3,301.97			
7	5	$3,742.91			
8	6	$4,242.72			
9	7	$4,809.29			
10	8	$5,451.51			
11	9	$6,179.49			
12					
13					

Excel

Values of x:	Values of f(x)
1	2267.074593
2	2569.813606
3	2912.979567
4	3301.970984
5	3742.907262
6	4242.724979
7	4809.287003
8	5451.506188
9	6179.485587
10	7004.677386

Website

Because the balance first exceeds $5000 in year 8, the answer is $t = 8$ years.

Carbon-14, an unstable isotope of carbon, decays exponentially to nitrogen. Because carbon-14 decay is extremely slow, it has important applications in the dating of fossils.

Example 5 Exponential Decay: Carbon Dating

The amount of carbon-14 remaining in a sample that originally contained A grams is approximately

$$C(t) = A(0.999879)^t$$

where t is time in years. A fossilized plant unearthed in an archaeological dig contains 0.50 gram of carbon-14 and is known to be 50,000 years old. How much carbon-14 did the plant originally contain?

Solution We are given the following information: $C = 0.50$, $A =$ the unknown, and $t = 50,000$. Substituting gives

$$0.50 = A(0.999879)^{50,000}$$

Solving for A gives

$$A = \frac{0.5}{0.999879^{50,000}} \approx 212 \text{ grams}$$

Thus, the plant originally contained 212 grams of carbon-14.

+*Before we go on...* The formula we used for A in Example 5 has the form

$$A(t) = \frac{C}{0.999879^t}$$

which gives the original amount of carbon-14 t years ago in terms of the amount C that is left now. A similar formula can be used in finance to find the present value, given the future value. ∎

The Number *e* and More Applications

In nature we find examples of growth that occurs *continuously*, as though "interest" is being added more often than every second or fraction of a second. To model this, we need to see what happens to our compound interest formula as we let m (the number of times interest is added per year) become extremely large. Something very interesting does happen: we end up with a more compact and elegant formula than we began with. To see why, let's look at a very simple situation.

Suppose we invest \$1 in the bank for 1 year at 100% interest, compounded m times per year. If $m = 1$, then 100% interest is added every year, and so our money doubles at the end of the year. In general, the accumulated capital at the end of the year is

$$A = 1\left(1 + \frac{1}{m}\right)^m = \left(1 + \frac{1}{m}\right)^m \qquad \text{(1+1/m)^m}$$

Now, we are interested in what A becomes for large values of m. Here is an Excel sheet showing the quantity $\left(1 + \frac{1}{m}\right)^m$ for larger and larger values of m:

	A	B	C
1	m	(1+1/m)^m	
2	1	2	
3	10	2.59374246	
4	100	2.704813829	
5	1000	2.716923932	
6	10000	2.718145927	
7	100000	2.718268237	
8	1000000	2.718280469	
9	10000000	2.718281694	
10	100000000	2.718281786	
11	1000000000	2.718282031	
12			
13			

Something interesting *does* seem to be happening! The numbers appear to be getting closer and closer to a specific value. In mathematical terminology, we say that the numbers **converge** to a fixed number, 2.71828 . . . , called the **limiting value**[12] of the quantities $\left(1 + \frac{1}{m}\right)^m$. This number, called e, is one of the most important in mathematics. The number e is irrational, just as the more familiar number π is, so we cannot write down its exact numerical value. To 20 decimal places,

$$e = 2.71828182845904523536\ldots.$$

We now say that, if \$1 is invested for 1 year at 100% interest **compounded continuously,** the accumulated money at the end of that year will amount to $\$e = \$2.72$ (to the nearest cent). But what about the following more general question?

Q: *What about a more general scenario: If we invest an amount $\$P$ for t years at an interest rate of r, compounded continuously, what will be the accumulated amount A at the end of that period?*

A: In the special case above, $(P, t,$ and r all equal to 1) we took the compound interest formula and let m get larger and larger. We do the same more generally, after a little preliminary work with the algebra of exponentials.

$$A = P\left(1 + \frac{r}{m}\right)^{mt}$$

$$= P\left(1 + \frac{1}{(m/r)}\right)^{mt} \qquad \text{Substituting } \frac{r}{m} = \frac{1}{(m/r)}.$$

$$= P\left(1 + \frac{1}{(m/r)}\right)^{(m/r)rt} \qquad \text{Substituting } m = \left(\frac{m}{r}\right)r.$$

$$= P\left[\left(1 + \frac{1}{(m/r)}\right)^{(m/r)}\right]^{rt} \qquad \text{Using the rule } a^{bc} = (a^b)^c$$

For continuous compounding of interest, we let m, and hence m/r, get very large. This affects only the term in brackets, which converges to e, and we get the formula

$$A = Pe^{rt} \quad \blacksquare$$

Q: *How do I obtain powers of e or e itself on a TI-83/84 or in Excel?*

A: On the TI-83/84, enter e^x as `e^(x)`, where `e^(` can be obtained by pressing `2nd` `LN`. Excel has a built-in function called `EXP`; `EXP(x)` gives the value of e^x. To obtain the number e on the TI-83/84, enter `e^(1)`. In Excel, enter `=EXP(1)`. ∎

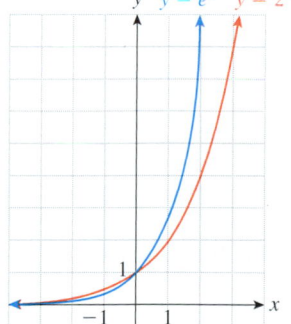

y $y = e^x$ $y = 2^x$

Technology formula: `e^(x)` or `EXP(x)`

Figure **12**

Figure 12 shows the graph of $y = e^x$ with that of $y = 2^x$ for comparison.

The Number e and Continuous Compounding

The number e is the limiting value of the quantities $\left(1 + \frac{1}{m}\right)^m$ as m gets larger and larger, and has the value $2.71828182845904523536\ldots$

   If $\$P$ is invested at an annual interest rate r compounded continuously, the accumulated amount after t years is

$$A(t) = Pe^{rt} \qquad \text{P*e^(r*t) or P*EXP(r*t)}$$

---

[12] See Chapter 3 for more on limits.

quick Examples

1. If $100 is invested in an account that bears 15% interest compounded continuously, at the end of 10 years the investment will be worth

$$A(10) = 100e^{(0.15)(10)} = \$448.17 \quad \texttt{100*e\^(0.15*10) or 100*EXP(0.15*10)}$$

2. If $1 is invested in an account that bears 100% interest compounded continuously, at the end of x years, the investment will be worth

$$A(x) = e^x \text{ dollars}$$

Example 6 Continuous Compounding

a. You invest $10,000 at Fastrack Savings & Loan, which pays 6% compounded continuously. Express the balance in your account as a function of the number of years t and calculate the amount of money you will have after five years.

b. Your friend has just invested $10,000 in Constant Growth Funds, whose stocks are continuously declining at a rate of 6% per year. How much will her investment be worth in five years?

Solution

a. We use the continuous growth formula with $P = 10,000$, $r = 0.06$, and t variable, getting

$$A(t) = Pe^{rt} = 10,000e^{0.06t}$$

In five years,

$$A(5) = 10,000e^{0.06(5)}$$
$$= 10,000e^{0.3}$$
$$\approx \$13,498.59$$

b. Because the investment is depreciating, we use a negative value for r and take $P = 10,000$, $r = -0.06$, and $t = 5$, getting

$$A(t) = Pe^{rt} = 10,000e^{-0.06t}$$
$$A(5) = 10,000e^{-0.06(5)}$$
$$= 10,000e^{-0.3}$$
$$\approx \$7408.18$$

+*Before we go on...*

Q: How does continuous compounding compare with monthly compounding?

A: To repeat the calculation in part (a) of Example 6 using monthly compounding instead of continuous compounding, we use the compound interest formula with $P = 10,000$, $r = 0.06$, $m = 12$, and $t = 5$ and find

$$A(5) = 10,000(1 + 0.06/12)^{60} \approx \$13,488.50$$

Thus, continuous compounding earns you approximately $10 more than monthly compounding on a five-year, $10,000 investment. This is little to get excited about. ∎

If we write the continuous compounding formula $A(t) = Pe^{rt}$ as $A(t) = P(e^r)^t$, we see that $A(t)$ is an exponential function of t, where the base is $b = e^r$, so we have really not introduced a new kind of function. In fact, exponential functions are often written in this way:

Exponential Functions: Alternative Form

We can write any exponential function in the following alternative form:

$$f(x) = Ae^{rx}$$

where A and r are constants. If r is positive, f models exponential growth; if r is negative, f models exponential decay.

quick Examples

1. $f(x) = 100e^{0.15x}$ Exponential growth $A = 100, r = 0.15$

2. $f(t) = Ae^{-0.000\,121\,01t}$ Exponential decay of carbon-14; $r = -0.000\,121\,01$

3. $f(t) = 100e^{0.15t} = 100\left(e^{0.15}\right)^t$
   $= 100(1.1618)^t$ Converting Ae^{rt} to the form Ab^t

We will see in Chapter 4 that the exponential function with base e exhibits some interesting properties when we measure its rate of change, and this is the real mathematical importance of e.

Exponential Regression

Starting with a set of data that suggests an exponential curve, we can use technology to compute the exponential regression curve in much the same way as we did for the quadratic regression curve in Example 5 of Section 9.1.

Example 7 Exponential Regression: Health Expenditures

The following table shows annual expenditure on health in the U.S. from 1980 through 2010 ($t = 0$ represents 1980).[*]

Year t	0	5	10	15	20	25	30
Expenditure ($ Billion)	246	427	696	990	1310	1920	2750

a. Find the exponential regression model

$$C(t) = Ab^t$$

for the annual expenditure.

b. Use the regression model to estimate the expenditure in 2002 ($t = 22$; the actual expenditure was approximately $1550 billion).

Solution

a. We use technology to obtain the exponential regression curve (See Figure 13):

$$C(t) \approx 282.33(1.0808)^t \quad \text{Coefficients rounded}$$

 using *Technology*

See the Technology Guides at the end of the chapter for detailed instructions on using a TI-83/84 or Excel to find exponential regression curves. Alternatively, go online and follow:

Chapter 9

→ Tools

→ Simple Regression

to find a utility for finding regression curves of various sorts.

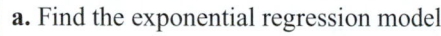

[*]Data are rounded. 2005 and 2010 figures are projections. SOURCE: Centers for Medicare and Medicaid Services, "National Health Expenditures," 2002 version, released January 2004; www.cms.hhs.gov/statistics/nhe/

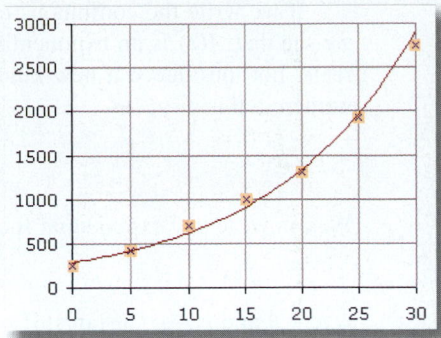

Figure **13**

b. Using the model $C(t) \approx 282.33(1.0808)^t$ we find that

$$C(22) \approx 282.33(1.0808)^{22} \approx \$1560 \, \text{billion}$$

which is close to the actual number of around $1550 billion.

+*Before we go on...* We said in the preceding section that the regression curve gives the smallest value of the sum-of-squares error, SSE (the sum of the squares of the residuals). However, exponential regression as computed via technology generally minimizes the sum of the squares of the residuals of the *logarithms* (logarithms are discussed in the next section). Using logarithms allows one easily to convert an exponential function into a linear one and then use linear regression formulas. However, in Section 9.4, we will discuss a way of using Excel's Solver to minimize SSE directly, which allows us to find the best-fit exponential curve directly without the need for devices to simplify the mathematics. If we do this, we obtain a very different equation:

$$C(t) \approx 316.79(1.0747^t)$$

If you plot this function, you will notice that it seems to fit the data more closely than the regression curve. ■

FAQs   **When to Use an Exponential Model for Data Points, and When to Use *e* in Your Model**

Q: Given a set of data points that appear to be curving upwards, how can I tell whether to use a quadratic model or an exponential model?

A: Here are some things to look for:

• Do the data values appear to double at regular intervals? (For example, do the values approximately double every five units?) If so, then an exponential model is appropriate. If it takes longer and longer to double, then a quadratic model may be more appropriate.
• Do the values first decrease to a low point and then increase? If so, then a quadratic model is more appropriate.

It is also helpful to use technology to graph both the regression quadratic and exponential curves and to visually inspect the graphs to determine which gives the closest fit to the data. ■

$Q:$ *We have two ways of writing exponential functions: $f(x) = Ab^x$ and $f(x) = Ae^{rx}$. How do we know which one to use?*

$A:$ The two forms are equivalent, and it is always possible to convert from one form to the other.* So, use whichever form seems to be convenient for a particular situation. For instance, $f(t) = A(3^t)$ conveniently models exponential growth that is tripling every unit of time, whereas $f(t) = Ae^{0.06t}$ conveniently models an investment with continuous compounding at 6%. ∎

---
*Quick Example 3 on p. 639 shows how to convert Ae^{rx} to Ab^x. Conversion from Ab^x to Ae^{rx} involves logarithms: $r = \ln b$.

9.2 EXERCISES

● denotes basic skills exercises

tech Ex indicates exercises that should be solved using technology

For each function in Exercises 1–12, compute the missing values in the following table and supply a valid technology formula for the given function: hint [see Quick Examples on p. 628]

x	-3	-2	-1	0	1	2	3
$f(x)$							

1. ● $f(x) = 4^x$ 2. ● $f(x) = 3^x$
3. ● $f(x) = 3^{-x}$ 4. ● $f(x) = 4^{-x}$
5. ● $g(x) = 2(2^x)$ 6. ● $g(x) = 2(3^x)$
7. ● $h(x) = -3(2^{-x})$ 8. ● $h(x) = -2(3^{-x})$
9. ● $r(x) = 2^x - 1$ 10. ● $r(x) = 2^{-x} + 1$
11. ● $s(x) = 2^{x-1}$ 12. ● $s(x) = 2^{1-x}$

Using a chart of values, graph each of the functions in Exercises 13–18. (Use $-3 \le x \le 3$.)

13. ● $f(x) = 3^{-x}$ 14. ● $f(x) = 4^{-x}$
15. ● $g(x) = 2(2^x)$ 16. ● $g(x) = 2(3^x)$
17. ● $h(x) = -3(2^{-x})$ 18. ● $h(x) = -2(3^{-x})$

In Exercises 19–24, the values of two functions, f and g, are given in a table. One, both, or neither of them may be exponential. Decide which, if any, are exponential, and give the exponential models for those that are. hint [see Example 1]

19. ●

x	-2	-1	0	1	2
$f(x)$	0.5	1.5	4.5	13.5	40.5
$g(x)$	8	4	2	1	$\frac{1}{2}$

20. ●

x	-2	-1	0	1	2
$f(x)$	$\frac{1}{2}$	1	2	4	8
$g(x)$	3	0	-1	0	3

21. ●

x	-2	-1	0	1	2
$f(x)$	22.5	7.5	2.5	7.5	22.5
$g(x)$	0.3	0.9	2.7	8.1	16.2

22. ●

x	-2	-1	0	1	2
$f(x)$	0.3	0.9	2.7	8.1	24.3
$g(x)$	3	1.5	0.75	0.375	0.1875

23. ●

x	-2	-1	0	1	2
$f(x)$	100	200	400	600	800
$g(x)$	100	20	4	0.8	0.16

24. ●

x	-2	-1	0	1	2
$f(x)$	0.8	0.2	0.1	0.05	0.025
$g(x)$	80	40	20	10	2

tech Ex *For each function in Exercises 25–30, supply a valid technology formula and then use technology to compute the missing values in the following table:*

x	-3	-2	-1	0	1	2	3
$f(x)$							

25. ● $f(x) = e^{-2x}$ 26. ● $g(x) = e^{x/5}$
27. ● $h(x) = 1.01(2.02^{-4x})$ 28. ● $h(x) = 3.42(3^{-x/5})$
29. ● $r(x) = 50\left(1 + \dfrac{1}{3.2}\right)^{2x}$ 30. ● $r(x) = 0.043\left(4.5 - \dfrac{5}{1.2}\right)^{-x}$

In Exercises 31–38, supply a valid technology formula for the given function.

31. ● 2^{x-1} 32. ● 2^{-4x} 33. ● $\dfrac{2}{1 - 2^{-4x}}$ 34. ● $\dfrac{2^{3-x}}{1 - 2^x}$

35. ● $\dfrac{(3+x)^{3x}}{x+1}$ 36. ● $\dfrac{20.3^{3x}}{1 + 20.3^{2x}}$ 37. ● $2e^{(1+x)/x}$ 38. ● $\dfrac{2e^{2/x}}{x}$

● basic skills   tech Ex technology exercise

tech Ex *On the same set of axes, use technology to graph the pairs of functions in Exercises 39–46 with $-3 \le x \le 3$. Identify which graph corresponds to which function.*

39. ● $f_1(x) = 1.6^x$, $f_2(x) = 1.8^x$

40. ● $f_1(x) = 2.2^x$, $f_2(x) = 2.5^x$

41. ● $f_1(x) = 300(1.1^x)$, $f_2(x) = 300(1.1^{2x})$

42. ● $f_1(x) = 100(1.01^{2x})$, $f_2(x) = 100(1.01^{3x})$

43. ● $f_1(x) = 2.5^{1.02x}$, $f_2(x) = e^{1.02x}$

44. ● $f_1(x) = 2.5^{-1.02x}$, $f_2(x) = e^{-1.02x}$

45. ● $f_1(x) = 1000(1.045^{-3x})$, $f_2(x) = 1000(1.045^{3x})$

46. ● $f_1(x) = 1202(1.034^{-3x})$, $f_2(x) = 1202(1.034^{3x})$

For Exercises 47–54, model the data using an exponential function $f(x) = Ab^x$. hint [see Example 1]

47. ●

x	0	1	2
$f(x)$	500	250	125

48. ●

x	0	1	2
$f(x)$	500	1000	2000

49. ●

x	0	1	2
$f(x)$	10	30	90

50. ●

x	0	1	2
$f(x)$	90	30	10

51. ●

x	0	1	2
$f(x)$	500	225	101.25

52. ●

x	0	1	2
$f(x)$	5	3	1.8

53. ●

x	1	2
$f(x)$	-110	-121

54. ●

x	1	2
$f(x)$	-41	-42.025

Find equations for exponential functions that pass through the pairs of points given in Exercises 55–62. (Round all coefficients to four decimal places when necessary.) hint [see Example 2]

55. ● Through (2, 36) and (4, 324)

56. ● Through (2, –4) and (4, –16)

57. ● Through (–2, –25) and (1, –0.2)

58. ● Through (1, 1.2) and (3, 0.108)

59. ● Through (1, 3) and (3, 6)

60. ● Through (1, 2) and (4, 6)

61. ● Through (2, 3) and (6, 2)

62. ● Through (−1, 2) and (3, 1)

Obtain exponential functions in the form $f(t) = Ae^{rt}$ in Exercises 63–66. hint [see Example 6]

63. ● $f(t)$ is the value after t years of a $5000 investment earning 10% interest compounded continuously.

64. ● $f(t)$ is the value after t years of a $2000 investment earning 5.3% interest compounded continuously.

65. ● $f(t)$ is the value after t years of a $1000 investment depreciating continuously at an annual rate of 6.3%.

66. ● $f(t)$ is the value after t years of a $10,000 investment depreciating continuously at an annual rate of 60%.

tech Ex *In Exercises 67–70, use technology to find the exponential regression function through the given points. (Round all coefficients to 4 decimal places.) hint [see Example 7]*

67. ● {(1, 2), (3, 5), (4, 9), (5, 20)}

68. ● {(−1, 2), (−3, 5), (−4, 9), (−5, 20)}

69. ● {(−1, 10), (−3, 5), (−4, 3)}

70. ● {(3, 3), (4, 5), (5, 10)}

Applications

71. ● *Bacteria* A bacteria culture starts with 1000 bacteria and doubles in size every three hours. Find an exponential model for the size of the culture as a function of time t in hours and use the model to predict how many bacteria there will be after two days. *hint [see Example 3]*

72. ● *Bacteria* A bacteria culture starts with 1000 bacteria. Two hours later there are 1500 bacteria. Find an exponential model for the size of the culture as a function of time t in hours, and use the model to predict how many bacteria there will be after two days.

73. ● *Investments* In 2004, the Scottish Widows Bank in the United Kingdom offered 4.39% interest on its savings accounts, with interest reinvested annually.[13] Find the associated exponential model for the value of a £5000 deposit after t years. Assuming this rate of return continued for five years, how much would a deposit of £5000 at the beginning of 2004 be worth at the start of 2009? (Answer to the nearest £1.) *hint [see Example 4]*

74. ● *Investments* In 2004, the Northern Rock Bank in the United Kingdom offered 4.76% interest on its online accounts, with interest reinvested annually.[14] Find the associated exponential model for the value of a £4000 deposit after t years. Assuming this rate of return continued for four years, how much would a deposit of £4000 made at the beginning of 2004 be worth at the start of 2008? (Answer to the nearest £1.)

75. ● **tech** Ex *Investments* Refer to Exercise 73. When will an investment of £5000 made at the beginning of 2004 first exceed £7500?

76. ● **tech** Ex *Investments* Refer to Exercise 74. In which year will an investment of £4000 made at the beginning of 2004 first exceed £6000?

77. ● *Carbon Dating* A fossil originally contained 104 grams of carbon-14. Refer to the formula for $C(t)$ in Example 5 and estimate the amount of carbon-14 left in the sample after 10,000 years, 20,000 years, and 30,000 years. *hint [see Example 5]*

---

[13] SOURCE: www.rate.co.uk, October, 2004.

[14] Ibid.

● basic skills   **tech** Ex technology exercise

78. ● *Carbon Dating* A fossil presently contains 4.06 grams of carbon-14. Refer to the formula for $A(t)$ in Example 5 and estimate the amount of carbon-14 in the sample 10,000 years, 20,000 years, and 30,000 years ago.

79. ● *Carbon Dating* A fossil presently contains 4.06 grams of carbon-14. It is estimated that the fossil originally contained 46 grams of carbon-14. By calculating the amount left after 5000 years, 10,000 years, . . . , 35,000 years, estimate the age of the sample to the nearest 5000 years. (Refer to the formula for $C(t)$ in Example 5.)

80. ● *Carbon Dating* A fossil presently contains 2.8 grams of carbon-14. It is estimated that the fossil originally contained 104 grams of carbon-14. By calculating the amount 5000 years, 10,000 years, . . . , 35,000 years ago, estimate the age of the sample to the nearest 5000 years. (Refer to the formula for $A(t)$ in Example 5.)

81. ● *Aspirin* Soon after taking an aspirin, a patient has absorbed 300 mg of the drug. If the amount of aspirin in the bloodstream decays exponentially, with half being removed every two hours, find the amount of aspirin in the bloodstream after five hours.

82. ● *Alcohol* After several drinks, a person has a blood alcohol level of 200 mg/dL (milligrams per deciliter). If the amount of alcohol in the blood decays exponentially, with one fourth being removed every hour, find the person's blood alcohol level after four hours.

83. ● *Profit* South African Breweries (SAB) reported profits of $360 million in 1997 ($t = 0$) and $480 million in 2000 ($t = 3$).[15] Use this information to find **a.** a linear model and **b.** an exponential model for SAB's profit P as a function of time t since 1997. (Round all coefficients to four decimal places.) Which, if either, of these models would you judge to be applicable to the data shown below?

t	0 (1997)	1	2	3	4 (2001)
Profit ($ million)	360	380	320	480	360

84. ● *Assets* South African Breweries (SAB) reported fixed assets of R9500 million in 1997 ($t = 0$) and R29,300 million in 2001 ($t = 4$).[16] Use this information to find **a.** a linear model and **b.** an exponential model for SAB's fixed assets as a function of time t since 1997. (Round all coefficients to four significant digits.) Which, if either, of these models would you judge to be applicable to the data shown below?

t	0 (1997)	1	2	3	4 (2001)
Fixed Assets (R million)	9500	11,100	16,100	22,900	29,300

85. *U.S. Population.* The U.S. population was 180 million in 1960 and 294 million in 2004.[17]

   a. Use these data to give an exponential growth model showing the U.S. population P as a function of time t in years since 1960. Round coefficients to five decimal places.

   b. By experimenting, determine the smallest number of decimal places to which you should round the coefficients in part (a) in order to obtain the correct 2004 population figure accurate to three significant digits.

   c. Using the model in part (a), predict the population in 2020.

86. *World Population.* World population was estimated at 2.56 billion people in 1950 and 6.40 billion people in 2004.[18]

   a. Use these data to give an exponential growth model showing the world population P as a function of time t in years since 1950. Round coefficients to five decimal places.

   b. By experimenting, determine the smallest number of decimal places to which you should round the coefficients in part (a) in order to obtain the correct 2004 population figure to three significant digits.

   c. Assuming the exponential growth model from part (a), estimate the world population in 1000 AD. Comment on your answer.

87. *Frogs* Frogs in Nassau County have been breeding like flies! Each year, the pledge class of Epsilon Delta is instructed by the brothers to tag all the frogs residing on the ESU campus (Nassau County Branch) as an educational exercise. Two years ago they managed to tag all 50,000 of them (with little Epsilon Delta Fraternity tags). This year's pledge class discovered that last year's tags had all fallen off, and they wound up tagging a total of 75,000 frogs.

   a. Find an exponential model for the frog population.

   b. Assuming exponential population growth, and that all this year's tags have fallen off, how many tags should Epsilon Delta order for next year's pledge class?

88. *Flies* Flies in Suffolk County have been breeding like frogs! Three years ago the Health Commission caught 4000 flies in a trap in one hour. This year it caught 7000 flies in one hour.

   a. Find an exponential model for the fly population.

   b. Assuming exponential population growth, how many flies should the commission expect to catch next year?

89. ● *Investments* Rock Solid Bank & Trust is offering a CD (certificate of deposit) that pays 4% compounded continuously. How much interest would a $1000 deposit earn over 10 years? *hint* [see Example 6]

90. ● *Savings* FlybynightSavings.com is offering a savings account that pays 31% interest compounded continuously. How much interest would a deposit of $2000 earn over 10 years?

[15] SOURCE: South African Breweries corporate website www.sab.co.za/investfr.asp, April, 2002.

[16] Ibid.

[17] Figures are rounded to three significant digits. SOURCE: *Statistical Abstract of the United States,* Population Reference Bureau, October 2004 www.prb.org

[18] Figures are rounded to three significant digits. SOURCE: U.S. Census Bureau, October 2004, www.census.gov/.

● basic skills   *tech* Ex technology exercise

91. ● *Global Warming* The most abundant greenhouse gas is carbon dioxide. According to a United Nations "worst-case scenario" prediction, the amount of carbon dioxide in the atmosphere (in parts of volume per million) can be approximated by

$$C(t) \approx 277e^{0.00353t} \text{ parts per million} \quad (0 \le t \le 350)$$

where t is time in years since 1750.[19]

a. Use the model to estimate the amount of carbon dioxide in the atmosphere in 1950, 2000, 2050, and 2100.

b. According to the model, when, to the nearest decade, will the level surpass 700 parts per million?

92. ● *Global Warming* Repeat Exercise 91 using the United Nations "midrange scenario" prediction:

$$C(t) \approx 277e^{0.00267t} \text{ parts per million} \quad (0 \le t \le 350)$$

where t is time in years since 1750.

93. `tech` Ex *New York City Housing Costs: Downtown* The following table shows the average price of a two-bedroom apartment in downtown New York City from 1994 to 2004.[20] *hint* [see Example 7]

t	0 (1994)	2	4	6	8	10 (2004)
Price (\$ million)	0.38	0.40	0.60	0.95	1.20	1.60

a. Use exponential regression to model the price $P(t)$ as a function of time t since 1994. Include a sketch of the points and the regression curve. (Round the coefficients to three decimal places.)

b. Extrapolate your model to estimate the cost of a two-bedroom downtown apartment in 2005.

94. `tech` Ex *New York City Housing Costs: Uptown* The following table shows the average price of a two-bedroom apartment in uptown New York City from 1994 to 2004.[21]

t	0 (1994)	2	4	6	8	10 (2004)
Price (\$ million)	0.18	0.18	1.19	0.2	0.35	0.4

a. Use exponential regression to model the price $P(t)$ as a function of time t since 1994. Include a sketch of the points and the regression curve. (Round the coefficients to three decimal places.)

b. Extrapolate your model to estimate the cost of a two-bedroom uptown apartment in 2005.

95. `tech` Ex *Grants* The following table shows the annual spending on grants by U.S. foundations from 1976 to 2001.[22]

t	0 (1976)	5	10	15	20	25 (2001)
Spending (\$ billion)	6	7	10	12	15	29

a. Use exponential regression to model the annual spending on grants by U.S. foundations as a function of time in years since 1976, and graph the data points and regression curve. (Round coefficients to four decimal places.)

b. According to your model, by what annual percentage has spending on grants by U.S. foundations been increasing over the period shown?

c. Use your model to estimate 1994 spending to the nearest \$1 billion.

96. `tech` Ex *Foundations* The following table shows the total number of active grant-making foundations in the U.S. from 1975 to 2000.[23]

t	0 (1975)	5	10	15	20	25 (2000)
Foundations (thousands)	22	22	25	32	40	57

a. Use exponential regression to model the number of active grant-making foundations in the U.S. as a function of time in years since 1975, and graph the data points and regression curve. (Round coefficients to four decimal places.)

b. According to your model, by what annual percentage has the number of grant-making foundations been increasing over the period shown?

c. Use your model to estimate, to the nearest 1000, the number of active grant-making foundations in 1994.

Communication and Reasoning Exercises

97. ● Which of the following three functions will be largest for large values of x?
(A) $f(x) = x^2$ **(B)** $r(x) = 2^x$ **(C)** $h(x) = x^{10}$

98. ● Which of the following three functions will be smallest for large values of x?
(A) $f(x) = x^{-2}$ **(B)** $r(x) = 2^{-x}$ **(C)** $h(x) = x^{-10}$

99. ● What limitations apply to using an exponential function to model growth in real-life situations? Illustrate your answer with an example.

100. ● Explain in words why 5% per year compounded continuously yields more interest than 5% per year compounded monthly.

101. ● Describe two real-life situations in which a linear model would be more appropriate than an exponential model, and two situations in which an exponential model would be more appropriate than a linear model.

102. ● Describe a real-life situation in which a quadratic model would be more appropriate than an exponential model and one in which an exponential model would be more appropriate than a quadratic model.

---

[19] Exponential regression based on the 1750 figure and the 2100 UN prediction. SOURCES: Tom Boden/Oak Ridge National Laboratory, Scripps Institute of Oceanography/University of California, International Panel on Climate Change/*The New York Times*, December 1, 1997, p. F1.

[20] Data are rounded and 2004 figure is an estimate. SOURCE: Miller Samuel/*New York Times*, March 28, 2004, p. RE 11.

[21] Ibid.

[22] Figures are rounded and adjusted for inflation. SOURCE: The Foundation Center/*New York Times*, April 2, 2002, p. A21.

[23] Figures are rounded. SOURCE: Ibid.

● basic skills   `tech` Ex technology exercise

103. ● How would you check whether data points of the form $(1, y_1)$, $(2, y_2)$, $(3, y_3)$ lie on an exponential curve?

104. You are told that the points $(1, y_1)$, $(2, y_2)$, $(3, y_3)$ lie on an exponential curve. Express y_3 in terms of y_1 and y_2.

105. Your local banker tells you that the reason his bank doesn't compound interest continuously is that it would be too de-manding of computer resources because the computer would need to spend a great deal of time keeping all accounts updated. Comment on his reasoning.

106. Your other local banker tells you that the reason *her* bank doesn't offer continuously compounded interest is that it is equivalent to offering a fractionally higher interest rate compounded daily. Comment on her reasoning.

● basic skills   **tech** Ex technology exercise

9.3 Logarithmic Functions and Models

Logarithms were invented by John Napier (1550–1617) in the late 16th century as a means of aiding calculation. His invention made possible the prodigious hand calculations of astronomer Johannes Kepler (1571–1630), who was the first to describe accurately the orbits and the motions of the planets. Today, computers and calculators have done away with that use of logarithms, but many other uses remain. In particular, the logarithm is used to model real-world phenomena in numerous fields, including physics, finance, and economics.

From the equation

$$2^3 = 8$$

we can see that the power to which we need to raise 2 in order to get 8 is 3. We abbreviate the phrase "the power to which we need to raise 2 in order to get 8" as "$\log_2 8$." Thus, another way of writing the equation $2^3 = 8$ is

$$\log_2 8 = 3 \qquad \text{The power to which we need to raise 2 in order to get 8 is 3.}$$

This is read "the base 2 logarithm of 8 is 3" or "the log, base 2, of 8 is 3."

Here is the general definition.

Base *b* Logarithm

The **base *b* logarithm of *x***, $\log_b x$, is the power to which we need to raise *b* in order to get *x*. Symbolically,

$$\log_b x = y \qquad \text{means} \qquad b^y = x$$
Logarithmic form   *Exponential form*

quick Examples

1. The following table lists some exponential equations and their equivalent logarithmic forms:

Exponential Form	$10^3 = 1000$	$4^2 = 16$	$3^3 = 27$	$5^1 = 5$	$7^0 = 1$	$4^{-2} = \frac{1}{16}$	$25^{1/2} = 5$
Logarithmic Form	$\log_{10} 1000 = 3$	$\log_4 16 = 2$	$\log_3 27 = 3$	$\log_5 5 = 1$	$\log_7 1 = 0$	$\log_4 \frac{1}{16} = -2$	$\log_{25} 5 = \frac{1}{2}$

2. $\log_3 9 =$ the power to which we need to raise 3 in order to get 9. Because $3^2 = 9$, this power is 2, so $\log_3 9 = 2$.

3. $\log_{10} 10{,}000 =$ the power to which we need to raise 10 in order to get 10,000. Because $10^4 = 10{,}000$, this power is 4, so $\log_{10} 10{,}000 = 4$.

4. $\log_3 \frac{1}{27}$ is the power to which we need to raise 3 in order to get $\frac{1}{27}$. Because $3^{-3} = \frac{1}{27}$ this power is -3, so $\log_3 \frac{1}{27} = -3$.

5. $\log_b 1 = 0$ for every positive number b other than 1 because $b^0 = 1$.

Note The number $\log_b x$ is defined only if b and x are both positive and $b \neq 1$. Thus, it is impossible to compute, say, $\log_3(-9)$ (because there is no power of 3 that equals -9), or $\log_1(2)$ (because there is no power of 1 that equals 2). ∎

Logarithms with base 10 and base e are frequently used, so they have special names and notations.

Common Logarithm, Natural Logarithm

The following are standard abbreviations.

		TI-83/84 & Excel Formula
Base 10: $\log_{10} x = \log x$ *Common Logarithm*		`log(x)`
Base e: $\log_e x = \ln x$ *Natural Logarithm*		`ln(x)`

quick Examples

Logarithmic Form	**Exponential Form**
1. $\log 10{,}000 = 4$	$10^4 = 10{,}000$
2. $\log 10 = 1$	$10^1 = 10$
3. $\log \dfrac{1}{10{,}000} = -4$	$10^{-4} = \dfrac{1}{10{,}000}$
4. $\ln e = 1$	$e^1 = e$
5. $\ln 1 = 0$	$e^0 = 1$
6. $\ln 2 = 0.69314718\ldots$	$e^{0.69314718\ldots} = 2$

Some technologies (such as calculators) do not permit direct calculation of logarithms other than common and natural logarithms. To compute logarithms with other bases with these technologies, we can use the following formula:

Change-of-Base Formula

$$\log_b a = \frac{\log a}{\log b} = \frac{\ln a}{\ln b}$$

Change-of-base formula[*]

quick Examples

1. $\log_{11} 9 = \dfrac{\log 9}{\log 11} \approx 0.91631$ `log(9)/log(11)`

2. $\log_{11} 9 = \dfrac{\ln 9}{\ln 11} \approx 0.91631$ `ln(9)/ln(11)`

3. $\log_{3.2}\left(\dfrac{1.42}{3.4}\right) \approx -0.75065$ `log(1.42/3.4)/log(3.2)`

Using Technology to Compute Logarithms

To compute $\log_b x$ using technology, use the following formulas:

TI-83:   `log(x)/log(b)`   Example: $\log_2(16)$ is `log(16)/log(2)`
Excel: `=LOG(x,b)`   Example: $\log_2(16)$ is `= LOG(16,2)`

---

[*] Here is a quick explanation of why this formula works: To calculate $\log_b a$, we ask, "to what power must we raise b to get a?" To check the formula, we try using $\log a/\log b$ as the exponent.

$$b^{\frac{\log a}{\log b}} = (10^{\log b})^{\frac{\log a}{\log b}} \quad \text{(because } b = 10^{\log b}\text{)}$$
$$= 10^{\log a} = a$$

so this exponent works!

One important use of logarithms is to solve equations in which the unknown is in the exponent.

Example 1 Solving Equations with Unknowns in the Exponent

Solve the following equations

a. $5^{-x} = 125$ **b.** $3^{2x-1} = 6$ **c.** $100(1.005)^{3x} = 200$

Solution

a. Write the given equation $5^{-x} = 125$ in logarithmic form:

$$-x = \log_5 125$$

This gives $x = -\log_5 125 = -3$.

b. In logarithmic form, $3^{2x-1} = 6$ becomes

$$2x - 1 = \log_3 6$$

$$2x = 1 + \log_3 6$$

giving $x = \dfrac{1 + \log_3 6}{2} \approx \dfrac{1 + 1.6309}{2} \approx 1.3155$

c. We cannot write the given equation, $100(1.005)^{3x} = 200$, directly in exponential form. We must first divide both sides by 100:

$$1.005^{3x} = \frac{200}{100} = 2$$

$$3x = \log_{1.005} 2$$

$$x = \frac{\log_{1.005} 2}{3} \approx \frac{138.9757}{3} \approx 46.3252$$

Now that we know what logarithms are, we can talk about functions based on logarithms:

Logarithmic Function

A **logarithmic function** has the form

$$f(x) = \log_b x + C \qquad \text{(}b \text{ and } C \text{ are constants with } b > 0, b \neq 1\text{)}$$

or, alternatively,

$$f(x) = A \ln x + C \qquad \text{(}A, C \text{ constants with } A \neq 0\text{)}$$

quick Examples

1. $f(x) = \log x$

2. $g(x) = \ln x - 5$

3. $h(x) = \log_2 x + 1$

Q: *What is the difference between the two forms of the logarithmic function?*

A: None, really, they're equivalent: We can start with an equation in the first form and use the change-of-base formula to rewrite it:

$$f(x) = \log_b x + C$$

$$= \frac{\ln x}{\ln b} + C \qquad \text{Change-of-base formula}$$

$$= \left(\frac{1}{\ln b}\right) \ln x + C$$

Our function now has the form $f(x) = A \ln x + C$, where $A = 1/\ln b$. We can go the other way as well, to rewrite $A \ln x + C$ in the form $\log_b x + C$. ∎

Example 2 Graphs of Logarithmic Functions

a. Sketch the graph of $f(x) = \log_2 x$ by hand.

b. Use technology to compare the graph in part (a) with the graphs of $\log_b x$ for $b = 1/4$, $1/2$, and 4.

Solution

a. To sketch the graph of $f(x) = \log_2 x$ by hand, we begin with a table of values. Because $\log_2 x$ is not defined when $x = 0$, we choose several values of x close to zero and also some larger values, all chosen so that their logarithms are easy to compute:

x	$\frac{1}{8}$	$\frac{1}{4}$	$\frac{1}{2}$	1	2	4	8
$f(x) = \log_2 x$	-3	-2	-1	0	1	2	3

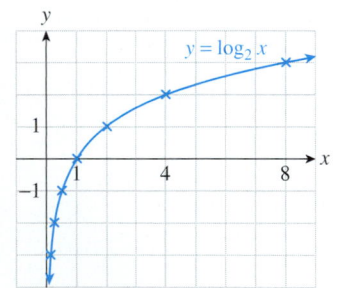

Figure 14

Graphing these points and joining them by a smooth curve gives us Figure 14.

b. We enter the logarithmic functions in graphing utilities as follows (note the use of the change-of-base formula in the TI-83 version):

TI-83	**Excel**
Y₁=log(X)/log(0.25)	=LOG(x,0.25)
Y₂=log(X)/log(0.5)	=LOG(x,0.5)
Y₃=log(X)/log(2)	=LOG(x,2)
Y₄=log(X)/log(4)	=LOG(x,4)

Figure 15 shows the resulting graphs.

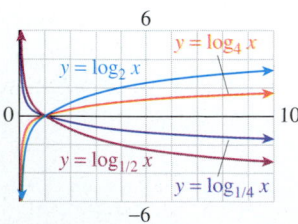

Figure 15

+*Before we go on...* Notice that the graphs of the logarithmic functions in Example 2 all pass through the point $(1, 0)$. (Why?) Notice further that the graphs of the logarithmic functions with bases less than 1 are upside-down versions of the others. Finally, how are these graphs related to the graphs of exponential functions? ∎

The following lists some important algebraic properties of logarithms.

Logarithm Identities

The following identities hold for all positive bases $a \neq 1$ and $b \neq 1$, all positive numbers x and y, and every real number r. These identities follow from the laws of exponents.

Identity

1. $\log_b(xy) = \log_b x + \log_b y$

2. $\log_b\left(\dfrac{x}{y}\right) = \log_b x - \log_b y$

3. $\log_b(x^r) = r \log_b x$

4. $\log_b b = 1; \log_b 1 = 0$

5. $\log_b\left(\dfrac{1}{x}\right) = -\log_b x$

6. $\log_b x = \dfrac{\log_a x}{\log_a b}$

Quick Examples

$\log_2 16 = \log_2 8 + \log_2 2$

$\log_2\left(\dfrac{5}{3}\right) = \log_2 5 - \log_2 3$

$\log_2(6^5) = 5 \log_2 6$

$\log_2 2 = 1; \ln e = 1; \log_{11} 1 = 0$

$\log_2\left(\dfrac{1}{3}\right) = -\log_2 3$

$\log_2 5 = \dfrac{\log_{10} 5}{\log_{10} 2} = \dfrac{\log 5}{\log 2}$

Relationship with Exponential Functions

The following two identities demonstrate that the operations of taking the base b logarithm and raising b to a power are "inverse" to each other.*

Identity

1. $\log_b(b^x) = x$

In words: The power to which you raise b in order to get b^x is x (!)

2. $b^{\log_b x} = x$

In words: Raising b to the power to which it must be raised to get x, yields x (!)

Quick Example

$\log_2(2^7) = 7$

$5^{\log_5 8} = 8$

---

* See the online topic on inverse functions mentioned in the margin.

Online, follow:

Chapter 9

→ Using and Deriving Algebraic Properties of Logarithms

to find a list of logarithmic identities and a discussion on where they come from. Follow:

Chapter 9

→ Inverse Functions

for a general discussion of inverse functions, including further discussion of the relationship between logarithmic and exponential functions.

Applications

Example 3 Investments: How Long?

Ten-year government bonds in Italy are yielding an average of 5.2% per year.* At that interest rate, how long will it take a €1000 investment to be worth €1500 if the interest is compounded monthly?

Solution Substituting $A = 1500$, $P = 1000$, $r = 0.052$, and $m = 12$ in the compound interest equation gives

$$A(t) = P\left(1 + \frac{r}{m}\right)^{mt}$$

$$1500 = 1000\left(1 + \frac{0.052}{12}\right)^{12t}$$

$$\approx 1000(1.004333)^{12t}$$

---

* In 2001. Source: *ECB, Reuters and BE.*

and we must solve for t. We first divide both sides by 1000, getting an equation in exponential form:

$$1.5 = 1.004333^{12t}$$

In logarithmic form, this becomes

$$12t = \log_{1.004333}(1.5)$$

We can now solve for t:

$$t = \frac{\log_{1.004333}(1.5)}{12} \approx 7.8 \text{ years} \qquad \texttt{log(1.5)/(log(1.004333)*12)}$$

Thus, it will take approximately 7.8 years for a €1000 investment to be worth €1500.

Example 4 Half-Life

a. The weight of carbon-14 that remains in a sample that originally contained A grams is given by

$$C(t) = A(0.999879)^t$$

where t is time in years. Find the **half-life,** the time it takes half of the carbon-14 in a sample to decay.

b. Repeat part (a) using the following alternative form of the exponential model in part (a):

$$C(t) = Ae^{-0.000\,121\,01t} \qquad \color{red}{\text{See p. 639.}}$$

c. Another radioactive material has a half-life of 7000 years. Find an exponential decay model in the form

$$R(t) = Ae^{-kt}$$

for the amount of undecayed material remaining. (The constant k is called the **decay constant.**)

d. How long will it take for 99.95% of the substance in a sample of the material in part (c) to decay?

Solution

a. We want to find the value of t for which $C(t) =$ the weight of undecayed carbon-14 left $=$ half the original weight $= 0.5A$. Substituting, we get

$$0.5A = A(0.999879)^t$$

Dividing both sides by A gives

$$0.5 = 0.999879^t \qquad \color{red}{\text{Exponential form}}$$
$$t = \log_{0.999879} 0.5 \approx 5728 \text{ years} \qquad \color{red}{\text{Logarithmic form}}$$

b. This is similar to part (a): We want to solve the equation

$$0.5A = Ae^{-0.000\,121\,01t}$$

for t. Dividing both sides by A gives

$$0.5 = e^{-0.000\,121\,01t}$$

Taking the natural logarithm of both sides gives

$$\ln(0.5) = \ln(e^{-0.000\,121\,01t}) = -0.000\,121\,01t \qquad \textcolor{red}{\ln(e^a) = a\ln e = a}$$

$$t = \frac{\ln(0.5)}{-0.000\,121\,01} \approx 5728 \text{ years}$$

as we obtained in part (a).

c. This time we are given the half-life, which we can use to find the exponential model $R(t) = Ae^{-kt}$. At time $t = 0$, the amount of radioactive material is

$$R(0) = Ae^0 = A$$

Because half of the sample decays in 7000 years, this sample will decay to $0.5A$ grams in 7000 years ($t = 7000$). Substituting this information gives

$$0.5A = Ae^{-k(7000)}$$

Canceling A and taking natural logarithms gives

$$\ln(0.5) = -7000k$$

so the decay constant k is

$$k = -\frac{\ln(0.5)}{7000} \approx 0.000\,099\,021$$

Therefore, the model is

$$R(t) = Ae^{-0.000\,099\,021t}$$

d. If 99.95% of the substance in a sample has decayed, then the amount of undecayed material left is 0.05% of the original amount $= 0.0005A$. We have

$$0.0005A = Ae^{-0.000\,099\,021t}$$

$$0.0005 = e^{-0.000\,099\,021t}$$

$$\ln(0.0005) = -0.000\,099\,021t$$

$$t = \frac{\ln(0.0005)}{-0.000\,099\,021} \approx 76{,}760 \text{ years}$$

+ *Before we go on...*

Q: In parts (a) and (b) of Example 4 we were given two different forms of the model for carbon-14 decay. How do we convert an exponential function in one form to the other?

A: We have already seen (See Quick Example 3 on p. 639) how to convert from the form $f(t) = Ae^{rt}$ in part (b) to the form $f(t) = Ab^t$ in part (a). To go the other way, start with the model in part (a), and equate it to the desired form:

$$C(t) = A(0.999\,879)^t = Ae^{rt}$$

To solve for r, cancel the As and take the natural logarithm of both sides:

$$t\ln(0.999\,879) = rt\ln e = rt$$

so $\qquad\qquad r = \ln(0.999\,879) \approx -0.000\,121\,01$

giving

$$C(t) = Ae^{-0.000\,121\,01t}$$

as in part (b). ∎

We can use the work we did in parts (b) and (c) of the above example to obtain a formula for the decay constant in an exponential decay model for any radioactive substance when we know its half-life. Write the half-life as t_h. Then the calculation in part (b) gives

$$k = -\frac{\ln(0.5)}{t_h} = \frac{\ln 2}{t_h} \qquad -\ln(0.5) = -\ln\left(\frac{1}{2}\right) = \ln 2$$

Multiplying both sides by t_h gives us the relationship $t_h k = \ln 2$.

Exponential Decay Model and Half-Life

An **exponential decay function** has the form

$$Q(t) = Q_0 e^{-kt} \qquad Q_0, k \text{ both positive}$$

Q_0 represents the value of Q at time $t = 0$, and k is the **decay constant.** The decay constant k and half-life t_h for Q are related by

$$t_h k = \ln 2$$

quick Examples

1. $Q(t) = Q_0 e^{-0.000\,121\,01t}$ is the decay function for carbon-14 (see Example 4b).

2. If $t_h = 10$ years, then $10k = \ln 2$, so $k = \dfrac{\ln 2}{10} \approx 0.06931$ so the decay model is

$$Q(t) = Q_0 e^{-0.06931t}$$

3. If $k = 0.0123$, then $t_h(0.0123) = \ln 2$, so the half-life is $t_h = \dfrac{\ln 2}{0.0123} \approx 56.35$ years.

We can repeat the analysis above for exponential growth models:

Exponential Growth Model and Doubling Time

An **exponential growth function** has the form

$$Q(t) = Q_0 e^{kt} \qquad Q_0, k \text{ both positive}$$

Q_0 represents the value of Q at time $t = 0$, and k is the **growth constant.** The growth constant k and doubling time t_d for Q are related by

$$t_d k = \ln 2$$

quick Examples

1. $P(t) = 1000e^{0.05t}$ $1000 invested at 5% annually with interest compounded continuously

2. If $t_d = 10$ years, then $10k = \ln 2$, so $k = \dfrac{\ln 2}{10} \approx 0.06931$ so the growth model is

$$Q(t) = Q_0 e^{0.06931t}$$

3. If $k = 0.0123$, then $t_d(0.0123) = \ln 2$, so the doubling time is $t_d = \dfrac{\ln 2}{0.0123} \approx 56.35$ years.

Logarithmic Regression

If we start with a set of data that suggests a logarithmic curve we can, by repeating the methods from previous sections, use technology to find the logarithmic regression curve $y = \log_b x + C$ approximating the data.

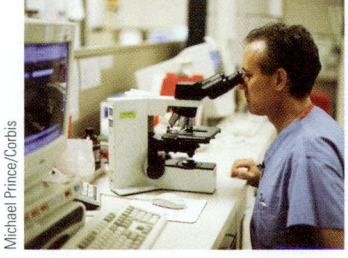

Michael Prince/Corbis

 using _Technology_

See the Technology Guides at the end of the chapter for detailed instructions on using a TI-83/84 or Excel to find logarithmic regression curves. Alternatively, go online and follow:

Chapter 9

→ Tools

→ Simple Regression

to find a utility for finding regression curves of various sorts.

techEx   Example 5 Research & Development

The following table shows the total spent on research and development in the U.S., in billions of dollars, for the period 1995–2005 ($t = 5$ represents 1995).[*]

Year t	5	6	7	8	9	10
Spending ($ billions)	187	197	208	219	232	248
Year t	11	12	13	14	15	
Spending ($ billions)	250	250	253	260	265	

Find the best-fit logarithmic model of the form

$$S(t) = A \ln t + C$$

and use the model to project total spending on research in 2010, assuming the trend continues.

Solution

We use technology to get the following regression model:

$$S(t) = 73.77 \ln t + 67.75 \qquad \text{Coefficients rounded}$$

Because 2010 is represented by $t = 20$, we have

$$S(20) = 73.77 \ln(20) + 67.75 \approx 289 \qquad \text{Why did we round the result to three significant digits?}$$

So, research and development spending is predicted to be around $289 billion in 2010.

_____

[*] Data are approximate and are given in constant 1996 dollars. 2004 and 2005 figures are projections. SOURCE: National Science Foundation, Division of Science Resource Statistics, National Patterns of R&D Resources. www.nsf.gov/sbe/srs/nprdr/start.hrm October 2004.

+ _Before we go on..._ The model in Example 5 seems to give reasonable estimates when we extrapolate forward, but extrapolating backward is quite another matter: The logarithm curve drops sharply to the left of the given range and becomes negative for small values of t (Figure 16).

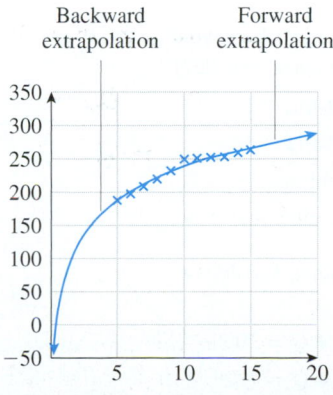

Figure **16**

9.3 EXERCISES

● denotes basic skills exercises

techEx indicates exercises that should be solved using technology

In Exercises 1–4, complete the given tables. hint [see Quick Example on p. 645]

1. ●

Exponential Form	$10^4 = 10{,}000$	$4^2 = 16$	$3^3 = 27$	$5^1 = 5$	$7^0 = 1$	$4^{-2} = \frac{1}{16}$
Logarithmic Form						

2. ●

Exponential form	$4^3 = 64$	$10^{-1} = 0.1$	$2^8 = 256$	$5^0 = 1$	$(0.5)^2 = 0.25$	$6^{-2} = \frac{1}{36}$
Logarithmic form						

3. ●

Exponential form						
Logarithmic form	$\log_{0.5} 0.25 = 2$	$\log_5 1 = 0$	$\log_{10} 0.1 = -1$	$\log_4 64 = 3$	$\log_2 256 = 8$	$\log_2 \frac{1}{4} = -2$

4. ●

Exponential form						
Logarithmic form	$\log_5 5 = 1$	$\log_4 \frac{1}{16} = -2$	$\log_4 16 = 2$	$\log_{10} 10{,}000 = 4$	$\log_3 27 = 3$	$\log_7 1 = 0$

In Exercises 5–12, use logarithms to solve the given equation. (Round answers to four decimal places.) hint [see Example 1]

5. ● $3^x = 5$ **6.** ● $4^x = 3$

7. ● $5^{-2x} = 40$ **8.** ● $6^{3x+1} = 30$

9. ● $4.16e^x = 2$ **10.** ● $5.3(10^x) = 2$

11. ● $5(1.06^{2x+1}) = 11$ **12.** ● $4(1.5^{2x-1}) = 8$

In Exercises 13–18, graph the given function. hint [see Example 2]

13. ● $f(x) = \log_4 x$ **14.** ● $f(x) = \log_5 x$

15. ● $f(x) = \log_4(x - 1)$ **16.** ● $f(x) = \log_5(x + 1)$

17. ● $f(x) = \log_{1/4} x$ **18.** ● $f(x) = \log_{1/5} x$

In Exercises 19–22 find the associated exponential decay or growth model. hint [see Quick Examples on p. 652]

19. ● $Q = 1000$ when $t = 0$; Half-life $= 1$

20. ● $Q = 2000$ when $t = 0$; Half-life $= 5$

21. ● $Q = 1000$ when $t = 0$; Doubling time $= 2$

22. ● $Q = 2000$ when $t = 0$; Doubling time $= 5$

In Exercises 23–26 find the associated half-life or doubling time.

23. ● $Q = 1000e^{0.5t}$ **24.** ● $Q = 1000e^{-0.025t}$

25. ● $Q = Q_0e^{-4t}$ **26.** ● $Q = Q_0e^t$

In Exercises 27–32 convert the given exponential function to the form indicated. Round all coefficients to four significant digits. hint [see Example 4 Before we go on]

27. ● $f(x) = 4e^{2x}$; $f(x) = Ab^x$

28. ● $f(x) = 2.1e^{-0.1x}$; $f(x) = Ab^x$

29. ● $f(t) = 2.1(1.001)^t$; $f(t) = Q_0e^{kt}$

30. ● $f(t) = 23.4(0.991)^t$; $f(t) = Q_0e^{-kt}$

31. ● $f(t) = 10(0.987)^t$; $f(t) = Q_0e^{-kt}$

32. ● $f(t) = 2.3(2.2)^t$; $f(t) = Q_0e^{kt}$

Applications

33. ● *Investments* How long will it take a $500 investment to be worth $700 if it is continuously compounded at 10% per year? (Give the answer to two decimal places.) hint [see Example 3]

34. ● *Investments* How long will it take a $500 investment to be worth $700 if it is continuously compounded at 15% per year? (Give the answer to two decimal places.)

35. ● *Investments* How long, to the nearest year, will it take an investment to triple if it is continuously compounded at 10% per year?

36. ● *Investments* How long, to the nearest year, will it take me to become a millionaire if I invest $1000 at 10% interest compounded continuously?

● basic skills   techEx technology exercise

37. ● *Investments* I would like my investment to double in value every three years. At what rate of interest would I need to invest it, assuming the interest is compounded continuously? *hint* [see Quick Examples on p. 652]

38. ● *Depreciation* My investment in OHaganBooks.com stocks is losing half its value every two years. Find and interpret the associated decay rate.

39. ● *Carbon Dating* The amount of carbon-14 remaining in a sample that originally contained A grams is given by

$$C(t) = A(0.999879)^t$$

where t is time in years. If tests on a fossilized skull reveal that 99.95% of the carbon-14 has decayed, how old, to the nearest 1000 years, is the skull?

40. ● *Carbon Dating* Refer back to Exercise 39. How old, to the nearest 1000 years, is a fossil in which only 30% of the carbon-14 has decayed?

Long-Term Investments Exercises 41–48 are based on the following table, which lists interest rates on long-term investments (based on 10-year government bonds) in several countries in 2004–2005.[24]

Country	U.S.	Japan	Canada	Korea	Australia
Yield	5.3%	1.5%	5.2%	5.4%	6.0%

41. ● Assuming that you invest $10,000 in the U.S., how long (to the nearest year) must you wait before your investment is worth $15,000 if the interest is compounded annually?

42. ● Assuming that you invest $10,000 in Japan, how long (to the nearest year) must you wait before your investment is worth $15,000 if the interest is compounded annually?

43. ● If you invest $10,400 in Canada and the interest is compounded monthly, how many months will it take for your investment to grow to $20,000?

44. ● If you invest $10,400 in the U.S., and the interest is compounded monthly, how many months will it take for your investment to grow to $20,000?

45. ● How long, to the nearest year, will it take an investment in Australia to double its value if the interest is compounded every six months?

46. ● How long, to the nearest year, will it take an investment in Korea to double its value if the interest is compounded every six months?

47. ● If the interest on a long-term U.S. investment is compounded continuously, how long will it take the value of an investment to double? (Give the answer correct to two decimal places.)

48. ● If the interest on a long-term Australia investment is compounded continuously, how long will it take the value of an investment to double? (Give an answer correct to two decimal places.)

---
[24] Approximate interest rates based on 10-year government bonds and similar investments. SOURCE: Organization for Economic Co-operation and Development, www.oecd.org.

49. ● *Half-life* The amount of radium-226 remaining in a sample that originally contained A grams is approximately

$$C(t) = A(0.999\ 567)^t$$

where t is time in years. Find the half-life to the nearest 100 years. *hint* [see Example 4a]

50. ● *Half-life* The amount of iodine-131 remaining in a sample that originally contained A grams is approximately

$$C(t) = A(0.9175)^t$$

where t is time in days. Find the half-life to two decimal places.

51. ● *Automobiles* The rate of auto thefts triples every six months.

  a. Determine, to two decimal places, the base b for an exponential model $y = Ab^t$ of the rate of auto thefts as a function of time in months.

  b. Find the doubling time to the nearest tenth of a month.

52. ● *Televisions* The rate of television thefts is doubling every four months.

  a. Determine, to two decimal places, the base b for an exponential model $y = Ab^t$ of the rate of television thefts as a function of time in months.

  b. Find the tripling time to the nearest tenth of a month.

53. ● *Half-life* The half-life of cobalt-60 is five years.

  a. Obtain an exponential decay model for cobalt-60 in the form $Q(t) = Q_0 e^{-kt}$. (Round coefficients to three significant digits.) *hint* [see Quick Examples on p. 652]

  b. Use your model to predict, to the nearest year, the time it takes one third of a sample of cobalt-50 to decay.

54. ● *Half-life* The half-life of strontium-90 is 28 years.

  a. Obtain an exponential decay model for strontium-90 in the form $Q(t) = Q_0 e^{-kt}$. (Round coefficients to three significant digits.)

  b. Use your model to predict, to the nearest year, the time it takes three-fifths of a sample of strontium-90 to decay.

55. ● *Radioactive Decay* Uranium-235 is used as fuel for some nuclear reactors. It has a half-life of 710 million years. How long will it take 10 grams of uranium-235 to decay to 1 gram? (Round your answer to three significant digits.)

56. ● *Radioactive Decay* Plutonium-239 is used as fuel for some nuclear reactors, and also as the fissionable material in atomic bombs. It has a half-life of 24,400 years. How long would it take 10 grams of Plutonium-239 to decay to 1 gram? (Round your answer to three significant digits.)

57. ● *Aspirin* Soon after taking an aspirin, a patient has absorbed 300 mg of the drug. If the amount of aspirin in the bloodstream decays exponentially, with half being removed every two hours, find, to the nearest 0.1 hour, the time it will take for the amount of aspirin in the bloodstream to decrease to 100 mg.

58. *Alcohol* After several drinks, a person has a blood alcohol level of 200 mg/dL (milligrams per deciliter). If the amount of

● basic skills   **tech** Ex technology exercise

alcohol in the blood decays exponentially, with one fourth being removed every hour, find the time it will take for the person's blood alcohol level to decrease to 80 mg/dL.

59. *Radioactive Decay* You are trying to determine the half-life of a new radioactive element you have isolated. You start with 1 gram, and two days later you determine that it has decayed down to 0.7 grams. What is its half-life? (Round your answer to three significant digits.)

60. *Radioactive Decay* You have just isolated a new radioactive element. If you can determine its half-life, you will win the Nobel Prize in physics. You purify a sample of 2 grams. One of your colleagues steals half of it, and three days later you find that 0.1 gram of the radioactive material is still left. What is the half-life? (Round your answer to three significant digits.)

61. `tech` Ex ***Population Aging*** The following table shows the percentage of U.S. residents over the age of 65 in 1950, 1960, . . . , 2010 (t is time in years since 1900):[25]

t (Year since 1900)	50	60	70	80	90	100	110
P (% over 65)	8.2	9.2	9.9	11.3	12.6	12.6	13

a. Find the logarithmic regression model of the form $P(t) = A \ln t + C$. (Round the coefficients to four significant digits).

b. In 1940, 6.9% of the population was over 65. To how many significant digits does the model reflect this figure?

c. Which of the following is correct? The model, if extrapolated into the indefinite future, predicts that

(A) The percentage of U.S. residents over the age of 65 will increase without bound.

(B) The percentage of U.S. residents over the age of 65 will level off at around 14.2%.

(C) The percentage of U.S. residents over the age of 65 will eventually decrease. *hint* [see Example 5]

62. `tech` Ex ***Population Aging*** The following table shows the percentage of U.S. residents over the age of 85 in 1950, 1960, . . . , 2010 (t is time in years since 1900):[26]

t (Year since 1900)	50	60	70	80	90	100	110
P (% over 85)	0.4	0.5	0.7	1	1.2	1.6	1.9

a. Find the logarithmic regression model of the form $P(t) = A \ln t + C$. (Round the coefficients to four significant digits).

[25] SOURCE: U.S. Census Bureau.
[26] Ibid.

b. In 2020, 2.1% of the population is projected to be over 85. To how many significant digits does the model reflect this figure?

c. Which of the following is correct? If you increase A by 0.1 and decrease C by 0.1 in the logarithmic model, then

(A) The new model predicts eventually lower percentages.

(B) The long-term prediction is essentially the same.

(C) The new model predicts eventually higher percentages.

63. `tech` Ex ***Market Share*** The following table shows the U.S. market share of sport utility vehicles (SUVs) (including "crossover utility vehicles"—smaller SUVs based on car designs) for the period 1994–2001.[27]

Year t	4 (1994)	5	6	7	8	9	10	11 (2001)
Market share	10%	11.5	12	15	17.5	19	19.5	21

Find the logarithmic regression model of the form $M(t) = A \ln t + C$. Comment on the long-term suitability of the model.

64. `tech` Ex ***Market Share*** The following table shows the U.S. market share of large size sport utility vehicles (SUVs) for the period 1994–2001.[28]

Year t	4 (1994)	5	6	7	8	9	10	11 (2001)
Market Share	1.5%	2	2.5	3.5	4.5	5	5	5

Find the logarithmic regression model of the form $M(t) = A \ln t + C$. Comment on the suitability of the model to estimate the market share during the period 1990–1993.

65. *Richter Scale* The **Richter scale** is used to measure the intensity of earthquakes. The Richter scale rating of an earthquake is given by the formula

$$R = \frac{2}{3}(\log E - 11.8)$$

where E is the energy released by the earthquake (measured in ergs[29]).

a. The San Francisco earthquake of 1906 registered $R = 8.2$ on the Richter scale. How many ergs of energy were released?

[27] Market share is given as a percentage of all vehicles sold in the U.S. 2001 figure is based on sales through April. SOURCE: Ward's Auto Infobank/ *New York Times,* May 19, 2001, p. C1.

[28] Ibid.

[29] An erg is a unit of energy. One erg is the amount of energy it takes to move a mass of one gram one centimeter in one second.

● basic skills   `tech` Ex technology exercise

b. In 1989 another San Francisco earthquake registered 7.1 on the Richter scale. Compare the two: The energy released in the 1989 earthquake was what percentage of the energy released in the 1906 quake?

c. Show that if two earthquakes registering R_1 and R_2 on the Richter scale release E_1 and E_2 joules of energy, respectively, then

$$\frac{E_2}{E_1} = 10^{1.5(R_2 - R_1)}$$

d. Fill in the blank: If one earthquake registers 2 points more on the Richter scale than another, then it releases ___ times the amount of energy.

66. *Sound Intensity* The loudness of a sound is measured in **decibels.** The decibel level of a sound is given by the formula

$$D = 10 \, \log \frac{I}{I_0}$$

where D is the decibel level (dB), I is its intensity in watts per square meter (W/m^2), and $I_0 = 10^{-12}$ W/m^2 is the intensity of a barely audible "threshold" sound. A sound intensity of 90 dB or greater causes damage to the average human ear.

a. Find the decibel levels of each of the following, rounding to the nearest decibel:

Whisper:	115×10^{-12} W/m^2
TV (average volume from 10 feet):	320×10^{-7} W/m^2
Loud music:	900×10^{-3} W/m^2
Jet aircraft (from 500 feet):	100 W/m^2

b. Which of the sounds above damages the average human ear?

c. Show that if two sounds of intensity I_1 and I_2 register decibel levels of D_1 and D_2 respectively, then

$$\frac{I_2}{I_1} = 10^{0.1(D_2 - D_1)}$$

d. Fill in the blank: If one sound registers one decibel more than another, then it is ___ times as intense.

67. *Sound Intensity* The decibel level of a TV set decreases with the distance from the set according to the formula

$$D = 10 \log \left(\frac{320 \times 10^7}{r^2} \right)$$

where D is the decibel level and r is the distance from the TV set in feet.

a. Find the decibel level (to the nearest decibel) at distances of 10, 20, and 50 feet.

b. Express D in the form $D = A + B \log r$ for suitable constants A and B. (Round A and D to two significant digits.)

c. How far must a listener be from a TV so that the decibel level drops to 0? (Round the answer to two significant digits.)

68. *Acidity* The acidity of a solution is measured by its pH, which is given by the formula

$$pH = -\log(H^+)$$

where H^+ measures the concentration of hydrogen ions in moles per liter.[30] The pH of pure water is 7. A solution is referred to as *acidic* if its pH is below 7 and as *basic* if its pH is above 7.

a. Calculate the pH of each of the following substances.

Blood:	3.9×10^{-8} moles/liter
Milk:	4.0×10^{-7} moles/liter
Soap solution:	1.0×10^{-11} moles/liter
Black coffee:	1.2×10^{-7} moles/liter

b. How many moles of hydrogen ions are contained in a liter of acid rain that has a pH of 5.0?

c. Complete the following sentence: If the pH of a solution increases by 1.0, then the concentration of hydrogen ions _____.

Communication and Reasoning Exercises

69. ● Why is the logarithm of a negative number not defined?

70. ● Of what use are logarithms, now that they are no longer needed to perform complex calculations?

71. ● If $y = 4^x$, then $x =$ _____.

72. ● If $y = \log_6 x$, then $x =$ _____.

73. ● Simplify: $2^{\log_2 8}$.

74. ● Simplify: $e^{\ln x}$.

75. ● Simplify: $\ln(e^x)$.

76. ● Simplify: $\ln \sqrt{a}$.

77. ● Your company's market share is undergoing steady growth. Explain why a logarithmic function is *not* appropriate for long-term future prediction of your market share.

78. ● Your company's market share is undergoing steady growth. Explain why a logarithmic function is *not* appropriate for long-term backward extrapolation of your market share.

79. If a town's population is increasing exponentially with time, how is time increasing with population? Explain.

80. If a town's population is increasing logarithmically with time, how is time increasing with population? Explain.

---

[30] A mole corresponds to about 6.0×10^{23} hydrogen ions. (This number is known as Avogadro's number.)

9.4 Logistic Functions and Models

Figure 17 shows the percentage of U.S. households with personal computers as a function of time t in years ($t = 0$ represents 1994).[31]

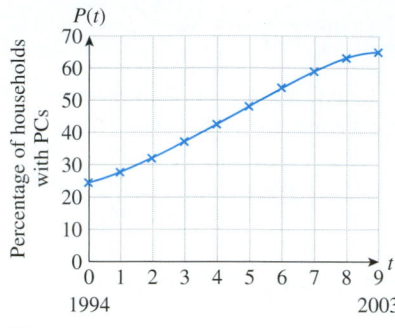

Figure **17**

The left-hand part of the curve in Figure 17, from $t = 0$ to, say, $t = 4$, looks like exponential growth: P behaves (roughly) like an exponential function, growing at a rate of around 15% per year. Then, as the market starts to become saturated, the growth of P slows and its value approaches a "saturation" point. **Logistic** functions have just this kind of behavior, growing exponentially at first and then leveling off. In addition to modeling the demand for a new technology or product, logistic functions are often used in epidemic and population modeling. In an epidemic, the number of infected people often grows exponentially at first and then slows when a significant proportion of the entire susceptible population is infected and the epidemic has "run its course." Similarly, populations may grow exponentially at first and then slow as they approach the capacity of the available resources.

Logistic Function

A **logistic function** has the form

$$f(x) = \frac{N}{1 + Ab^{-x}}$$

for some constants A, N, and b ($b > 0$ and $b \neq 1$).

quick Example

$N = 6$, $A = 2$, $b = 1.1$ gives $f(x) = \dfrac{6}{1 + 2(1.1^{-x})}$ 6/(1+2*1.1^-x)

$f(0) = \dfrac{6}{1 + 2} = 2$ The y-intercept is $N/(1 + A)$

$f(1000) = \dfrac{6}{1 + 2(1.1^{-1000})} \approx \dfrac{6}{1 + 0} = 6 = N$ When x is large, $f(x) \approx N$

---

[31] SOURCE: NTIA/Census Bureau/Pegasus Research International, LLC
www.pegasusresearch.com/metrics/growthus.htm

Graph of a Logistic Function

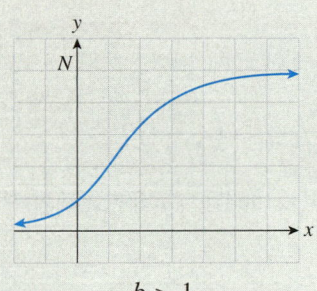

$b > 1$

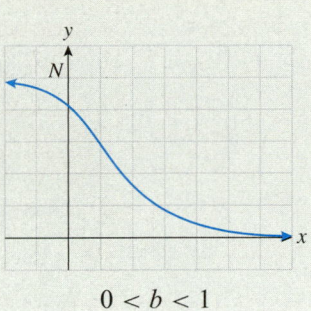

$0 < b < 1$

$$y = \frac{N}{1 + Ab^{-x}}$$

Properties of the Logistic Curve $y = \dfrac{N}{1 + Ab^{-x}}$

- The graph is an S-shaped curve sandwiched between the horizontal lines $y = 0$ and $y = N$. N is called the **limiting value** of the logistic curve.
- If $b > 1$ the graph rises; if $b < 1$, the graph falls.
- The y-intercept is $\dfrac{N}{1 + A}$.

Note If we write b^{-x} as e^{-kx} (where $k = \ln b$), we get the following alternative form of the logistic function:

$$f(x) = \frac{N}{1 + Ae^{-kx}}$$ ∎

Q: How does the constant b affect the graph*?*

A: To understand the role of b, we first rewrite the logistic function by multiplying top and bottom by b^x:

$$f(x) = \frac{N}{1 + Ab^{-x}}$$

$$= \frac{Nb^x}{(1 + Ab^{-x})b^x}$$

$$= \frac{Nb^x}{b^x + A} \qquad \text{Because } b^{-x}b^x = 1$$

For values of x close to 0, the quantity b^x is close to 1, so the denominator is approximately $1 + A$, giving

$$f(x) \approx \frac{Nb^x}{1 + A} = \left(\frac{N}{1 + A}\right)b^x$$ ∎

In other words, $f(x)$ is approximately exponential with base b for values of x close to 0. Put another way, if x represents time, then initially the logistic function behaves like an exponential function.

To summarize:

Logistic Function for Small x and the Role of b

For small values of x, we have

$$\frac{N}{1 + Ab^{-x}} \approx \left(\frac{N}{1 + A}\right) b^x$$

Thus, for small x, the logistic function grows approximately exponentially with base b.

quick Example   Let

$$f(x) = \frac{50}{1 + 24(3^{-x})} \qquad \color{red}{N = 50,\ A = 24,\ b = 3}$$

Then

$$f(x) \approx \left(\frac{50}{1 + 24}\right)(3^x) = 2(3^x)$$

for small values of x. The following figure compares their graphs:

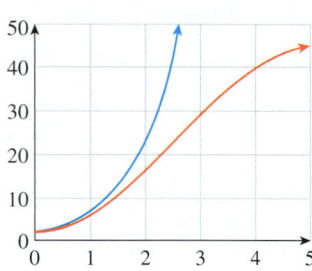 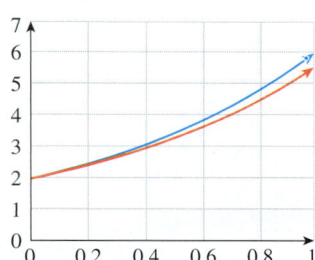

The upper curve is the exponential curve.

Modeling with the Logistic Function

Example 1 Epidemics

A flu epidemic is spreading through the U.S. population. An estimated 150 million people are susceptible to this particular strain, and it is predicted that all susceptible people will eventually become infected. There are 10,000 people already infected, and the number is doubling every two weeks. Use a logistic function to model the number of people infected. Hence predict when, to the nearest week, 1 million people will be infected.

Solution Let t be time in weeks, and let $P(t)$ be the total number of people infected at time t. We want to express P as a logistic function of t, so that

$$P(t) = \frac{N}{1 + Ab^{-t}}$$

We are told that, in the long run, 150 million people will be infected, so that

$$N = 150,000,000 \qquad \color{red}{\text{Limiting value of } P}$$

At the current time ($t = 0$), 10,000 people are infected, so

$$10,000 = \frac{N}{1 + A} = \frac{150,000,000}{1 + A} \qquad \color{red}{\text{Value of } P \text{ when } t = 0}$$

Solving for A gives

$$10,000(1 + A) = 150,000,000$$
$$1 + A = 15,000$$
$$A = 14,999$$

What about b? At the beginning of the epidemic (t near 0), P is growing approximately exponentially, doubling every two weeks. Using the technique of Section 9.2, we find that the exponential curve passing through the points (0, 10,000) and (2, 20,000) is

$$y = 10,000(\sqrt{2})^t$$

giving us $b = \sqrt{2}$. Now that we have the constants N, A, and b, we can write down the logistic model:

$$P(t) = \frac{150,000,000}{1 + 14,999(\sqrt{2})^{-t}}$$

The graph of this function is shown in Figure 18.

Now we tackle the question of prediction: When will 1 million people be infected? In other words: When is $P(t) = 1,000,000$?

$$1,000,000 = \frac{150,000,000}{1 + 14,999(\sqrt{2})^{-t}}$$

$$1,000,000[1 + 14,999(\sqrt{2})^{-t}] = 150,000,000$$

$$1 + 14,999(\sqrt{2})^{-t} = 150$$

$$14,999(\sqrt{2})^{-t} = 149$$

$$(\sqrt{2})^{-t} = \frac{149}{14,999}$$

$$-t = \log_{\sqrt{2}}\left(\frac{149}{14,999}\right) \approx -13.31 \qquad \text{Logarithmic form}$$

Thus, 1 million people will be infected by about the 13th week.

P versus Weeks graph

People infected (millions) on vertical axis with markings 50, 100, 150; Weeks on horizontal axis with markings 0, 10, 20, 30, 40, 50.

Figure **18**

Logistic Regression

Let us go back to the data on the percentage of households with PCs and try to estimate the percentage of households that will have PCs in the long term. In order to be able to make predictions such as this, we require a model for the data, so we will need to do some form of regression.

tech Ex

Example **2** Households with PCs

Here are the data graphed in Figure 17:

Year (t)	0	1	2	3	4	5	6	7	8	9
Households with PCs (%) (P)	24	28	32	37	42	48	54	59	63	65

Find a logistic regression curve of the form

$$P(t) = \frac{N}{1 + Ab^{-t}}$$

In the long term, what percentage of households with PCs does the model predict?

Solution We can use technology to obtain the following regression model:

$$P(t) \approx \frac{84.573}{1 + 2.6428(1.2845)^{-t}}$$

Its graph and the original data are shown in Figure 19.

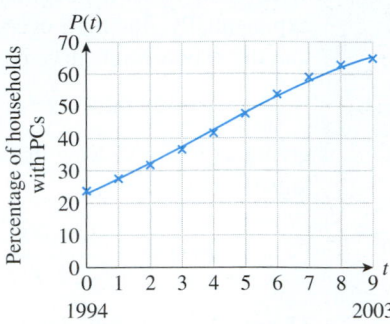

Figure **19**

 using *Technology*

See the Technology Guides at the end of the chapter for detailed instructions on using a TI-83/84 or Excel to find logistic regression curves.

Because $N \approx 85$, the model predicts that about 85% of all households will have PCs in the long term.

Note Logistic regression estimates all three constants N, A, and b for a model $y = \frac{N}{1 + Ab^{-x}}$. However, there are times, as in Example 1, when we already know the limiting value N and require estimates of only A and b. In such cases, we can use exponential regression to compute these estimates: First rewrite the logistic equation as

$$\frac{N}{y} = 1 + Ab^{-x}$$

so that

$$\frac{N}{y} - 1 = Ab^{-x} = A(b^{-1})^x$$

This equation gives $N/y - 1$ as an exponential function of x. Thus, if we do exponential regression using the data points $(x, N/y - 1)$, we can obtain estimates for A and b^{-1} (and hence b). This is done in Exercises 33 and 34. ∎

9.4 EXERCISES

● denotes basic skills exercises

◆ denotes challenging exercises

tech Ex indicates exercises that should be solved using technology

In Exercises 1–6, find N, A, and b, give a technology formula for the given function, and use technology to sketch its graph for the given range of values of x. *hint* [see Quick Example on p. 658]

1. ● $f(x) = \dfrac{7}{1 + 6(2^{-x})}$; $[0, 10]$

2. ● $g(x) = \dfrac{4}{1 + 0.333(4^{-x})}$; $[0, 2]$

3. ● $f(x) = \dfrac{10}{1 + 4(0.3^{-x})}$; $[-5, 5]$

4. ● $g(x) = \dfrac{100}{1 + 5(0.5^{-x})}$; $[-5, 5]$

5. ● $h(x) = \dfrac{2}{0.5 + 3.5(1.5^{-x})}$; $[0, 15]$

(First divide top and bottom by 0.5.)

● basic skills   ◆ challenging   *tech* Ex technology exercise

6. ● $k(x) = \dfrac{17}{2 + 6.5(1.05^{-x})}$; [0, 100]

(First divide top and bottom by 2.)

In Exercises 7–10, find the logistic function f with the given properties. hint [see Example 1]

7. ● $f(0) = 10$, f has limiting value 200, and for small values of x, f is approximately exponential and doubles with every increase of 1 in x.

8. ● $f(0) = 1$, f has limiting value 10, and for small values of x, f is approximately exponential and grows by 50% with every increase of 1 in x.

9. ● f has limiting value 6 and passes through (0, 3) and (1, 4).

10. ● f has limiting value 4 and passes through (0, 1) and (1, 2).

In Exercises 11–16, choose the logistic function that best approximates the given curve.

11. ●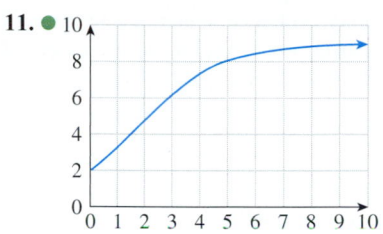

(A) $f(x) = \dfrac{6}{1 + 0.5(3^{-x})}$ **(B)** $f(x) = \dfrac{9}{1 + 3.5(2^{-x})}$

(C) $f(x) = \dfrac{9}{1 + 0.5(1.01)^{-x}}$

12. ●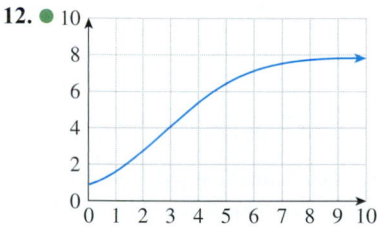

(A) $f(x) = \dfrac{8}{1 + 7(2)^{-x}}$ **(B)** $f(x) = \dfrac{8}{1 + 3(2)^{-x}}$

(C) $f(x) = \dfrac{6}{1 + 11(5)^{-x}}$

13. ●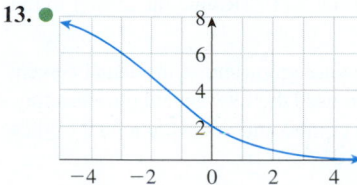

(A) $f(x) = \dfrac{8}{1 + 7(0.5)^{-x}}$ **(B)** $f(x) = \dfrac{8}{1 + 3(0.5)^{-x}}$

(C) $f(x) = \dfrac{8}{1 + 3(2)^{-x}}$

14. ●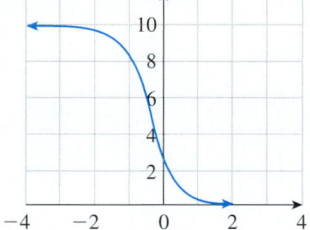

(A) $f(x) = \dfrac{10}{1 + 3(1.01)^{-x}}$ **(B)** $f(x) = \dfrac{8}{1 + 7(0.1)^{-x}}$

(C) $f(x) = \dfrac{10}{1 + 3(0.1)^{-x}}$

15. ●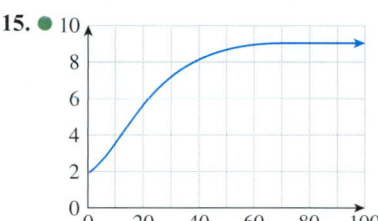

(A) $f(x) = \dfrac{18}{2 + 7(5)^{-x}}$ **(B)** $f(x) = \dfrac{18}{2 + 3(1.1)^{-x}}$

(C) $f(x) = \dfrac{18}{2 + 7(1.1)^{-x}}$

16. ●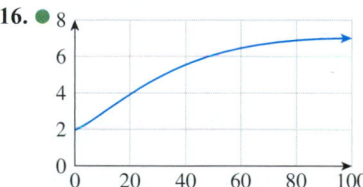

(A) $f(x) = \dfrac{14}{2 + 5(15)^{-x}}$ **(B)** $f(x) = \dfrac{14}{1 + 13(1.05)^{-x}}$

(C) $f(x) = \dfrac{14}{2 + 5(1.05)^{-x}}$

tech Ex *In Exercises 17–20, use technology to find a logistic regression curve* $y = \dfrac{N}{1 + Ab^{-x}}$ *approximating the given data. Draw a graph showing the data points and regression curve. (Round b to three significant digits and A and N to two significant digits.)* hint [see Example 2]

17. ●

x	0	20	40	60	80	100
y	2.1	3.6	5.0	6.1	6.8	6.9

18. ●

x	0	30	60	90	120	150
y	2.8	5.8	7.9	9.4	9.7	9.9

● basic skills   ◆ challenging   tech Ex technology exercise

19. ●

x	0	20	40	60	80	100
y	30.1	11.6	3.8	1.2	0.4	0.1

20. ●

x	0	30	60	90	120	150
y	30.1	20	12	7.2	3.8	2.4

Applications

21. ● *Scientific Research* The following chart shows the number of research articles in the prominent journal *Physics Review* that were written by researchers in Europe during 1983–2003 ($t = 0$ represents 1983).[32] *hint* [see Properties of Logistic Curves on p. 659.]

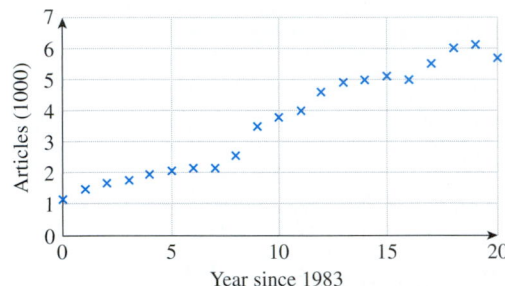

Year since 1983

a. Which of the following logistic functions best models the data? (t is the number of years since 1983.) Try to determine the correct model without actually computing data points.

(A) $A(t) = \dfrac{7.0}{1 + 5.4(1.2)^{-t}}$ **(B)** $A(t) = \dfrac{4.0}{1 + 3.4(1.2)^{-t}}$

(C) $A(t) = \dfrac{4.0}{1 + 3.4(0.8)^{-t}}$ **(D)** $A(t) = \dfrac{7.0}{1 + 5.4(0.8)^{-t}}$

b. According to the model you selected, at what percentage was the number of articles growing around 1985?

22. ● *Scientific Research* The following chart shows the percentage, above 25%, of research articles in the prominent journal *Physics Review* that were written by researchers in the U.S. during 1983–2003 ($t = 0$ represents 1983).[33]

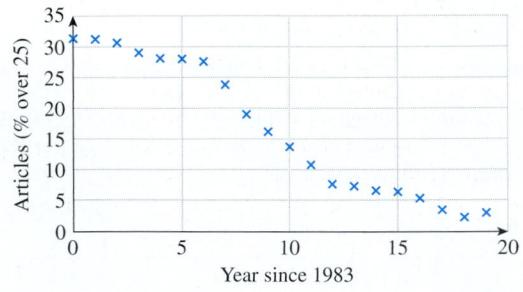

Year since 1983

a. Which of the following logistic functions best models the data? (t is the number of years since 1983.) Try to determine the correct model without actually computing data points.

(A) $P(t) = \dfrac{36}{1 + 0.06(1.7)^{-t}}$

(B) $P(t) = \dfrac{12}{1 + 0.06(1.7)^{-t}}$

(C) $P(t) = \dfrac{12}{1 + 0.06(0.7)^{-t}}$

(D) $P(t) = \dfrac{36}{1 + 0.06(0.7)^{-t}}$

b. According to the model you selected, how fast was the value of P declining around 1985?

23. ● *Computer Use* The following graph shows the actual percentage of U.S. households with a computer as a function of household income (the data points) and a logistic model of these data (the curve).[34]

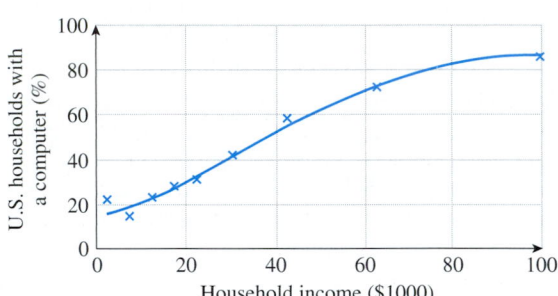

Household income ($1000)

The logistic model is

$$P(x) = \frac{91}{1 + 5.35(1.05)^{-x}} \text{ percent}$$

where x is the household income in thousands of dollars.
a. According to the model, what percentage of extremely wealthy households had computers?
b. For low incomes, the logistic model is approximately exponential. Which exponential model best approximates $P(x)$ for small x?
c. According to the model, 50% of households of what income had computers in 2000? (Round the answer to the nearest $1000.)

24. ● *Internet Use* The following graph shows the actual percentage of U.S. residents who used the Internet at home as a function of income (the data points) and a logistic model of these data (the curve).[35]

[32] SOURCE: The American Physical Society/*New York Times* May 3, 2003, p. A1.
[33] Ibid.

[34] Income levels are midpoints of income brackets. (The top income level is an estimate.) SOURCE: NTIA and ESA, U.S. Department of Commerce, using U.S. Bureau of the Census Current Population, 2000.
[35] Ibid.

● basic skills   ◆ challenging   **tech** Ex technology exercise

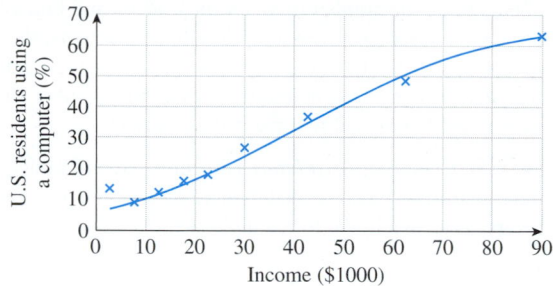

The logistic model is given by

$$P(x) = \frac{64.2}{1 + 9.6(1.06)^{-x}} \text{ percent}$$

where x is an individual's income in thousands of dollars.

a. According to the model, what percentage of extremely wealthy people used the Internet at home?

b. For low incomes, the logistic model is approximately exponential. Which exponential model best approximates $P(x)$ for small x?

c. According to the model, 50 percent of individuals with what income used the Internet at home in 2000? (Round the answer to the nearest $1000.)

25. ● **Epidemics** There are currently 1000 cases of Venusian flu in a total susceptible population of 10,000 and the number of cases is increasing by 25 percent each day. Find a logistic model for the number of cases of Venusian flu and use your model to predict the number of flu cases a week from now. *hint* [see Example 1]

26. ● **Epidemics** Last year's epidemic of Martian flu began with a single case in a total susceptible population of 10,000. The number of cases was increasing initially by 40 percent per day. Find a logistic model for the number of cases of Martian flu and use your model to predict the number of flu cases three weeks into the epidemic.

27. ● **Sales** You have sold 100 "I ♥ Calculus" T-shirts and sales appear to be doubling every five days. You estimate the total market for "I ♥ Calculus" T-shirts to be 3000. Give a logistic model for your sales and use it to predict, to the nearest day, when you will have sold 700 T-shirts.

28. ● **Sales** In Russia the average consumer drank two servings of Coca-Cola® in 1993. This amount appeared to be increasing exponentially with a doubling time of two years.[36] Given a long-range market saturation estimate of 100 servings per year, find a logistic model for the consumption of Coca-Cola in Russia and use your model to predict when, to the nearest year, the average consumption will be 50 servings per year.

29. tech Ex **Scientific Research** The following chart shows some the data shown in the graph in Exercise 21:

Year, t	0	5	10	15	20
Research Articles, A (1000)	1.2	2.1	3.8	5.1	5.7

($t = 0$ represents 1983.)[37]

a. What is the logistic regression model for the data? (Round all coefficients to two significant digits.) At what value does the model predict that the number of research articles will level off? *hint* [see Example 2]

b. According to the model, how many *Physics Review* articles were published by U.S. researchers in 2000 ($t = 17$)? (The actual number was about 5500 articles.)

30. tech Ex **Scientific Research** The following chart shows some of the data shown in the graph in Exercise 22:

Year, t	0	5	10	15	20
Percentage, P (Percentage over 25)	36	28	16	7	3

($t = 0$ represents 1983.)[38]

a. What is the logistic regression model for the data? (Round all coefficients to two significant digits.)

b. According to the model, what percentage of *Physics Review* articles were published by researchers in the U.S. in 2000 ($t = 17$)? (The actual figure was about 30.1%.)

31. tech Ex **Online Book Sales** The following table shows the number of books sold online in the U.S. in the period 1997–2000 ($t = 0$ represents 1997).[39]

t (Year)	0	1	2	3
Book sales (millions)	4.5	20.0	58.2	78.0

a. What is the logistic regression model for the data? (Round all coefficients to three significant digits.) At what value (to three significant digits) does the model predict that book sales will level off?

b. The 2000 figure ($t = 3$) represents approximately 7 percent of the total number of books sold. Some analysts predict that book sales will level off at around 15 percent of the market. Is this prediction consistent with the logistic regression model? Comment on the answer.

c. In what year does the model predict that book sales first exceed 80 million?

---

[36] The doubling time is based on retail sales of Coca-Cola products in Russia. Sales in 1993 were double those in 1991, and were expected to double again by 1995. SOURCE: *The New York Times*, September 26, 1994, p. D2.

[37] SOURCE: The American Physical Society/*New York Times* May 3, 2003, p. A1.

[38] Ibid.

[39] SOURCE: Ipsos-NPD Book Trends/*New York Times*, April 16, 2001, p. C1.

● basic skills   ◆ challenging   tech Ex technology exercise

32. `tech` Ex *South African Exports* The following table shows the value of South African exports to African nations for the period 1991–1999 ($t = 0$ represents 1990).[40]

t (Year)	1	2	3	4	5	6	7	8	9
Exports (billions of Rand)	5	6	7	9	13	18	20	20	22

a. What is the logistic regression model for the data? (Round all coefficients to three significant digits.) At what value does the model predict that the exports will level off?

b. In 2000, South Africa exported approximately R29 billion to African nations. Comment on this figure in light of your model.

`tech` Ex *Exercises 33 and 34 are based on the discussion following Example 2. If the limiting value N is known, then*

$$\frac{N}{y} - 1 = A(b^{-1})^x$$

and so $N/y - 1$ is an exponential function of x. In Exercises 33 and 34, use the given value of N and the data points $(x, N/y - 1)$ to obtain A and b, and hence a logistic model.

33. ◆ `tech` Ex *Education* The following chart shows the number of high-school graduates in the U.S. over the period 1993–1999.[41]

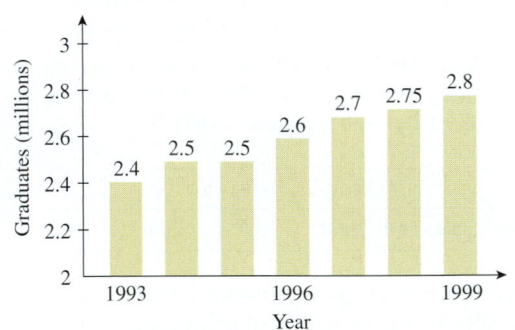

Take t to be the number of years since 1993, and find a logistic model based on the assumption that, eventually, the number of high school graduates will grow to 5 million per year. In what year does your model predict the number of high school graduates will first reach 3.5 million?

34. ◆ `tech` Ex *College Athletics* The percentage of college athletes who are women tended to increase over the preceding 20 years, as shown in the following chart.

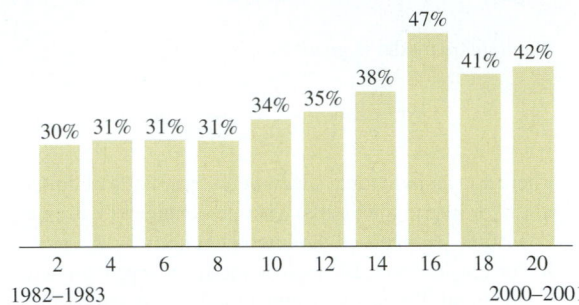

The long-term goal of many athletic policy makers is to attain a parity level of 50% women in college athletics. Using $N = 50$, find a logistic model for the percentage of women college athletes as a function of time t in years ($t = 2$ represents the 1982–1983 academic year). In which year does your model predict that 45% of all college athletes will be women?

Communication and Reasoning Exercises

35. ● Logistic functions are commonly used to model the spread of epidemics. Given this fact, explain why a logistic function is also useful to model the spread of a new technology.

36. ● Why is a logistic function more appropriate than an exponential function for modeling the spread of an epidemic?

37. ● Give one practical use for logistic regression.

38. ● What happens to the function $P(t) = \dfrac{N}{1 + Ab^{-t}}$ if $A = 0$? if $A < 0$?

---

[40] Data are approximate. SOURCE: ABSA: "South Africa's Foreign Trade, 2001 Edition"/*New York Times,* February 17, 2002, p. 14.

[41] Data are rounded. SOURCE: Western Interstate Commission for Higher Education and the College Board/*The New York Times,* February 17, 1999, p. B9.

Chapter **9** Review

KEY CONCEPTS

9.1 Quadratic Functions and Models

A **quadratic function** has the form
$f(x) = ax^2 + bx + c$ *p. 616*

The graph of $f(x) = ax^2 + bx + c$
$(a \neq 0)$ is a **parabola** *p. 617*

The x-coordinate of the **vertex** is $-\frac{b}{2a}$.
The y-coordinate is $f\left(-\frac{b}{2a}\right)$ *p. 617*

x-**intercepts** (if any) occur at
$$x = \frac{-b \pm \sqrt{b^2 - 4ac}}{2a} \quad p. 617$$

The y-**intercept** occurs at $y = c$ *p. 617*

The parabola is **symmetric** with respect to the vertical line through the vertex. *p. 617*

Sketching the graph of a quadratic function *p. 618*

Application to maximizing revenue *p. 619*

Application to maximizing profit *p. 620*

Finding the quadratic regression curve *p. 622*

9.2 Exponential Functions and Models

An **exponential function** has the form
$f(x) = Ab^x$ *p. 628*

Recognizing exponential data *p. 629*

Roles of the constants A and b in an exponential function $f(x) = Ab^x$ *p. 630*

Finding the exponential curve through two points *p. 632*

Application to exponential growth (epidemics) *p. 632*

Application to compound interest *p. 634*

Application to exponential decay (carbon dating) *p. 635*

The number e and continuous compounding *p. 636*

Alternative form of an exponential function: $f(x) = Ae^{rx}$ *p. 639*

Finding the exponential regression curve *p. 639*

9.3 Logarithmic Functions and Models

The **base b logarithm of x**: $y = \log_b x$
means $b^y = x$ *p. 645*

Common logarithm, $\log x = \log_{10} x$, and **natural logarithm,**
$\ln x = \log_e x$ *p. 646*

Change of base formula *p. 646*

Solving equations with unknowns in the exponent *p. 647*

A **logarithmic function** has the form
$f(x) = Ab^x$ *p. 647*

Graphs of logarithmic functions *p. 648*

Logarithm identities *p. 649*

Application to investments (how long?) *p. 649*

Application to half-life *p. 650*

Exponential growth and decay models and half-life *p. 652*

Finding the logarithmic regression curve *p. 652*

9.4 Logistic Functions and Models

A **logistic function** has the form
$$f(x) = \frac{N}{1 + Ab^{-x}} \quad p. 658$$

Properties of the logistic curve *p. 659*

Logistic function for small x, and the role of b *p. 660*

Application to epidemics *p. 660*

Finding the logistic regression curve *p. 661*

REVIEW EXERCISES

Sketch the graph of the quadratic functions in Exercises 1 and 2, indicating the coordinates of the vertex, the y-intercept, and the x-intercepts (if any).

1. $f(x) = x^2 + 2x - 3$ **2.** $f(x) = -x^2 - x - 1$

In Exercises 3 and 4, the values of two functions, f and g, are given in a table. One, both, or neither of them may be exponential. Decide which, if any, are exponential, and give the exponential models for those that are.

3.

x	-2	-1	0	1	2
$f(x)$	20	10	5	2.5	1.25
$g(x)$	8	4	2	1	0

4.

x	-2	-1	0	1	2
$f(x)$	8	6	4	2	1
$g(x)$	$\frac{3}{4}$	$\frac{3}{2}$	3	6	12

In Exercises 5 and 6 graph the given pairs of functions on the same set of axes with $-3 \leq x \leq 3$.

5. $f(x) = \frac{1}{2}(3^x); g(x) = \frac{1}{2}(3^{-x})$

6. $f(x) = 2(4^x); g(x) = 2(4^{-x})$

tech Ex *On the same set of axes, use technology to graph the pairs of functions in Exercises 7 and 8 for the given range of x. Identify which graph corresponds to which function.*

7. $f(x) = e^x; g(x) = e^{0.8x}; -3 \leq x \leq 3$

8. $f(x) = 2(1.01)^x; g(x) = 2(0.99)^x; -100 \leq x \leq 100$

In Exercises 9–14, compute the indicated quantity.

9. A \$3000 investment earns 3% interest, compounded monthly. Find its value after 5 years.

10. A \$10,000 investment earns 2.5% interest, compounded quarterly. Find its value after 10 years.

11. An investment earns 3% interest, compounded monthly and is worth \$5000 after 10 years. Find its initial value.

12. An investment earns 2.5% interest, compounded quarterly and is worth $10,000 after 10 years. Find its initial value.

13. A $3000 investment earns 3% interest, compounded continuously. Find its value after 5 years.

14. A $10,000 investment earns 2.5% interest, compounded continuously. Find its value after 10 years.

In Exercises 15–18, find a formula of the form $f(x) = Ab^x$ using the given information.

15. $f(0) = 4.5$; the value of f triples for every half-unit increase in x.

16. $f(0) = 5$; the value of f decreases by 75% for every one-unit increase in x.

17. $f(1) = 2$, $f(3) = 18$

18. $f(1) = 10$, $f(3) = 5$

In Exercises 19–22, use logarithms to solve the given equation for x.

19. $3^{-2x} = 4$

20. $2^{2x^2-1} = 2$

21. $300(10^{3x}) = 315$

22. $P(1+i)^{mx} = A$

On the same set of axes, graph the pairs of functions in Exercises 23 and 24.

23. $f(x) = \log_3 x$; $g(x) = \log_{(1/3)} x$

24. $f(x) = \log x$; $g(x) = \log_{(1/10)} x$

In Exercises 25–28, use the given information to find an exponential model of the form $Q = Q_0 e^{-kt}$ or $Q = Q_0 e^{kt}$, as appropriate. Round all coefficients to three significant digits when rounding is necessary.

25. Q is the amount of radioactive substance with a half-life of 100 years in a sample originally containing 5g (t is time in years).

26. Q is the number of cats on an island whose cat population was originally 10,000 but is being cut in half every five years (t is time in years).

27. Q is the diameter (in cm) of a circular patch of mold on your roommate's damp towel that you have been monitoring with morbid fascination. You measured the patch at 2.5 cm across four days ago, and have observed that it is doubling in diameter every two days (t is time in days).

28. Q is the population of cats on another island whose cat population was originally 10,000 but is doubling every 15 months (t is time in months).

In Exercises 29–32, find the time required, to the nearest 0.1 year, for the investment to reach the desired goal.

29. $2000 invested at 4% , compounded monthly; goal: $3000.

30. $2000 invested at 6.75%, compounded daily; goal: $3000.

31. $2000 invested at 3.75%, compounded continuously; goal: $3000.

32. $1000 invested at 100%, compounded quarterly; goal: $1200.

In Exercises 33–36, find equations for the logistic functions of x with the stated properties.

33. Through (0, 100), initially increasing by 50% per unit of x, and limiting value 900.

34. Initially exponential of the form $y = 5(1.1)^x$ with limiting value 25.

35. Passing through (0, 5) and decreasing from a limiting value of 20 to 0 at a rate of 20% per unit of x when x is near 0.

36. Initially exponential of the form $y = 2(0.8)^x$ with a value close to 10 when $x = -60$.

Applications

37. *Website Traffic* The daily traffic ("hits per day") at OHaganBooks.com seems to depend on the monthly expenditure on advertising through banner ads on well-known Internet portals. The following model, based on information you have collected over the past few months, shows the approximate relationship:

$$h = -0.000005c^2 + 0.085c + 1750$$

where h is the average number of hits per day at OHaganBooks.com, and c is the monthly advertising expenditure.

a. According to the model, what monthly advertising expenditure will result in the largest volume of traffic at OHaganBooks.com? What is that volume?

b. In addition to predicting a maximum volume of traffic, the model predicts that the traffic will eventually drop to zero if the advertising expenditure is increased too far. What expenditure (to the nearest dollar) results in no website traffic?

c. What feature of the formula for this quadratic model indicates that it will predict an eventual decline in traffic as advertising expenditure increases?

38. *Broadband Access* Pablo Pelogrande, a new summer intern at OHaganBooks.com, argues that broadband access to the Internet is increasing as a rate that justifies the website upgrades (video and audio content for broadband) that the company is planning: indeed the rate of growth of broadband was approximately $n(t) = 2t^2 - 2t + 8$ million new American adults with broadband per year, where t is time in years since the beginning of 2001.[42]

a. According to the model, when was the rate of growth at a minimum?

b. Does the model predict a zero rate of growth at any particular time? If so, when?

---

[42] Based on data for 2001–2003. Source for data: Pew Internet and American Life Project data memos dated May 18, 2003 and April 19, 2004, available at www.pewinternet.org.

c. What feature of the formula for this quadratic model indicates that the rate of growth eventually increases?

d. Does the fact that $n(t)$ decreases for $t \le 0.5$ suggest that the number of broadband users actually declined in early 2001? Explain.

39. Revenue Some time ago, you formulated the following linear model of demand:

$$q = -60p + 950$$

where q is the monthly demand for OHaganBooks.com's online novels at a price of p dollars per novel. Use this model to express the monthly revenue as a function of the unit price p, and hence determine the price you should charge for a maximum monthly revenue.

40. Profit Refer to the linear demand model in Exercise 39 for online novels. Author royalties and copyright fees cost the company an average of $4 per novel, and the monthly cost of operating and maintaining the online publishing service amounts to $900 per month. Express the monthly profit P as a function of the unit price p, and hence determine the unit price you should charge for a maximum monthly profit. What is the resulting profit (or loss)?

41. Lobsters Marjory Duffin is particularly fond of having steamed lobster at working lunches with executives from OHaganBooks.com, and is therefore alarmed by the news that the yearly lobster harvest from New York's Long Island Sound has been decreasing dramatically since 1997. Indeed, the size of the annual harvest can be approximated by

$$n(t) = 10(0.66^t) \text{ million pounds}$$

where t is time in years since June, 1997.[43]

a. The model tells us that the harvest was _____ million pounds in 1997 and decreasing by ___% each year.

b. What does the model predict for the 2005 harvest?

42. Stock Prices In the period immediately following its initial public offering (IPO), OHaganBooks.com's stock is doubling in value every three hours. If you bought $10,000 worth of the stock when it was first offered, how much was your stock worth after eight hours?

43. Lobsters (See Exercise 41.) Marjory Duffin has just left John O'Hagan, CEO of OHaganBooks.com, a frantic phone message to the effect that yearly lobster harvest from New York's Long Island Sound has just dipped below 100,000 pounds, making that planned lobster working lunch more urgent than ever. What year is it?

44. Stock Prices We saw in Exercise 42 that OHaganBooks.com's stock was doubling in value every three hours, following its

[43] Based on a regression model. Source for data: NY State Department of Environmental Conservation/*Newsday*, October 6, 2004, p. A4.

IPO. If you bought $10,000 worth of the stock when it was first offered, how long from the initial offering did it take your investment to reach $50,000?

45. Stock Prices We saw in Exercise 42 that OHaganBooks.com's stock was doubling in value every three hours, following its IPO. After 10 hours of trading, the stock turns around and starts losing one third of its value every four hours. How long (from the initial offering) will it be before your stock is once again worth $10,000?

46. tech Ex Lobsters The model in Exercise 41 was based on the data shown in the following chart:

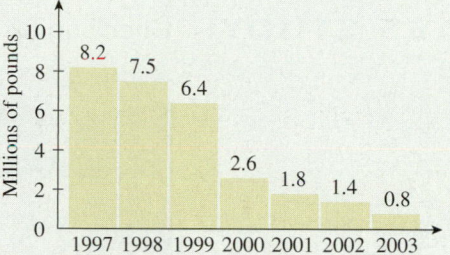

Yearly Lobster Harvest from Long Island Sound

Use the data to obtain an exponential regression curve of the form $n(t) = Ab^t$, with $t = 0$ corresponding to 1997 and coefficients rounded to two significant digits.

47. Hardware Life (*Based on a question from the GRE economics exam*) To estimate the rate at which new computer hard drives will have to be retired, OHaganBooks.com uses the "survivor curve":

$$L_x = L_0 e^{-x/t}$$

where

L_x = number of surviving hard drives at age x
L_0 = number of hard drives initially
t = average life in years

All of the following are implied by the curve *except*:

(A) Some of the equipment is retired during the first year of service.

(B) Some equipment survives three average lives.

(C) More than half the equipment survives the average life.

(D) Increasing the average life of equipment by using more durable materials would increase the number surviving at every age.

(E) The number of survivors never reaches zero.

48. Sales OHaganBooks.com modeled its weekly sales over a period of time with the function

$$s(t) = 6050 + \frac{4470}{1 + 14(1.73^{-t})}$$

as shown in the following graph (t is measured in weeks):

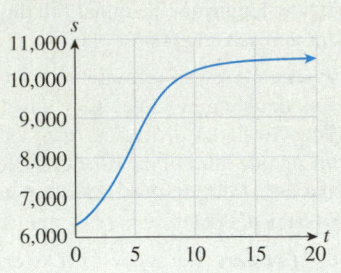

a. As time goes on, it appears that weekly sales are leveling off. At what value are they leveling off?

b. When did weekly sales rise above 10,000?

Ⓥ**Mentor** Do you need a live tutor for homework problems? Access vMentor on the ThomsonNOW! website at **www.thomsonedu.com** for one-on-one tutoring from a mathematics expert.

CASE STUDY: Checking up on Malthus

Park Street/PhotoEdit

In 1798 Thomas R. Malthus (1766–1834) published an influential pamphlet, later expanded into a book, titled *An Essay on the Principle of Population As It Affects the Future Improvement of Society*. One of his main contentions was that population grows geometrically (exponentially) while the supply of resources such as food grows only arithmetically (linearly). This led him to the pessimistic conclusion that population would always reach the limits of sub-sistence and precipitate famine, war, and ill-health, unless population could be checked by other means. He advocated "moral restraint," which includes the pattern of late marriage common in Western Europe at the time and is now common in most developed countries, and which leads to a lower reproduction rate.

Two hundred years later, you have been asked to check the validity of Malthus's con-tention. That population grows geometrically, at least over short periods of time, is com-monly assumed. That resources grow linearly is more questionable. You decide to check the actual production of a common crop, wheat, in the United States. Agricultural statis-tics like these are available from the U.S. government on the Internet, through the U.S. Department of Agriculture's National Agricultural Statistics Service (NASS). As of 2006, this service was available at http://www.nass.usda.gov/. Looking through this site, you locate data on the annual production of all wheat in the U.S. from 1866 through 2001.

A current link to NASS, as well as the data, are available online. Follow:

Chapter 9

→ Case Study

Year	1866	1867	. . .	2000	2001
Thousands of Bushels	169,703	210,878	. . .	2,232,460	1,957,643

Graphing these data (using Excel, for example), you obtain the graph in Figure 20.

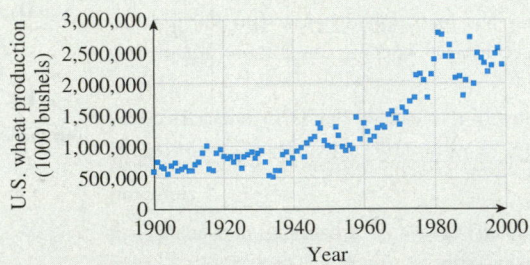

Figure **20**

This does not look very linear, particularly in the last half of the 20th century, but you continue checking the mathematics. Using Excel's built-in linear regression capabilities, you find that the line that best fits these data, shown in Figure 21, has $r^2 = 0.8002$. (Recall the discussion of the correlation coefficient r in Section 1.5. A similar statistic is available for other types of regression as well.)

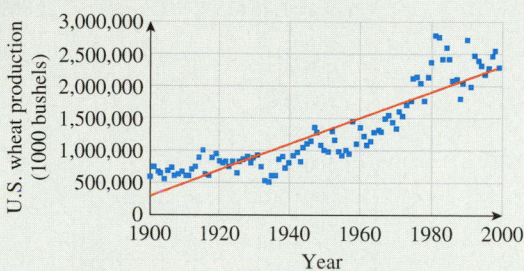

Figure **21**

Although that is a fairly high correlation, you notice that the residuals are not distributed randomly: The actual wheat production starts out higher than the line, is below the line from about 1920 to about 1970, and then is mostly above the line. This suggests that you would get a better fit from a function whose graph bends upwards. But what kind of curve? You decide to compare three different models: quadratic, exponential, and logistic.

Following is a comparison of the results of fitting the three proposed models. (Coefficients are rounded to six significant digits. SSE is the sum-of-squares error and r^2 is called the coefficient of determination, and measures how closely the regression curve fits the data.)

Model and Graph	SSE	r^2
Quadratic $P(t) \approx 245.067t^2 - 4319.47t + 711{,}703$ 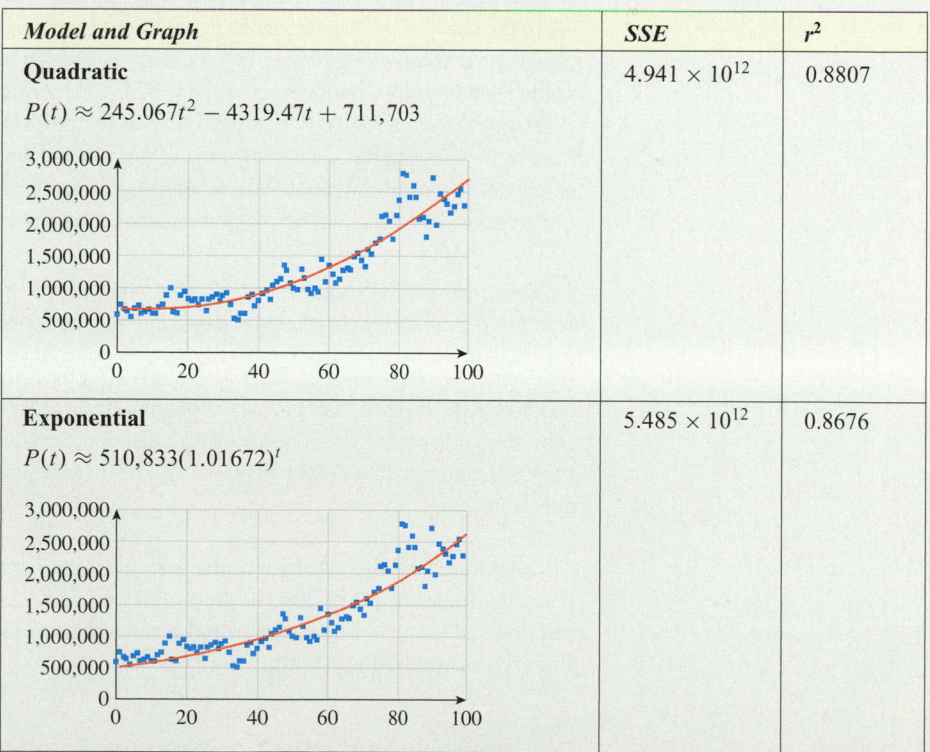	4.941×10^{12}	0.8807
Exponential $P(t) \approx 510{,}833(1.01672)^t$	5.485×10^{12}	0.8676

Model and Graph	SSE	r^2
Logistic $$P(t) \approx \frac{1.15135 \times 10^{15}}{1 + 2.25867 \times 10^{9} \times 1.01672^{-t}}$$ $(N \approx 1.15135 \times 10^{15}, \ A \approx 2.25867 \times 10^9, \ b \approx 1.01672)$	5.485×10^{12}	0.8676

Q: *Because the quadratic model gives a slightly higher value for r^2, it follows that the quadratic model is more appropriate than the exponential model, right?*

A: The quadratic model does give a better fit to the data than the exponential model (as evidenced by the values of both SSE and r^2). However, a better fit does not necessarily imply a more appropriate model. ∎

Q: *Why not?*

A: Suppose, for arguments' sake, that production of wheat *was* growing exponentially according to the above regression model *on average,* but that due to random fluctuations in market conditions and weather patterns, the actual output varied from the predicted output by a random amount. This randomness would explain the fact that the observed values do not lie exactly on the regression curve, but are instead scattered about in its vicinity. Moreover, because the scatter is due to random factors, any attempt to obtain a mathematical curve that fits the actual data more closely would be questionable, given our assumption. (For example, we *could* find a degree 99 polynomial whose graph passes through every data point!) ∎

Q: *Could you not then argue that the linear model might be the most appropriate of all three, and that the illusion of curvature is caused by those same random fluctuations?*

A: That could conceivably be happening, but two factors argue against it: First, the positive-negative-positive pattern of the residuals in the linear model (Figure 21) strongly suggest that something other than a linear relationship is going on. Second (and we will be vague here) there are rigorous statistical tests one can perform to support the hypothesis that there is curvature.[44] ∎

Although neither the quadratic nor the exponential model has a clear advantage over the other, the good fit of both models provides reasonable evidence that wheat production, at least, is better described as increasing quadratically or exponentially than linearly, contradicting Malthus.

---

[44] These hypothesis tests are generally discussed in statistics courses.

If you compare the exponential and logistic models, you get additional interesting information. The logistic model seems as though it *ought* to be the most appropriate, because wheat production cannot reasonably be expected to continue increasing exponentially forever; eventually resource limitations must lead to a leveling-off of wheat production. Such a leveling off, if it occurred before the population started to level off, would seem to vindicate Malthus's pessimistic predictions.[45]

However, the logistic regression model looks suspect: It predicts a leveling-off value (N) that is orders of magnitude higher than what seems reasonable. Moreover, the logistic model (as well as SSE and r^2) looks almost indistinguishable from the exponential model, suggesting that it is no better. You are forced to conclude that wheat production—even if it is logistic—is still in the early (exponential) stage of growth, and thus shows no sign, as yet, of leveling off. In general, for a logistic model to be reliable in its prediction of the leveling-off value N, we would need to see significant evidence of leveling-off in the data.

You tentatively conclude that wheat production for the past 100 years is better described as increasing quadratically or exponentially than linearly, contradicting Malthus, and moreover that it shows no sign of leveling off as yet.

Exercises

1. Find the best-fit line for wheat production for the period 1866–1940. How good a fit is the line?

2. Find the best-fit line for wheat production for the period 1940–1997. How good a fit is the line?

3. Find the production figures for another common crop grown in the U.S. Compare the linear, quadratic, exponential, and logistic models. What can you conclude?

4. Below are the census figures for the population of the U.S. (in thousands) from 1800 to 2000.[46] Compare the linear, quadratic, and exponential models. What can you conclude?

Population of the U.S. (1000)

1810	1820	1830	1840	1850	1860	1870	1880	1890	1900
7,240	9,638	12,861	17,063	23,192	31,443	38,558	50,189	62,980	76,212
1910	**1920**	**1930**	**1940**	**1950**	**1960**	**1970**	**1980**	**1990**	**2000**
92,228	106,022	123,203	132,165	151,326	179,323	203,302	226,542	248,710	281,422

---

[45] See the exercise set for data on population.

[46] SOURCE: Bureau of the Census, U.S. Department of Commerce.

Section 9.1

Example 2 Sketch the graph of each quadratic function, showing the location of the vertex and intercepts.

a. $f(x) = 4x^2 - 12x + 9$ **b.** $g(x) = -\frac{1}{2}x^2 + 4x - 12$

Solution with Technology We will do part (a).

1. Start by storing the coefficients a, b, c using

    4→A:-12→B:9→C

    STO> gives the arrow   ALPHA . gives the colon

2. Save your quadratic as Y₁ using the Y= screen:

    Y₁=AX^2+BX+C

3. To obtain the x-coordinate of the vertex, enter its formula as shown:

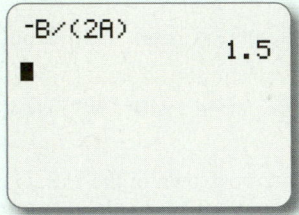

4. The y-coordinate of the vertex can be obtained from the table screen by entering $x = 1.5$ as shown. (If you can't enter values of x, press 2ND   TBLSET , and set Indpnt to Ask.)

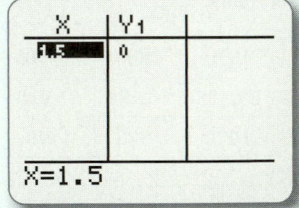

From the table, we see that the vertex is at the point $(1.5, 0)$.

5. To obtain the x-intercepts, enter the quadratic formula on the home screen as shown:

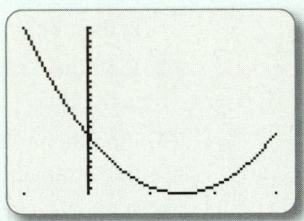

Because both intercepts agree, we conclude that the graph intersects the x-axis on a single point (at the vertex).

6. To graph the function, we need to select good values for Xmin and Xmax. In general, we would like our graph to show the vertex as well as all the intercepts. To see the vertex, make sure that its x-coordinate (1.5) is between Xmin and Xmax. To see the x-intercepts, make sure that they are also between Xmin and Xmax (also 1.5). To see the y-intercept, make sure that $x = 0$ is between Xmin and Xmax. Thus, to see everything, choose Xmin and Xmax so that the interval [xMin, xMax] contains the x-coordinate of the vertex, the x-intercepts, and 0. For this example, we can choose an interval like $[-1, 3]$.

7. Once xMin and xMax are chosen, you can obtain convenient values of yMin and yMax by pressing ZOOM and selecting the option ZoomFit. (Make sure that your quadratic equation is entered in the Y= screen before doing this!)

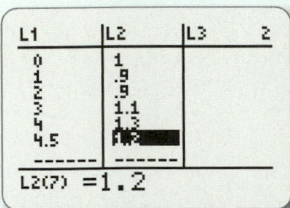

Example 5b The following table shows the value of the euro (€) in U.S. dollars since it began trading in January, 1999 ($t = 0$ represents January, 2000).[47]

Year t	−1	0	1	2	3	4	4.5
Value ($)	1.2	1	0.9	0.9	1.1	1.3	1.2

Find the quadratic regression model.

Solution with Technology

1. Using STAT EDIT enter the data in the TI-83/84 putting the x-coordinates (values of t) in L₁ and the y-coordinates (values of the euro) in L₂, just as in Section 1.5:

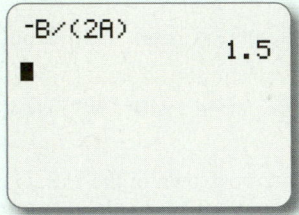

---

[47] Prices are rounded 20-day averages. Source for data: http://finance.yahoo.com/.

2. Press $\boxed{\text{STAT}}$, select CALC, and choose the option QuadReg. Pressing $\boxed{\text{ENTER}}$ gives the quadratic regression curve in the home screen:

$$y \approx 0.0399x^2 - 0.1150x + 1.0154$$

Coefficients rounded to four decimal places

3. To graph the points and regression line in the same window, turn Stat Plot on by pressing $\boxed{\text{2ND}}$ $\boxed{\text{STAT PLOT}}$, selecting 1 and turning PLOT1 on:

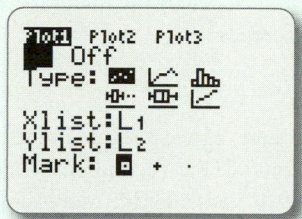

4. Next, enter the regression equation in the Y= screen by pressing $\boxed{\text{Y=}}$, clearing out whatever function is there, and pressing $\boxed{\text{VARS}}$ $\boxed{5}$ and selecting EQ (Option 1: RegEq):

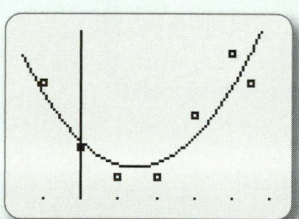

5. To obtain a convenient window showing all the points and the lines, press $\boxed{\text{ZOOM}}$ and choose option #9: ZoomStat:

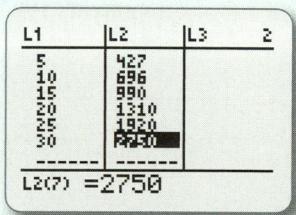

Note When you are done viewing the graph, it is a good idea to turn PLOT1 off again to avoid errors in graphing or data points showing up in your other graphs. ∎

Section **9.2**

Example 4b You invest $2000 in a mutual fund with an annual yield of 12.6% and the interest is reinvested each month. Use the model to estimate the year when the value of your investment will reach $5000.

Solution with Technology We need to find the value of t for which $A(t) = \$5000$, so we need to solve the equation $5000 = 2000(1.0105)^{12t}$ for t.

1. In the Y= screen, enter

Y$_1$=2000*1.0105^(12X)

2. Go to the Table screen, where you can list values of this function for various values of X:

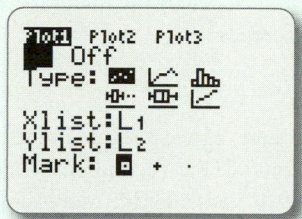

Because the balance first exceeds $5000 in year 8, the answer is $t = 8$ years.

Example 7a The following table shows annual expenditure on health in the U.S. from 1980 through 2010 ($t = 0$ represents 1980).[48]

Year t	0	5	10	15	20	25	30
Expenditure ($ Billion)	246	427	696	990	1310	1920	2750

Find the exponential regression model $C(t) = Ab^t$.

Solution with Technology This is very similar to Example 5 in Section 9.1 (see the Technology Guide for Section 9.1):

1. Use $\boxed{\text{STAT}}$ EDIT to enter the above table of values.

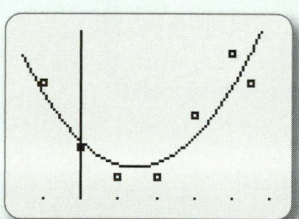

---

[48] Data are rounded. The 2005 and 2010 figures are projections.
SOURCE: Centers for Medicare and Medicaid Services, "National Health Expenditures," 2002 version, released January 2004; www.cms.hhs.gov/statistics/nhe/.

2. Press STAT, select CALC, and choose the option ExpReg. Pressing ENTER gives the exponential regression curve in the home screen:

$$C(t) \approx 282.33(1.0808)^t \quad \text{Coefficients rounded}$$

3. To graph the points and regression line in the same window, turn Stat Plot on (see the Technology Guide for Example 5 in Section 9.1) and enter the regression equation in the Y= screen by pressing Y=, clearing out whatever function is there, and pressing VARS 5 and selecting EQ (Option 1: RegEq). Then press ZOOM and choose option #9: ZoomStat to see the graph.

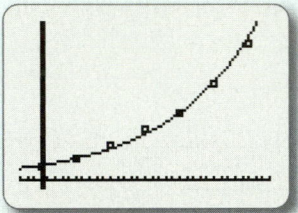

Note When you are done viewing the graph, it is a good idea to turn PLOT1 off again to avoid errors in graphing or data points showing up in your other graphs. ■

Section **9.3**

Example 5 The following table shows the total spent on research and development in the U.S., in billions of dollars, for the period 1995–2005 ($t = 5$ represents 1995).[49]

Year t	5	6	7	8	9	10
Spending ($ billions)	187	197	208	219	232	248
Year t	11	12	13	14	15	
Spending ($ billions)	250	250	253	260	265	

---

[49] Data are approximate and are given in constant 1996 dollars. 2004 and 2005 figures are projections. Source: National Science Foundation, Division of Science Resource Statistics, National Patterns of R&D Resources. www.nsf.gov/sbe/srs/nprdr/start.hrm October 2004.

Find the best-fit logarithmic model of the form:

$$S(t) = A \ln t + C$$

Solution with Technology This is very similar to Example 5 in Section 9.1 and Example 7 in Section 9.2 (see the Technology Guide for Section 9.1):

1. Use STAT EDIT to the enter above table of values.

2. Press STAT, select CALC, and choose the option LnReg. Pressing ENTER gives the quadratic regression curve in the home screen:

$$S(t) = 73.77 \ln t + 67.75 \quad \text{Coefficients rounded}$$

3. To graph the points and regression line in the same window, turn Stat Plot on (see the Technology Guide for Example 5 in Section 9.1) and enter the regression equation in the Y= screen by pressing Y=, clearing out whatever function is there, and pressing VARS 5 and selecting EQ (Option 1: RegEq). To see the graph, press ZOOM and choose option #9: ZoomStat:

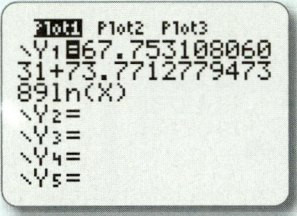

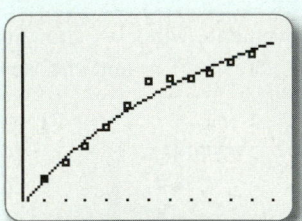

Section **9.4**

Example 2 Here are the data graphed in Figure 17:

Year (t)	0	1	2	3	4	5	6	7	8	9
Households with PCs (%) (P)	24	28	32	37	42	48	54	59	63	65

Find a logistic regression curve of the form

$$P(t) = \frac{N}{1 + Ab^{-t}}$$

TECHNOLOGY GUIDE

Solution with Technology This is very similar to Example 5 in Section 9.1 (see the Technology Guide for Section 9.1):

1. Use $\boxed{\text{STAT}}$ EDIT to enter the above table of values.

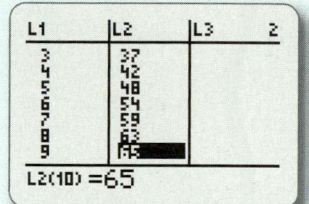

2. Press $\boxed{\text{STAT}}$, select CALC, and choose the option Logistic. Pressing $\boxed{\text{ENTER}}$ gives the exponential regression curve in the home screen:

$$P(t) \approx \frac{84.573}{1 + 2.6428e^{-0.250363t}}$$ Coefficients rounded

This is not exactly the form we are seeking, but we can convert it to that form by writing

$$e^{-0.250363t} = (e^{0.250363})^{-t} \approx 1.2845^{-t}$$

so

$$P(t) \approx \frac{84.573}{1 + 2.6428(1.2845)^{-t}}$$

3. To graph the points and regression line in the same window, turn Stat Plot on (see the Technology Guide for Example 5 in Section 9.1) and enter the regression equation

in the Y= screen by pressing $\boxed{\text{Y=}}$, clearing out whatever function is there, and pressing $\boxed{\text{VARS}}$ $\boxed{5}$ and selecting EQ (Option 1: RegEq):

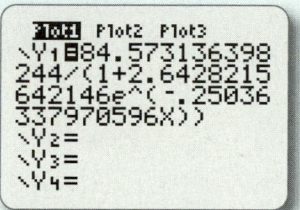

To obtain a convenient window showing all the points and the lines, press $\boxed{\text{ZOOM}}$ and choose option #9: Zoom-Stat:

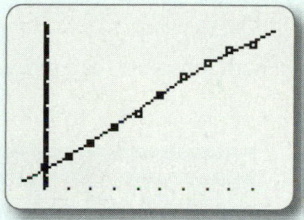

Note When you are done viewing the graph, it is a good idea to turn PLOT1 off again to avoid errors in graphing or data points showing up in your other graphs. ∎

EXCEL Technology Guide

Section 9.1

Example 2 Sketch the graph of each quadratic function, showing the location of the vertex and intercepts.

a. $f(x) = 4x^2 - 12x + 9$ **b.** $g(x) = -\frac{1}{2}x^2 + 4x - 12$

Solution with Technology We can set up a worksheet so that all we have to enter are the coefficients a, b, and c, and a range of x values for the graph. Here is a possible layout that will plot 100 points using the coefficients for part (a) (similar to the Excel Graphing Worksheet we used in Example 2 of Section 1.2).

1. First, we compute the x-coordinates:

	A	B	C	D	E
1	x	y		a	4
2	=$D$4			b	-12
3	=A2+$D$6			c	9
4				Xmin	-10
5				Xmax	10
6				Delta X	=(D5-D4)/100
7					
102					
103					

2. To add the y-coordinates, we use the technology formula

```
a*x^2+b*x+c
```

replacing a, b, and c with (absolute) references to the cells containing their values.

	A	B	C	D	E
1	x	y		a	4
2	-10	=$D$1*A2^2+$D$2*A2+$D$3		b	-12
3	-9.8			c	9
4	-9.6			Xmin	-10
5	-9.4			Xmax	10
6	-9.2			Delta X	0.2
7					
	9.8				
102	10				
103					
104					

3. Graphing the data in columns A and B gives the graph shown here:

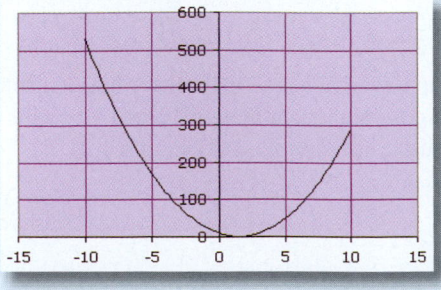

$$y = 4x^2 - 12x + 9$$

4. We can go further and compute the exact coordinates of the vertex and intercepts:

	A	B	C	D	E	F
1	x	y	a	4	Vertex (x)	=-D2/(2*D1)
2	-10	529	b	-12	Vertex (y)	=D1*F1^2+D2*F1+D3
3	-9.8	510.76	c	9	X Intercept 1	=(-D2-SQRT(D2^2-4*D1*D3))/(2*D1)
4	-9.6	492.84	Xmin	-10	X Intercept 2	=(-D2+SQRT(D2^2-4*D1*D3))/(2*D1)
5	-9.4	475.24	Xmax	10	Y Intercept	=D3
6	-9.2	457.96	Delta X	0.2		
7	-9	441				
8	-9.8	424.36				
9						

The completed sheet should look like this:

	A	B	C	D	E	F
1	x	y	a	4	Vertex (x)	1.5
2	-10	529	b	-12	Vertex (y)	0
3	-9.8	510.76	c	9	X Intercept 1	1.5
4	-9.6	492.84	Xmin	-10	X Intercept 2	1.5
5	-9.4	475.24	Xmax	10	Y Intercept	9
6	-9.2	457.96	Delta X	0.2		
7	-9	441				
8	-9.8	424.36				
9						

We can now save this sheet as a template to handle all quadratic functions. For instance, to do part (b), we just change the values of a, b, and c in column D to $a = -1/2$, $b = 4$, and $c = -12$.

Example 5b The following table shows the value of the euro (€) in U.S. dollars since it began trading in January, 1999 ($t = 0$ represents January, 2000).[50]

Year t	−1	0	1	2	3	4	4.5
Value ($)	1.2	1	0.9	0.9	1.1	1.3	1.2

Find the quadratic regression model.

Solution with Technology As in Section 1.5, Example 3, we start with a scatter plot of the original data, and add a trendline:

---

[50] Prices are rounded 20-day averages. Source for data: http://finance.yahoo.com/.

1. Start with the original data and a "Scatter plot."

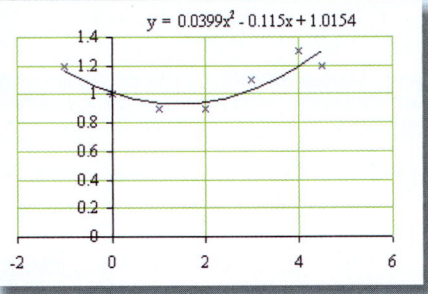

2. Click on the chart, and select "Add Trendline . . ." from the Chart menu. Then select a "Polynomial" type, set the order to 2, and, under "Options," check the option "Display equation on chart".

Section **9.2**

Example 4b You invest $2000 in a mutual fund with an annual yield of 12.6% and the interest is reinvested each month. Use the model to estimate the year when the value of your investment will reach $5000.

Solution with Technology We need to find the value of t for which $A(t) = \$5000$, so we need to solve the equation $5000 = 2000(1.0105)^{12t}$ for t. Here is a table you can obtain in Excel (to format a cell as currency, highlight the cell and use the Format menu: Format → Cells → Currency):

Because the balance first exceeds \$5000 in year 8, the answer is $t = 8$ years.

Example 7a The following table shows annual expenditure on health in the U.S. from 1980 through 2010 ($t = 0$ represents 1980).[51]

Year t	0	5	10	15	20	25	30
Expenditure (\$ Billion)	246	427	696	990	1310	1920	2750

Find the exponential regression model $C(t) = Ab^t$.

Solution with Technology This is very similar to Example 5 in Section 9.1 (see the Technology Guide for Section 9.1):

1. Start with a "Scatter plot" of the observed data, click on the chart, and select "Add Trendline . . ." from the Chart menu.
2. In the dialogue box that now appears, select the "Exponential" type and, under "Options", check the option "Display equation on chart":

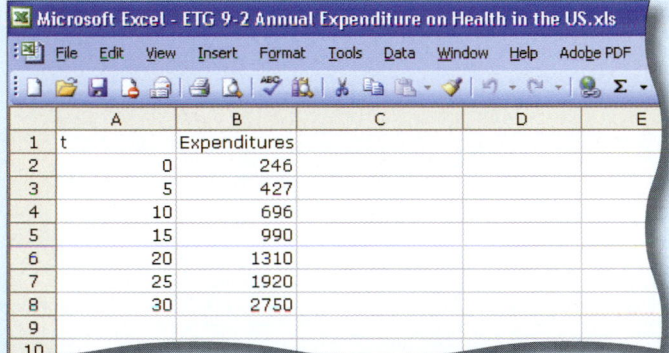
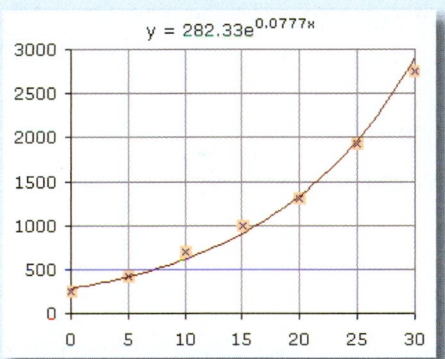

Notice that the regression curve is given in the form Ae^{kt} rather than Ab^t. To transform it, write

$$282.33e^{0.0777t} = 282.33(e^{0.0777})^t$$
$$\approx 282.33(1.0808)^t \qquad e^{0.0777} \approx 1.0808$$

Section 9.3

Example 5 The following table shows the total spent on research and development in the U.S., in billions of dollars, for the period 1995–2005 ($t = 5$ represents 1995).[52]

Year t	5	6	7	8	9	10
Spending (\$ billions)	187	197	208	219	232	248
Year t	11	12	13	14	15	
Spending (\$ billions)	250	250	253	260	265	

---

[51] Data are rounded. 2005 and 2010 figures are projections. SOURCE: Centers for Medicare and Medicaid Services, "National Health Expenditures," 2002 version, released January 2004; www.cms.hhs.gov/statistics/nhe/

[52] Data are approximate and are given in constant 1996 dollars. The 2004 and 2005 figures are projections. SOURCE: National Science Foundation, Division of Science Resource Statistics, National Patterns of R&D Resources. www.nsf.gov/sbe/srs/nprdr/start.hrm October 2004.

Find the best-fit logarithmic model of the form

$$S(t) = A \ln t + C$$

Solution with Technology This is very similar to Example 5 in Section 9.1 and Example 7 in Section 9.2 (see the Technology Guide for Section 9.1): We start, as usual, with a "Scatter plot" of the observed data and add a `Logarithmic` trendline. Here is the result:

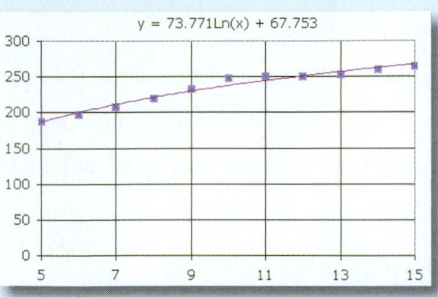

Section **9.4**

Example 2 Here are the data graphed in Figure 17:

Year (t)	0	1	2	3	4	5	6	7	8	9
Households with PCs (%) (P)	24	28	32	37	42	48	54	59	63	65

Find a logistic regression curve of the form

$$P(t) = \frac{N}{1 + Ab^{-t}}$$

Solution with Technology Excel does not have a built-in logistic regression calculation, so we use an alternative method that works for any type of regression curve.

1. First use rough estimates for N, A, and b, and compute the sum-of-squares error (SSE; see Section 1.5) directly:

	A	B	C	D	E	F	G
1	t	P (Observed)	P (Predicted)	Residual^2	N	A	b
2	0	24	=$E$2/(1+$F$2*$G$2^(-A2))	=(C2-B2)^2	70	2	1.15
3	1	28					
4	2	32					
5	3	37			SSE		
6	4	42			=SUM(D2:D11)		
7	5	48					
8	6	54					
9	7	59					
10	8	63					
11	9	65					
12							
13							

Microsoft Excel - ETG 9-4 Logistic Regression Calculation.xls

Cells E2:G2 contain our initial rough estimates of N, A, and b. For N, we used 70 (notice that the y-coordinates do appear to level off around 70). For A, we used the fact that the y-intercept is $N/(1 + A)$. In other words,

$$24 = \frac{70}{1 + A}$$

Because a very rough estimate is all we are after, using $A = 2$ will do just fine. For b, we chose 1.15 as the values of P appear to be increasing by around 15% per year initially (again this is rough—we could equally have use 1.1 or 1.2).

2. Cell C2 contains the formula for $P(t)$, and the square of the resulting residual is computed in D2.

3. Cell F6 will contain SSE. The completed spreadsheet should look like this:

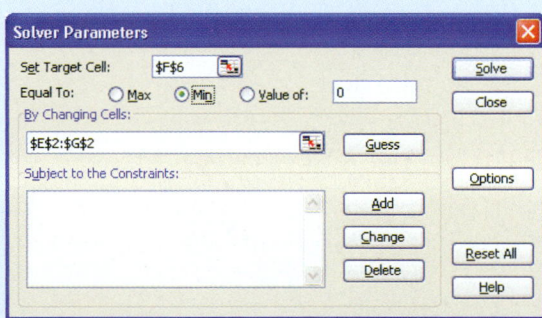

The best-fit curve will result from values of N, A, and b that give a minimum value for SSE. We shall use Excel's "Solver," found in the "Tools" menu, to find these values for us. (If "Solver" does not appear in the Tools menu, select Add-Ins in the Tools menu and install it.) Figure 22 shows the dialogue box with the necessary fields completed to solve the problem.

Figure **22**

- The Target Cell refers to the cell that contains SSE
- "Min" is selected because we are minimizing SSE.
- "Changing Cells" are obtained by selecting the cells that contain the current values of N, A, and b.

4. When you have filled in the values for the three items above, press "Solve" and tell Solver to Keep Solver Solution when done. You will find $N \approx 84.573$, $A \approx 2.6428$, and $b \approx 1.2845$ so that

$$P(t) \approx \frac{84.573}{1 + 2.6428(1.2845)^{-t}}$$

If you use a scatter plot to graph the data in columns A, B and C, you will obtain the following graph:

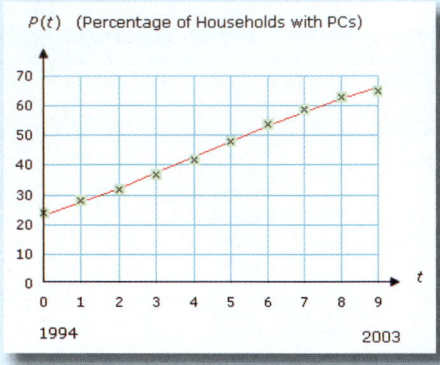

10

Introduction to the Derivative

Get a better Grade!
www.thomsonedu.com/login

Logon to your Personalized Study plan to find:

• Section by section tutorials

• A detailed chapter summary

• Additional review exercises

• Graphers, Excel tutorials, and other resources

• Optional sections:
Sketching the Graph of the Derivative
Proof of the Power Rule
Continuity and Differentiability

CASE STUDY Reducing Sulfur Emissions

The Environmental Protection Agency (EPA) wants to formulate a policy that will encourage utilities to reduce sulfur emissions. Its goal is to reduce annual emissions of sulfur dioxide by a total of 10 million tons from the current level of 25 million tons by imposing a fixed charge for every ton of sulfur released into the environment per year. The EPA has some data showing the marginal cost to utilities of reducing sulfur emissions. As a consultant to the EPA, you must determine the amount to be charged per ton of sulfur emissions in light of these data.

Creatas/Superstock

Introduction

In the world around us, everything is changing. The mathematics of change is largely about the rate of change: how fast and in which direction the change is occurring. Is the Dow Jones average going up, and if so, how fast? If I raise my prices, how many customers will I lose? If I launch this missile, how fast will it be traveling after two seconds, how high will it go, and where will it come down?

We have already discussed the concept of rate of change for linear functions (straight lines), where the slope measures the rate of change. But this works only because a straight line maintains a constant rate of change along its whole length. Other functions rise faster here than there—or rise in one place and fall in another—so that the rate of change varies along the graph. The first achievement of calculus is to provide a systematic and straightforward way of calculating (hence the name) these rates of change. To describe a changing world, we need a language of change, and that is what calculus is.

The history of calculus is an interesting story of personalities, intellectual movements, and controversy. Credit for its invention is given to two mathematicians: Isaac Newton (1642–1727) and Gottfried Leibniz (1646–1716). Newton, an English mathematician and scientist, developed calculus first, probably in the 1660s. We say "probably" because, for various reasons, he did not publish his ideas until much later. This allowed Leibniz, a German mathematician and philosopher, to publish his own version of calculus first, in 1684. Fifteen years later, stirred up by nationalist fervor in England and on the continent, controversy erupted over who should get the credit for the invention of calculus. The debate got so heated that the Royal Society (of which Newton and Leibniz were both members) set up a commission to investigate the question. The commission decided in favor of Newton, who happened to be president of the society at the time. The consensus today is that both mathematicians deserve credit because they came to the same conclusions working independently. This is not really surprising: Both built on well-known work of other people, and it was almost inevitable that someone would put it all together at about that time.

algebra Review

For this chapter, you should be familiar with the algebra reviewed in Chapter 0, Section 2.

10.1 Limits: Numerical and Graphical Approaches (OPTIONAL)

Rates of change are calculated by derivatives, but an important part of the definition of the derivative is something called a **limit.** Arguably, much of mathematics since the 18th century has revolved around understanding, refining, and exploiting the idea of the limit. The basic idea is easy, but getting the technicalities right is not.

Evaluating Limits Numerically

Start with a very simple example: Look at the function $f(x) = 2 + x$ and ask: What happens to $f(x)$ as x approaches 3? The following table shows the value of $f(x)$ for values of x close to and on either side of 3:

	x approaching 3 from the left →					← *x* approaching 3 from the right			
x	2.9	2.99	2.999	2.9999	**3**	3.0001	3.001	3.01	3.1
f(x) = 2 + x	4.9	4.99	4.999	4.9999		5.0001	5.001	5.01	5.1

We have left the entry under 3 blank to emphasize that when calculating the limit of $f(x)$ as x *approaches* 3, we are not interested in its value when x *equals* 3.

Notice from the table that the closer x gets to 3 from either side, the closer $f(x)$ gets to 5. We write this as

$$\lim_{x \to 3} f(x) = 5 \qquad \text{The limit of } f(x), \text{ as } x \text{ approaches 3, equals 5.}$$

Q: Why all the fuss? Can't we simply substitute $x = 3$ and avoid having to use a table*?*

A: This happens to work for *some* functions, but not for *all* functions. The following example illustrates this point. ∎

Example 1 Estimating a Limit Numerically

Use a table to estimate the following limits:

a. $\displaystyle\lim_{x \to 2} \frac{x^3 - 8}{x - 2}$ **b.** $\displaystyle\lim_{x \to 0} \frac{e^{2x} - 1}{x}$

Solution

a. We cannot simply substitute $x = 2$, because the function $f(x) = \dfrac{x^3 - 8}{x - 2}$ is not defined at $x = 2$. (Why?)[*] Instead, we use a table of values as we did above, with x approaching 2 from both sides.

	x approaching 2 from the left →					← x approaching 2 from the right			
x	1.9	1.99	1.999	1.9999	2	2.0001	2.001	2.01	2.1
$f(x) = \dfrac{x^3 - 8}{x - 2}$	11.41	11.9401	11.9940	11.9994		12.0006	12.0060	12.0601	12.61

We notice that as x approaches 2 from either side, $f(x)$ appears to be approaching 12. This suggests that the limit is 12, and we write

$$\lim_{x \to 2} \frac{x^3 - 8}{x - 2} = 12$$

b. The function $g(x) = \dfrac{e^{2x} - 1}{x}$ is not defined at $x = 0$ (nor can it even be simplified to one which *is* defined at $x = 0$). In the following table, we allow x to approach 0 from both sides:

	x approaching 0 from the left →					← x approaching 0 from the right			
x	-0.1	-0.01	-0.001	-0.0001	0	0.0001	0.001	0.01	0.1
$g(x) = \dfrac{e^{2x} - 1}{x}$	1.8127	1.9801	1.9980	1.9998		2.0002	2.0020	2.0201	2.2140

The table suggests that $\displaystyle\lim_{x \to 0} \frac{e^{2x} - 1}{x} = 2$.

 using *Technology*

We can automate these computations using a graphing calculator or Excel. See the Technology Guides at the end of the chapter to find out how to create tables like these using a TI-83/84 or Excel.

---

[*]However, if you factor $x^3 - 8$, you will find that $f(x)$ can be simplified to a function which *is* defined at $x = 2$. This point will be discussed (and this example redone) in Section 10.3. The function in part (b) cannot be simplified by factoring.

+ *Before we go on...* Although the table *suggests* that the limit in Example 1 part (b) is 2, it by no means establishes that fact conclusively. It is *conceivable* (though not in fact the case here) that putting $x = 0.000000087$ could result in $g(x) = 426$. Using a table can only suggest a value for the limit. In the next two sections we shall discuss algebraic techniques for finding limits. ■

Before we continue, let us make a more formal definition.

Definition of a Limit

If $f(x)$ approaches the number L as x approaches (but is not equal to) a from both sides, then we say that $f(x)$ **approaches L as $x \to a$** ("x approaches a") or that the **limit** of $f(x)$ as $x \to a$ is L. We write

$$\lim_{x \to a} f(x) = L$$

or

$$f(x) \to L \text{ as } x \to a$$

If $f(x)$ *fails* to approach *a single fixed number* as x approaches a from both sides, then we say that $f(x)$ **has no limit** as $x \to a$, or

$$\lim_{x \to a} f(x) \textbf{ does not exist.}$$

quick Examples

1. $\lim\limits_{x \to 3}(2 + x) = 5$ See discussion before Example 1.

2. $\lim\limits_{x \to -2}(3x) = -6$ As x approaches -2, $3x$ approaches -6.

3. $\lim\limits_{x \to 0}(x^2 - 2x + 1)$ exists.   In fact, the limit is 1.

4. $\lim\limits_{x \to 5} \dfrac{1}{x} = \dfrac{1}{5}$ As x approaches 5, $\dfrac{1}{x}$ approaches $\dfrac{1}{5}$.

5. $\lim\limits_{x \to 2} \dfrac{x^3 - 8}{x - 2} = 12$ See Example 1. (We cannot just put $x = 2$ here.)

(For examples where the limit does not exist, see Example 2.)

Notes

1. It is important that $f(x)$ approach the same number as x approaches a from either side. For instance, if $f(x)$ approaches 5 for $x = 1.9, 1.99, 1.999, \ldots$, but approaches 4 for $x = 2.1, 2.01, 2.001, \ldots$, then the limit as $x \to 2$ does not exist. (See Example 2 for such a situation.)

2. It may happen that $f(x)$ does not approach any fixed number at all as $x \to a$ from either side. In this case, we also say that the limit does not exist.

3. We are being deliberately vague as to exactly what we mean by the word "approaches"; instead, we trust your intuition. However, the following phrasing of the definition of the limit is close to the more technical one used by mathematicians. ■

Verbal Form of the Mathematical Definition of a Limit

We can make $f(x)$ be as close to L as we like by choosing any x sufficiently close to (but not equal to) a.

The following example gives instances in which a stated limit does not exist.

Example 2 Limits Do Not Always Exist

Do the following limits exist?

a. $\lim\limits_{x \to 0} \dfrac{1}{x^2}$ **b.** $\lim\limits_{x \to 0} \dfrac{|x|}{x}$ **c.** $\lim\limits_{x \to 2} \dfrac{1}{x - 2}$

Solution

a. Here is a table of values for $f(x) = \dfrac{1}{x^2}$, with x approaching 0 from both sides.

x approaching 0 from the left →   ← x approaching 0 from the right

x	-0.1	-0.01	-0.001	-0.0001	**0**	0.0001	0.001	0.01	0.1
$f(x) = \dfrac{1}{x^2}$	100	10,000	1,000,000	100,000,000		100,000,000	1,000,000	10,000	100

The table shows that as x gets closer to zero on either side, $f(x)$ gets larger and larger **without bound**—that is, if you name any number, no matter how large, $f(x)$ will be even larger than that if x is sufficiently close to 0. Because $f(x)$ is not approaching any real number, we conclude that $\lim\limits_{x \to 0} \dfrac{1}{x^2}$ does not exist. Because $f(x)$ is becoming arbitrarily large, we also say that $\lim\limits_{x \to 0} \dfrac{1}{x^2}$ **diverges to** $+\infty$, or just

$$\lim_{x \to 0} \frac{1}{x^2} = +\infty$$

Note This is not meant to imply that the limit exists; the symbol $+\infty$ does not represent any real number. We write $\lim_{x \to a} f(x) = +\infty$ to indicate two things: (1) the limit does not exist and (2) the function gets large without bound as x approaches a. ■

b. Here is a table of values for $f(x) = \dfrac{|x|}{x}$, with x approaching 0 from both sides.

x approaching 0 from the left →   ← x approaching 0 from the right

x	-0.1	-0.01	-0.001	-0.0001	**0**	0.0001	0.001	0.01	0.1
$f(x) = \dfrac{|x|}{x}$	-1	-1	-1	-1		1	1	1	1

The table shows that $f(x)$ does not approach the same limit as x approaches 0 from both sides. There appear to be two *different* limits: the limit as we approach 0 from the left and the limit as we approach from the right. We write

$$\lim_{x \to 0^-} f(x) = -1$$

read as "the limit as x approaches 0 from the left (or from below) is -1" and

$$\lim_{x \to 0^+} f(x) = 1$$

read as "the limit as x approaches 0 from the right (or from above) is 1." These are called the **one-sided limits** of $f(x)$. In order for f to have a **two-sided limit**, the two one-sided limits must be equal. Because they are not, we conclude that $\lim_{x \to 0} f(x)$ does not exist.

c. Near $x = 2$, we have the following table of values for $f(x) = \dfrac{1}{x - 2}$:

	x approaching 2 from the left →					← x approaching 2 from the right			
x	1.9	1.99	1.999	1.9999	**2**	2.0001	2.001	2.01	2.1
$f(x) = \dfrac{1}{x - 2}$	-10	-100	-1000	$-10{,}000$		$10{,}000$	1000	100	10

Because $\dfrac{1}{x - 2}$ is approaching no (single) real number as $x \to 2$, we see that $\displaystyle\lim_{x \to 2} \dfrac{1}{x - 2}$ does not exist. Notice also that $\dfrac{1}{x - 2}$ diverges to $+\infty$ as $x \to 2$ from the positive side (right half of the table) and to $-\infty$ as $x \to 2$ from the left (left half of the table). In other words,

$$\lim_{x \to 2^-} \frac{1}{x - 2} = -\infty$$

$$\lim_{x \to 2^+} \frac{1}{x - 2} = +\infty$$

$$\lim_{x \to 2} \frac{1}{x - 2} \text{ does not exist}$$

In another useful kind of limit, we let x approach either $+\infty$ or $-\infty$, by which we mean that we let x get arbitrarily large or let x become an arbitrarily large negative number. The next example illustrates this.

Example 3 Limits at Infinity

Use a table to estimate: **a.** $\displaystyle\lim_{x \to +\infty} \dfrac{2x^2 - 4x}{x^2 - 1}$ and **b.** $\displaystyle\lim_{x \to -\infty} \dfrac{2x^2 - 4x}{x^2 - 1}$.

Solution

a. By saying that x is "approaching $+\infty$," we mean that x is getting larger and larger without bound, so we make the following table:

			x approaching $+\infty$ →		
x	10	100	1,000	10,000	100,000
$f(x) = \dfrac{2x^2 - 4x}{x^2 - 1}$	1.6162	1.9602	1.9960	1.9996	2.0000

(Note that we are only approaching $+\infty$ from the left because we can hardly approach it from the right!) What seems to be happening is that $f(x)$ is approaching 2. Thus we write

$$\lim_{x \to +\infty} f(x) = 2.$$

b. Here, x is approaching $-\infty$, so we make a similar table, this time with x assuming negative values of greater and greater magnitude (read this table from right to left):

	← x approaching $-\infty$				
x	$-100{,}000$	$-10{,}000$	$-1{,}000$	-100	-10
$f(x) = \dfrac{2x^2 - 4x}{x^2 - 1}$	2.0000	2.0004	2.0040	2.0402	2.4242

Once again, $f(x)$ is approaching 2. Thus, $\lim_{x \to -\infty} f(x) = 2$.

Estimating Limits Graphically

We can often estimate a limit from a graph, as the next example shows.

Example 4 Estimating Limits Graphically

The graph of a function f is shown in Figure 1. (Recall that the solid dots indicate points on the graph, and the hollow dots indicate points not on the graph.)

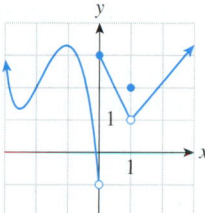

Figure **1**

From the graph, analyze the following limits.

a. $\lim\limits_{x \to -2} f(x)$ **b.** $\lim\limits_{x \to 0} f(x)$ **c.** $\lim\limits_{x \to 1} f(x)$ **d.** $\lim\limits_{x \to +\infty} f(x)$

Solution Since we are given only a graph of f, we must analyze these limits graphically.

a. Imagine that Figure 1 was drawn on a graphing calculator equipped with a trace feature that allows us to move a cursor along the graph and see the coordinates as we go. To simulate this, place a pencil point on the graph to the left of $x = -2$, and move it along the curve so that the x-coordinate approaches -2. (See Figure 2.) We evaluate the limit numerically by noting the behavior of the y-coordinates.[*]

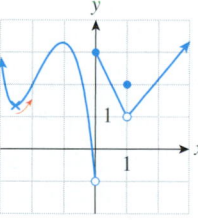   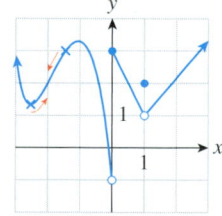

Figure **2**   Figure **3**

However, we can see directly from the graph that the y-coordinate approaches 2. Similarly, if we place our pencil point to the right of $x = -2$ and move it to the left, the y-coordinate will approach 2 from that side as well (Figure 3). Therefore, as x approaches -2 from either side, $f(x)$ approaches 2, so

$$\lim\limits_{x \to -2} f(x) = 2$$

b. This time we move our pencil point toward $x = 0$. Referring to Figure 4, if we start from the left of $x = 0$ and approach 0 (by moving right), the y-coordinate

---

[*] For a visual animation of this process, look at the online tutorial for this section.

approaches -1. However, if we start from the right of $x = 0$ and approach 0 (by moving left), the y-coordinate approaches 3. Thus (see Example 2),

$$\lim_{x \to 0^-} f(x) = -1$$

and

$$\lim_{x \to 0^+} f(x) = 3$$

Because these limits are not equal, we conclude that

$$\lim_{x \to 0} f(x) \text{ does not exist}$$

In this case there is a "break" in the graph at $x = 0$, and we say that the function is **discontinuous** at $x = 0$ (see Section 10.7).

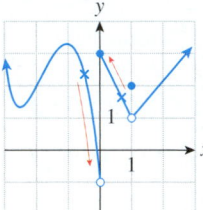   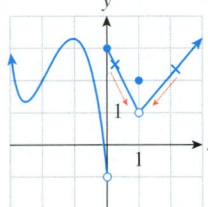   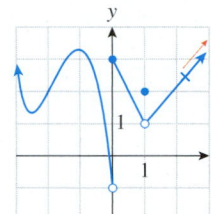

Figure **4**   Figure **5**   Figure **6**

c. Once more we think about a pencil point moving along the graph with the x-coordinate this time approaching $x = 1$ from the left and from the right (Figure 5). As the x-coordinate of the point approaches 1 from either side, the y-coordinate approaches 1 also. Therefore,

$$\lim_{x \to 1} f(x) = 1$$

d. For this limit, x is supposed to approach infinity. We think about a pencil point moving along the graph further and further to the right as shown in Figure 6.

As the x-coordinate gets larger, the y-coordinate also gets larger and larger without bound. Thus, $f(x)$ diverges to $+\infty$:

$$\lim_{x \to +\infty} f(x) = +\infty$$

Similarly,

$$\lim_{x \to -\infty} f(x) = +\infty$$

➕*Before we go on...* In Example 4(c) $\lim_{x \to 1} f(x) = 1$ but $f(1) = 2$ (why?). Thus, $\lim_{x \to 1} f(x) \neq f(1)$. In other words, the limit of $f(x)$ as x *approaches* 1 is not the same as the value of f at $x = 1$. Always keep in mind that when we evaluate a limit as $x \to a$, *we do not care about the value of the function at $x = a$*. We only care about the value of $f(x)$ as x *approaches* a. In other words, $f(a)$ may or may not equal $\lim_{x \to a} f(x)$. ∎

Here is a summary of the graphical method we used in Example 4, together with some additional information:

> ### Evaluating Limits Graphically
>
> To decide whether $\lim_{x \to a} f(x)$ exists and to find its value if it does:
>
> 1. Draw the graph of $f(x)$ by hand or with graphing technology.
> 2. Position your pencil point (or the Trace cursor) on a point of the graph to the right of $x = a$.
> 3. Move the point *along the graph* toward $x = a$ from the right and read the y-coordinate as you go. The value the y-coordinate approaches (if any) is the limit $\lim_{x \to a^+} f(x)$.
> 4. Repeat Steps 2 and 3, this time starting from a point on the graph to the left of $x = a$, and approaching $x = a$ along the graph from the left. The value the y-coordinate approaches (if any) is $\lim_{x \to a^-} f(x)$.
> 5. If the left and right limits both exist and have the same value L, then $\lim_{x \to a} f(x) = L$. Otherwise, the limit does not exist. The value $f(a)$ has no relevance whatsoever.
> 6. To evaluate $\lim_{x \to +\infty} f(x)$, move the pencil point toward the far right of the graph and estimate the value the y-coordinate approaches (if any). For $\lim_{x \to -\infty} f(x)$, move the pencil point toward the far left.
> 7. If $x = a$ happens to be an endpoint of the domain of f, then only a one-sided limit is possible at $x = a$. For instance, if the domain is $(-\infty,\ 4]$, then $\lim_{x \to 4^-} f(x)$ can be computed, but not $\lim_{x \to 4} f(x)$ or $\lim_{x \to 4^+} f(x)$.

In the next example we use both the numerical and graphical approaches.

Example 5 Infinite Limit

Does $\displaystyle\lim_{x \to 0^+} \frac{1}{x}$ exist?

Solution

Numerical Method Because we are asked for only the right-hand limit, we need only list values of x approaching 0 from the right.

← x approaching 0 from the right

x	0	0.0001	0.001	0.01	0.1
$f(x) = \dfrac{1}{x}$		10,000	1000	100	10

What seems to be happening as x approaches 0 from the right is that $f(x)$ is increasing without bound, as in Example 4(d). That is, if you name any number, no matter how large, $f(x)$ will be even larger than that if x is sufficiently close to zero. Thus, the limit diverges to $+\infty$, so

$$\lim_{x \to 0^+} \frac{1}{x} = +\infty$$

Graphical Method Recall that the graph of $f(x) = \dfrac{1}{x}$ is the standard hyperbola shown in Figure 7.

The figure also shows the pencil point moving so that its x-coordinate approaches 0 from the right. Because the point moves along the graph, it is forced to go higher and

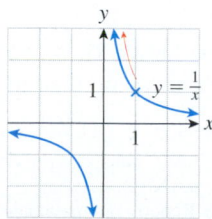

Figure **7**

higher. In other words, its y-coordinate becomes larger and larger, approaching $+\infty$. Thus, we conclude that

$$\lim_{x \to 0^+} \frac{1}{x} = +\infty$$

+*Before we go on...* In Example 5(a) you should also check that

$$\lim_{x \to 0^-} \frac{1}{x} = -\infty$$

We say that as x approaches 0 from the left, $\frac{1}{x}$ diverges to $-\infty$. Also, check that

$$\lim_{x \to +\infty} \frac{1}{x} = \lim_{x \to -\infty} \frac{1}{x} = 0$$ ■

Application

Example 6 Internet Connectivity

The number of U.S. households connected to the Internet can be modeled by[*]

$$N(t) = \frac{80}{1 + 2.2(3.68)^{-t}} \quad (t \geq 0)$$

where t is time in years since 1999.

a. Estimate $\lim_{t \to +\infty} N(t)$ and interpret the answer.

b. Estimate $\lim_{t \to 0^+} N(t)$ and interpret the answer.

Solution

a. Figure 8 shows a plot of $N(t)$ for $0 < t < 10$:

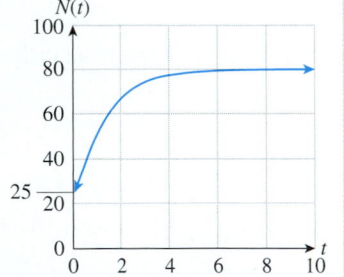

Figure **8**

Using either the numerical or graphical approach, we find

$$\lim_{t \to +\infty} N(t) = \lim_{t \to +\infty} \frac{80}{1 + 2.2(3.68)^{-t}} = 80$$

Thus, in the long term (as t gets larger and larger) the percentage of U.S. households connected to the Internet is expected to approach 80%.

b. The limit here is

$$\lim_{t \to 0^+} N(t) = \lim_{t \to 0^+} \frac{80}{1 + 2.2(3.68)^{-t}} = 25$$

(Notice that in this case, we can simply put $t = 0$ to evaluate this limit.) Thus, the closer t gets to 0 (representing 1999) the closer $N(t)$ gets to 25%, meaning that, in 1999, 25% of U.S. households were connected to the Internet.

---

[*] Based on a regression model by the authors. Sources for data: Telecommunications Reports International/ *New York Times,* May 21, 2001, p. C9.

FAQs   **Determining When a Limit Does or Does Not Exist**

Q: *If I substitute $x = a$ in the formula for a function and find that the function is not defined there, it means that $\lim_{x \to a} f(x)$ does not exist, right?*

A: Wrong. The limit may still exist, as in Example 1, or may not exist, as in Example 2. In general, whether or not $\lim_{x \to a} f(x)$ exists has nothing to do with $f(a)$, but rather the value of f when x is *very close to, but not equal to a.* ∎

Q: *Is there a quick and easy way of telling from a graph whether $\lim_{x \to a} f(x)$ exists?*

A: Yes. If you cover up the portion of the graph corresponding to $x = a$, and it appears as though the visible part of the graph could be made into a continuous line by filling in a suitable point at $x = a$, then the limit exists. (The "suitable point" need not be $(a, f(a))$.) Otherwise, it does not. Try this method with the curves in Example 4. ∎

10.1 | EXERCISES

● denotes basic skills exercises

◆ denotes challenging exercises

Estimate the limits in Exercises 1–18 numerically.

1. ● $\lim\limits_{x \to 0} \dfrac{x^2}{x + 1}$ *hint* [see Example 1]   **2.** ● $\lim\limits_{x \to 0} \dfrac{x - 3}{x - 1}$

3. ● $\lim\limits_{x \to 2} \dfrac{x^2 - 4}{x - 2}$ **4.** ● $\lim\limits_{x \to 2} \dfrac{x^2 - 1}{x - 2}$

5. ● $\lim\limits_{x \to -1} \dfrac{x^2 + 1}{x + 1}$

6. ● $\lim\limits_{x \to -1} \dfrac{x^2 + 2x + 1}{x + 1}$

7. ● $\lim\limits_{x \to +\infty} \dfrac{3x^2 + 10x - 1}{2x^2 - 5x}$ *hint* [see Example 3]

8. ● $\lim\limits_{x \to +\infty} \dfrac{6x^2 + 5x + 100}{3x^2 - 9}$

9. ● $\lim\limits_{x \to -\infty} \dfrac{x^5 - 1,000x^4}{2x^5 + 10,000}$

10. ● $\lim\limits_{x \to -\infty} \dfrac{x^6 + 3,000x^3 + 1,000,000}{2x^6 + 1,000x^3}$

11. ● $\lim\limits_{x \to +\infty} \dfrac{10x^2 + 300x + 1}{5x + 2}$ **12.** ● $\lim\limits_{x \to +\infty} \dfrac{2x^4 + 20x^3}{1,000x^6 + 6}$

13. ● $\lim\limits_{x \to +\infty} \dfrac{10x^2 + 300x + 1}{5x^3 + 2}$ **14.** ● $\lim\limits_{x \to +\infty} \dfrac{2x^4 + 20x^3}{1,000x^3 + 6}$

15. ● $\lim\limits_{x \to 2} e^{x-2}$ **16.** ● $\lim\limits_{x \to +\infty} e^{-x}$

17. ● $\lim\limits_{x \to +\infty} xe^{-x}$ **18.** ● $\lim\limits_{x \to -\infty} xe^{x}$

In each of Exercises 19–30, the graph of f is given. Use the graph to compute the quantities asked for. *hint* [see Example 4]

19. ●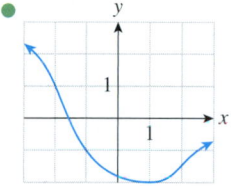

a. $\lim\limits_{x \to 1} f(x)$ **b.** $\lim\limits_{x \to -1} f(x)$

20. ●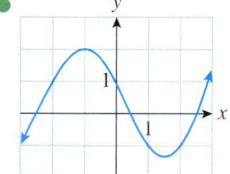

a. $\lim\limits_{x \to -1} f(x)$ **b.** $\lim\limits_{x \to 1} f(x)$

21. ●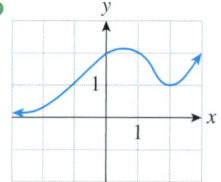

a. $\lim\limits_{x \to 0} f(x)$ **b.** $\lim\limits_{x \to 2} f(x)$

c. $\lim\limits_{x \to -\infty} f(x)$ **d.** $\lim\limits_{x \to +\infty} f(x)$

22. ●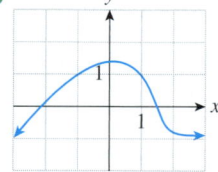

a. $\lim\limits_{x \to -1} f(x)$ **b.** $\lim\limits_{x \to 1} f(x)$

c. $\lim\limits_{x \to +\infty} f(x)$ **d.** $\lim\limits_{x \to -\infty} f(x)$

23. ●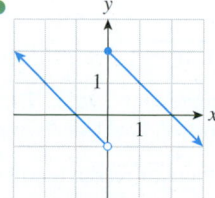

a. $\lim\limits_{x \to 2} f(x)$ **b.** $\lim\limits_{x \to 0^+} f(x)$

c. $\lim\limits_{x \to 0^-} f(x)$ **d.** $\lim\limits_{x \to 0} f(x)$

e. $f(0)$ **f.** $\lim\limits_{x \to -\infty} f(x)$

24. ●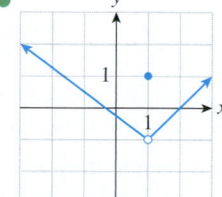

a. $\lim\limits_{x \to 3} f(x)$ **b.** $\lim\limits_{x \to 1^+} f(x)$

c. $\lim\limits_{x \to 1^-} f(x)$ **d.** $\lim\limits_{x \to 1} f(x)$

e. $f(1)$ **f.** $\lim\limits_{x \to +\infty} f(x)$

● basic skills   ◆ challenging

25. ●

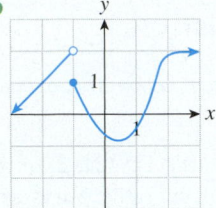

a. $\lim_{x \to -2} f(x)$ **b.** $\lim_{x \to -1^+} f(x)$

c. $\lim_{x \to -1^-} f(x)$ **d.** $\lim_{x \to -1} f(x)$

e. $f(-1)$ **f.** $\lim_{x \to +\infty} f(x)$

26. ●

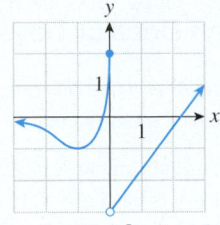

a. $\lim_{x \to -1} f(x)$ **b.** $\lim_{x \to 0^+} f(x)$

c. $\lim_{x \to 0^-} f(x)$ **d.** $\lim_{x \to 0} f(x)$

e. $f(0)$ **f.** $\lim_{x \to -\infty} f(x)$

27. ●

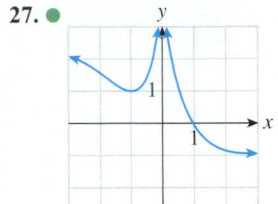

a. $\lim_{x \to -1} f(x)$ **b.** $\lim_{x \to 0^+} f(x)$

c. $\lim_{x \to 0^-} f(x)$ **d.** $\lim_{x \to 0} f(x)$

e. $f(0)$ **f.** $\lim_{x \to +\infty} f(x)$

28. ●

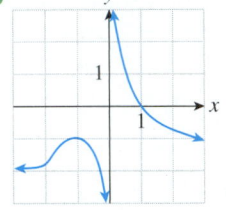

a. $\lim_{x \to 1} f(x)$ **b.** $\lim_{x \to 0^+} f(x)$

c. $\lim_{x \to 0^-} f(x)$ **d.** $\lim_{x \to 0} f(x)$

e. $f(0)$ **f.** $\lim_{x \to -\infty} f(x)$

29. ●

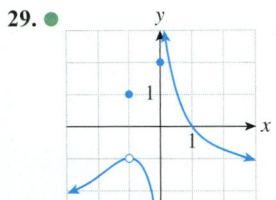

a. $\lim_{x \to -1} f(x)$ **b.** $\lim_{x \to 0^+} f(x)$

c. $\lim_{x \to 0^-} f(x)$ **d.** $\lim_{x \to 0} f(x)$

e. $f(0)$ **f.** $f(-1)$

30. ●

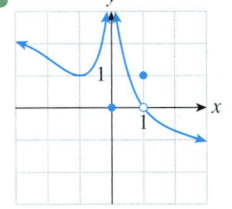

a. $\lim_{x \to 0^-} f(x)$ **b.** $\lim_{x \to 1^+} f(x)$

c. $\lim_{x \to 0} f(x)$ **d.** $\lim_{x \to 1} f(x)$

e. $f(0)$ **f.** $f(1)$

Applications

31. ● *Scientific Research* The number of research articles per year, in thousands, in the prominent journal *Physics Review* written by researchers in Europe can be modeled by

$$A(t) = \frac{7.0}{1 + 5.4(1.2)^{-t}}$$

where t is time in years ($t = 0$ represents 1983).[1] Estimate $\lim_{t \to +\infty} A(t)$ and interpret the answer. *hint* [see Example 6]

32. ● *Scientific Research* The percentage of research articles in the prominent journal *Physics Review* written by researchers

in the U.S. can be modeled by

$$A(t) = 25 + \frac{36}{1 + 0.6(0.7)^{-t}}$$

where t is time in years ($t = 0$ represents 1983).[2] Estimate $\lim_{t \to +\infty} A(t)$ and interpret the answer.

33. ● *SAT Scores by Income* The following bar graph shows U.S. verbal SAT scores as a function of parents' income level:[3]

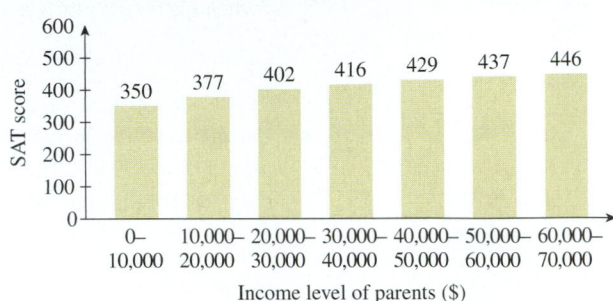

These data can be modeled by

$$S(x) = 470 - 136(0.974)^x$$

where $S(x)$ is the average SAT verbal score of a student whose parents' income is x thousand dollars per year. Evaluate $\lim_{x \to +\infty} S(x)$ and interpret the result.

34. ● *SAT Scores by Income* The following bar graph shows U.S. math SAT scores as a function of parents' income level:[4]

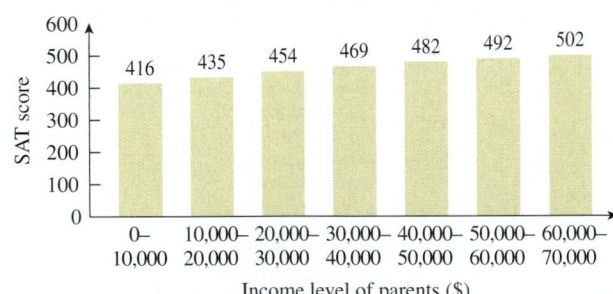

These data can be modeled by

$$S(x) = 535 - 136(0.979)^x$$

where $S(x)$ is the average math SAT score of a student whose parents' income is x thousand dollars per year. Evaluate $\lim_{x \to +\infty} S(x)$ and interpret the result.

35. ● *Electric Rates* The cost of electricity in Portland, Oregon, for residential customers increased suddenly on October 1, 2001, from around \$0.06 to around \$0.08 per kilowatt hour.[5]

[1] Based on data from 1983 to 2003. SOURCE: The American Physics Society/*New York Times,* May 3, 2003, p. A1.

[2] Based on data from 1983 to 2003. SOURCE: The American Physics Society/*New York Times,* May 3, 2003, p. A1.

[3] Based on 1994 data. SOURCE: The College Board/*New York Times,* March 5, 1995, p. E16.

[4] Ibid.

[5] SOURCE: Portland General Electric/*New York Times,* February 2, 2002, p. C1.

● basic skills   ◆ challenging

Let $C(t)$ be this cost at time t, and take $t = 1$ to represent October 1, 2001. What does the given information tell you about $\lim_{t \to 1} C(t)$? *hint* [see Example 4b]

36. ● *Airline Stocks* Prior to the September 11, 2001 attacks, United Airlines stock was trading at around \$35 per share. Immediately following the attacks, the share price dropped by \$15.[6] Let $U(t)$ be this cost at time t, and take $t = 11$ to represent September 11, 2001. What does the given information tell you about $\lim_{t \to 11} U(t)$?

Foreign Trade Annual U.S. imports from China in the years 1996 through 2003 could be approximated by

$$I(t) = t^2 + 3.5t + 50 \quad (1 \le t \le 9)$$

billion dollars, where t represents time in years since 1995. Annual U.S. exports to China in the same years could be approximated by

$$E(t) = 0.4t^2 - 1.6t + 14 \quad (0 \le t \le 10)$$

billion dollars.[7] Exercises 37 and 38 are based on these models.

37. Assuming the trends shown in the above models continue indefinitely, numerically estimate

$$\lim_{t \to +\infty} I(t) \text{ and } \lim_{t \to +\infty} \frac{I(t)}{E(t)}$$

interpret your answers, and comment on the results.

38. Repeat Exercise 37, this time calculating

$$\lim_{t \to +\infty} E(t) \text{ and } \lim_{t \to +\infty} \frac{E(t)}{I(t)}$$

39. *Online Sales* The following graph shows the approximate annual sales of books online in the U.S. for the period 1997–2004.[8]

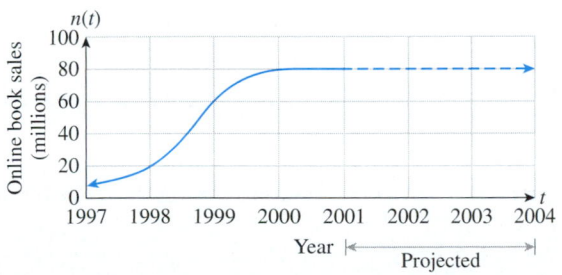

Estimate $\lim_{t \to +\infty} n(t)$ and interpret your answer.

40. *Employment* The following graph shows the number of new employees per year at Amerada Hess Corp. from 1984 $(t = 0)$.[9]

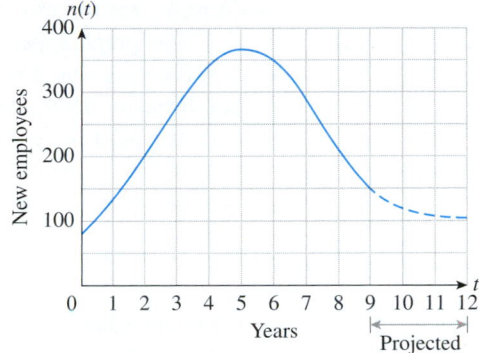

Estimate $\lim_{t \to +\infty} n(t)$ and interpret your answer.

Communication and Reasoning Exercises

41. ● Describe the method of evaluating limits numerically. Give at least one disadvantage of this method.

42. ● Describe the method of evaluating limits graphically. Give at least one disadvantage of this method.

43. What is wrong with the following statement? "Because $f(a)$ is not defined, $\lim_{x \to a} f(x)$ does not exist."

44. What is wrong with the following statement? "If $f(a)$ is defined, then $\lim_{x \to a} f(x)$ exists, and equals $f(a)$."

45. Your friend Dion, a business student, claims that the study of limits that do not exist is completely unrealistic and has nothing to do with the world of business. Give two examples from the world of business that might convince him that he is wrong.

46. If $D(t)$ is the Dow Jones Average at time t and $\lim_{t \to +\infty} D(t) = +\infty$, is it possible that the Dow will fluctuate indefinitely into the future?

47. ◆ Give an example of a function f with $\lim_{x \to 1} f(x) = f(2)$.

48. ◆ If $S(t)$ represents the size of the universe in billions of light years at time t years since the big bang and $\lim_{t \to +\infty} S(t) = 130,000$, is it possible that the universe will continue to expand forever?

---

[6] Stock prices are approximate.

[7] Based on quadratic regression using data from the U.S. Census Bureau Foreign Trade Division website www.census.gov/foreign-trade/sitc1/ as of December 2004.

[8] Source for 1997–2000 data: Ipsos-NPD Book Trends/*New York Times,* April 16, 2001, p. C1. (2001–2004 data were projections.)

[9] The projected part of the curve (from $t = 9$ on) is fictitious. The model is based on a best-fit logistic curve. Source for data: Hoover's Handbook Database (World Wide website), The Reference Press, Inc., Austin, Texas, 1995.

● basic skills   ◆ challenging

10.2 Limits and Continuity (OPTIONAL)

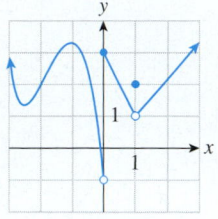

Figure 9

In Section 10.1 we saw examples of graphs that had various kinds of "breaks" or "jumps." For instance, in Example 4 we looked at the graph in Figure 9.

This graph appears to have breaks, or **discontinuities**, at $x = 0$ and at $x = 1$. At $x = 0$ we saw that $\lim_{x \to 0} f(x)$ does not exist because the left- and right-hand limits are not the same. Thus, the discontinuity at $x = 0$ seems to be due to the fact that the limit does not exist there. On the other hand, at $x = 1$, $\lim_{x \to 1} f(x)$ *does* exist (it is equal to 1), but is not equal to $f(1) = 2$.

Thus, we have identified two kinds of discontinuity:

(1) Points where the limit of the function does not exist. $x = 0$ in Figure 9 because $\lim_{x \to 0} f(x)$ does not exist.

(2) Points where the limit exists but does not equal the value of the function. $x = 1$ in Figure 9 because $\lim_{x \to 1} f(x) = 1 \neq f(1)$

On the other hand, there is no discontinuity at, say, $x = -2$, where we find that $\lim_{x \to -2} f(x)$ exists and equals 2 and $f(-2)$ is also equal to 2. In other words,

$$\lim_{x \to -2} f(x) = 2 = f(-2)$$

The point $x = -2$ is an example of a point where f is **continuous**. (Notice that you can draw the portion of the graph near $x = -2$ without lifting your pencil from the paper.) Similarly, f is continuous at *every* point other than $x = 0$ and $x = 1$. Here is the mathematical definition.

Continuous Function

Let f be a function and let a be a number in the domain of f. Then f is **continuous at a** if

a. $\lim_{x \to a} f(x)$ exists, and

b. $\lim_{x \to a} f(x) = f(a)$.

The function f is said to be **continuous on its domain** if it is continuous at each point in its domain.

If f is not continuous at a particular a in its domain, we say that f is **discontinuous** at a or that f has a **discontinuity** at a. Thus, a discontinuity can occur at $x = a$ if either

a. $\lim_{x \to a} f(x)$ does not exist, or

b. $\lim_{x \to a} f(x)$ exists but is not equal to $f(a)$.

quick Examples

1. The function shown in Figure 9 is continuous at $x = -1$ and $x = 2$. It is discontinuous at $x = 0$ and $x = 1$, and so is not continuous on its domain.

2. The function $f(x) = x^2$ is continuous on its domain. (Think of its graph, which contains no breaks.)

3. The function f whose graph is shown on the left in the following figure is continuous on its domain. (Although the graph breaks at $x = 2$, that is not a point of its domain.) The function g whose graph is shown on the right is not continuous on its

domain because it has a discontinuity at $x = 2$. (Here, $x = 2$ is a point of the domain of g.)

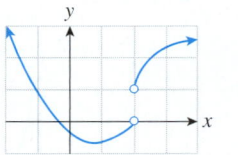

$y = f(x)$: Continuous on its domain

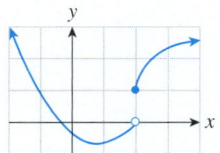

$y = g(x)$: Not continuous on its domain

Note If the number a is not in the domain of f—that is, if $f(a)$ is not defined—we will not consider the question of continuity at a. A function cannot be continuous at a point not in its domain, and it cannot be discontinuous there either. ∎

Example 1 Continuous and Discontinuous Functions

Which of the following functions are continuous on their domains?

a. $h(x) = \begin{cases} x + 3 & \text{if } x \leq 1 \\ 5 - x & \text{if } x > 1 \end{cases}$

b. $k(x) = \begin{cases} x + 3 & \text{if } x \leq 1 \\ 1 - x & \text{if } x > 1 \end{cases}$

c. $f(x) = \dfrac{1}{x}$

d. $g(x) = \begin{cases} \frac{1}{x} & \text{if } x \neq 0 \\ 0 & \text{if } x = 0 \end{cases}$

Solution

a. and **b.** The graphs of h and k are shown in Figure 10.

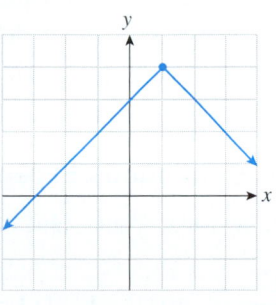

$y = h(x)$

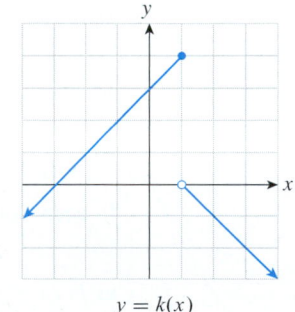

$y = k(x)$

Figure **10**

Even though the graph of h is made up of two different line segments, it is continuous at every point of its domain, including $x = 1$ because

$$\lim_{x \to 1} h(x) = 4 = h(1)$$

On the other hand, $x = 1$ is also in the domain of k, but $\lim_{x \to 1} k(x)$ does not exist. Thus, k is discontinuous at $x = 1$ and thus not continuous on its domain.

c. and **d.** The graphs of f and g are shown in Figure 11.

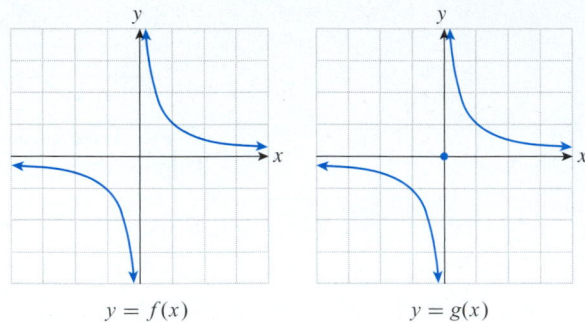

$$y = f(x) \qquad\qquad y = g(x)$$

Figure **11**

The domain of f consists of all real numbers except 0 and f is continuous at all such numbers. (Notice that 0 is not in the domain of f, so the question of continuity at 0 does not arise.) Thus, f is continuous on its domain.

The function g, on the other hand, has its domain expanded to include 0, so we now need to check whether g is continuous at 0. From the graph, it is easy to see that g is discontinuous there because $\lim_{x \to 0} g(x)$ does not exist. Thus, g is not continuous on its domain because it is discontinuous at 0.

+ Before we go on...

Q: *Wait a minute! How can a function like $f(x) = 1/x$ be continuous when its graph has a break in it?*

A: We are not claiming that f is continuous *at every real number*. What we are saying is that f is continuous *on its domain*; the break in the graph occurs at a point not in the domain of f. In other words, f is continuous on the set of all nonzero real numbers; it is not continuous on the set of *all* real numbers because it is not even defined on the set. ■

Example **2** Continuous Except at a Point

In each case, say what, if any, value of $f(a)$ would make f continuous at a.

a. $f(x) = \dfrac{x^3 - 8}{x - 2}$; $\; a = 2$ **b.** $f(x) = \dfrac{e^{2x} - 1}{x}$; $\; a = 0$ **c.** $f(x) = \dfrac{|x|}{x}$; $\; a = 0$

Solution

a. In Figure 12 we see the graph of $f(x) = \dfrac{x^3 - 8}{x - 2}$. The point corresponding to $x = 2$ is missing because f is not (yet) defined there. (Your graphing utility will probably miss this subtlety and render a continuous curve.)

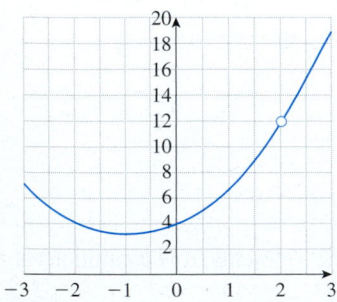

Figure **12**

To turn f into a function that is continuous at $x = 2$, we need to "fill in the gap" so as to obtain a continuous curve. Since the graph suggests that the missing point is $(2, 12)$, let us define $f(2) = 12$.

Does f now become continuous if we take $f(2) = 12$? From the graph, or Example 1(a) of Section 10.1,

$$\lim_{x \to 2} f(x) = \lim_{x \to 2} \frac{x^3 - 8}{x - 2} = 12$$

which is now equal to $f(2)$. Thus, $\lim_{x \to 2} f(x) = f(2)$, showing that f is now continuous at $x = 2$.

b. In Example 1(b) of the preceding section, we saw that

$$\lim_{x \to 0} f(x) = \lim_{x \to 0} \frac{e^{2x} - 1}{x} = 2$$

and so, as in part (a), we must define $f(0) = 2$. This is confirmed by the graph shown in Figure 13.

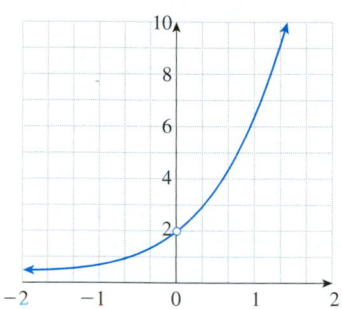

Figure **13**

c. We considered the function $f(x) = |x|/x$ in Example 2 in Section 10.1. Its graph is shown in Figure 14.

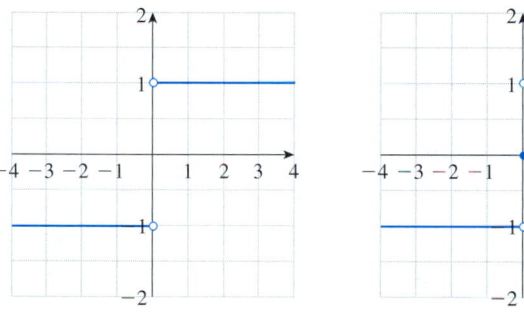

Figure **14**   Figure **15**

Now we encounter a problem: No matter how we try to fill in the gap at $x = 0$, the result will be a discontinuous function. For example, setting $f(0) = 0$ will result in the discontinuous function shown in Figure 15.

We conclude that it is impossible to assign any value to $f(0)$ to turn f into a function that is continuous at $x = 0$.

We can also see this result algebraically: In Example 2 of Section 10.1, we saw that $\lim_{x \to 0} \dfrac{|x|}{x}$ does not exist. Thus, the resulting function will fail to be continuous at 0, no matter how we define $f(0)$.

A function not defined at an isolated point is said to have a **singularity** at that point. The function in part (a) of Example 2 has a singularity at $x = 2$, and the functions in parts (b) and (c) have singularities at $x = 0$. The functions in parts (a) and (b) have *removable* singularities because we can make these functions continuous at $x = a$ by properly defining $f(a)$. The function in part (c) has an **essential singularity** because we cannot make f continuous at $x = a$ just by defining $f(a)$ properly.

10.2 EXERCISES

● denotes basic skills exercises

In Exercises 1–12, the graph of a function f is given. Determine whether f is continuous on its domain. If it is not continuous on its domain, say why. hint [see Quick Examples p. 698]

1. ●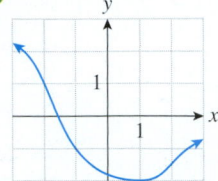

2. ●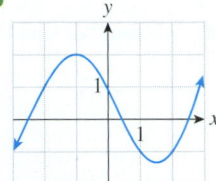

3. ●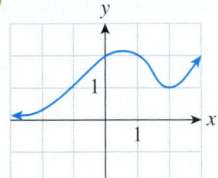

4. ●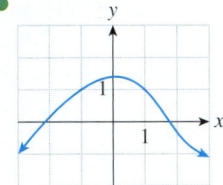

5. ●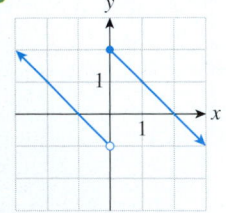

6. ●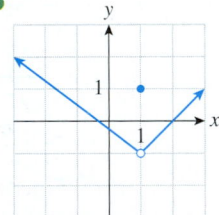

7. ●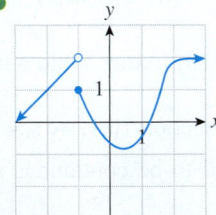

8. ●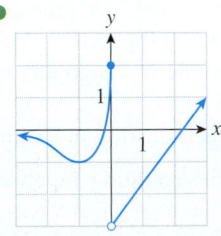

9. ●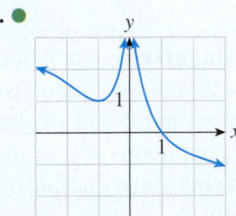

10. ●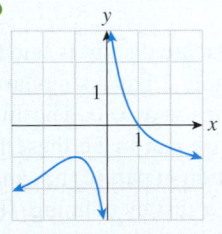

11. ●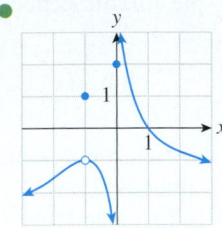

12. ●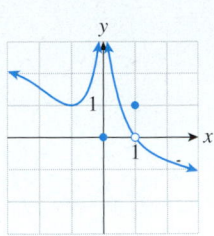

In Exercises 13 and 14, identify which (if any) of the given graphs represent functions continuous on their domains. hint [see Example 1]

13. ●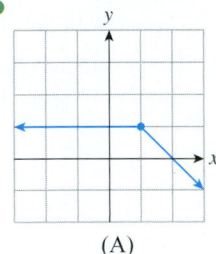
(A)

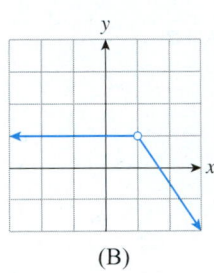
(B)

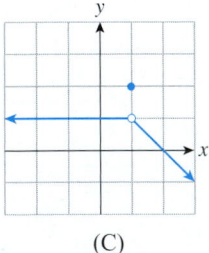
(C)

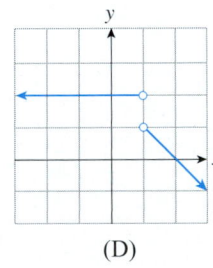
(D)

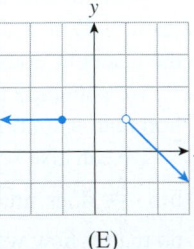
(E)

14. ●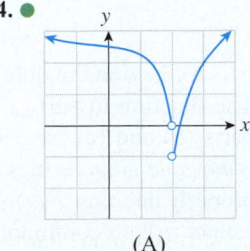
(A)

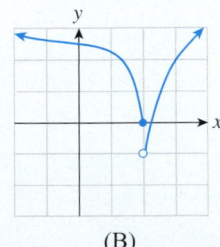
(B)

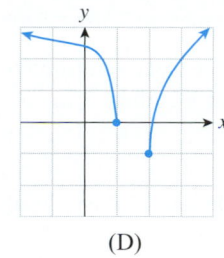

(C)   (D)

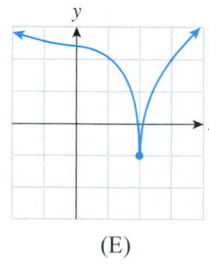

(E)

In Exercises 15–22, use a graph of f or some other method to determine what, if any, value to assign to $f(a)$ to make f continuous at $x = a$.

15. ● $f(x) = \dfrac{x^2 - 2x + 1}{x - 1}$; $a = 1$ *hint* [see Example 2]

16. ● $f(x) = \dfrac{x^2 + 3x + 2}{x + 1}$; $a = -1$

17. ● $f(x) = \dfrac{x}{3x^2 - x}$; $a = 0$ **18.** ● $f(x) = \dfrac{x^2 - 3x}{x + 4}$; $a = -4$

19. ● $f(x) = \dfrac{3}{3x^2 - x}$; $a = 0$ **20.** ● $f(x) = \dfrac{x - 1}{x^3 - 1}$; $a = 1$

21. ● $f(x) = \dfrac{1 - e^x}{x}$; $a = 0$ **22.** ● $f(x) = \dfrac{1 + e^x}{1 - e^x}$; $a = 0$

In Exercises 23–32, use a graph to determine whether the given function is continuous on its domain. If it is not continuous on its domain, list the points of discontinuity.

23. ● $f(x) = |x|$

24. ● $f(x) = \dfrac{|x|}{x}$

25. ● $g(x) = \dfrac{1}{x^2 - 1}$

26. ● $g(x) = \dfrac{x - 1}{x + 2}$

27. ● $f(x) = \begin{cases} x + 2 & \text{if } x < 0 \\ 2x - 1 & \text{if } x \geq 0 \end{cases}$

28. ● $f(x) = \begin{cases} 1 - x & \text{if } x \leq 1 \\ x - 1 & \text{if } x > 1 \end{cases}$

29. ● $h(x) = \begin{cases} \frac{|x|}{x} & \text{if } x \neq 0 \\ 0 & \text{if } x = 0 \end{cases}$

30. ● $h(x) = \begin{cases} \frac{1}{x^2} & \text{if } x \neq 0 \\ 2 & \text{if } x = 0 \end{cases}$

31. ● $g(x) = \begin{cases} x + 2 & \text{if } x < 0 \\ 2x + 2 & \text{if } x \geq 0 \end{cases}$

32. ● $g(x) = \begin{cases} 1 - x & \text{if } x \leq 1 \\ x + 1 & \text{if } x > 1 \end{cases}$

Communication and Reasoning Exercises

33. ● If a function is continuous on its domain, is it continuous at every real number? Explain.

34. ● True or false: The graph of a function that is continuous on its domain is a continuous curve with no breaks in it.

35. ● True or false: The graph of a function that is continuous at every real number is a continuous curve with no breaks in it.

36. Give an example of a function that is not continuous at $x = -1$ but is not discontinuous there either.

37. Draw the graph of a function that is discontinuous at every integer.

38. Draw the graph of a function that is continuous on its domain but not continuous at any integer,

39. Describe a real-life scenario in the stock market that can be modeled by a discontinuous function.

40. Describe a real-life scenario in your room that can be modeled by a discontinuous function.

● basic skills

10.3 Limits and Continuity: Algebraic Approach (OPTIONAL)

Although numerical and graphical estimation of limits is effective, the estimates these methods yield may not be perfectly accurate. The algebraic method, when it can be used, will always yield an exact answer. Moreover, algebraic analysis of a function often enables us to take a function apart and see "what makes it tick."

Let's start with the function $f(x) = 2 + x$ and ask: What happens to $f(x)$ as x approaches 3? To answer this algebraically, notice that as x gets closer and closer to 3, the quantity $2 + x$ must get closer and closer to $2 + 3 = 5$. Hence,

$$\lim_{x \to 3} f(x) = \lim_{x \to 3}(2 + x) = 2 + 3 = 5$$

Q: Is that all there is to the algebraic method? Just substitute $x = a$?

A: Under certain circumstances: Notice that by substituting $x = 3$ we *evaluated the function at* $x = 3$. In other words, we relied on the fact that

$$\lim_{x \to 3} f(x) = f(3)$$

In Section 10.2 we said that a function satisfying this equation is *continuous* at $x = 3$. ∎

Thus,

> *If we know that the function f is continuous at a point a, we can compute $\lim_{x \to a} f(x)$ by simply substituting $x = a$ into $f(x)$.*

To use this fact, we need to know how to recognize continuous functions when we see them. Geometrically, they are easy to spot: A function is continuous at $x = a$ if its graph has no break at $x = a$. Algebraically, a large class of functions are known to be continuous on their domains—those, roughly speaking, that are *specified by a single formula*.

We can be more precise: A **closed-form function** is any function that can be obtained by combining constants, powers of x, exponential functions, radicals, logarithms, and trigonometric functions (and some other functions we do not encounter in this text) into a *single* mathematical formula by means of the usual arithmetic operations and composition of functions. (They can be as complicated as we like.)

Closed-Form Functions

A **closed-form function** is any function that can be obtained by combining constants, powers of x, exponential functions, radicals, logarithms, and trigonometric functions (and some other functions we do not encounter in this text) into a *single* mathematical formula by means of the usual arithmetic operations and composition of functions.

quick Examples

1. $3x^2 - x + 1$, $\dfrac{\sqrt{x^2 - 1}}{6x - 1}$, $e^{\frac{-4x^2-1}{x}}$, and $\sqrt{\log_3(x^2 - 1)}$ are all closed form functions.

2. $f(x) = \begin{cases} -1 & \text{if } x \le -1 \\ x^2 + x & \text{if } -1 < x \le 1 \\ 2 - x & \text{if } 1 < x \le 2 \end{cases}$ is *not* a closed-form function because $f(x)$ is not specified by a *single* mathematical formula.

What is so special about closed-form functions is the following theorem:

Theorem: Continuity of Closed-Form Functions

Every closed-form function is continuous on its domain. Thus, if f is a closed-form function and $f(a)$ is defined, we have $\lim_{x \to a} f(x) = f(a)$.

quick Example

$f(x) = 1/x$ is a closed-form function, and its natural domain consists of all real numbers except 0. Thus, f is continuous at every nonzero real number. That is,

$$\lim_{x \to a} \frac{1}{x} = \frac{1}{a}$$

provided $a \neq 0$.

Mathematics majors spend a great deal of time studying the proof of this theorem. We ask you to accept it without proof.

Example 1 Limit of a Closed-Form Function

Evaluate $\lim\limits_{x \to 1} \dfrac{x^3 - 8}{x - 2}$ algebraically.

Solution

First, notice that $(x^3 - 8)/(x - 2)$ is a closed-form function because it is specified by a single algebraic formula. Also, $x = 1$ is in the domain of this function. Therefore,

$$\lim_{x \to 1} \frac{x^3 - 8}{x - 2} = \frac{1^3 - 8}{1 - 2} = 7$$

✛*Before we go on...* In Example 1, the point $x = 2$ is not in the domain of the function $(x^3 - 8)/(x - 2)$, so we cannot evaluate $\lim\limits_{x \to 2} \dfrac{x^3 - 8}{x - 2}$ by substituting $x = 2$. However— and this is the key to finding such limits—some preliminary algebraic simplification will allow us to obtain a closed-form function with $x = 2$ in its domain, as we shall see in Example 2. ∎

Example 2 Simplifying to Obtain the Limit

Evaluate $\lim\limits_{x \to 2} \dfrac{x^3 - 8}{x - 2}$ algebraically.

Solution

Again, although $(x^3 - 8)/(x - 2)$ is a closed-form function, $x = 2$ is not in its domain. Thus, we cannot obtain the limit by substitution. Instead, we first simplify $f(x)$ to obtain a new function with $x = 2$ in its domain. To do this, notice first that the numerator can be factored as

$$x^3 - 8 = (x - 2)(x^2 + 2x + 4)$$

Thus,

$$\frac{x^3 - 8}{x - 2} = \frac{(x - 2)(x^2 + 2x + 4)}{x - 2} = x^2 + 2x + 4$$

Once we have canceled the offending $(x - 2)$ in the denominator, we are left with a closed-form function *with 2 in its domain*. Thus,

$$\lim_{x \to 2} \frac{x^3 - 8}{x - 2} = \lim_{x \to 2}(x^2 + 2x + 4)$$
$$= 2^2 + 2(2) + 4 = 12 \quad \text{Substitute } x = 2.$$

This confirms the answer we found numerically in Example 1 in Section 10.1.

If the given function fails to simplify, we can always approximate the limit numerically. It may very well be that the limit does not exist in such a case.

Q: *There is something suspicious about Example 2. If 2 was not in the domain before simplifying but was in the domain after simplifying, we must have changed the function, right?*

A: Correct. In fact, when we said that

$$\frac{x^3 - 8}{x - 2} = x^2 + 2x + 4$$

Domain excludes 2   Domain includes 2

we were lying a little bit. What we really meant is that these two expressions are equal *where both are defined*. The functions $(x^3 - 8)/(x - 2)$ and $x^2 + 2x + 4$ are different functions. The difference is that $x = 2$ is not in the domain of $(x^3 - 8)/(x - 2)$ and is in the domain of $x^2 + 2x + 4$. Since $\lim_{x \to 2} f(x)$ explicitly *ignores* any value that f may have at 2, this does not affect the limit. From the point of view of the limit at 2, these functions *are* equal. In general we have the following rule. ∎

Functions with Equal Limits

If $f(x) = g(x)$ for all x except possibly $x = a$, then

$$\lim_{x \to a} f(x) = \lim_{x \to a} g(x).$$

quick Example

$$\frac{x^2 - 1}{x - 1} = x + 1 \text{ for all } x \text{ except } x = 1. \qquad \text{<span style="color:magenta">Write } \frac{x^2 - 1}{x - 1} \text{ as } \frac{(x + 1)(x - 1)}{x - 1} \text{ and cancel the } (x - 1).</span>$$

Therefore,

$$\lim_{x \to 1} \frac{x^2 - 1}{x - 1} = \lim_{x \to 1} (x + 1) = 1 + 1 = 2$$

We can also use algebraic techniques to analyze functions that are not given in closed form.

Example 3 Nonclosed-Form Function

For which values of x are the following piecewise defined functions continuous?

a. $f(x) = \begin{cases} x^2 + 2 & \text{if } x < 1 \\ 2x - 1 & \text{if } x \geq 1 \end{cases}$ **b.** $g(x) = \begin{cases} x^2 - x + 1 & \text{if } x \leq 0 \\ 1 - x & \text{if } 0 < x \leq 1 \\ x - 3 & \text{if } x > 1 \end{cases}$

Solution

a. The function $f(x)$ is given in closed form over the intervals $(-\infty, 1)$ and $[1, +\infty)$. At $x = 1$, $f(x)$ suddenly switches from one closed-form formula to another, so $x = 1$ is the only place where there is a potential problem with continuity. To investigate the continuity of $f(x)$ at $x = 1$, let's calculate the limit there:

$$\lim_{x \to 1^-} f(x) = \lim_{x \to 1^-} (x^2 + 2) \qquad \text{<span style="color:magenta">} f(x) = x^2 + 2 \text{ for } x < 1.</span>$$
$$= (1)^2 + 2 = 3 \qquad \text{<span style="color:magenta">} x^2 + 2 \text{ is closed-form.}</span>$$

$$\lim_{x \to 1^+} f(x) = \lim_{x \to 1^+} (2x - 1) \qquad \text{<span style="color:magenta">} f(x) = 2x - 1 \text{ for } x > 1.</span>$$
$$= 2(1) - 1 = 1 \qquad \text{<span style="color:magenta">} 2x - 1 \text{ is closed-form.}</span>$$

Because the left and right limits are different, $\lim_{x \to 1} f(x)$ does not exist, and so $f(x)$ is discontinuous at $x = 1$.

b. The only potential points of discontinuity for $g(x)$ occur at $x = 0$ and $x = 1$:

$$\lim_{x \to 0^-} g(x) = \lim_{x \to 0^-} x^2 - x + 1 = 1$$

$$\lim_{x \to 0^+} g(x) = \lim_{x \to 0^+} 1 - x = 1$$

Thus, $\lim_{x \to 0} g(x) = 1$. Further, $g(0) = 0^2 - 0 + 1 = 1$ from the formula, and so

$$\lim_{x \to 0} g(x) = g(0)$$

which shows that $g(x)$ is continuous at $x = 0$. At $x = 1$ we have

$$\lim_{x \to 1^-} g(x) = \lim_{x \to 1^-} 1 - x = 0$$

$$\lim_{x \to 1^+} g(x) = \lim_{x \to 1^+} x - 3 = -2$$

so that $\lim_{x \to 1} g(x)$ does not exist. Thus, $g(x)$ is discontinuous at $x = 1$. We conclude that $g(x)$ is continuous at every real number x except $x = 1$.

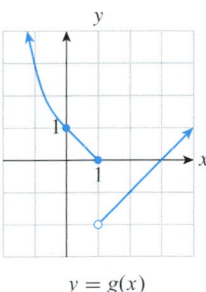

$y = g(x)$

Figure **16**

+ *Before we go on...* Figure 16 shows the graph of g from Example 3(b). Notice how the discontinuity at $x = 1$ shows up as a break in the graph, whereas at $x = 0$ the two pieces "fit together" at the point $(0, 1)$. ∎

Limits at Infinity

Let's look once again at Example 3 in Section 10.1 and some similar limits.

Example **4** Limits at Infinity

Compute the following limits, if they exist:

a. $\lim_{x \to +\infty} \dfrac{2x^2 - 4x}{x^2 - 1}$ **b.** $\lim_{x \to -\infty} \dfrac{2x^2 - 4x}{x^2 - 1}$

c. $\lim_{x \to +\infty} \dfrac{-x^3 - 4x}{2x^2 - 1}$ **d.** $\lim_{x \to +\infty} \dfrac{2x^2 - 4x}{5x^3 - 3x + 5}$

Solution

a. and **b.** While calculating the values for the tables used in Example 3 in Section 10.1, you might have noticed that the highest power of x in both the numerator and denominator dominated the calculations. For instance, when $x = 100,000$, the term $2x^2$ in the numerator has the value of 20,000,000,000, whereas the term $4x$ has the comparatively insignificant value of 400,000. Similarly, the term x^2 in the denominator overwhelms the term -1. In other words, for large values of x (or negative values with large magnitude),

$$\frac{2x^2 - 4x}{x^2 - 1} \approx \frac{2x^2}{x^2}$$ Use only the highest powers top & bottom.

$$= 2$$

Therefore,

$$\lim_{x \to \pm\infty} \frac{2x^2 - 4x}{x^2 - 1} = 2$$

The procedure of using only the highest powers of x to compute the limit is stated formally and justified after this example.

c. Applying the above technique of looking only at highest powers gives

$$\frac{-x^3 - 4x}{2x^2 - 1} \approx \frac{-x^3}{2x^2} \qquad \text{\color{magenta}Use only the highest powers top \& bottom.}$$

$$= \frac{-x}{2} \qquad \text{\color{magenta}Simplify.}$$

As x gets large, $-x/2$ gets large and negative, so

$$\lim_{x \to +\infty} \frac{-x^3 - 4x}{2x^2 - 1} = -\infty$$

d. $\dfrac{2x^2 - 4x}{5x^3 - 3x + 5} \approx \dfrac{2x^2}{5x^3} = \dfrac{2}{5x}$. As x gets large, $2/(5x)$ gets close to zero, so

$$\lim_{x \to +\infty} \frac{2x^2 - 4x}{5x^3 - 3x + 5} = 0$$

Let's look again at the limits (a) and (b) in Example 4. We say that the graph of f has a **horizontal asymptote** at $y = 2$ because of the limits we have just calculated. This means that the graph approaches the horizontal line $y = 2$ far to the right or left (in this case, to both the right and left). Figure 17 shows the graph of f together with the line $y = 2$.

The graph reveals some additional interesting information: as $x \to 1^+$, $f(x) \to -\infty$, and as $x \to 1^-$, $f(x) \to +\infty$. Thus,

$$\lim_{x \to 1} f(x) \text{ does not exist}$$

See if you can determine what happens as $x \to -1$.

If you graph the function in part (d) of Example 4, you will again see a horizontal asymptote at $y = 0$. Does the limit in part (c) show a horizontal asymptote?

In Example 4, $f(x)$ was a **rational function:** a quotient of polynomial functions. We calculated the limit of $f(x)$ at $\pm\infty$ by ignoring all powers of x in both the numerator and denominator except for the largest. Following is a theorem that justifies this procedure:

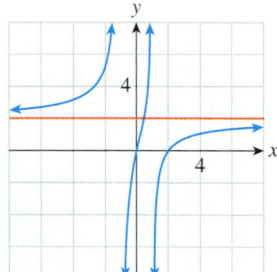

Figure 17

Theorem: Evaluating the Limit of a Rational Function at $\pm\infty$

If $f(x)$ has the form

$$f(x) = \frac{c_n x^n + c_{n-1} x^{n-1} + \cdots + c_1 x + c_0}{d_m x^m + d_{m-1} x^{m-1} \cdots + d_1 x + d_0}$$

with the c_i and d_i constants ($c_n \neq 0$ and $d_m \neq 0$), then we can calculate the limit of $f(x)$ as $x \to \pm\infty$ by ignoring all powers of x except the highest in both the numerator and denominator. Thus,

$$\lim_{x \to \pm\infty} f(x) = \lim_{x \to \pm\infty} \frac{c_n x^n}{d_m x^m}$$

quick Examples   (See Example 4)

1. $\displaystyle\lim_{x \to +\infty} \frac{2x^2 - 4x}{x^2 - 1} = \lim_{x \to +\infty} \frac{2x^2}{x^2} = \lim_{x \to +\infty} 2 = 2$

2. $\lim\limits_{x \to +\infty} \dfrac{-x^3 - 4x}{2x^2 - 1} = \lim\limits_{x \to +\infty} \dfrac{-x^3}{2x^2} = \lim\limits_{x \to +\infty} \dfrac{-x}{2} = -\infty$

3. $\lim\limits_{x \to +\infty} \dfrac{2x^2 - 4x}{5x^3 - 3x + 5} = \lim\limits_{x \to +\infty} \dfrac{2x^2}{5x^3} = \lim\limits_{x \to +\infty} \dfrac{2}{5x} = 0$

Proof Our function $f(x)$ is a polynomial of degree n divided by a polynomial of degree m. If n happens to be larger than m, then dividing the top and bottom by the largest power x^n of x gives

$$f(x) = \frac{c_n x^n + c_{n-1} x^{n-1} + \cdots + c_1 x + c_0}{d_m x^m + d_{m-1} x^{m-1} \cdots + d_1 x + d_0}$$

$$= \frac{c_n x^n / x^n + c_{n-1} x^{n-1} / x^n + \cdots + c_1 x / x^n + c_0 / x^n}{d_m x^m / x^n + d_{m-1} x^{m-1} / x^n + \cdots + d_1 x / x^n + d_0 / x^n}$$

Canceling powers of x in each term and remembering that $n > m$ leaves us with

$$f(x) = \frac{c_n + c_{n-1}/x + \cdots + c_1/x^{n-1} + c_0/x^n}{d_m/x^{n-m} + d_{m-1}/x^{n-m+1} + \cdots + d_1/x^{n-1} + d_0/x^n}$$

As $x \to \pm\infty$, all the terms shown in red approach 0, so we can ignore them in taking the limit. (The first term in the denominator happens to approach zero as well, but we retain it for convenience.) Thus,

$$\lim_{x \to \pm\infty} f(x) = \lim_{x \to \pm\infty} \frac{c_n}{d_m/x^{n-m}} = \lim_{x \to \pm\infty} \frac{c_n x^n}{d_m x^m}$$

as required. The cases when n is smaller than m and $m = n$ are proved similarly by dividing top and bottom by the largest power of x in each case.

FAQs **Strategy for Evaluating Limits Algebraically**

Q: *Is there a systematic way to evaluate a limit* $\lim_{x \to a} f(x)$ *algebraically?*

A: The following approach is often successful:

Case 1: a Is a Finite Number (Not $\pm\infty$)
1. Decide whether f is a closed-form function. If it is not, then find the left and right limits at the values of x where the function changes from one formula to another.
2. If f is a closed-form function, try substituting $x = a$ in the formula for $f(x)$. Then one of three things will happen:

  $f(a)$ is defined. Then $\lim_{x \to a} f(x) = f(a)$.
  $f(a)$ is not defined and has the form 0/0. Try to simplify the expression for f to cancel one of the terms that gives 0.
  $f(a)$ is not defined and has the form $k/0$ where k is not zero. Then the function diverges to $\pm\infty$ as x approaches a from each side. Check this graphically, as in Example 4 (see "Before we go on").

Case 2: $a = \pm\infty$
If the given function is a polynomial or ratio of polynomials, use the technique of Example 4: Focus only on the highest powers of x. ∎

There is another technique for evaluating certain difficult limits, called *L'Hospital's Rule,* but this uses derivatives, so we'll have to wait to discuss it until Section 10.7.

10.3 EXERCISES

● denotes basic skills exercises

In Exercises 1–4 complete the given sentence.

1. ● The closed-form function $f(x) = \dfrac{1}{x-1}$ is continuous for all x except _____. *hint* [see Quick Example p. 704]

2. ● The closed-form function $f(x) = \dfrac{1}{x^2-1}$ is continuous for all x except _____.

3. ● The closed-form function $f(x) = \sqrt{x+1}$ has $x = 3$ in its domain. Therefore, $\lim_{x\to 3} \sqrt{x+1} = $ ____. *hint* [see Example 1]

4. ● The closed-form function $f(x) = \sqrt{x-1}$ has $x = 10$ in its domain. Therefore, $\lim_{x\to 10} \sqrt{x-1} = $ ____.

Calculate the limits in Exercises 5–42 algebraically. If a limit does not exist, say why. *hint* [see Example 1]

5. ● $\lim\limits_{x\to 0} (x+1)$

6. ● $\lim\limits_{x\to 0} (2x-4)$

7. ● $\lim\limits_{x\to 2} \dfrac{2+x}{x}$

8. ● $\lim\limits_{x\to -1} \dfrac{4x^2+1}{x}$

9. ● $\lim\limits_{x\to -1} \dfrac{x+1}{x}$

10. ● $\lim\limits_{x\to 4} (x+\sqrt{x})$

11. ● $\lim\limits_{x\to 8} (x-\sqrt[3]{x})$

12. ● $\lim\limits_{x\to 1} \dfrac{x-2}{x+1}$

13. ● $\lim\limits_{h\to 1} (h^2+2h+1)$

14. ● $\lim\limits_{h\to 0} (h^3-4)$

15. ● $\lim\limits_{h\to 3} 2$

16. ● $\lim\limits_{h\to 0} -5$

17. ● $\lim\limits_{h\to 0} \dfrac{h^2}{h+h^2}$ *hint* [see Example 2]

18. ● $\lim\limits_{h\to 0} \dfrac{h^2+h}{h^2+2h}$

19. ● $\lim\limits_{x\to 1} \dfrac{x^2-2x+1}{x^2-x}$

20. ● $\lim\limits_{x\to -1} \dfrac{x^2+3x+2}{x^2+x}$

21. ● $\lim\limits_{x\to 2} \dfrac{x^3-8}{x-2}$

22. ● $\lim\limits_{x\to -2} \dfrac{x^3+8}{x^2+3x+2}$

23. ● $\lim\limits_{x\to 0^+} \dfrac{1}{x^2}$

24. ● $\lim\limits_{x\to 0^+} \dfrac{1}{x^2-x}$

25. ● $\lim\limits_{x\to -1} \dfrac{x^2+1}{x+1}$

26. ● $\lim\limits_{x\to -1^-} \dfrac{x^2+1}{x+1}$

27. ● $\lim\limits_{x\to +\infty} \dfrac{3x^2+10x-1}{2x^2-5x}$ *hint* [see Example 4]

28. ● $\lim\limits_{x\to +\infty} \dfrac{6x^2+5x+100}{3x^2-9}$

29. ● $\lim\limits_{x\to +\infty} \dfrac{x^5-1000x^4}{2x^5+10,000}$

30. ● $\lim\limits_{x\to +\infty} \dfrac{x^6+3000x^3+1,000,000}{2x^6+1000x^3}$

31. ● $\lim\limits_{x\to +\infty} \dfrac{10x^2+300x+1}{5x+2}$

32. ● $\lim\limits_{x\to +\infty} \dfrac{2x^4+20x^3}{1000x^3+6}$

33. ● $\lim\limits_{x\to +\infty} \dfrac{10x^2+300x+1}{5x^3+2}$

34. ● $\lim\limits_{x\to +\infty} \dfrac{2x^4+20x^3}{1000x^6+6}$

35. ● $\lim\limits_{x\to -\infty} \dfrac{3x^2+10x-1}{2x^2-5x}$

36. ● $\lim\limits_{x\to -\infty} \dfrac{6x^2+5x+100}{3x^2-9}$

37. ● $\lim\limits_{x\to -\infty} \dfrac{x^5-1000x^4}{2x^5+10,000}$

38. ● $\lim\limits_{x\to -\infty} \dfrac{x^6+3000x^3+1,000,000}{2x^6+1000x^3}$

39. ● $\lim\limits_{x\to -\infty} \dfrac{10x^2+300x+1}{5x+2}$

40. ● $\lim\limits_{x\to -\infty} \dfrac{2x^4+20x^3}{1000x^3+6}$

41. ● $\lim\limits_{x\to -\infty} \dfrac{10x^2+300x+1}{5x^3+2}$

42. ● $\lim\limits_{x\to -\infty} \dfrac{2x^4+20x^3}{1000x^6+6}$

In each of Exercises 43–50, find all points of discontinuity of the given function. *hint* [see Example 3]

43. ● $f(x) = \begin{cases} x+2 & \text{if } x < 0 \\ 2x-1 & \text{if } x \geq 0 \end{cases}$

44. ● $g(x) = \begin{cases} 1-x & \text{if } x \leq 1 \\ x-1 & \text{if } x > 1 \end{cases}$

45. ● $g(x) = \begin{cases} x+2 & \text{if } x < 0 \\ 2x+2 & \text{if } 0 \leq x < 2 \\ x^2+2 & \text{if } x \geq 2 \end{cases}$

46. ● $f(x) = \begin{cases} 1-x & \text{if } x \leq 1 \\ x+2 & \text{if } 1 < x < 3 \\ x^2-4 & \text{if } x \geq 3 \end{cases}$

47. $h(x) = \begin{cases} x+2 & \text{if } x < 0 \\ 0 & \text{if } x = 0 \\ 2x+2 & \text{if } x > 0 \end{cases}$

48. $h(x) = \begin{cases} 1-x & \text{if } x < 1 \\ 1 & \text{if } x = 1 \\ x+2 & \text{if } x > 1 \end{cases}$

49. $f(x) = \begin{cases} 1/x & \text{if } x < 0 \\ x & \text{if } 0 \leq x \leq 2 \\ 2^{x-1} & \text{if } x > 2 \end{cases}$

50. $f(x) = \begin{cases} x^3+2 & \text{if } x \leq -1 \\ x^2 & \text{if } -1 < x < 0 \\ x & \text{if } x \geq 0 \end{cases}$

● basic skills

Applications

51. ● *Movie Advertising* Movie expenditures, in billions of dollars, on advertising in newspapers from 1995 to 2004 could be approximated by

$$f(t) = \begin{cases} 0.04t + 0.33 & \text{if } t \leq 4 \\ -0.01t + 1.2 & \text{if } t > 4 \end{cases}$$

where t is time in years since 1995.[10]

a. Compute $\lim_{t \to 4^-} f(t)$ and $\lim_{t \to 4^+} f(t)$, and interpret each answer.

b. Is the function f continuous at $t = 4$? What does the answer tell you about movie advertising expenditures? *hint* [see Example 3]

52. ● *Movie Advertising* The percentage of movie advertising as a share of newspapers' total advertising revenue from 1995 to 2004 could be approximated by

$$p(t) = \begin{cases} -0.07t + 6.0 & \text{if } t \leq 4 \\ 0.3t + 17.0 & \text{if } t > 4 \end{cases}$$

where t is time in years since 1995.[11]

a. Compute $\lim_{t \to 4^-} p(t)$ and $\lim_{t \to 4^+} p(t)$, and interpret each answer.

b. Is the function p continuous at $t = 4$? What does the answer tell you about newspaper revenues?

53. ● *Law Enforcement* The cost of fighting crime in the U.S. increased steadily in the period 1982–1999. Total spending on police and courts can be approximated, respectively, by[12]

$$P(t) = 1.745t + 29.84 \text{ billion dollars} \quad (2 \leq t \leq 19)$$
$$C(t) = 1.097t + 10.65 \text{ billion dollars} \quad (2 \leq t \leq 19)$$

where t is time in years since 1980. Compute $\lim_{t \to +\infty} \dfrac{P(t)}{C(t)}$ to two decimal places and interpret the result. *hint* [see Example 4]

54. *Law Enforcement* Refer to Exercise 53. Total spending on police, courts, and prisons can be approximated, respectively, by[13]

$$P(t) = 1.745t + 29.84 \text{ billion dollars} \quad (2 \leq t \leq 19)$$
$$C(t) = 1.097t + 10.65 \text{ billion dollars} \quad (2 \leq t \leq 19)$$
$$J(t) = 1.919x + 12.36 \text{ billion dollars} \quad (2 \leq t \leq 19)$$

where t is time in years since 1980. Compute $\lim_{t \to +\infty} \dfrac{P(t)}{P(t) + C(t) + J(t)}$ to two decimal places and interpret the result.

Foreign Trade Annual U.S. imports from China in the years 1996 through 2003 could be approximated by

$$I(t) = t^2 + 3.5t + 50 \quad (1 \leq t \leq 8)$$

billion dollars, where t represents time in years since 1995. Annual U.S. exports to China in the same years could be approximated by

$$E(t) = 0.4t^2 + 1.6t + 14 \quad (0 \leq t \leq 10)$$

billion dollars.[14] Exercises 55 and 56 are based on these models.

55. Assuming that the trends shown in the above models continue indefinitely, calculate the limits

$$\lim_{t \to +\infty} I(t) \quad \text{and} \quad \lim_{t \to +\infty} \frac{I(t)}{E(t)}$$

algebraically, interpret your answers, and comment on the results.

56. Repeat Exercise 55, this time calculating

$$\lim_{t \to +\infty} E(t) \quad \text{and} \quad \lim_{t \to +\infty} \frac{E(t)}{I(t)}$$

57. *Acquisition of Language* The percentage $p(t)$ of children who can speak in at least single words by the age of t months can be approximated by the equation[15]

$$p(t) = 100\left(1 - \frac{12,200}{t^{4.48}}\right) \quad (t \geq 8.5)$$

Calculate $\lim_{t \to +\infty} p(t)$ and interpret the result.

58. *Acquisition of Language* The percentage $q(t)$ of children who can speak in sentences of five or more words by the age of t months can be approximated by the equation[16]

$$q(t) = 100\left(1 - \frac{5.27 \times 10^{17}}{t^{12}}\right) \quad (t \geq 30)$$

If p is the function referred to in the preceding exercise, calculate $\lim_{t \to +\infty}[p(t) - q(t)]$ and interpret the result.

59. *Television Advertising* The cost, in millions of dollars, of a 30-second television ad during Super Bowls from 1990 to 2001 can be approximated by the following piecewise linear function ($t = 0$ represents 1990):[17]

$$C(t) = \begin{cases} 0.08t + 0.6 & \text{if } 0 \leq t < 8 \\ 0.355t - 1.6 & \text{if } 8 \leq t \leq 11 \end{cases}$$

Is C a continuous function of t? Why?

---

[10] Model by the authors. Source for data: Newspaper Association of America Business Analysis and Research/*New York Times,* May 16, 2005.

[11] Model by the authors. Source for data: Newspaper Association of America Business Analysis and Research/*New York Times,* May 16, 2005.

[12] Spending is adjusted for inflation and shown in 1999 dollars. Models are based on a linear regression. Source for data: Bureau of Justice Statistics/*New York Times,* February 11, 2002, p. A14.

[13] Ibid.

[14] Based on quadratic regression using data from the U.S. Census Bureau Foreign Trade Division website www.census.gov/foreign-trade/sitc1/ as of December 2004.

[15] The model is the authors' and is based on data presented in the article *The Emergence of Intelligence* by William H. Calvin, *Scientific American,* October, 1994, pp. 101–107.

[16] Ibid.

[17] SOURCE: *New York Times,* January 26, 2001, p. C1.

60. *Internet Purchases* The percentage $p(t)$ of buyers of new cars who used the Internet for research or purchase since 1997 is given by the following function[18] ($t = 0$ represents 1997):

$$p(t) = \begin{cases} 10t + 15 & \text{if } 0 \le t < 1 \\ 15t + 10 & \text{if } 1 \le t \le 4 \end{cases}$$

Is p a continuous function of t? Why?

Communication and Reasoning Exercises

61. ● Describe the algebraic method of evaluating limits as discussed in this section and give at least one disadvantage of this method.

---

[18] Model is based on data through 2000 (the 2000 value is estimated). SOURCE: J.D. Power Associates/*New York Times,* January 25, 2000, p. C1.

62. ● What is a closed-form function? What can we say about such functions?

63. Your friend Karin tells you that $f(x) = 1/(x - 2)^2$ cannot be a closed-form function because it is not continuous at $x = 2$. Comment on her assertion.

64. Give an example of a function f specified by means of algebraic formulas such that the domain of f consists of all real numbers and f is not continuous at $x = 2$. Is f a closed-form function?

65. ● What is wrong with the following statement? If $f(x)$ is specified algebraically and $f(x)$ is defined, then $\lim_{x \to a} f(x)$ exists and equals $f(a)$.

66. ● What is wrong with the following statement? $\lim_{x \to -2} \dfrac{x^2 - 4}{x + 2}$ does not exist because substituting $x = -2$ yields 0/0, which is undefined.

67. Find a function that is continuous everywhere except at two points.

68. Find a function that is continuous everywhere except at three points.

● basic skills

10.4 | Average Rate of Change

Calculus is the mathematics of change, inspired largely by observation of continuously changing quantities around us in the real world. As an example, the New York metro area consumer confidence index C decreased from 100 points in January 2000 to 80 points in January 2002.[19] As we saw in Chapter 1, the **change** in this index can be measured as the difference:

$$\Delta C = \text{Second value} - \text{First value} = 80 - 100 = -20 \text{ points}$$

(The fact that the confidence index decreased is reflected in the negative sign of the change.) The kind of question we will concentrate on is *how fast* the confidence index was dropping. Because C decreased by 20 points in 2 years, we say it averaged a $20/2 = 10$ point drop each year. (It actually dropped less than 10 points the first year and more the second, giving an average drop of 10 points each year.)

Alternatively, we might want to measure this rate in points per month rather than points per year. Because C decreased by 20 points in 24 months, it went down at an average rate of $20/24 \approx 0.833$ points per month.

---

[19] Figures are approximate. SOURCE: Siena College Research Institute/*New York Times,* February 10, 2002, p. LI.

In both cases, we obtained the average rate of change by dividing the change by the corresponding length of time:

$$\text{Average rate of change} = \frac{\text{Change in } C}{\text{Change in time}} = \frac{-20}{2} = -10 \text{ points per year}$$

$$\text{Average rate of change} = \frac{\text{Change in } C}{\text{Change in time}} = \frac{-20}{24} \approx -0.833 \text{ points per month}$$

Average Rate of Change of a Function Using Numerical Data

Example 1 Average Rate of Change from a Table

The following table lists the approximate value of Standard and Poors 500 stock market index (S&P) during the 10-year period 1998–2004[*] ($t = 8$ represents 1998):

t (year)	8	9	10	11	12	13	14
$S(t)$ S&P 500 Index (points)	1200	1350	1450	1200	950	1000	1100

a. What was the average rate of change in the S&P over the two-year period 1998–2000 (the period $8 \le t \le 10$ or [8, 10] in interval notation); over the four-year period 2000–2004 (the period $10 \le t \le 14$ or [10, 14]; and over the period [8, 11]?

b. Graph the values shown in the table. How are the rates of change reflected in the graph?

Solution

a. During the two-year period [8, 10], the S&P changed as follows:

Start of the period ($t = 8$):	$S(8) = 1200$
End of the period ($t = 10$):	$S(10) = 1450$
Change during the period [8, 10]:	$S(10) - S(8) = 250$

Thus, the S&P increased by 250 points in 2 years, giving an average rate of change of $250/2 = 125$ points per year. We can write the calculation this way:

$$\begin{aligned}\text{Average rate of change of } S &= \frac{\text{Change in } S}{\text{Change in } t} \\ &= \frac{\Delta S}{\Delta t} \\ &= \frac{S(10) - S(8)}{10 - 8} \\ &= \frac{1450 - 1200}{10 - 8} = \frac{250}{2} = 125 \text{ points per year}\end{aligned}$$

[*] The values are approximate values midway through the given year. SOURCE: http://money.excite.com, May, 2004.

Interpreting the result: During the period [8, 10] (that is, 1998–2000), the S&P increased at an average rate of 125 points per year.

Similarly, the average rate of change during the period [10, 14] was

$$\text{Average rate of change of } S = \frac{\Delta S}{\Delta t} = \frac{S(14) - S(10)}{14 - 10} = \frac{1100 - 1450}{14 - 10}$$

$$= -\frac{350}{4} = -87.5 \text{ points per year}$$

Interpreting the result: During the period [10, 14] the S&P *decreased* at an average rate of 87.5 points per year.

Finally, during the period [8, 11], the average rate of change was

$$\text{Average rate of change of } S = \frac{\Delta S}{\Delta t} = \frac{S(11) - S(8)}{11 - 8} = \frac{1200 - 1200}{11 - 8}$$

$$= \frac{0}{3} = 0 \text{ points per year}$$

Interpreting the result: During the period [8, 11] the average rate of change of the S&P was zero points per year (even though its value did fluctuate during that period).

b. In Chapter 1, we saw that the rate of change of a quantity that changes linearly with time is measured by the slope of its graph. However, the S&P index does not change linearly with time. Figure 18 shows the data plotted two different ways: (a) as a bar chart and (b) as a piecewise linear graph. Bar charts are more commonly used in the media, but Figure 18b on the right illustrates the changing index price more clearly.

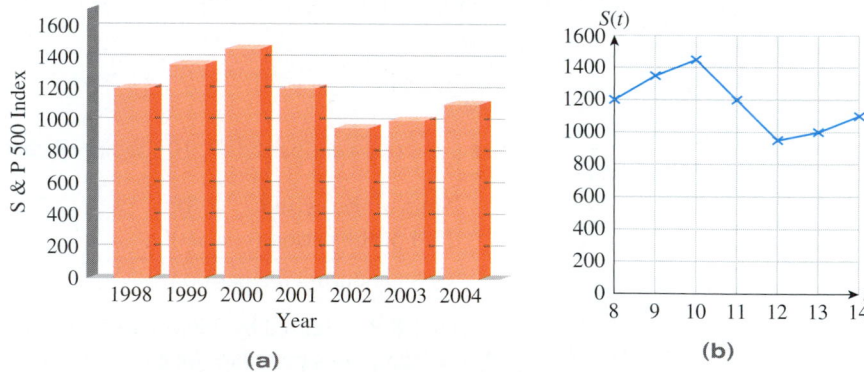

Figure 18

We saw in part (a) that the average rate of change of S over the interval [8, 10] is the ratio

$$\text{Rate of change of } S = \frac{\Delta S}{\Delta t} = \frac{\text{Change in } S}{\text{Change in } t} = \frac{S(10) - S(8)}{10 - 8}$$

Notice that this rate of change is also the slope of the line through P and Q shown in Figure 19.

Average Rate of Change as Slope: The average rate of change of the S&P over the interval [8, 10] is the slope of the line passing through the points on the graph where $t = 8$ and $t = 10$.

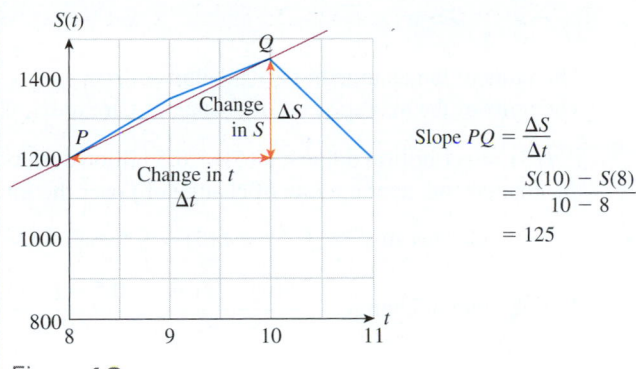

Figure **19**

Similarly, the average rates of change of the S&P over the intervals [10, 14] and [8, 11] are the slopes of the lines through pairs of corresponding points.

Here is the formal definition of the average rate of change of a function over an interval.

Change and Average Rate of Change of *f* over [*a*, *b*]: Difference Quotient

The **change** in $f(x)$ over the interval $[a, b]$ is

$$\text{Change in } f = \Delta f$$
$$= \text{Second value} - \text{First value}$$
$$= f(b) - f(a)$$

The **average rate of change** of $f(x)$ over the interval $[a, b]$ is

$$\text{Average rate of change of } f = \frac{\text{Change in } f}{\text{Change in } x}$$
$$= \frac{\Delta f}{\Delta x} = \frac{f(b) - f(a)}{b - a}$$
$$= \text{Slope of line through points } P \text{ and } Q$$
$$\text{(see figure)}$$

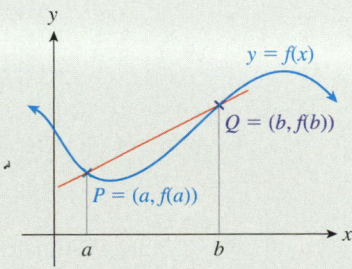

Average rate of change = Slope of *PQ*

We also call this average rate of change the **difference quotient** of f over the interval $[a, b]$. (It is the *quotient* of the *differences* $f(b) - f(a)$ and $b - a$.) A line through two points of a graph like P and Q is called a **secant line** of the graph.

> **Units**
>
> The units of the change in f are the units of $f(x)$.
> The units of the average rate of change of f are units of $f(x)$ per unit of x.

quick Example

If $f(3) = -1$ billion dollars, $f(5) = 0.5$ billion dollars, and x is measured in years, then the change and average rate of change of f over the interval $[3, 5]$ are given by

$$\text{Change in } f = f(5) - f(3) = 0.5 - (-1) = 1.5 \text{ billion dollars}$$

$$\text{Average rate of change} = \frac{f(5) - f(3)}{5 - 3} = \frac{0.5 - (-1)}{2} = 0.75 \text{ billion dollars/year}$$

> **Alternative Formula: Average Rate of Change of f over $[a, a + h]$**
>
> (Replace b above by $a + h$.) The average rate of change of f over the interval $[a, a + h]$ is
>
> $$\text{Average rate of change of } f = \frac{f(a + h) - f(a)}{h}$$

Average Rate of Change of a Function Using Graphical Data

In Example 1 we saw that the average rate of change of a quantity can be determined directly from a graph. Here is an example that further illustrates the graphical approach.

Example 2 Average Rate of Change from a Graph

Figure 20 shows the number of sports utility vehicles (SUVs) sold in the U.S. each year from 1990 through 2003 ($t = 0$ represents the year 1990, and $N(t)$ represents sales in year t in thousands of vehicles).[*]

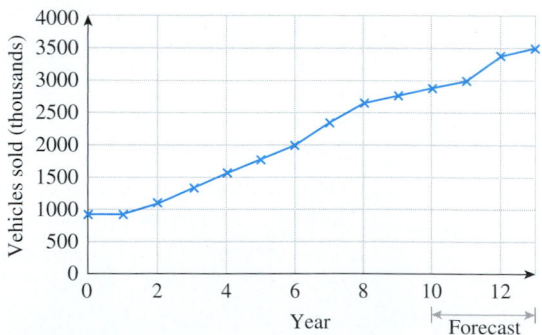

Figure **20**

a. Use the graph to estimate the average rate of change of $N(t)$ with respect to t over the interval $[6, 11]$ and interpret the result.

b. Over which one-year period(s) was the average rate of change of $N(t)$ the greatest?

---

[*] SOURCES: Ford Motor Company/*New York Times,* February 9, 1995, p. D17, Oak Ridge National Laboratory, Light Vehicle MPG and Market Shares System, AutoPacific, *The U.S. Car and Light Truck Market,* 1999, pp. 24, 120, 121.

Solution

a. The average rate of change of N over the interval $[6, 11]$ is given by the slope of the line through the points P and Q shown in Figure 21.

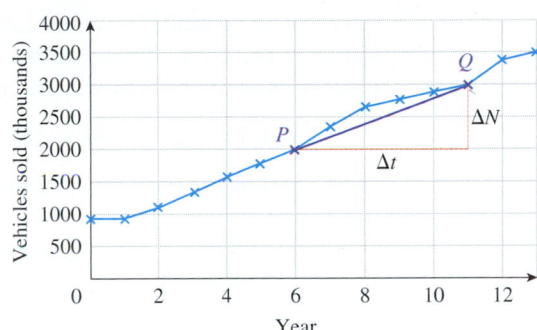

Figure **21**

From the figure,

$$\text{Average rate of change of } N = \frac{\Delta N}{\Delta t} = \text{slope } PQ \approx \frac{3000 - 2000}{11 - 6} = \frac{1000}{5} = 200$$

Thus, the rate of change of N over the interval $[6, 11]$ is approximately 200.

Q: How do we interpret the result?

A: A clue is given by the units of the average rate of change: units of N per unit of t. The units of N are thousands of SUVs and the units of t are years. Thus, the average rate of change of N is measured in thousands of SUVs per year, and we can now interpret the result as follows:

Interpreting the average rate of change: Sales of SUVs were increasing at an average rate of 200,000 SUVs per year from 1996 to 2001. ∎

b. The rates of change of annual sales over successive one-year periods are given by the slopes of the individual line segments that make up the graph in Figure 20. Thus, the greatest average rate of change over a single year corresponds to the segment(s) with the largest slope. If you look carefully at the figure, you will notice that the segment corresponding to $[11, 12]$ is the steepest. (The segment corresponding to $[6, 7]$ is slightly less steep.) Thus, the average rate of change of annual sales was largest over the one-year period from 2001 to 2002.

+*Before we go on...* Notice in Example 2 that we do not get exact answers from a graph; the best we can do is *estimate* the rates of change: Was the exact answer to part (a) closer to 201 or 202? Two people can reasonably disagree about results read from a graph, and you should bear this in mind when you check the answers to the exercises.

Perhaps the most sophisticated way to compute the average rate of change of a quantity is through the use of a mathematical formula or model for the quantity in question. ∎

Average Rate of Change of a Function Using Algebraic Data

Example 3 Average Rate of Change from a Formula

You are a commodities trader and you monitor the price of gold on the New York Spot Market very closely during an active morning. Suppose you find that the price of an ounce of gold can be approximated by the function

$$G(t) = -2t^2 + 36t + 228 \text{ dollars} \quad (7.5 \le t \le 10.5)$$

where t is time in hours. (See Figure 22. $t = 8$ represents 8:00 AM.)

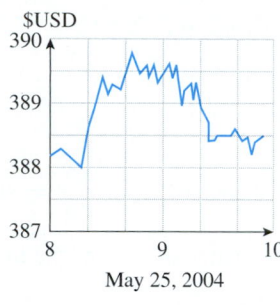

Source: www.kitco.com May 25, 2004

(a)

Figure **22**

Looking at the graph on the right, we can see that the price of gold rose rather rapidly at the beginning of the time period, but by $t = 8.5$ the rise had slowed, until the market faltered and the price began to fall more and more rapidly toward the end of the period. What was the average rate of change of the price of gold over the $1\frac{1}{2}$-hour period starting at 8:00 AM (the interval [8, 9.5] on the t-axis)?

Solution We have

$$\text{Average rate of change of } G \text{ over } [8. 9.5] = \frac{\Delta G}{\Delta t} = \frac{G(9.5) - G(8)}{9.5 - 8}$$

From the formula for $G(t)$, we find

$$G(9.5) = -2(9.5)^2 + 36(9.5) + 228 = 389.5$$
$$G(8) = -2(8)^2 + 36(8) + 228 = 388$$

Thus, the average rate of change of G is given by

$$\frac{G(9.5) - G(8)}{9.5 - 8} = \frac{389.5 - 388}{1.5} = \frac{1.5}{1.5} = \$1 \text{ per hour}$$

In other words, the price of gold was increasing at an average rate of $1 per hour over the given $1\frac{1}{2}$-hour period.

using *Technology*

See the Technology Guides at the end of the chapter to find out how we can use a TI-83/84 or Excel to compute average rates of change for a given function.

Example 4 Rates of Change over Shorter Intervals

Continuing with Example 3, use technology to compute the average rate of change of

$$G(t) = -2t^2 + 36t + 228 \qquad (7.5 \le t \le 10.5)$$

over the intervals $[8, 8 + h]$, where $h = 1, 0.1, 0.01, 0.001,$ and 0.0001. What do the answers tell you about the price of gold?

Solution

We use the alternative formula

$$\text{Average rate of change of } G \text{ over } [a, a + h] = \frac{G(a + h) - G(a)}{h}$$

so

$$\text{Average rate of change of } G \text{ over } [8, 8 + h] = \frac{G(8 + h) - G(8)}{h}$$

Let us calculate this average rate of change for some of the values of h listed:

$h = 1$: $G(8 + h) = G(8 + 1) = G(9) = -2(9)^2 + 36(9) + 228 = 390$

$G(8) = -2(8)^2 + 36(8) + 228 = 388$

$$\text{Average rate of change of } G = \frac{G(9) - G(8)}{1} = \frac{390 - 388}{1} = 2$$

$h = 0.1$: $G(8 + h) = G(8 + 0.1) = G(8.1) = -2(8.1)^2 + 36(8.1) + 228 = 388.38$

$G(8) = -2(8)^2 + 36(8) + 228 = 388$

$$\text{Average rate of change of } G = \frac{G(8.1) - G(8)}{0.1} = \frac{388.38 - 388}{0.1} = 3.8$$

$h = 0.01$: $G(8 + h) = G(8 + 0.01) = G(8.01) = -2(8.01)^2 + 36(8.01) + 288$

$= 388.0398$

$G(8) = -2(8)^2 + 36(8) + 228 = 388$

$$\text{Average rate of change of } G = \frac{G(8.01) - G(8)}{0.01} = \frac{388.0398 - 388}{0.01} = 3.98$$

Continuing in this way, we get the values in the following table:

h	1	0.1	0.01	0.001	0.0001
Ave. Rate of Change $\dfrac{G(8 + h) - G(8)}{h}$	2	3.8	3.98	3.998	3.9998

Each value is an average rate of change of G. For example, the value corresponding to $h = 0.01$ is 3.98, which tells us:

Over the interval $[8, 8.01]$ the price of gold was increasing at an average rate of $3.98 per hour.

In other words, during the first one-hundredth of an hour (or 36 seconds) starting at $t = 8$ AM, the price of gold was increasing at an average rate of $3.98 per hour. Put another way, in those 36 seconds, the price of gold increased at a rate that, if continued, would have produced an increase of $3.98 in the price of gold during the next hour. We will return to this example at the beginning of Section 10.5.

using *Technology*

This is the kind of example where the use of technology can make a huge difference. See the Technology Guides at the end of the chapter to find out how to do the above computations almost effortlessly.

FAQs   Recognizing When and How to Compute the Average Rate of Change and How to Interpret the Answer

Q: *How do I know, by looking at the wording of a problem, that it is asking for an average rate of change?*

A: If a problem does not ask for an average rate of change directly, it might do so indirectly, as in "On average, how fast is quantity *q* increasing?" ∎

Q: *If I know that a problem calls for computing an average rate of change, how should I compute it? By hand or using technology?*

A: All the computations can be done by hand, but when hand calculations are not called for, using technology might save time. ∎

Q: *Lots of problems ask us to "interpret" the answer. How do I do that for questions involving average rates of change?*

A: The *units* of the average rate of change are often the key to interpreting the results:

The units of the average rate of change of f(x) are units of f(x) per unit of x.

Thus, for instance, if $f(x)$ is the cost, in dollars, of a trip of x miles in length, and the average rate of change of f is calculated to be 10, then we can say that the cost of a trip rises an average of $10 for each additional mile. ∎

10.4 EXERCISES

● denotes basic skills exercises

◆ denotes challenging exercises

tech Ex indicates exercises that should be solved using technology

In Exercises 1–18, calculate the average rate of change of the given function over the given interval. Where appropriate, specify the units of measurement.

1. ●

x	0	1	2	3
$f(x)$	3	5	2	−1

Interval: [1, 3] *hint* [see Example 1]

2. ●

x	0	1	2	3
$f(x)$	−1	3	2	1

Interval: [0, 2]

3. ●

x	−3	−2	−1	0
$f(x)$	−2.1	0	−1.5	0

Interval: [−3, −1]

4. ●

x	−2	−1	0	1
$f(x)$	−1.5	−0.5	4	6.5

Interval: [−1, 1]

5. ●

t (months)	2	4	6
$R(t)$ ($ millions)	20.2	24.3	20.1

Interval: [2, 6]

6. ●

x (kilos)	1	2	3
$C(x)$ (£)	2.20	3.30	4.00

Interval: [1, 3]

7. ●

p ($)	5.00	5.50	6.00
$q(p)$ (items)	400	300	150

Interval: [5, 5.5]

8. ●

t (hours)	0	0.1	0.2
$D(t)$ (miles)	0	3	6

Interval: [0.1, 0.2]

● basic skills   ◆ challenging   tech Ex technology exercise

9. ● *hint* [see Example 2]

Apple Computer Stock Price ($)

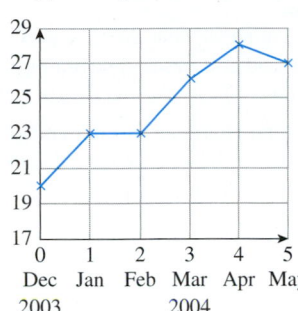

Dec   Jan   Feb   Mar   Apr   May
2003   2004

Interval: [2, 5]

10. ●

Cisco Systems Stock Price ($)

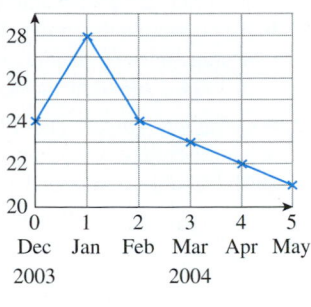

Dec   Jan   Feb   Mar   Apr   May
2003   2004

Interval: [1, 5]

11. ●

Unemployment (%)

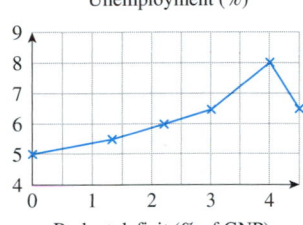

Budget deficit (% of GNP)

Interval: [0, 4]

12. ●

Inflation (%)

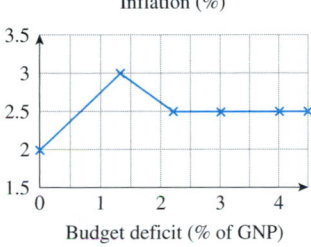

Budget deficit (% of GNP)

Interval: [0, 4]

13. ● $f(x) = x^2 - 3$; [1, 3] *hint* [see Example 3]

14. ● $f(x) = 2x^2 + 4$; [−1, 2]

15. ● $f(x) = 2x + 4$; [−2, 0]   **16. ●** $f(x) = \frac{1}{x}$; [1, 4]

17. ● $f(x) = \frac{x^2}{2} + \frac{1}{x}$; [2, 3]   **18. ●** $f(x) = 3x^2 - \frac{x}{2}$; [3, 4]

In Exercises 19–24, calculate the average rate of change of the given function f over the intervals [a, a + h] where h = 1, 0.1, 0.01, 0.001, and 0.0001. (Technology is recommended for the cases h = 0.01, 0.001, and 0.0001.)

19. ● $f(x) = 2x^2$; $a = 0$ *hint* [see Example 4]

20. ● $f(x) = \frac{x^2}{2}$; $a = 1$

21. ● $f(x) = \frac{1}{x}$; $a = 2$ **22. ●** $f(x) = \frac{2}{x}$; $a = 1$

23. ● $f(x) = x^2 + 2x$; $a = 3$ **24. ●** $f(x) = 3x^2 - 2x$; $a = 0$

Applications

25. ● *Employment* The following table lists the approximate number of people employed in the U.S. during the period 1998–2004, on July 1 of each year[20] ($t = 0$ represents 2000):

Year t	−2	−1	0	1	2	3	4
Employment $P(t)$ (millions)	126	130	132	132	130	130	131

Compute and interpret the average rate of change of $P(t)$ **a.** over the period 2000–2004 (that is, [0, 4]), and **b.** over the period [−1, 2]. Be sure to state the units of measurement. *hint* [see Example 1]

26. ● *Cell Phone Sales* The following table lists the net sales (after-tax revenue) at the Finnish cell phone company Nokia during the period 1997–2003[21] ($t = 0$ represents 2000):

Year t	−3	−2	−1	0	1	2	3
Nokia net sales $P(t)$ (€ Billion)	9	14	20	31	31	30	29

Compute and interpret the average rate of change of $P(t)$ **a.** over the period [−2, 3], and **b.** over the period [0, 1]. Be sure to state the units of measurement.

27. ● *Venture Capital* The following table shows the number of companies that invested in venture capital each year during the period 1995–2001[22] ($t = 5$ represents 1995):

Year t	5	6	7	8	9	10	11
Number of companies $N(t)$	100	150	300	400	1000	1700	900

During which two-year interval(s) was the average rate of change of $N(t)$ **a.** greatest **b.** least? Interpret your answers by referring to the rates of change.

28. ● *Venture Capital* The following table shows the amount of money that companies invested in venture capital during the period 1995–2001[23] ($t = 5$ represents 1995):

Year t	5	6	7	8	9	10	11
Investment $M(t)$ (billions)	$0.05	0.5	1	2	8	16	4

During which three-year interval(s) was the average rate of change of $M(t)$ **a.** greatest **b.** least? Interpret your answers by referring to the rates of change.

29. *Physics Research in the U.S.* The following table shows the number of research articles in the journal *Physics Review* authored by U.S researchers during the period 1993–2003[24] ($t = 3$ represents 1993):

---

[20] The given (approximate) values represent nonfarm employment. SOURCE: Bureau of Labor Statistics http://stats.bls.gov/, June 10, 2004.

[21] SOURCES: *New York Times*, February 6, 2002, p. A3, Nokia June 10, 2003 www.nokia.com.

[22] 2001 figure is a projection based on data through September. SOURCE: Venture Economics/National Venture Capital Association/*New York Times*, February 3, 2002, p. BU4.

[23] Ibid.

[24] SOURCE: The American Physical Society/*New York Times* May 3, 2003, p. A1.

● basic skills   ◆ challenging   **tech**Ex technology exercise

t (Year)	3	5	7	9	11	13
N(t) (Articles, thousands)	5.1	4.6	4.3	4.3	4.5	4.2

a. Find the interval(s) over which the average rate of change of N was the most negative. What was that rate of change? Interpret your answer.

b. The **percentage change of N over the interval $[a, b]$** is defined to be

$$\text{Percentage change of } N = \frac{\text{Change in } N}{\text{First value}} = \frac{N(b) - N(a)}{N(a)}$$

Compute the percentage change of N over the interval $[3, 13]$ and also the average rate of change. Interpret the answers.

30. *Physics Research in Europe* The following table shows the number of research articles in the journal *Physics Review* authored by researchers in Europe during the period 1993–2003[25] ($t = 3+$ represents 1993):

t (Year)	3	5	7	9	11	13
N(t) (Articles, thousands)	3.8	4.6	5.0	5.0	6.0	5.7

a. Find the interval(s) over which the average rate of change of N was the most positive. What was that rate of change? Interpret your answer.

b. The **percentage change of N over the interval $[a, b]$** is defined to be

$$\text{Percentage change of } N = \frac{\text{Change in } N}{\text{First value}} = \frac{N(b) - N(a)}{N(a)}$$

Compute the percentage change of N over the interval $[7, 13]$ and also the average rate of change. Interpret the answers.

31. ● *Collegiate Sports* The following chart shows the number of women's college soccer teams in the U.S. from 1982 to 2001[26] ($t = 1$ represents the 1981–1982 academic year):

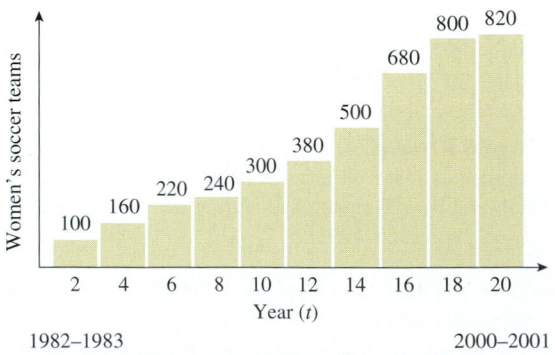

1982–1983   2000–2001

a. On average, how fast was the number of women's college soccer teams growing over the four-year period starting in the 1992–1993 academic year?

b. By inspecting the graph, determine whether the four-year average rates of change increased or decreased beginning in 1994–1995. *hint* [see Example 2]

32. ● *Collegiate Sports.* The following chart shows the number of men's college wrestling teams in the U.S. from 1982 to 2001[27] ($t = 1$ represents the 1981–1982 academic year):

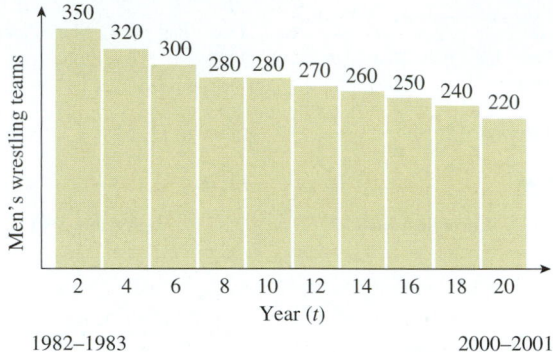

1982–1983   2000–2001

a. On average, how fast was the number of men's college wrestling teams decreasing over the eight-year period, starting in the 1984–1985 academic year?

b. By inspecting the graph, determine when the number of men's college wrestling teams was declining the fastest.

33. ● *Online Shopping* The following graph shows the annual number $N(t)$ of online shopping transactions in the U.S. for the period January 2000–January 2002[28] ($t = 0$ represents January, 2000):

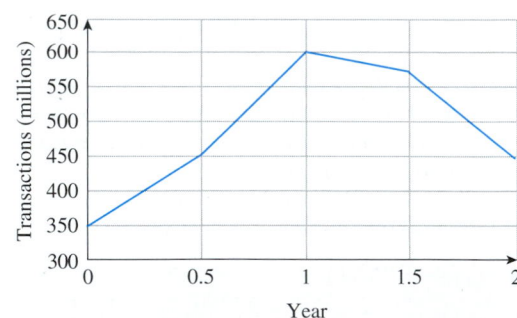

a. Estimate the average rate of change of $N(t)$ over the intervals $[0, 1]$, $[1, 2]$, and $[0, 2]$. Interpret your answers.

b. How can the average rate of change of $N(t)$ over $[0, 2]$ be obtained from the rates over $[0, 1]$ and $[1, 2]$?

34. ● *Online Shopping* The following graph shows the percentage of people in the U.S. who have ever purchased anything

[25] SOURCE: The American Physical Society/*New York Times* May 3, 2003, p. A1.
[26] SOURCE: N.C.A.A./*New York Times,* May 9, 2002, p. D4.
[27] SOURCE: N.C.A.A./*New York Times,* May 9, 2002, p. D4.
[28] Second half of 2001 data was an estimate. Source for data: Odyssey Research/*New York Times,* November 5, 2001, p. C1.

● basic skills   ◆ challenging   **tech** Ex technology exercise

online for the period January 2000–January 2002 ($t = 0$ represents January 2000):[29]

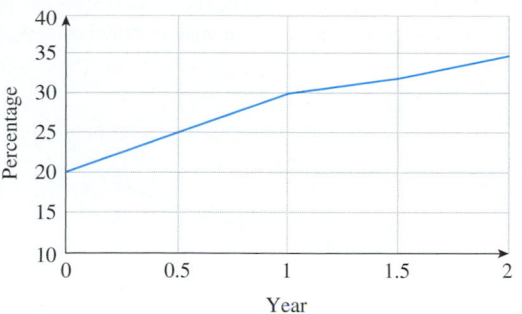

Year

a. Estimate the average rate of change of $P(t)$ over the interval [0, 2], and interpret your answer.

b. Are any of the one-year average rates of change greater than the two-year rate? Refer in your answer to the slopes of certain lines.

35. *Funding for the Arts* The following chart shows the total annual support for the arts in the U.S. by federal, state, and local government in 1995–2003 as a function of time in years ($t = 0$ represents 1995) together with the regression line:[30]

Government Funding for
the Arts

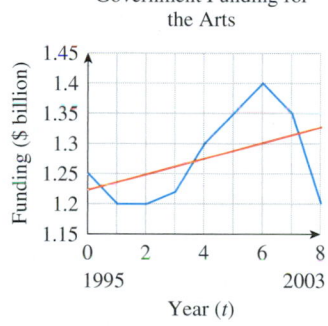

Year (t)

Multiple choice:

a. Over the period [0, 4] the average rate of change of government funding for the arts was
(A) less than   **(B)** greater than   **(C)** approximately equal to the rate predicted by the regression line.

b. Over the period [4, 8] the average rate of change of government funding for the arts was
(A) less than   **(B)** greater than   **(C)** approximately equal to the rate predicted by the regression line.

c. Over the period [3, 6] the average rate of change of government funding for the arts was
(A) less than   **(B)** greater than   **(C)** approximately equal to the rate predicted by the regression line.

d. Estimate, to two significant digits, the average rate of change of government funding for the arts over the period [0, 8]. (Be careful to state the units of measurement.) How does it compare to the slope of the regression line?

36. *Funding for the Arts* The following chart shows the total annual support for the arts in the U.S. by foundation endowments in 1995–2002 as a function of time in years ($t = 0$ represents 1995) together with the regression line:[31]

Foundation Funding for
the Arts

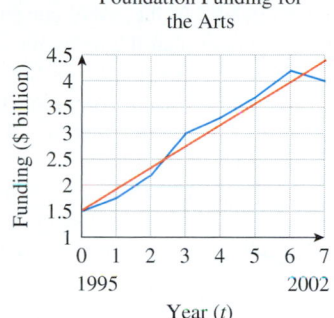

Year (t)

Multiple choice:

a. Over the period [0, 2.5] the average rate of change of government funding for the arts was
(A) less than   **(B)** greater than   **(C)** approximately equal to the rate predicted by the regression line.

b. Over the period [2, 6] the average rate of change of government funding for the arts was
(A) less than   **(B)** greater than   **(C)** approximately equal to the rate predicted by the regression line.

c. Over the period [3, 7] the average rate of change of government funding for the arts was
(A) less than   **(B)** greater than   **(C)** approximately equal to the rate predicted by the regression line.

d. Estimate, to two significant digits, the average rate of change of foundation funding for the arts over the period [0, 7]. (Be careful to state the units of measurement.) How does it compare to the slope of the regression line?

37. *Market Volatility* A volatility index generally measures the extent to which a market undergoes sudden changes in value. The volatility of the S&P 500 (as measured by one such index) was decreasing at an average rate of 0.2 points per year during 1991–1995, and was increasing at an average rate of about 0.3 points per year during 1995–1999. In 1995, the volatility of the S&P was 1.1.[32] Use this information to give a rough sketch of the volatility of the S&P 500 as a function of time, showing its values in 1991 and 1999.

---

[29] January 2002 data was estimated. Source for data: Odyssey Research/*New York Times,* November 5, 2001, p. C1.
[30] Figures are adjusted for inflation. SOURCES: Giving USA, The Foundation Center, Americans for the Arts/*New York Times,* June 19, 2004, p. B7.

[31] Figures are adjusted for inflation. SOURCES: Giving USA, The Foundation Center, Americans for the Arts/*New York Times,* June 19, 2004, p. B7.
[32] Source for data: Sanford C. Bernstein Company/*New York Times,* March 24, 2000, p. C1.

● basic skills   ◆ challenging   *tech* Ex technology exercise

38. *Market Volatility* The volatility (see the preceding exercise) of the NASDAQ had an average rate of change of 0 points per year during 1992–1995, and increased at an average rate of 0.2 points per year during 1995–1998. In 1995, the volatility of the NASDAQ was 1.1.[33] Use this information to give a rough sketch of the volatility of the NASDAQ as a function of time.

39. ● *Market Index* Joe Downs runs a small investment company from his basement. Every week he publishes a report on the success of his investments, including the progress of the "Joe Downs Index." At the end of one particularly memorable week, he reported that the index for that week had the value $I(t) = 1000 + 1500t - 800t^2 + 100t^3$ points, where t represents the number of business days into the week; t ranges from 0 at the beginning of the week to 5 at end of the week. The graph of I is shown below.

On average, how fast, and in what direction, was the index changing over the first two business days (the interval [0, 2])?
hint [see Example 3]

40. ● *Market Index* Refer to the Joe Downs Index in the preceding exercise. On average, how fast, and in which direction, was the index changing over the last three business days (the interval [2, 5])?

41. ● *Currency* The value of the euro (€) since its introduction in January, 1999 can be approximated by

$e(t) = 0.036t^2 - 0.10t + 1.0$ U.S. dollars $(-1 \le t \le 4.5)$

where $t = 0$ represents January, 2000.[34] Compute the average rate of change of $e(t)$ over the interval [0.5, 4.5] and interpret your answer.

42. ● *Interest Rates* The prime lending rate (the lowest short-term interest rate charged by commercial banks to their most creditworthy clients) for the period January 1982 – June 2004 can be approximated by

$p(t) = 0.028t^2 - 0.52t + 9.3$ percentage points $(-8 \le t \le 14)$

where $t = 0$ represents January, 1990.[35] Compute the average rate of change of $p(t)$ over the interval [−4, 4] and interpret your answer.

---

[33] Ibid.

[34] SOURCES: Bank of England, Reuters, July, 2004.

[35] SOURCES: Bloomberg News, www.icmarc.org July 6, 2004.

43. *Ecology* Increasing numbers of manatees ("sea sirens") have been killed by boats off the Florida coast. The following graph shows the relationship between the number of boats registered in Florida and the number of manatees killed each year:

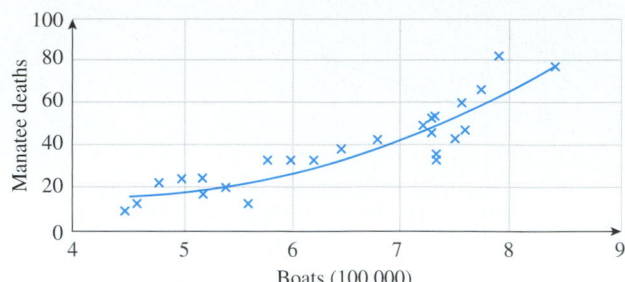

Boats (100,000)

The regression curve shown is given by
$f(x) = 3.55x^2 - 30.2x + 81$ manatees $(4.5 \le x \le 8.5)$

where x is the number of boats (in hundreds of thousands) registered in Florida in a particular year, and $f(x)$ is the number of manatees killed by boats in Florida that year.[36]

a. Compute the average rate of change of f over the intervals [5, 6] and [7, 8].

b. What does the answer to part (a) tell you about the manatee deaths per boat?

44. *Ecology* Refer to Exercise 43,

a. Compute the average rate of change of f over the intervals [5, 7] and [6, 8].

b. Had we used a linear model instead of a quadratic one, how would the two answers in part (a) be related to each other?

45. *Advertising Revenue* The following table shows the annual advertising revenue earned by America Online (AOL) during the last three years of the 1990s:[37]

Year	1997	1998	1999
Revenue ($ million)	150	360	760

These data can be modeled by

$R(t) = 95t^2 + 115t + 150$ million dollars $(0 \le t \le 2)$

where t is time in years since December 1997.

a. What was the average rate of change of R over the period 1997–1999? Interpret the result.

b. Which of the following is true? From 1997 to 1999, annual online advertising revenues
(A) increased at a faster and faster rate.
(B) increased at a slower and slower rate.

---

[36] Regression model is based on data from 1976 to 2000. Sources for data: Florida Department of Highway Safety & Motor Vehicles, Florida Marine Institute/*New York Times,* February 12, 2002, p. F4.

[37] Figures are rounded to the nearest $10 million. SOURCES: AOL; Forrester Research/*New York Times,* January 31, 2000, p. C1.

● basic skills   ◆ challenging   **tech** Ex technology exercise

(C) decreased at a faster and faster rate.

(D) decreased at a slower and slower rate.

c. Use the model to project the average rate of change of R over the one-year period ending December 2000. Interpret the result.

46. *Religion* The following table shows the population of Roman Catholic nuns in the U.S. during the last 25 years of the 1900s:[38]

Year	1975	1985	1995
Population	130,000	120,000	80,000

These data can be modeled by

$$P(t) = -0.15t^2 + 0.50t + 130 \quad \text{thousand nuns} \quad (0 \le t \le 20)$$

where t is time in years since December 1975.

a. What was the average rate of change of P over the period 1975–1995? Interpret the result.

b. Which of the following is true? From 1975 to 1995, the population of nuns

(A) increased at a faster and faster rate.

(B) increased at a slower and slower rate.

(C) decreased at a faster and faster rate.

(D) decreased at a slower and slower rate.

c. Use the model to project the average rate of change of P over the 10-year period ending December 2005. Interpret the result.

47. tech Ex *Poverty vs Income* Based on data from 1988 through 2003, the poverty rate (percentage of households with incomes below the poverty threshold) in the U.S. can be approximated by

$$p(x) = 0.092x^2 - 8.1x + 190 \text{ percentage points} \quad (38 \le x \le 44)$$

where x is the U.S. median household income in thousands of dollars.[39]

a. Use technology to complete the following table which shows the average rate of change of p over successive intervals of length $\frac{1}{2}$. (Round all answers to two decimal places.) *hint* [see Example 4]

Interval	[39, 39.5]	[39.5, 40]	[40, 40.5]	[40.5, 41]	[41, 41.5]	[41.5, 42]
Average Rate of change of p						

b. Interpret your answer for the interval [40, 40.5], being sure to indicate the direction of change and the units of measurement.

c. Multiple choice: As the median household income rises, the poverty rate

(A) Increases

(B) Decreases

(C) Increases, then decreases

(D) Decreases, then increases

d. Multiple choice: As the median income increases, the effect on the poverty rate is

(A) More pronounced

(B) Less pronounced

48. tech Ex *Poverty vs Unemployment* Based on data from 1988 through 2003, the poverty rate (percentage of households with incomes below the poverty threshold) in the U.S. can be approximated by

$$p(x) = -0.12x^2 + 2.4x + 3.2 \text{ percentage points} \quad (4 \le x \le 8)$$

where x is the unemployment rate in percentage points.[40]

a. Use technology to complete the following table which shows the average rate of change of p over successive intervals of length $\frac{1}{2}$. (Round all answers to two decimal places.)

Interval	[5.0, 5.5]	[5.5, 6.0]	[6.0, 6.5]	[6.5, 7.0]	[7.0, 7.5]	[7.5, 8.0]
Average Rate of change of p						

b. Interpret your answer for the interval [5.0, 5.5], being sure to indicate the direction of change and the units of measurement.

c. Multiple choice: As the median household income rises, the poverty rate

(A) Increases

(B) Decreases

(C) Increases, then decreases

(D) Decreases, then increases

d. Multiple choice: As the unemployment rate increases, the effect on the poverty rate is

(A) More pronounced

(B) Less pronounced

---

[38] Figures are rounded. SOURCE: Center for Applied Research in the Apostolate/*New York Times,* January 16, 2000, p. A1.

[39] The model is based on a quadratic regression. Household incomes are in constant 2002 dollars. The poverty threshold is approximately $18,000 for a family of four and $9200 for an individual. SOURCES: Census Bureau Current Population Survey/*New York Times,* Sept 27, 2003, p. A10, U.S. Department of Labor Bureau of Labor Statistics http://stats.bls.gov June 17, 2004.

[40] The model is based on a quadratic regression. Household incomes are in constant 2002 dollars. The poverty threshold is approximately $18,000 for a family of four and $9200 for an individual. SOURCES: Census Bureau Current Population Survey/*New York Times,* Sept 27, 2003, p. A10, U.S. Department of Labor Bureau of Labor Statistics http://stats.bls.gov June 17, 2004.

● basic skills   ◆ challenging   tech Ex technology exercise

Communication and Reasoning Exercises

49. ● Describe three ways we have used to determine the average rate of change of f over an interval $[a, b]$. Which of the three ways is *least* precise? Explain.

50. ● If f is a linear function of x with slope m, what is its average rate of change over any interval $[a, b]$?

51. ● Sketch the graph of a function whose average rate of change over $[0, 3]$ is negative but whose average rate of change over $[1, 3]$ is positive.

52. ● Sketch the graph of a function whose average rate of change over $[0, 2]$ is positive but whose average rate of change over $[0, 1]$ is negative.

53. If the rate of change of quantity A is 2 units of quantity A per unit of quantity B, and the rate of change of quantity B is 3 units of quantity B per unit of quantity C, what is the rate of change of quantity A with respect to quantity C?

54. If the rate of change of quantity A is 2 units of quantity A per unit of quantity B, what is the rate of change of quantity B with respect to quantity A?

55. A certain function has the property that its average rate of change over the interval $[1, 1 + h]$ (for positive h) increases as h decreases. Which of the following graphs could be the graph of f?

(A)   **(B)**

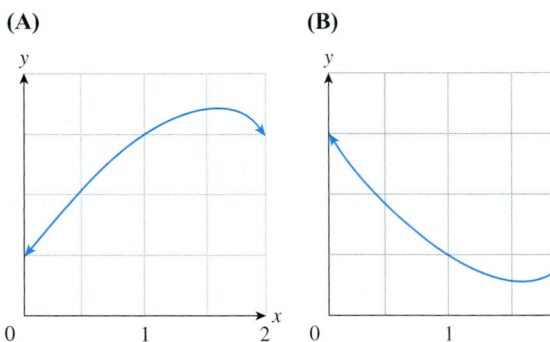

(C)

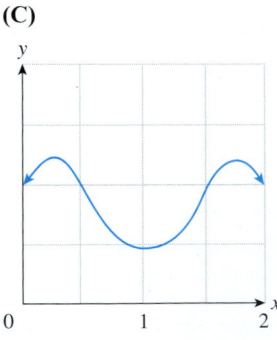

56. ● A certain function has the property that its average rate of change over the interval $[1, 1 + h]$ (for positive h) decreases as h decreases. Which of the following graphs could be the graph of f?

(A)   **(B)**

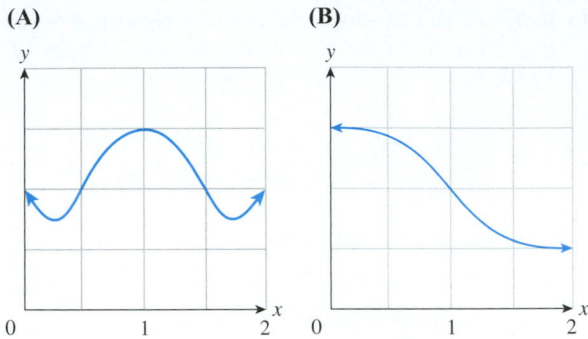

(C)

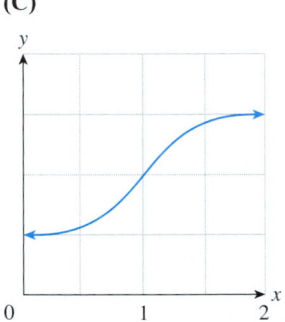

57. Is it possible for a company's revenue to have a negative three-year average rate of growth, but a positive average rate of growth in two of the three years? (If not, explain; if so, illustrate with an example.)

58. Is it possible for a company's revenue to have a larger two-year average rate of change than either of the one-year average rates of change? (If not, explain why with the aid of a graph; if so, illustrate with an example.)

59. ◆ The average rate of change of f over $[1, 3]$ is

(A) always equal to   (B) never equal to
(C) sometimes equal to the average of its average rates of change over $[1, 2]$ and $[2, 3]$.

60. ◆ The average rate of change of f over $[1, 4]$ is

(A) always equal to   (B) never equal to
(C) sometimes equal to the average of its average rates of change over $[1, 2]$, $[2, 3]$, and $[3, 4]$.

10.5 Derivatives: Numerical and Graphical Viewpoints

In Example 4 of Section 10.4, we looked at the average rate of change of the function $G(t) = -2t^2 + 36t + 228$, approximating the price of gold on the New York Spot Market, over the intervals $[8, 8 + h]$ for successively smaller values of h. Here are some values we got:

h getting smaller; interval $[8, 8 + h]$ getting smaller →

h	0.1	0.01	0.001	0.0001
Ave. rate of change over $[8, 8 + h]$	3.8	3.98	3.998	3.9998

Rate of change approaching \$4 per hour →

The average rate of change of the price of gold over smaller and smaller periods of time, starting at the instant $t = 8$ (8 AM), appears to be getting closer and closer to \$4 per hour. As we look at these shrinking periods of time, we are getting closer to looking at what happens at the *instant* $t = 8$. So it seems reasonable to say that the average rates of change are approaching the **instantaneous rate of change** at $t = 8$, which the table suggests is \$4 per hour. This is how fast the price of gold was changing *exactly* at 8 AM.

> At $t = 8$, *the instantaneous rate of change of $G(t)$ is* 4

We express this fact mathematically by writing $G'(8) = 4$ (which we read as "G prime of 8 equals 4"). Thus,

> $G'(8) = 4$ *means that, at $t = 8$, the instantaneous rate of change of $G(t)$ is* 4

The process of letting h get smaller and smaller is called taking the **limit** as h approaches 0 (as you recognize if you've done the sections on limits). We write $h \to 0$ as shorthand for "h approaches 0." Thus, taking the limit of the average rates of change as $h \to 0$ gives us the instantaneous rate of change.

Q: All these intervals $[8, 8 + h]$ are intervals to the right of 8. What about small intervals to the left of 8, such as $[7.9, 8]$?

A: We can compute the average rate of change of our function for such intervals by choosing h to be negative ($h = -0.1, -0.01$, etc.) and using the same difference quotient formula we used for positive h:

$$\text{Average rate of change of } G \text{ over } [8 + h, 8] = \frac{G(8) - G(8 + h)}{8 - (8 + h)}$$
$$= \frac{G(8 + h) - G(8)}{h}$$

Here are the results we get using negative h:

h getting closer to 0; interval $[8 + h, 8]$ getting smaller →

h	−0.1	−0.01	−0.001	−0.0001
Ave. rate of change over $[8 + h, 8]$	4.2	4.02	4.002	4.0002

Rate of change approaching \$4 per hour →

Notice that the average rates of change are again getting closer and closer to 4 as h approaches 0, suggesting once again that the instantaneous rate of change is $4 per hour. ∎

Instantaneous Rate of Change of $f(x)$ at $x = a$: Derivative

The **instantaneous rate of change** of $f(x)$ at $x = a$ is defined as

$$f'(a) = \lim_{h \to 0} \frac{f(a+h) - f(a)}{h}$$

f prime of *a* equals the limit as *h* approaches 0, of the ratio $\frac{f(a+h) - f(a)}{h}$

The quantity $f'(a)$ is also called the **derivative of $f(x)$ at $x = a$.** Finding the derivative of f is called **differentiating f.**

Units

The units of $f'(a)$ are the same as the units of the average rate of change: units of f per unit of x.

---

quick Example

If $f(x) = -2x^2 + 36x + 228$, then the two tables above suggest that

$$f'(8) = \lim_{h \to 0} \frac{f(8+h) - f(8)}{h} = 4$$

Important Notes

1. Sections 10.1–10.3 discuss limits in some detail. If you have not (yet) covered those sections, you can trust to your intuition.

2. The formula for the derivative tells us that the instantaneous rate of change is the limit of the average rates of change $[f(a+h) - f(a)]/h$ over smaller and smaller intervals. Thus, value of $f'(a)$ can be approximated by computing the average rate of change for smaller and smaller values of h, both positive and negative.

3. In this section we will only *approximate* derivatives. In Section 10.6 we will begin to see how we find the *exact* values of derivatives.

4. $f'(a)$ is a number we can calculate, or at least approximate, for various values of a, as we have done in the example above. Since $f'(a)$ depends on the value of a, we can think of f' as *a function of a*. (We return to this idea at the end of this section.) An old name for f' is "the function *derived from f*," which has been shortened to the *derivative* of f.

5. It is because f' is a function that we sometimes refer to $f'(a)$ as "the derivative of f evaluated at a," or the "derivative of $f(x)$ evaluated at $x = a$."

It may happen that the average rates of change $[f(a+h) - f(a)]/h$ do not approach any fixed number at all as h approaches zero, or that they approach one number on the intervals using positive h, and another on those using negative h. If this happens, we say that f is **not differentiable** at $x = a$, or $f'(a)$ **does not exist.** When the average rates of change *do* approach a fixed limit for both positive and negative h, we say that f is **differentiable** at the point $x = a$, or $f'(a)$ **exists.** It is comforting to know that all polynomials and exponential functions are differentiable at *every* point. On the other hand, certain functions are not differentiable. Examples are $f(x) = |x|$ and $f(x) = x^{1/3}$, neither of which is differentiable at $x = 0$ (see Section 10.7).

Example 1 Instantaneous Rate of Change: Numerically and Graphically

The air temperature one spring morning, t hours after 7:00 AM, was given by the function $f(t) = 50 + 0.1t^4 \,°F \ (0 \le t \le 4)$.

a. How fast was the temperature rising at 9:00 AM?

b. How is the instantaneous rate of change of temperature at 9:00 AM reflected in the graph of temperature vs. time?

Solution

a. We are being asked to find the instantaneous rate of change of the temperature at $t = 2$, so we need to find $f'(2)$. To do this we examine the average rates of change

$$\frac{f(2+h) - f(2)}{h} \qquad \text{\textcolor{red}{Average rate of change = difference quotient}}$$

for values of h approaching 0. Calculating the average rate of change over $[2, 2 + h]$ for $h = 1, 0.1, 0.01, 0.001,$ and 0.0001 we get the following values (rounded to four decimal places):[*]

h	1	0.1	0.01	0.001	0.0001
Ave. rate of change over $[2, 2 + h]$	6.5	3.4481	3.2241	3.2024	3.2002

Here are the values we get using negative values of h:

h	−1	−0.1	−0.01	−0.001	−0.0001
Ave. rate of change over $[2 + h, 2]$	1.5	2.9679	3.1761	3.1976	3.1998

The average rates of change are clearly approaching the number 3.2, so we can say that $f'(2) = 3.2$. Thus, at 9:00 in the morning, the temperature was rising at the rate of 3.2°F per hour.

b. We saw in Section 10.4 that the average rate of change of f over an interval is the slope of the secant line through the corresponding points on the graph of f. Figure 23 illustrates this for the intervals $[2, 2 + h]$ with $h = 1, 0.5,$ and 0.1.

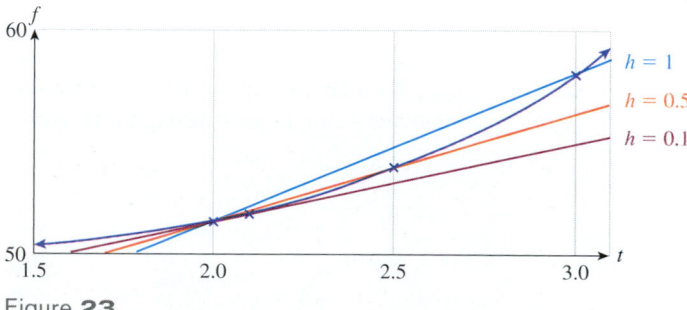

Figure 23

---

[*] We can quickly compute these values using technology as in Example 4 in Section 10.4 (see the Technology Guides at the end of the chapter).

All three secant lines pass though the point $(2, f(2)) = (2, 51.6)$ on the graph of f. Each of them passes through a second point on the curve (the second point is different for each secant line) and this second point gets closer and closer to $(2, 51.6)$ as h gets closer to 0. What seems to be happening is that the secant lines are getting closer and closer to a line that just touches the curve at $(2, 51.6)$: the **tangent line** at $(2, 51.6)$, shown in Figure 24.

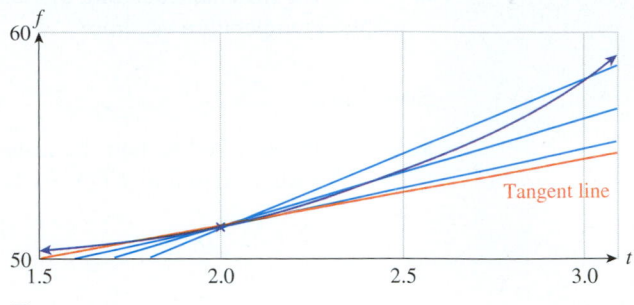

Figure **24**

Q: *What is the slope of this tangent line?*

A: Because the slopes of the secant lines are getting closer and closer to 3.2, and because the secant lines are approaching the tangent line, the tangent line must have slope 3.2. In other words,

At the point on the graph where $x = 2$, the slope of the tangent line is $f'(2)$. ∎

Be sure you understand the difference between $f(2)$ and $f'(2)$: Briefly, $f(2)$ is the *value of f* when $t = 2$, while $f'(2)$ is the *rate at which f is changing* when $t = 2$. Here,

$$f(2) = 50 + 0.1(2)^4 = 51.6°\text{F}$$

Thus, at 9:00 AM $(t = 2)$, the temperature was 51.6°F. On the other hand,

$$f'(2) = 3.2°\text{F/hour} \qquad \text{Units of slope are units of } f \text{ per unit of } t.$$

This means that, at 9:00 AM $(t = 2)$, the temperature was increasing at a rate of 3.2°F per hour.

Because we have been talking about tangent lines, we should say more about what they *are*. A tangent line to a *circle* is a line that touches the circle in just one point. A tangent line gives the circle "a glancing blow," as shown in Figure 25.

For a smooth curve other than a circle, a tangent line may touch the curve at more than one point, or pass through it (Figure 26).

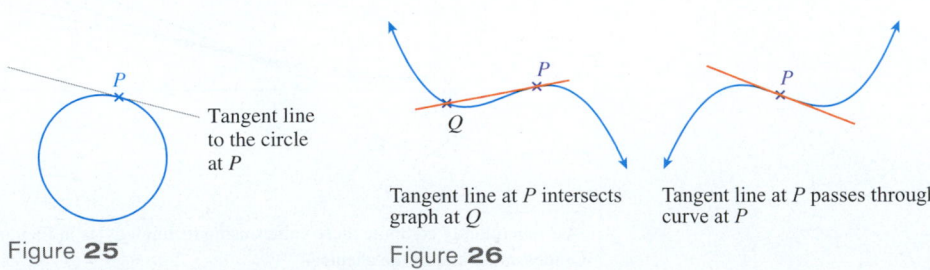

Figure **25**

Tangent line to the circle at P

Tangent line at P intersects graph at Q

Figure **26**

Tangent line at P passes through curve at P

However, all tangent lines have the following interesting property in common: If we focus on a small portion of the curve very close to the point *P*—in other words, if we "zoom in" to the graph near the point *P*—the curve will appear almost straight, and almost indistinguishable from the tangent line. (Figure 27).

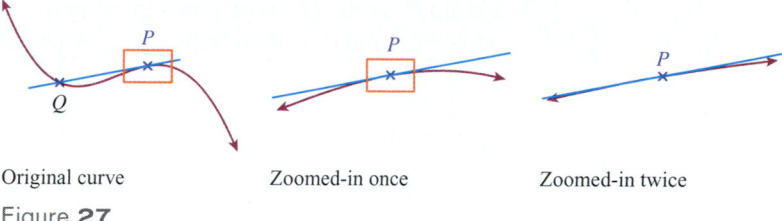

Original curve   Zoomed-in once   Zoomed-in twice

Figure **27**

You can check this property by zooming in on the curve shown in Figures 23 and 24 in the above example near the point where $x = 2$.

Secant and Tangent Lines

The *slope of the secant line* through the points on the graph of f where $x = a$ and $x = a + h$ is given by the average rate of change, or difference quotient,

$$m_{\text{sec}} = \text{slope of secant} = \text{average rate of change} = \frac{f(a + h) - f(a)}{h}$$

The *slope of the tangent line* through the point on the graph of f where $x = a$ is given by the instantaneous rate of change, or derivative

$$m_{\text{tan}} = \text{slope of tangent} = \text{derivative} = f'(a) = \lim_{h \to 0} \frac{f(a + h) - f(a)}{h}$$

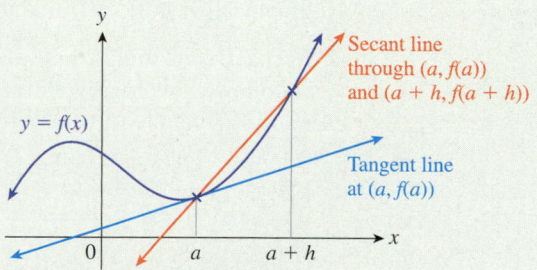

quick Example

In the following graph, the tangent line at the point where $x = 2$ has slope 3. Therefore, the derivative at $x = 2$ is 3. That is, $f'(2) = 3$.

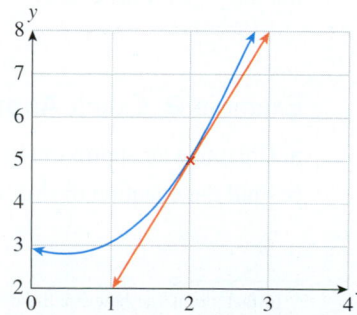

Note It might happen that the tangent line is vertical at some point or does not exist at all. These are the cases in which f is not differentiable at the given point. (See Sections 10.6 and 10.7.) ∎

We can now give a more precise definition of what we mean by the tangent line to a point P on the graph of f at a given point: The **tangent line** to the graph of f at the point $P(a, f(a))$ is the straight line passing through P with slope $f'(a)$.

Quick Approximation of the Derivative

Q: *Do we always need to make tables of difference quotients as above in order to calculate an approximate value for the derivative?*

A: We can usually *approximate* the value of the derivative by using a single, small value of h. In the example above, the value $h = 0.0001$ would have given a pretty good approximation. The problems with using a fixed value of h are that (1) we do not get an *exact* answer, only an *approximation* of the derivative, and (2) how good an approximation it is depends on the function we're differentiating.[41] However, with most of the functions we'll be considering, setting $h = 0.0001$ does give us a good approximation. ∎

Calculating a Quick Approximation of the Derivative

We can calculate an approximate value of $f'(a)$ by using the formula

$$f'(a) \approx \frac{f(a+h) - f(a)}{h} \qquad \text{Rate of change over } [a,\, a+h]$$

with a small value of h. The value $h = 0.0001$ often works (but see the next example for a graphical way of determining a good value to use).

Alternative Formula: the Balanced Difference Quotient
The following alternative formula, which measures the rate of change of f over the interval $[a - h, a + h]$, often gives a more accurate result, and is the one used in many calculators:

$$f'(a) \approx \frac{f(a+h) - f(a-h)}{2h} \qquad \text{Rate of change over } [a-h,\, a+h]$$

Note For the quick approximations to be valid, the function f must be differentiable; that is, $f'(a)$ must exist. ∎

Example 2 Quick Approximation of the Derivative

a. Calculate an approximate value of $f'(1.5)$ if $f(x) = x^2 - 4x$.

b. Find the equation of the tangent line at the point on the graph where $x = 1.5$.

---

[41] In fact, no matter how small the value we decide to use for h, it is possible to craft a function f for which the difference quotient at a is not even close to $f'(a)$.

Solution

a. We shall compute both the ordinary difference quotient and the balanced difference quotient. Using $h = 0.0001$, the ordinary difference quotient is:

$$f'(1.5) \approx \frac{f(1.5 + 0.0001) - f(1.5)}{0.0001} \qquad \text{\color{red}Usual difference quotient}$$

$$= \frac{f(1.5001) - f(1.5)}{0.0001}$$

$$= \frac{(1.5001^2 - 4 \times 1.5001) - (1.5^2 - 4 \times 1.5)}{0.0001} = -0.9999$$

This answer is accurate to 0.0001; in fact, $f'(1.5) = -1$.

Graphically, we can picture this approximation as follows: Zoom in on the curve using the window $1.5 \le x \le 1.5001$ and measure the slope of the secant line joining both ends of the curve segment. Figure 28 shows close-up views of the curve and tangent line near the point P in which we are interested, the third view being the zoomed-in view used for this approximation.

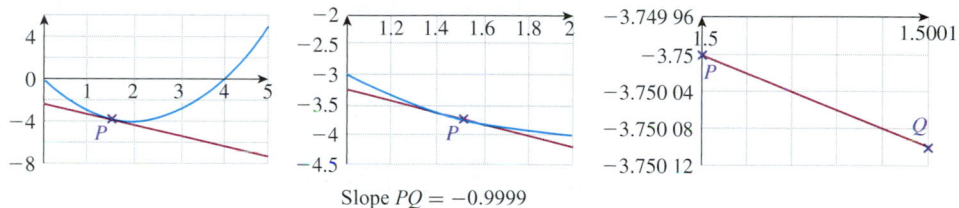

Slope $PQ = -0.9999$

Figure **28**

Notice that in the third window the tangent line and curve are indistinguishable. Also, the point P in which we are interested is on the left edge of the window.

Turning to the balanced difference quotient, we get

$$f'(1.5) \approx \frac{f(1.5 + 0.0001) - f(1.5 - 0.0001)}{2(0.0001)} \qquad \text{\color{red}Balanced difference quotient}$$

$$= \frac{f(1.5001) - f(1.4999)}{0.0002}$$

$$= \frac{(1.5001^2 - 4 \times 1.5001) - (1.4999^2 - 4 \times 1.4999)}{0.0002} = -1$$

This balanced difference quotient gives the exact answer in this case![*] Graphically, it is as though we have zoomed in using a window that puts the point P in the *center* of the screen (Figure 29) rather than at the left edge.

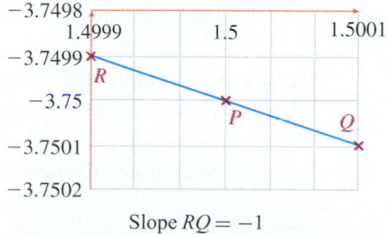

Slope $RQ = -1$

Figure **29**

---

[*]The balanced difference quotient always gives the exact derivative for a quadratic function.

using *Technology*

See the Technology Guides at the end of the chapter to find out how to compute the usual and balanced difference quotient very easily.

b. To find the equation of the tangent line, we use the point-slope formula from Chapter 1:

- **Point** $(1.5, f(1.5)) = (1.5, -3.75)$.
- **Slope** $m = f'(1.5) = -1$.   Slope of the tangent line = derivative

The equation is

$$y = mx + b$$

where $m = -1$ and $b = y_1 - mx_1 = -3.75 - (-1)(1.5) = -2.25$. Thus, the equation of the tangent line is

$$y = -x - 2.25$$

Q: *Why can't we simply put* $h = 0.000\,000\,000\,000\,000\,000\,01$ *for an incredibly accurate approximation to the instantaneous rate of change and be done with it?*

A: This approach would certainly work if you were patient enough to do the (thankless) calculation by hand! However, doing it with the help of technology—even an ordinary calculator—will cause problems: The issue is that calculators and spreadsheets represent numbers with a maximum number of significant digits (15 in the case of Excel). As the value of h gets smaller, the value of $f(a + h)$ gets closer and closer to the value of $f(a)$. For example, if $f(x) = 50 + 0.1x^4$, Excel might compute

$$f(2 + 0.000\,000\,000\,000\,1) - f(2)$$

$$= 51.600\,000\,000\,000\,3 - 51.6 \qquad \text{\color{red}Rounded to 15 digits}$$

$$= 0.000\,000\,000\,000\,3$$

and the corresponding difference quotient would be 3, not 3.2 as it should be. If h gets even smaller, Excel will not be able to distinguish between $f(a + h)$ and $f(a)$ at all, in which case it will compute 0 for the rate of change. This loss in accuracy when subtracting two very close numbers is called **subtractive error.**

Thus, there is a trade-off in lowering the value of h: smaller values of h yield *mathematically* more accurate approximations of the derivative, but if h gets too small, subtractive error becomes a problem and decreases the accuracy of computations that use technology. ∎

Leibniz *d* Notation

We introduced the notation $f'(x)$ for the derivative of f at x, but there is another interesting notation. We have written the average rate of change as

$$\text{Average rate of change} = \frac{\Delta f}{\Delta x} \qquad \color{red}\frac{\text{change in } f}{\text{change in } x}$$

As we use smaller and smaller values for Δx, we approach the instantaneous rate of change, or derivative, for which we also have the notation df/dx, due to Leibniz:

$$\text{Instantaneous rate of change} = \lim_{\Delta x \to 0} \frac{\Delta f}{\Delta x} = \frac{df}{dx}$$

That is, df/dx is just another notation for $f'(x)$. Do not think of df/dx as an actual quotient of two numbers: remember that we only use an actual quotient $\Delta f/\Delta x$ to *approximate* the value of df/dx.

In Example 3, we apply the quick approximation method of estimating the derivative.

Example 3 Velocity

My friend Eric, an enthusiastic baseball player, claims he can "probably" throw a ball upward at a speed of 100 feet per second (ft/sec).[*] Our physicist friends tell us that its height s (in feet) t seconds later would be $s = 100t - 16t^2$. Find its average velocity over the interval $[2, 3]$ and its instantaneous velocity exactly 2 seconds after Eric throws it.

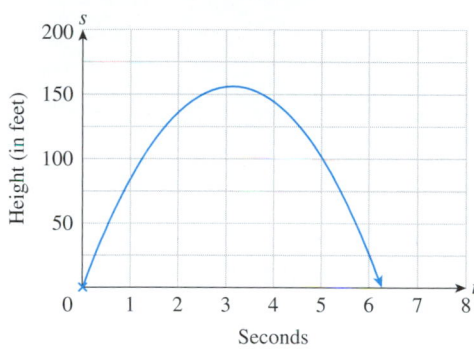

Figure 30

Solution The graph of the ball's height as a function of time is shown in Figure 30. Asking for the velocity is really asking for the rate of change of height with respect to time. (Why?) Consider average velocity first. To compute the **average velocity** of the ball from time 2 to time 3, we first compute the change in height:

$$\Delta s = s(3) - s(2) = 156 - 136 = 20 \text{ ft}$$

Since it rises 20 feet in $\Delta t = 1$ second, we use the defining formula *speed = distance/time* to get the average velocity:

$$\text{Average velocity} = \frac{\Delta s}{\Delta t} = \frac{20}{1} = 20 \text{ ft/sec}$$

from time $t = 2$ to $t = 3$. This is just the difference quotient, so

The average velocity is the average rate of change of height.

To get the **instantaneous velocity** at $t = 2$, we find the instantaneous rate of change of height. In other words, we need to calculate the derivative ds/dt at $t = 2$. Using the balanced quick approximation described above, we get

$$\frac{ds}{dt} \approx \frac{s(2 + 0.0001) - s(2 - 0.0001)}{2(0.0001)}$$

$$= \frac{s(2.0001) - s(1.9999)}{0.0002}$$

$$= \frac{100(2.0001) - 16(2.0001)^2 - (100(1.9999) - 16(1.9999)^2)}{0.0002}$$

$$= 36 \text{ ft/sec}$$

In fact, this happens to be the exact answer; the instantaneous velocity at $t = 2$ is exactly 36 ft/sec. (Try an even smaller value of h to persuade yourself.)

---

[*] Eric's claim is difficult to believe; 100 ft/sec corresponds to around 68 mph, and professional pitchers can throw *forward* at about 100 mph.

✚*Before we go on...* If we repeat the calculation in Example 3 at time $t = 5$, we get

$$\frac{ds}{dt} = -60 \text{ ft/s}$$

The negative sign tells us that the ball is *falling* at a rate of 60 feet per second at time $t = 5$. (How does the fact that it is falling at $t = 5$ show up on the graph?)

The preceding example gives another interpretation of the derivative. ∎

Average and Instantaneous Velocity

For an object moving in a straight line with position $s(t)$ at time t, the **average velocity** from time t to time $t + h$ is the average rate of change of position with respect to time:

$$v_{ave} = \frac{s(t + h) - s(t)}{h} = \frac{\Delta s}{\Delta t}$$

Average velocity =
Average rate of change of position

The **instantaneous velocity** at time t is

$$v = \lim_{h \to 0} \frac{s(t + h) - s(t)}{h} = \frac{ds}{dt}$$

Instantaneous velocity =
Instantaneous rate of change of position

In other words, *instantaneous velocity is the derivative of position with respect to time.*

Here is one last comment on Leibniz notation. In Example 3, we could have written the velocity either as s' or as ds/dt, as we chose to do. To write the answer to the question, that the velocity at $t = 2$ sec was 36 ft/sec, we can write either

$$s'(2) = 36$$

or

$$\left.\frac{ds}{dt}\right|_{t=2} = 36$$

The notation "$|_{t=2}$" is read "evaluated at $t = 2$." Similarly, if $y = f(x)$, we can write the instantaneous rate of change of f at $x = 5$ in either functional notation as

$$f'(5) \qquad \text{The derivative of } f \text{, evaluated at } x = 5$$

or in Leibniz notation as

$$\left.\frac{dy}{dx}\right|_{x=5} \qquad \text{The derivative of } y \text{, evaluated at } x = 5$$

The latter notation is obviously more cumbersome than the functional notation $f'(5)$, but the notation dy/dx has compensating advantages. You should practice using both notations.

The Derivative Function

The derivative $f'(x)$ is a number we can calculate, or at least approximate, for various values of x. Because $f'(x)$ depends on the value of x, we may think of f' as a function of x. This function is the **derivative function.**

Derivative Function

If f is a function, its **derivative function** f' is the function whose value $f'(x)$ is the derivative of f at x. Its domain is the set of all x at which f is differentiable. Equivalently, f' associates to each x the slope of the tangent to the graph of the function f at x, or the rate of change of f at x. The formula for the derivative function is

$$f'(x) = \lim_{h \to 0} \frac{f(x+h) - f(x)}{h} \qquad \text{Derivative function}$$

quick Examples

1. Let $f(x) = 3x - 1$. The graph of f is a straight line that has slope 3 everywhere. In other words, $f'(x) = 3$ for every choice of x; that is, f' is a constant function.

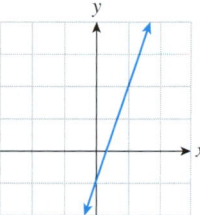 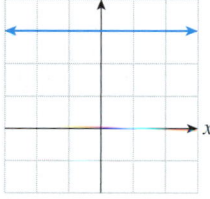

Original Function f Derivative Function f'
$f(x) = 3x - 1$ $f'(x) = 3$

2. Given the graph of a function f, we can get a rough sketch of the graph of f' by estimating the slope of the tangent to the graph of f at several points, as illustrated below.[*]

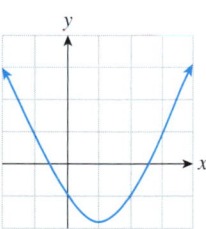 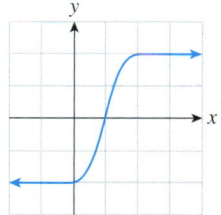

Original Function f Derivative Function f'
$y = f(x)$ $y = f'(x)$

For x between -2 and 0, the graph of f is linear with slope -2. As x increases from 0 to 2, the slope increases from -2 to 2. For x larger than 2, the graph of f is linear with slope 2. (Notice that, when $x = 1$, the graph of f has a horizontal tangent, so $f'(1) = 0$.)

---

[*] This method is discussed in detail online at Chapter 10→ Online Text → Sketching the Graph of the Derivative.

The following example shows how we can use technology to graph the (approximate) derivative of a function, where it exists.

Example 4 Tabulating and Graphing the Derivative with Technology

Use technology to obtain a table of values of and graph the derivative of $f(x) = -2x^2 + 6x + 5$ for values of x starting at -5.

Solution The TI-83 has a built-in function that approximates the derivative, and we can use it to graph the derivative of a function. In Excel, we need to create the approximation using one of the quick approximation formulas and we can then graph a table of its values. See the Technology Guides at the end of the chapter to find out how to graph the derivative (Figure 31) using the TI-83/84 and Excel.

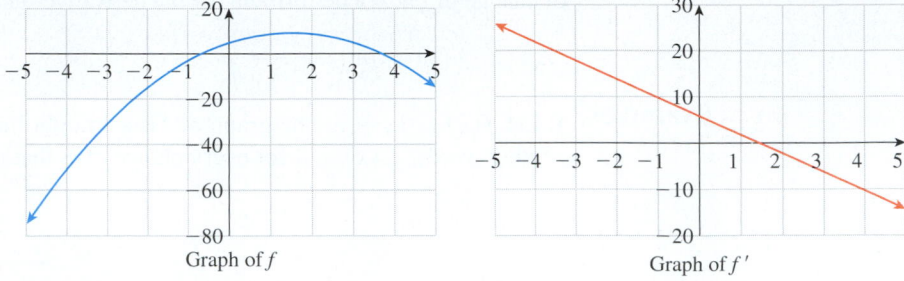

Graph of f Graph of f'

Figure **31**

We said that f' records the slope of (the tangent line to) the function f at each point. Notice that the graph of f' confirms that the slope of the graph of f is decreasing as x increases from -5 to 5. Note also that the graph of f reaches a high point at $x = 1.5$ (the vertex of the parabola). At that point, the slope of the tangent is zero; that is, $f'(1.5) = 0$ as we see in the graph of f'.

Example 5 An Application: Market Growth

The number N of U.S. households connected to the Internet from the start of 1999 could be modeled by the logistic function

$$N(t) = \frac{79.317}{1 + 2.2116(1.3854)^{-t}} \text{ million households} \qquad (0 \le t \le 9)$$

where t is the number of quarters since the start of 1999.[*] Graph both N and its derivative, and determine when Internet usage was growing most rapidly.

Solution Using one of the methods in Example 4, we obtain the graphs shown in Figure 32.

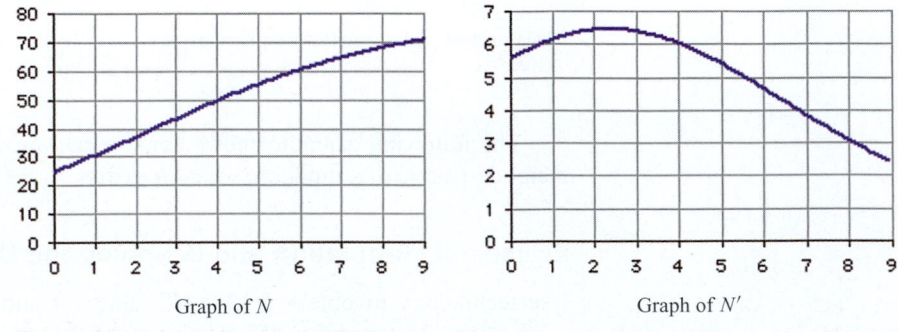

Graph of N Graph of N'

Figure **32**

---

[*] The model was obtained In Section 9.4 by logistic regression. Sources for data: Telecommunications Reports International/*New York Times,* May 21, 2001, p. C9.

From the graph on the right, we see that N' reaches a peak somewhere between $t = 2$ and $t = 3$ (sometime during the third quarter of 1999). Recalling that N' measures the *slope* of the graph of N, we can conclude that the graph of N is steepest between $t = 2$ and $t = 3$, indicating that, according to the model, the number of U.S. households connected to the Internet was growing most rapidly during the third quarter of 1999. Notice that this is not so easy to see directly on the graph of N.

To determine the point of maximum growth more accurately, we can zoom in on the graph of N' using the range $2 \leq t \leq 3$ (Figure 33).

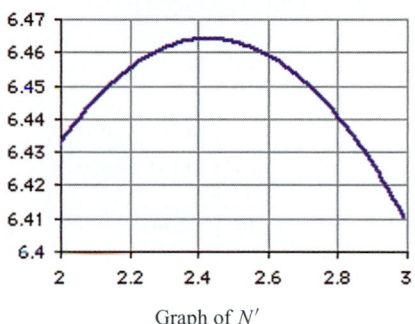

Graph of N'

Figure **33**

We can now see that N' reaches its highest point around $t = 2.4$, so we conclude that the number of U.S. households connected to the Internet was growing most rapidly shortly before the midpoint of the third quarter of 1999.

+*Before we go on...* Besides helping us to determine the point of maximum growth, the graph of N' in Example 5 gives us a great deal of additional information. As just one example, in Figure 32 we can see that the maximum value of N' is approximately 6.5, indicating that Internet usage in the U.S. never grew at a faster rate than about 6.5 million households per quarter. ∎

FAQs   Recognizing When and How to Compute the Instantaneous Rate of Change

Q: *How do I know, by looking at the wording of a problem, that it is asking for an instantaneous rate of change?*

A: If a problem does not ask for an instantaneous rate of change directly, it might do so indirectly, as in "How fast is quantity q increasing?" or "Find the rate of increase of q." ∎

Q: *If I know that a problem calls for estimating an instantaneous rate of change, how should I estimate it: with a table showing smaller and smaller values of h, or by using a quick approximation?*

A: For most practical purposes, a quick approximation is accurate enough. Use a table showing smaller and smaller values of h when you would like to check the accuracy. ∎

Q: *Which should I use in computing a quick approximation: the balanced difference quotient or the ordinary difference quotient?*

A: In general, the balanced difference quotient gives a more accurate answer. ∎

You can find the following optional sections:
Continuity and Differentiability
Sketching the Graph of the Derivative
Online, follow:
  Chapter 10
  → Online Text

10.5 EXERCISES

● denotes basic skills exercises

◆ denotes challenging exercises

tech Ex indicates exercises that should be solved using technology

In Exercises 1–4, estimate the derivative from the table of average rates of change. hint *[see discussion at the beginning of the section]*

1. ●

h	1	0.1	0.01	0.001	0.0001
Ave. rate of change of f over $[5, 5 + h]$	12	6.4	6.04	6.004	6.0004
h	−1	−0.1	−0.01	−0.001	−0.0001
Ave. rate of change of f over $[5 + h, 5]$	3	5.6	5.96	5.996	5.9996

Estimate $f'(5)$.

2. ●

h	1	0.1	0.01	0.001	0.0001
Ave. rate of change of g over $[7, 7 + h]$	4	4.8	4.98	4.998	4.9998
h	−1	−0.1	−0.01	−0.001	−0.0001
Ave. rate of change of g over $[7 + h, 7]$	5	5.3	5.03	5.003	5.0003

Estimate $g'(7)$.

3. ●

h	1	0.1	0.01	0.001	0.0001
Ave. rate of change of r over $[-6, -6 + h]$	−5.4	−5.498	−5.4998	−5.499982	−5.49999822
h	−1	−0.1	−0.01	−0.001	−0.0001
Ave. rate of change of r over $[-6 + h, -6]$	−7.52	−6.13	−5.5014	−5.5000144	−5.500001444

Estimate $r'(-6)$.

4. ●

h	1	0.1	0.01	0.001	0.0001
Ave. rate of change of s over $[0, h]$	−2.52	−1.13	0.6014	−0.6000144	−0.600001444
h	−1	−0.1	−0.01	−0.001	−0.0001
Ave. rate of change of s over $[h, 0]$	−0.4	−0.598	−0.5998	−0.599982	−0.59999822

Estimate $s'(0)$.

Consider the functions in Exercises 5–8 as representing the value of an ounce of silver in Indian rupees as a function of the time t in days.[42] *Find the average rates of change of R(t) over the time intervals $[t, t + h]$, where t is as indicated and $h = 1, 0.1,$ and 0.01 days. Hence, estimate the instantaneous rate of change of R at time t, specifying the units of measurement. (Use smaller values of h to check your estimates.)* hint *[see Example 1]*

5. ● $R(t) = 50t - t^2; t = 5$

6. ● $R(t) = 60t - 2t^2; t = 3$

7. ● $R(t) = 100 + 20t^3; t = 1$

8. ● $R(t) = 1000 + 50t - t^3; t = 2$

Each of the functions in Exercises 9–12 gives the cost to manufacture x items. Find the average cost per unit of manufacturing h more items (i.e., the average rate of change of the total cost) at a production level of x, where x is as indicated and $h = 10$ and 1. Hence, estimate the instantaneous rate of change of the total cost at the given production level x, specifying the units of measurement. (Use smaller values of h to check your estimates.)

9. ● $C(x) = 10,000 + 5x - \dfrac{x^2}{10,000}; \ x = 1000$

10. ● $C(x) = 20,000 + 7x - \dfrac{x^2}{20,000}; \ x = 10,000$

11. ● $C(x) = 15,000 + 100x + \dfrac{1000}{x}; \ x = 100$

12. ● $C(x) = 20,000 + 50x + \dfrac{10,000}{x}; x = 100$

In each of the graphs given in Exercises 13–18, say at which labeled point the slope of the tangent is (a) greatest and (b) least (in the sense that -7 is less than 1).

13. ●
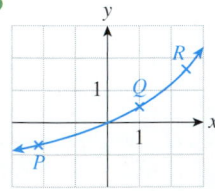

14. ●
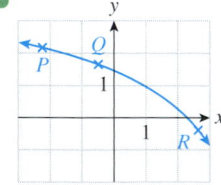

15. ●
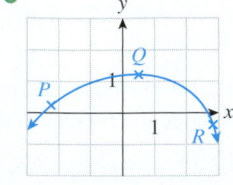

16. ●
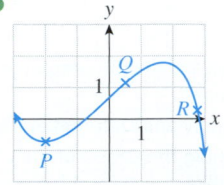

[42] Silver was trading at around 290 rupees in July, 2004.

17. ●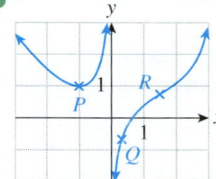

18. ●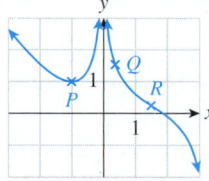

In Exercises 19–22, the graph of a function is shown together with the tangent line at a point P. Estimate the derivative of f at the corresponding x-value. hint [see Quick Example on p. 731]

19. ●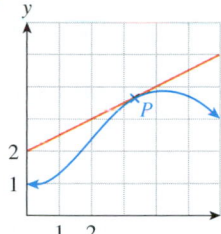

20. ●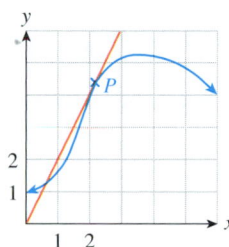

21. ●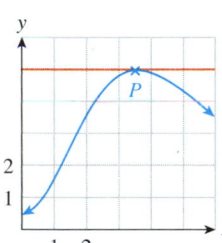

22. ●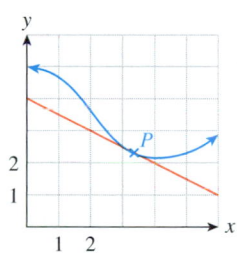

In each of Exercises 23–26, three slopes are given. For each slope, determine at which of the labeled points on the graph the tangent line has that slope.

23. ● **a.** 0   **b.** 4   **c.** −1

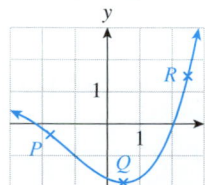

24. ● **a.** 0   **b.** 1   **c.** −1

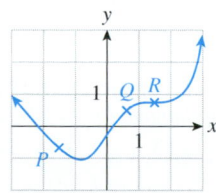

25. ● **a.** 0   **b.** 3   **c.** −3

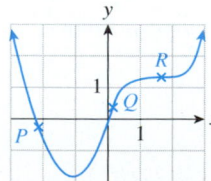

26. ● **a.** 0   **b.** 3   **c.** 1

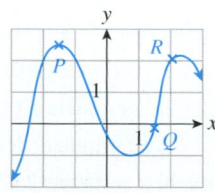

*In each of Exercises 27–30, find the approximate coordinates of all points (if any) where the slope of the tangent is: **(a)** 0, **(b)** 1, **(c)** −1.*

27. ●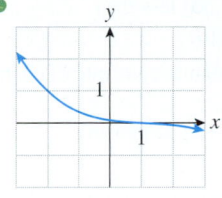

28. ●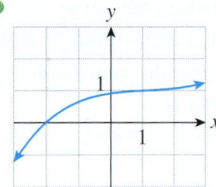

29. ●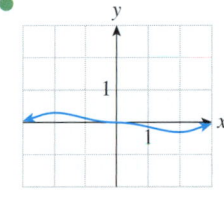

30. ●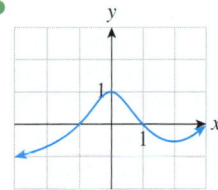

31. ● Complete the following: The tangent to the graph of the function f at the point where $x = a$ is the line passing through the point _____ with slope _____ .

32. ● Complete the following: The difference quotient for f at the point where $x = a$ gives the slope of the _____ line that passes through _____ .

33. ● Which is correct? The derivative function assigns to each value x
   (A) the average rate of change of f at x.
   (B) the slope of the tangent to the graph of f at $(x, f(x))$.
   (C) the rate at which f is changing over the interval $[x, x + h]$ for $h = 0.0001$.
   (D) the balanced difference quotient $[f(x + h) − f(x − h)]/(2h)$ for $h \approx 0.0001$.

34. ● Which is correct? The derivative function $f'(x)$ tells us
   (A) the slope of the tangent line at each of the points $(x, f(x))$.
   (B) the approximate slope of the tangent line at each of the points $(x, f(x))$.
   (C) the slope of the secant line through $(x, f(x))$ and $(x + h, f(x + h))$ for $h = 0.0001$.
   (D) the slope of a certain secant line through each of the points $(x, f(x))$.

35. Let f have the graph shown.

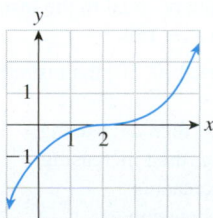

a. The average rate of change of f over the interval [2, 4] is
  (A) greater than   **(B)** less than
  (C) approximately equal to $f'(2)$.

b. The average rate of change of f over the interval $[-1, 1]$ is
  (A) greater than   **(B)** less than
  (C) approximately equal to $f'(0)$.

c. Over the interval [0, 2], the instantaneous rate of change of f is
  (A) increasing   **(B)** decreasing
  (C) neither.

d. Over the interval [0, 4], the instantaneous rate of change of f is
  (A) increasing, then decreasing.
  (B) decreasing, then increasing.
  (C) always increasing.
  (D) always decreasing.

e. When $x = 4$, $f(x)$ is
  (A) approximately 0, and increasing at a rate of about 0.7 units per unit of x.
  (B) approximately 0, and decreasing at a rate of about 0.7 unit per unit of x.
  (C) approximately 0.7, and increasing at a rate of about 1 unit per unit of x.
  (D) approximately 0.7, and increasing at a rate of about 3 units per unit of x.

36. A function f has the following graph.

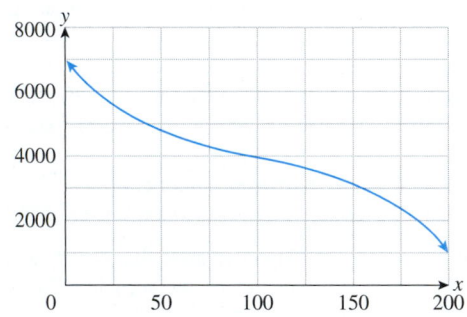

a. The average rate of change of f over [0, 200] is
  (A) greater than
  (B) less than
  (C) approximately equal to the instantaneous rate of change at $x = 100$.

b. The average rate of change of f over [0, 200] is
  (A) greater than
  (B) less than
  (C) approximately equal to the instantaneous rate of change at $x = 150$.

c. Over the interval [0, 50] the instantaneous rate of change of f is
  (A) increasing, then decreasing.
  (B) decreasing, then increasing.
  (C) always increasing.
  (D) always decreasing.

d. On the interval [0, 200], the instantaneous rate of change of f is
  (A) Always positive.   **(B)** Always negative.
  (C) Negative, positive, and then negative.

e. $f'(100)$ is
  (A) greater than   **(B)** less than
  (C) approximately equal to $f'(25)$.

In Exercises 37–40, use a quick approximation to estimate the derivative of the given function at the indicated point. hint [see Example 2a]

37. ● $f(x) = 1 - 2x$; $x = 2$ **38.** ● $f(x) = \dfrac{x}{3} - 1$; $x = -3$

39. ● $f(x) = \dfrac{x^2}{4} - \dfrac{x^3}{3}$; $x = -1$ **40.** ● $f(x) = \dfrac{x^2}{2} + \dfrac{x}{4}$; $x = 2$

In Exercises 41–48, estimate the indicated derivative by any method.

41. ● $g(t) = \dfrac{1}{t^5}$; estimate $g'(1)$

42. ● $s(t) = \dfrac{1}{t^3}$; estimate $s'(-2)$

43. ● $y = 4x^2$; estimate $\left.\dfrac{dy}{dx}\right|_{x=2}$

44. ● $y = 1 - x^2$; estimate $\left.\dfrac{dy}{dx}\right|_{x=-1}$

45. ● $s = 4t + t^2$; estimate $\left.\dfrac{ds}{dt}\right|_{t=-2}$

46. ● $s = t - t^2$; estimate $\left.\dfrac{ds}{dt}\right|_{t=2}$

47. ● $R = \dfrac{1}{p}$; estimate $\left.\dfrac{dR}{dp}\right|_{p=20}$

48. ● $R = \sqrt{p}$; estimate $\left.\dfrac{dR}{dp}\right|_{p=400}$

In Exercises 49–54, (a) use any method to estimate the slope of the tangent to the graph of the given function at the point with the given x-coordinate and, (b) find an equation of the tangent line in part (a). In each case, sketch the curve together with the appropriate tangent line. hint [see Example 2b]

49. ● $f(x) = x^3$; $x = -1$ **50.** ● $f(x) = x^2$; $x = 0$

51. ● $f(x) = x + \dfrac{1}{x}$; $x = 2$ **52.** ● $f(x) = \dfrac{1}{x^2}$; $x = 1$

53. ● $f(x) = \sqrt{x}$; $x = 4$ **54.** ● $f(x) = 2x + 4$; $x = -1$

In each of Exercises 55–58, estimate the given quantity.

55. ● $f(x) = e^x$; estimate $f'(0)$

56. ● $f(x) = 2e^x$; estimate $f'(1)$

57. ● $f(x) = \ln x$; estimate $f'(1)$

58. ● $f(x) = \ln x$; estimate $f'(2)$

● basic skills   ◆ challenging   **tech Ex** technology exercise

In Exercises 59–64, match the graph of f to the graph of f' (the graphs of f' are shown after Exercise 64).

59.

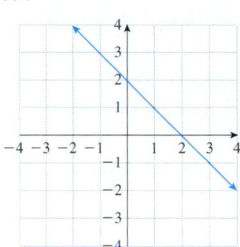

60.

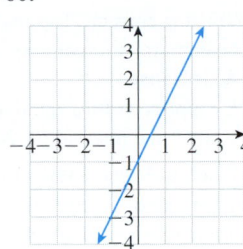

61.

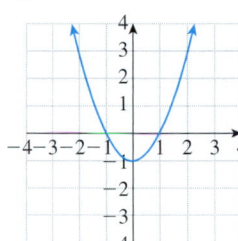

62.

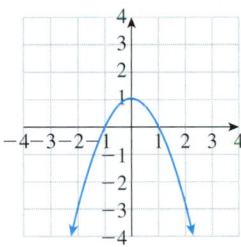

63.

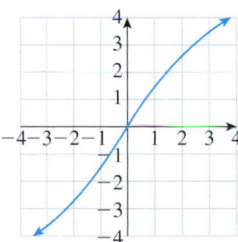

64.

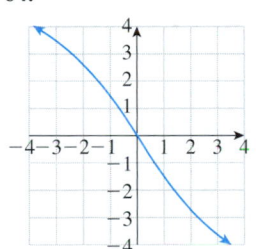

Graphs of derivatives for Exercises 59–64:

(A)

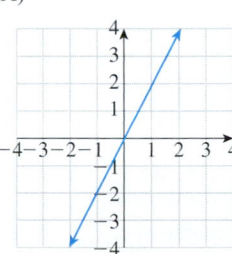

(B)

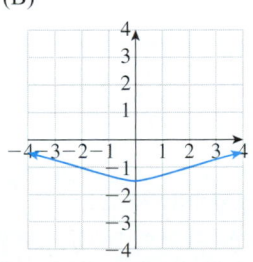

(C)

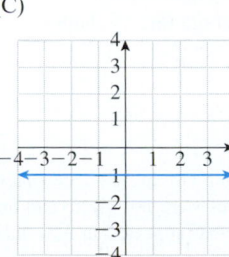

(D)

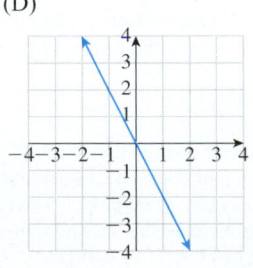

(E)

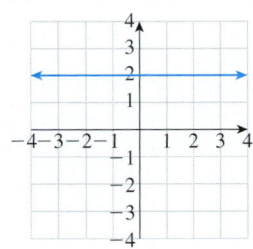

(F)

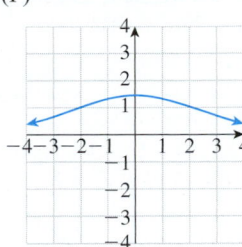

tech Ex *In Exercises 65 and 66, use technology to graph the derivative of the given function for the given range of values of x. Then use your graph to estimate all values of x (if any) where the tangent line to the graph of the given function is horizontal. Round answers to one decimal place.*

hint [see Example 1]

65. ● $f(x) = x^4 + 2x^3 - 1; -2 \le x \le 1$

66. ● $f(x) = -x^3 - 3x^2 - 1; -3 \le x \le 1$

tech Ex *In Exercises 67 and 68, use the method of Example 4 to list approximate values of f'(x) for x in the given range. Graph f(x) together with f'(x) for x the given range.*

67. ● $f(x) = \dfrac{x+2}{x-3}; 4 \le x \le 5$

68. ● $f(x) = \dfrac{10x}{x-2}; 2.5 \le x \le 3$

Applications

69. ● ***Demand*** Suppose the demand for a new brand of sneakers is given by

$$q = \frac{5,000,000}{p}$$

where p is the price per pair of sneakers, in dollars, and q is the number of pairs of sneakers that can be sold at price p. Find $q(100)$ and estimate $q'(100)$. Interpret your answers.

70. ● ***Demand*** Suppose the demand for an old brand of TV is given by

$$q = \frac{100,000}{p+10}$$

where p is the price per TV set, in dollars, and q is the number of TV sets that can be sold at price p. Find $q(190)$ and estimate $q'(190)$. Interpret your answers.

71. ● ***Swimming Pool Sales*** The following graph shows the approximate annual U.S. sales of in ground swimming pools.[43]

---

[43] Based on a regression model using 1996–2002 data. Source for data: PK Data/*New York Times,* July 5, 2001, p. C1, Industry reports, June 2004 www.poolspamarketing.com/statistics/

● basic skills   ◆ challenging   **tech Ex** technology exercise

Also shown is the tangent line (and its slope) at the point corresponding to year 2000.

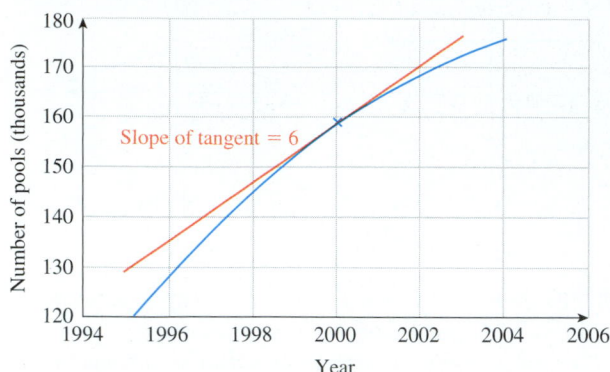

a. What does the graph tell you about swimming pool sales in 2000?

b. According to the graph, is the rate of change of swimming pool sales increasing or decreasing? Why?

72. ● *Swimming Pool Sales* Repeat Exercise 71 using the following graph showing approximate annual U.S. sales of aboveground swimming pools.[44]

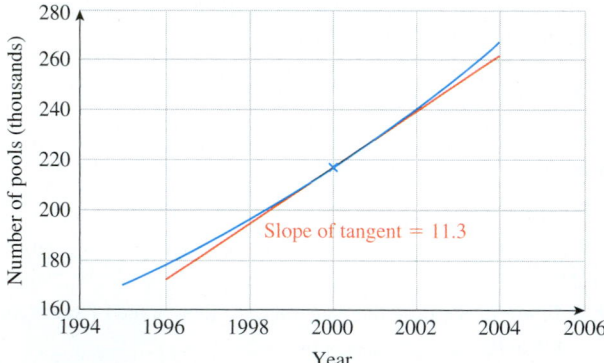

73. *Prison Population* The following curve is a model of the total population in state prisons as a function of time in years ($t = 0$ represents 1980).[45]

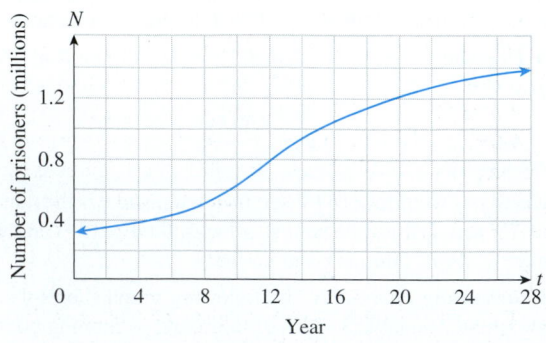

a. Which is correct? Over the period [16, 20] the instantaneous rate of change of N is
(**A**) increasing   (**B**) decreasing.

b. Which is correct? The instantaneous rate of change of prison population at $t = 12$ was
(**A**) less than   (**B**) greater than
(**C**) approximately equal to the average rate of change over the interval [0, 24].

c. Which is correct? Over the period [0, 28] the instantaneous rate of change of N is
(**A**) increasing, then decreasing
(**B**) decreasing, then increasing
(**C**) always increasing
(**D**) always decreasing

d. According to the model, the total state prison population was increasing fastest around what year?

e. Roughly estimate the instantaneous rate of change of N at $t = 16$ by using a balanced difference quotient with $h = 4$. Interpret the result.

74. *Demand for Freon* The demand for chlorofluorocarbon-12 (CFC-12)—the ozone-depleting refrigerant commonly known as freon[46]—has been declining significantly in response to regulation and concern about the ozone layer. The graph below represents a model for the projected demand for CFC-12 as a function of time in years ($t = 0$ represents 1990).[47]

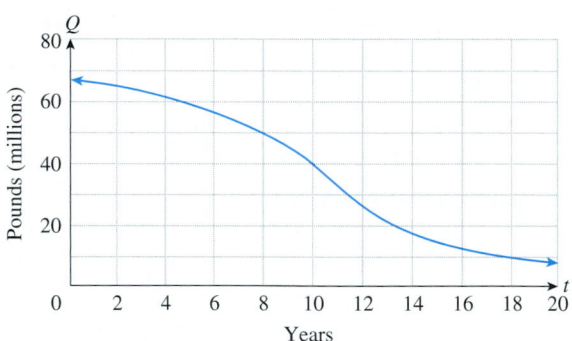

a. Which is correct? Over the period [12, 20] the instantaneous rate of change of Q is
(**A**) increasing   (**B**) decreasing.

b. Which is correct? The instantaneous rate of change of demand for freon at $t = 10$ was
(**A**) less than   (**B**) greater than
(**C**) approximately equal to the average rate of change over the interval [0, 20].

[44] Ibid.

[45] The prison population represented excludes federal prisons. Source for 1980–2000 data: Bureau of Justice Statistics/*New York Times,* June 9, 2001, p. A10.

[46] The name given to it by DuPont.

[47] Source for data: The Automobile Consulting Group/*New York Times,* December 26, 1993, p. F23. The exact figures were not given, and the chart is a reasonable facsimile of the chart that appeared in *New York Times.*

● basic skills   ◆ challenging   tech**Ex** technology exercise

c. Which is correct? Over the period [0, 20] the instantaneous rate of change of Q is
   (A) increasing, then decreasing
   (B) decreasing, then increasing
   (C) always increasing
   (D) always decreasing

d. According to the model, the demand for freon was decreasing most rapidly around what year?

e. Roughly estimate the instantaneous rate of change of Q at $t = 13$ by using a balanced difference quotient with $h = 5$. Interpret the result.

75. ● **Velocity** If a stone is dropped from a height of 400 feet, its height after t seconds is given by $s = 400 - 16t^2$. *hint* [see Example 3]

   a. Find its average velocity over the period [2, 4].
   b. Estimate its instantaneous velocity at time $t = 4$.

76. ● **Velocity** If a stone is thrown down at 120 ft/sec from a height of 1000 feet, its height after t seconds is given by $s = 1000 - 120t - 16t^2$.

   a. Find its average velocity over the period [1, 3].
   b. Estimate its instantaneous velocity at time $t = 3$.

77. ● **Currency** The value of the euro (€) since its introduction in January, 1999 can be approximated by
   $$e(t) = 0.036t^2 - 0.10t + 1.0 \text{ U.S. dollars} \quad (-1 \le t \le 4.5)$$
   where $t = 0$ represents January, 2000.[48]

   a. Compute the average rate of change of $e(t)$ over the interval [0, 4], and interpret your answer.
   b. Estimate the instantaneous rate of change of $e(t)$ at $t = 0$, and interpret your answer.
   c. The answers to part (a) and part (b) have opposite sign. What does this indicate about the value of the euro?

78. ● **Interest Rates** The prime lending rate (the lowest short-term interest rate charged by commercial banks to their most creditworthy clients) for the period January 1982–June 2004 can be approximated by
   $$p(t) = 0.028t^2 - 0.52t + 9.3 \text{ percentage points}$$
   $$(-8 \le t \le 14)$$
   where $t = 0$ represents January, 1990.[49]

   a. Compute the average rate of change of $p(t)$ over the interval [0, 5], and interpret your answer.
   b. Estimate the instantaneous rate of change of $p(t)$ at $t = 0$, and interpret your answer.
   c. The answer to part (b) has larger absolute value than the answer to part (a). What does this indicate about the prime lending rate?

79. ● **Advertising Revenue** The following table shows the annual advertising revenue earned by America Online (AOL) during the last three years of the 1990s.[50]

Year	1997	1998	1999
Revenue ($ million)	150	360	760

These data can be modeled by
$$R(t) = 95t^2 + 115t + 150 \text{ million dollars} \quad (0 \le t \le 2)$$
where t is time in years since December, 1997.

   a. How fast was AOL's advertising revenue increasing in December 1998?
   b. Which of the following is true? During 1997–1999, annual online advertising revenues
      (A) increased at a faster and faster rate.
      (B) increased at a slower and slower rate.
      (C) decreased at a faster and faster rate.
      (D) decreased at a slower and slower rate.
   c. Use the model to project the instantaneous rate of change of R in December, 2000. Interpret the result.

80. ● **Religion** The following table shows the population of Roman Catholic nuns in the U.S. at the end of the 20th century.[51]

Year	1975	1985	1995
Population	130,000	120,000	80,000

These data can be modeled by
$$P(t) = -0.15t^2 + 0.50t + 130 \text{ thousand nuns} \quad (0 \le t \le 20)$$
where t is time in years since December, 1975.

   a. How fast was the number of Roman Catholic nuns in the U.S. decreasing in December, 1995?
   b. Which of the following is true? During 1975–1995, the annual population of nuns
      (A) increased at a faster and faster rate.
      (B) increased at a slower and slower rate.
      (C) decreased at a faster and faster rate.
      (D) decreased at a slower and slower rate.
   c. Use the model to project the instantaneous rate of change of P in December, 2005. Interpret the result.

81. ● **Online Services** On January 1, 1996, America Online was the biggest online service provider, with 4.5 million subscribers, and was adding new subscribers at a rate of 60,000 per week.[52] If $A(t)$ is the number of America Online subscribers t weeks

---

[48] SOURCES: Bank of England, Reuters, July, 2004.

[49] SOURCE: Bloomberg News, www.icmarc.org July 6, 2004.

[50] Figures are rounded to the nearest $10 million. SOURCES: AOL; Forrester Research/*New York Times,* January 31, 2000, p. C1.

[51] Figures are rounded. SOURCE: Center for Applied Research in the Apostolate/*New York Times,* January 16, 2000, p. A1.

[52] SOURCE: Information and Interactive Services Report/*New York Times,* January 2, 1996, p. C14.

● basic skills   ◆ challenging   **tech** Ex technology exercise

after January 1, 1996, what do the given data tell you about values of the function A and its derivative?

82. ● **Online Services** On January 1, 1996, Prodigy was the third-biggest online service provider, with 1.6 million subscribers, but was losing subscribers.[53] If $P(t)$ is the number of Prodigy subscribers t weeks after January 1, 1996, what do the given data tell you about values of the function P and its derivative?

83. **Learning to Speak** Let $p(t)$ represent the percentage of children who are able to speak at the age of t months.

    a. It is found that $p(10) = 60$ and $\left.\dfrac{dp}{dt}\right|_{t=10} = 18.2$.[54] What does this mean?

    b. As t increases, what happens to p and $\dfrac{dp}{dt}$?

84. **Learning to Read** Let $p(t)$ represent the number of children in your class who learned to read at the age of t years.

    a. Assuming that everyone in your class could read by the age of 7, what does this tell you about $p(7)$ and $\left.\dfrac{dp}{dt}\right|_{t=7}$?

    b. Assuming that 25.0% of the people in your class could read by the age of 5, and that 25.3% of them could read by the age of 5 years and one month, estimate $\left.\dfrac{dp}{dt}\right|_{t=5}$. Remember to give its units.

85. ● **Sales** Weekly sales of a new brand of sneakers are given by

$$S(t) = 200 - 150e^{-t/10}$$

pairs sold per week, where t is the number of weeks since the introduction of the brand. Estimate $S(5)$ and $\left.\dfrac{dS}{dt}\right|_{t=5}$ and interpret your answers.

86. ● **Sales** Weekly sales of an old brand of TV are given by

$$S(t) = 100e^{-t/5}$$

sets per week, where t is the number of weeks after the introduction of a competing brand. Estimate $S(5)$ and $\left.\dfrac{dS}{dt}\right|_{t=5}$ and interpret your answers.

87. ● **Computer Use** The percentage of U.S. households with a computer in 2000 as a function of household income can be modeled by the logistic function[55]

$$P(x) = \frac{91}{1 + 5.35(1.05)^{-x}} \text{ percent} \quad 0 \le x \le 100$$

where x is the household income in thousands of dollars.

    a. Estimate $P(50)$ and $P'(50)$. What do the answers tell you about computer use in the U.S.? *hint* [see Example 5]

    b. Graph the function and its derivative for $0 \le x \le 100$ and use your graphs to describe how the derivative behaves for values of x approaching 100.

88. ● **Online Book Sales** The number of books sold online in the U.S. for the period 1997–2000 can be modeled by the logistic function[56]

$$N(t) = \frac{82.8}{1 + 21.8(7.14)^{-t}} \quad 0 \le t \le 3$$

where t is time in years since the start of 1997, and $N(t)$ is the number of books sold in the year beginning at time t.

    a. Estimate $N(1)$ and $N'(1)$. What do the answers tell you about online book sales?

    b. Graph the function and its derivative for $0 \le t \le 3$ and use your graphs to estimate, to the nearest 6 months, when N' was greatest.

89. **tech Ex** **Embryo Development** The oxygen consumption of a turkey embryo increases from the time the egg is laid through the time the turkey chick hatches. In a brush turkey, the oxygen consumption (in milliliters per hour) can be approximated by

$$c(t) = -0.0012t^3 + 0.12t^2 - 1.83t + 3.97 \quad (20 \le t \le 50)$$

where t is the time (in days) since the egg was laid.[57] (An egg will typically hatch at around $t = 50$.) Use technology to graph $c'(t)$ and use your graph to answer the following questions.

    a. Over the interval $[20, 32]$ the derivative c' is
      (A) increasing, then decreasing
      (B) decreasing, then increasing
      (C) decreasing   **(D)** increasing

    b. When, to the nearest day, is the oxygen consumption increasing at the fastest rate?

    c. When, to the nearest day, is the oxygen consumption increasing at the slowest rate?

90. **tech Ex** **Embryo Development** The oxygen consumption of a bird embryo increases from the time the egg is laid through the time the chick hatches. In a typical galliform bird, the oxygen consumption (in milliliters per hour) can be approximated by

$$c(t) = -0.0027t^3 + 0.14t^2 - 0.89t + 0.15 \quad (8 \le t \le 30)$$

where t is the time (in days) since the egg was laid.[58] (An egg will typically hatch at around $t = 28$.) Use technology to graph $c'(t)$ and use your graph to answer the following questions.

    a. Over the interval $[8, 30]$ the derivative c' is
      (A) increasing, then decreasing
      (B) decreasing, then increasing
      (C) decreasing   **(D)** increasing

[53] Ibid.

[54] Based on data presented in the article *The Emergence of Intelligence* by William H. Calvin, *Scientific American,* October, 1994, pp. 101–107.

[55] Source: NTIA and ESA, U.S. Department of Commerce, using U.S. Bureau of the Census Current Population, 2000.

[56] Source: Ipsos-NPD Book Trends/*New York Times,* April 16, 2001, p. C1.

[57] Ibid.

[58] The model approximates graphical data published in the article *The Brush Turkey* by Roger S. Seymour, *Scientific American,* December, 1991, pp. 108–114.

● basic skills   ◆ challenging   **tech Ex** technology exercise

b. When, to the nearest day, is the oxygen consumption increasing the fastest?

c. When, to the nearest day, is the oxygen consumption increasing at the slowest rate?

The next two exercises are applications of Einstein's Special Theory of Relativity and relate to objects that are moving extremely fast. In science fiction terminology, a speed of *warp 1* is the speed of light—about 3×10^8 meters per second. (Thus, for instance, a speed of warp 0.8 corresponds to 80% of the speed of light—about 2.4×10^8 meters per second.)

91. ◆ *Lorentz Contraction* According to Einstein's Special Theory of Relativity, a moving object appears to get shorter to a stationary observer as its speed approaches the speed of light. If a spaceship that has a length of 100 meters at rest travels at a speed of warp p, its length in meters, as measured by a stationary observer, is given by

$$L(p) = 100\sqrt{1 - p^2}$$

with domain [0, 1). Estimate $L(0.95)$ and $L'(0.95)$. What do these figures tell you?

92. ◆ *Time Dilation* Another prediction of Einstein's Special Theory of Relativity is that, to a stationary observer, clocks (as well as all biological processes) in a moving object appear to go more and more slowly as the speed of the object approaches that of light. If a spaceship travels at a speed of warp p, the time it takes for an onboard clock to register one second, as measured by a stationary observer, will be given by

$$T(p) = \frac{1}{\sqrt{1 - p^2}} \text{ seconds}$$

with domain [0, 1). Estimate $T(0.95)$ and $T'(0.95)$. What do these figures tell you?

Communication and Reasoning Exercises

93. ● Explain why we cannot put $h = 0$ in the approximation

$$f'(x) \approx \frac{f(x + h) - f(x)}{h}$$

for the derivative of f.

94. ● The balanced difference quotient

$$f'(a) \approx \frac{f(a + 0.0001) - f(a - 0.0001)}{0.0002}$$

is the average rate of change of f on what interval?

95. ● It is now eight months since the Garden City lacrosse team won the national championship, and sales of team paraphernalia, while still increasing, have been leveling off. What does this tell you about the derivative of the sales curve?

96. ● Having been soundly defeated in the national lacrosse championships, Brakpan High has been faced with decreasing sales of its team paraphernalia. However, sales, while still decreasing, appear to be bottoming out. What does this tell you about the derivative of the sales curve?

97. Company A's profits are given by $P(0) = \$1$ million and $P'(0) = -\$1$ million/month. Company B's profits are given by $P(0) = -\$1$ million and $P'(0) = \$1$ million/month. In which company would you rather invest? Why?

98. Company C's profits are given by $P(0) = \$1$ million and $P'(0) = \$0.5$ million/month. Company D's profits are given by $P(0) = \$0.5$ million and $P'(0) = \$1$ million/month. In which company would you rather invest? Why?

99. During the one-month period starting last January 1, your company's profits increased at an average rate of change of $4 million per month. On January 1, profits were increasing at an instantaneous rate of $5 million per month. Which of the following graphs could represent your company's profits? Why?

(A)

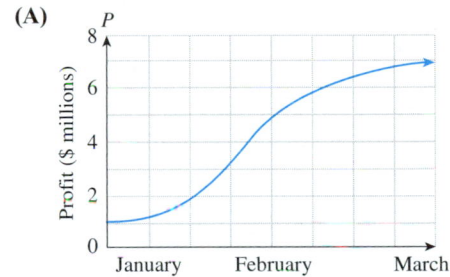

(B)

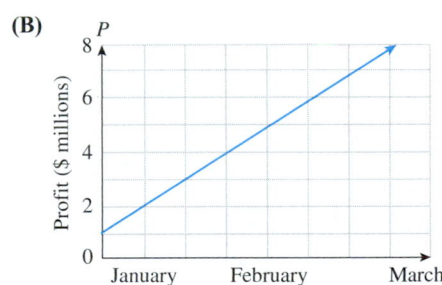

(C)

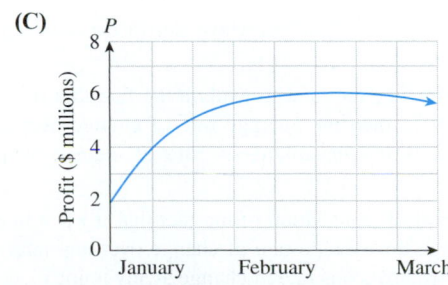

100. During the one-month period starting last January 1, your company's sales increased at an average rate of change of $3000 per month. On January 1, sales were changing at an instantaneous rate of −$1000 per month. Which of the following graphs could represent your company's sales? Why?

(A)

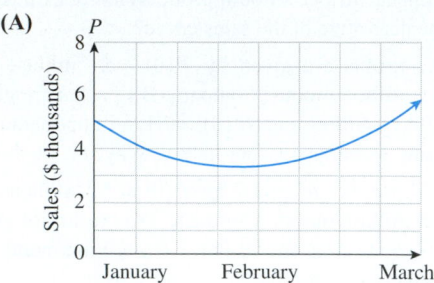

(B)

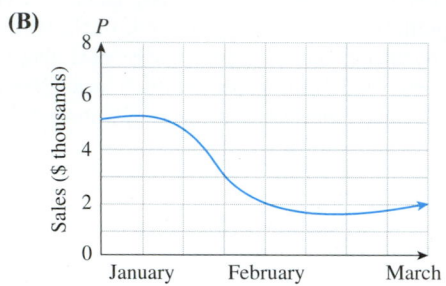

(C)

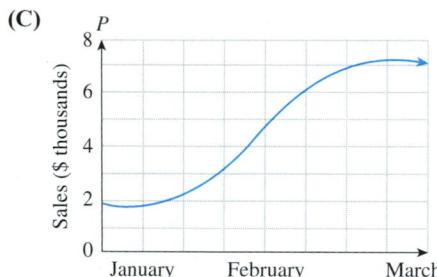

101. If the derivative of f is zero at a point, what do you know about the graph of f near that point?

102. Sketch the graph of a function whose derivative never exceeds 1.

103. Sketch the graph of a function whose derivative exceeds 1 at every point.

104. Sketch the graph of a function whose derivative is exactly 1 at every point.

105. Use the difference quotient to explain the fact that if f is a linear function, then the average rate of change over any interval equals the instantaneous rate of change at any point.

106. Give a numerical explanation of the fact that if f is a linear function, then the average rate of change over any interval equals the instantaneous rate of change at any point.

107. ◆ Consider the following values of the function f from Exercise 1.

h	0.1	0.01	0.001	0.0001
Ave. rate of change of f over $[5, 5+h]$	6.4	6.04	6.004	6.0004
h	−0.1	−0.01	−0.001	−0.0001
Ave. rate of change of f over $[5+h, 5]$	5.6	5.96	5.996	5.9996

Does the table suggests that the instantaneous rate of change of f is

(A) increasing   **(B)** decreasing

as x increases toward 5?

108. ◆ Consider the following values of the function g from Exercise 2.

h	0.1	0.01	0.001	0.0001
Ave. rate of change of g over $[7, 7+h]$	4.8	4.98	4.998	4.9998
h	−0.1	−0.01	−0.001	−0.0001
Ave. rate of change of g over $[7+h, 7]$	5.3	5.03	5.003	5.0003

Does the table suggests that the instantaneous rate of change of g is

(A) increasing   **(B)** decreasing

as x increases toward 7?

109. Sketch the graph of a function whose derivative is never zero but decreases as x increases.

110. Sketch the graph of a function whose derivative is never negative but is zero at exactly two points.

111. ◆ Here is the graph of the derivative f' of a function f. Give a rough sketch of the graph of f, given that $f(0) = 0$.

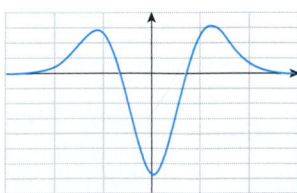

112. ◆ Here is the graph of the derivative f' of a function f. Give a rough sketch of the graph of f, given that $f(0) = 0$.

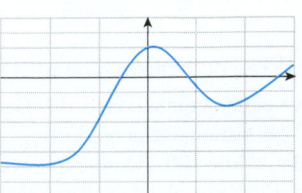

113. ◆ Professor Talker of the physics department drove a 60-mile stretch of road in exactly one hour. The speed limit along that stretch was 55 miles per hour. Which of the following must be correct:

(A) He exceeded the speed limit at no point of the journey.

(B) He exceeded the speed limit at some point of the journey.

(C) He exceeded the speed limit throughout the journey.

(D) He traveled slower than the speed limit at some point of the journey

114. ◆ Professor Silent, another physics professor, drove a 50-mile stretch of road in exactly one hour. The speed limit along that stretch was 55 miles per hour. Which of the following must be correct:

(A) She exceeded the speed limit at no point of the journey.

(B) She exceeded the speed limit at some point of the journey.

(C) She traveled slower than the speed limit throughout the journey.

(D) She traveled slower than the speed limit at some point of the journey.

115. ◆ Draw the graph of a function f with the property that the balanced difference quotient gives a more accurate approximation of $f'(1)$ than the ordinary difference quotient.

116. ◆ Draw the graph of a function f with the property that the balanced difference quotient gives a less accurate approximation of $f'(1)$ than the ordinary difference quotient.

● basic skills   ◆ challenging   **tech** Ex technology exercise

10.6 The Derivative: Algebraic Viewpoint

In Section 10.5 we saw how to estimate the derivative of a function using numerical and graphical approaches. In this section we use an algebraic approach that will give us the *exact value* of the derivative, rather than just an approximation, when the function is specified algebraically.

This algebraic approach is quite straightforward: Instead of subtracting numbers to estimate the average rate of change over smaller and smaller intervals, we subtract algebraic expressions. Our starting point is the definition of the derivative in terms of the difference quotient:

$$f'(a) = \lim_{h \to 0} \frac{f(a+h) - f(a)}{h}$$

Example 1 Calculating the Derivative at a Point Algebraically

Let $f(x) = x^2$. Use the definition of the derivative to compute $f'(3)$ algebraically.

Solution Substituting $a = 3$ into the definition of the derivative, we get:

$$f'(3) = \lim_{h \to 0} \frac{f(3+h) - f(3)}{h} \qquad \text{Formula for the derivative}$$

$$= \lim_{h \to 0} \frac{\overbrace{(3+h)^2}^{f(3+h)} - \overbrace{3^2}^{f(3)}}{h} \qquad \text{Substitute for } f(3) \text{ and } f(3+h)$$

$$= \lim_{h \to 0} \frac{(9 + 6h + h^2) - 9}{h} \qquad \text{Expand } (3+h)^2$$

$$= \lim_{h \to 0} \frac{6h + h^2}{h} \qquad \text{Cancel the 9}$$

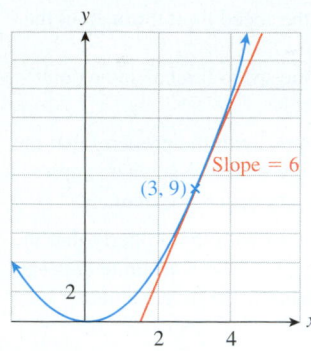

Figure **34**

$$= \lim_{h \to 0} \frac{h(6 + h)}{h} \qquad \text{Factor out } h$$

$$= \lim_{h \to 0} (6 + h) \qquad \text{Cancel the } h$$

Now we let h approach 0. As h gets closer and closer to 0, the sum $6 + h$ clearly gets closer and closer to $6 + 0 = 6$. Thus,

$$f'(3) = \lim_{h \to 0} (6 + h) = 6 \qquad \text{As } h \to 0, (6 + h) \to 6$$

(Calculations of limits like this are discussed and justified more fully in Sections 10.2 and 10.3.)

+ *Before we go on...* We did the following calculation in Example 1: If $f(x) = x^2$, then $f'(3) = 6$. In other words, the tangent to the graph of $y = x^2$ at the point $(3, 9)$ has slope 6 (Figure 34). ■

There is nothing very special about $a = 3$ in Example 1. Let's try to compute $f'(x)$ for general x.

Example **2** Calculating the Derivative Function Algebraically

Let $f(x) = x^2$. Use the definition of the derivative to compute $f'(x)$ algebraically.

Solution Once again, our starting point is the definition of the derivative in terms of the difference quotient:

$$f'(x) = \lim_{h \to 0} \frac{f(x + h) - f(x)}{h} \qquad \text{Formula for the derivative}$$

$$= \lim_{h \to 0} \frac{\overbrace{(x + h)^2}^{f(x+h)} - \overbrace{x^2}^{f(x)}}{h} \qquad \text{Substitute for } f(x) \text{ and } f(x + h)$$

$$= \lim_{h \to 0} \frac{(x^2 + 2xh + h^2) - x^2}{h} \qquad \text{Expand } (x + h)^2.$$

$$= \lim_{h \to 0} \frac{2xh + h^2}{h} \qquad \text{Cancel the } x^2.$$

$$= \lim_{h \to 0} \frac{h(2x + h)}{h} \qquad \text{Factor out } h.$$

$$= \lim_{h \to 0} (2x + h) \qquad \text{Cancel the } h.$$

Now we let h approach 0. As h gets closer and closer to 0, the sum $2x + h$ clearly gets closer and closer to $2x + 0 = 2x$. Thus,

$$f'(x) = \lim_{h \to 0} (2x + h) = 2x$$

This is the derivative function. Now that we have a *formula* for the derivative of f, we can obtain $f'(a)$ for any value of a we choose by simply evaluating f' there. For instance,

$$f'(3) = 2(3) = 6$$

as we saw in Example 1.

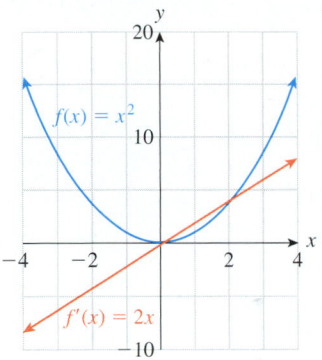

Figure 35

+*Before we go on...* The graphs of $f(x) = x^2$ and $f'(x) = 2x$ from Example 2 are familiar. Their graphs are shown in Figure 35.

When $x < 0$, the parabola slopes downward, which is reflected in the fact that the derivative $2x$ is negative there. When $x > 0$, the parabola slopes upward, which is reflected in the fact that the derivative is positive there. The parabola has a horizontal tangent line at $x = 0$, reflected in the fact that $2x = 0$ there. ■

Example 3 More Computations of Derivative Functions

Compute the derivative $f'(x)$ for each of the following functions:

a. $f(x) = x^3$ **b.** $f(x) = 2x^2 - x$ **c.** $f(x) = \dfrac{1}{x}$

Solution

a. $f'(x) = \lim_{h \to 0} \dfrac{f(x+h) - f(x)}{h}$ Derivative formula

$$= \lim_{h \to 0} \dfrac{\overbrace{(x+h)^3}^{f(x+h)} - \overbrace{x^3}^{f(x)}}{h}$$ Substitute for $f(x)$ and $f(x+h)$

$$= \lim_{h \to 0} \dfrac{(x^3 + 3x^2 h + 3xh^2 + h^3) - x^3}{h}$$ Expand $(x+h)^3$

$$= \lim_{h \to 0} \dfrac{3x^2 h + 3xh^2 + h^3}{h}$$ Cancel the x^3

$$= \lim_{h \to 0} \dfrac{h(3x^2 + 3xh + h^2)}{h}$$ Factor out h

$$= \lim_{h \to 0} (3x^2 + 3xh + h^2)$$ Cancel the h

$$= 3x^2$$ Let h approach 0

b. $f'(x) = \lim_{h \to 0} \dfrac{f(x+h) - f(x)}{h}$ Derivative formula

$$= \lim_{h \to 0} \dfrac{\overbrace{(2(x+h)^2 - (x+h))}^{f(x+h)} - \overbrace{(2x^2 - x)}^{f(x)}}{h}$$ Substitute for $f(x)$ and $f(x+h)$

$$= \lim_{h \to 0} \dfrac{(2x^2 + 4xh + 2h^2 - x - h) - (2x^2 - x)}{h}$$ Expand

$$= \lim_{h \to 0} \dfrac{4xh + 2h^2 - h}{h}$$ Cancel the $2x^2$ and x

$$= \lim_{h \to 0} \dfrac{h(4x + 2h - 1)}{h}$$ Factor out h

$$= \lim_{h \to 0} (4x + 2h - 1)$$ Cancel the h

$$= 4x - 1$$ Let h approach 0

c. $f'(x) = \lim_{h \to 0} \dfrac{f(x+h) - f(x)}{h}$ Derivative formula

$$= \lim_{h \to 0} \dfrac{\left[\overbrace{\dfrac{1}{x+h}}^{f(x+h)} - \overbrace{\dfrac{1}{x}}^{f(x)}\right]}{h}$$ Substitute for $f(x)$ and $f(x+h)$

$$= \lim_{h \to 0} \frac{\left[\dfrac{x - (x + h)}{(x + h)x} \right]}{h} \qquad \text{Subtract the fractions}$$

$$= \lim_{h \to 0} \frac{1}{h} \left[\frac{x - (x + h)}{(x + h)x} \right] \qquad \text{Dividing by } h = \text{Multiplying by } 1/h$$

$$= \lim_{h \to 0} \left[\frac{-h}{h(x + h)x} \right] \qquad \text{Simplify}$$

$$= \lim_{h \to 0} \left[\frac{-1}{(x + h)x} \right] \qquad \text{Cancel the } h$$

$$= \frac{-1}{x^2} \qquad \text{Let } h \text{ approach } 0$$

In Example 4, we redo Example 3 of Section 10.5, this time getting an exact, rather than approximate, answer.

Example 4 Velocity

My friend Eric, an enthusiastic baseball player, claims he can "probably" throw a ball upward at a speed of 100 feet per second (ft/sec). Our physicist friends tell us that its height s (in feet) t seconds later would be $s(t) = 100t - 16t^2$. Find the ball's instantaneous velocity function and its velocity exactly 2 seconds after Eric throws it.

Solution The instantaneous velocity function is the derivative ds/dt, which we calculate as follow:

$$\frac{ds}{dt} = \lim_{h \to 0} \frac{s(t + h) - s(t)}{h}$$

Let us compute $s(t + h)$ and $s(t)$ separately:

$$s(t) = 100t - 16t^2$$

$$\begin{aligned} s(t + h) &= 100(t + h) - 16(t + h)^2 \\ &= 100t + 100h - 16(t^2 + 2th + h^2) \\ &= 100t + 100h - 16t^2 - 32th - 16h^2 \end{aligned}$$

Therefore,

$$\begin{aligned} \frac{ds}{dt} &= \lim_{h \to 0} \frac{s(t + h) - s(t)}{h} \\ &= \lim_{h \to 0} \frac{100t + 100h - 16t^2 - 32th - 16h^2 - (100t - 16t^2)}{h} \\ &= \lim_{h \to 0} \frac{100h - 32th - 16h^2}{h} \\ &= \lim_{h \to 0} \frac{h(100 - 32t - 16h)}{h} \\ &= \lim_{h \to 0} 100 - 32t - 16h \\ &= 100 - 32t \text{ ft/sec} \end{aligned}$$

Thus, the velocity exactly 2 seconds after Eric throws it is

$$\frac{ds}{dt}\bigg|_{t=2} = 100 - 32(2) = 36 \text{ ft/sec}$$

This verifies the accuracy of the approximation we made in Section 10.5.

+*Before we go on...* From the derivative function in Example 4, we can now describe the behavior of the velocity of the ball: Immediately on release ($t = 0$) the ball is traveling at 100 feet per second upward. The ball then slows down; precisely, it loses 32 feet per second of speed every second. When, exactly, does the velocity become zero and what happens after that? ∎

Q: Do we always have to calculate the limit of the difference quotient to find a formula for the derivative function?

A: As it turns out, no. In Section 10.7 we will start to look at shortcuts for finding derivatives that allow us to bypass the definition of the derivative in many cases. ∎

A Function Not Differentiable at a Point

Recall from Section 10.5 that a function is **differentiable** at a point a if $f'(a)$ exists; that is, if the difference quotient $[f(a + h) - f(a)]/h$ approaches a fixed value as h approaches 0. In Section 10.5, we mentioned that the function $f(x) = |x|$ is not differentiable at $x = 0$. In Example 5, we find out why.

Example 5 A Function Not Differentiable at 0

Numerically, graphically, and algebraically investigate the differentiability of the function $f(x) = |x|$ at the points **a.** $x = 1$ and **b.** $x = 0$.

Solution

a. We compute

$$f'(1) = \lim_{h \to 0} \frac{f(1 + h) - f(1)}{h}$$

$$= \lim_{h \to 0} \frac{|1 + h| - 1}{h}$$

Numerically, we can make tables of the values of the average rate of change ($|1 + h| - 1)/h$ for h positive or negative and approaching 0:

h	1	0.1	0.01	0.001	0.0001
Ave. rate of change over [1, 1 + h]	1	1	1	1	1

h	−1	−0.1	−0.01	−0.001	−0.0001
Ave. rate of change over [1 + h, 1]	1	1	1	1	1

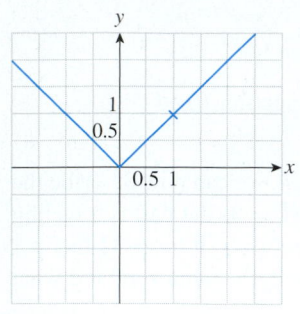

Figure **36**

From these tables it appears that $f'(1)$ is equal to 1. We can verify that algebraically: For h that is sufficiently small, $1 + h$ is positive (even if h is negative) and so

$$f'(1) = \lim_{h \to 0} \frac{1 + h - 1}{h}$$

$$= \lim_{h \to 0} \frac{h}{h} \qquad \text{Cancel the 1s}$$

$$= \lim_{h \to 0} 1 \qquad \text{Cancel the } h$$

$$= 1$$

Graphically, we are seeing the fact that the tangent line at the point $(1, 1)$ has slope 1 because the graph is a straight line with slope 1 near that point (Figure 36).

b. $f'(0) = \lim_{h \to 0} \dfrac{f(0 + h) - f(0)}{h}$

$$= \lim_{h \to 0} \frac{|0 + h| - 0}{h}$$

$$= \lim_{h \to 0} \frac{|h|}{h}$$

If we make tables of values in this case we get the following:

h	1	0.1	0.01	0.001	0.0001
Ave. rate of change over $[0, 0 + h]$	1	1	1	1	1

h	-1	-0.1	-0.01	-0.001	-0.0001
Ave. rate of change over $[0 + h, 0]$	-1	-1	-1	-1	-1

For the limit and hence the derivative $f'(0)$ to exist, the average rates of change should approach the same number for both positive and negative h. Because they do not, f is not differentiable at $x = 0$. We can verify this conclusion algebraically: If h is positive, then $|h| = h$, and so the ratio $|h|/h$ is 1, regardless of how small h is. Thus, according to the values of the difference quotients with $h > 0$, the limit should be 1. On the other hand if h is negative, then $|h| = -h$ (positive) and so $|h|/h = -1$, meaning that the limit should be -1. Because the limit cannot be both -1 and 1 (it must be a single number for the derivative to exist), we conclude that $f'(0)$ does not exist.

To see what is happening graphically, take a look at Figure 37, which shows zoomed-in views of the graph of f near $x = 0$.

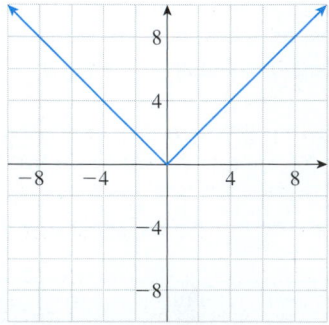 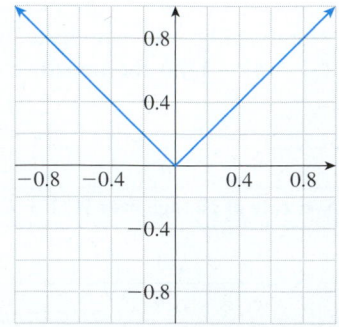 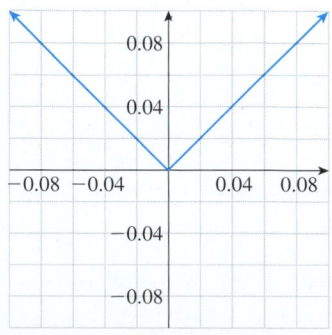

Figure **37**

No matter what scale we use to view the graph, it has a sharp corner at $x = 0$ and hence has no tangent line there. Since there is no tangent line at $x = 0$, the function is not differentiable there.

$+$*Before we go on...* If we repeat the computation in Example 5(a) using any nonzero value for a in place of 1, we see that f is differentiable there as well. If a is positive, we find that $f'(a) = 1$ and, if a is negative, $f'(a) = -1$. In other words, the derivative function is

$$f'(x) = \begin{cases} -1 & \text{if } x < 0 \\ 1 & \text{if } x > 0 \end{cases}$$

Of course, $f'(x)$ is not defined when $x = 0$. ∎

FAQs **Computing Derivatives Algebraically**

Q: *The algebraic computation of $f'(x)$ seems to require a number of steps. How do I remember what to do, and when?*

A: If you examine the computations in the examples above, you will find the following pattern:

1. Write out the formula for $f'(x)$ as the limit of the difference quotient, then substitute $f(x + h)$ and $f(x)$.
2. Expand and simplify the *numerator* of the expression, but not the denominator.
3. After simplifying the numerator, factor out an h to cancel with the h in the denominator. If h does not factor out of the numerator, you might have made an error. (A frequent error is a wrong sign.)
4. After canceling the h, you should be able to see what the limit is by letting $h \to 0$. ∎

10.6 EXERCISES

● denotes basic skills exercises

In Exercises 1–14, compute $f'(a)$ algebraically for the given value of a.

1. ● $f(x) = x^2 + 1; a = 2$ *hint* [see Example 1]

2. ● $f(x) = x^2 - 3; a = 1$

3. ● $f(x) = 3x - 4; a = -1$

4. ● $f(x) = -2x + 4; a = -1$

5. ● $f(x) = 3x^2 + x; a = 1$

6. ● $f(x) = 2x^2 + x; a = -2$

7. ● $f(x) = 2x - x^2; a = -1$

8. ● $f(x) = -x - x^2; a = 0$

9. ● $f(x) = x^3 + 2x; a = 2$

10. ● $f(x) = x - 2x^3; a = 1$

11. ● $f(x) = \dfrac{-1}{x}; a = 1$ *hint* [see Example 3]

12. ● $f(x) = \dfrac{2}{x}; a = 5$

13. $f(x) = mx + b; a = 43$

14. $f(x) = \dfrac{x}{k} - b\ (k \neq 0); a = 12$

In Exercises 15–28, compute the derivative function $f'(x)$ algebraically. (Notice that the functions are the same as those in Exercises 1–14.)

15. ● $f(x) = x^2 + 1$ *hint* [see Examples 2 and 3]

16. ● $f(x) = x^2 - 3; a = 1$ **17.** ● $f(x) = 3x - 4$

18. ● $f(x) = -2x + 4$ **19.** ● $f(x) = 3x^2 + x$

20. ● $f(x) = 2x^2 + x$ **21.** ● $f(x) = 2x - x^2$

● basic skills

22. ● $f(x) = -x - x^2$

23. ● $f(x) = x^3 + 2x$

24. ● $f(x) = x - 2x^3$

25. $f(x) = \dfrac{-1}{x}$

26. $f(x) = \dfrac{2}{x}$

27. $f(x) = mx + b$

28. $f(x) = \dfrac{x}{k} - b \ (k \neq 0)$

In Exercises 29–38, compute the indicated derivative.

29. ● $R(t) = -0.3t^2; R'(2)$

30. ● $S(t) = 1.4t^2; S'(-1)$

31. ● $U(t) = 5.1t^2 + 5.1; U'(3)$

32. ● $U(t) = -1.3t^2 + 1.1; U'(4)$

33. ● $U(t) = -1.3t^2 - 4.5t; U'(1)$

34. ● $U(t) = 5.1t^2 - 1.1t; U'(1)$

35. ● $L(r) = 4.25r - 5.01; L'(1.2)$

36. ● $L(r) = -1.02r + 5.7; L'(3.1)$

37. $q(p) = \dfrac{2.4}{p} + 3.1; q'(2)$

38. $q(p) = \dfrac{1}{0.5p} - 3.1; q'(2)$

In Exercises 39–44, find the equation of the tangent to the graph at the indicated point.

39. $f(x) = x^2 - 3; a = 2$

40. $f(x) = x^2 + 1; a = 2$

41. $f(x) = -2x - 4; a = 3$

42. $f(x) = 3x + 1; a = 1$

43. $f(x) = x^2 - x; a = -1$

44. $f(x) = x^2 + x; a = -1$

Applications

45. ● *Velocity* If a stone is dropped from a height of 400 feet, its height after t seconds is given by $s = 400 - 16t^2$. Find its instantaneous velocity function and its velocity at time $t = 4$. *hint* [see Example 4]

46. ● *Velocity* If a stone is thrown down at 120 feet per second from a height of 1000 feet, its height after t seconds is given by $s = 1000 - 120t - 16t^2$. Find its instantaneous velocity function and its velocity at time $t = 3$.

47. ● *Foreign Trade* Annual U.S. imports from China in the years 1996 through 2003 could be approximated by

$$I(t) = t^2 + 3.5t + 50 \text{ billion dollars} \quad (1 \leq t \leq 9)$$

where t represents time in years since 1995.[59] At what rate was this number increasing in 2000?

48. ● *Foreign Trade* Annual U.S. exports to China in the years 1995 through 2003 could be approximated by

$$E(t) = 0.4t^2 - 1.6t + 14 \text{ billion dollars} \quad (0 \leq t \leq 8)$$

where t represents time in years since 1995.[60] At what rate was this number increasing in 2000?

49. ● *Bottled Water Sales* Annual U.S. sales of bottled water rose through the period 1993–2003 as shown in the following chart.[61]

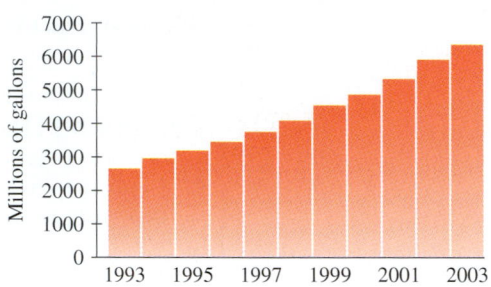

Bottled Water Sales in the U.S.

The function

$$R(t) = 17t^2 + 100t + 2300 \text{ million gallons} \quad (3 \leq t \leq 13)$$

gives a good approximation, where t is time in years since 1990. Find the derivative function $R'(t)$. According to the model, how fast were annual sales of bottled water increasing in 2000?

50. ● *Bottled Water Sales* Annual U.S. per capita sales of bottled water rose through the period 1993–2003 as shown in the following chart.[62]

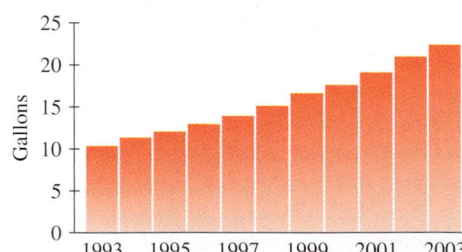

Per Capita Bottled Water Sales in the U.S.

---

[59] Based on quadratic regression using data from the U.S. Census Bureau Foreign Trade Division website www.census.gov/foreign-trade/sitc1/ as of December 2004.

[60] Based on quadratic regression using data from the U.S. Census Bureau Foreign Trade Division website www.census.gov/foreign-trade/sitc1/ as of December 2004.

[61] SOURCE: Beverage Marketing Corporation news release, "Bottled water now number-two commercial beverage in U.S., says Beverage Marketing Corporation," April 8, 2004, available at www.beveragemarketing.com.

[62] Ibid.

● basic skills

The function

$$Q(t) = 0.05t^2 + 0.4t + 9 \text{ gallons}$$

gives a good approximation, where t is the time in years since 1990. Find the derivative function $Q'(t)$. According to the model, how fast were annual per capita sales of bottled water increasing in 2000?

51. *Ecology* Increasing numbers of manatees ("sea sirens") have been killed by boats off the Florida coast. The following graph shows the relationship between the number of boats registered in Florida and the number of manatees killed each year.

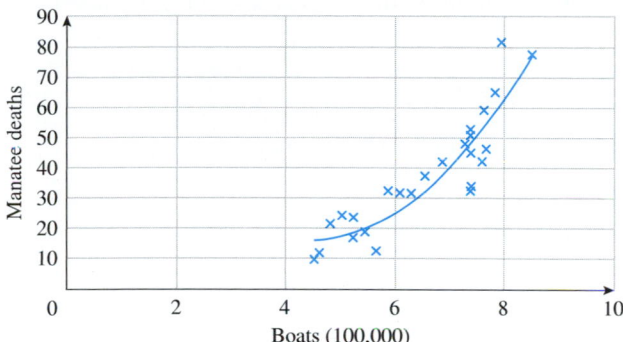

Boats (100,000)

The regression curve shown is given by

$$f(x) = 3.55x^2 - 30.2x + 81 \text{ manatee deaths}$$
$$(4.5 \le x \le 8.5)$$

where x is the number of boats (hundreds of thousands) registered in Florida in a particular year and $f(x)$ is the number of manatees killed by boats in Florida that year.[63] Compute and interpret $f'(8)$.

52. *SAT Scores by Income* The following graph shows U.S. verbal SAT scores as a function of parents' income level.[64]

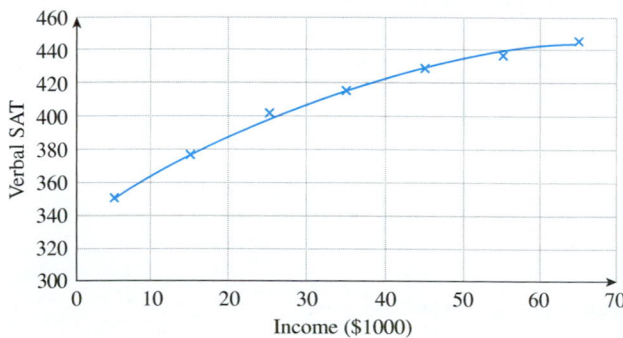

Income ($1000)

The regression curve shown is given by

$$f(x) = -0.021x^2 + 3.0x + 336 \quad (5 \le x \le 65)$$

where $f(x)$ is the average SAT verbal score of a student whose parents earn x thousand dollars per year.[65] Compute and interpret $f'(30)$.

Communication and Reasoning Exercises

53. ● Of the three methods (numerical, graphical, algebraic) we can use to estimate the derivative of a function at a given value of x, which is always the most accurate? Explain.

54. ● Explain why we cannot put $h = 0$ in the formula

$$f'(a) = \lim_{h \to 0} \frac{f(a+h) - f(a)}{h}$$

for the derivative of f.

55. ● Your friend Muffy claims that, because the balanced difference quotient is more accurate, it would be better to use that instead of the usual difference quotient when computing the derivative algebraically. Comment on this advice.

56. ● Use the balanced difference quotient formula,

$$f'(a) = \lim_{h \to 0} \frac{f(a+h) - f(a-h)}{2h}$$

to compute $f'(3)$ when $f(x) = x^2$. What do you find?

57. A certain function f has the property that $f'(a)$ does not exist. How is that reflected in the attempt to compute $f'(a)$ algebraically?

58. One cannot put $h = 0$ in the formula

$$f'(a) = \lim_{h \to 0} \frac{f(a+h) - f(a)}{h}$$

for the derivative of f. (See Exercise 54). However, in the last step of each of the computations in the text, we are effectively setting $h = 0$ when taking the limit. What is going on here?

[63] Regression model is based on data from 1976 to 2000. Sources for data: Florida Department of Highway Safety & Motor Vehicles, Florida Marine Institute/*New York Times,* February 12, 2002, p. F4.

[64] Based on 1994 data. SOURCE: The College Board/*New York Times,* March 5, 1995, p. E16.

[65] Regression model is based on 1994 data. SOURCE: The College Board/*New York Times,* March 5, 1995, p. E16.

10.7 Derivatives of Powers, Sums, and Constant Multiples

So far in this chapter we have approximated derivatives using difference quotients, and we have done exact calculations using the definition of the derivative as the limit of a difference quotient. In general, we would prefer to have an exact calculation, and it is also very useful to have a formula for the derivative function when we can find one. However, the calculation of a derivative as a limit is often tedious, so it would be nice to have a quicker method. We discuss the first of the shortcut rules in this section. By the end of Chapter 11, we will be able to find fairly quickly the derivative of almost any function we can write.

Shortcut Formula: The Power Rule

If you look again at Examples 2 and 3 in Section 10.6, you may notice a pattern:

$$f(x) = x^2 \quad \Rightarrow \quad f'(x) = 2x$$
$$f(x) = x^3 \quad \Rightarrow \quad f'(x) = 3x^2$$

Theorem: The Power Rule

If n is any constant and $f(x) = x^n$, then

$$f'(x) = nx^{n-1}$$

quick Examples

1. If $f(x) = x^2$, then $f'(x) = 2x^1 = 2x$.
2. If $f(x) = x^3$, then $f'(x) = 3x^2$.
3. If $f(x) = x$, rewrite as $f(x) = x^1$, so $f'(x) = 1x^0 = 1$.
4. If $f(x) = 1$, rewrite as $f(x) = x^0$, so $f'(x) = 0x^{-1} = 0$.

The proof of the power rule involves first studying the case when n is a positive integer, and then studying the cases of other types of exponents (negative integer, rational number, irrational number). You can find a proof at the website.

Find a proof of the power rule online.
Follow:

Chapter 10
→ Proof of the Power Rule

Example 1 Using the Power Rule for Negative and Fractional Exponents

Calculate the derivatives of the following:

a. $f(x) = \dfrac{1}{x}$ **b.** $f(x) = \dfrac{1}{x^2}$ **c.** $f(x) = \sqrt{x}$

Solution

a. Rewrite* as $f(x) = x^{-1}$. Then $f'(x) = (-1)x^{-2} = -\dfrac{1}{x^2}$.

b. Rewrite as $f(x) = x^{-2}$. Then $f'(x) = (-2)x^{-3} = -\dfrac{2}{x^3}$.

c. Rewrite as $f(x) = x^{0.5}$. Then $f'(x) = 0.5x^{-0.5} = \dfrac{0.5}{x^{0.5}}$.

Alternatively, rewrite $f(x)$ as $x^{1/2}$, so that $f'(x) = \dfrac{1}{2}x^{-1/2} = \dfrac{1}{2x^{1/2}} = \dfrac{1}{2\sqrt{x}}$.

---

* See the section on exponents in the algebra review to brush up on negative and fractional exponents.

By rewriting the given functions in Example 1 before taking derivatives, we converted them from **rational** or **radical form** (as in, say, $\dfrac{1}{x^2}$ or $\sqrt{x}$) to **exponent form** (as in x^{-2} and $x^{0.5}$) to enable us to use the power rule (see the caution below).

Caution

We cannot apply the power rule to terms in the denominators or under square roots. For example:

1. The derivative of $\dfrac{1}{x^2}$ is **NOT** $\dfrac{1}{2x}$; but is $-\dfrac{2}{x^3}$.   See Example 1(b)

2. The derivative of $\sqrt{x^3}$ is **NOT** $\sqrt{3x^2}$; but is $1.5x^{0.5}$.   Rewrite $\sqrt{x^3}$ as $x^{3/2}$ or $x^{1.5}$ and apply the power rule

---

Some of the derivatives in Example 1 are very useful to remember, so we summarize them in Table 1. We suggest that you add to this table as you learn more derivatives. It is *extremely* helpful to remember the derivatives of common functions such as $1/x$ and $\sqrt{x}$, even though they can be obtained using the power rule as in the above example.

Table 1 Table of Derivative Formulas

$f(x)$	$f'(x)$
1	0
x	1
x^2	$2x$
x^3	$3x^2$
x^n	nx^{n-1}
$\dfrac{1}{x}$	$-\dfrac{1}{x^2}$
$\dfrac{1}{x^2}$	$-\dfrac{2}{x^3}$
$\sqrt{x}$	$\dfrac{1}{2\sqrt{x}}$

Another Notation: Differential Notation

Here is a useful notation based on the "*d*-notation" we discussed in Section 10.5. **Differential notation** is based on an abbreviation for the phrase "the derivative with respect to x." For example, we learned that if $f(x) = x^3$, then $f'(x) = 3x^2$. When we say "$f'(x) = 3x^2$," we mean the following:

The derivative of x^3 with respect to x equals $3x^2$.

You may wonder why we sneaked in the words "with respect to x." All this means is that the variable of the function is x, and not any other variable.[66] Since we use the phrase "the derivative with respect to x" often, we use the following abbreviation.

Differential Notation; Differentiation

$\dfrac{d}{dx}$ means "the derivative with respect to x."

Thus, $\dfrac{d}{dx}[f(x)]$ is the same thing as $f'(x)$, the derivative of $f(x)$ with respect to x. If y is a function of x, then the derivative of y with respect to x is

$$\frac{d}{dx}(y) \qquad \text{or, more compactly,} \qquad \frac{dy}{dx}$$

---

[66] This may seem odd in the case of $f(x) = x^3$ because there are no other variables to worry about. But in expressions like st^3 that involve variables other than x, it is necessary to specify just what the variable of the function is. This is the same reason that we write "$f(x) = x^3$" rather than just "$f = x^3$."

To **differentiate** a function $f(x)$ with respect to x means to take its derivative with respect to x.

quick Examples

In Words	**Formula**
1. The derivative with respect to x of x^3 is $3x^2$.	$\dfrac{d}{dx}(x^3) = 3x^2$
2. The derivative with respect to t of $\dfrac{1}{t}$ is $-\dfrac{1}{t^2}$.	$\dfrac{d}{dt}\left(\dfrac{1}{t}\right) = -\dfrac{1}{t^2}$
3. If $y = x^4$, then $\dfrac{dy}{dx} = 4x^3$.	
4. If $u = \dfrac{1}{t^2}$, then $\dfrac{du}{dt} = -\dfrac{2}{t^3}$.	

Notes

1. $\dfrac{dy}{dx}$ is Leibniz notation for the derivative we discussed in Section 10.5 (see the discussion before Example 3 there).

2. Leibniz notation illustrates units nicely: units of $\dfrac{dy}{dx}$ are units of y per unit of x. ∎

The Rules for Sums and Constant Multiples

We can now find the derivatives of more complicated functions, such as polynomials, using the following rules:

Theorem: Derivatives of Sums, Differences, and Constant Multiples

If $f(x)$ and $g(x)$ are any two differentiable functions, and if c is any constant, then the functions $f(x) + g(x)$ and $cf(x)$ are differentiable, and

$$[f(x) \pm g(x)]' = f'(x) \pm g'(x) \qquad \text{Sum Rule}$$

$$[cf(x)]' = cf'(x) \qquad \text{Constant Multiple Rule}$$

In Words:
- The derivative of a sum is the sum of the derivatives, and the derivative of a difference is the difference of the derivatives.
- The derivative of c times a function is c times the derivative of the function.

Differential Notation:

$$\frac{d}{dx}[f(x) \pm g(x)] = \frac{d}{dx}f(x) \pm \frac{d}{dx}g(x)$$

$$\frac{d}{dx}[cf(x)] = c\frac{d}{dx}f(x)$$

quick Examples

1. $\dfrac{d}{dx}[x^2 - x^4] = \dfrac{d}{dx}[x^2] - \dfrac{d}{dx}[x^4] = 2x - 4x^3$

2. $\dfrac{d}{dx}[7x^3] = 7\dfrac{d}{dx}[x^3] = 7(3x^2) = 21x^2$

In other words, we multiply the coefficient (7) by the exponent (3), and then decrease the exponent by 1.

3. $\dfrac{d}{dx}[12x] = 12\dfrac{d}{dx}[x] = 12(1) = 12$

In other words, *the derivative of a constant times x is that constant.*

4. $\dfrac{d}{dx}[-x^{0.5}] = \dfrac{d}{dx}[(-1)x^{0.5}] = (-1)\dfrac{d}{dx}[x^{0.5}] = (-1)(0.5)x^{-0.5} = -0.5x^{-0.5}$

5. $\dfrac{d}{dx}[12] = \dfrac{d}{dx}[12(1)] = 12\dfrac{d}{dx}[1] = 12(0) = 0.$

In other words, *the derivative of a constant is zero.*

6. If my company earns twice as much (annual) revenue as yours and the derivative of your revenue function is the curve on the left, then the derivative of my revenue function is the curve on the right.

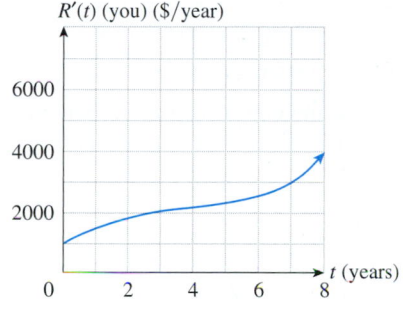
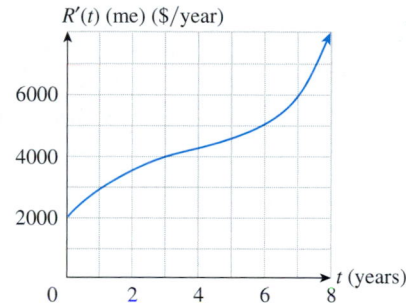

7. Suppose that a company's revenue R and cost C are changing with time. Then so is the profit, $P(t) = R(t) - C(t)$, and the rate of change of the profit is

$$P'(t) = R'(t) - C'(t)$$

In words: *The derivative of the profit is the derivative of revenue minus the derivative of cost.*

Proof of the Sum Rule

By the definition of the derivative of a function,

$$
\begin{aligned}
\frac{d}{dx}[f(x) + g(x)] &= \lim_{h \to 0} \frac{[f(x+h) + g(x+h)] - [f(x) + g(x)]}{h} \\
&= \lim_{h \to 0} \frac{[f(x+h) - f(x)] + [g(x+h) - g(x)]}{h} \\
&= \lim_{h \to 0} \left[\frac{f(x+h) - f(x)}{h} + \frac{g(x+h) - g(x)}{h} \right] \\
&= \lim_{h \to 0} \frac{f(x+h) - f(x)}{h} + \lim_{h \to 0} \frac{g(x+h) - g(x)}{h} \\
&= \frac{d}{dx}[f(x)] + \frac{d}{dx}[g(x)]
\end{aligned}
$$

The next-to-last step uses a property of limits: the limit of a sum is the sum of the limits. Think about why this should be true. The last step uses the definition of the derivative again (and the fact that the functions are differentiable).

The proof of the rule for constant multiples is similar.

> ### Example 2 Combining the Sum and Constant Multiple Rules, and Dealing with *x* in the Denominator
>
> Find the derivatives of the following:
>
> **a.** $f(x) = 3x^2 + 2x - 4$ **b.** $f(x) = \dfrac{2x}{3} - \dfrac{6}{x} + \dfrac{2}{3x^{0.2}} - \dfrac{x^4}{2}$
>
> **Solution**
>
> **a.** $\dfrac{d}{dx}(3x^2 + 2x - 4) = \dfrac{d}{dx}(3x^2) + \dfrac{d}{dx}(2x - 4)$ Rule for sums
>
> $\qquad\qquad\qquad\qquad = \dfrac{d}{dx}(3x^2) + \dfrac{d}{dx}(2x) - \dfrac{d}{dx}(4)$ Rule for differences
>
> $\qquad\qquad\qquad\qquad = 3(2x) + 2(1) - 0$ See Quick Example 2
>
> $\qquad\qquad\qquad\qquad = 6x + 2$
>
> **b.** Notice that f has x and powers of x in the denominator. We deal with these terms the same way we did in Example 1, by rewriting them in exponent form:
>
> $\qquad f(x) = \dfrac{2x}{3} - \dfrac{6}{x} + \dfrac{2}{3x^{0.2}} - \dfrac{x^4}{2}$ Rational form
>
> $\qquad\qquad = \dfrac{2}{3}x - 6x^{-1} + \dfrac{2}{3}x^{-0.2} - \dfrac{1}{2}x^4$ Exponent form
>
> We are now ready to take the derivative:
>
> $\qquad f'(x) = \dfrac{2}{3}(1) - 6(-1)x^{-2} + \dfrac{2}{3}(-0.2)x^{-1.2} - \dfrac{1}{2}(4x^3)$
>
> $\qquad\qquad = \dfrac{2}{3} + 6x^{-2} - \dfrac{0.4}{3}x^{-1.2} - 2x^3$ Exponent form
>
> $\qquad\qquad = \dfrac{2}{3} + \dfrac{6}{x^2} - \dfrac{0.4}{3x^{1.2}} - 2x^3$ Rational form

Notice that in Example 2(a) we had three terms in the expression for $f(x)$, not just two. By applying the rule for sums and differences twice, we saw that the derivative of a sum or difference of three terms is the sum or difference of the derivatives of the terms. (One of those terms had zero derivatives, so the final answer had only two terms.) In fact, the derivative of a sum or difference of any number of terms is the sum or difference of the derivatives of the terms. Put another way, to take the derivative of a sum or difference of any number of terms, we take derivatives term by term.

Note Nothing forces us to use only x as the independent variable when taking derivatives (although it is traditional to give x preference). For instance, part (a) in Example 2 can be rewritten as

$$\frac{d}{dt}(3t^2 + 2t - 4) = 6t + 2 \qquad \frac{d}{dt} \text{ means "derivative with respect to } t\text{".}$$

or $\qquad \dfrac{d}{du}(3u^2 + 2u - 4) = 6u + 2 \qquad \dfrac{d}{du}$ means "derivative with respect to u". ∎

In the examples above, we saw instances of the following important facts. (Think about these graphically to see why they must be true.)

The Derivative of a Constant Times *x* and the Derivative of a Constant

If c is any constant, then:

Rule

$$\frac{d}{dx}(cx) = c$$

$$\frac{d}{dx}(c) = 0$$

$$\frac{d}{dx}(6x) = 6 \qquad \frac{d}{dx}(-x) = -1$$

$$\frac{d}{dx}(5) = 0 \qquad \frac{d}{dx}(\pi) = 0$$

In Example 5 of Section 10.6 we saw that $f(x) = |x|$ fails to be differentiable at $x = 0$. In the next example we use the power rule and find more functions not differentiable at a point.

Example 3 Functions Not Differentiable at a Point

Find the natural domains of the derivatives of $f(x) = x^{1/3}$ and $g(x) = x^{2/3}$, and $h(x) = |x|$.

Solution Let's first look at the functions f and g. By the power rule,

$$f'(x) = \frac{1}{3}x^{-2/3} = \frac{1}{3x^{2/3}}$$

and

$$g'(x) = \frac{2}{3}x^{-1/3} = \frac{2}{3x^{1/3}}$$

$f'(x)$ and $g'(x)$ are defined only for nonzero values of x, and their natural domains consist of all real numbers except 0. Thus, the derivatives f' and g' do not exist at $x = 0$. In other words, f and g are not differentiable at $x = 0$. If we look at Figure 38, we notice why these functions fail to be differentiable at $x = 0$: The graph of f has a vertical tangent line at 0. Because a vertical line has undefined slope, the derivative is undefined at that point. The graph of g comes to a sharp point (called a **cusp**) at 0, so it is not meaningful to speak about a tangent line at that point; therefore, the derivative of g is not defined there. (Actually, there is a reasonable candidate for the tangent line at $x = 0$, but it is the vertical line again.)

We can also detect this nondifferentiability by computing some difference quotients numerically. In the case of $f(x) = x^{1/3}$, we get the following table:

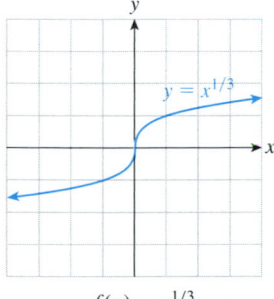

$f(x) = x^{1/3}$

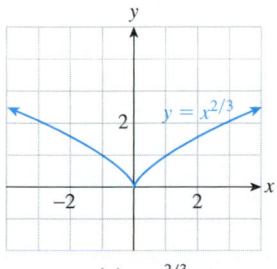

$g(x) = x^{2/3}$

Figure **38**

h	± 1	± 0.1	± 0.01	± 0.001	± 0.0001
$\dfrac{f(0+h) - f(0)}{h}$	1	4.6416	21.544	100	464.16

suggesting that the difference quotients $[f(0 + h) - f(0)]/h$ grow large without bound rather than approach any fixed number as h approaches 0. (Can you see how the behavior of the difference quotients in the table is reflected in the graph?)

using *Technology*

If you try to graph the function $f(x) = x^{2/3}$ using the format

X^(2/3)

you may get only the right-hand portion of Figure 38 because graphing utilities are (often) not programmed to raise negative numbers to fractional exponents. (However, many will handle X^(1/3) correctly, as a special case they recognize.) To avoid this difficulty, you can take advantage of the identity

$$x^{2/3} = (x^2)^{1/3}$$

so that it is always a nonnegative number that is being raised to a fractional exponent. Thus, use the format

(X^2)^(1/3)

to obtain both portions of the graph.

Now we return to the function $h(x) = |x|$ discussed in Example 5 of Section 10.6. We can write

$$|x| = \begin{cases} -x & \text{if } x < 0 \\ x & \text{if } x > 0 \end{cases}$$

Hence, by the power rule (think of x as x^1):

$$f'(x) = \begin{cases} -1 & \text{if } x < 0 \\ 1 & \text{if } x > 0 \end{cases}$$

Q: *So does that mean there is no single formula for the derivative of |x|?*

A: Actually, there *is* a convenient formula. Consider the ratio

$$\frac{|x|}{x} \quad \blacksquare$$

If x is positive then $|x| = x$, so $|x|/x = x/x = 1$. On the other hand, if x is negative then $|x| = -x$, so $|x|/x = -x/x = -1$. In other words,

$$\frac{|x|}{x} = \begin{cases} -1 & \text{if } x < 0 \\ 1 & \text{if } x > 0 \end{cases}$$

which is exactly the formula we obtained for $f'(x)$. In other words:

Derivative of |x|

$$\frac{d}{dx}|x| = \frac{|x|}{x}$$

Note that the derivative does not exist when $x = 0$.

quick Example

$$\frac{d}{dx}[3|x| + x] = 3\frac{|x|}{x} + 1$$

An Application to Limits: L'Hospital's Rule (Optional)

The limits that caused us some trouble in Sections 10.1–10.3 are those of the form $\lim_{x \to a} f(x)$ in which we cannot just substitute $x = a$, such as

$$\lim_{x \to 2} \frac{x^3 - 8}{x - 2} \qquad \text{Substituting } x = 2 \text{ yields } \frac{0}{0}.$$

$$\lim_{x \to +\infty} \frac{2x - 4}{x - 1} \qquad \text{Substituting } x = +\infty \text{ yields } \frac{\infty}{\infty}.$$

L'Hospital's rule gives us an alternate way of computing limits such as these without the need to do any preliminary simplification. It also allows us to compute some limits for which algebraic simplification does not work.[67]

---

[67] Guillame François Antoine Marquis de L'Hospital (1661–1704) wrote the first textbook on calculus, *Analyse des infiniment petits pour l'intelligence des lignes courbes,* in 1692. The rule now known as L'Hospital's Rule appeared first in this book.

Theorem: L'Hospital's Rule

If f and g are two differentiable functions such that substituting $x = a$ in the expression $\dfrac{f(x)}{g(x)}$ gives either $\dfrac{0}{0}$ or $\dfrac{\infty}{\infty}$, then

$$\lim_{x \to a} \frac{f(x)}{g(x)} = \lim_{x \to a} \frac{f'(x)}{g'(x)}$$

That is, we can replace $f(x)$ and $g(x)$ with their *derivatives* and try again to take the limit.

quick Examples

1. Substituting $x = 2$ in $\dfrac{x^3 - 8}{x - 2}$ yields $\dfrac{0}{0}$. Therefore, L'Hospital's rule applies and

$$\lim_{x \to 2} \frac{x^3 - 8}{x - 2} = \lim_{x \to 2} \frac{3x^2}{1} = \frac{3(2)^2}{1} = 12$$

2. Substituting $x = +\infty$ in $\dfrac{2x - 4}{x - 1}$ yields $\dfrac{\infty}{\infty}$. Therefore, L'Hospital's rule applies and

$$\lim_{x \to +\infty} \frac{2x - 4}{x - 1} = \lim_{x \to +\infty} \frac{2}{1} = 2$$

The proof of L'Hospital's rule is beyond the scope of this text.[68]

Example 4 Applying L'Hospital's Rule

Check whether L'Hospital's rule applies to each of the following limits. If it does, use it to evaluate the limit. Otherwise, use some other method to evaluate the limit.

a. $\displaystyle\lim_{x \to 1} \frac{x^2 - 2x + 1}{4x^3 - 3x^2 - 6x + 5}$ **b.** $\displaystyle\lim_{x \to +\infty} \frac{2x^2 - 4x}{5x^3 - 3x + 5}$

c. $\displaystyle\lim_{x \to 1} \frac{x - 1}{x^3 - 3x^2 + 3x - 1}$ **d.** $\displaystyle\lim_{x \to 1} \frac{x}{x^3 - 3x^2 + 3x - 1}$

Solution

a. Setting $x = 1$ yields

$$\frac{1 - 2 + 1}{4 - 3 - 6 + 5} = \frac{0}{0}$$

Therefore L'Hospital's rule applies and

$$\lim_{x \to 1} \frac{x^2 - 2x + 1}{4x^3 - 3x^2 - 6x + 5} = \lim_{x \to 1} \frac{2x - 2}{12x^2 - 6x - 6}$$

---

[68] A proof of L'Hospital's rule can be found in most advanced calculus textbooks.

We are left with a closed-form function. However, we cannot substitute $x = 1$ to find the limit because the function $(2x - 2)/(12x^2 - 6x - 6)$ is still not defined at $x = 1$. In fact, if we set $x = 1$, we again get $0/0$. Thus, L'Hospital's rule applies again, and

$$\lim_{x \to 1} \frac{2x - 2}{12x^2 - 6x - 6} = \lim_{x \to 1} \frac{2}{24x - 6}$$

Once again we have a closed-form function, but this time it is defined when $x = 1$, giving

$$\frac{2}{24 - 6} = \frac{1}{9}$$

Thus,

$$\lim_{x \to 1} \frac{x^2 - 2x + 1}{4x^3 - 3x^2 - 6x + 5} = \frac{1}{9}$$

b. Setting $x = +\infty$ yields $\dfrac{\infty}{\infty}$, so

$$\lim_{x \to +\infty} \frac{2x^2 - 4x}{5x^3 - 3x + 5} = \lim_{x \to +\infty} \frac{4x - 4}{15x^2 - 3}$$

Setting $x = +\infty$ again yields $\dfrac{\infty}{\infty}$, so we can apply the rule again to obtain

$$\lim_{x \to +\infty} \frac{4x - 4}{15x^2 - 3} = \lim_{x \to +\infty} \frac{4}{30x}$$

Note that we cannot apply L'Hospital's rule a third time because setting $x = +\infty$ yields $4/\infty$. However, we can easily see now that the limit is zero. (Refer to Example 4(d) in Section 10.3.)

c. Setting $x = 1$ yields $0/0$ so, by L'Hospital's rule,

$$\lim_{x \to 1} \frac{x - 1}{x^3 - 3x^2 + 3x - 1} = \lim_{x \to 1} \frac{1}{3x^2 - 6x + 3}$$

We are left with a closed-form function that is still not defined at $x = 1$. Further, L'Hospital's rule no longer applies because putting $x = 1$ yields $1/0$. To investigate this limit, we must either graph it or create a table of values. If we use either method, we find that

$$\lim_{x \to 1} \frac{x - 1}{x^3 - 3x^2 + 3x - 1} = +\infty$$

d. Setting $x = 1$ in the expression yields $1/0$, so L'Hospital's rule does not apply here. If we graph the function or create a table of values, we find that the limit does not exist.

FAQs Using the Rules and Recognizing When a Function Is Not Differentiable

Q: I would like to say that the derivative of $5x^2 - 8x + 4$ is just $10x - 8$ without having to go through all that stuff about derivatives of sums and constant multiples. Can I simply forget about all the rules and write down the answer?

A: We developed the rules for sums and constant multiples precisely for that reason: so that we could simply write down a derivative without having to think about it too hard. So, you are perfectly justified in simply writing down the derivative without going through the rules, but bear in mind that what you are really doing is applying the power rule, the rule for sums, and the rule for multiples over and over. ∎

Q: Is there a way of telling from its formula whether a function f is not differentiable at a point?

A: Here are some indicators to look for in the formula for f:

• The absolute value of some expression; f may not be differentiable at points where that expression is zero.

 Example: $f(x) = 3x^2 - |x - 4|$ is not differentiable at $x = 4$.

• A fractional power smaller than 1 of some expression; f may not be differentiable at points where that expression is zero.

 Example: $f(x) = (x^2 - 16)^{2/3}$ is not differentiable at $x = \pm 4$. ∎

10.7 | EXERCISES

● denotes basic skills exercises
◆ denotes challenging exercises
tech Ex indicates exercises that should be solved using technology

*In Exercises 1–10, use the shortcut rules to **mentally** calculate the derivative of the given function.*

1. ● $f(x) = x^5$ *hint* [see Quick Examples on p. 758]

2. ● $f(x) = x^4$ **3.** ● $f(x) = 2x^{-2}$

4. ● $f(x) = 3x^{-1}$ **5.** ● $f(x) = -x^{0.25}$

6. ● $f(x) = -x^{-0.5}$ **7.** ● $f(x) = 2x^4 + 3x^3 - 1$

8. ● $f(x) = -x^3 - 3x^2 - 1$

9. ● $f(x) = -x + \dfrac{1}{x} + 1$ *hint* [see Example 1]

10. ● $f(x) = \dfrac{1}{x} + \dfrac{1}{x^2}$

In Exercises 11–16, obtain the derivative dy/dx and state the rules that you use. *hint* [see Example 2]

11. ● $y = 10$ **12.** ● $y = x^3$

13. ● $y = x^2 + x$ **14.** ● $y = x - 5$

15. ● $y = 4x^3 + 2x - 1$ **16.** ● $y = 4x^{-1} - 2x - 10$

In Exercises 17–40, find the derivative of each function. *hint* [see Example 2]

17. ● $f(x) = x^2 - 3x + 5$ **18.** ● $f(x) = 3x^3 - 2x^2 + x$

19. ● $f(x) = x + x^{0.5}$ **20.** ● $f(x) = x^{0.5} + 2x^{-0.5}$

21. ● $g(x) = x^{-2} - 3x^{-1} - 2;$ **22.** ● $g(x) = 2x^{-1} + 4x^{-2}$

23. ● $g(x) = \dfrac{1}{x} - \dfrac{1}{x^2}$ **24.** ● $g(x) = \dfrac{1}{x^2} + \dfrac{1}{x^3}$

25. ● $h(x) = \dfrac{2}{x^{0.4}}$ **26.** ● $h(x) = -\dfrac{1}{2x^{0.2}}$

27. ● $h(x) = \dfrac{1}{x^2} + \dfrac{2}{x^3}$ **28.** ● $h(x) = \dfrac{2}{x} - \dfrac{2}{x^3} + \dfrac{1}{x^4}$

29. ● $r(x) = \dfrac{2}{3x} - \dfrac{1}{2x^{0.1}}$ **30.** ● $r(x) = \dfrac{4}{3x^2} + \dfrac{1}{x^{3.2}}$

31. ● $r(x) = \dfrac{2x}{3} - \dfrac{x^{0.1}}{2} + \dfrac{4}{3x^{1.1}} - 2$

32. ● $r(x) = \dfrac{4x^2}{3} + \dfrac{x^{3.2}}{6} - \dfrac{2}{3x^2} + 4$

33. ● $t(x) = |x| + \dfrac{1}{x}$ **34.** ● $t(x) = 3|x| - \sqrt{x}$

● basic skills   ◆ challenging   tech Ex technology exercise

35. $s(x) = \sqrt{x} + \dfrac{1}{\sqrt{x}}$ **36.** $s(x) = x + \dfrac{7}{\sqrt{x}}$

[Hint: For Exercises 37–40: First expand the given function]

37. $s(x) = x\left(x^2 - \dfrac{1}{x}\right)$ **38.** $s(x) = x^{-1}\left(x - \dfrac{2}{x}\right)$

39. $t(x) = \dfrac{x^2 - 2x^3}{x}$ **40.** $t(x) = \dfrac{2x + x^2}{x}$

In Exercises 41–46, evaluate the given expression.

41. $\dfrac{d}{dx}(2x^{1.3} - x^{-1.2})$ **42.** $\dfrac{d}{dx}(2x^{4.3} + x^{0.6})$

43. $\dfrac{d}{dx}[1.2(x - |x|)]$ **44.** $\dfrac{d}{dx}[4(x^2 + 3|x|)]$

45. $\dfrac{d}{dt}(at^3 - 4at)$; ($a$ constant)

46. $\dfrac{d}{dt}(at^2 + bt + c)$ (a, b, c constant)

In Exercises 47–52, find the indicated derivative.

47. $y = \dfrac{x^{10.3}}{2} + 99x^{-1}; \dfrac{dy}{dx}$ **48.** $y = \dfrac{x^{1.2}}{3} - \dfrac{x^{0.9}}{2}; \dfrac{dy}{dx}$

49. $s = 2.3 + \dfrac{2.1}{t^{1.1}} - \dfrac{t^{0.6}}{2}; \dfrac{ds}{dt}$ **50.** $s = \dfrac{2}{t^{1.1}} + t^{-1.2}; \dfrac{ds}{dt}$

51. $V = \dfrac{4}{3}\pi r^3; \dfrac{dV}{dr}$ **52.** $A = 4\pi r^2; \dfrac{dA}{dr}$

In Exercises 53–58, find the slope of the tangent to the graph of the given function at the indicated point.

53. $f(x) = x^3; (-1, -1)$ **54.** $g(x) = x^4; (-2, 16)$

55. $f(x) = 1 - 2x; (2, -3)$ **56.** $f(x) = \dfrac{x}{3} - 1; (-3, -2)$

57. $g(t) = \dfrac{1}{t^5}; (1, 1)$ **58.** $s(t) = \dfrac{1}{t^3}; \left(-2, -\dfrac{1}{8}\right)$

In Exercises 59–64, find the equation of the tangent line to the graph of the given function at the point with the indicated x-coordinate. In each case, sketch the curve together with the appropriate tangent line.

59. $f(x) = x^3; x = -1$ **60.** $f(x) = x^2; x = 0$

61. $f(x) = x + \dfrac{1}{x}; x = 2$ **62.** $f(x) = \dfrac{1}{x^2}; x = 1$

63. $f(x) = \sqrt{x}; x = 4$ **64.** $f(x) = 2x + 4; x = -1$

In Exercises 65–70, find all values of x (if any) where the tangent line to the graph of the given equation is horizontal.

65. $y = 2x^2 + 3x - 1$ **66.** $y = -3x^2 - x$

67. $y = 2x + 8$ **68.** $y = -x + 1$

69. $y = x + \dfrac{1}{x}$ **70.** $y = x - \sqrt{x}$

71. ◆ Write out the proof that $\dfrac{d}{dx}(x^4) = 4x^3$.

72. ◆ Write out the proof that $\dfrac{d}{dx}(x^5) = 5x^4$.

tech Ex *In Exercises 73–76, use technology to graph the derivative of the given function for the given range of values of x. Then use your graph to estimate all values of x (if any) where (a) the given function is not differentiable, and (b) the tangent line to the graph of the given function is horizontal. Round answers to one decimal place.*

73. ● $h(x) = |x - 3|; -5 \le x \le 5$

74. **tech** Ex $h(x) = 2x + (x - 3)^{1/3}; -5 \le x \le 5$

75. **tech** Ex $f(x) = x - 5(x - 1)^{2/5}; -4 \le x \le 6$

76. **tech** Ex $f(x) = |2x + 5| - x^2; -4 \le x \le 4$

tech Ex *In Exercises 77–80, investigate the differentiability of the given function at the given points numerically (that is, use a table of values). If $f'(a)$ exists, give its approximate value.* hint *[see Example 3]*

77. **tech** Ex $f(x) = x^{1/3}$ **a.** $a = 1$ **b.** $a = 0$

78. **tech** Ex $f(x) = x + |1 - x|$ **a.** $a = 1$ **b.** $a = 0$

79. **tech** Ex $f(x) = [x(1 - x)]^{1/3}$ **a.** $a = 1$ **b.** $a = 0$

80. **tech** Ex $f(x) = (1 - x)^{2/3}$ **a.** $a = -1$ **b.** $a = 1$

In Exercises 81–92 say whether L'Hospital's rule applies. It is does, use it to evaluate the given limit. If not, use some other method.

81. ● $\displaystyle\lim_{x \to 1} \dfrac{x^2 - 2x + 1}{x^2 - x}$ **82.** ● $\displaystyle\lim_{x \to -1} \dfrac{x^2 + 3x + 2}{x^2 + x}$

83. ● $\displaystyle\lim_{x \to 2} \dfrac{x^3 - 8}{x - 2}$ **84.** ● $\displaystyle\lim_{x \to 0} \dfrac{x^3 + 8}{x^2 + 3x + 2}$

85. ● $\displaystyle\lim_{x \to 1} \dfrac{x^2 + 3x + 2}{x^2 + x}$ **86.** ● $\displaystyle\lim_{x \to -2} \dfrac{x^3 + 8}{x^2 + 3x + 2}$

87. ● $\displaystyle\lim_{x \to -\infty} \dfrac{3x^2 + 10x - 1}{2x^2 - 5x}$ **88.** ● $\displaystyle\lim_{x \to -\infty} \dfrac{6x^2 + 5x + 100}{3x^2 - 9}$

89. ● $\displaystyle\lim_{x \to -\infty} \dfrac{10x^2 + 300x + 1}{5x + 2}$ **90.** ● $\displaystyle\lim_{x \to -\infty} \dfrac{2x^4 + 20x^3}{1000x^3 + 6}$

91. ● $\displaystyle\lim_{x \to -\infty} \dfrac{x^3 - 100}{2x^2 + 500}$ **92.** ● $\displaystyle\lim_{x \to -\infty} \dfrac{x^2 + 30x}{2x^6 + 10x}$

Applications

93. ● *Collegiate Sports* The number of women's college soccer teams in the U.S. from 1982 to 2001 can be modeled by

$$s(t) = 1.52t^2 + 9.45t + 82.7 \qquad 2 \le t \le 20$$

where t is time in years since the 1980–1981 academic year.[69]

---

[69] The model is based on a quadratic regression. Source for data: N.C.A.A./ *New York Times,* May 9, 2002, p. D4.

● basic skills   ◆ challenging   **tech** Ex technology exercise

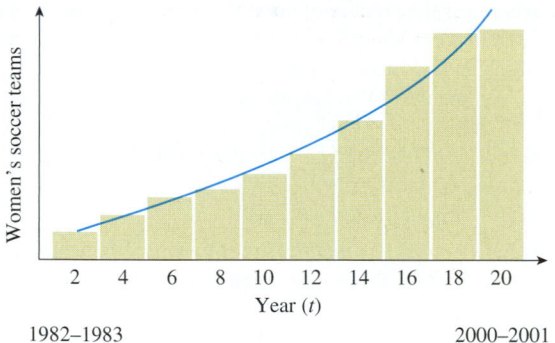

a. Find $s'(t)$.

b. How fast (to the nearest whole number) was the number of women's college soccer teams increasing in the 1994–1995 academic year?

94. ● **Collegiate Sports** The number of men's college wrestling teams in the U.S. from 1982 to 2001 can be modeled by

$$w(t) = 0.161t^2 - 9.75t + 360 \qquad 2 \leq t \leq 20$$

where t is time in years since the 1980–1981 academic year[70]

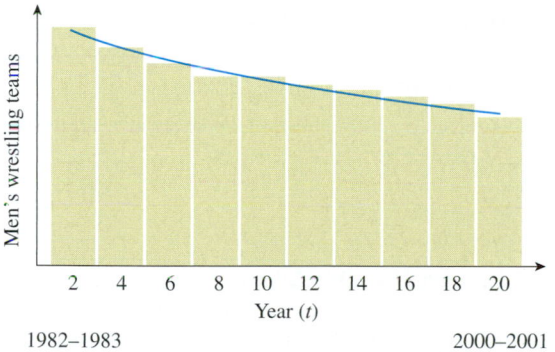

a. Find and graph $w'(t)$.

b. How fast (to the nearest whole number) was the number of men's college wrestling teams decreasing in the 1998–1999 academic year?

95. ● **Online Shopping** In January 2000–January 2002, the percentage of people in the U.S. who had ever purchased anything online can be approximated by

$$P(t) = -2.6t^2 + 13t + 19 \text{ percent} \quad (0 \leq t \leq 2)$$

where t is time in years since January 2000.[71] Find $P'(t)$. What does your answer tell you about online shopping transactions in January 2002?

96. ● **Online Shopping** The annual number of online shopping transactions in the U.S. for the period January 2000–January 2002 can be approximated by

$$N(t) = -180t^2 + 440t + 320 \text{ million transactions } (0 \leq t \leq 2)$$

where t is time in years since January 2000.[72] Find $N'(t)$. What does your answer tell you about online shopping transactions in January 2002?

97. ● **Food Versus Education** The following equation shows the approximate relationship between the percentage y of total personal consumption spent on food and the corresponding percentage x spent on education.[73]

$$y = \frac{35}{x^{0.35}} \text{ percentage points} \quad (6.5 \leq x \leq 17.5)$$

According to the model, spending on food is decreasing at a rate of _____ percentage points per one percentage point increase in spending on education when 10% of total consumption is spent on education. (Answer should be rounded to two significant digits.)

98. ● **Food Versus Recreation** The following equation shows the approximate relationship between the percentage y of total personal consumption spent on food and the corresponding percentage x spent on recreation.[74]

$$y = \frac{33}{x^{0.63}} \text{ percentage points} \quad (2.5 \leq x \leq 4.5)$$

According to the model, spending on food is decreasing at a rate of _____ percentage points per one percentage point increase in spending on recreation when 3% of total consumption is spent on recreation. (Answer should be rounded to two significant digits.)

99. ● **Velocity** If a stone is dropped from a height of 400 feet, its height s after t seconds is given by $s(t) = 400 - 16t^2$, with s in feet.

a. Compute $s'(t)$ and hence find its velocity at times $t = 0$, 1, 2, 3, and 4 seconds.

b. When does it reach the ground, and how fast is it traveling when it hits the ground?

100. ● **Velocity** If a stone is thrown down at 120 ft/sec from a height of 1000 feet, its height s after t seconds is given by $s(t) = 1000 - 120t - 16t^2$, with s in feet.

a. Compute $s'(t)$ and hence find its velocity at times $t = 0$, 1, 2, 3, and 4 seconds.

b. When does it reach the ground, and how fast is it traveling when it hits the ground?

---

[70] Ibid.

[71] Based on a regression model. (Second half of 2001 data was an estimate.) Source for data: Odyssey Research/*New York Times,* November 5, 2001, p. C1.

[72] Based on a regression model. (Second half of 2001 data was an estimate.) Source for data: Odyssey Research/*New York Times,* November 5, 2001, p. C1.

[73] Model based on historical and projected data from 1908–2010. SOURCES: Historical data, Bureau of Economic Analysis; projected data, Bureau of Labor Statistics/*New York Times,* December 1, 2003, p. C2.

[74] Ibid.

● basic skills   ◆ challenging   **tech** Ex technology exercise

101. ● *Currency* The value of the euro (€) since its introduction in January, 1999 can be approximated by

$$E(t) = 0.036t^2 - 0.10t + 1.0 \text{ U.S. dollars} \quad (-1 \le t \le 4.5)$$

where $t = 0$ represents January 2000.[75]

a. Compute $E'(t)$. How fast was the value of the euro changing in January, 2004?

b. According to the model, the value of the euro

   (A) increased at a faster and faster rate

   (B) increased at a slower and slower rate

   (C) decreased at a faster and faster rate

   (D) decreased at a slower and slower rate

from January 2000 to January 2001.

102. ● *Funding for the Arts* Total annual support for the arts in the U.S. by federal, state, and local government for the period January 1998 – January 2003 can be approximated by

$$f(t) = -0.028t^2 + 0.031t + 1.4 \text{ billion dollars} \quad (-2 \le t \le 3)$$

where $t = 0$ represents January 2000.[76]

a. Compute $f'(t)$. How fast was annual support for the arts changing in January, 2002?

b. According to the model, annual support for the arts

   (A) increased at a faster and faster rate

   (B) increased at a slower and slower rate

   (C) decreased at a faster and faster rate

   (D) decreased at a slower and slower rate

from January 1998 to January 2000.

103. ● *Ecology* Increasing numbers of manatees ("sea sirens") have been killed by boats off the Florida coast. The following graph shows the relationship between the number of boats registered in Florida and the number of manatees killed each year.

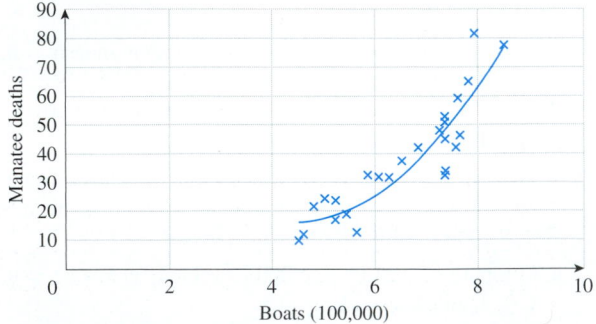

The regression curve shown is given by

$$f(x) = 3.55x^2 - 30.2x + 81 \quad (4.5 \le x \le 8.5)$$

where x is the number of boats (hundreds of thousands) registered in Florida in a particular year and $f(x)$ is the number of manatees killed by boats in Florida that year.[77]

a. Compute $f'(x)$. What are the units of measurement of $f'(x)$?

b. Is $f'(x)$ increasing or decreasing with increasing x? Interpret the answer.

c. Compute and interpret $f'(8)$.

104. ● *SAT Scores by Income* The following graph shows U.S. verbal SAT scores as a function of parents' income level.[78]

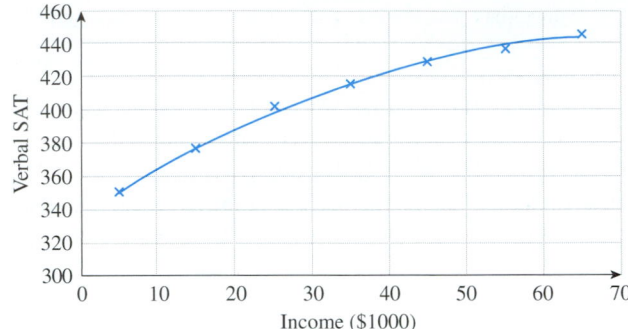

The regression curve shown is given by

$$f(x) = -0.021x^2 + 3.0x + 336 \quad (5 \le x \le 65)$$

where $f(x)$ is the average SAT verbal score of a student whose parents earn x thousand dollars per year.[79]

a. Compute $f'(x)$. What are the units of measurement of $f'(x)$?

b. Is $f'(x)$ increasing or decreasing with increasing x? Interpret the answer.

c. Compute and interpret $f'(30)$

105. *ISP Market Share* The following graph shows approximate market shares, in percentage points, of Microsoft's MSN Internet service provider, and the combined shares of MSN, Comcast, Earthlink, and AOL for the period 1999–2004.[80]

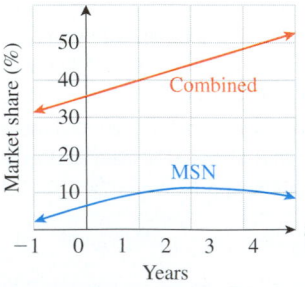

[75] SOURCES: Bank of England, Reuters, July, 2004.

[76] Based in a quadratic regression of original data. Figures are adjusted for inflation. SOURCES: Giving USA, The Foundation Center, Americans for the Arts/*New York Times,* June 19, 2004, p. B7.

[77] Regression model is based on data from 1976 to 2000. Sources for data: Florida Department of Highway Safety & Motor Vehicles, Florida Marine Institute/*New York Times,* February 12, 2002, p. F4.

[78] Based on 1994 data. SOURCE: The College Board/*New York Times,* March 5, 1995, p. E16.

[79] Ibid.

[80] The curves are regression models. Source for data: Solomon Research, Morgan Stanley/*New York Times,* July 19, 2004.

● basic skills   ◆ challenging   **tech Ex** technology exercise

Here, t is time in years since June, 2000. Let $c(t)$ be the combined market share at time t, and let $m(t)$ be MSN's share at time t.

a. What does the function $c(t) - m(t)$ measure? What does $c'(t) - m'(t)$ measure?

b. Based on the graphs shown, $c(t) - m(t)$ is
   (A) Increasing   **(B)** Decreasing
   (C) Increasing, then decreasing
   (D) Decreasing, then increasing
   on the interval [3, 4]

c. Based on the graphs shown, $c'(t) - m'(t)$ is
   (A) Positive   **(B)** Negative
   (C) Positive, then negative
   (D) Negative, then positive
   on the interval [3, 4].

d. The two market shares are approximated by
   MSN: $m(t) = -0.83t^2 + 3.8t + 6.8$ $(-1 \le t \le 4)$
   Combined: $c(t) = 4.2t + 36$ $(-1 \le t \le 4)$

   Compute $c'(2) - m'(2)$. Interpret your answer.

106. *ISP Revenue* The following graph shows the approximate total revenue, in millions of dollars, of Microsoft's MSN Internet service provider, as well as the portion of the revenue due to advertising for the period June, 2001–January, 2004.[81]

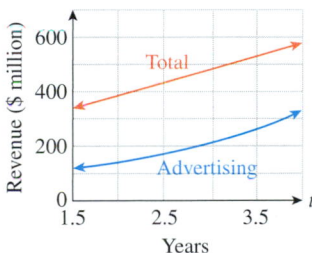

Years

Here, t is time in years since January, 2000. Let $s(t)$ be the total revenue at time t, and let $a(t)$ be revenue due to advertising at time t.

a. What does the function $s(t) - a(t)$ measure? What does $s'(t) - a'(t)$ measure?

b. Based on the graphs shown, $s(t) - a(t)$ is
   (A) Increasing   **(B)** Decreasing
   (C) Increasing, then decreasing
   (D) Decreasing, then increasing
   on the interval [2, 4].

c. Based on the graphs shown, $s'(t) - a'(t)$ is
   (A) Positive   **(B)** Negative
   (C) Positive, then negative
   (D) Negative, then positive
   on the interval [2, 4].

d. The two revenue curves are approximated by
   Advertising: $a(t) = 20t^2 - 27t + 120$ $(1.5 \le t \le 4)$
   Total: $s(t) = 96t + 190$ $(1.5 \le t \le 4)$

   Compute $a'(2)$, $s'(2)$, and hence $s'(2) - a'(2)$. Interpret your answer.

Communication and Reasoning Exercises

107. ● What instructions would you give to a fellow student who wanted to accurately graph the tangent line to the curve $y = 3x^2$ at the point $(-1, 3)$?

108. ● What instructions would you give to a fellow student who wanted to accurately graph a line at right angles to the curve $y = 4/x$ at the point where $x = 0.5$?

109. ● Consider $f(x) = x^2$ and $g(x) = 2x^2$. How do the slopes of the tangent lines of f and g at the same x compare?

110. ● Consider $f(x) = x^3$ and $g(x) = x^3 + 3$. How do the slopes of the tangent lines of f and g compare?

111. ● Suppose $g(x) = -f(x)$. How do the derivatives of f and g compare?

112. ● Suppose $g(x) = f(x) - 50$. How do the derivatives of f and g compare?

113. ● Following is an excerpt from your best friend's graded homework:

   $$3x^4 + 11x^5 = 12x^3 + 55x^4 \quad ✗ \quad \textbf{\textit{WRONG}} \quad \left(-8\right)$$

   Why was it marked wrong?

114. ● Following is an excerpt from your second best friend's graded homework:

   $$f(x) = \frac{3}{4x^2}; f'(x) = \frac{3}{8x} \quad ✗ \quad \textbf{\textit{WRONG}} \quad \left(-10\right)$$

   Why was it marked wrong?

115. ● Following is an excerpt from your worst enemy's graded homework:

   $$f(x) = 4x^2; f'(x) = (0)(2x) = 0 \quad ✗ \quad \textbf{\textit{WRONG}} \quad \left(-6\right)$$

   Why was it marked wrong?

116. How would you respond to an acquaintance who says, "I finally understand what the derivative is: it is nx^{n-1}! Why weren't we taught that in the first place instead of the difficult way using limits?"

117. Sketch the graph of a function whose derivative is undefined at exactly two points but which has a tangent line at all but one point.

118. Sketch the graph of a function that has a tangent line at each of its points, but whose derivative is undefined at exactly two points.

[81] The curves are regression models. Source for data: Solomon Research, Morgan Stanley/*New York Times,* July 19, 2004.

● basic skills   ◆ challenging   **tech** Ex technology exercise

10.8 A First Application: Marginal Analysis

In Chapter 1, we considered linear *cost functions* of the form $C(x) = mx + b$, where C is the total cost, x is the number of items, and m and b are constants. The slope m is the *marginal cost*. It measures the *cost of one more item*. Notice that the derivative of $C(x) = mx + b$ is $C'(x) = m$. In other words, for a linear cost function, *the marginal cost is the derivative of the cost function*.

In general, we make the following definition.

Marginal Cost

A **cost function** specifies the total cost C as a function of the number of items x. In other words, $C(x)$ is the total cost of x items. The **marginal cost function** is the derivative $C'(x)$ of the cost function $C(x)$. It measures the rate of change of cost with respect to x.

Units

The units of marginal cost are units of cost (dollars, say) per item.

Interpretation

We interpret $C'(x)$ as the approximate cost of one more item.[*]

quick Example   If $C(x) = 400x + 1000$ dollars, then the marginal cost function is $C'(x) = \$400$ per item (a constant).

---

[*] See Example 1 below.

Example 1 Marginal Cost

Suppose that the cost in dollars to manufacture portable CD players is given by

$$C(x) = 150{,}000 + 20x - 0.0001x^2$$

where x is the number of CD players manufactured.[†] Find the marginal cost function $C'(x)$ and use it to estimate the cost of manufacturing the 50,001st CD player.

Solution Since

$$C(x) = 150{,}000 + 20x - 0.0001x^2$$

[†] You might well ask where on earth this formula came from. There are two approaches to obtaining cost functions in real life: analytical and empirical. The analytical approach is to calculate the cost function from scratch. For example, in the above situation, we might have fixed costs of $150,000, plus a production cost of $20 per CD player. The term $0.0001x^2$ may reflect a cost saving for high levels of production, such as a bulk discount in the cost of electronic components. In the empirical approach, we first obtain the cost at several different production levels by direct observation. This gives several points on the (as yet unknown) cost versus production level graph. Then find the equation of the curve that best fits these points, usually using regression.

the marginal cost function is

$$C'(x) = 20 - 0.0002x$$

The units of $C'(x)$ are units of C (dollars) per unit of x (CD players). Thus, $C'(x)$ is measured in dollars per CD player.

The cost of the 50,001st CD player is the amount by which the total cost would rise if we increased production from 50,000 CD players to 50,001. Thus, we need to know the rate at which the total cost rises as we increase production. This rate of change is measured by the derivative, or marginal cost, which we just computed. At $x = 50,000$, we get

$$C'(50,000) = 20 - 0.0002(50,000) = \$10 \text{ per CD player}$$

In other words, we estimate that the 50,001st CD player will cost approximately \$10.

$+$ *Before we go on...* In Example 1, the marginal cost is really only an *approximation* to the cost of the 50,001st CD player:

$$C'(50,000) \approx \frac{C(50,001) - C(50,000)}{1} \qquad \text{Set } h = 1 \text{ in the definition of the derivative}$$

$$= C(50,001) - C(50,000)$$

$$= \text{Cost of the 50,001st CD player}$$

The exact cost of the 50,001st CD player is

$$C(50,001) - C(50,000) = [150,000 + 20(50,001) - 0.0001(50,001)^2]$$
$$- [150,000 + 20(50,000) - 0.0001(50,000)^2]$$
$$= \$9.9999$$

So, the marginal cost is a good approximation to the actual cost.

Graphically, we are using the tangent line to approximate the cost function near a production level of 50,000. Figure 39 shows the graph of the cost function together with the tangent line at $x = 50,000$. Notice that the tangent line is essentially indistinguishable from the graph of the function for some distance on either side of 50,000.

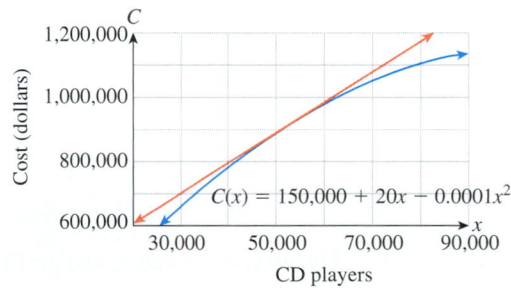

Figure **39**

Notes

1. In general, the difference quotient $[C(x + h) - C(x)]/h$ gives the **average cost per item** to produce h more items at a current production level of x items. (Why?)

2. Notice that $C'(x)$ is much easier to calculate than $[C(x + h) - C(x)]/h$. (Try it.) ∎

We can extend the idea of marginal cost to include other functions, like revenue and profit:

Marginal Revenue and Profit

A **revenue** or **profit function** specifies the total revenue R or profit P as a function of the number of items x. The derivatives, $R'(x)$ and $P'(x)$ of these functions are called the **marginal revenue** and **marginal profit** functions. They measure the rate of change of revenue and profit with respect to x.

Units
The units of marginal revenue and profit are the same as those of marginal cost: dollars (or euros, pesos, etc.) per item.

Interpretation
We interpret $R'(x)$ and $P'(x)$ as the approximate revenue and profit from the sale of one more item.

Example 2 Marginal Revenue and Profit

You operate an *iPod* customizing service (a typical customized iPod might have a custom color case with blinking lights and a personalized logo). The cost to refurbish x iPods in a month is calculated to be

$$C(x) = 0.25x^2 + 40x + 1000 \text{ dollars}$$

You charge customers \$80 per iPod for the work.

a. Calculate the marginal revenue and profit functions. Interpret the results.

b. Compute the revenue and profit, and also the marginal revenue and profit, if you have refurbished 20 units this month. Interpret the results.

c. For which value of x is the marginal profit is zero? Interpret your answer.

Solution

a. We first calculate the revenue and profit functions:

$$
\begin{aligned}
R(x) &= 80x && \text{Revenue} = \text{Price} \times \text{Quantity} \\
P(x) &= R(x) - C(x) && \text{Profit} = \text{Revenue} - \text{Cost} \\
&= 80x - (0.25x^2 + 40x + 1000) \\
P(x) &= -0.25x^2 + 40x - 1000
\end{aligned}
$$

The marginal revenue and profit functions are then the derivatives:

$$\text{Marginal revenue} = R'(x) = 80$$
$$\text{Marginal Profit} = P'(x) = -0.5x + 40$$

Interpretation: $R'(x)$ gives the approximate revenue from the refurbishing of one more item, and $P'(x)$ gives the approximate profit from the refurbishing of one more item. Thus, if x iPods have been refurbished in a month, you will earn a revenue of \$80 and make a profit of approximately \$$(-0.5x + 40)$ if you refurbish one more that month.

Notice that the marginal revenue is a constant, so you earn the same revenue ($80) for each iPod you refurbish. However, the marginal profit, $(-0.5x + 40)$, decreases as x increases, so your additional profit is about 50¢ less for each additional iPod you refurbish.

b. From part (a), the revenue, profit, marginal revenue, and marginal profit functions are

$$R(x) = 80x$$
$$P(x) = -0.25x^2 + 40x - 1000$$
$$R'(x) = 80$$
$$P'(x) = -0.5x + 40$$

Because you have refurbished $x = 20$ iPods this month, $x = 20$, so

$R(20) = 80(20) = \$1600$	Total revenue from 20 iPods
$P(20) = -0.25(20)^2 + 40(20) - 1000 = -\$300$	Total profit from 20 iPods
$R'(20) = \$80$ per unit	Approximate revenue from the 21st iPod
$P'(20) = -0.5(20) + 40 = \$30$ per unit	Approximate profit from the 21st iPod

Interpretation: If you refurbish 20 iPods in a month, you will earn a total revenue of $160 and a profit of $-$300 (indicating a loss of $300). Refurbishing one more iPod that month will earn you an additional revenue of $80 and an additional profit of about $30.

c. The marginal profit is zero when $P'(x) = 0$

$$-0.5x + 40 = 0$$
$$x = \frac{40}{0.5} = 80 \text{ iPods}$$

Thus, if you refurbish 80 iPods in a month, refurbishing one more will get you (approximately) zero additional profit. To understand this further, let us take a look at the graph of the profit function, shown in Figure 40.

Notice that the graph is a parabola (the profit function is quadratic) with vertex at the point $x = 80$, where $P'(x) = 0$, so the profit is a maximum at this value of x.

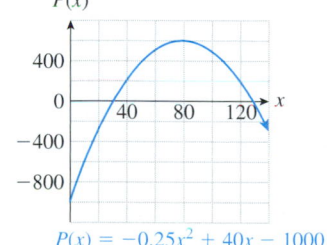

$P(x) = -0.25x^2 + 40x - 1000$

Figure **40**

$+$ *Before we go on...* In general, setting $P'(x) = 0$ and solving for x will always give the exact values of x for which the profit peaks as in Figure 40, assuming there is such a value. We recommend that you graph the profit function to check whether the profit is indeed a maximum at such a point. ∎

Example **3 Marginal Product**

A consultant determines that Precision Manufacturers' annual profit (in dollars) is given by

$$P(n) = -200{,}000 + 400{,}000n - 4600n^2 - 10n^3 \quad (10 \le n \le 50)$$

where n is the number of assembly-line workers it employs.

a. Compute $P'(n)$. $P'(n)$ is called the **marginal product** at the employment level of n assembly-line workers. What are its units?

b. Calculate $P(20)$ and $P'(20)$, and interpret the results.

c. Precision Manufacturers currently employs 20 assembly-line workers and is considering laying off some of them. What advice would you give the company's management?

Solution

a. Taking the derivative gives

$$P'(n) = 400{,}000 - 9200n - 30n^2$$

The units of $P'(n)$ are profit (in dollars) per worker.

b. Substituting into the formula for $P(n)$, we get

$$P(20) = -200{,}000 + 400{,}000(20) - 4600(20)^2 - 10(20)^3 = \$5{,}880{,}000$$

Thus, Precision Manufacturer will make an annual profit of $5,880,000 if it employs 20 assembly-line workers. On the other hand,

$$P'(20) = 400{,}000 - 9200(20) - 30(20)^2 = \$204{,}000/\text{worker}$$

Thus, at an employment level of 20 assembly-line workers, annual profit is increasing at a rate of $204,000 per additional worker. In other words, if the company were to employ one more assembly-line worker, its annual profit would increase by approximately $204,000.

c. Because the marginal product is positive, profits will increase if the company increases the number of workers and will decrease if it decreases the number of workers, so your advice would be to hire additional assembly-line workers. Downsizing their assembly-line workforce would reduce their annual profits.

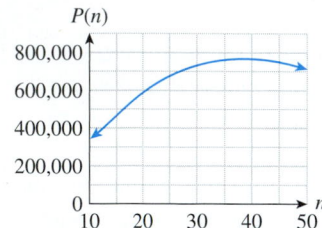

$P(n)$

Figure **41**

➕*Before we go on...* In Example 3, it would be interesting for Precision Manufacturers to ascertain how many additional assembly-line workers they should hire to obtain the *maximum* annual profit. Taking our cue from Example 2, we suspect that such a value of n would correspond to a point where $P'(n) = 0$. Figure 41 shows the graph of P, and on it we see that the highest point of the graph is indeed a point where the tangent line is horizontal; that is, $P'(n) = 0$, and occurs somewhere between $n = 35$ and 40.

To compute this value of n more accurately, set $P'(n) = 0$ and solve for n:

$$P'(n) = 400{,}000 - 9200n - 30n^2 = 0$$

or $\qquad 40{,}000 - 920n - 3n^2 = 0$

We can now obtain n using the quadratic formula:

$$n = \frac{-b \pm \sqrt{b^2 - 4ac}}{2a} = \frac{920 \pm \sqrt{920^2 - 4(-3)(40{,}000)}}{2(-3)}$$

$$= \frac{920 \pm \sqrt{1{,}326{,}400}}{-6} \approx -345.3 \text{ or } 38.6$$

The only meaningful solution is the positive one, $n \approx 38.6$ workers, and we conclude that the company should employ between 38 and 39 assembly-line workers for a maximum profit. To see which gives the larger profit, 38 or 39, we check:

$$P(38) = \$7,808,880$$

while

$$P(39) = \$7,810,210$$

This tells us that the company should employ 39 assembly-line workers for a maximum profit. Thus, instead of laying off any of its 20 assembly-line workers, the company should hire 19 additional assembly line workers for a total of 39. ∎

Average Cost

Example 4 Average Cost

Suppose the cost in dollars to manufacture portable CD players is given by

$$C(x) = 150,000 + 20x - 0.0001x^2$$

where x is the number of CD players manufactured. (This is the cost equation we saw in Example 1.)

a. Find the average cost per CD player if 50,000 CD players are manufactured.

b. Find a formula for the average cost per CD player if x CD players are manufactured. This function of x is called the **average cost function, $\bar{C}(x)$.**

Solution

a. The total cost of manufacturing 50,000 CD players is given by

$$
\begin{aligned}
C(50,000) &= 150,000 + 20(50,000) - 0.0001(50,000)^2 \\
&= \$900,000
\end{aligned}
$$

Because 50,000 CD players cost a total of $900,000 to manufacture, the average cost of manufacturing one CD player is this total cost divided by 50,000:

$$\bar{C}(50,000) = \frac{900,000}{50,000} = \$18.00 \text{ per CD player}$$

Thus, if 50,000 CD players are manufactured, each CD player costs the manufacturer an average of $18.00 to manufacture.

b. If we replace 50,000 by x, we get the general formula for the average cost of manufacturing x CD players:

$$
\begin{aligned}
\bar{C}(x) &= \frac{C(x)}{x} \\
&= \frac{1}{x}(150,000 + 20x - 0.0001x^2) \\
&= \frac{150,000}{x} + 20 - 0.0001x \qquad \text{Average cost function}
\end{aligned}
$$

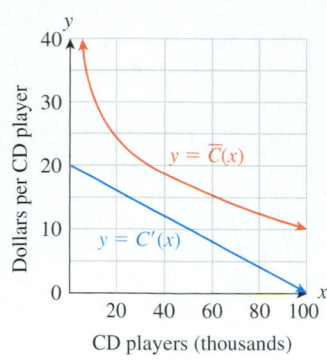

Figure **42**

+*Before we go on...* Average cost and marginal cost convey different but related information. The average cost $\bar{C}(50{,}000) = \$18$ that we calculated in Example 4 is the cost per item of manufacturing the first 50,000 CD players, whereas the marginal cost $C'(50{,}000) = \$10$ that we calculated in Example 1 gives the (approximate) cost of manufacturing the *next* CD player. Thus, according to our calculations, the first 50,000 CD players cost an average of \$18 to manufacture, but it costs only about \$10 to manufacture the next one. Note that the marginal cost at a production level of 50,000 CD players is lower than the average cost. This means that the average cost to manufacture CDs is going down with increasing volume. (Think about why.)

Figure 42 shows the graphs of average and marginal cost. Notice how the decreasing marginal cost seems to pull the average cost down with it. ∎

To summarize:

> ### Average Cost
>
> Given a cost function C, the **average cost** of the first x items is given by
>
> $$\bar{C}(x) = \frac{C(x)}{x}$$
>
> The average cost is distinct from the **marginal cost** $C'(x)$, which tells us the approximate cost of the *next* item.

quick Example

For the cost function $C(x) = 20x + 100$ dollars

$$\text{Marginal Cost} = C'(x) = \$20 \text{ per additional item}$$

$$\text{Average Cost} = \bar{C}(x) = \frac{C(x)}{x} = \frac{20x + 100}{x} = \$(20 + 100/x) \text{ per item}$$

10.8 EXERCISES

● denotes basic skills exercises

◆ denotes challenging exercises

In Exercises 1–4, for each cost function, find the marginal cost at the given production level x, and state the units of measurement. (All costs are in dollars.) hint [see Example 1]

1. ● $C(x) = 10{,}000 + 5x - 0.0001x^2;\ x = 1000$

2. ● $C(x) = 20{,}000 + 7x - 0.00005x^2;\ x = 10{,}000$

3. ● $C(x) = 15{,}000 + 100x + \dfrac{1000}{x};\ x = 100$

4. ● $C(x) = 20{,}000 + 50x + \dfrac{10{,}000}{x};\ x = 100$

In Exercises 5 and 6, find the marginal cost, marginal revenue, and marginal profit functions, and find all values of x for which the marginal profit is zero. Interpret your answer. hint [see Example 2]

5. ● $C(x) = 4x;\ R(x) = 8x - 0.001x^2$

6. ● $C(x) = 5x^2;\ R(x) = x^3 + 7x + 10$

7. A certain cost function has the following graph:

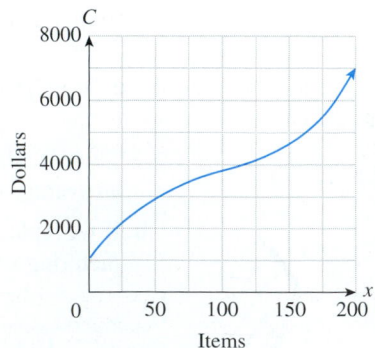

a. The associated marginal cost is
   (A) increasing, then decreasing.
   (B) decreasing, then increasing.
   (C) always increasing.
   (D) always decreasing.

b. The marginal cost is least at approximately
   (A) $x = 0$ **(B)** $x = 50$
   (C) $x = 100$ **(D)** $x = 150$
c. The cost of 50 items is
   (A) approximately $20, and increasing at a rate of about $3000 per item.
   (B) approximately $0.50, and increasing at a rate of about $3000 per item.
   (C) approximately $3000, and increasing at a rate of about $20 per item.
   (D) approximately $3000, and increasing at a rate of about $0.50 per item.

8. A certain cost function has the following graph:

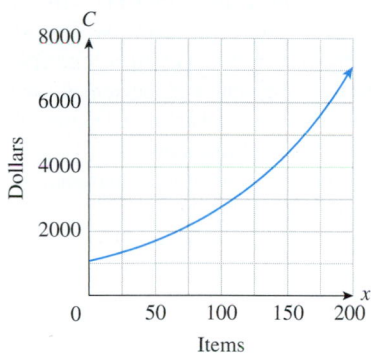

a. The associated marginal cost is
   (A) increasing, then decreasing.
   (B) decreasing, then increasing.
   (C) always increasing.
   (D) always decreasing.

b. When $x = 100$, the marginal cost is
   (A) greater than   **(B)** less than
   (C) approximately equal to the average cost.

c. The cost of 150 items is
   (A) approximately $4400, and increasing at a rate of about $40 per item.
   (B) approximately $40, and increasing at a rate of about $4400 per item.
   (C) approximately $4400, and increasing at a rate of about $1 per item.
   (D) approximately $1, and increasing at a rate of about $4400 per item.

Applications

9. ● *Advertising Costs* The cost, in thousands of dollars, of airing x television commercials during a Super Bowl game is given by[82]

$$C(x) = 150 + 2250x - 0.02x^2$$

[82] CBS charged an average of $2.25 million per 30-second television spot during the 2004 Super Bowl game. This explains the coefficient of x in the cost function. SOURCE: Advertising Age Research, www.AdAge.com/.

a. Find the marginal cost function and use it to estimate how fast the cost is increasing when $x = 4$. Compare this with the exact cost of airing the fifth commercial.
b. Find the average cost function $\bar{C}$, and evaluate $\bar{C}(4)$. What does the answer tell you? *hint* [see Example 1]

10. ● *Marginal Cost and Average Cost* The cost of producing x teddy bears per day at the Cuddly Companion Co. is calculated by their marketing staff to be given by the formula

$$C(x) = 100 + 40x - 0.001x^2$$

a. Find the marginal cost function and use it to estimate how fast the cost is going up at a production level of 100 teddy bears. Compare this with the exact cost of producing the 101st teddy bear.
b. Find the average cost function $\bar{C}$, and evaluate $\bar{C}(100)$. What does the answer tell you?

11. ● *Marginal Revenue and Profit* Your college newspaper, *The Collegiate Investigator,* sells for 90¢ per copy. The cost of producing x copies of an edition is given by

$$C(x) = 70 + 0.10x + 0.001x^2 \text{ dollars}$$

a. Calculate the marginal revenue and profit functions.
b. Compute the revenue and profit, and also the marginal revenue and profit, if you have produced and sold 500 copies of the latest edition. Interpret the results.
c. For which value of x is the marginal profit is zero? Interpret your answer. *hint* [see Example 2]

12. ● *Marginal Revenue and Profit* The Audubon Society at Enormous State University (ESU) is planning its annual fund-raising "Eatathon." The society will charge students $1.10 per serving of pasta. The society estimates that the total cost of producing x servings of pasta at the event will be

$$C(x) = 350 + 0.10x + 0.002x^2 \text{ dollars}$$

a. Calculate the marginal revenue and profit functions.
b. Compute the revenue and profit, and also the marginal revenue and profit, if you have produced and sold 200 servings of pasta. Interpret the results.
c. For which value of x is the marginal profit is zero? Interpret your answer.

13. ● *Marginal Profit* Suppose $P(x)$ represents the profit on the sale of x DVDs. If $P(1000) = 3000$ and $P'(1000) = -3$, what do these values tell you about the profit?

14. ● *Marginal Loss* An automobile retailer calculates that its loss on the sale of type M cars is given by $L(50) = 5000$ and $L'(50) = -200$, where $L(x)$ represents the loss on the sale of x type M cars. What do these values tell you about losses?

15. ● *Marginal Profit* Your monthly profit (in dollars) from selling magazines is given by

$$P = 5x + \sqrt{x}$$

where x is the number of magazines you sell in a month. If you are currently selling $x = 50$ magazines per month, find your profit and your marginal profit. Interpret your answers.

● basic skills   ◆ challenging

16. *Marginal Profit* Your monthly profit (in dollars) from your newspaper route is given by

$$P = 2n - \sqrt{n}$$

where n is the number of subscribers on your route. If you currently have 100 subscribers, find your profit and your marginal profit. Interpret your answers.

17. *Marginal Revenue: Pricing Tuna* Assume that the demand function for tuna in a small coastal town is given by

$$p = \frac{20,000}{q^{1.5}} \qquad (200 \le q \le 800)$$

where p is the price (in dollars) per pound of tuna, and q is the number of pounds of tuna that can be sold at the price p in one month.

a. Calculate the price that the town's fishery should charge for tuna in order to produce a demand of 400 pounds of tuna per month.

b. Calculate the monthly revenue R as a function of the number of pounds of tuna q.

c. Calculate the revenue and marginal revenue (derivative of the revenue with respect to q) at a demand level of 400 pounds per month, and interpret the results.

d. If the town fishery's monthly tuna catch amounted to 400 pounds of tuna, and the price is at the level in part (a), would you recommend that the fishery raise or lower the price of tuna in order to increase its revenue?

18. *Marginal Revenue: Pricing Tuna* Repeat Exercise 17, assuming a demand equation of

$$p = \frac{60}{q^{0.5}} \qquad (200 \le q \le 800)$$

19. ● *Marginal Product* A car wash firm calculates that its daily profit (in dollars) depends on the number n of workers it employs according to the formula

$$P = 400n - 0.5n^2$$

Calculate the marginal product at an employment level of 50 workers, and interpret the result. *hint* [see Example 3]

20. ● *Marginal Product* Repeat the preceding exercise using the formula

$$P = -100n + 25n^2 - 0.005n^4$$

21. ● *Average and Marginal Cost* The daily cost to manufacture generic trinkets for gullible tourists is given by the cost function

$$C(x) = -0.001x^2 + 0.3x + 500 \text{ dollars}$$

where x is the number of trinkets. *hint* [see Example 4]
a. As x increases, the marginal cost
   (A) increases   **(B)** decreases
   (C) increases, then decreases.
   (D) decreases, then increases.

b. As x increases, the average cost
   (A) increases   **(B)** decreases
   (C) increases, then decreases.
   (D) decreases, then increases.
c. The marginal cost is
   (A) greater than   **(B)** equal to   **(C)** less than
   the average cost when $x = 100$.

22. ● *Average and Marginal Cost* Repeat Exercise 21, using the following cost function for imitation oil paintings (x is the number of "oil paintings" manufactured):

$$C(x) = 0.1x^2 - 3.5x + 500 \text{ dollars}$$

23. ● *Advertising Cost* Your company is planning to air a number of television commercials during the ABC Television Network's presentation of the Academy Awards. ABC is charging your company $1.6 million per 30 second spot.[83] Additional fixed costs (development and personnel costs) amount to $500,000, and the network has agreed to provide a discount of $10,000\sqrt{x}$ for x television spots.

a. Write down the cost function C, marginal cost function C', and average cost function $\bar{C}$.

b. Compute $C'(3)$ and $\bar{C}(3)$. (Round all answers to three significant digits.) Use these two answers to say whether the average cost is increasing or decreasing as x increases.

24. ● *Housing Costs* The cost C of building a house is related to the number k of carpenters used and the number x of electricians used by the formula[84]

$$C = 15,000 + 50k^2 + 60x^2$$

a. Assuming that 10 carpenters are currently being used, find the cost function C, marginal cost function C', and average cost function $\bar{C}$, all as functions of x.

b. Use the functions you obtained in part (a) to compute $C'(15)$ and $\bar{C}(15)$. Use these two answers to say whether the average cost is increasing or decreasing as the number of electricians increases.

25. *Emission Control* The cost of controlling emissions at a firm rises rapidly as the amount of emissions reduced increases. Here is a possible model:

$$C(q) = 4000 + 100q^2$$

where q is the reduction in emissions (in pounds of pollutant per day) and C is the daily cost (in dollars) of this reduction.

a. If a firm is currently reducing its emissions by 10 pounds each day, what is the marginal cost of reducing emissions further?

---

[83] ABC charged an average of $1.6 million for a 30-second spot during the 2005 Academy Awards presentation. Source: CNN/Reuters, www.cnn.com/, February 9, 2005.

[84] Based on an exercise in *Introduction to Mathematical Economics* by A. L. Ostrosky, Jr., and J. V. Koch (Waveland Press, Prospect Heights, Illinois, 1979).

b. Government clean-air subsidies to the firm are based on the formula

$$S(q) = 500q$$

where q is again the reduction in emissions (in pounds per day) and S is the subsidy (in dollars). At what reduction level does the marginal cost surpass the marginal subsidy?

c. Calculate the net cost function, $N(q) = C(q) - S(q)$, given the cost function and subsidy above, and find the value of q that gives the lowest net cost. What is this lowest net cost? Compare your answer to that for part (b) and comment on what you find.

26. *Taxation Schemes* Here is a curious proposal for taxation rates based on income:

$$T(i) = 0.001i^{0.5}$$

where i represents total annual income in dollars and $T(i)$ is the income tax rate as a percentage of total annual income. (Thus, for example, an income of $50,000 per year would be taxed at about 22%, while an income of double that amount would be taxed at about 32%.)[85]

a. Calculate the after-tax (net) income $N(i)$ an individual can expect to earn as a function of income i.

b. Calculate an individual's marginal after-tax income at income levels of $100,000 and $500,000.

c. At what income does an individual's marginal after-tax income become negative? What is the after-tax income at that level, and what happens at higher income levels?

d. What do you suspect is the most anyone can earn after taxes? (See the footnote.)

27. *Fuel Economy* Your Porsche's gas mileage (in miles per gallon) is given as a function $M(x)$ of speed x in miles per hour. It is found that

$$M'(x) = \frac{3600x^{-2} - 1}{(3600x^{-1} + x)^2}$$

Estimate $M'(10)$, $M'(60)$, and $M'(70)$. What do the answers tell you about your car?

28. *Marginal Revenue* The estimated marginal revenue for sales of ESU soccer team T-shirts is given by

$$R'(p) = \frac{(8 - 2p)e^{-p^2 + 8p}}{10,000,000}$$

where p is the price (in dollars) that the soccer players charge for each shirt. Estimate $R'(3)$, $R'(4)$, and $R'(5)$. What do the answers tell you?

29. ◆ *Marginal Cost (from the GRE Economics Test)* In a multiplant firm in which the different plants have different and continuous cost schedules, if costs of production for a given output level are to be minimized, which of the following is essential?

(A) Marginal costs must equal marginal revenue.
(B) Average variable costs must be the same in all plants.
(C) Marginal costs must be the same in all plants.
(D) Total costs must be the same in all plants.
(E) Output per worker per hour must be the same in all plants.

30. ◆ *Study Time (from the GRE economics test)* A student has a fixed number of hours to devote to study and is certain of the relationship between hours of study and the final grade for each course. Grades are given on a numerical scale (e.g., 0 to 100), and each course is counted equally in computing the grade average. In order to maximize his or her grade average, the student should allocate these hours to different courses so that

(A) the grade in each course is the same.
(B) the marginal product of an hour's study (in terms of final grade) in each course is zero.
(C) the marginal product of an hour's study (in terms of final grade) in each course is equal, although not necessarily equal to zero.
(D) the average product of an hour's study (in terms of final grade) in each course is equal.
(E) the number of hours spent in study for each course are equal.

31. ◆ *Marginal Product (from the GRE economics test)* Assume that the marginal product of an additional senior professor is 50% higher than the marginal product of an additional junior professor and that junior professors are paid one-half the amount that senior professors receive. With a fixed overall budget, a university that wishes to maximize its quantity of output from professors should do which of the following?

(A) Hire equal numbers of senior professors and junior professors.
(B) Hire more senior professors and junior professors.
(C) Hire more senior professors and discharge junior professors.
(D) Discharge senior professors and hire more junior professors.
(E) Discharge all senior professors and half of the junior professors.

32. ◆ *Marginal Product (Based on a Question from the GRE Economics Test)* Assume that the marginal product of an additional senior professor is twice the marginal product of an additional junior professor and that junior professors are paid two-thirds the amount that senior professors receive. With a fixed overall budget, a university that wishes to maximize its quantity of output from professors should do which of the following?

(A) Hire equal numbers of senior professors and junior professors.
(B) Hire more senior professors and junior professors.
(C) Hire more senior professors and discharge junior professors.

---

[85] This model has the following interesting feature: an income of a million dollars per year would be taxed at 100%, leaving the individual penniless!

● basic skills   ◆ challenging

(D) Discharge senior professors and hire more junior professors.

(E) Discharge all senior professors and half of the junior professors.

Communication and Reasoning Exercises

33. ● The marginal cost of producing the 1001st item is

(A) Equal to
(B) Approximately equal to
(C) Always slightly greater than

the actual cost of producing the 1001st item.

34. ● For the cost function $C(x) = mx + b$, the marginal cost of producing the 1001st item is

(A) Equal to
(B) Approximately equal to
(C) Always slightly greater than

the actual cost of producing the 1001st item.

35. ● What is a cost function? Carefully explain the difference between *average cost* and *marginal cost* in terms of **a.** their mathematical definition, **b.** graphs, and **c.** interpretation.

36. ● The cost function for your grand piano manufacturing plant has the property that $\bar{C}(1000) = \$3000$ per unit and $C'(1000) = \$2500$ per unit. Will the average cost increase or decrease if your company manufactures a slightly larger number of pianos? Explain your reasoning.

37. ● If the average cost to manufacture one grand piano increases as the production level increases, which is greater, the marginal cost or the average cost?

38. ● If your analysis of a manufacturing company yielded positive marginal profit but negative profit at the company's current production levels, what would you advise the company to do?

39. If the marginal cost is decreasing, is the average cost necessarily decreasing? Explain.

40. If the average cost is decreasing, is the marginal cost necessarily decreasing?

41. ◆ If a company's marginal average cost is zero at the current production level, positive for a slightly higher production level, and negative for a slightly lower production level, what should you advise the company to do?

42. ◆ The **acceleration** of cost is defined as the derivative of the marginal cost function: that is, the derivative of the derivative—or *second derivative*—of the cost function. What are the units of acceleration of cost, and how does one interpret this measure?

● basic skills   ◆ challenging

Chapter **10** Review

KEY CONCEPTS

10.1 Limits: Numerical and Graphical Approaches

$\lim_{x \to a} f(x) = L$ means that $f(x)$ approaches L as x approaches a. *p. 688*

What it means for a limit to exist *p. 688*

Limits at infinity *p. 690*

Estimating limits graphically *p. 691*

Interpreting limits in real-world situations *p. 694*

10.2 Limits and Continuity

f is continuous at a if $\lim_{x \to a} f(x)$ exists and $\lim_{x \to a} f(x) = f(a)$. *p. 698*

Discontinuous, continuous on domain *p. 698*

Determining whether a given function is continuous *p. 699*

10.3 Limits and Continuity Algebraic Approach

Closed-form function *p. 704*

Limits of closed form functions *p. 705*

Simplifying to obtain limits *p. 705*

Limits of piecewise defined functions *p. 706*

Limits at infinity *p. 707*

10.4 Average Rate of Change

Average rate of change of $f(x)$ over

$[a, b]$: $\dfrac{\Delta f}{\Delta x} = \dfrac{f(b) - f(a)}{b - a}$ *p. 715*

Average rate of change as slope of the secant line *p. 715*

Computing the average rate of change from a graph *p. 716*

Computing the average rate of change from a formula *p. 718*

Computing the average rate of change over short intervals $[a, a + h]$ *p. 719*

10.5 The Derivative: Numerical and Graphical Viewpoints

Instantaneous rate of change of $f(x)$ (derivative of f at a);

$f'(a) = \lim_{h \to 0} \dfrac{f(a + h) - f(a)}{h}$ *p. 728*

The derivative as slope of the tangent line *p. 731*

Quick approximation of the derivative *p. 732*

Leibniz d notation *p. 734*

The derivative as velocity *p. 735*

Average and instantaneous velocity *p. 736*

The derivative function *p. 737*

Graphing the derivative function with technology *p. 737*

10.6 The Derivative: Algebraic Viewpoint

Derivative at the point $x = a$:

$f'(a) = \lim_{h \to 0} \dfrac{f(a + h) - f(a)}{h}$ *p. 749*

Derivative function:

$f'(x) = \lim_{h \to 0} \dfrac{f(x + h) - f(x)}{h}$ *p. 750*

Examples of the computation of $f'(x)$ *p. 750*

$f(x) = |x|$ is not differentiable at $x = 0$. *p. 753*

10.7 Derivatives of Powers, Sums and Constant Multiples

Power Rule: If n is any constant and $f(x) = x^n$, then $f'(x) = nx^{n-1}$ *p. 758*

Using the power rule for negative and fractional exponents *p. 758*

$\dfrac{d}{dx}$ Notation *p. 759*

Sums, Differences, and Constant Multiples *p. 760*

Combining the rules *p. 762*

$\dfrac{d}{dx}(cx) = c$, $\dfrac{d}{dx}(c) = 0$ *p. 763*

$f(x) = x^{1/3}$ and $g(x) = x^{2/3}$ are not differentiable at $x = 0$. *p. 763*

Derivative of $f(x) = |x|$:

$\dfrac{d}{dx}|x| = \dfrac{|x|}{x}$ *p. 764*

10.8 A First Application: Marginal Analysis

Marginal cost function $C'(x)$ *p. 772*

Marginal revenue and profit functions $R'(x)$ and $P'(x)$ *p. 774*

What it means when the marginal profit is zero *p. 774*

Marginal product *p. 775*

Average cost of the first x items:

$\bar{C}(x) = \dfrac{C(x)}{x}$ *p. 778*

REVIEW EXERCISES

Numerically *estimate whether the limits in Exercises 1–4 exist. If a limit does exist, give its approximate value.*

1. $\lim_{x \to 3} \dfrac{x^2 - x - 6}{x - 3}$

2. $\lim_{x \to 3} \dfrac{x^2 - 2x - 6}{x - 3}$

3. $\lim_{x \to -1} \dfrac{|x + 1|}{x^2 - x - 2}$

4. $\lim_{x \to -1} \dfrac{|x + 1|}{x^2 + x - 2}$

In Exercises 5 and 6, the graph of a function f is shown. Graphically determine whether the given limits exist. If a limit does exist, give its approximate value.

5.

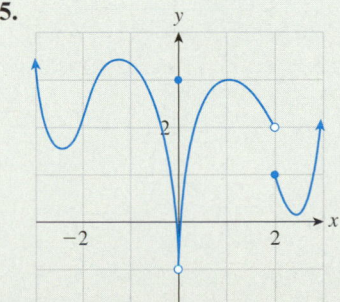

a. $\lim_{x \to 0} f(x)$

b. $\lim_{x \to 1} f(x)$

c. $\lim_{x \to 2} f(x)$

6.

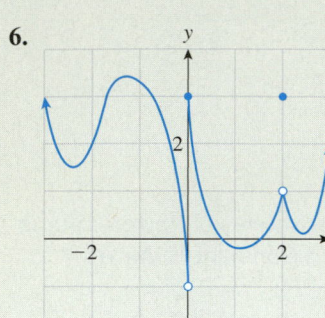

    a. $\lim\limits_{x\to 0} f(x)$

    b. $\lim\limits_{x\to -2} f(x)$

    c. $\lim\limits_{x\to 2} f(x)$

Calculate the limits in Exercises 7–12 algebraically. If a limit does not exist, say why.

7. $\lim\limits_{x\to -2} \dfrac{x^2}{x-3}$　　　　**8.** $\lim\limits_{x\to 3} \dfrac{x^2-9}{2x-6}$

9. $\lim\limits_{x\to 0} \dfrac{x}{2x^2-x}$　　　　**10.** $\lim\limits_{x\to 1} \dfrac{x^2-9}{x-1}$

11. $\lim\limits_{x\to -\infty} \dfrac{x^2-x-6}{x-3}$　　**12.** $\lim\limits_{x\to \infty} \dfrac{x^2-x-6}{4x^2-3}$

In Exercises 13–16, find the average rate of change of the given function over the interval [a, a + h] for h = 1, 0.01, and 0.001. (Round answers to four decimal places.) Then estimate the slope of the tangent line to the graph of the function at a.

13. $f(x) = \dfrac{1}{x+1}; a = 0$　　**14.** $f(x) = x^x; a = 2$

15. $f(x) = e^{2x}; a = 0$　　　**16.** $f(x) = \ln(2x); a = 1$

In Exercises 17–20 you are given the graph of a function with four points marked. Determine at which (if any) of these points the derivative of the function is: **(i)** -1 **(ii)** 0 **(iii)** 1, and **(iv)** 2.

17.
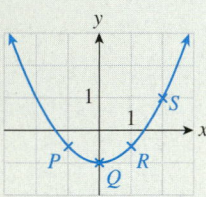

18.
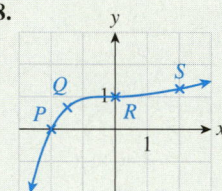

19.
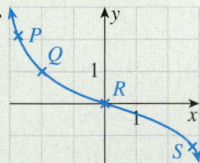

20.
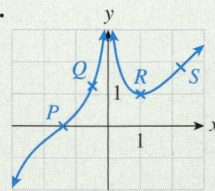

21. Let f have the graph shown.

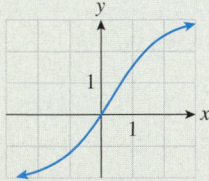

Select the correct answer.

    a. The average rate of change of f over the interval $[0, 2]$ is
      (A) greater than　　　　**(B)** less than
      (C) approximately equal to $f'(0)$.

    b. The average rate of change of f over the interval $[-1, 1]$ is
      (A) greater than　　　　**(B)** less than
      (C) approximately equal to $f'(0)$.

    c. Over the interval $[0, 2]$, the instantaneous rate of change of f is
      (A) increasing　　　　　　**(B)** decreasing
      (C) neither increasing nor decreasing

    d. Over the interval $[-2, 2]$, the instantaneous rate of change of f is
      (A) increasing, then decreasing
      (B) decreasing, then increasing
      (C) approximately constant

    e. When $x = 2$, $f(x)$ is
      (A) approximately 1 and increasing at a rate of about 2.5 units per unit of x
      (B) approximately 1.2 and increasing at a rate of about 1 unit per unit of x
      (C) approximately 2.5 and increasing at a rate of about 0.5 units per unit of x
      (D) approximately 2.5 and increasing at a rate of about 2.5 units per unit of x

22. Let f have the graph shown.

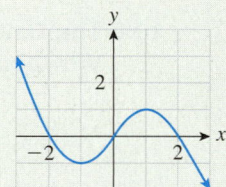

Select the correct answer.

    a. The average rate of change of f over the interval $[0, 1]$ is
      (A) greater than　　　　**(B)** less than
      (C) approximately equal to $f'(0)$.

    b. The average rate of change of f over the interval $[0, 2]$ is
      (A) greater than　　　　**(B)** less than
      (C) approximately equal to $f'(1)$.

    c. Over the interval $[-2, 0]$, the instantaneous rate of change of f is
      (A) increasing　　　　　　**(B)** decreasing
      (C) neither increasing nor decreasing

    d. Over the interval $[-2, 2]$, the instantaneous rate of change of f is
      (A) increasing, then decreasing
      (B) decreasing, then increasing
      (C) approximately constant

    e. When $x = 0$, $f(x)$ is
      (A) approximately 0 and increasing at a rate of about 1.5 units per unit of x

(B) approximately 0 and decreasing at a rate of about 1.5 units per unit of x

(C) approximately 1.5 and neither increasing nor decreasing

(D) approximately 0 and neither increasing nor decreasing

In Exercises 23–26, use the definition of the derivative to calculate the derivative of each of the given functions algebraically.

23. $f(x) = x^2 + x$

24. $f(x) = 3x^2 - x + 1$

25. $f(x) = 1 - \dfrac{2}{x}$

26. $f(x) = \dfrac{1}{x} + 1$

In Exercises 27–30, find the derivative of the given function.

27. $f(x) = 10x^5 + \dfrac{1}{2}x^4 - x + 2$

28. $f(x) = \dfrac{10}{x^5} + \dfrac{1}{2x^4} - \dfrac{1}{x} + 2$

29. $f(x) = 3x^3 + 3\sqrt[3]{x}$

30. $f(x) = \dfrac{2}{x^{2.1}} - \dfrac{x^{0.1}}{2}$

In Exercises 31–34, evaluate the given expressions.

31. $\dfrac{d}{dx}\left(x + \dfrac{1}{x^2}\right)$

32. $\dfrac{d}{dx}\left(2x - \dfrac{1}{x}\right)$

33. $\dfrac{d}{dx}\left(\dfrac{4}{3x} - \dfrac{2}{x^{0.1}} + \dfrac{x^{1.1}}{3.2} - 4\right)$

34. $\dfrac{d}{dx}\left(\dfrac{4}{x} + \dfrac{x}{4} - |x|\right)$

tech Ex *In Exercises 35–38, use technology to graph the derivative of the given function. In each case, choose a range of x-values and y-values that shows the interesting features of the graph.*

35. $f(x) = 10x^5 + \dfrac{1}{2}x^4 - x + 2$

36. $f(x) = \dfrac{10}{x^5} + \dfrac{1}{2x^4} - \dfrac{1}{x} + 2$

37. $f(x) = 3x^3 + 3\sqrt[3]{x}$

38. $f(x) = \dfrac{2}{x^{2.1}} - \dfrac{x^{0.1}}{2}$

Applications

39. OHaganBooks.com CEO John O'Hagan has terrible luck with stocks. The following graph shows the value of Fly-By-Night Airlines stock that he bought acting on a "hot tip" from Marjory Duffin (CEO of Duffin Press and a close business associate):

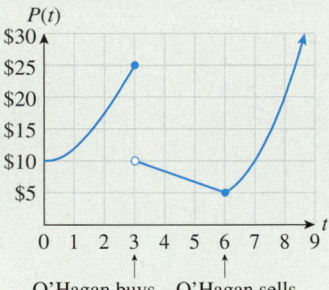

Fly-by-night stock

a. Compute $P(3)$, $\lim_{t\to 3^-} P(t)$ and $\lim_{t\to 3^+} P(t)$. Does $\lim_{t\to 3} P(t)$ exist? Interpret your answers in terms of Fly-By-Night stocks.

b. Is P continuous at $t = 6$? Is P differentiable at $t = 6$? Interpret your answers in terms of Fly-By-Night stocks.

40. *Advertising Costs* OHaganBooks.com has (on further advice from Marjory Duffin) mounted an aggressive online marketing strategy. The following graph shows the weekly cost of this campaign for the six-week period since the start of July (t is time in weeks):

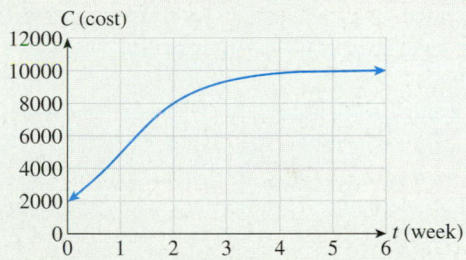

a. Assuming the trend shown in the graph were to continue indefinitely, estimate $\lim_{t\to 2} C(t)$ and $\lim_{t\to +\infty} C(t)$ and interpret the results.

b. Estimate $\lim_{t\to +\infty} C'(t)$ and interpret the result.

41. *Sales* Since the start of July, OHaganBooks.com has seen its weekly sales increase, as shown in the following table:

Week	1	2	3	4	5	6
Sales (Books)	6500	7000	7200	7800	8500	9000

a. What was the average rate of increase of weekly sales over this entire period?

b. During which 1-week interval(s) did the rate of increase of sales exceed the average rate?

c. During which 2-week interval(s) did the weekly sales rise at the highest average rate, and what was that average rate?

42. *Advertising Costs* The following graph (see Exercise 40) shows the weekly cost of OHaganBooks.com's online ad campaign for the six-week period since the start of July (*t* is time in weeks).

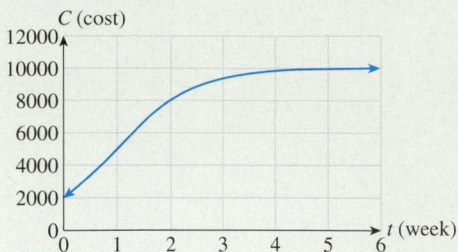

Use the graph to answer the following questions:

a. What was the average rate of change of cost over the entire six-week period?

b. What was the average rate of change of cost over the period [2, 6]?

c. Which of the following is correct? Over the period [2, 6],

    (A) The rate of change of cost increased and the cost increased

    (B) The rate of change of cost decreased and the cost increased

    (C) The rate of change of cost increased and the cost decreased

    (D) The rate of change of cost decreased and the cost decreased

43. *Sales* OHaganBooks.com fits the cubic curve

$$w(t) = -3.7t^3 + 74.6t^2 + 135.5t + 6300$$

to its weekly sales figures from Exercise 41, as shown in the following graph:

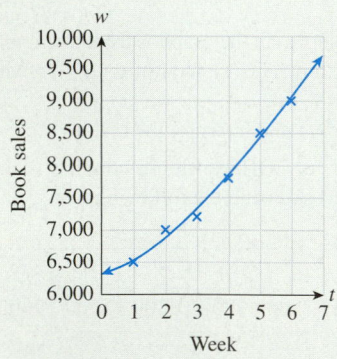

a. According to the cubic model, what was the rate of increase of sales at the beginning of the second week (*t* = 1)? (Round your answer to the nearest unit.)

b. If we extrapolate the model, what would be the rate of increase of weekly sales at the beginning of the 8th week (*t* = 7)?

c. Graph the function *w* for $0 \leq t \leq 20$. Would it be realistic to use the function to predict sales through week 20? Why?

44. tech Ex **Sales** OHaganBooks.com decided that the cubic curve in Exercise 43 was not suitable for extrapolation, so instead it tried

$$s(t) = 6053 + \frac{4474}{1 + e^{-0.55(t-4.8)}}$$

which is shown in the following graph:

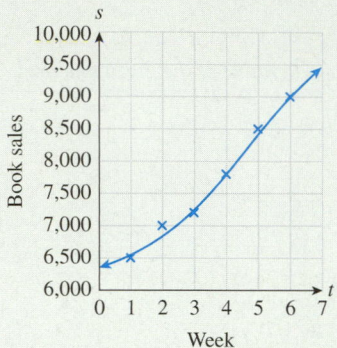

a. Using this function, estimate the rate of increase of weekly sales at the beginning of the 7th week (*t* = 6). (Round your answer to the nearest unit.)

b. If we extrapolate the model, what would be the rate of increase of weekly sales at the beginning of the 15th week (*t* = 14)?

c. Graph the function *s* for $0 \leq t \leq 20$. What is the long-term prediction for weekly sales? What is the long-term prediction for the rate of change of weekly sales?

45. As OHaganBooks.com's sales increase, so do its costs. If we take into account volume discounts from suppliers and shippers, the weekly cost of selling *x* books is

$$C(x) = -0.00002x^2 + 3.2x + 5400 \text{ dollars}$$

a. What is the marginal cost at a sales level of 8000 books per week?

b. What is the average cost per book at a sales level of 8000 books per week?

c. What is the marginal average cost at a sales level of 8000 books per week?

d. Interpret the results of parts (a)–(c).

CASE STUDY: Reducing Sulfur Emissions

The Environmental Protection Agency (EPA) wishes to formulate a policy that will encourage utilities to reduce sulfur emissions. Its goal is to reduce annual emissions of sulfur dioxide by a total of 10 million tons from the current level of 25 million tons by imposing a fixed charge for every ton of sulfur released into the environment per year. As a consultant to the EPA, you must determine the amount to be charged per ton of sulfur emissions.

You have the following data, which show the marginal cost to the utility industry of reducing sulfur emissions at several levels of reduction.[86]

Reduction (millions of tons)	8	10	12
Marginal Cost ($ per ton)	270	360	779

If $C(q)$ is the cost of removing q tons of sulfur dioxide, the table tells you that $C'(8,000,000) = \$270$ per ton, $C'(10,000,000) = \$360$ per ton, and $C'(12,000,000) = \$779$ per ton. Recalling that $C'(q)$ is the slope of the tangent to the graph of the cost function, you can see from the table that this slope is positive and increasing as q increases, so the graph of this cost function has the general shape shown in Figure 42.

(Notice that the slope is increasing as you move to the right.) Thus, the utility industry has no cost incentive to reduce emissions. What you would like to do—if the goal of reducing total emissions by 10 million tons is to be reached—is to alter this cost curve so that it has the general shape shown in Figure 43.

In this curve, the cost D to utilities is lowest at a reduction level of 10 million tons, so if the utilities act to minimize cost, they can be expected to reduce emissions by 10 million tons, which is the EPA goal. From the graph, you can see that $D'(10,000,000) = \$0$ per ton, whereas $D'(q)$ is negative for $q < 10,000,000$.

At first you are bothered by the fact that you were not given a cost function. Only the marginal costs were supplied, but you decide to work as best you can without knowing the original cost function $C(q)$.

You now assume that the EPA will impose an annual emission charge of $\$k$ per ton of sulfur released into the environment. It is your job to calculate k. Because you are working with q as the independent variable, you decide that it would be best to formulate the emission charge as a function of q, where q represents the amount by which sulfur emissions are *reduced*. The relationship between the annual sulfur emissions and the amount q by which emissions are reduced from the original 25 million tons is given by

$$\text{Annual sulfur emissions} = \text{original emissions} - \text{amount of reduction}$$

$$= 25,000,000 - q$$

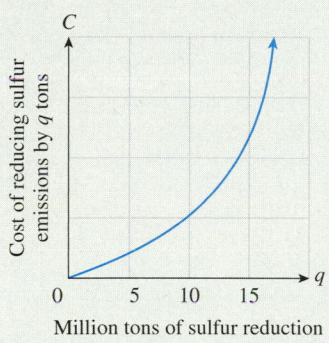

Figure 42

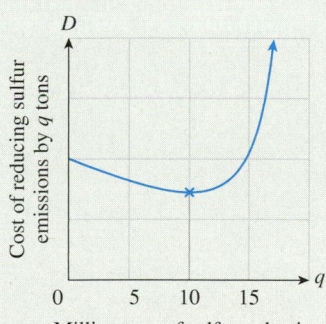

Figure 43

---

[86] These figures were produced in a computerized study of reducing sulfur emissions from the 1980 level by the given amounts. SOURCE: Congress of the United States, Congressional Budget Office, *Curbing Acid Rain: Cost, Budget and Coal Market Effects* (Washington, DC: Government Printing Office, 1986): xx, xxii, 23, 80.

Thus, the total annual emission charge to the utilities is

$$k(25,000,000 - q) = 25,000,000k - kq$$

This results in a total cost to the utilities of

Total cost = Cost of reducing emissions + emission charge

$$D(q) = C(q) + 25,000,000k - kq$$

Even though you have no idea of the form of $C(q)$, you remember that the derivative of a sum is the sum of the derivatives, so you differentiate both sides and obtain

$$D'(q) = C'(q) + 0 - k \qquad \text{The derivative of } kq \text{ is } k \text{ (the slope).}$$
$$= C'(q) - k$$

Remember that you want

$$D'(10,000,000) = 0$$

Thus,

$$C'(10,000,000) - k = 0$$

Referring to the table, you see that

$$360 - k = 0$$

so

$$k = \$360 \text{ per ton}$$

In other words, all you need to do is set the emission charge at $k = \$360$ per ton of sulfur emitted. Further, to ensure that the resulting curve will have the general shape shown in Figure 2, you would like to have $D'(q)$ negative for $q < 10,000,000$ and positive for $q > 10,000,000$. To check this, write

$$D'(q) = C'(q) - k$$
$$= C'(q) - 360$$

and refer to the table to obtain

$$D'(8,000,000) = 270 - 360 = -90 < 0 \quad ✔$$

and

$$D'(12,000,000) = 779 - 360 = 419 > 0 \quad ✔$$

Thus, based on the given data, the resulting curve will have the shape you require. You therefore inform the EPA that an annual emissions charge of \$360 per ton of sulfur released into the environment will create the desired incentive: to reduce sulfur emissions by 10 million tons per year.

One week later, you are informed that this charge would be unrealistic because the utilities cannot possibly afford such a cost. You are asked whether there is an alternative plan that accomplishes the 10-million-ton reduction goal and yet is cheaper to the utilities by \$5 billion per year. You then look at your expression for the emission charge

$$25,000,000k - kq$$

and notice that, if you decrease this amount by \$5 billion, the derivative will not change at all because the derivative of a constant is zero. Thus, you propose the following revised formula for the emission charge:

$$25,000,000k - kq - 5,000,000,000$$
$$= 25,000,000(360) - 360q - 5,000,000,000$$
$$= 4,000,000,000 - 360q$$

At the expected reduction level of 10 million tons, the total amount paid by the utilities will then be

$$4,000,000,000 - 360(10,000,000) = \$400,000,000$$

Thus, your revised proposal is the following: Impose an annual emissions charge of $360 per ton of sulfur released into the environment and hand back $5 billion in the form of subsidies. The effect of this policy will be to cause the utilities industry to reduce sulfur emissions by 10 million tons per year and will result in $400 million in annual revenues to the government.

Notice that this policy also provides an incentive for the utilities to search for cheaper ways to reduce emissions. For instance, if they lowered costs to the point where they could achieve a reduction level of 12 million tons, they would have a total emission charge of

$$4,000,000,000 - 360(12,000,000) = -\$320,000,000$$

The fact that this is negative means that the government would be paying the utilities industry $320 million more in annual subsidies than the industry is paying in per ton emission charges.

Exercises

1. Excluding subsidies, what should the annual emission charge be if the goal is to reduce sulfur emissions by 8 million tons?

2. Excluding subsidies, what should the annual emission charge be if the goal is to reduce sulfur emissions by 12 million tons?

3. What is the *marginal emission charge* in your revised proposal (as stated before the exercise set)? What is the relationship between the marginal cost of reducing sulfur emissions before emissions charges are implemented and the marginal emission charge, at the optimal reduction under your revised proposal?

4. We said that the revised policy provided an incentive for utilities to find cheaper ways to reduce emissions. How would $C(q)$ have to change to make 12 million tons the optimum reduction?

5. What change in $C(q)$ would make 8 million tons the optimum reduction?

6. If the scenario in Exercise 5 took place, what would the EPA have to do in order to make 10 million tons the optimal reduction once again?

7. Due to intense lobbying by the utility industry, you are asked to revise the proposed policy so that the utility industry will pay no charge if sulfur emissions are reduced by the desired 10 million tons. How can you accomplish this?

8. Suppose that instead of imposing a fixed charge per ton of emission, you decide to use a sliding scale, so that the total charge to the industry for annual emissions of x tons will be $\$kx^2$ for some k. What must k be to again make 10 million tons the optimum reduction? [The derivative of kx^2 is $2kx$.]

Section 10.1

Example 1 Use a table to estimate the following limits.

a. $\lim_{x \to 2} \dfrac{x^3 - 8}{x - 2}$ **b.** $\lim_{x \to 0} \dfrac{e^{2x} - 1}{x}$

Solution with Technology On the TI-83/84, use the table feature to automate these computations as follows:

1. Define `Y₁=(x^3-8)/(x-2)` for part (a) or `Y₁=(e^(2x)-1)/x` for part (b).

2. Press $\boxed{\text{2ND}}$ $\boxed{\text{TABLE}}$ to list its values for the given values of x. (If the calculator does not allow you to enter values of x, press $\boxed{\text{2ND}}$ $\boxed{\text{TBLSET}}$ and set Indpnt to Ask).

Here is the table showing some of the values for part (a):

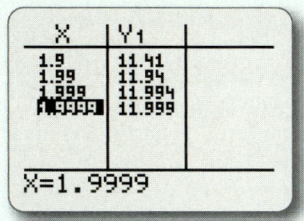

For part (b) use `Y₁=(e^(2*x)-1)/X` and use values of x approaching 0 from either side.

Section 10.4

Example 3 You are a commodities trader and you monitor the price of gold on the New York Spot Market very closely during an active morning. Suppose you find that the price of an ounce of gold can be approximated by the function

$$G(t) = -2t^2 + 36t + 228 \text{ dollars} \quad (7.5 \le t \le 10.5)$$

where t is time in hours. What was the average rate of change of the price of gold over the $1\frac{1}{2}$-hour period starting at 8:00 AM (the interval [8, 9.5] on the t-axis)?

Solution with Technology On the TI-83/84:

1. Enter the function G as Y_1 (using X for t):

`Y₁=-2*X^2+36*X+228`

2. Now find the average rate of change over [8, 9.5] by evaluating the following on the home screen:

`(Y₁(9.5)-Y₁(8))/(9.5-8)`

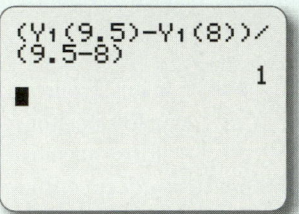

As shown on the screen, the average rate is of change is 1.

Example 4 Continuing with Example 3, use technology to compute the average rate of change of

$$G(t) = -2t^2 + 36t + 228 \quad (7.5 \le t \le 10.5)$$

over the intervals [8, 8 + h], where $h = 1, 0.1, 0.01, 0.001,$ and 0.0001.

Solution with Technology

1. As in Example 3, enter the function G as Y_1 (using X for t):

`Y₁=-2*X^2+36*X+228`

2. Now find the average rate of change for $h = 1$ by evaluating, on the home screen,

`(Y₁(8+1)-Y₁(8))/1`

which gives 2.

3. To evaluate for $h = 0.1$, recall the expression using $\boxed{\text{2nd}}$ $\boxed{\text{ENTER}}$ and then change the 1, both places it occurs, to 0.1, getting

`(Y₁(8+0.1)-Y₁(8))/0.1`

which gives 3.8.

4. Continuing, we can evaluate the average rate of change for all the desired values of h:

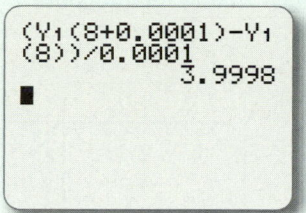

We get the values in the following table:

h	1	0.1	0.01	0.001	0.0001
Ave. Rate of Change $\dfrac{G(8+h) - G(8)}{h}$	2	3.8	3.98	3.998	3.9998

Section 10.5

Example 2 Calculate an approximate value of $f'(1.5)$ if $f(x) = x^2 - 4x$, and then find the equation of the tangent line at the point on the graph where $x = 1.5$.

Solution with Technology

1. In the TI-83/84, enter the function f as Y_1

   $Y_1=x^2-4*x$

2. Go to the home screen to compute the approximations:

   $(Y_1(1.5001)-Y_1(1.5))/0.0001$
   Usual difference quotient

   $(Y_1(1.5001)-Y_1(1.4999))/0.0002$
   Balanced difference quotient

 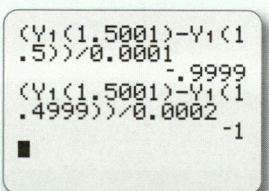

From the display on the right, we find that the difference quotient quick approximation is –0.9999 and the balanced difference quotient quick approximation is –1, which is in fact is the exact value of $f'(1.5)$. See the

discussion in the text for the calculation of the equation of the tangent line.

Example 4 Use technology to obtain a table of values of and graph the derivative of $f(x) = -2x^2 + 6x + 5$ for values of x in starting at -5.

Solution with Technology On the TI-83/84, the easiest way to obtain quick approximations of the derivative of a given function is to use the built-in `nDeriv` function, which calculates balanced difference quotients.

1. On the $Y=$ screen, first enter the function:

   $Y_1=-2x^2+6x+5$

2. Then set

   $Y_2=nDeriv(Y_1,X,X)$ For `nDeriv` press MATH 8

which is the TI-83's approximation of $f'(x)$.

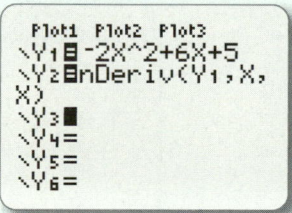

Alternatively, we can enter the balanced difference quotient directly:

   $Y_2=(Y_1(X+0.001)-Y_1(X-0.001))/0.002$

(The TI-83 uses $h = 0.001$ by default in the balanced difference quotient when calculating `nDeriv`, but this can be changed by giving a value of h as a fourth argument, like `nDeriv(Y₁,X,X,0.0001)`.) To see a table of approximate values of the derivative, we press 2ND TABLE and choose a collection of values for *x*:

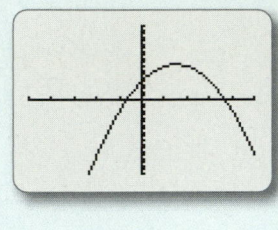

Here, Y₁ shows the value of *f* and Y₂ shows the values of *f'*.

To graph the function or its derivative, we can graph Y₁ or Y₂ in a window showing the given domain $[-5, 5]$:

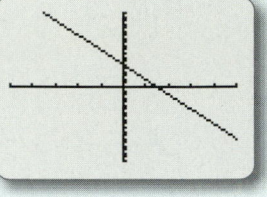

Graph of *f*   Graph of *f'*

EXCEL Technology Guide

Section 10.1

Example 1 Use a table to estimate the following limits.

a. $\lim_{x \to 2} \dfrac{x^3 - 8}{x - 2}$

b. $\lim_{x \to 0} \dfrac{e^{2x} - 1}{x}$

Solution with Technology Set up your spreadsheet to duplicate the table in part (a) as follows:

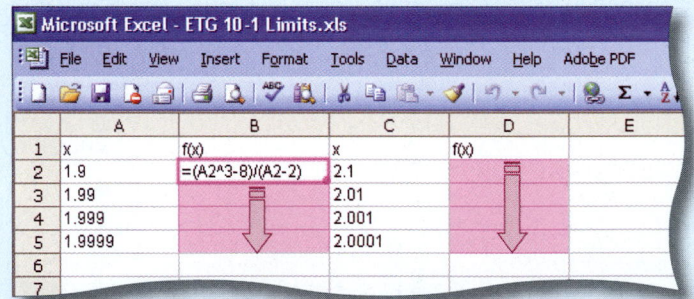

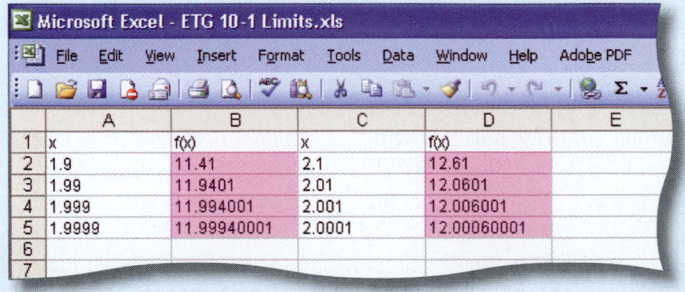

TECHNOLOGY GUIDE

(The formula in cell B2 is copied to columns B and D as indicated by the shading.) The values of $f(x)$ will be calculated in columns B and D.

For part (b), use the formula $= (\text{EXP}(2*\text{A2})-1)/\text{A2}$ in cell B2 and, in columns A and C, use values of x approaching 0 from either side.

Section 10.4

Example 3 You are a commodities trader and you monitor the price of gold on the New York Spot Market very closely during an active morning. Suppose you find that the price of an ounce of gold can be approximated by the function

$$G(t) = -2t^2 + 36t + 228 \text{ dollars} \quad (7.5 \le t \le 10.5)$$

where t is time in hours. What was the average rate of change of the price of gold over the $1\frac{1}{2}$-hour period starting at 8:00 AM (the interval $[8, 9.5]$ on the t-axis)?

Solution with Technology To use Excel to compute the average rate of change of G:

1. Start with two columns, one for values of t and one for values of $G(t)$, which you enter using the formula for G:

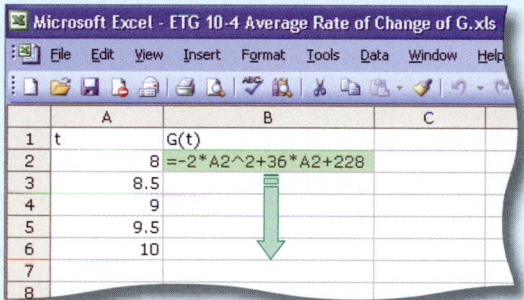

2. Next, calculate the average rate of change as shown here:

In Example 4, we describe another, more versatile Excel template for computing rates of change.

Example 4 Continuing with Example 3, use technology to compute the average rate of change of

$$G(t) = -2t^2 + 36t + 228 \quad (7.5 \le t \le 10.5)$$

over the intervals $[8, 8 + h]$, where $h = 1, 0.1, 0.01, 0.001,$ and 0.0001.

Solution with Technology The template we can use to compute the rates of change is an extension of what we used in Example 3:

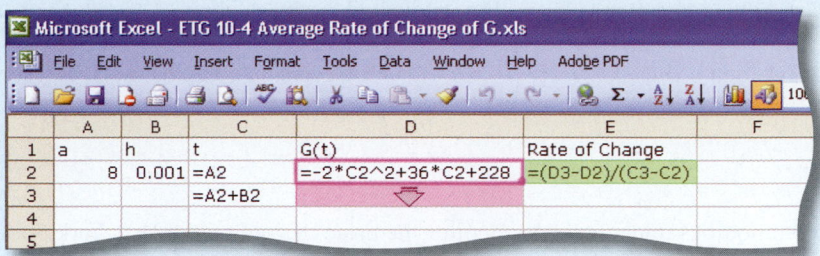

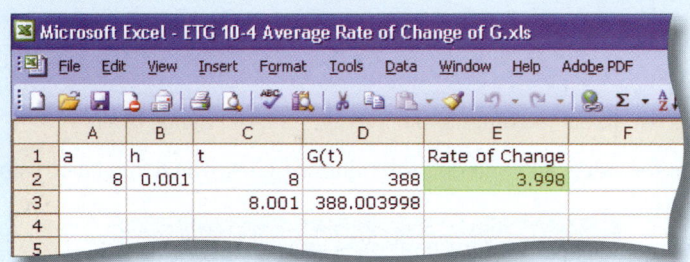

Column C contains the values $t = a$ and $t = a + h$ we are using for the independent variable. The formula in cell E2 is the average rate of change formula $\Delta G / \Delta t$. Entering the different values $h = 1, 0.1, 0.01, 0.001,$ and 0.0001 gives the results shown in the following table:

h	1	0.1	0.01	0.001	0.0001
Ave. Rate of Change $\dfrac{G(8 + h) - G(8)}{h}$	2	3.8	3.98	3.998	3.9998

Section 10.5

Example 2 Calculate an approximate value of $f'(1.5)$ if $f(x) = x^2 - 4x$, and then find the equation of the tangent line at the point on the graph where $x = 1.5$.

Solution with Technology You can compute both the difference quotient and the balanced difference quotient approximations in Excel using the following extension of the worksheet in Example 4 in Section 10.1.

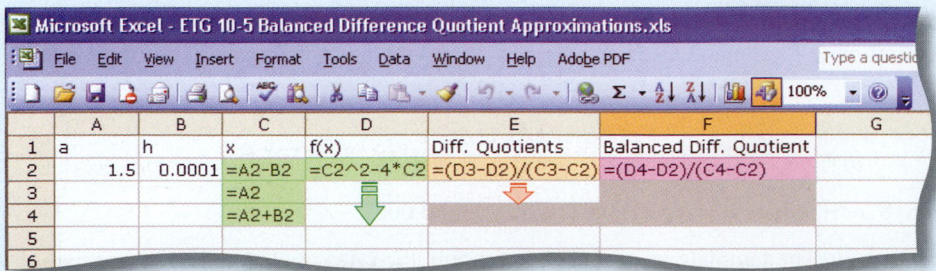

Notice that we get two difference quotients in column E. The first uses $h = -0.0001$ while the second uses $h = 0.0001$ and is the one we use for our quick approximation. The balanced quotient is their average (column F). The results are as follows.

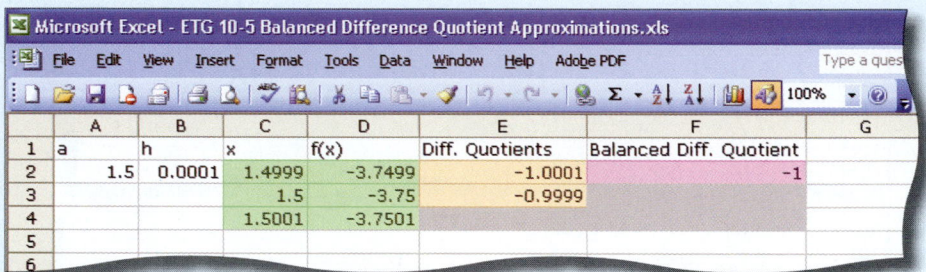

From the results shown above, we find that the difference quotient quick approximation is -0.9999 and that the balanced difference quotient quick approximation is -1, which is in fact it is the exact value of $f'(1.5)$. See the discussion in the text for the calculation of the equation of the tangent line.

Example 4 Use technology to obtain a table of values of and graph the derivative of $f(x) = -2x^2 + 6x + 5$ for values of x in starting at -5.

Solution with Technology

1. Start with a table of values for the function f, reminiscent of our graphing spreadsheet from Chapter 1:

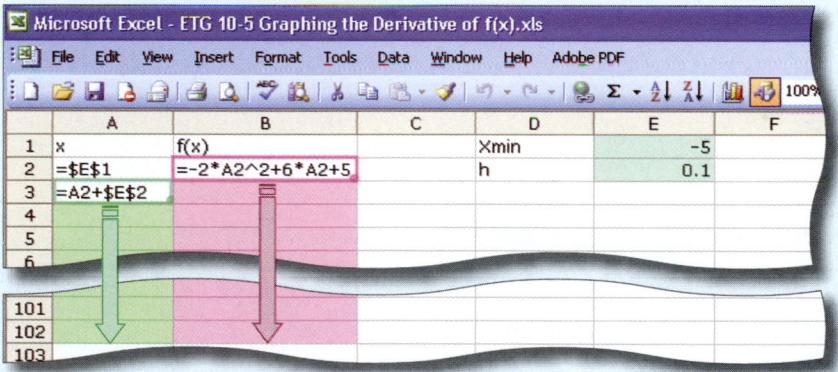

2. Next, compute approximate derivatives in Column C:

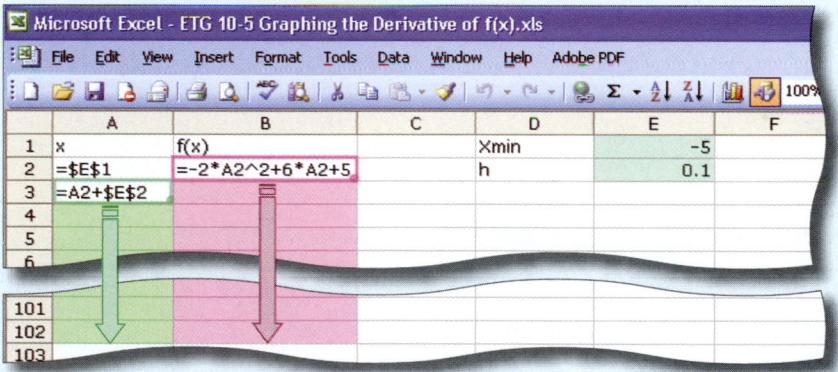

	A	B	C	D	E	F
1	x	f(x)	f'(x)	Xmin	−5	
2	−5	−75	25.8	h	0.1	
3	−4.9	−72.42	25.4			
4	−4.8	−69.88	25			
5	−4.7	−67.38	24.6			
6		−64.92	−100.0			
101	−4.9	−13.62	−13.8			
102	−5	−15				
103						
104						

You cannot paste the difference quotient formula into cell C102. (Why?) Notice that this worksheet uses the ordinary difference quotients, $[f(x + h) - f(x)]/h$. If you prefer, you can use balanced difference quotients $[f(x + h) - f(x - h)]/(2h)$, in which case cells C2 and C102 would both have to be left blank.

We then graph the function and the derivative on different graphs as follows:

1. First, graph the function f in the usual way, using Columns A and B.

2. Make a copy of this graph and click on it once. Columns A and B should be outlined, indicating that these are the columns used in the graph.

3. By dragging from the center of the bottom edge, move the Column B box over to Column C as shown:

	4.4	−7.32	
97	4.5	−8.5	−12.2
98	4.6	−9.72	−12.6
99	4.7	−10.98	−13
100	4.8	−12.28	−13.4
101	4.9	−13.62	−13.8
102	5	−15	
103			
104			

	4.4	−7.32	
97	4.5	−8.5	−12.2
98	4.6	−9.72	−12.6
99	4.7	−10.98	−13
100	4.8	−12.28	−13.4
101	4.9	−13.62	−13.8
102	5		−15
103			
104			

The graph will then show the derivative (Columns A and C):

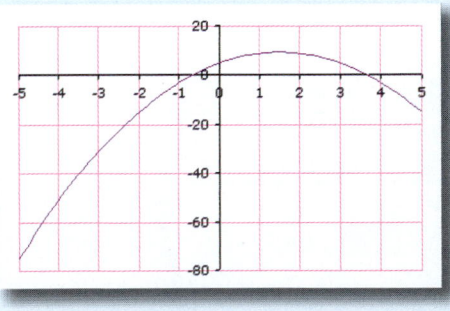

Graph of f

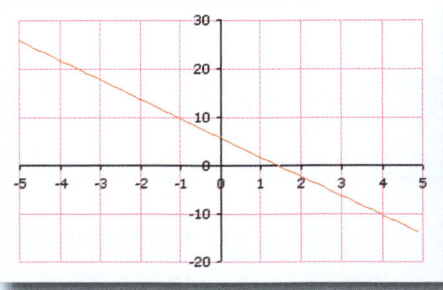

Graph of f'

11

Techniques of Differentiation

CASE STUDY Projecting Market Growth

You are on the board of directors at Fullcourt Academic Press. The sales director of the high school division has just burst into your office with a proposal for a major expansion strategy based on the assumption that the number of high school seniors in the U.S. will be growing at a rate of at least 20,000 per year through the year 2005. Because the figures actually appear to be leveling off, you are suspicious about this estimate. You would like to devise a model that predicts this trend before tomorrow's scheduled board meeting. How do you go about doing this?

John Giustina/Getty Images

Introduction

In Chapter 10 we studied the concept of the derivative of a function, and we saw some of the applications for which derivatives are useful. However, the only functions we could differentiate easily were sums of terms of the form ax^n, where a and n are constants.

In this chapter we develop techniques that help us differentiate any closed-form function—that is, any function, no matter how complicated, that can be specified by a formula involving powers, radicals, exponents, and logarithms. (In a later chapter, we will discuss how to add trigonometric functions to this list.) We also show how to find the derivatives of functions that are only specified *implicitly*—that is, functions for which we are not given an explicit formula for y in terms of x but only an equation relating x and y.

algebra Review

For this chapter, you should be familiar with the algebra reviewed in Chapter 0, Sections 3 and 4.

11.1 : The Product and Quotient Rules

We know how to find the derivatives of functions that are sums of powers, like polynomials. In general, if a function is a sum or difference of functions whose derivatives we know, then we know how to find its derivative. But what about *products and quotients* of functions whose derivatives we know? For instance, how do we calculate the derivative of something like $x^2/(x + 1)$? The derivative of $x^2/(x + 1)$ is not, as one might suspect, $2x/1 = 2x$. That calculation is based on an assumption that the derivative of a quotient is the quotient of the derivatives. But it is easy to see that this assumption is false: For instance, the derivative of $1/x$ is not $0/1 = 0$, but $-1/x^2$. Similarly, the derivative of a product is not the product of the derivatives: For instance, the derivative of $x = 1 \cdot x$ is not $0 \cdot 1 = 0$, but 1.

To identify the correct method of computing the derivatives of products and quotients, let's look at a simple example. We know that the daily revenue resulting from the sale of q items per day at a price of p dollars per item is given by the product, $R = pq$ dollars. Suppose you are currently selling wall posters on campus. At this time your daily sales are 50 posters, and sales are increasing at a rate of 4 per day. Furthermore, you are currently charging $10 per poster, and you are also raising the price at a rate of $2 per day. Let's use this information to estimate how fast your daily revenue is increasing. In other words, let us estimate the rate of change, dR/dt, of the revenue R.

There are two contributions to the rate of change of daily revenue: the increase in daily sales and the increase in the unit price. We have

$\dfrac{dR}{dt}$ due to increasing price: \$2 per day $\times$ 50 posters = \$100 per day

$\dfrac{dR}{dt}$ due to increasing sales: \$10 per poster $\times$ 4 posters per day = \$40 per day

Thus, we estimate the daily revenue to be increasing at a rate of \$100 + \$40 = \$140 per day. Let us translate what we have said into symbols:

$\dfrac{dR}{dt}$ due to increasing price: $\dfrac{dp}{dt} \times q$

$\dfrac{dR}{dt}$ due to increasing sales: $p \times \dfrac{dq}{dt}$

Thus, the rate of change of revenue is given by

$$\frac{dR}{dt} = \frac{dp}{dt} q + p \frac{dq}{dt}$$

Because $R = pq$, we have discovered the following rule for differentiating a product:

$$\frac{d}{dt}(pq) = \frac{dp}{dt}q + p\frac{dq}{dt}$$

The derivative of a product is the derivative of the first times the second, plus the first times the derivative of the second.

This rule and a similar rule for differentiating quotients are given next, and also a discussion of how these results are proved rigorously.

Product Rule

If $f(x)$ and $g(x)$ are differentiable functions of x, then so is their product $f(x)g(x)$, and

$$\frac{d}{dx}[f(x)g(x)] = f'(x)g(x) + f(x)g'(x)$$

Product Rule in Words

The derivative of a product is the derivative of the first times the second, plus the first times the derivative of the second.

quick Example $f(x) = x^2$ and $g(x) = 3x - 1$ are both differentiable functions of x, and so their product $x^2(3x - 1)$ is differentiable, and

$$\frac{d}{dx}[x^2(3x - 1)] = 2x \cdot (3x - 1) \quad + \quad x^2 \cdot (3)$$

Derivative of first   Second   First   Derivative of second

Quotient Rule

If $f(x)$ and $g(x)$ are differentiable functions of x, then so is their quotient $f(x)/g(x)$ (provided $g(x) \neq 0$), and

$$\frac{d}{dx}\left(\frac{f(x)}{g(x)}\right) = \frac{f'(x)g(x) - f(x)g'(x)}{[g(x)]^2}$$

Quotient Rule in Words

The derivative of a quotient is the derivative of the top times the bottom, minus the top times the derivative of the bottom, all over the bottom squared.

quick Example $f(x) = x^3$ and $g(x) = x^2 + 1$ are both differentiable functions of x, and so their quotient $x^3/(x^2 + 1)$ is differentiable, and

Derivative of top   Bottom   Top   Derivative of bottom

$$\frac{d}{dx}\left(\frac{x^3}{x^2 + 1}\right) = \frac{3x^2(x^2 + 1) \quad - \quad x^3 \cdot 2x}{(x^2 + 1)^2}$$

Bottom squared

Notes

1. Don't try to remember the rules by the symbols we have used, but remember them in words. (The slogans are easy to remember, even if the terms are not precise.)

2. One more time: *The derivative of a product is* NOT *the product of the derivatives, and the derivative of a quotient is* NOT *the quotient of the derivatives.* To find the derivative of a product, you must use the product rule, and to find the derivative of a quotient, you must use the quotient rule.[1]

Q: *Wait a minute! The expression $2x^3$ is a product, and we already know that its derivative is $6x^2$. Where did we use the product rule?*

A: To differentiate functions such as $2x^3$, we have used the rule from Section 10.4:
The derivative of c times a function is c times the derivative of the function.

However, the product rule gives us the same result:

Derivative of first   Second   First   Derivative of Second
  ↓   ↓   ↓   ↓

$$\frac{d}{dx}(2x^3) = (0)(x^3) \quad + \quad (2)(3x^2) = 6x^2 \qquad \text{Product rule}$$

$$\frac{d}{dx}(2x^3) = (2)(3x^2) = 6x^2 \qquad\qquad\qquad \text{Derivative of a constant times a function}$$

We do not recommend that you use the product rule to differentiate functions like $2x^3$; continue to use the simpler rule when one of the factors is a constant. ■

Derivation of the Product Rule

Before we look at more examples of using the product and quotient rules, let's see why the product rule is true. To calculate the derivative of the product $f(x)g(x)$ of two differentiable functions, we go back to the definition of the derivative:

$$\frac{d}{dx}[f(x)g(x)] = \lim_{h\to 0}\frac{f(x+h)g(x+h) - f(x)g(x)}{h}$$

We now rewrite this expression so that we can evaluate the limit: Notice that the numerator reflects a simultaneous change in f [from $f(x)$ to $f(x+h)$] and g [from $g(x)$ to $g(x+h)$]. To separate the two effects, we add and subtract a quantity in the numerator that reflects a change in only one of the functions:

$$\frac{d}{dx}[f(x)g(x)] = \lim_{h\to 0}\frac{f(x+h)g(x+h) - f(x)g(x)}{h}$$

$$= \lim_{h\to 0}\frac{f(x+h)g(x+h) - f(x)g(x+h) + f(x)g(x+h) - f(x)g(x)}{h} \qquad \text{We subtracted and added the quantity[2] } f(x)g(x+h)$$

$$= \lim_{h\to 0}\frac{[f(x+h) - f(x)]g(x+h) + f(x)[g(x+h) - g(x)]}{h} \qquad \text{Common factors}$$

$$= \lim_{h\to 0}\left(\frac{f(x+h) - f(x)}{h}\right)g(x+h) + \lim_{h\to 0}f(x)\left(\frac{g(x+h) - g(x)}{h}\right) \qquad \text{Limit of sum}$$

$$= \lim_{h\to 0}\left(\frac{f(x+h) - f(x)}{h}\right)\lim_{h\to 0}g(x+h) + \lim_{h\to 0}f(x)\lim_{h\to 0}\left(\frac{g(x+h) - g(x)}{h}\right) \qquad \text{Limit of product}$$

---

[1] Leibniz made this mistake at first, too, so you would be in good company if you forgot to use the product or quotient rule.

[2] Adding an appropriate form of zero is an age-old mathematical ploy.

For a proof of the fact that, if g is differentiable, it must be continuous, go online and follow:

Chapter 10

→ Continuity and Differentiability

The quotient rule can be proved in a very similar way. Go online and follow:

Chapter 11

→ Proof of Quotient Rule

to find a proof.

Now we already know the following four limits:

$$\lim_{h\to 0}\frac{f(x+h)-f(x)}{h}=f'(x)$$ Definition of derivative of f; f is differentiable

$$\lim_{h\to 0}\frac{g(x+h)-g(x)}{h}=g'(x)$$ Definition of derivative of g; g is differentiable

$$\lim_{h\to 0}g(x+h)=g(x)$$ If g is differentiable, it must be continuous.

$$\lim_{h\to 0}f(x)=f(x)$$ Limit of a constant

Putting these limits into the one we're calculating, we get

$$\frac{d}{dx}[f(x)g(x)]=f'(x)g(x)+f(x)g'(x)$$

which is the product rule.

Example 1 Using the Product Rule

Compute the following derivatives.

a. $\dfrac{d}{dx}[(x^{3.2}+1)(1-x)]$ Simplify the answer.

b. $\dfrac{d}{dx}[(x+1)(x^2+1)(x^3+1)]$ Do not expand the answer.

Solution

a. We can do the calculation in two ways.

	derivative of first	second	first	derivative of second
	↓	↓	↓	↓

Using the Product Rule:
$$\frac{d}{dx}[(x^{3.2}+1)(1-x)]=(3.2x^{2.2})(1-x)+(x^{3.2}+1)(-1)$$

$$=3.2x^{2.2}-3.2x^{3.2}-x^{3.2}-1$$ Expand the answer.

$$=-4.2x^{3.2}+3.2x^{2.2}-1$$

Not using the Product Rule: First, expand the given expression.

$$(x^{3.2}+1)(1-x)=-x^{4.2}+x^{3.2}-x+1$$

Thus,

$$\frac{d}{dx}[(x^{3.2}+1)(1-x)]=\frac{d}{dx}(-x^{4.2}+x^{3.2}-x+1)$$

$$=-4.2x^{3.2}+3.2x^{2.2}-1$$

In this example the product rule saves us little or no work, but in later sections we shall see examples that can be done in no other way. Learn how to use the product rule now!

b. Here we have a product of *three* functions, not just two. We can find the derivative by using the product rule twice:

$$\frac{d}{dx}[(x+1)(x^2+1)(x^3+1)]$$

$$= \frac{d}{dx}(x+1)\cdot[(x^2+1)(x^3+1)] + (x+1)\cdot\frac{d}{dx}[(x^2+1)(x^3+1)]$$

$$= (1)(x^2+1)(x^3+1) + (x+1)[(2x)(x^3+1) + (x^2+1)(3x^2)]$$

$$= (1)(x^2+1)(x^3+1) + (x+1)(2x)(x^3+1) + (x+1)(x^2+1)(3x^2)$$

We can see here a more general product rule:

$$(fgh)' = f'gh + fg'h + fgh'$$

Notice that every factor has a chance to contribute to the rate of change of the product. There are similar formulas for products of four or more functions.

Example 2   Using the Quotient Rule

Compute the derivatives   **a.** $\dfrac{d}{dx}\left[\dfrac{1-3.2x^{-0.1}}{x+1}\right]$ **b.** $\dfrac{d}{dx}\left[\dfrac{(x+1)(x+2)}{x-1}\right]$

Solution

Derivative of top   Bottom   Top   Derivative of bottom

a. $\dfrac{d}{dx}\left[\dfrac{1-3.2x^{-0.1}}{x+1}\right] = \dfrac{(0.32x^{-1.1})(x+1) - (1-3.2x^{-0.1})(1)}{(x+1)^2}$

Bottom squared

$$= \frac{0.32x^{-0.1} + 0.32x^{-1.1} - 1 + 3.2x^{-0.1}}{(x+1)^2}$$ Expand the numerator

$$= \frac{3.52x^{-0.1} + 0.32x^{-1.1} - 1}{(x+1)^2}$$

b. Here we have both a product and a quotient. Which rule do we use, the product or the quotient rule? Here is a way to decide. Think about how we would calculate, step by step, the value of $(x+1)(x+2)/(x-1)$ for a specific value of x—say $x = 11$. Here is how we would probably do it:

1. Calculate $(x+1)(x+2) = (11+1)(11+2) = 156$.

2. Calculate $x - 1 = 11 - 1 = 10$.

3. Divide 156 by 10 to get 15.6.

Now ask: *What was the last operation we performed?* The last operation we performed was division, so we can regard the whole expression as a *quotient*—that is, as $(x+1)(x+2)$ *divided by* $(x-1)$. Therefore, we should use the quotient rule.

The first thing the quotient rule tells us to do is take the derivative of the numerator. Now, the numerator is a product, so we must use the product rule to take its derivative.

Here is the calculation:

$$\frac{d}{dx}\left[\frac{(x+1)(x+2)}{x-1}\right] = \frac{\overbrace{[(1)(x+2)+(x+1)(1)]}^{\text{Derivative of top}}\overbrace{(x-1)}^{\text{Bottom}} - \overbrace{[(x+1)(x+2)]}^{\text{Top}}\overbrace{(1)}^{\text{Derivative of bottom}}}{\underset{\text{Bottom squared}}{(x-1)^2}}$$

$$= \frac{(2x+3)(x-1)-(x+1)(x+2)}{(x-1)^2}$$

$$= \frac{x^2-2x-5}{(x-1)^2}$$

What is important is to determine the *order of operations* and, in particular, to determine the last operation to be performed. Pretending to do an actual calculation reminds us of the order of operations; we call this technique the **calculation thought experiment.**

+ *Before we go on...* We used the quotient rule in Example 2 because the function was a quotient; we used the product rule to calculate the derivative of the numerator because the numerator was a product. Get used to this: Differentiation rules usually must be used in combination.

Here is another way we could have done this problem: Our calculation thought experiment could have taken the following form:

1. Calculate $(x+1)/(x-1) = (11+1)/(11-1) = 1.2$.

2. Calculate $x+2 = 11+2 = 13$.

3. Multiply 1.2 by 13 to get 15.6.

We would have then regarded the expression as a *product*—the product of the factors $(x+1)/(x-1)$ and $(x+2)$—and used the product rule instead. We can't escape the quotient rule however: We need to use it to take the derivative of the first factor, $(x+1)/(x-1)$. Try this approach for practice and check that you get the same answer. ∎

Calculation Thought Experiment

The **calculation thought experiment** is a technique to determine whether to treat an algebraic expression as a product, quotient, sum, or difference. Given an expression, consider the steps you would use in computing its value. If the last operation is multiplication, treat the expression as a product; if the last operation is division, treat the expression as a quotient; and so on.

quick Examples

1. $(3x^2 - 4)(2x+1)$ can be computed by first calculating the expressions in parentheses and then multiplying. Because the last step is multiplication, we can treat the expression as a product.

2. $\dfrac{2x-1}{x}$ can be computed by first calculating the numerator and denominator and then dividing one by the other. Because the last step is division, we can treat the expression as a quotient.

3. $x^2 + (4x - 1)(x + 2)$ can be computed by first calculating x^2, then calculating the product $(4x - 1)(x + 2)$, and finally adding the two answers. Thus, we can treat the expression as a sum.

4. $(3x^2 - 1)^5$ can be computed by first calculating the expression in parentheses and then raising the answer to the fifth power. Thus, we can treat the expression as a power. (We shall see how to differentiate powers of expressions in Section 11.2.)

It often happens that the same expression can be calculated in different ways. For example, $(x + 1)(x + 2)/(x - 1)$ can be treated as either a quotient or a product; see Example 2(b).

Example 3   Using the Calculation Thought Experiment

Find $\dfrac{d}{dx}\left[6x^2 + 5\left(\dfrac{x}{x - 1}\right)\right]$.

Solution

The calculation thought experiment tells us that the expression we are asked to differentiate can be treated as a *sum*. Because the derivative of a sum is the sum of the derivatives, we get

$$\frac{d}{dx}\left[6x^2 + 5\left(\frac{x}{x-1}\right)\right] = \frac{d}{dx}(6x^2) + \frac{d}{dx}\left[5\left(\frac{x}{x-1}\right)\right]$$

In other words, we must take the derivatives of $6x^2$ and $5\left(\frac{x}{x-1}\right)$ separately and then add the answers. The derivative of $6x^2$ is $12x$. There are two ways of taking the derivative of $5\left(\frac{x}{x-1}\right)$: we could either first multiply the expression $\left(\frac{x}{x-1}\right)$ by 5 to get $\left(\frac{5x}{x-1}\right)$ and then take its derivative using the quotient rule, or we could pull the 5 out, as we do next.

$$\frac{d}{dx}\left[6x^2 + 5\left(\frac{x}{x-1}\right)\right] = \frac{d}{dx}(6x^2) + \frac{d}{dx}\left[5\left(\frac{x}{x-1}\right)\right] \qquad \text{Derivative of sum}$$

$$= 12x + 5\frac{d}{dx}\left(\frac{x}{x-1}\right) \qquad \text{Constant} \times \text{Function}$$

$$= 12x + 5\left(\frac{(1)(x-1) - (x)(1)}{(x-1)^2}\right) \qquad \text{Quotient rule}$$

$$= 12x + 5\left(\frac{-1}{(x-1)^2}\right)$$

$$= 12x - \frac{5}{(x-1)^2}$$

Applications

In the next example, we return to a scenario similar to the one discussed at the start of this section.

Example 4 Applying the Product and Quotient Rules: Revenue and Average Cost

Sales of your newly launched miniature wall posters for college dorms, *iMiniPosters,* are really taking off. (Those old-fashioned large wall posters no longer fit in today's "downsized" college dorm rooms.) Monthly sales to students at the start of this year were 1500 iMiniPosters, and since that time, sales have been increasing by 300 posters each month, even though the price you charge has also been going up.

a. The price you charge for iMiniPosters is given by:

$$p(t) = 10 + 0.05t^2 \text{ dollars per poster}$$

where t is time in months since the start of January of this year. Find a formula for the monthly revenue, and then compute its rate of change at the beginning of March.

b. The number of students who purchase iMiniPosters in a month is given by

$$n(t) = 800 + 0.2t$$

where t is as in part (a). Find a formula for the average number of posters each student buys, and hence estimate the rate at which this number was growing at the beginning of March.

Solution

a. To compute monthly revenue as a function of time t, we use

$$R(t) = p(t)q(t) \qquad \text{Revenue} = \text{Price} \times \text{Quantity}$$

We already have a formula for $p(t)$. The function $q(t)$ measures sales, which were 1500 posters/month at time $t = 0$, and rising by 300 per month:

$$q(t) = 1500 + 300t$$

Therefore, the formula for revenue is

$$R(t) = p(t)q(t)$$
$$R(t) = (10 + 0.05t^2)(1500 + 300t)$$

Rather than expand this expression, we shall leave it as a product so that we can use the product rule in computing its rate of change:

$$R'(t) = p'(t)q(t) + p(t)q'(t)$$
$$= [0.10t][1500 + 300t] + [10 + 0.05t^2][300]$$

Because the beginning of March corresponds to $t = 2$, we have

$$R'(2) = [0.10(2)][1500 + 300(2)] + [10 + 0.05(2)^2][300]$$
$$= (0.2)(2100) + (10.2)(300) = \$3480 \text{ per month}$$

Therefore, your monthly revenue was increasing at a rate of $3480 per month at the beginning of March.

b. The average number of posters sold to each student is

$$k(t) = \frac{\text{Number of posters}}{\text{Number of students}}$$

$$k(t) = \frac{q(t)}{n(t)} = \frac{1500 + 300t}{800 + 0.2t}$$

The rate of change of $k(t)$ is computed with the quotient rule:

$$k'(t) = \frac{q'(t)n(t) - q(t)n'(t)}{n(t)^2}$$

$$= \frac{(300)(800 + 0.2t) - (1500 + 300t)(0.2)}{(800 + 0.2t)^2}$$

so that

$$k'(2) = \frac{(300)[800 + 0.2(2)] - [1500 + 300(2)](0.2)}{[800 + 0.2(2)]^2}$$

$$= \frac{(300)(800.4) - (2100)(0.2)}{800.4^2} \approx 0.37 \text{ posters/student per month}$$

Therefore, the average number of posters sold to each student was increasing at a rate of about 0.37 posters/student per month.

11.1 EXERCISES

● denotes basic skills exercises

◆ denotes challenging exercises

In Exercises 1–12:

a. Calculate the derivative of the given function without using either the product or quotient rule.

b. Use the product or quotient rule to find the derivative. Check that you obtain the same answer.

1. ● $f(x) = 3x$ **2.** ● $f(x) = 2x^2$ **3.** ● $g(x) = x \cdot x^2$

4. ● $g(x) = x \cdot x$ **5.** ● $h(x) = x(x+3)$ **6.** ● $h(x) = x(1+2x)$

7. ● $r(x) = 100x^{2.1}$ **8.** ● $r(x) = 0.2x^{-1}$ **9.** ● $s(x) = \dfrac{2}{x}$

10. ● $t(x) = \dfrac{x}{3}$ **11.** ● $u(x) = \dfrac{x^2}{3}$ **12.** ● $s(x) = \dfrac{3}{x^2}$

Calculate $\dfrac{dy}{dx}$ in Exercises 13–20. Simplify your answer.

hint [see Example 1]

13. ● $y = 3x(4x^2 - 1)$ **14.** ● $y = 3x^2(2x + 1)$

15. ● $y = x^3(1 - x^2)$ **16.** ● $y = x^5(1 - x)$

17. ● $y = (2x + 3)^2$ **18.** ● $y = (4x - 1)^2$

19. ● $x\sqrt{x}$ **20.** ● $x^2\sqrt{x}$

Calculate $\dfrac{dy}{dx}$ in Exercises 21–56. You need not expand your answers.

21. ● $y = (x + 1)(x^2 - 1)$

22. ● $y = (4x^2 + x)(x - x^2)$

23. ● $y = (2x^{0.5} + 4x - 5)(x - x^{-1})$

24. ● $y = (x^{0.7} - 4x - 5)(x^{-1} + x^{-2})$

25. ● $y = (2x^2 - 4x + 1)^2$

26. ● $y = (2x^{0.5} - x^2)^2$

27. ● $y = \left(\dfrac{x}{3.2} + \dfrac{3.2}{x}\right)(x^2 + 1)$

28. ● $y = \left(\dfrac{x^{2.1}}{7} + \dfrac{2}{x^{2.1}}\right)(7x - 1)$

29. ● $x^2(2x + 3)(7x + 2)$ *hint* [see Example 1b]

30. ● $x(x^2 - 3)(2x^2 + 1)$

31. ● $(5.3x - 1)(1 - x^{2.1})(x^{-2.3} - 3.4)$

32. ● $(1.1x + 4)(x^{2.1} - x)(3.4 - x^{-2.1})$

33. $y = (\sqrt{x} + 1)\left(\sqrt{x} + \dfrac{1}{x^2}\right)$

34. $y = (4x^2 - \sqrt{x})\left(\sqrt{x} - \dfrac{2}{x^2}\right)$

35. ● $y = \dfrac{2x + 4}{3x - 1}$ *hint* [see Example 2]

36. ● $y = \dfrac{3x - 9}{2x + 4}$

37. ● $y = \dfrac{2x^2 + 4x + 1}{3x - 1}$ **38.** ● $y = \dfrac{3x^2 - 9x + 11}{2x + 4}$

39. ● $y = \dfrac{x^2 - 4x + 1}{x^2 + x + 1}$ **40.** ● $y = \dfrac{x^2 + 9x - 1}{x^2 + 2x - 1}$

41. ● $y = \dfrac{x^{0.23} - 5.7x}{1 - x^{-2.9}}$ **42.** ● $y = \dfrac{8.43x^{-0.1} - 0.5x^{-1}}{3.2 + x^{2.9}}$

● basic skills   ◆ challenging

43. $y = \dfrac{\sqrt{x}+1}{\sqrt{x}-1}$

44. $y = \dfrac{\sqrt{x}-1}{\sqrt{x}+1}$

45. $y = \dfrac{\left(\dfrac{1}{x}+\dfrac{1}{x^2}\right)}{x+x^2}$

46. $y = \dfrac{\left(1-\dfrac{1}{x^2}\right)}{x^2-1}$

47. ● $y = \dfrac{(x+3)(x+1)}{3x-1}$ *hint* [see Example 2b]

48. ● $y = \dfrac{x}{(x-5)(x-4)}$

49. ● $y = \dfrac{(x+3)(x+1)(x+2)}{3x-1}$

50. ● $y = \dfrac{3x-1}{(x-5)(x-4)(x-1)}$

51. ● $y = x^4 - (x^2+120)(4x-1)$ *hint* [see Example 3]

52. ● $y = x^4 - \dfrac{x^2+120}{4x-1}$

53. ● $y = x + 1 + 2\left(\dfrac{x}{x+1}\right)$

54. ● $y = x + 2 - 4(x^2-x)\left(x+\dfrac{1}{x}\right)$ (Do not simplify the answer)

55. ● $y = (x+1)(x-2) - 2\left(\dfrac{x}{x+1}\right)$

56. ● $y = \dfrac{x+2}{x+1} + (x+1)(x-2)$

In Exercises 57–62, compute the derivatives.

57. ● $\dfrac{d}{dx}[(x^2+x)(x^2-x)]$

58. ● $\dfrac{d}{dx}[(x^2+x^3)(x+1)]$

59. ● $\dfrac{d}{dx}[(x^3+2x)(x^2-x)]\Big|_{x=2}$

60. ● $\dfrac{d}{dx}[(x^2+x)(x^2-x)]\Big|_{x=1}$

61. ● $\dfrac{d}{dt}[(t^2-t^{0.5})(t^{0.5}+t^{-0.5})]\Big|_{t=1}$

62. ● $\dfrac{d}{dt}[(t^2+t^{0.5})(t^{0.5}-t^{-0.5})]\Big|_{t=1}$

In Exercises 63–68, find the equation of the line tangent to the graph of the given function at the point with the indicated x-coordinate.

63. ● $f(x) = (x^2+1)(x^3+x); \ x = 1$

64. ● $f(x) = (x^{0.5}+1)(x^2+x); \ x = 1$

65. ● $f(x) = \dfrac{x+1}{x+2}; \ x = 0$

66. ● $f(x) = \dfrac{\sqrt{x}+1}{\sqrt{x}+2}; \ x = 4$

67. ● $f(x) = \dfrac{x^2+1}{x}; \ x = -1$

68. ● $f(x) = \dfrac{x}{x^2+1}; \ x = 1$

● basic skills ◆ challenging

Applications

69. ● **Revenue** The monthly sales of Sunny Electronics' new sound system are given by $q(t) = 2000t - 100t^2$ units per month, t months after its introduction. The price Sunny charges is $p(t) = 1000 - t^2$ dollars per sound system, t months after introduction. Find the rate of change of monthly sales, the rate of change of the price, and the rate of change of monthly revenue five months after the introduction of the sound system. Interpret your answers. *hint* [see Example 4a]

70. ● **Revenue** The monthly sales of Sunny Electronics' new *iSun* walkman is given by $q(t) = 2000t - 100t^2$ units per month, t months after its introduction. The price Sunny charges is $p(t) = 100 - t^2$ dollars per *iSun, t* months after introduction. Find the rate of change of monthly sales, the rate of change of the price, and the rate of change of monthly revenue six months after the introduction of the *iSun*. Interpret your answers.

71. ● **Saudi Oil Revenues** The spot price of crude oil during the period 2000–2005 can be approximated by

$$P(t) = 5t + 25 \text{ dollars per barrel} \quad (0 \le t \le 5)$$

in year t, where $t = 0$ represents 2000. Saudi Arabia's crude oil production over the same period can be approximated by

$$Q(t) = 0.082t^2 - 0.22t + 8.2 \text{ million barrels per day}[3]$$
$$(0 \le t \le 5)$$

Use these models to estimate Saudi Arabia's daily oil revenue and also its rate of change in 2001. (Round your answers to the nearest \$1 million.)

72. ● **Russian Oil Revenues** Russia's crude oil production during the period 2000–2005 can be approximated by

$$Q(t) = -0.066t^2 + 0.96t + 6.1 \text{ million barrels per day}[4]$$
$$(0 \le t \le 5)$$

in year t, where $t = 0$ represents 2000. Use the model for the spot price in Exercise 71 to estimate Russia's daily oil revenue and also its rate of change in 2001.

73. ● **Revenue** Dorothy Wagner is currently selling 20 "I ♥ Calculus" T-shirts per day, but sales are dropping at a rate of 3 per day. She is currently charging \$7 per T-shirt, but to compensate for dwindling sales, she is increasing the unit price by \$1 per day. How fast, and in what direction is her daily revenue currently changing?

74. ● **Pricing Policy** Let us turn Exercise 73 around a little: Dorothy Wagner is currently selling 20 "I ♥ Calculus" T-shirts per day, but sales are dropping at a rate of 3 per day. She is currently charging \$7 per T-shirt, and she wishes to increase

---

[3] Source for data: EIA/Saudi British Bank (www.sabb.com). 2004 figures are based on mid-year data, and 2005 data are estimates.

[4] Source for data: Energy Information Administration (http://www.eia. doe.gov), Pravda (http://english.pravda.ru). 2004 figures are based on mid-year data, and 2005 data are estimates.

her daily revenue by $10 per day. At what rate should she increase the unit price to accomplish this (assuming that the price increase does not affect sales)?

75. ● **Bus Travel** Thoroughbred Bus Company finds that its monthly costs for one particular year were given by $C(t) = 10,000 + t^2$ dollars after t months. After t months the company had $P(t) = 1000 + t^2$ passengers per month. How fast is its cost per passenger changing after 6 months? *hint* [see Example 4b]

76. ● **Bus Travel** Thoroughbred Bus Company finds that its monthly costs for one particular year were given by $C(t) = 100 + t^2$ dollars after t months. After t months, the company had $P(t) = 1000 + t^2$ passengers per month. How fast is its cost per passenger changing after 6 months?

77. ● **Fuel Economy** Your muscle car's gas mileage (in miles per gallon) is given as a function $M(x)$ of speed x in mph, where

$$M(x) = \frac{3000}{x + 3600x^{-1}}$$

Calculate $M'(x)$, and then $M'(10)$, $M'(60)$, and $M'(70)$. What do the answers tell you about your car?

78. ● **Fuel Economy** Your used Chevy's gas mileage (in miles per gallon) is given as a function $M(x)$ of speed x in mph, where

$$M(x) = \frac{4000}{x + 3025x^{-1}}$$

Calculate $M'(x)$ and hence determine *the sign* of each of the following: $M'(40)$, $M'(55)$, $M'(60)$. Interpret your results.

79. **Military Spending** The annual cost per active-duty armed service member in the U.S. increased from $80,000 in 1995 to a projected $120,000 in 2007. In 1995, there were 1.5 million armed service personnel, and this number was projected to decrease to 1.4 million in 2003.[5] Use linear models for annual cost and personnel to estimate, to the nearest $10 million, the rate of change of total military personnel costs in 2002.

80. **Military Spending in the 1990s** The annual cost per active-duty armed service member in the U.S. increased from $80,000 in 1995 to $90,000 in 2000. In 1990, there were 2 million armed service personnel and this number decreased to 1.5 million in 2000.[6] Use linear models for annual cost and personnel to estimate, to the nearest $10 million, the rate of change of total military personnel costs in 1995.

81. **Biology—Reproduction** The Verhulst model for population growth specifies the reproductive rate of an organism as a function of the total population according to the following formula:

$$R(p) = \frac{r}{1 + kp}$$

where p is the total population in thousands of organisms, r and k are constants that depend on the particular circumstances and the organism being studied, and $R(p)$ is the reproduction rate in thousands of organisms per hour.[7] If $k = 0.125$ and $r = 45$, find $R'(p)$ and then $R'(4)$. Interpret the result.

82. **Biology—Reproduction** Another model, the predator satiation model for population growth, specifies that the reproductive rate of an organism as a function of the total population varies according to the following formula:

$$R(p) = \frac{rp}{1 + kp}$$

where p is the total population in thousands of organisms, r and k are constants that depend on the particular circumstances and the organism being studied, and $R(p)$ is the reproduction rate in new organisms per hour.[8] Given that $k = 0.2$ and $r = 0.08$, find $R'(p)$ and $R'(2)$. Interpret the result.

83. **Embryo Development** Bird embryos consume oxygen from the time the egg is laid through the time the chick hatches. For a typical galliform bird egg, the total oxygen consumption (in milliliters) t days after the egg was laid can be approximated by[9]

$$C(t) = -0.016t^4 + 1.1t^3 - 11t^2 + 3.6t \quad (15 \le t \le 30)$$

(An egg will usually hatch at around $t = 28$.) Suppose that at time $t = 0$ you have a collection of 30 newly laid eggs and that the number of eggs decreases linearly to zero at time $t = 30$ days. How fast is the total oxygen consumption of your collection of embryos changing after 25 days? (Round your answers to 2 significant digits.) Comment on the result.

84. **Embryo Development** Turkey embryos consume oxygen from the time the egg is laid through the time the chick hatches. For a brush turkey, the total oxygen consumption (in milliliters) t days after the egg was laid can be approximated by[10]

$$C(t) = -0.0071t^4 + 0.95t^3 - 22t^2 + 95t \quad (25 \le t \le 50)$$

(An egg will typically hatch at around $t = 50$.) Suppose that at time $t = 0$ you have a collection of 100 newly laid eggs and that the number of eggs decreases linearly to zero at time $t = 50$ days. How fast is the total oxygen consumption of your collection of embryos changing after 40 days? (Round your answer to 2 significant digits.) Interpret the result.

85. **ISP Market Share** The following graphs show approximate market shares, in percentage points, of Microsoft's MSN

[5] Annual costs are adjusted for inflation. SOURCES: Department of Defense, Stephen Daggett, military analyst, Congressional Research Service/*New York Times,* April 19, 2002, p. A21.

[6] Ibid.

[7] SOURCE: *Mathematics in Medicine and the Life Sciences* by F. C. Hoppensteadt and C. S. Peskin (Springer-Verlag, New York, 1992) pp. 20–22.

[8] Ibid.

[9] The model is derived from graphical data published in the article "The Brush Turkey" by Roger S. Seymour, *Scientific American,* December, 1991, pp. 108–114.

[10] Ibid.

Internet service provider, and the combined shares of MSN, Comcast, Earthlink, and AOL for the period 1999–2004.[11]

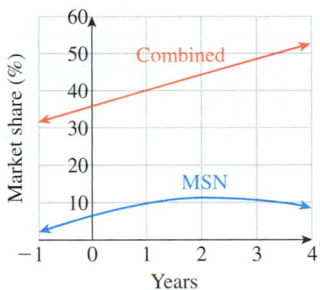

Market share (%) vs. Years

Formulas for the curves are:

MSN: $m(t) = -0.83t^2 + 3.8t + 6.8$ $(-1 \leq t \leq 4)$

Combined: $c(t) = 4.2t + 36$ $(-1 \leq t \leq 4)$

t is time in years since June, 2000.

a. What are represented by the functions $c(t) - m(t)$ and $m(t)/c(t)$?

b. Compute $\dfrac{d}{dt}\left(\dfrac{m(t)}{c(t)}\right)\Big|_{t=3}$ to two significant digits. What does the answer tell you about MSN?

86. _ISP Revenue_ The following graphs show the approximate total revenue, in millions of dollars, of Microsoft's MSN Internet service, as well as the portion of that revenue due to advertising for the period June, 2001–January, 2004.[12]

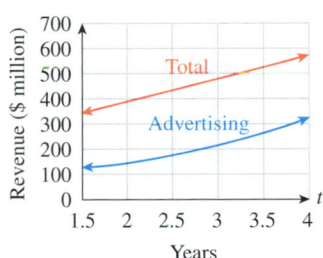

Revenue ($ million) vs. Years

Formulas for the curves are:

Advertising: $a(t) = 20t^2 - 27t + 120$ $(1.5 \leq t \leq 4)$

Total: $s(t) = 96t + 190$ $(1.5 \leq t \leq 4)$

a. What are represented by the functions $s(t) - a(t)$ and $a(t)/s(t)$?

b. Compute $\dfrac{d}{dt}\left(\dfrac{a(t)}{s(t)}\right)\Big|_{t=2}$ to two significant digits. What does the answer tell you about MSN?

---

[11] The curves are regression models. Source for data: Solomon Research, Morgan Stanley/*New York Times,* July 19, 2004.

[12] Ibid.

Communication and Reasoning Exercises

87. ● You have come across the following in a newspaper article: "Revenues of HAL Home Heating Oil Inc. are rising by $4.2 million per year. This is due to an annual increase of 70¢ per gallon in the price HAL charges for heating oil and an increase in sales of 6 million gallons of oil per year." Comment on this analysis.

88. ● Your friend says that because average cost is obtained by dividing the cost function by the number of units x, it follows that the derivative of average cost is the same as marginal cost because the derivative of x is 1. Comment on this analysis.

89. Find a demand function $q(p)$ such that, at a price per item of $p = \$100$, revenue will rise if the price per item is increased.

90. What must be true about a demand function $q(p)$ so that, at a price per item of $p = \$100$, revenue will decrease if the price per item is increased?

91. You and I are both selling a steady 20 T-shirts per day. The price I am getting for my T-shirts is increasing twice as fast as yours, but your T-shirts are currently selling for twice the price of mine. Whose revenue is increasing faster: yours, mine, or neither? Explain.

92. You and I are both selling T-shirts for a steady $20 per shirt. Sales of my T-shirts are increasing at twice the rate of yours, but you are currently selling twice as many as I am. Whose revenue is increasing faster: yours, mine, or neither? Explain.

93. ◆ **_Marginal Product_** *(From the GRE Economics Test)* Which of the following statements about average product and marginal product is correct?
 (A) If average product is decreasing, marginal product must be less than average product.
 (B) If average product is increasing, marginal product must be increasing.
 (C) If marginal product is decreasing, average product must be less than marginal product.
 (D) If marginal product is increasing, average product must be decreasing.
 (E) If marginal product is constant over some range, average product must be constant over that range.

94. ◆ **_Marginal Cost_** *(Based on a Question from the GRE Economics Test)* Which of the following statements about average cost and marginal cost is correct?
 (A) If average cost is increasing, marginal cost must be increasing.
 (B) If average cost is increasing, marginal cost must be decreasing.
 (C) If average cost is increasing, marginal cost must be more than average cost.
 (D) If marginal cost is increasing, average cost must be increasing.
 (E) If marginal cost is increasing, average cost must be larger than marginal cost.

● basic skills   ◆ challenging

11.2 The Chain Rule

We can now find the derivatives of expressions involving powers of x combined using addition, subtraction, multiplication, and division, but we still cannot take the derivative of an expression like $(3x + 1)^{0.5}$. For this we need one more rule. The function $h(x) = (3x + 1)^{0.5}$ is not a sum, difference, product, or quotient. We can use the calculation thought experiment to find the last operation we would perform in calculating $h(x)$.

1. Calculate $3x + 1$.

2. Take the 0.5 power (square root) of the answer.

Thus, the last operation is "take the 0.5 power." We do not yet have a rule for finding the derivative of the 0.5 power of a quantity other than x.

There is a way to build $h(x) = (3x + 1)^{0.5}$ out of two simpler functions: $u(x) = 3x + 1$ (the function that corresponds to the first step in the calculation above) and $f(x) = x^{0.5}$ (the function that corresponds to the second step):

$$h(x) = (3x + 1)^{0.5}$$
$$= [u(x)]^{0.5} \qquad u(x) = 3x + 1$$
$$= f(u(x)) \qquad f(x) = x^{0.5}$$

We say that h is the **composite** of f and u. We read $f(u(x))$ as "f of u of x."

To compute $h(1)$, say, we first compute $3 \cdot 1 + 1 = 4$ and then take the square root of 4, giving $h(1) = 2$. To compute $f(u(1))$ we follow exactly the same steps: First compute $u(1) = 4$ and then $f(u(1)) = f(4) = 2$. We always compute $f(u(x))$ numerically from the inside out: Given x, first compute $u(x)$ and then $f(u(x))$.

Now, f and u are functions *whose derivatives we know*. The *chain rule* allows us to use our knowledge of the derivatives of f and u to find the derivative of $f(u(x))$. For the purposes of stating the rule, let us avoid some of the nested parentheses by abbreviating $u(x)$ as u. Thus, we write $f(u)$ instead of $f(u(x))$ and remember that u is a function of x.

Chain Rule

If f is a differentiable function of u and u is a differentiable function of x, then the composite $f(u)$ is a differentiable function of x, and

$$\frac{d}{dx}[f(u)] = f'(u)\frac{du}{dx} \qquad \text{Chain Rule}$$

In words *The derivative of f(quantity) is the derivative of f, evaluated at that quantity, times the derivative of the quantity.*

quick Examples

1. Take $f(u) = u^2$. Then

$$\frac{d}{dx}[u^2] = 2u\frac{du}{dx} \qquad \text{Since } f'(u) = 2u$$

The derivative of a quantity squared is two times the quantity, times the derivative of the quantity.

2. Take $f(u) = u^{0.5}$. Then

$$\frac{d}{dx}[u^{0.5}] = 0.5u^{-0.5}\frac{du}{dx} \qquad \text{Since } f'(u) = 0.5u^{-0.5}.$$

The derivative of a quantity raised to the 0.5 is 0.5 times the quantity raised to the *-0.5, times the derivative of the quantity.*

As the quick examples illustrate, for every power of a function u whose derivative we know, we now get a "generalized" differentiation rule. The following table gives more examples.

Original Rule	Generalized Rule	In Words
$\dfrac{d}{dx}[x^2] = 2x$	$\dfrac{d}{dx}[u^2] = 2u\dfrac{du}{dx}$	The derivative of a quantity squared is twice the quantity, times the derivative of the quantity.
$\dfrac{d}{dx}[x^3] = 3x^2$	$\dfrac{d}{dx}[u^3] = 3u^2\dfrac{du}{dx}$	The derivative of a quantity cubed is 3 times the quantity squared, times the derivative of the quantity.
$\dfrac{d}{dx}\left(\dfrac{1}{x}\right) = -\dfrac{1}{x^2}$	$\dfrac{d}{dx}\left(\dfrac{1}{u}\right) = -\dfrac{1}{u^2}\dfrac{du}{dx}$	The derivative of 1 over a quantity is negative 1 over the quantity squared, times the derivative of the quantity.
Power Rule	**Generalized Power Rule**	**In Words**
$\dfrac{d}{dx}[x^n] = nx^{n-1}$	$\dfrac{d}{dx}[u^n] = nu^{n-1}\dfrac{du}{dx}$	The derivative of a quantity raised to the n is n times the quantity raised to the $n-1$, times the derivative of the quantity.

To motivate the chain rule, let us see why it is true in the special case when $f(u) = u^3$, where the chain rule tells us that

$$\frac{d}{dx}[u^3] = 3u^2\frac{du}{dx} \qquad \textcolor{red}{\text{Generalized Power Rule with } n = 3}$$

But we could have done this using the product rule instead:

$$\frac{d}{dx}[u^3] = \frac{d}{dx}[u \cdot u \cdot u] = \frac{du}{dx}u \cdot u + u\frac{du}{dx}u + u \cdot u\frac{du}{dx} = 3u^2\frac{du}{dx}$$

which gives us the same result. A similar argument works for $f(u) = u^n$ where $n = 2, 3, 4, \ldots$ We can then use the quotient rule and the chain rule for positive powers to verify the generalized power rule for *negative* powers as well. For the case of a general differentiable function f, the proof of the chain rule is beyond the scope of this book, but see the note in the margin.

For the proof of the chain rule for a general differentiable function f, go online and follow:

Chapter 11

→ Proof of Chain Rule

Example 1 Using the Chain Rule

Compute the following derivatives.

a. $\dfrac{d}{dx}[(2x^2 + x)^3]$ **b.** $\dfrac{d}{dx}[(x^3 + x)^{100}]$ **c.** $\dfrac{d}{dx}\sqrt{3x + 1}$

Solution

a. Using the calculation thought experiment, we see that the last operation we would perform in calculating $(2x^2 + x)^3$ is that of *cubing*. Thus we think of $(2x^2 + x)^3$ as *a quantity cubed*. There are two similar methods we can use to calculate its derivative.

Method 1: Using the formula We think of $(2x^2 + x)^3$ as u^3, where $u = 2x^2 + x$. By the formula,

$$\frac{d}{dx}[u^3] = 3u^2\frac{du}{dx} \qquad \text{\textcolor{red}{Generalized Power Rule}}$$

Now substitute for u:

$$\frac{d}{dx}[(2x^2 + x)^3] = 3(2x^2 + x)^2\frac{d}{dx}(2x^2 + x)$$
$$= 3(2x^2 + x)^2(4x + 1)$$

Method 2: Using the verbal form If we prefer to use the verbal form, we get:

The derivative of $(2x^2 + x)$ cubed is three times $(2x^2 + x)$ squared, times the derivative of $(2x^2 + x)$.

In symbols,

$$\frac{d}{dx}[(2x^2 + x)^3] = 3(2x^2 + x)^2(4x + 1)$$

as we obtained above.

b. First, the calculation thought experiment: If we were computing $(x^3 + x)^{100}$, the last operation we would perform is *raising a quantity to the power* 100. Thus we are dealing with *a quantity raised to the power* 100, and so we must again use the generalized power rule. According to the verbal form of the generalized power rule, the derivative of a quantity raised to the power 100 is 100 times that quantity to the power 99, times the derivative of that quantity. In symbols,

$$\frac{d}{dx}[(x^3 + x)^{100}] = 100(x^3 + x)^{99}(3x^2 + 1)$$

c. We first rewrite the expression $\sqrt{3x + 1}$ as $(3x + 1)^{0.5}$ and then use the generalized power rule as in parts (a) and (b):

The derivative of a quantity raised to the 0.5 is 0.5 times the quantity raised to the −0.5, times the derivative of the quantity.

Thus,

$$\frac{d}{dx}[(3x + 1)^{0.5}] = 0.5(3x + 1)^{-0.5} \cdot 3 = 1.5(3x + 1)^{-0.5}$$

+ *Before we go on...* The following are examples of common errors in solving Example 1(b):

$$\text{``}\frac{d}{dx}[(x^3+x)^{100}] = 100(3x^2+1)^{99}\text{''} \quad \text{✗ } \textit{WRONG!}$$

$$\text{``}\frac{d}{dx}[(x^3+x)^{100}] = 100(x^3+x)^{99}\text{''} \quad \text{✗ } \textit{WRONG!}$$

Remember that the generalized power rule says that the derivative of a quantity to the power 100 is 100 times *that same quantity* raised to the power 99, *times the derivative of that quantity.* ∎

Q: *It seems that there are now two formulas for the derivative of an nth power:*

(1) $\dfrac{d}{dx}[x^n] = nx^{n-1}$

(2) $\dfrac{d}{dx}[u^n] = nu^{n-1}\dfrac{du}{dx}$

Which one do I use?

A: Formula 1 is the original power rule, which applies only to a power of *x*. For instance, it applies to x^{10}, but it does not apply to $(2x+1)^{10}$ because the quantity that is being raised to a power is not *x*. Formula 2 applies to a power of any *function of x*, such as $(2x+1)^{10}$. It can even be used in place of the original power rule. For example, if we take $u = x$ in Formula 2, we obtain

$$\frac{d}{dx}[x^n] = nx^{n-1}\frac{dx}{dx}$$

$$= nx^{n-1} \qquad \text{\color{red}{The derivative of } } x \text{ \color{red}{with respect to } } x \text{ \color{red}{is 1.}}$$

Thus, the generalized power rule really *is* a generalization of the original power rule, as its name suggests. ∎

Example 2 More Examples Using the Chain Rule

Find: **a.** $\dfrac{d}{dx}[(2x^5+x^2-20)^{-2/3}]$ **b.** $\dfrac{d}{dx}\left[\dfrac{1}{\sqrt{x+2}}\right]$ **c.** $\dfrac{d}{dx}\left[\dfrac{1}{x^2+x}\right]$.

Solution

Each of the given functions is, or can be rewritten as, a power of a function whose derivative we know. Thus, we can use the method of Example 1.

a. $\dfrac{d}{dx}[(2x^5+x^2-20)^{-2/3}] = -\dfrac{2}{3}(2x^5+x^2-20)^{-5/3}(10x^4+2x)$

b. $\dfrac{d}{dx}\left[\dfrac{1}{\sqrt{x+2}}\right] = \dfrac{d}{dx}(x+2)^{-1/2} = -\dfrac{1}{2}(x+2)^{-3/2}\cdot 1 = -\dfrac{1}{2(x+2)^{3/2}}$

c. $\dfrac{d}{dx}\left[\dfrac{1}{x^2+x}\right] = \dfrac{d}{dx}(x^2+x)^{-1} = -(x^2+x)^{-2}(2x+1) = -\dfrac{2x+1}{(x^2+x)^2}$

+ *Before we go on...* In Example 2(c), we could have used the quotient rule instead of the generalized power rule. We can think of the quantity $1/(x^2 + x)$ in two different ways using the calculation thought experiment:

1. As 1 divided by something—in other words, as a quotient

2. As something raised to the -1 power

Of course, we get the same derivative using either approach. ∎

We now look at some more complicated examples.

Example 3 Harder Examples Using the Chain Rule

Find $\dfrac{dy}{dx}$ in each case. **a.** $y = [(x + 1)^{-2.5} + 3x]^{-3}$ **b.** $y = (x + 10)^3 \sqrt{1 - x^2}$

Solution

a. The calculation thought experiment tells us that the last operation we would perform in calculating y is raising the quantity $[(x + 1)^{-2.5} + 3x]$ to the power -3. Thus, we use the generalized power rule.

$$\frac{dy}{dx} = -3[(x + 1)^{-2.5} + 3x]^{-4} \frac{d}{dx}[(x + 1)^{-2.5} + 3x]$$

We are not yet done; we must still find the derivative of $(x + 1)^{-2.5} + 3x$. Finding the derivative of a complicated function in several steps helps to keep the problem manageable. Continuing, we have

$$\frac{dy}{dx} = -3[(x + 1)^{-2.5} + 3x]^{-4} \frac{d}{dx}[(x + 1)^{-2.5} + 3x]$$

$$= -3[(x + 1)^{-2.5} + 3x]^{-4} \left[\frac{d}{dx}[(x + 1)^{-2.5}] + \frac{d}{dx}(3x) \right] \quad \text{Derivative of a sum}$$

Now we have two derivatives left to calculate. The second of these we know to be 3, and the first is the derivative of a quantity raised to the -2.5 power. Thus

$$\frac{dy}{dx} = -3[(x + 1)^{-2.5} + 3x]^{-4}[-2.5(x + 1)^{-3.5} \cdot 1 + 3]$$

b. The expression $(x + 10)^3 \sqrt{1 - x^2}$ is a product, so we use the product rule:

$$\frac{d}{dx}\left[(x + 10)^3 \sqrt{1 - x^2}\right] = \left(\frac{d}{dx}[(x + 10)^3]\right) \sqrt{1 - x^2} + (x + 10)^3 \left(\frac{d}{dx}\sqrt{1 - x^2}\right)$$

$$= 3(x + 10)^2 \sqrt{1 - x^2} + (x + 10)^3 \frac{1}{2\sqrt{1 - x^2}}(-2x)$$

$$= 3(x + 10)^2 \sqrt{1 - x^2} - \frac{x(x + 10)^3}{\sqrt{1 - x^2}}$$

Applications

The next example is a new treatment of Example 3 from Section 10.8.

Example 4 Marginal Product

Precision Manufacturers is informed by a consultant that its annual profit is given by

$$P = -200{,}000 + 4000q - 0.46q^2 - 0.00001q^3$$

where q is the number of surgical lasers it sells each year. The consultant also informs Precision that the number of surgical lasers it can manufacture each year depends on the number n of assembly line workers it employs according to the equation

$$q = 100n \qquad \text{Each worker contributes 100 lasers per year}$$

Use the chain rule to find the marginal product $\dfrac{dP}{dn}$.

Solution

We could calculate the marginal product by substituting the expression for q in the expression for P to obtain P as a function of n (as given in Chapter 10) and then finding dP/dn. Alternatively—and this will simplify the calculation—we can use the chain rule. To see how the chain rule applies, notice that P is a function of q, where q in turn is given as a function of n. By the chain rule,

$$\frac{dP}{dn} = P'(q)\frac{dq}{dn} \qquad \text{Chain Rule}$$

$$= \frac{dP}{dq}\frac{dq}{dn} \qquad \text{Notice how the ``quantities'' } dq \text{ appear to cancel}$$

Now we compute

$$\frac{dP}{dq} = 4000 - 0.92q - 0.00003q^2$$

and $\qquad \dfrac{dq}{dn} = 100$

Substituting into the equation for $\dfrac{dP}{dn}$ gives

$$\frac{dP}{dn} = (4000 - 0.92q - 0.00003q^2)(100)$$

$$= 400{,}000 - 92q - 0.003q^2$$

Notice that the answer has q as a variable. We can express dP/dn as a function of n by substituting $100n$ for q:

$$\frac{dP}{dn} = 400{,}000 - 92(100n) - 0.003(100n)^2$$

$$= 400{,}000 - 9200n - 30n^2$$

The equation

$$\frac{dP}{dn} = \frac{dP}{dq}\frac{dq}{dn}$$

in the example above is an appealing way of writing the chain rule because it suggests that the "quantities" dq cancel. In general, we can write the chain rule as follows.

Chain Rule in Differential Notation

If y is a differentiable function of u, and u is a differentiable function of x, then

$$\frac{dy}{dx} = \frac{dy}{du}\frac{du}{dx}$$

Notice how the units cancel:

$$\frac{\text{Units of } y}{\text{Units of } x} = \frac{\text{Units of } y}{\text{Units of } u}\frac{\text{Units of } u}{\text{Units of } x}$$

quick Example   If $y = u^3$, where $u = 4x + 1$, then

$$\frac{dy}{dx} = \frac{dy}{du}\frac{du}{dx} = 3u^2 \cdot 4 = 12u^2 = 12(4x + 1)^2$$

You can see one of the reasons we still use Leibniz differential notation: The chain rule looks like a simple "cancellation" of du terms.

Example 5 Marginal Revenue

Suppose a company's weekly revenue R is given as a function of the unit price p, and p in turn is given as a function of weekly sales q (by means of a demand equation). If

$$\left.\frac{dR}{dp}\right|_{q=1000} = \$40 \text{ per } \$1 \text{ increase in price}$$

and

$$\left.\frac{dp}{dq}\right|_{q=1000} = -\$20 \text{ per additional item sold per week}$$

find the marginal revenue when sales are 1000 items per week.

Solution

The marginal revenue is $\dfrac{dR}{dq}$. By the chain rule, we have

$$\frac{dR}{dq} = \frac{dR}{dp}\frac{dp}{dq} \qquad \text{Units: Revenue per item =}$$
Revenue per $1 price increase × price increase per additional item

Because we are interested in the marginal revenue at a demand level of 1000 items per week, we have

$$\left.\frac{dR}{dq}\right|_{q=1000} = (40)(-20) = -\$800 \text{ per additional item sold}$$

Thus, if the price is lowered to increase the demand from 1000 to 1001 items per week, the weekly revenue will drop by approximately $800.

Look again at the way the terms "du" appeared to cancel in the differential formula $\dfrac{dy}{dx} = \dfrac{dy}{du}\dfrac{du}{dx}$. In fact, the chain rule tells us more:

Manipulating Derivatives in Differential Notation

1. Suppose y is a function of x. Then, thinking of x as a function of y (as, for instance, when we can solve for x)[*] one has

$$\frac{dx}{dy} = \frac{1}{\left(\dfrac{dy}{dx}\right)}, \text{ provided } \frac{dy}{dx} \neq 0 \qquad \text{Notice again how } \frac{dy}{dx} \text{ behaves like a fraction.}$$

quick Example

In the demand equation $q = -0.2p - 8$, we have $\dfrac{dq}{dp} = -0.2$. Therefore,

$$\frac{dp}{dq} = \frac{1}{\left(\dfrac{dq}{dp}\right)} = \frac{1}{-0.2} = -5$$

2. Suppose x and y are functions of t. Then, thinking of y as a function of x (as, for instance, when we can solve for t as a function of x, and hence obtain y as a function of x) one has

$$\frac{dy}{dx} = \frac{dy/dt}{dx/dt} \qquad \text{The terms } dt \text{ appear to cancel.}$$

quick Example

If $x = 3 - 0.2t$ and $y = 6 + 6t$, then

$$\frac{dy}{dx} = \frac{dy/dt}{dx/dt} = \frac{6}{-0.2} = -30$$

_____

[*] The notion of "thinking of x as a function of y" will be made more precise in Section 11.4.

To see why the above formulas work, notice that the second formula,

$$\frac{dy}{dx} = \frac{\left(\dfrac{dy}{dt}\right)}{\left(\dfrac{dx}{dt}\right)}$$

can be written as

$$\frac{dy}{dx}\frac{dx}{dt} = \frac{dy}{dt} \qquad \text{Multiply both sides by } \frac{dx}{dt}$$

which is just the differential form of the chain rule. For the first formula, use the second formula with y playing the role of t:

$$\frac{dy}{dx} = \frac{dy/dy}{dx/dy}$$

$$= \frac{1}{dx/dy} \qquad \frac{dy}{dy} = \frac{d}{dy}[y] = 1$$

FAQs **Using the Chain Rule**

Q: *How do I decide whether or not to use the chain rule when taking a derivative?*

A: Use the Calculation Thought Experiment (Section 11.1): Given an expression, consider the steps you would use in computing its value.

- If the last step is *raising a quantity to a power*, as in $\left(\dfrac{x^2-1}{x+4}\right)^4$, then the first step to use is the chain rule (in the form of the generalized power rule):

$$\frac{d}{dx}\left(\frac{x^2-1}{x+4}\right)^4 = 4\left(\frac{x^2-1}{x+4}\right)^3 \frac{d}{dx}\left(\frac{x^2-1}{x+4}\right)$$

Then use the appropriate rules to finish the computation. You may need to again use the Calculation Thought Experiment to decide on the next step (here the quotient rule):

$$= 4\left(\frac{x^2-1}{x+4}\right)^3 \frac{(2x)(x+4)-(x^2-1)(1)}{(x+4)^2}$$

- If the last step is *division*, as in $\dfrac{(x^2-1)}{(3x+4)^4}$, then the first step to use is the quotient rule:

$$\frac{d}{dx}\frac{(x^2-1)}{(3x+4)^4} = \frac{(2x)(3x+4)^4-(x^2-1)\dfrac{d}{dx}(3x+4)^4}{(3x+4)^8}$$

Then use the appropriate rules to finish the computation (here the chain rule):

$$= \frac{(2x)(3x+4)^4-(x^2-1)4(3x+4)^3(3)}{(3x+4)^8}$$

- If the last step is *multiplication, addition, subtraction, or multiplication by a constant*, then the first rule to use is the product rule, or the rule for sums, differences or constant multiples as appropriate. ■

Q: *Every time I compute the derivative, I leave something out. How do I make sure I am really done when taking the derivative of a complicated-looking expression?*

A: Until you are an expert at taking derivatives, the key is to use one rule at a time and write out each step, rather than trying to compute the derivative in a single step. To illustrate this, try computing the derivative of $(x+10)^3\sqrt{1-x^2}$ in Example 3(b) in two ways: First try to compute it in a single step, and then compute it by writing out each step as shown in the example. How do your results compare? For more practice, try Exercises 83 and 84 below. ■

11.2 EXERCISES

- ● denotes basic skills exercises
- ◆ denotes challenging exercises
- tech Ex indicates exercises that should be solved using technology

Calculate the derivatives of the functions in Exercises 1–46.
hint [see Example 1]

1. ● $f(x) = (2x+1)^2$
2. ● $f(x) = (3x-1)^2$
3. ● $f(x) = (x-1)^{-1}$
4. ● $f(x) = (2x-1)^{-2}$
5. ● $f(x) = (2-x)^{-2}$
6. ● $f(x) = (1-x)^{-1}$
7. ● $f(x) = (2x+1)^{0.5}$
8. ● $f(x) = (-x+2)^{1.5}$

9. ● $f(x) = (4x-1)^{-1}$
10. ● $f(x) = (x+7)^{-2}$
11. ● $f(x) = \dfrac{1}{3x-1}$
12. ● $f(x) = \dfrac{1}{(x+1)^2}$
13. ● $f(x) = (x^2+2x)^4$
14. ● $f(x) = (x^3-x)^3$
15. ● $f(x) = (2x^2-2)^{-1}$
16. ● $f(x) = (2x^3+x)^{-2}$
17. ● $g(x) = (x^2-3x-1)^{-5}$
18. ● $g(x) = (2x^2+x+1)^{-3}$
19. ● $h(x) = \dfrac{1}{(x^2+1)^3}$ hint [see Example 2]

● basic skills   ◆ challenging   tech Ex technology exercise

20. ● $h(x) = \dfrac{1}{(x^2 + x + 1)^2}$

21. ● $r(x) = (0.1x^2 - 4.2x + 9.5)^{1.5}$

22. ● $r(x) = (0.1x - 4.2x^{-1})^{0.5}$

23. ● $r(s) = (s^2 - s^{0.5})^4$ **24.** ● $r(s) = (2s + s^{0.5})^{-1}$

25. ● $f(x) = \sqrt{1 - x^2}$ **26.** ● $f(x) = \sqrt{x + x^2}$

27. ● $h(x) = 2[(x + 1)(x^2 - 1)]^{-1/2}$ *hint* [see Example 3]

28. ● $h(x) = 3[(2x - 1)(x - 1)]^{-1/3}$

29. ● $h(x) = (3.1x - 2)^2 - \dfrac{1}{(3.1x - 2)^2}$

30. ● $h(x) = \left[3.1x^2 - 2 - \dfrac{1}{3.1x - 2}\right]^2$

31. ● $f(x) = [(6.4x - 1)^2 + (5.4x - 2)^3]^2$

32. ● $f(x) = (6.4x - 3)^{-2} + (4.3x - 1)^{-2}$

33. ● $f(x) = (x^2 - 3x)^{-2}(1 - x^2)^{0.5}$

34. ● $f(x) = (3x^2 + x)(1 - x^2)^{0.5}$

35. ● $s(x) = \left(\dfrac{2x + 4}{3x - 1}\right)^2$ **36.** ● $s(x) = \left(\dfrac{3x - 9}{2x + 4}\right)^3$

37. ● $g(z) = \left(\dfrac{z}{1 + z^2}\right)^3$ **38.** ● $g(z) = \left(\dfrac{z^2}{1 + z}\right)^2$

39. ● $f(x) = [(1 + 2x)^4 - (1 - x)^2]^3$

40. ● $f(x) = [(3x - 1)^2 + (1 - x)^5]^2$

41. ● $t(x) = [2 + (x + 1)^{-0.1}]^{4.3}$

42. ● $t(x) = [(x + 1)^{0.1} - 4x]^{-5.1}$

43. $r(x) = \left(\sqrt{2x + 1} - x^2\right)^{-1}$

44. $r(x) = \left(\sqrt{x + 1} + \sqrt{x}\right)^3$

45. $f(x) = \left(1 + \left(1 + (1 + 2x)^3\right)^3\right)^3$

46. $f(x) = 2x + \left(2x + (2x + 1)^3\right)^3$

Find the indicated derivatives in Exercises 47–54. In each case, the independent variable is a (unspecified) function of t.

47. ● $y = x^{100} + 99x^{-1}$. Find $\dfrac{dy}{dt}$.

48. ● $y = x^{0.5}(1 + x)$. Find $\dfrac{dy}{dt}$.

49. ● $s = \dfrac{1}{r^3} + r^{0.5}$. Find $\dfrac{ds}{dt}$.

50. ● $s = r + r^{-1}$. Find $\dfrac{ds}{dt}$.

51. ● $V = \dfrac{4}{3}\pi r^3$. Find $\dfrac{dV}{dt}$.

52. ● $A = 4\pi r^2$. Find $\dfrac{dA}{dt}$.

53. $y = x^3 + \dfrac{1}{x}$, $x = 2$ when $t = 1$, $\left.\dfrac{dx}{dt}\right|_{t=1} = -1$.

Find $\left.\dfrac{dy}{dt}\right|_{t=1}$.

54. $y = \sqrt{x} + \dfrac{1}{\sqrt{x}}$, $x = 9$ when $t = 1$, $\left.\dfrac{dx}{dt}\right|_{t=1} = -1$.

Find $\left.\dfrac{dy}{dt}\right|_{t=1}$.

In Exercises 55–60, compute the indicated derivative using the chain rule. *hint* [see Quick Examples on p. 817]

55. ● $y = 3x - 2;\ \dfrac{dx}{dy}$

56. ● $y = 8x + 4;\ \dfrac{dx}{dy}$

57. ● $x = 2 + 3t,\ y = -5t;\ \dfrac{dy}{dx}$

58. ● $x = 1 - t/2,\ y = 4t - 1;\ \dfrac{dy}{dx}$

59. ● $y = 3x^2 - 2x;\ \left.\dfrac{dx}{dy}\right|_{x=1}$

60. ● $y = 3x - \dfrac{2}{x};\ \left.\dfrac{dx}{dy}\right|_{x=2}$

Applications

61. ● *Food Versus Education* The percentage y (of total personal consumption) an individual spends on food is approximately

$$y = 35x^{-0.25} \text{ percentage points} \quad (6.5 \le x \le 17.5)$$

where x is the percentage the individual spends on education.[13] An individual finds that she is spending

$$x = 7 + 0.2t$$

percent of her personal consumption on education, where t is time in months since January 1. Use direct substitution to express the percentage y as a function of time t (do not simplify the expression) and then use the chain rule to estimate how fast the percentage she spends on food is changing on November 1. Be sure to specify the units.

62. ● *Food Versus Recreation* The percentage y (of total personal consumption) an individual spends on food is approximately

$$y = 33x^{-0.63} \text{ percentage points} \quad (2.5 \le x \le 4.5)$$

where x is the percentage the individual spends on recreation.[14] A college student finds that he is spending

$$x = 3.5 + 0.1t$$

percent of his personal consumption on recreation, where t is time in months since January 1. Use direct substitution to

---

[13] Model based on historical and projected data from 1908–2010. SOURCES: Historical data, Bureau of Economic Analysis; projected data, Bureau of Labor Statistics/*New York Times,* December 1, 2003, p. C2.

[14] Ibid.

● basic skills   ◆ challenging   *tech* Ex technology exercise

express the percentage y as a function of time t (do not simplify the expression) and then use the chain rule to estimate how fast the percentage he spends on food is changing on November 1. Be sure to specify the units.

63. ● **Marginal Product** Paramount Electronics has an annual profit given by

$$P = -100,000 + 5000q - 0.25q^2$$

where q is the number of laptop computers it sells each year. The number of laptop computers it can make and sell each year depends on the number n of electrical engineers Paramount employs, according to the equation

$$q = 30n + 0.01n^2$$

Use the chain rule to find $\left.\dfrac{dp}{dn}\right|_{n=10}$ and interpret the result. *hint* [see Example 4]

64. ● **Marginal Product** Refer back to Exercise 63. The average profit $\bar{P}$ per computer is given by dividing the total profit P by q:

$$\bar{P} = -\frac{100,000}{q} + 5000 - 0.25q$$

Determine the **marginal average product**, $d\bar{P}/dn$ at an employee level of 10 engineers. Interpret the result.

65. ● **Marginal Revenue** The weekly revenue from the sale of rubies at Royal Ruby Retailers (RRR) is increasing at a rate of $40 per $1 increase in price, and the price is decreasing at a rate of $0.75 per additional ruby sold. What is the marginal revenue? (Be sure to state the units of measurement.) Interpret the result. *hint* [see Example 5]

66. ● **Marginal Revenue** The weekly revenue from the sale of emeralds at Eduardo's Emerald Emporium (EEE) is decreasing at a rate of €500 per €1 increase in price, and the price is decreasing at a rate of €0.45 per additional emerald sold. What is the marginal revenue? (Be sure to state the units of measurement.) Interpret the result.

67. ● **Crime Statistics** The murder rate in large cities (over 1 million residents) can be related to that in smaller cities (500,000–1,000,000 residents) by the following linear model:[15]

$$y = 1.5x - 1.9 \quad (15 \le x \le 25)$$

where y is the murder rate (in murders per 100,000 residents each year) in large cities and x is the murder rate in smaller cities. During the period 1991–1998, the murder rate in small cities was decreasing at an average rate of 2 murders per 100,000 residents each year. Use the chain rule to estimate how fast the murder rate was changing in larger cities during that period. (Show how you used the chain rule in your answer.)

68. ● **Crime Statistics** Following is a quadratic model relating the murder rates described in the preceding exercise:

$$y = 0.1x^2 - 3x + 39 \quad (15 \le x \le 25)$$

In 1996, the murder rate in smaller cities was approximately 22 murders per 100,000 residents each year and was decreasing at a rate of approximately 2.5 murders per 100,000 residents each year. Use the chain rule to estimate how fast the murder rate was changing for large cities. (Show how you used the chain rule in your answer.)

69. ● **Ecology** Manatees are grazing sea mammals sometimes referred to as "sea sirens." Increasing numbers of manatees have been killed by boats off the Florida coast, as shown in the following chart:

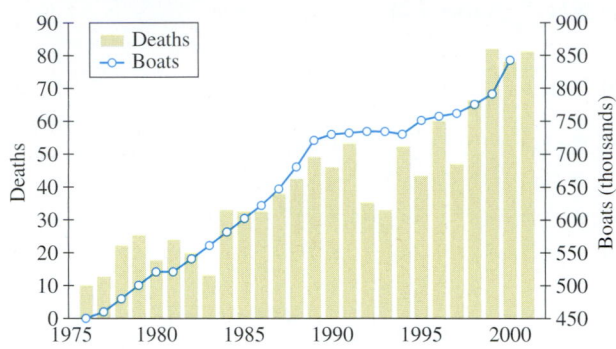

Since 1976 the number M of manatees killed by boats each year is roughly linear, with

$$M(t) = 2.48t + 6.87 \text{ manatees} \quad (1 \le t \le 26)$$

where t is the number of years since 1975.[16] Over the same period, the total number B of boats registered in Florida has also been increasing at a roughly linear rate, given by

$$B(t) = 15,700t + 444,000 \text{ boats} \quad (1 \le t \le 25)$$

Use the chain rule to give an estimate of dM/dB. What does the answer tell you about manatee deaths? *hint* [see Quick Examples on p. 817]

70. ● **Ecology** Refer to Exercise 69. If we use only the data from 1990 on, we obtain the following linear models:

$$M(t) = 3.87t + 35.0 \text{ manatees} \quad (0 \le t \le 11)$$

$$B(t) = 9020t + 712,000 \text{ boats} \quad (0 \le t \le 10)$$

where t is the number of years since 1990.[17] Repeat Exercise 69 using these models.

71. Pollution An offshore oil well is leaking oil and creating a circular oil slick. If the radius of the slick is growing at a rate of 2 miles/hour, find the rate at which the area is increasing when the radius is 3 miles. (The area of a disc of radius r is $A = \pi r^2$.)

---

[15] The model is a linear regression model. Source for data: Federal Bureau of Investigation, Supplementary Homicide Reports/*New York Times,* May 29, 2000, p. A12.

[16] Rounded regression model. Sources for data: Florida Department of Highway Safety & Motor Vehicles, Florida Marine Institute/*New York Times,* February 12, 2002, p. F4.

[17] Ibid.

● basic skills   ◆ challenging   **tech** Ex technology exercise

72. Mold A mold culture in a dorm refrigerator is circular and growing. The radius is growing at a rate of 0.3 cm/day. How fast is the area growing when the culture is 4 centimeters in radius? (The area of a disc of radius r is $A = \pi r^2$.)

73. Budget Overruns The Pentagon is planning to build a new, spherical satellite. As is typical in these cases, the specifications keep changing, so that the size of the satellite keeps growing. In fact, the radius of the planned satellite is growing 0.5 feet per week. Its cost will be $1000 per cubic foot. At the point when the plans call for a satellite 10 feet in radius, how fast is the cost growing? (The volume of a solid sphere of radius r is $V = \frac{4}{3}\pi r^3$.)

74. Soap Bubbles The soap bubble I am blowing has a radius that is growing at a rate of 4 cm/s. How fast is the surface area growing when the radius is 10 cm? (The surface area of a sphere of radius r is $S = 4\pi r^2$.)

75. `tech`Ex **Revenue Growth** The demand for the Cyberpunk II arcade video game is modeled by the logistic curve

$$q(t) = \frac{10,000}{1 + 0.5e^{-0.4t}}$$

where $q(t)$ is the total number of units sold t months after its introduction.

a. Use technology to estimate $q'(4)$.

b. Assume that the manufacturers of Cyberpunk II sell each unit for $800. What is the company's marginal revenue dR/dq?

c. Use the chain rule to estimate the rate at which revenue is growing 4 months after the introduction of the video game.

76. `tech`Ex **Information Highway** The amount of information transmitted each month in the early years of the Internet (1988 to 1994) can be modeled by the equation

$$q(t) = \frac{2e^{0.69t}}{3 + 1.5e^{-0.4t}} \quad (0 \le t \le 6)$$

where q is the amount of information transmitted each month in billions of data packets and t is the number of years since the start of 1988.[18]

a. Use technology to estimate $q'(2)$.

b. Assume that it costs $5 to transmit a million packets of data. What is the marginal cost $C'(q)$?

c. How fast was the cost increasing at the start of 1990?

Money Stock *Exercises 77–80 are based on the following demand function for money (taken from a question on the GRE economics test):*

$$M_d = 2 \times y^{0.6} \times r^{-0.3} \times p$$

where

M_d = demand for nominal money balances (money stock)
y = real income

---

[18] This is the authors' model, based on figures published in *New York Times,* November 3, 1993.

r = an index of interest rates
p = an index of prices

*These exercises also use the idea of **percentage rate of growth:***

$$\text{Percentage Rate of Growth of } M = \frac{\text{Rate of Growth of } M}{M}$$
$$= \frac{dM/dt}{M}$$

77. ◆ (From the GRE economics test) If the interest rate and price level are to remain constant while real income grows at 5 percent per year, the money stock must grow at what percent per year?

78. ◆ (From the GRE economics test) If real income and price level are to remain constant while the interest rate grows at 5 percent per year, the money stock must change by what percent per year?

79. ◆ (From the GRE economics test) If the interest rate is to remain constant while real income grows at 5 percent per year and the price level rises at 5 percent per year, the money stock must grow at what percent per year?

80. ◆ (From the GRE economics test) If real income grows by 5 percent per year, the interest rate grows by 2 percent per year, and the price level drops by 3 percent per year, the money stock must change by what percent per year?

Communication and Reasoning Exercises

81. ● Complete the following: The derivative of one over a glob is -1 over ____.

82. ● Complete the following: The derivative of the square root of a glob is 1 over ____.

83. ● Why was the following marked wrong?

$$\frac{d}{dx}[(3x^3 - x)^3] = 3(9x^2 - 1)^2 \qquad \text{✗ \quad WRONG!}$$

84. ● Why was the following marked wrong?

$$\frac{d}{dx}\left[\left(\frac{3x^2 - 1}{2x - 2}\right)^3\right] = 3\left(\frac{3x^2 - 1}{2x - 2}\right)^2\left(\frac{6x}{2}\right) \quad \text{✗ \quad WRONG!}$$

85. Formulate a simple procedure for deciding whether to apply first the chain rule, the product rule, or the quotient rule when finding the derivative of a function.

86. Give an example of a function f with the property that calculating $f'(x)$ requires use of the following rules in the given order: (1) the chain rule, (2) the quotient rule, and (3) the chain rule.

87. ◆ Give an example of a function f with the property that calculating $f'(x)$ requires use of the chain rule five times in succession.

88. ◆ What can you say about composites of linear functions?

11.3 Derivatives of Logarithmic and Exponential Functions

At this point, we know how to take the derivative of any algebraic expression in x (involving powers, radicals, and so on). We now turn to the derivatives of logarithmic and exponential functions.

Derivative of the Natural Logarithm

$$\frac{d}{dx}[\ln x] = \frac{1}{x} \qquad \text{Recall that } \ln x = \log_e x$$

quick Examples

1. $\dfrac{d}{dx}[3 \ln x] = 3 \cdot \dfrac{1}{x} = \dfrac{3}{x}$ Derivative of a constant times a function

2. $\dfrac{d}{dx}[x \ln x] = 1 \cdot \ln x + x \cdot \dfrac{1}{x}$ Product rule, because $x \ln x$ is a product

$\qquad\qquad = \ln x + 1$

The above simple formula works only for the natural logarithm (the logarithm with base e). For logarithms with bases other than e, we have the following:

Derivative of the Logarithm with Base b

$$\frac{d}{dx}[\log_b x] = \frac{1}{x \ln b} \qquad \text{Notice that, if } b = e, \text{ we get the same formula as above}$$

quick Examples

1. $\dfrac{d}{dx}[\log_3 x] = \dfrac{1}{x \ln 3} \approx \dfrac{1}{1.0986x}$

2. $\dfrac{d}{dx}[\log_2(x^4)] = \dfrac{d}{dx}(4 \log_2 x)$ We used the logarithm identity $\log_b(x^r) = r \log_b x$

$\qquad\qquad = 4 \cdot \dfrac{1}{x \ln 2} \approx \dfrac{4}{0.6931x}$

Derivation of the Formulas $\frac{d}{dx}[\ln x] = \frac{1}{x}$ and $\frac{d}{dx}[\log_b x] = \frac{1}{x \ln b}$

To compute $\dfrac{d}{dx}[\ln x]$, we need to use the definition of the derivative. We also use properties of the logarithm to help evaluate the limit.

$$\frac{d}{dx}[\ln x] = \lim_{h \to 0} \frac{\ln(x + h) - \ln x}{h} \qquad \text{Definition of the derivative}$$

$$= \lim_{h \to 0} \frac{1}{h}[\ln(x + h) - \ln x] \qquad \text{Algebra}$$

$$= \lim_{h \to 0} \frac{1}{h} \ln\left(\frac{x + h}{x}\right) \qquad \text{Properties of the logarithm}$$

$$= \lim_{h \to 0} \frac{1}{h} \ln\left(1 + \frac{h}{x}\right) \qquad \text{Algebra}$$

$$= \lim_{h \to 0} \ln\left(1 + \frac{h}{x}\right)^{1/h} \qquad \text{Properties of the logarithm}$$

which we rewrite as

$$\lim_{h \to 0} \ln\left[\left(1 + \frac{1}{(x/h)}\right)^{x/h}\right]^{1/x}$$

As $h \to 0^+$, the quantity x/h is getting large and positive, and so the quantity in brackets is approaching e (see the definition of e in Section 9.2), which leaves us with

$$\ln[e]^{1/x} = \frac{1}{x} \ln e = \frac{1}{x}$$

which is the derivative we are after.[19] What about the limit as $h \to 0^-$? We will glide over that case and leave it for the interested reader to pursue.[20]

The rule for the derivative of $\log_b x$ follows from the fact that $\log_b x = \ln x / \ln b$.

If we were to take the derivative of the natural logarithm of a *quantity* (a function of x), rather than just x, we would need to use the chain rule:

Derivatives of Logarithms of Functions

Original Rule	*Generalized Rule*	*In Words*
$\dfrac{d}{dx}[\ln x] = \dfrac{1}{x}$	$\dfrac{d}{dx}[\ln u] = \dfrac{1}{u}\dfrac{du}{dx}$	The derivative of the natural logarithm of a quantity is 1 over that quantity, times the derivative of that quantity.
$\dfrac{d}{dx}[\log_b x] = \dfrac{1}{x \ln b}$	$\dfrac{d}{dx}[\log_b u] = \dfrac{1}{u \ln b}\dfrac{du}{dx}$	The derivative of the log to base b of a quantity is 1 over the product of $\ln b$ and that quantity, times the derivative of that quantity.

quick Examples

1. $\dfrac{d}{dx} \ln[x^2 + 1] = \dfrac{1}{x^2 + 1}\dfrac{d}{dx}(x^2 + 1) \qquad u = x^2 + 1$ (see the footnote[*])

$$= \frac{1}{x^2 + 1}(2x) = \frac{2x}{x^2 + 1}$$

2. $\dfrac{d}{dx} \log_2[x^3 + x] = \dfrac{1}{(x^3 + x) \ln 2}\dfrac{d}{dx}(x^3 + x) \qquad u = x^3 + x$

$$= \frac{1}{(x^3 + x) \ln 2}(3x^2 + 1) = \frac{3x^2 + 1}{(x^3 + x) \ln 2}$$

---

[*] If we were to evaluate $\ln(x^2 + 1)$, the last operation we would perform would be to take the natural logarithm of a quantity. Thus, the calculation thought experiment tells us that we are dealing with $\ln$ *of a quantity*, and so we need the generalized logarithm rule as stated above.

---

[19] We actually used the fact that the logarithm function is continuous when we took the limit.

[20] Here is an outline of the argument for negative h. Since x must be positive for $\ln x$ to be defined, we find that $x/h \to -\infty$ as $h \to 0^-$, and so we must consider the quantity $(1 + 1/m)^m$ for large *negative* m. It turns out the limit is still e (check it numerically!) and so the computation above still works.

Example 1 Derivative of Logarithmic Function

Find $\dfrac{d}{dx}[\ln\sqrt{x+1}]$

Solution The calculation thought experiment tells us that we have the natural logarithm of a quantity, so

$$\frac{d}{dx}[\ln\sqrt{x+1}] = \frac{1}{\sqrt{x+1}}\frac{d}{dx}\sqrt{x+1} \qquad \frac{d}{dx}\ln u = \frac{1}{u}\frac{du}{dx}$$

$$= \frac{1}{\sqrt{x+1}}\cdot\frac{1}{2\sqrt{x+1}} \qquad \frac{d}{dx}\sqrt{u} = \frac{1}{2\sqrt{u}}\frac{du}{dx}$$

$$= \frac{1}{2(x+1)}$$

+**Before we go on...** What happened to the square root in Example 1? As with many problems involving logarithms, we could have done this one differently and with less bother if we had simplified the expression $\ln\sqrt{x+1}$ using the properties of logarithms *before* differentiating. Doing this, we get

$$\ln\sqrt{x+1} = \ln(x+1)^{1/2} = \frac{1}{2}\ln(x+1) \qquad \text{Simplify the logarithm first.}$$

Thus,

$$\frac{d}{dx}[\ln\sqrt{x+1}] = \frac{d}{dx}\left(\frac{1}{2}\ln(x+1)\right)$$

$$= \frac{1}{2}\left(\frac{1}{x+1}\right)\cdot 1 = \frac{1}{2(x+1)}$$

the same answer as above. ∎

Example 2 Derivative of a Logarithmic Function

Find $\dfrac{d}{dx}[\ln[(1+x)(2-x)]]$.

Solution This time, we simplify the expression $\ln[(1+x)(2-x)]$ before taking the derivative.

$$\ln[(1+x)(2-x)] = \ln(1+x) + \ln(2-x) \qquad \text{Simplify the logarithm first.}$$

Thus,

$$\frac{d}{dx}[\ln[(1+x)(2-x)]] = \frac{d}{dx}[\ln(1+x)] + \frac{d}{dx}[\ln(2-x)]$$

$$= \frac{1}{1+x} - \frac{1}{2-x} \qquad \text{Because } \frac{d}{dx}\ln(2-x) = -\frac{1}{2-x}$$

$$= \frac{1-2x}{(1+x)(2-x)}$$

$+$*Before we go on...* For practice, try doing Example 2 without simplifying first. What other differentiation rule do you need to use? ∎

Example 3 Logarithm of an Absolute Value

Find $\dfrac{d}{dx}[\ln |x|]$.

Solution Before we start, we note that $\ln x$ is defined only for positive values of x, so its domain is the set of positive real numbers. The domain of $\ln |x|$, on the other hand, is the set of *all* nonzero real numbers. For example, $\ln |-2| = \ln 2 \approx 0.6931$. For this reason, $\ln |x|$ often turns out to be more useful than the ordinary logarithm function.

Now we'll get to work. The calculation thought experiment tells us that $\ln |x|$ is the natural logarithm of a quantity, so we use the chain rule:

$$
\begin{aligned}
\frac{d}{dx}[\ln |x|] &= \frac{1}{|x|}\frac{d}{dx}|x| \qquad &&\color{red} u = |x| \\[2mm]
&= \frac{1}{|x|}\frac{|x|}{x} \qquad &&\color{red} \text{Recall that } \frac{d}{dx}|x| = \frac{|x|}{x} \\[2mm]
&= \frac{1}{x}
\end{aligned}
$$

$+$*Before we go on...* Figure 1a shows the graphs of $y = \ln |x|$ and $y = 1/x$. Figure 1b shows the graphs of $y = \ln |x|$ and $y = 1/|x|$. You should be able to see from these graphs why the derivative of $\ln |x|$ is $1/x$ and not $1/|x|$.

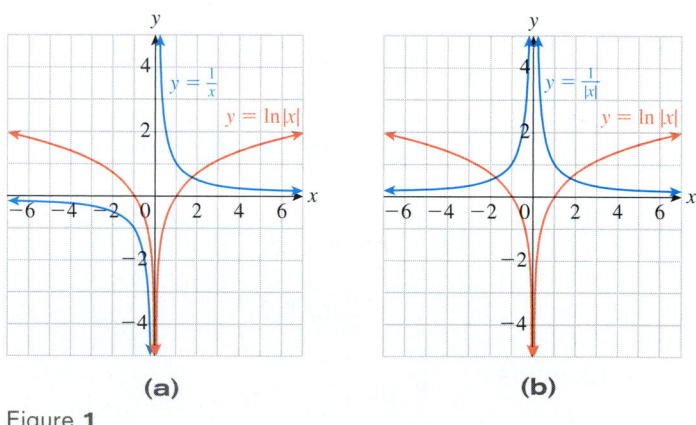

(a)   **(b)**

Figure **1**   ∎

This last example, in conjunction with the chain rule, gives us the following formulas.

Derivative of Logarithms of Absolute Values

Original Rule	*Generalized Rule*	*In Words*
$\dfrac{d}{dx}[\ln|x|] = \dfrac{1}{x}$	$\dfrac{d}{dx}[\ln|u|] = \dfrac{1}{u}\dfrac{du}{dx}$	The derivative of the natural logarithm of the absolute value of a quantity is 1 over that quantity, times the derivative of that quantity.
$\dfrac{d}{dx}[\log_b|x|] = \dfrac{1}{x\ln b}$	$\dfrac{d}{dx}[\log_b|u|] = \dfrac{1}{u\ln b}\dfrac{du}{dx}$	The derivative of the log to base b of the absolute value of a quantity is 1 over the product of $\ln b$ and that quantity, times the derivative of that quantity.

quick Examples

1. $\dfrac{d}{dx}[\ln|x^2 + 1|] = \dfrac{1}{x^2 + 1}\dfrac{d}{dx}(x^2 + 1)$ $u = x^2 + 1$

$\qquad\qquad\qquad = \dfrac{1}{x^2 + 1}(2x) = \dfrac{2x}{x^2 + 1}$

2. $\dfrac{d}{dx}[\log_2|x^3 + x|] = \dfrac{1}{(x^3 + x)\ln 2}\dfrac{d}{dx}(x^3 + x)$ $u = x^3 + x$

$\qquad\qquad\qquad = \dfrac{1}{(x^3 + x)\ln 2}(3x^2 + 1) = \dfrac{3x^2 + 1}{(x^3 + x)\ln 2}$

In other words, when taking the derivative of the logarithm of the absolute value of a quantity, we can simply ignore the absolute value!

We now turn to the derivatives of *exponential* functions—that is, functions of the form $f(x) = b^x$. We begin by showing how *not* to differentiate them.

Caution The derivative of b^x is *not* xb^{x-1}. The power rule applies only to *constant* exponents. In this case the exponent is decidedly *not* constant, and so the power rule does not apply.

The following shows the correct way of differentiating b^x.

Derivative of e^x

$$\frac{d}{dx}[e^x] = e^x$$

quick Examples

1. $\dfrac{d}{dx}[3e^x] = 3\dfrac{d}{dx}[e^x] = 3e^x$

2. $\dfrac{d}{dx}\left[\dfrac{e^x}{x}\right] = \dfrac{e^x x - e^x(1)}{x^2}$ Quotient rule

$\qquad\qquad\quad = \dfrac{e^x(x - 1)}{x^2}$

Thus, e^x has the amazing property that its derivative is itself![21] For bases other than e, we have the following generalization:

Derivative of b^x

If b is any positive number, then

$$\frac{d}{dx}[b^x] = b^x \ln b$$

Note that if $b = e$, we obtain the previous formula,

quick Example

$$\frac{d}{dx}[3^x] = 3^x \ln 3$$

Derivation of the Formula $\frac{d}{dx}[e^x] = e^x$

To find the derivative of e^x we use a shortcut.[22] Write $g(x) = e^x$. Then

$$\ln g(x) = x$$

Take the derivative of both sides of this equation to get

$$\frac{g'(x)}{g(x)} = 1$$

or

$$g'(x) = g(x) = e^x$$

In other words, the exponential function with base e is its own derivative. The rule for exponential functions with other bases follows from the equality $b^x = e^{x \ln b}$ (why?) and the chain rule. (Try it.)

If we were to take the derivative of e raised to a *quantity,* not just x, we would need to use the chain rule, as follows.

Derivatives of Exponentials of Functions

Original Rule	*Generalized Rule*	*In Words*
$\dfrac{d}{dx}[e^x] = e^x$	$\dfrac{d}{dx}[e^u] = e^u \dfrac{du}{dx}$	The derivative of e raised to a quantity is e raised to that quantity, times the derivative of that quantity.
$\dfrac{d}{dx}[b^x] = b^x \ln b$	$\dfrac{d}{dx}[b^u] = b^u \ln b \dfrac{du}{dx}$	The derivative of b raised to a quantity is b raised to that quantity, times $\ln b$, times the derivative of that quantity.

quick Examples

1. $\dfrac{d}{dx}[e^{x^2+1}] = e^{x^2+1} \dfrac{d}{dx}[x^2 + 1]$ $u = x^2 + 1$ (see note*)

$$= e^{x^2+1}(2x) = 2x\, e^{x^2+1}$$

---
* The calculation thought experiment tells us that we have e raised to a quantity.

---

[21] There is another—very simple—function that is its own derivative. What is it?

[22] This shortcut is an example of a technique called *logarithmic differentiation,* which is occasionally useful. We will see it again in the next section.

2. $\dfrac{d}{dx}[2^{3x}] = 2^{3x} \ln 2 \dfrac{d}{dx}[3x]$ $u = 3x$

$\qquad\qquad = 2^{3x}(\ln 2)(3) = (3 \ln 2)2^{3x}$

3. If \$1000 is invested in an account earning 5% per year compounded continuously, then the rate of change of the account balance after t years is

$$\frac{d}{dt}[1000e^{0.05t}] = 1000(0.05)e^{0.05t} = 50e^{0.05t} \text{ dollars/year}$$

Applications

Example 4 Epidemics

In the early stages of the AIDS epidemic during the 1980s, the number of cases in the U.S. was increasing by about 50% every 6 months. By the start of 1983, there were approximately 1600 AIDS cases in the United States.[*] Had this trend continued, how many new cases per year would have been occurring by the start of 1993?

Solution To find the answer, we must first model this exponential growth using the methods of Chapter 9. Referring to Example 3 in Section 9.2, we find that t years after the start of 1983 the number of cases is

$$A = 1600(2.25^t)$$

We are asking for the number of new cases each year. In other words, we want the rate of change, dA/dt:

$$\frac{dA}{dt} = 1600(2.25)^t \ln 2.25 \text{ cases per year}$$

At the start of 1993, $t = 10$, so the number of new cases per year is

$$\left.\frac{dA}{dt}\right|_{t=10} = 1600(2.25)^{10} \ln 2.25 \approx 4{,}300{,}000 \text{ cases per year}$$

---

[*] Data based on regression of 1982–1986 figures. Source for data: Centers for Disease Control and Prevention. HIV/AIDS Surveillance Report, 2000;12 (No. 2).

+ *Before we go on...* In Example 4, the figure for the number of new cases per year is so large because we assumed that exponential growth—the 50% increase every six months—would continue. A more realistic model for the spread of a disease is the logistic model. (See Section 9.4, as well as the next example.) ∎

Example 5 Sales Growth

The sales of the Cyberpunk II video game can be modeled by the logistic curve

$$q(t) = \frac{10{,}000}{1 + 0.5e^{-0.4t}}$$

where $q(t)$ is the total number of units sold t months after its introduction. How fast is the game selling two years after its introduction?

Solution We are asked for $q'(24)$. We can find the derivative of $q(t)$ using the quotient rule, or we can first write

$$q(t) = 10,000(1 + 0.5e^{-0.4t})^{-1}$$

and then use the generalized power rule:

$$q'(t) = -10,000(1 + 0.5e^{-0.4t})^{-2}(0.5e^{-0.4t})(-0.4)$$

$$= \frac{2000e^{-0.4t}}{(1 + 0.5e^{-0.4t})^2}$$

Thus,

$$q'(24) = \frac{2000e^{-0.4(24)}}{(1 + 0.5e^{-0.4(24)})^2} \approx 0.135 \text{ units per month}$$

So, after 2 years, sales are quite slow.

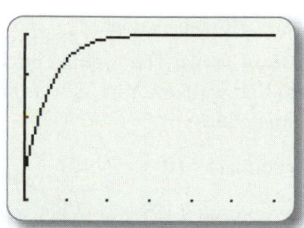

Figure 2

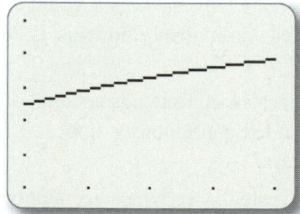

Figure 3

$+$*Before we go on...* We can check the answer in Example 5 graphically. If we plot the total sales curve for $0 \le t \le 30$ and $6000 \le q \le 10,000$, on a TI-83/84, for example, we get the graph shown in Figure 2. Notice that total sales level off at about 10,000 units.[23] We computed $q'(24)$, which is the slope of the curve at the point with t-coordinate 24. If we zoom in to the portion of the curve near $t = 24$, we obtain the graph shown in Figure 3, with $23 \le t \le 25$ and $9999 \le q \le 10,000$. The curve is almost linear in this range. If we use the two endpoints of this segment of the curve, $(23, 9999.4948)$ and $(25, 9999.7730)$, we can approximate the derivative as

$$\frac{9999.7730 - 9999.4948}{25 - 23} = 0.1391$$

which is accurate to two decimal places. ■

_____
[23] We can also say this using limits: $\lim_{t \to +\infty} q(t) = 10,000$.

11.3 EXERCISES

● denotes basic skills exercises

◆ denotes challenging exercises

tech Ex indicates exercises that should be solved using technology

Find the derivatives of the functions in Exercises 1–76.

1. ● $f(x) = \ln(x - 1)$ *hint* [see Quick Examples on p. 826]

2. ● $f(x) = \ln(x + 3)$

3. ● $f(x) = \log_2 x$

4. ● $f(x) = \log_3 x$

5. ● $g(x) = \ln|x^2 + 3|$

6. ● $g(x) = \ln|2x - 4|$

7. ● $h(x) = e^{x+3}$ *hint* [see Quick Examples on pp. 827, 828]

8. ● $h(x) = e^{x^2}$

9. ● $f(x) = e^{-x}$

10. ● $f(x) = e^{1-x}$

11. ● $g(x) = 4^x$

12. ● $g(x) = 5^x$

13. ● $h(x) = 2^{x^2-1}$

14. ● $h(x) = 3^{x^2-x}$

15. ● $f(x) = x \ln x$

16. ● $f(x) = 3 \ln x$

17. ● $f(x) = (x^2 + 1) \ln x$

18. ● $f(x) = (4x^2 - x) \ln x$

19. ● $f(x) = (x^2 + 1)^5 \ln x$

20. ● $f(x) = (x + 1)^{0.5} \ln x$

● basic skills   ◆ challenging   **tech** Ex technology exercise

21. ● $g(x) = \ln|3x - 1|$

22. ● $g(x) = \ln|5 - 9x|$

23. ● $g(x) = \ln|2x^2 + 1|$

24. ● $g(x) = \ln|x^2 - x|$

25. ● $g(x) = \ln(x^2 - 2.1x^{0.3})$

26. ● $g(x) = \ln(x - 3.1x^{-1})$

27. ● $h(x) = \ln[(-2x + 1)(x + 1)]$

28. ● $h(x) = \ln[(3x + 1)(-x + 1)]$

29. ● $h(x) = \ln\left(\dfrac{3x + 1}{4x - 2}\right)$

30. ● $h(x) = \ln\left(\dfrac{9x}{4x - 2}\right)$

31. ● $r(x) = \ln\left|\dfrac{(x + 1)(x - 3)}{-2x - 9}\right|$

32. ● $r(x) = \ln\left|\dfrac{-x + 1}{(3x - 4)(x - 9)}\right|$

33. ● $s(x) = \ln(4x - 2)^{1.3}$

34. ● $s(x) = \ln(x - 8)^{-2}$

35. ● $s(x) = \ln\left|\dfrac{(x + 1)^2}{(3x - 4)^3(x - 9)}\right|$

36. ● $s(x) = \ln\left|\dfrac{(x + 1)^2(x - 3)^4}{2x + 9}\right|$

37. ● $h(x) = \log_2(x + 1)$

38. ● $h(x) = \log_3(x^2 + x)$

39. ● $r(t) = \log_3(t + 1/t)$

40. ● $r(t) = \log_3(t + \sqrt{t})$

41. ● $f(x) = (\ln|x|)^2$

42. ● $f(x) = \dfrac{1}{\ln|x|}$

43. ● $r(x) = \ln(x^2) - [\ln(x - 1)]^2$

44. ● $r(x) = (\ln(x^2))^2$

45. ● $f(x) = xe^x$

46. ● $f(x) = 2e^x - x^2e^x$

47. ● $r(x) = \ln(x + 1) + 3x^3e^x$

48. ● $r(x) = \ln|x + e^x|$

49. ● $f(x) = e^x \ln|x|$

50. ● $f(x) = e^x \log_2|x|$

51. ● $f(x) = e^{2x+1}$

52. ● $f(x) = e^{4x-5}$

53. ● $h(x) = e^{x^2-x+1}$

54. ● $h(x) = e^{2x^2-x+1/x}$

55. ● $s(x) = x^2e^{2x-1}$

56. ● $s(x) = \dfrac{e^{4x-1}}{x^3 - 1}$

57. ● $r(x) = (e^{2x-1})^2$

58. ● $r(x) = (e^{2x^2})^3$

59. ● $t(x) = 3^{2x-4}$

60. ● $t(x) = 4^{-x+5}$

61. ● $v(x) = 3^{2x+1} + e^{3x+1}$

62. ● $v(x) = e^{2x}4^{2x}$

63. ● $u(x) = \dfrac{3^{x^2}}{x^2 + 1}$

64. ● $u(x) = (x^2 + 1)4^{x^2-1}$

65. ● $g(x) = \dfrac{e^x + e^{-x}}{e^x - e^{-x}}$

66. ● $g(x) = \dfrac{1}{e^x + e^{-x}}$

67. ● $g(x) = e^{3x-1}e^{x-2}e^x$

68. ● $g(x) = e^{-x+3}e^{2x-1}e^{-x+11}$

69. ● $f(x) = \dfrac{1}{x \ln x}$

70. ● $f(x) = \dfrac{e^{-x}}{xe^x}$

71. ● $f(x) = [\ln(e^x)]^2 - \ln[(e^x)^2]$

72. ● $f(x) = e^{\ln x} - e^{2\ln(x^2)}$

73. ● $f(x) = \ln|\ln x|$

74. ● $f(x) = \ln|\ln|\ln x||$

75. ● $s(x) = \ln\sqrt{\ln x}$

76. ● $s(x) = \sqrt{\ln(\ln x)}$

● basic skills   ◆ challenging   **tech** Ex technology exercise

---

Find the equations of the straight lines described in Exercises 77–82. Use graphing technology to check your answers by plotting the given curve together with the tangent line.

77. ● Tangent to $y = e^x \log_2 x$ at the point $(1, 0)$

78. ● Tangent to $y = e^x + e^{-x}$ at the point $(0, 2)$

79. ● Tangent to $y = \ln\sqrt{2x + 1}$ at the point where $x = 0$

80. ● Tangent to $y = \ln\sqrt{2x^2 + 1}$ at the point where $x = 1$

81. ● At right angles to $y = e^{x^2}$ at the point where $x = 1$

82. ● At right angles to $y = \log_2(3x + 1)$ at the point where $x = 1$

Applications

83. ● **New York City Housing Costs: Downtown** The average price of a two-bedroom apartment in downtown New York City from 1994 to 2004 could be approximated by

$$p(t) = 0.33e^{0.16t} \text{ million dollars} \quad (0 \leq t \leq 10)$$

where t is time in years ($t = 0$ represents 1994).[24] What was the average price of a two-bedroom apartment in downtown New York City in 2003, and how fast was it increasing? (Round your answers to two significant digits.) *hint* [see Example 4]

84. ● **New York City Housing Costs: Uptown** The average price of a two-bedroom apartment in uptown New York City from 1994 to 2004 could be approximated by

$$p(t) = 0.14e^{0.10t} \text{ million dollars} \quad (0 \leq t \leq 10)$$

where t is time in years ($t = 0$ represents 1994).[25] What was the average price of a two-bedroom apartment in uptown New York City in 2002, and how fast was it increasing? (Round your answers to two significant digits.)

85. ● **Investments** If $10,000 is invested in a savings account offering 4% per year, compounded continuously, how fast is the balance growing after 3 years?

86. ● **Investments** If $20,000 is invested in a savings account offering 3.5% per year, compounded continuously, how fast is the balance growing after 3 years?

87. ● **Investments** If $10,000 is invested in a savings account offering 4% per year, compounded semiannually, how fast is the balance growing after 3 years?

88. ● **Investments** If $20,000 is invested in a savings account offering 3.5% per year, compounded semiannually, how fast is the balance growing after 3 years?

89. ● **Scientific Research** The number of research articles in the research journal *Physics Review* that were written by researchers in Europe during 1983–2003 can be approximated by

$$A(t) = \dfrac{7.0}{1 + 5.4e^{-0.18t}} \text{ thousand articles} \quad (0 \leq t \leq 10)$$

---

[24] Model is based on a exponential regression. Source for data: Miller Samuel/*New York Times*, March 28, 2004, p. RE 11.

[25] Ibid.

($t = 0$ represents 1983).[26] How fast was this number increasing in 2000 ($t = 7$)? (Round your answer to two significant digits.) *hint* [see Example 5]

90. ● *Personal Computers* The percentage of U.S households with personal computers can be approximated by

$$P(t) = \frac{85}{1 + 2.6e^{0.26t}} \quad (0 \le t \le 9)$$

where t is time in years ($t = 0$ represents 1994).[27] How fast was this percentage increasing in 2000 ($t = 6$)? (Round your answer to two significant digits.)

91. ● *Epidemics* A flu epidemic described in Example 1 in Section 9.4 approximately followed the curve

$$P = \frac{150}{1 + 15,000e^{-0.35t}} \text{ million people}$$

where P is the number of people infected and t is the number of weeks after the start of the epidemic. How fast is the epidemic growing (that is, how many new cases are there each week) after 20 weeks? After 30 weeks? After 40 weeks? (Round your answers to two significant digits.)

92. ● *Epidemics* Another epidemic follows the curve

$$P = \frac{200}{1 + 20,000e^{-0.549t}} \text{ million people}$$

where t is in years. How fast is the epidemic growing after 10 years? After 20 years? After 30 years? (Round your answers to two significant digits.)

93. ● *Scientific Research* (Compare Exercise 89.) The number of research articles in the research journal *Physics Review* that were written by researchers in Europe during 1983–2003 can be approximated by

$$A(t) = \frac{7.0}{1 + 5.4(1.2)^{-t}} \text{ thousand articles}$$

($t = 0$ represents 1983).[28] How fast was this number increasing in 2000 ($t = 7$)? (Round your answer to two significant digits.)

94. ● *Personal Computers* (Compare Exercise 90.) The percentage of U.S households with personal computers can be approximated by

$$P(t) = \frac{85}{1 + 2.6(1.3)^{-t}} \quad (0 \le t \le 9)$$

where t is time in years ($t = 0$ represents 1994).[29] How fast was this percentage increasing in 2000 ($t = 6$)? (Round your answer to two significant digits.)

95. *Population Growth* The population of Lower Anchovia was 4,000,000 at the start of 1995 and was doubling every 10 years.

How fast was it growing per year at the start of 1995? (Round your answer to 3 significant digits.)

96. *Population Growth* The population of Upper Anchovia was 3,000,000 at the start of 1996 and doubling every 7 years. How fast was it growing per year at the start of 1996? (Round your answer to 3 significant digits.)

97. *Radioactive Decay* Plutonium-239 has a half-life of 24,400 years. How fast is a lump of 10 grams decaying after 100 years?

98. *Radioactive Decay* Carbon-14 has a half-life of 5730 years. How fast is a lump of 20 grams decaying after 100 years?

99. *SAT Scores by Income* The following chart shows U.S. verbal SAT scores as a function of parents' income level:[30]

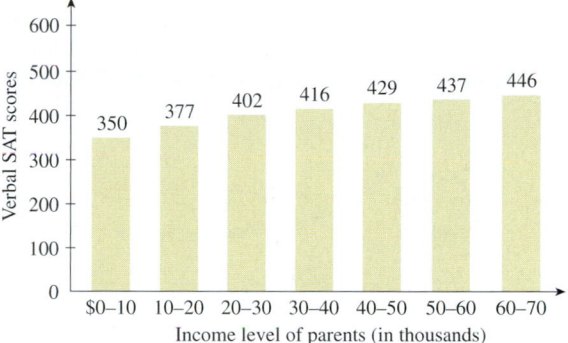

a. The data can best be modeled by which of the following?
   (A) $S(x) = 470 - 136e^{-0.0000264x}$
   (B) $S(x) = 136e^{-0.0000264x}$
   (C) $S(x) = 355(1.000004^x)$
   (D) $S(x) = 470 - 355(1.000004^x)$

   ($S(x)$ is the average verbal SAT score of students whose parents earn $\$x$ per year.)

b. Use $S'(x)$ to predict how a student's verbal SAT score is affected by a $\$1000$ increase in parents' income for a student whose parents earn $\$45,000$.

c. Does $S'(x)$ increase or decrease as x increases? Interpret your answer.

100. *SAT Scores by Income* The following chart shows U.S. average math SAT scores as a function of parents' income level:[31]

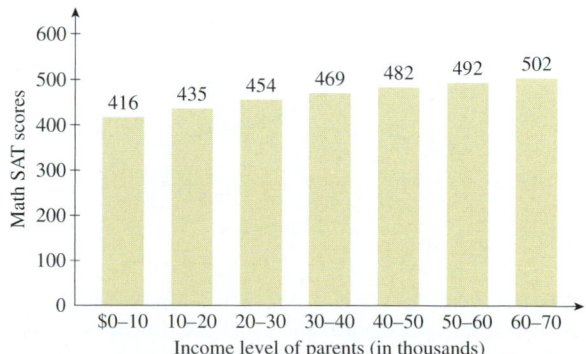

[26] SOURCE: The American Physical Society/*New York Times* May 3, 2003, p. A1.

[27] SOURCE: NTIA/Census Bureau/Pegasus Research International, LLC http://www.pegasusresearch.com/metrics/growthus.htm.

[28] SOURCE: The American Physical Society/*New York Times* May 3, 2003, p. A1.

[29] SOURCE: NTIA/Census Bureau/Pegasus Research International, LLC http://www.pegasusresearch.com/metrics/growthus.htm.

[30] SOURCE: The College Board/*New York Times*, March 5, 1995, p. E16.
[31] Ibid.

● basic skills   ◆ challenging   **tech** Ex technology exercise

a. The data can best be modeled by which of the following?

(A) $S(x) = 535 - 415(1.000003^x)$

(B) $S(x) = 535 - 136e^{0.0000213x}$

(C) $S(x) = 535 - 136e^{-0.0000213x}$

(D) $S(x) = 415(1.000003^x)$

($S(x)$ is the average math SAT score of students whose parents earn \$$x$ per year.)

b. Use $S'(x)$ to predict how a student's math SAT score is affected by a \$1000 increase in parents' income for a student whose parents earn \$45,000.

c. Does $S'(x)$ increase or decrease as x increases? Interpret your answer.

101. ***Demographics: Average Age and Fertility*** The following graph shows a plot of average age of a population versus fertility rate (the average number of children each woman has in her lifetime) in the U.S. and Europe over the period 1950–2005.[32]

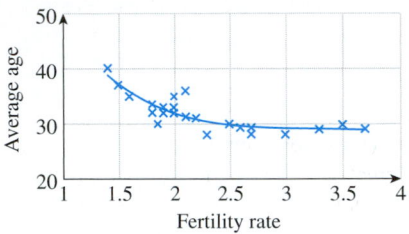

The equation of the accompanying curve is

$$a = 28.5 + 120(0.172)^x \quad (1.4 \le x \le 3.7)$$

where a is the average age (in years) of the population and x is the fertility rate.

a. Compute $a'(2)$. What does the answer tell you about average age and fertility rates?

b. Use the answer to part (a) to estimate how much the fertility rate would need to increase from a level of 2 children per woman to lower the average age of a population by about 1 year.

102. ***Demographics: Average Age and Fertility*** The following graph shows a plot of average age of a population versus fertility rate (the average number of children each woman has in her lifetime) in Europe over the period 1950–2005.[33]

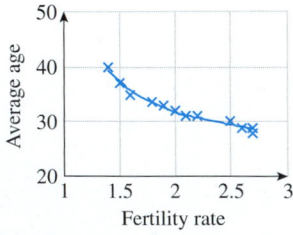

The equation of the accompanying curve is

$$g = 27.6 + 128(0.181)^x \quad (1.4 \le x \le 3.7)$$

where g is the average age (in years) of the population and x is the fertility rate.

a. Compute $g'(2.5)$. What does the answer tell you about average age and fertility rates?

b. Referring to the model that combines the data for Europe and the U.S. in Exercise 101, which population's average age is affected more by a changing fertility rate at the level of 2.5 children per woman?

103. ***Big Brother*** The following chart shows the number of wiretaps authorized each year by U.S. courts from 1990 to 2000 ($t = 0$ represents 1990):[34]

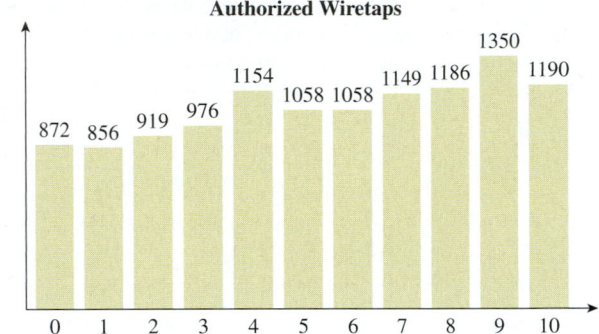

These data can be approximated with the logistic model

$$W(t) = \frac{1500}{1 + 0.77(1.16)^{-t}} \quad (0 \le t \le 10)$$

where $W(t)$ is the number of authorized wiretaps in year $1990 + t$.

a. Calculate $W'(t)$ and use it to approximate $W'(6)$. To how many significant digits should we round the answer? Why? What does the answer tell you?

b. *tech* Ex Graph the function $W'(t)$. Based on the graph, the number of wiretaps authorized each year (choose one)

(A) increased at a decreasing rate

(B) decreased at an increasing rate

(C) increased at an increasing rate

(D) decreased at a decreasing rate

from 1990 to 2000.

104. ***Big Brother*** The following chart shows the number of wiretaps authorized each year by U.S. federal courts from 1990 to 2000 ($t = 0$ represents 1990):[35]

---

[32] The separate data for Europe and the U.S. are collected in the same graph. 2005 figures are estimates. SOURCE: United Nations World Population Division/*New York Times*, June 29, 2003, p. 3.

[33] All European countries including the Russian Federation. 2005 figures are estimates. SOURCE: United Nations World Population Division/*New York Times*, June 29, 2003, p. 3.

[34] SOURCE: 2000 Wiretap Report, Administrative Office of the United States Courts www.epic.org/privacy/wiretap/stats/2000_report/default.html.

[35] Ibid.

● basic skills   ◆ challenging   *tech* Ex technology exercise

Authorized Wiretaps: Federal

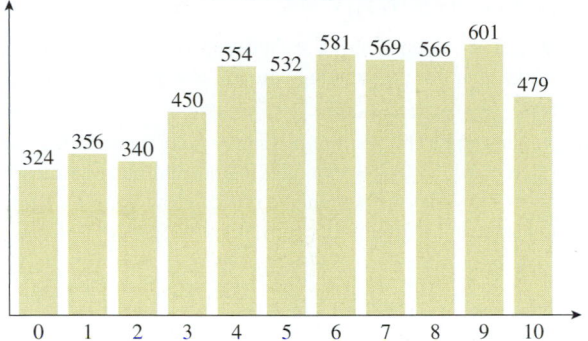

These data can be approximated with the logistic model

$$W(t) = \frac{600}{1 + 1.00(1.62)^{-t}} \qquad (0 \le t \le 10)$$

where $W(t)$ is the number of authorized wiretaps in year $1990 + t$.

a. Calculate $W'(t)$ and use it to approximate $W'(6)$. To how many significant digits should we round the answer? Why? What does the answer tell you?

b. tech Ex Graph the function $W'(t)$. Based on the graph, the number of wiretaps authorized each year (choose one)
  (A) increased at an increasing rate
  (B) decreased at a decreasing rate
  (C) increased at a decreasing rate
  (D) decreased at an increasing rate

from 1990 to 2000.

105. tech Ex *Diffusion of New Technology* Numeric control is a technology whereby the operation of machines is controlled by numerical instructions on disks, tapes, or cards. In a study, E. Mansfield et al[36] modeled the growth of this technology using the equation

$$p(t) = \frac{0.80}{1 + e^{4.46 - 0.477t}}$$

where $p(t)$ is the fraction of firms using numeric control in year t.

a. Graph this function for $0 \le t \le 20$ and estimate $p'(10)$ graphically. Interpret the result.

b. Use your graph to estimate $\lim_{t \to +\infty} p(t)$ and interpret the result.

c. Compute $p'(t)$, graph it, and again find $p'(10)$.

d. Use your graph to estimate $\lim_{t \to +\infty} p'(t)$ and interpret the result.

106. tech Ex *Diffusion of New Technology* Repeat Exercise 105 using the revised formula

$$p(t) = \frac{0.90e^{-0.1t}}{1 + e^{4.50 - 0.477t}}$$

which takes into account that in the long run this new technology will eventually become outmoded and will be replaced by a newer technology. Draw your graphs using the range $0 \le t \le 40$.

107. ◆ *Cell Phone Revenues* The number of cell phone subscribers in China for the period 2000–2005 was projected to follow the equation[37]

$$N(t) = 39t + 68 \text{ million subscribers}$$

in year t ($t = 0$ represents 2000). The average annual revenue per cell phone user was \$350 in 2000. Assuming that, due to competition, the revenue per cell phone user decreases continuously at an annual rate of 10%, give a formula for the annual revenue in year t. Hence, project the annual revenue and its rate of change in 2002. Round all answers to the nearest billion dollars or billion dollars per year.

108. ◆ *Cell Phone Revenues* The annual revenue for cell phone use in China for the period 2000–2005 was projected to follow the equation[38]

$$R(t) = 14t + 24 \text{ billion dollars}$$

in year t ($t = 0$ represents 2000). At the same time, there were approximately 68 million subscribers in 2000. Assuming that the number of subscribers increases continuously at an annual rate of 10%, give a formula for the annual revenue per subscriber in year t. Hence, project to the nearest dollar the annual revenue per subscriber and its rate of change in 2002. (Be careful with units!)

Communication and Reasoning Exercises

109. ● Complete the following: The derivative of e raised to a glob is ____.

110. ● Complete the following: The derivative of the natural logarithm of a glob is ____.

111. ● Complete the following: The derivative of 2 raised to a glob is ____.

112. ● Complete the following: The derivative of the base 2 logarithm of a glob is ____.

113. ● What is wrong with the following?

$$\frac{d}{dx} 3^{2x} = (2x)3^{2x-1} \quad ✗ \quad \textit{WRONG!}$$

114. ● What is wrong with the following?

$$\frac{d}{dx} \ln(3x^2 - 1) = \frac{1}{6x} \quad ✗ \quad \textit{WRONG!}$$

115. ● The number N of music downloads on campus is growing exponentially with time. Can $N'(t)$ grow linearly with time? Explain.

[36] SOURCE: "The Diffusion of a Major Manufacturing Innovation," in *Research and Innovation in the Modern Corporation* (W.W. Norton and Company, Inc., New York, 1971, pp. 186–205).

[37] Based on a regression of projected figures (coefficients are rounded). SOURCE: Intrinsic Technology/*New York Times*, Nov. 24, 2000, p. C1.

[38] Not allowing for discounting due to increased competition. SOURCE: Ibid.

● basic skills   ◆ challenging   tech Ex technology exercise

116. The number N of graphing calculators sold on campus is decaying exponentially with time. Can $N'(t)$ grow with time? Explain.

*The **percentage rate of change** or **fractional rate of change** of a function is defined to be the ratio $f'(x)/f(x)$. (It is customary to express this as a percentage when speaking about percentage rate of change.)*

117. ◆ Show that the fractional rate of change of the exponential function e^{kx} is equal to k, which is often called its **fractional growth rate.**

118. ◆ Show that the fractional rate of change of $f(x)$ is the rate of change of $\ln(f(x))$.

119. ◆ Let $A(t)$ represent a quantity growing exponentially. Show that the percentage rate of change, $A'(t)/A(t)$, is constant.

120. ◆ Let $A(t)$ be the amount of money in an account that pays interest which is compounded some number of times per year. Show that the percentage rate of growth, $A'(t)/A(t)$, is constant. What might this constant represent?

● basic skills   ◆ challenging   **tech** Ex technology exercise

---

11.4 | Implicit Differentiation (OPTIONAL)

Consider the equation $y^5 + y + x = 0$, whose graph is shown in Figure 4.

How did we obtain this graph? We did not solve for y as a function of x; that is impossible. In fact, we solved for x in terms of y to find points to plot. Nonetheless, the graph in Figure 4 is the graph of a function because it passes the vertical line test: Every vertical line crosses the graph no more than once, so for each value of x there is no more than one corresponding value of y. Because we cannot solve for y explicitly in terms of x, we say that the equation $y^5 + y + x = 0$ determines y as an **implicit function** of x.

Now, suppose we want to find the slope of the tangent line to this curve at, say, the point $(2, -1)$ (which, you should check, is a point on the curve). In the following example we find, surprisingly, that it is possible to obtain a formula for dy/dx without having to first solve the equation for y.

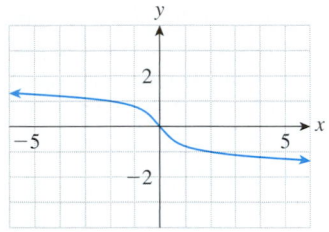

Figure 4

Example 1 Implicit Differentiation

Find $\dfrac{dy}{dx}$, given that $y^5 + y + x = 0$

Solution

We use the chain rule and a little cleverness. Think of y as a function of x and take the derivative with respect to x of both sides of the equation:

$$y^5 + y + x = 0 \qquad \text{\color{magenta}{Original equation}}$$

$$\frac{d}{dx}[y^5 + y + x] = \frac{d}{dx}[0] \qquad \text{\color{magenta}{Derivative with respect to x of both sides}}$$

$$\frac{d}{dx}[y^5] + \frac{d}{dx}[y] + \frac{d}{dx}[x] = 0 \qquad \text{\color{magenta}{Derivative rules}}$$

Now we must be careful. The derivative *with respect to x* of y^5 is *not* $5y^4$. Rather, because y is a function of x, we must use the chain rule, which tells us that

$$\frac{d}{dx}[y^5] = 5y^4 \frac{dy}{dx}$$

Thus, we get

$$5y^4 \frac{dy}{dx} + \frac{dy}{dx} + 1 = 0$$

We want to find dy/dx, so we *solve for it:*

$$(5y^4 + 1)\frac{dy}{dx} = -1 \qquad \textcolor{red}{\text{Isolate } dy/dx \text{ on one side.}}$$

$$\frac{dy}{dx} = -\frac{1}{5y^4 + 1} \qquad \textcolor{red}{\text{Divide both sides by } 5y^4 + 1.}$$

+ *Before we go on...* Note that we should not expect to obtain dy/dx as an explicit function of x if y was not an explicit function of x to begin with. For example, the formula we found in Example 1 for dy/dx is not a function of x because there is a y in it. However, the result is still useful because we can evaluate the derivative at any point on the graph. For instance, at the point $(2, -1)$ on the graph, we get

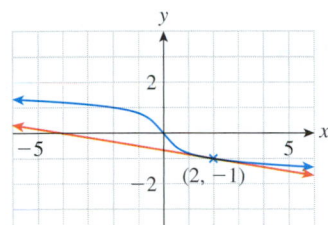

Figure **5**

$$\frac{dy}{dx} = -\frac{1}{5y^4 + 1} = -\frac{1}{5(-1)^4 + 1} = -\frac{1}{6}$$

Thus, the slope of the tangent line to the curve $y^5 + y + x = 0$ at the point $(2, -1)$ is $-1/6$. Figure 5 shows the graph and this tangent line. ∎

This procedure we just used—differentiating an equation to find dy/dx without first solving the equation for y—is called **implicit differentiation.**

In Example 1 we were given an equation in x and y that determined y as an (implicit) function of x, even though we could not solve for y. But an equation in x and y need not always determine y, as a function of x. Consider, for example, the equation

$$2x^2 + y^2 = 2$$

Solving for y yields $y = \pm\sqrt{2 - 2x^2}$. The $\pm$ sign reminds us that for some values of x there are two corresponding values for y. We can graph this equation by superimposing the graphs of

$$y = \sqrt{2 - 2x^2} \quad \text{and } y = -\sqrt{2 - 2x^2}$$

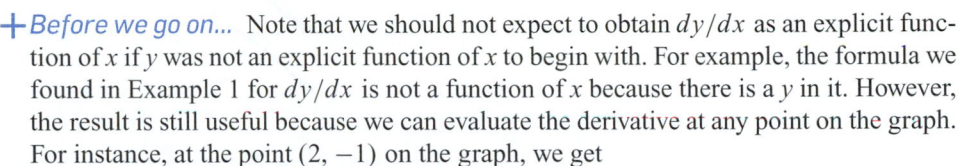

The graph, an *ellipse,* is shown in Figure 6.

The graph of $y = \sqrt{2 - 2x^2}$ constitutes the top half of the ellipse, and the graph of $y = -\sqrt{2 - 2x^2}$ constitutes the bottom half.

Figure **6**

Example **2** Slope of Tangent Line

Refer to Figure 6. Find she slope of the tangent line to the ellipse $2x^2 + y^2 = 2$ at the point $(1/\sqrt{2}, 1)$.

Solution Because $(1/\sqrt{2}, 1)$ is on the top half of the ellipse in Figure 6, we *could* differentiate the function $y = \sqrt{2 - 2x^2}$, to obtain the result, but it is actually easier to apply implicit differentiation to the original equation.

$$2x^2 + y^2 = 2 \qquad \textcolor{red}{\text{Original equation}}$$

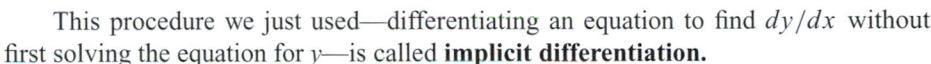

$$\frac{d}{dx}[2x^2 + y^2] = \frac{d}{dx}[2] \qquad \textcolor{red}{\text{Derivative with respect to } x \text{ of both sides}}$$

$$4x + 2y\frac{dy}{dx} = 0$$

$$2y\frac{dy}{dx} = -4x \qquad \text{Solve for } dy/dx$$

$$\frac{dy}{dx} = -\frac{4x}{2y} = -\frac{2x}{y}$$

To find the slope at $(1/\sqrt{2},\ 1)$ we now substitute for x and y:

$$\frac{dy}{dx}\bigg|_{(1/\sqrt{2},1)} = -\frac{2/\sqrt{2}}{2} = -\sqrt{2}$$

Thus, the slope of the tangent to the ellipse at the point $(1/\sqrt{2},\ 1)$ is $-\sqrt{2} \approx -1.414$.

Example 3 Tangent Line for an Implicit Function

Find the equation of the tangent line to the curve $\ln y = xy$ at the point where $y = 1$.

Solution First, we use implicit differentiation to find dy/dx:

$$\frac{d}{dx}[\ln y] = \frac{d}{dx}[xy] \qquad \text{Take } d/dx \text{ of both sides}$$

$$\frac{1}{y}\frac{dy}{dx} = (1)y + x\frac{dy}{dx} \qquad \text{Chain rule on left, product rule on right}$$

To solve for dy/dx, we bring all the terms containing dy/dx to the left-hand side and all terms not containing it to the right-hand side:

$$\frac{1}{y}\frac{dy}{dx} - x\frac{dy}{dx} = y \qquad \text{Bring the terms with } dy/dx \text{ to the left.}$$

$$\frac{dy}{dx}\left(\frac{1}{y} - x\right) = y \qquad \text{Factor out } dy/dx.$$

$$\frac{dy}{dx}\left(\frac{1 - xy}{y}\right) = y$$

$$\frac{dy}{dx} = y\left(\frac{y}{1 - xy}\right) = \frac{y^2}{1 - xy} \qquad \text{Solve for } dy/dx.$$

The derivative gives the slope of the tangent line, so we want to evaluate the derivative at the point where $y = 1$. However, the formula for dy/dx requires values for both x and y. We get the value of x by substituting $y = 1$ in the original equation:

$$\ln y = xy$$
$$\ln(1) = x \cdot 1$$

But $\ln(1) = 0$, and so $x = 0$ for this point. Thus,

$$\frac{dy}{dx}\bigg|_{(0,1)} = \frac{1^2}{1 - (0)(1)} = 1$$

Therefore, the tangent line is the line through $(x, y) = (0, 1)$ with slope 1, which is

$$y = x + 1$$

+*Before we go on...* Example 3 presents an instance of an implicit function in which it is simply not possible to solve for y. Try it. ∎

Sometimes, it is easiest to differentiate a complicated function of x by first taking the logarithm and then using implicit differentiation—a technique called **logarithmic differentiation.**

Example 4 Logarithmic Differentiation

Find $\dfrac{d}{dx}\left[\dfrac{(x+1)^{10}(x^2+1)^{11}}{(x^3+1)^{12}}\right]$ without using the product or quotient rules.

Solution Write

$$y = \frac{(x+1)^{10}(x^2+1)^{11}}{(x^3+1)^{12}}$$

and then take the natural logarithm of both sides:

$$\ln y = \ln\left[\frac{(x+1)^{10}(x^2+1)^{11}}{(x^3+1)^{12}}\right]$$

We can use properties of the logarithm to simplify the right-hand side:

$$\ln y = \ln(x+1)^{10} + \ln(x^2+1)^{11} - \ln(x^3+1)^{12}$$

$$= 10\ln(x+1) + 11\ln(x^2+1) - 12\ln(x^3+1)$$

Now we can find $\dfrac{dy}{dx}$ using implicit differentiation:

$$\frac{1}{y}\frac{dy}{dx} = \frac{10}{x+1} + \frac{22x}{x^2+1} - \frac{36x^2}{x^3+1} \qquad \text{Take } d/dx \text{ of both sides.}$$

$$\frac{dy}{dx} = y\left(\frac{10}{x+1} + \frac{22x}{x^2+1} - \frac{36x^2}{x^3+1}\right) \qquad \text{Solve for } dy/dx.$$

$$= \frac{(x+1)^{10}(x^2+1)^{11}}{(x^3+1)^{12}}\left(\frac{10}{x+1} + \frac{22x}{x^2+1} - \frac{36x^2}{x^3+1}\right) \qquad \text{Substitute for } y.$$

+*Before we go on...* Redo Example 4 using the product and quotient rules (and the chain rule) instead of logarithmic differentiation and compare the answers. Compare also the amount of work involved in both methods. ∎

Application

Productivity usually depends on both labor and capital. Suppose, for example, you are managing a surfboard manufacturing company. You can measure its productivity by counting the number of surfboards the company makes each year. As a measure of labor, you can use the number of employees, and as a measure of capital you can use its operating budget. The so-called *Cobb-Douglas* model uses a function of the form:

$$P = Kx^a y^{1-a} \qquad \text{Cobb-Douglas model for productivity}$$

David Samuel Robbins/Corbis

where P stands for the number of surfboards made each year, x is the number of employees, and y is the operating budget. The numbers K and a are constants that depend on the particular situation studied, with a between 0 and 1.

Example 5 Cobb-Douglas Production Function

The surfboard company you own has the Cobb-Douglas production function

$$P = x^{0.3}y^{0.7}$$

where P is the number of surfboards it produces per year, x is the number of employees, and y is the daily operating budget (in dollars). Assume that the production level P is constant.

a. Find $\dfrac{dy}{dx}$.

b. Evaluate this derivative at $x = 30$ and $y = 10,000$, and interpret the answer.

Solution

a. We are given the equation $P = x^{0.3}y^{0.7}$, in which P is constant. We find $\dfrac{dy}{dx}$ by implicit differentiation

$$0 = \frac{d}{dx}[x^{0.3}y^{0.7}] \qquad\qquad d/dx \text{ of both sides}$$

$$0 = 0.3x^{-0.7}y^{0.7} + x^{0.3}(0.7)y^{-0.3}\frac{dy}{dx} \qquad\qquad \text{Product \& chain rules}$$

$$-0.7x^{0.3}y^{-0.3}\frac{dy}{dx} = 0.3x^{-0.7}y^{0.7} \qquad\qquad \text{Bring term with } dy/dx \text{ to left}$$

$$\frac{dy}{dx} = -\frac{0.3x^{-0.7}y^{0.7}}{0.7x^{0.3}y^{-0.3}} \qquad\qquad \text{Solve for } dy/dx$$

$$= -\frac{3y}{7x} \qquad\qquad \text{Simplify}$$

b. Evaluating this derivative at $x = 30$ and $y = 10,000$ gives

$$\left.\frac{dy}{dx}\right|_{x=30,\ y=10,000} = -\frac{3(10,000)}{7(30)} \approx -143$$

To interpret this result, first look at the units of the derivative: We recall that the units of dy/dx are units of y per unit of x. Because y is the daily budget, its units are dollars; because x is the number of employees, its units are employees. Thus,

$$\left.\frac{dy}{dx}\right|_{x=30,\ y=10,000} \approx -\$143 \text{ per employee}$$

Next, recall that dy/dx measures the rate of change of y as x changes. Because the answer is negative, the daily budget to maintain production at the fixed level is decreasing by approximately \$143 per additional employee at an employment level of 30 employees and a daily operating budget of \$10,000. In other words, increasing the workforce by one worker will result in a savings of approximately \$143 per day. Roughly speaking, *a new employee is worth \$143 per day* at the current levels of employment and production.

11.4 EXERCISES

● denotes basic skills exercises

◆ denotes challenging exercises

In Exercises 1–10, find dy/dx, using implicit differentiation. In each case, compare your answer with the result obtained by first solving for y as a function of x and then taking the derivative. hint [see Example 1]

1. ● $2x + 3y = 7$

2. ● $4x - 5y = 9$

3. ● $x^2 - 2y = 6$

4. ● $3y + x^2 = 5$

5. ● $2x + 3y = xy$

6. ● $x - y = xy$

7. ● $e^x y = 1$

8. ● $e^x y - y = 2$

9. ● $y \ln x + y = 2$

10. ● $\dfrac{\ln x}{y} = 2 - x$

In Exercises 11–30, find the indicated derivative using implicit differentiation.

11. ● $x^2 + y^2 = 5; \dfrac{dy}{dx}$

12. ● $2x^2 - y^2 = 4; \dfrac{dy}{dx}$

13. ● $x^2 y - y^2 = 4; \dfrac{dy}{dx}$

14. ● $xy^2 - y = x; \dfrac{dy}{dx}$

15. ● $3xy - \dfrac{y}{3} = \dfrac{2}{x}; \dfrac{dy}{dx}$

16. ● $\dfrac{xy}{2} - y^2 = 3; \dfrac{dy}{dx}$

17. ● $x^2 - 3y^2 = 8; \dfrac{dx}{dy}$

18. ● $(xy)^2 + y^2 = 8; \dfrac{dx}{dy}$

19. ● $p^2 - pq = 5p^2 q^2; \dfrac{dp}{dq}$

20. ● $q^2 - pq = 5p^2 q^2; \dfrac{dp}{dq}$

21. ● $xe^y - ye^x = 1; \dfrac{dy}{dx}$

22. ● $x^2 e^y - y^2 = e^x; \dfrac{dy}{dx}$

23. ● $e^{st} = s^2; \dfrac{ds}{dt}$

24. ● $e^{s^2 t} - st = 1; \dfrac{ds}{dt}$

25. ● $\dfrac{e^x}{y^2} = 1 + e^y; \dfrac{dy}{dx}$

26. ● $\dfrac{x}{e^y} + xy = 9y; \dfrac{dy}{dx}$

27. ● $\ln(y^2 - y) + x = y; \dfrac{dy}{dx}$

28. ● $\ln(xy) - x \ln y = y; \dfrac{dy}{dx}$

29. ● $\ln(xy + y^2) = e^y; \dfrac{dy}{dx}$

30. ● $\ln(1 + e^{xy}) = y; \dfrac{dy}{dx}$

In Exercises 31–42, use implicit differentiation to find **(a)** *the slope of the tangent line, and* **(b)** *the equation of the tangent line at the indicated point on the graph. (Round answers to 4 decimal places as needed.) If only the x-coordinate is given, you must also find the y-coordinate.)* hint [see Examples 2, 3]

31. ● $4x^2 + 2y^2 = 12, (1, -2)$

32. ● $3x^2 - y^2 = 11, (-2, 1)$

33. ● $2x^2 - y^2 = xy, (-1, 2)$

34. ● $2x^2 + xy = 3y^2, (-1, -1)$

35. ● $x^2 y - y^2 + x = 1, (1, 0)$

36. ● $(xy)^2 + xy - x = 8, (-8, 0)$

37. ● $xy - 2000 = y, x = 2$

38. ● $x^2 - 10xy = 200, x = 10$

39. ● $\ln(x + y) - x = 3x^2, x = 0$

40. ● $\ln(x - y) + 1 = 3x^2, x = 0$

41. ● $e^{xy} - x = 4x, x = 3$

42. ● $e^{-xy} + 2x = 1, x = -1$

In Exercises 43–52, use logarithmic differentiation to find dy/dx. Do not simplify the result. hint [see Example 4]

43. ● $y = \dfrac{2x + 1}{4x - 2}$

44. ● $y = (3x + 2)(8x - 5)$

45. ● $y = \dfrac{(3x + 1)^2}{4x(2x - 1)^3}$

46. ● $y = \dfrac{x^2(3x + 1)^2}{(2x - 1)^3}$

47. ● $y = (8x - 1)^{1/3}(x - 1)$

48. ● $y = \dfrac{(3x + 2)^{2/3}}{3x - 1}$

49. ● $y = (x^3 + x)\sqrt{x^3 + 2}$

50. ● $y = \sqrt{\dfrac{x - 1}{x^2 + 2}}$

51. $y = x^x$

52. $y = x^{-x}$

Applications

53. ● *Productivity* The number of CDs per hour that Snappy Hardware can manufacture at its plant is given by

$$P = x^{0.6} y^{0.4}$$

where x is the number of workers at the plant and y is the monthly budget (in dollars). Assume P is constant, and compute $\dfrac{dy}{dx}$ when $x = 100$ and $y = 200,000$. Interpret the result. hint [see Example 5]

54. ● *Productivity* The number of cell-phone accessory kits (neon lights, matching covers and earpods) per day that USA Cellular Makeover Inc. can manufacture at its plant in Cambodia is given by

$$P = x^{0.5} y^{0.5}$$

where x is the number of workers at the plant and y is the monthly budget (in dollars). Assume P is constant, and compute $\dfrac{dy}{dx}$ when $x = 200$ and $y = 100,000$. Interpret the result.

55. ● *Demand* The demand equation for soccer tournament T-shirts is

$$xy - 2000 = y$$

where y is the number of T-shirts the Enormous State University soccer team can sell at a price of $x per shirt. Find $\dfrac{dy}{dx}\bigg|_{x=5}$, and interpret the result.

56. ● *Cost Equations* The cost y (in cents) of producing x gallons of Ectoplasm hair gel is given by the cost equation

$$y^2 - 10xy = 200$$

Evaluate $\dfrac{dy}{dx}$ at $x = 1$ and interpret the result.

57. ● *Housing Costs*[39] The cost C (in dollars) of building a house is related to the number k of carpenters used and the number e of electricians used by the formula

$$C = 15{,}000 + 50k^2 + 60e^2$$

If the cost of the house is fixed at \$200,000, find $\left.\dfrac{dk}{de}\right|_{e=15}$ and interpret your result.

58. ● *Employment* An employment research company estimates that the value of a recent MBA graduate to an accounting company is

$$V = 3e^2 + 5g^3$$

where V is the value of the graduate, e is the number of years of prior business experience, and g is the graduate school grade-point average. If V is fixed at 200, find $\dfrac{de}{dg}$ when $g = 3.0$ and interpret the result.

59. *Grades*[40] A productivity formula for a student's performance on a difficult English examination is

$$g = 4tx - 0.2t^2 - 10x^2 \quad (t < 30)$$

where g is the score the student can expect to obtain, t is the number of hours of study for the examination, and x is the student's grade-point average.

a. For how long should a student with a 3.0 grade-point average study in order to score 80 on the examination?

b. Find $\dfrac{dt}{dx}$ for a student who earns a score of 80, evaluate it when $x = 3.0$, and interpret the result.

60. *Grades* Repeat the preceding exercise using the following productivity formula for a basket-weaving examination:

$$g = 10tx - 0.2t^2 - 10x^2 \quad (t < 10)$$

Comment on the result.

---

[39] Based on an Exercise in *Introduction to Mathematical Economics* by A. L. Ostrosky Jr., and J. V. Koch (Waveland Press, Springfield, Illinois, 1979).

[40] Ibid.

Exercises 61 and 62 are based on the following demand function for money (taken from a question on the GRE economics test):

$$M_d = (2) \times (y)^{0.6} \times (r)^{-0.3} \times (p)$$

where

M_d = demand for nominal money balances (money stock)
y = real income
r = an index of interest rates
p = an index of prices.

61. ◆ *Money Stock* If real income grows while the money stock and the price level remain constant, the interest rate must change at what rate? (First find dr/dy, then dr/dt; your answers will be expressed in terms of r, y, and $\dfrac{dy}{dt}$.)

62. ◆ *Money Stock* If real income grows while the money stock and the interest rate remain constant, the price level must change at what rate? (See hint for Exercise 61.)

Communication and Reasoning Exercises

63. ● Fill in the missing terms: The equation $x = y^3 + y - 3$ specifies ___ as a function of ___, and ___ as an implicit function of ___.

64. ● Fill in the missing terms: When $x \neq 0$ in the equation $xy = x^3 + 4$, it is possible to specify ___ as a function of ___. However, ___ is only an implicit function of ___.

65. Use logarithmic differentiation to give another proof of the product rule.

66. Use logarithmic differentiation to give a proof of the quotient rule.

67. If y is given explicitly as a function of x by an equation $y = f(x)$, compare finding dy/dx by implicit differentiation to finding it explicitly in the usual way.

68. Explain why one should not expect dy/dx to be a function of x if y is not a function of x.

69. ◆ If y is a function of x and $dy/dx \neq 0$ at some point, regard x as an implicit function of y and use implicit differentiation to obtain the equation

$$\frac{dx}{dy} = \frac{1}{dy/dx}$$

70. ◆ If you are given an equation in x and y such that dy/dx is a function of x only, what can you say about the graph of the equation?

● basic skills   ◆ challenging

Chapter **11** Review

KEY CONCEPTS

11.1 The Product and Quotient Rules

Product rule: $\dfrac{d}{dx}[f(x)g(x)] =$

$\quad f'(x)g(x) + f(x)g'(x)$ *p. 799*

Quotient rule: $\dfrac{d}{dx}\left(\dfrac{f(x)}{g(x)}\right) =$

$\quad \dfrac{f'(x)g(x) - f(x)g'(x)}{[g(x)]^2}$ *p. 799*

Using the product rule *p. 801*
Using the quotient rule *p. 802*
Calculation thought experiment *p. 803*
Application to revenue and average cost *p. 805*

11.2 The Chain Rule

Chain rule: $\dfrac{d}{dx}[f(u)] = f'(u)\dfrac{du}{dx}$
p. 810

Generalized power rule:

$\quad \dfrac{d}{dx}[u^n] = nu^{n-1}\dfrac{du}{dx}$ *p. 811*

Using the chain rule *p. 812*
Application to marginal product *p. 815*
Chain rule in differential notation:

$$\frac{dy}{dx} = \frac{dy}{du}\frac{du}{dx} \quad p.\ 816$$

Manipulating derivatives in differential notation *p. 817*

11.3 Derivatives of Logarithmic and Exponential Functions

Derivative of the natural logarithm:

$$\frac{d}{dx}[\ln x] = \frac{1}{x} \quad p.\ 822$$

Derivative of logarithm with base

b: $\dfrac{d}{dx}[\log_b x] = \dfrac{1}{x\ln b}$ *p. 822*

Derivatives of logarithms of functions:

$$\frac{d}{dx}[\ln u] = \frac{1}{u}\frac{du}{dx}$$

$$\frac{d}{dx}[\log_b u] = \frac{1}{u\ln b}\frac{du}{dx} \quad p.\ 823$$

Derivative of logarithms of absolute values:

$$\frac{d}{dx}[\ln|x|] = \frac{1}{x}$$

$$\frac{d}{dx}[\ln|u|] = \frac{1}{u}\frac{du}{dx}$$

$$\frac{d}{dx}[\log_b|x|] = \frac{1}{x\ln b}$$

$$\frac{d}{dx}[\log_b|u|] = \frac{1}{u\ln b}\frac{du}{dx} \quad p.\ 825$$

Derivative of e^x: $\dfrac{d}{dx}[e^x] = e^x$ *p. 826*

Derivative of b^x: $\dfrac{d}{dx}[b^x] = b^x\ln b$ *p. 827*

Derivatives of exponential functions *p. 827*
Application to epidemics *p. 828*
Application to sales growth (logistic function) *p. 828*

11.4 Implicit Differentiation

Implicit function of x *p. 834*
Implicit differentiation *p. 834*
Using implicit differentiation *p. 835*
Finding a tangent line *p. 836*
Logarithmic differentiation *p. 837*

REVIEW EXERCISES

In Exercises 1–12 find the derivative of the given function.

1. $f(x) = e^x(x^2 - 1)$

2. $f(x) = \dfrac{x^2 + 1}{x^2 - 1}$

3. $f(x) = (x^2 - 1)^{10}$

4. $f(x) = \dfrac{1}{(x^2 - 1)^{10}}$

5. $f(x) = e^x(x^2 + 1)^{10}$

6. $f(x) = \left[\dfrac{x-1}{3x+1}\right]^3$

7. $f(x) = \dfrac{3^x}{x-1}$

8. $f(x) = 4^{-x}(x+1)$

9. $f(x) = e^{x^2 - 1}$

10. $f(x) = (x^2 + 1)e^{x^2 - 1}$

11. $f(x) = \ln(x^2 - 1)$

12. $f(x) = \dfrac{\ln(x^2 - 1)}{x^2 - 1}$

In Exercises 13–16 find all values of x (if any) where the tangent line to the graph of the given equation is horizontal.

13. $y = x - e^{2x-1}$

14. $y = e^{x^2}$

15. $y = \dfrac{x}{x+1}$

16. $y = \sqrt{x}(x-1)$

In Exercises 17–22, find dy/dx for the given equation.

17. $x^2 - y^2 = x$

18. $2xy + y^2 = y$

19. $e^{xy} + xy = 1$

20. $\ln\left(\dfrac{y}{x}\right) = y$

21. $y = \dfrac{(2x-1)^4(3x+4)}{(x+1)(3x-1)^3}$

22. $y = x^{x-1}3^x$

In Exercises 23 and 24 find the equation of the tangent line to the graph of the given equation at the specified point.

23. $xy - y^2 = x^2 - 3;\ (-1, 1)$

24. $\ln(xy) + y^2 = 1;\ (-1, -1)$

Applications

25. Revenue At the moment, OHaganBooks.com is selling 1000 books per week and its sales are rising at a rate of 200 books per week. Also, it is now selling all its books for $20 each, but its price is dropping at a rate of $1 per week. At what rate is OHaganBooks.com's revenue rising or falling?

26. Revenue Refer to Exercise 25. John O'Hagan would like to see the company's revenue increase at a rate of $5000 per week. At what rate would sales have to have been increasing to accomplish that goal, assuming all the other information is as given in Exercise 25?

27. Percentage Rate of Change of Revenue The percentage rate of change of a quantity Q is Q'/Q. Why is the percentage rate of change of revenue always equal to the sum of the percentage rates of change of unit price and weekly sales?

28. P/E Ratios At the beginning of last week, OHaganBooks.com stock was selling for $100 per share, rising at a rate of $50 per year. Its earnings amounted to $1 per share, rising at a rate of $0.10 per year. At what rate was its price-to-earnings (P/E) ratio, the ratio of its stock price to its earnings per share, rising or falling?

29. P/E Ratios Refer to Exercise 28. Curt Hinrichs, who recently invested in OHaganBooks.com stock, would have liked to see the P/E ratio increase at a rate of 100 points per year. How fast would the stock have to have been rising, assuming all the other information is as given in Exercise 28?

30. Percentage Rate of Change of P/E Ratios The percentage rate of change of a quantity Q is Q'/Q. Why is the percentage rate of change of P/E always equal to the percentage rate of change of unit price minus the percentage rate of change of earnings?

31. Sales OHaganBooks.com modeled its weekly sales over a period of time with the function

$$s(t) = 6053 + \frac{4474}{1 + e^{-0.55(t-4.8)}}$$

as shown in the following graph:

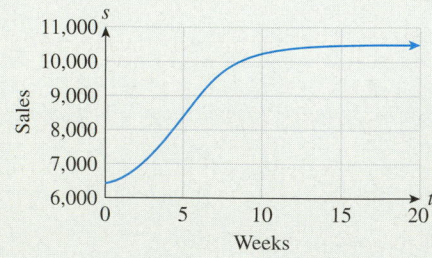

Compute $s'(t)$ and use the answer to compute the rate of increase of weekly sales at the beginning of the 7th week ($t = 6$). (Round your answer to the nearest unit.)

32. Sales refer to Exercise 31. Find the rate of increase of weekly sales at the beginning of the 15th week ($t = 14$).

33. Website Activity The number of "hits" on OHaganBooks.com's website was 1000 per day at the beginning of the year, growing at a rate of 5% per week. If this growth rate continued for the whole year (52 weeks), find the rate of increase (in hits per day per week) at the end of the year.

34. Demand and Revenue The price p that OHaganBooks.com charges for its latest leather-bound gift edition of *The Lord of the Rings* is related to the demand q in weekly sales by the equation

$$100pq + q^2 = 5,000,000$$

Suppose the price is set at $40, which would make the demand 1000 copies per week.

a. Using implicit differentiation, compute the rate of change of demand with respect to price, and interpret the result. (Round the answer to two decimal places.)

b. Use the result of part (a) to compute the rate of change of revenue with respect to price. Should the price be raised or lowered to increase revenue?

vMentor Do you need a live tutor for homework problems? Access vMentor on the ThomsonNOW! website at **www.thomsonedu.com** for one-on-one tutoring from a mathematics expert.

CASE STUDY: Projecting Market Growth

You are on the board of directors at Fullcourt Academic Press, and TJM, the sales director of the high school division, has just burst into your office with data showing the number of high school graduates each year over the past decade (Figure 7).[41]

TJM is pleased that the figures appear to support a basic premise of his recent proposal for a major expansion strategy: The number of high school seniors in the U.S. will be growing at a rate of at least 20,000 per year through the year 2005. The rate of increase, as he points out, has averaged around 50,000 per year since 1994, so it would not be overly optimistic to assume that the trend will continue—at least for the next 5 years.

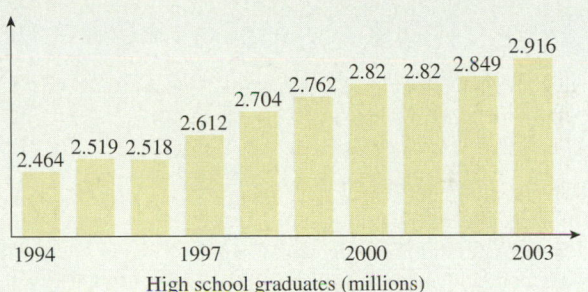

Figure **7**

Although you are tempted to support TJM's proposal at the next board meeting, you would like to estimate first whether the 20,000 figure is a realistic expectation, especially because the graph suggests that the number of graduates began to "level off" (in the language of calculus, the *derivative appears to be decreasing*) during the second half of the period. Moreover, you recall reading somewhere that the numbers of students in the lower grades have also begun to level off, so it is safe to predict that the slowing of growth in the senior class will continue over the next few years. You really need precise data about numbers in the lower grades in order to make a meaningful prediction, but TJM's report is scheduled to be presented tomorrow and you would like a quick and easy way of "extending the curve to the right" by then.

It would certainly be helpful if you had a mathematical model of the data in Figure 7 that you could use to project the current trend. But what kind of model should you use? A linear model would be no good because it would not show any change in the derivative (the derivative of a linear function is constant). In addition, best-fit polynomial and exponential functions do not accurately reflect the leveling off, as you realize after trying to fit a few of them (Figure 8).

You then recall that a logistic curve can model the leveling-off property you desire, and so you try fitting a curve of the form

$$y = \frac{N}{1 + Ab^{-t}}$$

---

[41] Data starting in 2000 are projections. SOURCE: U.S. Department of Education
http://nces.ed.gov/pubs2001/proj01/tables/table23.asp.

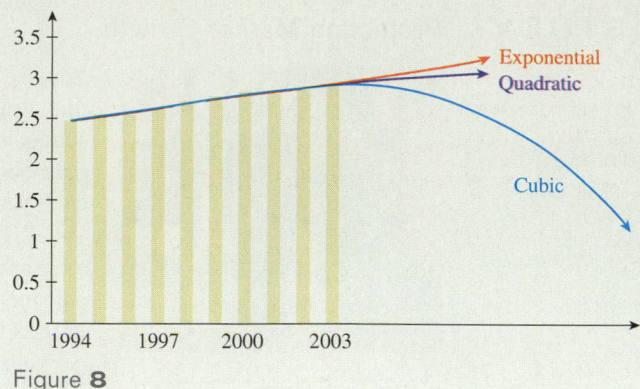

Figure **8**

Figure 9 shows the best-fit logistic curve, which eventually levels off at around $N = 3.3$.

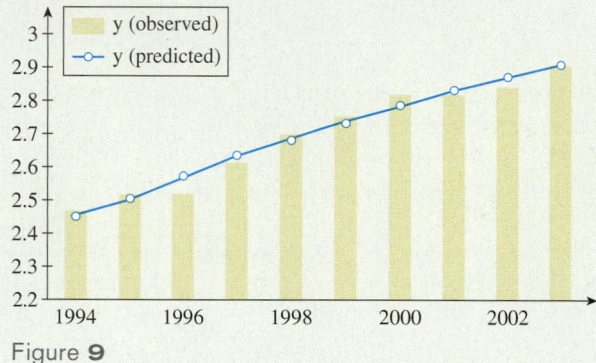

Figure **9**

The leveling-off prediction certainly seems reasonable, but you are slightly troubled by the shape of the regression curve: It doesn't seem to "follow the s-shape" of the data very convincingly. Moreover, the curve doesn't appear to fit the data significantly more snugly than the quadratic or cubic models.[42] To reassure yourself, you decide to look for another kind of s-shaped model as a backup.

After flipping through a calculus book, you stumble across a function whose graph looks rather like the one you have (Figure 10).

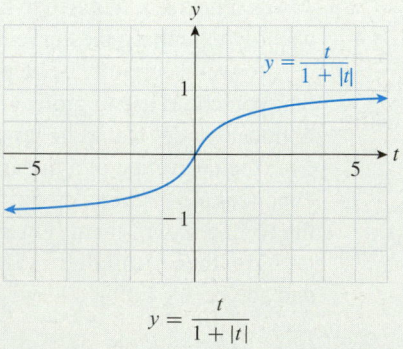

$$y = \frac{t}{1 + |t|}$$

Figure **10**

---

[42] There is another, more mathematical, reason for not using logistic regression to predict long-term leveling off: The regression value of the long-term level N is extremely sensitive to the values of the other coefficients. As a result, there can be good fits to the same set of data with a wide variety of values of N.

Online, follow:

Chapter 1

→ New Functions from Old:
   Scaled and Shifted Functions

to find a detailed treatment of
scaled and shifted functions.

Curves of this form are sometimes called predator satiation curves, and they are used to model the population of predators in an environment with limited prey. Although the curve does seem to have the proper shape, you realize that you will need to shift and scale the function in order to fit the actual data. The most general scaled and shifted version of this curve has the form

$$y = c + b\frac{a(t-m)}{1 + a|t-m|} \qquad (a, b, c, m \text{ constant})$$

and you decide to try a model of this form, where y will represent the number of high school seniors (in millions), and t will represent years since 1994.

For the moment, you postpone the question of finding the best values for the constants a, b, c, and m, and decide first to calculate the derivative of the model in terms of the given constants. The derivative, dy/dt, will represent the rate of increase of high school graduates, which is exactly what you wish to estimate.

The main part of the function is a quotient, so you start with the quotient rule:

$$\frac{dy}{dt} = b\frac{a(1 + a|t-m|) - a(t-m)\frac{d}{dt}(1 + a|t-m|)}{(1 + a|t-m|)^2}$$

You now recall the formula for the derivative of an absolute value:

$$\frac{d}{dt}[|t|] = \frac{|t|}{t} = \begin{cases} 1 & \text{if } t > 0 \\ -1 & \text{if } t \le 0 \end{cases}$$

which, you are disturbed to notice, is not defined at $t = 0$. Undaunted, you make a mental note and press on. Because you need the derivative of the absolute value of a quantity other than t, you use the chain rule, which tells you that

$$\frac{d}{dt}[|u|] = \frac{|u|}{u}\frac{du}{dt}$$

Thus,

$$\frac{d}{dt}[|t-m|] = \frac{|t-m|}{t-m} \cdot 1 \qquad u = t - m; \quad m = \text{constant}$$

(This is not defined when $t = m$.) Substituting into the formula for dy/dt, you find:

$$\frac{dy}{dt} = b\frac{a(1 + a|t-m|) - a(t-m)\cdot a\dfrac{|t-m|}{(t-m)}}{(1 + a|t-m|)^2}$$

$$= b\frac{a(1 + a|t-m|) - a^2|t-m|}{(1 + a|t-m|)^2} \qquad \text{Cancel } (t-m)$$

$$= \frac{ab}{(1 + a|t-m|)^2} \qquad a^2|t-m| - a^2|t-m| = 0$$

It is interesting that, although the derivative of $|t - m|$ is not defined when $t = m$, the offending term $t - m$ was canceled, so that dy/dt seems to be defined[43] at $t = m$.

Now you have a simple-looking expression for dy/dt, which will give you an estimate of the rate of change of the high school senior population. However, you still need

[43] In fact, it is defined and has the value given by the formula just derived: ab. To show this takes a bit more work. How might you do it?

values for the constants a, b, c, and m. (You don't really need the value of c to compute the derivative—where has it gone?—but it is a part of the model.) How do you find the values of a, b, c, and m that result in the curve that best fits the given data?

Turning once again to your calculus book (see the discussion of logistic regression in Section 9.4), you see that a best-fit curve is one that minimizes the sum-of-squares error. Here is an Excel spreadsheet showing the errors for $a = 1$, $b = 1$, $c = 1$, and $m = 1$.

	A	B	C	D	E	F	G
	t (year)	y (Observed)	y (Predicted)	Residue^2	Constants		
1	0	2.464	0.5	3.857296			
2	1	2.519	1	2.307361	a	1	
3	2	2.518	1.5	1.036324	b	1	
4	3	2.612	1.666666667	0.89365511	c	1	
5	4	2.704	1.75	0.910116	m	1	
6	5	2.762	1.8	0.925444			
7	6	2.82	1.833333333	0.97351111	SSE:	13.8344343	
8	7	2.82	1.857142857	0.92709388			
9	8	2.849	1.875	0.948676			
10	9	2.916	1.888888889	1.05495723			

(Microsoft Excel - CS 11 High School Graduates in 1994.xls)

The first two columns show the observed data (t = year since 1994, y = number of high school graduates in millions). The formula for y (Predicted) is our model

$$ y = c + b\frac{a(t - m)}{1 + a|t - m|} $$

entered in cell C2 as

```
=$F$4 + $F$3*$F$2*(A2 - $F$5) / (1 + $F$2*ABS(A2 - $F$5))
   c  +  b *  a  * (t - m)      / (1 +   a *    |t - m|)
```

and then copied into the cells below it. Since the square error (Residue^2) is defined as the square of the difference between y and y (Predicted), we enter

```
= (C2-B2)^2
```

in cell D2 and then copy into the cells below it. The sum-of-squares error, SSE (sum of the entries in D2–D11), is then placed in cell F7.

The values of a, b, c, and m shown are initial values and don't matter too much (but see below); you will have Excel change these values in order to improve your model. The smaller the SSE is, the better your model. (For a perfect fit, the y (Observed) column would equal the y (Observed) column, and SSE would be zero.) Hence, the goal is now to find values of a, b, c, and m that make the value of SSE as small as possible. (See the discussion in Section 1.4.) Finding these values analytically is an extremely difficult mathematical problem. However, there is software, such as Excel's built-in "Solver" routine,[44] that can be used to find *numerical* solutions.

Figure 11 shows how to set up Solver to find the best values for a, b, c, and m for the setup used in this spreadsheet.

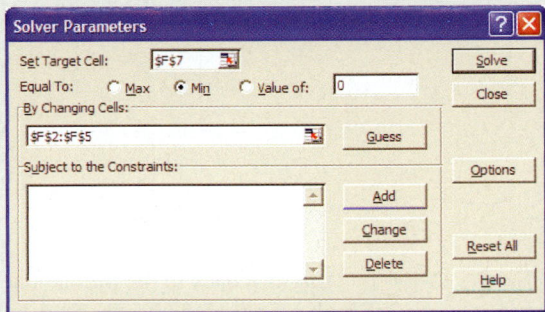

Figure **11**

The Target Cell, $F$7, contains the value of SSE, which is to be minimized. The Changing Cells are the cells containing the values of the constants a, b, c, and m that we want to change. That's it.

Now press "Solve." After thinking about it for a few seconds, Excel gives the optimal values of a, b, c, and m in cells F2–F5, and the minimum value of SSE in cell D12.[45] You find

$$a = 0.24250175, \; b = 0.4102711, \; c = 2.65447873, \; m = 3.5211956$$

with SSE ≈ 0.00288, which is a better fit than the logistic regression curve (SSE ≈ 0.00653)

Figure 12 shows that not only does this choice of model and constants give an excellent fit, but that the curve seems to follow the "s-shape" more convincingly than the logistic curve.

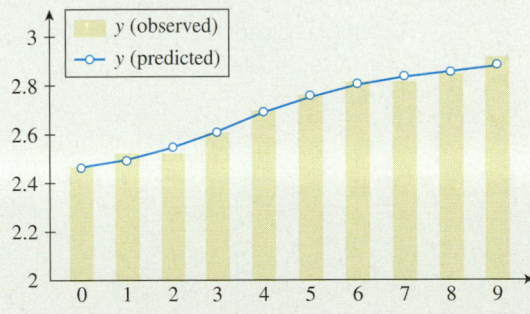

Figure **12**

---

[44] See the similar discussion in Section 9.4. If "Solver" does not appear in the "Tools" menu, you should first install it using your Excel installation software. (Solver is one of the "Excel Add-Ins.")

[45] Depending on the settings in Solver, you may need to run the utility twice in succession to reach the minimum value of SSE.

Figure 13 shows how the model predicts the long-term leveling-off phenomenon you were looking for.

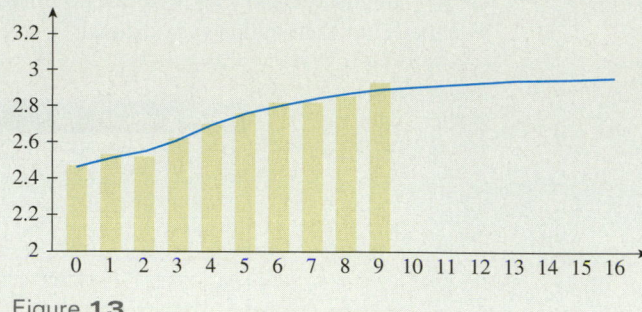

Figure **13**

You turn back to the problem at hand: projecting the rate of increase of the number of high school graduates in 2005. You have the formula

$$\frac{dy}{dt} = \frac{ab}{(1 + a|t - m|)^2}$$

and also values for the constants. So you compute:

$$\frac{dy}{dt} = \frac{(0.24250175)(0.4102711)}{(1 + 0.24250175[11 - 3.5211956])^2} \qquad t = 11 \text{ in } 2005$$

$$\approx 0.0126 \text{ million students per year}$$

or 12,600 students per year—far less than the optimistic estimate of 20,000 in the proposal!

You now conclude that TJM's prediction is suspect and that further research will have to be done before the board can support the proposal.

Q: *How accurately does the model predict the number of high school graduates?*

A: Using a regression curve-fitting model to make long-term predictions is always risky. A more accurate model would have to take into account such factors as the birth rate and current school populations at all levels. The U.S. Department of Education has used more sophisticated models to make the projections shown below, which we compare with those predicted by our model.

Year	U.S. Dept. of Ed. Projections (Millions)	Model Predictions (Millions)
2004	2.921	2.91
2005	2.929	2.92
2006	2.986	2.93
2007	3.054	2.94
2008	3.132	2.95
2009	3.127	2.96
2010	3.103	2.96
2011	3.063	2.97

Q: Which values of the constants should I use as starting values when using Excel to find the best-fit curve?

A: If the starting values of the constants are far from the optimal values, Solver may find a nonoptimal solution. Thus, you need to obtain some rough initial estimate of the constants by examining the graph. Figure 14 shows some important features of the curve that you can use to obtain estimates of *a*, *b*, *c*, and *m* by inspecting the graph.

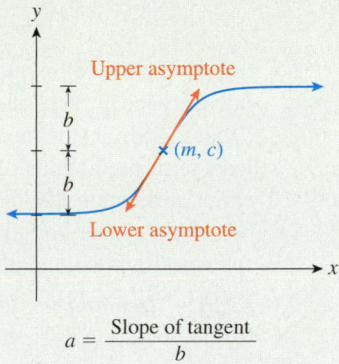

$$a = \frac{\text{Slope of tangent}}{b}$$

Figure **14**

From the graph, *m* and *c* are the coordinates of the point on the curve where it is steepest, and *b* is the vertical distance from that point to the upper or lower asymptote (where the curve "levels off"). To estimate *a*, first estimate the slope of the tangent at the point of steepest inclination, then divide by *b*. If *b* is negative (and *a* is positive), we obtain an "upside-down" version of the curve (Figure 15).

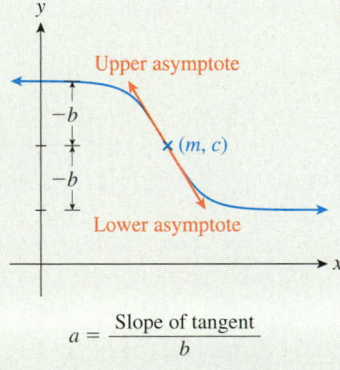

$$a = \frac{\text{Slope of tangent}}{b}$$

Figure **15**

Exercises

1. In 1993 there were 2.49 million high school graduates. What does the regression model "predict" for 1993? What is the residue ($y_{\text{predicted}} - y_{\text{observed}}$)? (Round answers to the nearest 0.01 million.)

2. What is the long-term prediction of the model?

3. Find $\lim\limits_{t \to \infty} \dfrac{dy}{dt}$ and interpret the result.

4. tech Ex You receive a memo to the effect that the 1994 and 1995 figures are not accurate. Use Excel Solver to re-estimate the best-fit constants a, b, c, and m in the absence of this data and obtain new estimates for the 1994 and 1995 data. What does the new model predict the rate of change in the number of high school seniors will be in 2005?

5. tech Ex **Shifted Logistic Model** Using the original data, find the best-fit shifted logistic curve of the form

$$f(t) = c + \frac{N}{1 + Ab^{-t}}$$

(Start with the following values: $c = 0$, $N = 3$, and $A = b = 1$. You might have to run Solver twice in succession to minimize SSE.) Graph the data together with the model. What is SSE? Is the model as accurate a fit as the model used in the text? How do the long-term predictions of the two models compare with the U.S. Department of Education projections? What does this model predict will be the growth rate of the number of high school graduates in 2005? Round the coefficients in the model and all answers to four decimal places.

6. tech Ex **Demand for Freon** The demand for chlorofluorocarbon-12 (CFC-12)—the ozone-depleting refrigerant commonly known as freon[46]—has been declining significantly in response to regulation and concern about the ozone layer. The chart below shows the projected demand for CFC-12 for the period 1994–2005.[47]

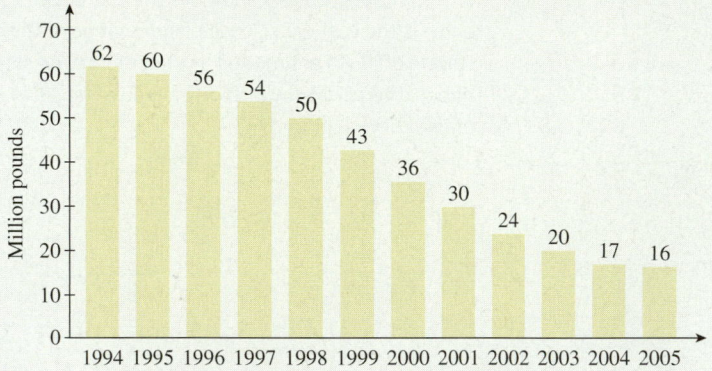

a. Use Excel Solver to obtain the best-fit equation of the form

$$f(t) = c + b\frac{a(t - m)}{1 + a|t - m|}$$

where $t =$ years since 1990. Use your function to estimate the total demand for CFC-12 from the start of the year 2000 to the start of 2010. [Start with the following values: $a = 1$, $b = -25$, $c = 35$, and $m = 10$, and round your answers to four decimal places.]

b. According to your model, how fast is the demand for freon declining in 2000?

---

[46] The name given to it by Du Pont.

[47] SOURCE: The Automobile Consulting Group (*New York Times*, December 26, 1993, p. F23). The exact figures were not given, and the chart is a reasonable facsimile of the chart that appeared in the *New York Times*.

12

Applications of the Derivative

CASE STUDY Production Lot Size Management

Your publishing company is planning the production of its latest best seller, which it predicts will sell 100,000 copies each month over the coming year. The book will be printed in several batches of the same number, evenly spaced throughout the year. Each print run has a setup cost of $5000, a single book costs $1 to produce, and monthly storage costs for books awaiting shipment average 1¢ per book. To meet the anticipated demand at minimum total cost to your company, how many printing runs should you plan?

Jeff Greenberg/PhotoEdit

Introduction

In this chapter we begin to see the power of calculus as an optimization tool. In Chapter 9 we saw how to price an item in order to get the largest revenue when the demand function is linear. Using calculus, we can handle nonlinear functions, which are much more general. In Section 12.1 we show how calculus can be used to solve the problem of finding the values of a variable that lead to a maximum or minimum value of a given function. In Section 12.2 we show how this helps us in various real-world applications.

Another theme in this chapter is that calculus can help us to draw and understand the graph of a function. By the time you have completed the material in Section 12.1, you will be able to locate and sketch some of the important features of a graph. In Section 12.3 we discuss further how to explain what you see in a graph (drawn, for example, using graphing technology) and to locate its most important points.

algebra Review

For this chapter, you should be familiar with the algebra reviewed in Chapter 0, sections 5 and 6.

We also include sections on related rates and elasticity of demand. The first of these (Section 12.4) examines further the concept of the derivative as a rate of change. The second (Section 12.5) returns to the problem of optimizing revenue based on the demand equation, looking at it in a new way that leads to an important idea in economics—elasticity.

12.1 | Maxima and Minima

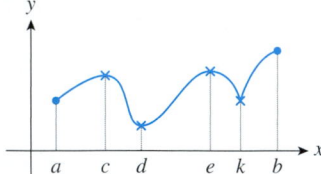

Figure **1**

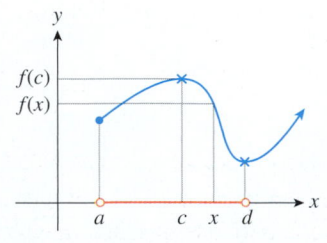

Figure **2**

Figure 1 shows the graph of a function f whose domain is the closed interval $[a, b]$. A mathematician sees lots of interesting things going on here. There are hills and valleys, and even a small chasm (called a *cusp*) toward the right. For many purposes, the important features of this curve are the highs and lows. Suppose, for example, you know that the price of the stock of a certain company will follow this graph during the course of a week. Although you would certainly make a handsome profit if you bought at time a and sold at time b, your best strategy would be to follow the old adage to "buy low and sell high," buying at all the lows and selling at all the highs.

Figure 2 shows the graph once again with the highs and lows marked. Mathematicians have names for these points: the highs (at the x-values c, e, and b) are referred to as **relative maxima,** and the lows (at the x-values a, d, and k) are referred to as **relative minima.** Collectively, these highs and lows are referred to as **relative extrema.** (A point of language: The singular forms of the plurals *minima, maxima,* and *extrema* are *minimum, maximum,* and *extremum.*)

Why do we refer to these points as relative extrema? Take a look at the point corresponding to $x = c$. It is the highest point of the graph *compared to other points nearby*. If you were an extremely nearsighted mountaineer standing at point c, you would *think* that you were at the highest point of the graph, not being able to see the distant peaks at $x = e$ and $x = b$.

Let's translate into mathematical terms. We are talking about the heights of various points on the curve. The height of the curve at $x = c$ is $f(c)$, so we are saying that $f(c)$ is greater than $f(x)$ for every x near c. For instance, $f(c)$ *is the greatest value that $f(x)$ has for all choices of x between a and d* (see Figure 3).

We can phrase the formal definition as follows.

Relative Extrema

f has a **relative maximum** at *c* if there is some interval (r, s) (even a very small one) containing *c* for which $f(c) \geq f(x)$ for all *x* between *r* and *s* for which $f(x)$ is defined.

f has a **relative minimum** at *c* if there is some interval (r, s) (even a very small one) containing *c* for which $f(c) \leq f(x)$ for all *x* between *r* and *s* for which $f(x)$ is defined.

quick Examples

In Figure 2, *f* has the following relative extrema:

1. A relative maximum at *c*, as shown by the interval (a, d)

2. A relative maximum at *e*, as shown by the interval (d, k)

3. A relative maximum at *b*, as shown by the interval $(k, b + 1)$

Note that $f(x)$ is not defined for $x > b$. However, $f(b) \geq f(x)$ for every *x* in the interval $(k, b + 1)$ *for which $f(x)$ is defined*—that is, for every *x* in $(k, b]$.

4. A relative minimum at *d*, as shown by the interval (c, e)

5. A relative minimum at *k*, as shown by the interval (e, b)

6. A relative minimum at *a*, as shown by the interval $(a - 1, c)$ (See Quick Example 3)

Note Our definition of relative extremum allows *f* to have a relative extremum at an endpoint of its domain; the definitions used in some books do not. In view of examples like our stock-market investing strategy, we find it very useful to count endpoints as extrema. ■

Looking carefully at Figure 2, we can see that the lowest point on the whole graph is where $x = d$ and the highest point is where $x = b$. This means that $f(d)$ is the smallest value of *f* on the whole domain of *f* (the interval $[a, b]$) and $f(b)$ is the largest value. We call these the *absolute* minimum and maximum.

Absolute Extrema

f has an **absolute maximum** at *c* if $f(c) \geq f(x)$ for every *x* in the domain of *f*.

f has an **absolute minimum** at *c* if $f(c) \leq f(x)$ for every *x* in the domain of *f*.

quick Examples

1. In Figure 2, *f* has an absolute maximum at *b* and an absolute minimum at *d*.

2. If $f(x) = x^2$ then $f(x) \geq f(0)$ for every real number *x*. Therefore, $f(x) = x^2$ has an absolute minimum at $x = 0$ (see the figure).

3. Generalizing (2), every quadratic function $f(x) = ax^2 + bx + c$ has an absolute extremum at its vertex $x = -b/(2a)$; it is an absolute minimum if $a > 0$ and an absolute maximum if $a < 0$.

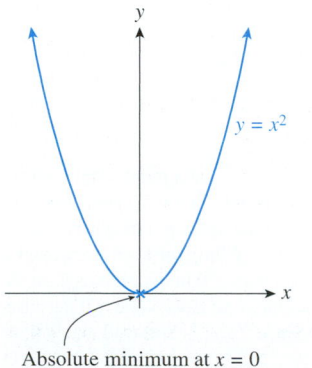

y

$y = x^2$

x

Absolute minimum at $x = 0$

Some graphs have no absolute extrema at all (think of the graph of $y = x$), while others might have an absolute minimum but no absolute maximum (like $y = x^2$), or vice versa. When *f* does have an absolute maximum, there is only one absolute maximum *value* of *f*, but this value may occur at different values of *x*. (see Figure 4).

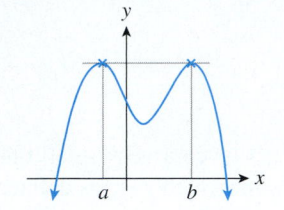

Absolute maxima at $x = a$ and $x = b$

Figure **4**

Q: At how many different values of x can f take on its absolute maximum value *?*

A: An extreme case is that of a constant function; because we use $\geq$ in the definition of absolute maximum, a constant function has an absolute maximum (and minimum) at every point in its domain. ∎

Now, how do we go about locating extrema? In many cases we can get a good idea by using graphing technology to zoom in on a maximum or minimum and approximate its coordinates. However, calculus gives us a way to find the exact locations of the extrema and at the same time to understand why the graph of a function behaves the way it does. In fact, it is often best to combine the powers of graphing technology with those of calculus, as we shall see.

In Figure 5 we see the graph from Figure 1 once more, but we have labeled each extreme point as one of three types.

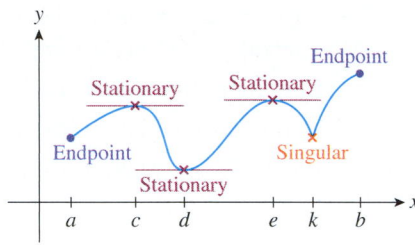

Figure **5**

At the points labeled "stationary," the tangent lines to the graph are horizontal, and so have slope 0, so f' (which gives the slope) is 0. Any time $f'(x) = 0$, we say that f has a **stationary point** at x because the rate of change of f is zero there. We call an *extremum* that occurs at a stationary point a **stationary extremum.** In general, to find the exact location of each stationary point, we need to solve the equation $f'(x) = 0$.

There is a relative minimum in Figure 5 at $x = k$, but there is no horizontal tangent there. In fact, there is no tangent line at all; $f'(k)$ is not defined. (Recall a similar situation with the graph of $f(x) = |x|$ at $x = 0$.) When $f'(x)$ does not exist, we say that f has a **singular point** at x. We shall call an extremum that occurs at a singular point a **singular extremum.** The points that are either stationary or singular we call collectively the **critical points** of f.

The remaining two extrema are at the **endpoints** of the domain.[1] As we see in the figure, they are (almost) always either relative maxima or relative minima.

Q: Are there any other types of relative extrema *?*

A: No; relative extrema of a function always occur at critical points or endpoints (a rigorous proof is beyond the scope of this book).[2] ∎

---

[1] Remember that we do allow relative extrema at endpoints.

[2] Here is an outline of the argument. Suppose f has a relative maximum, say, at $x = a$, at a point other than an endpoint of the domain. Then either f is differentiable there, or it is not. If it is not, then we have a singular point. If f is differentiable at $x = a$, then consider the slope of the secant line through the points where $x = a$ and $x = a + h$ for small positive h. Since f has a relative maximum at $x = a$, it is falling (or level) to the right of $x = a$, and so the slope of this secant line must be ≤ 0. Thus we must have $f'(a) \leq 0$ in the limit as $h \to 0$. On the other hand, if h is small and *negative,* then the corresponding secant line must have slope ≥ 0 because f is also falling (or level) as we move left from $x = a$, and so $f'(a) \geq 0$. Since $f'(a)$ is both ≥ 0 and ≤ 0, it must be zero, and so we have a stationary point at $x = a$.

Locating Candidates for Relative Extrema

If f is a real valued function, then its relative extrema occur among the following types of points:

1. **Stationary Points:** f has a stationary point at x if x is in the domain and $f'(x) = 0$. To locate stationary points, set $f'(x) = 0$ and solve for x.

2. **Singular Points:** f has a singular point at x if x is in the domain and $f'(x)$ is not defined. To locate singular points, find values of x where $f'(x)$ is *not* defined, but $f(x)$ *is* defined.

3. **Endpoints:** The x-coordinates of endpoints are endpoints of the domain, if any. Recall that closed intervals contain endpoints, but open intervals do not.

Once we have the x-coordinates of a candidate for a relative extremum, we find the corresponding y-coordinate using $y = f(x)$.

quick Examples

1. **Stationary Points:** Let $f(x) = x^3 - 12x$. Then to locate the stationary points, set $f'(x) = 0$ and solve for x. This gives $3x^2 - 12 = 0$, so f has stationary points at $x = \pm 2$. Thus, the stationary points are $(-2, f(-2)) = (-2, 16)$ and $(2, f(2)) = (2, -16)$.

2. **Singular points:** Let $f(x) = 3(x-1)^{1/3}$. Then $f'(x) = (x-1)^{-2/3} = 1/(x-1)^{2/3}$. $f'(1)$ is not defined, although $f(1)$ *is* defined. Thus, the (only) singular point occurs at $x = 1$. Its coordinates are $(1, f(1)) = (1, 0)$.

3. **Endpoints:** Let $f(x) = 1/x$, with domain $(-\infty, 0) \cup [1, +\infty)$. Then the only endpoint in the domain of f occurs when $x = 1$ and has coordinates $(1, 1)$. The natural domain of $1/x$ has no endpoints.

Remember, though, that these are only *candidates* for relative extrema. It is quite possible, as we shall see, to have a stationary point (or singular point) that is neither a relative maximum nor a relative minimum.

Now let's look at some examples of finding maxima and minima. In all of these examples, we will use the following procedure: First, we find the derivative, which we examine to find the stationary points and singular points. Next, we make a table listing the x-coordinates of the critical points and endpoints, together with their y-coordinates. We use this table to make a rough sketch of the graph. From the table and rough sketch, we usually have enough data to be able to say where the extreme points are and what kind they are.

Example 1 Maxima and Minima

Find the relative and absolute maxima and minima of
$$f(x) = x^2 - 2x$$
on the interval $[0, 4]$.

Solution We first calculate $f'(x) = 2x - 2$. We use this derivative to locate the stationary and singular points.

Stationary Points To locate the stationary points, we solve the equation $f'(x) = 0$, or
$$2x - 2 = 0,$$
getting $x = 1$. The domain of the function is $[0, 4]$, so $x = 1$ is in the domain. Thus, the only candidate for a stationary relative extremum occurs when $x = 1$.

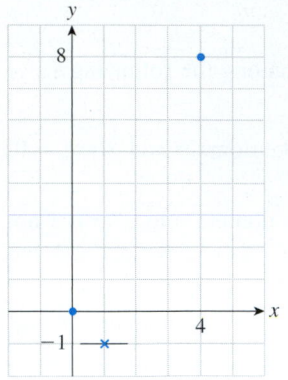

Figure **6**

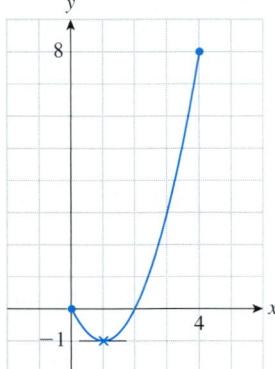

Figure **7**

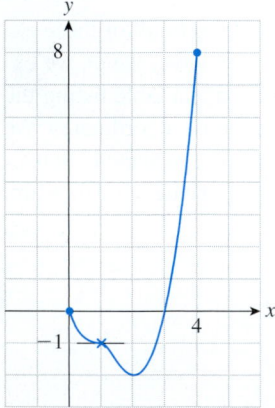

Figure **8**

Singular Points We look for points where the derivative is not defined. However, the derivative is $2x - 2$, which is defined for every x. Thus, there are no singular points and hence no candidates for singular relative extrema.

Endpoints The domain is [0, 4], so the endpoints occur when $x = 0$ and $x = 4$.

We record these values of x in a table, together with the corresponding y-coordinates (values of f):

x	0	1	4
$f(x) = x^2 - 2x$	0	−1	8

This gives us three points on the graph, $(0, 0)$, $(1, -1)$, and $(4, 8)$, which we plot in Figure 6.

We remind ourselves that the point $(1, -1)$ is a stationary point of the graph by drawing in a part of the horizontal tangent line. Connecting these points must give us a graph something like that in Figure 7.

Notice that the graph has a horizontal tangent line at $x = 1$ but not at either of the endpoints because the endpoints are not stationary points.

From Figure 7 we can see that f has the following extrema:

x	$y = x^2 - 2x$	Classification
0	0	Relative maximum (endpoint)
1	−1	Absolute minimum (stationary point)
4	8	Absolute maximum (endpoint)

Q: *How can we be sure that the graph in Example 1 doesn't look like Figure 8?*

A: If it did, there would be another critical point somewhere between $x = 1$ and $x = 4$. But we already know that there aren't any other critical points. The table we made listed all of the possible extrema; there can be no more. ∎

In Example 1 we found that $f'(1) = 0$; f has a stationary point at $x = 1$. It is also useful to consider values of $f'(x)$ to the left and right of the critical point. Here is a table with some values to the left and right of the critical point $x = 1$ in the above example:

		Critical Point	
x	0	1	2
$f'(x) = 2x - 2$	−2	0	2
Direction of Graph	↘	→	↗

At $x = 0$, $f'(0) = -2 < 0$, so the graph has negative slope and f is **decreasing;** its values are going down as x increases. We note this with the downward pointing arrow in the chart. At $x = 2$, $f'(2) = 2 > 0$, so the graph has positive slope and f is **increasing;** its values are going up as x increases. In fact, because $f'(x) = 0$ only at $x = 1$, we know that $f'(x) < 0$ for all x in [0, 1), and we can say that f is decreasing on the interval [0, 1].

Similarly, f is increasing on $[1, 4]$. So, starting at $x = 0$, the graph of f goes down until we reach $x = 1$ and then it goes back up. Notice that the arrows suggest exactly this type of graph. This is another way of checking that a critical point is a relative minimum and is known as the **first derivative test.**[3]

Note Here is some terminology: If the point (a, b) is a maximum (or minimum) of f, we sometimes say that f **has a maximum (or minimum) value of b at $x = a$.** Thus, in the above example, we could have said the following:

• f has a relative maximum value of 0 at $x = 0$.

• f has an absolute minimum value of -1 at $x = 1$.

• f has an absolute maximum value of 8 at $x = 4$. ∎

Example 2 Unbounded Interval

Find all extrema of $f(x) = 3x^4 - 4x^3$ on $[-1, \infty)$.

Solution We first calculate $f'(x) = 12x^3 - 12x^2$.

Stationary points We solve the equation $f'(x) = 0$, which is

$$12x^3 - 12x^2 = 0 \text{ or}$$
$$12x^2(x - 1) = 0$$

There are two solutions, $x = 0$ and $x = 1$, and both are in the domain. These are our candidates for the x-coordinates of stationary relative extrema.

Singular points There are no points where $f'(x)$ is not defined, so there are no singular points.

Endpoints The domain is $[-1, \infty)$, so there is one endpoint, at $x = -1$.

We record these points in a table with the corresponding y-coordinates:

x	-1	0	1
$f(x) = 3x^4 - 4x^3$	7	0	-1

We will illustrate three methods we can use to determine which are minima, which are maxima, and which are neither:

1. Plot these points and sketch the graph by hand.

2. Use the First Derivative Test.

3. Use technology to help us.

Use the method you find most convenient.

Using a Hand Plot: If we plot these points by hand, we obtain Figure 9(a), which suggests Figure 9(b).

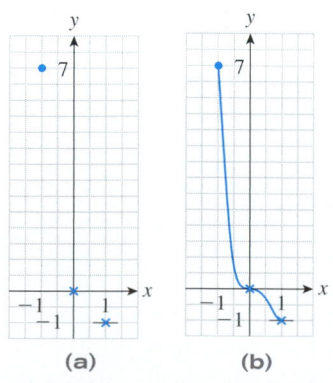

(a)   (b)

Figure 9

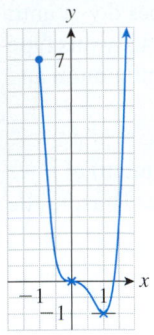

Figure **10**

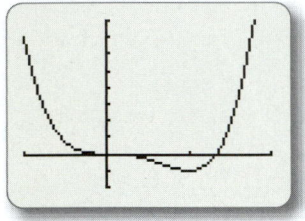

Figure **11**

 using *Technology*

We can't be sure what happens to the right of $x = 1$. Does the curve go up, or does it go down? To find out, let's plot a "test point" to the right of $x = 1$. Choosing $x = 2$, we obtain $y = 3(2)^4 - 4(2)^3 = 16$, so $(2, 16)$ is another point on the graph. Thus, it must turn upwards to the right of $x = 1$, as shown in Figure 10.

From the graph, we find that f has the following extrema:

A relative (endpoint) maximum at $(-1, 7)$

An absolute (stationary) minimum at $(1, -1)$

Using the First Derivative Test: List the critical and endpoints in a table, and add additional points as necessary so that each critical point has a noncritical point on either side. Then compute the derivative at each of these points, and draw an arrow to indicate the direction of the graph.

	Endpoint	Critical point		Critical point	
x	-1	0	0.5	1	2
$f'(x) = 12x^3 - 12x^2$	-24	0	-1.5	0	48
Direction of Graph	↘	→	↘	→	↗

Notice that the arrows now suggest the shape of the curve in Figure 10, and hence permit us to determine that the function has a maximum at $x = -1$, neither a maximum nor a minimum at $x = 0$, and a minimum at $x = 1$. Deciding which of these extrema are absolute and which are relative requires us to compute y-coordinates and plot the corresponding points on the graph by hand, as we did in the first method.

If we use technology to show the graph, we should choose the viewing window so that it contains the three interesting points we found: $x = -1$, $x = 0$, and $x = 1$. Again, we can't be sure yet what happens to the right of $x = 1$; does the graph go up or down from that point? If we set the viewing window to an interval of $[-1, 2]$ for x and $[-2, 8]$ for y, we will leave enough room to the right of $x = 1$ and below $y = -1$ to see what the graph will do. The result will be something like Figure 11.

Now we can tell what happens to the right of $x = 1$: the function increases. We know that it cannot later decrease again because if it did, there would have to be another critical point where it turns around, and we found that there are no other critical points. ∎

+*Before we go on...* Notice that the stationary point at $x = 0$ in Example 2 is neither a relative maximum nor a relative minimum. It is simply a place where the graph of f flattens out for a moment before it continues to fall. Notice also that f has no absolute maximum because $f(x)$ increases without bound as x gets large. ∎

Example **3 Singular Point**

Find all extrema of $f(t) = t^{2/3}$ on $[-1, 1]$.

Solution First, $f'(t) = \dfrac{2}{3}t^{-1/3}$

Stationary points We need to solve

$$\frac{2}{3}t^{-1/3} = 0$$

We can rewrite this equation without the negative exponent:

$$\frac{2}{3t^{1/3}} = 0$$

Now, the only way that a fraction can equal 0 is if the numerator is 0, so this fraction can never equal 0. Thus, there are no stationary points.

Singular points The derivative

$$f'(t) = \frac{2}{3t^{1/3}}$$

is not defined for $t = 0$. However, f itself *is* defined at $t = 0$, so 0 is in the domain. Thus, f has a singular point at $t = 0$.

Endpoints There are two endpoints, -1 and 1.

We now put these three points in a table with the corresponding y-coordinates:

t	-1	0	1
$f(t)$	1	0	1

Using a Hand Plot: The derivative, $f'(t) = 2/(3t^{1/3})$, is not defined at the singular point $t = 0$. To help us sketch the graph, let's use limits to investigate what happens to the derivative as we approach 0 from either side:

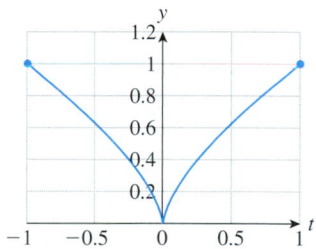

Figure **12**

$$\lim_{t \to 0^-} f'(t) = \lim_{t \to 0^-} \frac{2}{3t^{1/3}} = -\infty$$

$$\lim_{t \to 0^+} f'(t) = \lim_{t \to 0^+} \frac{2}{3t^{1/3}} = +\infty$$

Thus, the graph decreases very steeply, approaching $t = 0$ from the left, and then rises very steeply as it leaves to the right. It would make sense to say that the tangent line at $x = 0$ is vertical, as shown in Figure 12.

From this graph, we find the following extrema for f:

An absolute (endpoint) maximum at $(-1, 1)$

An absolute (singular) minimum at $(0, 0)$

An absolute (endpoint) maximum at $(1, 1)$

Notice that the absolute maximum value of f is achieved at two values of t: $t = -1$ and $t = 1$.

First Derivative Test: Here is the corresponding table for the first derivative test.

	t	-1	0	1
$f'(t) = \dfrac{2}{3t^{1/3}}$		$-\dfrac{2}{3}$	Undefined	$\dfrac{2}{3}$
Direction of graph		↘	↕	↗

We drew a vertical arrow at $t = 0$ to indicate a vertical tangent. Again, notice how the arrows suggest the shape of the curve in Figure 12.

 using *Technology*

Because there is only one critical point, at $t = 0$, it is clear from this table that f must decrease from $t = -1$ to $t = 0$ and then increase from $t = 0$ to $t = 1$. To graph f using technology, choose a viewing window with an interval of $[-1, 1]$ for t and $[0, 1]$ for y. The result will be something like Figure 12.[*]

---

[*] Many graphing calculators will give you only the right-hand half of the graph shown in Figure 12 because fractional powers of negative numbers are not, in general, real numbers. To obtain the whole curve, enter the formula as `Y=(x^2)^(1/3)`, a fractional power of the nonnegative function x^2.

In Examples 1 and 3, we could have found the absolute maxima and minima without doing any graphing. In Example 1, after finding the critical points and endpoints, we created the following table:

x	0	1	4
$f(x)$	0	−1	8

From this table we can see that f must decrease from its value of 0 at $x = 0$ to −1 at $x = 1$, and then increase to 8 at $x = 4$. The value of 8 must be the largest value it takes on, and the value of −1 must be the smallest, on the interval $[0, 4]$. Similarly, in Example 3 we created the following table:

t	−1	0	1
$f(t)$	1	0	1

From this table we can see that the largest value of f on the interval $[-1, 1]$ is 1 and the smallest value is 0. We are taking advantage of the following fact, the proof of which uses some deep and beautiful mathematics (alas, beyond the scope of this book):

Absolute Extrema on a Closed Interval

If f is *continuous* on a closed interval $[a, b]$, then it will have an absolute maximum and an absolute minimum value on that interval. Each absolute extremum must occur either at an endpoint or a critical point. Therefore, the absolute maximum is the largest value in a table of the values of f at the endpoints and critical points, and the absolute minimum is the smallest value.

quick Example

The function $f(x) = 3x - x^3$ on the interval $[0, 2]$, has one critical point, at $x = 1$. The values of f at the critical point and the endpoints of the interval are given in the following table:

	Endpoint	Critical point	Endpoint
x	0	1	2
$f(x)$	0	2	−2

From this table we can say that the absolute maximum value of f on $[0, 2]$ is 2, which occurs at $x = 1$, and the absolute minimum value of f is −2, which occurs at $x = 2$.

As we can see in Example 2 and the following examples, if the domain is not a closed interval then f may not have an absolute maximum and minimum, and a table of values as above is of little help in determining whether it does.

Example 4 Domain Not a Closed Interval

Find all extrema of $f(x) = x + \dfrac{1}{x}$.

Solution Because no domain is specified, we take the domain to be as large as possible. The function is not defined at $x = 0$ but is at all other points, so we take its domain to be $(-\infty, 0) \cup (0, +\infty)$. We calculate

$$f'(x) = 1 - \frac{1}{x^2}$$

Stationary Points Setting $f'(x) = 0$, we solve

$$1 - \frac{1}{x^2} = 0$$

to find $x = \pm 1$. Calculating the corresponding values of f, we get the two stationary points $(1, 2)$ and $(-1, -2)$.

Singular Points The only value of x for which $f'(x)$ is not defined is $x = 0$, but then f is not defined there either, so there are no singular points in the domain.

Endpoints The domain, $(-\infty, 0) \cup (0, +\infty)$, has no endpoints.

From this scant information, it is hard to tell what f does. If we are sketching the graph by hand, or using the first derivative test, we will need to plot additional "test points" to the left and right of the stationary points $x = \pm 1$.

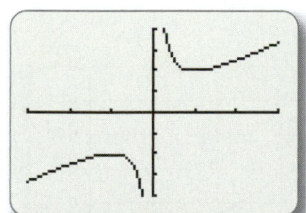 using *Technology*

For the technology approach, let's choose a viewing window with an interval of $[-3, 3]$ for x and $[-4, 4]$ for y, which should leave plenty of room to see how f behaves near the stationary points. The result is something like Figure 13.

From this graph we can see that f has

A relative (stationary) maximum at $(-1, -2)$

A relative (stationary) minimum at $(1, 2)$

Curiously, the relative maximum is lower than the relative minimum! Notice also that, because of the break in the graph at $x = 0$, the graph did not need to rise to get from $(-1, -2)$ to $(1, 2)$. ∎

Figure **13**

So far we have been solving the equation $f'(x) = 0$ to obtain our candidates for stationary extrema. However, it is often not easy—or even possible—to solve equations analytically. In the next example, we show a way around this problem by using graphing technology.

 tech Ex

Example 5 Finding Approximate Extrema Using Technology

Graph the function $f(x) = (x - 1)^{2/3} - \dfrac{x^2}{2}$ with domain $[-2, +\infty)$. Also graph its derivative and hence locate and classify all extrema of f, with coordinates accurate to two decimal places.

Solution In Example 4 of Section 10.5, we saw how to draw the graphs of f and f' using technology. Note that the technology formula to use for the graph of f is

```
((x-1)^2)^(1/3)-0.5*x^2
```

instead of

```
(x-1)^(2/3)-0.5*x^2
```

(Why?)

Figure 14 shows the resulting graphs of f and f'.

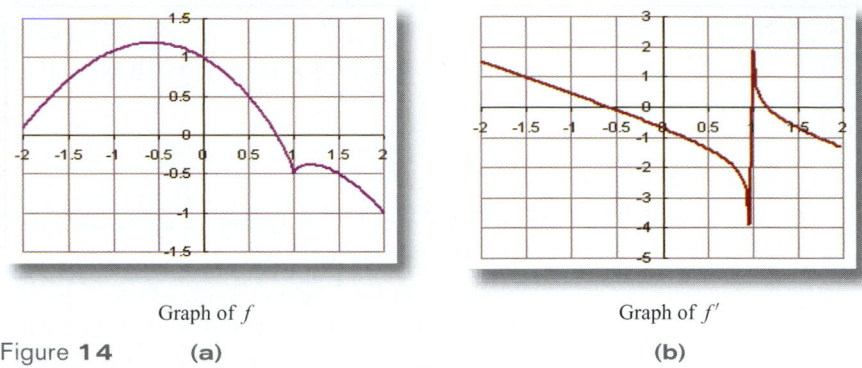

Graph of f Graph of f'

Figure **14**   **(a)**   **(b)**

If we extend Xmax beyond $x = 2$, we find that the graph continues downward, apparently without any further interesting behavior.

Stationary Points The graph of f shows two stationary points, both maxima, at around $x = -0.6$ and $x = 1.2$. Notice that the graph of f' is zero at precisely these points. Moreover, it is easier to locate these values accurately on the graph of f' because it is easier to pinpoint where a graph crosses the x-axis than to locate a stationary point. Zooming in to the stationary point at $x \approx -0.6$ results in Figure 15.

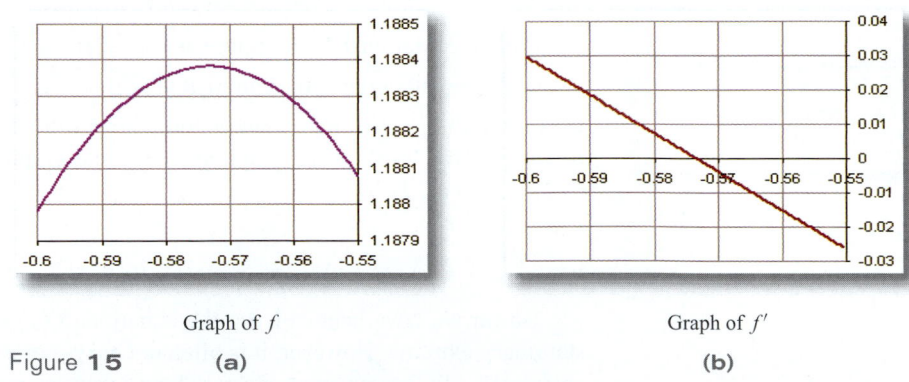

Graph of f Graph of f'

Figure **15**   **(a)**   **(b)**

From the graph of f, we can see that the stationary point is somewhere between -0.58 and -0.57. The graph of f' shows more clearly that the zero of f', hence the stationary point of f lies somewhat closer to -0.57 than to -0.58. Thus, the stationary point occurs at $x \approx -0.57$, rounded to two decimal places.

In a similar way, we find the second stationary point at $x \approx 1.18$.

Singular Points Going back to Figure 14, we notice what appears to be a cusp (singular point) at the relative minimum around $x = 1$, and this is confirmed by a glance at the graph of f', which seems to take a sudden jump at that value. Zooming in closer suggests that the singular point occurs at exactly $x = 1$. In fact, we can calculate

$$f'(x) = \frac{2}{3(x-1)^{1/3}} - x$$

From this formula we see clearly that $f'(x)$ is defined everywhere except at $x = 1$.
Endpoints The only endpoint in the domain is $x = -2$, which gives a relative minimum.

Thus, we have found the following approximate extrema for f:

A relative (endpoint) minimum at $(-2, 0.08)$

An absolute (stationary) maximum at $(-0.57, 1.19)$

A relative (singular) minimum at $(1, -0.5)$

A relative (stationary) maximum at $(1.18, -0.38)$

12.1 EXERCISES

● denotes basic skills exercises

tech Ex indicates exercises that should be solved using technology

In Exercises 1–12, locate and classify all extrema in each graph. (By classifying the extrema, we mean listing whether each extremum is a relative or absolute maximum or minimum.) Also, locate any stationary points or singular points that are not relative extrema.

1. ●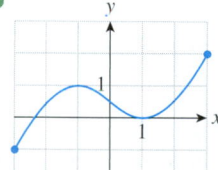

2. ●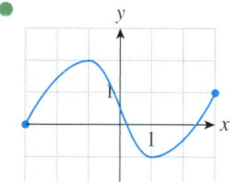

3. ●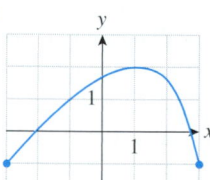

4. ●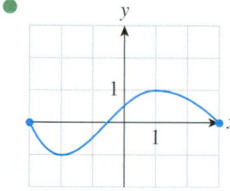

5. ●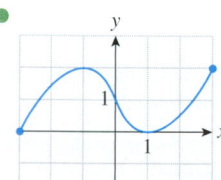

6. ●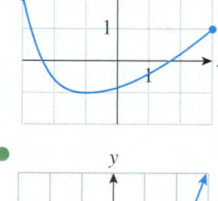

7. ●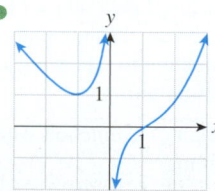

8. ●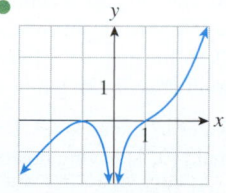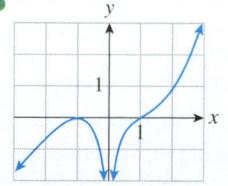

9. ●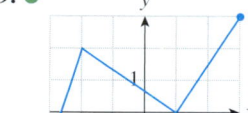

10. ●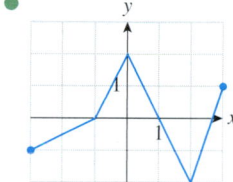

11. ●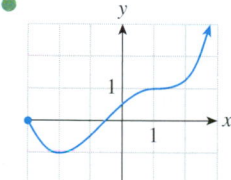

12. ●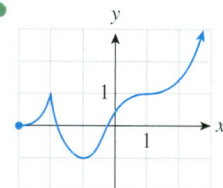

Find the exact location of all the relative and absolute extrema of each function in Exercises 13–44.

13. ● $f(x) = x^2 - 4x + 1$ with domain $[0, 3]$ *hint* [see Example 1]

14. ● $f(x) = 2x^2 - 2x + 3$ with domain $[0, 3]$

15. ● $g(x) = x^3 - 12x$ with domain $[-4, 4]$

16. ● $g(x) = 2x^3 - 6x + 3$ with domain $[-2, 2]$

17. ● $f(t) = t^3 + t$ with domain $[-2, 2]$

18. ● $f(t) = -2t^3 - 3t$ with domain $[-1, 1]$

19. ● $h(t) = 2t^3 + 3t^2$ with domain $[-2, +\infty)$ *hint* [see Example 2]

20. ● $h(t) = t^3 - 3t^2$ with domain $[-1, +\infty)$

21. ● $f(x) = x^4 - 4x^3$ with domain $[-1, +\infty)$

22. ● $f(x) = 3x^4 - 2x^3$ with domain $[-1, +\infty)$

23. ● $g(t) = \frac{1}{4}t^4 - \frac{2}{3}t^3 + \frac{1}{2}t^2$ with domain $(-\infty, +\infty)$

24. ● $g(t) = 3t^4 - 16t^3 + 24t^2 + 1$ with domain $(-\infty, +\infty)$

25. ● $h(x) = (x-1)^{2/3}$ with domain $[0, 2]$ *hint* [see Example 3]

● basic skills   **tech** Ex technology exercise

26. ● $h(x) = (x + 1)^{2/5}$ with domain $[-2, 0]$

27. ● $k(x) = \dfrac{2x}{3} + (x + 1)^{2/3}$ with domain $(-\infty, 0]$

28. ● $k(x) = \dfrac{2x}{5} - (x - 1)^{2/5}$ with domain $[0, +\infty)$

29. $f(t) = \dfrac{t^2 + 1}{t^2 - 1}$; $-2 \le t \le 2, t \ne \pm 1$

30. $f(t) = \dfrac{t^2 - 1}{t^2 + 1}$ with domain $[-2, 2]$

31. $f(x) = \sqrt{x}(x - 1)$; $x \ge 0$

32. $f(x) = \sqrt{x}(x + 1)$; $x \ge 0$

33. $g(x) = x^2 - 4\sqrt{x}$

34. $g(x) = \dfrac{1}{x} - \dfrac{1}{x^2}$ **35.** $g(x) = \dfrac{x^3}{x^2 + 3}$

36. $g(x) = \dfrac{x^3}{x^2 - 3}$

37. $f(x) = x - \ln x$ with domain $(0, +\infty)$

38. $f(x) = x - \ln x^2$ with domain $(0, +\infty)$

39. $g(t) = e^t - t$ with domain $[-1, 1]$

40. $g(t) = e^{-t^2}$ with domain $(-\infty, +\infty)$

41. $f(x) = \dfrac{2x^2 - 24}{x + 4}$ **42.** $f(x) = \dfrac{x - 4}{x^2 + 20}$

43. $f(x) = xe^{1-x^2}$

44. $f(x) = x \ln x$ with domain $(0, +\infty)$

In Exercises 45–48, use graphing technology and the method in Example 5 to find the x-coordinates of the critical points, accurate to two decimal places. Find all relative and absolute maxima and minima.

45. **tech** Ex $y = x^2 + \dfrac{1}{x - 2}$ with domain $(-3, 2) \cup (2, 6)$

   hint [see Example 5]

46. **tech** Ex $y = x^2 - 10(x - 1)^{2/3}$ with domain $(-4, 4)$

47. **tech** Ex $f(x) = (x - 5)^2(x + 4)(x - 2)$ with domain $[-5, 6]$

48. **tech** Ex $f(x) = (x + 3)^2(x - 2)^2$ with domain $[-5, 5]$

In Exercises 49–56, the graph of the derivative of a function f is shown. Determine the x-coordinates of all stationary and singular points of f, and classify each as a relative maximum, relative minimum, or neither. (Assume that $f(x)$ is defined and continuous everywhere in $[-3, 3]$.) *hint* [see Example 5]

49.
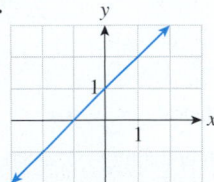

50.
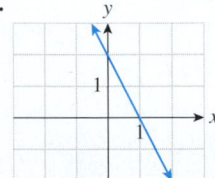

51.
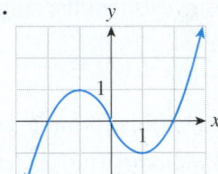

52.
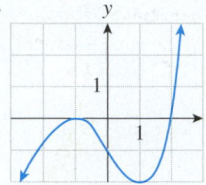

53.
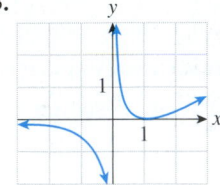

54.
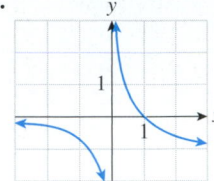

55.
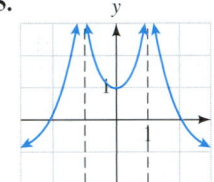

56.
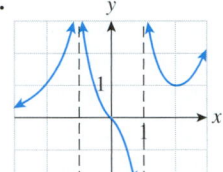

Communication and Reasoning Exercises

57. ● Draw the graph of a function f with domain the set of all real numbers, such that f is not linear and has no relative extrema.

58. ● Draw the graph of a function g with domain the set of all real numbers, such that g has a relative maximum and minimum but no absolute extrema.

59. ● Draw the graph of a function that has stationary and singular points but no relative extrema.

60. ● Draw the graph of a function that has relative, not absolute, maxima and minima, but has no stationary or singular points.

61. ● If a stationary point is not a relative maximum, then must it be a relative minimum? Explain your answer.

62. ● If one endpoint is a relative maximum, must the other be a relative minimum? Explain your answer.

63. We said that if f is continuous on a closed interval $[a, b]$, then it will have an absolute maximum and an absolute minimum. Draw the graph of a function with domain $[0, 1]$ having an absolute maximum but no absolute minimum.

64. Refer to Exercise 63. Draw the graph of a function with domain $[0, 1]$ having no absolute extrema.

12.2 | Applications of Maxima and Minima

In many applications we would like to find the largest or smallest possible value of some quantity—for instance, the greatest possible profit or the lowest cost. We call this the *optimal* (best) value. In this section we consider several such examples and use calculus to find the optimal value in each.

In all applications the first step is to translate a written description into a mathematical problem. In the problems we look at in this section, there are *unknowns* that we are asked to find, there is an expression involving those unknowns that must be made as large or as small as possible—the **objective function**—and there may be **constraints**—equations or inequalities relating the variables.[4]

Example 1 Minimizing Average Cost

Gymnast Clothing manufactures expensive hockey jerseys for sale to college bookstores in runs of up to 500. Its cost (in dollars) for a run of x hockey jerseys is

$$C(x) = 2000 + 10x + 0.2x^2$$

How many jerseys should Gymnast produce per run in order to minimize average cost?[*]

Solution Here is the procedure we will follow to solve problems like this.

1. ***Identify the unknown(s).*** There is one unknown: x, the number of hockey jerseys Gymnast should produce per run. (We know this because the question is, How many jerseys. . . ?)

2. ***Identify the objective function.*** The objective function is the quantity that must be made as small (in this case) as possible. In this example it is the average cost, which is given by

$$\bar{C}(x) = \frac{C(x)}{x} = \frac{2000 + 10x + 0.2x^2}{x}$$

$$= \frac{2000}{x} + 10 + 0.2x \text{ dollars/jersey}$$

3. ***Identify the constraints (if any).*** At most 500 jerseys can be manufactured in a run. Also, $\bar{C}(0)$ is not defined. Thus, x is constrained by

$$0 < x \leq 500$$

Put another way, the domain of the objective function $\bar{C}(x)$ is $(0, 500]$.

4. ***State and solve the resulting optimization problem.*** Our optimization problem is:

$$\text{Minimize } \bar{C}(x) = \frac{2000}{x} + 10 + 0.2x \qquad \textcolor{red}{\text{Objective function}}$$

$$\text{subject to } 0 < x \leq 500 \qquad \textcolor{red}{\text{Constraint}}$$

---

[*] Why don't we seek to minimize total cost? The answer would be uninteresting; to minimize total cost, we would make *no* jerseys at all. Minimizing the average cost is a more practical objective.

---

[4] If you have studied linear programming, you will notice a similarity here, but unlike the situation in linear programming, neither the objective function nor the constraints need be linear.

We now solve this problem as in Section 12.1. We first calculate

$$\bar{C}'(x) = -\frac{2000}{x^2} + 0.2$$

We solve $\bar{C}'(x) = 0$ to find $x = \pm 100$. We reject $x = -100$ because -100 is not in the domain of $\bar{C}$ (and makes no sense), so we have one stationary point, at $x = 100$. There, the average cost is $\bar{C}(100) = \$50$ per jersey.

The only point at which the formula for $\bar{C}'$ is not defined is $x = 0$, but that is not in the domain of $\bar{C}$, so we have no singular points. We have one endpoint in the domain, at $x = 500$. There, the average cost is $\bar{C}(500) = \$114$.

using *Technology*

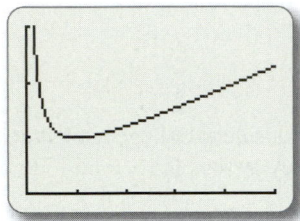

Figure **16**

Let's plot $\bar{C}$ in a viewing window with the intervals $[0, 500]$ for x and $[0, 150]$ for y, which will show the whole domain and the two interesting points we've found so far. The result is Figure 16.

From the graph of $\bar{C}$, we can see that the stationary point at $x = 100$ gives the absolute minimum. We can therefore say that Gymnast Clothing should produce 100 jerseys per run, for a lowest possible average cost of $\$50$ per jersey. ∎

Example **2** Maximizing Area

Slim wants to build a rectangular enclosure for his pet rabbit, Killer, against the side of his house, as shown in Figure 17. He has bought 100 feet of fencing. What are the dimensions of the largest area that he can enclose?

Figure **17**

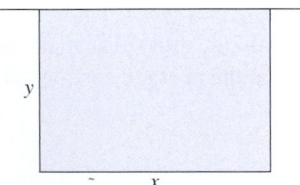

Figure **18**

Solution

1. ***Identify the unknown(s).*** To identify the unknown(s), we look at the question: What are the *dimensions* of the largest area he can enclose? Thus, the unknowns are the dimensions of the fence. We call these x and y, as shown in Figure 18.

2. ***Identify the objective function.*** We look for what it is that we are trying to maximize (or minimize). The phrase "largest area" tells us that our object is to *maximize the area,* which is the product of length and width, so our objective function is

$$A = xy \quad \text{where } A \text{ is the area of the enclosure}$$

3. ***Identify the constraints (if any).*** What stops Slim from making the area as large as he wants? He has only 100 feet of fencing to work with. Looking again at Figure 18, we see that the sum of the lengths of the three sides must equal 100, so

$$x + 2y = 100$$

One more point: Because x and y represent the lengths of the sides of the enclosure, neither can be a negative number.

4. ***State and solve the resulting optimization problem.*** Our mathematical problem is:

Maximize $A = xy$ Objective function
subject to $x + 2y = 100$, $x \geq 0$, and $y \geq 0$ Constraints

We know how to find maxima and minima of a function of one variable, but A appears to depend on two variables. We can remedy this by using a constraint to express one variable in terms of the other. Let's take the constraint $x + 2y = 100$ and solve for x in terms of y:

$$x = 100 - 2y$$

Substituting into the objective function gives

$$A = xy = (100 - 2y)y = 100y - 2y^2$$

and we have eliminated x from the objective function. What about the inequalities? One says that $x \geq 0$, but we want to eliminate x from this as well. We substitute for x again, getting

$$100 - 2y \geq 0$$

Solving this inequality for y gives $y \leq 50$. The second inequality says that $y \geq 0$. Now, we can restate our problem with x eliminated:

Maximize $A(y) = 100y - 2y^2$ subject to $0 \leq y \leq 50$

We now proceed with our usual method of solving such problems. We calculate $A'(y) = 100 - 4y$. Solving $100 - 4y = 0$, we get one stationary point at $y = 25$. There, $A(25) = 1250$. There are no points at which $A'(y)$ is not defined, so there are no singular points. We have two endpoints, at $y = 0$ and $y = 50$. The corresponding areas are $A(0) = 0$ and $A(50) = 0$. We record the three points we found in a table:

y	0	25	50
$A(y)$	0	1250	0

It's clear now how A must behave: It increases from 0 at $y = 0$ to 1250 at $y = 25$ and then decreases back to 0 at $y = 50$. Thus, the largest possible value of A is 1250 square feet, which occurs when $y = 25$. To completely answer the question that was asked, we need to know the corresponding value of x. We have $x = 100 - 2y$, so $x = 50$ when $y = 25$. Thus, Slim should build his enclosure 50 feet across and 25 feet deep (with the "missing" 50-foot side being formed by part of the house).

+*Before we go on...* Notice that the problem in Example 2 came down to finding the absolute maximum value of A on the closed and bounded interval [0, 50]. As we noted in the preceding section, the table of values of A at its critical points and the endpoints of the interval gives us enough information to find the absolute maximum. ∎

Let's stop for a moment and summarize the steps we've taken in these two examples.

Solving an Optimization Problem

1. Identify the unknown(s), possibly with the aid of a diagram. These are usually the quantities asked for in the problem.

2. Identify the objective function. This is the quantity you are asked to maximize or minimize. You should name it explicitly, as in "Let S = surface area."

3. Identify the constraint(s). These can be equations relating variables or inequalities expressing limitations on the values of variables.

> **4. State the optimization problem.** This will have the form "Maximize [minimize] the objective function subject to the constraint(s)."
>
> **5. Eliminate extra variables.** If the objective function depends on several variables, solve the constraint equations to express all variables in terms of one particular variable. Substitute these expressions into the objective function to rewrite it as a function of a single variable. Substitute the expressions into any inequality constraints to help determine the domain of the objective function.
>
> **6. Find the absolute maximum (or minimum) of the objective function.** Use the techniques of the preceding section.

Now for some further examples.

Example 3 Maximizing Revenue

Cozy Carriage Company builds baby strollers. Using market research, the company estimates that if it sets the price of a stroller at p dollars, then it can sell $q = 300{,}000 - 10p^2$ strollers per year. What price will bring in the greatest annual revenue?

Solution The question we are asked identifies our main unknown, the price p. However, there is another quantity that we do not know, q, the number of strollers the company will sell per year. The question also identifies the objective function, revenue, which is

$$R = pq$$

Including the equality constraint given to us, that $q = 300{,}000 - 10p^2$, and the "reality" inequality constraints $p \geq 0$ and $q \geq 0$, we can write our problem as

$$\text{Maximize } R = pq \text{ subject to } q = 300{,}000 - 10p^2, \; p \geq 0, \text{ and } q \geq 0$$

We are given q in terms of p, so let's substitute to eliminate q:

$$R = pq = p(300{,}000 - 10p^2) = 300{,}000p - 10p^3$$

Substituting in the inequality $q \geq 0$, we get

$$300{,}000 - 10p^2 \geq 0$$

Thus, $p^2 \leq 30{,}000$, which gives $-100\sqrt{3} \leq p \leq 100\sqrt{3}$. When we combine this with $p \geq 0$, we get the following restatement of our problem:

$$\text{Maximize } R(p) = 300{,}000p - 10p^3 \text{ such that } 0 \leq p \leq 100\sqrt{3}$$

We solve this problem in much the same way we did the preceding one. We calculate $R'(p) = 300{,}000 - 30p^2$. Setting $300{,}000 - 30p^2 = 0$, we find one stationary point at $p = 100$. There are no singular points and we have the endpoints $p = 0$ and $p = 100\sqrt{3}$. Putting these points in a table and computing the corresponding values of R, we get the following:

p	0	100	$100\sqrt{3}$
$R(p)$	0	20,000,000	0

Thus, Cozy Carriage should price its strollers at $100 each, which will bring in the largest possible revenue of $20,000,000.

Figure **19**

Example **4** Optimizing Resources

The Metal Can Company has an order to make cylindrical cans with a volume of 250 cubic centimeters. What should be the dimensions of the cans in order to use the least amount of metal in their production?

Solution We are asked to find the dimensions of the cans. It is traditional to take as the dimensions of a cylinder the height h and the radius of the base r, as in Figure 19.

We are also asked to minimize the amount of metal used in the can, which is the area of the surface of the cylinder. We can look up the formula or figure it out ourselves: Imagine removing the circular top and bottom and then cutting vertically and flattening out the hollow cylinder to get a rectangle, as shown in Figure 20.

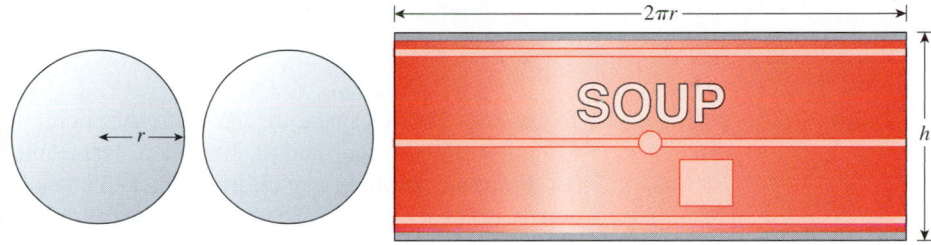

Figure **20**

Our objective function is the (total) surface area S of the can. The area of each disc is πr^2, while the area of the rectangular piece is $2\pi rh$. Thus, our objective function is

$$S = 2\pi r^2 + 2\pi rh$$

As usual, there is a constraint: The volume must be exactly 250 cubic centimeters. The formula for the volume of a cylinder is $V = \pi r^2 h$, so

$$\pi r^2 h = 250$$

It is easiest to solve this constraint for h in terms of r:

$$h = \frac{250}{\pi r^2}$$

Substituting in the objective function, we get

$$S = 2\pi r^2 + 2\pi r \frac{250}{\pi r^2} = 2\pi r^2 + \frac{500}{r}$$

Now r cannot be negative or 0, but it can become very large (a very wide but very short can could have the right volume). We therefore take the domain of $S(r)$ to be $(0, +\infty)$, so our mathematical problem is as follows:

$$\text{Minimize } S(r) = 2\pi r^2 + \frac{500}{r} \text{ subject to } r > 0$$

Now we calculate

$$S'(r) = 4\pi r - \frac{500}{r^2}$$

To find stationary points, we set this equal to 0 and solve:

$$4\pi r - \frac{500}{r^2} = 0$$

$$4\pi r = \frac{500}{r^2}$$

$$4\pi r^3 = 500$$

$$r^3 = \frac{125}{\pi}$$

So

$$r = \sqrt[3]{\frac{125}{\pi}} = \frac{5}{\sqrt[3]{\pi}} \approx 3.41$$

The corresponding surface area is approximately $S(3.41) \approx 220$. There are no singular points or endpoints in the domain.

 using *Technology*

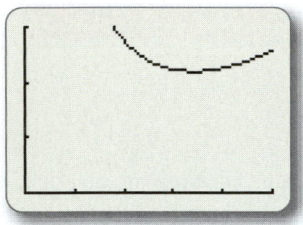

Figure **21**

To see how S behaves near the one stationary point, let's graph it in a viewing window with interval $[0, 5]$ for r and $[0, 300]$ for S. The result is Figure 21.

From the graph we can clearly see that the smallest surface area occurs at the stationary point at $r \approx 3.41$. The height of the can will be

$$h = \frac{250}{\pi r^2} \approx 6.83$$

Thus, the can that uses the least amount of metal has a height of approximately 6.83 centimeters and a radius of approximately 3.41 centimeters. Such a can will use approximately 220 square centimeters of metal.

+*Before we go on...* We obtained the value of r in Example 4 by solving the equation

$$4\pi r = \frac{500}{r^2}$$

This time, let us do things differently: divide both sides by 4π to obtain

$$r = \frac{500}{4\pi r^2} = \frac{125}{\pi r^2}$$

and compare what we got with the expression for h:

$$h = \frac{250}{\pi r^2}$$

which we see is exactly twice the expression for r. Put another way, the height is exactly equal to the diameter so that the can looks square when viewed from the side. Have you ever seen cans with that shape? Why do you think most cans do not have this shape? ■

Example 5 Allocation of Labor

The Gym Sock Company manufactures cotton athletic socks. Production is partially automated through the use of robots. Daily operating costs amount to $50 per laborer and $30 per robot. The number of pairs of socks the company can manufacture in a day is given by a Cobb-Douglas[*] production formula

$$q = 50n^{0.6}r^{0.4}$$

where q is the number of pairs of socks that can be manufactured by n laborers and r robots. Assuming that the company wishes to produce 1000 pairs of socks per day at a minimum cost, how many laborers and how many robots should it use?

Solution The unknowns are the number of laborers n and the number of robots r. The objective is to minimize the daily cost:

$$C = 50n + 30r$$

The constraints are given by the daily quota

$$1000 = 50n^{0.6}r^{0.4}$$

and the fact that n and r are nonnegative. We solve the constraint equation for one of the variables; let's solve for n:

$$n^{0.6} = \frac{1000}{50r^{0.4}} = \frac{20}{r^{0.4}}$$

Taking the $1/0.6$ power of both sides gives

$$n = \left(\frac{20}{r^{0.4}}\right)^{1/0.6} = \frac{20^{1/0.6}}{r^{0.4/0.6}} = \frac{20^{5/3}}{r^{2/3}} \approx \frac{147.36}{r^{2/3}}$$

Substituting in the objective equation gives us the cost as a function of r:

$$C(r) \approx 50\left(\frac{147.36}{r^{2/3}}\right) + 30r$$

$$= 7368r^{-2/3} + 30r$$

The only remaining constraint on r is that $r > 0$. To find the minimum value of $C(r)$, we first take the derivative:

$$C'(r) \approx -4912r^{-5/3} + 30$$

Setting this equal to zero, we solve for r:

$$r^{-5/3} \approx 0.006107$$

$$r \approx (0.006107)^{-3/5} \approx 21.3$$

The corresponding cost is $C(21.3) \approx \$1600$. There are no singular points or endpoints in the domain of C.

using *Technology*

To see how C behaves near its stationary point, let's draw its graph in a viewing window with an interval of $[0, 40]$ for r and $[0, 2000]$ for C. The result is Figure 22.

---

[*] Cobb-Douglas production formulas were discussed in Section 11.4.

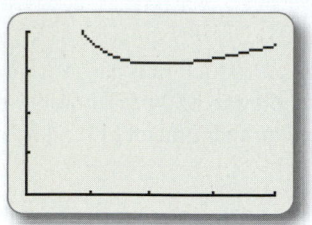

Figure **22**

From the graph we can see that C does have its minimum at the stationary point. The corresponding value of n is

$$n \approx \frac{147.36}{r^{2/3}} \approx 19.2$$

■

At this point, our solution appears to be this: Use (approximately) 19.2 laborers and (approximately) 21.3 robots to meet the manufacturing quota at a minimum cost. However, we are not interested in fractions of robots or people, so we need to find integer solutions for n and r. If we round these numbers, we get the solution $(n, r) = (19, 21)$. However, a quick calculation shows that

$$q = 50(19)^{0.6}(21)^{0.4} \approx 989 \text{ pairs of socks}$$

which fails to meet the quota of 1000. Thus we need to round at least one of the quantities n or r *upward* in order to meet the quota. The three possibilities, with corresponding values of q and C, are as follows:

$(n, r) = (20, 21)$, with $q \approx 1020$ and $C = \$1630$

$(n, r) = (19, 22)$, with $q \approx 1007$ and $C = \$1610$

$(n, r) = (20, 22)$, with $q \approx 1039$ and $C = \$1660$

Of these, the solution that meets the quota at a minimum cost is $(n, r) = (19, 22)$. Thus, the Gym Sock Co. should use 19 laborers and 22 robots, at a cost of $50 \times 19 + 30 \times 22 = \$1610$, to manufacture $50 \times 19^{0.6} \times 22^{0.4} \approx 1007$ pairs of socks.

12.2 EXERCISES

● denotes basic skills exercises

◆ denotes challenging exercises

tech Ex indicates exercises that should be solved using technology

Solve the optimization problems in Exercises 1–8.

1. ● Maximize $P = xy$ with $x + y = 10$. *hint* [see Example 2]

2. ● Maximize $P = xy$ with $x + 2y = 40$.

3. ● Minimize $S = x + y$ with $xy = 9$ and both x and $y > 0$.

4. ● Minimize $S = x + 2y$ with $xy = 2$ and both x and $y > 0$.

5. ● Minimize $F = x^2 + y^2$ with $x + 2y = 10$.

6. ● Minimize $F = x^2 + y^2$ with $xy^2 = 16$.

7. ● Maximize $P = xyz$ with $x + y = 30$ and $y + z = 30$, and x, y, and $z \geq 0$.

8. ● Maximize $P = xyz$ with $x + z = 12$ and $y + z = 12$, and x, y, and $z \geq 0$.

9. ● For a rectangle with perimeter 20 to have the largest area, what dimensions should it have?

10. ● For a rectangle with area 100 to have the smallest perimeter, what dimensions should it have?

Applications

11. ● *Average Cost* The cost function for the manufacture of portable MP3 players is given by

$$C(x) = 25{,}000 + 20x + 0.001x^2 \text{ dollars}$$

where x is the number of MP3 players manufactured. How many MP3 players should be manufactured in order to minimize average cost? What is the resulting average cost of an MP3 player? (Give your answer to the nearest dollar.) *hint* [see Example 1]

12. ● *Average Cost* Repeat the preceding exercise using the revised cost function

$$C(x) = 6400 + 10x + 0.01x^2$$

13. ● *Pollution Control* The cost of controlling emissions at a firm rises rapidly as the amount of emissions reduced increases. Here is a possible model:

$$C(q) = 4000 + 100q^2$$

where q is the reduction in emissions (in pounds of pollutant per day) and C is the daily cost to the firm (in dollars) of this

● basic skills   ◆ challenging   tech Ex technology exercise

reduction. What level of reduction corresponds to the lowest average cost per pound of pollutant, and what would be the resulting average cost to the nearest dollar?

14. ● *Pollution Control* Repeat the preceding exercise using the following cost function:

$$C(q) = 2000 + 200q^2$$

15. ● *Pollution Control* (Compare Exercise 13.) The cost of controlling emissions at a firm is given by

$$C(q) = 4000 + 100q^2$$

where q is the reduction in emissions (in pounds of pollutant per day) and C is the daily cost to the firm (in dollars) of this reduction. Government clean-air subsidies amount to $500 per pound of pollutant removed. How many pounds of pollutant should the firm remove each day in order to minimize *net* cost (cost minus subsidy)?

16. ● *Pollution Control* (Compare Exercise 14.) Repeat the preceding exercise, using the following cost function:

$$C(q) = 2000 + 200q^2$$

with government subsidies amounting to $100 per pound of pollutant removed per day.

17. ● *Fences* I want to fence in a rectangular vegetable patch. The fencing for the east and west sides costs $4 per foot, and the fencing for the north and south sides costs only $2 per foot. I have a budget of $80 for the project. What is the largest area I can enclose? *hint* [see Example 2]

18. ● *Fences* My orchid garden abuts my house so that the house itself forms the northern boundary. The fencing for the southern boundary costs $4 per foot, and the fencing for the east and west sides costs $2 per foot. If I have a budget of $80 for the project, what is the largest area I can enclose?

19. ● *Revenue* Hercules Films is deciding on the price of the video release of its film *Son of Frankenstein*. Its marketing people estimate that at a price of p dollars, it can sell a total of $q = 200,000 - 10,000p$ copies. What price will bring in the greatest revenue? *hint* [see Example 3]

20. ● *Profit* Hercules Films is also deciding on the price of the video release of its film *Bride of the Son of Frankenstein*. Again, marketing estimates that at a price of p dollars, it can sell $q = 200,000 - 10,000p$ copies, but each copy costs $4 to make. What price will give the greatest *profit*?

21. ● *Revenue* The demand for rubies at Royal Ruby Retailers (RRR) is given by the equation

$$q = -\frac{4}{3}p + 80$$

where p is the price RRR charges (in dollars) and q is the number of rubies RRR sells per week. At what price should RRR sell its rubies in order to maximize its weekly revenue?

22. ● *Revenue* The consumer demand curve for tissues is given by

$$q = (100 - p)^2 \qquad (0 \le p \le 100)$$

where p is the price per case of tissues and q is the demand in weekly sales. At what price should tissues be sold in order to maximize revenue?

23. ● *Revenue* Assume that the demand for tuna in a small coastal town is given by

$$p = \frac{500,000}{q^{1.5}}$$

where q is the number of pounds of tuna that can be sold in a month at p dollars per pound. Assume that the town's fishery wishes to sell at least 5000 pounds of tuna per month.

  a. How much should the town's fishery charge for tuna in order to maximize monthly revenue?

  b. How much tuna will it sell per month at that price?

  c. What will be its resulting revenue?

24. ● *Revenue* Economist Henry Schultz devised the following demand function for corn:

$$p = \frac{6,570,000}{q^{1.3}}$$

where q is the number of bushels of corn that could be sold at p dollars per bushel in one year.[5] Assume that at least 10,000 bushels of corn per year must be sold.

  a. How much should farmers charge per bushel of corn to maximize annual revenue?

  b. How much corn can farmers sell per year at that price?

  c. What will be the farmers' resulting revenue?

25. ● *Revenue* The wholesale price for chicken in the United States fell from 25¢ per pound to 14¢ per pound, while per capita chicken consumption rose from 22 pounds per year to 27.5 pounds per year.[6] Assuming that the demand for chicken depends linearly on the price, what wholesale price for chicken maximizes revenues for poultry farmers, and what does that revenue amount to?

26. ● *Revenue* Your underground used book business is booming. Your policy is to sell all used versions of *Calculus and You* at the same price (regardless of condition). When you set the price at $10, sales amounted to 120 volumes during the first week of classes. The following semester, you set the price at $30 and sold not a single book. Assuming that the demand for books depends linearly on the price, what price gives you the maximum revenue, and what does that revenue amount to?

27. ● *Profit* The demand for rubies at Royal Ruby Retailers (RRR) is given by the equation

$$q = -\frac{4}{3}p + 80$$

[5] Based on data for the period 1915–1929. SOURCE: Henry Schultz, *The Theory and Measurement of Demand*, (as cited in *Introduction to Mathematical Economics* by A. L. Ostrosky, Jr., and J. V. Koch (Waveland Press, Prospect Heights, Illinois, 1979).

[6] Data are provided for the years 1951–1958. SOURCE: U.S. Department of Agriculture, *Agricultural Statistics*.

● basic skills   ◆ challenging   *tech* Ex technology exercise

where p is the price RRR charges (in dollars) and q is the number of rubies RRR sells per week. Assuming that due to extraordinary market conditions, RRR can obtain rubies for $25 each, how much should it charge per ruby to make the greatest possible weekly profit, and what will that profit be?

28. ● **Profit** The consumer demand curve for tissues is given by

$$q = (100 - p)^2 \qquad (0 \le p \le 100)$$

where p is the price per case of tissues and q is the demand in weekly sales. If tissues cost $30 per case to produce, at what price should tissues be sold for the largest possible weekly profit? What will that profit be?

29. ● **Profit** The demand equation for your company's virtual reality video headsets is

$$p = \frac{1000}{q^{0.3}}$$

where q is the total number of headsets that your company can sell in a week at a price of p dollars. The total manufacturing and shipping cost amounts to $100 per headset.

   a. What is the greatest profit your company can make in a week, and how many headsets will your company sell at this level of profit? (Give answers to the nearest whole number.)

   b. How much, to the nearest $1, should your company charge per headset for the maximum profit?

30. ● **Profit** Due to sales by a competing company, your company's sales of virtual reality video headsets have dropped, and your financial consultant revises the demand equation to

$$p = \frac{800}{q^{0.35}}$$

where q is the total number of headsets that your company can sell in a week at a price of p dollars. The total manufacturing and shipping cost still amounts to $100 per headset.

   a. What is the greatest profit your company can make in a week, and how many headsets will your company sell at this level of profit? (Give answers to the nearest whole number.)

   b. How much, to the nearest $1, should your company charge per headset for the maximum profit?

31. **Box Design** Chocolate Box Company is going to make open-topped boxes out of 6" × 16" rectangles of cardboard by cutting squares out of the corners and folding up the sides. What is the largest volume box it can make this way?

32. **Box Design** Vanilla Box Company is going to make open-topped boxes out of 12" × 12" rectangles of cardboard by cutting squares out of the corners and folding up the sides. What is the largest volume box it can make this way?

33. **Box Design** A packaging company is going to make closed boxes, with square bases, that hold 125 cubic centimeters. What are the dimensions of the box that can be built with the least material?

34. **Box Design** A packaging company is going to make open-topped boxes, with square bases, that hold 108 cubic centimeters. What are the dimensions of the box that can be built with the least material?

35. **Luggage Dimensions** American Airlines requires that the total outside dimensions (length + width + height) of a checked bag not exceed 62 inches.[7] Suppose you want to check a bag whose height equals its width. What is the largest volume bag of this shape that you can check on an American flight?

36. **Luggage Dimensions** American Airlines requires that the total outside dimensions (length + width + height) of a carry-on bag not exceed 45 inches.[8] Suppose you want to carry on a bag whose length is twice its height. What is the largest volume bag of this shape that you can carry on an American flight?

37. **Luggage Dimensions** Fly-by-Night Airlines has a peculiar rule about luggage: The length and width of a bag must add up to at most 45 inches, and the width and height must also add up to at most 45 inches. What are the dimensions of the bag with the largest volume that Fly-by-Night will accept?

38. **Luggage Dimensions** Fair Weather Airlines has a similar rule. It will accept only bags for which the sum of the length and width is at most 36 inches, while the sum of length, height, and twice the width is at most 72 inches. What are the dimensions of the bag with the largest volume that Fair Weather will accept?

39. **Package Dimensions** The U.S. Postal Service (USPS) will accept packages only if the length plus girth is no more than 108 inches.[9] (See the figure.)

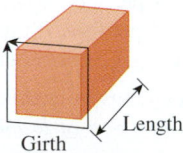

Girth | Length

Assuming that the front face of the package (as shown in the figure) is square, what is the largest volume package that the USPS will accept?

40. **Package Dimensions** United Parcel Service (UPS) will only accept packages with a length of no more than 108 inches and length plus girth of no more than 165 inches.[10] (See figure for the preceding exercise.) Assuming that the front face of the package (as shown in the figure) is square, what is the largest volume package that UPS will accept?

---

[7] According to information on its website (http://www.aa.com) as of April, 2005.

[8] Ibid.

[9] The requirement for packages sent other than Parcel Post, as of April 2005 (www.usps.com).

[10] The requirement as of April, 2005 (www.ups.com).

● basic skills   ◆ challenging   **tech** Ex technology exercise

41. Cell Phone Revenues The number of cell phone subscribers in China in the years 2000–2005 was projected to follow the equation $N(t) = 39t + 68$ million subscribers in year t ($t = 0$ represents January 2000). The average annual revenue per cell phone user was $350 in 2000.[11] If we assume that due to competition the revenue per cell phone user decreases continuously at an annual rate of 30%, we can model the annual revenue as

$$R(t) = 350(39t + 68)e^{-0.3t} \text{ million dollars}$$

Determine **a.** when to the nearest 0.1 year the revenue is projected to peak and **b.** the revenue, to the nearest $1 million, at that time.

42. Cell Phone Revenues Refer to Exercise 41. If we assume instead that the revenue per cell phone user decreases continuously at an annual rate of 20%, we obtain the revenue model

$$R(t) = 350(39t + 68)e^{-0.2t} \text{ million dollars}$$

Determine **a.** when to the nearest 0.1 year the revenue is projected to peak and **b.** the revenue, to the nearest $1 million, at that time.

43. Research and Development Spending on research and development by drug companies in the U.S. t years after 1970 can be modeled by

$$S(t) = 2.5e^{0.08t} \text{ billion dollars} \qquad (0 \le t \le 31)$$

The number of new drugs approved by the FDA over the same period can be modeled by

$$D(t) = 10 + t \text{ drugs per year}^{12} \qquad (0 \le t \le 31)$$

When was the function $D(t)/S(t)$ at a maximum? What is the maximum value of $D(t)/S(t)$? What does the answer tell you about the cost of developing new drugs?

44. Research and Development Refer to Exercise 43. If the number of new drugs approved by the FDA had been $10 + 2t$ new drugs each year, when would the function $D(t)/S(t)$ have reached a maximum? What does the answer tell you about the cost of developing new drugs?

45. Asset Appreciation As the financial consultant to a classic auto dealership, you estimate that the total value (in dollars) of its collection of 1959 Chevrolets and Fords is given by the formula

$$v = 300{,}000 + 1000t^2 \quad (t \ge 5)$$

where t is the number of years from now. You anticipate a continuous inflation rate of 5% per year, so that the discounted (present) value of an item that will be worth v in t years' time is

$$p = ve^{-0.05t}$$

When would you advise the dealership to sell the vehicles to maximize their discounted value?

46. Plantation Management The value of a fir tree in your plantation increases with the age of the tree according to the formula

$$v = \frac{20t}{1 + 0.05t}$$

where t is the age of the tree in years. Given a continuous inflation rate of 5% per year, the discounted (present) value of a newly planted seedling is

$$p = ve^{-0.05t}$$

At what age (to the nearest year) should you harvest your trees in order to ensure the greatest possible discounted value?

47. Marketing Strategy FeatureRich Software Company has a dilemma. Its new program, Doors-X 10.27, is almost ready to go on the market. However, the longer the company works on it, the better it can make the program and the more it can charge for it. The company's marketing analysts estimate that if it delays t days, it can set the price at $100 + 2t$ dollars. On the other hand, the longer it delays, the more market share they will lose to their main competitor (see the next exercise) so that if it delays t days it will be able to sell $400{,}000 - 2500t$ copies of the program. How many days should FeatureRich delay the release in order to get the greatest revenue?

48. Marketing Strategy FeatureRich Software's main competitor (see previous exercise) is Moon Systems, and Moon is in a similar predicament. Its product, Walls-Y 11.4, could be sold now for $200, but for each day Moon delays, it could increase the price by $4. On the other hand, it could sell 300,000 copies now, but each day it waits will cut sales by 1500. How many days should Moon delay the release in order to get the greatest revenue?

49. Average Profit The FeatureRich Software Company sells its graphing program, Dogwood, with a volume discount. If a customer buys x copies, then he pays[13] $500\sqrt{x}$. It cost the company $10,000 to develop the program and $2 to manufacture each copy. If a single customer were to buy all the copies of Dogwood, how many copies would the customer have to buy for FeatureRich Software's average profit per copy to be maximized? How are average profit and marginal profit related at this number of copies?

[11] Based on a regression of projected figures (coefficients are rounded). SOURCE: Intrinsic Technology/*New York Times*, Nov. 24, 2000, p. C1.

[12] The exponential model for R&D is based on the 1970 and 2001 spending in constant 2001 dollars, while the linear model for new drugs approved is based on the six-year moving average from data from 1970–2000. Source for data: Pharmaceutical Research and Manufacturers of America, FDA/*New York Times*, April 19, 2002, p. C1.

[13] This is similar to the way site licenses have been structured for the program Maple®.

● basic skills   ◆ challenging   **tech Ex** technology exercise

50. *Average Profit* Repeat the preceding exercise with the charge to the customer $600\sqrt{x}$ and the cost to develop the program $9000.

51. *Prison Population* The prison population of the U.S. followed the curve

$$N(t) = -145t^3 + 5300t^2 + 1300t + 350{,}000 \quad (0 \le t \le 23)$$

in the years 1980–2003. Here t is the number of years since 1980 and N is the number of prisoners.[14] When, to the nearest year, was the prison population increasing most rapidly? When was it increasing least rapidly?

52. *Test Scores* Combined SAT scores in the U.S. in the years 1985–2003 could be approximated by

$$T(t) = -0.015t^3 + 0.75t^2 - 10t + 1040 \quad (5 \le t \le 23)$$

where t is the number of years since 1980 and T is the combined SAT score average.[15] Based on this model, when (to the nearest year) was the average SAT score decreasing most rapidly? When was it increasing most rapidly?

53. *Embryo Development* The oxygen consumption of a bird embryo increases from the time the egg is laid through the time the chick hatches. In a typical galliform bird, the oxygen consumption can be approximated by

$$c(t) = -0.065t^3 + 3.4t^2 - 22t + 3.6 \text{ milliliters per day}$$
$$(8 \le t \le 30)$$

where t is the time (in days) since the egg was laid.[16] (An egg will typically hatch at around $t = 28$.) When, to the nearest day, is $c'(t)$ a maximum? What does the answer tell you?

54. *Embryo Development* The oxygen consumption of a turkey embryo increases from the time the egg is laid through the time the chick hatches. In a brush turkey, the oxygen consumption can be approximated by

$$c(t) = -0.028t^3 + 2.9t^2 - 44t + 95 \text{ milliliters per day}$$
$$(20 \le t \le 50)$$

where t is the time (in days) since the egg was laid.[17] (An egg will typically hatch at around $t = 50$.) When, to the nearest day, is $c'(t)$ a maximum? What does the answer tell you?

55. *Minimizing Resources* Basic Buckets, Inc., has an order for plastic buckets holding 5000 cubic centimeters. The buckets

are open-topped cylinders, and the company want to know what dimensions will use the least plastic per bucket. (The volume of an open-topped cylinder with height h and radius r is $\pi r^2 h$, while the surface area is $\pi r^2 + 2\pi r h$.)

56. *Optimizing Capacity* Basic Buckets would like to build a bucket with a surface area of 1000 square centimeters. What is the volume of the largest bucket they can build? (See the preceding exercise.)

57. ◆ *Agriculture* The fruit yield per tree in an orchard containing 50 trees is 100 pounds per tree each year. Due to crowding, the yield decreases by 1 pound per season for every additional tree planted. How may additional trees should be planted for a maximum total annual yield?

58. ◆ *Agriculture* Two years ago your orange orchard contained 50 trees and the total yield was 75 bags of oranges. Last year you removed ten of the trees and noticed that the total yield increased to 80 bags. Assuming that the yield per tree depends linearly on the number of trees in the orchard, what should you do this year to maximize your total yield?

59. ● *Resource Allocation* Your automobile assembly plant has a Cobb-Douglas production function given by

$$q = x^{0.4} y^{0.6}$$

where q is the number of automobiles it produces per year, x is the number of employees, and y is the daily operating budget (in dollars). Annual operating costs amount to an average of $20,000 per employee plus the operating budget of $365y. Assume that you wish to produce 1000 automobiles per year at a minimum cost. How many employees should you hire? *hint* [see Example 5]

60. ● *Resource Allocation* Repeat the preceding exercise using the production formula

$$q = x^{0.5} y^{0.5}$$

The use of technology is recommended for Exercises 61–66.

61. `tech` Ex *iPod Sales* The quarterly sales of Apple iPods from the fourth quarter of 2002 through the third quarter of 2004 could be roughly approximated by the function

$$N(t) = \frac{1100}{1 + 9(1.8)^{-t}} \text{ thousand iPods} \quad (-1 \le t \le 6)$$

where t is time in quarters since the first quarter of 2003.[18] During which quarter were iPod sales increasing most rapidly? How fast, to the nearest 10 thousand iPods per quarter, were iPod sales increasing at that time?

62. `tech` Ex *Grants* The annual spending on grants by U.S. foundations in the period 1993 to 2003 was approximately

$$s(t) = 11 + \frac{20}{1 + 1800(2.5)^{-t}} \text{ billion dollars} \quad (3 \le t \le 13)$$

---

[14] The authors' model from data obtained from *Sourcebook of Criminal Justice Statistics Online,* http://www.albany.edu/sourcebook/, as of April, 2005.

[15] The model is the authors'. Source for data: *Digest of Educational Statistics, 2003,* National Center for Education Statistics, U.S. Dept. of Education, http://nces.ed.gov/programs/digest/.

[16] The model approximates graphical data published in the article "The Brush Turkey" by Roger S. Seymour, *Scientific American,* December, 1991, pp. 108–114.

[17] Ibid.

[18] Based on a logistic regression. Source for data: Apple Computer, Inc., quarterly earnings reports, available at www.apple.com.

● basic skills   ◆ challenging   `tech` Ex technology exercise

where t is the number of years since 1990.[19] When, to the nearest year, was grant spending increasing most rapidly? How fast was it increasing at that time?

63. `tech` Ex *Bottled Water Sales* Annual sales of bottled water in the U.S. (including sparkling water) in the period 1993–2003 could be approximated by

$$R(t) = 17t^2 + 100t + 2300 \text{ million gallons} \quad (3 \le t \le 13)$$

where t is time in years since 1990.[20] Sales of sparkling water could be approximated by[21]

$$S(t) = -1.3t^2 + 4t + 160 \text{ million gallons} \quad (3 \le t \le 13)$$

Graph the derivative of $S(t)/R(t)$ for $3 \le t \le 13$. Determine when, to the nearest half year, this derivative had an absolute minimum and find its approximate value at that time. What does the answer tell you?

64. `tech` Ex *Bottled Water versus Coffee* Annual per capita consumption of bottled water in the U.S. for the period 1990–2003 could be approximated by

$$W(t) = 0.05t^2 + 0.4t + 9 \text{ gallons}[22]$$

where t is the time in years since 1990. During the same period, annual per capita consumption of coffee could be approximated by

$$C(t) = -0.05t^2 - 0.03t + 26 \text{ gallons}[23]$$

Graph the derivative of $W(t)/C(t)$ for $3 \le t \le 13$. Determine when, to the nearest half year, this derivative had an absolute maximum, and find its approximate value at that time. What does the answer tell you?

65. `tech` Ex *Asset Appreciation* You manage a small antique company that owns a collection of Louis XVI jewelry boxes. Their value v is increasing according to the formula

$$v = \frac{10,000}{1 + 500e^{-0.5t}}$$

where t is the number of years from now. You anticipate an inflation rate of 5% per year, so that the present value of an item that will be worth $\$v$ in t years' time is given by

$$p = v(1.05)^{-t}$$

When (to the nearest year) should you sell the jewelry boxes to maximize their present value? How much (to the nearest constant dollar) will they be worth at that time?

66. `tech` Ex *Harvesting Forests* The following equation models the approximate volume in cubic feet of a typical Douglas fir tree of age t years.[24]

$$V = \frac{22,514}{1 + 22,514t^{-2.55}}$$

The lumber will be sold at $10 per cubic foot, and you do not expect the price of lumber to appreciate in the foreseeable future. On the other hand, you anticipate a general inflation rate of 5% per year, so that the present value of an item that will be worth $\$v$ in t years' time is given by

$$p = v(1.05)^{-t}$$

At what age (to the nearest year) should you harvest a Douglas fir tree in order to maximize its present value? How much (to the nearest constant dollar) will a Douglas fir tree be worth at that time?

67. ◆ *Revenue (based on a question on the GRE economics test[25])* If total revenue (TR) is specified by $TR = a + bQ - cQ^2$, where Q is quantity of output and a, b, and c are positive parameters, then TR is maximized for this firm when it produces Q equal to:
(A) $b/2ac$ (B) $b/4c$ (C) $(a + b)/c$ (D) $b/2c$ (E) $c/2b$

68. ◆ *Revenue (based on a question on the GRE economics test)* If total demand (Q) is specified by $Q = -aP + b$, where P is unit price and a and b are positive parameters, then total revenue is maximized for this firm when it charges P equal to:
(A) $b/2a$ (B) $b/4a$ (C) a/b (D) $a/2b$ (E) $-b/2a$

Communication and Reasoning Exercises

69. ● Explain why the following problem is uninteresting: A packaging company wishes to make cardboard boxes with open tops by cutting square pieces from the corners of a square sheet of cardboard and folding up the sides. What is the box with the least surface area it can make this way?

70. ● Explain why finding the production level that minimizes a cost function is frequently uninteresting. What would a more interesting objective be?

71. ● Your friend Margo claims that all you have to do to find the absolute maxima and minima in applications is set the

---

[19] Based on a logistic regression. Source for data: The Foundation Center, *Foundation Growth and Giving Estimates*, 2004, downloaded from the Center's website, http://fdncenter.org.

[20] SOURCE: Beverage Marketing Corporation news release, "Bottled water now number-two commercial beverage in U.S., says Beverage Marketing Corporation," April 8, 2004, available at www.beveragemarketing.com.

[21] Source for data: Beverage Marketing Corporation of New York/*New York Times*, June 21, 2001, p. C1.

[22] Ibid.

[23] Extrapolated from data through 2000. Source for data: www.fas.usda.gov/htp/tropical/2002/06-02/coffusco.pdf

---

[24] The model is the authors' and is based on data in *Environmental and Natural Resource Economics* by Tom Tietenberg, Third Edition, (New York: HarperCollins, 1992), p. 282.

[25] SOURCE: GRE Economics Test, by G. Gallagher, G. E. Pollock, W. J. Simeone, G. Yohe (Piscataway, NJ: Research and Education Association, 1989).

● basic skills   ◆ challenging   `tech` Ex technology exercise

derivative equal to zero and solve. "All that other stuff about endpoints and so-on is a waste of time just to make life hard for us," according to Margo. Explain why she is wrong, and find at least one exercise in this exercise set to illustrate your point.

72. ● You are having a hard time persuading your friend Marco that maximizing revenue is not the same as maximizing profit. "How on earth can you expect to obtain the largest profit if you are not taking in the largest revenue?" Explain why he is wrong, and find at least one exercise in this exercise set to illustrate your point.

73. If demand q decreases as price p increases, what does the minimum value of dq/dp measure?

74. Explain how you would solve an optimization problem of the following form: Maximize $P = f(x, y, z)$ subject to $z = g(x, y)$ and $y = h(x)$.

● basic skills   ◆ challenging   *tech* Ex technology exercise

12.3 The Second Derivative and Analyzing Graphs

The **second derivative** is simply the derivative of the derivative function. To explain why we would be interested in such a thing, we start by discussing one of its interpretations.

Acceleration

Recall that if $s(t)$ represents the position of a car at time t, then its velocity is given by the derivative: $v(t) = s'(t)$. But one rarely drives a car at a constant speed; the velocity itself is changing. The rate at which the velocity is changing is the **acceleration.** Since the derivative measures the rate of change, acceleration is the derivative of velocity: $a(t) = v'(t)$. Because v is the derivative of s, we can express the acceleration in terms of s:

$$a(t) = v'(t) = (s')'(t) = s''(t)$$

That is, a is the derivative of the derivative of s, in other words, the second derivative of s which we write as s''. (In this context you will often hear the derivative s' referred to as the **first derivative.)**

Second Derivative, Acceleration

The **second derivative** of a function f is the derivative of the derivative of f, written as f''.

quick Examples

1. If $f(x) = x^3 - x$, then $f'(x) = 3x^2 - 1$, so $f''(x) = 6x$.

2. If $f(x) = 3x + 1$, then $f'(x) = 3$, so $f''(x) = 0$.

The **acceleration** of a moving object is the derivative of its velocity—that is, the second derivative of the position function.

quick Example

If t is time in hours and the position of a car at time t is $s(t) = t^3 + 2t^2$ miles, then the car's velocity is $v(t) = s'(t) = 3t^2 + 4t$ miles per hour and its acceleration is $a(t) = s''(t) = v'(t) = 6t + 4$ miles per hour per hour.

Differential Notation for the Second Derivative

We have written the second derivative of $f(x)$ as $f''(x)$. We could also use differential notation:

$$f''(x) = \frac{d^2 f}{dx^2}$$

This notation comes from writing the second derivative as the derivative of the derivative in differential notation:

$$f''(x) = \frac{d}{dx}\left[\frac{df}{dx}\right] = \frac{d^2 f}{dx^2}$$

Similarly, if $y = f(x)$, we write $f''(x)$ as $\frac{d}{dx}\left[\frac{dy}{dx}\right] = \frac{d^2 y}{dx^2}$. For example, if $y = x^3$, then $\frac{d^2 y}{dx^2} = 6x$.

An important example of acceleration is the acceleration due to gravity.

Example 1 Acceleration Due to Gravity

According to the laws of physics, the height of an object near the surface of the earth falling in a vacuum from an initial rest position 100 feet above the ground under the influence of gravity is approximately

$$s(t) = 100 - 16t^2 \text{ feet}$$

in t seconds. Find its acceleration.

Solution The velocity of the object is

$$v(t) = s'(t) = -32t \text{ ft/s} \qquad \text{Differential notation: } v = \frac{ds}{dt} = -32t \text{ ft/s}$$

The reason for the negative sign is that the height of the object is decreasing with time, so its velocity is negative. Hence, the acceleration is

$$a(t) = s''(t) = -32 \text{ ft/s}^2 \qquad \text{Differential notation: } a = \frac{d^2 s}{dt^2} = -32 \text{ ft/s}^2$$

(We write ft/s^2 as an abbreviation for feet/second/second—that is, feet per second per second. It is often read "feet per second squared.") Thus, the *downward* velocity is increasing by 32 ft/s every second. We say that 32 ft/s^2 is the **acceleration due to gravity.** If we ignore air resistance, all falling bodies near the surface of the earth, no matter what their weight, will fall with this acceleration.[*]

──────────

[*] On other planets the acceleration due to gravity is different. For example, on Jupiter, it is about three times as large as on Earth.

+*Before we go on...* In very careful experiments using balls rolling down inclined planes, Galileo made one of his most important discoveries—that the acceleration due to gravity is constant and does not depend on the weight or composition of the object falling.[26] A famous, though probably apocryphal, story has him dropping cannonballs of different weights off the Leaning Tower of Pisa to prove his point.[27] ∎

Example 2 Acceleration of Sales

For the first 15 months after the introduction of a new video game, the total sales can be modeled by the curve

$$S(t) = 20e^{0.4t} \text{ units sold}$$

where t is the time in months since the game was introduced. After about 25 months total sales follow more closely the curve

$$S(t) = 100,000 - 20e^{17-0.4t}$$

How fast are total sales accelerating after 10 months? How fast are they accelerating after 30 months? What do these numbers mean?

Solution By acceleration we mean the rate of change of the rate of change, which is the second derivative. During the first 15 months, the first derivative of sales is

$$\frac{dS}{dt} = 8e^{0.4t}$$

and so the second derivative is

$$\frac{d^2S}{dt^2} = 3.2e^{0.4t}$$

Thus, after 10 months the acceleration of sales is

$$\left.\frac{d^2S}{dt^2}\right|_{t=10} = 3.2e^4 \approx 175 \text{ units/month/month, or units/month}^2$$

We can also compute total sales

$$S(10) = 20e^4 \approx 1092 \text{ units}$$

and the rate of change of sales

$$\left.\frac{dS}{dt}\right|_{t=10} = 8e^4 \approx 437 \text{ units/month}$$

What do these numbers mean? By the end of the tenth month, a total of 1092 video games have been sold. At that time the game is selling at the rate of 437 units per month. This rate of sales is increasing by 175 units per month per month. More games will be sold each month than the month before.

To analyze the sales after 30 months is similar, using the formula

$$S(t) = 100,000 - 20e^{17-0.4t}$$

---

[26] An interesting aside: Galileo's experiments depended on getting extremely accurate timings. Since the time-pieces of his day were very inaccurate, he used the most accurate time measurement he could: He sang and used the beat as his stopwatch.

[27] A true story: The point was made again during the Apollo 15 mission to the moon (July, 1971) when astronaut David R. Scott dropped a feather and a hammer from the same height. The moon has no atmosphere, so the two hit the surface of the moon simultaneously.

The derivative is

$$\frac{dS}{dt} = 8e^{17-0.4t}$$

and the second derivative is

$$\frac{d^2S}{dt^2} = -3.2e^{17-0.4t}$$

After 30 months,

$$S(30) = 100{,}000 - 20e^{17-12} \approx 97{,}032 \text{ units}$$

$$\left.\frac{dS}{dt}\right|_{t=30} = 8e^{17-12} \approx 1187 \text{ units/month}$$

$$\left.\frac{d^2S}{dt^2}\right|_{t=30} = -3.2e^{17-12} \approx -475 \text{ units/month}^2$$

By the end of the 30th month, 97,032 video games have been sold, the game is selling at a rate of 1187 units per month, and the rate of sales is *decreasing* by 475 units per month per month. Fewer games are sold each month than the month before.

Concavity

The first derivative of f tells us where the graph of f is rising [where $f'(x) > 0$] and where it is falling [where $f'(x) < 0$]. The second derivative tells in what direction the graph of f *curves* or *bends*. Consider the graphs in Figures 23 and 24.

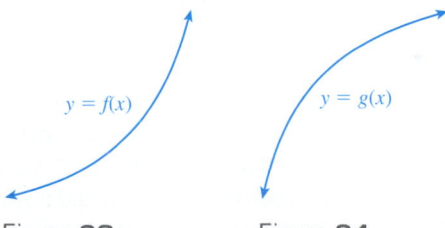

Figure **23**   Figure **24**

Think of a car driving from left to right along each of the roads shown in the two figures. A car driving along the graph of f in Figure 23 will turn to the left (upward); a car driving along the graph of g in Figure 24 will turn to the right (downward). We say that the graph of f is **concave up** while the graph of g is **concave down.** Now think about the derivatives of f and g. The derivative $f'(x)$ starts small but *increases* as the graph gets steeper. Since $f'(x)$ is increasing, its derivative $f''(x)$ must be positive. On the other hand, $g'(x)$ *decreases* as we go to the right. Since $g'(x)$ is decreasing, its derivative $g''(x)$ must be negative. Summarizing, we have the following.

Concavity and the Second Derivative

A curve is **concave up** if its slope is increasing, in which case the second derivative is positive. A curve is **concave down** if its slope is decreasing, in which case the second derivative is negative. A point where the graph of f changes concavity, from concave up to concave down or vice versa, is called a **point of inflection.** At a point of inflection, the second derivative is either zero or undefined.

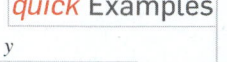

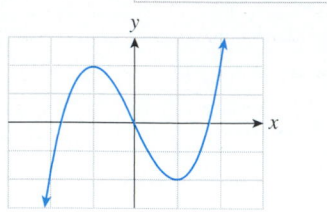

Figure 26

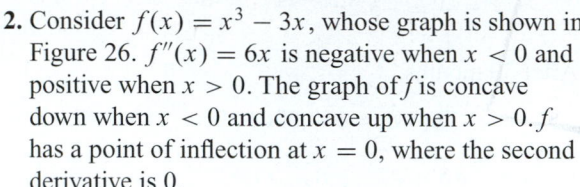

Locating Points of Inflection
To locate possible points of inflection, list points where $f''(x) = 0$ and also points where $f''(x)$ is not defined.

1. The function f whose graph is shown in Figure 25 has points of inflection at approximately $x = 1$ and $x = 3$.

2. Consider $f(x) = x^3 - 3x$, whose graph is shown in Figure 26. $f''(x) = 6x$ is negative when $x < 0$ and positive when $x > 0$. The graph of f is concave down when $x < 0$ and concave up when $x > 0$. f has a point of inflection at $x = 0$, where the second derivative is 0.

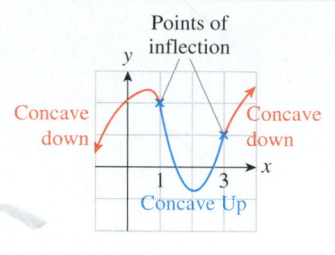

Figure 25

The following example shows one of the reasons it's useful to look at concavity.

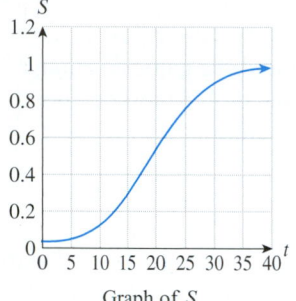

Graph of S

(a)

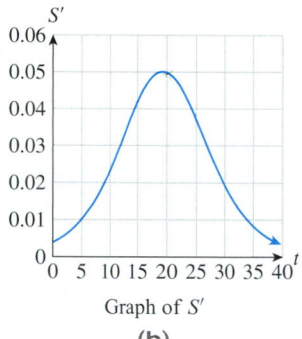

Graph of S′

(b)

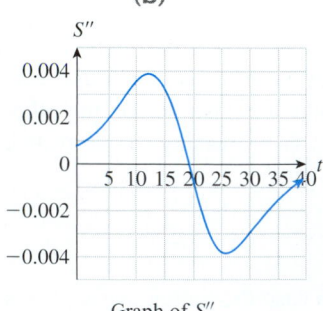

Graph of S″

(c)

Figure 27

Example 3 The Point of Diminishing Returns

After the introduction of a new video game, the worldwide sales are modeled by the curve

$$S(t) = \frac{1}{1 + 50e^{-0.2t}} \text{ million units sold}$$

where t is the time in months since the game was introduced (compare Example 2). The graphs of $S(t)$, $S'(t)$ and $S''(t)$ are shown in Figure 27.

Where is the graph of S concave up, and where is it concave down? Where are any points of inflection? What does this all mean?

Solution Look at the graph of S. We see that the graph of S is concave up in the early months and then becomes concave down later. The point of inflection, where the concavity changes, is somewhere between 15 and 25 months.

Now look at the graph of S''. This graph crosses the t-axis very close to $t = 20$, is positive before that point, and negative after that point. Since positive values of S'' indicate S is concave up and negative values concave down, we conclude that the graph of S is concave up for about the first 20 months; that is, for $0 < t < 20$ and concave down for $20 < t < 40$. The concavity switches at the point of inflection which occurs at about $t = 20$ (when $S''(t) = 0$; a more accurate answer is $t \approx 19.56$).

What does this all mean? Look at the graph of S', which shows sales per unit time, or monthly sales. From this graph we see that monthly sales are increasing for $t < 20$: more units are being sold each month than the month before. Monthly sales reach a peak of 0.05 million = 50,000 games per month at the point of inflection $t = 20$ and then begin to drop off. Thus, the point of inflection is the time when sales stop increasing and start to fall off. This is sometimes called the **point of diminishing returns.** Although the total sales figure continues to rise (see the graph of S: game units continue to be sold), the *rate* at which units are sold starts to drop.

Analyzing Graphs

We now have the tools we need to find the most interesting points on the graph of a function. It is easy to use graphing technology to draw the graph, but we need to use calculus to understand what we are seeing. The most interesting features of a graph are the following.

Features of a Graph

1. ***The x- and y-intercepts:*** If $y = f(x)$, find the x-intercept(s) by setting $y = 0$ and solving for x; find the y-intercept by setting $x = 0$ and solving for y.

2. ***Relative extrema:*** Use the technique of Section 12.1 to locate the relative extrema.

3. ***Points of inflection:*** Use the technique of this section to find the points of inflection.

4. ***Behavior near points where the function is not defined:*** If $f(x)$ is not defined at $x = a$, consider $\lim_{x \to a^-} f(x)$ and $\lim_{x \to a^+} f(x)$ to see how the graph of f behaves as x approaches a.

5. ***Behavior at infinity:*** Consider $\lim_{x \to -\infty} f(x)$ and $\lim_{x \to +\infty} f(x)$ if appropriate, to see how the graph of f behaves far to the left and right.

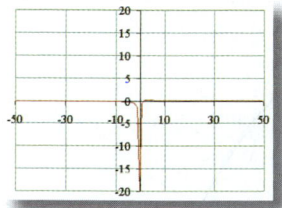

$-50 \le x \le 50, \ -20 \le y \le 20$

(a)

Note It is sometimes difficult or impossible to solve all of the equations that come up in Steps 1, 2, and 3 of the above analysis. As a consequence, we might not be able to say exactly where the x-intercept, extrema, or points of inflection are. When this happens, we will use graphing technology to assist us in determining accurate numerical approximations. ∎

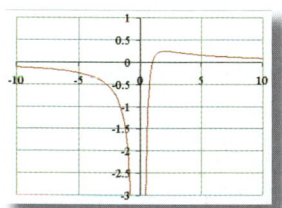

$-10 \le x \le 10, \ -3 \le y \le 1$

(b)

Figure 28

Example 4 Analyzing a Graph

Analyze the graph of $f(x) = \dfrac{1}{x} - \dfrac{1}{x^2}$.

Solution The graph, as drawn using graphing technology, is shown in Figure 28, using two different viewing windows. (Note that $x = 0$ is not in the domain of f.)

The window in Figure 28(b) seems to show the features of the graph better than the window in Figure 28(a). Does the viewing window in Figure 28(b) include *all* the interesting features of the graph? Or are there perhaps some interesting features to the right of $x = 10$ or to the left of $x = -10$? Also, where exactly do features like maxima, minima, and points of inflection occur? In our five-step process of analyzing the interesting features of the graph, we will be able to sketch the curve by hand, and also answer these questions.

1. ***The x- and y-intercepts:*** We consider $y = \dfrac{1}{x} - \dfrac{1}{x^2}$. To find the x-intercept(s), we set $y = 0$ and solve for x:

$$0 = \frac{1}{x} - \frac{1}{x^2}$$

$$\frac{1}{x} = \frac{1}{x^2}$$

Multiplying both sides by x^2 (we know that x cannot be zero, so we are not multiplying both sides by 0) gives

$$x = 1$$

Thus, there is one x-intercept (which we can see in Figure 28) at $x = 1$.

For the y-intercept, we would substitute $x = 0$ and solve for y. However, we cannot substitute $x = 0$; because $f(0)$ is not defined, the graph does not meet the y-axis.

We add features to our freehand sketch as we go. Figure 29 shows what we have so far.

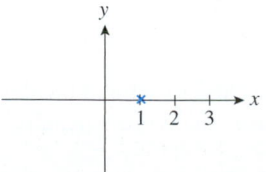

Figure 29

2. *Relative extrema:* We calculate $f'(x) = -\dfrac{1}{x^2} + \dfrac{2}{x^3}$. To find any stationary points, we set the derivative equal to 0 and solve for x:

$$-\frac{1}{x^2} + \frac{2}{x^3} = 0$$

$$\frac{1}{x^2} = \frac{2}{x^3}$$

$$x = 2$$

Thus, there is one stationary point, at $x = 2$. We can use a test point to the right to determine that this stationary point is a relative maximum:

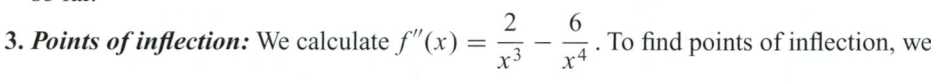

x	1 (Intercept)	2	3 (Test point)
$y = \dfrac{1}{x} - \dfrac{1}{x^2}$	0	$\dfrac{1}{4}$	$\dfrac{2}{9}$

The only possible singular point is at $x = 0$ because $f'(0)$ is not defined. However, $f(0)$ is not defined either, so there are no singular points. Figure 30 shows our graph so far.

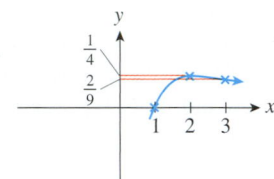

Figure 30

3. *Points of inflection:* We calculate $f''(x) = \dfrac{2}{x^3} - \dfrac{6}{x^4}$. To find points of inflection, we set the second derivative equal to 0 and solve for x:

$$\frac{2}{x^3} - \frac{6}{x^4} = 0$$

$$\frac{2}{x^3} = \frac{6}{x^4}$$

$$2x = 6$$

$$x = 3$$

Figure 28 confirms that the graph of f changes from being concave down to being concave up at $x = 3$, so this is a point of inflection. $f''(x)$ is not defined at $x = 0$, but that is not in the domain, so there are no other points of inflection. For example, the graph must be concave down in the whole region $(-\infty, \, 0)$.

Figure 31 shows our graph so far (we extended the curve near $x = 3$ to suggest a point of inflection at $x = 3$).

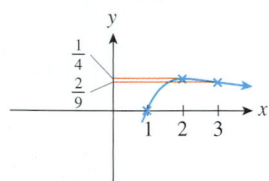

Figure 31

4. *Behavior near points where f is not defined:* The only point where $f(x)$ is not defined is $x = 0$. From the graph, $f(x)$ appears to go to $-\infty$ as x approaches 0 from either side. To calculate these limits, we rewrite $f(x)$:

$$f(x) = \frac{1}{x} - \frac{1}{x^2} = \frac{x - 1}{x^2}$$

Now, if x is close to 0 (on either side), the numerator $x - 1$ is close to -1 and the denominator is a very small but positive number. The quotient is therefore a negative number of very large magnitude. Therefore,

$$\lim_{x \to 0^-} f(x) = -\infty$$

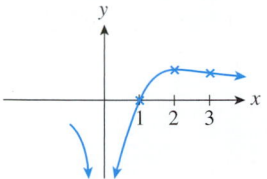

Figure 32

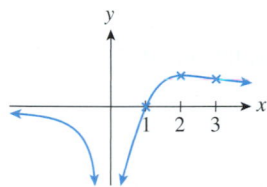

Figure 33

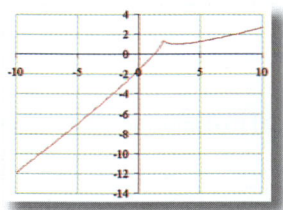

Technology:
2*x/3-((x-2)^2)^(1/3)

Figure 34

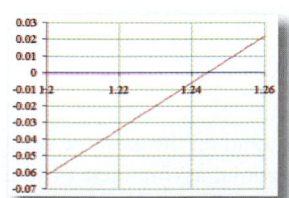

Figure 35

and

$$\lim_{x \to 0^+} f(x) = -\infty$$

From these limits, we see the following:

(1) Immediately to the *left* of $x = 0$, the graph plunges down toward $-\infty$.

(2) Immediately to the *right* of $x = 0$, the graph also plunges down toward $-\infty$. Figure 32 shows our graph with these features added.

We say that f has a **vertical asymptote** at $x = 0$, meaning that the graph approaches the line $x = 0$ without touching it.

5. *Behavior at infinity:* Both $1/x$ and $1/x^2$ go to 0 as x goes to $-\infty$ or $+\infty$; that is,

$$\lim_{x \to -\infty} f(x) = 0$$

and

$$\lim_{x \to +\infty} f(x) = 0$$

Thus, on the extreme left and right of our picture, the height of the curve levels off toward zero. Figure 33 shows the completed freehand sketch of the graph.

We say that f has a **horizontal asymptote** at $y = 0$. (Notice another thing: we haven't plotted a single point to the left of the y-axis, and yet we have a pretty good idea of what the curve looks like there! Compare the technology-drawn curve in Figure 28).

In summary, there is one x-intercept at $x = 1$; there is one relative maximum (which, we can now see, is also an absolute maximum) at $x = 2$; there is one point of inflection at $x = 3$, where the graph changes from being concave down to concave up. There is a vertical asymptote at $x = 0$, on both sides of which the graph goes down toward $-\infty$, and a horizontal asymptote at $y = 0$.

Example 5 Analyzing a Graph

Analyze the graph of $f(x) = \dfrac{2x}{3} - (x - 2)^{2/3}$.

Solution

Figure 34 shows a technology-generated version of the graph. (Note that in the technology formulation $(x - 2)^{2/3}$ is written as $[(x - 2)^2]^{1/3}$ to avoid problems with some graphing calculators and Excel.)

Let us now recreate this graph by hand, and in the process identify the features we see in Figure 34.

1. *The x- and y-intercepts:* We consider $y = \dfrac{2x}{3} - (x - 2)^{2/3}$. For the y-intercept, we set $x = 0$ and solve for y:

$$y = \frac{2(0)}{3} - (0 - 2)^{2/3} = -2^{2/3} \approx -1.59$$

To find the x-intercept(s), we set $y = 0$ and solve for x. However, if we attempt this, we will find ourselves with a cubic equation that is hard to solve. (Try it!) Following the advice in the note on p. 883, we use graphing technology to locate the x-intercept we see in Figure 34 by zooming in (Figure 35). From Figure 35, we find $x \approx 1.24$. We shall see in the discussion to follow that there can be no other x-intercepts.

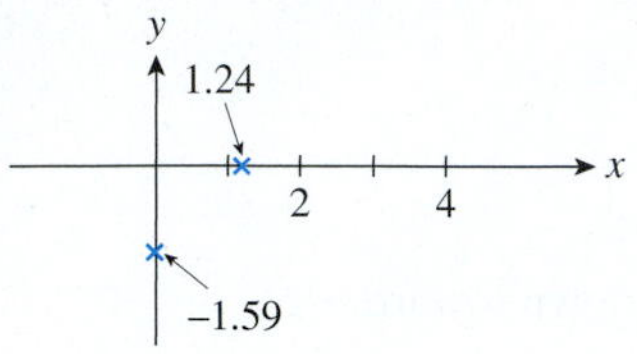

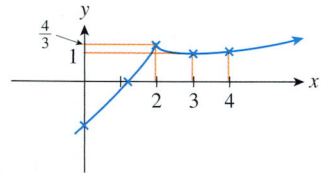

Figure 36

Figure 36 shows our freehand sketch so far.

2. Relative extrema: We calculate

$$f'(x) = \frac{2}{3} - \frac{2}{3}(x-2)^{-1/3}$$

$$= \frac{2}{3} - \frac{2}{3(x-2)^{1/3}}$$

To find any stationary points, we set the derivative equal to 0 and solve for x:

$$\frac{2}{3} - \frac{2}{3(x-2)^{1/3}} = 0$$

$$(x-2)^{1/3} = 1$$

$$x - 2 = 1^3 = 1$$

$$x = 3$$

To check for singular points, look for points where $f(x)$ is defined and $f'(x)$ is not defined. The only such point is $x = 2$: $f'(x)$ is not defined at $x = 2$, whereas $f(x)$ is defined there, so we have a singular point at $x = 2$.

x	2 (Singular point)	3 (Stationary point)	4 (Test point)
$y = \dfrac{2x}{3} - (x-2)^{2/3}$	$\dfrac{4}{3}$	1	1.079

Figure 37 shows our graph so far.

We see that there is a singular relative maximum at $(2, \ 4/3)$ (we will confirm that the graph eventually gets higher on the right) and a stationary relative minimum at $x = 3$.

3. Points of inflection: We calculate

$$f''(x) = \frac{2}{9(x-2)^{4/3}}$$

To find points of inflection, we set the second derivative equal to 0 and solve for x. But the equation

$$0 = \frac{2}{9(x-2)^{4/3}}$$

has no solution for x, so there are no points of inflection on the graph.

4. Behavior near points where f is not defined: Because $f(x)$ is defined everywhere, there are no such points to consider. In particular, there are no vertical asymptotes.

5. Behavior at infinity: We estimate the following limits numerically:

$$\lim_{x \to -\infty} \left[\frac{2x}{3} - (x-2)^{2/3} \right] = -\infty$$

and

$$\lim_{x \to +\infty} \left[\frac{2x}{3} - (x-2)^{2/3} \right] = +\infty$$

Thus, on the extreme left the curve goes down toward $-\infty$, and on the extreme right the curve rises toward $+\infty$. In particular, there are no horizontal asymptotes. (There can also be no other x-intercepts.)

Figure 38 shows the completed graph.

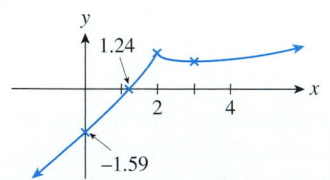

Figure 38

Figure 37

12.3 EXERCISES

● denotes basic skills exercises

tech Ex indicates exercises that should be solved using technology

In Exercises 1–10, calculate $\dfrac{d^2y}{dx^2}$ *hint* [see Quick Examples on p. 878]

1. ● $y = 3x^2 - 6$

2. ● $y = -x^2 + x$

3. ● $y = \dfrac{2}{x}$

4. ● $y = -\dfrac{2}{x^2}$

5. ● $y = 4x^{0.4} - x$

6. ● $y = 0.2x^{-0.1}$

7. ● $y = e^{-(x-1)} - x$

8. ● $y = e^{-x} + e^{x}$

9. ● $y = \dfrac{1}{x} - \ln x$

10. ● $y = x^{-2} + \ln x$

*In Exercises 11–16, the position s of a point (in feet) is given as a function of time t (in seconds). Find **(a)** its acceleration as a function of t and **(b)** its acceleration at the specified time.*

11. ● $s = 12 + 3t - 16t^2; t = 2$

12. ● $s = -12 + t - 16t^2; t = 2$

13. ● $s = \dfrac{1}{t} + \dfrac{1}{t^2}; t = 1$

14. ● $s = \dfrac{1}{t} - \dfrac{1}{t^2}; t = 2$

15. ● $s = \sqrt{t} + t^2; t = 4$

16. ● $s = 2\sqrt{t} + t^3; t = 1$

In Exercises 17–24, the graph of a function is given. Find the approximate coordinates of all points of inflection of each function (if any). hint [see Quick Examples on p. 882]

17. ●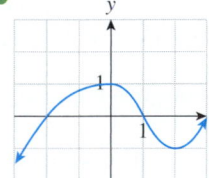

18. ●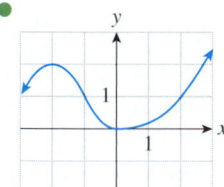

19. ●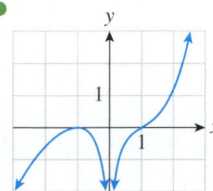

20. ●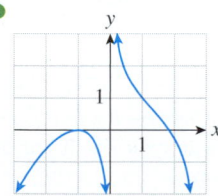

21. ●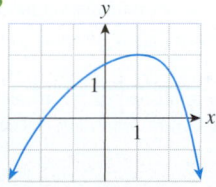

22. ●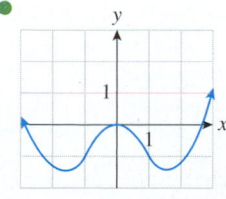

23. ●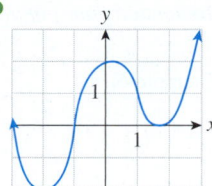

24. ●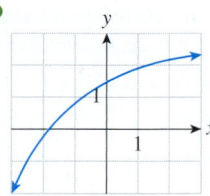

In Exercises 25–28, the graph of the derivative, $f'(x)$, *is given. Determine the x-coordinates of all points of inflection of* $f(x)$, *if any. (Assume that* $f(x)$ *is defined and continuous everywhere in* $[-3, 3]$.) *hint* [see *Before we go on* discussion in Example 3]

25. ●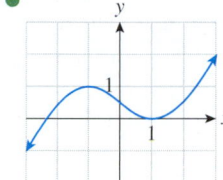

26. ●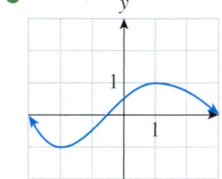

27. ●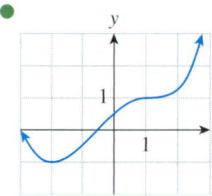

28. ●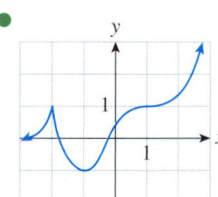

In Exercises 29–32, the graph of the second derivative, $f''(x)$, *is given. Determine the x-coordinates of all points of inflection of* $f(x)$, *if any. (Assume that* $f(x)$ *is defined and continuous everywhere in* $[-3, 3]$.)*

29.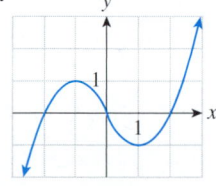

30.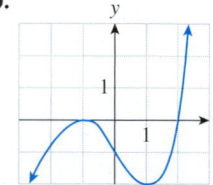

31.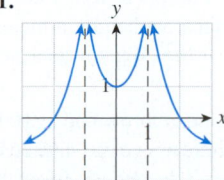

32.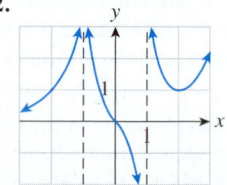

● basic skills   tech Ex technology exercise

In Exercises 33–58, sketch the graph of the given function, labeling all relative and absolute extrema and points of inflection, and vertical and horizontal asymptotes. Check your graph using technology. In the marked exercises, use graphing technology to approximate the coordinates of the extrema and points of inflection to two decimal places.

33. ● $f(x) = x^2 + 2x + 1$ *hint* [see Example 4]

34. ● $f(x) = -x^2 - 2x - 1$

35. ● $f(x) = 2x^3 + 3x^2 - 12x + 1$

36. ● $f(x) = 4x^3 + 3x^2 + 2$

37. ● $g(x) = x^3 - 12x$, domain $[-4, 4]$

38. ● $g(x) = 2x^3 - 6x$, domain $[-4, 4]$

39. ● $g(t) = \frac{1}{4}t^4 - \frac{2}{3}t^3 + \frac{1}{2}t^2$

40. ● $g(t) = 3t^4 - 16t^3 + 24t^2 + 1$

41. ● $f(t) = \frac{t^2 + 1}{t^2 - 1}$, domain $[-2, 2]$, $t \neq \pm 1$

42. ● $f(t) = \frac{t^2 - 1}{t^2 + 1}$, domain $[-2, 2]$

43. ● $f(x) = x + \frac{1}{x}$

44. ● $f(x) = x^2 + \frac{1}{x^2}$

45. ● $k(x) = \frac{2x}{3} + (x+1)^{2/3}$ *hint* [see Example 5]

46. ● $k(x) = \frac{2x}{5} - (x-1)^{2/5}$

47. ● $g(x) = x^3/(x^2 + 3)$ **48.** ● $g(x) = x^3/(x^2 - 3)$

49. ● $f(x) = x - \ln x$, domain $(0, +\infty)$

50. ● $f(x) = x - \ln x^2$, domain $(0, +\infty)$

51. ● $f(x) = x^2 + \ln x^2$ **52.** ● $f(x) = 2x^2 + \ln x$

53. ● $g(t) = e^t - t$, domain $[-1, 1]$

54. ● $g(t) = e^{-t^2}$

55. `tech` Ex $f(x) = x^4 - 2x^3 + x^2 - 2x + 1$

56. `tech` Ex $f(x) = x^4 + x^3 + x^2 + x + 1$

57. `tech` Ex $f(x) = e^x - x^3$

58. `tech` Ex $f(x) = e^x - \frac{x^4}{4}$

Applications

59. ● **Acceleration on Mars** If a stone is dropped from a height of 40 meters above the Martian surface, its height in meters after t seconds is given by $s = 40 - 1.9t^2$. What is its acceleration? *hint* [see Example 1]

60. ● **Acceleration on the Moon** If a stone is thrown up at 10 m per second from a height of 100 meters above the surface of the Moon, its height in meters after t seconds is given by $s = 100 + 10t - 0.8t^2$. What is its acceleration?

61. ● **Motion in a Straight Line** The position of a particle moving in a straight line is given by $s = t^3 - t^2$ ft after t seconds. Find an expression for its acceleration after a time t. Is its velocity increasing or decreasing when $t = 1$?

62. ● **Motion in a Straight Line** The position of a particle moving in a straight line is given by $s = 3e^t - 8t^2$ ft after t seconds. Find an expression for its acceleration after a time t. Is its velocity increasing or decreasing when $t = 1$?

63. ● **Bottled Water Sales** Annual sales of bottled water in the U.S. in the period 1993–2003 could be approximated by

$$R(t) = 17t^2 + 100t + 2300 \text{ million gallons} \quad (3 \le t \le 13)$$

where t is time in years since 1990.[28] Were sales of bottled water accelerating or decelerating in 2000? How fast? *hint* [see Example 2]

64. ● **Sparkling Water Sales** Annual sales of sparkling water in the U.S. in the period 1993–2003 could be approximated by

$$S(t) = -1.3t^2 + 4t + 160 \text{ million gallons} \quad (3 \le t \le 13)$$

where t is time in years since 1990.[29] Were sales of sparkling water accelerating or decelerating in 2002? How fast?

65. ● **Embryo Development** The daily oxygen consumption of a bird embryo increases from the time the egg is laid through the time the chick hatches. In a typical galliform bird, the oxygen consumption can be approximated by

$$c(t) = -0.065t^3 + 3.4t^2 - 22t + 3.6 \text{ ml} \quad (8 \le t \le 30)$$

where t is the time (in days) since the egg was laid.[30] (An egg will typically hatch at around $t = 28$.) Use the model to estimate the following (give the units of measurement for each answer and round all answers to two significant digits):

a. The daily oxygen consumption 20 days after the egg was laid

b. The rate at which the oxygen consumption is changing 20 days after the egg was laid

c. The rate at which the oxygen consumption is accelerating 20 days after the egg was laid

66. ● **Embryo Development** The daily oxygen consumption of a turkey embryo increases from the time the egg is laid through the time the chick hatches. In a brush turkey, the oxygen consumption can be approximated by

$$c(t) = -0.028t^3 + 2.9t^2 - 44t + 95 \text{ ml} \quad (20 \le t \le 50)$$

where t is the time (in days) since the egg was laid.[31] (An egg will typically hatch at around $t = 50$.) Use the model to

---

[28] SOURCE: Beverage Marketing Corporation news release, "Bottled water now number-two commercial beverage in U.S., says Beverage Marketing Corporation," April 8, 2004, available at www.beveragemarketing.com.

[29] Source for data: Beverage Marketing Corporation of New York/*New York Times,* June 21, 2001, p. C1.

[30] The model approximates graphical data published in the article "The Brush Turkey" by Roger S. Seymour, *Scientific American,* December, 1991, pp. 108–114.

[31] Ibid.

● basic skills   `tech` Ex technology exercise

estimate the following (give the units of measurement for each answer and round all answers to two significant digits):

a. The daily oxygen consumption 40 days after the egg was laid

b. The rate at which the oxygen consumption is changing 40 days after the egg was laid

c. The rate at which the oxygen consumption is accelerating 40 days after the egg was laid

67. ● *Epidemics* The following graph shows the total number n of people (in millions) infected in an epidemic as a function of time t (in years):

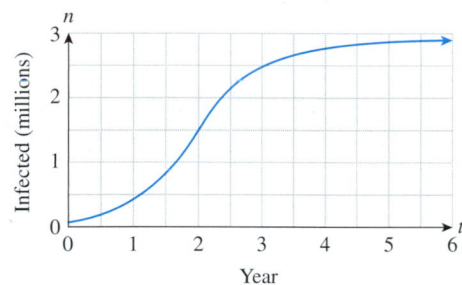

a. When to the nearest year was the rate of new infection largest?

b. When could the Centers for Disease Control and Prevention announce that the rate of new infection was beginning to drop?

68. ● *Sales* The following graph shows the total number of Pomegranate Q4 computers sold since their release (t is in years):

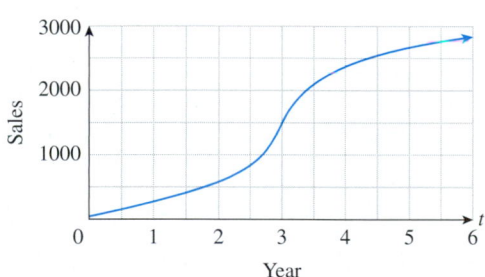

a. When were the computers selling fastest?

b. Explain why this graph might look as it does.

69. ● *Industrial Output* The following graph shows the yearly industrial output (measured in billions of dollars) of the Republic of Mars over a seven-year period:

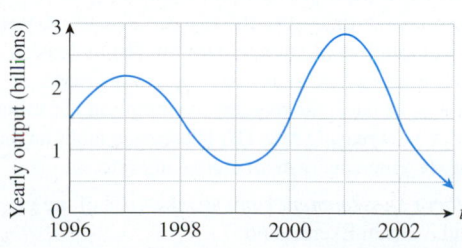

a. When to the nearest year did the rate of change of yearly industrial output reach a maximum?

b. When to the nearest year did the rate of change of yearly industrial output reach a minimum?

c. When to the nearest year did the rate of change of yearly industrial output first start to increase?

70. ● *Profits* The following graph shows the yearly profits of Gigantic Conglomerate, Inc. (GCI) from 1990 to 2005:

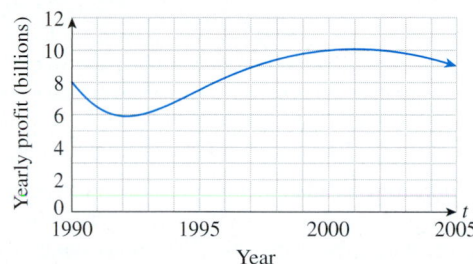

a. When were the profits rising most rapidly?

b. When were the profits falling most rapidly?

c. When could GCI's board of directors legitimately tell stockholders that they had "turned the company around"?

71. ● *Scientific Research* The percentage of research articles in the prominent journal *Physics Review* that were written by researchers in the U.S. during the years 1983–2003 can be modeled by

$$P(t) = 25 + \frac{36}{1 + 0.06(0.7)^{-t}}$$

where t is time in years since 1983.[32] The graphs of P, P' and P'' are shown here:

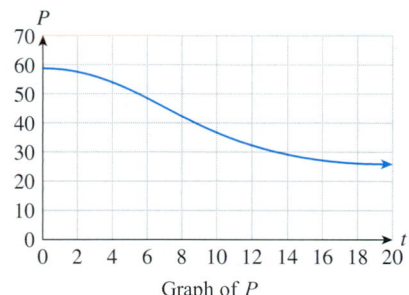

Graph of P

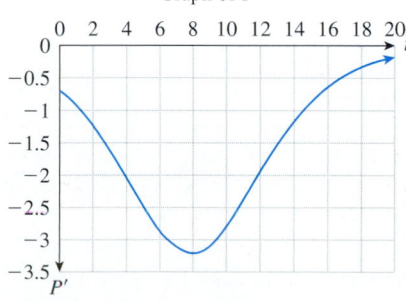

Graph of P'

[32] Source: The American Physical Society/*New York Times*, May 3, 2003, p. A1.

● basic skills   **tech** Ex technology exercise

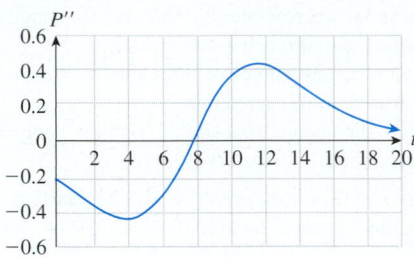

Graph of P''

Determine, to the nearest whole number, the values of t for which the graph of P is concave up, where it is concave down, and locate any points of inflection. What does the point of inflection tell you about science articles? *hint* [see Example 3]

72. ● *Scientific Research* The number of research articles in the prominent journal *Physics Review* that were written by researchers in Europe during the years 1983–2003 can be modeled by

$$P(t) = \frac{7.0}{1 + 5.4(1.2)^{-t}}$$

where t is time in years since 1983.[33] The graphs of P, P', and P'' are shown here:

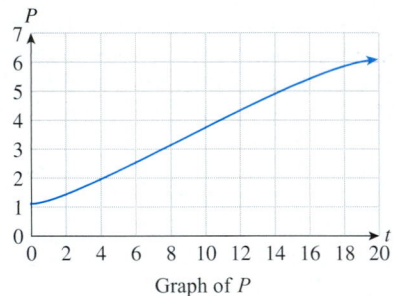

Graph of P

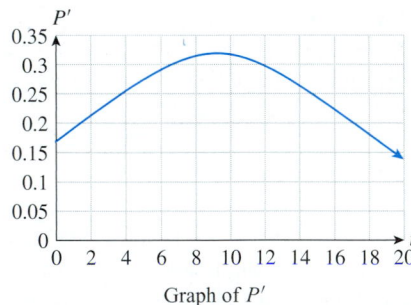

Graph of P'

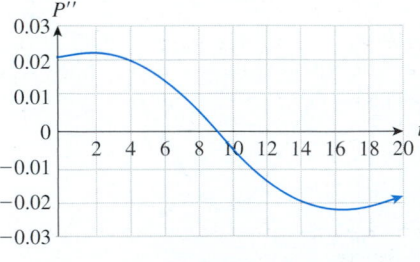

Graph of P''

[33] Source: The American Physical Society/*New York Times,* May 3, 2003, p. A1.

Determine, to the nearest whole number, the values of t for which the graph of P is concave up, where it is concave down, and locate any points of inflection. What does the point of inflection tell you about science articles?

73. ● *Embryo Development* Here are sketches of the graphs of c, c', and c'' from Exercise 65:

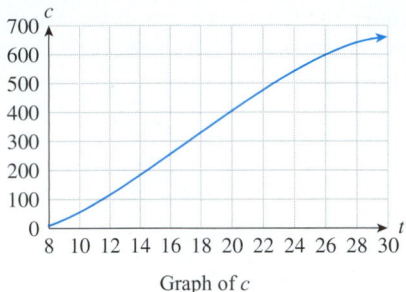

Graph of c

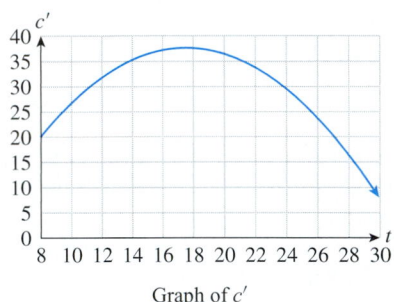

Graph of c'

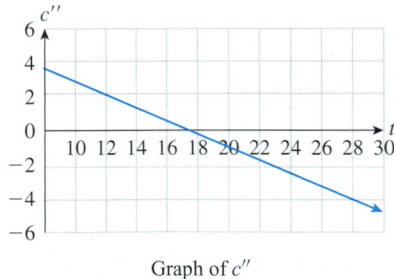

Graph of c''

Multiple choice:

a. The graph of c' **(A)** has a point of inflection. **(B)** has no points of inflection in the range shown.

b. At around 18 days after the egg is laid, daily oxygen consumption is **(A)** at a maximum. **(B)** increasing at a maximum rate. **(C)** just beginning to decrease.

c. For $t > 18$ days, the oxygen consumption is **(A)** increasing at a decreasing rate. **(B)** decreasing at an increasing rate. **(C)** increasing at an increasing rate.

74. ● *Embryo Development* Here are sketches of the graphs of c, c', and c'' from Exercise 66:

● basic skills   **tech** Ex technology exercise

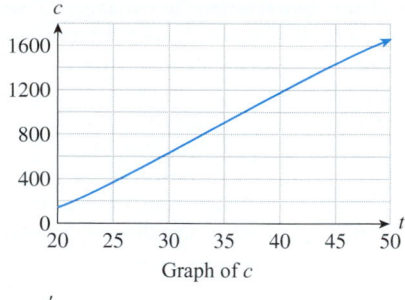

Graph of c

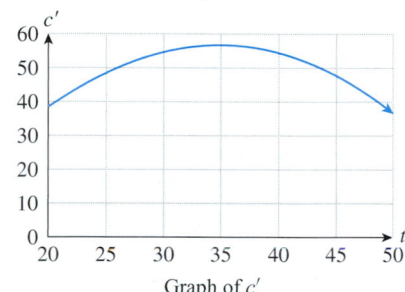

Graph of c'

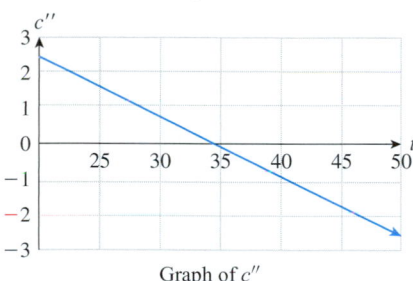

Graph of c''

Multiple choice:

a. The graph of c **(A)** has points of inflection. **(B)** has no points of inflection. **(C)** impossible to say from the graphs.

b. At around 35 days after the egg is laid, the rate of change of daily oxygen consumption is **(A)** at a maximum. **(B)** increasing at a maximum rate. **(C)** just becoming negative.

c. For $t < 35$ days, the oxygen consumption is **(A)** increasing at an increasing rate. **(B)** increasing at a decreasing rate. **(C)** decreasing at an increasing rate.

75. *Education and Crime* The following graph shows a striking relationship between the total prison population and the average combined SAT score in the U.S.:

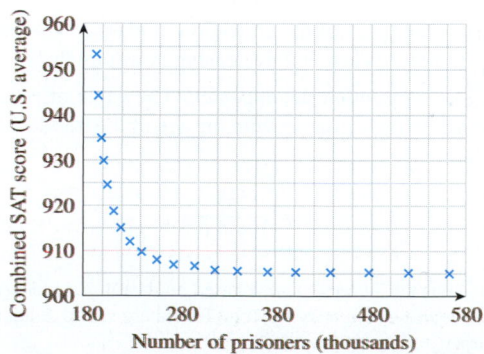

These data can be accurately modeled by

$$S(n) = 904 + \frac{1326}{(n - 180)^{1.325}} \quad (192 \leq n \leq 563)$$

Here, $S(n)$ is the combined U.S. average SAT score at a time when the total U.S. prison population was n thousand.[34]

a. Are there any points of inflection on the graph of S?

b. What does the concavity of the graph of S tell you about prison populations and SAT scores?

76. *Education and Crime* Refer back to the model in the preceding exercise,

a. Are there any points of inflection on the graph of S'?

b. When is S'' a maximum? Interpret your answer in terms of prisoners and SAT scores.

77. *Patents* In 1965, the economist F.M. Scherer modeled the number, n, of patents produced by a firm as a function of the size, s, of the firm (measured in annual sales in millions of dollars). He came up with the following equation based on a study of 448 large firms:[35]

$$n = -3.79 + 144.42s - 23.86s^2 + 1.457s^3$$

a. Find $\left.\dfrac{d^2n}{ds^2}\right|_{s=3}$. Is the rate at which patents are produced as the size of a firm goes up increasing or decreasing with size when $s = 3$? Comment on Scherer's words, ". . . we find diminishing returns dominating."

b. Find $\left.\dfrac{d^2n}{ds^2}\right|_{s=7}$ and interpret the answer.

c. Find the s-coordinate of any points of inflection and interpret the result.

78. *Returns on Investments* A company finds that the number of new products it develops per year depends on the size of its annual R&D budget, x (in thousands of dollars), according to the formula

$$n(x) = -1 + 8x + 2x^2 - 0.4x^3$$

a. Find $n''(1)$ and $n''(3)$, and interpret the results.

b. Find the size of the budget that gives the largest rate of return as measured in new products per dollar (again, called the point of diminishing returns).

79. techEx *Ecology* Manatees are grazing sea mammals sometimes referred to as sea sirens. Increasing numbers of manatees have been killed by boats off the Florida coast, as shown in the following chart:

[34] The model is the authors' based on data for the years 1967–1989. SOURCES: *Sourcebook of Criminal Justice Statistics*, 1990, p. 604/Educational Testing Service.

[35] SOURCE: F. M. Scherer, "Firm Size, Market Structure, Opportunity, and the Output of Patented Inventions," *American Economic Review*, 55 (December 1965): pp. 1097–1125.

● basic skills   techEx technology exercise

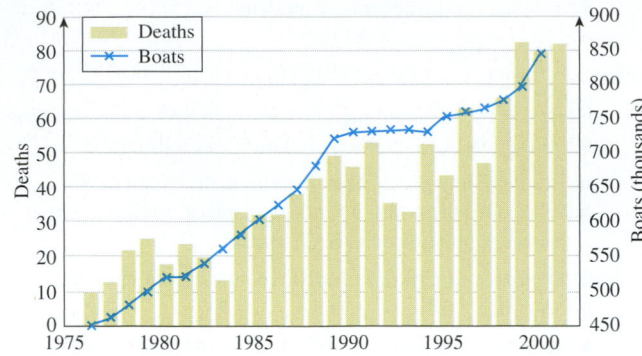

Let $M(t)$ be the number of manatees killed and let $B(t)$ be the number of boats registered in Florida in year t since 1975. These functions can be approximated by the following linear functions:[36]

$$M(t) = 2.48t + 6.87 \text{ manatees}$$
$$B(t) = 15{,}700t + 444{,}000 \text{ boats}$$

Graph the function $M(t)/B(t)$. Is the graph of $M(t)/B(t)$ concave up or concave down? The concavity of $M(t)/B(t)$ tells you that:

(A) Fewer manatees are killed per boat each year.

(B) More manatees are killed per boat each year.

(C) The number of manatees killed per boat is increasing at a decreasing rate.

(D) The number of manatees killed per boat is decreasing at an increasing rate.

80. `tech` Ex *Saudi Oil Revenues* The spot price of crude oil during the period 2000–2005 can be approximated by

$$P(t) = 5t + 25 \text{ dollars per barrel} \quad (0 \le t \le 5)$$

in year t, where $t = 0$ represents 2000. Saudi Arabia's crude oil production over the same period can be approximated by

$$Q(t) = 0.082t^2 - 0.22t + 8.2 \text{ million barrels per day}^{37}$$
$$(0 \le t \le 5)$$

Graph the revenue function $R(t) = P(t)Q(t)$. Is the graph of $R(t)$ concave up or concave down? The concavity of $R(t)$ tells you that:

(A) Saudi crude oil production was increasing at an accelerating rate.

(B) Revenues from Saudi crude oil production were increasing at an increasing rate.

(C) The acceleration of revenues from Saudi crude oil production was increasing.

(D) The rate of change of revenues from Saudi crude oil production was accelerating.

<hr />

[36] Rounded regression model. Sources for data: Florida Department of Highway Safety & Motor Vehicles, Florida Marine Institute/*New York Times,* February 12, 2002, p. F4.

[37] Source for data: EIA/Saudi British Bank (www.sabb.com). 2004 figures are based on mid-year data, and 2005 data are estimates.

81. `tech` Ex *Asset Appreciation* You manage a small antique store that owns a collection of Louis XVI jewelry boxes. Their value v is increasing according to the formula

$$v = \frac{10{,}000}{1 + 500e^{-0.5t}}$$

where t is the number of years from now. You anticipate an inflation rate of 5% per year, so that the present value of an item that will be worth $\$v$ in t years' time is given by

$$p = v(1.05)^{-t}$$

What is the greatest rate of increase of the value of your antiques, and when is this rate attained?

82. `tech` Ex *Harvesting Forests* The following equation models the approximate volume in cubic feet of a typical Douglas fir tree of age t years.[38]

$$V = \frac{22{,}514}{1 + 22{,}514t^{-2.55}}$$

The lumber will be sold at $10 per cubic foot, and you do not expect the price of lumber to appreciate in the foreseeable future. On the other hand, you anticipate a general inflation rate of 5% per year, so that the present value of an item that will be worth $\$v$ in t years time is given by

$$p = v(1.05)^{-t}$$

What is the largest rate of increase of the value of a fir tree, and when is this rate attained?

83. `tech` Ex *Asset Appreciation* As the financial consultant to a classic auto dealership, you estimate that the total value of its collection of 1959 Chevrolets and Fords is given by the formula

$$v = 300{,}000 + 1000t^2$$

where t is the number of years from now. You anticipate a continuous inflation rate of 5% per year, so that the discounted (present) value of an item that will be worth $\$v$ in t years' time is given by

$$p = ve^{-0.05t}$$

When is the value of the collection of classic cars increasing most rapidly? When is it decreasing most rapidly?

84. `tech` Ex *Plantation Management* The value of a fir tree in your plantation increases with the age of the tree according to the formula

$$v = \frac{20t}{1 + 0.05t}$$

<hr />

[38] The model is the authors', and is based on data in *Environmental and Natural Resource Economics* by Tom Tietenberg, Third Edition, (New York: HarperCollins, 1992), p. 282.

● basic skills   `tech` Ex technology exercise

where t is the age of the tree in years. Given a continuous inflation rate of 5% per year, the discounted (present) value of a newly planted seedling is

$$p = ve^{-0.05t}$$

When is the discounted value of a tree increasing most rapidly? Decreasing most rapidly?

Communication and Reasoning Exercises

85. ● Complete the following: If the graph of a function is concave up on its entire domain, then its second derivative is _____ on the domain.

86. ● Complete the following: If the graph of a function is concave up on its entire domain, then its first derivative is _____ on the domain.

87. ● Daily sales of Kent's Tents reached a maximum in January, 2002 and declined to a minimum in January, 2003 before starting to climb again. The graph of daily sales shows a point of inflection at June, 2002. What is the significance of the point of inflection?

88. ● The graph of daily sales of Luddington's Wellington boots is concave down, although sales continue to increase. What properties of the graph of daily sales versus time are reflected in the following behaviors: **a.** a point of inflection next year **b.** a horizontal asymptote?

89. ● Company A's profits satisfy $P(0) = \$1$ million, $P'(0) = \$1$ million per year, and $P''(0) = -\$1$ million per year per year. Company B's profits satisfy $P(0) = \$1$ million, $P'(0) = -\$1$ million per year, and $P''(0) = \$1$ million per year per year. There are no points of inflection in either company's profit curve. Sketch two pairs of profit curves: one in which Company A ultimately outperforms Company B and another in which Company B ultimately outperforms Company A.

90. ● Company C's profits satisfy $P(0) = \$1$ million, $P'(0) = \$1$ million per year, and $P''(0) = -\$1$ million per year per year. Company D's profits satisfy $P(0) = \$0$ million, $P'(0) = \$0$ million per year, and $P''(0) = \$1$ million per year per year. There are no points of inflection in either company's profit curve. Sketch two pairs of profit curves: one in which Company C ultimately outperforms Company D and another in which Company D ultimately outperforms Company C.

91. Explain geometrically why the derivative of a function has a relative extremum at a point of inflection, if it is defined there. Which points of inflection give rise to relative maxima in the derivative?

92. If we regard position, s, as a function of time, t, what is the significance of the *third* derivative, $s'''(t)$? Describe an everyday scenario in which this arises.

● basic skills   **tech** Ex technology exercise

12.4 Related Rates

We start by recalling some basic facts about the rate of change of a quantity:

Rate of Change of Q
If Q is a quantity changing over time t, then the derivative dQ/dt is the rate at which Q changes over time.

quick Examples

1. If A is the area of an expanding circle, then dA/dt is the rate at which the area is increasing.

2. *Words:* The radius r of a sphere is currently 3 cm and increasing at a rate of 2 cm/s.
Symbols: $r = 3$ cm and $dr/dt = 2$ cm/s.

In this section we are concerned with what are called **related rates** problems. In such a problem we have two (sometimes more) related quantities, we know the rate at which one is changing, and we wish to find the rate at which another is changing. A typical example is the following.

Example 1 The Expanding Circle

The radius of a circle is increasing at a rate of 10 cm/s. How fast is the area increasing at the instant when the radius has reached 5 cm?

Solution We have two related quantities: the radius of the circle, r, and its area, A. The first sentence of the problem tells us that r is increasing at a certain rate. When we see a sentence referring to speed or change, it is very helpful to rephrase the sentence using the phrase "the rate of change of." Here, we can say

The rate of change of r is 10 cm/s.

Because the rate of change is the derivative, we can rewrite this sentence as the equation

$$\frac{dr}{dt} = 10$$

Similarly, the second sentence of the problem asks how A is changing. We can rewrite that question:

What is the rate of change of A when the radius is 5 *cm?*

Using mathematical notation, the question is:

What is $\dfrac{dA}{dt}$ *when* $r = 5$?

Thus, knowing one rate of change, dr/dt, we wish to find a related rate of change, dA/dt. To find exactly how these derivatives are related, we need the equation relating the variables, which is

$$A = \pi r^2$$

To find the relationship between the derivatives, we take the derivative of both sides of this equation *with respect to t.* On the left we get dA/dt. On the right we need to re-member that r is a function of t and use the chain rule. We get

$$\frac{dA}{dt} = 2\pi r \frac{dr}{dt}$$

Now we substitute the given values $r = 5$ and $dr/dt = 10$. This gives

$$\left. \frac{dA}{dt} \right|_{r=5} = 2\pi(5)(10) = 100\pi \approx 314 \, \text{cm}^2/\text{s}$$

Thus, the area is increasing at the rate of 314 cm²/s when the radius is 5 cm.

We can organize our work as follows:

Solving a Related Rates Problem

A. The Problem

   1. List the related, changing quantities.

   2. Restate the problem in terms of rates of change. Rewrite the problem using mathematical notation for the changing quantities and their derivatives.

B. The Relationship

   1. Draw a diagram, if appropriate, showing the changing quantities.

   2. Find an equation or equations relating the changing quantities.

   3. Take the derivative with respect to time of the equation(s) relating the quantities to get the **derived equation(s),** which relate the rates of change of the quantities.

C. The Solution

   1. Substitute into the derived equation(s) the given values of the quantities and their derivatives.

   2. Solve for the derivative required.

We can illustrate the procedure with the "ladder problem" found in almost every calculus textbook.

Example 2 The Falling Ladder

Jane is at the top of a 5-foot ladder when it starts to slide down the wall at a rate of 3 feet per minute. Jack is standing on the ground behind her. How fast is the base of the ladder moving when it hits him if Jane is 4 feet from the ground at that instant?

Solution The first sentence talks about (the top of) the ladder sliding down the wall. Thus, one of the changing quantities is the height of the top of the ladder. The question asked refers to the motion of the base of the ladder, so another changing quantity is the distance of the base of the ladder from the wall. Let's record these variables and follow the outline above to obtain the solution.

A. The Problem

1. The changing quantities are

     h = height of the top of the ladder

     b = distance of the base of the ladder from the wall

2. We rephrase the problem in words, using the phrase "rate of change":

The rate of change of the height of the top of the ladder is −3 feet per minute. What is the rate of change of the distance of the base from the wall when the top of the ladder is 4 feet from the ground?

We can now rewrite the problem mathematically:

$$\frac{dh}{dt} = -3. \text{ Find } \frac{db}{dt} \text{ when } h = 4$$

B. The Relationship

1. Figure 39 shows the ladder and the variables h and b. Notice that we put in the figure the fixed length, 5, of the ladder, but any changing quantities, like h and b, we leave as variables. We shall not use any specific values for h or b until the very end.

2. From the figure, we can see that h and b are related by the Pythagorean theorem:

$$h^2 + b^2 = 25$$

Figure 39

3. Taking the derivative with respect to time of the equation above gives us the derived equation:

$$2h\frac{dh}{dt} + 2b\frac{db}{dt} = 0$$

C. The Solution

1. We substitute the known values $dh/dt = -3$ and $h = 4$ into the derived equation:

$$2(4)(-3) + 2b\frac{db}{dt} = 0$$

We would like to solve for db/dt, but first we need the value of b, which we can determine from the equation $h^2 + b^2 = 25$, using the value $h = 4$:

$$16 + b^2 = 25$$
$$b^2 = 9$$
$$b = 3$$

Substituting into the derived equation, we get

$$-24 + 2(3)\frac{db}{dt} = 0$$

2. Solving for db/dt gives

$$\frac{db}{dt} = \frac{24}{6} = 4$$

Thus, the base of the ladder is sliding away from the wall at 4 ft/min when it hits Jack.

Example 3 Average Cost

The cost to manufacture x cell phones in a day is

$$C(x) = 10{,}000 + 20x + \frac{x^2}{10{,}000} \text{ dollars}$$

The daily production level is currently $x = 5000$ cell phones and is increasing at a rate of 100 units per day. How fast is the average cost changing?

Solution

A. The Problem

1. The changing quantities are the production level x and the average cost, $\bar{C}$.

2. We rephrase the problem as follows:

The daily production level is $x = 5000$ units and the rate of change of x is 100 units/day. What is the rate of change of the average cost, $\bar{C}$?

In mathematical notation,

$$x = 5000 \text{ and } \frac{dx}{dt} = 100. \text{ Find } \frac{d\bar{C}}{dt}$$

B. The Relationship

1. In this example the changing quantities cannot easily be depicted geometrically.

2. We are given a formula for the *total* cost. We get the *average* cost by dividing the total cost by x:

$$\bar{C} = \frac{C}{x}$$

So,

$$\bar{C} = \frac{10{,}000}{x} + 20 + \frac{x}{10{,}000}$$

3. Taking derivatives with respect to t of both sides, we get the derived equation:

$$\frac{d\bar{C}}{dt} = \left(-\frac{10{,}000}{x^2} + \frac{1}{10{,}000} \right) \frac{dx}{dt}$$

C. The Solution

Substituting the values from part A into the derived equation, we get

$$\frac{d\bar{C}}{dt} = \left(-\frac{10{,}000}{5{,}000^2} + \frac{1}{10{,}000} \right) 100$$

$$= -0.03 \text{ dollars/day}$$

Thus, the average cost is decreasing by 3¢ per day.

The scenario in the following example is similar to Example 5 in Section 12.2.

Example 4 Allocation of Labor

The Gym Sock Company manufactures cotton athletic socks. Production is partially automated through the use of robots. The number of pairs of socks the company can manufacture in a day is given by a Cobb-Douglas production formula:

$$q = 50n^{0.6}r^{0.4}$$

where q is the number of pairs of socks that can be manufactured by n laborers and r robots. The company currently produces 1000 pairs of socks each day and employs 20 laborers. It is bringing one new robot on line every month. At what rate are laborers being laid off, assuming that the number of socks produced remains constant?

Solution

A. The Problem

1. The changing quantities are the number of laborers n and the number of robots r.

2. $\dfrac{dr}{dt} = 1$. Find $\dfrac{dn}{dt}$ when $n = 20$

B. The Relationship

1. No diagram is appropriate here.

2. The equation relating the changing quantities:

$$1000 = 50n^{0.6}r^{0.4}$$

or

$$20 = n^{0.6}r^{0.4}$$

(Productivity is constant at 1000 pairs of socks each day.)

3. The derived equation is

$$0 = 0.6n^{-0.4}\left(\frac{dn}{dt}\right)r^{0.4} + 0.4n^{0.6}r^{-0.6}\left(\frac{dr}{dt}\right)$$

$$= 0.6\left(\frac{r}{n}\right)^{0.4}\left(\frac{dn}{dt}\right) + 0.4\left(\frac{n}{r}\right)^{0.6}\left(\frac{dr}{dt}\right)$$

We solve this equation for dn/dt because we shall want to find dn/dt below and because the equation becomes simpler when we do this:

$$0.6\left(\frac{r}{n}\right)^{0.4}\left(\frac{dn}{dt}\right) = -0.4\left(\frac{n}{r}\right)^{0.6}\left(\frac{dr}{dt}\right)$$

$$\frac{dn}{dt} = -\frac{0.4}{0.6}\left(\frac{n}{r}\right)^{0.6}\left(\frac{n}{r}\right)^{0.4}\left(\frac{dr}{dt}\right)$$

$$= -\frac{2}{3}\left(\frac{n}{r}\right)\left(\frac{dr}{dt}\right)$$

C. The Solution

Substituting the numbers in A into the last equation in B, we get

$$\frac{dn}{dt} = -\frac{2}{3}\left(\frac{20}{r}\right)(1)$$

We need to compute r by substituting the known value of n in the original formula:

$$20 = n^{0.6}r^{0.4}$$

$$20 = 20^{0.6}r^{0.4}$$

$$r^{0.4} = \frac{20}{20^{0.6}} = 20^{0.4}$$

$$r = 20$$

Thus,

$$\frac{dn}{dt} = -\frac{2}{3}\left(\frac{20}{20}\right)(1) = -\frac{2}{3} \text{ laborers per month}$$

The company is laying off laborers at a rate of $2/3$ per month, or two every three months.

We can interpret this result as saying that, at the current level of production and number of laborers, one robot is as productive as $2/3$ of a laborer, or 3 robots are as productive as 2 laborers.

12.4 EXERCISES

● denotes basic skills exercises

◆ denotes challenging exercises

Rewrite the statements and questions in Exercises 1–8 in mathematical notation.

1. ● The population P is currently 10,000 and growing at a rate of 1000 per year. *hint* [see Quick Examples on p. 893]

2. ● There are presently 400 cases of Bangkok flu, and the number is growing by 30 new cases every month.

3. ● The annual revenue of your tie-dye T-shirt operation is currently $7000 but is decreasing by $700 each year. How fast are annual sales changing?

4. ● A ladder is sliding down a wall so that the distance between the top of the ladder and the floor is decreasing at a rate of

● basic skills　◆ challenging

3 feet per second. How fast is the base of the ladder receding from the wall?

5. ● The price of shoes is rising $5 per year. How fast is the demand changing?

6. ● Stock prices are rising $1000 per year. How fast is the value of your portfolio increasing?

7. ● The average global temperature is 60°F and rising by 0.1°F per decade. How fast are annual sales of Bermuda shorts increasing?

8. ● The country's population is now 260,000,000 and is increasing by 1,000,000 people per year. How fast is the annual demand for diapers increasing?

Applications

9. ● **Sun Spots** The area of a circular sun spot is growing at a rate of 1200 km²/sec. *hint* [see Example 1]

   a. How fast is the radius growing at the instant when it equals 10,000 km?

   b. How fast is the radius growing at the instant when the sun spot has an area of 640,000 km²?

10. ● **Puddles** The radius of a circular puddle is growing at a rate of 5 cm/sec.

    a. How fast is its area growing at the instant when the radius is 10 cm?

    b. How fast is the area growing at the instant when it equals 36 cm²?

11. ● **Balloons** A spherical party balloon is being inflated with helium pumped in at a rate of 3 cubic feet per minute. How fast is the radius growing at the instant when the radius has reached 1 foot? (The volume of a sphere of radius r is $V = \frac{4}{3}\pi r^3$.)

12. ● **More Balloons** A rather flimsy spherical balloon is designed to pop at the instant its radius has reached 10 centimeters. Assuming the balloon is filled with helium at a rate of 10 cubic centimeters per second, calculate how fast the radius is growing at the instant it pops. (The volume of a sphere of radius r is $V = \frac{4}{3}\pi r^3$.)

13. ● **Sliding Ladders** The base of a 50-foot ladder is being pulled away from a wall at a rate of 10 feet per second. How fast is the top of the ladder sliding down the wall at the instant when the base of the ladder is 30 feet from the wall? *hint* [see Example 2]

14. ● **Sliding Ladders** The top of a 5-foot ladder is sliding down a wall at a rate of 10 feet per second. How fast is the base of the ladder sliding away from the wall at the instant when the top of the ladder is 3 feet from the ground?

15. ● **Average Cost** The average cost function for the weekly manufacture of portable CD players is given by

$$\bar{C}(x) = 150,000x^{-1} + 20 + 0.0001x \text{ dollars per player}$$

where x is the number of CD players manufactured that week. Weekly production is currently 3000 players and is increasing at a rate of 100 players per week. What is happening to the average cost? *hint* [see Example 3]

16. ● **Average Cost** Repeat the preceding exercise, using the revised average cost function

$$\bar{C}(x) = 150,000x^{-1} + 20 + 0.01x \text{ dollars per player}$$

17. ● **Demand** Demand for your tie-dyed T-shirts is given by the formula

$$q = 500 - 100p^{0.5}$$

where q is the number of T-shirts you can sell each month at a price of p dollars. If you currently sell T-shirts for $15 each and you raise your price by $2 per month, how fast will the demand drop? (Round your answer to the nearest whole number.)

18. ● **Supply** The number of portable CD players you are prepared to supply to a retail outlet every week is given by the formula

$$q = 0.1p^2 + 3p$$

where p is the price it offers you. The retail outlet is currently offering you $40 per CD player. If the price it offers decreases at a rate of $2 per week, how will this affect the number you supply?

19. ● **Revenue** You can now sell 50 cups of lemonade per week at 30¢ per cup, but demand is dropping at a rate of 5 cups per week each week. Assuming that raising the price does not affect demand, how fast do you have to raise your price if you want to keep your weekly revenue constant?

20. ● **Revenue** You can now sell 40 cars per month at $20,000 per car, and demand is increasing at a rate of 3 cars per month each month. What is the fastest you could drop your price before your monthly revenue starts to drop?

21. ● **Production** The automobile assembly plant you manage has a Cobb-Douglas production function given by

$$P = 10x^{0.3}y^{0.7}$$

where P is the number of automobiles it produces per year, x is the number of employees, and y is the daily operating budget (in dollars). You maintain a production level of 1000 automobiles per year. If you currently employ 150 workers and are hiring new workers at a rate of 10 per year, how fast is your daily operating budget changing? *hint* [see Example 4]

22. ● **Production** Refer back to the Cobb-Douglas production formula in the preceding exercise. Assume that you maintain a constant work force of 200 workers and wish to increase production in order to meet a demand that is increasing by 100 automobiles per year. The current demand is 1000 automobiles per year. How fast should your daily operating budget be increasing?

● basic skills   ◆ challenging

23. ● *Demand* Assume that the demand equation for tuna in a small coastal town is

$$pq^{1.5} = 50,000$$

where q is the number of pounds of tuna that can be sold in one month at the price of p dollars per pound. The town's fishery finds that the demand for tuna is currently 900 pounds per month and is increasing at a rate of 100 pounds per month each month. How fast is the price changing?

24. ● *Demand* The demand equation for rubies at Royal Ruby Retailers is

$$q + \frac{4}{3}p = 80$$

where q is the number of rubies RRR can sell per week at p dollars per ruby. RRR finds that the demand for its rubies is currently 20 rubies per week and is dropping at a rate of one ruby per week. How fast is the price changing?

25. **Ships Sailing Apart** The H.M.S. Dreadnaught is 40 miles north of Montauk and steaming due north at 20 miles/hour, while the U.S.S. Mona Lisa is 50 miles east of Montauk and steaming due east at an even 30 miles/hour. How fast is their distance apart increasing?

26. *Near Miss* My aunt and I were approaching the same intersection, she from the south and I from the west. She was traveling at a steady speed of 10 miles/hour, while I was approaching the intersection at 60 miles/hour. At a certain instant in time, I was one-tenth of a mile from the intersection, while she was one-twentieth of a mile from it. How fast were we approaching each other at that instant?

27. *Baseball* A baseball diamond is a square with side 90 ft.

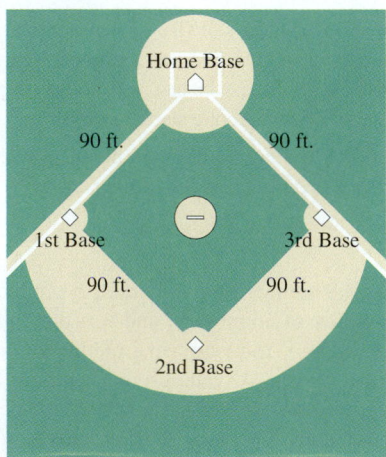

A batter at home base hits the ball and runs toward first base with a speed of 24 ft/sec. At what rate is his distance from third base increasing when he is halfway to first base?

28. *Baseball* Refer to Exercise 27. Another player is running from third base to home at 30 ft/sec. How fast is her distance from second base increasing when she is 60 feet from third base?

29. *Movement along a Graph* A point on the graph of $y = 1/x$ is moving along the curve in such a way that its x-coordinate is increasing at a rate of 4 units per second. What is happening to the y-coordinate at the instant the y-coordinate is equal to 2?

30. *Motion around a Circle* A point is moving along the circle $x^2 + (y-1)^2 = 8$ in such a way that its x-coordinate is decreasing at a rate of 1 unit per second. What is happening to the y-coordinate at the instant when the point has reached $(-2, 3)$?

31. *Education* In 1991, the expected income of an individual depended on his or her educational level according to the following formula:

$$I(n) = 2928.8n^3 - 115,860n^2 + 1,532,900n - 6,760,800$$
$$(12 \leq n \leq 15)$$

Here, n is the number of school years completed and $I(n)$ is the individual's expected income.[39] You have completed 13 years of school and are currently a part-time student. Your schedule is such that you will complete the equivalent of one year of college every three years. Assuming that your salary is linked to the above model, how fast is your income going up? (Round your answer to the nearest $1.)

32. *Education* Refer back to the model in the preceding exercise. Assume that someone has completed 14 years of school and that her income is increasing by $10,000 per year. How much schooling per year is this rate of increase equivalent to?

33. *Employment* An employment research company estimates that the value of a recent MBA graduate to an accounting company is

$$V = 3e^2 + 5g^3$$

where V is the value of the graduate, e is the number of years of prior business experience, and g is the graduate school grade point average. A company that currently employs graduates with a 3.0 average wishes to maintain a constant employee value of $V = 200$, but finds that the grade point average of its new employees is dropping at a rate of 0.2 per year. How fast must the experience of its new employees be growing in order to compensate for the decline in grade point average?

34. *Grades*[40] A production formula for a student's performance on a difficult English examination is given by

$$g = 4hx - 0.2h^2 - 10x^2$$

where g is the grade the student can expect to obtain, h is the number of hours of study for the examination, and x is the student's grade point average. The instructor finds that students'

---

[39] The model is a best-fit cubic based on Table 358, U.S. Department of Education, *Digest of Education Statistics, 1991,* Washington, DC: Government Printing Office, 1991.

[40] Based on an Exercise in *Introduction to Mathematical Economics* by A.L. Ostrosky Jr. and J.V. Koch (Waveland Press, Illinois, 1979.)

● basic skills   ◆ challenging

grade point averages have remained constant at 3.0 over the years, and that students currently spend an average of 15 hours studying for the examination. However, scores on the examination are dropping at a rate of 10 points per year. At what rate is the average study time decreasing?

35. **Cones** A right circular conical vessel is being filled with green industrial waste at a rate of 100 cubic meters per second. How fast is the level rising after 200π cubic meters have been poured in? The cone has a height of 50 m and a radius 30 m at its brim. (The volume of a cone of height h and cross-sectional radius r at its brim is given by $V = \frac{1}{3}\pi r^2 h$.)

36. **More Cones** A circular conical vessel is being filled with ink at a rate of 10 cm³/sec. How fast is the level rising after 20 cm³ have been poured in? The cone has height 50 cm and radius 20 cm at its brim. (The volume of a cone of height h and cross-sectional radius r at its brim is given by $V = \frac{1}{3}\pi r^2 h$.)

37. **Cylinders** The volume of paint in a right cylindrical can is given by $V = 4t^2 - t$ where t is time in seconds and V is the volume in cm³. How fast is the level rising when the height is 2 cm? The can has a height of 4 cm and a radius of 2 cm. [Hint: To get h as a function of t, first solve the volume $V = \pi r^2 h$ for h.]

38. **Cylinders** A cylindrical bucket is being filled with paint at a rate of 6 cm³ per minute. How fast is the level rising when the bucket starts to overflow? The bucket has a radius of 30 cm and a height of 60 cm.

39. **Computers vs. Income** The demand for personal computers in the home goes up with household income. For a given community, we can approximate the average number of computers in a home as

$$q = 0.3454 \ln x - 3.047 \qquad 10{,}000 \le x \le 125{,}000$$

where x is mean household income.[41] Your community has a mean income of $30,000, increasing at a rate of $2,000 per year. How many computers per household are there, and how fast is the number of computers in a home increasing? (Round your answer to four decimal places.)

40. **Computers vs. Income** Refer back to the model in the preceding exercise. The average number of computers per household in your town is 0.5 and is increasing at a rate of 0.02 computers per household per year. What is the average household income in your town, and how fast is it increasing? (Round your answers to the nearest $10).

Education and Crime *The following graph shows a striking relationship between the total prison population and the average combined SAT score in the U.S. Exercises 41 and 42 are based*

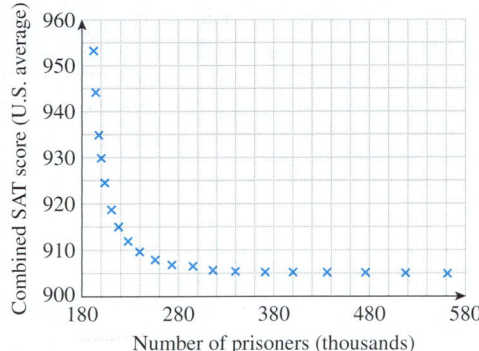

on the following model for these data:

$$S(n) = 904 + \frac{1326}{(n - 180)^{1.325}} \qquad (192 \le n \le 563)$$

Here, $S(n)$ is the combined average SAT score at a time when the total prison population is n thousand.[42]

41. In 1985, the U.S. prison population was 475,000 and increasing at a rate of 35,000 per year. What was the average SAT score, and how fast, and in what direction, was it changing? (Round your answers to two decimal places.)

42. In 1970, the U.S. combined SAT average was 940 and dropping by 10 points per year. What was the U.S. prison population, and how fast, and in what direction, was it changing? (Round your answers to the nearest 100.)

Divorce Rates *A study found that the divorce rate d (given as a percentage) appears to depend on the ratio r of available men to available women.[43] This function can be approximated by*

$$d(r) = \begin{cases} -40r + 74 & \text{if } r \le 1.3 \\ \dfrac{130r}{3} - \dfrac{103}{3} & \text{if } r > 1.3 \end{cases}$$

Exercises 43 and 44 are based on this model.

43. ◆ There are currently 1.1 available men per available woman in Littleville, and this ratio is increasing by 0.05 per year. What is happening to the divorce rate?

44. ◆ There are currently 1.5 available men per available woman in Largeville, and this ratio is decreasing by 0.03 per year. What is happening to the divorce rate?

[41] The model is a regression model. Source for data: Income distribution: Luxembourg Income Study/*New York Times,* August 14, 1995, p. A9. Computer data: Forrester Research/*The New York Times,* August 8, 1999, p. BU4.

[42] The model is the authors' based on data for the years 1967–1989. SOURCES: Sourcebook of Criminal Justice Statistics, 1990, p. 604/Educational Testing Service.

[43] The cited study, by Scott J. South and associates, appeared in the *American Sociological Review* (February, 1995). Figures are rounded. SOURCE: *The New York Times,* February 19, 1995, p. 40.

● basic skills   ◆ challenging

Communication and Reasoning Exercises

45. ● Why is this section titled "related rates"?

46. ● If you know how fast one quantity is changing and need to compute how fast a second quantity is changing, what kind of information do you need?

47. ● In a related rates problem, there is no limit to the number of changing quantities we can consider. Illustrate this by creating a related rates problem with four changing quantities.

48. ● If three quantities are related by a single equation, how would you go about computing how fast one of them is changing based on a knowledge of the other two?

49. The demand and unit price for your store's checkered T-shirts are changing with time. Show that the percentage rate of change of revenue equals the sum of the percentage rates of change of price and demand. (The percentage rate of change of a quantity Q is $Q'(t)/Q(t)$.)

50. The number N of employees and the total floor space S of your company are both changing with time. Show that the percentage rate of change of square footage per employee equals the percentage rate of change of S minus the percentage rate of change of N. (The percentage rate of change of a quantity Q is $Q'(t)/Q(t)$.)

51. In solving a related rates problem, a key step is solving the derived equation for the unknown rate of change (once we have substituted the other values into the equation). Call the unknown rate of change X. The derived equation is what kind of equation in X?

52. On a recent exam, you were given a related rates problem based on an algebraic equation relating two variables x and y. Your friend told you that the correct relationship between dx/dt and dy/dt was given by

$$\left(\frac{dx}{dt}\right) = \left(\frac{dy}{dt}\right)^2$$

Could he be correct?

53. Transform the following into a mathematical statement about derivatives: If my grades are improving at twice the speed of yours, then your grades are improving at half the speed of mine.

54. If two quantities x and y are related by a linear equation, how are their rates of change related?

● basic skills ◆ challenging

12.5 | Elasticity

You manufacture an extremely popular brand of sneakers and want to know what will happen if you increase the selling price. Common sense tells you that demand will drop as you raise the price. But will the drop in demand be enough to cause your revenue to fall? Or will it be small enough that your revenue will rise because of the higher selling price? For example, if you raise the price by 1%, you might suffer only a 0.5% loss in sales. In this case, the loss in sales will be more than offset by the increase in price and your revenue will rise. In such a case, we say that the demand is **inelastic,** because it is not very sensitive to the increase in price. On the other hand, if your 1% price increase results in a 2% drop in demand, then raising the price will cause a drop in revenues. We then say that the demand is **elastic** because it reacts strongly to a price change.

We can use calculus to measure the response of demand to price changes if we have a demand equation for the item we are selling.[44] We need to know the *percentage drop in demand per percentage increase in price*. This ratio is called the **elasticity of demand,** or **price elasticity of demand,** and is usually denoted by E. Let's derive a formula for E in terms of the demand equation.

---

[44] Coming up with a good demand equation is not always easy. We saw in Chapter 1 that it is possible to find a linear demand equation if we know the sales figures at two different prices. However, such an equation is only a first approximation. To come up with a more accurate demand equation, we might need to gather data corresponding to sales at several different prices and use curve-fitting techniques like regression. Another approach would be an analytic one, based on mathematical modeling techniques that an economist might use.

Assume that we have a demand equation

$$q = f(p)$$

where q stands for the number of items we would sell (per week, per month, or what have you) if we set the price per item at p. Now suppose we increase the price p by a very small amount, Δp. Then our percentage increase in price is $(\Delta p/p) \times 100\%$. This increase in p will presumably result in a decrease in the demand q. Let's denote this corresponding decrease in q by $-\Delta q$ (we use the minus sign because, by convention, Δq stands for the *increase* in demand). Thus, the percentage decrease in demand is $(-\Delta q/q) \times 100\%$.

Now E is the ratio

$$E = \frac{\text{Percentage decrease in demand}}{\text{Percentage increase in price}}$$

so

$$E = \frac{-\frac{\Delta q}{q} \times 100\%}{\frac{\Delta p}{p} \times 100\%}$$

Canceling the 100%s and reorganizing, we get

$$E = -\frac{\Delta q}{\Delta p} \cdot \frac{p}{q}$$

Q: *What small change in price will we use for Δp?*

A: It should probably be pretty small. If, say, we increased the price of sneakers to $1 million per pair, the sales would likely drop to zero. But knowing this tells us nothing about how the market would respond to a modest increase in price. In fact, we'll do the usual thing we do in calculus and let Δp approach 0. ∎

In the expression for E, if we let Δp go to 0, then the ratio $\Delta q/\Delta p$ goes to the derivative dq/dp. This gives us our final and most useful definition of the elasticity.

Price Elasticity of Demand

The **price elasticity of demand E** is the percentage rate of decrease of demand per percentage increase in price. E is given by the formula

$$E = -\frac{dq}{dp} \cdot \frac{p}{q}$$

We say that the demand is **elastic** if $E > 1$, is **inelastic** if $E < 1$, and has **unit elasticity** if $E = 1$.

quick Example

Suppose that the demand equation is $q = 20{,}000 - 2p$ where p is the price in dollars. Then

$$E = -(-2)\frac{p}{20{,}000 - 2p} = \frac{p}{10{,}000 - p}$$

If $p = \$2000$, then $E = 1/4$, and demand is inelastic at this price.
If $p = \$8000$, then $E = 4$, and demand is elastic at this price.
If $p = \$5000$, then $E = 1$, and the demand has unit elasticity at this price.

Jeffery Olson

Patrick Farace

TITLE Network Manager, Clinical Network Services
INSTITUTION United Behavioral Health

As a Network Manager for United Behavioral Health (UBH), my objective is to enable UBH members to obtain high-quality, affordable behavioral health care by providing an outstanding panel of clinicians and facilities with sufficient availability to meet member needs.

To determine the adequacy of our clinical network we first determine a utilization rate. A utilization rate is the extent to which insured members utilize mental health/substance abuse services. We use this number to determine the number of clinicians and facilities needed to meet the needs of our membership. An analysis is completed to review the existing network against urban, suburban, and rural access standards. Clinician-to-member and facility-to-member ratios are assessed, along with the linguistic and cultural competency of the network as compared to U.S. Census data. If a network is found to be below standard on any of these measures, an action plan is developed to enhance the network via targeted recruitment in order to close any potential network gap as quickly as possible.

Clinicians who are contracted with a managed care organization agree to accept a negotiated rate for their services. This rate is an important factor when recruiting and contracting with clinicians. The managed care model is built on the principle that the managed care organization provides volume in exchange for a discounted fee. These negotiated rates are created using many factors. These factors consist of values based on human and financial resources consumed in the delivery of each procedure as well as the cost of malpractice insurance. This value is then multiplied by a conversion factor to take into account regional differences in service locations.

In addition to maintaining and recruiting clinicians for our insured members, I must also report many statistics required for state and federal regulatory bodies as well as other accrediting organizations.

Throughout my 13 year career in managed care I have worked with both medical and mental health providers and have also held positions in sales and accounting. All of these functions within managed care rely heavily on mathematics. As I progress in my field I realize more and more the important role that mathematics plays in not only managed care, but life.

We are generally interested in the price that maximizes revenue and, in ordinary cases, the price that maximizes revenue must give unit elasticity. One way of seeing this is as follows:[45] If the demand is inelastic (which ordinarily occurs at a low unit price) then raising the price by a small percentage—1% say—results in a smaller percentage drop in demand. For example, in the Quick Example above, if $p = \$2000$, then the demand would drop by only $\frac{1}{4}$% for every 1% increase in price. To see the effect on revenue, we use the fact[46] that, for small changes in price,

Percentage change in revenue $\approx$ Percentage change in price

$$+ \text{ Percentage change in demand}$$

$$= 1 + \left(-\frac{1}{4}\right) = \frac{3}{4}\%$$

Thus, the revenue will increase by about 3/4%. Put another way:

If the demand is inelastic, raising the price increases revenue.

On the other hand, if the price is elastic (which ordinarily occurs at a high unit price), then increasing the price slightly will lower the revenue, so:

If the demand is elastic, lowering the price increases revenue.

The price that results in the largest revenue must therefore be at unit elasticity.

---

[45] For another—more rigorous—argument, see Exercise 27.

[46] See, for example, Exercise 49 in Section 12.4.

Example 1 Price Elasticity of Demand: Dolls

Suppose that the demand equation for Bobby Dolls is given by $q = 216 - p^2$, where p is the price per doll in dollars and q is the number of dolls sold per week.

a. Compute the price elasticity of demand when $p = \$5$ and $p = \$10$, and interpret the results.

b. Find the ranges of prices for which the demand is elastic and the range for which the demand is inelastic.

c. Find the price at which the weekly revenue is maximized. What is the maximum weekly revenue?

Solution

a. The price elasticity of demand is

$$E = -\frac{dq}{dp} \cdot \frac{p}{q}$$

Taking the derivative and substituting for q gives

$$E = 2p \cdot \frac{p}{216 - p^2} = \frac{2p^2}{216 - p^2}$$

When $p = \$5$,

$$E = \frac{2(5)^2}{216 - 5^2} = \frac{50}{191} \approx 0.26$$

Thus, when the price is set at $5, the demand is dropping at a rate of 0.26% per 1% increase in the price. Because $E < 1$, the demand is inelastic at this price, so raising the price will increase revenue.
    When $p = \$10$,

$$E = \frac{2(10)^2}{216 - 10^2} = \frac{200}{116} \approx 1.72$$

Thus, when the price is set at $10, the demand is dropping at a rate of 1.72% per 1% increase in the price. Because $E > 1$, demand is elastic at this price, so raising the price will decrease revenue; lowering the price will increase revenue.

b. and c. We answer part (c) first. Setting $E = 1$, we get

$$\frac{2p^2}{216 - p^2} = 1$$

$$p^2 = 72$$

 using *Technology*

See the Technology Guide at the end of the chapter to find out how to automate computations such as those in part (a) using a TI-83/84 or Excel.

Thus, we conclude that the maximum revenue occurs when $p = \sqrt{72} \approx \$8.49$. We can now answer part (b): The demand is elastic when $p > \$8.49$ (the price is too high), and the demand is inelastic when $p < \$8.49$ (the price is too low). Finally, we calculate the maximum weekly revenue, which equals the revenue corresponding to the price of $8.49:

$$R = qp = (216 - p^2)p = (216 - 72)\sqrt{72} = 144\sqrt{72} \approx \$1222$$

The concept of elasticity can be applied in other situations. In the following example we consider *income* elasticity of demand—the percentage increase in demand for a particular item per percentage increase in personal income.

Example 2 Income Elasticity of Demand: Porsches

You are the sales director at Suburban Porsche and have noticed that demand for Porsches depends on income according to

$$q = 0.005e^{-0.05x^2+x} \quad (1 \le x \le 10)$$

Here, x is the income of a potential customer in hundreds of thousands of dollars and q is the probability that the person will actually purchase a Porsche.[*] The **income elasticity of demand** is

$$E = \frac{dq}{dx}\frac{x}{q}$$

Compute and interpret E for $x = 2$ and 9.

Solution

Q: *Why is there no negative sign in the formula?*

A: Because we anticipate that the demand will increase as income increases, the ratio

$$\frac{\text{Percentage increase in demand}}{\text{Percentage increase in income}}$$

will be positive, so there is no need to introduce a negative sign. ∎

Turning to the calculation, since $q = 0.005e^{-0.05x^2+x}$,

$$\frac{dq}{dx} = 0.005e^{-0.05x^2+x}(-0.1x + 1)$$

and so

$$E = \frac{dq}{dx}\frac{x}{q}$$
$$= 0.005e^{-0.05x^2+x}(-0.1x + 1)\frac{x}{0.005e^{-0.05x^2+x}}$$
$$= x(-0.1x + 1)$$

When $x = 2$, $E = 2[-0.1(2) + 1)] = 1.6$. Thus, at an income level of $200,000, the probability that a customer will purchase a Porsche increases at a rate of 1.6% per 1% increase in income.

When $x = 9$, $E = 9[-0.1(9) + 1)] = 0.9$. Thus, at an income level of $900,000, the probability that a customer will purchase a Porsche increases at a rate of 0.9% per 1% increase in income.

---

[*] In other words, q is the fraction of visitors to your showroom having income x who actually purchase a Porsche.

12.5 EXERCISES

● denotes basic skills exercises

◆ denotes challenging exercises

Applications

1. ● **Demand for Oranges** The weekly sales of Honolulu Red Oranges is given by $q = 1000 - 20p$. Calculate the price elasticity of demand when the price is $30 per orange (yes, $30 per orange[47]). Interpret your answer. Also, calculate the price that gives a maximum weekly revenue, and find this maximum revenue. *hint* [see Example 1]

2. ● **Demand for Oranges** Repeat the preceding exercise for weekly sales of $1000 - 10p$.

3. ● **Tissues** The consumer demand equation for tissues is given by $q = (100 - p)^2$, where p is the price per case of tissues and q is the demand in weekly sales.

   a. Determine the price elasticity of demand E when the price is set at $30, and interpret your answer.
   b. At what price should tissues be sold in order to maximize the revenue?
   c. Approximately how many cases of tissues would be demanded at that price?

4. ● **Bodybuilding** The consumer demand curve for Professor Stefan Schwarzenegger dumbbells is given by $q = (100 - 2p)^2$, where p is the price per dumbbell, and q is the demand in weekly sales. Find the price Professor Schwarzenegger should charge for his dumbbells in order to maximize revenue.

5. ● **T-Shirts** The Physics Club sells $E = mc^2$ T-shirts at the local flea market. Unfortunately, the club's previous administration has been losing money for years, so you decide to do an analysis of the sales. A quadratic regression based on old sales data reveals the following demand equation for the T-shirts:

$$q = -2p^2 + 33p \qquad (9 \le p \le 15)$$

   Here, p is the price the club charges per T shirt, and q is the number it can sell each day at the flea market.

   a. Obtain a formula for the price elasticity of demand for $E = mc^2$ T-shirts.
   b. Compute the elasticity of demand if the price is set at $10 per shirt. *Interpret the result.*
   c. How much should the Physics Club charge for the T-shirts in order to obtain the maximum daily revenue? What will this revenue be?

6. ● **Comics** The demand curve for original *Iguanawoman* comics is given by

$$q = \frac{(400 - p)^2}{100} \qquad (0 \le p \le 400)$$

   where q is the number of copies the publisher can sell per week if it sets the price at $p.

   a. Find the price elasticity of demand when the price is set at $40 per copy.
   b. Find the price at which the publisher should sell the books in order to maximize weekly revenue.
   c. What, to the nearest $1, is the maximum weekly revenue the publisher can realize from sales of *Iguanawoman* comics?

7. ● **College Tuition** A study of about 1800 U.S. colleges and universities resulted in the demand equation $q = 9900 - 2.2p$, where q is the enrollment at a college or university, and p is the average annual tuition (plus fees) it charges.[48]

   a. The study also found that the average tuition charged by universities and colleges was $2900. What is the corresponding price elasticity of demand? Is the price elastic or inelastic? Should colleges charge more or less on average to maximize revenue?
   b. Based on the study, what would you advise a college to charge its students in order to maximize total revenue, and what would the revenue be?

8. ● **Demand for Fried Chicken** A fried chicken franchise finds that the demand equation for its new roast chicken product, "Roasted Rooster," is given by

$$p = \frac{40}{q^{1.5}}$$

   where p is the price (in dollars) per quarter-chicken serving and q is the number of quarter-chicken servings that can be sold per hour at this price. Express q as a function of p and find the price elasticity of demand when the price is set at $4 per serving. Interpret the result.

9. ● **Paint-By-Number** The estimated monthly sales of *Mona Lisa* paint-by-number sets is given by the formula $q = 100e^{-3p^2+p}$, where q is the demand in monthly sales and p is the retail price in yen.

   a. Determine the price elasticity of demand E when the retail price is set at ¥3 and interpret your answer.
   b. At what price will revenue be a maximum?
   c. Approximately how many paint-by-number sets will be sold per month at the price in part (b)?

---

[47] They are very hard to find, and their possession confers considerable social status.

[48] Based on a study by A.L. Ostrosky Jr. and J.V. Koch, as cited in their book, *Introduction to Mathematical Economics* (Waveland Press, Illinois, 1979) p. 133.

● basic skills   ◆ challenging

10. ● *Paint-By-Number* Repeat the previous exercise using the demand equation $q = 100e^{p-3p^2/2}$.

11. *Linear Demand Functions* A general linear demand function has the form $q = mp + b$ (*m* and *b* constants, $m \neq 0$).

a. Obtain a formula for the price elasticity of demand at a unit price of p.

b. Obtain a formula for the price that maximizes revenue.

12. *Exponential Demand Functions* A general exponential demand function has the form $q = Ae^{-bp}$ (*A* and *b* nonzero constants).

a. Obtain a formula for the price elasticity of demand at a unit price of p.

b. Obtain a formula for the price that maximizes revenue.

13. *Hyperbolic Demand Functions* A general hyperbolic demand function has the form $q = \dfrac{k}{p^r}$ (*r* and *k* nonzero constants).

a. Obtain a formula for the price elasticity of demand at unit price p.

b. How does E vary with p?

c. What does the answer to part (b) say about the model?

14. *Quadratic Demand Functions* A general quadratic demand function has the form $q = ap^2 + bp + c$ (*a*, *b*, and *c* constants with $a \neq 0$).

a. Obtain a formula for the price elasticity of demand at a unit price p.

b. Obtain a formula for the price or prices that could maximize revenue.

15. *Modeling Linear Demand* You have been hired as a marketing consultant to Johannesburg Burger Supply, Inc., and you wish to come up with a unit price for its hamburgers in order to maximize its weekly revenue. To make life as simple as possible, you assume that the demand equation for Johannesburg hamburgers has the linear form $q = mp + b$, where p is the price per hamburger, q is the demand in weekly sales, and m and b are certain constants you must determine.

a. Your market studies reveal the following sales figures: When the price is set at \$2.00 per hamburger, the sales amount to 3000 per week, but when the price is set at \$4.00 per hamburger, the sales drop to zero. Use these data to calculate the demand equation.

b. Now estimate the unit price that maximizes weekly revenue and predict what the weekly revenue will be at that price.

16. *Modeling Linear Demand* You have been hired as a marketing consultant to Big Book Publishing, Inc., and you have been approached to determine the best selling price for the hit calculus text by Whiner and Istanbul entitled *Fun with Derivatives*. You decide to make life easy and assume that the demand equation for *Fun with Derivatives* has the linear form $q = mp + b$, where p is the price per book, q is the demand in annual sales, and m and b are certain constants you'll have to figure out.

a. Your market studies reveal the following sales figures: when the price is set at \$50.00 per book, the sales amount

to 10,000 per year; when the price is set at \$80.00 per book, the sales drop to 1000 per year. Use these data to calculate the demand equation.

b. Now estimate the unit price that maximizes annual revenue and predict what Big Book Publishing, Inc.'s annual revenue will be at that price.

17. ● *Income Elasticity of Demand: Live Drama* The likelihood that a child will attend a live theatrical performance can be modeled by

$$q = 0.01(-0.0078x^2 + 1.5x + 4.1) \quad (15 \leq x \leq 100)$$

Here, q is the fraction of children with annual household income x thousand dollars who will attend a live dramatic performance at a theater during the year.[49] Compute the income elasticity of demand at an income level of \$20,000 and interpret the result. (Round your answer to two significant digits.) *hint* [see Example 2]

18. ● *Income Elasticity of Demand: Live Concerts* The likelihood that a child will attend a live musical performance can be modeled by

$$q = 0.01(0.0006x^2 + 0.38x + 35) \quad (15 \leq x \leq 100)$$

Here, q is the fraction of children with annual household income x who will attend a live musical performance during the year.[50] Compute the income elasticity of demand at an income level of \$30,000 and interpret the result.

19. ● *Income Elasticity of Demand: Computer Usage* The demand for personal computers in the home goes up with household income. The following graph shows some data on computer usage together with the logarithmic model $q = 0.3454 \ln(x) - 3.047$, where q is the probability that a household with annual income x will have a computer.[51]

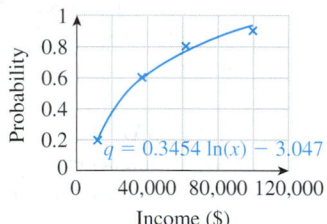

a. Compute the income elasticity of demand for computers, to two decimal places, for a household income of \$60,000 and interpret the result.

[49] Based on a quadratic regression of data from a 2001 survey. Source for data: New York Foundation of the Arts; www.nyfa.org/culturalblueprint/.

[50] Ibid.

[51] All figures are approximate. The model is a regression model, and x measures the probability that a given household will have one or more computers. Source for data: Income distribution: Luxembourg Income Study/*New York Times*, August 14, 1995, p. A9. Computer data: Forrester Research/ *The New York Times*, August 8, 1999, p. BU4.

● basic skills   ◆ challenging

b. As household income increases, how is income elasticity of demand affected?

c. How reliable is the given model of demand for incomes well above $120,000? Explain.

d. What can you say about E for incomes much larger than those shown?

20. ● *Income Elasticity of Demand: Internet Usage* The demand for Internet connectivity also goes up with household income. The following graph shows some data on Internet usage, together with the logarithmic model $q = 0.2802 \ln(x) - 2.505$, where q is the probability that a home with annual household income x will have an Internet connection.[52]

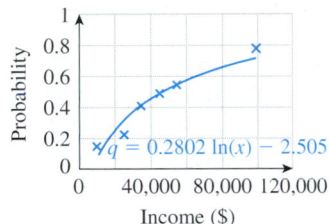

Income ($)

a. Compute the income elasticity of demand to two decimal places for a household income of $60,000 and interpret the result.

b. As household income increases, how is income elasticity of demand affected?

c. The logarithmic model shown above is not appropriate for incomes well above $100,000. Suggest a model that might be more appropriate.

d. In the model you propose, how does E behave for very large incomes?

21. *Income Elasticity of Demand (Based on a question on the GRE economics test)* If $Q = aP^\alpha Y^\beta$ is the individual's demand function for a commodity, where P is the (fixed) price of the commodity, Y is the individual's income, and $a, \alpha,$ and β are parameters, explain why β can be interpreted as the income elasticity of demand.

22. *College Tuition (From the GRE economics test)* A time-series study of the demand for higher education, using tuition charges as a price variable, yields the following result:

$$\frac{dq}{dp} \cdot \frac{p}{q} = -0.4$$

where p is tuition and q is the quantity of higher education. Which of the following is suggested by the result?

(A) As tuition rises, students want to buy a greater quantity of education.

(B) As a determinant of the demand for higher education, income is more important than price.

(C) If colleges lowered tuition slightly, their total tuition receipts would increase.

(D) If colleges raised tuition slightly, their total tuition receipts would increase.

(E) Colleges cannot increase enrollments by offering larger scholarships.

23. *Modeling Exponential Demand* As the new owner of a supermarket, you have inherited a large inventory of unsold imported Limburger cheese, and you would like to set the price so that your revenue from selling it is as large as possible. Previous sales figures of the cheese are shown in the following table:

Price per pound, p	$3.00	$4.00	$5.00
Monthly sales in pounds, q	407	287	223

a. Use the sales figures for the prices $3 and $5 per pound to construct a demand function of the form $q = Ae^{-bp}$, where A and b are constants you must determine. (Round A and b to two significant digits.)

b. Use your demand function to find the price elasticity of demand at each of the prices listed.

c. At what price should you sell the cheese in order to maximize monthly revenue?

d. If your total inventory of cheese amounts to only 200 pounds, and it will spoil one month from now, how should you price it in order to receive the greatest revenue? Is this the same answer you got in part (c)? If not, give a brief explanation.

24. *Modeling Exponential Demand* Repeat the preceding exercise, but this time use the sales figures for $4 and $5 per pound to construct the demand function.

Communication and Reasoning Exercises

25. ● Complete the following: When demand is inelastic, revenue will decrease if ____.

26. ● Complete the following: When demand has unit elasticity, revenue will decrease if ____ .

27. Given that the demand q is a differentiable function of the unit price p, show that the revenue $R = pq$ has a stationary point when

$$q + p\frac{dq}{dp} = 0$$

Deduce that the stationary points of R are the same as the points of unit price elasticity of demand. (Ordinarily, there is only one such stationary point, corresponding to the

---

[52] All figures are approximate, and the model is a regression model. The Internet connection figures were actually quoted as "share of consumers who use the Internet, by household income." SOURCES: Luxembourg Income Study/*The New York Times*, August 14, 1995, p. A9, Commerce Department, Deloitte & Touche Survey/*The New York Times,* November 24, 1999, p. C1.

absolute maximum of R.) [Hint: Differentiate R with respect to p.]

28. Given that the demand q is a differentiable function of income x, show that the quantity $R = q/x$ has a stationary point when

$$q - x\frac{dq}{dx} = 0$$

Deduce that stationary points of R are the same as the points of unit income elasticity of demand. [Hint: Differentiate R with respect to x.]

29. ◆ Your calculus study group is discussing price elasticity of demand, and a member of the group asks the following question: "Since elasticity of demand measures the response of demand to change in unit price, what is the difference between elasticity of demand and the quantity $-dq/dp$?" How would you respond?

30. ◆ Another member of your study group claims that unit price elasticity of demand need not always correspond to maximum revenue. Is he correct? Explain your answer.

● basic skills   ◆ challenging

Chapter **12** Review

KEY CONCEPTS

12.1 Maxima and Minima

Relative maximum, relative minimum *p. 853*

Absolute maximum, absolute minimum *p. 853*

Stationary points, singular points, endpoints *p. 854*

Finding and classifying maxima and minima *p. 855*

Finding absolute extrema on a closed interval *p. 860*

Using technology to locate approximate extrema *p. 861*

12.2 Applications of Maxima and Minima

Minimizing average cost *p. 865*

Maximizing area *p. 866*

Steps in solving optimization problems *p. 867*

Maximizing revenue *p. 868*

Optimizing resources *p. 869*

Allocation of labor *p. 871*

12.3 The Second Derivative and Analyzing Graphs

The second derivative of a function f is the derivative of the derivative of f, written as f''. *p. 878*

The acceleration of a moving object is the second derivative of the position function. *p. 878*

Acceleration due to gravity *p. 879*

Acceleration of sales *p. 880*

Concave up, concave down, point of inflection *p. 881*

Locating points of inflection *p. 882*

The point of diminishing returns *p. 882*

Features of a graph: x- and y-intercepts, relative extrema, points of inflection; behavior near points where the function is not defined, behavior at infinity *p. 883*

Analyzing a graph *p. 883*

12.4 Related Rates

If Q is a quantity changing over time t, then the derivative dQ/dt is the rate at which Q changes over time. *p. 893*

The expanding circle *p. 894*

Steps in solving related rates problems *p. 894*

The falling ladder *p. 895*

Average cost *p. 896*

Allocation of labor *p. 897*

12.5 Elasticity

Price elasticity of demand

$$E = -\frac{dq}{dp} \cdot \frac{p}{q}; \text{ demand is elastic}$$

if $E > 1$, inelastic if $E < 1$, has unit elasticity if $E = 1$ *p. 903*

Computing and interpreting elasticity, and maximizing revenue *p. 904*

Using technology to compute elasticity *p. 905*

Income elasticity of demand *p. 906*

REVIEW EXERCISES

In Exercises 1–8, find all the relative and absolute extrema of the given functions on the given domain (if supplied) or on the largest possible domain (if no domain is supplied).

1. $f(x) = 2x^3 - 6x + 1$ on $[-2, +\infty)$

2. $f(x) = x^3 - x^2 - x - 1$ on $(-\infty, \infty)$

3. $g(x) = x^4 - 4x$ on $[-1, 1]$

4. $f(x) = \dfrac{x+1}{(x-1)^2}$ on $[-2, 1) \cup (1, 2]$

5. $g(x) = (x-1)^{2/3}$

6. $g(x) = x^2 + \ln x$ on $(0, +\infty)$

7. $h(x) = \dfrac{1}{x} + \dfrac{1}{x^2}$

8. $h(x) = e^{x^2} + 1$

In Exercises 9–12, the graph of the function f or its derivative is given. Find the approximate x-coordinates of all relative extrema and points of inflection of the original function f (if any).

9. Graph of f:

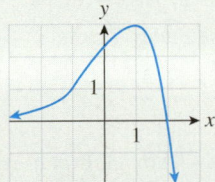

10. Graph of f:

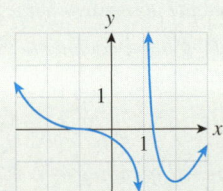

11. Graph of f':

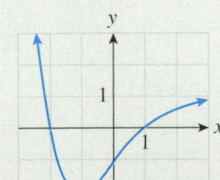

12. Graph of f':

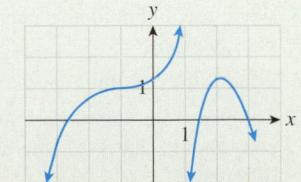

In Exercises 13 and 14, the graph of the second derivative of a function f is given. Find the approximate x-coordinates of all points of inflection of the original function f (if any).

13. Graph of f''

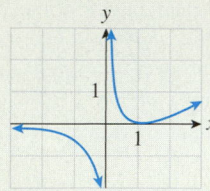

14. Graph of f''

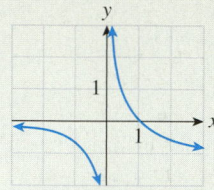

*In Exercises 15 and 16, the position s of a point (in meters) is given as a function of time t (in seconds). Find **(a)** its acceleration as a function of t and **(b)** its acceleration at the specified time.*

15. $s = \dfrac{2}{3t^2} - \dfrac{1}{t}; t = 1$

16. $s = \dfrac{4}{t^2} - \dfrac{3t}{4}; t = 2$

In Exercises 17–22, sketch the graph of the given function, indicating all relative and absolute extrema and points of inflection. Find the coordinates of these points exactly, where possible. Also indicate any horizontal and vertical asymptotes.

17. $f(x) = x^3 - 12x$ on $[-2, +\infty)$

18. $g(x) = x^4 - 4x$ on $[-1, 1]$

19. $f(x) = \dfrac{x^2 - 3}{x^3}$

20. $f(x) = (x - 1)^{2/3} + \dfrac{2x}{3}$

21. $g(x) = (x - 3)\sqrt{x}$

22. $g(x) = (x + 3)\sqrt{x}$

Applications

23. Revenue Demand for the latest best-seller at OHaganBooks.com, *A River Burns Through It,* is given by

$$q = -p^2 + 33p + 9 \qquad (18 \le p \le 28)$$

copies sold per week when the price is *p* dollars. What price should the company charge to obtain the largest revenue?

Profit *Taking into account storage and shipping, it costs OHaganBooks.com*

$$C = 9q + 100$$

dollars to sell q copies of A River Burns Through It *in a week.*

24. If demand is as in the preceding exercise, express the weekly profit earned by OHaganBooks.com from the sale of *A River Burns Through It* as a function of unit price *p*.

25. What price should the company charge to get the largest weekly profit?

26. What is the maximum possible weekly profit?

27. Compare your answer to Exercise 25 with the price the company should charge to obtain the largest revenue. Explain any difference.

28. Box Design The sales department at OHaganBooks.com, which has decided to send chocolate lobsters to each of its customers, is trying to design a shipping box with a square base. It has a roll of cardboard 36 inches wide from which to make the boxes. Each box will be obtained by cutting out corners from a rectangle of cardboard as shown in the following diagram:

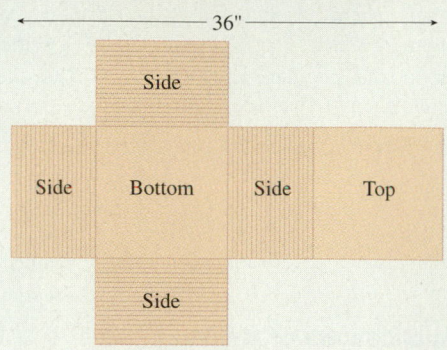

(Notice that the top and bottom of each box will be square, but the sides will not necessarily be square.) What are the dimensions of the boxes with the largest volume that can be made in this way? What is the maximum volume?

Elasticity of Demand *(Compare Exercise 23). Demand for the latest best-seller at OHaganBooks.com,* A River Burns Through It, *is given by*

$$q = -p^2 + 33p + 9 \qquad (18 \le p \le 28)$$

copies sold per week when the price is p dollars.

29. Find the price elasticity of demand as a function of *p*.

30. Find the elasticity of demand for this book at a price of $20 and at a price of $25. (Round your answers to two decimal places.) Interpret the answers.

31. What price should the company charge to obtain the largest revenue?

Elasticity of Demand *Last year OHaganBooks.com experimented with an online subscriber service, Red On Line (ROL) for its electronic book service. The consumer demand for ROL was modeled by the equation*

$$q = 1000e^{-p^2 + p}$$

where p was the monthly access charge, and q is the number of subscribers.

32. Obtain a formula for the price elasticity of demand, *E*, for ROL services.

33. Compute the elasticity of demand if the monthly access charge is set at $2 per month. Interpret the result.

34. How much should the company have charged in order to obtain the maximum monthly revenue? What would this revenue have been?

Sales OHaganBooks.com modeled its weekly sales over a period of time with the function

$$s(t) = 6053 + \frac{4474}{1 + e^{-0.55(t-4.8)}}$$

where t is the time in weeks. Following are the graphs of s, s′, and s″:

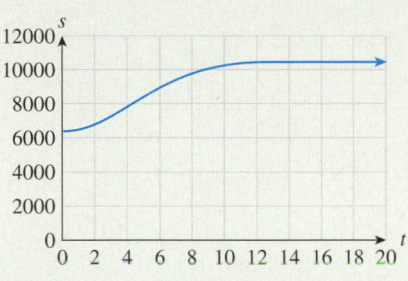

Graph of s

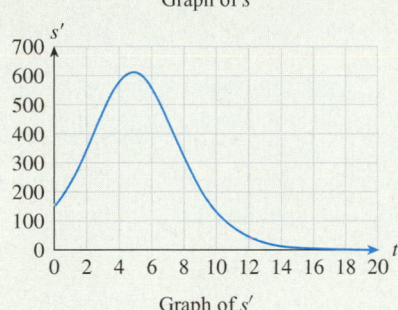

Graph of s′

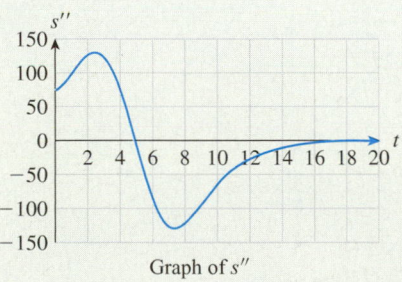

Graph of s″

35. Estimate when, to the nearest week, the weekly sales were growing fastest.

36. To what features on the graphs of s, s′, and s″ does your answer to part (a) correspond?

37. The graph of s has a horizontal asymptote. What is the approximate value (s-coordinate) of this asymptote, and what is its significance in terms of weekly sales at OHaganBooks.com?

38. The graph of s′ has a horizontal asymptote. What is the value (s′-coordinate) of this asymptote, and what is its significance in terms of weekly sales at OHaganBooks.com?

39. *Chance Encounter* Marjory Duffin is walking north towards the corner entrance of OHaganBooks.com company headquarters at 5 ft/sec, while John O'Hagan is walking west toward the same entrance, also at 5 ft/sec. How fast is their distance apart decreasing when:

   a. Each of them is 2 ft from the corner?
   b. Each of them is 1 ft. from the corner?
   c. Each of them is h ft. from the corner?
   d. They collide on the corner?

40. *Company Logos* OHaganBooks.com's website has an animated graphic with its name in a rectangle whose height and width change; on either side of the rectangle are semicircles, as in the figure, whose diameters are the same as the height of the rectangle.

For reasons too complicated to explain, the designer wanted the combined area of the rectangle and semicircles to remain constant. At one point during the animation, the width of the rectangle is 1 inch, growing at a rate of 0.5 inches per second, while the height is 3 inches. How fast is the height changing?

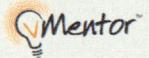 Do you need a live tutor for homework problems? Access vMentor on the ThomsonNOW! website at **www.thomsonedu.com** for one-on-one tutoring from a mathematics expert.

CASE STUDY: Production Lot Size Management

Your publishing company, Knockem Dead Paperbacks, Inc., is about to release its next best-seller, *Henrietta's Heaving Heart* by Celestine A. Lafleur. The company expects to sell 100,000 books each month in the next year. You have been given the job of scheduling print runs to meet the anticipated demand and minimize total costs to the company. Each print run has a setup cost of $5000, each book costs $1 to produce, and monthly storage costs for books awaiting shipment average 1¢ per book. What will you do?

If you decide to print all 1,200,000 books (the total demand for the year, 100,000 books per month for 12 months) in a single run at the start of the year and sales run as predicted, then the number of books in stock would begin at 1,200,000 and decrease to zero by the end of the year, as shown in Figure 40.

Figure **40**

On average, you would be storing 600,000 books for 12 months at 1¢ per book, giving a total storage cost of $600,000 \times 12 \times .01 = \$72,000$. The setup cost for the single print run would be $5000. When you add to these the total cost of producing 1,200,000 books at $1 per book, your total cost would be $1,277,000.

If, on the other hand, you decide to cut down on storage costs by printing the books in two runs of 600,000 each, you would get the picture shown in Figure 41.

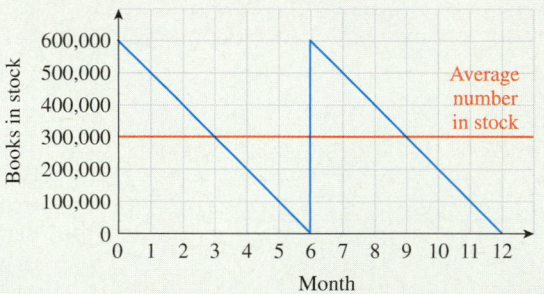

Figure **41**

Now, the storage cost would be cut in half because on average there would be only 300,000 books in stock. Thus, the total storage cost would be $36,000, and the setup cost would double to $10,000 (because there would now be two runs). The production costs would be the same: 1,200,000 books @ $1 per book. The total cost would therefore be reduced to $1,246,000, a savings of $31,000 compared to your first scenario.

"Aha!" you say to yourself, after doing these calculations. "Why not drastically cut costs by setting up a run every month?" You calculate that the setup costs alone would be $12 \times \$5000 = \$60,000$, which is already more than the setup plus storage costs for two runs, so a run every month will cost too much. Perhaps, then, you should investigate three runs, four runs, and so on, until you find the lowest cost. This strikes you as too laborious a process, especially considering that you will have to do it all over again when planning for Lafleur's sequel, *Lorenzo's Lost Love,* due to be released next year. Realizing that this is an optimization problem, you decide to use some calculus to help you come up with a *formula* that you can use for all future plans. So you get to work.

Instead of working with the number 1,200,000, you use the letter N so that you can be as flexible as possible. (What if *Lorenzo's Lost Love* sells more copies?) Thus, you have a total of N books to be produced for the year. You now calculate the total cost of using x print runs per year. Because you are to produce a total of N books in x print runs, you will have to produce N/x books in each print run. N/x is called the **lot size.** As you can see from the diagrams above, the average number of books in storage will be half that amount, $N/(2x)$.

Now you can calculate the total cost for a year. Write P for the setup cost of a single print run ($P = \$5000$ in your case) and c for the *annual* cost of storing a book (to convert all of the time measurements to years; $c = \$0.12$ here). Finally, write b for the cost of producing a single book ($b = \$1$ here). The costs break down as follows.

Setup Costs: x print runs @ P dollars per run: $\qquad\qquad\qquad Px$

Storage Costs: $N/(2x)$ books stored @ c dollars per year: $\quad cN/(2x)$

Production Costs: N books @ b dollars per book: $\qquad\qquad \dfrac{Nb}{}$

$$\textbf{Total Cost:}\quad Px + \frac{cN}{2x} + Nb$$

Remember that P, N, c, and b are all constants and x is the only variable. Thus, your cost function is

$$C(x) = Px + \frac{cN}{2x} + Nb$$

and you need to find the value of x that will minimize $C(x)$. But that's easy! All you need to do is find the relative extrema and select the absolute minimum (if any).

The domain of $C(x)$ is $(0, +\infty)$ because there is an x in the denominator and x can't be negative. To locate the extrema, you start by locating the critical points:

$$C'(x) = P - \frac{cN}{2x^2}$$

The only singular point would be at $x = 0$, but 0 is not in the domain. To find stationary points, you set $C'(x) = 0$ and solve for x:

$$P - \frac{cN}{2x^2} = 0$$

$$2x^2 = \frac{cN}{P}$$

so

$$x = \sqrt{\frac{cN}{2P}}$$

There is only one stationary point, and there are no singular points or endpoints. To graph the function you will need to put in numbers for the various constants. Substituting $N = 1{,}200{,}000$, $P = 5000$, $c = 0.12$, and $b = 1$, you get

$$C(x) = 5000x + \frac{72{,}000}{x} + 1{,}200{,}000$$

with the stationary point at

$$x = \sqrt{\frac{(0.12)(1{,}200{,}000)}{2(5000)}} \approx 3.79$$

The total cost at the stationary point is

$$C(3.79) \approx 1{,}240{,}000$$

You now graph $C(x)$ in a window that includes the stationary point, say, $0 \le x \le 12$ and $1{,}100{,}000 \le C \le 1{,}500{,}000$, getting Figure 42.

From the graph, you can see that the stationary point is an absolute minimum. In the graph it appears that the graph is always concave up, which also tells you that your

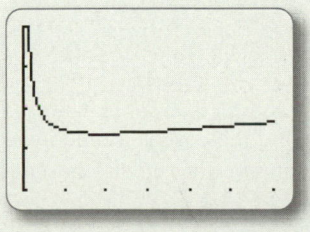

Figure **42**

stationary point is a minimum. You can check the concavity by computing the second derivative:

$$C''(x) = \frac{cN}{x^3} > 0$$

The second derivative is always positive because c, N, and x are all positive numbers, so indeed the graph is always concave up. Now you also know that it works regardless of the particular values of the constants.

So now you are practically done! You know that the absolute minimum cost occurs when you have $x \approx 3.79$ print runs per year. Don't be disappointed that the answer is not a whole number; whole number solutions are rarely found in real scenarios. What the answer (and the graph) do indicate is that either 3 or 4 print runs per year will cost the least money. If you take $x = 3$, you get a total cost of

$$C(3) = \$1{,}239{,}000$$

If you take $x = 4$, you get a total cost of

$$C(4) = \$1{,}238{,}000$$

So, four print runs per year will allow you to minimize your total costs.

Exercises

1. *Lorenzo's Lost Love* will sell 2,000,000 copies in a year. The remaining costs are the same. How many print runs should you use now?

2. In general, what happens to the number of runs that minimizes cost if both the setup cost and the total number of books are doubled?

3. In general, what happens to the number of runs that minimizes cost if the setup cost increases by a factor of 4?

4. Assuming that the total number of copies and storage costs are as originally stated, find the setup cost that would result in a single print run.

5. Assuming that the total number of copies and setup cost are as originally stated, find the storage cost that would result in a print run each month.

6. In Figure 41 we assumed that all the books in each run were manufactured in a very short time; otherwise the figure might have looked more like Figure 43, which shows the inventory, assuming a slower rate of production.

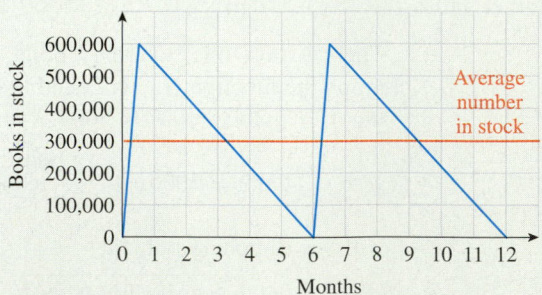

Figure **43**

How would this affect the answer?

7. Referring to the general situation discussed in the text, find the cost as a function of the total number of books produced, assuming that the number of runs is chosen to minimize total cost. Also find the average cost per book.

8. Let $\bar{C}$ be the average cost function found in the preceding exercise. Calculate $\lim_{N \to +\infty} \bar{C}(N)$ and interpret the result.

TI-83/84 Technology Guide

Section 12.5

Example 1 (a) Suppose that the demand equation for Bobby Dolls is given by $q = 216 - p^2$, where p is the price per doll in dollars and q is the number of dolls sold per week. Compute the price elasticity of demand when $p = \$5$ and $p = \$10$, and interpret the results.

Solution with Technology The TI-83/84 function `nDeriv` can be used to compute approximations of E at various prices.

1. Set

```
Y₁=216-X²          Demand equation
Y₂=-nDeriv(Y₁,X,X)*X/Y₁    Formula for E
```

2. Use the table feature to list the values of elasticity for a range of prices, For part (a) we chose values of X close to 5:

X	Y₁	Y₂
4.7	193.91	.22784
4.8	192.96	.23881
4.9	191.99	.25012
5	191	.26178
5.1	189.99	.2738
5.2	188.96	.2862
5.3	187.91	.29897

Y₂=.261780104712

EXCEL Technology Guide

Section 12.5

Example 1 (a) Suppose that the demand equation for Bobby Dolls is given by $q = 216 - p^2$, where p is the price per doll in dollars and q is the number of dolls sold per week. Compute the price elasticity of demand when $p = \$5$ and $p = \$10$, and interpret the results.

Solution with Technology To approximate E in Excel, we can use the following approximation of E.

$$E \approx \frac{\text{Percentage decrease in demand}}{\text{Percentage increase in price}} \approx -\frac{\left(\dfrac{\Delta q}{q}\right)}{\left(\dfrac{\Delta p}{p}\right)}$$

The smaller Δp is, the better the approximation. Let's use $\Delta p = 1\cent$, or 0.01 (which is small compared with the typical prices we consider—around $5 to $10).

1. We start by setting up our worksheet to list a range of prices, in increments of Δp, on either side of a price in which we are interested, such as $p_0 = \$5$:

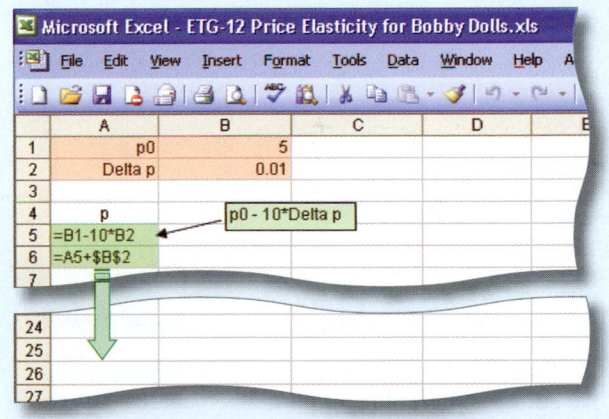

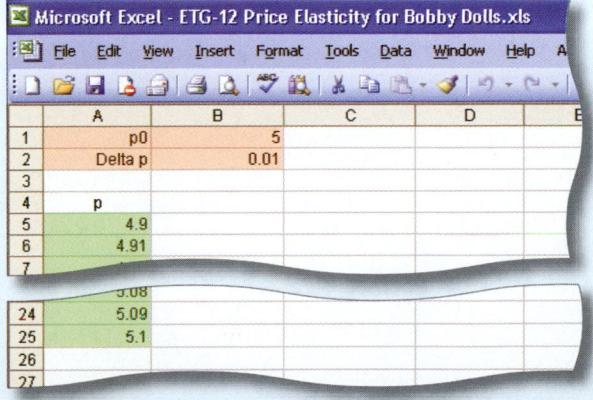

We start in cell A5 with the formula for $p_0 - 10\Delta p$ and then successively add Δp going down column A. You will find that the value $p_0 = 5$ appears midway down the list.

2. Next, we compute the corresponding values for the demand q in Column B.

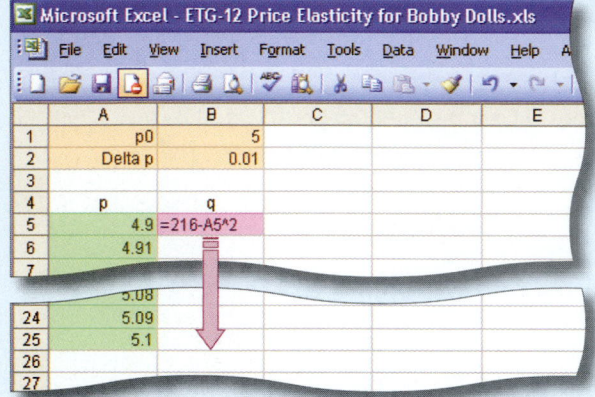

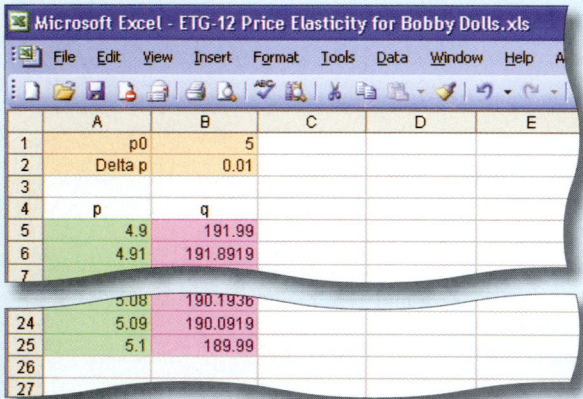

3. We add two new columns for the percentage changes in p and q. The formula shown in cell C5 is copied down columns C and D, to Row 24. (Why not Row 25?)

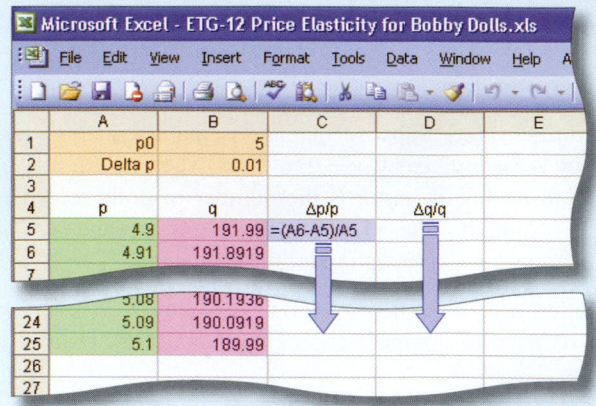

4. The elasticity can now be computed in column E as shown:

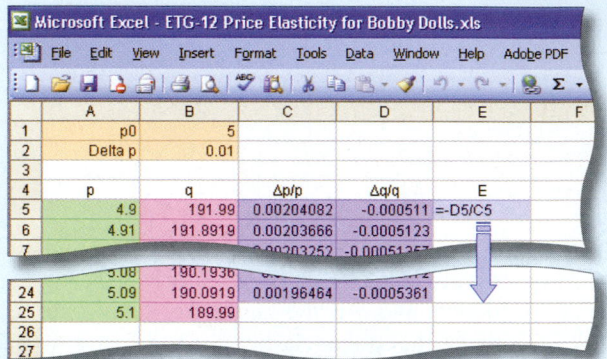

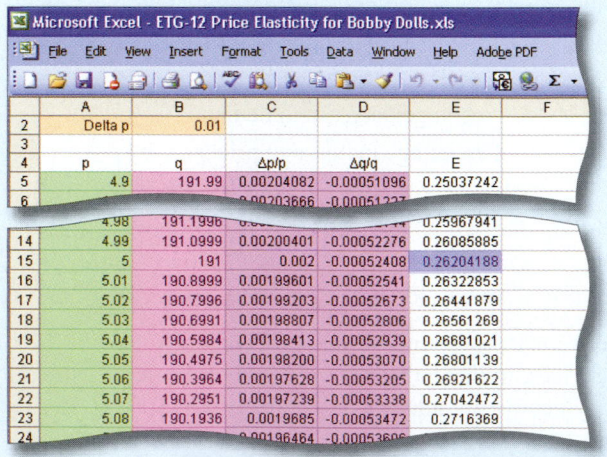

13

The Integral

CASE STUDY Wage Inflation

As assistant personnel manager for a large corporation, you have been asked to estimate the average annual wage earned by a worker in your company, from the time the worker is hired to the time the worker retires. You have data about wage increases. How will you estimate this average?

Digital Vision/Getty Images

Introduction

Roughly speaking, calculus is divided into two parts: **differential calculus** (the calculus of derivatives) and **integral calculus,** which is the subject of this chapter and the next. Integral calculus is concerned with problems that are in some sense the reverse of the problems seen in differential calculus. For example, where differential calculus shows how to compute the rate of change of a quantity, integral calculus shows how to find the quantity if we know its rate of change. This idea is made precise in the **Fundamental Theorem of Calculus.** Integral calculus and the Fundamental Theorem of Calculus allow us to solve many problems in economics, physics, and geometry, including one of the oldest problems in mathematics—computing areas of regions with curved boundaries.

13.1 The Indefinite Integral

Suppose that we knew the marginal cost to manufacture an item and we wanted to reconstruct the cost function. We would have to *reverse* the process of differentiation, to go from the derivative (the marginal cost function) back to the original function (the total cost). We'll first discuss how to do that and then look at some applications.

Here is an example: If the derivative of $F(x)$ is $4x^3$, what was $F(x)$? We recognize $4x^3$ as the derivative of x^4. So, we might have $F(x) = x^4$. However, $F(x) = x^4 + 7$ works just as well. In fact, $F(x) = x^4 + C$ works for any number C. Thus, there are *infinitely many* possible answers to this question.

In fact, we will see shortly that the formula $F(x) = x^4 + C$ covers *all* possible answers to the question. Let's give a name to what we are doing.

Antiderivative

An **antiderivative** of a function f is a function F such that $F' = f$.

quick Examples

1. An antiderivative of $4x^3$ is x^4.   Because the derivative of x^4 is $4x^3$

2. Another antiderivative of $4x^3$ is $x^4 + 7$.   Because the derivative of $x^4 + 7$ is $4x^3$

3. An antiderivative of $2x$ is $x^2 + 12$.   Because the derivative of $x^2 + 12$ is $2x$

Thus,

If the derivative of A(x) is B(x), then an antiderivative of B(x) is A(x).

We call the set of *all* antiderivatives of a function the **indefinite integral** of the function.

Indefinite Integral

$$\int f(x)\,dx$$

is read "the **indefinite integral** of $f(x)$ with respect to x" and stands for the set of all antiderivatives of f. Thus, $\int f(x)\,dx$ is a *collection of functions;* it is not a single function,

---

or a number. The function *f* that is being **integrated** is called the **integrand,** and the variable *x* is called the **variable of integration.**

quick Examples

1. $\int 4x^3\, dx = x^4 + C$ Every possible antiderivative of $4x^3$ has the form $x^4 + C$.

2. $\int 2x\, dx = x^2 + C$ Every possible antiderivative of $2x$ has the form $x^2 + C$.

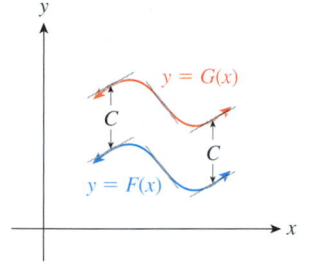

Figure **1**

The **constant of integration** C reminds us that we can add any constant and get a different antiderivative.

Q: *If $F(x)$ is one antiderivative of $f(x)$, why must all other antiderivatives have the form $F(x) + C$?*

A: Suppose $F(x)$ and $G(x)$ are both antiderivatives of $f(x)$, so that $F'(x) = G'(x)$. Consider what this means by looking at Figure 1. ∎

If $F'(x) = G'(x)$ for all x, then F and G have the *same slope* at each value of x. This means that their graphs must be *parallel* and hence remain exactly the same vertical distance apart. But that is the same as saying that the functions differ by a constant—that is, that $G(x) = F(x) + C$ for some constant C.[1]

Example **1** Indefinite Integral

Check that $\displaystyle\int x\, dx = \frac{x^2}{2} + C$.

Solution We check our answer by taking the derivative of the right-hand side:

$$\frac{d}{dx}\left(\frac{x^2}{2} + C\right) = \frac{2x}{2} + 0 = x \qquad ✔$$

Because the derivative of the right-hand side is the integrand x, we can conclude that $\displaystyle\int x\, dx = \frac{x^2}{2} + C$, as claimed.

Now, we would like to make the process of finding indefinite integrals (antiderivatives) more mechanical. For example, it would be nice to have a power rule for indefinite integrals similar to the one we already have for derivatives. Two cases suggested by the examples above are:

$$\int x\, dx = \frac{x^2}{2} + C \qquad\qquad \int x^3\, dx = \frac{x^4}{4} + C$$

You should check the last equation by taking the derivative of its right-hand side. These cases suggest the following general statement:

---

[1] This argument can be turned into a more rigorous proof—that is, a proof that does not rely on geometric concepts such as "parallel graphs." We should also say that the result requires that F and G have the same derivative *on an interval* $[a, b]$.

Power Rule for the Indefinite Integral, Part I

$$\int x^n \, dx = \frac{x^{n+1}}{n+1} + C \qquad (\text{if } n \neq -1)$$

In Words

To find the integral of x^n, add 1 to the exponent, and then divide by the new exponent. This rule works provided that n is not -1.

quick Examples

1. $\displaystyle\int x^{55} \, dx = \frac{x^{56}}{56} + C$

2. $\displaystyle\int \frac{1}{x^{55}} \, dx = \int x^{-55} \, dx$ Exponent form

   $$= \frac{x^{-54}}{-54} + C \qquad \text{When we add 1 to } -55, \text{ we get } -54, \textit{not } -56.$$

   $$= -\frac{1}{54x^{54}} + C \qquad \text{See below}^*$$

3. $\displaystyle\int 1 \, dx = x + C$ Because $1 = x^0$. This is an important special case.

---

* We are glossing over a subtlety here: The constant of integration C can be different for $x < 0$ and $x > 0$ because the graph breaks at $x = 0$. In general, our understanding will be that the constant of integration may be different on disconnected intervals of the domain.

Notes

1. The integral $\displaystyle\int 1 \, dx$ is commonly written as $\displaystyle\int dx$. Similarly, the integral $\displaystyle\int \frac{1}{x^{55}} \, dx$ may be written as $\displaystyle\int \frac{dx}{x^{55}}$.

2. We can easily check the power rule formula by taking the derivative of the right-hand side:

$$\frac{d}{dx}\left(\frac{x^{n+1}}{n+1} + C\right) = \frac{(n+1)x^n}{n+1} = x^n \qquad ✔ \qquad ■$$

Q: *What is the reason for the restriction $n \neq -1$?*

A: The right-hand side of the power rule formula has $n + 1$ in the denominator, and thus makes no sense if $n = -1$. This leaves us not yet knowing how to compute

$$\int x^{-1} \, dx = \int \frac{1}{x} \, dx$$

Computing this derivative amounts to finding a function whose derivative is $1/x$. Prodding our memories a little, we recall that $\ln x$ has derivative $1/x$. In fact, as we pointed out when we first discussed it, $\ln|x|$ also has derivative $1/x$, but it has the advantage that its domain is the same as that of $1/x$. Thus, we can fill in the missing case as follows:

$$\int x^{-1} \, dx = \ln |x| + C \qquad\qquad ■$$

This and two other indefinite integrals that come from corresponding formulas for differentiation are summarized here:

Power Rule for the Indefinite Integral, Part II

$$\int x^{-1}\, dx = \ln |x| + C \qquad \text{Equivalently, } \int \frac{1}{x}\, dx = \ln |x| + C.$$

Indefinite Integral of e^x and b^x

$$\int e^x\, dx = e^x + C \qquad \text{Because } \frac{d}{dx}(e^x) = e^x$$

If b is any positive number other than 1, then

$$\int b^x\, dx = \frac{b^x}{\ln b} + C \qquad \text{Because } \frac{d}{dx}\left(\frac{b^x}{\ln b}\right) = \frac{b^x \ln b}{\ln b} = b^x$$

quick Example

$$\int 2^x\, dx = \frac{2^x}{\ln 2} + C$$

For more complicated functions, like $2x^3 + 6x^5 - 1$, we need the following rules for integrating sums, differences, and constant multiples.

Sums, Differences, and Constant Multiples

Sum and Difference Rules

$$\int [f(x) \pm g(x)]\, dx = \int f(x)\, dx \pm \int g(x)\, dx$$

In Words: The integral of a sum is the sum of the integrals, and the integral of a difference is the difference of the integrals.

Constant Multiple Rule

$$\int k f(x)\, dx = k \int f(x)\, dx \qquad (k \text{ constant})$$

In Words: The integral of a constant times a function is the constant times the integral of the function. (In other words, the constant "goes along for the ride.")

quick Examples

Sum Rule: $\displaystyle\int (x^3 + 1)\, dx = \int x^3\, dx + \int 1\, dx = \frac{x^4}{4} + x + C \qquad f(x) = x^3;\ g(x) = 1$

Constant Multiple Rule: $\displaystyle\int 5x^3\, dx = 5 \int x^3\, dx = 5\frac{x^4}{4} + C \qquad k = 5;\ f(x) = x^3$

Constant Multiple Rule: $\displaystyle\int 4\, dx = 4 \int 1\, dx = 4x + C \qquad k = 4;\ f(x) = 1$

Constant Multiple Rule: $\displaystyle\int 4e^x\, dx = 4 \int e^x\, dx = 4e^x + C \qquad k = 4;\ f(x) = e^x$

Proof of the Sum Rule

We saw above that if two functions have the same derivative, they differ by a (possibly zero) constant. Look at the rule for sums:

$$\int [f(x) + g(x)]\, dx = \int f(x)\, dx + \int g(x)\, dx$$

If we take the derivative of the left-hand side with respect to x, we get the integrand, $f(x) + g(x)$. If we take the derivative of the right-hand side, we get

$$\frac{d}{dx}\left[\int f(x)\, dx + \int g(x)\, dx\right] = \frac{d}{dx}\left[\int f(x)\, dx\right] + \frac{d}{dx}\left[\int g(x)\, dx\right]$$

Derivative of a sum = Sum of derivatives

$$= f(x) + g(x)$$

Because the left- and right-hand sides have the same derivative, they differ by a constant. But, because both expressions are indefinite integrals, adding a constant does not affect their value, so they are the same as indefinite integrals.

Notice that a key step in the proof was the fact that the derivative of a sum is the sum of the derivatives.

A similar proof works for the difference and constant multiple rule.

Example 2 Using the Sum and Difference Rules

Find the integrals

a. $\displaystyle\int (x^3 + x^5 - 1)\, dx$ **b.** $\displaystyle\int \left(x^{2.1} + \frac{1}{x^{1.1}} + \frac{1}{x} + e^x\right) dx$ **c.** $\displaystyle\int (e^x + 3^x - 1)\, dx$

Solution

a. $\displaystyle\int (x^3 + x^5 - 1)\, dx = \int x^3\, dx + \int x^5\, dx - \int 1\, dx$ Sum/difference rule

$$= \frac{x^4}{4} + \frac{x^6}{6} - x + C$$ Power rule

b. $\displaystyle\int \left(x^{2.1} + \frac{1}{x^{1.1}} + \frac{1}{x} + e^x\right) dx$

$$= \int (x^{2.1} + x^{-1.1} + x^{-1} + e^x)\, dx$$ Exponent form

$$= \int x^{2.1}\, dx + \int x^{-1.1}\, dx + \int x^{-1}\, dx + \int e^x\, dx$$ Sum rule

$$= \frac{x^{3.1}}{3.1} + \frac{x^{-0.1}}{-0.1} + \ln|x| + e^x + C$$ Power rule and exponential rule

$$= \frac{x^{3.1}}{3.1} - \frac{10}{x^{0.1}} + \ln|x| + e^x + C$$

c. $\displaystyle\int (e^x + 3^x - 1)\, dx = \int e^x\, dx + \int 3^x\, dx - \int 1\, dx$ Sum/difference rule

$$= e^x + \frac{3^x}{\ln 3} - x + C$$ Power rule and exponential rule

+*Before we go on...* You should check each of the answers in Example 2 by differentiating.

Q: *Why is there only a single arbitrary constant C in each of the answers?*

A: We could have written the answer to part (a) as

$$\frac{x^4}{4} + D + \frac{x^6}{6} + E - x + F$$

where D, E, and F are all arbitrary constants. Now suppose, for example, we set $D = 1$, $E = -2$, and $F = 6$. Then the particular antiderivative we get is $x^4/4 + x^6/6 - x + 5$, which has the form $x^4/4 + x^6/6 - x + C$. Thus, we could have chosen the single constant C to be 5 and obtained the same answer. In other words, the answer $x^4/4 + x^6/6 - x + C$ is just as general as the answer $x^4/4 + D + x^6/6 + E - x + F$, but simpler. ∎

In practice we do not explicitly write the integral of a sum as a sum of integrals but just "integrate term by term," much as we learned to differentiate term by term.

Example 3 Combining the Rules

Find the integrals

a. $\int (10x^4 + 2x^2 - 3e^x)\, dx$ **b.** $\int \left(\frac{2}{x^{0.1}} + \frac{x^{0.1}}{2} - \frac{3}{4x} \right) dx$ **c.** $\int [3e^x - 2(1.2^x)]\, dx$

Solution

a. We need to integrate separately each of the terms $10x^4$, $2x^2$, and $3e^x$. To integrate $10x^4$ we use the rules for constant multiples and powers:

$$\int 10x^4\, dx = 10 \int x^4\, dx = 10\frac{x^5}{5} + C = 2x^5 + C$$

The other two terms are similar. We get

$$\int (10x^4 + 2x^2 - 3e^x)\, dx = 10\frac{x^5}{5} + 2\frac{x^3}{3} - 3e^x + C = 2x^5 + \frac{2}{3}x^3 - 3e^x + C$$

b. We first convert to exponent form and then integrate term by term:

$$\int \left(\frac{2}{x^{0.1}} + \frac{x^{0.1}}{2} - \frac{3}{4x} \right) dx = \int \left(2x^{-0.1} + \frac{1}{2}x^{0.1} - \frac{3}{4}x^{-1} \right) dx \qquad \text{Exponent form}$$

$$= 2\frac{x^{0.9}}{0.9} + \frac{1}{2}\frac{x^{1.1}}{1.1} - \frac{3}{4}\ln|x| + C \qquad \text{Integrate term by term.}$$

$$= \frac{20x^{0.9}}{9} + \frac{x^{1.1}}{2.2} - \frac{3}{4}\ln|x| + C \qquad \text{Back to rational form}$$

c. $\int [3e^x - 2(1.2^x)]\, dx = 3e^x - 2\frac{1.2^x}{\ln(1.2)} + C$

Example 4 Different Variable Name

Find $\displaystyle\int \left(\frac{1}{u} + \frac{1}{u^2} \right) du$.

Solution

This integral may look a little strange because we are using the letter u instead of x, but there is really nothing special about x. Using u as the variable of integration, we get

$$\int \left(\frac{1}{u} + \frac{1}{u^2} \right) du = \int (u^{-1} + u^{-2})\, du \qquad \text{Exponent form.}$$

$$= \ln |u| + \frac{u^{-1}}{-1} + C \qquad \text{Integrate term by term.}$$

$$= \ln |u| - \frac{1}{u} + C \qquad \text{Simplify the result.}$$

+ *Before we go on...* When we compute an indefinite integral, we want the independent variable in the answer to be the same as the variable of integration. Thus, if the integral in Example 4 had been written in terms of x rather than u, we would have written

$$\int \left(\frac{1}{x} + \frac{1}{x^2} \right) dx = \ln |x| - \frac{1}{x} + C \qquad \blacksquare$$

Application: Cost and Marginal Cost

Rudi Von Briel/PhotoEdit

Example 5 Finding Cost from Marginal Cost

The marginal cost to produce baseball caps at a production level of x caps is $4 - 0.001x$ dollars per cap, and the cost of producing 100 caps is \$500. Find the cost function.

Solution We are asked to find the cost function $C(x)$, given that the *marginal* cost function is $4 - 0.001x$. Recalling that the marginal cost function is the derivative of the cost function, we can write

$$C'(x) = 4 - 0.001x$$

and must find $C(x)$. Now $C(x)$ must be an antiderivative of $C'(x)$, so

$$C(x) = \int (4 - 0.001x)\, dx$$

$$= 4x - 0.001\frac{x^2}{2} + K \qquad \text{K is the constant of integration.}^{*}$$

$$= 4x - 0.0005x^2 + K$$

Now, unless we have a value for K, we don't really know what the cost function is. However, there is another piece of information we have ignored: The cost of producing 100 baseball caps is \$500. In symbols

$$C(100) = 500$$

---

* We used K and not C for the constant of integration because we are using C for cost.

Substituting in our formula for $C(x)$, we have

$$C(100) = 4(100) - 0.0005(100)^2 + K$$
$$500 = 395 + K$$
$$K = 105$$

Now that we know what K is, we can write down the cost function:

$$C(x) = 4x - 0.0005x^2 + 105$$

+ *Before we go on...* Let us consider the significance of the constant term 105 in Example 5. If we substitute $x = 0$ into the cost function, we get

$$C(0) = 4(0) - 0.0005(0)^2 + 105 = 105$$

Thus, $105 is the cost of producing zero items; in other words, it is the **fixed cost.** ■

Application: Motion in a Straight Line

An important application of the indefinite integral is to the study of motion. The application of calculus to problems about motion is an example of the intertwining of mathematics and physics. We begin by bringing together some facts, scattered through the last several chapters, that have to do with an object moving in a straight line, and then restating them in terms of antiderivatives.

Position, Velocity, and Acceleration: Derivative Form

If $s = s(t)$ is the **position** of an object at time t, then its **velocity** is given by the derivative

$$v = \frac{ds}{dt}$$

In Words: Velocity is the derivative of position.
The **acceleration** of an object is given by the derivative

$$a = \frac{dv}{dt}$$

In Words: Acceleration is the derivative of velocity.

Position, Velocity, and Acceleration: Integral Form

$$s(t) = \int v(t)\, dt \qquad \text{Because } v = \frac{ds}{dt}$$

$$v(t) = \int a(t)\, dt \qquad \text{Because } a = \frac{dv}{dt}$$

quick Example

If the velocity of a particle moving in a straight line is given by $v(t) = 4t + 1$, then its position after t seconds is given by $s(t) = \int v(t)\, dt = \int (4t + 1)\, dt = 2t^2 + t + C$.

Example 6 Motion in a Straight Line

a. The velocity of a particle moving along in a straight line is given by $v(t) = 4t + 1$ m/sec. Given that the particle is at position $s = 2$ meters at time $t = 1$, find an expression for s in terms of t.

b. For a freely falling body experiencing no air resistance and zero initial velocity, find an expression for the velocity v in terms of t. [Note: On Earth, a freely falling body experiencing no air resistance accelerates downward at approximately 9.8 meters per second per second, or 9.8 m/s^2 (or 32 ft/s^2).]

Solution

a. As we saw in the Quick Example above, the position of the particle after t seconds is given by

$$s(t) = \int v(t)\, dt$$

$$= \int (4t + 1)\, dt = 2t^2 + t + C$$

But what is the value of C? Now, we are told that the particle is at position $s = 2$ at time $t = 1$. In other words, $s(1) = 2$. Substituting this into the expression for $s(t)$ gives

$$2 = 2(1)^2 + 1 + C$$

So $C = -1$

Hence the position after t seconds is given by

$$s(t) = 2t^2 + t - 1 \text{ meters.}$$

b. Let's measure heights above the ground as positive, so that a rising object has positive velocity and the acceleration due to gravity is negative (it causes the upward velocity to decrease in value). Thus, the acceleration of the stone is given by

$$a(t) = -9.8 \text{ m/s}^2$$

We wish to know the velocity, which is an antiderivative of acceleration, so we compute

$$v(t) = \int a(t)\, dt = \int (-9.8)\, dt = -9.8t + C$$

To find the value of C, we use the given information that at time $t = 0$ the velocity is $0 : v(0) = 0$. Substituting this into the expression for $v(t)$ gives

$$0 = -9.8(0) + C$$

so $C = 0$

Hence the velocity after t seconds is given by

$$v(t) = -9.8t \text{ m/s}$$

Example 7 Motion in a Straight Line Under Gravity

You are standing on the edge of a cliff and toss a stone upward at a speed of 30 feet per second.

a. Find the stone's velocity as a function of time. How fast and in what direction is it going after 5 seconds? (Neglect the effects of air resistance.)

b. Find the position of the stone as a function of time. Where will it be after 5 seconds?

c. When and where will the stone reach its zenith, its highest point?

Solution

a. This is similar to Example 6(b): Measuring height above the ground as positive, the acceleration of the stone is given by $a(t) = -32$ ft/s^2, and so

$$v(t) = \int (-32)\, dt = -32t + C$$

To obtain C, we use the fact that you tossed the stone upward at 30 ft/s; that is, when $t = 0$, $v = 30$, or $v(0) = 30$. Thus,

$$30 = v(0) = -32(0) + C$$

so $C = 30$ and the formula for velocity is $v(t) = -32t + 30$ ft/sec. In particular, after 5 seconds the velocity will be

$$v(5) = -32(5) + 30 = -130 \text{ ft/s}$$

After 5 seconds the stone is *falling* with a speed of 130 ft/s.

b. We wish to know the position, but position is an antiderivative of velocity. Thus,

$$s(t) = \int v(t)\, dt = \int (-32t + 30)\, dt = -16t^2 + 30t + C$$

Now to find C, we need to know the initial position $s(0)$. We are not told this, so let's measure heights so that the initial position is zero. Then

$$0 = s(0) = C$$

and $s(t) = -16t^2 + 30t$. In particular, after 5 seconds the stone has a height of

$$s(5) = -16(5)^2 + 30(5) = -250 \text{ ft}$$

In other words, the stone is now 250 ft *below* where it was when you first threw it, as shown in Figure 2.

c. The stone reaches its zenith when its height $s(t)$ is at its maximum value, which occurs when $v(t) = s'(t)$ is zero. So we solve

$$v(t) = -32t + 30 = 0$$

getting $t = 30/32 = 15/16 = 0.9375$ s. This is the time when the stone reaches its zenith. The height of the stone at that time is

$$s(15/16) = -16(15/16)^2 + 30(15/16) = 14.0625 \text{ ft}$$

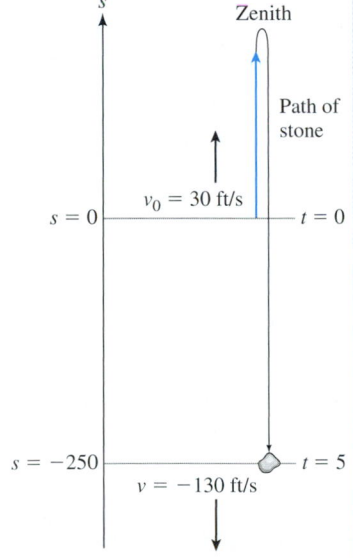

Figure 2

13.1 EXERCISES

● denotes basic skills exercises

Evaluate the integrals in Exercises 1–40.

1. ● $\int x^5\, dx$ *hint* [see Quick Examples p. 922]

2. ● $\int x^7\, dx$ 3. ● $\int 6\, dx$ 4. ● $\int (-5)\, dx$

5. ● $\int x\, dx$ 6. ● $\int (-x)\, dx$

7. ● $\int (x^2 - x)\, dx$ *hint* [see Example 2]

8. ● $\int (x + x^3)\, dx$ 9. ● $\int (1 + x)\, dx$

10. ● $\int (4 - x)\, dx$ 11. ● $\int x^{-5}\, dx$

12. ● $\int x^{-7}\, dx$ 13. ● $\int (x^{2.3} + x^{-1.3})\, dx$

14. ● $\int (x^{-0.2} - x^{0.2})\, dx$

15. ● $\int (u^2 - 1/u)\, du$ *hint* [see Example 4]

16. ● $\int (1/v^2 + 2/v)\, dv$

17. ● $\int \sqrt{x}\, dx$ 18. ● $\int \sqrt[3]{x}\, dx$

19. ● $\int (3x^4 - 2x^{-2} + x^{-5} + 4)\, dx$ *hint* [see Example 3]

20. ● $\int (4x^7 - x^{-3} + 1)\, dx$

21. ● $\int \left(\dfrac{2}{u} + \dfrac{u}{4}\right) du$ 22. ● $\int \left(\dfrac{2}{u^2} + \dfrac{u^2}{4}\right) du$

23. ● $\int \left(\dfrac{1}{x} + \dfrac{2}{x^2} - \dfrac{1}{x^3}\right) dx$ 24. ● $\int \left(\dfrac{3}{x} - \dfrac{1}{x^5} + \dfrac{1}{x^7}\right) dx$

25. ● $\int (3x^{0.1} - x^{4.3} - 4.1)\, dx$ 26. ● $\int \left(\dfrac{x^{2.1}}{2} - 2.3\right) dx$

27. ● $\int \left(\dfrac{3}{x^{0.1}} - \dfrac{4}{x^{1.1}}\right) dx$ 28. ● $\int \left(\dfrac{1}{x^{1.1}} - \dfrac{1}{x}\right) dx$

29. ● $\int \left(5.1t - \dfrac{1.2}{t} + \dfrac{3}{t^{1.2}}\right) dt$

30. ● $\int \left(3.2 + \dfrac{1}{t^{0.9}} + \dfrac{t^{1.2}}{3}\right) dt$

31. ● $\int (2e^x + 5/x + 1/4)\, dx$

32. ● $\int (-e^x + x^{-2} - 1/8)\, dx$

33. ● $\int \left(\dfrac{6.1}{x^{0.5}} + \dfrac{x^{0.5}}{6} - e^x\right) dx$

34. ● $\int \left(\dfrac{4.2}{x^{0.4}} + \dfrac{x^{0.4}}{3} - 2e^x\right) dx$

35. ● $\int (2^x - 3^x)\, dx$ 36. ● $\int (1.1^x + 2^x)\, dx$

37. ● $\int 100(1.1^x)\, dx$ 38. ● $\int 1000(0.9^x)\, dx$

39. $\int \dfrac{x + 2}{x^3}\, dx$ 40. $\int \dfrac{x^2 - 2}{x}\, dx$

41. ● Find $f(x)$ if $f(0) = 1$ and the tangent line at $(x, f(x))$ has slope x. *hint* [see Example 5]

42. ● Find $f(x)$ if $f(1) = 1$ and the tangent line at $(x, f(x))$ has slope $\dfrac{1}{x}$.

43. ● Find $f(x)$ if $f(0) = 0$ and the tangent line at $(x, f(x))$ has slope $e^x - 1$.

44. ● Find $f(x)$ if $f(1) = -1$ and the tangent line at $(x, f(x))$ has slope $2e^x + 1$.

Applications

45. ● *Marginal Cost* The marginal cost of producing the xth box of light bulbs is $5 - \dfrac{x}{10{,}000}$ and the fixed cost is $20{,}000$. Find the cost function $C(x)$. *hint* [see Example 5]

46. ● *Marginal Cost* The marginal cost of producing the xth box of Zip disks is $10 + \dfrac{x^2}{100{,}000}$ and the fixed cost is $100{,}000$. Find the cost function $C(x)$.

47. ● *Marginal Cost* The marginal cost of producing the xth roll of film is $5 + 2x + \dfrac{1}{x}$. The total cost to produce one roll is $1000. Find the cost function $C(x)$.

48. ● *Marginal Cost* The marginal cost of producing the xth box of CDs is $10 + x + \dfrac{1}{x^2}$. The total cost to produce 100 boxes is $10{,}000. Find the cost function $C(x)$.

49. ● *Motion in a Straight Line* The velocity of a particle moving in a straight line is given by $v(t) = t^2 + 1$.

   a. Find an expression for the position s after a time t.
   b. Given that $s = 1$ at time $t = 0$, find the constant of integration C, and hence find an expression for s in terms of t without any unknown constants. *hint* [see Example 6]

50. ● *Motion in a Straight Line* The velocity of a particle moving in a straight line is given by $v = 3e^t + t$.

● basic skills

a. Find an expression for the position s after a time t.

b. Given that $s = 3$ at time $t = 0$, find the constant of integration C, and hence find an expression for s in terms of t without any unknown constants.

51. ● ***Motion in a Straight Line*** If a stone is dropped from a rest position above the ground, how fast (in feet per second) and in what direction will it be traveling after 10 seconds? (Neglect the effects of air resistance.)

52. ● ***Motion in a Straight Line*** If a stone is thrown upward at 10 feet per second, how fast (in feet per second) and in what direction will it be traveling after 10 seconds? (Neglect the effects of air resistance.)

53. ● ***Motion in a Straight Line*** Your name is Galileo Galilei and you toss a weight upward at 16 feet per second from the top of the Leaning Tower of Pisa (height 185 ft).

a. Neglecting air resistance, find the weight's velocity as a function of time t in seconds.

b. Find the height of the weight above the ground as a function of time. Where and when will it reach its zenith? *hint* [see Example 7]

54. ● ***Motion in a Straight Line*** Your name is Francesca Dragonetti (an assistant of Galileo Galilei) and, to impress your boss, you toss a weight upward at 24 feet per second from the top of the Leaning Tower of Pisa (height 185 ft).

a. Neglecting air resistance, find the weight's velocity as a function of time t in seconds.

b. Find the height of the weight above the ground as a function of time. Where and when will it reach its zenith?

Vertical Motion *In Exercises 55–62, neglect the effects of air resistance.*

55. Show that if a projectile is thrown upward with a velocity of v_0 ft/s, then it will reach its highest point after $v_0/32$ ds.

56. Use the result of the preceding exercise to show that if a projectile is thrown upward with a velocity of v_0 ft/s, its highest point will be $v_0^2/64$ feet above the starting point.

Exercises 57–62 use the results in the preceding two exercises.

57. I threw a ball up in the air to a height of 20 feet. How fast was the ball traveling when it left my hand?

58. I threw a ball up in the air to a height of 40 feet. How fast was the ball traveling when it left my hand?

59. A piece of chalk is tossed vertically upward by Prof. Schwarzenegger and hits the ceiling 100 feet above with a *BANG*.

a. What is the minimum speed the piece of chalk must have been traveling to enable it to hit the ceiling?

b. Assuming that Prof. Schwarzenegger in fact tossed the piece of chalk up at 100 ft/sec, how fast was it moving when it struck the ceiling?

c. Assuming that Prof. Schwarzenegger tossed the chalk up at 100 ft/sec, and that it recoils from the ceiling with the same speed it had at the instant it hit, how long will it take the chalk to make the return journey and hit the ground?

60. A projectile is fired vertically upward from ground level at 16,000 feet per second.

a. How high does the projectile go?

b. How long does it take to reach its zenith (highest point)?

c. How fast is it traveling when it hits the ground?

61. ***Strength*** Prof. Strong can throw a 10-pound dumbbell twice as high as Professor Weak can. How much faster can Prof. Strong throw it?

62. ***Weakness*** Prof. Weak can throw a computer disk three times as high as Professor Strong can. How much faster can Prof. Weak throw it?

63. ***Household Income*** From 1990 to 2003, median household income in the U.S. rose by an average of approximately $1000 per year.[2] Given that the median household income in 1990 was approximately $30,000, use an indefinite integral to find a formula for median household income I as a function of time t since 1990 ($t = 0$ represents 1990), and use your formula to find the median household income in 2003.

64. ***Household Income*** From 1990 to 2003, mean household income in the U.S. rose by an average of approximately $1700 per year.[3] Given that the mean household income in 2003 was approximately $59,000, find a formula for mean household income A as a function of time t since 1990 ($t = 0$ represents 1990), and use your formula to find the mean household income in 2000.

65. ***Health Care Spending*** Write $H(t)$ for the amount spent in the U.S. on health care in year t, where t is measured in years since 1990. The rate of increase of $H(t)$ was approximately $65 billion per year in 1990 and rose to $100 billion per year in 2000.[4]

a. Find a linear model for the rate of change $H'(t)$.

b. Given that $700 billion was spent on health care in the U.S. in 1990, find the function $H(t)$.

66. ***Health Care Spending*** Write $H(t)$ for the amount spent in the U.S. on health care in year t, where t is measured in years since 2000. The rate of increase of $H(t)$ was projected to rise from $100 billion per year in 2000 to approximately $190 billion per year in 2010.[5]

a. Find a linear model for the rate of change $H'(t)$.

b. Given that $1300 billion was spent on health care in the U.S. in 2000, and using your model from (a), find the function $H(t)$.

---

[2] In current dollars, unadjusted for inflation. SOURCE: U.S. Census Bureau; "Table H-5. Race and Hispanic Origin of Householder—Households by Median and Mean Income: 1967 to 2003;" published August 27, 2004; www.census.gov/hhes/income.

[3] Ibid.

[4] SOURCE: Centers for Medicare and Medicaid Services, "National Health Expenditures," 2002 version, released January 2004; www.cms.hhs.gov/statistics/nhe/.

[5] SOURCE: Centers for Medicare and Medicaid Services, "National Health Expenditures 1965–2013, History and Projections," www.cms.hhs.gov/statistics/nhe/.

● basic skills

67. Bottled-Water Sales The rate of U.S. sales of bottled water for the period 1993–2003 could be approximated by

$$R(t) = 17t^2 + 100t + 2300 \text{ million gallons per year}$$
$$(3 \le t \le 13)$$

where t is time in years since 1990.[6] Use an indefinite integral to approximate the total sales $S(t)$ of bottled water since 1993. Approximately how much bottled water was sold from 1993 to 2003?

68. Bottled-Water Sales The rate of U.S. per capita sales of bottled water for the period 1993–2003 could be approximated by

$$Q(t) = 0.05t^2 + 0.4t + 9 \text{ gallons per year} \quad (3 \le t \le 13)$$

where t is the time in years since 1990.[7] Use an indefinite integral to approximate the total per capita sales $P(t)$ of bottled water since 1993. Approximately how much bottled water was sold, per capita, from 1993 to 2003?

Communication and Reasoning Exercises

69. ● If $F(x)$ and $G(x)$ are both antiderivatives of $f(x)$, how are $F(x)$ and $G(x)$ related?

70. ● Your friend Marco claims that once you have one antiderivative of $f(x)$, you have all of them. Explain what he means.

---

[6] The authors' regression model, based on data in the Beverage Marketing Corporation news release, "Bottled water now number-two commercial beverage in U.S., says Beverage Marketing Corporation," April 8, 2004, available at www.beveragemarketing.com.

[7] Ibid.

71. ● Complete the following: The total cost function is an _____ of the _____ cost function.

72. ● Complete the following: The distance covered is an antiderivative of the _____ function, and the velocity is an antiderivative of the _____ function.

73. ● If x represents the number of items manufactured and $f(x)$ represents dollars per item, what does $\int f(x)\,dx$ represent? In general, how are the units of $f(x)$ and the units of $\int f(x)\,dx$ related?

74. ● Complete the following: $-\dfrac{1}{x}$ is an _____ of $\dfrac{1}{x^2}$, whereas $\ln x^2$ is not. Also, $-\dfrac{1}{x} + C$ is the _____ of $\dfrac{1}{x^2}$, because the _____ of $-\dfrac{1}{x} + C$ is _____.

75. Give an argument for the rule that the integral of a sum is the sum of the integrals.

76. Is it true that $\displaystyle\int \frac{1}{x^3}\,dx = \ln(x^3) + C$? Give a reason for your answer.

77. Give an example to show that the integral of a product is not the product of the integrals.

78. Give an example to show that the integral of a quotient is not the quotient of the integrals.

79. Complete the following: If you take the _____ of the _____ of $f(x)$, you obtain $f(x)$ back. On the other hand, if you take the _____ of the _____ of $f(x)$, you obtain $f(x) + C$.

80. If a Martian told you that the Institute of Alien Mathematics, after a long and difficult search, has announced the discovery of a new antiderivative of $x - 1$ called $M(x)$ [the formula for $M(x)$ is classified information and cannot be revealed here], how would you respond?

● basic skills

13.2 Substitution

The chain rule for derivatives gives us an extremely useful technique for finding antiderivatives. This technique is called **change of variables** or **substitution.**

Recall that to differentiate a function like $(x^2 + 1)^6$, we first think of the function as $g(u)$ where $u = x^2 + 1$ and $g(u) = u^6$. We then compute the derivative, using the chain rule, as

$$\frac{d}{dx}g(u) = g'(u)\frac{du}{dx}$$

Any rule for derivatives can be turned into a technique for finding antiderivatives. The chain rule turns into the following formula:

$$\int g'(u)\frac{du}{dx}\,dx = g(u) + C$$

But, if we write $g(u) + C = \int g'(u)du$, we get the following interesting equation:

$$\int g'(u)\frac{du}{dx}dx = \int g'(u)du$$

This equation is the one usually called the change of variables formula. We can turn it into a more useful integration technique as follows. Let $f = g'(u)(du/dx)$. We can rewrite the above change of variables formula using f:

$$\int f\,dx = \int \left(\frac{f}{du/dx}\right)du$$

In essence, we are making the formal substitution

$$dx = \frac{1}{du/dx}du$$

Here's the technique:

Substitution Rule

If u is a function of x, then we can use the following formula to evaluate an integral:

$$\int f\,dx = \int \left(\frac{f}{du/dx}\right)du$$

Rather than use the formula directly, we use the following step-by-step procedure:

1. Write u as a function of x.

2. Take the derivative du/dx and solve for the quantity dx in terms of du.

3. Use the expression you obtain in Step 2 to substitute for dx in the given integral and substitute u for its defining expression.

Now let's see how this procedure works in practice.

Example 1 Substitution

Find $\int 4x(x^2 + 1)^6 dx$.

Solution To use substitution we need to choose an expression to be u. There is no hard and fast rule, but here is one hint that often works:

Take u to be an expression that is being raised to a power.

In this case, let's set $u = x^2 + 1$. Continuing the procedure above, we place the calculations for step (2) in a box.

$$u = x^2 + 1 \qquad \text{Write } u \text{ as a function of } x.$$

$$\frac{du}{dx} = 2x \qquad \text{Take the derivative of } u \text{ with respect to } x.$$

$$dx = \frac{1}{2x}du \qquad \text{Solve for } dx\text{: } dx = \frac{1}{du/dx}du.$$

Now we *substitute u for its defining expression, and substitute for dx* in the original integral:

$$\int 4x(x^2 + 1)^6 \, dx = \int 4xu^6 \frac{1}{2x} du \qquad \text{Substitute}^* \text{ for } u \text{ and } dx.$$

$$= \int 2u^6 du \qquad \text{Cancel the } xs \text{ and simplify.}$$

We have boiled the given integral down to the much simpler integral $\int 2u^6 du$, and we can now write down the solution:

$$2\frac{u^7}{7} + C = \frac{2(x^2 + 1)^7}{7} + C \qquad \text{Substitute } (x^2 + 1) \text{ for } u \text{ in the answer.}$$

---

* This step is equivalent to using the formula stated in the Substitution Rule box. If it should bother you that the integral contains both x and u, note that x is now a function of u.

+ *Before we go on...* There are two points to notice in Example 1. First, before we can actually integrate with respect to u, *we must eliminate all xs from the integrand.* If we cannot, we may have chosen the wrong expression for u. Second, after integrating, we must substitute back to obtain an expression involving x.

It is easy to check our answer. We differentiate:

$$\frac{d}{dx}\left[\frac{2(x^2 + 1)^7}{7}\right] = \frac{2(7)(x^2 + 1)^6(2x)}{7} = 4x(x^2 + 1)^6 \qquad ✔$$

Notice how we used the chain rule to check the result obtained by substitution. ∎

When we use substitution, the first step is always to decide what to take as u. Again, there are no set rules, but we see some common cases in the examples.

Example 2 More Substitution

Calculate $\displaystyle\int x^2(x^3 + 1)^2 dx.$

Solution

As we said in Example 1, it often works to take u to be an expression that is being raised to a power. We usually also want to see the derivative of u as a factor in the integrand so that we can cancel terms involving x. In this case, $x^3 + 1$ is being raised to a power, so let's set $u = x^3 + 1$. Its derivative is $3x^2$; in the integrand is x^2, which is missing the factor 3, but missing or incorrect constant factors are not a problem.

$$u = x^3 + 1 \qquad \text{Write } u \text{ as a function of } x.$$

$$\frac{du}{dx} = 3x^2 \qquad \text{Take the derivative of } u \text{ with respect to } x.$$

$$dx = \frac{1}{3x^2} du \qquad \text{Solve for } dx: \ dx = \frac{1}{du/dx} du.$$

$$\int x^2(x^3+1)^2dx \quad = \int x^2u^2\frac{1}{3x^2}du \qquad \text{Substitute for } u \text{ and } dx.$$

$$= \int \frac{1}{3}u^2du \qquad \text{Cancel the terms with } x.$$

$$= \frac{1}{9}u^3 + C \qquad \text{Take the antiderivative.}$$

$$= \frac{1}{9}(x^3+1)^3 + C \qquad \text{Substitute for } u \text{ in the answer.}$$

Example 3 An Expression in the Exponent

Evaluate $\int 3xe^{x^2}\,dx$.

Solution When we have an exponential with an expression in the exponent, it often works to substitute u for that expression. In this case, let's set $u = x^2$.

$$u = x^2$$
$$\frac{du}{dx} = 2x$$
$$dx = \frac{1}{2x}du$$

Substituting into the integral, we have

$$\int 3xe^{x^2}dx = \int 3xe^u\frac{1}{2x}du = \int \frac{3}{2}e^u du$$

$$= \frac{3}{2}e^u + C = \frac{3}{2}e^{x^2} + C$$

Example 4 A Special Power

Evaluate $\int \frac{1}{2x+5}dx$.

Solution We begin by rewriting the integrand as a power.

$$\int \frac{1}{2x+5}dx = \int (2x+5)^{-1}dx$$

Now we take our earlier advice and set u equal to the expression that is being raised to a power:

$$u = 2x+5$$
$$\frac{du}{dx} = 2$$
$$dx = \frac{1}{2}du$$

Substituting into the integral, we have

$$\int \frac{1}{2x+5}dx = \int \frac{1}{2}u^{-1}du = \frac{1}{2}\ln|u| + C$$

$$= \frac{1}{2}\ln|2x+5| + C$$

Example 5 Choosing *u*

Evaluate $\int (x+3)\sqrt{x^2+6x}\,dx$.

Solution There are two parenthetical expressions. Notice however, that the derivative of the expression (x^2+6x) is $2x+6$, which is twice the term $(x+3)$ in front of the radical. Recall that we would like the derivative of u to appear as a factor. Thus, let's take $u = x^2 + 6x$.

$$u = x^2 + 6x$$
$$\frac{du}{dx} = 2x + 6 = 2(x+3)$$
$$dx = \frac{1}{2(x+3)}du$$

Substituting into the integral, we have

$$\int (x+3)\sqrt{x^2+6x}\,dx = \int (x+3)\sqrt{u}\left(\frac{1}{2(x+3)}\right)du$$

$$= \int \frac{1}{2}\sqrt{u}\,du = \frac{1}{2}\int u^{1/2}du$$

$$= \frac{1}{2}\frac{2}{3}u^{3/2} + C = \frac{1}{3}(x^2+6x)^{3/2} + C$$

Some cases require a little more work.

Example 6 When the *x* Terms Do Not Cancel

Evaluate $\int \dfrac{2x}{(x-5)^2}dx$.

Solution We first rewrite

$$\int \frac{2x}{(x-5)^2}dx = \int 2x(x-5)^{-2}dx$$

This suggests that we should set $u = x - 5$.

$$u = x - 5$$
$$\frac{du}{dx} = 1$$
$$dx = du$$

Substituting, we have

$$\int \frac{2x}{(x-5)^2}dx = \int 2xu^{-2}du$$

Now, there is nothing in the integrand to cancel the x that appears. If, as here, there is still an x in the integrand after substituting, we go back to the expression for u, solve for x, and substitute the expression we obtain for x in the integrand. So, we take $u = x - 5$ and solve for $x = u + 5$. Substituting, we get

$$\int 2xu^{-2}du = \int 2(u+5)u^{-2}du$$

$$= 2\int (u^{-1} + 5u^{-2})du$$

$$= 2\ln|u| - \frac{10}{u} + C$$

$$= 2\ln|x-5| - \frac{10}{x-5} + C$$

Example 7 Application: iPod Sales

The rate of sales of Apple iPods from the fourth quarter of 2002 through the third quarter of 2004 could be roughly approximated by the logistic function

$$r(t) = \frac{1100e^{0.6t}}{9 + e^{0.6t}} \text{ thousand iPods per quarter} \quad (-1 \leq t \leq 6)$$

where t is time in quarters since the first quarter of 2003.[*]

a. Find an expression for the total number of iPods sold since the first quarter of 2003 ($t = 0$).

b. Roughly how many iPods were sold in 2003?

Solution

a. If we write the total number of iPods sold from the first quarter of 2003 to time t as $N(t)$, then the information we are given says that

$$N'(t) = r(t) = \frac{1100e^{0.6t}}{9 + e^{0.6t}}$$

Thus,
$$N(t) = \int \frac{1100e^{0.6t}}{9 + e^{0.6t}} \, dt$$

is the function we are after. To integrate the expression, take u to be the denominator of the integrand:

$$u = 9 + e^{0.6t}$$

$$\frac{du}{dt} = 0.6e^{0.6t}$$

$$dt = \frac{1}{0.6e^{0.6t}} \, du$$

$$N(t) = \int \frac{1100e^{0.6t}}{9 + e^{0.6t}} \, dt$$

$$= \int \frac{1100e^{0.6t}}{u} \cdot \frac{1}{0.6e^{0.6t}} \, du$$

$$= \frac{1100}{0.6} \int \frac{1}{u} \, du$$

$$\approx 1800 \ln |u| + C = 1800 \ln(9 + e^{0.6t}) + C$$

(Why could we drop the absolute value in the last step?)

Now what is C? Because $N(t)$ represents the total number of iPods sold *since* time $t = 0$, we have $N(0) = 0$ (because that is when we started counting). Thus,

$$0 = 1800 \ln(9 + e^{0.6(0)}) + C$$

$$= 1800 \ln(10) + C$$

$$C = -1800 \ln(10) \approx -4100$$

Therefore, the total number of iPods sold since the first quarter of 2003 is approximately

$$N(t) = 1800 \ln(9 + e^{0.6t}) - 4100 \text{ thousand iPods}$$

b. Since the last quarter of 2003 ended when $t = 4$, the total number of iPods sold in 2003 was approximately

$$N(4) = 1800 \ln(9 + e^{0.6(4)}) - 4100 \approx 1300 \text{ thousand iPods}[†]$$

---

[*] Based on a logistic regression. Source for data: Apple Computer, Inc., quarterly earnings reports, available at www.apple.com.

[†] The actual number of iPods sold in 2003 was 939 thousand. This discrepancy results from a less than perfect fit of the logistic model to the actual data.

+ *Before we go on...* You might wonder why we are writing a logistic function in the form we used in Example 7 rather than in one of the "standard" forms $\dfrac{N}{1 + Ab^{-t}}$ or $\dfrac{N}{1 + Ae^{-kt}}$. Our only reason for doing this is to make the substitution work. To convert from the second "standard" form to the form we used in the example, multiply top and bottom by e^{kt}. (See Exercises 67 and 68 in Section 13.4 for further discussion.) ∎

Shortcuts

The following shortcuts allow us to simply write down the antiderivative in cases where we would otherwise need the substitution $u = ax + b$, as in Example 4. (a and b are constants with $a \neq 0$.) All of the shortcuts can be obtained using the substitution $u = ax + b$. Their derivation will appear in the exercises.

Shortcuts: Integrals of Expressions Involving ($ax + b$)

Rule

$$\int (ax + b)^n \, dx = \frac{(ax + b)^{n+1}}{a(n + 1)} + C$$
$$(\text{if } n \neq -1)$$

$$\int (ax + b)^{-1} dx = \frac{1}{a} \ln |ax + b| + C$$

$$\int e^{ax+b} dx = \frac{1}{a} e^{ax+b} + C$$

$$\int c^{ax+b} \, dx = \frac{1}{a \ln c} c^{ax+b} + C$$

quick Example

$$\int (3x - 1)^2 \, dx = \frac{(3x - 1)^3}{3(3)} + C$$
$$= \frac{(3x - 1)^3}{9} + C$$

$$\int (3 - 2x)^{-1} dx = \frac{1}{(-2)} \ln |3 - 2x| + C$$
$$= -\frac{1}{2} \ln |3 - 2x| + C$$

$$\int e^{-x+4} dx = \frac{1}{(-1)} e^{-x+4} + C$$
$$= -e^{-x+4} + C$$

$$\int 2^{-3x+4} \, dx = \frac{1}{(-3 \ln 2)} 2^{-3x+4} + C$$
$$= -\frac{1}{3 \ln 2} 2^{-3x+4} + C$$

FAQs   When to Use Substitution and What to Use for u

Q: *If I am asked to calculate an antiderivative, how do I know when to use a substitution and when not to use one?*

A: Do *not* use substitution when integrating sums, differences, and/or constant multiples of powers of x and exponential functions, such as $2x^3 - \dfrac{4}{x^2} + \dfrac{1}{2x} + 3^x + \dfrac{2^x}{3}$.

To recognize when you should try a substitution, pretend that you are *differentiating* the given expression instead of integrating it. If differentiating the expression would require use of the chain rule, then integrating that expression may well require a substitution, as in, say, $x(3x^2 - 4)^3$ or $(x + 1)e^{x^2+2x-1}$. (In the first we have a *quantity* cubed, and in the second we have *e* raised to a *quantity*.) ■

Q: *If an integral seems to call for a substitution, what should I use for u?*

A: There are no set rules for deciding what to use for *u*, but the preceding examples show some common patterns:

- If you see a linear expression raised to a power, try setting *u* equal to that linear expression. For example, in $(3x - 2)^{-3}$, set $u = 3x - 2$. (Alternatively, try using the shortcuts above.)
- If you see a constant raised to a linear expression, try setting *u* equal to that linear expression. For example, in $3^{(2x+1)}$, set $u = 2x + 1$. (Alternatively, try a shortcut.)
- If you see an expression raised to a power multiplied by the derivative of that expression (or a constant multiple of the derivative), try setting *u* equal to that expression. For example, in $x^2(3x^3 - 4)^{-1}$, set $u = 3x^3 - 4$.
- If you see a constant raised to an expression, multiplied by the derivative of that expression (or a constant multiple of its derivative), try setting *u* equal to that expression. For example, in $5(x + 1)e^{x^2+2x-1}$, set $u = x^2 + 2x - 1$.
- If you see an expression in the denominator and its derivative (or a constant multiple of its derivative) in the numerator, try setting *u* equal to that expression. For example, in $\dfrac{2^{3x}}{3 - 2^{3x}}$, set $u = 3 - 2^{3x}$.

Persistence often pays off: if a certain substitution does not work, try another approach or a different substitution. ■

13.2 EXERCISES

● denotes basic skills exercises

In Exercises 1–10, evaluate the given integral using the substitution (or method) indicated.

1. ● $\int (3x - 5)^3 \, dx$; $u = 3x - 5$

2. ● $\int (2x + 5)^{-2} \, dx$; $u = 2x + 5$

3. ● $\int (3x - 5)^3 \, dx$; Shortcut p. 938

4. ● $\int (2x + 5)^{-2} \, dx$; Shortcut p. 938

5. ● $\int e^{-x} \, dx$; $u = -x$ **6.** ● $\int e^{x/2} \, dx$; $u = x/2$

7. ● $\int e^{-x} \, dx$; Shortcut p. 938   **8.** ● $\int e^{x/2} \, dx$; Shortcut p. 938

9. ● $\int (x + 1)e^{(x+1)^2} \, dx$; $u = (x + 1)^2$

10. ● $\int (x - 1)^2 e^{(x-1)^3} \, dx$; $u = (x - 1)^3$

In Exercises 11–48, decide on what substitution to use, and then evaluate the given integral using a substitution.

11. ● $\int (3x + 1)^5 \, dx$ *hint* [see Example 1]

12. ● $\int (-x - 1)^7 \, dx$ **13.** ● $\int (-2x + 2)^{-2} \, dx$

14. ● $\int (2x)^{-1} \, dx$ **15.** ● $\int 7.2\sqrt{3x - 4} \, dx$

16. ● $\int 4.4e^{(-3x+4)} \, dx$ **17.** ● $\int 1.2e^{(0.6x+2)} \, dx$

18. ● $\int 8.1\sqrt{-3x + 4} \, dx$ **19.** ● $\int x(3x^2 + 3)^3 \, dx$

20. ● $\int x(-x^2 - 1)^3 \, dx$ **21.** ● $\int x(x^2 + 1)^{1.3} \, dx$

22. ● $\int \dfrac{x}{(3x^2 - 1)^{0.4}} \, dx$ **23.** ● $\int (1 + 9.3e^{3.1x-2}) \, dx$

24. ● $\int (3.2 - 4e^{1.2x-3}) \, dx$ **25.** ● $\int 2x\sqrt{3x^2 - 1} \, dx$

● basic skills

26. $\bullet$ $\displaystyle\int 3x\sqrt{-x^2+1}\,dx$ **27.** $\bullet$ $\displaystyle\int xe^{-x^2+1}\,dx$

28. $\bullet$ $\displaystyle\int xe^{2x^2-1}\,dx$

29. $\bullet$ $\displaystyle\int (x+1)e^{-(x^2+2x)}\,dx$ *hint* [see Example 5]

30. $\bullet$ $\displaystyle\int (2x-1)e^{2x^2-2x}\,dx$

31. $\bullet$ $\displaystyle\int \frac{-2x-1}{(x^2+x+1)^3}\,dx$ **32.** $\bullet$ $\displaystyle\int \frac{x^3-x^2}{3x^4-4x^3}\,dx$

33. $\bullet$ $\displaystyle\int \frac{x^2+x^5}{\sqrt{2x^3+x^6-5}}\,dx$ **34.** $\bullet$ $\displaystyle\int \frac{2(x^3-x^4)}{(5x^4-4x^5)^5}\,dx$

35. $\bullet$ $\displaystyle\int x(x-2)^5\,dx$ *hint* [see Example 6]

36. $\bullet$ $\displaystyle\int x(x-2)^{1/3}\,dx$ **37.** $\bullet$ $\displaystyle\int 2x\sqrt{x+1}\,dx$

38. $\bullet$ $\displaystyle\int \frac{x}{\sqrt{x+1}}\,dx$

39. $\bullet$ $\displaystyle\int \frac{e^{-0.05x}}{1-e^{-0.05x}}\,dx$ *hint* [see Example 7]

40. $\bullet$ $\displaystyle\int \frac{3e^{1.2x}}{2+e^{1.2x}}\,dx$ **41.** $\displaystyle\int \frac{3e^{-1/x}}{x^2}\,dx$

42. $\displaystyle\int \frac{2e^{2/x}}{x^2}\,dx$ **43.** $\displaystyle\int \frac{e^x+e^{-x}}{2}\,dx$

44. $\displaystyle\int e^{x/2}+e^{-x/2}\,dx$ **45.** $\displaystyle\int \frac{e^x-e^{-x}}{e^x+e^{-x}}\,dx$

46. $\displaystyle\int \frac{e^{x/2}+e^{-x/2}}{e^{x/2}-e^{-x/2}}\,dx$

47. $\displaystyle\int \left((2x-1)e^{2x^2-2x}+xe^{x^2}\right)dx$

48. $\displaystyle\int \left(xe^{-x^2+1}+e^{2x}\right)dx$

In Exercises 49–52, derive the given equation, where a and b are constants with $a \neq 0$.

49. $\bullet$ $\displaystyle\int (ax+b)^n\,dx = \frac{(ax+b)^{n+1}}{a(n+1)} + C$ (if $n \neq -1$)

50. $\bullet$ $\displaystyle\int (ax+b)^{-1}\,dx = \frac{1}{a}\ln|ax+b| + C$

51. $\bullet$ $\displaystyle\int \sqrt{ax+b}\,dx = \frac{2}{3a}(ax+b)^{3/2} + C$

52. $\bullet$ $\displaystyle\int e^{ax+b}\,dx = \frac{1}{a}e^{ax+b} + C$

In Exercises 53–66, use the shortcut formulas on p. 938 and Exercises 49–52 to calculate the given integral.

53. $\bullet$ $\displaystyle\int e^{-x}\,dx$ **54.** $\bullet$ $\displaystyle\int e^{x-1}\,dx$

55. $\bullet$ $\displaystyle\int e^{2x-1}\,dx$ **56.** $\bullet$ $\displaystyle\int e^{-3x}\,dx$

57. $\bullet$ $\displaystyle\int (2x+4)^2\,dx$ **58.** $\bullet$ $\displaystyle\int (3x-2)^4\,dx$

59. $\bullet$ $\displaystyle\int \frac{1}{5x-1}\,dx$ **60.** $\bullet$ $\displaystyle\int (x-1)^{-1}\,dx$

61. $\bullet$ $\displaystyle\int (1.5x)^3\,dx$ **62.** $\bullet$ $\displaystyle\int e^{2.1x}\,dx$

63. $\bullet$ $\displaystyle\int 1.5^{3x}\,dx$ **64.** $\bullet$ $\displaystyle\int 4^{-2x}\,dx$

65. $\bullet$ $\displaystyle\int (2^{3x+4}+2^{-3x+4})\,dx$ **66.** $\bullet$ $\displaystyle\int (1.1^{-x+4}+1.1^{x+4})\,dx$

67. $\bullet$ Find $f(x)$ if $f(0)=0$ and the tangent line at $(x, f(x))$ has slope $x(x^2+1)^3$.

68. $\bullet$ Find $f(x)$ if $f(1)=0$ and the tangent line at $(x, f(x))$ has slope $\dfrac{x}{x^2+1}$.

69. $\bullet$ Find $f(x)$ if $f(1)=1/2$ and the tangent line at $(x, f(x))$ has slope xe^{x^2-1}.

70. $\bullet$ Find $f(x)$ if $f(2)=1$ and the tangent line at x has slope $(x-1)e^{x^2-2x}$.

Applications

71. $\bullet$ *Cost* The marginal cost of producing the xth roll of film is given by $5 + 1/(x+1)^2$. The total cost to produce one roll is $1000. Find the total cost function $C(x)$.

72. $\bullet$ *Cost* The marginal cost of producing the xth box of CDs is given by $10 - x/(x^2+1)^2$. The total cost to produce 2 boxes is $1000. Find the total cost function $C(x)$.

73. $\bullet$ *Scientific Research* The number of research articles in the prominent journal *Physics Review* written by researchers in Europe can be approximated by

$$E(t) = \frac{7e^{0.2t}}{5+e^{0.2t}} \text{ thousand articles per year}\quad (t \geq 0)$$

where t is time in years ($t = 0$ represents 1983).[8]

a. Find an (approximate) expression for the total number of articles written by researchers in Europe since 1983 ($t = 0$).

b. Roughly how many articles were written by researchers in Europe from 1983 to 2003? *hint* [see Example 7]

74. $\bullet$ *Scientific Research* The number of research articles in the prominent journal *Physics Review* written by researchers in the U.S. can be approximated by

$$U(t) = \frac{4.6e^{0.6t}}{0.4+e^{0.6t}} \text{ thousand articles per year}\quad (t \geq 0)$$

where t is time in years ($t = 0$ represents 1983).[9]

---

[8] Based on data from 1983 to 2003. SOURCE: The American Physical Society/*New York Times*, May 3, 2003, p. A1.

[9] Ibid.

$\bullet$ basic skills

a. Find an (approximate) expression for the total number of articles written by researchers in the U.S. since 1983 ($t = 0$).

b. Roughly how many articles were written by researchers in the U.S. from 1983 to 2003?

75. ● *Motion in a Straight Line* The velocity of a particle moving in a straight line is given by $v = t(t^2 + 1)^4 + t$.

a. Find an expression for the position s after a time t.

b. Given that $s = 1$ at time $t = 0$, find the constant of integration C and hence an expression for s in terms of t without any unknown constants.

76. ● *Motion in a Straight Line* The velocity of a particle moving in a straight line is given by $v = 3te^{t^2} + t$.

a. Find an expression for the position s after a time t.

b. Given that $s = 3$ at time $t = 0$, find the constant of integration C and hence an expression for s in terms of t without any unknown constants.

77. ● *Bottled-Water Sales* The rate of U.S. sales of bottled water for the period 1993–2003 could be approximated by

$R(t) = 17(t - 1990)^2 + 100(t - 1990) + 2300$
    million gallons per year   ($1993 \le t \le 2003$)

where t is the year.[10] Use an indefinite integral to approximate the total sales $S(t)$ of bottled water since 1993 ($t = 1993$). Approximately how much bottled water was sold from 1993 to 2003?

78. ● *Bottled-Water Sales* The rate of U.S. per capita sales of bottled water for the period 1993–2003 could be approximated by

$Q(t) = 0.05(t - 1990)^2 + 0.4(t - 1990) + 9$
    gallons per year   ($1993 \le t \le 2003$)

where t is the year.[11] Use an indefinite integral to approximate the total per capita sales $P(t)$ of bottled water since 1993.

---

[10] The authors' regression model, based on data in the Beverage Marketing Corporation news release, "Bottled water now number-two commercial beverage in U.S., says Beverage Marketing Corporation," April 8, 2004, available at www.beveragemarketing.com.

[11] Ibid.

Approximately how much bottled water was sold, per capita, from 1993 to 2003?

Communication and Reasoning Exercises

79. ● Are there any circumstances in which you should use the substitution $u = x$? Illustrate your answer by giving an example that shows the effect of this substitution.

80. ● At what stage of a calculation using a u substitution should you substitute back for u in terms of x: before or after taking the antiderivative?

81. ● You are asked to calculate $\int \dfrac{u}{u^2 + 1}\, du$. What is wrong with the substitution $u = u^2 + 1$?

82. ● What is wrong with the following "calculation" of $\int \dfrac{1}{x^2 - 1}\, dx$?

$$\int \frac{1}{x^2 - 1} = \int \frac{1}{u} \qquad \textcolor{red}{\text{Using the substitution } u = x^2 - 1}$$
$$= \ln|u| + C$$
$$= \ln|x^2 - 1| + C$$

83. ● Give an example of an integral that can be calculated by using the substitution $u = x^2 + 1$, and then carry out the calculation.

84. Give an example of an integral that can be calculated either by using the power rule for antiderivatives or by using the substitution $u = x^2 + x$, and then carry out the calculations.

85. Show that *none* of the following substitutions work for $\int e^{-x^2}\, dx$: $u = -x$, $u = x^2$, $u = -x^2$. (The antiderivative of e^{-x^2} involves the *error function* erf(x).)

86. Show that *none* of the following substitutions work for $\int \sqrt{1 - x^2}\, dx$: $u = 1 - x^2$, $u = x^2$, and $u = -x^2$. (The antiderivative of $\sqrt{1 - x^2}$ involves inverse trigonometric functions, discussion of which is beyond the scope of this book.)

● basic skills

13.3 The Definite Integral: Numerical and Graphical Approaches

In Sections 13.1 and 13.2, we discussed the indefinite integral. There is an older, related concept called the **definite integral.** Let's introduce this new idea with an example. (We'll drop hints now and then about how the two types of integral are related. In Section 13.4 we discuss the exact relationship, which is one of the most important results in calculus.)

In Section 13.1, we used antiderivatives to answer questions of the form "Given the marginal cost, compute the total cost" (see Example 5 in Section 13.1). In this section we approach such questions more directly, and we will forget about antiderivatives for now.

Example 1 Total Cost

Your cell phone company offers you an innovative pricing scheme. When you make a call, the *marginal* cost is

$$c(t) = \frac{5}{10t + 1} \text{ dollars per hour}$$

Use a numerical calculation to estimate the total cost of a two-hour phone call.

Solution The graph of $c(t)$ is shown in Figure 3.

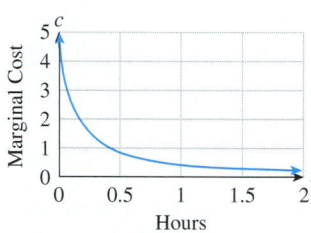

Figure **3**

Let's start with a very crude estimate of the total cost, using the graph as a guide. The marginal cost at the beginning of your call is $c(0) = 5/(0 + 1) = \$5$ per hour. If this cost were to remain constant for the length of your call, the total cost of the call would be

Cost of Call = Cost per hour × Number of hours = $5 \times 2 = \$10$

Figure 4 shows how we can represent this calculation on the graph of $c(t)$.

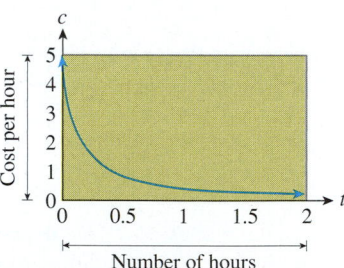

Figure **4**

The cost per hour based on $c(0) = 5$ is represented by the *y*-coordinate of the graph at its left edge, while the number of hours is represented by the width of the interval [0, 2] on the *x*-axis. Therefore, computing the area of the shaded rectangle in the figure gives the same calculation:

Area of rectangle = Cost per hour × Number of hours

$$= 5 \times 2 = \$10 = \text{Cost of Call}$$

But, as we see in the graph, the marginal cost does not remain constant, but goes down quite dramatically over the course of the call. We can obtain a somewhat more accurate estimate of the total cost by looking at the call hour by hour—that is, by dividing the length of the call into two equal intervals, or subdivisions. We estimate the cost of each one-hour subdivision, using the marginal cost at the beginning of that hour.

$$\text{Cost of first hour} = \text{Cost per hour} \times \text{Number of hours}$$
$$= c(0) \times 1 = 5 \times 1 = \$5$$
$$\text{Cost of second hour} = \text{Cost per hour} \times \text{Number of hours}$$
$$= c(1) \times 1 = 5/11 \times 1 \approx \$0.45$$

Adding these costs gives us the more accurate estimate

$$c(0) \times 1 + c(1) \times 1 = \$5.45 \qquad \textcolor{red}{\text{Calculation using 2 subdivisions}}$$

In Figure 5 we see that we are computing the combined area of two rectangles, each of whose heights is determined by the height of the graph at its left edge:

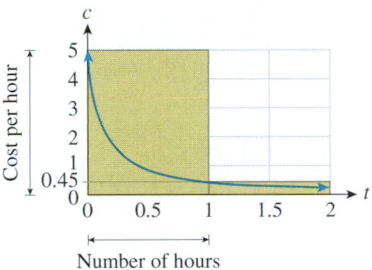

The areas of the rectangles are estimates of the costs for successive one-hour periods.

Figure **5**

$$\text{Area of first rectangle} = \text{Cost per hour} \times \text{Number of hours}$$
$$= c(0) \times 1 = \$5 = \text{Cost of first hour}$$
$$\text{Area of second rectangle} = \text{Cost per hour} \times \text{Number of hours}$$
$$= c(1) \times 1 \approx \$0.45 = \text{Cost of second hour}$$

If we assume that the phone company is honest about $c(t)$ being the marginal cost and is actually calculating your cost more than once an hour, we get an even better estimate of the cost by looking at the call by using four divisions of a half-hour each. The calculation for this estimate is

$$c(0) \times 0.5 + c(0.5) \times 0.5 + c(1) \times 0.5 + c(1.5) \times 0.5$$
$$\approx 2.500 + 0.417 + 0.227 + 0.156 = \$3.30 \qquad \textcolor{red}{\text{Calculation using 4 subdivisions}}$$

As we see in Figure 6, we have now computed the combined area of *four* rectangles each of whose heights is again determined by the height of the graph at its left edge.

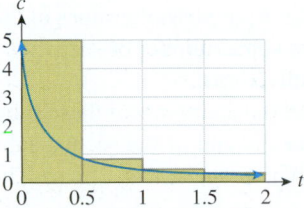

Estimated Cost Using 4 Subdivisions.

The areas of the rectangles are estimates of the costs for successive half-hour periods.

Figure **6**

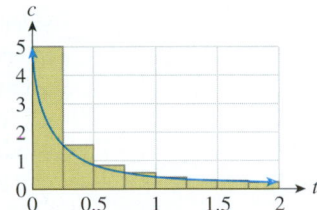

Estimated Cost Using 8 Subdivisions.

The areas of the rectangles are estimates of the costs for successive quarter-hour periods.

Figure **7**

Notice how the cost seems to be decreasing as we use more subdivisions. More importantly, total cost seems to be getting closer to the area under the graph. Figure 7 illustrates the calculation for 8 equal subdivisions. The approximate total cost using 8 subdivisions is the total area of the shaded region in Figure 7:

$$c(0) \times 0.25 + c(0.25) \times 0.25 + c(0.5) \times 0.25 + \cdots + c(1.75) \times 0.25 \approx \$2.31$$

Calculation using 8 subdivisions

Looking at Figure 7, one still gets the impression that we are being overcharged, especially for the first period. If the phone company wants to be *really* honest about $c(t)$ being the marginal cost, it should really be calculating your cost *continuously,* minute by minute or, better yet, second-by-second, as illustrated in Figure 8.

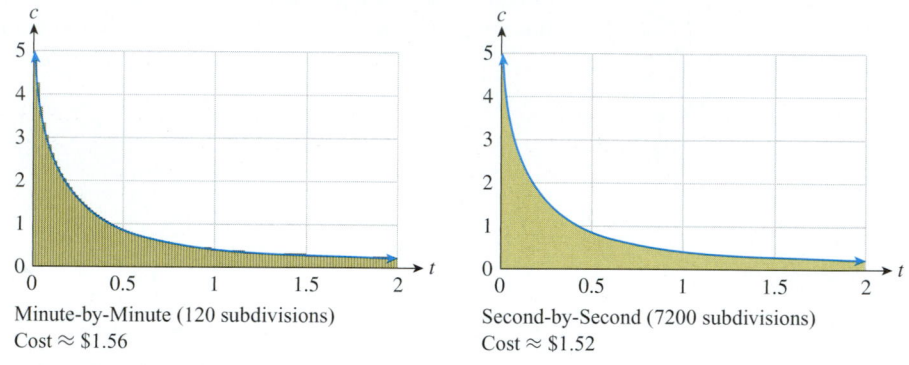

Minute-by-Minute (120 subdivisions)
Cost ≈ $1.56

Second-by-Second (7200 subdivisions)
Cost ≈ $1.52

Figure **8**

Figure 8 strongly suggests that the more accurately we calculate the total cost, the closer the answer gets to the exact area under the portion of the graph of $c(t)$ with $0 \leq t \leq 2$, and leads us to the conclusion that the *exact* total cost is the exact area under the marginal cost curve for $0 \leq t \leq 2$. In other words, we have made the following remarkable discovery:

Total cost is the area under the marginal cost curve!

+*Before we go on...* The minute-by-minute calculation in Example 1 is tedious to do by hand, and no one in their right mind would even *attempt* to do the second-by-second calculation by hand! Below we discuss ways of doing these calculations with the aid of technology. ■

The type of calculation done in Example 1 is useful in many applications. Let's look at the general case and give the result a name.

In general, we have a function f (such as the function c in the example) and we consider an interval $[a, b]$ of possible values of the independent variable x. We subdivide the interval $[a, b]$ into some number of segments of equal length. Write n for the number of segments, or **subdivisions.**

Next, we label the endpoints of these subdivisions x_0 for a, x_1 for the end of the first subdivision, x_2 for the end of the second subdivision, and so on until we get to x_n, the end of the nth subdivision, so that $x_n = b$. Thus,

$$a = x_0 < x_1 < \cdots < x_n = b$$

The first subdivision is the interval $[x_0, x_1]$, the second subdivision is $[x_1, x_2]$, and so on until we get to the last subdivision, which is $[x_{n-1}, x_n]$. We are dividing the interval

$[a, b]$ into n subdivisions of equal length, so each segment has length $(b - a)/n$. We write Δx for $(b - a)/n$ (Figure 9).

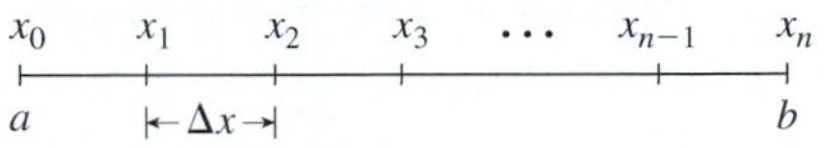

$$x_0 \quad x_1 \quad x_2 \quad x_3 \quad \cdots \quad x_{n-1} \quad x_n$$

$$a \quad \vdash \Delta x \dashv \qquad\qquad\qquad b$$

Figure **9**

Having established this notation, we can write the calculation that we want to do as follows: For each subdivision $[x_{k-1}, x_k]$, compute $f(x_{k-1})$, the value of the function f at the left endpoint. Multiply this value by the length of the interval, which is Δx. Then add together all n of these products to get the number

$$f(x_0)\Delta x + f(x_1)\Delta x + \cdots + f(x_{n-1})\Delta x$$

This sum is called a **(left) Riemann[12] sum** for f. In Example 1 we computed several different Riemann sums. Here is the computation for $n = 4$ we used in the cell phone example (see Figure 10):

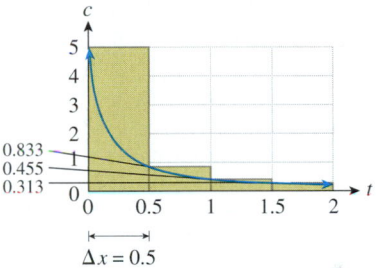

Figure **10**

Left Riemann Sum $= f(x_0)\Delta x + f(x_1)\Delta x + \cdots + f(x_{n-1})\Delta x$

$$= f(0)(0.5) + f(0.5)(0.5) + f(1)(0.5) + f(1.5)(0.5)$$

$$\approx (5)(0.5) + (0.833)(0.5) + (0.455)(0.5) + (0.313)(0.5) \approx 3.30$$

Because sums are often used in mathematics, mathematicians have developed a shorthand notation for them. We write

$$f(x_0)\Delta x + f(x_1)\Delta x + \cdots + f(x_{n-1})\Delta x \quad \text{as} \quad \sum_{k=0}^{n-1} f(x_k)\Delta x$$

The symbol $\sum$ is the Greek letter sigma and stands for **summation.** The letter k here is called the index of summation, and we can think of it as counting off the segments. We read the notation as "the sum from $k = 0$ to $n - 1$ of the quantities $f(x_k)\Delta x$." Think of it as a set of instructions:

Set $k = 0$, and calculate $f(x_0)\Delta x$. $f(0)(0.5)$ in the above calculation
Set $k = 1$, and calculate $f(x_1)\Delta x$. $f(0.5)(0.5)$ in the above calculation
$\cdots$
Set $k = n - 1$, and calculate $f(x_{n-1})\Delta x$. $f(1.5)(0.5)$ in the above calculation

Then sum all the quantities so calculated.

---

[12] After Georg Friedrich Bernhard Riemann (1826–1866).

Riemann Sum

If f is a continuous function, the **left Riemann sum** with n equal subdivisions for f over the interval $[a, b]$ is defined to be

$$\text{Left Riemann sum} = \sum_{k=0}^{n-1} f(x_k)\Delta x$$

$$= f(x_0)\Delta x + f(x_1)\Delta x + \cdots + f(x_{n-1})\Delta x$$

$$= [f(x_0) + f(x_1) + \cdots + f(x_{n-1})]\Delta x$$

where $a = x_0 < x_1 < \cdots < x_n = b$ are the subdivisions, and $\Delta x = (b - a)/n$.

Interpretation of the Riemann Sum

If f is the rate of change of a quantity F (that is, $f = F'$), then the Riemann sum of f approximates the total change of F from $x = a$ to $x = b$. The approximation improves as the number of subdivisions increases toward infinity.

quick Example

If $f(t)$ is the rate of change in the number of bats in a bell tower and $[a, b] = [2, 3]$, then the Riemann sum approximates the total change in the number of bats in the tower from time $t = 2$ to time $t = 3$.

Visualizing a Left Riemann Sum (Nonnegative Function)

Graphically, we can represent a left Riemann sum of a nonnegative function as an approximation of the area under a curve:

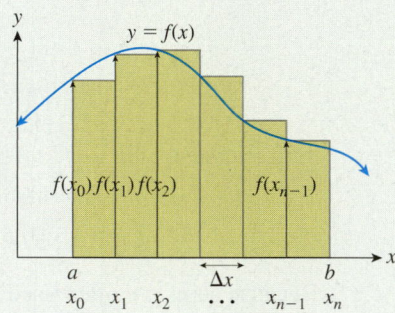

Riemann Sum = Shaded Area = Area of first rectangle + Area of second rectangle + $\cdots$ +
Area of nth rectangle = $f(x_0)\Delta x + f(x_1)\Delta x + f(x_2)\Delta x + \cdots + f(x_{n-1})\Delta x$

quick Example

In Example 1 we computed several Riemann sums, including these:

$n = 1$:   Riemann sum $= c(0)\Delta t = 5 \times 2 = \$10$

$n = 2$:   Riemann sum $= [c(t_0) + c(t_1)]\Delta t$
$$= [c(0) + c(1)] \cdot (1) \approx \$5.45$$

$n = 4$:   Riemann sum $= [c(t_0) + c(t_1) + c(t_2) + c(t_3)]\Delta t$
$$= [c(0) + c(0.5) + c(1) + c(1.5)] \cdot (0.5) \approx \$3.30$$

$n = 8$:   Riemann sum $= [c(t_0) + c(t_1) + \cdots + c(t_7)]\Delta t$
$$= [c(0) + c(0.25) + \cdots + c(1.75)] \cdot (0.25) \approx \$2.31$$

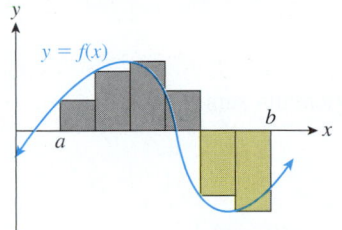

Riemann sum = Area above x-axis
$\qquad\qquad$ − Area below x-axis

Figure 11

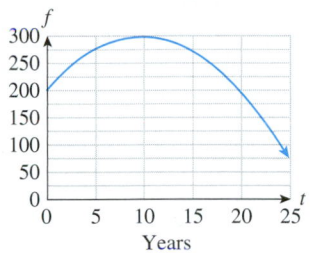

Figure 12

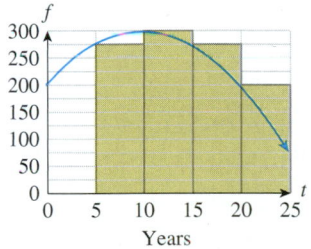

Figure 13

Note To visualize the Riemann sum of a function that is negative, look again at the formula $f(x_0)\Delta x + f(x_1)\Delta x + f(x_2)\Delta x + \cdots + f(x_{n-1})\Delta x$ for the Riemann sum. Each term $f(x_i)\Delta x_i$ represents the area of one rectangle in the figure above. So, the areas of the rectangles with negative values of $f(x_i)$ are automatically counted as negative. They appear as gold rectangles in Figure 11. ∎

Example 2 Computing a Riemann Sum From a Graph

Figure 12 shows the approximate rate f at which convicts have been given the death sentence in the U.S. since 1980 (t is time in years since 1980.)[*]

Use a left Riemann sum with 4 subdivisions to estimate the total number of death sentences handed down from 1985 to 2005.

Solution Let us represent the total number of death sentences handed down since the institution of the death sentence up to time t by $F(t)$. The total number of death sentences handed down from 1985 to 2005 is then the total change in F over the interval $[5, 25]$. In view of the above discussion, we can approximate the total change in F using a Riemann sum of its rate of change f. Because $n = 4$ subdivisions are specified, the width of each subdivision is

$$\Delta x = \frac{b-a}{n} = \frac{25-5}{4} = 5$$

We can therefore represent the left Riemann sum by the shaded area shown in Figure 13.
From the graph,

$$\text{Left Sum} = f(5)\Delta t + f(10)\Delta t + f(15)\Delta t + f(20)\Delta t$$
$$= (275)(5) + (300)(5) + (275)(5) + (200)(5) = 5250$$

So, we estimate that a total of 5250 death sentences were handed down during the given period.

———————

[*] The death penalty was reinstated by the U.S. Supreme Court in 1976. Source for data through 2003: Bureau of Justice Statistics, NAACP Defense Fund Inc./*New York Times,* September 15, 2004, p. A16.

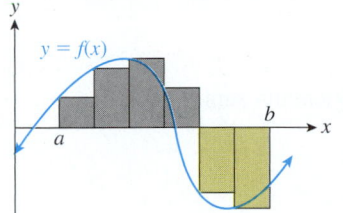

Right Riemann Sum = (300)(5)
$\quad$ + (275)(5) + (200)(5) + (75)(5)
$\quad$ = 4250

Height of each rectangle is determined by height of graph at right edge.

Figure 14

+*Before we go on...* Although in this section we focus primarily on left Riemann sums, we could also approximate the total in Example 2 using a **right Riemann sum,** as shown Figure 14.
$\qquad$ For continuous functions, the distinction between these two types of Riemann sums approaches zero as the number of subdivisions approaches infinity (see below). ∎

Example 3 Computing a Riemann Sum From a Formula

Compute the left Riemann sum for $f(x) = x^2 + 1$ over the interval $[-1, 1]$, using $n = 5$ subdivisions.

Solution Because the interval is $[a, b] = [-1, 1]$ and $n = 5$, we have

$$\Delta x = \frac{b-a}{n} = \frac{1-(-1)}{5} = 0.4 \qquad \text{Width of subdivisions}$$

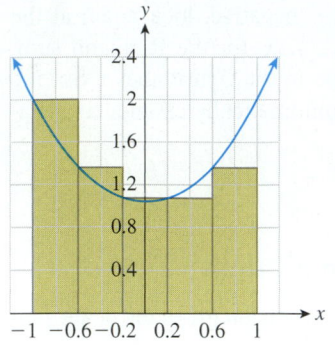

Figure 15

Thus, the subdivisions of $[-1, 1]$ are given by

$$-1 < -0.6 < -0.2 < 0.2 < 0.6 < 1 \qquad \text{Start with } -1 \text{ and keep adding } \Delta x = 0.4.$$

Figure 15 shows the graph with a representation of the Riemann sum.

The Riemann sum we want is

$$[f(x_0) + f(x_1) + \cdots + f(x_4)]\Delta x$$
$$= [f(-1) + f(-0.6) + f(-0.2) + f(0.2) + f(0.6)]0.4$$

We can conveniently organize this calculation in a table as follows:

x	-1	-0.6	-0.2	0.2	0.6	**Total**
$f(x) = x^2 + 1$	2	1.36	1.04	1.04	1.36	6.8

The Riemann sum is therefore

$$6.8\Delta x = 6.8 \times 0.4 = 2.72$$

As in Example 1, we're most interested in what happens to the Riemann sum when we let n get very large. When f is continuous,[13] its Riemann sums will always approach a limit as n goes to infinity. (This is not meant to be obvious. Proofs may be found in advanced calculus texts.) We give the limit a name.

> ### The Definite Integral
>
> If f is a continuous function, the **definite integral of f from a to b** is defined to be the limit of the Riemann sums as the number of partitions approaches infinity:
>
> $$\int_a^b f(x)\, dx = \lim_{n \to \infty} \sum_{k=0}^{n-1} f(x_k)\, \Delta x$$
>
> **In Words:** The integral, from a to b, of $f(x)\, dx$ equals the limit, as $n \to \infty$, of the Riemann Sum with a partition of n subdivisions.
>
> The function f is called the **integrand,** the numbers a and b are the **limits of integration,** and the variable x is the **variable of integration.** A Riemann sum with a large number of subdivisions may be used to approximate the definite integral.
>
> **Interpretation of the Definite Integral**
> If f is the rate of change of a quantity F (that is, $f = F'$), then $\int_a^b f(x)\, dx$ is the (exact) total change of F from $x = a$ to $x = b$.
>
> *quick* Examples
> 1. If $f(t)$ is the rate of change in the number of bats in a bell tower and $[a, b] = [2, 3]$, then $\int_2^3 f(t)\, dt$ is the total change in the number of bats in the tower from time $t = 2$ to time $t = 3$.
> 2. If, at time t hours, you are selling wall posters at a rate of $s(t)$ posters per hour, then
>
>   Total number of posters sold from hour 3 to hour 5 $= \int_3^5 s(t)\, dt$

---

[13] And for some other functions as well.

Visualizing the Definite Integral

Nonnegative Functions: If $f(x) \geq 0$ for all x in $[a, b]$, then $\int_a^b f(x)\, dx$ is the area under the graph of f over the interval $[a, b]$, as shaded in the figure.

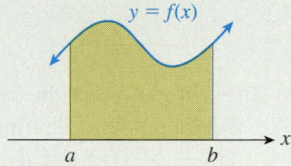

General Functions: $\int_a^b f(x)\, dx$ is the area between $x = a$ and $x = b$ that is above the x-axis and below the graph of f, minus the area that is below the x-axis and above the graph of f:

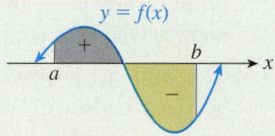

$$\int_a^b f(x)\, dx = \text{Area above } x\text{-axis} - \text{Area below } x\text{-axis}$$

quick Examples

1.

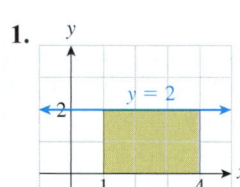

$$\int_1^4 2\, dx = \text{Area of rectangle} = 6$$

2.

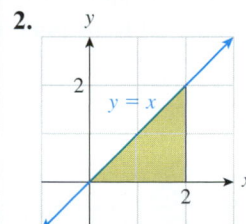

$$\int_0^2 x\, dx = \text{Area of triangle} = \frac{1}{2} \text{ base} \times \text{height} = 2$$

3.

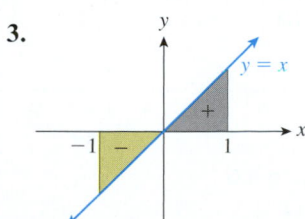

$$\int_{-1}^1 x\, dx = 0 \qquad \textcolor{red}{\text{The areas above and below the } x\text{-axis are equal}}$$

Notes

1. Remember that $\int_a^b f(x)\,dx$ stands for a number that depends on f, a, and b. The variable x that appears is called a **dummy variable** because it has no effect on the answer. In other words,

$$\int_a^b f(x)\,dx = \int_a^b f(t)\,dt \qquad \text{\textcolor{red}{x or t is just a name we give the variable.}}$$

2. The notation for the definite integral (due to Leibniz) comes from the notation for the Riemann sum. The integral sign $\int$ is an elongated S, the Roman equivalent of the Greek $\sum$. The d in dx is the lowercase Roman equivalent of the Greek Δ.

3. The definition above is adequate for continuous functions, but more complicated definitions are needed to handle other functions. For example, we broke the interval $[a, b]$ into n subdivisions of equal length, but other definitions allow a **partition** of the interval into subdivisions of possibly unequal lengths. We have evaluated f at the left endpoint of each subdivision, but we could equally well have used the right endpoint or any other point in the subdivision. All of these variations lead to the same answer when f is continuous.

4. The similarity between the notations for the definite integral and the indefinite integral is no mistake. We will discuss the exact connection in the next section. ∎

Computing Definite Integrals

In some cases, we can compute the definite integral directly from the graph (see the quick examples above and the next example below). In general, the only method of computing definite integrals we have discussed so far is numerical estimation: compute the Riemann sums for larger and larger values of n and then estimate the number it seems to be approaching as we did in Example 1. (In the next section we will discuss an algebraic method for computing them.)

Example 4   Estimating a Definite Integral From a Graph

Figure 16 shows the graph of the (approximate) rate $f'(t)$, in billions of barrels per year, at which the U.S. has been consuming oil from 1995 through 2004. (t is time in years since 1995).[*]

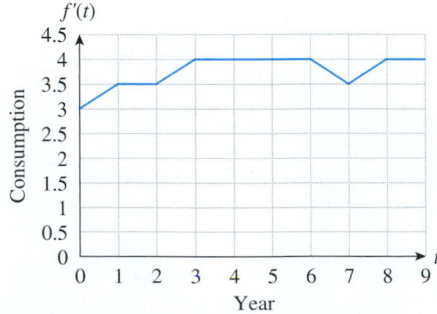

Figure **16**

[*] Source: Energy Information Administration: Standard & Poors/*New York Times,* April 23, 2005, p. C3.

Use the graph to estimate the total U.S. consumption of oil over the period shown.

Solution The derivative $f'(t)$ represents the rate of change of the U.S. consumption of oil, and so the total U.S. consumption of oil over the given period [0, 9] is given by the definite integral:

$$\text{Total U.S. consumption of oil} = \text{Total change in } f'(t) = \int_0^9 f'(t)\, dt$$

and is given by the area under the graph (Figure 17).

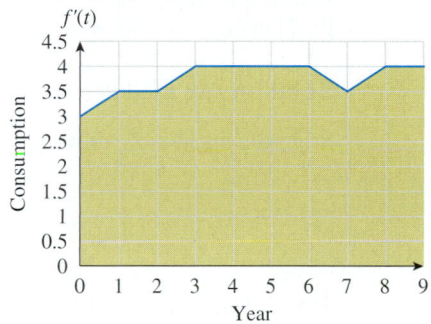

Figure **17**

One way to determine the area is to count the number of filled rectangles as defined by the grid. Each rectangle has an area of $1 \times 0.5 = 0.5$ units (so that the half-rectangles determined by diagonal portions of the graph have half that area). Counting rectangles, we find a total of 68 complete rectangles, so

$$\text{Total area} = 34$$

Because $f'(t)$ is in billions of barrels per year, we conclude that the total U.S. consumption of oil over the given period was about 34 billion barrels.

While counting rectangles might seem easy, it becomes awkward in cases involving large numbers of rectangles or partial rectangles whose area is not easy to determine. Rather than counting rectangles, we can get the area by averaging the left and right Riemann sums whose subdivisions are determined by the grid:

$$\text{Left Sum} = (3 + 3.5 + 3.5 + 4 + 4 + 4 + 4 + 3.5 + 4)(1) = 33.5$$

$$\text{Right Sum} = (3.5 + 3.5 + 4 + 4 + 4 + 4 + 3.5 + 4 + 4)(1) = 34.5$$

$$\text{Average} = \frac{33.5 + 34.5}{2} = 34$$

To see why this works, look at the single interval [0, 1]. The left sum contributes $3 \times 1 = 3$ and the right sum contributes $3.5 \times 1 = 3.5$. The exact area is their average, 3.25 (Figure 18).

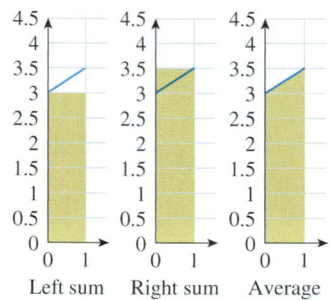

Left sum   Right sum   Average

Figure **18**

✚ *Before we go on...* It is important to check that the units we are using in Example 4 match up correctly: t is given in *years* and $f'(t)$ is given in billions of barrels per *year*. The integral is then given in

$$\text{Years} \times \frac{\text{Billions of barrels}}{\text{Year}} = \text{Billions of barrels}$$

If we had specified $f'(t)$ in, say, billions of barrels per *day* but t in years, then we would have needed to convert either t or $f'(t)$ so that the units of time match. ∎

The next example illustrates the use of technology in estimating definite integrals using Riemann sums.

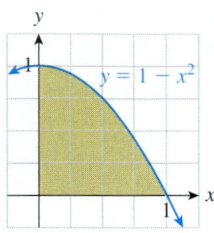

y

$y = 1 - x^2$

x

1

Figure 19

Example 5 Using Technology to Approximate the Definite Integral

Use technology to estimate the area under the graph of $f(x) = 1 - x^2$ over the interval $[0, 1]$ using $n = 100$, $n = 200$, and $n = 500$ partitions.

Solution We need to estimate the area under the parabola shown in Figure 19.

From the discussion above,

$$\text{Area} = \int_0^1 (1 - x^2)\, dx$$

The Riemann sum with $n = 100$ has $\Delta x = (b - a)/n = (1 - 0)/100 = 0.01$ and is given by

$$\sum_{k=0}^{99} f(x_k)\Delta x = [f(0) + f(0.01) + \cdots + f(0.99)](0.01)$$

Similarly, the Riemann sum with $n = 200$ has $\Delta x = (b - a)/n = (1 - 0)/200 = 0.005$ and is given by

$$\sum_{k=0}^{199} f(x_k)\Delta x = [f(0) + f(0.005) + \cdots + f(0.995)](0.005)$$

For $n = 500$, $x = (b - a)/n = (1 - 0)/500 = 0.002$ and the Riemann sum is

$$\sum_{k=0}^{499} f(x_k)\Delta x = [f(0) + f(0.002) + \cdots + f(0.998)](0.002)$$

See the Technology Guides at the end of the chapter to find out how to compute these sums on a TI-83/84 and Excel.

From the numerical results, we find the following Riemann sums:

$$n = 100: \sum_{k=0}^{99} f(x_k)\Delta x = 0.67165$$

$$n = 200: \sum_{k=0}^{199} f(x_k)\Delta x = 0.6691625$$

$$n = 500: \sum_{k=0}^{499} f(x_k)\Delta x = 0.667666$$

so we estimate that the area under the curve is about 0.667. (The exact answer is 2/3, as we will be able to verify using the techniques in the next section.)

Online, follow:

Chapter 13
→ **Tools**
  → **Numerical Integration Utility**

to obtain a utility that computes left and right Riemann sums. In the utilities, there is also a downloadable Excel spreadsheet that computes and also graphs Riemann sums (Riemann Sum Grapher).

Example 6 Motion

A fast car has velocity $v(t) = 6t^2 + 10t$ ft/s after t seconds (as measured by a radar gun). Use several values of n to find the distance covered by the car from time $t = 3$ seconds to time $t = 4$ seconds.

Solution Because the velocity $v(t)$ is rate of change of position, the total change in position over the interval [3, 4] is

$$\text{Distance covered} = \text{Total change in position} = \int_3^4 v(t)\,dt = \int_3^4 (6t^2 + 10t)\,dt$$

As in Examples 1 and 5, we can subdivide the one-second interval [3, 4] into smaller and smaller pieces to get more and more accurate approximations of the integral. By computing Riemann sums for various values of n, we get the following results.

$$n = 10: \sum_{k=0}^{9} v(t_k)\Delta t = 106.41 \qquad n = 100: \sum_{k=0}^{99} v(t_k)\Delta t \approx 108.740$$

$$n = 1000: \sum_{k=0}^{999} v(t_k)\Delta t \approx 108.974 \quad n = 10{,}000: \sum_{k=0}^{9999} v(t_k)\Delta t \approx 108.997$$

These calculations suggest that the total distance covered by the car, the value of the definite integral, is approximately 109 feet.

Online, find optional section:
Numerical Integration

$+$*Before we go on...* Do Example 6 using antiderivatives instead of Riemann sums, as in Section 13.1. Do you notice a relationship between antiderivatives and definite integrals? This will be explored in the next section. ∎

13.3 EXERCISES

● denotes basic skills exercises

tech Ex indicates exercises that should be solved using technology

In Exercises 1–8, use the given graph to estimate the left Riemann sum for the given interval with the stated number of subdivisions. hint [see Example 2]

1. ● [0, 5], $n = 5$

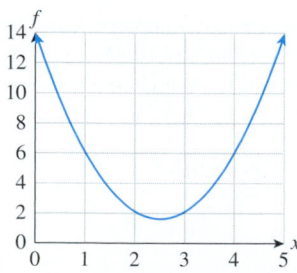

2. ● [0, 8], $n = 4$

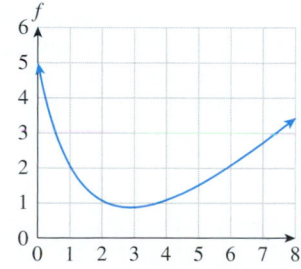

3. ● [1, 9], $n = 4$

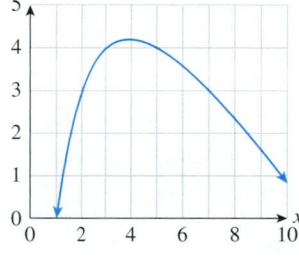

4. ● [0.5, 2.5], $n = 4$

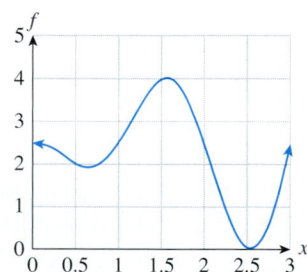

5. ● [1, 3.5], $n = 5$ **6.** ● [0.5, 3.5], $n = 3$

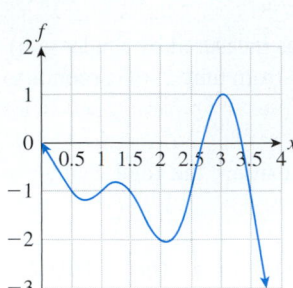   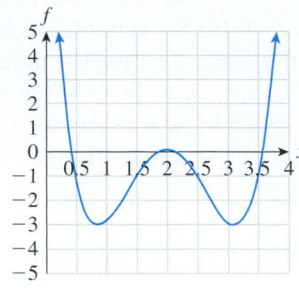

7. ● [0, 3]; $n = 3$ **8.** ● [0.5, 3]; $n = 5$

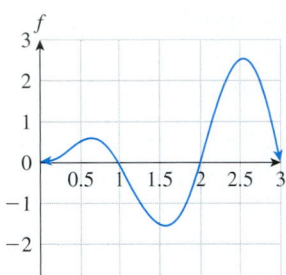   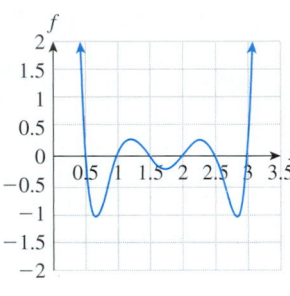

Calculate the left Riemann sums for the given functions over the given interval in Exercises 9–18, using the given values of n. (When rounding, round answers to four decimal places.) *hint* [see Example 3]

9. ● $f(x) = 4x - 1$ over [0, 2], $n = 4$

10. ● $f(x) = 1 - 3x$ over [−1, 1], $n = 4$

11. ● $f(x) = x^2$ over [−2, 2], $n = 4$

12. ● $f(x) = x^2$ over [1, 5], $n = 4$

13. ● $f(x) = \dfrac{1}{1 + x}$ over [0, 1], $n = 5$

14. ● $f(x) = \dfrac{x}{1 + x^2}$ over [0, 1], $n = 5$

15. ● $f(x) = e^{-x}$ over [0, 10], $n = 5$

16. ● $f(x) = e^{-x}$ over [−5, 5], $n = 5$

17. ● $f(x) = e^{-x^2}$ over [0, 10], $n = 4$

18. ● $f(x) = e^{-x^2}$ over [0, 100], $n = 4$

Use geometry (not Riemann sums) to compute the integrals in Exercises 19–28. *hint* [see Quick Examples p. 949]

19. ● $\displaystyle\int_0^1 1\,dx$ **20.** ● $\displaystyle\int_0^2 5\,dx$

21. ● $\displaystyle\int_0^1 x\,dx$ **22.** ● $\displaystyle\int_1^2 x\,dx$

23. ● $\displaystyle\int_0^1 \frac{x}{2}\,dx$ **24.** ● $\displaystyle\int_1^2 \frac{x}{2}\,dx$

25. ● $\displaystyle\int_2^4 (x - 2)\,dx$ **26.** ● $\displaystyle\int_3^6 (x - 3)\,dx$

27. ● $\displaystyle\int_{-1}^1 x^3\,dx$ **28.** ● $\displaystyle\int_{-2}^2 \frac{x}{2}\,dx$

In Exercises 29–34, the graph of the derivative $f'(t)$ of $f(t)$ is shown. Compute the total change of $f(t)$ over the given interval. *hint* [see Example 4]

29. ● [1, 5]   **30.** ● [2, 6]

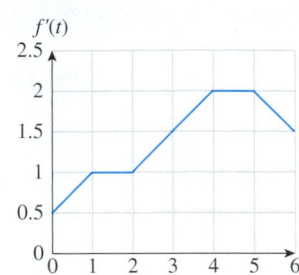   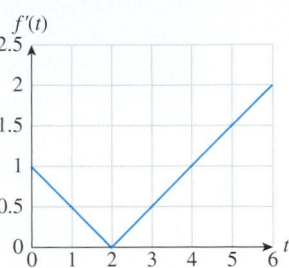

31. ● [2, 6]   **32.** ● [0, 5]

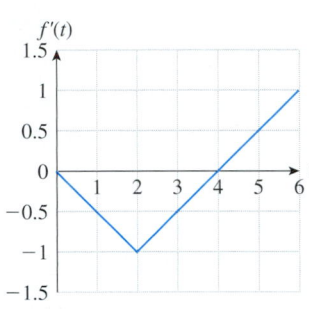   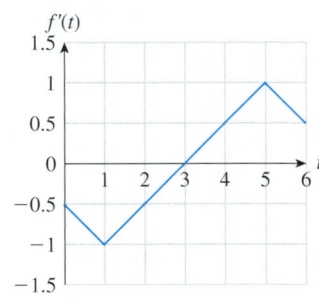

33. ● [−1, 2]   **34.** ● [−1, 2]

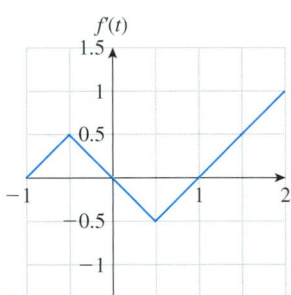   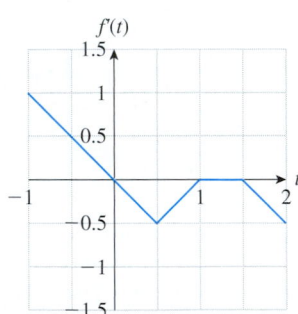

tech Ex *In Exercises 35–38, use technology to approximate the given integrals with Riemann sums, using (a) $n = 10$, (b) $n = 100$, and (c) $n = 1000$. Round all answers to four decimal places.* *hint* [see Example 5]

35. *tech* Ex $\displaystyle\int_0^1 4\sqrt{1 - x^2}\,dx$ **36.** *tech* Ex $\displaystyle\int_0^1 \frac{4}{1 + x^2}\,dx$

37. *tech* Ex $\displaystyle\int_2^3 \frac{2x^{1.2}}{1 + 3.5x^{4.7}}\,dx$

38. *tech* Ex $\displaystyle\int_3^4 3xe^{1.3x}\,dx$

● basic skills   *tech* Ex technology exercise

Applications

39. ● Cost The marginal cost function for the manufacture of portable MP3 players is given by

$$C'(x) = 20 - \frac{x}{200}$$

where x is the number of MP3 players manufactured. Use a Riemann sum with $n = 5$ to estimate the cost of producing the first 5 MP3 players. *hint* [see Example 1]

40. ● Cost Repeat the preceding exercise using the marginal cost function

$$C'(x) = 25 - \frac{x}{50}$$

41. ● Bottled-Water Sales The rate of U.S. sales of bottled water for the period 1993–2003 could be approximated by

$$R(t) = 17t^2 + 100t + 2300 \text{ million gallons per year}$$

$$(3 \le t \le 13)$$

where t is time in years since 1990.[14] Use a Riemann sum with $n = 5$ to estimate the total U.S. sales of bottled water from 1995 to 2000. (Round your answer to the nearest billion gallons.)

42. ● Bottled-Water Sales The rate of U.S. per capita sales of bottled water for the period 1993–2003 could be approximated by

$$Q(t) = 0.05t^2 + 0.4t + 9 \text{ gallons per year} \quad (3 \le t \le 13)$$

where t is the time in years since 1990.[15] Use a Riemann sum with $n = 5$ to estimate the total U.S. per capita sales of bottled water from 1995 to 2000. (Round your answer to the nearest gallon.)

43. ● Online Auctions: U.S. The following graph shows the approximate rate of change $s(t)$ of the total value, in billions of dollars, of goods sold in the U.S. through eBay. (t is time in years since 1998)

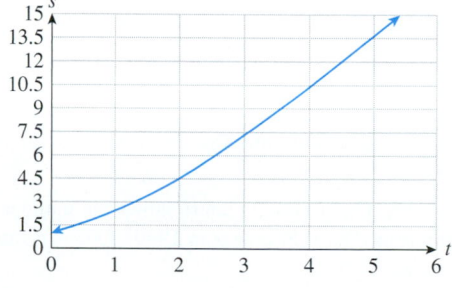

Use the graph to estimate the total value of goods sold in the U.S. through eBay for 2000–2003. (Use a left Riemann sum with 3 subdivisions).[16] *hint* [see Example 2]

44. ● Online Auctions: Global The following graph shows the approximate rate of change $s(t)$ of the total value, in billions of dollars, of goods sold through eBay throughout the world. (t is time in years since 1998).

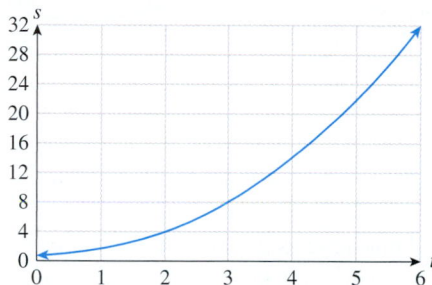

Use the graph to estimate the total value of goods sold globally through eBay for 1999–2004. (Use a left Riemann sum with 5 subdivisions).[17]

45. ● Scientific Research The rate of change $r(t)$ of the total number of research articles in the prominent journal *Physics Review* written by researchers in Europe is shown in the following graph:

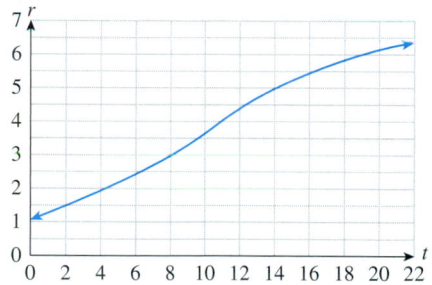

Here, t is time in years ($t = 0$ represents the start of 1983).[18]

a. Use both left and right Riemann sums with 8 subdivisions to estimate the total number of articles in *Physics Review* written by researchers in Europe during the 16-year period beginning at the start of 1983. (Estimate each value of $r(t)$ to the nearest 0.5.)

b. Use the answers from part (a) to obtain an estimate of $\int_0^{16} r(t) \, dt$. (See the "Before we go on" discussion at the end of Example 4.) Interpret the result.

---

[14] The authors' regression model, based on data in the Beverage Marketing Corporation news release, "Bottled water now number-two commercial beverage in U.S., says Beverage Marketing Corporation," April 8, 2004, available at www.beveragemarketing.com.

[15] Ibid.

[16] Source: Company Reports/Bloomberg Financial Market/*New York Times*, March 6, 2005, p. BU4.

[17] Ibid.

[18] Based on data from 1983 to 2003. Source: The American Physical Society/*New York Times*, May 3, 2003, p. A1.

46. ● *Scientific Research* The rate of change $r(t)$ of the total number of research articles in the prominent journal *Physics Review* written by researchers in the U.S. is shown in the following graph:

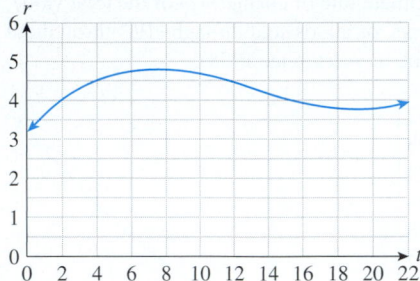

Here, t is time in years ($t = 0$ represents the start of 1983).[19]

a. Use both left and right Riemann sums with 6 subdivisions to estimate the total number of articles in *Physics Review* written by researchers in Europe during the 12-year period beginning at the start of 1993. (Estimate each value of $r(t)$ to the nearest 0.25.)

b. Use the answers from part (a) to obtain an estimate of $\int_{10}^{22} r(t)\,dt$. (See the "Before we go on" discussion at the end of Example 4.) Interpret the result.

47. ● *Visiting Students* The following graph shows the approximate rate of change $c'(t)$ in the number of students from China who have taken the GRE exam required for admission to U.S. universities (t is time in years since 2000):[20]

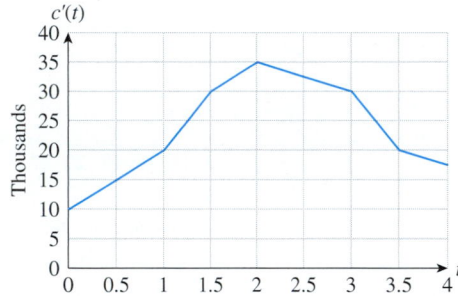

Use the graph to estimate, to the nearest 1000, the total number of students from China who took the GRE exams during the period 2002–2004. *hint* [see Example 4]

48. ● *Visiting Students* Repeat Exercise 47, using the following graph for students from India:[21]

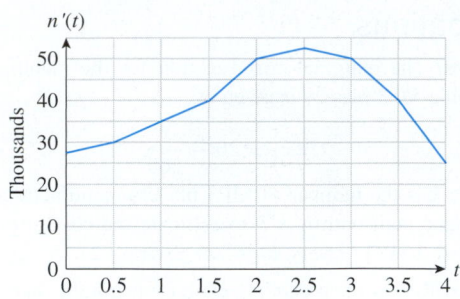

49. ● *Downsizing* The following graph shows the approximate rate of change $p'(t)$ in the total General Motors payroll for hourly employees in 2000–2004. t is time in years since 2000, and $p'(t)$ is in billions of dollars per year.[22]

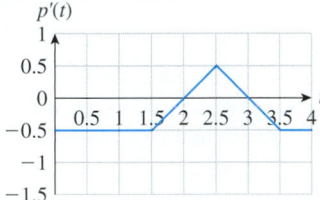

Use the graph to estimate the total change in the General Motors payroll for hourly employees for the given period.

50. ● *Downsizing* The following graph shows the approximate rate of change $n'(t)$ in the total number of General Motors employees in the U.S. in 2000–2004; t is time in years since 2000, and $n'(t)$ is thousands of employees per year.[23]

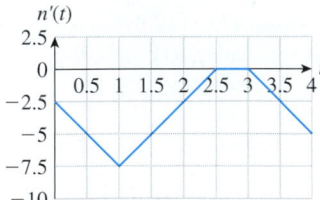

Use the graph to estimate the total change in the total number of General Motors employees in the U.S for the given period.

51. *Big Brother* The following chart shows the number of wiretaps authorized by U.S. courts in 1990–2003 (t is the number of years since 1990):[24]

[19] Based on data from 1983 to 2003. Source: The American Physical Society/*New York Times,* May 3, 2003, p. A1.

[20] Source: Educational Testing Services/Shanghai and Jiao Tong University/*New York Times,* December 21, 2004, p. A25.

[21] Ibid.

[22] Source: G.M./Automotive News/*New York Times,* June 8, 2005, p, C1.

[23] Ibid.

[24] Source: 2000 & 2003 Wiretap Reports, Administrative Office of the United States Courts, www.uscourts.gov/library/wiretap.

● basic skills   **tech**Ex technology exercise

Authorized Wiretaps

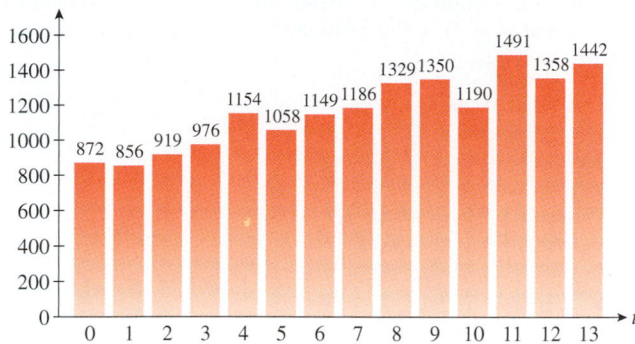

Let $W(t)$ be the number of wiretaps in year t. Estimate $\int_8^{14} W(t)\,dt$ using a Riemann sum with $n = 6$. What does this number represent?

52. Profit The following chart shows the annual profits of SABMiller, in millions of dollars, for the fiscal years ending March 31, 1997 through 2004:[25]

SABMiller Profits ($ million)

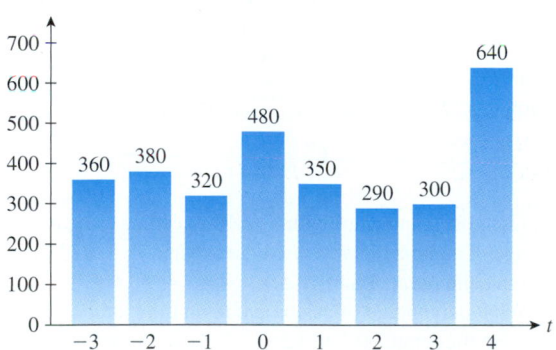

Let $p(t)$ be the annual profit of SABMiller in the year beginning at time t, where $t = 0$ represents March 31, 1999. Estimate $\int_1^5 p(t)\,dt$ using a Riemann sum with $n = 4$. What does this number represent?

53. ● Motion A model rocket has upward velocity $v(t) = 40t^2$ ft/s, t seconds after launch. Use a Riemann sum with $n = 10$ to estimate how high the rocket is 2 seconds after launch. *hint* [see Example 6]

54. ● Motion A race car has a velocity of $v(t) = 600(1 - e^{-0.5t})$ ft/s, t seconds after starting. Use a Riemann sum with $n = 10$ to estimate how far the car travels in the first 4 seconds. (Round your answer to the nearest whole number.)

55. `tech` Ex **Household Income** In the period 1967 to 2003, median household income in the U.S. increased at a rate of approximately

$$R(t) = -1.5t^2 - 0.9t + 1200 \text{ dollars per year}$$

where t is the number of years since 1990.[26] Estimate $\int_{-10}^{10} R(t)\,dt$ using a Riemann sum with $n = 100$. (Round your answer to the nearest \$1000.) Interpret the answer.

56. `tech` Ex **Health-Care Spending** In the period 1965 to 2003, the rate of increase of health-care spending in the U.S. was approximately

$$K(t) = 17t^2 + 2600t + 55{,}000 \text{ million dollars per year}$$

where t is the number of years since 1990.[27] Estimate $\int_{-10}^{10} K(t)\,dt$ using a Riemann sum with $n = 100$. (Round your answer to the nearest \$100 billion.) Interpret the answer.

57. Surveying My uncle intends to build a kidney-shaped swimming pool in his small yard, and the town zoning board will approve the project only if the total area of the pool does not exceed 500 square feet. The accompanying figure shows a diagram of the planned swimming pool, with measurements of its width at the indicated points. Will my uncle's plans be approved? Use a (left) Riemann sum to approximate the area.

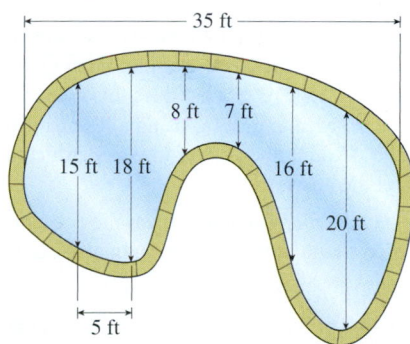

58. Pollution An aerial photograph of an ocean oil spill shows the pattern in the accompanying diagram. Assuming that the oil slick has a uniform depth of 0.01 m, how many cubic meters of oil would you estimate to be in the spill? (Volume = Area × Thickness. Use a (left) Riemann sum to approximate the area.)

[25] SOURCE: 2004 Annual Report, obtained from www.SABMiller.com/. In 2002 South African Breweries acquired Miller Brewing and changed its name to SABMiller.

[26] In current dollars, unadjusted for inflation. SOURCE: U.S. Census Bureau; "Table H-5. Race and Hispanic Origin of Householder—Households by Median and Mean Income: 1967 to 2003;" published August 27, 2004; www.census.gov/hhes/income.

[27] SOURCE: Centers for Medicare and Medicaid Services, "National Health Expenditures," 2002 version, released January 2004; www.cms.hhs.gov/statistics/nhe/.

● basic skills   `tech` Ex technology exercise

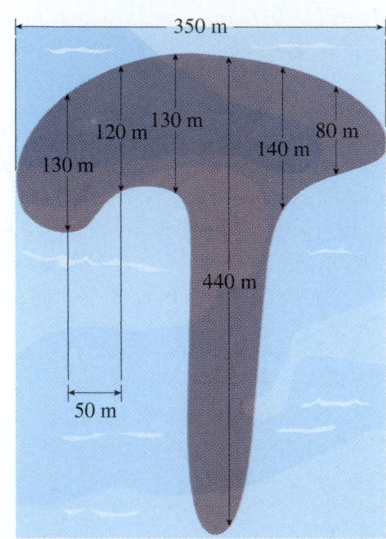

59. tech Ex *Household Income* In the period 1967 to 2003, mean annual household income in the U.S. could be approximated by

$$I(t) = -0.6t^3 + 9t^2 + 1800t + 37{,}000 \text{ dollars per household}$$

where t is the number of years since 1990.[28] Over the same period, the number of households could be approximated by

$$P(t) = 1.4t + 94 \text{ million households}$$

a. Graph the annual household income function $A(t) = I(t)P(t)$ for $0 \le t \le 14$, indicating the area that represents $\int_{10}^{14} A(t)\, dt$. What does this area mean?
b. Estimate the area in part (a) using a Riemann sum with $n = 200$. (Round the answer to two significant digits.) Interpret the answer.

60. tech Ex *Household Income* Repeat the preceding exercise, using the constant dollar median annual household income function

$$I(t) = 0.1t^3 + 10t^2 + 680t + 51{,}000 \text{ dollars per household}[29]$$

The Normal Curve The *normal distribution* curve, which models the distributions of data in a wide range of applications, is given by the function

$$p(x) = \frac{1}{\sqrt{2\pi}\,\sigma} e^{-(x-\mu)^2/2\sigma^2}$$

---

[28] In current dollars, unadjusted for inflation. SOURCE: U.S. Census Bureau; "Table H-5. Race and Hispanic Origin of Householder—Households by Median and Mean Income: 1967 to 2003;" published August 27, 2004; www.census.gov/hhes/income.

[29] In constant 2003 dollars. SOURCE: U.S. Census Bureau; "Table H-5. Race and Hispanic Origin of Householder—Households by Median and Mean Income: 1967 to 2003;" published August 27, 2004; www.census.gov/hhes/income.

where $\pi = 3.14159265\ldots$ and σ and μ are constants called the **standard deviation** and the **mean,** respectively. Its graph (when $\sigma = 1$ and $\mu = 2$) is shown in the figure. Exercises 61 and 62 illustrate its use.

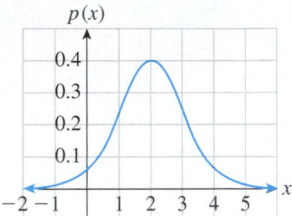

61. tech Ex *Test Scores* Enormous State University's Calculus I test scores are modeled by a normal distribution with $\mu = 72.6$ and $\sigma = 5.2$. The percentage of students who obtained scores between a and b on the test is given by

$$\int_a^b p(x)\, dx$$

a. Use a Riemann sum with $n = 40$ to estimate the percentage of students who obtained between 60 and 100 on the test.
b. What percentage of students scored less than 30?

62. tech Ex *Consumer Satisfaction* In a survey, consumers were asked to rate a new toothpaste on a scale of 1–10. The resulting data are modeled by a normal distribution with $\mu = 4.5$ and $\sigma = 1.0$. The percentage of consumers who rated the toothpaste with a score between a and b on the test is given by

$$\int_a^b p(x)\, dx$$

a. Use a Riemann sum with $n = 10$ to estimate the percentage of customers who rated the toothpaste 5 or above. (Use the range 4.5 to 10.5.)
b. What percentage of customers rated the toothpaste 0 or 1? (Use the range –0.5 to 1.5.)

Communication and Reasoning Exercises

63. ● If $f(x) = 6$, then the left Riemann sum _____ (increases/decreases/stays the same) as n increases.

64. ● If $f(x) = -1$, then the left Riemann sum _____ (increases/decreases/stays the same) as n increases.

65. ● If f is an increasing function of x, then the left Riemann sum _____ (increases/decreases/stays the same) as n increases.

66. ● If f is a decreasing function of x, then the left Riemann sum _____ (increases/decreases/stays the same) as n increases.

67. ● If $\int_a^b f(x)\, dx = 0$, what can you say about the graph of f?

68. ● Sketch the graphs of two (different) functions $f(x)$ and $g(x)$ such that $\int_a^b f(x)\, dx = \int_a^b g(x)\, dx$.

69. ● The definite integral counts the area under the x-axis as negative. Give an example that shows how this can be useful in applications.

● basic skills   tech Ex technology exercise

70. Sketch the graph of a nonconstant function whose Riemann sum with $n = 1$ gives the exact value of the definite integral.

71. Sketch the graph of a nonconstant function whose Riemann sums with $n = 1, 5,$ and 10 are all zero.

72. Besides left and right Riemann sums, another approximation of the integral is the **midpoint** approximation, in which we compute the sum

$$\sum_{k=1}^{n} f(\bar{x}_k) \, \Delta x$$

where $\bar{x}_k = (x_{k-1} + x_k)/2$ is the point midway between the left and right endpoints of the interval $[x_{k-1}, x_k]$. Why is it true that the midpoint approximation is exact if f is linear? (Draw a picture.)

73. Your cell phone company charges you $c(t) = \dfrac{20}{t + 100}$ dollars for the tth minute. You make a 60-minute phone call.

What kind of (left) Riemann sum represents the total cost of the call? Explain.

74. Your friend's cell phone company charges her $c(t) = \dfrac{20}{t + 100}$ dollars for the $(t + 1)$st minute. Your friend makes a 60-minute phone call. What kind of (left) Riemann sum represents the total cost of the call? Explain.

75. Give a formula for the **right Riemann Sum** with n equal subdivisions $a = x_0 < x_1 < \cdots < x_n = b$ for f over the interval $[a, b]$.

76. Refer to Exercise 75. If f is continuous, what happens to the difference between the left and right Riemann sums as $n \to \infty$? Explain.

77. When approximating a definite integral by computing Riemann sums, how might you judge whether you have chosen n large enough to get your answer accurate to, say, three decimal places?

● basic skills   *tech* Ex technology exercise

13.4 The Definite Integral: Algebraic Approach and the Fundamental Theorem of Calculus

In Section 13.3 we saw that the definite integral of the marginal cost function gives the total cost. However, in Section 13.1 we used antiderivatives to recover the cost function from the marginal cost function, so we *could* use antiderivatives to compute total cost. The following example, based on Example 5 in Section 13.1, compares these two approaches.

Example 1 Finding Cost from Marginal Cost

The marginal cost of producing baseball caps at a production level of x caps is $4 - 0.001x$ dollars per cap. Find the total change of cost if production is increased from 100 to 200 caps.

Solution

Method 1: Using an Antiderivative (based on Example 5 in Section 13.1): Let $C(x)$ be the cost function. Because the marginal cost function is the derivative of the cost function, we have $C'(x) = 4 - 0.001x$ and so

$$C(x) = \int (4 - 0.001x) \, dx$$

$$= 4x - 0.001\frac{x^2}{2} + K \qquad \textcolor{red}{K \text{ is the constant of integration.}}$$

$$= 4x - 0.0005x^2 + K$$

Although we do not know what to use for the value of the constant K, we can say:

Cost at production level of 100 caps $= C(100)$

$$= 4(100) - 0.0005(100)^2 + K = \$395 + K$$

Cost at production level of 200 caps $= C(200)$

$$= 4(200) - 0.0005(200)^2 + K = \$780 + K$$

Therefore,

Total change in cost $= C(200) - C(100)$

$$= (\$780 + K) - (\$395 + K) = \$385$$

Notice how the constant of integration simply canceled out! So, we could choose any value for K that we wanted (such as $K = 0$) and still come out with the correct total change. Put another way, we could use *any antiderivative* of $C'(x)$, such as

$$F(x) = 4x - 0.0005x^2$$

$F(x)$ is *any* antiderivative of $C'(x)$

or $\qquad F(x) = 4x - 0.0005x^2 + 4$

whereas $C(x)$ is the actual cost function

compute $F(200) - F(100)$, and obtain the total change, \$385.

Summarizing this method: To compute the total change of $C(x)$ over the interval [100, 200], use any antiderivative $F(x)$ of $C'(x)$, and compute $F(200) - F(100)$.

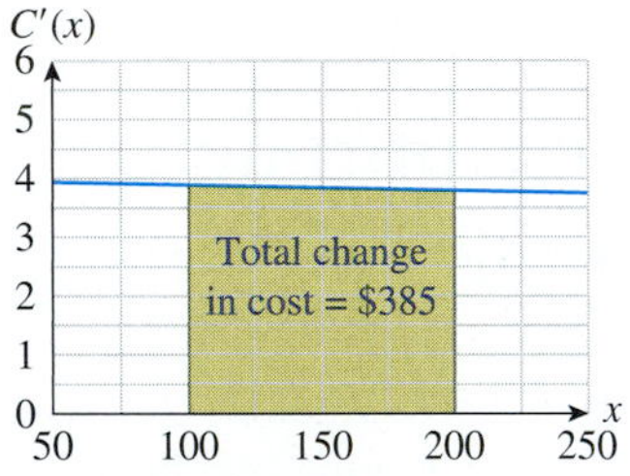

Method 2: Using a Definite Integral (based on Example 1 in Section 13.3): Because the marginal cost $C(x)$ is the rate of change of the total cost function $C(x)$, the total change in $C(x)$ over the interval [100, 200] is given by

Total change in cost $=$ Area under the marginal cost function curve

$$= \int_{100}^{200} C'(x)\, dx$$

$$= \int_{100}^{200} (4 - 0.001x)\, dx \qquad \text{See Figure 20.}$$

$$= \$385 \qquad \text{Using geometry or Riemann sums}$$

$C'(x)$

Total change in cost = \$385

Figure **20**

Putting these two methods together gives us the following surprising result:

$$\int_{100}^{200} C'(x)\, dx = F(200) - F(100)$$

where $F(x)$ is any antiderivative of $C'(x)$.

Now, there is nothing special in Example 1 about the specific function $C'(x)$ or the choice of endpoints of integration. So if we replace $C'(x)$ by a general continuous function $f(x)$, we can write

$$\int_a^b f(x)\, dx = F(b) - F(a)$$

where $F(x)$ is any antiderivative of $f(x)$. This result is known as the **Fundamental Theorem of Calculus.**

The Fundamental Theorem of Calculus (FTC)

Let f be a continuous function defined on the interval $[a, b]$ and if F is *any* antiderivative of f and is defined on $[a, b]$, we have

$$\int_a^b f(x)\,dx = F(b) - F(a)$$

Moreover, such an antiderivative is guaranteed to exist.

In Words

Every continuous function has an antiderivative. To compute the definite integral of $f(x)$ over $[a, b]$, first find an antiderivative $F(x)$, then evaluate it at $x = b$, evaluate it at $x = a$, and subtract the two answers.

quick Example

Because $F(x) = x^2$ is an antiderivative of $f(x) = 2x$,

$$\int_0^1 2x\,dx = F(1) - F(0) = 1^2 - 0^2 = 1$$

Example 2 Using the FTC to Calculate a Definite Integral

Calculate $\displaystyle\int_0^1 (1 - x^2)\,dx$

Solution To use the FTC, we need to find an antiderivative of $1 - x^2$. But we know that

$$\int (1 - x^2)\,dx = x - \frac{x^3}{3} + C$$

We need only one antiderivative, so let's take $F(x) = x - x^3/3$. The FTC tells us that

$$\int_0^1 (1 - x^2)\,dx = F(1) - F(0) = \left(1 - \frac{1}{3}\right) - (0) = \frac{2}{3}$$

which is the value we estimated in Section 13.4.

+*Before we go on...* A useful piece of notation is often used here. We write[30]

$$\left[F(x)\right]_a^b = F(b) - F(a)$$

Thus, we can rewrite the computation in Example 2 as

$$\int_0^1 (1 - x^2)\,dx = \left[x - \frac{x^3}{3}\right]_0^1 = \left(1 - \frac{1}{3}\right) - (0) = \frac{2}{3}$$

■

---

[30] There seem to be several notations in use, actually. Another common notation is $F(x)\Big|_a^b$.

Example 3 More Use of the FTC

Compute the following definite integrals.

a. $\displaystyle\int_0^1 (2x^3 + 10x + 1)\,dx$ **b.** $\displaystyle\int_1^5 \left(\frac{1}{x^2} + \frac{1}{x}\right) dx$

Solution

a. $\displaystyle\int_0^1 (2x^3 + 10x + 1)\,dx = \left[\frac{1}{2}x^4 + 5x^2 + x\right]_0^1$

$$= \left(\frac{1}{2} + 5 + 1\right) - (0)$$

$$= \frac{13}{2}$$

b. $\displaystyle\int_1^5 \left(\frac{1}{x^2} + \frac{1}{x}\right) dx = \int_1^5 (x^{-2} + x^{-1})\,dx$

$$= \left[-x^{-1} + \ln|x|\right]_1^5$$

$$= \left(-\frac{1}{5} + \ln 5\right) - (-1 + \ln 1)$$

$$= \frac{4}{5} + \ln 5$$

When calculating a definite integral, we may have to use substitution to find the necessary antiderivative. We could substitute, evaluate the indefinite integral with respect to u, express the answer in terms of x, and then evaluate at the limits of integration. However, there is a shortcut, as we shall see in the next example.

Example 4 Using the FTC with Substitution

Evaluate $\displaystyle\int_1^2 (2x - 1)e^{2x^2 - 2x}\,dx$.

Solution The shortcut we promised is to put *everything* in terms of u, including the limits of integration.

$u = 2x^2 - 2x$

$\dfrac{du}{dx} = 4x - 2$

$dx = \dfrac{1}{4x - 2}\,du$

When $x = 1,\, u = 0$ Substitute $x = 1$ in the formula for u

When $x = 2,\, u = 4$ Substitute $x = 2$ in the formula for u

We get the value $u = 0$, for example, by substituting $x = 1$ in the equation $u = 2x^2 - 2x$. We can now rewrite the integral.

$$\int_1^2 (2x - 1)e^{2x^2 - 2x}\, dx = \int_0^4 (2x - 1)e^u \frac{1}{4x - 2}\, du$$

$$= \int_0^4 \frac{1}{2} e^u\, du$$

$$= \left[\frac{1}{2} e^u\right]_0^4 = \frac{1}{2} e^4 - \frac{1}{2}$$

+ *Before we go on...* The alternative, longer calculation in Example 4 is first to calculate the indefinite integral:

$$\int (2x - 1)e^{2x^2 - 2x}\, dx = \int \frac{1}{2} e^u\, du$$

$$= \frac{1}{2} e^u + C = \frac{1}{2} e^{2x^2 - 2x} + C$$

Then we can say that

$$\int_1^2 (2x - 1)e^{2x^2 - 2x}\, dx = \left[\frac{1}{2} e^{2x^2 - 2x}\right]_1^2 = \frac{1}{2} e^4 - \frac{1}{2}$$ ∎

Applications

Example 5 Total Cost

In Section 13.3 we considered the following example. Your cell phone company offers you an innovative pricing scheme. When you make a call, the marginal cost is

$$c(t) = \frac{5}{10t + 1} \text{ dollars per hour}$$

Compute the total cost of a two-hour phone call.

Solution We calculate

$$\text{Total Cost} = \int_0^2 \frac{5}{10t + 1}\, dt = 5\int_0^2 \frac{1}{10t + 1}\, dt$$

$$= 5\left[\frac{1}{10} \ln(10t + 1)\right]_0^2 \qquad \text{See the shortcuts on p. 938}$$

$$= \frac{5}{10}[\ln(21) - \ln(1)]$$

$$= \frac{1}{2} \ln 21 \approx \$1.52$$

Compare this with Example 1 of Section 13.3, where we found the same answer by approximating with Riemann sums.

mathematics At Work

Drew Taylor

Patrick Farace

TITLE Geographic Information Systems Analyst
INSTITUTION RECON Environmental

As a Geographic Information Systems (GIS) Analyst for an environmental consulting firm in San Diego, I'm helping to create a higher quality of life for Southern Californians. My company, RECON Environmental, Inc. is dedicated to balancing the demands of industry and a growing population with the hope of preserving our cultural landscape and protecting our environment. My job is a combination of cartography and information systems; I provide spatial statistics and geographic resources to archeologists and biologists. The data I provide are presented both as hard numbers and graphically as maps and figures. These are presented in documents such as Environmental Impact Reports.

GIS is a powerful tool used to answer spatial questions that help manage our environment. Even though I'm not solving equations day to day, it is my responsibility to input the correct variables and to know what to "tell" the GIS software to do with the data. Without a background in calculus I would not be able evaluate the computer's output. My background in applied mathematics allows me to understand the methodology and to review my work by making sure the final data are in fact realistic and accurate. This ensures a quality product to my clients.

GIS can quickly calculate the area of a site and determine the distance to other geographic features such as bodies of water, developed/urban areas, or competing habitats. Like other information systems, GIS organizes and manipulates information, but what makes GIS different is that all the data it works with are spatially referenced—either by X- or Y- (and sometimes Z-elevation) coordinates or in some cases a street address. By assigning each record of a GIS database a coordinate value—essentially representing everything in space with numbers, analysts can then input these data into software such as ArcGIS or ArcView that is capable of running rigorous algorithms in order to calculate area and distance. This allows us to understand the relationship of geographic features in a less abstract, and more quantitative way, resulting in more informed decisions about the way we manage our world.

In the end, my work—with the help of understanding applied mathematics—allows policy makers, city and regional planners, and environmentally conscious individuals the ability to make intelligent and informed land use decisions that increase the quality of life for all of us.

Example 6 Computing Area

Find the total area of the region enclosed by the graph of $y = xe^{x^2}$, the x-axis, and the vertical lines $x = -1$ and $x = 1$.

Solution The region whose area we want is shown in Figure 21. Notice the symmetry of the graph. Also, half the region we are interested in is above the x-axis, while the other half is below. If we calculated the integral $\int_{-1}^{1} xe^{x^2}\, dx$, the result would be

$$\text{Area above } x\text{-axis} - \text{Area below } x\text{-axis} = 0$$

which does not give us the total area. To prevent the area below the x-axis from being combined with the area above the axis, we do the calculation in two parts, as illustrated in Figure 22.

(In Figure 22 we broke the integral at $x = 0$ because that is where the graph crosses the x-axis.) These integrals can be calculated using the substitution $u = x^2$:

$$\int_{-1}^{0} xe^{x^2}\, dx = \frac{1}{2}\left[e^{x^2}\right]_{-1}^{0} = \frac{1}{2}(1 - e) \approx -0.85914 \qquad \text{Why is it negative?}$$

$$\int_{0}^{1} xe^{x^2}\, dx = \frac{1}{2}\left[e^{x^2}\right]_{0}^{1} = \frac{1}{2}(e - 1) \approx 0.85914$$

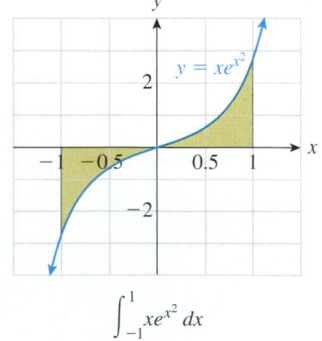

$$\int_{-1}^{1} xe^{x^2}\, dx$$

Figure 21

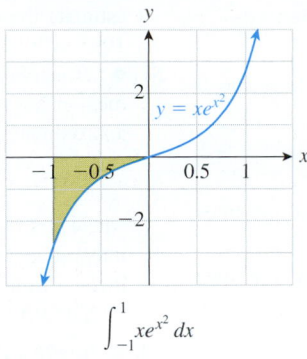

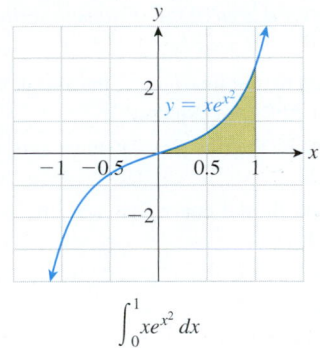

$$\int_{-1}^{1} xe^{x^2}\, dx$$

$$\int_{0}^{1} xe^{x^2}\, dx$$

Figure **22**

To obtain the total area, we should add the *absolute* values of these answers because we don't wish to count any area as negative. Thus,

$$\text{Total area} \approx 0.85914 + 0.85914 = 1.71828.$$

13.4 EXERCISES

● denotes basic skills exercises

◆ denotes challenging exercises

tech Ex indicates exercises that should be solved using technology

Evaluate the integrals in Exercises 1–42.

1. ● $\int_{-1}^{1} (x^2 + 2)\, dx$ *hint* [see Example 2]

2. ● $\int_{-2}^{1} (x - 2)\, dx$

3. ● $\int_{0}^{1} (12x^5 + 5x^4 - 6x^2 + 4)\, dx$

4. ● $\int_{0}^{1} (4x^3 - 3x^2 + 4x - 1)\, dx$

5. ● $\int_{-2}^{2} (x^3 - 2x)\, dx$ 6. ● $\int_{-1}^{1} (2x^3 + x)\, dx$

7. ● $\int_{1}^{3} \left(\frac{2}{x^2} + 3x\right) dx$ 8. ● $\int_{2}^{3} \left(x + \frac{1}{x}\right) dx$

9. ● $\int_{0}^{1} (2.1x - 4.3x^{1.2})\, dx$ 10. ● $\int_{-1}^{0} (4.3x^2 - 1)\, dx$

11. ● $\int_{0}^{1} 2e^x\, dx$ 12. ● $\int_{-1}^{0} 3e^x\, dx$

13. ● $\int_{0}^{1} \sqrt{x}\, dx$ 14. ● $\int_{-1}^{1} \sqrt[3]{x}\, dx$

15. ● $\int_{0}^{1} 2^x\, dx$ 16. ● $\int_{0}^{1} 3^x\, dx$

17. ● $\int_{0}^{1} 18(3x + 1)^5\, dx$ 18. ● $\int_{0}^{1} 8(-x + 1)^7\, dx$

19. ● $\int_{-1}^{1} e^{2x-1}\, dx$ *hint* [see Example 4]

20. ● $\int_{0}^{2} e^{-x+1}\, dx$

21. ● $\int_{0}^{2} 2^{-x+1}\, dx$ 22. ● $\int_{-1}^{1} 3^{2x-1}\, dx$

23. ● $\int_{0}^{50} e^{-0.02x-1}\, dx$ 24. ● $\int_{-20}^{0} 3e^{2.2x}\, dx$

25. ● $\int_{-1.1}^{1.1} e^{x+1}\, dx$ 26. ● $\int_{0}^{\sqrt{2}} x\sqrt{2x^2 + 1}\, dx$

27. ● $\int_{-\sqrt{2}}^{\sqrt{2}} 3x\sqrt{2x^2 + 1}\, dx$ 28. $\int_{-1.2}^{1.2} e^{-x-1}\, dx$

29. ● $\int_{0}^{1} 5xe^{x^2+2}\, dx$ 30. ● $\int_{0}^{2} \frac{3x}{x^2 + 2}\, dx$

31. ● $\int_{2}^{3} \frac{x^2}{x^3 - 1}\, dx$ 32. ● $\int_{2}^{3} \frac{x}{2x^2 - 5}\, dx$

33. ● $\int_{0}^{1} x(1.1)^{-x^2}\, dx$ 34. ● $\int_{0}^{1} x^2(2.1)^{x^3}\, dx$

35. $\int_{1}^{2} \frac{e^{1/x}}{x^2}\, dx$ 36. $\int_{1}^{2} \frac{\sqrt{\ln x}}{x}\, dx$

37. $\int_{0}^{2} \frac{x}{x + 1}\, dx$ 38. $\int_{-1}^{1} \frac{2x}{x + 2}\, dx$

39. $\int_{1}^{2} x(x - 2)^5\, dx$ 40. $\int_{1}^{2} x(x - 2)^{1/3}\, dx$

41. $\int_{0}^{1} x\sqrt{2x + 1}\, dx$ 42. $\int_{-1}^{0} 2x\sqrt{x + 1}\, dx$

● basic skills   ◆ challenging   tech Ex technology exercise

Calculate the total area of the regions described in Exercises 43–50. Do not count area beneath the x-axis as negative.

43. ● Bounded by the line $y = x$, the x-axis, and the lines $x = 0$ and $x = 1$ *hint* [see Example 6]

44. ● Bounded by the line $y = 2x$, the x-axis, and the lines $x = 1$ and $x = 2$

45. ● Bounded by the curve $y = \sqrt{x}$, the x-axis, and the lines $x = 0$ and $x = 4$

46. ● Bounded by the curve $y = 2\sqrt{x}$, the x-axis, and the lines $x = 0$ and $x = 16$

47. Bounded by the curve $y = x^2 - 1$, the x-axis, and the lines $x = 0$ and $x = 4$

48. Bounded by the curve $y = 1 - x^2$, the x-axis, and the lines $x = -1$ and $x = 2$

49. Bounded by the x-axis, the curve $y = xe^{x^2}$, and the lines $x = 0$ and $x = (\ln 2)^{1/2}$

50. Bounded by the x-axis, the curve $y = xe^{x^2-1}$ and the lines $x = 0$ and $x = 1$

Applications

51. ● **Cost** The marginal cost of producing the xth box of light bulbs is $5 + x^2/1000$ dollars. Determine how much is added to the total cost by a change in production from $x = 10$ to $x = 100$ boxes. *hint* [see Example 5]

52. ● **Revenue** The marginal revenue of the xth box of zip disks sold is $100e^{-0.001x}$ dollars. Find the revenue generated by selling items 101 through 1000.

53. ● **Displacement** A car traveling down a road has a velocity of $v(t) = 60 - e^{-t/10}$ mph at time t hours. Find the total distance it travels from time $t = 1$ hour to time $t = 6$. (Round your answer to the nearest mile.)

54. ● **Displacement** A ball thrown in the air has a velocity of $v(t) = 100 - 32t$ ft/s at time t seconds. Find the total displacement of the ball between times $t = 1$ second and $t = 7$ seconds, and interpret your answer.

55. ● **Bottled-Water Sales** The rate of U.S. sales of bottled water for the period 1993–2003 could be approximated by

$$R(t) = 17t^2 + 100t + 2300 \text{ million gallons per year}$$
$$(3 \leq t \leq 13)$$

where t is time in years since 1990.[31] Use the FTC to estimate the total U.S. sales of bottled water from 1995 to 2000. (Round your answer to the nearest billion gallons.)

56. ● **Bottled-Water Sales** The rate of U.S. per capita sales of bottled water for the period 1993–2003 could be approximated by

$$Q(t) = 0.05t^2 + 0.4t + 9 \text{ gallons per year} \quad (3 \leq t \leq 13)$$

where t is the time in years since 1990.[32] Use the FTC to

estimate the total U.S. per capita sales of bottled water from 1995 to 2000. (Round your answer to the nearest gallon.)

57. ● **Household Income** In the period 1967 to 2003, median household income in the U.S. increased at a rate of approximately

$$R(t) = -1.5t^2 - 0.9t + 1200 \text{ dollars per year}$$

where t is the number of years since 1990.[33] Use a definite integral to estimate the total change in median household income from 1980 to 2000. (Round your answer to the nearest $1000.)

58. ● **Health-Care Spending** In the period 1965 to 2003, the rate of increase of health-care spending in the U.S. was approximately

$$K(t) = 17t^2 + 2600t + 55{,}000 \text{ million dollars per year}$$

where t is the number of years since 1990.[34] Use a definite integral to estimate the total change in health-care spending from 1980 to 2000. (Round your answer to two significant digits.)

59. **Embryo Development** The oxygen consumption of a bird embryo increases from the time the egg is laid through the time the chick hatches. In a typical galliform bird, the oxygen consumption can be approximated by

$$c(t) = -0.065t^3 + 3.4t^2 - 22t + 3.6 \text{ milliliters per day}$$
$$(8 \leq t \leq 30)$$

where t is the time (in days) since the egg was laid.[35] (An egg will typically hatch at around $t = 28$.) Find the total amount of oxygen consumed during the ninth and tenth days ($t = 8$ to $t = 10$). Round your answer to the nearest milliliter.

60. **Embryo Development** The oxygen consumption of a turkey embryo increases from the time the egg is laid through the time the chick hatches. In a brush turkey, the oxygen consumption can be approximated by

$$c(t) = -0.028t^3 + 2.9t^2 - 44t + 95 \text{ milliliters per day}$$
$$(20 \leq t \leq 50)$$

where t is the time (in days) since the egg was laid [36] (An egg will typically hatch at around $t = 50$.) Find the total amount of oxygen consumed during the 21st and 22nd days ($t = 20$ to $t = 22$). Round your answer to the nearest 10 milliliters.

---

[31] The authors' regression model, based on data in the Beverage Marketing Corporation news release, "Bottled water now number-two commercial beverage in U.S., says Beverage Marketing Corporation," April 8, 2004, available at www.beveragemarketing.com.

[32] Ibid.

[33] In current dollars, unadjusted for inflation. SOURCE: U.S. Census Bureau; "Table H-5. Race and Hispanic Origin of Householder—Households by Median and Mean Income: 1967 to 2003;" published August 27, 2004; www.census.gov/hhes/income.

[34] SOURCE: Centers for Medicare and Medicaid Services, "National Health Expenditures," 2002 version, released January 2004; www.cms.hhs.gov/statistics/nhe/.

[35] The model approximates graphical data published in the article "The Brush Turkey" by Roger S. Seymour, *Scientific American,* December, 1991, pp. 108–114.

[36] Ibid.

● basic skills   ◆ challenging   **tech** Ex technology exercise

61. Sales Weekly sales of your *Lord of the Rings* T-shirts have been falling by 5% per week. Assuming you are now selling 50 T-shirts per week, how many shirts will you sell during the coming year? (Round your answer to the nearest shirt.)

62. Sales Annual sales of fountain pens in Littleville are presently 4000 per year and are increasing by 10% per year. How many fountain pens will be sold over the next five years?

63. Fuel Consumption The way Professor Waner drives, he burns gas at the rate of $1 - e^{-t}$ gallons each hour, t hours after a fill-up. Find the number of gallons of gas he burns in the first 10 hours after a fill-up.

64. Fuel Consumption The way Professor Costenoble drives, he burns gas at the rate of $1/(t + 1)$ gallons each hour, t hours after a fill-up. Find the number of gallons of gas he burns in the first 10 hours after a fill-up.

65. Total Cost Use the Fundamental Theorem of Calculus to show that if $m(x)$ is the marginal cost at a production level of x items, then the cost function $C(x)$ is given by

$$C(x) = C(0) + \int_0^x m(t)\, dt$$

What term do we use for $C(0)$?

66. Total Sales The total cost of producing x items is given by

$$C(x) = 246.76 + \int_0^x 5t\, dt$$

Find the fixed cost and the marginal cost of producing the 10th item.

67. ◆ The Logistic Function and iPod Sales

a. Show that the logistic function $f(x) = \dfrac{N}{1 + Ab^{-x}}$ can be written in the form

$$f(x) = \frac{Nb^x}{A + b^x}$$

b. Use the result of part (a) and a suitable substitution to show that

$$\int \frac{N}{1 + Ab^{-x}}\, dx = \frac{N \ln(A + b^x)}{\ln b} + C$$

c. The rate of sales of Apple iPods from the fourth quarter of 2002 through the third quarter of 2004 could be roughly approximated by the function

$$R(t) = \frac{1100}{1 + 18(1.9)^{-t}} \text{ thousand iPods per quarter}$$

$$(0 \le t \le 7)$$

where t is time in quarters since the start of 2003.[37] Use the result of part (b) to estimate, to the nearest hundred thousand, the total number of iPods sold from the start of 2003 to the middle of 2004.

68. ◆ The Logistic Function and Grants

a. Show that the logistic function $f(x) = \dfrac{N}{1 + Ae^{-kx}}$ can be written in the form

$$f(x) = \frac{Ne^{kx}}{A + e^{kx}}$$

b. Use the result of part (a) and a suitable substitution to show that

$$\int \frac{N}{1 + Ae^{-kx}}\, dx = \frac{N \ln(A + e^{kx})}{k} + C$$

c. The rate of spending on grants by U.S. foundations in the period 1993 to 2003 was approximately

$$s(t) = 11 + \frac{20}{1 + 1800e^{-0.9t}} \text{ billion dollars per year}$$

$$(3 \le t \le 13)$$

where t is the number of years since 1990.[38] Use the result of part (b) to estimate, to the nearest \$10 billion, the total spending on grants from 1998 to 2003.

69. ◆ Big Brother The number of wiretaps authorized each year by U.S. courts from 1990 to 2003 can be approximated by

$$W(t) = 620 + \frac{900e^{0.25t}}{3 + e^{0.25t}} \quad (0 \le t \le 14)$$

where t is the number of years since the start of 1990.[39]

a. Use a definite integral to estimate the total number of wiretaps from the start of 1998 to the end of 2003. (Round the answer to two significant digits.)

b. The following graph shows the actual number of authorized wiretaps.

Authorized Wiretaps

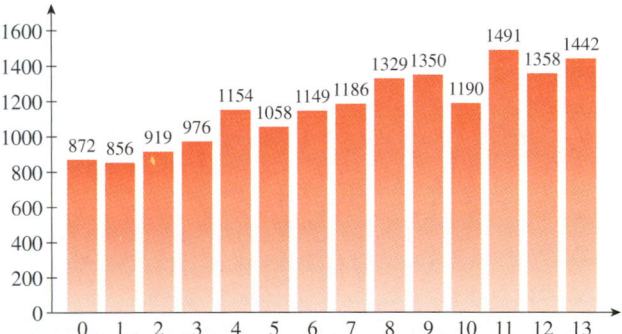

Does the integral in part (a) give an accurate estimate of the actual number to two significant digits? Explain.

[37]Based on a logistic regression. Source for data: Apple Computer, Inc., quarterly earnings reports, available at www.apple.com.

[38]Based on a logistic regression. Source for data: The Foundation Center, *Foundation Growth and Giving Estimates,* 2004, downloaded from the Center's website, http://fdncenter.org.

[39]SOURCE: 2000 & 2003 Wiretap Reports, Administrative Office of the United States Courts, www.uscourts.gov/library/wiretap.

● basic skills   ◆ challenging   **tech** Ex technology exercise

70. ◆ *Big Brother* The total number of wiretaps authorized each year by U.S. federal courts from 1990 to 2003 can be approximated by

$$W(t) = 340 + \frac{200e^{5t}}{3{,}000{,}000 + e^{5t}} \quad 0 \le t \le 13$$

where t is the number of years since the start of 1990.[40]

a. Use a definite integral to estimate the total number of federal wiretaps from the start of 1998 to the end of 2003. (Round the answer to two significant digits.)

b. The following graph shows the actual number of authorized wiretaps.

Authorized Wiretaps: Federal

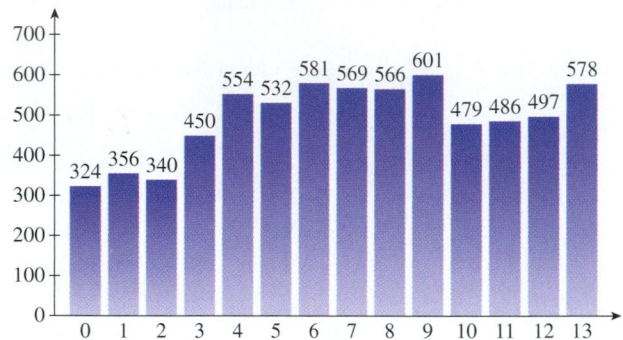

Does the integral in part (a) give an accurate estimate of the actual number to two significant digits? Explain.

71. ◆ *Kinetic Energy* The work done in accelerating an object from velocity v_0 to velocity v_1 is given by

$$W = \int_{v_0}^{v_1} v \frac{dp}{dv} dv$$

where p is its momentum, given by $p = mv$ (m = mass). Assuming that m is a constant, show that

$$W = \frac{1}{2}mv_1^2 - \frac{1}{2}mv_0^2$$

The quantity $\frac{1}{2}mv^2$ is referred to as the **kinetic energy** of the object, so the work required to accelerate an object is given by its change in kinetic energy.

72. ◆ *Einstein's Energy Equation* According to the special theory of relativity, the apparent mass of an object depends on its velocity according to the formula

$$m = \frac{m_0}{\left(1 - \dfrac{v^2}{c^2}\right)^{1/2}}$$

where v is its velocity, m_0 is the "rest mass" of the object (that is, its mass when $v = 0$), and c is the velocity of light: approximately 3×10^8 meters per second.

a. Show that, if $p = mv$ is the momentum,

$$\frac{dp}{dv} = \frac{m_0}{\left(1 - \dfrac{v^2}{c^2}\right)^{3/2}}$$

---

[40] Ibid.

b. Use the integral formula for W in the preceding exercise, together with the result in part (a), to show that the work required to accelerate an object from a velocity of v_0 to v_1 is given by

$$W = \frac{m_0c^2}{\sqrt{1 - \dfrac{v_1^2}{c^2}}} - \frac{m_0c^2}{\sqrt{1 - \dfrac{v_0^2}{c^2}}}$$

We call the quantity $\dfrac{m_0c^2}{\sqrt{1 - \dfrac{v^2}{c^2}}}$ the **total relativistic energy** of an object moving at velocity v. Thus, the work to accelerate an object from one velocity to another is given by the change in its total relativistic energy.

c. Deduce (as Albert Einstein did) that the total relativistic energy E of a body at rest with rest mass m is given by the famous equation

$$E = mc^2$$

Communication and Reasoning Exercises

73. ● Explain how the indefinite integral and the definite integral are related.

74. ● What is "definite" about the definite integral?

75. ● Complete the following: The total sales from time a to time b are obtained from the marginal sales by taking its _____ _____ from _____ to _____ .

76. ● What does the Fundamental Theorem of Calculus permit one to do?

77. Give an example of a nonzero velocity function that will produce a displacement of 0 from time $t = 0$ to time $t = 10$.

78. Give an example of a nonzero function whose definite integral over the interval [4, 6] is zero.

79. Give an example of a decreasing function $f(x)$ with the property that $\int_a^b f(x)\,dx$ is positive for every choice of a and $b > a$.

80. Explain why, in computing the total change of a quantity from its rate of change, it is useful to have the definite integral subtract area below the x-axis.

81. ◆ If $f(x)$ is a continuous function defined for $x \ge a$, define a new function $F(x)$ by the formula

$$F(x) = \int_a^x f(t)\,dt$$

Use the Fundamental Theorem of Calculus to deduce that $F'(x) = f(x)$. What, if anything, is interesting about this result?

82. ◆ tech Ex Use the result of Exercise 81 and technology to compute a table of values for $x = 1, 2, 3$ for an antiderivative of e^{-x^2} with the property that $A(0) = 0$. (Round answers to two decimal places.)

● basic skills   ◆ challenging   tech Ex technology exercise

KEY CONCEPTS

13.1 The Indefinite Integral

An antiderivative of a function f is a function F such that $F' = f$. *p. 920*

Indefinite integral $\int f(x)\,dx$ *p. 920*

Power rule for the indefinite integral:

$$\int x^n dx = \frac{x^{n+1}}{n+1} + C \qquad (\text{if } n \neq -1) \ \ p.\ 922$$

$$\int x^{-1} dx = \ln|x| + C \ \ p.\ 922$$

Indefinite Integral of e^x and b^x:

$$\int e^x dx = e^x + C$$

$$\int b^x dx = \frac{b^x}{\ln b} + C \ \ p.\ 923$$

Sums, differences, and constant multiples:

$$\int [f(x) \pm g(x)]\,dx = \int f(x)\,dx \pm \int g(x)\,dx$$

$$\int k f(x)\,dx = k \int f(x)\,dx \quad (k \ constant) \ p.\ 923$$

Combining the rules *p. 925*

Position, velocity, and acceleration:

$$v = \frac{ds}{dt} \qquad s(t) = \int v(t)\,dt$$

$$a = \frac{dv}{dt} \qquad v(t) = \int a(t)\,dt \ \ p.\ 927$$

Motion in a straight line *p. 928*
Motion in a straight line under gravity *p. 929*

13.2 Substitution

Substitution rule: $\displaystyle\int f\,dx = \int \left(\frac{f}{du/dx}\right) du$ *p. 933*

Using the substitution rule *pp. 933–934*

Shortcuts: integrals of expressions involving $(ax + b)$:

$$\int (ax+b)^n dx = \frac{(ax+b)^{n+1}}{a(n+1)} + C \quad (\text{if } n \neq -1)$$

$$\int (ax+b)^{-1} dx = \frac{1}{a} \ln|ax+b| + C$$

$$\int e^{ax+b} dx = \frac{1}{a} e^{ax+b} + C$$

$$\int c^{ax+b} dx = \frac{1}{a \ln c} c^{ax+b} + C \ \ p.\ 938$$

13.3 The Definite Integral: Numerical and Graphical Approaches

Left Riemann sum:

$$\sum_{k=0}^{n-1} f(x_k)\Delta x = [f(x_0) + f(x_1) + \cdots + f(x_{n-1})]\Delta x$$

p. 946

Computing the Riemann sum from a graph *p. 947*
Computing the Riemann sum from a formula *p. 947*
Definite integral of f from a to b:

$$\int_a^b f(x)\,dx = \lim_{n\to\infty} \sum_{k=0}^{n-1} f(x_k)\Delta x \ \ p.\ 948$$

Estimating the definite integral from a graph *p. 950*
Estimating the definite integral using technology *p. 952*
Application to motion in a straight line *p. 953*

13.4 The Definite Integral: An Algebraic Approach and The Fundamental Theorem of Calculus

The Fundamental Theorem of Calculus (FTC) *p. 961*
Using the FTC to compute definite integrals *pp. 961–962*
Computing total cost from marginal cost *p. 963*
Computing area *p. 964*

REVIEW EXERCISES

Evaluate the indefinite integrals in Exercises 1–12.

1. $\displaystyle\int (x^2 - 10x + 2)\,dx$

2. $\displaystyle\int (e^x + \sqrt{x})\,dx$

3. $\displaystyle\int \left(\frac{4x^2}{5} - \frac{4}{5x^2}\right) dx$

4. $\displaystyle\int \left(\frac{3x}{5} - \frac{3}{5x}\right) dx$

5. $\displaystyle\int e^{-2x+11}\,dx$

6. $\displaystyle\int \frac{dx}{(4x-3)^2}$

7. $\displaystyle\int x(x^2+4)^{10}\,dx$

8. $\displaystyle\int \frac{x^2+1}{(x^3+3x+2)^2}\,dx$

9. $\int 5e^{-2x}\,dx$

10. $\int xe^{-x^2/2}\,dx$

11. $\int \dfrac{x+1}{x+2}\,dx$

12. $\int x\sqrt{x-1}\,dx$

In Exercises 13 and 14, use the given graph to estimate the left Riemann sum for the given interval with the stated number of subdivisions.

13. $[0, 3]$, $n = 6$

14. $[1, 3]$, $n = 4$

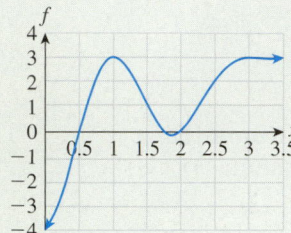
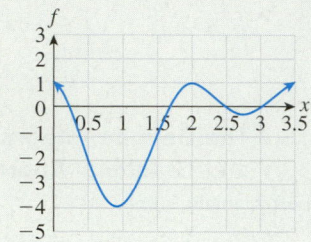

Calculate the left Riemann sums for the given functions over the given interval in Exercises 15–18, using the given values of n. (When rounding, round answers to four decimal places.)

15. $f(x) = x^2 + 1$ over $[-1, 1]$, $n = 4$

16. $f(x) = (x - 1)(x - 2) - 2$ over $[0, 4]$, $n = 4$

17. $f(x) = x(x^2 - 1)$ over $[0, 1]$, $n = 5$

18. $f(x) = \dfrac{x - 1}{x - 2}$ over $[0, 1.5]$, $n = 3$

tech Ex *In Exercises 19 and 20, use technology to approximate the given definite integrals using left Riemann sums with $n = 10, 100,$ and 1000. (Round answers to four decimal places.)*

19. $\int_0^1 e^{-x^2}\,dx$

20. $\int_1^3 x^{-x}\,dx$

In Exercises 21 and 22 the graph of the derivative $f'(x)$ of $f(x)$ is shown. Compute the total change of $f(x)$ over the given interval.

21. $[-1, 2]$

22. $[0, 2]$

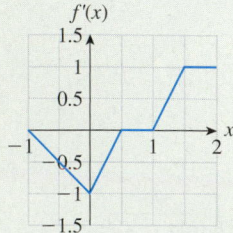
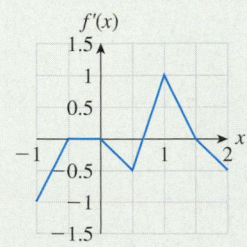

Evaluate the definite integrals in Exercises 23–30, using the Fundamental Theorem of Calculus.

23. $\int_0^1 (x - x^3)\,dx$

24. $\int_0^9 \dfrac{1}{x+1}\,dx$

25. $\int_{-1}^1 (1 + e^x)\,dx$

26. $\int_0^9 (x + \sqrt{x})\,dx$

27. $\int_0^2 x^2\sqrt{x^3 + 1}\,dx$

28. $\int_{-1}^1 3^{2x-2}\,dx$

29. $\int_0^{\ln 2} \dfrac{e^{-2x}}{1 + 4e^{-2x}}\,dx$

30. $\int_0^1 3xe^{-x^2}\,dx$

In Exercises 31–34, find the areas of the specified regions. (Do not count area below the x-axis as negative.)

31. The area bounded by $y = 4 - x^2$, the x-axis, and the lines $x = -2$ and $x = 2$.

32. The area bounded by $y = 4 - x^2$, the x-axis, and the lines $x = 0$ and $x = 5$.

33. The area bounded by $y = xe^{-x^2}$, the x-axis, and the lines $x = 0$ and $x = 5$.

34. The area bounded by $y = |2x|$, the x-axis, and the lines $x = -1$ and $x = 1$.

Applications

35. *Demand* If OHaganBooks.com were to give away its latest bestseller, *A River Burns Through It*, the demand would be 100,000 books. The marginal demand for the book is $-20p$ at a price of p dollars.

a. What is the demand function for this book?
b. At what price does demand drop to zero?

36. *Motion Under Gravity* An overworked employee at OHaganBooks.com goes to the top of the company's 100 foot tall headquarters building and flings a book up into the air at a speed of 60 feet per second.

a. When will the book hit the ground 100 feet below? (Neglect air resistance.)
b. How fast will it be traveling when it hits the ground?
c. How high will the book go?

37. *Sales* Sales at the OHaganBooks.com website of *Larry Potter Episode V: Return of the Headmasters* fluctuated rather wildly in first 5 months of last year as the following graph shows:

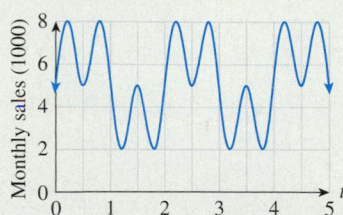

Puzzled by the graph, CEO John O'Hagan asks Jimmy Duffin[41] to estimate the total sales over the entire 5-month period shown. Jimmy decides to use a left Riemann sum with 10 partitions to estimate the total sales. What does he find?

38. *Promotions* Unlike *Larry Potter and the Riemann Sum,* sales at OHaganBooks.com of the special leather-bound gift editions of *Lord of the Rings* have been suffering lately, as shown in the following graph (negative sales indicate returns by dissatisfied customers; t is time in months since January 1 of this year):

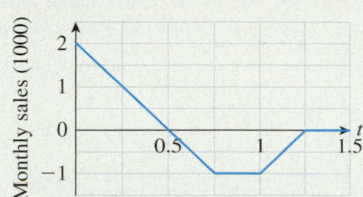

Use the graph to compute the total (net) sales over the period shown.

39. *Website Activity* The number of "hits" on the OHaganBooks. com website has been steadily increasing over the past month in response to recent publicity over a software glitch that caused the company to pay customers for buying books online. The activity can be modeled by

$$n(t) = 1000t - 10t^2 + t^3 \text{ hits per day}$$

where t is time in days since news about the software glitch was first publicized on GrungeReport.com. Use a left Riemann sum with 5 partitions to estimate the total number of hits during the first 10 days of the period.

―――――――――
[41] Marjory Duffin's nephew, currently at OHaganBooks.com on a summer internship.

40. *Legal Costs* The legal team maintained by OHaganBooks. com to handle the numerous lawsuits brought against the company by disgruntled clients may have to be expanded. The marginal monthly cost to maintain a team of x lawyers is estimated (by a method too complicated to explain) as

$$c(x) = (x - 2)^2 [8 - (x - 2)^3]^{3/2} \text{ thousand dollars per additional lawyer}$$

Compute, to the nearest $1000, the total monthly cost if O'HaganBooks goes ahead with a proposal to increase the size of the legal team from 2 to 4.

41. *Projected Sales* When OHaganBooks.com was about to go online, it estimated that its weekly sales would begin at about 6400 books per week, with sales increasing at such a rate that weekly sales would double about every 2 weeks. If these estimates had been correct, how many books would the company have sold in the first 5 weeks?

42. *Actual Sales* In fact, OHaganBooks.com modeled its weekly sales over a period of time after it went online with the function

$$s(t) = 6053 + \frac{4474e^{0.55t}}{e^{0.55t} + 14.01}$$

where t is the time in weeks after it went online. According to this model, how many books did it actually sell in the first five weeks?

vMentor Do you need a live tutor for homework problems? Access vMentor on the ThomsonNOW! website at **www.thomsonedu.com** for one-on-one tutoring from a mathematics expert.

CASE STUDY: Wage Inflation

You are the assistant personnel manager at ABC Development Enterprises, a large corporation, and yesterday you received the following memo from the Personnel Manager.

TO: SW
FROM: SC
SUBJECT: Cost of labor

Yesterday, the CEO asked me to find some mathematical formulas to estimate (1) the trend in annual wage increases and (2) the average annual wage of assembly line workers from the time they join the company to the time they retire. (She needs the information for next week's stockholder meeting.) So far, I have had very little luck: All I have been able to find is a table giving annual percentage wage increases (attached). Also, I know that the average wage for an assembly-line worker in 1981 was $25,000 per year. Do you have any ideas?

ATTACHMENT*

Date	'81	'82	'83	'84	'85	'86	'87	'88
Annual Change (%)	9.3	8.0	5.3	5.0	4.2	4.0	3.2	3.4

Date	'89	'90	'91	'92	'93	'94	'95
Annual Change (%)	4.2	4.2	4.0	3.4	2.8	3.0	2.8

---
*The data show approximate year-to-year percentage change in U.S. wages.
SOURCE: DataStream/*The New York Times,* August 13, 1995, p. 26.

(So, for example, the wages increased 9.3% from 1981 to 1982.)

Getting to work, you decide that the first thing to do is fit these data to a mathematical curve that you can use to project future changes in wages. You graph the data to get a sense of what mathematical models might be appropriate (Figure 1).

Annual Change in Wages

Figure **1**

The graph suggests a decreasing trend, leveling off at about 3%. You recall that there are a variety of curves that behave this way. One of the simplest is the curve

$$y = \frac{a}{t} + b \qquad (t \geq 1)$$

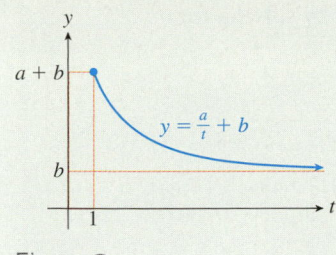

Figure **2**

where a and b are constants (Figure 2).[42]

You let $t = 1$ correspond to the time of the first data point, 1981, and convert all the percentages to decimals, giving the following table of data:

t	1	2	3	4	5	6	7	8
y	0.093	0.080	0.053	0.050	0.042	0.040	0.032	0.034
t	9	10	11	12	13	14	15	
y	0.042	0.042	0.040	0.034	0.028	0.030	0.028	

You then find the values of a and b that best fit the given data.[43] This gives you the following model for wage inflation (with figures rounded to five significant digits):

$$y = \frac{0.071813}{t} + 0.028647$$

(It is interesting that the model predicts wage inflation leveling off to about 2.86%.) Figure 2 shows the graph of y superimposed on the data.

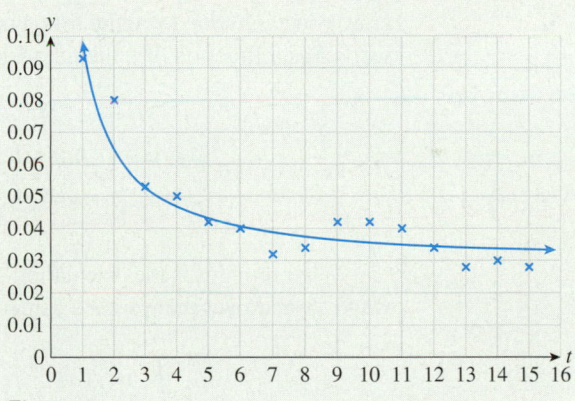

Figure **3**

Now that you have a model for wage inflation, you must use it to find the annual wage. First, you realize that the model gives the *fractional rate of increase* of wages (because it is specified as a percentage, or fraction, of the total wage). In other words, if $w(t)$ represents a worker's annual wage at time t, Then

$$y = \frac{dw/dt}{w} = \frac{d}{dt}(\ln w) \qquad \text{By the chain rule for derivatives.}$$

---

[42] There is a good mathematical reason for choosing a curve of this form: it is a first approximation (for $t \geq 1$) to a general rational function that approaches a constant as $t \to +\infty$.

[43] To do this, you note that y is a linear function of $1/t$, namely, $y = a(1/t) + b$. Then you run a linear regression on the data $(1/t, \ y)$.

You find an equation for a worker's annual wage at time t by solving for w:

$$\ln w = \int y\, dt$$

$$= \int \left(\frac{a}{t} + b\right) dt$$

$$= a \ln t + bt + C$$

where a and b are as above and C is the constant of integration, so

$$w = e^{a \ln t + bt + C}$$

To compute C, you substitute the initial data from the memo: $w(1) = 25{,}000$. Thus,

$$25{,}000 = e^{a \ln 1 + b + C} = e^{b+C} = e^{0.028647 + C}$$

Thus,

$$\ln(25{,}000) = 0.028647 + C$$

which gives

$$C = \ln(25{,}000) - 0.028647 \approx 10.098 \qquad \text{(to 5 significant digits).}$$

Now you can write down the following formula for the annual wage of an assembly-line worker as a function of t, the number of years since 1980:

$$w(t) = e^{a \ln t + bt + C} = e^{0.071813 \ln t + 0.028647t + 10.098}$$

$$= e^{0.071813 \ln t}\, e^{0.028647t}\, e^{10.098}$$

$$= t^{0.071813}\, e^{0.028647t}\, e^{10.098}$$

What remains is the calculation of the average annual wage. The average is the total wage earned over the worker's career divided by the number of years worked:

$$\bar{w} = \frac{1}{s - r} \int_{r}^{s} w(t)\, dt$$

where r is the time an employee begins working at the company and s is the time he or she retires. Substituting the formula for $w(t)$ gives

$$\bar{w} = \frac{1}{s - r} \int_{r}^{s} t^{0.071813}\, e^{0.028647t}\, e^{10.098}\, dt$$

$$= \frac{e^{10.098}}{s - r} \int_{r}^{s} t^{0.071813}\, e^{0.028647t}\, dt$$

You cannot find an explicit antiderivative for the integrand, so you decide that the only way to compute it is numerically. You send the following memo to SC.

TO: SC
FROM: SW
SUBJECT: The formula you wanted

The average annual salary of an assembly-line worker here at ABC Development Enterprises is given by the formula

$$\bar{w} = \frac{e^{10.098}}{s-r} \int_{r}^{s} t^{0.071813} e^{0.028647t} \, dt$$

where r is the time in years after 1980 that a worker joins ABC and s is the time (in years after 1980) the worker retires. (The formula is valid only from 1981 on.) To calculate it easily (and impress the board members), I suggest you enter the following on your graphing calculator:

```
Y₁=(e^(10.098)/(S-R))fnInt(T^0.071813e^(0.028647T),T,R,S)
```

Then suppose, for example, that a worker joined the company in 1983 ($r = 3$) and retired in 2001 ($s = 21$). All you do is enter

```
3→R
21→S
Y₁
```

and your calculator will give you the result: The average salary of the worker is $41,307.16.

Have a nice day.
SW

Exercises

1. Use the model developed above to compute the average annual income of a worker who joined the company in 1998 and left 3 years later.

2. What was the total amount paid by ABC Enterprises to the worker in Exercise 1?

3. What (if any) advantages are there to using a model for the annual wage inflation rate when the actual annual wage inflation rates are available?

4. The formula in the model was based on a 1981 salary of $25,000. Change the model to allow for an arbitrary 1981 salary of $\$w_0$.

5. **tech** Ex If we had used exponential regression to model the wage inflation data, we would have obtained

$$y = 0.07142 e^{-0.067135t}$$

Graph this equation along with the actual wage data and the earlier model for y. Is this a better model or a worse model?

6. Use the actual data in the table to calculate the average salary of an assembly-line worker for the 6-year period from 1981 through the end of 1986, and compare it with the figure predicted by the model in the text.

TECHNOLOGY GUIDE

Section 13.3

Example 5 Estimate the area under the graph of $f(x) = 1 - x^2$ over the interval $[0, 1]$ using $n = 100$, $n = 200$, and $n = 500$ partitions.

Solution with Technology   There are several ways to compute Riemann sums with a graphing calculator. We illustrate one method. For $n = 100$, we need to compute the sum

$$\sum_{k=0}^{99} f(x_k)\Delta x = [f(0) + f(0.01) + \cdots + f(0.99)](0.01) \quad \text{See discussion in Example 5}$$

Thus, we first need to calculate the numbers $f(0)$, $f(0.01)$, and so on, and add them up. The TI-83/84 has a built-in sum function (available in the LIST MATH menu), which, like the SUM function in a spreadsheet, sums the entries in a list.

1. To generate a list that contains the numbers we want to add together, use the seq function (available in the LIST OPS menu). If we enter

$$\texttt{seq(1-X\^2,X,0,0.99,0.01)} \quad \text{seq:} \boxed{\text{2ND}} \boxed{\text{LIST}} \text{ OPS } \boxed{5}$$

the calculator will calculate a list by evaluating $\texttt{1-X\^2}$ for values of X from 0 to 0.99 in steps of 0.01.

2. To take the sum of all these numbers, we wrap the seq function in a call to sum:

$$\texttt{sum(seq(1-X\^2,X,0,0.99,0.01))} \quad \text{sum:} \boxed{\text{2ND}} \boxed{\text{LIST}} \text{ MATH } \boxed{5}$$

This gives the sum

$$f(0) + f(0.01) + \cdots + f(0.99) = 67.165$$

3. To obtain the Riemann sum, we need to multiply this sum by $\Delta x = 0.01$, and we obtain the estimate of $67.165 \approx 0.67165$ for the Riemann sum:

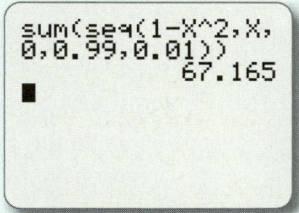

We obtain the other Riemann sums similarly, as shown here:

$n = 200$ $n = 500$

One disadvantage of this method is that the TI-83/84 can generate and sum a list of at most 999 entries. The TI-83/84 also has a built-in function `fnInt`, which finds a very accurate approximation of a definite integral, but it uses a more sophisticated technique than the one we are discussing here.

The LEFTSUM Program for the TI-83/84

The following program calculates (left) Riemann sums for any n. The latest version of this program (and others) is available at the website.

```
PROGRAM: LEFTSUM
:Input "LEFT ENDPOINT? ",A          Prompts for the left end-point a
:Input "RIGHT ENDPOINT? ",B         Prompts for the right end-point b
:Input "N? ",N                      Prompts for the number of rectangles
:(B-A)/N→D                          D is Δx = (b − a)/n
:Ø→L                                L will eventually be the left sum
:A→X                                X is the current x-coordinate
:For(I,1,N)                         Start of a loop—recall the sigma notation
:L+Y₁→L                             Add f(xᵢ₋₁) to L
:A+I*D→X                            Uses formula xᵢ = a + iΔx
:End                                End of loop
:L*D→L                              Multiply by Δx
:Disp "LEFT SUM IS ",L
:Stop
```

EXCEL Technology Guide

Section 13.3

Example 5 Estimate the area under the graph of $f(x) = 1 - x^2$ over the interval $[0, 1]$ using $n = 100$, $n = 200$, and $n = 500$ partitions.

Solution with Technology We need to compute various sums

$$\sum_{k=0}^{99} f(x_k)\Delta x = [f(0) + f(0.01) + \cdots + f(0.99)](0.01) \qquad \text{See discussion in Example 5}$$

$$\sum_{k=0}^{199} f(x_k)\Delta x = [f(0) + f(0.005) + \cdots + f(0.995)](0.005)$$

$$\sum_{k=0}^{499} f(x_k)\Delta x = [f(0) + f(0.002) + \cdots + f(0.998)](0.002)$$

Here is how you can compute them all on same spreadsheet.

1. Enter the values for the endpoints a and b, the number of subdivisions n, and the formula $\Delta x = (b - a)/n$:

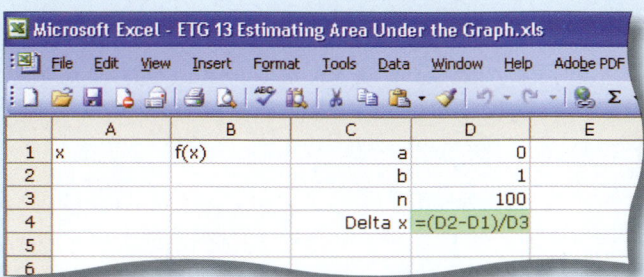

2. Next, we compute all the x-values we might need in column A. Because the largest value of n that we will be using is 500, we will need a total of 501 values of x. Note that the value in each cell below A3 is obtained from the one above by adding Δx.

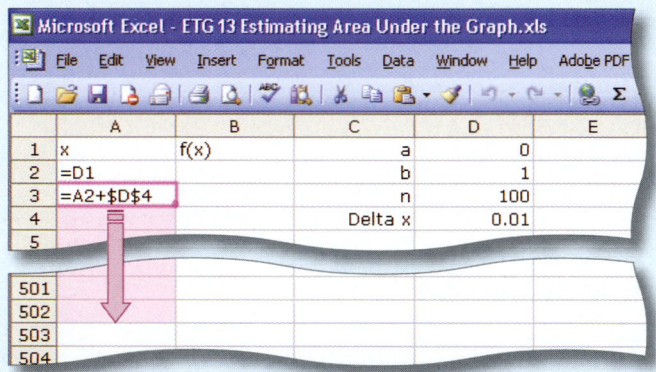

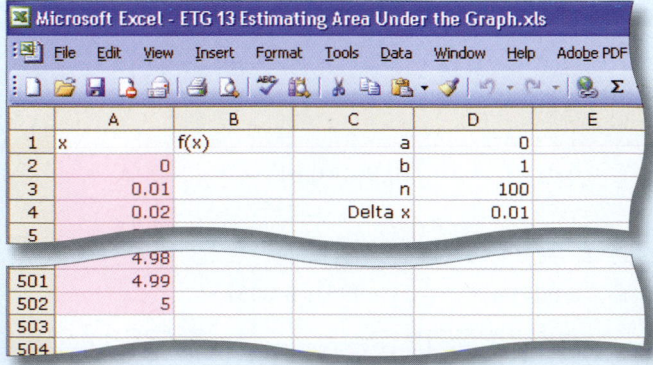

(The fact that the values of x presently go too far will be corrected in the next step.)

3. We need to calculate the numbers $f(0)$, $f(0.01)$, and so on, but only those for which the corresponding x-value is less than b. To do this, we use a logical formula as we did with piecewise-defined functions in Chapter 1:

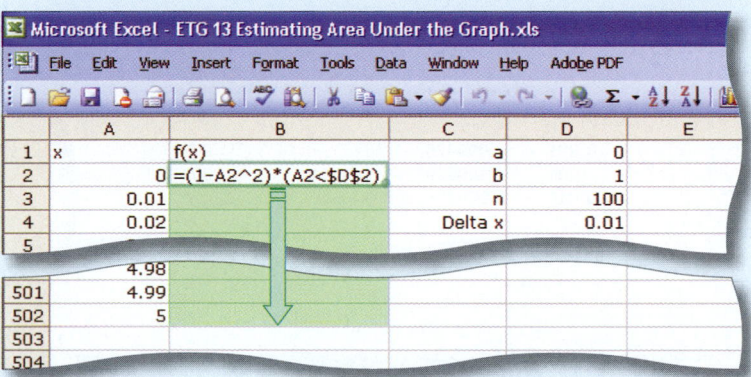

When the value of x is b or above, the function will evaluate to zero, because we do not want to count it.

4. Finally, we compute the Riemann sum by adding up everything in Column B and multiplying by Δx:

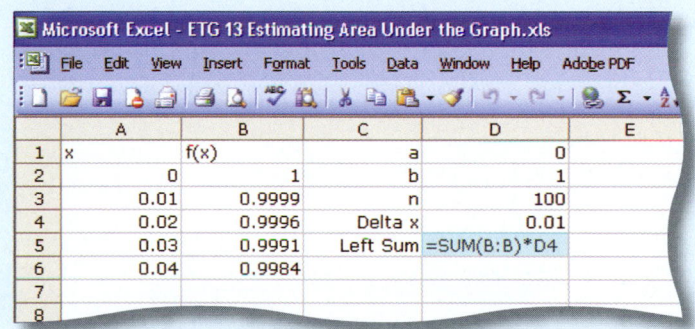

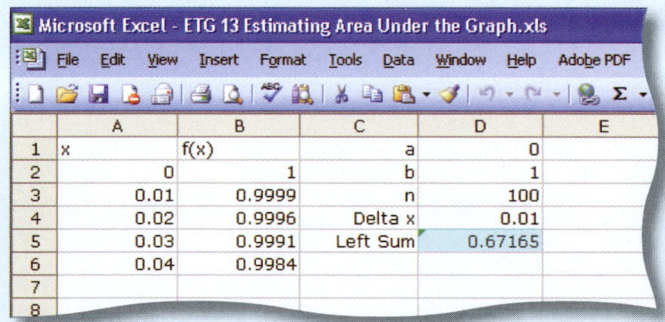

Now it is easy to obtain the sums for $n = 200$ and $n = 500$: Simply change the value of n in cell D3:

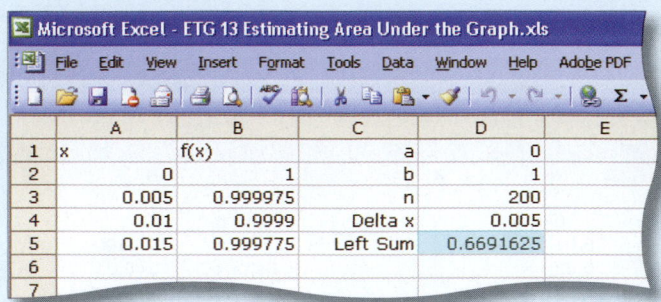

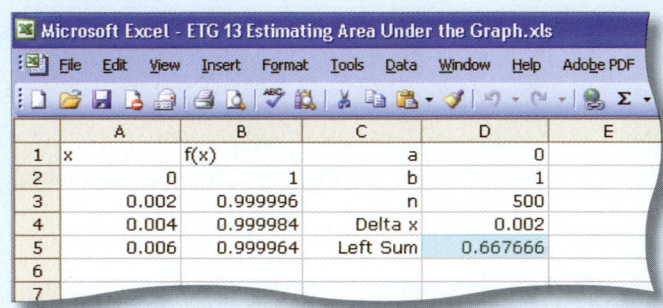

14

Further Integration Techniques and Applications of the Integral

CASE STUDY Estimating Tax Revenues

You have just been hired by the incoming administration to coordinate national tax policy, and the so-called experts on your staff can't seem to agree on which of two tax proposals will result in more revenue for the government. The data you have are the two income tax proposals (graphs of tax vs. income) and the distribution of incomes in the country. How do you use this information to decide which tax policy will result in more revenue?

Bob Daemmrich/The Image Works

Introduction

In the preceding chapter, we learned how to compute many integrals and saw some of the applications of the integral. In this chapter, we look at some further techniques for computing integrals and then at more applications of the integral. We also see how to extend the definition of the definite integral to include integrals over infinite intervals, and show how such integrals can be used for long-term forecasting. Finally, we introduce the beautiful theory of differential equations and some of its numerous applications.

14.1   Integration by Parts

Integration by parts is an integration technique that comes from the product rule for derivatives. The tabular method we present here has been around for some time and makes integration by parts quite simple, particularly in problems where it has to be used several times.[1]

We start with a little notation to simplify things while we introduce integration by parts (we use this notation only in the next few pages). If u is a function, denote its derivative by $D(u)$ and an antiderivative by $I(u)$. Thus, for example, if $u = 2x^2$, then

$$D(u) = 4x$$

and

$$I(u) = \frac{2x^3}{3}$$

[If we wished, we could instead take $I(u) = \frac{2x^3}{3} + 46$, but we usually opt to take the simplest antiderivative.]

Integration by Parts

If u and v are continuous functions of x, and u has a continuous derivative, then

$$\int u \cdot v \, dx = u \cdot I(v) - \int D(u)I(v) \, dx$$

quick Example   (The following will be discussed more fully in Example 1 below.)

$$\int x \cdot e^x \, dx = x I(e^x) - \int D(x)I(e^x) \, dx$$

$$= xe^x - \int 1 \cdot e^x \, dx \qquad I(e^x) = e^x; \ D(x) = 1$$

$$= xe^x - e^x + C \qquad \int e^x \, dx = e^x + C$$

As the Quick Example shows, although we could not immediately integrate $u \cdot v = x \cdot e^x$, we could easily integrate $D(u)I(v) = 1 \cdot e^x = e^x$.

---

[1] The version of the tabular method we use was developed and taught to us by Dan Rosen at Hofstra University.

Derivation of Integration by Parts Formula

As we mentioned, the integration-by-parts formula comes from the product rule for derivatives. We apply the product rule to the function $uI(v)$:

$$D[u \cdot I(v)] = D(u)I(v) + uD(I(v))$$
$$= D(u)I(v) + uv$$

because $D(I(v))$ is the derivative of an antiderivative of v, which is v. Integrating both sides gives

$$u \cdot I(v) = \int D(u)I(v)\, dx + \int uv\, dx$$

A simple rearrangement of the terms now gives us the integration-by-parts formula.

The integration-by-parts formula is easiest to use via the tabular method illustrated in the following example, where we repeat the calculation we did in the Quick Example above.

Example 1 Integration by Parts: Tabular Method

Calculate $\displaystyle\int xe^x\, dx$.

Solution First, the reason we *need* to use integration by parts to evaluate this integral is that none of the other techniques of integration that we've talked about up to now will help us. Furthermore, we cannot simply find antiderivatives of x and e^x and multiply them together. [You should check that $(x^2/2)e^x$ is *not* an antiderivative for xe^x.] However, as we saw above, this integral can be found by integration by parts. We want to find the integral of the *product* of x and e^x. We must make a decision: Which function will play the role of u and which will play the role of v in the integration-by-parts formula? Because the derivative of x is just 1, differentiating makes it simpler, so we try letting x be u and letting e^x be v. We need to calculate $D(u)$ and $I(v)$, which we record in the following table.

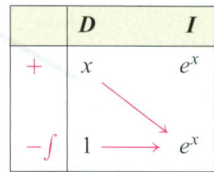

$$+\, x \cdot e^x - \int 1 \cdot e^x\, dx$$

Below x in the D column, we put $D(x) = 1$; below e^x in the I column, we put $I(e^x) = e^x$. The arrow at an angle connecting x and $I(e^x)$ reminds us that the product $xI(e^x)$ will appear in the answer; the plus sign on the left of the table reminds us that it is $+xI(e^x)$ that appears. The integral sign and the horizontal arrow connecting $D(x)$ and $I(e^x)$ remind us that the *integral* of the product $D(x)I(e^x)$ also appears in the answer; the minus sign on the left reminds us that we need to subtract this integral. Combining these two contributions, we get

$$\int xe^x\, dx = xe^x - \int e^x\, dx$$

The integral that appears on the right is much easier than the one we began with, so we can complete the problem:

$$\int xe^x\, dx = xe^x - \int e^x\, dx = xe^x - e^x + C$$

+*Before we go on...* In Example 1, what if we had made the opposite decision and put e^x in the D column and x in the I column? Then we would have had the following table:

	D	I
$+$	e^x	x
$-\int$	$e^x \longrightarrow$	$x^2/2$

This gives

$$\int x e^x \, dx = \frac{x^2}{2} e^x - \int \frac{x^2}{2} e^x \, dx$$

The integral on the right is harder than the one we started with, not easier! How do we know beforehand which way to go? We don't. We have to be willing to do a little trial and error: We try it one way, and if it doesn't make things simpler, we try it another way. *Remember, though, that the function we put in the I column must be one that we can integrate.* ∎

Example 2 Repeated Integration by Parts

Calculate $\displaystyle\int x^2 e^{-x} \, dx$.

Solution Again, we have a product—the integrand is the product of x^2 and e^{-x}. Because differentiating x^2 makes it simpler, we put it in the D column and get the following table:

	D	I
$+$	x^2	e^{-x}
$-\int$	$2x \longrightarrow$	$-e^{-x}$

This table gives us

$$\int x^2 e^{-x} \, dx = x^2(-e^{-x}) - \int 2x(-e^{-x}) \, dx$$

The last integral is simpler than the one we started with, but it still involves a product. It's a good candidate for another integration by parts. The table we would use would start with $2x$ in the D column and $-e^{-x}$ in the I column, which is exactly what we see in the last row of the table we've already made. Therefore, we *continue the process,* elongating the table above:

	D	I
$+$	x^2	e^{-x}
$-$	$2x$	$-e^{-x}$
$+\int$	$2 \longrightarrow$	e^{-x}

(Notice how the signs on the left alternate. Here's why: To compute $-\int 2x(-e^{-x})\,dx$ we use the negative of the following table:

	D	I
$+$	$2x$	$-e^{-x}$
$-\int$	$2 \longrightarrow$	e^{-x}

so we reverse all the signs.)

Now, we still have to compute an integral (the integral of the product of the functions in the bottom row) to complete the computation. But why stop here? Let's continue the process one more step:

	D	I
$+$	x^2	e^{-x}
$-$	$2x$	$-e^{-x}$
$+$	2	e^{-x}
$-\int$	$0 \longrightarrow$	$-e^{-x}$

In the bottom line we see that all that is left to integrate is $0(-e^{-x}) = 0$. Since the indefinite integral of 0 is C, we can read the answer from the table as

$$\int x^2 e^{-x}\,dx = x^2(-e^{-x}) - 2x(e^{-x}) + 2(-e^{-x}) + C$$

$$= -x^2 e^{-x} - 2xe^{-x} - 2e^{-x} + C$$

$$= -e^{-x}(x^2 + 2x + 2) + C$$

In Example 2 we saw a technique that we can summarize as follows:

Integrating a Polynomial Times a Function

If one of the factors in the integrand is a polynomial and the other factor is a function that can be integrated repeatedly, put the polynomial in the D column and keep differentiating until you get zero. Then complete the I column to the same depth, and read off the answer.

For practice, redo Example 1 using this technique.

It is not always the case that the integrand is a polynomial times something easy to integrate, so we can't always expect to end up with a zero in the D column. In that case we hope that at some point we will be able to integrate the product of the functions in the last row. Here are some examples.

Example 3 Polynomial Times a Logarithm

Calculate: **a.** $\displaystyle\int x \ln x \, dx$ **b.** $\displaystyle\int (x^2 - x) \ln x \, dx$ **c.** $\displaystyle\int \ln x \, dx$

Solution

a. This is a product and therefore a good candidate for integration by parts. Our first impulse is to differentiate x, but that would mean integrating $\ln x$, and we do not (yet) know how to do that. So we try it the other way around and hope for the best.

	D	I
$+$	$\ln x$	x
$-\int$	$\dfrac{1}{x}$ $\longrightarrow$	$\dfrac{x^2}{2}$

Why did we stop? If we continued the table, both columns would get more complicated. However, if we stop here we get

$$\int x \ln x \, dx = (\ln x)\left(\frac{x^2}{2}\right) - \int \left(\frac{1}{x}\right)\left(\frac{x^2}{2}\right) dx$$

$$= \frac{x^2}{2} \ln x - \frac{1}{2} \int x \, dx$$

$$= \frac{x^2}{2} \ln x - \frac{x^2}{4} + C$$

b. We can use the same technique we used in part (a) to integrate any polynomial times the logarithm of x:

	D	I
$+$	$\ln x$	$x^2 - x$
$-\int$	$\dfrac{1}{x}$ $\longrightarrow$	$\dfrac{x^3}{3} - \dfrac{x^2}{2}$

$$\int (x^2 - x) \ln x \, dx = (\ln x)\left(\frac{x^3}{3} - \frac{x^2}{2}\right) - \int \left(\frac{1}{x}\right)\left(\frac{x^3}{3} - \frac{x^2}{2}\right) dx$$

$$= \left(\frac{x^3}{3} - \frac{x^2}{2}\right) \ln x - \int \left(\frac{x^2}{3} - \frac{x}{2}\right) dx$$

$$= \left(\frac{x^3}{3} - \frac{x^2}{2}\right) \ln x - \frac{x^3}{9} + \frac{x^2}{4} + C$$

c. The integrand $\ln x$ is not a product. We can, however, *make* it into a product by thinking of it as $1 \cdot \ln x$. Because this is a polynomial times $\ln x$, we proceed as in parts (a) and (b):

	D	I
$+$	$\ln x$	1
$-\int$	$1/x$ $\longrightarrow$	x

We notice that the product of $1/x$ and x is just 1, which we know how to integrate, so we can stop here:

$$\int \ln x \, dx = x \ln x - \int \left(\frac{1}{x}\right) x \, dx$$

$$= x \ln x - \int 1 \, dx$$

$$= x \ln x - x + C$$

FAQs   Whether to Use Integration by Parts, and What Goes in the D and I Columns

Q: *Will integration by parts always work to integrate a product?*

A: No. While integration by parts often works for products in which one factor is a polynomial, it will almost *never* work in the examples of products we saw when discussing substitution in Section 13.2. For example, although integration by parts can be used to compute $\int (x^2 - x)e^{2x-1} \, dx$ (put $x^2 - x$ in the D column and e^{2x-1} in the I column), it *cannot* be used to compute $\int (2x - 1)e^{x^2-x} \, dx$ (put $u = x^2 - x$). Recognizing when to use integration by parts is best learned by experience. ■

Q: *When using integration by parts, which expression goes in the D column, and which in the I column?*

A: Although there is no general rule, the following guidelines are useful:

- To integrate a product in which one factor is a polynomial and the other can be integrated several times, put the polynomial in the D column and the other factor in the I column. Then differentiate the polynomial until you get zero.
- If one of the factors is a polynomial but the other factor cannot be integrated easily, put the polynomial in the I column and the other factor in the D column. Stop when the product of the functions in the bottom row can be integrated.
- If neither factor is a polynomial, put the factor that seems easier to integrate in the I column and the other factor in the D column. Again, stop the table as soon as the product of the functions in the bottom row can be integrated.
- If your method doesn't work, try switching the functions in the D and I columns or try breaking the integrand into a product in a different way. If none of this works, maybe integration by parts isn't the technique to use on this problem. ■

14.1 EXERCISES

- ● denotes basic skills exercises
- ◆ denotes challenging exercises

Evaluate the integrals in Exercises 1–38.

1. ● $\int 2xe^x \, dx$ *hint* [see Example 1]

2. ● $\int 3xe^{-x} \, dx$

3. ● $\int (3x - 1)e^{-x} \, dx$

4. ● $\int (1 - x)e^x \, dx$

5. ● $\int (x^2 - 1)e^{2x} \, dx$

6. ● $\int (x^2 + 1)e^{-2x} \, dx$

7. ● $\int (x^2 + 1)e^{-2x+4} \, dx$

8. ● $\int (x^2 + 1)e^{3x+1} \, dx$

9. ● $\int (2 - x)2^x \, dx$

10. ● $\int (3x - 2)4^x \, dx$

11. ● $\int (x^2 - 1)3^{-x} \, dx$

12. ● $\int (1 - x^2)2^{-x} \, dx$

● basic skills   ◆ challenging

13. $\displaystyle\int \frac{x^2 - x}{e^x}\, dx$ 14. $\displaystyle\int \frac{2x + 1}{e^{3x}}\, dx$

15. ● $\displaystyle\int x(x + 2)^6\, dx$ 16. ● $\displaystyle\int x^2(x - 1)^6\, dx$

17. $\displaystyle\int \frac{x}{(x - 2)^3}\, dx$ 18. $\displaystyle\int \frac{x}{(x - 1)^2}\, dx$

19. ● $\displaystyle\int x^3 \ln x\, dx$ *hint* [see Example 3]

20. ● $\displaystyle\int x^2 \ln x\, dx$

21. ● $\displaystyle\int (t^2 + 1) \ln(2t)\, dt$ 22. ● $\displaystyle\int (t^2 - t) \ln(-t)\, dt$

23. ● $\displaystyle\int t^{1/3} \ln t\, dt$ 24. ● $\displaystyle\int t^{-1/2} \ln t\, dt$

25. ● $\displaystyle\int \log_3 x\, dx$ 26. ● $\displaystyle\int x \log_2 x\, dx$

27. $\displaystyle\int (xe^{2x} - 4e^{3x})\, dx$ 28. $\displaystyle\int (x^2 e^{-x} + 2e^{-x+1})\, dx$

29. $\displaystyle\int (x^2 e^x - xe^{x^2})\, dx$ 30. $\displaystyle\int [(2x + 1)\, e^{x^2 + x} - x^2 e^{2x+1}]\, dx$

31. ● $\displaystyle\int_0^1 (x + 1)e^x\, dx$ 32. ● $\displaystyle\int_{-1}^1 (x^2 + x)e^{-x}\, dx$

33. ● $\displaystyle\int_0^1 x^2(x + 1)^{10}\, dx$ 34. ● $\displaystyle\int_0^1 x^3(x + 1)^{10}\, dx$

35. ● $\displaystyle\int_1^2 x \ln(2x)\, dx$ 36. ● $\displaystyle\int_1^2 x^2 \ln(3x)\, dx$

37. ● $\displaystyle\int_0^1 x \ln(x + 1)\, dx$ 38. ● $\displaystyle\int_0^1 x^2 \ln(x + 1)\, dx$

39. ● Find the area bounded by the curve $y = xe^{-x}$, the x-axis, and the lines $x = 0$ and $x = 10$.

40. ● Find the area bounded by the curve $y = x \ln x$, the x-axis, and the lines $x = 1$ and $x = e$.

41. ● Find the area bounded by the curve $y = (x + 1) \ln x$, the x-axis, and the lines $x = 1$ and $x = 2$.

42. ● Find the area bounded by the curve $y = (x - 1)e^x$, the x-axis, and the lines $x = 0$ and $x = 2$.

Applications

43. ● *Displacement* A rocket rising from the ground has a velocity of $2000te^{-t/120}$ ft/s, after t seconds. How far does it rise in the first two minutes?

44. ● *Sales* Weekly sales of graphing calculators can be modeled by the equation

$$s(t) = 10 - te^{-t/20}$$

where s is the number of calculators sold per week after t weeks. How many graphing calculators (to the nearest unit) will be sold in the first 20 weeks?

45. ● *Total Cost* The marginal cost of the xth box of light bulbs is $10 + [\ln(x + 1)]/(x + 1)^2$, and the fixed cost is $5000. Find the total cost to make x boxes of bulbs.

46. ● *Total Revenue* The marginal revenue for selling the xth box of light bulbs is $10 + 0.001x^2 e^{-x/100}$. Find the total revenue generated by selling 200 boxes of bulbs.

47. *Revenue* You have been raising the price of your *Lord of the Rings®* T-shirts by 50¢ per week, and sales have been falling continuously at a rate of 2% per week. Assuming you are now selling 50 T-shirts per week and charging $10 per T-shirt, how much revenue will you generate during the coming year? (Round your answer to the nearest dollar.) [Hint: Weekly revenue = weekly sales × price per T-shirt.]

48. *Revenue* Luckily, sales of your *Star Wars®* T-shirts are now 50 T-shirts per week and increasing continuously at a rate of 5% per week. You are now charging $10 per T-shirt and are decreasing the price by 50¢ per week. How much revenue will you generate during the next six weeks?

49. *Bottled-Water Revenue* U.S. sales of bottled water for the period 1993–2003 could be approximated by

$$q(t) = 17t^2 + 100t + 2300 \text{ million gallons per year}$$
$$(3 \le t \le 13)$$

where t is time in years since 1990.[2] Assume that the average price of a gallon of bottled water was $3.00 in 1990 and that this price rose continuously at 3% per year. How much revenue was generated by sales of bottled water in the U.S. in the period 1993 to 2003? (Give your answer to the nearest $10,000 million.)

50. *Bottled-Water Revenue* U.S. per capita sales of bottled water for the period 1993–2003 could be approximated by

$$Q(t) = 0.05t^2 + 0.4t + 9 \text{ gallons}$$

where t is the time in years since 1990.[3] Assume that the average price of a gallon of bottled water was $3.00 in 1990 and that this price rose continuously at 3% per year. How much total revenue was generated per capita in the period 1993 to 2003? (Give your answer to the nearest $100.)

Communication and Reasoning Exercises

51. ● Your friend Janice claims that integration by parts allows one to integrate any product of two functions. Prove her wrong by giving an example of a product of two functions that cannot be integrated using integration by parts.

52. ● Complete the following sentence in words: The integral of $u\, v$ is the first times the integral of the second minus the integral of _____.

---

[2] The authors' regression model, based on data in the Beverage Marketing Corporation news release, "Bottled water now number-two commercial beverage in U.S., says Beverage Marketing Corporation," April 8, 2004, available at www.beveragemarketing.com.
[3] Ibid.

● basic skills   ◆ challenging

53. If $p(x)$ is a polynomial of degree n and $f(x)$ is some function of x, how many times do we generally have to integrate $f(x)$ to compute $\int p(x) f(x)\, dx$?

54. Use integration by parts to show that $\int (\ln x)^2\, dx = x(\ln x)^2 - 2x \ln x + 2x + C$.

55. ◆ *Hermite's Identity* If $f(x)$ is a polynomial of degree n, show that

$$\int_0^b f(x) e^{-x}\, dx = F(0) - F(b) e^{-b}$$

where $F(x) = f(x) + f'(x) + f''(x) + \cdots + f^{(n)}(x)$ (this is the sum of f and all of its derivatives).

56. ◆ Write down a formula similar to Hermite's identity for $\int_0^b f(x) e^x\, dx$ when $f(x)$ is a polynomial of degree n.

● basic skills   ◆ challenging

14.2 Area Between Two Curves and Applications

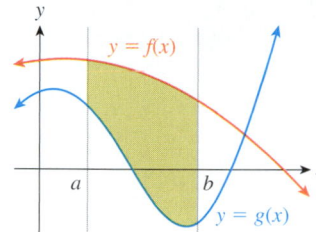

Figure 1

As we saw in the preceding chapter, we can use the definite integral to calculate the area between the graph of a function and the x-axis. With only a little more work, we can use it to calculate the area between two graphs. Figure 1 shows the graphs of two functions, $f(x)$ and $g(x)$, with $f(x) \geq g(x)$ for every x in the interval $[a, b]$.

To find the shaded area between the graphs of the two functions, we use the following formula:

Area Between Two Graphs

If $f(x) \geq g(x)$ for all x in $[a, b]$, then the area of the region between the graphs of f and g and between $x = a$ and $x = b$ is given by

$$A = \int_a^b [f(x) - g(x)]\, dx$$

Let's look at an example and then discuss why the formula works.

Example 1 The Area Between Two Curves

Find the area of the region between $f(x) = -x^2 - 3x + 4$ and $g(x) = x^2 - 3x - 4$ and between $x = -1$ and $x = 1$.

Solution The area in question is shown in Figure 2. Because the graph of f lies above the graph of g in the interval $[-1, 1]$, we have $f(x) \geq g(x)$ for all x in $[-1, 1]$. Therefore, we can use the formula given above and calculate the area as follows:

$$A = \int_{-1}^{1} [f(x) - g(x)]\, dx$$

$$= \int_{-1}^{1} [(-x^2 - 3x + 4) - (x^2 - 3x - 4)]\, dx$$

$$= \int_{-1}^{1} (8 - 2x^2)\, dx$$

$$= \left[8x - \frac{2}{3} x^3 \right]_{-1}^{1}$$

$$= \frac{44}{3}$$

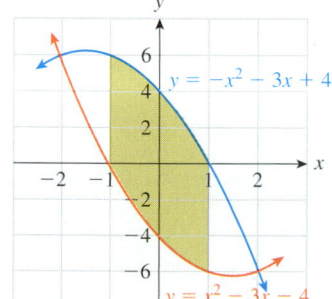

Figure 2

Q: *Why does the formula for the area between two curves work?*

A: Let's go back once again to the general case illustrated in Figure 1, where we were given two functions f and g with $f(x) \geq g(x)$ for every x in the interval $[a, b]$. To avoid complicating the argument by the fact that the graph of g, or f, or both, may dip below the x-axis in the interval $[a, b]$ (as occurs in Figure 1 and also in Example 1), we shift both graphs vertically upward by adding a big enough constant M to lift them both above the x-axis in the interval $[a, b]$, as shown in Figure 3.

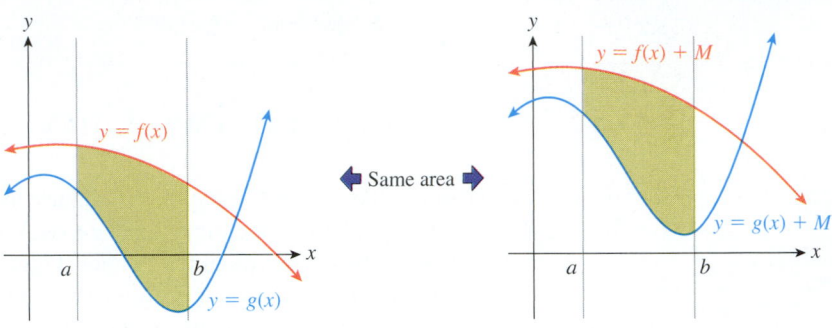

Figure **3**

As the figure illustrates, the area of the region between the graphs is not affected, so we will calculate the area of the region shown on the right of Figure 3. That calculation is shown in Figure 4.

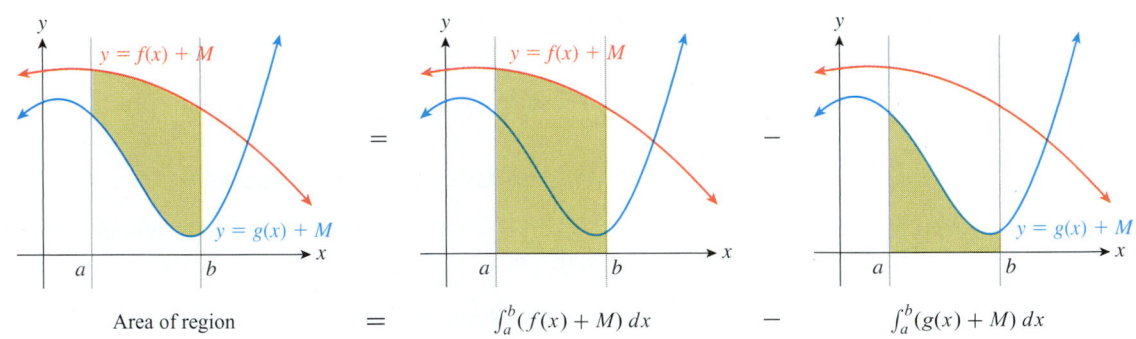

Figure **4**

From the figure, the area we want is

$$\int_a^b (f(x) + M)dx - \int_a^b (g(x) + M)dx = \int_a^b [(f(x) + M) - (g(x) + M)]dx$$

$$= \int_a^b [f(x) - g(x)]dx$$

which is the formula we gave originally. ∎

So far, we've been assuming that $f(x) \geq g(x)$, so that the graph of f never dips below the graph of g and so the graphs cannot cross (although they can touch). Example 2 shows what happens when the two graphs *do* cross.

Example 2 Regions Enclosed by Crossing Curves

Find the area of the region between $y = 3x^2$ and $y = 1 - x^2$ and between $x = 0$ and $x = 1$.

Solution The area we wish to calculate is shown in Figure 5. From the figure, we can see that neither graph lies above the other over the whole interval. To get around this, we break the area into the two pieces on either side of the point at which the graphs cross and then compute each area separately. To do this, we need to know exactly where that crossing point is. The crossing point is where $3x^2 = 1 - x^2$, so we solve for x:

$$3x^2 = 1 - x^2$$
$$4x^2 = 1$$
$$x^2 = \frac{1}{4}$$
$$x = \pm\frac{1}{2}$$

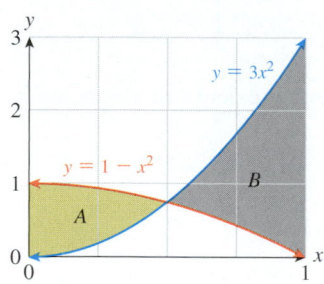

Figure **5**

Because we are interested only in the interval $[0, 1]$, the crossing point we're interested in is at $x = 1/2$.

Now, to compute the areas A and B, we need to know which graph is on top in each of these areas. We can see that from the figure, but what if the functions were more complicated and we could not easily draw the graphs? To be sure, we can test the values of the two functions at some point in each region. But we really need not worry. If we make the wrong choice for the top function, the integral will yield the negative of the area (why?), so we can simply take the absolute value of the integral to get the area of the region in question. For this example, we have

$$A = \int_0^{1/2} [(1 - x^2) - 3x^2]\, dx = \int_0^{1/2} (1 - 4x^2)\, dx$$
$$= \left[x - \frac{4x^3}{3} \right]_0^{1/2}$$
$$= \left(\frac{1}{2} - \frac{1}{6} \right) - (0 - 0) = \frac{1}{3}$$

and

$$B = \int_{1/2}^1 [3x^2 - (1 - x^2)]\, dx = \int_{1/2}^1 (4x^2 - 1)\, dx$$
$$= \left[\frac{4x^3}{3} - x \right]_{1/2}^1$$
$$= \left(\frac{4}{3} - 1 \right) - \left(\frac{1}{6} - \frac{1}{2} \right) = \frac{2}{3}$$

This gives a total area of $A + B = \frac{1}{3} + \frac{2}{3} = 1$.

+*Before we go on...* What would have happened in Example 2 if we had not broken the area into two pieces but had just calculated the integral of the difference of the two functions? We would have calculated

$$\int_0^1 [(1 - x^2) - 3x^2]\, dx = \int_0^1 [1 - 4x^2]\, dx = \left[x - \frac{4x^3}{3} \right]_0^1 = -\frac{1}{3}$$

which is not even close to the right answer. What this integral calculated was actually $A - B$ rather than $A + B$. Why? ■

Example 3 The Area Enclosed by Two Curves

Find the area enclosed by $y = x^2$ and $y = x^3$.

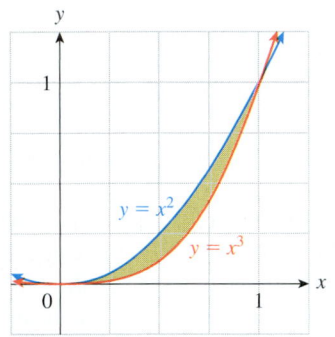

Figure **6**

Solution This example has a new wrinkle: we are not told what interval to use for x. However, if we look at the graph in Figure 6, we see that the question can have only one meaning.

We are being asked to find the area of the shaded sliver, which is the only region that is actually *enclosed* by the two graphs. This sliver is bounded on either side by the two points where the graphs cross, so our first task is to find those points. They are the points where $x^2 = x^3$, so we solve for x:

$$x^2 = x^3$$
$$x^3 - x^2 = 0$$
$$x^2(x - 1) = 0$$
$$x = 0 \quad \text{or} \quad x = 1$$

Thus, we must integrate over the interval $[0, 1]$. Although we see from the diagram (or by substituting $x = 1/2$) that the graph of $y = x^2$ is above that of $y = x^3$, if we didn't notice that we might calculate

$$\int_0^1 (x^3 - x^2)\, dx = \left[\frac{x^4}{4} - \frac{x^3}{3} \right]_0^1 = -\frac{1}{12}$$

This tells us that the required area is $1/12$ square units and also that we had our integral reversed. Had we calculated $\int_0^1 (x^2 - x^3)\, dx$ instead, we would have found the correct answer, $1/12$, directly.

We can summarize the procedure we used in the preceding two examples.

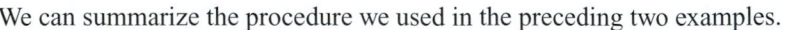

Finding the Area Between the Graphs of $f(x)$ and $g(x)$

1. Find all points of intersection by solving $f(x) = g(x)$ for x. This either determines the interval over which you will integrate or breaks up a given interval into regions between the intersection points.

2. Determine the area of each region you found by integrating the difference of the larger and the smaller function. (If you accidentally take the smaller minus the larger, the integral will give the negative of the area, so just take the absolute value.)

3. Add together the areas you found in Step 2 to get the total area.

Q: Is there any quick and easy method to find the area between two graphs without having to find all points of intersection? What if it is hard or impossible to find out where the curves intersect?

A: We *can* use technology to give the approximate area between two graphs. First recall that, if $f(x) \geq g(x)$ for all x in $[a, b]$, then the area between their graphs over $[a, b]$ is given by $\int_a^b [f(x) - g(x)]dx$, whereas if $g(x) \geq f(x)$, the area is given by $\int_a^b [g(x) - f(x)]dx$. Notice that both expressions are equal to

$$\int_a^b |f(x) - g(x)|\,dx$$

telling us that we can use the same formula in both cases:

---

tech Ex

Area Between Two Graphs: Approximation using Technology

The area of the region between the graphs of f and g and between $x = a$ and $x = b$ is given by

$$A = \int_a^b |f(x) - g(x)|\,dx$$

quick Example

To approximate the area of the region between $y = 3x^2$ and $y = 1 - x^2$ and between $x = 0$ and $x = 1$ we calculated in Example 2, use technology to compute

$$\int_a^b |3x^2 - (1 - x^2)|\,dx = 1 \qquad \text{TI-83/84: } \texttt{fnInt(abs(3x^2-(1-x^2)),X,0,1)}$$

14.2 | EXERCISES

● denotes basic skills exercises

tech Ex indicates exercises that should be solved using technology

Find the areas of the indicated regions in Exercises 1–24. (We suggest you use technology to check your answers.)

1. ● Between $y = x^2$ and $y = -1$ for x in $[-1, 1]$ *hint* [see Example 1]

2. ● Between $y = x^3$ and $y = -1$ for x in $[-1, 1]$

3. ● Between $y = -x$ and $y = x$ for x in $[0, 2]$

4. ● Between $y = -x$ and $y = x/2$ for x in $[0, 2]$

5. ● Between $y = x$ and $y = x^2$ for x in $[-1, 1]$ *hint* [see Example 2]

6. ● Between $y = x$ and $y = x^3$ for x in $[-1, 1]$

7. ● Between $y = e^x$ and $y = x$ for x in $[0, 1]$

8. ● Between $y = e^{-x}$ and $y = -x$ for x in $[0, 1]$

9. ● Between $y = (x - 1)^2$ and $y = -(x - 1)^2$ for x in $[0, 1]$

10. ● Between $y = x^2(x^3 + 1)^{10}$ and $y = -x(x^2 + 1)^{10}$ for x in $[0, 1]$

11. ● Enclosed by $y = x$ and $y = x^4$ *hint* [see Example 3]

12. ● Enclosed by $y = x$ and $y = -x^4$

13. ● Enclosed by $y = x^3$ and $y = x^4$

14. ● Enclosed by $y = x$ and $y = x^3$

15. ● Enclosed by $y = x^2$ and $y = x^4$

16. ● Enclosed by $y = x^4 - x^2$ and $y = x^2 - x^4$

17. ● Enclosed by $y = e^x$, $y = 2$, and the y-axis

18. ● Enclosed by $y = e^{-x}$, $y = 3$, and the y-axis

19. ● Enclosed by $y = \ln x$, $y = 2 - \ln x$, and $x = 4$

● basic skills   **tech** Ex technology exercise

20. ● Enclosed by $y = \ln x$, $y = 1 - \ln x$, and $x = 4$

21. `tech` Ex Enclosed by $y = e^x$, $y = 2x + 1$, $x = -1$, and $x = 1$ (Round answer to four significant digits.) *hint* [see Quick Example p. 993]

22. `tech` Ex Enclosed by $y = 2^x$, $y = x + 2$, $x = -2$, and $x = 2$ (Round answer to four significant digits.)

23. `tech` Ex Enclosed by $y = \ln x$ and $y = \dfrac{x}{2} - \dfrac{1}{2}$ (Round answer to four significant digits.) [First use technology to determine approximately where the graphs cross.]

24. `tech` Ex Enclosed by $y = \ln x$ and $y = x - 2$ (Round answer to four significant digits.) [First use technology to determine approximately where the graphs cross.]

Applications

25. ● **Revenue and Cost** Suppose your daily revenue from selling used DVDs is

$$R(t) = 100 + 10t \qquad (0 \le t \le 5)$$

dollars per day, where t represents days from the beginning of the week, while your daily costs are

$$C(t) = 90 + 5t \qquad (0 \le t \le 5)$$

dollars per day. Find the area between the graphs of $R(t)$ and $C(t)$ for $0 \le t \le 5$. What does your answer represent?

26. ● **Income and Expenses** Suppose your annual income is

$$I(t) = 50{,}000 + 2000t \qquad (0 \le t \le 3)$$

dollars per year, where t represents the number of years since you began your job, while your annual expenses are

$$E(t) = 45{,}000 + 1500t \qquad (0 \le t \le 3)$$

dollars per year. Find the area between the graphs of $I(t)$ and $E(t)$ for $0 \le t \le 3$. What does your answer represent?

27. **Foreign Trade** Annual U.S. imports from China in the years 1996–2004 could be approximated by

$$I = t^2 + 3.5t + 50 \qquad (1 \le t \le 9)$$

billion dollars per year, where t represents time in years since the start of 1995. During the same period, annual U.S. exports to China could be approximated by

$$E = 0.4t^2 - 1.6t + 14 \qquad (1 \le t \le 9)$$

billion dollars per year.[4]

a. What does the area between the graphs of I and E over the interval [1, 9] represent?

b. Compute the area in part (a), and interpret your answer. (Round your answer to the nearest $10 billion.)

28. ● **Revenue** Apple Computer, Inc.'s total revenue from the fourth quarter of 2002 through the third quarter of 2004 was flowing in at a rate of approximately

$$R_t = 5t^2 + 70t + 1500 \qquad (-1 \le t \le 6)$$

million dollars per quarter, where t is time in quarters since the first quarter of 2003. During the same period, revenue from sales of iPods alone flowed in at a rate of approximately

$$R_i = 2.6t^2 + 22t + 60 \qquad (-1 \le t \le 6)$$

million dollars per quarter.[5]

a. What does the area between the graphs of R_t and R_i over the interval [0, 6] represent?

b. Compute the area in part (a) and interpret your answer. (Round your answer to the nearest $100 million.)

29. **Health-Care Spending** The rate of private spending (including from insurance) on health care in the U.S. from 1965 through 2002 was approximately

$$P(t) = 20t^3 + 1000t^2 + 28{,}000t + 360{,}000$$
$$(-25 \le t \le 12)$$

million dollars per year, where t is the number of years since 1990. The rate of spending on health care in the U.S. from 1965 through 2002 by private insurance only was approximately

$$I(t) = 15t^3 + 800t^2 + 19{,}000t + 200{,}000$$
$$(-25 \le t \le 12)$$

million dollars per year.[6] Take these functions as projections of spending rates through 2010.

a. Use these models to project the total spent on health care from private funds other than insurance from 2000 to 2010. (Round your answer to the nearest $100 billion.)

b. How does your answer to part (a) relate to the area between two curves?

30. **Health-Care Spending** The rate of public spending (including by the federal government) on health care in the U.S. from 1965 through 2002 was approximately

$$G(t) = 17t^3 + 900t^2 + 24{,}000t + 270{,}000$$
$$(-25 \le t \le 12)$$

million dollars per year, where t is the number of years since 1990. The rate of spending on health care by the federal government only was approximately

$$F(t) = 11t^3 + 600t^2 + 17{,}000t + 190{,}000$$
$$(-25 \le t \le 12)$$

---

[4] The models are based on quadratic regression using data from the U.S. Census Bureau Foreign Trade Division website www.census.gov/foreign-trade/sitc1 as of December 2004.

[5] Based on a quadratic regression. Source for data: Apple Computer, Inc., quarterly earnings reports, available at www.apple.com.

[6] SOURCE: Centers for Medicare and Medicaid Services, "National Health Expenditures," 2002 version, released January 2004; www.cms.hhs.gov/statistics/nhe.

● basic skills   `tech` Ex technology exercise

million dollars per year.[7] Take these functions as projections of spending rates through 2010.

a. Use these models to project the total public spending on health care, excluding spending by the federal government, from 2000 to 2010. (Round your answer to the nearest $100 billion.)

b. How does your answer to part (a) relate to the area between two curves?

Communication and Reasoning Exercises

31. ● The following graph shows Canada's annual exports and imports for the period 1997–2001.[8]

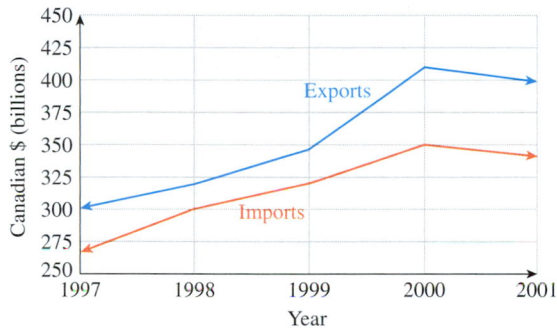

What does the area between the export and import curves represent?

32. ● The following graph shows a fictitious country's monthly exports and imports for the period 1997–2001.

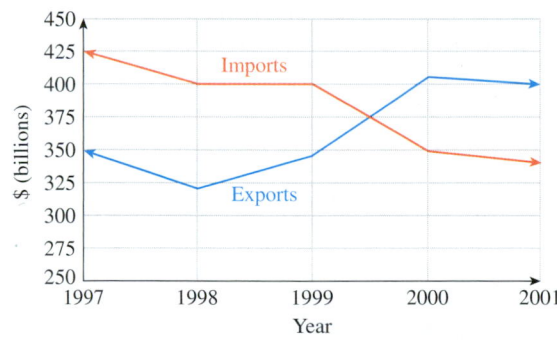

What does the total area enclosed by the export and import curves represent, and what does the definite integral of the difference, Exports − Imports, represent?

33. The following graph shows the daily revenue and cost in your Adopt-a-Chia operation *t* days from its inception:

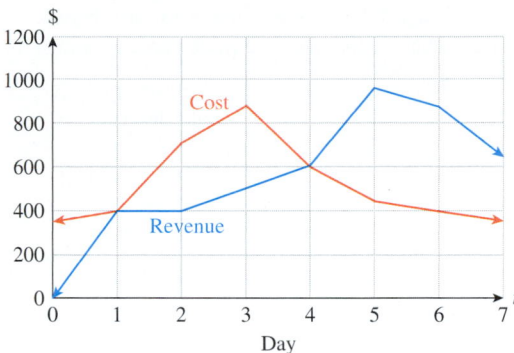

Multiple Choice: The area between the cost and revenue curves represents:

(A) The accumulated loss through day 4 plus the accumulated profit for days 5 through 7

(B) The accumulated profit for the week

(C) The accumulated loss for the week

(D) The accumulated cost through day 4 plus the accumulated revenue for days 5 through 7

34. The following graph shows daily orders and inventory (stock on hand) for your Mona Lisa paint-by-number sets *t* days into last week.

a. Multiple Choice: Which is greatest:

(A) $\int_0^7 (\text{Orders} - \text{Inventory})\, dt$

(B) $\int_0^7 (\text{Inventory} - \text{Orders})\, dt$

(C) The area between the Orders and Inventory curves.

b. Multiple Choice: The answer to part (a) measures

(A) The accumulated gap between orders and inventory

(B) The accumulated surplus through day 3 minus the accumulated shortage for days 3 through 5 plus the accumulated surplus through days 6 through 7

(C) The total net surplus

(D) The total net loss

[7] SOURCE: Centers for Medicare and Medicaid Services, "National Health Expenditures," 2002 version, released January 2004; www.cms.hhs.gov/statistics/nhe.

[8] SOURCE: http://strategis.ic.gc.ca, August, 2002.

● basic skills   **tech Ex** technology exercise

35. What is wrong with the following claim: "I purchased Consolidated Edison shares for $40 at the beginning of January, 2002. My total profit per share from this investment from January through July is represented by the area between the stock price curve and the purchase price curve as shown on the following graph."[9]

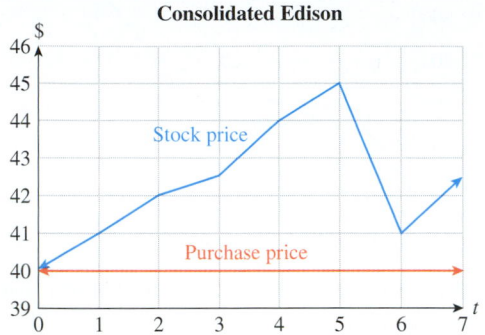

Consolidated Edison

---

[9] SOURCE: http://money.excite.com, August 5, 2002.

36. Your pharmaceutical company monitors the amount of medication in successive batches of 100 mg Tetracycline capsules, and obtains the following graph.

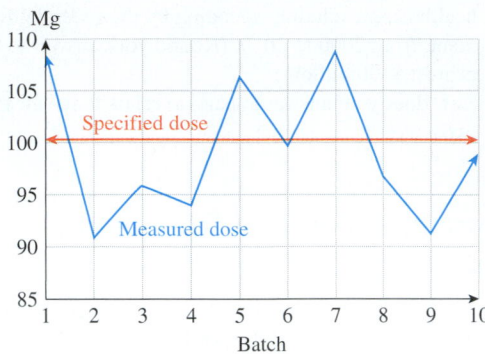

Your production manager claims that the batches of tetracycline conform to the exact dosage requirement because half of the area between the graphs is above the Specified Dose line and half is below it. Comment on this reasoning.

● basic skills   tech Ex technology exercise

14.3 Averages and Moving Averages

Averages

To find the average of, say, 20 numbers, we simply add them up and divide by 20. More generally, if we want to find the **average,** or **mean,** of the n numbers $y_1, y_2, y_3, \ldots, y_n$, we add them up and divide by n. We write this average as $\bar{y}$ ("y-bar").

Average, or Mean, of a Collection of Values

$$\bar{y} = \frac{y_1 + y_2 + \cdots + y_n}{n}$$

quick Example

The average of $\{0, 2, -1, 5\}$ is $\bar{y} = \dfrac{0 + 2 - 1 + 5}{4} = \dfrac{6}{4} = 1.5$

But, we also use the word *average* in other senses. For example, we speak of the average speed of a car during a trip.

Example 1 Average Speed

Over the course of 2 hours, my speed varied from 50 miles per hour to 60 miles per hour, following the function $v(t) = 50 + 2.5t^2$, $0 \le t \le 2$. What was my average speed over those two hours?

Solution Recall that average speed is simply the total distance traveled divided by the time it took. Recall, also, that we can find the distance traveled by integrating the speed:

$$
\begin{aligned}
\text{Distance traveled} &= \int_0^2 v(t)\,dt \\
&= \int_0^2 (50 + 2.5t^2)\,dt \\
&= \left[50t + \frac{2.5}{3} t^3 \right]_0^2 \\
&= 100 + \frac{20}{3} \\
&\approx 106.67 \text{ miles}
\end{aligned}
$$

It took 2 hours to travel this distance, so the average speed was

$$
\text{Average speed} \approx \frac{106.67}{2} \approx 53.3 \text{ mph}
$$

In general, if we travel with velocity $v(t)$ from time $t = a$ to time $t = b$, we will travel a distance of $\int_a^b v(t)\,dt$ in time $b - a$, which gives an average velocity of

$$
\text{Average velocity} = \frac{1}{b - a} \int_a^b v(t)\,dt
$$

Thinking of this calculation as finding the average value of the velocity function, we generalize and make the following definition:

Average Value of a Function

The **average,** or **mean,** of a function $f(x)$ on an interval $[a, b]$ is

$$
\bar{f} = \frac{1}{b - a} \int_a^b f(x)\,dx
$$

quick Example

The average of $f(x) = x$ on $[1, 5]$ is

$$
\begin{aligned}
\bar{f} &= \frac{1}{b - a} \int_a^b f(x)\,dx \\
&= \frac{1}{5 - 1} \int_1^5 x\,dx \\
&= \frac{1}{4} \left[\frac{x^2}{2} \right]_1^5 = \frac{1}{4} \left(\frac{25}{2} - \frac{1}{2} \right) = 3
\end{aligned}
$$

Interpreting the Average of a Function Geometrically

The average of a function has a geometric interpretation. Referring to the Quick Example above, we can compare the graph of $y = f(x)$ with the graph of $y = 3$, both over the interval $[1, 5]$ (Figure 7).

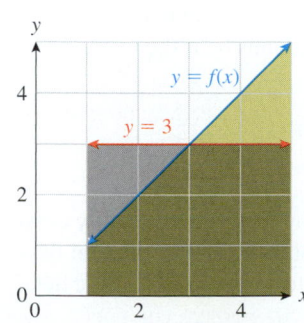

Figure 7

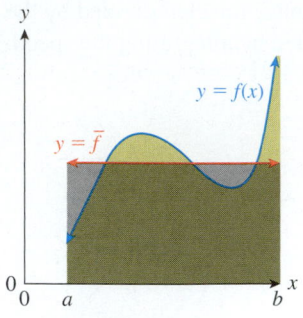

Figure 8

We can find the area under the graph of $f(x) = x$ by geometry or by calculus; it is 12. The area in the rectangle under $y = 3$ is also 12.

In general, the average $\bar{f}$ of a positive function over the interval $[a, b]$ gives the height of the rectangle over the interval $[a, b]$ that has the same area as the area under the graph of $f(x)$ as illustrated in Figure 8. The equality of these areas follows from the equation

$$(b - a)\bar{f} = \int_a^b f(x)\, dx$$

Example 2 Average Balance

A savings account at the People's Credit Union pays 3% interest, compounded continuously, and at the end of the year you get a bonus of 1% of the average balance in the account during the year. If you deposit $10,000 at the beginning of the year, how much interest and how large a bonus will you get?

Solution We can use the continuous compound interest formula to calculate the amount of money you have in the account at time t:

$$A(t) = 10,000e^{0.03t}$$

where t is measured in years. At the end of 1 year, the account will have

$$A(1) = \$10,304.55$$

so you will have earned $304.55 interest. To compute the bonus, we need to find the average amount in the account, which is the average of $A(t)$ over the interval $[0, 1]$. Thus,

$$\bar{A} = \frac{1}{b - a} \int_a^b A(t)\, dt$$

$$= \frac{1}{1 - 0} \int_0^1 10,000e^{0.03t}\, dt = \frac{10,000}{0.03} \left[e^{0.03t} \right]_0^1$$

$$\approx \$10,151.51$$

The bonus is 1% of this, or $101.52.

✛ *Before we go on...* The 1% bonus in Example 2 was one-third of the total interest. Why did this happen? What fraction of the total interest would the bonus be if the interest rate was 4%, 5%, or 10%? ∎

Moving Averages

Suppose you follow the performance of a company's stock by recording the daily closing prices. The graph of these prices may seem jagged or "jittery" due to random day-to-day fluctuations. To see any trends, you would like a way to "smooth out" these data. The **moving average** is one common way to do that.

Example 3 Stock Prices

The following table shows Colossal Conglomerate's closing stock prices for 20 consecutive trading days:

Day	1	2	3	4	5	6	7	8	9	10
Price	20	22	21	24	24	23	25	26	20	24
Day	11	12	13	14	15	16	17	18	19	20
Price	26	26	25	27	28	27	29	27	25	24

Plot these prices and the 5-day moving average.

Solution The 5-day moving average is the average of each day's price together with the prices of the preceding 4 days. We can compute the 5-day moving averages starting on the fifth day. We get these numbers:

Day	1	2	3	4	5	6	7	8	9	10
Moving Average					22.2	22.8	23.4	24.4	23.6	23.6
Day	11	12	13	14	15	16	17	18	19	20
Moving Average	24.2	24.4	24.2	25.6	26.4	26.6	27.2	27.6	27.2	26.4

The closing stock prices and moving averages are plotted in Figure 9.

Figure 9

As you can see, the moving average is less volatile than the closing price. Because the moving average incorporates the stock's performance over 5 days at a time, a single day's fluctuation is smoothed out. Look at day 9 in particular. The moving average also tends to lag behind the actual performance because it takes past history into account. Look at the downturns at days 6 and 18 in particular.

The period of 5 days for a moving average, as used in Example 3, is arbitrary. Using a longer period of time would smooth the data more but increase the lag. For data used as economic indicators, such as housing prices or retail sales, it is common to compute the 4-quarter moving average to smooth out seasonal variations.

using *Technology*

We can automate these computations using a graphing calculator or Excel. See the Technology Guides at the end of the chapter to find out how to tabulate and graph moving averages.

It is also sometimes useful to compute moving averages of continuous functions. We may want to do this if we use a mathematical model of a large collection of data. Also, some physical systems have the effect of converting an input function (an electrical signal, for example) into its moving average. By an **n-unit moving average** of a function $f(x)$ we mean the function $\bar{f}$ for which $\bar{f}(x)$ is the average of the value of $f(x)$ on $[x - n, x]$. Using the formula for the average of a function, we get the following formula.

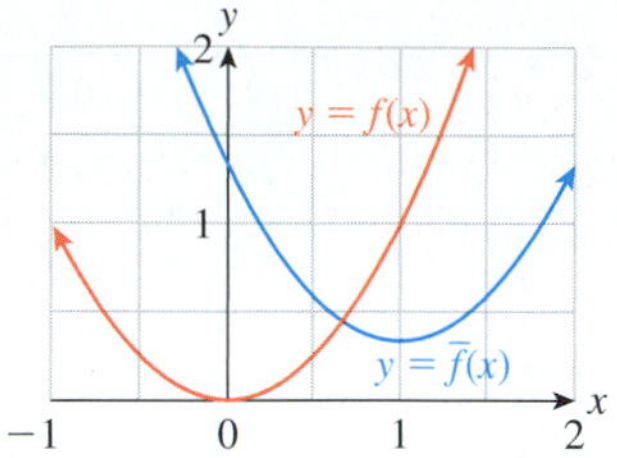

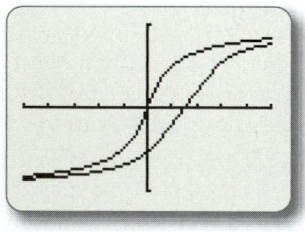

Figure 10

n-Unit Moving Average of a Function

The n-unit moving average of a function f is

$$\bar{f}(x) = \frac{1}{n} \int_{x-n}^{x} f(t)\, dt$$

quick Example

The 2-unit moving average of $f(x) = x^2$ is

$$\bar{f}(x) = \frac{1}{2} \int_{x-2}^{x} t^2\, dt = \frac{1}{6}\left[t^3\right]_{x-2}^{x} = x^2 - 2x + \frac{4}{3}$$

The graphs of $f(x)$ and $\bar{f}(x)$ are shown in Figure 10.

Example 4 Moving Average via Technology

Use technology to plot the 3-unit moving average of

$$f(x) = \frac{x}{1 + |x|} \qquad (-5 \le x \le 5)$$

Solution This function is a little tricky to integrate analytically because $|x|$ is defined differently for positive and negative values of x, so instead we use technology to approximate the integral. Figure 11 shows the output we can obtain using a TI-83/84 graphing calculator (the lower curve is the moving average). See the TI 83/84 Technology Guide at the end of the chapter to find out how to obtain this output.

Figure 11

14.3 EXERCISES

● denotes basic skills exercises

tech Ex indicates exercises that should be solved using technology

Find the averages of the functions in Exercises 1–6 over the given intervals. Plot each function and its average on the same graph (as in Figure 7). hint [see Quick Example p. 997]

1. ● $f(x) = x^3$ over $[0, 2]$ **2.** ● $f(x) = x^3$ over $[-1, 1]$

3. ● $f(x) = x^3 - x$ over $[0, 2]$ **4.** ● $f(x) = x^3 - x$ over $[0, 1]$

5. ● $f(x) = e^{-x}$ over $[0, 2]$ **6.** ● $f(x) = e^x$ over $[-1, 1]$

In Exercises 7 and 8, complete the given table with the values of the 3-unit moving average of the given function. hint [see Example 3]

7. ●

x	0	1	2	3	4	5	6	7
$r(x)$	3	5	10	3	2	5	6	7
$\bar{r}(x)$								

● basic skills   tech Ex technology exercise

8. ●

x	0	1	2	3	4	5	6	7
$s(x)$	2	9	7	3	2	5	7	1
$\bar{s}(x)$								

In Exercises 9 and 10, some values of a function and its 3-unit moving average are given. Supply the missing information.

9. ●

x	0	1	2	3	4	5	6	7
$r(x)$	1	2			11		10	2
$\bar{r}(x)$			3	5		11		

10. ●

x	0	1	2	3	4	5	6	7
$s(x)$	1	5		1				
$\bar{s}(x)$			5		5	2	3	2

Calculate the 5-unit moving average of each function in Exercises 11–18. Plot each function and its moving average on the same graph, as in Example 4. (You may use graphing technology for these plots, but you should compute the moving averages analytically.) hint [see Quick Example p. 1000]

11. ● $f(x) = x^3$

12. ● $f(x) = x^3 - x$

13. ● $f(x) = x^{2/3}$

14. ● $f(x) = x^{2/3} + x$

15. ● $f(x) = e^{0.5x}$

16. ● $f(x) = e^{-0.02x}$

17. ● $f(x) = \sqrt{x}$

18. ● $f(x) = x^{1/3}$

In Exercises 19–22, use graphing technology to plot the given functions together with their 3-unit moving averages. hint [see Example 4]

19. tech Ex $f(x) = \dfrac{10x}{1 + 5|x|}$ **20.** tech Ex $f(x) = \dfrac{1}{1 + e^x}$

21. tech Ex $f(x) = \ln(1 + x^2)$ **22.** tech Ex $f(x) = e^{1-x^2}$

Applications

23. ● *Employment* The following table shows the approximate number of people employed each year in the U.S. during the period 1995–2004[10]:

Year t	1995	1996	1997	1998	1999	2000	2001	2002	2003	2004
Employment (millions)	117	120	123	126	129	132	132	130	130	131

What was the average number of people employed in the U.S. for the years 1995 through 2004? hint [see Quick Example p. 996]

24. ● *Cell Phone Sales* The following table shows the net sales (revenue) of the Finnish cell phone company Nokia for each year in the period 1999–2003[11]:

Year t	1999	2000	2001	2002	2003
Nokia net sales (billions of euros)	20	30	31	30	29

What were Nokia's average net sales for the years 1999 through 2003?

25. ● *Television Advertising* The cost, in millions of dollars, of a 30-second television ad during the Super Bowl in the years 1998 to 2001 can be approximated by

$$C(t) = 0.355t - 1.6 \text{ million dollars} \quad (8 \le t \le 11)$$

($t = 8$ represents 1998).[12] What was the average cost of a Super Bowl ad during the given period? hint [see Example 1]

26. ● *Television Advertising* The cost, in millions of dollars, of a 30-second television ad during the Super Bowl in the years 1990 to 1998 can be approximated by

$$C(t) = 0.08t + 0.6 \text{ million dollars} \quad (0 \le t \le 8)$$

($t = 0$ represents 1990).[13] What was the average cost of a Super Bowl ad during the given period?

27. ● *Investments* If you invest $10,000 at 8% interest compounded continuously, what is the average amount in your account over one year? hint [see Example 2]

28. ● *Investments* If you invest $10,000 at 12% interest compounded continuously, what is the average amount in your account over one year?

29. ● *Average Balance* Suppose you have an account (paying no interest) into which you deposit $3000 at the beginning of each month. You withdraw money continuously so that the amount in the account decreases linearly to 0 by the end of the month. Find the average amount in the account over a period of several months. (Assume that the account starts at $0 at $t = 0$ months.)

30. ● *Average Balance* Suppose you have an account (paying no interest) into which you deposit $4000 at the beginning of each month. You withdraw $3000 during the course of each month, in such a way that the amount decreases linearly. Find the average amount in the account in the first two months. (Assume that the account starts at $0 at $t = 0$ months.)

31. ● *Employment* Refer back to Exercise 23. Complete the following table by computing the 4-year moving average of employment in the U.S. Round each average to the nearest whole number.

Year t	1995	1996	1997	1998	1999	2000	2001	2002	2003	2004
Employment (millions)	117	120	123	126	129	132	132	130	130	131
Moving average (millions)										

How do the year-by-year changes in the moving average compare with those in the employment figures? hint [see Example 3]

---

[10] The values represent nonfarm employment. SOURCE: U.S. Bureau of Labor Statistics, Division of Current Employment Statistics, http://bls.gov/ces, January 13, 2005.

[11] SOURCE: Nokia financial statements downloaded from www.nokia.com, January 13, 2005.

[12] SOURCE: *New York Times,* January 26, 2001, p. C1.

[13] Ibid.

● basic skills   tech Ex technology exercise

32. ● *Cell Phone Sales* Refer back to Exercise 24. Complete the following table by computing the 3-year moving average of Nokia's net sales. Round each average to the nearest whole number.

Year t	1999	2000	2001	2002	2003
Nokia net sales (billions of euros)	20	30	31	30	29
Moving Average (billions of euros)					

Is the average of the moving averages the same as the overall average? Explain.

33. techEx *Health-Care Spending* The following table shows approximate public spending on health care in the U.S. in the years 1981–2000, in billions of dollars.[14]

1981	1982	1983	1984	1985	1986	1987	1988	1989	1990
121	134	147	161	175	190	209	226	252	282
1991	**1992**	**1993**	**1994**	**1995**	**1996**	**1997**	**1998**	**1999**	**2000**
321	359	390	427	457	482	504	521	553	595

a. Use technology to compute and plot the 5-year moving average of these data.
b. The graph of the moving average will appear almost linear over the range 1991–1997. Use the 1991 and 1997 moving average figures to give an estimate (to the nearest billion dollars per year) of the rate of change of public spending on health care in the U.S. for the period 1991–1997. Interpret the result.

34. techEx *Health-Care Spending* The following table shows approximate private spending on health care in the U.S. in the years 1981–2000, in billions of dollars.[15]

1981	1982	1983	1984	1985	1986	1987	1988	1989	1990
160	190	210	230	250	270	290	330	370	410
1991	**1992**	**1993**	**1994**	**1995**	**1996**	**1997**	**1998**	**1999**	**2000**
440	470	500	510	530	560	590	630	670	710

a. Use technology to compute and plot the 5-year moving average of these data.
b. The graph of the moving average will appear almost linear over the range 1993–2000. Use the 1993 and 2000 moving average figures to give an estimate (to the nearest 10 billion dollars per year) of the rate of change of private spending on health care in the U.S. for the period 1993–2000. Interpret the result.

35. ● *Bottled-Water Sales* The rate of U.S. sales of bottled water for the period 1993–2003 could be approximated by

$$R(t) = 17t^2 + 100t + 2300 \text{ million gallons per year}$$
$$(3 \le t \le 13)$$

where t is time in years since 1990.[16]

a. Compute the average annual sales of bottled water over the period 1993–2003, to the nearest 100 million gallons per year.
b. Compute the two-year moving average of R. (You need not simplify the answer.)
c. Without simplifying the answer in part (b), say what kind of function the moving average is.

hint [see Quick Examples p. 997, 1000]

36. ● *Bottled-Water Sales* The rate of U.S. per capita sales of bottled water for the period 1993–2003 could be approximated by

$$Q(t) = 0.05t^2 + 0.4t + 9 \text{ gallons per year}$$

where t is the time in years since 1990.[17] Repeat the preceding exercise as applied to per capita sales. (Give your answer to (a) to the nearest gallon per year.)

37. ● *Medicare Spending* Annual federal spending on Medicare (in constant 2000 dollars) was projected to increase from $240 billion in 2000 to $600 billion in 2025.[18]

a. Use this information to express s, the annual spending on Medicare (in billions of dollars), as a linear function of t, the number of years since 2000.
b. Find the 4-year moving average of your model.
c. What can you say about the slope of the moving average?

38. ● *Pasta Imports* In 1990, the United States imported 290 million pounds of pasta. From 1990 to 2001 imports increased by an average of 40 million pounds per year.[19]

a. Use these data to express q, the annual U.S. imports of pasta (in millions of pounds), as a linear function of t, the number of years since 1990.

[14] SOURCE: Centers for Medicare and Medicaid Services, "National Health Expenditures," 2002 version, released January 2004; www.cms.hhs.gov/statistics/nhe.

[15] Ibid.

[16] The authors' regression model, based on data in the Beverage Marketing Corporation news release, "Bottled water now number-two commercial beverage in U.S., says Beverage Marketing Corporation," April 8, 2004, available at www.beveragemarketing.com.

[17] Ibid.

[18] Data are rounded. SOURCE: The Urban Institute's Analysis of the 1999 Trustee's Report www.urban.org.

[19] Data are rounded. SOURCES: Department of Commerce/*New York Times*, September 5, 1995, p. D4, International Trade Administration (www.ita.doc.gov) March 31, 2002.

● basic skills   techEx technology exercise

b. Find the 4-year moving average of your model.

c. What can you say about the slope of the moving average?

39. *Moving Average of a Linear Function* Find a formula for the a-unit moving average of a general linear function $f(x) = mx + b$.

40. *Moving Average of an Exponential Function* Find a formula for the a-unit moving average of a general exponential function $f(x) = Ae^{kx}$.

41. *Fair Weather*[20] The Cancun Royal Hotel's advertising brochure features the following chart, showing the year-round temperature:

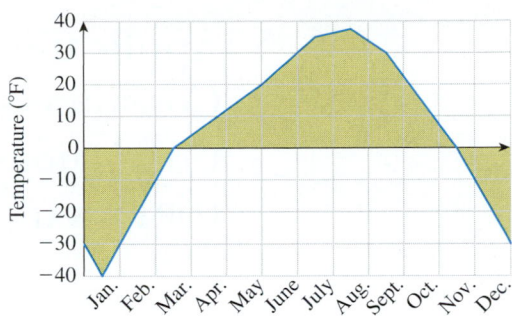

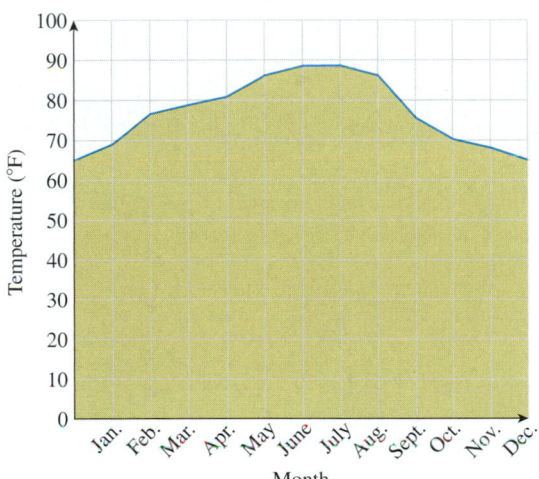

Month

a. Estimate and plot the two- and three-month moving averages. (Use graphing technology, if available.)

b. What can you say about the 24-month moving average?

c. Comment on the limitations of a quadratic model for these data.

42. *Foul Weather* Repeat the preceding exercise, using the following data from the Tough Traveler Lodge in Frigidville:

---

[20] Inspired by an exercise in the Harvard Consortium Calculus project.

Communication and Reasoning Exercises

43. ● Explain why it is sometimes more useful to consider the moving average of a stock price rather than the stock price itself.

44. ● Your monthly salary has been increasing steadily for the past year, and your average monthly salary over the past year was x dollars. Would you have earned more money if you had been paid x dollars per month? Explain your answer.

45. What property does a (nonconstant) function have if its average value over an interval is zero? Sketch a graph of such a function.

46. Can the average value of a function f on an interval be greater than its value at *every* point in that interval? Explain.

47. Criticize the following claim: The average value of a function on an interval is midway between its highest and lowest value.

48. Your manager tells you that 12-month moving averages gives at least as much information as shorter-term moving averages and very often more. How would you argue that he is wrong?

49. Which of the following most closely approximates the original function, (A) its 10-unit moving average, (B) its 1-unit moving average, or (C) its 0.8-unit moving average? Explain your answer.

50. Is an increasing function larger or smaller than its 1-unit moving average? Explain.

● basic skills   **tech** Ex technology exercise

14.4 Applications to Business and Economics: Consumers' and Producers' Surplus and Continuous Income Streams (OPTIONAL)

Consumers' Surplus

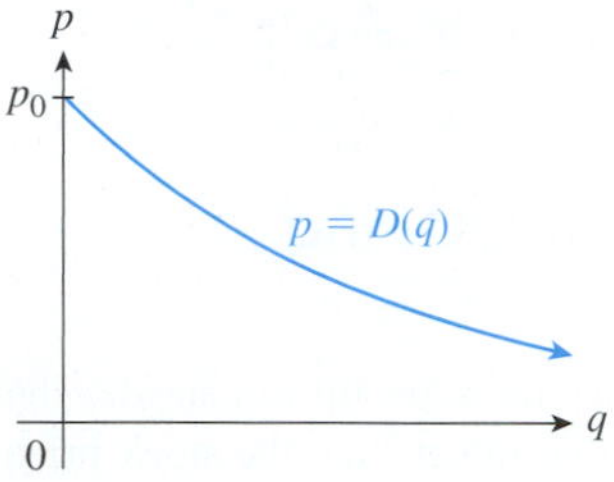

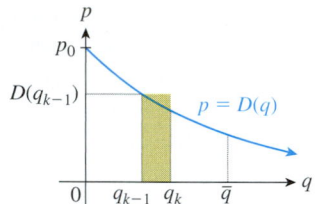

Figure **12**

Consider a general demand curve presented, as is traditional in economics, as $p = D(q)$, where p is unit price and q is demand measured, say, in annual sales (Figure 12). Thus, $D(q)$ is the price at which the demand will be q units per year. The price p_0 shown on the graph is the highest price that customers are willing to pay.

Suppose, for example, that the graph is the demand curve for a particular new model of computer. When the computer first comes out and supplies are low (q is small), "early adopters" will be willing to pay a high price. This is the part of the graph on the left, near the p axis. As supplies increase and the price drops, more consumers will be willing to pay and more computers will be sold. We can ask the following question: How much are consumers willing to spend for the first $\bar{q}$ units?

Consumers' Willingness to Spend

We can approximate consumers' willingness to spend on the first $\bar{q}$ units as follows. We partition the interval $[0, \bar{q}\,]$ into n subintervals of equal length, as we did when discussing Riemann sums. Figure 13 shows a typical subinterval, $[q_{k-1}, q_k]$.

The price consumers are willing to pay for each of units q_{k-1} through q_k is approximately $D(q_{k-1})$, so the total that consumers are willing to spend for these units is approximately $D(q_{k-1})(q_k - q_{k-1}) = D(q_{k-1})\Delta q$, the area of the shaded region in Figure 13. Thus, the total amount that consumers are willing to spend for items 0 through $\bar{q}$ is

Figure **13**

$$W \approx D(q_0)\Delta q + D(q_1)\Delta q + \cdots + D(q_{n-1})\Delta q = \sum_{k=0}^{n-1} D(q_k)\Delta q$$

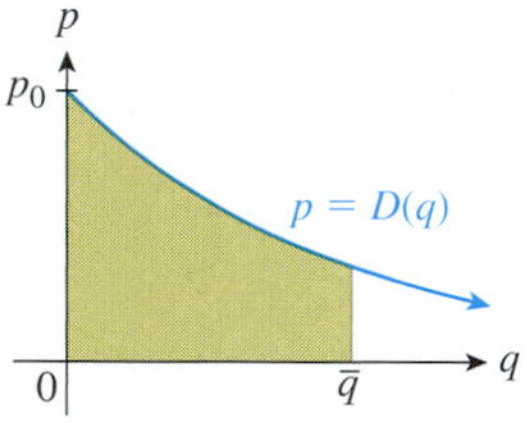

which is a Riemann sum. The approximation becomes better the larger n becomes, and in the limit the Riemann sums converge to the integral

$$W = \int_0^{\bar{q}} D(q)\, dq$$

Figure **14**

This quantity, the area shaded in Figure 14, is the total consumers' willingness to spend to buy the first $\bar{q}$ units.

Consumers' Expenditure

Now suppose that the manufacturer simply sets the price at $\bar{p}$, with a corresponding demand of $\bar{q}$, so $D(\bar{q}) = \bar{p}$. Then the amount that consumers will actually spend to buy these $\bar{q}$ is $\bar{p}\bar{q}$, the product of the unit price and the quantity sold. This is the area of the rectangle shown in Figure 15. Notice that we can write $\bar{p}\bar{q} = \int_0^{\bar{q}} \bar{p}\, dq$, as suggested by the figure.

The difference between what consumers are willing to pay and what they actually pay is money in their pockets and is called the **consumers' surplus.**

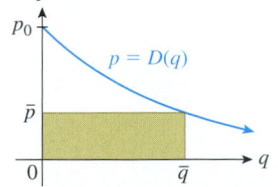

Figure **15**

Consumers' Surplus

If demand for an item is given by $p = D(q)$, the selling price is $\bar{p}$, and $\bar{q}$ is the corresponding demand [so that $D(\bar{q}) = \bar{p}$], then the **consumers' surplus** is the difference between their willingness to spend and their actual expenditure:

$$CS = \int_0^{\bar{q}} D(q)\, dq - \bar{p}\bar{q} = \int_0^{\bar{q}} (D(q) - \bar{p})\, dq$$

Graphically, it is the area between the graphs of $p = D(q)$ and $p = \bar{p}$, as shown in the figure.

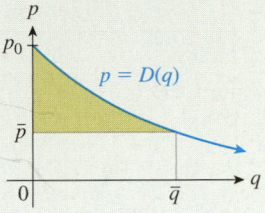

Example 1 Consumers' Surplus

Your used-CD store has an exponential demand equation of the form

$$p = 15e^{-0.01q}$$

where q represents daily sales of used CDs and p is the price you charge per CD. Calculate the daily consumers' surplus if you sell your used CDs at $5 each.

Solution We are given $D(q) = 15e^{-0.01q}$ and $\bar{p} = 5$. We also need $\bar{q}$. By definition,

$$D(\bar{q}) = \bar{p}$$

or $15e^{-0.01\bar{q}} = 5$

which we must solve for $\bar{q}$:

$$e^{-0.01\bar{q}} = \frac{1}{3}$$

$$-0.01\bar{q} = \ln\left(\frac{1}{3}\right) = -\ln 3$$

$$\bar{q} = \frac{\ln 3}{0.01} \approx 109.8612$$

We now have

$$CS = \int_0^{\bar{q}} (D(q) - \bar{p})\, dq$$

$$= \int_0^{109.8612} (15e^{-0.01q} - 5)\, dq$$

$$= \left[\frac{15}{-0.01}e^{-0.01q} - 5q\right]_0^{109.8612}$$

$$\approx (-500 - 549.31) - (-1500 - 0)$$

$$= \$450.69 \text{ per day}$$

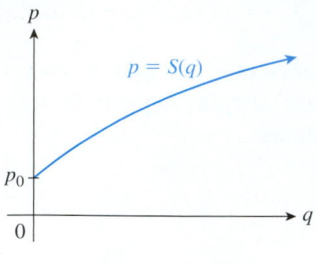

Figure **16**

Producers' Surplus

We can also calculate extra income earned by producers. Consider a supply equation of the form $p = S(q)$, where $S(q)$ is the price at which a supplier is willing to supply q items (per time period). Because a producer is generally willing to supply more units at a higher price per unit, a supply curve usually has a positive slope, as shown in Figure 16. The price p_0 is the lowest price that a producer is willing to charge.

Arguing as before, we see that the minimum amount of money producers are willing to receive in exchange for $\bar{q}$ items is $\int_0^{\bar{q}} S(q)\, dq$. On the other hand, if the producers charge $\bar{p}$ per item for $\bar{q}$ items, their actual revenue is $\bar{p}\bar{q} = \int_0^{\bar{q}} \bar{p}\, dq$.

The difference between the producers' actual revenue and the minimum they would have been willing to receive is the **producers' surplus.**

Producers' Surplus

The **producers' surplus** is the extra amount earned by producers who were willing to charge less than the selling price of $\bar{p}$ per unit and is given by

$$PS = \int_0^{\bar{q}} [\bar{p} - S(q)]\, dq$$

where $S(\bar{q}) = \bar{p}$. Graphically, it is the area of the region between the graphs of $p = \bar{p}$ and $p = S(q)$ for $0 \le q \le \bar{q}$, as in the figure.

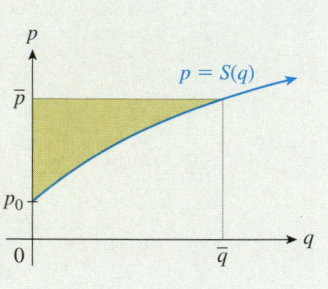

Example **2** Producers' Surplus

My tie-dye T-shirt enterprise has grown to the extent that I am now able to produce T-shirts in bulk, and several campus groups have begun placing orders. I have informed one group that I am prepared to supply $20\sqrt{p - 4}$ T-shirts at a price of p dollars per shirt. What is my total surplus if I sell T-shirts to the group at $8 each?

Solution We need to calculate the producers' surplus when $\bar{p} = 8$. The supply equation is

$$q = 20\sqrt{p - 4}$$

but in order to use the formula for producers' surplus, we need to express p as a function of q. First, we square both sides to remove the radical sign:

$$q^2 = 400(p - 4)$$

so

$$p - 4 = \frac{q^2}{400}$$

giving

$$p = S(q) = \frac{q^2}{400} + 4$$

We now need the value of $\bar{q}$ corresponding to $\bar{p} = 8$. Substituting $p = 8$ in the original equation, gives

$$\bar{q} = 20\sqrt{8 - 4} = 20\sqrt{4} = 40$$

Thus,

$$
\begin{aligned}
PS &= \int_0^{\bar{q}} (\bar{p} - S(q))\, dq \\
&= \int_0^{40} \left[8 - \left(\frac{q^2}{400} + 4 \right) \right] dq \\
&= \int_0^{40} \left(4 - \frac{q^2}{400} \right) dq \\
&= \left[4q - \frac{q^3}{1200} \right]_0^{40} \approx \$106.67
\end{aligned}
$$

Thus, I earn a surplus of \$106.67 if I sell T-shirts to the group at \$8 each.

Example 3 Equilibrium

To continue the preceding example: A representative informs me that the campus group is prepared to order only $\sqrt{200(16 - p)}$ T-shirts at p dollars each. I would like to produce as many T-shirts for them as possible but avoid being left with unsold T-shirts. Given the supply curve from the preceding example, what price should I charge per T-shirt, and what are the consumers' and producers' surpluses at that price?

Solution The price that guarantees neither a shortage nor a surplus of T-shirts is the **equilibrium price,** the price where supply equals demand. We have

$$
\begin{aligned}
\text{Supply:} \quad & q = 20\sqrt{p - 4} \\
\text{Demand:} \quad & q = \sqrt{200(16 - p)}
\end{aligned}
$$

Equating these gives

$$20\sqrt{p - 4} = \sqrt{200(16 - p)}$$

$$400(p - 4) = 200(16 - p)$$

$$400p - 1600 = 3200 - 200p$$

$$600p = 4800$$

$$p = \$8 \text{ per T-shirt}$$

We therefore take $\bar{p} = 8$ (which happens to be the price we used in the preceding example). We get the corresponding value for q by substituting $p = 8$ into either the demand or supply equation:

$$\bar{q} = 20\sqrt{8 - 4} = 40$$

Thus, $\bar{p} = 8$ and $\bar{q} = 40$.

We must now calculate the consumers' surplus and the producers' surplus. We calculated the producers' surplus for $\bar{p} = 8$ in the preceding example:

$$PS = \$106.67$$

For the consumers' surplus, we must first express p as a function of q for the demand equation. Thus, we solve the demand equation for p as we did for the supply equation and we obtain

Demand: $D(q) = 16 - \dfrac{q^2}{200}$

Therefore,

$$
\begin{aligned}
CS &= \int_0^{\bar{q}} (D(q) - \bar{p})\, dq \\
&= \int_0^{40} \left[\left(16 - \frac{q^2}{200} \right) - 8 \right] dq \\
&= \int_0^{40} \left(8 - \frac{q^2}{200} \right) dq \\
&= \left[8q - \frac{q^3}{600} \right]_0^{40} \approx \$213.33
\end{aligned}
$$

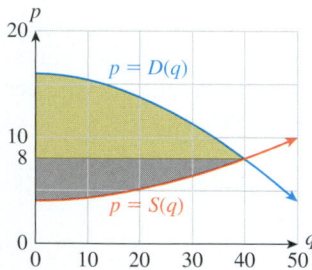

Figure **17**

+ *Before we go on...* Figure 17 shows both the consumers' surplus (top portion) and the producers' surplus (bottom portion) from Example 3. Because extra money in people's pockets is a Good Thing, the total of the consumers' and the producers' surpluses is called the **total social gain.** In this case it is

Social gain $= CS + PS = 213.33 + 106.67 = \$320.00$

As you can see from the figure, the total social gain is also the area between two curves and equals

$$\int_0^{40} (D(q) - S(q))\, dq$$

Continuous Income Streams

For purposes of calculation, it is often convenient to assume that a company with a high sales volume receives money continuously. In such a case, we have a function $R(t)$ that represents the rate at which money is being received by the company at time t.

Example **4** **Continuous Income**

An ice cream store's business peaks in late summer; the store's summer revenue is approximated by

$$R(t) = 300 + 4.5t - 0.05t^2 \text{ dollars per day}   (0 \le t \le 92)$$

where t is measured in days after June 1. What is its total revenue for the months of June, July, and August?

TITLE **Director**
INSTITUTION **Kroll Zolfo Cooper**

Patrick Farace

Kroll Zolfo Cooper is one of the world's leading restructuring firms and works with companies experiencing critical financial or operational problems. We are retained as interim management or advisors to assist in stabilizing, rehabilitating and reorganizing the business, whether through performance improvement changes or Chapter 11 (bankruptcy) reorganization. We then make recommendations based on the financial analysis to the company's senior management team and use the analysis to develop concrete action plans to help get the company back on track. As a Director at Kroll Zolfo Cooper, I use mathematics when I perform financial analysis on companies in financial distress and/or bankruptcy.

At a large merchant power company we built a highly complex mathematical model to analyze the financial capa-

bilities of the corporation and to project the company's financial performance into the future. The model was used by the company to forecast performance and by the lenders, investors and other key stakeholders to determine whether or not to lend money, invest and/or partner with the company.

At a major jewelry manufacturer, we determined that the company was losing money on most of its smaller orders. We used our mathematical analysis to work with the company to eliminate a series of costly activities that were unimportant to the smaller customers, determined the true cost of handling smaller orders, revised their computer pricing models, improved the company's pricing, and created incentives for customers to place larger orders. Almost immediately our financial analysis resulted in higher profitability for the department and higher overall sales results.

While these are just a few examples, the bulk of the work performed in the financial world has its basis in the same mathematical principals taught in your college coursework.

Solution Let's approximate the total revenue by breaking up the interval $[0, 92]$ representing the three months into n subintervals $[t_{k-1}, t_k]$, each with length Δt. In the interval $[t_{k-1}, t_k]$ the store receives money at a rate of approximately $R(t_{k-1})$ dollars per day for Δt days, so it will receive a total of $R(t_{k-1})\Delta t$ dollars. Over the whole summer, then, the store will receive approximately

$$R(t_0)\Delta t + R(t_1)\Delta t + \cdots + R(t_{n-1})\Delta t \text{ dollars}$$

As we let n become large to better approximate the total revenue, this Riemann sum approaches the integral

$$\text{Total revenue} = \int_0^{92} R(t)\, dt$$

Substituting the function we were given, we get

$$\text{Total revenue} = \int_0^{92} (300 + 4.5t - 0.05t^2)\, dt$$

$$= \left[300t + 2.25t^2 - \frac{0.05}{3}t^3 \right]_0^{92}$$

$$\approx \$33,666$$

+*Before we go on...* We could approach the calculation in Example 4 another way: $R(t) = S'(t)$ where $S(t)$ is the total revenue earned up to day t. By the Fundamental Theorem of Calculus,

$$\text{Total revenue} = S(92) - S(0) = \int_0^{92} R(t)\, dt$$

We did the calculation using Riemann sums mainly as practice for the next example. ∎

Generalizing Example 4, we can say the following:

Total Value of a Continuous Income Stream

If the rate of receipt of income is $R(t)$ dollars per unit of time, then the total income received from time $t = a$ to $t = b$ is

$$\text{Total value} = TV = \int_a^b R(t)\, dt$$

Example 5 Future Value

Suppose the ice cream store in the preceding example deposits its receipts in an account paying 5% interest per year compounded continuously. How much money will it have in its account at the end of August?

Solution Now we have to take into account not only the revenue but also the interest it earns in the account. Again, we break the interval $[0, 92]$ into n subintervals. During the interval $[t_{k-1}, t_k]$, approximately $R(t_{k-1})\Delta t$ dollars are deposited in the account. That money will earn interest until the end of August, a period of $92 - t_{k-1}$ days, or $(92 - t_{k-1})/365$ years. The formula for continuous compounding tells us that by the end of August, those $R(t_{k-1})\Delta t$ dollars will have turned into

$$R(t_{k-1})\Delta t e^{0.05(92-t_{k-1})/365} = R(t_{k-1})e^{0.05(92-t_{k-1})/365}\Delta t \text{ dollars}$$

(Recall that 5% is the *annual* interest rate.) Adding up the contributions from each subinterval, we see that the total in the account at the end of August will be approximately

$$R(t_0)e^{0.05(92-t_0)/365}\Delta t + R(t_1)e^{0.05(92-t_1)/365}\Delta t + \cdots + R(t_{n-1})e^{0.05(92-t_{n-1})/365}\Delta t$$

This is a Riemann sum; as n gets large the sum approaches the integral

$$\text{Future value} = FV = \int_0^{92} R(t)e^{0.05(92-t)/365}\, dt$$

Substituting $R(t) = 300 + 4.5t - 0.05t^2$, we obtain

$$FV = \int_0^{92} (300 + 4.5t - 0.05t^2)e^{0.05(92-t)/365}\, dt$$

$$\approx \$33{,}880 \qquad \text{\color{red}{Using technology or integration by parts}}$$

+ *Before we go on...* The interest earned in the account in Example 5 was fairly small (compare this answer to that in Example 4). Not only was the money in the account for only three months, but much of it was put in the account towards the end of that period, so had very little time to earn interest. ∎

Generalizing again, we have the following:

Future Value of a Continuous Income Stream

If the rate of receipt of income from time $t = a$ to $t = b$ is $R(t)$ dollars per unit of time and the income is deposited as it is received in an account paying interest at rate r per unit of time, compounded continuously, then the amount of money in the account at time $t = b$ is

$$\text{Future value} = FV = \int_a^b R(t)e^{r(b-t)}\,dt$$

Example 6 Present Value

You are thinking of buying the ice cream store discussed in the preceding two examples. What is its income stream worth to you on June 1? Assume that you have access to the same account paying 5% per year compounded continuously.

Solution The value of the income stream on June 1 is the amount of money that, if deposited June 1, would give you the same future value as the income stream will. If we let PV denote this "present value," its value after 92 days will be

$$PV e^{0.05 \times 92/365}$$

We equate this with the future value of the income stream to get

$$PV e^{0.05 \times 92/365} = \int_0^{92} R(t)e^{0.05(92-t)/365}\,dt$$

so

$$PV = \int_0^{92} R(t)e^{-0.05t/365}\,dt$$

Substituting the formula for $R(t)$ and integrating using technology or integration by parts, we get

$$PV \approx \$33,455$$

The general formula is the following:

Present Value of a Continuous Income Stream

If the rate of receipt of income from time $t = a$ to $t = b$ is $R(t)$ dollars per unit of time and the income is deposited as it is received in an account paying interest at rate r per unit of time, compounded continuously, then the value of the income stream at time $t = a$ is

$$\text{Present value} = PV = \int_a^b R(t)e^{r(a-t)}\,dt$$

We can derive this formula from the relation

$$FV = PV e^{r(b-a)}$$

because the present value is the amount that would have to be deposited at time $t = a$ to give a future value of FV at time $t = b$.

Note These formulas are more general than we've said. They still work when $R(t) < 0$ if we interpret negative values as money flowing *out* rather than in. That is, we can use these formulas for income we receive or for payments that we make, or for situations where we sometimes receive money and sometimes pay it out. These formulas can also be used for flows of quantities other than money. For example, if we use an exponential model for population growth and we let $R(t)$ represent the rate of immigration $[R(t) > 0]$ or emigration $[R(t) < 0]$, then the future value formula gives the future population. ■

14.4 EXERCISES

● denotes basic skills exercises

Calculate the consumers' surplus at the indicated unit price $\bar{p}$ for each of the demand equations in Exercises 1–12. hint [see Example 1]

1. ● $p = 10 - 2q$; $\bar{p} = 5$ 2. ● $p = 100 - q$; $\bar{p} = 20$

3. ● $p = 100 - 3\sqrt{q}$; $\bar{p} = 76$ 4. ● $p = 10 - 2q^{1/3}$; $\bar{p} = 6$

5. ● $p = 500e^{-2q}$; $\bar{p} = 100$ 6. ● $p = 100 - e^{0.1q}$; $\bar{p} = 50$

7. ● $q = 100 - 2p$; $\bar{p} = 20$ 8. ● $q = 50 - 3p$; $\bar{p} = 10$

9. ● $q = 100 - 0.25p^2$; $\bar{p} = 10$

10. ● $q = 20 - 0.05p^2$; $\bar{p} = 5$

11. ● $q = 500e^{-0.5p} - 50$; $\bar{p} = 1$

12. ● $q = 100 - e^{0.1p}$; $\bar{p} = 20$

Calculate the producers' surplus for each of the supply equations in Exercises 13–24 at the indicated unit price $\bar{p}$. hint [see Example 2]

13. ● $p = 10 + 2q$; $\bar{p} = 20$

14. ● $p = 100 + q$; $\bar{p} = 200$

15. ● $p = 10 + 2q^{1/3}$; $\bar{p} = 12$

16. ● $p = 100 + 3\sqrt{q}$; $\bar{p} = 124$

17. ● $p = 500e^{0.5q}$; $\bar{p} = 1000$

18. ● $p = 100 + e^{0.01q}$; $\bar{p} = 120$

19. ● $q = 2p - 50$; $\bar{p} = 40$

20. ● $q = 4p - 1000$; $\bar{p} = 1000$

21. ● $q = 0.25p^2 - 10$; $\bar{p} = 10$

22. ● $q = 0.05p^2 - 20$; $\bar{p} = 50$

23. ● $q = 500e^{0.05p} - 50$; $\bar{p} = 10$

24. ● $q = 10(e^{0.1p} - 1)$; $\bar{p} = 5$

In Exercises 25–30, find the total value of the given income stream and also find its future value (at the end of the given interval) using the given interest rate. hint [see Examples 4 & 5]

25. ● $R(t) = 30,000$, $0 \le t \le 10$, at 7%

26. ● $R(t) = 40,000$, $0 \le t \le 5$, at 10%

27. ● $R(t) = 30,000 + 1000t$, $0 \le t \le 10$, at 7%

28. ● $R(t) = 40,000 + 2000t$, $0 \le t \le 5$, at 10%

29. ● $R(t) = 30,000e^{0.05t}$, $0 \le t \le 10$, at 7%

30. ● $R(t) = 40,000e^{0.04t}$, $0 \le t \le 5$, at 10%

In Exercises 31–36, find the total value of the given income stream and also find its present value (at the beginning of the given interval) using the given interest rate. hint [see Examples 4 & 6]

31. ● $R(t) = 20,000$, $0 \le t \le 5$, at 8%

32. ● $R(t) = 50,000$, $0 \le t \le 10$, at 5%

33. ● $R(t) = 20,000 + 1000t$, $0 \le t \le 5$, at 8%

34. ● $R(t) = 50,000 + 2000t$, $0 \le t \le 10$, at 5%

35. ● $R(t) = 20,000e^{0.03t}$, $0 \le t \le 5$, at 8%

36. ● $R(t) = 50,000e^{0.06t}$, $0 \le t \le 10$, at 5%

Applications

37. ● **College Tuition** A study of U.S. colleges and universities resulted in the demand equation $q = 20,000 - 2p$, where q is the enrollment at a public college or university and p is the average annual tuition (plus fees) it charges.[21] Officials at Enormous State University have developed a policy whereby the number of students it will accept per year at a tuition level of p dollars is given by $q = 7500 + 0.5p$. Find the equilibrium tuition price $\bar{p}$ and the consumers' and producers' surpluses at this tuition level. What is the total social gain at the equilibrium price? hint [see Example 3]

38. ● **Fast Food** A fast-food outlet finds that the demand equation for its new side dish, "Sweetdough Tidbit," is given by

$$p = \frac{128}{(q + 1)^2}$$

where p is the price (in cents) per serving and q is the number of servings that can be sold per hour at this price. At the same time, the franchise is prepared to sell $q = 0.5p - 1$ servings per hour at a price of p cents. Find the equilibrium price $\bar{p}$ and

---

[21] Idea based on a study by A. L. Ostrosky Jr. and J. V. Koch, as cited in their book, *Introduction to Mathematical Economics* (Waveland Press, Illinois, 1979. p. 133). The data used here are fictitious, however.

the consumers' and producers' surpluses at this price level. What is the total social gain at the equilibrium price?

39. **Linear Demand** Given a linear demand equation of the form $q = -mp + b$ $(m > 0)$, find a formula for the consumers' surplus at a price level of $\bar{p}$ per unit.

40. **Linear Supply** Given a linear supply equation of the form $q = mp + b$ $(m > 0)$, find a formula for the producers' surplus at a price level of $\bar{p}$ per unit.

41. ● **Revenue** The annual net sales (revenue) earned by the Finnish cell phone company Nokia from January 1999 to January 2004 can be approximated by

$$R(t) = -1.7t^2 + 5t + 28 \text{ billion euros per year}$$
$$(-1 \le t \le 4)$$

where t is time in years ($t = 0$ represents January 2000).[22] Estimate, to the nearest €10 billion, Nokia's total revenue from January 1999 through December 2003. *hint* [see Example 4]

42. ● **Revenue** The annual net sales (revenue) earned by Nintendo Co., Ltd., in the fiscal years ending March 31, 1995 to March 31, 2004, can be approximated by

$$R(t) = -4t^2 + 10t + 530 \text{ billion yen per year}$$
$$(-6 \le t \le 4)$$

where t is time in years ($t = 0$ represents March 31, 2000).[23] Estimate, to the nearest ¥100 billion, Nintendo's total revenue from April 1, 1994, through March 31, 2004.

43. **Revenues** The annual revenue earned by Wal-Mart in the fiscal years ending January 31, 1994 to January 31, 2004 can be approximated by

$$R(t) = 150e^{0.14t} \text{ billion dollars per year} \quad (-7 \le t \le 4)$$

where t is time in years ($t = 0$ represents January 31, 2000).[24] Estimate, to the nearest $10 billion, Wal-Mart's total revenue from January 31, 1999 through January 31, 2004.

44. **Revenues** The annual revenue earned by Target for fiscal years 1998 through 2003 can be approximated by

$$R(t) = 37e^{0.09t} \text{ billion dollars per year} \quad (-2 \le t \le 4)$$

where t is time in years ($t = 0$ represents the beginning of fiscal year 2000).[25] Estimate, to the nearest $10 billion, Target's total revenue in fiscal years 1998 through 2003.

45. **Revenue** Refer back to Exercise 41. Suppose that, from January 1999 on, Nokia invested its revenue in an investment yielding 4% compounded continuously. What, to the nearest €10 billion, would the total value of Nokia's revenues have been at the end of 2003? *hint* [see Example 5]

46. **Revenue** Refer back to Exercise 42. Suppose that, from April 1994 on, Nintendo invested its revenue in an investment yielding 5% compounded continuously. What, to the nearest ¥100 billion, would the total value of Nintendo's revenue have been by the end of March 2004?

47. **Revenue** Refer back to Exercise 43. Suppose that, from January 1999 on, Wal-Mart invested its revenue in an investment that depreciated continuously at a rate of 5% per year. What, to the nearest $10 billion, would the total value of Wal-Mart's revenues have been by the end of January 2004?

48. **Revenue** Refer back to Exercise 44. Suppose that, from fiscal year 1998 on, Target invested its revenue in an investment that depreciated continuously at a rate of 3% per year. What, to the nearest $10 billion, would the total value of Target's revenue have been by the end of fiscal year 2003?

49. **Saving for Retirement** You are saving for your retirement by investing $700 per month in an annuity with a guaranteed interest rate of 6% per year. With a continuous stream of investment and continuous compounding, how much will you have accumulated in the annuity by the time you retire in 45 years?

50. **Saving for College** When your first child is born, you begin to save for college by depositing $400 per month in an account paying 12% interest per year. With a continuous stream of investment and continuous compounding, how much will you have accumulated in the account by the time your child enters college 18 years later?

51. **Saving for Retirement** You begin saving for your retirement by investing $700 per month in an annuity with a guaranteed interest rate of 6% per year. You increase the amount you invest at the rate of 3% per year. With continuous investment and compounding, how much will you have accumulated in the annuity by the time you retire in 45 years?

52. **Saving for College** When your first child is born, you begin to save for college by depositing $400 per month in an account paying 12% interest per year. You increase the amount you save by 2% per year. With continuous investment and compounding, how much will have accumulated in the account by the time your child enters college 18 years later?

53. **Bonds** The U.S. Treasury issued a 30-year bond on October 15, 2001, paying 3.375% interest.[26] Thus, if you bought $100,000 worth of these bonds you would receive $3375 per year in interest for 30 years. An investor wishes to buy the

---

[22] SOURCE: Nokia financial statements downloaded from www.nokia.com, January 13, 2005.

[23] The model is based on a quadratic regression. Source for data: Nintendo Co., Ltd. annual reports, downloaded from www.nintendo.com, January 23, 2005.

[24] The model is based on an exponential regression. Source for data: WalMart 2004 annual report, downloaded from www.walmartstores.com.

[25] The model is based on an exponential regression. Source for data: Target 2003 annual report, downloaded from www.targetcorp.com.

[26] The U.S. Treasury suspended selling Treasury Bonds after October 2001 but resumed selling them in February 2006. SOURCE: The Bureau of the Public Debt's website: www.publicdebt.treas.gov.

● basic skills

rights to receive the interest on $100,000 worth of these bonds. The amount the investor is willing to pay is the present value of the interest payments, assuming a 4% rate of return. Assuming (incorrectly, but approximately) that the interest payments are made continuously, what will the investor pay? *hint* [see Example 6]

54. **Bonds** The Megabucks Corporation is issuing a 20-year bond paying 7% interest (see the preceding exercise). An investor wishes to buy the rights to receive the interest on $50,000 worth of these bonds, and seeks a 6% rate of return. Assuming that the interest payments are made continuously, what will the investor pay?

55. **Valuing Future Income** Inga was injured and can no longer work. As a result of a lawsuit, she is to be awarded the present value of the income she would have received over the next 20 years. Her income at the time she was injured was $100,000 per year, increasing by $5000 per year. What will be the amount of her award, assuming continuous income and a 5% interest rate?

56. **Valuing Future Income** Max was injured and can no longer work. As a result of a lawsuit, he is to be awarded the present value of the income he would have received over the next 30 years. His income at the time he was injured was $30,000 per year, increasing by $1500 per year. What will be the amount of his award, assuming continuous income and a 6% interest rate?

Communication and Reasoning Exercises

57. ● Complete the following: The future value of a continuous income stream earning 0% interest is the same as the _____ value.

58. ● Complete the following: The present value of a continuous income stream earning 0% interest is the same as the _____ value.

59. Your study group friend says that the future value of a continuous stream of income is always greater than the total value, assuming a positive rate of return. Is she correct? Why?

60. Your other study group friend says that the present value of a continuous stream of income can sometimes be greater than the total value, depending on the (positive) interest rate. Is he correct? Explain.

61. Arrange from smallest to largest: Total Value, Future Value, Present Value of a continuous stream of income (assuming a positive income and positive rate of return).

62. **a.** Arrange the following functions from smallest to largest: $R(t)$, $R(t)e^{r(b-t)}$, $R(t)e^{r(a-t)}$, where $a \le t \le b$, and r and $R(t)$ are positive.

   b. Use the result from part (a) to justify your answers in Exercises 59–61.

● basic skills

14.5 Improper Integrals and Applications

All the definite integrals we have seen so far have had the form $\int_a^b f(x)\,dx$, with a and b finite and $f(x)$ piecewise continuous on the closed interval $[a, b]$. If we relax one or both of these requirements somewhat, we obtain what are called **improper integrals.** There are various types of improper integrals.

Integrals in Which a Limit of Integration is Infinite

Integrals in which one or more limits of integration are infinite can be written as

$$\int_a^{+\infty} f(x)\,dx, \quad \int_{-\infty}^b f(x)\,dx, \quad \text{or} \quad \int_{-\infty}^{+\infty} f(x)\,dx$$

Let's concentrate for a moment on the first form, $\int_a^{+\infty} f(x)\,dx$. What does the $+\infty$ mean here? As it often does, it means that we are to take a limit as something gets large. Specifically, it means the limit as the upper bound of integration gets large.

Improper Integral with an Infinite Limit of Integration

We define

$$\int_a^{+\infty} f(x)\,dx = \lim_{M \to +\infty} \int_a^M f(x)\,dx$$

provided the limit exists. If the limit exists, we say that $\int_a^{+\infty} f(x)\,dx$ **converges.** Otherwise, we say that $\int_a^{+\infty} f(x)\,dx$ **diverges.** Similarly, we define

$$\int_{-\infty}^b f(x)\,dx = \lim_{M \to -\infty} \int_M^b f(x)\,dx$$

provided the limit exists. Finally, we define

$$\int_{-\infty}^{+\infty} f(x)\,dx = \int_{-\infty}^a f(x)\,dx + \int_a^{+\infty} f(x)\,dx$$

for some convenient a, provided *both* integrals on the right converge.

quick Examples

1. $\displaystyle\int_1^{+\infty} \frac{dx}{x^2} = \lim_{M \to +\infty} \int_1^M \frac{dx}{x^2} = \lim_{M \to +\infty} \left[-\frac{1}{x} \right]_1^M = \lim_{M \to +\infty} \left(-\frac{1}{M} + 1 \right) = 1$ Converges

2. $\displaystyle\int_1^{+\infty} \frac{dx}{x} = \lim_{M \to +\infty} \int_1^M \frac{dx}{x} = \lim_{M \to +\infty} \left[\ln|x| \right]_1^M = \lim_{M \to +\infty} (\ln M - \ln 1) = +\infty$ Diverges

3. $\displaystyle\int_{-\infty}^{-1} \frac{dx}{x^2} = \lim_{M \to -\infty} \int_M^{-1} \frac{dx}{x^2} = \lim_{M \to -\infty} \left[-\frac{1}{x} \right]_M^{-1} = \lim_{M \to -\infty} \left(1 + \frac{1}{M} \right) = 1$ Converges

4. $\displaystyle\int_{-\infty}^{+\infty} e^{-x}\,dx = \int_{-\infty}^0 e^{-x}\,dx + \int_0^{+\infty} e^{-x}\,dx$

$$= \lim_{M \to -\infty} \int_M^0 e^{-x}\,dx + \lim_{M \to +\infty} \int_0^M e^{-x}\,dx$$

$$= \lim_{M \to -\infty} -[e^{-x}]_M^0 + \lim_{M \to +\infty} -[e^{-x}]_0^M$$

$$= \lim_{M \to -\infty} (e^{-M} - 1) + \lim_{M \to +\infty} (1 - e^{-M})$$

$$= +\infty + 1 \qquad\qquad \text{<span style="color:#c00">Diverges</span>}$$

5. $\displaystyle\int_{-\infty}^{+\infty} xe^{-x^2}\,dx = \int_{-\infty}^0 xe^{-x^2}\,dx + \int_0^{+\infty} xe^{-x^2}\,dx$

$$= \lim_{M \to -\infty} \int_M^0 xe^{-x^2}\,dx + \lim_{M \to +\infty} \int_0^M xe^{-x^2}\,dx$$

$$= \lim_{M \to -\infty} \left[-\frac{1}{2} e^{-x^2} \right]_M^0 + \lim_{M \to +\infty} \left[-\frac{1}{2} e^{-x^2} \right]_0^M$$

$$= \lim_{M \to -\infty} \left(-\frac{1}{2} + \frac{1}{2} e^{-M^2} \right) + \lim_{M \to +\infty} \left(-\frac{1}{2} e^{-M^2} + \frac{1}{2} \right)$$

$$= -\frac{1}{2} + \frac{1}{2} = 0 \qquad\qquad \text{<span style="color:#c00">Converges</span>}$$

Q: *We learned that the integral can be interpreted as the area under the curve. Is this still true for improper integrals* **?**

A: Yes. Figure 18 illustrates how we can represent an improper integral as the area of an infinite region.

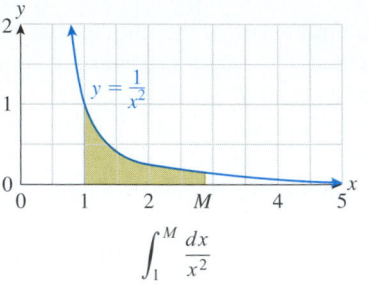

$$\int_1^M \frac{dx}{x^2}$$

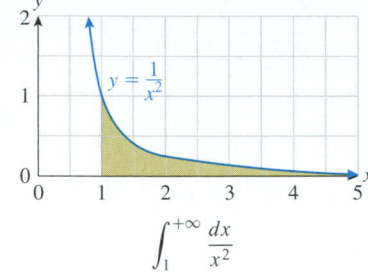

$$\int_1^{+\infty} \frac{dx}{x^2}$$

Figure **18**

On the left we see the area represented by $\int_1^M dx/x^2$. As M gets larger, the integral approaches $\int_1^{+\infty} dx/x^2$. In the picture, think of M being moved farther and farther along the x-axis in the direction of increasing x, resulting in the region shown on the right. ■

Q: *Wait! We calculated $\int_1^{+\infty} dx/x^2 = 1$. Does this mean that the infinitely long area in Figure 18 has an area of only 1 square unit* **?**

A: That is exactly what it means. If you had enough paint to cover 1 square unit, you would never run out of paint while painting the region in Figure 18. This is one of the places where mathematics seems to contradict common sense. But common sense is notoriously unreliable when dealing with infinities. ■

Example **1** **Future Sales of VCRs**

In 2001, sales of DVD players were starting to make inroads into the sales of VCRs, but VCRs were still selling well.[*] Approximately 15 million VCRs were expected to be sold in 2001. Suppose that sales of VCRs decrease by 15% per year from 2001 on. How many VCRs, total, will be sold from 2001 on?

Solution Recall that the total sales between two dates can be computed as the definite integral of annual sales. So, if we wanted the sales between the year 2001 and a year far in the future, we would compute $\int_0^M s(t)\, dt$ with a large M, where $s(t)$ is the annual sales t years after 2001. Because we want to know the *total* number of VCRs sold from 2001 on, we let $M \to +\infty$; that is, we compute $\int_0^{+\infty} s(t)\, dt$.

Because sales of VCRs are decreasing by 15% per year, we can model $s(t)$ by

$$s(t) = 15(0.85)^t \text{ million VCRs}$$

where t is the number of years since 2001.

$$\text{Total sales from 2001 on} = \int_0^{+\infty} 15(0.85)^t\, dt$$

$$= \lim_{M \to +\infty} \int_0^M 15(0.85)^t\, dt$$

---

[*] SOURCE: "VCRs outsell DVD players over holidays," *USA Today,* Jan. 24, 2001.

$$= \frac{15}{\ln 0.85} \lim_{M \to +\infty} [0.85^t]_0^M$$

$$= \frac{15}{\ln 0.85} \lim_{M \to +\infty} (0.85^M - 0.85^0)$$

$$= \frac{15}{\ln 0.85}(-1) \approx 92.3 \text{ million VCRs}$$

Integrals in Which the Integrand Becomes Infinite

We can sometimes compute integrals $\int_a^b f(x)\,dx$ in which $f(x)$ becomes infinite. As we'll see in Example 4, the Fundamental Theorem of Calculus does not work for such integrals. The first case to consider is when $f(x)$ approaches $\pm\infty$ at either a or b.

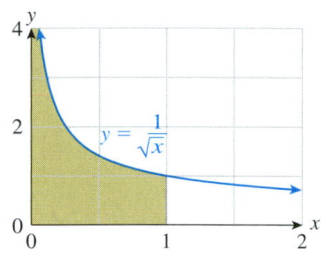

Figure **19**

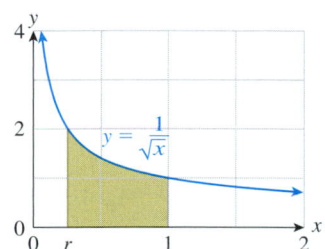

Figure **20**

Example 2 Integrand Infinite at One Endpoint

Calculate $\displaystyle\int_0^1 \frac{1}{\sqrt{x}}\,dx$.

Solution Notice that the integrand approaches $+\infty$ as x approaches 0 from the right and is not defined at 0. This makes the integral an improper integral. Figure 19 shows the region whose area we are trying to calculate; it extends infinitely vertically rather than horizontally.

Now, if $0 < r < 1$, the integral $\int_r^1 (1/\sqrt{x})\,dx$ is a proper integral because we avoid the bad behavior at 0. This integral gives the area shown in Figure 20. If we let r approach 0 from the right, the area in Figure 20 will approach the area in Figure 19. So, we calculate

$$\int_0^1 \frac{1}{\sqrt{x}}\,dx = \lim_{r \to 0^+} \int_r^1 \frac{1}{\sqrt{x}}\,dx$$

$$= \lim_{r \to 0^+} [2\sqrt{x}]_r^1$$

$$= \lim_{r \to 0^+} (2 - 2\sqrt{r})$$

$$= 2$$

Thus, we again have an infinitely long region with finite area.

Generalizing, we make the following definition.

Improper Integral in Which the Integrand Becomes Infinite

If $f(x)$ is defined for all x with $a < x \le b$ but approaches $\pm\infty$ as x approaches a, we define

$$\int_a^b f(x)\,dx = \lim_{r \to a^+} \int_r^b f(x)\,dx$$

provided the limit exists. Similarly, if $f(x)$ is defined for all x with $a \le x < b$ but approaches $\pm\infty$ as x approaches b, we define

$$\int_a^b f(x)\,dx = \lim_{r \to b^-} \int_a^r f(x)\,dx$$

provided the limit exists. In either case, if the limit exists, we say that $\int_a^b f(x)\,dx$ **converges**. Otherwise, we say that $\int_a^b f(x)\,dx$ **diverges**.

Example 3 Testing for Convergence

Does $\displaystyle\int_{-1}^{3} \frac{x}{x^2 - 9}\, dx$ converge? If so, to what?

Solution We first check to see where, if anywhere, the integrand approaches $\pm\infty$. That will happen where the denominator becomes 0, so we solve $x^2 - 9 = 0$.

$$x^2 - 9 = 0$$
$$x^2 = 9$$
$$x = \pm 3$$

The solution $x = -3$ is outside of the range of integration, so we ignore it. The solution $x = 3$ is, however, the right endpoint of the range of integration, so the integral is improper. We need to investigate the following limit:

$$\int_{-1}^{3} \frac{x}{x^2 - 9}\, dx = \lim_{r \to 3^-} \int_{-1}^{r} \frac{x}{x^2 - 9}\, dx$$

Now, to calculate the integral we use a substitution:

$$u = x^2 - 9$$
$$\frac{du}{dx} = 2x$$
$$dx = \frac{1}{2x}\, du$$
when $x = r$, $u = r^2 - 9$
when $x = -1$, $u = (-1)^2 - 9 = -8$

Thus,

$$\int_{-1}^{r} \frac{x}{x^2 - 9}\, dx = \int_{-8}^{r^2-9} \frac{1}{2u}\, du$$

$$= \frac{1}{2}[\ln |u|]_{-8}^{r^2-9}$$

$$= \frac{1}{2}(\ln |r^2 - 9| - \ln 8)$$

Now we take the limit:

$$\int_{-1}^{3} \frac{x}{x^2 - 9}\, dx = \lim_{r \to 3^-} \int_{-1}^{r} \frac{x}{x^2 - 9}\, dx$$

$$= \lim_{r \to 3^-} \frac{1}{2}(\ln |r^2 - 9| - \ln 8)$$

$$= -\infty$$

because, as $r \to 3$, $r^2 - 9 \to 0$, and so $\ln |r^2 - 9| \to -\infty$. Thus, this integral diverges.

Example 4 Integrand Infinite Between the Endpoints

Does $\displaystyle\int_{-2}^{3} \frac{1}{x^2}\, dx$ converge? If so, to what?

Solution Again we check to see if there are any points at which the integrand approaches $\pm\infty$. There is such a point, at $x = 0$. This is between the endpoints of the range of integration. To deal with this we break the integral into two integrals:

$$\int_{-2}^{3} \frac{1}{x^2}\, dx = \int_{-2}^{0} \frac{1}{x^2}\, dx + \int_{0}^{3} \frac{1}{x^2}\, dx$$

Each integral on the right is an improper integral with the integrand approaching $\pm\infty$ at an endpoint. If both of the integrals on the right converge, we take the sum as the value of the integral on the left. So now we compute

$$\int_{-2}^{0} \frac{1}{x^2}\, dx = \lim_{r \to 0^-} \int_{-2}^{r} \frac{1}{x^2}\, dx$$

$$= \lim_{r \to 0^-} \left[-\frac{1}{x} \right]_{-2}^{r}$$

$$= \lim_{r \to 0^-} \left(-\frac{1}{r} - \frac{1}{2} \right)$$

which diverges to $+\infty$. There is no need now to check $\int_{0}^{3}(1/x^2)\, dx$; because one of the two pieces of the integral diverges, we simply say that $\int_{-2}^{3}(1/x^2)\, dx$ diverges.

+*Before we go on...* What if we had been sloppy in Example 4 and had not checked first whether the integrand approached $\pm\infty$ somewhere? Then we probably would have applied the Fundamental Theorem of Calculus and done the following:

$$\int_{-2}^{3} \frac{1}{x^2}\, dx = \left[-\frac{1}{x} \right]_{-2}^{3} = \left(-\frac{1}{3} - \frac{1}{2} \right) = -\frac{5}{6} \quad \text{✗ \textit{WRONG!}}$$

Notice that the answer this "calculation" gives is patently ridiculous. Because $1/x^2 > 0$ for all x for which it is defined, any definite integral of $1/x^2$ must give a positive answer. *Moral:* Always check to see whether the integrand blows up anywhere in the range of integration. If it does, the FTC does not apply, and we must use the methods of this example. ∎

We end with an example of what to do if an integral is improper for more than one reason.

Example 5 An Integral Improper in Two Ways

Does $\displaystyle\int_{0}^{+\infty} \frac{1}{\sqrt{x}}\, dx$ converge? If so, to what?

Solution This integral is improper for two reasons. First, the range of integration is infinite. Second, the integrand blows up at the endpoint 0. In order to separate these two problems, we break up the integral at some convenient point:

$$\int_0^{+\infty} \frac{1}{\sqrt{x}}\,dx = \int_0^1 \frac{1}{\sqrt{x}}\,dx + \int_1^{+\infty} \frac{1}{\sqrt{x}}\,dx$$

We chose to break the integral at 1. Any positive number would have sufficed, but 1 is generally easier to use in calculations.

The first piece, $\int_0^1 (1/\sqrt{x})\,dx$, we discussed in Example 2; it converges to 2. For the second piece we have:

$$\int_1^{+\infty} \frac{1}{\sqrt{x}}\,dx = \lim_{M\to+\infty} \int_1^M \frac{1}{\sqrt{x}}\,dx$$

$$= \lim_{M\to+\infty} [2\sqrt{x}]_1^M$$

$$= \lim_{M\to+\infty} (2\sqrt{M} - 2)$$

which diverges to $+\infty$. Because the second piece of the integral diverges, we conclude that $\int_0^{+\infty}(1/\sqrt{x})\,dx$ diverges.

14.5 EXERCISES

● denotes basic skills exercises

◆ denotes challenging exercises

tech Ex indicates exercises that should be solved using technology

For some of the exercises in this section you need to assume the fact that $\lim_{M\to+\infty} M^n e^{-M} = 0$ *for all n.*

Decide whether each integral in Exercises 1–26 converges. If the integral converges, compute its value.

1. ● $\int_1^{+\infty} x\,dx$ *hint* [see Quick Examples p. 1015]

2. ● $\int_0^{+\infty} e^{-x}\,dx$

3. ● $\int_{-2}^{+\infty} e^{-0.5x}\,dx$ 4. ● $\int_1^{+\infty} \frac{1}{x^{1.5}}\,dx$

5. ● $\int_{-\infty}^2 e^x\,dx$ 6. ● $\int_{-\infty}^{-1} \frac{1}{x^{1/3}}\,dx$

7. ● $\int_{-\infty}^{-2} \frac{1}{x^2}\,dx$ 8. ● $\int_{-\infty}^0 e^{-x}\,dx$

9. ● $\int_0^{+\infty} x^2 e^{-6x}\,dx$ 10. ● $\int_0^{+\infty} (2x-4)e^{-x}\,dx$

11. ● $\int_0^5 \frac{2}{x^{1/3}}\,dx$ *hint* [see Example 2]

12. ● $\int_0^2 \frac{1}{x^2}\,dx$

13. ● $\int_{-1}^2 \frac{3}{(x+1)^2}\,dx$ 14. ● $\int_{-1}^2 \frac{3}{(x+1)^{1/2}}\,dx$

hint [see Example 3]

15. ● $\int_{-1}^2 \frac{3x}{x^2-1}\,dx$ *hint* [see Example 4]

16. ● $\int_{-1}^2 \frac{3}{x^{1/3}}\,dx$

17. ● $\int_{-2}^2 \frac{1}{(x+1)^{1/5}}\,dx$ 18. ● $\int_{-2}^2 \frac{2x}{\sqrt{4-x^2}}\,dx$

19. ● $\int_{-1}^1 \frac{2x}{x^2-1}\,dx$ 20. ● $\int_{-1}^2 \frac{2x}{x^2-1}\,dx$

21. ● $\int_{-\infty}^{+\infty} xe^{-x^2}\,dx$ 22. ● $\int_{-\infty}^{\infty} xe^{1-x^2}\,dx$

23. ● $\int_0^{+\infty} \frac{1}{x\ln x}\,dx$ *hint* [see Example 5]

24. ● $\int_0^{+\infty} \ln x\,dx$

25. $\int_0^{+\infty} \frac{2x}{x^2-1}\,dx$ 26. $\int_{-\infty}^0 \frac{2x}{x^2-1}\,dx$

Applications

27. ● *Advertising Revenue* From June 2001 to June 2002, *GQ Magazine's* advertising revenues could be approximated by

$$R(t) = 91.7(0.90)^t \text{ million dollars per year} \quad (0 \le t \le 1)$$

● basic skills   ◆ challenging   tech Ex technology exercise

where t is time in years since June 2001.[27] By extrapolating this model into the indefinite future, project *GQ's* total advertising revenue from June 2001 on. (Round your answer to the nearest \$1 million.) *hint* [see Example 1]

28. ● *Advertising Revenue* From June 2001 to June 2002, *Esquire Magazine's* advertising revenues could be approximated by

$$R(t) = 57.0(0.927)^t \text{ million dollars per year} \quad (0 \le t \le 1)$$

where t is time in years since June 2001.[28] By extrapolating this model into the indefinite future, project *Esquire's* total advertising revenue from June 2001 on. (Round your answer to the nearest \$1 million.)

29. *Cigarette Sales* According to the Federal Trade Commission, the number of cigarettes sold domestically in 2002 decreased by 5.5% from the 2001 total of approximately 400 billion cigarettes.[29] Use an exponential model to forecast the total number of cigarettes sold from 2001 on. (Round your answer to the nearest 100 billion cigarettes.)

30. *Sales* Sales of the text *Calculus and You* have been declining continuously at a rate of 5% per year. Assuming that *Calculus and You* currently sells 5000 copies per year and that sales will continue this pattern of decline, calculate total future sales of the text.

31. *Sales* My financial adviser has predicted that annual sales of Frodo T-shirts will continue to decline by 10% each year. At the moment, I have 3200 of the shirts in stock and am selling them at a rate of 200 per year. Will I ever sell them all?

32. *Revenue* Alarmed about the sales prospects for my Frodo T-shirts (see the preceding exercise), I will try to make up lost revenues by increasing the price by \$1 each year. I now charge \$10 per shirt. What is the total amount of revenue I can expect to earn from sales of my T-shirts, assuming the sales levels described in the previous exercise? (Give your answer to the nearest \$1000.)

33. *Education* Let $N(t)$ be the number of high school students graduated in the U.S. in year t. This number is projected to change at a rate of about

$$N'(t) = 0.214t^{-0.91} \text{ million graduates per year}$$
$$(0 \le t \le 21)$$

where t is time in years since 1990.[30] In 1991, there were about 2.5 million high school students graduated. By extrapolating the model, what can you say about the number of high school students graduated in a year far in the future?

34. *Education, Martian* Let $M(t)$ be the number of high school students graduated in the Republic of Mars in year t. This number is projected to change at a rate of about

$$M'(t) = 0.321t^{-1.10} \text{ thousand graduates per year}$$
$$(0 \le t \le 50)$$

where t is time in years since 2020. In 2021, there were about 1300 high school students graduated. By extrapolating the model, what can you say about the number of high school students graduated in a year far in the future?

35. *Cell Phone Revenues* The number of cell phone subscribers in China for the period 2000–2005 was projected to follow the equation[31]

$$N(t) = 39t + 68 \text{ million subscribers}$$

in year t ($t = 0$ represents 2000). The average annual revenue per cell phone user was \$350 in 2000.

  a. Assuming that, due to competition, the revenue per cell phone user decreases continuously at an annual rate of 10%, give a formula for the annual revenue in year t.

  b. Using the model you obtained in part (a) as an estimate of the rate of change of total revenue, estimate the total revenue from 2000 into the indefinite future.

36. *Vid Phone Revenues* The number of vid phone subscribers in the Republic of Mars for the period 2200–2300 was projected to follow the equation

$$N(t) = 18t - 10 \text{ thousand subscribers}$$

in year t ($t = 0$ represents 2200). The average annual revenue per vid phone user was $\bar{\bar{Z}}$ 40 in 2200.[32]

  a. Assuming that, due to competition, the revenue per vid phone user decreases continuously at an annual rate of 20%, give a formula for the annual revenue in year t.

  b. Using the model you obtained in part (a) as an estimate of the rate of change of total revenue, estimate the total revenue from 2200 into the indefinite future.

37. *Foreign Investments* According to data published by the World Bank, foreign direct investment in low income countries from 1999 through 2002 was approximately

$$q(t) = 1.7t^2 - 0.5t + 8 \text{ billion dollars per year}$$

where t is time in years since 2000.[33] Assuming a worldwide inflation rate of 5% per year, find the value of all foreign direct investment in low income countries from 2000 on in constant dollars. (The constant dollar value of $q(t)$ dollars t years from now is given by $q(t)e^{-rt}$, where r is the fractional

---

[27] Based on six-month advertising revenue figures for June, 2001 and June, 2002. SOURCE: *New York Times*, July, 29, 2002, p. C1.

[28] Ibid.

[29] SOURCE: Federal Trade Commission Cigarette Report for 2002, issued 2004, available at www.ftc.gov/reports/cigarette/041022cigaretterpt.pdf.

[30] Based on a regression model. Source for data: U.S. Department of Education, 2002; http://nces.ed.gov.

[31] Based on a regression of projected figures (coefficients are rounded). SOURCE: Intrinsic Technology/*New York Times*, Nov. 24, 2000, p. C1.

[32] $\bar{\bar{Z}}$ designates Zonars, the designated currency for the city-state of Utarek, Mars. SOURCE: www.marsnext.com/comm/zonars.html.

[33] The authors' approximation, based on data from the World Bank, obtained from www.worldbank.org.

● basic skills   ◆ challenging   **tech** Ex technology exercise

rate of inflation. Give your answer to the nearest $1000 billion.)

38. **Foreign Aid** Repeat the preceding exercise, using the following model for per capita aid to the least developed countries.[34] (Give your answer to the nearest $1000.)

$$q(t) = t^2 + 0.7t + 19 \text{ dollars per year}$$

39. **Online Book Sales** The number of books per year sold online in the U.S. in the period 1997–2000 can be approximated by

$$N(t) = \frac{82.8(7.14)^t}{21.8 + (7.14)^t} \text{ million books per year}$$

($t = 0$ represents 1997).[35] Investigate the integrals $\int_0^{+\infty} N(t)\, dt$ and $\int_{-\infty}^0 N(t)\, dt$ and interpret your answers.

40. **Mousse Sales** The weekly demand for your company's Lo-Cal Mousse is modeled by the equation

$$q(t) = \frac{50e^{2t-1}}{1 + e^{2t-1}}$$

where t is time from now in weeks and $q(t)$ is the number of gallons sold per week. Investigate the integrals $\int_0^{+\infty} q(t)\, dt$ and $\int_{-\infty}^0 q(t)\, dt$ and interpret your answers.

tech Ex *The Normal Curve Exercises 41–44 require the use of a graphing calculator or computer programmed to do numerical integration. The* normal distribution curve, *which models the distributions of data in a wide range of applications, is given by the function*

$$p(x) = \frac{1}{\sqrt{2\pi}\sigma} e^{-(x-\mu)^2/2\sigma^2}$$

where $\pi = 3.14159265\ldots$ *and* σ *and* μ *are constants called the standard deviation and the mean, respectively. Its graph (for* $\sigma = 1$ *and* $\mu = 2$*) is shown in the figure.*

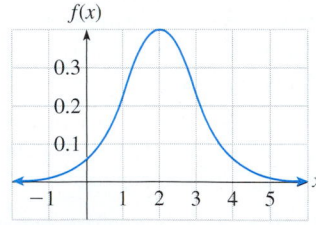

41. tech Ex With $\sigma = 4$ and $\mu = 1$, approximate $\int_{-\infty}^{+\infty} p(x)\, dx$.
42. tech Ex With $\sigma = 1$ and $\mu = 0$, approximate $\int_0^{+\infty} p(x)\, dx$.
43. tech Ex With $\sigma = 1$ and $\mu = 0$, approximate $\int_1^{+\infty} p(x)\, dx$.
44. tech Ex With $\sigma = 1$ and $\mu = 0$, approximate $\int_{-\infty}^1 p(x)\, dx$.
45. ◆ **Variable Sales** The value of your Chateau Petit Mont Blanc 1963 vintage burgundy is increasing continuously at an annual rate of 40%, and you have a supply of 1000 bottles worth

$85 each at today's prices. In order to ensure a steady income, you have decided to sell your wine at a diminishing rate—starting at 500 bottles per year, and then decreasing this figure continuously at a fractional rate of 100% per year. How much income (to the nearest dollar) can you expect to generate by this scheme? [*Hint:* Use the formula for continuously compounded interest.]

46. ◆ **Panic Sales** Unfortunately, your large supply of Chateau Petit Mont Blanc is continuously turning to vinegar at a fractional rate of 60% per year! You have thus decided to sell off your Petit Mont Blanc at $50 per bottle, but the market is a little thin, and you can only sell 400 bottles per year. Because you have no way of knowing which bottles now contain vinegar until they are opened, you shall have to give refunds for all the bottles of vinegar. What will your net income be before all the wine turns to vinegar?

47. ◆ **Meteor Impacts** The frequency of meteor impacts on earth can be modeled by

$$n(k) = \frac{1}{5.6997k^{1.081}}$$

where $n(k) = N'(k)$, and $N(k)$ is the average number of meteors of energy less than or equal to k megatons that will hit the earth in one year.[36] (A small nuclear bomb releases on the order of one megaton of energy.)

a. How many meteors of energy at least $k = 0.2$ hit the earth each year?
b. Investigate and interpret the integral $\int_0^1 n(k)\, dk$.

48. ◆ **Meteor Impacts** (continuing the previous exercise)

a. Explain why the integral

$$\int_a^b k \cdot n(k)\, dk$$

computes the total energy released each year by meteors with energies between a and b megatons.

b. Compute and interpret

$$\int_0^1 k \cdot n(k)\, dk$$

c. Compute and interpret

$$\int_1^{+\infty} k \cdot n(k)\, dk$$

49. ◆ **The Gamma Function** The gamma function is defined by the formula

$$\Gamma(x) = \int_0^{+\infty} t^{x-1} e^{-t}\, dt$$

---

[34] Ibid.

[35] The model is a logistic regression. Source for data: Ipsos-NPD Book Trends/*New York Times,* April 16, 2001, p. C1.

[36] The authors' model, based on data published by NASA International Near-Earth-Object Detection Workshop (*The New York Times,* Jan. 25, 1994, p. C1).

● basic skills ◆ challenging tech Ex technology exercise

a. Find $\Gamma(1)$ and $\Gamma(2)$.

b. Use integration by parts to show that for every positive integer n, $\Gamma(n+1) = n\Gamma(n)$.

c. Deduce that $\Gamma(n) = (n-1)!$ $[= (n-1)(n-2)\ldots 2 \cdot 1]$ for every positive integer n.

50. ◆ **Laplace Transforms** The Laplace Transform $F(x)$ of a function $f(t)$ is given by the formula

$$F(x) = \int_0^{+\infty} f(t)e^{-xt}\, dt \quad (x > 0)$$

a. Find $F(x)$ for $f(t) = 1$ and for $f(t) = t$.

b. Find a formula for $F(x)$ if $f(t) = t^n$ $(n = 1, 2, 3, \ldots)$.

c. Find a formula for $F(x)$ if $f(t) = e^{at}$ (a constant).

Communication and Reasoning Exercises

51. ● Why can't the Fundamental Theorem of Calculus be used to evaluate $\int_{-1}^{1} \dfrac{1}{x}\, dx$?

52. ● Why can't the Fundamental Theorem of Calculus be used to evaluate $\int_{1}^{+\infty} \dfrac{1}{x^2}\, dx$?

53. ● It sometimes happens that the Fundamental Theorem of Calculus gives the correct answer for an improper integral.

Does the FTC give the correct answer for improper integrals of the form

$$\int_{-a}^{a} \frac{1}{x^{1/r}}\, dx$$

if $r = 3, 5, 7, \ldots$?

54. ● Does the FTC give the correct answer for improper integrals of the form

$$\int_{-a}^{a} \frac{1}{x^r}\, dx$$

if $r = 3, 5, 7, \ldots$?

55. tech Ex How could you use technology to approximate improper integrals? (Your discussion should refer to each type of improper integral.)

56. tech Ex Use technology to approximate the integrals $\int_0^{M} e^{-(x-10)^2}\, dx$ for larger and larger values of M, using Riemann sums with 500 subdivisions. What do you find? Comment on the answer.

57. Make up an interesting application whose solution is $\int_{10}^{+\infty} 100te^{-0.2t}\, dt = \$1{,}015.01$.

58. Make up an interesting application whose solution is $\int_{100}^{+\infty} \dfrac{1}{r^2}\, dr = 0.01$.

● basic skills   ◆ challenging   tech Ex technology exercise

14.6 Differential Equations and Applications

A **differential equation** is an equation that involves a derivative of an unknown function. A **first-order differential equation** involves only the first derivative of the unknown function. A **second-order differential equation** involves the second derivative of the unknown function (and possibly the first derivative). Higher-order differential equations are defined similarly. In this book, we will deal only with first-order differential equations.

To **solve** a differential equation means to find the unknown function. Many of the laws of science and other fields describe how things change. When expressed mathematically, these laws take the form of equations involving derivatives—that is, differential equations. The field of differential equations is a large and very active area of study in mathematics, and we shall see only a small part of it in this section.

Example 1 Motion

A dragster accelerates from a stop so that its speed t seconds after starting is $40t$ ft/s. How far will the car go in 8 seconds?

Solution We wish to find the car's position function $s(t)$. We are told about its speed, which is ds/dt. Precisely, we are told that

$$\frac{ds}{dt} = 40t$$

This is the differential equation we have to solve to find $s(t)$. But we already know how to solve this kind of differential equation; we integrate:

$$s(t) = \int 40t \, dt = 20t^2 + C$$

We now have the **general solution** to the differential equation. By letting C take on different values, we get all the possible solutions. We can specify the one **particular solution** that gives the answer to our problem by imposing the **initial condition** that $s(0) = 0$. Substituting into $s(t) = 20t^2 + C$, we get

$$0 = s(0) = 20(0)^2 + C = C$$

so $C = 0$ and $s(t) = 20t^2$. To answer the question, the car travels $20(8)^2 = 1280$ feet in 8 seconds.

We did not have to work hard to solve the differential equation in Example 1. In fact, any differential equation of the form $dy/dx = f(x)$ can (in theory) be solved by integrating. (Whether we can actually carry out the integration is another matter!)

Simple Differential Equations

A **simple** differential equation has the form

$$\frac{dy}{dx} = f(x)$$

Its general solution is

$$y = \int f(x) \, dx$$

quick Example   The differential equation

$$\frac{dy}{dx} = 2x^2 - 4x^3$$

is simple and has general solution

$$y = \int f(x) \, dx = \frac{2x^3}{3} - x^4 + C$$

Not all differential equations are simple, as the next example shows.

Example 2 Separable Differential Equation

Consider the differential equation $\dfrac{dy}{dx} = \dfrac{x}{y^2}$.

a. Find the general solution.

b. Find the particular solution that satisfies the initial condition $y(0) = 2$.

Solution

a. This is not a simple differential equation because the right-hand side is a function of both x and y. We cannot solve this equation by just integrating; the solution to this problem is to "separate" the variables.

Step 1 *Separate the variables algebraically.* We rewrite the equation as

$$y^2 \, dy = x \, dx$$

Step 2 *Integrate both sides.*

$$\int y^2 \, dy = \int x \, dx$$

giving

$$\frac{y^3}{3} = \frac{x^2}{2} + C$$

Step 3 *Solve for the dependent variable.* We solve for y:

$$y^3 = \frac{3}{2}x^2 + 3C = \frac{3}{2}x^2 + D$$

(rewriting $3C$ as D, an equally arbitrary constant), so

$$y = \left(\frac{3}{2}x^2 + D\right)^{1/3}$$

This is the general solution of the differential equation.

b. We now need to find the value for D that will give us the solution satisfying the condition $y(0) = 2$. Substituting 0 for x and 2 for y in the general solution, we get

$$2 = \left(\frac{3}{2}(0)^2 + D\right)^{1/3} = D^{1/3}$$

so

$$D = 2^3 = 8$$

Thus, the particular solution we are looking for is

$$y = \left(\frac{3}{2}x^2 + 8\right)^{1/3}$$

+*Before we go on...* We can check the general solution in Example 2 by calculating both sides of the differential equation and comparing.

$$\frac{dy}{dx} = \frac{d}{dx}\left(\frac{3}{2}x^2 + D\right)^{1/3} = x\left(\frac{3}{2}x^2 + D\right)^{-2/3}$$

$$\frac{x}{y^2} = \frac{x}{\left(\frac{3}{2}x^2 + 8\right)^{2/3}} = x\left(\frac{3}{2}x^2 + D\right)^{-2/3} \qquad ✔$$

Q: *In Example 2, we wrote $y^2\,dy$ and $x\,dx$. What do they mean* ?

A: Although it is possible to give meaning to these symbols, for us they are just a notational convenience. We could have done the following instead:

$$y^2 \frac{dy}{dx} = x$$

Now we integrate both sides with respect to x.

$$\int y^2 \frac{dy}{dx}\,dx = \int x\,dx$$

We can use substitution to rewrite the left-hand side:

$$\int y^2 \frac{dy}{dx}\,dx = \int y^2\,dy$$

which brings us back to the equation

$$\int y^2\,dy = \int x\,dx$$ ∎

We were able to separate the variables in the preceding example because the right-hand side, x/y^2, was a *product* of a function of x and a function of y—namely,

$$\frac{x}{y^2} = x\left(\frac{1}{y^2}\right)$$

In general, we can say the following:

Separable Differential Equation

A **separable** differential equation has the form

$$\frac{dy}{dx} = f(x)g(y)$$

We solve a separable differential equation by separating the xs and the ys algebraically, writing

$$\frac{1}{g(y)}\,dy = f(x)\,dx$$

and then integrating:

$$\int \frac{1}{g(y)}\,dy = \int f(x)\,dx$$

Example 3 Rising Medical Costs

Spending on Medicare from 2000 to 2025 was projected to rise continuously at an instantaneous rate of 3.7% per year.[*] Find a formula for Medicare spending y as a function of time t in years since 2000.

Solution When we say that Medicare spending y was going up continuously at an instantaneous rate of 3.7% per year, we mean that

the instantaneous rate of increase of y was 3.7% of its value

---

[*] Spending is in constant 2000 dollars. Source for projected data: The Urban Institute's Analysis of the 1999 Trustee's Report; www.urban.org.

or $\dfrac{dy}{dt} = 0.037y$

This is a separable differential equation. Separating the variables gives

$$\frac{1}{y}dy = 0.037\,dt$$

Integrating both sides, we get

$$\int \frac{1}{y}dy = \int 0.037\,dt$$

so $\ln y = 0.037t + C$

(We should write $\ln |y|$, but we know that the medical costs are positive.) We now solve for y.

$$y = e^{0.037t+C} = e^C e^{0.037t} = Ae^{0.037t}$$

where A is a positive constant. This is the formula we used before for continuous percentage growth.

+ *Before we go on...* To determine A in Example 3 we need to know, for example, Medicare spending at time $t = 0$. The source cited estimates Medicare spending as $239.6 billion in 2000. Substituting $t = 0$ in the equation above gives

$$239.6 = Ae^0 = A$$

Thus, projected Medicare spending is

$$y = 239.6e^{0.037t} \text{ billion dollars}$$

t years after 2000. ∎

Example 4 Newton's Law of Cooling

Newton's Law of Cooling states that a hot object cools at a rate proportional to the difference between its temperature and the temperature of the surrounding environment (the **ambient temperature**). If a hot cup of coffee, at 170°F, is left to sit in a room at 70°F, how will the temperature of the coffee change over time?

Solution We let $H(t)$ denote the temperature of the coffee at time t. Newton's Law of Cooling tells us that $H(t)$ *decreases* at a rate proportional to the difference between $H(t)$ and 70°F, the ambient temperature. In other words,

$$\frac{dH}{dt} = -k(H - 70)$$

where k is some positive constant.[*] Note that $H \geq 70$: The coffee will never cool to less than the ambient temperature.

---

[*] When we say that a quantity Q is *proportional* to a quantity R, we mean that $Q = kR$ for some constant k. The constant k is referred to as the **constant of proportionality.**

The variables here are H and t, which we can separate as follows:

$$\frac{dH}{H - 70} = -k\,dt$$

Integrating, we get

$$\int \frac{dH}{H - 70} = \int (-k)\,dt$$

so $\ln(H - 70) = -kt + C$

(Note that $H - 70$ is positive, so we don't need absolute values.) We now solve for H:

$$H - 70 = e^{-kt + C}$$
$$= e^C e^{-kt}$$
$$= A e^{-kt}$$

so

$$H(t) = 70 + A e^{-kt}$$

where A is some positive constant. We can determine the constant A using the initial condition $H(0) = 170$:

$$170 = 70 + A e^0 = 70 + A$$

so $A = 100$

Therefore,

$$H(t) = 70 + 100 e^{-kt}$$

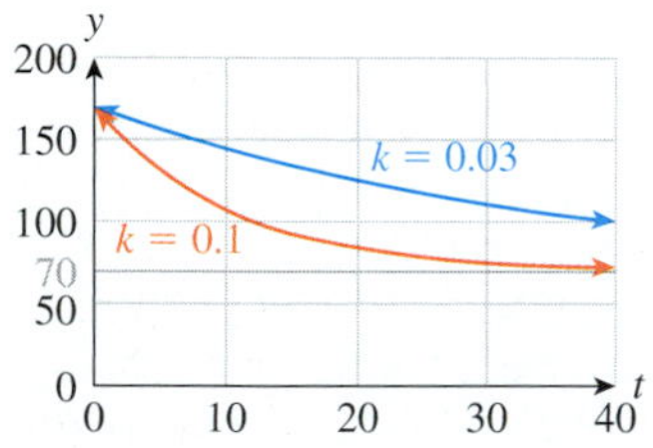

Q: But what is k?

A: The constant k determines the rate of cooling. Its value depends on the units of time we are using, on the substance cooling—in this case the coffee—and its container. Figure 21 shows two possible graphs, one with $k = 0.1$ and the other with $k = 0.03$ ($k \approx 0.03$ for a cup of coffee in a Styrofoam container with t measured in minutes).

In any case, we can see from the graph or the formula for $H(t)$ that the temperature of the coffee will approach the ambient temperature exponentially. ■

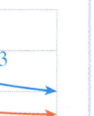

$k = 0.03$
$k = 0.1$

Figure 21

14.6 EXERCISES

● denotes basic skills exercises

tech Ex indicates exercises that should be solved using technology

Find the general solution of each differential equation in Exercises 1–10. Where possible, solve for y as a function of x.
hint [see Quick Examples p. 1024]

1. ● $\dfrac{dy}{dx} = x^2 + \sqrt{x}$

2. ● $\dfrac{dy}{dx} = \dfrac{1}{x} + 3$

3. ● $\dfrac{dy}{dx} = \dfrac{x}{y}$ hint [see Example 2a]

4. ● $\dfrac{dy}{dx} = \dfrac{y}{x}$

5. ● $\dfrac{dy}{dx} = xy$

6. ● $\dfrac{dy}{dx} = x^2 y$

7. ● $\dfrac{dy}{dx} = (x + 1)y^2$

8. ● $\dfrac{dy}{dx} = \dfrac{1}{(x + 1)y^2}$

9. ● $x\dfrac{dy}{dx} = \dfrac{1}{y}\ln x$

10. ● $\dfrac{1}{x}\dfrac{dy}{dx} = \dfrac{1}{y}\ln x$

For each differential equation in Exercises 11–20, find the particular solution indicated. hint [see Example 2b]

11. ● $\dfrac{dy}{dx} = x^3 - 2x$; $y = 1$ when $x = 0$

● basic skills   tech Ex technology exercise

12. ● $\dfrac{dy}{dx} = 2 - e^{-x}$; $y = 0$ when $x = 0$

13. ● $\dfrac{dy}{dx} = \dfrac{x^2}{y^2}$; $y = 2$ when $x = 0$

14. ● $\dfrac{dy}{dx} = \dfrac{y^2}{x^2}$; $y = \dfrac{1}{2}$ when $x = 1$

15. ● $x\dfrac{dy}{dx} = y$; $y(1) = 2$ **16.** ● $x^2\dfrac{dy}{dx} = y$; $y(1) = 1$

17. ● $\dfrac{dy}{dx} = x(y + 1)$; $y(0) = 0$ **18.** ● $\dfrac{dy}{dx} = \dfrac{y + 1}{x}$; $y(1) = 2$

19. ● $\dfrac{dy}{dx} = \dfrac{xy^2}{x^2 + 1}$; $y(0) = -1$

20. ● $\dfrac{dy}{dx} = \dfrac{xy}{(x^2 + 1)^2}$; $y(0) = 1$

Applications

21. ● **Sales** Your monthly sales of Green Tea Ice Cream are falling at an instantaneous rate of 5% per month. If you currently sell 1000 quarts per month, find the differential equation describing your change in sales, and then solve it to predict your monthly sales. *hint* [see Example 3]

22. ● **Profit** Your monthly profit on sales of Avocado Ice Cream is rising at an instantaneous rate of 10% per month. If you currently make a profit of $15,000 per month, find the differential equation describing your change in profit, and solve it to predict your monthly profits.

23. ● **Cooling** A bowl of clam chowder at 190°F is placed in a room whose air temperature is 75°F. After 10 minutes, the soup has cooled to 150°F. Find the temperature of the chowder as a function of time. (Refer to Example 4 for Newton's Law of Cooling.)

24. ● **Heating** Newton's Law of Heating is just the same as his Law of Cooling: The rate of change of temperature is proportional to the difference between the temperature of an object and its surroundings, whether the object is hotter or colder than its surroundings. Suppose that a pie, at 20°F, is put in an oven at 350°F. After 15 minutes, its temperature has risen to 80°F. Find the temperature of the pie as a function of time.

25. ● **Market Saturation** You have just introduced a new flatscreen monitor to the market. You predict that you will eventually sell 100,000 monitors and that your monthly rate of sales will be 10% of the difference between the saturation value and the total number you have sold up to that point. Find a differential equation for your total sales (as a function of the month) and solve. (What are your total sales at the moment when you first introduce the monitor?)

26. ● **Market Saturation** Repeat the preceding exercise, assuming that monthly sales will be 5% of the difference between the saturation value (of 100,000 monitors) and the total sales to that point, and assuming that you sell 5000 monitors to corporate customers before placing the monitor on the open market.

27. **Approach to Equilibrium** The Extrasoft Toy Co. has just released its latest creation, a plush platypus named "Eggbert." The demand function for Eggbert dolls is $D(p) = 50,000 - 500p$ dolls per month when the price is p dollars. The supply function is $S(p) = 30,000 + 500p$ dolls per month when the price is p dollars. This makes the equilibrium price $20. The Evans price adjustment model assumes that if the price is set at a value other than the equilibrium price, it will change over time in such a way that its rate of change is proportional to the shortage $D(p) - S(p)$.

  a. Write the differential equation given by the Evans price adjustment model for the price p as a function of time.

  b. Find the general solution of the differential equation you wrote in (a). (You will have two unknown constants, one being the constant of proportionality.)

  c. Find the particular solution in which Eggbert dolls are initially priced at $10 and the price rises to $12 after one month.

28. **Approach to Equilibrium** Spacely Sprockets has just released its latest model, the Dominator. The demand function is $D(p) = 10,000 - 1000p$ sprockets per year when the price is p dollars. The supply function is $S(p) = 8000 + 1000p$ sprockets per year when the price is p dollars.

  a. Using the Evans price adjustment model described in the preceding exercise, write the differential equation for the price $p(t)$ as a function of time.

  b. Find the general solution of the differential equation you wrote in (a).

  c. Find the particular solution in which Dominator sprockets are initially priced at $5 each but fall to $3 each after one year.

29. **Determining Demand** Nancy's Chocolates estimates that the elasticity of demand for its dark chocolate truffles is $E = 0.05p - 1.5$ where p is the price per pound. Nancy's sells 20 pounds of truffles per week when the price is $20 per pound. Find the formula expressing the demand q as a function of p. Recall that the elasticity of demand is given by

$$E = -\dfrac{dq}{dp} \times \dfrac{p}{q}$$

30. **Determining Demand** Nancy's Chocolates estimates that the elasticity of demand for its chocolate strawberries is $E = 0.02p - 0.5$ where p is the price per pound. It sells 30 pounds of chocolate strawberries per week when the price is $30 per pound. Find the formula expressing the demand q as a function of p. Recall that the elasticity of demand is given by

$$E = -\dfrac{dq}{dp} \times \dfrac{p}{q}$$

31. **Logistic Equation** There are many examples of growth in which the rate of growth is slow at first, becomes faster, and then slows again as a limit is reached. This pattern can be described by the differential equation

$$\dfrac{dy}{dt} = ay(L - y)$$

● basic skills   **tech** Ex technology exercise

where a is a constant and L is the limit of y. Show by substitution that

$$y = \frac{CL}{e^{-aLt} + C}$$

is a solution of this equation, where C is an arbitrary constant.

32. *Logistic Equation* Using separation of variables and integration with a table of integrals or a symbolic algebra program, solve the differential equation in the preceding exercise to derive the solution given there.

tech Ex *Exercises 33–36 require the use of technology.*

33. tech Ex *Market Saturation* You have just introduced a new model of DVD player. You predict that the market will saturate at 2,000,000 DVD players and that your total sales will be governed by the equation

$$\frac{dS}{dt} = \frac{1}{4}S(2 - S)$$

where S is the total sales in millions of DVD players and t is measured in months. If you give away 1000 DVD players when you first introduce them, what will S be? Sketch the graph of S as a function of t. About how long will it take to saturate the market? (See Exercise 31.)

34. tech Ex *Epidemics* A certain epidemic of influenza is predicted to follow the function defined by

$$\frac{dA}{dt} = \frac{1}{10}A(20 - A)$$

where A is the number of people infected in millions and t is the number of months after the epidemic starts. If 20,000 cases are reported initially, find $A(t)$ and sketch its graph. When is A growing fastest? How many people will eventually be affected? (See Exercise 31.)

35. tech Ex *Growth of Tumors* The growth of tumors in animals can be modeled by the Gompertz equation:

$$\frac{dy}{dt} = -ay \ln\left(\frac{y}{b}\right)$$

where y is the size of a tumor, t is time, and a and b are constants that depend on the type of tumor and the units of measurement.

a. Solve for y as a function of t.
b. If $a = 1$, $b = 10$, and $y(0) = 5$ cm^3 (with t measured in days), find the specific solution and graph it.

36. tech Ex *Growth of Tumors* Refer back to the preceding exercise. Suppose that $a = 1$, $b = 10$, and $y(0) = 15$ cm^3. Find the specific solution and graph it. Comparing its graph to the one obtained in the preceding exercise, what can you say about tumor growth in these instances?

Communication and Reasoning Exercises

37. ● What is the difference between a particular solution and the general solution of a differential equation? How do we get a particular solution from the general solution?

38. ● Why is there always an arbitrary constant in the general solution of a differential equation? Why are there not two or more arbitrary constants in a first-order differential equation?

39. Show by example that a **second-order** differential equation, one involving the second derivative y'', usually has two arbitrary constants in its general solution.

40. Find a differential equation that is not separable.

41. Find a differential equation whose general solution is $y = 4e^{-x} + 3x + C$.

42. Explain how, knowing the elasticity of demand as a function of either price or demand, you may find the demand equation (see Exercise 29).

Chapter **14** Review

KEY CONCEPTS

14.1 Integration by Parts

Integration by parts formula:

$$\int u \cdot v \, dx = u \cdot I(v) - \int D(u)I(v) \, dx$$
p. 982

Tabular method for integration by parts p. 983

Integrating a polynomial times a logarithm p. 986

14.2 Area Between Two Curves and Applications

If $f(x) \geq g(x)$ for all x in $[a, b]$, then the area of the region between the graphs of f and g and between $x = a$ and $x = b$ is given by

$$A = \int_a^b [f(x) - g(x)] \, dx \quad p. 989$$

Regions enclosed by crossing curves p. 991

Area enclosed by two curves p. 992

General instructions for finding the area between the graphs of $f(x)$ and $g(x)$ p. 992

Approximating the area between two curves using technology:

$$A = \int_a^b |f(x) - g(x)| \, dx \quad p. 993$$

14.3 Averages and Moving Averages

Average, or mean, of a collection of values

$$\bar{y} = \frac{y_1 + y_2 + \cdots + y_n}{n} \quad p. 996$$

The *average, or mean,* of a function $f(x)$ on an interval $[a, b]$ is

$$\bar{f} = \frac{1}{b-a} \int_a^b f(x) \, dx \quad p. 997$$

Average balance p. 998

Computing the moving average of a set of data p. 999

n-Unit moving average of a function:

$$\bar{f}(x) = \frac{1}{n} \int_{x-n}^x f(t) \, dt \quad p. 1000$$

Computing moving average using technology p. 1000

14.4 Applications to Business and Economics: Consumers' and Producers' Surplus and Continuous Income Streams

Consumers' surplus:

$$CS = \int_0^{\bar{q}} (D(q) - \bar{p}) \, dq \quad p. 1004$$

Producers' surplus:

$$PS = \int_0^{\bar{q}} [\bar{p} - S(q)] \, dq \quad p. 1006$$

Equilibrium price p. 1007

Social gain $= CS + PS$ p. 1008

Total value of a continuous income stream: $TV = \int_a^b R(t) \, dt$ p. 1010

Future value of a continuous income stream: $FV = \int_a^b R(t)e^{r(b-t)} \, dt$ p. 1010

Present value of a continuous income stream:

$$PV = \int_a^b R(t)e^{r(a-t)} \, dt \quad p. 1011$$

14.5 Improper Integrals and Applications

Improper integral with an infinite limit of integration:

$$\int_a^{+\infty} f(x) \, dx, \int_{-\infty}^b f(x) \, dx,$$

$$\int_{-\infty}^{+\infty} f(x) \, dx \quad p. 1015$$

Improper integral in which the integrand becomes infinite p. 1017

Testing for convergence p. 1018

Integrand infinite between the endpoints p. 1019

Integral improper in two ways p. 1019

14.6 Differential Equations and Applications

Simple differential equations:

$$\frac{dy}{dx} = f(x) \quad p. 1024$$

Separable differential equations:

$$\frac{dy}{dx} = f(x)g(y) \quad p. 1024$$

Newton's Law of Cooling p. 1027

REVIEW EXERCISES

Evaluate the integrals in Exercises 1–10 that converge. (Some of the integrals are improper and may diverge.)

1. $\int (x^2 + 2)e^x \, dx$

2. $\int (x^2 - x)e^{-3x+1} \, dx$

3. $\int x^2 \ln(2x) \, dx$

4. $\int \log_5 x \, dx$

5. $\int_{-2}^2 (x^3 + 1)e^{-x} \, dx$

6. $\int_1^e x^2 \ln x \, dx$

7. $\int_1^\infty \frac{1}{x^5} \, dx$

8. $\int_0^1 \frac{1}{x^5} \, dx$

9. $\int_{-2}^2 \frac{1}{(x+1)^{1/3}} \, dx$

10. $\int_0^1 \frac{1}{\sqrt{1-x}} \, dx$

In Exercises 11–14, find the areas of the given regions.

11. Between $y = x^3$ and $y = 1 - x^3$ for x in $[0, 1]$

12. Between $y = e^x$ and $y = e^{-x}$ for x in $[0, 2]$

13. Enclosed by $y = 1 - x^2$ and $y = x^2$

14. Between $y = x$ and $y = xe^{-x}$ for x in $[0, 2]$

In Exercises 15–18, find the average value of the given function over the indicated interval.

15. $f(x) = x^3 - 1$ over $[-2, 2]$

16. $f(x) = \dfrac{x}{x^2 + 1}$ over $[0, 1]$

17. $f(x) = x^2 e^x$ over $[0, 1]$

18. $f(x) = (x + 1) \ln x$ over $[1, 2e]$

In Exercises 19–22, find the 2-unit moving averages of the given function.

19. $f(x) = 3x + 1$ **20.** $f(x) = 6x^2 + 12$

21. $f(x) = x^{4/3}$ **22.** $f(x) = \ln x$

In Exercises 23 and 24, calculate the consumers' surplus at the indicated unit price $\bar{p}$ for the given demand equation.

23. $p = 50 - \dfrac{1}{2}q;\ \bar{p} = 10$

24. $p = 10 - q^{1/2};\ \bar{p} = 4$

In Exercises 25 and 26, calculate the producers' surplus at the indicated unit price $\bar{p}$ for the given supply equation.

25. $p = 50 + \dfrac{1}{2}q;\ \bar{p} = 100$

26. $p = 10 + q^{1/2};\ \bar{p} = 40$

Solve the differential equations in Exercises 27–30.

27. $\dfrac{dy}{dx} = x^2 y^2$

28. $\dfrac{dy}{dx} = xy + 2x$

29. $xy\dfrac{dy}{dx} = 1;\ y(1) = 1$

30. $y(x^2 + 1)\dfrac{dy}{dx} = xy^2;\ y(0) = 2$

Applications

31. *Investments* OHaganBooks.com keeps its cash reserves in a bank account paying 6% compounded continuously. It starts a year with $1 million in reserves and does not withdraw or deposit any money.

a. What is the average amount it will have in the account over the course of two years?

b. Find the one-month moving average of the amount it has in the account.

32. *Consumers' and Producers' Surplus* OHaganBooks.com is about to start selling a new coffee table book, *Computer Designs of the Late Twentieth Century*. It estimates the demand curve to be $q = 1000\sqrt{200 - 2p}$, and its willingness to order books from the publisher is given by the supply curve $q = 1000\sqrt{10p - 400}$.

a. Find the equilibrium price and demand.

b. Find the consumers' and producers' surpluses at the equilibrium price.

33. *Revenue* Sales of the bestseller *A River Burns Through It* are dropping at OHaganBooks.com. To try to bolster sales, the company is dropping the price of the book, now $40, at a rate of $2 per week. As a result, this week OHaganBooks.com will sell 5000 copies, and it estimates that sales will fall continuously at a rate of 10% per week. How much revenue will it earn on sales of this book over the next 8 weeks?

34. *Investments* OHaganBooks.com CEO John O'Hagan has started a gift account for the Marjory Duffin Foundation. The account pays 6% compounded continuously and is initially empty. OHaganBooks.com deposits money continuously into it, starting at the rate of $100,000 per month and increasing continuously by $10,000 per month.

a. How much money will the company have in the account at the end of two years?

b. How much of the amount you found in part (a) was principal deposited and how much was interest earned?

35. *Acquisitions* The Megabucks Corporation is considering buying OHaganBooks.com. They estimate OHaganBooks.com's revenue stream at $50 million per year, growing continuously at a 10% rate. Assuming interest rates of 6%, how much is OHaganBooks.com's revenue for the next year worth now?

36. *Incompetence* OHaganBooks.com is shopping around for a new bank. A junior executive at one bank offers them the following interesting deal: The bank will pay them interest continuously at a rate equal to 0.01% of the square of the amount of money they have in the account at any time. By considering what would happen if $10,000 was deposited in such an account, explain why the junior executive was fired shortly after this offer was made.

CASE STUDY: Estimating Tax Revenues

You have just been hired by the incoming administration of your country as chief consultant for national tax policy, and you have been getting conflicting advice from the finance experts on your staff. Several of them have come up with plausible suggestions for new tax structures, and your job is to choose the plan that results in more revenue for the government.

Before you can evaluate their plans, you realize that it is essential to know your country's income distribution—that is, how many people earn how much money.[37] One might think that the most useful way of specifying income distribution would be to use a function that gives the exact number $f(x)$ of people who earn a given salary x. This would necessarily be a discrete function—it only makes sense if x happens to be a whole number of cents. There is, after all, no one earning a salary of exactly $22,000.142567! Furthermore, this function would behave rather erratically, because there are, for example, probably many more people making a salary of exactly $30,000 than exactly $30,000.01. Given these problems, it is far more convenient to start with the function defined by

$$N(x) = \textit{the total number of people earning between } 0 \textit{ and } x \textit{ dollars}$$

Actually, you would want a "smoothed" version of this function. The graph of $N(x)$ might look like the one shown in Figure 22.

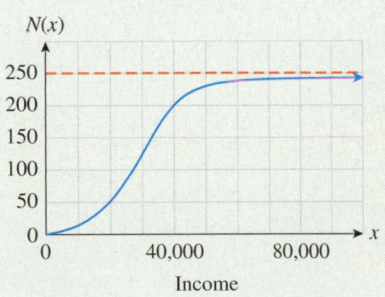

Figure **22**

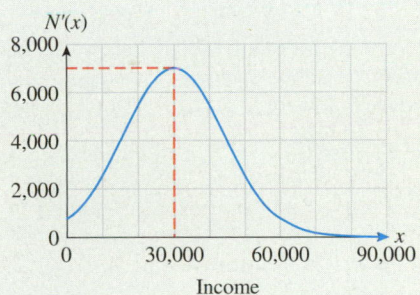

Figure **23**

If we take the *derivative* of $N(x)$, we get an income distribution function. Its graph might look like the one shown in Figure 23. Because the derivative measures the rate of change, its value at x is the additional number of taxpayers per $1 increase in salary. Thus, the fact that $N'(20,000) \approx 5500$ tells us that approximately 5500 people are earning a salary of between $20,000 and $20,001. In other words, N' shows the distribution of incomes among the population—hence, the name "distribution function."[38]

You thus send a memo to your experts requesting the income distribution function for the nation. After much collection of data, they tell you that the income distribution

---

[37] To simplify our discussion, we are assuming that (1) all tax revenues are based on earned income and that (2) everyone in the population we consider earns some income.

[38] A very similar idea is used in probability. See the optional chapter on Calculus Applied to Probability and Statistics at the website.

function is

$$N'(x) = 7000e^{-(x-30,000)^2/400,000,000}$$

This is in fact the function whose graph is shown in Figure 23 and is an example of a **normal distribution.** Notice that the curve is symmetric around the median income of \$30,000 and that about 7000 people are earning between \$30,000 and \$30,001 annually.[39]

Given this income distribution, your financial experts have come up with the two possible tax policies illustrated in Figures 24 and 25.

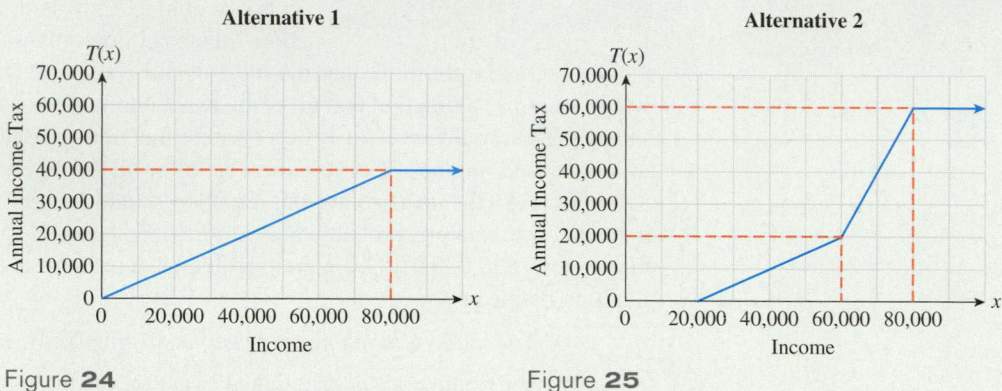

Figure **24**   Figure **25**

In the first alternative, all taxpayers pay half of their income in taxes, except that no one pays more than \$40,000 in taxes. In the second alternative, there are four tax brackets, described by the following table:

Income	Marginal tax rate
\$0–20,000	0%
\$20,000–60,000	50%
\$60,000–80,000	200%
Above \$80,000	0%

Now you must determine which alternative will generate more annual tax revenue.

Each of Figures 24 and 25 is the graph of a function, T. Rather than using the formulas for these particular functions, you begin by working with the general situation. You have an income distribution function N' and a tax function T, both functions of annual income. You need to find a formula for total tax revenues. First you decide to use a cutoff so that you need to work only with incomes in some finite bracket $[0, M]$; you might use, for example, $M = \$10$ million (later you will let M approach $+\infty$). Next you subdivide the interval $[0, M]$ into a large number of intervals of small width, Δx. If $[x_{k-1}, x_k]$ is a typical such interval, you wish to calculate the approximate tax revenue from people whose total incomes lie between x_{k-1} and x_k. You will then sum over k to get the total revenue.

You need to know how many people are making incomes between x_{k-1} and x_k. Because $N(x_k)$ people are making incomes *up to* x_k and $N(x_{k-1})$ people are making

---

[39] You might find it odd that you weren't given the original function N, but it will turn out that you don't need it. How would you compute it?

incomes up to x_{k-1}, the number of people making incomes between x_{k-1} and x_k is $N(x_k) - N(x_{k-1})$. Because x_k is very close to x_{k-1}, the incomes of these people are all approximately equal to x_{k-1} dollars, so each of these taxpayers is paying an annual tax of about $T(x_{k-1})$. This gives a tax revenue of

$$[N(x_k) - N(x_{k-1})]T(x_{k-1})$$

Now you do a clever thing. You write $x_k - x_{k-1} = \Delta x$ and replace $N(x_k) - N(x_{k-1})$ by

$$\frac{N(x_k) - N(x_{k-1})}{\Delta x} \Delta x$$

This gives you a tax revenue of about

$$\frac{N(x_k) - N(x_{k-1})}{\Delta x} T(x_{k-1}) \Delta x$$

from wage-earners in the bracket $[x_{k-1}, x_k]$. Summing over k gives an approximate total revenue of

$$\sum_{k=1}^{n} \frac{N(x_k) - N(x_{k-1})}{\Delta x} T(x_{k-1}) \Delta x$$

where n is the number of subintervals. The larger n is, the more accurate your estimate will be, so you take the limit of the sum as $n \to \infty$. When you do this, two things happen. First, the quantity

$$\frac{N(x_k) - N(x_{k-1})}{\Delta x}$$

approaches the derivative, $N'(x_{k-1})$. Second, the sum, which you recognize as a Riemann sum, approaches the integral

$$\int_0^M N'(x)T(x)\,dx$$

You now take the limit as $M \to +\infty$ to get

$$\text{Total tax revenue} = \int_0^{+\infty} N'(x)T(x)\,dx$$

This improper integral is fine in theory, but the actual calculation will have to be done numerically, so you stick with the upper limit of $10 million for now. You will have to check that it is reasonable at the end (notice that, by the graph of N', it appears that extremely few, if any, people earn that much). Now you already have a formula for $N'(x)$, but you still need to write formulas for the tax functions $T(x)$ for both alternatives.

Alternative 1 The graph in Figure 24 rises linearly from 0 to 40,000 as x ranges from 0 to 80,000, and then stays constant at 40,000. The slope of the first part is $40,000/80,000 = 1/2$. The taxation function is therefore

$$T(x) = \begin{cases} \frac{x}{2} & \text{if } 0 \le x \le 80,000 \\ 40,000 & \text{if } x \ge 80,000 \end{cases}$$

To perform the integration, you will therefore need to break the integral into two pieces, the first from 0 to 80,000 and the second from 80,000 to 10,000,000. In other words,

$$R_1 = \int_0^{80,000} (7000\, e^{-(x-30,000)^2/400,000,000})\frac{x}{2}\, dx$$

$$+ \int_{80,000}^{10,000,000} (7000\, e^{-(x-30,000)^2/400,000,000})40,000\, dx$$

You decide not to attempt this by hand![40] You use numerical integration software to obtain a grand total of $R_1 = \$3{,}732{,}760{,}000{,}000$, or \$3.73276 trillion (rounded to six significant digits).

Alternative 2 The graph in Figure 25 rises linearly from 0 to 20,000 as x ranges from 20,000 to 60,000, then rises from 20,000 to 60,000 as x ranges from 60,000 to 80,000, and then stays constant at 60,000. The slope of the first incline is $1/2$ and the slope of the second incline is 2 (this is why the *marginal* tax rates are 50% and 200% respectively). The taxation function is therefore

$$T(x) = \begin{cases} 0 & \text{if } 0 \le x \le 20,000 \\ \dfrac{x-20,000}{2} & \text{if } 20,000 \le x \le 60,000 \\ 20,000 + 2(x-60,000) & \text{if } 60,000 \le x \le 80,000 \\ 60,000 & \text{if } x \ge 80,000 \end{cases}$$

Values of x between 0 and 20,000 do not contribute to the integral, so

$$R_2 = \int_{20,000}^{60,000} (7000\, e^{-(x-30,000)^2/400,000,000}) \left(\frac{x-20,000}{2} \right) dx$$

$$+ \int_{60,000}^{80,000} (7000\, e^{-(x-30,000)^2/400,000,000}) [20,000 + 2(x-60,000)]\, dx$$

$$+ \int_{80,000}^{10,000,000} (7000\, e^{-(x-30,000)^2/400,000,000})60,000\, dx$$

Numerical integration software gives $R_2 = \$1.52016$ trillion—considerably less than Alternative 1. Thus, even though Alternative 2 taxes the wealthy more heavily, it yields less total revenue.

Now what about the cutoff at \$10 million annual income? If you try either integral again with an upper limit of \$100 million, you will see no change in either result to six significant digits. There simply are not enough taxpayers earning an income above \$10,000,000 to make a difference. You conclude that your answers are sufficiently accurate and that the first alternative provides more tax revenue.

Exercises

In Exercises 1–6, calculate the total tax revenue for a country with the given income distribution and tax policies (all currency in dollars).

1. `tech` Ex $N'(x) = 3000e^{-(x-10,000)^2/10,000}$; 25% tax on all income
2. `tech` Ex $N'(x) = 3000e^{-(x-10,000)^2/10,000}$; 45% tax on all income

---

[40] In fact, these integrals cannot be done in elementary terms at all.

3. `tech` Ex $N'(x) = 5000e^{-(x-30,000)^2/100,000}$; no tax on an income below \$30,000, \$10,000 tax on any income of \$30,000 or above

4. `tech` Ex $N'(x) = 5000e^{-(x-30,000)^2/100,000}$; no tax on an income below \$50,000, \$20,000 tax on any income of \$50,000 or above

5. `tech` Ex $N'(x) = 7000\,e^{-(x-30,000)^2/400,000,000}$; $T(x)$ with the following graph:

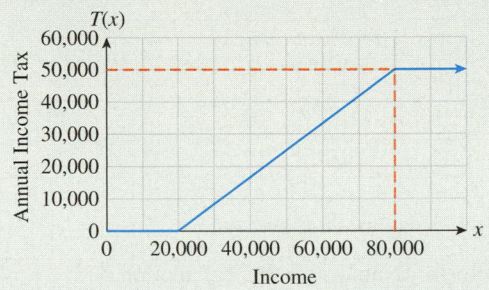

6. `tech` Ex $N'(x) = 7000\,e^{-(x-30,000)^2/400,000,000}$; $T(x)$ with the following graph:

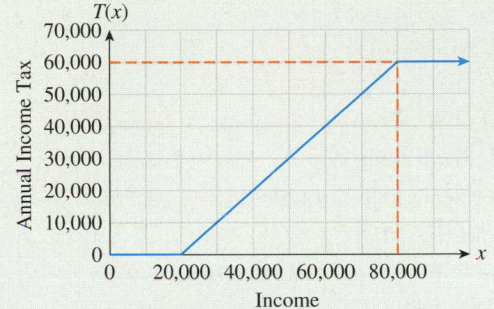

7. Let $P(x)$ be the number of people earning more than x dollars.

a. What is $N(x) + P(x)$?

b. Show that $P'(x) = -N'(x)$.

c. Use integration by parts to show that, if $T(0) = 0$, then the total tax revenue is

$$\int_0^{+\infty} P(x)T'(x)\,dx$$

[Note: You may assume that $T'(x)$ is continuous, but the result is still true if we assume only that $T(x)$ is continuous and piecewise continuously differentiable.]

8. Income tax functions T are most often described, as in the text, by tax brackets and marginal tax rates.

a. If one tax bracket is $a < x \le b$, show that $\int_a^b P(x)\,dx$ is the total income earned in the country that falls into that bracket (P as in the preceding exercise).

b. Use (a) to explain directly why $\int_0^{+\infty} P(x)T'(x)\,dx$ gives the total tax revenue in the case where T is described by tax brackets and constant marginal tax rates in each bracket.

Section 14.3

Example 3 The following table shows Colossal Conglomerate's closing stock prices for 20 consecutive trading days:

Day	1	2	3	4	5	6	7	8	9	10
Price	20	22	21	24	24	23	25	26	20	24
Day	11	12	13	14	15	16	17	18	19	20
Price	26	26	25	27	28	27	29	27	25	24

Plot these prices and the 5-day moving average.

Solution with Technology To automate this calculation on a TI-83/84:

1. Use

```
seq(X,X,1,20)→L₁
```
2ND STAT → OPS → 5
STO 2ND STAT → L₁

to enter the sequence of numbers 1 through 20 into the list L_1, representing the trading days.

2. Using the list editor accessible through the STAT menu, enter the daily stock prices in list L_2.

3. You can now calculate the list of 5-day moving averages by using the following command:

```
seq((L₂(X)+L₂(X-1)+L₂(X-2)+L₂
  (X-3)+L₂(X-4))/5,X,5,20)→L₃
```

This has the effect of putting the moving averages into elements 1 through 15 of list L_3.

4. If you wish to plot the moving average on the same graph as the daily prices, you will want the averages in L_3 to match up with the prices in L_2. One way to do this is to put four more entries at the beginning of L_3—say, copies of the first four entries of L_2. The following command accomplishes this:

```
augment(seq(L₂(X),X,1,4),L₃)→L₃
```
2ND STAT → OPS → 9

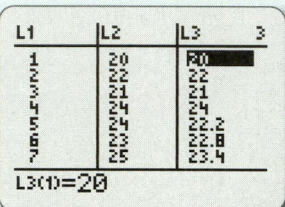

5. You can now graph the prices and moving averages by creating an xyLine scatter plot through the STAT PLOT menu, with L_1 being the Xlist and L_2 being the Ylist for `Plot1`, and L_1 being the Xlist and L_3 the Ylist for `Plot2`:

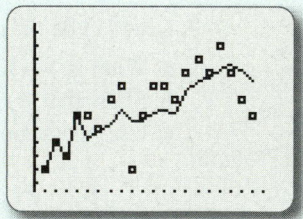

Example 4 Use technology to plot the 3-unit moving average of

$$f(x) = \frac{x}{1+|x|} \qquad (-5 \le x \le 5)$$

Solution with Technology

1. We enter the following:

```
Y₁ = X/(1+abs(X))
Y₂ = (1/3)fnInt(Y₁(T),T,X-3,X)
```

The Y_1 entry is $f(x)$, and the Y_2 entry is a numerical approximation of the 3-unit moving average of $f(x)$:

$$\bar{f}(x) = \frac{1}{3}\int_{x-3}^{x} \frac{t}{1+|t|}\,dt$$

2. We set the viewing window ranges to $-5 \leq x \leq 5$ and $-1 \leq y \leq 1$, and plot these curves. (Be patient—the calculator has to do a numerical integration to obtain each point on the graph of the moving average.) The result is shown here (the lower curve is the moving average):

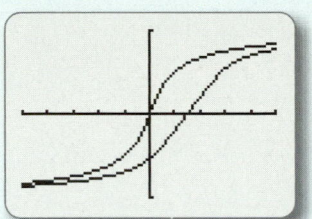

3. Of course, once we've entered $\mathtt{Y_1}$ and $\mathtt{Y_2}$ as above, we can use the calculator to evaluate the moving average at any value of x. For instance, to calculate $\bar{f}(1.2)$ we enter $\mathtt{Y_2(1.2)}$ on the home screen and find

$$\bar{f}(1.2) \approx -0.1196$$

EXCEL Technology Guide

Section 14.3

Example 3 The following table shows Colossal Conglomerate's closing stock prices for 20 consecutive trading days:

Day	1	2	3	4	5	6	7	8	9	10
Price	20	22	21	24	24	23	25	26	20	24
Day	11	12	13	14	15	16	17	18	19	20
Price	26	26	25	27	28	27	29	27	25	24

Plot these prices and the 5-day moving average.

Solution with Technology

1. Compute the moving averages in a column next to the daily prices, as shown here:

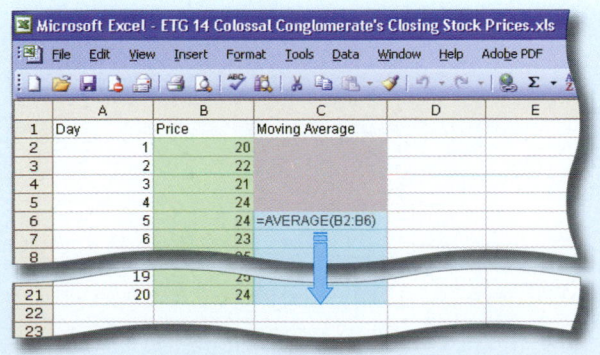 $\rightarrow$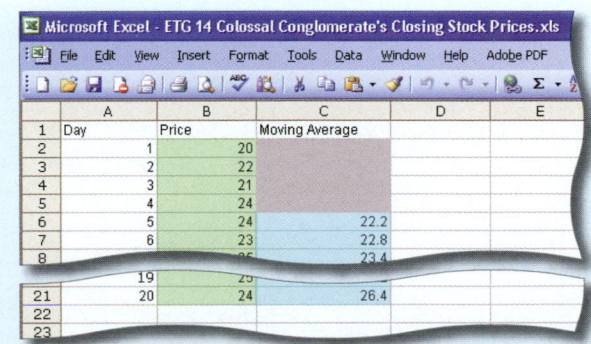

2. You can then graph the average and moving average using a scatter plot.

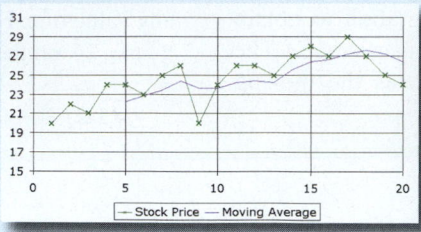

15

Functions of Several Variables

CASE STUDY Modeling Household Income

The Millennium Real Estate Development Corporation is interested in developing housing projects for medium-sized families that have high household incomes. To decide which income bracket to target, the company has asked you, a paid consultant, for an analysis of the relationship of household size to household income and the effect of increasing household size on household income. How can you analyze the relevant data?

Bill Varie/Corbis

Introduction

We have studied functions of a single variable extensively. But not every useful function is a function of only one variable. In fact, most are not. For example, if you operate an online bookstore in competition with Amazon.com, BN.com, and Booksamillion.com, your sales may depend on those of your competitors. Your company's daily revenue might be modeled by a function such as

$$R(x, y, z) = 10{,}000 - 0.01x - 0.02y - 0.01z + 0.00001yz$$

where x, y, and z are the online daily revenues of Amazon.com, BN.com, and Booksamillion.com, respectively. Here, R is a function of three variables because it *depends on* x, y, and z. As we shall see, the techniques of calculus extend readily to such functions. Among the applications we shall look at is optimization: finding, where possible, the maximum or minimum of a function of two or more variables.

15.1 Functions of Several Variables from the Numerical and Algebraic Viewpoints

Recall that a function of one variable is a rule for manufacturing a new number $f(x)$ from a single independent variable x. A function of two or more variables is similar, but the new number now depends on more than one independent variable.

Function of Several Variables

A **real-valued function**, f, **of** $x, y, z, \ldots$ is a rule for manufacturing a new number, written $f(x, y, z, \ldots)$, from the values of a sequence of independent variables $x, y, z, \ldots$. The function f is called a **real-valued function of two variables** if there are two independent variables, a **real-valued function of three variables** if there are three independent variables, and so on.

quick Examples

1. $f(x, y) = x - y$ Function of two variables

    $f(1, 2) = 1 - 2 = -1$ Substitute 1 for x and 2 for y.

    $f(2, -1) = 2 - (-1) = 3$ Substitute 2 for x and -1 for y.

    $f(y, x) = y - x$ Substitute y for x and x for y.

2. $g(x, y) = x^2 + y^2$ Function of two variables

    $g(-1, 3) = (-1)^2 + 3^2 = 10$ Substitute -1 for x and 3 for y.

3. $h(x, y, z) = x + y + xz$ Function of three variables

    $h(2, 2, -2) = 2 + 2 + 2(-2) = 0$ Substitute 2 for x, 2 for y, and -2 for z.

Figure 1 illustrates the concept of a function of two variables: In goes a pair of numbers and out comes a single number.

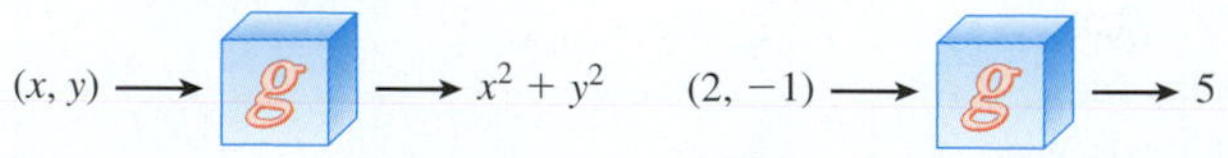

$(x, y) \longrightarrow \boxed{g} \longrightarrow x^2 + y^2 \qquad (2, -1) \longrightarrow \boxed{g} \longrightarrow 5$

Figure **1**

As with functions of one variable, functions of several variables can be represented numerically (using a table of values), algebraically (using a formula as in the above examples), and sometimes graphically[1] (using a graph).

Let's now look at a number of examples of interesting functions of several variables.

Example 1 Cost Function

You own a company that makes two models of speakers: the Ultra Mini and the Big Stack. Your total monthly cost (in dollars) to make x Ultra Minis and y Big Stacks is given by

$$C(x, y) = 10,000 + 20x + 40y$$

What is the significance of each term in this formula?

Solution The terms have meanings similar to those we saw for linear cost functions of a single variable. Let us look at the terms one at a time.

Constant Term Consider the monthly cost of making no speakers at all ($x = y = 0$). We find

$$C(0, 0) = 10,000 \qquad \text{Cost of making no speakers is \$10,000.}$$

Thus, the constant term 10,000 is the **fixed cost,** the amount you have to pay each month even if you make no speakers.

Coefficients of x and y Suppose you make a certain number of Ultra Minis and Big Stacks one month and the next month you increase production by one Ultra Mini. The costs are

$$C(x, y) = 10,000 + 20x + 40y \qquad \text{First Month}$$
$$C(x + 1, y) = 10,000 + 20(x + 1) + 40y \qquad \text{Second Month}$$
$$= 10,000 + 20x + 20 + 40y$$
$$= C(x, y) + 20$$

Thus, each Ultra Mini adds $20 to the total cost. We say that $20 is the **marginal cost** of each Ultra Mini. Similarly, because of the term $40y$, each Big Stack adds $40 to the total cost. The marginal cost of each Big Stack is $40.

This is an example of a linear function of two variables. The coefficients of x and y play roles similar to that of the slope of a line. In particular, they give the rates of change of the function as each variable increases while the other stays constant (think about it). We shall say more about linear functions below.

+ *Before we go on...* In Example 1, which values of x and y may we substitute into $C(x, y)$? Certainly we must have $x \geq 0$ and $y \geq 0$ because it makes no sense to speak of manufacturing a negative number of speakers. Also, there is certainly some upper bound to the number of speakers that can be made in a month. The bound might take one of several forms. The number of each model may be bounded—say $x \leq 100$ and $y \leq 75$. The inequalities $0 \leq x \leq 100$ and $0 \leq y \leq 75$ describe the region in the plane shaded in Figure 2.

using *Technology*

See the Technology Guides at the end of the chapter to find out how you can use a TI-83/84 and Excel to display various values of $C(x, y)$.

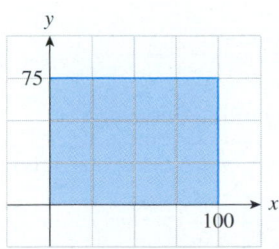

Figure **2**

---

[1] See the next section.

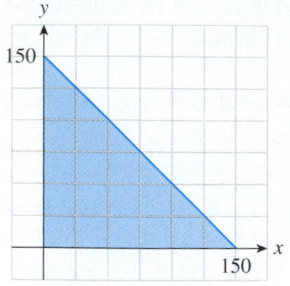

Figure 3

Another possibility is that the *total* number of speakers is bounded—say, $x + y \leq 150$. This, together with $x \geq 0$ and $y \geq 0$, describes the region shaded in Figure 3.

In either case, the region shown represents the pairs (x, y) for which $C(x, y)$ is defined. Just as with a function of one variable, we call this region the **domain** of the function. As before, when the domain is not given explicitly, we agree to take the largest domain possible. ∎

Example 2 Faculty Salaries

David Katz came up with the following function for the salary of a professor with 10 years of teaching experience in a large university.

$$S(x, y, z) = 13{,}005 + 230x + 18y + 102z$$

Here, S is the salary in 1969–1970 in dollars per year, x is the number of books the professor has published, y is the number of articles published, and z is the number of "excellent" articles published.[*] What salary do you expect that a professor with 10 years' experience earned in 1969–1970 if she published two books, 20 articles, and 3 "excellent" articles?

Solution All we need to do is calculate

$$S(2, 20, 3) = 13{,}005 + 230(2) + 18(20) + 102(3)$$
$$= \$14{,}131$$

---

[*] David A. Katz, "Faculty Salaries, Promotions and Productivity at a Large University," *American Economic Review,* June 1973, pp. 469–477. Prof. Katz's equation actually included other variables, such as the number of dissertations supervised; our equation assumes that all of these are zero.

✦ *Before we go on...* In Example 1, we gave a linear function of two variables. In Example 2 we have a linear function of three variables. Katz came up with his model by surveying a large number of faculty members and then finding the linear function "best" fitting the data. Such models are called **multiple linear regression** models. In the Case Study at the end of this chapter, we shall see a spreadsheet method of finding the coefficients of a multiple regression model from a set of observed data.

What does this model say about the value of a single book or a single article? If a book takes 15 times as long to write as an article, how would you recommend that a professor spend her writing time? ∎

Here are two simple kinds of functions of several variables.

Linear Function

A **linear function** of the variables $x_1, x_2, \ldots, x_n$ is a function of the form

$$f(x_1, x_2, \ldots, x_n) = a_0 + a_1 x_1 + \cdots + a_n x_n \qquad (a_0, a_1, a_2, \ldots, a_n \text{ constants})$$

quick Examples

1. $f(x, y) = 3x - 5y$ Linear function of x and y
2. $C(x, y) = 10{,}000 + 20x + 40y$ Example 1
3. $S(x, y, z) = 13{,}005 + 230x + 18y + 102z$ Example 2

Interaction Function

If we add to a linear function one or more terms of the form $bx_i x_j$ (b a nonzero constant and $i \neq j$), we get a **second-order interaction function.**

quick Examples

1. $C(x, y) = 10{,}000 + 20x + 40y + 0.1xy$

2. $R(x, y, z) = 10{,}000 - 0.01x - 0.02y - 0.01z + 0.00001yz$

So far, we have been specifying functions of several variables **algebraically**—by using algebraic formulas. If you have ever studied statistics, you are probably familiar with statistical tables. These tables may also be viewed as representing functions **numerically,** as the next example shows.

Example 3 Function Represented Numerically: Body Mass Index

The following table lists some values of the "body mass index," which gives a measure of the massiveness of your body, taking height into account.[*] The variable w represents

---

[*] It is interesting that weight-lifting competitions are usually based on weight, rather than body mass index. As a consequence, taller people are at a significant disadvantage in weight-lifting competitions because they must compete with shorter, stockier people of the same weight. (An extremely thin, very tall person can weigh as much as a muscular short person, although his body mass index would be significantly lower.) SOURCES: *Shape Up America*/National Institute of Health/*The New York Times,* May 2, 1999, p. WK3.

your weight in pounds, and h represents your height in inches. An individual with a body mass index of 25 or above is generally considered overweight.

$w \rightarrow$

$h \downarrow$		130	140	150	160	170	180	190	200	210
	60	25.2	27.1	29.1	31.0	32.9	34.9	36.8	38.8	40.7
	61	24.4	26.2	28.1	30.0	31.9	33.7	35.6	37.5	39.4
	62	23.6	25.4	27.2	29.0	30.8	32.7	34.5	36.3	38.1
	63	22.8	24.6	26.4	28.1	29.9	31.6	33.4	35.1	36.9
	64	22.1	23.8	25.5	27.2	28.9	30.7	32.4	34.1	35.8
	65	21.5	23.1	24.8	26.4	28.1	29.7	31.4	33.0	34.7
	66	20.8	22.4	24.0	25.6	27.2	28.8	30.4	32.0	33.6
	67	20.2	21.8	23.3	24.9	26.4	28.0	29.5	31.1	32.6
	68	19.6	21.1	22.6	24.1	25.6	27.2	28.7	30.2	31.7
	69	19.0	20.5	22.0	23.4	24.9	26.4	27.8	29.3	30.8
	70	18.5	19.9	21.4	22.8	24.2	25.6	27.0	28.5	29.9
	71	18.0	19.4	20.8	22.1	23.5	24.9	26.3	27.7	29.1
	72	17.5	18.8	20.2	21.5	22.9	24.2	25.6	26.9	28.3
	73	17.0	18.3	19.6	20.9	22.3	23.6	24.9	26.2	27.5
	74	16.6	17.8	19.1	20.4	21.7	22.9	24.2	25.5	26.7
	75	16.1	17.4	18.6	19.8	21.1	22.3	23.6	24.8	26.0
	76	15.7	16.9	18.1	19.3	20.5	21.7	22.9	24.2	25.4

As the table shows, the value of the body mass index depends on two quantities: w and h. Let us write $M(w, h)$ for the body mass index function. What are $M(140, 62)$ and $M(210, 63)$?

Solution We can read the answers from the table:

$$M(140, 62) = 25.4 \qquad w = 140\,\text{lb}, h = 62\,\text{in.}$$

and $$M(210, 63) = 36.9 \qquad w = 210\,\text{lb}, h = 63\,\text{in.}$$

The function $M(w, h)$ is actually given by the formula

using *Technology*

See the Technology Guides at the end of the chapter to find out how you can use Excel to create the table in this example.

$$M(w, h) = \frac{0.45w}{(0.0254h)^2}$$

[The factor 0.45 converts the weight to kilograms, and 0.0254 converts the height to meters. If w is in kilograms and h is in meters, the formula is simpler: $M(w, h) = w/h^2$.]

Distance and Related Functions

Newton's Law of Gravity states that the gravitational force exerted by one particle on another depends on their masses and the distance between them. The distance between two particles in the xy-plane can be expressed as a function of their coordinates, as follows:

Distance Formula

The distance between the points $P(x_1, y_1)$ and $Q(x_2, y_2)$ is

$$d = \sqrt{(x_2 - x_1)^2 + (y_2 - y_1)^2} = \sqrt{(\Delta x)^2 + (\Delta y)^2}$$

Derivation

The distance d is shown in the figure below.

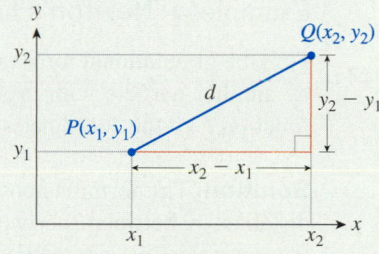

By the Pythagorean theorem applied to the right triangle shown, we get

$$d^2 = (x_2 - x_1)^2 + (y_2 - y_1)^2$$

Taking square roots (d is a distance, so we take the positive square root), we get the distance formula. Notice that if we switch x_1 with x_2 or y_1 with y_2, we get the same result.

quick Examples

1. The distance between the points $(3, -2)$ and $(-1, 1)$ is

$$d = \sqrt{(-1 - 3)^2 + (1 + 2)^2} = \sqrt{25} = 5$$

2. The distance from (x, y) to the origin $(0, 0)$ is

$$d = \sqrt{(x - 0)^2 + (y - 0)^2} = \sqrt{x^2 + y^2} \qquad \text{Distance to the origin}$$

The set of all points (x, y) whose distance from the origin $(0, 0)$ is a fixed quantity r is a circle centered at the origin with radius r. From the second Quick Example, we get the following equation for the circle centered at the origin with radius r:

$$\sqrt{x^2 + y^2} = r \qquad \text{Distance from the origin} = r$$

Squaring both sides gives the following equation, which we use in later sections:

Equation of the Circle of Radius *r* Centered at the Origin

$$x^2 + y^2 = r^2$$

quick Examples

1. The circle of radius 1 centered at the origin has equation $x^2 + y^2 = 1$.
2. The circle of radius 2 centered at the origin has equation $x^2 + y^2 = 4$.
3. The circle of radius 3 centered at the origin has equation $x^2 + y^2 = 9$.

Now, let's return to Newton's Law of Gravity. According to Newton's Law, the gravitational force exerted on a particle with mass m by another particle with mass M is given by the following function of distance:

$$F(r) = G\frac{Mm}{r^2}$$

Here, r is the distance in meters between the two particles, the masses M and m are given in kilograms, $G \approx 6.67 \times 10^{-11}$, and the resulting force is measured in newtons.[2]

Example 4 Newton's Law of Gravity

Find the gravitational force exerted on a particle with mass m situated at the point (x, y) by another particle with mass M situated at the point (a, b). Express the answer as a function F of the coordinates of the particle with mass m.

Solution The formula above for gravitational force is expressed as a function of the distance, r, between the two particles. Because we are given the coordinates of the two particles, we can express r in terms of these coordinates using the formula for distance:

$$r = \sqrt{(x - a)^2 + (y - b)^2}$$

Substituting for r, we get

$$F(x, y) = G\frac{Mm}{(x - a)^2 + (y - b)^2}$$

+*Before we go on...* In Example 4, notice that $F(a, b)$ is not defined because substituting $x = a$ and $y = b$ makes the denominator equal 0. Thus, the largest possible domain of F excludes the point (a, b). Because (a, b) is the only value of (x, y) for which F is not defined, we deduce that *the domain of F consists of all points (x, y) except for (a, b).* In other words, the domain of F is the whole xy-plane with the single point (a, b) missing.[3] ■

Q: Why have we expressed F as a function of x and y only, and not also as a function of a and b?

A: It's a matter of interpretation. When we write F as a function of x and y, we are thinking of a and b as *constants*. For example, (a, b) could be the coordinates of the sun—which we often assume to be fixed in space—while (x, y) could be the coordinates of the earth—which is moving around the sun. In that case it is most natural to think of x and y as variables and a and b as constants. In another context we may indeed want to consider F as a function of four variables, $x, y, a,$ and b. ■

---

[2] A newton is the force that will cause a 1-kilogram mass to accelerate at 1 m/sec^2.

[3] Mathematicians often refer to this as a "punctured plane."

15.1 EXERCISES

● denotes basic skills exercises

tech Ex indicates exercises that should be solved using technology

For each function in Exercises 1–4, evaluate **(a)** $f(0, 0)$;
(b) $f(1, 0)$; **(c)** $f(0, -1)$; **(d)** $f(a, 2)$; **(e)** $f(y, x)$;
(f) $f(x + h, y + k)$ hint [see Quick Examples p. 1042]

1. ● $f(x, y) = x^2 + y^2 - x + 1$

2. ● $f(x, y) = x^2 - y - xy + 1$

3. ● $f(x, y) = 0.2x + 0.1y - 0.01xy$

4. ● $f(x, y) = 0.4x - 0.5y - 0.05xy$

For each function in Exercises 5–8, evaluate **(a)** $g(0, 0, 0)$;
(b) $g(1, 0, 0)$; **(c)** $g(0, 1, 0)$; **(d)** $g(z, x, y)$;
(e) $g(x + h, y + k, z + l)$, provided such a value exists.

5. ● $g(x, y, z) = e^{x+y+z}$ **6.** ● $g(x, y, z) = \ln(x + y + z)$

7. ● $g(x, y, z) = \dfrac{xyz}{x^2 + y^2 + z^2}$

8. ● $g(x, y, z) = \dfrac{e^{xyz}}{x + y + z}$

9. ● Let $f(x, y, z) = 1.5 + 2.3x - 1.4y - 2.5z$. Complete the following sentences.

  a. f ___ by ___ units for every 1 unit of increase in x.

  b. f ___ by ___ units for every 1 unit of increase in y.

  c. _____ by 2.5 units for every _____.
  hint [see Example 1]

10. ● Let $g(x, y, z) = 0.01x + 0.02y - 0.03z - 0.05$. Complete the following sentences.

  a. g ___ by ___ units for every 1 unit of increase in z.

  b. g ___ by ___ units for every 1 unit of increase in x.

  c. _____ by 0.02 units for every _____.

In Exercises 11–18, classify each function as linear, interaction, or neither. hint [see Quick Examples pp. 1044, 1045]

11. ● $L(x, y) = 3x - 2y + 6xy - 4y^2$

12. ● $L(x, y, z) = 3x - 2y + 6xz$

13. ● $P(x_1, x_2, x_3) = 0.4 + 2x_1 - x_3$

14. ● $Q(x_1, x_2) = 4x_2 - 0.5x_1 - x_1^2$

15. ● $f(x, y, z) = \dfrac{x + y - z}{3}$

16. ● $g(x, y, z) = \dfrac{xz - 3yz + z^2}{4z}$ $(z \neq 0)$

17. ● $g(x, y, z) = \dfrac{xz - 3yz + z^2 y}{4z}$ $(z \neq 0)$

18. ● $f(x, y) = x + y + xy + x^2 y$

In Exercises 19 and 20, use the given tabular representation of the function f to compute the quantities asked for. hint [see Example 3]

19. ●

$x \rightarrow$	10	20	30	40
$y\downarrow$ 10	−1	107	162	−3
20	−6	194	294	−14
30	−11	281	426	−25
40	−16	368	558	−36

  a. $f(20, 10)$ **b.** $f(40, 20)$

  c. $f(10, 20) - f(20, 10)$

20. ●

$x \rightarrow$	10	20	30	40
$y\downarrow$ 10	162	107	−5	−7
20	294	194	−22	−30
30	426	281	−39	−53
40	558	368	−56	−76

  a. $f(10, 30)$ **b.** $f(20, 10)$

  c. $f(10, 40) + f(10, 20)$

tech Ex In Exercises 21 and 22, use a spreadsheet or some other method to complete the given tables.

21. ● $P(x, y) = x - 0.3y + 0.45xy$

$x \rightarrow$	10	20	30	40
$y\downarrow$ 10				
20				
30				
40				

22. ● $Q(x, y) = 0.4x + 0.1y - 0.06xy$

$x \rightarrow$	10	20	30	40
$y\downarrow$ 10				
20				
30				
40				

23. tech Ex The following statistical table lists some values of the "Inverse F distribution" ($\alpha = 0.5$):

$n \rightarrow$

d		1	2	3	4	5	6	7	8	9	10
$\downarrow$	1	161.4	199.5	215.7	224.6	230.2	234.0	236.8	238.9	240.5	241.9
	2	18.51	19.00	19.16	19.25	19.30	19.33	19.35	19.37	19.39	19.40
	3	10.13	9.552	9.277	9.117	9.013	8.941	8.887	8.812	8.812	8.785
	4	7.709	6.944	6.591	6.388	6.256	6.163	6.094	5.999	5.999	5.964
	5	6.608	5.786	5.409	5.192	5.050	4.950	4.876	4.772	4.772	4.735
	6	5.987	5.143	4.757	4.534	4.387	4.284	4.207	4.099	4.099	4.060
	7	5.591	4.737	4.347	4.120	3.972	3.866	3.787	3.677	3.677	3.637
	8	5.318	4.459	4.066	3.838	3.688	3.581	3.500	3.388	3.388	3.347
	9	5.117	4.256	3.863	3.633	3.482	3.374	3.293	3.179	3.179	3.137
	10	4.965	4.103	3.708	3.478	3.326	3.217	3.135	3.020	3.020	2.978

In Excel, you can compute the value of this function at (n, d) by the formula

$=$ FINV(0.05, n, d)   The 0.05 is the value of alpha (α).

Use Excel to re-create this table.

24. tech Ex The formula for the body mass index $M(w, h)$, if w is given in kilograms and h is given in meters, is

$$M(w, h) = \frac{w}{h^2}$$ See Example 3.

Use this formula to complete the following table in Excel:

$w \rightarrow$

h		70	80	90	100	110	120	130
$\downarrow$	1.8							
	1.85							
	1.9							
	1.95							
	2							
	2.05							
	2.1							
	2.15							
	2.2							
	2.25							
	2.3							

tech Ex *In Exercises 25–28, use either a graphing calculator or a spreadsheet to complete each table. Express all your answers as decimals rounded to four decimal places.*

25. ●

x	y	$f(x, y) = x^2\sqrt{1 + xy}$
3	1	
1	15	
0.3	0.5	
56	4	

26. ●

x	y	$f(x, y) = x^2 e^y$
0	2	
-1	5	
1.4	2.5	
11	9	

27. ●

x	y	$f(x, y) = x \ln(x^2 + y^2)$
3	1	
1.4	-1	
e	0	
0	e	

28. ●

x	y	$f(x, y) = \dfrac{x}{x^2 - y^2}$
-1	2	
0	0.2	
0.4	2.5	
10	0	

29. Brand Z's annual sales are affected by the sales of related products X and Y as follows: Each $1 million increase in sales of brand X causes a $2.1 million decline in sales of brand Z, whereas each $1 million increase in sales of brand Y results in an increase of $0.4 million in sales of brand Z. Currently, brands X, Y, and Z are each selling $6 million per year. Model the sales of brand Z using a linear function.

30. Let $f(x, y, z) = 43.2 - 2.3x + 11.3y - 4.5z$. Complete the following: An increase of 1 in the value of y causes the value of f to ___ by ___, whereas increasing the value of x by 1 and ___ the value of z by ___ causes a decrease of 11.3 in the value of f.

In Exercises 31–34, find the distance between the given pairs of points.

31. ● $(1, -1)$ and $(2, -2)$ **32.** ● $(1, 0)$ and $(6, 1)$

33. ● $(a, 0)$ and $(0, b)$ **34.** ● (a, a) and (b, b)

35. Find the value of k such that $(1, k)$ is equidistant from $(0, 0)$ and $(2, 1)$.

36. Find the value of k such that (k, k) is equidistant from $(-1, 0)$ and $(0, 2)$.

● basic skills   tech Ex technology exercise

37. Describe the set of points (x, y) such that
$(x - 2)^2 + (y + 1)^2 = 9$.

38. Describe the set of points (x, y) such that
$(x + 3)^2 + (y - 1)^2 = 4$.

Applications

39. ● *Marginal Cost* Your weekly cost (in dollars) to manufacture x cars and y trucks is

$$C(x, y) = 240{,}000 + 6000x + 4000y$$

What is the marginal cost of a car? Of a truck?

40. ● *Marginal Cost* Your weekly cost (in dollars) to manufacture x bicycles and y tricycles is

$$C(x, y) = 24{,}000 + 60x + 20y$$

What is the marginal cost of a bicycle? Of a tricycle?

41. ● *Marginal Cost* Your sales of online video and audio clips are booming. Your Internet provider, Moneydrain.com, wants to get in on the action and has offered you unlimited technical assistance and consulting if you agree to pay Moneydrain 3¢ for every video clip and 4¢ for every audio clip you sell on the site. Further, Moneydrain agrees to charge you only $10 per month to host your site. Set up a (monthly) cost function for the scenario, and describe each variable.

42. ● *Marginal Cost* Your Cabaret nightspot "Jazz on Jupiter" has become an expensive proposition: You are paying monthly costs of $50,000 just to keep the place running. On top of that, your regular cabaret artist is charging you $3000 per performance, and your jazz ensemble is charging $1000 per hour. Set up a (monthly) cost function for the scenario, and describe each variable.

Scientific Research In 2004, physics research in the U.S appeared to be losing ground to Europe and other countries as evidenced in the following graph. The graph shows the number of articles published in the prominent physics research journal Physical Review:[4]

Articles in *Physical Review* (Thousands)

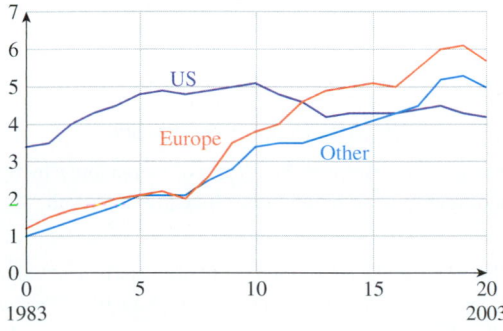

---

[4] SOURCE: The American Physical Society/*New York Times,* May 3, 2003, p. A1.

Exercises 43 and 44 are based on mathematical models derived from the graphical data shown above.

43. ● In each year from 1983 to 2003, the percentage y of research articles in *Physical Review* written by researchers in the U.S. can be approximated by

$$y = 82 - 0.78t - 1.02x \text{ percentage points} \quad (0 \le t \le 20)$$

where t is the year since 1983 and x is the percentage of articles written by researchers in Europe.

a. In 2003, researchers in Europe wrote 38% of the articles published by the journal that year. What percentage was written by researchers in the U.S.?

b. In 1983, researchers in the U.S. wrote 61% of the articles published that year. What percentage was written by researchers in Europe?

c. What are the units of measurement of the coefficient of t?

44. ● The number z of research articles in *Physical Review* that were written by researchers in the U.S. from 1993 through 2003 can be approximated by

$$z = 5960 - 0.71x + 0.50y \quad (3000 \le x, y \le 6000)$$

articles each year, where x is the number of articles written by researchers in Europe and y is the number written by researchers in other countries (excluding Europe and the U.S.).

a. In the year 2000, approximately 5500 articles were written by researchers in Europe, and 4500 by researchers in other countries. How many articles (to the nearest 100) were written by researchers in the U.S.?

b. According to the model, if 5000 articles were written in Europe and an equal number by researchers in the U.S. and other countries, what would that number be?

c. What is the significance of the fact that the coefficient of x is negative?

45. ● *Career Choices* Graduating MBAs who do not become consultants often join investment banking, venture capital, or high-technology companies. The following linear model is based on data on Harvard Business School MBAs:[5]

$$c(x, y, z) = 48.4 + 0.06x - 0.40y - 1.3z$$

Here, c is the percentage of MBAs who become consultants, x is the percentage who join high technology companies, y is the percentage who join investment banking companies, and z is the percentage who join venture capital companies.

a. In 1999, approximately 18% of Harvard MBAs joined high technology companies, 12% joined investment banking companies, and 12% joined venture capital companies. Use the model to estimate the percentage who became consultants. (Round to the nearest one percent.)

---

[5] The model is based on a regression of the data from 1995 to 1999. SOURCE: Harvard Business School/*The New York Times,* January 19, 2000, p. C1.

● basic skills   **tech** Ex technology exercise

b. In 1997, approximately 32% of Harvard MBAs became consultants, 16% joined investment banking companies, and 8% joined venture capital companies. Use the model to estimate the percentage who joined high technology companies. (Round to the nearest one percent.)

c. Complete the following: For every 1-point rise in the percentage of Harvard MBAs who join investment banking companies, there is a ___-point ___ in the percentage who become consultants, assuming the number joining high technology companies and venture capital companies are unchanged.

46. ● ***Career Choices*** Refer to the preceding exercise. An alternative, interaction model based on Harvard Business School MBAs is

$$c(y, z) = 44.3 - 0.28y + 0.04yz - 1.6z$$

Here, c is the percentage of MBAs who become consultants, y is the percentage who join investment banking companies, and z is the percentage who join venture capital companies.[6]

a. In 1999, approximately 12% joined investment banking companies and 12% joined venture capital companies. Use the model to estimate the percentage who became consultants. (Round to the nearest one percent.)

b. In 1997, approximately 32% of Harvard MBAs became consultants and 8% joined venture capital companies. Use the model to estimate the percentage who joined investment banking companies. (Round to the nearest one percent.)

47. ***Online Revenue*** Let us look once again at the example we used to introduce the chapter. Your major online bookstore is in direct competition with Amazon.com, BN.com, and Borders.com. Your company's daily revenue in dollars is given by

$$R(x, y, z) = 10,000 - 0.01x - 0.02y - 0.01z + 0.00001yz$$

where $x, y,$ and z are the online daily revenues of Amazon.com, BN.com, and Borders.com, respectively.

a. If, on a certain day, Amazon.com shows revenue of $12,000, while BN.com and Borders.com each show $5000, what does the model predict for your company's revenue that day?

b. If Amazon.com and BN.com each show daily revenue of $5000, give an equation showing how your daily revenue depends on that of Borders.com.

48. ***Online Revenue*** Repeat the preceding exercise, using the revised revenue function

$$R(x, y, z) = \$20,000 - 0.02x - 0.04y - 0.01z + 0.00001yz$$

49. ***Modeling the Growth of Wireless with a Linear Function*** The following table shows the approximate number of

wireless phone subscribers and the number of cell sites in the U.S. in 1997, 2002, and 2005.[7]

	1997	**2002**	**2005**
Subscribers (millions)	60	110	200
Cell sites (thousands)	50	100	180

Model the number of subscribers as a function of the number of cell sites and time, using a linear function of the form

$$s(c, t) = Ac + Bt + C \quad (A, B, C \text{ constants})$$

where s represents the number of subscribers (in millions), c represents the number of cell sites (in thousands), and t is time in years since 1997.

50. ***Modeling Sales with a Linear Function*** The following table shows Toyota's sales, in millions of vehicles, in the U.S. and Japan in 1991, 1996, and 2001.[8]

	1991	**1996**	**2001**
U.S.	1.0	1.2	1.7
Japan	2.3	2.0	1.7

Model Toyota sales in Japan as a function of sales in the U.S. and time, using a linear function of the form

$$j(u, t) = Au + Bt + C \quad (A, B, C \text{ constants})$$

where j represents annual Toyota sales (in millions of vehicles) in Japan, u represents sales in the U.S., and t is time in years since 1991.

51. ***Utility*** Suppose your newspaper is trying to decide between two competing desktop publishing software packages, Macro Publish and Turbo Publish. You estimate that if you purchase x copies of Macro Publish and y copies of Turbo Publish, your company's daily productivity will be

$$U(x, y) = 6x^{0.8}y^{0.2} + x$$

where $U(x, y)$ is measured in pages per day (U is called a *utility function*). If $x = y = 10$, calculate the effect of increasing x by one unit, and interpret the result.

52. ***Housing Costs***[9] The cost C (in dollars) of building a house is related to the number k of carpenters used and the number e of electricians used by

$$C(k, e) = 15,000 + 50k^2 + 60e^2$$

If $k = e = 10$, compare the effects of increasing k by one unit and of increasing e by one unit. Interpret the result.

---

[7] 2002 figures are estimates. SOURCES: Cellular Telecommunications and Internet association/*New York Times,* February 14, 2002, p. G1, *Wired.com,* 2005.

[8] SOURCE: Toyota Motor North America/*New York Times,* February 17, 2002, p. BU1.

[9] Based on an Exercise in *Introduction to Mathematical Economics* by A. L. Ostrosky Jr. and J. V. Koch (Waveland Press, Illinois, 1979).

---

[6] Ibid.

● basic skills   *tech* Ex technology exercise

53. *Volume* The volume of an ellipsoid with cross-sectional radii a, b, and c is $V(a, b, c) = \frac{4}{3}\pi abc$.

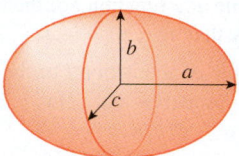

a. Find at least two sets of values for a, b and c such that $V(a, b, c) = 1$.

b. Find the value of a such that $V(a, a, a) = 1$, and describe the resulting ellipsoid.

54. *Volume* The volume of a right elliptical cone with height h and radii a and b of its base is $V(a, b, h) = \frac{1}{3}\pi abh$.

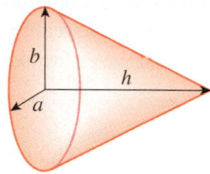

a. Find at least two sets of values for a, b and h such that $V(a, b, h) = 1$.

b. Find the value of a such that $V(a, a, a) = 1$, and describe the resulting cone.

Exercises 55–58 involve "Cobb-Douglas" productivity functions. These functions have the form

$$P(x, y) = Kx^a y^{1-a}$$

where P stands for the number of items produced per year, x is the number of employees, and y is the annual operating budget. (The numbers K and a are constants that depend on the situation we are looking at, with $0 \leq a \leq 1$.)

55. ● *Productivity* How many items will be produced per year by a company with 100 employees and an annual operating budget of $500,000 if $K = 1000$ and $a = 0.5$? (Round your answer to one significant digit.)

56. ● *Productivity* How many items will be produced per year by a company with 50 employees and an annual operating budget of $1,000,000 if $K = 1000$ and $a = 0.5$? (Round your answer to one significant digit.)

57. *Modeling Production with Cobb-Douglas* Two years ago my piano manufacturing plant employed 1000 workers, had an operating budget of $1 million, and turned out 100 pianos. Last year I slashed the operating budget to $10,000, and production dropped to 10 pianos.

a. Use the data for each of the two years and the Cobb-Douglas formula to obtain two equations in K and a.

b. Take logs of both sides in each equation and obtain two linear equations in a and $\log K$.

c. Solve these equations to obtain values for a and K.

d. Use these values in the Cobb-Douglas formula to predict production if I increase the operating budget back to $1 million but lay off half the work force.

58. *Modeling Production with Cobb-Douglas* Repeat the preceding exercise using the following data: Two years ago—1000 employees, $1 million operating budget, 100 pianos; Last year—1000 employees, $100,000 operating budget, 10 pianos.

59. *Pollution* The burden of man-made aerosol sulfate in the earth's atmosphere, in grams per square meter, is

$$B(x, n) = \frac{xn}{A}$$

where x is the total weight of aerosol sulfate emitted into the atmosphere per year and n is the number of years it remains in the atmosphere. A is the surface area of the earth, approximately 5.1×10^{14} square meters.[10]

a. Calculate the burden, given the 1995 estimated values of $x = 1.5 \times 10^{14}$ grams per year, and $n = 5$ days.

b. What does the function $W(x, n) = xn$ measure?

60. *Pollution* The amount of aerosol sulfate (in grams) was approximately 45×10^{12} grams in 1940 and has been increasing exponentially ever since, with a doubling time of approximately 20 years.[11] Use the model from the preceding exercise to give a formula for the atmospheric burden of aerosol sulfate as a function of the time t in years since 1940 and the number of years n it remains in the atmosphere.

61. *Alien Intelligence* Frank Drake, an astronomer at the University of California at Santa Cruz, devised the following equation to estimate the number of planet-based civilizations in our Milky Way galaxy willing and able to communicate with Earth:[12]

$$N(R, f_p, n_e, f_l, f_i, f_c, L) = Rf_p n_e f_l f_i f_c L$$

$R =$ the number of new stars formed in our galaxy each year

$f_p =$ the fraction of those stars that have planetary systems

$n_e =$ the average number of planets in each such system that can support life

$f_l =$ the fraction of such planets on which life actually evolves

$f_i =$ the fraction of life-sustaining planets on which intelligent life evolves

$f_c =$ the fraction of intelligent-life-bearing planets on which the intelligent beings develop the means and the will to communicate over interstellar distances

$L =$ the average lifetime of such technological civilizations (in years)

---

[10] SOURCE: Robert J. Charlson and Tom M. L. Wigley, "Sulfate Aerosol and Climatic Change," *Scientific American,* February, 1994, pp. 48–57.

[11] Ibid.

[12] SOURCE: "First Contact" (Plume Books/Penguin Group)/*The New York Times,* October 6, 1992, p. C1.

● basic skills   *tech* Ex technology exercise

a. What would be the effect on N if any one of the variables were doubled?

b. How would you modify the formula if you were interested only in the number of intelligent-life-bearing planets in the galaxy?

c. How could one convert this function into a linear function?

d. (For discussion) Try to come up with an estimate of N.

62. *More Alien Intelligence* The formula given in the preceding exercise restricts attention to planet-based civilizations in our galaxy. Give a formula that includes intelligent planet-based aliens from the galaxy Andromeda. (Assume that all the variables used in the formula for the Milky Way have the same values for Andromeda.)

63. tech Ex *Level Curves* The height of each point in a hilly region is given as a function of its coordinates by the formula

$$f(x, y) = y^2 - x^2$$

a. Use technology to plot the curves on which the height is 0, 1, and 2 on the same set of axes. These are called **level curves of** f.

b. Sketch the curve $f(x, y) = 3$ *without* using technology.

c. Sketch the curves $f(y, x) = 1$ and $f(y, x) = 2$ without using technology.

64. tech Ex *Isotherms* The temperature (in degrees Fahrenheit) at each point in a region is given as a function of the coordinates by the formula

$$T(x, y) = 60.5(x - y^2)$$

a. Use technology to sketch the curves on which the temperature is $0°$, $30°$, and $90°$. These curves are called **isotherms.**

b. Sketch the isotherms corresponding to $20°$, $50°$, and $100°$ *without* using technology.

c. What do the isotherms corresponding to negative temperatures look like?

Communication and Reasoning Exercises

65. ● Let $f(x, y) = \dfrac{x}{y}$. How are $f(x, y)$ and $f(y, x)$ related?

66. ● Let $f(x, y) = x^2 y^3$. How are $f(x, y)$ and $f(-x, -y)$ related?

67. ● Give an example of a function of the two variables x and y with the property that interchanging x and y has no effect.

68. ● Give an example of a function f of the two variables x and y with the property that $f(x, y) = -f(y, x)$.

69. ● Give an example of a function f of the three variables x, y, and z with the property that $f(x, y, z) = f(y, x, z)$ and $f(-x, -y, -z) = -f(x, y, z)$.

70. ● Give an example of a function f of the three variables x, y, and z with the property that $f(x, y, z) = f(y, x, z)$ and $f(-x, -y, -z) = f(x, y, z)$.

71. ● Illustrate by means of an example how a real-valued function of the two variables x and y gives different real-valued functions of one variable when we restrict y to be different constants.

72. ● Illustrate by means of an example how a real-valued function of one variable x gives different real-valued functions of the two variables y and z when we substitute for x suitable functions of y and z.

73. If f is a linear function of x and y, show that if we restrict y to be a fixed constant, then the resulting function of x is linear. Does the slope of this linear function depend on the choice of y?

74. If f is an interaction function of x and y, show that if we restrict y to be a fixed constant, then the resulting function of x is linear. Does the slope of this linear function depend on the choice of y?

75. Suppose that $C(x, y)$ represents the cost of x CDs and y cassettes. If $C(x, y + 1) < C(x + 1, y)$ for every $x \geq 0$ and $y \geq 0$, what does this tell you about the cost of CDs and cassettes?

76. Suppose that $C(x, y)$ represents the cost of renting x DVDs and y video games. If $C(x + 2, y) < C(x, y + 1)$ for every $x \geq 0$ and $y \geq 0$, what does this tell you about the cost of renting DVDs and video games?

● basic skills   tech Ex technology exercise

15.2 Three-Dimensional Space and the Graph of a Function of Two Variables

Just as functions of a single variable have graphs, so do functions of two or more variables. Recall that the graph of $f(x)$ consists of all points $(x, f(x))$ in the xy-plane. By analogy, we would like to say that the graph of a function of *two* variables, $f(x, y)$, consists of all points of the form $(x, y, f(x, y))$. Thus, we need three axes: the x-, y-, and z-axes. In other words, our graph will live in **three-dimensional space, or 3-space.**[13]

Just as we had two mutually perpendicular axes in two-dimensional space (the xy-plane; see Figure 4a), so we have three mutually perpendicular axes in three-dimensional space (Figure 4b).

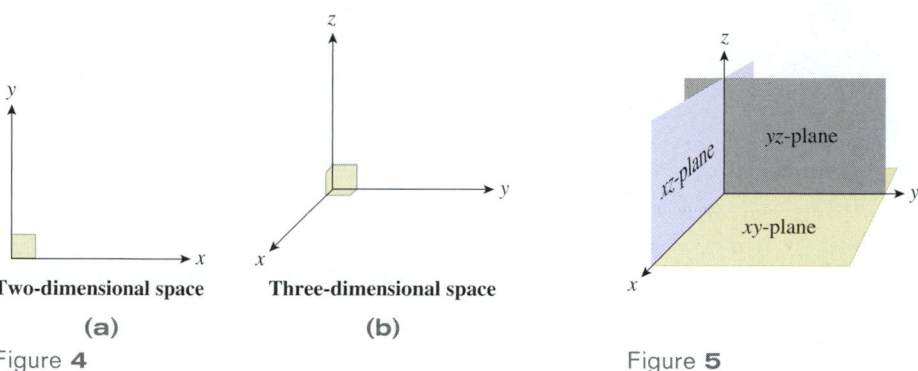

Two-dimensional space

(a)

Three-dimensional space

(b)

Figure **4**

Figure **5**

In both 2-space and 3-space, the axis labeled with the last letter goes up. Thus, the z-direction is the "up" direction in 3-space, rather than the y-direction.

Three important planes are associated with these axes: the xy-plane, the yz-plane, and the xz-plane. These planes are shown in Figure 5. Any two of these planes intersect in one of the axes (for example, the xy- and xz-planes intersect in the x-axis) and all three meet at the origin. Notice that the xy-plane consists of all points with z-coordinate zero, the xz-plane consists of all points with $y = 0$, and the yz-plane consists of all points with $x = 0$.

In 3-space, each point has *three* coordinates, as you might expect: the x-coordinate, the y-coordinate, and the z-coordinate. To see how this works, look at the following examples.

Example **1** Plotting Points in Three Dimensions

Locate the points $P(1, 2, 3)$, $Q(-1, 2, 3)$, $R(1, -1, 0)$, and $S(1, 2, -2)$ in 3-space.

Solution To locate P, the procedure is similar to the one we used in 2-space: Start at the origin, proceed 1 unit in the x direction, then proceed 2 units in the y direction, and finally, proceed three units in the z direction. We wind up at the point P shown in Figures 6a and 6b.

Here is another, extremely useful way of thinking about the location of P. First, look at the x- and y-coordinates, obtaining the point $(1, 2)$ in the xy-plane. The point we want is then three units vertically above the point $(1, 2)$ because the z-coordinate of a point is just its height above the xy-plane. This strategy is shown in Figure 6c.

---

[13] If we were dealing instead with a function of *three* variables, then we would need to go to *four-dimensional* space. Here we run into visualization problems (to say the least!) so we won't discuss the graphs of functions of three or more variables in this text.

The z-coordinate of a point is its height above the xy-plane.

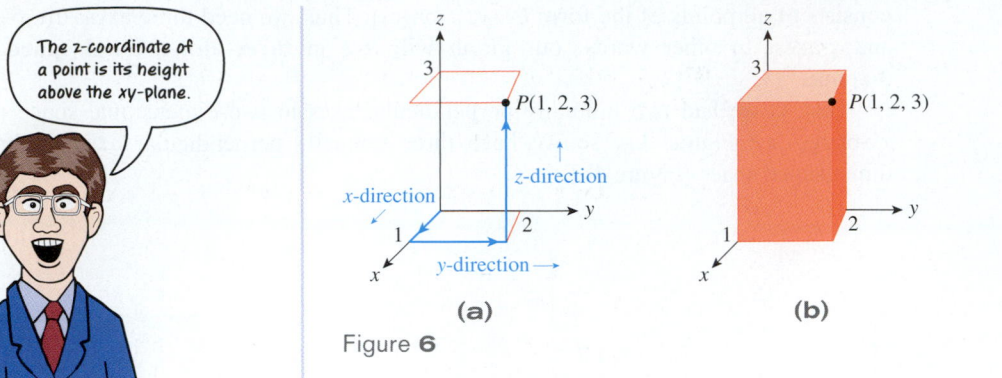

Figure 6

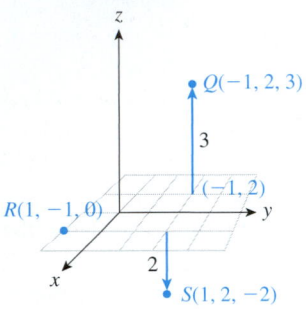

Figure 7

Plotting the points Q, R and S is similar, using the convention that negative coordinates correspond to moves back, left, or down (see Figure 7).

Our next task is to describe the graph of a function $f(x, y)$ of two variables.

Graph of a Function of Two Variables

The **graph of the function f of two variables** is the set of all points $(x, y, f(x, y))$ in three-dimensional space, where we restrict the values of (x, y) to lie in the domain of f. In other words, the graph is the set of all the points (x, y, z) with $z = f(x, y)$.

For *every* point (x, y) in the domain of f, the z-coordinate of the corresponding point on the graph is given by evaluating the function at (x, y). Thus, there will be a point on the graph above *every* point in the domain of f, so that the graph is usually a *surface* of some sort.

Example 2 Graph of a Function of Two Variables

Describe the graph of $f(x, y) = x^2 + y^2$.

Solution Your first thought might be to make a table of values. You could choose some values for x and y and then, for each such pair, calculate $z = x^2 + y^2$. For example, you might get the following table:

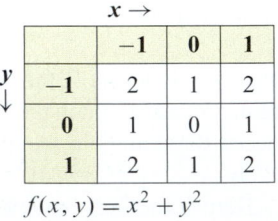

		$x \rightarrow$	
$y \downarrow$	**−1**	**0**	**1**
−1	2	1	2
0	1	0	1
1	2	1	2

$f(x, y) = x^2 + y^2$

This gives the following nine points on the graph of f: $(-1, -1, 2), (-1, 0, 1),\backslash$ $(-1, 1, 2), (0, -1, 1), (0, 0, 0), (0, 1, 1), (1, -1, 2), (1, 0, 1),$ and $(1, 1, 2)$. These points are shown in Figure 8.

The points on the xy-plane we chose for our table are the grid points in the xy-plane, and the corresponding points on the graph are marked with solid dots. The problem is that this small number of points hardly tells us what the surface looks like, and even if we plotted more points, it is not clear that we would get anything more than a mass of dots on the page.

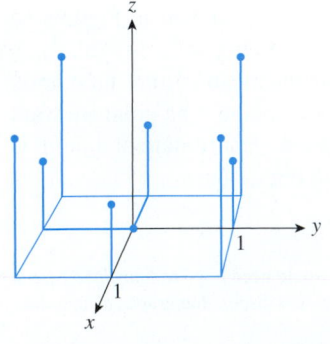

Figure 8

What can we do? There are several alternatives. One place to start is to use technology to draw the graph.[*] We then obtain something like Figure 9. This particular surface is called a **paraboloid.**

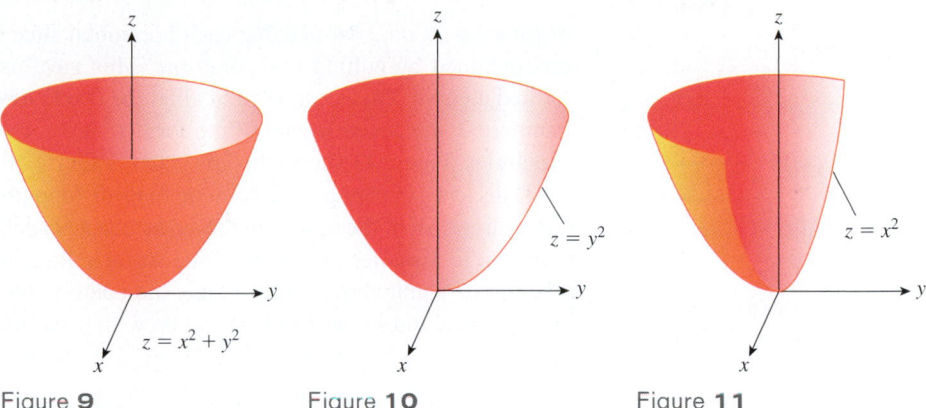

Figure **9**   Figure **10**   Figure **11**

If we slice vertically through this surface along the yz-plane, we get the picture in Figure 10. The shape of the front edge, where we cut, is a parabola. To see why, note that the yz-plane is the set of points where $x = 0$. To get the intersection of $x = 0$ and $z = x^2 + y^2$, we substitute $x = 0$ in the second equation, getting $z = y^2$. This is the equation of a parabola in the yz-plane.

Similarly, we can slice through the surface with the xz-plane by setting $y = 0$. This gives the parabola $z = x^2$ in the xz-plane (Figure 11).

We can also look at horizontal slices through the surface, that is, slices by planes parallel to the xy-plane. These are given by setting $z = c$ for various numbers c. For example, if we set $z = 1$, we will see only the points with height 1. Substituting in the equation $z = x^2 + y^2$ gives the equation

$$1 = x^2 + y^2$$

which is the equation of a circle of radius 1. If we set $z = 4$, we get the equation of a circle of radius 2:

$$4 = x^2 + y^2$$

In general, if we slice through the surface at height $z = c$, we get a circle (of radius $\sqrt{c}$). Figure 12 shows several of these circles.

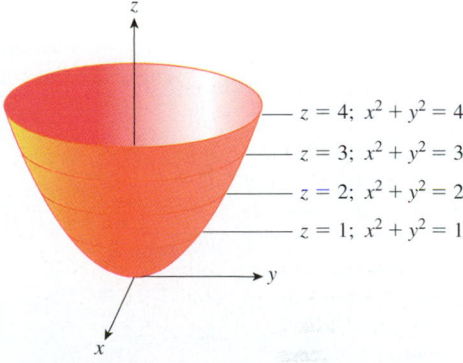

Figure **12**

---

[*] See Example 3 for a discussion of the use of a spreadsheet to draw a surface.

Looking at these circular slices, we see that this surface is the one we get by taking the parabola $z = x^2$ and spinning it around the z-axis. This is an example of what is known as a **surface of revolution.**

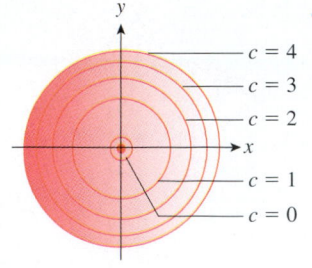

Level curves of the paraboloid
$z = x^2 + y^2$

Figure **13**

+Before we go on... Notice that each horizontal slice through the surface in Example 2 was obtained by putting $z = constant$. This gave us an equation in x and y that described a curve. These curves are called the **level curves** of the surface $z = f(x, y)$. In Example 2, the equations are of the form $x^2 + y^2 = constant$, and so the level curves are circles. Figure 13 shows the level curves for $c = 0, 1, 2, 3,$ and 4.

The level curves give a contour map or topographical map of the surface. Each curve shows all of the points on the surface at a particular height c. You can use this contour map to visualize the shape of the surface. Imagine moving the contour at $c = 1$ to a height of 1 unit above the xy-plane, the contour at $c = 2$ to a height of 2 units above the xy-plane, and so on. You will end up with something like Figure 12. ∎

The following summary includes the techniques we have just used plus some additional ones:

Analyzing the Graph of a Function of Two Variables

If possible, use technology to render the graph of a given function $z = f(x, y)$. Given the function $z = f(x, y)$, you can analyze its graph as follows:

Step 1 Obtain the **x-, y-, and z-intercepts** (the places where the surface crosses the coordinate axes).

x-Intercept(s): Set $y = 0$ and $z = 0$ and solve for x.

y-Intercept(s): Set $x = 0$ and $z = 0$ and solve for y.

z-Intercept: Set $x = 0$ and $y = 0$ and compute z.

Step 2 Slice the surface along planes parallel to the xy-, yz-, and xz-planes.

$z = constant$ Set $z = constant$ and analyze the resulting curves.
(level curves) These are the curves resulting from horizontal slices.

$x = constant$ Set $x = constant$ and analyze the resulting curves.
These are the curves resulting from slices parallel to the yz-plane.

$y = constant$ Set $y = constant$ and analyze the resulting curves.
These are the curves resulting from slices parallel to the xz-plane.

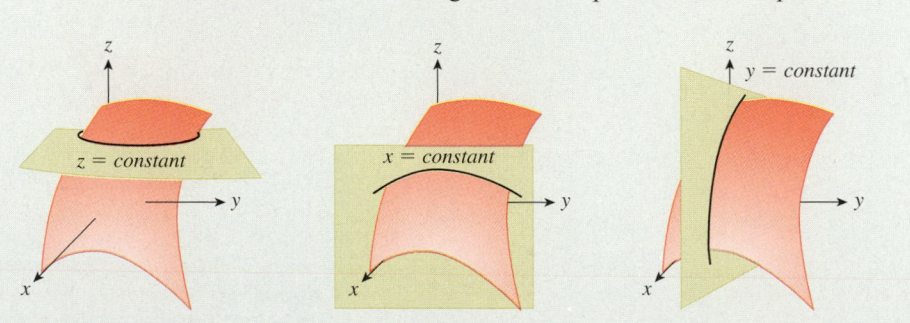

Spreadsheets often have built-in features to render surfaces such as the paraboloid in Example 2. In the following example, we use Excel to graph another surface and then analyze it as above.

 **tech** Ex | ## Example 3 Analyzing a Surface

Describe the graph of $f(x, y) = x^2 - y^2$.

Solution First we obtain a picture of the graph using technology. Figure 14 was obtained using the three-dimensional Excel graphing utility you can find online by following:

Chapter 15 → Tools → Excel Surface Graphing Utility

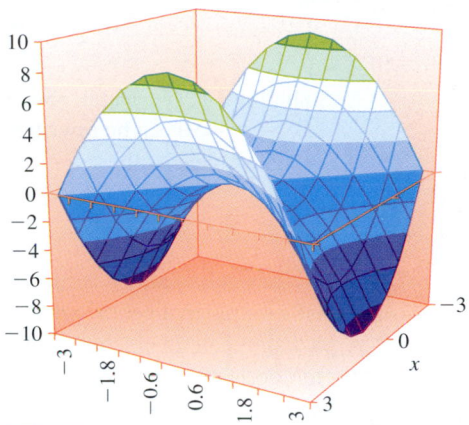

Figure **14**

See the Technology Guides at the end of the chapter to find out how to obtain a similar graph from scratch on an ordinary Excel sheet.

The graph shows an example of a "saddle point" at the origin (we return to this idea in a Section 15.4). To analyze the graph for the features shown in the box above, replace $f(x, y)$ by z to obtain

$$z = x^2 - y^2$$

Step 1 Intercepts Setting any two of the variables x, y, and z equal to zero results in the third also being zero, so the x-, y-, and z-intercepts are all 0. In other words, the surface touches all three axes in exactly one point, the origin.

Step 1 Slices Slices in various directions show more interesting features.

Slice by $x = c$ This gives $z = c^2 - y^2$, which is the equation of a parabola that opens downward. You can see two of these slices ($c = -3$, $c = 3$) as the front and back edges of the surface in Figure 14. [More are shown in Figure 15a.]

Slice by $y = c$ This gives $z = x^2 - c^2$, which is the equation of a parabola once again—this time, opening upward. You can see two of these slices ($c = -3$, $c = 3$) as the left and right edges of the surface in Figure 14. [More are shown in Figure 15b.]

Slice by $z = c$ This gives $x^2 - y^2 = c$, which is a hyperbola. The level curves for various values of c are visible in Figure 14 as the boundaries between the different

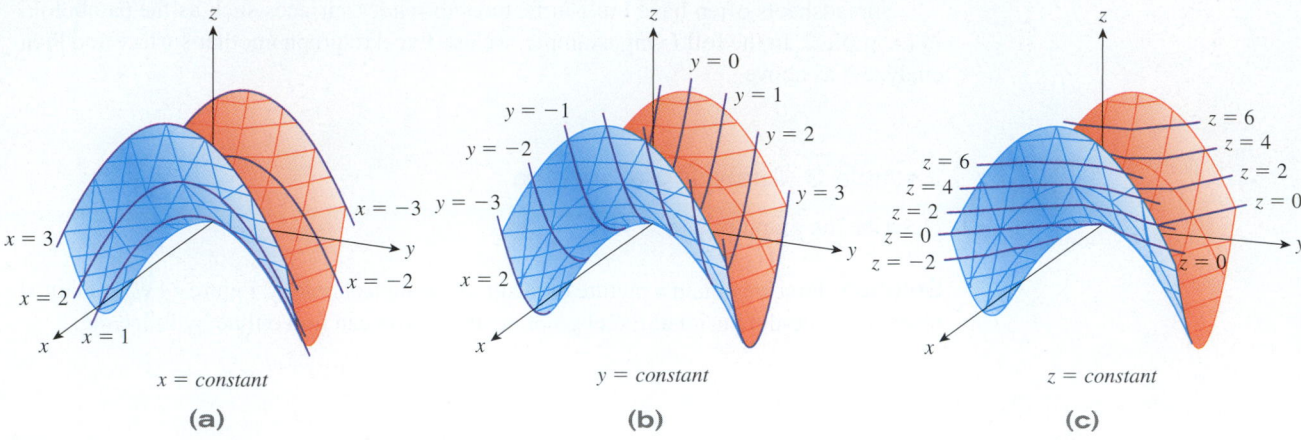

$x = constant$

(a)

$y = constant$

(b)

$z = constant$

(c)

Figure **15**

shadings in the graph. [See Figure 15c.] The case $c = 0$ is interesting: The equation $x^2 - y^2 = 0$ can be rewritten as $x = \pm y$ (why?), which represents two lines at right-angles to each other.

To obtain really beautiful renderings of surfaces, you could use one of the commercial computer algebra software packages, such as Mathematica® or Maple®. These packages can do much more than render surfaces and can be used, for example, to compute derivatives and antiderivatives, to solve equations algebraically, and to perform a variety of algebraic computations.

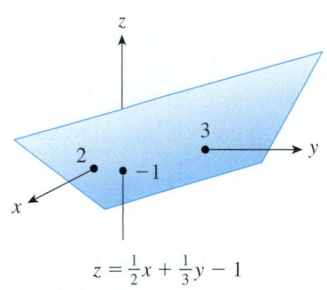

$z = \frac{1}{2}x + \frac{1}{3}y - 1$

Figure **16**

Example 4 Graph of a Linear Function

Describe the graph of $g(x, y) = \frac{1}{2}x + \frac{1}{3}y - 1$.

Solution Notice first that g is a linear function of x and y. Figure 16 shows a portion of the graph, which is a plane.

We can get a good idea of what plane this is by looking at the x-, y-, and z-intercepts.

x-intercept Set $y = z = 0$, which gives $x = 2$.

y-intercept Set $x = z = 0$, which gives $y = 3$.

z-intercept Set $x = y = 0$, which gives $z = -1$.

Three points are enough to define a plane, so we can say that the plane is the one passing through the three points $(2, 0, 0)$, $(0, 3, 0)$, and $(0, 0, -1)$.

Note It can be shown that the graph of every linear function of two variables is a plane. What do the level curves look like? ∎

15.2 EXERCISES

● denotes basic skills exercises

tech Ex indicates exercises that should be solved using technology

1. ● Sketch the cube with vertices $(0, 0, 0)$, $(1, 0, 0)$, $(0, 1, 0)$, $(0, 0, 1)$, $(1, 1, 0)$, $(1, 0, 1)$, $(0, 1, 1)$, and $(1, 1, 1)$. *hint* [see Example 1]

2. ● Sketch the cube with vertices $(-1, -1, -1)$, $(1, -1, -1)$, $(-1, 1, -1)$, $(-1, -1, 1)$, $(1, 1, -1)$, $(1, -1, 1)$, $(-1, 1, 1)$, and $(1, 1, 1)$.

3. ● Sketch the pyramid with vertices $(1, 1, 0)$, $(1, -1, 0)$, $(-1, 1, 0)$, $(-1, -1, 0)$, and $(0, 0, 2)$.

4. ● Sketch the solid with vertices $(1, 1, 0)$, $(1, -1, 0)$, $(-1, 1, 0)$, $(-1, -1, 0)$, $(0, 0, -1)$, and $(0, 0, 1)$.

Sketch the planes in Exercises 5–10.

5. ● $z = -2$

6. ● $z = 4$

7. ● $y = 2$

8. ● $y = -3$

9. ● $x = -3$

10. ● $x = 2$

Match each equation in Exercises 11–18 with one of the graphs below. (If necessary, use technology to render the surfaces.) hint [see Examples 2, 3, 4]

11. ● $f(x, y) = 1 - 3x + 2y$

12. ● $f(x, y) = 1 - \sqrt{x^2 + y^2}$

13. ● $f(x, y) = 1 - (x^2 + y^2)$

14. ● $f(x, y) = y^2 - x^2$

15. ● $f(x, y) = -\sqrt{1 - (x^2 + y^2)}$

16. ● $f(x, y) = 1 + (x^2 + y^2)$

17. ● $f(x, y) = \dfrac{1}{x^2 + y^2}$

18. ● $f(x, y) = 3x - 2y + 1$

(A)

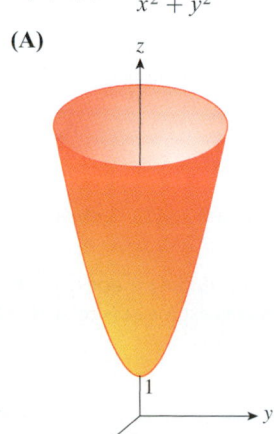

(B)

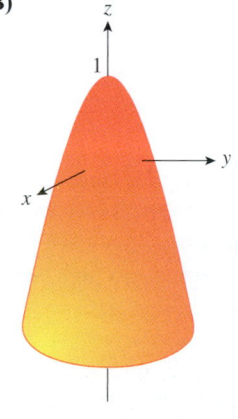

(C) **(D)**

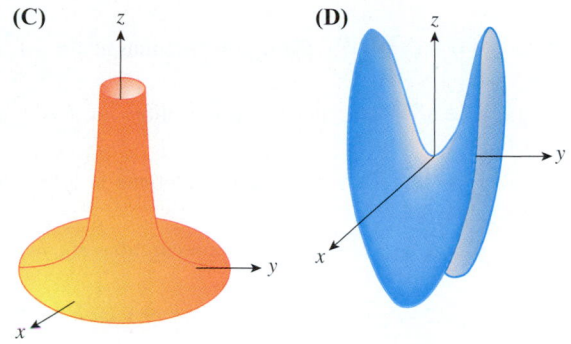

(E) **(F)**

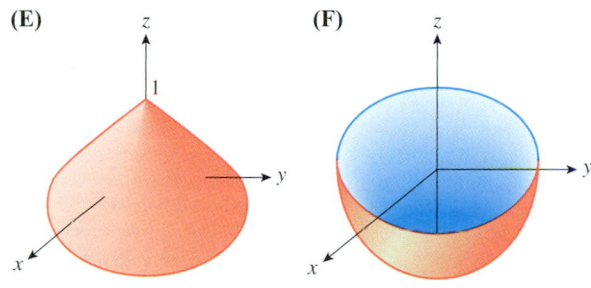

(G) **(H)**

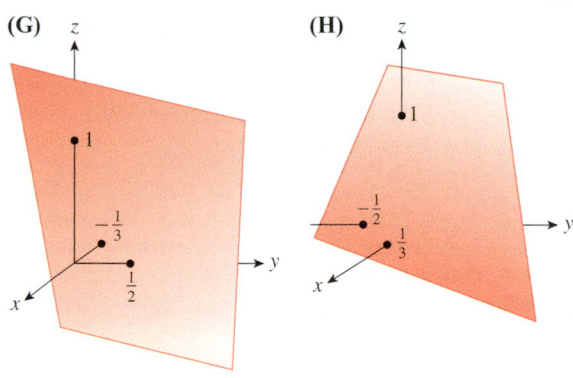

Sketch the graphs of the functions in Exercises 19–40. hint [see Example 4]

19. ● $f(x, y) = 1 - x - y$

20. ● $f(x, y) = x + y - 2$

21. ● $g(x, y) = 2x + y - 2$

22. ● $g(x, y) = 3 - x + 2y$

23. ● $h(x, y) = x + 2$

24. ● $h(x, y) = 3 - y$

25. ● $r(x, y) = x + y$

26. ● $r(x, y) = x - y$

tech Ex *Use of technology is suggested in Exercises 27–40.*
hint [see Example 3]

27. ● $s(x, y) = 2x^2 + 2y^2$. Show cross sections at $z = 1$ and $z = 2$.

28. ● $s(x, y) = -(x^2 + y^2)$. Show cross sections at $z = -1$ and $z = -2$.

29. ● $t(x, y) = x^2 + 2y^2$. Show cross sections at $x = 0$ and $z = 1$.

30. ● $t(x, y) = \frac{1}{2}x^2 + y^2$. Show cross sections at $x = 0$ and $z = 1$.

31. ● $f(x, y) = 2 + \sqrt{x^2 + y^2}$. Show cross sections at $z = 3$ and $y = 0$.

32. ● $f(x, y) = 2 - \sqrt{x^2 + y^2}$. Show cross sections at $z = 0$ and $y = 0$.

33. ● $f(x, y) = -2\sqrt{x^2 + y^2}$. Show cross sections at $z = -4$ and $y = 1$.

34. ● $f(x, y) = 2 + 2\sqrt{x^2 + y^2}$. Show cross sections at $z = 4$ and $y = 1$.

35. ● $f(x, y) = y^2$

36. ● $g(x, y) = x^2$

37. ● $h(x, y) = \frac{1}{y}$

38. ● $k(x, y) = e^y$

39. ● $f(x, y) = e^{-(x^2+y^2)}$

40. ● $g(x, y) = \frac{1}{\sqrt{x^2 + y^2}}$

Applications

41. ● *Marginal Cost (Linear Model)* Your weekly cost (in dollars) to manufacture x cars and y trucks is

$$C(x, y) = 240{,}000 + 6000x + 4000y$$

   a. Describe the graph of the cost function C.
   b. Describe the slice $x = 10$. What cost function does this slice describe?
   c. Describe the level curve $z = 480{,}000$. What does this curve tell you about costs?

42. ● *Marginal Cost (Linear Model)* Your weekly cost (in dollars) to manufacture x bicycles and y tricycles is

$$C(x, y) = 24{,}000 + 60x + 20y$$

   a. Describe the graph of the cost function C.
   b. Describe the slice by $y = 100$. What cost function does this slice describe?
   c. Describe the level curve $z = 72{,}000$. What does this curve tell you about costs?

43. ● *Market Share (Cars and Light Trucks)* Based on data in the 1980s and 1990s, the relationship between the domestic market shares of three major U.S. manufacturers of cars and light trucks could be modeled by

$$x_3 = 0.66 - 2.2x_1 - 0.02x_2$$

where x_1, x_2, and x_3 are, respectively, the fractions of the market held by Chrysler, Ford, and General Motors.[14] Thinking of General Motors' market share as a function of the shares of the other two manufacturers, describe the graph of the resulting function. How are the different slices by $x_1 = constant$ related to one another? What does this say about market share?

44. ● *Market Share (Cereals)* Based on data in the 1980s and 1990s, the relationship among the domestic market shares of three major manufacturers of breakfast cereal is

$$x_1 = -0.4 + 1.2x_2 + 2x_3$$

where x_1, x_2, and x_3 are, respectively, the fractions of the market held by Kellogg, General Mills, and General Foods.[15] Thinking of Kellogg's market share as a function of shares of the other two manufacturers, describe the graph of the resulting function. How are the different slices by $x_2 = constant$ related to one another? What does this say about market share?

45. ● *Marginal Cost (Interaction Model)* Your weekly cost (in dollars) to manufacture x cars and y trucks is

$$C(x, y) = 240{,}000 + 6000x + 4000y - 20xy$$

(Compare with Exercise 41.)

   a. Describe the slices $x = $ constant and $y = $ constant.
   b. Is the graph of the cost function a plane? How does your answer relate to part (a)?
   c. What are the slopes of the slices $x = 10$ and $x = 20$? What does this say about cost?

46. ● *Marginal Cost (Interaction Model)* Repeat the preceding exercise using the weekly cost to manufacture x bicycles and y tricycles given by

$$C(x, y) = 24{,}000 + 60x + 20y + 0.3xy$$

(Compare with Exercise 42.)

47. ● *Housing Costs*[16] The cost C of building a house is related to the number k of carpenters used and the number e of electricians used by

$$C(k, e) = 15{,}000 + 50k^2 + 50e^2$$

Describe the level curves $C = 30{,}000$ and $C = 40{,}000$. What do these level curves represent?

48. ● *Housing Costs*[17] The cost C of building a house (in a different area from that in the previous exercise) is related to the

---

[14] The model is based on a linear regression. Source of data: Ward's AutoInfoBank/*The New York Times*, July 29, 1998, p. D6.

[15] The models are based on a linear regression. Source of data: Bloomberg Financial Markets/*The New York Times*, November 28, 1998, p. C1.

[16] Based on an exercise in *Introduction to Mathematical Economics* by A. L. Ostrosky Jr. and J. V. Koch (Waveland Press, Illinois, 1979).

[17] Ibid.

number k of carpenters used and the number e of electricians used by

$$C(k, e) = 15,000 + 70k^2 + 40e^2$$

Describe the slices by the planes $k = 2$ and $e = 2$. What do these slices represent?

49. **Area** The area of a rectangle of height h and width w is $A(h, w) = hw$. Sketch a few level curves of A. If the perimeter $h + w$ of the rectangle is constant, which h and w give the largest area? (We suggest you draw in the line $h + w = c$ for several values of c.)

50. **Area** The area of an ellipse with semiminor axis a and semimajor axis b is $A(a, b) = \pi ab$. Sketch the graph of A. If $a^2 + b^2$ is constant, what a and b give the largest area?

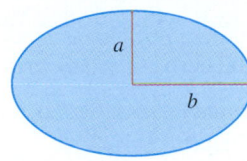

Graphing technology is suggested for Exercises 51–54.

51. **tech Ex** **Production (Cobb-Douglas Model)** Graph the level curves at $z = 0, 1, 2$ and 3 of $P(x, y) = Kx^a y^{1-a}$ if $K = 1$ and $a = 0.5$. Here, x is the number of workers, y is the operating budget, and $P(x, y)$ is the productivity. Interpret the level curve at $z = 3$.

52. **tech Ex** **Production (Cobb-Douglas Model)** Graph the level curves at $z = 0, 1, 2$ and 3 of $P(x, y) = Kx^a y^{1-a}$ if $K = 1$ and $a = 0.25$. Here, x is the number of workers, y is the operating budget, and $P(x, y)$ is the productivity. Interpret the level curve at $z = 0$.

53. **tech Ex** **Utility** Suppose that your newspaper is trying to decide between two competing desktop publishing software packages, Macro Publish and Turbo Publish. You estimate that if you purchase x copies of Macro Publish and y copies of Turbo Publish, your company's daily productivity will be

$$U(x, y) = 6x^{0.8}y^{0.2} + x$$

where $U(x, y)$ is measured in pages per day (U is called a *utility function*). Graph the level curves at $z = 0, 10, 20$, and 30. What does the level curve at $z = 0$ tell you?

54. **tech Ex** **Utility** Suppose that your small publishing company is trying to decide between two competing desktop publishing software packages, Macro Publish and Turbo Publish. You estimate that if you purchase x copies of Macro Publish and y copies of Turbo Publish, your company's daily productivity will be given by

$$U(x, y) = 5x^{0.2}y^{0.8} + x$$

where $U(x, y)$ is measured in pages per day. Graph the level curves at $z = 0, 10, 20$ and 30. Give a formula for the level curve at $z = 30$ specifying y as a function of x. What does this curve tell you?

Communication and Reasoning Exercises

55. ● Complete the following: The graph of a linear function of two variables is a _____ .

56. ● Complete the following: The level curves of a linear function of two variables are _____ .

57. ● Your study partner Slim claims that because the surface $z = f(x, y)$ you have been studying is a plane, it follows that all the slices $x = constant$ and $y = constant$ are straight lines. Do you agree or disagree? Explain.

58. ● Your other study partner Shady just told you that the surface $z = xy$ you have been trying to graph must be a plane because you've already found that the slices $x = constant$ and $y = constant$ are all straight lines. Do you agree or disagree? Explain.

59. Why do we not sketch the graphs of functions of three or more variables?

60. The surface of a mountain can be thought of as the graph of what function?

61. Show that the distance between the points (x, y, z) and (a, b, c) is given by the following **three-dimensional distance formula:**

$$d = \sqrt{(x - a)^2 + (y - b)^2 + (z - c)^2}$$

or

$$d = \sqrt{(\Delta x)^2 + (\Delta y)^2 + (\Delta z)^2}$$

The following diagram should be of assistance:

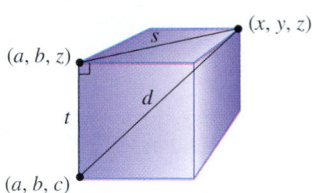

62. Use the result of the preceding exercise to show that the sphere of radius r centered at the origin has equation

$$x^2 + y^2 + z^2 = r^2$$

63. Why is three-dimensional space used to represent the graph of a function of two variables?

64. Why is it that we can sketch the graphs of functions of two variables on the two-dimensional flat surfaces of these pages?

15.3 Partial Derivatives

Recall that if f is a function of x, then the derivative df/dx measures how fast f changes as x increases. If f is a function of two or more variables, we can ask how fast f changes as each variable increases while the others remain fixed. These rates of change are called the "partial derivatives of f," and they measure how each variable contributes to the change in f. Here is a more precise definition.

Partial Derivatives

The **partial derivative of f with respect to x** is the derivative of f with respect to x, when all other variables are treated as constant. Similarly, the **partial derivative of f with respect to y** is the derivative of f with respect to y, with all other variables treated as constant, and so on for other variables. The partial derivatives are written as $\dfrac{\partial f}{\partial x}, \dfrac{\partial f}{\partial y}$, and so on. The symbol ∂ is used (instead of d) to remind us that there is more than one variable and that we are holding the other variables fixed.

quick Examples

1. Let $f(x, y) = x^2 + y^2$.

$$\frac{\partial f}{\partial x} = 2x + 0 = 2x \qquad \text{Because } y^2 \text{ is treated as a constant}$$

$$\frac{\partial f}{\partial y} = 0 + 2y = 2y \qquad \text{Because } x^2 \text{ is treated as a constant}$$

2. Let $z = x^2 + xy$.

$$\frac{\partial z}{\partial x} = 2x + y \qquad \frac{\partial}{\partial x}(xy) = \frac{\partial}{\partial x}(x \cdot \text{constant}) = \text{constant} = y$$

$$\frac{\partial z}{\partial y} = 0 + x = x \qquad \frac{\partial}{\partial y}(xy) = \frac{\partial}{\partial x}(\text{constant} \cdot y) = \text{constant} = x$$

3. Let $f(x, y) = x^2 y + y^2 x - xy + y$.

$$\frac{\partial f}{\partial x} = 2xy + y^2 - y \qquad y \text{ is treated as a constant}$$

$$\frac{\partial f}{\partial y} = x^2 + 2xy - x + 1 \qquad x \text{ is treated as a constant}$$

Interpretation

$\dfrac{\partial f}{\partial x}$ is the rate at which f changes as x changes, for a fixed (constant) y.

$\dfrac{\partial f}{\partial y}$ is the rate at which f changes as y changes, for a fixed (constant) x.

Example 1 Marginal Cost: Linear Model

We return to Example 1 from Section 15.1. Suppose that you own a company that makes two models of speakers, the Ultra Mini and the Big Stack. Your total monthly cost (in dollars) to make x Ultra Minis and y Big Stacks is given by

$$C(x, y) = 10{,}000 + 20x + 40y$$

What is the significance of $\dfrac{\partial C}{\partial x}$ and of $\dfrac{\partial C}{\partial y}$?

Solution First we compute these partial derivatives:

$$\frac{\partial C}{\partial x} = 20$$

$$\frac{\partial C}{\partial y} = 40$$

We interpret the results as follows: $\dfrac{\partial C}{\partial x} = 20$ means that the cost is increasing at a rate of \$20 per additional Ultra Mini (if production of Big Stacks is held constant); $\dfrac{\partial C}{\partial y} = 40$ means that the cost is increasing at a rate of \$40 per additional Big Stack (if production of Ultra Minis is held constant). In other words, these are the **marginal costs** of each model of speaker.

+ *Before we go on...* How much does the cost rise if you increase x by Δx and y by Δy? In Example 1, the change in cost is given by

$$\Delta C = 20\Delta x + 40\Delta y = \frac{\partial C}{\partial x}\Delta x + \frac{\partial C}{\partial y}\Delta y$$

This suggests the **chain rule for several variables.** Part of this rule says that if x and y are both functions of t, then C is a function of t through them, and the rate of change of C with respect to t can be calculated as

$$\frac{dC}{dt} = \frac{\partial C}{\partial x}\cdot\frac{dx}{dt} + \frac{\partial C}{\partial y}\cdot\frac{dy}{dt}$$

We shall not have a chance to use this interesting result in this book. ▪

Example 2 Marginal Cost: Interaction Model

Another possibility for the cost function in the preceding example is the interaction model

$$C(x, y) = 10{,}000 + 20x + 40y + 0.1xy$$

a. Now what are the marginal costs of the two models of speakers?

b. What is the marginal cost of manufacturing Big Stacks at a production level of 100 Ultra Minis and 50 Big Stacks per month?

Solution

a. We compute the partial derivatives:

$$\frac{\partial C}{\partial x} = 20 + 0.1y$$

$$\frac{\partial C}{\partial y} = 40 + 0.1x$$

Thus, the marginal cost of manufacturing Ultra Minis increases by \$0.1 or 10¢ for each Big Stack that is manufactured. Similarly, the marginal cost of manufacturing Big Stacks increases by 10¢ for each Ultra Mini that is manufactured.

b. From part (a), the marginal cost of manufacturing Big Stacks is

$$\frac{\partial C}{\partial y} = 40 + 0.1x$$

At a production level of 100 Ultra Minis and 50 Big Stacks per month, we have $x = 100$ and $y = 50$. Thus, the marginal cost of manufacturing Big Stacks at these production levels is

$$\left.\frac{\partial C}{\partial y}\right|_{(100,50)} = 40 + 0.1(100) = \$50 \text{ per Big Stack}$$

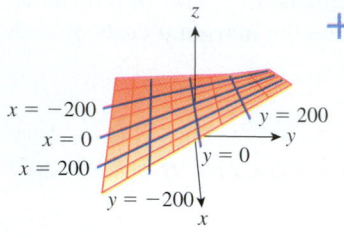

Figure **17**

+ *Before we go on...* A portion of the surface $z = 10{,}000 + 20x + 40y + 0.1xy$ from Example 2 is shown in Figure 17, together with some of the slices through $x = constant$ and $y = constant$.

These slices are straight lines whose slopes are given by the partial derivatives (which represent the marginal costs when x and y are nonnegative):

$$y = -200: \quad \text{Slope} = \left.\frac{\partial z}{\partial x}\right|_{(x,-200)} = 20 + 0.1(-200) = 0$$

$$y = 0: \quad \text{Slope} = \left.\frac{\partial z}{\partial x}\right|_{(x,0)} = 20 + 0.1(0) = 20$$

$$y = 200: \quad \text{Slope} = \left.\frac{\partial z}{\partial x}\right|_{(x,200)} = 20 + 0.1(200) = 40$$

Notice that these slopes increase as y increases, as we can confirm in Figure 17.

$$x = -200 \quad \text{Slope} = \left.\frac{\partial z}{\partial y}\right|_{(-200,y)} = 40 + 0.1(-200) = 20$$

$$x = 0 \quad \text{Slope} = \left.\frac{\partial z}{\partial y}\right|_{(0,y)} = 40 + 0.1(0) = 40$$

$$x = 200 \quad \text{Slope} = \left.\frac{\partial z}{\partial y}\right|_{(200,y)} = 40 + 0.1(200) = 60$$

Notice that these slopes increase as x increases, as we can confirm in Figure 17. ∎

Partial derivatives of functions of three variables are obtained in the same way as those for functions of two variables, as the following example shows:

Example 3 Function of Three Variables

Calculate $\dfrac{\partial f}{\partial x}, \dfrac{\partial f}{\partial y}$ and $\dfrac{\partial f}{\partial z}$ if $f(x, y, z) = xy^2z^3 - xy$.

Solution Although we now have three variables, the calculation remains the same: $\partial f/\partial x$ is the derivative of f with respect to x, with *both* other variables, y and z, held constant:

$$\frac{\partial f}{\partial x} = y^2 z^3 - y$$

Similarly, $\partial f/\partial y$ is the derivative of f with respect to y, with both x and z held constant:

$$\frac{\partial f}{\partial y} = 2xyz^3 - x$$

Finally, to find $\partial f/\partial z$, we hold both x and y constant and take the derivative with respect to z.

$$\frac{\partial f}{\partial z} = 3xy^2 z^2$$

Note The procedure for finding a partial derivative is the same for any number of variables: To get the partial derivative with respect to any one variable, we treat all the others as constants. ■

Geometric Interpretation of Partial Derivatives

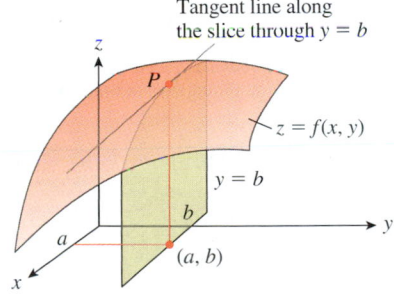

Tangent line along
the slice through $y = b$

$z = f(x, y)$

$y = b$

(a, b)

$\left. \dfrac{\partial f}{\partial x} \right|_{(a,b)}$ is the slope of the tangent line
at the point $P(a, b, f(a, b))$
along the slice through $y = b$.

Figure **18**

Recall that if f is a function of one variable x, then the derivative df/dx gives the slopes of the tangent lines to its graph. Now, suppose that f is a function of x and y. By definition, $\partial f/\partial x$ is the derivative of the function of x we get by holding y fixed. If we evaluate this derivative at the point (a, b), we are holding y fixed at the value b, taking the ordinary derivative of the resulting function of x, and evaluating this at $x = a$. Now, holding y fixed at b amounts to slicing through the graph of f along the plane $y = b$, resulting in a curve. Thus, the partial derivative is the slope of the tangent line to this curve at the point where $x = a$ and $y = b$, along the plane $y = b$ (Figure 18).

This fits with our interpretation of $\partial f/\partial x$ as the rate of increase of f with increasing x when y is held fixed at b.

The other partial derivative, $\partial f/\partial y|_{(a, b)}$ is, similarly, the slope of the tangent line at the same point $P(a, b, f(a, b))$ but along the slice by the plane $x = a$. You should draw the corresponding picture for this on your own.

Second-Order Partial Derivatives

Just as for functions of a single variable, we can calculate second derivatives. Suppose, for example, that we have a function of x and y, say, $f(x, y) = x^2 - x^2 y^2$. We know that

$$\frac{\partial f}{\partial x} = 2x - 2xy^2$$

If we take the partial derivative with respect to x once again, we obtain

$$\frac{\partial}{\partial x}\left(\frac{\partial f}{\partial x}\right) = 2 - 2y^2 \qquad \text{Take } \frac{\partial}{\partial x} \text{ of } \frac{\partial f}{\partial x}$$

(The symbol $\partial/\partial x$ means "the partial derivative with respect to x," just as d/dx stands for "the derivative with respect to x.") This is called the **second-order partial derivative** and is written $\dfrac{\partial^2 f}{\partial x^2}$. We get the following derivatives similarly:

$$\frac{\partial f}{\partial y} = -2x^2 y$$

$$\frac{\partial^2 f}{\partial y^2} = -2x^2 \qquad\qquad \text{Take } \frac{\partial}{\partial y} \text{ of } \frac{\partial f}{\partial y}$$

Now what if we instead take the partial derivative with respect to y of $\partial f/\partial x$?

$$\frac{\partial^2 f}{\partial y \partial x} = \frac{\partial}{\partial y}\left(\frac{\partial f}{\partial x}\right) \qquad\qquad \text{Take } \frac{\partial}{\partial y} \text{ of } \frac{\partial f}{\partial x}$$

$$= \frac{\partial}{\partial y}[2x - 2xy^2] = -4xy$$

Here, $\dfrac{\partial^2 f}{\partial y \partial x}$ means "first take the partial derivative with respect to x and then with respect to y," and is called a **mixed partial derivative.** If we differentiate in the opposite order, we get

$$\frac{\partial^2 f}{\partial x \partial y} = \frac{\partial}{\partial x}\left(\frac{\partial f}{\partial y}\right) = \frac{\partial}{\partial x}[-2x^2 y] = -4xy$$

the same expression as $\dfrac{\partial^2 f}{\partial y \partial x}$. This is no coincidence: The mixed partial derivatives $\dfrac{\partial^2 f}{\partial x \partial y}$ and $\dfrac{\partial^2 f}{\partial y \partial x}$ are always the same as long as the first partial derivatives are both differentiable functions of x and y and the mixed partial derivatives are continuous. Because all the functions we shall use are of this type, we can take the derivatives in any order we like when calculating mixed derivatives.

Here is another notation for partial derivatives that is especially convenient for second-order partial derivatives:

$$f_x \text{ means } \frac{\partial f}{\partial x}$$

$$f_y \text{ means } \frac{\partial f}{\partial y}$$

$$f_{xy} \text{ means } (f_x)_y = \frac{\partial^2 f}{\partial y \partial x} \quad \text{(Note the order in which the derivatives are taken.)}$$

$$f_{yx} \text{ means } (f_y)_x = \frac{\partial^2 f}{\partial x \partial y}$$

15.3 EXERCISES

● denotes basic skills exercises

In Exercises 1–18, calculate $\dfrac{\partial f}{\partial x}, \dfrac{\partial f}{\partial y}, \dfrac{\partial f}{\partial x}\Big|_{(1,-1)}$, *and* $\dfrac{\partial f}{\partial y}\Big|_{(1,-1)}$

when defined. hint [see Quick Examples p. 1064]

1. ● $f(x, y) = 10{,}000 - 40x + 20y$

2. ● $f(x, y) = 1000 + 5x - 4y$

3. ● $f(x, y) = 3x^2 - y^3 + x - 1$

4. ● $f(x, y) = x^{1/2} - 2y^4 + y + 6$

5. ● $f(x, y) = 10{,}000 - 40x + 20y + 10xy$

6. ● $f(x, y) = 1000 + 5x - 4y - 3xy$

7. ● $f(x, y) = 3x^2 y$

8. ● $f(x, y) = x^4 y^2 - x$

● basic skills

9. ● $f(x, y) = x^2 y^3 - x^3 y^2 - xy$

10. ● $f(x, y) = x^{-1} y^2 + xy^2 + xy$

11. ● $f(x, y) = (2xy + 1)^3$ 12. ● $f(x, y) = \dfrac{1}{(xy + 1)^2}$

13. $f(x, y) = e^{x+y}$ 14. $f(x, y) = e^{2x+y}$

15. $f(x, y) = 5x^{0.6} y^{0.4}$ 16. $f(x, y) = -2x^{0.1} y^{0.9}$

17. $f(x, y) = e^{0.2xy}$ 18. $f(x, y) = xe^{xy}$

In Exercises 19–28, find $\dfrac{\partial^2 f}{\partial x^2}, \dfrac{\partial^2 f}{\partial y^2}, \dfrac{\partial^2 f}{\partial x\, \partial y}$, *and* $\dfrac{\partial^2 f}{\partial y\, \partial x}$, *and evaluate them all at* (1, −1) *if possible.* hint [see Discussion on pp. 1067–1068]

19. ● $f(x, y) = 10{,}000 - 40x + 20y$

20. ● $f(x, y) = 1000 + 5x - 4y$

21. ● $f(x, y) = 10{,}000 - 40x + 20y + 10xy$

22. ● $f(x, y) = 1000 + 5x - 4y - 3xy$

23. ● $f(x, y) = 3x^2 y$ 24. ● $f(x, y) = x^4 y^2 - x$

25. $f(x, y) = e^{x+y}$ 26. $f(x, y) = e^{2x+y}$

27. $f(x, y) = 5x^{0.6} y^{0.4}$ 28. $f(x, y) = -2x^{0.1} y^{0.9}$

In Exercises 29–40, find $\dfrac{\partial f}{\partial x}, \dfrac{\partial f}{\partial y}, \dfrac{\partial f}{\partial z}$, *and their values at* (0, −1, 1) *if possible.* hint [see Example 3]

29. ● $f(x, y, z) = xyz$

30. ● $f(x, y, z) = xy + xz - yz$

31. $f(x, y, z) = -\dfrac{4}{x + y + z^2}$

32. $f(x, y, z) = \dfrac{6}{x^2 + y^2 + z^2}$

33. $f(x, y, z) = xe^{yz} + ye^{xz}$

34. $f(x, y, z) = xye^z + xe^{yz} + e^{xyz}$

35. $f(x, y, z) = x^{0.1} y^{0.4} z^{0.5}$

36. $f(x, y, z) = 2x^{0.2} y^{0.8} + z^2$

37. $f(x, y, z) = e^{xyz}$ 38. $f(x, y, z) = \ln(x + y + z)$

39. $f(x, y, z) = \dfrac{2000z}{1 + y^{0.3}}$ 40. $f(x, y, z) = \dfrac{e^{0.2x}}{1 + e^{-0.1y}}$

Applications

41. ● *Marginal Cost (Linear Model)* Your weekly cost (in dollars) to manufacture x cars and y trucks is

$$C(x, y) = 240{,}000 + 6000x + 4000y$$

Calculate and interpret $\dfrac{\partial C}{\partial x}$ and $\dfrac{\partial C}{\partial y}$. hint [see Example 1]

42. ● *Marginal Cost (Linear Model)* Your weekly cost (in dollars) to manufacture x bicycles and y tricycles is

$$C(x, y) = 24{,}000 + 60x + 20y$$

Calculate and interpret $\dfrac{\partial C}{\partial x}$ and $\dfrac{\partial C}{\partial y}$.

43. ● *Scientific Research* In each year from 1983 to 2003, the percentage y of research articles in *Physical Review* written by researchers in the U.S. can be approximated by

$$y = 82 - 0.78t - 1.02x \text{ percentage points} \quad (0 \le t \le 20)$$

where t is the year since 1983 and x is the percentage of articles written by researchers in Europe.[18] Calculate and interpret $\dfrac{\partial y}{\partial t}$ and $\dfrac{\partial y}{\partial x}$.

44. ● *Scientific Research* The number z of research articles in *Physical Review* that were written by researchers in the U.S. from 1993 through 2003 can be approximated by

$$z = 5960 - 0.71x + 0.50y \quad (3000 \le x, y \le 6000)$$

articles each year, where x is the number of articles written by researchers in Europe and y is the number written by researchers in other countries (excluding Europe and the U.S.).[19] Calculate and interpret $\dfrac{\partial z}{\partial x}$ and $\dfrac{\partial z}{\partial y}$.

45. ● *Marginal Cost (Interaction Model)* Your weekly cost (in dollars) to manufacture x cars and y trucks is

$$C(x, y) = 240{,}000 + 6000x + 4000y - 20xy$$

(Compare with Exercise 41.) Compute the marginal cost of manufacturing cars at a production level of 10 cars and 20 trucks. hint [see Example 2]

46. ● *Marginal Cost (Interaction Model)* Your weekly cost (in dollars) to manufacture x bicycles and y tricycles is

$$C(x, y) = 24{,}000 + 60x + 20y + 0.3xy$$

(Compare with Exercise 42.) Compute the marginal cost of manufacturing tricycles at a production level of 10 bicycles and 20 tricycles.

47. ● *Brand Loyalty* The fraction of Mazda car owners who chose another new Mazda can be modeled by the following function:[20]

$$M(c, f, g, h, t) = 1.1 - 3.8c + 2.2f + 1.9g - 1.7h - 1.3t$$

Here, c is the fraction of Chrysler car owners who remained loyal to Chrysler, f is the fraction of Ford car owners remaining loyal to Ford, g the corresponding figure for General Motors, h the corresponding figure for Honda, and t for Toyota.

a. Calculate $\dfrac{\partial M}{\partial c}$ and $\dfrac{\partial M}{\partial f}$ and interpret the answers.

[18] SOURCE: The American Physical Society/*New York Times*, May 3, 2003, p. A1.

[19] Ibid.

[20] The model is an approximation of a linear regression based on data from the period 1988–1995. Source for data: Chrysler, Maritz Market Research, Consumer Attitude Research, and Strategic Vision/*The New York Times*, November 3, 1995, p. D2.

● basic skills

b. In 1995 it was observed that $c = 0.56$, $f = 0.56$, $g = 0.72$, $h = 0.50$, and $t = 0.43$. According to the model, what percentage of Mazda owners remained loyal to Mazda? (Round your answer to the nearest percentage point.)

48. ● **Brand Loyalty** The fraction of Mazda car owners who chose another new Mazda can be modeled by the following function:[21]

$$M(c, f) = 9.4 + 7.8c + 3.6c^2 - 38f - 22cf + 43f^2$$

where c is the fraction of Chrysler car owners who remained loyal to Chrysler and f is the fraction of Ford car owners remaining loyal to Ford.

a. Calculate $\dfrac{\partial M}{\partial c}$ and $\dfrac{\partial M}{\partial f}$ evaluated at the point $(0.7, 0.7)$, and interpret the answers.

b. In 1995, it was observed that $c = 0.56$, and $f = 0.56$. According to the model, what percentage of Mazda owners remained loyal to Mazda? (Round your answer to the nearest percentage point.)

49. **Family Income** The following model is based on statistical data on the median family incomes of black and white families in the U.S. for the period 1950–2000:[22]

$$z = 13{,}000 + 350t + 9900x + 220xt$$

where

$z = $ median family income

$t = $ year ($t = 0$ represents 1950)

$x = \begin{cases} 0 & \text{if the income was for a black family} \\ 1 & \text{if the income was for a white family} \end{cases}$

a. Use the model to estimate the median income of a black family in 1960.

b. Use the model to estimate the median income of a white family in 1960.

c. According to the model, how fast was the median income for a black family increasing in 1960?

d. According to the model, how fast was the median income for a white family increasing in 1960?

e. Do the answers to parts (c) and (d) suggest that the income gap between white and black families was widening or narrowing in the second half of the twentieth century?

50. **Life Expectancy** The following model is based on life expectancy for men and women in the U.S. for the period 1900–2000:[23]

$$z = 50.9 + 0.325t - 1.95x - 0.055xt$$

where:

$z = $ life expectancy of a person born in the U.S.

$t = $ year of birth ($t = 0$ represents 1900)

$x = \begin{cases} 0 & \text{if the person was a female} \\ 1 & \text{if the person was a male} \end{cases}$

a. Use the model to estimate, to the nearest year, the life expectancy of a female born in 1950.

b. Use the model to estimate, to the nearest year, the life expectancy of a male born in 1950.

c. According to the model, how fast was the life expectancy for a female increasing in 1950?

d. According to the model, how fast was the life expectancy for a male increasing in 1950?

e. Do the answers you have given suggest that the gap between male and female life expectancy was widening or narrowing in the twentieth century?

51. **Marginal Cost** Your weekly cost (in dollars) to manufacture x cars and y trucks is

$$C(x, y) = 200{,}000 + 6000x$$
$$+ 4000y - 100{,}000e^{-0.01(x+y)}$$

What is the marginal cost of a car? Of a truck? How do these marginal costs behave as total production increases?

52. **Marginal Cost** Your weekly cost (in dollars) to manufacture x bicycles and y tricycles is

$$C(x, y) = 20{,}000 + 60x + 20y + 50\sqrt{xy}$$

What is the marginal cost of a bicycle? Of a tricycle? How do these marginal costs behave as x and y increase?

53. **Average Cost** If you average your costs over your total production, you get the **average cost,** written $\bar{C}$:

$$\bar{C}(x, y) = \frac{C(x, y)}{x + y}$$

Find the average cost for the cost function in Exercise 51. Then find the marginal average cost of a car and the marginal average cost of a truck at a production level of 50 cars and 50 trucks. Interpret your answers.

54. **Average Cost** Find the average cost for the cost function in Exercise 52 (see the preceding exercise). Then find the marginal average cost of a bicycle and the marginal average cost of a tricycle at a production level of 5 bicycles and 5 tricycles. Interpret your answers.

---

[21] The model is an approximation of a second-order regression based on data from the period 1988–1995. Source for data: Ibid.

[22] Incomes are in constant 1997 dollars. The model is a multiple regression model and coefficients are rounded to two significant digits. Source for data: Statistical Abstract of the United States, 1999, U.S. Census Bureau/ *The New York Times,* December 19, 1999, p. WK5.

---

[23] The model is a multiple regression model and coefficients are rounded to 3 significant digits. Source for data: Ibid.

● basic skills

55. *Marginal Revenue* As manager of an auto dealership, you offer a car rental company the following deal: You will charge $15,000 per car and $10,000 per truck, but you will then give the company a discount of $5000 times the square root of the total number of vehicles it buys from you. Looking at your marginal revenue, is this a good deal for the rental company?

56. *Marginal Revenue* As marketing director for a bicycle manufacturer, you come up with the following scheme: You will offer to sell a dealer x bicycles and y tricycles for

$$R(x, y) = 3500 - 3500e^{-0.02x - 0.01y} \text{dollars}$$

Find your marginal revenue for bicycles and for tricycles. Are you likely to be fired for your suggestion?

57. *Research Productivity* Here we apply a variant of the Cobb-Douglas function to the modeling of research productivity. A mathematical model of research productivity at a particular physics laboratory is

$$P = 0.04x^{0.4}y^{0.2}z^{0.4}$$

where P is the annual number of groundbreaking research papers produced by the staff, x is the number of physicists on the research team, y is the laboratory's annual research budget, and z is the annual National Science Foundation subsidy to the laboratory. Find the rate of increase of research papers per government-subsidy dollar at a subsidy level of $1,000,000 per year and a staff level of 10 physicists if the annual budget is $100,000.

58. *Research Productivity* A major drug company estimates that the annual number P of patents for new drugs developed by its research team is best modeled by the formula

$$P = 0.3x^{0.3}y^{0.4}z^{0.3}$$

where x is the number of research biochemists on the payroll, y is the annual research budget, and z is the size of the bonus awarded to discoverers of new drugs. Assuming that the company has 12 biochemists on the staff, has an annual research budget of $500,000 and pays $40,000 bonuses to developers of new drugs, calculate the rate of growth in the annual number of patents per new research staff member.

59. *Utility* Your newspaper is trying to decide between two competing desktop publishing software packages, Macro Publish and Turbo Publish. You estimate that if you purchase x copies of Macro Publish and y copies of Turbo Publish, your company's daily productivity will be

$$U(x, y) = 6x^{0.8}y^{0.2} + x$$

$U(x, y)$ is measured in pages per day.

a. Calculate $\dfrac{\partial U}{\partial x}\Big|_{(10, 5)}$ and $\dfrac{\partial U}{\partial y}\Big|_{(10, 5)}$ to two decimal places, and interpret the results.

b. What does the ratio $\dfrac{\partial U}{\partial x}\Big|_{(10, 5)}\Big/\dfrac{\partial U}{\partial y}\Big|_{(10, 5)}$ tell about the usefulness of these products?

60. *Grades*[24] A production formula for a student's performance on a difficult English examination is given by

$$g(t, x) = 4tx - 0.2t^2 - x^2$$

where g is the grade the student can expect to get, t is the number of hours of study for the examination, and x is the student's grade point average.

a. Calculate $\dfrac{\partial g}{\partial t}\Big|_{(10, 3)}$ and $\dfrac{\partial g}{\partial x}\Big|_{(10, 3)}$ and interpret the results.

b. What does the ratio $\dfrac{\partial g}{\partial t}\Big|_{(10, 3)}\Big/\dfrac{\partial g}{\partial x}\Big|_{(10, 3)}$ tell about the relative merits of study and grade point average?

61. *Electrostatic Repulsion* If positive electric charges of Q and q coulombs are situated at positions (a, b, c) and (x, y, z) respectively, then the force of repulsion they experience is given by

$$F = K\frac{Qq}{(x - a)^2 + (y - b)^2 + (z - c)^2}$$

where $K \approx 9 \times 10^9$, F is given in newtons, and all positions are measured in meters. Assume that a charge of 10 coulombs is situated at the origin, and that a second charge of 5 coulombs is situated at $(2, 3, 3)$ and moving in the y-direction at one meter per second. How fast is the electrostatic force it experiences decreasing? (Round the answer to one significant digit.)

62. *Electrostatic Repulsion* Repeat the preceding exercise, assuming that a charge of 10 coulombs is situated at the origin and that a second charge of 5 coulombs is situated at $(2, 3, 3)$ and moving in the negative z direction at one meter per second. (Round the answer to one significant digit.)

63. *Investments* Recall that the compound interest formula for annual compounding is

$$A(P, r, t) = P(1 + r)^t$$

where A is the future value of an investment of P dollars after t years at an interest rate of r.

a. Calculate $\dfrac{\partial A}{\partial P}, \dfrac{\partial A}{\partial r}$, and $\dfrac{\partial A}{\partial t}$, all evaluated at $(100, 0.10, 10)$. (Round your answers to two decimal places.) Interpret your answers.

_____

[24] Based on an exercise in *Introduction to Mathematical Economics* by A. L. Ostrosky Jr. and J. V. Koch (Waveland Press, Illinois, 1979).

● basic skills

b. What does the function $\left.\dfrac{\partial A}{\partial P}\right|_{(100,\,0.10,\,t)}$ of t tell about your investment?

64. *Investments* Repeat the preceding exercise, using the formula for continuous compounding:

$$A(P, r, t) = Pe^{rt}$$

65. *Modeling with the Cobb-Douglas Production Formula* Assume you are given a production formula of the form

$$P(x, y) = Kx^a y^b \quad (a + b = 1)$$

a. Obtain formulas for $\dfrac{\partial P}{\partial x}$ and $\dfrac{\partial P}{\partial y}$, and show that $\dfrac{\partial P}{\partial x} = \dfrac{\partial P}{\partial y}$ precisely when $x/y = a/b$.

b. Let x be the number of workers a firm employs and let y be its monthly operating budget in thousands of dollars. Assume that the firm currently employs 100 workers and has a monthly operating budget of $200,000. If each additional worker contributes as much to productivity as each additional $1000 per month, find values of a and b that model the firm's productivity.

66. *Housing Costs*[25] The cost C of building a house is related to the number k of carpenters used and the number e of electricians used by

$$C(k, e) = 15{,}000 + 50k^2 + 60e^2$$

If three electricians are currently employed in building your new house and the marginal cost per additional electrician is the same as the marginal cost per additional carpenter, how many carpenters are being used? (Round your answer to the nearest carpenter.)

67. *Nutrient Diffusion* Suppose that one cubic centimeter of nutrient is placed at the center of a circular petri dish filled with water. We might wonder how the nutrient is distributed after a time of t seconds. According to the classical theory of diffusion, the concentration of nutrient (in parts of nutrient per part of water) after a time t is given by

$$u(r, t) = \frac{1}{4\pi Dt} e^{-\frac{r^2}{4Dt}}$$

Here D is the *diffusivity,* which we will take to be 1, and r is the distance from the center in centimeters. How fast is the concentration increasing at a distance of 1 cm from the center 3 seconds after the nutrient is introduced?

68. *Nutrient Diffusion* Refer back to the preceding exercise. How fast is the concentration increasing at a distance of 4 cm from the center 4 seconds after the nutrient is introduced?

Communication and Reasoning Exercises

69. ● Given that $f(a, b) = r$, $f_x(a, b) = s$, and $f_y(a, b) = t$, complete the following: ___ is increasing at a rate of ___ units per unit of x, ___ is increasing at a rate of ___ units per unit of y, and the value of ___ is ___ when $x = $ ___ and $y = $ ___.

70. ● A firm's productivity depends on two variables, x and y. Currently, $x = a$ and $y = b$, and the firm's productivity is 4000 units. Productivity is increasing at a rate of 400 units per unit *decrease* in x, and is decreasing at a rate of 300 units per unit increase in y. What does all of this information tell you about the firm's productivity function $g(x, y)$?

71. ● Complete the following: Let $f(x, y, z)$ be the cost to build a development of x cypods (one-bedroom units) in the city-state of Utarek, Mars, y argaats (two-bedroom units), and z orbici (singular: orbicus; three-bedroom units) in $\overline{\overline{Z}}$ (zonars, the designated currency in Utarek).[26] Then $\dfrac{\partial f}{\partial z}$ measures _____ and has units _____ .

72. ● Complete the following: Let $f(t, x, y)$ be the projected number of citizens of the Principality State of Voodice, Luna[27] in year t since its founding, assuming the presence of x lunar vehicle factories and y domed settlements. Then $\dfrac{\partial f}{\partial x}$ measures _____ and has units _____ .

73. ● Give an example of a function $f(x, y)$ with $f_x(1, 1) = -2$ and $f_y(1, 1) = 3$.

74. ● Give an example of a function $f(x, y, z)$ that has all of its partial derivatives nonzero constants.

75. The graph of $z = b + mx + ny$ ($b, m,$ and n constants) is a plane.

a. Explain the geometric significance of the numbers $b, m,$ and n.

b. Show that the equation of the plane passing through (h, k, l) with slope m in the x direction (in the sense of $\partial/\partial x$) and slope n in the y direction is

$$z = l + m(x - h) + n(y - k)$$

76. The **tangent plane** to the graph of $f(x, y)$ at $P(a, b, f(a, b))$ is the plane containing the lines tangent to the slice through the graph by $y = b$ (as in Figure 18) and the slice through the graph by $x = a$. Use the result of the preceding exercise to show that the equation of the tangent plane is

$$z = f(a, b) + f_x(a, b)(x - a) + f_y(a, b)(y - b)$$

---

[25] Based on an Exercise in *Introduction to Mathematical Economics* by A.L. Ostrosky Jr. and J.V. Koch (Waveland Press, Illinois, 1979).

[26] SOURCE: www.marsnext.com/comm/zonars.html.

[27] SOURCE: www.voodice.info.

15.4 Maxima and Minima

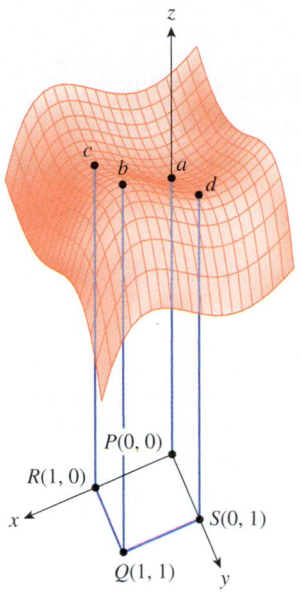

Figure **19**

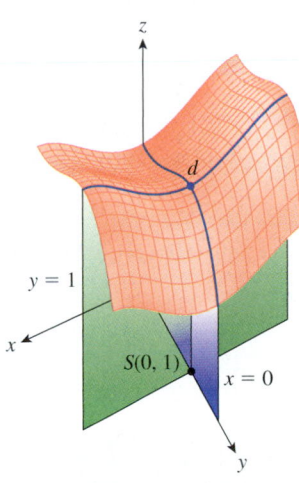

Figure **20**

In Chapter 12 on applications of the derivative, we saw how to locate relative extrema of a function of a single variable. In this section we extend our methods to functions of two variables. Similar techniques work for functions of three or more variables.

Figure 19 shows a portion of the graph of the function $f(x, y) = 2(x^2 + y^2) - (x^4 + y^4) + 3$. The graph resembles a "flying carpet," and several interesting points, marked a, b, c, and d are shown.

1. The point a has coordinates $(0, 0, f(0, 0))$, is directly above the origin $(0, 0)$, and is the lowest point in its vicinity; water would puddle there. We say that f has a **relative minimum** at $(0, 0)$ because $f(0, 0)$ is smaller than $f(x, y)$ for any (x, y) near $(0, 0)$.

2. Similarly, the point b is higher than any point in its vicinity. Thus, we say that f has a **relative maximum** at $(1, 1)$.

3. The points c and d represent a new phenomenon and are called **saddle points.** They are neither relative maxima nor relative minima but seem to be a little of both.

To see more clearly what features a saddle point has, look at Figure 20, which shows a portion of the graph near the point d.

If we slice through the graph along $y = 1$, we get a curve on which d is the *lowest* point. Thus, d looks like a relative minimum along this slice. On the other hand, if we slice through the graph along $x = 0$, we get another curve, on which d is the *highest* point, so d looks like a relative maximum along this slice. This kind of behavior characterizes a saddle point: f has a **saddle point** at (r, s) if f has a relative minimum at (r, s) along some slice through that point and a relative maximum along another slice through that point. If you look at the other saddle point, c, in Figure 19, you see the same characteristics.

While numerical information can help us locate the approximate position of relative extrema and saddle points, calculus permits us to locate these points accurately as we did for functions of a single variable. Look once again at Figure 19, and notice the following:

• The points P, Q, R and S are all in the **interior** of the domain of f; that is, none of them lies on the boundary of the domain. Said another way, we can move some distance in any direction from any of these points without leaving the domain of f.

• The tangent lines along the slices through these points parallel to the x- and y-axes are *horizontal.* Thus, the partial derivatives $\partial f/\partial x$ and $\partial f/\partial y$ are zero when evaluated at any of the points P, Q, R and S. This gives us a way of locating candidates for relative extrema and saddle points.

The following summary generalizes and also expands on some of what we have just said:

Relative and Absolute Maxima and Minima

The function $f(x, y, \ldots)$ has a **relative maximum** at $(r, s, \ldots)$ if $f(r, s, \ldots) \geq f(x, y, \ldots)$ for every point $(x, y, \ldots)$ near $(r, s, \ldots)$. in the domain of f. We say that $f(x, y, \ldots)$ has an **absolute maximum** at $(r, s, \ldots)$ if $f(r, s, \ldots) \geq f(x, y, \ldots)$ for every point $(x, y, \ldots)$ in the domain of f. The terms **relative minimum** and **absolute minimum** are defined in a similar way.

Locating Candidates for Relative Extrema and Saddle Points in the Interior of the Domain of f:

- Set $\dfrac{\partial f}{\partial x} = 0$, $\dfrac{\partial f}{\partial y} = 0$, ... simultaneously, and solve for $x, y, \ldots$.

- Check that the resulting points $(x, y, \ldots)$ are in the interior of the domain of f.

Points at which all the partial derivatives of f are zero are called **critical points.** Thus, the critical points are the only candidates for relative extrema and saddle points in the interior of the domain of f.[*]

quick Examples

1. Let $f(x, y) = x^3 + (y - 1)^2$. Then $\dfrac{\partial f}{\partial x} = 3x^2$ and $\dfrac{\partial f}{\partial y} = 2(y - 1)$. Thus, we solve the system

$$3x^2 = 0 \quad \text{and} \quad 2(y - 1) = 0$$

The first equation gives $x = 0$, and the second gives $y = 1$. Thus, the only critical point is $(0, 1)$. Because the domain of f is the whole Cartesian plane, the point $(0, 1)$ is interior, and hence a candidate for a relative extremum or saddle point.[†]

2. Let $f(x, y) = e^{-(x^2 + y^2)}$. Taking partial derivatives and setting them equal to zero gives

$$-2xe^{-(x^2 + y^2)} = 0 \qquad \text{We set } \frac{\partial f}{\partial x} = 0.$$

$$-2ye^{-(x^2 + y^2)} = 0 \qquad \text{We set } \frac{\partial f}{\partial y} = 0.$$

The first equation implies that $x = 0$,[‡] and the second implies that $y = 0$. Thus, the only critical point is $(0, 0)$. This point is interior, and hence a candidate for a relative extremum or saddle point.

---

[*] We'll be looking at extrema on the *boundary* of the domain of a function in the next section. What we are calling critical points correspond to the *stationary* points of a function of one variable. We shall not consider the analogs of the singular points.

[†] In fact, it is a saddle point. (Can you see why?)

[‡] Recall that if a product of two numbers is zero, then one or the other must be zero. In this case the number $e^{-(x^2 + y^2)}$ can't be zero (since e^u is never zero), which gives the result claimed.

In the next example we first locate all critical points, and then classify each one as a relative maximum, minimum, saddle point, or none of these.

Example 1 Locating and Classifying Critical Points

Locate all critical points of $f(x, y) = x^2 y - x^2 - 2y^2$. Graph the function to classify the critical points as relative maxima, minima, saddle points, or none of these.

Solution The partial derivatives are

$$f_x = 2xy - 2x = 2x(y - 1)$$
$$f_y = x^2 - 4y$$

Setting these equal to zero gives

$$x = 0 \quad \text{or} \quad y = 1$$
$$x^2 = 4y$$

We get a solution by choosing either $x = 0$ or $y = 1$ and substituting into $x^2 = 4y$.

Case 1: $x = 0$ Substituting into $x^2 = 4y$ gives $0 = 4y$ and hence $y = 0$. Thus, the critical point for this case is $(x, y) = (0, 0)$.

Case 2: $y = 1$ Substituting into $x^2 = 4y$ gives $x^2 = 4$ and hence $x = \pm 2$. Thus, we get two critical points for this case: $(2, 1)$ and $(-2, 1)$.

We now have three critical points altogether: $(0, 0)$, $(2, 1)$, and $(-2, 1)$. We get the corresponding points on the graph by substituting for x and y in the equation for f to get the z-coordinates. The points are $(0, 0, 0)$, $(2, 1, -2)$, and $(-2, 1, -2)$.

Classifying the Critical Points Graphically To classify the critical points graphically, we look at the graph of f shown in Figure 21.

Examining the graph carefully, we see that the point $(0, 0, 0)$ is a relative maximum. As for the other two critical points, are they saddle points or are they relative maxima? They *seem* to be relative maxima along the y-direction, but the slice in the x-direction (through $x = 1$) seems to be horizontal. However, a diagonal slice (along $x = \pm y$) shows these two points as minima and so they are saddle points. (If you don't believe this, we will get more evidence below and in a later example.)

Classifying the Critical Points Numerically We can use a tabular representation of the function to classify the critical points numerically. The following tabular representation of the function can be obtained using Excel. (See the Excel Technology Guide discussion of Section 15.1 Example 3 at the end of the chapter for information on using Excel to generate such a table.)

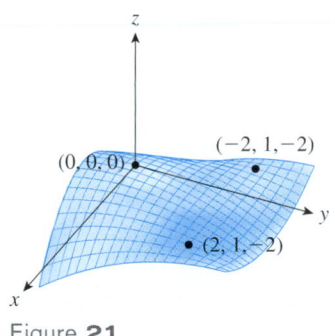

Figure **21**

		$x \rightarrow$						
		−3	**−2**	**−1**	**0**	**1**	**2**	**3**
y ↓	**−3**	−54	−34	−22	−18	−22	−34	−54
	−2	−35	−20	−11	−8	−11	−20	−35
	−1	−20	−10	−4	−2	−4	−10	−20
	0	−9	−4	−1	0	−1	−4	−9
	1	−2	−2	−2	−2	−2	−2	−2
	2	1	−4	−7	−8	−7	−4	1
	3	0	−10	−16	−18	−16	−10	0

The shaded and colored cells show rectangular neighborhoods of the three critical points $(0, 0)$, $(2, 1)$, and $(-2, 1)$. (Notice that they overlap.) The values of f at the points are at the centers of these rectangles. Looking at the gray neighborhood of $(x, y) = (0, 0)$, we see that $f(0, 0) = 0$ is the largest value of f in the shaded cells, suggesting that f has a maximum at $(0, 0)$. The shaded neighborhood of $(2, 1)$ on the right shows $f(2, 1) = -2$ as the maximum along some slices (e.g., the vertical slice), and a minimum along the diagonal slice from top left to bottom right. This is what results in a saddle point on the graph. The point $(-2, 1)$ is similar, and thus f also has a saddle point at $(-2, 1)$.

Q: *Is there an algebraic way of deciding whether a given point is a relative maximum, relative minimum, or saddle point?*

A: There is a "second derivative test" for functions of two variables, stated as follows. ∎

Second Derivative Test for Functions of Two Variables

Suppose (a, b) is a critical point in the interior of the domain of the function f of two variables. Let H be the quantity

$$H = f_{xx}(a, b) f_{yy}(a, b) - [f_{xy}(a, b)]^2 \qquad \text{H is called the } Hessian.$$

Then, if H is *positive*,

- f has a relative minimum at (a, b) if $f_{xx}(a, b) > 0$.
- f has a relative maximum at (a, b) if $f_{xx}(a, b) < 0$.

If H is *negative*,

- f has a saddle point at (a, b).

If $H = 0$, the test tells us nothing, so we need to look at the graph or a numerical table to see what is going on.

quick Examples

1. Let $f(x, y) = x^2 - y^2$. Then

$$f_x = 2x \quad \text{and} \quad f_y = -2y$$

which gives $(0, 0)$ as the only critical point. Also,

$$f_{xx} = 2, f_{xy} = 0, \quad \text{and} \quad f_{yy} = -2 \qquad \text{Note that these are constant}$$

which gives $H = (2)(-2) - 0^2 = -4$. Because H is negative, we have a saddle point at $(0, 0)$.

2. Let $f(x, y) = x^2 + 2y^2 + 2xy + 4x$. Then

$$f_x = 2x + 2y + 4 \quad \text{and} \quad f_y = 2x + 4y$$

Setting these equal to zero gives a system of two linear equations in two unknowns:

$$x + y = -2$$
$$x + 2y = 0$$

This system has solution $(-4, 2)$, so this is our only critical point. The second partial derivatives are $f_{xx} = 2$, $f_{xy} = 2$, and $f_{yy} = 4$, so $H = (2)(4) - 2^2 = 4$. Since $H > 0$ and $f_{xx} > 0$, we have a relative minimum at $(-4, 2)$.

Note There is a second derivative test for functions of three or more variables, but it is considerably more complicated. We stick with functions of two variables for the most part in this book. The justification of the second derivative test is beyond the scope of this book. ∎

Example 2 Using the Second Derivative Test

Use the second derivative test to analyze the function $f(x, y) = x^2 y - x^2 - 2y^2$ discussed in Example 1, and confirm the results we got there.

Solution We saw in Example 1 that the first-order derivatives are

$$f_x = 2xy - 2x = 2x(y - 1)$$
$$f_y = x^2 - 4y$$

and the critical points are $(0, 0)$, $(2, 1)$, and $(-2, 1)$. We also need the second derivatives:

$$f_{xx} = 2y - 2$$
$$f_{xy} = 2x$$
$$f_{yy} = -4$$

The point (0, 0): $f_{xx}(0, 0) = -2$, $f_{xy}(0, 0) = 0$, $f_{yy}(0, 0) = -4$, so $H = 8$. Because $H > 0$ and $f_{xx}(0, 0) < 0$, the second derivative test tells us that f has a relative maximum at $(0, 0)$.

The point (2, 1): $f_{xx}(2, 1) = 0$, $f_{xy}(2, 1) = 4$ and $f_{yy}(2, 1) = -4$, so $H = -16$. Since $H < 0$, we know that f has a saddle point at $(2, 1)$.

The point (−2, 1): $f_{xx}(-2, 1) = 0$, $f_{xy}(-2, 1) = -4$ and $f_{yy}(-2, 1) = -4$, so once again $H = -16$, and f has a saddle point at $(-2, 1)$.

Deriving the Regression Formulas

Back in Section 1.5, we presented the following set of formulas for the **regression** or **best-fit** line associated with a given set of data points $(x_1, y_1), (x_2, y_2), \ldots, (x_n, y_n)$.

Regression Line

The line that best fits the n data points $(x_1, y_1), (x_2, y_2), \ldots, (x_n, y_n)$ has the form

$$y = mx + b$$

where

$$m = \frac{n\left(\sum xy\right) - \left(\sum x\right)\left(\sum y\right)}{n\left(\sum x^2\right) - \left(\sum x\right)^2}$$

$$b = \frac{\sum y - m\left(\sum x\right)}{n}$$

n = number of data points

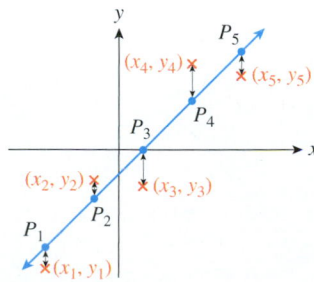

Figure **22**

We now show how to derive these formulas. Recall that the regression line is defined to be the line that minimizes the sum of the squares of the **residuals,** measured by the vertical distances shown in Figure 22, which shows a regression line associated with $n = 5$ data points. In the figure, the points $P_1, \ldots, P_n$ on the regression line have coordinates

$(x_1, mx_1 + b)$, $(x_2, mx_2 + b)$, ..., $(x_n, mx_n + b)$. The residuals are the quantities $y_{\text{Observed}} - y_{\text{Predicted}}$:

$$y_1 - (mx_1 + b), \ y_2 - (mx_2 + b), \ldots, y_n - (mx_n + b)$$

The sum of the squares of the residuals is therefore

$$S(m, b) = [y_1 - (mx_1 + b)]^2 + [y_2 - (mx_2 + b)]^2 + \cdots + [y_n - (mx_n + b)]^2$$

and this is the quantity we must minimize by choosing m and b. Because we reason that there is a line that minimizes this quantity, there must be a relative minimum at that point. We shall see in a moment that the function S has at most one critical point, which must therefore be the desired absolute minimum. To obtain the critical points of S, we set the partial derivatives equal to zero and solve:

$$S_m = 0: \quad -2x_1[y_1 - (mx_1 + b)] - \cdots - 2x_n[y_n - (mx_n + b)] = 0$$
$$S_b = 0: \quad -2[y_1 - (mx_1 + b)] - \cdots - 2[y_n - (mx_n + b)] = 0$$

Dividing by -2 and gathering terms allows us to rewrite the equations as

$$m(x_1^2 + \cdots + x_n^2) + b(x_1 + \cdots + x_n) = x_1 y_1 + \cdots + x_n y_n$$
$$m(x_1 + \cdots + x_n) + nb \qquad\qquad = y_1 + \cdots + y_n$$

We can rewrite these equations more neatly using $\sum$-notation:

$$m\left(\sum x^2\right) + b\left(\sum x\right) = \sum xy$$
$$m\left(\sum x\right) + nb \qquad = \sum y$$

This is a system of two linear equations in the two unknowns m and b. It may or may not have a unique solution. When there is a unique solution, we can conclude that the best fit line is given by solving these two equations for m and b. Alternatively, there is a general formula for the solution of any system of two equations in two unknowns, and if we apply this formula to our two equations, we get the regression formulas above.

15.4 EXERCISES

● denotes basic skills exercises

◆ denotes challenging exercises

tech Ex indicates exercises that should be solved using technology

In Exercises 1–4, classify each labeled point on the graph as one of the following:

(A) a relative maximum
(B) a relative minimum
(C) a saddle point
(D) a critical point but neither a relative extremum nor a saddle point
(E) none of the above

1. ● *hint* [see Example 1]

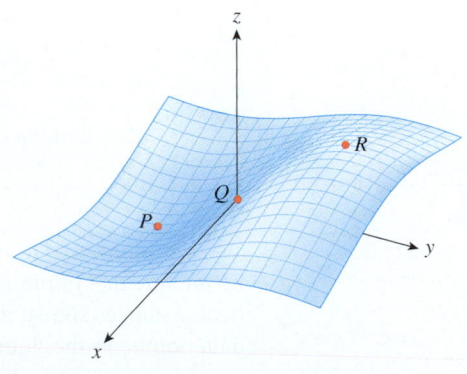

2. ●

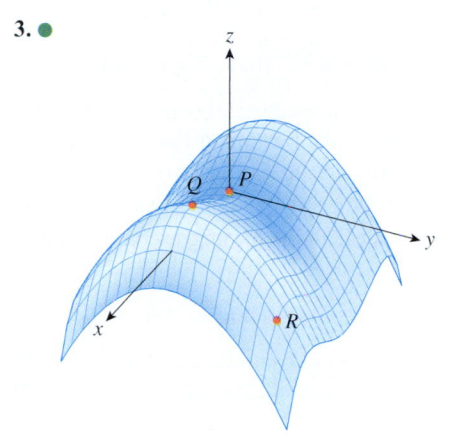

3. ●

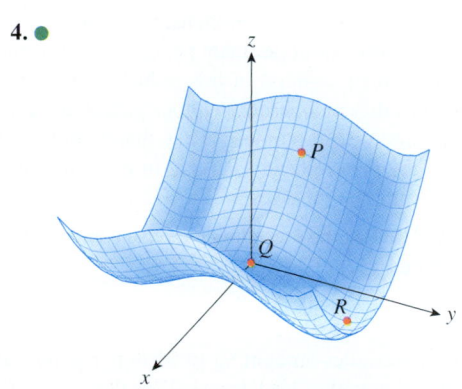

4. ●

In Exercises 5–10, classify the shaded value in each table as one of the following:

(A) a relative maximum
(B) a relative minimum
(C) a saddle point
(D) neither a relative extremum nor a saddle point

5. ●

$x \rightarrow$

y↓		-3	-2	-1	0	1	2
	-3	10	5	2	1	2	5
	-2	9	4	1	0	1	4
	-1	10	5	2	1	2	5
	0	13	8	5	4	5	8
	1	18	13	10	9	10	13
	2	25	20	17	16	17	20
	3	34	29	26	25	26	29

6. ●

$x \rightarrow$

y↓		-3	-2	-1	0	1	2
	-3	5	0	-3	-4	-3	0
	-2	8	3	0	-1	0	3
	-1	9	4	1	0	1	4
	0	8	3	0	-1	0	3
	1	5	0	-3	-4	-3	0
	2	0	-5	-8	-9	-8	-5
	3	-7	-12	-15	-16	-15	-12

7. ●

$x \rightarrow$

y↓		-3	-2	-1	0	1	2
	-3	5	0	-3	-4	-3	0
	-2	8	3	0	-1	0	3
	-1	9	4	1	0	1	4
	0	8	3	0	-1	0	3
	1	5	0	-3	-4	-3	0
	2	0	-5	-8	-9	-8	-5
	3	-7	-12	-15	-16	-15	-12

8. ●

$x \rightarrow$

y↓		-3	-2	-1	0	1	2
	-3	2	3	2	-1	-6	-13
	-2	3	4	3	0	-5	-12
	-1	2	3	2	-1	-6	-13
	0	-1	0	-1	-4	-9	-16
	1	-6	-5	-6	-9	-14	-21
	2	-13	-12	-13	-16	-21	-28
	3	-22	-21	-22	-25	-30	-37

9. ●

	$x \rightarrow$					
	−3	**−2**	**−1**	**0**	**1**	**2**
y ↓						
−3	4	5	4	1	−4	−11
−2	3	4	3	0	−5	−12
−1	4	5	4	1	−4	−11
0	7	8	7	4	−1	−8
1	12	13	12	9	4	−3
2	19	20	19	16	11	4
3	28	29	28	25	20	13

10. ●

	$x \rightarrow$					
	−3	**−2**	**−1**	**0**	**1**	**2**
y ↓						
−3	100	101	100	97	92	85
−2	99	100	99	96	91	84
−1	98	99	98	95	90	83
0	91	92	91	88	83	76
1	72	73	72	69	64	57
2	35	36	35	32	27	20
3	−26	−25	−26	−29	−34	−41

Locate and classify all the critical points of the functions in Exercises 11–28. *hint* [see Example 2]

11. ● $f(x, y) = x^2 + y^2 + 1$

12. ● $f(x, y) = 4 - (x^2 + y^2)$

13. ● $g(x, y) = 1 - x^2 - x - y^2 + y$

14. ● $g(x, y) = x^2 + x + y^2 - y - 1$

15. ● $h(x, y) = x^2y - 2x^2 - 4y^2$

16. ● $h(x, y) = x^2 + y^2 - y^2x - 4$

17. ● $s(x, y) = e^{x^2+y^2}$

18. ● $s(x, y) = e^{-(x^2+y^2)}$

19. ● $t(x, y) = x^4 + 8xy^2 + 2y^4$

20. ● $t(x, y) = x^3 - 3xy + y^3$

21. ● $f(x, y) = x^2 + y - e^y$

22. ● $f(x, y) = xe^y$

23. ● $f(x, y) = e^{-(x^2+y^2+2x)}$

24. ● $f(x, y) = e^{-(x^2+y^2-2x)}$

25. $f(x, y) = xy + \dfrac{2}{x} + \dfrac{2}{y}$

26. $f(x, y) = xy + \dfrac{4}{x} + \dfrac{2}{y}$

27. $g(x, y) = x^2 + y^2 + \dfrac{2}{xy}$

28. $g(x, y) = x^3 + y^3 + \dfrac{3}{xy}$

29. Refer back to Exercise 11. Which (if any) of the critical points of $f(x, y) = x^2 + y^2 + 1$ are absolute extrema?

30. Refer back to Exercise 12. Which (if any) of the critical points of $f(x, y) = 4 - (x^2 + y^2)$ are absolute extrema?

31. *tech Ex* Refer back to Exercise 15. Which (if any) of the critical points of $h(x, y) = x^2y - 2x^2 - 4y^2$ are absolute extrema?

32. *tech Ex* Refer back to Exercise 16. Which (if any) of the critical points of $h(x, y) = x^2 + y^2 - y^2x - 4$ are absolute extrema?

Applications

33. ● ***Brand Loyalty*** Suppose the fraction of Mazda car owners who chose another new Mazda can be modeled by the following function:[28]

$$M(c, f) = 11 + 8c + 4c^2 - 40f - 20cf + 40f^2$$

where c is the fraction of Chrysler car owners who remained loyal to Chrysler and f is the fraction of Ford car owners remaining loyal to Ford. Locate and classify all the critical points and interpret your answer. *hint* [see Example 2]

34. ● ***Brand Loyalty*** Repeat the preceding exercise using the function:

$$M(c, f) = -10 - 8f - 4f^2 + 40c + 20fc - 40c^2$$

35. ***Pollution Control*** The cost of controlling emissions at a firm goes up rapidly as the amount of emissions reduced goes up. Here is a possible model:

$$C(x, y) = 4000 + 100x^2 + 50y^2$$

where x is the reduction in sulfur emissions, y is the reduction in lead emissions (in pounds of pollutant per day), and C is the daily cost to the firm (in dollars) of this reduction. Government clean-air subsidies amount to $500 per pound of sulfur and $100 per pound of lead removed. How many pounds of pollutant should the firm remove each day in order to minimize *net* cost (cost minus subsidy)?

36. ***Pollution Control*** Repeat the preceding exercise using the following information:

$$C(x, y) = 2000 + 200x^2 + 100y^2$$

with government subsidies amounting to $100 per pound of sulfur and $500 per pound of lead removed per day.

37. ***Revenue*** Your company manufactures two models of speakers, the Ultra Mini and the Big Stack. Demand for each depends partly on the price of the other. If one is expensive, then more people will buy the other. If p_1 is the price of the

---

[28] This model is not accurate, although it was inspired by an approximation of a second-order regression based on data from the period 1988–1995. Source for original data: Chrysler, Maritz Market Research, Consumer Attitude Research, and Strategic Vision/*The New York Times*, November 3, 1995, p. D2.

● basic skills   ◆ challenging   *tech Ex* technology exercise

Ultra Mini, and p_2 is the price of the Big Stack, demand for the Ultra Mini is given by

$$q_1(p_1, p_2) = 100,000 - 100p_1 + 10p_2$$

where q_1 represents the number of Ultra Minis that will be sold in a year. The demand for the Big Stack is given by

$$q_2(p_1, p_2) = 150,000 + 10p_1 - 100p_2$$

Find the prices for the Ultra Mini and the Big Stack that will maximize your total revenue.

38. *Revenue* Repeat the preceding exercise, using the following demand functions:

$$q_1(p_1, p_2) = 100,000 - 100p_1 + p_2$$
$$q_2(p_1, p_2) = 150,000 + p_1 - 100p_2$$

39. *Luggage Dimensions* American Airlines requires that the total outside dimensions (length + width + height) of a checked bag not exceed 62 inches.[29] What are the dimensions of the largest volume bag that you can check on an American flight?

40. *Luggage Dimensions* American Airlines requires that the total outside dimensions (length + width + height) of a carry-on bag not exceed 45 inches.[30] What are the dimensions of the largest volume bag that you can carry on an American flight?

41. *Package Dimensions* The U.S. Postal Service (USPS) will accept only packages with length plus girth of no more than 108 inches.[31] (See the figure.)

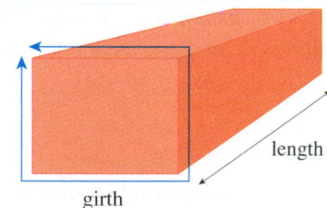

length

girth

What are the dimensions of the largest volume package that the USPS will accept? What is its volume?

42. *Package Dimensions* United Parcel Service (UPS) will accept only packages with length no more than 108 inches and length plus girth no more than 130 inches.[32] (See figure for

---

[29] According to information on its website (www.aa.com) as of August, 2002.

[30] Ibid.

[31] The requirement for priority Mail as of July, 2002.

[32] The requirement as of July, 2002.

the preceding exercise.) What are the dimensions of the largest volume package that UPS will accept? What is its volume?

Communication and Reasoning Exercises

43. ● Sketch the graph of a function that has one extremum and no saddle points.

44. ● Sketch the graph of a function that has one saddle point and one extremum.

45. Sketch the graph of a function that has one relative extremum, no absolute extrema, and no saddle points.

46. Sketch the graph of a function that has infinitely many absolute maxima.

47. ● Let $H = f_{xx}(a, b)f_{yy}(a, b) - [f_{xy}(a, b)]^2$. What condition on H guarantees that f has a relative extremum at the point (a, b)?

48. ● Let H be as in the preceding exercise. Give an example to show that it is possible to have $H = 0$ and a relative minimum at (a, b).

49. Suppose that when the graph of $f(x, y)$ is sliced by a vertical plane through (a, b) parallel to either the xz-plane or the yz-plane, the resulting curve has a relative maximum at (a, b). Does this mean that f has a relative maximum at (a, b)? Explain your answer.

50. Suppose that f has a relative maximum at (a, b). Does it follow that, if the graph of f is sliced by a vertical plane parallel to either the xz-plane or the yz-plane, the resulting curve has a relative maximum at (a, b)? Explain your answer.

51. *Average Cost* Let $C(x, y)$ be any cost function. Show that when the average cost is minimized, the marginal costs C_x and C_y both equal the average cost. Explain why this is reasonable.

52. *Average Profit* Let $P(x, y)$ be any profit function. Show that when the average profit is maximized, the marginal profits P_x and P_y both equal the average profit. Explain why this is reasonable.

53. ◆ The tangent plane to a graph was introduced in Exercise 76 in the preceding section. Use the equation of the tangent plane given there to explain why the tangent plane is parallel to the xy-plane at a relative maximum or minimum of $f(x, y)$.

54. ◆ Use the equation of the tangent plane given in Exercise 76 in the preceding section to explain why the tangent plane is parallel to the xy-plane at a saddle point of $f(x, y)$.

15.5 Constrained Maxima and Minima and Applications

So far we have looked only at the relative extrema of functions with no constraints. However, in Section 12.2 we saw examples in which we needed to find the maximum or minimum of an objective function subject to one or more constraints on the independent variables. For instance, consider the following problem:

$$\text{Minimize } S = xy + 2xz + 2yz \quad \text{subject to } xyz = 4 \text{ with } x > 0, \ y > 0, z > 0$$

Our strategy for solving such problems is essentially the same as the strategy we used earlier. First, we use the constraint equations to eliminate variables. In the examples in this section, we are able to reduce our objective function to a function of only two variables. Next, we locate any critical points, and then determine whether they are maxima, minima, or neither.

An alternative method, called the *method of Lagrange Multipliers,* can be used even when it is impossible to eliminate variables using the constraint equations.

First, we see how to solve the above constrained minimization problem using the first method.

Example 1 Constrained Minimization Problem

Minimize $S = xy + 2xz + 2yz$ subject to $xyz = 4$ with $x > 0, y > 0, z > 0$.

Solution   As suggested in the above discussion, we proceed as follows:

Solve the constraint equation for one of the variables and then substitute in the objective function.   The constraint equation is $xyz = 4$. Solving for z gives

$$z = \frac{4}{xy}$$

The objective function is $S = xy + 2xz + 2yz$, so substituting $z = 4/xy$ gives

$$S = xy + 2x\frac{4}{xy} + 2y\frac{4}{xy}$$

$$= xy + \frac{8}{y} + \frac{8}{x}$$

Minimize the resulting function of two variables.   We use the method in Section 15.4 to find the minimum of $S = xy + \dfrac{8}{y} + \dfrac{8}{x}$ for $x > 0$ and $y > 0$: We look for critical points:

$$S_x = y - \frac{8}{x^2} \qquad S_y = x - \frac{8}{y^2}$$

$$S_{xx} = \frac{16}{x^3} \qquad S_{xy} = 1 \qquad S_{yy} = \frac{16}{y^3}$$

We now equate the first partial derivatives to zero:

$$y = \frac{8}{x^2} \qquad \text{and} \qquad x = \frac{8}{y^2}$$

To solve for x and y, we substitute the first of these equations in the second, getting

$$x = \frac{x^4}{8}$$
$$x^4 - 8x = 0$$
$$x(x^3 - 8) = 0$$

The two solutions are $x = 0$, which we reject because x cannot be zero, and $x = 2$. Substituting $x = 2$ in $y = 8/x^2$ gives $y = 2$ also. Thus, the only critical point is $(2, 2)$. To apply the second derivative test, we compute

$$S_{xx}(2, 2) = 2 \qquad S_{xy}(2, 2) = 1 \qquad S_{yy}(2, 2) = 2$$

and find that $H = 3 > 0$, so we have a relative minimum at $(2, 2)$.

The corresponding value of z is given by the constraint equation:

$$z = \frac{4}{xy} = \frac{4}{4} = 1$$

The corresponding value of the objective function is

$$S = xy + \frac{8}{y} + \frac{8}{x} = 4 + \frac{8}{2} + \frac{8}{2} = 12$$

Figure 23 shows a portion of the graph of $S = xy + \dfrac{8}{y} + \dfrac{8}{x}$ for positive x and y (drawn using the Excel 3-D Grapher in the Chapter 15 utilities available online), and suggests that there is a single absolute minimum, which must be at our only candidate point $(2, 2)$.

We conclude that the minimum of S is 12 and occurs at $(2, 2, 1)$.

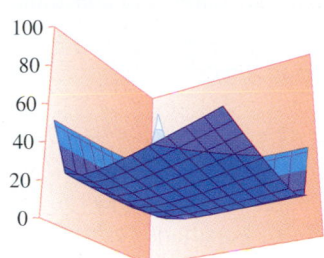

Graph of $S = xy + \frac{8}{y} + \frac{8}{x}$
$(0.2 \le x \le 5,\ 0.2 \le y \le 5)$

Figure **23**

Example **2** Minimizing Area

Find the dimensions of an open-top rectangular box that has a volume of 4 cubic feet and the smallest possible surface area.

Solution Our first task is to rephrase this request as a mathematical optimization problem. Figure 24 shows a picture of the box with dimensions x, y, and z.

We want to minimize the total surface area, which is given by

$$A = xy + 2xz + 2yz \qquad \text{Base + Sides + Front and Back}$$

This is our objective function. We can't simply choose x, y, and z to all be zero; however, because the enclosed volume must be 4 cubic feet. So,

$$xyz = 4 \qquad \text{Constraint}$$

This is our constraint equation. Other unstated constraints are $x > 0$, $y > 0$, and $z > 0$, because the dimensions of the box must be positive. We now restate the problem as follows:

$$\text{Minimize } A = xy + 2xz + 2yz \qquad \text{subject to } xyz = 4,\ x > 0,\ y > 0,\ z > 0$$

But this is exactly the problem in Example 1, which has a solution $x = 2$, $y = 2$, $z = 1$, $A = 12$. Thus, the required dimensions of the box are

$$x = 2 \text{ ft},\ y = 2 \text{ ft},\ z = 1 \text{ ft}$$

requiring a total surface area of 12 ft^2.

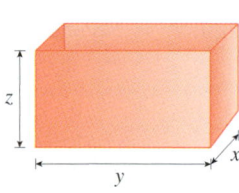

Figure **24**

Q: *In Example 1 we checked that we had a relative minimum at* $(x, y) = (2, 2)$ *and we were persuaded graphically that this was probably an absolute minimum. Can we be sure that this relative minimum is an absolute minimum?*

A: Yes. There must be a least surface area among all boxes that hold 4 cubic feet. (Why?) Because this would give a relative minimum of A and because the only possible relative minimum of A occurs at $(2, 2)$, this is the absolute minimum. ■

The Method of Lagrange Multipliers

Suppose we have a constrained optimization problem in which it is difficult or impossible to solve a constraint equation for one of the variables. Then we can use the method of **Lagrange multipliers** to avoid this difficulty. We restrict attention to the case of a single constraint equation, although the method generalizes to any number of constraint equations.

Locating Relative Extrema Using the Method of Lagrange Multipliers

To locate the candidates for relative extrema of a function $f(x, y, \ldots)$ subject to the constraint $g(x, y, \ldots) = 0$, we solve the following system of equations for $x, y, \ldots$ and λ:

$$f_x = \lambda g_x$$
$$f_y = \lambda g_y$$
$$\ldots$$
$$g = 0$$

The unknown λ is called a **Lagrange multiplier.** The points $(x, y, \ldots)$ that occur in solutions are then the candidates for the relative extrema of f subject to $g = 0$.

Although the justification for the method of Lagrange multipliers is beyond the scope of this text, we will demonstrate by example how it is used.

Example 3 Using Lagrange Multipliers

Use the method of Lagrange multipliers to find the maximum value of $f(x, y) = 2xy$ subject to $x^2 + 4y^2 = 32$.

Solution We start by rewriting the problem in standard form:

$$\text{Maximize } f(x, y) = 2xy \quad \text{subject to } x^2 + 4y^2 - 32 = 0$$

Here, $g(x, y) = x^2 + 4y^2 - 32$, and the system of equations we need to solve is thus

$$f_x = \lambda g_x \quad \text{or} \quad 2y = 2\lambda x$$
$$f_y = \lambda g_y \quad \text{or} \quad 2x = 8\lambda y$$
$$g = 0 \quad \text{or} \quad x^2 + 4y^2 - 32 = 0$$

A convenient way to solve such a system is to solve one of the equations for λ and then substitute in the remaining equations. Thus, we start by solving the first equation to obtain

$$\lambda = \frac{y}{x}$$

(A word of caution: Because we divided by x, we made the implicit assumption that $x \neq 0$, so before continuing we should check what happens if $x = 0$. But if $x = 0$, then the first equation, $2y = 2\lambda x$, tells us that $y = 0$ as well, and this contradicts the third equation: $x^2 + 4y^2 - 32 = 0$. Thus, we can rule out the possibility that $x = 0$.) Substituting in the remaining equations gives

$$x = 4\lambda y = \frac{4y^2}{x} \quad \text{or} \quad x^2 = 4y^2$$
$$x^2 + 4y^2 - 32 = 0$$

Notice how we have reduced the number of unknowns and also the number of equations by one. We can now substitute $x^2 = 4y^2$ in the last equation, obtaining

$$4y^2 + 4y^2 - 32 = 0$$
$$8y^2 = 32$$
$$y = \pm 2$$

We now substitute back to obtain

$$x^2 = 4y^2 = 16$$

or $\qquad x = \pm 4$

We don't need the value of λ, so we won't solve for it. Thus, the candidates for relative extrema are given by $x = \pm 4$ and $y = \pm 2$, that is, the four points $(-4, -2)$, $(-4, 2)$, $(4, -2)$, and $(4, 2)$. Recall that we are seeking the values of x and y that give the maximum value for $f(x, y) = 2xy$. Because we now have only four points to choose from, we compare the values of f at these four points and conclude that the maximum value of f occurs when $(x, y) = (-4, -2)$ or $(4, 2)$.

Something is suspicious in Example 3. We didn't check to see whether these candidates were relative extrema to begin with, let alone absolute extrema! How do we justify this omission? One of the difficulties with using the method of Lagrange Multipliers is that it does not provide us with a test analogous to the second derivative test for functions of several variables. However, if you grant that the function in question does have an absolute maximum, then we require no test, because one of the candidates must give this maximum.

Q: *But how do we know that the given function has an absolute maximum?*

A: The best way to see this is by giving a geometric interpretation. The constraint $x^2 + 4y^2 = 32$ tells us that the point (x, y) must lie on the ellipse shown in Figure 25. The function $f(x, y) = 2xy$ gives the area of the rectangle shaded in Figure 25.

Because there must be a largest such rectangle, the function f must have an absolute maximum for at least one pair of coordinates (x, y). ∎

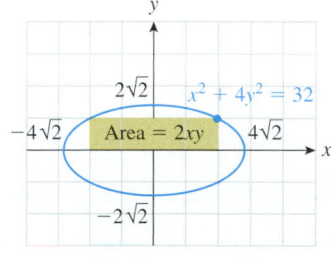

Figure 25

As a last example, we show how to use Lagrange Multipliers to solve the minimization problem in Example 1:

Example 4 Using Lagrange Multipliers: Function of Three Variables

Use the method of Lagrange multipliers to find the minimum value of $S = xy + 2xz + 2yz$ subject to $xyz = 4$ with $x > 0$, $y > 0$, $z > 0$.

Solution We start by rewriting the problem in standard form:

$$\text{Maximize } f(x, y, z) = xy + 2xz + 2yz$$
$$\text{subject to } xyz - 4 = 0 \text{ (with } x > 0, y > 0, z > 0)$$

Here, $g(x, y, z) = xyz - 4$, and the system of equations we need to solve is thus

$$f_x = \lambda g_x \quad \text{or} \quad y + 2z = \lambda yz$$
$$f_y = \lambda g_y \quad \text{or} \quad x + 2z = \lambda xz$$
$$f_z = \lambda g_z \quad \text{or} \quad 2x + 2y = \lambda xy$$
$$g = 0 \quad \text{or} \quad xyz - 4 = 0$$

As in the last example, we solve one of the equations for λ and substitute in the others. The first equation gives

$$\lambda = \frac{1}{z} + \frac{2}{y}$$

Substituting this into the second and third equations gives

$$x + 2z = x + \frac{2xz}{y}$$

or $\qquad 2 = \dfrac{2x}{y}$ \qquad *Subtract x from both sides and then divide by z.*

giving $\qquad y = x$

Substituting the expression for λ into the third equation gives

$$2x + 2y = \frac{xy}{z} + 2x$$

or $\qquad 2 = \dfrac{x}{z}$ \qquad *Subtract $2x$ from both sides and then divide by y.*

giving $\qquad z = \dfrac{x}{2}$

Now we have both y and z in terms of x. We substitute these values in the last (constraint) equation:

$$x(x)\left(\frac{x}{2}\right) - 4 = 0$$
$$x^3 = 8$$
$$x = 2$$

Thus, $y = x = 2$, and $z = \dfrac{x}{2} = 1$. Therefore, the only critical point occurs at $(2, 2, 1)$ as we found in Example 1, and the corresponding value of S is

$$S = xy + 2xz + 2yz = (2)(2) + 2(2)(1) + 2(2)(1) = 12$$

+ *Before we go on...* Again, the method of Lagrange Multipliers does not tell us whether the critical point in Example 4 is a maximum, minimum, or neither. However, if you grant that the function in question does have an absolute minimum, then the values we found must give this minimum value. ∎

FAQs When to Use Lagrange Multipliers

Q: When can I use the method of Lagrange multipliers? When should I use it?

A: We have discussed the method only when there is a single equality constraint. There is a generalization, which we do not discuss, that works when there are more equality constraints (we need to introduce one multiplier for each constraint). So, if you have a problem with more than one equality constraint, or with any inequality constraints, you must use the method we discussed earlier in this section. On the other hand, if you have one equality constraint, and it would be difficult to solve it for one of the variables, then you should use Lagrange multipliers. We have noted in the exercises where you might use Lagrange multipliers. ∎

15.5 EXERCISES

● denotes basic skills exercises

◆ denotes challenging exercises

In Exercises 1–6, use substitution to solve the given optimization problem. hint [see Example 1]

1. ● Find the maximum value of $f(x, y, z) = 1 - x^2 - y^2 - z^2$ subject to $z = 2y$. Also find the corresponding point(s) (x, y, z).

2. ● Find the minimum value of $f(x, y, z) = x^2 + y^2 + z^2 - 2$ subject to $x = y$. Also find the corresponding point(s) (x, y, z).

3. ● Find the maximum value of $f(x, y, z) = 1 - x^2 - x - y^2 + y - z^2 + z$ subject to $3x = y$. Also find the corresponding point(s) (x, y, z).

4. ● Find the minimum value of $f(x, y, z) = 2x^2 + 2x + y^2 - y + z^2 - z - 1$ subject to $z = 2y$. Also find the corresponding point(s) (x, y, z).

5. ● Minimize $S = xy + 4xz + 2yz$ subject to $xyz = 1$ with $x > 0, y > 0, z > 0$.

6. ● Minimize $S = xy + xz + yz$ subject to $xyz = 2$ with $x > 0$, $y > 0, z > 0$.

In Exercises 7–18, use Lagrange Multipliers to solve the given optimization problem. hint [see Example 3]

7. ● Find the maximum value of $f(x, y) = xy$ subject to $x + 2y = 40$. Also find the corresponding point(s) (x, y).

8. ● Find the maximum value of $f(x, y) = xy$ subject to $3x + y = 60$. Also find the corresponding point(s) (x, y).

9. ● Find the maximum value of $f(x, y) = 4xy$ subject to $x^2 + y^2 = 8$. Also find the corresponding point(s) (x, y).

10. ● Find the maximum value of $f(x, y) = xy$ subject to $y = 3 - x^2$. Also find the corresponding point(s) (x, y).

11. ● Find the minimum value of $f(x, y) = x^2 + y^2$ subject to $x + 2y = 10$. Also find the corresponding point(s) (x, y).

12. ● Find the minimum value of $f(x, y) = x^2 + y^2$ subject to $xy^2 = 16$. Also find the corresponding point(s) (x, y).

13. ● The problem in Exercise 1. *hint* [see Example 4]

14. ● The problem in Exercise 2.

15. ● The problem in Exercise 3.

16. ● The problem in Exercise 4.

17. ● The problem in Exercise 5.

18. ● The problem in Exercise 6.

Applications

Exercises 19–22 were solved in Section 12.2. This time, use the method of Lagrange Multipliers to solve them.

19. ● *Fences* I want to fence in a rectangular vegetable patch. The fencing for the east and west sides costs $4 per foot, and the fencing for the north and south sides costs only $2 per foot. I have a budget of $80 for the project. What is the largest area I can enclose?

20. ● *Fences* My orchid garden abuts my house so that the house itself forms the northern boundary. The fencing for the southern boundary costs $4 per foot, and the fencing for the east and west sides costs $2 per foot. If I have a budget of $80 for the project, what is the largest area I can enclose this time?

21. ● *Revenue* Hercules Films is deciding on the price of the video release of its film *Son of Frankenstein*. Its marketing people estimate that at a price of p dollars, it can sell a total of $q = 200,000 - 10,000p$ copies. What price will bring in the greatest revenue?

22. ● *Profit* Hercules Films is also deciding on the price of the video release of its film *Bride of the Son of Frankenstein*. Again, marketing estimates that at a price of p dollars it can sell $q = 200,000 - 10,000p$ copies, but each copy costs $4 to make. What price will give the greatest *profit*?

● basic skills   ◆ challenging

23. ● *Geometry* At what points on the sphere $x^2 + y^2 + z^2 = 1$ is the product xyz a maximum? (The method of Lagrange multipliers can be used.)

24. ● *Geometry* At what point on the surface $z = (x^2 + x + y^2 + 4)^{1/2}$ is the quantity $x^2 + y^2 + z^2$ a minimum? (The method of Lagrange multipliers can be used.)

25. *Geometry* What point on the surface $z = x^2 + y - 1$ is closest to the origin? [*Hint:* Minimize the square of the distance from (x, y, z) to the origin.]

26. *Geometry* What point on the surface $z = x + y^2 - 3$ is closest to the origin? [*Hint:* Minimize the square of the distance from (x, y, z) to the origin.]

27. *Geometry* Find the point on the plane $-2x + 2y + z - 5 = 0$ closest to $(-1, 1, 3)$. [*Hint:* Minimize the square of the distance from the given point to a general point on the plane.]

28. *Geometry* Find the point on the plane $2x - 2y - z + 1 = 0$ closest to $(1, 1, 0)$.

29. ● *Construction Cost* A closed rectangular box is made with two kinds of materials. The top and bottom are made with heavy-duty cardboard costing 20¢ per square foot, and the sides are made with lightweight cardboard costing 10¢ per square foot. Given that the box is to have a capacity of 2 cubic feet, what should its dimensions be if the cost is to be minimized? *hint* [see Example 2]

30. ● *Construction Cost* Repeat the preceding exercise assuming that the heavy-duty cardboard costs 30¢ per square foot, the lightweight cardboard costs 5¢ per square foot, and the box is to have a capacity of 6 cubic feet.

31. ● *Package Dimensions* The U.S. Postal Service (USPS) will accept only packages with length plus girth no more than 108 inches.[33] (See the figure.)

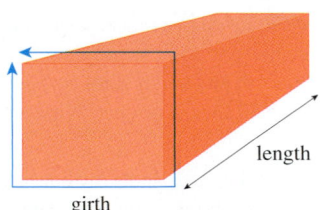

girth

length

What are the dimensions of the largest volume package the USPS will accept? What is its volume? (This exercise is the same as Exercise 41 in the preceding section. This time, solve it using Lagrange Multipliers.)

32. ● *Package Dimensions* United Parcel Service (UPS) will accept only packages with length no more than 108 inches and length plus girth no more than 130 inches.[34] (See the figure for the preceding exercise.) What are the dimensions of the largest volume package UPS will accept? What is its volume? (This exercise is the same as Exercise 42 in the preceding section. This time, solve it using Lagrange Multipliers.)

[33] The requirement for priority Mail as of July, 2002.

[34] The requirement as of July, 2002.

33. ● *Construction Cost* My company wishes to manufacture boxes similar to those described in Exercise 29 as cheaply as possible, but unfortunately the company that manufactures the cardboard is unable to give me price quotes for the heavy-duty and lightweight cardboard. Find formulas for the dimensions of the box in terms of the price per square foot of heavy-duty and lightweight cardboard.

34. ● *Construction Cost* Repeat the preceding exercise, assuming that only the bottoms of the boxes are to be made using heavy-duty cardboard.

35. *Geometry* Find the dimensions of the rectangular box with largest volume that can be inscribed above the xy-plane and under the paraboloid $z = 1 - (x^2 + y^2)$.

36. *Geometry* Find the dimensions of the rectangular box with largest volume that can be inscribed above the xy-plane and under the paraboloid $z = 2 - (2x^2 + y^2)$.

Communication and Reasoning Exercises

37. ● Outline two methods of solution of the problem "*Maximize $f(x, y, z)$ subject to $g(x, y, z) = 0$*" and give an advantage and disadvantage of each.

38. ● Suppose we know that $f(x, y)$ has both partial derivatives in its domain $D: x > 0, y > 0$, and that (a, b) is the only point in D such that $f_x(a, b) = f_y(a, b) = 0$. Must it be the case that f has an absolute maximum at (a, b)? Explain.

39. ● Under what circumstances would it be necessary to use the method of Lagrange multipliers?

40. ● Under what circumstances would the method of Lagrange multipliers not apply?

41. ● Restate the following problem as a maximization problem of the form "*Maximize $f(x, y)$ subject to $g(x, y) = 0$*":

Find the maximum value of $h(x) = 1 - 2x^2$

42. ● Restate the following problem as a maximization problem of the form "*Maximize $f(x, y, z)$ subject to $g(x, y, z) = 0$*":

Find the maximum value of $h(x, y) = 1 - 2(x^2 + y^2)$

43. If the partial derivatives of a function of several variables are never 0, is it possible for the function to have relative extrema on some domain? Explain your answer.

44. ◆ A **linear programming problem in two variables** is a problem of the form: "*Maximize (or minimize) $f(x, y)$ subject to constraints of the form $C(x, y) \geq 0$ or $C(x, y) \leq 0$.*" Here, the objective function f and the constraints C are linear functions. There may be several linear constraints in one problem. Explain why the solution cannot occur in the interior of the domain of f.

45. ◆ Refer back to Exercise 44. Explain why the solution will actually be at a corner of the domain of f (where two or more of the line segments that make up the boundary meet). This result—or rather a slight generalization of it—is known as the Fundamental Theorem of Linear Programming.

● basic skills   ◆ challenging

15.6 Double Integrals and Applications

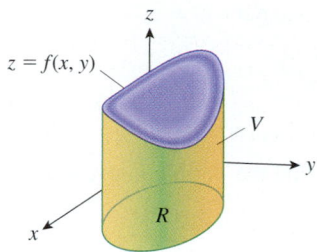

Figure 26

When discussing functions of one variable, we computed the area under a graph by integration. The analog for the graph of a function of two variables is the *volume V* under the graph, as in Figure 26.

Think of the region R in the xy-plane as the "shadow" under the portion of the surface $z = f(x, y)$ shown.

By analogy with the definite integral of a function of one variable, we make the following definition:

Geometric Definition of the Double Integral

The **double integral of $f(x, y)$ over the region R in the xy-plane** is defined as

(Volume *above* the region R and under the graph of f)

$\qquad$ − (Volume *below* the region R and above the graph of f)

We denote the double integral of $f(x, y)$ over the region R by $\iint_R f(x, y) \, dx \, dy$.

quick Example $\quad$ Take $f(x, y) = 2$ and take R to be the rectangle $0 \leq x \leq 1$, $0 \leq y \leq 1$. Then the graph of f is a flat horizontal surface $z = 2$, and

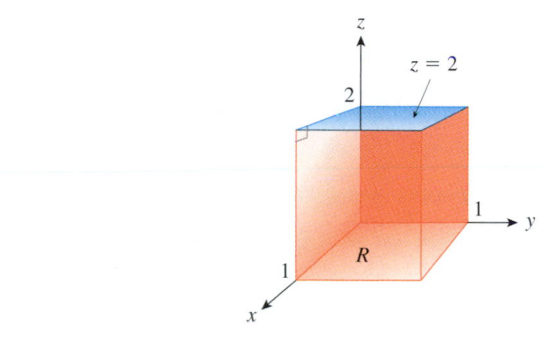

$$\iint_R f(x, y) \, dx \, dy = \text{Volume of box}$$
$$= \text{Width} \times \text{Length} \times \text{Height} = 1 \times 1 \times 2 = 2$$

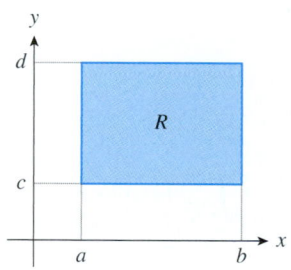

Figure 27

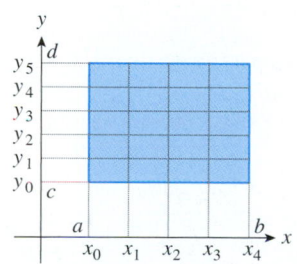

Figure 28

As we saw in the case of the definite integral of a function of one variable, we also desire *numerical* and *algebraic* definitions for two reasons: (1) to make the mathematical definition more precise, so as not to rely on the notion of "volume," and (2) for direct computation of the integral using technology or analytical tools.

We start with the simplest case, when the region R is a rectangle $a \leq x \leq b$ and $c \leq y \leq d$ (see Figure 27). To compute the volume over R, we mimic what we did to find the area under the graph of a function of one variable. We break up the interval $[a, b]$ into m intervals all of width $\Delta x = (b - a)/m$, and we break up $[c, d]$ into n intervals all of width $\Delta y = (d - c)/n$. Figure 28 shows an example with $m = 4$ and $n = 5$.

This gives us mn rectangles defined by $x_{i-1} \leq x \leq x_i$ and $y_{j-1} \leq y \leq y_j$. Over one of these rectangles, f is approximately equal to its value at one corner—say $f(x_i, y_j)$.

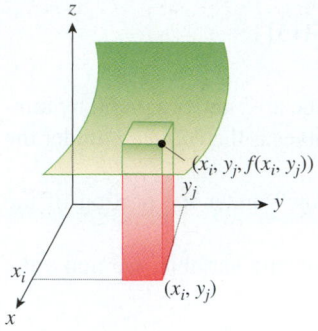

Figure **29**

The volume under f over this small rectangle is then approximately the volume of the rectangular brick (size exaggerated) shown in Figure 29. This brick has height $f(x_i, y_j)$, and its base is Δx by Δy. Its volume is therefore $f(x_i, y_j)\Delta x \Delta y$. Adding together the volumes of all of the bricks over the small rectangles in R, we get

$$\iint_R f(x, y)\, dx\, dy \approx \sum_{j=1}^{n} \sum_{i=1}^{m} f(x_i, y_j)\Delta x\, \Delta y$$

This double sum is called a **double Riemann sum.** We define the double integral to be the limit of the Riemann sums as m and n go to infinity.

Algebraic Definition of the Double Integral

$$\iint_R f(x, y)\, dx\, dy = \lim_{n \to \infty}\lim_{m \to \infty} \sum_{j=1}^{n} \sum_{i=1}^{m} f(x_i, y_j)\Delta x\, \Delta y$$

Note This definition is adequate (the limit exists) when f is continuous. More elaborate definitions are needed for general functions. ∎

This definition also gives us a clue about how to compute a double integral. The innermost sum is $\sum_{i=1}^{m} f(x_i, y_j)\Delta x$, which is a Riemann sum for $\int_a^b f(x, y_j)\, dx$. The innermost limit is therefore

$$\lim_{m \to \infty}\sum_{i=1}^{m} f(x_i, y_j)\Delta x = \int_a^b f(x, y_j)\, dx$$

The outermost limit is then also a Riemann sum, and we get the following way of calculating double integrals:

Computing the Double Integral over a Rectangle

If R is the rectangle $a \le x \le b$ and $c \le y \le d$, then

$$\iint_R f(x, y)\, dx\, dy = \int_c^d \left(\int_a^b f(x, y)\, dx \right) dy = \int_a^b \left(\int_c^d f(x, y)\, dy \right) dx$$

The second formula comes from switching the order of summation in the double sum.

quick Example   If R is the rectangle $1 \le x \le 2$ and $1 \le y \le 3$, then

$$\iint_R 1\, dx\, dy = \int_1^3 \left(\int_1^2 1\, dx \right) dy$$

$$= \int_1^3 [x]_{x=1}^2\, dy \qquad \text{Evaluate the inner integral}$$

$$= \int_1^3 1\, dy \qquad [x]_{x=1}^2 = 2 - 1 = 1$$

$$= [y]_{y=1}^3 = 3 - 1 = 2$$

The Quick Example used a constant function for the integrand. Here is an example in which the integrand is not constant.

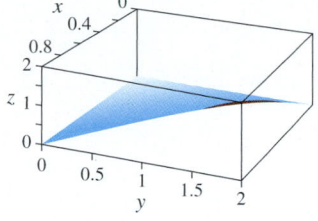

$z = xy, 0 \leq x \leq 1$ and $0 \leq y \leq 2$

Figure **30**

Example **1** **Double Integral over a Rectangle**

Let R be the rectangle $0 \leq x \leq 1$ and $0 \leq y \leq 2$. Compute $\iint_R xy \, dx \, dy$. This integral gives the volume of the part of the boxed region under the surface $z = xy$ shown in Figure 30.

Solution

$$\iint_R xy \, dx \, dy = \int_0^2 \int_0^1 xy \, dx \, dy$$

(We usually drop the parentheses around the inner integral like this.) As in the Quick Example, we compute this **iterated integral** from the inside out. First we compute

$$\int_0^1 xy \, dx$$

To do this computation, we do as we did when finding partial derivatives: We treat y as a constant. This gives

$$\int_0^1 xy \, dx = \left[\frac{x^2}{2} \cdot y\right]_{x=0}^1 = \frac{1}{2}y - 0 = \frac{y}{2}$$

We can now calculate the outer integral.

$$\int_0^2 \int_0^1 xy \, dx \, dy = \int_0^2 \frac{y}{2} \, dy = \left[\frac{y^2}{4}\right]_0^2 = 1$$

+*Before we go on...* We could also reverse the order of integration in Example 1.

$$\int_0^1 \int_0^2 xy \, dy \, dx = \int_0^1 \left(\left[x \cdot \frac{y^2}{2}\right]_{y=0}^2\right) dx = \int_0^1 2x \, dx = \left[x^2\right]_0^1 = 1 \quad \blacksquare$$

Often we need to integrate over regions R that are not rectangular. There are two cases that come up. The first is a region like the one shown in Figure 31.

In this region, the bottom and top sides are defined by functions $y = c(x)$ and $y = d(x)$, respectively, so that the whole region can be described by the inequalities $a \leq x \leq b$ and $c(x) \leq y \leq d(x)$. To evaluate a double integral over such a region, we have the following formula:

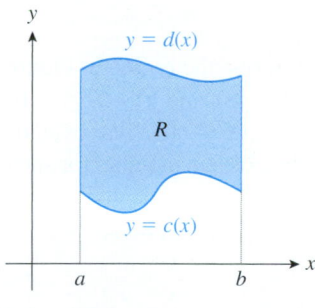

Figure **31**

Computing the Double Integral over a Nonrectangular Region

If R is the region $a \leq x \leq b$ and $c(x) \leq y \leq d(x)$ (Figure 31), then we integrate over R according to the following equation:

$$\iint_R f(x, y) \, dx \, dy = \int_a^b \int_{c(x)}^{d(x)} f(x, y) \, dy \, dx$$

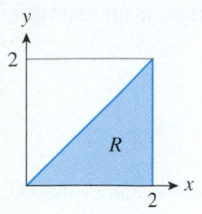

Figure 32

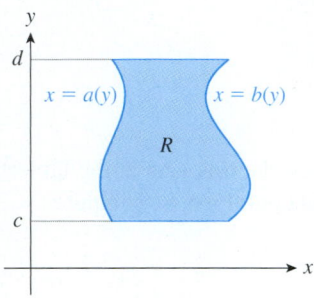

Figure 33

Example 2 Double Integral over a Nonrectangular Region

R is the triangle shown in Figure 32. Compute $\iint_R x \, dx \, dy$.

Solution R is the region described by $0 \le x \le 2, 0 \le y \le x$. We have

$$\iint_R x \, dx \, dy = \int_0^2 \int_0^x x \, dy \, dx$$

$$= \int_0^2 [xy]_{y=0}^x \, dx$$

$$= \int_0^2 x^2 \, dx$$

$$= \left[\frac{x^3}{3}\right]_0^2 = \frac{8}{3}$$

The second type of region is shown in Figure 33.
This is the region described by $c \le y \le d$ and $a(y) \le x \le b(y)$. To evaluate a double integral over such a region, we have the following formula:

Double Integral over a Nonrectangular Region (continued)

If R is the region $c \le y \le d$ and $a(y) \le x \le b(y)$ (Figure 33), then we integrate over R according to the following equation:

$$\iint_R f(x, y) \, dx \, dy = \int_c^d \int_{a(y)}^{b(y)} f(x, y) \, dx \, dy$$

Example 3 Double Integral over a Nonrectangular Region

Redo Example 2, integrating in the opposite order.

Solution We can integrate in the opposite order if we can describe the region in Figure 32 in the way shown in Figure 33. In fact, it is the region $0 \le y \le 2$ and $y \le x \le 2$. To see this, we draw a horizontal line through the region, as in Figure 34. The line extends from $x = y$ on the left to $x = 2$ on the right, so $y \le x \le 2$. The possible heights for such a line are $0 \le y \le 2$. We can now compute the integral:

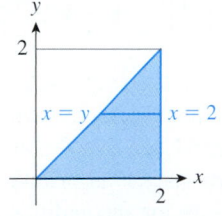

Figure 34

$$\iint_R x \, dx \, dy = \int_0^2 \int_y^2 x \, dx \, dy$$

$$= \int_0^2 \left[\frac{x^2}{2}\right]_{x=y}^2 \, dy$$

$$= \int_0^2 \left(2 - \frac{y^2}{2}\right) dy$$

$$= \left[2y - \frac{y^3}{6}\right]_0^2 = \frac{8}{3}$$

Note Many regions can be described in two different ways, as we saw in Examples 2 and 3. Sometimes one description will be much easier to work with than the other, so it pays to consider both. ∎

Applications

There are many applications of double integrals besides finding volumes. For example, we can use them to find *averages*. Remember that the average of $f(x)$ on $[a, b]$ is given by $\int_a^b f(x)\,dx$ divided by $(b - a)$, the length of the interval.

Average of a Function of Two Variables

The average of $f(x, y)$ on the region R is

$$\bar{f} = \frac{1}{A} \iint_R f(x, y)\,dx\,dy$$

Here, A is the area of R. We can compute the area A geometrically, or by using the techniques from the chapter on applications of the integral, or by computing

$$A = \iint_R 1\,dx\,dy$$

quick Example

The average value of $f(x, y) = xy$ on the rectangle given by $0 \le x \le 1$ and $0 \le y \le 2$ is

$$\bar{f} = \frac{1}{2} \iint_R xy\,dx\,dy \qquad \text{The area of the rectangle is 2.}$$

$$= \frac{1}{2} \int_0^2 \int_0^1 xy\,dx\,dy$$

$$= \frac{1}{2} \cdot 1 = \frac{1}{2} \qquad \text{We calculated the integral in Example 1.}$$

Example 4 Average Revenue

Your company is planning to price its new line of subcompact cars at between \$10,000 and \$15,000. The marketing department reports that if the company prices the cars at p dollars per car, the demand will be between $q = 20,000 - p$ and $q = 25,000 - p$ cars sold in the first year. What is the average of all the possible revenues your company could expect in the first year?

Solution Revenue is given by $R = pq$ as usual, and we are told that

$$10,000 \le p \le 15,000$$

and $\qquad 20,000 - p \le q \le 25,000 - p$

This domain D of prices and demands is shown in Figure 35.

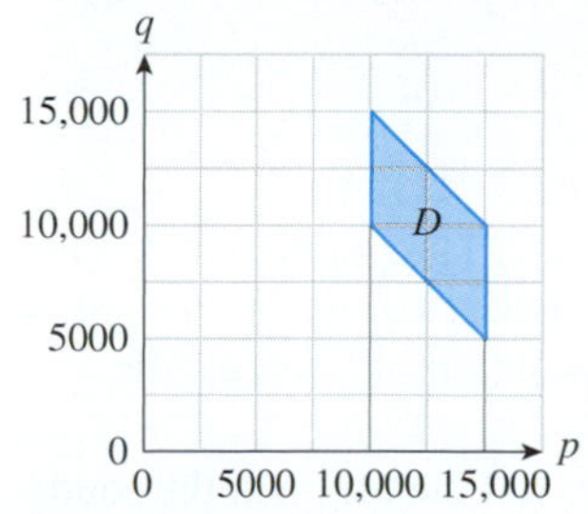

To average the revenue R over the domain D, we need to compute the area A of D. Using either calculus or geometry, we get $A = 25,000,000$. We then need to integrate R over D:

Figure **35**

$$\iint_D pq \, dp \, dq = \int_{10,000}^{15,000} \int_{20,000-p}^{25,000-p} pq \, dq \, dp$$

$$= \int_{10,000}^{15,000} \left[\frac{pq^2}{2} \right]_{q=20,000-p}^{25,000-p} dp$$

$$= \frac{1}{2} \int_{10,000}^{15,000} [p(25,000 - p)^2 - p(20,000 - p)^2] \, dp$$

$$= \frac{1}{2} \int_{10,000}^{15,000} [225,000,000p - 10,000p^2] \, dp$$

$$\approx 3,072,900,000,000,000$$

The average of all the possible revenues your company could expect in the first year is therefore

$$\bar{R} = \frac{3,072,900,000,000,000}{25,000,000} \approx \$122,900,000$$

+**Before we go on...** To check that the answer obtained in Example 4 is reasonable, notice that the revenues at the corners of the domain are $100,000,000 per year, $150,000,000 per year (at two corners), and $75,000,000 per year. Some of these are smaller than the average and some larger, as we would expect. ∎

Another useful application of the double integral comes about when we consider density. For example, suppose that $P(x, y)$ represents the population density (in people per square mile, say) in the city shown in Figure 36.

If we break the city up into small rectangles (for example, city blocks), then the population in the small rectangle $x_{i-1} \leq x \leq x_i$ and $y_{j-1} \leq y \leq y_j$ is approximately $P(x_i, y_j) \Delta x \Delta y$. Adding up all of these population estimates, we get

$$\text{Total population} \approx \sum_{j=1}^{n} \sum_{i=1}^{m} P(x_i, y_j) \, \Delta x \, \Delta y$$

Since this is a double Riemann sum, when we take the limit as m and n go to infinity, we get the following calculation of the population of the city:

$$\text{Total population} = \iint_{\text{City}} P(x, y) \, dx \, dy$$

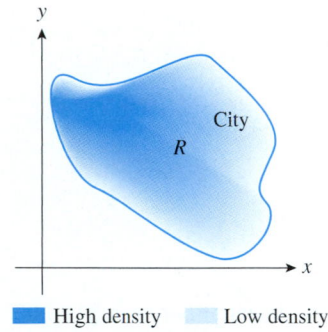

High density ■ Low density ■

Figure 36

Example 5 Population

Squaresville is a city in the shape of a square 5 miles on a side. The population density at a distance of x miles east and y miles north of the southwest corner is $P(x, y) = x^2 + y^2$ thousand people per square mile. Find the total population of Squaresville.

Solution Squaresville is pictured in Figure 37, in which we put the origin in the southwest corner of the city.

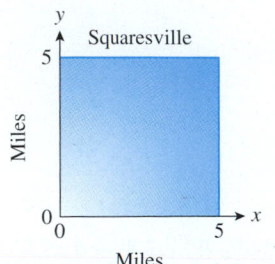

Figure 37

To compute the total population, we integrate the population density over the city S.

$$\text{Total population} = \iint_{\text{Squaresville}} P(x, y)\, dx\, dy$$

$$= \int_0^5 \int_0^5 (x^2 + y^2)\, dx\, dy$$

$$= \int_0^5 \left[\frac{x^3}{3} + xy^2 \right]_{x=0}^5 dy$$

$$= \int_0^5 \left[\frac{125}{3} + 5y^2 \right] dy$$

$$= \frac{1250}{3} \approx 417 \text{ thousand people}$$

✛*Before we go on...* Note that the average population density is the total population divided by the area of the city, which is about 17,000 people per square mile. Compare this calculation with the calculations of averages in the previous two examples. ∎

15.6 EXERCISES

● denotes basic skills exercises

Compute the integrals in Exercises 1–16.

1. ● $\int_0^1 \int_0^1 (x - 2y)\, dx\, dy$ *hint* [see Example 1]

2. ● $\int_{-1}^1 \int_0^2 (2x + 3y)\, dx\, dy$

3. ● $\int_0^1 \int_0^2 (ye^x - x - y)\, dx\, dy$

4. ● $\int_1^2 \int_2^3 \left(\frac{1}{x} + \frac{1}{y} \right) dx\, dy$

5. ● $\int_0^2 \int_0^3 e^{x+y}\, dx\, dy$ **6.** ● $\int_0^1 \int_0^1 e^{x-y}\, dx\, dy$

7. ● $\int_0^1 \int_0^{2-y} x\, dx\, dy$ **8.** ● $\int_0^1 \int_0^{2-y} y\, dx\, dy$

9. ● $\int_{-1}^1 \int_{y-1}^{y+1} e^{x+y}\, dx\, dy$ **10.** ● $\int_0^1 \int_y^{y+2} \frac{1}{\sqrt{x+y}}\, dx\, dy$

11. ● $\int_0^1 \int_{-x^2}^{x^2} x\, dy\, dx$ **12.** ● $\int_1^4 \int_{-\sqrt{x}}^{\sqrt{x}} \frac{1}{x}\, dy\, dx$

13. ● $\int_0^1 \int_0^x e^{x^2}\, dy\, dx$ **14.** ● $\int_0^1 \int_0^{x^2} e^{x^3+1}\, dy\, dx$

15. ● $\int_0^2 \int_{1-x}^{8-x} (x + y)^{1/3}\, dy\, dx$

16. ● $\int_1^2 \int_{1-2x}^{x^2} \frac{x+1}{(2x+y)^3}\, dy\, dx$

In Exercises 17–24, find $\iint_R f(x, y)\, dx\, dy$, where R is the indicated domain. (Remember that you often have a choice as to the order of integration.)

17. ● $f(x, y) = 2$ **18.** ● $f(x, y) = x$

hint [see Example 2]

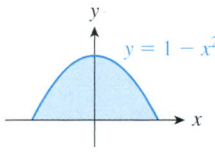   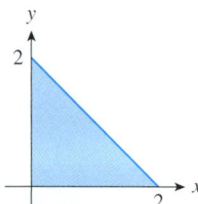

19. ● $f(x, y) = 1 + y$ **20.** ● $f(x, y) = e^{x+y}$

hint [see Example 3]

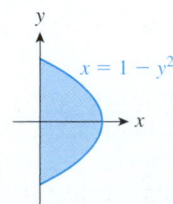   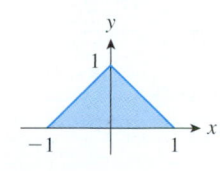

● basic skills

21. ● $f(x, y) = xy^2$

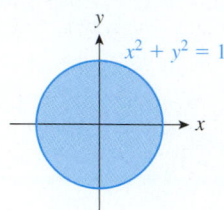

$x^2 + y^2 = 1$

22. ● $f(x, y) = xy^2$

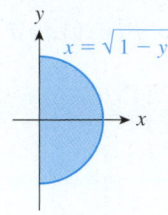

$x = \sqrt{1 - y^2}$

23. ● $f(x, y) = x^2 + y^2$

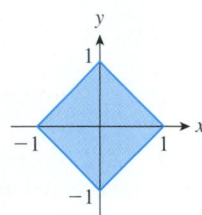

24. ● $f(x, y) = x^2$

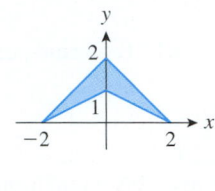

In Exercises 25–30, find the average value of the given function over the indicated domain. hint [see Quick Examples p. 1093]

25. ● $f(x, y) = y$

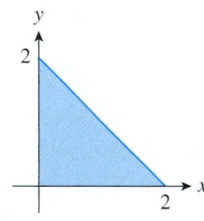

26. ● $f(x, y) = 2 + x$

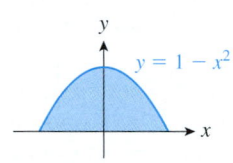

$y = 1 - x^2$

27. ● $f(x, y) = e^y$

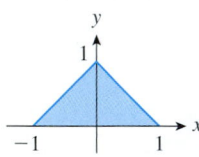

28. ● $f(x, y) = y$

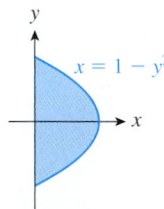

$x = 1 - y^2$

29. ● $f(x, y) = x^2$

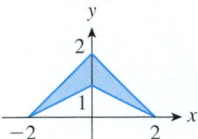

30. ● $f(x, y) = x^2 + y^2$

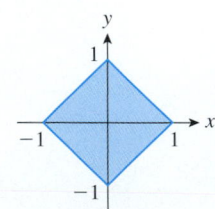

In Exercises 31–36, sketch the region over which you are integrating, then write down the integral with the order of integration reversed (changing the limits of integration as necessary).

31. $\displaystyle\int_0^1 \int_0^{1-y} f(x, y)\, dx\, dy$

32. $\displaystyle\int_{-1}^1 \int_0^{1+y} f(x, y)\, dx\, dy$

33. $\displaystyle\int_{-1}^1 \int_0^{\sqrt{1+y}} f(x, y)\, dx\, dy$

34. $\displaystyle\int_{-1}^1 \int_0^{\sqrt{1-y}} f(x, y)\, dx\, dy$

35. $\displaystyle\int_1^2 \int_1^{4/x^2} f(x, y)\, dy\, dx$

36. $\displaystyle\int_1^{e^2} \int_0^{\ln x} f(x, y)\, dy\, dx$

37. ● Find the volume under the graph of $z = 1 - x^2$ over the region $0 \le x \le 1$ and $0 \le y \le 2$.

38. ● Find the volume under the graph of $z = 1 - x^2$ over the triangle $0 \le x \le 1$ and $0 \le y \le 1 - x$.

39. ● Find the volume of the tetrahedron shown in the figure. Its corners are $(0, 0, 0)$, $(1, 0, 0)$, $(0, 1, 0)$, and $(0, 0, 1)$.

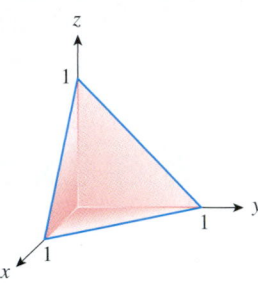

40. Find the volume of the tetrahedron with corners at $(0, 0, 0)$, $(a, 0, 0)$, $(0, b, 0)$, and $(0, 0, c)$.

Applications

41. ● **Productivity** A productivity model at the Handy Gadget Company is

$$P = 10,000x^{0.3}y^{0.7}$$

where P is the number of gadgets the company turns out per month, x is the number of employees at the company, and y is the monthly operating budget in thousands of dollars. Because the company hires part-time workers, it uses anywhere between 45 and 55 workers each month, and its operating budget varies from $8000 to $12,000 per month. What is the average of the possible numbers of gadgets it can turn out per month? (Round the answer to the nearest 1000 gadgets.) hint [see Quick Examples p. 1093]

42. ● **Productivity** Repeat the preceding exercise using the productivity model

$$P = 10,000x^{0.7}y^{0.3}$$

43. ● **Revenue** Your latest CD-ROM of clip art is expected to sell between $q = 8000 - p^2$ and $q = 10,000 - p^2$ copies if priced at p dollars. You plan to set the price between $40 and

● basic skills

$50. What is the average of all the possible revenues you can make? *hint* [see Example 4]

44. ● *Revenue* Your latest DVD drive is expected to sell between $q = 180,000 - p^2$ and $q = 200,000 - p^2$ units if priced at p dollars. You plan to set the price between $300 and $400. What is the average of all the possible revenues you can make?

45. ● *Revenue* Your self-published novel has demand curves between $p = 15,000/q$ and $p = 20,000/q$. You expect to sell between 500 and 1000 copies. What is the average of all the possible revenues you can make?

46. ● *Revenue* Your self-published book of poetry has demand curves between $p = 80,000/q^2$ and $p = 100,000/q^2$. You expect to sell between 50 and 100 copies. What is the average of all the possible revenues you can make?

47. ● *Population Density* The town of West Podunk is shaped like a rectangle 20 miles from west to east and 30 miles from north to south (see the figure). It has a population density of $P(x, y) = e^{-0.1(x+y)}$ hundred people per square mile x miles east and y miles north of the southwest corner of town. What is the total population of the town? *hint* [see Example 5]

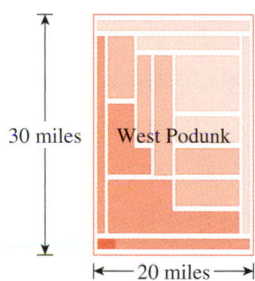

30 miles   West Podunk

←— 20 miles —→

48. ● *Population Density* The town of East Podunk is shaped like a triangle with an east-west base of 20 miles and a north-south height of 30 miles (see the figure). It has a population density of $P(x, y) = e^{-0.1(x+y)}$ hundred people per square mile x miles east and y miles north of the southwest corner of town. What is the total population of the town?

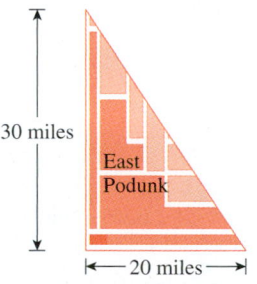

30 miles   East Podunk

←— 20 miles —→

49. ● *Temperature* The temperature at the point (x, y) on the square with vertices $(0, 0)$, $(0, 1)$, $(1, 0)$ and $(1, 1)$ is given by $T(x, y) = x^2 + 2y^2$. Find the average temperature on the square.

50. ● *Temperature* The temperature at the point (x, y) on the square with vertices $(0, 0)$, $(0, 1)$, $(1, 0)$ and $(1, 1)$ is given by $T(x, y) = x^2 + 2y^2 - x$. Find the average temperature on the square.

Communication and Reasoning Exercises

51. ● Explain how double integrals can be used to compute the area between two curves in the xy plane.

52. ● Explain how double integrals can be used to compute the volume of solids in 3-space.

53. ● Complete the following: The first step in calculating an integral of the form $\int_a^b \int_{r(x)}^{s(x)} f(x, y) \, dy \, dx$ is to evaluate the integral _____, obtained by holding ___ constant and integrating with respect to ___ .

54. ● If the units of $f(x, y)$ are zonars per square meter, and x and y are given in meters, what are the units of $\int_a^b \int_{r(x)}^{s(x)} f(x, y) \, dy \, dx$?

55. ● If the units of $\int_a^b \int_{r(x)}^{s(x)} f(x, y) \, dy \, dx$ are paintings, the units of x are picassos, and the units of y are dalis, what are the units of $f(x, y)$?

56. ● Complete the following: If the region R is bounded on the left and right by vertical lines and on the top and bottom by the graphs of functions of x, then we integrate over R by first integrating with respect to _____ and then with respect to ___,

57. Show that if a, b, c and d are constant, then $\int_a^b \int_c^d f(x) g(y) \, dx \, dy = \int_c^d f(x) \, dx \int_a^b g(y) \, dy$. Test this result on the integral $\int_0^1 \int_1^2 y e^x \, dx \, dy$.

58. Refer to Exercise 57. If a, b, c, and d are constants, can $\int_a^b \int_c^d \frac{f(x)}{g(y)} \, dx \, dy$ be expressed as a product of two integrals? Explain.

Chapter **15** Review

KEY CONCEPTS

REVIEW EXERCISES

1. Let $g(x, y, z) = xy(x + y - z) + x^2$. Evaluate $g(0, 0, 0)$, $g(1, 0, 0)$, $g(0, 1, 0)$, $g(x, x, x)$ and $g(x, y + k, z)$.

2. Let $f(x, y, z) = 2.72 - 0.32x - 3.21y + 12.5z$. Complete the following: f ___ by ___ units for every 1 unit of increase in x, and ___ by ___ units for every unit of increase in z.

3. Let $h(x, y) = 2x^2 + xy - x$. Complete the following table of values.

$x \to$

	−1	0	1
$y \downarrow$ −1			
0			
1			

4. Give a formula for a (single) function f with the property that $f(x, y) = -f(y, x)$ and $f(1, -1) = 3$.

In Exercises 5–10, compute the partial derivatives shown for the given function.

5. $f(x, y) = x^2 + xy$; find f_x, f_y, and f_{yy}.

6. $f(x, y) = \dfrac{6}{xy} + \dfrac{xy}{6}$; find f_x, f_y, and f_{yy}.

7. $f(x, y) = 4x + 5y - 6xy$; find $f_{xx}(1, 0) - f_{xx}(3, 2)$

8. $f(x, y) = e^{xy} + e^{3x^2 - y^2}$; find $\dfrac{\partial f}{\partial x}$ and $\dfrac{\partial^2 f}{\partial x\, \partial y}$.

9. $f(x, y, z) = \dfrac{x}{x^2 + y^2 + z^2}$; find $\dfrac{\partial f}{\partial x}$, $\dfrac{\partial f}{\partial y}$, $\dfrac{\partial f}{\partial z}$ and $\left.\dfrac{\partial f}{\partial x}\right|_{(0,1,0)}$.

10. $f(x, y, z) = x^2 + y^2 + z^2 + xyz$; find $f_{xx} + f_{yy} + f_{zz}$.

In Exercises 11–15, locate and classify all critical points.

11. $f(x, y) = (x - 1)^2 + (2y - 3)^2$

12. $g(x, y) = (x - 1)^2 - 3y^2 + 9$

13. $h(x, y) = e^{xy}$

14. $j(x, y) = xy + x^2$

15. $f(x, y) = \ln(x^2 + y^2) - (x^2 + y^2)$

In Exercises 16–19, solve the following constrained optimization problems by using substitution to eliminate a variable. (Do not use Lagrange Multipliers.)

16. Find the minimum value of $f(x, y, z) = x^2 + y^2 + z^2 - 1$ subject to $x = y + z$. Also find the corresponding point(s) (x, y, z).

17. Find the largest value of xyz subject to $x + y + z = 1$ with $x > 0, y > 0, z > 0$. Also find the corresponding point(s) (x, y, z).

18. Minimize $S = xy + x^2z^2 + 4yz$ subject to $xyz = 1$ with $x > 0, y > 0, z > 0$.

19. Find the point on the surface $z = \sqrt{x^2 + 2(y - 3)^2}$ closest to the origin.

In Exercises 20–24, use Lagrange Multipliers to solve the given optimization problem.

20. Find the maximum value of $f(x, y) = xy$ subject to $y = e^{-x}$. Also find the corresponding point(s) (x, y).

21. Find the minimum value of $f(x, y) = x^2 + y^2$ subject to $xy = 2$. Also find the corresponding point(s) (x, y).

22. The problem in Exercise 16.

23. The problem in Exercise 18.

24. The problem in Exercise 19.

In Exercises 25–30, compute the given quantities.

25. $\displaystyle\int_0^1 \int_0^2 (2xy)\, dx\, dy$

26. $\displaystyle\int_1^2 \int_0^1 xye^{x+y}\, dx\, dy$

27. $\displaystyle\int_0^2 \int_0^{2x} \frac{1}{x^2 + 1}\, dy\, dx$

28. The average value of xye^{x+y} over the rectangle $0 \le x \le 1$, $1 \le y \le 2$.

29. $\iint_R (x^2 - y^2)\, dx\, dy$, where R is the region shown in the figure

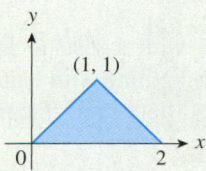

30. The volume under the graph of $z = 1 - y$ over the region in the xy plane between the parabola $y = 1 - x^2$ and the x-axis.

Applications

31. **Website Traffic** OHaganBooks.com has two principal competitors: JungleBooks.com and FarmerBooks.com. Current website traffic at OHaganBooks.com is estimated at 5000 hits per day. This number is predicted to decrease by 0.8 for every new customer of JungleBooks.com and by 0.6 for every new customer of FarmerBooks.com.

   a. Use this information to model the daily website traffic at OHaganBooks.com as a linear function of the new customers of its two competitors.

   b. According to the model, if Junglebooks.com gets 100 new customers and OHaganBooks.com traffic drops to 4770 hits per day, how many new customers has FarmerBooks.com obtained?

32. **Website Traffic** Refer to the model in Exercise 31.

   a. The model in Exercise 31 did not take into account the growth of the total online consumer base. OHaganBooks.com expects to get approximately one additional hit per day for every 10,000 new Internet shoppers. Modify your model in part (a) so as to include this information using a new independent variable.

   b. How many new Internet shoppers would it take to offset the effects on traffic at OHaganBooks.com of 100 new customers at each of its competitor sites?

33. **Internet Advertising** To increase business at OHaganBooks.com, you have purchased banner ads at well-known Internet portals and have advertised on television. The following interaction model shows the average number h of hits per day as a function of monthly expenditures x on banner ads and y on television advertising (x and y are in dollars).

$$h(x, y) = 1800 + 0.05x + 0.08y + 0.00003xy$$

   a. Based on your model, how much traffic can you anticipate if you spend $2000 per month for banner ads and $3000 per month on television advertising?

   b. Evaluate $\dfrac{\partial h}{\partial y}$, specify its units of measurement, and indicate whether it increases or decreases with increasing x.

   c. How much should the company spend on banner ads to obtain 1 hit per day for each $5 spent per month on television advertising?

34. **Internet Advertising** Refer to the model in Exercise 33. One or more of the following five statements is correct. Identify which one(s).

   (A) If nothing is spent on television advertising, one more dollar spent per month in banner ads will buy approximately 0.05 hits per day at OHaganBooks.com

   (B) If nothing is spent on television advertising, one more hit per day at OHaganBooks.com will cost the company about 5¢ per month in banner ads.

   (C) If nothing is spent on banner ads, one more hit per day at OHaganBooks.com will cost the company about 5¢ per month in banner ads.

(D) If nothing is spent on banner ads, one more dollar spent per month in banner ads will buy approximately 0.05 hits per day at OHaganBooks.com

(E) Hits at OHaganBooks.com cost approximately 5¢ per month spent on banner ads, and this cost increases at a rate of 0.003¢ per month, per hit.

35. *Productivity* The holiday season is now at its peak and OHaganBooks.com has been understaffed and swamped with orders. The current backlog (orders unshipped for two or more days) has grown to a staggering 50,000, and new orders are coming in at a rate of 5000 per day. Research based on productivity data at OHaganBooks.com results in the following model:

$$P(x, y) = 1000x^{0.9}y^{0.1} \text{ additional orders filled per day}$$

where x is the number of additional personnel hired and y is the daily budget (excluding salaries) allocated to eliminating the backlog. How many additional orders will be filled per day if the company hires 10 additional employees and budgets an additional \$1000 per day? (Round the answer to the nearest 100.)

36. *Productivity* Refer to the model in Exercise 35. In addition to the daily budget, extra staffing costs the company \$150 per day for every new staff member hired. In order to fill at least 15,000 orders per day at a minimum total daily cost, how many new staff members should the company hire? (Use the method of Lagrange Multipliers).

37. *Profit* If OHaganBooks.com sells x paperback books and y hardcover books per week, it will make an average weekly profit of

$$P(x, y) = 3x + 10y \text{ dollars}$$

If it sells between 1200 and 1500 paperback books and between 1800 and 2000 hardcover books per week, what is the average of all its possible weekly profits?

vMentor Do you need a live tutor for homework problems? Access vMentor on the ThomsonNOW! website at **www.thomsonedu.com** for one-on-one tutoring from a mathematics expert.

CASE STUDY: Modeling Household Income

Bill Varie/Corbis

The Millennium Real Estate Development Corporation is interested in developing housing projects for medium-sized families that have high household incomes. To decide which income bracket to target, the company has asked you, a paid consultant, for an analysis of household income and household size in the United States up to the year 2000. In particular, Millennium is interested in three issues:

- The relationship between household size and household income and the effect of increasing household size on household income.
- The household size that corresponds to the highest household income.
- The change in the relationship between household size and household income over time.
- Some near-term projections of household income vs. household size following the year 2000 (to, say, 2005).

You decide that a good place to start would be with a visit to the Census Bureau's website at http://www.census.gov. After some time battling with search engines, you discover detailed information on household size vs. household income[35] for the period

---

[35] Household income is adjusted for inflation and given in 2000 dollars. SOURCE: Bureau of the Census, 2002; www.census.gov/hhes/income/histinc.

1967–2000. The following table summarizes the information on median household income:

Median Household Income by Household Size and Year

Household Size →

		1	2	3	4	5	6	7
Year ↓	1967	10,321	27,507	36,680	39,520	40,241	39,630	36,658
	1968	11,364	29,055	37,756	41,598	41,889	41,032	39,400
	1969	11,947	30,507	38,959	43,345	44,300	43,903	41,364
	1970	11,993	30,346	38,826	43,181	44,661	44,572	41,241
	1971	12,104	30,058	38,330	43,209	44,378	44,060	40,738
	1972	12,816	31,489	40,562	46,308	47,280	46,634	42,622
	1973	13,730	32,465	41,111	46,674	48,569	48,653	45,718
	1974	13,930	32,317	40,067	46,041	48,008	47,181	45,064
	1975	13,640	31,704	39,968	44,980	46,796	45,811	41,500
	1976	14,400	32,761	40,634	45,986	47,569	47,368	44,093
	1977	14,898	32,981	41,294	47,072	48,098	48,131	43,562
	1978	16,132	34,962	43,963	49,497	50,803	50,334	48,902
	1979	16,259	35,607	44,773	49,708	51,718	50,427	49,969
	1980	16,240	34,833	43,249	48,570	49,457	48,580	46,423
	1981	16,719	34,394	42,929	47,882	47,531	49,012	44,672
	1982	17,159	34,717	42,013	47,464	46,706	46,732	40,762
	1983	17,734	34,758	42,419	48,161	46,251	44,368	38,650
	1984	18,266	35,884	44,316	49,243	48,911	44,924	41,714
	1985	18,238	36,628	45,514	50,188	48,737	47,714	43,370
	1986	18,329	38,144	46,755	52,267	51,523	49,508	41,979
	1987	18,584	38,679	47,300	54,039	52,109	48,824	45,452
	1988	19,675	39,429	47,767	54,442	50,689	51,939	44,573
	1989	19,997	40,268	48,919	54,942	52,970	47,607	44,020
	1990	19,701	40,263	47,205	53,250	50,428	48,995	46,362
	1991	19,125	38,670	47,368	53,326	50,524	45,696	42,280
	1992	18,606	38,389	46,580	53,111	50,853	44,760	40,010
	1993	18,896	38,150	46,360	53,032	49,685	48,336	38,957
	1994	18,672	39,082	47,241	53,817	50,800	49,128	42,152
	1995	19,159	40,085	47,433	55,615	51,325	49,700	43,805
	1996	19,564	40,756	48,988	56,194	52,298	46,391	44,095
	1997	20,075	42,096	50,412	56,885	53,934	49,716	45,306
	1998	21,267	43,804	51,778	58,971	56,671	51,790	49,221
	1999	21,791	44,798	52,910	61,776	56,269	53,630	53,898
	2000	21,468	44,530	54,196	61,847	60,295	54,841	54,663

You notice that the table is actually a numerical representation of the median household income I as a function of two variables, the household size n and the year t.

The numbers are a bit overwhelming, so you decide to use Excel to graph the data as a surface (Figure 38).

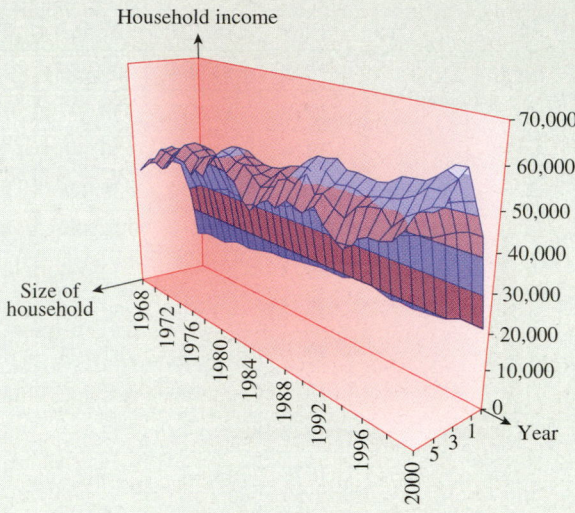

Figure **38**

Now you definitely see two trends. First, the household income peaks at around 5 or 6 people per household, and then drops off at both ends. In fact, the slices through $t = constant$ look parabolic. Second, the household income for all household sizes seems to increase more or less linearly with time (the slices through $n = constant$ are approximately linear).

At this point you realize that a mathematical model of these data would be useful; not only would it "smooth out the bumps" but it would give you a way to complete the project for Millennium. Although technology can give you a regression model for data such as this, it is up to you to decide on the form of the model. It is in choosing an appropriate model that your analysis of the graph comes in handy. Because I appears to vary quadratically with the household size, you would like a general quadratic of the form

$$I = a + bn + cn^2$$

for each value of time t. Also, because I should vary linearly with time t, you would like

$$I = mt + k$$

for each value of n. Putting these together, you get the following candidate model

$$I(n, t) = a_1 + a_2n + a_3n^2 + a_4t$$

where $a_1, a_2, a_3,$ and a_4 are constants you need to determine.

You decide to use Excel to generate your model. The specific software tool you need is called the "Analysis Toolpack" which comes with Excel. (It is found in the Tools menu as "Data Analysis." If it is missing, select Add-Ins from the Tools menu and check "Analysis Toolpack.")

Now you are set to do regression analysis. However, the data as shown in the table are not in a form Excel can use for regression; the data need to be organized into columns, as shown below.

	A	B	C	D	E
1	n	n^2	t	i	
2	1	1	0	10,321	
3	1	1	1	11,364	
4	1	1	2	11,947	
5			3	11,993	
	1	1			
35	1	1	33	21,468	
36	2	4	0	27,507	
37	2	4	1	29,055	
38			1	30,507	
	2	4			
69	2	4	33	44,530	
70	3	9	0	36,680	
71	3	9	1	37,756	
72			2	38,959	
	7	49			
237	7	49	31	49,221	
238	7	49	32	53,898	
239	7	49	33	54,663	
240					
241					

The headings of each column show the variables n and t, with the income i in column D. (Instead of using the calendar year for t, we have represented 1967 by $t = 0$.) Notice that the columns of the original table are in column D, one beneath the other. Thus, Columns A–C show the independent variables (and n^2, which we will treat as an independent variable for the regression), and column D contains the dependent variable.

You now select Data Analysis from the Tools menu. Under "Type of Analysis" you select "Regression," identify where the dependent and independent variables are (D1–D239 for the Y range, and A1–C239 for the X range), check "Labels," and hit "OK."

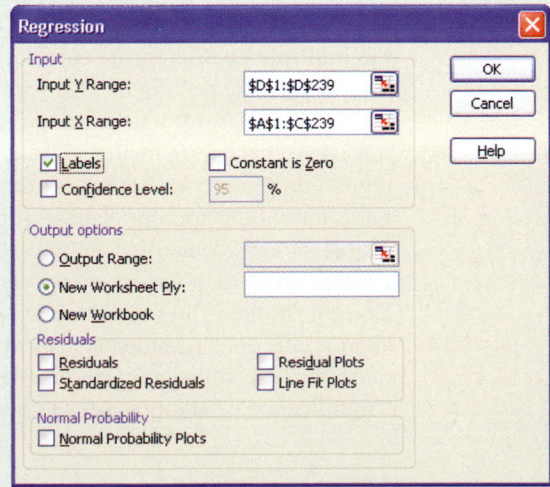

A portion of the output is shown below, with some of the important statistics highlighted.

The desired constants a_1, a_2, a_3, and a_4 appear in the coefficients column at the bottom left, in the correct order: a_1 is the "intercept,", a_2 is the coefficient of n, and so on. Thus, if we round to 5 significant digits, we have

$$a_1 = -5778.0 \quad a_2 = 21{,}008 \quad a_3 = -2139.1 \quad a_4 = 352.06$$

which gives our regression model:

$$I(n, t) = -5778.0 + 21{,}008n - 2139.1n^2 + 352.06t$$

Fine, you say to yourself, now you have the model, but how good a fit is it to the data? That is where the "Multiple R" at the top of the data analysis comes in. R is called the **multiple coefficient of correlation,** and generalizes the coefficient of correlation discussed in the section on regression in Chapter 1: The closer R is to 1, the better the fit. We can interpret its square, given in the table as "R Square" with value 0.938, as indicating that approximately 94% of the variation in median income is explained by the regression model, indicating an excellent fit. The "P-values" at bottom right are also indicators of the appropriateness of the model; a P-value close to zero indicates a high degree of confidence that the corresponding coefficient is really nonzero, whereas a P-value close to 1 indicates low confidence (there is a P-value for each coefficient). Because all the values are extremely tiny, you are confident indeed that the model is an appropriate one. Another statistical indicator is the value of "F" on the right—an indicator of confidence in the model as a whole. The fact that it too is large and its "Significance F" is tiny is yet another good sign.[36]

---

[36] We are being deliberately vague about the exact meaning of these statistics, which are discussed fully in many applied statistics texts.

As comforting as these statistics are, nothing can be quite as persuasive as a graph. You turn to the graphing software of your choice and notice that the graph of the model appears to be a faithful representation of the data. (See Figure 2.)

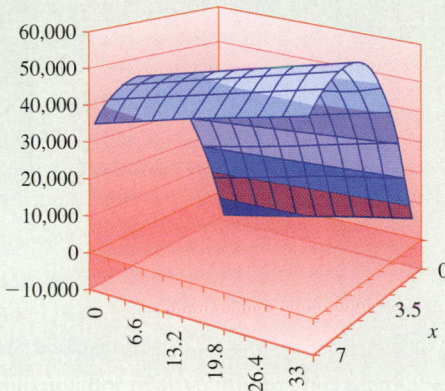

Figure **2**

Now you get to work, using the model to address the questions posed by Millennium.

1. *The relationship between household size and household income and the effect of increasing household size on household income.* You already have a quantitative relationship in the form of the regression model. As for the second part of the question, the rate of change of median household income with respect to household size is given by the partial derivative:

$$\frac{\partial I}{\partial n} = 21{,}008 - 4278.2n \text{ dollars per additional family member}$$

Thus, for example, in a household of 4 people,

$$\frac{\partial I}{\partial n} = 21{,}008 - 4278.2(4) \approx \$3895 \text{ per additional family member}$$

On the other hand, when $n = 5$, one has

$$\frac{\partial I}{\partial n} = 21{,}008 - 4278.2(5) = -\$383 \text{ per additional family member}$$

Notice that the derivative is independent of time: the rate of change of average family income with respect to household size is independent of the date (according to your model).

2, 3. *The household size corresponding to the highest household income and the change in the relationship over time.* Although a glance at the graph shows you that there are no relative maxima, holding *t* constant (that is, on any given year) gives a relative maximum along the corresponding slice when

$$\frac{\partial I}{\partial n} = 0$$

or $$21{,}008 - 4278.2n = 0,$$

which gives $$n = \frac{21{,}008}{4278.2} \approx 4.91$$

In other words, households of 5 tend to have the highest household incomes. Again, $\partial I/\partial n$ does not depend on t, so that this optimal household size seems independent of time t.

4. *Some near-term projections of household income vs. household size.* As we have seen throughout the book, extrapolation can be a risky venture; however, *near-term* extrapolation from a good model can be reasonable. You enter the model in an Excel spreadsheet to obtain the following predicted median household incomes for the years 2001–2005:

Household Size →

Year ↓		1	2	3	4	5	6	7
	2001	25,061	39,652	49,964	55,998	57,755	55,232	48,432
	2002	25,413	40,004	50,316	56,351	58,107	55,585	48,784
	2003	25,765	40,356	50,668	56,703	58,459	55,937	49,136
	2004	26,117	40,708	51,020	57,055	58,811	56,289	49,488
	2005	26,469	41,060	51,372	57,407	59,163	56,641	49,840

Exercises

1. Use Excel to obtain an interaction model of the form

$$I(n, t) = a_1 + a_2 n + a_3 t + a_4 nt$$

Compare the fit of this model with that of the quadratic model above. Comment on the result.

2. How much is there to be gained by including a term of the form $a_5 t^2$ in the original model? (Perform the regression and analyze the result by referring to the P-value for the resulting coefficient of t^2.)

3. The following table shows some data on U.S. population (in thousands) vs. age and year. (This table can be found as an Excel file online by following

Chapter 15 → Case Study Excel Data)

Year $(0 = 1990) \rightarrow$

		0	2	4	6	8	8.5
Age	**2.5**	18,851	19,489	19,694	19,324	19,020	18,974
↓	**7.5**	18,058	18,285	18,742	19,425	19,912	19,931
	12.5	17,191	18,065	18,666	18,949	19,184	19,291
	17.5	17,763	17,170	17,707	18,644	19,460	19,554
	22.5	19,137	19,085	18,451	17,562	17,685	17,796
	27.5	21,233	20,152	19,142	18,993	18,621	18,513
	32.5	21,909	22,237	22,141	21,328	20,163	19,965
	37.5	19,980	21,092	21,973	22,550	22,600	22,589
	42.5	17,793	18,806	19,714	20,809	21,875	22,014
	47.5	13,823	15,362	16,685	18,438	18,850	19,007
	52.5	11,370	12,059	13,199	13,931	15,727	15,973
	57.5	10,474	10,487	10,937	11,362	12,408	12,631
	62.5	10,619	10,440	10,079	9,997	10,256	10,358
	67.5	10,076	9,973	9,963	9,895	9,575	9,515
	72.5	8,022	8,467	8,733	8,778	8,781	8,780
	77.5	6,146	6,392	6,575	6,873	7,195	7,238
	82.5	3,934	4,135	4,350	4,559	4,712	4,748
	87.5	2,050	2,170	2,287	2,395	2,533	2,560
	92.5	765	860	956	1,024	1,094	1,108
	97.5	206	231	249	287	317	324
	102.5	37	44	50	57	63	62

Use multiple regression to construct **(a)** a linear model and **(b)** an interaction model for the data (round all coefficients to four significant digits). Does the interaction model give a significantly better fit in terms of the multiple regression coefficient? Referring to the linear model, does the P-value for the coefficient of y provide strong evidence that the population profile has been changing with time? (A P-value of α indicates that we can be certain with a confidence level of $1 - \alpha$ that the associated coefficient is nonzero.)

4. Graph the data from the preceding exercise and decide whether the linear model gives a faithful representation of the actual data. If not, propose and construct an alternative model. How is the confidence level in the coefficient of y changed?

5. According to your model in the preceding question, why does the age-group with maximum population not change over time? Propose a model in which it does. Construct such a model, and test the additional coefficient(s).

Section 15.1

Example 1 You own a company that makes two models of speakers: the Ultra Mini and the Big Stack. Your total monthly cost (in dollars) to make x Ultra Minis and y Big Stacks is given by

$$C(x, y) = 10,000 + 20x + 40y$$

Compute several values of this function.

Solution with Technology You can have a TI-83/84 compute $C(x, y)$ numerically as follows:

1. In the "Y=" screen, enter

$Y_1 = 10000 + 20X + 40Y$

2. To evaluate, say, $C(10, 30)$ (the cost to make 10 Ultra Minis and 30 Big Stacks), enter

$10 \rightarrow X$
$30 \rightarrow Y$
Y_1

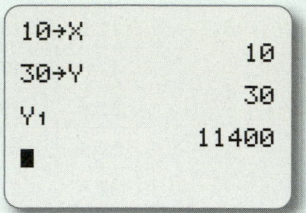

and the calculator will evaluate the function and give the answer, $C(10, 30) = 11,400$.

This procedure is too laborious if you want to calculate $f(x, y)$ for a large number of different values of x and y.

EXCEL Technology Guide

Section 15.1

Example 1 You own a company that makes two models of speakers: the Ultra Mini and the Big Stack. Your total monthly cost (in dollars) to make x Ultra Minis and y Big Stacks is given by

$$C(x, y) = 10,000 + 20x + 40y$$

Compute several values of this function.

Solution with Technology Spreadsheets like Excel handle functions of several variables easily. The following setup shows how a table of values of C can be created, using values of x and y you enter:

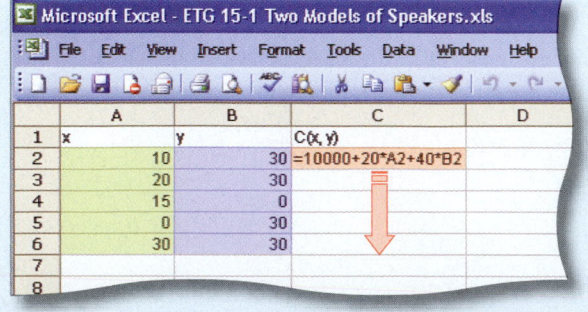 →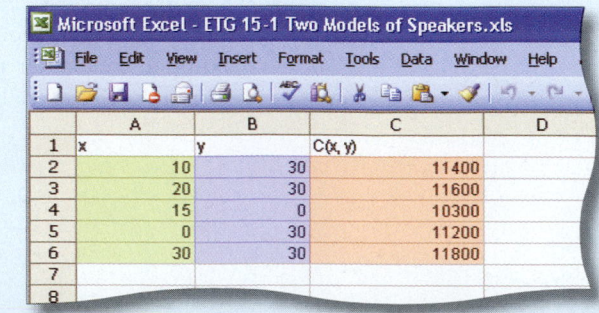

A disadvantage of this layout is that it's not easy to enter values of *x* and *y* systematically in two columns. Can you find a way to remedy this? (See Example 3 for one method.)

Example **3** Use technology to create a table of values of the body mass index

$$M(w, h) = \frac{0.45w}{(0.0254h)^2}$$

Solution with Technology We can use this formula to recreate a table in Excel, as follows:

In the formula in cell B2 we have used B$1 instead of B1 for the *w*-coordinate because we want all references to *w* to use the same row (1). Similarly, we want all references to *h* to refer to the same column (A), so we used $A2 instead of A2.

We copy the formula in cell B2 to all of the red shaded area to obtain the desired table:

Section **15.2**

Example **3** Describe the graph of $f(x, y) = x^2 - y^2$.

Solution with Technology

1. Set up a table showing a range of values of *x* and *y* and the corresponding values of the function (see Example 3 in Section 15.1):

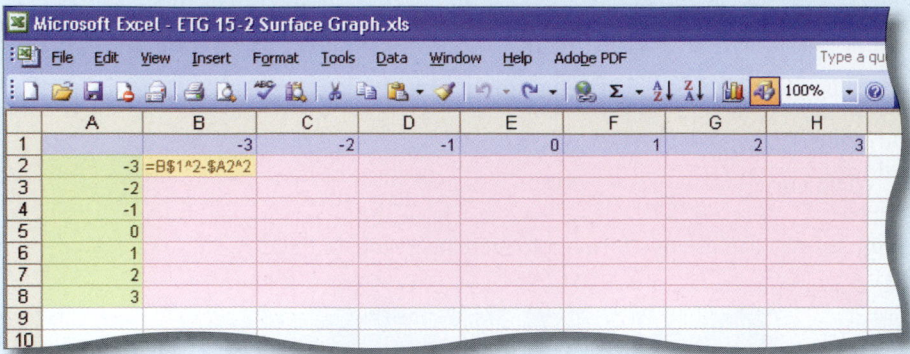

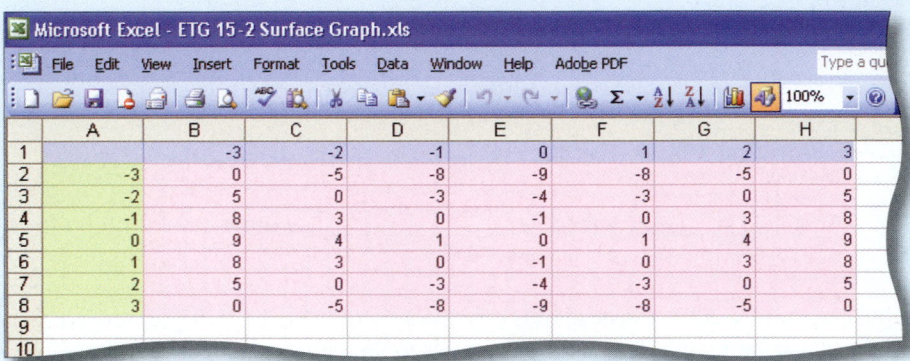

2. Select the cells with the values (B2: H8) and insert a chart, with the "Surface" option selected and "Series in Columns" selected as the data option, to obtain a graph like the following:

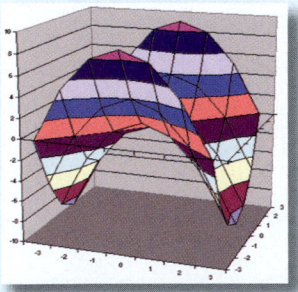

16

Trigonometric Models

CASE STUDY Predicting Cocoa Inventories

As a consultant to the Chocoholic Cocoa Company, you have been asked to model the world-wide production and consumption of cocoa to determine how production cycles are affected by consumption, to estimate the trend in average world cocoa inventories, and to advise your cocoa producers whether to increase or decrease production. You have data showing the production and consumption of cocoa for the past 10 years. How will you analyze these data to prepare your report?

Stephen Studd/Getty Images

Introduction

Cyclical behavior is common in the business world: There are seasonal fluctuations in the demand for surfing equipment, swim wear, snow shovels, and many other items. The nonlinear functions we have studied up to now cannot model this kind of behavior. To model cyclical behavior, we need the **trigonometric** functions.

In the first section, we study the basic trigonometric functions—especially the **sine** and **cosine** functions from which all the trigonometric functions are built—and see how to model various kinds of periodic behavior using these functions. The rest of the chapter is devoted to the calculus of the trigonometric functions—their derivatives and integrals—and to its numerous applications.

16.1 Trigonometric Functions, Models, and Regression

The Sine Function

Figure 1 shows the approximate average daily high temperatures in New York's Central Park.[1] If we draw the graph for several years, we get the repeating pattern shown in Figure 2 where the *x*-coordinate represents time in years, with $x = 0$ corresponding to August 1, and where the *y*-coordinate represents the temperature in °F. This is an example of **cyclical** or **periodic** behavior.

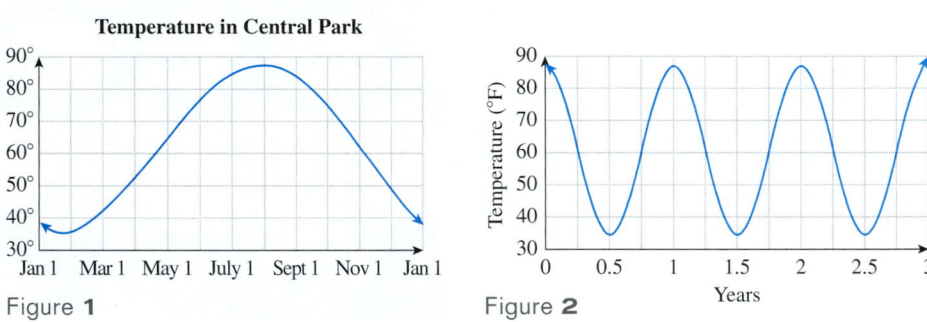

Figure **1**   Figure **2**

Cyclical behavior is also common in the business world. The graph in Figure 3 suggests cyclical behavior in the U.S. unemployment level.[2]

Figure **3**

---

[1] SOURCE: National Weather Service/*The New York Times,* January 7, 1996, p. 36.

[2] SOURCE: Bureau of Labor Statistics, data.bls.gov January, 2006.

From a mathematical point of view, the simplest models of cyclical behavior are the **sine** and **cosine** functions. An easy way to describe these functions is as follows. Imagine a bicycle wheel whose radius is one unit, with a marker attached to the rim of the rear wheel, as shown in Figure 4.

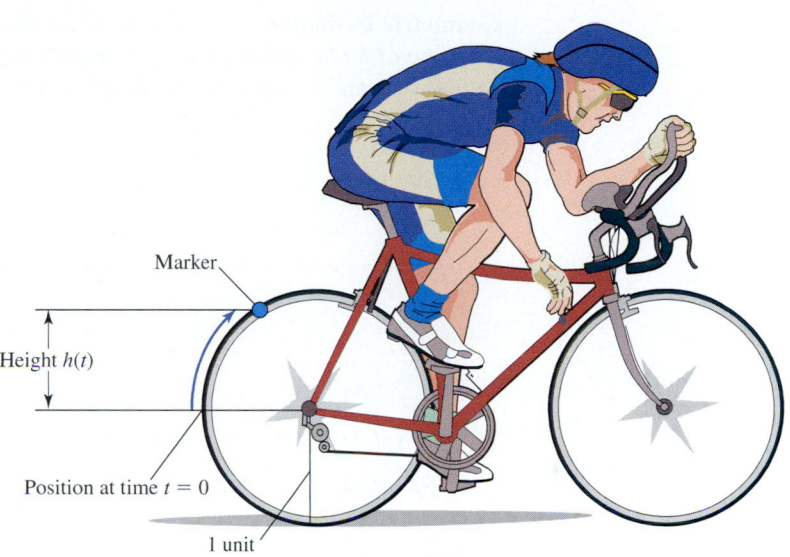

Marker

Height $h(t)$

Position at time $t = 0$

1 unit

Figure **4**

Now, we can measure the height $h(t)$ of the marker above the center of the wheel. As the wheel rotates, $h(t)$ fluctuates between -1 and $+1$. Suppose that, at time $t = 0$ the marker was at height zero as shown in the diagram, so $h(0) = 0$. Because the wheel has a radius of one unit, its circumference (the distance all around) is 2π, where $\pi = 3.14159265\ldots$. If the cyclist happens to be moving at a speed of 1 unit per second, it will take the bicycle wheel 2π seconds to make one complete revolution. During the time interval $[0, 2\pi]$, the marker will first rise to a maximum height of $+1$, drop to a low point of -1, and then return to the starting position of 0 at $t = 2\pi$. This function $h(t)$ is called the **sine function,** denoted by $\sin(t)$. Figure 5 shows its graph.

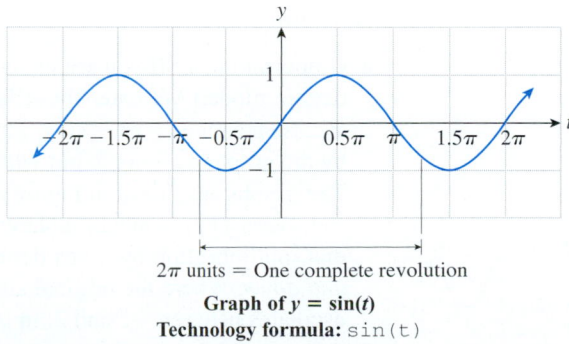

2π units = One complete revolution

Graph of $y = \sin(t)$
Technology formula: `sin(t)`

Figure **5**

Sine Function

"Bicycle Wheel" Definition

If a wheel of radius 1 unit rolls forward at a speed of 1 unit per second, then $\sin(t)$ is the height after t seconds of a marker on the rim of the wheel, starting in the position shown in Figure 4.

Geometric Definition

The **sine** of a real number t is the y-coordinate (height) of the point P in the following diagram, where $|t|$ is the length of the arc shown.

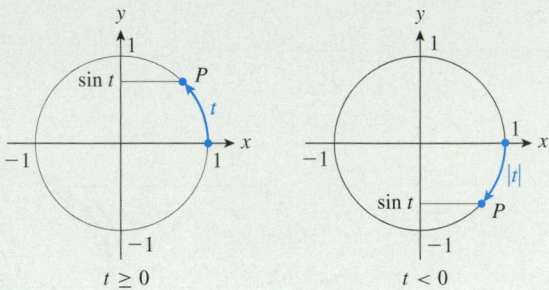

$\sin(t) = y$-coordinate of the point P

quick Examples

From the graph, we see that

1. $\sin(\pi) = 0$

    Graphing Calculator: `sin(π)`
    Excel: `sin(PI())`
      π is `PI()` in Excel

2. $\sin\left(\dfrac{\pi}{2}\right) = 1$

    Graphing Calculator: `sin(π/2)`
    Excel: `sin(PI()/2)`

3. $\sin\left(\dfrac{3\pi}{2}\right) = -1$

    Graphing Calculator: `sin(3π/2)`
    Excel: `sin(3*PI()/2)`

 Ex

Example 1 Some Trigonometric Functions

Use technology to plot the following pairs of graphs on the same set of axes.

a. $f(x) = \sin(x)$; $g(x) = 2\sin(x)$

b. $f(x) = \sin(x)$; $g(x) = \sin(x + 1)$

c. $f(x) = \sin(x)$; $g(x) = \sin(2x)$

Solution

a. (Important note: If you are using a calculator, make sure it is set to *radian mode,* not degree mode.) We enter these functions as `sin(x)` and `2*sin(x)`, respectively. We use the range $-2\pi \le x \le 2\pi$ (approximately $-6.28 \le x \le 6.28$) for x suggested by the graph in Figure 5, but with larger range of y-coordinates (why?): $-3 \le y \le 3$. The graphs are shown in Figure 6.

    Here, $f(x) = \sin(x)$ is shown in orange, and $g(x) = 2\sin(x)$ in turquoise. Notice that multiplication by 2 has doubled the **amplitude,** or *distance that it oscillates up and down.* Where the original sine curve oscillates between -1 and 1, the new curve oscillates between -2 and 2. In general:

    The graph of $A\sin(x)$ *has amplitude* A

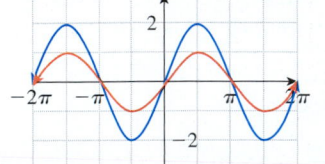

Figure **6**

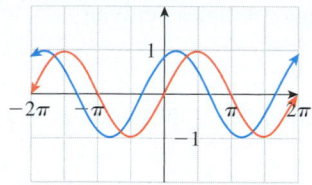

Figure 7

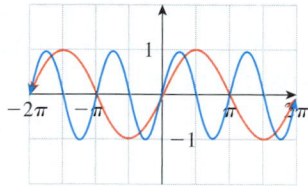

Figure 8

b. We enter these functions as `sin(x)` and `sin(x+1)`, respectively, and we get Figure 7.

Once again, $f(x) = \sin(x)$ is shown in orange and $g(x) = \sin(x + 1)$ is in turquoise. The addition of 1 to the argument has shifted the graph to the left by 1 unit. In general:

Replacing x by x + c shifts the graph to the left c units

(How would we shift the graph to the *right* 1 unit?)

c. We enter these functions as `sin(x)` and `sin(2*x)`, respectively, and get the graph in Figure 8.

The graph of $\sin(2x)$ oscillates twice as fast as the graph of $\sin(x)$. In other words, the graph of $\sin(2x)$ makes two complete cycles on the interval $[0, 2\pi]$, whereas the graph of $\sin x$ completes only one cycle. In general:

Replacing x by bx multiplies the rate of oscillation by b

We can combine the operations in Example 1, and a vertical shift as well, to obtain the following:

Online, follow:

Chapter 16

→ New Functions from Old: Scaled and Shifted Functions

for more discussion of the operations we just used to modify the graph of the sine function.

The General Sine Function

The general sine function is

$$f(x) = A \sin[\omega(x - \alpha)] + C$$

Its graph is shown here.

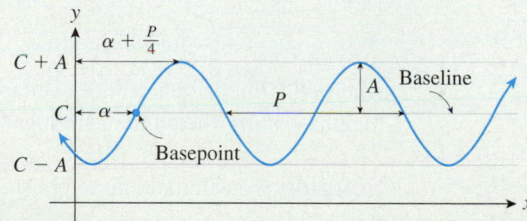

- A is the **amplitude** (the height of each peak above the baseline).
- C is the **vertical offset** (height of the baseline).
- P is the **period** or **wavelength** (the length of each cycle) and is related to ω by

$$P = 2\pi/\omega \quad \text{or} \quad \omega = 2\pi/P$$

- ω is the **angular frequency** (the number of cycles in every interval of length 2π).
- α is the **phase shift.**

Example 2 Electrical Current

The typical voltage V supplied by an electrical outlet in the U.S. is a sinusoidal function that oscillates between -165 volts and $+165$ volts with a frequency of 60 cycles per second. Find an equation for the voltage as a function of time t.

Solution What we are looking for is a function of the form:

$$V(t) = A \sin[\omega(t - \alpha)] + C$$

Referring to the figure above, we can determine the constants.

Amplitude A and Vertical Offset C: Because the voltage oscillates between -165 volts and $+165$ volts, we see that $A = 165$ and $C = 0$.

Period P: Because the electric current completes 60 cycles in one second, the length of time it takes to complete one cycle is $1/60$ second. Thus, the period is $P = 1/60$.

Angular Frequency ω: This is given by the formula:

$$\omega = \frac{2\pi}{P} = 2\pi(60) = 120\pi$$

Phase Shift α: The phase shift α tells us when the curve first crosses the t-axis as it ascends. Because we are free to specify what time $t = 0$ represents, let us say that the curve crosses 0 when $t = 0$, so $\alpha = 0$.

Thus, the equation for the voltage at time t is

$$V(t) = A\,\sin[\omega(t - \alpha)] + C$$
$$= 165\,\sin(120\pi t)$$

where t is time in seconds.

Example 3 Cyclical Employment Patterns

An economist consulted by your employment agency indicates that the demand for temporary employment (measured in thousands of job applications per week) in your county can be roughly approximated by the function

$$d = 4.3\,\sin(0.82t - 0.3) + 7.3$$

where t is time in years since January 2000. Calculate the amplitude, the vertical offset, the phase shift, the angular frequency, and the period, and interpret the results.

Solution To calculate these constants, we write:

$$d = A\,\sin[\omega(t - \alpha)] + C = A\,\sin[\omega t - \omega\alpha] + C$$
$$= 4.3\,\sin(0.82t - 0.3) + 7.3$$

and we see right away that $A = 4.3$ (the amplitude), $C = 7.3$ (vertical offset), and $\omega = 0.82$ (angular frequency). We also have

$$\omega\alpha = 0.3$$

so that $\qquad \alpha = \dfrac{0.3}{\omega} = \dfrac{0.3}{0.82} \approx 0.37$

(rounding to two significant digits; notice that all the constants are given to two digits). Finally, we get the period using the formula:

$$P = \frac{2\pi}{\omega} = \frac{2\pi}{0.82} \approx 7.7$$

We can interpret these numbers as follows: The demand for temporary employment fluctuates in cycles of 7.7 years about a baseline of 7300 job applications per week. Every cycle, the demand peaks at 11,600 applications per week (4300 above the baseline) and dips to a low of 3000. In May, 2000, ($t = 0.37$) the demand for employment was at the baseline level and rising.

The Cosine Function

Closely related to the sine function is the cosine function, defined as follows (refer to the definition of the sine function for comparison):

Cosine Function

Geometric Definition

The **cosine** of a real number t is the x-coordinate of the point P in the following diagram, in which $|t|$ is the length of the arc shown.

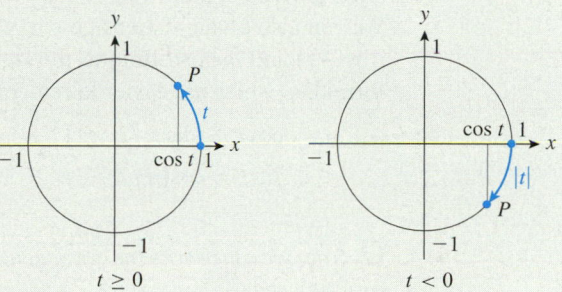

$$\cos(t) = x\text{-coordinate of the point } P$$

Graph of the Cosine Function

The graph of the cosine function is identical to the graph of the sine function, except that it is shifted $\pi/2$ units to the left.

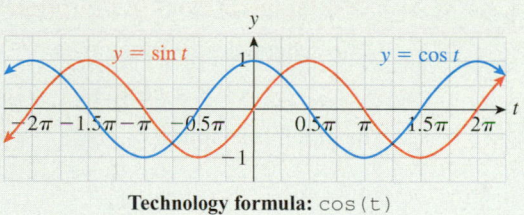

Technology formula: `cos(t)`

Notice that the coordinates of the point P in the diagram above are $(\cos t, \sin t)$ and that the distance from P to the origin is 1 unit. It follows from the Pythagorean Theorem that the distance from a point (x, y) to the origin is $\sqrt{x^2 + y^2}$. Thus:

Square of the distance from P to $(0, 0) = 1$

$$(\sin t)^2 + (\cos t)^2 = 1$$

We often write $(\sin t)^2$ as $\sin^2 t$ and similarly for the cosine, so we can rewrite the equation as

$$\sin^2 t + \cos^2 t = 1$$

This equation is one of the important relationships between the sine and cosine functions.

Fundamental Trigonometric Identities: Relationships between Sine and Cosine

The sine and cosine of a number t are related by

$$\sin^2 t + \cos^2 t = 1$$

We can obtain the cosine curve by shifting the sine curve to the left a distance of $\pi/2$. [See Example 1(b) for a shifted sine function.] Conversely, we can obtain the sine curve from the cosine curve by shifting it $\pi/2$ units to the right. These facts can be expressed as

$$\cos t = \sin(t + \pi/2)$$
$$\sin t = \cos(t - \pi/2)$$

Alternative Formulation

We can also obtain the cosine curve by first inverting the sine curve vertically (replace t by $-t$) and then shifting to the *right* a distance of $\pi/2$. This gives us two alternative formulas (which are easier to remember):

$$\cos t = \sin(\pi/2 - t)$$ Cosine is the sine of the <u>complementary</u> angle
$$\sin t = \cos(\pi/2 - t)$$

Q: *Since we can rewrite the cosine function in terms of the sine function, do we really need the cosine function?*

A: Technically, we don't need the cosine function and could get by with only the sine function. On the other hand, it is convenient to have the cosine function because it starts at its highest point rather than at zero. These two functions and their relationship play important roles throughout mathematics. ∎

The General Cosine Function

The general cosine function is

$$f(x) = A \cos[\omega(x - \alpha)] + C$$

Its graph is as follows:

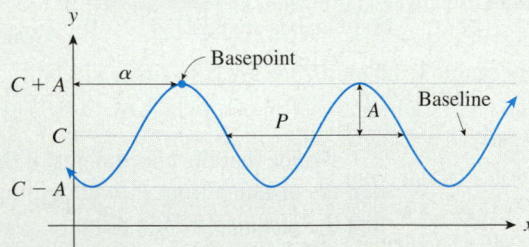

Note that the basepoint of the cosine curve is at the highest point of the curve. All the constants have the same meaning as for the general sine curve:

- A is the amplitude (the height of each peak above the baseline).
- C is the vertical offset (height of the baseline).
- P is the **period** or **wavelength** (the length of each cycle) and is related to ω by

$$P = 2\pi/\omega \quad \text{or} \quad \omega = 2\pi/P$$

- ω is the **angular frequency** (the number of cycles in every interval of length 2π).
- α is the phase shift.

Example 4 Cash Flows into Stock Funds

The annual cash flow into stock funds (measured as a percentage of total assets) has fluctuated in cycles of approximately 40 years since 1955, when it was at a high point. The highs were roughly +15% of total assets, whereas the lows were roughly −10% of total assets.[*]

a. Model this cash flow with a cosine function of the time t in years, with $t = 0$ representing 1955.

b. Convert the answer in part (a) to a sine function model.

Solution

a. Cosine modeling is similar to sine modeling; we are seeking a function of the form

$$P(t) = A \cos[\omega(t - \alpha)] + C$$

Amplitude A and Vertical Offset C: The cash flow fluctuates between −10% and +15%. We can express this as a fluctuation of $A = 12.5$ about the average $C = 2.5$.

Period P: This is given as $P = 40$.

Angular Frequency ω: We find ω from the formula

$$\omega = \frac{2\pi}{P} = \frac{2\pi}{40} = \frac{\pi}{20} \approx 0.157$$

Phase Shift α: The basepoint is at the high point of the curve, and we are told that cash flow was at its high point at $t = 0$. Therefore, the basepoint occurs at $t = 0$, and so $\alpha = 0$.

Putting the model together gives

$$P(t) = A \cos[\omega(t - \alpha)] + C$$
$$\approx 12.5 \cos(0.157t) + 2.5$$

where t is time in years since 1955.

b. To convert between a sine and cosine model, we can use one of the relationships given earlier. Let us use the formula

$$\cos x = \sin(x + \pi/2)$$

Therefore,

$$P(t) \approx 12.5 \cos(0.157t) + 2.5$$
$$= 12.5 \sin(0.157t + \pi/2) + 2.5$$

---

[*] SOURCE: Investment Company Institute/*The New York Times,* February 2, 1997. p. F8.

The Other Trigonometric Functions

The ratios and reciprocals of sine and cosine are given their own names.

Tangent, Cotangent, Secant, Cosecant

Tangent: $\tan x = \dfrac{\sin x}{\cos x}$

$$\textbf{Cotangent:} \quad \cot x = \cot x = \frac{\cos x}{\sin x} = \frac{1}{\tan x}$$

$$\textbf{Secant:} \quad \sec x = \frac{1}{\cos x}$$

$$\textbf{Cosecant:} \quad \csc x = \cosec x = \frac{1}{\sin x}$$

Trigonometric Regression

In the examples so far, we were given enough information to obtain a sine (or cosine) model directly. Often, however, we are given data that only *suggest* a sine curve. In such cases we can use regression to find the best-fit generalized sine (or cosine) curve.

Example 5 Spam

The authors of this book tend to get inundated with spam e-mail. One of us has been systematically documenting the number of spam e-mails arriving at his e-mail account, and noticed a curious cyclical pattern in the average number of e-mails arriving each week.[*] Figure 9 shows the daily spam for the 16-week period beginning June 6, 2005. (Each point is a one-week average):

Daily Spam

Week	0	1	2	3	4	5	6	7	8	9	10	11	12	13	14	15
Number	107	163	170	176	167	140	149	137	158	157	185	151	122	132	134	182

a. Use technology to find the best-fit sine curve of the form $S(t) = A \sin[\omega(t - \alpha)] + C$.

b. Use your model to estimate the period of the cyclical pattern in spam mail, and also to predict the daily spam average for week 23.

Solution

a. Following are the models obtained by using the TI-83/84 and Excel. (See the Technology Guide at the end of the chapter to find out how to obtain these models.)

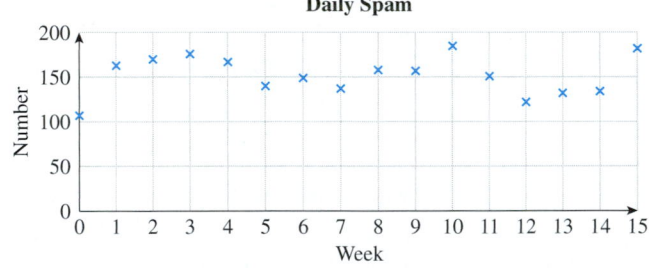

TI-83/84 : $S(t) \approx 11.6 \sin[0.910(t - 1.63)] + 155$

Excel : $S(t) \approx 25.8 \sin[0.96(t - 1.22)] + 153$

---

[*] Confirming the notion that academics have little else to do but fritter away their time in pointless pursuits.

Q: Why do the models from TI-83/84 and Excel differ so drastically?

A: Not all regression algorithms are identical, and it seems that the TI-83/84's algorithm is not very efficient at finding the best-fit sine curve. Indeed, the value for the sum of squares error (SSE) for the TI-83/84 regression curve is around 5030, whereas it is around 2148 for the Excel curve, indicating a far better fit.* Notice another thing: the sine curve does not appear to fit the data well in either graph. In general, we can expect better agreement between the different forms of technology for data that follow a sine curve more closely. ∎

b. This model gives a period of approximately

$$\text{TI-83/84:} \quad P = \frac{2\pi}{\omega} \approx \frac{2\pi}{0.910} \approx 6.9 \text{ weeks}$$

$$\text{Excel:} \quad P = \frac{2\pi}{\omega} \approx \frac{2\pi}{0.96} \approx 6.5 \text{ weeks}$$

So, both models predict a very similar period.
    In week 23, we obtain the following predictions:

$$\text{TI-83/84:} \quad S(23) \approx 11.6 \sin\left[0.910\left(23 - 1.63\right)\right] + 155 \approx 162 \text{ spam e-mails per day}$$

$$\text{Excel:} \quad S(23) \approx 25.8 \sin\left[0.96\left(23 - 1.22\right)\right] + 153 \approx 176 \text{ spam e-mails per day}$$

Note: The actual figure for week 23 was 213 spam e-mails per day. The discrepancy illustrates the danger of using regression models to extrapolate. ∎

_____

* This comparison is actually unfair: The method using Excel's Solver starts with an initial guess of the coefficients, so the TI-83/84 algorithm, which does not require an initial guess, is starting at a significant disadvantage. An initial guess that is way off can result in Solver coming up with a very different result! On the other hand, the TI-83/84 algorithm seems problematic and tends to fail (giving an error message) on many sets of data.

For further discussion of the graphs of the trigonometric functions, their relationship to right triangles, and some exercises, go to the online version of this chapter:

Chapter 16
→ Trigonometric Functions and Calculus
    → The Six Trigonometric Functions

16.1 EXERCISES

● denotes basic skills exercises

tech Ex indicates exercises that should be solved using technology

In Exercises 1–12, graph the given functions or pairs of functions on the same set of axes.

a. Sketch the curves without any technological help by consulting the discussion in Example 1.

b. tech Ex *Use technology to check your sketches.*

1. ● $f(t) = \sin(t)$; $g(t) = 3\sin(t)$ *hint* [see Example 1]

2. ● $f(t) = \sin(t)$; $g(t) = 2.2\sin(t)$

3. ● $f(t) = \sin(t)$; $g(t) = \sin(t - \pi/4)$

4. ● $f(t) = \sin(t)$; $g(t) = \sin(t + \pi)$

5. ● $f(t) = \sin(t)$; $g(t) = \sin(2t)$

6. ● $f(t) = \sin(t)$; $g(t) = \sin(-t)$

7. ● $f(t) = 2\sin(3\pi(t - 0.5)) - 3$

8. ● $f(t) = 2\sin(3\pi(t + 1.5)) + 1.5$

9. ● $f(t) = \cos(t)$; $g(t) = 5\cos(3(t - 1.5\pi))$

10. ● $f(t) = \cos(t)$; $g(t) = 3.1\cos(3t)$

11. ● $f(t) = \cos(t)$; $g(t) = -2.5\cos(t)$

12. ● $f(t) = \cos(t)$; $g(t) = 2\cos(t - \pi)$

In Exercises 13–18, model each curve with a sine function. (Note that not all are drawn with the same scale on the two axes.)

13. ● *hint* [see Example 2]   14. ●

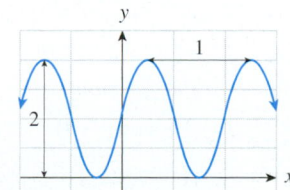

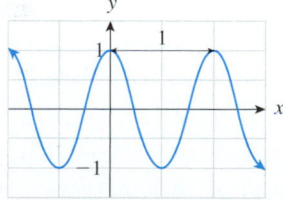

● basic skills   tech Ex technology exercise

15. ●

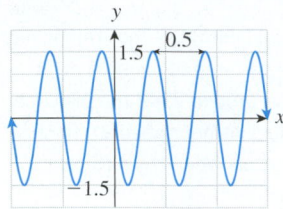

16. ●

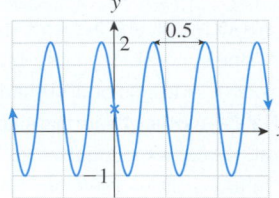

17. ●

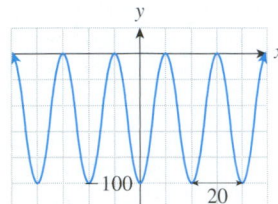

18. ●

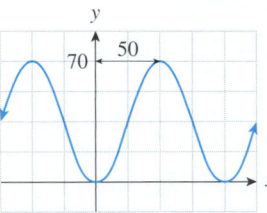

In Exercises 19–24, model each curve with a cosine function. (Note that not all are drawn with the same scale on the two axes.)

19. ● *hint* [see Example 4]

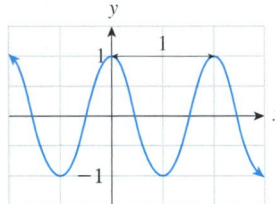

20. ●

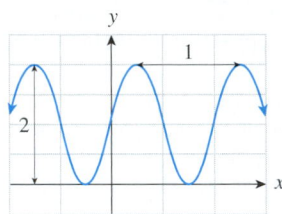

21. ●

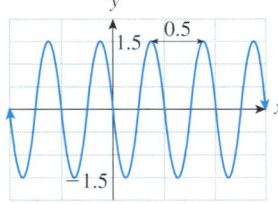

22. ●

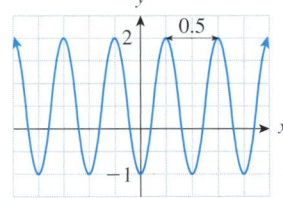

23. ●

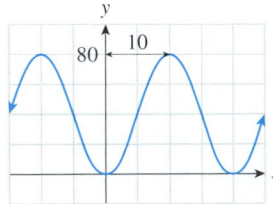

24. ●

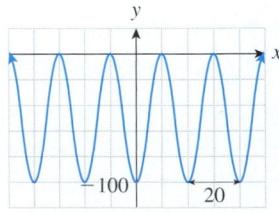

In Exercises 25–28, use the conversion formula $\cos x = \sin(\pi/2 - x)$ to replace each expression by a sine function.

25. $f(t) = 4.2\cos(2\pi t) + 3$ **26.** $f(t) = 3 - \cos(t - 4)$

27. $g(x) = 4 - 1.3\cos[2.3(x - 4)]$

28. $g(x) = 4.5\cos[2\pi(3x - 1)] + 7$

Some Identities Starting with the identity $\sin^2 x + \cos^2 x = 1$ and then dividing both sides of the equation by a suitable trigonometric function, derive the trigonometric identities in Exercises 29 and 30.

29. $\sec^2 x = 1 + \tan^2 x$ **30.** $\csc^2 x = 1 + \cot^2 x$

*Exercises 31–38 are based on the **addition formulas:***

$$\sin(x + y) = \sin x \cos y + \cos x \sin y$$
$$\sin(x - y) = \sin x \cos y - \cos x \sin y$$
$$\cos(x + y) = \cos x \cos y - \sin x \sin y$$
$$\cos(x - y) = \cos x \cos y + \sin x \sin y$$

31. Calculate $\sin(\pi/3)$, given that $\sin(\pi/6) = 1/2$ and $\cos(\pi/6) = \sqrt{3}/2$.

32. Calculate $\cos(\pi/3)$, given that $\sin(\pi/6) = 1/2$ and $\cos(\pi/6) = \sqrt{3}/2$.

33. Use the formula for $\sin(x + y)$ to obtain the identity $\sin(t + \pi/2) = \cos t$.

34. Use the formula for $\cos(x + y)$ to obtain the identity $\cos(t - \pi/2) = \sin t$.

35. Show that $\sin(\pi - x) = \sin x$.

36. Show that $\cos(\pi - x) = -\cos x$.

37. Use the addition formulas to express $\tan(x + \pi)$ in terms of $\tan(x)$.

38. Use the addition formulas to express $\cot(x + \pi)$ in terms of $\cot(x)$.

Applications

39. ● ***Sunspot Activity*** The activity of the sun (sunspots, solar flares, and coronal mass ejection) fluctuates in cycles of around 10–11 years. Sunspot activity can be modeled by the following function:[6]

$$N(t) = 57.7\sin[0.602(t - 1.43)] + 58.8$$

where t is the number of years since January 1, 1997, and $N(t)$ is the number of sunspots observed at time t.

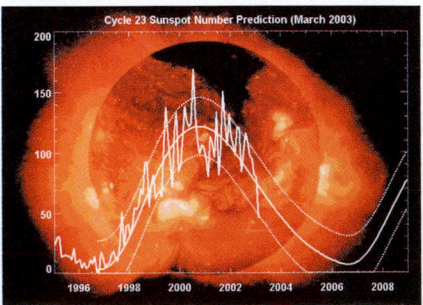

---

[6] The model is based on a regression obtained from predicted data for 1997–2006 and the mean historical period of sunspot activity from 1755 to 1995. SOURCE: NASA Science Directorate; Marshall Space Flight Center http://solarscience.msfc.nasa.gov/predict.shtml, August 2002.

● basic skills   tech Ex technology exercise

a. What is the period of sunspot activity according to this model? (Round your answer to the nearest 0.1 year.)

b. What is the maximum number of sunspots observed? What is the minimum number? (Round your answers to the nearest sunspot.)

c. When, to the nearest year, is sunspot activity next expected to reach a high point? *hint* [see Example 3]

40. ● *Solar Emissions* The following model gives the flux of radio emissions from the sun:

$$F(t) = 49.6 \sin[0.602(t - 1.48)] + 111$$

where t is the number of years since January 1, 1997, and $F(t)$ is the flux of solar emissions of a specified wavelength at time t.[7]

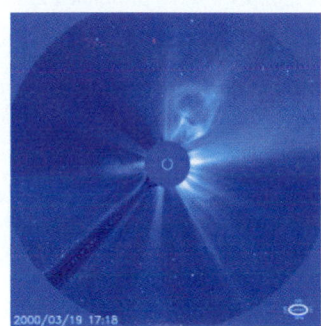

2000/03/19 17:18

a. What is the period of radio activity, according to this model? (Round your answer to the nearest 0.1 year.)

b. What is the maximum flux of radio emissions? What is the minimum flux? (Round your answers to the nearest whole number.)

c. When, to the nearest year, is radio activity next expected to reach a low point?

41. tech Ex *Computer Sales* Sales of computers are subject to seasonal fluctuations. Computer City's sales of computers in 1995 and 1996 can be approximated by the function:

$$s(t) = 0.106 \sin(1.39t + 1.61) + 0.455 \quad (1 \le t \le 8)$$

where t is time in quarters ($t = 1$ represents the end of the first quarter of 1995) and $s(t)$ is computer sales (quarterly revenue) in billions of dollars.[8]

a. Use technology to plot sales versus time from the end of the first quarter of 1995 through the end of the last quarter of 1996. Then use your graph to estimate the value of t and the quarter during which sales were lowest and highest.

b. Estimate Computer City's maximum and minimum quarterly revenue from computer sales.

c. Indicate how the answers to part (b) can be obtained directly from the equation for $s(t)$.

42. tech Ex *Computer Sales* Repeat the preceding exercise using the following model for CompUSA's quarterly sales of computers:[9]

$$s(t) = 0.0778 \sin(1.52t + 1.06) + 0.591$$

43. ● *Computer Sales* (Based on Exercise 41, but no technology required) Computer City's sales of computers in 1995 and 1996 can be approximated by the function

$$s(t) = 0.106 \sin(1.39t + 1.61) + 0.455 \quad (1 \le t \le 8)$$

where t is time in quarters ($t = 1$ represents the end of the first quarter of 1995) and $s(t)$ is computer sales (quarterly revenue) in billions of dollars.[10] Calculate the amplitude, the vertical offset, the phase shift, the angular frequency, and the period, and interpret the results.

44. ● *Computer Sales* Repeat the preceding exercise using the following model for CompUSA's quarterly sales of computers:

$$s(t) = 0.0778 \sin(1.52t + 1.06) + 0.591$$

45. ● *Biology* Sigatoka leaf spot is a plant disease that affects bananas. In an infected plant, the percentage of leaf area affected varies from a low of around 5% at the start of each year to a high of around 20% at the middle of each year.[11] Use the sine function to model the percentage of leaf area affected by Sigatoka leaf spot t weeks since the start of a year. *hint* [see Example 2]

46. ● *Biology* Apple powdery mildew is an epidemic that affects apple shoots. In a new infection, the percentage of apple shoots infected varies from a low of around 10% at the start of May to a high of around 60% six months later.[12] Use the sine function to model the percentage of apple shoots affected by apple powdery mildew t months since the start of a year.

47. ● *Sales Fluctuations* Sales of General Motors cars and light trucks in 1996 fluctuated from a high of $95 billion in October ($t = 0$) to a low of $80 billion in April ($t = 6$).[13] Construct a sinusoidal model for the monthly sales $s(t)$ of General Motors.

48. ● *Sales Fluctuations* Sales of cypods (one-bedroom units) in the city-state of Utarek, Mars[14] fluctuate from a low of 5 units per week each February 1 ($t = 1$) to a high of 35 units per week each August 1 ($t = 7$). Use a sine function to model the weekly sales $s(t)$ of cypods, where t is time in months.

---

[7] Flux measured at a wavelength of 10.7 cm. Ibid.

[8] The model is based on a regression of data that appeared in *The New York Times,* January 8, 1997, p. D1. Constants are rounded to three significant digits.

[9] Ibid.

[10] The model is based on a regression of data which appeared in *The New York Times,* January 8, 1997, p. D1. Constants are rounded to three significant digits.

[11] Based on graphical data. SOURCE: American Phytopathological Society. www.apsnet.org/education/AdvancedPlantPath/Topics/ Epidemiology/CyclicalNature.htm, July 2002.

[12] Ibid.

[13] These are rough figures based on the percentage of the market held by GM as published in *The New York Times,* January 9, 1997, p. D4.

[14] See www.marsnext.com/comm/zonars.html.

● basic skills   tech Ex technology exercise

49. ● *Sales Fluctuations* Repeat Exercise 47, but this time use a cosine function for your model. *hint* [see Example 4]

50. ● *Sales Fluctuations* Repeat Exercise 48, but this time use a cosine function for your model.

51. ● *Tides* The depth of water at my favorite surfing spot varies from 5 to 15 feet, depending on the time. Last Sunday, high tide occurred at 5:00 A.M. and the next high tide occurred at 6:30 P.M. Use a sine function model to describe the depth of water as a function of time t in hours since midnight on Sunday morning.

52. ● *Tides* Repeat Exercise 51 using data from the depth of water at my second favorite surfing spot, where the tide last Sunday varied from a low of 6 feet at 4:00 A.M. to a high of 10 feet at noon.

53. *Inflation* The uninflated cost of Dugout brand snow shovels currently varies from a high of $10 on January 1 ($t = 0$) to a low of $5 on July 1 ($t = 0.5$).

  a. Assuming this trend continues indefinitely, calculate the uninflated cost $u(t)$ of Dugout snow shovels as a function of time t in years. (Use a sine function.)

  b. Assuming a 4% annual rate of inflation in the cost of snow shovels, the cost of a snow shovel t years from now, adjusted for inflation, will be 1.04^t times the uninflated cost. Find the cost $c(t)$ of Dugout snow shovels as a function of time t.

54. *Deflation* Sales of my exclusive 1997 vintage Chateau Petit Mont Blanc vary from a high of 10 bottles per day on April 1 ($t = 0.25$) to a low of four bottles per day on October 1.

  a. Assuming this trend continues indefinitely, find the undeflated sales $u(t)$ of Chateau Petit Mont Blanc as a function of time t in years. (Use a sine function.)

  b. Regrettably, ever since that undercover exposé of my wine-making process, sales of Chateau Petit Mont Blanc have been declining at an annual rate of 12%. Using the preceding exercise as a guide, write down a model for the deflated sales $s(t)$ of Chateau Petit Mont Blanc t years from now.

55. tech Ex *Consumer Spending* The following table shows the annual percentage change in consumer spending compared with the preceding year for even-numbered years. ($t = 0$ represents 1990).[15] *hint* [see Example 5]

t (year)	0	2	4	6	8	10
Change in spending	7.3%	8.0%	7%	4.5%	5.5%	7.2%

  a. Plot the data and *roughly* estimate the period P and the parameters C, A, and α.

  b. Find the best-fit sine curve approximating the given data. (You may have to use your estimates from part (a) as initial guesses if you are using Solver.) Plot the given data together with the regression curve (round coefficients to 3 decimal places).

  c. Complete the following: Based on the regression model, one might argue that the annual percentage change in consumer spending shows a pattern that repeats itself every _____ years, from a low of __% to a high of __%. (Round answers to 1 decimal place.)

56. tech Ex *Consumer Spending* Repeat Exercise 55 using the following data for the odd-numbered years (but not using the data for the even-numbered years).[16]

t (year)	1	3	5	7	9	11
Change in spending	5.0%	7.0%	6.0%	4.8%	8.0%	6.1%

Music Musical sounds exhibit the same kind of periodic behavior as the trigonometric functions. High-pitched notes have short periods (less than 1/1000 second), while the lowest audible notes have periods of about 1/100 second. Some electronic synthesizers work by superimposing (adding) sinusoidal functions of different frequencies to create different textures. Exercises 57–60 show some examples of how superposition can be used to create interesting periodic functions.

57. tech Ex *Sawtooth Wave*

  a. Graph the following functions in a window with $-7 \le x \le 7$ and $-1.5 \le y \le 1.5$.

$$y_1 = \frac{2}{\pi} \cos x$$

$$y_3 = \frac{2}{\pi} \cos x + \frac{2}{3\pi} \cos 3x$$

$$y_5 = \frac{2}{\pi} \cos x + \frac{2}{3\pi} \cos 3x + \frac{2}{5\pi} \cos 5x$$

  b. Following the pattern established above, give a formula for y_{11} and graph it in the same window.

  c. How would you modify y_{11} to approximate a saw-tooth wave with an amplitude of 3 and a period of 4π?

58. tech Ex *Square Wave* Repeat the preceding exercise using sine functions in place of cosine functions in order to approximate a square wave.

59. tech Ex *Harmony* If we add two sinusoidal functions with frequencies that are simple ratios of each other, the result is a pleasing sound. The following function models two notes an octave apart together with the intermediate fifth:

$$y = \cos(x) + \cos(1.5x) + \cos(2x)$$

Graph this function in the window $0 \le x \le 20$ and $-3 \le y \le 3$ and estimate the period of the resulting wave.

60. tech Ex *Discord* If we add two sinusoidal functions with similar, but unequal, frequencies, the result is a function that "pulsates," or exhibits "beats." (Piano tuners and guitar players use this phenomenon to help them tune an instrument.) Graph the function

$$y = \cos(x) + \cos(0.9x)$$

---

[15] Figures are approximate, and are based on purchases of general merchandise. SOURCE: *New York Times*, February 6, 2001, p. C1.

[16] Some figures were from the second or third quarter of the stated year. SOURCE: Ibid.

● basic skills   tech Ex technology exercise

in the window $-50 \le x \le 50$ and $-2 \le y \le 2$ and estimate the period of the resulting wave.

Communication and Reasoning Exercises

61. ● What are the seasonal highs and lows for sales of a commodity modeled by a function of the form $s(t) = A \sin(2\pi t) + B$ (A, B constants)?

62. ● Your friend has come up with the following model for choral society Tupperware stock inventory: $r(t) = 4 \sin(2\pi(t - 2)/3) + 2.3$, where t is time in weeks and $r(t)$ is the number of items in stock. Why is the model not realistic?

63. ● Your friend is telling everybody that all six trigonometric functions can be obtained from the single function $\sin x$. Is he correct? Explain your answer.

64. ● Another friend claims that all six trigonometric functions can be obtained from the single function $\cos x$. Is she correct? Explain your answer.

65. ● If weekly sales of sodas at a movie theater are given by $s(t) = A + B \cos(\omega t)$, what is the largest B can be? Explain your answer.

66. ● Complete the following: If the cost of an item is given by $c(t) = A + B \cos[\omega(t - \alpha)]$, then the cost fluctuates by _____ with a period of _____ about a base of _____, peaking at time $t = $ _____.

● basic skills   **tech** Ex technology exercise

16.2 Derivatives of Trigonometric Functions and Applications

We start with the derivatives of the sine and cosine functions.

Theorem: Derivatives of the Sine and Cosine Functions

The sine and cosine functions are differentiable with

$$\frac{d}{dx} \sin x = \cos x$$

$$\frac{d}{dx} \cos x = -\sin x \qquad \text{Notice the sign change.}$$

quick Examples

1. $\dfrac{d}{dx}[x \cos x] = 1 \cdot \cos x + x \cdot (-\sin x)$ Product rule: $x \cos x$ is a product.[*]

$\qquad\qquad\qquad = \cos x - x \sin x$

2. $\dfrac{d}{dx}\left[\dfrac{x^2 + x}{\sin x}\right] = \dfrac{(2x + 1)(\sin x) - (x^2 + x)(\cos x)}{\sin^2 x}$ Quotient rule

---

[*] Apply the calculation thought experiment: If we were to compute $x \cos x$, the last operation we would perform is the multiplication of x and $\cos x$. Hence, $x \cos x$ is a product.

Before deriving these formulas, we can see right away that they are plausible by examining Figure 10, which shows the graphs of the sine and cosine functions together with their derivatives.

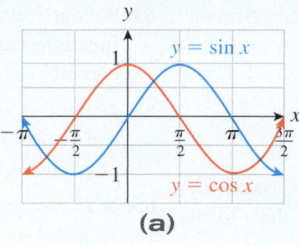

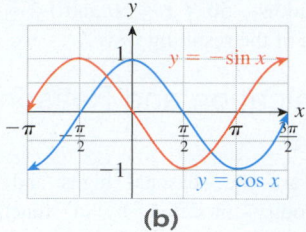

(a)   **(b)**

Figure **10**

Notice, for instance, that in Figure 10(a), the graph of $\sin x$ is rising most rapidly when $x = 0$, corresponding to the maximum value of its derivative, $\cos x$. When $x = \pi/2$, the graph of $\sin x$ levels off, so that its derivative, $\cos x$, is 0. Another point to notice: Because periodic functions (such as sine and cosine) repeat their behavior, their derivatives must also be periodic.

Derivation of Formulas for Derivatives of the Sine and Cosine Functions

We first calculate the derivative of $\sin x$ from scratch, using the definition of the derivative:

$$\frac{d}{dx} f(x) = \lim_{h \to 0} \frac{f(x+h) - f(x)}{h}$$

Thus,

$$\frac{d}{dx} \sin x = \lim_{h \to 0} \frac{\sin(x+h) - \sin x}{h}$$

We now use the addition formula given in Exercise Set 16.1:

$$\sin(x + h) = \sin x \cos h + \cos x \sin h$$

Substituting this expression for $\sin(x + h)$ gives

$$\frac{d}{dx} \sin x = \lim_{h \to 0} \frac{\sin x \cos h + \cos x \sin h - \sin x}{h}$$

Grouping the first and third terms together and factoring out the term $\sin x$ gives

$$\frac{d}{dx} \sin x = \lim_{h \to 0} \frac{\sin x (\cos h - 1) + \cos x \sin h}{h}$$

$$= \lim_{h \to 0} \frac{\sin x (\cos h - 1)}{h} + \lim_{h \to 0} \frac{\cos x \sin h}{h} \qquad \text{Limit of a sum}$$

$$= \sin x \lim_{h \to 0} \frac{\cos h - 1}{h} + \cos x \lim_{h \to 0} \frac{\sin h}{h}$$

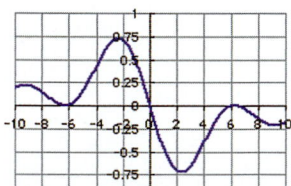

Figure **11**

and we are left with two limits to evaluate. Calculating these limits analytically requires a little trigonometry.[17] Alternatively, we can get a good idea of what these two limits are by estimating them numerically or graphically. Figures 11 and 12 show the graphs of $(\cos h - 1)/h$ and $(\sin h)/h$, respectively.

We find that:

$$\lim_{h \to 0} \frac{\cos h - 1}{h} = 0$$

Figure **12**

---

[17] You can find these calculations online. Follow:

Chapter 16 → Proof of Some Trigonometric Limits

and

$$\lim_{h \to 0} \frac{\sin h}{h} = 1$$

Therefore,

$$\frac{d}{dx} \sin x = (\sin x)(0) + (\cos x)(1) = \cos x$$

This is the required formula for the derivative of $\sin x$.

Turning to the derivative of the cosine function, we use the identity

$$\cos x = \sin(\pi/2 - x)$$

from Section 16.1. If $y = \cos x = \sin(\pi/2 - x)$, then, using the chain rule, we have

$$\frac{dy}{dx} = \cos(\pi/2 - x)\frac{d}{dx}(\pi/2 - x)$$
$$= (-1)\cos(\pi/2 - x)$$
$$= -\sin x \qquad \text{Using the identity } \cos(\pi/2 - x) = \sin x$$

This is the required formula for the derivative of $\cos x$.

Just as with logarithmic and exponential functions, the chain rule can be used to find more general derivatives.

Derivatives of Sines and Cosines of Functions

Original Rule	Generalized Rule	In Words
$\dfrac{d}{dx} \sin x = \cos x$	$\dfrac{d}{dx} \sin u = \cos u \dfrac{du}{dx}$	The derivative of the sine of a quantity is the cosine of that quantity, times the derivative of that quantity.
$\dfrac{d}{dx} \cos x = -\sin x$	$\dfrac{d}{dx} \cos u = -\sin u \dfrac{du}{dx}$	The derivative of the cosine of a quantity is negative sine of that quantity, times the derivative of that quantity.

quick Examples

1. $\dfrac{d}{dx} \sin(3x^2 - 1) = \cos(3x^2 - 1)\dfrac{d}{dx}(3x^2 - 1)$ $u = 3x^2 - 1$ (see footnote*)

    $= 6x \cos(3x^2 - 1)$ We placed the $6x$ in front—see Note below.

2. $\dfrac{d}{dx} \cos(x^3 + x) = -\sin(x^3 + x)\dfrac{d}{dx}(x^3 + x)$ $u = x^3 + x$

    $= -(3x^2 + 1)\sin(x^3 + x)$

---
* If we were to evaluate $\sin(3x^2 - 1)$, the last operation we would perform is taking the sine of a quantity. Thus, the calculation thought experiment tells us that we are dealing with the *sine of a quantity,* and we use the generalized rule.

Note: Avoid writing ambiguous expressions like $\cos(3x^2 - 1)(6x)$. Does this mean

$$\cos[(3x^2 - 1)(6x)] \qquad \text{The cosine of the quantity } (3x^2 - 1)(6x)$$

or does it mean

$$[\cos(3x^2 - 1)](6x)? \qquad \text{The product of } \cos(3x^2 - 1) \text{ and } 6x$$

To avoid the ambiguity, place the $6x$ in front of the cosine expression and write

$$6x \cos(3x^2 - 1) \qquad \text{The product of } 6x \text{ and } \cos(3x^2 - 1)$$

Example 1 Derivatives of Trigonometric Functions

Find the derivatives of the following functions:

a. $f(x) = \sin^2 x$ **b.** $g(x) = \sin^2(x^2)$ **c.** $h(x) = e^{-x}\cos(2x)$

Solution

a. Recall that $\sin^2 x = (\sin x)^2$. The calculation thought experiment tells us that $f(x)$ is the square of a quantity.[*] Therefore, we use the chain rule (or generalized power rule) for differentiating the square of a quantity:

$$\frac{d}{dx}(u^2) = 2u\frac{du}{dx}$$

$$\frac{d}{dx}(\sin x)^2 = 2(\sin x)\frac{d(\sin x)}{dx} \qquad u = \sin x$$

$$= 2\sin x \cos x$$

Thus, $f'(x) = 2\sin x \cos x$.

b. We rewrite the function $g(x) = \sin^2(x^2)$ as $[\sin(x^2)]^2$. Because $g(x)$ is the square of a quantity, we have

$$\frac{d}{dx}\sin^2(x^2) = \frac{d}{dx}[\sin(x^2)]^2 \qquad \text{Rewrite } \sin^2(-) \text{ as } [\sin(-)]^2.$$

$$= 2\sin(x^2)\frac{d[\sin(x^2)]}{dx} \qquad \frac{d}{dx}[u^2] = 2u\frac{du}{dx} \text{ with } u = \sin(x^2)$$

$$= 2\sin(x^2)\cdot\cos(x^2)\cdot 2x \qquad \frac{d}{dx}\sin u = \cos u\frac{du}{dx} \text{ with } u = x^2$$

Thus, $g'(x) = 4x\sin(x^2)\cos(x^2)$.

c. Because $h(x)$ is the product of e^{-x} and $\cos(2x)$, we use the product rule:

$$h'(x) = (-e^{-x})\cos(2x) + e^{-x}\frac{d}{dx}[\cos(2x)]$$

$$= (-e^{-x})\cos(2x) - e^{-x}\sin(2x)\frac{d}{dx}[2x] \qquad \frac{d}{dx}\cos u = -\sin u\frac{du}{dx}$$

$$= -e^{-x}\cos(2x) - 2e^{-x}\sin(2x)$$

$$= -e^{-x}[\cos(2x) + 2\sin(2x)]$$

---

[*] Notice the difference between $\sin^2 x$ and $\sin(x^2)$. The first is the square of $\sin x$, whereas the second is the sin of the quantity x^2.

Derivatives of Other Trigonometric Functions

Because the remaining trigonometric functions are ratios of sines and cosines, we can use the quotient rule to find their derivatives. For example, we can find the derivative of tan x as follows:

$$\frac{d}{dx} \tan x = \frac{d}{dx}\left(\frac{\sin x}{\cos x}\right)$$

$$= \frac{(\cos x)(\cos x) - (\sin x)(-\sin x)}{\cos^2 x}$$

$$= \frac{\cos^2 x + \sin^2 x}{\cos^2 x}$$

$$= \frac{1}{\cos^2 x}$$

$$= \sec^2 x$$

We ask you to derive the other three derivatives in the exercises. Here is a list of the derivatives of all six trigonometric functions and their chain rule variants.

Derivatives of the Trigonometric Functions

Original Rule	*Generalized Rule*
$\dfrac{d}{dx}\sin x = \cos x$	$\dfrac{d}{dx}\sin u = \cos u\,\dfrac{du}{dx}$
$\dfrac{d}{dx}\cos x = -\sin x$	$\dfrac{d}{dx}\cos u = -\sin u\,\dfrac{du}{dx}$
$\dfrac{d}{dx}\tan x = \sec^2 x$	$\dfrac{d}{dx}\tan u = \sec^2 u\,\dfrac{du}{dx}$
$\dfrac{d}{dx}\cot x = -\csc^2 x$	$\dfrac{d}{dx}\cot u = -\csc^2 u\,\dfrac{du}{dx}$
$\dfrac{d}{dx}\sec x = \sec x \tan x$	$\dfrac{d}{dx}\sec u = \sec u \tan u\,\dfrac{du}{dx}$
$\dfrac{d}{dx}\csc x = -\csc x \cot x$	$\dfrac{d}{dx}\csc u = -\csc u \cot u\,\dfrac{du}{dx}$

quick Examples

1. $\dfrac{d}{dx}\tan(x^2 - 1) = \sec^2(x^2 - 1)\dfrac{d(x^2 - 1)}{dx}$ $u = x^2 - 1$

$$= 2x \sec^2(x^2 - 1)$$

2. $\dfrac{d}{dx}\csc(e^{3x}) = -\csc(e^{3x})\cot(e^{3x})\dfrac{d(e^{3x})}{dx}$ $u = e^{3x}$

$$= -3e^{3x}\csc(e^{3x})\cot(e^{3x})$$ The derivative of e^{3x} is $3e^{3x}$.

Figure **13**

Example **2** Gas Heating Demand

In the preceding section, we saw that seasonal fluctuations in temperature suggested a sine function. For instance, we can use the function

$$T = 60 + 25 \sin\left[\frac{\pi}{6}(x - 4)\right]$$ t = temperature in °F; x = months since Jan 1

to model a temperature that fluctuates between 35°F on Feb. 1 ($x = 1$) and 85°F on Aug. 1 ($x = 7$) (see Figure 13).

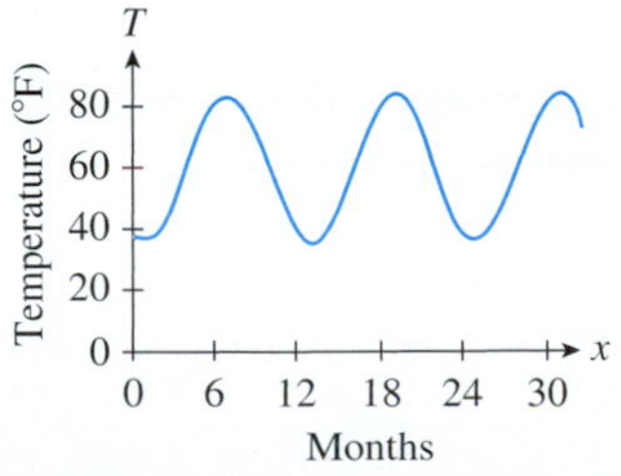

The demand for gas at a utility company can be expected to fluctuate in a similar way because demand grows with increased heating requirements. A reasonable model might therefore be

$$G = 400 - 100 \sin\left[\frac{\pi}{6}(x - 4)\right]$$ Why did we subtract the sine term?

where G is the demand for gas in cubic yards per day. Find and interpret $G'(10)$.

Solution First, we take the derivative of G:

$$G'(x) = -100 \cos\left[\frac{\pi}{6}(x - 4)\right] \cdot \frac{\pi}{6}$$

$$= -\frac{50\pi}{3} \cos\left[\frac{\pi}{6}(x - 4)\right] \text{ cubic yards per day, per month}$$

Thus,

$$G'(10) = -\frac{50\pi}{3} \cos\left[\frac{\pi}{6}(10 - 4)\right]$$

$$= -\frac{50\pi}{3} \cos(\pi) = \frac{50\pi}{3}$$ Since $\cos \pi = -1$

Because the units of $G'(10)$ are cubic yards per day per month, we interpret the result as follows: On November 1 ($x = 10$) the daily demand for gas is increasing at a rate of $50\pi/3 \approx 52$ cubic yards per day, per month. This is consistent with Figure 13, which shows the temperature decreasing on that date.

16.2 EXERCISES

● denotes basic skills exercises

In Exercises 1–28, find the derivatives of the given functions.
hint [see Quick Examples on p. 1125 & 1129]

1. ● $f(x) = \sin x - \cos x$

2. ● $f(x) = \tan x - \sin x$

3. ● $g(x) = (\sin x)(\tan x)$

4. ● $g(x) = (\cos x)(\cot x)$

5. ● $h(x) = 2 \csc x - \sec x + 3x$

6. ● $h(x) = 2 \sec x + 3 \tan x + 3x$

7. ● $r(x) = x \cos x + x^2 + 1$

8. ● $r(x) = 2x \sin x - x^2$

9. ● $s(x) = (x^2 - x + 1)\tan x$ **10.** ● $s(x) = \dfrac{\tan x}{x^2 - 1}$

11. ● $t(x) = \dfrac{\cot x}{1 + \sec x}$

12. ● $t(x) = (1 + \sec x)(1 - \cos x)$

13. ● $k(x) = \cos^2 x$ **14.** ● $k(x) = \tan^2 x$

15. ● $j(x) = \sec^2 x$ **16.** ● $j(x) = \csc^2 x$

17. ● $p(x) = 2 + 5 \sin\left[\dfrac{\pi}{5}(x - 4)\right]$

18. ● $p(x) = 10 - 3 \cos\left[\dfrac{\pi}{6}(x + 3)\right]$

19. ● $u(x) = \cos(x^2 - x)$

20. ● $u(x) = \sin(3x^2 + x - 1)$

21. ● $v(x) = \sec(x^{2.2} + 1.2x - 1)$

22. ● $v(x) = \tan(x^{2.2} + 1.2x - 1)$

23. ● $w(x) = \sec x \tan(x^2 - 1)$

24. ● $w(x) = \cos x \sec(x^2 - 1)$

25. ● $y(x) = \cos(e^x) + e^x \cos x$

26. ● $y(x) = \sec(e^x)$

27. ● $z(x) = \ln|\sec x + \tan x|$

28. ● $z(x) = \ln|\csc x + \cot x|$

In Exercises 29–32, derive the given formulas from the derivatives of sine and cosine. hint [see Discussion on pp. 1128–1129]

29. $\dfrac{d}{dx} \sec x = \sec x \tan x$ **30.** $\dfrac{d}{dx} \cot x = -\csc^2 x$

31. $\dfrac{d}{dx} \csc x = -\csc x \cot x$

32. $\dfrac{d}{dx} \ln|\sec x| = \tan x$

● basic skills

Calculate the derivatives in Exercises 33–40.

33. $\dfrac{d}{dx}[e^{-2x}\sin(3\pi x)]$ **34.** $\dfrac{d}{dx}[e^{5x}\sin(-4\pi x)]$

35. $\dfrac{d}{dx}[\sin(3x)]^{0.5}$ **36.** $\dfrac{d}{dx}\cos\left(\dfrac{x^2}{x-1}\right)$

37. $\dfrac{d}{dx}\sec\left(\dfrac{x^3}{x^2-1}\right)$ **38.** $\dfrac{d}{dx}\left(\dfrac{\tan x}{2+e^x}\right)^2$

39. $\dfrac{d}{dx}([\ln|x|][\cot(2x-1)])$

40. $\dfrac{d}{dx}\ln|\sin x - 2xe^{-x}|$

In Exercises 41 and 42, investigate the differentiability of the given functions at the given points. If $f'(a)$ exists, give its approximate value.

41. $f(x) = |\sin x|$ **a.** $a = 0$ **b.** $a = 1$

42. $f(x) = |\sin(1-x)|$ **a.** $a = 0$ **b.** $a = 1$

*Estimate the limits in Exercises 43–48 **(a)** numerically and **(b)** using L'Hospital's rule.*

43. $\displaystyle\lim_{x\to 0}\dfrac{\sin^2 x}{x}$ **44.** $\displaystyle\lim_{x\to 0}\dfrac{\sin x}{x^2}$

45. $\displaystyle\lim_{x\to 0}\dfrac{\sin(2x)}{x}$ **46.** $\displaystyle\lim_{x\to 0}\dfrac{\sin x}{\tan x}$

47. $\displaystyle\lim_{x\to 0}\dfrac{\cos x - 1}{x^3}$ **48.** $\displaystyle\lim_{x\to 0}\dfrac{\cos x - 1}{x^2}$

In Exercises 49–52, find the indicated derivative using implicit differentiation.

49. $x = \tan y$; find $\dfrac{dy}{dx}$

50. $x = \cos y$; find $\dfrac{dy}{dx}$

51. $x + y + \sin(xy) = 1$; find $\dfrac{dy}{dx}$

52. $xy + x\cos y = x$; find $\dfrac{dy}{dx}$

Applications

53. ● ***Cost*** The cost in dollars of Dig-It brand snow shovels is given by

$$c(t) = 3.5\sin[2\pi(t - 0.75)]$$

where t is time in years since January 1, 2002. How fast, in dollars per week, is the cost increasing each October 1? *hint* [see Example 2]

54. ● ***Sales*** Daily sales of Doggy brand cookies can be modeled by

$$s(t) = 400\cos[2\pi(t-2)/7]$$

cartons, where t is time in days since Monday morning. How fast are sales changing on Thursday morning?

55. ● ***Sunspot Activity*** The activity of the sun can be approximated by the following model of sunspot activity:[18]

$$N(t) = 57.7\sin[0.602(t - 1.43)] + 58.8$$

where t is the number of years since January 1, 1997, and $N(t)$ is the number of sunspots observed at time t. Compute and interpret $N'(6)$.

56. ● ***Solar Emissions*** The following model gives the flux of radio emissions from the sun:

$$F(t) = 49.6\sin[0.602(t - 1.48)] + 111$$

where t is the number of years since January 1, 1997, and $F(t)$ is the average flux of solar emissions of a specified wavelength at time t.[19] Compute and interpret $F'(5.5)$.

57. ● ***Inflation*** Taking a 3.5% rate of inflation into account, the cost of DigIn brand snow shovels is given by

$$c(t) = 1.035^t[0.8\sin(2\pi t) + 10.2]$$

where t is time in years since January 1, 2002. How fast, in dollars per week, is the cost of DigIn shovels increasing on January 1, 2003?

58. ● ***Deflation*** Sales, in bottles per day, of my exclusive mass-produced 2002 vintage Chateau Petit Mont Blanc follow the function

$$s(t) = 4.5e^{-0.2t}\sin(2\pi t)$$

where t is time in years since January 1, 2002. How fast were sales rising or falling on January 1, 2003?

59. ***Tides*** The depth of water at my favorite surfing spot varies from 5 to 15 feet, depending on the time. Last Sunday, high tide occurred at 5:00 A.M. and the next high tide occurred at 6:30 P.M.

   a. Obtain a cosine model describing the depth of water as a function of time t in hours since 5:00 A.M. on Sunday morning.

   b. How fast was the tide rising (or falling) at noon on Sunday?

60. ***Tides*** Repeat Exercise 59, using data from the depth of water at my other favorite surfing spot, where the tide last Sunday varied from a low of 6 feet at 4:00 A.M. to a high of 10 feet at noon. (As in Exercise 59, take t as time in hours since 5:00 A.M.)

61. ***Tilt of the Earth's Axis*** The tilt of the earth's axis from its plane of rotation about the sun oscillates between approximately

---

[18] The model is based on a regression obtained from predicted data for 1997–2006 and the mean historical period of sunspot activity from 1755 to 1995. SOURCE: NASA Science Directorate; Marshall Space Flight Center http://solarscience.msfc.nasa.gov/predict.shtml, August 2002.

[19] Flux measured at a wavelength of 10.7 cm. Ibid.

● basic skills

22.5° and 24.5° with a period of approximately 40,000 years.[20] We know that 500,000 years ago, the tilt of the earth's axis was 24.5°.

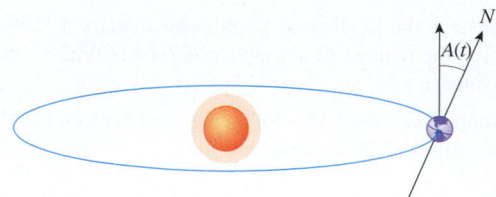

a. Which of the following functions best models the tilt of the earth's axis?

(I) $A(t) = 23.5 + 2\sin\left(\dfrac{2\pi t + 500}{40}\right)$

(II) $A(t) = 23.5 + \cos\left(\dfrac{t + 500}{80\pi}\right)$

(III) $A(t) = 23.5 + \cos\left(\dfrac{2\pi(t + 500)}{40}\right)$

where $A(t)$ is the tilt in degrees and t is time in thousands of years, with $t = 0$ being the present time.

b. Use the model you selected in part (a) to estimate the rate at which the tilt was changing 150,000 years ago. (Round your answer to three decimal places, and be sure to give the units of measurement.)

62. *Eccentricity of the Earth's Orbit* The eccentricity of the earth's orbit (that is, the deviation of the earth's orbit from a perfect circle) can be modeled by[21]

$$E(t) = 0.025\left[\cos\left(\frac{2\pi(t + 200)}{400}\right) + \cos\left(\frac{2\pi(t + 200)}{100}\right)\right]$$

where $E(t)$ is the eccentricity and t is time in thousands of years, with $t = 0$ being the present time. What was the value of the eccentricity 200,000 years ago, and how fast was it changing?

Communication and Reasoning Exercises

63. ● Complete the following: The rate of change of $f(x) = 3\sin(2x - 1) + 3$ oscillates between ____ and ____ .

64. ● Complete the following: The rate of change of $g(x) = -3\cos(-x + 2) + 2x$ oscillates between ____ and ____ .

65. ● Give two examples of a function $f(x)$ with the property that $f''(x) = -f(x)$.

66. ● Give two examples of a function $f(x)$ with the property that $f''(x) = -4f(x)$.

67. ● Give two examples of a function $f(x)$ with the property that $f'(x) = -f(x)$.

68. ● Give four examples of a function $f(x)$ with the property that $f^{(4)}(x) = f(x)$.

69. By referring to the graph of $f(x) = \cos x$, explain why $f'(x) = -\sin x$, rather than $\sin x$.

70. If A and B are constants, what is the relationship between $f(x) = A\cos x + B\sin x$ and its second derivative?

71. At what angle does the graph of $f(x) = \sin x$ depart from the origin?

72. At what angle does the graph of $f(x) = \cos x$ depart from the point (0, 1)?

---

[20] SOURCE: Dr. David Hodell, University of Florida/Juan Valesco/ *The New York Times,* February 16, 1999, p. F1.

[21] This is a rough model based on the actual data. SOURCE: Ibid.

● basic skills

16.3 Integrals of Trigonometric Functions and Applications

We saw in Section 13.1 that every calculation of a derivative also gives us a calculation of an antiderivative. For instance, since we know that $\cos x$ is the derivative of $\sin x$, we can say that an antiderivative of $\cos x$ is $\sin x$:

$$\int \cos x \, dx = \sin x + C \qquad \text{An antiderivative of } \cos x \text{ is } \sin x$$

The rules for the derivatives of sine, cosine, and tangent give us the following antiderivatives.

Indefinite Integrals of Some Trig Functions

$$\int \cos x \, dx = \sin x + C \qquad \text{Because } \frac{d}{dx}(\sin x) = \cos x$$

$$\int \sin x \, dx = -\cos x + C \qquad \text{Because } \frac{d}{dx}(-\cos x) = \sin x$$

$$\int \sec^2 x \, dx = \tan x + C \qquad \text{Because } \frac{d}{dx}(\tan x) = \sec^2 x$$

quick Examples

1. $\int (\sin x + \cos x) \, dx = -\cos x + \sin x + C$ Integral of sum = Sum of integrals

2. $\int (4 \sin x - \cos x) \, dx = -4 \cos x - \sin x + C$ Integral of constant multiple

3. $\int (e^x - \sin x + \cos x) \, dx = e^x + \cos x + \sin x + C$

Example 1 Substitution

Evaluate $\int (x + 3) \sin(x^2 + 6x) \, dx$.

Solution There are two parenthetical expressions that we might replace with u. Notice, however, that the derivative of the expression $(x^2 + 6x)$ is $2x + 6$, which is twice the term $(x + 3)$ in front of the sine. Recall that we would like the derivative of u to appear as a factor. Thus, let us take $u = x^2 + 6x$.

$$u = x^2 + 6x$$
$$\frac{du}{dx} = 2x + 6 = 2(x + 3)$$
$$dx = \frac{1}{2(x + 3)} \, du$$

Substituting into the integral we get

$$\int (x + 3) \sin(x^2 + 6x) \, dx = \int (x + 3) \sin u \left(\frac{1}{2(x + 3)} \right) du$$
$$= \int \frac{1}{2} \sin u \, du$$
$$= -\frac{1}{2} \cos u + C = -\frac{1}{2} \cos(x^2 + 6x) + C$$

Example 2 Definite Integrals

Compute the following:

a. $\int_0^\pi \sin x \, dx$ **b.** $\int_0^\pi x \sin(x^2) \, dx$

Solution

a. $\int_0^\pi \sin x \, dx = \left[-\cos x \right]_0^\pi = (-\cos \pi) - (-\cos 0)$

$$= -(-1) - (-1) = 2$$

Thus, the area under one "arch" of the sine curve is exactly 2 square units!

b. $\displaystyle\int_0^\pi x\,\sin(x^2)\,dx = \int_0^{\pi^2} \frac{1}{2}\sin u\,du$ After substituting $u = x^2$

$$= \left[-\frac{1}{2}\cos u\right]_0^{\pi^2}$$

$$= \left[-\frac{1}{2}\cos(\pi^2)\right] - \left[-\frac{1}{2}\cos(0)\right]$$

$$= -\frac{1}{2}\cos(\pi^2) + \frac{1}{2}$$ $\cos(0) = 1$

We can approximate $\frac{1}{2}\cos(\pi^2)$ by a decimal or leave it in the above form, depending on what we want to do with the answer.

Antiderivatives of the Six Trigonometric Functions

The following table gives the indefinite integrals of the six trigonometric functions. (The first two we have already seen.)

Integrals of the Trigonometric Functions

$$\int \sin x\,dx = -\cos x + C$$

$$\int \cos x\,dx = \sin x + C$$

$$\int \tan x\,dx = -\ln|\cos x| + C$$ Shown below

$$\int \cot x\,dx = \ln|\sin x| + C$$ See the Exercise Set

$$\int \sec x\,dx = \ln|\sec x + \tan x| + C$$ Shown below

$$\int \csc x\,dx = -\ln|\csc x + \cot x| + C$$ See the Exercise Set

Derivations of Formulas for Antiderivatives of Trigonometic Functions

To show that $\int \tan x\,dx = -\ln|\cos x| + C$, we first write $\tan x$ as $\dfrac{\sin x}{\cos x}$ and put $u = \cos x$ in the integral:

$$\int \tan x\,dx = \int \frac{\sin x}{\cos x}\,dx$$

$$= -\int \frac{\sin x}{u}\frac{du}{\sin x}$$

$$= -\int \frac{du}{u}$$

$$= -\ln|u| + C$$

$$= -\ln|\cos x| + C$$

$u = \cos x$

$\dfrac{du}{dx} = -\sin x$

$dx = -\dfrac{du}{\sin x}$

To show that $\int \sec x\, dx = \ln|\sec x + \tan x| + C$, first use a little "trick": write $\sec x$ as $\sec x \left(\dfrac{\sec x + \tan x}{\sec x + \tan x} \right)$ and put u equal to the denominator:

$$\int \sec x\, dx = \int \sec x \left(\frac{\sec x + \tan x}{\sec x + \tan x} \right) dx$$

$$= \int \sec x \frac{\sec x + \tan x}{u} \frac{du}{\sec x(\tan x + \sec x)}$$

$$= \int \frac{du}{u}$$

$$= \ln|u| + C$$

$$= \ln|\sec x + \tan x| + C$$

$$u = \sec x + \tan x$$

$$\frac{du}{dx} = \sec x \tan x + \sec^2 x$$

$$= \sec x(\tan x + \sec x)$$

$$dx = \frac{du}{\sec x(\tan x + \sec x)}$$

Shortcuts

If a and b are constants with $a \neq 0$, then we have the following formulas. (All of them can be obtained using the substitution $u = ax + b$. They will appear in the exercises.)

quick Example

Shortcuts: Integrals of Expressions Involving (*ax* + *b*)

Rule

$$\int \sin(ax + b)\, dx = -\frac{1}{a}\cos(ax + b) + C$$

$$\int \sin(-4x)\, dx = \frac{1}{4}\cos(-4x) + C$$

$$\int \cos(ax + b)\, dx = \frac{1}{a}\sin(ax + b) + C$$

$$\int \cos(x + 1)\, dx = \sin(x + 1) + C$$

$$\int \tan(ax + b)\, dx = -\frac{1}{a}\ln|\cos(ax + b)| + C$$

$$\int \tan(-2x)\, dx = \frac{1}{2}\ln|\cos(-2x)| + C$$

$$\int \cot(ax + b)\, dx = \frac{1}{a}\ln|\sin(ax + b)| + C$$

$$\int \cot(3x - 1)\, dx = \frac{1}{3}\ln|\sin(3x - 1)| + C$$

$$\int \sec(ax + b)\, dx = \frac{1}{a}\ln|\sec(ax + b)$$
$$+ \tan(ax + b)| + C$$

$$\int \sec(9x)\, dx = \frac{1}{9}\ln|\sec(9x)$$
$$+ \tan(9x)| + C$$

$$\int \csc(ax + b)\, dx = -\frac{1}{a}\ln|\csc(ax + b)$$
$$+ \cot(ax + b)| + C$$

$$\int \csc(x + 7)\, dx = -\ln|\csc(x + 7)$$
$$+ \cot(x + 7)| + C$$

Example 3 Sales

The rate of sales of cypods (one-bedroom units) in the city-state of Utarek, Mars[*] can be modeled by

$$s(t) = 7.5\cos(\pi t/6) + 87.5 \text{ units per month}$$

where t is time in months since January 1. How many cypods are sold in a calendar year?

[*] See www.marsnext.com/comm/zonars.html

Solution Total sales over one calendar year are given by

$$\int_0^{12} s(t)\,dt = \int_0^{12} [7.5\cos(\pi t/6) + 87.5]\,dt$$

$$= \left[7.5\frac{6}{\pi}\sin(\pi t/6) + 87.5t \right]_0^{12} \qquad \text{We used a shortcut on the first term}$$

$$= \left[7.5\frac{6}{\pi}\sin(2\pi) + 87.5(12) \right] - \left[7.5\frac{6}{\pi}\sin(0) + 87.5(0) \right]$$

$$= 87.5(12) \qquad\qquad\qquad \sin(2\pi) = \sin(0) = 0$$

$$= 1050\ \text{cypods}$$

+*Before we go on...* Would it have made any difference in Example 3 if we had computed total sales over the period $[12, 24]$, $[6, 18]$, or any interval of the form $[a, a + 12]$? ∎

Using Integration by Parts with Trig Functions

Example 4 Integrating a Polynomial Times Sine or Cosine

Calculate $\int (x^2 + 1)\sin(x + 1)\,dx$.

Solution We use the column method of integration by parts described in Section 14.1. Because differentiating $x^2 + 1$ makes it simpler, we put it in the D column and get the following table:

	D	I
$+$	$x^2 + 1$	$\sin(x + 1)$
$-$	$2x$	$-\cos(x + 1)$
$+$	2	$-\sin(x + 1)$
$-\int$	0	$\cos(x + 1)$

[Notice that we used the shortcut formulas to repeatedly integrate $\sin(x + 1)$.] We can now read the answer from the table:

$$\int (x^2 + 1)\sin(x + 1)\,dx = (x^2 + 1)[-\cos(x + 1)] - 2x[-\sin(x + 1)]$$

$$+ 2[\cos(x + 1)] + C$$

$$= (-x^2 - 1 + 2)\cos(x + 1) + 2x\sin(x + 1) + C$$

$$= (-x^2 + 1)\cos(x + 1) + 2x\sin(x + 1) + C$$

Example 5 Integrating an Exponential Times Sine or Cosine

Calculate $\int e^x \sin x \, dx$.

Solution The integrand is the product of e^x and $\sin x$, so we put one in the D column and the other in the I column. For this example, it doesn't matter much which we put where.

	D	I
$+$	$\sin x$	e^x
$-$	$\cos x$	e^x
$+\int$	$-\sin x$	e^x

It looks as if we're just spinning our wheels. Let's stop and see what we have:

$$\int e^x \sin x \, dx = e^x \sin x - e^x \cos x - \int e^x \sin x \, dx$$

At first glance, it appears that we are back where we started, still having to evaluate $\int e^x \sin x \, dx$. However, if we add this integral to both sides of the equation above, we can solve for it:

$$2 \int e^x \sin x \, dx = e^x \sin x - e^x \cos x + C$$

(Why $+ C$?) So,

$$\int e^x \sin x \, dx = \frac{1}{2} e^x \sin x - \frac{1}{2} e^x \cos x + \frac{C}{2}$$

Because $C/2$ is just as arbitrary as C, we write C instead of $C/2$, and obtain

$$\int e^x \sin x \, dx = \frac{1}{2} e^x \sin x - \frac{1}{2} e^x \cos x + C$$

16.3 EXERCISES

● denotes basic skills exercises

techEx indicates exercises that should be solved using technology

Evaluate the integrals in Exercises 1–28.

1. ● $\int (\sin x - 2 \cos x) \, dx$ *hint* [see Quick Examples p. 1133]

2. ● $\int (\cos x - \sin x) \, dx$

3. ● $\int (2 \cos x - 4.3 \sin x - 9.33) \, dx$

4. ● $\int (4.1 \sin x + \cos x - 9.33/x) \, dx$

5. ● $\int \left(3.4 \sec^2 x + \frac{\cos x}{1.3} - 3.2 e^x \right) dx$

6. ● $\int \left(\frac{3 \sec^2 x}{2} + 1.3 \sin x - \frac{e^x}{3.2} \right) dx$

7. ● $\int 7.6 \cos(3x - 4) \, dx$ *hint* [see Example 1]

8. ● $\int 4.4 \sin(-3x + 4) \, dx$ **9.** ● $\int x \sin(3x^2 - 4) \, dx$

10. ● $\int x \cos(-3x^2 + 4) \, dx$ **11.** ● $\int (4x + 2) \sin(x^2 + x) \, dx$

● basic skills   techEx technology exercise

11~

12. $\bullet \int (x+1)[\cos(x^2+2x)+(x^2+2x)]\,dx$

13. $\bullet \int (x+x^2)\sec^2(3x^2+2x^3)\,dx$

14. $\bullet \int (4x+2)\sec^2(x^2+x)\,dx$

15. $\bullet \int (x^2)\tan(2x^3)\,dx$ **16.** $\bullet \int (4x)\tan(x^2)\,dx$

17. $\bullet \int 6\sec(2x-4)\,dx$ **18.** $\bullet \int 3\csc(3x)\,dx$

19. $\bullet \int e^{2x}\cos(e^{2x}+1)\,dx$ **20.** $\bullet \int e^{-x}\sin(e^{-x})\,dx$

21. $\bullet \int_{-\pi}^{0} \sin x\,dx$ *hint* [see Example 2]

22. $\bullet \int_{\pi/2}^{\pi} \cos x\,dx$ **23.** $\bullet \int_{0}^{\pi/3} \tan x\,dx$

24. $\bullet \int_{\pi/6}^{\pi/2} \cot x\,dx$ **25.** $\bullet \int_{1}^{\sqrt{\pi+1}} x\cos(x^2-1)\,dx$

26. $\bullet \int_{0.5}^{(\pi+1)/2} \sin(2x-1)\,dx$

27. $\int_{1/\pi}^{2/\pi} \dfrac{\sin(1/x)}{x^2}\,dx$ **28.** $\int_{0}^{\pi/3} \dfrac{\sin x}{\cos^2 x}\,dx$

In Exercises 29–32, derive each equation, where a and b are constants with $a \neq 0$.

29. $\int \cos(ax+b)\,dx = \dfrac{1}{a}\sin(ax+b)+C$

30. $\int \sin(ax+b)\,dx = -\dfrac{1}{a}\cos(ax+b)+C$

31. $\int \cot x\,dx = \ln|\sin x|+C$

32. $\int \csc x\,dx = -\ln|\csc x + \cot x|+C$

Use the shortcut formulas given before Example 3 to calculate the integrals in Exercises 33–40 mentally.

33. $\bullet \int \sin(4x)\,dx$ **34.** $\bullet \int \cos(5x)\,dx$

35. $\bullet \int \cos(-x+1)\,dx$ **36.** $\bullet \int \sin\left(\dfrac{1}{2}x\right)\,dx$

37. $\bullet \int \sin(-1.1x-1)\,dx$ **38.** $\bullet \int \cos(4.2x-1)\,dx$

39. $\bullet \int \cot(-4x)\,dx$ **40.** $\bullet \int \tan(6x)\,dx$

Use geometry (not antiderivatives) to compute the integrals in Exercises 41–44. [Hint: First draw the graph.]

41. $\bullet \int_{-\pi/2}^{\pi/2} \sin x\,dx$ **42.** $\bullet \int_{0}^{\pi} \cos x\,dx$

43. $\int_{0}^{2\pi} (1+\sin x)\,dx$ **44.** $\int_{0}^{2\pi} (1+\cos x)\,dx$

Use integration by parts to evaluate the integrals in Exercises 45–52. hint [see Example 4]

45. $\bullet \int x\sin x\,dx$ **46.** $\bullet \int x^2\cos x\,dx$

47. $\bullet \int x^2\cos(2x)\,dx$ **48.** $\bullet \int (2x+1)\sin(2x-1)\,dx$

49. $\int e^{-x}\sin x\,dx$ **50.** $\int e^{2x}\cos x\,dx$

51. $\int_{0}^{\pi} x^2\sin x\,dx$ **52.** $\int_{0}^{\pi/2} x\cos x\,dx$

Recall from Section 14.3 that the average of a function $f(x)$ on an interval $[a, b]$ is

$$\bar{f} = \frac{1}{b-a}\int_{a}^{b} f(x)\,dx$$

Find the averages of the functions in Exercises 53 and 54 over the given intervals. Plot each function and its average on the same graph.

53. $f(x) = \sin x$ over $[0, \pi]$

54. $f(x) = \cos(2x)$ over $[0, \pi/4]$

Decide whether each integral in Exercises 55–58 converges. (See Section 14.5.) If the integral converges, compute its value.

55. $\bullet \int_{0}^{+\infty} \sin x\,dx$ **56.** $\bullet \int_{0}^{+\infty} \cos x\,dx$

57. $\int_{0}^{+\infty} e^{-x}\cos x\,dx$ **58.** $\int_{0}^{+\infty} e^{-x}\sin x\,dx$

Applications

59. $\bullet$ *Varying Cost* The cost of producing a bottle of suntan lotion is changing at a rate of $0.04 - 0.1\sin\left[\dfrac{\pi}{26}(t-25)\right]$ dollars per week, t weeks after January 1. If it cost $1.50 to produce a bottle 12 weeks into the year, find the cost $C(t)$ at time t.

60. $\bullet$ *Varying Cost* The cost of producing a box of holiday tree decorations is changing at a rate of $0.05 + 0.4\cos\left[\dfrac{\pi}{6}(t-11)\right]$ dollars per month, t months after January 1. If it cost $5 to produce a box on June 1, find the cost $C(t)$ at time t.

61. *Pets* My dog Miranda is running back and forth along a 12-foot stretch of garden in such a way that her velocity t seconds after she began is

$$v(t) = 3\pi\cos\left[\frac{\pi}{2}(t-1)\right] \text{ feet per second}$$

How far is she from where she began 10 seconds after starting the run? *hint* [see Example 3]

62. *Pets* My cat, Prince Sadar, is pacing back and forth along his favorite window ledge in such a way that his velocity

$\bullet$ basic skills   **tech** Ex technology exercise

t seconds after he began is

$$v(t) = -\frac{\pi}{2} \sin\left[\frac{\pi}{4}(t-2)\right] \text{ feet per second}$$

How far is he from where he began 10 seconds after starting to pace?

For Exercises 63–68, recall from Section 14.3 that the average of a function $f(x)$ on an interval $[a, b]$ is

$$\bar{f} = \frac{1}{b-a}\int_a^b f(x)\,dx$$

63. ● **Sunspot Activity** The activity of the sun (sunspots, solar flares, and coronal mass ejection) fluctuates in cycles of around 10–11 years. Sunspot activity can be modeled by the following function:[22]

$$N(t) = 57.7 \sin[0.602(t-1.43)] + 58.8$$

where *t* is the number of years since January 1, 1997, and $N(t)$ is the number of sunspots observed at time *t*. Estimate the average number of sunspots visible over the two-year period 2002–2003. (Round your answer to the nearest whole number.)

64. ● **Solar Emissions** The following model gives the flux of radio emissions from the sun:

$$F(t) = 49.6 \sin[0.602(t-1.48)] + 111$$

where *t* is the number of years since January 1, 1997, and $F(t)$ is the flux of solar emissions of a specified wavelength at time *t*.[23] Estimate the average flux of radio emissions over the five-year period 2001–2005. (Round your answer to the nearest whole number.)

65. **Biology** Sigatoka leaf spot is a plant disease that affects bananas. In an infected plant, the percentage of leaf area affected varies from a low of around 5% at the start of each year to a high of around 20% at the middle of each year.[24] Use a sine function model of the percentage of leaf area affected by Sigatoka leaf spot *t* weeks since the start of a year to estimate, to the nearest 0.1%, the average percentage of leaf area affected in the first quarter (13 weeks) of a year.

66. **Biology** Apple powdery mildew is an epidemic that affects apple shoots. In a new infection, the percentage of apple shoots infected varies from a low of around 10% at the start of May to a high of around 60% six months later.[25] Use a sine

function model of the percentage of apple shoots affected by apple powdery mildew *t* months since the start of a year to estimate, to the nearest 0.1%, the average percentage of apple shoots affected in the first two months of a year.

67. `tech` Ex **Electrical Current** The typical voltage *V* supplied by an electrical outlet in the United States is given by

$$V(t) = 165 \cos(120\pi t)$$

where *t* is time in seconds.

a. Find the average voltage over the interval $[0, 1/6]$. How many times does the voltage reach a maximum in one second? (This is referred to as the number of **cycles per second**.)

b. Plot the function $S(t) = (V(t))^2$ over the interval $[0, 1/6]$.

c. The **root mean square** voltage is given by the formula

$$V_{rms} = \sqrt{\bar{S}}$$

where $\bar{S}$ is the average value of $S(t)$ over one cycle. Estimate V_{rms}.

68. ● **Tides** The depth of water at my favorite surfing spot varies from 5 to 15 feet, depending on the time. Last Sunday, high tide occurred at 5:00 A.M. and the next high tide occurred at 6:30 P.M. Use the cosine function to model the depth of water as a function of time *t* in hours since midnight on Sunday morning. What was the average depth of the water between 10:00 A.M. and 2:00 P.M.?

Income Streams Recall from Section 14.4 that the total income received from time $t = a$ to time $t = b$ from a continuous income stream of $R(t)$ dollars per year is

$$\text{Total value} = TV = \int_a^b R(t)\,dt$$

In Exercises 69 and 70, find the total value of the given income stream over the given period.

69. ● $R(t) = 50{,}000 + 2000\pi \sin(2\pi t),\ 0 \le t \le 1$

70. ● $R(t) = 100{,}000 - 2000\pi \sin(\pi t),\ 0 \le t \le 1.5$

Communication and Reasoning Exercises

71. ● What can you say about the definite integral of a sine or cosine function over a whole number of periods?

72. ● How are the derivative and antiderivative of $\sin x$ related?

73. What is the average value of $1 + 2\cos x$ over a large interval?

74. What is the average value of $3 - \cos x$ over a large interval?

75. The acceleration of an object is given by $a = K\sin(\omega t - \alpha)$. What can you say about its displacement at time *t*?

76. Write down a function whose derivative is −2 times its antiderivative.

---

[22] The model is based on a regression obtained from predicted data for 1997–2006 and the mean historical period of sunspot activity from 1755 to 1995. SOURCE: NASA Science Directorate; Marshall Space Flight Center http://solarscience.msfc.nasa.gov/predict.shtml, August 2002.

[23] Flux measured at a wavelength of 10.7 cm. Ibid.

[24] Based on graphical data. SOURCE: American Phytopathological Society. www.apsnet.org/education/AdvancedPlantPath/Topics/Epidemiology/CyclicalNature.htm

[25] Ibid.

● basic skills   `tech` Ex technology exercise

Chapter 16 Review

KEY CONCEPTS

16.1 Trigonometric Functions, Models, and Regression

The **sine** of a real number *p. 1112*

Plotting the graphs of functions based on $\sin x$ *p. 1114*

The general sine function: $f(x) = A \sin[\omega(x - \alpha)] + C$ *p. 1115*

Modeling with the general sine function *p. 1115*

The **cosine** of a real number *p. 1117*

Fundamental trigonometric identities:

$$\sin^2 t + \cos^2 t = 1$$

$$\cos t = \sin(t + \pi/2) \qquad \cos t = \sin(\pi/2 - t)$$

$$\sin t = \cos(t - \pi/2) \qquad \sin t = \cos(\pi/2 - t) \quad p.\ 1118$$

The general cosine function:

$$f(x) = A \cos[\omega(x - \alpha)] + C \quad p.\ 1118$$

Modeling with the general cosine function *p. 1119*

Other trig functions:

$$\tan x = \frac{\sin x}{\cos x} \qquad \cot x = \cotan x = \frac{\cos x}{\sin x} = \frac{1}{\tan x}$$

$$\sec x = \frac{1}{\cos x} \qquad \csc x = \cosec x = \frac{1}{\sin x} \quad p.\ 1119$$

16.2 Derivatives of Trigonometric Functions and Applications

Derivatives of sine and cosine:

$$\frac{d}{dx} \sin x = \cos x \qquad \frac{d}{dx} \cos x = -\sin x \quad p.\ 1125$$

Derivatives of sines and cosines of functions

$$\frac{d}{dx} \sin u = \cos u \frac{du}{dx} \qquad \frac{d}{dx} \cos u = -\sin u \frac{du}{dx} \quad p.\ 1127$$

Derivatives of the other trigonometric functions

$$\frac{d}{dx} \tan x = \sec^2 x \qquad \frac{d}{dx} \cot x = -\csc^2 x$$

$$\frac{d}{dx} \sec x = \sec x \tan x \qquad \frac{d}{dx} \csc x = -\csc x \cot x \quad p.\ 1129$$

Some trigonometric limits

$$\lim_{h \to 0} \frac{\sin h}{h} = 1 \qquad \lim_{h \to 0} \frac{\cos h - 1}{h} = 0 \quad pp.\ 1126\text{--}1127$$

16.3 Integrals of Trigonometric Functions and Applications

$$\int \cos x\, dx = \sin x + C \qquad \int \sin x\, dx = -\cos x + C$$

$$\int \sec^2 x\, dx = \tan x + C \quad p.\ 1133$$

Substitution in integrals involving trig functions *p. 1133*

Definite integrals involving trig functions *p. 1133*

Antiderivatives of the other trigonometric functions:

$$\int \tan x\, dx = -\ln|\cos x| + C$$

$$\int \cot x\, dx = \ln|\sin x| + C$$

$$\int \sec x\, dx = \ln|\sec x + \tan x| + C$$

$$\int \csc x\, dx = -\ln|\csc x + \cot x| + C \quad p.\ 1134$$

Shortcuts: Integrals of expressions involving $(ax + b)$ *p. 1135*

Using integration by parts with trig functions *pp. 1136, 1137*

REVIEW EXERCISES

In Exercises 1–4, model the given curve with a sine function. (The scales on the two axes may not be the same.)

1.

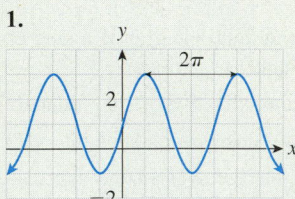

2.

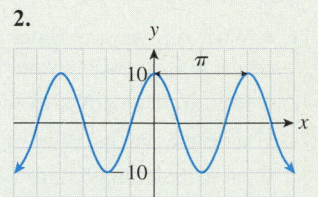

3.

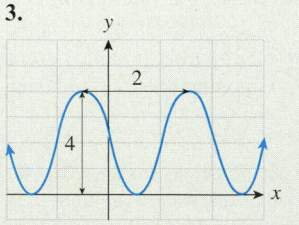

4.

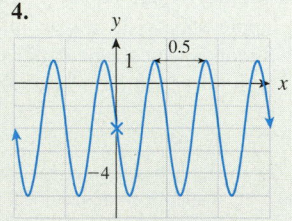

In Exercises 5–8, model the curves in Exercises 1–4 with cosine functions.

5. The curve in Exercise 1.

6. The curve in Exercise 2.

7. The curve in Exercise 3.

8. The curve in Exercise 4.

In Exercises 9–14, find the derivative of the given function.

9. $f(x) = \cos(x^2 - 1)$ **10.** $f(x) = \sin(x^2 + 1)\cos(x^2 - 1)$

11. $f(x) = \tan(2e^x - 1)$ **12.** $f(x) = \sec\sqrt{x^2 - x}$

13. $f(x) = \sin^2(x^2)$ **14.** $f(x) = \cos^2[1 - \sin(2x)]$

In Exercises 15–22, evaluate the given integral.

15. $\displaystyle\int 4\cos(2x - 1)\, dx$ **16.** $\displaystyle\int (x - 1)\sin(x^2 - 2x + 1)\, dx$

17. $\displaystyle\int 4x\,\sec^2(2x^2 - 1)\, dx$ **18.** $\displaystyle\int \frac{\cos\left(\dfrac{1}{x}\right)}{x^2\sin\left(\dfrac{1}{x}\right)}\, dx$

19. $\displaystyle\int x\,\tan(x^2 + 1)\, dx$ **20.** $\displaystyle\int_0^\pi \cos(x + \pi/2)\, dx$

21. $\displaystyle\int_{\ln(\pi/2)}^{\ln(\pi)} e^x\sin(e^x)\, dx$ **22.** $\displaystyle\int_\pi^{2\pi} \tan(x/6)\, dx$

Use integration by parts to evaluate the integrals in Exercises 23 and 24.

23. $\displaystyle\int x^2\sin x\, dx$ **24.** $\displaystyle\int e^x\sin 2x\, dx$

Applications

25. *Sales* After several years in the business, OHaganBooks.com noticed that its sales showed seasonal fluctuations, so that weekly sales oscillated in a sine wave from a low of 9000 books per week to a high of 12,000 books per week, with the high point of the year being three quarters of the way through the year, in October. Model OHaganBooks.com's weekly sales as a function of t, the number of weeks into the year.

26. *Elvish for Dummies* OHaganBooks.com recently hired a new sales manager, Mary Beth O'Connell, who has predicted that the revenue from sales of the latest blockbuster "Elvish for Dummies" will vary in accordance with annual releases of episodes of the movie series "Lord of the Rings Episodes 9–12." She has come up with the following model (which includes the effect of diminishing sales):

$$R(t) = 20{,}000 + 15{,}000e^{-0.12t}\cos\left[\frac{\pi}{6}(t - 4)\right] \text{ dollars}$$
$$(0 \le t \le 72)$$

where t is time in months from now and $R(t)$ is the monthly revenue. How fast, to the nearest dollar, will the revenue be changing 10 months from now?

27. *Revenue* Refer back to Question 26. Use technology or integration by parts to estimate, to the nearest $100, the total revenue from sales of "Elvish for Dummies" over the next 10 months.

28. *Mars Missions* Having completed his doctorate in biophysics, Billy Sean O'Hagan will be accompanying the first manned mission to Mars. For reasons too complicated to explain (but having to do with the continuation of his doctoral research project and the timing of messages from his fiancée), during the voyage he will be consuming protein at a rate of

$$P(t) = 150 + 50\sin\left[\frac{\pi}{2}(t - 1)\right] \text{ grams per day}$$

t days into the voyage. Find the total amount of protein he will consume as a function of time t.

🔅**Mentor** Do you need a live tutor for homework problems? Access vMentor on the ThomsonNOW! website at **www.thomsonedu.com** for one-on-one tutoring from a mathematics expert.

CASE STUDY: Predicting Cocoa Inventories

Stephen Studd/Getty Images

As a consultant to the Chocoholic Cocoa company, you have been asked to model the worldwide production and consumption of cocoa to determine how production cycles are affected by consumption, to estimate the trend in average world cocoa inventories, and to advise your cocoa producers whether to increase or decrease production. You have data showing the production and consumption of cocoa for the past 12 years (see Figure 14):[26]

---

[26] Approximate figures are in thousands of tonnes. 2001 supply figure is an estimate.

Source: EDF Man Cocoa Report, 2001/United Nations Conference on Trade and Development www.unctad.org/infocomm/Diversification/bangkok/Zemek.PDF.

	1990	*1991*	*1992*	*1993*	*1994*	*1995*	*1996*	*1997*	*1998*	*1999*	*2000*	*2001*
Demand (Consumption)	2200	2350	2300	2400	2450	2500	2600	2700	2800	2750	2950	2950
Supply (Production)	2400	2500	2250	2390	2450	2600	2900	2700	2650	2750	3000	3150

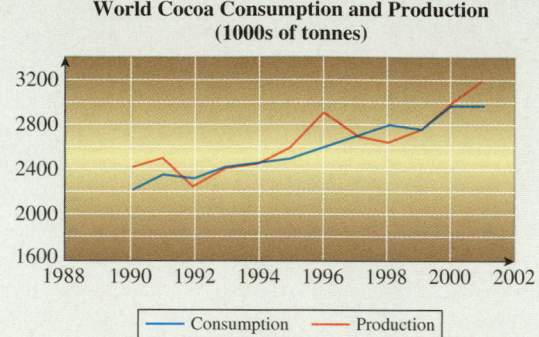

Figure **14**

From the graph, you deduce that the demand is roughly linear, but the supply tends to oscillate above and below the demand graph fairly regularly. To see this oscillation more clearly, you reason that it would be best to subtract the demand from the supply, and so you enter the data in your worksheet, and compute and graph the differences between the supply and demand (Figure 15).

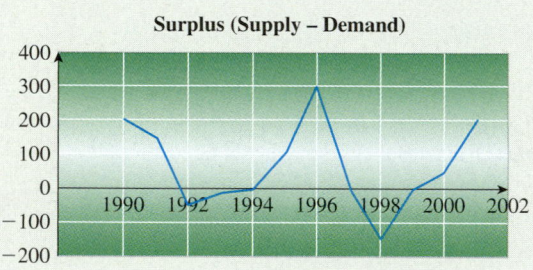

Figure **15**

The graph suggests a rough sine curve with amplitude around 150 and period around 6 years, so you use the techniques from Section 16.1 to construct a regression sine curve. Figure 16 shows the demand with its linear regression line and the surplus (supply–demand) with its sine regression curve ($t = 0$ represents 1990).

Thus, your models for cocoa production and consumption (in thousands of tonnes) are

Consumption $= 68.71t + 2201$

Production = Demand + Surplus $= 148.2 \sin[1.153(t + 1.170)] + 68.71t + 2254$

These models allow you to make the following observations:

- Consumption of cocoa is increasing at an approximately constant rate of 68,710 tonnes per year.

- Production of cocoa fluctuates from about 95,000 tonnes below consumption to 201,000 tonnes above it every 5.4 years. (See the exercises.)

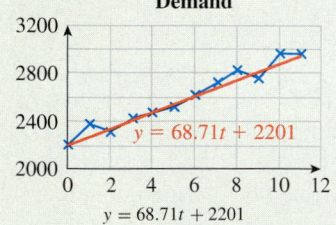

$y = 68.71t + 2201$

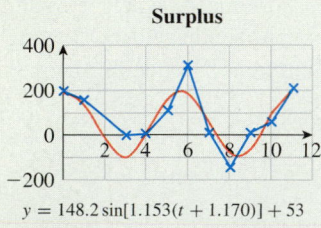

$y = 148.2 \sin[1.153(t + 1.170)] + 53$

Figure **16**

Interestingly—and this can also be seen in the raw data—production of cocoa exceeds consumption on average. This can be explained by the phenomenon that production anticipates a rising demand.

You now address the question: How does this pattern of production and consumption affect cocoa inventories? The accumulated cocoa inventory from 1990 to time t can be modeled by

$$A(t) = \int_0^t (148.2 \sin[1.153(x + 1.170)] + 53) \, dx$$

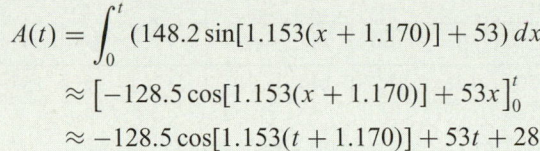

$$\approx \left[-128.5 \cos[1.153(x + 1.170)] + 53x \right]_0^t$$

$$\approx -128.5 \cos[1.153(t + 1.170)] + 53t + 28$$

The graph of $A(t)$ is shown in Figure 17.

The graph clearly indicates a rising world inventory of cocoa and strongly suggests that production must be cut back substantially in order to reduce the surplus.

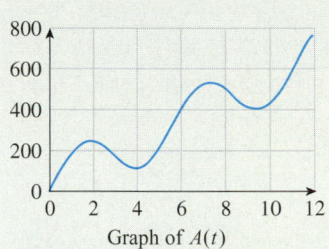

Graph of $A(t)$

Figure **17**

Exercises

1. Use the observed data to compute the actual accumulated cocoa inventory from the start of 1990 through 2001 and compare it with the value predicted by the model. (Use t to represent the time since the start of 1990 and round to the nearest 1000 tonnes.)

2. Justify the claim made above that production of cocoa fluctuates from 95,000 tonnes below consumption to 201,000 tonnes above it every 5.4 years.

3. What would the maximum accumulated inventory be if the production was given instead by the following formula?

$$P(t) = 148.2 \sin[1.153(t + 1.170)] + 68.71t + 2201$$

4. If the consumption of cocoa increased exponentially, but the surplus was still as shown above, would you still advise cocoa producers to cut back on production? Explain.

5. **tech** Ex Redo the entire analysis for the following (hypothetical) data on the sale of Fruit Punch on the planet Dune (in millions of tons).

	4000	4001	4002	4003	4004	4005	4006	4007	4008	4009	4010	4011
Demand (Consumption)	2210	2350	2310	2410	2450	2510	2610	2700	2810	2760	2210	2350
Supply (Production)	2340	2440	2190	2330	2390	2540	2840	2640	2590	2690	2340	2440

TI-83/84 Technology Guide

Section 16.1

Example 5(a) The following data shows the daily spam for the 16-week period beginning June 6, 2005. (Each figure is a one-week average):

Week	0	1	2	3	4	5	6	7	8	9	10	11	12	13	14	15
Number	107	163	170	176	167	140	149	137	158	157	185	151	122	132	134	182

Use technology to find the best-fit sine curve of the form $S(t) = A \sin[\omega(t - \alpha)] + C$.

Solution with Technology The TI-83/84 has a built-in sine regression utility.

1. As with the other forms of regression discussed in Chapter 9, we start by entering the coordinates of the data points in the lists L_1 and L_2, as shown:

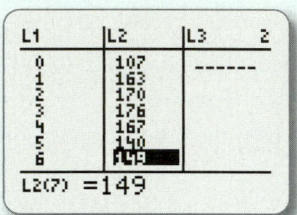

2. Press STAT, select CALC, and choose option #C: SinReg.

3. Pressing ENTER gives the sine regression equation in the home screen (we have rounded the coefficients):

$$S(t) \approx 11.57 \sin(0.9099t - 1.487) + 154.7$$

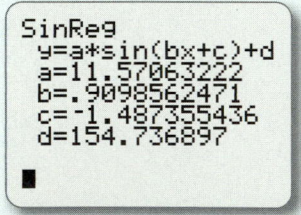

Although this is not exactly in the form we want, we can rewrite it:

$$S(t) \approx 11.57 \sin\left[0.9099\left(t - \frac{1.487}{0.9099}\right)\right] + 154.7$$
$$\approx 11.6 \sin[0.910(t - 1.63)] + 155$$

4. To graph the points and regression line in the same window, turn Stat Plot on by pressing 2nd STAT PLOT, selecting 1 and turning PLOT1 on:

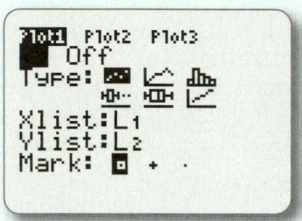

5. Enter the regression equation in the Y= screen by pressing Y=, clearing out whatever function is there, and pressing VARS 5 and selecting EQ (Option 1: RegEq.

6. To obtain a convenient window showing all the points and the lines, press ZOOM and choose option #9: Zoom-Stat, and you will obtain the following output:

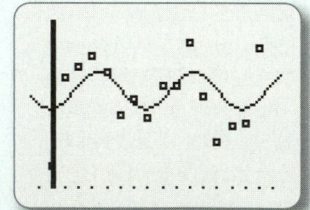

Section **16.1**

Example 5(a) The following data shows the daily spam for the 16-week period beginning June 6, 2005. (Each figure is a one-week average):

Week	0	1	2	3	4	5	6	7	8	9	10	11	12	13	14	15
Number	107	163	170	176	167	140	149	137	158	157	185	151	122	132	134	182

Use technology to find the best-fit sine curve of the form $S(t) = A \sin[\omega(t - \alpha)] + C$.

Solution with Technology We set up our worksheet as we did in Section 11.4, as shown below.

1. For our initial guesses, let us roughly estimate the parameters from the graph. The amplitude is around $A = 30$, and the vertical offset is roughly $C = 150$. The period seems to be around 7 weeks, so let us choose $P = 7$. This gives $\omega = 2\pi/P \approx 0.9$. Finally, let us take $\alpha = 0$ to begin with.

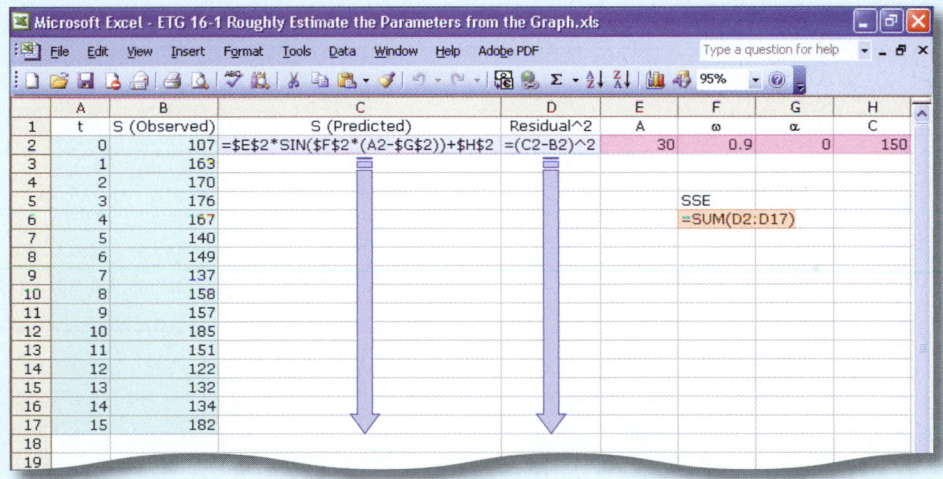

2. We then use Solver to minimize SSE by changing cells E2 through H2 as in Section 11.4:

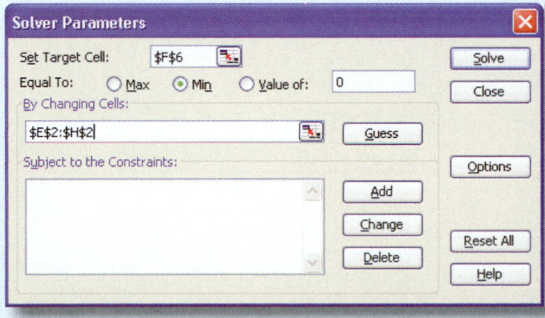

Solver gives us the following model (we have rounded the coefficients to 3 decimal places):

$$S(t) = 25.8 \sin[0.96(t - 1.22)] + 153$$

with SSE $\approx$ 2148. Plotting the observed and predicted values of S gives the following graph:

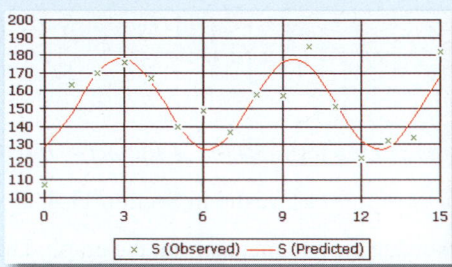

Note that Solver estimated the period for us. However, in many situations, we know what to use for the period beforehand: For example, we expect sales of snow shovels to fluctuate according to an annual cycle. Thus, if we were using regression to fit a regression sine or cosine curve to snow shovel sales data, we would set $P = 1$ year, compute ω, and have Solver estimate only the remaining coefficients: A, C, and α.

Appendix A   Logic

Introduction

Logic is the underpinning of all reasoned argument. The ancient Greeks recognized its role in mathematics and philosophy, and studied it extensively. Aristotle, in his *Organon,* wrote the first systematic treatise on logic. His work had a heavy influence on philosophy, science and religion through the Middle Ages.

But Aristotle's logic was expressed in ordinary language, so was subject to the ambiguities of ordinary language. Philosophers came to want to express logic more formally and symbolically, more like the way that mathematics is written (Leibniz, in the 17th century, was probably the first to envision and call for such a formalism). It was with the publication in 1847 of G. Boole's *The Mathematical Analysis of Logic* and A. DeMorgan's *Formal Logic* that **symbolic logic** came into being, and logic became recognized as part of mathematics. Since Boole and DeMorgan, logic and mathematics have been inextricably intertwined. Logic is part of mathematics, but at the same time it is the language of mathematics.

The study of symbolic logic is usually broken into several parts. The first and most fundamental is the **propositional logic.** Built on top of this is the **predicate logic,** which is the language of mathematics. In this appendix we give an introduction to propositional logic.

A.1   Statements and Logical Operators

Propositional logic is the study of *propositions.* A **statement,** or **proposition,** is any declarative sentence which is either true (T) or false (F). We refer to T or F as the **truth value** of the statement.

Example 1   Statements

a. "$2 + 2 = 4$" is a statement because it can be either true or false.[*] Because it happens to be a true statement, its truth value is T.

b. "$1 = 0$" is also a statement, but its truth value is F.

c. "It will rain tomorrow" is a statement. To determine its truth value, we shall have to wait for tomorrow.

d. "Solve the following equation for x" is not a statement, because it cannot be assigned any truth value whatsoever. It is an imperative, or command, rather than a declarative sentence.

e. "The number 5" is not a statement, because it is not even a complete sentence.

For a much more extensive interactive treatment of logic, including discussion of proofs, rules of inference, and an introduction to the predicate calculus, go online and follow:

Online Text

→ Online Topics in Finite Mathematics

　→ Introduction to Logic

---

[*] Is "$2 + 2 = 4$" a sentence? Read it aloud: "Two plus two equals four," is a perfectly respectable English sentence.

f. "This statement is false" gets us into a bind: If it were true, then, because it is declaring itself to be false, it must be false. On the other hand, if it were false, then its declaring itself false is a lie, so it is true! In other words, if it is true, then it is false, and if it is false, then it is true, and we go around in circles. We get out of this bind by saying that because the sentence cannot be either true or false, we refuse to call it a statement. An equivalent pseudo-statement is: "I am lying," so this sentence is known as **the liar's paradox.**

Note Sentences that refer to themselves, or *self-referential sentences,* as illustrated in Example 1(f), are not permitted to be statements. This eliminates the liar's paradox and several similar problems. ∎

We shall use letters like p, q, r and so on to stand for statements. Thus, for example, we might decide that p should stand for the statement "the moon is round." We write

p: "the moon is round"   *p is the statement that the moon is round*

to express this.

We can form new statements from old ones in several different ways. For example, starting with p: "I am an Anchovian," we can form the **negation** of p: "It is not the case that I am an Anchovian" or simply "I am not an Anchovian."

Negation of a Statement

If p is a statement, then its **negation** is the statement "not p" and is denoted by $\sim p$. We mean by this that, if p is true, then $\sim p$ is false, and vice versa.

quick Examples

1. If p: "$2 + 2 = 4$," then $\sim p$: "It is not the case that $2 + 2 = 4$," or, more simply, $\sim p$: "$2 + 2 \neq 4$."
2. If q: "$1 = 0$," then $\sim q$: "$1 \neq 0$."
3. If r: "Diamonds are a pearl's best friend," then $\sim r$: "Diamonds are not a pearl's best friend."
4. If s: "All politicians are crooks," then $\sim s$: "Not all politicians are crooks."
5. **Double Negation:** If p is any statement, then the negation of $\sim p$ is $\sim(\sim p)$: "not (not p)," or, in other words, p. Thus $\sim(\sim p)$ has the same meaning as p.

Notes

1. Notice in Quick Example 1 above that $\sim p$ is false, because p is true. However, in Quick Example 2, $\sim q$ is true, because q is false. A statement of the form $\sim q$ can very well be true; it is a common mistake to think it must be false.

2. Saying that not all politicians are crooks is not the same as saying that no politicians are crooks, but is the same as saying that some (meaning one or more) politicians are not crooks.

3. The symbol $\sim$ is our first example of a **logical operator.**

4. When we say in Quick Example 5 above that $\sim(\sim p)$ has the same meaning as p, we mean that they are *logically equivalent*—a notion we will make precise below. ∎

Here is another way we can form a new statement from old ones. Starting with p: "I am wise," and q: "I am strong," we can form the statement "I am wise and I am strong." We denote this new statement by $p \wedge q$, read "p and q." In order for $p \wedge q$ to be true, *both p and q* must be true. Thus, for example, if I am wise but not strong, then $p \wedge q$ is false. The symbol $\wedge$ is another logical operator. The statement $p \wedge q$ is called the **conjunction** of p and q.

Conjunction

The **conjunction** of p and q is the statement $p \wedge q$, which we read "p and q." It can also be said in a number of different ways, such as "p even though q." The statement $p \wedge q$ is true when both p and q are true and false otherwise.

quick Examples

1. If p: "This galaxy will ultimately disappear into a black hole" and q: "$2 + 2 = 4$," then $p \wedge q$ is the statement "Not only will this galaxy ultimately disappear into a black hole, but $2 + 2 = 4$!"
2. If p: "$2 + 2 = 4$" and q: "$1 = 0$," then $p \wedge q$: "$2 + 2 = 4$ and $1 = 0$." Its truth value is F because q is F.
3. With p and q as in Quick Example 1, the statement $p \wedge (\sim q)$ says: "This galaxy will ultimately disappear into a black hole and $2 + 2 \neq 4$," or, more colorfully, as "Contrary to your hopes, this galaxy is doomed to disappear into a black hole; moreover, two plus two is decidedly *not* equal to four!"

Notes

1. We sometimes use the word "but" as an emphatic form of "and." For instance, if p: "It is hot," and q: "It is not humid," then we can read $p \wedge q$ as "It is hot but not humid." There are always many ways of saying essentially the same thing in a natural language; one of the purposes of symbolic logic is to strip away the verbiage and record the underlying logical structure of a statement.

2. A **compound statement** is a statement formed from simpler statements via the use of logical operators. Examples are $\sim p$, $(\sim p) \wedge (q \wedge r)$ and $p \wedge (\sim p)$. A statement that cannot be expressed as a compound statement is called an **atomic statement**.[1] For example, "I am clever" is an atomic statement. In a compound statement such as $(\sim p) \wedge (q \wedge r)$, we refer to p, q and r as the **variables** of the statement. Thus, for example, $\sim p$ is a compound statement in the single variable p. ∎

Before discussing other logical operators, we pause for a moment to talk about **truth tables,** which give a convenient way to analyze compound statements.

Truth Table

The **truth table** for a compound statement shows, for each combination of possible truth values of its variables, the corresponding truth value of the statement.

quick Examples

1. The truth table for negation, that is, for $\sim p$, is:

p	$\sim p$
T	F
F	T

Each row shows a possible truth value for p and the corresponding value of $\sim p$.

---

[1] "Atomic" comes from the Greek for "not divisible." Atoms were originally thought to be the indivisible components of matter, but the march of science proved that wrong. The name stuck, though.

2. The truth table for conjunction, that is, for $p \wedge q$, is:

p	q	$p \wedge q$
T	T	T
T	F	F
F	T	F
F	F	F

Each row shows a possible combination of truth values of p and q and the corresponding value of $p \wedge q$.

Example 2 Construction of Truth Tables

Construct truth tables for the following compound statements.

a. $\sim(p \wedge q)$ **b.** $(\sim p) \wedge q$

Solution

a. Whenever we encounter a complex statement, we work from the inside out, just as we might do if we had to evaluate an algebraic expression like $-(a + b)$. Thus, we start with the p and q columns, then construct the $p \wedge q$ column, and finally, the $\sim(p \wedge q)$ column.

p	q	$p \wedge q$	$\sim(p \wedge q)$
T	T	T	F
T	F	F	T
F	T	F	T
F	F	F	T

Notice how we get the $\sim(p \wedge q)$ column from the $p \wedge q$ column: we reverse all the truth values.

b. Because there are two variables, p and q, we again start with the p and q columns. We then evaluate $\sim p$, and finally take the conjunction of the result with q.

p	q	$\sim p$	$(\sim p) \wedge q$
T	T	F	F
T	F	F	F
F	T	T	T
F	F	T	F

Because we are "and-ing" $\sim p$ with q, we look at the values in the $\sim p$ and q columns and combine these according to the instructions for "and." Thus, for example, in the first row we have $F \wedge T = F$ and in the third row we have $T \wedge T = T$.

Here is a third logical operator. Starting with p: "You are over 18" and q: "You are accompanied by an adult," we can form the statement "You are over 18 or are accompanied by an adult," which we write symbolically as $p \vee q$, read "p or q." Now in English the word "or" has several possible meanings, so we have to agree on which one we want here. Mathematicians have settled on the **inclusive or:** $p \vee q$ means p is true or q is true *or both are true*[2]. With p and q as above, $p \vee q$ stands for "You are over 18 or are accompanied by an adult, or both." We shall sometimes include the phrase "or both" for emphasis, but even if we leave it off we still interpret "or" as inclusive.

Disjunction

The **disjunction** of p and q is the statement $p \vee q$, which we read "*p* or *q*." Its truth value is defined by the following truth table.

p	q	$p \vee q$
T	T	T
T	F	T
F	T	T
F	F	F

This is the **inclusive** or, so $p \vee q$ is true when p is true or q is true *or both* are true.

quick Examples

1. Let p: "The butler did it" and let q: "The cook did it." Then $p \vee q$: "Either the butler or the cook did it."

2. Let p: "The butler did it," and let q: "The cook did it," and let r: "The lawyer did it." Then $(p \vee q) \wedge (\sim r)$: "Either the butler or the cook did it, but not the lawyer."

Note The only way for $p \vee q$ to be false is for *both* p and q to be false. For this reason, we can say that $p \vee q$ also means "p and q are not both false." ∎

To introduce our next logical operator, we ask you to consider the following statement: "If you earn an A in logic, then I'll buy you a new car." It seems to be made up out of two simpler statements,

p: "You earn an A in logic," and
q: "I will buy you a new car."

The original statement says: *if p is true, then q is true*, or, more simply, **if** p, **then** q. We can also phrase this as p **implies** q, and we write the statement symbolically as $p \rightarrow q$.

Now let us suppose for the sake of argument that the original statement: "If you earn an A in logic, then I'll buy you a new car," is true. This does *not* mean that you *will* earn an A in logic. All it says is that *if* you do so, then I will buy you that car. Thinking of this as a promise, the only way that it can be broken is if you *do* earn an A and I do *not* buy you a new car. With this in mind, we define the logical statement $p \rightarrow q$ as follows.

Conditional

The **conditional** $p \rightarrow q$, which we read "if p, then q" or "p implies q," is defined by the following truth table:

p	q	$p \rightarrow q$
T	T	T
T	F	F
F	T	T
F	F	T

[2] There is also the **exclusive or:** "p or q but not both." This can be expressed as $(p \vee q) \wedge \sim(p \wedge q)$. Do you see why?

The arrow "$\rightarrow$" is the **conditional** operator, and in $p \rightarrow q$ the statement p is called the **antecedent,** or **hypothesis,** and q is called the **consequent,** or **conclusion.** A statement of the form $p \rightarrow q$ is also called an **implication.**

quick Examples

1. "If $1 + 1 = 2$ then the sun rises in the east" has the form $p \rightarrow q$ where p: "$1 + 1 = 2$" is true and q: "the sun rises in the east." is also true. Therefore the statement is true.

2. "If the moon is made of green cheese, then I am Arnold Schwartzenegger" has the form $p \rightarrow q$ where p is false. From the truth table, we see that $p \rightarrow q$ is therefore true, regardless of whether or not I am Arnold Schwartzenegger.

3. "If $1 + 1 = 2$ then $0 = 1$" has the form $p \rightarrow q$ where this time p is true but q is false. Therefore, by the truth table, the given statement is false.

Notes

1. The only way that $p \rightarrow q$ can be false is if p is true and q is false—this is the case of the "broken promise" in the car example above.

2. If you look at the last two rows of the truth table, you see that we say that "$p \rightarrow q$" is true when p is false, *no matter what the truth value of q.* Think again about the promise—if you don't get that A, then whether or not I buy you a new car, I have not broken my promise. It may seem strange at first to say that $F \rightarrow T$ is T and $F \rightarrow F$ is also T, but, as they did in choosing to say that "or" is always inclusive, mathematicians agreed that the truth table above gives the most useful definition of the conditional. ∎

It is usually misleading to think of "if p then q" as meaning that p *causes* q. For instance, tropical weather conditions cause hurricanes, but one cannot claim that if there are tropical weather conditions, then there are (always) hurricanes. Here is a list of some English phrases that *do* have the same meaning as $p \rightarrow q$.

Some Phrasings of the Conditional

We interpret each of the following as equivalent to the conditional $p \rightarrow q$.

If p then q.	p implies q.
q follows from p.	Not p unless q.
q if p.	p only if q.
Whenever p, q.	q whenever p.
p is sufficient for q.	q is necessary for p.
p is a sufficient condition for q.	q is a necessary condition for p.

quick Example

"If it's Tuesday, this must be Belgium" can be rephrased in several ways as follows:

"Its being Tuesday implies that this is Belgium."
"This is Belgium if it's Tuesday."
"It's Tuesday only if this is Belgium."
"It can't be Tuesday unless this is Belgium."
"Its being Tuesday is sufficient for this to be Belgium."
"That this is Belgium is a necessary condition for its being Tuesday."

Notice the difference between "if" and "only if." We say that "p only if q" means $p \to q$ because, assuming that $p \to q$ is true, p can be true only if q is also. In other words, the only line of the truth table that has $p \to q$ true and p true also has q true. The phrasing "p is a sufficient condition for q" says that it suffices to know that p is true to be able to conclude that q is true. For example, it is sufficient that you get an A in logic for me to buy you a new car. Other things might induce me to buy you the car, but an A in logic would suffice. The phrasing "q is necessary for p" says that for p to be true, q must be true (just as we said for "p only if q").

Q: Does the commutative law hold for the conditional? In other words, is $p \to q$ the same as $q \to p$?

A: No, as we can see in the following truth table:

p	q	$p \to q$	$q \to p$
T	T	T	T
T	F	F	T
F	T	T	F
F	F	T	T

not the same

Converse and Contrapositive

The statement $q \to p$ is called the **converse** of the statement $p \to q$. A conditional and its converse are *not* the same.

The statement $\sim q \to \sim p$ is the **contrapositive** of the statement $p \to q$. A conditional and its contrapositive are logically equivalent in the sense we define below: they have the same truth value for all possible values of p and q.

Example 3 Converse and Contrapositive

Give the converse and contrapositive of the statement "If you earn an A in logic, then I'll buy you a new car."

Solution This statement has the form $p \to q$ where p: "you earn an A" and q: "I'll buy you a new car." The converse is $q \to p$. In words, this is "If I buy you a new car then you earned an A in logic."

The contrapositive is $(\sim q) \to (\sim p)$. In words, this is "If I don't buy you a new car, then you didn't earn an A in logic."

Assuming that the original statement is true, notice that the converse is not necessarily true. There is nothing in the original promise that prevents me from buying you a new car if you do not earn the A. On the other hand, the contrapositive is true. If I don't buy you a new car, it must be that you didn't earn an A; otherwise I would be breaking my promise.

It sometimes happens that we do want both a conditional and its converse to be true. The conjunction of a conditional and its converse is called a **biconditional.**

Biconditional

The **biconditional,** written $p \leftrightarrow q$, is defined to be the statement $(p \rightarrow q) \wedge (q \rightarrow p)$. Its truth table is the following:

p	q	$p \leftrightarrow q$
T	T	T
T	F	F
F	T	F
F	F	T

Phrasings of the Biconditional
We interpret each of the following as equivalent to $p \leftrightarrow q$.

p if and only if q.
p is necessary and sufficient for q.
p is equivalent to q.

quick Example

"I teach math if and only if I am paid a large sum of money." can be rephrased in several ways as follows:

"I am paid a large sum of money if and only if I teach math."
"My teaching math is necessary and sufficient for me to be paid a large sum of money."
"For me to teach math, it is necessary and sufficient that I be paid a large sum of money."

A.2 Logical Equivalence

We mentioned above that we say that two statements are **logically equivalent** if for all possible truth values of the variables involved the two statements always have the same truth values. If s and t are equivalent, we write $s \equiv t$. This is *not* another logical statement. It is simply the claim that the two statements s and t are logically equivalent. Here are some examples.

Example 4 Logical Equivalence

Use truth tables to show the following:

a. $p \equiv \sim(\sim p)$. This is called **double negation.**

b. $\sim(p \wedge q) \equiv (\sim p) \vee (\sim q)$. This is one of **DeMorgan's Laws.**

Solution

a. To demonstrate the logical equivalence of these two statements, we construct a truth table with columns for both p and $\sim(\sim p)$.

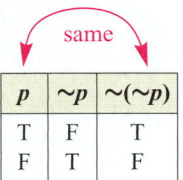

same

p	$\sim p$	$\sim(\sim p)$
T	F	T
F	T	F

Because the p and $\sim(\sim p)$ columns contain the same truth values in all rows, the two statements are logically equivalent.

b. We construct a truth table showing both $\sim(p \wedge q)$ and $(\sim p) \vee (\sim q)$.

same

p	q	$p \wedge q$	$\sim(p \wedge q)$	$\sim p$	$\sim q$	$(\sim p) \vee (\sim q)$
T	T	T	F	F	F	F
T	F	F	T	F	T	T
F	T	F	T	T	F	T
F	F	F	T	T	T	T

Because the $\sim(p \wedge q)$ column and $(\sim p) \vee (\sim q)$ column agree, the two statements are equivalent.

✛ *Before we go on...* The statement $\sim(p \wedge q)$ can be read as "It is not the case that both p and q are true" or "p and q are not both true." We have just shown that this is equivalent to "Either p is false or q is false." ∎

Here are the two equivalences known as DeMorgan's Laws.

DeMorgan's Laws

If p and q are statements, then

$$\sim(p \wedge q) \equiv (\sim p) \vee (\sim q)$$
$$\sim(p \vee q) \equiv (\sim p) \wedge (\sim q)$$

quick Example   Let p: "the President is a Democrat," and q: "the President is a Republican." Then the following two statements say the same thing:

$\sim(p \wedge q)$: "the President is not both a Democrat and a Republican."

$(\sim p) \vee (\sim q)$: "either the President is not a Democrat, or he is not a Republican (or he is neither)."

Here is a list of some important logical equivalences, some of which we have already encountered. All of them can be verified using truth tables as in Example 4. (The verifications of some of these are in the exercise set.)

Important Logical Equivalences

$\sim(\sim p) \equiv p$	the Double Negative Law
$p \wedge q \equiv q \wedge p$	the Commutative Law for Conjunction.
$p \vee q \equiv q \vee p$	the Commutative Law for Disjunction.
$(p \wedge q) \wedge r \equiv p \wedge (q \wedge r)$	the Associative Law for Conjunction.
$(p \vee q) \vee r \equiv p \vee (q \vee r)$	the Associative Law for Disjunction.
$\sim(p \vee q) \equiv (\sim p) \wedge (\sim q)$	DeMorgan's Laws
$\sim(p \wedge q) \equiv (\sim p) \vee (\sim q)$	
$p \wedge (q \vee r) \equiv (p \wedge q) \vee (p \wedge r)$	the Distributive Laws
$p \vee (q \wedge r) \equiv (p \vee q) \wedge (p \vee r)$	
$p \wedge p \equiv p$	Absorption Laws
$p \vee p \equiv p$	
$p \rightarrow q \equiv (\sim q) \rightarrow (\sim p)$	Contrapositive Law

Note that these logical equivalences apply to *any* statement. The ps, qs and rs can stand for atomic statements or compound statements, as we see in the next example.

Example 5 Applying Logical Equivalences

a. Apply DeMorgan's law (once) to the statement $\sim([p \wedge (\sim q)] \wedge r)$.

b. Apply the distributive law to the statement $(\sim p) \wedge [q \vee (\sim r)]$.

c. Consider: "You will get an A if either you are clever and the sun shines, or you are clever and it rains." Rephrase the condition more simply using the distributive law.

Solution

a. We can analyze the given statement from the outside in. It is first of all a negation, but further, it is the negation $\sim(A \wedge B)$, where A is the compound statement $[p \wedge (\sim q)]$ and B is r:

$$\sim(\quad \underbrace{A \quad\quad \wedge \quad B}\quad)$$
$$\sim(\ [p \wedge (\sim q)] \wedge \quad r\ )$$

Now one of DeMorgan's laws is

$$\sim(A \wedge B) \equiv (\sim A) \vee (\sim B)$$

Applying this equivalence gives

$$\sim([p \wedge (\sim q)] \wedge r) \equiv (\sim[p \wedge (\sim q)]) \vee (\sim r)$$

b. The given statement has the form $A \wedge [B \vee C]$, where $A = (\sim p)$, $B = q$, and $C = (\sim r)$. So, we apply the distributive law $A \wedge [B \vee C], \equiv [A \wedge B] \vee [A \wedge C]$:

$$(\sim p) \wedge [q \vee (\sim r)] \equiv [(\sim p) \wedge q] \vee [(\sim p) \wedge (\sim r)]$$

(We need not stop here: The second expression on the right is just begging for an application of DeMorgan's law. . .)

c. The condition is "either you are clever and the sun shines, or you are clever and it rains." Let's analyze this symbolically: Let p: "You are clever," q: "The sun shines," and r: "It rains." The condition is then $(p \wedge q) \vee (p \wedge r)$. We can "factor out" the p using one of the distributive laws in reverse, getting

$$(p \wedge q) \vee (p \wedge r) \equiv p \wedge (q \vee r)$$

We are taking advantage of the fact that the logical equivalences we listed can be read from right to left as well as from left to right. Putting $p \wedge (q \vee r)$ back into English, we can rephrase the sentence as "You will get an A if you are clever and either the sun shines or it rains."

+*Before we go on...* In part (a) of Example 5 we could, if we wanted, apply DeMorgan's law again, this time to the statement $\sim[p \wedge (\sim q)]$ that is part of the answer. Doing so gives

$$\sim[p \wedge (\sim q)] \equiv (\sim p) \vee \sim(\sim q) \equiv (\sim p) \vee q$$

Notice that we've also used the double negative law. Therefore, the original expression can be simplified as follows:

$$\sim([p \wedge (\sim q)] \wedge r) \equiv (\sim[p \wedge (\sim q)]) \vee (\sim r) \equiv ((\sim p) \vee q) \vee (\sim r)$$

which we can write as

$$(\sim p) \vee q \vee (\sim r)$$

because the associative law tells us that it does not matter which two expressions we "or" first. ∎

A.3 | Tautologies, Contradictions, and Arguments

Tautologies and Contradictions

A compound statement is a **tautology** if its truth value is always T, regardless of the truth values of its variables. It is a **contradiction** if its truth value is always F, regardless of the truth values of its variables.

quick Examples

1. $p \vee (\sim p)$ has truth table

p	$\sim p$	$p \vee (\sim p)$
T	F	T
F	T	T

— all T's

and is therefore a tautology.

2. $p \wedge (\sim p)$ has truth table

p	$\sim p$	$p \wedge (\sim p)$
T	F	F
F	T	F

and is therefore a contradiction.

When a statement is a tautology, we also say that the statement is **tautological.** In common usage this sometimes means simply that the statement is self-evident. In logic it means something stronger: that the statement is always true, under all circumstances. In contrast, a contradiction, or **contradictory** statement, is *never* true, under any circumstances.

Some of the most important tautologies are the **tautological implications,** tautologies that have the form of implications. We look at two of them: Direct Reasoning, and Indirect Reasoning:

Modus Ponens or Direct Reasoning

The following tautology is called *modus ponens* or **direct reasoning:**

$$[(p \rightarrow q) \wedge p] \rightarrow q$$

In Words
If an implication and its antecedent (p) are both true, then so is its consequent (q).

quick Example

If my loving math implies that I will pass this course, and if I do love math, then I will pass this course.

Note You can check that the statement $[(p \rightarrow q) \wedge p] \rightarrow q$ is a tautology by drawing its truth table. ∎

Tautological implications are useful mainly because they allow us to check the validity of **arguments.**

Argument

An **argument** is a list of statements called **premises** followed by a statement called the **conclusion.** If the premises are $P_1, P_2, \ldots, P_n$ and the conclusion is C, then we say that the argument is **valid** if the statement $(P_1 \wedge P_2 \wedge \ldots \wedge P_n) \rightarrow C$ is a tautology. In other words, an argument is valid if the truth of all its premises logically implies the truth of its conclusion.

quick Examples

1. The following is a valid argument:

$$p \rightarrow q$$
$$\underline{p \qquad\quad}$$
$$\therefore \quad q$$

(This is the traditional way of writing an argument: We list the premises above a line and then put the conclusion below; the symbol "∴" stands for the word "therefore.") This argument is valid because the statement $[(p \rightarrow q) \wedge p] \rightarrow q$ is a tautology, namely *modus ponens*.

2. The following is an invalid argument:

$$p \rightarrow q$$
$$\underline{q \qquad\quad}$$
$$\therefore \quad p$$

The argument is invalid because the statement $[(p \rightarrow q) \wedge q] \rightarrow p$ is not a tautology. In fact, if p is F and q is T, then the whole statement is F.

The argument in Quick Example 2 above is known as the *fallacy of affirming the consequent*. It is a common invalid argument and not always obviously flawed at first sight, so is often exploited by advertisers. For example, consider the following claim: All Olympic athletes drink Boors, so you should too. The suggestion is that, if you drink Boors, you will be an Olympic athlete:

If you are an Olympic Athlete you drink Boors.	Premise (Let's pretend this is True)
You drink Boors.	Premise (True)
∴ You are an Olympic Athlete.	Conclusion (May be false!)

This is an error that Boors hopes you will make!

There is, however, a correct argument in which we *deny* the consequent:

Modus Tollens or Indirect Reasoning

The following tautology is called *modus tollens* or **indirect reasoning:**

$$[(p \to q) \land (\sim q)] \to (\sim p)$$

In Words
If an implication is true but its consequent (q) is false, then its antecedent (p) is false.

In Argument Form

$$p \to q$$
$$\underline{\sim q}$$
$$\therefore \quad \sim q$$

quick Example

If my loving math implies that I will pass this course, and if I do not pass the course, then it must be the case that I do not love math.
In argument form:

If I love math, then I will pass this course.
I will not pass the course.
Therefore, I do not love math.

Note This argument is not as direct as *modus ponens;* it contains a little twist: "If I loved math I would pass this course. However, I will not pass this course. Therefore, it must be that I don't love math (else I *would* pass this course)." Hence the name "indirect reasoning."

Note that, again, there is a similar, but fallacious argument to avoid, for instance: "If I were an Olympic athlete then I would drink Boors ($p \to q$). However, I am not an Olympic athlete ($\sim p$). Therefore, I won't drink Boors. ($\sim q$)." This is a mistake Boors certainly hopes you do *not* make! ■

There are other interesting tautologies that we can use to justify arguments. We mention one more and refer the interested reader to the website for more examples and further study.

For an extensive list of tautologies go online and follow:

Chapter L Logic

→ List of Tautologies and Tautological Implications

Disjunctive Syllogism or "One or the Other"

The following tautologies are both known as the **disjunctive syllogism** or **one-or-the-other:**

$$[(p \vee q) \wedge (\sim p)] \to q \qquad\qquad [(p \vee q) \wedge (\sim q)] \to p$$

In Words

If one or the other of two statements is true, but one is known to be false, then the other must be true.

In Argument Form

$$p \vee q$$
$$\underline{\sim p}$$
$$\therefore \quad q$$

$$p \vee q$$
$$\underline{\sim q}$$
$$\therefore \quad p$$

quick Example   *The butler or the cook did it. The butler didn't do it. Therefore, the cook did it.*
In argument form:

The butler or the cook did it.

The butler did not do it.

Therefore, the cook did it.

A EXERCISES

Which of Exercises 1–10 are statements? Comment on the truth values of all the statements you encounter. If a sentence fails to be a statement, explain why. hint [see Example 1]

1. All swans are white.   **2.** The fat cat sat on the mat.

3. Look in thy glass and tell whose face thou viewest.[3]

4. My glass shall not persuade me I am old.[4]

5. There is no largest number.

6. 1,000,000,000 is the largest number.

7. Intelligent life abounds in the universe.

8. There may or may not be a largest number.

9. This is exercise number 9.   **10.** This sentence no verb.[5]

Let p: "Our mayor is trustworthy," q: "Our mayor is a good speller," and r = "Our mayor is a patriot." Express each of the statements in Exercises 11–16 in logical form: hint [see Quick Examples on pp. A2, A3, A5]

11. Although our mayor is not trustworthy, he is a good speller.

12. Either our mayor is trustworthy, or he is a good speller.

13. Our mayor is a trustworthy patriot who spells well.

14. While our mayor is both trustworthy and patriotic, he is not a good speller.

15. It may or may not be the case that our mayor is trustworthy.

16. Our mayor is either not trustworthy or not a patriot, yet he is an excellent speller.

Let p: "Willis is a good teacher," q: "Carla is a good teacher," r: "Willis' students hate math," s: "Carla's students hate math." Express the statements in Exercises 17–24 in words.

17. $p \wedge (\sim r)$ **18.** $(\sim p) \wedge (\sim q)$

19. $q \vee (\sim q)$ **20.** $((\sim p) \wedge (\sim s)) \vee q$

21. $r \wedge (\sim r)$ **22.** $(\sim s) \vee (\sim r)$

23. $\sim(q \vee s)$ **24.** $\sim(p \wedge r)$

Assume that it is true that "Polly sings well," it is false that "Quentin writes well," and it is true that "Rita is good at math." Determine the truth of each of the statements in Exercises 25–32.

25. Polly sings well and Quentin writes well.

26. Polly sings well or Quentin writes well.

27. Polly sings poorly and Quentin writes well.

28. Polly sings poorly or Quentin writes poorly.

29. Either Polly sings well and Quentin writes poorly, or Rita is good at math.

---

[3] William Shakespeare Sonnet 3

[4] Ibid., Sonnet 22.

[5] From *Metamagical Themas: Questing for the Essence of Mind and Pattern* by Douglas R. Hofstadter (Bantam Books, New York 1986)

30. Either Polly sings well and Quentin writes poorly, or Rita is not good at math.

31. Either Polly sings well or Quentin writes well, or Rita is good at math.

32. Either Polly sings well and Quentin writes well, or Rita is bad at math.

Find the truth value of each of the statements in Exercises 33–48. *hint* [see Quick Examples on p. A6]

33. "If $1 = 1$, then $2 = 2$."   **34.** "If $1 = 1$, then $2 = 3$."

35. "If $1 \neq 0$, then $2 \neq 2$."   **36.** "If $1 = 0$, then $1 = 1$."

37. "A sufficient condition for 1 to equal 2 is $1 = 3$."

38. "$1 = 1$ is a sufficient condition for 1 to equal 0."

39. "$1 = 0$ is a necessary condition for 1 to equal 1."

40. "$1 = 1$ is a necessary condition for 1 to equal 2."

41. "If I pay homage to the great Den, then the sun will rise in the east."

42. "If I fail to pay homage to the great Den, then the sun will still rise in the east."

43. "In order for the sun to rise in the east, it is necessary that it sets in the west."

44. "In order for the sun to rise in the east, it is sufficient that it sets in the west."

45. "The sun rises in the west only if it sets in the west."

46. "The sun rises in the east only if it sets in the east."

47. "In order for the sun to rise in the east, it is necessary and sufficient that it sets in the west."

48. "In order for the sun to rise in the west, it is necessary and sufficient that it sets in the east."

Construct the truth tables for the statements in Exercises 49–62. *hint* [see Example 2]

49. $p \wedge (\sim q)$ **50.** $p \vee (\sim q)$

51. $\sim(\sim p) \vee p$ **52.** $p \wedge (\sim p)$

53. $(\sim p) \wedge (\sim q)$ **54.** $(\sim p) \vee (\sim q)$

55. $(p \wedge q) \wedge r$ **56.** $p \wedge (q \wedge r)$

57. $p \wedge (q \vee r)$ **58.** $(p \wedge q) \vee (p \wedge r)$

59. $p \rightarrow (q \vee p)$ **60.** $(p \vee q) \rightarrow \sim p$

61. $p \leftrightarrow (p \vee q)$ **62.** $(p \wedge q) \leftrightarrow \sim p$

Use truth tables to verify the logical equivalences given in Exercises 63–72.

63. $p \wedge p \equiv p$ **64.** $p \vee p \equiv p$

65. $p \vee q \equiv q \vee p$ **66.** $p \wedge q \equiv q \wedge p$
(Commutative law for disjunction)   (Commutative law for conjunction)

67. $\sim(p \vee q) \equiv (\sim p) \wedge (\sim q)$ **68.** $\sim(p \wedge (\sim q)) \equiv (\sim p) \vee q$

69. $(p \wedge q) \wedge r \equiv p \wedge (q \wedge r)$ **70.** $(p \vee q) \vee r \equiv p \vee (q \vee r)$
(Associative law for conjunction)   (Associative law for disjunction)

71. $p \rightarrow q \equiv (\sim q) \rightarrow (\sim p)$ **72.** $\sim(p \rightarrow q) \equiv p \wedge (\sim q)$

In Exercises 73–78, use truth tables to check whether the given statement is a tautology, a contradiction, or neither. *hint* [see Quick Examples on p. A11]

73. $p \wedge (\sim p)$ **74.** $p \wedge p$

75. $p \wedge \sim(p \vee q)$ **76.** $p \vee \sim(p \vee q)$

77. $p \vee \sim(p \wedge q)$ **78.** $q \vee \sim(p \wedge (\sim p))$

Apply the stated logical equivalence to the given statement in Exercises 79–84. *hint* [see Example 5a, b]

79. $p \vee (\sim p)$; the commutative law

80. $p \wedge (\sim q)$; the commutative law

81. $\sim(p \wedge (\sim q))$; DeMorgan's law

82. $\sim(q \vee (\sim q))$; DeMorgan's law

83. $p \vee ((\sim p) \wedge q)$; the distributive law

84. $(\sim q) \wedge ((\sim p) \vee q)$; the distributive law.

In Exercises 85–88, use the given logical equivalence to rewrite the given sentence. *hint* [see Example 5c]

85. It is not true that both I am Julius Caesar and you are a fool. DeMorgan's law.

86. It is not true that either I am Julius Caesar or you are a fool. DeMorgan's law.

87. Either it is raining and I have forgotten my umbrella, or it is raining and I have forgotten my hat. The distributive law.

88. I forgot my hat or my umbrella, and I forgot my hat or my glasses. The distributive law.

Give the contrapositive and converse of each of the statements in Exercises 89 and 90, phrasing your answers in words.

89. "If I think, then I am."

90. "If these birds are of a feather, then they flock together."

Exercises 91 and 92 are multiple choice. Indicate which statement is equivalent to the given statement, and say why that statement is equivalent to the given one.

91. "In order for you to worship Den, it is necessary for you to sacrifice beasts of burden."
 (A) "If you are not sacrificing beasts of burden, then you are not worshiping Den."
 (B) "If you are sacrificing beasts of burden, then you are worshiping Den."
 (C) "If you are not worshiping Den, then you are not sacrificing beasts of burden."

92. "In order to read the Tarot, it is necessary for you to consult the Oracle."
 (A) "In order to consult the Oracle, it is necessary to read the Tarot."
 (B) "In order not to consult the Oracle, it is necessary not to read the Tarot."
 (C) "In order not to read the Tarot, it is necessary not to read the Oracle."

In Exercises 93–102, write the given argument in symbolic form (use the underlined letters to represent the statements containing them), then decide whether it is valid or not, If it is valid, name the validating tautology. hint [see Quick Examples on pp. A12, A13, A14]

93. If I am <u>h</u>ungry, I am also <u>t</u>hirsty, If I am thirsty, I drink <u>m</u>ango juice. Therefore, every time I am hungry, I drink mango juice.

94. If I am not <u>h</u>ungry, then I certainly am not <u>t</u>hirsty either. I am not thirsty, and so I cannot be hungry.

95. For me to bring my <u>u</u>mbrella, it's sufficient that it <u>r</u>ain. It is not raining. Therefore, I will not bring my umbrella.

96. For me to bring my <u>u</u>mbrella, it's necessary that it <u>r</u>ain. But it is not raining. Therefore, I will not bring my umbrella.

97. For me to pass <u>m</u>ath, it is sufficient that I have a <u>g</u>ood teacher. I will not pass math. Therefore, I have a bad teacher.

98. For me to pass <u>m</u>ath, it is necessary that I have a <u>g</u>ood teacher. I will pass math. Therefore, I have a good teacher.

99. I will either pass <u>m</u>ath or I have a <u>b</u>ad teacher. I have a good teacher. Therefore, I will pass math.

100. Either <u>r</u>oses are not red or <u>v</u>iolets are not blue. But roses are red. Therefore, violets are not blue.

101. I am either <u>s</u>mart or <u>a</u>thletic, and I am athletic. So I must not be smart.

102. The president is either <u>w</u>ise or <u>s</u>trong. She is strong. Therefore, she is not wise.

In Exercises 103–108, use the stated tautology to complete the argument.

103. If John is a swan, it is necessary that he is green. John is indeed a swan. Therefore, _____. (*Modus ponens.*)

104. If Jill had been born in Texas, then she would be able to ride horses. But Jill cannot ride horses. Therefore, ___. (*Modus tollens.*)

105. If John is a swan, it is necessary that he is green. But John is not green. Therefore, _____. (*Modus tollens.*)

106. If Jill had been born in Texas, then she would be able to ride horses. Jill was born in Texas. Therefore, ___ (*Modus ponens.*)

107. Peter is either a scholar or a gentleman. He is not, however, a scholar. Therefore, ___. (Disjunctive syllogism.)

108. Pam is either a plumber or an electrician. She is not, however, an electrician. Therefore, ___ (Disjunctive syllogism.)

Communication and Reasoning Exercises

109. If two statements are logically equivalent, what can be said about their truth tables?

110. If a proposition is neither a tautology nor a contradiction, what can be said about its truth table?

111. If A and B are two compound statements such that $A \vee B$ is a contradiction, what can you say about A and B?

112. If A and B are two compound statements such that $A \wedge B$ is a tautology, what can you say about A and B?

113. Give an example of an instance where $p \rightarrow q$ means that q causes p.

114. Complete the following. If $p \rightarrow q$, then its converse, ___ , is the statement that ___ and (<u>is/is not</u>) logically equivalent to $p \rightarrow q$.

115. Give an instance of a biconditional $p \leftrightarrow q$ where neither one of p or q causes the other. Answers may vary.

Appendix B Area Under a Normal Curve

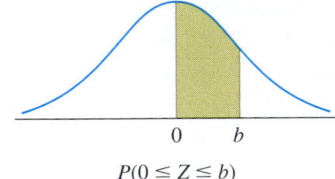

$P(0 \leq Z \leq b)$

The table below gives the probabilities $P(0 \leq Z \leq b)$ where Z is a standard normal variable. For example, to find $P(0 \leq Z \leq 2.43)$, write 2.43 as $2.4 + 0.03$, and read the entry in the row labeled 2.4 and the column labeled 0.03. From the portion of the table shown at left, you will see that $P(0 \leq Z \leq 2.43) = .4925$.

Z	0.00	0.01	0.02	0.03
2.3	.4893	.4896	.4898	.4901
2.4	.4918	.4920	.4922	.4925
2.5	.4938	.4940	.4941	.4943

Z	0.00	0.01	0.02	0.03	0.04	0.05	0.06	0.07	0.08	0.09
0.0	.0000	.0040	.0080	.0120	.0160	.0199	.0239	.0279	.0319	.0359
0.1	.0398	.0438	.0478	.0517	.0557	.0596	.0636	.0675	.0714	.0753
0.2	.0793	.0832	.0871	.0910	.0948	.0987	.1026	.1064	.1103	.1141
0.3	.1179	.1217	.1255	.1293	.1331	.1368	.1406	.1443	.1480	.1517
0.4	.1554	.1591	.1628	.1664	.1700	.1736	.1772	.1808	.1844	.1879
0.5	.1915	.1950	.1985	.2019	.2054	.2088	.2123	.2157	.2190	.2224
0.6	.2257	.2291	.2324	.2357	.2389	.2422	.2454	.2486	.2517	.2549
0.7	.2580	.2611	.2642	.2673	.2704	.2734	.2764	.2794	.2823	.2852
0.8	.2881	.2910	.2939	.2967	.2995	.3023	.3051	.3078	.3106	.3133
0.9	.3159	.3186	.3212	.3238	.3264	.3289	.3315	.3340	.3365	.3389
1.0	.3413	.3438	.3461	.3485	.3508	.3531	.3554	.3577	.3599	.3621
1.1	.3643	.3665	.3686	.3708	.3729	.3749	.3770	.3790	.3810	.3830
1.2	.3849	.3869	.3888	.3907	.3925	.3944	.3962	.3980	.3997	.4015
1.3	.4032	.4049	.4066	.4082	.4099	.4115	.4131	.4147	.4162	.4177
1.4	.4192	.4207	.4222	.4236	.4251	.4265	.4279	.4292	.4306	.4319
1.5	.4332	.4345	.4357	.4370	.4382	.4394	.4406	.4418	.4429	.4441
1.6	.4452	.4463	.4474	.4484	.4495	.4505	.4515	.4525	.4535	.4545
1.7	.4554	.4564	.4573	.4582	.4591	.4599	.4608	.4616	.4625	.4633
1.8	.4641	.4649	.4656	.4664	.4671	.4678	.4686	.4693	.4699	.4706
1.9	.4713	.4719	.4726	.4732	.4738	.4744	.4750	.4756	.4761	.4767
2.0	.4772	.4778	.4783	.4788	.4793	.4798	.4803	.4808	.4812	.4817
2.1	.4821	.4826	.4830	.4834	.4838	.4842	.4846	.4850	.4854	.4857
2.2	.4861	.4864	.4868	.4871	.4875	.4878	.4881	.4884	.4887	.4890
2.3	.4893	.4896	.4898	.4901	.4904	.4906	.4909	.4911	.4913	.4916
2.4	.4918	.4920	.4922	.4925	.4927	.4929	.4931	.4932	.4934	.4936
2.5	.4938	.4940	.4941	.4943	.4945	.4946	.4948	.4949	.4951	.4952
2.6	.4953	.4955	.4956	.4957	.4959	.4960	.4961	.4962	.4963	.4964
2.7	.4965	.4966	.4967	.4968	.4969	.4970	.4971	.4972	.4973	.4974
2.8	.4974	.4975	.4976	.4977	.4977	.4978	.4979	.4979	.4980	.4981
2.9	.4981	.4982	.4982	.4983	.4984	.4984	.4985	.4985	.4986	.4986
3.0	.4987	.4987	.4987	.4988	.4988	.4989	.4989	.4989	.4990	.4990

Answers to Selected Exercises

Chapter 0

Section 0.1

1. -48 **3.** $2/3$ **5.** -1 **7.** 9 **9.** 1 **11.** 33 **13.** 14
15. $5/18$ **17.** 13.31 **19.** 6 **21.** $43/16$ **23.** 0
25. `3*(2-5)` **27.** `3/(2-5)` **29.** `(3-1)/(8+6)`
31. `3-(4+7)/8` **33.** `2/(3+x)-x*y^2`
35. `3.1x^3-4x^(-2)-60/(x^2-1)` **37.** `(2/3)/5`
39. `3^(4-5)*6` **41.** `3*(1+4/100)^(-3)`
43. `3^(2*x-1)+4^x-1` **45.** `2^(2x^2-x+1)`
47. `4*e^(-2*x)/(2-3e^(-2*x))` or `4*(e^(-2*x))/(2-3e^(-2*x))` **49.** `3(1-(-1/2)^2)^2+1`

Section 0.2

1. 27 **3.** -36 **5.** $4/9$ **7.** $-1/8$ **9.** 16 **11.** 2 **13.** 32
15. 2 **17.** x^5 **19.** $-\dfrac{y}{x}$ **21.** $\dfrac{1}{x}$ **23.** $x^3 y$ **25.** $\dfrac{z^4}{y^3}$ **27.** $\dfrac{x^6}{y^6}$
29. $\dfrac{x^4 y^6}{z^4}$ **31.** $\dfrac{3}{x^4}$ **33.** $\dfrac{3}{4x^{2/3}}$ **35.** $1 - 0.3x^2 - \dfrac{6}{5x}$ **37.** 2
39. $1/2$ **41.** $4/3$ **43.** $2/5$ **45.** 7 **47.** 5 **49.** -2.668
51. $3/2$ **53.** 2 **55.** 2 **57.** ab **59.** $x + 9$ **61.** $x\sqrt[3]{a^3 + b^3}$
63. $\dfrac{2y}{\sqrt{x}}$ **65.** $3^{1/2}$ **67.** $x^{3/2}$ **69.** $(xy^2)^{1/3}$ **71.** $x^{3/2}$
73. $\dfrac{3}{5}x^{-2}$ **75.** $\dfrac{3}{2}x^{-1.2} - \dfrac{1}{3}x^{-2.1}$ **77.** $\dfrac{2}{3}x - \dfrac{1}{2}x^{0.1} + \dfrac{4}{3}x^{-1.1}$
79. $(x^2 + 1)^{-3} - \dfrac{3}{4}(x^2 + 1)^{-1/3}$ **81.** $\sqrt[3]{2^2}$ **83.** $\sqrt[3]{x^4}$
85. $\sqrt[5]{\sqrt{x}\,\sqrt[3]{y}}$ **87.** $-\dfrac{3}{2\sqrt[4]{x}}$ **89.** $\dfrac{0.2}{\sqrt[3]{x^2}} + \dfrac{3\sqrt{x}}{7}$
91. $\dfrac{3}{4\sqrt{(1-x)^5}}$ **93.** 64 **95.** $\sqrt{3}$ **97.** $1/x$ **99.** xy
101. $\left(\dfrac{y}{x}\right)^{1/3}$ **103.** ± 4 **105.** $\pm 2/3$ **107.** $-1, -1/3$
109. -2 **111.** 16 **113.** ± 1 **115.** $33/8$

Section 0.3

1. $4x^2 + 6x$ **3.** $2xy - y^2$ **5.** $x^2 - 2x - 3$
7. $2y^2 + 13y + 15$ **9.** $4x^2 - 12x + 9$ **11.** $x^2 + 2 + 1/x^2$
13. $4x^2 - 9$ **15.** $y^2 - 1/y^2$ **17.** $2x^3 + 6x^2 + 2x - 4$
19. $x^4 - 4x^3 + 6x^2 - 4x + 1$ **21.** $y^5 + 4y^4 + 4y^3 - y$
23. $(x + 1)(2x + 5)$ **25.** $(x^2 + 1)^5(x + 3)^3(x^2 + x + 4)$
27. $-x^3(x^3 + 1)\sqrt{x + 1}$ **29.** $(x + 2)\sqrt{(x + 1)^3}$
31. a. $x(2 + 3x)$ **b.** $x = 0, -2/3$ **33. a.** $2x^2(3x - 1)$
b. $x = 0, 1/3$ **35. a.** $(x - 1)(x - 7)$ **b.** $x = 1, 7$
37. a. $(x - 3)(x + 4)$ **b.** $x = 3, -4$ **39. a.** $(2x + 1)(x - 2)$
b. $x = -1/2, 2$ **41. a.** $(2x + 3)(3x + 2)$
b. $x = -3/2, -2/3$ **43. a.** $(3x - 2)(4x + 3)$
b. $x = 2/3, -3/4$ **45. a.** $(x + 2y)^2$ **b.** $x = -2y$
47. a. $(x^2 - 1)(x^2 - 4)$ **b.** $x = \pm 1, \pm 2$

Section 0.4

1. $\dfrac{2x^2 - 7x - 4}{x^2 - 1}$ **3.** $\dfrac{3x^2 - 2x + 5}{x^2 - 1}$ **5.** $\dfrac{x^2 - x + 1}{x + 1}$
7. $\dfrac{x^2 - 1}{x}$ **9.** $\dfrac{2x - 3}{x^2 y}$ **11.** $\dfrac{(x + 1)^2}{(x + 2)^4}$ **13.** $\dfrac{-1}{\sqrt{(x^2 + 1)^3}}$
15. $\dfrac{-(2x + y)}{x^2(x + y)^2}$

Section 0.5

1. -1 **3.** 5 **5.** $13/4$ **7.** $43/7$ **9.** -1 **11.** $(c - b)/a$
13. $x = -4, 1/2$ **15.** No solutions **17.** $\pm\sqrt{\dfrac{5}{2}}$ **19.** -1
21. $-1, 3$ **23.** $\dfrac{1 \pm \sqrt{5}}{2}$ **25.** 1 **27.** $\pm 1, \pm 3$
29. $\pm\sqrt{\dfrac{-1 \pm \sqrt{5}}{2}}$ **31.** $-1, -2, -3$ **33.** -3 **35.** 1
37. -2 **39.** $1, \pm\sqrt{5}$ **41.** $\pm 1, \pm\dfrac{1}{\sqrt{2}}$ **43.** $-2, -1, 2, 3$

Section 0.6

1. $0, 3$ **3.** $\pm\sqrt{2}$ **5.** $-1, -5/2$ **7.** -3 **9.** $0, -1, 1$
11. $x = -1$ ($x = -2$ is not a solution.) **13.** $-2, -3/2, -1$
15. -1 **17.** $\pm\sqrt[4]{2}$ **19.** ± 1 **21.** ± 3 **23.** $2/3$ **25.** $-4, -1/4$

Chapter 1

Section 1.1

1. a. 2 **b.** 0.5 **3. a.** -1.5 **b.** 8 **c.** -8 **5. a.** -7 **b.** -3
c. 1 **d.** $4y - 3$ **e.** $4(a + b) - 3$ **7. a.** 3 **b.** 6 **c.** 2 **d.** 6
e. $a^2 + 2a + 3$ **f.** $(x + h)^2 + 2(x + h) + 3$ **9. a.** 2
b. 0 **c.** $65/4$ **d.** $x^2 + 1/x$ **e.** $(s + h)^2 + 1/(s + h)$
f. $(s + h)^2 + 1/(s + h) - (s^2 + 1/s)$ **11. a.** 1 **b.** 1 **c.** 0
d. 27 **13. a.** Yes; $f(4) = 63/16$ **b.** Not defined
c. Not defined **15. a.** Not defined **b.** Not defined
c. Yes, $f(-10) = 0$ **17. a.** $h(2x + h)$ **b.** $2x + h$
19. a. $-h(2x + h)$ **b.** $-(2x + h)$
21. `0.1*x^2-4*x+5`

x	0	1	2	3	4	5	6	7	8	9	10
$f(x)$	5	1.1	-2.6	-6.1	-9.4	-12.5	-15.4	-18.1	-20.6	-22.9	-25

23. `(x^2-1)/(x^2+1)`

x	0.5	1.5	2.5	3.5	4.5	5.5	6.5	7.5	8.5	9.5	10.5
$h(x)$	-0.6000	0.3846	0.7241	0.8491	0.9059	0.9360	0.9538	0.9651	0.9727	0.9781	0.9820

25. a. $P(5) = 117$, $P(10) = 132$, and $P(9.5) \approx 131$. Approximately 117 million people were employed in the U.S. on July 1, 1995, 132 million people on July 1, 2000, and 131 million people

on January 1, 2000. **b.** [5, 11].   **27. a.** [0, 10]. $t \geq 0$ is not an appropriate domain because it would predict U.S. trade with China into the indefinite future with no basis. **b.** $280 billion; U.S. trade with China in 2004 was valued at approximately $280 billion. **29. a.** (2) **b.** $36.8 billion   **31. a.** 358,600 **b.** 361,200 **c.** $6.00   **33. a.** $P(0) = 200$: At the start of 1995, the processor speed was 200 megahertz. $P(4) = 500$: At the start of 1999, the processor speed was 500 megahertz. $P(5) = 1100$: At the start of 2000, the processor speed was 1100 megahertz. **b.** Midway through 2001
c:

t	0	1	2	3	4	5	6	7	8	9
$P(t)$	200	275	350	425	500	1100	1700	2300	2900	3500

35. a. `(0.08*t+0.6)*(t<8)+(0.355*t-1.6)*(t>=8)`
b:

t	0	1	2	3	4	5	6	7	8	9	10	11
$C(t)$	0.6	0.68	0.76	0.84	0.92	1	1.08	1.16	1.24	1.595	1.95	2.305

37. $T(26,000) = \$730 + 0.15(26,000 - 7300) = \$3535$; $T(65,000) = \$4090 + 0.25(65,000 - 29,700) = \$12,915$
39. a. $12,000 **b.** $N(q) = 2000 + 100q^2 - 500q$; $N(20) = \$32,000$ **41. a.** `100*(1-12200/t^4.48)`
b:

t	9	10	11	12	13	14	15	16	17	18	19	20
$p(t)$	35.2	59.6	73.6	82.2	87.5	91.1	93.4	95.1	96.3	97.1	97.7	98.2

c. 82.2% **d.** 14 months   **43.** t; m **45.** $y(x) = 4x^2 - 2$ (or $f(x) = 4x^2 - 2$)   **47.** $N(t) = 200 + 10t$ (N = number of sound files, t = time in days)   **49.** As the text reminds us: to evaluate f of a quantity (such as $x + h$) replace x everywhere by the *whole quantity* $x + h$, getting $f(x + h) = (x + h)^2 - 1$. **51.** False: Functions with infinitely many points in their domain (such as $f(x) = x^2$) cannot be specified numerically.

Section 1.2
1. a. 20 **b.** 30 **c.** 30 **d.** 20 **e.** 0   **3. a.** -1 **b.** 1.25 **c.** 0 **d.** 1 **e.** 0   **5. a.** (I) **b.** (IV) **c.** (V) **d.** (VI) **e.** (III) **f.** (II)
7.

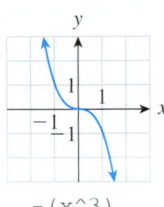

`-(x^3)`

9.

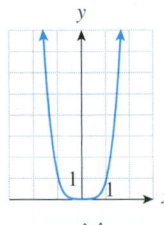

`x^4`

11.

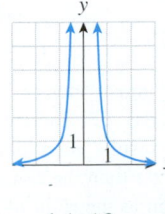

`1/x^2`

13. a. -1 **b.** 2 **c.** 2

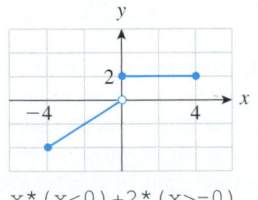

`x*(x<0)+2*(x>=0)`

15. a. 1 **b.** 0 **c.** 1

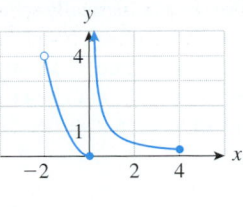

`(x^2)*(x<=0)+(1/x)*(0<x)`

17. a. 0 **b.** 2 **c.** 3 **d.** 3

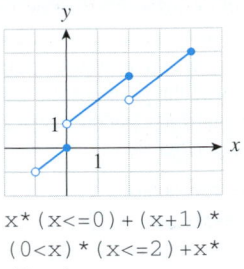

`x*(x<=0)+(x+1)*(0<x)*(x<=2)+x*(2<x)`

19. $f(6) \approx 2000$, $f(9) \approx 2800$, $f(7.5) \approx 2500$. In 1996, 2,000,000 SUVs were sold. In 1999, 2,800,000 were sold, and in the year beginning July, 1997, 2,500,000 were sold.
21. $f(6) - f(5)$; SUV sales increased more from 1995 to 1996 than from 1999 to 2000.   **23. a.** $[-1.5, 1.5]$
b. $N(-0.5) \approx 131$, $N(0) \approx 132$, $N(1) \approx 132$. In July 1999, approximately 131 million people were employed. In January 2000 and January 2001, approximately 132 million people were employed. **c.** [0.5, 1.5]; Employment was falling during the period July 2000–July 2001.   **25. a.** (C) **b.** $20.80 per shirt if the team buys 70 shirts. Graph:

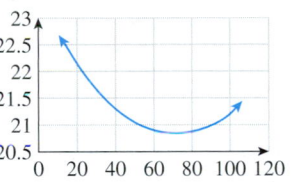

27. A quadratic model (B) is the best choice; the other models either predict perpetually increasing value of the euro or perpetually decreasing value of the euro.
29. a. `100*(1-12200/t^4.48)` **b.** Graph:

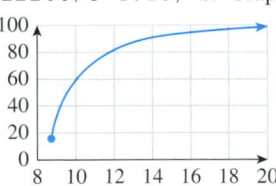

c. 82% **d.** 14 months
31. Midway through 2001

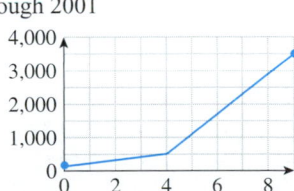

33. a. `(0.08*t+0.6)*(t<8)+(0.355*t-1.6)*(t>=8)` Graph:

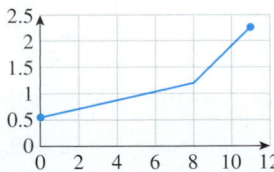

b. 2001

35. True. We can construct a table of values from any graph by reading off a set of values.   **37.** False. In a numerically specified function, only certain values of the function are specified, giving only certain points on the graph.   **39.** They are different portions of the graph of the associated equation $y = f(x)$.   **41.** The graph of $g(x)$ is the same as the graph of $f(x)$, but shifted 5 units to the right.

Section 1.3

1. Missing value: 11; $m = 3$ **3.** Missing value: -4; $m = -1$
5. Missing value: 7; $m = 3/2$ **7.** $f(x) = -x/2 - 2$
9. $f(0) = -5$, $f(x) = -x - 5$ **11.** f is linear: $f(x) = 4x + 6$
13. g is linear: $g(x) = 2x - 1$ **15.** $-3/2$ **17.** $1/6$
19. Undefined   **21.** 0   **23.** $-4/3$

25. 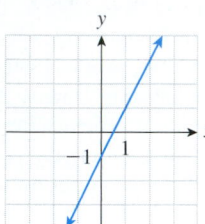   **27.**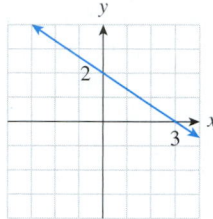

29. 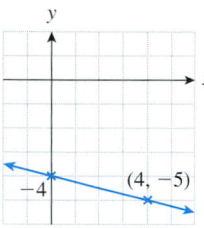   **31.**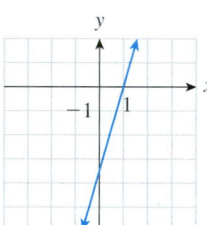

33. 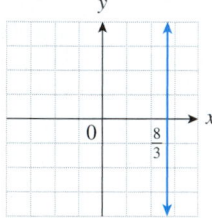   **35.**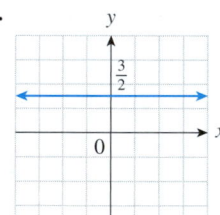

37.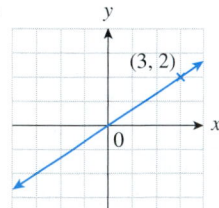

39. 2   **41.** 2   **43.** -2 **45.** Undefined   **47.** 1.5   **49.** -0.09
51. $1/2$ **53.** $(d - b)/(c - a)$ **55. a.** 1 **b.** $1/2$ **c.** 0 **d.** 3
e. $-1/3$ **f.** -1 **g.** Undefined **h.** $-1/4$ **i.** -2 **57.** $y = 3x$
59. $y = \dfrac{1}{4}x - 1$ **61.** $y = 10x - 203.5$ **63.** $y = -5x + 6$
65. $y = -3x + 2.25$ **67.** $y = -x + 12$ **69.** $y = 2x + 4$
71. Compute the corresponding successive changes Δx in x and

Δy in y, and compute the ratios $\Delta y/\Delta x$. If the answer is always the same number, then the values in the table come from a linear function.   **73.** $f(x) = -\dfrac{a}{b}x + \dfrac{c}{b}$. If $b = 0$, then $\dfrac{a}{b}$ is undefined, and y cannot be specified as a function of x. (The graph of the resulting equation would be a vertical line.)   **75.** slope, 3
77. If m is positive then y will increase as x increases; if m is negative then y will decrease as x increases; if m is zero then y will not change as x changes.   **79.** The slope increases, since an increase in the y-coordinate of the second point increases Δy while leaving Δx fixed.

Section 1.4

1. $C(x) = 1500x + 1200$ per day **a.** \$5700 **b.** \$1500
c. \$1500   **3.** Fixed cost $= \$8000$, marginal cost $= \$25$ per bicycle   **5. a.** $C(x) = 0.4x + 70$, $R(x) = 0.5x$,
$P(x) = 0.1x - 70$ **b.** $P(500) = -20$; a loss of \$20
c. 700 copies   **7.** $q = -40p + 2000$ **9. a.** $q = -p + 156.4$;
53.4 million phones **b.** \$1, 1 million   **11. a.** Demand:
$q = -60p + 150$; supply: $q = 80p - 60$ **b.** \$1.50 each
13. a. (1996, 125) and (1997, 135) or (1998, 140) and (1999, 150). **b.** The number of new in-ground pools increased most rapidly during the periods 1996–1997 and 1998–1999, when it rose by 10,000 new pools in a year.   **15.** $N = 400 + 50t$ million transactions. The slope gives the additional number of online shopping transactions per year, and is measured in (millions of) transactions per year.   **17. a.** $s = 14.4t + 240$; Medicare spending is predicted to rise at a rate of \$14.4 billion per year **b.** \$816 billion
19. a. 2.5 ft/sec **b.** 20 feet along the track **c.** after 6 seconds
21. a. 130 miles per hour **b.** $s = 130t - 1300$ **c.** After 5 seconds   **23.** $F = 1.8C + 32$; 86°F; 72°F; 14°F; 7°F   **25.**
$I(N) = 0.05N + 50,000$; $N = \$1,000,000$; marginal income is $m = 5$¢ per dollar of net profit.   **27.** $w = 2n - 58$; 42 billion pounds   **29.** $c = 0.075m - 1.5$; 0.75 pounds   **31.** $T(r) = (1/4)r + 45$; $T(100) = 70°F$ **33.** $P(x) = 100x - 5132$, with domain $[0, 405]$. For profit, $x \geq 52$ **35.** 5000 units   **37.**
$FC/(SP - VC)$ **39.** $P(x) = 579.7x - 20,000$, with domain $x \geq 0$; $x = 34.50$ g per day for break even   **41.** Increasing by \$355,000 per year   **43. a.** $y = -30t + 200$ **b.** $y = 50t - 200$

c. $y = \begin{cases} -30t + 200 & \text{if } 0 \leq t \leq 5 \\ 50t - 200 & \text{if } 5 < t \leq 9 \end{cases}$ **d.** 150

45. $C(t) = \begin{cases} -1{,}400t + 30{,}000 & \text{if } 0 \leq t \leq 5 \\ 7{,}400t - 14{,}000 & \text{if } 5 < t \leq 10 \end{cases}$

$C(3) = 25{,}800$ students

47. $d(r) = \begin{cases} -40r + 74 & \text{if } 1.1 \leq r \leq 1.3 \\ \dfrac{130r}{3} - \dfrac{103}{3} & \text{if } 1.3 < r \leq 1.6 \end{cases}$; $d(1) = 34\%$

49. Bootlags per zonar; bootlags   **51.** It must increase by 10 units each day, including the third.   **53.** (B)   **55.** Increasing the number of items from the breakeven results in a profit: Because the slope of the revenue graph is larger than the slope of the cost graph, it is higher than the cost graph to the right of the point of intersection, and hence corresponds to a profit.

Section 1.5

1. 6   **3.** 86   **5. a.** 0.5 (better fit) **b.** 0.75   **7. a.** 27.42
b. 27.16 (better fit)

9. $y = 1.5x - 0.6667$

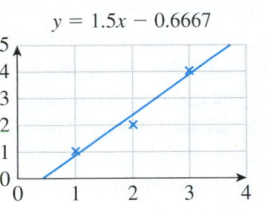

11. $y = 0.4118x + 0.9706$

13. a. $r = 0.9959$ (best, not perfect) **b.** $r = 0.9538$
c. $r = 0.3273$ (worst)

15.

x	y	xy	x^2
3	500	1500	9
5	600	3000	25
7	800	5600	49
Totals 15	1900	10100	83

$y = 75x + 258.33$; 858.33 million
17. $y = 2.5t + 5.67$; \$13.17 billion   **19.** $y = 0.135x + 0.15$;
6.9 million jobs   **21. a.** $y = 1.62x - 23.87$. Graph:

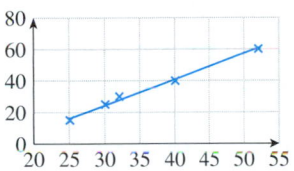

b. Each acre of cultivated land produces about 1.62 tons of
soybeans   **23. a.** Regression line: $y = -0.40x + 29$. Graph:

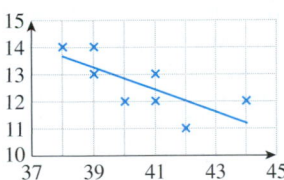

The graph suggests a relationship between x and y. **b.** The
poverty rate declines by 0.40% for each \$1000 increase in the
median household income. **c.** $r \approx -0.7338$; not a strong correlation   **25. a.** $p = 0.13t + 0.22$. Graph:

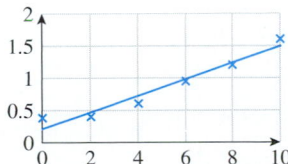

b. Yes; the first and last points lie above the regression line, while
the central points lie below it, suggesting a curve.

c.

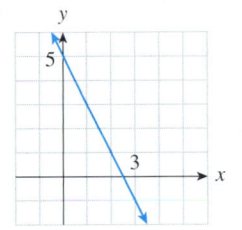

	A	B	C	D	E
1	t	p (Observed)	p (predicted)	Residual	
2	0	0.38	0.22	0.16	
3	2	0.4	0.48	-0.08	
4	4	0.6	0.74	-0.14	
5	6	0.95	1	-0.05	
6	8	1.2	1.26	-0.06	
7	10	1.6	1.52	0.08	
8					
9					

Notice that the residuals are positive at first, then become negative, and then become positive, confirming the impression from
the graph.   **27.** The line that passes through (a, b) and (c, d)
gives a sum-of-squares error SSE = 0, which is the smallest value
possible.   **29.** The regression line is the line passing through
the given points.   **31.** 0   **33.** No. The regression line through
$(-1, 1)$, $(0, 0)$, and $(1, 1)$ passes through none of these points.

Chapter 1 Review

1. a. 1   **b.** −2 **c.** 0 **d.** −1   **3. a.** 1 **b.** 0 **c.** 0 **d.** −1
5.

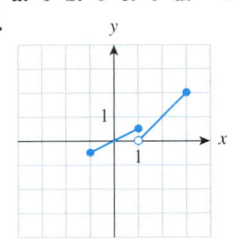

7.

9. Absolute value   **11.** Linear   **13.** Quadratic
15. $y = -x + 1$ **17.** $y = (1/2)x - 1$ **19.** The   first   line,
$y = x + 1$, is the better fit.   **21.** $y \approx 0.857x + 1.24$, $r \approx 0.92$
23. a. (A) **b.** (A) Leveling off (B) Rising (C) Rising; they begin
to fall after 7 months (D) Rising   **25. a.** 2080 hits per day
b. Probably not. This model predicts that Web site traffic will
start to decrease as advertising increases beyond \$8500 per
month, and then drop toward zero.   **27. a.** $q = -60p + 950$
b. 50 novels per month **c.** \$10, for a profit of \$1200.

Chapter 2

Section 2.1

1. (2, 2)   **3.** (3, 1) **5.** (6, 6)   **7.** (5/3, −4/3)   **9.** (0, −2)
11. $(x, (1 - 2x)/3)$ or $\left(\frac{1}{2}(1 - 3y), y\right)$ **13.** No solution
15. (5, 0)   **17.** (0.3, −1.1)   **19.** (116.6, −69.7)   **21.** (3.3, 1.8)
23. (3.4, 1.9)   **25.** 200 quarts of vanilla and 100 quarts of mocha
27. 2 servings of Mixed Cereal and 1 serving of Mango Tropical
Fruit   **29. a.** 4 servings of beans and 5 slices of bread **b.** No.
One of the variables in the solution of the system has a negative
value.   **31.** Mix 5 servings of Cell-Tech and 6 servings of Ribo-
Force HP for a cost of \$20.60.   **33.** 100 CSCO, 150 NOK
35. 100 ED, 200 KSE   **37.** 242 in favor and 193 against

39. 5 soccer games and 7 football games   **41.** 7   **43.** $1.50 each   **45.** 55 widgets   **47.** Demand: $q = -4p + 47$; supply: $q = 4p - 29$; equilibrium price: $9.50   **49.** 33 pairs of dirty socks and 11 T-shirts   **51.** $1200   **53.** A system of three equations in two unknowns will have a unique solution if either (1) the three corresponding lines intersect in a single point, or (2) two of the equations correspond to the same line, and the third line intersects it in a single point.   **55.** Yes. Even if two lines have negative slope, they will still intersect if the slopes differ.   **57.** You cannot round both of them up, because there will not be sufficient eggs and cream. Rounding both answers down will ensure that you will not run out of ingredients. It may be possible to round one answer down and the other up, and this should be tried.   **59.** (B)   **61.** (B)   **63.** Answers will vary.   **65.** It is very likely. Two randomly chosen straight lines are unlikely to be parallel.

Section 2.2

1. $(3, 1)$ **3.** $(6, 6)$ **5.** $\left(\frac{1}{2}(1 - 3y), y\right)$, y arbitrary   **7.** No solution   **9.** $(1/4, 3/4)$ **11.** No solution   **13.** $(10/3, 1/3)$ **15.** $(4, 4, 4)$ **17.** $\left(-1, -3, \frac{1}{2}\right)$ **19.** (z, z, z), z arbitrary   **21.** No solution   **23.** $(-1, 1, 1)$ **25.** $(1, z - 2, z)$, z arbitrary   **27.** $(4 + y, y, -1)$, y arbitrary   **29.** $(4 - y/3 + z/3, y, z)$, y arbitrary, z arbitrary   **31.** $(-17, 20, -2)$ **33.** $\left(-\frac{3}{2}, 0, \frac{1}{2}, 0\right)$ **35.** $(-3z, 1 - 2z, z, 0)$, z arbitrary   **37.** $\left(\frac{1}{5}(7 - 17z + 8w), \frac{1}{5}(1 - 6z - 6w), z, w\right)$, z, w arbitrary   **39.** $(1, 2, 3, 4, 5)$ **41.** $(-2, -2 + z - u, z, u, 0)$ z, u arbitrary   **43.** $(16, 12/7, -162/7, -88/7)$ **45.** $(-8/15, 7/15, 7/15, 7/15, 7/15)$ **47.** $(1.0, 1.4, 0.2)$ **49.** $(-5.5, -0.9, -7.4, -6.6)$ **51.** A pivot is an entry in a matrix that is selected to "clear a column;" that is, use the row operations of a certain type to obtain zeros everywhere above and below it. "Pivoting" is the procedure of clearing a column using a designated pivot.   **53.** $2R_1 + 5R_4$, or $6R_1 + 15R_4$ (which is less desirable).   **55.** It will include a row of zeros.   **57.** The claim is wrong. If there are more equations than unknowns, there can be a unique solution as well as row(s) of zeros in the reduced matrix, as in Example 6.   **59.** Two   **61.** The number of pivots must equal the number of variables, because no variable will be used as a parameter.   **63.** A simple example is: $x = 1$; $y - z = 1$; $x + y - z = 2$.

Section 2.3

1. 100 batches of vanilla, 50 batches of mocha, and 100 batches of strawberry   **3.** 3 sections of Finite Math, 2 sections of Applied Calculus and 1 section of Computer Methods   **5.** 5 of each   **7.** 22 tons from Cheesy Cream, 56 tons from Super Smooth & Sons, and 22 tons from Bagel's Best Friend   **9.** 10 evil sorcerers, 50 trolls, and 500 orcs   **11.** $3.6 billion for rock music, $1.8 billion for rap music, and $0.4 billion for classical music.   **13.** It donated $600 to each of the MPBF and the SCN, and $1200 to the Jets.   **15.** United: 120; American: 40; SouthWest: 50   **17.** $5000 in PNF, $2000 in FDMMX, $2000 in FFLIX   **19.** 100 APPL, 20 HPQ, 80 DELL   **21.** Microsoft: 88 million, Time Warner: 79 million, Yahoo: 75 million, Google: 42 million

23. The third equation is $x + y + z + w = 1$. General Solution: $x = -1.58 + 3.89w$, $y = 1.63 - 2.99w$, $z = 0.95 - 1.9w$, w arbitrary. State Farm is most impacted by Other.   **25. a.** Brooklyn to Manhattan: 500 books; Brooklyn to Long Island: 500 books; Queens to Manhattan: 1000 books; Queens to Long Island: 1000 books. **b.** Brooklyn to Manhattan: 1000 books; Brooklyn to Long Island: none; Queens to Manhattan: 500 books; Queens to Long Island: 1500 books, giving a total cost of $8000.   **27. a.** The associated system of equations has infinitely many solutions. **b.** No; the associated system of equations still has infinitely many solutions. **c.** Yes; North America to Australia: 440,000, North America to South Africa: 190,000, Europe to Australia: 950,000, Europe to South Africa: 950,000.   **29. a.** $x + y = 14,000$; $z + w = 95,000$; $x + z = 63,550$; $y + w = 45,450$. The system does not have a unique solution, indicating that the given data are insufficient to obtain the missing data. **b.** $(x, y, z, w) = (5600, 8400, 57,950, 37,050)$ **31. a.** No; The general solution is: Eastward Blvd.: $S + 200$; Northwest La.: $S + 50$; Southwest La.: S, where S is arbitrary. Thus it would suffice to know the traffic along Southwest La. **b.** Yes, as it leads to the solution Eastward Blvd.: 260; Northwest La.: 110; Southwest La.: 60 **c.** 50 vehicles per day   **33. a.** No; the corresponding system of equations is underdetermined. The net flow of traffic along any of the three stretches of Broadway would suffice. **b.** West   **35.** $10 billion   **37.** $x = $ Water, $y = $ Gray matter, $z = $ Tumor   **39.** $x = $ Water, $y = $ Bone, $z = $ Tumor, $u = $ Air   **41.** Tumor   **43.** 200 Democrats, 20 Republicans, 13 of other parties   **45.** Yes; $20m in Company X; $5m in Company Y, $10m in Company Z, and $30m in Company W   **47.** It is not realistic to expect to use exactly all of the ingredients. Solutions of the associated system may involve negative numbers or not exist. Only solutions with nonnegative values for all the unknowns correspond to being able to use up all of the ingredients.   **49.** Yes; $x = 100$ **51.** Yes; $0.3x - 0.7y + 0.3z = 0$ is one form of the equation.   **53.** No; represented by an inequality rather than an equation.   **55.** Answers will vary.

Chapter 2 Review

1. One solution

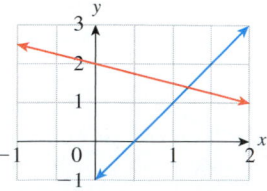

3. Infinitely many solutions

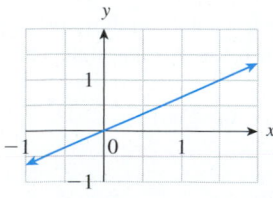

5. One solution.

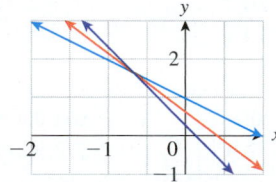

7. $(6/5, 7/5)$ **9.** $(3y/2, y)$, y arbitrary   **11.** $(-0.7, 1.7)$
13. $(-1, -1, -1)$ **15.** $(z - 2, 4(z - 1), z)$, z arbitrary
17. No solution   **19. a.** $-40°$ **b.** $320°F$ ($160°C$)   **c.** It is impossible; setting $F = 1.8C$ leads to an inconsistent system of equations.   **21.** 550 packages from Duffin House, 350 from Higgins Press   **23.** 600 packages from Duffin House, 200 from Higgins Press   **25.** \$40   **27.** 7 of each   **29.** 5000 hits per day at OHaganBooks.com, 1250 at JungleBooks.com, 3750 at FarmerBooks.com   **31.** DHS: 1000 shares, HPR: 600 shares, SPUB: 400 shares.   **33. a.** $x = 100, y = 100 + w, z = 300 - w$, w arbitrary   **b.** 100 book orders per day   **c.** 300 book orders per day   **d.** $x = 100, y = 400, z = 0, w = 300$ **e.** 100 book orders per day   **35.** New York to OHaganBooks.com: 450 packages, New York to FantasyBooks.com: 50 packages, Illinois to OHaganBooks.com: 150 packages, Illinois to FantasyBooks.com: 150 packages.

Chapter 3

Section 3.1

1. $1 \times 4; 0$ **3.** $4 \times 1; 5/2$ **5.** $p \times q; e_{22}$ **7.** $2 \times 2; 3$
9. $1 \times n; d_r$ **11.** $x = 1, y = 2, z = 3, w = 4$

13. $\begin{bmatrix} 0.25 & -2 \\ 1 & 0.5 \\ -2 & 5 \end{bmatrix}$ **15.** $\begin{bmatrix} -0.75 & -1 \\ 0 & -0.5 \\ -1 & 6 \end{bmatrix}$

17. $\begin{bmatrix} -1 & -1 \\ 1 & -1 \\ -1 & 5 \end{bmatrix}$ **19.** $\begin{bmatrix} 0 & 2 & -2 \\ -2 & 0 & 4 \end{bmatrix}$

21. $\begin{bmatrix} 4 & -1 & -1 \\ 5 & 1 & 0 \end{bmatrix}$ **23.** $\begin{bmatrix} -2 + x & 0 & 1 + w \\ -5 + z & 3 + r & 2 \end{bmatrix}$

25. $\begin{bmatrix} -1 & -2 & 1 \\ -5 & 5 & -3 \end{bmatrix}$ **27.** $\begin{bmatrix} 9 & 15 \\ 0 & -3 \\ -3 & 3 \end{bmatrix}$

29. $\begin{bmatrix} -8.5 & -22.35 & -24.4 \\ 54.2 & 20 & 42.2 \end{bmatrix}$

31. $\begin{bmatrix} 1.54 & 8.58 \\ 5.94 & 0 \\ 6.16 & 7.26 \end{bmatrix}$ **33.** $\begin{bmatrix} 7.38 & 76.96 \\ 20.33 & 0 \\ 29.12 & 39.92 \end{bmatrix}$

35. $\begin{bmatrix} -19.85 & 115.82 \\ -50.935 & 46 \\ -57.24 & 94.62 \end{bmatrix}$ **37. a.** $[720 \quad 680 \quad 350]$

b. $[760 \quad 800 \quad 300]$ **39.** Sales $= \begin{bmatrix} 700 & 1300 & 2000 \\ 400 & 300 & 500 \end{bmatrix}$

Inventory $-$ Sales $= \begin{bmatrix} 300 & 700 & 3000 \\ 600 & 4700 & 1500 \end{bmatrix}$

41. Profit = Revenue − Cost;

	2004	2005	2006
Full Boots	\$8000	\$7200	\$8800
Half Boots	\$5600	\$5760	\$7040
Sandals	\$2800	\$3500	\$4000

43. 1980 distribution $= A = [49.1 \quad 58.9 \quad 75.4 \quad 43.2]$; 1990 distribution $= B = [50.8 \quad 59.7 \quad 85.4 \quad 52.8]$; Net change 1980 to 1990 $= B - A = [1.7 \quad 0.8 \quad 10 \quad 9.6]$ (all net increases)
45. Total Bankruptcy Filings = Filings in Manhattan + Filings in Brooklyn + Filings in Newark $= [150 \quad 250 \quad 150 \quad 100 \quad 150] + [300 \quad 400 \quad 300 \quad 200 \quad 250] + [250 \quad 400 \quad 250 \quad 200 \quad 200] = [700 \quad 1050 \quad 700 \quad 500 \quad 600]$ **47.** Filings in Brooklyn − Filings in Newark $= [300 \quad 400 \quad 300 \quad 200 \quad 250] - [250 \quad 400 \quad 250 \quad 200 \quad 200] = [50 \quad 0 \quad 50 \quad 0 \quad 50]$. The difference was greatest in January 01, July 01, and January 02.

49. a. Use $= \begin{array}{c} \\ \text{Pom II} \\ \text{Pom Classic} \end{array} \begin{array}{ccc} \text{Proc} & \text{Mem} & \text{Tubes} \\ \end{array}$ $\begin{bmatrix} 2 & 16 & 20 \\ 1 & 4 & 40 \end{bmatrix}$

Inventory $= \begin{bmatrix} 500 & 5000 & 10{,}000 \\ 200 & 2000 & 20{,}000 \end{bmatrix}$

Inventory $- 100 \cdot$ Use $= \begin{bmatrix} 300 & 3400 & 8000 \\ 100 & 1600 & 16{,}000 \end{bmatrix}$

b. After 4 months.

51. a. $A = \begin{bmatrix} 440 & 190 \\ 950 & 950 \\ 1790 & 200 \end{bmatrix}$ $D = \begin{bmatrix} -20 & 40 \\ 50 & 50 \\ 0 & 100 \end{bmatrix}$

2008 Tourism $= A + D = \begin{bmatrix} 420 & 230 \\ 1000 & 1000 \\ 1790 & 300 \end{bmatrix}$

b. $\frac{1}{2}(A + B)$; $\begin{bmatrix} 430 & 210 \\ 975 & 975 \\ 1790 & 250 \end{bmatrix}$

53. The ijth entry of the sum $A + B$ is obtained by adding the ijth entries of A and B.   **55.** It would have zeros down the main

diagonal: $A = \begin{bmatrix} 0 & \# & \# & \# & \# \\ \# & 0 & \# & \# & \# \\ \# & \# & 0 & \# & \# \\ \# & \# & \# & 0 & \# \\ \# & \# & \# & \# & 0 \end{bmatrix}$ The symbols # indicate arbi-

trary numbers.   **57.** $(A^T)_{ij} = A_{ji}$ **59.** Answers will vary.

a. $\begin{bmatrix} 0 & -4 \\ 4 & 0 \end{bmatrix}$ **b.** $\begin{bmatrix} 0 & -4 & 5 \\ 4 & 0 & 1 \\ -5 & -1 & 0 \end{bmatrix}$ **61.** The associativity of

matrix addition is a consequence of the associativity of addition of numbers, since we add matrices by adding the corresponding entries (which are real numbers).   **63.** Answers will vary.

Section 3.2

1. [13]   **3.** [5/6]   **5.** $[-2y+z]$ **7.** Undefined
9. $[3 \quad 0 \quad -6 \quad -2\ ]$ **11.** $[-6 \quad 37 \quad 7]$

13. $\begin{bmatrix} -4 & -7 & -1 \\ 9 & 17 & 0 \end{bmatrix}$ **15.** $\begin{bmatrix} 0 & 1 \\ 0 & 0 \end{bmatrix}$ **17.** $\begin{bmatrix} 1 & -1 \\ 1 & -1 \end{bmatrix}$ **19.** $\begin{bmatrix} 0 & 0 \\ 0 & 0 \end{bmatrix}$

21. Undefined   **23.** $\begin{bmatrix} 1 & -5 & 3 \\ 0 & 0 & 9 \\ 0 & 4 & 1 \end{bmatrix}$ **25.** $\begin{bmatrix} 3 \\ -4 \\ 0 \\ 3 \end{bmatrix}$

27. $\begin{bmatrix} 0.23 & 5.36 & -21.65 \\ -13.18 & -5.82 & -16.62 \\ -11.21 & -9.9 & 0.99 \\ -2.1 & 2.34 & 2.46 \end{bmatrix}$

29. $A^2 = \begin{bmatrix} 0 & 0 & 1 & 2 \\ 0 & 0 & 0 & 1 \\ 0 & 0 & 0 & 0 \\ 0 & 0 & 0 & 0 \end{bmatrix}$ $A^3 = \begin{bmatrix} 0 & 0 & 0 & 1 \\ 0 & 0 & 0 & 0 \\ 0 & 0 & 0 & 0 \\ 0 & 0 & 0 & 0 \end{bmatrix}$

$A^4 = \begin{bmatrix} 0 & 0 & 0 & 0 \\ 0 & 0 & 0 & 0 \\ 0 & 0 & 0 & 0 \\ 0 & 0 & 0 & 0 \end{bmatrix}$ $A^{100} = \begin{bmatrix} 0 & 0 & 0 & 0 \\ 0 & 0 & 0 & 0 \\ 0 & 0 & 0 & 0 \\ 0 & 0 & 0 & 0 \end{bmatrix}$

31. $\begin{bmatrix} 4 & -1 \\ -1 & -7 \end{bmatrix}$ **33.** $\begin{bmatrix} 4 & -1 \\ -12 & 2 \end{bmatrix}$ **35.** $\begin{bmatrix} -2 & 1 & -2 \\ 10 & -2 & 2 \\ -10 & 2 & -2 \end{bmatrix}$

37. $\begin{bmatrix} -2+x-z & 2-r & -6+w \\ 10+2z & -2+2r & 10 \\ -10-2z & 2-2r & -10 \end{bmatrix}$

39. a.–d. $P^2 = P^4 = P^8 = P^{1000} = \begin{bmatrix} 0.2 & 0.8 \\ 0.2 & 0.8 \end{bmatrix}$

41. a. $P^2 = \begin{bmatrix} 0.01 & 0.99 \\ 0 & 1 \end{bmatrix}$ **b.** $P^4 = \begin{bmatrix} 0.0001 & 0.9999 \\ 0 & 1 \end{bmatrix}$

c. and d. $P^8 \approx P^{1000} \approx \begin{bmatrix} 0 & 1 \\ 0 & 1 \end{bmatrix}$

43. a.–d. $P^2 = P^4 = P^8 = P^{1000} = \begin{bmatrix} 0.25 & 0.25 & 0.50 \\ 0.25 & 0.25 & 0.50 \\ 0.25 & 0.25 & 0.50 \end{bmatrix}$

45. $2x - y + 4z = 3;\ -4x + 3y/4 + z/3 = -1;\ -3x = 0$
47. $x - y + w = -1;\ x + y + 2z + 4w = 2$

49. $\begin{bmatrix} 1 & -1 \\ 2 & -1 \end{bmatrix} \begin{bmatrix} x \\ y \end{bmatrix} = \begin{bmatrix} 4 \\ 0 \end{bmatrix}$

51. $\begin{bmatrix} 1 & 1 & -1 \\ 2 & 1 & 1 \\ \frac{3}{4} & 0 & \frac{1}{2} \end{bmatrix} \begin{bmatrix} x \\ y \\ z \end{bmatrix} = \begin{bmatrix} 8 \\ 4 \\ 1 \end{bmatrix}$

53. Revenue = Price × Quantity =

$[15 \quad 10 \quad 12] \begin{bmatrix} 50 \\ 40 \\ 30 \end{bmatrix} = [1510]$

55. Price: $\begin{matrix} \text{Hard} \\ \text{Soft} \\ \text{Plastic} \end{matrix} \begin{bmatrix} 30 \\ 10 \\ 15 \end{bmatrix}$;

$\begin{bmatrix} 700 & 1300 & 2000 \\ 400 & 300 & 500 \end{bmatrix} \begin{bmatrix} 30 \\ 10 \\ 15 \end{bmatrix} = \begin{bmatrix} \$64,000 \\ \$22,500 \end{bmatrix}$

57. Number of books = Number of books per editor × Number of

editors = $[3 \quad 3.5 \quad 5 \quad 5.2] \begin{bmatrix} 16,000 \\ 15,000 \\ 12,500 \\ 13,000 \end{bmatrix} = 230,600$ new books

59. \$4300 billion (or \$4.3 trillion)   **61.** $D = N(F - M)$ where N is the income per person, and F and M are, respectively, the female and male populations in 2005; \$140 billion.   **63.** [1.2 1.0], which represents the amount, in billions of pounds, by which cheese production in north central states exceeded that in western states.
65. Number of bankruptcy filings handled by firm = Percentage handled by firm × Total number =

$[0.10 \quad 0.05 \quad 0.20] \begin{bmatrix} 150 & 150 & 150 \\ 300 & 300 & 250 \\ 250 & 250 & 200 \end{bmatrix} = [80 \quad 80 \quad 67.5]$

67. The number of filings in Manhattan and Brooklyn combined in each of the months shown.

69. $[1 \quad -1 \quad 1] \begin{bmatrix} 150 & 150 & 150 \\ 300 & 300 & 250 \\ 250 & 250 & 200 \end{bmatrix} \begin{bmatrix} 1 \\ 1 \\ 1 \end{bmatrix} = [300]$

71. $\begin{bmatrix} 2 & 16 & 20 \\ 1 & 4 & 40 \end{bmatrix} \begin{bmatrix} 100 & 150 \\ 50 & 40 \\ 10 & 15 \end{bmatrix} = \begin{bmatrix} \$1200 & \$1240 \\ \$700 & \$910 \end{bmatrix}$

73. $AB = \begin{bmatrix} 29.6 \\ 85.5 \\ 97.5 \end{bmatrix}$ $AC = \begin{bmatrix} 22 & 7.6 \\ 47.5 & 38 \\ 89.5 & 8 \end{bmatrix}$ The entries of AB

give the number of people from each of the three regions who settle in Australia or South Africa, while the entries in AC break those figures down further into settlers in South Africa and settlers in Australia.   **75.** Distribution in 2003 = $A = [53.3 \quad 64.0$ $101.6 \quad 65.4]$; Distribution in 2004 = $A \cdot P \approx [53.1 \quad 63.9$ $102.0 \quad 65.3]$ **77.** Answers will vary. One example:

$A = [1 \quad 2], \quad B = \begin{bmatrix} 1 & 2 & 3 \\ 4 & 5 & 6 \end{bmatrix}$. Another example: $A = [1]$,

$B = [1 \quad 2]$.   **79.** Multiplication of 1×1 matrices is just ordinary multiplication of the single entries: $[a][b] = [ab]$.
81. The claim is correct. Every matrix equation represents the equality of two matrices. Equating the corresponding entries gives a system of equations.   **83.** Answers will vary. Here is a possible scenario: costs of items A, B and C in $1995 = [10 \quad 20 \quad 30]$, percentage increases in these costs in $1996 = [0.5 \quad 0.1 \quad 0.20]$, actual increases in costs = $[10 \times 0.5$ $20 \times 0.1 \quad 30 \times 0.20]$ **85.** It produces a matrix whose ij entry is the product of the ij entries of the two matrices.

Section 3.3

1. Yes **3.** Yes **5.** No **7.** $\begin{bmatrix} -1 & 1 \\ 2 & -1 \end{bmatrix}$ **9.** $\begin{bmatrix} 0 & 1 \\ 1 & 0 \end{bmatrix}$

11. $\begin{bmatrix} 1 & -1 \\ -1 & 2 \end{bmatrix}$ **13.** Singular **15.** $\begin{bmatrix} 1 & -1 & 0 \\ 0 & 1 & -1 \\ 0 & 0 & 1 \end{bmatrix}$

17. $\begin{bmatrix} 1 & -1 & 1 \\ \frac{1}{2} & 0 & -\frac{1}{2} \\ -\frac{1}{2} & 1 & -\frac{1}{2} \end{bmatrix}$ **19.** $\begin{bmatrix} 1 & \frac{1}{3} & -\frac{1}{3} \\ 1 & -\frac{2}{3} & -\frac{1}{3} \\ -1 & \frac{1}{3} & \frac{2}{3} \end{bmatrix}$

21. Singular **23.** $\begin{bmatrix} 0 & 1 & -2 & 1 \\ 0 & 1 & -1 & 0 \\ 1 & -1 & 2 & -1 \\ 0 & 1 & -1 & 1 \end{bmatrix}$

25. $\begin{bmatrix} 1 & -2 & 1 & 0 \\ 0 & 1 & -2 & 1 \\ 0 & 0 & 1 & -2 \\ 0 & 0 & 0 & 1 \end{bmatrix}$ **27.** $-2;\ \begin{bmatrix} \frac{1}{2} & \frac{1}{2} \\ \frac{1}{2} & -\frac{1}{2} \end{bmatrix}$

29. $-2;\ \begin{bmatrix} -2 & 1 \\ \frac{3}{2} & -\frac{1}{2} \end{bmatrix}$ **31.** $1/36;\ \begin{bmatrix} 6 & 6 \\ 0 & 6 \end{bmatrix}$ **33.** 0; Singular

35. $\begin{bmatrix} 0.38 & 0.45 \\ 0.49 & -0.41 \end{bmatrix}$ **37.** $\begin{bmatrix} 0.00 & -0.99 \\ 0.81 & 2.87 \end{bmatrix}$ **39.** Singular

41. $\begin{bmatrix} 91.35 & -8.65 & 0 & -71.30 \\ -0.07 & -0.07 & 0 & 2.49 \\ 2.60 & 2.60 & -4.35 & 1.37 \\ 2.69 & 2.69 & 0 & -2.10 \end{bmatrix}$ **43.** $(5/2, 3/2)$

45. $(6, -4)$ **47.** $(6, 6, 6)$ **49. a.** $(10, -5, -3)$ **b.** $(6, 1, 5)$
c. $(0, 0, 0)$ **51. a.** 10/3 servings of beans, and 5/6 slices of bread
b. $\begin{bmatrix} -1/2 & 1/6 \\ 7/8 & -5/24 \end{bmatrix} \begin{bmatrix} A \\ B \end{bmatrix} = \begin{bmatrix} -A/2 + B/6 \\ 7A/8 - 5B/24 \end{bmatrix}$; that is,
$-A/2 + B/6$ servings of beans and $7A/8 - 5B/24$ slices of
bread **53. a.** 100 batches of vanilla, 50 batches of mocha,
100 batches of strawberry **b.** 100 batches of vanilla, no mocha,
200 batches of strawberry **c.** $\begin{bmatrix} 1 & -1/3 & -1/3 \\ -1 & 0 & 1 \\ 0 & 2/3 & -1/3 \end{bmatrix} \begin{bmatrix} A \\ B \\ C \end{bmatrix}$, or
$A - B/3 - C/3$ batches of Vanilla, $-A + C$ batches of mocha,
and $2B/3 - C/3$ batches of strawberry **55.** $5000 in PNF,
$2000 in FDMMX, $2000 in FFLIX **57.** 100 APPL, 20 HPQ,
80 DELL **59.** Distribution in 2003 $= A = [53.3\ 64.0\ 101.6$
$65.4]$; Distribution in 2002 $= A \cdot P^{-1} \approx [53.5\ 64.1\ 101.2\ 65.5]$
61. a. $(-0.7071, 3.5355)$ **b.** R^2, R^3 **c.** R^{-1}
63. $[37\ 81\ 40\ 80\ 15\ 45\ 40\ 96\ 29\ 59\ 4\ 8]$ **65.** CORRECT
ANSWER **67.** (A) **69.** The inverse does not exist—the matrix is singular. (If two rows of a matrix are the same, then row reducing it will lead to a row of zeros, and so it cannot be reduced
to the identity.) **73.** When one or more of the d_i are zero. If that
is the case, then the matrix $[D \mid I]$ easily reduces to a matrix that
has a row of zeros on the left-hand portion, so that D is singular,
Conversely, if none of the d_i are zero, then $[D \mid I]$ easily reduces
to a matrix of the form $[I \mid E]$, showing that D is invertible.

75. $(AB)(B^{-1}A^{-1}) = A(BB^{-1})A^{-1} = AIA^{-1} = AA^{-1} = I$
77. If A has an inverse, then every system of equations $AX = B$
has a unique solution, namely $X = A^{-1}B$. But if A reduces to a
matrix with a row of zeros, then such a system has either infinitely
many solutions or no solution at all.

Section 3.4

1.
$$\begin{array}{c c} & \mathbf{B} \\ \mathbf{A} \begin{array}{c} a \\ b \end{array} & \begin{array}{c} p \quad r \\ \begin{bmatrix} 1 & 10 \\ 2 & -4 \end{bmatrix} \end{array} \end{array}$$
3.
$$\begin{array}{c c} & \mathbf{B} \\ \mathbf{A}\ 3 & \begin{array}{c} c \\ [-1] \end{array} \end{array}$$
5.
$$\begin{array}{c c} & \mathbf{B} \\ \mathbf{A}\ q & \begin{array}{c} b \\ [0] \end{array} \end{array}$$

7. Strictly determined. A's optimal strategy is a; B's optimal strategy is q; value: 1 **9.** Not strictly determined **11.** Not strictly
determined **13.** -1 **15.** -0.25 **17.** $[0\ 0\ 1\ 0]$;
$e = 2.25$ **19.** $[1\ 0\ 0]^T$ or $[0\ 1\ 0]^T$; $e = 1/4$
21. $R = [1/4\ 3/4], C = [3/4\ 1/4]^T, e = -1/4$
23. $R = [3/4\ 1/4], C = [3/4\ 1/4]^T, e = -5/4$

25.
$$\begin{array}{c c} & \text{Friend} \\ \text{You} \begin{array}{c} H \\ T \end{array} & \begin{array}{c} H \quad T \\ \begin{bmatrix} -1 & 1 \\ 1 & -1 \end{bmatrix} \end{array} \end{array}$$

27. F = France; S = Sweden; N = Norway;

$$\begin{array}{c c} & \text{Your Opponent Defends} \\ \text{You Invade} \begin{array}{c} F \\ S \\ N \end{array} & \begin{array}{c} F \quad S \quad N \\ \begin{bmatrix} -1 & 1 & 1 \\ 1 & -1 & 1 \\ 1 & 1 & -1 \end{bmatrix} \end{array} \end{array}$$

29. B = Brakpan; N = Nigel; S = Springs;

$$\begin{array}{c c} & \text{Your Opponent} \\ \text{You} \begin{array}{c} B \\ N \\ S \end{array} & \begin{array}{c} B \quad N \quad S \\ \begin{bmatrix} 0 & 0 & 1000 \\ 0 & 0 & 1000 \\ -1000 & -1000 & 0 \end{bmatrix} \end{array} \end{array}$$

31. P = PleasantTap; T = Thunder Rumble; S = Strike the Gold,
N = None;

$$\begin{array}{c c} & \text{Winner} \\ \text{You Bet} \begin{array}{c} P \\ T \\ S \end{array} & \begin{array}{c} P \quad T \quad S \quad N \\ \begin{bmatrix} 25 & -10 & -10 & -10 \\ -10 & 35 & -10 & -10 \\ -10 & -10 & 40 & -10 \end{bmatrix} \end{array} \end{array}$$

33. a. CE should charge $1000 and GCS should charge $900;
15% gain in market share for CE **b.** CE should charge $1200
(the more CE can charge for the same market, the better!)
35. Pablo vs. Noto; evenly matched **37.** Both commanders
should use the northern route; 1 day **39.** Confess

41. a.
$$\begin{array}{c c} & \text{Kerry} \\ \text{Bush} \begin{array}{c} F \\ O \end{array} & \begin{array}{c} F \quad O \\ \begin{bmatrix} 24 & 21 \\ 25 & 24 \end{bmatrix} \end{array} \end{array}$$
b. Both candidates should visit
Ohio, leaving Bush with a 21% chance of winning the election.
43. You can expect to lose 39 customers. **45.** Option 2: move to
the suburbs. **47. a.** About 66% **b.** Yes; spend the whole night

studying game theory; 75% **c.** Game theory; 57.5%   **49. a.** Lay off 10 workers; Cost: $40,000 **b.** 60 inches of snow, costing $350,000 **c.** Lay off 15 workers   **51.** Allocate 1/7 of the budget to WISH and the rest (6/7) to WASH. Softex will lose approximately $2860.   **53.** Like a saddle point in a payoff matrix, the center of a saddle is a low point (minimum height) in one direction and a high point (maximum) in a perpendicular direction. **55.** Although there is a saddle point in the 2,4 position, you would be wrong to use saddle points (based on the minimax criterion) to reach the conclusion that row strategy 2 is best. One reason is that the entries in the matrix do not represent payoffs, because high numbers of employees in an area do not necessarily represent benefit to the row player. Another reason for this is that there is no opponent deciding what your job will be in such a way as to force you into the least populated job.   **57.** If you strictly alternate the two strategies the column player will know which pure strategy you will play on each move, and can choose a pure strategy

accordingly. For example, consider the game $A \begin{matrix} a & b \\ \begin{bmatrix} 1 & 0 \\ 0 & 1 \end{bmatrix} \end{matrix}$. By the analysis of Example 3 (or the symmetry of the game), the best strategy for the row player is [0.5 0.5] and the best strategy for the column player is $[0.5 \ 0.5]^T$. This gives an expected value of 0.5 for the game. However, suppose that the row player alternates A and B strictly and that the column player catches on to this. Then, whenever the row player plays A the column player will play b and whenever the row player plays B the column player will play a. This gives a payoff of 0 each time, worse for the row player than the expected value of 0.5.

Section 3.5

1. a. 0.8 **b.** 0.2 **c.** 0.05   **3.** $\begin{bmatrix} 0.2 & 0.1 \\ 0.5 & 0 \end{bmatrix}$

5. $[52,000 \quad 40,000]^T$ **7.** $[50,000 \quad 50,000]^T$
9. $[2560 \quad 2800 \quad 4000]^T$
11. $[27,000 \quad 28,000 \quad 17,000]^T$ **13.** Increase of 100 units in each sector.   **15.** Increase of $[1.5 \quad 0.2 \quad 0.1]^T$; the ith column of $(I - A)^{-1}$ gives the change in production necessary to meet an increase in external demand of one unit for the product of Sector i.

17. $A = \begin{bmatrix} 0.2 & 0.4 & 0.5 \\ 0 & 0.8 & 0 \\ 0 & 0.2 & 0.5 \end{bmatrix}$ **19.** Main DR: $80,000, Bits &

Bytes: $38,000   **21.** Equipment Sector production approximately $86,000 million, Components Sector production approximately $140,000 million   **23. a.** 0.006 **b.** textiles; clothing and footwear

25. Columns of $\begin{bmatrix} 1140.99 & 2.05 & 13.17 & 20.87 \\ 332.10 & 1047.34 & 26.05 & 111.18 \\ 0.12 & 0.13 & 1031.19 & 1.35 \\ 93.88 & 95.69 & 215.50 & 1016.15 \end{bmatrix}$

(in millions of dollars)   **27. a.** $0.78 **b.** Other food products **29.** It would mean that all of the sectors require neither their own product or the product of any other sector.   **31.** It would mean that all of the output of that sector was used internally in the

economy; none of the output was available for export and no importing was necessary.   **33.** It means that an increase in demand for one sector (the column sector) has no effect on the production of another sector (the row sector).   **35.** Usually, to produce one unit of one sector requires less than one unit of input from another. We would expect then that an increase in demand of one unit for one sector would require a smaller increase in production in another sector.

Chapter 3 Review

1. Undefined   **3.** $\begin{bmatrix} 1 & 8 \\ 5 & 11 \\ 6 & 13 \end{bmatrix}$ **5.** $\begin{bmatrix} 1 & 3 \\ 2 & 3 \\ 3 & 3 \end{bmatrix}$ **7.** $\begin{bmatrix} 1 & -2 \\ 0 & 1 \end{bmatrix}$

9. $\begin{bmatrix} 2 & 4 \\ 1 & 12 \end{bmatrix}$ **11.** $\begin{bmatrix} 1 & 1 \\ 0 & 1 \end{bmatrix}$ **13.** $\begin{bmatrix} 1 & -1/2 & -5/2 \\ 0 & 1/4 & -1/4 \\ 0 & 0 & 1 \end{bmatrix}$

15. Singular   **17.** $\begin{bmatrix} 1 & 2 \\ 3 & 4 \end{bmatrix} \begin{bmatrix} x \\ y \end{bmatrix} = \begin{bmatrix} 0 \\ 2 \end{bmatrix}$; $\begin{bmatrix} x \\ y \end{bmatrix} = \begin{bmatrix} 2 \\ -1 \end{bmatrix}$

19. $\begin{bmatrix} 1 & 1 & 1 \\ 1 & 2 & 1 \\ 1 & 1 & 2 \end{bmatrix} \begin{bmatrix} x \\ y \\ z \end{bmatrix} = \begin{bmatrix} 2 \\ 3 \\ 1 \end{bmatrix}$; $\begin{bmatrix} x \\ y \\ z \end{bmatrix} = \begin{bmatrix} 2 \\ 1 \\ -1 \end{bmatrix}$

21. $R = [1 \quad 0 \quad 0], C = [0 \quad 1 \quad 0]^T, e = 1$
23. $R = [0 \quad 0.8 \quad 0.2], C = [0.2 \quad 0 \quad 0.8], e = -0.2$

25. $\begin{bmatrix} 1100 \\ 700 \end{bmatrix}$ **27.** $\begin{bmatrix} 48,125 \\ 22,500 \\ 10,000 \end{bmatrix}$

29. Inventory $-$ Sales $= \begin{bmatrix} 2500 & 4000 & 3000 \\ 1500 & 3000 & 1000 \end{bmatrix} -$

$\begin{bmatrix} 300 & 500 & 100 \\ 100 & 450 & 200 \end{bmatrix} = \begin{bmatrix} 2200 & 3500 & 2900 \\ 1400 & 2550 & 800 \end{bmatrix}$

31. Revenue = Quantity × Price

$= \begin{bmatrix} 280 & 550 & 100 \\ 50 & 500 & 120 \end{bmatrix} \begin{bmatrix} 5 \\ 6 \\ 5.5 \end{bmatrix} = \begin{bmatrix} 5250 \\ 3910 \end{bmatrix} \begin{matrix} \text{Texas} \\ \text{Nevada} \end{matrix}$

33. $[2000 \quad 4000 \quad 4000] \begin{bmatrix} 0.8 & 0.1 & 0.1 \\ 0.4 & 0.6 & 0 \\ 0.2 & 0 & 0.8 \end{bmatrix} =$

$[4000 \quad 2600 \quad 3400]$ **35.** Here are three. (1) It is possible for someone to be a customer at two different enterprises. (2) Some customers may stop using all three of the companies. (3) New customers can enter the field.   **37.** Loss = Number of shares × (Purchase price − Dividends − Selling price) =

$[1000 \quad 2000 \quad 2000] \left(\begin{bmatrix} 20 \\ 10 \\ 5 \end{bmatrix} - \begin{bmatrix} 0.10 \\ 0.10 \\ 0 \end{bmatrix} - \begin{bmatrix} 3 \\ 1 \\ 1 \end{bmatrix} \right)$

$= [42,700]$ **39.** Go with the "3 for 1" promotion and gain 20,000 customers from JungleBooks   **41.** $A = \begin{bmatrix} 0.1 & 0.5 \\ 0.01 & 0.05 \end{bmatrix}$

43. $1190 worth of paper, $1802 worth of books.

Chapter 4

Section 4.1

1.
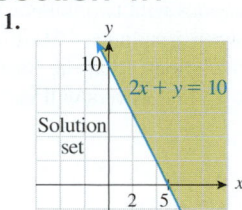
Unbounded

3.
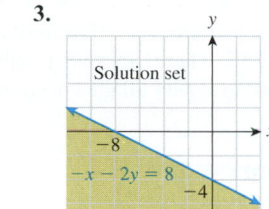
Unbounded

5.
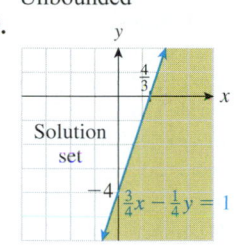
Unbounded

7.
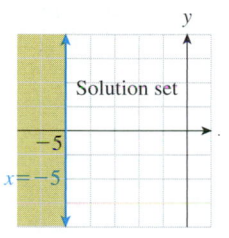
Unbounded

9.
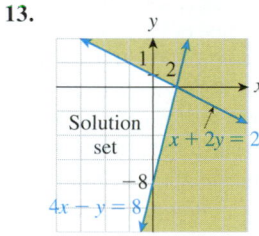
Unbounded

11.
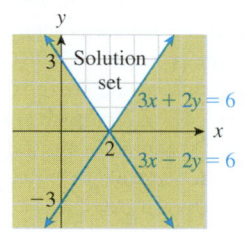
Unbounded

13.
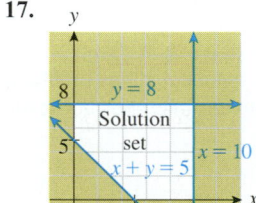
Unbounded;
Corner point: (2, 0)

15.
Unbounded;
Corner points: (2, 0), (0, 3)

17.
Bounded; Corner points:
(5, 0), (10, 0), (10, 8),
(0, 8), (0, 5)

19.
Bounded; Corner points:
(0, 0), (5, 0), (0, 5),
(2, 4), (4, 2)

21.
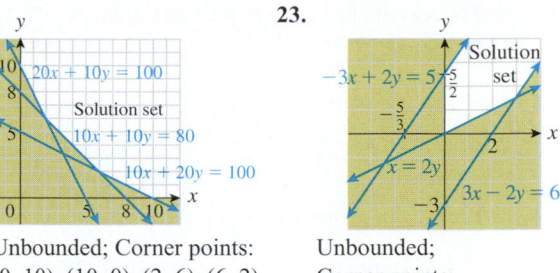
Unbounded; Corner points:
(0, 10), (10, 0), (2, 6), (6, 2)

23.
Unbounded;
Corner points:
(0, 0), (0, 5/2), (3, 3/2)

25.
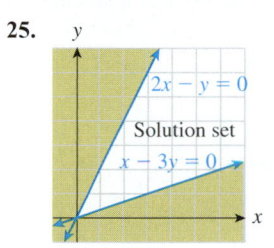
Unbounded; Corner point: (0, 0)

27.
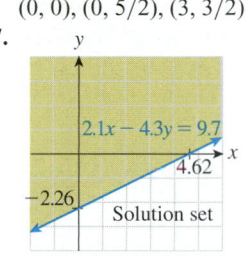

29.
Corner point: (−7.74, 2.50)

31.
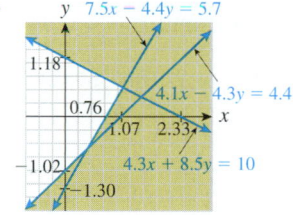
Corner points:
(0.36, −0.68), (1.12, 0.61)

33. x = # quarts of Creamy Vanilla,
y = # quarts of Continental Mocha

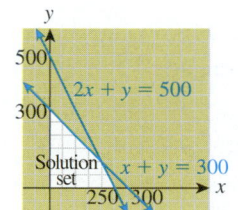
Corner points:
(0, 0), (250, 0), (0, 300), (200, 100)

35. $x =$ # ounces of chicken, $y =$ # ounces of grain

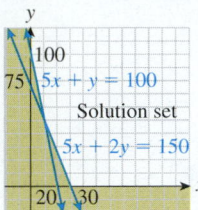

Corner points: (30, 0), (10, 50), (0, 100)

37. $x =$ # servings of Mixed Cereal for Baby, $y =$ # servings of Mango Tropical Fruit Dessert

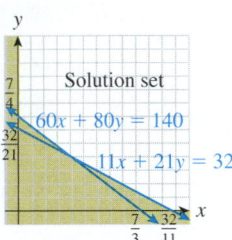

Corner points:
(0, 7/4), (1, 1), (32/11, 0)

39. $x =$ # dollars in PNF, $y =$ # dollars in FDMMX

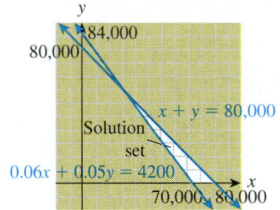

Corner points:
(70,000, 0), (80,000, 0), (20,000, 60,000)

41. $x =$ # shares of MO, $y =$ # shares of RAI

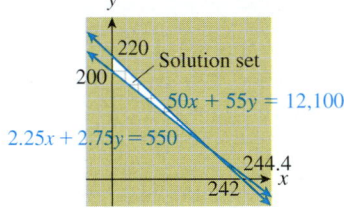

Corner points:
(0, 200), (0, 220), (220, 20)

43. $x =$ # full-page ads in Sports Illustrated, $y =$ # full-page ads in GQ

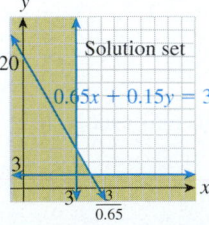

Corner points: (3, 7), (4, 3) (Rounded)

45. An example is $x \geq 0$, $y \geq 0$, $x + y \geq 1$ **47.** The given triangle can be described as the solution set of the system $x \geq 0$, $y \geq 0$, $x + 2y \leq 2$.   **49.** Answers may vary. One limitation is that the method is only suitable for situations with two unknown quantities. Accuracy is also limited when graphing.   **51.** (C)   **53.** (B)   **55.** There are no feasible solutions; that is, it is impossible to satisfy all the constraints.   **57.** Answers will vary.

Section 4.2

1. $p = 6$, $x = 3$, $y = 3$ **3.** $c = 4$, $x = 2$, $y = 2$ **5.** $p = 24$, $x = 7$, $y = 3$ **7.** $p = 16$, $x = 4$, $y = 2$ **9.** $c = 1.8$, $x = 6$, $y = 2$ **11.** Max: $p = 16$, $x = 4$, $y = 6$. Min: $p = 2$, $x = 2$, $y = 0$ **13.** No optimal solution; objective function unbounded   **15.** $c = 28$; $(x, y) = (14, 0)$ and $(6, 4)$ and the line connecting them   **17.** $c = 3$, $x = 3$, $y = 2$ **19.** No solution; feasible region empty   **21.** You should make 200 quarts of vanilla and 100 quarts of mocha.   **23.** Ruff, Inc., should use 100 oz of grain and no chicken.   **25.** Feed your child 1 serving of cereal and 1 serving of dessert.   **27.** Purchase 60 compact fluorescent light bulbs and 960 square feet of insulation for a saving of $312 per year in energy costs.   **29.** Mix 5 servings of Cell-Tech and 6 servings of RiboForce HP for a cost of $20.60.   **31.** Make 200 Dracula Salamis and 400 Frankenstein Sausages, for a profit of $1400.   **33.** Buy no shares of IBM and 500 shares of HPQ for maximum company earnings of $600.   **35.** Buy 220 shares of MO and 20 shares of RAI for a minimum total risk index is $c = 500$.   **37.** Purchase 20 spots on "Becker" and 20 spots on "The Simpsons."   **39.** He should instruct in diplomacy for 10 hours per week and in battle for 40 hours per week, giving a weekly profit of 2400 ducats.   **41.** Gillian could expend a minimum of 360,000 pico-shirleys of energy by using 480 sleep spells and 160 shock spells. (There is actually a whole line of solutions joining the one above with $x = 2880/7$, $y = 1440/7$.)   **43.** 100 hours per week for new customers and 60 hours per week for old customers.   **45.** (A)   **47.** Every point along the line connecting them is also an optimal solution.   **49.** Answers will vary.   **51.** Answers will vary.   **53.** Answers will vary. A simple example is the following: Maximize profit $p = 2x + y$ subject to $x \geq 0$, $y \geq 0$. Then p can be made as large as we like by choosing large values of x and/or y. Thus there is no optimal solution to the problem.   **55.** Mathematically, this means that there are infinitely many possible solutions: one for each point along the line joining the two corner points in question. In practice, select those points with integer solutions (because x and y must be whole numbers in this problem) that are in the feasible region and close to this line, and choose the one that gives the largest profit.

Section 4.3

1. $p = 8$; $x = 4$, $y = 0$ **3.** $p = 4$; $x = 4$, $y = 0$ **5.** $p = 80$; $x = 10$, $y = 0$, $z = 10$ **7.** $p = 53$; $x = 5$, $y = 0$, $z = 3$ **9.** $z = 14,500$; $x_1 = 0$, $x_2 = 500/3$, $x_3 = 5000/3$ **11.** $p = 6$; $x = 2$, $y = 1$, $z = 0$, $w = 3$ **13.** $p = 7$; $x = 1$, $y = 0$, $z = 2$, $w = 0$, $v = 4$ (or: $x = 1$, $y = 0$, $z = 2$, $w = 1$, $v = 3$.)

15. $p = 21$; $x = 0$, $y = 2.27$, $z = 5.73$ **17.** $p = 4.52$; $x = 1$, $y = 0$, $z = .67$, $w = 1.52$ **19.** $p = 7.7$; $x = 1.1$, $y = 0$, $z = 2.2$, $w = 0$, $v = 4$ **21.** You should purchase 500 calculus texts, no history texts and no marketing texts. The maximum profit is \$5000 per semester. **23.** The company can make a maximum profit of \$650 by making 100 gallons of PineOrange, 200 gallons of PineKiwi, and 150 gallons of OrangeKiwi. **25.** The department should offer no Ancient History, 30 sections of Medieval History, and 15 sections of Modern History, for a profit of \$1,050,000. There will be 500 students without classes, but all sections and professors are used. **27.** Plant 80 acres of tomatoes and leave the other 20 acres unplanted. This will give you a profit of \$160,000. **29.** It can make a profit of \$10,000 by selling 1000 servings of Granola, 500 servings of Nutty Granola and no Nuttiest Granola. It is left with 2000 oz. almonds. **31.** Allocate 5 million gals to process A and 45 million gals to process C. Another solution: Allocate 10 million gals to process B and 40 million gals to process C. **33.** Use 15 servings of RiboForce HP and none of the others for a maximum of 75g creatine. **35.** She is wrong; you should buy 125 shares of IBM and no others. **37.** Allocate \$2,250,000 to automobile loans, \$500,000 to signature loans, and \$2,250,000 to any combination of furniture loans and other secured loans. **39.** Invest \$75,000 in Universal, none in the rest. Another optimal solution is: Invest \$18,750 in Universal, and \$75,000 in EMI. **41.** Tucson to Honolulu: 290 boards; Tucson to Venice Beach: 330 boards; Toronto to Honolulu: 0 boards; Toronto to Venice Beach: 200 boards, giving 820 boards shipped. **43.** Fly 10 people from Chicago to Los Angeles, 5 people from Chicago to New York, and 10 people from Denver to New York. **45.** Yes; the given problem can be stated as: Maximize $p = 3x - 2y$ subject to $-x + y - z \le 0$, $x - y - z \le 6$. **47.** The graphical method applies only to LP problems in two unknowns, whereas the simplex method can be used to solve LP problems with any number of unknowns. **49.** She is correct. Because there are only two constraints, there can only be two active variables, giving two or fewer nonzero values for the unknowns at each stage. **51.** A basic solution to a system of linear equations is a solution in which all the non-pivotal variables are taken to be zero; that is, all variables whose values are arbitrary are assigned the value zero. To obtain a basic solution for a given system of linear equations, one can row reduce the associated augmented matrix, write down the general solution, and then set all the parameters (variables with "arbitrary" values) equal to zero. **53.** No. Let us assume for the sake of simplicity that all the pivots are 1's. (They may certainly be changed to 1's without affecting the value of any of the variables.) Because the entry at the bottom of the pivot column is negative, the bottom row gets replaced by itself plus a positive multiple of the pivot row. The value of the objective function (bottom-right entry) is thus replaced by itself plus a positive multiple of the nonnegative rightmost entry of the pivot row. Therefore, it cannot decrease.

Section 4.4

1. $p = 20/3$; $x = 4/3$, $y = 16/3$ **3.** $p = 850/3$; $x = 50/3$, $y = 25/3$ **5.** $p = 750$; $x = 0$, $y = 150$, $z = 0$ **7.** $p = 135$; $x = 0$, $y = 25$, $z = 0$, $w = 15$ **9.** $c = 80$; $x = 20/3$, $y = 20/3$

11. $c = 100$; $x = 0$, $y = 100$, $z = 0$ **13.** $c = 111$; $x = 1$, $y = 1, z = 1$ **15.** $c = 200$; $x = 200$, $y = 0, z = 0, w = 0$ **17.** $p = 136.75$; $x = 0$, $y = 25.25$, $z = 0$, $w = 15.25$ **19.** $c = 66.67$; $x = 0$, $y = 66.67, z = 0$ **21.** $c = -250$; $x = 0$, $y = 500$, $z = 500$; $w = 1500$ **23.** Plant 100 acres of tomatoes and no other crops. This will give you a profit of \$200,000. (You will be using all 100 acres of your farm.) **25.** 10 mailings to the East Coast, none to the Midwest, 10 to the West Coast. Cost: \$900. Another solution resulting in the same cost is no mailings to the East Coast, 15 to the Midwest, none to the West Coast. **27.** 10,000 quarts of orange juice and 2000 quarts of orange concentrate **29.** Stock 10,000 rock CDs, 5000 rap CDs, and 5000 classical CDs for a maximum retail value of \$255,000. **31.** One serving of cereal, one serving of juice, and no dessert! **33.** 15 bundles from Nadir, 5 from Sonny, and none from Blunt. Cost: \$70,000. Another solution resulting in the same cost is 10 bundles from Nadir, none from Sonny, and 10 from Blunt. **35.** Mix 6 servings of Riboforce HP and 10 servings of Creatine Transport for a cost of \$15.60. **37. a.** Build 1 convention-style hotel, 4 vacation-style hotels and 2 small motels. The total cost will amount to \$188 million. **b.** Because 20% of this is \$37.6 million, you will still be covered by the subsidy. **39.** Tucson to Honolulu: 500 boards/week; Tucson to Venice Beach: 120 boards/week; Toronto to Honolulu: 0 boards/week; Toronto to Venice Beach: 410 boards/week. Minimum weekly cost is \$9700. **41.** \$2500 from Congressional Integrity Bank, \$0 from Citizens' Trust, \$7500 from Checks R Us. **43.** Fly 10 people from Chicago to LA, 5 from Chicago to New York, none from Denver to LA, 10 from Denver to NY at a total cost of \$4520. **45.** Hire no more cardiologists, 12 rehabilitation specialists, and 5 infectious disease specialists. **47.** The solution $x = 0$, $y = 0, \ldots$, represented by the initial tableau may not be feasible. In phase I we use pivoting to arrive at a basic solution that is feasible. **49.** The basic solution corresponding to the initial tableau has all the unknowns equal to zero, and this is not a feasible solution because it does not satisfy the given inequality. **51.** (C) **53.** Answers may vary. Examples are Exercises 1 and 2. **55.** Answers may vary. A simple example is: Maximize $p = x + y$ subject to $x + y \le 10, x + y \ge 20, x \ge 0, y \ge 0$.

Section 4.5

1. Minimize $c = 6s + 2t$ subject to $s - t \ge 2$, $2s + t \ge 1$, $s \ge 0, t \ge 0$ **3.** Maximize $p = 100x + 50y$ subject to $x + 2y \le 2, x + y \le 1, x \le 3, x \ge 0, y \ge 0$. **5.** Minimize $c = 3s + 4t + 5u + 6v$ subject to $s + u + v \ge 1, s + t + v \ge 1$, $s + t + u \ge 1, t + u + v \ge 1, s \ge 0, t \ge 0, u \ge 0, v \ge 0$. **7.** Maximize $p = 1000x + 2000y + 500z$ subject to $5x + z \le 1, -x + z \le 3, y \le 1, x - y \le 0, x \ge 0, y \ge 0$, $z \ge 0$. **9.** $c = 4$; $s = 2, t = 2$ **11.** $c = 80$; $s = 20/3$, $t = 20/3$ **13.** $c = 1.8$; $s = 6, t = 2$ **15.** $c = 25$; $s = 5$, $t = 15$ **17.** $c = 30$; $s = 30, t = 0, u = 0$ **19.** $c = 100$; $s = 0, t = 100, u = 0$ **21.** $c = 30$; $s = 10, t = 10, u = 10$ **23.** $R = [3/5 \quad 2/5], C = [2/5 \quad 3/5 \quad 0]^T, e = 1/5$ **25.** $R = [1/4 \quad 0 \quad 3/4], C = [1/2 \quad 0 \quad 1/2]^T, e = 1/2$ **27.** $R = [0 \quad 3/11 \quad 3/11 \quad 5/11]$, $C = [8/11 \quad 0 \quad 2/11 \quad 1/11]^T, e = 9/11$

29. 4 ounces each of fish and cornmeal, for a total cost of 40¢ per can; 5/12¢ per gram of protein, 5/12¢ per gram of fat. **31.** 100 oz of grain and no chicken, for a total cost of $1; 1/2¢ per gram of protein, 0¢ per gram of fat. **33.** One serving of cereal, one serving of juice, and no dessert! for a total cost of 37¢; 1/6¢ per calorie and 17/120¢ per % U.S. RDA of Vitamin C. **35.** 10 mailings to the East coast, none to the Midwest, 10 to the West Coast. Cost: $900; 20¢ per Democrat and 40¢ per Republican. OR 15 mailings to the Midwest and no mailing to the coasts. Cost: $900; 20¢ per Democrat and 40¢ per Republican. **37.** Gillian should use 480 sleep spells and 160 shock spells, costing 360,000 pico-shirleys of energy OR 2880/7 sleep spells and 1440/7 shock spells. **39.** T. N. Spend should spend about 73% of the days in Littleville, 27% in Metropolis, and skip Urbantown. T. L. Down should spend about 91% of the days in Littleville, 9% in Metropolis, and skip Urbantown. The expected outcome is that T. L. Down will lose about 227 votes per day of campaigning. **41.** Each player should show one finger with probability 1/2, two fingers with probability 1/3, and three fingers with probability 1/6. The expected outcome is that player A will win 2/3 point per round, on average. **43.** Write moves as (x, y) where x represents the number of regiments sent to the first location and y represents the number sent to the second location. Colonel Blotto should play (0, 4) with probability 4/9, (2, 2) with probability 1/9, and (4, 0) with probability 4/9. Captain Kije has several optimal strategies, one of which is to play (0, 3) with probability 1/30, (1, 2) with probability 8/15, (2, 1) with probability 16/45, and (3, 0) with probability 7/90. The expected outcome is that Colonel Blotto will win 14/9 points on average. **45.** The dual of a standard minimization problem satisfying the nonnegative objective condition is a standard maximization problem, which can be solved using the standard simplex algorithm, thus avoiding the need to do Phase I. **47.** Answers will vary. An example is: Minimize $c = x - y$ subject to $x - y \geq 100$, $x + y \geq 200$, $x \geq 0$, $y \geq 0$. This problem can be solved using the techniques in Section 4.4. **49.** Build 1 convention-style hotel, 4 vacation-style hotels and 2 small motels. **51.** Answers will vary.

Chapter 4 Review

1.

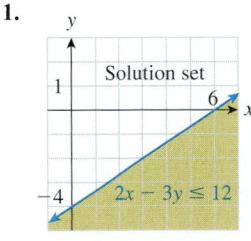

Unbounded

3.

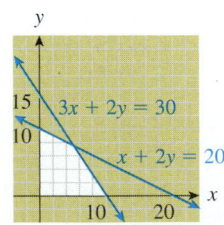

Bounded; Corner points: (0, 0), (0, 10), (5, 15/2), (10, 0)

5. $p = 21; x = 9, y = 3$ **7.** $c = 22; x = 8, y = 6$
9. $p = 45; x = 0, y = 15, z = 15$
11. $p = 220; x = 20, y = 20, z = 60$
13. $c = 30; x = 30, y = 0, z = 0$
15. $c = 50; x = 20, y = 10, z = 0, w = 20$

17. $c = 60; x = 24, y = 12$ **19.** $c = 20; x = 0, y = 20$
21. $R = [1/2 \quad 1/2 \quad 0], C = [0 \quad 1/3 \quad 2/3]^T, e = 0$
23. $R = [1/27 \quad 7/9 \quad 5/27], C = [8/27 \quad 5/27 \quad 14/27]^T$,
$e = 8/27$ **25.** (A) **27.** 35 **29.** (B), (D) **31.** 450 packages from Duffin House, and 375 from Higgins Press for a minimum cost of $52,500. **33.** $c = 90,000; x = 0, y = 600, z = 0$
35. Billy Sean should take the following combination: Sciences: 24 credits, Fine Arts: no credits, Liberal Arts: 48 credits, Mathematics: 48 credits, for a total cost of $26,400. **37.** Fantasy-Books should choose between "2 for 1" and "3 for 2" with probabilities 20% and 80%, respectively. O'HaganBooks should choose between "3 for 1" and "Finite Math" with probabilities 60% and 40%, respectively. O'HaganBooks expects to gain 12,000 customers from FantasyBooks.

Chapter 5

Section 5.1

1. $INT = \$120, FV = \$2120$ **3.** $INT = \$505, FV = \$20,705$ **5.** $INT = \$250, FV = \$10,250$ **7.** $PV = \$9090.91$
9. $PV = \$966.18$ **11.** $PV = \$14,457.83$ **13.** $5200
15. $787.40 **17.** 5% **19.** In 2 years **21.** 3.775% **23.** 65%
25. 10% **27.** 168.85% **29.** 85.28% if you sold in February, 2005 **31.** No. Simple interest increase is linear. The graph is visibly not linear in that time period. **33.** 9.2% **35.** 3,260,000
37. $P = 500 + 46t$ ($t = $ time in years since 1950) Graph:

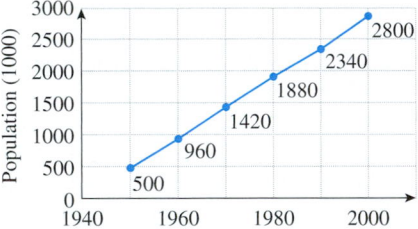

39. Graph (A) is the only possible choice, because the equation $FV = PV(1 + rt) = PV + PVrt$ gives the future value as a linear function of time. **41.** Wrong. In simple interest growth, the change each year is a fixed percentage of the *starting* value, and not the preceding year's value. (Also see Exercise 42.)
43. Simple interest is always calculated on a constant amount, PV. If interest is paid into your account, then the amount on which interest is calculated does not remain constant.

Section 5.2

1. $13,439.16 **3.** $11,327.08 **5.** $19,154.30 **7.** $12,709.44
9. $613.91 **11.** $810.65 **13.** $1227.74 **15.** 5.09%
17. 10.47% **19.** 10.52% **21.** $268.99 **23.** $2491.75
25. $2927.15 **27.** $21,161.79 **29.** $163,414.56
31. $55,526.45 per year **33.** $174,110 **35.** $750.00
37. $27,171.92 **39.** $111,678.96 **41.** $1039.21 **43.** The one earning 11.9% compounded monthly **45.** Yes. The investment will have grown to about $150,281 million **47.** 147 reals
49. 744 pesos **51.** 1224 pesos **53.** The Ecuadorian investment is better: it is worth 1.01614 units of currency (in constant units) per

unit invested as opposed to 1.01262 units for Chile.   **55.** 41.02%
57. 51.90% if you sold in February, 2005   **59.** No. Compound
interest increase is exponential. The graph looks roughly exponential in that period, but to really tell we can compare interest
rates between marked points to see if the rate remained roughly
constant: From December 1997 to August 1999 the rate was
$(16.31/3.28)^{12/20} - 1 = 1.6179$ or 161.79%, while from August
1999 to March 2000 the rate was $(33.95/16.31)^{12/7} - 1 = 2.5140$
or 251.40%. These rates are quite different.   **61.** 31 years;
about $26,100   **63.** 2.3 years   **65. a.** $1510.31 **b.** $54,701.29
c. 23.51%   **67.** The function $y = P(1 + r/m)^{mx}$ is not a linear
function of x, but an exponential function. Thus, its graph is not a
straight line.   **69.** Wrong. Its growth is exponential and can be
modeled by $0.01(1.10)^t$.   **71.** The graphs are the same because
the formulas give the same function of x; a compound-interest investment behaves as though it was being compounded once a year
at the effective rate.   **73.** The effective rate exceeds the nominal
rate when the interest is compounded more than once a year because
then interest is being paid on interest accumulated during each year,
resulting in a larger effective rate. Conversely, if the interest is compounded less often than once a year, the effective rate is less than the
nominal rate.   **75.** Compare their future values in constant dollars.
The investment with the larger future value is the better investment.
77. The graphs are approaching a particular curve as m gets larger,
approximately the curve given by the largest two values of m.

Section 5.3

1. $15,528.23   **3.** $171,793.82   **5.** $23,763.28   **7.** $147.05
9. $491.12   **11.** $105.38   **13.** $90,155.46   **15.** $69,610.99
17. $95,647.68   **19.** $554.60   **21.** $1366.41   **23.** $524.14
25. $248.85   **27.** $1984.65   **29.** $999.61   **31.** $998.47
33. 3.617%   **35.** 3.059%   **37.** $973.54   **39.** $7451.49
41. You should take the loan from Solid Savings & Loan: it will
have payments of $248.85 per month. The payments on the other
loan would be more than $300 per month.   **43.** Answers using
correctly rounded intermediate results:

Year	Interest	Payment on Principal
1	$3934.98	$1798.98
2	$3785.69	$1948.27
3	$3623.97	$2109.99
4	$3448.84	$2285.12
5	$3259.19	$2474.77
6	$3053.77	$2680.19
7	$2831.32	$2902.64
8	$2590.39	$3143.57
9	$2329.48	$3404.48
10	$2046.91	$3687.05
11	$1740.88	$3993.08
12	$1409.47	$4324.49
13	$1050.54	$4683.42
14	$661.81	$5072.15
15	$240.84	$5491.80

45. First five years: $402.62/month; last 25 years: $601.73
47. Original monthly payments were $824.79. The new monthly
payments will be $613.46. You will save $36,481.77 in interest.
49. 10.81%   **51.** 13 years   **53.** 4.5 years   **55.** 24 years
57. He is wrong because his estimate ignores the interest that
will be earned by your annuity—both while it is increasing and
while it is decreasing. Your payments will be considerably smaller
(depending on the interest earned).   **59.** He is not correct. For
instance, the payments on a $100,000 10-year mortgage at 12%
are $1434.71, while for a 20-year mortgage at the same rate,
they are $1101.09, which is a lot more than half the 10-year
mortgage payment.   **61.** $PV = FV(1 + i)^{-n} =$

$$PMT\frac{(1+i)^n - 1}{i}(1+i)^{-n} = PMT\frac{1 - (1+i)^{-n}}{i}$$

Chapter 5 Review

1. $7425.00   **3.** $7604.88   **5.** $6757.41   **7.** $4848.48
9. $4733.80   **11.** $5331.37   **13.** $177.58   **15.** $112.54
17. $187.57   **19.** $9584.17   **21.** 5.346%   **23.** 14.0 years
25. 10.8 years   **27.** 7.0 years   **29.** 2003

Year	2000	2001	2002	2003	2004
Revenue	$180,000	$216,000	$259,200	$311,040	$373,248

31. At least 52,515 shares   **33.** $224,111   **35.** $420,275
37. $1453.06   **39.** $53,055.66   **41.** 5.99%

Chapter 6

Section 6.1

1. $F = \{$spring, summer, fall, winter$\}$ **3.** $I = \{1, 2, 3, 4, 5, 6\}$
5. $A = \{1, 2, 3\}$ **7.** $B = \{2, 4, 6, 8\}$ **9. a.** $S = \{(H, H),$
$(H, T), (T, H), (T, T)\}$ **b.** $S = \{(H, H), (H, T), (T, T)\}$
11. $S = \{(1, 5), (2, 4), (3, 3), (4, 2), (5, 1)\}$
13. $S = \{(1, 5), (2, 4), (3, 3)\}$ **15.** $S = \varnothing$

17.

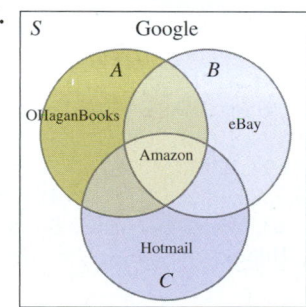

19.

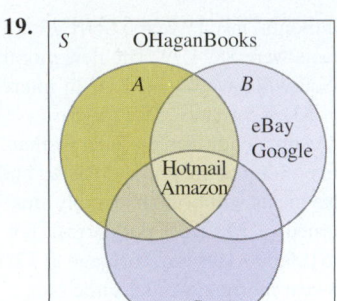

21. A **23.** A **25.** {June, Janet, Jill, Justin, Jeffrey, Jello, Sally, Solly, Molly, Jolly}   **27.** {Jello}   **29.** $\varnothing$ **31.** {Jello}
33. {Janet, Justin, Jello, Sally, Solly, Molly, Jolly}   **35.** {(small, triangle), (small, square), (medium, triangle), (medium, square), (large, triangle), (large, square)}   **37.** {(small, blue), (small, green), (medium, blue), (medium, green), (large, blue), (large, green)}
39.

Microsoft Excel - Ch 6-1 Exercise 37 Answer.xls

File Edit View Insert Format Tools Data Window Help Adobe PD

	A	B	C	D
1		**Triangle**	**Square**	
2	**Blue**	Blue Triangle	Blue Square	
3	**Green**	Green Triangle	Green Square	
4				
5				

41.

Microsoft Excel - Ch 6-1 Exercise 41 Answer.xls

File Edit View Insert Format Tools Data Window Help Adobe PD

	A	B	C	D
1		**Blue**	**Green**	
2	**Small**	Small Blue	Small Green	
3	**Medium**	Medium Blue	Medium Green	
4	**Large**	Large Blue	Large Green	
5				
6				

43. $B \times A = \{$1H, 1T, 2H, 2T, 3H, 3T, 4H, 4T, 5H, 5T, 6H, 6T$\}$
45. $A \times A \times A = \{$HHH, HHT, HTH, HTT, THH, THT, TTH, TTT$\}$ **47.** {(1,1), (1,3), (1,5), (3,1), (3,3), (3,5), (5,1), (5,3), (5,5)}   **49.** $\varnothing$ **51.** {(1,1), (1,3), (1,5), (3,1), (3,3), (3,5), (5,1), (5,3), (5,5), (2,2), (2,4), (2,6), (4,2), (4,4), (4,6), (6,2), (6,4), (6,6)}
61. $A \cap B = \{$Acme, Crafts$\}$ **63.** $B \cup C = \{$Acme, Brothers, Crafts, Dion, Effigy, Global, Hilbert$\}$ **65.** $A' \cap C = \{$Dion, Hilbert$\}$ **67.** $A \cap B' \cap C' = \varnothing$

69.

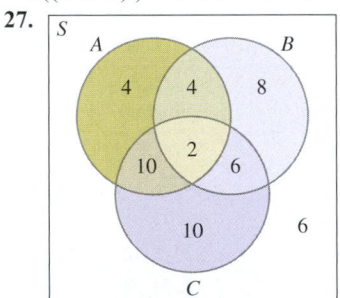

Microsoft Excel - Ch 6-1 Exercise 69 Answer.xls

File Edit View Insert Format Tools Data Window Help Adobe PDF

	A	B	C	D	E
1		**Sail Boats**	**Motor Boats**	**Yachts**	
2	**2003**	(2003 Sail Boats)	(2003 Motor Boats)	(2003 Yachts)	
3	**2004**	(2004 Sail Boats)	(2004 Motor Boats)	(2004 Yachts)	
4	**2005**	(2005 Sail Boats)	(2005 Motor Boats)	(2005 Yachts)	
5	**2006**	(2006 Sail Boats)	(2006 Motor Boats)	(2006 Yachts)	
6					
7					

71. $I \cup J$ **73.** (B)   **75.** Answers may vary. Let $A = \{1\}$, $B = \{2\}$, and $C = \{1, 2\}$. Then $(A \cap B) \cup C = \{1, 2\}$ but $A \cap (B \cup C) = \{1\}$. In general, $A \cap (B \cup C)$ must be a subset of A, but $(A \cap B) \cup C$ need not be; also, $(A \cap B) \cup C$ must contain C as a subset, but $A \cap (B \cup C)$ need not.   **77.** A universal set is a set containing all "things" currently under consideration. When discussing sets of positive integers, the universe might be the set of all positive integers, or the set of all integers (positive, negative, and 0), or any other set containing the set of all positive integers.
79. A is the set of suppliers who deliver components on time, B is the set of suppliers whose components are known to be of high quality, and C is the set of suppliers who do not promptly replace defective components.   **81.** Let $A = \{$movies that are violent$\}$, $B = \{$movies that are shorter than two hours$\}$, $C = \{$movies that have a tragic ending$\}$, and $D = \{$movies that have an unexpected ending$\}$. The given sentence can be rewritten as "She prefers movies in $A' \cap B \cap (C \cup D)'$." It can also be rewritten as "She prefers movies in $A' \cap B \cap C' \cap D'$."

Section 6.2

1. 9   **3.** 7   **5.** 4   **7.** $n(A \cup B) = 7, n(A) + n(B) - n(A \cap B) = 4 + 5 - 2 = 7$ **9.** 4   **11.** 18   **13.** 72   **15.** 60   **17.** 20   **19.** 6
21. 9   **23.** 4   **25.** $n((A \cap B)') = 9, n(A') + n(B') - n((A \cup B)') = 6 + 7 - 4 = 9$

27.

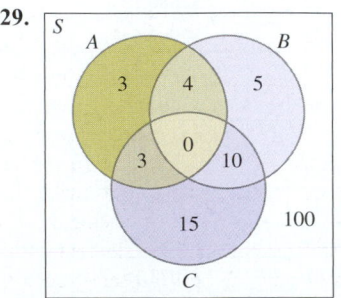

31. 76,000 **33.** 2 **35.** $C \cap N$ is the set of authors who are both successful and new. $C \cup N$ is the set of authors who are either successful or new (or both). $n(C) = 30$; $n(N) = 20$; $n(C \cap N) = 5$; $n(C \cup N) = 45$; $45 = 30 + 20 - 5$
37. $C \cap N'$ is the set of authors who are successful but not new. $n(C \cap N') = 25$ **39.** 31.25%; 83.33% **41.** $N \cap C$; $n(N \cap C) = 8$ billion **43.** $C \cap N'$; $n(C \cap N') = 13$ billion
45. $A \cap (N \cup U)$; $n(A \cap (N \cup U)) = 14$ billion
47. $V \cap I'$; $n(V \cap I') = 15$ **49.** 80; The number of stocks that were either not pharmaceutical stocks, or were unchanged in value after a year (or both). **51.** 3/8; the fraction of Internet stocks that increased in value **53. a.** 931 **b.** 382
55. a.

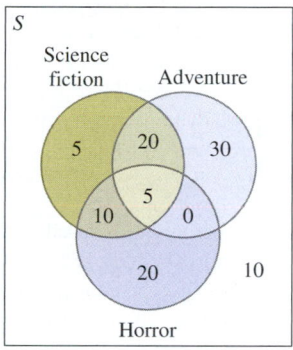

b. 37.5% **57.** 17 **59.** The number of elements in the Cartesian product of two finite sets is the product of the number of elements in the two sets. **61.** Answers will vary.
63. When $A \cap B \neq \emptyset$ **65.** When $B \subseteq A$
67. $n(A \cup B \cup C) = n(A) + n(B) + n(C) - n(A \cap B) - n(B \cap C) - n(A \cap C) + n(A \cap B \cap C)$

Section 6.3
1. 10 **3.** 30 **5.** 6 outcomes **7.** 15 outcomes
9. 13 outcomes **11.** 25 outcomes **13.** 4 **15.** 93
17. 16 **19.** 30 **21.** 13 **23.** 18 **25.** 25,600 **27.** 3381
29. a. 288 **b.** 288 **31.** 256 **33.** 10 **35.** 286 **37.** 4
39. a. 8,000,000 **b.** 30,000 **c.** 4,251,528 **41. a.** $4^3 = 64$
b. 4^n **c.** $4^{2.1 \times 10^{10}}$ **43. a.** $16^6 = 16,777,216$
b. $16^3 = 4096$ **c.** $16^2 = 256$ **d.** 766
45. $(10 \times 9 \times 8 \times 7 \times 6 \times 5 \times 4) \times (8 \times 7 \times 6 \times 5)$ $= 1,016,064,000$ possible casts **47. a.** $26^3 \times 10^3 = 17,576,000$
b. $26^2 \times 23 \times 10^3 = 15,548,000$ **c.** $15,548,000 - 3 \times 10^3 = 15,545,000$ **49. a.** 4 **b.** 4 **c.** There would be an infinite number of routes. **51. a.** 72 **b.** 36 **53.** 96 **55. a.** 36
b. 37 **57.** Step 1: Choose a day of the week on which Jan 1 will fall: 7 choices. Step 2: Decide whether or not it is a leap year: 2 choices. Total: $7 \times 2 = 14$ possible calendars. **59.** 1900
61. Step 1: choose a position in the Left-Right direction: m choices. Step 2: choose a position in the Front-Back direction: n choices. Step 3: choose a position in the Up-Down direction: r choices. Hence there are $m \cdot n \cdot r$ possible outcomes. **63.** 4 **65.** Cartesian product
67. The decision algorithm produces every pair of shirts twice, first in one order and then in the other. **69.** Think of placing the five squares in a row of five empty slots. Step 1: choose a slot for the blue square, 5 choices. Step 2: choose a slot for the green square, 4 choices.

Step 3: choose the remaining 3 slots for the yellow squares, 1 choice. Hence there are 20 possible five-square sequences.

Section 6.4
1. 720 **3.** 56 **5.** 360 **7.** 15 **9.** 3 **11.** 45 **13.** 20
15. 4950 **17.** 360 **19.** 35 **21.** 120 **23.** 120 **25.** 20
27. 60 **29.** 210 **31.** 7 **33.** 35 **35.** 24 **37.** 126
39. 196 **41.** 105 **43.** $\dfrac{C(30, 5) \times 5^{25}}{6^{30}} \approx 0.192$
45. $\dfrac{C(30, 15) \times 3^{15} \times 3^{15}}{6^{30}} \approx 0.144$ **47.** 24
49. $C(13, 2)C(4, 2)C(4, 2) \times 44 = 123,552$
51. $13 \times C(4, 2)C(12, 3) \times 4 \times 4 \times 4 = 1,098,240$
53. 10,200 **55. a.** 252 b. 20 c. 26 **57. a.** 300 **b.** 3 **c.** 1 in 100 or .01 **59. a.** 210 **b.** 77 **c.** No **61. a.** 23! **b.** 18!
c. $19 \times 18!$ **63.** $C(11, 1)C(10, 4)C(6, 4)C(2, 2)$
65. $C(11, 2)C(9, 1)C(8, 1)C(7, 3)C(4, 1)C(3, 1)C(2, 1)C(1, 1)$
67. $C(10, 2)C(8, 4)C(4, 1)C(3, 1)C(2, 1)C(1, 1)$
69. (A) **71.** (D) **73. a.** 9880 **b.** 1560 **c.** 11,480
75. a. $C(20, 2) = 190$ **b.** $C(n, 2)$ **77.** The multiplication principle; it can be used to solve all problems that use the formulas for permutations. **79.** Urge your friend not to focus on formulas, but instead learn to formulate decision algorithms and use the principles of counting. **81.** It is ambiguous on the following point: are the three students to play different characters, or are they to play a group of three, such as "three guards." This should be made clear in the exercise.

Chapter 6 Review
1. $N = \{-3, -2, -1\}$ **3.** $S = \{(1, 2), (1, 3), (1, 4), (1, 5), (1, 6),$ $(2, 1), (2, 3), (2, 4), (2, 5), (2, 6), (3, 1), (3, 2), (3, 4), (3, 5), (3, 6),$ $(4, 1), (4, 2), (4, 3), (4, 5), (4, 6), (5, 1), (5, 2), (5, 3), (5, 4), (5, 6),$ $(6, 1), (6, 2), (6, 3), (6, 4), (6, 5)\}$ **5.** $A \cup B' = \{a, b, d\}$,
$A \times B' = \{(a, a), (a, d), (b, a), (b, d)\}$ **7.** $A \times B$
9. $(P \cap E' \cap Q)'$ or $P' \cup E \cup Q'$ **11.** $n(A \cup B) = n(A) + n(B) - n(A \cap B), n(C') = n(S) - n(C)$; 100
13. $n(A \times B) = n(A)n(B), n(A \cup B) = n(A) + n(B) - n(A \cap B), n(A') = n(S) - n(A)$, 21
15. $C(12, 1)C(4, 2)C(11, 3)$ $C(4, 1)C(4, 1)C(4, 1)$
17. $C(4, 1)C(10, 1)$ **19.** $C(4, 4)C(8, 1) = 8$
21. $C(3, 2)C(9, 3) + C(3, 3)C(9, 2) = 288$ **23.** The set of books that are either sci-fi or stored in Texas (or both); $n(S \cup T) = 112,000$ **25.** The set of books that are either stored in California or not sci-fi; $n(C \cup S') = 175,000$
27. The romance books that are also horror books or stored in Texas; $n(R \cap (T \cup H)) = 20,000$ **29.** 1000
31. FarmerBooks.com; 1800 **33.** JungleBooks.com; 3500
35. 15,600 **37.** 2 letters, 4 digits; 2,948,400 **39.** 28,000

Chapter 7

Section 7.1
1. $S = \{HH, HT, TH, TT\}$; $E = \{HH, HT, TH\}$ **3.** $S = \{HHH, HHT, HTH, HTT, THH, THT, TTH, TTT\}$; $E = \{HTT, THT, TTH, TTT\}$

5. $S = \left\{ \begin{array}{llllll} (1,1) & (1,2) & (1,3) & (1,4) & (1,5) & (1,6) \\ (2,1) & (2,2) & (2,3) & (2,4) & (2,5) & (2,6) \\ (3,1) & (3,2) & (3,3) & (3,4) & (3,5) & (3,6) \\ (4,1) & (4,2) & (4,3) & (4,4) & (4,5) & (4,6) \\ (5,1) & (5,2) & (5,3) & (5,4) & (5,5) & (5,6) \\ (6,1) & (6,2) & (6,3) & (6,4) & (6,5) & (6,6) \end{array} \right\}$;

$E = \{(1,4), (2,3), (3,2), (4,1)\}$

7. $S = \left\{ \begin{array}{llllll} (1,1) & (1,2) & (1,3) & (1,4) & (1,5) & (1,6) \\ & (2,2) & (2,3) & (2,4) & (2,5) & (2,6) \\ & & (3,3) & (3,4) & (3,5) & (3,6) \\ & & & (4,4) & (4,5) & (4,6) \\ & & & & (5,5) & (5,6) \\ & & & & & (6,6) \end{array} \right\}$;

$E = \{(1,3), (2,2)\}$

9. S as in Exercise 7; $E = \{(2,2), (2,3), (2,5), (3,3),$ $(3,5), (5,5)\}$ **11.** $S = \{m, o, z, a, r, t\}$: $E = \{o, a\}$
13. $S = \{(s,o), (s,r), (s,e), (o,s), (o,r), (o,e), (r,s), (r,o),$ $(r,e), (e,s), (e,o), (e,r)\}$; $E = \{(o,s), (o,r), (o,e), (e,s),$ $(e,o), (e,r)\}$ **15.** $S = \{01, 02, 03, 04, 10, 12, 13, 14, 20, 21, 23,$ $24, 30, 31, 32, 34, 40, 41, 42, 43\}$; $E = \{10, 20, 21, 30, 31, 32,$ $40, 41, 42, 43\}$ **17.** $S = \{$domestic car, imported car, van, antique car, antique truck$\}$; $E = \{$van, antique truck$\}$
19. a. all sets of 4 gummy candies chosen from the packet of 12
b. all sets of 4 gummy candies in which two are strawberry and two are blackcurrant **21. a.** all lists of 14 people chosen from 20 **b.** all lists of 14 people chosen from 20, in which Colin Powell occupies the first position **23.** $A \cap B$; $n(A \cap B) = 1$ **25.** B'; $n(B') = 33$ **27.** $B' \cap D'$; $n(B' \cap D') = 2$
29. $C \cup B$; $n(C \cup B) = 12$ **31.** $W \cap I$ **33.** $E \cup I'$
35. $I \cup (W \cap E')$ **37.** $(I \cup W) \cap E'$ **39.** $E = \{$New England, Pacific, Middle Atlantic$\}$ **41.** $E \cup F$ is the event that you choose a region that saw an increase in housing prices of 15% or more or is on the east coast. $E \cup F = \{$Pacific, New England, Middle Atlantic, South Atlantic$\}$. $E \cap F$ is the event that you choose a region that saw an increase in housing prices of 15% or more and is on the east coast. $E \cap F = \{$New England, Middle Atlantic$\}$. **43. a.** Mutually exclusive **b.** Not mutually exclusive **45.** $S \cap N$ is the event that an author is successful and new. $S \cup N$ is the event that an author is either successful or new; $n(S \cap N) = 5$; $n(S \cup N) = 45$ **47.** N and E **49.** $S \cap N'$ is the event that an author is successful but not a new author. $n(S \cap N') = 25$ **51.** 31.25%; 83.33%
53. $V \cap I'$; $n(V \cap I') = 15$ **55.** 80; The number of stocks that were either not pharmaceutical stocks, or were unchanged in value after a year (or both). **57.** P and E, P and I, E and I, N and E, V and N, V and D, N and D **59.** 3/8; the fraction of Internet stocks that increased in value **61. a.** $E' \cap H$
b. $E \cup H$ **c.** $(E \cup G)' = E' \cap G'$ **63. a.** $\{9\}$ **b.** $\{6\}$
65. a. The dog's "fight" drive is weakest. **b.** The dog's "fight" and "flight" drives are either both strongest or both weakest.
c. Either the dog's "fight" drive is strongest, or its "flight" drive

is strongest. **67.** $C(6,4) = 15$; $C(1,1)C(5,3) = 10$
69. a. $n(S) = P(7,3) = 210$ **b.** $E \cap F$ is the event that Celera wins and Electoral College is in second or third place. In other words, it is the set of all lists of three horses in which Celera is first and Electoral College is second or third. $n(E \cap F) = 10$.
71. $C(8,3) = 56$ **73.** $C(4,1)C(2,1)C(2,1) = 16$
75. Subset of the sample space **77.** E and F do not both occur
79. True; Consider the following experiment: Select an element of the set S at random. **81.** Answers may vary. Cast a die and record the remainder when the number facing up is divided by 2.
83. Yes. For instance, $E = \{(2,5), (5,1)\}$ and $F = \{(4,3)\}$ are two such events.

Section 7.2
1. .4 **3.** .8

5.

Outcome	HH	HT	TH	TT
Probability	.275	.2375	.3	.1875

7. .575 **9.** The second coin *seems* slightly biased in favor of heads, because heads comes up approximately 58% of the time. On the other hand, it is conceivable that the coin is fair and that heads came up 58% of the time purely by chance. Deciding which conclusion is more reasonable requires some knowledge of inferential statistics. **15.** $P(E) = 1/4$ **17.** $P(E) = 1$
19. $P(E) = 3/4$ **21.** $P(E) = 3/4$ **23.** $P(E) = 1/2$
25. $P(E) = 1/9$ **27.** $P(E) = 0$ **29.** $P(E) = 1/4$
31. $1/12$; $\{(4,4), (2,3), (3,2)\}$

33.

Outcome	1	2	3	4	5	6
Probability	1/9	2/9	1/9	2/9	1/9	2/9

$P(\{1, 2, 3\}) = 4/9$

35.

Outcome	1	2	3	4
Probability	8/15	4/15	2/15	1/15

37. a. .04 **b.** .98

39. a.

Test Rating	3	2	1	0
Probability	.1	.4	.4	.1

b. 0.5 **41. a.** Dial-up: .63, Cable Modem: .21, DSL: .15, Other: .01 **b.** .36

43.

Outcome	Low	Middle	High
Probability	.5	.3	.2

45. .25 **47.** .2 **49.** .7 **51.** 5/6 **53.** 5/16

55.

Outcome	U	C	R
Probability	.2	.64	.16

57.

Conventional	No pesticide	Single pesticide	Multiple pesticide
Probability	.27	.13	.60
Organic	No pesticide	Single pesticide	Multiple pesticide
Probability	.77	.13	.10

59. $P(\text{false negative}) = 10/400 = .025$, $P(\text{false positive}) = 10/200 = .05$ **63.** .70   **65.** .86   **67.** .86
69.

Outcome	Hispanic or Latino	White (not Hispanic)	African American	Asian	Other
Probability	.42	.37	.09	.08	.04

$P(\text{Neither White nor Asian}) = .55$
71. a. $S = \{\text{Stock market success, Sold to other concern, Fail}\}$

b.

Outcome	Stock market success	Sold to other concern	Fail
Probability	.2	.3	.5

c. .5

73.

Outcome	SUV	Pickup	Passenger Car	Minivan
Probability	.25	.15	.50	.10

75. $P(1) = 0$, $P(6) = 0$; $P(2) = P(3) = P(4) = P(5) = 1/4 = .25$ **77.** $P(1) = P(6) = 1/10$; $P(2) = P(3) = P(4) = P(5) = 1/5$, $P(\text{odd}) = 1/2$ **79.** $P(1, 1) = P(2, 2) = \ldots = P(6, 6) = 1/66$; $P(1, 2) = \ldots = P(6, 5) = 1/33$, $P(\text{odd sum}) = 6/11$ **81.** $P(2) = 15/38$; $P(4) = 3/38$, $P(1) = P(3) = P(5) = P(6) = 5/38$, $P(\text{odd}) = 15/38$
83. The fraction of times E occurs   **85.** Wrong. For a pair of fair dice, the theoretical probability of a pair of matching numbers is 1/6, as Ruth says. However, it is quite possible, although not very likely, that if you cast a pair of fair dice 20 times, you will never obtain a matching pair (in fact, there is approximately a 2.6% chance that this will happen). In general, a nontrivial claim about theoretical probability can never be absolutely validated or refuted experimentally. All we can say is that the evidence suggests that the dice are not fair.   **87.** For a (large) number of days, record the temperature prediction for the next day and then check the actual temperature the next day. Record whether the prediction was accurate (within, say, $2°F$ of the actual temperature). The fraction of times the prediction was accurate is the estimated probability.   **89.** He is wrong. It is possible to have a run of losses of any length. Tony may have grounds to *suspect* that the game is rigged, but no proof.

Section 7.3
1. .65   **3.** .1   **5.** .7   **7.** .4   **9.** .25   **11.** 1.0   **13.** .3
15. 1.0   **17.** No; $P(A \cup B)$ should be $\leq P(A) + P(B)$.

19. Yes   **21.** No; $P(A \cup B)$ should be $\geq P(B)$.
23. $P(e) = .2$ **a.** .9   **b.** .95   **c.** .1   **d.** .8   **25.** 5/6
27. .39   **29.** .54   **31.** .24   **33.** .00   **35.** .76   **37.** .46
39. .54   **41.** .01   **43.** .56   **45.** .43   **47.** 22%; 43%
49. All of them   **51.** .884   **53.** They are mutually exclusive.
55. Wrong. For example, the theoretical probability of winning a state lotto is small but nonzero. However, the vast majority of people who play lotto day of their lives never win, no matter how frequently they play.   **57.** When $A \cap B = \emptyset$ we have $P(A \cap B) = P(\emptyset) = 0$, so $P(A \cup B) = P(A) + P(B) - P(A \cap B) = P(A) + P(B) - 0 = P(A) + P(B)$.   **59.** Zero. According to the assumption, no matter how many thunderstorms occur, lightning cannot strike your favorite spot more than once, and so, after n trials the estimated probability will never exceed $1/n$, and so will approach zero as the number of trials gets large.
61. $P(A \cup B \cup C) = P(A) + P(B) + P(C) - P(A \cap B) - P(A \cap C) - P(B \cap C) + P(A \cap B \cap C)$

Section 7.4
1. 1/42   **3.** 7/9   **5.** 1/7   **7.** 1/2   **9.** 41/42   **11.** 1/15
13. 4/15   **15.** 1/5   **17.** $1/(2^8 \times 5^5 \times 5!)$ **19.** .4226
21. .0475   **23.** .0020   **25.** $1/27^{39}$ **27.** 1/7   **29.** Probability of being a big winner $= 1/2,118,760 \approx .000000472$. Probability of being a small-fry winner $= 225/2,118,760 \approx .000106194$. Probability of being either a winner or a small-fry winner $= 226/2,118,760 \approx .000106666$.
31. a. $C(600,300)/C(700,400)$ **b.** $C(699,399)/C(700,400)$ or 400/700   **33.** $P(10, 3)/10^3 = 18/25 = .72$ **35.** $8!/8^8$
37. 1/8   **39.** 1/8   **41.** 37/10,000   **43. a.** 90,720 **b.** 25,200
c. $25,200/90,720 = 25/90 \approx .28$ **45.** The four outcomes listed are not equally likely; for example, (red, blue) can occur in four ways. The methods of this section yield a probability for (red, blue) of $C(2, 2)/C(4, 2) = 1/6$ **47.** No. If we do not pay attention to order, the probability is $C(5, 2)/C(9, 2) = 10/36 = 5/18$. If we do pay attention to order, the probability is $P(5, 2)/P(9, 2) = 20/72 = 5/18$ again. The difference between permutations and combinations cancels when we compute the probability.   **49.** Answers will vary.

Section 7.5
1. .4   **3.** .08   **5.** .75   **7.** .2   **9.** .5
11.

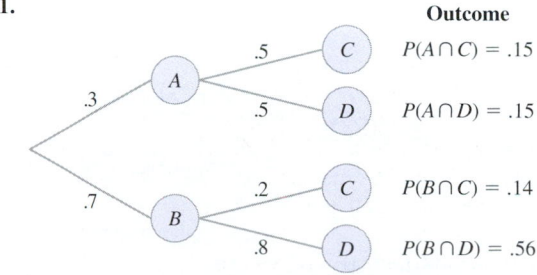

13.

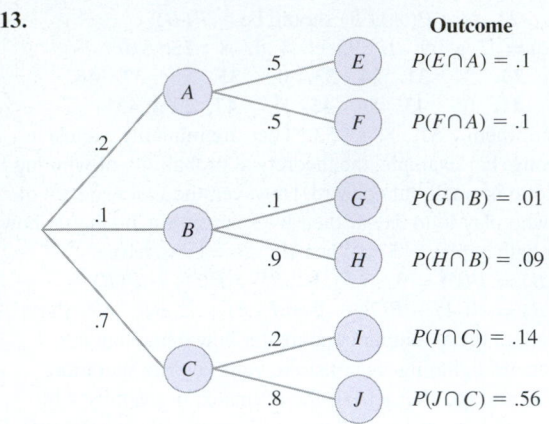

Outcome

$P(E \cap A) = .1$

$P(F \cap A) = .1$

$P(G \cap B) = .01$

$P(H \cap B) = .09$

$P(I \cap C) = .14$

$P(J \cap C) = .56$

15. 1/10 **17.** 1/5 **19.** 2/9 **21.** 1/84 **23.** 5/21

25. 24/175 **27.** (B) **29.** (C) **31.** $\frac{1}{2} \cdot \frac{1}{2} = \frac{1}{4}$ Independent

33. $\frac{5}{18} \cdot \frac{1}{2} \neq \frac{1}{9}$ Dependent **35.** $\frac{25}{36} \cdot \frac{5}{18} \neq \frac{2}{9}$ Dependent

37. $(1/2)^{11} = 1/2048$ **39.** .8 **41.** .43 **43.** .34

45. Not independent; $P(\text{giving up} \mid \text{used Brand X}) = 0.1$ is larger than $P(\text{giving up})$ **47.** .00015 **49.** 5/6 **51.** 3/4

53. 11/16 **55.** 11/14

57.

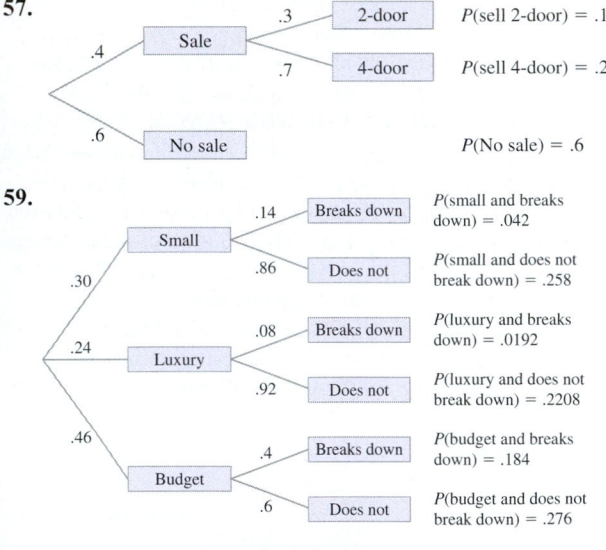

$P(\text{sell 2-door}) = .12$

$P(\text{sell 4-door}) = .28$

$P(\text{No sale}) = .6$

59.

$P(\text{small and breaks down}) = .042$

$P(\text{small and does not break down}) = .258$

$P(\text{luxury and breaks down}) = .0192$

$P(\text{luxury and does not break down}) = .2208$

$P(\text{budget and breaks down}) = .184$

$P(\text{budget and does not break down}) = .276$

61.

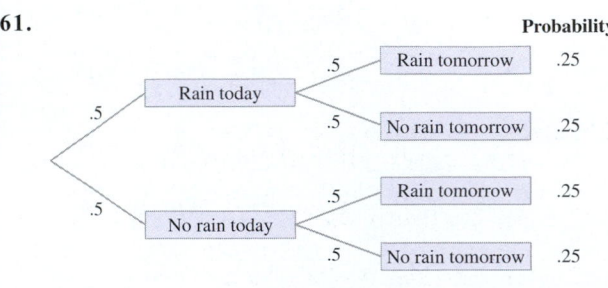

Probability

.25

.25

.25

.25

63. .76 **65.** .33 **67.** .37 **69.** .98 **71.** The claim is correct. The probability that an unemployed person has a high school diploma only is .35, while the corresponding figure for an employed person is .30. **73.** $P(K \mid D) = 1.31 P(K \mid D')$

75. a. $P(I \mid T) > P(I)$ **b.** It was ineffective. **77. a.** .59

b. \$35,000 or more: $P(\text{Internet user} \mid < \$35,000) \approx .27 <$ $P(\text{Internet user} \mid \geq \$35,000) \approx .59$. **79.** $P(R \mid J)$ **81.** (D)

83. a. .000057 **b.** .015043 **85.** 11% **87.** 106 **89.** .631

91. Answers will vary. Here is a simple one: E: the first toss is a head, F: the second toss is a head, G: the third toss is a head.

93. The probability you seek is $P(E \mid F)$, or should be. If, for example, you were going to place a wager on whether E occurs or not, it is crucial to know that the sample space has been reduced to F (you know that F did occur). If you base your wager on $P(E)$ rather than $P(E \mid F)$ you will misjudge your likelihood of winning. **95.** If $A \subseteq B$ then $A \cap B = A$, so $P(A \cap B) = P(A)$ and $P(A \mid B) = P(A \cap B)/P(B) = P(A)/P(B)$. **97.** Your friend is correct. If A and B are mutually exclusive then $P(A \cap B) = 0$. On the other hand, if A and B are independent then $P(A \cap B) = P(A)P(B)$. Thus, $P(A)P(B) = 0$. If a product is 0 then one of the factors must be 0, so either $P(A) = 0$ or $P(B) = 0$. Thus, it cannot be true that A and B are mutually exclusive, have nonzero probabilities, and are independent all at the same time. **99.** $P(A' \cap B') = 1 - P(A \cup B) = 1 - [P(A) + P(B) - P(A \cap B)] = 1 - [P(A) + P(B) - P(A)P(B)] = (1 - P(A))(1 - P(B)) = P(A')P(B')$.

Section 7.6

1. .4 **3.** .7887 **5.** .7442 **7.** .1163 **9.** 26.8% **11.** .1724

13. .73 **15.** .71 **17.** .1653 **19.** 88% **21.** 12%

23. a. 14.43%; **b.** 19.81% of single homeowners have pools. Thus they should go after the single homeowners. **25.** 9

27. .9310 **29.** 1.76% **31.** .20 **33.** K: child killed; D: Airbag deployed; $P(K \mid D) = 1.31 P(K \mid D')$; $P(D \mid K) = 1.31(.25)/[1.31(.25) + .75] = .30$ **35.** Show him an example like Example 1 of this section, where $P(T \mid A) = .95$ but $P(A \mid T) \approx .64$. **37.** Suppose that the steroid test gives 10% false negatives and that only 0.1% of the tested population uses steroids. Then the probability that an athlete uses steroids, given that he or she has tested positive, is

$$\frac{(.9)(.001)}{(.9)(.001) + (.01)(.999)} \approx .083.$$ **39.** Draw a tree in which the first branching shows which of R_1, R_2, or R_3 occurred, and the second branching shows which of T or T' then occurred. There are three final outcomes in which T occurs: $P(R_1 \cap T) = P(T \mid R_1) P(R_1)$, $P(R_2 \cap T) = P(T \mid R_2)P(R_2)$, and $P(R_3 \cap T) = P(T \mid R_3)P(R_3)$. In only one of these, the first, does R_1 occur. Thus,

$$P(R_1 \mid T) = \frac{P(R_1 \cap T)}{P(T)}$$

$$= \frac{P(T \mid R_1)P(R_1)}{P(T \mid R_1)P(R_1) + P(T \mid R_2)P(R_2) + P(T \mid R_3)P(R_3)}$$

41. The reasoning is flawed. Let A be the event that a Democrat agrees with Safire's column, and let F and M be the events that a Democrat reader is female and male respectively. Then A. D. makes the following argument:

$P(M \mid A) = 0.9, \ P(F \mid A') = 0.9.$ Therefore, $P(A \mid M) = 0.9.$

According to Bayes' Theorem we cannot conclude anything about $P(A \mid M)$ unless we know $P(A)$, the percentage of all Democrats who agreed with Safire's column. This was not given.

Section 7.7

1. $\begin{bmatrix} 1/4 & 3/4 \\ 1/2 & 1/2 \end{bmatrix}$ **3.** $\begin{bmatrix} 0 & 1 \\ 1/6 & 5/6 \end{bmatrix}$ **5.** $\begin{bmatrix} 0 & .8 & .2 \\ .9 & 0 & .1 \\ 0 & 0 & 1 \end{bmatrix}$

7. $\begin{bmatrix} 1 & 0 & 0 \\ 0 & 1 & 0 \\ 0 & 0 & 1 \end{bmatrix}$ **9.** $\begin{bmatrix} 1 & 0 & 0 & 0 & 0 & 0 \\ 2/3 & 0 & 1/3 & 0 & 0 & 0 \\ 0 & 2/3 & 0 & 1/3 & 0 & 0 \\ 0 & 0 & 2/3 & 0 & 1/3 & 0 \\ 0 & 0 & 0 & 2/3 & 0 & 1/3 \\ 0 & 0 & 0 & 0 & 1 & 0 \end{bmatrix}$

11. a. $\begin{bmatrix} .25 & .75 \\ 0 & 1 \end{bmatrix}$ **b.** distribution after one step: [.5   .5]; after two steps: [.25   .75]; after three steps: [.125   .875]

13. a. $\begin{bmatrix} .36 & .64 \\ .32 & .68 \end{bmatrix}$ **b.** distribution after one step: [.3   .7]; after two steps: [.34   .66]; after three steps: [.332   .668]

15. a. $\begin{bmatrix} 3/4 & 1/4 \\ 1/2 & 1/2 \end{bmatrix}$ **b.** distribution after one step: [2/3   1/3]; after two steps: [2/3   1/3]; after three steps: [2/3   1/3]

17. a. $\begin{bmatrix} 3/4 & 1/4 \\ 3/4 & 1/4 \end{bmatrix}$ **b.** distribution after one step: [3/4   1/4]; after two steps: [3/4   1/4]; after three steps: [3/4   1/4]

19. a. $\begin{bmatrix} .25 & .75 & 0 \\ 0 & 1 & 0 \\ 0 & .75 & .25 \end{bmatrix}$ **b.** distribution after one step: [.5   .5   0]; after two steps: [.25   .75   0]; after three steps: [.125   .875   0]   **21. a.** $\begin{bmatrix} 1/3 & 1/3 & 1/3 \\ 4/9 & 4/9 & 1/9 \\ 0 & 1 & 0 \end{bmatrix}$ **b.** distribution after one step: [1/2   1/2   0]; after two steps: [1/6   2/3   1/6]; after three steps: [7/18   7/18   2/9]

23. a. $\begin{bmatrix} .01 & .99 & 0 \\ 0 & 1 & 0 \\ 0 & .36 & .64 \end{bmatrix}$ **b.** distribution after one step: [.05   .55   .4]; after two steps: [.005   .675   .32]; after three steps: [.0005   .7435   .256]   **25.** [2/3   1/3]   **27.** [3/7   4/7]   **29.** [2/5   3/5]   **31.** [2/5   1/5   2/5]   **33.** [1/3   1/2   1/6]

35. [0   1   0]   **37.** 1 = Sorey state, 2 = C&T; $P = \begin{bmatrix} \frac{1}{2} & \frac{1}{2} \\ \frac{1}{4} & \frac{3}{4} \end{bmatrix}$;

$3/8 = .375$ **39. a.** 1 = not checked in; 2 = checked in

$P = \begin{bmatrix} .4 & .6 \\ 0 & 1 \end{bmatrix}, P^2 = \begin{bmatrix} .16 & .84 \\ 0 & 1 \end{bmatrix},$

$P^3 = \begin{bmatrix} .064 & .936 \\ 0 & 1 \end{bmatrix}$ **b.** 1 hour: .6; 2 hours: .84; 3 hours: .936

c. Eventually, all the roaches will have checked in.   **41.** 16.67% fall into the high-risk category and 83.33% into the low-risk category.   **43. a.** $47/300 \approx .156667$ **b.** 3/13   **45.** 41.67% of the customers will be in the Paid up category, 41.67% in the 0–90 days category, and 16.67% in the bad debt category.

47. a. $P = \begin{bmatrix} .729 & .271 & 0 \\ .075 & .84 & .085 \\ 0 & .304 & .696 \end{bmatrix}$ **b.** 2.3%

c. Affluent: 17.8%; Middle class: 64.3%; Poor: 18.0%
49. Long-term income distribution (top to bottom): [8.43%, 41.57%, 41.57%, 8.43%]

51. a. $P = \begin{bmatrix} .981 & .005 & .005 & .009 \\ .01 & .972 & .006 & .012 \\ .01 & .006 & .973 & .011 \\ .008 & .006 & .005 & .981 \end{bmatrix}$

b. Verizon: 29.6%, Cingular: 19.3%, AT&T: 18.1%, Other: 32.8%
c. Verizon: 30.3%, Cingular: 18.6%, AT&T: 17.6%, Other: 33.5%. The biggest gainers are Verizon and Other, each gaining 0.6%.   **53.** [1/5   1/5   1/5   1/5   1/5]
55. Answers will vary.   **57.** There are two assumptions made by Markov systems that may not be true about the stock market: the assumption that the transition probabilities do not change over time, and the assumption that the transition probability depends only on the current state.   **59.** If q is a row of Q, then by assumption $qP = q$. Thus, when we multiply the rows of Q by P, nothing changes, and $QP = Q$.   **61.** At each step, only 0.4 of the population in state 1 remains there, and nothing enters from any other state. Thus, when the first entry in the steady-state distribution vector is multiplied by 0.4 it must remain unchanged. The only number for which this true is 0.
63. An example is

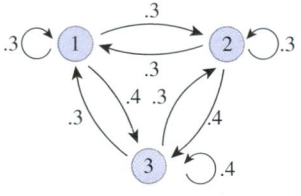

65. If $vP = v$ and $wP = w$, then $\frac{1}{2}(v + w)P = \frac{1}{2}vP + \frac{1}{2}wP = \frac{1}{2}v + \frac{1}{2}w = \frac{1}{2}(v + w)$. Further, if the entries of v and w add up to 1, then so do the entries of $(v + w)/2$.

Chapter 7 Review

1. $n(S) = 8$, $E = $ {HHT, HTH, HTT, THH, THT, TTH, TTT}, $P(E) = 7/8$ **3.** $n(S) = 36$; $E = $ {(1, 6), (2, 5), (3, 4), (4, 3), (5, 2), (6, 1)}; $P(E) = 1/6$ **5.** $n(S) = 6$; $E = $ {2}; $P(E) = 1/8$ **7.** .76   **9.** .25   **11.** .5   **13.** 7/15   **15.** 8/792   **17.** 48/792   **19.** 288/792   **21.** $C(8, 5)/C(52, 5)$ **23.** $C(4, 3)C(1, 1)C(3, 1)/C(52, 5)$ **25.** $C(9, 1)C(8, 1)C(4, 3)C(4, 2)/C(52, 5)$ **27.** 1/5; dependent   **29.** 1/6; independent   **31.** 1; dependent

33. $P = \begin{bmatrix} 1/2 & 1/2 \\ 1/4 & 3/4 \end{bmatrix}$

35. Brand A: $65/192 \approx .339$, Brand B: $127/192 \approx .661$
37. 14/25 **39.** 15/94 **41.** 79/167 **43.** 98%
45. .931 **47.** 0.75% **49.** .0049 **51.** 40% for
OHaganBooks.com, 26% for JungleBooks.com, and 34% for
FarmerBooks.com **53.** Here are three: (1) it is possible for
someone to be a customer at two different enterprises; (2) some
customers may stop using all three of the companies; (3) new
customers can enter the field.

Chapter 8

Section 8.1

1. Finite; $\{2, 3, \ldots, 12\}$ **3.** Discrete infinite; $\{0, 1, -1,$
$2, -2, \ldots\}$ (negative profits indicate loss) **5.** Continuous; X
can assume any value between 0 and 60. **7.** Finite; $\{0, 1,$
$2, \ldots, 10\}$ **9.** Discrete infinite $\{k/1, k/4, k/9, k/16, \ldots\}$
11. a. $S = \{HH, HT, TH, TT\}$
b. X is the rule that assigns to each outcome the number of tails.
c.

Outcome	HH	HT	TH	TT
Value of X	0	1	1	2

13. a. $S = \{(1, 1), (1, 2), \ldots, (1, 6), (2, 1), (2, 2), \ldots, (6, 6)\}$
b. X is the rule that assigns to each outcome the sum of the two
numbers.
c.

Outcome	(1, 1)	(1, 2)	(1, 3)	...	(6, 6)
Value of X	2	3	4	...	12

15. a. $S = \{(4, 0), (3, 1), (2, 2)\}$ (listed in order (red, green))
b. X is the rule that assigns to each outcome the number of red
marbles.
c.

Outcome	(4, 0)	(3, 1)	(2, 2)
Value of X	4	3	2

17. a. $S =$ the set of students in the study group
b. X is the rule that assigns to each student his or her final exam
score. **c.** The values of X, in the order given, are 89%, 85%, 95%,
63%, 92%, 80%. **19. a.** $P(X = 8) = P(X = 6) = .3$ **b.** .7

21.

x	1	2	3	4	5	6
$P(X = x)$	$\frac{1}{6}$	$\frac{1}{6}$	$\frac{1}{6}$	$\frac{1}{6}$	$\frac{1}{6}$	$\frac{1}{6}$

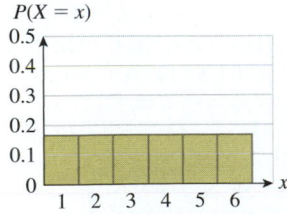

23.

x	0	1	4	9
$P(X = x)$	$\frac{1}{8}$	$\frac{3}{8}$	$\frac{3}{8}$	$\frac{1}{8}$

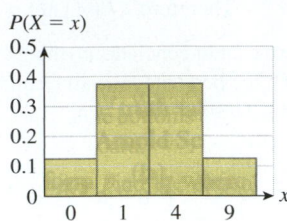

25.

x	2	3	4	5	6	7	8	9	10	11	12
$P(X = x)$	$\frac{1}{36}$	$\frac{2}{36}$	$\frac{3}{36}$	$\frac{4}{36}$	$\frac{5}{36}$	$\frac{6}{36}$	$\frac{5}{36}$	$\frac{4}{36}$	$\frac{3}{36}$	$\frac{2}{36}$	$\frac{1}{36}$

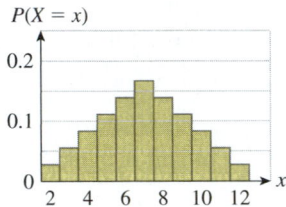

27.

x	1	2	3	4	5	6
$P(X = x)$	$\frac{1}{36}$	$\frac{3}{36}$	$\frac{5}{36}$	$\frac{7}{36}$	$\frac{9}{36}$	$\frac{11}{36}$

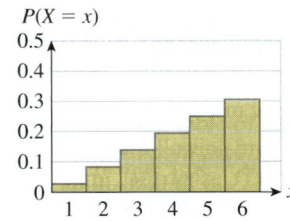

29. a. 2000, 3000, 4000, 5000, 6000, 7000, 8000 (7000 is
optional)
b.

x	2000	3000	4000	5000	6000	7000	8000
Freq.	2	1	1	1	2	0	3
$P(X = x)$	.2	.1	.1	.1	.2	.0	.3

c. $P(X \le 5000) = .5$ **31.** The random variable is $X =$ mold
count on a given day

X	750	2250	3750	5250
$P(X = x)$	11/16	2/16	1/16	2/16

33. a.

x	10,000	30,000	50,000	70,000	90,000
$P(X = x)$	.27	.28	.20	.15	.10

b. .25 Histogram:

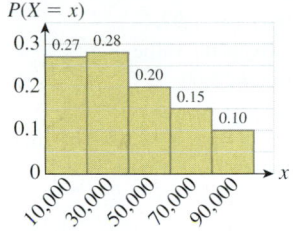

$P(X = x)$

35.

Class	1.1 − 2.0	2.1 − 3.0	3.1 − 4.0
Freq.	4	7	9

x	1.5	2.5	3.5
$P(X = x)$	.20	.35	.45

37. 95.5%

39.

x	3	2	1	0
$P(X = x)$	.0625	.6875	.125	.125

41. .75 The probability that a randomly selected small car is rated Good or Acceptable is .75.   **43.** $P(Y \geq 2) = .50$, $P(Z \geq 2) \approx .53$, suggesting that medium SUVs are safer than small SUVs in frontal crashes   **45.** Small cars   **47.** .375

49.

x	1	2	3	4
$P(X = x)$	$\frac{4}{35}$	$\frac{18}{35}$	$\frac{12}{35}$	$\frac{1}{35}$

$P(X \geq 2) = 31/35 \approx .886$

51. Answers will vary.   **53.** No; for instance, if X is the number of times you must toss a coin until heads comes up, then X is infinite but not continuous.   **55.** By measuring the values of X for a large number of outcomes, and then using the estimated probability (relative frequency)   **57.** Here is an example: let X be the number of days a diligent student waits before beginning to study for an exam scheduled in 10 days' time.   **59.** Answers may vary. If we are interested in exact page-counts, then the number of possible values is very large and the values are (relatively speaking) close together, so using a continuous random variable might be advantageous. In general, the finer and more numerous the measurement classes, the more likely it becomes that a continuous random variable could be advantageous.

Section 8.2

1. .0729   **3.** .59049   **5.** .00001   **7.** .99144   **9.** .00856
11. .27648   **13.** .54432   **15.** .04096   **17.** .77414

19. $P(X = x)$

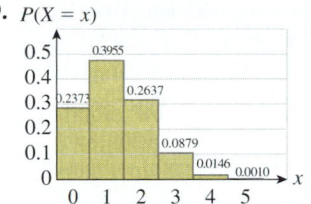

21. $P(X = x)$

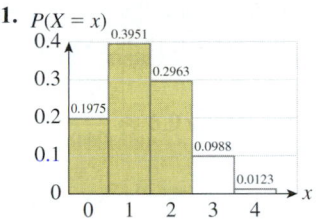

$P(X \leq 2) = .8889$

23. .2637   **25.** .8926   **27.** .875   **29. a.** .0081 **b.** .08146
31. .41   **33. a.** .0519 **b.** Probability distribution (entries rounded to 4 decimal places):

x	0	1	2	3	4
$P(X = x)$	.6648	.2770	.0519	.0058	.0004
5	6	7	8	9	10
.0000	.0000	.0000	.0000	.0000	.0000

c. None   **35.** .000298   **37.** .8321   **39. a.** 21 **b.** 20 **c.** The graph for $n = 50$ trials is more widely distributed than the graph for $n = 20$.   **41.** 69 trials   **43.** $.562 \times 10^{-5}$ **45.** .0266 Because there is only a 2.66% chance of detecting the disease in a given year, the government's claim seems dubious.   **47.** No; in a sequence of Bernoulli trials, the occurrence of one success does not affect the probability of success on the next attempt.   **49.** No; if life is a sequence of Bernoulli trials, then the occurrence of one misfortune ("success") does not affect the probability of a misfortune on the next trial. Hence, misfortunes may very well not "occur in threes."   **51.** The probability of selecting a red marble changes after each selection, as the number of marbles left in the bag decreases. This violates the requirement that, in a sequence of Bernoulli trials, the probability of "success" does not change.

Section 8.3

1. $\bar{x} = 6$, median = 5, mode = 5
3. $\bar{x} = 3$, median = 3.5, mode = −1
5. $\bar{x} = -0.1875$, median = 0.875, every value is a mode
7. $\bar{x} = 0.2$, median = −0.1, mode = −0.1   **9.** Answers may vary. Two examples are: 0, 0, 0, 0, 0, 6 and 0, 0, 0, 1, 2, 3
11. 0.9   **13.** 21   **15.** −0.1   **17.** 3.5   **19.** 1   **21.** 4.472
23. 2.667   **25.** 2   **27.** 0.385   **29.** $\bar{x} = 5000$, $m = 5500$; 5500
31. $\bar{x} = \$426$, median = \$425.50, mode = \$425. Over the 10-business day period sampled, the price of gold averaged \$426 per ounce. It was above \$425.50 as many times as it was below that, and stood at \$425 per ounce for most of the days sampled.

33. a. 6.5; There were an average of 6.5 checkout lanes in a supermarket that was surveyed. **b.** $P(X < \mu) = .42$; $P(X > \mu) = .58$, and is thus larger. Most supermarkets have more than the average number of checkout lanes.

35.

X	5	10	15	20	25	35
P(X)	.17	.33	.21	.19	.03	.07

$E(X) = 14.3$;

The average age of a school goer in 1998 was 14.3.   **37.** 1687.5
39. $41,000

41.

x	3	2	1	0
P(X = x)	.0625	.6875	.125	.125

$E(X) = 1.6875$

y	3	2	1	0
P(Y = y)	.1	.4	.4	.1

$E(Y) = 1.5$

Small cars
43. Large cars   **45.** Expect to lose 5.3¢.   **47.** 25.2 students
49. a. 2 defective airbags **b.** 120 airbags

51.

x	1	2	3	4
P(X = x)	$\frac{4}{35}$	$\frac{18}{35}$	$\frac{12}{35}$	$\frac{1}{35}$

$E(X) = 16/7 \approx 2.2857$ tents

53. FastForward: 3.97%; SolidState: 5.51%; SolidState gives the higher expected return.   **55.** A loss of $29,390   **57.** (A)   **59.** He is wrong; for example, the collection 0, 0, 300 has mean 100 and median 0.   **61.** No. The expected number of times you will hit the dart-board is the average number of times you will hit the bull's eye per 50 shots; the average of a set of whole numbers need not be a whole number.   **63.** Wrong. It might be the case that only a small fraction of people in the class scored better than you but received exceptionally high scores that raised the class average. Suppose, for instance, that there are 10 people in the class. Four received 100%, you received 80%, and the rest received 70%. Then the class average is 83%, 5 people have lower scores than you, but only four have higher scores.   **65.** No; the mean of a very large sample is only an *estimate* of the population mean. The means of larger and larger samples *approach* the population mean as the sample size increases.   **67.** Wrong. The statement attributed to President Bush asserts that the mean tax refund would be $1000, whereas the statements referred to as "The Truth" suggest that the *median* tax refund would be close to $100 (and that the 31st percentile would be zero).   **69.** Select a U.S. household at random, and let X be the income of that household. The expected value of X is then the population mean of all U.S. household incomes.

Section 8.4

1. $s^2 = 29$; $s = 5.39$ **3.** $s^2 = 12.4$; $s = 3.52$
5. $s^2 = 6.64$; $s = 2.58$ **7.** $s^2 = 13.01$; $s = 3.61$ **9.** 1.04
11. 9.43   **13.** 3.27   **15.** Expected value $= 3.5$, variance $= 2.918$, standard deviation $= 1.71$ **17.** Expected value $= 1$, variance $= 0.5$, standard deviation $= 0.71$
19. Expected value $= 4.47$, variance $= 1.97$, standard deviation $= 1.40$ **21.** Expected value $= 2.67$, variance $= 0.36$, standard deviation $= 0.60$
23. Expected value $= 2$, variance $= 1.8$, standard deviation $= 1.34$ **25. a.** $\bar{x} = 3$, $s = 3.54$ **b.** $[0, 6.54]$ We must assume that the population distribution is bell-shaped and symmetric.   **27. a.** $\bar{x} = 5.0$, $s = 0.6$ **b.** 3.8, 6.2
29. a. $\bar{x} = 5000$, $s \approx 2211$ **b.** 2789, 7211, 60%   **31. a.** 2.18 **b.** $[11.22, 24.28]$ **c.** 100%; Empirical rule
33. $\mu = 1.5$, $\sigma = 1.43$; 100%   **35.** $\mu = 40.6$, $\sigma \approx 26$; $52,000   **37. a.** $\mu \approx 30.2$ yrs. old, $\sigma = 11.78$ years **b.** 18–42
39. At most 6.25%   **41.** At most; 12.5%   **43. a.** $\mu = 25.2$, $\sigma = 3.05$ **b.** 31   **45. a.** $\mu = 780$, $\sigma \approx 13.1$ **b.** 754, 806
47. a. $\mu = 6.5$, $\sigma^2 = 4.0$, $\sigma = 2.0$ **b.** $[2.5, 10.5]$; 3 checkout lanes   **49.** $10,700 or less   **51.** $65,300 or more   **53.** U.S.
55. U.S.   **57.** 16%   **59.** 0–$76,000   **61.** $\mu = 12.56\%$, $\sigma \approx 1.8885\%$ **63.** 78%; The empirical rule predicts 68%. The associated probability distribution is roughly bell-shaped but not symmetric.   **65.** 96%. Chebyshev's rule is valid, since it predicts that *at least* 75% of the scores are in this range.   **67.** (B), (D)
69. The sample standard deviation is bigger; the formula for sample standard deviation involves division by the smaller term $n - 1$ instead of n, which makes the resulting number larger.
71. The grades in the first class were clustered fairly close to 75. By Chebyshev's inequality, at least 88% of the class had grades in the range 60–90. On the other hand, the grades in the second class were widely dispersed. The second class had a much wider spread of ability than did the first class.   **73.** The variable must take on only the value 10, with probability 1.   **75.** $(y - x)/2$

Section 8.5

1. .1915   **3.** .5222   **5.** .6710   **7.** .2417   **9.** .8664
11. .8621   **13.** .2286   **15.** .3830   **17.** .5028   **19.** .35   **21.** .05
23. .3830   **25.** .6687   **27.** 26%   **29.** 29,600,000   **31.** 0
33. About 6680   **35.** 28%   **37.** 5%   **39.** The U.S.   **41.** Wechsler. Because this test has a smaller standard deviation, a greater percentage of scores fall within 20 points of the mean.   **43.** This is surprising, because the time between failures was more than 5 standard deviations away from the mean, which happens with an extremely small probability.   **45.** .6103   **47.** .6103 × .5832 ≈ .3559
49. .6255   **51.** .7257   **53.** .8708   **55.** .0029   **57.** Probability that a person will say Goode $= .54$. Probability that Goode polls more than 52% $\approx .8925$.   **59.** 23.4   **61.** When the distribution is normal   **63.** Neither. They are equal.   **65.** $1/(b - a)$
67. A normal distribution with standard deviation 0.5, because it is narrower near the mean, but must enclose the same amount of area as the standard curve, and so it must be higher.

Chapter 8 Review

1.

x	0	1	2
$P(X = x)$	1/4	1/2	1/4

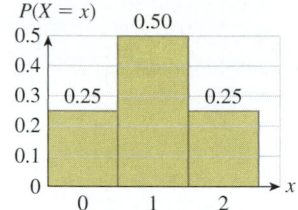

3.

x	15	25	35	45
$P(X = x)$	.482	.386	.116	.016

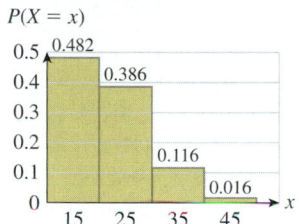

5.

x	0	1	2
$P(X = x)$	9/16	6/16	1/16

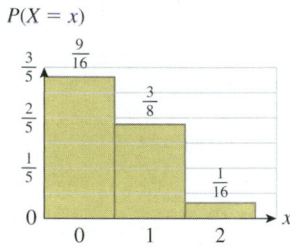

7. Two examples are: 0, 0, 0, 4 and −1, −1, 1, 5   **9.** An example is −1, −1, −1, 1, 1, 1   **11.** .4165   **13.** .3232   **15.** .7330

17.

x	−3	−2	−1	0	1	2	3
$P(X = x)$	1/16	2/16	3/16	4/16	3/16	2/16	1/16

$\mu = 0$, $\sigma = 1.5811$; within 1.3 standard deviations of the mean.
19. [40, 160]   **21.** .0668   **23.** .7888   **25.** .0000
27. a. $27,210 **b.** False; let X = price and Y = weekly sales. Then weekly Revenue = XY. However, $27,210 \neq 12.15 \times 2,620$. In other words, $E(XY) \neq E(X)E(Y)$.   **29.** Between 2.431 and 7.569 orders per million residents. The empirical rule does not apply because the distribution is not symmetric.   **31.** .190   **33.** .060   **35.** 2.5   **37.** .873   **39.** Using normal distribution table: 364,000 people. More accurate answer: 378,000 people   **41.** 148

Chapter 9

Section 9.1

1. Vertex: $(-3/2, -1/4)$; y-intercept: 2; x-intercepts: $-2, -1$

3. Vertex: $(2,0)$; y-intercept: -4; x-intercept: 2

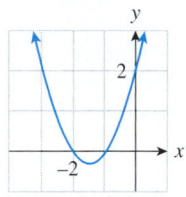

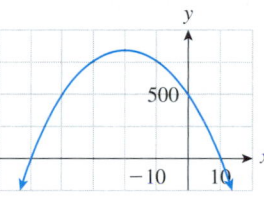

5. Vertex: $(-20, 900)$; y-intercept: 500; x-intercepts: $-50, 10$

7. Vertex: $(-1/2, -5/4)$; y-intercept: -1; x-intercepts: $-1/2 \pm \sqrt{5}/2$

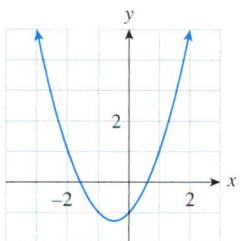

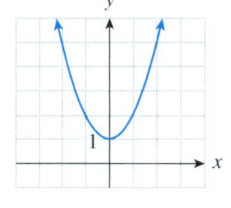

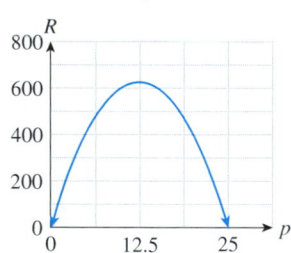

9. Vertex: $(0, 1)$; y-intercept: 1; No x-intercepts

11. $R = -4p^2 + 100p$; Maximum revenue when $p = \$12.50$

13. $R = -2p^2 + 400p$; Maximum revenue when $p = \$100$

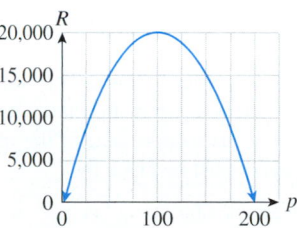

15. $y = -0.7955x^2 + 4.4591x - 1.6000$
17. $y = -1.1667x^2 - 6.1667x - 3.0000$
19. a. Positive because the data suggest a curve that is concave up.   **b.** (C)   **c.** 1995. The parabola rises to the left of the vertex and thus predicts increasing trade as we go back in time, contradicting history.

21. 1985 ($t = 15$); 3525 pounds

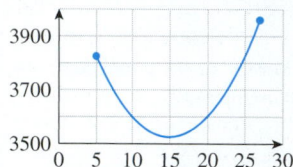

23. 5000 pounds. The model is not trustworthy for vehicle weights larger than 5000 pounds, because it predicts increasing fuel economy with increasing weight, and 5000 is close to the upper limit of the domain of the function.
25. Maximum revenue when $p = \$140$, $R = \$9800$
27. Maximum revenue with 70 houses, $R = \$9,800,000$
29. a. $q = -560x + 1400$; $R = -560x^2 + 1400x$
b. $P = -560x^2 + 1400x - 30$; $x = \$1.25$; $P = \$845$ per month
31. $C = -200x + 620$; $P = -400x^2 + 1400x - 620$
$x = \$1.75$ per log-on; $P = \$605$ per month
33. a. $q = -10p + 400$ **b.** $R = -10p^2 + 400p$
c. $C = -30p + 4200$
d. $P = -10p^2 + 430p - 4200$; $p = \$21.50$
35. $C(t) = 2.7t^2 - 4.5t + 50$; $\$120.2$ billion, which agrees with the actual value to the nearest $1 billion.
37. a. $S(t) = -12.27t^2 + 227.23t + 64.39$

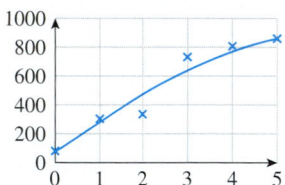

b. 986,000 units **c.** Mathematical regression cannot reliably be used to make predictions about sales. (Answers will vary.)
39. The x-coordinate of the vertex represents the unit price that leads to the maximum revenue, the y-coordinate of the vertex gives the maximum possible revenue, the x-intercepts give the unit prices that result in zero revenue, and the y-intercept gives the revenue resulting from zero unit price (which is obviously zero).
41. Graph the data to see whether the points suggest a curve rather than a straight line. If the curve suggested by the graph is concave up or concave down, then a quadratic model would be a likely candidate.   **43.** If $q = mp + b$ (with $m < 0$), then the revenue is given by $R = pq = mp^2 + bp$. This is the equation of a parabola with $a = m < 0$, and so is concave down. Thus the vertex is the highest point on the parabola, showing that there is a single highest value for R, namely, the y-coordinate of the vertex.
45. Because $R = pq$, the demand must be given by
$$q = \frac{R}{p} = \frac{-50p^2 + 60p}{p} = -50p + 60.$$

Section 9.2

1. `4^x`

x	-3	-2	-1	0	1	2	3
$f(x)$	$\frac{1}{64}$	$\frac{1}{16}$	$\frac{1}{4}$	1	4	16	64

3. `3^(-x)`

x	-3	-2	-1	0	1	2	3
$f(x)$	27	9	3	1	$\frac{1}{3}$	$\frac{1}{9}$	$\frac{1}{27}$

5. `2*2^x or 2*(2^x)`

x	-3	-2	-1	0	1	2	3
$g(x)$	$\frac{1}{4}$	$\frac{1}{2}$	1	2	4	8	16

7. `-3*2^(-x)`

x	-3	-2	-1	0	1	2	3
$h(x)$	-24	-12	-6	-3	$-\frac{3}{2}$	$-\frac{3}{4}$	$-\frac{3}{8}$

9. `2^x-1`

x	-3	-2	-1	0	1	2	3
$r(x)$	$-\frac{7}{8}$	$-\frac{3}{4}$	$-\frac{1}{2}$	0	1	3	7

11. `2^(x-1)`

x	-3	-2	-1	0	1	2	3
$s(x)$	$\frac{1}{16}$	$\frac{1}{8}$	$\frac{1}{4}$	$\frac{1}{2}$	1	2	4

13.   **15.**   **17.**

 $y = 3^{-x}$ $y = 2(2^x)$ $y = -3(2^{-x})$

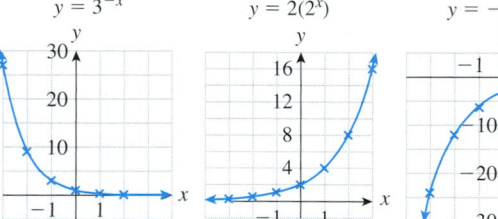

19. Both; $f(x) = 4.5(3^x)$, $g(x) = 2(1/2)^x$, or $2(2^{-x})$
21. Neither   **23.** g; $g(x) = 4(0.2)^x$
25. `e^(-2*x) or EXP(-2*x)`

x	-3	-2	-1	0	1	2	3
$f(x)$	403.4	54.60	7.389	1	0.1353	0.01832	0.002479

27. `1.01*2.02^(-4*x)`

x	-3	-2	-1	0	1	2	3
$h(x)$	4662	280.0	16.82	1.01	0.06066	0.003643	0.0002188

29. `50*(1+1/3.2)^(2*x)`

x	-3	-2	-1	0	1	2	3
$r(x)$	9.781	16.85	29.02	50	86.13	148.4	255.6

31. `2^(x-1)`; **not** `2^x-1`   **33.** `2/(1-2^(-4*x))`;
not `2/1-2^-4*x`; **not** `2/1-2^(-4*x)`
35. `(3+x)^(3*x)/(x+1) or ((3+x)^(3*x))/(x+1)`;
not `(3+x)^(3*x)/x+1`; **not** `(3+x^(3*x))/(x+1)`
37. `2*e^((1+x)/x) or 2*EXP((1+x)/x)`; **not**
`2*e^1+x/x`; **not** `2*e^(1+x)/x`; **not** `2*EXP(1+x)/x`

39.

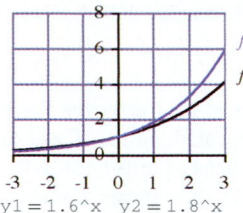

-3 -2 -1 0 1 2 3
y1 = 1.6^x   y2 = 1.8^x

41.

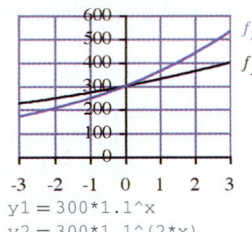

-3 -2 -1 0 1 2 3
y1 = 300*1.1^x
y2 = 300*1.1^(2*x)

43.

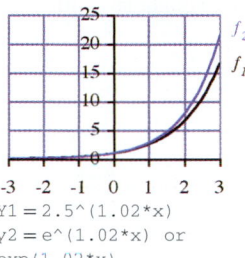

-3 -2 -1 0 1 2 3
Y1 = 2.5^(1.02*x)
y2 = e^(1.02*x) or
exp(1.02*x)

45.

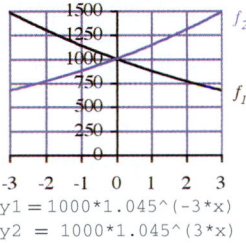

-3 -2 -1 0 1 2 3
y1 = 1000*1.045^(-3*x)
y2 = 1000*1.045^(3*x)

47. $f(x) = 500(0.5)^x$ **49.** $f(x) = 10(3)^x$
51. $f(x) = 500(0.45)^x$ **53.** $f(x) = -100(1.1)^x$
55. $y = 4(3^x)$ **57.** $y = -1(0.2^x)$ **59.** $y = 2.1213(1.4142^x)$
61. $y = 3.6742(0.9036^x)$ **63.** $f(t) = 5000e^{0.10t}$
65. $f(t) = 1000e^{-0.063t}$ **67.** $y = 1.0442(1.7564)^x$
69. $y = 15.1735(1.4822)^x$ **71.** $y = 1000(2^{t/3})$;
65,536,000 bacteria after 2 days   **73.** $A(t) = 5000(1.0439)^t$;
£6198   **75.** At the beginning of 2014   **77.** 31.0 grams,
9.25 grams, 2.76 grams   **79.** 20,000 years   **81.** 53 mg
83. a. $P = 40t + 360$ **b.** $P = 360(1.1006)^t$. Neither model is
applicable.   **85. a.** $P = 180(1.01121)^t$ million **b.** 4 decimal
places **c.** 351 million   **87. a.** $y = 50,000(1.5^{t/2})$, t = time in
years since two years ago **b.** 91,856 tags   **89.** $491.82

91. a.

Year	1950	2000	2050	2100
$C(t)$ parts per million	561	669	799	953

b. 2010 ($t = 260$)
93. a. $P(t) = 0.339(1.169)^t$.   **95. a.** $y = 5.4433(1.0609)^t$.
Graph: Graph:

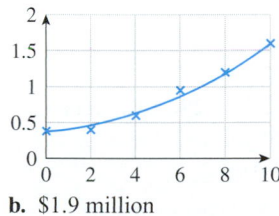

b. $1.9 million

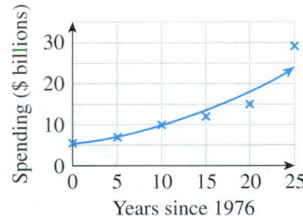

b. 609%   **c.** $16 billion

97. (B)   **99.** Exponential functions of the form $f(x) = A(b^x)$ ($b > 0$) increase rapidly for large values of x. In real-life situations, such as population growth, this model is reliable only for relatively short periods of growth. Eventually, population growth tapers off because of pressures such as limited resources and overcrowding.   **101.** Linear functions better: cost models where there is a fixed cost and a variable cost; simple interest, where interest is paid on the original amount invested. Exponential models better: compound interest, population growth. (In both of these, the rate of growth depends on the present number of items, rather than on some fixed quantity.)   **103.** Take the ratios y_2/y_1 and y_3/y_2. If they are the same, the points fit on an exponential curve.   **105.** This reasoning is suspect—the bank need not use its computer resources to update all the accounts every minute, but can instead use the continuous compounding formula to calculate the balance in any account at any time.

Section 9.3

1.

Logarithmic Form	$\log_{10} 10,000 = 4$	$\log_4 16 = 2$	$\log_3 27 = 3$	$\log_5 5 = 1$	$\log_7 1 = 0$	$\log_4 \frac{1}{16} = -2$

3.

Exponential Form	$(0.5)^2 = 0.25$	$5^0 = 1$	$10^{-1} = 0.1$	$4^3 = 64$	$2^8 = 256$	$2^{-2} = \frac{1}{4}$

5. 1.4650   **7.** -1.1460 **9.** -0.7324 **11.** 6.2657
13.   **15.**

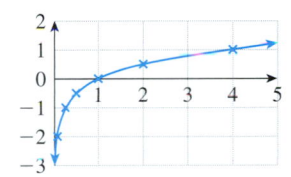   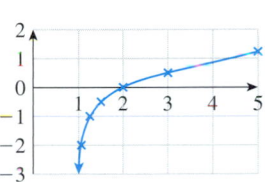

17.

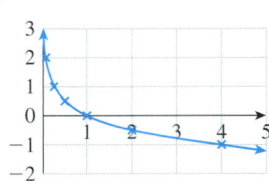

19. $Q = 1000e^{-t \ln 2}$ **21.** $Q = 1000e^{t(\ln 2)/2}$ **23.** Doubling
time $= 2 \ln 2$ **25.** Half-life $= (\ln 2)/4$
27. $f(x) = 4(7.389)^x$ **29.** $f(t) = 2.1e^{0.0009995t}$
31. $f(t) = 10e^{-0.01309}$ **33.** 3.36 years   **35.** 11 years
37. 23.1%   **39.** 63,000 years old   **41.** 8 years
43. 151 months   **45.** 12 years   **47.** 13.08 years
49. 1600 years   **51. a.** $b = 3^{1/6} \approx 1.20$ **b.** 3.8 months
53. a. $Q(t) = Q_0e^{-0.139t}$ **b.** 3 years   **55.** 2360 million years
57. 3.2 hours   **59.** 3.89 days   **61. a.** $P(t) =$
$6.591 \ln t - 17.69$ **b.** 1 digit **c.** (A)   **63.** $M(t) =$
$11.622 \ln t - 7.1358$. The model is unsuitable for large values
of t since, for sufficiently large values of t, $M(t)$ will eventually
become larger than 100%.   **65. a.** About 1.259×10^{24} ergs
b. about 2.24% **d.** 1000   **67. a.** 75 dB, 69 dB, 61 dB
b. $D = 95 - 20 \log r$ **c.** 57,000 feet   **69.** The logarithm of a
negative number, were it defined, would be the power to which a
base must be raised to give that negative number. But raising a

base to a power never results in a negative number, so there can be no such real number as the logarithm of a negative number.
71. $\log_4 y$ **73.** 8   **75.** x **77.** Any logarithmic curve $y = \log_b t + C$ will eventually surpass 100%, and hence not be suitable as a long-term predictor of market share.   **79.** Time is increasing logarithmically with population; Solving $P = Ab^t$ for t gives $t = \log_b(P/A) = \log_b P - \log_b A$, which is of the form $t = \log_b P + C$.

Section 9.4

1. $N = 7, A = 6, b = 2$;
    7/(1+6*2^-x)

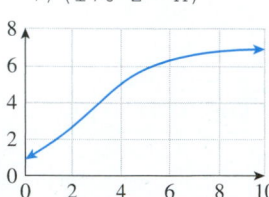

3. $N = 10, A = 4, b = 0.3$;
    10/(1+4*0.3^-x)

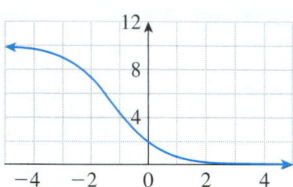

5. $N = 4, A = 7, b = 1.5$;
    4/(1+7*1.5^-x)

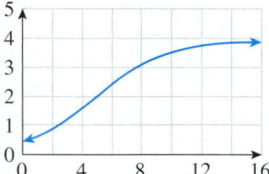

7. $f(x) = \dfrac{200}{1 + 19(2^{-x})}$ **9.** $f(x) = \dfrac{6}{1 + 2^{-x}}$

11. (B)   **13.** (B)   **15.** (C)

17. $y = \dfrac{7.2}{1 + 2.4(1.05)^{-x}}$ **19.** $y = \dfrac{97}{1 + 2.2(0.942)^{-x}}$

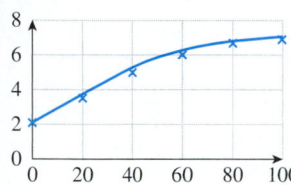

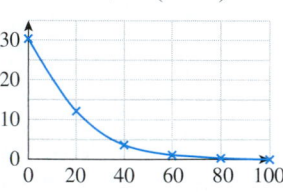

21. a. (A) **b.** 20% per year   **23. a.** 91% **b.** $P(x) \approx$ 14.33(1.05)^x **c.** \$38,000   **25.** $N(t) = \dfrac{10,000}{1 + 9(1.25)^{-t}}$;
$N(7) \approx 3463$ cases   **27.** $N(t) = \dfrac{3000}{1 + 29(2^{1/5})^{-t}}$;
$t = 16$ days   **29. a.** $A(t) = \dfrac{6.3}{1 + 4.8(1.2)^{-t}}$; 6300 articles
b. 5200 articles   **31. a.** $N(t) = \dfrac{82.8}{1 + 21.8(7.14)^{-t}}$. The model predicts that book sales will level off at around 82.8 million books per year. **b.** Not consistent; 15% of the market is represented by more than double the predicted value. This shows the

difficulty in making long-term predictions from regression models obtained from a small amount of data. **c.** 2001
33. $N(t) = \dfrac{5}{1 + 1.080(1.056)^{-t}}$; $t = 17$, or 2010.   **35.** Just as diseases are communicated via the spread of a pathogen (such as a virus), new technology is communicated via the spread of information (such as advertising and publicity). Further, just as the spread of a disease is ultimately limited by the number of susceptible individuals, so the spread of a new technology is ultimately limited by the size of the potential market.   **37.** It can be used to predict where the sales of a new commodity might level off.

Chapter 9 Review

1.

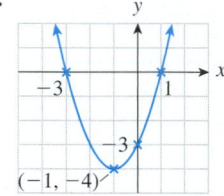

3. $f: f(x) = 5(1/2)^x$, or $5(2^{-x})$

5.   **7.**

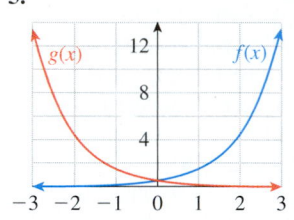

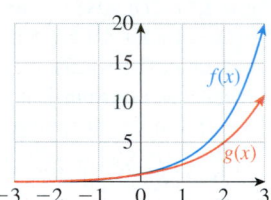

9. \$3484.85   **11.** \$3705.48   **13.** \$3485.50
15. $f(x) = 4.5(9^x)$ **17.** $f(x) = \dfrac{2}{3}3^x$ **19.** $-\dfrac{1}{2}\log_3 4$
21. $\dfrac{1}{3}\log 1.05$

23.

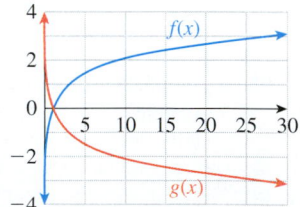

25. $Q = 5e^{-0.00693t}$ **27.** $Q = 2.5e^{0.347t}$ **29.** 10.2 years
31. 10.8 years   **33.** $f(x) = \dfrac{900}{1 + 8(1.5)^{-x}}$
35. $f(x) = \dfrac{20}{1 + 3(0.8)^{-x}}$ **37. a.** \$8500 per month; an average of approximately 2100 hits per day **b.** \$29,049 per month **c.** The fact that -0.000005, the coefficient of c^2, is negative.   **39.** $R = -60p^2 + 950p$; $p = \$7.92$ per novel, Monthly revenue $= \$3760.42$ **41. a.** 10, 34 **b.** About 360,000 pounds   **43.** 2008   **45.** 32.8 hours   **47.** (C)

Chapter 10

Section 10.1

1. 0 **3.** 4 **5.** Does not exist **7.** 1.5 **9.** 0.5 **11.** Diverges to $+\infty$ **13.** 0 **15.** 1 **17.** 0 **19. a.** -2 **b.** -1 **21. a.** 2 **b.** 1 **c.** 0 **d.** $+\infty$ **23. a.** 0 **b.** 2 **c.** -1 **d.** Does not exist **e.** 2 **f.** $+\infty$ **25. a.** 1 **b.** 1 **c.** 2 **d.** Does not exist **e.** 1 **f.** 2 **27. a.** 1 **b.** $+\infty$ **c.** $+\infty$ **d.** $+\infty$ **e.** not defined **f.** -1 **29. a.** -1 **b.** $+\infty$ **c.** $-\infty$ **d.** Does not exist **e.** 2 **f.** 1 **31.** 7.0; In the long term, the number of research articles in *Physics Review* written by researchers in Europe approaches 7000 per year. **33.** 470. This suggests that students whose parents earn an exceptionally large income score an average of 470 on the SAT verbal test. **35.** $\lim_{t\to 1^-} C(t) = 0.06$, $\lim_{t\to 1^+} C(t) = 0.08$, so $\lim_{t\to 1} C(t)$ does not exist. **37.** $\lim_{t\to+\infty} I(t) = +\infty$, $\lim_{t\to+\infty}(I(t)/E(t)) \approx 2.5$. In the long term, U.S. imports from China will rise without bound and be 2.5 times U.S. exports to China. In the real world, imports and exports cannot rise without bound. Thus, the given models should not be extrapolated far into the future. **39.** $\lim_{t\to+\infty} n(t) \approx 80$. Online book sales can be expected to level off at 80 million per year in the long term. **41.** To approximate $\lim_{x\to a} f(x)$ numerically, choose values of x closer and closer to, and on either side of $x = a$, and evaluate $f(x)$ for each of them. The limit (if it exists) is then the number that these values of $f(x)$ approach. A disadvantage of this method is that it may never give the exact value of the limit, but only an approximation. (However, we can make this as accurate as we like.) **43.** It is possible for $\lim_{x\to a} f(x)$ to exist even though $f(a)$ is not defined. An example is $\lim_{x\to 1} \dfrac{x^2 - 3x + 2}{x - 1}$.

45. Any situation in which there is a sudden change can be modeled by a function in which $\lim_{t\to a^+} f(t)$ is not the same as $\lim_{t\to a^-} f(t)$ One example is the value of a stock market index before and after a crash: $\lim_{t\to a^-} f(t)$ is the value immediately before the crash at time $t = a$, while $\lim_{t\to a^+} f(t)$ is the value immediately after the crash. Another example might be the price of a commodity that has suddenly increased from one level to another. **47.** An example is $f(x) = (x - 1)(x - 2)$.

Section 10.2

1. Continuous on its domain **3.** Continuous on its domain **5.** Discontinuous at $x = 0$ **7.** Discontinuous at $x = -1$ **9.** Continuous on its domain **11.** Discontinuous at $x = -1$ and 0 **13.** (A), (B), (D), (E) **15.** 0 **17.** -1 **19.** No value possible **21.** -1 **23.** Continuous on its domain **25.** Continuous on its domain **27.** Discontinuity at $x = 0$ **29.** Discontinuity at $x = 0$ **31.** Continuous on its domain **33.** Not unless the domain of the function consists of all real numbers. (It is impossible for a function to be continuous at points not in its domain.) For example, $f(x) = 1/x$ is continuous on its domain—the set of nonzero real numbers—but not at $x = 0$. **35.** True. If the graph of a function has a break in its graph at any point a, then it cannot be continuous at the point a.

37. Answers may vary.

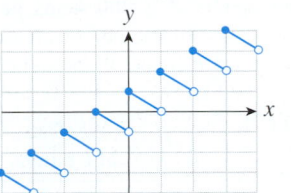

39. Answers may vary. The price of OHaganBooks.com stock suddenly drops by $10 as news spreads of a government investigation. Let $f(x) = $ Price of OHaganBooks.com stock.

Section 10.3

1. $x = 1$ **3.** 2 **5.** 1 **7.** 2 **9.** 0 **11.** 6 **13.** 4 **15.** 2 **17.** 0 **19.** 0 **21.** 12 **23.** Diverges to $+\infty$ **25.** Does not exist; left and right (infinite) limits differ **27.** 3/2 **29.** 1/2 **31.** Diverges to $+\infty$ **33.** 0 **35.** 3/2 **37.** 1/2 **39.** Diverges to $-\infty$ **41.** 0 **43.** Discontinuity at $x = 0$ **45.** Continuous everywhere **47.** Discontinuity at $x = 0$ **49.** Discontinuity at $x = 0$ **51. a.** 0.49, 1.16. Shortly before 1999, annual advertising expenditures were close to $0.49 billion. Shortly after 1999, annual advertising expenditures were close to $1.16 billion. **b.** Not continuous; Movie advertising expenditures jumped suddenly in 1999. **53.** 1.59; If the trend continues indefinitely, the annual spending on police will be 1.59 times the annual spending on courts in the long run. **55.** $\lim_{t\to+\infty} I(t) = +\infty$, $\lim_{t\to+\infty}(I(t)/E(t)) = 2.5$. In the long term, U.S. imports from China will rise without bound and be 2.5 times U.S. exports to China. In the real world, imports and exports cannot rise without bound. Thus, the given models should not be extrapolated far into the future. **57.** $\lim_{t\to+\infty} p(t) = 100$. The percentage of children who learn to speak approaches 100% as their age increases. **59.** Yes; $\lim_{t\to 8^-} C(t) = \lim_{t\to 8^+} C(t) = 1.24$. **61.** To evaluate $\lim_{x\to a} f(x)$ algebraically, first check whether $f(x)$ is a closed-form function. Then check whether $x = a$ is in its domain. If so, the limit is just $f(a)$; that is, it is obtained by substituting $x = a$. If not, then try to first simplify $f(x)$ in such a way as to transform it into a new function such that $x = a$ is in its domain, and then substitute. A disadvantage of this method is that it is sometimes extremely difficult to evaluate limits algebraically, and rather sophisticated methods are often needed. **63.** She is wrong. Closed-form functions are continuous only at points in their domains, and $x = 2$ is not in the domain of the closed-form function $f(x) = 1/(x - 2)^2$. **65.** The statement may not be true, for instance, if $f(x) = \begin{cases} x + 2 & \text{if } x < 0 \\ 2x - 1 & \text{if } x \geq 0 \end{cases}$, then $f(0)$ is defined and equals -1, and yet $\lim_{x\to 0} f(x)$ does not exist. The statement can be corrected by requiring that f be a closed-form function: "If f is a closed form function, and $f(a)$ is defined, then $\lim_{x\to a} f(x)$ exists and equals $f(a)$." **67.** Answers may vary, for example

$$f(x) = \begin{cases} 0 & \text{if } x \text{ is any number other than 1 or 2} \\ 1 & \text{if } x = 1 \text{ or 2} \end{cases}$$

Section 10.4

1. −3 **3.** 0.3 **5.** −$25,000 per month **7.** −200 items per dollar **9.** $1.33 per month **11.** 0.75 percentage point increase in unemployment per 1 percentage point increase in the deficit
13. 4 **15.** 2 **17.** 7/3
19.

	Ave. Rate of Change
h	
1	2
0.1	0.2
0.01	0.02
0.001	0.002
0.0001	0.0002

21.

	Ave. Rate of Change
h	
1	−0.1667
0.1	−0.2381
0.01	−0.2488
0.001	−0.2499
0.0001	−0.24999

23.

h	**Ave. Rate of Change**
1	9
0.1	8.1
0.01	8.01
0.001	8.001
0.0001	8.0001

25. a. −0.25 million people per year. During the period 2000–2004, employment in the U.S. decreased at an average rate of 0.25 million people per year. **b.** Zero people per year. During the period 1999–2002 the average rate of change of employment in the U.S. was zero people per year. **27. a.** 1998–2000. The number of companies that invested in venture capital each year was increasing most rapidly during the period 1998–2000, when it grew at an average rate of 650 companies per year. **b.** 1999–2001. The number of companies that invested in venture capital each year was decreasing most rapidly during the period 1999–2001, when it decreased at an average rate of 50 companies per year.
29. a. [3, 5]; −0.25 thousand articles per year. During the period 1993–1995, the number of articles authored by U.S. researchers decreased at an average rate of 250 articles per year. **b.** Percentage rate ≈ −0.1765, Average rate = −0.09 thousand articles/year. Over the period 1993–2003, the number of articles authored by U.S. researchers decreased at an average rate of 90 per year, representing a 17.65% decrease over that period. **31. a.** 75 teams per year **b.** Decreased **33. a.** 250 million transactions per year, −150 million transactions per year, 50 million transactions per year. Over the period January 2000–January 2001, the (annual) number of online shopping transactions in the U.S. increased at an average rate of 250 million per year. From January 2001 to January 2002, this number decreased at an average rate of 150 million per year. From January 2000 to January 2002, this number increased at an average rate of 50 million per year. **b.** The average rate of change of $N(t)$ over [0, 2] is the average of the rates of change over [0, 1] and [1, 2]. **35. a.** (C) **b.** (A) **c.** (B)
d. Approximately −0.0063 (to two significant digits) billion dollars per year, (−$6,300,000 per year). This is much less than the (positive) slope of the regression line, 0.0125 ≈ 0.013 billion dollars per year, ($13,000,000 per year).

37. Answers may vary

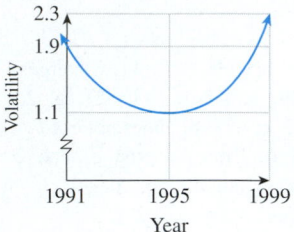

39. The index was increasing at an average rate of 300 points per day. **41.** $0.08 per year. The value of the euro in U.S. dollars was growing at an average rate of about $0.08 per year over the period June 2000–June 2004. **43. a.** 8.85 manatee deaths per 100,000 boats; 23.05 manatee deaths per 100,000 boats **b.** More boats result in more manatee deaths per additional boat. **45. a.** $305 million per year; Over the period 1997–1999, annual advertising revenues increased at an average rate of $305 million per year. **b.** (A) **c.** $590 million per year; The model projects annual advertising revenues to increase by $590 million per year in 2000.
47. a. −0.88, −0.79, −0.69, −0.60, −0.51, −0.42 **b.** For household incomes between $40,000 and $40,500, the poverty rate decreases at an average rate of 0.69 percentage points per $1000 increase in the median household income. **c.** (B) **d.** (B).
49. The average rate of change of f over an interval [a, b] can be determined numerically, using a table of values; graphically, by measuring the slope of the corresponding line segment through two points on the graph; or algebraically, using an algebraic formula for the function. Of these, the least precise is the graphical method, because it relies on reading coordinates of points on a graph.
51. Answers will vary.

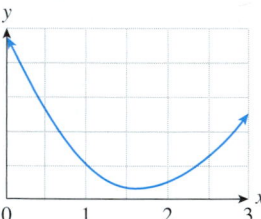

53. 6 units of quantity A per unit of quantity C **55.** (A)
57. Yes. Here is an example:

Year	2000	2001	2002	2003
Revenue ($ billion)	10	20	30	5

59. (A)

Section 10.5

1. 6 **3.** −5.5
5.

h	1	0.1	0.01
Ave. rate	39	39.9	39.99

Instantaneous Rate = 40 rupees per day

7.

h	1	0.1	0.01
Ave. rate	140	66.2	60.602

Instantaneous Rate = 60 rupees per day

9.

h	10	1
C_{ave}	4.799	4.7999

$C'(1,000) = \$4.8$ per item

11.

h	10	1
C_{ave}	99.91	99.90

$C'(100) = \$99.90$ per item

13. a. R **b.** P **15. a.** P **b.** R **17. a.** Q **b.** P **19.** $1/2$
21. 0 **23. a.** Q **b.** R **c.** P **25. a.** R **b.** Q **c.** P
27. a. $(1, 0)$ **b.** None **c.** $(-2, 1)$ **29. a.** $(-2, 0.3)$, $(0, 0)$,
$(2, -0.3)$ **b.** None **c.** None **31.** $(a, f(a))$; $f'(a)$ **33.** (B)
35. a. (A) **b.** (C) **c.** (B) **d.** (B) **e.** (C) **37.** -2 **39.** -1.5
41. -5 **43.** 16 **45.** 0 **47.** -0.0025

49. a. 3 **b.** $y = 3x + 2$ **51. a.** $\dfrac{3}{4}$ **b.** $y = \dfrac{3}{4}x + 1$

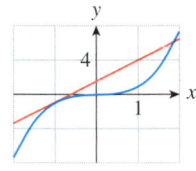   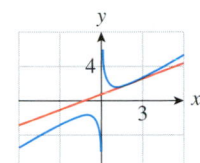

53. a. $\dfrac{1}{4}$ **b.** $y = \dfrac{1}{4}x + 1$

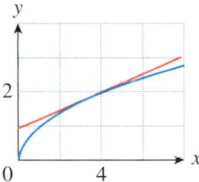

55. 1.000 **57.** 1.000 **59.** (C) **61.** (A) **63.** (F)
65.

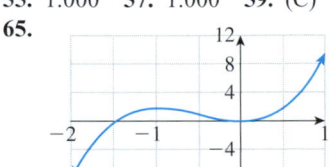

$x = -1.5, x = 0$

67. Note: Answers depend on the form of technology used.
Excel ($h = 0.1$):

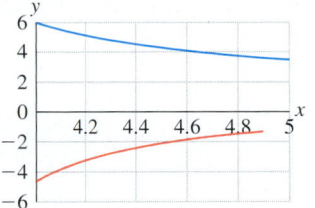

Graphs:

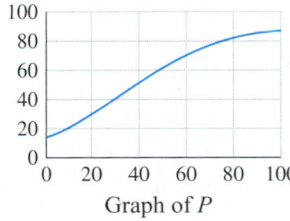

The top curve is $y = f(x)$; the bottom curve is $y = f'(x)$.

69. $q(100) = 50,000$, $q'(100) = -500$. A total of 50,000 pairs of sneakers can be sold at a price of $100, but the demand is decreasing at a rate of 500 pairs per $1 increase in the price.
71. a. Sales in 2000 were approximately 160,000 pools per year, and increasing at a rate of 6000 per year. **b.** Decreasing because the slope is decreasing. **73. a.** (B) **b.** (B) **c.** (A) **d.** 1992 **e.** 0.05. In 1996, the total number of state prisoners was increasing at a rate of approximately 50,000 prisoners per year.
75. a. -96 ft/sec **b.** -128 ft/sec **77. a.** $\$0.044$ per year. The value of the euro was increasing at an average rate of about $0.044 per year over the period January 2000–January 2004. **b.** $-\$0.10$ per year. In January, 2000, the value of the euro was decreasing at a rate of about $0.10 per year. **c.** The value of the euro was decreasing in January 2000, and then began to increase.
79. a. $305 million per year **b.** (A) **c.** $685 million/year. In December 2000, AOL's advertising revenue was projected to be increasing at a rate of $685 million per year. **81.** $A(0) = 4.5$ million; $A'(0) = 60,000$ **83. a.** 60% of children can speak at the age of 10 months. At the age of 10 months, this percentage is increasing by 18.2 percentage points per month. **b.** As t increases, p approaches 100 percentage points (all children eventually learn to speak), and dp/dt approaches zero because the percentage stops increasing. **85.** $S(5) \approx 109$, $\left.\dfrac{dS}{dt}\right|_{t=5} \approx 9.1$.
After 5 weeks, sales are 109 pairs of sneakers per week, and sales are increasing at a rate of 9.1 pairs per week each week.
87. a. $P(50) \approx 62$, $P'(50) \approx 0.96$; 62% of U.S. households with an income of $50,000 have a computer. This percentage is increasing at a rate of 0.96 percentage points per $1000 increase in household income. **b.** P' decreases toward zero.

Graphs:

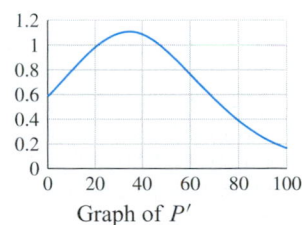

Graph of P Graph of P'

89. a. (D) **b.** 33 days after the egg was laid **c.** 50 days after the egg was laid. Graph:

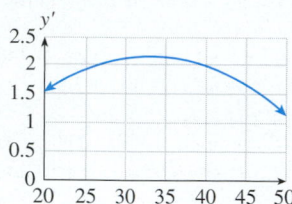

91. $L(0.95) = 31.2$ meters and $L'(0.95) = -304.2$ meters/warp. Thus, at a speed of warp 0.95, the spaceship has an observed length of 31.2 meters and its length is decreasing at a rate of 304.2 meters per unit warp, or 3.042 meters per increase in speed of 0.01 warp. **93.** The difference quotient is not defined when $h = 0$ because there is no such number as $0/0$. **95.** The derivative is positive and decreasing toward zero. **97.** Company B. Although the company is currently losing money, the derivative is positive, showing that the profit is increasing. Company A, on the other hand, has profits that are declining. **99.** (C) is the only graph in which the instantaneous rate of change on January 1 is greater than the one-month average rate of change. **101.** The tangent to the graph is horizontal at that point, and so the graph is almost horizontal near that point.

103. Answers may vary.

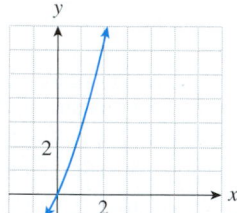

105. If $f(x) = mx + b$, then its average rate of change over any interval $[x, x + h]$ is $\dfrac{m(x + h) + b - (mx + b)}{h} = m$. Because this does not depend on h, the instantaneous rate is also equal to m. **107.** Increasing because the average rate of change appears to be rising as we get closer to 5 from the left (see the bottom row).

109. Answers may vary

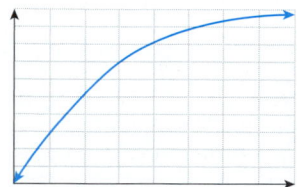

111. Answers may vary

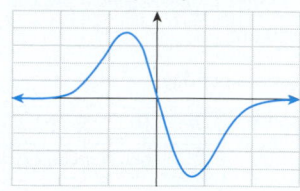

113. (B)
115. Answers may vary.

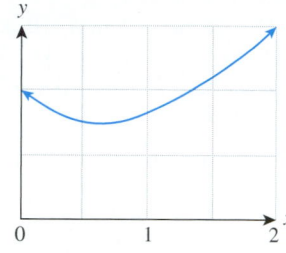

Section 10.6

1. 4 **3.** 3 **5.** 7 **7.** 4 **9.** 14 **11.** 1 **13.** m **15.** $2x$
17. 3 **19.** $6x + 1$ **21.** $2 - 2x$ **23.** $3x^2 + 2$ **25.** $1/x^2$
27. m **29.** -1.2 **31.** 30.6 **33.** -7.1 **35.** 4.25 **37.** -0.6
39. $y = 4x - 7$ **41.** $y = -2x - 4$ **43.** $y = -3x - 1$
45. $s'(t) = -32t$; $s'(4) = -128$ ft/sec **47.** Annual U.S. imports from China were increasing by \$13.5 billion per year in 2000. **49.** $R'(t) = 34t + 100$. Annual U.S. sales of bottled water were increasing by 440 million gallons per year in 2000. **51.** $f'(8) = 26.6$ manatee deaths per 100,000 boats. At a level of 800,000 boats, the number of manatee deaths is increasing at a rate of 26.6 manatees per 100,000 additional boats. **53.** The algebraic method because it gives the exact value of the derivative. The other two approaches give only approximate values (except in some special cases). **55.** Because the algebraic computation of $f'(a)$ is exact and not an approximation, it makes no difference whether one uses the balanced difference quotient or the ordinary difference quotient in the algebraic computation. **57.** The computation results in a limit that cannot be evaluated.

Section 10.7

1. $5x^4$ **3.** $-4x^{-3}$ **5.** $-0.25x^{-0.75}$
7. $8x^3 + 9x^2$ **9.** $-1 - 1/x^2$

11. $\dfrac{dy}{dx} = 10(0) = 0$ (constant multiple and power rule)

13. $\dfrac{dy}{dx} = \dfrac{d}{dx}(x^2) + \dfrac{d}{dx}(x)$ (sum rule) $= 2x + 1$ (power rule)

15. $\dfrac{dy}{dx} = \dfrac{d}{dx}(4x^3) + \dfrac{d}{dx}(2x) - \dfrac{d}{dx}(1)$ (sum and difference)

$= 4\dfrac{d}{dx}(x^3) + 2\dfrac{d}{dx}(x) - \dfrac{d}{dx}(1)$ (constant multiples)

$= 12x^2 + 2$ (power rule)

17. $f'(x) = 2x - 3$ **19.** $f'(x) = 1 + 0.5x^{-0.5}$

21. $g'(x) = -2x^{-3} + 3x^{-2}$ **23.** $g'(x) = -\dfrac{1}{x^2} + \dfrac{2}{x^3}$

25. $h'(x) = -\dfrac{0.8}{x^{1.4}}$ **27.** $h'(x) = -\dfrac{2}{x^3} - \dfrac{6}{x^4}$

29. $r'(x) = -\dfrac{2}{3x^2} + \dfrac{0.1}{2x^{1.1}}$ **31.** $r'(x) = \dfrac{2}{3} - \dfrac{0.1}{2x^{0.9}} - \dfrac{4.4}{3x^{2.1}}$

33. $t'(x) = |x|/x - 1/x^2$ **35.** $s'(x) = \dfrac{1}{2\sqrt{x}} - \dfrac{1}{2x\sqrt{x}}$

37. $s'(x) = 3x^2$ **39.** $t'(x) = 1 - 4x$ **41.** $2.6x^{0.3} + 1.2x^{-2.2}$

43. $1.2(1 - |x|/x)$ **45.** $3at^2 - 4a$ **47.** $5.15x^{9.3} - 99x^{-2}$

49. $-\dfrac{2.31}{t^{2.1}} - \dfrac{0.3}{t^{0.4}}$ **51.** $4\pi r^2$ **53.** 3   **55.** -2 **57.** -5

59. $y = 3x + 2$ **61.** $y = \dfrac{3}{4}x + 1$

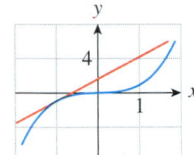   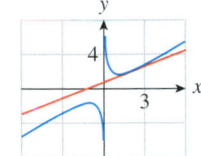

63. $y = \dfrac{1}{4}x + 1$

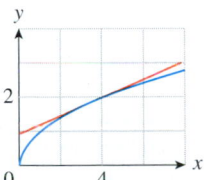

65. $x = -3/4$ **67.** No such values   **69.** $x = 1, -1$
73.   **75.**

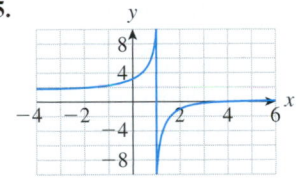

a. $x = 3$ **b.** None   **a.** $x = 1$ **b.** $x = 4.2$
77. a. $f'(1) = 1/3$ **b.** Not differentiable at 0   **79. a.** Not differentiable at 1   **b.** Not differentiable at 0   **81.** Yes; 0
83. Yes; 12   **85.** No; 3   **87.** Yes; 3/2   **89.** Yes; Diverges to $-\infty$ **91.** Yes; Diverges to $-\infty$ **93. a.** $s'(t) = 3.04t + 9.45$
b. 52 teams/year   **95.** $P'(t) = -5.2t + 13$; increasing at a rate of 2.6 percentage points per year   **97.** 0.55   **99. a.** $s'(t) = -32t$; 0, -32, -64, -96, -128 ft/sec   **b.** 5 seconds; downward at 160 ft/sec   **101. a.** $E'(t) = 0.072t - 0.10$. In January 2004 the value of the euro was increasing at a rate of $0.188 per year.
b. (D)   **103. a.** $f'(x) = 7.1x - 30.2$ manatees per 100,000 boats.   **b.** Increasing; the number of manatees killed per additional 100,000 boats increases as the number of boats increases.   **c.** $f'(8) = 26.6$ manatees per 100,000 additional boats. At a level of 800,000 boats, the number of manatee deaths is increasing at a rate of 26.6 manatees per 100,000 additional

boats.   **105. a.** $c(t) - m(t)$ measures the combined market share of the other three providers (Comcast, Earthlink, and AOL); $c'(t) - m'(t)$ measures the rate of change of the combined market share of the other three providers.   **b.** (A)   **c.** (A)
d. 3.72% per year. In 1992, the combined market share of the other three providers was increasing at a rate of about 3.72 percentage points per year.   **107.** After graphing the curve $y = 3x^2$, draw the line passing through $(-1, 3)$ with slope -6.   **109.** The slope of the tangent line of g is twice the slope of the tangent line of f.   **111.** $g'(x) = -f'(x)$ **113.** The left-hand side is not equal to the right-hand side. The *derivative* of the left-hand side is equal to the right-hand side, so your friend should have written

$$\dfrac{d}{dx}(3x^4 + 11x^5) = 12x^3 + 55x^4$$ **115.** The derivative of a

constant times a function is the constant times the derivative of the function, so that $f'(x) = (2)(2x) = 4x$. Your enemy mistakenly computed the *derivative* of the constant times the derivative of the function. (The derivative of a product of two functions is not the product of the derivative of the two functions. The rule for taking the derivative of a product is discussed in the next chapter.).
117. Answers may vary.

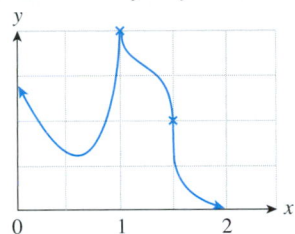

Section 10.8

1. $C'(1000) = \$4.80$ per item   **3.** $C'(100) = \$99.90$ per item
5. $C'(x) = 4$; $R'(x) = 8 - x/500$; $P'(x) = 4 - x/500$;
$P'(x) = 0$ when $x = 2000$. Thus, at a production level of 2000, the profit is stationary (neither increasing nor decreasing) with respect to the production level. This may indicate a maximum profit at a production level of 2000.   **7. a.** (B)   **b.** (C)   **c.** (C)
9. a. $C'(x) = 2250 - 0.04x$. The cost is going up at a rate of $2,249,840 per television commercial. The exact cost of airing the fifth television commercial is $C(5) - C(4) = \$2,249,820$.
b. $\bar{C}(x) = 150/x + 2250 - 0.02x$; $\bar{C}(4) = \$2,287,420$ per television commercial. The average cost of airing the first four television commercials is $2,287,420.   **11. a.** $R'(x) = 0.90$, $P'(x) = 0.80 - 0.002x$ **b.** Revenue: $450, Profit: $80, Marginal revenue: $0.90, Marginal profit: $-\$0.20$. The total revenue from the sale of 500 copies is $450. The profit from production and sale of 500 copies is $80. Approximate revenue from the sale of the 501st copy is 90¢. Approximate loss from the sale of the 501st copy is 20¢.   **c.** $x = 400$. The profit is a maximum when you produce and sell 400 copies.   **13.** The profit on the sale of 1000 DVDs is $3000, and is decreasing at a rate of $3 per additional DVD sold.   **15.** $P \approx \$257.07$ and $dP/dx \approx 5.07$. Your current profit is $257.07 per month, and this would increase at a rate of $5.07 per additional magazine in sales.
17. a. $2.50 per pound   **b.** $R(q) = 20{,}000/q^{0.5}$

c. $R(400) = \$1000$. This is the monthly revenue that will result from setting the price at \$2.50 per pound. $R'(400) = -\$1.25$ per pound of tuna. Thus, at a demand level of 400 pounds per month, the revenue is decreasing at a rate of \$1.25 per pound.   **d.** The fishery should raise the price (to reduce the demand).

19. $P'(50) = \$350$. This means that, at an employment level of 50 workers, the firm's daily profit will increase at a rate of \$350 per additional worker it hires.   **21. a.** (B) **b.** (B) **c.** (C)
23. a. $C(x) = 500{,}000 + 1{,}600{,}000x - 100{,}000\sqrt{x}$;

$C'(x) = 1{,}600{,}000 - \dfrac{50{,}000}{\sqrt{x}}$; $\bar{C}(x) = \dfrac{500{,}000}{x} +$

$1{,}600{,}000 - \dfrac{100{,}000}{\sqrt{x}}$ **b.** $C'(3) \approx \$1{,}570{,}000$ per spot,

$\bar{C}(3) \approx \$1{,}710{,}000$ per spot. The average cost will decrease as x increases.   **25. a.** $C'(q) = 200q$; $C'(10) = \$2000$ per one-pound reduction in emissions.   **b.** $S'(q) = 500$. Thus $S'(q) = C'(q)$ when $500 = 200q$, or $q = 2.5$ pounds per day reduction.   **c.** $N(q) = C(q) - S(q) = 100q^2 - 500q + 4000$. This is a parabola with lowest point (vertex) given by $q = 2.5$. The net cost at this production level is $N(2.5) = \$3375$ per day. The value of q is the same as that for part (b). The net cost to the firm is minimized at the reduction level for which the cost of controlling emissions begins to increase faster than the subsidy. This is why we get the answer by setting these two rates of increase equal to each other.   **27.** $M'(10) \approx 0.0002557$ mpg/mph. This means that, at a speed of 10 mph, the fuel economy is increasing at a rate of 0.0002557 miles per gallon per 1-mph increase in speed. $M'(60) = 0$ mpg/mph. This means that, at a speed of 60 mph, the fuel economy is neither increasing nor decreasing with increasing speed. $M'(70) \approx -0.00001799$. This means that, at 70 mph, the fuel economy is decreasing at a rate of 0.00001799 miles per gallon per 1-mph increase in speed. Thus 60 mph is the most fuel-efficient speed for the car.   **29.** (C)   **31.** (D)   **33.** (B)
35. Cost is often measured as a function of the number of items x. Thus, $C(x)$ is the cost of producing (or purchasing, as the case may be) x items.   **a.** The average cost function $\bar{C}(x)$ is given by $\bar{C}(x) = C(x)/x$. The marginal cost function is the derivative, $C'(x)$, of the cost function.   **b.** The average cost $\bar{C}(r)$ is the slope of the line through the origin and the point on the graph where $x = r$. The marginal cost of the rth unit is the slope of the tangent to the graph of the cost function at the point where $x = r$.   **c.** The average cost function $\bar{C}(x)$ gives the average cost of producing the first x items. The marginal cost function $C'(x)$ is the rate at which cost is changing with respect to the number of items x, or the incremental cost per item, and approximates the cost of producing the $(x + 1)$st item.   **37.** The marginal cost   **39.** Not necessarily. For example, it may be the case that the marginal cost of the 101st item is larger than the average cost of the first 100 items (even though the marginal cost is decreasing). Thus, adding this additional item will *raise* the average cost.   **41.** The circumstances described suggest that the average cost function is at a relatively low point at the current production level, and so it would be appropriate to advise the company to maintain current production levels; raising or lowering the production level will result in increasing average costs.

Chapter 10 Review

1. 5   **3.** Does not exist   **5. a.** -1 **b.** 3 **c.** Does not exist
7. $-4/5$ **9.** -1 **11.** Diverges to $-\infty$
13.

h	1	0.01	0.001
Ave. Rate of Change	-0.5	-0.9901	-0.9990

Slope ≈ -1

15.

h	1	0.01	0.001
Avg. Rate of Change	6.3891	2.0201	2.0020

Slope ≈ 2

17. a. (i) P **(ii)** Q **(iii)** R **(iv)** S **19. (i)** Q **(ii)** None
(iii) None **(iv)** None   **21. a.** (B) **b.** (B) **c.** (B) **d.** (A)
e. (C)   **23.** $2x + 1$ **25.** $2/x^2$ **27.** $50x^4 + 2x^3 - 1$
29. $9x^2 + x^{-2/3}$ **31.** $1 - 2/x^3$
33. $-4/(3x^2) + 0.2/x^{1.1} + 1.1x^{0.1}/3.2$
35. `50*x^4 + 2*x^3 - 1`   **37.** `9*x^2 + 1/(x^2)^(1/3)`

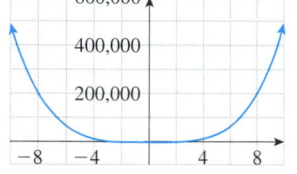
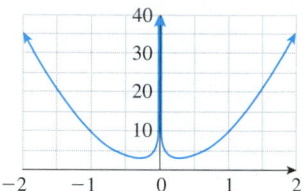

39. a. $P(3) = 25$: O'Hagan purchased the stock at \$25. $\lim_{t \to 3^-} P(t) = 25$: The value of the stock had been approaching \$25 up to the time he bought it. $\lim_{t \to 3^+} P(t) = 10$: The value of the stock dropped to \$10 immediately after he bought it.
b. Continuous but not differentiable. Interpretation: the stock price changed continuously but suddenly reversed direction (and started to go up) the instant O'Hagan sold it.   **41. a.** 500 books per week **b.** [3, 4], [4, 5] **c.** [3, 5]; 650 books per week
43. a. 274 books per week **b.** 636 books per week **c.** No; the function w begins to decrease after $t = 14$. Graph:

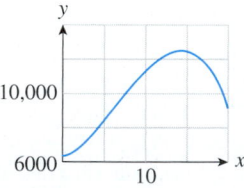

45. a. \$2.88 per book **b.** \$3.715 per book **c.** Approximately $-\$0.000104$ per book, per additional book sold. **d.** At a sales level of 8000 books per week, the cost is increasing at a rate of \$2.88 per book (so that the 8001st book costs approximately \$2.88 to sell), and it costs an average of \$3.715 per book to sell the first 8000 books. Moreover, the average cost is decreasing at a rate of \$0.000104 per book, per additional book sold.

Chapter 11

Section 11.1

1. 3   **3.** $3x^2$ **5.** $2x + 3$ **7.** $210x^{1.1}$ **9.** $-2/x^2$ **11.** $2x/3$
13. $3(4x^2 - 1) + 3x(8x) = 36x^2 - 3$ **15.** $3x^2(1 - x^2) +$
$x^3(-2x) = 3x^2 - 5x^4$ **17.** $2(2x + 3) + (2x + 3)(2) =$
$8x + 12$ **19.** $3\sqrt{x}/2$ **21.** $(x^2 - 1) + 2x(x + 1) =$
$(x + 1)(3x - 1)$ **23.** $(x^{-0.5} + 4)(x - x^{-1}) + (2x^{0.5} +$
$4x - 5)(1 + x^{-2})$ **25.** $8(2x^2 - 4x + 1)(x - 1)$
27. $(1/3.2 - 3.2/x^2)(x^2 + 1) + 2x(x/3.2 + 3.2/x)$
29. $2x(2x + 3)(7x + 2) + 2x^2(7x + 2) + 7x^2(2x + 3)$
31. $5.3(1 - x^{2.1})(x^{-2.3} - 3.4) - 2.1x^{1.1}(5.3x - 1) \cdot$
$(x^{-2.3} - 3.4) - 2.3x^{-3.3}(5.3x - 1)(1 - x^{2.1})$

33. $\dfrac{1}{2\sqrt{x}}\left(\sqrt{x} + \dfrac{1}{x^2}\right) + (\sqrt{x} + 1)\left(\dfrac{1}{2\sqrt{x}} - \dfrac{2}{x^3}\right)$

35. $\dfrac{2(3x - 1) - 3(2x + 4)}{(3x - 1)^2} = -14/(3x - 1)^2$

37.

$\dfrac{(4x + 4)(3x - 1) - 3(2x^2 + 4x + 1)}{(3x - 1)^2} = (6x^2 - 4x - 7)/(3x - 1)^2$

39. $\dfrac{(2x - 4)(x^2 + x + 1) - (x^2 - 4x + 1)(2x + 1)}{(x^2 + x + 1)^2}$

$= (5x^2 - 5)/(x^2 + x + 1)^2$

41. $\dfrac{(0.23x^{-0.77} - 5.7)(1 - x^{-2.9}) - 2.9x^{-3.9}(x^{0.23} - 5.7x)}{(1 - x^{-2.9})^2}$

43. $\dfrac{\frac{1}{2}x^{-1/2}(x^{1/2} - 1) - \frac{1}{2}x^{-1/2}(x^{1/2} + 1)}{(x^{1/2} - 1)^2} = \dfrac{-1}{\sqrt{x}(\sqrt{x} - 1)^2}$

45. $-3/x^4$

47. $\dfrac{[(x + 1) + (x + 3)](3x - 1) - 3(x + 3)(x + 1)}{(3x - 1)^2}$

$= (3x^2 - 2x - 13)/(3x - 1)^2$

49. $\dfrac{[(x+1)(x+2)+(x+3)(x+2)+(x+3)(x+1)](3x-1)-3(x+3)(x+1)(x+2)}{(3x-1)^2}$

51. $4x^3 - 12x^2 + 2x - 480$ **53.** $1 + 2/(x + 1)^2$
55. $2x - 1 - 2/(x + 1)^2$ **57.** $4x^3 - 2x$ **59.** 64   **61.** 3
63. $y = 12x - 8$ **65.** $y = x/4 + 1/2$ **67.** $y = -2$
69. $q'(5) = 1000$ units/month (sales are increasing at a rate of 1000 units per month); $p'(5) = -\$10$/month (the price of a sound system is dropping at a rate of \$10 per month); $R'(5) = 900{,}000$ (revenue is increasing at a rate of \$900,000 per month)   **71.** \$242 million; increasing at a rate of \$39 million per year   **73.** Decreasing at a rate of \$1 per day   **75.** Decreasing at a rate of approximately \$0.10 per month

77. $M'(x) = \dfrac{3000(3600x^{-2} - 1)}{(x + 3600x^{-1})^2}$; $M'(10) \approx 0.7670$ mpg/mph.

This means that, at a speed of 10 mph, the fuel economy is increasing at a rate of 0.7670 miles per gallon per one mph increase in speed. $M'(60) = 0$ mpg/mph. This means that, at a speed of 60 mph, the fuel economy is neither increasing nor decreasing with increasing speed. $M'(70) \approx -0.0540$. This means that, at 70 mph, the fuel economy is decreasing at a rate of

0.0540 miles per gallon per one mph increase in speed. 60 mph is the most fuel-efficient speed for the car. (In the next chapter we shall discuss how to locate largest values in general.)   **79.** Increasing at a rate of about \$3420 million per year.

81. $R'(p) = -\dfrac{5.625}{(1 + 0.125p)^2}$; $R'(4) = -2.5$ thousand

organisms per hour, per 1000 organisms. This means that the reproduction rate of organisms in a culture containing 4000 organisms is declining at a rate of 2500 organisms per hour, per 1000 additional organisms.   **83.** Oxygen consumption is decreasing at a rate of 1600 milliliters per day. This is due to the fact that the number of eggs is decreasing, because $C'(25)$ is positive.   **85. a.** $c(t) - m(t)$ represents the combined market share of the other three providers (Comcast, Earthlink, and AOL). $m(t)/c(t)$ represents MSN's market share as a fraction of

the four providers considered.   **b.** $\dfrac{d}{dt}\left(\dfrac{m(t)}{c(t)}\right)\Big|_{t=3} \approx -0.043$

(or -4.3 percentage points) per year. In June, 2003, MSN's market share as a fraction of the four providers considered was decreasing at a rate of about 0.043 (or 4.3 percentage points) per year.   **87.** The analysis is suspect, because it seems to be asserting that the annual increase in revenue, which we can think of as dR/dt, is the product of the annual increases, dp/dt in price, and dq/dt in sales. However, because $R = pq$, the product rule implies that dR/dt is not the product of dp/dt and dq/dt, but is

instead $\dfrac{dR}{dt} = \dfrac{dp}{dt} \cdot q + p \cdot \dfrac{dq}{dt}$.   **89.** Answers will vary;

$q = -p + 1000$ is one example.   **91.** Mine; it is increasing twice as fast as yours. The rate of change of revenue is given by $R'(t) = p'(t)q(t)$ because $q'(t) = 0$ Thus, $R'(t)$ does not depend on the selling price $p(t)$.   **93.** (A)

Section 11.2

1. $4(2x + 1)$ **3.** $-(x - 1)^{-2}$ **5.** $2(2 - x)^{-3}$
7. $(2x + 1)^{-0.5}$ **9.** $-4(4x - 1)^{-2}$ **11.** $-3/(3x - 1)^2$
13. $4(x^2 + 2x)^3(2x + 2)$ **15.** $-4x(2x^2 - 2)^{-2}$
17. $-5(2x - 3)(x^2 - 3x - 1)^{-6}$ **19.** $-6x/(x^2 + 1)^4$
21. $1.5(0.2x - 4.2)(0.1x^2 - 4.2x + 9.5)^{0.5}$
23. $4(2s - 0.5s^{-0.5})(s^2 - s^{0.5})^3$ **25.** $-x/\sqrt{1 - x^2}$
27. $-[(x + 1)(x^2 - 1)]^{-3/2}(3x - 1)(x + 1)$
29. $6.2(3.1x - 2) + 6.2/(3.1x - 2)^3$
31. $2[(6.4x - 1)^2 + (5.4x - 2)^3][12.8(6.4x - 1) + 16.2(5.4x - 2)^2]$
33. $-2(x^2 - 3x)^{-3}(2x - 3)(1 - x^2)^{0.5} - x(x^2 - 3x)^{-2}(1 - x^2)^{-0.5}$
35. $-56(x + 2)/(3x - 1)^3$ **37.** $3z^2(1 - z^2)/(1 + z^2)^4$
39. $3[(1 + 2x)^4 - (1 - x)^2]^2[8(1 + 2x)^3 + 2(1 - x)]$
41. $-0.43(x + 1)^{-1.1}[2 + (x + 1)^{-0.1}]^{3.3}$

43. $-\dfrac{\left(\dfrac{1}{\sqrt{2x+1}} - 2x\right)}{\left(\sqrt{2x + 1} - x^2\right)^2}$

45. $54(1 + 2x)^2\left(1 + (1 + 2x)^3\right)^2\left(1 + \left(1 + (1 + 2x)^3\right)^3\right)^2$
47. $(100x^{99} - 99x^{-2})dx/dt$ **49.** $(-3r^{-4} + 0.5r^{-0.5})dr/dt$
51. $4\pi r^2 dr/dt$ **53.** $-47/4$ **55.** $1/3$ **57.** $-5/3$ **59.** $1/4$
61. $y = 35(7 + 0.2t)^{-0.25}$; -0.11 percentage points per month.

63. $\left.\dfrac{dP}{dn}\right|_{n=10} = 146{,}454.9$. At an employment level of 10 engineers, Paramount will increase its profit at a rate of \$146,454.90 per additional engineer hired.   **65.** −\$30 per additional ruby sold. The revenue is decreasing at a rate of \$30 per additional ruby sold.   **67.** $\dfrac{dy}{dt} = \dfrac{dy}{dx}\dfrac{dx}{dt} = (1.5)(-2) = -3$ murders per 100,000 residents/yr each year.   **69.** 0.000158 manatees per boat, or 15.8 manatees per 100,000 boats. Approximately 15.8 more manatees are killed each year for each additional 100,000 registered boats.   **71.** 12π mi²/h   **73.** \$200,000π/week ≈ \$628,000/week   **75. a.** $q'(4) \approx 333$ units per month **b.** $dR/dq = \$800$/unit **c.** $dR/dt \approx \$267{,}000$ per month   **77.** 3% per year   **79.** 8% per year   **81.** The glob squared, times the derivative of the glob.   **83.** The derivative of a quantity cubed is three times the (original) quantity squared, times the derivative of the quantity. Thus, the correct answer is $3(3x^3 - x)^2(9x^2 - 1)$.   **85.** Following the calculation thought experiment, pretend that you are evaluating the function at a specific value of x. If the last operation you would perform is addition or subtraction, look at each summand separately. If the last operation is multiplication, use the product rule first; if it is division, use the quotient rule first; if it is any other operation (such as raising a quantity to a power or taking a radical of a quantity), then use the chain rule first.   **87.** An example is

$$f(x) = \sqrt{x + \sqrt{x + \sqrt{x + \sqrt{x + \sqrt{x + 1}}}}}\,.$$

Section 11.3

1. $1/(x - 1)$ **3.** $1/(x \ln 2)$ **5.** $2x/(x^2 + 3)$ **7.** e^{x+3}
9. $-e^{-x}$ **11.** $4^x \ln 4$ **13.** $2^{x^2-1} 2x \ln 2$ **15.** $1 + \ln x$
17. $2x \ln x + (x^2 + 1)/x$ **19.** $10x(x^2 + 1)^4 \ln x + (x^2 + 1)^5/x$
21. $3/(3x - 1)$ **23.** $4x/(2x^2 + 1)$
25. $(2x - 0.63x^{-0.7})/(x^2 - 2.1x^{0.3})$
27. $-2/(-2x + 1) + 1/(x + 1)$ **29.** $3/(3x + 1) - 4/(4x - 2)$
31. $1/(x + 1) + 1/(x - 3) - 2/(2x + 9)$ **33.** $5.2/(4x - 2)$
35. $2/(x + 1) - 9/(3x - 4) - 1/(x - 9)$ **37.** $\dfrac{1}{(x + 1) \ln 2}$
39. $\dfrac{1 - 1/t^2}{(t + 1/t) \ln 3}$ **41.** $\dfrac{2 \ln |x|}{x}$ **43.** $\dfrac{2}{x} - \dfrac{2 \ln(x - 1)}{x - 1}$
45. $e^x(1 + x)$ **47.** $1/(x + 1) + 3e^x(x^3 + 3x^2)$
49. $e^x(\ln|x| + 1/x)$ **51.** $2e^{2x+1}$ **53.** $(2x - 1)e^{x^2-x+1}$
55. $2xe^{2x-1}(1 + x)$ **57.** $4(e^{2x-1})^2$ **59.** $2 \cdot 3^{2x-4} \ln 3$
61. $2 \cdot 3^{2x+1} \ln 3 + 3e^{3x+1}$ **63.** $\dfrac{2x3^{x^2}[(x^2 + 1)\ln 3 - 1]}{(x^2 + 1)^2}$
65. $-4/(e^x - e^{-x})^2$ **67.** $5e^{5x-3}$ **69.** $-\dfrac{\ln x + 1}{(x \ln x)^2}$
71. $2(x - 1)$ **73.** $\dfrac{1}{x \ln x}$ **75.** $\dfrac{1}{2x \ln x}$
77. $y = (e/\ln 2)(x - 1) \approx 3.92(x - 1)$ **79.** $y = x$
81. $y = -[1/(2e)](x - 1) + e$ **83.** Average price: \$1.4 million; increasing at a rate of about \$220,000 per year.   **85.** \$451.00 per year   **87.** \$446.02 per year   **89.** 300 articles per year   **91.** 3,300,000 cases/week; 11,000,000 cases/week; 640,000 cases/week   **93.** 310 articles per year   **95.** 277,000 people/yr   **97.** 0.000283 g/yr   **99. a.** (A) **b.** The verbal SAT increases by

approximately 1 point.   **c.** $S'(x)$ decreases with increasing x, so that as parental income increases, the effect on SAT scores decreases.   **101. a.** −6.25 years/child; When the fertility rate is 2 children per woman, the average age of a population is dropping at a rate of 6.25 years per one-child increase in the fertility rate. **b.** 0.160
103. a. $W'(t) = -\dfrac{1500(0.77)(1.16)^{-t}(-1) \ln(1.16)}{(1 + 0.77(1.16)^{-t})^2}$

$\approx \dfrac{171.425(1.16)^{-t}}{(1 + 0.77(1.16)^{-t})^2}$; $W'(6) \approx 40.624 \approx 41$

to two significant digits. The constants in the model are specified to two and three significant digits, so we cannot expect the answer to be accurate to more than two digits. In other words, all digits from the third on are probably meaningless. The answer tells one that in 1996, the number of authorized wiretaps was increasing at a rate of approximately 41 wiretaps per year.
b. (A) Graph: **105. a.**

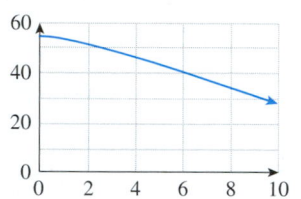
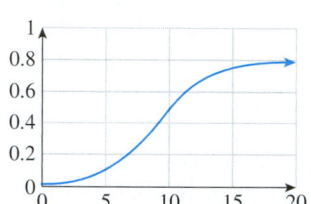

$p'(10) \approx 0.09$, so the percentage of firms using numeric control is increasing at a rate of 9 percentage points per year after 10 years. **b.** 0.80. Thus, in the long run, 80% of all firms will be using numeric control.
c. $p'(t) = 0.3816e^{4.46-0.477t}/(1 + e^{4.46-0.477t})^2$.
$p'(10) = 0.0931$. Graph:

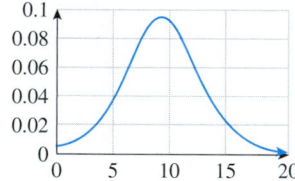

d. 0. Thus, in the long run, the percentage of firms using numeric control will stop increasing.
107. $R(t) = 350e^{-0.1t}(39t + 68)$ million dollars; $R(2) \approx \$42$ billion; $R'(2) \approx \$7$ billion per year   **109.** e raised to the glob, times the derivative of the glob.   **111.** 2 raised to the glob, times the derivative of the glob, times the natural logarithm of 2.
113. The power rule does not apply when the exponent is not constant. The derivative of 3 raised to a quantity is 3 raised to the quantity, times the derivative of the quantity, times $\ln 3$. Thus, the correct answer is $3^{2x} 2 \ln 3$.   **115.** No. If $N(t)$ is exponential, so is its derivative.   **117.** If $f(x) = e^{kx}$, then the fractional rate of change is $\dfrac{f'(x)}{f(x)} = \dfrac{ke^{kx}}{e^{kx}} = k$, the fractional growth rate.   **119.** If $A(t)$ is growing exponentially, then $A(t) = A_0 e^{kt}$ for constants A_0 and k. Its percentage rate of change is then

$\dfrac{A'(t)}{A(t)} = \dfrac{kA_0 e^{kt}}{A_0 e^{kt}} = k$, a constant.

Section 11.4

1. $-2/3$ **3.** x **5.** $(y-2)/(3-x)$ **7.** $-y$

9. $-\dfrac{y}{x(1+\ln x)}$ **11.** $-x/y$ **13.** $-2xy/(x^2-2y)$

15. $-(6+9x^2y)/(9x^3-x^2)$ **17.** $3y/x$

19. $(p+10p^2q)/(2p-q-10pq^2)$

21. $(ye^x-e^y)/(xe^y-e^x)$ **23.** $se^{st}/(2s-te^{st})$

25. $ye^x/(2e^x+y^3e^y)$ **27.** $(y-y^2)/(-1+3y-y^2)$

29. $-y/(x+2y-xye^y-y^2e^y)$ **31. a.** 1 **b.** $y=x-3$

33. a. -2 **b.** $y=-2x$ **35. a.** -1 **b.** $y=-x+1$

37. a. -2000 **b.** $y=-2000x+6000$ **39. a.** 0 **b.** $y=1$

41. a. -0.1898 **b.** $y=-0.1898x+1.4721$

43. $\dfrac{2x+1}{4x-2}\left[\dfrac{2}{2x+1}-\dfrac{4}{4x-2}\right]$

45. $\dfrac{(3x+1)^2}{4x(2x-1)^3}\left[\dfrac{6}{3x+1}-\dfrac{1}{x}-\dfrac{6}{2x-1}\right]$

47. $(8x-1)^{1/3}(x-1)\left[\dfrac{8}{3(8x-1)}+\dfrac{1}{x-1}\right]$

49. $(x^3+x)\sqrt{x^3+2}\left[\dfrac{3x^2+1}{x^3+x}+\dfrac{1}{2}\dfrac{3x^2}{x^3+2}\right]$

51. $x^x(1+\ln x)$ **53.** $-\$3000$ per worker. The monthly budget to maintain production at the fixed level P is decreasing by approximately $\$3000$ per additional worker at an employment level of 100 workers and a monthly operating budget of $\$200,000$.
55. -125 T-shirts per dollar; when the price is set at $\$5$, the demand is dropping by 125 T-shirts per $\$1$ increase in price.

57. $\left.\dfrac{dk}{de}\right|_{e=15}=-0.307$ carpenters per electrician. This means that, for a $\$200,000$ house whose construction employs 15 electricians, adding one more electrician would cost as much as approximately 0.307 additional carpenters. In other words, one electrician is worth approximately 0.307 carpenters.
59. a. 22.93 hours. (The other root is rejected because it is larger

than 30.) **b.** $\dfrac{dt}{dx}=\dfrac{4t-20x}{0.4t-4x}$; $\left.\dfrac{dt}{dx}\right|_{x=3.0}\approx-11.2$ hours per

grade point. This means that, for a 3.0 student who scores 80 on the examination, 1 grade point is worth approximately 11.2 hours.

61. $\dfrac{dr}{dy}=2\dfrac{r}{y}$, so $\dfrac{dr}{dt}=2\dfrac{r}{y}\dfrac{dy}{dt}$ by the chain rule.

63. $x,\ y,\ y,\ x$

65. Then $\ln y=\ln f(x)+\ln g(x)$, and $\dfrac{1}{y}\dfrac{dy}{dx}=\dfrac{f'(x)}{f(x)}+\dfrac{g'(x)}{g(x)}$,

so $\dfrac{dy}{dx}=y\left(\dfrac{f'(x)}{f(x)}+\dfrac{g'(x)}{g(x)}\right)=f(x)g(x)\left(\dfrac{f'(x)}{f(x)}+\dfrac{g'(x)}{g(x)}\right)=$

$f'(x)g(x)+f(x)g'(x)$.
67. Writing $y=f(x)$ specifies y as an explicit function of x. This can be regarded as an equation giving y as an *implicit* function of x. The procedure of finding dy/dx by implicit differentiation is then the same as finding the derivative of y as an explicit function of x: we take d/dx of both sides.   **69.** Differentiate both sides of the equation $y=f(x)$ with respect to y to get

$1=f'(x)\cdot\dfrac{dx}{dy}$, giving $\dfrac{dx}{dy}=\dfrac{1}{f'(x)}=\dfrac{1}{dy/dx}$.

Chapter 11 Review

1. $e^x(x^2+2x-1)$ **3.** $20x(x^2-1)^9$

5. $e^x(x^2+1)^9(x^2+20x+1)$

7. $3^x[(x-1)\ln 3-1]/(x-1)^2$ **9.** $2xe^{x^2-1}$

11. $2x/(x^2-1)$ **13.** $x=(1-\ln 2)/2$

15. None   **17.** $\dfrac{2x-1}{2y}$ **19.** $-y/x$

21.

$\dfrac{(2x-1)^4(3x+4)}{(x+1)(3x-1)^3}\left[\dfrac{8}{2x-1}+\dfrac{3}{3x+4}-\dfrac{1}{x+1}-\dfrac{9}{3x-1}\right]$

23. $y=x+2$

25. $R'(0)=p'(0)q(0)+p(0)q'(0)$
$\quad\quad=(-1)(1000)+20(200)=\$3000$ per week (rising)

27. $R=pq$ gives $R'=p'q+pq'$. Thus,
$R'/R=R'/(pq)=(p'q+pq')/pq=p'/p+q'/q$

29. $\$110$ per year   **31.** $s'(t)=\dfrac{2460.7e^{-0.55(t-4.8)}}{(1+e^{-0.55(t-4.8)})^2}$;

553 books per week   **33.** 616.8 hits per day per week

Chapter 12

Section 12.1

1. Absolute min.: $(-3,-1)$, relative max: $(-1,1)$, relative min: $(1,0)$, absolute max: $(3,2)$ **3.** Absolute min: $(3,-1)$ and $(-3,-1)$, absolute max: $(1,2)$ **5.** Absolute min: $(-3,0)$ and $(1,0)$, absolute max: $(-1,2)$ and $(3,2)$ **7.** Relative min: $(-1,1)$ **9.** Absolute min: $(-3,-1)$, relative max: $(-2,2)$, relative min: $(1,0)$, absolute max: $(3,3)$ **11.** Relative max: $(-3,0)$, absolute min: $(-2,-1)$, stationary non-extreme point: $(1,1)$ **13.** Absolute max: $(0,1)$, absolute min: $(2,-3)$, relative max: $(3,-2)$ **15.** Absolute min: $(-4,-16)$, absolute max: $(-2,16)$, absolute min: $(2,-16)$, absolute max: $(4,16)$
17. Absolute min: $(-2,-10)$, absolute max: $(2,10)$
19. Absolute min: $(-2,-4)$, relative max: $(-1,1)$, relative min: $(0,0)$ **21.** Relative max: $(-1,5)$, absolute min: $(3,-27)$
23. Absolute min: $(0,0)$ **25.** Absolute maxima at $(0,1)$ and $(2,1)$, absolute min at $(1,0)$ **27.** Relative maximum at $(-2,-1/3)$, relative minimum at $(-1,-2/3)$, absolute maximum at $(0,1)$ **29.** Relative min: $(-2,5/3)$, relative max: $(0,-1)$, relative min: $(2,5/3)$ **31.** Relative max: $(0,0)$; absolute min: $(1/3,-2\sqrt{3}/9)$ **33.** Relative max: $(0,0)$, absolute min: $(1,-3)$ **35.** No relative extrema   **37.** Absolute min: $(1,1)$ **39.** Relative max: $(-1,1+1/e)$, absolute min: $(0,1)$, absolute max: $(1,e-1)$ **41.** Relative max: $(-6,-24)$, relative min: $(-2,-8)$ **43.** Absolute max $(1/\sqrt{2},\sqrt{e/2})$, absolute min: $(-1/\sqrt{2},-\sqrt{e/2})$ **45.** Relative min at $(0.15,-0.52)$ and $(2.45,8.22)$, relative max at $(1.40,0.29)$ **47.** Absolute max at $(-5,700)$, relative max at $(3.10,28.19)$ and $(6,40)$, absolute min at $(-2.10,-392.69)$ and relative min at $(5,0)$.   **49.** Stationary minimum at $x=-1$ **51.** Stationary minima at $x=-2$ and $x=2$, stationary maximum at $x=0$ **53.** Singular minimum at $x=0$, stationary non-extreme point at $x=1$ **55.** Stationary minimum at $x=-2$, singular non-extreme points at $x=-1$ and $x=1$, stationary maximum at $x=2$

57. Answers will vary.   **59.** Answers will vary.

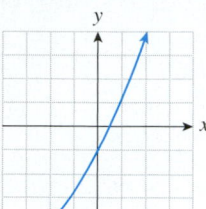   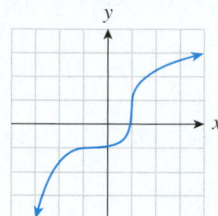

61. Not necessarily; it could be neither a relative maximum nor a relative minimum, as in the graph of $y = x^3$ at the origin.
63. Answers will vary.

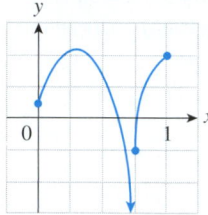

Section 12.2

1. $x = y = 5; P = 25$ **3.** $x = y = 3; S = 6$ **5.** $x = 2, y = 4;$ $F = 20$ **7.** $x = 20,$ $y = 10,$ $z = 20;$ $P = 4000$
9. 5×5 **11.** 5000 MP3 players, giving an average cost of $30 per MP3 player   **13.** $\sqrt{40} \approx 6.32$ pounds of pollutant per day, for an average cost of about $1265 per pound   **15.** 2.5 lb
17. $5 \times 10 = 50$ square feet   **19.** $10   **21.** $30
23. a. $1.41 per pound **b.** 5000 pounds **c.** $7071.07 per month
25. 34.5¢ per pound, for an annual (per capita) revenue of $5.95
27. $42.50 per ruby, for a weekly profit of $408.33   **29. a.** 656 headsets, for a profit of $28,120 **b.** $143 per headset
31. $13\frac{1}{3}$ in $\times 3\frac{1}{3}$ in $\times 1\frac{1}{3}$ in for a volume of $1600/27 \approx$ 59 cubic inches   **33.** $5 \times 5 \times 5$ cm   **35.** $l = w = h \approx 20.67$ in, volume ≈ 8827 in^3 **37.** $l = 30$ in, $w = 15$ in, $h = 30$ in
39. $l = 36$ in, $w = h = 18$ in, $V = 11,664$ in^3 **41. a.** 1.6 years, or year 2001.6; **b.** $R_{max} = \$28,241$ million   **43.** $t = 2.5$ or midway through 1972.; $D(2.5)/S(2.5) \approx 4.09$ The number of new (approved) drugs per $1 billion of spending on research and development reached a high of around 4 approved drugs per $1 billion midway through 1972.   **45.** 30 years from now
47. 55 days   **49.** 1600 copies. At this value of x, average profit equals marginal profit; beyond this the marginal profit is smaller than the average.   **51.** Increasing most rapidly in 1992; increasing least rapidly in 1980   **53.** Maximum when $t = 17$ days. This means that the embryo's oxygen consumption is increasing most rapidly 17 days after the egg is laid.   **55.** $h = r \approx 11.7$ cm
57. 25 additional trees   **59.** 71 employees   **61.** Fourth quarter of 2003 ($t \approx 3.7$); 160 thousand iPods per quarter
63. Graph of derivative:

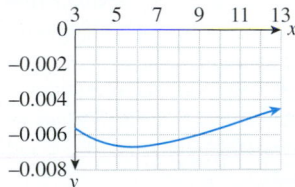

Absolute minimum occurs at approximately $t = 6$, with value approximately -0.0067. The fraction of bottled water sales due to sparkling water was decreasing most rapidly in 1996. At that time it was decreasing at a rate of 0.67 percentage points per year.
65. You should sell them in 17 years' time, when they will be worth approximately $3960.   **67.** (D) **69.** The problem is uninteresting because the company can accomplish the objective by cutting away the entire sheet of cardboard, resulting in a box with surface area zero.   **71.** Not all absolute extrema occur at stationary points; some may occur at an endpoint or singular point of the domain, as in Exercises 23, 24, 51 and 52.   **73.** The minimum of dq/dp is the fastest that the demand is dropping in response to increasing price.

Section 12.3

1. 6   **3.** $4/x^3$ **5.** $-0.96x^{-1.6}$ **7.** $e^{-(x-1)}$ **9.** $2/x^3 + 1/x^2$
11. a. $a = -32$ ft/sec^2 **b.** $a = -32$ ft/sec^2
13. a. $a = 2/t^3 + 6/t^4$ ft/sec^2 **b.** $a = 8$ ft/sec^2
15. a. $a = -1/(4t^{3/2}) + 2$ ft/sec^2
b. $a = 63/32$ ft/sec^2 **17.** $(1, 0)$ **19.** $(1, 0)$ **21.** None
23. $(-1, 0)$, $(1, 1)$ **25.** Points of inflection at $x = -1$ and $x = 1$ **27.** One point of inflection, at $x = -2$ **29.** Points of inflection at $x = -2, x = 0, x = 2$ **31.** Points of inflection at $x = -2$ and $x = 2$
33. Absolute min at $(-1, 0)$; no points of inflection   **35.** Relative max at $(-2, 21)$; relative min at $(1, -6)$; point of inflection at $(-1/2, 15/2)$

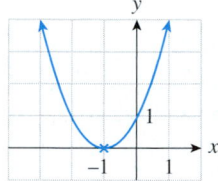   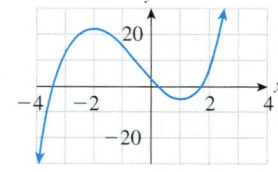

37. Absolute min at $(-4, -16)$ and $(2, -16)$; absolute max at $(-2, 16)$ and $(4, 16)$; point of inflection at $(0, 0)$ **39.** Absolute min at $(0, 0)$; points of inflection at $(1/3, 11/324)$ and $(1, 1/12)$

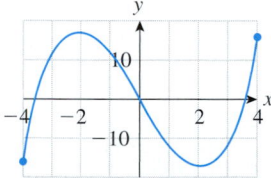   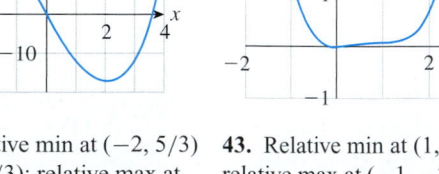

41. Relative min at $(-2, 5/3)$ and $(2, 5/3)$; relative max at $(0, -1)$; vertical asymptotes: $x = \pm 1$ **43.** Relative min at $(1, 2)$; relative max at $(-1, -2)$; vertical asymptote: $y = 0$

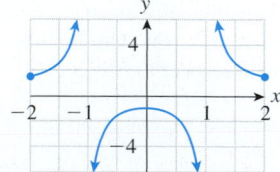   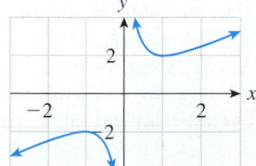

45. Relative maximum at $(-2, -1/3)$; relative minimum at $(-1, -2/3)$; no points of inflection

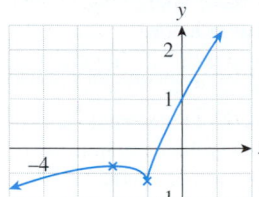

47. No extrema; points of inflection at $(0, 0)$, $(-3, -9/4)$, and $(3, 9/4)$

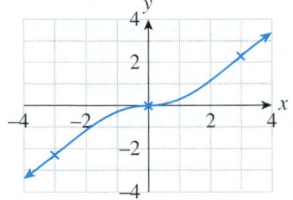

49. Absolute min at $(1, 1)$; vertical asymptote at $x = 0$

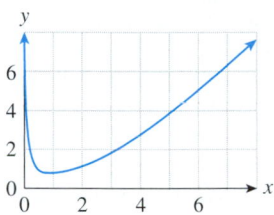

51. No relative extrema; point of inflection at $(1, 1)$ and $(-1, 1)$; vertical asymptote at $x = 0$

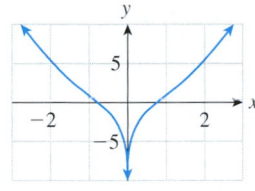

53. Absolute min at $(0, 1)$, absolute max at $(1, e - 1)$, relative max at $(-1, 1 + e^{-1})$

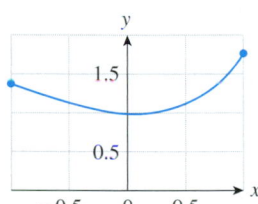

55. Absolute min at $(1.40, -1.49)$; points of inflection: $(0.21, 0.61)$, $(0.79, -0.55)$

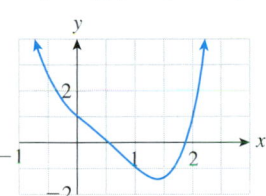

57. Relative min at $(-0.46, 0.73)$; relative max at $(0.91, 1.73)$; absolute min at $(3.73, -10.22)$; points of inflection at $(0.20, 1.22)$ and $(2.83, -5.74)$

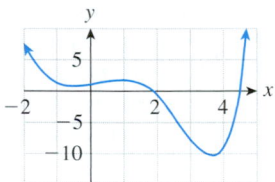

59. -3.8 m/s^2 **61.** $6t - 2$ ft/s^2; increasing **63.** Accelerating by 34 million gals/yr^2 **65. a.** 400 ml **b.** 36 ml/day **c.** -1 ml/day^2 **67. a.** Two years into the epidemic **b.** Two years into the epidemic **69. a.** 2000 **b.** 2002 **c.** 1998 **71.** Concave up for $8 < t < 20$, concave down for $0 < t < 8$, point of inflection around $t = 8$. The percentage of articles written by researchers in the U.S. was decreasing most rapidly at around $t = 8$ (1991). **73. a.** (B) **b.** (B) **c.** (A) **75. a.** There are no points of inflection in the graph of S. **b.** Because the graph is concave up, the derivative of S is increasing, and so the rate of *decrease* of SAT scores with increasing numbers of prisoners is diminishing. In other words, the apparent effect of more prisoners is diminishing.

77. a. $\left.\dfrac{d^2n}{ds^2}\right|_{s=3} = -21.494$. Thus, for a firm with annual sales of \$3 million, the rate at which new patents are produced decreases with increasing firm size. This means that the returns (as measured in the number of new patents per increase of \$1 million in sales) are diminishing as the firm size increases.

b. $\left.\dfrac{d^2n}{ds^2}\right|_{s=7} = 13.474$. Thus, for a firm with annual sales of \$7 million, the rate at which new patents are produced increases with increasing firm size by 13.474 new patents per \$1 million increase in annual sales. **c.** There is a point of inflection when $s \approx 5.4587$, so that in a firm with sales of \$5,458,700 per year, the number of new patents produced per additional \$1 million in sales is a minimum. **79.** Concave down; (C). Graph:

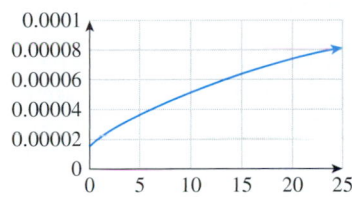

81. About \$570 per year, after about 12 years **83.** Increasing most rapidly in 17.64 years, decreasing most rapidly now (at $t = 0$) **85.** Nonnegative **87.** Daily sales were decreasing most rapidly in June, 2002.

89.

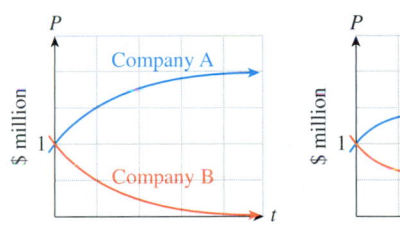
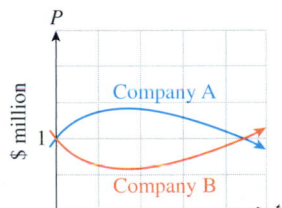

91. At a point of inflection, the graph of a function changes either from concave up to concave down, or vice versa. If it changes from concave up to concave down, then the derivative changes from increasing to decreasing, and hence has a relative maximum. Similarly, if it changes from concave down to concave up, the derivative has a relative minimum.

Section 12.4

1. $P = 10,000;$ $\dfrac{dP}{dt} = 1000$ **3.** Let R be the annual revenue of my company, and let q be annual sales. $R = 7000$ and $\dfrac{dR}{dt} = -700$. Find $\dfrac{dq}{dt}$. **5.** Let p be the price of a pair of shoes, and let q be the demand for shoes. $\dfrac{dp}{dt} = 5$. Find $\dfrac{dq}{dt}$. **7.** Let T be the average global temperature, and let q be the number of Bermuda shorts sold per year. $T = 60$ and $\dfrac{dT}{dt} = 0.1$. Find $\dfrac{dq}{dt}$. **9. a.** $6/(100\pi) \approx 0.019$ km/sec **b.** $6/(8\sqrt{\pi}) \approx 0.4231$ km/sec **11.** $3/(4\pi) \approx 0.24$ ft/min

13. 7.5 ft/sec   **15.** Decreasing at a rate of $1.66 per player per week   **17.** Monthly sales will drop at a rate of 26 T-shirts per month.   **19.** Raise the price by 3¢ per week.   **21.** The daily operating budget is dropping at a rate of $2.40 per year.
23. The price is decreasing at a rate of approximately 31¢ per pound per month.   **25.** $2300/\sqrt{4100} \approx 36$ miles/hour.
27. 10.7 ft/sec   **29.** The y-coordinate is decreasing at a rate of 16 units per second.   **31.** $1814 per year   **33.** Their prior experience must increase at a rate of approximately 0.97 years every year.

35. $\dfrac{2500}{9\pi}\left(\dfrac{3}{5000}\right)^{2/3} \approx 0.63$ m/sec

37. $\sqrt{\dfrac{1+128\pi}{4\pi}} \approx 1.6$ cm/sec   **39.** 0.5137 computers per household, and increasing at a rate of 0.0230 computers per household per year.   **41.** The average SAT score was 904.71 and decreasing at a rate of 0.11 per year.   **43.** Decreasing by 2 percentage points per year   **45.** The section is called "related rates" because the goal is to compute the rate of change of a quantity based on a knowledge of the rate of change of a related quantity.
47. Answers may vary: A rectangular solid has dimensions 2 cm × 5 cm × 10 cm, and each side is expanding at a rate of 3 cm/second. How fast is the volume increasing?   **51.** Linear
53. Let $x =$ my grades and $y =$ your grades. If $dx/dt = 2\,dy/dt$, then $dy/dt = (1/2)\,dx/dt$.

Section 12.5

1. $E = 1.5$; the demand is going down 1.5% per 1% increase in price at that price level; revenue is maximized when $p = \$25$; weekly revenue at that price is $12,500.   **3. a.** $E = 6/7$; the demand is going down 6% per 7% increase in price at that price level; thus a price increase is in order. **b.** Revenue is maximized when $p = 100/3 \approx \$33.33$ **c.** Demand would be $(100 - 100/3)^2 = (200/3)^2 \approx 4444$ cases per week.
5. a. $E = (4p - 33)/(-2p + 33)$ **b.** 0.54; The demand for $E = mc^2$ T-shirts is going down by about 0.54% per 1% increase in the price. **c.** $11 per shirt for a daily revenue of $1331
7. a. $E = 1.81$. Thus, the demand is elastic at the given tuition level, showing that a decrease in tuition will result in an increase in revenue. **b.** They should charge an average of $2250 per student, and this will result in an enrollment of about 4950 students, giving a revenue of about $11,137,500.   **9. a.** $E = 51$; the demand is going down 51% per 1% increase in price at that price level; thus a large price decrease is advised.
b. Revenue is maximized when $p = ¥0.50$. **c.** Demand would be $100e^{-3/4+1/2} \approx 78$ paint-by-number sets per month.
11. a. $E = -\dfrac{mp}{mp + b}$ **b.** $p = -\dfrac{b}{2m}$ **13. a.** $E = r$
b. E is independent of p. **c.** If $r = 1$, then the revenue is not affected by the price. If $r > 1$, then the revenue is always elastic, while if $r < 1$, the revenue is always inelastic. This is an unrealistic model because there should always be a price at which the revenue is a maximum.   **15. a.** $q = -1500p + 6000$. **b.** $2 per hamburger, giving a total weekly revenue of $6000.
17. $E \approx 0.77$. At a family income level of $20,000, the fraction of children attending a live theatrical performance is increasing by

0.77% per 1% increase in household income.   **19. a.** $E \approx 0.46$. The demand for computers is increasing by 0.46% per one percent increase in household income. **b.** E decreases as income increases. **c.** Unreliable; it predicts a likelihood greater than 1 at incomes of $123,000 and above. In a more appropriate model, one would expect the curve to level off at or below 1.
d. $E \approx 0$ **21.** $\dfrac{Y}{Q} \cdot \dfrac{dQ}{dY} = \beta$. An increase in income of x% will result in an increase in demand of βx%. (Note that we do *not* take the negative here, because we expect an increase in income to produce an *increase* in demand.)   **23. a.** $q = 1000e^{-0.30p}$
b. At $p = \$3$, $E = 0.9$; at $p = \$4$, $E = 1.2$; at $p = \$5$, $E = 1.5$ **c.** $p = \$3.33$ **d.** $p = \$5.36$. Selling at a lower price would increase demand, but you cannot sell more than 200 pounds anyway. You should charge as much as you can and still be able to sell all 200 pounds.   **25.** The price is lowered.
27. Start with $R = pq$, and differentiate with respect to p to obtain $\dfrac{dR}{dp} = q + p\dfrac{dq}{dp}$. For a stationary point, $dR/dp = 0$, and so $q + p\dfrac{dq}{dp} = 0$. Rearranging this result gives $p\dfrac{dq}{dp} = -q$, and hence $-\dfrac{dq}{dp} \cdot \dfrac{p}{q} = 1$, or $E = 1$, showing that stationary points of R correspond to points of unit elasticity.   **29.** The distinction is best illustrated by an example. Suppose that q is measured in weekly sales and p is the unit price in dollars. Then the quantity $-dq/dp$ measures the drop in weekly sales per $1 increase in price. The elasticity of demand E, on the other hand, measures the *percentage* drop in sales per *one percent* increase in price. Thus, $-dq/dp$ measures absolute change, while E measures fractional, or percentage, change.

Chapter 12 Review

1. Relative max: $(-1, 5)$, Absolute min: $(-2, -3)$ and $(1, -3)$
3. Absolute max: $(-1, 5)$, Absolute min: $(1, -3)$ **5.** Absolute min: $(1, 0)$ **7.** Absolute min: $(-2, -1/4)$ **9.** Relative max at $x = 1$, point of inflection at $x = -1$ **11.** Relative max at $x = -2$, relative min at $x = 1$, point of inflection at $x = -1$ **13.** One point of inflection, at $x = 0$ **15. a.** $a = 4/t^4 - 2/t^3$ m/sec² **b.** 2 m/sec²
17. Relative max: $(-2, 16)$; absolute min: $(2, -16)$; point of inflection: $(0, 0)$; no horizontal or vertical asymptotes   **19.** Relative min: $(-3, -2/9)$; relative max: $(3, 2/9)$; inflection: $(-3\sqrt{2}, -5\sqrt{2}/36)$, $(3\sqrt{2}, 5\sqrt{2}/36)$; vertical asymptote: $x = 0$; horizontal asymptote: $y = 0$

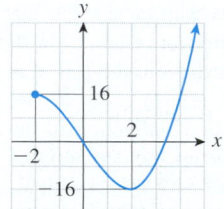

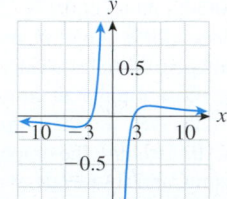

21. Relative max at $(0,0)$, absolute min at $(1, -2)$, no asymptotes

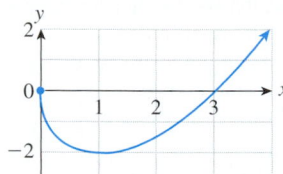

23. \$22.14 per book   **25.** \$24 per copy   **27.** For maximum revenue, the company should charge \$22.14 per copy. At this price, the cost is decreasing at a linear rate with increasing price, while the revenue is not decreasing (its derivative is zero). Thus, the profit is increasing with increasing price, suggesting that the maximum profit will occur at a higher price.

29. $E = \dfrac{2p^2 - 33p}{-p^2 + 33p + 9}$ **31.** \$22.14 per book   **33.** $E = 6$;

The demand is dropping at a rate of 6% per 1% increase in the price.   **35.** Week 5   **37.** 10,500; If weekly sales continue as predicted by the model, they will level off at around 10,500 books per week in the long-term.   **39. a–d.** $10/\sqrt{2}$ ft/sec

Chapter 13

Section 13.1

1. $x^6/6 + C$ **3.** $6x + C$ **5.** $x^2/2 + C$ **7.** $x^3/3 - x^2/2 + C$
9. $x + x^2/2 + C$ **11.** $-x^{-4}/4 + C$
13. $x^{3.3}/3.3 - x^{-0.3}/0.3 + C$ **15.** $u^3/3 - \ln|u| + C$
17. $\dfrac{2x^{3/2}}{3} + C$ **19.** $3x^5/5 + 2x^{-1} - x^{-4}/4 + 4x + C$
21. $2\ln|u| + u^2/8 + C$ **23.** $\ln|x| - \dfrac{2}{x} + \dfrac{1}{2x^2} + C$
25. $3x^{1.1}/1.1 - x^{5.3}/5.3 - 4.1x + C$
27. $\dfrac{x^{0.9}}{0.3} + \dfrac{40}{x^{0.1}} + C$ **29.** $2.55t^2 - 1.2\ln|t| - \dfrac{15}{t^{0.2}} + C$
31. $2e^x + 5\ln|x| + x/4 + C$ **33.** $12.2x^{0.5} + x^{1.5}/9 - e^x + C$
35. $\dfrac{2^x}{\ln 2} - \dfrac{3^x}{\ln 3} + C$ **37.** $\dfrac{100(1.1^x)}{\ln(1.1)} + C$
39. $-1/x - 1/x^2 + C$ **41.** $f(x) = x^2/2 + 1$
43. $f(x) = e^x - x - 1$
45. $C(x) = 5x - x^2/20{,}000 + 20{,}000$
47. $C(x) = 5x + x^2 + \ln x + 994$ **49. a.** $s = t^3/3 + t + C$
b. $C = 1; s = t^3/3 + t + 1$ **51.** 320 ft/s downwards
53. a. $v(t) = -32t + 16$ **b.** $s(t) = -16t^2 + 16t + 185$;
zenith at $t = 0.5$ sec $s = 189$ feet, 4 feet above the top of the tower.   **57.** $(1280)^{1/2} \approx 35.78$ ft/sec   **59. a.** 80 ft/sec
b. 60 ft/sec   **c.** 1.25 seconds   **61.** $\sqrt{2} \approx 1.414$ times as fast
63. $I(t) = 30{,}000 + 1000t$; $I(13) = \$43{,}000$
65. a. $H'(t) = 3.5t + 65$ billion dollars per year
b. $H(t) = 1.75t^2 + 65t + 700$ billion dollars
67. $S(t) \approx \dfrac{17}{3}t^3 + 50t^2 + 2300t - 7503$ million gallons.
Approximately 43,000 million gallons.   **69.** They differ by a constant, $G(x) - F(x) = $ Constant   **71.** Antiderivative, marginal   **73.** $\int f(x)\,dx$ represents the total cost of manufacturing

x items. The units of $\int f(x)\,dx$ are the product of the units of $f(x)$ and the units of x.   **75.** $\int (f(x) + g(x))\,dx$ is, by definition, an antiderivative of $f(x) + g(x)$. Let $F(x)$ be an antiderivative of $f(x)$ and let $G(x)$ be an antiderivative of $g(x)$. Then, because the derivative of $F(x) + G(x)$ is $f(x) + g(x)$ (by the rule for sums of derivatives), this means that $F(x) + G(x)$ is an antiderivative of $f(x) + g(x)$. In symbols, $\int (f(x) + g(x))\,dx = F(x) + G(x) + C = \int f(x)\,dx + \int g(x)\,dx$, the sum of the indefinite integrals.   **77.** $\int x \cdot 1\,dx = \int x\,dx = x^2/2 + C$, whereas $\int x\,dx \cdot \int 1\,dx = (x^2/2 + D) \cdot (x + E)$, which is not the same as $x^2/2 + C$, no matter what values we choose for the constants C, D and E.   **79.** If you take the *derivative* of the *indefinite integral* of $f(x)$, you obtain $f(x)$ back. On the other hand, if you take the *indefinite integral* of the *derivative* of $f(x)$, you obtain $f(x) + C$.

Section 13.2

1. $(3x - 5)^4/12 + C$ **3.** $(3x - 5)^4/12 + C$ **5.** $-e^{-x} + C$
7. $-e^{-x} + C$ **9.** $\dfrac{1}{2}e^{(x+1)^2} + C$ **11.** $(3x + 1)^6/18 + C$
13. $(-2x + 2)^{-1}/2 + C$ **15.** $1.6(3x - 4)^{3/2} + C$
17. $2e^{(0.6x+2)} + C$ **19.** $(3x^2 + 3)^4/24 + C$
21. $(x^2 + 1)^{2.3}/4.6 + C$ **23.** $x + 3e^{3.1x-2} + C$
25. $2(3x^2 - 1)^{3/2}/9 + C$ **27.** $-(1/2)e^{-x^2+1} + C$
29. $-(1/2)e^{-(x^2+2x)} + C$ **31.** $(x^2 + x + 1)^{-2}/2 + C$
33. $(2x^3 + x^6 - 5)^{1/2}/3 + C$
35. $(x - 2)^7/7 + (x - 2)^6/3 + C$
37. $4[(x + 1)^{5/2}/5 - (x + 1)^{3/2}/3] + C$
39. $20\ln|1 - e^{-0.05x}| + C$ **41.** $3e^{-1/x} + C$
43. $(e^x - e^{-x})/2 + C$ **45.** $\ln(e^x + e^{-x}) + C$
47. $(e^{2x^2-2x} + e^{x^2})/2 + C$ **53.** $-e^{-x} + C$
55. $(1/2)e^{2x-1} + C$ **57.** $(2x + 4)^3/6 + C$
59. $(1/5)\ln|5x - 1| + C$ **61.** $(1.5x)^4/6 + C$
63. $\dfrac{1.5^{3x}}{3\ln(1.5)} + C$ **65.** $\dfrac{2^{3x+4} - 2^{-3x+4}}{3\ln 2} + C$
67. $f(x) = (x^2 + 1)^4/8 - 1/8$ **69.** $f(x) = (1/2)e^{x^2-1}$
71. $C(x) = 5x - 1/(x + 1) + 995.5$
73. a. $N(t) = 35\ln(5 + e^{0.2t}) - 63$ **b.** 80,000 articles
75. a. $s = (t^2 + 1)^5/10 + t^2/2 + C$
b. $C = 9/10; s = (t^2 + 1)^5/10 + t^2/2 + 9/10$
77. $S(t) = \dfrac{17}{3}(t - 1990)^3 + 50(t - 1990)^2 +$

$2300(t - 1900) - 7503$ million gallons. $S(2003) \approx 43{,}000$ million gallons.   **79.** None; the substitution $u = x$ simply replaces the letter x throughout by the letter u, and thus does not change the integral at all. For instance, the integral $\int x(3x^2 + 1)\,dx$ becomes $\int u(3u^2 + 1)\,du$ if we substitute $u = x$.   **81.** The purpose of substitution is to introduce a new variable that is defined in terms of the variable of integration. One cannot say $u = u^2 + 1$, because u is not a new variable. Instead, define $w = u^2 + 1$ (or any other letter different from u).   **83.** The integral $\int x(x^2 + 1)\,dx$ can be solved by the substitution $u = x^2 + 1$,

because it leads to $\dfrac{1}{2}\displaystyle\int u\,du = \dfrac{1}{4}u^2 + C = \dfrac{(x^2 + 1)^2}{4} + C$.

Section 13.3

1. 30 **3.** 22 **5.** −2 **7.** 0 **9.** 4 **11.** 6 **13.** 0.7456
15. 2.3129 **17.** 2.5048 **19.** 1 **21.** 1/2 **23.** 1/4 **25.** 2
27. 0 **29.** 6 **31.** 0 **33.** 0.5 **35.** 3.3045, 3.1604, 3.1436
37. 0.0275, 0.0258, 0.0256 **39.** $99.95 **41.** 19 billion gallons
43. $22.5 billion **45. a.** Left sum: about 46,000 articles, Right
sum: about 55,000 articles **b.** 50.5; A total of about 50,500 articles in *Physics Review* were written by researchers in Europe in
the 16-year period beginning 1983. **47.** 54,000 students
49. −$1 billion. **51.** 8160; This represents the total number of
wiretaps authorized by U.S. courts from 1998 through 2003.
53. 91.2 ft **55.** $\int_{-10}^{10} R(t)\, dt \approx \$23,000$. The median household
income rose a total of approximately $23,000 from 1980 to 2000.
57. Yes. The Riemann sum gives an estimated area of 420 square
feet. **59. a.** The area represents the total amount earned by
households in the period 2000 through 2003, in millions of
dollars. **b.** $\int_{10}^{14} A(t)\, dt \approx 26{,}000{,}000$. The total amount earned
by households from 2000 through 2003 was approximately
$26 trillion.

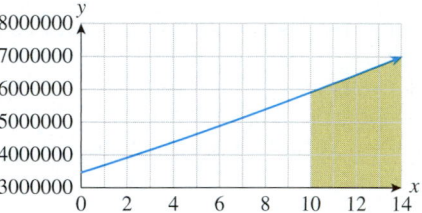

61. a. 99.4% **b.** 0 (to at least 15 decimal places) **63.** Stays
the same. **65.** Increases. **67.** The area under the curve and
above the x-axis equals the area above the curve and below the
x-axis. **69.** Answers will vary. One example: Let $r(t)$ be
the rate of change of net income at time t. If $r(t)$ is negative, then
the net income is decreasing, so the change in net income, represented by the definite integral of $r(t)$, is negative.
71. Answers may vary:

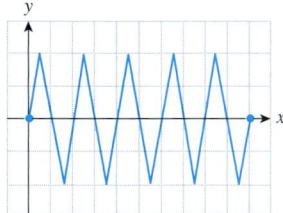

73. The total cost is $c(1) + c(2) + \cdots + c(60)$, which is
represented by the Riemann sum approximation of $\int_1^{61} c(t)\, dt$
with $n = 60$.
75. $[f(x_1) + f(x_2) + \cdots + f(x_n)]\Delta x = \sum_{k=1}^{n} f(x_k)\Delta x$
77. There is no simple way of knowing for certain how accurate
your answer is, but here is a rule of thumb: If increasing n does
not change the value of the answer when rounded to three decimal
places, then the answer can be taken to be accurate to three
decimal places.

Section 13.4

1. 14/3 **3.** 5 **5.** 0 **7.** 40/3 **9.** −0.9045 **11.** $2(e - 1)$
13. 2/3 **15.** $1/\ln 2$ **17.** $4^6 - 1 = 4095$ **19.** $(e^1 - e^{-3})/2$
21. $3/(2\ln 2)$ **23.** $50(e^{-1} - e^{-2})$ **25.** $e^{2.1} - e^{-0.1}$ **27.** 0
29. $(5/2)(e^3 - e^2)$ **31.** $(1/3)[\ln 26 - \ln 7]$ **33.** $\dfrac{0.1}{2.2\ln(1.1)}$
35. $e - e^{1/2}$ **37.** $2 - \ln 3$ **39.** $-4/21$
41. $3^{5/2}/10 - 3^{3/2}/6 + 1/15$ **43.** 1/2 **45.** 16/3 **47.** 56/3
49. 1/2 **51.** $783 **53.** 296 miles **55.** 20 billion gallons
57. $23,000 **59.** 68 milliliters **61.** 907 T-shirts **63.** 9 gallons
67. c. 2,100,000 iPods **69. a.** 8200 wiretaps **b.** The actual
number of wiretaps is 8160, which agrees with the answer in part
(a) to two significant digits. Therefore, the integral in part (a) does
give an accurate estimate. **73.** They are related by the Fundamental Theorem of Calculus, which states (summarized briefly)
that the definite integral of a suitable function can be calculated
by evaluating the indefinite integral at the two endpoints and
subtracting. **75.** The total sales from time a to time b are obtained from the marginal sales by taking its *definite integral* from
a to b. **77.** An example is $v(t) = t - 5$. **79.** An example is
$f(x) = e^{-x}$. **81.** By the FTC, $\int_a^x f(t)\, dt = G(x) - G(a)$
where G is an antiderivative of f. Hence, $F(x) = G(x) - G(a)$.
Taking derivatives of both sides, $F'(x) = G'(x) + 0 = f(x)$, as
required. The result gives us a formula, in terms of area, for an
antiderivative of any continuous function.

Chapter 13 Review

1. $\dfrac{x^3}{3} - 5x^2 + 2x + C$ **3.** $4x^3/15 + 4/(5x) + C$
5. $-e^{-2x+11}/2 + C$ **7.** $\dfrac{1}{22}(x^2 + 4)^{11} + C$ **9.** $-\dfrac{5}{2}e^{-2x} + C$
11. $(x + 2) + \ln|x + 2| + C$ or $x + \ln|x + 2| + C$
13. 1 **15.** −4 **17.** $5/12 \approx 0.4167$
19. 0.7778, 0.7500, 0.7471 **21.** 0 **23.** 1/4
25. $2 + e - e^{-1}$ **27.** 52/9
29. $[\ln 5 - \ln 2]/8 = \ln(2.5)/8$ **31.** 32/3 **33.** $(1 - e^{-25})/2$
35. a. $100{,}000 - 10p^2$; **b.** $100 **37.** 25,000 copies
39. 39,200 hits **41.** About 86,000 books

Chapter 14

Section 14.1

1. $2e^x(x - 1) + C$ **3.** $-e^{-x}(2 + 3x) + C$
5. $e^{2x}(2x^2 - 2x - 1)/4 + C$
7. $-e^{-2x+4}(2x^2 + 2x + 3)/4 + C$
9. $2^x[(2 - x)/\ln 2 + 1/(\ln 2)^2] + C$
11. $-3^{-x}[(x^2 - 1)/\ln 3 + 2x/(\ln 3)^2 + 2/(\ln 3)^3] + C$
13. $-e^{-x}(x^2 + x + 1) + C$
15. $\dfrac{1}{7^x}(x + 2)^7 - \dfrac{1}{56}(x + 2)^8 + C$
17. $-\dfrac{x}{2(x - 2)^2} - \dfrac{1}{2(x - 2)} + C$
19. $(x^4 \ln x)/4 - x^4/16 + C$
21. $(t^3/3 + t)\ln(2t) - t^3/9 - t + C$

23. $(3/4)t^{4/3}(\ln t - 3/4) + C$ **25.** $x \log_3 x - x/\ln 3 + C$
27. $e^{2x}(x/2 - 1/4) - 4e^{3x}/3 + C$
29. $e^x(x^2 - 2x + 2) - e^{x^2}/2 + C$ **31.** e **33.** $38229/286$
35. $(7/2)\ln 2 - 3/4$ **37.** $1/4$ **39.** $1 - 11e^{-10}$
41. $4\ln 2 - 7/4$ **43.** $28{,}800{,}000(1 - 2e^{-1})$ ft.
45. $5001 + 10x - 1/(x + 1) - [\ln(x + 1)]/(x + 1)$
47. \$33,598   **49.** \$170,000 million   **51.** Answers will vary.
Examples are xe^{x^2} and $e^{x^2} = 1 \cdot e^{x^2}$ **53.** $n + 1$ times

Section 14.2
1. $8/3$ **3.** 4 **5.** 1 **7.** $e - 3/2$ **9.** $2/3$ **11.** $3/10$
13. $1/20$ **15.** $4/15$ **17.** $2\ln 2 - 1$ **19.** $8\ln 4 + 2e - 16$
21. 0.9138 **23.** 0.3222 **25.** 112.5. This represents your total
profit for the week, \$112.50.   **27. a.** The area represents the ac-
cumulated U.S. trade deficit with China (total excess value of imports over exports) for the 8-year period 1996–2004. **b.** 640. The
U.S. accumulated a \$640 billion trade deficit with China over the
period 1996–2004.   **29. a.** \$3600 billion. **b.** This is the area of
the region between the graphs of $P(t)$ and $I(t)$ for $10 \le t \le 20$.
31. The area between the export and import curves represents
Canada's accumulated trade surplus (that is, the total excess of
exports over imports) from January, 1997 to January, 2001.
33. (A)   **35.** The claim is wrong because the area under a curve
can only represent income if the curve is a graph of income *per
unit time*. The value of a stock price is not income per unit time—
the income can only be realized when the stock is sold, and it
amounts to the current market price. The total net income (per
share) from the given investment would be the stock price on the
date of sale minus the purchase price of \$40.

Section 14.3
1. Average $= 2$

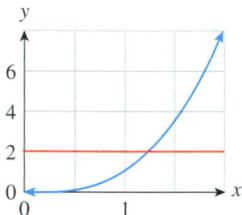

3. Average $= 1$

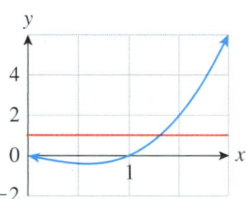

5. Average $= (1 - e^{-2})/2$

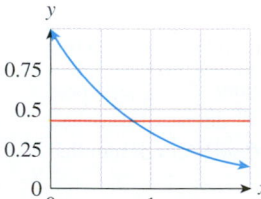

7.

x	0	1	2	3	4	5	6	7
$r(x)$	3	5	10	3	2	5	6	7
$\bar{r}(x)$			6	6	5	10/3	13/3	6

9.

x	0	1	2	3	4	5	6	7
$r(x)$	1	2	6	7	11	15	10	2
$\bar{r}(x)$			3	5	8	11	12	9

11. Moving average:
$\bar{f}(x) = x^3 - (15/2)x^2 + 25x - 125/4$

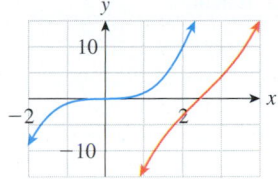

13. Moving average:
$\bar{f}(x) = (3/25)[x^{5/3} - (x - 5)^{5/3}]$

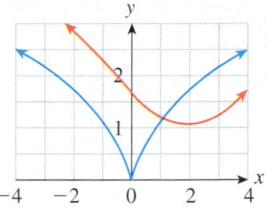

15. $\bar{f}(x) = \dfrac{2}{5}(e^{0.5x} - e^{0.5(x-5)})$

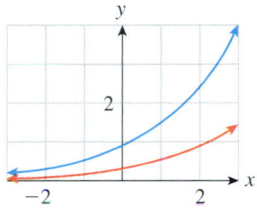

17. $\bar{f}(x) = \dfrac{2}{15}(x^{3/2} - (x - 5)^{3/2})$

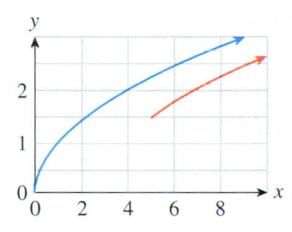

19.

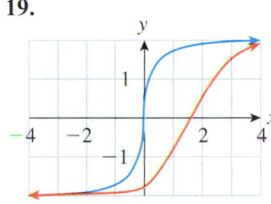

21.

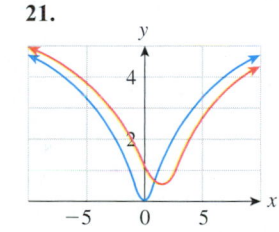

23. 127 million people   **25.** \$1.7345 million   **27.** \$10,410.88
29. \$1500
31.

Year t	1995	1996	1997	1998	1999	2000	2001	2002	2003	2004
Employment (millions)	117	120	123	126	129	132	132	130	130	131
Moving average (millions)				122	125	128	130	131	131	131

Some changes are larger, and others are smaller.

33. a.

— Spending   — Moving Average

b. $31 billion per year; Public spending on health care in the U.S. was increasing at a rate of approximately $31 billion per year during the given period.

35. a. 4300 million gallons per year
b. $\dfrac{1}{2}\left[\dfrac{17}{3}[t^3 - (t-2)^3] + 50[t^2 - (t-2)^2] + 4600\right]$
c. Quadratic   **37. a.** $s = 14.4t + 240$
b. $\bar{s}(t) = 14.4t + 211.2$ **c.** The slope of the moving average is the same as the slope of the original function.
39. $\bar{f}(x) = mx + b - \dfrac{ma}{2}$
41. a.

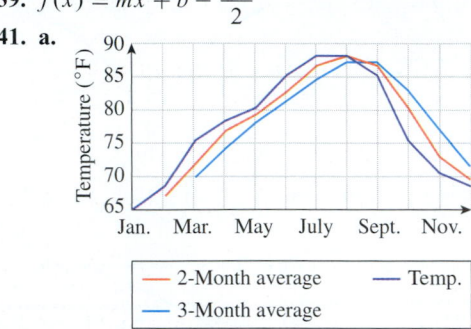

b. The 24-month moving average is constant and equal to the year-long average of approximately 77°. **c.** A quadratic model could not be used to predict temperatures beyond the given 12-month period, since temperature patterns are periodic, whereas parabolas are not.   **43.** The moving average "blurs" the effects of short-term oscillations in the price, and shows the longer-term trend of the stock price.   **45.** The area above the x-axis equals the area below the x-axis. Example: $y = x$ on $[-1, 1]$ **47.** This need not be the case; for instance, the function $f(x) = x^2$ on $[0, 1]$ has average value $1/3$, whereas the value midway between the maximum and minimum is $1/2$.
49. (C). A shorter term moving average most closely approximates the original function, since it averages the function over a shorter period, and continuous functions change by only a small amount over a small period.

Section 14.4

1. $6.25   **3.** $512   **5.** $119.53   **7.** $900   **9.** $416.67
11. $326.27   **13.** $25   **15.** $0.50   **17.** $386.29   **19.** $225
21. $25.50   **23.** $12,684.63   **25.** $TV = \$300,000$,
$FV = \$434,465.45$ **27.** $TV = \$350,000$, $FV = \$498,496.61$
29. $TV = \$389,232.76$, $FV = \$547,547.16$
31. $TV = \$100,000$, $PV = \$82,419.99$
33. $TV = \$112,500$, $PV = \$92,037.48$
35. $TV = \$107,889.50$, $PV = \$88,479.69$ **37.** $\bar{p} = \$5000$,
$\bar{q} = 10,000$, $CS = \$25$ million, $PS = \$100$ million. The total social gain is $125 million.   **39.** $CS = \dfrac{1}{2m}(b - m\bar{p})^2$
41. €140 billion   **43.** $940 billion   **45.** €160 billion
47. $850 billion   **49.** $1,943,162.44   **51.** $3,086,245.73
53. $58,961.74   **55.** $1,792,723.35   **57.** Total   **59.** She is correct, provided there is a positive rate of return, in which case the future value (which includes interest) is greater than the total value (which does not).   **61.** $PV < TV < FV$

Section 14.5

1. Diverges   **3.** Converges to $2e$ **5.** Converges to e^2
7. Converges to $1/2$ **9.** Converges to $1/108$ **11.** Converges to $3 \times 5^{2/3}$ **13.** Diverges   **15.** Diverges   **17.** Converges to $\dfrac{5}{4}(3^{4/5} - 1)$.   **19.** Diverges   **21.** Converges to 0
23. Diverges   **25.** Diverges   **27.** $870 million   **29.** 7100 billion cigarettes   **31.** No; You will not sell more than 2000 of them.   **33.** The integral diverges, and so the number of graduates each year will rise without bound.   **35. a.** $R(t) = 350e^{-0.1t}(39t + 68)$ million dollars/yr;   **b.** $1,603,000 million
37. $27,000 billion   **39.** $\int_0^{+\infty} N(t)\,dt$ diverges, indicating that there is no bound to the expected future total online sales of books. $\int_{-\infty}^0 N(t)\,dt$ converges to approximately 1.889, indicating that total online sales of books prior to 1997 amounted to approximately 1.889 million books   **41.** 1   **43.** 0.1587
45. $70,833   **47. a.** 2.468 meteors on average **b.** The integral diverges. We can interpret this as saying that the number of impacts by meteors smaller than 1 megaton is very large. (This makes sense because, for example, this number includes meteors no larger than a grain of dust.)   **49. a.** $\Gamma(1) = 1$; $\Gamma(2) = 1$
51. The integral does not converge, so the number given by the FTC is meaningless.   **53.** Yes; the integrals converge to 0, and the FTC also gives 0.   **55.** In all cases, you need to rewrite the improper integral as a limit and use technology to evaluate the integral of which you are taking the limit. Evaluate for several values of the endpoint approaching the limit. In the case of an integral in which one of the limits of integration is infinite, you may have to instruct the calculator or computer to use more subdivisions as you approach $+\infty$.

Section 14.6

1. $y = \dfrac{x^3}{3} + \dfrac{2x^{3/2}}{3} + C$ **3.** $\dfrac{y^2}{2} = \dfrac{x^2}{2} + C$ **5.** $y = Ae^{x^2/2}$
7. $y = -\dfrac{2}{(x+1)^2 + C}$ **9.** $y = \pm\sqrt{(\ln x)^2 + C}$
11. $y = \dfrac{x^4}{4} - x^2 + 1$ **13.** $y = (x^3 + 8)^{1/3}$ **15.** $y = 2x$
17. $y = e^{x^2/2} - 1$ **19.** $y = -\dfrac{2}{\ln(x^2 + 1) + 2}$
21. With $s(t) =$ monthly sales after t months, $\dfrac{ds}{dt} = -0.05s$;
$s = 1000$ when $t = 0$. Solution: $s = 1000e^{-0.05t}$ quarts per month.   **23.** $H(t) = 75 + 115e^{-0.04274t}$ degrees Fahrenheit after t minutes.   **25.** With $S(t) =$ total sales after t months,
$\dfrac{dS}{dt} = 0.1(100,000 - S)$; $S(0) = 0$.
Solution: $S = 100,000(1 - e^{-0.1t})$ monitors after t months.
27. a. $\dfrac{dp}{dt} = k(D(p) - S(p)) = k(20,000 - 1,000p)$
b. $p = 20 - Ae^{-kt}$ **c.** $p = 20 - 10e^{-0.2231t}$ dollars after t months.   **29.** $q = 0.6078e^{-0.05p}p^{1.5}$

33. $S = \dfrac{2/1999}{e^{-0.5t} + 1/1999}$

It will take about 27 months to saturate the market. Graph:

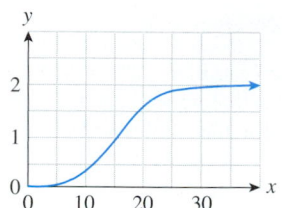

35. a. $y = be^{Ae^{-at}}$, $A = $ constant
b. $y = 10e^{-0.69315e^{-t}}$ Graph:

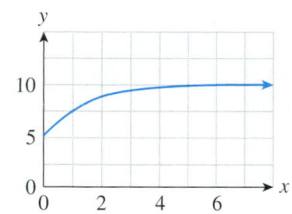

37. A general solution gives all possible solutions to the equation, using at least one arbitrary constant. A particular solution is one specific function that satisfies the equation. We obtain a particular solution by substituting specific values for any arbitrary constants in the general solution.   **39.** Example: $y'' = x$ has general solution $y = \dfrac{1}{6}x^3 + Cx + D$ (integrate twice).

41. $y' = -4e^{-x} + 3$

Chapter 14 Review

1. $(x^2 - 2x + 4)e^x + C$ **3.** $(1/3)x^3 \ln 2x - x^3/9 + C$

5. $-e^2 - 39/e^2$ **7.** $1/4$ **9.** $3(3^{2/3} - 1)/2$ **11.** $\dfrac{3}{2 \cdot 2^{1/3}} - \dfrac{1}{2}$

13. $\dfrac{2\sqrt{2}}{3}$ **15.** -1 **17.** $e - 2$ **19.** $3x - 2$

21. $\dfrac{3}{14}[x^{7/3} - (x-2)^{7/3}]$ **23.** $1600   **25.** $2500

27. $y = -\dfrac{3}{x^3 + C}$ **29.** $y = \sqrt{2\ln|x| + 1}$

31. a. $1,062,500; **b.** $997,500e^{0.06t}$ **33.** Approximately $910,000   **35.** $51 million

Chapter 15

Section 15.1

1. a. 1 **b.** 1 **c.** 2 **d.** $a^2 - a + 5$ **e.** $y^2 + x^2 - y + 1$
f. $(x + h)^2 + (y + k)^2 - (x + h) + 1$ **3. a.** 0 **b.** 0.2
c. -0.1 **d.** $0.18a + 0.2$ **e.** $0.1x + 0.2y - 0.01xy$
f. $0.2(x + h) + 0.1(y + k) - 0.01(x + h)(y + k)$
5. a. 1 **b.** e **c.** e **d.** e^{x+y+z} **e.** $e^{x+h+y+k+z+l}$
7. a. Does not exist **b.** 0 **c.** 0 **d.** $xyz/(x^2 + y^2 + z^2)$
e. $(x + h)(y + k)(z + l)/[(x + h)^2 + (y + k)^2 + (z + l)^2]$
9. a. Increases; 2.3 **b.** Decreases; 1.4 **c.** Decreases; 1 unit increase in z **11.** Neither   **13.** Linear   **15.** Linear
17. Interaction   **19. a.** 107 **b.** -14 **c.** -113
21.

$x \rightarrow$	10	20	30	40
y ↓ 10	52	107	162	217
20	94	194	294	394
30	136	281	426	571
40	178	368	558	748

25. 18, 4, 0.0965, 47,040   **27.** 6.9078, 1.5193, 5.4366, 0
29. Let $z = $ annual sales of Z (in millions of dollars), $x = $ annual sales of X, and $y = $ annual sales of Y. The model is $z = -2.1x + 0.4y + 16.2$ **31.** $\sqrt{2}$ **33.** $\sqrt{a^2 + b^2}$
35. $1/2$ **37.** Circle with center $(2, -1)$ and radius 3
39. The marginal cost of cars is $6000 per car. The marginal cost of trucks is $4000 per truck.
41. $C(x, y) = 10 + 0.03x + 0.04y$ where C is the cost in dollars, $x = $ # video clips sold per month, $y = $ # audio clips sold per month   **43. a.** 28% **b.** 21% **c.** Percentage points per year
45. a. 29% **b.** 7% **c.** 0.4-point drop   **47. a.** $9980
b. $R(z) = 9850 + 0.04z$ **49.** $s(c, t) = 1.2c - 2t$
51. $U(11, 10) - U(10, 10) \approx 5.75$. This means that, if your company now has 10 copies of Macro Publish and 10 copies of Turbo Publish, then the purchase of one additional copy of Macro Publish will result in a productivity increase of approximately 5.75 pages per day.   **53. a.** Answers will vary. $(a, b, c) = (3, 1/4, 1/\pi); (a, b, c) = (1/\pi, 3, 1/4)$.

b. $a = \left(\dfrac{3}{4\pi}\right)^{1/3}$. The resulting ellipsoid is a sphere with radius a.

55. 7,000,000   **57. a.** $100 = K(1,000)^a(1,000,000)^{1-a}$; $10 = K(1,000)^a(10,000)^{1-a}$ **b.** $\log K - 3a = -4$; $\log K - a = -3$ **c.** $a = 0.5, K \approx 0.003162$
d. $P = 71$ pianos (to the nearest piano)   **59. a.** 4×10^{-3} gram per square meter **b.** The total weight of sulfates in the Earth's atmosphere   **61. a.** The value of N would be doubled.
b. $N(R, f_p, n_e, f_l, f_i, L) = Rf_pn_ef_lf_iL$, where here L is the average lifetime of an intelligent civilization **c.** Take the logarithm of both sides, since this would yield the linear function $\ln(N) = \ln(R) + \ln(f_p) + \ln(n_e) + \ln(f_l) + \ln(f_i) + \ln(f_c) + \ln(L)$.
63. a.   **b.**   **c.**

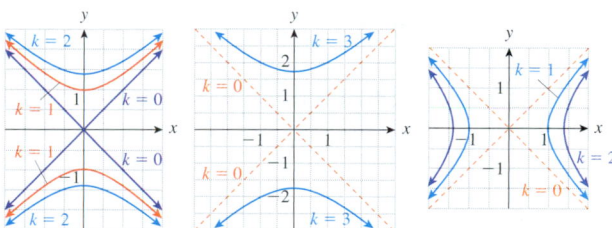

65. They are reciprocals of each other.   **67.** For example, $f(x, y) = x^2 + y^2$.   **69.** For example, $f(x, y, z) = xyz$
71. For example, take $f(x, y) = x + y$. Then setting $y = 3$ gives $f(x, 3) = x + 3$. This can be viewed as a function of the single variable x. Choosing other values for y gives other functions of x.   **73.** If $f = ax + by + c$, then fixing $y = k$ gives $f = ax + (bk + c)$, a linear function with slope a and intercept $bk + c$. The slope is independent of the choice of $y = k$.
75. CDs cost more than cassettes.

Section 15.2

1.

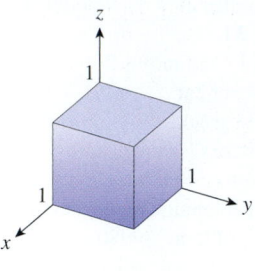

3.

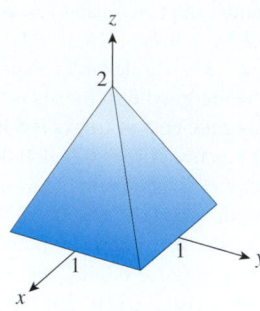

5.

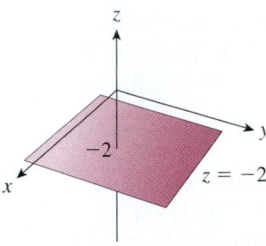

7.

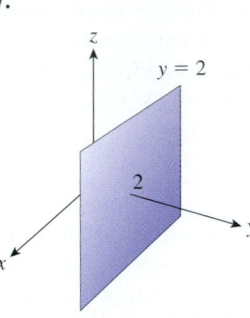

9.

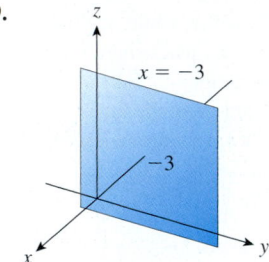

11. (H)   **13.** (B)   **15.** (F)   **17.** (C)

19.

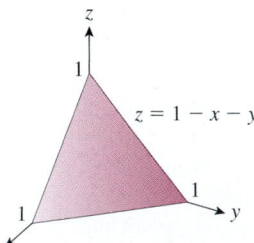

21.

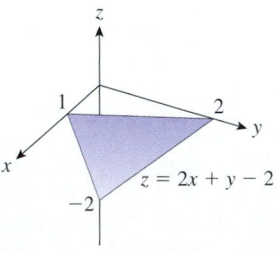

23.

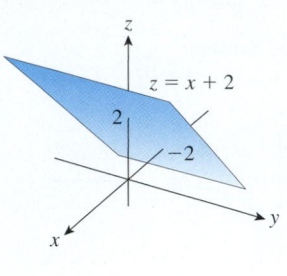

25.

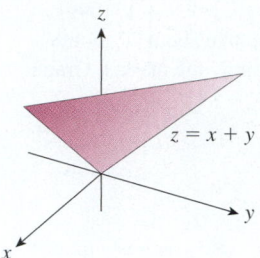

27.

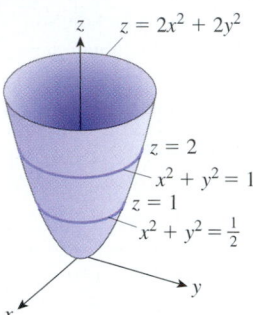

29.

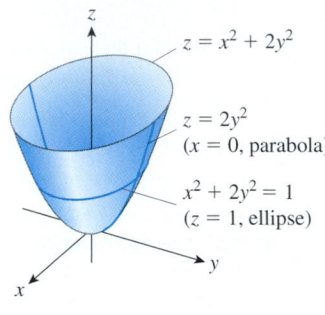

31.

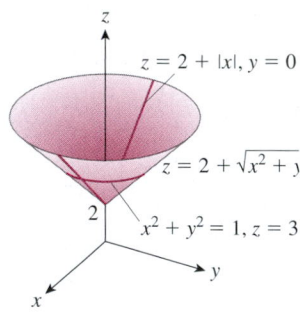

33.

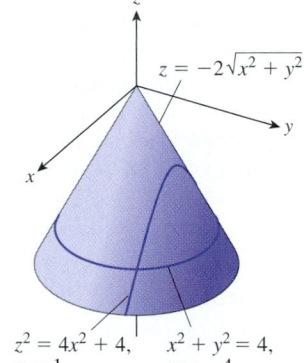

35.

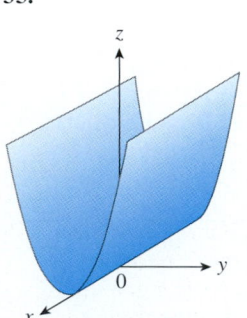

37.

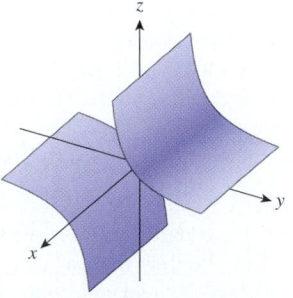

39.

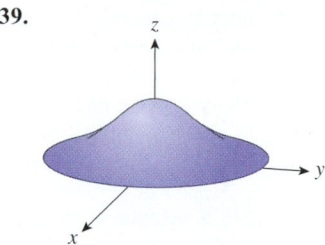

41. a. The graph is a plane with x-intercept -40, y-intercept -60, and z-intercept $240{,}000$. **b.** The slice $x = 10$ is the straight line with equation $z = 300{,}000 + 4000y$. It describes the cost function for the manufacture of trucks if car production is held fixed at 10 cars per week. **c.** The level curve $z = 480{,}000$ is the straight line $6000x + 4000y = 240{,}000$. It describes the number of cars and trucks you can manufacture to maintain weekly costs at $480{,}000. **43.** The graph is a plane with x_1-intercept 0.3, x_2-intercept 33, and x_3-intercept 0.66. The slices by $x_1 =$ constant are straight lines that are parallel to each other. Thus, the rate of change of General Motors' share as a function of Ford's share does not depend on Chrysler's share. Specifically, GM's share decreases by 0.02 percentage points per 1 percentage-point increase in Ford's market share, regardless of Chrysler's share. **45. a.** The slices $x =$ constant and $y =$ constant are straight lines. **b.** No. Even though the slices $x =$ constant and $y =$ constant are straight lines, the level curves are not, and so the surface is not a plane. **c.** The slice $x = 10$ has a slope of 3800. The slice $x = 20$ has a slope of 3600. Manufacturing more cars lowers the marginal cost of manufacturing trucks. **47.** Both level curves are quarter-circles. (We see only the portion in the first quadrant because $e \geq 0$ and $k \geq 0$.) The level curve $C = 30{,}000$ represents the relationship between the number of electricians and the number of carpenters used in building a home that costs $30{,}000. Similarly for the level curve $C = 40{,}000$. **49.** The following figure shows several level curves together with several lines of the form $h + w = c$.

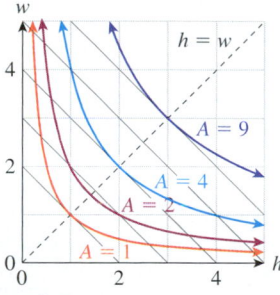

From the figure, thinking of the curves as contours on a map, we see that the largest value of A anywhere along any of the lines $h + w = c$ occurs midway along the line, when $h = w$. Thus, the largest area rectangle with a fixed perimeter occurs when $h = w$ (that is, when the rectangle is a square).

51. The level curve at $z = 3$ has the form $3 = x^{0.5}y^{0.5}$, or $y = 9/x$, and shows the relationship between the number of workers and the operating budget at a production level of 3 units.

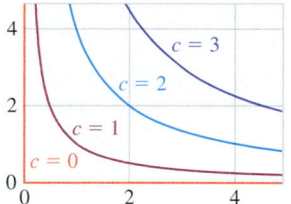

53. The level curve at $z = 0$ consists of the nonnegative y-axis ($x = 0$) and tells us that zero utility corresponds to zero copies of Macro Publish, regardless of the number of copies of Turbo Publish. (Zero copies of Turbo Publish does not necessarily result in zero utility, according to the formula.)

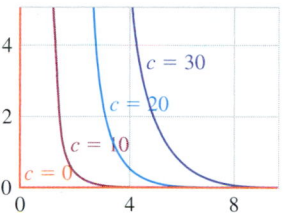

55. Plane **57.** Agree: any slice through a plane is a straight line. **59.** The graph of a function of three or more variables lives in four-dimensional (or higher) space, which makes it difficult to draw and visualize. **63.** We need one dimension for each of the variables plus one dimension for the value of the function.

Section 15.3

1. $f_x(x, y) = -40$; $f_y(x, y) = 20$; $f_x(1, -1) = -40$; $f_y(1, -1) = 20$ **3.** $f_x(x, y) = 6x + 1$; $f_y(x, y) = -3y^2$; $f_x(1, -1) = 7$; $f_y(1, -1) = -3$ **5.** $f_x(x, y) = -40 + 10y$; $f_y(x, y) = 20 + 10x$; $f_x(1, -1) = -50$; $f_y(1, -1) = 30$
7. $f_x(x, y) = 6xy$; $f_y(x, y) = 3x^2$; $f_x(1, -1) = -6$; $f_y(1, -1) = 3$ **9.** $f_x(x, y) = 2xy^3 - 3x^2y^2 - y$; $f_y(x, y) = 3x^2y^2 - 2x^3y - x$; $f_x(1, -1) = -4$; $f_y(1, -1) = 4$
11. $f_x(x, y) = 6y(2xy + 1)^2$; $f_y(x, y) = 6x(2xy + 1)^2$; $f_x(1, -1) = -6$; $f_y(1, -1) = 6$ **13.** $f_x(x, y) = e^{x+y}$; $f_y(x, y) = e^{x+y}$; $f_x(1, -1) = 1$; $f_y(1, -1) = 1$
15. $f_x(x, y) = 3x^{-0.4}y^{0.4}$; $f_y(x, y) = 2x^{0.6}y^{-0.6}$; $f_x(1, -1)$ undefined; $f_y(1, -1)$ undefined **17.** $f_x(x, y) = 0.2ye^{0.2xy}$; $f_y(x, y) = 0.2xe^{0.2xy}$; $f_x(1, -1) = -0.2e^{-0.2}$; $f_y(1, -1) = 0.2e^{-0.2}$ **19.** $f_{xx}(x, y) = 0$; $f_{yy}(x, y) = 0$; $f_{xy}(x, y) = f_{yx}(x, y) = 0$; $f_{xx}(1, -1) = 0$; $f_{yy}(1, -1) = 0$; $f_{xy}(1, -1) = f_{yx}(1, -1) = 0$ **21.** $f_{xx}(x, y) = 0$; $f_{yy}(x, y) = 0$; $f_{xy}(x, y) = f_{yx}(x, y) = 10$; $f_{xx}(1, -1) = 0$; $f_{yy}(1, -1) = 0$; $f_{xy}(1, -1) = f_{yx}(1, -1) = 10$
23. $f_{xx}(x, y) = 6y$; $f_{yy}(x, y) = 0$; $f_{xy}(x, y) = f_{yx}(x, y) = 6x$; $f_{xx}(1, -1) = -6$; $f_{yy}(1, -1) = 0$; $f_{xy}(1, -1) = f_{yx}(1, -1) = 6$ **25.** $f_{xx}(x, y) = e^{x+y}$; $f_{yy}(x, y) = e^{x+y}$; $f_{xy}(x, y) = f_{yx}(x, y) = e^{x+y}$; $f_{xx}(1, -1) = 1$; $f_{yy}(1, -1) = 1$; $f_{xy}(1, -1) = f_{yx}(1, -1) = 1$

27. $f_{xx}(x, y) = -1.2x^{-1.4}y^{0.4}$; $f_{yy}(x, y) = -1.2x^{0.6}y^{-1.6}$; $f_{xy}(x, y) = f_{yx}(x, y) = 1.2x^{-0.4}y^{-0.6}$; $f_{xx}(1, -1)$ undefined; $f_{yy}(1, -1)$ undefined; $f_{xy}(1, -1)$ & $f_{yx}(1, -1)$ undefined   **29.** $f_x(x, y, z) = yz$; $f_y(x, y, z) = xz$; $f_z(x, y, z) = xy$; $f_x(0, -1, 1) = -1$; $f_y(0, -1, 1) = 0$; $f_z(0, -1, 1) = 0$ **31.** $f_x(x, y, z) = 4/(x + y + z^2)^2$; $f_y(x, y, z) = 4/(x + y + z^2)^2$; $f_z(x, y, z) = 8z/(x + y + z^2)^2$; $f_x(0, -1, 1)$ undefined; $f_y(0, -1, 1)$ undefined; $f_z(0, -1, 1)$ undefined   **33.** $f_x(x, y, z) = e^{yz} + yze^{xz}$; $f_y(x, y, z) = xze^{yz} + e^{xz}$; $f_z(x, y, z) = xy(e^{yz} + e^{xz})$; $f_x(0, -1, 1) = e^{-1} - 1$; $f_y(0, -1, 1) = 1$; $f_z(0, -1, 1) = 0$ **35.** $f_x(x, y, z) = 0.1x^{-0.9}y^{0.4}z^{0.5}$; $f_y(x, y, z) = 0.4x^{0.1}y^{-0.6}z^{0.5}$; $f_z(x, y, z) = 0.5x^{0.1}y^{0.4}z^{-0.5}$; $f_x(0, -1, 1)$ undefined; $f_y(0, -1, 1)$ undefined, $f_z(0, -1, 1)$ undefined   **37.** $f_x(x, y, z) = yze^{xyz}$, $f_y(x, y, z) = xze^{xyz}$, $f_z(x, y, z) = xye^{xyz}$; $f_x(0, -1, 1) = -1$; $f_y(0, -1, 1) = f_z(0, -1, 1) = 0$ **39.** $f_x(x, y, z) = 0$; $f_y(x, y, z) = -\dfrac{600z}{y^{0.7}(1 + y^{0.3})^2}$; $f_z(x, y, z) = \dfrac{2{,}000}{1 + y^{0.3}}$; $f_x(0, -1, 1)$ undefined; $f_y(0, -1, 1)$ undefined; $f_z(0, -1, 1)$ undefined   **41.** $\partial C/\partial x = 6000$, the marginal cost to manufacture each car is $6000. $\partial C/\partial y = 4000$, the marginal cost to manufacture each truck is $4000.   **43.** $\partial y/\partial t = -0.78$. The number of articles written by researchers in the U.S. was decreasing at a rate of 0.78 percentage points per year. $\partial y/\partial x = -1.02$. The number of articles written by researchers in the U.S. was decreasing at a rate of 1.02 percentage points per one percentage point increase in articles written in Europe.   **45.** $5600 per car   **47. a.** $\partial M/\partial c = -3.8$, $\partial M/\partial f = 2.2$. For every 1 point increase in the percentage of Chrysler owners who remain loyal, the percentage of Mazda owners who remain loyal decreases by 3.8 points. For every 1 point increase in the percentage of Ford owners who remain loyal, the percentage of Mazda owners who remain loyal increases by 2.2 points.   **b.** 16%   **49. a.** $16,500 **b.** $28,600 **c.** $350 per year **d.** $570 per year **e.** Widening   **51.** The marginal cost of cars is $6000 + 1{,}000e^{-0.01(x+y)}$ per car. The marginal cost of trucks is $4000 + 1{,}000e^{-0.01(x+y)}$ per truck. Both marginal costs decrease as production rises.   **53.**
$$\bar{C}(x, y) = \frac{200{,}000 + 6{,}000x + 4{,}000y - 100{,}000e^{-0.01(x+y)}}{x + y};$$
$\bar{C}_x(50, 50) = -$2.64$ per car. This means that at a production level of 50 cars and 50 trucks per week, the average cost per vehicle is decreasing by $2.64 for each additional car manufactured. $\bar{C}_y(50, 50) = -$22.64$ per truck. This means that at a production level of 50 cars and 50 trucks per week, the average cost per vehicle is decreasing by $22.64 for each additional truck manufactured.   **55.** No; your marginal revenue from the sale of cars is $\$15{,}000 - \dfrac{2{,}500}{\sqrt{x + y}}$ per car and $\$10{,}000 - \dfrac{2{,}500}{\sqrt{x + y}}$ per truck from the sale of trucks. These increase with increasing x and y. In other words, you will earn more revenue per vehicle with increasing sales, and so the rental company will pay more for each additional vehicle it buys.

57. $P_z(10,100{,}000, 1{,}000{,}000) \approx 0.0001010$ papers/$ **59.** $U_x(10, 5) = 5.18$, $U_y(10, 5) = 2.09$. This means that, if 10 copies of Macro Publish and 5 copies of Turbo Publish are purchased, the company's daily productivity is increasing at a rate of 5.18 pages per day for each additional copy of Macro purchased and by 2.09 pages per day for each additional copy of Turbo purchased. **b.** $\dfrac{U_x(10, 5)}{U_y(10, 5)} \approx 2.48$ is the ratio of the usefulness of one additional copy of Macro to one of Turbo. Thus, with 10 copies of Macro and 5 copies of Turbo, the company can expect approximately 2.48 times the productivity per additional copy of Macro compared to Turbo.   **61.** 6×10^9 N/sec   **63. a.** $A_P(100, 0.1, 10) = 2.59$; $A_r(100, 0.1, 10) = 2{,}357.95$; $A_t(100, 0.1, 10) = 24.72$. Thus, for a $100 investment at 10% interest, after 10 years the accumulated amount is increasing at a rate of $2.59 per $1 of principal, at a rate of $2,357.95 per increase of 1 in r (note that this would correspond to an increase in the interest rate of 100%), and at a rate of $24.72 per year.   **b.** $A_P(100, 0.1, t)$ tells you the rate at which the accumulated amount in an account bearing 10% interest with a principal of $100 is growing per $1 increase in the principal, t years after the investment.   **65. a.** $P_x = Ka\left(\dfrac{y}{x}\right)^b$ and $P_y = Kb\left(\dfrac{x}{y}\right)^a$. They are equal precisely when $\dfrac{a}{b} = \left(\dfrac{x}{y}\right)^b \left(\dfrac{x}{y}\right)^a$. Substituting $b = 1 - a$ now gives $\dfrac{a}{b} = \dfrac{x}{y}$.   **b.** The given information implies that $P_x(100, 200) = P_y(100, 200)$. By part (a), this occurs precisely when $a/b = x/y = 100/200 = 1/2$. But $b = 1 - a$, so $a/(1 - a) = 1/2$, giving $a = 1/3$ and $b = 2/3$.   **67.** Decreasing at 0.0075 parts of nutrient per part of water/sec   **69.** f is increasing at a rate of s units per unit of x, f is increasing at a rate of t units per unit of y, and the value of f is r when $x = a$ and $y = b$ **71.** The marginal cost of building an additional orbicus; zonars per unit.   **73.** Answers will vary. One example is $f(x, y) = -2x + 3y$. Others are $f(x, y) = -2x + 3y + 9$ and $f(x, y) = xy - 3x + 2y + 10$.   **75. a.** b is the z-intercept of the plane. m is the slope of the intersection of the plane with the xz-plane. n is the slope of the intersection of the plane with the yz-plane.   **b.** Write $z = b + rx + sy$. We are told that $\partial z/\partial x = m$, so $r = m$. Similarly, $s = n$. Thus, $z = b + mx + ny$. We are also told that the plane passes through (h, k, l). Substituting gives $l = b + mh + nk$. This gives b as $l - mh - nk$. Substituting in the equation for z therefore gives $z = l - mh - nk + mx + ny = l + m(x - h) + n(y - k)$, as required.

Section 15.4

1. P: relative minimum; Q: none of the above; R: relative maximum   **3.** P: saddle point; Q: relative maximum; R: none of the above   **5.** Relative minimum   **7.** Neither   **9.** Saddle point   **11.** Relative minimum at $(0, 0, 1)$ **13.** Relative maximum at $(-1/2, 1/2, 3/2)$ **15.** Relative maximum at $(0, 0, 0)$, saddle points at $(\pm 4, 2, -16)$ **17.** Relative minimum at $(0, 0, 1)$ **19.** Relative minimum at $(-2, \pm 2, -16)$, $(0, 0)$ a critical point that is not a relative extremum   **21.** Saddle point at $(0, 0, -1)$ **23.** Relative maximum at $(-1, 0, e)$ **25.** Relative minimum at $(2^{1/3}, 2^{1/3}, 3(2^{2/3}))$ **27.** Relative minimum at $(1, 1, 4)$ and $(-1, -1, 4)$ **29.** Absolute minimum at $(0, 0, 1)$ **31.** None; the relative maximum at $(0, 0, 0)$ is not absolute. (look at, say, $(10, 10)$).   **33.** Minimum of $1/3$ at $(c, f) = (2/3, 2/3)$. Thus, at least $1/3$ of all Mazda owners would choose another new Mazda, and this lowest loyalty occurs when $2/3$ of Chrysler and Ford owners remain loyal to their brands.   **35.** It should remove 2.5 pounds of sulfur and 1 pound of lead per day.   **37.** You should charge $580.81 for the Ultra Mini and $808.08 for the Big Stack.   **39.** $l = w = h \approx 20.67$ in, volume ≈ 8827 cubic inches.   **41.** 18 in $\times$ 18 in $\times$ 36 in, volume $= 11,664$ cubic inches

43.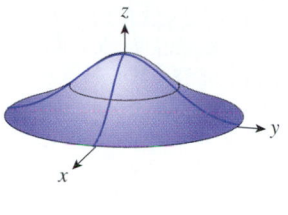

45.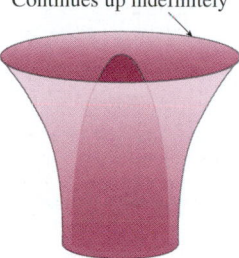

Continues up indefinitely

Continues down indefinitely
Function not defined on circle

47. H must be positive.   **49.** No. In order for there to be a relative maximum at (a, b), *all* vertical planes through (a, b) should yield a curve with a relative maximum at (a, b). It could happen that a slice by another vertical plane through (a, b) (such as $x - a = y - b$) does not yield a curve with a relative maximum at (a, b). [An example is $f(x, y) = x^2 + y^2 - \sqrt{xy}$, at the point $(0, 0)$. Look at the slices through $x = 0$, $y = 0$ and $y = x$.]   **51.** $\bar{C}_x = \dfrac{\partial}{\partial x}\left(\dfrac{C}{x+y}\right) = \dfrac{(x+y)C_x - C}{(x+y)^2}$. If this is zero, then $(x+y)C_x = C$, or $C_x = \dfrac{C}{x+y} = \bar{C}$. Similarly, if $\bar{C}_y = 0$ then $C_y = \bar{C}$. This is reasonable because if the average cost is decreasing with increasing x, then the average cost is greater than the marginal cost C_x. Similarly, if the average cost is increasing with increasing x, then the average cost is less than the marginal cost C_x. Thus, if the average cost is stationary with increasing x, then the average cost equals the marginal cost C_x. (The situation is similar for the case of increasing y.)   **53.** The equation of the tangent plane at the point (a, b) is $z = f(a, b) + f_x(a, b)(x - a) + f_y(a, b)(y - b)$. If f has a relative extremum at (a, b), then $f_x(a, b) = 0 = f_y(a, b)$. Substituting these into the equation of the tangent plane gives $z = f(a, b)$, a constant. But the graph of $z = constant$ is a plane parallel to the xy-plane.

Section 15.5

1. 1; $(0, 0, 0)$ **3.** 1.35; $(1/10, 3/10, 1/2)$
5. Minimum value $= 6$ at $(1, 2, 1/2)$ **7.** 200; $(20, 10)$
9. 16; $(2, 2)$ and $(-2, -2)$ **11.** 20; $(2, 4)$ **13.** 1; $(0, 0, 0)$

15. 1.35; $(1/10, 3/10, 1/2)$ **17.** Minimum value $= 6$ at $(1, 2, 1/2)$ **19.** $5 \times 10 = 50$ sq. ft.   **21.** $10
23. $(1/\sqrt{3}, 1/\sqrt{3}, 1/\sqrt{3})$, $(-1/\sqrt{3}, -1/\sqrt{3}, 1/\sqrt{3})$, $(1/\sqrt{3}, -1/\sqrt{3}, -1/\sqrt{3})$, $(-1/\sqrt{3}, 1/\sqrt{3}, -1/\sqrt{3})$ **25.** $(0, 1/2, -1/2)$ **27.** $(-5/9, 5/9, 25/9)$
29. $l \times w \times h = 1 \times 1 \times 2$ **31.** 18 in $\times$ 18 in $\times$ 36 in, volume $= 11,664$ cubic inches
33. $(2l/h)^{1/3} \times (2l/h)^{1/3} \times 2^{1/3}(h/l)^{2/3}$, where $l =$ cost of lightweight cardboard and $h =$ cost of heavy-duty cardboard per square foot   **35.** $1 \times 1 \times 1/2$ **37.** Method 1: Solve $g(x, y, z) = 0$ for one of the variables and substitute in $f(x, y, z)$. Then find the maximum value of the resulting function of 2 variables. Advantage (Answers may vary): We can use the second derivative test to check whether the resulting critical points are maxima, minima, saddle points, or none of these. Disadvantage (Answers may vary): We may not be able to solve $g(x, y, z) = 0$ for one of the variables. Method 2: Use the method of Lagrange Multipliers. Advantage (Answers may vary): We do not need to solve the constraint equation for one of the variables. Disadvantage (Answers may vary): The method does not tell us whether the critical points obtained are maxima, minima, saddle points, or none of these.   **39.** If the only constraint is an equality constraint, and if it is impossible to eliminate one of the variables in the objective function by substitution (solving the constraint equation for a variable or some other method).   **41.** Answers may vary: Maximize $f(x, y) = 1 - x^2 - y^2$ subject to $x = y$.   **43.** Yes. There may be relative extrema at points on the boundary of the domain of the function. The partial derivatives of the function need not be 0 at such points.   **45.** If the solution were located in the interior of one of the line segments making up the boundary of the domain of f, then the derivative of a certain function would be 0. This function is obtained by substituting the linear equation $C(x, y) = 0$ in the linear objective function. But because the result would again be a linear function, it is either constant, or its derivative is a nonzero constant. In either event, extrema lie on the boundary of that line segment; that is, at one of the corners of the domain.

Section 15.6

1. $-1/2$ **3.** $e^2/2 - 7/2$ **5.** $(e^3 - 1)(e^2 - 1)$ **7.** $7/6$
9. $(e^3 - e - e^{-1} + e^{-3})/2$ **11.** $1/2$ **13.** $(e - 1)/2$
15. $45/2$ **17.** $8/3$ **19.** $4/3$ **21.** 0   **23.** $2/3$ **25.** $2/3$
27. $2(e - 2)$ **29.** $2/3$

31. $\displaystyle\int_0^1 \int_0^{1-x} f(x, y)\, dy\, dx$ **33.** $\displaystyle\int_0^{\sqrt{2}} \int_{x^2-1}^1 f(x, y)\, dy\, dx$

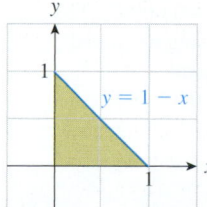

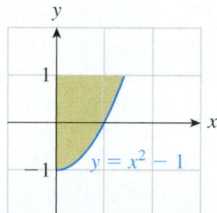

35. $\displaystyle\int_1^4 \int_1^{2/\sqrt{y}} f(x, y)\, dx\, dy$

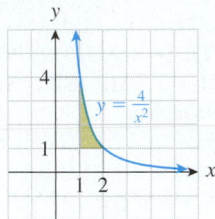

37. 4/3   **39.** 1/6   **41.** 162,000 gadgets   **43.** Average revenue is $312,750.   **45.** Average revenue is $17,500.   **47.** 8216
49. 1 degree   **51.** The area between the curves $y = r(x)$ and $y = s(x)$ and the vertical lines $x = a$ and $x = b$ is given by

$\int_a^b \int_{r(x)}^{s(x)} dy\, dx$ assuming that $r(x) \le s(x)$ for $a \le x \le b$.

53. The first step in calculating an integral of the form

$\int_a^b \int_{r(x)}^{s(x)} f(x, y)\, dy\, dx$ is to evaluate the integral $\int_{r(x)}^{s(x)} f(x, y)\, dy$, obtained by holding x constant and integrating with respect to y.
55. Paintings per picasso per dali   **57.** Left-hand side is
$\int_a^b \int_c^d f(x)g(y)\, dx\, dy = \int_a^b \left(g(y) \int_c^d f(x)\, dx \right) dy$ (since $g(y)$ is treated as a constant in the inner integral) $=$
$\left(\int_c^d f(x)\, dx \right) \int_a^b g(y)\, dy$ (since $\int_c^d f(x)\, dx$ is a constant and can therefore be taken outside the integral).

$\int_0^1 \int_1^2 ye^x\, dx\, dy = \dfrac{1}{2}(e^2 - e)$ no matter how we compute it.

Chapter 15 Review

1. 0; 1; 0; $x^3 + x^2$; $x(y + k)(x + y + k - z) + x^2$
3. Reading left to right, starting at the top: 4, 0, 0, 3, 0, 1, 2, 0, 2
5. $f_x = 2x + y$, $f_y = x$, $f_{yy} = 0$ **7.** 0
9. $\dfrac{\partial f}{\partial x} = \dfrac{-x^2 + y^2 + z^2}{(x^2 + y^2 + z^2)^2}$, $\dfrac{\partial f}{\partial y} = -\dfrac{2xy}{(x^2 + y^2 + z^2)^2}$,
$\dfrac{\partial f}{\partial z} = -\dfrac{2xz}{(x^2 + y^2 + z^2)^2}$, $\dfrac{\partial f}{\partial x}\bigg|_{(0,1,0)} = 1$

11. Absolute minimum at $(1, 3/2)$ **13.** Saddle point at $(0, 0)$
15. Absolute maximum at each point on the circle $x^2 + y^2 = 1$
17. 1/27 at $(1/3, 1/3, 1/3)$ **19.** $(0, 2, \sqrt{2})$ **21.** 4;
$(\sqrt{2}, \sqrt{2})$ and $(-\sqrt{2}, -\sqrt{2})$ **23.** Minimum value $= 5$ at
$(2, 1, 1/2)$ **25.** 2   **27.** ln 5   **29.** 4/5
31. a. $h(x, y) = 5000 - 0.8x - 0.6y$ hits per day
($x =$ number of new customers at JungleBooks.com,
$y =$ number of new customers at FarmerBooks.com)
b. 250   **33. a.** 2320 hits per day   **b.** $0.08 + 0.00003x$ hits
(daily) per dollar spent on television advertising per month;
increases with increasing x **c.** $4000 per month
35. About 15,800 orders per day   **37.** $23,050

Chapter 16

Section 16.1

1.

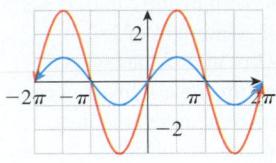

3.

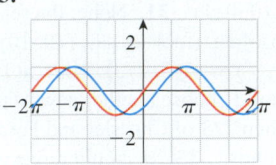

5.

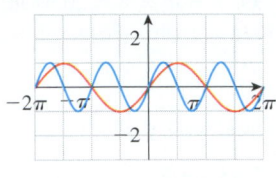

7.

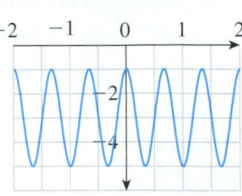

9.

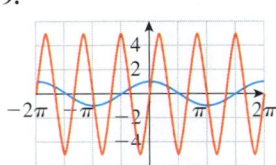

11.

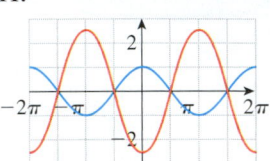

13. $f(x) = \sin(2\pi x) + 1$ **15.** $f(x) = 1.5 \sin(4\pi(x - 0.25))$
17. $f(x) = 50 \sin(\pi(x - 5)/10) - 50$ **19.** $f(x) = \cos(2\pi x)$
21. $f(x) = 1.5 \cos(4\pi(x - 0.375))$
23. $f(x) = 40 \cos(\pi(x - 10)/10) + 40$
25. $f(t) = 4.2 \sin(\pi/2 - 2\pi t) + 3$
27. $g(x) = 4 - 1.3 \sin[\pi/2 - 2.3(x - 4)]$ **31.** $\sqrt{3}/2$
37. $\tan(x + \pi) = \tan(x)$ **39. a.** $2\pi/0.602 \approx 10.4$ years.
b. Maximum: $58.8 + 57.7 = 116.5 \approx 117$;
minimum: $58.8 - 57.7 = 1.1 \approx 1$
c. $1.43 + P/4 + P = 1.43 + 13.05 \approx 14.5$ years, or midway
through 2011   **41. a.** Maximum sales occurred when $t \approx 4.5$
(during the first quarter of 1996). Minimum sales occurred when
$t \approx 2.2$ (during the third quarter of 1995) and $t \approx 6.8$ (during
the third quarter of 1996).   **b.** Maximum quarterly revenues
were $0.561 billion; minimum quarterly revenues were
$0.349 billion.   **c.** maximum: $0.455 + 0.106 = 0.561$; mini-
mum: $0.455 - 0.106 = 0.349$ **43.** Amplitude $= 0.106$, vertical offset $= 0.455$, phase shift $= -1.61/1.39 \approx -1.16$ angular
frequency $= 1.39$, period $= 4.52$. In 1995 and 1996, quarterly
revenue from the sale of computers at Computer City fluctuated
in cycles of 4.52 quarters about a baseline of $0.455 billion.
Every cycle, quarterly revenue peaked at $0.561 billion
($0.106 billion above the baseline) and dipped to a low of
$0.349 billion. Revenue peaked early in the middle of the first
quarter of 1996 (at $t = -1.16 + (5/4) \times 4.52 = 4.49$).
45. $P(t) = 7.5 \sin[\pi(t - 13)/26] + 12.5$
47. $s(t) = 7.5 \sin(\pi(t - 9)/6) + 87.5$
49. $s(t) = 7.5 \cos(\pi t/6) + 87.5$
51. $d(t) = 5 \sin(2\pi(t - 1.625)/13.5) + 10$

53. a. $u(t) = 2.5 \sin(2\pi(t - 0.75)) + 7.5$
b. $c(t) = 1.04^t[2.5 \sin(2\pi(t - 0.75)) + 7.5]$
55. a. $P \approx 8$, $C \approx 6$, $A \approx 2$, $\alpha \approx 8$ (Answers may vary)

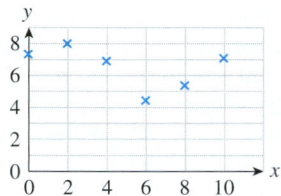

b. $C(t) = 1.755 \sin[0.636(t - 9.161)] + 6.437$

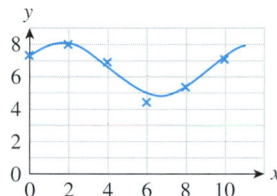

c. 9.9, 4.7%, 8.2%

57. a.

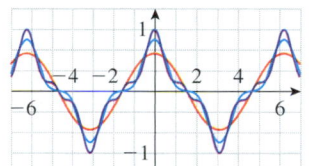

b. $y_{11} = \dfrac{2}{\pi} \cos x + \dfrac{2}{3\pi} \cos 3x + \dfrac{2}{5\pi} \cos 5x + \dfrac{2}{7\pi} \cos 7x$
$+ \dfrac{2}{9\pi} \cos 9x + \dfrac{2}{11\pi} \cos 11x$

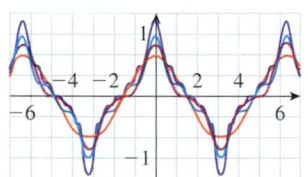

c. $y_{11} = \dfrac{6}{\pi} \cos \dfrac{x}{2} + \dfrac{6}{3\pi} \cos \dfrac{3x}{2} + \dfrac{6}{5\pi} \cos \dfrac{5x}{2} + \dfrac{6}{7\pi} \cos \dfrac{7x}{2}$
$+ \dfrac{6}{9\pi} \cos \dfrac{9x}{2} + \dfrac{6}{11\pi} \cos \dfrac{11x}{2}$

59. The period is approximately 12.6 units

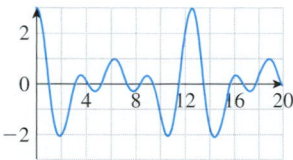

61. Lows: $B - A$; Highs: $B + A$ **63.** He is correct. The other trig functions can be obtained from the sine function

by first using the formula $\cos x = \sin(x + \pi/2)$ to obtain cosine, and then using the formulas

$$\tan x = \frac{\sin x}{\cos x}, \quad \cot x = \frac{\cos x}{\sin x}, \quad \sec x = \frac{1}{\cos x}, \quad \csc x = \frac{1}{\sin x}$$

to obtain the rest.   **65.** The largest B can be is A. Otherwise, if B is larger than A, the low figure for sales would have the negative value of $A - B$.

Section 16.2

1. $\cos x + \sin x$ **3.** $(\cos x)(\tan x) + (\sin x)(\sec^2 x)$
5. $-2 \csc x \cot x - \sec x \tan x + 3$ **7.** $\cos x - x \sin x + 2x$
9. $(2x - 1) \tan x + (x^2 - x + 1) \sec^2 x$
11. $-[\csc^2 x(1 + \sec x) + \cot x \sec x \tan x]/(1 + \sec x)^2$

13. $-2 \cos x \sin x$ **15.** $2 \sec^2 x \tan x$ **17.** $\pi \cos\left[\dfrac{\pi}{5}(x - 4)\right]$
19. $-(2x - 1) \sin(x^2 - x)$
21. $(2.2x^{1.2} + 1.2) \sec(x^{2.2} + 1.2x - 1) \tan(x^{2.2} + 1.2x - 1)$
23. $\sec x \tan x \tan(x^2 - 1) + 2x \sec x \sec^2(x^2 - 1)$
25. $e^x[-\sin(e^x) + \cos x - \sin x]$ **27.** $\sec x$
33. $e^{-2x}[-2 \sin(3\pi x) + 3\pi \cos(3\pi x)]$
35. $1.5[\sin(3x)]^{-0.5} \cos(3x)$

37. $\dfrac{x^4 - 3x^2}{(x^2 - 1)^2} \sec\left(\dfrac{x^3}{x^2 - 1}\right) \tan\left(\dfrac{x^3}{x^2 - 1}\right)$

39. $\dfrac{\cot(2x - 1)}{x} - 2 \ln|x| \csc^2(2x - 1)$

41. a. Not differentiable at 0   **b.** $f'(1) \approx 0.5403$
43. 0   **45.** 2   **47.** Does not exist   **49.** $1/\sec^2 y$
51. $-[1 + y \cos(xy)]/[1 + x \cos(xy)]$
53. $c'(t) = 7\pi \cos[2\pi(t - 0.75)]$; $c'(0.75) \approx$
$\$21.99$ per *year* $\approx \$0.42$ per week   **55.** $N'(6) \approx -32.12$
On January 1, 2003, the number of sunspots was decreasing at a rate of 32.12 sunspots per year.
57. $c'(t) = 1.035^t[\ln(1.035)(0.8 \sin(2\pi t) + 10.2) + 1.6\pi \cos(2\pi t)]$; $c'(1) = 1.035[10.2 \ln|1.035| + 1.6\pi] \approx$
$\$5.57$ per year, or $\$0.11$ per week.
59. a. $d(t) = 5 \cos(2\pi t/13.5) + 10$
b. $d'(t) = -(10\pi/13.5) \sin(2\pi t/13.5)$; $d'(7) \approx 0.270$. At noon, the tide was rising at a rate of 0.270 feet per hour.
61. a. (III)   **b.** Increasing at a rate of 0.157 degrees per thousand years   **63.** -6; 6   **65.** Answers will vary. Examples: $f(x) = \sin x$; $f(x) = \cos x$ **67.** Answers will vary. Examples: $f(x) = e^{-x}$; $f(x) = -2e^{-x}$ **69.** The graph of $\cos x$ slopes down over the interval $(0, \pi)$, so that its derivative is negative over that interval. The function $-\sin x$, and not $\sin x$, has this property.   **71.** The derivative of $\sin x$ is $\cos x$. When $x = 0$, this is $\cos(0) = 1$. Thus, the tangent to the graph of $\sin x$ at the point $(0, 0)$ has slope 1, which means it slopes upward at $45°$.

Section 16.3

1. $-\cos x - 2 \sin x + C$ **3.** $2 \sin x + 4.3 \cos x - 9.33x + C$
5. $3.4 \tan x + (\sin x)/1.3 - 3.2e^x + C$
7. $(7.6/3) \sin(3x - 4) + C$ **9.** $-(1/6) \cos(3x^2 - 4) + C$
11. $-2 \cos(x^2 + x) + C$ **13.** $(1/6) \tan(3x^2 + 2x^3) + C$
15. $-(1/6) \ln|\cos(2x^3)| + C$

17. $3\ln|\sec(2x-4)+\tan(2x-4)|+C$
19. $(1/2)\sin(e^{2x}+1)+C$ **21.** -2 **23.** $\ln(2)$ **25.** 0
27. 1 **33.** $-\dfrac{1}{4}\cos(4x)+C$ **35.** $-\sin(-x+1)+C$
37. $[\cos(-1.1x-1)]/1.1+C$ **39.** $-\dfrac{1}{4}\ln|\sin(-4x)|+C$
41. 0 **43.** 2π **45.** $-x\cos x+\sin x+C$
47. $\left[\dfrac{x^2}{2}-\dfrac{1}{4}\right]\sin(2x)+\dfrac{x}{2}\cos(2x)+C$
49. $-\dfrac{1}{2}e^{-x}\cos x-\dfrac{1}{2}e^{-x}\sin x+C$ **51.** π^2-4
53. Average $=2/\pi$

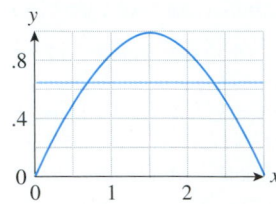

55. Diverges   **57.** Converges to $1/2$
59. $C(t)=0.04t+\dfrac{2.6}{\pi}\cos\left[\dfrac{\pi}{26}(t-25)\right]+1.02$
61. 12 feet   **63.** 79 sunspots
65. $P(t)=7.5\sin[(\pi/26(t-13)]+12.5;\,7.7\%$
67. a. Average voltage over $[0,1/6]$ is zero; 60 cycles per second.
b.

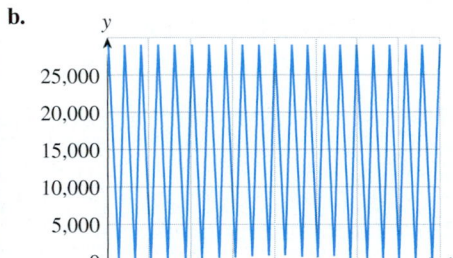

c. 116.673 volts.   **69.** \$50,000   **71.** It is always zero.
73. 1 **75.** $s=-\dfrac{K}{\omega^2}\sin(\omega t-\alpha)+Lt+M$ for constants L and M

Chapter 16 Review

1. $f(x)=1+2\sin x$
3. $f(x)=2+2\sin[\pi(x-1)]=2+2\sin[\pi(x+1)]$
5. $f(x)=1+2\cos(x-\pi/2)$
7. $f(x)=2+2\cos[\pi(x+1/2)]=2+2\cos[\pi(x-3/2)]$
9. $-2x\sin(x^2-1)$ **11.** $2e^x\sec^2(2e^x-1)$
13. $4x\sin(x^2)\cos(x^2)$ **15.** $2\sin(2x-1)+C$
17. $\tan(2x^2-1)+C$ **19.** $-\dfrac{1}{2}\ln|\cos(x^2+1)|+C$ **21.** 1
23. $-x^2\cos x+2x\sin x+2\cos x+C$
25. $s(t)=10{,}500+1500\sin[(2\pi/52)t-\pi]\approx$
$10{,}500+1500\sin(0.12083t-3.14159)$ **27.** \$222,300

Appendix A

1. False statement   **3.** Not a statement, because it is not a declarative sentence   **5.** True statement   **7.** True (we hope!) statement
9. Not a statement, because it is self-referential.   **11.** $(\sim p)\wedge q$
13. $(p\wedge r)\wedge q$ or just $p\wedge q\wedge r$ **15.** $p\vee(\sim p)$ **17.** Willis is a good teacher and his students do not hate math.   **19.** Either Carla is a good teacher, or she is not.   **21.** Willis' students both hate and do not hate math.   **23.** It is not true that either Carla is a good teacher or her students hate math.   **25.** F   **27.** F   **29.** T   **31.** T
33. T   **35.** F   **37.** T   **39.** F   **41.** T   **43.** T   **45.** T   **47.** T

49.

p	q	$\sim q$	$p\wedge\sim q$
T	T	F	F
T	F	T	T
F	T	F	F
F	F	T	F

51.

p	$\sim p$	$\sim(\sim p)$	$\sim(\sim p)\vee p$
T	F	T	T
F	T	F	F

53.

p	q	$\sim p$	$\sim q$	$(\sim p)\wedge(\sim q)$
T	T	F	F	F
T	F	F	T	F
F	T	T	F	F
F	F	T	T	T

55.

p	q	r	$p\wedge q$	$(p\wedge q)\wedge r$
T	T	T	T	T
T	T	F	T	F
T	F	T	F	F
T	F	F	F	F
F	T	T	F	F
F	T	F	F	F
F	F	T	F	F
F	F	F	F	F

57.

p	q	r	$q\vee r$	$p\wedge(q\vee r)$
T	T	T	T	T
T	T	F	T	T
T	F	T	T	T
T	F	F	F	F
F	T	T	T	F
F	T	F	T	F
F	F	T	T	F
F	F	F	F	F

59.

p	q	$q \vee p$	$p \to (q \vee p)$
T	T	T	T
T	F	T	T
F	T	T	T
F	F	F	T

61.

p	q	$p \vee q$	$p \leftrightarrow (p \vee q)$
T	T	T	T
T	F	T	T
F	T	T	F
F	F	F	T

63.

p	$p \wedge p$
T	T
F	F

same

65.

p	q	$p \vee q$	$q \vee p$
T	T	T	T
T	F	T	T
F	T	T	T
F	F	F	F

same

67.

p	q	$p \vee q$	$\sim(p \vee q)$	$\sim p$	$\sim q$	$(\sim p) \wedge (\sim q)$
T	T	T	F	F	F	F
T	F	T	F	F	T	F
F	T	T	F	T	F	F
F	F	F	T	T	T	T

same

69.

p	q	r	$p \wedge q$	$(p \wedge q) \wedge r$	$q \wedge r$	$p \wedge (q \wedge r)$
T	T	T	T	T	T	T
T	T	F	T	F	F	F
T	F	T	F	F	F	F
T	F	F	F	F	F	F
F	T	T	F	F	T	F
F	T	F	F	F	F	F
F	F	T	F	F	F	F
F	F	F	F	F	F	F

same

71.

p	q	$p \to q$	$\sim p$	$\sim q$	$(\sim q) \to (\sim p)$
T	T	T	F	F	T
T	F	F	F	T	F
F	T	T	T	F	T
F	F	T	T	T	T

same

73. Contradiction   **75.** Contradiction   **77.** Tautology
79. $(\sim p) \vee p$ **81.** $(\sim p) \vee \sim(\sim q)$
83. $(p \vee (\sim p)) \wedge (p \vee q)$ **85.** Either I am not Julius Caesar or you are no fool.   **87.** It is raining and I have forgotten either my umbrella or my hat.   **89.** Contrapositive: "If I do not exist, then I do not think."Converse: "If I am, then I think."
91. (A) It is the contrapositive of the given statement.

93. $h \to t$
$\underline{\hspace{1cm} h \hspace{1cm}}$
$\therefore t$
Valid; Modus
Ponens

95. $r \to u$
$\underline{\hspace{1cm} \sim r \hspace{1cm}}$
$\therefore \sim u$
Invalid

97. $g \to m$
$\underline{\hspace{1cm} \sim m \hspace{1cm}}$
$\therefore \sim g$
Valid: Modus
Tollens

99. $m \vee b$
$\underline{\hspace{1cm} \sim b \hspace{1cm}}$
$\therefore m$
Valid: Disjunc-
tive Syllogism

101. $s \vee a$
$\underline{\hspace{1cm} a \hspace{1cm}}$
$\therefore \sim s$
Invalid

103. John is green.   **105.** John is not a swan.   **107.** He is a gentleman.   **109.** Their truth tables have the same truth values for corresponding values of the variables.   **111.** A and B are both contradictions.   **113.** Answers may vary. Let p: "You have smoker's cough," and q: "You smoke."   **115.** Let p: "It is summer in New York," and q: "It is summer in Seattle."

Index

(continued)